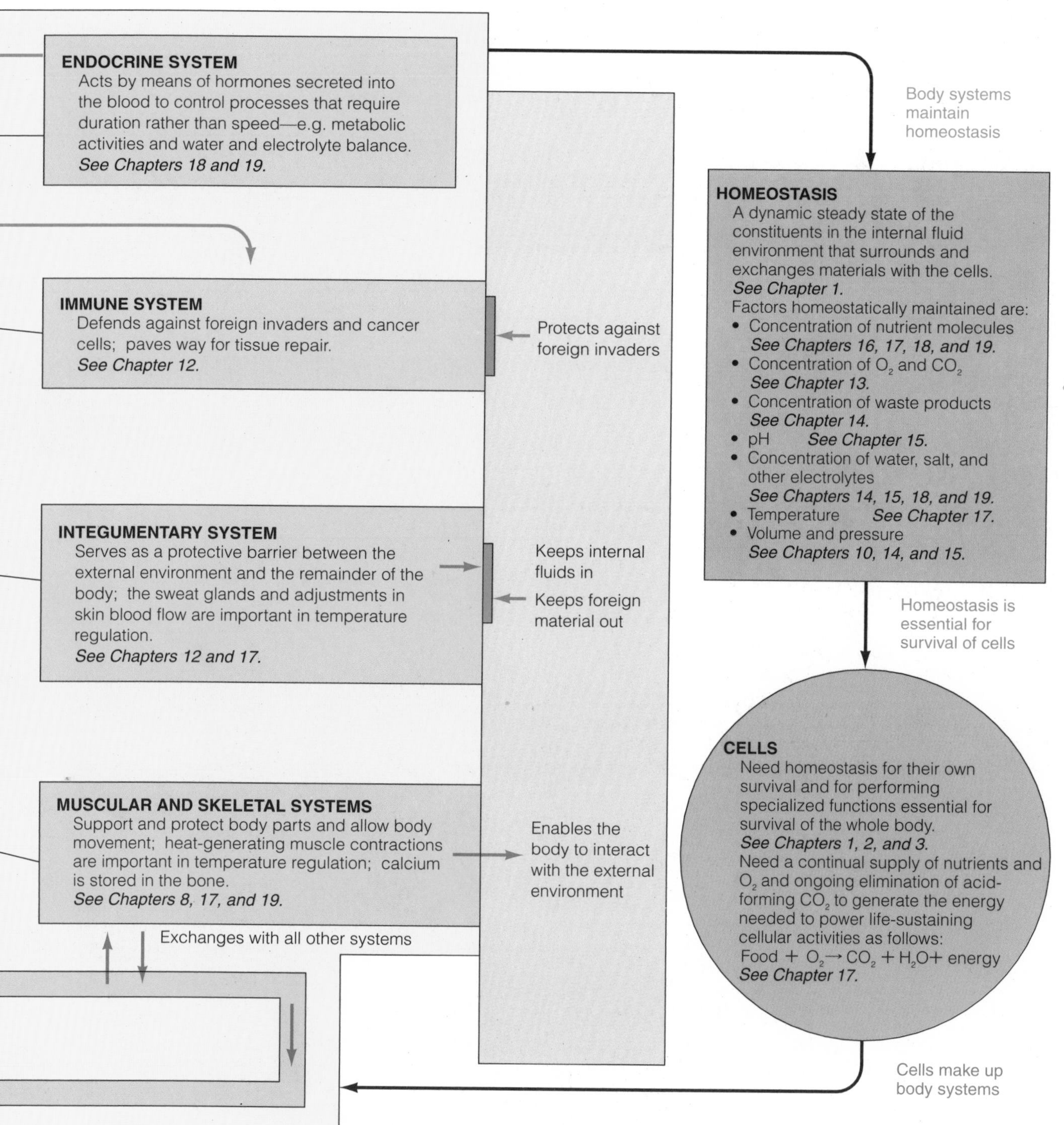
ENDOCRINE SYSTEM
Acts by means of hormones secreted into the blood to control processes that require duration rather than speed—e.g. metabolic activities and water and electrolyte balance.
See Chapters 18 and 19.
IMMUNE SYSTEM
Defends against foreign invaders and cancer cells; paves way for tissue repair.
See Chapter 12.
Protects against foreign invaders
INTEGUMENTARY SYSTEM
Serves as a protective barrier between the external environment and the remainder of the body; the sweat glands and adjustments in skin blood flow are important in temperature regulation.
See Chapters 12 and 17.
Keeps internal fluids in
Keeps foreign material out
MUSCULAR AND SKELETAL SYSTEMS
Support and protect body parts and allow body movement; heat-generating muscle contractions are important in temperature regulation; calcium is stored in the bone.
See Chapters 8, 17, and 19.
Enables the body to interact with the external environment
Exchanges with all other systems
Body systems maintain homeostasis
HOMEOSTASIS
A dynamic steady state of the constituents in the internal fluid environment that surrounds and exchanges materials with the cells.
See Chapter 1.
Factors homeostatically maintained are:
• Concentration of nutrient molecules See Chapters 16, 17, 18, and 19.
• Concentration of O_2 and CO_2 See Chapter 13.
• Concentration of waste products See Chapter 14.
• pH See Chapter 15.
• Concentration of water, salt, and other electrolytes See Chapters 14, 15, 18, and 19.
• Temperature See Chapter 17.
• Volume and pressure See Chapters 10, 14, and 15.
Homeostasis is essential for survival of cells
CELLS
Need homeostasis for their own survival and for performing specialized functions essential for survival of the whole body.
See Chapters 1, 2, and 3.
Need a continual supply of nutrients and O_2 and ongoing elimination of acid-forming CO_2 to generate the energy needed to power life-sustaining cellular activities as follows:
Food + $O_2 \rightarrow CO_2 + H_2O$ + energy
See Chapter 17.
Cells make up body systems

Human Physiology

FROM CELLS TO SYSTEMS

Third Edition

Lauralee Sherwood

Department of Physiology
School of Medicine
West Virginia University

Wadsworth Publishing Company

I(T)P® An International Thomson Publishing Company

Belmont, CA • Albany, NY • Bonn • Boston • Cincinnati • Detroit • Johannesburg
London • Madrid • Melbourne • Mexico City • New York • Paris • Singapore
Tokyo • Toronto • Washington

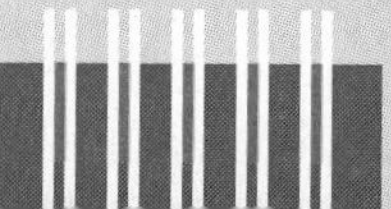

Production Credits

COPYEDITING
Patricia Lewis

INTERIOR DESIGN
Diane Beasley

COMPOSITION
Parkwood Composition, Inc.

ARTWORK
Wayne Clark, Darwen and Vally Hennings, Carlyn Iverson, Sandra McMahon, Elizabeth Morales, Precision Graphics, Publication Services, Rolin Graphics, John and Judy Waller, and Cyndie C. H.-Wooley

ELECTRONIC PAGE LAYOUT
Diane Beasley

COVER DESIGN
Diane Beasley

COVER IMAGE
© Linda S. Nye/Phototake

PRODUCTION, PREPRESS, PRINTING, AND BINDING
West Publishing Company

Printed In the United States of America

4 5 6 7 8 9 10

Library of Congress Cataloging-in-Publication Data

Sherwood, Lauralee.
Human physiology : from cells to systems / Lauralee Sherwood. — 3rd ed.
p. cm.
Includes bibliographical references and index.
ISBN 0-314-09245-5 (hard : alk. paper)
1. Human physiology. I. Title.
QP34.5.S48 1997
612--dc20

96-42161

For more information, contact Wadsworth Publishing Company, 10 Davis Drive, Belmont, CA 94002, or electronically at http://www.thomson.com/wadsworth.html

International Thomson Publishing Europe
Berkshire House 168-173
High Holborn
London, WC1V 7AA, England

Thomas Nelson Australia
102 Dodds Street
South Melbourne 3205
Victoria, Australia

Nelson Canada
1120 Birchmount Road
Scarborough, Ontario
Canada MIK 5G4

International Thomson Publishing GmbH
Königswinterer Strasse 418
53227 Bonn, Germany

International Thomson Editores
Campos Eliseos 385, Piso 7
Col. Polanco
11560 México D.F. México

International Thomson Publishing Asia
221 Henderson Road
#05-10 Henderson Building
Singapore 0315

International Thomson Publishing Japan
Hirakawacho Kyowa Building, 3F
2-2-1 Hirakawacho
Chiyoda-ku, Tokyo 102, Japan

International Thomson Publishing Southern Africa
Building 18, Constantia Park
240 Old Pretoria Road
Halfway House, 1685 South Africa

To my family,
in celebration of the body functions that make life possible
and with love and appreciation for the sharing and caring
that make life worthwhile:

my grandparents (in memorium),
George and Lottie Wonch
Clarence and Amy Sherwood

my parents, Larry and Lee Sherwood

my husband, Peter Marshall

my mother- and (in memorium) father-in-law,
Sarah and Harold Marshall

my daughter, Allison Marshall

my daughter and son-in-law, Melinda and Mark Marple

and my first grandchild, Lindsay Marple

Brief Contents

Contents

CHAPTER 6

The Peripheral Nervous System: Afferent Division; Special Senses 155

CHAPTER 7

The Peripheral Nervous System: Efferent Division 203

CHAPTER 8

Muscle Physiology 221

CHAPTER 9

Cardiac Physiology 265

CHAPTER 10

The Blood Vessels and Blood Pressure 305

CHAPTER 11

The Blood 352

CHAPTER *12*

The Body Defenses 373

CHAPTER *13*

The Respiratory System 417

CHAPTER *14*

The Urinary System 468

CHAPTER *15*

Fluid and Acid-Base Balance 515

CHAPTER *16*

The Digestive System 546

CHAPTER *17*

Energy Balance and Temperature Regulation 601

CHAPTER *18*

Principles of Endocrinology; The Central Endocrine Glands 620

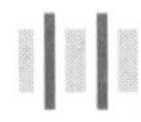

CHAPTER *19*

The Peripheral Endocrine Glands 654

CHAPTER *20*

The Reproductive System 700

Appendices

Preface

Philosophy, Goals, and Theme

When a baby first discovers that it can control its own hands, it will be fascinated and spend many hours manipulating them in front of its face. Most of us, even infants, have a natural curiosity about how our bodies work. Our bodies are quite miraculous. No machine has been constructed that can take over even a portion of a natural body function as effectively. By capitalizing on students' natural curiosity about themselves, I have strived to make physiology a subject that they can enjoy learning.

Even the most tantalizing subject matter, however, can be drudgery to study and difficult to comprehend if not effectively presented. Therefore, this book has a logical, understandable format that is unencumbered by unnecessary details and emphasizes how each concept is an integral part of the whole subject matter. Too often, students view isolated sections of a physiology course as separate entities; by understanding how each component of the body depends on other components, a student can appreciate the integrated functioning of the human body. The text focuses on the mechanisms of body function from cells to systems and is organized around the central theme of homeostasis—how the body meets changing demands while maintaining the internal constancy necessary for all cells and organs to function.

The text is written with its primary target audience—undergraduate students preparing for health-related careers—in mind. Its approach and depth are appropriate, however, for other undergraduate student populations. Because it is intended to serve as an introductory text and, for most students, may be their only exposure to a formal physiology text, all aspects of physiology receive broad coverage, yet depth, where needed, is not sacrificed. The scope of this text has been limited by judicious selection of pertinent content that a student can reasonably be expected to assimilate in a one-semester physiology course.

This edition of *Human Physiology: From Cells to Systems* has been revised with the goal of continuing to offer instructors two choices in length and depth of physiology textbooks. Also available is a carefully condensed version of the original parent book for those students who do not have a math and physics background, *Fundamentals of Physiology: A Human Perspective, Second Edition* (L. Sherwood, West Publishing Company, 1995). The briefer text is more suitable for lower-level undergraduate courses of shorter duration and intensity. This third edition of the parent book is designed primarily for upper-level undergraduate students who have more background in math and science and need more breadth and depth of coverage in physiology. Accordingly, this edition is slightly more quantitative than the first two editions while largely retaining the scope, length, and depth of the preceding books. The extent of quantitative coverage within the chapters remains unchanged, but quantitative exercises are now provided within the end-of-chapter pedagogy, and a new appendix reviewing the principles of quantitative reasoning has been added. Chemistry and genetic appendixes are also provided for those who need a review or a quick reference source. Furthermore, because anatomy is not a prerequisite course, enough relevant anatomy is integrated within the text to make the inseparable relation between structure and function meaningful.

To keep pace with today's rapid advances in the health sciences, students in the health professions must be able to draw on their conceptual understanding of physiology instead of merely recalling isolated facts that soon may be outdated. Therefore, this text is designed to promote understanding of the basic principles and concepts of physiology rather than memorization of details. The text is written in simple, straightforward language, and every effort has been made to assure smooth reading through good transitions, logical reasoning, and integration of ideas throughout the text.

In consideration of the clinical orientation of the target group, research methodologies and data are not emphasized, although the material is based on up-to-date evidence. New information based on recent discoveries has been added to all chapters. Redundant and extraneous points were carefully pared from the preceding edition so that addition of cutting-edge content and helpful new pedagogy has not increased the length of the book for the third edition. Some controversial ideas and hypotheses are presented to illustrate that physiology is a dynamic, changing discipline.

Changes in this Edition

Organizational Changes This edition retains the same number and sequence of chapters as its forerunner. However, the content within some chapters has been slightly reorganized as the following examples illustrate:

- A new section on "Intercellular Communication and Signal Transduction" has been added to chapter 3, involving the following organizational changes and new coverage:
  1. The section on cell-to-cell communication, which compares paracrines, neurotransmitters, hormones, and neurohormones, has been moved from chapter 18 to this new section, providing early introduction to the types of chemical messengers.
  2. Added to this section is the recent discovery of caveolae, membranous structures believed to play roles in membrane transport and signal transduction.
  3. Much of the material previously presented in chapter 3 under the title "Membrane Receptors and Postreceptor Events" has been incorporated into this new section.
  4. Coverage of channel regulation by extracellular messengers has been expanded and updated. The concept of gating has been moved from chapter 4 to this new chapter 3 section.
  5. Discussion of the concept of signal transduction has been significantly expanded.
  6. The role of G proteins is discussed in considerably greater detail than in the preceding edition.
- Unlike the preceding edition in which the coverage of endocytosis was split evenly between chapters 2 and 3, this topic is now discussed primarily in chapter 2, with only a brief reference in chapter 3. The expanded discussion of endocytosis incorporates new knowledge about the role of SNARES in vesicle fusion with a membrane and the recently discovered role of dynamin in pinching off the necks of the endocytotic vesicles.
- Coverage of countercurrent exchange, urea recycling in the renal medulla, and vasopressin-controlled water reabsorption in chapter 14 has been reorganized for smoother flow of the material.
- The description of visceral and sensory afferents in chapter 6 has been reorganized to eliminate redundancy and improve clarity.
- A new section on the pineal gland was added to chapter 18, including a discussion of the multiple proposed roles of its hormonal product, melatonin, that have recently been identified.

Updated Content To keep this edition timely and relevant, all chapters have been updated based on current research findings. Following are examples of the more than 75 recent findings that have been incorporated as new or expanded topics in this edition:

- The emerging concept of apoptosis, programmed cell death.
- New evidence that adaptation in the Pacinian corpuscle involves an electrochemical component of adaptation in addition to the classic mechanical component.
- A flurry of new findings on odor transduction and odor discrimination, such as the role of the olfactory-bulb glomeruli as "smell files."
- Recent discoveries in muscle research. Among these are the roles of the dihydropyridine receptors and ryanodine receptors in the various types of muscle and the newly identified ATP binding cleft in skeletal muscle's myosin head, including a discussion of the widening and narrowing of the cleft in relation to the power stroke.
- The latest research on the ionic basis of the heart's pacemaker potential.
- An updated and expanded discussion of atherosclerosis, including the pathogenic sequence and possible contributing factors.
- The recent discovery of a new hormone, thrombopoietin, which controls platelet production.
- More extensive coverage of helper T cells, including the newly identified subsets of T helper 1 (TH1) cells and T helper 2 (TH2) cells and the patterns of response they elicit.
- Expanded discussion of the generation of respiratory rhythm to include the recently discovered role of the ventromedial medulla.
- The newly identified water channels, or aquaporins, in kidney tubular cells.
- New findings regarding the identity and location of the pacesetter cells of the digestive tract (the interstitial cells of Cajal).
- The discovery of the Ob gene and its product leptin and their possible roles, along with other factors, in obesity.
- New data that dispel the traditional view that "normal" body temperature is 98.6°F.
- Expanded coverage of the postreceptor events for lipophilic hormones to include new research regarding the association of steroid receptors with molecular chaperone proteins.
- Updated discussion of insulin-dependent glucose transport to include the role of various GLUT molecules and expanded coverage of transporter recruitment.
- New information on the advantages of breast-feeding, including the multiple immunoprotective components of breast milk, many of which have been recently identified.

In some instances, new findings that are introduced challenge traditional views, reinforcing the notion that physiology is a dynamic discipline. The following are illustrative:

- New data that suggest that many membrane proteins are restricted in their mobility by confinement zones and other means and are thus not as mobile as proposed by the fluid mosaic model of membrane structure.
- Evidence that questions the accuracy of the classic somatotopic map of the primary motor cortex.
- New findings that challenge the traditional view of a three-cone system of color vision.

Revised Coverage of Difficult Topics Every chapter has been carefully evaluated for ways to improve the presentation. A particular effort was made to further clarify topic areas that traditionally give students the most difficulty. Further explanations, new analogies, new tables, and new or reconceptualized art have been incorporated to help students better understand troublesome concepts, as exemplified by the following:

- Ion movements responsible for membrane electrical properties.
- Comparison of the impact of presynaptic inhibition and inhibitory postsynaptic potentials.
- Excitation-contraction coupling.
- Distinction between the equation for *calculating* mean arterial pressure (mean arterial pressure = diastolic pressure + ⅓ pulse pressure) and the equation indicating the *determinants* of mean arterial pressure (mean arterial pressure = cardiac output × total peripheral resistance).
- How the kidneys are able to excrete urine of varying concentration, depending on the body's needs, including the roles of the countercurrent system and vasopressin-controlled water reabsorption.
- Table summarizing changes in $[CO_2]$, $[HCO_3^-]$, and pH during uncompensated and compensated acid-base disorders.
- Significance of plasma clearance rates for substances handled in different ways by the kidneys.

New Features and Pedagogical Aids In addition to these organizational and content changes, new features and pedagogical aids have been added to complement those in the preceding edition, as will be described in the next section.

Features and Pedagogical Aids New to this Edition

Improved Illustrations Almost every figure has been improved by using color in new ways to help students sort out the components of the figure. Colored screens, boxed labels, colored type, and reverse type help to delineate key features of the figures with the added bonus of making the art more aesthetically pleasing. In flow diagrams, the corners of all physical entities, such as body structures or chemicals, have been rounded to distinguish them from the square corners of all actions. New figures have been added, and many others have been reconceptualized for improved clarity.

New Layout Design The static two-column format has been replaced by a dynamic new layout that not only offers more visual appeal but more importantly permits art to be positioned in closer proximity to its text reference.

New Boxed Features Fifteen percent of the **Concepts, Challenges, and Controversies** boxed features have been deleted and replaced with more current, high-interest topics such as xenotransplantation (transplantation of animal organs in humans) and environmental estrogens. The majority of the retained boxed features have been updated to keep them current and relevant.

Expanded End-of-Chapter Learning Activities The Review Exercises at the end of each chapter now include **Quantitative Exercises** in addition to the previous edition's objective questions and essay questions. The Quantitative Exercises provide the students with an opportunity to practice calculations that will enhance their understanding of complex relationships. The **Points to Ponder** section, which features thought-provoking questions related to material covered in the chapter, has been expanded to include a **Clinical Consideration.** This mini case study challenges students to apply their physiology knowledge to a patient's specific symptoms. These new learning activities could be assigned or used as a basis for lecture or small-group discussions.

New Appendixes A new appendix, **Principles of Quantitative Reasoning,** is designed to help students become more comfortable working with equations and translating back and forth between words, concepts, and equations.

Answers to all end-of-chapter learning activities, including solutions to the Quantitative Exercises and explanations for the Points to Ponder and Clinical Consideration, are provided in another appendix.

Glossary with Phonetic Pronunciations The glossary, which enables students to quickly review key terms when they occur later in the book, has been upgraded to include phonetic pronunciations of the entries.

Retained and Improved Features and Pedagogical Aids

Homeostatic Model and Chapter Opening A revised, unique, easy-to-follow, pictorial homeostatic model depicting the relationship among cells, systems, and homeostasis is developed in the introductory chapter and presented on the inside front cover as a quick reference. Each chapter begins with (1) a revised specialized version of this model to emphasize how each body system functionally fits in with the body as a whole, (2) an accompanying didactic introduction, and (3) a **Chapter Contents at a Glance,** which lists the principal topics to be covered in that chapter. These opening features are designed to orient the student and help put the material that follows in perspective.

Chapter Closing Focusing on Homeostasis Each chapter concludes with a narrative, **Chapter in Perspective: Focus on Homeostasis,** which helps the students put into perspective how the part of the body just discussed contributes to homeostasis. This pedagogical feature, the opening homeostatic model, and the introductory comments are designed to work together to facilitate the students' comprehension of the interactions and interdependency of body systems, even though each system is discussed separately.

Narrative Chapter Summaries A concise, section-by-section, narrative **Chapter Summary** at each chapter's end enables students to focus on the main concepts before moving on. Furthermore, an instructor can assign the summary (or part of it) in lieu of the full chapter (or section) for a topic that needs only a concise overview in a particular course. Thus, the instructor has maximum flexibility to cover topics in depth or superficially.

Boxed Features Each chapter in this edition has at least two boxed features, one entitled **Concepts, Challenges, and Controversies** and the other **A Closer Look at Exercise Physiology.** The Concepts, Challenges, and Controversies boxes expose students to high-interest, tangentially relevant information on such diverse topics as environmental impact on the body, aging, ethical issues, new discoveries regarding common diseases, historical perspectives, and body responses to new environments such as those encountered in space flight and deep-sea diving.

Current concepts related to exercise physiology are included in the other boxed feature for three reasons: increasing national awareness of the importance of physical fitness; increasing recognition of the value of prescribed therapeutic exercise programs for a variety of conditions; and growing career opportunities related to fitness and exercise.

Analogies Many analogies and frequent references to everyday experiences are included to help students relate to the physiology concepts presented.

Pathophysiology Another effective way to keep students' interest is to help them realize that they are learning worthwhile and applicable material. Because most students using this text will have health-related careers, frequent references to pathophysiology and clinical physiology demonstrate the contents' relevance to their professional goals. Even more topics of clinical interest have been incorporated in this edition.

Full-Color Illustrations A full-color art program is used as a functional tool to learning. Anatomical illustrations, schematic representations, photographs, tables, and graphs are designed to complement and reinforce the written material. Flow diagrams are used extensively to help the students integrate the didactic information presented. Thorough figure captions are provided to improve understanding of the figures.

A colored symbol precedes each figure ▬ and table number and title ▮ and also precedes the first reference to the figure or table in the text. This feature enables students to easily find the text description of a figure or table and enables them to return quickly to the text they were reading before they referred to the learning aid.

Integrated Color-Coded Figure/Table Combinations Figure/table combinations enable students to better visualize what part of the body is responsible for what activities. For example, an anatomical depiction of the brain is integrated with a table of the functions of the major brain components, with each component shown in the same color in the figure and the table.

Diversity of Human Models A unique feature of this book is that the people depicted in the various illustrations are realistic representatives of a cross section of humanity (they were drawn from photographs of real people). Sensitivity to the various races, sexes, and ages of undergraduate students should enable all students to identify with the material being presented.

Feedforward Statements as Subsection Titles Instead of traditional topic titles for each subsection (for example, **Heart valves**), feedforward statements alert the student to the main point of the subsection to come (for example, **Heart valves ensure the proper direction of blood flow through the heart.**).

Cross-References Cross-references to related material in earlier chapters enable students to quickly refresh their memories and also give instructors more flexibility in organizing the presentation of materials.

Key Terms Key terms are defined as they appear in the text. Word derivations are provided as necessary to enhance understanding of new words.

Cellular Approach Even though the text is primarily organized according to body systems, it provides extensive coverage of cellular physiology at the beginning and incorporates explanations of function at the cellular and molecular level throughout as a basis for understanding organ function. The cellular/molecular approach is slighted in most general undergraduate physiology textbooks, yet this approach is at the cutting edge of much of physiology research.

Supplemental Reading List An appendix of suggested current readings is available as a guide for students who wish to explore specific topics in greater detail.

Appendixes In addition to the new appendix on quantitative reasoning and the appendix of answers to the end-of-chapter learning activities, three appendixes from the preceding edition have been retained in this book: a chemistry appendix, a genetics appendix, and an immune effectors appendix.

Most undergraduate physiology texts have a chapter on chemistry, yet physiology instructors rarely teach basic chemistry concepts. The decision was made, therefore, to reserve valuable text space for physiological concepts and to provide instead an appendix entitled **A Review of Chemical Principles** as a handy reference for students who need a review of basic chemistry concepts that are essential to understanding physiology.

Likewise, an appendix entitled **Storage, Replication, and Expression of Genetic Information** serves as a reference for students or as assigned material if the instructor deems appropriate. It includes a discussion of DNA and chromosomes, protein synthesis, cell division, and mutations.

The remaining appendix entitled **Control, Functions, and Interactions of Immune-Response Effectors** is a handy reference for sorting out at a glance the complex actions and interactions of the immune effector cells and chemical mediators.

Organization

There is no ideal organization of physiological processes into a logical sequence. With the sequence chosen, most chapters build on material presented in immediately preceding chapters, yet each chapter is designed to stand on its own to allow the instructor flexibility in curriculum design. The general flow is from introductory background information to cells to excitable tissue to organ systems. Every attempt has been made to provide logical transitions from one chapter to the next. For example, chapter 8, *Muscle Physiology,* ends with a discussion of cardiac muscle, which is carried forward into chapter 9, *Cardiac Physiology.* Even topics that seem unrelated in sequence, such as chapter 12, *Defense Mechanisms of the Body,* and chapter 13, *Respiratory System,* are linked together, in this case by ending chapter 12 with a discussion of respiratory defense mechanisms.

Several organizational features warrant specific mention. The most difficult decision in organizing this text was the placement of the chapters on the endocrine system. Intermediary metabolism of absorbed nutrient molecules is largely under endocrine control, providing a link from digestion (chapter 16) and energy balance (chapter 17) to the endocrine system (chapters 18 and 19). There is merit in placing the chapters on the nervous and endocrine systems in close proximity because of these systems' roles as the body's major control systems. Placing the endocrine system chapters earlier, immediately after the discussion of the nervous system (chapters 4 through 7), however, would have created two problems. First, it would have disrupted the logical flow of material related to excitable tissue. Second, the endocrine system could not have been covered at the level of depth its importance warrants if it had been discussed before the students were provided the background essential to understanding this system's roles in maintaining homeostasis. Placing the endocrine system chapters late in the book does not mean, however, that students are not exposed to endocrine function or hormones until near the book's completion. Endocrine control and hormones are defined in chapter 1, are revisited again in chapter 3 in the discussion of intercellular communication, and are compared with nervous control in chapter 5. Specific hormones are introduced in appropriate chapters, such as vasopressin and aldosterone in the chapters on kidney and fluid balance. Chapters 18 and 19 explore the basic characteristics of endocrine glands and hormones as well as the control and functions of specific endocrine secretions.

Unique to this book, the skin is covered in the chapter on defense mechanisms of the body in consideration of the skin's newly recognized immune functions. Bone is also covered more extensively in the endocrine chapters than in most undergraduate physiology texts, especially with regard to hormonal control of bone growth and bone's dynamic role in calcium metabolism.

Departure from traditional groupings of material in several important instances has permitted more independent and more extensive coverage of topics that are frequently omitted or buried within chapters concerned with other subject matter. For example, a separate chapter is devoted to fluid balance and acid-base regulation, topics often tucked within the kidney chapter. The grouping of the autonomic nervous system, motor neurons, and the neuromuscular junction in an independent chapter on the efferent division of the peripheral nervous system, which serves as a link between the nervous system chapters and the muscle chapter, is another example.

Although there is a rationale for covering the various aspects of physiology in the order given here, it is by no means the only logical way of presenting the topics. Each chapter is able to stand on its own, especially with the cross-references provided, so that the sequence of presentation can be varied at the instructor's discretion. Some chapters may even be omitted, depending on the students' needs and interests and the time constraints of the course. For example, a cursory explanation of the defense role of the leukocytes is covered in the chapter on blood, so an instructor could choose to omit the more detailed explanations of immune defense in chapter 12. Similarly, the in-depth coverage of topics in chapters 2, 6, 15, 17, and 19 could selectively be omitted without sacrificing a student's general appreciation of systems-approach physiology.

As an alternative to total omission of certain chapters, the **Chapter Summary** could be used as a supplement for less comprehensive coverage of these topics.

Ancillaries

Instructor's Manual and Test Bank This comprehensive teaching aid by Jay Templin and Lauralee Sherwood includes a brief chapter outline, lecture hints and suggestions, and a thoroughly revised and expanded test bank of true/false, fill-in-the-blank, matching, and multiple choice questions.

Computerized Test Service The publisher provides the entire test bank on diskette along with WESTEST, a computerized testing package. Using WESTEST 3.1, instructors can generate examinations containing questions they select or have questions randomly generated by the computer. Instructors can also use the WESTEST 3.2 edit function to modify these questions, add new questions, or delete existing questions. Classroom management software is also available from the publisher for recording, storing, and preparing reports using students' data. Contact your local sales representative for details.

Acetates A full-color transparency acetate set provides clear, effective illustrations of 225 of the most important artwork and diagrams from the text.

Clinical Health Issues Handbook, Second Edition Revised and updated by Karil Bellah, M.D., this handbook discusses pathophysiological conditions of interest to students. The handbook is organized by body systems. Each chapter opens with a case history. Questions concerning the case and the body system are followed by discussions of clinical conditions associated with that system. The chapters conclude with a review of the case study. The publisher can make this supplement available with the text or it can be purchased separately.

Readings in Human Physiology, 1997 Edition Edited by Lauralee Sherwood and Linda Vona-Davis, West Virginia University, this is a collection of approximately fifty very current articles chosen from general interest and science magazines to supplement material students will encounter in their course work. Each article begins with an introduction by the editors and ends with a set of questions that help the students test their understanding of the material. Answers are provided at the end of the book. The publisher can make this supplement available with the text or it can be purchased separately.

Learning Resource Manual Written by David Shepherd, Southeastern Louisiana University, this study guide provides an overview and detailed outline of each chapter of the main text; a list of key terms; open-ended questions that help the student synthesize the broader concepts of the chapter; simple and safe physiology experiments; and a full set of review exercises consisting of true/false, fill-in-blank, matching, and multiple choice questions.

Software The *Human Physiology Interactive Tutorial and Workbook* software series, developed by J. M. Yochim, University of Kansas, and Y. J. Dori, The Technion-Israel Institute of Technology, provides nine hypercard modules for Macintosh computers on the endocrine system, neurophysiology 1 and 2, digestive physiology, the brain, the kidney, skeletal muscle, cardiac physiology, and pregnancy and lactation. This series is a self-paced learning environment that can be used by students as an introduction, a study guide, a progress monitor, an evaluation tool, or a review of material already presented in the text or lecture. The programs follow a learning path that branches off to provide increasing depth with each level. The software and workbooks can be purchased by nonadopters and students. Contact your local sales representative for information on site licenses and further detail. Other software is available to qualified adopters.

Videotapes Free selections from a publisher-based video library are available to qualified adopters. Contact your local sales representative for more information.

Acknowledgments

I gratefully acknowledge the many people who helped with the first two editions or this edition of the textbook. A special thank-you goes to four individuals who contributed substantially to the content of the book: Rachel Yeater (professor and chairwoman, Sports Exercise Program, and director, Human Performance Laboratory, School of Medicine, West Virginia University), who contributed the material for the boxed features **A Closer Look at Exercise Physiology;** Spencer Seager (chairman, Chemistry Department, Weber State College) who prepared Appendix A, **A Review of Chemical Principles;** and Kim Cooper (Assistant Professor), and John Nagy (Instructor) Department of Zoology, Arizona State University, who provided the **Quantitative Exercises** for the end-of-chapter pedagogy and prepared Appendix C, **Principles of Quantitative Reasoning.**

During the book's creation and revision, many colleagues provided assistance. George Hedge and Robert Goodman, Department of Physiology, West Virginia University, deserve a special note of gratitude for their willingness to share materials used from their publication, **Clinical Endocrine Physiology** (Saunders, 1987), and for their thoughtful reviews of this book's chapters on endocrinology and reproduction. Appreciation is also extended to Elizabeth Walker and Dennis Overman, Department of Anatomy, West Virginia University, who provided many custom-made light and electron micrographs for the book.

Others at West Virginia University deserving of recognition for providing resource materials or countless clarifications include James Culberson and William Beresford, Department of Anatomy; Marta Henderson, Division of Medical Technology; Ronald Gaskins, Department of Medicine; Sidney Schochet, Jr., and Val Vallyathan, Department of Pathology; Karen Curto and Robert Stitzel, Department of Pharmacology; and Christine Baylis, Paul Brown, John Connors, Gunter Franz, Wilbert Gladfelter, Linda Huffman, Michael Johnson, Ping Lee, Philip Miles, Ronald Millechia, and William Stauber, Department of Physiology.

In addition to the 46 reviewers who carefully evaluated the preceding two editions for accuracy, clarity, and relevance, I express sincere appreciation to the following individuals who served as reviewers for this book. Their comments and suggestions were very helpful while I was considering ways to improve this edition.

Ronald J. Adkins
Washington State University

Lee B. Astheimer
University of Wollongong

Sunny K. Boyd
University of Notre Dame

David Busath
Brown University

Carl M. Christenson
Indiana University-Southeast

Kim Cooper
Arizona State University

Alan Donnelly
University of Wolverhampton

Charles A. Fuller
University of California, Davis

Jack M. Goldberg
University of California, Davis

Eric Hall
Northeastern University

Jon F. Hunter
Texas A & M University

Graham Huxham
University of Queensland

John Keener
University of Minnesota

Suzanne M. Kelley
Harvard School of Public Health

David Kendall
University of Nottingham

Janice M. Lapsansky
Western Washington University

Randall Packer
George Washington University

Gary Ritchison
Eastern Kentucky University

Kevin T. Strang
University of Wisconsin-Madison

Roger Thies
University of Oklahoma

Martyn Ward
University of Dundee

David W. Washington
University of Central Florida

It was my personal and professional pleasure to work with the same two highly competent editors at West Publishing Company who oversaw the development and production of my previous editions—Jerry Westby, executive editor, who served as the developmental editor, and Barbara Fuller, assistant manager, who served as the production editor. It is more difficult to find the words to express my appreciation to Jerry and Barb than to explain a difficult physiology concept! For over twelve years and through five textbooks, Jerry has offered invaluable guidance, inspiration, and moral support. He continues to amaze me with his knack for coming up with fresh ideas for improving an already successful book. I am grateful for the personal attention he pays to all aspects of the development and production process, despite the multitude of other responsibilities he has, and for his ability to smooth over the inevitable rough spots. It has also been my good fortune to have Barb as the production editor for all the editions of this book. I have the greatest confidence in her ability to get things done right and in timely fashion. Barb, as usual, through her competency, efficiency, resourcefulness, painstaking attention to detail, and unflagging dedication, brought this edition smoothly and punctually to fruition. Besides being a great facilitator, Barb offers regular, welcome doses of good cheer, words of encouragement, and understanding. Both Jerry and Barb went "above and beyond the call of duty" to help make this the best book it could possibly be.

A warm thank-you is also extended to others who helped make this book a reality: Dean DeChambeau, the developmental editor who oversaw the review process and helped with the development of ancillary materials; Ellen Stanton, the promotion manager for the book; Patricia Lewis, copyeditor; Diane Beasley, interior designer, layout designer, and cover designer; and Parkwood Composition, typesetter. I further wish to thank the text's artists for their excellent contributions: John and Judy Waller, Darwen and Vally Hennings, Cyndie C. H.-Wooley, Wayne Clark, Sandy McMahon, Carlyn Iverson, Elizabeth Morales, Rolin Graphics, Publication Services, and Precision Graphics.

Finally, my love and gratitude go to my family for another year of sacrifices in family life as this third edition was being developed and produced. I want to thank them for their patience, understanding, and support during the times I was working on the book instead of being there with them or for them.

1 Homeostasis: The Foundation of Physiology

During the minute that it will take you to read this page:

Your eyes will convert the image from this page into electrical signals (nerve impulses) that will transmit the information to your brain for processing.

Besides receiving and processing information such as visual input, your brain will also provide output to your muscles to help maintain your posture, move your eyes across the page as you read, and turn the page as needed. Chemical messengers will carry signals between your nerves and muscles to trigger appropriate muscle contraction.

Your heart will beat seventy times, pumping 5 liters (about 5 quarts) of blood to your lungs and another 5 liters to the rest of your body.

You will breathe in and out about twelve times, exchanging 6 liters of air between the atmosphere and your lungs.

More than 1 liter of blood will flow through your kidneys, which will act on the blood to conserve the "wanted" materials and eliminate the "unwanted" materials in the urine. Your kidneys will produce 1 ml (about a thimbleful) of urine during this minute.

Your cells will consume 250 ml (about a cup) of oxygen and produce 200 ml of carbon dioxide.

Your digestive system will be processing your last meal for transfer into your bloodstream for delivery to your cells.

You will use about 2 calories of energy derived from food to support your body's "cost of living," and your contracting muscles will burn additional calories.

Chapter Contents At a Glance

Introduction

The activities described on the preceding page are a sampling of the various processes that occur in our bodies all the time just to keep us alive. We usually take these life-sustaining activities of the body for granted and don't really think about "what makes us tick," but that's what physiology is all about. **Physiology** is the study of the functions of the body, or how the body works.

Physiologists view the body as a machine whose mechanisms of action can be explained in terms of cause-and-effect sequences of physical and chemical processes—the same types of processes that occur in other components of the universe. It is important to distinguish between the mechanistic and teleological approaches to explaining the various events that occur in the body. With the **mechanistic approach,** which is employed by physiologists, the *mechanisms* of action are emphasized. That is, physiologists explain the "how" of events that occur in the body. With a **teleological approach,** phenomena that occur in the body are explained in terms of their particular purpose in fulfilling a bodily *need,* without considering how this outcome is accomplished. That is, the teleological approach emphasizes the "why" of body processes.

A simple example will help you to distinguish between these approaches. A teleological explanation of why you shiver when you are cold is "to keep warm," because shivering generates heat. A physiologist's mechanistic explanation of why you shiver is that when temperature-sensitive nerve cells detect a fall in body temperature, they signal the hypothalamus, the part of the brain responsible for temperature regulation. In response, the hypothalamus activates nerve pathways that ultimately bring about involuntary, oscillating muscle contractions (that is, shivering).

Because those mechanisms of action that are most beneficial to survival have prevailed throughout evolutionary history, it is helpful when studying physiology to predict what mechanistic process would be teleologically useful to the body under a particular circumstance. Chances are such a mechanism will exist. This does not mean that the discipline of physiology is an elaborate guessing game. Predictions must be verified by facts ascertained by careful scientific investigation. However, the fact that most bodily mechanisms do serve a useful purpose (having been naturally selected throughout evolutionary time) allows you to apply a certain amount of logical reasoning to each new situation you encounter in your study of physiology. If you always try to find the thread of logic in what you are studying, you can avoid a good deal of pure memorization, and, more importantly, you will better understand and appreciate the concepts being presented.

Physiology is closely interrelated with **anatomy,** the study of the structure of the body. Just as the functioning of an automobile depends on the shapes, organization, and interactions of its various parts, the structure and function of the human body are inseparable. Therefore, as we tell the story of how the body works, we will provide sufficient anatomical background for you to understand the function of the body part being discussed.

Levels of Organization in the Body

Cells are the basic units of life.

The basic unit of both structure and function is the **cell,** the foundation of all living organisms. The cell is the smallest unit capable of carrying out the processes associated with life. In fact, simple life-forms include **unicellular** (single-celled) **organisms** such as bacteria and amoebae. Humans are **multicellular** (many-celled) **organisms**—the adult human body is an aggregate of trillions of cells.

All cells, whether they exist as solitary cells or as part of a multicellular organism, perform certain basic functions essential for survival of the cell and, in turn, survival of the organism. These basic cell functions include:

1. Obtaining food (nutrients) and oxygen (O_2) from the environment surrounding the cell
2. Performing various chemical reactions that use nutrients and O_2 to provide energy for the cells as follows:

$$\text{Food} + O_2 \rightarrow CO_2 + H_2O + \text{energy}$$

3. Eliminating to the cell's surrounding environment carbon dioxide (CO_2) and other by-products or wastes produced during these chemical reactions
4. Synthesizing proteins and other components needed for cellular structure, for growth, and for carrying out particular cell functions

5. Being sensitive and responsive to changes in the environment surrounding the cell
6. Controlling to a large extent the exchange of materials between the cell and its surrounding environment
7. Moving materials from one part of the cell to another in carrying out cellular activities, with some cells even being able to move in entirety through their surrounding environment
8. In the case of most cells, reproducing. Some body cells, such as nerve cells and muscle cells, have lost the ability to reproduce. When these cells are destroyed through trauma or disease processes, they cannot be replaced. With other body cells, replacement of damaged or old cells is possible.

Cells are remarkable in the similarity with which they carry out these functions. Thus, all cells share many common characteristics. In multicellular organisms, each cell also performs a specialized function, which is usually a modification or elaboration of one of the basic cell functions. The following are a few examples:

- By taking special advantage of their protein-synthesizing ability, the gland cells of the digestive system secrete digestive enzymes, which are all proteins.
- Capitalizing on the basic ability of cells to respond to changes in their surrounding environment, nerve cells generate and transmit to other regions of the body electrical impulses that relay information about changes to which the nerve cells are responsive. For example nerve cells in the ear can relay information to the brain about sound in the external environment.
- The ability of kidney cells to selectively retain the substances needed by the body while eliminating unwanted substances in the urine depends on these cells' highly specialized ability to control exchange of materials between the cell and its environment.
- Muscle contraction, which involves selective movement of internal structures to bring about shortening of muscle cells, is an elaboration of the inherent capability of these cells to produce intracellular ("within the cell") movement.

It is important to recognize that each cell performs these specialized activities in addition to carrying on the unceasing, fundamental activities required of all cells. The fundamental cellular activities are essential for the survival of each individual cell, whereas the specialized contributions and interactions among the cells of a multicellular organism are essential for the survival of the whole organism.

Cells are progressively organized into tissues, organs, systems, and finally the whole body.

Just as a machine does not function unless all its various parts are properly assembled, the cells of the body must be specifically organized to carry out the life-sustaining processes of the body as a whole, such as digestion, respiration, and circulation. The body is made up of four levels of organization: cells, tissues, organs, and systems.

Cells of similar structure and function are organized into **tissues,** of which there are four primary types: *muscle, nervous, epithelial,* and *connective tissue.* Each tissue consists of cells of a single specialized type, along with varying amounts of extracellular ("outside the cell") material.

- **Muscle tissue** is composed of cells specialized for contraction and force generation. There are three types of muscle tissue: *skeletal muscle,* which accomplishes movement of the skeleton; *cardiac muscle,* which is responsible for pumping blood out of the heart; and *smooth muscle,* which encloses and controls movement of contents through hollow tubes and organs, such as movement of food through the digestive tract.
- **Nervous tissue** consists of cells specialized for initiation and transmission of electrical impulses, sometimes over long distances. These electrical impulses act as signals that relay information from one part of the body to another. Nervous tissue is found in (1) the brain; (2) the spinal cord; (3) the nerves that signal information about the external environment and about the status of various internal factors in the body that are subject to regulation, such as blood pressure; and (4) the nerves that influence muscle contraction or gland secretion.
- **Epithelial tissue** is made up of cells specialized in the exchange of materials between the cell and its environment. This tissue is organized into two general types of structures: epithelial sheets and secretory glands. Epithelial cells are joined together very tightly to form sheets of tissue that cover and line various parts of the body. For example, the outer layer of the skin is epithelial tissue, as is the lining of the digestive tract. In general, these epithelial sheets serve as boundaries that separate the body from the external environment and from the contents of cavities that communicate with the external environment, such as the digestive tract lumen. (A **lumen** is the cavity within a hollow organ or tube.) Only selected transfer of materials is permitted between the regions separated by an epithelial barrier. The type and extent of controlled exchange vary, depending on the location and function of the epithelial tissue. For example, very little can be exchanged between the body and external environment across the skin, whereas the epithelial cells lining the digestive tract are specialized for absorption of nutrients.

  Glands are epithelial tissue derivatives that are specialized for secretion. **Secretion** is the release from a cell, in response to appropriate stimulation, of specific products that have in large part been synthesized by the cell. Glands are formed during embryonic development by pockets of epithelial tissue that dip inward from the surface. There are two categories of glands: *exocrine* and *endocrine* (— Fig. 1-1). If during development the connecting cells between the epithelial surface cells and the secretory gland cells within the depths of the invagination remain intact as a duct between the gland and the surface, an exocrine gland is formed. **Exocrine glands** (*exo* means "external"; *crine* means "secretion") secrete through ducts to the outside of the body (or into a cavity that communicates with the out-

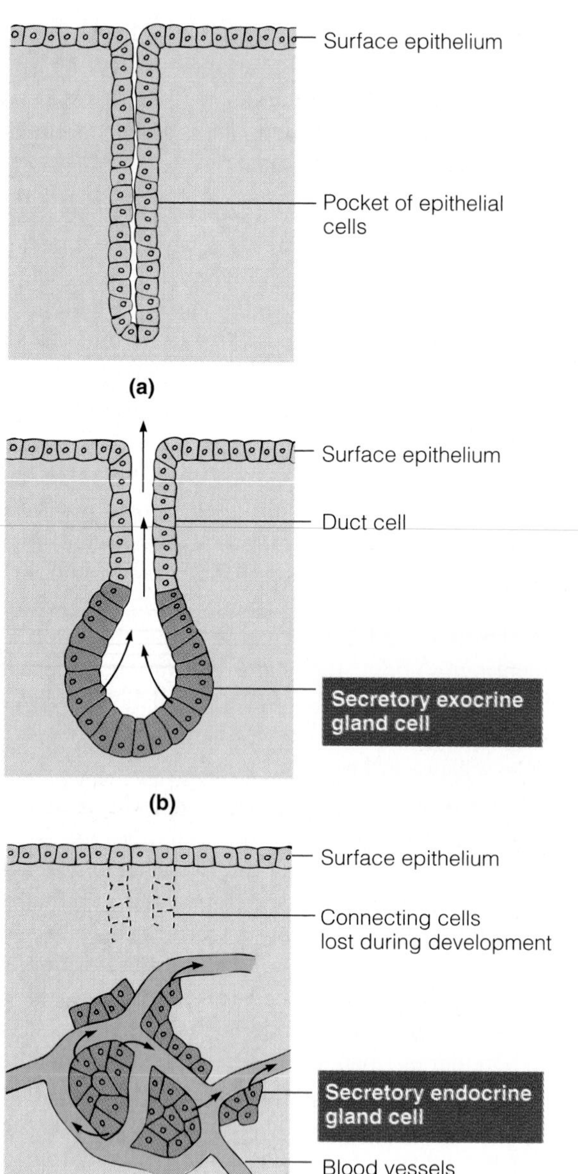

Figure 1-1 Exocrine and Endocrine Gland Formation
(a) Glands arise during development from the formation of pocketlike invaginations of surface epithelial cells. (b) If the cells at the deepest part of the invagination become secretory and release their product through the connecting duct to the surface, an exocrine gland is formed. (c) If the connecting cells are lost and the deepest secretory cells release their product into the blood, an endocrine gland is formed.

side). Examples are sweat glands and glands that secrete digestive juices. If, on the other hand, the connecting cells disappear during development and the secretory gland cells are isolated from the surface, an endocrine gland is formed. **Endocrine glands** (*endo* means "internal") lack ducts and release their secretory products, known as *hormones,* internally into the blood within the body. For example, the parathyroid gland secretes parathyroid hormone into the blood, which transports this hormone to its sites of action at the bones and kidneys.

- **Connective tissue** is distinguished by having relatively few cells dispersed within an abundance of extracellular material. As its name implies, connective tissue connects, supports, and anchors various body parts. It includes such diverse structures as the loose connective tissue that attaches epithelial tissue to underlying structures; tendons, which attach skeletal muscles to bones; bone, which gives the body shape, support, and protection; and blood, which transports materials from one part of the body to another. Except for blood, the cells within connective tissue produce specific molecules that they release into the extracellular spaces between the cells. One such molecule is the rubber-band–like protein fiber **elastin,** whose presence facilitates the stretching and recoiling of structures such as the lungs, which alternately inflate and deflate during breathing.

Muscle, nervous, epithelial, and connective tissue are the primary tissues in a classical sense; that is, each is an integrated collection of cells of the same specialized structure and function. The term *tissue* is also frequently used to refer to the aggregate of various cellular and extracellular components that make up a particular organ (for example, lung tissue or liver tissue).

Organs are composed of two or more types of primary tissue organized to perform a particular function or functions. The stomach is an example of an organ made up of all four primary tissue types. The tissues that compose the stomach function collectively to store the ingested food and move it forward into the remainder of the digestive tract as well as to begin the digestion of protein. The stomach is lined with epithelial tissue that restricts the transfer of harsh digestive chemicals and undigested food from the stomach lumen into the blood. Epithelially derived gland cells in the stomach include exocrine cells, which secrete protein-digesting juices into the lumen, and endocrine cells, which secrete a hormone that helps regulate the stomach's exocrine secretion and muscle contraction. The walls of the stomach contain smooth muscle tissue whose contraction mixes ingested food with the digestive juices and propels the mixture forward into the intestine. Also within the walls is nervous tissue, which, along with hormones, controls muscle contraction and gland secretion. These various tissues are all bound together by connective tissue.

Organs are further organized into **body systems,** each of which is a collection of organs that perform related functions and interact to accomplish a common activity that is essential for survival of the whole body. For example, the digestive system consists of the mouth, pharynx (throat), esophagus, stomach, small intestine, large intestine, salivary glands, pancreas, liver, and gallbladder. These digestive organs cooperate to accomplish the breakdown of dietary food into small nutrient molecules that can be absorbed into the blood.

The **total body**—a single, independently living individual—is composed of the various organ systems structurally and functionally linked together as an entity that is separate from the **external** (outside the body) **environment.** Thus, the body is made up of living cells organized into life-sustaining systems. (See the boxed feature on p. 6, ◆ Concepts, Challenges, and Controversies.)

Concept of Homeostasis

Body cells are in contact with a privately maintained internal environment instead of with the external environment that surrounds the body.

If each cell possesses basic survival skills, why can't the body cells live without performing specialized tasks and being organized according to specialization into systems that accomplish functions essential for the whole body's survival? The cells in a multicellular organism must contribute to the survival of the organism as a whole and cannot live and function without contributions from the other body cells because the vast majority of cells are not in direct contact with the external environment in which the organism lives. A unicellular organism such as an amoeba can directly obtain nutrients and O_2 from its immediate external surroundings and eliminate wastes back into those surroundings. A muscle cell or any other cell in a multicellular organism has the same need for life-supporting nutrient and O_2 uptake and waste elimination, yet the muscle cell cannot directly make these exchanges with the environment surrounding the body because the cell is isolated from this external environment.

How is it possible for a muscle cell to make vital exchanges with the external environment with which it has no contact? The key is the presence of an aqueous **internal environment** with which the body cells are in direct contact. This internal environment is *outside* the cells but *inside* the body. It consists of the **extracellular** (*extra* means "outside of") **fluid,** which is made up of **plasma,** the fluid portion of the blood, and **interstitial fluid,** which surrounds and bathes the cells (▬ Fig. 1-2). Various body systems accomplish exchanges between the external environment and the internal environment. For example, the digestive system transfers the nutrients required by all body cells from the external environment into the plasma. Likewise, the respiratory system transfers O_2 from the external environment into the plasma. The circulatory system distributes these nutrients and O_2 throughout the body. Materials are thoroughly mixed and exchanged between the plasma and the interstitial fluid across the thin, pore-lined walls of the capillaries, the smallest of the blood vessels. As a result, the nutrients and O_2 originally obtained from the external environment are delivered to the interstitial fluid that surrounds the cells. The body cells, in turn, make life-sustaining exchanges with the internal environment. No matter how remote a cell is from the body surface, it is able to take in from the internal environment the nutrients and O_2 needed to support its own existence. Similarly, wastes produced by the cells are extruded into the interstitial fluid, picked up by the plasma, and transported to the organs that specialize in eliminating these wastes from the internal environment to the external environment. The lungs remove CO_2 from the plasma, and the kidneys remove other wastes for elimination in the urine.

▬ *Figure 1-2* **Components of the Extracellular Fluid (Internal Environment)**

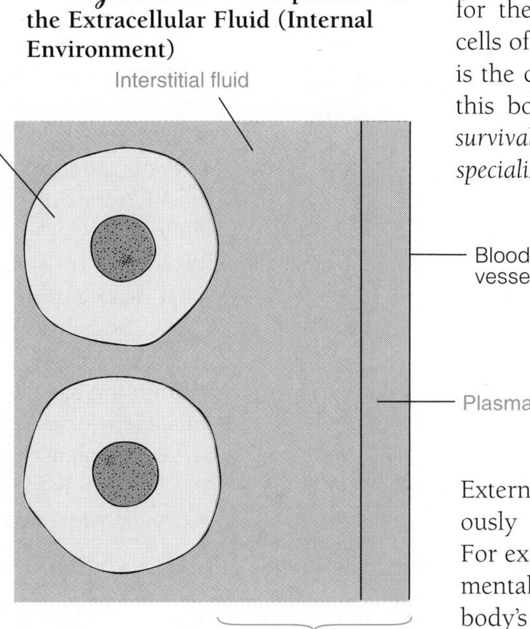

Thus, a body cell takes in essential nutrients from and eliminates wastes into its watery surroundings, just as an amoeba does. The major difference is that each body cell must help maintain the composition of the internal environment so that this fluid continuously remains suitable to support the existence of all the body cells. In contrast, an amoeba does nothing to regulate its surroundings.

Homeostasis is essential for cell survival, and each cell, as part of an organized system, contributes to homeostasis.

The body cells can live and function only when they are bathed by extracellular fluid that is compatible with their survival; thus, the chemical composition and physical state of the internal environment can be allowed to deviate only within narrow limits. As cells remove nutrients and O_2 from the internal environment, these essential materials must constantly be replenished in order for the cells' ongoing maintenance of life processes to continue. Likewise, wastes must constantly be removed from the internal environment so that they do not reach toxic levels. Other elements in the internal environment that are important for the maintenance of life also must be kept relatively constant. Maintenance of a relatively stable internal environment is termed **homeostasis** (*homeo* means "the same"; *stasis* means "to stand or stay").

The functions performed by each body system contribute to homeostasis, thereby maintaining within the body the environment required for the survival and function of all the cells of which the body is composed. This is the central theme of physiology and of this book: *homeostasis is essential for the survival of each cell, and each cell, through its specialized activities, contributes as part of a body system to the maintenance of the internal environment shared by all cells* (▬ Fig. 1-3).

The fact that the internal environment must be kept relatively stable does not mean that its composition, temperature, and other characteristics are absolutely unchanging. External and internal factors continuously threaten to disrupt homeostasis. For example, exposure to a cold environmental temperature tends to reduce the body's internal temperature. Likewise, addition of CO_2 into the internal environ-

Xenotransplantation and Tissue Engineering: A Quest for Replacement Parts

Kidney failure, extensive burns, surgical removal of a cancerous breast, an arm severed and mangled in an accident—though our bodies are remarkable and normally serve us well, sometimes a body part is defective, injured beyond repair, or lost in situations such as these. Ideally, when the body suffers an irreparable loss, new, permanent replacement parts could be substituted to restore normal function and appearance. As the turn of the century approaches, this possibility is moving rapidly from the realm of science fiction to the reality of scientific progress.

One solution for replacing damaged or lost parts is implantation of synthetic devices, such as artificial joints or artificial heart valves. Even though remarkable advances have been made in this field, no synthetic device manufactured thus far can perform the functions of a body part as effectively as the normal, healthy original part. Indeed, some synthetic devices, such as silicone breast implants, have raised medical and public concerns because of their implicated role in triggering health problems.

Another solution to organ failure is transplantation of organs such as kidneys, hearts, and livers from donors (recently deceased in the case of most donated organs) to patients whose survival depends on a healthy replacement. Over the last two decades, organ transplantation has become almost commonplace. The major obstacle limiting the number of transplantations is not a lack of technical capability but a shortage of donor organs. Waiting lists for replacement organs outstrip the number of donor organs available by as much as tenfold.

Because of the limitations of implanting synthetic devices and transplanting donor organs, some researchers are exploring two alternative methods for replacing body parts that are no longer serviceable: xenotransplantation and tissue engineering. Interest in *xenotransplantation,* transplantation of organs from one species to another (*xenos* means "strange" or "foreign"), is growing, not only because of the shortage of human donors but because our technical capabilities have improved to the point that this technique is now becoming feasible. One of the major hurdles to xenotransplantation has been the swift rejection of the animal organ by the human recipient's body. The patient's immune system, recognizing the donated animal organ as "foreign," immediately launches a destructive attack on the transplanted organ.

Pioneering efforts on several fronts are breaking down this barrier to xenotransplantation. First, researchers now have a more thorough understanding of the underlying immune mechanisms that cause transplanted tissue to be rejected. This knowledge has led to better *immunosuppressive drugs* (drugs that suppress or interfere with the immune system's attack on the transplanted organ). Even recipients of human donor organs must be treated with immunosuppressive drugs because of the differences in "self-identity markers" found on the surfaces of the donor's and the recipient's cells. Because the differences in these cell surface markers are much greater between humans and animals, more powerful drugs must be used to prevent the rejection of xenotransplants.

The second advance toward xenotransplantation is the development of *transgenic animals* through genetic engineering. A transgenic animal possesses not only its own genes but also

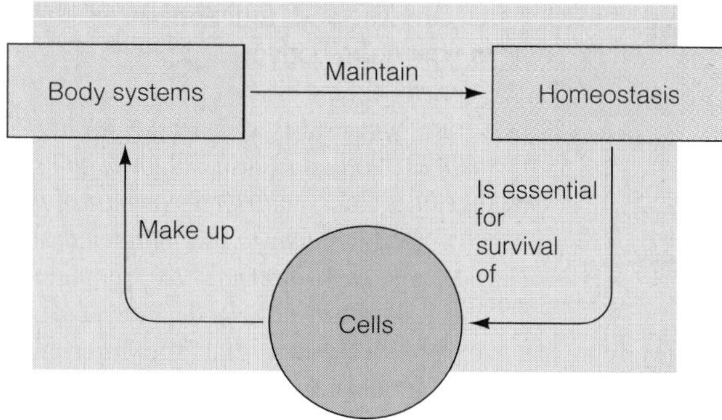

Figure 1-3 Interdependent Relationship of Cells, Body Systems, and Homeostasis The depicted interdependent relationship serves as the foundation for modern-day physiology: *homeostasis is esssential for the survival of cells, body systems maintain homeostasis, and cells make up body systems.*

ment as a result of energy-generating chemical reactions tends to raise the concentration of this gas within the body. When any factor starts to move the internal environment away from optimal conditions, appropriate counterreactions are initiated to restore these conditions. When the body temperature starts to fall on a cold day, compensatory shivering is initiated. This shivering internally generates heat that restores the body temperature to normal. Similarly, a rise in the CO_2 levels within the internal environment triggers an increase in breathing. As a result, the extra CO_2 is blown off to the external environment, restoring the CO_2 concentration in the extracellular fluid to normal. Thus, homeostasis should be viewed not as a fixed state but as a dynamic steady state in which the changes that do occur are minimized by compensatory physiological responses. For each factor in the internal environment, the small fluctuations around the optimal level are normally kept within the narrow limits compatible with life by carefully regulated mechanisms. (See the boxed feature on p. 10, A Closer Look at Exercise Physiology.)

The factors of the internal environment that must be homeostatically maintained include the following (Fig. 1-4):

1. *Concentration of nutrient molecules.* Cells need a constant supply of nutrient molecules to serve as a metabolic fuel for energy production. Energy, in turn, is needed to support life-sustaining and specialized cellular activities.

some specific genes of another species that have been artificially introduced. Scientists have bred transgenic pigs that carry human genes for use as organ donors for xenotransplantation. Pigs were chosen because they are easy to breed and raise and because their organs are anatomically and physiologically similar to human organs. The goal is to trick the human xenotransplant recipient's immune system into viewing the pig organ as human, thus reducing the extent of the immunological attack.

Even as the technical barriers to xenotransplantation are crumbling, however, several obstacles to continued progress in this area are emerging. Scientists are vigorously debating the likelihood that xenotransplantation will introduce new infectious agents from an animal source into the human population. Opponents of xenotransplants fear that an undetected disease-causing microorganism lurking in a donor animal could cause infection in a human recipient and spread from there into the general population. Furthermore, animal rights advocates are launching strong opposition to what they view as exploitation of animals as sources of spare parts for humans.

Besides xenotransplantation, the second frontier in replacement parts is *tissue engineering*—growing in the laboratory new tissues and organs that can be implanted in humans to serve as permanent replacements for body parts that cannot be repaired. The era of tissue engineering is being ushered in by advances in cell biology, plastic manufacturing, and computer graphics.

Most human cells can already be *cultured;* that is, when removed from the body, they will continue to thrive and reproduce in laboratory dishes when supplied with appropriate nutrients and other supportive materials. Now researchers are working toward growing specific tissues and even whole organs in the laboratory for use as replacement parts. Using computer-aided designs, scientists plan to shape highly pure, dissolvable plastics into molds or scaffoldings that mimic the structure of a particular tissue or organ. The plastic mold will then be "seeded" with the desired cell types, which will be coaxed into multiplying and assembling into the desired tissue. After the biodegradable plastic scaffolding dissolves, only the newly generated tissue will remain, ready to be implanted into a patient as a permanent, living replacement part. To prevent rejection by the immune system, without the necessity of lifelong immunosuppressive drugs, the plastic mold could be seeded, if possible, with appropriate cells from the recipient, or perhaps through genetic engineering "universal" seed cells could be produced that would be immunologically acceptable to any recipient.

Following are some of the tissue engineers' early accomplishments and future predictions:

- Engineered skin patches have already been used to treat victims of severe burns and are anticipated to be available worldwide during the next decade.
- Laboratory-grown cartilage and bone have already been successfully implanted in experimental animals. Examples include artificial heart valves, ears, and noses.
- Completely natural replacement breasts will be grown for implantation following surgical removal of a cancerous breast.
- Engineered joints will be used as a living, more satisfactory alternative to the plastic and metal devices used as replacements today.
- Whole organs like kidneys, livers, and hearts will be grown and transferred to patients as these organs wear out.
- Ultimately, complex body parts such as arms and hands will be produced in the laboratory for attachment as needed.

Thus, tissue engineering holds the promise that damaged or lost body parts can be replaced with the best alternative, a laboratory-grown version of "the real thing."

2. *Concentration of O_2 and CO_2.* Cells need O_2 to perform chemical reactions that extract from nutrient molecules the most energy possible for use by the cell. The CO_2 produced during these chemical reactions must be balanced by CO_2 removal from the lungs so that acid-forming CO_2 does not increase the acidity of the internal environment.
3. *Concentration of waste products.* Various chemical reactions produce end products that exert a toxic effect on the body's cells if these wastes are allowed to accumulate beyond a certain limit.
4. *pH.* Among the most pronounced effects of changes in the pH (acidity) of the internal fluid environment are alterations in the electrical signaling mechanism of nerve cells and in the enzyme activity of all cells.
5. *Concentration of water, salt, and other electrolytes.* Because the relative concentrations of salt (NaCl) and water in the extracellular fluid (internal environment) influence how much water enters or leaves the cells, these concentrations are carefully regulated to maintain the proper volume of the cells. Cells do not function normally when they are swollen or shrunken. Other electrolytes perform a variety of vital functions. For example, the rhythmic beating of the heart depends on a relatively constant concentration of potassium (K^+) in the extracellular fluid.
6. *Temperature.* Body cells function optimally within a narrow temperature range. Cells slow down too much if they are too cold, and, worse yet, their structural and enzymatic proteins are impaired if they get too hot.
7. *Volume and pressure.* The circulating component of the internal environment, the plasma, must be maintained at adequate volume and blood pressure to ensure bodywide distribution of this important link between the external environment and the cells.

There are eleven major body systems (▮ Table 1-1 on p. 10 and Fig. 1-4); their most important contributions to homeostasis are as follows:

1. The *circulatory system* is the transport system that carries materials such as nutrients, O_2, CO_2, wastes, electrolytes, and hormones from one part of the body to another.
2. The *digestive system* breaks down dietary food into small nutrient molecules that can be absorbed into the plasma

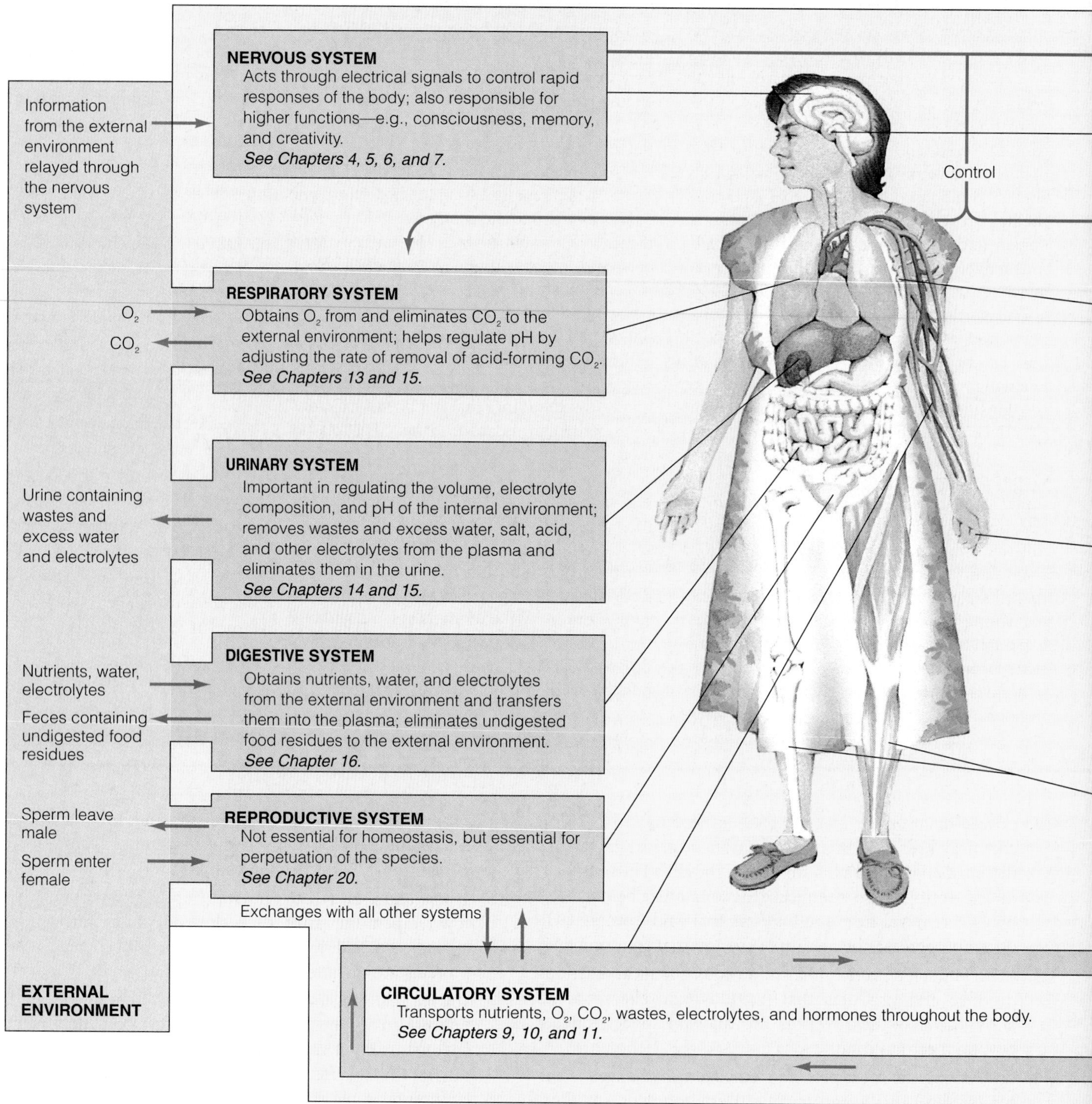

BODY SYSTEMS
Made up of cells organized according to specialization to maintain homeostasis.
See Chapter 1.
NERVOUS SYSTEM
Acts through electrical signals to control rapid responses of the body; also responsible for higher functions—e.g., consciousness, memory, and creativity.
See Chapters 4, 5, 6, and 7.
Information from the external environment relayed through the nervous system
Control
RESPIRATORY SYSTEM
Obtains O_2 from and eliminates CO_2 to the external environment; helps regulate pH by adjusting the rate of removal of acid-forming CO_2.
See Chapters 13 and 15.
O_2
CO_2
URINARY SYSTEM
Important in regulating the volume, electrolyte composition, and pH of the internal environment; removes wastes and excess water, salt, acid, and other electrolytes from the plasma and eliminates them in the urine.
See Chapters 14 and 15.
Urine containing wastes and excess water and electrolytes
DIGESTIVE SYSTEM
Obtains nutrients, water, and electrolytes from the external environment and transfers them into the plasma; eliminates undigested food residues to the external environment.
See Chapter 16.
Nutrients, water, electrolytes
Feces containing undigested food residues
REPRODUCTIVE SYSTEM
Not essential for homeostasis, but essential for perpetuation of the species.
See Chapter 20.
Sperm leave male
Sperm enter female
Exchanges with all other systems
EXTERNAL ENVIRONMENT
CIRCULATORY SYSTEM
Transports nutrients, O_2, CO_2, wastes, electrolytes, and hormones throughout the body.
See Chapters 9, 10, and 11.

Figure 1-4 **Role of the Body Systems in Maintaining Homeostasis**

for distribution to the body cells. It also transfers water and electrolytes from the external environment into the internal environment. It eliminates undigested food residues to the external environment in the feces.

3. The *respiratory system* obtains O_2 from and eliminates CO_2 to the external environment. By adjusting the rate of removal of acid-forming CO_2, the respiratory system is also important in maintaining the proper pH of the internal environment.
4. The *urinary system* removes excess water, salt, acid, and other electrolytes from the plasma and eliminates them in the urine, along with waste products other than CO_2.

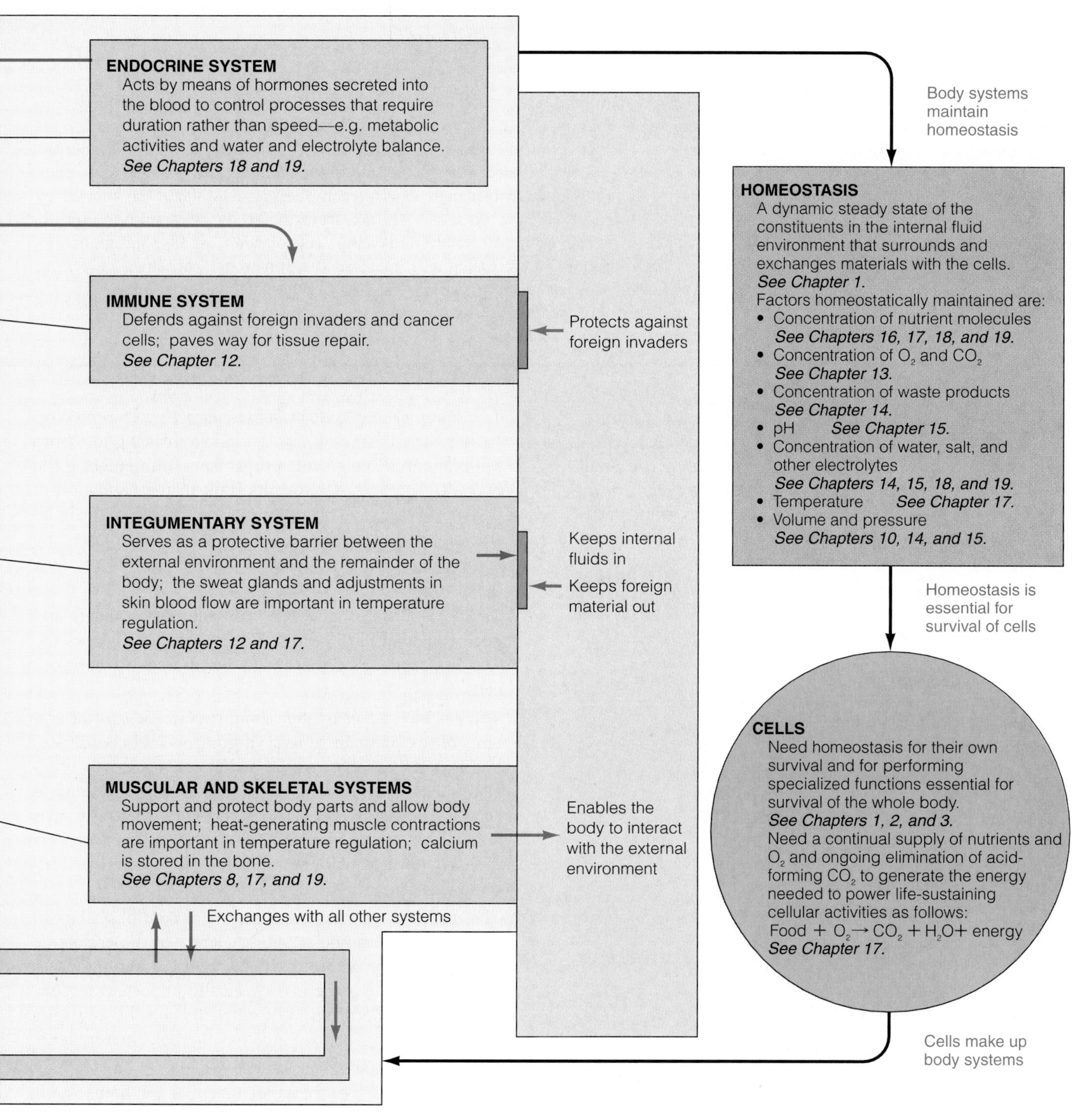
ENDOCRINE SYSTEM
Acts by means of hormones secreted into the blood to control processes that require duration rather than speed—e.g. metabolic activities and water and electrolyte balance.
See Chapters 18 and 19.
Body systems maintain homeostasis
HOMEOSTASIS
A dynamic steady state of the constituents in the internal fluid environment that surrounds and exchanges materials with the cells.
See Chapter 1.
Factors homeostatically maintained are:
• Concentration of nutrient molecules See Chapters 16, 17, 18, and 19.
• Concentration of O2 and CO2 See Chapter 13.
• Concentration of waste products See Chapter 14.
• pH See Chapter 15.
• Concentration of water, salt, and other electrolytes See Chapters 14, 15, 18, and 19.
• Temperature See Chapter 17.
• Volume and pressure See Chapters 10, 14, and 15.
IMMUNE SYSTEM
Defends against foreign invaders and cancer cells; paves way for tissue repair.
See Chapter 12.
Protects against foreign invaders
INTEGUMENTARY SYSTEM
Serves as a protective barrier between the external environment and the remainder of the body; the sweat glands and adjustments in skin blood flow are important in temperature regulation.
See Chapters 12 and 17.
Keeps internal fluids in
Keeps foreign material out
Homeostasis is essential for survival of cells
CELLS
Need homeostasis for their own survival and for performing specialized functions essential for survival of the whole body.
See Chapters 1, 2, and 3.
Need a continual supply of nutrients and O2 and ongoing elimination of acid-forming CO2 to generate the energy needed to power life-sustaining cellular activities as follows:
Food + O2 → CO2 + H2O + energy
See Chapter 17.
MUSCULAR AND SKELETAL SYSTEMS
Support and protect body parts and allow body movement; heat-generating muscle contractions are important in temperature regulation; calcium is stored in the bone.
See Chapters 8, 17, and 19.
Enables the body to interact with the external environment
Exchanges with all other systems
Cells make up body systems

5. The *skeletal system* provides support and protection for the soft tissues and organs. It also serves as a storage reservoir for calcium (Ca^{2+}), an electrolyte whose plasma concentration must be maintained within very narrow limits. Together with the muscular system, the skeletal system also enables movement of the body and its parts.

6. The *muscular system* moves the bones to which the skeletal muscles are attached. From a purely homeostatic view, this system enables an individual to move toward food or away from harm. Furthermore, the heat generated by muscle contraction is important in temperature regulation. In addition, because skeletal muscles are under vol-

A CLOSER LOOK AT
EXERCISE PHYSIOLOGY

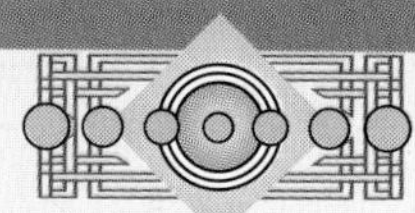

What Is Exercise Physiology?

Exercise physiology is the study of both the functional changes that occur in response to a single session of exercise and the adaptations that occur as a result of regular, repeated exercise sessions. Exercise initially disrupts homeostasis. The changes that occur in response to exercise are the body's attempt to meet the challenge of maintaining homeostasis when increased demands are placed on the body.

Heart rate is one of the easiest factors to monitor that shows both an immediate response to exercise and long-term adaptation to a regular exercise program. When a person begins to exercise, the active muscle cells use more O_2 to support their increased energy demands. Heart rate increases to deliver more oxygenated blood to the exercising muscles. The heart adapts to regular exercise of sufficient intensity and duration by increasing its strength and efficiency so that it pumps more blood per beat. Because of increased pumping ability, the heart does not have to beat as rapidly to pump a given quantity of blood as it did before physical training.

Exercise physiologists study the mechanisms responsible for the changes that occur as a result of exercise. Much of the knowledge gained from the study of exercise is used to develop appropriate exercise programs to increase the functional capacities of people ranging from athletes to the infirm.

untary control, a person is able to use them to accomplish myriad other movements of his or her own choice. These movements, which range from the fine motor skills required for delicate needlework to the powerful movements involved in weight lifting, are not necessarily directed toward maintaining homeostasis.

Table 1-1
Components of Body Systems

System	*Components*
Circulatory system	Heart, blood vessels, blood
Digestive system	Mouth, pharnyx, esophagus, stomach, small intestine, large intestine, salivary glands, exocrine pancreas, liver, gallbladder
Respiratory system	Nose, pharnyx, larynx, trachea, bronchi, lungs
Urinary system	Kidneys, ureters, urinary bladder, urethra
Skeletal system	Bones, cartilage, joints
Muscular system	Skeletal muscles
Integumentary system	Skin, hair, nails
Immune system	White blood cells, thymus, bone marrow, tonsils, adenoids, lymph nodes, spleen, appendix, gut-associated lymphoid tissue, skin-associated lymphoid tissue
Nervous system	Brain, spinal cord, peripheral nerves, special sense organs
Endocrine system	All hormone-secreting tissues, including hypothalamus, pituitary, thyroid, adrenals, endocrine pancreas, parathyroids, gonads, kidneys, intestine, heart, thymus, pineal, and skin
Reproductive system	Male: testes, penis, prostate gland, seminal vesicles, bulbourethral glands, and associated ducts Female: ovaries, oviducts, uterus, vagina, breasts

7. The *integumentary system* serves as an outer protective barrier that prevents internal fluid from being lost from the body and foreign microorganisms from entering the body. This system is also important in the regulation of body temperature. The amount of heat lost from the body surface to the external environment can be adjusted by controlling sweat production and by regulating the flow of warm blood through the skin.
8. The *immune system* defends against foreign invaders and body cells that have become cancerous. It also paves the way for repair or replacement of injured or worn-out cells.
9. The *nervous system* is one of the two major control systems of the body. In general, it controls and coordinates bodily activities that require swift responses. It is especially important in detecting and initiating reactions to changes in the external environment. Furthermore, it is responsible for higher functions that are not entirely directed toward maintaining homeostasis, such as consciousness, memory, and creativity.
10. The *endocrine system* is the other major control system. In general, the hormone-secreting glands of the endocrine system regulate activities that require duration rather than speed. This system is especially important in controlling the concentration of nutrients and, by adjusting kidney function, controlling the internal environment's volume and electrolyte composition.
11. The *reproductive system* is not essential for homeostasis and therefore is not essential for survival of the individual. It is essential, however, for perpetuation of the species.

As we examine each of these systems in greater detail, always keep in mind that the body is a coordinated whole even though each system provides its own special contributions. It is easy to forget that all of the body parts actually fit together into a functioning, interdependent whole body. Accordingly, each chapter begins with a figure and discussion that will help you focus on how the body system to be described fits into the body as a whole. In addition, each chapter concludes with a brief overview of the homeostatic contributions of the body system. As a further tool to help you keep track of how all the pieces fit together, Figure 1-4 is duplicated on the inside cover as a handy reference.

Keep another point in mind as you read through the book from cells to systems: The functioning whole is greater than the sum of its separate parts. Through specialization, cooperation, and interdependence, cells combine to form a coordinated, unique, single living organism with more diverse and complex capabilities than is possessed by any of the cells that make it up. For humans, these capabilities go far beyond the processes needed to maintain life. Obviously a cell, or even a random combination of cells, cannot create an artistic masterpiece or design a spacecraft, but body cells working together permit those capabilities in an individual.

Negative feedback is a common regulatory mechanism for maintaining homeostasis.

To maintain homeostasis, the body must be able to detect deviations in the internal environmental factors that need to be held within narrow limits, and it must be able to control the various body systems responsible for adjusting these factors. For example, to maintain the concentration of CO_2 in the extracellular fluid at an optimal value, the body must be able to detect a change in CO_2 concentration and then appropriately alter respiratory activity so that CO_2 concentration is returned to the desirable level.

Control systems that operate to maintain homeostasis can be grouped into two classes—intrinsic and extrinsic controls. **Intrinsic (local) controls** (*intrinsic* means "within") are built in or inherent to an organ. For example, as an exercising muscle rapidly uses up O_2 and produces CO_2 to generate energy to support its contractile activity, the O_2 concentration falls and the CO_2 concentration increases within the muscle. By acting directly on the smooth muscle in the walls of the blood vessels that supply the exercising muscle, these local chemical changes cause the smooth muscle to relax and the vessels to open widely to accommodate increased blood flow into the exercising muscle. This local mechanism contributes to the maintenance of an optimal level of O_2 and CO_2 in the internal fluid environment surrounding the exercising muscle's cells.

Most factors in the internal environment are maintained, however, by **extrinsic controls** (*extrinsic* means "outside of"), which are regulatory mechanisms initiated outside an organ to alter the activity of the organ. Extrinsic control of the various organs and systems is accomplished by the nervous and endocrine systems, the two major control systems of the body. Extrinsic control permits coordinated regulation of several organs toward a common goal; in contrast, intrinsic controls are self-serving for the organ in which they occur. Coordinated, overall regulatory mechanisms are critical for maintaining the dynamic steady state in the internal environment as a whole. For example, to restore blood pressure to the proper level when it falls too low, the nervous system simultaneously acts on the heart and the blood vessels throughout the body to increase the blood pressure to normal.

The body's homeostatic control mechanisms primarily operate on the principle of negative feedback. **Negative feedback** occurs when a change in a controlled variable triggers a response that opposes the change, driving the variable in the opposite direction of the initial change.

A common example of negative feedback is control of room temperature (the **controlled variable**) by a **control system** that includes a furnace, a thermostatic device, and all their electrical connections. The room temperature is determined by the activity of the furnace, a heat source that can be turned on or off. To switch on or off appropriately, the control system as a whole, must "know" what the *actual* room temperature is, "compare" it with the *desired* room temperature, and "adjust" the output of the heat source to bring the actual temperature to the desired level. Information about the actual room temperature is provided by a thermometer in the thermostat; the thermometer is the **sensor** that monitors the magnitude of the controlled variable. The desired temperature level, or **set point**, is provided by the thermostat setting. The thermostat acts as an **integrator:** it compares the sensor's input with the set point and adjusts the heat output of the **effector**, the furnace, to bring about the appropriate effect or response to oppose the deviation from the set point.

For example, if the room temperature falls below the set-point level during cold weather (■ Fig. 1-5a), the thermostat, through connecting circuitry, activates the furnace, which produces heat to increase the room temperature. Once the room temperature reaches the set point, the activating mechanism in the thermostat and consequently the furnace are switched off. Thus, the heat produced by the furnace counteracts or is "negative" to the original fall in the temperature. If heat production were to continue unabated, the room temperature would be increased above the set point. Overshooting beyond the set point does not occur because the heat "feeds back" to shut off the thermostat that triggered its output, thereby limiting its own production by controlling the signal that initiated the heat production. Thus, the control system takes corrective actions to prevent the controlled variable from drifting too far below or too far above the set point.

What if the original deviation is a rise in room temperature above the set point because it is hot outside? A heat-producing furnace is of no use in returning the room temperature to the desired level. In this case, the thermostat, through connecting circuitry, can activate the air conditioner, which cools the room air, the opposite effect of the furnace. In negative-feedback fashion, once the set point is reached, the air conditioner is turned off to prevent the room from becoming too cold. Note that if the controlled variable can be deliberately adjusted to oppose a change in one direction only, the variable can move in uncontrolled fashion in the opposite direction. For example, if the house is equipped only with a furnace that produces heat to oppose a fall in room temperature, no mechanism is available to prevent the house from getting too hot in warm weather. However, the room temperature can be kept relatively constant through two opposing mechanisms, one

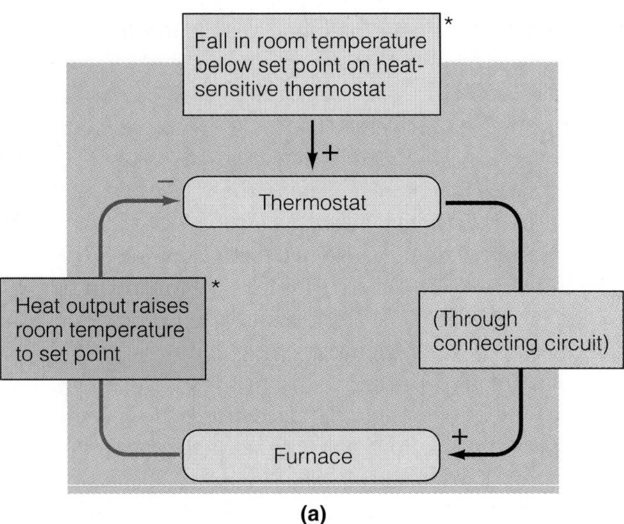

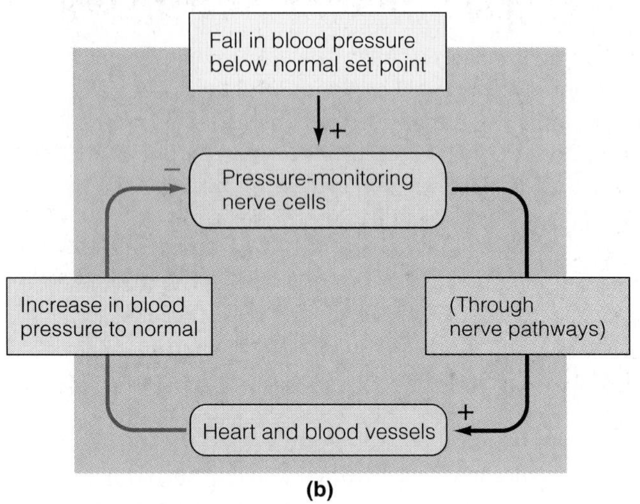

For flow diagrams throughout the text:
+ = Stimulates or activates
– = Inhibits or turns off
▭ = Physical entity, such as a body structure or a chemical
□ = Actions
* Note that lighter and darker shades of the same color are used to denote respectively a decrease or an increase in a controlled variable.

Figure 1-5 **Negative Feedback** (a) Negative-feedback control of a home heating system. (b) Negative-feedback control of blood pressure.

that heats and one that cools the room, despite wide variations in the temperature of the external environment.

Homeostatic negative-feedback systems operate in the same way to maintain a controlled factor in a relatively steady state. For example, when the pressure-monitoring nerve cells detect a *decrease* in blood pressure below the desired level, they bring about a sequence of events that culminates in nerve-controlled changes in the circulatory system to *increase* blood pressure to the proper level (Fig. 1-5b). When the blood pressure increases to the set point, the stimulatory signal to the heart and blood vessels arising from the pressure-monitoring nerve cells is turned off. As a result, the blood pressure does not continue to increase above the set point. The opposite events occur when the original change is an elevation in blood pressure above normal. The pressure-monitoring nerve cells bring about a reduction in blood pressure to normal by triggering compensatory responses in the circulatory system. The blood pressure does not fall too low, because the pressure-monitoring nerve cells cease triggering the pressure-reducing responses when the blood pressure reaches the right level.

Positive feedback occurs less frequently in the body. In negative feedback, a control system's output is regulated to resist change, so that the regulated variable is maintained at a relatively steady set point. With positive feedback, however, the output is continually enhanced so that the controlled variable continues to be moved in the direction of the initial change. Instead of bringing about a response that counteracts the initial change, positive feedback reinforces the change in the same direction. Such action would be comparable to the heat generated by a furnace triggering the thermostat to call for even more heat output from the furnace so that the room temperature would continuously rise.

Because positive feedback moves the controlled variable even farther from a steady state, it does not occur very often in the body, where the major goal is maintenance of stable, homeostatic conditions. Positive feedback does occur in certain instances, however, such as during the birth of a baby. The hormone oxytocin causes powerful contractions of the uterus. As uterine contractions push the baby against the cervix (the exit from the uterus), the resultant stretching of the cervix triggers a sequence of events that brings about the release of even more oxytocin, which causes even stronger uterine contractions, triggering the release of more oxytocin, and so on. The positive-feedback cycle does not cease until the baby is finally expelled.

In addition to feedback mechanisms, which bring about a response in *reaction* to a change in a regulated variable, the body less frequently employs **feedforward** mechanisms, which bring about a response in *anticipation of* a change in a regulated variable. For example, when a meal is still in the digestive tract, a feedforward mechanism increases the secretion of a hormone that will promote the cellular uptake and storage of ingested nutrients after they have been absorbed from the digestive tract. This anticipatory response helps limit the rise in blood nutrient concentration that occurs following nutrient absorption.

Disruptions in homeostasis can lead to illness and death

When one or more of the body's systems fail to function properly, homeostasis is disrupted, and all of the cells suffer because they no longer have an optimal environment in which to live and function. Various pathophysiological states ensue, depending on the type and extent of homeostatic disruption. **Pathophysiology** refers to the abnormal functioning of the body (altered physiology) associated with disease. When a homeostatic disruption becomes so severe that it is no longer compatible with survival, death results.

Chapter in Perspective: Focus on Homeostasis

In this chapter, you have learned what homeostasis is: a dynamic steady state of the constituents in the internal fluid environment (the extracellular fluid) that surrounds and exchanges materials with the cells. Maintenance of homeostasis is essential for the survival and normal functioning of cells, and each cell, through its specialized activities, contributes as part of a body system to the maintenance of homeostasis. We have already described how cells are organized according to specialization into body systems. How homeostasis is essential for cell survival and how body systems maintain this internal constancy are the topics covered in the remainder of this book.

Chapter Summary

Levels of Organization in the Body

The human body is composed of an interactive society of cells, which are the basic units of both structure and function. Each cell performs basic functions essential for its own survival, such as obtaining O_2 and nutrients, which the cell uses to acquire energy; eliminating wastes; synthesizing needed cellular components; reacting to changes in the surrounding environment; controlling movement of materials within the cell and between the cell and its environment; and reproducing.

In multicellular organisms, each cell performs, in addition to these fundamental cell functions, a specialized activity that is usually an elaboration of one of the basic cell functions. The body's cells are highly organized into functional groupings, with cells of similar structure and specialized activity organized into tissues. There are four primary types of tissue: (1) muscle tissue, which is specialized for contraction and force generation; (2) nerous tissue, which is specialized for initiation and transmission of electrical impulses; (3) epithelial tissue, which lines and covers various body surfaces and cavities and also forms secretory glands; and (4) connective tissue, which connects, supports, and anchors various body parts. Tissues are further organized into organs, which are structures composed of several types of primary tissue that act together to perform one or more functions. Organs make up body systems, which are collections of organs that perform related functions and interact to accomplish a common activity essential for survival of the whole body. Organ systems, in turn, compose the whole body.

Concept of Homeostasis

Homeostasis refers to the maintenance of a dynamic steady state within the internal fluid environment that bathes all of the body's cells. Because the body's cells are not in direct contact with the external environment, cell survival depends on maintenance of a stable internal fluid environment with which the cells directly make exchanges. For example, O_2 and nutrients must constantly be replenished in the internal environment to keep pace with the rate at which the cells use these materials for energy production. The factors of the internal environment that must be homeostatically maintained are its (1) concentration of nutrient molecules, (2) concentration of O_2 and CO_2, (3) concentration of waste products, (4) pH, (5) concentration of water, salt, and other electrolytes, (6) temperature, and (7) volume and pressure.

The functions performed by each of the eleven body systems are directed toward maintaining homeostasis. Each body system's functions ultimately depend on the specialized activities of the cells composing the system. Thus, homeostasis is essential for each cell's survival, and each cell contributes to homeostasis.

Control systems that regulate the body systems' various activities to maintain homeostasis can be classified as (1) intrinsic controls, which are inherent compensatory responses of an organ to a change, and (2) extrinsic controls, which are responses of an organ that are triggered by factors external to the organ, namely, by the nervous and endocrine systems. Both intrinsic and extrinsic control systems generally operate on the principle of negative feedback: a change in a regulated variable triggers a response that drives the variable in the opposite direction of the initial change, thus opposing the change.

Pathophysiological states ensue when one or more of the body systems fail to function properly so that an optimal internal environment can no longer be maintained. Serious homeostatic disruption leads to death.

Review Exercises

Objective Questions (Answers on p. E–1.)

1. Which of the following activities is *not* carried out by every cell in the body?
   a. obtaining O_2 and nutrients
   b. performing chemical reactions to acquire energy for the cell's use
   c. eliminating wastes
   d. controlling to a large extent exchange of materials between the cell and its external environment
   e. reproducing
2. Which of the following is the proper progression of the levels of organization in the body?
   a. cells, organs, tissues, body systems, whole body
   b. cells, tissues, organs, body systems, whole body
   c. cells, tissues, organs, whole body, body systems
   d. cells, organs, tissues, whole body, body systems
   e. cells, tissues, body systems, organs, whole body
3. Which of the following is *not* a type of connective tissue?
   a. bone
   b. blood
   c. the spinal cord
   d. tendons
   e. the tissue that attaches epithelial tissue to underlying structures
4. The term *tissue* can apply either to one of the four primary tissue types or to a particular organ's aggregate of cellular and extracellular components. (True or false?)
5. Cells in a multicellular organism have specialized to such an extent that they have little in common with single-celled organisms. (True or false?)
6. Cellular specializations are usually a modification or elaboration of one of the basic cell functions (True or false?)
7. The four primary types of tissue are ____________, ____________, ____________, and ____________.
8. ____________ refers to the release from a cell, in response to appropriate stimulation, of specific products that have in large part been synthesized by the cell.
9. ____________ glands secrete through ducts to the outside of the body, whereas ____________ release their secretory products, known as ____________, internally into the blood.
10. ____________ controls are inherent to an organ, whereas ____________ controls are regulatory mechanisms initiated outside of an organ that alter the activity of the organ.
11. Match the following:

___ 1. circulatory system	(a) obtains O_2 and eliminates CO_2
___ 2. digestive system	(b) support and protect body parts and allow movement
___ 3. respiratory system	(c) controls, via hormones it secretes, processes that require duration
___ 4. urinary system	(d) transport system
___ 5. muscular and skeletal systems	(e) removes wastes and excess water, salt, and other electrolytes
___ 6. integumentary system	(f) essential for perpetuation of species
___ 7. immune system	(g) obtains nutrients, water, and electrolytes
___ 8. nervous system	(h) defends against foreign invaders and cancer
___ 9. endocrine system	(i) acts through electrical signals to control body's rapid responses
___ 10. reproductive system	(j) serves as protective barrier between body and external environment

Essay Questions

1. Define physiology.
2. What are the basic cell functions?
3. Distinguish between the external environment and the internal environment.
4. Of what fluid compartments is the internal environment composed?
5. Define homeostasis.
6. Describe the interrelationship among cells, body systems, and homeostasis.
7. What factors of the internal environment must be homeostatically maintained?
8. Compare negative and positive feedback.

Points to Ponder

(Explanations on p. E–1.)

1. Considering the nature of negative-feedback control and the function of the respiratory system, what effect do you predict that a decrease in CO_2 in the internal environment would have on how rapidly and deeply a person breathes?
2. Would the O_2 levels in the blood be (a) normal, (b) below normal, or (c) elevated in a patient with severe pneumonia in whom exchange of O_2 and CO_2 between the air and blood in the lungs is impaired? Would the CO_2 levels in the same patient's blood be (a) normal, (b) below normal, or (c) elevated? Because CO_2 reacts with H_2O to form carbonic acid (H_2CO_3), would the patient's blood (a) have a normal pH, (b) be too acidic, or (c) not be acidic enough (that is, be too alkaline), assuming that other compensatory measures have not yet had time to act?
3. The hormone insulin enhances the transport of glucose (sugar) from the blood into most of the body's cells. Its secretion is controlled by a negative-feedback system between the concentration of glucose in the blood and the insulin-secreting cells. Therefore, which of the following statements is correct?
   a. A decrease in blood glucose concentration stimulates insulin secretion, which in turn further lowers the blood glucose concentration.
   b. An increase in blood glucose concentration stimulates insulin secretion, which in turn lowers the blood glucose concentration.
   c. A decrease in blood glucose concentration stimulates insulin secretion, which in turn increases the blood glucose concentration.
   d. An increase in blood glucose concentration stimulates insulin secretion, which in turn further increases the blood glucose concentration.
   e. None of the above are correct.
4. Given that most AIDS victims die from overwhelming infections or rare types of cancer, what body system do you think is impaired by the AIDS virus?
5. Body temperature is homeostatically regulated around a set point. Based on your knowledge of negative feedback and homeostatic control systems, predict whether narrowing or widening of the blood vessels of the skin will occur when a person is engaged in strenuous exercise. (*Hints:* Muscle contraction generates heat. Narrowing of the vessels supplying an organ decreases blood flow through the organ, whereas widening of the vessels increases blood flow through the organ. The more warm blood flowing through the skin, the greater the loss of heat from the skin to the surrounding environment.)
6. ***Clinical Consideration*** Jennifer R. has the "stomach flu" that is going around campus and has been vomiting profusely for the past twenty-four hours. Not only has she been unable to keep down fluids or food that she has consumed, but she has also lost the acidic digestive juices secreted by the stomach that are normally reabsorbed back into the blood farther down the digestive tract. In what ways might this condition threaten to disrupt the homeostatic maintenance of Jennifer's internal environment? That is, what homeostatically maintained factors will be moved away from normal as a result of her profuse vomiting? What organ systems will respond to resist these changes?

2 Cellular Physiology

Cells are the body's living building blocks. Just as the body as a whole is highly organized, so too is a cell's interior. A cell is made up of three major parts: a **plasma membrane** that encloses the cell; the **nucleus,** which houses the cell's genetic material; and the **cytoplasm,** which is organized into discrete highly specialized *organelles* dispersed throughout a gelatin-like liquid, the *cytosol.* The cytosol is pervaded by a protein scaffolding, the *cytoskeleton,* that serves as the "bone and muscle" of the cell.

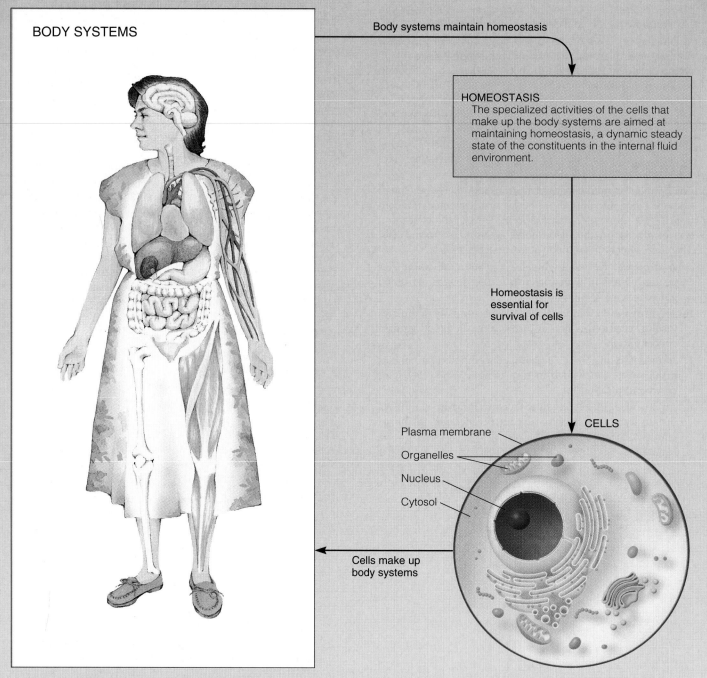

Through the coordinated action of each of these cellular components, every cell is capable of performing certain basic functions essential to its own survival and a specialized task that contributes to the maintenance of homeostasis. Cells are organized according to their specialization into body systems that maintain the stable internal environment essential for the whole body's survival. All body functions ultimately depend on the activities of the individual cells that compose the body.

Chapter Contents At a Glance

Introduction

Cells are the bridge between molecules and humans.

Even though researchers have analyzed the chemicals of which cells are made, it has not been possible to organize these chemicals into a living cell in a laboratory. It is not merely the presence of various molecules that confers the unique characteristics of life, but rather the complex organization and interaction of these molecules within the cell. Within each cell, inanimate chemical molecules are organized into a living entity. Cells, in turn, serve as the living building blocks for the immensely complicated whole body. Modern physiologists are unraveling many of the broader mysteries of how the body works by probing deeper into the molecular structure and organization of the cells that make up the body.

Increasingly better tools are revealing the complexity of cells.

The cells that compose the human body are so small that they cannot be seen by the unaided eye. The smallest visible particle is about five to ten times larger than a typical human cell, which averages about 10 to 20 micrometers (μm) in diameter. (1μm = 1 millionth of a meter, 1m = 39.37 inches. Refer to the inside back cover for a comparison of metric units and their English equivalents.) About 100 average-sized cells lined up side by side would stretch a distance of only 1 mm.

Not until the microscope was invented in the middle of the seventeenth century was the existence of cells revealed. In the early part of the nineteenth century, with the development of better light microscopes, researchers learned that all plant and animal tissues are composed of individual cells. Another important discovery was that cells are filled with a fluid, which, with the microscopic capabilities of the time, appeared to be a rather homogeneous, soupy mixture believed to be the elusive "stuff of life." Not until the 1940s, when the technique of electron microscopy was first employed to observe living matter, did an understanding of the great diversity and complexity of the internal structure of cells begin to emerge. (Electron microscopes are about 100 times more powerful than light microscopes.) Now, with the availability of even more sophisticated microscopes, biochemical techniques, cell-culture technology (see the boxed feature on p. 20, ◆Concepts, Challenges, and Controversies), and genetic engineering, the concept of the cell as a microscopic bag of amorphous fluid has given way to our present-day knowledge of the cell as a complex, highly organized, compartmentalized structure.

Most cells are subdivided into the plasma membrane, nucleus, and cytoplasm.

Even though there is no such thing as a "typical" cell because of diverse structural and functional specializations, different cells share many common features. Most cells have three major subdivisions: the *plasma membrane,* the *nucleus,* and the *cytoplasm* (▮ Table 2-1). The **plasma membrane,** or **cell membrane,** is a very thin membranous structure that encloses each cell, separating the cell's contents from its surroundings. The fluid contained within all of the cells of the body is known collectively as **intracellular fluid (ICF),** and the fluid outside the cells is referred to as **extracellular fluid (ECF).** The plasma membrane does not merely serve as a mechanical barrier to hold in the contents of the cells; it has the ability to selectively control movement of molecules between the ICF and ECF.

The two major parts of the cell's interior are the nucleus and the cytoplasm. The **nucleus,** which is typically the largest single organized cellular component, can be seen as a distinct spherical or oval structure, usually located near the center of the cell. It is surrounded by a double-layered membrane, which separates the nucleus from the remainder of the cell. Sequestered within the nucleus is the cell's genetic material, **deoxyribonucleic acid (DNA),** which has two important functions: directing protein synthesis and serving as a genetic blueprint during cell replication. DNA provides codes, or "instructions," for directing synthesis of specific structural and enzymatic proteins within the cell. By directing the kinds and amounts of various enzymes and other proteins

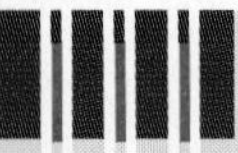

Table 2-1 Summary of Cell Structures and Functions

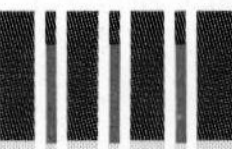

Cell Part	*Number Per Cell*	*Structure*	*Function*
Plasma membrane	1	Lipid bilayer studded with proteins and small amounts of carbohydrate	Acts as selective barrier between cellular contents and extracellular fluid; controls traffic in and out of the cell
Nucleus	1	DNA and specialized proteins enclosed by a double-layered membrane	Acts as control center of the cell, providing storage of genetic information Provides codes for the synthesis of structural and enzymatic proteins that determine the cell's specific nature Serves as blueprint for cell replication
Cytoplasm			
Organelles			
Endoplasmic reticulum	1	Extensive, continuous membranous network of fluid-filled tubules and flattened sacs, partially studded with ribosomes	Forms new cell membrane and other cell components and manufactures products for secretion
Golgi complex	1 to several hundred	Sets of stacked, flattened membranous sacs	Modifies, packages, and distributes newly synthesized proteins
Lysosomes	300	Membranous sacs containing hydrolytic enzymes	Serve as digestive system of the cell, destroying unwanted material, such as foreign substances and cellular debris
Peroxisomes	200	Membranous sacs containing oxidative enzymes	Perform detoxification activities
Mitochondria	100–2,000	Rod- or oval-shaped bodies enclosed by two membranes, with the inner membrane folded into cristae that project into the interior matrix	Act as energy organelles; major site of ATP production; contain enzymes for citric acid cycle and electron transport chain
Vaults	Thousands	Shaped like octagonal barrels	Unclear; may transport messenger RNA from nucleus to cytoplasm; may be important in cellular contractile systems
Cytostol			
Intermediary metabolism enzymes	Many	Sequential arrangement within the cytoskeleton	Facilitate intracellular reactions involving the degradation, synthesis, and transformation of small organic molecules
Ribosomes	Many	Granules of RNA and proteins—some attached to rough endoplasmic reticulum, some free in the cytoplasm	Serve as workbenches for protein synthesis
Secretory vesicles	Varies	Membrane-enclosed packages of secretory products	Store secretory products until signaled to empty to the outside
Inclusions	Varies	Glycogen granules, fat droplets	Store excess nutrients

that are produced, the nucleus indirectly governs most cellular activities and serves as the cell's control center. Three types of **ribonucleic acid (RNA)** play a role in this protein synthesis. First, DNA's genetic code for a particular protein is transcribed into a **messenger RNA** molecule, which exits the nucleus through the nuclear pores that pierce the nuclear membrane. Within the cytoplasm, messenger RNA delivers the coded message to **ribosomal RNA,** which "reads" the code and translates it into the appropriate amino acid sequence for the designated protein being synthesized.

Table 2-1 **Summary of Cell Structures and Functions (continued)**

Cell Part	*Number Per Cell*	*Structure*	*Function*
Cytoskeleton			
Microtubules	Many	Long, slender, hollow tubes composed of tubulin molecules	Maintain asymmetrical cell shapes Coordinate complex cell movements Facilitate transport of secretory vesicles within cell, such as axonal transport Serve as dominant structural and functional component of cilia and flagella Form mitotic spindle during cell division
Microfilaments	Many	Helically intertwined chains of actin molecules; microfilaments composed of myosin molecules also present in muscle cells	Play a vital role in various cellular contractile systems Play a dominant role in muscle contraction Form nonmuscle contractile assemblies, such as in white blood cells during amoeboid movement Serve as a mechanical stiffener for microvilli Increase the surface area available for absorption in small intestine and kidneys Are specialized to detect sound and positional changes in ear
Intermediate filaments	Many	Irregular, threadlike proteins	Play a structural role in parts of the cell subject to mechanical stress
Microtrabecular lattice	1	Meshwork of exceedingly fine interlinked filaments	Suspends and functionally links larger cytoskeletal elements and various organelles Organizes cytosolic enzymes *Integrated Functions of Entire Cytoskeleton:* Responsible for the shape, rigidity, and spatial geometry of each type of cell; the "bone" of the cell Responsible for directing intracellular transport and for regulating cellular movements; the "muscle" of the cell Appears to play a role in regulating growth and division of cells

Finally, **transfer RNA** transfers the appropriate amino acids within the cytoplasm to their designated site in the protein under construction.

In addition to providing codes for protein synthesis, DNA also serves as a genetic blueprint during cell replication to ensure that the cell produces additional cells just like itself, thus continuing the identical type of cell line within the body. Furthermore, in the reproductive cells, the DNA blueprint serves to pass on genetic characteristics to future generations. (See Appendix B for further details of DNA function.)

HeLa Cells: Problems in a "Growing" Industry

Many basic advances in cell physiology and related fields such as genetics and cancer research have come about through the use of cells grown outside the body in what is known as *in vitro* culture (*in vitro* is a Latin phrase that means "in glass"). In the early 1950s, many attempts were made to culture human cells using tissues obtained from biopsies or surgical procedures. These early attempts usually met with failure; the cells died after a few days or weeks in culture, mostly without undergoing cell replication. These difficulties continued until February 1951, when a researcher at Johns Hopkins University received a sample of a cervical cancer from a woman named Henrietta Lacks. Following convention, the culture was named HeLa by combining the first two letters of the donor's first and last names. This cell line not only grew but prospered under culture conditions and represented one of the earliest cell lines successfully grown outside the body.

Researchers were eager to have human cells available on demand to study the effects of drugs, toxic chemicals, radiation, and viruses on human tissue. For example, it was found that polio virus reproduced well in HeLa cells, providing a breakthrough in the development of a polio vaccine. As cell-culture techniques improved, human cell lines were started from other cancers and normal tissues, including heart, kidney, and liver tissues. By the early 1960s, a central collection of cell lines had been established in Washington, D.C., and cultured human cells were an important tool in many areas of biological research.

The first clouds in this happy picture began to gather in 1966, when Stanley Gartler, a geneticist at the University of Washington, discovered that eighteen different human cell lines he had analyzed had all been contaminated and taken over by HeLa cells. Over the next two years, it was confirmed that twenty-four of the thirty-four cell lines in the central repository were actually HeLa cells. Researchers who had spent years studying what they thought were heart or kidney cells had in reality been working with a cervical cancer cell instead. Gartler's discovery meant that hundreds of thousands of experiments performed in laboratories around the world were invalid.

As painful as this lesson was, scientists started over, preparing new cell lines and using new, stricter rules to prevent contamination with HeLa cells. Unfortunately, the problem did not end. In 1974, Walter Nelson-Rees published a paper demonstrating that five cell lines extensively used in cancer research were in fact HeLa cells. In 1976, another paper announced that eleven additional cell lines, each widely used in research, were also HeLa cells; and in 1981, Nelson-Rees listed twenty-two more cell lines that were contaminated with HeLa. In all, one-third of all cell lines used in cancer research were apparently really HeLa cells. The result was an enormous waste of dollars and resources. Clearly, not enough care was taken to prevent the spread of HeLa cells to other cultured lines. This spread can be attributed to poor record keeping, mislabeled cultures, and sloppy laboratory techniques. The response of the scientific community was interesting. Some scientists quickly stepped forward to acknowledge the problem and retract their conclusions. Others engaged in denials and steadfastly refused to acknowledge that their work was invalid. Henrietta Lacks provided the scientific world with two valuable gifts: a cell line and proof that science is a human endeavor, subject to human traits that include not only honesty, candor, and veracity, but also ego, fear, and denial.

The **cytoplasm** is that portion of the cell interior not occupied by the nucleus. It contains a number of distinct, highly organized, membrane-enclosed structures—the **organelles**—dispersed within a complex, gel-like mass called the **cytosol.** Nearly all cells contain six main types of organelles—the endoplasmic reticulum, Golgi complex, lysosomes, peroxisomes, mitochondria, and vaults. (■ Fig. 2-1). These organelles are similar in all cells, although there are some variations depending on the specialized capabilities of each cell type. Organelles are like intracellular "specialty shops." Each is a separate internal compartment that contains a specific set of chemicals for carrying out a particular cellular function. This compartmentalization is advantageous because it permits chemical activities that would not be compatible with each other to occur simultaneously within the cell. For example, the enzymes that destroy unwanted proteins in the cell do so within the protective confines of the lysosomes without the risk of destroying essential cellular proteins. About half of the total cell volume is occupied by organelles.

The remainder of the cytoplasm (the part not occupied by organelles) consists of cytosol, a semiliquid mass laced with an elaborate protein network that constitutes the cytoskeleton. Many of the chemical reactions that are compatible with each other are carried on in the cytosol. The cytoskeletal network gives the cell its shape, provides for its internal organization, and regulates its various movements. In this chapter, we will examine each of the cytoplasmic components in more detail, concentrating first on the organelles.

Organelles

The endoplasmic reticulum is a synthesizing factory.

The **endoplasmic reticulum (ER)** is an elaborate fluid-filled membranous system distributed extensively throughout the cytosol. Two distinct types of endoplasmic reticulum—the smooth ER and the rough ER—can be distinguished. The **smooth ER** is a meshwork of tiny interconnected tubules, whereas the **rough ER** projects outward from the smooth ER as stacks of relatively flattened sacs (■ Fig. 2-2). Even though these two regions differ considerably in appearance and function, they are thought to be continuous with each other. In other words, the ER is one continuous organelle with many

Figure 2-1 Schematic Three-Dimensional Illustration of Cell Structures Visible Under an Electron Microscope

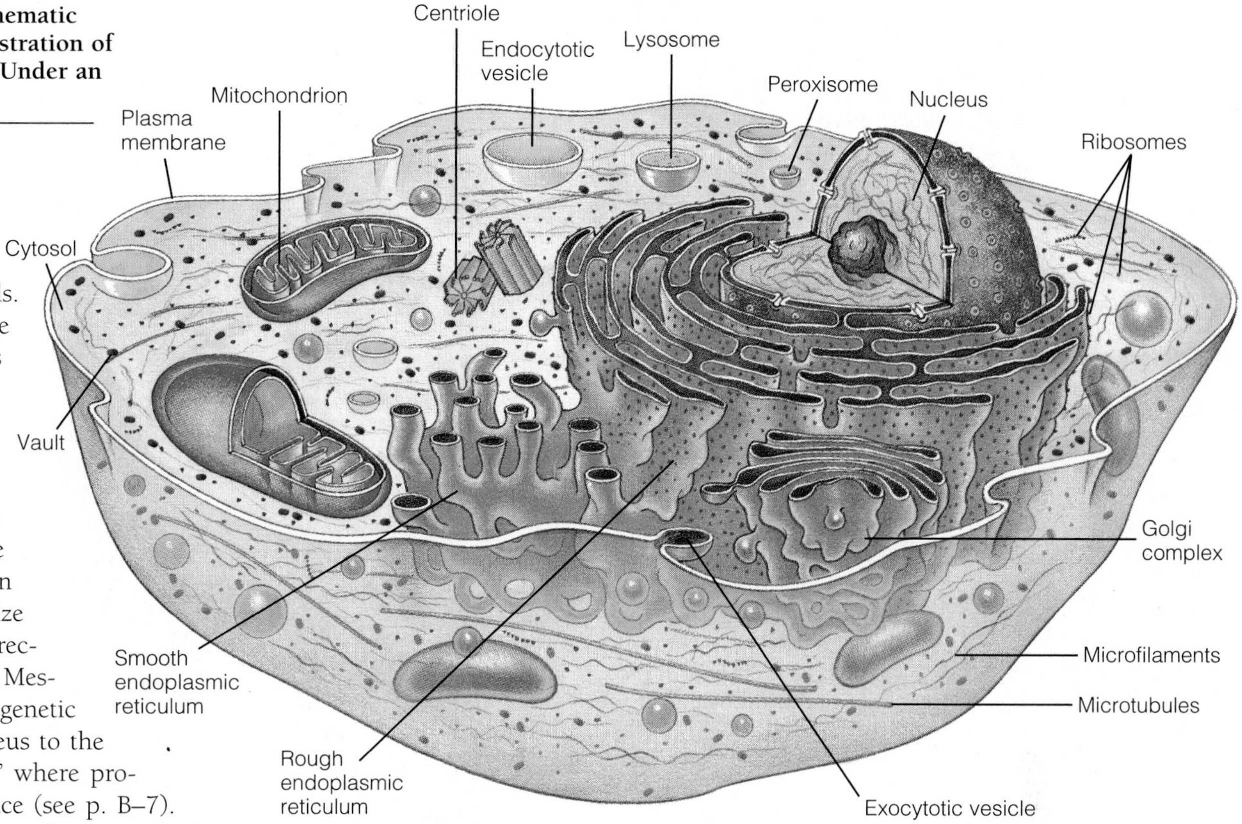

interconnected channels. The outer surface of the rough ER membrane is studded with small, dark-staining particles that give it a "rough" or granular appearance. These particles are **ribosomes,** which are ribosomal RNA-protein complexes that synthesize proteins under the direction of nuclear DNA. Messenger RNA carries the genetic message from the nucleus to the ribosome "workbench," where protein synthesis takes place (see p. B–7). Not all ribosomes in the cell are attached to the rough ER. Unattached or "free" ribosomes are dispersed throughout the cytosol. The relative amount of smooth and rough ER varies between cells, depending on the activity of the cell.

Rough endoplasmic reticulum The rough ER, in association with its ribosomes, synthesizes and releases a variety of new proteins into the ER lumen, the fluid-filled space enclosed by the ER membrane. These proteins serve one of two purposes: (1) Some proteins are destined for export to the cell's exterior as secretory products, such as proteinaceous hormones or enzymes. (All enzymes are proteins—see p. A–6). (2) Other proteins are transported to sites within the cell for use in the construction of new cellular membrane (either new plasma membrane or new organelle membrane) or other protein components of organelles. Cellular membranes consist predominantly of lipids (fats) and proteins. The membranous wall of the ER also contains enzymes essential for the synthesis of nearly all the lipids needed for the production of new membranes. These newly synthesized lipids enter the ER lumen along with the proteins. Predictably, the rough ER is most abundant in cells specialized for protein secretion (for example, cells that secrete digestive enzymes) or in cells that require extensive membrane synthesis (for example, rapidly growing cells such as immature egg cells.

Once assembled, each ribosome participates in the synthesis of only one type of protein. The free and ER-bound ribosomes differ only in the proteins that they help to synthesize. In contrast to the rough ER ribosomes, the free ribosomes

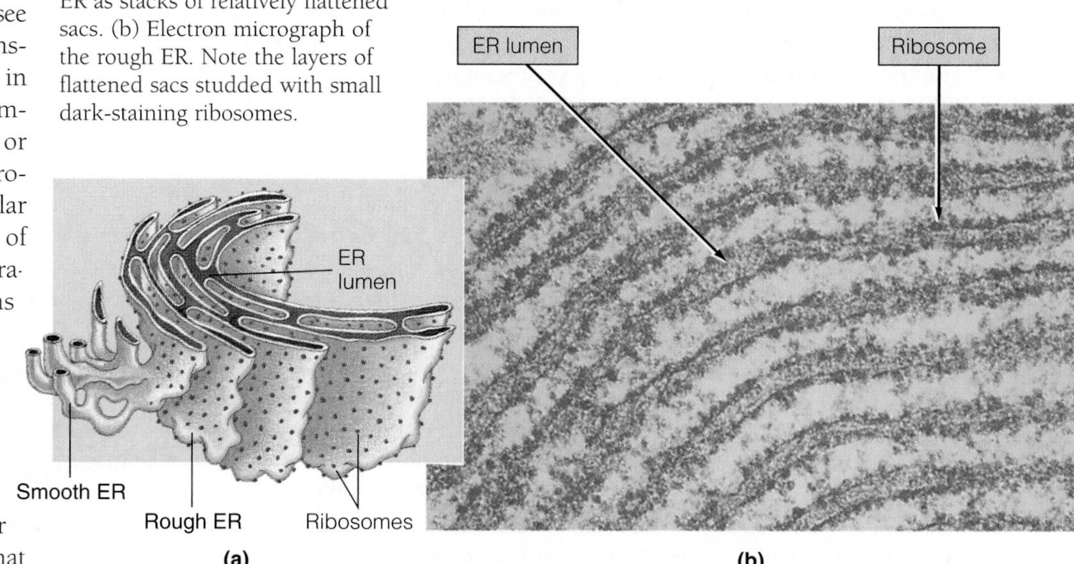

Figure 2-2 Endoplasmic Reticulum (ER) (a) Schematic three-dimensional representation of the relationship between the rough and the smooth ER. The smooth ER is a meshwork of tiny interconnected tubules. The rough ER, which is studded with ribosomes, projects outward from the smooth ER as stacks of relatively flattened sacs. (b) Electron micrograph of the rough ER. Note the layers of flattened sacs studded with small dark-staining ribosomes.

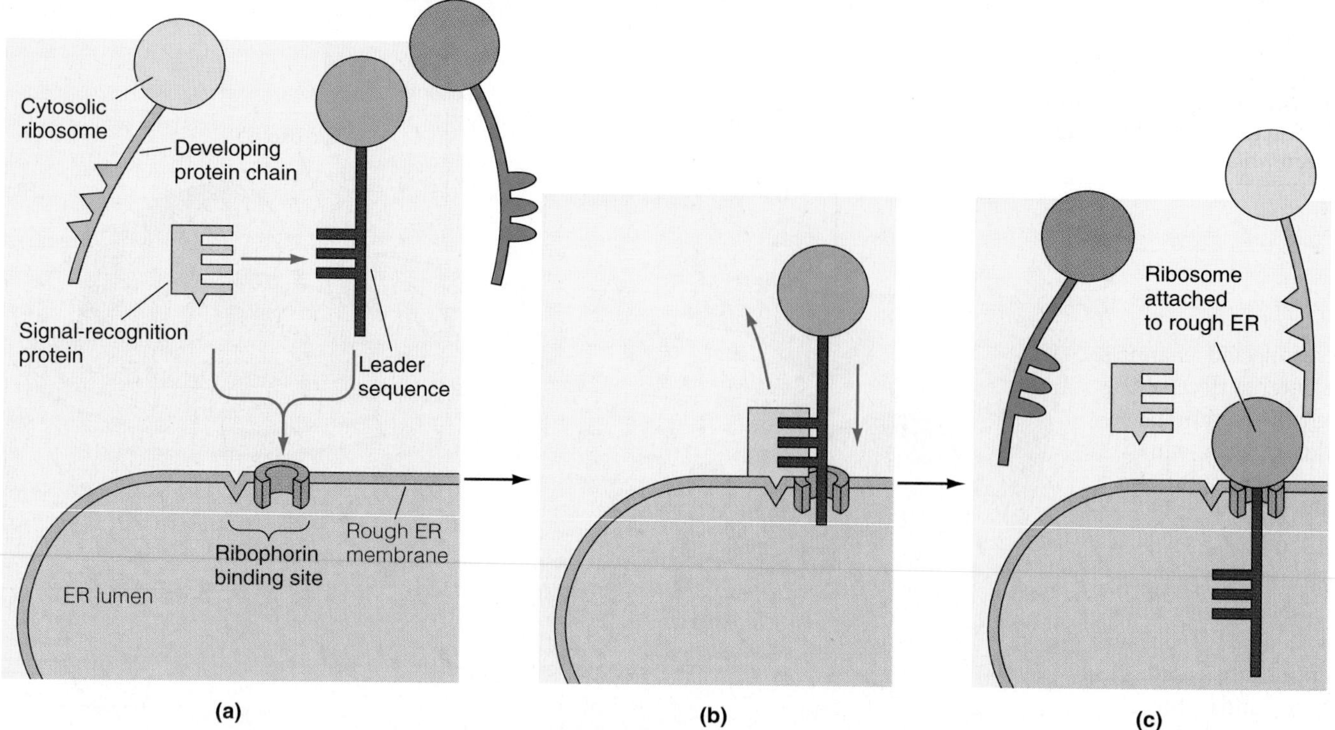

Figure 2-3 Delivery of Ribosomes to the Rough Endoplasmic Reticulum (a) The signal-recognition protein "recognizes" the correct leader sequence on the developing protein chain of a cytosolic ribosome destined to become attached to the rough ER. (b) The signal-recognition protein carries the ribosome to the correct ribophorin "address" on the ER membrane. (c) The signal-recognition protein departs, and the protein chain threads through the ER membrane into the lumen; the proper ribosome is now attached. Other ribosomes synthesizing cytosolic proteins are not recognized by the signal-recognition protein.

synthesize enzymatic proteins that are used intracellularly within the cytosol. All ribosomes are produced in the cell's nucleus under the direction of DNA, with each being "programmed" at any given time to facilitate the synthesis of only one specific protein needed by that particular cell.

Currently, it is thought that newly formed ribosomes destined to become attached to the rough ER start to synthesize a **leader sequence** that acts as a signal, much like an address on a letter (Fig. 2-3). Another protein present in the cytosol, the **signal-recognition protein,** binds to one of these ribosomes upon recognizing the newly synthesized leader sequence. Subsequently, the signal-recognition protein, acting much like a mail carrier, delivers the ribosome to the proper "address" on the ER membrane. How does the signal-recognition protein know the proper address? Special proteins called **ribophorins** found exclusively in the rough regions of the ER membrane act as the "house address" on this membrane. The ribophorins serve as binding sites for preferential ribosomal attachment. The signal-recognition protein (the "mail carrier") can recognize both the leader-sequence signal on the ribosome (the "address on the envelope") and the ribophorin binding site on the ER (the "house address") and therefore delivers the proper ribosome to the proper site on the rough ER for binding.

Once a ribosome is attached to the ER membrane and starts directing synthesis of a specific protein, the developing protein threads its way across the ER membrane into the lumen. Here the leader sequence is removed and the protein chain is folded into its final conformation. The protein may also be modified in other ways, such as being "pruned" (Fig. 2-4) or having sugar molecules attached to it. After this processing within the ER lumen, a new protein is unable to pass through the ER membrane and therefore becomes permanently separated from the cytosol as soon as it has been synthesized. In this way, the ER provides cells with a mechanism for separating the newly produced molecules that are

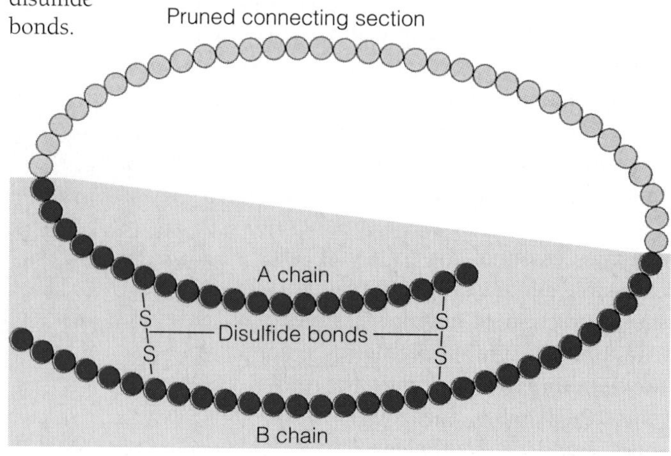

Figure 2-4 Formation of an Insulin Molecule within the Endoplasmic Reticulum Insulin is originally formed as a single long polypeptide chain that is folded back on itself, with the two overlapping ends being joined by two disulfide (sulfur-to-sulfur) bonds. The connecting piece (in gray) between the two overlapping ends is then pruned away, leaving the insulin molecule (in blue) consisting of an A chain and a B chain connected by the disulfide bonds.

Each circle represents a specific amino acid.

destined for export out of the cell or for synthesis of new cellular components (those synthesized by the ER) from those that belong in the cytosol (those produced by the free ribosomes).

How do the newly synthesized molecules within the ER lumen get to their destinations at other intracellular sites or to the exterior of the cell if these molecules cannot pass out through the ER membrane? The smooth ER is important in accomplishing this feat.

Smooth endoplasmic reticulum The smooth ER does not contain ribosomes; hence it is "smooth." Lacking ribosomes, it is not involved in protein synthesis. Instead it serves a variety of other purposes that vary in different cell types.

In the majority of cells, the smooth ER is rather sparse and serves primarily as a central packaging and discharge site for molecules that are to be transported from the ER. Newly synthesized proteins and lipids pass from the rough ER to gather in the smooth ER. Portions of the smooth ER then "bud off" (that is, are pinched off), giving rise to **transport vesicles** that contain the new molecules enclosed in a membrane layer derived from the smooth ER membrane (Fig. 2-5). Newly synthesized membrane components are rapidly incorporated into the ER membrane itself to replace the membrane that was used to "wrap" the transport vesicle. Transport vesicles move to the Gogli complex for further processing of their cargo.

In contrast to the sparseness of the smooth ER in most cells, some specialized types of cells have an extensive smooth ER, which has additional responsibilities as follows:

- The smooth ER is abundant in cells that specialize in lipid metabolism—for example, cells that secrete steroid hormones. (A steroid is a special type of lipid derived from cholesterol.) The membranous wall of the smooth ER, like that of the rough ER, contains enzymes for synthesis of lipids. The lipid-producing enzymes in the membranous wall of the rough ER alone are insufficient to carry out the extensive lipid synthesis necessary to maintain adequate steroid-hormone secretion levels. These cells have an expanded smooth ER compartment to house the additional enzymes necessary to keep pace with demands for hormone secretion.
- In liver cells, the smooth ER has a special capability. It contains enzymes that are involved in detoxifying harmful substances produced within the body by metabolism or substances that enter the body from the outside in the form of drugs or other foreign compounds. These detoxification enzymes alter toxic substances so that the latter can be eliminated more readily in the urine. The amount of smooth ER available in liver cells for the task of detoxification can vary dramatically, depending on the need. For example, if phenobarbital (a barbiturate drug used as a sedative) is administered in large quantities, the amount of smooth ER with its associated detoxification enzymes doubles within a few days, only to return to normal within five days after drug administration ceases. The mechanisms involved in regulating these changes are not well understood. Unfortunately, in some cases the same enzymes transform otherwise harmless substances into carcinogenic ("cancer-producing") substances that play a role in cancer development.
- Muscle cells have developed another specialized use for the smooth ER. They have an elaborate, modified smooth ER known as the *sarcoplasmic reticulum,* which stores calcium and plays an important role in the process of muscle contraction (see p. 229).

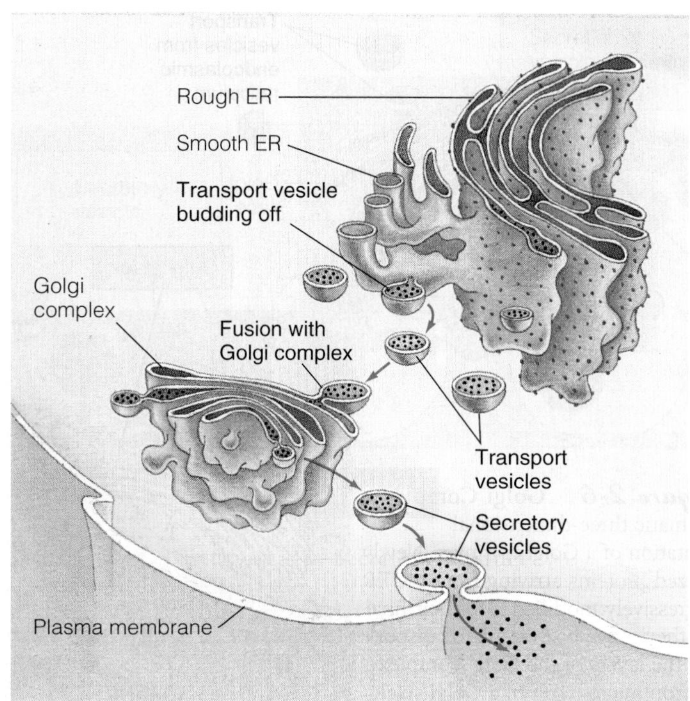

Figure 2-5 **Movement of Transport Vesicles from the Endoplasmic Reticulum to the Golgi Complex and Release of Secretory Vesicles from the Golgi Complex after Further Processing**

The Golgi complex is a refining plant and directs molecular traffic.

Closely associated with the endoplasmic reticulum is the **Golgi complex,** which consists of sets of flattened, slightly curved, membrane-enclosed sacs, or cisternae, stacked in layers (Fig. 2-6). Note that the flattened sacs are thin in the middle but have dilated edges. The number of Golgi stacks varies, depending on the cell type. Some cells have only one stack, whereas cells highly specialized for protein secretion may have hundreds of stacks.

The majority of the newly synthesized molecules that have just budded off from the smooth ER enter a Golgi stack. When a transport vesicle carrying its newly synthesized cargo reaches a Golgi stack, the vesicle membrane fuses with the membrane of the sac closest to the center of the cell. The vesicle membrane opens up and becomes a new part of the Golgi membrane, and the contents of the vesicle are released to the interior of the sac (Fig. 2-5).

These newly synthesized raw materials from the ER travel by means of vesicle formation through the layers of the Golgi stack, where two important interrelated functions take place:

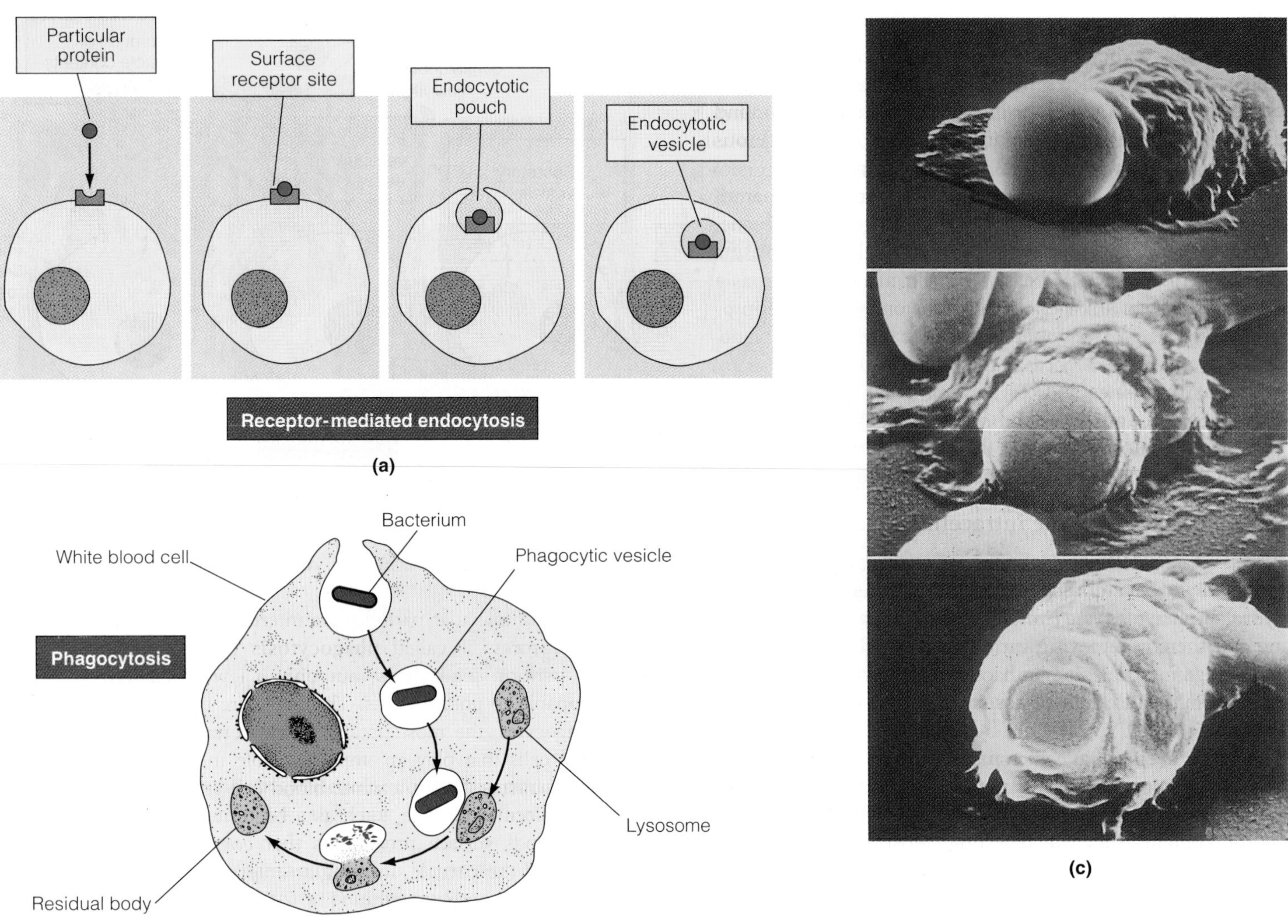

Figure 2-9 **Endocytosis** (a) Receptor-mediated endocytosis. When a large molecule such as a protein attaches to a specific surface receptor site, the membrane dips inward to form a pouch, then seals the surface to internalize the molecule along with a bit of extracellular fluid within an intracellular vesicle. (b) Phagocytosis. White blood cells perform a special form of endocytosis known as phagocytosis. They internalize multimolecular particles such as bacteria by extending surface projections that seal in the targeted material. A lysosome fuses with the internalized vesicle, releasing enzymes that attack the engulfed material within the confines of the vesicle. (c) Scanning-electron-micrograph series of a white blood cell phagocytizing an old, worn-out red blood cell.

In specific instances, lysosomes cause intentional self-destruction of healthy cells. This happens as a normal part of embryonic development when certain unwanted tissues that form are programmed for destruction. For example, embryonic ducts capable of forming a male reproductive tract are deliberately destroyed during the development of a female fetus. Lysosomes also play an important role in tissue regression, such as during the normal reduction in the uterine lining following pregnancy. Such programmed cell death is termed **apoptosis.** The mechanisms that control lyosomal activity under such circumstances are still unknown.

An inherent danger even in a healthy, intact cell is that lysosomal membranes may inadvertently rupture. Because these organelles have the potential ability to self-destruct the cell, their discoverer named them "suicide bags." However, two factors make this threat less severe than imagined. First, the lysosomal hydrolytic enzymes function best in an acid environment. The lysosomal membrane transports hydrogen ions (acid formers) into the lysosome, making it considerably more acidic than the remainder of the cell. Should the enzymes accidentally leak into the cytosol, they would be less potent than they are in their acidic home. Second, in most instances, cells could tolerate the limited damage that would occur if only one or two lysosomes inadvertently ruptured because most parts of the cell are renewable. The biggest danger is accidental digestion of part of the irreplaceable DNA molecule within the nucleus. Such nuclear damage would alter the cell's genetic properties, and the defect would be perpetuated to all of the cell's progeny.

Some individuals lack the ability to synthesize one or more of the lysosomal enzymes. The result is massive accumulation within the lysosomes of the specific compound that is normally digested by the missing enzyme. Clinical manifestations often accompany such disorders because the engorged lysosomes interfere with normal cell activity. The nature and severity of the symptoms depend on the type of substance that is accumulating, which in turn depends on what lysosomal enzyme is missing. Among these so-called storage diseases is **Tay-Sachs disease.** It is characterized by abnormal accumulation of gangliosides, which are complex molecules found in nerve cells. Profound symptoms of progressive nervous system degeneration result as the accumulation continues.

Peroxisomes house oxidative enzymes that detoxify various wastes.

Typically, several hundred small **peroxisomes** about one-third to one-half the average size of lysosomes are present in a cell. Peroxisomes are similar to lysosomes in that they are membrane-enclosed sacs containing enzymes, but unlike the lysosomes, which contain hydrolytic enzymes, peroxisomes house several powerful oxidative enzymes and contain most of the cell's catalase. **Oxidative enzymes,** as the name implies, use oxygen (O_2), in this case to strip hydrogen from specific molecules. Such a reaction is important in detoxifying various wastes produced within the cell or foreign compounds that have entered the cell, such as the ethanol consumed in alcoholic beverages. The major product generated in the peroxisome is **hydrogen peroxide** (H_2O_2), which is formed by molecular oxygen and the hydrogen atoms stripped from the waste.

Hydrogen peroxide, itself a powerful oxidant, is potentially destructive if it is allowed to accumulate or escape from the confines of the peroxisome. However, peroxisomes also contain an abundance of **catalase,** an antioxidant enzyme that decomposes potent H_2O_2 into harmless H_2O and O_2. This latter reaction is an important safety mechanism that destroys the potentially deadly peroxide at the site of its production, thereby preventing its possible devastating escape into the cytosol.

Mitochondria are the energy organelles.

Mitochondria are the energy organelles or "powerhouses" of the cell; they extract energy from the nutrients in food and transform it into a usable form to energize cellular activities. The number of mitochondria per cell varies greatly, depending on the energy needs of each particular cell type. A single cell may contain as few as a hundred or as many as several thousand mitochondria. In some cell types, the mitochondria are densely compacted in cellular regions that use most of the cell's energy. For example, mitochondria are packed between the contractile units in the muscle cells of the heart.

Mitochondria are rod- or oval-shaped structures about the size of bacteria. In fact, considerable evidence indicates that mitochondria are descendants of bacteria that invaded or were engulfed by primitive cells early in evolutionary history. It is generally believed that a symbiotic ("living together") relationship subsequently evolved, with the bacteria eventually becoming permanent organelles.

Each mitochondrion is enclosed by a double membrane—a smooth outer membrane that surrounds the mitochondrion itself and an inner membrane that forms a series of infoldings or shelves called **cristae,** which project into an inner cavity filled with a gel-like solution known as the **matrix** (Fig. 2-10). These cristae contain crucial proteins that ultimately are responsible for converting much of the energy in food into a usable form (the electron transport proteins, to be described shortly). The generous folds of the inner membrane greatly increase the surface area available for housing these important proteins. (This is a common theme in physiology: increasing the surface area of various body structures by means of surface projections or infoldings makes more surface area available to participate in the structure's functions.) The matrix consists of a concentrated mixture of hundreds of different dissolved enzymes (the citric acid cycle enzymes, soon to be described) that are important in preparing nutrient molecules

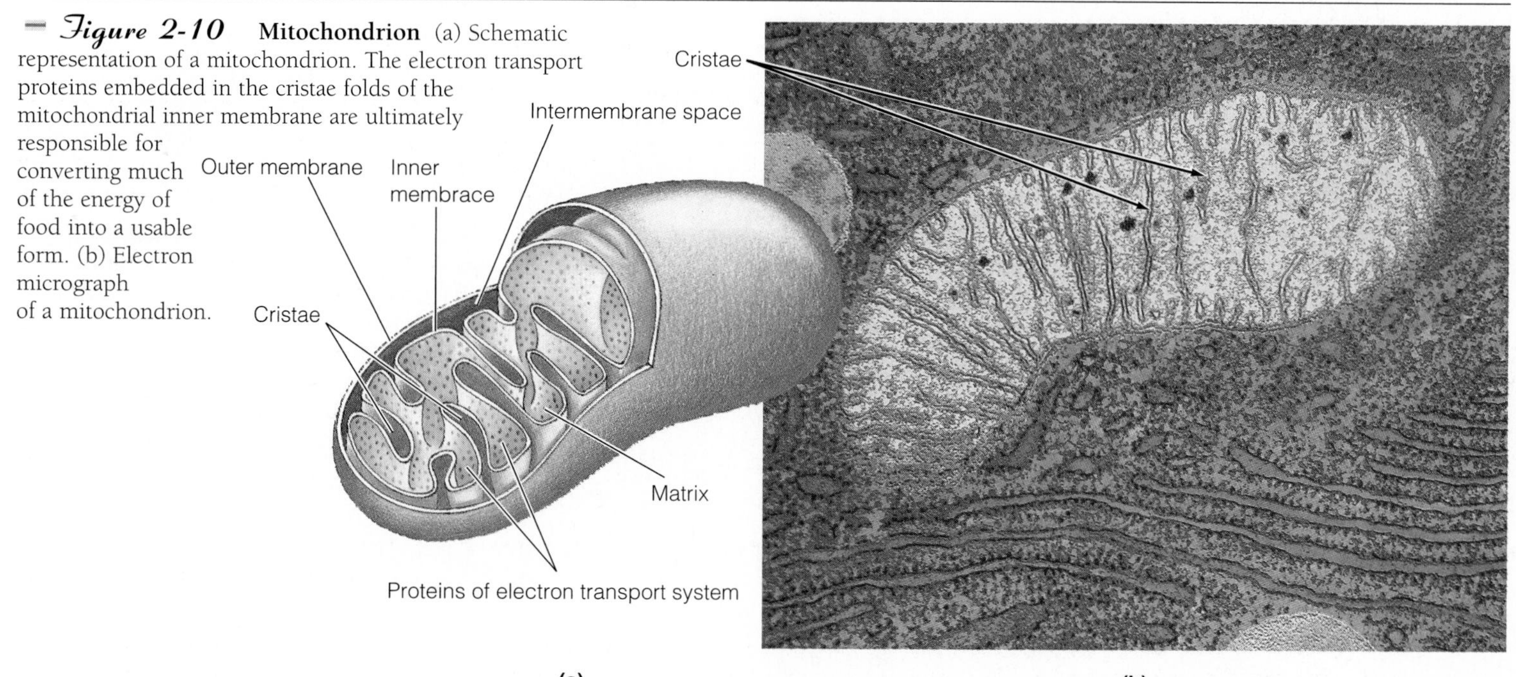

Figure 2-10 **Mitochondrion** (a) Schematic representation of a mitochondrion. The electron transport proteins embedded in the cristae folds of the mitochondrial inner membrane are ultimately responsible for converting much of the energy of food into a usable form. (b) Electron micrograph of a mitochondrion.

for the final extraction of usable energy by the cristae proteins.

Energy derived from food is stored in ATP.

The source of energy for the body is the chemical energy stored in the carbon-hydrogen bonds in ingested food. Body cells are not equipped to use this energy directly, however. Instead they must extract energy from food nutrients and convert it into an energy form that they can use—namely, the high-energy phosphate bonds of **adenosine triphosphate, (ATP),** which consists of adenosine with three phosphate groups attached (see p. A–14). When a high-energy bond such as that binding the terminal phosphate to adenosine is split, a substantial amount of energy is released. Adenosine triphosphate is the universal energy carrier—the common energy "currency" of the body. Cells can "cash in" ATP to pay the energy "price" for running the cellular machinery. To obtain immediate usable energy, cells split the terminal phosphate bond of ATP, which yields **adenosine diphosphate (ADP)**—adenosine with two phosphate groups attached—plus inorganic phosphate (Pi) plus energy:

$$\text{ATP} \xrightarrow{\text{splitting}} \text{ADP} + \text{P}_i + \text{energy for use by the cell}$$

In this energy scheme, food might be thought of as the "crude fuel," whereas ATP is the "refined fuel" for operating the body's machinery. Let us elaborate on this fuel conversion process (Table 2-2). Dietary food is digested, or broken down, by the digestive system into smaller absorbable units that can be transferred from the digestive tract lumen into the circulatory system (chapter 16). For example, dietary carbohydrates are broken down primarily into glucose, which can be absorbed into the blood. No usable energy is released during the digestion of food. When delivered to the cells by the blood, the nutrient molecules are transported across the plasma membrane into the cytosol. Among the thousands of enzymes within the cytosol are those responsible for **glycolysis,** a chemical process involving nine separate sequential reactions that break down the simple six-carbon sugar molecule, glucose, into two pyruvic acid molecules, each of which contains three carbons. During this process, some of the energy stored in the chemical bonds of glucose is used to convert ADP into ATP (Fig. 2-11). However, glycolysis is not very efficient in

Figure 2-11 **A Simplified Summary of Glycolysis** Glycolysis involves the breakdown of glucose into pyruvic acid, with a net yield of two molecules of ATP for every glucose molecule processed.

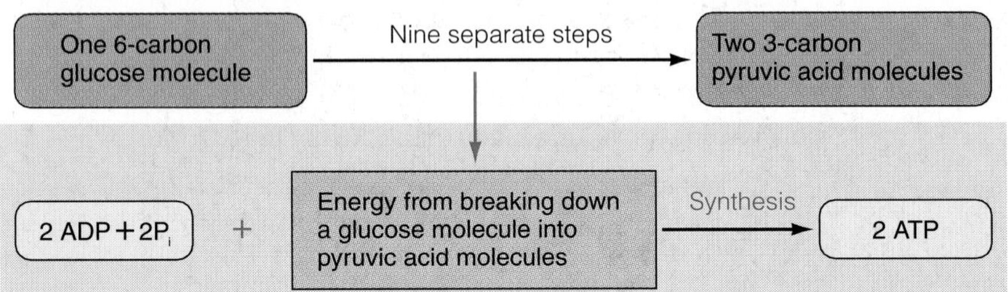

Table 2-2 **Overview of Cellular Energy Production from Glucose**

Reaction	*Substance Processed*	*Location*	*Energy Yield (per glucose molecule (processed)*	*End Products Available for Further Energy Extraction (per glucose molecule processed)*	*Need for Oxygen*
Glycolysis	Glucose	Cytosol	Two molecules of ATP	Two pyruvic acid molecules	No; anaerobic
Citric acid cycle	Acetyl CoA, which is derived from pyruvic acid, the end product of glycolysis. Two acetyl CoA molecules result from the processing of one glucose molecule	Mitochondrial matrix	Two molecules of ATP	Eight NADH and two $FADH_2$ hydrogen carrier molecules	Yes, derived from molecules involved in citric acid cycle reactions
Electron transport chain	High-energy electrons stored in hydrogen atoms in the hydrogen carrier molecules NADH and $FADH_2$ derived from citric acid cycle reactions	Mitochondrial inner-membrane cristae	Thirty-two molecules of ATP	None	Yes, derived from molecular oxygen acquired from breathing

terms of energy extraction; one molecule of glucose has a net yield of only two molecules of ATP. Much of the energy originally contained in the glucose molecule is still locked in the chemical bonds of the pyruvic acid molecules. The low-energy yield of glycolysis is insufficient to support the body's demand for ATP. This is where the mitochondria come into play.

The pyruvic acid produced by glycolysis in the cytosol can be selectively transported into the mitochondrial matrix. Here it is further broken down into a two-carbon molecule, acetic acid, by enzymatic removal of one of the carbons in the form of carbon dioxide (CO_2), which eventually is eliminated from the body as an end product, or waste (Fig. 2-12). During this

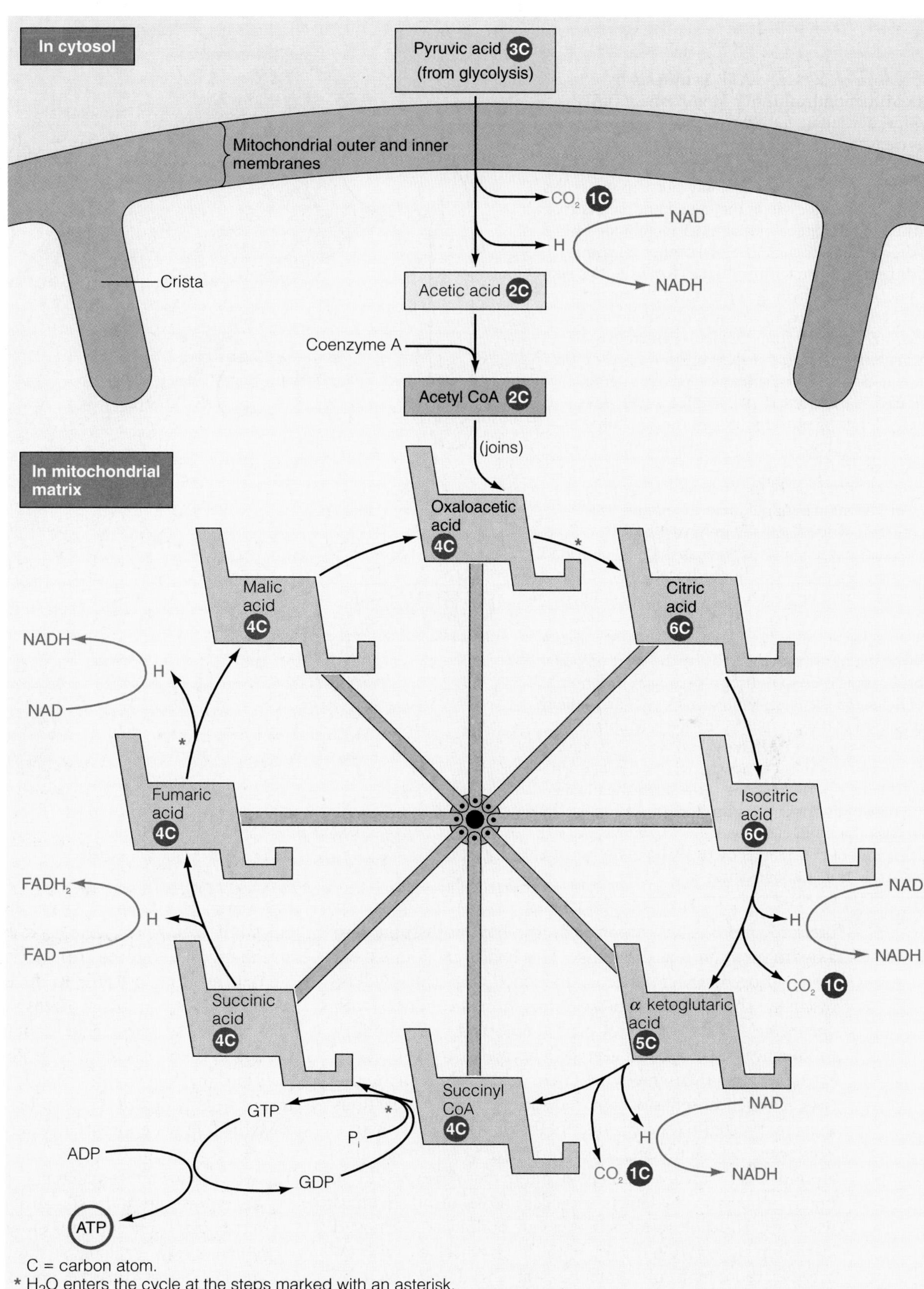

Figure 2-12

Citric Acid Cycle A simplified version of the citric acid cycle, showing how the two carbons entering the cycle by means of acetyl CoA are eventually converted to CO_2, with oxaloacetic acid, which accepts the acetyl CoA, being regenerated at the end of the cyclical pathway. Also denoted is the release of hydrogen atoms at specific points along the pathway, with these hydrogens binding to the hydrogen carrier molecules NAD and FAD for further processing. One molecule of ATP is generated for each molecule of acetyl CoA that enters the citric acid cycle, for a total of two molecules of ATP for each molecule of processed glucose.

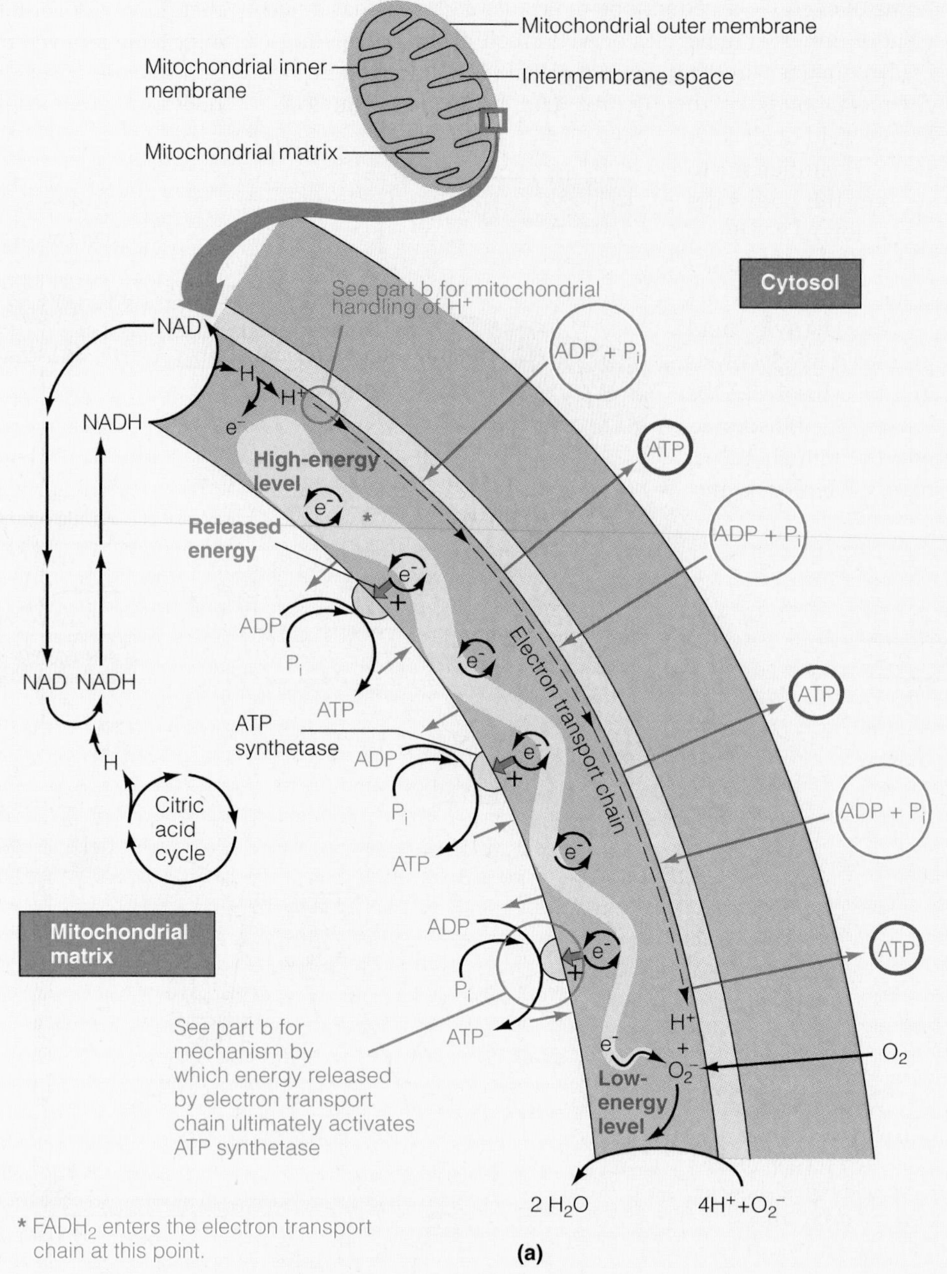

Figure 2-13 ATP Synthesis by the Mitochondrial Inner Membrane (a) ATP synthesis resulting from the passage of high-energy electrons through the mitochondrial electron transport chain. Hydrogen (H) that is released during the degradation of carbon-containing nutrient molecules by the citric acid cycle in the mitochondrial matrix is carried to the mitochondrial inner membrane by hydrogen carriers such as NADH. After releasing hydrogen at the inner membrane, NAD shuttles back to pick up more hydrogen generated by the citric acid cycle in the matrix. Meanwhile, high-energy electrons extracted from the hydrogen are passed through the electron transport chain located on the mitochondrial inner membrane. Energy is gradually released as the electrons fall to successively lower energy levels by moving through the electron transport chain of reactions. The released energy triggers a sequence of steps (shown in part b) that ultimately results in activation of the enzyme ATP synthetase within the mitochondrial inner membrane. ATP synthetase synthesizes ATP from ADP and P_i. Molecular oxygen, after serving as the final electron acceptor, combines with the hydrogen ions (H^+) generated from hydrogen upon extraction of high-energy electrons to produce water. (b) Activation of ATP synthetase by movement of H^+. Energy released during the transfer of electrons by the electron transport chain is used to transport hydrogen ions from the matrix to the intermembrane space. The resultant buildup of hydrogen ions in the intermembrane space brings about the flow of hydrogen ions from the intermembrane space to the matrix through special channels in the mitochondrial inner membrane. The flow of hydrogen ions through the channel activates ATP synthetase, which is located at the matrix end of the channel. Activation of ATP synthetase brings about synthesis of ATP.

breakdown process, a carbon-hydrogen bond is disrupted, so a hydrogen atom is also released. This hydrogen atom is held by a hydrogen carrier molecule, the function of which will be discussed shortly. The acetic acid thus formed combines with coenzyme A, a derivative of pantothenic acid (a B vitamin), producing the compound acetyl coenzyme A (acetyl CoA).

Acetyl CoA then enters the **citric acid cycle,** which consists of a cyclical series of eight separate biochemical reactions that are directed by the enzymes of the mitochondrial matrix. This cycle of reactions can be compared to one revolution around a ferris wheel. (Keep in mind that Figure 2-12 is highly schematic. It depicts a cyclical series of biochemical reactions. The molecules themselves are not physically moved around in a cycle.) On the top of the ferris wheel, acetyl CoA, a two-carbon molecule, enters a seat already occupied by oxaloacetic acid, a four-carbon molecule. These two molecules link together to form a six-carbon citric acid molecule, and the trip around the citric acid cycle begins. (This cycle is alternatively known as the **Krebs cycle** in honor of its principal discoverer, Sir Hans Krebs, or the **tricarboxylic acid cycle,** because citric acid contains three carboxylic acid groups.) As the seat moves around the cycle, at each new position, matrix enzymes modify the passenger molecule to form a slightly different molecule. These molecular alterations have the following important consequences:

1. Two carbons are sequentially "kicked off the ride" as they are removed from the six-carbon citric acid molecule, converting it back into the four-carbon oxaloacetic acid, which is now available at the top of the cycle to pick up another acetyl CoA for another revolution through the cycle.

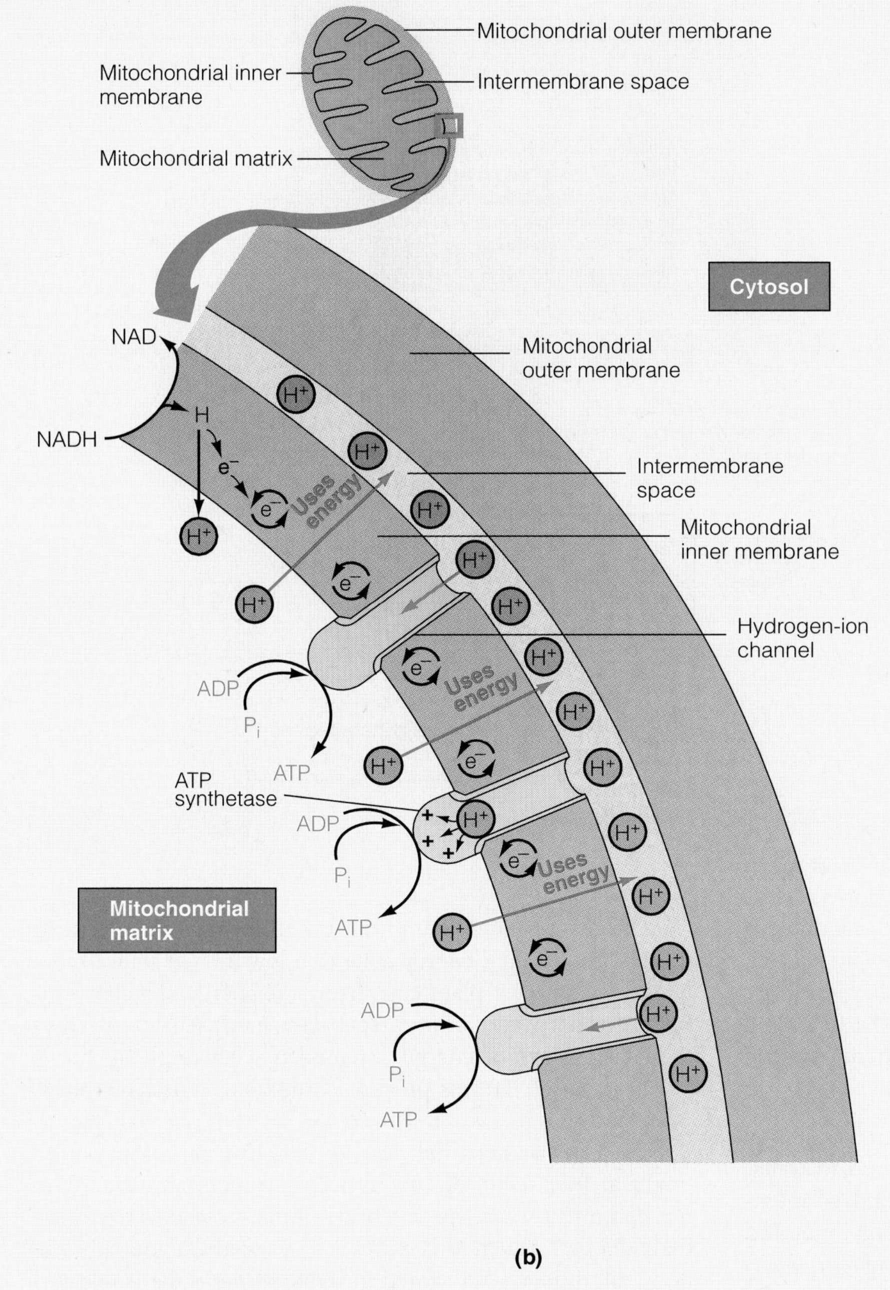

(NAD), a derivative of the B vitamin niacin, and **flavine adenine dinucleotide (FAD)**, a derivative of the B vitamin riboflavin. These compounds are converted by the transfer of hydrogen to NADH and $FADH_2$, respectively.

4. One more molecule of ATP is produced for each molecule of acetyl CoA processed. Actually, ATP is not directly produced by the citric acid cycle. The released energy is used to directly link inorganic phosphate to **guanosine diphosphate (GDP)** to form **guanosine triphosphate (GTP)**, a high-energy molecule similar to ATP. The energy from GTP can then be transferred to ATP as follows:

$$ADP + GTP \rightleftharpoons ATP + GDP$$

Because each glucose molecule is converted into two acetic acid molecules, thus permitting two turns of the citric acid cycle, two more ATP molecules are produced from each glucose molecule.

These two additional ATPs are still not much of an energy profit. However, the citric acid cycle is important in preparing the hydrogen carrier molecules for their entry into the **electron transport chain,** which produces far more energy than the sparse amount of ATP produced by the cycle itself. Considerable untapped energy is still stored in the released hydrogen atoms, which contain electrons at high energy levels. The "big payoff" comes when NADH and $FADH_2$ enter the electron transport chain, which consists of electron carrier molecules located in the inner mitochondrial membrane lining the cristae (Fig. 2-13a). The high energy electrons are extracted from the hydrogens held in NADH and $FADH_2$ and are transferred sequentially to the electron carrier molecules, freeing NAD and FAD to pick up more hydrogen atoms. The electron transport molecules are arranged in a specifically ordered fashion on the inner membrane so that the high-energy electrons are progressively transferred through a chain of reactions, with the electrons falling to successively lower energy levels with each step.

This electron transport chain is also called the **respiratory chain** because it is critical to cellular respiration, which refers to the intracellular oxidation of nutrient derivatives. Ultimately, the electrons are passed to molecular oxygen (O_2) derived from the air we breathe. Electrons bound to O_2 are in their lowest energy state. Oxygen breathed in from the atmosphere enters the mitochondria to serve as the final electron acceptor of the electron transport chain. This negatively charged oxygen (negative because it has acquired additional electrons) then combines with the positively charged hydro-

2. The released carbon atoms, which were originally present in the acetyl CoA that entered the cycle, are converted into two molecules of CO_2. This CO_2, as well as the CO_2 produced during the formation of acetic acid from pyruvic acid, passes out of the mitochondrial matrix and subsequently out of the cell to enter the blood. In turn, the blood carries it to the lungs, where it is finally eliminated into the atmosphere through the process of breathing. The oxygen used to make CO_2 from these released carbon atoms is derived from the molecules that were involved in the reactions, not from free molecular oxygen supplied by breathing.
3. Hydrogen atoms are also "bumped off" during the cycle at four of the chemical conversion steps. These hydrogens are "caught" by two other compounds that act as hydrogen carrier molecules—**nicotinamide adenine dinucleotide**

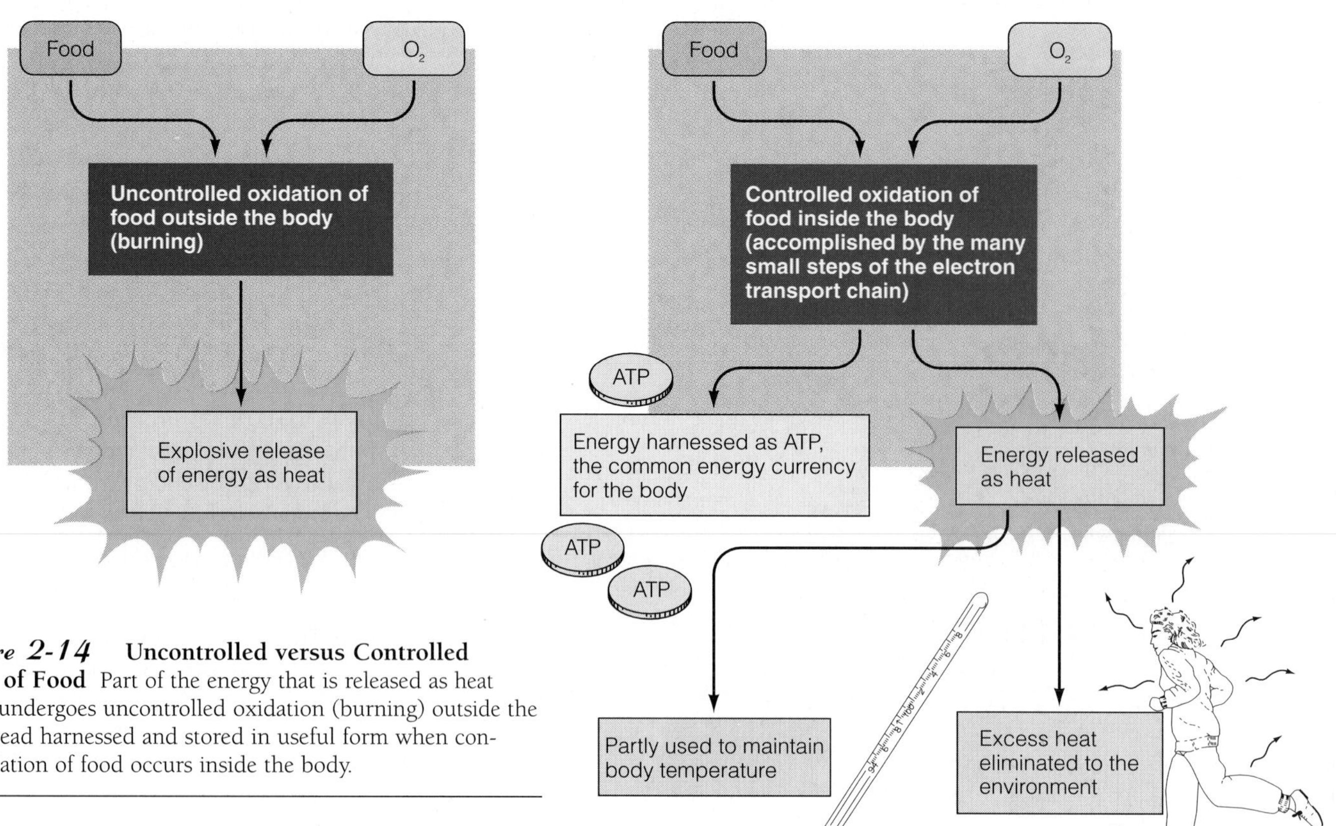

Figure 2-14 Uncontrolled versus Controlled Oxidation of Food Part of the energy that is released as heat when food undergoes uncontrolled oxidation (burning) outside the body is instead harnessed and stored in useful form when controlled oxidation of food occurs inside the body.

gen ions (positive because they have donated the electrons at the beginning of the electron transport chain) to form water.

As the electrons move through this chain of reactions to ever lower energy levels, they release energy. Part of the released energy is lost as heat, but some of it is harnessed by the mitochondrion to synthesize ATP through the following steps, which are collectively known as the **chemiosmotic mechanism:**

1. At three sites in the electron transport chain, the energy released during the transfer of electrons is used to transport hydrogen ions across the inner mitochondrial membrane from the matrix to the space between the inner and outer mitochondrial membranes (the *intermembrane space*) (Fig. 2-13b).
2. As a result of this transport process, hydrogen ions are in greater concentration in the mitochondrial intermembrane space than in the matrix. Due to this difference in concentration, the transported hydrogen ions have a strong tendency to flow back into the matrix through channels or passageways formed by special proteins within the inner mitochondrial membrane.
3. The channels through which the transported hydrogen ions return to the matrix bear the enzyme **ATP synthetase,** which is activated by the flow of hydrogen ions from the intermembrane space to the matrix.
4. Upon activation, ATP synthetase converts ADP + P_i to ATP, providing a rich yield of thirty-two more ATP molecules for each glucose molecule thus processed. ATP is subsequently transported out of the mitochondrion into the cytosol for use as the cell's energy source.

The harnessing of energy into a useful form as the electrons tumble from a high-energy state to a low-energy state can be likened to a power plant converting the energy of water tumbling down a waterfall into electricity. Because O_2 is used in these final steps of energy conversion when a phosphate is added to form ATP, this process is known as **oxidative phosphorylation.**

The series of steps that lead to oxidative phosphorylation might at first seem like an unnecessary complication. Why not just directly oxidize, or "burn," food molecules to release their energy? When this process is carried out outside the body, all of the energy stored in the food molecule is released explosively in the form of heat (Fig. 2-14). In the body, oxidation of food molecules occurs in many small, controlled steps so that the food molecule's chemical energy is gradually made available for convenient packaging in a storage form that is useful to the cell. The cell, by means of its mitochondria, can more efficiently capture the energy from the food molecules within ATP bonds when it is released in small quantities. In this way, much less of the energy is converted to heat. The heat that is produced is not completely wasted energy; it is used to help maintain body temperature, with any excess heat being eliminated to the environment.

The cell is a much more efficient energy converter when oxygen is available (Fig. 2-15). In an **anaerobic** ("lack of air," specifically "lack of O_2") condition, the degradation of glucose cannot proceed beyond glycolysis. Recall that glycolysis takes place in the cytosol and involves the breakdown of glucose into pyruvic acid, producing a low yield of two mol-

A CLOSER LOOK AT EXERCISE PHYSIOLOGY

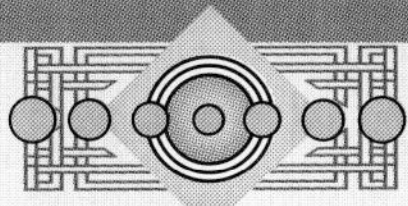

Aerobic Exercise: What For and How Much?

Aerobic ("with O_2") **exercise** involves large muscle groups and is performed at a low enough intensity and for a long enough period of time that fuel sources can be converted to ATP by using the citric acid cycle and electron transport chain as the predominant metabolic pathway. Aerobic exercise can be sustained for from fifteen to twenty minutes to several hours at a time. Short-duration, high-intensity activities, such as weight training and the 100 meter dash, which last for a matter of seconds and rely solely on energy stored in the muscles and on glycolysis, are forms of **anaerobic** ("without O_2") **exercise.**

Inactivity is associated with increased risk of developing both hypertension (high blood pressure) and coronary artery disease (blockage of the arteries that supply the heart). To reduce the risk of hypertension and coronary artery disease and to improve physical work capacity, the American College of Sports Medicine recommends that an individual participate in aerobic exercise a minimum of three times per week for twenty to sixty minutes. The intensity of the exercise should be based on a percentage of the individual's maximal capacity to work. The easiest way to establish the proper intensity of exercise and to monitor intensity levels is by checking the heart rate. The estimated maximal heart rate is determined by subtracting the person's age from 220. Significant benefits can be derived from aerobic exercise performed between 70% and 80% of maximal heart rate. For example, the estimated maximal heart rate for a twenty-year-old is 200 beats per minute. If this person exercised three times per week for twenty to sixty minutes at an intensity that increased the heart rate to 140 to 160 beats per minute, the participant should significantly improve his or her aerobic work capacity and reduce the risk of cardiovascular disease.

ecules of ATP per molecule of glucose. The untapped energy of the glucose molecule remains locked in the bonds of the pyruvic acid molecules, which are eventually converted to lactic acid if they do not enter the pathway that ultimately leads to oxidative phosphorylation. When sufficient O_2 is present— an **aerobic** ("with air" or "with O_2") condition— mitochondrial processing (that is, the citric acid cycle in the matrix and the electron transport chain on the cristae) harnesses sufficient energy to generate thirty-four more molecules of ATP, for a total net yield of thirty-six ATPs per molecule of glucose processed. (For a description of aerobic exercise, see the accompanying boxed feature, A Closer Look at Exercise Physiology.) The overall reaction for the oxidation of food molecules to yield energy is as follows:

food	+	O_2	→	CO_2	+	H_2O	+	ATP
		(necessary for oxidative phosphorylation)		(produced primarily by the citric acid cycle)		(produced by the electron transport chain)		(produced primarily by the electron chain)

Glucose, the principal nutrient derived from dietary carbohydrates, is the fuel preference of most cells. However,

Figure 2-15 Comparison of Energy Yield and Products under Anaerobic and Aerobic Conditions In anaerobic conditions, only two ATPs are produced for every glucose molecule processed, but in aerobic conditions, a total of thirty-six ATPs are produced per glucose molecule.

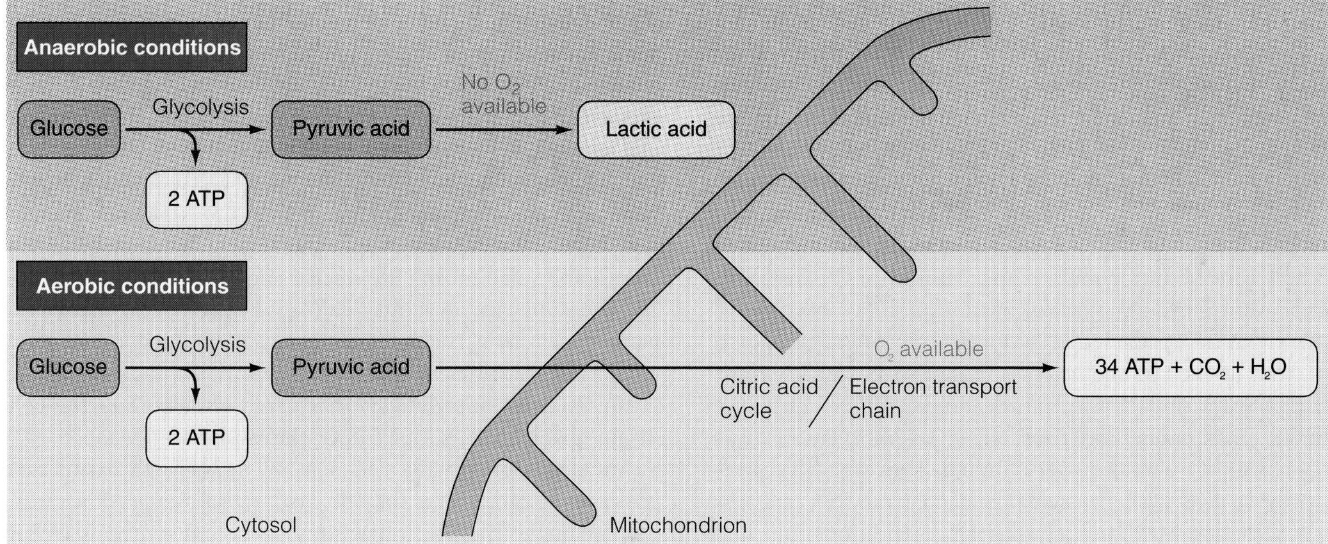

nutrient molecules derived from fats (fatty acids) and, if necessary, from protein (amino acids) can also participate at specific points in this overall chemical reaction to eventually produce energy. Amino acids are usually used for protein synthesis instead of energy production, but they can be used as fuel if insufficient glucose and fat are available (chapter 17).

Note that the oxidative reactions within the mitochondria generate energy, unlike the oxidative reactions controlled by the peroxisome enzymes. Both organelles use O_2, but for different purposes.

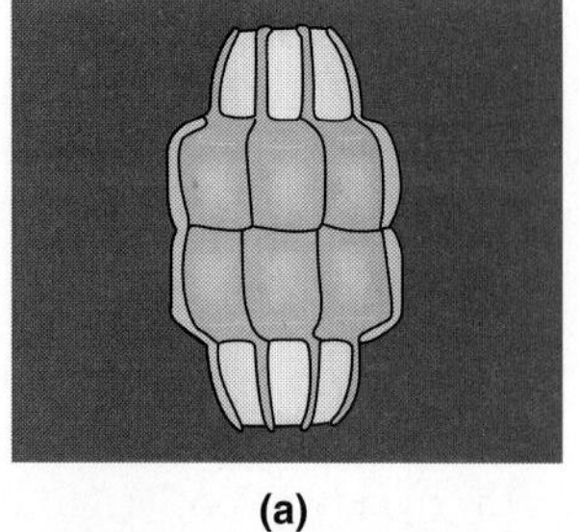

(a)

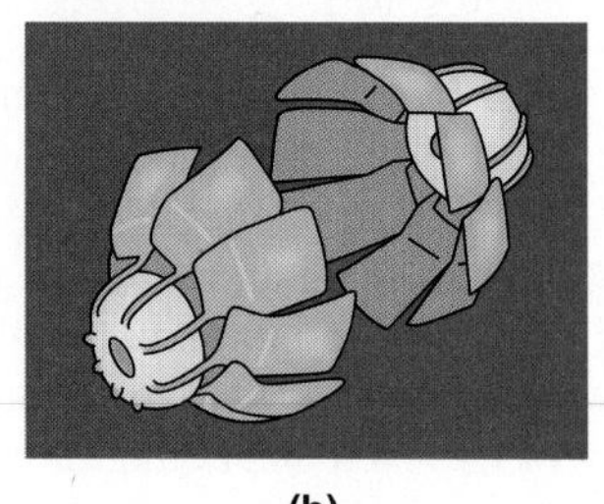

(b)

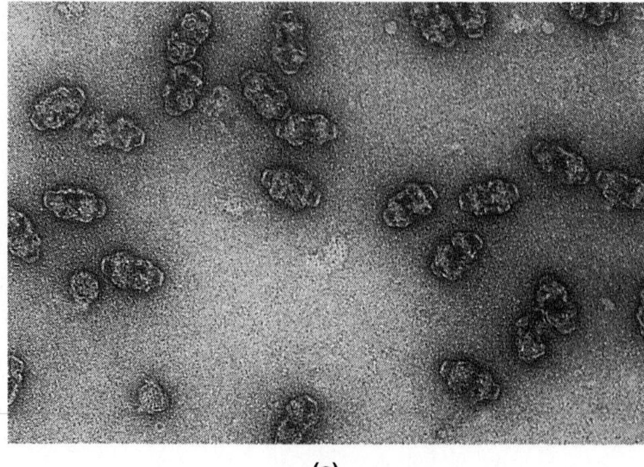

(c)

Figure 2-16 **Vaults** (a) Schematic three-dimensional representation of a vault, an octagonal barrel-shaped organelle believed to transport messenger RNA from the nucleus to the cytoplasmic ribosomes. (b) Schematic representation of an opened vault, showing its octagonal structure. (c) Electron micrograph of vaults.

The energy stored within ATP is used for synthesis, transport, and mechanical work.

Once formed, ATP is transported out of the mitochondria and is then available as an energy source as needed within the cell. Cellular activities that require energy expenditure fall into three main categories:

1. *Synthesis of new chemical compounds,* such as protein synthesis by the endoplasmic reticulum. Some cells, especially cells with a high rate of secretion and cells in the growth phase, use up to 75% of the ATP they generate just to synthesize new chemical compounds.
2. *Membrane transport,* such as the selective transport of molecules across the kidney tubules during the process of urine formation. Kidney cells can expend as much as 80% of their ATP currency to operate their selective membrane-transport mechanisms.
3. *Mechanical work,* such as contraction of the heart muscle to pump blood or contraction of skeletal muscles to lift an object. These activities require tremendous quantities of ATP.

As a result of cellular energy expenditure to support these various activities, large quantities of ATP are produced. These energy-depleted ADP molecules enter the mitochondria for "recharging" and then cycle back into the cytosol as energy-rich ATP molecules after participating in oxidative phosphorylation. A single ADP/ATP molecule may shuttle back and forth between the mitochondria and cytosol for this recharging/expenditure cycle thousands of times per day.

The high demands for ATP render glycolysis alone an insufficient as well as inefficient supplier of power for most cells. If it were not for the mitochondria, which house the metabolic machinery for oxidative phosphorylation, our energy capability would be very limited. However, glycolysis does provide cells with a sustenance mechanism that can produce at least some ATP under anaerobic conditions. Skeletal muscle cells in particular take advantage of this ability during short bursts of strenuous exercise, when energy demands for contractile activity outstrip the body's ability to bring adequate O_2 to the exercising muscles to support oxidative phosphorylation. Also, red blood cells, which are the only cells that do not contain any mitochondria, rely solely on glycolysis for their limited energy production. The energy needs of red blood cells are low, however, because they also lack a nucleus and therefore are not capable of synthesizing new substances, the biggest energy expenditure for most noncontractile cells.

Vaults are a newly discovered organelle.

In addition to the five well-documented organelles, in the early 1990s researchers identified a sixth type of organelle—**vaults.** Vaults, which are three times as large as ribosomes, are shaped like octagonal barrels (Fig. 2-16). Their name comes from their multiple arches, which reminded their discoverers of vaulted or cathedral ceilings. A cell may contain thousands of vaults. Why would the presence of these numerous, relatively large organelles have been elusive until recently? The reason is that they do not show up with ordinary staining techniques. One clue to the function of vaults may be their octagonal shape. Interestingly, the pores in the membrane surrounding the nucleus are also octagonal shaped and the same size as vaults, leading to speculation that vaults may be cellular "trucks." According to this proposal, vaults would dock at nuclear pores, pick up molecules synthesized in the nucleus, and deliver their cargo elsewhere in the cell. At any given time, about 5% of the vaults are localized near the nuclear pores while the rest are in the cytoplasm. One possibility is that vaults may be carrying messenger RNA from the nucleus to the ribosomal sites of protein synthesis within the cytoplasm. Also, because vaults are especially abundant in regions of the cell where actin is being assembled, they may somehow be involved with cellular contractile systems. Furthermore, vaults may play an undesirable role in bringing about the multidrug resistance sometimes displayed by cancer cells. Researchers have shown that some cancer cells resistant to chemotherapy produce more than normal quantities of a particular protein, which, surprisingly, turns out to be the major vault protein. If further investigation confirms that vaults play a role in drug resistance—perhaps by transporting the drugs to sites for exocytosis from the cancer cells—the exciting possibility exists that interference with this vault activity could improve the sensitivity of cancer cells to chemotherapeutic drugs.

Cytosol and Cytoskeleton

Occupying about 55% of the total cell volume, the cytosol is the semiliquid portion of the cytoplasm that surrounds the organelles. Its amorphous appearance under an electron microscope belies the fact that the cytosol is not a uniform liquid mixture but is actually more like a highly organized, gelatinous mass with differences in composition and consistency between various regions of the cell.

The cytosol is important in intermediary metabolism, ribosomal protein synthesis, and storage of fat and glycogen.

Three general categories of activities are associated with the cytosol: (1) enzymatic regulation of intermediary metabolism; (2) ribosomal protein synthesis; and (3) storage of fat, carbohydrate, and secretory vesicles. Dispersed throughout the cytosol is a cytoskeleton that gives shape to the cell, provides an intracellular organizational framework, and is responsible for various cell movements.

Enzymatic regulation of intermediary metabolism **Intermediary metabolism** refers collectively to the large set of intracellular chemical reactions that involve the degradation, synthesis, and transformation of small organic molecules such as simple sugars, amino acids, and fatty acids. These reactions are critical for ultimately capturing energy to be used for cellular activities and for providing the raw materials needed for maintenance of the cell's structure and function and for the cell's growth. All intermediary metabolism occurs in the cytoplasm, with most of it being accomplished in the cytosol. Thousands of enzymes involved in glycolysis and other intermediary biochemical reactions are found in the cytosol.

Ribosome protein synthesis Also dispersed throughout the cytosol are the free ribosomes, which synthesize proteins for use in the cytosol itself. In contrast, rough ER ribosomes synthesize proteins for secretion and for construction of new cellular components. Often, cytosolic ribosomes that are synthesizing identical proteins are clustered together in "assembly lines" known as **polyribosomes.**

Storage of fat and glycogen Excess nutrients not immediately used for ATP production are converted in the cytosol into storage forms that are readily visible even under a light microscope. Such nonpermanent masses of stored material are known as **inclusions.** The largest and most important storage product is fat. Small fat droplets can be seen within the cytosol in various cells. In **adipose tissue,** the tissue specialized for fat storage, the stored fat molecules can occupy almost the entire cytosol, coalescing to form one large fat droplet (Fig. 2-17a). The other visible storage product is **glycogen,** the storage form of glucose, which appears as aggregates or clusters dispersed throughout the cell (Fig. 2-17b). Cells vary in their ability to store glycogen, with liver and muscle cells having the greatest stores. When food is not

Figure 2-17 **Inclusions** (a) Light micrograph depicting fat storage in an adipose cell. Note that the fat droplet occupies almost the entire cytosol. (b) Light micrograph depicting glycogen storage in a liver cell. The red-staining granules throughout the liver cell's cytosol are glycogen deposits.

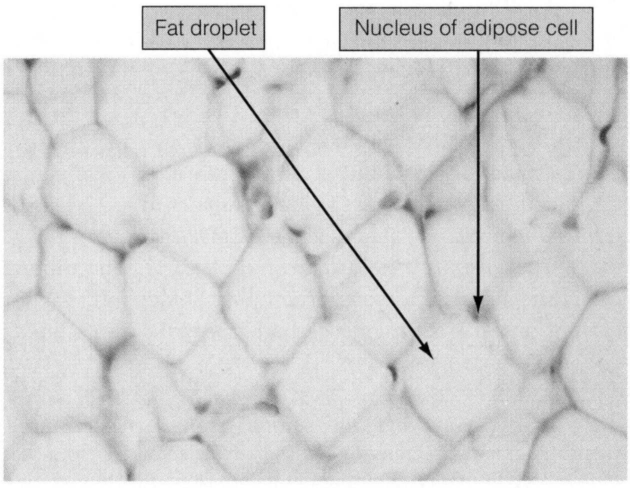

(a)

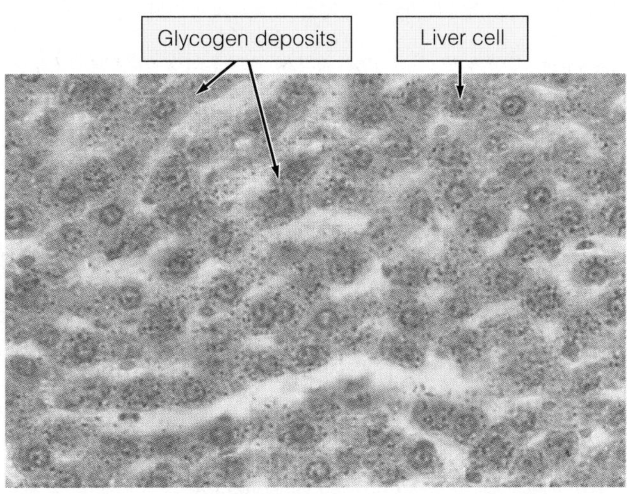

(b)

available to provide fuel for the citric acid cycle and electron transport chain, stored glycogen and fat are broken down to release glucose and fatty acids, respectively, which can feed the mitochondrial energy-producing machinery. An average adult has enough glycogen stored to provide sufficient energy for about a day of normal activities, and typically enough fat is stored to provide energy for two months.

Secretory vesicles that have been processed and packaged by the endoplasmic reticulum and Golgi complex also remain in the cytosol, where they are stored until signaled to empty their contents to the outside.

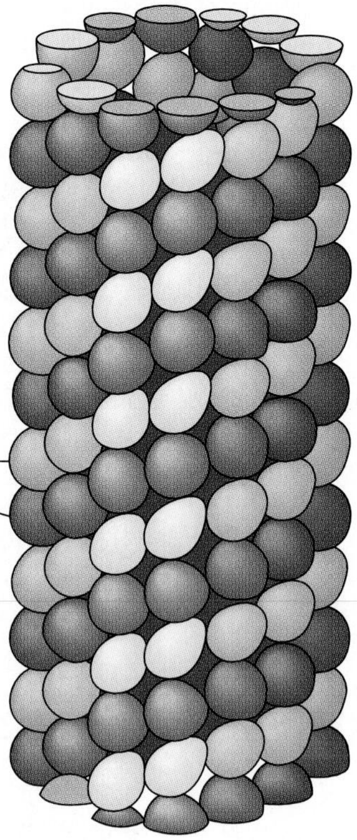

Figure 2-18 **Arrangement of Tubulin Molecules in a Microtubule** Microtubules, the largest of the cytoskeletal elements, are long, hollow tubes formed by globular-shaped tubulin molecules.

Presence of the cytoskeleton Permeating the cytosol is the **cytoskeleton,** a complex protein network that acts as the "bone and muscle" of the cell. The distinct shape, size, complexity, and intracellular specialization of the various body cells necessitate intracellular scaffolding to support and organize the cellular components into an appropriate arrangement and to control their movements. These functions are performed by the cytoskeleton. This elaborate network has at least four distinct elements: (1) *microtubules,* (2) *microfilaments,* (3) *intermediate filaments,* and (4) the *microtrabecular lattice* (Table 2-1). The different parts of the cytoskeleton are structurally linked and functionally coordinated to provide certain integrated functions for the cell. Because of the complexity of this network and the variety of functions it serves, its elements will be addressed separately.

Microtubules are essential for maintaining asymmetrical cell shapes and are important in complex cell movements.

The **microtubules** are the largest of the cytoskeletal elements. They are very slender [22 nanometer (nm) diameter, 1 nm = 1 billionth of a meter], long, hollow, unbranched tubes composed primarily of **tubulin,** a small, globular protein molecule (6 nm diameter) (Fig. 2-18). Microtubules are essential for maintaining an asymmetrical cell shape, such as that of a nerve cell, whose elongated axon may extend up to a meter in length from the origin of the cell body in the spinal cord to the termination of the axon at a muscle (Fig. 2-19). Microtubules, along with specialized intermediate filaments, stabilize this asymmetrical axonal extension. In cells whose mature shape and organization do not depend on the presence of microtubules, these cytoskeletal structures are present at an earlier developmental stage of the cell. Here they function as temporary scaffolding, creating a shape and organization that are later maintained by other cellular elements. For example, spiral rings of microtubules wrap around the base of the tail of a developing sperm and subsequently disappear. Before their disappearance, the microtubules serve as a transient "mold" for arranging the mitochondria spirally

Figure 2-19 **Two-Way Vesicular Axonal Transport Facilitated by the Microtubular "Highway" in a Nerve Cell** Schematic illustration of a neuron depicting secretory vesicles being transported from the site of production in the cell body along a microtubule "highway" to the terminal end for secretion. Vesicles containing debris are being transported in the opposite direction for degradation in the cell body. The enlargement depicts kinesin, a motor protein, carrying a secretory vesicle down the microtubule by using its "feet" to "step" on one tubulin molecule after another.

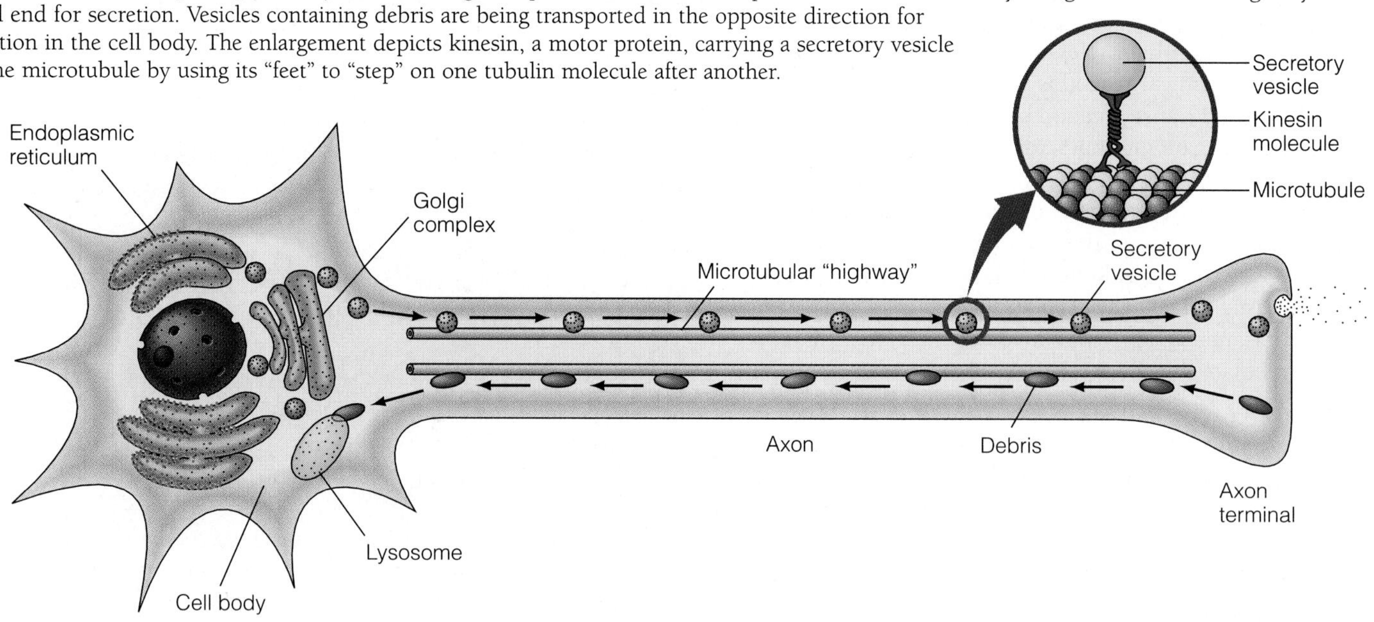

around the tail, where these energy organelles will be readily available to meet the cell's energy needs for motility.

Microtubules also play an important role in coordinating numerous complex cell movements, including (1) transport of secretory vesicles from one region of the cell to another; (2) movement of specialized cell projections such as cilia and flagella; and (3) distribution of chromosomes during cell division through formation of a mitotic spindle.

Transport of secretory vesicles Axonal transport provides a good example of the importance of an organized system for moving secretory vesicles. In a nerve cell, specific chemicals are released from the terminal end of the elongated axon to influence a muscle or another structure that the nerve cell controls. These chemicals are largely produced within the cell body, where the nuclear DNA blueprint, endoplasmic reticular factory, and Golgi packaging and distribution outlet are located. Yet these chemicals ultimately function at the end of the axon, which may be a meter away. If these chemicals had to diffuse on their own from the cell body to a distant axon terminal, it would take them about fifty years to get there—obviously an impractical solution. The microtubules provide a "highway" for vesicular traffic along the axon, with the driving force depending on ATP (Fig. 2-19). Specific protein molecules attach to the particle to be transported and "walk" along the microtubule, powered by ATP. **Kinesin** is one such transport, or motor (*motor* means "movement"), protein. It consists of a long tail and two globular heads. Kinesin's tail binds to the particle to be moved while its globular heads act like little feet, alternately attaching to one tubulin molecule on the microtubule, bending and pushing forward, letting go, then recocking and reattaching to the next tubulin molecule farther down the microtubule. The process is repeated over and over as kinesin moves its cargo to the end of the axon by using each of the tubulin molecules as a "stepping stone."

Reverse vesicular traffic also occurs along these microtubular highways. Vesicles that contain debris are transported by different motor proteins from the axon terminal to the cell body for degradation by lysosomes, which are confined within the cell body. Coincidentally, these microtubular highways may also serve as pathways for the movement of such infectious agents as herpes virus, poliomyelitis virus, and rabies virus, which travel retrograde (backward) along nerves from their surface site of contamination to the central nervous system (brain and spinal cord).

Movement of cilia and flagella Microtubules are also the dominant structural and functional components of cilia and flagella. These specialized protrusions from the cell surface allow a cell to move materials across its surface (in the case of a stationary cell) or to propel itself through its environment (in the case of a motile cell). **Cilia** are numerous tiny, hairlike protrusions, whereas a **flagellum** is a single, long, whiplike appendage. Even though they project from the surface of the cell, cilia and flagella are both intracellular structures covered by the plasma membrane.

Cilia beat or stroke in unison, much like the coordinated efforts of a rowing team. Each cilium exerts a rapid active stroke, which moves material on the cell surface forward. This stroke is followed by a recovery phase in which the cilium more slowly returns to its original position with a kind of unrolling, backward movement that does not exert much force. In this way, the material just pushed forward is not futilely pushed backward but instead is moved only in the direction of the forward stroke.

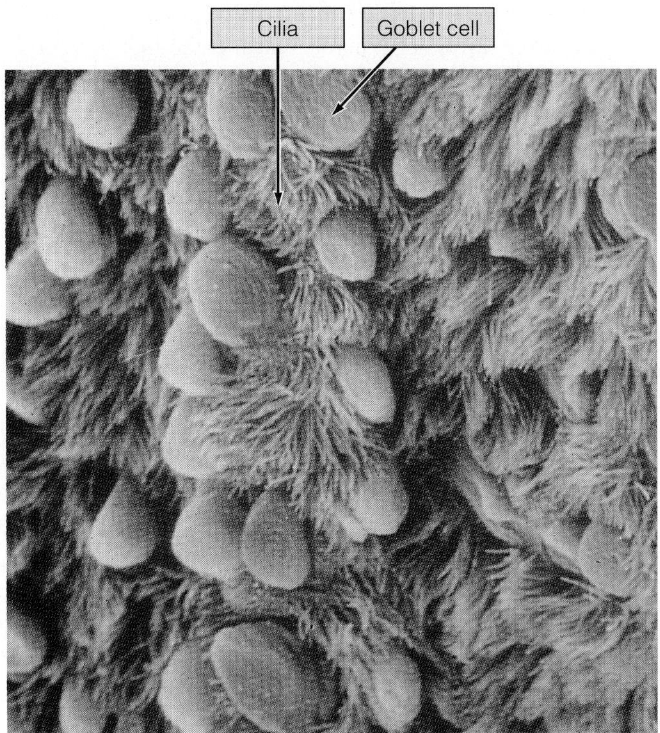

Figure 2-20 **Scanning Electron Micrograph of Cilia on Cells Lining the Respiratory Tract in Humans** The respiratory airways are lined by goblet cells, which secrete a sticky mucus that traps inspired particles, and epithelial cells that bear numerous hairlike cilia. The cilia beat in unison to sweep inspired particles up and out of the airways.

In humans, ciliated cells are found in the stationary cells that line the respiratory tract and the oviduct of the female reproductive tract. Respiratory cilia help keep foreign particles out of the lungs (Fig. 2-20). The thousands of cilia lining the respiratory airways project into a layer of sticky mucus that traps dust and other inspired particles. The coordinated stroking action of these cilia sweeps this dust-laden mucus up to the throat, where it can be expectorated (spit out) or swallowed and eventually eliminated in the feces. In the female reproductive tract, the sweeping action of the cilia that line the oviduct draws the egg (ovum) released from the ovary during ovulation into the oviduct and then guides it toward the uterus (womb).

The only human cells that bear flagella are sperm. The whiplike motion of the flagellum or "tail" enables a sperm to move through its environment (see Fig. 20-8, p. 712). This ability is particularly useful when the sperm maneuvers for final penetration of the ovum during fertilization.

Cilia and flagella have the same basic internal structure. Both consist of nine fused pairs of microtubules (doublets) arranged in an outer ring around two single unfused microtubules in the center (Fig. 2-21). This characteristic "nine plus two" array of microtubules extends throughout the length of the motile appendage. A cilium or flagellum originates from a specialized cytoplasmic structure, the **basal body**, which is located inside the main part of the cell and provides a base from which the microtubules grow to form the appendage. Each basal body is a short cylinder composed of a parallel microtubular symmetry similar to that of the cilum or flagellum.

Associated with these microtubules are accessory proteins that maintain the microtubules' organization and play an essential part in the microtubular movement that causes the entire structure to bend. The most important of these accessory proteins is **dynein**, which forms a set of armlike projections from each doublet of microtubules (Fig. 2-21a). The bending movements of cilia and flagella are produced by the sliding of adjacent microtubule doublets past each other. Sliding is accomplished by the dynein arms, which are motor proteins like kinesin. The dynein arms have the capability of splitting ATP and then using the released energy to "crawl" along the neighboring microtubule doublet (similar to kinesin's movement) to cause relative displacement of the doublets (Fig. 2-22). Groups of cilia working together are oriented to beat in the same direction and contract in a synchronized manner through controlling mechanisms that are poorly understood. These mechanisms appear to involve the single microtubules at the cilium's center and their surrounding accessory proteins, the **inner sheath** (Fig. 2-21a).

Formation of the mitotic spindle Cell division involves two discrete but related activities: *mitosis* (nuclear division) and *cytokinesis* (cytoplasmic division). **Mitosis** involves replication of the DNA-containing chromosomes, which are then evenly distributed in the two halves of the cell (see p. B–11). In **cytokinesis**, the plasma membrane is constricted in the middle of the cell, and the two halves separate into two new daughter cells, each with a full complement of chromosomes (Fig. 2-23).

Microtubules are transiently assembled to form a **mitotic spindle**, which organizes and directs the movement of the replicated chromosomes away from each other toward opposite ends of the cell so that the genetic material is evenly distributed when the cell divides (see Fig. B–10). The mitotic spindle is formed by the **centrioles**, a pair of short cylindrical structures that lie at right angles to each other near the nucleus (Fig. 2-1). The centrioles also duplicate during cell division. After self-replication, the centriole pairs move toward opposite ends of the cell and form the spindle apparatus between them through a precisely organized assemblage of microtubules.

Besides their role in mitotic-spindle formation, the centrioles and surrounding complex of densely staining proteinaceous material together assemble the many microtubules that normally radiate throughout the cytoskeleton. The centrioles are identical in structure to basal bodies. In fact, under some circumstances the centrioles and basal bodies are intercon-

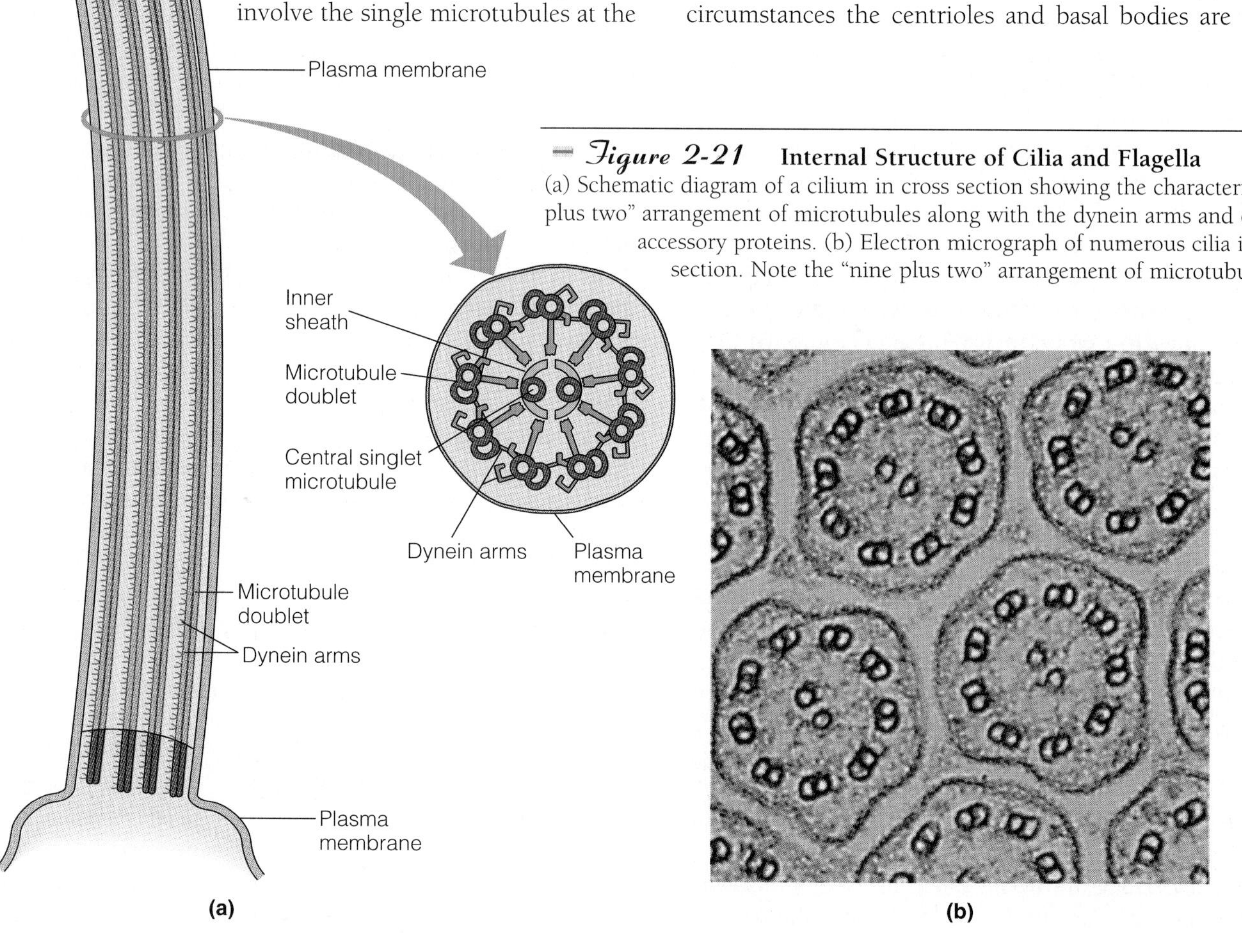

Figure 2-21 **Internal Structure of Cilia and Flagella**
(a) Schematic diagram of a cilium in cross section showing the characteristic "nine plus two" arrangement of microtubules along with the dynein arms and other accessory proteins. (b) Electron micrograph of numerous cilia in cross section. Note the "nine plus two" arrangement of microtubules.

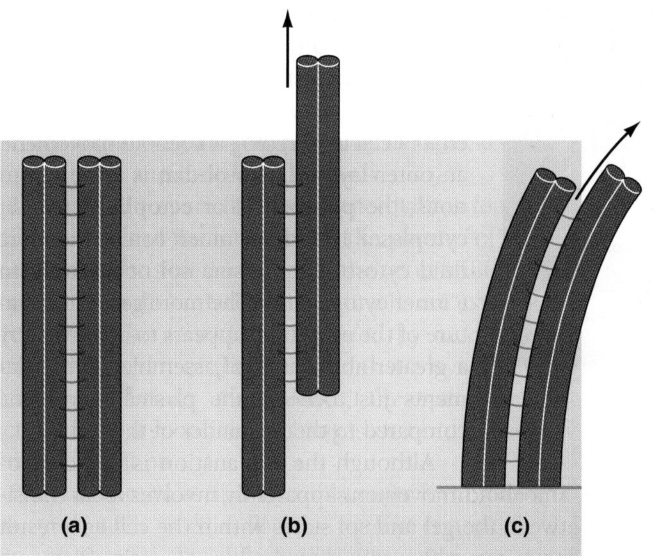

Figure 2-22 **Bending of Cilium Accomplished by Relative Sliding of Anchored Microtubule Doublets**
(a) Dynein arms extend between the microtubule doublets within a cilium. Relative sliding of adjacent microtubule doublets is believed to be accomplished by ATP-powered dynein "crawling." (b) If the microtubule doublets were not anchored at their base, they would freely slide past each other as a result of this dynein activity. (c) Because the doublets are anchored at their base, bending occurs when the doublets slide with respect to each other as a result of dynein activity.

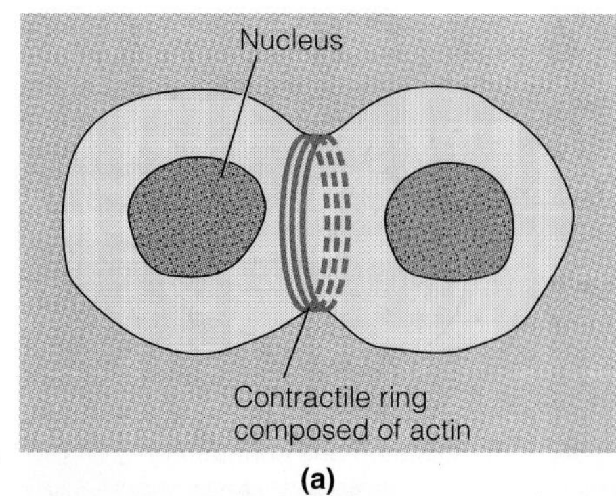

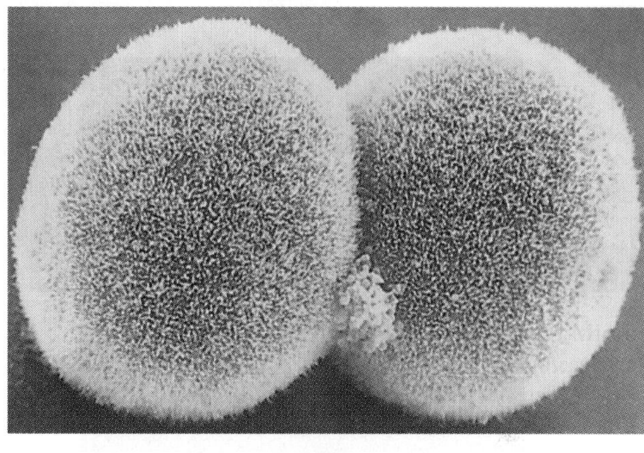

Figure 2-23 **Cytokinesis** (a) Schematic illustration of the actin contractile ring squeezing apart the two duplicate cell halves during cytokinesis. (b) Photograph of a cell undergoing cytokinesis.

vertible. During development of ciliated human cells, the centriole pair migrates to the region of the cell where the cilia will be formed and duplicates itself to produce the many basal bodies that will form the cilia.

Microfilaments are important to cellular contractile systems and as mechanical stiffeners.

The **microfilaments** are the smallest (6 nm diameter) elements of the cytoskeleton visible with a conventional electron microscope. The most obvious microfilaments in most cells are those composed of **actin,** a protein molecule that has a globular shape similar to tubulin. Unlike tubulin, which forms a hollow tube, actin is assembled into two twisted strands, much like two strings of pearls twisted into a helix (spiral) to form a microfilament (Fig. 2-24). In muscle cells, another protein called **myosin** forms a different kind of microfilament (see Fig. 8–5, p. 225). In most cells, myosin is not as abundant and does not form such distinct filaments.

Microfilaments serve at least two functions: (1) they play a vital role in various cellular contractile systems, and (2) they act as mechanical stiffeners for several specific cellular projections.

Microfilaments in cellular contractile systems Actin-based assemblies are involved in muscle contraction, cell division, and cell locomotion. The most obvious, best organized, and most clearly understood cellular contractile system is that found in muscle. Muscle contains an abundance of actin and myosin filaments, which are organized so that when they are triggered by electrically induced ATP splitting to slide past each other, they generate a contractile force (chapter 8).

Surprisingly, nonmuscle cells may also contain "muscle-like" assemblies. Some of these microfilament contractile systems are transiently assembled to perform a specific function when needed. A good example is the contractile ring that

Figure 2-24 **Arrangement of Actin Molecules in a Microfilament**

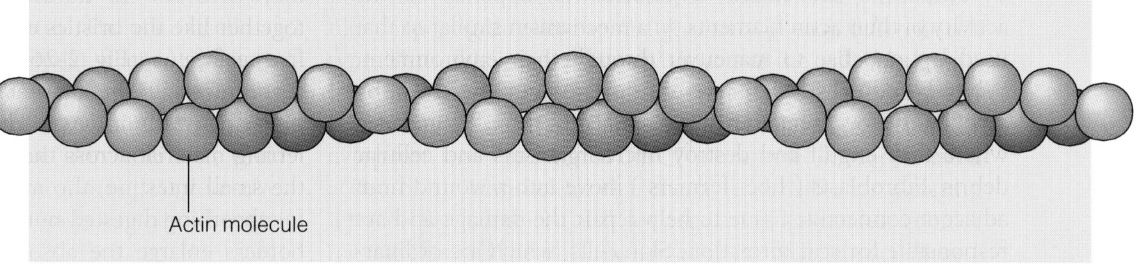

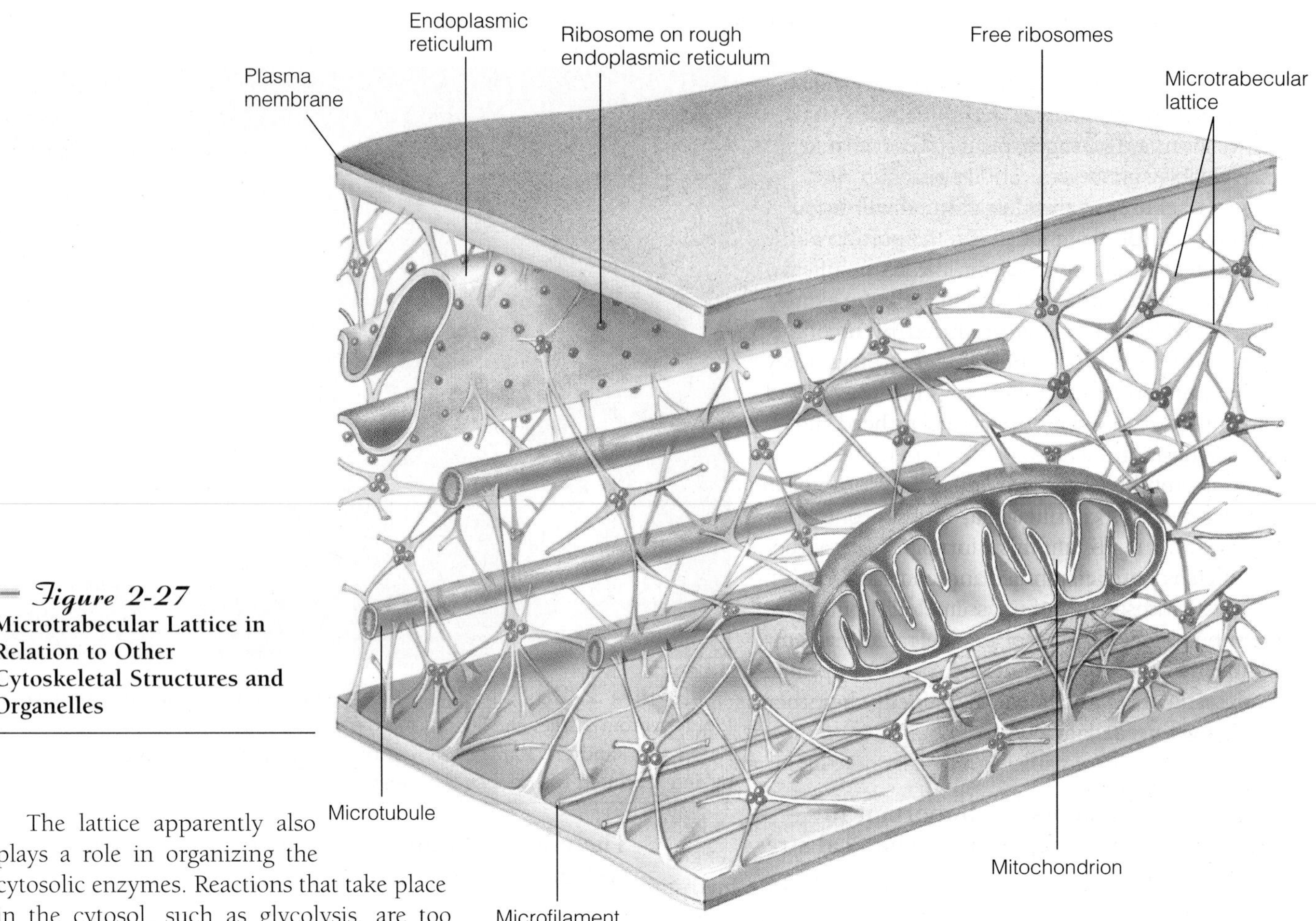

Figure 2-27
Microtrabecular Lattice in Relation to Other Cytoskeletal Structures and Organelles

The lattice apparently also plays a role in organizing the cytosolic enzymes. Reactions that take place in the cytosol, such as glycolysis, are too well choreographed and too rapid to occur by random contacts between enzymes and their substrates (the substances acted on by the enzymes). There is convincing evidence that these enzymes are somehow incorporated into the lattice, probably in some sort of sequential alignment that guides glucose through the steps of the glycolytic pathway. Furthermore, the lattice, by sequestering the free ribosomes at its intersections, plays an important role in governing the dispensation of newly synthesized cytosolic proteins. Once proteins have been synthesized, they are available in a controlled, nonrandom fashion for enzymatic activities and for the assembly of microtubules, microfilaments, and other cytosolic components.

The microtrabecular lattice is not a rigid, static structure. Its structure has been observed to vary reversibly when the cell changes shape or when intracellular movements are occurring. The lattice changes constantly through local contractions, expansions, and deformations, so that the organelles and cytoskeletal fibers are continually redistributed and reoriented as the cell carries on its various activities. Recent observations suggest that signals received via an organized cytoskeleton normally play a role in regulating cell growth and division, a system that might be disrupted in cancer cells. Conspicuous cytoskeletal changes accompanied by abnormal growth behavior are often present in cancer cells. Very little is known about the interactions among the components of the cytoskeleton in carrying out these and other integrated activities.

Chapter in Perspective: Focus on Homeostasis

The ability of cells to perform functions essential for their own survival as well as specialized tasks that contribute to the maintenance of homeostasis within the body ultimately depends on the successful, cooperative operation of the intracellular components. For example, to support life-sustaining activities, all cells must generate energy in a usable form from nutrient molecules. Energy is generated intracellularly by chemical reactions that take place within the cytosol and mitochondria.

In addition to being essential for basic cell survival, the organelles and cytoskeleton also participate in many cells' specialized tasks that contribute to homeostasis. Following are several examples:

- Nerve and endocrine cells both release proteinaceous chemical messengers that are important in regulatory activities aimed at maintaining homeostasis—for example, chemical messengers released from nerve cells stimulate the respiratory muscles, which accomplish life-sustaining exchanges of O_2 and CO_2 between the body and atmosphere through breathing. These proteinaceous chemical messengers (neurotransmitters in nerve cells and hormones in endocrine cells) are all produced by the endoplasmic reticulum and Golgi complex and released by exocytosis from the cell when needed.

- The ability of muscle cells to contract depends on their highly developed cytoskeletal microfilaments sliding past each other. Muscle contraction is responsible for many homeostatic activities, including (1) contraction of the heart muscle, which pumps life-supporting blood throughout the body; (2) contraction of the muscles attached to bones, which enables the body to procure food; and (3) contraction of the muscle in the walls of the stomach and intestine, which moves the food along the digestive tract so that ingested nutrients can be progressively broken down into a form that can be absorbed into the blood for delivery to the cells.
- White blood cells help the body resist infection by making extensive use of lysosomal destruction of engulfed particles as they police the body for microbial invaders.

As we begin to examine the various organs and systems, keep in mind that proper cellular functioning is the foundation of all organ activities.

Chapter Summary

Introduction

The complex organization and interaction of the various chemicals within a cell confer the unique characteristics of life. Cells, in turn, are the living building blocks of the body.

Body cells, which are too small to be seen by the unaided eye, have been shown by microscopic techniques to consist of three major subdivisions: (1) the plasma membrane, which encloses the cell and separates the intracellular and extracellular fluid; (2) the nucleus, which contains deoxyribonucleic acid (DNA), the cell's genetic material; and (3) the cytoplasm, the portion of the cell's interior not occupied by the nucleus. The cytoplasm consists of cytosol, a complex gelatinlike mass, and organelles, which are highly organized, membrane-enclosed structures dispersed within the cytosol. Compartmentalization of specific sets of chemicals within the organelles permits chemical activities that would not be compatible with each other to occur simultaneously within separate organelle compartments.

Organelles

Six types of organelles are found in most cells: endoplasmic reticulum, Golgi complex, lysosomes, peroxisomes, mitochondria, and vaults. The endoplasmic reticulum (ER) is a single, complex membranous network that encloses a fluid-filled lumen. The primary function of the ER is to serve as a factory for synthesizing proteins and lipids to be used for (1) the production of new cellular components, particularly cell membranes, and (2) the secretion of special products such as enzymes and hormones to the exterior of the cell. There are two types of endoplasmic reticulum: rough endoplasmic reticulum, which is studded with ribosomes, and smooth endoplasmic reticulum, which lacks ribosomes. The rough endoplasmic reticular ribosomes synthesize proteins, which are released into the ER lumen so that they are separated from the cytosol. Also entering the lumen are lipids produced within the membranous walls of the ER. Synthesized products move from the rough ER to the smooth ER, where they are packaged and discharged as transport vesicles. Transport vesicles are formed as a portion of the smooth ER "buds off," containing a collection of newly synthesized proteins and lipids wrapped in smooth ER membrane.

The Golgi complex, which consists of stacks of flattened, membrane-enclosed sacs, serves a twofold function: (1) to act as a refining plant for modifying into a finished product the newly synthesized molecules delivered to it in crude form from the endoplasmic reticular factory; and (2) to sort, package, and direct molecular traffic to appropriate intracellular and extracellular destinations.

Each cell contains several hundred lysosomes, which are membrane-enclosed sacs that contain powerful hydrolytic (digestive) enzymes. Serving as the intracellular digestive system, lysosomes destroy phagocytized foreign material such as bacteria, demolish worn-out cell parts to make way for new replacement parts, and eliminate the entire cell if it is severely damaged or dead.

Peroxisomes, small membrane-enclosed sacs containing powerful oxidative enzymes, are specialized for carrying out particular oxidative reactions, including certain detoxification activities.

The rod-shaped mitochondria are the energy organelles of the cell. They house the enzymes of the citric acid cycle and electron transport chain, which efficiently convert the energy in food molecules to the usable energy stored in ATP molecules. During this process, which is known as oxidative phosphorylation, the mitochondria utilize molecular oxygen and produce carbon dioxide and water as by-products. The body's cells use ATP as an energy source for synthesis of new chemical compounds, for membrane transport, and for mechanical work.

Vaults, which are shaped like octagonal barrels, are the same shape and size as the nuclear pores. It is speculated that vaults may transport messenger RNA from the nucleus to the cytoplasmic sites of protein synthesis. They are especially abundant at sites of actin assembly.

Cytosol and Cytoskeleton

The cytosol contains the enzymes involved in intermediary metabolism and the ribosomal machinery essential for synthesis of these enzymes as well as other cytosolic proteins. Furthermore, many cells store unused nutrients within the cytosol in the form of glycogen granules or fat droplets. Also present in the cytosol are secretory vesicles containing products that are to be discharged from the cell on appropriate stimulation. Pervading the cytosol is the cytoskeleton, which serves as the "bone and muscle" of the cell. The four types of cytoskeletal elements—microtubules, microfilaments, intermediate filaments, and microtrabecular lattice—are each composed of different proteins and perform various roles. Collectively, the cytoskeletal elements give the cell shape and support, enable it to organize and move its internal structures as needed, and, in some cells, allow movement between the cell and its environment.

Review Exercises

Objective Questions (Answers on p. E–1.)

1. The barrier that separates and controls movement between the cellular contents and the extracellular fluid is the ____________.
2. The chemical that directs protein synthesis and serves as a genetic blueprint is ____________, which is found in the ____________ of the cell.
3. The cytoplasm consists of ____________, which are specialized, membrane-enclosed intracellular compartments, and a gel-like mass known as ____________, which contains an elaborate protein network called the ____________.
4. Transport vesicles from the ____________ fuse with and enter the ____________ for modification and sorting.
5. The (what kind of) ____________ enzymes within the peroxisomes primarily detoxify various wastes produced within the cell or foreign compounds that have entered the cell, generating ____________ in the process.
6. The universal energy carrier of the body is ____________.
7. The largest cells in the human body can be seen by the unaided eye. (True or false?)
8. Inadvertent rupture of lysosomal membranes inevitably leads to complete self-destruction of the cell. (True or false?)
9. Amoeboid movement is accomplished by transitions of the cytosol between a gel and a sol state as a result of alternate assembly and disassembly, respectively, of actin filaments. (True or false?)
10. Using the answer code below, indicate which type of ribosome is being described:

    (a) free ribosome
    (b) rough ER–bound ribosome

    ____1. synthesizes proteins used to construct new cell membrane
    ____2. synthesizes proteins used intracellularly within the cytosol
    ____3. synthesizes secretory proteins such as enzymes or hormones
11. Using the answer code below, indicate which form of energy production is being described:

    (a) glycolysis
    (b) citric acid cycle
    (c) electron transport chain

    ____1. takes place in the mitochondrial matrix
    ____2. produces H_2O as a by-product
    ____3. rich yield of ATP
    ____4. takes place in the cytosol
    ____5. processes acetyl CoA
    ____6. located in the mitrochondrial inner-membrane cristae
    ____7. converts glucose into two pyruvic acid molecules
    ____8. utilizes molecular oxygen

Essay Questions

1. What are a cell's three major subdivisions?
2. Distinguish between intracellular and extracellular fluid.
3. State an advantage of organelle compartmentalization.
4. List the six types of organelles.
5. Describe the structure of the endoplasmic reticulum, distinguishing between the rough and the smooth ER. What is the function of each?
6. Compare exocytosis and endocytosis. Define secretion, pinocytosis, and phagocytosis.
7. Which organelles serve as the intracellular digestive system? What type of enzymes do they contain? What functions do these organelles serve?
8. Compare lysosomes with peroxisomes.
9. Describe the structure of mitochondria and explain their role in oxidative phosphorylation.
10. Distinguish between the oxidative enzymes found in peroxisomes and those found in mitochondria.
11. What three categories of cellular activities require energy expenditure?
12. List and describe the functions of each of the components of the cytoskeleton.

Quantitative Exercises (Solutions on p. E–2.) (See Appendix C, *Principles of Quantitative Reasoning*)

1. Each "turn" of the Krebs cycle
   a. generates 3 NAD^+, 1 FADH, and 2 CO_2
   b. generates 1 GTP, 2 CO_2, and 1 $FADH_2$
   c. consumes 1 pyruvate and 1 oxaloacetate
   d. consumes an amino acid
2. Let's consider how much ATP you synthesize in a day. Assume that you consume one mole of O_2 per hour or 24 moles/day (a mole is the number of grams of a chemical equal to its molecular weight). About 6 moles of ATP are produced per mole of O_2 consumed. The molecular weight of ATP is 507. How many grams of ATP do you produce per day at this rate?

   Given that 1,000 gm equal 2.2 pounds, how many pounds of ATP do you produce per day at this rate? (This is under relatively inactive conditions!)
3. Under resting circumstances a person produces about 144 moles ATP per day (73,000 g ATP/day). The amount of *free energy* represented by this amount of ATP can be calculated as follows. Cleavage of the terminal phosphate bond from ATP results in a decrease of free energy of approximately 7,300 cal/mole. This is a crude measure of the energy available to do work that is contained in the terminal phosphate bond of the ATP molecule. How many calories, in the form of ATP, are produced per day by a resting individual, crudely speaking?
4. Calculate the number of cells in the body of an average 68-kg (150 pound) adult. (This will only be accurate to about one part in ten but should give you an idea how this commonly

quoted number is arrived at). Assume all cells are spheres 20 µm in diameter. The volume of a sphere can be determined by the equation: $v = \frac{4}{3} \pi r^3$. *Hint:* We know that about 2/3 of the water in the body is intracellular, and the density of cells is nearly 1 g/ml. The proportion of mass made up of water is about 60%.

5. If sucrose is injected into the bloodstream, it tends to stay out of the cells (cells do not use sucrose directly). If it doesn't go into cells, where does it go? In other words, how much "space" is in the body that is not inside some cell? Sucrose can be used to determine this space. Suppose 150 mg of sucrose is injected into a 55 kg woman. If the concentration of sucrose in her blood is 0.015 mg/ml, what is the volume of her extracellular space, assuming that no metabolism is occurring and that the blood sucrose concentration is equal to the sucrose concentration throughout the extracellular space?

Points to Ponder

(Explanations on p. E–2.)

1. After a mother stops breast-feeding her infant, the highly developed milk-secreting glands and supportive structures in the breasts gradually diminish. What organelle do you think is responsible for diminution of the breast tissue upon cessation of lactation (milk production)?
2. The poison *cyanide* acts by binding irreversibly to one of the components of the electron transport chain, blocking its action. As a result, the entire electron transport process comes to a screeching halt and the cells lose over 94% of their ATP-producing capacity. Considering the types of cellular activities that depend on energy expenditure, what would be the consequences of cyanide poisoning?
3. Hydrogen peroxide, which belongs to a class of very unstable compounds known as *free radicals,* can bring about drastic, detrimental changes in a cell's structure and function by reacting with almost any molecule with which it comes in contact, including DNA. The resultant cellular changes can lead to genetic mutations, cancer, or other serious consequences. Furthermore, some researchers speculate that cumulative effects of more subtle cellular damage resulting from free radical reactions over a period of time might contribute to the gradual deterioration associated with aging. Related to this speculation, studies have shown that longevity decreases in fruit flies in direct proportion to a decrease in a specific chemical found in one of the cellular organelles. Based on your knowledge of how the body rids itself of dangerous hydrogen peroxide, what do you think this chemical in the organelle is?
4. Why do you think a person is able to perform anaerobic exercise (such as lifting and holding a heavy weight) only briefly but can sustain aerobic exercise (such as walking or swimming) for long periods? (*Hint:* muscles have limited energy stores.)
5. One type of the affliction *epidermolysis bullosa* is caused by a genetic defect that results in production of abnormally weak keratin. Based on your knowledge of the role of keratin, what part of the body do you think would be affected by this condition?
6. ***Clinical Consideration*** Kevin S. and his wife have been trying to have a baby for the past three years. On seeking the help of a fertility specialist, Kevin learned that he has a hereditary form of male sterility involving nonmotile sperm. His condition can be traced to defects in the cytoskeletal components of the sperm's flagella. Based on this finding, the physician suspected that Kevin also has a long history of recurrent respiratory tract disease. Kevin confirmed that indeed he has had colds, bronchitis, and influenza more frequently than his friends. Why would the physician suspect that Kevin probably had a history of frequent respiratory disease based on his diagnosis of sterility due to nonmotile sperm?

3 The Plasma Membrane and Membrane Potential

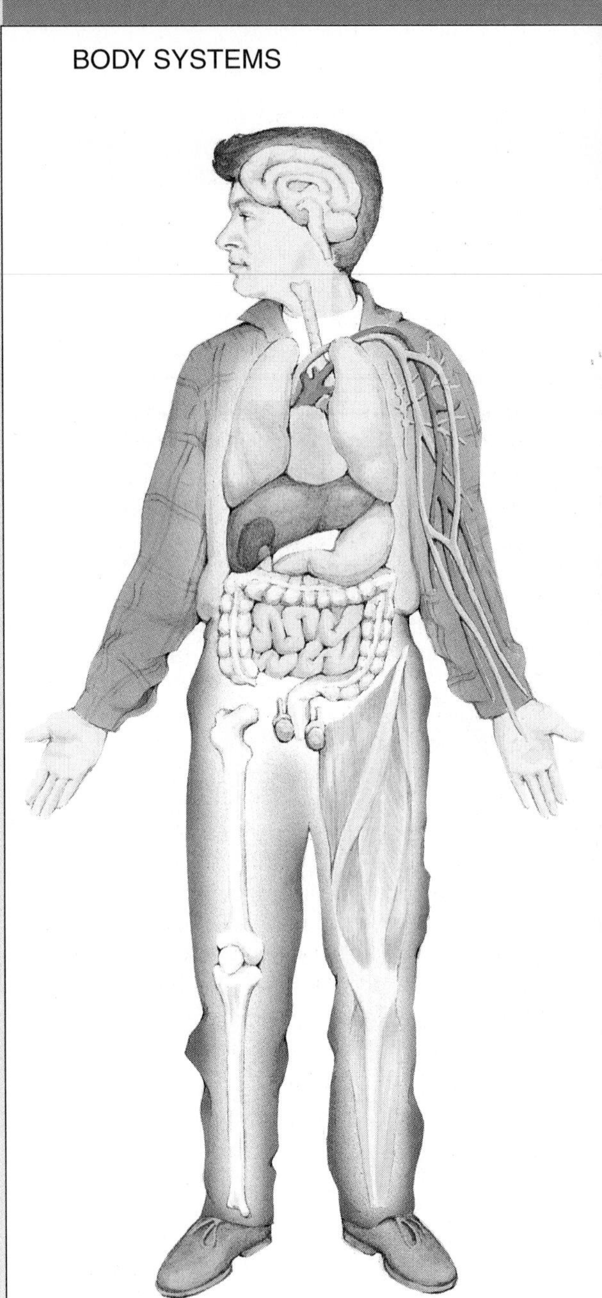

BODY SYSTEMS
Body systems maintain homeostasis
HOMEOSTASIS
The plasma membranes of the cells that make up the body systems play a dynamic role in exchanges and interactions between constituents in the intracellular and extracellular fluid. Many of these plasma membrane activities, including controlled changes in membrane potential, are important in maintaining homeostasis.
Homeostasis is essential for survival of cells
Plasma membrane
Membrane potential
CELLS
Cells make up body systems

All cells are enveloped by a **plasma membrane,** a thin, flexible, lipid barrier that separates the contents of the cell from its surroundings. To carry on life-sustaining and specialized activities, each cell must exchange materials across its plasma membrane with the homeostatically maintained internal fluid environment that surrounds it. This discriminating barrier contains specific proteins, some of which enable selective passage of materials through the membrane. Other proteins in the plasma membrane serve as receptor sites for interaction with specific chemical messengers in the cell's environment. These chemical messengers control many cellular activities critical to the maintenance of homeostasis.

Cells have a **membrane potential,** which refers to a slight excess of negative charges lined up along the inside of the membrane and a slight excess of positive charges on the outside. The specialization of nerve and muscle cells depends on the ability of these cells to alter their potential on appropriate stimulation.

Chapter Contents At a Glance

Membrane Structure and Composition

The plasma membrane separates the intracellular and extracellular fluid.

The survival of every cell depends on the maintenance of intracellular contents unique for that cell type despite the remarkably different composition of the extracellular fluid surrounding it. This difference in fluid composition inside and outside a cell is maintained by the **plasma membrane,** an extremely thin layer of lipids and proteins that forms the outer boundary of every cell and encloses the intracellular contents. In addition to serving as a mechanical barrier that traps needed molecules within the cell, the plasma membrane plays an active role in determining the composition of the cell by selectively permitting specific substances to pass between the cell and its environment. Many of the functional differences between cell types are due to subtle variations in the composition of their plasma membranes, which in turn enable the cells to interact in different ways with essentially the same extracellular fluid environment.

The plasma membrane is a fluid lipid bilayer embedded with proteins.

The plasma membrane is too thin to be seen under an ordinary light microscope, but with an electron microscope it appears as a **trilaminar** (three-layered) **structure** consisting of two dark layers separated by a light middle layer (▬ Fig. 3-1). The specific arrangement of the molecules that make up the plasma membrane is believed to be responsible for this three-layered "sandwich" appearance.

All plasma membranes consist mostly of lipids (fats) and proteins plus small amounts of carbohydrate. The most abundant membrane lipids are phospholipids, with lesser amounts of cholesterol. **Phospholipids** have a polar (electrically charged; see p. A–11) head containing a negatively charged phosphate group and two nonpolar (electrically neutral) fatty acid tails (▬ Fig. 3-2a). The polar end is hydrophilic ("water-loving") because it can interact with water molecules, which are also polar; the nonpolar end is hydrophobic ("water-fearing") and will not mix with water. Such two-sided molecules self-assemble into a **lipid bilayer,** a double layer of lipid molecules, when in contact with water (Fig. 3-2b). The hydrophobic tails bury themselves in the center away from the water, while the hydrophilic heads line up on both sides in contact with the water. The outer surface of the layer is exposed to extracellular fluid (ECF), whereas the inner surface is in contact with the intracellular fluid (ICF).

This lipid bilayer is not a rigid structure but instead is fluid in nature, with a consistency more like liquid cooking oil than solid shortening. The phospholipids, which are not held together by chemical bonds, are able to twirl around rapidly as well as move about within their own half of the layer, much like skaters on a crowded skating rink.

Also contributing to the fluidity as well as the stability of the membrane is **cholesterol.** By being tucked in between the phospholipid molecules, the cholesterol molecules prevent the fatty acid chains from packing together and crystallizing, a process that would drastically reduce membrane fluidity. Recent studies also suggest that choles-

▬ *Figure 3-1* **Trilaminar Appearance of a Plasma Membrane in an Electron Micrograph** Depicted are the plasma membranes of two adjacent cells. Note the trilaminar structure (that is, two dark layers separated by a light middle layer) of each membrane.

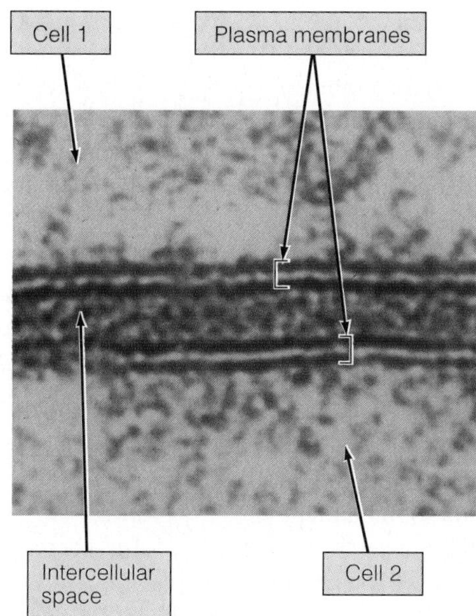

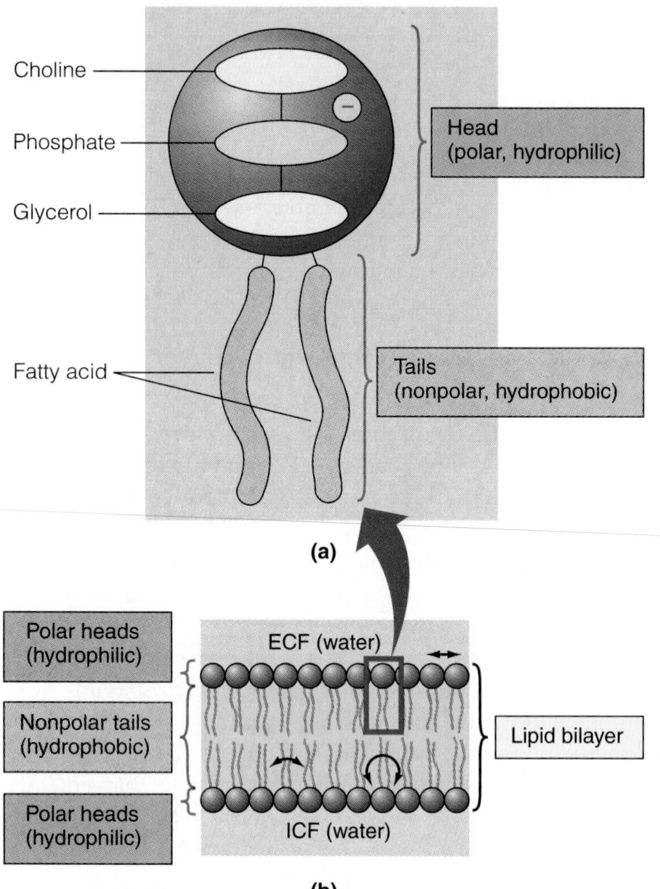

Figure 3-2 **Structure and Organization of Phospholipid Molecules in a Lipid Bilayer** (a) Phospholipid molecule. (b) When in contact with water, phospholipid molecules organize themselves into a lipid bilayer with the polar heads interacting with the polar water molecules at each surface and the nonpolar tails all facing the interior of the bilayer.

terol exerts a regulatory role on some of the membrane proteins.

The fluid nature of the membrane permits it to be flexible, enabling the cell to change its shape. Red blood cells, for example, must change shape considerably as they squeeze their way single file through the capillaries, the tiniest of blood vessels. It is suspected that other essential membrane functions, such as transport processes, are also dependent on the fluidity of the lipid bilayer.

Attached to or inserted within the lipid bilayer are the **membrane proteins** (Fig. 3-3). Some of these proteins extend through the entire thickness of the membrane; they have polar regions at both ends joined by a nonpolar central portion. Other proteins stud only the outer or inner surface; they are anchored by interactions with a protein that spans the membrane or by attachment to the lipid bilayer. The fluidity of the lipid bilayer enables many membrane proteins to float freely like "icebergs" in a moving "sea" of lipid, although the mobility of proteins that perform a specialized function in a specific area of the cell is restricted. This view of membrane structure is known as the **fluid mosaic model,** in reference to the membrane fluidity and the ever-changing mosaic pattern of the proteins embedded within the lipid bilayer.

The small amount of **membrane carbohydrate** is located only at the outer surface. Short-chain carbohydrates, which protrude from the outer surface, are bound primarily to membrane proteins and to a lesser extent to lipids, forming *glycoproteins* and *glycolipids,* respectively (Fig. 3-3).

This proposed structure can account for the trilaminar appearance of the plasma membrane. The two dark lines are believed to be caused by the preferential staining of the hydrophilic polar regions of the lipid and protein molecules, whereas the light space between corresponds to the hydrophobic core formed by the nonpolar regions of these molecules.

Even though the outer and inner layers have the same appearance when viewed with an electron microscope, it is obvious from our description that the plasma membrane actually is asymmetrical; that is, the surface of the membrane facing the extracellular fluid is strikingly different from the surface facing the cytoplasm. Carbohydrate is located only on the outer surface; the outer and inner surfaces bear different amounts and types of protein; and even the lipid composition of the outer half of the bilayer varies somewhat from the inner half. This distinct sidedness of the membrane is related to the different functions carried out at the outer and inner surfaces.

Furthermore, considerable heterogeneity exists within each half of the membrane. For example, recent information suggests that not all membrane proteins are as freely mobile and uniformly distributed as proposed by the fluid mosaic model. The mobility of proteins that perform a specialized function in a specific area of the membrane is restricted by several means. While some proteins do drift randomly within the membrane (similar to a dog who freely roams the neighborhood), other proteins are believed to be tethered to the cytoskeleton (see p. 36) (much like a dog restricted by a leash). Still other proteins appear to be restricted to *confinement zones* by fine cytoskeletal meshwork fences within the membrane (similar to a dog confined by a fence to its own backyard). Sometimes these barriers to mobility temporarily open so that proteins are able to move between confinement zones. Finally, a few proteins undergo rapid, highly directed transport toward a specific region of the membrane, perhaps being carried by motor proteins similar to kinesin (see p. 37). (This is analogous to a dog being carried by its owner to a particular destination.) Thus, the emerging view of the plasma membrane is that it is a highly complex, dynamic, regionally differentiated structure.

The lipid bilayer forms the primary barrier to diffusion, whereas proteins perform most of the specific membrane functions.

The various components of the plasma membrane are responsible for carrying out a variety of different functions.

Lipid bilayer The lipid bilayer serves at least three important functions:

1. It forms the basic structure of the membrane. (The phospholipids can be visualized as the "pickets" that form the "fence" around the cell.)

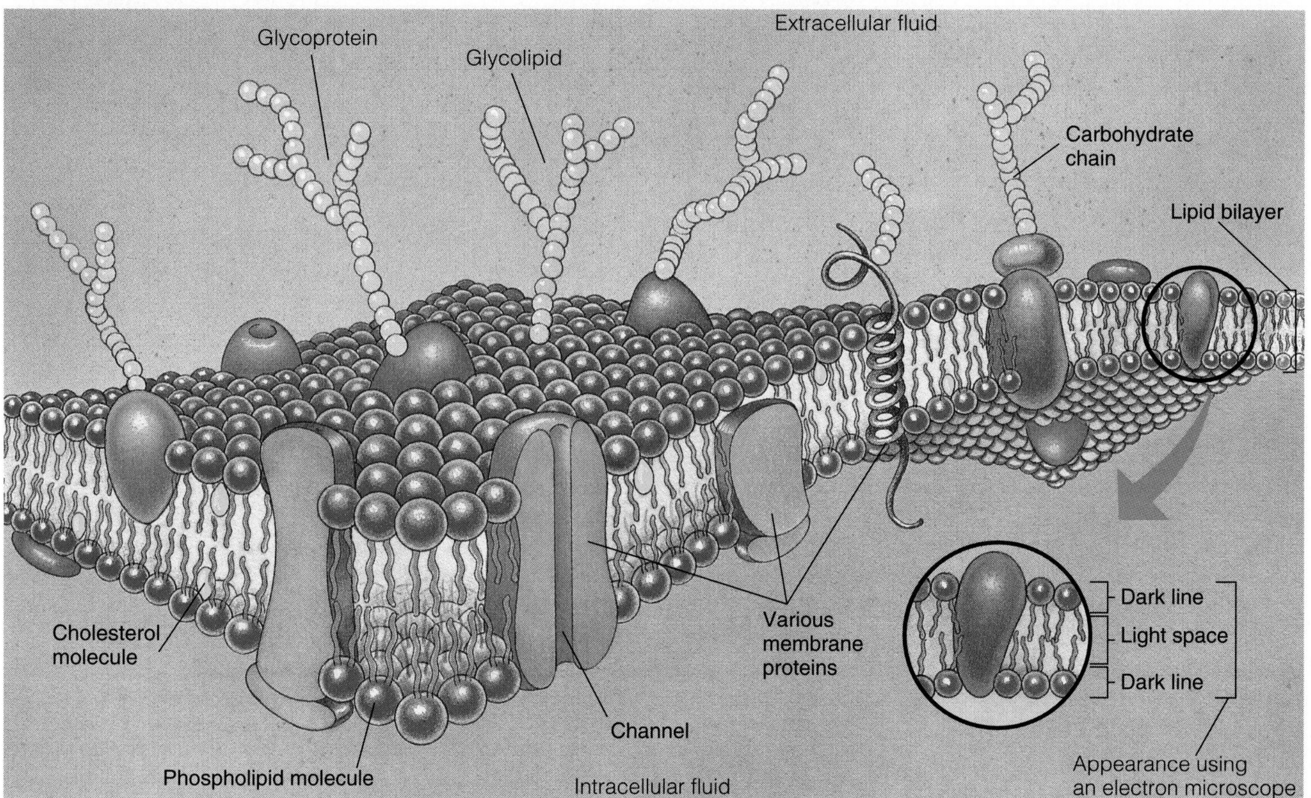

Figure 3-3 Fluid Mosaic Model of Plasma Membrane Structure The plasma membrane is composed of a lipid bilayer embedded with proteins. Some of these proteins extend through the thickness of the membrane, some are partially submerged in the membrane, and others are loosely attached to the surface of the membrane. Short carbohydrate chains are attached to proteins or lipids on the outer surface only.

2. Its hydrophobic interior serves as a barrier to passage of water-soluble substances between the ICF and ECF. Water-soluble substances cannot dissolve in and pass through the lipid bilayer. (However, water molecules themselves are small enough to pass between the molecules that form this barrier.)
3. It is responsible for the fluidity of the membrane.

Membrane proteins A variety of different proteins within the plasma membrane serve the following specialized functions:

1. Some proteins that span the membrane form water-filled pathways, or **channels,** across the lipid bilayer. Their presence enables water-soluble substances that are small enough to enter a channel, such as ions, to pass through the membrane without coming into direct contact with the hydrophobic lipid interior (Fig. 3-3). The channels are highly selective. Not only does their small diameter preclude passage of particles greater than 0.8 nanometer (nm) in diameter (1 nm = 1 billionth of a meter or 40 billionths of an inch), but a given channel can also selectively attract or repel particular ions. For example, sodium (Na^+) channels can accommodate the passage of only Na^+ whereas only K^+ can pass through potassium (K^+) channels. This selectivity is believed to be due to specific arrangements of charged amino acid groups on the interior surfaces of the proteins that form the channel walls. Cells vary in the number, kind, and activity of channels they possess. It is even possible for a given channel to be *open* or *closed* to its specific ion as a result of changes in channel shape in response to a controlling mechanism. (See the boxed feature on p. 50, ♦ Concepts, Challenges, and Controversies.)
2. Other proteins serve as **carrier molecules** that transfer specific substances unable to cross the membrane on their own. Thus, channels and carrier molecules are both important in the transport of substances between the ECF and ICF. Each carrier can transport only a particular molecule or closely related molecules. Variation in the kinds of carriers different cells possess permits them to selectively transport different substances across their membranes. For example, the thyroid gland requires iodine for the synthesis of thyroid hormone. Accordingly, the plasma membranes of thyroid gland cells uniquely possess carriers for iodine, enabling this essential element to be transported from the blood into thyroid gland cells, a capability not present in other body cells.
3. Many of the proteins on the outer surface serve as **receptor sites** that "recognize" and bind with specific molecules in the environment of the cell. This binding initiates a series of membrane and intracellular events that alter the activity of the particular cell. (The postreceptor pathways involved in altering cell function are discussed in a later

CONCEPTS, CHALLENGES, AND CONTROVERSIES

Cystic Fibrosis: A Fatal Defect in Membrane Transport

Cystic **fibrosis (CF),** the most common fatal genetic disease in the United States, strikes 1 in every 2,000 Caucasian children. With cystic fibrosis, the body's exocrine glands (see p. 3) secrete an abnormally thick, sticky mucus. Researchers have recently found that cystic fibrosis is caused by any one of several different genetic defects that lead to production of a flawed version of a protein known as *cystic fibrosis transmembrane conductance regulator (CFTR)*. CFTR normally helps form and regulate the chloride (Cl^-) channels in the plasma membrane of exocrine gland cells. With CF, the defective CFTR "gets stuck" in the endoplasmic reticulum/Golgi system, which normally manufactures and processes this product and ships it to the plasma membrane (see p. 20); that is, in CF patients, the mutated version of CFTR is only partially processed and never makes it to the cell surface. The resultant absence of CFTR protein in the plasma membrane's Cl^- channels leads to membrane impermeability to Cl^-. Because Cl^- transport across the membrane is closely linked to Na^+ transport, and, in turn, water transport across the membrane is closely linked to salt (NaCl) transport, CF patients are unable to secrete sufficient salt and water to dilute their mucus secretions to a normal consistency.

Most dramatically affected are the respiratory airways and the pancreas. The presence of the thick, sticky mucus in the respiratory airways makes it difficult to get adequate air in and out of the lungs. Also, because bacteria thrive in the accumulated mucus, CF patients suffer from repeated respiratory infections. Gradually, the involved lung tissue becomes scarred (fibrotic), making the lungs harder to inflate. This complication increases the work of breathing beyond the extra work required to move air through the clogged airways.

Similarly, the pancreatic duct, which carries secretions from the pancreas to the small intestine, becomes plugged with thick mucus in CF patients. Because the pancreas produces enzymes important in the digestion of food, malnourishment eventually results. Furthermore, as the pancreatic digestive secretions accumulate behind the blocked pancreatic duct, fluid-filled cysts form in the pancreas, with the affected pancreatic tissue gradually degenerating and becoming fibrotic. The name "cystic fibrosis" aptly describes long-term changes that occur in the pancreas and lungs as the result of a single genetic flaw in the plasma membrane Cl^- channels.

Treatment consists of physical therapy to help clear the airways of the excess mucus and antibiotic therapy to combat respiratory infections, plus special diets and administration of supplemental pancreatic enzymes to maintain adequate nutrition. Despite this supportive treatment, most CF victims do not survive beyond their early twenties, with most dying from lung complications.

With the recent discovery of the genetic defect responsible for the majority of CF cases, investigators are hopeful of developing a means to correct or compensate for the defective gene. For example, researchers recently reported success in inserting a healthy human CFTR gene into a disabled cold virus that was unable to cause disease but could still penetrate the cells lining the respiratory airways. When this gene-carrying virus was deposited in the lungs of CF patients in a limited clinical study, the stowaway gene produced human CFTR, although in insufficient quantity to remedy the problem. Though of limited success, these preliminary results lead scientists to believe that gene therapy could one day be a potential cure for CF. Scientists are currently searching for a better means of shuttling the healthy gene into the patient's respiratory airways. The treatment would have to be repeated several times per year as old respiratory airway cells die and are replaced by new cells.

Another potential cure being studied is development of a drug that induces the mutated CFTR to be "finished off" and inserted in the plasma membrane. Furthermore, several promising new drug therapy approaches, such as a mucus-thinning aerosol drug that can be inhaled, offer hope of reducing the number of lung infections and extending the life span of CF victims until a cure can be found.

---

section.) In this way, chemical messengers in the blood, such as hormones, are able to influence only the specific cells that possess receptors for the messenger while having no effect on other cells, even though every cell is exposed to the same messenger via its widespread distribution by the blood. To illustrate, the anterior pituitary gland secretes into the blood thyroid-stimulating hormone (TSH), which can attach only to the surface of thyroid gland cells to stimulate secretion of thyroid hormone. No other cells have receptor sites for TSH, so only thyroid cells are influenced by TSH despite its widespread distribution.

4. Another group of proteins function as **membrane-bound enzymes** that control specific chemical reactions at either the inner or the outer cell surface. Cells display specialization in the types of enzymes embedded within their plasma membranes. For example, the outer layer of the plasma membrane of skeletal muscle cells contains an enzyme that destroys the chemical messenger that triggers muscle contraction, thus enabling the muscle to relax.
5. Some proteins are arranged in a **filamentous meshwork** on the inner surface of the membrane and are secured to certain internal protein elements of the cytoskeleton. These membrane proteins appear to be structurally important in maintaining cell shape and probably participate in surface changes accompanying cell movements.
6. Other proteins serve as **cell adhesion molecules (CAMs).** These molecules protrude from the membrane surface and form loops or other appendages that the cells use to grip each other and to grasp the connective tissue fibers that interlace between cells. For example, *cadherins,* one type of CAM, on the surface of adjacent cells interdigitate in zipper fashion to help hold tissues and organs together. Still other CAMs enable white blood cells to adhere to and exit

blood vessels within infected or injured tissues so that they can perform their defense and cleanup responsibilities instead of being swept along in the bloodstream.

7. Finally, still other proteins, especially in conjunction with carbohydrates, are important in the cells' ability to recognize "self" (that is, cells of the same type) and in cell-to-cell interactions.

Membrane carbohydrates The short sugar chains on the outer-membrane surface serve as self-identity markers that enable cells to identify and interact with each other in the following ways:

1. The complexity and diversity of these carbohydrate chains as well as their location on the external surface enable them to play an important role in recognition of "self" and in cell-to-cell interactions. Cells are able to recognize other cells of the same type and join together to form tissues. This is especially important during embryonic development. If cultures of embryonic cells of two different types, such as nerve cells and muscle cells, are mixed together, the cells will sort themselves into separate aggregates of nerve cells and muscle cells. Apparently, the unique combination of sugar chains projecting from the surface membrane proteins serves as the "trademark" of a particular cell type, enabling a cell to recognize others of its own kind in tissue formation.
2. Carbohydrate-containing surface markers also appear to be involved in tissue growth, which is normally held within certain limits of cell density. Cells do not "trespass" across the boundaries of neighboring tissues; that is, they do not overgrow their own territory. Abnormal surface carbohydrate markers have been identified in certain tumor cells, suggesting that this abnormality might underlie the uncontrolled growth of tumor cells.
3. Some CAMs bear carbohydrates on their outermost tip, where these sugary chains participate in cell adhesion activities.

Cell-to-Cell Adhesions

In multicellular organisms such as humans, plasma membranes not only serve as the outer boundaries of all cells but also participate in cell-to-cell adhesions, allowing groups of cells to bind together into tissues and to be packaged further into organs. Organization of cells into appropriate groupings may be at least partially attributable to the carbohydrate chains on the membrane surface. Once arranged, cells are held together by three different means: (1) cell adhesion molecules in the cells' plasma membranes, (2) the extracellular matrix, and (3) specialized cell junctions.

The extracellular matrix serves as the biological "glue."

Many cells within a tissue are not in direct physical contact with neighboring cells. Instead, they are held together by the **extracellular matrix,** an intricate meshwork of fibrous proteins embedded in a watery, gel-like substance composed of complex carbohydrates. The watery gel provides a pathway for diffusion of nutrients, wastes, and other water-soluble traffic between the blood and tissue cells. Interwoven within this gel are three major types of protein fibers: collagen, elastin, and fibronectin.

1. **Collagen** forms cablelike fibers or sheets that provide tensile strength (resistance to longitudinal stress). In *scurvy,* a condition caused by vitamin C deficiency, these fibers are not properly formed. As a result, the tissues, especially those of the skin and blood vessels, become very fragile. This leads to bleeding in the skin and mucous membranes, which is especially noticeable in the gums.
2. **Elastin** is a rubberlike protein fiber most abundant in tissues that must be capable of easily stretching and then recoiling after the stretching force is removed. It is found, for example, in the lungs, which stretch and recoil as air moves in and out.
3. **Fibronectin** promotes cell adhesion and holds cells in position. Reduced amounts of this protein have been found within certain types of cancerous tissue, possibly accounting for the fact that cancer cells do not adhere well to each other but tend to break loose and metastasize (spread elsewhere in the body).

The extracellular matrix is secreted by local cells, most commonly by **fibroblasts** ("fiber formers") present in the matrix. Often the matrix and the cells within it are known collectively as *connective tissue* because they connect cells together into tissues and tissues into organs. The exact composition of extracellular matrix components varies for different tissues, thus providing distinct local environments for the various cell types in the body. In some tissues, the matrix becomes highly specialized to form such structures as cartilage or tendons or, upon appropriate calcification, the hardened structures of bones and teeth.

New information suggests that the extracellular matrix is not just a passive scaffolding for cellular attachment but also helps regulate the behavior and functions of the cells with which it interacts. Investigators have shown that cells are able to function normally only when in association with their normal matrix components. The matrix is especially influential in cell growth and differentiation.

Some cells are directly linked together by specialized cell junctions.

In addition to the tissue cohesion provided by CAMs and by the extracellular matrix, some cells are directly linked together by one of three types of specialized cell junctions: (1) *desmosomes* (adhering junctions), (2) *tight junctions* (impermeable junctions), or (3) *gap junctions* (communicating junctions).

At a **desmosome,** filaments of unknown composition extend between the plasma membranes of two closely adjacent but nontouching cells and act as "spot rivets" anchoring the cells together (— Fig. 3-4). Desmosomes are distributed

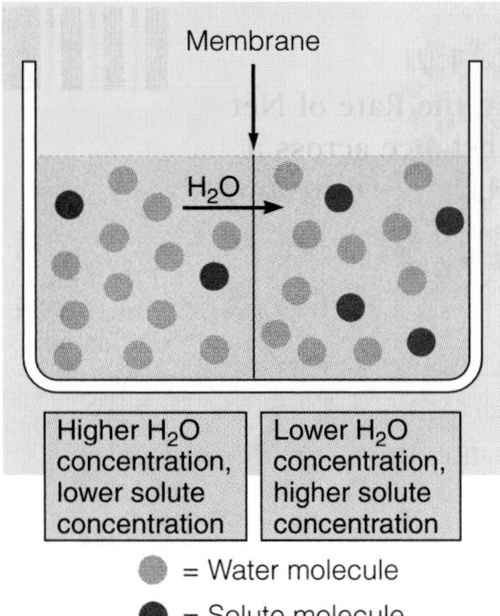

Figure 3-11 **Osmosis**
Osmosis is the net diffusion of water down its own concentration gradient (to the area of higher solute concentration).

container in (a) is full of pure water, so the water concentration is 100% while the solute concentration is 0%. In part (b), 10% of the water molecules have been replaced by solute. The water concentration is now 90%, and the solute concentration is 10%—a lower water concentration and a higher solute concentration than in (a). Note that as the solute concentration increases, the water concentration decreases correspondingly.

If solutions of unequal solute concentration (and hence unequal water concentration) are separated by a membrane that permits passage of water, such as the plasma membrane (Fig. 3-11), water will diffuse down its own concentration gradient from the area of higher water concentration (lower solute concentration) to the area of lower water concentration (higher solute concentration). This net diffusion of water is known as **osmosis.** Because solutions are always referred to in terms of concentration of solute, *water moves by osmosis to the area of higher solute concentration.* Very loosely, then, the solute can be thought of as "drawing," or attracting, water, but in reality, osmosis is nothing more than the diffusion of water down its own concentration gradient.

Thus far in our discussion of osmosis, we have ignored any solute movement. Let us compare the results of osmosis in the two cases when the solute (1) can and (2) cannot permeate the membrane:

1. If the membrane is permeable to the solute as well as to water, the solute is able to move down its own concentration gradient in the opposite direction of the net water movement (Fig. 3-12). This movement continues until both the solute and water are evenly distributed across the membrane. Even though some transitory differences in volume between the two compartments may occur because of differences in the speed with which water and the solute diffuse through the membrane, the final volume of the compartments when the steady state is achieved is the same as at the onset. Water and solute molecules have merely exchanged places between the two compartments until their distributions have equalized; that is, an equal number of water molecules have moved from side 1 to side 2 as solute molecules have moved from side 2 to side 1. With all concentration gradients abolished, osmosis ceases.
2. If the membrane is impermeable to the solute, the solute is not able to cross the membrane down its concentration gradient (Fig. 3-13). At the onset, the concentration gradients are identical to those in the previous example. However, even though net diffusion of water takes place from side 1 to side 2, the solute cannot move. As a result of water movement alone, side 2's volume increases while the volume of side 1 correspondingly decreases. Loss of water from side 1 increases the solute concentration on side 1, whereas addition of water to side 2 reduces the solute concentration on that side. Eventually, the concentrations of water and solute on the two sides of the membrane become equal, and net diffusion of water ceases. Unlike the situation in which the solute can also permeate, diffusion of water alone has resulted in a change in the final volumes of the two compartments. The side originally containing the greater solute concentration has a larger volume, having gained water.

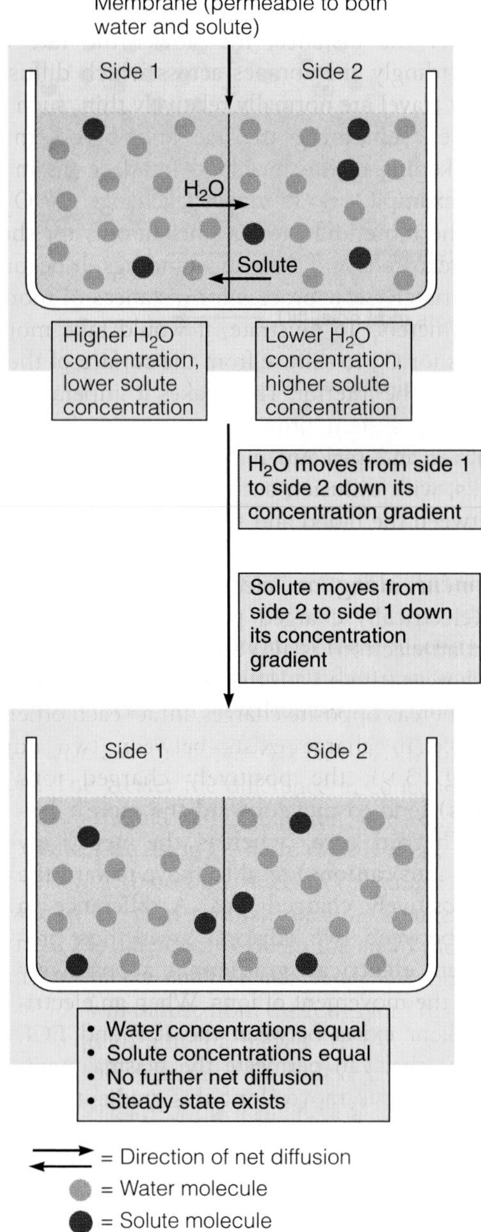

Figure 3-12 **Movement of Water and a Penetrating Solute Unequally Distributed across a Membrane**

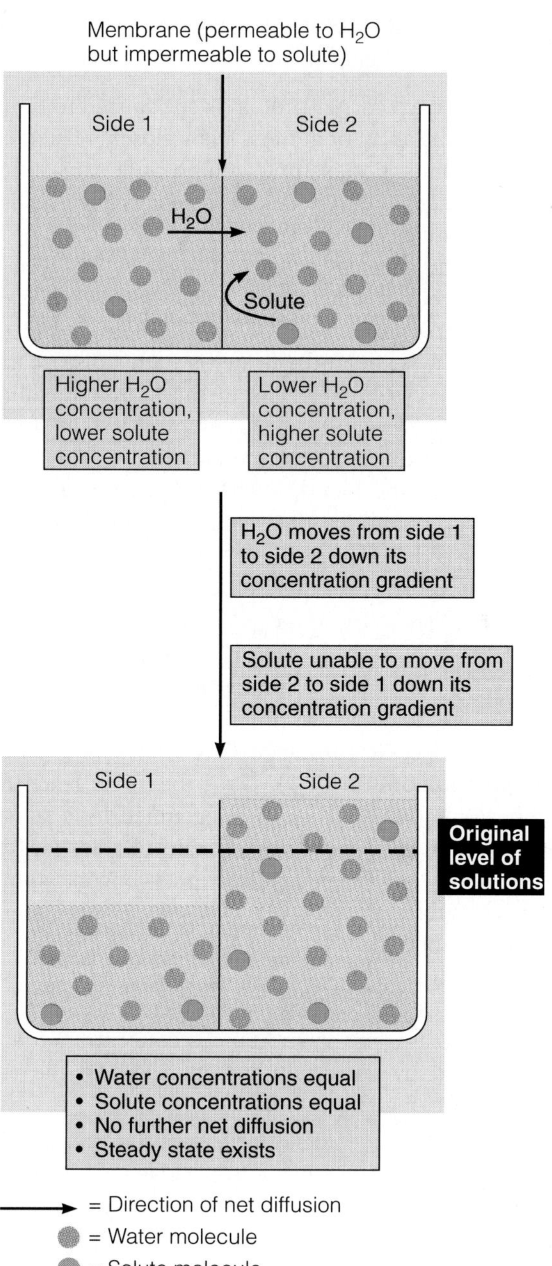

Figure 3-13 **Osmosis in the Presence of an Unequally Distributed Nonpenetrating Solute**

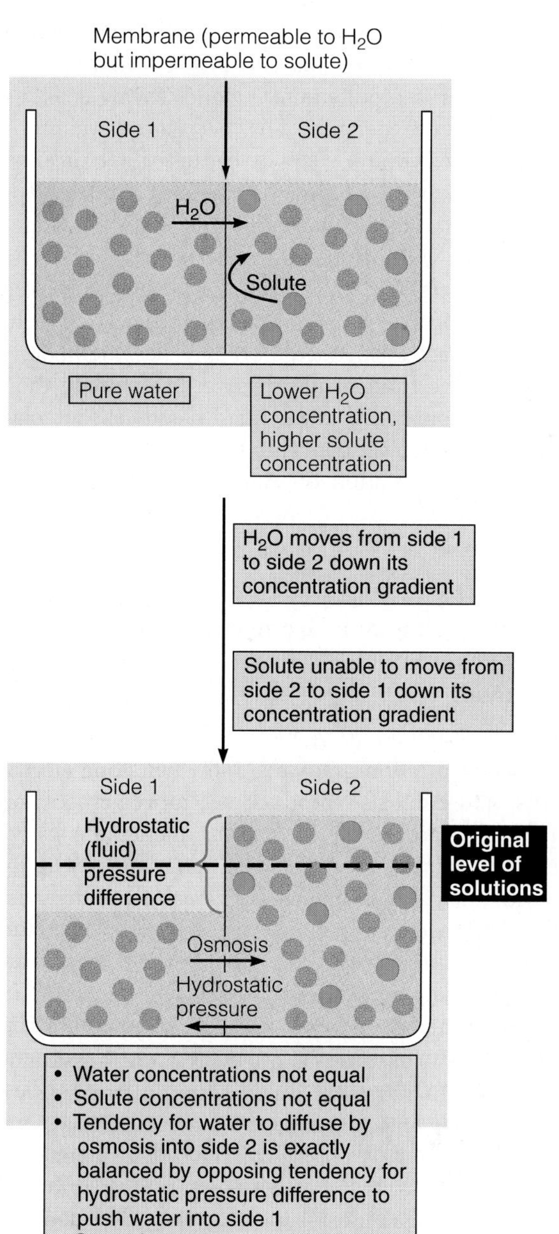

Figure 3-14 **Osmosis When Pure Water is Separated from a Solution Containing a Nonpenetrating Solute**

What will happen if a nonpenetrating solute is present on side 2 and pure water is present on side 1 (Fig. 3-14)? Osmosis occurs from side 1 to side 2, but the concentrations between the two compartments can never become equal. No matter how dilute side 2 becomes because of water diffusing into it, it can never become pure water, nor can side 1 ever acquire any solute. Since equilibrium is impossible to achieve, does net diffusion of water (osmosis) continue unabated until all the water has left side 1? No. As the volume expands in compartment 2, a difference in **hydrostatic (fluid) pressure** between the two compartments is created, and it opposes osmosis. The magnitude of opposing pressure necessary to completely stop osmosis is equal to the **osmotic pressure** of the solution on side 2. The osmotic pressure can be related directly to the concentration of nonpenetrating solute. The greater the concentration of nonpenetrating solute → the lower the concentration of water → the greater the drive for water to move by osmosis from pure water into the solution → the greater the opposing pressure required to stop the osmotic flow → the greater the osmotic pressure of the solution. Therefore, a solution with a high solute concentration exerts greater osmotic pressure than does a solution with a lower solute concentration.

Approximately one hundred times the volume of water in a cell crosses the plasma membrane every second. Even though water is rapidly entering and leaving cells, they normally do not experience any net gain (swelling) or loss (shrinking) of volume. This is because the concentration of solutes in the ECF is normally carefully regulated (primarily by the kidneys) to maintain the same level of osmotic activity as is present within the cells; thus, no net diffusion of water occurs. (See pp. 521–528 for further details on the regulation of ECF osmolarity.) For example, the plasma in which the red blood cells are suspended normally has the same osmotic activity as the fluid inside these cells, enabling the cells to maintain a constant volume. If red blood cells are placed in a solution with a lower concentration of solutes (and, therefore, a higher concentration of water), water enters the cells by osmosis, causing them to swell, perhaps to the point of rupturing or *lysing*.

Special mechanisms are used to transport selected molecules unable to cross the plasma membrane on their own.

All kinds of transport we have discussed thus far—diffusion down concentration gradients, diffusion along electrical gradients, and osmosis—produce net movement of molecules capable of permeating the plasma membrane by virtue of their lipid solubility or small size. Large, poorly lipid-soluble molecules such as proteins, glucose, and amino acids cannot cross the plasma membrane on their own no matter what forces are acting on them. This impermeability ensures that the large polar intracellular proteins cannot escape from the cell. It also means, however, that the cell must provide mechanisms for transporting into the cell essential nutrients, such as glucose for energy and amino acids for the synthesis of proteins, as well as for transporting out of the cell metabolic wastes and secretory products, such as proteinaceous hormones and enzymes. Furthermore, passive diffusion alone cannot always account for the movement of small ions. Cells utilize two different mechanisms to accomplish these selective transport processes: *carrier-mediated transport* and *vesicular transport*.

Carrier-mediated transport All carrier proteins span the thickness of the plasma membrane and are able to undergo reversible changes in shape so that specific binding sites can alternately be exposed at either side of the membrane. The details of the conformational changes that carriers undergo are unknown, but ▬ Figure 3-15 is a schematic representation of how this **carrier-mediated transport** process might take place. As the molecule to be transported attaches to a binding site on the carrier on one side of the membrane (step 1), it triggers a change in the carrier's shape that causes the same site to be exposed to the other side of the membrane (step 2). Then, having been moved in this way from one side of the membrane to the other, the bound molecule detaches from the carrier (step 3).

Carrier-mediated transport systems display three important characteristics that determine the kind and amount of material that can be transferred across the membrane: *specificity, saturation,* and *competition.*

1. **Specificity.** Each carrier protein is specialized to transport a specific substance, or at most a few closely related chemical compounds. For example, amino acids cannot bind to glucose carriers, although several similar amino acids may be able to utilize the same carrier. Cells vary in the types of carriers they possess, thus permitting transport selectivity among cells. A number of inherited diseases involve defects in transport systems for a particular substance. *Cysteinuria* (cysteine in the urine) is such a disease involving defective cysteine carriers in the kidney membranes. This transport system normally removes cysteine from the fluid destined to become urine and returns this essential amino acid to the blood. When this carrier is malfunctional, large quantities of cysteine remain in the urine, where it is relatively insoluble and tends to precipitate. This is one cause of urinary stones.
2. **Saturation.** There is a limit to the amount of a substance that can be transported across the membrane via a carrier in a given time; that is, a limited number of carrier binding sites are available within a particular plasma membrane for a specific substance. This limit is known as the **transport maximum (T_m).** Until the T_m is reached, the number of carrier binding sites occupied by a substance and, accordingly, the substance's rate of transport across the membrane are directly related to its concentration. The more of a substance available to be transported, the more will be transported. When the T_m is reached, the carrier is saturated (all binding sites are occupied), and the rate of the substance's transport across the membrane is maximal. Further increases in the substance's concentration are not accompanied by corresponding increases in the rate of transport (▬ Fig. 3-16).

   As an analogy, consider a ferry boat that can maximally carry 100 people at a time across a river in an hour-long trip. If 25 people are on hand to board the ferry, 25 will be transported that hour. Doubling the number of people on hand to 50 will double the rate of transport to 50 people per hour. Such a direct relationship will exist between the number of people waiting to board (the concentration) and the rate of transport until the ferry is fully occupied (its T_m is reached). The ferry can maximally transport 100 people per hour. Even if 150 people are waiting to board, still only 100 will be transported per hour.

   Saturation of carriers is a critical rate-limiting factor in the transport of selected substances across the kidney membranes during urine formation and across the intestinal membranes during absorption of digested foods. Furthermore, it is sometimes possible to regulate (for example, by hormones) the rate of carrier-mediated transport by varying the affinity (attraction) of the binding site for its passenger or by varying the number of binding sites. For example, the hormone insulin greatly increases the carrier-mediated transport of glucose into most cells of the body by promoting an increase in the number of glucose carriers in the cell's plasma membranes. Deficiency of insulin *(diabetes mellitus)* drastically impairs the body's ability to utilize glucose as the primary energy source.

3. **Competition.** Several closely related compounds may compete for a ride across the membrane on the same carrier. If a given binding site can be occupied by more than one type of molecule, the rate of transport of each substance is less when both molecules are present than when either is present by itself. To illustrate, assume the ferry has 100 seats (binding sites) that can be occupied by either men or women. If only men are waiting to board, up to 100 men can be transported during each trip; the same holds true if only women are waiting to board. If, however, both men and women are waiting to board, they will compete for the available seats so that fewer men and fewer women will be transported than when either group is present alone. Fifty of each might make the trip, although the total number of people transported will still be the same, 100 people. In other words, when a carrier is able to transport two closely related substances, such as the amino acids glycine and alanine, the presence of both diminishes the rate of transfer of either.

Carrier-mediated transport takes two forms, depending on whether energy must be supplied to complete the process: facilitated diffusion (not requiring energy) and active transport (requiring energy). **Facilitated diffusion** uses a carrier to facilitate (assist) the transfer of a particular substance across the membrane "downhill" from high to low concentration. This process is passive and does not require energy

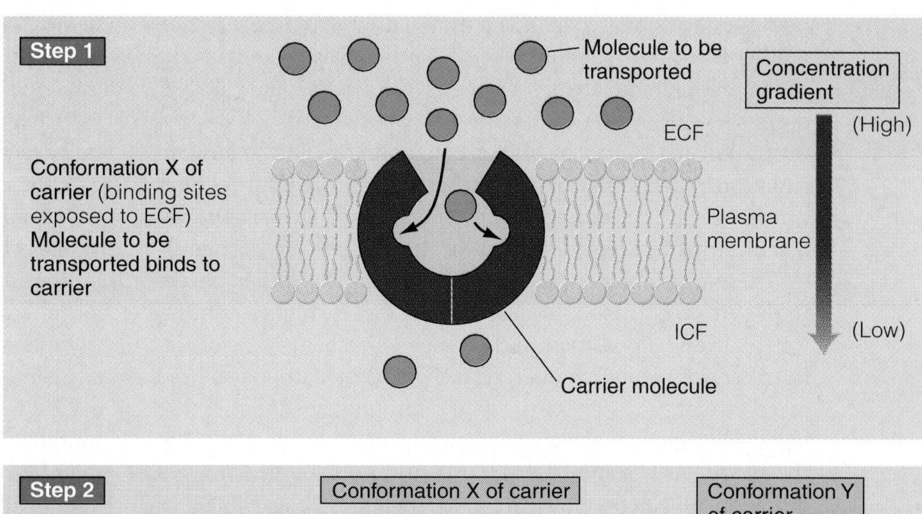

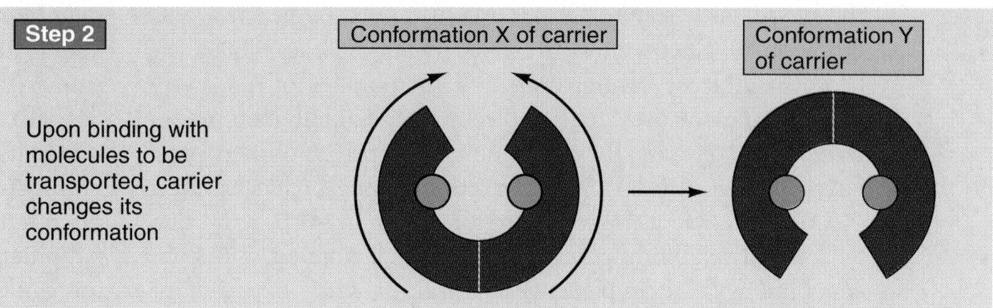

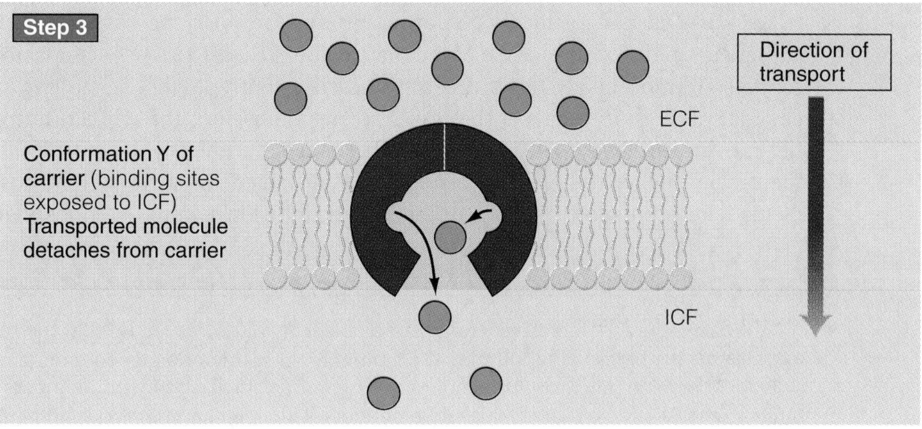

Figure 3-15 **Schematic Representation of Carrier-Mediated Transport: Facilitated Diffusion**

Figure 3-16 **Comparison of Carrier-Mediated Transport and Simple Diffusion down a Concentration Gradient** With simple diffusion of a molecule down its concentration gradient, the rate of transport of the molecule into the cell is directly proportional to the extracellular concentration of the molecule. With carrier-mediated transport of a molecule down its concentration gradient, the rate of transport of the molecule into the cell is directly proportional to the extracellular concentration of the molecule until the carrier is saturated, at which time the rate of transport reaches a maximal value (transport maximum, or T_m). The rate of transport does not increase with further increases in the ECF concentration of the molecule.

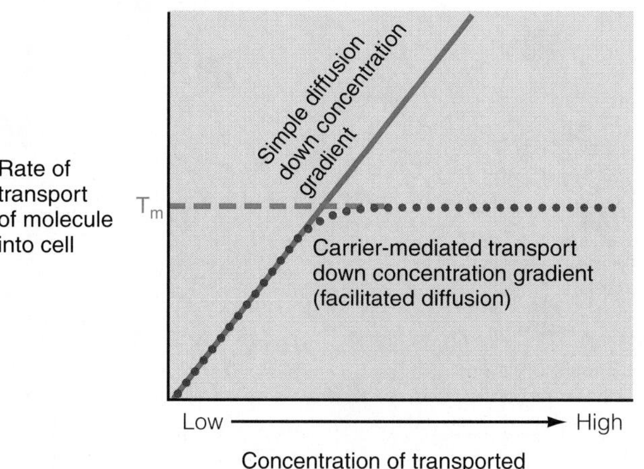

because movement occurs naturally down a concentration gradient. **Active transport,** on the other hand, requires the carrier to expend energy to transfer its passenger "uphill" *against* a concentration gradient, from an area of lower concentration to an area of higher concentration. An analogous situation is a car on a hill. To move the car downhill requires no energy; it will coast from the top down. Driving the car uphill, however, requires the utilization of energy (gasoline).

The most notable example of *facilitated diffusion* is the transport of glucose into cells. Glucose is in higher concentration in the blood than in the tissues. Fresh supplies of this nutrient are regularly added to the blood by eating and from reserve energy stores in the body. Simultaneously, the cells metabolize glucose almost as rapidly as it enters the cells from the blood. As a result, there is a continuous gradient for net diffusion of glucose into the cells. Being a polar molecule, however, glucose cannot cross cell membranes on its own. Without the glucose carrier molecules to facilitate membrane transport of glucose, the cells would be deprived of their preferred source of fuel. (See the accompanying boxed feature, A Closer Look at Exercise Physiology.)

The carrier binding sites involved in facilitated diffusion can bind with their passenger molecules when exposed to either side of the membrane (Fig. 3-15). Passenger binding triggers the carrier to flip its conformation and drop off the passenger on the opposite side of the membrane. Because passengers are more likely to bind with the carrier on the high-concentration side than on the low-concentration side, the net movement always proceeds down the concentration gradient from higher to lower concentration. As is characteristic of mediated transport, the rate of facilitated diffusion is limited by saturation of the carrier binding sites, unlike the rate of simple diffusion, which is always directly proportional to the concentration gradient (Fig. 3-16).

Active transport also involves the use of a protein carrier to transfer a specific substance across the membrane, but in this case the carrier transports the substance against its concentration gradient. For example, the uptake of iodine by thyroid gland cells necessitates active transport because 99% of the iodine in the body is concentrated in the thyroid. To move iodine from the blood, where its concentration is low, into the thyroid, where its concentration is high, requires expenditure of energy to drive the carrier. Specifically, energy in the form of ATP is required in active transport to vary the affinity of the binding site when exposed on opposite sides of the plasma membrane. In contrast, the affinity of the binding site in facilitated diffusion is the same when exposed to either the outside or the inside of the cell.

With active transport, the binding site has a greater affinity for its passenger on the low-concentration side as a result of **phosphorylation** of the carrier on this side (Fig. 3-17a). The carrier exhibits ATPase activity in that it splits the terminal phosphate from an ATP molecule to yield ADP plus a free inorganic phosphate (see p. 28). Phosphorylation involves the binding of this phosphate group to the carrier. Phosphorylation and the binding of the passenger on the low-concentration side induce a conformational change in the carrier protein so that the passenger is now exposed to the high-concentration side of the membrane (Fig. 3-17b). The change in carrier shape is accompanied by **dephosphorylation**; that is, the phosphate group detaches from the carrier. Removal of phosphate reduces the affinity of the binding site for the pas-

Figure 3-17 **Active Transport** The energy of ATP is required in the phosphorylation-dephosphorylation cycle of the carrier to transport the molecule uphill from a region of low concentration to a region of high concentration. (a) On the low-concentration side, the phosphorylated conformation of the carrier has high-affinity binding sites for the molecule to be transported so the molecule binds to the carrier. (b) On the high-concentration side, the dephosphorylated conformation of the carrier has low-affinity binding sites for the molecule being transported so the molecule is released from the carrier.

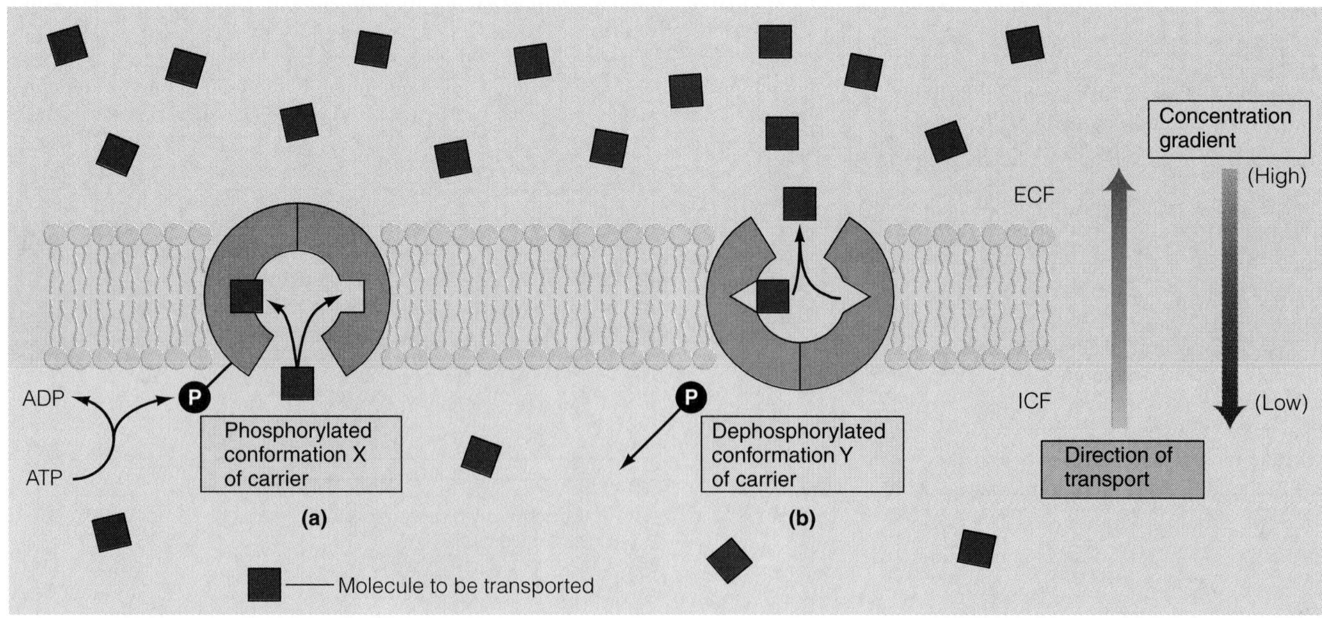

A CLOSER LOOK AT EXERCISE PHYSIOLOGY

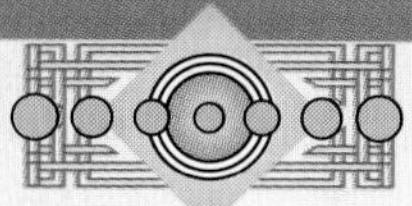

Exercising Muscles Have a "Sweet Tooth"

During exercise, muscle cells use more glucose and other nutrient fuels than usual to power their increased contractile activity. The rate of glucose transport into exercising muscle may increase more than tenfold during moderate or intense physical activity. Glucose uptake by cells is accomplished by glucose carriers in the plasma membrane. Cells maintain an intracellular pool of additional carriers that can be inserted into the plasma membrane as the need for glucose uptake increases. In many cells, including resting muscle cells, facilitated diffusion of glucose into the cells depends on the hormone insulin. Insulin promotes the insertion of glucose carriers in the plasma membranes of insulin-dependent cells. Because plasma insulin levels fall during exercise, however, insulin is not responsible for the increased transport of glucose into exercising muscles. Instead, researchers have shown that muscle cells insert more glucose carriers in their plasma membranes in response to exercise. This has been demonstrated in rats that have undergone physical training.

Exercise influences glucose transport into cells in yet another way. Regular aerobic exercise (see p. 33) has been shown to increase both the affinity (degree of attraction) and number of plasma membrane receptor sites that bind specifically with insulin. This adaptation results in an increase in insulin sensitivity; that is, the cells are more responsive than normal to a given level of circulating insulin.

Because insulin enhances the facilitated diffusion of glucose into most cells, an exercise-induced increase in insulin sensitivity is one of the factors that makes exercise a beneficial therapy for controlling diabetes mellitus, a disorder characterized by insulin deficiency (see chapter 19). As a result of inadequate insulin action, glucose entry into most cells is impaired. Plasma levels of glucose become elevated because glucose remains in the plasma instead of being transported into the cells. In the Type II form of the disease, insulin is being produced, but not in sufficient quantities to meet the body's need for glucose uptake. By increasing the cells' responsiveness to the limited amount of insulin available, regular aerobic exercise helps drive glucose into the cells, where it can be used for energy production, instead of remaining in the plasma, where it leads to detrimental consequences for the body.

senger, so the passenger is released on the high-concentration side. The carrier then returns to its original conformation. Thus, ATP energy is used in the phosphorylation-dephosphorylation cycle of the carrier. It alters the affinity of the carrier's binding sites on opposite sides of the membrane so that transported particles are moved uphill from an area of low concentration to an area of higher concentration. These active transport mechanisms are frequently called **pumps** analogous to water pumps that require energy to lift water against the downward pull of gravity.

The simplest active transport systems pump a single type of passenger. An example is the **hydrogen-ion (H^+) pump** used by specialized stomach cells to transport H^+ into the stomach lumen in association with the secretion of hydrochloric acid during digestion of a meal. This pump moves H^+ against a gradient three to four million times more concentrated.

Other more complicated active transport mechanisms involve the transfer of two different passengers, either simultaneously in the same direction or sequentially in opposite directions. For example, the plasma membrane of all cells contains a sequentially active **Na^+-K^+ ATPase pump (Na^+-K^+ pump** for short). This carrier transports Na^+ out of the cell, concentrating it in the ECF, and picks up K^+ from the outside, concentrating it in the ICF (■ Fig. 3-18). Splitting of ATP through ATPase activity and the subsequent phosphorylation of the carrier on the intracellular side increases the carrier's affinity for Na^+ and induces a change in carrier shape, leading to deposition of Na^+ on the exterior. The subsequent dephosphorylation of the carrier increases its affinity for K^+ on the extracellular side and restores the original carrier conformation, thereby transferring K^+ into the cytoplasm. There is not a direct exchange of Na^+ for K^+, however. The Na^+-K^+ pump moves three Na^+ out of the cell for every two K^+ it pumps in. (To appreciate the magnitude of active Na^+-K^+ pumping that takes place, consider that a single nerve cell membrane contains perhaps 1 million Na^+-K^+ pumps capable of transporting about 200 million ions per second.) The Na^+-K^+ pump plays three important roles:

1. It establishes Na^+ and K^+ concentration gradients across the plasma membrane of all cells; these gradients are critically important in the ability of nerve and muscle cells to generate electrical impulses essential to their functioning (a topic that will soon be discussed more thoroughly).
2. It helps regulate cell volume by controlling the concentrations of solutes inside the cell and thus minimizing osmotic effects that would induce swelling or shrinking of the cell.
3. The energy used to run the Na^+-K^+ pump also indirectly serves as the energy source for the cotransport of glucose and amino acids across intestinal and kidney cells.

Unlike most cells of the body, the intestinal and kidney cells actively transport glucose and amino acids by moving them uphill from low to high concentration. The intestinal cells transport these nutrients from inside the intestinal lumen into the blood, concentrating them in the blood until none of these molecules are left in the lumen to be lost in the feces. The kidney cells save these nutrient molecules for the body by transporting them out of the fluid that is to become urine, moving them against a concentration gradient into the blood.

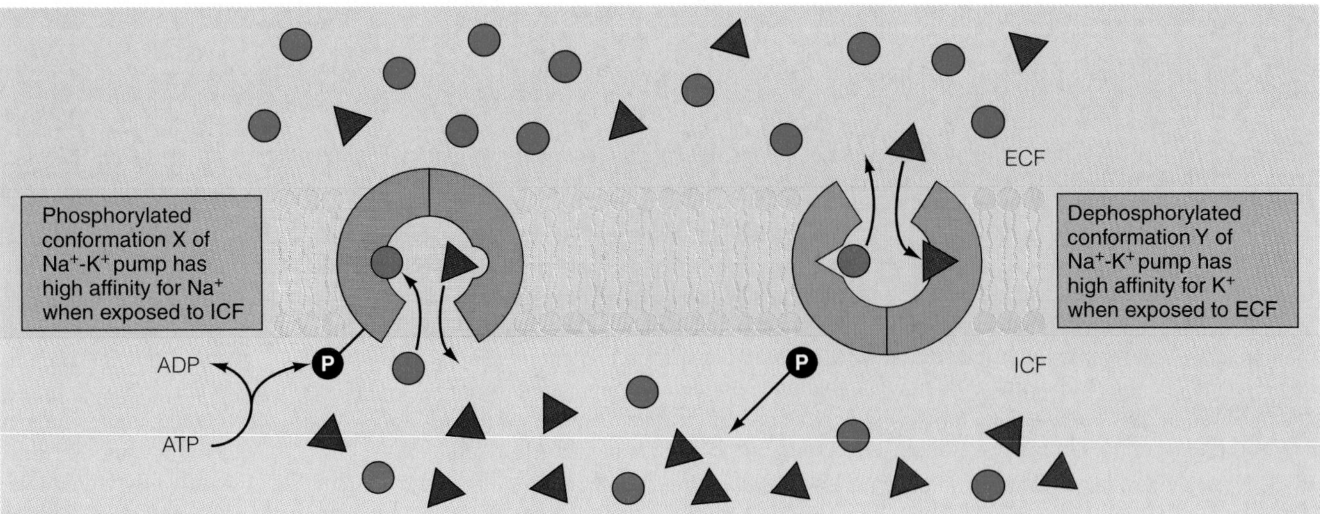

Figure 3-18 **Na^+-K^+ ATPase Pump** The plasma membrane of all cells contains an active transport carrier, the Na^+-K^+ ATPase pump, which uses energy in the carrier's phosphyorylation-dephosphyorylation cycle to sequentially transport Na^+ out of the cell and K^+ into the cell against these ions' concentration gradients.

However, energy is not directly supplied to the carrier in these instances. The carriers that transport glucose against its concentration gradient from the lumen in the intestine and kidneys are distinct from the glucose facilitated-diffusion carriers. The luminal carriers in intestinal and kidney cells are *cotransport carriers* in that they have two binding sites, one for Na^+ and one for the nutrient molecule. The Na^+-K^+ pumps in these cells are located in the basolateral membrane (the membrane at the base of the cell opposite the lumen and along the lateral edge of the cell below the tight junction; see Fig. 3-5). Since more Na^+ is present in the lumen than inside the cells because of the energy-requiring Na^+-K^+ pump that transports Na^+ out of the cell, more Na^+ binds to the luminal cotransport carrier when it is exposed to the outside (Fig. 3-19). Binding of Na^+ to the cotransport carrier increases the carrier's affinity for its other passenger (for example, glucose), so the carrier has a high affinity for glucose when exposed to the outside. When both Na^+ and glucose are bound to the carrier, it undergoes a conformational change and opens to the inside of the cell. Both Na^+ and glucose are released to the interior, Na^+ because of the lower intracellular Na^+ concentration, and glucose because of the reduced affinity of the binding site upon release of Na^+.

The movement of Na^+ into the cell by this cotransport carrier is downhill because the intracellular Na^+ concentration is low, but the movement of glucose is uphill because glucose becomes concentrated in the cell. The released Na^+ is quickly pumped out by the active Na^+-K^+ transport mechanism, keeping the level of intracellular Na^+ low. The energy expended in this process is not directly utilized to run the cotransport carrier, because phosphorylation is not required to alter the affinity of the binding site to glucose. Instead, the establishment of a Na^+ concentration gradient by a primary active transport mechanism (the Na^+-K^+ pump) drives this secondary active transport mechanism (Na^+-glucose cotransport carrier) to move glucose against its concentration gradient. With **primary active transport,** energy is directly required to move a substance uphill. With **secondary active transport,** energy is required in the entire process, but it is not directly required to run the pump. Rather it uses "second hand" energy stored in the form of an **ion concentration gradient** (for example, a Na^+ gradient) to move the cotransported molecule uphill. This is a very efficient interaction. The cotransported molecule is essentially getting a free ride, because Na^+ must be pumped out anyway to maintain the electrical and osmotic integrity of the cell.

The glucose that is carried into the cell across the luminal border by secondary active transport then passively moves out of the cell across the basolateral border down its concentration gradient and enters the blood. This movement is accomplished by facilitated diffusion, mediated by another carrier in the plasma membrane. This passive carrier is identical to the one that transports glucose into other cells, but in intestinal and kidney cells, it transports glucose out of the cell. The difference depends on the direction of the glucose concentration gradient. In the case of intestinal and kidney cells, glucose is in higher concentration inside the cells.

Vesicular transport: endocytosis/exocytosis The special carrier-mediated transport systems embedded in the plasma membrane can selectively transport ions and small polar molecules. But what about large polar molecules or even multimolecular materials that must leave or enter the cell, such as during secretion of protein hormones by endocrine cells or during ingestion of invading bacteria by white blood cells? These materials are unable to cross the plasma membrane, even with assistance. These large particles are transferred between the ICF and ECF not by crossing the membrane but by being

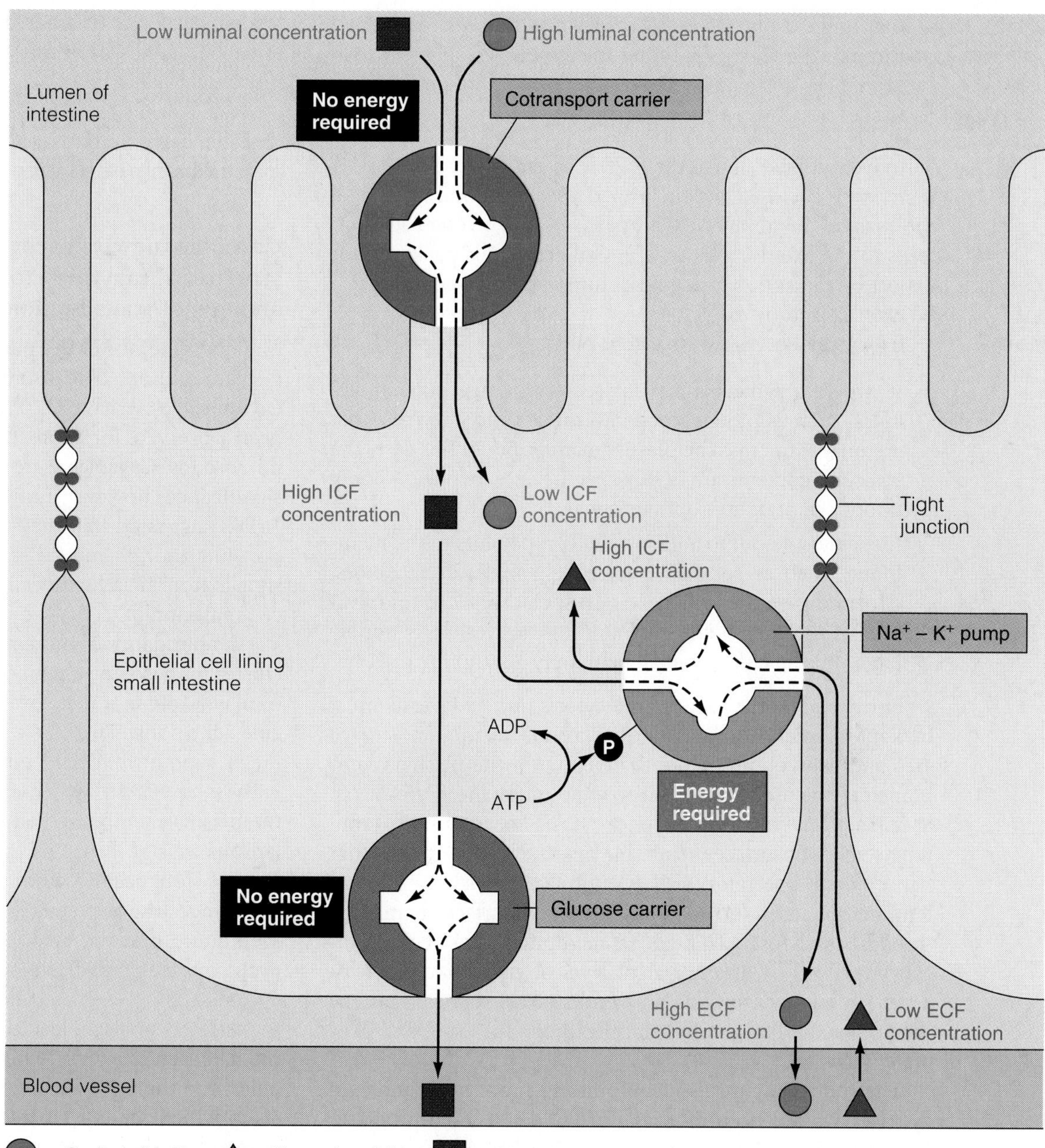

Figure 3-19
Secondary Active Transport
Glucose (as well as amino acids) is transported across intestinal and kidney cells against its concentration gradient by means of secondary active transport. A cotransport carrier at the luminal border simultaneously transfers glucose against a concentration gradient and Na^+ down a concentration gradient from the lumen into the cell. No energy is directly used by the cotransport carrier to move glucose uphill. Instead, operation of the cotransport carrier is driven by the Na^+ concentration gradient established by the energy-using Na^+-K^+pump, which actively transports Na^+ out of the cell at the basolateral border, keeping the ICF Na^+ concentration lower than the luminal Na^+ concentration. After entering the cell by secondary active transport, glucose is transported down its concentration gradient from the cell into the blood by facilitated diffusion, mediated by a passive glucose carrier.

wrapped in membrane. This process of a substance being transported into or out of the cell in a membrane-enclosed vesicle is known as **vesicular transport.** Transport into the cell in this manner is termed *endocytosis*, whereas transport out of the cell is called *exocytosis*.

In **endocytosis**, the plasma membrane surrounds the substance to be ingested, then fuses over the surface, pinching off a membrane-enclosed vesicle so that the engulfed material is trapped within the cell (see p. 26). The transported material has not actually passed through the surface membrane but has gained entrance to the interior of the cell by being wrapped in a piece of the membrane. Endocytosis of fluid is called *pinocytosis* (cell drinking), whereas endocytosis of a large multimolecular particle is known as *phagocytosis* ("cell eating").

Once inside the cell, an engulfed vesicle has two possible destinies:

1. In most instances, lysosomes fuse with the vesicle to degrade and release its contents into the intracellular fluid.
2. In some cells, the endocytotic vesicle bypasses the lysosomes and travels to the opposite side of the cell, where it releases its contents by exocytosis. This provides a pathway to shuttle intact particles through the cell. Such vesic-

ular traffic is one means by which materials are transferred through the thin cells lining the capillaries, across which exchanges are made between the blood and surrounding tissues.

In **exocytosis**, almost the reverse of endocytosis occurs. A membrane-enclosed vesicle formed within the cell fuses with the plasma membrane, then opens up and releases its contents to the exterior (see p. 25). Only materials packaged for export by the endoplasmic reticulum and Golgi complex can be externalized by exocytosis.

Exocytosis serves two different purposes:

1. It provides a mechanism for secreting large polar molecules, such as proteinaceous hormones and enzymes, that are unable to cross the plasma membrane. In this case, the vesicular contents are highly specific and are released only upon receipt of appropriate signals.
2. It enables the cell to add specific components to the membrane, such as selected carriers, channels, or receptors, depending on the cell's needs. In such cases, the composition of the membrane surrounding the vesicle is important, and the contents may be merely a sampling of ICF.

Furthermore, the rate of endocytosis and exocytosis must be kept in balance to maintain a constant membrane surface area and cell volume. More than 100% of the plasma membrane may be used in an hour to wrap internalized vesicles in a cell actively involved in endocytosis, necessitating rapid replacement of surface membrane by exocytosis. On the other hand, when a secretory cell is stimulated to secrete, it may temporarily insert up to thirty times its surface membrane through exocytosis. This added membrane must be specifically retrieved by an equivalent level of endocytotic activity. Thus, through exocytosis and endocytosis, portions of the membrane are constantly being restored, retrieved, and generally recycled.

The controlling mechanisms guiding endocytosis and exocytosis have not been thoroughly elucidated. Both processes are known to require energy and are considered active mechanisms. In some instances of endocytosis, receptor sites on the surface membrane recognize and bind specific molecules in the environment of the cell. This combination triggers a localized invagination process that selectively traps the bound material. One of the primary ways in which antibodies help defend us against invading organisms is by attaching to the bacterial surface to form a coat that can be recognized by specific receptor sites in the plasma membrane of the phagocytic white blood cells. Such "marked" bacteria are quickly engulfed and destroyed.

Exocytosis of secretory products also appears to be a triggered event. In most instances, a specific nervous or hormonal stimulus brings about opening of Ca^{2+} channels in the plasma membrane of the secretory cell. As Ca^{2+} enters the cell down its concentration gradient, the resultant rise in cytosolic Ca^{2+} triggers fusion of the exocytotic vesicle with the plasma membrane and the subsequent release of its secretory product. For example, nerve cells controlling skeletal muscles release a specific chemical via Ca^{2+}-induced exocytosis in response to a nerve impulse. This chemical, in turn, initiates a series of events to bring about contraction of the muscle.

This completes our discussion of membrane transport. ▮ Table 3-2 summarizes the established pathways by which materials can pass between the ECF and ICF.

Caveolae, newly discovered membranous structures, may play roles in membrane transport and signal transduction.

The outer surface of the plasma membrane is not smooth but instead is dimpled with tiny cavelike indentations known as **caveolae** ("tiny caves"). These small flask-shaped pits have been observed for more than forty years through electron microscopy. Caveolae were not considered to have functional significance, however, until further investigation in the mid-1990s suggested that they (1) provide a new route for transport into the cell and (2) serve as a "switchboard" for relaying signals from many extracellular chemical messengers into the cell's interior.

An abundance of proteins, including a variety of receptors, cluster in these tiny chambers. Some of these receptors appear to play a role in a new form of cellular uptake of small molecules and ions. The best-studied example is the transport of the B vitamin folic acid into the cell. When folic acid binds with its receptors, which are concentrated in the portions of the plasma membrane that form the caveolae, the extracellular openings of these tiny caves close off. The high concentration of folic acid within a closed caveolar compartment encourages the movement of this vitamin across the caveolar membrane into the cytoplasm. Cellular uptake through the cyclic opening and closing of caveolae has been termed **potocytosis.** Potocytosis is believed to be an uptake mechanism for selected small molecules and ions, in contrast to receptor-mediated endocytosis, which transports selected large molecules into the cell (see p. 25). Furthermore, unlike endocytosis in which the invaginated pouch pinches off at the surface and breaks free, forming a membrane-enclosed vesicle that enters the cytoplasm, in potocytosis the caveolae are thought to remain attached to the plasma membrane.

Besides serving as uptake vehicles, caveolae also appear to be important sites for **signal transduction,** a complicated process in which incoming signals (instructions from extracellular chemical messengers) are conveyed to the cell's interior for execution. Extracellular chemical messengers that cannot gain entry to the cell, such as proteinaceous hormones and other regulatory molecules, signal the cell to perform a given response by first binding with membrane receptors specific for the extracellular messenger. This binding brings about a series of intracellular events that result in the desired cellular response. Interestingly, many membrane receptors important in signal transduction are concentrated in the caveolae, leading one investigator to call these tiny caves "signaling organelles." Caveolae are thought to be important in cell-to-cell communication because they gather, process, and transmit into the cell signals carried by chemical messengers released by other cells.

Table 3-2 Characteristics of the Methods of Membrane Transport

Methods of Transport	Substances Involved	Energy Requirements and Force-Producing Movement	Limit to Transport
Diffusion			
Through lipid bilayer	Nonpolar molecules of any size (e.g., O_2, CO_2, fatty acids)	Passive; molecules move down concentration gradient (from high to low concentration)	Continues until the gradient is abolished (steady state with no net diffusion)
Through protein channel	Specific small ions (e.g., Na^+, K^+, Ca^{2+}, Cl^-)	Passive; ions move down electrochemical gradient (from high to low concentration and attraction of ion to area of opposite charge)	Continues until there is no net movement and a steady state is established
Special case of osmosis	Water only	Passive; water moves down its own concentration gradient (water moves to area of lower water concentration, i.e., higher solute concentration)	Continues until concentration difference is abolished or until stopped by an opposing hydrostatic pressure or until cell is destroyed
Carrier-mediated transport			
Facilitated diffusion	Specific polar molecules for which a carrier is available (e.g., glucose)	Passive; molecules move down concentration gradient (from high to low concentration)	Displays a transport maximum (T_m); carrier can become saturated
Primary active transport	Specific ions or polar molecules for which carriers are available (e.g., Na^+, K^+, amino acids)	Active; ions move against concentration gradient (from low to high concentration); requires ATP	Displays a transport maximum; carrier can become saturated
Secondary active transport	Specific polar molecules and ions for which cotransport carriers are available (e.g., glucose, amino acids, some ions, including Ca^{2+})	Active; molecules move against concentration gradient (from low to high concentration); driven directly by ion gradient (usually Na^+) established by ATP-requiring primary pump	Displays a transport maximum; cotransport carrier can become saturated
Vesicular transport			
Endocytosis			
Pinocytosis	Small volume of ECF fluid, perhaps with specific bound molecules (e.g., protein); also important in membrane recycling	ATP required	Control poorly understood
Phagocytosis	Multimolecular particles (e.g., bacteria and cellular debris)	ATP required	Necessitates binding to specific receptor site on membrane surface
Exocytosis	Secretory products (e.g., hormones and enzymes) as well as large molecules passing through cell intact; also important in membrane recycling	ATP required; increase in cytosolic Ca^{2+} induces fusion of vesicle with plasma membrane	Secretion triggered by specific neural or hormonal stimuli; other controls involved in transcellular traffic and membrane recycling not known

Intercellular Communication and Signal Transduction

Communication between cells is largely orchestrated by four types of extracellular chemical messengers.

Communication is critical for the survival of the society of cells that collectively compose the body. The ability of cells to communicate with each other is essential for coordination of their diverse activities to maintain homeostasis as well as to control growth and development of the body as a whole. There are three types of intercellular ("between cell") communication (Fig. 3-20):

1. The most intimate means of intercellular communication is through gap junctions, which are minute tunnels that bridge the cytoplasm of neighboring cells in some types of tissue. Through these specialized anatomical arrangements (Fig. 3-6), small molecules and ions are directly exchanged between interacting cells without ever entering the ECF. Gap junctions are especially important in permit-

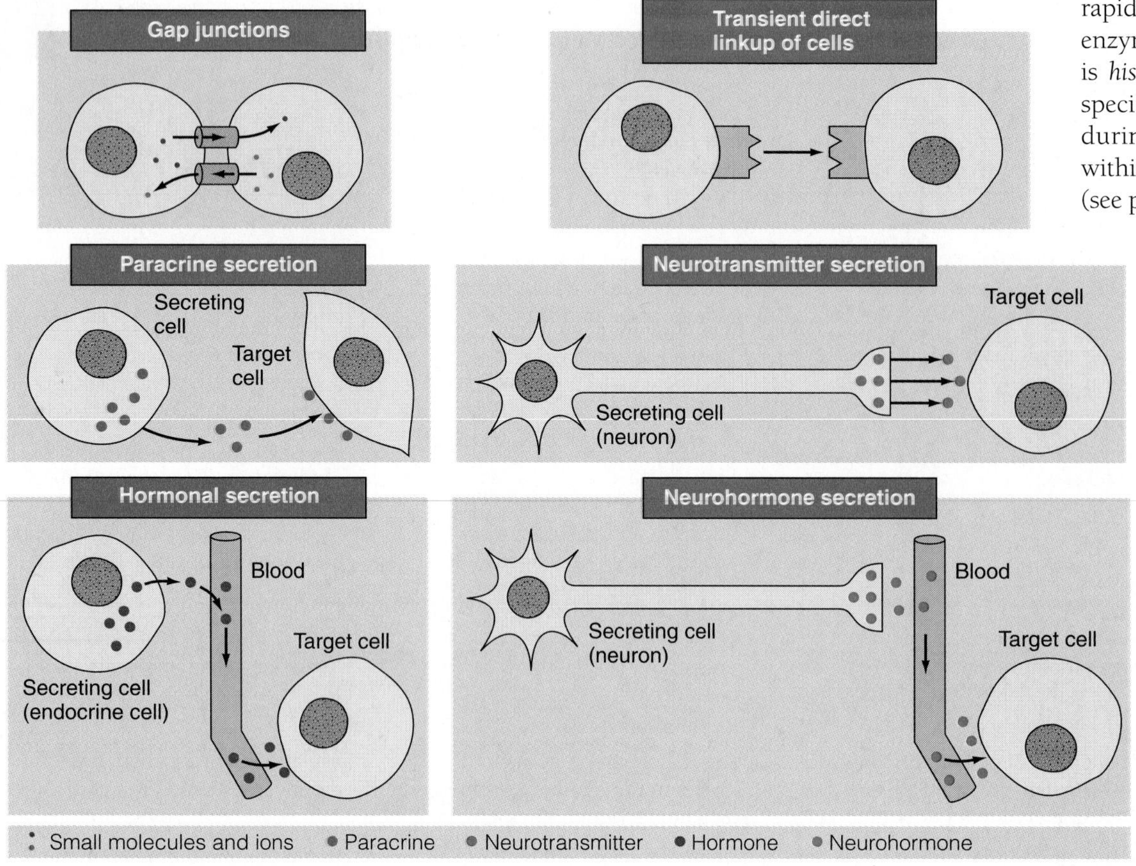

Figure 3-20 **Types of Intercellular Communication** Paracrines, neurotransmitters, hormones, and neurohormones are all intercellular chemical messengers that accomplish indirect communication between cells.

ting electrical signals to spread from one cell to the next in cardiac and smooth muscle.

2. The presence of signaling molecules on the surface membrane of some cells permits them to directly link up and interact with certain other cells in a specialized way. This is the means by which the phagocytes of the body's defense system specifically recognize and selectively destroy only undesirable cells, such as microbial invaders, while leaving the body's own cells alone.
3. The most common means by which cells communicate with each other is through intercellular chemical messengers, of which there are four types: *paracrines, neurotransmitters, hormones,* and *neurohormones.* In each case, a specific chemical messenger is synthesized by specialized cells to serve a designated purpose. On being released into the ECF by appropriate stimulation, these signaling agents act on other particular cells, the messenger's **target cells**, in a prescribed manner. These four types of chemical messengers differ in their source and the distance and means by which they get to their site of action.

   - **Paracrines** are local chemical messengers whose effect is exerted only on neighboring cells in the immediate environment of their site of secretion. Since paracrines are distributed by simple diffusion, their action is restricted to short distances. They do not gain entry to the blood in any significant quantity because they are rapidly inactivated by locally existing enzymes. One example of a paracrine is *histamine*, which is released from a specific type of connective tissue cell during an inflammatory response within an invaded or injured tissue (see p. 376). Among other things, histamine dilates (opens more widely) the blood vessels in the vicinity to increase blood flow to the tissue. This action brings additional blood-borne combat supplies into the affected area.

     Paracrines must be distinguished from chemicals that influence neighboring cells after being nonspecifically released during the course of cellular activity. For example, an increased local concentration of CO_2 in an exercising muscle is among the factors that promote local dilation of the blood vessels supplying the muscle. The resultant increased blood flow helps to meet the more active tissue's increased metabolic demands. However, CO_2 is produced by all cells and is not specifically released to accomplish this particular response, so it and similar nonspecifically released chemicals are not considered paracrines.
   - Neurons (nerve cells) communicate directly with the cells they innervate (their target cells) by releasing **neurotransmitters,** which are very short-range chemical messengers, in response to electrical signals (see p. 95). Like paracrines, neurotransmitters diffuse from their site of release across a narrow extracellular space to act locally on only an adjoining target cell, which is either another neuron, a muscle, or a gland.
   - **Hormones** are long-range chemical messengers that are specifically secreted into the blood by endocrine (ductless) glands (see p. 4) in response to an appropriate signal. The blood carries the messengers to other sites in the body, where they exert their effects on their target cells some distance away from their site of release.
   - **Neurohormones** are hormones released into the blood specifically by neurosecretory neurons. Like ordinary neurons, **neurosecretory neurons** can respond to and conduct electrical signals. Instead of directly innervating target cells, however, a neurosecretory neuron releases its chemical messenger, a neurohormone, into the blood upon appropriate stimulation. The neurohormone is then distributed through the blood to the target cells. Thus, like endocrine cells, neurosecretory neurons release blood-borne chemical messengers, whereas ordi-

nary neurons secrete short-range neurotransmitters into a confined space.

Binding of chemical messengers to membrane receptors brings about a wide range of responses in different cells though only a few remarkably similar pathways are used.

Dispersed within the outer surface of the plasma membrane are specialized protein receptors that bind with the selected chemical messengers that come into contact with the cell—for example, hormones delivered by the blood or neurotransmitters released from nerve endings. This combination of messenger with receptor triggers a sequence of cellular events that ultimately controls a particular cellular activity important in the maintenance of homeostasis, such as membrane transport, secretion, metabolism, or contraction.

In spite of the wide range of possible responses, there are only two general means by which binding of the receptor with the extracellular chemical messenger (the **first messenger**) brings about the desired intracellular response: (1) by opening or closing specific channels in the membrane to regulate the movement of particular ions into or out of the cell or (2) by transferring the signal to an intracellular chemical messenger (the **second messenger**), which in turn triggers a preprogrammed series of biochemical events within the cell. Because of the universal nature of these postreceptor (after receptor binding) events, let us examine each more closely.

Channel regulation by extracellular messengers Membrane channels behave as if they have "gates" that can be opened or closed depending on the circumstances. Changes in conformation (shape) of the proteins that form the channels can alternatively block the channel (channel closed) or permit passage through it (channel open). Channel opening and closing can be triggered in one of three ways: (1) by binding of an extracellular chemical messenger to a specific membrane receptor that is in close association with the channel, (2) by changes in electrical current in the plasma membrane, and (3) by stretching or other mechanical deformation of the channel. For now we will concentrate on channel regulation by means of chemical messengers and defer the other two methods of controlling channels until later (see p. 84).

An extracellular messenger can alter a channel through one of two mechanisms. For some channels, the receptor binding site on the plasma membrane is part of the channel itself. When an extracellular messenger binds with its receptor site, the channel opens (pathway 1 in ▬ Figure 3-21). For other channels, the receptor is a separate protein located near the

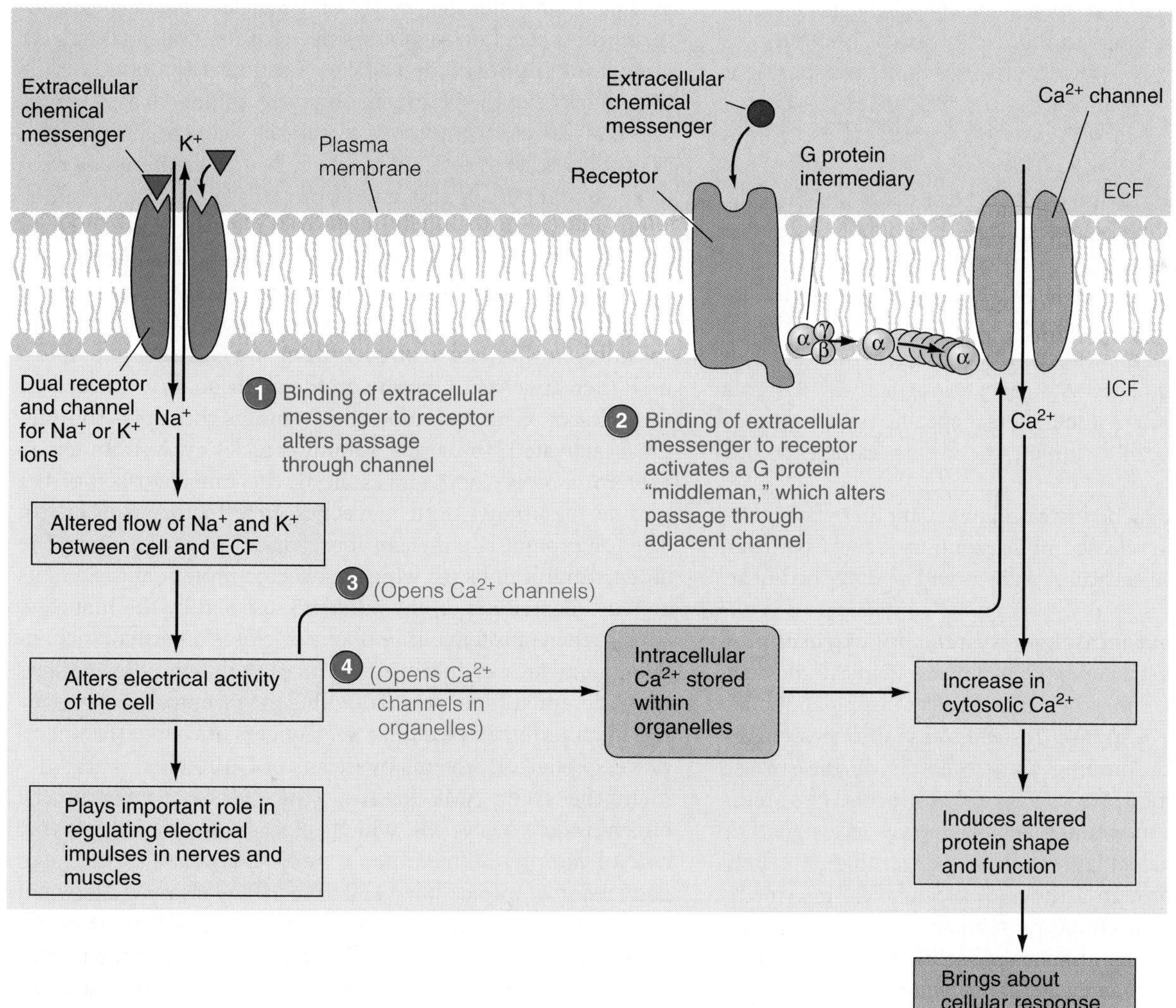

▬ *Figure 3-21* **Some Ways by Which Extracellular Messengers Regulate Channel Function** Binding of an extracellular chemical messenger with a surface membrane receptor can regulate ionic movement through channels to bring about the desired cellular response in the following ways: Binding of an extracellular messenger to a dual receptor/channel brings about a short-lived opening or closing of ion channels, such as Na^+ or K^+ channels, which generates electrical impulses ①. A transient opening of membrane Ca^{2+} channels occurs when binding of an extracellular messenger to a receptor activates a G protein intermediary, which alters a nearby ion channel, such as a Ca^{2+} channel ②. A transient opening of Ca^{2+} channels also occurs indirectly in response to electrical impulses produced by extracellular messenger-induced changes in Na^+ and K^+ channels ③. Release of Ca^{2+} from intracellular stores results when Ca^{2+} channels in organelles open in response to electrical impulses ④. An increase in cytosolic Ca^{2+} arising from pathways 2, 3, or 4 causes changes in the shape and function of specific intracellular proteins to produce the desired cellular response.

channel. In this case, binding of the extracellular messenger with its receptor activates membrane-bound intermediaries known as *G proteins,* which in turn open (or in some instances close) the appropriate adjacent channel (pathway 2). (The mechanism of action of G proteins will be discussed shortly.)

By opening or closing specific channels, extracellular messengers can regulate the flow of particular ions across the membrane. This ionic movement can be responsible for two different cellular events:

1. A small, short-lived movement of Na^+, K^+, or both across the membrane (pathway 1 in Fig. 3-21) alters the electrical activity of cells that are capable of generating electrical signals (or impulses), such as nerve and muscle cells.
2. A transient flow of calcium (Ca^{2+}) into the cell through opened Ca^{2+} channels (pathway 2) triggers an alteration in shape and function of specific intracellular proteins, which leads to the cell's response. Illustrative is the increase in cytosolic Ca^{2+} responsible for triggering the release of secretory product from many gland cells.

Upon completion of the response, the ions that moved across the membrane through opened channels to trigger the response are quickly returned to their original location by special carrier mechanisms in the membrane.

In some instances, the chemical messenger (or another stimulus) acts indirectly to open the Ca^{2+} channels by altering Na^+ and K^+ channels to induce an electrical impulse in the cell. The electrical impulse, in turn, is directly responsible for opening the Ca^{2+} channels (pathway 3). Release of chemicals from nerve cells in response to a nerve impulse is one such example.

In some cells, a rise in cytosolic Ca^{2+} can be brought about by release of Ca^{2+} from intracellular stores instead of Ca^{2+} entry through membrane channels. For example, large amounts of Ca^{2+} are stored within a modified endoplasmic reticulum (the sarcoplasmic reticulum) in skeletal muscle cells. An electrical impulse in these cells triggers the release of Ca^{2+} from this organelle into the cytosol (pathway 4). This increased cytosolic Ca^{2+} then alters a specific protein within the skeletal muscle cell to initiate the events leading to contraction.

Cytosolic Ca^{2+} may be increased in yet another way; intracellular Ca^{2+} can be released in a second messenger pathway independent of any electrical events, as will be described next.

Activation of second messenger systems by extracellular (first) messengers Many extracellular chemical messengers cannot actually enter their target cells to bring about the desired intracellular response. Instead, these first messengers issue their orders by binding with receptors on the surface membrane, triggering a "Psst, pass it on" process. First messenger binding to a membrane receptor serves as a signal for activation of an intracellular second messenger that ultimately relays the orders through a series of biochemical intermediaries to particular intracellular proteins that carry out the dictated response, such as changes in cellular metabolism or secretory activity. The intracellular pathways activated by a second messenger in response to binding of the first messenger to a surface receptor are remarkably similar among different cells despite the diversity of ultimate responses to that signal. The variability in response depends on the specialization of the cell, not on the mechanism utilized.

There are two major second messenger pathways: one utilizes **cyclic adenosine monophosphate** (**cyclic AMP** or **cAMP**) as a second messenger, and the other employs Ca^{2+} in this role. The two pathways have much in common. The initial stages are similar in that binding of a chemical messenger to a receptor on the outer surface of the membrane leads, via a series of biochemical steps, to activation of an enzyme on the cytoplasmic side of the membrane. This enzyme, in turn, activates intracellular second messengers that diffuse throughout the cell to trigger the appropriate cellular response. In both pathways, the cellular response is accomplished by altering the structure and subsequent function of particular cell proteins. For example, a particular enzymatic protein regulating a specific metabolic event may be modified so that its activity is increased or decreased.

In the cyclic AMP pathway, binding of an appropriate extracellular messenger to its surface receptor eventually activates (or in some instances inhibits) the enzyme **adenylyl cyclase** (step 1 in Fig. 3-22) on the inner surface of the membrane. A membrane-bound "middleman," a **G protein,** acts as an intermediary between the receptor and adenylyl cyclase. G proteins are so named because they are bound to guanine nucleotides—**guanosine triphosphate (GTP)** or **guanosine diphosphate (GDP).** G proteins are found on the inner surface of the plasma membrane. An unactivated G protein consists of a complex of alpha (α), beta (β), and gamma (γ) subunits, with a GDP molecule bound to the alpha subunit. A number of different G proteins with varying alpha subunits have been identified. The different G proteins are activated in response to binding of various first messengers to surface receptors. When a first messenger binds with its receptor, the receptor attaches to the appropriate G protein, resulting in the release of GDP from the G protein complex. GTP then attaches to the site on the alpha subunit vacated by the released GDP, an action that activates the alpha subunit. Once activated, the alpha subunit breaks away from the G protein complex and moves along the inner surface of the plasma membrane until it reaches an **effector protein.** An effector protein is either an ion channel or an enzyme. The alpha subunit links up with the effector protein and alters its activity. Depending on the outcome signaled by the first messenger, the G protein either opens or closes a particular channel or activates or inhibits a particular enzyme. Researchers have identified more than 100 different receptors that convey instructions of extracellular messengers through the membrane to effector proteins by means of G proteins.

In the cyclic AMP pathway, the effector protein is the enzyme adenylyl cyclase, which is located on the cytoplasmic side of the plasma membrane. Adenylyl cyclase induces the conversion of intracellular ATP to cAMP by cleaving off two of the phosphates (step 2). (This is the same ATP used as the common energy currency in the body.) The extracellular messenger cannot gain entry into the cell to "personally" deliver its message to the proteins that carry out the desired response.

Instead, it initiates membrane events that "arouse" an intracellular messenger, cAMP, which carries the signal beyond the membrane, triggering a sequence of intracellular events to bring about the response dictated by the extracellular messenger. To fulfill this function, cAMP activates a specific intracellular enzyme, **cAMP-dependent protein kinase** (step 3). Protein kinase in turn phosphorylates a specific intracellular protein (step 4), such as an enzyme important in a particular metabolic pathway. Phosphorylation refers to the transfer of a phosphate group from ATP to the protein at the expense of degrading ATP to ADP. Attachment of a phosphate group to the protein induces the protein to change its shape and function (either activating or inhibiting it) to bring about the desired response (step 5). For example, a particular enzymatic protein regulating a specific metabolic event may be modified so that its activity is increased or decreased. After the response is accomplished, the alpha subunit converts GTP to GDP by cleaving off a phosphate, in essence shutting itself off, then rejoins the beta and gamma subunits to restore the inactive G protein complex. Intracellular enzymes inactivate the other participating chemicals so that the response can be terminated. Otherwise, once triggered, the response would go on indefinitely until the cell ran out of necessary supplies.

Note that in this signal transduction pathway, the steps involving the extracellular first messenger, the receptor, the G protein complex, and the effector protein occur *in the plasma membrane* and lead to activation of the second messenger. The second messenger then triggers a chain reaction of biochemical events *inside the cell* that leads to the cellular response dictated by the extracellular messenger.

It is important to recognize that different types of cells have different proteins available for phosphorylation and modification by protein kinase. Therefore, a *common second messenger, cAMP, can induce widely differing responses in different cells,* depending on what proteins are modified. Cyclic AMP can be thought of as a commonly used molecular "switch" that can "turn on" (or "turn off") different cellular events depending on the unique specialization of a particular cell type. The variable responsiveness once the switch is turned on is due to the genetically programmed differences in the sets of proteins within different cells. For example, activation of the cAMP system brings about modification of heart rate in the heart, stimulation of the formation of female sex hormones in the ovaries, breakdown of stored glucose in the liver, and control of water conservation during urine formation in the kidneys.

Instead of cAMP, some cells use Ca^{2+} as a second messenger. In such cases, binding of the first messenger to the surface receptor eventually leads by means of G proteins to activation of the enzyme **phospholipase C,** a protein effector that is bound to the inner side of the membrane (step 1 in Fig. 3-23). This enzyme breaks down **phosphatidylinositol biphosphate** (abbreviated **PIP_2**), a component of the tails of the phospholipid molecules within the membrane itself. The products of PIP_2 breakdown are **diacylglycerol (DAG)** and **inositol triphosphate (IP_3)** (step 2). IP_3 is the fragment responsible for mobilizing intracellular Ca^{2+} stores to increase cytosolic Ca^{2+} (step 3). Calcium then takes over the role of second messenger, ultimately bringing about the response dictated by the first messenger. Many of the Ca^{2+}-dependent cellular events are triggered by activation of **calmodulin**, an intracellular Ca^{2+}-binding protein (step 4). Activation of calmodulin by Ca^{2+} is similar to activation of protein kinase by cAMP. From here the patterns of the two pathways are similar. Once bound to Ca^{2+}, activated calmodulin alters other cellular proteins (step 5), either activating or inhibiting them, to bring about the ultimate desired cellular response.

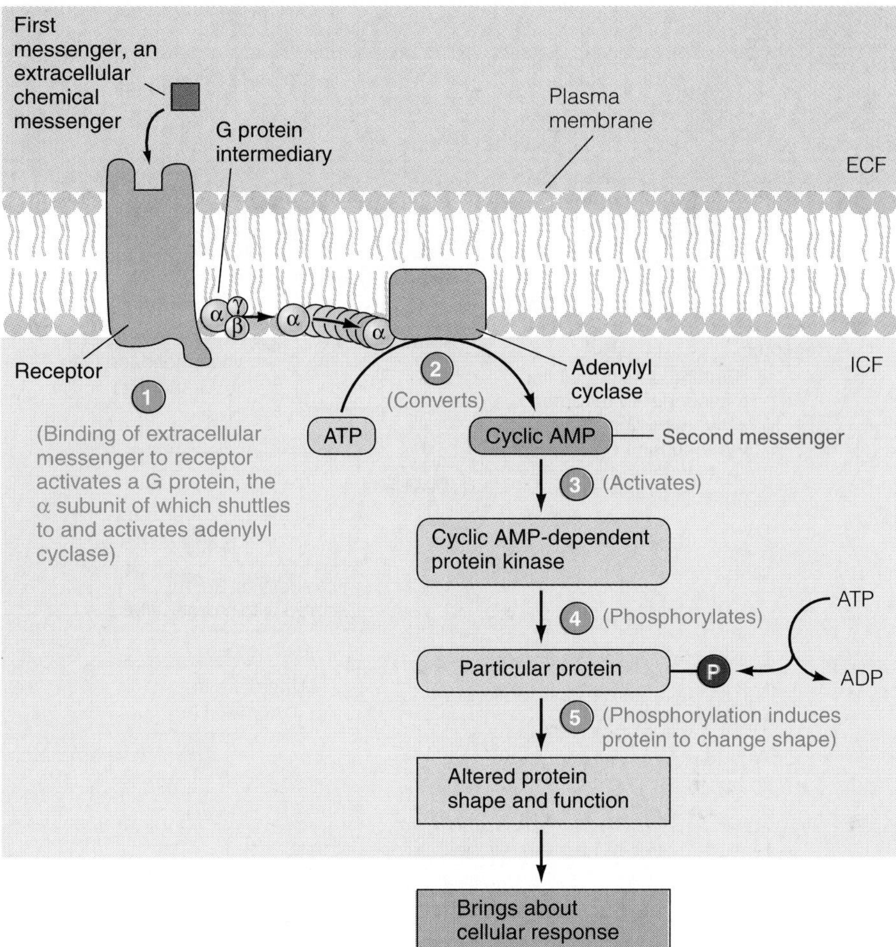

Figure 3-22 **Activation of the Cyclic AMP Second Messenger System by an Extracellular Messenger** Binding of an extracellular chemical messenger, the *first messenger,* to a surface membrane receptor activates by means of a G protein intermediary the membrane-bound enzyme adenylyl cyclase ①, which in turn converts intracellular ATP into cyclic AMP ②. Cyclic AMP acts as an intracellular *second messenger,* triggering the desired cellular response by activating cAMP-dependent protein kinase ③, which in turn phosphorylates ④ and thereby modifies ⑤ a particular intracellular protein. The altered protein then accomplishes the cellular response dictated by the extracellular messenger.

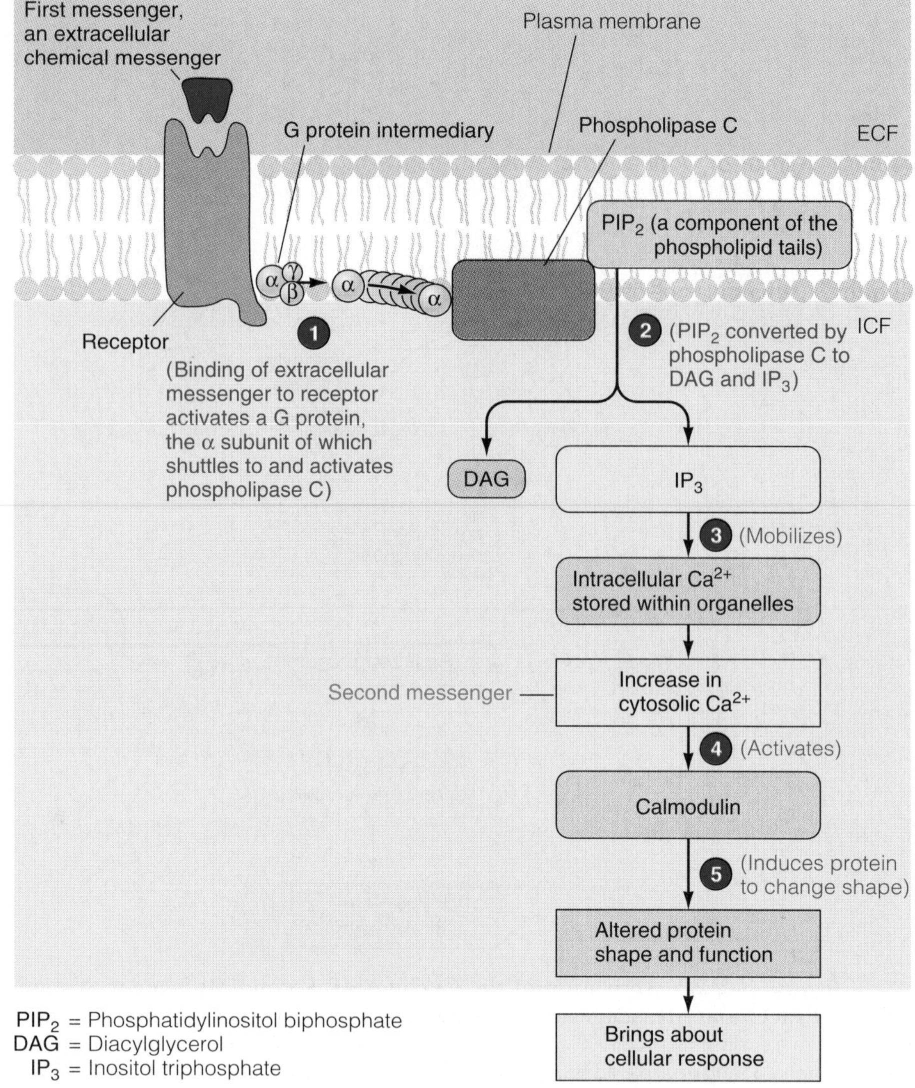

Figure 3-23 **Activation of the Calcium Second Messenger System by an Extracellular Messenger** Binding of an extracellular chemical messenger to a surface membrane receptor activates by means of a G protein intermediary the membrane-bound enzyme phospholipase C ①. Phospholipase C converts PIP_2, a membrane component, into DAG and IP_3 ②. IP_3 in turn mobilizes Ca^{2+} stored within organelles ③. Ca^{2+}, acting as a second messenger, activates calmodulin ④, which induces a change in the shape and function of a particular intracellular protein to produce the desired cellular response ⑤.

The cAMP and Ca^{2+} pathways frequently overlap in bringing about a particular cellular activity. For example, cAMP and Ca^{2+} can influence each other. Calcium-activated calmodulin can regulate adenylyl cyclase and thus influence cAMP, whereas cAMP-dependent kinase may phosphorylate and thereby change the activity of Ca^{2+} channels or carriers. Furthermore, in some instances, both Ca^{2+} and cAMP regulate the same intracellular protein. Also, remember that some Ca^{2+} channels are opened in response to changes in electrical current in the plasma membrane unrelated to second messenger systems.

Even though the Ca^{2+} and cAMP effects can become complexly intertwined, it is still notable that a great many diverse cellular events can be traced to a surprisingly small number of pathways: channel effects, one of the two major second messenger pathways, or some combination of these. It would not be correct, however, to leave the impression that these are the only possible pathways. For example, in a few cells **cyclic guanosine monophosphate (cyclic GMP)** serves as a second messenger in a system analogous to the cAMP system. An example is the signal transduction pathway involved in vision.

Several remaining points about receptor activation and the ensuing events merit attention. First, a multiplying (cascading) effect of these pathways greatly amplifies the initial signal (Fig. 3-24). Binding of one chemical messenger molecule to a receptor activates a number of adenylyl cyclase molecules (let us arbitrarily say 10), each of which activates many (in our example, 100) cAMP molecules. Each cAMP molecule then acts on a single protein kinase, which phosphorylates and thereby influences many (again, less us say 100) specific proteins, such as enzymes. Each enzyme, in turn, is responsible for producing many (perhaps 100) molecules of a particular product, such as a secretory product. The result of this cascade of events, with one event triggering the next event in sequence, is a tremendous amplification of the initial signal; in other words, the magnitude of the output is much greater than the input. In the hypothetical example in Figure 3-24, one chemical messenger molecule has been responsible for inducing a yield of 10 million molecules of a secretory product. In this way, very low concentrations of hormones and other chemical messengers can trigger pronounced cellular responses.

While membrane receptors serve as links between extracellular first messengers and intracellular second messengers in the regulation of specific cellular activities, the receptors themselves are also frequently subject to regulation. In many instances, the number and affinity (attraction of a receptor for its chemical messenger) can be altered, depending on the circumstances.

Many disease processes can be linked to malfunctioning receptors or defects in one of the components of the ensuing signal transduction pathways. For example, defective receptors are responsible for the extreme muscular weakness that characterizes *myasthenia gravis.* With this disease, affected skeletal muscle receptors are unable to respond to the chemical messenger released by nerves that normally triggers muscle contraction. On the other hand, in the intestinal disease *cholera,* the receptors are all functional, but the toxin produced by the cholera pathogen prevents the α subunit of the G protein from converting GTP to GDP. The toxin thus prevents the α subunit from turning itself off. This inability prevents the normal inactivation of cAMP in intestinal cells. In these cells, cAMP induces fluid secretion into the lumen. The severe diarrhea characteristic of cholera is caused by the continued secretion of this fluid into the gut, triggered by the continual presence of cAMP.

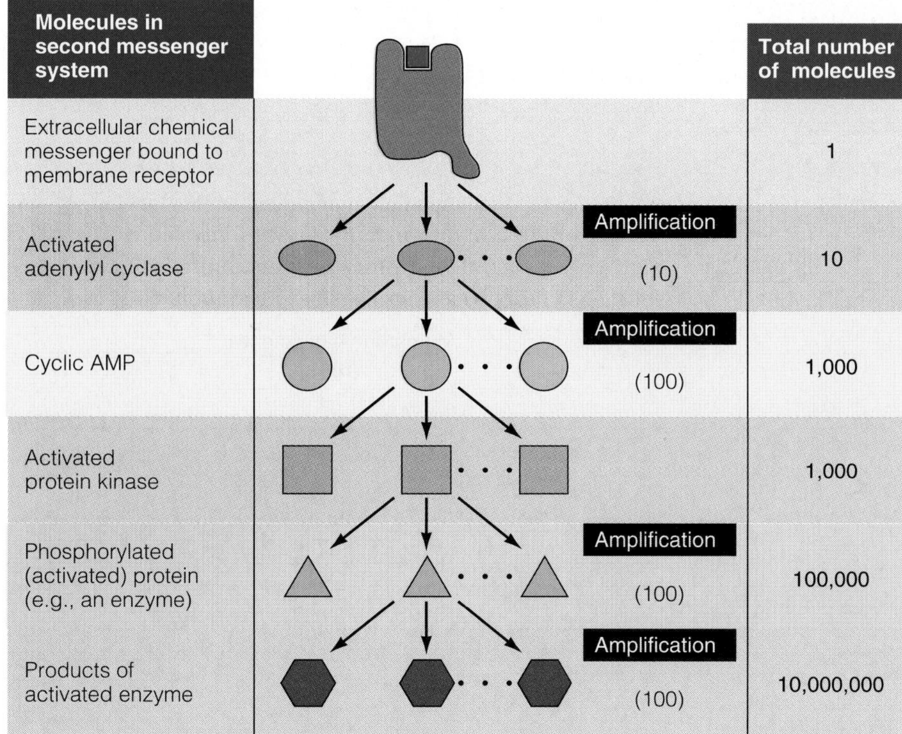

Figure 3-24 **Amplification of the Initial Signal by a Second Messenger System** Through amplification, very low concentrations of extracellular chemical messengers, such as hormones, can trigger pronounced cellular responses.

Membrane Potential

Membrane potential refers to a separation of opposite charges across the plasma membrane.

The unequal distribution of a few key ions between the ICF and ECF and their selective movement through the plasma membrane are responsible for the electrical properties of the membrane. All plasma membranes have a membrane potential or are polarized electrically. **Membrane potential** refers to a separation of charges across the membrane, or to a difference in the relative number of cations and anions in the ICF and ECF. Recall that opposite charges tend to attract each other and like charges tend to repel each other. Work must be performed (energy expended) to separate opposite charges after they have come together. Conversely, when oppositely charged particles have been separated, the electrical force of attraction between them can be harnessed to perform work when the charges are permitted to come together again. This is the basic principle underlying electrically powered devices. Because separated charges have the "potential" to do work, a separation of charges across the membrane is referred to as a *membrane potential.* Potential is measured in units of volts (the same unit used for the voltage in electrical devices), but because the membrane potential is relatively low, the unit used is **millivolts** (mV) (1 mV = 1/1,000 volt).

Since the concept of potential is fundamental to understanding nerve and muscle physiology, it is important to understand clearly what this term means. The membrane in Fig. 3-25a is electrically neutral. An equal number of positive (+) and negative (−) charges are on each side of the membrane, so no membrane potential exists. In Figure 3-25b, some of the + charges from the right side have been moved to the left. Now the left has an excess of + charges, leaving an excess of − charges on the right. In other words, there is a separation of opposite charges across the membrane, or a difference in the relative number of + and − charges between the two sides (i.e., a membrane potential exists). The attractive force between these separated charges will cause them to accumulate in a thin layer along the outer and inner surfaces of the plasma membrane (Fig. 3-25c). These separated charges represent only a small fraction of the total number of charged particles (ions) present in the ICF and ECF. The vast majority of the fluid inside and outside the cells is electrically neutral (Fig. 3-25d). The electrically balanced ions can be ignored, because they do not contribute to membrane potential. Thus, an almost insignificant fraction of the total number of charged particles present in the body fluids is responsible for the membrane potential.

The magnitude of the potential depends on the degree of separation of the opposite charges: the greater the number of charges separated, the larger the potential. Therefore, in Figure 3-25e, membrane B has more potential than A and less potential than C.

Membrane potential is primarily due to differences in the distribution and membrane permeability of sodium, potassium, and large intracellular anions.

All living cells have a membrane potential characterized by a slight excess of positive charges outside and a corresponding slight excess of negative charges on the inside. In the body, electrical charges are carried by ions. The ions primarily responsible for the generation of membrane potential are Na^+, K^+, and A^-. The latter refers to the large, negatively charged (anionic) intracellular proteins. Other ions (calcium, magnesium, chloride, bicarbonate, and phosphate, to name a few) do not make a direct contribution to the electrical properties of the plasma membrane in most cells, even though they play other important roles in the body.

The concentrations and relative permeabilities of the ions critical to membrane electrical activity are compared in Table 3-3. Note that *Na^+ is in greater concentration in the extracellular fluid and K^+ is in much higher concentration in the intracellular fluid.* These concentration differences are maintained by the Na^+-K^+ pump at the expense of energy. Because the plasma membrane is virtually impermeable to A^-,

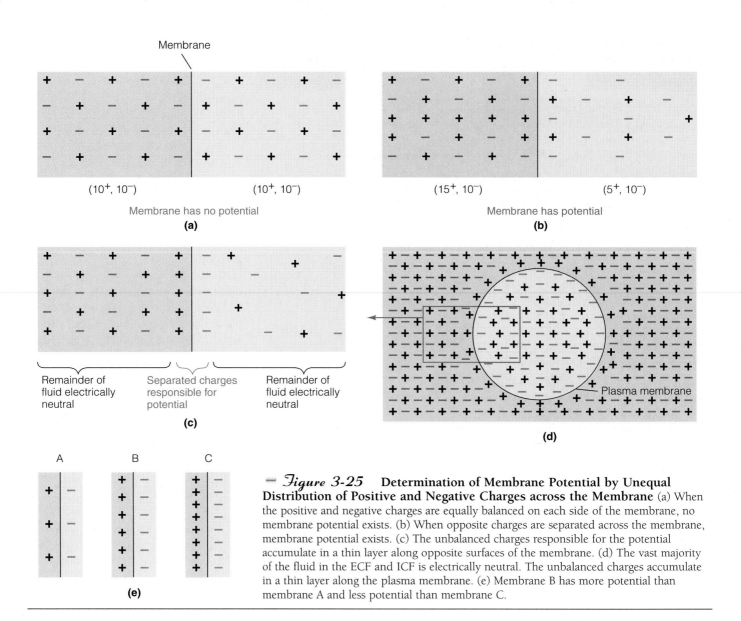

Figure 3-25 **Determination of Membrane Potential by Unequal Distribution of Positive and Negative Charges across the Membrane** (a) When the positive and negative charges are equally balanced on each side of the membrane, no membrane potential exists. (b) When opposite charges are separated across the membrane, membrane potential exists. (c) The unbalanced charges responsible for the potential accumulate in a thin layer along opposite surfaces of the membrane. (d) The vast majority of the fluid in the ECF and ICF is electrically neutral. The unbalanced charges accumulate in a thin layer along the plasma membrane. (e) Membrane B has more potential than membrane A and less potential than membrane C.

these large, negatively charged proteins are found *only inside* the cell. After they have been synthesized from amino acids transported into the cell, they remain trapped within the cell. In addition to the active carrier mechanism, Na^+ and K^+ can passively cross the membrane through protein channels specific for them. It is usually much easier for K^+ than for Na^+ to get through the membrane because typically more K^+ channels than Na^+ channels are open. In a nerve cell at rest (i.e., when it is not conducting a nerve impulse), the membrane is about fifty to seventy-five times more permeable to K^+ than to Na^+.

Table 3-3 Concentration and Permeability of Ions Responsible for Membrane Potential in a Resting Nerve Cell

Ion	*Concentration (Millimoles/Liter)* Extracellular	Intracellular	*Relative Permeability*
Na^+	150	15	1
K^+	5	150	50–75
A^-	0	65	0

Armed with a knowledge of the relative concentrations and permeabilities of these ions, we can now analyze the forces acting across the plasma membrane. This analysis will be broken down as follows: first, we will consider the direct contributions of the Na^+-K^+ pump to membrane potential; second, the effect the movement of K^+ alone would have on membrane potential; third, the effect of Na^+ alone; and finally the situation that exists in the cells when both K^+ and Na^+ effects are taking place concurrently. Table 3-4 summarizes the concentration and electrical gradients that exist for K^+ and for Na^+ under various conditions. Remember that the concentration gradient for K^+ will always be outward and the concentration gradient for Na^+ will always be inward because the Na^+-K^+ pump maintains a higher concentration of K^+ inside the cell and a higher concentration of Na^+ outside the cell. Also, note that because K^+ and Na^+ are both cations (positively charged), the electrical gradient for both will always be toward the negatively charged side of the membrane.

Effect of sodium-potassium pump on membrane potential About 20% of the membrane potential is directly generated by the Na^+-K^+ pump. This active transport mechanism pumps three Na^+ out for every two K^+ it transports in. Because Na^+ and K^+ are both positive ions, this unequal transport generates a membrane potential, with the outside becoming relatively more positive than the inside as more positive ions are transported out than in. However, most of the membrane potential—the remaining 80%—is caused by the passive diffusion of K^+ and Na^+ down concentration gradients. Thus, most of the Na^+-K^+ pump's role in producing membrane potential is indirect through its critical contribution to maintaining the concentration gradients directly responsible for the ion movements that generate most of the potential.

Effect of the movement of potassium alone on membrane potential Let's consider a hypothetical situation characterized by (1) the concentrations that exist for K^+ and A^- across the plasma membrane, (2) free permeability of the membrane to K^+ but not to A^-, and (3) no potential as yet present. The concentration gradient for K^+ would tend to move this ion out of the cell (Fig. 3-26). Because the membrane is permeable to K^+, this ion would readily pass through. As potassium ions moved to the outside, they would carry their positive charge with them, so more positive charges would be on the outside whereas negative charges in the form of A^- would be left behind on the inside, similar to the situation shown in Figure 3-25b. (Remember that the large protein anions cannot diffuse out, despite a tremendous concentration gradient.) A membrane potential would now exist. Because an electrical gradient would also be present, K^+, being a positively charged ion, would be attracted toward the negatively charged interior and repelled by the positively charged exterior. Thus, two opposing forces would now be acting on K^+: the concentration gradient tending to move K^+ out of the cell and the electrical gradient tending to move these same ions into the cell.

Initially, the concentration gradient would be stronger than the electrical gradient, so net diffusion of K^+ out of the cell would continue and the membrane potential would increase. As more and more K^+ moved down its concentration gradient and out of the cell, however, the opposing electrical gradient would also become greater as the outside became increasingly more positive and the inside more negative. Net outward diffusion would gradually be reduced as the strength of the electrical gradient approached that of the concentration gradient. Finally, when these two forces exactly balanced each other, no further net movement of K^+ would occur. The potential that would exist at this equilibrium is known as the **equilibrium potential** for **K^+ (E_{K^+}).** At this point, a large concentration gradient for K^+ would still exist, but no more K^+ would move out down this concentration gradient because of the exactly equal opposing electrical gradient (Fig. 3-26).

Table 3-4

Concentration and Electrical Gradients for K^+ and Na^+ at Equilibrium Potential (E) and Resting Potential

Ion	*Condition*	*Gradient*	*Direction of Gradient*
K^+	E_{K^+} (−90 mV)	Concentration gradient Electrical gradient	Outward Inward
Na^+	E_{Na^+} (+60 mV)	Concentration gradient Electrical gradient	Inward Outward
K^+	Resting potential (−70 mV)	Concentration gradient Electrical gradient	Outward Inward
Na^+	Resting potential (−70 mV)	Concentration gradient Electrical gradient	Inward Inward

The membrane potential at E_{K^+} is −90 mV. It is not really a negative potential. By convention, *the sign always designates the polarity of the excess charge on the inside of the membrane.* A membrane potential of −90 mV means that the potential is of a magnitude of 90 mV, with the inside being negative relative to the outside. A potential of +90 mV would have the same strength, but in this case the inside would be more positive than the outside.

The equilibrium potential for a given ion of differing concentrations across a membrane can be calculated by means of the **Nernst equation** as follows:

$$E = 61 \log \frac{C_o}{C_i}$$

Figure 3-26 Equilibrium Potential for K^+ At the equilibrium potential for K^+ (E_{K^+}), the outward concentration gradient is exactly counterbalanced by the inward electrical gradient. The membrane potential at this point is −90 mV.

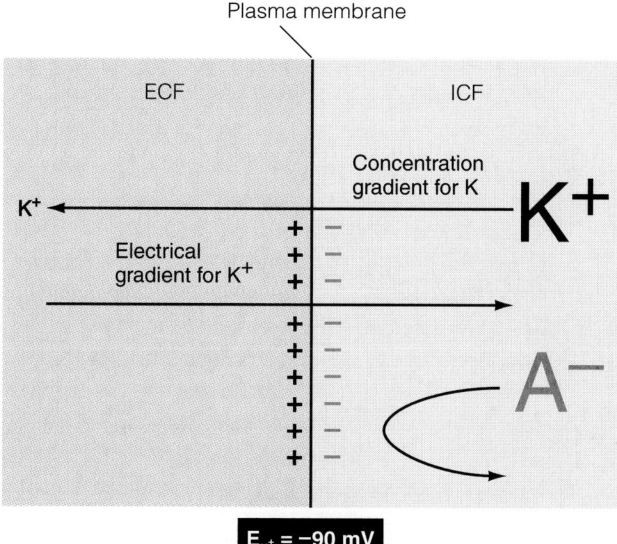

where

E = equilibrium potential for ion in mV

61 = a constant that incorporates the universal gas constant (R), absolute temperature (T), the ion's valence (z), and an electrical constant known as Faraday (F); 61 = RT/zF

C_o = concentration of the ion outside the cell in millimoles/liter (millimolars;mM)

C_i = concentration of the ion inside the cell in mM

Given that the ECF concentration of K^+ is 5 mM and the ICF concentration is 150 mM,

$$E_{K^+} = 61 \log \frac{5 \text{ mM}}{150 \text{ mM}}$$

$$= 61 \log \frac{1}{30}$$

Since the log of $\frac{1}{30} = -1.477$,

$$E_{K^+} = 61(-1.477) = -90 \text{ mV}$$

Since 61 is a constant, the equilibrium potential is essentially a measure of the membrane potential (that is, the magnitude of the electrical gradient) that exactly counterbalances the concentration gradient that exists for the ion (that is, the ratio between the ion's concentration outside and inside the cell). Note that the larger the concentration gradient for an ion, the greater the ion's equilibrium potential. A comparably greater opposing electrical gradient would be required to counterbalance the larger concentration gradient.

Effect of movement of sodium alone on membrane potential A similar hypothetical situation could be developed for Na^+ alone (Fig. 3-27). The concentration gradient for Na^+ would move this ion into the cell, producing a buildup of positive charges on the interior of the membrane and leaving negative charges unbalanced outside (primarily in the form of chloride, Cl^-; Na^+ and Cl^- are the predominant ECF ions). Net diffusion inward would continue until equilibrium was established by the development of an opposing electrical gradient that exactly counterbalanced the concentration gradient. At this point, given the concentrations for Na^+, the **Na^+ equilibrium potential (E_{Na^+})** would be +60 mV. In this case the inside of the cell would be positive, in contrast to the equilibrium potential for K^+. The magnitude of E_{Na^+} is somewhat less than for E_{K^+} (60 mV compared to 90 mV) because the concentration gradient for Na^+ is not as large (Table 3-3); thus, the opposing electrical gradient (membrane potential) is not as great at equilibrium.

Figure 3-27 **Equilibrium Potential for Na^+** At the equilibrium potential for Na^+ (E_{Na^+}), the inward concentration gradient is exactly counterbalanced by the outward electrical gradient. The membrane potential at this point is +60 mV.

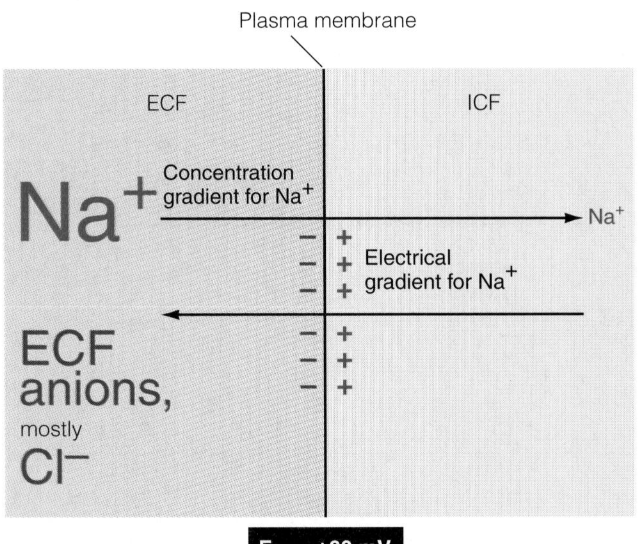

Concurrent potassium and sodium effects on membrane potential Neither K^+ nor Na^+ exists alone in the body fluids, so equilibrium potentials are not present in the body cells. They exist only in hypothetical or experimental conditions. In a living cell, the effects of both K^+ and Na^+ must be taken into account. Because the membrane at rest is fifty to seventy-five times more permeable to K^+ than to Na^+, K^+ passes through more readily than Na^+; thus, K^+ influences the resting membrane potential to a much greater extent than does Na^+. Recall that K^+ acting alone would establish an equilibrium potential of −90 mV. The membrane is somewhat permeable to Na^+, however, so some Na^+ enters the cell in a limited attempt to reach its equilibrium potential. This Na^+ influx neutralizes, or cancels, some of the potential produced by K^+ alone.

To facilitate an understanding of this concept, assume that each separated pair of charges in Figure 3-28 represents 10 mV of potential. (This is not technically correct because in reality, many separated charges must be present to account for a potential of 10 mV.) In this simplified example, nine separated pluses and minuses, with the minuses on the inside, would represent the E_{K^+} of −90 mV. Superimposing the slight influence of Na^+ on this K^+-dominated membrane, assume that two sodium ions enter the cell down the Na^+ concentration and electrical gradients. (Note that the electrical gradient for Na^+ is now inward in contrast to the outward electrical gradient for Na^+ at E_{Na^+}. At E_{Na^+}, the inside of the cell is positive as a result of the outward movement of Na^+ down its concentration gradient. In a resting nerve cell, however, the inside is negative because of the dominant influence of K^+ on membrane potential. Thus, both the concentration and electrical gradients now favor the inward movement of Na^+.) The inward movement of these two positively charged sodium ions neutralizes some of the potential established by K^+, so now only seven pairs of charges are separated, and the potential is −70 mV. This is the **resting membrane potential** of a typical nerve cell. The resting potential is much closer to E_{K^+} than to E_{Na^+} because of the greater permeability of the membrane to K^+, but it is slightly less than E_{K^+} (−70 mV is a lower potential than −90 mV) because of the weak influence of Na^+.

Figure 3-28 Effect of Concurrent K^+ and Na^+ Movement on Establishing the Resting Membrane Potential Given the concentration gradients that exist across the plasma membrane, K^+ tends to drive the membrane potential to K^+'s equilibrium potential (−90 mV), whereas Na^+ tends to drive the membrane potential to Na^+'s equilibrium potential (+60 mV). However, K^+ exerts the dominant effect on the resting membrane potential because the membrane is more permeable to K^+. As a result, the resting potential (−70 mV) is much closer to E_{K^+} than to E_{Na^+}. During the establishment of resting potential, the relatively large net diffusion of K^+ outward does not produce a potential of −90 mV because the resting membrane is slightly permeable to Na^+ and the relatively small net diffusion of Na^+ inward neutralizes some of the potential that would be created by K^+ alone, bringing the resting potential to −70 mV, slightly less than E_{K^+}.

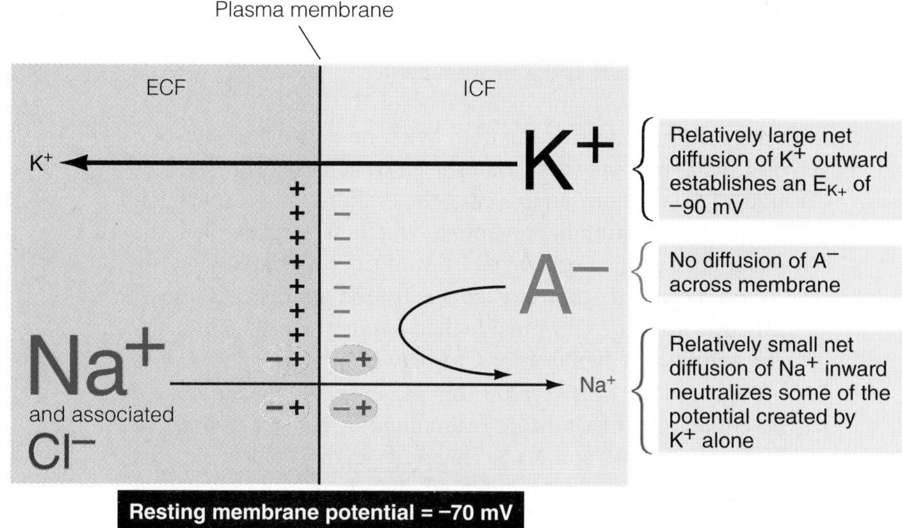

At resting potential, neither K^+ nor Na^+ is at equilibrium. A potential of −70 mV does not exactly counterbalance the concentration gradient for K^+; it takes a potential of −90 mV to do that. Thus, there is a continual tendency for K^+ to passively *leak* out through its channels. In the case of Na^+, the concentration and electrical gradients do not even oppose each other; they both favor the inward movement of Na^+. Therefore, Na^+ continually leaks inward down its electrochemical gradient, but only slowly because of its low permeability.

Since such leaking goes on all the time, why doesn't the intracellular concentration of K^+ continue to fall and the concentration of Na^+ inside the cell progressively increase? This does not happen because of the Na^+−K^+ pump. This active transport mechanism counterbalances the rate of leakage (Fig. 3-29). At resting potential, the pump transports back into the cell essentially the same number of potassium ions that have leaked out and simultaneously transports to the outside the sodium ions that have leaked in. Through this balance between leak and pump, the concentration gradients for K^+ and Na^+ remain constant across the membrane. Thus, not only is the Na^+-K^+ pump initially responsible for the Na^+ and K^+ concentration differences across the membrane, but it also maintains these differences.

As just discussed, it is the presence of these concentration gradients, together with the difference in permeability of the membrane to these ions, that accounts for the resting membrane potential. In this resting state, the potential remains constant. There is no net movement of any ions. All passive forces are exactly balanced by active forces. A steady state exists, even though there is still a strong concentration gradient for both K^+ and Na^+ in opposite directions, as well as a slight excess of positive charges in the ECF accompanied by a corresponding slight excess of negative charges in the ICF (enough to account for a potential of the magnitude of 70 mV). At this point, although movement across the membrane is taking place by means of passive leaks and active pumping, the exchange of charges between the ICF and ECF is exactly balanced, with the potential that has been established by these forces remaining constant.

Thus far, we have largely ignored one other ion present in high concentration in the ECF—Cl^-. Chloride is the principal ECF anion. Its equilibrium potential is −70 mV, exactly the same as the resting membrane potential. Movement alone of negatively charged Cl^- into the cell down its concentration gradient would produce an opposing electrical gradient, with the inside negative compared to the outside. When physiologists were first examining the ionic effects that could account for the membrane potential, it was tempting to think that Cl^- movements and establishment of the Cl^- equilibrium potential could be solely responsible for producing the identical

Figure 3-29 Counterbalance between Passive Na^+ and K^+ Leaks and the Active Na^+-K^+ Pump At resting membrane potential, the passive leaks of Na^+ and K^+ down their electrochemical gradients are exactly counterbalanced by the active Na^+-K^+ pump, so that there is no net movement of Na^+ and K^+ and the membrane potential remains constant.

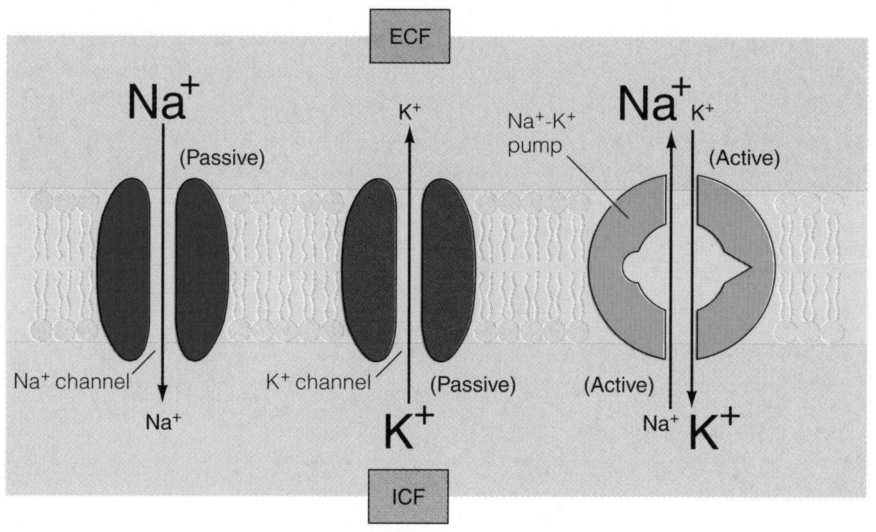

resting membrane potential. Actually, the reverse is the case. The membrane potential is responsible for driving the distribution of Cl^- across the membrane.

Most cells are highly permeable to Cl^- but have no active transport mechanisms for Cl^-. With no active forces acting on it, Cl^- passively distributes itself to achieve an individual state of equilibrium. In this case, the established electrical gradient (the membrane potential, which is negative on the inside) drives the negative Cl^- out of the cell until an exactly counterbalanced opposing concentration gradient is established, with more Cl^- outside than inside the cells. Thus, the concentration difference for Cl^- between the ECF and ICF is brought about passively by the presence of the membrane potential, rather than being maintained by an active pump, as is the case for K^+ and Na^+. Therefore, in most cells Cl^- does not influence membrane potential; instead, membrane potential passively influences the Cl^- distribution. (Some specialized cells have an active Cl^- pump, with subsequent movement of Cl^- accounting for part of the potential.)

Nerve and muscle cells have developed a specialized use for membrane potential. They are able to rapidly and transiently alter their membrane permeabilities to the involved ions in response to appropriate stimulation, thereby bringing about fluctuations in membrane potential. The rapid fluctuations in potential are responsible for producing nerve impulses in nerve cells and for triggering contraction in muscle cells. These activities will be the focus of the next five chapters. Even though all cells display a membrane potential, its significance in other cells is uncertain, although the potential of some secretory cells may somehow be linked to their level of secretory activity.

Chapter in Perspective: Focus on Homeostasis

All cells of the body must obtain vital materials such as nutrients and O_2 from the surrounding ECF and must transfer to the ECF wastes to be eliminated as well as secretory products such as chemical messengers and digestive enzymes. Thus, transport of materials across the plasma membrane between the ECF and ICF is essential for cell survival, and the constituents in the ECF must be homeostatically maintained in order to support these life-sustaining exchanges.

Many cell types utilize membrane transport to carry out their specialized activities geared toward maintaining homeostasis. Following are several examples:

1. Absorption of nutrients from the digestive tract lumen involves the transport of these energy-giving molecules across the membranes of the cells lining the tract.
2. Exchange of O_2 and CO_2 between the air and blood in the lungs involves the transport of these gases across the membranes of the cells lining the lungs' air sacs and blood vessels.
3. Urine formation is accomplished by the selective transfer of materials between the blood and the fluid within the kidney tubules across the membranes of the cells lining the tubules.
4. The beating of the heart is triggered by cyclical changes in the transport of Na^+, K^+, and Ca^{2+} across the heart cells' membranes.
5. Secretion of chemical messengers such as neurotransmitters from nerve cells and hormones from endocrine cells involves the transport of these regulatory products to the ECF on appropriate stimulation.

In addition to providing selective transport of materials between the ECF and ICF, the plasma membrane contains receptor sites for binding with specific chemical messengers that regulate various cell activities, many of which are specialized activities aimed toward maintaining homeostasis. For example, the hormone vasopressin, which is secreted in response to a water deficit in the body, binds with receptor sites in the plasma membrane of a specific type of kidney cell. This binding triggers these cells to conserve water during urine formation, thus helping alleviate the water deficit that initiated the response.

All living cells have a membrane potential, with the cell's interior being slightly more negative than the fluid surrounding the cell when the cell is electrically at rest. The specialized activities of nerve and muscle cells depends on these cells' ability to change their membrane potential rapidly on appropriate stimulation. These transient, rapid changes in potential in nerve cells serve as electrical signals or nerve impulses, which provide a means to transmit information along nerve pathways. This information is used to accomplish homeostatic adjustments, such as restoring blood pressure to normal when signaled that it has fallen to low.

Rapid changes in membrane potential in muscle cells trigger muscle contraction, the specialized activity of muscle. Muscle contraction contributes to homeostasis in many ways, including the pumping of blood by the heart and moving food through the digestive tract.

Chapter Summary

Membrane Structure and Composition

All cells are bounded by a plasma membrane, a thin lipid bilayer in which proteins are interspersed and to which carbohydrates are attached on the outer surface. The electron microscopic appearance of the plasma membrane as a trilaminar structure (two dark lines separated by a light interspace) is believed to be caused by the arrangement of the molecules composing it. The phospholipids orient themselves to form a bilayer with a hydrophobic interior (light interspace) sandwiched between the hydrophilic outer and inner surfaces (dark lines). This lipid bilayer forms the structural boundary of the cell, serving as a barrier for water-soluble substances and being responsible for the fluid nature of the membrane. Cholesterol molecules tucked between the phospholipids contribute to the fluidity and stability of the membrane.

Membrane proteins, which vary in type and distribution among cells, serve as channels for passage of small ions across the membrane; carriers for transport of specific substances in or out of the cell; receptor sites for detecting and responding to chemical messengers that alter cell function; membrane-bound enzymes that govern specific chemical reactions; cell adhesion molecules that help hold cells together; and a support meshwork on the inner-membrane surface to help maintain cell shape in association with the cytoskeleton.

The membrane carbohydrates, short sugar chains that project from the outer surface only, serve as self-identity markers. They are important in recognition of "self" in cell-to-cell interactions such as tissue formation and tissue growth.

Cell-to-Cell Adhesions

Special cells locally secrete a complex extracellular matrix, which serves as a biological "glue" between the cells of a tissue. Many cells are further joined by specialized cell junctions. Desmosomes serve as adhering junctions to hold cells together mechanically and are especially important in tissues subject to a great deal of stretching. Tight junctions actually fuse cells together to seal off passage between cells, thereby permitting only regulated passage of materials through the cells. These impermeable junctions are found in the epithelial sheets that separate compartments with very different chemical compositions. Cells joined by gap junctions are connected by small tunnels that permit exchange of ions and small molecules between the cells. Such movement of ions plays a key role in the spread of electrical activity to synchronize contraction in heart and smooth muscle.

Membrane Transport

Materials can pass between the ECF and ICF by the following pathways. Nonpolar (lipid-soluble) molecules of any size can dissolve in and pass through the lipid bilayer. Small ions traverse through protein channels specific for them. Particles move passively through these pathways down electrochemical gradients. Osmosis is a special case of water moving down its own concentration gradient.

Other substances can be selectively transferred across the membrane by specific carrier proteins, being moved either down a concentration gradient without the need for energy expenditure (facilitated diffusion) or moved against a concentration gradient at the expense of cellular energy (active transport). Carriers can move a single substance in one direction, two substances in opposite directions, or two substances in the same direction. Primary active transport requires the direct utilization of ATP to drive the pump, whereas secondary active transport is driven by an ion concentration gradient established by a primary active transport system. Carrier mechanisms are important for transfer of small polar molecules and for selected movement of ions.

Large polar molecules and multimolecular particles can leave or enter the cell by being wrapped in a piece of membrane to form vesicles that can be internalized (endocytosis) or externalized (exocytosis). Cells are differentially selective in what enters or leaves because they possess varying numbers and kinds of channels, carriers, and mechanisms for vesicular transport. Large polar molecules (too large for channels and not lipid-soluble) for which there are no special transport mechanisms are unable to permeate.

Intercellular Communication and Signal Transduction

Cells communicate with each other to carry out various coordinated activities largely by dispatching extracellular chemical messengers, which act on particular target cells to bring about the desired response. Transferral of the signal carried by the extracellular messenger into the cell for execution is known as signal transduction. Attachment of an extracellular chemical messenger such as a hormone (the first messenger) to a membrane receptor initiates one of several related intracellular pathways to bring about the desired response. Chemical messengers trigger cellular responses by two major methods: (1) opening or closing specific channels or (2) activating an intracellular messenger (the second messenger). Two commonly employed second messengers are cyclic AMP and Ca^{2+}. Once activated, these second messengers initiate a similar cascade of intracellular events that ultimately lead to a change in the shape and function of particular proteins to cause the appropriate cellular response. Despite the often widespread distribution of a single chemical messenger and the similarity of intracellular pathways employed, cells vary in their response because (1) different cell types are equipped with different sets of receptors that can bind with only selected types of messengers from among the many that might come into contact with each cell; and (2) various cell types contain different intracellular proteins, each of which responds uniquely to an identical second messenger.

Membrane Potential

All cells have a membrane potential, which is a separation of opposite charges across the plasma membrane. The Na^+-K^+ pump makes a small direct contribution to membrane potential through its unequal transport of positive ions; it transports more Na^+ ions out than K^+ ions in. The primary role of the Na^+-K^+ pump, however, is to actively maintain a greater concentration of Na^+ outside the cell and a greater concentration of K^+ inside the cell. These concentration gradients tend to passively move K^+ out of the cell and Na^+ into the cell. Because the resting membrane is much more permeable to K^+ than to Na^+, substantially more K^+ leaves the cell than Na^+ enters. This results in an excess of positive charges outside the cell and leaves an unbalanced excess of negative charges inside in the form of large protein anions (A^-) that are trapped within the cell. When the resting membrane potential of -70 mV is achieved, no further net movement of K^+ and Na^+ takes place because any further leaking of these ions down their concentration gradients is quickly reversed by the Na^+-K^+ pump. The distribution of Cl^- across the membrane is passively driven by the established membrane potential so that Cl^- is concentrated in the ECF.

Review Exercises

Objective Questions (Answers on p. E–2.)

1. The nonpolar tails of the phospholipid molecules bury themselves in the interior of the plasma membrane. (True or false?)
2. The hydrophobic regions of the molecules composing the plasma membrane correspond to the two dark layers of this structure visible under an electron microscope. (True or false?)
3. Second messenger systems ultimately bring about the desired cellular response by inducing a change in the shape and function of particular intracellular proteins. (True or false?)
4. Through its unequal pumping, the Na^+-K^+ pump is directly responsible for separating sufficient charges to establish a resting membrane potential of -70 mV. (True or false?)
5. Engulfment of a small volume of fluid by a cell is known __________, whereas cellular ingestion of a multimolecular particle is referred to as __________. Collectively, these processes are termed __________.
6. The two general ways in which interaction of a chemical messenger with a membrane receptor can bring about the desired intracellular responses are __________ and __________.
7. A common membrane-bound intermediary between the receptor and the effector protein within the plasma membrane is the __________.
8. At resting membrane potential, there is a slight excess of __________ (positive/negative) charges on the inside of the membrane, with a corresponding slight excess of __________ charges on the outside.
9. Using the answer code below, indicate which membrane component is responsible for the function in question:

   (a) lipid bilayer
   (b) proteins
   (c) carbohydrates

   ____1. channel formation
   ____2. barrier to passage of water-soluble substances
   ____3. receptor sites
   ____4. membrane fluidity
   ____5. recognition of "self"
   ____6. membrane-bound enzymes
   ____7. structural boundary
   ____8. carriers

10. Using the answer code below, indicate the direction of net movement in each case:

    (a) movement from high to low concentration
    (b) movement from low to high concentration

    ____1. simple passive diffusion
    ____2. facilitated diffusion
    ____3. primary active transport
    ____4. Na^+ during secondary active transport
    ____5. cotransported molecule during secondary active transport
    ____6. water with regard to the water concentration gradient during osmosis
    ____7. water with regard to the solute concentration gradient during osmosis

11. Using the answer code below, indicate the type of cell junction described:

    (a) gap junction
    (b) tight junction
    (c) desmosome

    ____1. adhering junction
    ____2. impermeable junction
    ____3. communicating junction
    ____4. consists of connexons, which permit passage of ions and small molecules between cells
    ____5. consists of interconnecting fibers, which spot rivet adjacent cells
    ____6. consists of an actual fusion of proteins on the outer surfaces of two interacting cells
    ____7. important in tissues subject to mechanical stretching
    ____8. important in synchronizing contractions within heart and smooth muscle by allowing spread of electrical activity between the cells composing the muscle mass
    ____9. important in preventing passage between cells in epithelial sheets that separate compartments of two different chemical compositions

Essay Questions

1. Describe the fluid mosaic model of membrane structure.
2. What are the functions of the three major types of protein fibers in the extracellular matrix?
3. What two properties of a particle influence whether it can permeate the plasma membrane?
4. List and describe the methods of membrane transport. Indicate what types of substances are transported by each method, and state whether each is a passive or active means of transport.
5. As stated by Fick's law of diffusion, what factors influence the rate of net diffusion across a membrane?
6. State three important roles of the Na^+-K^+ pump.
7. List and describe the types of intercellular communication.
8. Compare the cAMP and Ca^{2+} second messenger pathways.
9. Describe the contribution of each of the following to the establishment and maintenance of membrane potential: (a) the Na^+-K^+ pump; (b) passive movement of K^+ across the membrane; (c) passive movement of Na^+ across the membrane; (d) the large intracellular anions.

Quantitative Exercises (Solutions on p. E–2.) (See Appendix C, *Principles of Quantitative Reasoning*)

1. When using the Nernst equation for an ion that has a valence other than 1, one must divide the potential by the valence. Thus, for Ca^{2+}, the Nernst equation becomes:

$$E = \frac{61\text{mV}}{z} \log \frac{C_o}{C_i}$$

Use this equation to calculate the Nernst (equilibrium) potentials from the following sets of data:

a. Given $[Ca^{2+}]_o = 1$ mM, $[Ca^{2+}]_i = 100$ mM, find $E_{Ca^{2+}}$

b. Given $[Cl^-]_o = 110$ mM, $[Cl^-]_i = 10$ mM, find E_{Cl^-}

2. One of the important uses of the Nernst equation is in describing the flow of ions across cell membranes. Ions move under the influence of two forces, the concentration gradient (given in electrical units by the Nernst equation) and the electrical gradient (given by the membrane voltage). This is summarized by *Ohm's law* as follows:

$$I_x = G_x(V_m - E_x)$$

This equation describes the movement of ion x across the membrane. I is the current in amperes (A); G is the conductance, a measure of the permeability of x, in Siemens (S), which is $\Delta I/\Delta V$; V_m is the membrane voltage; and E_x is the equilibrium potential of ion x. Not only does this equation tell how large the current is, it also tells what direction the current is flowing. By convention, a negative value of the current represents either a positive ion entering the cell or a negative ion leaving the cell. The opposite is true of a positive value of the current.

a. Using the following information, calculate the magnitude of I_{Na^+}.

$$[Na^+]_o = 145 \text{ mM}, \; [Na^+]_i = 15\text{mM}$$
$$G_{Na^+} = 1 \text{ nS}, \; V_m = -70\text{mV}$$

b. Is Na^+ entering or leaving the cell?

c. Is Na^+ moving with or against the concentration gradient? Is it moving with or against the electrical gradient?

3. Another important use of the Nernst equation is in determining the resting membrane potential of a cell. The cell resting membrane potential is a weighted average of the equilibrium potentials of all permeant ions. The weighting factor is the relative permeability (conductance) to that ion. For a cell permeable only to Na^+ and K^+, this equation is:

$$V_m = \{G_{Na^+}/G_T\}E_{Na^+} + \{G_{K^+}/G_T\}E_{K^+}$$

In this equation, G_T is the total conductance (in this case, $G_T = G_{Na^+} + G_{K^+}$).

a. Given the following information, calculate V_m:

$$G_{Na^+} = 1 \text{ nS}, \; G_{K^+} = 5.3 \text{ nS}, \; E_{Na^+} = 59.1 \text{ mV},$$
$$E_{K^+} = -94.4 \text{ mV}$$

b. What would happen to V_m if $[K^+]_o$ were increased to 150 mM?

Points to Ponder

(Explanations on p. E–3.)

1. Assume that a membrane permeable to Na^+ but not to Cl^- separates two solutions. The concentration of sodium chloride on side 1 is much higher than on side 2. Which of the following ionic movements would occur?
   a. Na^+ would move until its concentration gradient is dissipated (i.e., until the concentration of Na^+ on side 2 is the same as the concentration of Na^+ on side 1).
   b. Cl^- would move down its concentration gradient from side 1 to side 2.
   c. A membrane potential, negative on side 1, would develop.
   d. A membrane potential, positive on side 1, would develop.
   e. None of the above are correct.
2. Compared to resting potential, would the membrane potential become more negative or more positive if the membrane were more permeable to Na^+ than to K^+?
3. Which of the following methods of transport is being utilized to transfer the substance into the cell in the accompanying graph?
   a. diffusion down a concentration gradient
   b. osmosis
   c. facilitated diffusion
   d. active transport
   e. vesicular transport
   f. It is impossible to tell with the information provided.

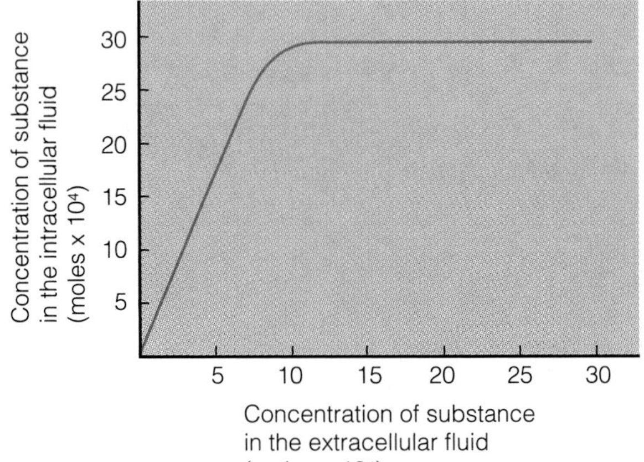

4. Colostrum, the first milk that a mother produces, contains an abundance of antibodies. These maternal antibodies help protect breast-fed infants from infections until the babies are capable of producing their own antibodies. By what means would you suspect these maternal antibodies are transported across the cells lining a newborn's digestive tract into the bloodstream?
5. The rate at which the Na^+-K^+ pump operates is not constant but is controlled by a combined effect of changes in ICF Na^+ concentration and ECF K^+ concentration. Do you think an increase in both ICF Na^+ and ECF K^+ concentrations would accelerate or slow down the Na^+-K^+ pump? What would be the benefit of this response? Before you reply, consider the following additional information about Na^+ and K^+ movement across the membrane. Not only do Na^+ and K^+ slowly and passively leak through their channels in a resting cell, but during an electrical impulse, known as an action potential, Na^+ rapidly and passively enters the cell; this movement is followed by a rapid, passive outflow of K^+. (These ion movements, which result from rapid changes in membrane permeability, bring about rapid, pronounced changes in membrane potential. This sequence of rapid potential changes—an action potential—serves as an electrical signal for conveying information along a nerve pathway.)
6. ***Clinical Consideration*** When William H. was helping victims following a devastating earthquake in a region that was not prepared to swiftly set up adequate temporary shelter, he developed severe diarrhea. He was diagnosed as having cholera, a disease transmitted through unsanitary water supplies that have been contaminated by fecal material from infected individuals. In this condition, an increase in cAMP in the intestinal cells opens the Cl^- channels in the luminal membranes of these cells, thereby increasing the secretion of Cl^- from the cells into the intestinal tract lumen. By what mechanisms would Na^+ and water be secreted into the lumen in accompaniment with Cl^- secretion? How does this secretory response account for the severe diarrhea that is characteristic of cholera? (See p. 70 for the underlying defect in cholera.)

Neuronal Physiology

4

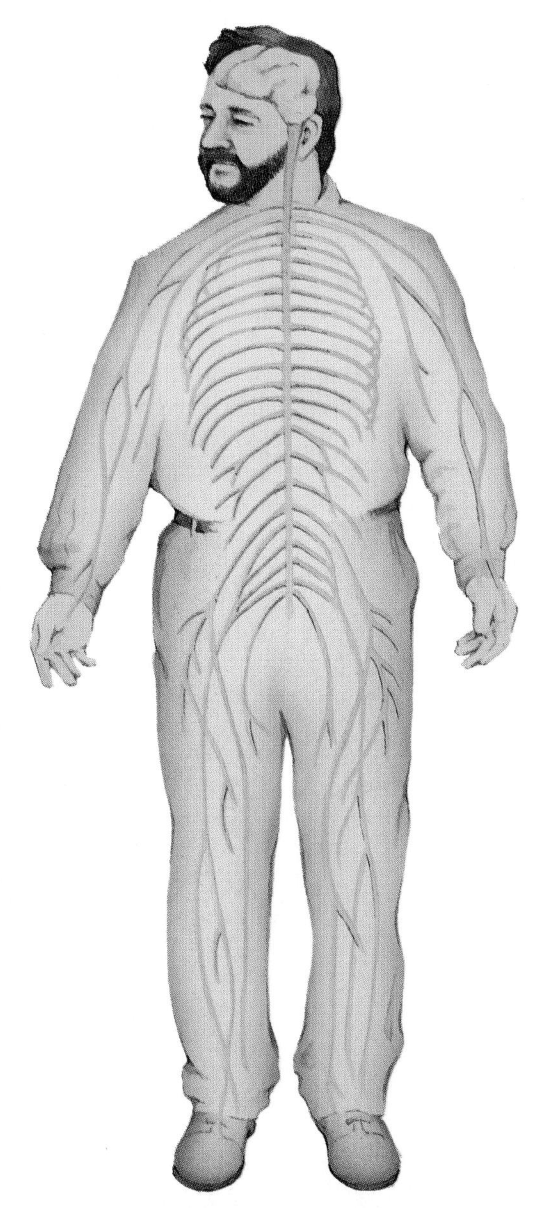

Body systems maintain homeostasis

HOMEOSTASIS
The nervous system, as one of the body's two major control systems, regulates many body activities aimed at maintaining a stable internal fluid environment.

Homeostasis is essential for survival of cells

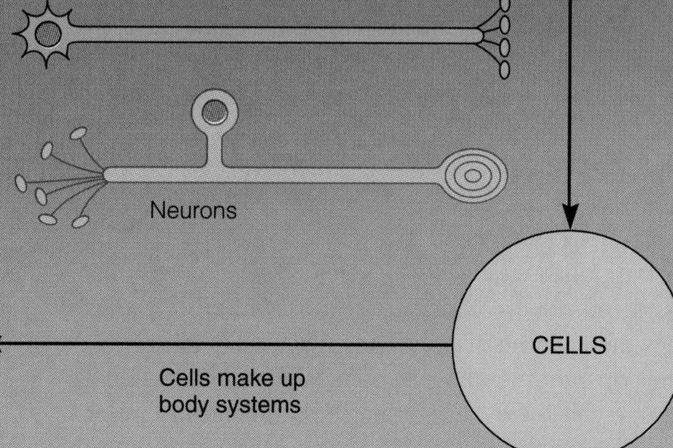

CELLS

Cells make up body systems

Nerve cells, or **neurons,** make up the nervous system, one of the two major control systems of the body. The nervous system exerts control over much of the body's muscular and glandular activities, most of which are directed toward maintaining homeostasis. Neurons are specialized for rapid electrical and chemical signaling. They are able to process, initiate, code, and conduct changes in their membrane potential as a means of rapidly transmitting a message throughout their length. Moreover, neurons have developed chemical means of passing this information through intricate nerve pathways from neuron to neuron as well as to muscles and glands.

Chapter Contents At a Glance

Electrical Signals: Graded Potentials and Action Potentials

Nerve and muscle are excitable tissues.

All cells of the body possess a membrane potential related to the nonuniform distribution of and differential permeability to Na^+, K^+, and large intracellular anions (chapter 3). Two types of cells, *nerve cells* and *muscle cells,* have developed a specialized use for this membrane potential. They are able to undergo transient, rapid changes in their membrane potentials. These fluctuations in potential, which serve as electrical signals, take two basic forms: (1) *graded potentials,* which serve as short-distance signals, and (2) *action potentials,* which signal over long distances.

Nerve and muscle are considered to be **excitable tissues** because they are capable of producing electrical signals when excited. The constant membrane potential that exists when an excitable tissue cell is not displaying rapid changes in potential is referred to as the *resting membrane potential* (although the membrane is far from "resting" because of the balanced leak-pump activity constantly going on). Recall that the resting potential of a typical nerve cell is −70 mV. Let us examine the types of electrical signals in more detail.

Graded potentials die out over short distances.

Graded potentials are local changes in membrane potential that occur in varying grades or degrees of magnitude or strength. For example, membrane potential could change from −70 mV to −60 mV (a 10 mV graded potential change) or from −70 mV to −50 mV (a 20 mV graded potential change). The magnitude of a graded potential is related to the magnitude of the triggering event that brings about the potential change; that is, *the stronger the triggering event, the larger the graded potential* (Fig. 4-1). Depending on the location or function of the graded potential, a triggering event might be (1) a stimulus, such as light stimulating specialized nerve cells in the eye; (2) an interaction of a chemical messenger with a surface receptor on a nerve or muscle cell membrane; or (3) a spontaneous change of potential caused by imbalances in the leak-pump cycle.

When a graded potential occurs locally in a nerve or muscle cell membrane, a different potential exists in this area than in the remainder of the membrane, which is still at resting potential. Because opposite charges attract each other, current (movement of charges) passively flows between the involved area and the adjacent resting regions on both the inside and outside of the membrane. For example, assume that a triggering event has temporarily reversed the charges in a particular region of an excitable tissue membrane, a region now called an *active area* (Fig. 4-2). Current will flow on both sides of the membrane between the active and neighboring *inactive* (still at resting potential) *areas.* By convention, the direction of current flow is always designated by the movement of the positive charges.

The charges in the active area do not actually have to reverse for local current flow to occur. A reduction in potential in the active area compared with that in the remainder of the membrane (the usual case with graded potentials) will also initiate current flow between the active and neighboring

Figure 4-1 **Graded Potentials** The greater the magnitude of a triggering event such as a stimulus, the larger the graded potential.

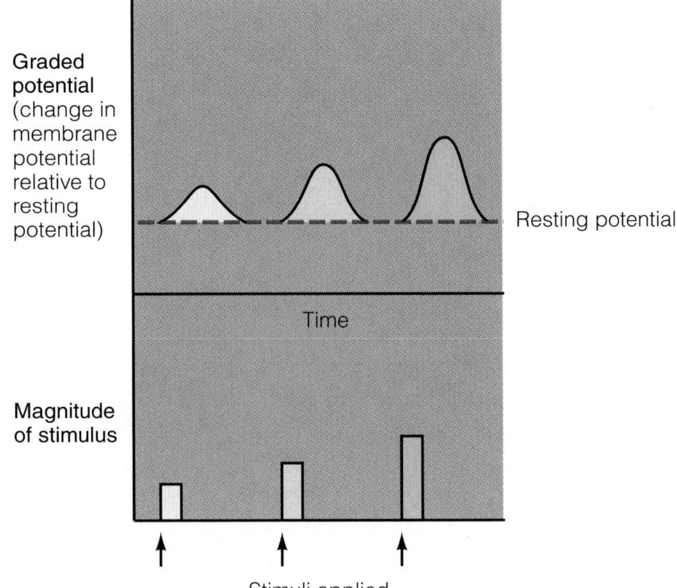

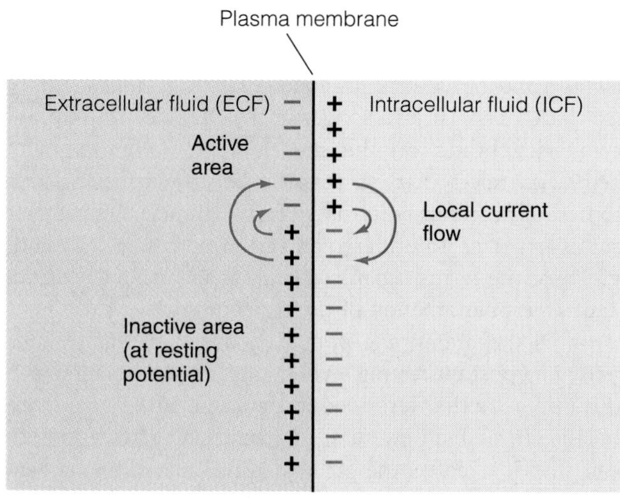

Figure 4-2 Local Current Flow between Active and Adjacent Inactive Areas of a Membrane Because opposite charges attract, current flows locally on both sides of the membrane between the active area that is undergoing a potential change and the adjacent inactive area that is still at resting potential.

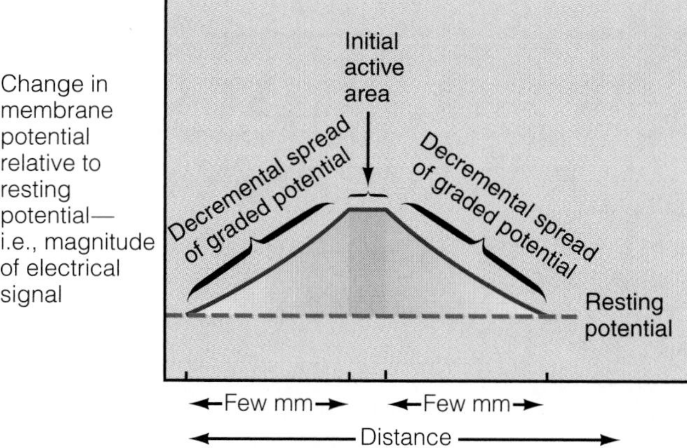

Figure 4-3 Decremental Spread of Graded Potentials Due to leakage of current, the magnitude of a graded potential continues to decrease as it passively spreads from the initial active area. The potential dies out altogether within a few mm of its site of initiation.

inactive areas. Current flow is easier to visualize, however, if we assume a reversal of charges. This local flow of current alters the potential in the previously inactive area, so this area's potential now differs from that of the region immediately next to it on the other side, inducing further current flow at this next site, and so on.

This passive current flow is similar to the means by which current is carried through electrical wires. We know from experience that current leaks out of an electrical wire unless the wire is covered with an insulating material such as rubber. The regions on excitable tissues at which graded potentials take place are not wrapped in insulating material. Because current from the activated region of the membrane leaks into the surrounding extracellular fluid (ECF), the magnitude of the graded potential continues to decrease the farther it moves away from the initial active area; that is, the spread of a graded potential is *decremental* (Fig. 4-3). In fact, these local currents die out within a few millimeters from the initial site of potential change and consequently can function as signals only for very short distances. This situation is similar to two people speaking directly to each other. Over short distances, the voices are clearly understood, but as the two people move farther apart, the sounds progressively diminish until further communication by this means becomes impossible. Just as telephones enable people to communicate over long distances, action potentials provide a means of transmitting an electrical signal long distances through the body. The limited signaling distance of graded potentials does not mean that they are of no value, however. The following graded potentials are critically important to the body: postsynaptic potentials, receptor potentials, end-plate potentials, pacemaker potentials, and slow-wave potentials. These terms are unfamiliar to you now, but you will become well acquainted with them as we continue discussing nerve and muscle physiology in the next several chapters.

Action potentials are brief reversals of membrane potential brought about by rapid changes in membrane permeability.

Because the passive current flow accompanying a graded potential fades very quickly as it moves away from its site of initiation, there must be another mechanism by which an electrical signal can be transmitted over long distances, with the strength of the signal being maintained as it travels away from its site of initiation. When appropriately triggered, nerve and muscle cell membranes undergo brief, rapid reversals of membrane potential; these reversals, known as **action potentials,** are able to spread throughout the membrane in nondecremental fashion. To understand the processes that occur during an action potential, it is necessary to be familiar with the following terms (Fig. 4-4):

1. **Polarization:** The membrane has potential; there is a separation of opposite charges.
2. **Depolarization:** The membrane potential is reduced from resting potential; it has decreased or moved toward 0 mV (that is, the potential is less than −70 mV, such as at −60 mV); fewer charges are separated than at resting potential.

Figure 4-4 Types of Changes in Membrane Potential

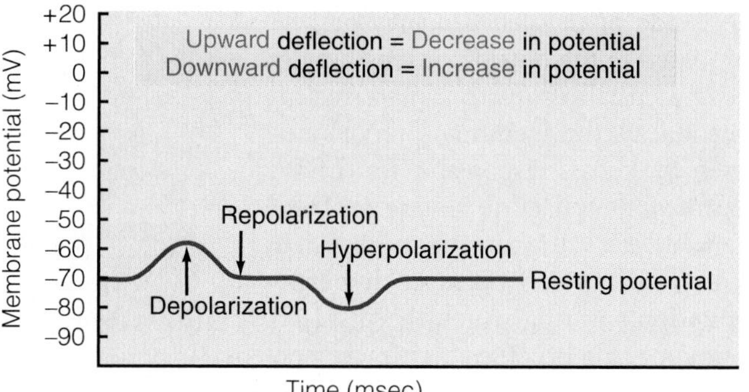

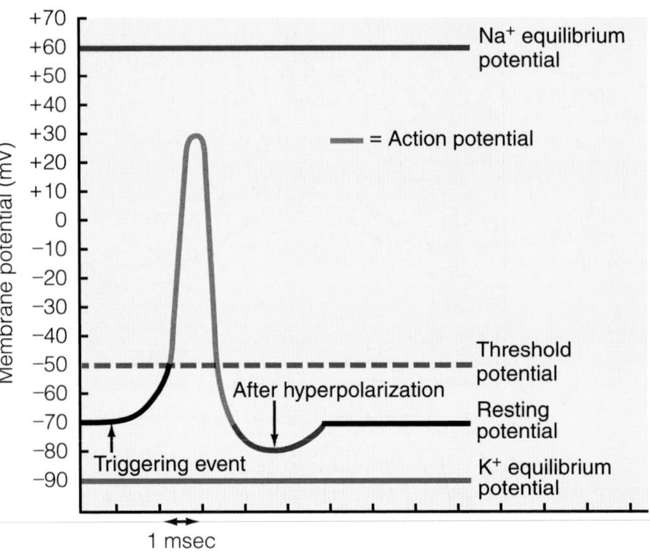

Figure 4-5 **Changes in Membrane Potential during an Action Potential**

3. **Hyperpolarization:** The potential is greater than resting potential; it has increased or become even more negative; (that is, the potential is greater than −70 mV, such as at −80 mV); more charges are separated than at resting potential.
4. **Repolarization:** The membrane returns to resting potential after having been depolarized.

One possibly confusing point should be clarified. On the device used for recording rapid changes in potential, a *decrease* in potential is represented as an *upward* deflection whereas an *increase* in potential is represented as a *downward* deflection.

With these terms in mind, let us consider the changes that occur in the membrane potential during an action potential (Fig. 4-5). To initiate an action potential, a triggering event causes the membrane to depolarize from the resting potential of −70 mV. Depolarization proceeds slowly at first until it reaches a critical level known as **threshold potential,** typically between −50 and −55 mV. At threshold potential, an explosive depolarization takes place. A recording of the potential at this time shows a sharp upward deflection to +30 mV as the potential rapidly decreases toward 0 mV, then reverses itself so that the inside of the cell becomes positive compared to the outside. Just as rapidly, the potential drops back to resting potential as the membrane repolarizes. Often the forces responsible for driving the membrane back to resting potential push it too far, causing a transient hyperpolarization (the **after hyperpolarization**), during which the inside of the membrane becomes even more negative than normal (for example −80 mV). The entire rapid change in potential from threshold to peak reversal and then back to resting is called the *action potential.* In a nerve cell, an action potential lasts for only 1 msec (0.001 sec). It lasts longer in muscle, with the duration varying depending on the muscle type. Often an action potential is referred to as a **spike** because of its spikelike recorded appearance. Alternatively, when an excitable membrane is triggered to undergo an action potential, it is said to **fire.** Thus, the terms *action potential, spike,* and *firing* all refer to the same phenomenon of rapid potential reversal.

How is the membrane potential, which is usually maintained at a constant resting level by the counterbalancing leak and pump activities described in the last chapter, thrown out of balance to such an extent as to produce an action potential? Recall that K^+ makes the greatest contribution to the establishment of the resting potential because the membrane at rest is considerably more permeable to K^+ than to Na^+. During an action potential, marked changes in membrane permeability to Na^+ and K^+ take place, permitting rapid fluxes of these ions down their electrochemical gradients. These ion movements carry the current responsible for the potential changes that occur during an action potential.

Recall that membrane channels behave as if they have "gates" that can be open or closed, depending on the circumstances (see p. 67). The channels are formed by proteins that span the thickness of the membrane. Changes in conformation (shape) of these proteins are believed to alternatively block the channel or permit passage through it. Channels apparently are able to exist in at least three different conformations (Fig. 4-6): (1) gates closed but capable of opening; (2) gates open (activated); and (3) gates closed and not capable of opening (inactivated).

There are three kinds of channels, depending on the factor that induces the change in channel conformation: (1) **voltage-gated channels,** which open or close in response to changes in membrane potential; (2) **chemical messenger-gated channels,** which change conformation in response to the binding of a specific chemical messenger with a membrane receptor that is in close association with the channel, as in channels involved in signal transduction (see p. 67); and (3) **mechanically gated channels,** which respond to stretching or other mechanical deformation. Voltage-gated channels are the ones

Figure 4-6 **Conformations of a Voltage-Gated Na^+ Channel**

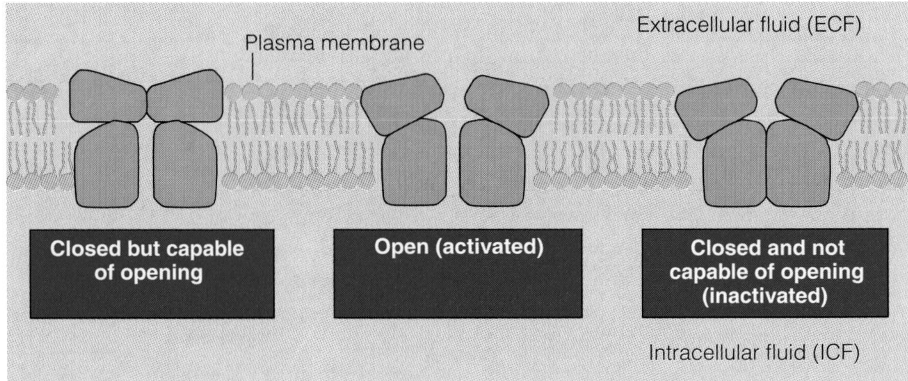

involved in action potentials. (The role of mechanically gated channels will be described later.)

Because proteins that compose the voltage-gated channels contain a number of charged groups, the electric field (potential) surrounding them can exert a distorting force on the channel structure; that is, various charged portions of the channel protein are electrically attracted or repelled by an unequal distribution of cations and anions in the fluids surrounding the membrane. Unlike the majority of membrane proteins, which remain stable in spite of fluctuations in membrane potential, the voltage-gated channel proteins are especially sensitive to voltage changes. Small distortions in shape induced by potential changes can cause them to flip to another conformation.

At resting potential (−70 mV), many K^+ channels are open but most of the Na^+ channels are closed; thus, the resting membrane is 50 to 75 times more permeable to K^+ (Table 4-1). When a membrane starts to depolarize toward threshold as a result of a triggering event, some of its voltage-gated Na^+ channels open. Since both the concentration and electrical gradients for Na^+ favor its movement into the cell, Na^+ starts to move in, carrying its positive charge with it. This depolarizes the membrane further, thereby opening more voltage-gated Na^+ channels and allowing more Na^+ to enter. As a result, still further depolarization occurs, opening more Na^+ channels, and so on, in a positive-feedback cycle (Fig. 4-7).

At threshold potential, there is an explosive increase in Na^+ permeability **(P Na^+)** as the membrane becomes 600 times more permeable to Na^+ than to K^+. Each individual channel is either closed or open and cannot be partially open.

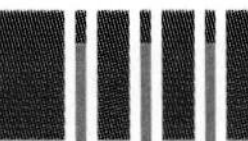

Table 4-1 Permeabilities and Gradients for Na^+ and K^+ before, during, and after an Action Potential

Ion	*Condition*	*Permeability Compared to Resting P Na^+*	*Gradient*	*Direction of Gradient*
Na^+	Resting potential (−70 mV) before an action potential	1	Concentration gradient	Inward
			Electrical gradient	Inward
K^+	Resting potential (−70 mV) before an action potential	50–75×	Concentration gradient	Outward
			Electrical gradient	Inward
Na^+	Threshold potential (−50 mV)	600×	Concentration gradient	Inward
			Electrical gradient	Inward
K^+	Peak of action potential (+30 mV)	300×	Concentration gradient	Outward
			Electrical gradient	Outward
Na^+	Resting potential (−70 mV) after an action potential	1	Concentration gradient	Inward
			Electrical gradient	Intward
K^+	Resting potential (−70 mV) after an action potential	50–75×	Concentration gradient	Outward
			Electrical gradient	Inward

Figure 4-7 Positive-Feedback Cycle Responsible for Opening Na^+ Channels at Threshold

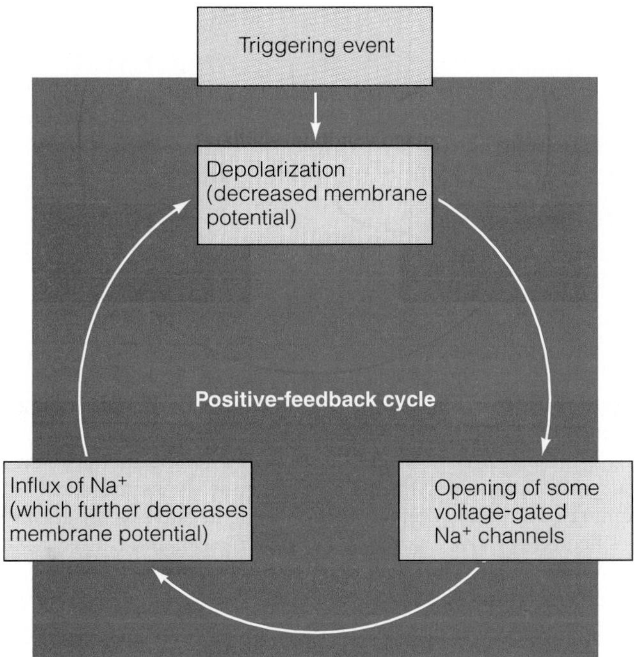

However, the delicately poised gating mechanisms of the various Na^+ channels are jolted open by slightly different voltage changes. During the early depolarizing phase, Na^+ channels are gradually opened as the potential progressively decreases. By threshold the gates of all the Na^+ channels have swung open so that Na^+ permeability now dominates the membrane, in contrast to the K^+ domination of the resting potential. Thus, at threshold Na^+ rushes into the cell, rapidly eliminating the internal negativity and even making the inside of the cell more positive than the outside (Fig. 4-8). The potential reaches +30 mV, close to the Na^+ equilibrium potential. The potential does not become any more positive because, at the peak of the action potential, the Na^+ channels close to the inactivated state, and P Na^+ falls to its low resting value. What causes the Na^+ channels to close? When the membrane potential reaches threshold, two simultaneous events are believed to take place in the gates of each Na^+ channel. First, the outer gates adjacent to the ECF are triggered to *open rapidly*, converting the channel to its open (activated) conformation (Fig. 4-6). Simultaneously, the same potential change triggers the inner gates adjacent to the intra-

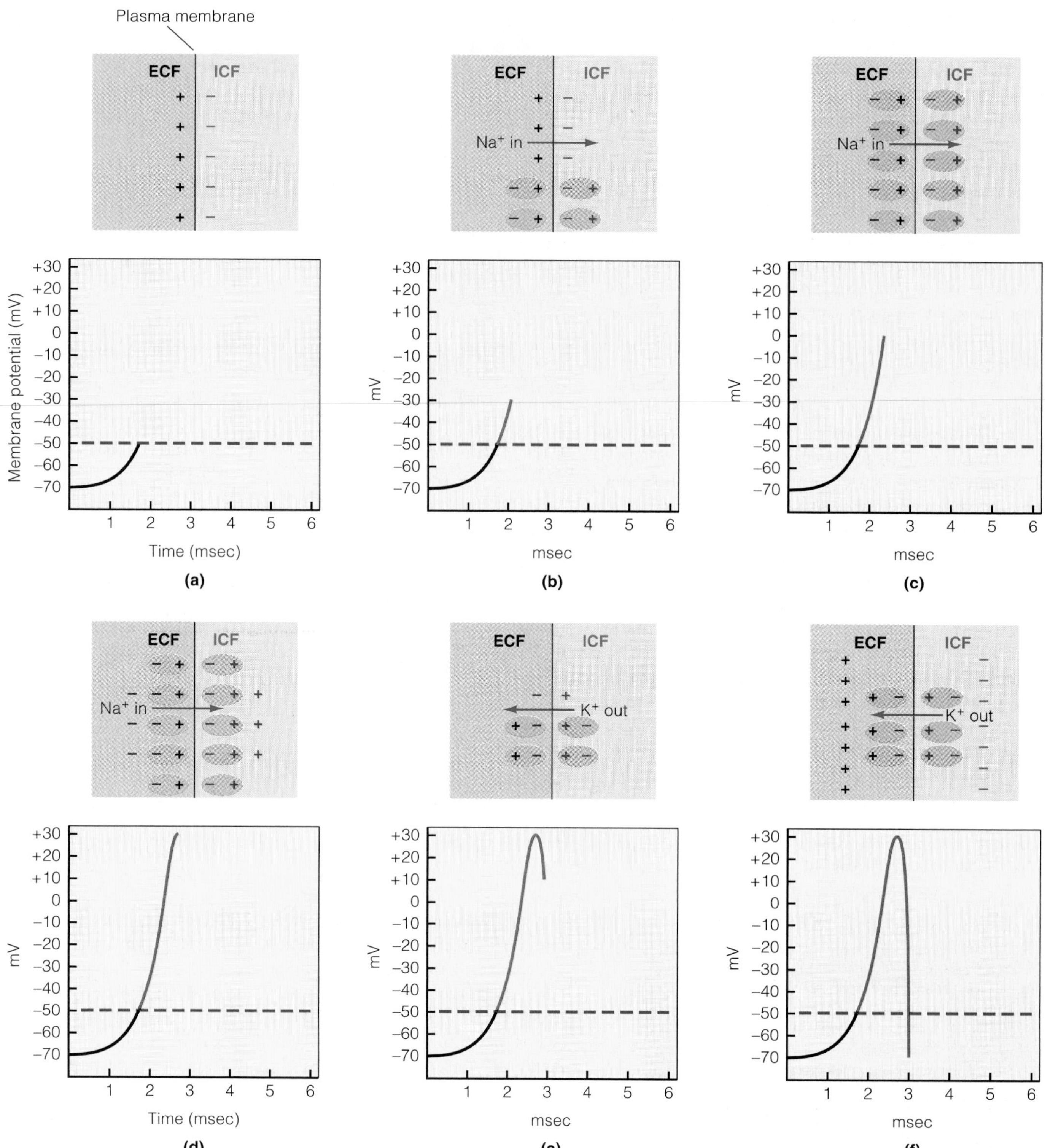

Figure 4-8 **Ionic Movements Responsible for Changes in Membrane Potential during an Action Potential** After the threshold potential of −50 mV is reached (a), movement of Na^+ inward neutralizes negative charges inside the cell, progressively decreasing the internal negativity (b) until 0 mV is reached (c). Inward movement of Na^+ leaves the negative charges (primarily Cl^-) with which Na^+ had been paired behind on the outside. These negative charges neutralize positive charges that had been contributing to membrane potential, making the outside progressively less positive (b) and (c). Continued inward movement of Na^+ reverses the potential, with the inside becoming positive and the outside becoming negative as the action potential peaks (d). After the peak of the action potential, outward movement of K^+ leaves behind negative charges (A^-) inside the cell and neutralizes negative charges outside; as a consequence, the inside becomes progressively less positive and the outside less negative (e) until 0 mV is reached. Continued outward movement of K^+ restores the resting membrane potential, with the potential reversing back, so that the inside is once again negative and the outside positive (f).

cellular fluid (ICF) to *close slowly.* Consequently, after the channel opens, there is a time delay before it is converted to its inactivated conformation. Meanwhile, the channel has remained open for about 0.5 msec, and Na^+ has rushed into the cell, bringing the action potential to its peak of +30 mV. Then the inner gates slam closed and remain shut in an inactivated state until the membrane potential has been restored to its negative resting value.

Simultaneous with inactivation of the Na^+ channels, K^+ permeability **(P K^+)** greatly increases to about 300 times the resting P Na^+. This opening of even more K^+ channels is also a delayed voltage-gated response triggered by the initial depolarization to threshold. The marked increase in P K^+ causes K^+ to rush out of the cell down its concentration and electrical gradients, carrying positive charges back to the outside. Note that at the peak of the action potential, the internal positivity of the cell tends to repel the positive K^+ ions, so the electrical gradient for K^+ is outward, unlike at resting potential (Table 4-1). The outward movement of K^+ rapidly restores the internal negativity and returns the potential to resting (Fig. 4-8).

To review (Fig. 4-9), *the rising phase of the action potential (depolarization) is due to Na^+ influx* (Na^+ entering the cell) induced by an explosive increase in P Na^+ at threshold. *The falling phase (repolarization) is brought about by K^+ efflux* (K^+ leaving the cell) caused by the marked increase in P K^+ occurring simultaneously with the inactivation of the Na^+ channels at the peak of the action potential. As the potential returns to resting, the changing voltage shifts the Na^+ channels to their "closed but capable of opening" conformation. The newly opened K^+ channels also close, so the membrane returns to the resting number of open K^+ channels. Sometimes more K^+ leaves than is necessary to bring the potential to resting because the K^+ channels do not close quickly enough. This slight excessive K^+ efflux makes the interior of the cell transiently even more negative than resting potential, causing the after hyperpolarization.

The Na^+-K^+ pump gradually restores the concentration gradients disrupted by action potentials.

At the completion of an action potential, the membrane potential has been restored to its resting condition, but the ion distribution has been altered slightly. Sodium has entered the cell during the rising phase, and a comparable amount of K^+ has left during the falling phase. It is the task of the Na^+-K^+ pump to restore these ions to their original locations in the long run, but not after each action potential.

The active pumping process takes much longer to restore Na^+ and K^+ to their original locations than it takes for the passive fluxes of these ions during an action potential. However, the membrane does not need to wait until the Na^+-K^+ pump slowly restores the concentration gradients before it can undergo another action potential. Actually, the movement of only relatively few of the total number of Na^+ and K^+ ions present is responsible for the dramatic swings in potential that occur during an action potential. Only about 1

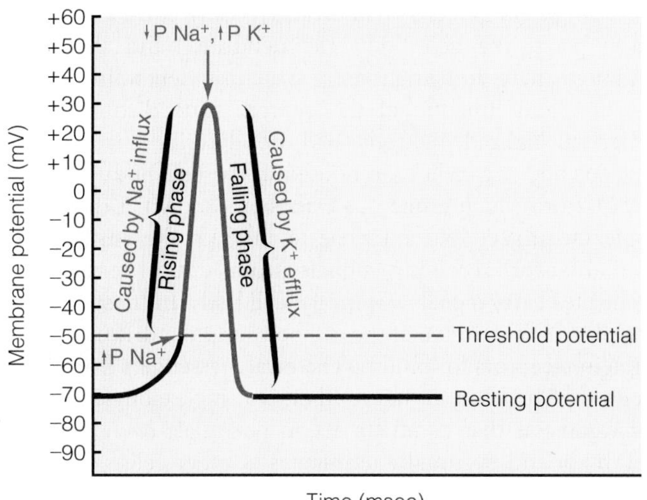

Figure 4-9 **Permeability Changes and Ionic Fluxes during an Action Potential**

out of 100,000 K^+ ions present in the cell leaves during an action potential, while a comparable number of Na^+ ions enter from the ECF. The shifting locations of these extremely small percentages of the total Na^+ and K^+ during a single action potential produce only infinitesimal changes in their ICF and ECF concentrations. There is still much more K^+ inside the cell than outside, and Na^+ is still predominantly an extracellular cation (Table 4-1). Consequently, the Na^+ and K^+ concentration gradients still exist, so repeated action potentials can occur without the pump having to keep pace to restore the gradients.

Of course, were it not for the pump, even tiny fluxes accompanying repeated action potentials would eventually "run down" the concentration gradients so that further action potentials would be impossible. If the concentrations of Na^+ and K^+ were equal between the ECF and ICF, changes in permeability to these ions would not bring about ionic fluxes, so no change in potential would occur. Thus, the Na^+-K^+ pump is critical to maintaining the concentration gradients in the long run. However, it does not have to perform its role between action potentials, nor is it directly involved in the ion fluxes or potential changes that occur during an action potential.

Once initiated, action potentials are propagated throughout an excitable cell.

A single action potential involves only a small patch of the total surface membrane of an excitable cell. If action potentials are to serve as long-distance signals, obviously they cannot be merely isolated events occurring in a limited area of a nerve or muscle cell membrane. Mechanisms must exist to conduct or spread the action potential throughout the entire cell membrane. Furthermore, the signal must be transmitted from one cell to the next cell (for example, along specific nerve pathways). Let us first examine how an action potential (nerve impulse) is conducted throughout a nerve cell before turning our attention to how the impulse is passed to another cell.

A single nerve cell, or **neuron,** typically consists of three basic parts: the cell body, the dendrites, and the axon, although there are variations in structure, depending on the location and function of the neuron. (The distinctions of other specialized neurons will be described later.) The nucleus and organelles are housed in the **cell body** (▬ Fig. 4-10), from which numerous extensions known as **dendrites** typically project like antennae to increase the surface area available for receiving signals from other nerve cells. Dendrites carry signals *toward* the cell body. In most neurons the plasma membrane of the cell body and dendrites contains protein receptors for binding chemical messengers from other neurons. The **axon,** or **nerve fiber,** is a single elongated tubular extension that conducts action potentials *away from* the cell body and eventually terminates at other cells. The axon frequently gives off side branches, or **collaterals,** along its course. The first portion of the axon plus the region of the cell body from which the axon leaves is known as the **axon hillock.** It is the site where action potentials are initiated in a neuron (with the exception of neurons specialized to carry sensory information, a topic described in a later chapter). The impulses are then propagated along the axon to its typically highly branched ending at the **axon terminals.** These terminals release chemical messengers that simultaneously influence numerous other cells with which they come into close association.

Axons vary in length from less than a millimeter in neurons that communicate only with neighboring cells to longer than a meter in neurons that communicate with distant parts of the nervous system or with peripheral organs. For example, the axon of the nerve cell innervating your big toe must traverse the distance between the origin of its cell body within

▬ *Figure 4-10* **Anatomy of a Neuron (Nerve Cell)** (a) Most but not all neurons consist of the basic parts schematically represented in the figure. (b) An electron micrograph highlighting the cell body, dendrites, and part of the axon of a neuron within the central nervous system.

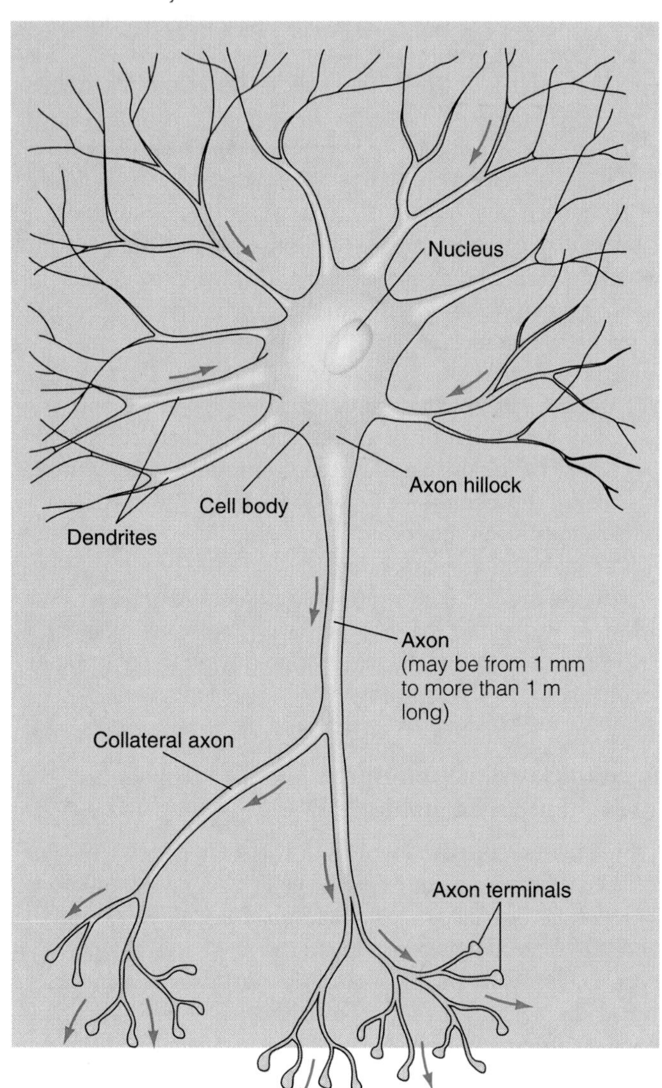

Arrows indicate direction in which nerve signals are conveyed.

(a)

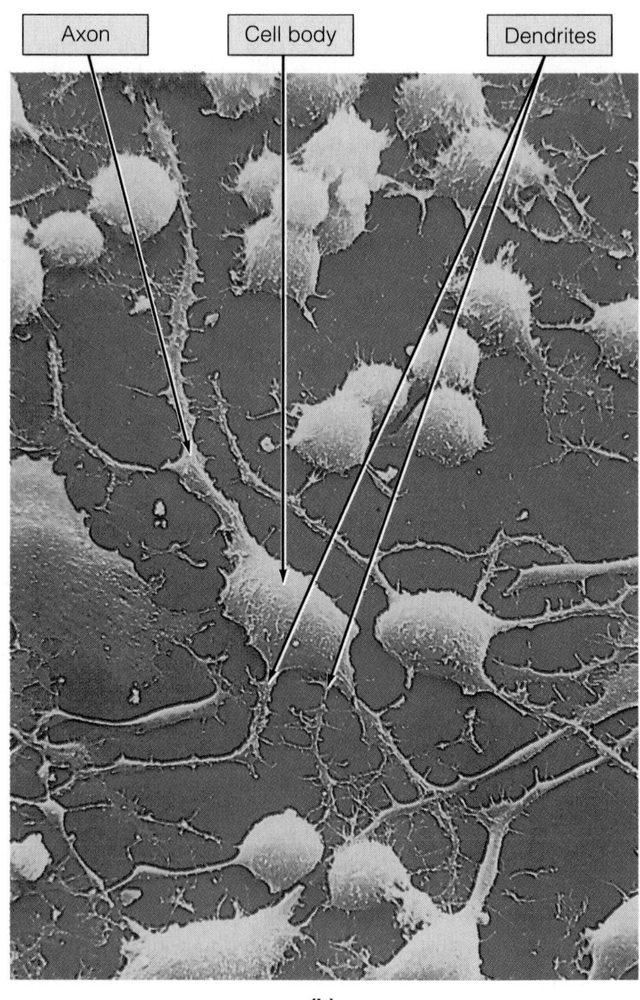

(b)

the spinal cord in the lower region of your back all the way down your leg to your toe.

Once an action potential is initiated at the axon hillock, no further triggering event is necessary to activate the remainder of the nerve fiber. The impulse is automatically conducted throughout the neuron without further stimulation by one of two methods of propagation: *conduction by local current flow* or *saltatory conduction.*

Figure 4-11 illustrates **conduction by local current flow.** You are viewing a schematic representation of a longitudinal section of the axon hillock and the portion of the axon immediately beyond it. The axon hillock is at the peak of an action potential. The inside of the cell is positive in this active area because Na^+ has already entered the nerve cell at this point. The remainder of the axon, still at resting potential and negative inside, is considered to be inactive. For the action potential to spread from the active to the inactive areas, the inactive areas must somehow be depolarized to threshold before they can undergo an action potential. This depolarization is accomplished by local current flow between the area already undergoing an action potential (positive inside, negative outside) and the adjacent inactive area (negative inside, positive outside), similar to the current flow responsible for the spread of graded potentials. Because opposite charges attract, current is able to flow locally between the active area and the neighboring inactive area on both the inside and the

Figure 4-11 **Conduction by Local Current Flow** Local current flow between the active area at the peak of an action potential and the adjacent inactive area still at resting potential reduces the potential in the inactive area to threshold, which triggers an action potential in the previously inactive area. The original active area returns to resting potential, and the new active area induces an action potential in the next adjacent inactive area by local current flow as the cycle repeats itself down the length of the axon.

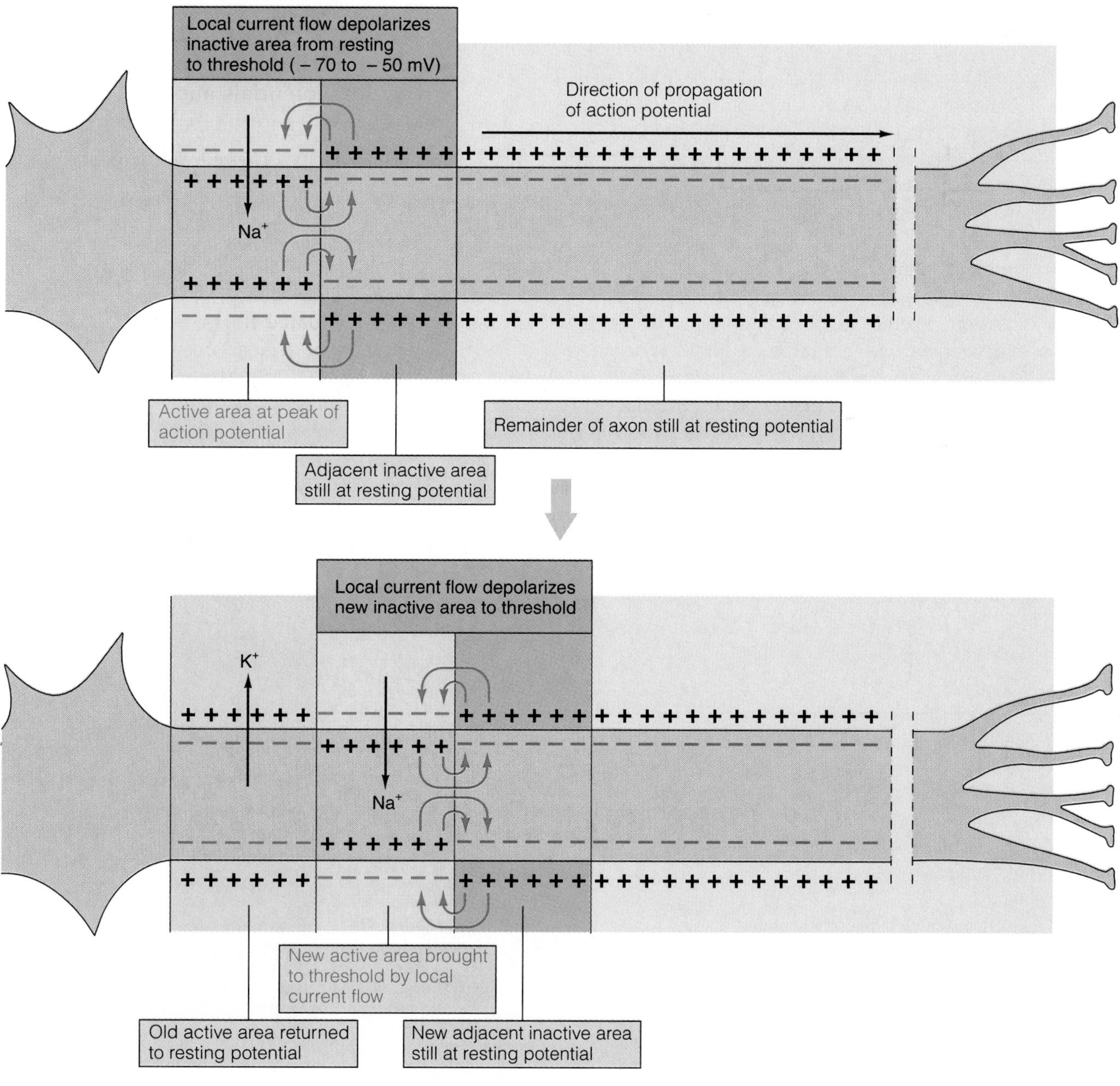

outside of the membrane. This local current flow in effect neutralizes or eliminates some of the unbalanced charges in the inactive area; that is, it reduces the number of opposite charges separated across the membrane or reduces the potential in this area. This depolarizing effect quickly brings the involved inactive area to threshold, at which time the voltage-gated Na^+ channels in this region of the membrane are all thrown open, leading to an action potential in this previously inactive area.

Meanwhile, the original active area returns to resting potential as a result of K^+ efflux. (Local current flow does *not* make a significant contribution to returning the original active area to resting. Repolarization is accomplished primarily by a concurrent drop in P Na^+ and rise in P K^+, as described earlier.) In turn, beyond the new active area is another inactive area, so the same thing happens again. Local current flow brings this next inactive area to threshold, causing it to fire and become a new active area. This cycle repeats itself until the action potential has spread to the end of the axon. *Once an action potential is initiated in one part of a nerve cell membrane, a self-perpetuating cycle is initiated so that the action potential is propagated throughout the rest of the fiber automatically.* In this way, the axon is similar to a firecracker fuse that needs to be lit at only one end. Once ignited, the fire spreads down the fuse; it is not necessary to hold a match to every separate section of the fuse.

Note that the original action potential does not travel along the membrane. Instead, it triggers an identical new action potential in the adjacent area of the membrane, with this process being repeated along the axon's length. Since each new action potential in the conduction process is a fresh event dependent on the induced permeability changes and electrochemical gradients, which are virtually identical down the length of the axon, the last action potential at the end of the axon is identical to the original one, no matter how long the axon. Thus, an action potential is spread throughout the axon in undiminished fashion. It always goes to maximal amplitude, rather than getting progressively smaller as it moves down the axon. In this way, action potentials can serve as faithful long-distance signals without attenuation or distortion.

The nondecremental propagation of an action potential is in contrast to the decremental spread of a graded potential, which dies out over a very short distance because it is not able to regenerate itself. Typically, regions of excitable cells where graded potentials take place do not have an achievable threshold for undergoing action potentials because of a sparsity of voltage-gated Na^+ channels. (There is no threshold phenomenon for the occurrence of graded potentials themselves.) Therefore, sites specialized for graded potentials do not undergo action potentials, even though they might be depolarized considerably. However, as will be elaborated on later, graded potentials can, before dying out, trigger action potentials in adjacent portions of the membrane by bringing these more sensitive regions to threshold through local current flow spreading from the site of the graded potential. ▮ Table 4-2 summarizes the differences between graded potentials and action potentials, some of which are yet to be discussed.

Myelination increases the speed of conduction of action potentials and conserves energy in the process.

The velocity, or speed, with which an action potential travels down the axon depends on two factors: (1) whether the fiber is myelinated and (2) the diameter of the fiber. Conduction by local current flow occurs in unmyelinated fibers. A faster method of propagation, *saltatory conduction,* takes place in myelinated fibers.

Myelinated fibers, as the name implies, are covered with **myelin** at regular intervals along the length of the axon (▬ Fig. 4-12a). Myelin is composed primarily of lipids. Because the water-soluble ions responsible for carrying current across the membrane cannot permeate this thick lipid barrier, the myelin coating acts as an insulator, just like rubber around an electrical wire, to prevent current leakage across the myelinated portion of the membrane. Myelin is not

Table 4-2 **Comparison of Graded Potentials and Action Potentials**

Graded Potentials	*Action Potentials*
Graded potential change; magnitude varies with magnitude of triggering event	All-or-none membrane response; magnitude of triggering event coded in frequency rather than amplitude of action potentials
Decremental conduction; magnitude diminishes with distance from initial site	Propagated throughout membrane in undiminishing fashion
Passive spread to neighboring inactive areas of membrane	Self-regeneration in neighboring inactive areas of membrane
No refractory period	Refractory period
Can be summed	Summation impossible
Can be depolarization or hyperpolarization	Always depolarization and reversal of charges
Triggered by stimulus, by combination of neurotransmitter with receptor, or by spontaneous shifts in leak-pump cycle	Triggered by depolarization to threshold, usually through spread of graded potential
Occurs in specialized regions of membrane designed to respond to triggering event	Occurs in regions of membrane with abundance of voltage-gated Na^+ channels

Figure 4-12 Myelinated Fibers (a) A myelinated fiber is surrounded by myelin at regular intervals. The intervening unmyelinated regions are known as nodes of Ranvier. (b) In the peripheral nervous system, each patch of myelin is formed by a separate Schwann cell that wraps itself jelly-roll fashion around the nerve fiber. In the central nervous system, each of the several processes of a myelin-forming oligodendrocyte forms a patch of myelin around a separate nerve fiber. (c) An electron micrograph of a myelinated fiber in cross section. (d) Membrane potential exists only at the nodes of Ranvier where the bare axon is exposed to the ECF. No charges exist across the insulated myelinated regions.

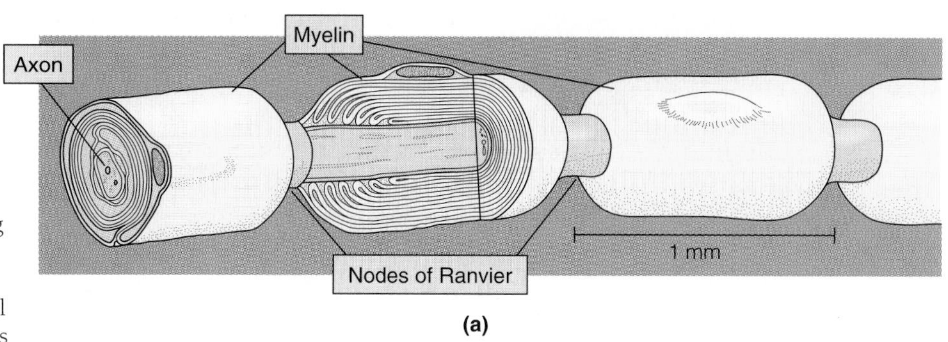

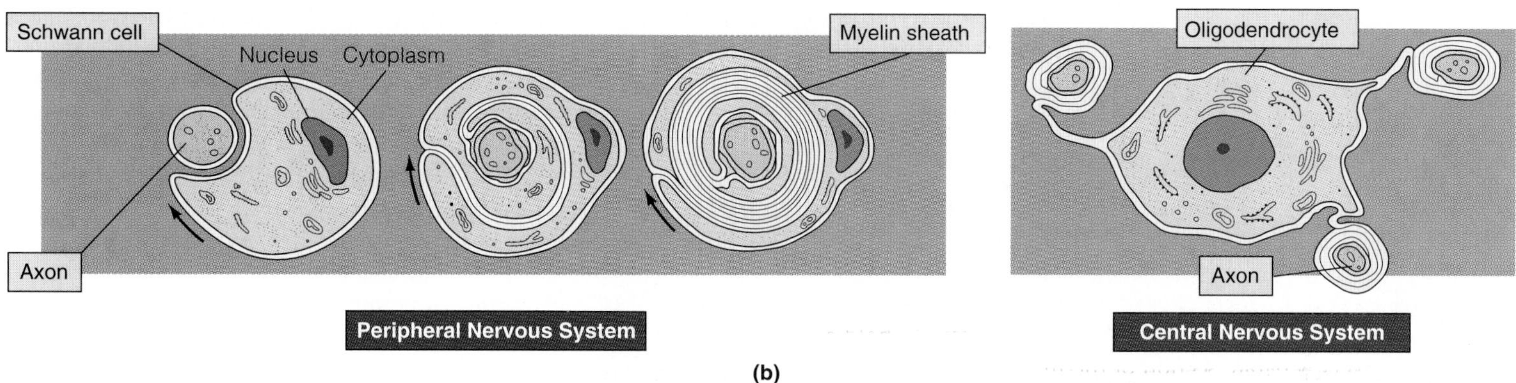

actually a part of the nerve cell but consists of separate myelin-forming cells that wrap themselves around the axon in jelly-roll fashion (Fig. 4-12b and c). These myelin-forming cells are **oligodendrocytes** in the central nervous system (the brain and spinal cord) and **Schwann cells** in the peripheral nervous system (the nerves running between the central nervous system and the various regions of the body). The lipid composition of myelin is due to the presence of layer upon layer of the lipid bilayer that composes the plasma membrane of these myelin-forming cells. Between the myelinated regions, the axonal membrane is bare and exposed to the ECF. It is only at these bare spaces, called **nodes of Ranvier,** that membrane potential can exist and current can flow across the membrane (Fig. 4-12d). Sodium channels are concentrated at the nodal areas; the myelin-covered regions are almost devoid of these special passageways.

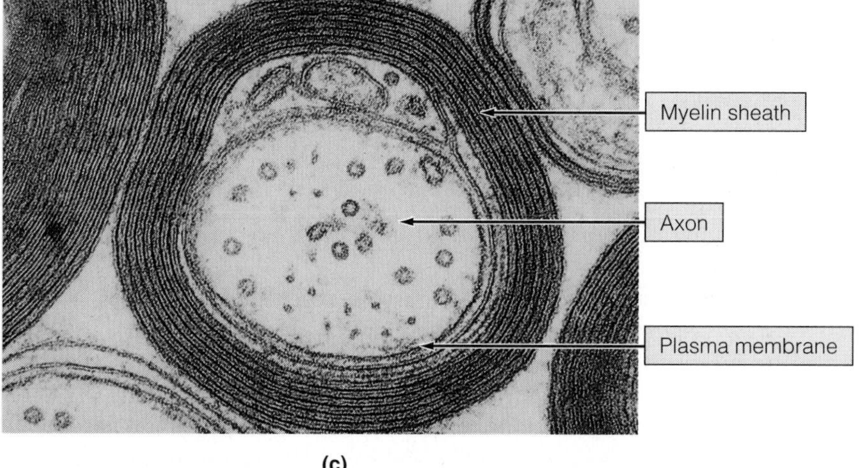

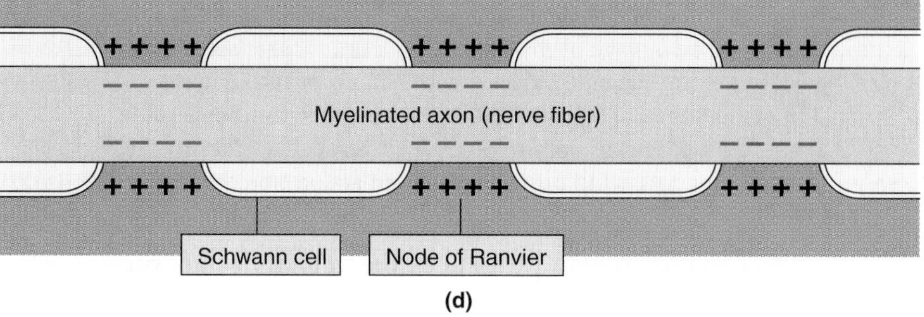

The nodes are usually about 1 mm apart, a distance short enough that local current from an active node can reach an adjacent node before dying off. When an action potential occurs at one of the nodes, opposite charges attract from the adjacent inactive node, reducing its potential to threshold so that it undergoes an action potential, and so on. Consequently, in a myelinated fiber, the impulse "jumps" from node to node, skipping over the myelinated sections of the axon (Fig. 4-13); this process is called **saltatory conduction** (*saltere* means "to jump or leap"). Saltatory conduction propagates action potentials more rapidly than does conduction by local current flow, because the action potential leaps

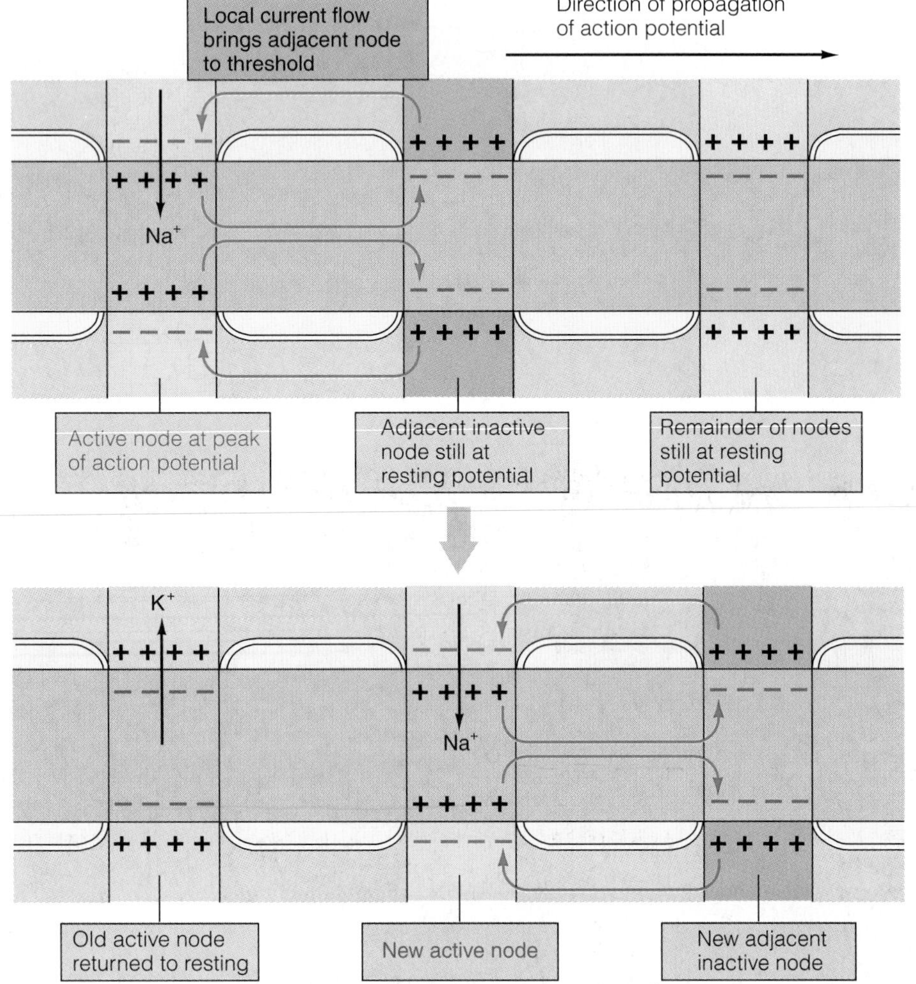

Figure 4-13 **Saltatory Conduction** The impulse "jumps" from node to node in a myelinated fiber.

over myelinated sections but must be regenerated within every section of an unmyelinated axonal membrane from beginning to end. Myelinated fibers conduct impulses about 50 times faster than unmyelinated fibers of comparable size. As a general rule, the most urgent types of information are transmitted via myelinated fibers, whereas the nervous pathways carrying less urgent information are unmyelinated.

In addition to permitting action potentials to travel faster, a second advantage of myelination is that it conserves energy. Since the ion fluxes associated with action potentials are confined to the nodal regions, the energy-consuming Na^+-K^+ pump must restore fewer ions to their respective sides of the membrane following propagation of an action potential.

Fiber diameter also influences the velocity of action potential propagation.

Besides the effect of myelination, the *diameter* of the fiber also influences the speed with which an axon can conduct action potentials. The magnitude of current flow (that is, the amount of charge that moves) depends not only on the difference in potential between two adjacent electrically charged regions but also on the *resistance* or hindrance to electrical charge movement between the two regions. When fiber diameter increases, the resistance decreases. Thus, the larger the diameter of the nerve fiber, the faster it can propagate action potentials.

Large myelinated fibers, such as those supplying skeletal muscles, can conduct action potentials at a speed of up to 120 meters (m)/sec (360 miles/hr), compared with a conduction velocity of 0.7 m/sec (2 miles/hr) in small unmyelinated fibers such as those supplying the digestive tract. This variability in speed of propagation of action potentials is related to the urgency of the information being conveyed. A signal to skeletal muscles to execute a particular movement (for example, to prevent you from falling as you trip on something) must be transmitted more rapidly than a signal to modify a slow-acting digestive process. Were it not for myelination, axon diameters within urgent nerve pathways would have to be very large and cumbersome to achieve the necessary conduction velocities. Indeed, this is the case in many invertebrates. Vertebrates have escaped the necessity of very large fibers by wrapping the axons in myelin to permit economic, rapid long-distance signaling.

Myelin plays a central role in several disorders.

Multiple sclerosis (MS) is a pathophysiological condition in which nerve fibers in various locations throughout the nervous system become demyelinated (lose their myelin). MS is an autoimmune (*auto* means "self"; *immune* means "defense against") disease in which the body's defense system erroneously attacks the myelin sheath surrounding myelinated nerve fibers. The symptoms vary considerably, depending on the extent and location of the myelin damage. Loss of myelin slows transmission of impulses in the affected neurons. Also, scarring associated with myelin damage can injure the underlying axons, further interfering with action potential propagation.

Myelin defects may be the culprit in MS, but the presence of myelinating cells can be of tremendous benefit when an axon is cut. In the case of a cut axon in a peripheral nerve, the portion of the axon farthest from the cell body degenerates, and the surrounding Schwann cells phagocytize the debris (Fig. 4-14a). The Schwann cells themselves remain and form a **regeneration tube** to guide the regenerating nerve fiber to its proper destination (Fig. 4-14b). The remaining portion of the axon connected to the cell body starts to grow and move forward within the Schwann cell column by ameoboid movement (see p. 40). It is believed that the growing axon tip "sniffs" its way forward in the proper direction, guided by a chemical secreted into the regeneration tube by the Schwann cells. Successful fiber regeneration (Fig. 4-14c) is responsible for the return of sensation and movement after a period of time following traumatic peripheral nerve injuries, although regeneration is not always successful.

Fibers in the central nervous system, which are myelinated by oligodendrocytes rather than Schwann cells, do not have this same regenerative ability. Therefore, damaged neuronal fibers in the brain and spinal cord never regenerate. In the future, however, with the help of exciting new findings, it may be possible to induce significant regeneration of damaged fibers in the central nervous system by chemical means. Nerve growth in the brain and spinal cord appears to be controlled by a delicate balance between *nerve growth–enhancing and nerve growth–inhibiting proteins.* During fetal development, nerve growth in the central nervous system is possible as the brain and spinal cord are being formed. Researchers speculate that the nerve growth inhibitors, which appear late in fetal development in the myelin sheaths surrounding central nerve fibers, may normally serve as "guardrails" to keep new nerve endings from "straying outside their proper paths." Scientists have been able for the first time to induce significant nerve regeneration in rats with severed spinal cords by chemically blocking these nerve growth inhibitors, thereby allowing the nerve growth enhancers to promote abundant sprouting of new nerve fibers at the site of injury.

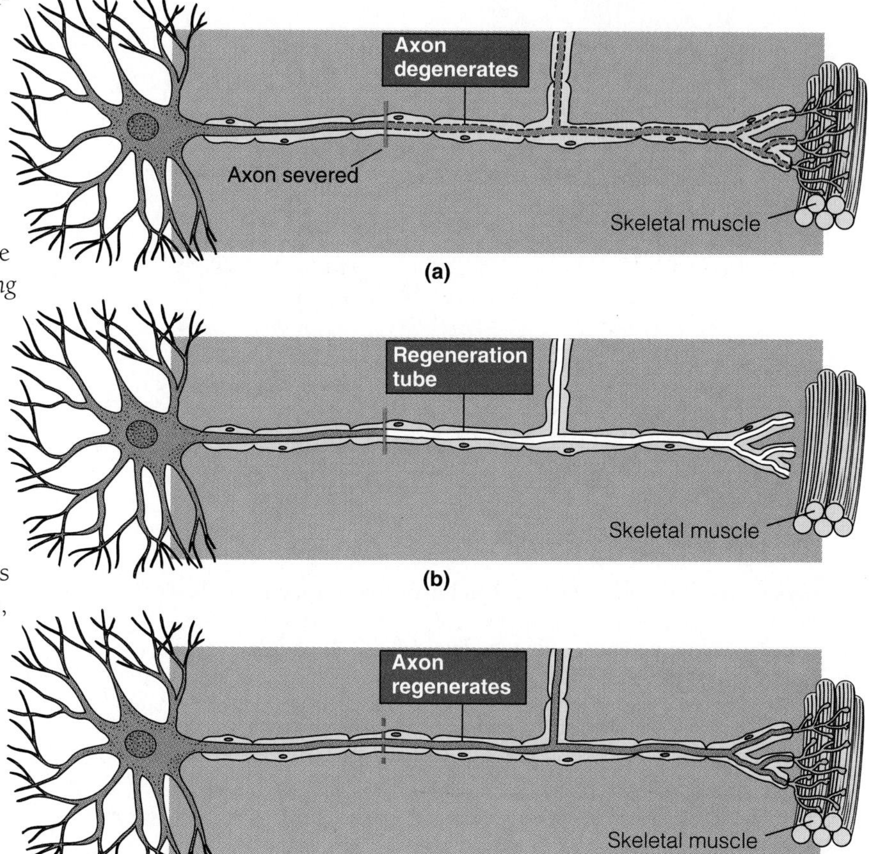

Figure 4-14 Regeneration of a Peripheral Nerve Fiber A severed axon can regenerate in the peripheral nervous system. (a) First, the segment from the cut to the terminal end degenerates. (b) The myelin sheath remains and forms a regeneration tube to guide the regenerating axon to its proper destination. (c) The remaining portion of the axon, which is connected to the cell body, regenerates the missing portion, often reestablishing previous contacts and restoring the neuron's function.

The refractory period ensures unidirectional propagation of the action potential and limits the frequency of action potentials.

What ensures the one-way propagation of an action potential away from the initial site of activation? Note in Fig. 4-15 that once the action potential has been regenerated at a new neighboring site (now positive inside) and the original active area has returned to resting (once again negative inside), the close proximity of opposite charges between these two areas is conducive to local current flow taking place in the backward direction as well as in the forward direction (into as yet unexcited portions of the membrane). If such backward current flow were able to bring the just inactivated area to threshold, another action potential would be initiated here, which would spread both forward and backward, initiating another action potential, and so forth. The situation would be chaotic, with numerous action potentials bouncing back and forth along the axon until the nerve cell eventually fatigued. Fortunately, neurons are saved from this fate of oscillating action potentials by the existence of the **refractory period,** which has two components: the absolute refractory period and the relative refractory period.

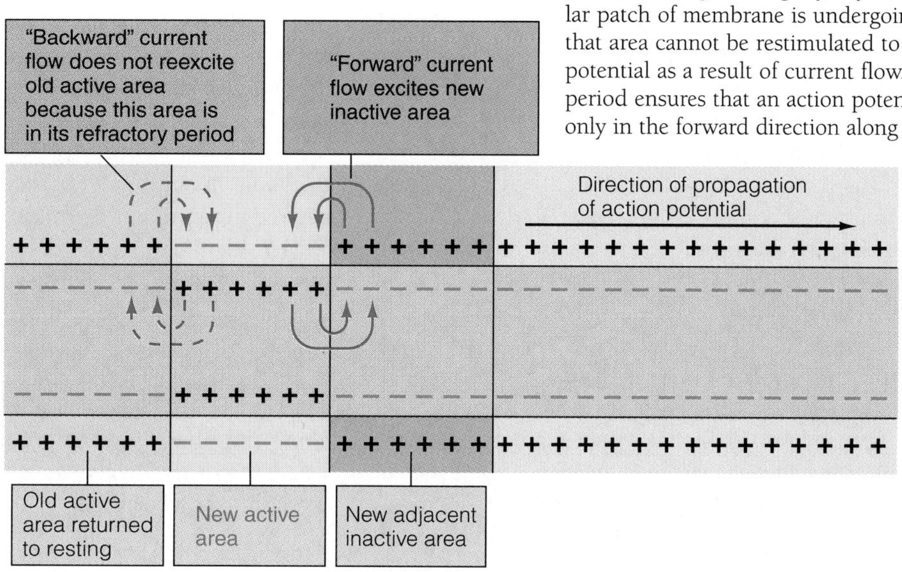

Figure 4-15 Value of the Refractory Period "Backward" current flow is prevented by the refractory period. During and slightly beyond the time when a particular patch of membrane is undergoing an action potential, that area cannot be restimulated to undergo another action potential as a result of current flow. Thus, the refractory period ensures that an action potential can be propagated only in the forward direction along the axon.

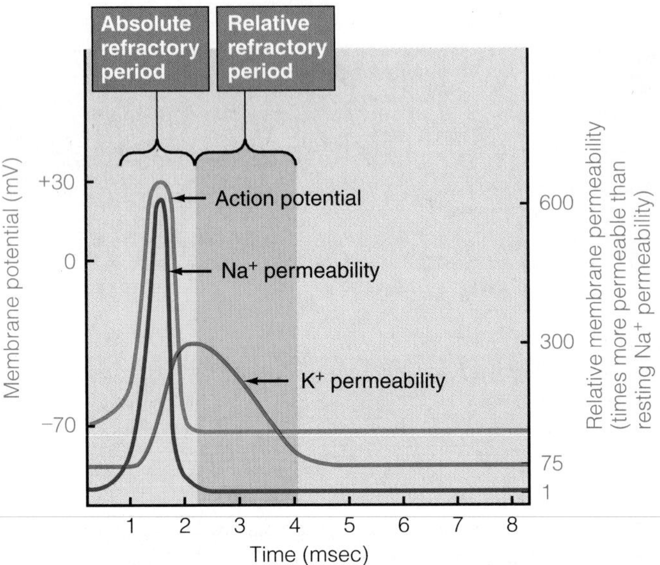

Figure 4-16 **Absolute and Relative Refractory Periods** During the absolute refractory period, the portion of the membrane that has just undergone an action potential cannot be restimulated. It corresponds to the time during which the Na^+ gates are not in their resting conformation. During the relative refractory period, the membrane can be restimulated only by a stronger stimulus than is usually necessary. It corresponds to the time during which the K^+ gates that were opened during the action potential are in their closed and inactivated state.

During the time that a particular patch of axonal membrane is undergoing an action potential, it is incapable of initiating another action potential, no matter how strongly it is stimulated. This time period when a recently activated patch of membrane is completely refractory (unresponsive) to further stimulation is known as the *absolute refractory period* (Fig. 4-16). It corresponds to the time period during which the Na^+ gates are first opened and then closed and inactivated. Not until the potential has returned to resting and the voltage-dependent Na^+ channels are restored to their "closed but capable of opening" conformation can they respond to another depolarization with an explosive increase in P Na^+ to initiate another action potential.

Following the absolute refractory period is a *relative refractory period* during which a second action potential can be produced only by a stimulus considerably stronger than is usually necessary. During this time, the K^+ gates (those which had opened at the peak of the action potential to bring about repolarization) are closed and in their inactivated state. Only when all channels have been restored to their resting conformation is the patch of membrane that has just undergone an action potential ready to respond again in a normal fashion. Meanwhile, the impulse has continued to be rapidly propagated in the forward direction only. By the time the original site has recovered from its refractory period and is capable of being restimulated by normal current flow, the action potential is so far away that it can no longer influence the original site. Thus, *the refractory period ensures the unidirectional propagation of the action potential down the axon away from the initial site of activation.*

The refractory period is also responsible for setting an upper limit on the frequency of action potentials; that is, it determines the maximum number of new action potentials that can be initiated and propagated along the fiber in a given period of time. The original site must recover from its refractory period before a new impulse can be triggered to follow the first impulse. The length of the refractory period varies for different types of neurons. The longer the refractory period, the greater the delay before the new action potential can be initiated and the lower the frequency with which a nerve cell can respond to repeated or ongoing stimulation.

Action potentials occur in all-or-none fashion.

If any portion of the neuronal membrane is depolarized to threshold, an action potential is initiated and relayed throughout the membrane in undiminished fashion. Furthermore, once threshold has been reached, the resultant action potential always goes to maximal height, because the changes in voltage during an action potential are due to ion movements down concentration and electrical gradients, which are not affected by stimulus strength. A stimulus stronger than one necessary to bring the membrane to threshold does not produce a larger action potential. On the other hand, a stimulus that fails to depolarize the membrane to threshold does not trigger an action potential at all. Thus, *an excitable membrane either responds to a stimulus with a maximal action potential that spreads nondecrementally throughout the membrane, or it does not respond with an action potential at all.* This is called the **all-or-none law.**

This all-or-none concept is analogous to firing a gun. Either the trigger is not pulled sufficiently to fire the bullet at all (threshold is not reached), or it is pulled hard enough to elicit the full firing response of the gun (threshold is reached). Squeezing the trigger harder does not produce a greater explosion. Just as it is not possible to fire a gun halfway, it is not possible to have a halfway action potential.

The threshold phenomenon is a means by which some discrimination can take place between important and unimportant stimuli. Stimuli too weak to bring the membrane to threshold do not initiate an action potential and therefore do not clutter up the nervous system with transmission of insignificant signals. How is it possible, however, to differentiate between two stimuli of varying strengths if both bring the membrane to threshold and generate action potentials of the same magnitude? For example, how can one distinguish between touching a warm object or a very hot object if both trigger identical action potentials in a nerve fiber relaying information about skin temperature to the central nervous system? The answer lies in the *frequency* with which the action potentials are generated. A stronger stimulus does not produce a larger action potential, but it does trigger a greater number of action potentials per second to be propagated along the fiber. In addition, a stronger stimulus in a region will cause the thresholds of more neurons to be reached, increasing the total information sent to the central nervous system.

Synapses and Neuronal Integration

A neurotransmitter carries the signal across a synapse.

What happens once an action potential reaches the end of an axon? A neuron may terminate at one of three structures: a muscle, a gland, or another neuron. The junctions between nerves and the muscles and glands that they innervate will be described later. For now we will concentrate on the junction between two neurons—a **synapse.**

Typically, a neuron-to-neuron synapse involves a junction between an axon terminal of one neuron and the dendrites or cell body of a second neuron. Less frequently, axon-to-axon and dendrite-to-dendrite connections occur. Most neuronal cell bodies and associated dendrites receive thousands of synaptic inputs, which are axon terminals from many other neurons. It has been estimated that some neurons within the central nervous system receive as many as 100,000 synaptic inputs (▬ Fig. 4-17a and b).

The anatomy of one of these thousands of synapses is shown in Fig. 4-17c and d. The axon terminal of the **presynaptic neuron,** which conducts its action potentials *toward* the synapse, ends in a slight swelling, the **synaptic knob.** The synaptic knob contains **synaptic vesicles,** which store a specific chemical messenger, a **neurotransmitter,** that has been synthesized and packaged by the presynaptic neuron. The synaptic knob comes into close proximity to, but does not actually directly contact, the **postsynaptic neuron,** the neuron whose action potentials are propagated *away* from the synapse. The space between the presynaptic and postsynaptic neurons, the **synaptic cleft,** is too wide for the direct spread of current from one cell to the other and therefore prevents action potentials from electrically passing between the neurons. The portion of the postsynaptic membrane immediately underlying the synaptic knob is referred to as the **subsynaptic membrane.**

Synapses operate in one direction only; that is, the presynaptic neuron brings about changes in membrane potential of the postsynaptic neuron, but the postsynaptic neuron does not influence the potential of the presynaptic neuron. The reason for this becomes readily apparent when one examines the events that occur at a synapse.

When an action potential in a presynaptic neuron has been propagated to the axon terminal (step 1 in Fig. 4-17c), this change in potential triggers the opening of voltage-gated Ca^{2+} channels in the synaptic knob. Because Ca^{2+} is in much higher concentration in the ECF, this ion flows into the synaptic knob (step 2). Here it induces the release by exocytosis (see p. 25) of a neurotransmitter from some of the synaptic vesicles into the synaptic cleft (step 3). The released neurotransmitter diffuses across the cleft and combines with specific protein receptor sites on the subsynaptic membrane (step 4). This binding triggers the opening of specific ion channels in the subsynaptic membrane, thereby altering the permeability of the postsynaptic neuron (step 5). This is an example of chemical messenger–gated channels, in contrast to the voltage-gated channels responsible for the action potential and for the Ca^{2+} influx into the synaptic knob. Because only the presynaptic terminal can release a neurotransmitter and only the subsynaptic membrane of the postsynaptic neuron has receptor sites for the neurotransmitter, the synapse can operate only in the direction from presynaptic to postsynaptic neuron.

Some synapses excite the postsynaptic neuron whereas others inhibit it.

There are two types of synapses, depending on the permeability changes induced in the postsynaptic neuron by the combination of neurotransmitter with receptor sites: *excitatory synapses* and *inhibitory synapses.* At an **excitatory synapse,** the response to the neurotransmitter-receptor combination is an opening of Na^+ and K^+ channels within the subsynaptic membrane, thus increasing permeability to both of these ions. Both the concentration and electrical gradients for Na^+ favor its movement into the postsynaptic neuron at resting potential, whereas only the concentration gradient for K^+ favors its movement outward (Table 4-1). Therefore, the permeability change induced at an excitatory synapse results in the simultaneous movement of a few K^+ ions out of the postsynaptic neuron while a relatively larger number of Na^+ ions enter this neuron. The result is a net movement of positive ions into the cell. This makes the inside of the membrane slightly less negative than at resting potential, thus producing a *small depolarization* of the postsynaptic neuron. Activation of one excitatory synapse can rarely depolarize the postsynaptic membrane sufficiently to bring it to threshold. Too few channels are involved at a single subsynaptic membrane to permit adequate depolarizing fluxes to reduce the potential to threshold. This small depolarization, however, does bring the membrane of the postsynaptic neuron closer to threshold, increasing the likelihood that threshold will be reached (in response to further excitatory input) and an action potential will occur. That is, the membrane is more excitable (easier to bring to threshold) than when at rest. Accordingly, such a postsynaptic potential change occurring at an excitatory synapse is called an **excitatory postsynaptic potential,** or **EPSP** (▬ Fig. 4-18a).

At an **inhibitory synapse,** the combination of the released chemical messenger with its receptor sites increases the permeability of the subsynaptic membrane to either K^+ or Cl^- by altering these ions' respective channel conformations. In either case, the resulting ion movements bring about a *small hyperpolarization* of the postsynaptic neuron (greater internal negativity). In the case of increased $P\ K^+$, more positive charges leave the cell via K^+ efflux, leaving more negative charges behind on the inside; in the case of increased $P\ Cl^-$, negative charges enter the cell in the form of Cl^- ions because Cl^- concentration is higher outside the cell. This small hyperpolarization moves the membrane potential even farther away from threshold (Fig. 4-18b), lessening the likelihood that the postsynaptic neuron will reach threshold and undergo an action potential. That is, the membrane would be harder to bring to threshold by excitatory input than when it is at resting potential. The membrane is said to be inhibited under

Figure 4-17 Synaptic Structure and Function (a) Schematic representation of presynaptic inputs to a single neuron. (b) Electron micrograph showing multiple presynaptic inputs to a single postsynaptic cell body. (c) Schematic representation of the structure of a single synapse. Propagation of an action potential to the axon terminal of a presynaptic neuron ① triggers the opening of voltage-gated Ca^{2+} channels and the subsequent entry of Ca^{2+} into the synaptic knob ②. Calcium induces the release by exocytosis of neurotransmitter from synaptic vesicles into the snyaptic cleft ③. After diffusing across the cleft, the neurotransmitter binds with its receptor sites on the subsynaptic membrane ④. This binding triggers the opening of specific ion channels in the subsynaptic membrane ⑤, which alters the permeability of the postsynaptic neuron.

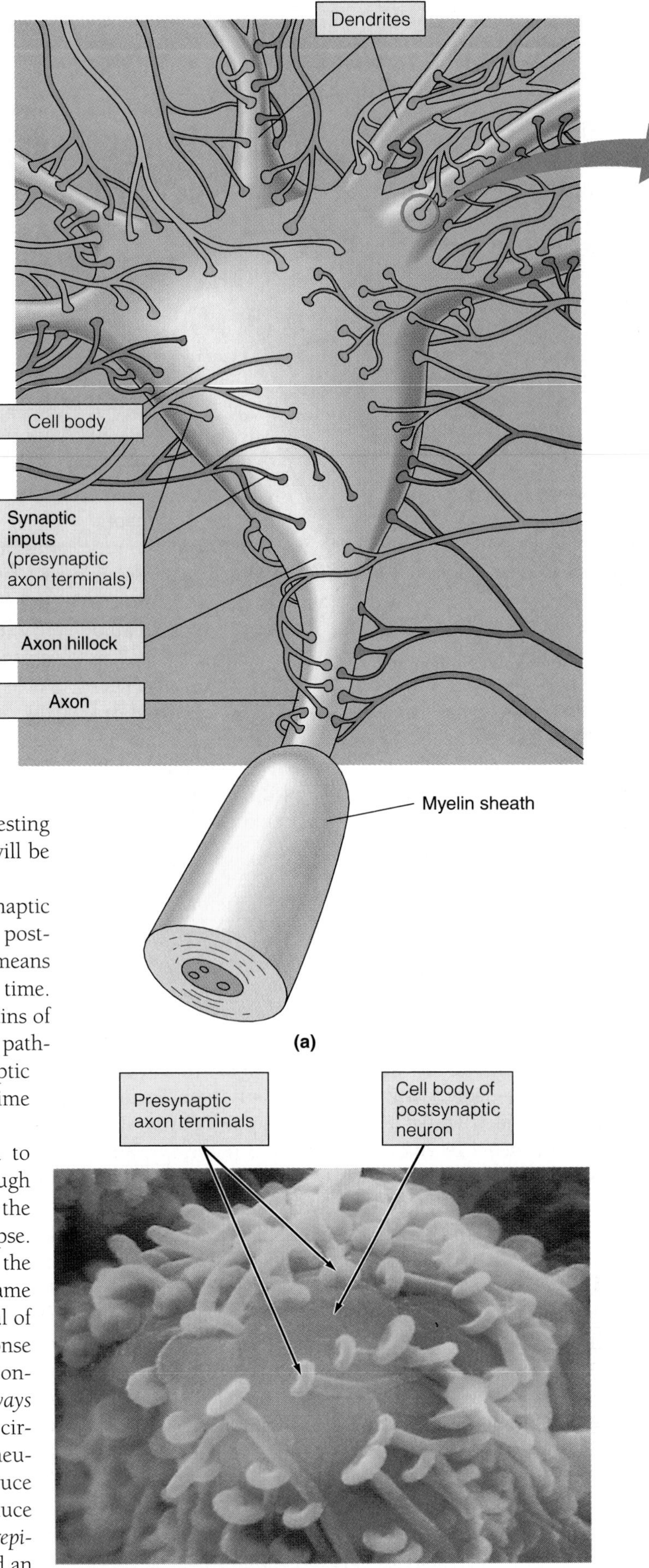

these circumstances, and the small hyperpolarization of the postsynaptic cell is called an **inhibitory postsynaptic potential**, or **IPSP.** (In cells where the equilibrium potential for Cl^- exactly equals the resting potential, an increased P Cl^- does not result in a hyperpolarization because there is no driving force to produce Cl^- movement. Opening of Cl^- channels in these cells tends to hold the membrane at resting potential, thus reducing the likelihood that threshold will be reached.)

This conversion of the electrical signal in the presynaptic neuron (an action potential) to an electrical signal in the postsynaptic neuron (either an EPSP or IPSP) by chemical means (via the neurotransmitter-receptor combination) takes time. This **synaptic delay** is usually about 0.5 to 1 msec. Chains of neurons often must be traversed along a specific neural pathway. The more complex the pathway, the more synaptic delays, and the longer the *total reaction time* (the time required to respond to a particular event).

Many different chemicals are known or suspected to serve as neurotransmitters (Table 4-3). Even though transmitter substances vary from synapse to synapse, the same transmitter is always released at a given synapse. Furthermore, binding of a specific neurotransmitter with the appropriate subsynaptic receptors always leads to the same change in permeability and resultant change in potential of a particular postsynaptic membrane. That is, the response to a given transmitter-receptor combination is always constant. *A given synapse is either always excitatory or always inhibitory.* It does not give rise to an EPSP under one circumstance and produce an IPSP at another time. Some neurotransmitters (for example, *glutamate*) always induce EPSPs whereas others (for example, *glycine*) always induce IPSPs. Still other neurotransmitters (for example, *norepinephrine*) may even produce an EPSP at one synapse and an IPSP at another synapse.

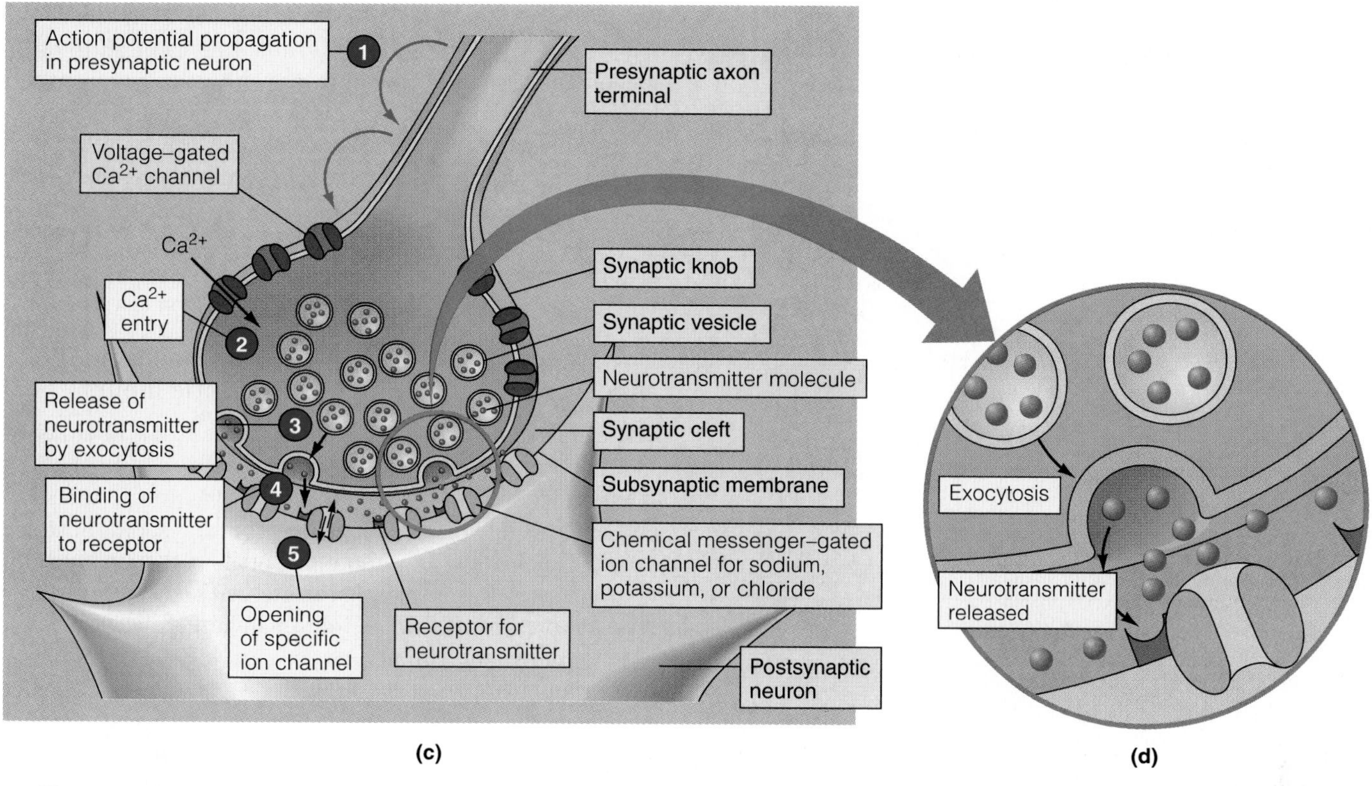

Figure 4-17 **Synaptic Structure and Function (continued)**

Neurotransmitters are quickly removed from the synaptic cleft to wipe the postsynaptic slate clean.

As long as the neurotransmitter remains bound to the receptor sites, the alteration in membrane permeability responsible for the EPSP or IPSP continues. It is important that the neurotransmitter be inactivated or removed after it has produced the appropriate response in the postsynaptic neuron so that the postsynaptic "slate" is "wiped clean," leaving it ready to receive additional messages from the same or other presynaptic inputs. Thus, after combining with the postsynaptic receptor, chemical transmitters are removed and the response is terminated. For example, the transmitter may diffuse away from the synaptic cleft, be inactivated by specific enzymes within the subsynaptic membrane, or be actively taken back up into the axon terminal by transport mechanisms in the presynaptic membrane. Once it is within the synaptic knob,

Figure 4-18 **Postsynaptic Potentials** (a) Excitatory synapse. An excitatory postsynaptic potential (EPSP) brought about by activation of an excitatory presynaptic input brings the postsynaptic neuron closer to threshold potential. (b) Inhibitory synapse. An inhibitory postsynaptic potential (IPSP) brought about by activation of an inhibitory presynaptic input moves the postsynaptic neuron farther from threshold potential.

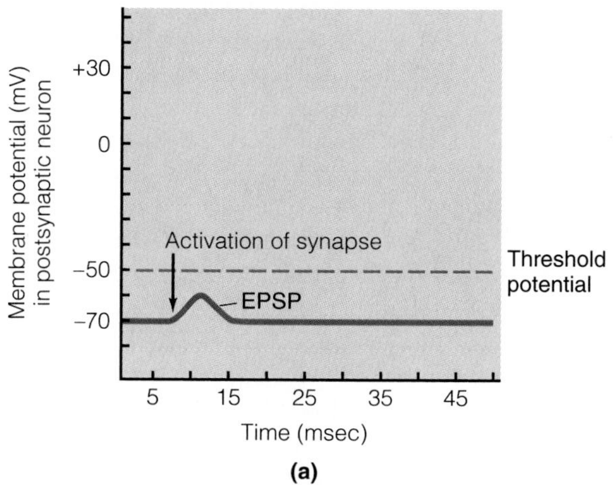

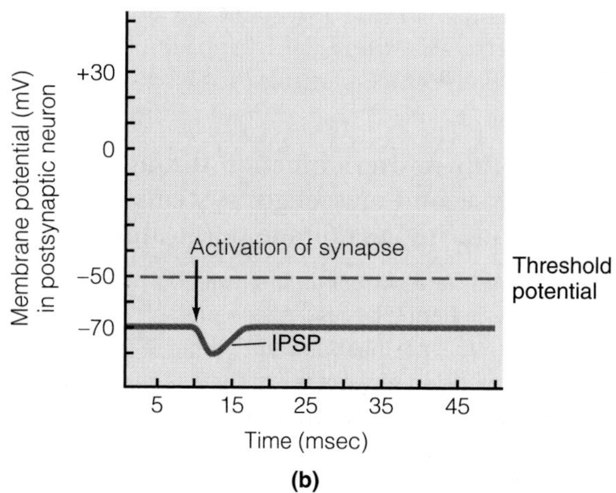

Table 4-3 Some Known or Suspected Neurotransmitters and Neuropeptides

Classical Neurotransmitters (Small, Rapid-Acting Molecules)
Acetylcholine
Dopamine
Norepinephrine
Epinephrine
Serotonin
Histamine
Glycine
Glutamate
Aspartate
Gamma-aminobutyric acid (GABA)

Neuropeptides (Large, Slow-Acting Molecules)
β-endorphin
Adrenocorticotropic hormone (ACTH)
α-melanocyte-stimulating hormone (MSH)
Thyrotropin-releasing hormone (TRH)
Gonadotropin-releasing hormone (GnRH)
Somatostatin
Vasoactive intestinal polypeptide (VIP)
Cholecystokinin (CCK)
Gastrin
Substance P
Neurotensin
Leucine enkephalin
Methionine enkephalin
Motilin
Insulin
Glucagon
Angiotensin II
Bradykinin
Vasopressin
Oxytocin
Carnosine
Bombesin

the transmitter can be stored and released another time (recycled) in response to a subsequent action potential or destroyed by enzymes within the synaptic knob. The method employed depends on the particular synapse.

Some neurotransmitters function through intracellular second messenger systems rather than by directly altering membrane permeability.

Most, but not all, neurotransmitters function by changing the conformation of chemical messenger–gated channels, thereby altering membrane permeability and ionic fluxes across the postsynaptic membrane. Another mode of synaptic transmission utilized by some neurotransmitters, for example *serotonin,* involves the activation of second messengers (see p. 67) within the postsynaptic neuron. Activation of the intracellular messenger, cyclic AMP, can induce both short- and long-term effects. In the short term, cAMP can lead to opening of ionic gates, a task that the other neurotransmitter-receptor combinations do directly without involving the intracellular messenger. The gating effects can be either excitatory or inhibitory. In addition, cAMP may trigger more long-term changes in the postsynaptic cell, even to the extent of possibly altering the cell's genetic expression. Such long-term cellular changes may play a role in learning and memory.

The grand postsynaptic potential depends on the sum of the activities of all presynaptic inputs.

The events that occur at a single synapse result in either an EPSP or an IPSP at the postsynaptic neuron. If a single EPSP is inadequate to bring the postsynaptic neuron to threshold and an IPSP moves it even farther from threshold, how is it possible to initiate an action potential in the postsynaptic neuron? Recall that a typical neuronal cell body receives thousands of presynaptic inputs from many other neurons. Some of these presynaptic inputs may be carrying sensory information brought from the environment; some may be signaling internal changes in homeostatic balance; others may be transmitting signals from control centers in the brain; and still others may arrive carrying other bits of information. At any given time, any number of these presynaptic neurons (probably hundreds) may be firing and thus influencing the postsynaptic neuron's level of activity. The total potential in the postsynaptic neuron, the **grand postsynaptic potential (GPSP),** is a composite of all EPSPs and IPSPs occurring at approximately the same time.

The postsynaptic neuron can be brought to threshold in two ways: (1) *temporal summation* and (2) *spatial summation.* To illustrate these methods of summation, we will examine the possible interactions of three presynaptic inputs—two excitatory (Ex1 and Ex2) and one inhibitory (In1)—on a hypothetical postsynaptic neuron (■ Fig. 4-19). The recording represents the potential in the postsynaptic cell. Bear in mind during our discussion of this simplified version that many thousands of synapses are actually interacting in the same way on a single cell body.

Now suppose that Ex1 has an action potential that causes the release of neurotransmitter from a few of its synaptic vesicles to bring about an EPSP in the postsynaptic neuron. Because EPSPs (as well as IPSPs) are graded potentials, the EPSP spreads only a short distance before dying off. If another action potential subsequently occurs in Ex1, an EPSP of the same magnitude takes place (panel A in Fig. 4-19). Next assume that Ex1 has two action potentials in close succession to each other (panel B). The first action potential in Ex1 produces an EPSP in the postsynaptic neuron. While the postsynaptic membrane is still partially depolarized from this first EPSP (before it has returned to resting), the second presynaptic action potential produces a second EPSP in the postsynaptic neuron. The second EPSP will add on to the first EPSP, bringing the membrane to threshold, so that an action potential can occur in the postsynaptic neuron. Graded potentials do not have a refractory period, so this additive effect is possible.

The summing of several EPSPs occurring very close together in time because of successive firing of a single presynaptic neuron is known as **temporal summation** (*tempus* means "time"). The actual situation is much more complex than the one just described. Up to fifty EPSPs might have to sum to bring the postsynaptic membrane to threshold. Each action potential in a presynaptic neuron triggers the emptying of a certain number of synaptic vesicles. The amount of neurotransmitter released and the resultant magnitude of the change in postsynaptic potential are thus directly related to the frequency of presynaptic action potentials. One way, then, in which the postsynaptic membrane can be brought to threshold is through rapid, repetitive excitation from a single persistent input.

Let us now see what will happen in the postsynaptic neuron if both excitatory inputs are stimulated simultaneously (panel C). An action potential in either Ex1 or Ex2 will produce an EPSP in the postsynaptic neuron; however, neither of these alone will bring the membrane to threshold to elicit a postsynaptic action potential. Simultaneous action potentials in Ex1 and Ex2, however, will produce EPSPs that add to each other, bringing the postsynaptic membrane to threshold, so that an action potential occurs. Such summation of EPSPs originating simultaneously from several different presynaptic inputs (i.e., from different points in "space") is known as **spatial summation.** A second way, therefore, to elicit an action potential in a postsynaptic cell is through concurrent activation of several excitatory inputs. Again, in reality up to fifty EPSPs arriving simultaneously on the postsynaptic membrane are required to bring it to threshold. Similarly, IPSPs can undergo temporal and spatial summation. As IPSPs add together, however, they progressively move the potential farther from threshold.

If an excitatory and an inhibitory input are simultaneously activated, the concurrent EPSP and IPSP more or less cancel each other out (the extent of cancellation depends on their respective magnitudes). In most cases, the postsynaptic membrane potential remains close to resting (panel D).

Thus, the grand postsynaptic potential depends on the sum of activity in all presynaptic inputs. The following oversimplified real-life example demonstrates the benefits of this neuronal integration. The explanation is not completely accurate technically, but the principles of summation are accurate. Assume for simplicity's sake that urination is controlled by a postsynaptic neuron supplying the urinary bladder. (Actually, voluntary control of urination is accomplished by postsynaptic integration at the neuron controlling the external urethral sphincter rather than the bladder itself). As the bladder starts to fill with urine and becomes stretched, a reflex is initiated that ultimately produces EPSPs in the postsynaptic neuron responsible for causing bladder contraction. Partial filling of the bladder does not cause sufficient excitation to bring the neuron to threshold, so urination does not take place (panel A of Fig. 4-19). As the bladder becomes progressively filled, the frequency of action potentials is progressively increased in the presynaptic neuron that signals the postsynaptic neuron of the extent of bladder filling (Ex1 in panel B of Fig. 4-19). When the frequency becomes great enough that the EPSPs are temporally summed to threshold, the postsynaptic neuron has an action potential that stimulates bladder contraction.

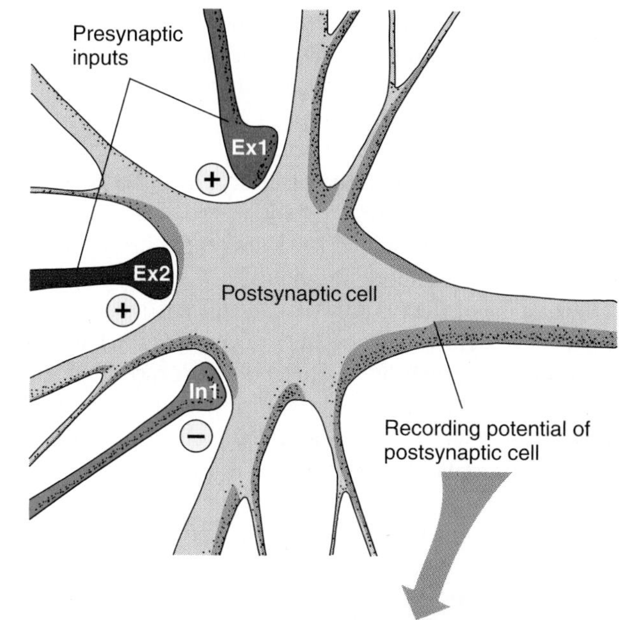

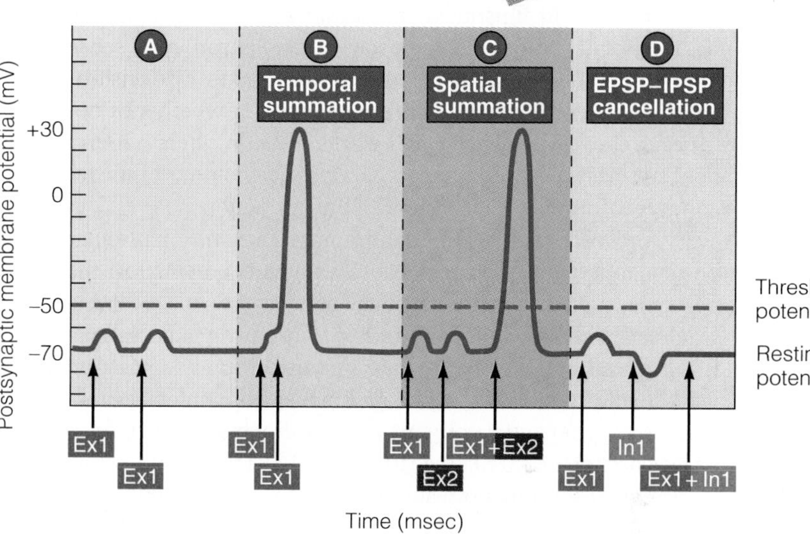

Figure 4-19 **Determination of the Grand Postsynaptic Potential by the Sum of Activity in the Presynaptic Inputs** Two excitatory (Ex1 and Ex2) and one inhibitory (In1) presynaptic inputs terminate on this hypothetical postsynaptic neuron. The potential of the postsynaptic neuron is being recorded. If an excitatory presynaptic input (Ex1) is stimulated a second time after the first EPSP in the postsynaptic cell has died off, a second EPSP of the same magnitude will occur (panel A). If, however, Ex1 is stimulated a second time before the first EPSP has died off, the second EPSP will add onto, or sum with, the first EPSP, resulting in temporal summation, which may bring the postsynaptic cell to threshold (panel B). The postsynaptic cell may also be brought to threshold by spatial summation of EPSPs that are initiated by simultaneous activation of two (Ex1 and Ex2) or more excitatory presynaptic inputs (panel C). Simultaneous activation of an excitatory (Ex1) and inhibitory (In1) presynaptic input does not change the postsynaptic potential because the resultant EPSP and IPSP cancel each other out (panel D).

What if the time is inopportune for urination to take place? IPSPs can be produced at the bladder postsynaptic neuron by presynaptic inputs originating in higher levels of the brain responsible for voluntary control (In1 in panel D of Fig. 4-19). These "voluntary" IPSPs in effect cancel out the "reflex" EPSPs triggered by stretching of the bladder. Thus, the postsynaptic neuron remains at resting potential and does not have an action potential, so the bladder is prevented from contracting and emptying even though it is full.

What if the bladder is only partially filled, so that the presynaptic input originating from this source is insufficient to bring the postsynaptic neuron to threshold to cause bladder contraction, yet the person needs to supply a urine specimen for laboratory analysis? The person can voluntarily activate an excitatory presynaptic neuron (Ex2 in panel C of Fig. 4-19), which spatially summates with the reflex-activated presynaptic neuron (Ex1) to bring the postsynaptic neuron to threshold. This achieves the action potential necessary to stimulate bladder contraction, even though the bladder is not full.

This example illustrates the importance of postsynaptic neuronal integration. Each postsynaptic neuron in a sense "computes" all the input it receives and makes a "decision" about whether to pass the information on (i.e., whether or not to reach threshold). Each postsynaptic neuron filters out and does not pass on information it receives that is not significant enough to bring it to threshold. If every action potential in every presynaptic neuron that impinges upon a particular postsynaptic neuron were to cause an action potential in the postsynaptic neuron, the neuronal pathways would be overwhelmed with trivia. Only if an excitatory presynaptic signal is reinforced by other supporting signals through summation will the information be passed on. Furthermore, interaction of postsynaptic potentials provide a way for one set of signals to offset another set (IPSPs negating EPSPs). This allows a fine degree of discrimination and control in determining what information will be passed on. (See the accompanying boxed feature, A Closer Look at Exercise Physiology.)

Action potentials are initiated at the axon hillock because it has the lowest threshold.

Threshold potential is not uniform throughout the postsynaptic neuron. The lowest threshold is present at the axon hillock because this region has an abundance of voltage-gated Na^+ channels, making it considerably more sensitive to changes in potential than the remainder of the cell body and dendrites. The latter regions have a significantly higher threshold than the axon hillock. Because of local current flow, changes in membrane potential (EPSPs or IPSPs) occurring anywhere on the cell body or dendrites spread throughout the cell body, dendrites, and axon hillock. When summation of EPSPs takes place, the lower threshold of the axon hillock is reached first, whereas the cell body and dendrites at the same potential are still considerably below their own much higher thresholds. Therefore, an action potential originates in the axon hillock and is propagated from there throughout the rest of the neuron.

Neuropeptides, which are also released from axon terminals, act primarily as neuromodulators.

Researchers have recently discovered that in addition to the classical neurotransmitters just described, some neurons also release **neuropeptides** (Table 4-3). Neuropeptides differ from classical neurotransmitters in several important ways (Table 4-4). Classical neurotransmitters are small, rapid-acting molecules that typically bring about a response in the postsynaptic neuron within a few milliseconds or less, usually a change in postsynaptic potential (an EPSP or IPSP) as a result of triggering the opening of specific ion channels. Classical neurotransmitters are synthesized and packaged locally in synaptic vesicles in the cytosol of the axon terminal. These chemical messengers are primarily amino acids or closely related compounds.

Neuropeptides are larger molecules made up of anywhere from two to about forty amino acids. They are synthesized in

Table 4-4 Comparison of Classical Neurotransmitters and Neuropeptides

Characteristic	*Classical neurotransmitters*	*Neuropeptides*
Size	Small; one amino acid or similar chemical	Large; 2 to 40 amino acids in length
Site of synthesis	Cytosol of synaptic knob	Endoplasmic reticulum and Golgi complex in cell body; travel to synaptic knob by axonal transport
Site of storage	In small synaptic vesicles in axon terminal	In large dense-core vesicles in axon terminal
Site of release	Axon terminal	Axon terminal; may be cosecreted with neurotransmitter
Speed and duration of action	Rapid, brief response	Slow, prolonged response
Site of action	Subsynaptic membrane of postsynaptic cell	Nonsynaptic sites on either presynaptic or postsynaptic cell at much lower concentrations than classical neurotransmitters
Effect	Usually alter potential of postsynaptic cell by opening specific ion channels	Usually enhance or suppress synaptic effectiveness by long-term changes in neurotransmitter synthesis or postsynaptic receptor sites

A CLOSER LOOK AT EXERCISE PHYSIOLOGY

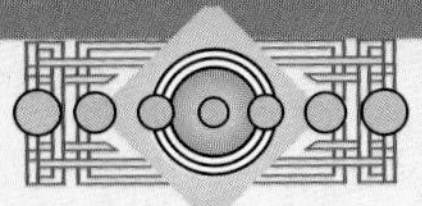

The Yells and Grunts of an Athlete May Serve a Phsyiological Purpose

Skeletal muscles are innervated by a special type of neuron called motor neurons (see chapter 7). Action potentials in motor neurons initiate the contractile process in the muscle on which they terminate. Both excitatory and inhibitory presynaptic inputs from a variety of sources impinge on the cell body and dendrites of the motor neurons. These neurons fire, thereby initiating muscle contraction, when their membrane is brought to threshold by summation of EPSPs. A person can voluntarily contract a given skeletal muscle by consciously activating nerve pathways that generate EPSPs at the appropriate motor neurons.

Among the inhibitory inputs to motor neurons are those arising from the Golgi tendon organ, a sensory device located within the tendon that attaches a muscle to the skeleton (see chapter 8). The Golgi tendon organ is sensitive to stretch imposed on it by the muscle's contraction. When stretched, the Golgi tendon organ initiates a nerve pathway that generates IPSPs on the motor neurons that innervate the very same contracting muscle in which the sensory device is located. If the grand postsynaptic potential of the involved motor neurons is reduced below threshold by the inhibitory input from the Golgi tendon organ, the muscle is no longer stimulated and it relaxes. This reflex relaxation is protective in nature; its purpose is to prevent the muscle from contracting hard enough to rupture the muscle or tendon or break a bone.

Weight training requires the voluntary control of this natural muscular inhibitory reflex. When persons begin to weight train, their ability to lift dramatically increases in the first two weeks even though no physiologic change has yet occurred in the muscle. This initial improvement in strength is thought to be brought about by the body learning to voluntarily produce more EPSPs, thus voluntarily overriding the muscles' inhibition by their stretched Golgi tendon organs. As weight training progresses, muscles get physiologically stronger as the muscle cells enlarge, the tendons get thicker, and the Golgi tendon organs become less sensitive.

Power lifters sometimes yell as they lift to try to further increase the excitatory stimuli during heavy lifting. Similarly, tennis players and other athletes often grunt during the execution of powerful maneuvers. In sports such as arm wrestling, the inhibitory reflex is overridden by voluntary control to the point that participants have ruptured muscles or tendons and have even broken the bones in the arm during the excitement of competition.

the neuronal cell body in the endoplasmic reticulum and Golgi complex (see p. 20) and are subsequently moved by axonal transport along the microtubular highways to the axon terminal (see p. 37). Neuropeptides are not stored within the small synaptic vesicles with the classical neurotransmitter but instead are packaged in large **dense-core vesicles,** which are also present in the axon terminal. The dense-core vesicles undergo Ca^{2+}-induced exocytosis and release neuropeptides at the same time that the neurotransmitter is released from the synaptic vesicles. A given axon terminal releases only a single classical neurotransmitter, but the same terminal may also contain one or more neuropeptides that are secreted along with the classical neurotransmitter.

Even though neuropeptides are currently the subject of intense investigation, our knowledge about their functions and control is still sketchy. They are known to diffuse locally and act on other adjacent neurons at much lower concentrations than classical neurotransmitters, and they bring about slower, more prolonged responses. Some neuropeptides released at synapses may function as true neurotransmitters, but most are believed to function as **neuromodulators.** Neuromodulators are chemical messengers that bind to neuronal receptors at nonsynaptic sites (that is, not at the subsynaptic membrane). They do not cause the formation of EPSPs and IPSPs but instead bring about long-term changes that subtly depress or enhance synaptic effectiveness, often by activating second messenger systems. Neuromodulators may act at either presynaptic or postsynaptic sites. For example, a neuromodulator may influence the level of an enzyme critical in the synthesis of a neurotransmitter by a presynaptic neuron, or it may alter the sensitivity of the postsynaptic neuron to a particular neurotransmitter by causing long-term changes in the number of postsynaptic receptor sites for the neurotransmitter. Thus, neuromodulators delicately fine-tune the synaptic response. The effect may last for days or even months or years. Whereas neurotransmitters serve as chemical messengers for rapid communication between neurons, neuromodulators are involved with more long-term neural events such as learning.

Interestingly, the synaptically released neuromodulators include many substances that also have distinctly different roles in other body regions as hormones released into the blood from neural and nonneural tissues. For example, *cholecystokinin (CCK)* is a well-known hormone released from the small intestine that causes the gallbladder to contract and release bile into the intestine, among other digestive activities. CCK has also been found in axon-terminal vesicles in diffuse areas of the brain, where it is believed to mediate satiety (the feeling of no longer being hungry) and perhaps decrease the perception of painful stimuli, among other unknown activities. In fact, in many instances neuropeptides are named for their first-discovered role as hormones, as is the case with cholecystokinin (*chole* means "bile"; *cysto* means "bladder"; *kinin* means "contraction"). It appears that a number of chemical messengers are quite versatile and can assume different roles, depending on their source, distribution, and interaction with a particular type of cell.

Presynaptic inhibition or facilitation can selectively alter the effectiveness of a given presynaptic input.

Besides neuromodulation, presynaptic inhibition or facilitation is another means of depressing or enhancing synaptic effectiveness. Sometimes a third neuron influences activity between a presynaptic ending and a postsynaptic neuron. A presynaptic axon terminal (labeled A in ▬ Fig. 4-20) may itself be innervated by another axon terminal (labeled B) that can modify the amount of transmitter released from presynaptic terminal A. When the neurotransmitter released from terminal B binds with receptor sites on terminal A, the amount of transmitter released from terminal A in response to action potentials is altered. If the amount of transmitter released from A is reduced, the phenomenon is known as **presynaptic inhibition.** If the release of transmitter is enhanced, the effect is called **presynaptic facilitation.** The mechanism for these effects is unclear, but it most likely involves Ca^{2+} because the amount of cytosolic Ca^{2+} in the synaptic knob determines the extent to which synaptic vesicles release neurotransmitter.

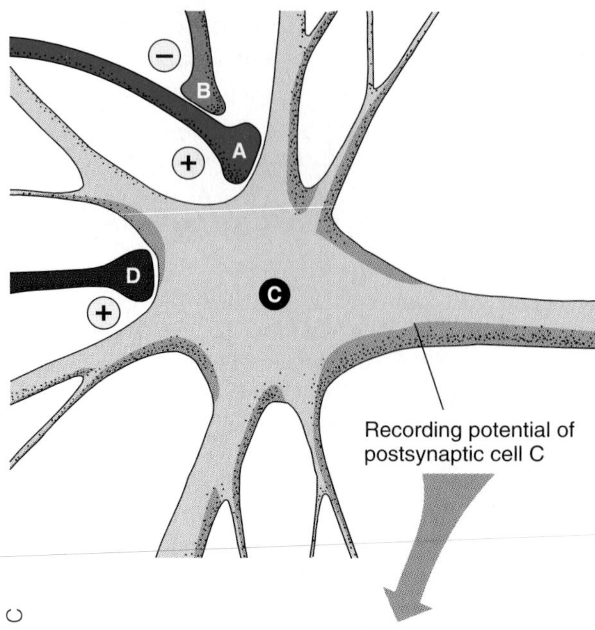

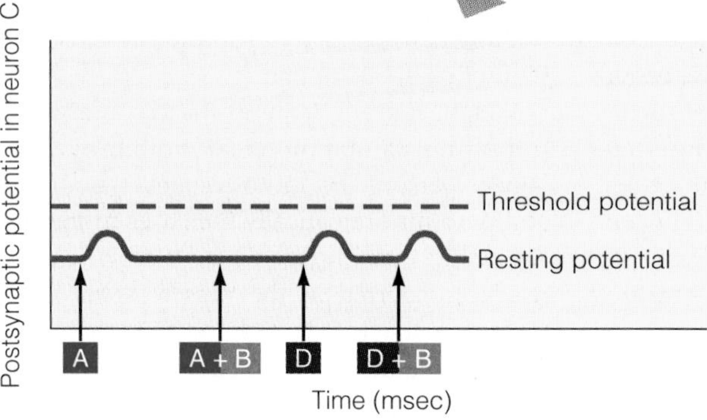

▬ *Figure 4-20* **Presynaptic Inhibition** A, an excitatory terminal ending on postsynaptic cell C, is itself innervated by inhibitory terminal B. Stimulation of terminal A alone produces an EPSP in cell C, but simultaneous stimulation of terminal B prevents the release of excitatory transmitter from terminal A. Consequently, no EPSP is produced in cell C despite the fact that terminal A has been stimulated. Such presynaptic inhibition selectively depresses activity from terminal A without suppressing any other excitatory input to cell C. Stimulation of excitatory terminal D produces an EPSP in cell C even though inhibitory terminal B is simultaneously stimulated because terminal B only inhibits terminal A.

What is the importance of this presynaptic modulation? The amount of transmitter released from presynaptic terminal A influences the potential in the postsynaptic neuron with which it is interacting (labeled C on Fig. 4-20). For example, if A, an excitatory input to C, is prevented from releasing its transmitter by presynaptic inhibition via B, the formation of EPSPs on postsynaptic membrane C will specifically be prevented from input A. As a result, no change in the potential of the postsynaptic neuron will occur in spite of action potentials in A. Could the same thing be accomplished by a simultaneous production of an IPSP through activation of an inhibitory input to negate an EPSP produced by activation of A? Not quite. The entire postsynaptic membrane is hyperpolarized by IPSPs, thereby negating excitatory information fed into any part of the cell from any presynaptic input. In other words, an IPSP would not only cancel out an EPSP from excitatory presynaptic input A but could just as well cancel out an EPSP from any other excitatory input, such as the one labeled D on Figure 4-20. Presynaptic inhibition (or presynaptic facilitation), on the other hand, provides a means by which certain inputs to the postsynaptic neuron can be *selectively* altered without affecting the contributions of any other inputs. For example, firing of B specifically prevents the formation of an EPSP in the postsynaptic neuron from excitatory presynaptic neuron A but does not have any influence on other excitatory presynaptic inputs. Excitatory input D can still produce an EPSP in the postsynaptic neuron even when B is firing. This type of neuronal integration is another means by which electrical signaling between nerve cells can be carefully fine-tuned.

The effectiveness of synaptic transmission can be modified by drugs and diseases.

Numerous opportunities exist for influencing synaptic transmission by drugs. In fact, the vast majority of drugs that influence the nervous system perform their function by altering synaptic mechanisms. Synaptic drugs may block an undesirable effect or enhance a desirable effect. Possible drug actions include (1) altering the synthesis, axonal transport, storage, or release of a neurotransmitter; (2) modifying neurotransmitter interaction with the postsynaptic receptor; (3) influencing neurotransmitter reuptake or destruction; or (4) replacing a deficient neurotransmitter with a substitute transmitter.

For example, the illegal drug *cocaine* blocks the reuptake of the neurotransmitter *dopamine* at presynaptic terminals by competitively binding with the dopamine reuptake transporter. The transporter is a protein molecule that serves the ferrying function of picking up released dopamine from the synaptic cleft and shuttling it back to the axon terminal for reuptake. With cocaine occupying the dopamine transporter, dopamine remains in the synaptic cleft longer than usual and continues to interact with its postsynaptic receptor sites. The result is prolonged activation of neural pathways that use this chemical as a neurotransmitter. Among these pathways are those believed to play a role in emotional responses, especially feelings of pleasure.

Whereas cocaine abuse leads to excessive dopamine activity, **Parksinson's disease** is attributable to a deficiency of dopamine in a particular region of the brain involved in controlling complex movements. (See the boxed feature on p. 104, ♦ Concepts, Challenges, and Controversies.) Treatment of Parkinson's disease is an example of a deficient neurotransmitter being replaced with a substitute transmitter. When patients with this disease are given *levodopa (L-dopa)*, a compound closely related to dopamine, the L-dopa can gain access to the brain and be taken up by the dopamine deficient synaptic knobs, thereby substituting for the lacking "home-grown variety" of this neurotransmitter. This greatly alleviates the symptoms associated with the deficit in most patients. Dopamine itself cannot be administered because it is unable to cross the blood-brain barrier (see p.119), whereas L-dopa can enter the brain from the blood.

Synaptic transmission is also vulnerable to a number of disease processes, including defects at both presynaptic and postsynaptic sites. For example, two different neural poisons, strychnine and tetanus toxin, act at different synaptic sites to block inhibitory impulses while leaving the excitatory inputs unchecked. **Strychnine** competes with one of the inhibitory neurotransmitters, glycine, at the postsynaptic receptor site. This poison combines with the receptor but does not directly alter the potential of the postsynaptic cell in any way. Instead, strychnine blocks the receptor so that it is not available for interaction with glycine when the latter is released from the inhibitory presynaptic ending. Thus, postsynaptic inhibition (formation of IPSPs) is abolished in nerve pathways that use glycine as an inhibitory transmitter. Unchecked excitatory pathways lead to convulsions, muscle spasticity, and death. **Tetanus toxin,** on the other hand, prevents the release of another inhibitory transmitter, *gamma-aminobutyric acid (GABA),* from inhibitory presynaptic inputs terminating on neurons that supply skeletal muscles. Unchecked excitatory inputs to these neurons result in uncontrolled muscle spasms. These spasms occur especially in the jaw muscles early in the disease, giving rise to the common name of *lockjaw* for this condition. Later they progress to the muscles responsible for breathing, at which point the disease proves fatal. Strychnine and tetanus toxin poisoning have similar outcomes, but one poison (strychnine) blocks specific postsynaptic inhibitory receptors, whereas the other (tetanus toxin) prevents the presynaptic release of a specific inhibitory neurotransmitter. Other drugs and diseases that influence synaptic transmission are too numerous to mention, but as these examples illustrate, any site along the synaptic pathway is vulnerable to interference, either phamacological (drug-induced) or pathological (disease-induced).

Neurons are linked to each other through convergence and divergence to form vast and complex nerve pathways.

Two important relationships exist between neurons: convergence and divergence. A given neuron may have many other neurons synapsing on it. Such a relationship is known as **convergence** (▬ Fig. 4-21). Through this converging input, a single cell is influenced by thousands of other cells. This single cell, in turn, influences the level of activity in many other cells by divergence of output. **Divergence** refers to the branching of axon terminals so that a single cell synapses with and influences many other cells.

▬ *Figure 4-21* **Convergence and Divergence**

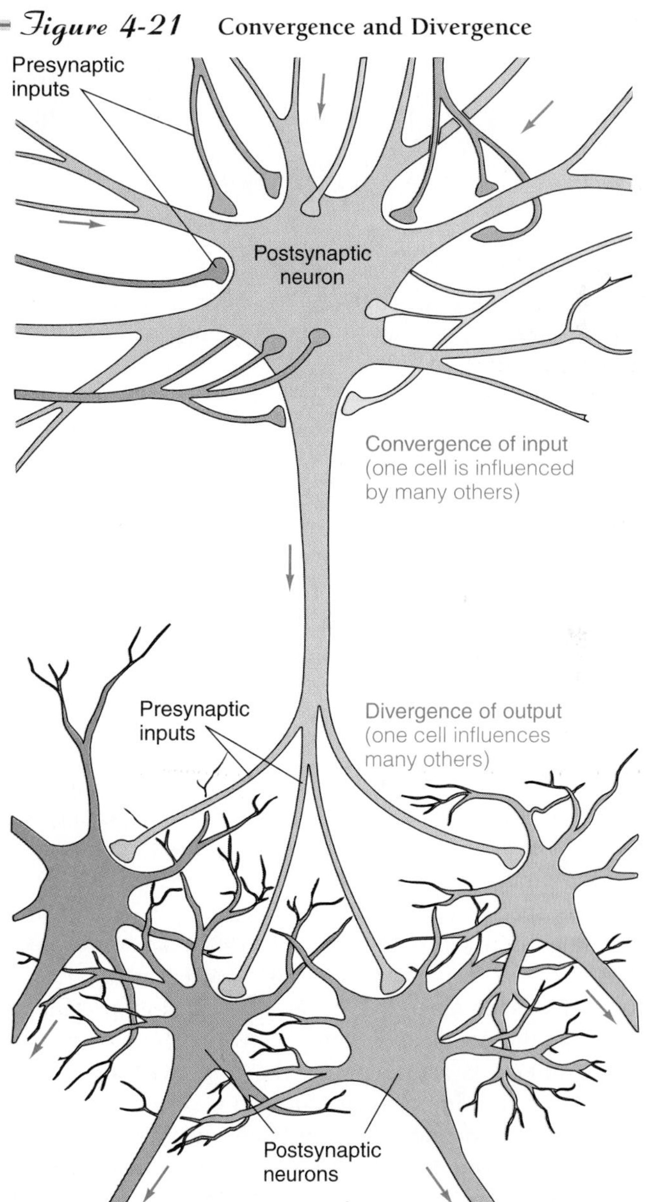

Arrows indicate direction in which information is being conveyed.

CONCEPTS, CHALLENGES, AND CONTROVERSIES

Parkinson's Disease, Pollution, Ethical Problems, and Politics

In 1983, Dr. J. W. Langston, a California neurologist, was confronted with a medical puzzle unlike any he had ever seen. Several young men and women were admitted to the hospital in a rigid, stuporous state as if they had been frozen in their tracks. Confined to their beds, the patients lay immobile day after day. They could neither talk nor feed themselves and could not move their limbs.

Langston began an intensive study of the victims and found that they all were drug addicts. Each of them had injected *meperidine (MPPP)*, a synthetic form of heroin. The MPPP was contaminated with a slightly different drug, MPTP, which is converted by a naturally occurring enzyme in the body into a paralyzing chemical, MPP^+. Made in a basement by one of California's small-time drug pushers, the synthetic heroin is one of the "designer drugs" available on the black market today. Sloppy chemical technique resulted in the MPPP being contaminated with MPTP.

Researchers have shown that MPP^+ destroys dopamine-secreting cells in a part of the brain called the *substantia nigra*. Axons from these cells terminate in the *basal nuclei*, another region of the brain involved in the coordination of slow, sustained movements, inhibition of muscle tone, and suppression of useless patterns of movement. Reduced dopamine activity in the basal nuclei resulting from destruction of these substantia nigra cells was responsible for the ensuing symptoms. Most prominently, loss of the basal nuclei's inhibitory influence on muscle tone led to the pronounced muscular rigidity experienced by these patients.

A gradual destruction of these same dopamine-secreting cells in the substantia nigra and the resultant loss of basal nuclei function are responsible for *Parkinson's disease*. The symptoms, which develop slowly as dopamine activity in the basal nuclei gradually diminishes, begin with involuntary tremors at rest, such as involuntary rhythmic shaking of the hands or head. As the disease worsens, patients speak in slow monotone, become increasingly stiff, and walk with a shuffling, stooped gait. In later stages, memory and thinking become severely impaired. Ultimately, debilitating rigidity ensues, similar to that seen in the MPTP victims.

Interestingly, MPP^+ is very similar to *paraquat*, a commonly used agricultural pesticide. This observation led to the hypothesis that Parkinson's disease may be linked to pesticide use. Indeed, epidemiological studies have shown a remarkable correlation between the use of pesticides and the incidence of Parkinson's disease. Other chemical pollutants are also suspected of contributing to the destruction of the dopamine-secreting brain cells. Researchers note that Parkinson's disease was unheard of before the Industrial Revolution. As the Industrial Revolution and environmental pollution spread, the incidence of the disease rose sharply, reaching a plateau in the early 1900s. Currently, an estimated 1.5 million people in the United States have Parkinson's disease. Some investigators warn that increased use of paraquat and other pesticides and a rise in industrial pollution could

Note that a particular neuron is postsynaptic to the neurons converging on it but presynaptic to the other cells on which it terminates. Thus, the terms *presynaptic* and *postsynaptic* refer only to a single synapse. Most neurons are presynaptic to one group of neurons and postsynaptic to another group.

Note that most neurons have three different functional parts: (1) The cell body/dendrite region is specialized to serve as the postsynaptic component that binds with and responds to neurotransmitters released from other neurons. (2) The axon is specialized for initiation and propagation of action potentials in response to graded potential changes induced by binding of a neurotransmitter with receptors on the cell body/dendrite region. (3) The axon terminal is specialized to serve as the presynaptic component that releases a neurotransmitter that influences other postsynaptic cells in response to action potential propagation down the axon.

There are an estimated 100 billion neurons in the brain alone. When you consider the vast and intricate interconnections possible between these neurons through converging and diverging pathways, you can begin to imagine how complex the wiring mechanism of our nervous system really is. Even the most sophisticated computers are far less complex than the human brain.

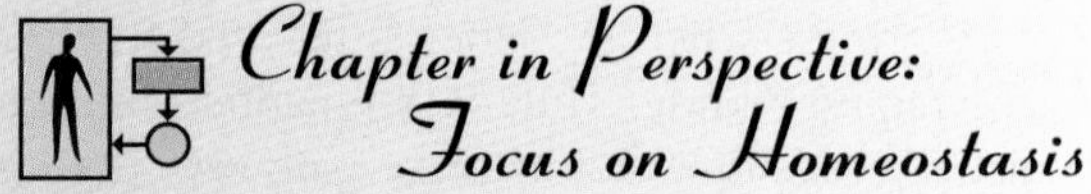

Chapter in Perspective: Focus on Homeostasis

All cells depend on maintenance of a stable internal environment for their survival, and all cells contribute to this homeostatic condition by means of their specialized activities. The specialization of nerve cells (neurons) and muscle cells depends on their ability to rapidly alter their membrane potential and thus produce electrical signals in response to appropriate triggering events.

Nerve cells are specialized to receive, process, encode, and rapidly transmit information from one part of the body to another. The information is transmitted over intricate nerve pathways by propagation of action potentials along the nerve cell's length as well as by chemical transmission of the signal from neuron to neuron and from neuron to muscles and glands through neurotransmitter-receptor interactions at synapses.

Collectively, the nerve cells make up the nervous system, one of the two major control systems of the body. Many of the activities controlled by the nervous system are geared toward maintaining homeostasis. Some neuronal electrical signals convey information about changes to which the body must substantially increase the incidence of the disease in years to come, underscoring the fact that to protect our health we must also protect the quality of our environment.

A standard treatment for Parkinson's disease has been administration of levodopa *(L-dopa)*, a precursor of dopamine. Dopamine itself is unable to cross the blood-brain barrier to get into the brain, but L-dopa can. By serving as a substitute neurotransmitter for dopamine, L-dopa has been an effective form of therapy for many patients. Unfortunately, however, with prolonged use, the beneficial effects of L-dopa tend to diminish and troublesome side effects develop. Therefore, researchers have been seeking other means to manage this disabling condition. One new drug that holds promise is *selegiline*, which enhances the effectiveness of dopamine at synapses and hopefully can forestall the symptoms of Parkinson's disease. Also attracting considerable interest is a newly discovered brain protein *GDNF*, a neurotropic ("nerve nourishing") factor that maintains and nourishes neurons. Researchers are hopeful that one day injection of GDNF into the brain of a Parkinson's disease patient may halt or even reverse the relentless, progressive neural degeneration characteristic of the disease.

One alternative approach to drug therapy for Parkinson's disease has stirred considerable controversy—harvesting dopamine-secreting cells from aborted fetuses and transplanting them into the brains of Parkinson's patients. The legal and ethical issues encumbering fetal cell transplants came into focus in 1988 when a group of researchers sought permission to transplant fetal brain cells into a patient with Parkinson's disease. This was the first request to perform a fetal cell transplant in humans in the United States, although the procedure had been performed in other countries. Opponents of the procedure were concerned that, among other issues, the medical use of aborted fetuses would encourage and justify the practice of abortion. Proponents argued that fetal tissue obtained from a perfectly legal procedure was being wasted when the tissue could be used to potentially treat or cure a number of diseases, including Parkinson's disease. During the Reagan and Bush presidencies, this politically heated debate led to a ban on federally funded research using cells from aborted fetuses, a ban that was lifted by President Clinton during his first week in office.

Early clinical studies have now been completed, but their mixed results have led to new debates among scientists over the most appropriate methodologies for administering the fetal cells. Although mounting evidence indicates that fetal cells can set up shop and replace damaged dopamine-secreting tissue in the adult brain, the outcomes have been inconsistent. A third of the patients experience remarkable improvement following surgery, while another third show only moderate gains and the remaining third have no long-lasting benefit.

Some investigators who doubt that enough tissue will be available from aborted fetuses to meet the needs of all potential patients are searching for alternative sources of tissue. (Cells from five to nine fetuses appear to be required to treat one Parkinson's patient.) Some researchers are using fetal pig brain cells for transplantation into patients—the first xenotransplantation experiment approved by the Food and Drug Administration (see p.6). Others are exploring using cells from the patient's own skin that have been genetically engineered to produce dopamine when multiplied in tissue culture and transplanted into the brain.

respond in order to maintain homeostasis—for example, these signals convey information about a fall in blood pressure. Other neuronal electrical signals convey messages to muscles and glands to stimulate appropriate responses to counteract these changes—for example, the nervous system through its electrical signals initiates adjustments in heart and blood vessel activity to restore blood pressure to normal when it starts to fall.

The specialization of muscle cells, contraction, also depends on these cells' ability to undergo action potentials. Action potentials trigger muscle contractions, many of which are important in maintaining homeostasis. For example, beating of the heart, mixing of ingested food with digestive enzymes, and shivering to generate heat when the body is cold are all accomplished by action potential–induced muscle contractions.

Chapter Summary

Electrical Signals: Graded Potentials and Action Potentials

Nerve and muscle cells are known as excitable tissues because they can rapidly alter their membrane permeabilities and thus undergo transient membrane potential changes when excited. There are two kinds of potential change: (1) graded potentials, which serve as short-distance signals that quickly die off within a close range of the small patch of membrane where they are first triggered, and (2) action potentials, the long-distance signals.

During an action potential, depolarization of the membrane to threshold potential triggers sequential changes in permeability caused by conformational changes in voltage-gated channels. These permeability changes bring about a brief reversal of membrane potential, with Na^+ influx being responsible for the rising phase (from −70 mV to +30 mV), followed by K^+ efflux during the falling phase (from peak back to resting potential). Before an action potential returns to resting, it regenerates an identical new action potential in the area next to it by means of current flow that brings the previously inactive area to threshold. This self-perpetuating cycle continues until the action potential has spread throughout the cell membrane in undiminished fashion. There are two types of action potential propagation: (1) conduction by local current flow in unmyelinated fibers, in which the action potential spreads along every portion of the membrane; and (2) the more rapid saltatory conduction in myelinated fibers, where the impulse jumps over the sections of the fiber covered with insulating myelin. The Na^+-K^+ pump gradually restores the ions that moved during propagation of the action potential to their original location to maintain the concentration gradients.

It is impossible to restimulate the portion of the membrane where the impulse has just passed until it has recovered from its refractory period. The refractory period ensures the unidirectional propagation of action potentials away from the original site of activation.

Action potentials occur either maximally in response to stimulation or not at all. Variable strengths of stimuli are coded by varying the frequency of action potentials, not their magnitude.

Synapses and Neuronal Integration

The primary means by which one neuron directly interacts with another neuron is through a synapse. An action potential in the presynaptic neuron triggers the release of a neurotransmitter, which combines with receptor sites on the postsynaptic neuron. This combination alters the postsynaptic cell in one of two ways. (1) The most typical response is the opening of chemical messenger–gated channels. If both Na^+ and K^+ channels are opened, the resultant ionic fluxes cause an EPSP, a small depolarization that brings the postsynaptic cell closer to threshold. On the other hand, the likelihood that the postsynaptic neuron will reach threshold is diminished when an IPSP, a small hyperpolarization, is produced as a result of the opening of either K^+ or Cl^- channels, or both. (2) In an alternate synaptic mechanism, an intracellular second messenger system, such as cyclic AMP, is activated by the neurotransmitter-receptor combination. Cyclic AMP can bring about channel opening or can have more prolonged effects within the cell, including alterations of the cell's genetic expression. Even though there are a number of different neurotransmitters, each synapse always releases the same transmitter to produce a given response when combined with a particular receptor. The response is terminated when the neurotransmitter is removed from the synaptic cleft by methods specific for the synapse.

Many neurons cosecrete larger, more slowly acting neuropeptides along with the classical neurotransmitter. The neuropeptides largely function as neuromodulators at nonsynaptic sites on either the presynaptic or the postsynaptic neuron to enhance or suppress synaptic effectiveness.

The interconnecting synaptic pathways between various neurons are incredibly complex due to convergence of neuronal input and divergence of its output. Usually, many presynaptic inputs converge upon a single neuron and jointly control its level of excitability. This same neuron, in turn, diverges to synapse with and influence the excitability of many other cells. Each neuron thus has the task of computing an output to numerous other cells from a complex set of inputs to itself. Depending on the combination of signals it is receiving from its various presynaptic inputs, at any given time a neuron may react by (1) firing action potentials along its axon, (2) remaining at rest and not passing any signals along, or (3) having its level of excitability reduced.

If the dominant activity is in its excitatory inputs, the postsynaptic cell is likely to be brought to threshold and have an action potential. This can be accomplished by either temporal summation (EPSPs from a single, repetitively firing presynaptic input occurring so close together in time that they add together) or spatial summation (adding of EPSPs occurring simultaneously from several different presynaptic inputs). Because the axon hillock has the lowest threshold, action potentials are initiated there. The frequency of action protentials reflects the magnitude of EPSP summation. If inhibitory inputs dominate, the postsynaptic potential will be brought farther than usual away from threshold. If excitatory and inhibitory activity to the postsynaptic neuron is balanced, the membrane will remain close to resting.

Numerous factors may alter synaptic effectiveness: some are built-in mechanisms to fine-tune neural responsiveness, some are deliberate pharmacological manipulations to achieve a desired result, and some are accidents caused by poisons or disease processes.

Review Exercises

Objective Questions (Answers on p. E–3.)

1. Conformational changes in channel proteins brought about by voltage changes are believed to be responsible for opening and closing Na^+ and K^+ gates during the generation of an action potential. (True or false?)
2. The Na^+-K^+ pump restores the membrane to resting potential after it reaches the peak of an action potential. (True or false?)
3. Following an action potential, there is more K^+ outside the cell than inside because of the K^+ efflux. (True or false?)
4. Postsynaptic neurons can either excite or inhibit presynaptic neurons. (True or false?)
5. The unidirectional propagation of action potentials away from the original site of activation is ensured by the ____________.
6. The ____________ is the site of action potential initiation in most neurons because it has the lowest threshold.
7. A junction in which electrical activity in one neuron influences the electrical activity in another neuron by means of a neurotransmitter is called a ____________.
8. Summing of EPSPs occurring very close together in time as a result of repetitive firing of a single presynaptic input is known as ____________.
9. Summing of EPSPs occurring simultaneously from several different presynaptic inputs is known as ____________.

10. The neuronal relationship where synapses from many presynaptic inputs act on a single postsynaptic cell is called ______________, whereas the relationship in which a single presynaptic neuron synapses with and thereby influences the activity of many postsynaptic cells is known as ______________.

11. Using the answer code below, indicate which potential is being described:

(a) graded potential
(b) action potential

____1. behaves in all-or-none fashion
____2. magnitude of the potential change varies with the magnitude of the triggering response
____3. decremental spread away from the original site
____4. spreads throughout the membrane in nondiminishing fashion
____5. serves as a long-distance signal
____6. serves as a short-distance signal

Essay Questions

1. What are the two types of excitable tissue?
2. Define the following terms: polarization, depolarization, hyperpolarization, repolarization, resting membrane potential, threshold potential, action potential, refractory period, all-or-none law.
3. Compare the three kinds of channels in terms of the factor that induces the change in channel conformation.
4. Describe the permeability changes and ionic fluxes that occur during an action potential.
5. Compare conduction by local current flow and saltatory conduction.
6. Compare the events that occur at excitatory and inhibitory synapses.
7. Distinguish between classical neurotransmitters and neuropeptides. Compare neurotransmitters and neuromodulators.
8. Discuss the possible outcomes of the grand postsynaptic potential brought about by interactions between EPSPs and IPSPs.
9. Distinguish between presynaptic inhibition and an inhibitory postsynaptic potential.

Quantitative Exercises (Solutions on p. E–3.) (See Appendix C, *Principles of Quantitative Reasoning*)

1. The following calculations give some insight into action potential conduction.
   a. How long would it take for an action potential to travel 0.6 m along the axon of an unmyelinated neuron of the digestive tract?
   b. How long would it take for an action potential to travel the same distance along the axon of a large myelinated neuron innervating a skeletal muscle?
   c. Suppose there were two synapses in a 0.6 m nerve tract and the delay at each synapse is 1 msec. How long would it take an action potential/chemical signal to travel the 0.6 m now, for both the myelinated and unmyelinated neurons?
   d. What if there were 5 synapses?
2. Suppose point A is 1 m from point B. Compare the following situations: i) A single axon spans the distance from A to B, and its conduction velocity is 60 m/sec. ii) three neurons span the distance from A to B, all three neurons have the same conduction velocity, and the synaptic delay at both synapses (draw a picture) is 1 msec. What are the conduction velocities of the three neurons in the second situation if the total conduction time in both cases is the same?
3. One can predict what the Na^+ current produced by the Na^+-K^+ pump is with the following equation[1]:

$$p = \frac{kT}{q}\left(\frac{G_{Na^+} G_{K^+}}{G_{Na^+} + G_{K^+}}\right)\log\frac{G_{K^+}[Na^+]_o}{G_{Na^+}[K^+]_i}$$

where p is the sodium pump current, G is the conductance of the membrane to the indicated ion, $[x]_o$ and $[x]_i$ are the concentrations of ion x outside and inside the cell respectively, k is Boltzmann's constant, T is the temperature in Kelvins, and q is the elementary charge constant. Suppose $kT/q = 25$ mV, $G_{Na^+} = 3.3\ \mu S/cm^2$, $G_{K^+} = 240\ \mu S/cm^2$, $[Na^+]_o = 145$ mM, and $[K^+]_i = 4$ mM. What is the pump current for sodium, in $\mu A/cm^2$?

[1]Hoppensteadt, F. C. and C. S. Peskin. *Mathematics in Medicine and the Life Sciences* (equation 7.4.35, p. 178). Springer-Verlag 1992.

Points to Ponder

(Explanations on p. E–4.)

1. Which of the following would occur if a neuron were experimentally stimulated simultaneously at both ends?
   a. The action potentials would pass in the middle and travel to the opposite ends.
   b. The action potentials would meet in the middle and then be propagated back to their starting positions.
   c. The action potentials would stop as they met in the middle.
   d. The stronger action potential would override the weaker action potential.
   e. Summation would occur when the action potentials met in the middle, resulting in a larger action potential.
2. Compare the expected changes in membrane potential of a neuron stimulated with a *subthreshold stimulus* (a stimulus not suffficient to bring a membrane to threshold), a *threshold stimulus* (a stimulus just sufficient to bring the membrane to threshold), and a *suprathreshold stimulus* (a stimulus larger than that necessary to bring the membrane to threshold).
3. Assume you touched a hot stove with your finger. Contraction of the biceps muscle causes flexion (bending) of the elbow, whereas contraction of the triceps muscle causes extension (straightening) of the elbow. What pattern of postsynaptic potentials (EPSPs and IPSPs) would you expect to be initiated reflexly in the cell bodies of the neurons controlling these muscles to pull your hand away from the painful stimulus?

Now assume your finger is being pricked to obtain a blood sample. The same *withdrawal reflex* would be initiated. What pattern of postsynaptic potentials would you voluntarily produce in the neurons controlling the biceps and triceps to keep your arm extended in spite of the painful stimulus?

4. *Schizophrenia* is believed to be caused by excessive dopamine activity in a particular region of the brain. Explain why symptoms of schizophrenia sometimes occur as a side effect in patients being treated for Parkinson's disease.
5. Assume presynaptic excitatory neuron A terminates on a postsynaptic cell near the axon hillock and presynaptic excitatory neuron B terminates on the same postsynaptic cell on a dendrite located on the side of the cell body opposite the axon hillock. Explain why rapid firing of presynaptic neuron A could bring the postsynaptic neuron to threshold through temporal summation, thus initiating an action potential, whereas firing of presynaptic neuron B at the same frequency and the same magnitude of EPSPs may not bring the postsynaptic neuron to threshold.
6. ***Clinical Consideration*** Becky N. was apprehensive as she sat in the dentist's chair awaiting the placement of her first silver amalgam (the "filling" in a cavity in a tooth). Before preparing the tooth for the amalgam by drilling away the decayed portion of the tooth, the dentist injected a local anesthetic in the nerve pathway supplying the region. As a result, Becky, much to her relief, did not feel any pain during the drilling and filling procedure. Local anesthetics block Na^+ channels. Explain how this action prevents the transmission of pain impulses to the brain.

5 The Central Nervous System

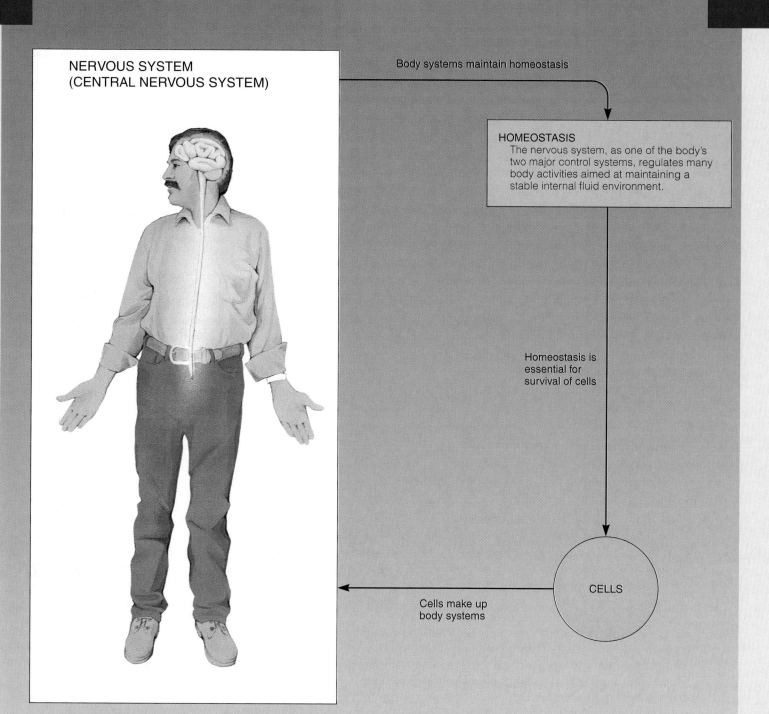

NERVOUS SYSTEM
(CENTRAL NERVOUS SYSTEM)

Body systems maintain homeostasis

HOMEOSTASIS
The nervous system, as one of the body's two major control systems, regulates many body activities aimed at maintaining a stable internal fluid environment.

Homeostasis is essential for survival of cells

CELLS

Cells make up body systems

The **nervous system** is one of the two major control systems of the body, the other being the endocrine system. A complex interactive network of three basic types of nerve cells—afferent neurons, efferent neurons, and interneurons—constitutes the nervous system. The **central nervous system (CNS)** is composed of the brain and spinal cord, which receive input about the external and internal environment from the afferent neurons. The CNS sorts and processes this input, then initiates appropriate directions in the efferent neurons, which carry the instructions to glands or muscles to bring about the desired response—some type of secretion or movement. Many of these neurally controlled activities are directed toward maintaining homeostasis. In general, the nervous system acts by means of its electrical signals (action potentials) to control the rapid responses of the body.

Chapter Contents At a Glance

Introduction

The way humans act and react depends on complex, organized, discrete neuronal processing. Many of the basic life-supporting neuronal patterns, such as those controlling respiration and circulation, are similar in all individuals. However, there must be subtle differences in neuronal integration between someone who is a talented composer and someone who cannot carry a tune, or between someone who is a math wizard and someone who struggles with long division. Some differences in the nervous systems of individuals are genetically endowed. The rest, however, are due to environmental encounters and experiences. When the immature nervous system develops according to its genetic plan, an overabundance of neurons and synapses is formed. Depending on external stimuli and the extent these pathways are used, some are retained, firmly established, and even enhanced, whereas others are eliminated. A case in point is **amblyopia (lazy eye)**, in which the weaker of the two eyes is not used for vision. A lazy eye that does not get appropriate visual stimulation during a critical developmental period will almost completely and permanently lose the power of vision. The blind eye itself is completely normal; the defect lies in the lost neuronal connections in the brain's visual pathways. If, however, the weak eye is forced to work by covering the stronger eye with a patch during the sensitive developmental period, its vision will be retained. The maturation of the nervous system truly does involve instances of "use it or lose it." Once the nervous system has matured, ongoing modifications still occur as we continue to learn from our unique set of experiences. For example, the act of reading this page is somehow altering the neuronal activity of your brain as you (it is hoped) tuck the information away in your memory.

Comparison of the Nervous and Endocrine Systems

The nervous system is a "wired" system, and the endocrine system is a "wireless" system, even though both systems influence their target cells by means of chemical messengers.

The nervous and endocrine systems are the two main control systems of the body. The **nervous system,** through its swift transmission of electrical impulses, generally coordinates the rapid activities of the body, such as muscle movements. It is especially important in the body's interactions with the external environment. The **endocrine system,** which secretes *hormones* into the blood for delivery to distant sites of action, primarily controls metabolic and other activities that require duration rather than speed, such as maintenance of blood glucose levels. Although these two systems differ in many respects, they have much in common (▮ Table 5-1). They both ultimately alter their **target cells** (their sites of action) by releasing chemical messengers (neurotransmitters in the case of nerve cells, hormones in the case of endocrine cells), which interact in particular ways with specific receptors (particular plasma membrane proteins) of the target cells.

Anatomically, the nervous and endocrine systems are quite different. In the nervous system, each nerve cell terminates directly on its specific target cells; that is, the nervous system is "wired" in a very specific way into highly organized, distinct anatomical pathways for transmission of signals from one part of the body to another. Information is carried along chains of neurons to the desired destination through action potential propagation coupled with synaptic transmission (chapter 4). In contrast, the endocrine system is a "wireless" system in that the endocrine glands are not anatomically linked with their target cells. Instead, the endocrine chemical messengers are secreted into the blood and delivered to distant target sites. In fact, the components of the endocrine system itself are not anatomically interconnected; the endocrine glands are scattered throughout the body (see Fig. 18-1, p. 621). These glands constitute a system in a functional sense, however, because they all secrete hormones and many interactions take place between various endocrine glands.

As a result of these anatomical differences, the two systems accomplish specificity of action by distinctly different means. Specificity of neural communication depends on nerve cells having a close anatomical relationship with their target cells, so that each neuron has a very narrow range of influence. A neurotransmitter is released for restricted distribution only to specific adjacent target cells, then is swiftly inactivated by enzymes at the nerve–target cell juncture or is taken back up by the nerve terminal before it is able to gain access to the blood. The target cells for a particular neuron have receptors for the neurotransmitter, but so do many other cells in other locations, and they could respond to this same mediator if it were delivered to them. For example, the entire system of nerve cells supplying all of your body's skeletal muscles (motor neurons) use the same neurotransmitter, *acetylcholine (ACh),* and all of your skeletal muscles bear complementary ACh receptors (chapter 7). Yet specific muscles are able to be discretely activated to contract because of precise structural arrangements, or wiring patterns, between motor neurons and muscle cells. You are able to specifically wiggle your big toe without influencing any of your other muscles because ACh can be discretely released from the motor neurons that are specifically wired to the muscles controlling your toe. If ACh were indiscriminately released into the blood, as are the hormones of the wireless endocrine system, all of the skeletal muscles would simultaneously respond by contracting because they all have identical receptors for ACh. This does not happen, of course, due to the direct lines of communication between neurons and their target cells.

This specificity is in sharp contrast to the way specificity of communication is built into the endocrine system. Because hormones travel in the blood, they are able to reach virtually all tissues. Yet despite this ubiquitous distribution, only specific target cells are able to respond to each hormone. Specificity of hormonal action depends on specialization of target cell receptors. For a hormone to exert its effect, the essential first step is the binding of the hormone with receptors specific for it that are located only on or in the hormone's target cells. Target cell receptors are highly discerning in their binding function. They will recognize and bind only a certain hormone, even though they are exposed simultaneously to many other blood-borne hormones, some of which are structurally very similar to the one that they discriminately bind. A receptor recognizes a specific hormone because the conformation of a portion of the receptor molecule matches a unique portion of its binding hormone in "lock-and-key" fashion. Binding of a hormone with target cell receptors initiates a reaction (or series of reactions) that culminates in the hormone's final effect. The hormone cannot influence any other cells because they lack the right binding receptors. This lock-and-key type of interaction between a hormone and its target cells ensures that the regulation of a particular tissue will be governed by the appropriate hormone.

Note that in contrast to neural communication, specificity of endocrine communication is built into the receiving end, that is, at the target cells. At the transmitting end of endocrine control systems, the original hormonal messenger is widely dispersed instead of being directed specifically toward the target cells. Radio transmission and reception provide an analogy. Radio waves (a specific hormone) are nondiscriminately dispersed in all directions from their point of origin (an endocrine gland). Even though radio waves are present in the air surrounding all of your appliances (the hormone circulating throughout the body in the blood), only your radio (the target cell) is able to respond to them because it alone has receiving components specific for radio waves (receptors). Thus, even though your radio and coffee pot are being equally

Table 5-1 Comparison of the Nervous System and the Endocrine System

Property	*Nervous System*	*Endocrine System*
Anatomical arrangement	A "wired" system; specific structural arrangement between neurons and their target cells; structural continuity in the system	A "wireless" system; endocrine glands widely dispersed and not structurally related to one another or to their target cells
Type of chemical messenger	Neurotransmitters released into synaptic cleft	Hormones released into blood
Distance of action of chemical messenger	Very short distance (diffuses across synaptic cleft)	Long distance (carried by blood)
Means of specificity of action on target cell	Dependent on close anatomical relationship between nerve cells and their target cells	Dependent on specificity of target cell binding and responsiveness to a particular hormone
Speed of response	Rapid (milliseconds)	Slow (minutes to hours)
Duration of action	Brief (milliseconds)	Long (minutes to days or longer)
Major functions	Coordinates rapid, precise responses	Controls activities that require long duration rather than speed
Influence on other major control system?	Yes	Yes

bombarded by radio waves, only your radio can respond. Similarly, even though all tissues are equally exposed to a given hormone through the blood, only the target cells with the appropriate receptors for that hormone are able to respond to it.

Even though the nervous and endocrine systems have their own realms of authority, they are functionally interconnected.

The nervous and endocrine systems are specialized for controlling different types of activities. In general, the nervous system is responsible for coordinating rapid, precise responses. Neural signals in the form of action potentials are rapidly propagated along nerve cell fibers, resulting in the release at the nerve terminal of a neurotransmitter that has to diffuse only a microscopic distance to its target cell before a response is effected. A neurally mediated response is not only rapid but brief; the action is quickly brought to a halt as the neurotransmitter is swiftly removed from the target site. This permits either termination of the response, almost immediate repetition of the response, or rapid initiation of an alternate response, depending on the circumstances (for example, the swift changes in commands to muscle groups needed to coordinate walking). This mode of action makes neural communication extremely rapid and precise. The target tissues of the nervous system are the muscles and glands, especially exocrine glands, of the body.

The endocrine system, in contrast, is specialized to control activities that require duration rather than speed, such as regulating organic metabolism and H_2O and electrolyte balance, promoting smooth, sequential growth and development, controlling reproduction, and regulating red blood cell production. The endocrine system responds more slowly to its triggering stimuli than the nervous system does for several reasons. First, the endocrine system must depend on blood flow to convey its hormonal messengers over long distances. Second, hormones' mechanism of action at their target cells is more complex than that of neurotransmitters and thus requires more time before a response occurs. The ultimate effect of some hormones cannot be detected until a few hours after they bind with target cell receptors. Also, due to the receptors' high affinity for their respective hormone, hormones often remain bound to receptors for some time, thus prolonging their biological effectiveness. Furthermore, unlike the brief, neurally induced responses that come to a halt almost immediately after the neurotransmitter is removed, endocrine effects usually last for some time after the hormone's withdrawal. Neural responses to a single burst of neurotransmitter release usually last only milliseconds to seconds, whereas the alterations in target cells induced by hormones range from minutes to days or, in the case of growth-promoting effects, even a lifetime. Thus, hormonal action is relatively slow and prolonged, making endocrine control particularly suitable for the regulation of metabolic activities that require long-term stability.

Although the endocrine and nervous systems have their own areas of specialization, they are intimately interconnected functionally. Some nerve cells do not release neurotransmitters at synapses but instead terminate at blood vessels and release their chemical messengers (neurohormones, see p. 66) into the blood, where these chemicals act as hormones. A given messenger may even be a neurotransmitter when released from a nerve ending and a hormone when secreted by an endocrine cell. The nervous system directly or indirectly controls the secretion of many hormones. At the same time, many hormones act as neuromodulators (see p. 101), altering synaptic effectiveness and thereby influencing the excitability of the nervous system. The presence of certain key hormones is even essential for the proper development and maturation of the brain during fetal life. Furthermore, in many instances the nervous and endocrine systems both influence the same target cells in supplementary fashion. For example, these two major control systems both contribute to the regulation of the circulatory and digestive systems. Thus, many important regulatory interfaces exist between the nervous and endocrine systems; the study of these relationships is known as **neuroendocrinology.**

For now we will concentrate on the nervous system and will examine the endocrine system in more detail in later chapters. Throughout the text we will continue to point out the numerous ways these two control systems interact, so that the body is a coordinated whole, even though each system has its own realm of authority.

Organization of the Nervous System

The nervous system is organized into the central nervous system and the peripheral nervous system.

The nervous system is organized into the **central nervous system (CNS),** consisting of the **brain** and **spinal cord,** and the **peripheral nervous system (PNS),** consisting of nerve fibers that carry information between the CNS and other parts of the body (the periphery) (■ Fig. 5-1). The PNS is further subdivided into afferent and efferent divisions. The **afferent division** (*a* is from *ad,* meaning "toward," as in advance; *ferent* means "carrying"; thus, *afferent* means "carrying toward") carries information *to* the CNS, apprising it of the external environment and providing status reports on internal activities being regulated by the nervous system. Instructions *from* the CNS are transmitted via the **efferent division** (*e* is from *ex,* meaning "from," as in exit; thus, *efferent* means "carrying from") to **effector organs**—the muscles or glands that carry out the orders to bring about the desired effect. The efferent nervous system is divided into the **somatic nervous system,** which consists of the fibers of the **motor neurons** that supply the skeletal muscles, and the **autonomic nervous system** fibers, which innervate smooth muscle, cardiac muscle, and glands. The latter system is further subdivided into the **sympathetic nervous system** and the **parasympathetic nervous system,** both of which innervate most of the organs supplied by the autonomic system.

Brain
Central nervous system (CNS)
Spinal cord
(Input to CNS from periphery)
(Output from CNS to periphery
Afferent division
Peripheral nervous system (PNS)
Efferent division
Sensory stimuli
Visceral stimuli
Somatic nervous system
Autonomic nervous system
Motor neurons
Sympathetic nervous system
Parasympathetic nervous system
Skeletal muscle
Smooth muscle
Cardiac muscle
Glands
Effector organs (made up of muscle and gland tissue)

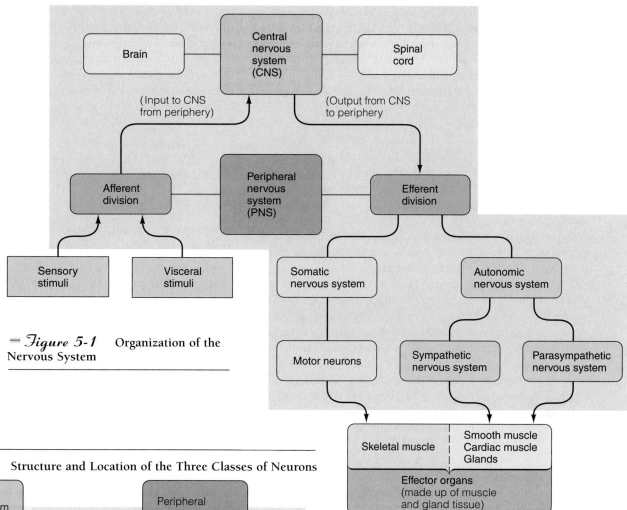

Figure 5-1 Organization of the Nervous System

Figure 5-2 Structure and Location of the Three Classes of Neurons

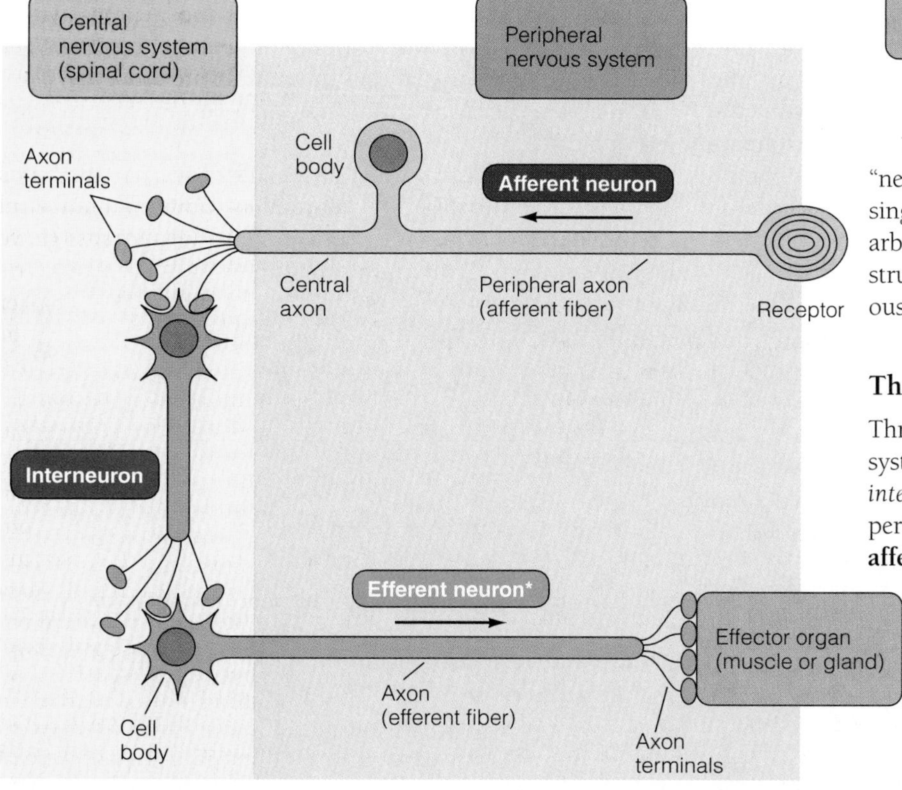

* Efferent autonomic nerve pathways consist of a two-neuron chain between the CNS and the effector organ.

It is important to recognize that all of these "nervous systems" are really subdivisions of a single, integrated nervous system. They are arbitrary divisions based on differences in the structure, location, and functions of the various diverse parts of the whole nervous system.

There are three classes of neurons.

Three classes of neurons make up the nervous system: *afferent neurons, efferent neurons,* and *interneurons.* The afferent division of the peripheral nervous system is composed of **afferent neurons,** which are shaped differently than efferent neurons and interneurons (Fig. 5-2). At its peripheral ending, an afferent neuron has a **sensory receptor** that generates action potentials in response to a particular type of stimulus. (This stimulus-sensitive afferent neuronal receptor should not be confused with the special protein receptors that bind chemical messengers and are found in the plasma membrane of all cells.) The afferent neuron cell body, which is devoid of dendrites and presynaptic inputs, is located adjacent to the spinal cord. A long *peripheral axon,* commonly called the *afferent fiber,* extends from the receptor to the cell body, and a short *central axon* passes from the cell body into the spinal cord. Action potentials are initiated at the receptor end of the peripheral axon in response to a stimulus and are propagated along the peripheral axon and central axon toward the spinal cord. The terminals of the central axon diverge and synapse with other neurons within the spinal cord, thus disseminating information about the stimulus. Afferent neurons lie primarily within the peripheral nervous system. Only a small portion of their central axon endings project into the spinal cord to relay peripheral signals.

Efferent neurons also lie primarily in the peripheral nervous system (Fig. 5-2). The cell bodies of efferent neurons originate in the CNS, where many centrally located presynaptic inputs converge on them to influence their outputs to the effector organs. Efferent axons *(efferent fibers)* leave the CNS to course their way to the muscles or glands they innervate, conveying their integrated output for the effector organs to put into effect. (An autonomic nerve pathway actually consists of a two-neuron chain between the CNS and the effector organ.)

Interneurons lie entirely within the CNS. About 99% of all neurons belong to this category. They serve two main roles. First, as their name implies, they lie between (*inter* means "between") the afferent and efferent neurons and are important in the integration of peripheral responses to peripheral information. For example, on receiving information through afferent neurons that you are touching a hot object, appropriate interneurons signal efferent neurons that transmit to your hand and arm muscles the message, "Pull the hand away from the hot object!" The more complex the required action, the greater the number of interneurons interposed between the afferent message and efferent response. Second, interconnections between interneurons themselves are responsible for the abstract phenomena associated with the "mind," such as thoughts, emotions, memory, creativity, intellect, and motivation. These activities are the least understood functions of the nervous system. (See the boxed feature on p. 116, ◆ Concepts, Challenges, and Controversies.)

With this brief introduction to the types of neurons and their location in the various divisions of the nervous system, we will now turn our attention to the central nervous system, followed in the next two chapters by a discussion of the two divisions of the peripheral nervous system.

Protection and Nourishment of the Brain

Neuroglia physically support the interneurons and help sustain them metabolically.

About 90% of the cells within the CNS are not neurons but **glial cells** or **neuroglia.** Despite their large numbers, the glial cells occupy only about half the volume of the brain because they do not branch as extensively as the neurons do. Unlike neurons, glial cells do not initiate or conduct nerve impulses. They are important in the viability of the CNS, however. For much of the time since their discovery in the nineteenth century, the glial cells were thought to be passive "mortar" that physically supported the functionally important neurons. In the last decade, however, the varied and important roles of these dynamic cells have become apparent. The glial cells serve as the connective tissue of the CNS and as such help support the neurons both physically and metabolically. They homeostatically maintain the composition of the specialized extracellular environment surrounding the neurons within the narrow limits optimal for normal neuronal function. The four major types of glial cells in the CNS are *astrocytes, oligodendrocytes, ependymal cells,* and *microglia* (Fig. 5-3 and Table 5-2).

Named for their starlike shape (*astro* means "star," *cyte* means "cell") (Fig. 5-4), **astrocytes** provide a number of critical functions. First, as the main "glue" (*glia* means "glue")

Table 5-2 Functions of Glial Cells

Type of Glial Cell	*Functions*
Astrocytes	Physically support neurons in proper spatial relationships
	Serve as scaffold during fetal brain development
	Induce formation of blood-brain barrier
	Form neural scar tissue
	Take up and degrade released neurotransmitters into raw materials for synthesis of more neurotransmitters by neurons
	Take up excess K^+ to help maintain proper brain ECF ion concentration and normal neural excitability
	Possess receptors for neurotransmitters, which may be important in a chemical signaling system
Oligodendrocytes	Form myelin sheaths in CNS
Ependymal cells	Line internal cavities of brain and spinal cord
	Contribute to formation of cerebrospinal fluid
Microglia	Play a role in defense of brain as phagocytic scavengers

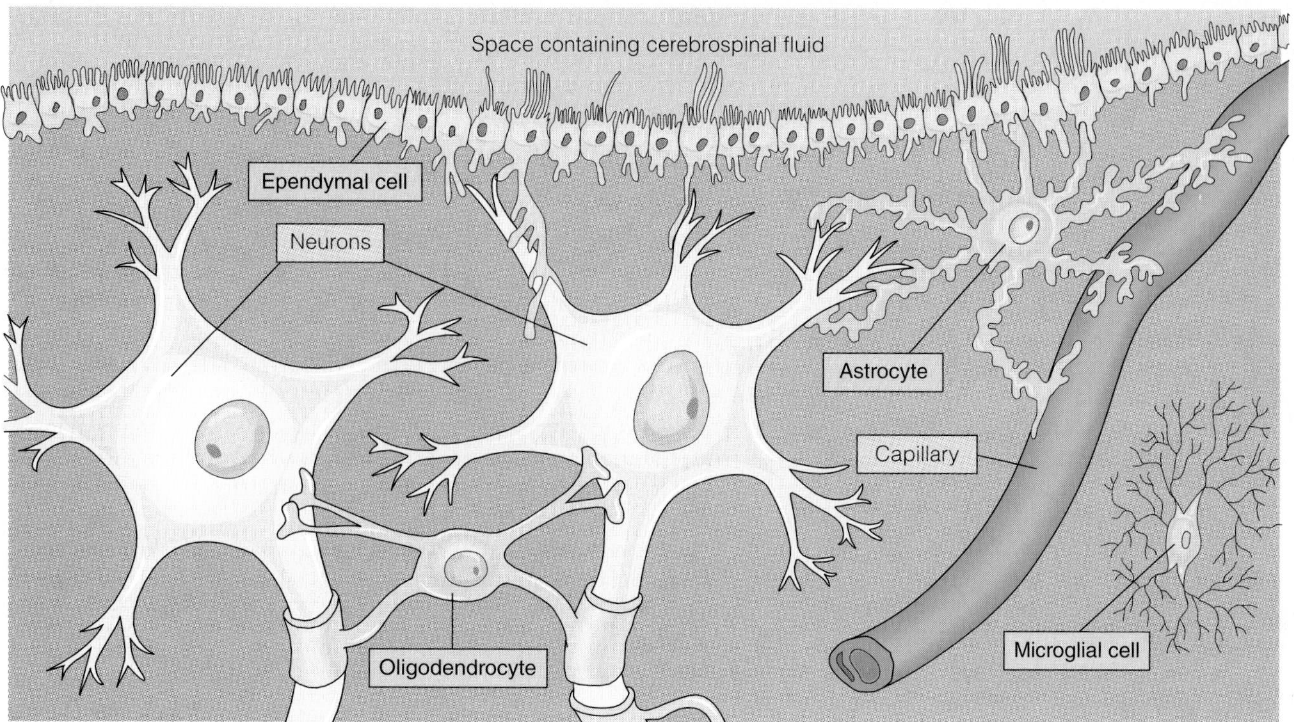

Figure 5-3 **Glial Cells of the Central Nervous System** Astrocytes, oligodendrocytes, microglia, and ependymal cells protect, nourish, and otherwise support neurons in the brain and spinal cord.

of the CNS, they hold the neurons together in proper spatial relationships. Second, astrocytes serve as a scaffold to guide neurons to their proper final destination during fetal brain development. Third, astrocytes induce the small blood vessels of the brain to undergo the anatomical and functional changes that are responsible for the establishment of the blood-brain barrier, a highly selective barricade between the blood and brain that will soon be described in greater detail. Fourth, astrocytes are important in the repair of brain injuries and in neural scar formation. Fifth, they play a role in neurotransmitter activity. Astrocytes take up glutamate and gamma-aminobutyric acid (GABA), excitatory and inhibitory neurotransmitters, respectively, thus bringing the actions of these chemical messengers to a halt. Furthermore, they degrade the sopped-up chemical messengers into raw materials for the neurons to use in making more of these neurotransmitters. Sixth, the astrocytes take up excess K^+ from the brain ECF when high action potential activity outpaces the ability of the Na^+-K^+ pump to return the effluxed K^+ to the neurons. (Recall that K^+ leaves a neuron during the falling phase of an action potential; see p. 87.) By taking up excess K^+, the astrocytes help maintain the proper brain ECF ion concentration to sustain normal neural excitability. If brain ECF K^+ levels were allowed to rise, the resultant lower K^+ concentration gradient between the neuronal ICF and surrounding ECF would reduce the neuronal membrane closer to threshold, even at rest. This would increase the excitability of the brain. In fact, an elevation in brain ECF K^+ concentration may be one of the factors responsible for the brain cells' explosive convulsive discharge that occurs during epileptic seizures. Finally, astrocytes possess receptors for the same neurotransmitters that neurons do. Researchers have shown that binding of the common neurotransmitter glutamate to astrocyte receptors causes these cells to release stored calcium ions. The purpose of this calcium unleashing is unclear, but it may rep-

Figure 5-4 **Astrocytes** Note the starlike shape of these astrocytes that have been grown in tissue culture.

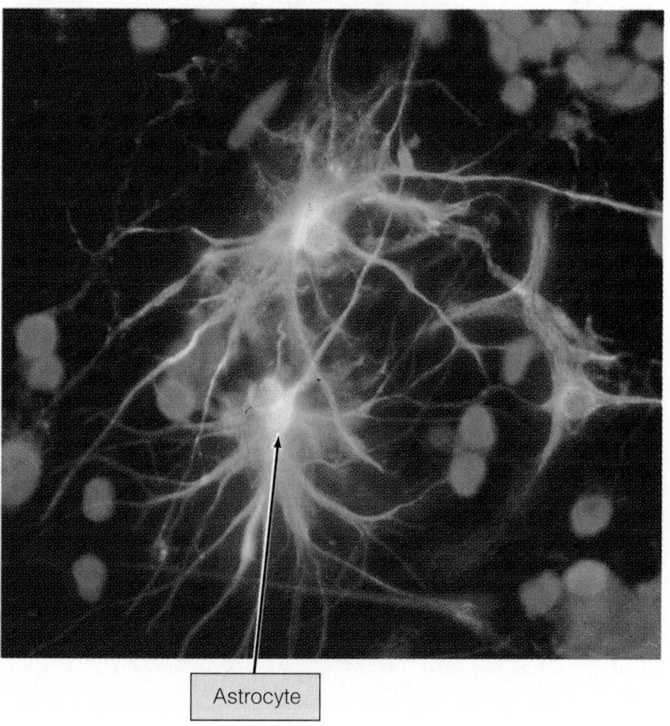

The Human Brain: In Search of Self-Understanding

The responsibilities and mechanisms of function of much of the human brain are still very poorly understood. Part of our lack of understanding derives from the complexity of the brain. It is the most complicated, mysterious, awesome organ on earth. Considering that myriad interconnections are possible among the estimated 100 billion neurons in the brain, it is a wonder that scientists have been able to unravel even the bits of information that they have.

Furthermore, unlike other organs of the body, the human brain is so unique that no experimental animal models are available for studying its most complex and sophisticated functions, such as language and creativity. In contrast, because human hearts are essentially the same as dog hearts, investigations probing canine heart function provide valuable insights into human heart function. Many aspects of brain function involve subjective feelings that cannot be measured or even communicated by nonverbal animals. We do not know to what extent, if any, animals experience what we know of as happiness, sadness, love, fear, jealousy, and so on. We can only observe in them behaviors that we associate with certain emotional states (for example, a dog wagging its tail when it is "happy").

In the past, knowledge about the brain was gleaned from (1) stimulation during brain surgery; (2) analysis of changes in electrical activity reaching the surface of the skull when a person is engaged in various activities; (3) observation of deficits in function associated with diseases or destruction of particular areas of the brain; (4) detailed anatomical studies in cadavers; and (5) inferences from animal studies. The latter, of course, are limited to characteristics shared by these species and humans and therefore leave most of the very characteristics that confer upon us our uniqueness as humans largely unexplored.

Now, however, several exciting new noninvasive techniques of imaging the living human brain, as well as new methods of probing the cellular and molecular functions of neurons, are contributing pieces of knowledge to our efforts to bridge the gap between our understanding of the biology of nerve cells and the behavior and intellect of the human brain. In fact, it is the human brain itself that is developing techniques for unraveling the secrets of its own complex nature.

These new imaging techniques, which are also becoming useful diagnostic tools, include **computerized axial tomography (CAT)**, **magnetic resonance imaging (MRI)**, and **positron emission tomography (PET)**. All three of these techniques depend on a computerized reconstruction of an image built from a series of "snapshots" that reveal specific differences in the imaged tissue. In CAT scans, images are reconstructed from differences in X-ray absorption by various areas in the brain or other body parts being examined. With MRI scans, an externally applied magnetic field perturbs hydrogen ions (protons), which behave like little compass needles. The magnetic perturbation causes the hydrogen ions to transiently emit their own detectable signal. Differences in signal patterns in various areas of the brain are reconstructed into a computerized image. CAT and standard MRI scans are especially useful in detecting physical (structural) abnormalities, such as tumors and intracranial hemorrhages. The functional MRI, a recent modification, detects functionally induced changes in regional cerebral blood flow and oxygen use by taking advantage of the fact that the magnetic properties of hemoglobin, the O_2-carrying molecule in the blood, is affected by the amount of O_2 it is carrying. Similarly, PET scans are also useful in detecting blood flow in the brain. PET scans depend on the injection of short-lived radioactively labeled water. As this tracer circulates in the blood, it emits positrons, which collide with electrons and produce a tiny burst of gamma-ray energy. The PET equipment detects this gamma radiation. Increased gamma activity means that more blood is flowing through that area. A computer then reconstructs an image of blood flow in the brain from a rapid series of PET snapshots. Because more blood flows into a particular region of the brain when it is more active, neuroscientists are able to "take pictures" of the brain at work. For example, as a person performs different tasks, different areas of the brain "light up" on PET scans, indicative of increased activity in these areas (see the accompanying figure).

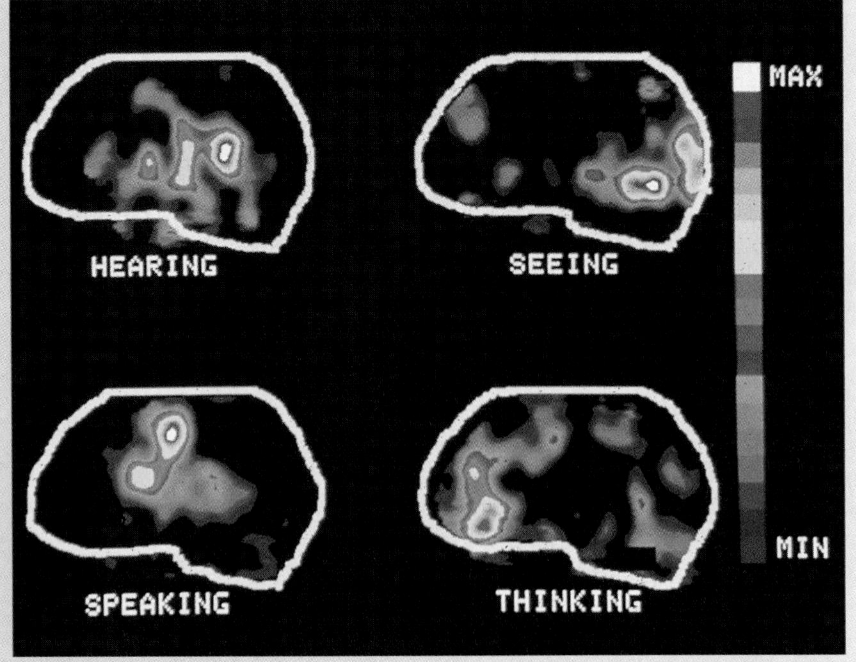

resent a type of chemical signaling system between cells in the brain, that is, a way in which these glial cells "talk" to each other or to neurons.

Oligodendrocytes form the insulative myelin sheaths around axons in the CNS. An oligodendrocyte has several elongated projections, each of which is wrapped jelly-roll fashion around a section of an interneuronal axon to form a patch of myelin (see Fig. 4-12b, p. 91 and Fig. 5-3).

Ependymal cells line the internal cavities of the CNS. As the nervous system develops embryonically from a hollow neural tube, the original central cavity of this tube is maintained and modified to form the **ventricles** of the brain and the **central canal** of the spinal cord. The ependymal cells lining the ventricles contribute to the formation of cerebrospinal fluid, a topic to be discussed shortly.

Microglia are the immune defense cells of the CNS. These scavengers are "cousins" of monocytes, a type of white blood cell that exits the blood and sets up residence as front-line defense agents in various tissues throughout the body. Microglia are derived from the same tissue that gives rise to monocytes, and during embryonic development, they migrate to the CNS. There they remain stationary until activated by an infection or injury. In the resting state, microglia are wispy cells with many long branches that radiate outward. Recent evidence suggests that resting microglia are not just waiting watchfully. They are thought to release low levels of growth factors, such as *nerve growth factor,* which help neurons and other glial cells survive and thrive. When trouble occurs in the CNS, microglia retract their branches, round up, and become highly mobile, moving toward the affected area to remove any foreign invaders or tissue debris. Activated microglia release destructive chemicals for assault against their target. Researchers are increasingly suspicious that excessive release of these chemicals from overzealous microglia may damage the neurons they are meant to protect, thus contributing to the neuronal damage seen in stroke, Alzheimer's disease, multiple sclerosis, the dementia (mental failing) of AIDS, and other neurodegenerative diseases.

Unlike neurons, glial cells do not lose the ability to undergo cell division, so most brain tumors of neural origin consist of glial cells **(gliomas).** Neurons themselves do not form tumors because they are unable to divide and multiply. Brain tumors of nonneural origin are of two types: (1) those that metastasize (spread) to the brain from other sites and (2) **meningiomas,** which originate from the meninges, the protective membranes covering the central nervous system.

The delicate central nervous tissue is well protected.

Central nervous tissue is very delicate. This characteristic, coupled with the fact that damaged nerve cells cannot be replaced because neurons are unable to divide, makes it imperative that this fragile, irreplaceable tissue be well protected. Four major features help protect the CNS from injury:

1. It is enclosed by hard, bony structures. The **cranium (skull)** encases the brain, and the **vertebral column** surrounds the spinal cord.
2. Three protective and nourishing membranes, the **meninges,** lie between the bony covering and the nervous tissue.
3. The brain "floats" in a special cushioning fluid, the **cerebrospinal fluid (CSF).**
4. A highly selective **blood-brain barrier** limits access of blood-borne materials into the vulnerable brain tissue.

The role of the first of these protective devices, the bony covering, is self-evident. The latter three protective mechanisms warrant further discussion.

Three meningeal "mothers" provide protection and nourishment for the central nervous system.

The meninges, the three membranes that wrap the central nervous system are, from the outermost to the innermost layer, the dura mater, the arachnoid mater, and the pia mater (■ Fig. 5-5). (*Mater* means "mother," indicative of the protective and supportive role played by these membranes.)

The **dura mater** (*dura* means "tough") is a tough, inelastic covering consisting of two layers. Usually, these layers adhere closely, but in some regions they are separated to form blood-filled cavities, **dural sinuses,** or in the case of the larger cavities, **venous sinuses.** Venous blood draining from the brain empties into these sinuses to be returned to the heart. Cerebrospinal fluid also reenters the blood at these sinus sites.

The **arachnoid mater** (*arachnoid* means "spiderlike") is a delicate, richly vascularized layer with a "cobwebby" appearance. The space between the arachnoid layer and the underlying pia mater, the **subarachnoid space,** is filled with CSF. Protrusions of arachnoid tissue, the **arachnoid villi,** penetrate through gaps in the overlying dura and project into the dural sinuses. It is across the surfaces of these villi that CSF is reabsorbed into the blood circulating within the sinuses.

The innermost meningeal layer, the **pia mater** (*pia* means "gentle"), is the most fragile. It is highly vascular and closely adheres to the surfaces of the brain and spinal cord, following every ridge and valley. In certain areas it dips deeply into the brain to bring a rich blood supply into close contact with the ependymal cells lining the ventricles. This relationship is important in the formation of CSF, a topic to which we now turn our attention.

The central nervous system is suspended in its own special cerebrospinal fluid.

Cerebrospinal fluid is formed primarily by the **choroid plexuses** found in particular regions of the ventricle cavities of the brain. Choroid plexuses consist of richly vascularized, cauliflower-like masses of pia mater tissue that dip into pockets formed by ependymal cells. Once CSF is formed, it flows through the four interconnected ventricles within the interior of the brain and through the spinal cord's narrow central canal, which is continuous with the last ventricle. Cerebrospinal fluid escapes through small openings from the fourth ventricle at the base of the brain to enter the subarachnoid space and subsequently flows between the meningeal layers over the entire surface of the brain and

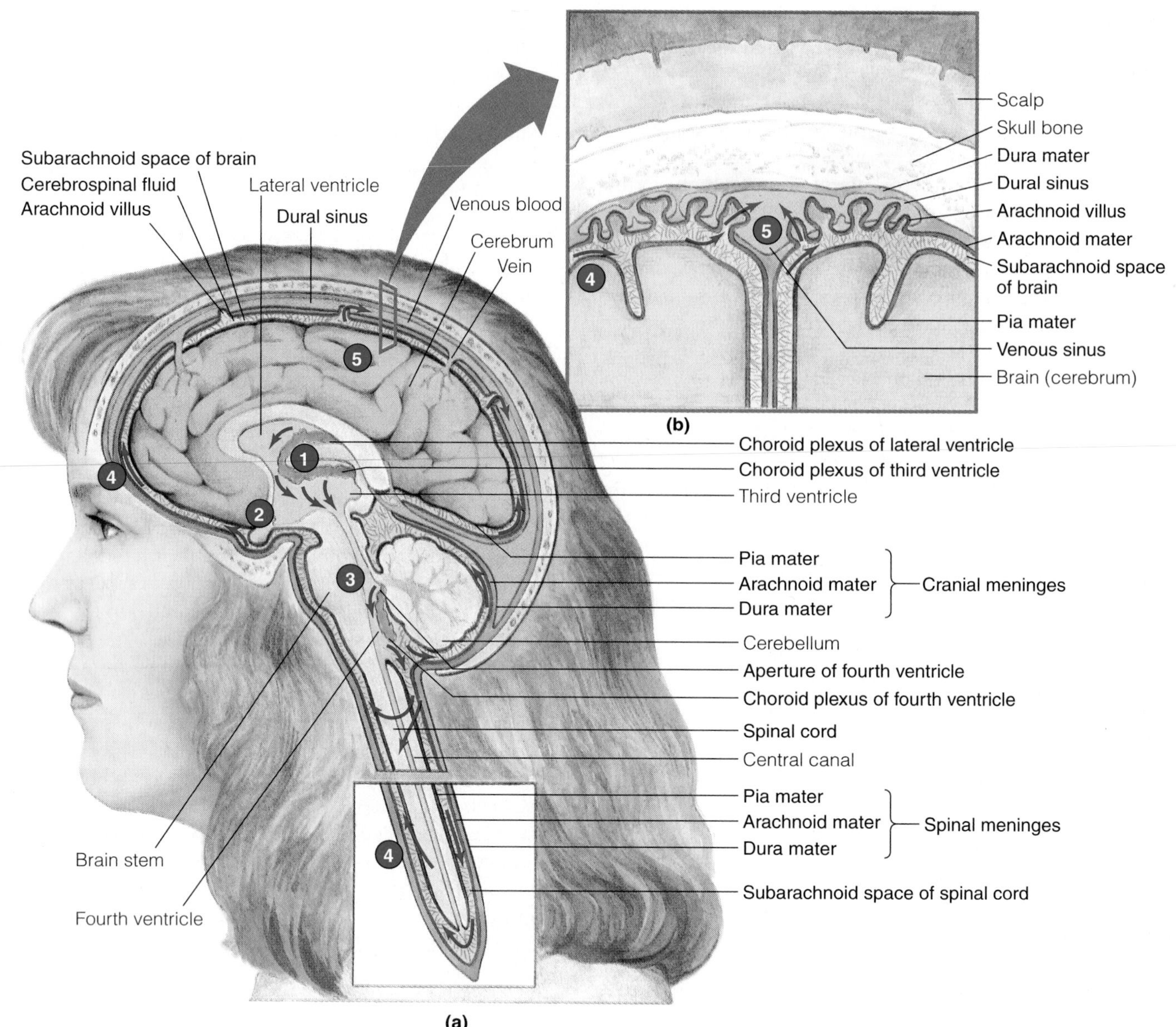

Figure 5-5 **Relationship of the Meninges and Cerebrospinal Fluid to the Brain and Spinal Cord** (a) Brain, spinal cord, and meninges in sagittal section. The arrows depict the direction of flow of cerebrospinal fluid (in yellow), which is produced by the choroid plexuses ①, circulates throughout the ventricles ②, exits the fourth ventricle at the base of the brain ③, flows in the subarachnoid space between the meningeal layers ④, and is finally reabsorbed from the subarachnoid space into the venous blood across the arachnoid villi ⑤ (b) Frontal section in the region between the two cerebral hemispheres of the brain, depicting the meninges in greater detail.

spinal cord (Fig. 5-5). When the CSF reaches the upper regions of the brain, it is reabsorbed from the subarachnoid space into the venous blood through the arachnoid villi.

Flow of CSF through this system is facilitated by circulatory and postural factors that result in a CSF pressure of about 10 mm Hg. Reduction of this pressure by removal of even a few milliliters (ml) of CSF during a spinal tap for laboratory analysis may produce severe headaches.

Through the ongoing processes of formation, circulation, and reabsorption, the entire volume of CSF of about 125 to 150 ml is replaced more than three times a day. **Hydrocephalus** ("water on the brain") occurs if any one of these processes is defective so that excess CSF accumulates. The resulting increase in CSF pressure can lead to brain damage and mental retardation if untreated. Treatment consists of surgically shunting the excess CSF to veins elsewhere in the body.

The CSF has about the same density as the brain itself, so the brain is essentially suspended in its special fluid environment. The major function of CSF is to serve as a shock-absorbing (cushioning) fluid to prevent the brain from bumping against the interior of the hard skull when the head is subjected to sudden, jarring movements.

In addition to protecting the delicate brain from mechanical trauma, the CSF performs an important role related to the exchange of materials between the body fluids and the brain. In all body tissues, the spaces between the cells are filled with a type of extracellular fluid known as interstitial fluid. It is the interstitial fluid of the brain—and not the blood plasma or the

CSF—that comes into direct contact with the neuronal and glial cells (▬ Fig. 5-6). Since the brain interstitial fluid directly bathes the neural cells, its composition is critical. The composition of brain interstitial fluid is influenced more by changes in the composition of the CSF than by alterations in blood plasma because there is fairly free exchange of materials between the CSF and brain interstitial fluid but only limited exchange between the blood and the brain interstitial fluid. Accordingly, it is essential that the composition of the CSF be regulated within very narrow limits.

Cerebrospinal fluid is formed as a result of selective transport mechanisms across the membranes of the choroid plexuses. The composition of CSF differs from that of plasma. For example, CSF is lower in K^+ and higher in Na^+, making it an ideal environment for the movement of these ions down concentration gradients, a process essential for conduction of nerve impulses (see pp. 87–92).

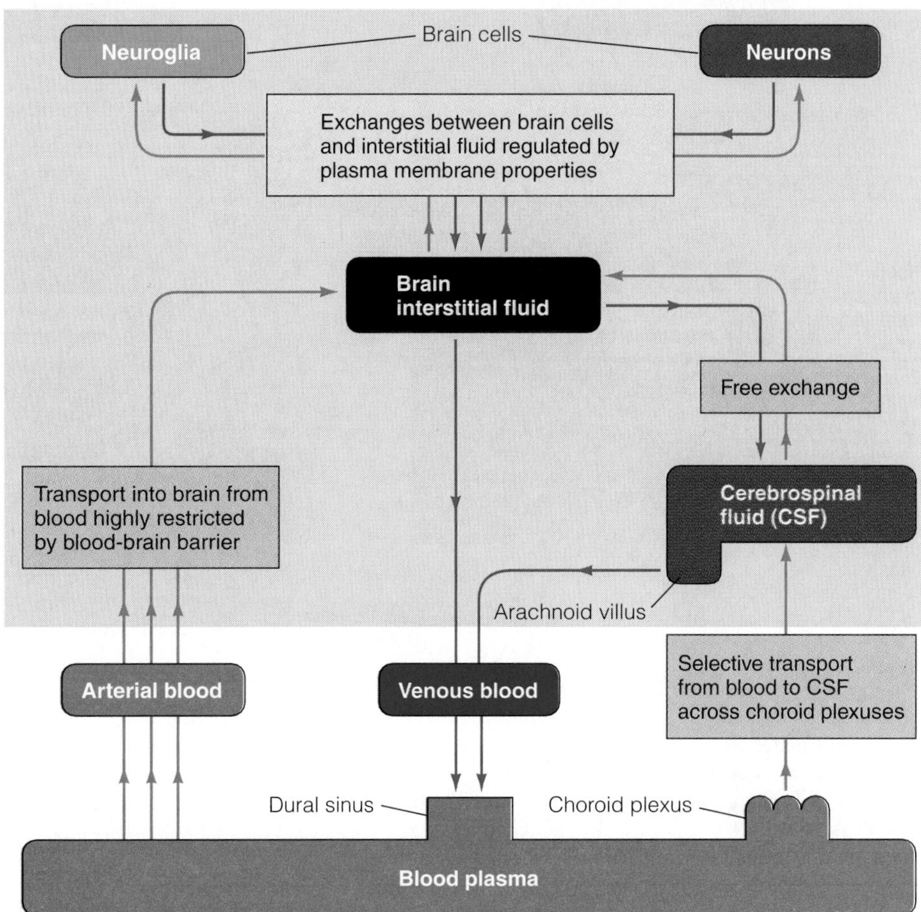

▬ *Figure 5-6* **Exchanges between the Blood and Brain Cells** The brain cells (neurons and neuroglia) exchange materials with the surrounding brain interstitial fluid. The composition of the brain interstitial fluid is influenced primarily by the composition of the cerebrospinal fluid (CSF), not by the composition of the blood, because there is free exchange between the CSF and brain interstitial fluid whereas exchange between the blood and brain interstitial fluid is restricted by the highly discriminating blood-brain barrier. The composition of CSF, in turn, is determined by selective transport processes from the blood to the CSF across the choroid plexuses of the brain.

A highly selective blood-brain barrier carefully regulates exchanges between the blood and brain.

The brain is carefully shielded from harmful changes in the blood by the blood-brain barrier. Exchange of materials between the blood and the surrounding interstitial fluid can take place only across the walls of capillaries, the smallest of blood vessels. Unlike the rather free exchange across capillaries elsewhere in the body, permissible exchanges across brain capillaries are strictly limited. Changes in most plasma constituents do not easily influence the composition of brain interstitial fluid because only selected exchanges can be made. For example, even if the K^+ level in the blood is doubled, little change occurs in the K^+ concentration of the fluid bathing the central neurons. This is beneficial because alterations in interstitial fluid K^+ would be detrimental to neuronal function.

The blood-brain barrier consists of both anatomical and physiological factors. Capillary walls throughout the body are formed by a single layer of cells. Usually, holes or pores between the cells making up the capillary wall permit free exchange of all plasma components, except the large plasma proteins, with the surrounding interstitial fluid. In the brain capillaries, however, the cells are joined by **tight junctions,** which completely seal the capillary wall so that nothing can be exchanged across the wall by passing *between* the cells (▬ Fig. 5-7). The only possible exchanges are *through* the capillary cells themselves. Lipid-soluble substances such as O_2, CO_2, alcohol, and steroid hormones penetrate these cells easily by dissolving in their lipid plasma membrane. Small water molecules also diffuse through readily, apparently by passing between the phospholipid molecules that compose the plasma membrane. All other substances exchanged between the blood and brain interstitial fluid, including such essential materials as glucose, amino acids, and ions, are transported by highly selective membrane-bound carriers. Accordingly, transport across the capillary walls between the cells is anatomically prevented, and transport through the cells is physiologically restricted.

The brain capillaries are surrounded by astrocyte processes, which at one time were thought to be physically responsible for the blood-brain barrier. Evidence now indicates that the astrocytes have two roles regarding the blood-brain barrier: (1) They appear to signal the cells forming the brain capillaries to "get tight." Capillary cells do not have an inherent ability to form tight junctions; they do so only at the command of a signal within their neural environment. (2) Astrocytes are believed to participate in the cross-cellular transport of some substances, such as K^+.

The blood-brain barrier protects the delicate brain and spinal cord from chemical fluctuations in the blood and min-

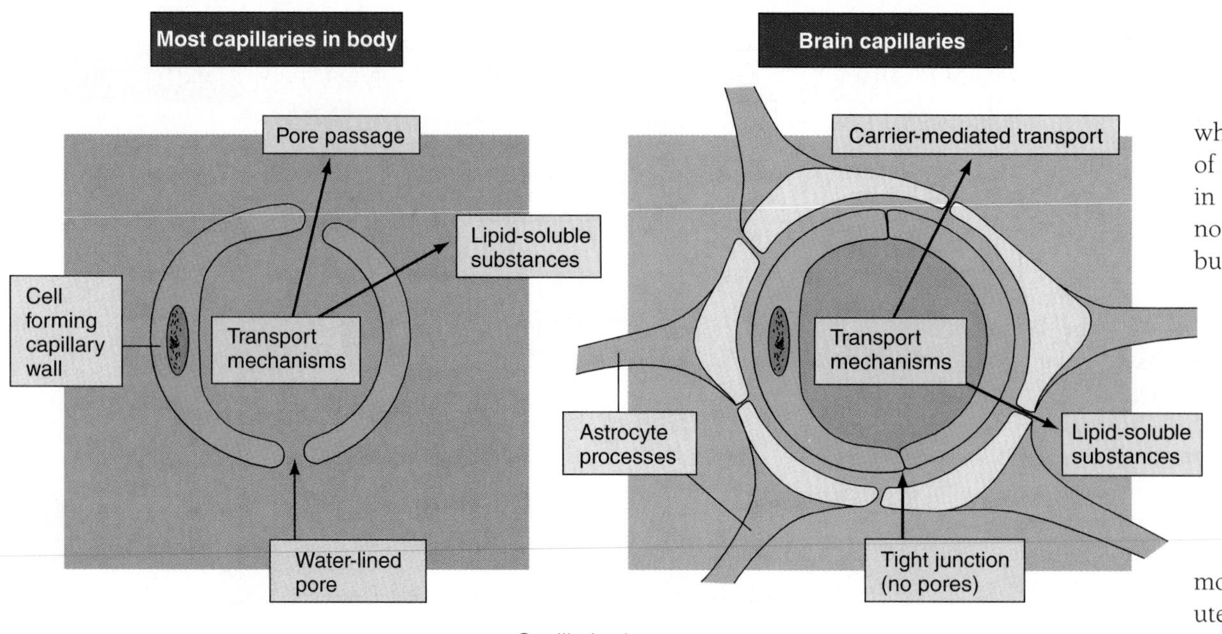

Figure 5-7 **Blood-Brain Barrier** Unlike most capillaries in the body, the cells forming the walls of brain capillaries are joined by tight junctions that prevent materials from passing between the cells. The only passage across brain capillaries is through the cells that form the capillary walls. With the exception of lipid-soluble substances and water, passage of all other materials through these cells is physiologically regulated by carrier-mediated systems, which are not present in capillaries elsewhere.

imizes the possibility that potentially harmful blood-borne substances might reach the central neural tissue. It further prevents certain circulating hormones that could also act as neurotransmitters from reaching the brain, where they could produce uncontrolled nervous activity. On the negative side, the blood-brain barrier limits the use of drugs for the treatment of brain and spinal cord disorders because many drugs are unable to penetrate this barrier.

Certain areas of the brain are not subject to the blood-brain barrier, most notably a portion of the hypothalamus. Functioning of the hypothalamus depends on its "sampling" the blood and adjusting its controlling output accordingly in order to maintain homeostasis. Part of this output is in the form of hormones that must enter hypothalamic capillaries to be transported to their sites of action. Appropriately, these hypothalamic capillaries are not sealed by tight junctions.

The brain depends on constant delivery of oxygen and glucose by the blood.

Even though many substances in the blood never actually come in contact with the brain tissue, the brain, more than any other tissue, is highly dependent on a constant blood supply. Unlike most tissues, which can resort to anaerobic metabolism to produce ATP in the absence of O_2 for at least short periods (see p. 32), the brain cannot produce ATP in the absence of O_2. Furthermore, in contrast to most tissues, which can use other sources of fuel for energy production in lieu of glucose, the brain normally uses only glucose but does not store any of this nutrient. Therefore, the brain is absolutely dependent on a continuous, adequate blood supply of O_2 and glucose. Accordingly, brain damage results if this organ is deprived of its critical O_2 supply for more than four to five minutes or if its glucose supply is cut off for more than ten to fifteen minutes.

Brain damage may occur in spite of protective mechanisms.

Even though the brain is carefully protected in its cushiony vault, traumatic head injuries may damage the delicate brain tissue. Direct brain damage occurs if the brain is violently shaken or jarred by a forceful impact, such as a fall or a blow, or if crushed cranial bones are pushed against the underlying neural tissue. Further indirect brain damage may occur following a head injury as a consequence of swelling or hemorrhaging within the enclosed confines of the cranium. The resultant increase in intracranial pressure may cause compression and damage of brain tissue. Brain damage may also occur as a result of infectious or degenerative neural disorders or brain tumors, all of which can lead to destruction of brain tissue.

The most common cause of brain damage, however, is **cerebrovascular accidents (strokes).** When a brain (cerebral) blood vessel ruptures or is blocked by a clot, the brain tissue being supplied by that vessel is deprived of its vital O_2 and glucose supply. The result is damage and usually death of the deprived tissue. Recently, researchers have learned that neural damage (and the subsequent loss of neural function) extends well beyond the blood-deprived area as a result of a toxic release of glutamate, a common excitatory neurotransmitter, from the O_2-starved neurons. Glutamate or other neurotransmitters are normally released in small amounts from neurons as a means of chemical communication between brain cells (see p. 95). Damaged brain cells release excessive amounts of glutamate, which binds with and overexcites surrounding neurons. The excitatory overdose of glutamate subsequently destroys these surrounding cells by triggering damaging chemical reactions within them. These doomed neighbors then spew out more glutamate, which in turn kills even more neuronal victims as the damaging cascade of chemical reactions spreads from the initial site of O_2 deprivation. This glutamate cascade is believed to be responsible for the majority of neuron deaths following a stroke.

Armed with this new knowledge, scientists are currently investigating new drugs, such as glutamate-receptor blockers, that will halt this domino effect. The goal, of course, is to limit the extent of neuronal damage and thus minimize or even prevent clinical symptoms such as paralysis. In the near future, this new therapeutic approach, coupled with recent successes in restoring blood flow through blocked cerebral vessels by administration of clot-dissolving drugs, holds much promise for treating strokes, which are the most prevalent cause of adult disability and the third leading cause of death in the United States. Until recently, treatment of strokes had been limited to rehabilitative therapy after the damage was already complete.

No matter what the cause of the destruction of brain tissue, the nature of the ensuing loss of neurological function depends on the area of the brain involved and the extent of permanent damage. Possible outcomes fall into the following range: (1) death if an area responsible for maintenance of a vital function is destroyed (for example, the brain center controlling respiration); (2) severe or mild loss of specific sensory awareness; (3) motor (movement) disorders of varying severity; (4) language impairment; (5) impaired mental abilities; or (6) full functional recovery.

Because mature neurons are unable to divide, destroyed CNS neurons cannot be replaced through cell division. Furthermore, damaged axons cannot regenerate within the CNS as they can in the peripheral nervous system due to the release of nerve growth–inhibiting proteins by the myelin-forming oligodendrocytes of the CNS. However, the brain displays a degree of **plasticity,** that is, an ability to change or be functionally remodeled in response to the demands placed on it. This ability is more pronounced in the early developmental years, but even adults retain some plasticity. When an area of the brain associated with a particular activity is destroyed, other areas of the brain may gradually assume some or all of the responsibilities of the damaged region. The underlying molecular mechanisms responsible for the brain's plasticity are only beginning to be unraveled. Current evidence suggests that the formation of new neural pathways (not new neurons, but new connections between existing neurons) in response to changes in experience are mediated in part by alterations in dendritic shape resulting from modifications in certain cytoskeletal elements (see p. 36). As its dendrites become more branched and elongated, a neuron is able to receive and integrate more signals from other neurons.

Headaches are seldom due to brain damage.

Headaches are the most common form of pain. Almost everyone experiences headaches at least occasionally. Fortunately, most headaches are not associated with brain damage. They usually result from tension or increased pressure within pain-sensitive structures inside the cranium. The following are among the causes of headaches:

1. *Tension* associated with sustained tightening of muscles in the neck, scalp, and forehead in conjunction with anxiety, stress, or fatigue.
2. *Swelling of the mucous membranes* lining the sinuses in response to respiratory infections or allergies.
3. *Eye disorders* accompanied by straining of eye muscles.
4. *Dilation of cerebral blood vessels* in association with high blood pressure, hangovers, or migraine headaches.
5. *Increased intracranial pressure* accompanying brain tumors or intracranial hemorrhaging.
6. *Inflammation and swelling* in association with meningeal infections **(meningitis)** or infection of the brain itself **(encephalitis).**

Cerebral Cortex

Newer, more sophisticated regions of the brain are piled on top of older, more primitive regions.

Although the brain is a functional whole, it is organized into several different regions. The parts of the brain can be arbitrarily grouped in various ways based on anatomical distinctions, functional specialization, and evolutionary development. We will use the following grouping (Table 5-3):

1. Brain stem
2. Cerebellum
3. Forebrain
   a. Diencephalon
      1. Hypothalamus
      2. Thalamus
   b. Cerebrum
      1. Basal nuclei
      2. Cerebral cortex

The order in which these components are listed generally represents both their anatomical location (from bottom to top) and their complexity and sophistication of function (from the least specialized, oldest level to the newest, most specialized level).

A primitive nervous system consists of comparatively few interneurons interspersed between afferent and efferent neurons. During evolutionary development, the interneuronal component progressively expanded, formed more complex interconnections, and became localized at the head end of the nervous system, forming the brain. Newer, more sophisticated layers of the brain were added on to the older, more primitive layers. The human brain represents the present peak of development.

The *brain stem,* the oldest and smallest region of the brain, is continuous with the spinal cord. It controls many of the life-sustaining processes, such as breathing, circulation, and digestion, that are common to many of the lower vertebrate forms. These processes are often referred to as "vegetative" functions because, with the loss of higher brain functions, these lower brain levels, in accompaniment with appropriate

Table 5-3 Overview of Structures and Functions of the Major Components of the Brain

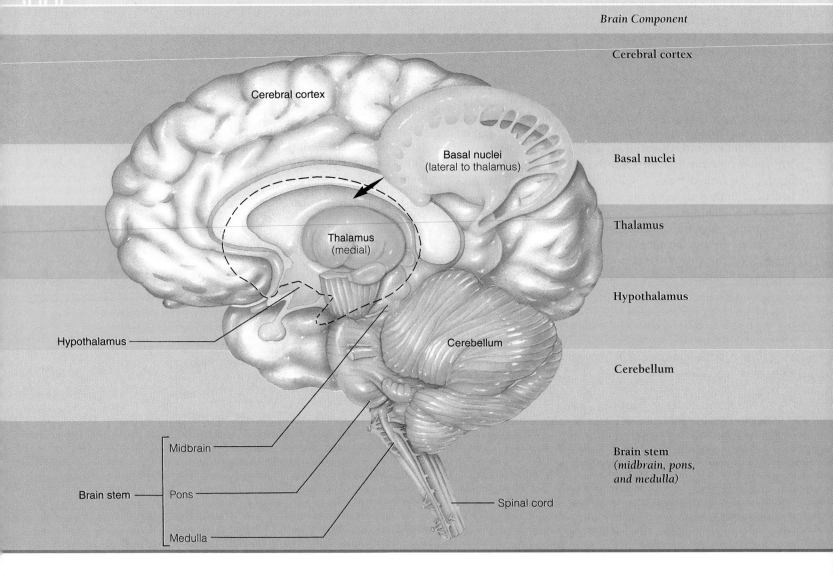

Brain Component
Cerebral cortex
Basal nuclei
Thalamus
Hypothalamus
Cerebellum
Brain stem *(midbrain, pons, and medulla)*

supportive therapy such as intravenous feeding, can still sustain the functions essential for survival. Because the person has no awareness or control of that life, however, the condition is sometimes referred to as "being a vegetable."

Attached at the top rear portion of the brain stem is the *cerebellum,* which is concerned with maintaining proper position of the body in space and subconscious coordination of motor activity (movement). On top of the brain stem, tucked within the interior of the cerebrum is the *diencephalon.* It houses two brain components: the *hypothalamus,* which controls many homeostatic functions important in maintaining stability of the internal environment, and the *thalamus,* which performs some primitive sensory processing. On top of this "cone" of lower brain regions is the *cerebrum,* whose "scoop" gets progressively larger and more highly convoluted (that is, has tortuous ridges delineated by deep grooves or folds) the more advanced the vertebrate species is. The cerebrum is most highly developed in humans, where it constitutes about 80% of the total brain weight. The outer layer of the cerebrum is the highly convoluted *cerebral cortex,* which caps an inner core that houses the *basal nuclei.* The cerebral cortex plays a key role in the most sophisticated neural functions, such as voluntary initiation of movement, final sensory perception, conscious thought, language, personality traits, and other factors we associate with the mind or intellect. It is the highest, most complex integrating area of the brain. Each of these regions of the brain will be discussed in turn, starting with the cerebral cortex.

The cerebral cortex is an outer shell of gray matter covering an inner core of white matter.

The **cerebrum,** by far the largest portion of the human brain, is divided into two halves, the right and left **cerebral hemispheres.** They are connected to each other by the **corpus callosum,** a thick band consisting of an estimated 300 million neuronal axons traversing between the two hemispheres (see Fig. 5-15, p. 131).

Major Functions
1. Sensory perception 2. Voluntary control of movement 3. Language 4. Personality traits 5. Sophisticated mental events, such as thinking, memory decision making, creativity, and self-consciousness
1. Inhibition of muscle tone 2. Coordination of slow, sustained movements 3. Suppression of useless patterns of movement
1. Relay station for all synaptic input 2. Crude awareness of sensation 3. Some degree of consciousness 4. Role in motor control
1. Regulation of many homeostatic functions, such as temperature control, thirst, urine output, and food intake 2. Important link between nervous and endocrine systems 3. Extensive involvement with emotion and basic behavioral patterns
1. Maintenance of balance 2. Enhancement of muscle tone 3. Coordination and planning of skilled voluntary muscle activity
1. Origin of majority of peripheral cranial nerves 2. Cardiovascular, respiratory, and digestive control centers 3. Regulation of muscle reflexes involved with equilibrium and posture 4. Reception and integration of all synaptic input from spinal cord; arousal and activation of cerebral cortex 5. Sleep centers

Each hemisphere is composed of a thin outer shell of *gray matter,* the **cerebral cortex,** covering a thick central core of *white matter* (see Fig. 5-15, p. 131). Located deep within the white matter is another region of gray matter, the basal nuclei. Throughout the entire CNS, **gray matter** consists predominantly of densely packaged cell bodies and their dendrites as well as glial cells. Bundles or tracts of myelinated nerve fibers (axons) constitute the **white matter;** its white appearance is due to the lipid (fat) composition of the myelin. The gray matter can be viewed as the "computers" of the CNS and the white matter as the "wires" that connect the computers to each other. The fiber tracts in the white matter transmit signals from one part of the cerebral cortex to another or between the cortex and other regions of the CNS. Such communication between different areas of the cortex and elsewhere facilitates integration of their activity. This integration is essential for even a relatively simple task such as picking a flower. Vision of the flower is received by one area of the cortex, reception of its fragrance takes place in another area, and movement is initiated by still another area. More subtle neuronal responses, such as appreciation of the flower's beauty and the urge to pick it, are poorly understood but undoubtedly extensively involve interconnecting fibers between different cortical regions.

The cerebral cortex is organized into layers and functional columns.

The cerebral cortex is organized into six well-defined layers based on varying distributions of the cell bodies and locally associated fibers of several distinctive cell types. These layers are organized into functional vertical columns that extend perpendicularly from the surface down through the depths of the cortex to the underlying white matter. The neurons within a given column are believed to function as a "team," with each cell being involved in different aspects of the same specific activity—for example, perceptual processing of the same stimulus from the same location.

The functional differences between various areas of the cortex result from different layering patterns within the columns and from different input-output connections, not from the presence of unique cell types or different neuronal mechanisms. For example, those regions of the cortex responsible for perception of senses have an expanded layer 4, a layer rich in **stellate cells,** which are responsible for initial processing of sensory input to the cortex. In contrast, the cortical areas that control output to skeletal muscles have a thickened layer 5, which contains an abundance of large **pyramidal cells.** These cells send fibers down the spinal cord from the cortex to terminate on the efferent motor neurons that innervate the skeletal muscles.

The four pairs of lobes in the cerebral cortex are specialized for different activities.

It is important to recognize that even though a discrete activity is ultimately attributed to a particular region of the brain, no part of the brain functions in isolation. Each part depends on complex interplay among numerous other regions for both incoming and outgoing messages. With this in mind, let us now consider the locations of the major functional areas of the brain.

The anatomical landmarks used in cortical mapping are certain deep folds that divide each half of the cortex into four major lobes: the *occipital, temporal, parietal,* and *frontal lobes* (■ Fig. 5-8). Refer to the basic functional map of the cortex in ■ Figure 5-9 during the following discussion of the major activities attributed to various regions of these lobes.

Occipital and temporal lobes The **occipital lobes,** which are located posteriorly (at the back of the head), are responsible for initially processing visual input. Sound sensation is initially received by the **temporal lobes,** located laterally (on the sides of the head).

Parietal lobes The parietal lobes and frontal lobes, located on the top of the head, are separated by a deep infolding, the **central sulcus,** which runs roughly down the middle of the lateral surface of each hemisphere. The parietal lobes lie to the rear of the central sulcus on each side, and the frontal lobes lie in front of it.

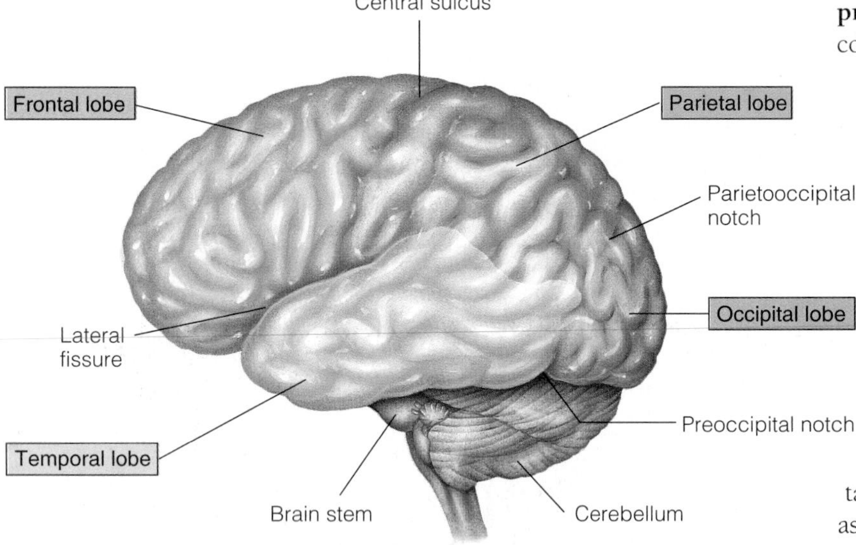

Figure 5-8 **Cortical Lobes** Each half of the cerebral cortex is divided into the occipital, temporal, parietal, and frontal lobes, as depicted in this schematic lateral view of the brain.

The **parietal lobes** are primarily responsible for receiving and processing sensory input such as touch, pressure, heat, cold, and pain from the surface of the body. These sensations are collectively known as **somesthetic sensations** (*somesthetic* means "body feelings"). The parietal lobes also perceive awareness of body position, a phenomenon referred to as **proprioception.** The **somatosensory cortex,** the site for initial cortical processing of this somesthetic and proprioceptive input, is located at the front of each parietal lobe immediately behind the central sulcus (Figs. 5-9 and 5-10a). Each region within the somatosensory cortex receives sensory input from a specific area of the body. This distribution of cortical sensory processing is depicted in Figure 5-10b. Note that on this so-called **sensory homunculus** (*homunculus* means "little man"), the body is represented upside down on the somatosensory cortex and, more importantly, *different parts of the body are not equally represented.* The size of each body part in this homunculus is indicative of the relative proportion of the somatosensory cortex devoted to that area. The exaggerated size of the face, tongue, hands, and genitalia is indicative of the high degree of sensory perception associated with these body parts.

The somatosensory cortex on each side of the brain for the most part receives sensory input from the opposite side of the

Figure 5-9 **Functional Areas of the Cerebral Cortex** Various regions of the cerebral cortex are primarily responsible for various aspects of neural processing, as indicated in this schematic lateral view of the brain.

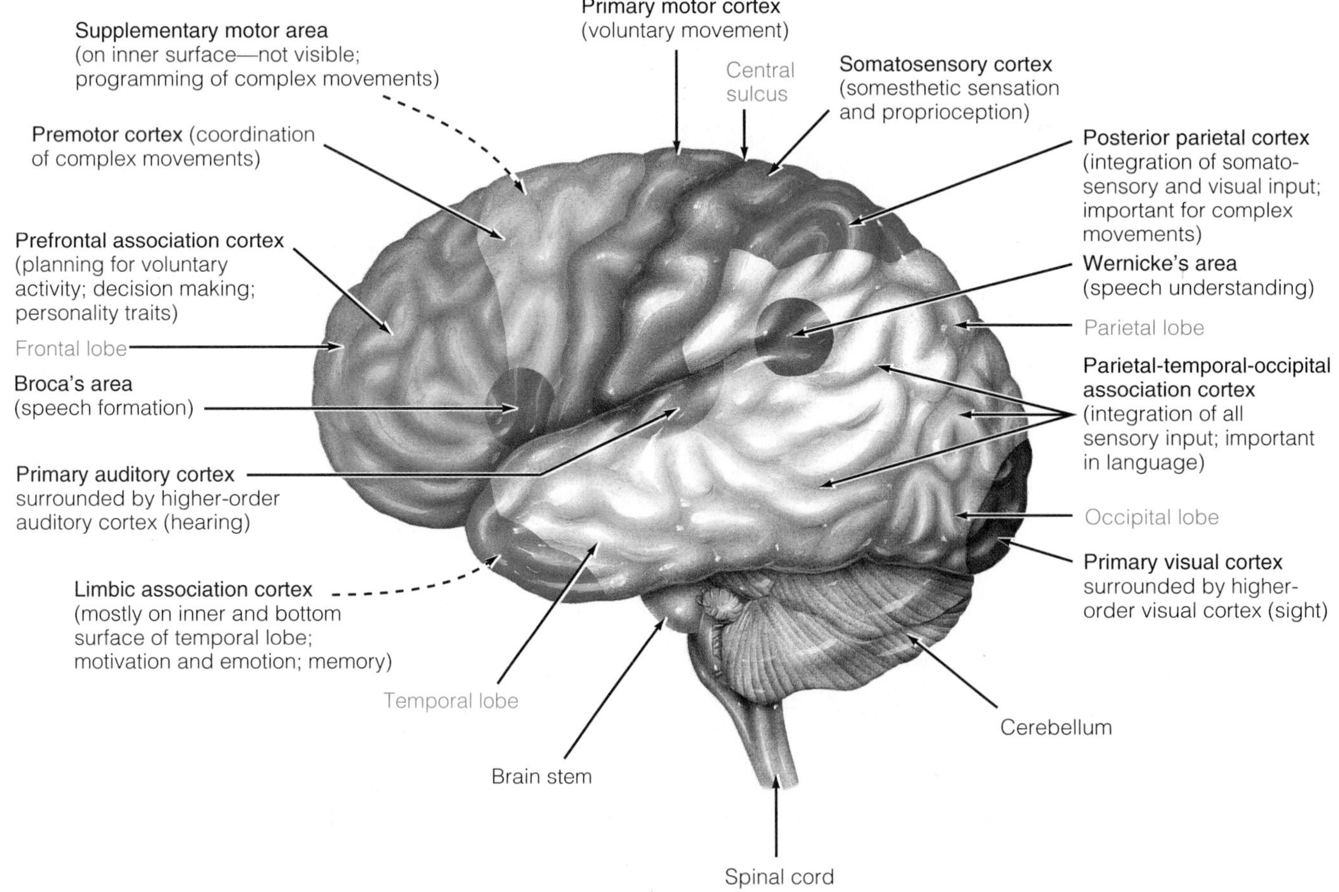

body, because most of the ascending pathways carrying sensory information up the spinal cord cross over to the opposite side before eventually terminating in the cortex. Thus, damage to the somatosensory cortex in the left hemisphere produces sensory deficits on the right side of the body, whereas sensory losses on the left side are associated with damage to the right half of the cortex.

Simple awareness of touch, pressure, or temperature is detected by the thalamus, a lower level of the brain, but the somatosensory cortex goes beyond pure recognition of sensations to fuller sensory perception. The thalamus makes you aware that something hot versus something cold is touching your body, but it does not tell you where or of what intensity. The somatosensory cortex localizes the source of sensory input and perceives the level of intensity of the stimulus. It also is capable of spatial discrimination, so it can discern shapes of objects being held and can distinguish subtle differences in similar objects that come into contact with the skin.

The somatosensory cortex, in turn, projects this sensory input via white matter fibers to adjacent higher sensory areas for even further elaboration, analysis, and integration of sensory information. These higher areas are important in the perception of complex patterns of somatosensory stimulation—for example, simultaneous appreciation of the texture, firmness, temperature, shape, position, and location of an object you are holding.

Frontal lobes The **frontal lobes,** lying at the front of the cortex, are responsible for three main functions: (1) voluntary motor activity, (2) speaking ability, and (3) elaboration of thought. The area at the rear of the frontal lobe immediately in front of the central sulcus and adjacent to the somatosensory cortex is the **primary motor cortex** (Figs. 5-9 and 5-11a). It confers voluntary control over movement produced by skeletal muscles. As in sensory processing, the motor cortex on each side of the brain primarily controls muscles on the opposite side of the body. Neuronal tracts originating in the motor cortex of the left hemisphere cross over before passing down the spinal cord to terminate on efferent motor neurons that trigger skeletal muscle contraction on the right side of the body. Accordingly, damage to the motor cortex on the left side of the brain produces paralysis on the right side of the body, and the converse is also true.

Stimulation of different areas of the primary motor cortex brings about movement in different regions of the body. Like the sensory homunculus for the somatosensory cortex, the **motor homunculus,** which depicts the location and relative amount of motor cortex devoted to output to the muscles of each body part, is upside down and distorted (Fig. 5-11b). The fingers, thumbs, and muscles important in speech, especially those of the lips and tongue, are grossly exaggerated, indicative of the fine degree of motor control with which these body parts are endowed. Compare this to how little brain tissue is devoted to the trunk, arms, and lower extremities, which are not capable of such complex movements. Thus, the extent of representation in the motor cortex is proportional to the precision and complexity of motor skills required of the respective part.

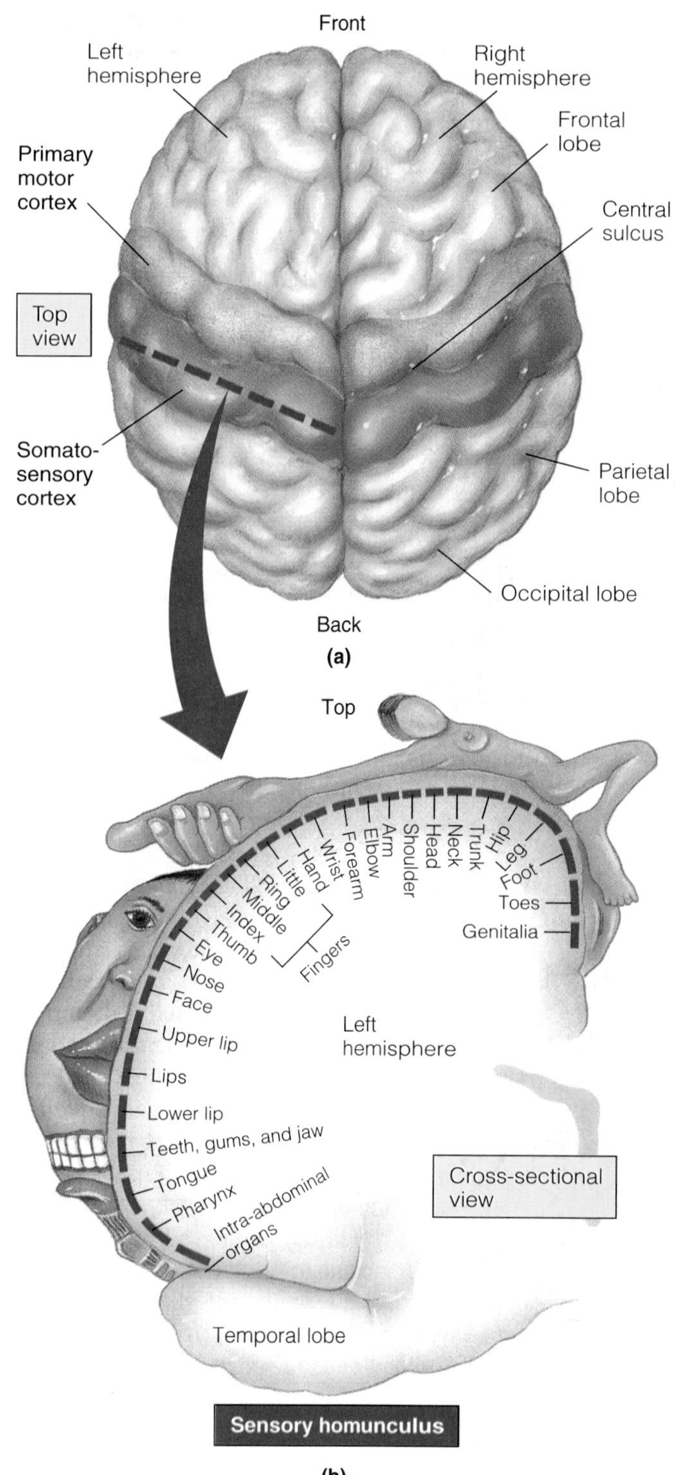

Figure 5-10 **Somatotopic Map of the Somatosensory Cortex**
(a) Top view of cerebral hemispheres. (b) Sensory homunculus showing the distribution of sensory input to the somatosensory cortex from different parts of the body. The distorted graphic representation of the body parts is indicative of the relative proportion of the somatosensory cortex devoted to reception of sensory input from each area.

Recent evidence suggests that the "map" of the primary motor cortex (the depiction of which cortical areas control the muscles in which parts of the body) is not as orderly as the motor homunculus suggests; that is, there is not a neat one-

Chapter Contents At a Glance

Introduction

The peripheral nervous system consists of nerve fibers that carry information between the CNS and other parts of the body. The afferent division of the peripheral nervous system sends information about the internal and external environment to the CNS. Afferent information about the internal environment, such as the blood pressure and the concentration of CO_2 in the body fluids, never reaches the level of conscious awareness, but this input is essential for determining the appropriate efferent output to maintain homeostasis. The incoming pathway for subconscious information derived from the internal viscera is called a **visceral afferent.** Afferent input that does reach the level of conscious awareness is known as **sensory information,** and the incoming pathway is considered to be a **sensory afferent.** Sensory information is categorized as either (1) **somatic** (body sense) **sensation** arising from the body surface, including *somesthetic sensation* from the skin and *proprioception* from the muscles, joints, skin, and inner ear (see p. 124), or (2) **special senses,** including *vision, hearing, taste,* and *smell.* (See the accompanying boxed feature, A Closer Look at Exercise Physiology.) Final processing of sensory input by the CNS not only is essential for interaction with the environment for basic survival (for example, food procurement and defense from danger) but also adds immeasurably to the richness of life.

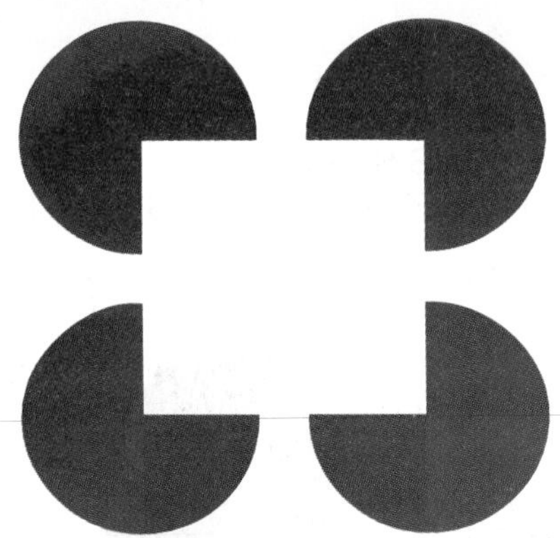

Figure 6-1 **Do You "See" a White Square That Is Not Really There?**

Perception is our conscious interpretation of the external world as created by the brain from a pattern of nerve impulses delivered to it from sensory receptors. Is the world as we perceive it reality? The answer is a resounding no. Our perception is different from what is really "out there" for several reasons. First, humans have receptors that detect only a limited number of existing energy forms. We perceive sounds, colors, shapes, textures, smells, tastes, and temperature but are not informed of magnetic forces, polarized light waves, radio waves, or X rays because we do not have receptors to respond to the latter energy forms. Our response range is limited even for the energy forms for which we do have receptors. For example, dogs can hear a dog whistle whose pitch is above our level of detection. Second, the information channels to our brains are not high-fidelity recorders. During precortical processing of sensory input, some features of stimuli are accentuated and others are suppressed or ignored. Third, the cerebral cortex further manipulates the data, comparing the sensory input with other incoming information as well as with memories of past experiences to extract the significant features—for example, sifting out a friend's words from the hubbub of sound in a school cafeteria. In the process, the cortex often fills in or distorts the information to abstract a logical perception; that is, it "completes the picture." As a simple example, you "see" a white square in Figure 6-1 even though there is no white square but merely right-angle wedges taken out of four purple circles. Optical illusions illustrate how the brain interprets reality according to its own rules. Do you see two faces in profile or a wineglass in Fig. 6-2? You can alternately see one or the other out of identical visual input. Thus, our perceptions do not replicate reality. Other species equipped with different types of receptors and sensitivities and with different neural processing perceive a markedly different world than we do.

A CLOSER LOOK AT EXERCISE PHYSIOLOGY

Back Swings and Pre-Jump Crouches: What Do They Share in Common?

Proprioception, the sense of the body's position in space, is critical to any movement and is especially important in athletic performance, whether it be a figure skater performing triple jumps on ice, a gymnast performing a difficult floor routine, or a football quarterback throwing perfectly to a spot 60 yards down field. In order to control skeletal muscle contraction to achieve the desired movement, the CNS must be continuously apprised of the results of its actions by means of sensory feedback information.

A number of receptors provide proprioceptive input. Muscle proprioceptors provide feedback information on muscle tension and length. Joint proprioceptors provide feedback on joint acceleration, angle, and direction of movement. Skin proprioceptors inform the CNS of weight-bearing pressure on the skin. Proprioceptors in the inner ear, along with those in neck muscles, provide information about head and neck position so that the CNS can orient the head correctly. For example, neck reflexes facilitate essential trunk and limb movements during somersaults, and divers and tumblers use strong movements of the head to maintain spins.

The most complex and probably one of the most important proprioceptors is the muscle spindle (see p. 250). Muscle spindles are found throughout a muscle but tend to be concentrated in its center. Each spindle lies parallel to the muscle fibers within the muscle. The spindle is sensitive to both the muscle's rate of change in length and the final length achieved. If a muscle is stretched, each muscle spindle within the muscle is also stretched, and the afferent neuron whose peripheral axon terminates on the muscle spindle is stimulated. The afferent fiber passes into the spinal cord and synapses directly on the motor neurons that supply the same muscle. Stimulation of the stretched muscle as a result of this stretch reflex causes the muscle to contract sufficiently to relieve the stretch.

Older persons or those with weak quadriceps (thigh) muscles unknowingly take advantage of the muscle spindle by pushing on the center of the thighs when they get up from a sitting position. Contraction of the quadriceps muscle extends the knee joint, thus straightening the leg. The act of pushing on the center of the thighs when getting up slightly stretches the quadriceps muscle in both limbs, stimulating the muscle spindles. The resultant stretch reflex aids in contraction of the quadriceps muscles and helps the person to assume a standing position.

In sports, people use the muscle spindle to advantage all the time. To jump high, as in basketball jumpball, an athlete starts by crouching down. This action stretches the quadriceps muscles and increases the firing rate of their spindles, thus triggering the stretch reflex that reinforces the quadriceps muscles' contractile response so that these extensor muscles of the legs gain additional power. The same is true for crouch starts in running events. The back swing in tennis, golf, and baseball similarly provides increased muscular excitation through reflex activity initiated by stretched muscle spindles.

Receptor Physiology

Receptors have differential sensitivities to various stimuli.

At their peripheral endings, afferent neurons have **receptors** that apprise the CNS of detectable changes, or **stimuli,** in both the external world and the internal environment by generating action potentials in response to the stimuli. These action potentials are transmitted via the afferent fibers to the CNS. Stimuli exist in a variety of energy forms, or **modalities,** such as heat, light, sound, pressure, and chemical changes. Because the only way that afferent neurons can transmit information to the CNS is via action potential propagation, receptors must convert these other forms of energy into electrical energy (action potentials). This energy conversion process is known as **transduction.**

Each type of receptor is specialized to respond more readily to one type of stimulus, its **adequate stimulus,** than to other stimuli. For example, receptors in the eye are most sensitive to light, receptors in the ear to sound waves, and warmth receptors in the skin to heat energy. We cannot "see" with our ears or "hear" with our eyes because of this differential sensitivity of receptors, a principle known as the **law of specific nerve energies.** Some receptors can respond weakly to stimuli other than their adequate stimulus, but even when activated by a different stimulus, a receptor still gives rise to the sensation usually detected by that receptor type. As an example, the adequate stimulus for eye receptors (photoreceptors) is light, to which they are exquisitely sensitive, but these receptors can also be activated to a lesser degree by mechanical stimulation. When hit in the eye, a person often "sees stars" because the mechanical pressure stimulates the photoreceptors. Thus, the sensation perceived depends on the type of receptor stimulated rather than on the type of stimulus. However, because receptors typically are activated by their adequate stimulus, the sensation usually corresponds to the stimulus modality.

Figure 6-2 Variable Perceptions from the Same Visual Input Do you see two faces in profile or a wineglass?

Depending on the type of energy to which they ordinarily respond, receptors are categorized as follows:

- **Photoreceptors** are responsive to light.
- **Mechanoreceptors** are sensitive to mechanical energy. Examples include skeletal muscle receptors sensitive to stretch, the receptors in the ear containing fine hair cells that are bent as a result of sound waves, and blood pressure–monitoring baroreceptors.
- **Thermoreceptors** are sensitive to heat and cold.
- **Osmoreceptors** detect changes in the concentration of solutes in the body fluids and the resultant changes in osmotic activity (see p. 55).
- **Chemoreceptors** are sensitive to specific chemicals. Chemoreceptors include the receptors for smell and taste, as well as those located deeper within the body that detect O_2 and CO_2 concentrations in the blood or the chemical content of the digestive tract.
- **Nociceptors,** or **pain receptors,** are sensitive to tissue damage such as pinching or burning or to distortion of tissue. Intense stimulation of any receptor is also perceived as painful.

Some sensations are compound sensations in that their perception arises from central integration of several simultaneously activated primary sensory inputs. For example, the perception of wetness comes from touch, pressure, and thermal receptor input; there is no such thing as a "wet receptor."

The information detected by receptors is conveyed via afferent neurons to the CNS, where it is used for a variety of purposes:

- First, afferent input is essential for the control of efferent output, both for regulation of motor behavior in accordance with external circumstances and for coordination of internal activities directed toward maintenance of homeostasis. At the most basic level, afferent input provides information (of which the individual may or may not be consciously aware) for the CNS to use in directing activities necessary for survival.
- Second, processing of sensory input by the reticular activating system in the brain stem is critical for cortical arousal and consciousness (see p. 143).
- Third, central processing of sensory information gives rise to our perceptions of the world around us.
- Finally, selected information delivered to the CNS may be stored for future reference.

Altered membrane permeability of receptors in response to a stimulus produces a graded receptor potential.

A receptor may be either a specialized ending of the afferent neuron or a separate cell closely associated with the peripheral ending of the neuron. Stimulation of a receptor alters its membrane permeability, usually by causing a nonselective opening of all small ion channels. The means by which this permeability change takes place is individualized for each receptor type. Because the electrochemical driving force is greater for Na^+ than for other small ions at resting potential, the predominant effect is an inward flux of Na^+, which depolarizes the receptor membrane. (There are exceptions; for example, photoreceptors are hyperpolarized upon stimulation.)

This local depolarizing change in potential is known as a **receptor potential** in the case of a separate receptor or as a **generator potential** if the receptor is a specialized ending of an afferent neuron. The receptor (or generator) potential is a graded potential (see p. 82) whose amplitude and duration can vary, depending on the strength and the rate of application or removal of the stimulus. The stronger the stimulus, the greater the permeability change and the larger the receptor potential. As is true of all graded potentials, receptor potentials have no refractory period, so summation in response to rapidly successive stimuli is possible. Because the receptor region has a very high threshold, action potentials do not take place at the receptor itself. For long-distance transmission, the receptor potential must be converted into action potentials that can be propagated along the afferent fiber. This conversion is accomplished by the opening of Na^+ channels in the afferent neuron membrane adjacent to the receptor in response to the presence of a receptor (or generator) potential. If the resulting Na^+ influx is sufficient to bring this region adjacent to the receptor to threshold, an action potential is initiated that is self-propagated along the afferent fiber to the CNS.

The means by which the Na^+ channels in the adjacent afferent membrane are opened differ depending on whether the receptor is a separate cell or a specialized afferent ending. In the case of a separate receptor, a receptor potential triggers the release of a chemical messenger that diffuses across the small space separating the receptor from the ending of the afferent neuron, similar to a synapse (— Fig. 6-3a). Binding of the chemical messenger with specific protein receptor sites on the afferent neuron opens chemical messenger–gated Na^+ channels (see p. 84). If the magnitude of the resulting ionic flux is sufficient to bring the adjacent membrane to threshold, a self-propagating action potential is initiated in the afferent neuron. In the case of a specialized afferent ending (Fig. 6-3b), local current flow between the activated receptor ending undergoing a generator potential and the cell membrane adjacent to the receptor brings about opening of voltage-gated Na^+ channels in this adacent region. Once threshold is reached, an action potential is initiated that is propagated along the afferent fiber. (For convenience, from here on we will refer to both receptor potentials and generator potentials as receptor potentials.)

The intensity of the stimulus is reflected by the magnitude of the receptor potential. In turn, the larger the receptor potential, the greater the frequency of action potentials generated in the afferent neuron. A larger receptor potential cannot bring about a larger action potential (all-or-none law), but it can induce more rapid firing of action potentials. This is one way stimulus strength is coded; the stronger the stimulus, the greater the frequency of action potentials. Stimulus strength is also reflected by the size of the area stimulated. Stronger stimuli usually affect larger areas, so a correspondingly greater

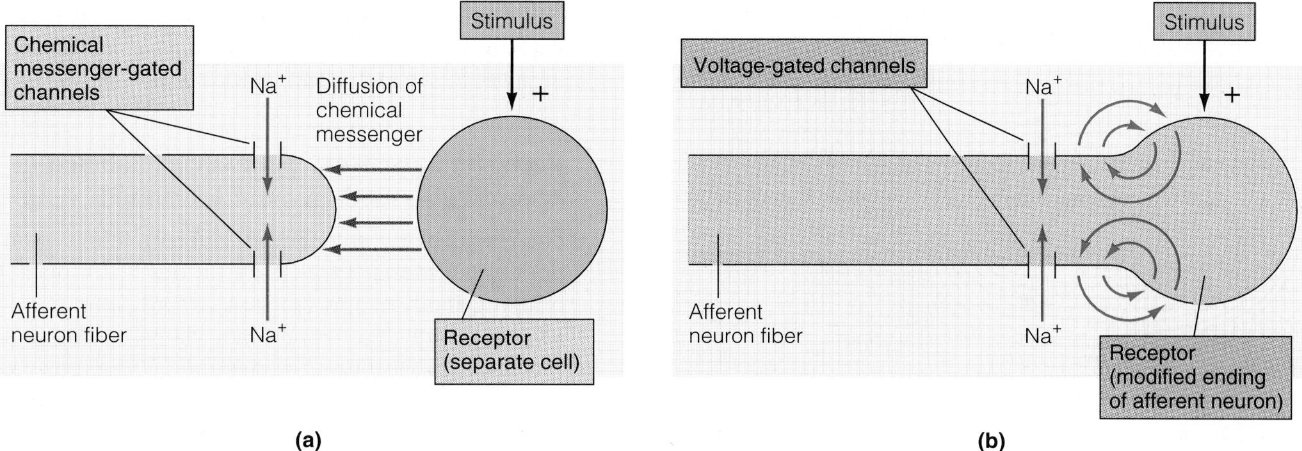

Figure 6-3 **Conversion of Receptor and Generator Potentials into Action Potentials** (a) Receptor potential. The chemical messenger released from a separate receptor initiates an action potential in the fiber by opening chemical messenger–gated Na^+ channels. (b) Generator potential. The local current flow between the depolarized receptor ending and the afferent fiber initiates an action potential in the fiber by opening voltage-gated Na^+ channels.

does not activate as many pressure receptors in the skin as a more forceful touch applied to the same area. Stimulus intensity is therefore distinguished both by the frequency of action potentials generated in the afferent neuron **(frequency code)** and by the number of receptors activated within the area **(population code).**

Receptors may adapt slowly or rapidly to sustained stimulation.

Because of **adaptation,** stimuli of the same intensity do not always elicit receptor potentials of the same magnitude from the same receptor. Some receptors have the ability to diminish the extent of their depolarization in spite of a sustained stimulus strength, with a subsequent decrease in the frequency of action potentials generated in the afferent neuron. The receptor "adapts" to the stimulus by no longer responding to it to the same degree.

There are two types of receptors—*tonic receptors* and *phasic receptors*—based on their speed of adaptation. **Tonic receptors** do not adapt at all or adapt slowly (Fig. 6-4a). These receptors are important in situations where maintained information about a stimulus is valuable. Examples are muscle stretch receptors, which monitor muscle length, and joint proprioceptors, which measure the degree of joint flexion. To maintain posture and balance, the CNS must be continually apprised of the degree of muscle length and joint position. It is important, therefore, that these receptors do not adapt to a stimulus but continue to generate action potentials to relay this information to the CNS.

Phasic receptors, on the other hand, are rapidly adapting receptors. These receptors often exhibit an **off response** (Fig. 6-4b). The receptor rapidly adapts by no longer responding to a maintained stimulus, but when the stimulus is removed, the receptor responds with a slight depolarization, the off response. Phasic receptors are useful in situations where it is important to signal a *change* in stimulus intensity rather than to relay status quo information. Rapidly adapting receptors

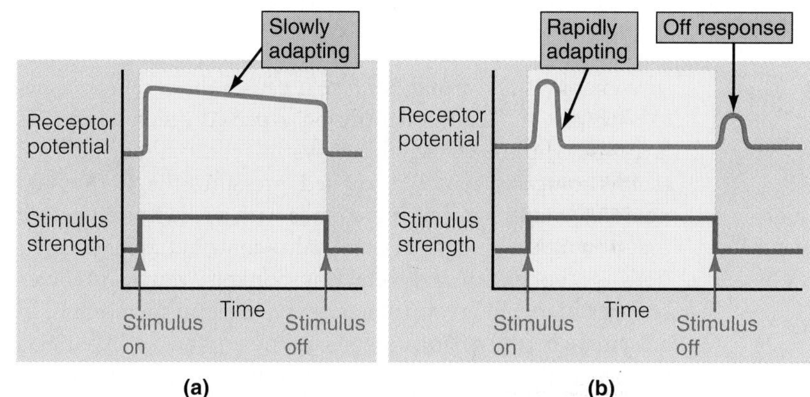

Figure 6-4 **Tonic and Phasic Receptors** (a) Tonic receptor. This receptor type does not adapt at all or adapts slowly to a sustained stimulus and thus provides continuous information about the stimulus. (b) Phasic receptor. This receptor type adapts rapidly to a sustained stimulus and frequently exhibits an off response when the stimulus is removed. Thus, the receptor signals changes in stimulus intensity rather than relaying status quo information.

include *tactile (touch)* receptors in the skin that signal changes in pressure on the skin surface. Because these receptors adapt rapidly, you are not continually conscious of wearing your watch, rings, and clothing. When you put something on, you soon become accustomed to it because of these receptors' rapid adaptation. When you take the item off, you are aware of its removal because of the off response.

The mechanism by which adaptation is accomplished varies for different receptors and has not been fully elucidated for all receptor types. One receptor type that has been extensively studied is the **Pacinian corpuscle,** a rapidly adapting skin receptor that detects pressure and vibration. Adaptation in a Pacinian corpuscle is believed to involve both mechanical and electrochemical components. The mechanical component depends on the physical properties of this receptor. A Pacinian corpuscle is a specialized receptor ending that consists of concentric layers of connective tissue resembling lay-

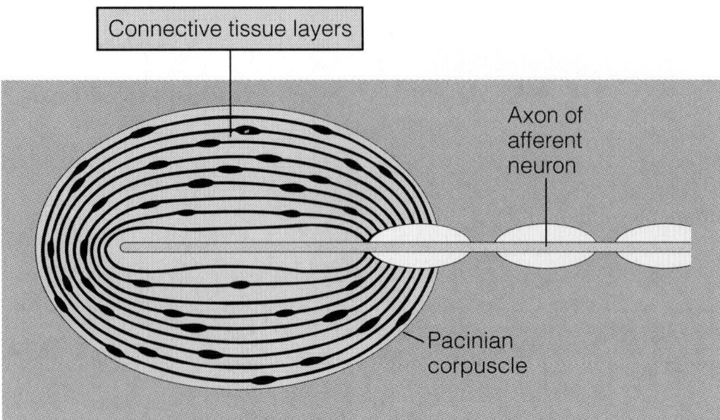

Figure 6-5 **Pacinian Corpuscle** The Pacinian corpuscle, a skin receptor that signals a change in pressure and detects vibration, consists of concentric layers of connective tissue wrapped around the peripheral terminal of an afferent neuron.

ers of an onion wrapped around the peripheral terminal of an afferent neuron (Fig. 6-5). When pressure is first applied to the Pacinian corpuscle, the underlying terminal responds with a receptor potential of a magnitude that reflects the intensity of the stimulus. As the stimulus continues, the pressure energy is dissipated because it causes the receptor layers to slip (just as steady pressure on a peeled onion causes its layers to slip). Because this physical effect filters out the steady component of the applied pressure, the underlying neuronal ending no longer responds with a receptor potential; that is, adaptation has occurred. Also contributing to adaptation is the electrochemical component, which involves changes in ionic movement across the receptor membrane. In a Pacinian corpuscle, for reasons unknown, the Na^+ channels that opened in response to the stimulus are slowly inactivated, thus reducing the inward flow of Na^+ ions that was largely responsible for the depolarizing receptor potential.

Adaptation should not be confused with habituation (see p. 137). Although both these phenomena involve decreased neural responsiveness to repetitive stimuli, they operate at different points in the neural pathway. Adaptation is a receptor adjustment in the PNS, whereas habituation involves a modification in synaptic effectiveness in the CNS.

Table 6-1
Coding of Sensory Information

Stimulus Property	*Mechanism of Coding*
Type of stimulus (stimulus modality)	Distinguished by the type of receptor activated and the specific pathway over which this information is transmitted to a particular area of the cerebral cortex
Location of stimulus	Distinguished by the location of the activated receptor field and the pathway that is subsequently activated to transmit this information to the area of the somatosensory cortex representing that particular location
Intensity of stimulus (stimulus strength)	Distinguished by the frequency of action potentials initiated in an activated afferent neuron and the number of receptors (and afferent neurons) activated

Each somatosensory pathway is "labeled" according to modality and location.

On reaching the spinal cord, afferent information has two possible destinies: (1) it may become part of a reflex arc, bringing about an appropriate effector response, or (2) it may be relayed upward to the brain via ascending pathways for further processing and possible conscious awareness. Pathways conveying conscious somatic sensation, the **somatosensory pathways,** consist of discrete chains of neurons synaptically interconnected in a particular sequence to accomplish progressively more sophisticated processing of the sensory information. The afferent neuron with its peripheral receptor that first detects the stimulus is known as a **first-order sensory neuron.** It synapses on a **second-order sensory neuron,** either in the spinal cord or the medulla, depending on which sensory pathway is involved. This neuron then synapses on a **third-order sensory neuron** in the thalamus, and so on. With each step, the input is processed further. A particular sensory modality detected by a specialized receptor type is sent over a specific afferent and ascending pathway (a neural pathway committed to that modality) to excite a defined area in the somatosensory cortex. This process is known as **projection;** that is, a particular sensory input is "projected" to a specific region of the cortex. Thus, information is kept separated within specific **labeled lines** between the periphery and the cortex. In this way, even though all information is propagated to the CNS via the same type of signal (action potentials), the brain can decode the type and location of the stimulus. Table 6-1 summarizes how the CNS is informed of the type (what?), location (where?), and intensity (how much?) of a stimulus.

Activation of a sensory pathway at any point gives rise to the same sensation that would be produced by stimulation of the receptors in the body part itself. This phenomenon has served as the traditional explanation for **phantom pain**—for example, pain perceived as originating in the foot by a person whose leg has been amputated at the knee. Irritation of the severed endings of the afferent pathways in the stump can trigger action potentials that, upon reaching the foot region of the somatosensory cortex, are interpreted as pain in the missing foot. New evidence suggests that additionally the sensation of phantom pain might arise from the recently documented extensive remodeling of the brain region that originally handled sensations from the severed limb. This "remapping" of the "vacated" area of the brain is speculated to somehow lead to signals from elsewhere being misinterpreted as pain arising from the missing extremity.

Acuity is influenced by receptive field size and lateral inhibition.

Each sensory neuron responds to stimulus information only within a circumscribed region of the skin surface surrounding it; this region is known as its **receptive field.** The size of a receptive field varies inversely with the density of receptors in the region; the more closely receptors of a particular type are spaced, the smaller the area of skin each monitors. The smaller the receptive field in a region, the greater its **acuity** or **discriminative ability.** Compare the tactile (touch) discrimination in your fingertips with that in your elbow by "feeling" the same object with both. You are able to discern more precise information about the object with your richly innervated fingertips because the receptive fields there are small; as a result, each neuron signals information about small, discrete portions of the object's surface. An estimated 17,000 tactile mechanoreceptors are present in the fingertips and palm of each hand. In contrast, the skin over the elbow is served by relatively few sensory endings with larger receptive fields. Subtle differences within each large receptive field cannot be detected (Fig. 6-6). The distorted cortical representation of various body parts in the sensory homunculus (see p. 125) corresponds precisely with the innervation density; more cortical space is allotted for sensory reception from areas with smaller receptive fields and, accordingly, greater tactile discriminative ability.

Besides receptor density, a second factor influencing acuity is **lateral inhibition.** You can appreciate the importance of this phenomenon by slightly indenting the surface of your skin with the point of a pencil (Fig. 6-7a). The receptive

Figure 6-7 **Lateral Inhibition** (a) The receptor at the site of most intense stimulation is activated to the greatest extent. Surrounding receptors are also stimulated but to a lesser degree. (b) The most intensely activated receptor pathway halts transmission of impulses in the less intensely stimulated pathways through lateral inhibition. This process facilitates localization of the site of stimulation.

Figure 6-6 **Comparison of Discriminative Ability of Regions with Small versus Large Receptive Fields** The relative tactile acuity of a given region can be determined by the *two-point threshold of discrimination test*. If the two points of a pair of calipers applied to the surface of the skin stimulate two different receptive fields, two separate points will be felt. If the two points touch the same receptive field, they will be perceived as only one point. By adjusting the distance between the caliper points, one can determine the minimal distance at which the two points can be recognized as two rather than one, which is a reflection of the size of the receptive fields in the region. With this technique, it is possible to plot the discriminative ability of the body surface. The two-point threshold ranges from 2 mm in the fingertip (enabling one to read Braille, where the raised dots are spaced 2.5 mm apart) to 48 mm in the poorly discriminative skin of the calf. (a) Region with small receptive fields. (b) Region with large receptive fields.

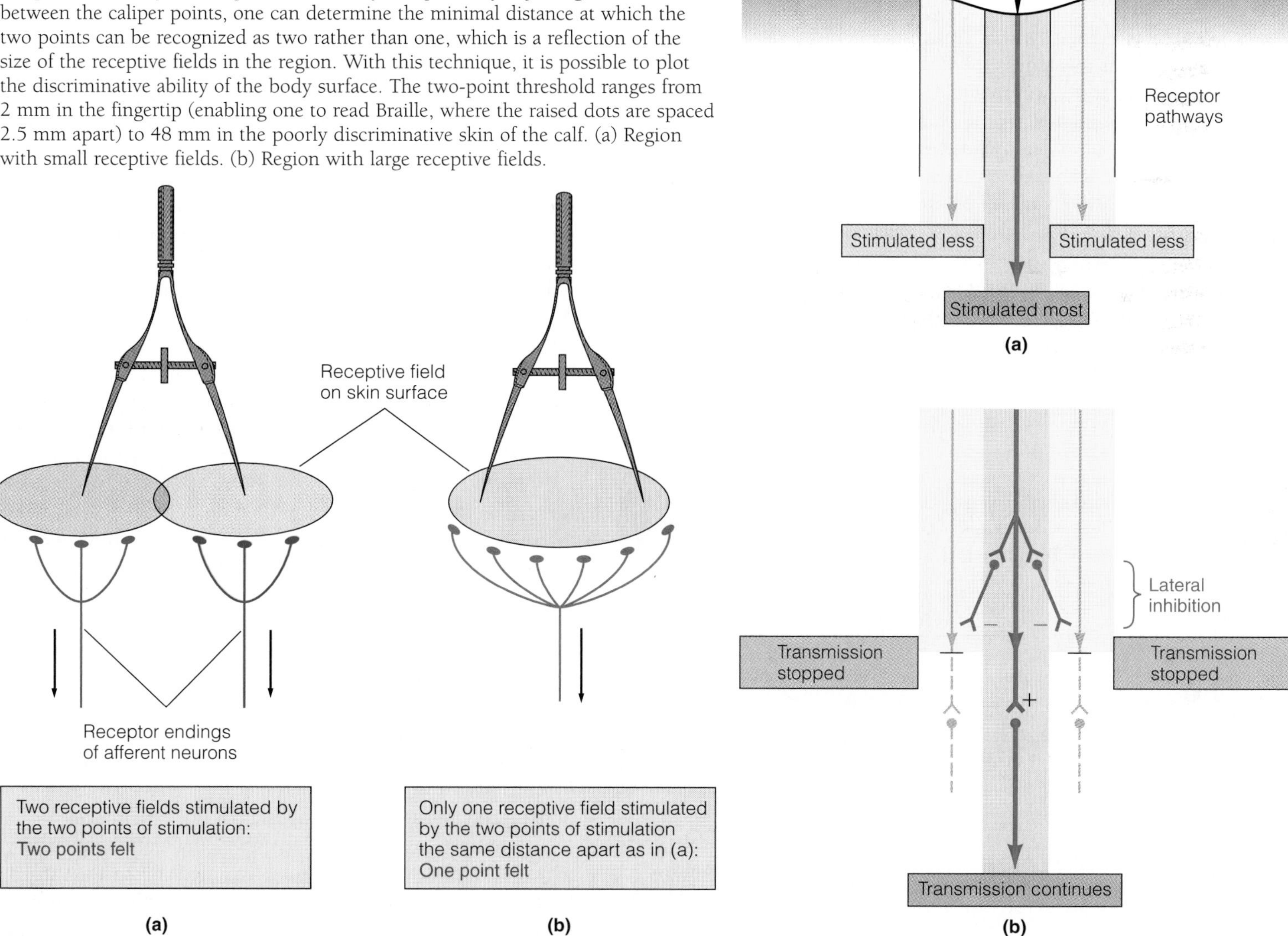

field is excited immediately under the center of the pencil point where the stimulus is most intense, but the surrounding receptive fields are also stimulated, only to a lesser extent because they are less distorted. If information from these marginally excited afferent fibers in the fringe of the stimulus area were to reach the cortex, localization of the pencil point would be blurred. To facilitate localization and sharpen contrast, lateral inhibition occurs within the CNS (Fig. 6-7b). The most strongly activated signal pathway originating from the center of the stimulus area inhibits the less excited pathways from the fringe areas. This occurs via inhibitory interneurons that pass laterally between ascending fibers serving neighboring receptive fields. Blockage of further transmission in the weaker inputs increases the contrast between wanted and unwanted information so that the pencil point can be precisely localized. The extent of lateral inhibitory connections within sensory pathways varies for different modalities. Those with the most lateral inhibition—touch and vision—bring about the most accurate localization.

Pain

Stimulation of nociceptors elicits the perception of pain plus motivational and emotional responses.

Pain is primarily a protective mechanism meant to bring to conscious awareness the fact that tissue damage is occurring or is about to occur. Unlike other somatosensory modalities, it is accompanied by motivated behavioral responses (such as withdrawal or defense) as well as emotional reactions (such as crying or fear). Also, unlike other sensations, the subjective perception of pain can be influenced by other past or present experiences (for example, heightened pain perception accompanying fear of the dentist or lowered pain perception in an injured athlete during a competitive event).

There are three categories of pain receptors: **mechanical nociceptors** respond to mechanical damage such as cutting, crushing, or pinching; **thermal nociceptors** respond to temperature extremes, especially heat; and **polymodal nociceptors** respond equally to all kinds of damaging stimuli, including irritating chemicals released from injured tissues. None of the nociceptors have specialized receptor structures; they are all naked nerve endings. Because of their value to survival, nociceptors do not adapt to sustained or repetitive stimulation. On the other hand, all nociceptors can be sensitized by the presence of *prostaglandins,* which greatly enhance the receptor response to noxious stimuli (that is, it hurts more when prostaglandins are present). Prostaglandins are a special group of fatty acid derivatives that are cleaved from the lipid bilayer of the plasma membrane and act locally on being released (see p. 716). Aspirinlike drugs inhibit the synthesis of prostaglandins, accounting at least in part for the **analgesic** (pain-relieving) properties of these drugs.

Pain impulses originating at nociceptors are transmitted to the CNS via one of two types of afferent fibers (▮ Table 6-2). Signals arising from mechanical and thermal nociceptors are transmitted over large myelinated **A-delta fibers** at rates of up to 30 meters/sec (the **fast pain pathway**). Impulses from polymodal nociceptors are carried by small unmyelinated **C fibers** at a much slower rate of 12 meters/sec (the **slow pain pathway**). Think about the last time you cut or burned your finger. You undoubtedly felt a sharp twinge of pain at first, with a more diffuse, disagreeable pain commencing shortly thereafter. Pain typically is perceived initially as a brief, sharp, prickling sensation that is easily localized; this is the fast pain pathway originating from specific mechanical or heat nociceptors. This feeling is followed by a dull, aching, poorly localized sensation that persists for a longer time and is more unpleasant; this is the slow pain pathway, which is activated by chemicals, especially **bradykinin,** a normally inactive substance that is activated by enzymes released into the ECF from damaged tissue. Bradykinin and related compounds not only provoke pain, presumably by stimulating the polymodal nociceptors, but they also contribute to the inflammatory response to tissue injury (chapter 12). The persistence of these chemicals might explain the long-lasting, aching pain that continues after removal of the mechanical or thermal stimulus that caused the tissue damage.

The primary afferent pain fibers synapse with specific second-order interneurons in the dorsal horn of the spinal cord (▬ Fig. 6-8a). One of the neurotransmitters released from these afferent pain terminals is **substance P,** which is believed to be unique to pain fibers. Ascending pain pathways have poorly understood destinations in the *somatosensory cortex,* the *thalamus,* and the *reticular formation.* The role of the cortex in pain perception is not clear, although it is probably important at least in localizing the pain. Pain can still be perceived in the absence of the cortex, presumably at the level of the thalamus. The reticular formation increases the level of alertness associated with the noxious encounter. Interconnections from the thalamus and reticular formation to the *hypothalamus* and *limbic system* elicit the behavioral and emotional responses accompanying the painful experience.

In contrast to the pain accompanying peripheral injury, which serves as a normal protective mechanism to warn of impending or actual damage to the body, abnormal chronic pain states are speculated to result from damage within the

Table 6-2
Characteristics of Pain

Fast Pain	*Slow Pain*
Occurs upon stimulation of mechanical and thermal nociceptors	Occurs upon stimulation of polymodal nociceptors
Carried by large myelinated A-delta fibers	Carried by small unmyelinated C fibers
Produces sharp, prickling sensation	Produces dull, aching, burning sensation
Easily localized	Poorly localized
Occurs first	Occurs second; persists for longer time; more unpleasant

pain pathways in the peripheral nerves or the CNS. That is, pain is perceived because of abnormal signaling within the pain pathways in the absence of peripheral injury or typical painful stimuli. For example, strokes that damage ascending pathways can lead to an abnormal, persistent sensation of pain.

The brain has a built-in analgesic system.

In addition to the chain of neurons connecting peripheral nociceptors with higher CNS structures for pain perception, the CNS also contains a neuronal system that suppresses pain. Our knowledge about this built-in **analgesic system** is still fragmentary. It appears that there are neural mechanisms that suppress transmission in the pain pathways as they enter the spinal cord. Electrical stimulation of the **periaqueductal gray matter** (gray matter surrounding the cerebral aqueduct, a narrow canal that connects the third and fourth ventricular cavities) results in profound analgesia, as does stimulation of the reticular formation within the brain stem. These two regions are thought to be part of a descending analgesic pathway (Fig. 6-8b) that blocks by presynaptic inhibition (see p. 102) the release of substance P from afferent pain fiber terminals.

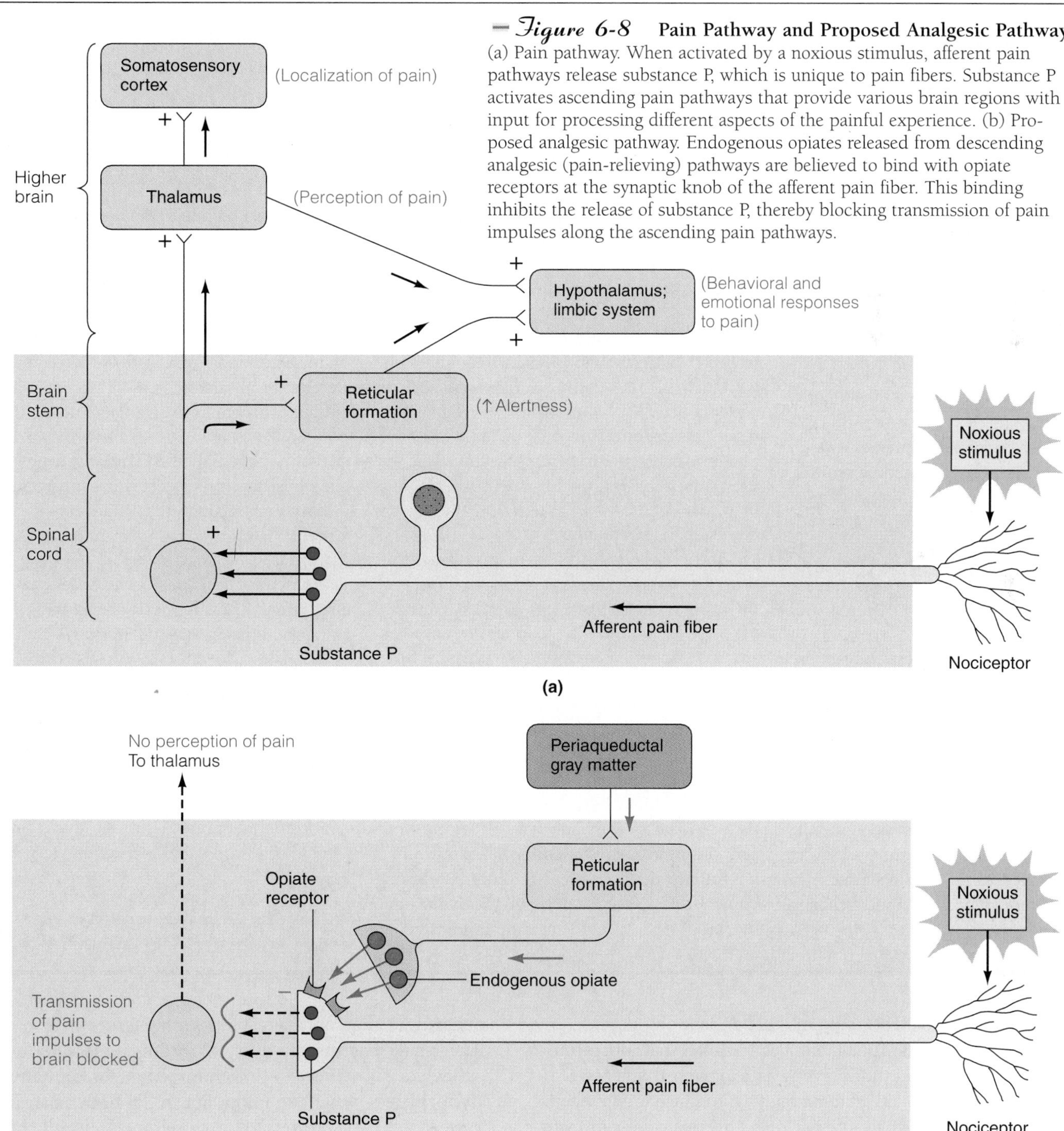

Figure 6-8 **Pain Pathway and Proposed Analgesic Pathway** (a) Pain pathway. When activated by a noxious stimulus, afferent pain pathways release substance P, which is unique to pain fibers. Substance P activates ascending pain pathways that provide various brain regions with input for processing different aspects of the painful experience. (b) Proposed analgesic pathway. Endogenous opiates released from descending analgesic (pain-relieving) pathways are believed to bind with opiate receptors at the synaptic knob of the afferent pain fiber. This binding inhibits the release of substance P, thereby blocking transmission of pain impulses along the ascending pain pathways.

CONCEPTS, CHALLENGES, AND CONTROVERSIES

Acupuncture: Is It For Real?

It sounds like science fiction. How can a needle inserted in the hand relieve a toothache? **Acupuncture analgesia (AA)**, the technique of relieving pain by the insertion and manipulation of threadlike needles at key points, has been practiced in China for over 2,000 years but is relatively new to Western medicine and still remains controversial in our country. Many Western scientists were skeptical because, until recently, the phenomenon could not be explained on the basis of any known, logical, physiological principles, although a tremendous body of anecdotal evidence in support of the effectiveness of AA existed in China.

According to a leading expert in acupuncture, the technique was not embraced in Western culture because of a clash in philosophies between West and East:

> Western medical science is quick to reject a phenomenon if it does not fit the current scientific theories. Chinese Taoism had a distaste for explanatory theories and chose instead merely to observe phenomena in order to be in harmony with mother nature. If a needle in the hand cured a toothache, that was sufficient for Chinese Taoism. For Western medicine acupuncture was impossible and hence was relegated to the wastebasket of placebo effects.*

The *placebo effect* refers to a chemical or technique that brings about a desired response through the power of suggestion or distraction rather than through any direct action. The placebo effect was first documented in 1945 when a physician injected patients with what they thought was morphine for pain relief, but some received sugar (a placebo) instead. Pain was relieved in 70% of those who actually received morphine, but, surprisingly, 35% of those who received sugar, but believed they were receiving morphine, also reported pain relief.

Because the Chinese were content with anecdotal evidence for the success of AA, this phenomenon did not come under close scientific scrutiny until the last two decades when European and American scientists started studying it. As a result of these efforts, an impressive body of rigorous scientific investigation supports the contention that AA really works (that is, by a physiological rather than a placebo/psychological effect). Furthermore, its mechanisms of action have become apparent. Indeed, more is known about the underlying physiological mechanisms of AA than of many conventional medical techniques, such as gas anesthesia.

AA has been proven to be effective in treating chronic pain and to exert a real physical effect in that it is more effective than placebo controls. In fact, AA compares favorably with morphine for treating chronic pain. In controlled clinical studies, 55% to 85% of patients were helped by AA. (By comparison, 70% of patients benefit from morphine therapy.) Pain relief was reported by only 30% to 35% of placebo controls (individuals who thought they were receiving proper AA treatment, but in whom needles were inserted in the wrong places or not deep enough).

The overwhelming body of evidence supports the *acupuncture endorphin hypothesis* as the primary mechanism of AA's action. According to this hypothesis, acupuncture needles activate specific afferent nerve fibers, which send impulses to the central nervous system. Here the incoming impulses activate three centers (a spinal cord center, a mid-brain center, and a hormonal center, the hypothalamus/anterior pituitary unit) to cause analgesia. All three centers have been shown to block pain transmission through use of endorphins and closely related compounds. Several other neurotransmitters, such as serotonin and norepinephrine, as well as cortisol, the major hormone released during stress, are implicated as well. (Pain relief in placebo controls is believed to occur as a result of the persons subconsciously activating their own built-in analgesic system.)

In the United States, AA is not used in mainstream medicine, even by physicians who have been convinced by scientific evidence that the technique is valid. AA methodology is not taught in our medical colleges, and the techniques take time to learn. Also, AA is much more time-consuming than using drugs. Western physicians who have been trained to use drugs to solve most pain problems are generally reluctant to scrap their known methods for an unfamiliar, time-consuming technique. In a limited way, however, acupuncture is gaining favor as an alternative treatment for relief of chronic pain, especially since analgesic drugs can have troublesome side effects. The number of U.S. physicians who use acupuncture has increased sixfold in the past decade from about 500 to 3,000. AA has caught on more extensively in Europe than in the United States. For example, in a recent survey of pain clinics in Germany, over 90% of the physicians reported using acupuncture.

Because acupuncture is relatively new in the United States, the laws governing its use vary from state to state. About half the states have acupuncture licensing agencies. Some states permit only trained physicians to perform AA, whereas others register nonphysician acupuncturists. Eleven schools in the United States are currently offering four-year training programs in AA for nonphysicians, an indication that AA will come to be used more commonly for pain relief in our country, if not by physicians then by others trained in this now scientifically legitimate technique. Of the 10,000 practicing acupuncturists in the United States today, over two-thirds are nonphysicians.

*Gabriel Stux and Bruce Pomeranz, *Basics of Acupuncture*, 2nd ed. (Springer Verlag, 1994), p. 1.

The built-in analgesic system is dependent on the presence of **opiate receptors.** It has long been known that **morphine**, a derivative of the opium poppy, is a powerful analgesic. It seemed highly unlikely that the body would be endowed with opiate receptors only to interact with chemicals derived from a flower! A search was therefore undertaken to discover the substances that normally bind with these opiate receptors. The result was the discovery of **endogenous opiates** (morphinelike substances)—the **endorphins, enkephalins,** and **dynorphin**—which are important in the body's natural analgesic system. According to a proposed model for the analgesic system, these endogenous opiates serve as analgesic neurotransmitters; they are released from a descending analgesic pathway and bind with opiate receptors on the afferent pain fiber terminal. This binding suppresses the release of substance P, thereby blocking further transmission of the pain signal. Morphine binds to these same opiate receptors, which accounts for its analgesic properties.

It is unclear how the natural pain-suppressing mechanisms are activated. Factors known to modulate pain include exercise (endorphins are believed to be released during prolonged exercise and presumably are responsible for the "runner's high"), acupuncture, hypnosis, and stress. (See the accompanying boxed feature, ◆ Concepts, Challenges, and Controversies.) There is evidence that some types of stress induce analgesia via the opiate pathway and other less understood nonopiate mechanisms. It is sometimes disadvantageous for a stressed organism to display the normal reaction to pain. For example, when two males are fighting for dominance of the herd, withdrawing, escaping, or resting when injured would mean certain defeat.

Pain can frequently be managed by drugs that suppress transmitter activity at some point along the pain pathway. Surgical intervention is occasionally necessary for relief of intractable pain, as may occur in terminally ill cancer patients. Surgical relief from pain entails either interrupting the ascending pain pathways within the spinal cord or severing certain pathways in the brain to modify emotional response to the pain. Recall that a painful experience includes both the sensation of pain and an emotional and behavioral reaction to it. These two components can be dissociated. The person still feels pain but does not mind as much.

Eye: Vision

Somatic sensation is detected by widely distributed receptors that provide information about the body's interactions with the environment in general. In contrast, each of the special senses has highly localized, extensively specialized receptors that respond to unique environmental stimuli. The special senses include **vision, hearing, taste** and **smell,** to which we now turn our attention, starting with vision.

The eye is a fluid-filled sphere enclosed by three specialized tissue layers.

The eyes capture the patterns of illumination in the environment as an "optical picture" on a layer of light-sensitive cells, the *retina,* much as a camera captures an image on film. Just as film can be developed into a visual likeness of the original image, the coded image on the retina is transmitted through a series of progressively more complex steps of visual processing until it is finally consciously perceived as a visual likeness of the original image.

Each **eye** (▮ Table 6-3) is a spherical, fluid-filled structure enclosed by three layers. From outermost to innermost, these are (1) the sclera/cornea; (2) the choroid/ciliary body/iris; and (3) the retina (▬ Fig. 6-9a). Most of the eyeball is covered by a tough outer layer of connective tissue, the **sclera,** which forms the visible white part of the eye (Fig. 6-9b). Anteriorly (toward the front), the outer layer consists of the transparent **cornea** through which light rays pass into the interior of the

▬ *Figure 6-9* **Structure of the Eye** (a) Internal sagittal view. (b) External front view.

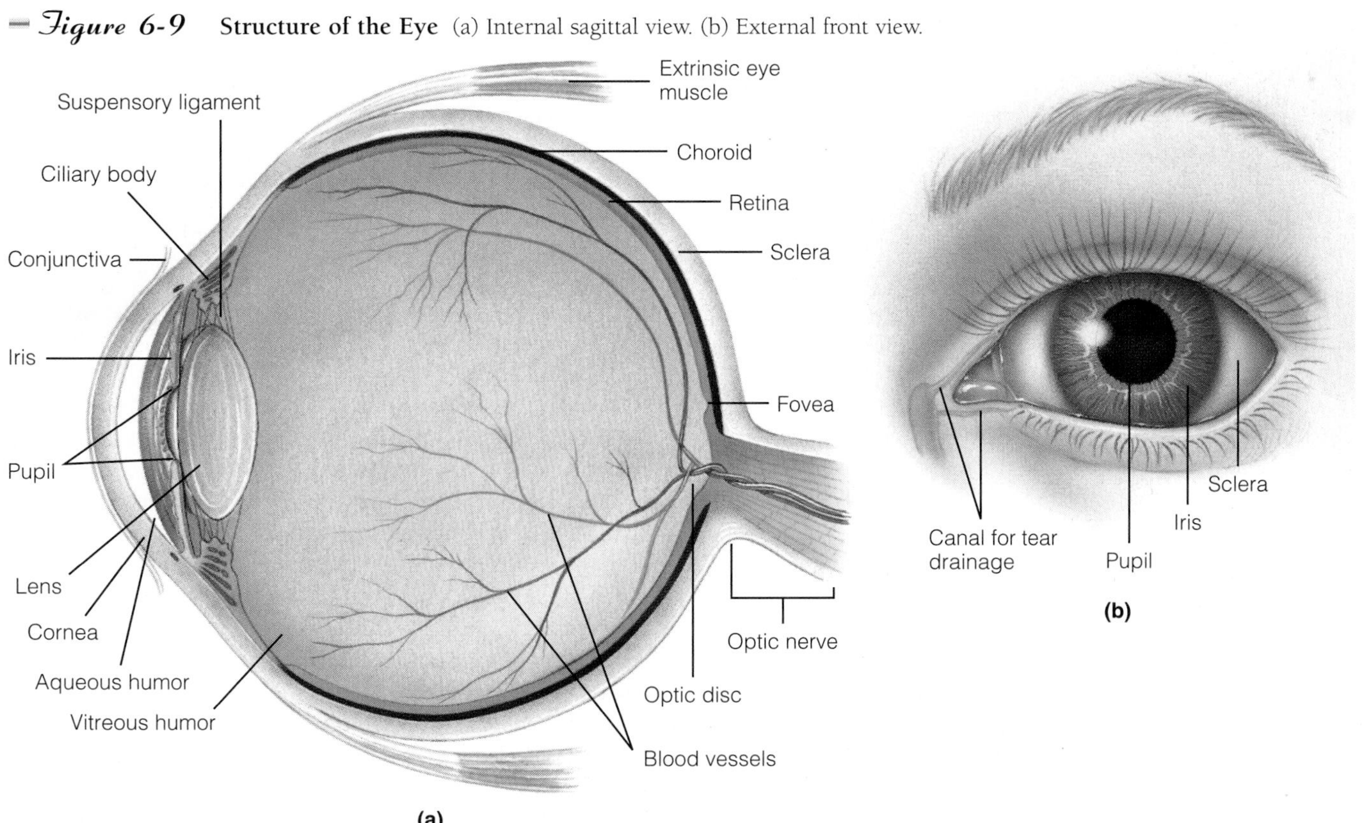

Table 6-3 Functions of the Major Components of the Eye

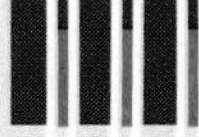

Structure (in alphabetical order)	*Location*	*Function*
Aqueous humor	Anterior cavity between cornea and lens	Clear watery fluid that is continually formed and carries nutrients to the cornea and lens
Bipolar neurons	Middle layer of nerve cells in retina	Important in retinal processing of light stimulus
Blind spot	Point slightly off-center on retina that is devoid of photoreceptors (also known as optic disc)	Route for passage of optic nerve and blood vessels
Choroid	Middle layer of eye	Pigmented to prevent scattering of light rays in eye; contains blood vessels that nourish retina; anteriorly specialized to form ciliary body and iris
Ciliary body	Specialized anterior derivative of the choroid layer; forms a ring around the outer edge of the lens	Produces aqueous humor and contains ciliary muscle
Ciliary muscle	Circular muscular component of ciliary body; attaches to lens by means of suspensory ligaments	Important in accommodation
Cones	Photoreceptors in outermost layer of retina	Responsible for high acuity, color, and day vision
Cornea	Anterior clear outermost layer of eye	Contributes most extensively to eye's refractive ability
Fovea	Exact center of retina	Region with greatest acuity
Ganglion cells	Inner layer of retina	Important in retinal processing of light stimulus; form optic nerve
Iris	Visible pigmented ring of muscle within aqueous humor	Varies size of pupil by variable contraction; responsible for eye color
Lens	Between aqueous humor and vitreous humor; attaches to ciliary muscle by suspensory ligaments	Provides variable refractive ability during accommodation
Macula lutea	Area immediately surrounding the fovea	Has high acuity because of abundance of cones
Optic disc	(*see* blind spot)	
Optic nerve	Leaves each eye at optic disc (blind spot)	First part of visual pathway to the brain
Pupil	Anterior round opening in middle of iris	Permits variable amounts of light to enter eye
Retina	Innermost layer of eye	Contains the photoreceptors (rods and cones)
Rods	Photoreceptors in outermost layer of retina	Responsible for high sensitivity, black-and-white, and night vision
Sclera	Tough outer layer of eye	Protective connective tissue coat; forms visible white part of eye; anteriorly specialized to form cornea
Suspensory ligaments	Suspended between ciliary muscle and lens	Important in accommodation
Vitreous humor	Between lens and retina	Semifluid, jellylike substance that helps maintain spherical shape of eye

eye. The middle layer underneath the sclera is the highly pigmented **choroid**, which contains many blood vessels that nourish the retina. The choroid layer becomes specialized anteriorly to form the *ciliary body* and *iris,* which will be described shortly. The innermost coat under the choroid is the **retina**, which consists of an outer pigmented layer and an inner nervous tissue layer. The latter contains the **rods** and **cones**, the photoreceptors that convert light energy into nerve impulses. Like the black walls of a photographic studio, the pigment in the choroid and retina absorbs light after it strikes the retina to prevent reflection or scattering of light within the eye.

The interior of the eye consists of two fluid-filled cavities, separated by a **lens**, all of which are transparent to permit light to pass through the eye from the cornea to the retina. The anterior (front) cavity between the cornea and lens contains a clear watery fluid, the **aqueous humor**, and the larger posterior (rear) cavity between the lens and retina contains a semifluid, jellylike substance, the **vitreous humor.**

The vitreous humor is important in maintaining the spherical shape of the eyeball. The aqueous humor carries nutrients for the cornea and lens, both of which lack a blood supply. Blood vessels in these structures would impede the passage of light to the photoreceptors. Aqueous humor is produced at a rate of about 5 ml/day by a capillary network within the **ciliary body**, a specialized anterior derivative of the choroid layer. This fluid drains into a canal at the edge of the cornea and eventually enters the blood (■ Fig. 6-10). If aqueous humor is not drained as rapidly as it forms (for example,

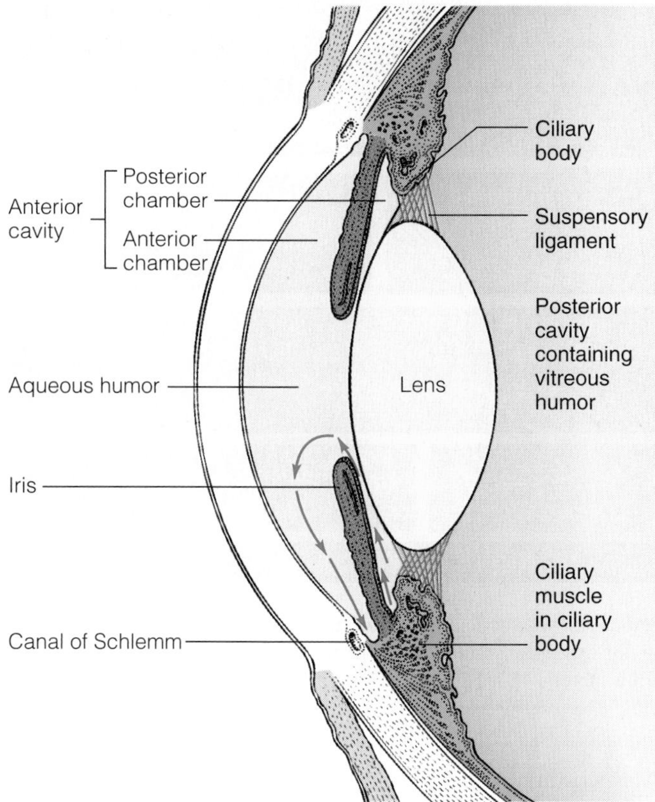

Figure 6-10 Formation and Drainage of Aqueous Humor Aqueous humor is formed by a capillary network in the ciliary body, then drains into the canal of Schlemm, and eventually enters the blood.

because of a blockage in the drainage canal), the excess will accumulate in the anterior cavity, causing the intraocular ("within the eye") pressure to rise. This condition is known as **glaucoma.** The excess aqueous humor pushes the lens backward into the vitreous humor, which in turn is pushed against the inner neural layer of the retina. This compression causes retinal and optic nerve damage that can lead to blindness if the condition is not treated.

The amount of light entering the eye is controlled by the iris.

Not all of the light passing through the cornea reaches the light-sensitive photoreceptors because of the presence of the **iris,** a thin, pigmented smooth muscle that forms a visible ringlike structure within the aqueous humor (Fig. 6-9a and b). The pigment in the iris is responsible for eye color. The round opening in the center of the iris through which light enters the interior portions of the eye is the **pupil.** The size of this opening can be adjusted by variable contraction of the iris muscles to admit more or less light as needed, much as the shutter controls the amount of light entering a camera. The iris contains two sets of smooth muscle networks, one *circular* (the muscle fibers run in a ringlike fashion within the iris) and the other *radial* (the fibers project outward from the pupillary margin like bicycle spokes) (Fig. 6-11). Because muscle fibers shorten when they contract, the pupil gets

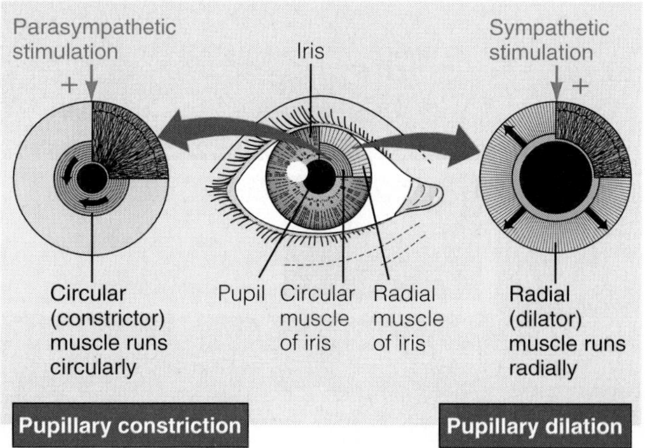

Figure 6-11 Control of Pupillary Size

smaller when the **circular** (or **constrictor**) **muscle** contracts and forms a smaller ring. This reflex pupillary constriction occurs in bright light to decrease the amount of light entering the eye. When the **radial** (or **dilator**) **muscle** shortens, the size of the pupil increases. Such pupillary dilation occurs in dim light to allow the entrance of more light.

Iris muscles are controlled by the autonomic nervous system. Parasympathetic nerve fibers innervate the circular muscle, and sympathetic fibers supply the radial muscle. Acting via the autonomic nervous system, conditions other than light can induce changes in pupillary size. For example, dilation of the pupils accompanies generalized discharge of the sympathetic nervous system in response to actual or impending danger.

The eye refracts the entering light to focus the image on the retina.

Light is a form of electromagnetic radiation composed of particlelike individual packets of energy called **photons** that travel in wavelike fashion. The distance between two wave peaks is known as the **wavelength** (Fig. 6-12). The pho-

Figure 6-12 Properties of an Electromagnetic Wave A wavelength is the distance between two wave peaks. Intensity refers to the amplitude of the wave.

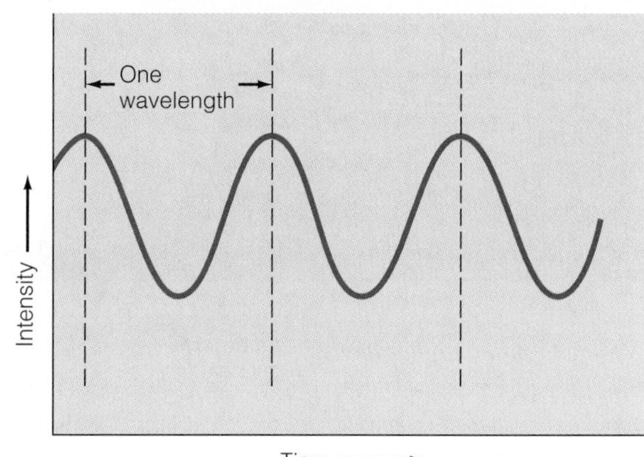

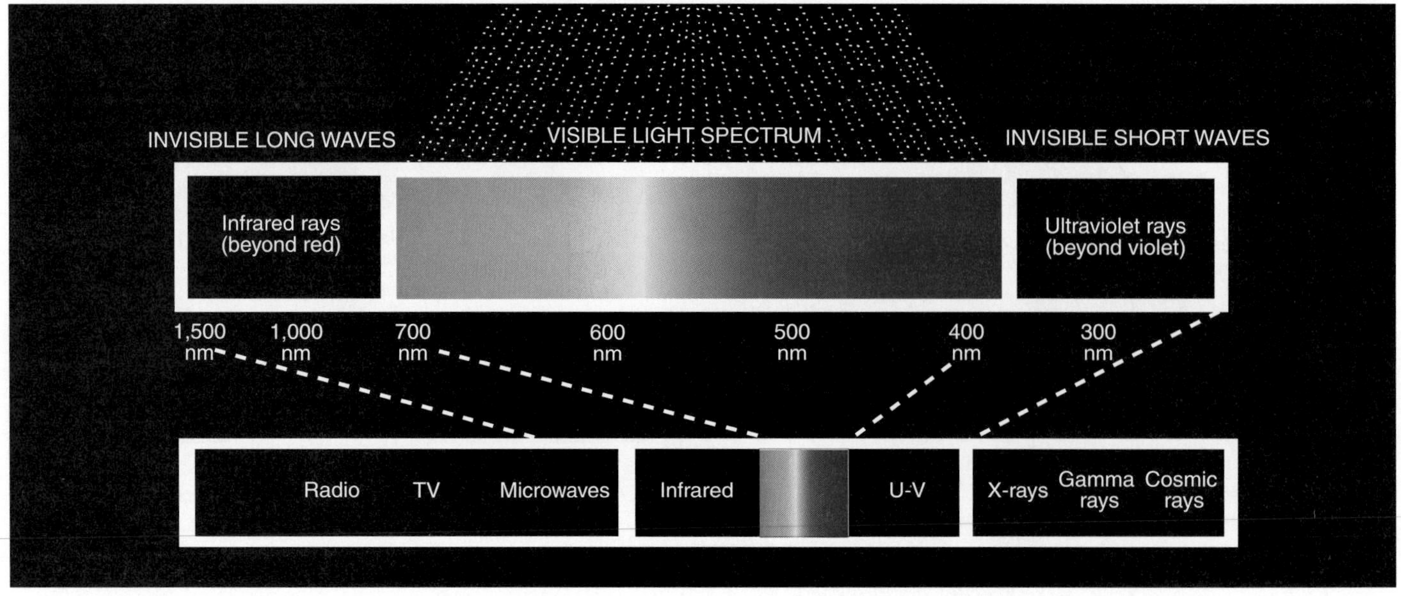

Figure 6-13 Electromagnetic Spectrum The wavelengths in the electromagnetic spectrum range from 10^4m (10 km—for example, long radio waves) to less than 10^{-14}m (quadrillionths of a meter—for example, gamma and cosmic rays). The visible spectrum includes wavelengths ranging from 400 to 700 nanometers (nm; billionths of a meter).

toreceptors in the eye are sensitive only to wavelengths between 400 and 700 nanometers (nm; billionths of a meter between peaks). This **visible light** is only a small portion of the total electromagnetic spectrum (Fig. 6-13). Light of different wavelengths in this visible band is perceived as different color sensations. Short wavelengths are sensed as violet and blue; long wavelengths are interpreted as orange and red.

In addition to having variable wavelengths, light energy also varies in **intensity;** that is, the amplitude, or height, of the wave (Fig. 6-12). Dimming a bright red light does not change its color; it just becomes less intense or less bright.

Light waves *diverge* (radiate outward) in all directions from every point of a light source. The forward movement of a light wave in a particular direction is known as a **light ray.** Divergent light rays reaching the eye must be bent inward to be focused back into a point on the light-sensitive retina to provide an accurate image of the light source (Fig. 6-14).

Figure 6-14 Focusing of Diverging Light Rays Diverging light rays must be bent inward to be focused.

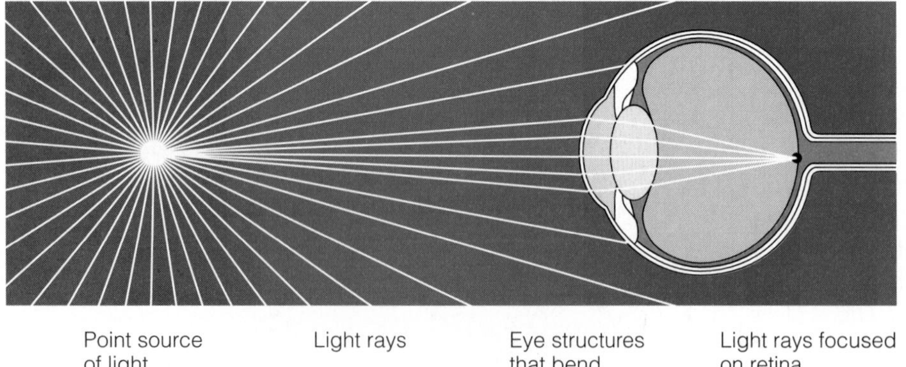

The bending of a light ray **(refraction)** occurs when the ray passes from a medium of one density into a medium of a different density (Fig. 6-15). Light travels faster through air than through other transparent media such as water and glass. When a light ray enters a medium of greater density, it is slowed down (the converse is also true). The course of direction of the ray changes if it strikes the surface of the new medium at any angle other than perpendicular.

Two factors contribute to the degree of refraction: the comparative densities of the two media (the greater the difference in density, the greater the degree of bending) and the angle at which the light strikes the second medium (the greater the angle, the greater the refraction).

With a curved surface such as a lens, the greater the curvature, the greater the degree of bending and the stronger the lens. When a light ray strikes the curved surface of any object of greater density, the direction of refraction depends on the angle of the curvature (Fig. 6-16). A lens with **convex** surfaces converges light rays, bringing them closer together, a requirement for bringing an image to a focal point. Refractive surfaces of the eye are therefore convex. A lens with **concave** surfaces diverges light rays (spreads them farther apart). A concave lens is useful for correcting certain refractive errors of the eye, such as nearsightedness.

The two structures most important in the eye's refractive ability are the *cornea* and the *lens.* The curved corneal surface, the first structure light passes through as it enters the eye, contributes most extensively to the eye's total refractive ability because the

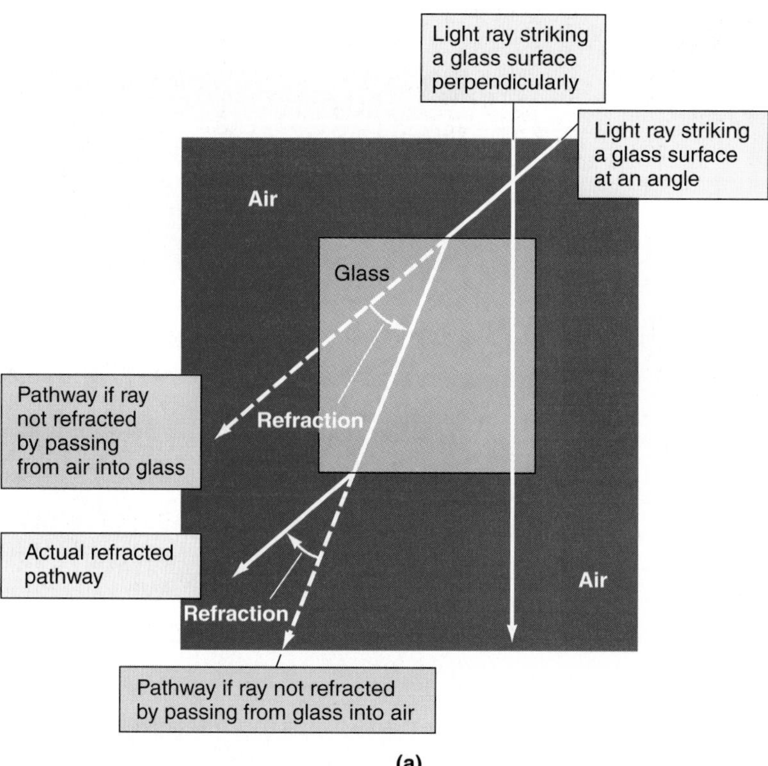

Figure 6-15 **Refraction** (a) A light ray is bent (refracted) when it strikes the surface of a medium of different density than the one in which it had been traveling (for example, moving from air into glass) at any angle other than perpendicular to the new medium's surface. (b) The pencil in the glass of water appears to bend. What is happening, though, is that the light rays coming to the camera (or your eyes) are bent as they pass through the water, then the glass, and then the air. Consequently, the pencil appears distorted.

difference in density at the air/corneal interface is much greater than the differences in density between the lens and the fluids surrounding it. In **astigmatism,** the curvature of the cornea is uneven, so light rays are unequally refracted. The refractive ability of a person's cornea remains constant because the curvature of the cornea never changes. In contrast, the refractive ability of the lens can be adjusted by changing its curvature as needed for near or far vision.

The refractive structures of the eye must bring light images into focus on the retina for clear vision. If an image is focused

Figure 6-16 **Refraction by Convex and Concave Lenses** (a) A lens with a convex surface, which converges the rays (brings them closer together). (b) A lens with a concave surface, which diverges the rays (spreads them farther apart).

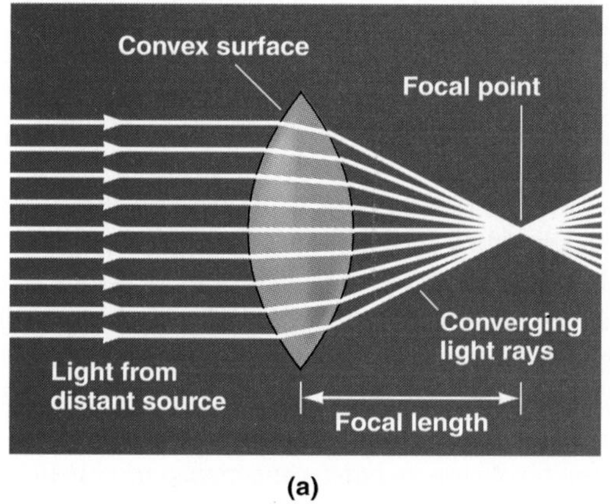

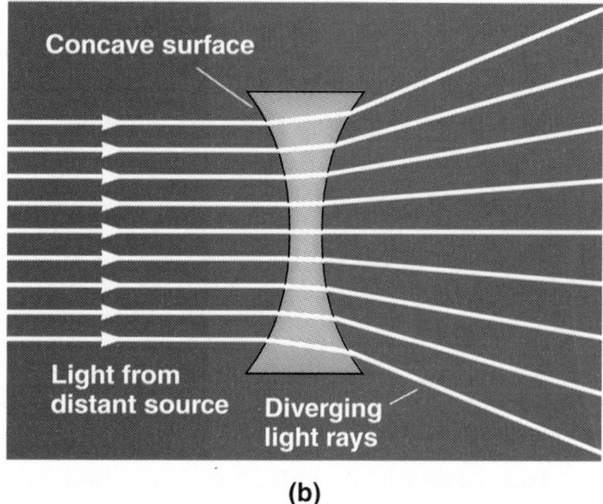

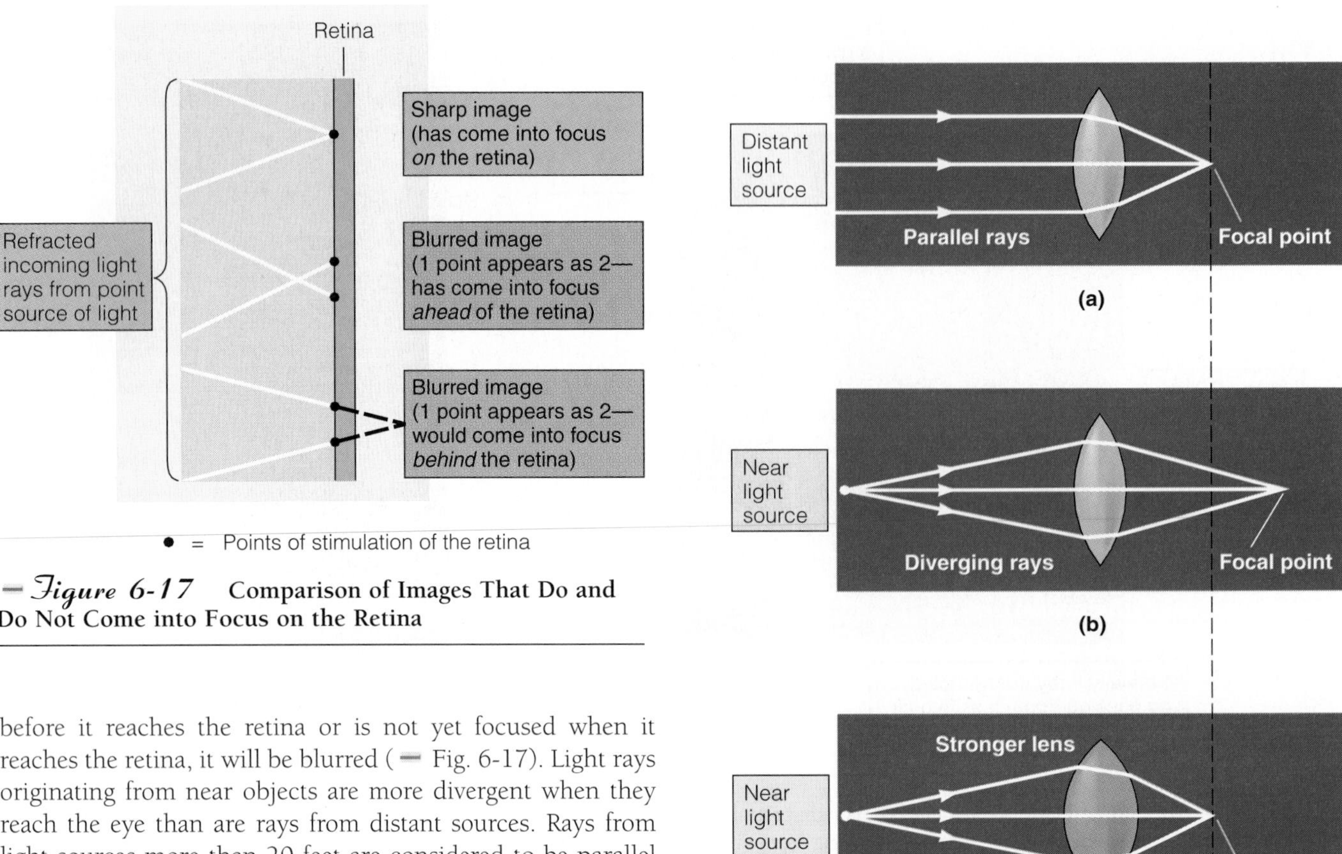

Figure 6-17 Comparison of Images That Do and Do Not Come into Focus on the Retina

Figure 6-18 Focusing of Distant and Near Sources of Light (a) A distant (far) light source is a light source more than 20 feet from the eye. By the time the rays reach the eye from a distant source, they are considered to be parallel. (b) The rays from a near light source (a light source less than 20 feet from the eye) are still diverging when they reach the eye. A longer distance is required for a lens of a given strength to bend the diverging rays from a near light source into focus (compared to the parallel rays from a distant light source). (c) To focus both a distant and a near light source in the same distance (the distance between the lens and retina), a stronger lens must be used for the near source. A stronger lens is able to focus a near image in the same distance as a weaker lens focuses a distant image.

before it reaches the retina or is not yet focused when it reaches the retina, it will be blurred (Fig. 6-17). Light rays originating from near objects are more divergent when they reach the eye than are rays from distant sources. Rays from light sources more than 20 feet are considered to be parallel by the time they reach the eye. For a given refractive ability of the eye, a near source of light requires a greater distance behind the lens for focusing than a far source does, because the near source rays are still diverging when they reach the eye (Fig. 6-18a and b).

In a particular eye, the distance between the lens and the retina always remains the same. To bring both near and far light sources into focus on the retina (that is, in the same distance), a stronger lens must be used for the near source (Fig. 6-18c). The strength of the lens can be adjusted through the process of accommodation, to which we now turn our attention.

Accommodation increases the strength of the lens for near vision.

The ability to adjust the strength of the lens so that both near and far sources can be focused on the retina is known as **accommodation.** The strength of the lens depends on its shape, which in turn is regulated by the ciliary muscle.

The **ciliary muscle** is part of the ciliary body, an anterior specialization of the choroid layer. The ciliary body has two major components: the ciliary muscle and the capillary network that produces the aqueous humor. The ciliary muscle is a circular ring of smooth muscle attached to the lens by **suspensory ligaments** (Fig. 6-19).

When the ciliary muscle is relaxed, the suspensory ligaments are taut, and they pull the lens into a flattened, weakly refractive shape (Fig. 6-19c). As the muscle contracts, its circumference decreases, slackening the tension in the suspensory ligaments (Fig. 6-19d). When the lens is subjected to less tension by the suspensory ligaments, it assumes a more spherical shape because of its inherent elasticity. The greater curvature of the more rounded lens increases its strength, causing greater bending of light rays.

In the normal eye, the ciliary muscle is relaxed and the lens is flat for far vision, but the muscle contracts to allow the lens to become more convex and stronger for near vision. The ciliary muscle is controlled by the autonomic nervous system. Sympathetic nerve fibers induce relaxation of the ciliary muscle for far vision, whereas the parasympathetic nervous system causes the muscle's contraction for near vision.

The lens is an elastic structure consisting of transparent fibers. Occasionally, these fibers become opaque so that light rays cannot pass through, a condition known as a **cataract.**

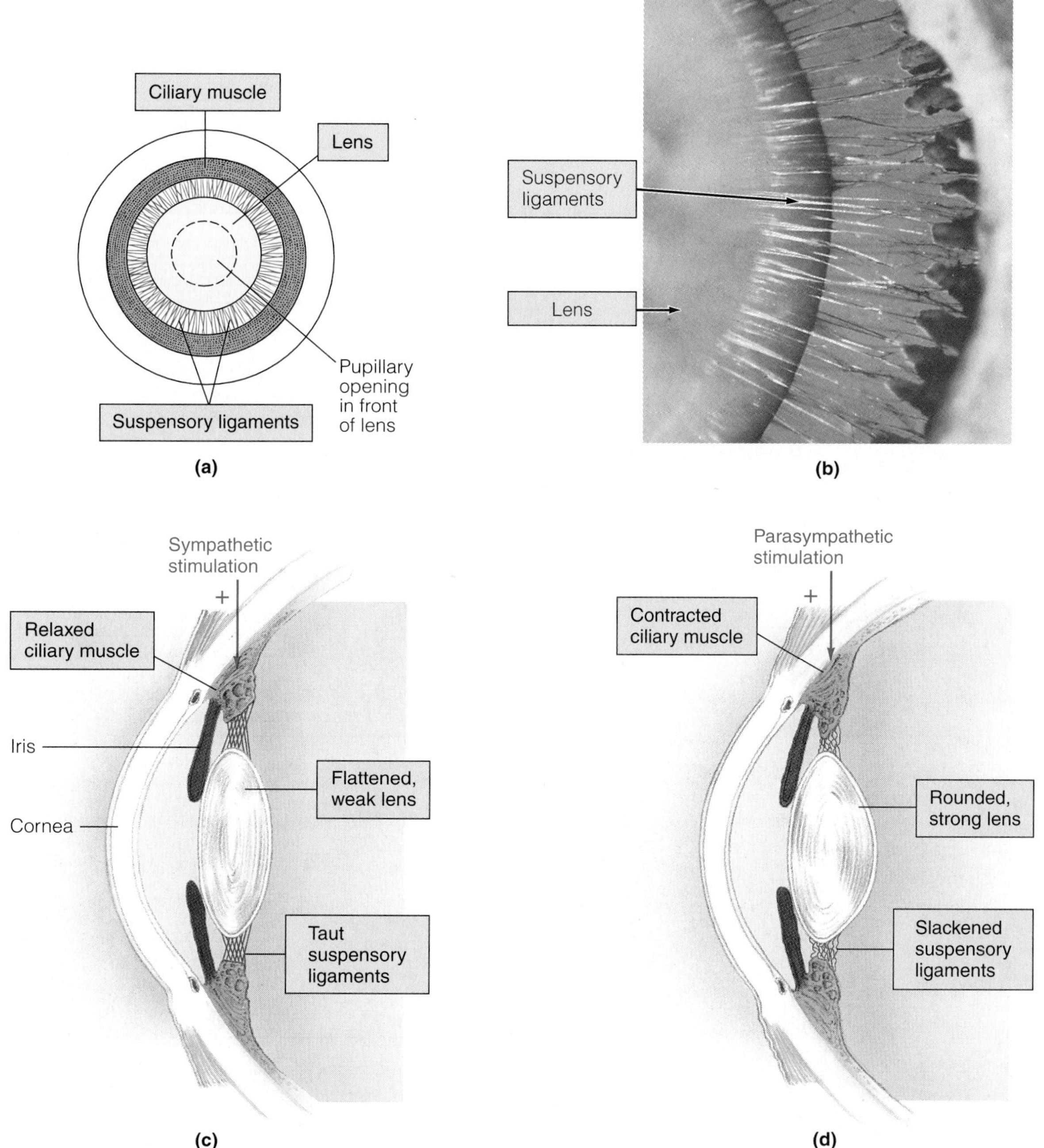

Figure 6-19 **Mechanism of Accommodation** (a) Schematic representation of suspensory ligaments extending from the ciliary muscle to the outer edge of the lens. (b) Scanning electron micrograph showing the suspensory ligaments attached to the lens. (c) When the ciliary muscle is relaxed, the suspensory ligaments are taut, putting tension on the lens so that it is flat and weak. (d) When the ciliary muscle is contracted, the suspensory ligaments become slack, reducing the tension on the lens. The lens can then assume a stronger, rounder shape because of its elasticity.

The defective lens can usually be surgically removed and vision restored by an implanted artificial lens or compensating eyeglasses.

Throughout life, only cells at the outer edges of the lens are replaced. Cells in the center of the lens are in double jeopardy. Not only are they oldest, but they also are the farthest away from the aqueous humor, the lens' nutrient source. With advancing age, these nonrenewable central cells die and become stiff. With loss of elasticity, the lens is no longer able to assume the spherical shape required to accommodate for near vision. This age-related reduction in accommodative ability, **presbyopia**, affects most people by middle age (45 to 50), requiring them to resort to corrective lens for near vision (reading).

Other common vision disorders are *nearsightedness (myopia)* and *farsightedness (hyperopia)*. In a normal eye

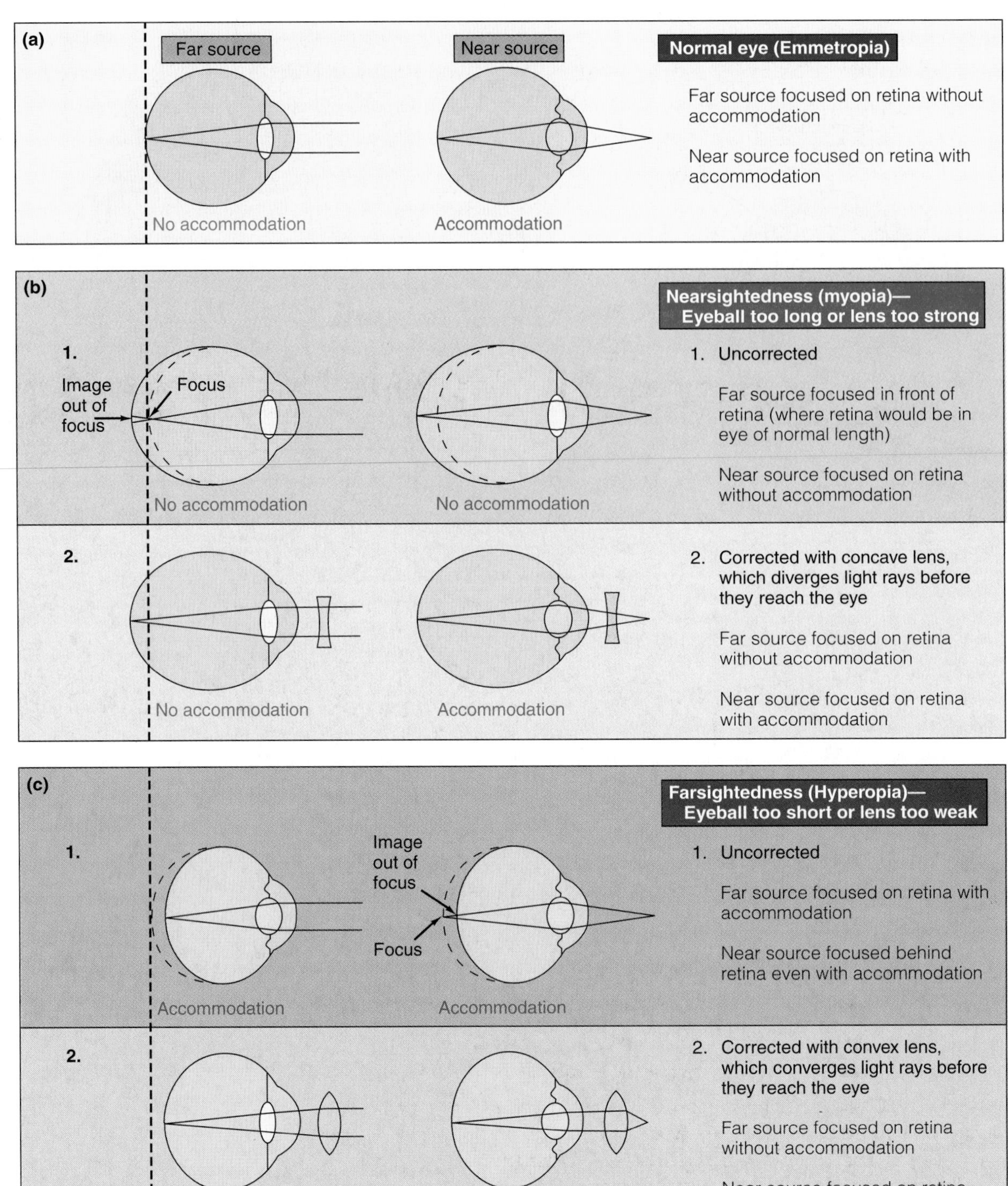

Figure 6-20 **Emmetropia, Myopia, and Hyperopia**

emmetropia (Fig. 6-20a), a far light source is focused on the retina without accommodation, whereas the strength of the lens is increased by accommodation to bring a near source into focus. In **myopia** (Fig. 6-20b1), because the eyeball is too long or the lens is too strong, a near light source is brought into focus on the retina without accommodation (even though accommodation is normally used for near vision), whereas a far light source is focused in front of the retina and is blurry. Thus, a myopic individual has better near vision than far vision, a condition that can be corrected by a concave lens (Fig. 6-20b2). With **hyperopia** (Fig. 6-20c1), either the eyeball is too short or the lens is too weak. Far objects are focused on the retina only with accommodation, whereas near objects are focused behind the retina even with accommodation and, accordingly, are blurry. Thus, a hyperopic individual has better far vision than near vision, a condi-

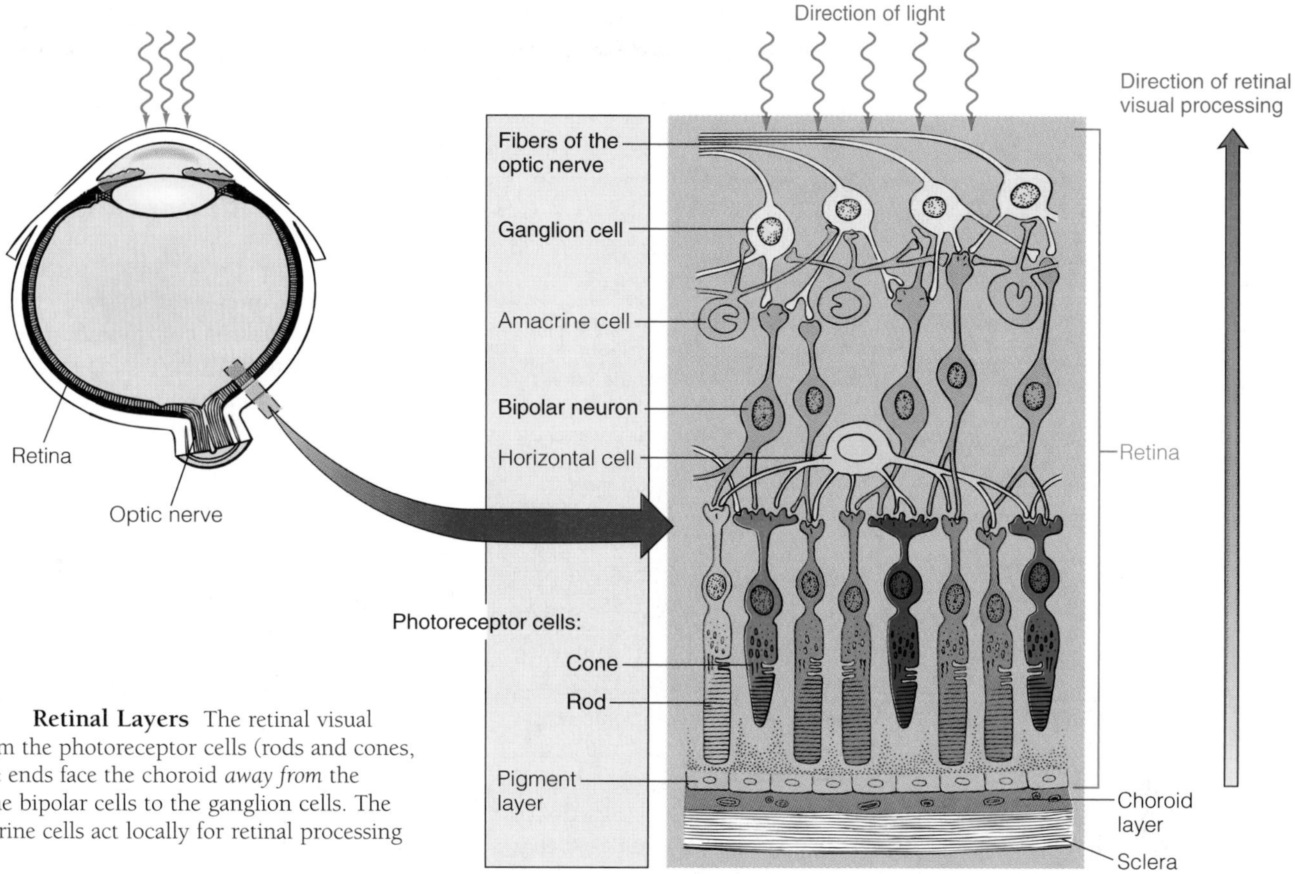

Direction of light
Direction of retinal visual processing
Fibers of the optic nerve
Ganglion cell
Amacrine cell
Bipolar neuron
Horizontal cell
Retina
Optic nerve
Photoreceptor cells:
Cone
Rod
Pigment layer
Choroid layer
Sclera

Figure 6-21 Retinal Layers The retinal visual pathway extends from the photoreceptor cells (rods and cones, whose light-sensitive ends face the choroid *away from* the incoming light) to the bipolar cells to the ganglion cells. The horizontal and amacrine cells act locally for retinal processing of visual input.

tion that can be corrected by a convex lens (Fig. 6-20c2). Such vision tends to get worse as the person gets older because of loss of accommodative ability with the onset of presbyopia.

Light must pass through several retinal layers before reaching the photoreceptors.

The major function of the eye is to focus light rays from the environment on the *rods* and *cones,* the photoreceptor cells of the retina. The photoreceptors then transform the light energy into electrical signals for transmission to the CNS.

The receptor-containing portion of the retina is actually an extension of the CNS and not a separate peripheral organ. During embryonic development, the retinal cells "back out" of the nervous system, so the retinal layers, surprisingly, are facing backward! The neural portion of the retina consists of three layers (Fig. 6-21): (1) the outermost layer (closest to the choroid) containing the **rods** and **cones,** whose light-sensitive ends face the choroid (away from the incoming light); (2) a middle layer of **bipolar neurons;** and (3) an inner layer of **ganglion cells.** Axons of the ganglion cells join together to form the **optic nerve,** which leaves the retina slightly off-center. The point on the retina at which the optic nerve leaves and through which blood vessels pass is the **optic disc** (Figs. 6-9a and 6-22). This region is often referred to as the **blind spot;** no image can be detected in this area because it is devoid of rods and cones. We are normally not aware of the blind spot because central processing somehow "fills in" the

Figure 6-22 View of the Retina Seen through an Ophthalmoscope With an ophthalmoscope, a lighted viewing instrument, it is possible to view the optic disc (blind spot) and macula within the retina of the rear of the eye.

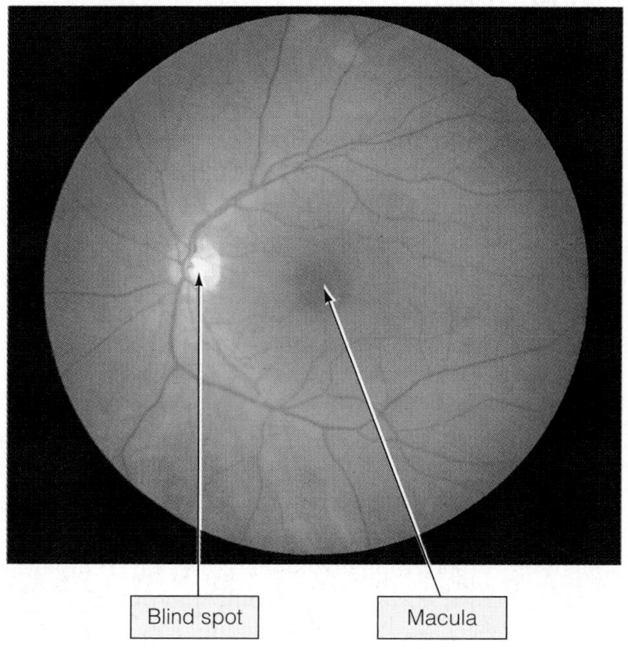

Figure 6-23 **Demonstration of the Blind Spot** Discover the blind spot in your right eye by closing your left eye and holding the book about 4 inches from your face. While focusing on the circle, gradually move the book away from you until the cross vanishes from view. At this time, the image of the cross is striking the blind spot of your right eye. You can similarly discover the blind spot in your left eye by closing your right eye and focusing on the cross. The circle will disappear when its image strikes the blind spot of your left eye.

missing spot. You can discover the existence of your own blind spot by a simple demonstration (Fig. 6-23).

Light must pass through the ganglion and bipolar layers before reaching the photoreceptors in all areas of the retina except the **fovea.** In the fovea, which is a pinhead-sized depression located in the exact center of the retina (Fig. 6-9a), the bipolar and ganglion cell layers are pulled aside so that light strikes the photoreceptors directly. This feature, coupled with the fact that only cones (which have greater acuity or discriminative ability than the rods) are found here, makes the fovea the point of most distinct vision. Thus, we turn our eyes so that the object at which we are looking is focused on the fovea. The area immediately surrounding the fovea, the **macula lutea,** also has a high concentration of cones and fairly high acuity (Fig. 6-22). Macular acuity, however, is less than that of the fovea because of the overlying ganglion and bipolar cells in the macula.

Phototransduction by retinal cells converts light stimuli into neural signals.

Photoreceptors consist of three parts (Fig. 6-24a): (1) an *outer segment,* which lies closest to the eye's exterior, facing the choroid, and detects the light stimulus; (2) an *inner segment,* which lies in the middle of the photoreceptor's length and contains the metabolic machinery of the cell; and (3) a *synaptic terminal,* which lies closest to the eye's interior, facing the bipolar neurons, and transmits the signal generated in the photoreceptor upon light stimulation to these next cells in the visual pathway. The outer segment, which is rod-shaped in rods and cone-shaped in cones (Fig. 6-24b and c), is composed of stacked, flattened, membranous discs containing an abundance of **photopigment** molecules. Over a billion of these molecules may be packed into the outer segment of each photoreceptor. Photopigments undergo chemical alterations when activated by light. A photopigment consists of an enzymatic protein called **opsin** combined with **retinene,** a derivative of vitamin A. There are four different photopigments, one in the rods and one in each of three types of cones. Retinene is identical in all four photopigments, but the photoreceptors' opsins vary slightly, enabling the photopigments to differentially absorb various wavelengths of light. **Rhodopsin,** the rod photopigment, cannot discriminate between various wavelengths in the visible spectrum; it absorbs all visible wavelengths. Therefore, rods provide vision only in shades of gray by detecting different intensities, not different colors. The photopigments in the three types of cones—**red, green,** and **blue cones**—respond selectively to various wavelengths of light, making **color vision** possible.

Phototransduction, the mechanism of excitation, is basically the same for all photoreceptors (Fig. 6-25a). When a photopigment molecule absorbs light it dissociates into its retinene and opsin components, and its retinene portion changes shape, thus triggering the enzymatic activity of opsin. Through a series of steps, this light-induced breakdown and subsequent activation of the photopigment bring about a hyperpolarizing receptor potential that influences transmitter release from the synaptic terminal of the photoreceptor.

Specifically, the light-induced biochemical change in the photopigment leads to closing of chemical messenger-gated Na^+ channels in the outer segment's membrane. Unlike other chemical-gated channels that respond to external chemical messengers, these channels respond to an internal second messenger, **cyclic GMP** (cyclic guanosine monophosphate), which links photopigment light absorption with Na^+ channel closing. Intracellular cyclic GMP, which is in high concentration in the dark, keeps the Na^+ channels in the outer segment's plasma membrane open (Fig. 6-25b). Therefore, unlike most receptors, the Na^+ channels of a photoreceptor are open in the absence of stimulation, that is, in the dark. The resultant passive inward Na^+ leak depolarizes the photoreceptor and keeps the Ca^{2+} channels in its synaptic terminal open, thereby permitting release of transmitter while in the dark.

Upon exposure to light, the concentration of cyclic GMP is decreased through a series of biochemical steps triggered by photopigment activation (Fig. 6-25a). Rod and cone cells contain a G protein (see p. 68) called **transducin.** Upon exposure to light, the activated photopigment activates transducin, which in turn activates the enzyme phosphodiesterase. This enzyme degrades cyclic GMP, thus decreasing the concentration of this second messenger in the photoreceptor. During the light-excitation process, the reduction in cyclic GMP brings about closure of Na^+ channels, which stops the depolarizing Na^+ leak and causes membrane hyperpolarization. This hyperpolarization, which is the receptor potential, passively spreads from the outer segment to the synaptic terminal of the photoreceptor. Here the potential changes leads to a reduction in transmitter release from the synaptic terminal. Thus, photoreceptors are inhibited by their adequate stimulus (hyperpolarized by light) and excited in the absence of stimulation (depolarized by darkness). The hyperpolarizing potential and subsequent decrease in transmitter release are graded according to the intensity of light. The brighter the light, the greater the hyperpolarizing response and the greater the reduction in transmitter release.

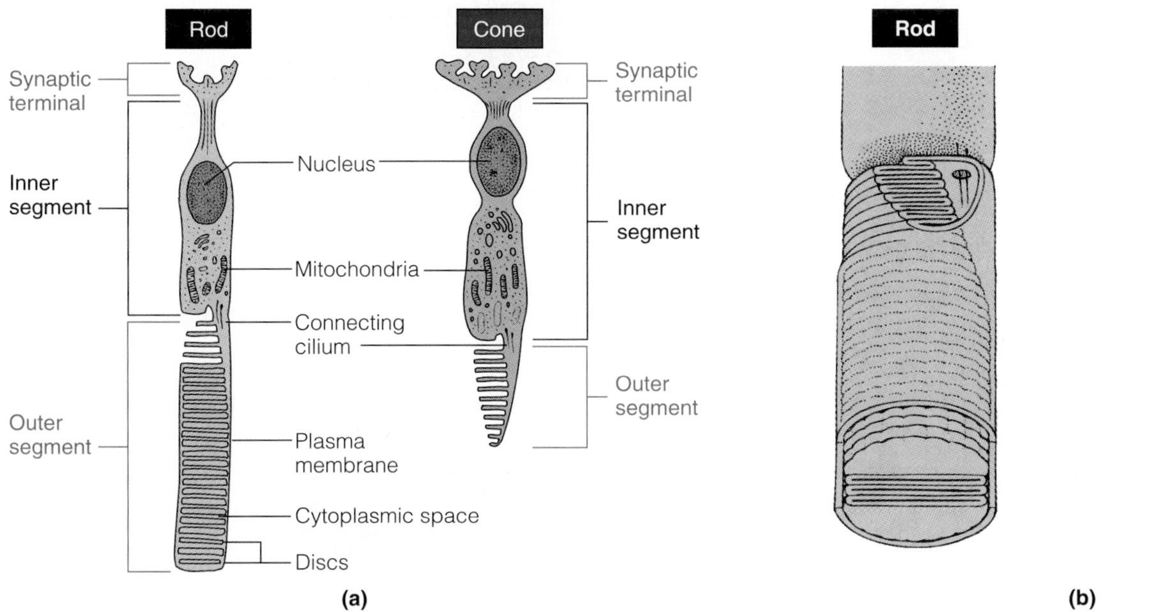

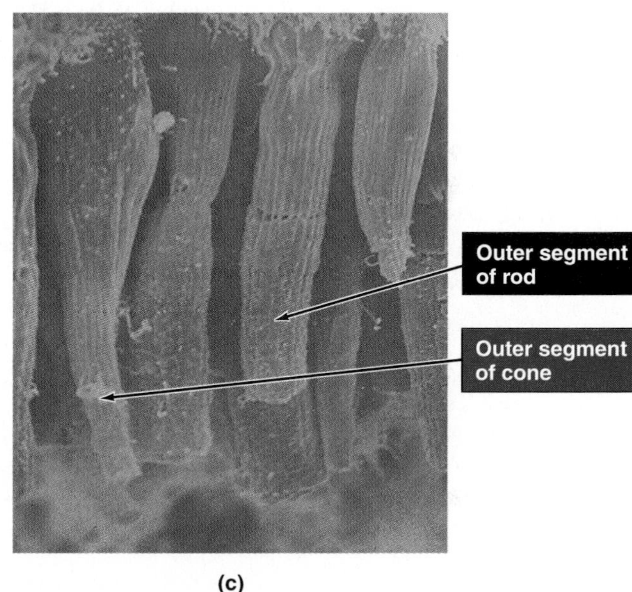

Figure 6-24 **Photoreceptors** (a) Schematic representation of the three parts of the rods and cones, the eye's photoreceptors (oriented in the same direction as in Fig. 6-21). (b) Schematic three-dimensional representation of the outer segments of the rods and cones. Note the stacked, flattened, membranous discs, which contain an abundance of photopigment molecules. (c) Colorized scanning electron micrograph of the outer segments of the rods and cones. Note the rod shape in rods and the cone shape in cones.

How does the retina signal the brain about light stimulation through such an inhibitory response? The photoreceptors synapse with bipolar cells. These cells in turn terminate on the ganglion cells, whose axons form the optic nerve for transmission of signals to the brain. The answer to our seeming paradox lies in the fact that the transmitter released from the photoreceptors' synaptic terminal has an *inhibitory* action on many of the bipolar cells. The reduction in transmitter release that accompanies light-induced receptor hyperpolarization decreases this inhibitory action on the bipolar cells. Removal of inhibition has the same effect as direct excitation of the bipolar cells. The greater the illumination on the receptor cells, the greater the removal of inhibition from the bipolar cells and the greater in effect the excitation of these next cells in the visual pathway to the brain.

Bipolar cells display graded potentials similar to the photoreceptors. Action potentials do not originate until the ganglion cells, the first neurons in the chain that must propagate the visual message over long distances to the brain.

The altered photopigments are restored to their original conformation in the dark by enzyme-mediated mechanisms. Thereupon, the membrane potential and rate of transmitter release of the photoreceptor are returned to their unexcited state, and no action potentials are transmitted to the visual cortex.

Researchers are currently working on an ambitious and still highly speculative microelectronic chip that would serve as a partial substitute retina. Their hope is that the device will be able to restore at least some sight in those who are blinded because of loss of photoreceptor cells but whose ganglion cells and optic pathways remain healthy. For example, if successful, the chip could benefit persons with **macular degeneration,** the leading cause of blindness in the western hemisphere. This condition is characterized by loss of photoreceptors in the macula in association with advancing age. The envisioned "vision chip" would bypass the photoreceptor step altogether: images received by means of a camera mounted on eyeglasses would be translated by the chip into electrical signals detectable by the ganglion cells and transmitted on for further optical processing.

Rods provide indistinct gray vision at night whereas cones provide sharp color vision during the day.

The retina contains more than 30 times more rods than cones (100 million rods compared to 3 million cones per eye). Cones are most abundant in the center of the retina in the

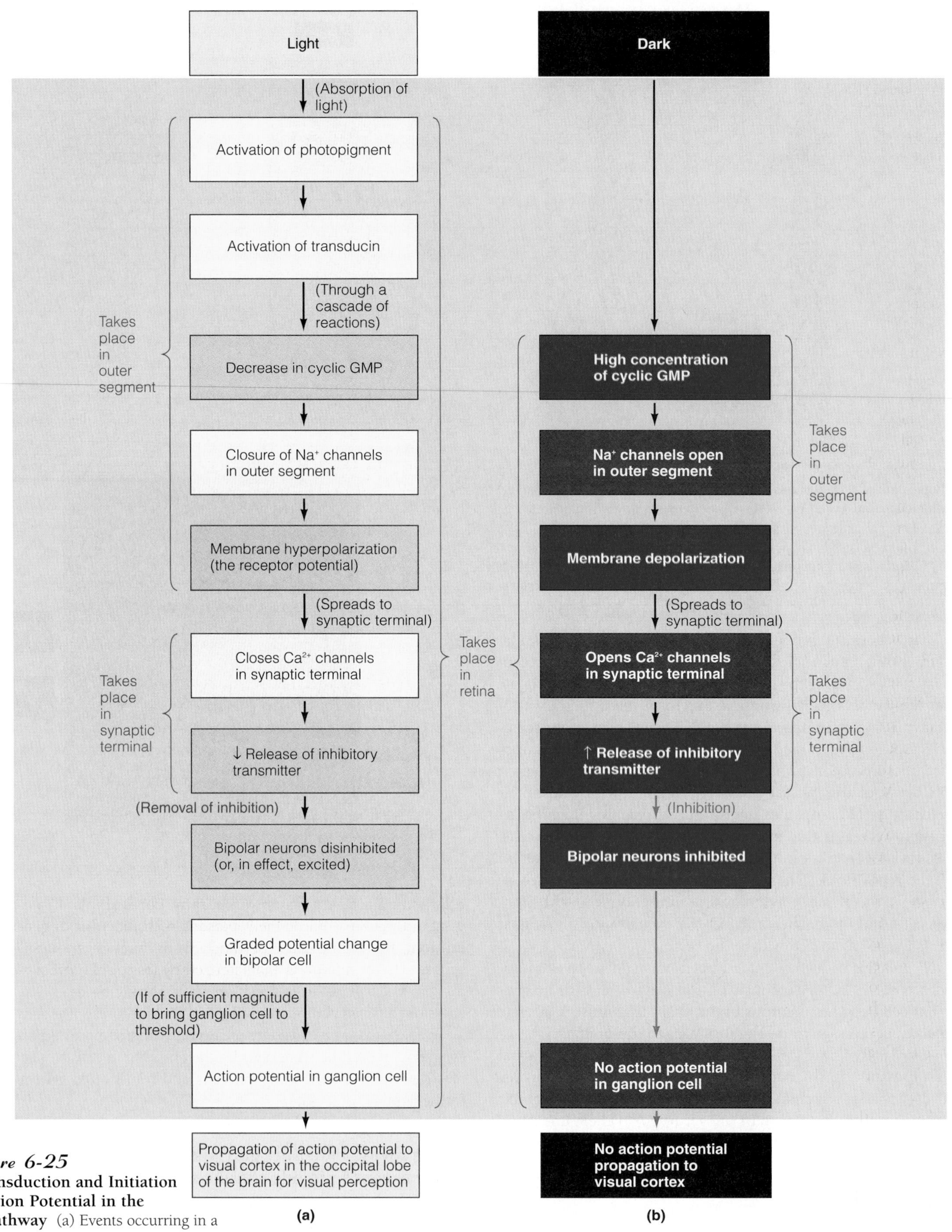

Figure 6-25
Phototransduction and Initiation of an Action Potential in the Visual Pathway (a) Events occurring in a photoreceptor in response to a light stimulus that initiate an action potential in the visual pathway (phototransduction). (b) Events occurring in a photoreceptor in response to the dark that prevent action potentials from being initiated in the visual pathway.

macula. From this point outward, the concentration of cones decreases and the concentration of rods increases. Rods are most abundant in the periphery. Because of differential absorption of various wavelengths of light, cones provide color vision whereas rods provide vision only in shades of gray. The capabilities of the rods and cones also differ in other respects due to a difference in the "wiring patterns" between these photoreceptor types and other retinal neuronal layers (▮ Table 6-4). Cones have low sensitivity to light, being "turned on" only by bright daylight, but they have high acuity (sharpness; ability to distinguish between two nearby points). Thus, cones provide sharp vision with high resolution for fine detail. Humans use cones for day vision, which is in color and distinct. Rods, on the other hand, have low acuity but high sensitivity, so they respond to the dim light of night. We are able to see at night with our rods but at the expense of color and distinctness. Let us see how wiring patterns influence sensitivity and acuity.

There is little convergence of neurons (see p. 103) in the retinal pathways for cone output. Each cone generally has a private line connecting it to a particular ganglion cell. In contrast, there is much convergence in rod pathways. Output from more than one hundred rods may converge via bipolar cells on a single ganglion cell.

Before a ganglion cell can have an action potential that will be propagated to the CNS, the cell must be brought to threshold through influence of the graded potentials in the receptors to which it is wired. Because a single-cone ganglion cell is influenced by only one cone, only bright daylight is intense enough to induce a sufficient receptor potential in the cone to ultimately bring the ganglion cell to threshold. The abundant convergence in the rod visual pathways, in contrast, offers good opportunities for summation of subthreshold events in a rod ganglion cell (see p. 99). Whereas a small receptor potential induced by dim light in a single cone would not be sufficient to bring its ganglion cell to threshold, similar small receptor potentials induced by the same dim light in multiple rods converging on a single ganglion cell would have an additive effect to bring the rod ganglion cell to threshold. Because rods can bring about action potentials in response to small amounts of light, they are much more sensitive than cones. However, because cones have private lines into the optic nerve, each cone transmits information about an extremely small receptive field on the retinal surface. Cones are thus able to provide highly detailed vision at the expense of sensitivity. With rod vision, acuity is sacrificed for sensitivity. Since many rods share a single ganglion cell, once an action potential is initiated, it is impossible to discern which of the multiple rod inputs were activated to bring the ganglion cell to threshold. Objects appear fuzzy when rod vision is used because of this poor ability to distinguish between two nearby points.

Table 6-4

Properties of Rod and Cone Vision

Rods	*Cones*
100 million per retina	3 million per retina
Vision in shades of gray	Color vision
High sensitivity	Low sensitivity
Low acuity	High acuity
Night vision	Day vision
Much convergence in retinal pathways	Little convergence in retinal pathways
More numerous in periphery	Concentrated in fovea

The sensitivity of the eyes can vary markedly through dark and light adaptation.

The eyes' sensitivity to light depends on the amount of photopigment present in the rods and cones. When you go from bright sunlight into darkened surroundings, you cannot see anything at first, but gradually you begin to distinguish objects as a result of the process of **dark adaptation.** Breakdown of photopigments during exposure to sunlight tremendously decreases photoreceptor sensitivity. For example, a reduction in rhodopsin content of only 0.6% from its maximum value decreases rod sensitivity approximately 3,000 times. In the dark, the photopigments broken down during light exposure are gradually regenerated. As a result, the sensitivity of your eyes gradually increases so that you can begin to see in the darkened surroundings. However, only the highly sensitive, rejuvenated rods are "turned on" by the dim light.

Conversely, when you move from the dark to the light (for example, leaving a movie theater and entering the bright sunlight), your eyes are very sensitive to the dazzling light at first. With little contrast between lighter and darker parts, the entire image appears bleached. As some of the photopigments are rapidly broken down by the intense light, the sensitivity of the eyes decreases and normal contrasts can once again be detected, a process known as **light adaptation.** The rods are so sensitive to light that sufficient rhodopsin is broken down to essentially "burn out" the rods in bright light; that is, the rod photopigments, having already been broken down by the bright light, are no longer able to respond to the light. Furthermore, a central neural adaptive mechanism switches the eye from the rod system to the cone system on exposure to bright light. Therefore, only the less-sensitive cones are used for day vision.

It is estimated that our eyes' sensitivity can change as much as 1 million times as they adjust to various levels of illumination through dark and light adaptation. These adaptative measures are also enhanced by pupillary reflexes that adjust the amount of available light permitted to enter the eye.

Since retinene, one of the photopigment components, is a derivative of vitamin A, adequate amounts of this nutrient must be available for the ongoing resynthesis of photopigments. **Night blindness** occurs because of dietary deficiencies of vitamin A. Although photopigment concentrations in both rods and cones are reduced in this condition, there is still sufficient cone photopigment to respond to the intense stimulation of bright light, except in the most severe cases. However, even modest reductions in rhodopsin content can decrease the sensitivity of rods so much that they are unable to respond to dim light. The person can see in the day using cones but cannot see at night because the rods are no longer functional. Thus, carrots are "good for your eyes" because they are rich in vitamin A.

Color vision depends on the ratios of stimulation of the three cone types.

Vision depends on stimulation of retinal photoreceptors by light. Certain objects in the environment such as the sun, fire, and lightbulbs, emit light. But how do we see objects like chairs, trees, and people, which do not emit light? The pigments in various objects selectively absorb particular wavelengths of light transmitted to them from light-emitting sources, and the unabsorbed wavelengths are reflected from the objects' surfaces. These reflected light rays enable us to see the objects. An object perceived as blue absorbs the longer red and green wavelengths of light and reflects the shorter blue wavelengths, which can be absorbed by the photopigment in the eyes' blue cones, thereby activating them.

Each cone type is most effectively activated by a particular wavelength of light in the range of color indicated by its name—blue, green, or red. However, cones also respond in varying degrees to other wavelengths (Fig. 6-26). Our perception of the many colors of the world depends on the three cone types' various *ratios of stimulation* in response to different wavelengths. A wavelength perceived as blue does not stimulate red or green cones at all but excites blue cones maximally (the percentage of maximal stimulation for red, green and blue cones, respectively, is 0:0:100). The sensation of yellow, in comparison, arises from a stimulation ratio of 83:83:0, red and green cones each being stimulated 83% of maximum while blue cones are not excited at all. The ratio for green is 31:67:36, and so on, with various combinations giving rise to the sensation of all the different colors. White is a mixture of all wavelengths of light, whereas black is the absence of light.

The extent that each of the cone types is excited is coded and transmitted in separate parallel pathways to the brain. A distinct color vision center in the primary visual cortex has recently been identified. This center combines and processes these inputs to generate the perception of color, taking into consideration the object in comparison with its background. The concept of color is therefore in the mind of the beholder. Most of us agree on what color we see because we have the same types of cones and use similar neural pathways for comparing their output. Occasionally, however, individuals lack a particular cone type, so their color vision is a product of the differential sensitivity of only two types of cones, a condition known as **color blindness.** Not only do color-defective individuals perceive certain colors differently, but they are also unable to distinguish as many varieties of colors (Fig. 6-27). For example, people with certain color defects are unable to distinguish between red and green. At a traffic light they can tell which light is "on" by its intensity, but they must rely on the position of the bright light to know whether to stop or go.

Although the three-cone system has been accepted as the standard model of color vision for more than two centuries, new evidence suggests that perception of color may be more complex. DNA studies have shown that men with normal color vision have a variable number of genes coding for cone pigments. For example, many had multiple genes (from 2 up to 4) for red-light detection and could distinguish more subtle differences in colors in this long-wavelength range than those with single copies of red-cone genes. This finding will undoubtedly lead to a reevaluation of how the various photopigments contribute to color vision.

Figure 6-26 **Sensitivity of the Three Types of Cones to Different Wavelengths** The ratios of stimulation of the three cone types are shown for three sample colors.

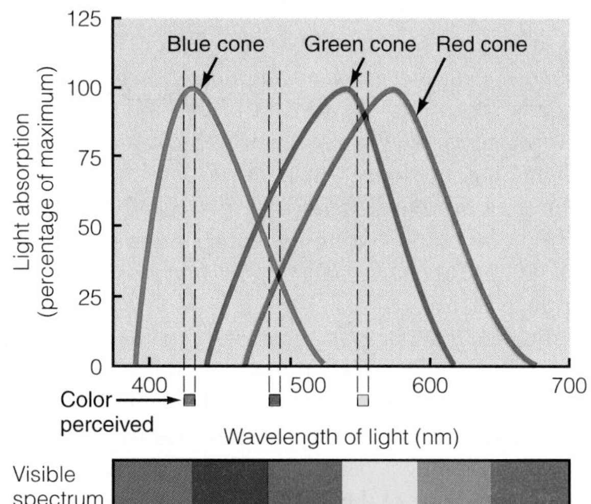

Color perceived	Percent of maximum stimulation		
	Red cones	Green cones	Blue cones
■	0	0	100
■	31	67	36
□	83	83	0

Visual information is separated and modified within the visual pathway before it is integrated into a perceptual image of the visual field by the cortex.

The field of view that can be seen without moving the head is known as the **visual field.** The information that reaches the visual cortex in the occipital lobe is not a replica of the visual field for several reasons:

1. The image detected on the retina at the onset of visual processing is upside down and backward because of bending of the light rays (Fig. 6-28). Once it is projected to the brain, the inverted image is interpreted as being in its correct orientation.
2. The information transmitted from the retina to the brain is not merely a point-to-point record of photoreceptor activation. Before the information reaches the brain, the retinal neuronal layers beyond the rods and cones reinforce

selected information and suppress other information to enhance contrast. One mechanism of retinal processing is lateral inhibition, by which strongly excited cone pathways suppress activity in surrounding pathways of weakly stimulated cones. This increases the dark-bright contrast to enhance the sharpness of boundaries.

Another mechanism of retinal processing involves differential activation of two types of ganglion cells, **on-center** and **off-center ganglion cells.** The receptive field of a cone ganglion cell is determined by the field of light detection by the cone with which it is linked. On-center and off-center ganglion cells respond in opposite ways, depending on the relative comparison of illumination between the center and periphery of their receptive fields. Think of the receptive field as a donut. An on-center ganglion cell increases its rate of firing when light is most intense at the center of its receptive field (that is, when the donut hole is lit up). In contrast, an off-center cell increases its firing rate when the periphery of its receptive field is most intensely illuminated (that is, when the donut itself is lit up). This is useful for enhancing the difference in light level between one small area at the center of a receptive field and the illumination immediately around it. By emphasizing differences in relative brightness, this mechanism helps define contours of images, but in so doing, information about absolute brightness is sacrificed (Fig. 6-29).

3. Various aspects of visual information such as form, color, depth, and movement are separated and projected in parallel pathways to different regions of the cortex. Only when these separate bits of processed information are integrated by higher visual regions is a reassembled picture of the visual scene perceived. This is similar to the blobs of paint on an artist's palette versus the finished portrait; the separate pigments do not represent a portrait of a face until they are appropriately integrated on a canvas.

Figure 6-27 **Color Blindness Chart** People with red-green color blindness cannot detect the number 29 in this chart.

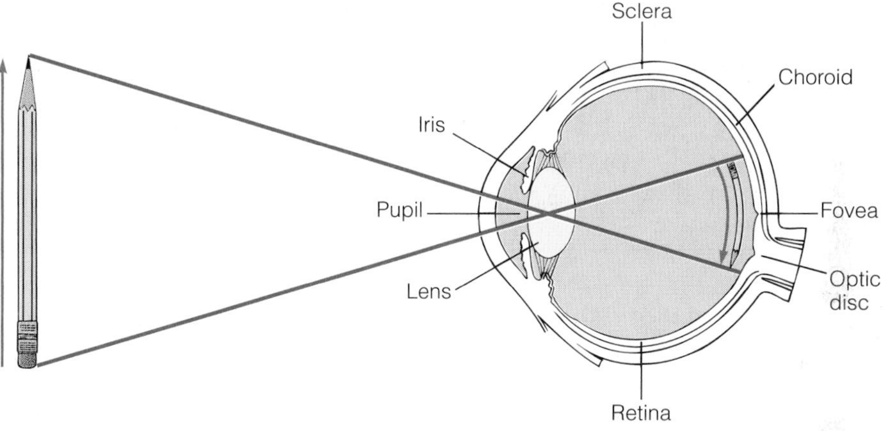

Figure 6-28 **Inversion of the Image on the Retina**

Figure 6-29 **Example of the Outcome of Retinal Processing by On-Center and Off-Center Ganglion Cells** Note that the gray circle surrounded by black appears brighter than the one surrounded by white, even though the two circles are identical (same shade and size). Retinal processing by on-center and off-center ganglion cells is largely responsible for enhancing differences in relative (rather than absolute) brightness, which helps to define contours.

Patients with lesions in specific visual processing regions of the brain may be unable to completely combine components of a visual impression. For example, a person may be unable to discern movement of an object but have reasonably good vision for shape, pattern, and color. Sometimes the defect can be remarkably specific, like being unable to recognize familiar faces while retaining the ability to recognize inanimate objects.

4. Because of the pattern of wiring between the eyes and the visual cortex, the left half of the cortex receives information only from the right half of the visual field as detected by both eyes, and the right half receives input only from the left half of the visual field of both eyes.

As a result of refraction, light rays from the left half of the visual field fall on the right half of the retina of both eyes (the medial or inner half of the left retina and the lateral or outer half of the right retina) (Fig. 6-30a). Similarly, rays from the right half of the visual field reach the left half of each retina (the lateral half of the left retina

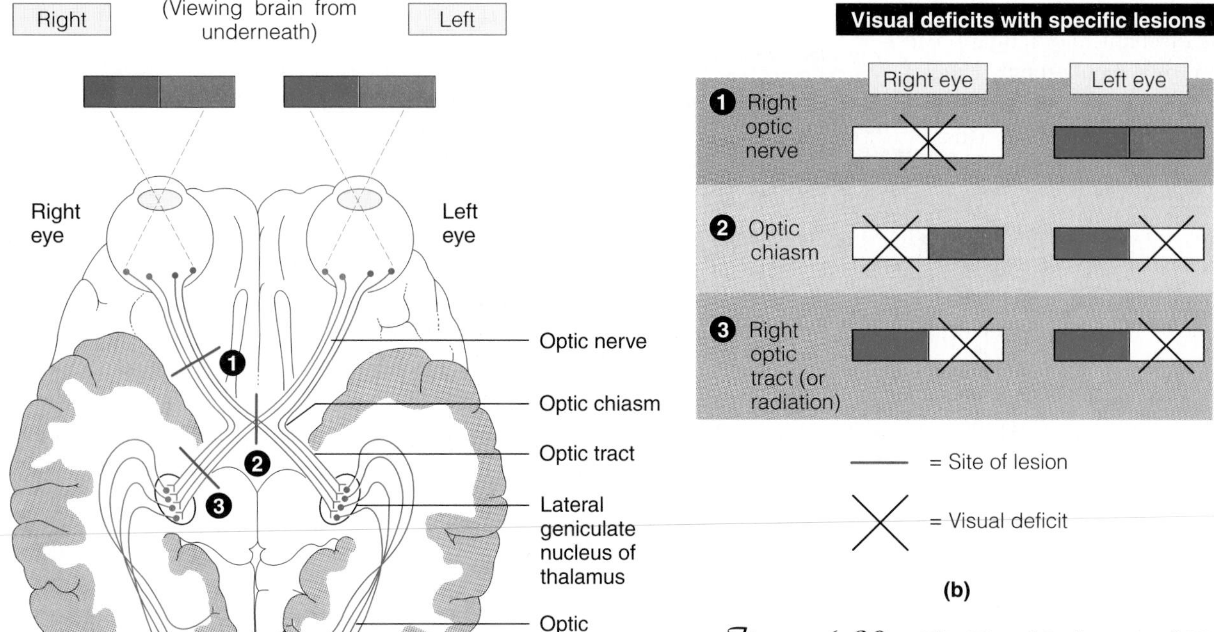

Figure 6-30 The Visual Pathway and Visual Deficits Associated with Lesions in the Pathway (a) Visual pathway. Note that the left half of the visual cortex in the occipital lobe receives information from the right half of the visual field of both eyes (in blue), and the right half of the cortex receives information from the left half of the visual field of both eyes (in red). (b) Visual deficits with specific lesions in the visual pathway. Each visual deficit illustrated is associated with a lesion at the corresponding numbered point in the visual pathway in part (a).

and the medial half of the right retina). Each optic nerve exiting the retina carries information from both halves of the retina it serves. This information is separated as the optic nerves meet at the **optic chiasm** (*chiasm* means "cross") located underneath the hypothalamus. Within the optic chiasm, the fibers from the medial half of each retina cross to the opposite side, but those from the lateral half remain on the original side. The reorganized bundles of fibers leaving the optic chiasm are known as **optic tracts.** Each optic tract carries information from the lateral half of one retina and the medial half of the other retina. Therefore, this partial crossover brings together from the two eyes fibers that carry information from the same half of the visual field. Each optic tract, in turn, delivers to the half of the brain on its same side information about the opposite half of the visual field. A knowledge of these pathways can facilitate diagnosis of visual defects arising from interruption of the visual pathway at various points (Fig. 6-30b).

The thalamus and visual cortices elaborate the visual message.

The first stop in the brain for information in the visual pathway is the **lateral geniculate nucleus** in the thalamus (Fig. 6-30a). It separates information received from the eyes and relays it via fiber bundles known as **optic radiations** to different zones in the cortex, each of which processes different aspects of the visual stimulus (for example, color, form, depth, movement). This sorting process is no small task, because each optic nerve contains more than 1 million fibers carrying information from the photoreceptors in one retina. This is more than all the afferent fibers carrying somatosensory input from all the regions of the body! Researchers estimate that hundreds of millions of neurons occupying about 30% of the cortex participate in visual processing, compared to 8% devoted to touch perception and 3% to hearing. Yet the connections in the visual pathways are precise. The lateral geniculate nucleus and each of the zones in the cortex that processes visual information have a topographical map representing the retina point-for-point. As with the somastosensory cortex, the neural maps of the retina are distorted. The fovea, the retinal region capable of greatest acuity, has much greater representation in the neural map than do the more peripheral regions of the retina.

Although each half of the visual cortex receives information simultaneously from the same part of the visual field as received by both eyes, the messages from the two eyes are not identical. Each eye views an object from a slightly different vantage point, even though there is a tremendous area of overlap (Fig. 6-31). The overlapping area seen by both eyes at the same time is known as the **binocular** ("two-eyed") field of vision, which is important for **depth perception.** Like other areas of the cortex, the primary visual cortex is organized into functional columns, each processing information from a small region of the retina. Independent alternating columns are devoted to information about the same point in the visual field from the right and left eyes. The brain uses the slight disparity in the information received from the two eyes to estimate distance, allowing us to perceive three-dimensional objects in spatial depth. Some depth perception is possible using only one eye, based on experience and comparison with other cues. For example, if your one-eyed view includes a car and a building and the car is much larger, you correctly interpret that the car must be closer to you than the building.

Occasionally, the two views are not successfully merged. This condition may occur for two reasons: (1) the eyes are not both focused on the same object simultaneously due to defects of the external eye muscles that make fusion of the two eyes' visual fields impossible, or (2) the binocular information is improperly integrated during visual processing. The result is double vision or **diplopia,** a condition in which the disparate views from both eyes are seen simultaneously.

Within the cortex, visual information is first processed in the primary visual cortex, then is projected to higher-order visual areas for even more complex processing and abstraction. The cortex contains a hierarchy of visual cells that respond to increasingly complex stimuli. Three types of visual cortical neurons have been identified based on the complexity of stimulus requirements needed for the cell to respond; these are called **simple, complex,** and **hypercomplex cells.** Simple and complex cells are stacked on top of each other within the cortical columns of the primary visual cortex, whereas hypercomplex cells are found in the higher visual processing areas. Unlike a retinal cell, which responds to the amount of light, a cortical cell fires only when it receives a particular pattern of illumination for which it is programmed. These patterns are built up by converging connections that originate from closely aligned photoreceptor cells in the retina. For example, some simple cells fire only when a bar is viewed vertically in a specific location, others when a bar is horizontal, and others at various oblique orientations. Movement of a critical axis of orientation becomes important for response by some of the complex cells. Hypercomplex cells add a new dimension to visual processing by responding only to particular edges, corners, and curves. Each level of cortical visual neurons has increasingly greater capacity for abstraction of information built up from the increasing convergence of input from lower-order neurons. In this way, the dotlike pattern of photoreceptors stimulated to varying degrees by varying light intensities in the retinal image is transformed in the cortex into information about depth, position, orientation, movement, contour, and length.

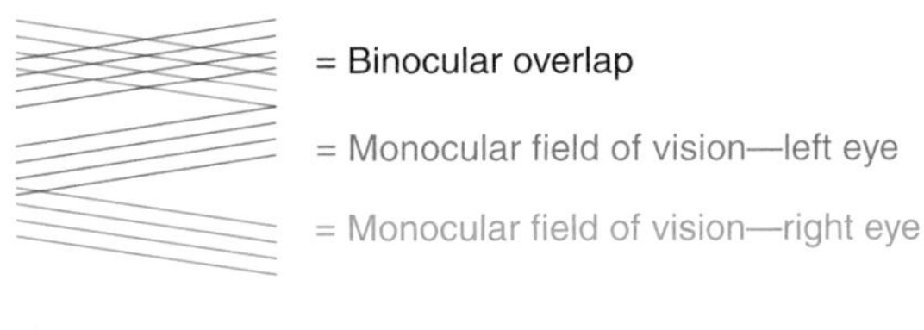

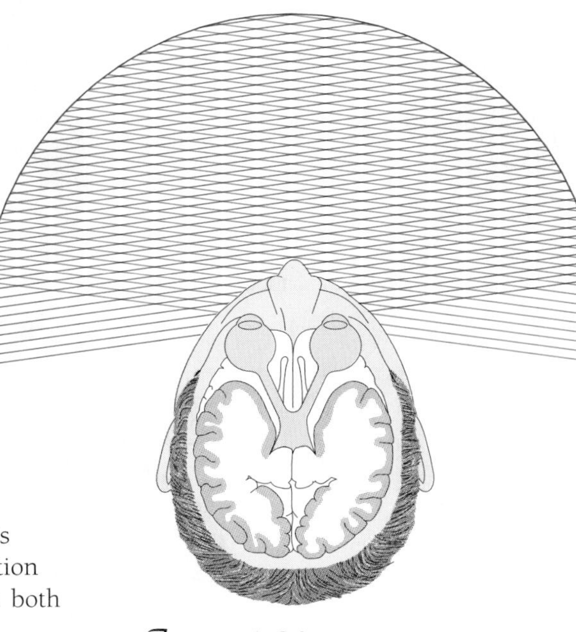

Figure 6-31 **Visual Fields**

Other aspects of this information, such as color perception, are processed simultaneously through a similar hierarchical organization. How and where the entire image is finally put together is still unresolved.

Visual input goes to other areas of the brain not involved in vision perception.

Not all fibers in the visual pathway terminate in the visual cortices. Some are projected to other regions of the brain for purposes other than direct vision perception. Following are examples of nonsight activities dependent on input from the retina:

1. Control of pupil size.
2. Synchronization of biological clocks to cyclical variations in light intensity (for example, the sleep-wake cycle synchronized to the night-day cycle).
3. Contribution to cortical alertness and attention.
4. Control of eye movements.

In the latter regard, each eye is equipped with a set of six **external eye muscles** that position and move the eye so that it can better locate, see, and track objects. Eye movements are among the fastest, most discretely controlled movements of the body.

Protective mechanisms help prevent eye injuries.

Several mechanisms help protect the eyes from injury. Except for its anterior portion, the eyeball is sheltered by the bony socket in which it is positioned. The **eyelids** act like shutters to protect the anterior portion of the eye from environmental insults. They close reflexly to cover the eye under threatening circumstances, such as rapidly approaching objects, dazzling light, and instances when the cornea or eyelashes are touched. Frequent spontaneous blinking of the eyelids helps disperse the lubricating, cleansing, bactericidal ("germ-killing") tears. **Tears** are produced continuously by the **lacrimal gland** in the upper lateral corner under the eyelid. This eye-washing fluid flows across the surface of the cornea and drains into tiny canals in the corner of each eye (Fig. 6-9b), eventually emptying into the back of the nasal passageway. This drainage system cannot handle the profuse tear production during crying, so the tears overflow from the eyes. The eyes are also equipped with protective **eyelashes,** which trap fine airborne debris such as dust before it can fall into the eye.

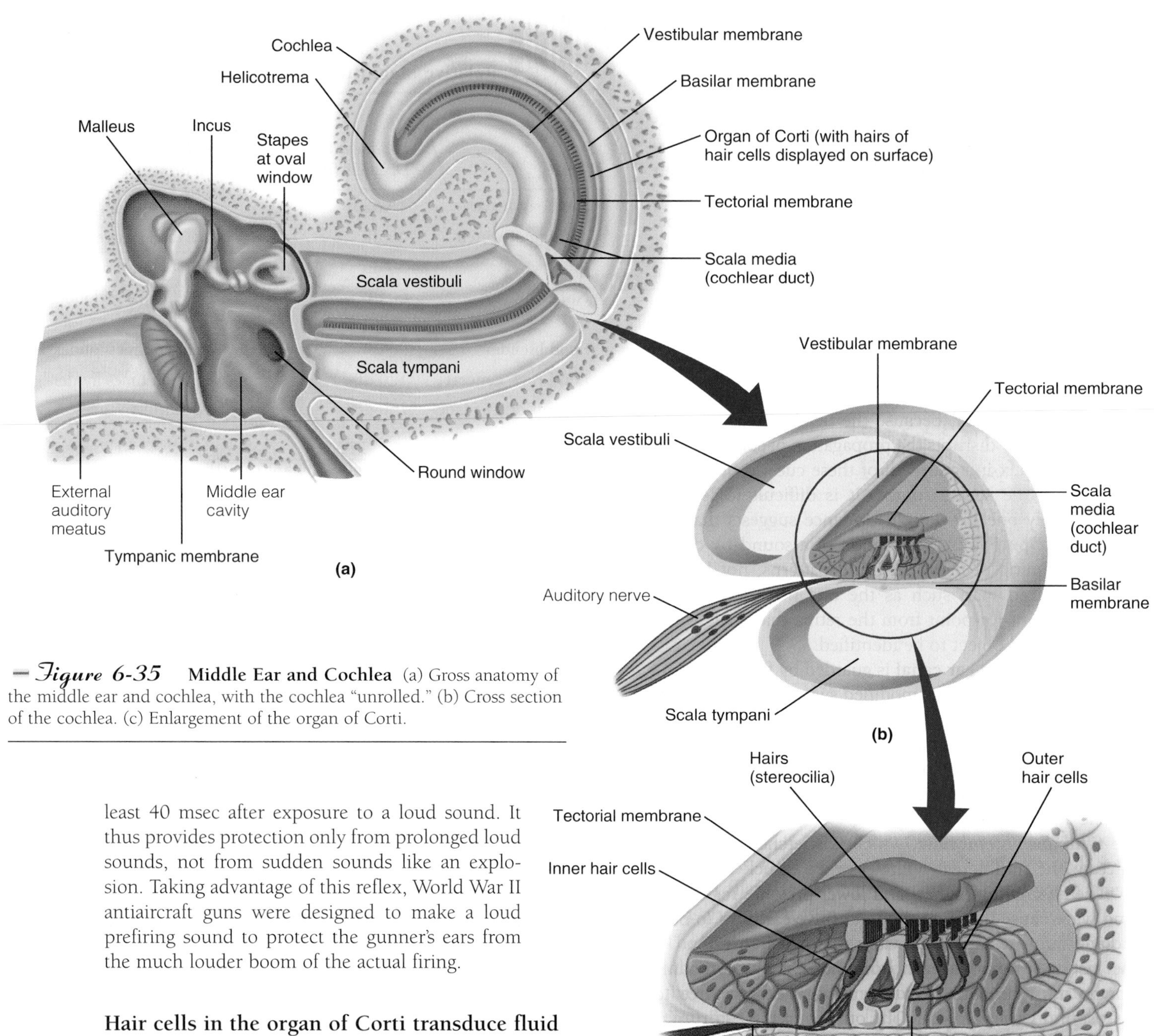

Figure 6-35 Middle Ear and Cochlea (a) Gross anatomy of the middle ear and cochlea, with the cochlea "unrolled." (b) Cross section of the cochlea. (c) Enlargement of the organ of Corti.

least 40 msec after exposure to a loud sound. It thus provides protection only from prolonged loud sounds, not from sudden sounds like an explosion. Taking advantage of this reflex, World War II antiaircraft guns were designed to make a loud prefiring sound to protect the gunner's ears from the much louder boom of the actual firing.

Hair cells in the organ of Corti transduce fluid movements into neural signals.

The pea-sized, snail-shaped cochlear portion of the **inner ear** is a coiled tubular system lying deep within the temporal bone (Fig. 6-32). It is easier to understand the functional components of the **cochlea** by "unrolling" it, as shown in Figure 6-35a). The cochlea is divided throughout most of its length into three fluid-filled longitudinal compartments. A blind-ended **cochlear duct,** which is also known as the **scala media,** constitutes the middle compartment. It tunnels lengthwise through the center of the cochlea, almost but not quite reaching its end. The upper compartment, the **scala vestibuli** (Figs. 6-35a and b), follows the inner contours of the spiral, and the **scala tympani,** the lower compartment, follows the outer contours. The fluid within the cochlear duct is called **endolymph** (Fig. 6-36a). The scala vestibuli and scala tympani both contain a slightly different fluid, the **perilymph.** The region beyond the tip of the cochlear duct where the fluid in the upper and lower compartments is continuous is called the **helicotrema.** The scala vestibuli is sealed from the middle ear cavity by the oval window, to which the stapes is attached. Another small membrane-covered opening, the **round window,** seals the scala tympani from the middle ear. The thin **vestibular membrane** forms the ceiling of the cochlear duct and separates it from the scala vestibuli. The **basilar membrane** forms the floor of the cochlear duct, separating it from the scala tympani. The basilar membrane is

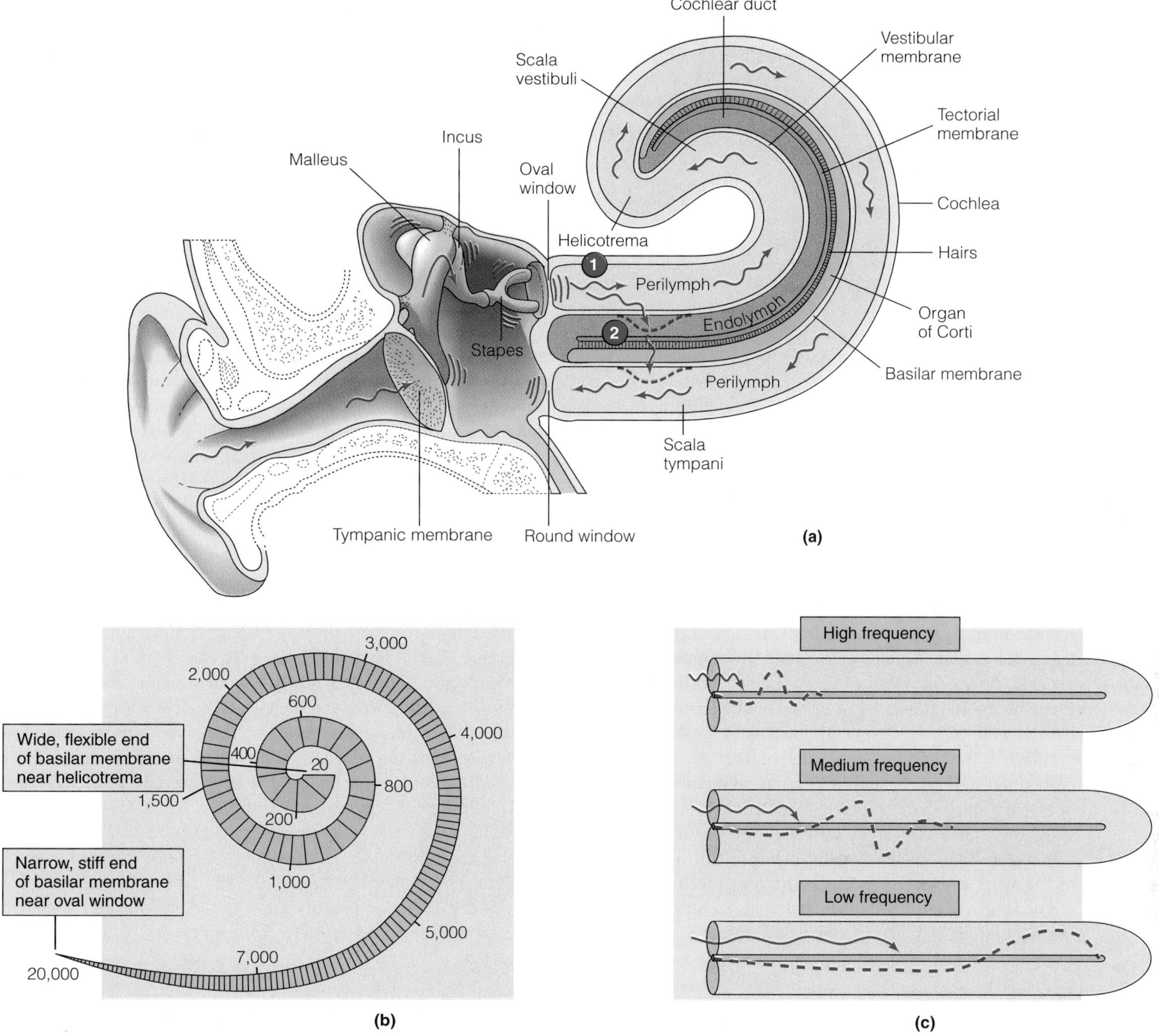

The numbers indicate the frequencies with which different regions of the basilar membrane maximally vibrate.

Figure 6-36 **Transmission of Sound Waves** (a) Fluid movement within the perilymph set up by vibration of the oval window follows two pathways: (1) through the scala vestibuli, around the helicotrema, and through the scala tympani, causing the round window to vibrate; and (2) a "shortcut" from the scala vestibuli through the basilar membrane to the scala tympani. The first pathway just dissipates sound energy, but the second pathway triggers activation of the receptors for sound by bending the hairs of the hair cells as the organ of Corti on top of the vibrating basilar membrane is displaced in relation to the overlying tectorial membrane. (b) Different regions of the basilar membrane vibrate maximally at different frequencies. (c) The narrow, stiff end of the basilar membrane nearest the oval window vibrates best with high-frequency pitches. The wide, flexible end of the basilar membrane near the helicotrema vibrates best with low-frequency pitches.

especially important because it bears the **organ of Corti,** the sense organ for hearing.

The organ of Corti, which rests on top of the basilar membrane throughout its full length, contains **hair cells** that are the receptors for sound. The 16,000 hair cells within each cochlea are arranged in four parallel rows along the length of the basilar membrane: one row of **inner hair cells** and three rows of **outer hair cells** (Fig. 6-35c). Protruding from the surface of each hair cell are about 100 hairs known as **stereocilia,** which are actin-stiffened microvilli (see p. 40). Hair cells generate neural signals when their surface hairs are mechanically deformed in association with fluid movements in the inner ear. These stereocilia are mechanically embedded in the **tectorial membrane,** an awninglike projection overhanging the organ of Corti throughout its length (Fig. 6-35b and c).

The pistonlike action of the stapes against the oval window sets up pressure waves in the upper compartment. Because fluid is incompressible, pressure is dissipated in two ways as the stapes causes the oval window to bulge inward: (1) displacement of the round window and (2) deflection of the basilar membrane (Fig. 6-36a). In the first of these pathways, the pressure wave pushes the perilymph forward in the upper compartment, then around the helicotrema, and into the lower compartment, where it causes the round window to bulge outward into the middle ear cavity to compensate for the pressure increase. As the stapes rocks backward and pulls the oval window outward toward the middle ear, the perilymph shifts in the opposite direction, displacing the round window inward. This pathway does not result in sound reception; it just dissipates pressure.

Pressure waves of frequencies associated with sound reception take a "shortcut." Pressure waves in the upper compartment are transferred through the thin vestibular membrane, into the cochlear duct, and then through the basilar membrane into the lower compartment, where they cause the round window to alternately bulge outward and inward. The main difference in this pathway is that transmission of pressure waves through the basilar membrane causes this membrane to move up and down, or vibrate, in synchrony with the pressure wave. Since the organ of Corti rides on the basilar membrane, the hair cells also move up and down as the basilar membrane oscillates. Because the stereocilia of the receptor cells are embedded in the stiff, stationary tectorial membrane, they are bent back and forth when the oscillating basilar membrane shifts their position in relationship to the tectorial membrane (— Fig. 6-37). This back-and-forth mechanical deformation of the hairs alternately opens and closes mechanically gated ion channels (see p. 84) in the hair cell, resulting in alternating depolarizing and hyperpolarizing potential changes—the receptor potential—at the same frequency as the original sound stimulus.

— *Figure 6-37* **Bending of Hair Cells upon Deflection of the Basilar Membrane**

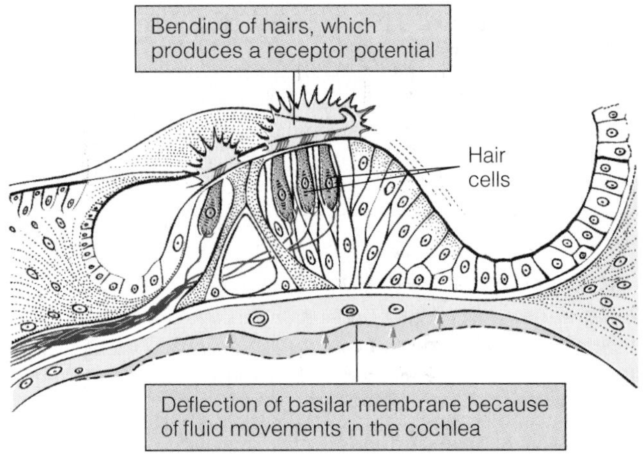

The inner and outer hair cells also differ in function, a more important difference than their geographical locations. The inner hair cells are the ones that transform the mechanical forces of sound (cochlear fluid vibration) into the electrical impulses of hearing (action potentials propagating auditory messages to the cerebral cortex). The inner hair cells are specialized receptor cells that communicate via a chemical synapse with the terminals of afferent nerve fibers making up the **auditory (cochlear) nerve.** Depolarization of these hair cells (when the basilar membrane is deflected upward) increases their rate of transmitter release, which steps up the rate of firing in the afferent fibers. Conversely, the firing rate decreases as these hair cells release less transmitter when they are hyperpolarized upon displacement in the opposite direction.

Thus, the ear converts sound waves in the air into oscillating movements of the basilar membrane that bends the hairs of the receptor cells back and forth. This shifting mechanical deformation of the hairs alternately opens and closes the receptor cells' channels, bringing about graded potential changes in the receptor that lead to changes in the rate of action potentials propagated to the brain. In this way, sound waves are translated into neural signals that can be perceived by the brain as sound sensations (— Fig. 6-38).

Whereas the inner hair cells primarily send auditory signals *to* the brain over afferent fibers, the outer hair cells, in addition to conveying meager afferent information, primarily receive signals *from* the brain over efferent fibers. Instead of vibrating in sync with the basilar membrane, the outer hair cells have been observed to bounce up and down at a frantic pace, like hyperactive kids on a trampoline. This so-called *fast motility,* most likely triggered by neural messages carried over the efferent fibers, is believed to be important in accelerating the motion of the basilar membrane. Such modification of basilar membrane movement is speculated to improve and tune the stimulation of the inner hair cells. Thus, the outer hair cells enhance the response of the inner hair cells, the real auditory sensory receptors, making them exquisitely sensitive to sound intensity and highly discriminatory between various pitches of sound.

Pitch discrimination depends on the region of the basilar membrane that vibrates; loudness discrimination depends on the amplitude of the vibration.

Pitch discrimination (that is, the ability to distinguish between various frequencies of incoming sound waves) depends on the shape and properties of the basilar membrane, which is narrow and stiff at its oval-window end and wide and flexible at its helicotrema end (Fig. 6-36b). Different regions of the basilar membrane naturally vibrate maximally at different frequencies; that is, each frequency displays peak vibration at a different position along the membrane. The narrow end nearest the oval window vibrates best with high-frequency pitches, whereas the wide end nearest the helicotrema vibrates maximally with low-frequency tones (Fig. 6-36c). The pitches in between are sorted out along the length of the membrane from higher to lower frequency. As a sound wave of a particular frequency is set up in the cochlea by oscillation of the stapes, the wave travels to the region of the basilar membrane that naturally responds maximally to that frequency. The energy of the pressure wave is dissipated with this vigorous membrane oscillation, so the wave dies out at the region of maximal displacement.

The hair cells in the region of peak vibration of the basilar membrane undergo the most mechanical deformation and, accordingly, are the most excited. This information is propagated to the CNS, which interprets the pattern of hair cell stimulation as a sound of a particular frequency. Modern techniques have determined that the basilar membrane is so fine-tuned that the peak membrane response to a single pitch probably extends no more than the width of a few hair cells.

Overtones of varying frequencies cause many points along the basilar membrane to vibrate simultaneously but less intensely than the fundamental tone, enabling the CNS to distinguish the timbre of the sound **(timbre discrimination).**

Intensity (loudness) discrimination depends on the amplitude of vibration. As sound waves originating from louder sound sources strike the eardrum, they cause it to vibrate more vigorously (that is, bulge in and out to a greater extent) but at the same frequency as a softer sound of the same pitch. The greater tympanic membrane deflection is converted into a greater amplitude of basilar membrane movement in the region of peak responsiveness. The CNS interprets greater basilar membrane oscillation as a louder sound.

The auditory system is so sensitive and can detect sounds so faint that the distance of basilar membrane deflection is comparable to only a fraction of the diameter of a hydrogen atom, the smallest of atoms. It is no wonder that very loud sounds, which cannot be sufficiently attenuated by protective middle ear reflexes (for example, the sounds of a typical rock concert), can set up such violent vibrations of the basilar membrane that irreplaceable hair cells are actually sheared off or permanently distorted, leading to partial hearing loss (Fig. 6-39).

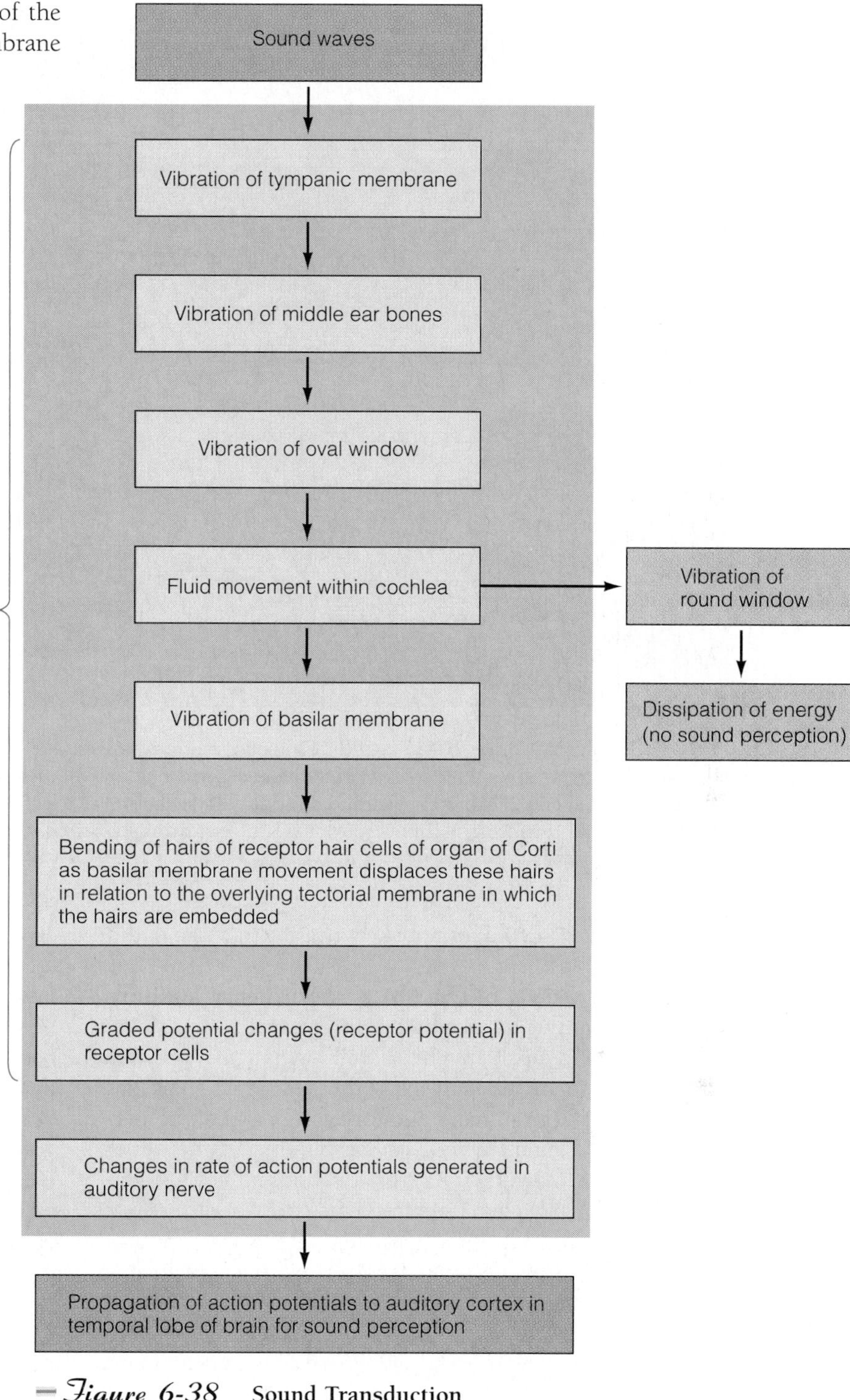

Figure 6-38 **Sound Transduction**

The auditory cortex is mapped according to tone.

Just as various regions of the basilar membrane are associated with particular tones, the auditory cortex is also *tonotopically* organized. Each region of the basilar membrane is linked to a specific region of the auditory cortex in the temporal lobe. Accordingly, specific cortical neurons are activated only by particular tones; each region of the auditory cortex becomes excited only in response to a specific tone detected by a selected portion of the basilar membrane.

The afferent neurons that pick up the auditory signals from the inner hair cells exit the cochlea via the auditory nerve. The neural pathway between the organ of Corti and the auditory cortex involves several synapses en route, the most notable of which are in the brain stem and medial geniculate nucleus of the thalamus. The brain stem uses the auditory input for alertness and arousal. The thalamus sorts and relays the signals upward. Unlike the visual pathways, auditory signals from each ear are transmitted to both temporal lobes because the fibers partially cross over in the brain stem. For this reason, a disruption of the auditory pathways on one side

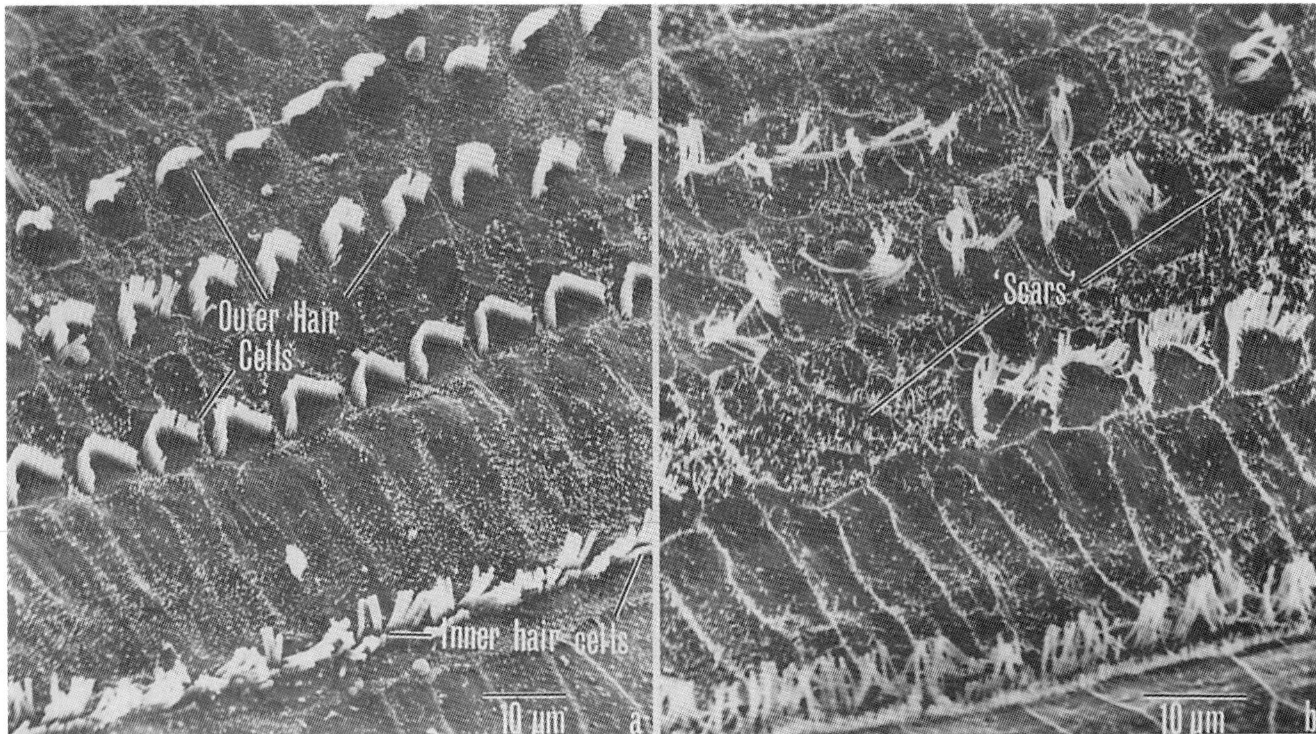

Figure 6-39 **Loss of Hair Cells Caused by Loud Noises** Injury and loss of hair cells caused by intense noise. Portions of the organ of Corti, with its three rows of outer hair cells and one row of inner hair cells, from the inner ear of (a) a normal guinea pig and (b) a guinea pig after a 24-hour exposure to noise at 120 decibels SPL, a level approached by loud rock music. (Scanning electron micrographs by R. S. Preston and J. E. Hawkins, Kresge Hearing Research Institute, University of Michigan.)

beyond the brain stem does not affect hearing in either ear to any extent.

Like other regions of the cortex, the **primary auditory cortex** is organized into columns. In contrast to the visual system, however, no hierarchy of cells within the columns that respond to increasingly complex auditory signals has been identified. The primary auditory cortex appears to perceive discrete sounds while the surrounding higher-order auditory cortex integrates the separate sounds into a coherent, meaningful pattern. Think about the complexity of the task accomplished by your auditory system. When you are at a concert, your organ of Corti responds to the simultaneous mixture of the instruments, the applause and hushed talking of the audience, and the background noises in the theater. You are able to distinguish these separate parts of the many sound waves reaching your ears and pay attention to those of importance to you.

Deafness is caused by defects either in conduction or neural processing of sound waves.

Loss of hearing, or **deafness,** may be temporary or permanent, partial or complete. Deafness is classified into two types—*conductive deafness* and *sensorineural deafness*—depending on the part of the hearing mechanism that fails to function adequately. **Conductive deafness** occurs when sound waves are not adequately conducted through the external and middle portions of the ear to set the fluids in the inner ear in motion. Possible causes include physical blockage of the ear canal with earwax, rupture of the eardrum, middle ear infections with accompanying fluid accumulation, or restriction of the ossicular movement because of bony adhesions between the stapes and the oval window. In **sensorineural deafness,** the sound waves are transmitted to the inner ear, but they are not translated into nerve signals that are interpreted by the brain as sound sensations. The defect can lie in the organ of Corti or the auditory nerves or, rarely, in the ascending auditory pathways or auditory cortex.

One of the most common causes of partial hearing loss, **neural presbycusis,** is a degenerative, age-related process that occurs as hair cells "wear out" with use. Over time, exposure to even ordinary modern-day sounds eventually damages hair cells, so that on average a male has lost more than 40% of his cochlear hair cells by age 65. The hair cells that process high-frequency sounds are the most vulnerable to destruction.

Hearing aids are helpful in conductive deafness but are less beneficial for sensorineural deafness. These devices increase the intensity of air-borne sounds and may modify the sound spectrum and tailor it to the patient's particular pattern of hearing loss at higher or lower frequencies. For the sound to be perceived, however, the receptor cell–neural pathway system must still be intact.

Within the past decade, **cochlear implants** have become available. These electronic devices, which are surgically implanted, transduce sound signals into electrical signals that can directly stimulate the auditory nerve, thus bypassing a

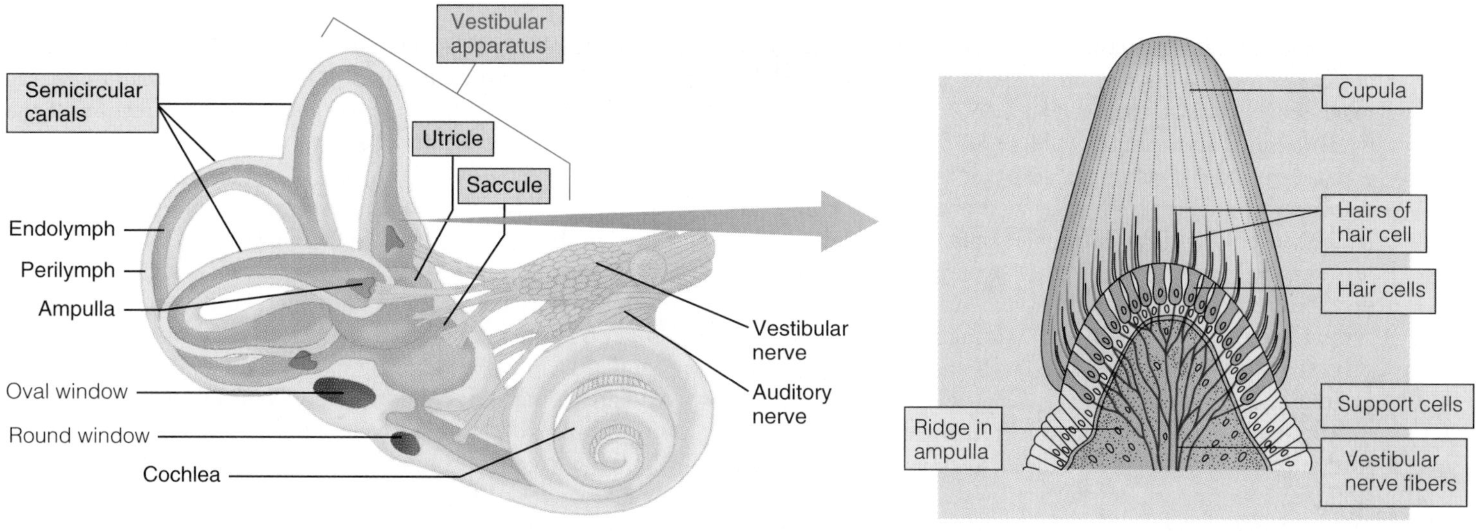

Figure 6-40

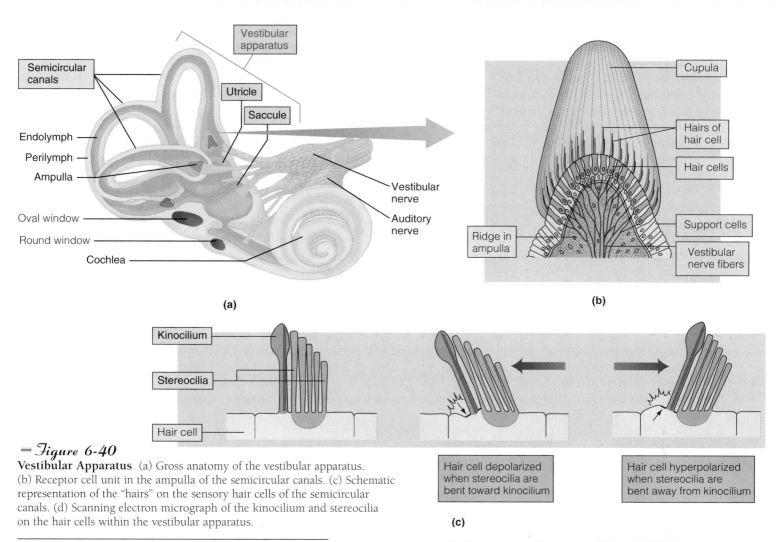

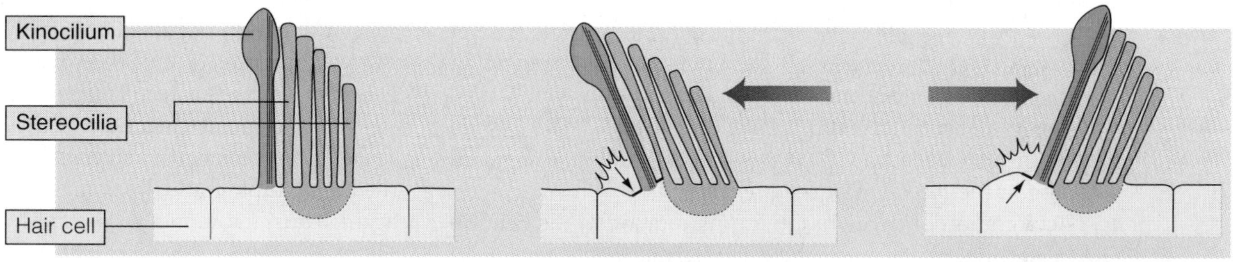

Vestibular Apparatus (a) Gross anatomy of the vestibular apparatus. (b) Receptor cell unit in the ampulla of the semicircular canals. (c) Schematic representation of the "hairs" on the sensory hair cells of the semicircular canals. (d) Scanning electron micrograph of the kinocilium and stereocilia on the hair cells within the vestibular apparatus.

defective cochlear system. Cochlear implants cannot restore normal hearing, but they do permit recipients to recognize sounds. Success ranges from an ability to "hear" a phone ringing to being able to carry on a conversation over the telephone (without visual cues, such as reading lips).

Exciting new findings suggest that in the future it may be possible to restore hearing by stimulating an injured inner ear to repair itself. Scientists have long considered the hair cells of the inner ear irreplaceable. Thus, hearing loss resulting from hair cell damage due to the aging process or exposure to loud noises is considered permanent. Encouraging new studies suggest, to the contrary, that hair cells in the inner ear have the latent ability to regenerate in response to an appropriate chemical signal. Researchers are currently trying to develop a drug that will spur regrowth of hair cells, thus repairing inner ear damage and hopefully restoring hearing.

(d)

The vestibular apparatus detects position and motion of the head and is important for equilibrium and coordination of head, eye, and body movements.

In addition to its cochlear-dependent role in hearing, the inner ear has another specialized component, the **vestibular apparatus,** which provides information essential for the sense of equilibrium and for coordinating head movements with eye and postural movements. The vestibular apparatus consists of two sets of structures lying within a tunneled-out region of the temporal bone near the cochlea—the *semicircular canals* and the *otolith organs,* namely, the *utricle* and *saccule* (Fig. 6-40a).

The vestibular apparatus detects changes in position and motion of the head. As in the cochlea, all components of the vestibular apparatus contain endolymph and are surrounded by perilymph. Also, similar to the organ of Corti, the vestibular components each contain hair cells that respond to mechanical deformation triggered by specific movements of the endolymph. Also, like the auditory hair cells, the vestibular receptors may be either depolarized or hyperpolarized, depending on the direction of the fluid movement. Unlike the auditory system, however, much of the information provided by the vestibular apparatus does not reach the level of conscious awareness.

The **semicircular canals** detect rotational or angular acceleration or deceleration of the head, such as when starting or stopping spinning, somersaulting, or turning the head. Each ear contains three semicircular canals arranged three-dimensionally in planes that lie at right angles to each other. The receptive hair cells of each semicircular canal are situated on top of a ridge located in the **ampulla,** a swelling at the base of the canal (Figs. 6-40a and b). The hairs are embedded in an overlying caplike, gelatinous layer, the **cupula,** which protrudes into the endolymph within the ampulla. The cupula sways in the direction of fluid movement, much like seaweed leaning in the direction of the prevailing tide.

Acceleration or deceleration during rotation of the head in any direction causes endolymph movement in at least one of the semicircular canals because of their three-dimensional arrangement. As the head starts to move, the bony canal and the ridge of hair cells embedded in the cupula move with the head. Initially, however, the fluid within the canal, not being attached to the skull, does not move in the direction of the rotation but lags behind because of its inertia. (Because of inertia, a resting object remains at rest, and a moving object continues to move in the same direction unless the object is acted on by some external force that induces change.) When the endolymph is left behind as the head starts to rotate, the endolymph that is in the same plane as the head movement is in effect shifted in the opposite direction from the movement (similar to your body tilting to the right as the car in which you are riding suddenly tuns to the left) (■ Fig. 6-41). This fluid movement causes the cupula to lean in the opposite direction from the head movement, bending the sensory hairs embedded in it. If the head movement continues at the same rate in the same direction, the endolymph catches up and moves in unison with the head so that the hairs return to their unbent position. When the head slows down and stops, the reverse situation occurs. The endolymph briefly continues to move in the direction of the rotation while the head decelerates to a stop. As a result, the cupula and its hairs are transiently bent in the direction of the preceding spin, which is opposite to the way they were bent during acceleration. When the endolymph gradually comes to a halt, the hairs straighten again. Thus, the semicircular canals detect changes in the rate of rotational movement of the head. They do not respond when the head is motionless or during circular motion at a constant speed.

The hairs of a vestibular hair cell consist of 20 to 50 microvilli—the stereocilia—and one cilium (see p. 37), the **kinocilium** (Fig. 6-40c and d). Each hair cell is oriented so that it depolarizes when its stereocilia are bent toward the kinocilium; bending in the opposite direction hyperpolarizes the cell. The hair cells form a chemically mediated synapse with terminal endings of afferent neurons whose axons join

■ *Figure 6-41* **Activation of the Hair Cells in the Semicircular Canals**

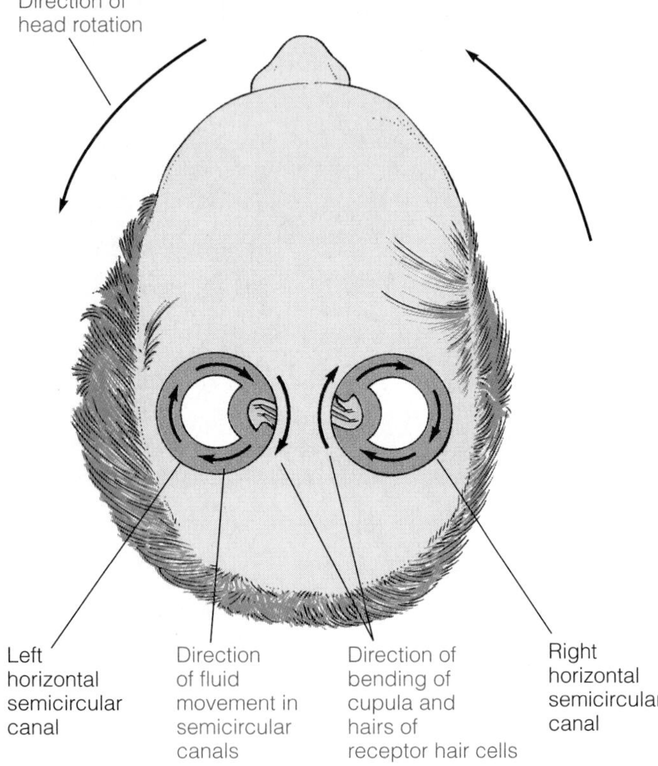

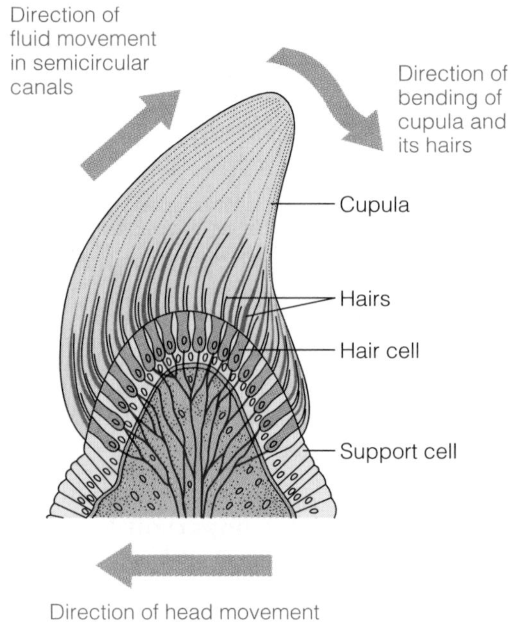

with those of the other vestibular structures to form the **vestibular nerve.** This nerve unites with the auditory nerve from the cochlea to form the **vestibulocochlear nerve.** Depolarization of the hair cells increases the rate of firing in the afferent fibers; conversely, when the hair cells are hyperpolarized, the frequency of action potentials in the afferent fibers is reduced.

Whereas the semicircular canals provide the CNS with information about rotational changes in head movement, the **otolith organs** provide information about the position of the head relative to gravity and also detect changes in the rate of linear motion (moving in a straight line regardless of direction). The **utricle** and **saccule** are saclike structures housed within a bony chamber situated between the semicircular canals and the cochlea (Fig. 6-40a). The hairs of the receptive hair cells in these sense organs also protrude into an overlying gelatinous sheet, whose movement displaces the hairs and results in changes in hair cell potential. Many tiny crystals of calcium carbonate—the **otoliths** ("ear stones")—are suspended within the gelatinous layer, making it heavier and giving it more inertia than the surrounding fluid (Fig. 6-42). When a person is in an upright position, the hairs within the utricles are oriented vertically and the saccule hairs are lined up horizontally.

Let's look at the utricle as an example. Its otolith-embedded, gelatinous mass shifts positions and bends the hairs in two ways:

1. When the head is tilted in any direction other than vertical (that is, other than straight up and down), the hairs are bent in the direction of the tilt because of the gravitational force exterted on the top-heavy gelatinous layer (Fig. 6-42b). Within the utricles on each side of the head, some of the hair cell bundles are oriented to depolarize and others to hyperpolarize when the head is in any position other than upright. The CNS thus receives different patterns of neural activity depending on head position with respect to gravity.
2. The utricle hairs are also displaced by any change in horizontal linear motion (such as moving straight forward, backward, or to the side). As a person starts to walk for-

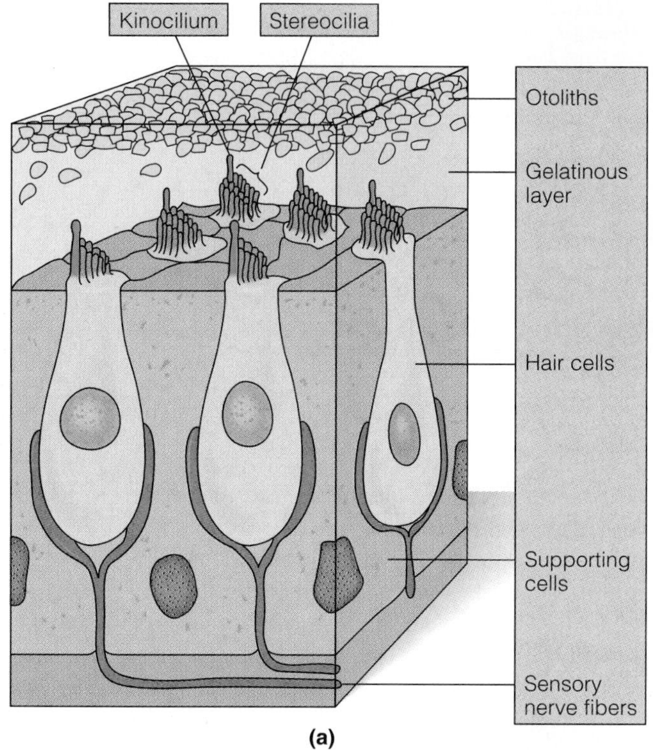

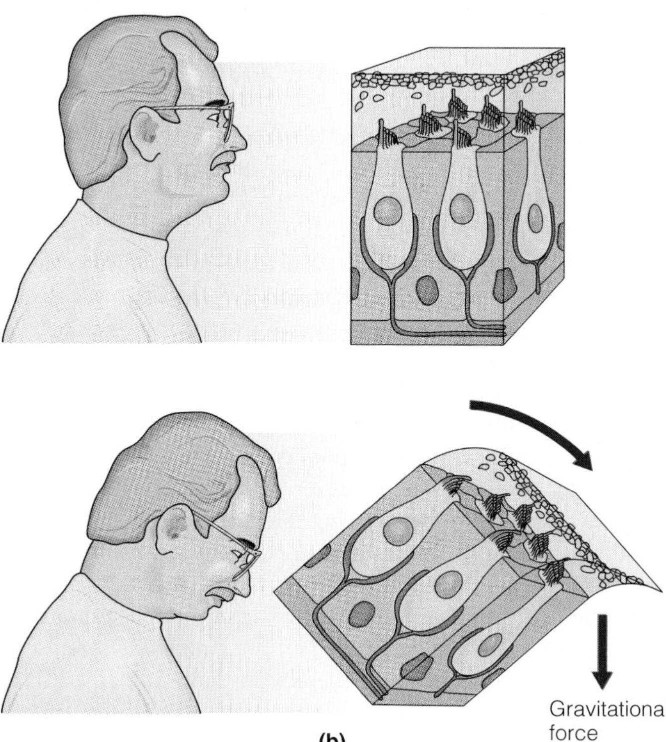

Figure 6-42 **Utricle** (a) Receptor unit in utricle. (b) Activation of the utricle by a change in head position. (c) Activation of the utricle by horizontal linear accleration.

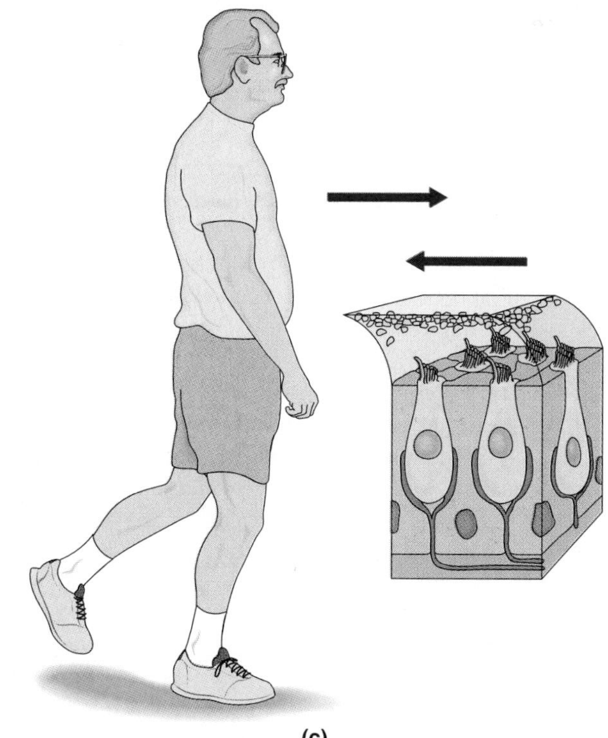

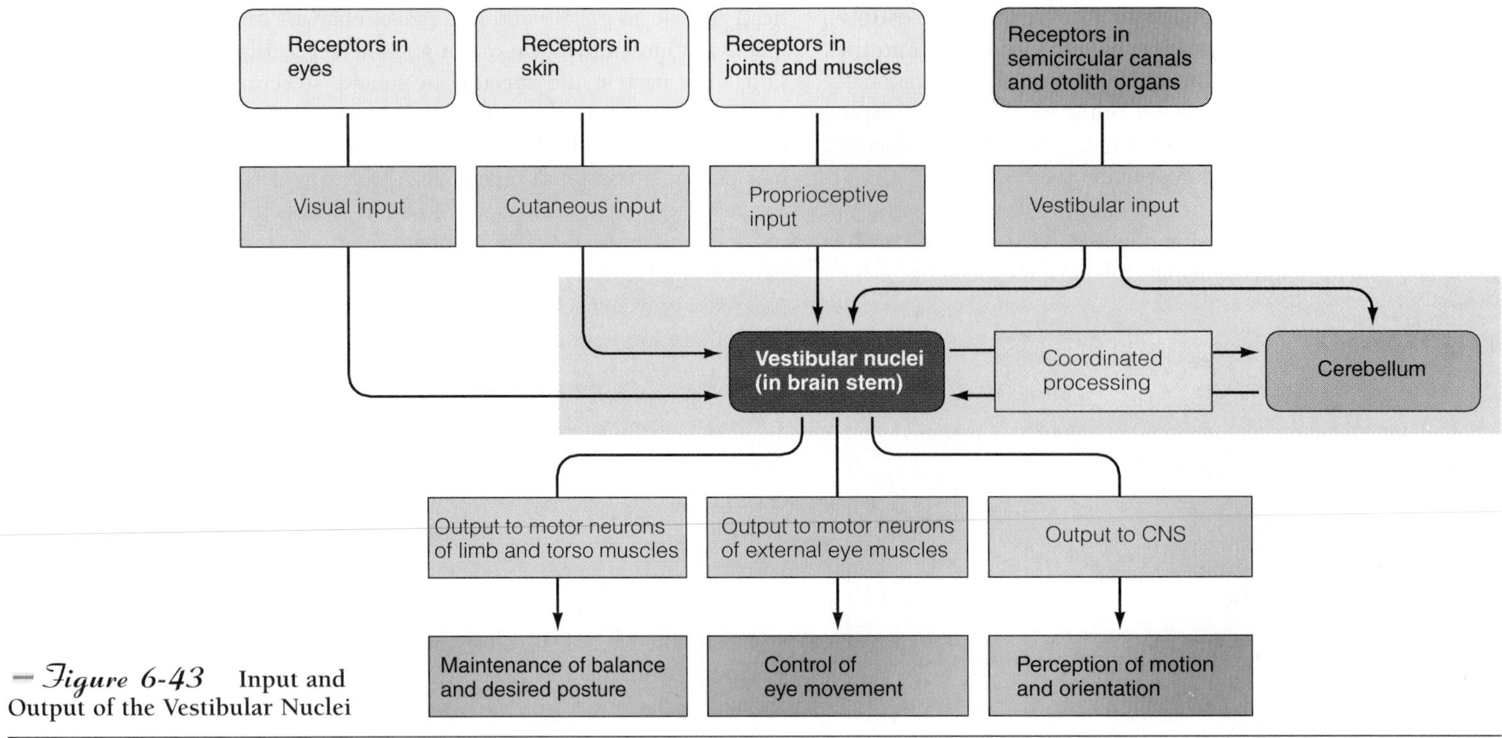

Figure 6-43 **Input and Output of the Vestibular Nuclei**

ward (Fig. 6-42c), the top-heavy otolith membrane at first lags behind the endolymph and hair cells because of its greater inertia. The hairs are thus bent to the rear, in the opposite direction of the forward movement of the head. If the walking pace is maintained, the gelatinous layer soon catches up and moves at the same rate as the head so that the hairs are no longer bent. When the person stops walking, the otolith sheet continues to move forward briefly as the head slows and stops, bending the hairs toward the front. Thus, the hair cells of the utricle detect horizontally directed linear acceleration and deceleration, but they do not provide information about movement in a straight line at constant speed.

The saccule functions similarly to the utricle, except that it responds selectively to the tilting of the head away from a horizontal position (such as getting up from bed) and to vertically directed linear acceleration and deceleration (such as jumping up and down or riding in an elevator).

Signals arising from the various components of the vestibular apparatus are carried through the vestibulocochlear nerve to the **vestibular nuclei**, a cluster of neuronal cell bodies in the brain stem, and to the cerebellum. Here the vestibular information is integrated with input from the skin surface, eyes, joints, and muscles for (1) maintaining balance and desired posture; (2) controlling the external eye muscles so that the eyes remain fixed on the same point, despite movement of the head; and (3) perceiving motion and orientation (— Fig. 6-43).

Some individuals, for poorly understood reasons, are especially sensitive to particular motions that activate the vestibular apparatus and cause symptoms of dizziness and nausea; this sensitivity is called **motion sickness.** Occasionally, fluid imbalances within the inner ear lead to **Meniere's disease.** Not surprisingly, because both the vestibular apparatus and cochlea contain the same inner ear fluids, both vestibular and auditory symptoms occur. An afflicted individual suffers transient attacks of severe **vertigo** (dizziness) accompanied by pronounced ringing in the ears and some loss of hearing. During these episodes, the person cannot stand upright and reports feeling as though self or surrounding objects in the room are spinning around.

Chemical Senses: Taste and Smell

Unlike the photoreceptors of the eye and the mechanoreceptors of the ear, the receptors for taste and smell are chemoreceptors, which generate neural signals upon binding with particular chemicals in their environment. The sensations of taste and smell in association with food intake influence the flow of digestive juices and affect appetite. Furthermore, stimulation of taste or smell receptors induces pleasurable or objectionable sensations and signals the presence of something to seek (a nutritionally useful, good-tasting food) or to avoid (a potentially toxic, bad-tasting substance). Thus, the chemical senses provide a "quality control" checkpoint for substances available for ingestion. In lower animals, smell also plays a major role in finding direction, in seeking prey or avoiding predators, and in sexual attraction to a mate. The sense of smell is less sensitive in humans and much less important in influencing our behavior (although millions of dollars are spent annually on perfumes and deodorants to make us "smell" better and thereby be more socially attractive). We will first examine the mechanism of taste **(gustation)** and then turn our attention to smell **(olfaction).**

Figure 6-44 Location and Structure of the Taste Buds Taste buds are located primarily along the edges of moundlike papillae on the upper surface of the tongue. The receptor cells and supporting cells of a taste bud are arranged like slices of an orange.

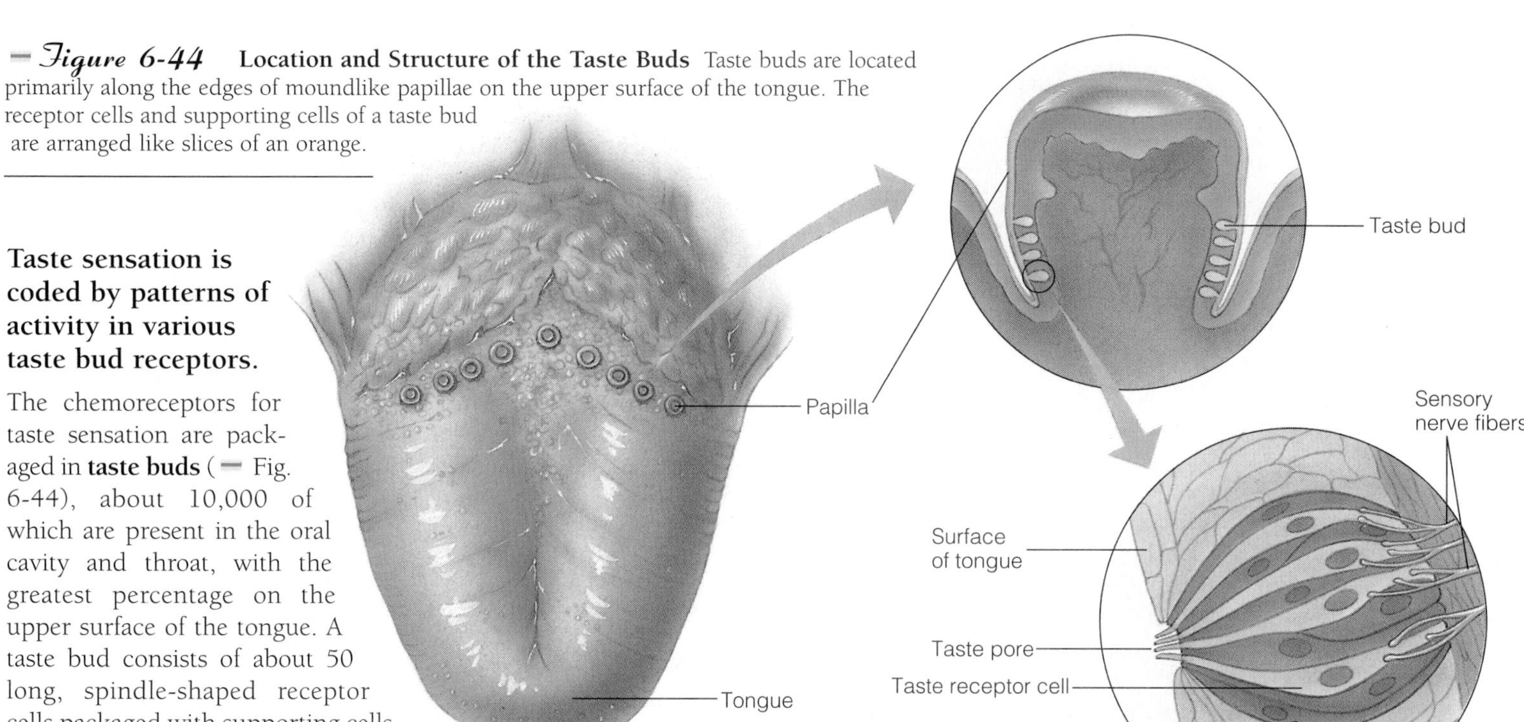

Taste sensation is coded by patterns of activity in various taste bud receptors.

The chemoreceptors for taste sensation are packaged in **taste buds** (Fig. 6-44), about 10,000 of which are present in the oral cavity and throat, with the greatest percentage on the upper surface of the tongue. A taste bud consists of about 50 long, spindle-shaped receptor cells packaged with supporting cells in an arrangement like slices of an orange. Each taste bud has a small opening, the **taste pore,** through which fluids in the mouth come into contact with the surface of its receptor cells.

Taste receptor cells are modified epithelial cells with many surface folds, or microvilli, that protrude slightly through the taste pore, greatly increasing the surface area exposed to the oral contents. The plasma membrane of the microvilli contains receptor sites that bind selectively with chemical molecules in the environment. Only chemicals in solution—either ingested liquids or solids that have been dissolved in saliva—can attach to receptor cells and evoke the sensation of taste. Binding of a taste-provoking chemical, a **tastant,** with a receptor cell alters the cell's ionic channels to produce a depolarizing receptor potential. This receptor potential, in turn, initiates action potentials within terminal endings of afferent nerve fibers with which the receptor cell synapses.

Most receptors are carefully sheltered from direct exposure to the environment, but the taste receptor cells, by virtue of their task, frequently come into contact with potent chemicals. Unlike the eye or ear receptors, which are irreplaceable, taste receptors have a life span of about 10 days. Epithelial cells surrounding the taste bud differentiate first into supporting cells and then into receptor cells to constantly renew the taste bud components.

Terminal afferent endings of several cranial nerves synapse with taste buds in various regions of the mouth. Signals in these sensory inputs are conveyed via synaptic stops in the brain stem and thalamus to the **cortical gustatory area,** a region in the parietal lobe adjacent to the "tongue" area of the somatosensory cortex. Unlike most sensory input, the gustatory pathways are primarily uncrossed. The brain stem also projects fibers to the hypothalamus and limbic system, presumably to add affective dimensions, such as whether the taste is pleasant or unpleasant, and to process behavioral aspects associated with taste and smell.

We can discriminate among thousands of different taste sensations, yet all tastes are varying combinations of four **primary tastes**: *salty, sour, sweet,* and *bitter.* A growing number of sensory physiologists suggest a fifth primary taste—*umami,* an amino acid and peptide (a short chain of amino acids) taste—should be added to the list, but this decision awaits further investigation.

Each receptor cell responds in varying degrees to all four primary tastes but is generally preferentially responsive to one of the taste modalities (Fig. 6-45). The richness of fine taste discrimination beyond the four primary tastes depends on subtle differences in the stimulation patterns of all the taste buds in response to various substances, similar to the variable stimulation of the three cone types that gives rise to the range of color sensations. Recent research has shown that each

Figure 6-45 Relative Responsiveness of Different Taste Buds to Different Stimuli Each taste bud responds in varying degrees to the four primary tastes. The pattern of stimulation for four different taste buds is depicted.

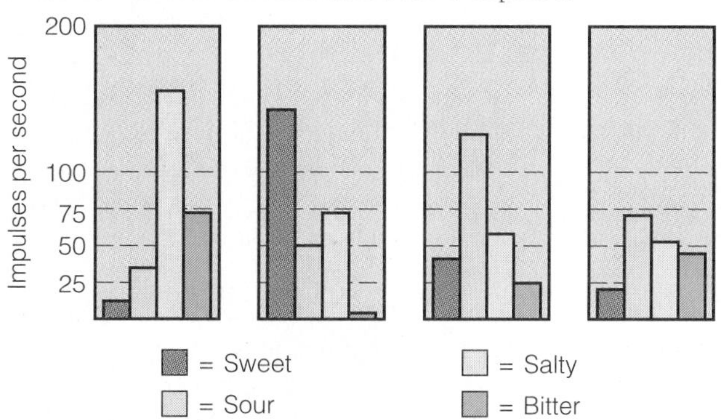

receptor cell uses different signaling pathways to bring about a depolarizing receptor potential in response to each of the four primary tastant categories:

- **Salt taste** is stimulated by chemical salts, especially NaCl (table salt). Direct entry of positively charged Na^+ ions through specialized Na^+ channels in the receptor cell membrane, a movement that reduces the cell's internal negativity, is believed to be responsible for receptor depolarization in response to salt.
- **Sour taste** is caused by acids, which contain a free hydrogen ion, H^+. The citric acid content of lemons, for example, accounts for their distinctly sour taste. Depolarization of the receptor cell by sour tastants is thought to occur because H^+ blocks K^+ channels in the receptor cell membrane. The resultant decrease in the passive movement of positively charged K^+ ions out of the cell reduces the internal negativity, producing a depolarizing receptor potential.
- **Sweet taste** is evoked by the particular configuration of glucose. Other organic molecules with similar structures, such as saccharin, aspartame, and other artificial sweeteners, can also interact with sweet receptor binding sites. Binding of glucose or another chemical with the taste cell receptor activates a G protein, which turns on the cAMP second messenger pathway in the taste cell (see p. 68). The second messenger pathway ultimately results in phosphorylation and blockage of K^+ channels in the receptor cell membrane, leading to a depolarizing receptor potential.
- **Bitter taste** is elicited by a more chemically diverse group of tastants than the other taste sensations. For example, alkaloids (such as caffeine, nicotine, strychnine, morphine, and other toxic plant derivatives) as well as poisonous substances all taste bitter, presumably as a protective mechanism to discourage ingestion of these potentially dangerous compounds. Studies reveal that different bitter substances employ various signaling pathways to bring about a receptor potential. Using a variety of mechanisms for perception of bitterness expands the ability of the taste receptor to detect a wide range of potentially harmful chemicals. Investigators are currently trying to sort out these signaling strategies for bitterness. The first G protein in taste—**gustducin**—has been identified in one of the bitter signaling pathways. Interestingly, this G protein, which sets off a second messenger pathway in the taste cell, is very similar to the visual G protein, transducin.

Taste perception is also influenced by information derived from other receptors, especially odor. When you temporarily lose your sense of smell because of swollen nasal passageways during a cold, your sense of taste is also markedly reduced, even though your taste receptors are unaffected by the cold. Other factors affecting taste include temperature and texture of the food as well as psychological factors associated with past experiences with the food. How the cortex accomplishes the complex perceptual processing of taste sensation is currently not known.

During smell perception, various parts of an odor are detected by different olfactory receptors and sorted into "smell files."

The **olfactory (smell) mucosa**, located in the ceiling of the nasal cavity, contains three cell types: *olfactory receptors, supporting cells,* and *basal cells* (Fig. 6-46). The supporting

Figure 6-46
Location and Structure of the Olfactory Receptors

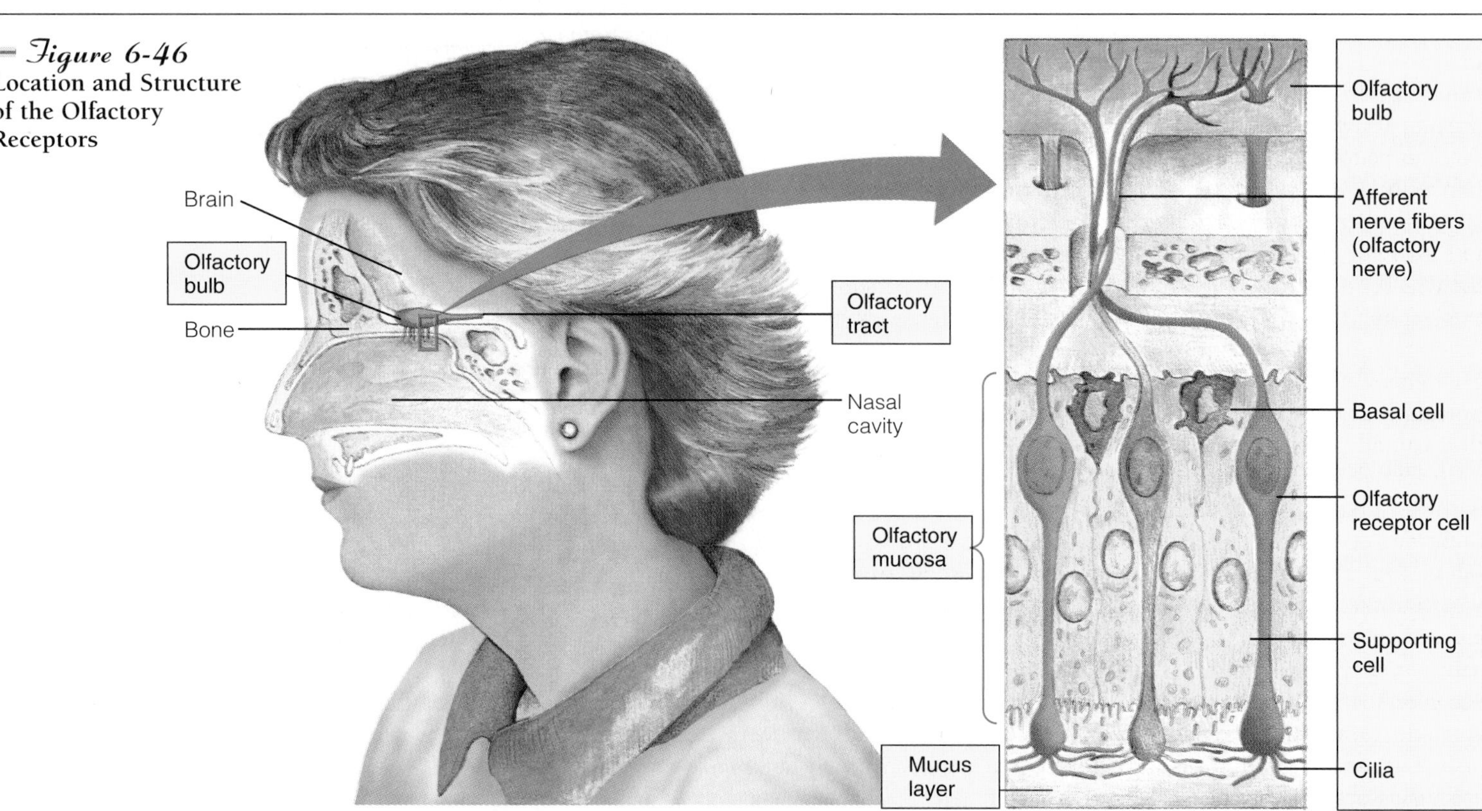

cells secrete **mucus,** which coats the nasal passages. The basal cells are precursors for new olfactory receptor cells, which are replaced about every two months. This is remarkable because, unlike the other special sense receptors, **olfactory receptors** are specialized endings of afferent neurons, not separate cells. The entire neuron, including its afferent axon projecting into the brain, is replaced. These are the only neurons that undergo cell division. The axons of the receptor cells collectively form the **olfactory nerve.** The receptor portion of the olfactory receptor cells consists of an enlarged knob bearing several long cilia that extend like a tassle to the surface of the mucosa. These cilia contain the binding sites for attachment of **odorants,** molecules that can be smelled. During quiet breathing, odorants typically reach the sensitive receptors only by diffusion because the olfactory mucosa is above the normal path of air flow. The act of sniffing enhances this process by drawing the air currents upward within the nasal cavity so that a greater percentage of the odoriferous molecules in the air come into contact with the olfactory mucosa. Odorants also reach the olfactory mucosa during eating by wafting up to the nose from the mouth through the pharynx (back of the throat).

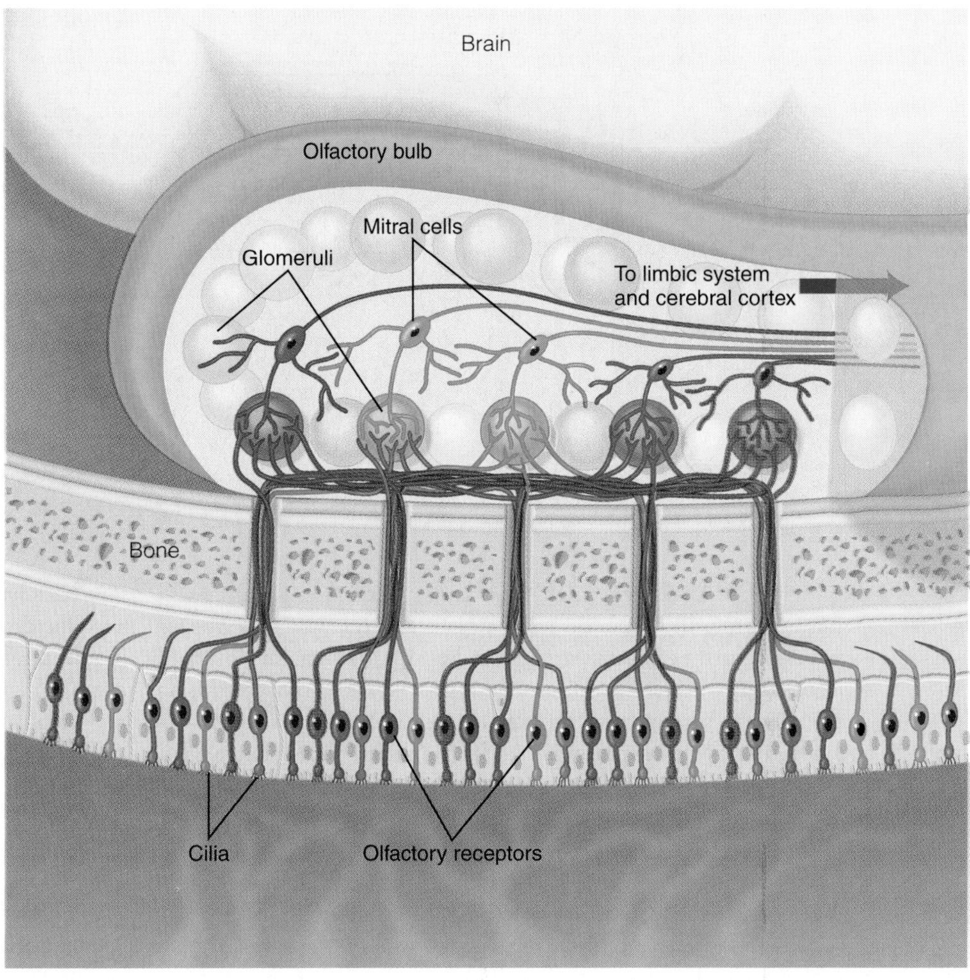

Figure 6-47 **Processing of Scents in the Olfactory Bulb** Each of the glomeruli lining the olfactory bulb receives synaptic input from only one type of olfactory receptor, which, in turn, responds to only one discrete component of an odorant. Thus, the glomeruli sort and file the various components of an odoriferous molecule before relaying the smell signal to the mitral cells and higher brain levels for further processing.

To be smelled, a substance must be (1) sufficiently volatile (easily vaporized) that some of its molecules can enter the nose in the inspired air and (2) sufficiently water-soluble that it can dissolve in the mucus layer coating the olfactory mucosa. As with taste receptors, molecules must be dissolved in order to be detected by olfactory receptors.

The human nose contains 5 million olfactory receptors, of which there are 1,000 different types, far more than the three receptors that code for color vision and the four that code for taste. During smell detection, an odor is "dissected" into various components. Each receptor responds to only one discrete component of an odor rather than to the whole odorant molecule. Accordingly, the various parts of an odor are detected by different receptors, and a given receptor can respond to a particular odor component shared in common by different scents.

Binding of an appropriate scent signal to an olfactory receptor activates a G protein, triggering a cascade of intracellular reactions that leads to opening of Na^+ channels. The resultant ion movement brings about a depolarizing receptor potential that generates action potentials in the afferent fiber. The frequency of the action potentials depends on the concentration of the stimulating chemical molecules.

The afferent fibers arising from the receptor endings in the nose pass through tiny holes in the flat bone plate separating the olfactory mucosa from the overlying brain tissue (Fig. 6-46). They immediately synapse in the **olfactory bulb,** a complex neural structure containing several different layers of cells that are functionally similar to the retinal layers of the eye. The twin olfactory bulbs, one on each side, are about the size of small grapes. Each olfactory bulb is lined by small ball-like neural junctions known as **glomeruli** ("little balls") (Fig. 6-47). Within each glomerulus, the terminals of receptor cells carrying information about a particular scent component synapse with the next cells in the olfactory pathway, the **mitral cells.** Because each glomerulus receives signals only from receptors that detect a particular odor component, the glomeruli serve as "smell files." The separate components of an odor are sorted into different glomeruli, one component per file. Thus, the glomeruli, which serve as the first relay station in the brain for processing olfactory information, play a key role in organizing scent perception.

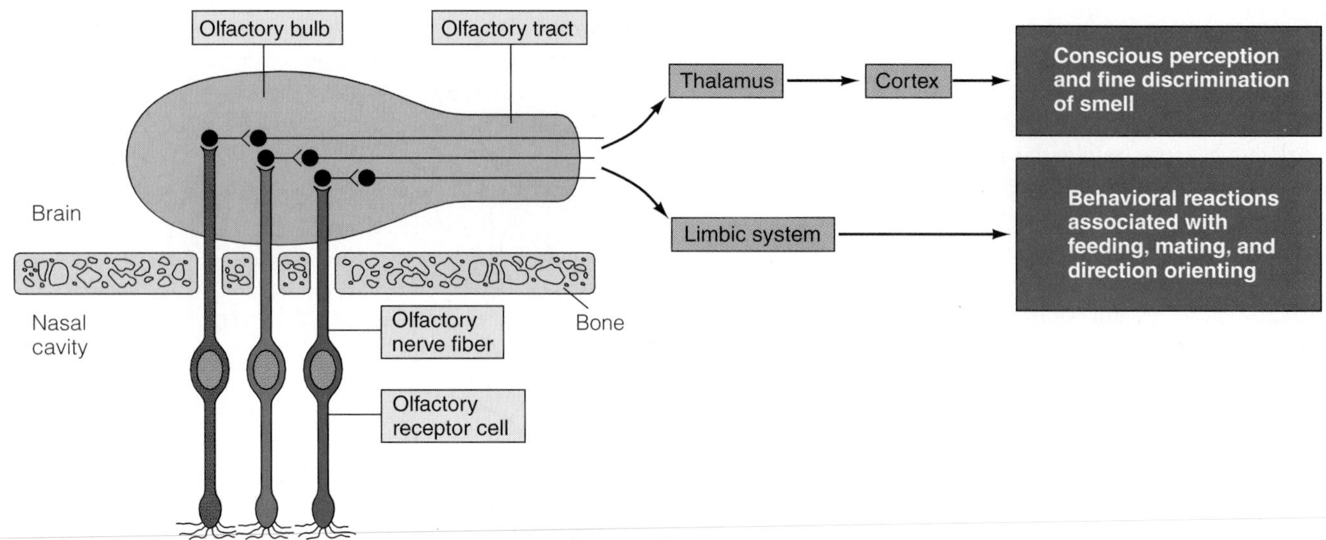

Figure 6-48 **Olfactory Pathways** See Figures 6-46 and 6-47 for a location reference point.

The mitral cells on which the olfactory receptors terminate in the glomeruli refine the smell signals and relay them to the brain for further processing. Fibers leaving the olfactory bulb travel in two different routes (Fig. 6-48): (1) a *subcortical route* going primarily to regions of the limbic system, especially the lower medial sides of the temporal lobes (considered to be the **primary olfactory cortex**), and (2) a *thalamic-cortical route*. Until recently, the subcortical route was thought to be the only olfactory pathway. This route, which includes hypothalamic involvement, permits close coordination between smell and behavioral reactions associated with feeding, mating, and direction orienting. As with the other senses, the thalamic-cortical route is important for conscious perception and fine discrimination of smell.

Since each given odorant activates multiple receptors and glomeruli in response to its various odor components, odor discrimination is based on different patterns of glomeruli activated by various scents. In this way, the cortex is able to distinguish about 10,000 different scents. This mechanism for sorting out and distinguishing different odors is very effective, even in humans, who have a poor sense of smell compared to other species. (By comparison, dogs' sense of smell is hundreds of times more sensitive than that of humans). A noteworthy example is our ability to detect methyl mercaptan (garlic odor) at a concentration of 1 molecule per 50 billion molecules in the air! This substance is added to odorless natural gas to enable us to detect potentially lethal gas leaks.

Although the olfactory system is sensitive and highly discriminating, it is also quickly adaptive. Our sensitivity to a new odor diminishes rapidly after a short period of exposure to it, even though the odor source continues to be present. This reduced sensitivity does not involve receptor adaptation, as has been thought for years; actually, the olfactory receptors themselves adapt slowly. It apparently involves some sort of adaptation process in the CNS. Adaptation is specific for a particular odor, and responsiveness to other odors remains unchanged.

What clears the odorants away from their binding sites on the olfactory receptors so that the sensation of smell doesn't "linger" after the source of the odor is removed? Several "odor-eating" enzymes have recently been discovered in the olfactory mucosa that could serve as molecular janitors, clearing away the odoriferous molecules so that they do not continue to stimulate the olfactory receptors. Interestingly, these odorant-clearing enzymes are very similar chemically to detoxification enzymes found in the liver. (These liver enzymes inactivate potential toxins absorbed from the digestive tract; see p. 23). This resemblance may not be coincidental. Researchers speculate that these enzymes may serve the dual purpose of clearing the olfactory mucosa of old odorants and transforming potentially harmful chemicals into harmless molecules. Such detoxification would serve a very useful purpose, considering the open passageway between the olfactory mucosa and the brain.

In addition to the olfactory mucosa, the nose contains another sense organ, the **vomeronasal organ (VNO)**, which is common in animals but until recently was thought to be nonexistent in humans. The VNO is located about half an inch inside the human nose next to the vomer bone, hence its name. It detects **pheromones**, chemical signals passed subconsciously from one individual to another. In animals, the VNO is known as the "sexual nose" for its role in governing reproductive and social behaviors, such as identifying and attracting a mate. Although the role of the VNO in human behavior has not been validated, researchers suspect that it is responsible for spontaneous "feelings" between people, either "good chemistry," such as "love at first sight," or "bad chemistry," such as "getting bad vibes" from someone you just met. Because messages conveyed by the VNO are believed to bypass cortical consciousness, the response to the largely odorless pheromones is not a distinct, discrete perception, such as smelling a favorite fragrance, but more like an inexplicable impression.

Chapter in Perspective: Focus on Homeostasis

To maintain a life-sustaining stable internal environment, adjustments must constantly be made to compensate for the myriad external and internal factors that continuously threaten to disrupt homeostasis, such as external exposure to cold or internal acid production. Many of these adjustments are directed by the nervous system, one of the body's two major control systems. The central nervous system (CNS), the integrating and decision-making component of the nervous system, must continuously be informed of "what's happening" in both the internal and the external environment so that it can command appropriate responses in the organ systems to maintain the body's viability. In other words, the CNS must know what changes are taking place before it can respond to these changes.

The afferent division of the peripheral nervous system is the communication link by which the CNS is informed about the internal and external environment. The afferent division detects, encodes, and transmits peripheral signals to the CNS for processing. Afferent input is necessary for arousal, perception, and determination of efferent output.

Afferent information about the internal environment, such as the CO_2 level in the blood, never reaches the level of conscious awareness, but this input to the controlling centers of the CNS is essential for the maintenance of homeostasis. Afferent input reaching the level of conscious awareness is known as sensory information and includes somesthetic and proprioceptive sensation (body sense) and special senses (vision, hearing, taste, and smell).

The body-sense receptors are distributed over the entire body surface as well as throughout the joints and muscles. Afferent signals from these receptors provide information about what's happening directly to each specific body part in relation to the external environment (that is, the "what," "where," and "how much" of stimulatory inputs to the body's surface and the momentary position of the body in space). In contrast, each special-sense organ is restricted to a single site in the body. Rather than providing information about a specific body part, a special-sense organ provides a specific type of information about the external environment that is useful to the body as a whole. For example, through their ability to detect, extensively analyze, and integrate patterns of illumination in the external environment, the eyes and visual processing system enable us to "see" our surroundings. The same integrative effect could not be achieved if photoreceptors were scattered over the entire body surface as are the touch receptors.

Sensory input (both body sense and special senses) enables a complex multicellular organism such as a human to interact in meaningful ways with the external environment in procuring food, defending against danger, and engaging in other behavioral actions geared toward maintaining homeostasis. In addition to providing information essential for interactions with the external environment for basic survival, the perceptual processing of sensory input adds immeasurably to the richness of life, such as enjoyment of a good book, concert, or meal.

Chapter Summary

Receptor Physiology

Receptors are specialized peripheral endings of afferent neurons; they respond to particular stimuli, translating the energy forms of the stimuli into electrical signals, the language of the nervous system. There are discrete labeled-line pathways from the receptors to the CNS so that information about the type and location of the stimuli can be deciphered by the CNS, even though all the information arrives in the form of action potentials. What the brain perceives from its input, however, is an abstraction and not reality. The only stimuli that can be detected are those for which receptors are present. Furthermore, as sensory signals ascend through progressively more complex processing, some of the information may be suppressed, whereas other parts of it may be enhanced.

Stimulation of a receptor produces a graded receptor potential. The strength and rate of change of the stimulus are reflected in the magnitude of the receptor potential, which in turn determines the frequency of action potentials generated in the afferent neuron. The magnitude of the receptor potential is also influenced by the extent of receptor adaptation, which refers to a reduction in receptor potential in spite of sustained stimulation. Tonic receptors adapt slowly or not at all and thus provide continuous information about the stimuli they monitor. Phasic receptors adapt rapidly and frequently exhibit off responses, thereby providing information about changes in the energy form they monitor.

Pain

Painful experiences are elicited by noxious mechanical, thermal, or chemical stimuli and consist of two components: the perception of pain coupled with emotional and behavioral responses to it. Pain signals are transmitted over two afferent pathways: a fast pathway that carries sharp, prickling pain signals, and a slow pathway that carries dull, aching, persistent pain signals. Afferent pain fibers terminate in the spinal cord on ascending pathways that transmit the signal to the brain for processing. Descending pathways from the brain use endogenous opiates to suppress the release of substance P, the neurotransmitter from the afferent pain fiber terminal. Thus, these descending pathways block further transmission of the pain signal and serve as a built-in analgesic system.

Eye: Vision

The eye is a specialized structure housing the light-sensitive receptors essential for vision perception—namely, the rods and cones found in its retinal layer. The iris controls the size of the pupil, thereby adjusting the amount of light permitted to enter the eye. The cornea and lens are the primary refractive structures that bend the incoming light rays to focus the image on the retina. The cornea contributes most to the total refractive ability of the eye. The strength of the lens can be adjusted through action of the ciliary muscle to accommodate for differences in near and far vision.

Rods and cones are activated when the photopigments they contain differentially absorb various wavelengths of light. Light absorption causes a biochemical change in the photopigment that is ultimately converted into a change in the rate of action potential propagation in the visual pathway leaving the retina. The visual message is transmitted to the visual cortex in the brain for perceptual processing.

Cones display high acuity but can be used only for day vision because of their low sensitivity to light. Different ratios of stimulation of three cone types by varying wavelengths of light lead to color vision. Rods provide only indistinct vision in shades of gray, but because they are very sensitive to light, they can be used for night vision.

Ear: Hearing and Equilibrium

The ear performs two unrelated functions: (1) hearing, which involves the external ear, middle ear, and cochlea of the inner ear; and (2) sense of equilibrium, which involves the vestibular apparatus of the inner ear. In contrast to the photoreceptors of the eye, the ear receptors located in the inner ear—the hair cells in the cochlea and vestibular apparatus—are mechanoreceptors. Hearing depends on the ear's ability to convert airborne sound waves into mechanical deformations of receptive hair cells, thereby initiating neural signals. Sound waves consist of high-pressure regions of compression alternating with low-pressure regions of rarefaction of air molecules. The pitch (tone) of a sound is determined by the frequency of its waves and the loudness (intensity) by the amplitude of the waves. Sound waves are funneled through the external ear canal to the tympanic membrane, which vibrates in synchrony with the waves. Middle ear bones bridging the gap between the tympanic membrane and the inner ear amplify the tympanic movements and transmit them to the oval window, whose movement sets up traveling waves in the cochlear fluid. These waves, which are at the same frequency as the original sound waves, set the basilar membrane in motion. Various regions of this membrane selectively vibrate more vigorously in response to different frequencies of sound. On top of the basilar membrane are the receptive hair cells of the organ of Corti, whose hairs are bent as the basilar membrane is deflected up and down in relation to the overhanging stationary tectorial membrane in which the hairs are embedded. This mechanical deformation of specific hair cells in the region of maximal basilar membrane vibration is transduced into neural signals that are transmitted to the auditory cortex in the brain for sound perception.

The vestibular apparatus in the inner ear consists of (1) the semicircular canals, which detect rotational acceleration or deceleration in any direction, and (2) the utricle and saccule, which detect changes in the rate of linear movement in any direction and provide information important for determining head position in relation to gravity. Neural signals are generated in response to mechanical deformation of hair cells caused by specific movement of fluid and related structures within these sense organs. This information is important for the sense of equilibrium and for maintaining posture.

Chemical Senses: Taste and Smell

Taste and smell are chemical senses. In both cases, attachment of specific dissolved molecules to binding sites on the receptor membrane causes receptor potentials that, in turn, set up neural impulses that signal the presence of the chemical. Taste receptors are housed in taste buds on the tongue; olfactory receptors are located in the mucosa in the upper part of the nasal cavity. Both sensory pathways include two routes: one to the limbic system for emotional and behavioral processing and one through the thalamus to the cortex for conscious perception and fine discrimination. Taste and olfactory receptors are continuously renewed, unlike visual and hearing receptors, which are irreplaceable.

Review Exercises

Objective Questions (Answers on p. E–5.)

1. Conversion of the energy forms of stimuli into electrical energy by the receptors is known as ____________.
2. The type of stimulus to which a particular receptor is most responsive is called its ____________.
3. The sensation perceived depends on the type of receptor stimulated rather than on the type of stimulus. (True or false?)
4. All afferent information is sensory information. (True or false?)
5. Off-center ganglion cells increase their rate of firing when a beam of light strikes the periphery of their respective field. (True or false?)
6. During dark adaptation, rhodopsin is gradually regenerated to increase the sensitivity of the eyes. (True or false?)
7. An optic nerve carries information from the lateral and medial halves of the same eye, whereas an optic tract carries information from the lateral half of one eye and the medial half of the other. (True or false?)
8. Displacement of the round window generates neural impulses that are perceived as sound sensations. (True or false?)
9. Hair cells in different regions of the organ of Corti and neurons in different regions of the auditory cortex are activated by different tones. (True or false?)
10. Each taste receptor responds to just one of the four primary tastes. (True or false?)
11. Rapid adaptation to odors results from adaptation of the olfactory receptors. (True or false?)

12. Using the answer code below, indicate which properties apply to taste and/or smell:

(a) applies to taste
(b) applies to smell
(c) applies to both taste and smell

___1. Receptors are separate cells that synapse with terminal endings of afferent neurons.
___2. Receptors are specialized endings of afferent neurons.
___3. Receptors are regularly replaced.
___4. Specific chemicals in the environment attach to special binding sites on the receptor surface, leading to a depolarizing receptor potential.
___5. Has two processing pathways: a limbic system route and a thalamic-cortical route.
___6. Uses four different receptor types.
___7. Uses 1,000 different receptor types.
___8. Information from receptor cells is filed and sorted by neural junctions called glomeruli.

13. Match the following:

___ 1. layer that contains the photoreceptors
___ 2. point from which the optic nerve leaves the retina
___ 3. forms the white part of the eye
___ 4. thalamic structure that processes visual input
___ 5. colored diaphragm of muscle that controls the amount of light entering the eye
___ 6. contributes most to the refractive ability of the eye
___ 7. supplies nutrients to the lens and cornea
___ 8. produces aqueous humor
___ 9. contains the vascular supply for the retina and a pigment that minimizes scattering of light within the eye
___10. has adjustable refractive ability
___11. portion of the retina with greatest acuity
___12. point at which fibers from the medial half of each retina cross to the opposite side

(a) choroid
(b) aqueous humor
(c) fovea
(d) lateral geniculate nucleus
(e) cornea
(f) retina
(g) lens
(h) optic disc; blind spot
(i) iris
(j) ciliary body
(k) optic chiasm
(l) sclera

Essay Questions

1. List and describe the receptor types according to their adequate stimulus.
2. Compare tonic and phasic receptors.
3. Explain how acuity is influenced by receptive field size and by lateral inhibition.
4. Compare the fast and slow pain pathways.
5. Describe the built-in analgesic system of the brain.
6. Describe the process of phototransduction.
7. Compare the functional characteristics of rods and cones.
8. What are sound waves? What is responsible for the pitch, intensity, and timbre of a sound?
9. Describe the function of each of the following parts of the ear: pinna, ear canal, tympanic membrane, ossicles, oval window, and the various parts of the cochlea. Include a discussion of how sound waves are transduced into action potentials.
10. Discuss the functions of the semicircular canals, the utricle, and the saccule.
11. Describe the location, structure, and activation of the receptors for taste and smell.
12. Compare the processes of color vision, hearing, taste, and smell discrimination.

Quantitative Exercises (Solutions on p. E–5.)

1. Calculate the difference in the time it takes for an action potential to travel 1.3 m between the slow (12 m/sec) and fast (30 m/sec) pain pathways.
2. Have you ever noticed that humans have circular pupils while cats' pupils are more rectangular? The following calculations will help you understand the implication of this difference. For simplicity, assume a constant intensity of light.
   a. If the diameter of a human's circular pupil were decreased by half on contraction of the constrictor muscle of the iris, by what percentage would the amount of light allowed into the eye be decreased?
   b. If a cat's rectangular pupil were decreased by half along one axis only, by what percentage would the amount of light allowed into the eye be decreased?
   c. Comparing these calculations, do humans or cats have more precise control over the amount of light falling on the retina?
3. A decibel is the unit of *sound level*, β, defined as follows:

$$\beta = (10\text{ dB}) \log_{10} (I/I_o),$$

where I is *sound intensity*, or the rate at which sound waves transmit energy per unit area. The units of I are watts per square meter (W/m^2). I_o is a constant intensity close to the human hearing threshold, namely 10^{-12} W/m^2.

   a. For the following sound levels, calculate the corresponding sound intensities:
      i. 20 dB (a whisper)
      ii. 70 dB (a car horn)
      iii. 120 dB (a low-flying jet)
      iv. 170 dB (a space shuttle launch)
   b. Explain why the sound levels of these sounds increase by the same increment (that is, each sound is 50 dB higher than the one preceding it), yet the incremental increases in sound intensities that you calculated are so different. What implications does this have for the performance of the human ear?

Points to Ponder

(Explanations on p. E–6.)

1. Patients with certain nerve disorders are unable to feel pain. Why is this disadvantageous?
2. Ophthamologists frequently instill eye drops in their patients' eyes to bring about pupillary dilation, which makes it easier for the physician to view the eye's interior. In what way would the drug in the eye drops affect autonomic nervous system activity in the eye in order to cause the pupils to dilate?
3. A patient complains of not being able to see the right half of the visual field with either eye. At what point in the patient's visual pathway is the defect?
4. Explain how middle ear infections interfere with hearing. Of what value are the "tubes" that are sometimes surgically placed in the eardrums of patients with a history of repeated middle ear infections accompanied by chronic fluid accumulation?
5. Explain why your sense of smell is reduced when you have a cold, even though the cold virus does not directly adversely affect the olfactory receptor cells.
6. ***Clinical Consideration*** Suzanne J. complained to her physician of bouts of dizziness. The physician asked her whether by "dizziness" she meant a feeling of lightheadedness, as if she felt she were going to faint (a condition known as *syncope*), or a feeling that she or surrounding objects in the room were spinning around (a condition known as *vertigo*). Why is this distinction important in the differential diagnosis of her condition? What are some possible causes of each of these symptoms?

7 The Peripheral Nervous System: Efferent Division

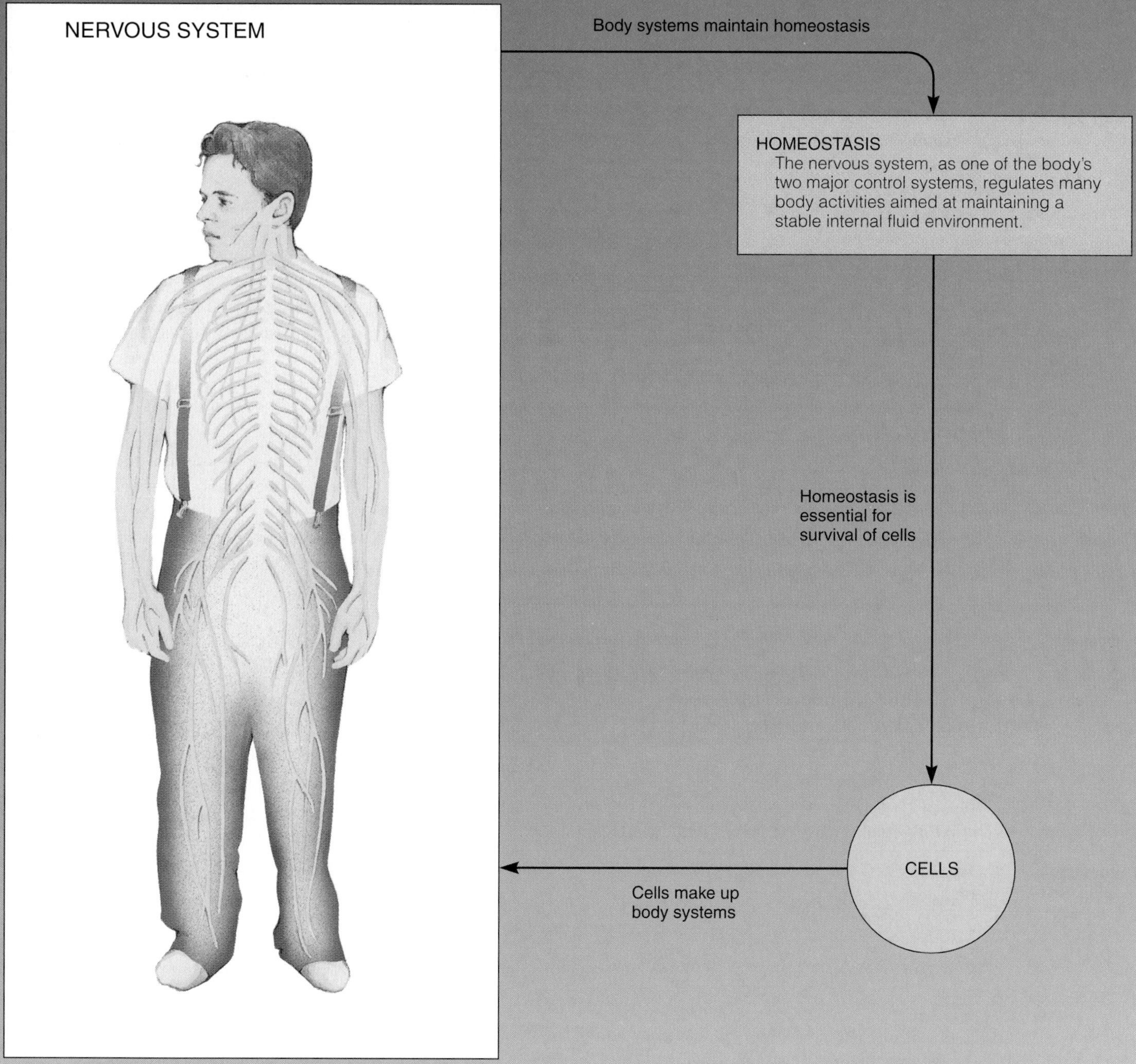

NERVOUS SYSTEM

Body systems maintain homeostasis

HOMEOSTASIS
The nervous system, as one of the body's two major control systems, regulates many body activities aimed at maintaining a stable internal fluid environment.

Homeostasis is essential for survival of cells

CELLS

Cells make up body systems

The nervous system, one of the two major control systems of the body, consists of the central nervous system (CNS), composed of the brain and spinal cord, and the **peripheral nervous system,** composed of the afferent and efferent fibers that relay signals between the CNS and periphery (other parts of the body).

Once informed by the afferent division of the peripheral nervous system that a change in the internal or external environment is threatening homeostasis, the CNS makes appropriate adjustments to maintain homeostasis. The CNS makes these adjustments by controlling the activities of effector organs (muscles and glands) by transmitting signals *from* the CNS to these organs through the **efferent division** of the peripheral nervous system.

Chapter Contents At a Glance

Introduction

The efferent division of the peripheral nervous system is the communication link by which the central nervous system controls the activities of muscles and glands. The CNS regulates these effector organs (see p. 112) by initiating action potentials in the cell bodies of efferent neurons whose axons terminate on these organs. Cardiac muscle, smooth muscle, most exocrine glands, and some endocrine glands are innervated by the **autonomic nervous system,** which is considered the involuntary branch of the peripheral efferent division. Skeletal muscle is innervated by the **somatic nervous system,** which is the voluntary branch of the efferent division. Efferent output typically influences either movement or secretion, as illustrated by the examples of the effects of neural control on various types of effector organs in Table 7-1. Much of this efferent output is directed toward maintaining homeostasis. The efferent output to skeletal muscles is also directed toward voluntarily controlled nonhomeostatic activities, such as riding a bicycle. (It is important to realize that many effector organs are also subject to hormonal control and/or intrinsic control mechanisms; see p. 11.)

How many different neurotransmitters would you guess are released from the various efferent neuronal terminals to elicit essentially all the neurally controlled effector organ responses? Only two—acetylcholine and norepinephrine! Acting independently, these neurotransmitters bring about such diverse effects as salivary secretion, bladder contraction, and voluntary motor movements. These effects are a prime example of how the same chemical messenger may elicit a multiplicity of responses from various tissues, depending on specialization of the effector organs.

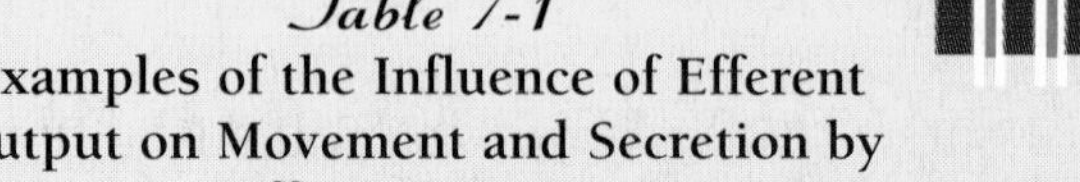

Table 7-1
Examples of the Influence of Efferent Output on Movement and Secretion by Effector Organs

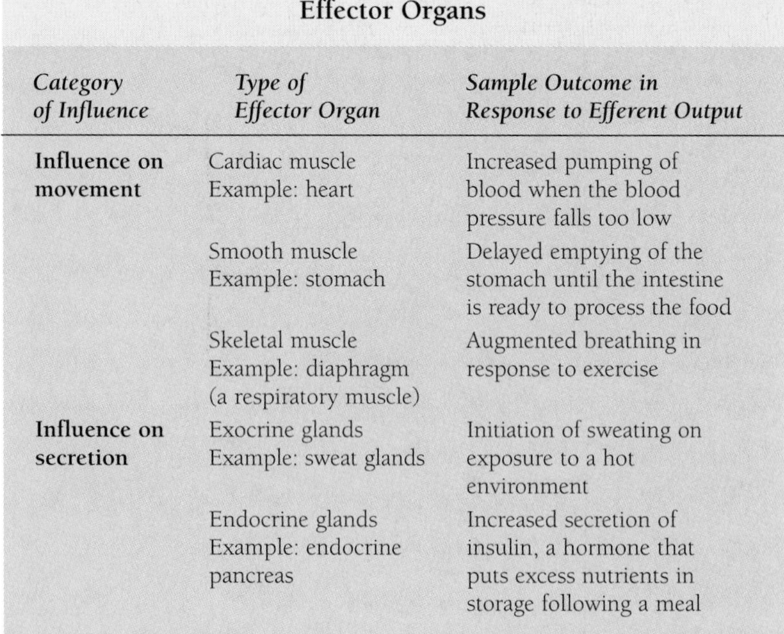

Category of Influence	*Type of Effector Organ*	*Sample Outcome in Response to Efferent Output*
Influence on movement	Cardiac muscle Example: heart	Increased pumping of blood when the blood pressure falls too low
	Smooth muscle Example: stomach	Delayed emptying of the stomach until the intestine is ready to process the food
	Skeletal muscle Example: diaphragm (a respiratory muscle)	Augmented breathing in response to exercise
Influence on secretion	Exocrine glands Example: sweat glands	Initiation of sweating on exposure to a hot environment
	Endocrine glands Example: endocrine pancreas	Increased secretion of insulin, a hormone that puts excess nutrients in storage following a meal

Autonomic Nervous System

An autonomic nerve pathway consists of a two-neuron chain, with the terminal neurotransmitter differing between sympathetic and parasympathetic nerves.

Each autonomic nerve pathway extending from the CNS to an innervated organ consists of a two-neuron chain. The cell body of the first neuron in the series is located in the CNS. Its axon, the **preganglionic fiber,** synapses with the cell body of the second neuron, which lies within a ganglion. (Recall that a ganglion is a cluster of neuronal cell bodies located outside the CNS.) The axon of the second neuron, the **postganglionic fiber,** innervates the effector organ.

The autonomic nervous system consists of two subdivisions—the **sympathetic** and the **parasympathetic nervous systems** (Fig. 7-1). Sympathetic nerve fibers originate in the thoracic and lumbar regions of the spinal cord. Most sympathetic preganglionic fibers are very short, synapsing with cell bodies of postganglionic neurons within ganglia that lie in a **sympathetic ganglion chain** (the **sympathetic trunk**) located along either side of the spinal cord. Long postganglionic fibers originating in the ganglion chain terminate on the effector organs. Some preganglionic fibers pass through

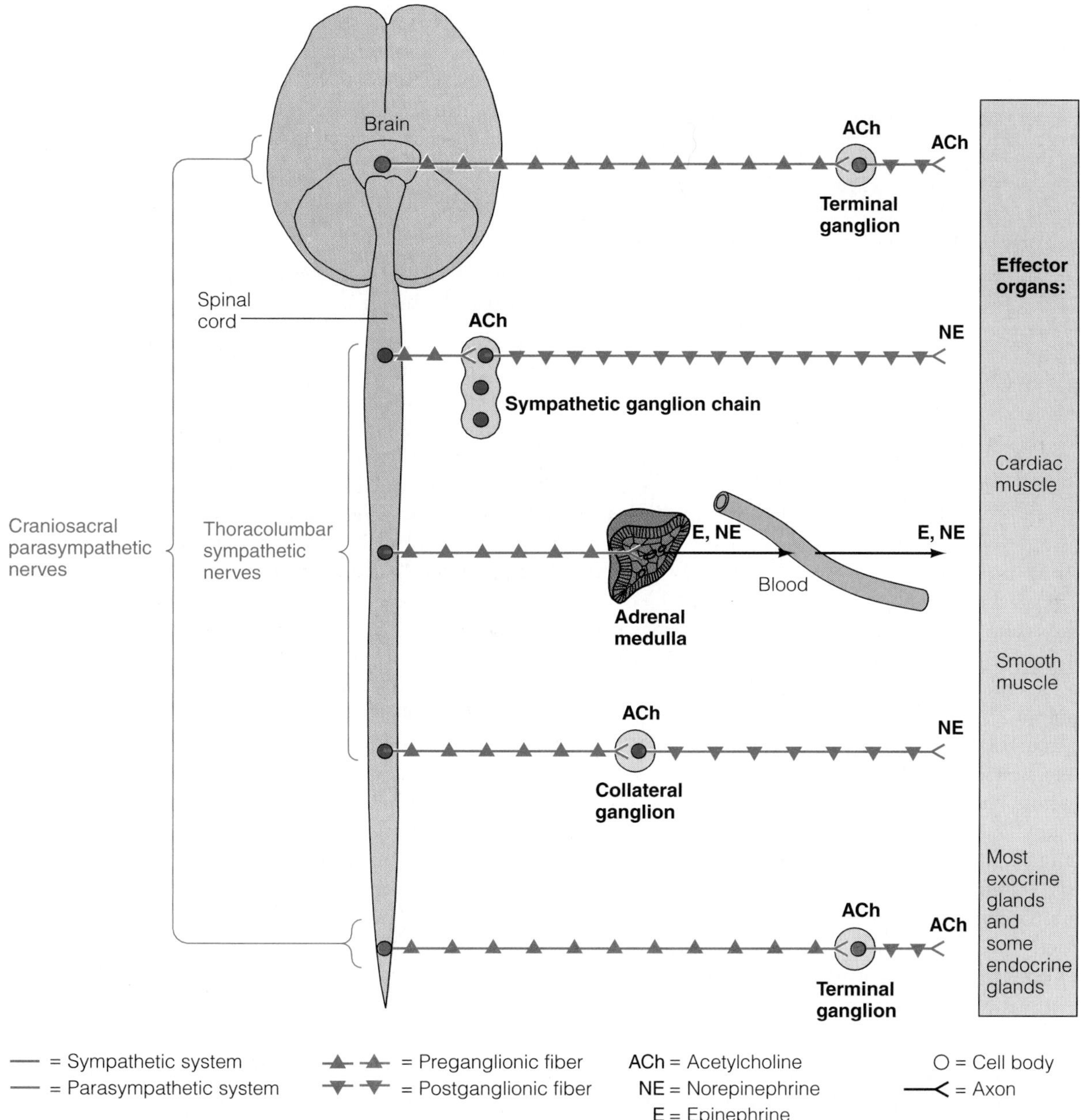

Figure 7-1 **Autonomic Nervous System** The sympathetic nervous system, which originates in the thoracolumbar regions of the spinal cord, has short cholinergic (acetylcholine-releasing) preganglionic fibers and long adrenergic (norepinephrine-releasing) postganglionic fibers. The parasympathetic nervous system, which originates in the brain and sacral region of the spinal cord, has long cholinergic preganglionic fibers and short cholinergic postganglionic fibers. In most instances, sympathetic and parasympathetic postganglionic fibers both innervate the same effector organs. The adrenal medulla is a modified sympathetic ganglion, which releases epinephrine and norepinephrine into the blood.

the ganglion chain without synapsing and terminate later in sympathetic **collateral ganglia** located about halfway between the CNS and the innervated organs, with postganglionic fibers traveling the remainder of the distance.

Parasympathetic preganglionic fibers arise from the cranial and sacral areas of the CNS. (Some cranial nerves contain parasympathetic fibers.) These fibers are long in comparison to sympathetic preganglionic fibers because they do not end until they reach **terminal ganglia** that lie in or near the effector organs. Very short postganglionic fibers terminate on the cells of an organ itself.

Sympathetic and parasympathetic preganglionic fibers release the same neurotransmitter, **acetylcholine (ACh),** but the postganglionic endings of these two systems release different neurotransmitters (the neurotransmitters that influence the effector organs). Parasympathetic postganglionic fibers release acetylcholine. Accordingly, they, along with all autonomic preganglionic fibers, are called **cholinergic fibers.**

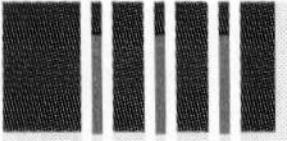

Table 7-2

Sites of Release for Acetylcholine and Norepinephrine

Acetylcholine	*Norepinephrine*
All preganglionic terminals of the autonomic nervous system	Most sympathetic postganglionic terminals
All parasympathetic postganglionic terminals	Adrenal medulla
Sympathetic postganglionic terminals at sweat glands and some blood vessels in skeletal muscle	Central nervous system
Terminals of efferent neurons supplying skeletal muscle (motor neurons)	
Central nervous system	

Most sympathetic postganglionic fibers, in contrast, are called **adrenergic fibers** because they release **noradrenaline**, commonly known as **norepinephrine.**[1] Both acetylcholine and norepinephrine also serve as chemical messengers elsewhere in the body (Table 7-2).

Postganglionic autonomic fibers do not end in a single terminal swelling like a synaptic knob. Instead, the terminal branches of autonomic fibers contain numerous swellings, or **varicosities,** that simultaneously release neurotransmitter over a large area of the innervated organ rather than on single cells. This diffuse release of neurotransmitter, coupled with the fact that any resulting change in electrical activity is spread throughout a smooth or cardiac muscle mass via gap junctions (see p. 52), means that whole organs instead of discrete cells are typically influenced by autonomic activity.

The autonomic nervous system controls involuntary visceral organ activities.

The autonomic nervous system regulates visceral activities normally outside the realm of consciousness and voluntary control, such as circulation, digestion, sweating, and pupillary size. It is not entirely true, however, that an individual has no control over activities governed by the autonomic system. Visceral afferent information usually does not reach the conscious level, so individuals have no way of consciously controlling the resultant efferent output. With the technique of **biofeedback,** however, people are provided with a conscious signal regarding visceral afferent information. This signal, which may be a sound, a light, or a graphic display on a computer screen, enables them to exert some voluntary control over events that are normally considered subsconscious activities. For example, individuals have learned to consciously lower their blood pressure when they "hear" that it is elevated via special devices that convert blood pressure levels into sound signals. Such biofeedback techniques are gaining wider acceptance and usage.

[1]*Noradrenaline (norepinephrine)* is chemically very similar to *adrenaline (epinephrine),* the primary hormone product secreted by the adrenal medulla gland. Because a U.S. pharmaceutical company marketed this product for use as a drug under the trade name "adrenaline," the scientific community in this country prefers the alternate name "epinephrine" as a generic term for this chemical messenger, and accordingly, "noradrenaline" is known as "norepinephrine." In most other English-speaking countries, however, "adrenaline" and "noradrenaline" are the terms of choice.

The sympathetic and parasympathetic nervous systems dually innervate most visceral organs.

Most visceral organs are innervated by both sympathetic and parasympathetic nerve fibers (Fig. 7-2). Table 7-3 summarizes the major effects of these autonomic branches. Although the details of this wide array of autonomic responses are described more fully in later chapters that discuss the individual organs involved, several general concepts can be derived now. As can be seen from the table, the sympathetic and parasympathetic nervous systems generally exert opposite effects in a particular organ. Sympathetic stimulation increases the heart rate, whereas parasympathetic stimulation decreases it; sympathetic stimulation slows down movement within the digestive tract, whereas parasympathetic stimulation enhances digestive motility. Note that one system is not always excitatory and the other always inhibitory. Both systems increase the activity of some organs and reduce the activity of others.

Rather than memorizing a list such as that presented in the table, it is better to logically deduce the actions of the two systems based on an understanding of the circumstances under which each system dominates. Usually, both systems are partially active; that is, normally some level of action potential activity exists in both the sympathetic and the parasympathetic fibers supplying a particular organ. This ongoing activity is called **sympathetic** or **parasympathetic tone** or **tonic activity.** Under given circumstances, activity of one division can dominate the other. *Sympathetic dominance* to a particular organ exists when the sympathetic fibers' rate of firing to that organ increases above tonic level, coupled with a simultaneous decrease below tonic level in the parasympathetic fibers' frequency of action potentials to the same organ. The reverse situation is true for *parasympathetic dominance.* Shifts in balance between sympathetic and parasympathetic activity can be accomplished discretely for individual organs to meet specific demands (for example, sympathetically induced dilation of the pupil in dim light; see p. 167), or a more generalized, widespread discharge of one autonomic system in favor of the other can be elicited to control bodywide functions. Massive widespread discharges take place more frequently in the sympathetic system. The value of this potential for massive sympathetic discharge is evident considering the circumstances during which this system usually dominates.

The sympathetic system promotes responses that prepare the body for strenuous physical activity in the face of emergency or stressful situations, such as a physical threat from the outside environment. This response is typically referred to as

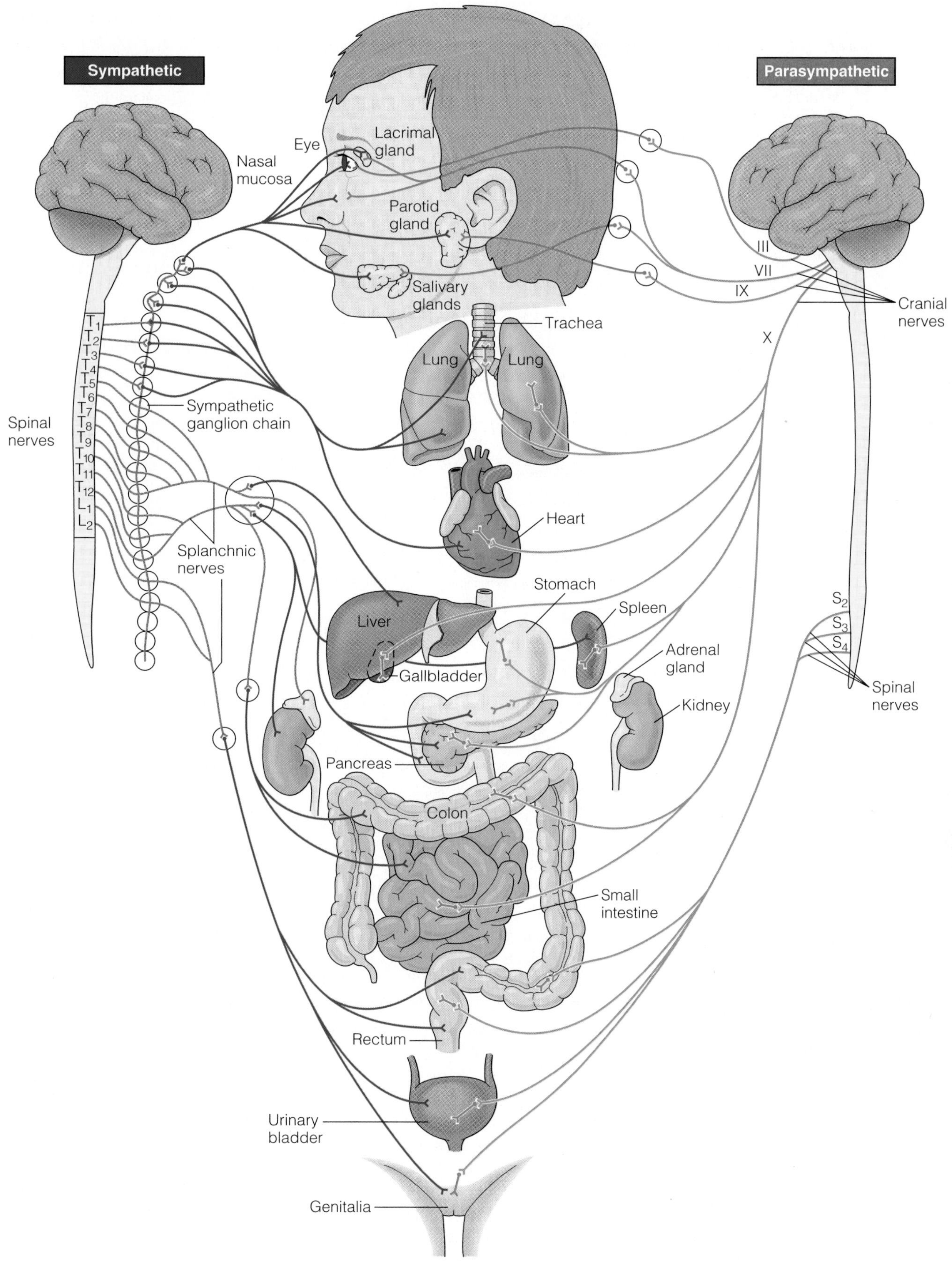

Figure 7-2 **Schematic Representation of the Structures Innervated by the Sympathetic and Parasympathetic Nervous Systems**

Table 7-3 Effects of the Autonomic Nervous System on Various Organs

Organ	*Predominant Type of Sympathetic Receptor*	*Effect of Sympathetic Stimulation*	*Effect of Parasympathetic Stimulation*
Heart	β_1	Increased rate, increased force of contraction (of whole heart)	Decreased rate, decreased force of contraction (of atria only)
Blood vessels	α (most organs)	Constriction	Dilation of vessels supplying the penis and clitoris only
	β_2 (heart and skeletal muscle vessels)	Dilation	
	*Cholinergic (skeletal muscle vessels)	Dilation	
Lungs	β_2 (airways)	Dilation of bronchioles (airways)	Constriction of bronchioles
	? (gland cells)	Inhibition (?) of mucus secretion	Stimulation of mucus secretion
Digestive tract	α, β_2 (organs)	Decreased motility (movement)	Increased motility
	α (sphincters)	Contraction of sphincters (to prevent forward movement of contents)	Relaxation of sphincters (to permit forward movement of contents)
	? (gland cells)	Inhibition (?) of digestive secretions	Stimulation of digestive secretions
Gallbladder	?	Relaxation	Contraction (emptying)
Urinary bladder	β_2	Relaxation	Contraction (emptying)
Eye	α (iris)	Dilation of pupil	Constriction of pupil
	β_2 (ciliary muscle)	Adjustment of eye for far vision	Adjustment of eye for near vision
Liver (glycogen stores)	β_2	Glycogenolysis (glucose released)	None
Adipose cells (fat stores)	β_2	Lipolysis (fatty acids released)	None
Exocrine glands			
Exocrine pancreas	α	Inhibition of pancreatic exocrine secretion	Stimulation of pancreatic exocrine secretion (important for digestion)
Sweat glands	α, cholinergic	Stimulation of secretion by sweat glands	None
Salivary glands	α	Stimulation of small volume of thick saliva rich in mucus	Stimulation of large volume of watery saliva rich in enzymes
Endocrine glands			
Adrenal medulla	Cholinergic	Stimulation of epinephrine and norepinephrine secretion	None
Endocrine pancreas	α	Inhibition of insulin secretion; stimulation of glucagon secretion	Stimulation of insulin and glucagon secretion
Genitals	α	Ejaculation and orgasmic contractions (males); orgasmic contractions (females)	Erection (caused by dilation of blood vessels in penis [male] and clitoris [female])
Brain activity	?	Increased alertness	None

*Uncertain as to whether cholinergic sympathetic receptors are present in skeletal muscle vessels in humans, although they have been demonstrated in other species.
?Receptor type unknown.

a **fight-or-flight response** because the sympathetic system readies the body to fight against or flee from the threat. Think about the body resources needed in such circumstances. The heart beats more rapidly and more forcefully; blood pressure is elevated because of generalized constriction of the blood vessels; the respiratory airways open wide to permit maximal air flow; glycogen (stored sugar) and fat stores are broken down to release extra fuel into the blood; and blood vessels supplying skeletal muscles dilate (open more widely). All of these responses are aimed at providing increased flow of oxygenated, nutrient-rich blood to the skeletal muscles in anticipation of strenuous physical activity. Furthermore, the pupils dilate and the eyes adjust for far vision, enabling the person to make a quick visual assessment of the entire threatening scene. Sweating is promoted in anticipation of excess heat production by the physical exertion. Because digestive and urinary activities are unessential in meeting the threat, the sympathetic system inhibits these activities.

The parasympathetic system, on the other hand, dominates in quiet, relaxed situations. Under such nonthreatening circumstances, the body can be concerned with its own "general housekeeping" activities, such as digestion and emptying of the urinary bladder. The parasympathetic system promotes these types of bodily functions while slowing down those activities that are enhanced by the sympathetic system. There is no need, for example, to have the heart beating rapidly and forcefully when the person is in a tranquil setting.

What is the advantage of dual innervation of organs with nerve fibers whose actions oppose each other? It enables precise control over an organ's activity, similar to having both an accelerator and a brake to control the speed of a car. If an animal suddenly darts across the road as you are driving, you could eventually stop if you simply took your foot off the accelerator, but you might stop too slowly to avoid hitting the animal. If you simultaneously apply the brake as you lift up on the accelerator, however, you can come to a more rapid, controlled stop. In a similar manner, a sympathetically accelerated heart rate could gradually be reduced to normal following a stressful situation by decreasing the rate of firing in the cardiac sympathetic nerve (letting up on the accelerator), but the heart rate can be reduced more rapidly by simultaneously increasing activity in the parasympathetic supply to the heart (applying the brake). Indeed, the two divisions of the autonomic nervous system are usually reciprocally controlled; increased activity in one division is accompanied by a corresponding decrease in the other. Thus, dual innervation of an organ by the two branches of the autonomic system permits more precise control over the organ.

Inhibition of the parasympathetic nervous system by *cocaine,* an illegal addictive drug, may be a major contributing factor to sudden death in cocaine overdose. If cocaine blocks the protective parasympathetic brakes, as it appears to do, the sympathetic nervous system could proceed unchecked in accelerating the heartbeat. Sudden death results if the heartbeat becomes too rapid and irregular to adequately pump blood.

There are several exceptions to the general rule of dual reciprocal innervation by the two branches of the autonomic nervous system (Table 7-3); the most notable are the following:

- *Innervated blood vessels* (most arterioles and veins are innervated, arteries and capillaries are not) receive only sympathetic nerve fibers. Regulation is accomplished by increasing or decreasing the firing rate above or below the tonic level in these sympathetic fibers. The only blood vessels to receive both sympathetic and parasympathetic fibers are those supplying the penis and clitoris. The precise vascular control this dual innervation affords these organs is important in accomplishing erection.
- *Sweat glands* are innervated only by sympathetic nerves. The postganglionic fibers of these nerves are unusual because they secrete acetylcholine rather than norepinephrine.
- *Salivary glands* are innervated by both autonomic divisions, but unlike elsewhere, sympathetic and parasympathetic activity is not antagonistic. Both stimulate salivary secretion, but the saliva's volume and composition differ, depending on which autonomic branch is dominant.

A wide variety of autonomic malfunctions accompany aging. Preliminary studies are providing clues to the age-related decline in autonomic control. (See the boxed feature on p. 210, ♦ Concepts, Challenges, and Controversies.)

The adrenal medulla, an endocrine gland, is a modified part of the sympathetic nervous system.

The *adrenal gland,* which lies above the kidney on each side (*ad* means "next to"; *renal* means "kidney"), is an endocrine gland consisting of an outer portion, the *adrenal cortex,* and an inner portion, the *adrenal medulla.* The **adrenal medulla** is considered a modified sympathetic ganglion that does not give rise to postganglionic fibers. Instead, it secretes hormones into the blood upon stimulation by the preganglionic fiber that originates in the CNS (Figs. 7-1 and 7-3). Not surprisingly, the

Figure 7-3 **Comparison of the Release and Binding to Receptors of Epinephrine and Norepinephrine** Norepinephrine is released both as a neurotransmitter from sympathetic postganglionic fibers and as a hormone from the adrenal medulla. Alpha (α) and beta$_1$ (β_1) receptors bind with both norepinephrine and epinephrine, whereas beta$_2$ (β_2) receptors bind primarily with epinephrine.

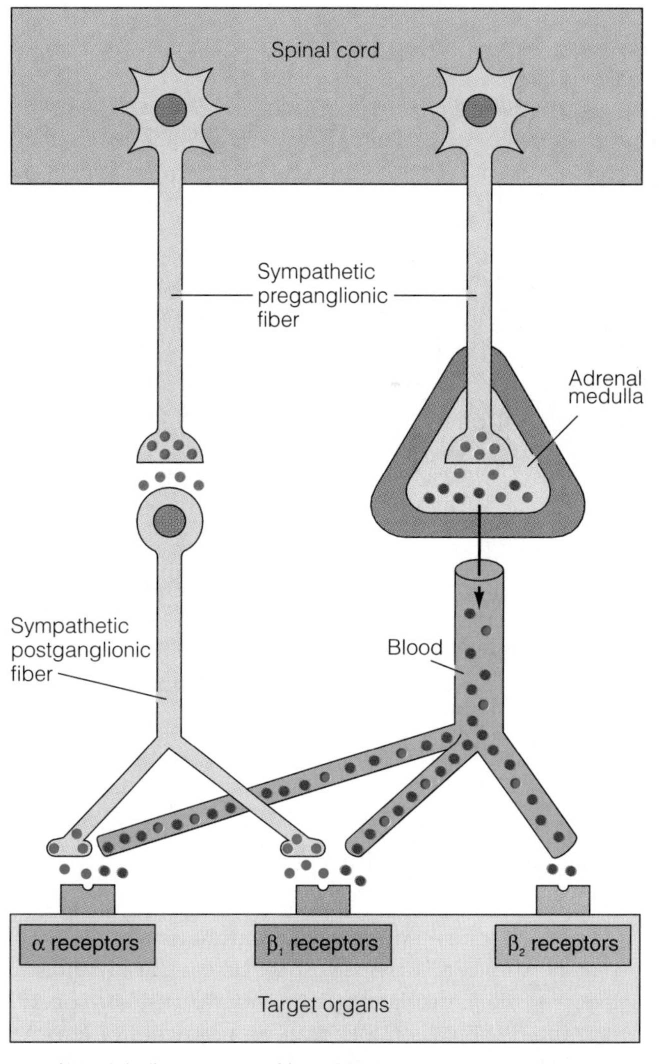

CONCEPTS, CHALLENGES, AND CONTROVERSIES

The Autonomic Nervous System and Aging: A Fortuitous Find

Often in scientific research, an investigator unexpectedly comes across an important finding while studying a totally different phenomenon. This happened with Robert Schmidt, a neuropathologist, who accidentally stumbled across the first evidence of the effects of aging on the autonomic nervous system while investigating a different problem—what might be responsible for the intestinal disorders that often accompany diabetes mellitus. Since nerve damage is known to be one of the long-term complications of diabetes, Schmidt suspected that damage to the autonomic nerves that help regulate the intestine might be the cause of the intestinal symptoms. Upon microscopic examination of the sympathetic ganglia supplying the intestine in diabetic rats, Schmidt was astonished to see that some of the preganglionic fibers (axons) near their terminal endings were swollen to more than 30 times their normal size. Thinking this dramatic pathology must somehow be linked to diabetes, he examined the same ganglia in nondiabetic control rats for comparison, fully expecting to find normal preterminal axons. To his surprise, some of the axon endings in these normal rats were as swollen as those in the diabetic rats.

Puzzling over his results, Schmidt finally thought of one thing that the two groups had in common. They were both old. Most studies on rats involve young specimens, but Schmidt's animals were old because he was investigating a long-term effect of diabetes.

In 1990, Schmidt published results in the *American Journal of Pathology* demonstrating comparable evidence of this autonomic aging in humans. In autopsy examinations of sympathetic ganglia in 56 people who had died between the ages of 15 and 93 from various causes, Schmidt found very few enlarged preterminal axons in persons younger than 60, whereas the number of swollen axons increased dramatically in those over 60. In fact, abnormal preterminal axons were significantly more frequent in older humans than in aging rats.

Further study revealed that the abnormal sympathetic axons contain an overabundance of disoriented neurofilaments, cytoskeletal components that normally serve as internal scaffolding to help maintain a nerve cell's shape (see p. 41). Neurofilaments are normally degraded in the axon terminals for recycling, but this process is apparently impaired in connection with the age-related autonomic damage.

Not all of the preganglionic fibers in a sympathetic ganglion are swollen and packed with neurofilaments. Only terminals that contain a recently discovered neuromodulator (see p. 101), **neuropeptide Y (NPY)**, are affected. The role that NPY plays in sympathetic ganglia is unknown, but other investigators have found evidence that NPY is a cotransmitter in sympathetic control of vascular (blood vessel) smooth muscle.

The autonomic nervous system helps control many body functions such as regulation of blood pressure and movement of food through the digestive tract via its influence on smooth muscle. Aging is often accompanied by autonomic malfunctions, such as becoming lightheaded when standing up too quickly because of inadequate vascular control or constipation due to sluggish bowel motility. These age-related autonomic dysfunctions are assuming greater significance as life expectancy and consequently the number of elderly are increasing.

Normal function of the smooth muscle of the intestine and blood vessels depends on complex, incompletely understood control mechanisms involving the interaction of multiple neural elements and hormones. One can speculate that an abnormality in NPY neuromodulation of these activities might contribute to age-related autonomic malfunctions, although this remains to be confirmed. According to Schmidt, "We predict that when the function of NPY-containing processes in human prevertebral (sympathetic) ganglia is understood, that function will be compromised in the elderly."

hormones are identical or similar to postganglionic sympathetic neurotransmitters. About 20% of the adrenal medullary hormone output is norepinephrine, and the remaining 80% is the closely related substance **epinephrine (adrenaline)** (see footnote 1, p. 206). These hormones, in general, reinforce activity of the sympathetic nervous system.

There are several different types of membrane receptor proteins for each autonomic neurotransmitter.

Because each autonomic neurotransmitter and medullary hormone stimulate activity in some tissues but inhibit activity in others, the particular responses must depend on specialization of the tissue cells rather than on properties of the chemicals themselves. Responsive tissue cells possess one or more of several different types of plasma membrane receptor proteins for these chemical messengers. Binding of a neurotransmitter to a receptor induces the tissue-specific response by means of a second messenger system within the cell (see p. 68).

Two types of acetylcholine (cholinergic) receptors—*nicotinic* and *muscarinic* receptors—have been identified on the basis of their response to particular drugs (▮Table 7-4). **Nicotinic receptors** (activated by the tobacco plant derivative nicotine) are found on the postganglionic cell bodies in all autonomic ganglia. They respond to acetylcholine released from both sympathetic and parasympathetic preganglionic fibers. **Muscarinic receptors** (activated by the mushroom poison muscarine) are found on effector cell membranes (smooth muscle, cardiac muscle, and glands). They bind with acetylcholine released from parasympathetic postganglionic fibers.

There are two major classes of adrenergic receptors for norepinephrine and epinephrine based on the ability of various drugs to either initiate or prevent responses in the effector organ. These receptors are designated as **alpha** (α) and

beta (β) receptors, with the latter further subclassified into β_1 and β_2 **receptors.**[2] These various receptor types are distinctly distributed among the effector organs (Table 7-3). Receptors of the β_2 type bind primarily with epinephrine, whereas β_1 and α receptors have about equal affinities for norepinephrine and epinephrine (Fig. 7-3). Activation of α receptors usually brings about an excitatory response in the effector organ—for example, arteriolar constriction caused by increased contraction of the smooth muscle in the walls of these blood vessels. Stimulation of β_1 receptors, which are found primarily in the heart, also causes an excitatory response, namely, increased rate and force of cardiac contraction. The response to β_2 receptor activation is generally inhibitory, such as arteriolar or bronchiolar (respiratory airway) dilation caused by relaxation of the smooth muscle in the walls of these tubular structures.

Because activation of various receptor types brings about different responses to the same autonomic messenger, these receptors can be manipulated fairly selectively by drugs. Drugs are available that selectively enhance or mimic **(agonists)** or block **(antagonists)** autonomic responses at each of the receptor types. Some are only of experimental interest, but others are very important therapeutically. For example, **atropine** blocks the effect of acetylcholine at muscarinic receptors but does not affect nicotinic receptors. Since the acetylcholine released at both parasympathetic and sympathetic preganglionic fibers combines with nicotinic receptors, blockage at nicotinic synapses would knock out both of these autonomic branches. By acting selectively to interfere with acetylcholine action only at muscarinic junctions, which are the sites of parasympathetic postganglionic action, atropine effectively blocks parasympathetic effects but does not influence sympathetic activity at all. This principle is used to suppress salivary and bronchial secretions before surgery to reduce the risk of a patient inhaling these secretions into the lungs.

Likewise, drugs that act selectively at α and β adrenergic receptor sites to either activate or block specific sympathetic effects are widely used. **Salbutamol** is an excellent example. It selectively activates β_2 adrenergic receptors at low doses, making it possible to dilate the bronchioles in the treatment of asthma without undesirably stimulating the heart (the heart has mostly β_1 receptors). Other drugs that act selectively at α and β receptors are beneficial in manipulating blood pressure and heart rate in the treatment of hypertension and cardiac arrhythmias.

Many regions of the central nervous system are involved in the control of autonomic activities.

Messages from the CNS are delivered to cardiac muscle, smooth muscle, and glands via the autonomic nerves, but what regions of the CNS regulate autonomic output?

[2]Alpha receptors are also subclassified into α_1 and α_2 receptors, but these categories are less distinct than the β receptor subclassifications. Many of the α receptors have not been designated as α_1 or α_2, so we are grouping them together as a single category.

Table 7-4
Location of Nicotinic and Muscarinic Cholinergic Receptors

Type of Receptor	*Site of Receptor*	*Respond to Acetylcholine Released From:*
Nicotinic receptors	All autonomic ganglia	Sympathetic and parasympathetic preganglionic fibers
	Motor end plates of skeletal muscle fibers	Motor neurons
	Some CNS cell bodies and dendrites	Some CNS presynaptic terminals
Muscarinic receptors	Effector cells (cardiac muscle, smooth muscle, glands)	Parasympathetic postganglionic fibers
	Some CNS cell bodies and dendrites	Some CNS presynaptic terminals

- Some autonomic reflexes, such as urination, defecation, and erection, are integrated at the spinal cord level, but all of these spinal reflexes are subject to control by higher levels of consciousness.
- The medulla within the brain stem is the region most directly responsible for autonomic output. Centers for controlling cardiovascular, respiratory, and digestive activity via the autonomic system are located there.
- The hypothalamus plays an important role in integrating the autonomic, somatic, and endocrine responses that automatically accompany various emotional and behavioral states. For example, the increased heart rate, blood pressure, and respiratory activity associated with anger or fear are brought about by the hypothalamus acting through the medulla.
- Autonomic activity can also be influenced by the prefrontal association cortex through its involvement with emotional expression characteristic of the individual's personality. An example is blushing when embarrassed, which is caused by dilation of blood vessels supplying the skin of the cheeks. Such responses are mediated through hypothalamic-medullary pathways.

Table 7-5 summarizes the main distinguishing features of the sympathetic and parasympathetic nervous systems.

Somatic Nervous System

Motor neurons supply skeletal muscle.

Skeletal muscle is innervated by **motor neurons,** the axons of which constitute the **somatic nervous system.** The cell bodies of these motor neurons are located within the ventral horn of the spinal cord. Unlike the two-neuron chain of autonomic nerve fibers, the axon of a motor neuron is continuous from

Table 7-5 Distinguishing Features of the Sympathetic and Parasympathetic Nervous System

Feature	*Sympathetic System*	*Parasympathetic System*
Origin of preganglionic fiber	Thoracic and lumbar regions of spinal cord	Brain and sacral region of spinal cord
Origin of postganglionic fiber (location of ganglion)	Sympathetic ganglion chain (near spinal cord) or collateral ganglia (about halfway between spinal cord and effector organs)	Terminal ganglia (in or near effector organs)
Length and type of fiber	Short cholinergic preganglionic fibers Long adrenergic postganglionic fibers (most) Long cholinergic postganglionic fibers (few)	Long cholinergic preganglionic fibers Short cholinergic postganglionic fibers
Effector organs innervated	Cardiac muscle, almost all smooth muscle, most exocrine glands, and some endocrine glands	Cardiac muscle, most smooth muscle, most exocrine glands, and some endocrine glands
Types of receptors for neurotransmitters	α, β_1, β_2	Nicotinic, muscarinic
Dominance	Dominates in emergency "fight-or-flight" situations; prepares body for strenuous physical activity	Dominates in quiet, relaxed situations; promotes "general housekeeping" activities such as digestion
Types of discharge	Frequently mass discharge of whole system; may involve only discrete organs	Normally involves discrete organs rather than mass discharge

its origin in the spinal cord to its termination on skeletal muscle. Motor neuron axon terminals release acetylcholine, which brings about excitation and contraction of the innervated muscle fibers. Motor neurons can only stimulate skeletal muscles, in contrast to autonomic fibers, which can either stimulate or inhibit their effector organs. Inhibition of skeletal muscle activity can be accomplished only within the CNS through activation of inhibitory synaptic input to the cell bodies and dendrites of the motor neurons supplying that particular muscle.

Motor neurons are the final common pathway.

Motor neurons are influenced by many converging presynaptic inputs, both excitatory and inhibitory. Some of these inputs are part of spinal-reflex pathways originating with peripheral sensory receptors. Others are part of descending pathways originating within the brain. Areas of the brain that exert control over skeletal muscle movements include the motor regions of the cortex, the basal nuclei, the cerebellum, and the brain stem (see p. 247).

Motor neurons are considered the **final common pathway** since the only way any other parts of the nervous system can influence skeletal muscle activity is by acting on these motor neurons. The level of activity in a motor neuron and its subsequent output to the skeletal muscle fibers it innervates depend on the relative balance of EPSPs and IPSPs (see p. 98) brought about by its presynaptic inputs originating from these diverse sites in the brain.

The somatic system is considered to be under voluntary control, but much of the skeletal muscle activity involving posture, balance, and stereotypical movements is subconsciously controlled. You may decide you want to start walking, but you do not have to consciously bring about the alternate contraction and relaxation of the involved muscles because these movements are involuntarily coordinated by lower brain centers.

The cell bodies of the crucial motor neurons may be selectively destroyed by **poliovirus.** The result is paralysis of the muscles innervated by the affected neurons. **Amyotropic lateral sclerosis (ALS),** more familiarly known as **Lou Gehrig's disease,** is the most common motor neuron disease. ALS is characterized by progressive degeneration and death of motor neurons. This adult-onset condition leads to gradual loss of motor control and ultimately death, as it did in baseball legend Lou Gehrig. During the last few years, several different mechanisms have been implicated as possible causes of motor neuron loss, such as abnormalities in glutamate receptors on the motor neuron cell bodies, leading to excitotoxicity similar to much of the neuronal death in the brain during a stroke (see p. 120), and pathological changes in neurofilaments, which block axonal transport of crucial materials (see p. 37).

Table 7-6 summarizes the features of the two divisions of the efferent nervous system discussed in this chapter. Table 7-7 compares the three types of neurons that have been examined in the last three chapters.

Neuromuscular Junction

Acetylcholine chemically links electrical activity in motor neurons with electrical activity in skeletal muscle cells.

An action potential in a motor neuron is rapidly propagated from the CNS to the skeletal muscle along the large myelinated fiber (axon) of the neuron. As the axon approaches a muscle, it divides into many terminal branches and loses its myelin sheath. Each of these axon terminals forms a special junction, a **neuromuscular junction,** with one of the many muscle cells that compose the whole muscle (Fig. 7-4). A single muscle cell, referred to as a **muscle fiber,** is long and cylindrical in shape. The axon terminal is enlarged into a knoblike structure, the **terminal button,** which fits into a shallow depression, or groove, in the underlying muscle fiber (Fig. 7-5). Some scientists alternatively call the neuromuscular junction a motor end plate. However, we will reserve the term **motor end plate** for the specialized portion of the muscle cell membrane immediately under the terminal button.

Nerve and muscle cells do not actually come into direct contact at a neuromuscular junction. The space, or cleft, between these two structures is too large to permit electrical transmission of an impulse between them (that is, the action potential cannot "jump" that far). Therefore, just as at a neuronal synapse (see p. 95), a chemical messenger is used to carry the signal between the neuron terminal and the muscle fiber. Each terminal button contains thousands of vesicles that store the chemical transmitter acetylcholine (ACh). Propagation of an action potential to the axon terminal (step 1, Fig. 7-5) triggers the opening of voltage-gated Ca^{2+} channels (see p. 84) in the terminal button. Opening of Ca^{2+} channels permits Ca^{2+} to diffuse into the terminal button from its higher extracellular concentration (step 2), which in turn causes the release of ACh by exocytosis from several hundred of the vesicles into the cleft (step 3).

The released ACh diffuses across the cleft and binds with specific receptor sites, which are specialized membrane proteins unique to the motor end-plate portion of the muscle fiber membrane (step 4). (These cholinergic receptors are of the nicotinic type.) Binding of ACh with these receptor sites induces the opening of chemical messenger–gated channels in the motor end plate. These channels permit a small amount of cation traffic through them (both Na^+ and K^+) but no

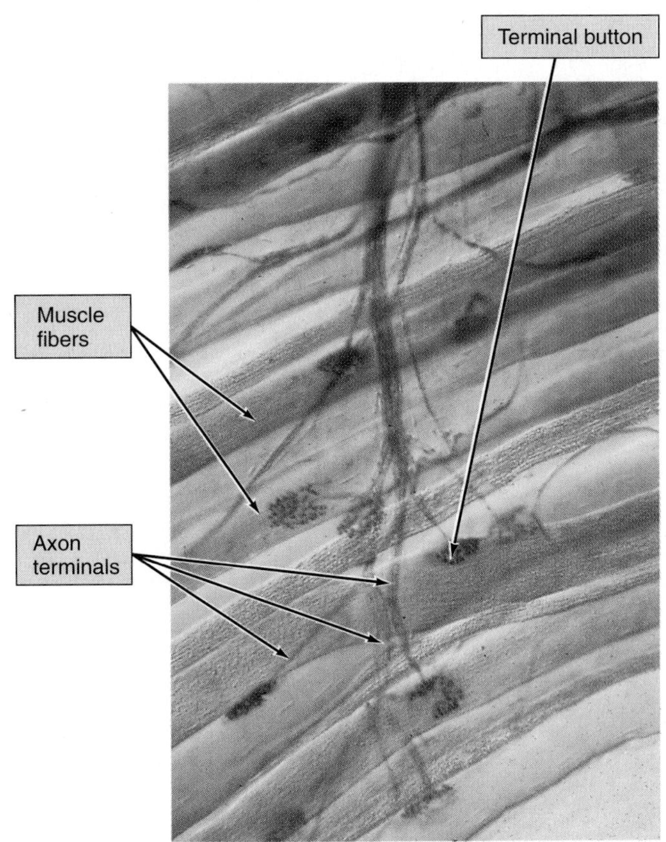

Figure 7-4 **Motor Neuron Innervating Skeletal Muscle Cells** When a motor neuron reaches a skeletal muscle, it divides into many terminal branches, each of which forms a neuromuscular junction with a single muscle cell (muscle fiber).

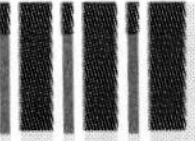

Table 7-6 **Comparison of the Autonomic Nervous System and the Somatic Nervous System**

Feature	*Autonomic Nervous System*	*Somatic Nervous System*
Site of origin	Brain or lateral horn of spinal cord	Ventral horn of spinal cord
Number of neurons from origin in CNS to effector organ	Two-neuron chain (preganglionic and postganglionic)	Single neuron (motor neuron)
Organs innervated	Cardiac muscle, smooth muscle, exocrine and some endocrine glands	Skeletal muscle
Type of innervation	Most effector organs dually innervated by the two antagonistic branches of this system (sympathetic and parasympathetic)	Effector organs innervated only by motor neurons
Neurotransmitter at effector organs	May be acetylcholine (parasympathetic terminals) or norepinephrine (sympathetic terminals)	Only acetylcholine
Effects on effector organs	Either stimulation or inhibition (antagonistic actions of two branches)	Stimulation only (inhibition possible only centrally through IPSPs on cell body of motor neuron)
Types of control	Under involuntary control; may be voluntarily controlled with biofeedback techniques and training	Subject to voluntary control; much activity subconsciously coordinated
Higher centers involved in control	Spinal cord, medulla, hypothalamus, prefrontal association cortex	Spinal cord, motor cortex, basal nuclei, cerebellum, brain stem

Table 7-7 Comparison of Types of Neurons

Feature	*Afferent Neuron*	*Efferent Neuron* — *Autonomic Nervous System*	*Efferent Neuron* — *Somatic Nervous System*	*Interneuron*
Origin, structure, location	Receptor at peripheral ending; elongated peripheral axon, which travels in peripheral nerve; cell body located in dorsal root ganglion; short central axon entering spinal cord	Two-neuron chain; first neuron (preganglionic fiber) originating in CNS and terminating on a ganglion; second neuron (postganglionic fiber) originating in the ganglion and terminating on the effector organ	Cell body of motor neuron lying in spinal cord; long axon traveling in peripheral nerve and terminating on the effector organ	Various shapes; lying entirely within CNS; some cell bodies originating in brain, with long axons traveling down the spinal cord in descending pathways; some originating in spinal cord, with long axons traveling up the cord to the brain in ascending pathways; others forming short local connections
Termination	Interneurons*	Effector organs (cardiac muscle, smooth muscle, exocrine and some endocrine glands)	Effector organs (skeletal muscle)	Other interneurons and efferent neurons
Function	Carries information about the external and internal environment to CNS	Carries instructions from CNS to effector organs	Carries instructions from CNS to effector organs	Processes and integrates afferent input; initiates and coordinates efferent output; responsible for thought and other higher mental functions
Convergence of input on cell body	No (only input is through receptor)	Yes	Yes	Yes
Effect of input to neuron	Can only be excited (through receptor potential induced by stimulus; must reach threshold for action potential)	Can be excited or inhibited (through EPSPs and IPSPs at first neuron; must reach threshold for action potential)	Can be excited or inhibited (through EPSPs and IPSPs; must reach threshold for action potential)	Can be excited or inhibited (through EPSPs and IPSPs; must reach threshold for action potential)
Site of action potential initiation	First excitable portion of membrane adjacent to receptor	Axon hillock	Axon hillock	Axon hillock
Divergence of output	Yes	Yes	Yes	Yes
Effect of output on structure on which it terminates	Only excites	Postganglionic fiber either excites or inhibits	Only excites	Either excites or inhibits

*The afferent neuron terminates directly on the alpha motor neuron in the case of the monosynaptic stretch reflex (see p. 251).

anions (step 5). Because the permeability of the end-plate membrane to Na^+ and K^+ on opening of these channels is essentially equal, the relative movement of these ions through the channels depends on their electrochemical driving forces. Recall that at resting potential, the net driving force for Na^+ is much greater than that for K^+, because the resting potential is much closer to the K^+ equilibrium potential. Both the concentration and electrical gradients for Na^+ are inward, whereas the outward concentration gradient for K^+ is almost, but not quite, balanced by the opposing inward electrical gradient. As a result, when ACh triggers the opening of these channels, considerably more Na^+ moves inward than K^+ outward, bringing about a depolarization of the motor end plate. This potential change is known as the **end-plate potential (EPP).** It is similar to an EPSP (excitatory postsynaptic potential; see p. 95), except that the magnitude of an EPP is much larger because (1) more transmitter is released from a terminal button than from a presynaptic knob in response to an action potential; (2) the motor end plate has a larger surface area and, accordingly, more transmitter receptor sites than a subsynaptic membrane; and (3) many more ion channels are opened in response to the transmitter-receptor complex in the motor end plate. This permits a greater net influx of positive ions and a larger depolarization. As with an EPSP, an EPP is a

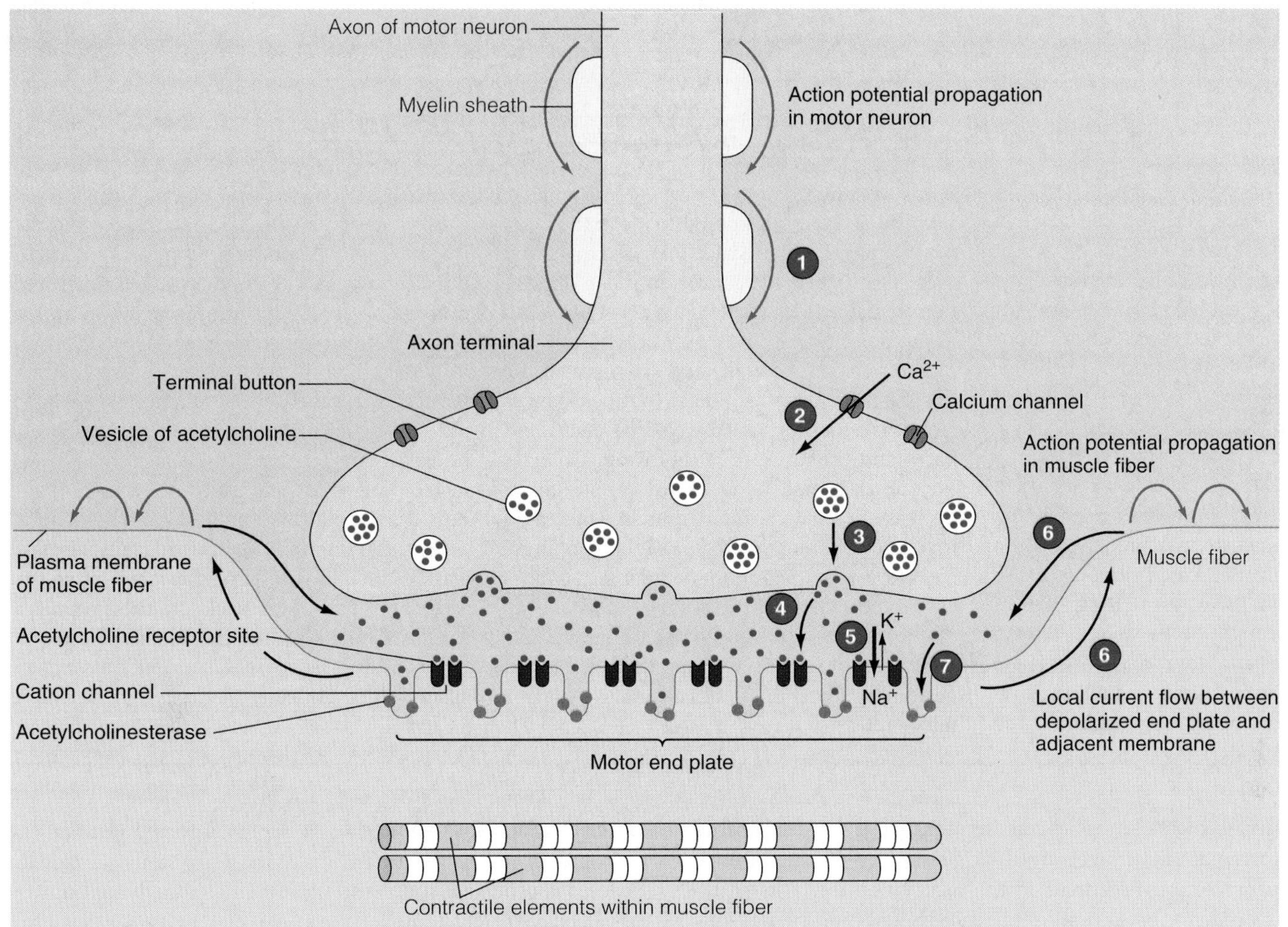

Figure 7-5 Events at a Neuromuscular Junction Propagation of an action potential to the terminal button of a motor neuron ① triggers the opening of voltage-gated Ca^{2+} channels and the subsequent entry of Ca^{2+} into the terminal button ②. Ca^{2+} triggers the release of acetylcholine by exocytosis from a portion of the vesicles ③. Acetylcholine diffuses across the space separating the nerve and muscle cells and binds with receptor sites specific for it on the motor end plate of the muscle cell membrane ④. This binding brings about the opening of cation channels, leading to a relatively large movement of Na^+ into the muscle cell compared to a smaller movement of K^+ outward ⑤. The result is an end-plate potential. Local current flow between the depolarized end plate and adjacent membrane initiates an action potential, which is propagated throughout the muscle fiber ⑥. Acetylcholine is subsequently destroyed by acetylcholinesterase, an enzyme located in the muscle cell membrane, terminating the muscle cell's response ⑦.

graded potential, whose magnitude depends on the amount and duration of ACh at the end plate.

The motor end-plate region itself does not have a threshold potential, so an action potential cannot be initiated at this site. However, an EPP brings about an action potential in the rest of the muscle fiber, as follows. The neuromuscular junction is usually located in the middle of the long cylindrical muscle fiber. When an EPP takes place at the motor end plate, local current flow occurs between the depolarized end plate and the adjacent, resting cell membrane in both directions, reducing the potential to threshold in the adjacent areas (step 6). The subsequent action potential initiated at these sites is propagated throughout the muscle fiber membrane by conduction by local current flow (see p. 89). The spread occurs in both directions, away from the motor end plate toward both ends of the fiber. This electrical activity triggers contraction of the muscle fiber. Thus, by means of ACh, an action potential in a motor neuron brings about an action potential and subsequent contraction in the muscle fiber. (See the boxed feature on p.216, A Closer Look at Exercise Physiology.)

Unlike synaptic transmission, the magnitude of an EPP is normally sufficient to cause an action potential in the muscle cell. Typically, therefore, one-to-one transmission of an action potential occurs at a neuromuscular junction; one action potential in a nerve cell triggers one action potential in a muscle cell that it innervates. Other comparisons of neuromuscular junctions with synapses can be found in Table 7-8.

Acetylcholinesterase terminates acetylcholine activity at the neuromuscular junction.

If ACh continued to remain in contact with the motor end plate, the channels would remain open and the EPP would continue to exist. Because this persistent end-plate depolarization would continue to initiate action potentials in the

A CLOSER LOOK AT
EXERCISE PHYSIOLOGY

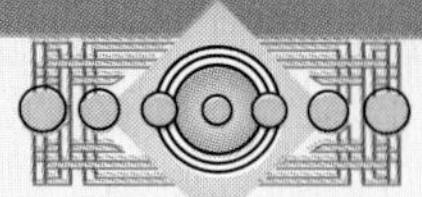

Loss of Muscle Mass: A Plight of Space Flight

Skeletal muscles are a case of "use it or lose it." Stimulation of skeletal muscles by motor neurons is essential not only to induce the muscles to contract but also to maintain their size and strength. Muscles that are not routinely stimulated gradually atrophy, or diminish in size and strength.

When humans entered the weightlessness of space, it became apparent that the muscular system required the stress of work or gravity to maintain its size and strength. In 1991, the space shuttle *Columbia* was launched for a nine-day mission dedicated among other things to comprehensive research on physiological changes brought on by weightlessness. The three female and four male astronauts aboard suffered a dramatic and significant 25% reduction of mass in their weight-bearing muscles. The effort required to move the body is remarkably less in space than on earth, and there is no need for active muscular opposition to gravity. The result is what some refer to as *functional atrophy.*

The muscles most affected are those in the lower extremities, the gluteal (buttocks) muscles, the extensor muscles of the neck and back, and the muscles of the trunk. Changes include a decrease in muscle volume and mass, decrease in strength and endurance, increased breakdown of muscle protein, and loss of muscle nitrogen (an important component of muscle protein). The exact biological mechanisms that induce muscle atrophy are unknown, but a majority of scientists believe that the lack of customary forcefulness of contraction plays a major role.

Space programs in the United States and the former Soviet Union have employed intervention techniques that emphasize both diet and exercise in an attempt to prevent muscle atrophy. Faithful performance of vigorous, carefully designed physical exercise has helped reduce the severity of functional atrophy. Studies of nitrogen and mineral balances, however, suggest that muscle atrophy continues to progress during exposure to weightlessness despite efforts to prevent it.

Furthermore, only half of the muscle mass was restored in the *Columbia* crew after the astronauts had been back on the ground a length of time equal to their flight. These and other findings suggest that extended stays in space will be difficult.

remainder of the muscle cell membrane, the muscle fiber would remain contracted until fatigued, even in the absence of further action potentials in the motor neuron. This situation would not be desirable, because there could be no controlled, purposeful alterations in movement. To control muscle contraction, electrical activity in the muscle fiber must be switched off promptly when there is no longer a signal from its motor neuron. The muscle's electrical response is turned off by an enzyme present in the motor end-plate membrane, **acetylcholinesterase** (AChE), which inactivates ACh.

Acetylcholine binds very briefly (for about 1 billionth of a second) with a receptor site, then detaches. Some of the ACh molecules quickly rebind with receptor sites, keeping the channels open, but some diffuse deeper into the folds of the motor end plate, where AChE is located (step 7). As this process is repeated, more and more ACh is inactivated until it has been virtually removed from the cleft within a few milliseconds after its release. Removal of ACh terminates the EPP so that no more action potentials are initiated.

The neuromuscular junction is vulnerable to several chemical agents and diseases.

Several chemical agents and diseases are known to affect the neuromuscular junction by acting at different sites in the transmission process (Table 7-9). Two well-known toxins—

Table 7-8 Comparison of a Synapse and a Neuromuscular Junction

Similarities	*Differences*
Both consist of two excitable cells separated by a narrow cleft that prevents direct transmission of electrical activity between them.	A synapse is a junction between two neurons. A neuromuscular junction exists between a motor neuron and a skeletal muscle fiber.
The axon terminals of both store chemical messengers (neurotransmitters) that are released by the Ca^{2+}-induced exocytosis of storage vesicles when an action potential reaches the terminal.	There is a one-to-one transmission of action potentials at a neuromuscular junction, whereas one action potential in a presynaptic neuron cannot by itself bring about an action potential in a postsynaptic neuron. An action potential in a postsynaptic neuron occurs only when the summation of EPSPs brings the membrane to threshold.
In both, binding of the neurotransmitter with receptor sites in the membrane of the cell underlying the axon terminal opens specific channels in the membrane, permitting ionic movements that alter the membrane potential of the cell.	A neuromuscular junction is always excitatory (an EPP); a synapse may be either excitatory (an EPSP) or inhibitory (an IPSP). The inhibition of skeletal muscles cannot be accomplished at the neuromuscular junction; it can take place only in the CNS through IPSPs at the cell body of the motor neuron.
The resultant change in membrane potential in both cases is a graded potential.	

black widow spider venom and *botulinum toxin*—alter the release of ACh, but in opposite directions. The venom of black widow spiders exerts its deadly effect by causing an explosive release of ACh from the storage vesicles, not only at neuromuscular junctions but at all cholinergic sites. All cholinergic sites undergo prolonged depolarization, the most detrimental consequence of which is respiratory failure. Breathing is accomplished by alternate contraction and relaxation of skeletal muscles, particularly the diaphragm. Inability to relax the diaphragm in the presence of a prolonged EPP caused by excessive amounts of ACh results in respiratory paralysis.

Botulinum toxin, on the other hand, exerts its lethal blow by blocking the release of ACh from the terminal button in response to an action potential in the motor neuron. *Clostridium botulinum* toxin is responsible for **botulism,** a form of food poisoning. It most frequently results from improperly canned foods contaminated with clostridial bacteria that survive and multiply, producing their toxin in the process. When this toxin is consumed, it prevents muscles from responding to nerve impulses. Death is due to respiratory failure caused by the inability to contract the diaphragm. Botulinum toxin is one of the most lethal poisons known; ingestion of less than 0.0001 mg can kill an adult.

Other chemicals interfere with neuromuscular junction activity by blocking the effect of released ACh. The best known example is **curare,** which reversibly binds to the ACh receptor sites on the motor end plate. Unlike ACh, however, curare does not alter membrane permeability, nor is it inactivated by AChE. When ACh receptor sites are occupied by curare, ACh cannot combine with these sites to open the channels that would permit the ionic movement responsible for an EPP. Consequently, because muscle action potentials cannot occur in response to nerve impulses to these muscles, paralysis ensues. When sufficient curare is present to effectively block a significant number of ACh receptor sites, the person dies from respiratory paralysis caused by an inability to contract the diaphragm. Curare was used in the past as a deadly arrowhead poison. Curare and related drugs have also been used medically during surgery to help achieve more complete skeletal muscle relaxation with less anesthetic. Under these circumstances, the amount of the agent is carefully administered, and facilities are available to maintain respiration artificially, if necessary, until the effects of the drug wear off.

Organophosphates are a group of chemicals that modify neuromuscular junction activity in yet another way—namely, by irreversibly inhibiting AChE. Inhibition of AChE prevents the inactivation of released ACh. Death from organophosphates is also due to respiratory failure because the diaphragm is unable to repolarize and return to resting conditions, then contract again to bring in a fresh breath of air. These toxic agents are found in some pesticides and military nerve gases.

One disease known to involve the neuromuscular junction is **myasthenia gravis** (*myasthenia* means "muscular weakness"; *gravis* means "severe"), a condition characterized by extreme muscular weakness. It is an autoimmune condition (*autoimmune* means "immunity against self") in which the body erroneously produces antibodies against its own motor end-plate ACh receptors. Consequently, not all of the released ACh molecules are able to find a functioning receptor site with which to bind. As a result, much of the ACh is destroyed by AChE without ever having an opportunity to interact with a receptor site and contribute to the EPP. Treatment consists of administration of a drug that inhibits AChE temporarily (in contrast to the toxic organophosphates, which irreversibly block this enzyme). **Neostigmine** is such a short-term anti-acetylcholinesterase drug. In myasthenia gravis patients, this drug prolongs the action of ACh at the neuromuscular junction by permitting it to build up. The resultant EPP is of sufficient magnitude to initiate action potentials and subsequent contraction in the muscle fiber, as it normally would.

Table 7-9
Examples of Chemical Agents and Diseases That Affect the Neuromuscular Junction

Mechanism	*Chemical Agent or Disease*
Alters Release of Acetylcholine	
Causes explosive release of acetylcholine	Black widow spider venom
Blocks release of acetylcholine	*Clostridium botulinum* toxin
Blocks Acetylcholine Receptor Sites	
Reversibly binds with acetylcholine receptor sites	Curare
Self-produced antibodies inactivate acetylcholine receptor sites	Myasthenia gravis
Prevents Inactivation of Acetylcholine	
Irreversibly inhibits acetylcholinesterase	Organophosphates (certain pesticides and military nerve gases)

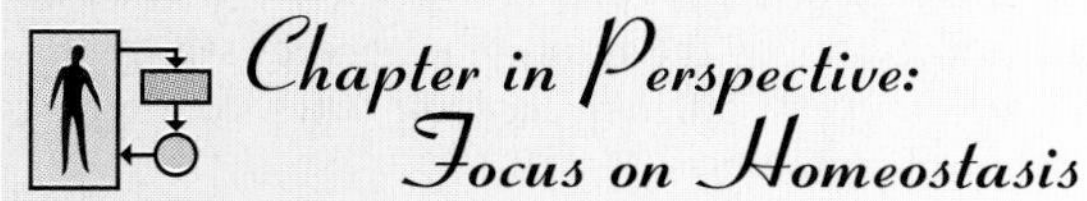

Chapter in Perspective: Focus on Homeostasis

The nervous system, along with the other major control system, the endocrine system, is responsible for controlling most of the body's muscular and glandular activities. Whereas the afferent division of the peripheral nervous system detects and carries information to the central nervous system (CNS) for processing and decision making, the efferent division of the peripheral nervous system carries directives from the CNS to the effector organs (muscles and glands), which carry out the intended response. Much of this efferent output is directed toward maintaining homeostasis.

The autonomic nervous system, which is the efferent branch that innervates smooth muscle, cardiac muscle, and glands, plays a major role in the following range of homeostatic activities:

- Regulation of blood pressure.
- Control of digestive juice secretion and of digestive tract contractions that mix ingested food with the digestive juices.
- Control of sweating to help maintain body temperature.

The somatic nervous system, the efferent branch that innervates skeletal muscle, contributes to homeostasis by stimulating the following activities:

- Skeletal muscle contractions that enable the body to move in relation to the external environment contribute to homeostasis by moving the body toward food or away from harm.
- Skeletal muscle contraction also accomplishes breathing to maintain appropriate levels of O_2 and CO_2 in the body.
- Shivering is a skeletal muscle activity important in the maintenance of body temperature.

Additionally, efferent output to skeletal muscles accomplishes many movements that are not aimed at maintaining a stable internal environment but nevertheless enrich our lives and enable us to contribute to society, such as dancing, building bridges, or performing surgery.

Chapter Summary

Introduction

The CNS controls muscles and glands by transmitting signals to these effector organs through the efferent division of the peripheral nervous system. There are two types of efferent output: the autonomic nervous system, which is under involuntary control and supplies cardiac and smooth muscle as well as most exocrine and some endocrine glands, and the somatic nervous system, which is subject to voluntary control and supplies skeletal muscle.

Autonomic Nervous System

The autonomic nervous system consists of two subdivisions—the sympathetic and parasympathetic nervous systems. An autonomic nerve pathway consists of a two-neuron chain. The preganglionic fiber originates in the CNS and synapses with the cell body of the postganglionic fiber in a ganglion outside the CNS. The postganglionic fiber terminates on the effector organ. All preganglionic fibers and parasympathetic postganglionic fibers release acetylcholine. Sympathetic postganglionic fibers release norepinephrine. The same neurotransmitter elicits various responses from different tissues. Thus, the response depends on specialization of the tissue cells, not on the properties of the messenger. Tissues innervated by the autonomic nervous system possess one or more of several different receptor types for the postganglionic chemical messengers.

A given autonomic fiber either excites or inhibits activity in the organ it innervates. Most visceral organs are innervated by both sympathetic and parasympathetic nerve fibers, which in general produce opposite effects in a particular organ. Dual innervation of visceral organs by both branches of the autonomic nervous system permits precise control over an organ's activity. The sympathetic system dominates in emergency or stressful situations and promotes responses that prepare the body for strenuous physical activity (for "fight" or "flight"). The parasympathetic system dominates in quiet, relaxed situations and promotes body maintenance activities such as digestion.

Somatic Nervous System

The somatic nervous system consists of the axons of motor neurons, which originate in the spinal cord and terminate on skeletal muscle. Acetylcholine, the neurotransmitter released from a motor neuron, stimulates muscle contraction. Motor neurons are the final common pathway by which various regions of the CNS exert control over skeletal muscle activity.

Neuromuscular Junction

Each axon terminal of a motor neuron forms a neuromuscular junction with a single muscle cell (fiber). Because these structures do not make direct contact, signals are passed between the nerve terminal and muscle fiber by means of the chemical messenger acetylcholine (ACh). An action potential in the axon terminal causes the release of ACh from its storage vesicles. The released ACh diffuses across the space separating the nerve and muscle cell and binds to special receptor sites on the underlying motor end plate of the muscle cell membrane. This combination of ACh with the receptor sites triggers the opening of specific channels in the motor end plate. The subsequent ion movements depolarize the motor end plate, producing the end plate potential (EPP). Local current flow between the depolarized end plate and adjacent muscle cell membrane brings these adjacent areas to threshold, initiating an action potential that is propagated throughout the muscle fiber. This muscle action potential triggers muscle contraction. Acetylcholinesterase inactivates ACh, terminating the EPP and, subsequently, the action potential.

Review Exercises

Objective Questions (Answers on p. E–6.)

1. Sympathetic preganglionic fibers originate in the thoracic and lumbar segments of the spinal cord. (True or false?)
2. Action potentials are transmitted on a one-to-one basis at both a neuromuscular junction and a synapse. (True or false?)
3. The sympathetic nervous system:
   a. is always excitatory.
   b. innervates only tissues concerned with protecting the body against challenges from the outside environment.
   c. has short preganglionic and long postganglionic fibers.
   d. is part of the afferent division of the peripheral nervous system.
   e. is part of the somatic nervous system.
4. Acetylcholinesterase:
   a. is stored in vesicles in the terminal button.
   b. combines with receptor sites on the motor end plate to bring about an end-plate potential.
   c. is inhibited by organophosphates.
   d. is the chemical transmitter at the neuromuscular junction.
   e. paralyzes skeletal muscle by strongly binding with acetylcholine receptor sites.
5. The two divisions of the autonomic nervous system are the ____________ nervous system, which dominates in fight-or-flight situations, and the ____________ nervous system, which dominates in quiet, relaxed situations.
6. The ____________ is a modified sympathetic ganglion that does not give rise to postganglionic fibers but instead secretes hormones similar or identical to sympathetic postganglionic neurotransmitters into the blood.
7. Using the answer code below, identify the autonomic transmitter being described:

   (a) acetylcholine
   (b) norepinephrine

   ___1. secreted by all preganglionic fibers
   ___2. secreted by sympathetic postganglionic fibers
   ___3. secreted by parasympathetic postganglionic fibers
   ___4. secreted by the adrenal medulla
   ___5. secreted by motor neurons
   ___6. binds to muscarinic or nicotinic receptors
   ___7. binds to α or β receptors
8. Using the answer code below, indicate which type of efferent output is being described:

   (a) characteristic of the somatic nervous system
   (b) characteristic of the autonomic nervous system

   ___1. composed of two-neuron chains
   ___2. innervates cardiac muscle, smooth muscle, and glands
   ___3. innervates skeletal muscle
   ___4. consists of the axons of motor neurons
   ___5. exerts either an excitatory or an inhibitory effect on its effector organs
   ___6. dually innervates its effector organs
   ___7. exerts only an excitatory effect on its effector organs

Essay Questions

1. Distinguish between preganglionic and postganglionic fibers.
2. Compare the origin, preganglionic and postganglionic fiber length, and neurotransmitters of the sympathetic and parasympathetic nervous systems.
3. What is the advantage of dual innervation of many organs by both branches of the autonomic nervous system?
4. Distinguish among the following types of receptors: nicotinic receptors, muscarinic receptors, α receptors, β_1 receptors, and β_2 receptors.
5. What regions of the CNS regulate autonomic output?
6. Why are motor neurons called the final common pathway?
7. Describe the sequence of events that occurs at a neuromuscular junction.
8. Discuss the effect each of the following has at the neuromuscular junction: black widow spider venom, botulinum toxin, curare, organophosphates, myasthenia gravis, and neostigmine.

Quantitative Exercises (Solutions on p. E–6.)

1. When a muscle fiber is activated at the neuromuscular junction, tension does not begin to rise until about 1 msec after initiation of the action potential in the muscle fiber. Many things are occurring during this delay, one time-consuming event being the diffusion of acetylcholine across the neuromuscular junction. The following equation can be used to calculate how long this diffusion takes:

   $$t = x^2/2D$$

   In this equation, x is the distance covered, D is the diffusion coefficient, and t is the time it takes for diffusion of the substance across the distance x. In this example, x is the width of the cleft between the neuronal axon terminal and the muscle fiber at the neuromuscular junction (assume 200 nm) and D is the diffusion coefficient of acetylcholine (assume 1×10^{-5} cm^2/sec). How long does it take the acetylcholine to diffuse across the neuromuscular junction?

Points to Ponder

(Explanations on p. E–6.)

1. Explain why epinephrine, which causes arteriolar constriction (narrowing) in most tissues, is frequently administered in conjunction with local anesthetics.
2. Would skeletal muscle activity be affected by atropine? Why or why not?
3. Considering that you can voluntarily control the emptying of your urinary bladder by contracting (preventing emptying) or relaxing (permitting emptying) your external urethral sphincter, a ring of muscle that guards the exit from the bladder, of what type of muscle is this sphincter composed and what branch of the nervous system supplies it?
4. The venom of certain poisonous snakes contains α bungarotoxin, which binds tenaciously to acetylcholine receptor sites on the motor end-plate membrane. What would the resultant symptoms be?
5. Explain how destruction of motor neurons by poliovirus or amyotropic lateral sclerosis can be fatal.
6. ***Clinical Consideration*** Christopher K. experienced chest pains when he climbed the stairs to his fourth-floor office or played tennis but was symptom-free when not physically exerting himself. His condition was diagnosed as *angina pectoris* (*angina* means "pain"; *pectoris* means "chest"), heart pain that occurs whenever the blood supply to the heart muscle cannot meet the muscle's need for oxygen delivery. This condition usually is caused by narrowing of the blood vessels supplying the heart by cholesterol-containing deposits. Most persons with this condition do not have any pain at rest but experience bouts of pain whenever the heart's need for oxygen increases, such as during exercise or emotionally stressful situations that increase sympathetic nervous activity. Christopher obtains immediate relief of angina attacks by promptly taking a vasodilator drug such as *nitroglycerin,* which relaxes the smooth muscle in the walls of his narrowed heart vessels. Consequently, the vessels open more widely and more blood can flow through them. For prolonged treatment, Christopher's doctor has indicated that he will experience fewer and less severe angina attacks if he takes a β_1-blocker drug, such as *metoprolol,* on a regular basis. Explain why.

8 Muscle Physiology

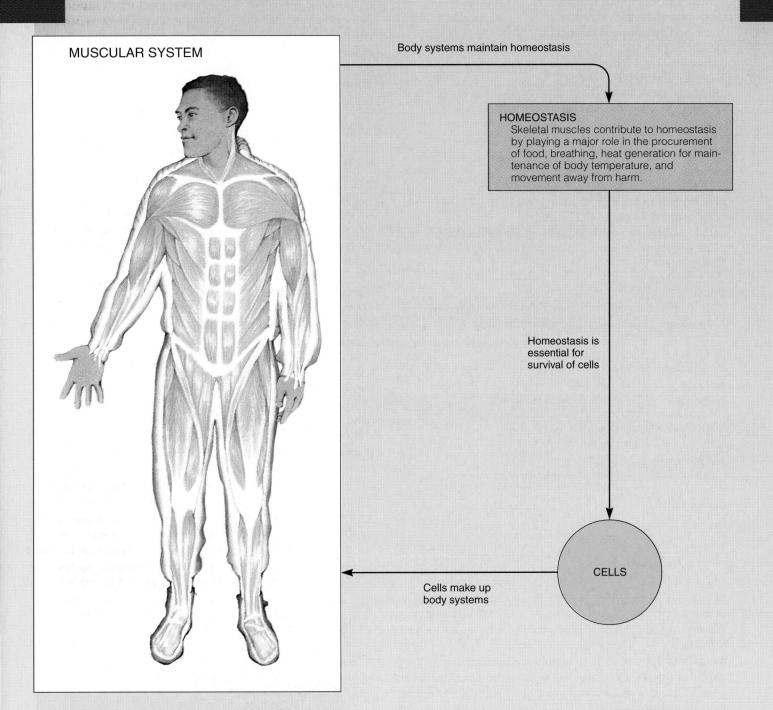

Muscles are the contraction specialists of the body. **Skeletal muscle** attaches to the skeleton. Contraction of skeletal muscles causes the bones to which they are attached to move, allowing the body to perform a variety of motor activities. Skeletal muscles that support homeostasis include those important in the acquisition, chewing, and swallowing of food and those essential for breathing. Skeletal muscle contraction is also used to move the body away from harm. Heat-generating muscle contractions are important in temperature regulation. Skeletal muscles are also used for nonhomeostatic activities, such as dancing or operating a computer. **Smooth muscle** is found in the walls of hollow organs and tubes. Controlled contraction of smooth muscle is responsible for regulating movement of blood through the blood vessels, food through the digestive tract, air through the respiratory airways, and urine to the exterior. **Cardiac muscle** is found only in the walls of the heart, whose contraction pumps life-sustaining blood throughout the body.

Chapter Contents At a Glance

Introduction

Almost all living cells possess rudimentary intracellular machinery for producing such movement as redistributing various components of the cell during cell division. White blood cells use intracellular contractile proteins to propel themselves through their environment. The contraction specialists of the body, however, are the muscle cells. Recall that there are three types of muscle: *skeletal muscle, cardiac muscle,* and *smooth muscle* (see p. 3). Through their highly developed ability to contract, muscle cells are capable of shortening and developing tension, which enables them to produce movement and to do work. In contrast to sensory systems, which transform other forms of energy in the environment into electrical signals, muscles in response to electrical signals convert the chemical energy of ATP into mechanical energy that can act on the environment. Controlled contraction of muscles allows (1) purposeful movement of the whole body or parts of the body in relation to the environment (such as walking or waving your hand); (2) manipulation of external objects (such as driving a car or moving a piece of furniture); (3) propulsion of contents through various hollow internal organs (such as circulation of blood or movement of materials through the digestive tract); and (4) emptying the contents of certain organs to the external environment (such as urination or giving birth).

Muscle comprises the largest group of tissues in the body, accounting for approximately half of the body's weight. Skeletal muscle alone makes up about 40% of body weight in men and 32% in women, with smooth and cardiac muscle making up another 10% of the total weight. Although the three muscle types are structurally and functionally distinct, they can be classified in two different ways according to their common characteristics (Fig. 8-1). First, muscles are categorized as *striated* (skeletal and cardiac muscle) or *unstriated* (smooth muscle), depending on whether alternating dark and light bands, or striations, can be seen when the muscle is viewed under a light microscope. Second, muscles are categorized as *voluntary* (skeletal muscle) or *involuntary* (cardiac and smooth muscle), depending respectively on whether they are innervated by the somatic nervous system and are subject to voluntary control or are innervated by the autonomic nervous system and are not subject to voluntary control (see p. 206).

Most of this chapter will be devoted to a detailed examination of how the most abundant and best understood muscle, skeletal muscle, works. The chapter will conclude with a discussion of the unique properties of smooth and cardiac muscle in comparison to skeletal muscle.

Figure 8-1 Categorization of Muscle

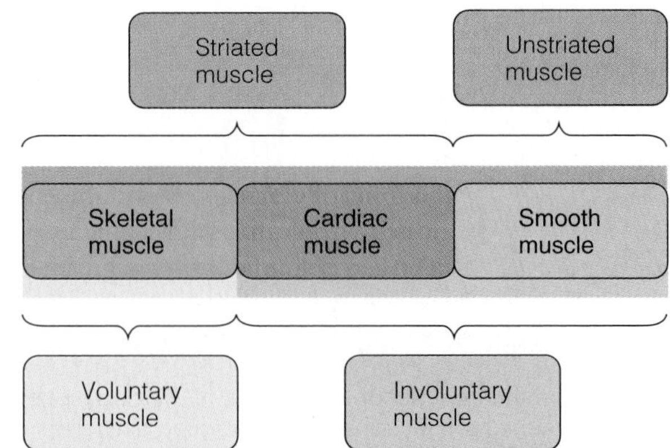

Structure of Skeletal Muscle

Skeletal muscle fibers have a highly organized internal arrangement that creates a striated appearance.

Skeletal muscles are stimulated to contract via release of acetylcholine (ACh) at neuromuscular junctions between motor neuron terminals and muscle cells (see p. 212). A basic understanding of the structural components of a skeletal muscle fiber is essential to understanding how the muscle action potential initiated by ACh brings about contraction.

A single skeletal muscle cell, known as a **muscle fiber,** is relatively large, elongated, and cylinder-shaped, measuring from 10 to 100 micrometers (1 μ = 1 millionth of a meter) in diameter and up to 750,000 μ, or 2.5 feet, in length. A skeletal muscle consists of a number of muscle fibers lying parallel to each other and bundled together by connective tissue (Fig. 8-2a). The fibers usually extend the entire length of the muscle. During embryonic development, the huge skeletal muscle fibers are formed by the fusion of many smaller cells; thus, one striking feature is the presence of multiple nuclei in a single muscle cell. Another feature is the abundance of mitochondria, the energy-generating organelles, as would be expected with the high energy demands of a tissue as active as skeletal muscle.

The most predominant structural feature of a skeletal muscle fiber is the presence of numerous **myofibrils.** These specialized contractile elements, which constitute 80% of the volume of the muscle fiber, are cylinder-shaped intracellular structures 1 μm in diameter that extend the entire length of the muscle fiber (Fig. 8-2b). Each myofibril consists of a regular arrangement of highly organized cytoskeletal elements—the thick and thin filaments (Fig. 8-2c). The **thick filaments,** which are 12 to 18 nm in diameter and 1.6 μm in length, are special assemblies of the protein *myosin,* whereas the **thin filaments,** which are 5 to 8 nm in diameter and 1.0 μm long, are made up primarily of the protein *actin* (Fig. 8-2d). These same proteins are found in all other cells of the body but in a less-organized fashion. The levels of organization in a skeletal muscle can be summarized as follows:

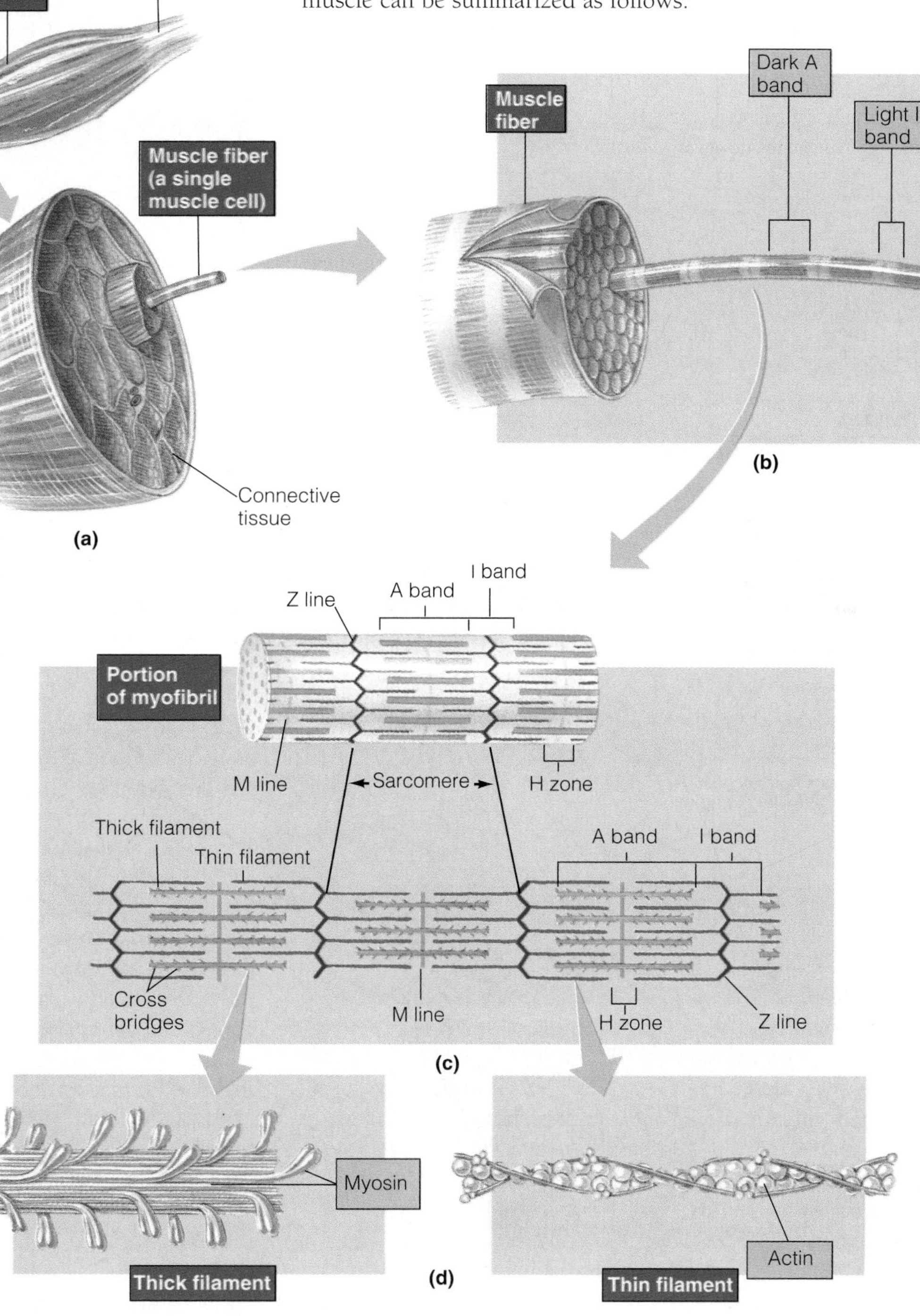

Figure 8-2 Levels of Organization in a Skeletal Muscle
(a) Enlargement of a cross section of a whole muscle. (b) Enlargement of a myofibril within a muscle fiber. (c) Cytoskeletal components of a myofibril. (d) Protein components of thick and thin filaments.

whole muscle →	muscle fiber →	myofibril →	thick and thin filaments →	myosin and actin
(an organ)	(a cell)	(a specialized intracellular structure)	(cytoskeletal elements)	(proteins)

Viewed with a light microscope, a relaxed myofibril (Fig. 8-3a) displays alternating dark bands (the A bands) and light bands (the I bands). The bands of all the myofibrils lined up parallel to each other collectively lead to the *striated* appearance of a skeletal muscle fiber (Fig. 8-3b). Alternate stacked sets of thick and thin filaments that slightly overlap each other are responsible for the A and I bands (Fig. 8-2c). An **A band** consists of a stacked set of thick filaments along with the portions of the thin filaments that overlap on both ends of the thick filaments. The thick filaments are found only within the A band and extend its entire width. The lighter area within the middle of the A band, where the thin filaments do not reach, is known as the **H zone.** Only the central portions of the thick filaments are found in this region. The **I band** consists of the remaining portion of the thin filaments that do not project into the A band. Thus, the I band contains only thin filaments but not their entire length.

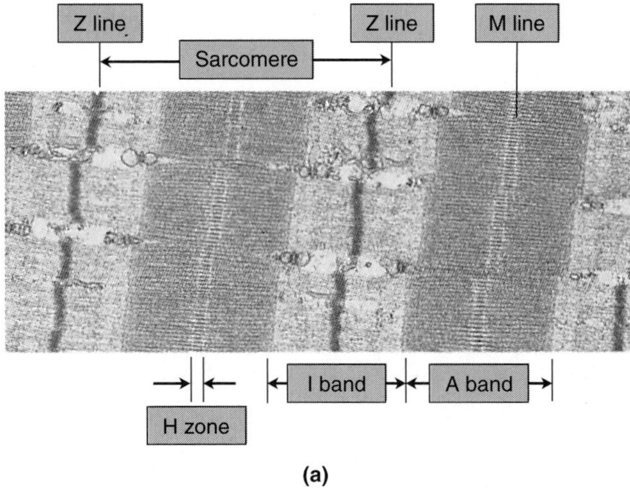

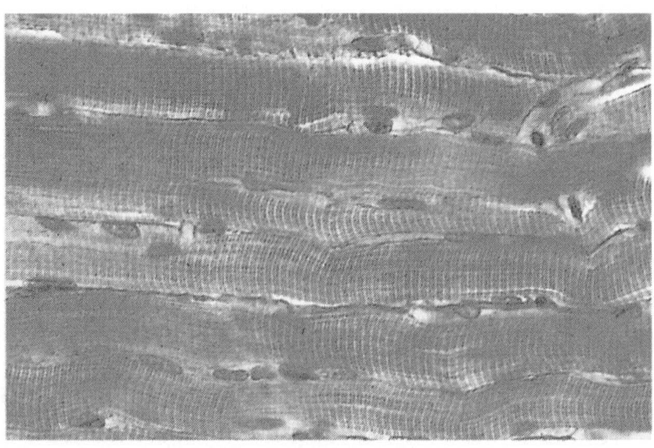

Figure 8-3 **Light-Microscope View of Skeletal Muscle Components** (a) High-power light-microscope view of a myofibril. (b) Low-power light-microscope view of skeletal muscle fibers. Note striated appearance. [Source: Reprinted with permission from Sydney Schochet Jr., M.D., Professor, Department of Pathology, School of Medicine, West Virginia University: *Diagnostic Pathology of Skeletal Muscle and Nerve* (Stamford, Connecticut: Appleton & Lange, 1986), Figure 1-13.]

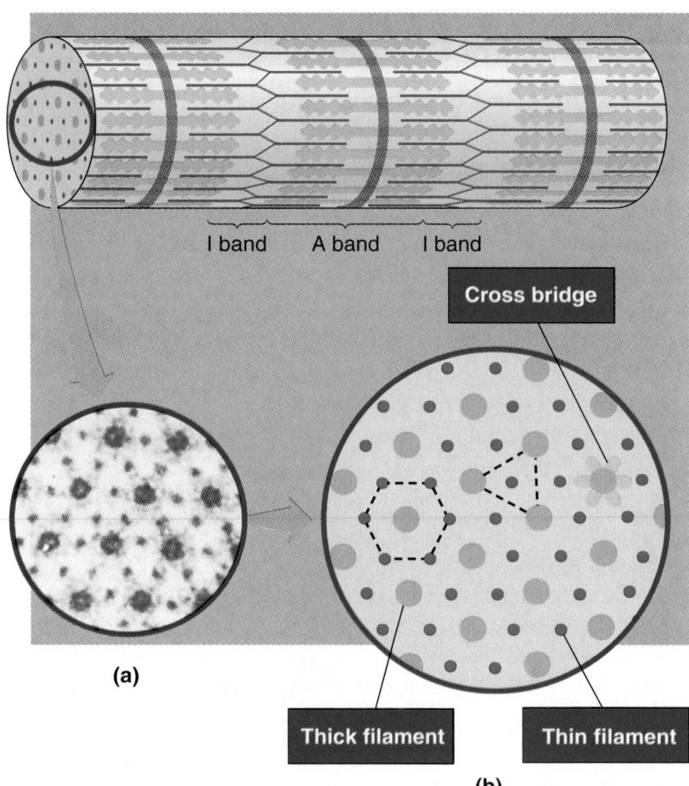

Figure 8-4 **Cross-Sectional Arrangement of Thick and Thin Filaments** (a) Electron micrograph cross section through the A band in the region of thick and thin filament overlap. Note the fine cross bridges extending from the thick filaments. (b) Schematic representation of the geometric relation among thick and thin filaments and cross bridges.

Visible in the middle of each I band is a dense, vertical **Z line.** The area between two Z lines is called a **sarcomere,** which is the functional unit of skeletal muscle. A **functional unit** of any organ is the smallest component that can perform all the functions of that organ. Accordingly, a sarcomere is the smallest component of a muscle fiber that is capable of contraction. The Z line is actually a flattened disc-like cytoskeletal protein that connects the thin filaments of two adjoining sarcomeres. Each relaxed sarcomere is about 2.5 µm in width and consists of one whole A band and half of each of the two I bands located on either side. During growth, a muscle increases in length by adding new sarcomeres, not by increasing the size of each sarcomere. Just as the Z lines hold the sarcomeres together in a chain along the myofibril's length, another system of supporting proteins holds the thick filaments together vertically within each stack. These proteins can be seen as the **M line,** which extends vertically down the middle of the A band within the center of the H zone.

With an electron microscope, fine **cross bridges** can be seen extending from each thick filament toward the surrounding thin filaments in the regions where the thick and thin filaments overlap (Figs. 8-2c and 8-4a). Three-dimensionally, the thin filaments are arranged hexagonally

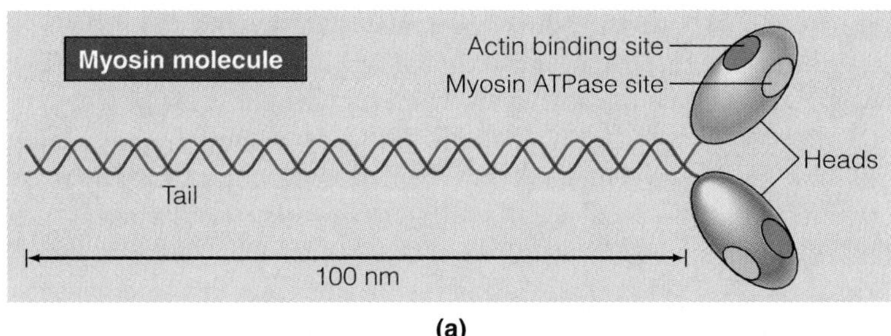

(a)

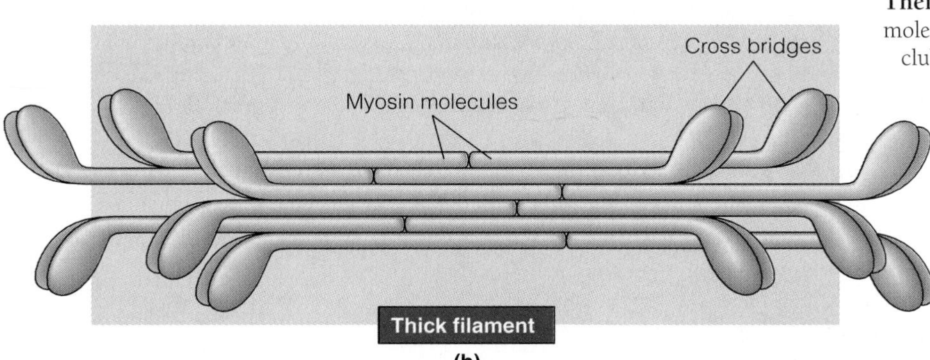

(b)

Figure 8-5 Structure of Myosin Molecules and Their Organization within a Thick Filament (a) Myosin molecule. Each myosin molecule consists of two identical golf-club–shaped subunits with their tails intertwined and their globular heads, each of which contains an actin binding site and a myosin ATPase site, projecting out at one end. (b) Thick filament. A thick filament is made up of myosin molecules lying lengthwise parallel to each other. Half are oriented in one direction and half in the opposite direction so that the tails from the two halves line up end-to-end in the middle of the filament. The globular heads, which protrude at regular intervals along the thick filament, form the cross bridges.

around the thick filaments. Cross bridges project from each thick filament in all six directions toward the surrounding thin filaments. Each thin filament, in turn, is surrounded by three thick filaments (Fig. 8-4b). To give you an idea of the magnitude of these filaments, a single muscle fiber may contain an estimated 16 billion thick and 32 billion thin filaments, all arranged in this very precise pattern within the myofibrils.

Myosin forms the thick filaments, whereas actin is the main structural component of the thin filaments.

Each thick filament is composed of several hundred myosin molecules packed together in a specific arrangement. A **myosin** molecule is a protein consisting of two identical subunits, each shaped somewhat like a golf club (Fig. 8-5a). The protein's tail ends are intertwined around each other, with the two globular heads projecting out at one end. The two halves of each thick filament are mirror images made up of myosin molecules lying lengthwise in a regular staggered array, with their tails oriented toward the center of the filament and their globular heads protruding outward at regularly spaced intervals (Fig. 8-5b). These heads form the cross bridges between the thick and thin filaments. Each cross bridge has two important sites crucial to the contractile process: an *actin binding site* and a *myosin ATPase (ATP-splitting) site.*

Thin filaments are composed of three proteins: *actin, tropomyosin,* and *troponin* (Fig. 8-6).

Figure 8-6 Composition of a Thin Filament The main structural component of a thin filament is two chains of spherical-shaped actin molecules that are twisted together. Troponin molecules, which consist of three small spherical subunits, and threadlike tropomyosin molecules are arranged to form a ribbon that lies alongside the groove of the actin helix and physically covers the binding sites on actin molecules for attachment with myosin cross bridges. (The thin filaments shown here are not drawn in proportion to the thick filaments in Figure 8-5. Thick filaments are two to three times larger in diameter than thin filaments.)

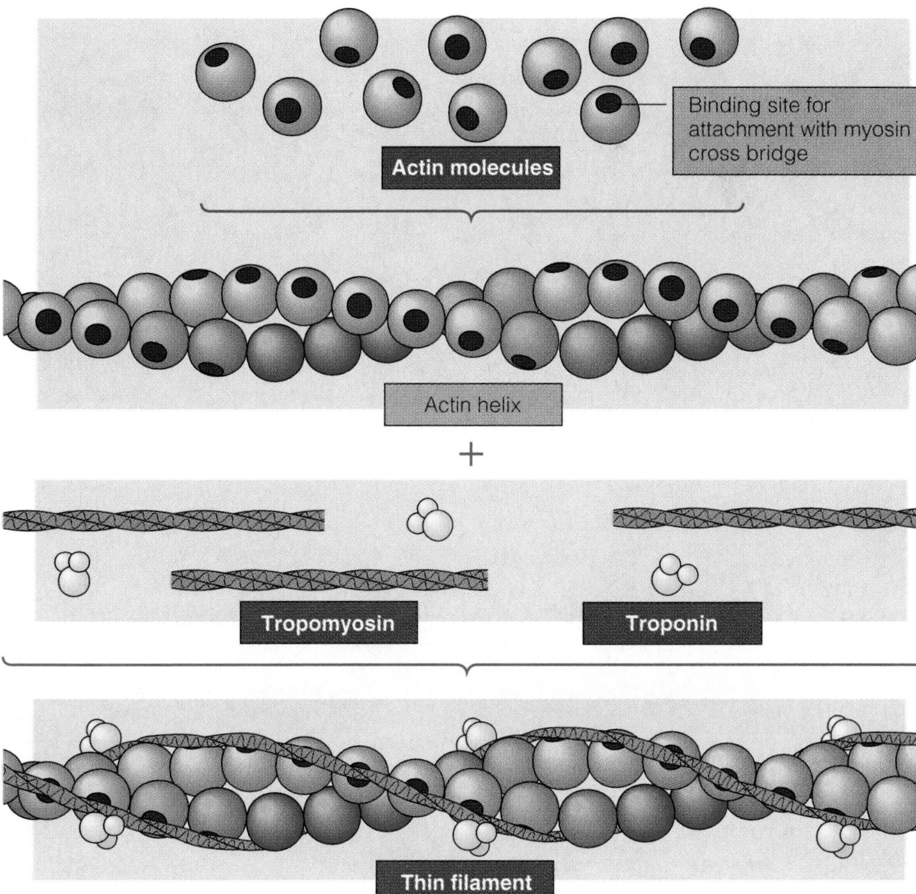

Actin molecules, the primary structural proteins of the thin filament, are spherical in shape. The backbone of a thin filament is formed by actin molecules joined into two strands and twisted together, like two chains of pearls wrapped around each other. Each actin molecule has a special binding site for attachment with a myosin cross bridge. By a mechanism to be described shortly, binding of actin and myosin molecules at the cross bridges results in energy-consuming contraction of the muscle fiber. Accordingly, actin and myosin are often referred to as **contractile proteins**, even though, as we will see, neither myosin nor actin actually contracts.

In a relaxed muscle fiber, contraction does not take place; actin is not able to bind with cross bridges because of the position of the two other types of protein within the thin filament—tropomyosin and troponin. **Tropomyosin** molecules are threadlike proteins that lie end-to-end alongside the groove of the actin spiral. In this position, tropomyosin covers the actin sites that bind with the cross bridges, thus blocking the interaction that leads to muscle contraction. Tropomyosin is stabilized in this blocking position by **troponin** molecules, which fasten down the ends of each tropomyosin molecule. Troponin is a protein complex consisting of three polypeptide units: one that binds to tropomyosin, one that binds to actin, and a third that can bind with Ca^{2+}. When Ca^{2+} binds to troponin, the shape of this protein is changed in such a way that tropomyosin is allowed to slide away from its blocking position (Fig. 8-7). With tropomyosin out of the way, actin and myosin can bind and interact at the cross bridges, resulting in muscle contraction. Tropomyosin and troponin are often referred to as **regulatory proteins** because

Figure 8-7 **Role of Calcium in Turning on Cross Bridges**

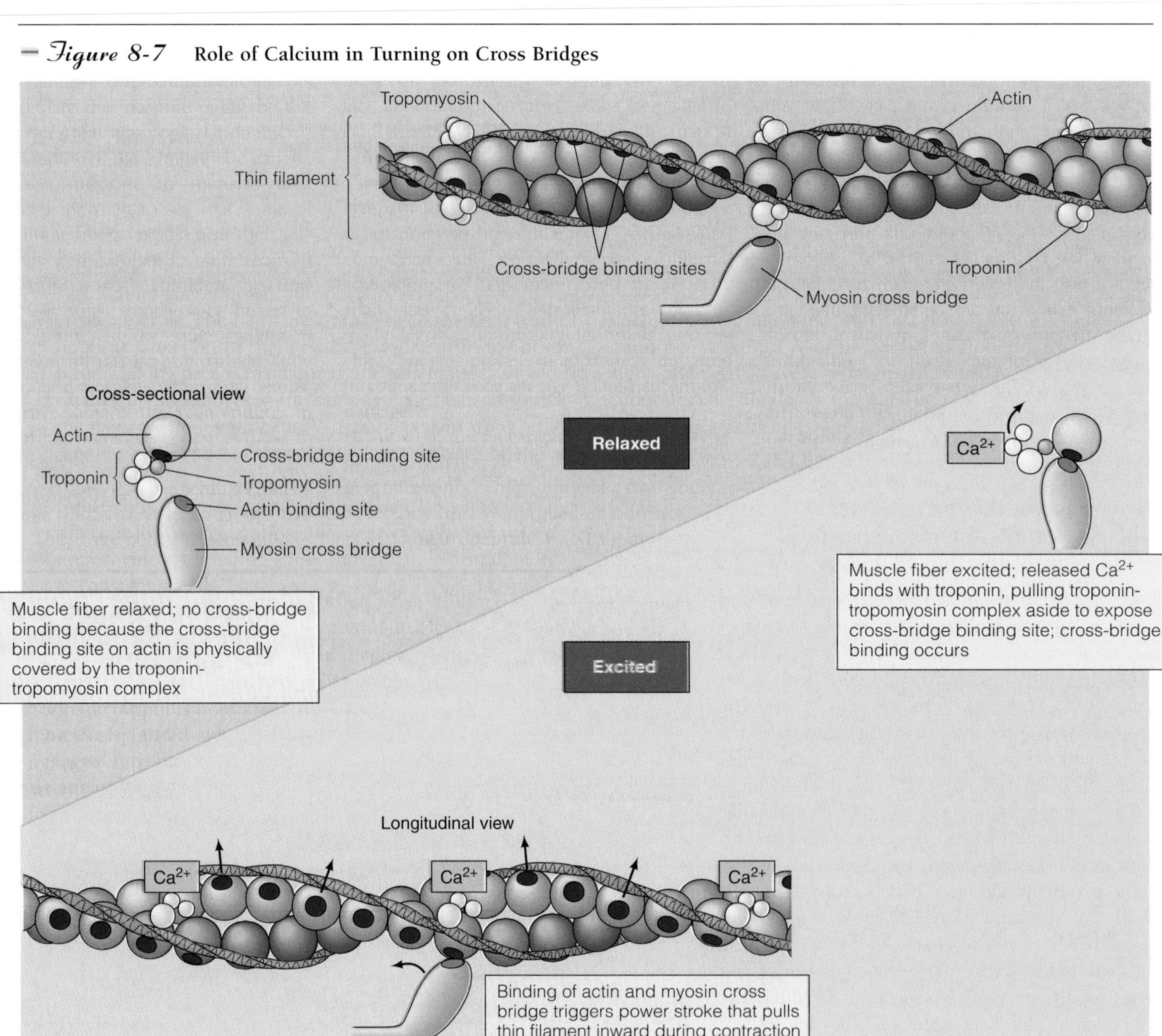

of their role in covering (preventing contraction) or exposing (permitting contraction) the binding sites for cross-bridge interaction between actin and myosin.

Several important links in the contractile process remain to be discussed. How does cross-bridge interaction between actin and myosin bring about muscle contraction? How does a muscle action potential trigger this contractile process? What is the source of the Ca^{2+} that physically repositions troponin and tropomyosin to permit cross-bridge binding? We will turn our attention to these topics in the next section.

Molecular Basis of Skeletal Muscle Contraction

Cycles of cross-bridge binding and bending pull the thin filaments closer together between the thick filaments during contraction.

The thin filaments on each side of a sarcomere slide inward toward the A band's center during contraction (Fig. 8-8). As they slide inward, the thin filaments pull the Z lines to which they are attached closer together, so the sarcomere shortens. As all the sarcomeres throughout the muscle fiber's length shorten simultaneously, the entire fiber becomes shorter. This is known as the **sliding-filament mechanism** of muscle contraction. The H zone, the region in the center of the A band where the thin filaments do not reach, becomes smaller as the thin filaments approach each other when they slide more deeply inward. The H zone may even disappear if the thin filaments meet in the middle of the A band. The I band, which consists of the portions of the thin filaments that do not overlap with the thick filaments, decreases in width as the thin filaments further overlap the thick filaments during their inward slide. The thin filaments themselves do not change length during muscle fiber shortening. The width of the A band remains unchanged during contraction, because its width is determined by the length of the thick filaments, and the thick filaments do not change length during the shortening process. Note that neither the thick nor thin filaments decrease in length to shorten the sarcomere. Instead, contraction is accomplished by the thin filaments sliding closer together between the thick filaments.

The thin filaments are pulled inward relative to the stationary thick filaments by cross-bridge activity. During contraction, with the tropomyosin and troponin "chaperones" pulled out of the way by Ca^{2+}, the myosin cross bridges from a thick filament are able to bind with the actin molecules in the surrounding thin filaments. Let's concentrate on a single cross-bridge interaction (Fig. 8-9a). When myosin and actin make contact at a cross bridge, the conformation of the bridge is altered so that it bends inward as if it were on a hinge, "stroking" toward the center of the sarcomere, similar to the stroking of a boat oar. This so-called **power stroke** of a cross bridge pulls the thin filament to which it is attached inward. A single power stroke pulls the thin filament inward only a small percentage of the total shortening distance. Complete shortening is accomplished by repeated cycles of cross-bridge binding and bending. At the end of one cross-bridge cycle, the link between the myosin cross bridge and actin molecule is broken. The cross bridge returns to its original conformation and binds to the next actin molecule posi-

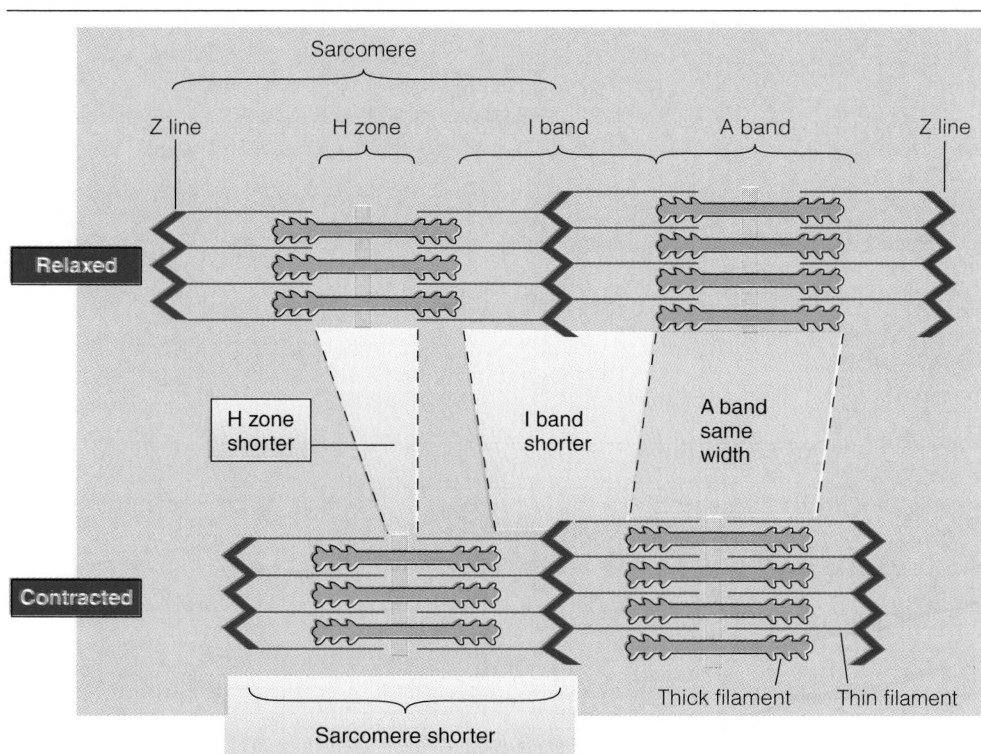

Figure 8-8 **Changes in Banding Pattern during Shortening** During muscle contraction, each sarcomere shortens as the thin filaments slide closer together between the thick filaments so that the Z lines are pulled closer together. The width of the A bands does not change as a muscle fiber shortens, but the I bands and H zones become shorter.

tioned behind its previous actin partner. The cross bridge bends once again to pull the thin filament in further, then detaches and repeats the cycle. Repeated cycles of cross-bridge binding and bending successively pull in the thin filaments, much like pulling in a rope hand over hand. Because of the orientation of the myosin molecules within a thick filament (Fig. 8-9b), all of the cross bridges' power strokes are directed toward the center, so that all six of the surrounding thin filaments are pulled inward simultaneously (Fig. 8-9c). The cross bridges aligned with given thin filaments do not all stroke in unison, however. At any time during contraction, part of the cross bridges are attached to the thin filaments and are stroking while others are returning to their original conformation in preparation for binding with another actin molecule. Thus, some cross bridges are "holding on" to the thin filaments, while others "let go" to bind with new actin. If it were not for this asynchronous cycling of the cross bridges, the thin filaments would be able to slip back toward their resting position between strokes.

Calcium is the link between excitation and contraction.

How is this cross-bridge cycling switched on by muscle excitation? **Excitation-contraction coupling** refers to the series of events linking muscle excitation (the presence of an action potential in a muscle fiber) to muscle contraction (cross-bridge activity that causes the thin filaments to slide closer together to produce sarcomere shortening) (▮ Table 8-1).

Skeletal muscles are stimulated to contract by release of acetylcholine (ACh) at neuromuscular junctions between motor neuron terminals and muscle fibers.

Actin molecules in thin myofilament

Myosin cross bridge

Z line

BINDING Myosin cross bridge binds to actin molecule.

POWER STROKE Cross bridge bends, pulling thin myofilament inward.

DETACHMENT Cross bridge detaches at end of power stroke and returns to original conformation.

BINDING Cross bridge binds to more distal actin molecule; cycle repeated.

(a)

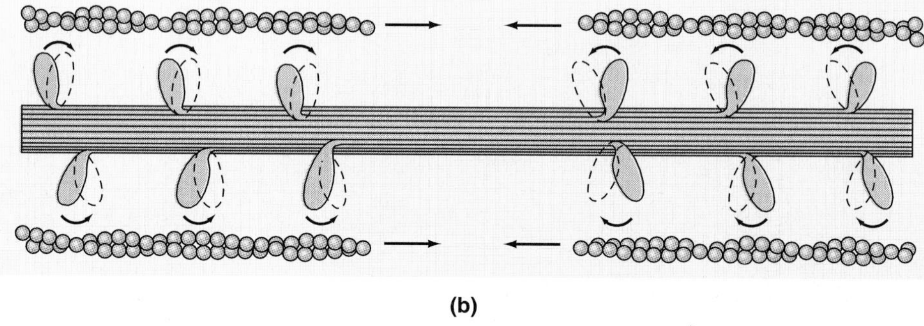

(b)

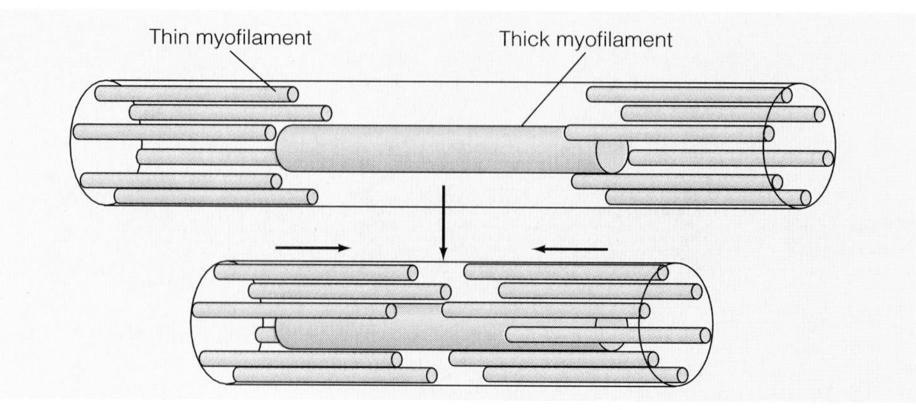

(c)

Figure 8-9 **Cross-Bridge Activity** (a) During each cross-bridge cycle, the cross bridge binds with an actin molecule, bends to pull the thin filament inward during the power stroke, then detaches and returns to its resting conformation, ready to repeat the cycle. (b) The power strokes of all cross bridges extending from a thick filament are directed toward the center of the thick filament. (c) Each thick filament is surrounded by six thin filaments, all of which are pulled inward simultaneously through cross-bridge cycling during muscle contraction.

Recall that the binding of ACh with the motor end plate of a muscle fiber brings about permeability changes in the muscle fiber that result in an action potential that is conducted over the entire surface of the muscle cell membrane (see p. 215).

At each junction of an A band and I band, the surface membrane dips into the muscle fiber to form a **transverse tubule (T tubule)**, which runs perpendicularly from the surface of the muscle cell membrane into the central portions of the muscle fiber (Fig. 8-10). Because the T tubule membrane is continuous with the surface membrane, an action potential on the surface membrane also spreads down into the T tubule, providing a means of rapidly transmitting the surface electric activity into the central portions of the fiber. The presence of a local action potential in the T tubules induces permeability changes in a separate membranous network within the muscle fiber, the sarcoplasmic reticulum.

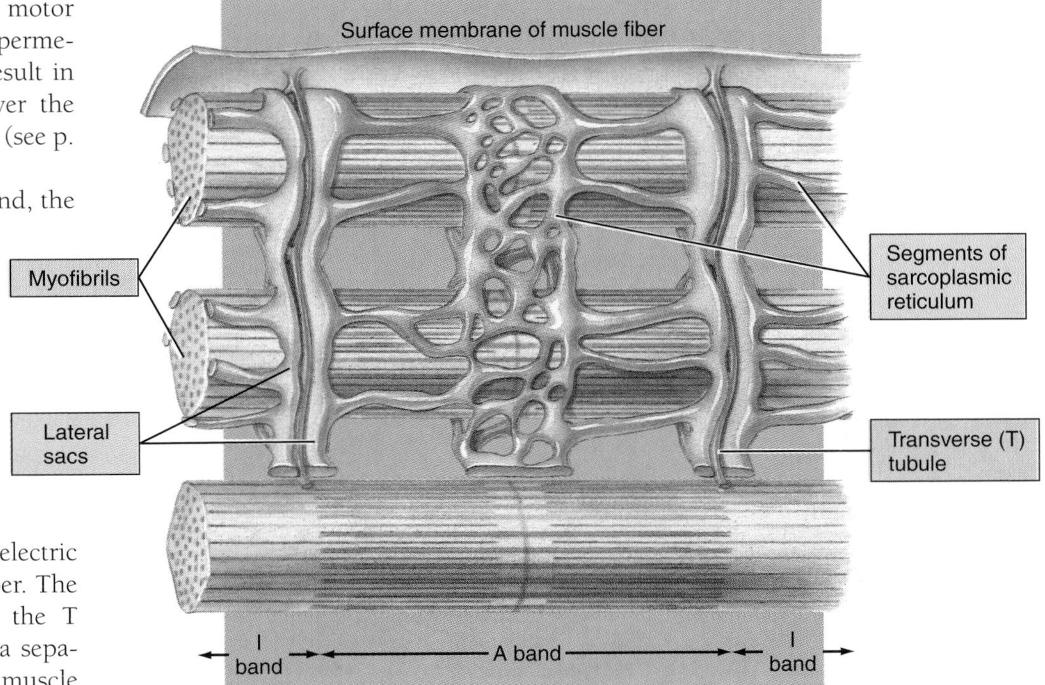

Figure 8-10 **The T Tubules and Sarcoplasmic Reticulum in Relationship to the Myofibrils** The transverse (T) tubules are membranous perpendicular extensions of the surface membrane that dip deep into the muscle fiber at the junctions between the A and I bands of the myofibrils. The sarcoplasmic reticulum is a fine membranous network that runs longitudinally and surrounds each myofibril, with separate segments encircling each A band and I band. The ends of each segment are expanded to form lateral sacs that lie next to the adjacent T tubules.

The **sarcoplasmic reticulum** is a modified endoplasmic reticulum (see p. 23) that consists of a fine network of interconnected tubules surrounding each myofibril like a mesh sleeve (Fig. 8-10). This membranous network runs longitudinally down the myofibril (that is, encircles the myofibril throughout its length), but is not continuous. Separate segments of sarcoplasmic reticulum are wrapped around each A band and each I band. The ends of each segment expand to form saclike regions, the **lateral sacs** (alternatively known as **terminal cisternae**), which are separated from the adjacent T tubules by a slight gap (Figs.

Table 8-1 Steps of Excitation-Contraction Coupling and Relaxation

Steps
Acetylcholine released from the terminal of a motor neuron initiates an action potential in the muscle cell that is propagated over the entire surface of the muscle cell membrane.
The surface electric activity is carried into the central portions of the muscle fiber by the T tubules.
Spread of the action potential down the T tubules triggers the release of stored Ca^{2+} from the adjacent lateral sacs of the sarcoplasmic reticulum.
Released Ca^{2+} binds with troponin and changes its shape so that the troponin-tropomyosin complex is physically pulled aside, uncovering actin's cross-bridge binding sites.
Exposed actin sites bind with myosin cross bridges, which have previously been energized by the splitting of ATP into ADP + P_i + energy by the myosin ATPase site on the cross bridges.
Binding of actin and myosin at a cross bridge causes the cross bridge to bend, producing a power stroke that pulls the thin filament inward. Inward sliding of all the thin filaments surrounding a thick filament shortens the sarcomere (causes muscle contraction).
ADP and P_i are released from the cross bridge during the power stroke.
Attachment of a new molecule of ATP permits detachment of the cross bridge, which returns to its original conformation.
Splitting of the fresh ATP molecule by myosin ATPase energizes the cross bridge once again.
If Ca^{2+} is still present so that the troponin-tropomyosin complex remains pulled aside, the cross bridges go through another cycle of binding and bending, pulling the thin filament in even further.
When there is no longer a local action potential and Ca^{2+} has been actively returned to its storage site in the sarcoplasmic reticulum's lateral sacs, the troponin-tropomyosin complex slips back into its blocking position, actin and myosin no longer bind at the cross bridges, and the thin filaments slide back to their resting position as relaxation takes place.

8-10 and 8-11). The sarcoplasmic reticulum's lateral sacs store Ca^{2+}. Spread of an action potential down a T tubule triggers release of Ca^{2+} from the sarcoplasmic reticulum into the cytosol.

Recent studies have revealed how a change in T tubule potential is linked with the release of Ca^{2+} from the sarcoplasmic reticulum's lateral sacs. Investigators have discovered an orderly arrangement of **foot proteins,** which extend from the sarcoplasmic reticulum and span the gap between the lateral sac and T tubule. These foot proteins not only bridge the gap but also serve as Ca^{2+}-release channels. These foot protein Ca^{2+} channels are known as **ryanodine receptors** because they are locked in the open position by the plant chemical ryanodine. The T tubule membrane also bears receptors at the points where it contacts the foot proteins protruding from the sarcoplasmic reticulum. These T tubule receptors, known as **dihydropyridine receptors** because they are blocked by the drug dihydropyridine, are voltage-gated sensors. When an action potential is propagated down the T tubule, the local depolarization activates the voltage-gated dihydropyridine receptors. These activated T tubule receptors in turn trigger the opening of the directly abutting Ca^{2+}-release channels (alias ryanodine receptors alias foot proteins) in the adjacent lateral sacs of the sarcoplasmic reticulum. Consequently, Ca^{2+} is released from the lateral sacs. By slightly repositioning the troponin and tropomyosin molecules, this released Ca^{2+} exposes the binding sites on the actin molecules so that they can link with the myosin cross bridges at their complementary binding sites (Fig. 8-11).

Recall that a myosin cross bridge has two special sites, an actin binding site and an ATPase site. The ATPase site has recently been identified as an open cleft or pocket formed by

Figure 8-11 **Calcium Release in Excitation-Contraction Coupling** Steps ① through ⑤ depict the events that couple neurotransmitter release and subsequent electrical excitation of the muscle cell with muscle contraction. Note that calcium (Ca^{2+}) released from the lateral sacs of the sarcoplasmic reticulum in response to a local action potential in the adjacent T tubule binds with troponin on the thin filaments to permit cross-bridge binding and power stroking. Steps ⑥ and ⑦ depict events associated with muscle relaxation.

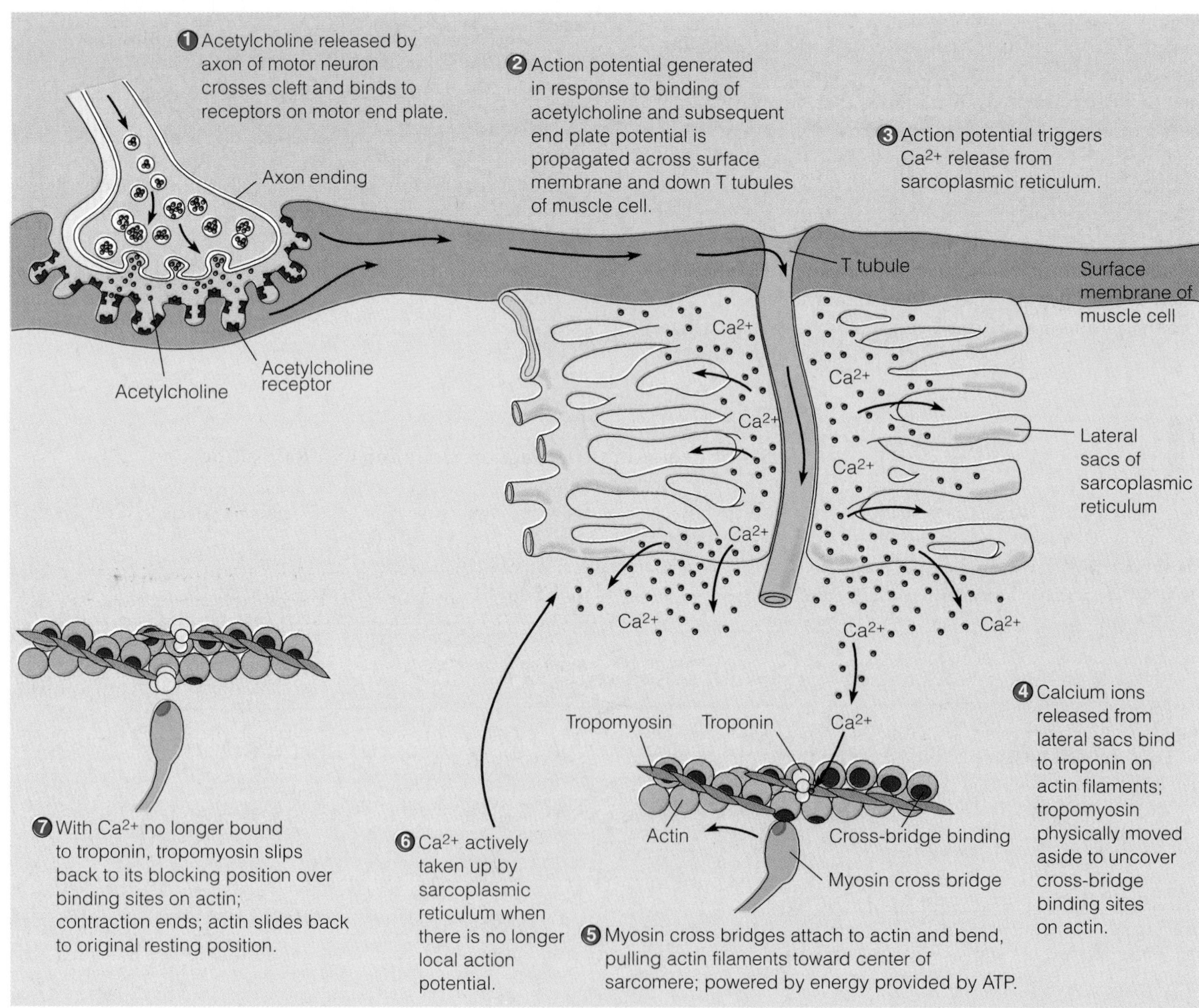

amino acid chains located on the opposite side of the myosin head from the actin binding site. This cleft is an enzymatic site that can bind the energy carrier *adenosine triphosphate (ATP)* and split it into *adenosine diphosphate (ADP)* and *inorganic phosphate* (P_i), yielding energy in the process. In skeletal muscle, magnesium (Mg^{2+}) must be attached to ATP before myosin ATPase can split the ATP. The breakdown of ATP occurs on the myosin cross bridge before the bridge ever links with an actin molecule (step 1 in Fig. 8-12). The ADP and P_i remain tightly bound to the myosin, and the generated energy is stored within the cross bridge to produce a high-energy form of myosin. To use an analogy, the cross bridge is "cocked" like a gun, ready to be fired when the trigger is pulled. When the muscle fiber is excited, Ca^{2+} pulls the troponin-tropomyosin complex out of its blocking position so that the energized (cocked) myosin cross bridge can bind with an actin molecule (step 2a). This contact between myosin and actin "pulls the trigger," causing the cross-bridge

Figure 8-12
Cross-Bridge Cycle

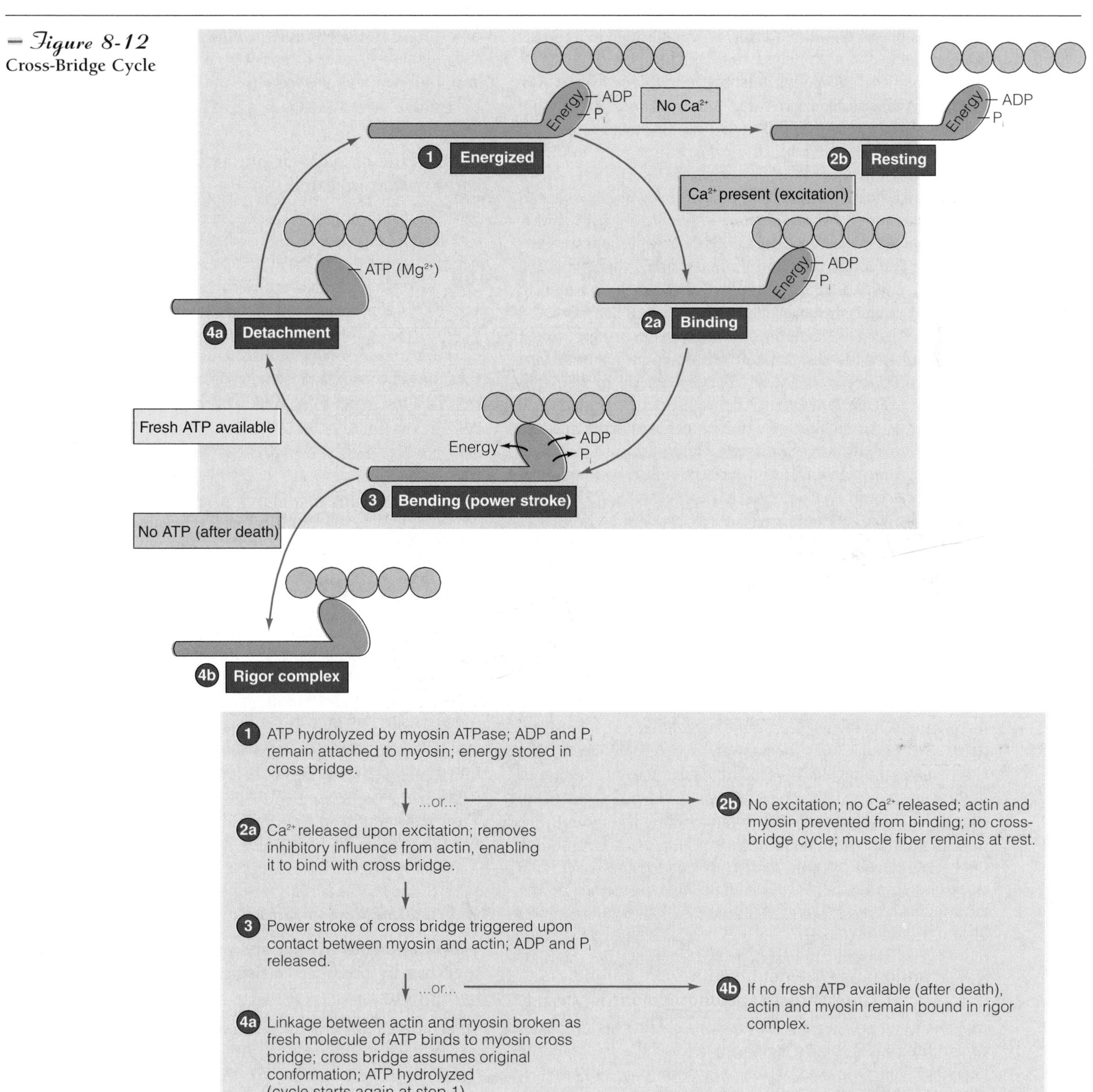
Energy
ADP
P_i
No Ca^{2+}
1 Energized
2b Resting
Ca^{2+} present (excitation)
ATP (Mg^{2+})
4a Detachment
2a Binding
Fresh ATP available
Energy
3 Bending (power stroke)
No ATP (after death)
4b Rigor complex
1 ATP hydrolyzed by myosin ATPase; ADP and P_i remain attached to myosin; energy stored in cross bridge.
...or...
2a Ca^{2+} released upon excitation; removes inhibitory influence from actin, enabling it to bind with cross bridge.
2b No excitation; no Ca^{2+} released; actin and myosin prevented from binding; no cross-bridge cycle; muscle fiber remains at rest.
3 Power stroke of cross bridge triggered upon contact between myosin and actin; ADP and P_i released.
...or...
4a Linkage between actin and myosin broken as fresh molecule of ATP binds to myosin cross bridge; cross bridge assumes original conformation; ATP hydrolyzed (cycle starts again at step 1).
4b If no fresh ATP available (after death), actin and myosin remain bound in rigor complex.

bending responsible for the power stroke that pulls the thin actin filament inward toward the center of the sarcomere (step 3). When the muscle is not excited and Ca^{2+} is not released, troponin and tropomyosin remain in their blocking position, so that actin and the myosin cross bridges do not bind and no power stroking takes place (step 2b).

The changes in shape of the myosin cross bridge during the cross-bridge cycle have recently been traced to the alternate widening and narrowing of the ATP binding cleft. The narrow cleft widens as the cross bridge first binds with ATP and narrows again as the cross bridge binds with actin. When the cleft is open on ATP binding, the cross bridge is "unbent" and strained, like a stretched rubber band. When the cleft closes again on binding with actin, the products of the ATP splitting, ADP and P_i, are released from the cross bridge (step 3). Loss of P_i triggers a conformational (shape) change that releases the strain, allowing the cross bridge to rebound or "bend," which produces the power stroke.

When ADP and inorganic phosphate are rapidly released from myosin upon contact with actin during a power stroke, the myosin ATPase site is free for attachment of another ATP molecule. The actin and myosin remain linked together at the cross bridge until a fresh molecule of ATP attaches to myosin at the end of the power stroke. Attachment of the new ATP molecule permits detachment of the cross bridge, which returns to its open-cleft, unbent conformation, ready to start another cycle (step 4a). The newly attached ATP is then split by myosin ATPase, energizing the myosin cross bridge once again (step 1). Upon binding with another actin molecule, the energized cross bridge again bends, and so on, successively pulling the thin filament inward to accomplish contraction.

Note that fresh ATP must attach to myosin to permit the cross-bridge link between myosin and actin to be broken at the end of a cycle, even though the ATP is not split during this dissociation process. The necessity for ATP in the separation of myosin and actin is amply demonstrated by the phenomenon of **rigor mortis.** This "stiffness of death" is a generalized locking in place of the skeletal muscles that begins three to four hours after death and becomes complete in about twelve hours. Following death, the cytosolic concentration of Ca^{2+} begins to rise, most likely because the inactive muscle cell membrane is unable to keep out extracellular Ca^{2+} and perhaps also because Ca^{2+} leaks out of the lateral sacs. This Ca^{2+} moves the regulatory proteins aside, permitting actin to bind with the myosin cross bridges, which were already charged with ATP before death. Since dead cells cannot produce any more ATP, actin and myosin, once bound, are unable to detach because of the absence of fresh ATP. The thick and thin filaments thus remain linked together by the immobilized cross bridges, resulting in the stiffened condition of dead muscles (step 4b). During the next several days, rigor mortis gradually subsides as the proteins involved in the rigor complex begin to degrade.

How is **relaxation** normally accomplished in a living muscle? Just as an action potential in a muscle fiber turns on the contractile process by triggering the release of Ca^{2+} from the lateral sacs into the cytosol, the contractile process is turned off when Ca^{2+} is returned to the lateral sacs upon cessation of local electrical activity. The sarcoplasmic reticulum possesses an energy-consuming carrier, a Ca^{2+}-ATPase pump, which actively transports Ca^{2+} from the cytosol and concentrates it in the lateral sacs. When acetylcholinesterase removes ACh from the neuromuscular junction, the muscle fiber action potential ceases. When there is no longer a local action potential in the T tubules to trigger the release of Ca^{2+}, the ongoing activity of the sarcoplasmic reticulum's Ca^{2+} pump returns the released Ca^{2+} back into its lateral sacs. Removal of cytosolic Ca^{2+} allows the troponin-tropomyosin complex to slip back into its blocking position, so that actin and myosin are no longer able to bind at the cross bridges. The thin filaments, freed from cycles of cross-bridge attachment and pulling, are able to return to their resting position. Relaxation has occurred.

Contractile activity far outlasts the electrical activity that initiated it.

A single action potential in a skeletal muscle fiber lasts only 1 to 2 msec. The onset of the resultant contractile response lags behind the action potential because the entire excitation-contraction coupling process must take place before cross-bridge activity begins. In fact, the action potential is completed before the contractile apparatus even becomes operational. This time delay of a few milliseconds between stimulation and the onset of contraction is known as the **latent period** (Fig. 8-13). Time is also required for the generation of tension within the muscle fiber produced by means of the sliding interactions between the thick and thin filaments through cross-bridge activity. The time from the onset of contraction until peak tension is developed—the **contraction time**—averages about 50 msec, although this time varies, depending on the type of muscle fiber. The contractile response does not cease until the lateral sacs have taken up all of the Ca^{2+} released in response to the action potential. This reuptake of Ca^{2+} is also time-consuming. Even after Ca^{2+} is removed, it takes time for the filaments to return to their resting positions. The time from peak tension until relaxation is complete, the **relaxation time,** usually lasts slightly longer than contraction time, another 50 msec or more. Consequently, the entire contractile response to a single action potential may last up to 100 msec or more; this is considerably longer than the duration of the action potential that initiated it (100 msec compared to 1 to 2 msec). This fact is important in the body's ability to produce muscle contractions of variable strength, as you will discover in the next section.

Skeletal Muscle Mechanics

Whole muscles are groups of muscle fibers bundled together by connective tissue and attached to bones by tendons.

Thus far we have described the contractile response in a single muscle fiber. In the body, groups of muscle fibers are organized into whole muscles. We will now turn our attention to contraction of whole muscles. Each person has about 600 skeletal muscles, which range in size from the delicate external eye muscles that control eye movements and contain only a few hundred fibers to the large, powerful leg muscles that contain several hundred thousand fibers.

Each muscle is covered by a sheath of connective tissue that penetrates from the surface into the muscle to envelop each individual fiber and divide the muscle into columns or bundles. The connective tissue extends beyond the ends of the muscle to form tough, collagenous **tendons** that attach the muscle to bones. A tendon may be quite long, attaching to a bone some distance from the fleshy portion of the muscle. For example, some of the muscles involved in finger movement are found in the forearm, with long tendons extending down to attach to the bones of the fingers. (You can readily observe the movement of these tendons on the top of your hand when you wiggle your fingers.) This arrangement permits greater dexterity; the fingers would be much thicker and more awkward if all the muscles involved in finger movement were actually located in the fingers.

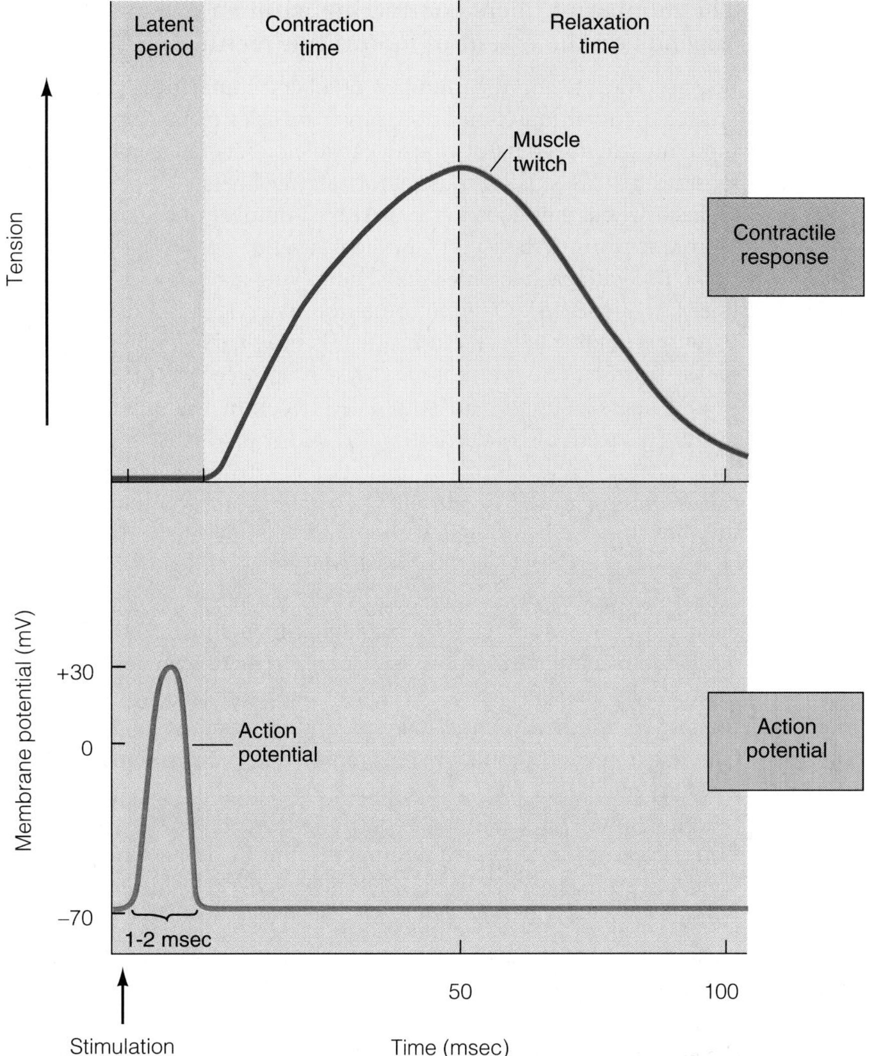

The duration of the action potential is not drawn to scale but is exaggerated.

Figure 8-13 **Relationship of an Action Potential to the Resultant Muscle Twitch**

Contractions of a whole muscle can be of varying strength.

A single action potential in a muscle fiber produces a brief, weak contraction known as a **twitch,** which is too short and too weak to be useful and normally does not take place in the body. Muscle fibers are arranged into whole muscles where they can function cooperatively to produce contractions of variable grades of strength stronger than a twitch. In other words, the force exerted by the same muscle can be made to vary, depending on whether the person is picking up a piece of paper, a book, or a 50-pound weight. Two primary factors can be adjusted to accomplish gradation of whole-muscle tension: (1) *the number of muscle fibers contracting within a muscle* and (2) *the tension developed by each contracting fiber* (Table 8-2). We will discuss each of these factors in turn.

Table 8-2 **Determinants of Whole-Muscle Tension in Skeletal Muscle**

Number of Fibers Contracting	*Tension Developed By Each Contracting Fiber*
Number of motor units recruited*	Frequency of stimulation (twitch summation and tetanus)*
Number of muscle fibers per motor unit	Length of fiber at onset of contraction (length-tension relationship)
Number of muscle fibers available to contract	Extent of fatigue
Size of muscle (number of muscle fibers in muscle)	Duration of activity
Presence of disease (e.g., muscular dystrophy)	Amount of asynchronous recruitment of motor units
Extent of recovery from traumatic losses	Type of fiber (fatigue-resistant oxidative or fatigue-prone glycolytic)
	Thickness of fiber
	Type of fiber (small-diameter oxidative or large-diameter glycolytic)
	Pattern of neural activity (hypertrophy, atrophy)
	Amount of testosterone (larger fibers in males than females)

*Factors controlled to accomplish gradation of contraction.

The number of fibers contracting within a muscle depends on the extent of motor unit recruitment.

Because the greater the number of fibers contracting, the greater the total muscle tension, larger muscles consisting of more muscle fibers are obviously capable of generating more tension than are smaller muscles with fewer fibers.

Each whole muscle is innervated by a number of different motor neurons. When a motor neuron enters a muscle, it branches, with each axon terminal supplying a single muscle fiber (■ Fig. 8-14). One motor neuron innervates a number of muscle fibers, but each muscle fiber is supplied by only one motor neuron. When a motor neuron is activated, all of the muscle fibers it supplies are stimulated to contract simultaneously. This team of concurrently activated components—one motor neuron plus all of the muscle fibers it innervates—is called a **motor unit.** The muscle fibers that compose a motor unit are dispersed throughout the whole muscle; thus, their simultaneous contraction results in an evenly distributed, although weak, contraction of the whole muscle. Each muscle consists of a number of intermingled motor units. For a weak contraction of the whole muscle, only one or a few of its motor units are activated. For stronger and stronger contractions, more and more motor units are *recruited,* or stimulated to contract, a phenomenon known as **motor unit recruitment.**

How much stronger the contraction will be with the recruitment of each additional motor unit depends on the size of the motor units (that is, the number of muscle fibers controlled by a single motor neuron) (■ Fig. 8-15). The number of muscle fibers per motor unit and the number of motor units per muscle vary widely, depending on the specific function of the muscle. For muscles that produce precise, delicate movements, such as the external eye muscles and the hand muscles, a single motor unit may contain as few as a dozen muscle fibers. Because so few muscle fibers are involved with each motor unit, recruitment of each additional motor unit results in only a small additional increment in the whole muscle's strength of contraction. These small motor units allow a very fine degree of control over muscle tension. In contrast, in muscles designed for powerful, coarsely controlled movement, such as those of the legs, a single motor unit may contain 1,500 to 2,000 muscle fibers. Recruitment of motor units in these muscles results in large incremental increases in whole-muscle tension. More powerful contractions occur at the expense of less precisely controlled gradations. Thus, the number of muscle fibers participating in the whole muscle's total contractile effort depends on the number of motor units recruited and the number of muscle fibers per motor unit in that muscle.

To delay or prevent **fatigue** (inability to maintain muscle tension at a given level) during a sustained contraction involving only a portion of a muscle's motor units, as is necessary in muscles supporting the weight of the body against the force of gravity, **asynchronous recruitment of motor units** takes place. The body alternates motor unit activity, like shifts at a factory, to give motor units that have been active an opportunity to rest while others take over. Changing of the shifts is carefully coordinated, so that the sustained contraction is smooth rather than jerky. Asynchronous motor unit recruitment is possible only for submaximal contractions, during which only some of the motor units are required to maintain the desired level of tension. During maximal contractions, when participation of all the muscle fibers is essential, it is impossible to alternate motor unit activity to prevent fatigue. This is one reason why you cannot support a heavy object as long as one that is light.

Furthermore, the type of muscle fiber that is activated varies with the extent of gradation. Most muscles consist of a mixture of fiber types that differ metabolically, some being more resistant to fatigue than others. During weak or moderate endurance-type activities (aerobic exercise), the motor units most resistant to fatigue are recruited first. The last fibers to be called into play in the face of demands for further increases in tension are those that fatigue rapidly. An individual can therefore engage in endurance activities for prolonged periods of time but can only briefly maintain bursts of all-out, powerful effort. Of course, even the muscle fibers most resistant to fatigue will eventually fatigue if required to maintain a certain level of sustained tension.

■ *Figure 8-14* **Schematic Representation of Motor Units in a Skeletal Muscle**

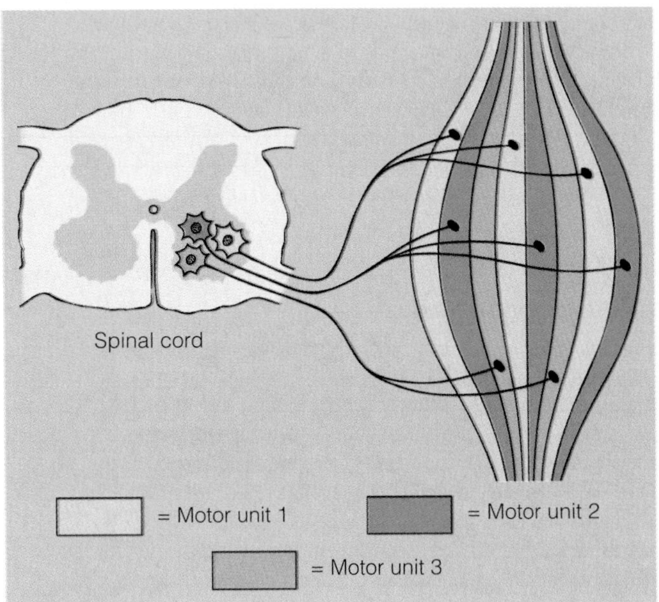

The frequency of stimulation can influence the tension developed by each muscle fiber.

Whole-muscle tension depends not only on the number of muscle fibers contracting but also on the tension developed by each contracting fiber. Various factors influence the extent to which tension can be developed. These factors include:

1. The frequency of stimulation
2. The length of the fiber at the onset of contraction
3. The extent of fatigue
4. The thickness of the fiber

We will now examine the effect of frequency of stimulation. (The other factors will be discussed in later sections.)

Even though a single action potential in a muscle fiber produces only a twitch, contractions with longer duration and greater tension can be achieved by repetitive stimulation of the fiber. Let us see what happens when a second action potential occurs in a muscle fiber. If the muscle fiber has completely relaxed before the next action potential takes place, a second twitch of the same magnitude as the first occurs (Fig. 8-16a). The same excitation-contraction events take place each time, resulting in identical twitch responses. If, however, the muscle fiber is stimulated a second time before it has completely relaxed from the first twitch, a second action potential occurs that causes a second contractile response, which is added "piggyback" on top of the first twitch (Fig. 8-16b). The two twitches resulting from the two action potentials add together, or sum, to produce greater tension in the fiber than that produced by a single action potential. This **twitch summation** is similar to temporal summation of EPSPs at the postsynaptic neuron (see p. 99). Twitch summation is possible only because the duration of the action potential (1 to 2 msec) is much shorter than the duration of the resultant twitch (100 msec). Remember that once an action potential has been initiated, a brief refractory period occurs during which another action potential cannot be initiated (see p. 93). It is therefore impossible to achieve summation of action potentials. The membrane must return to resting

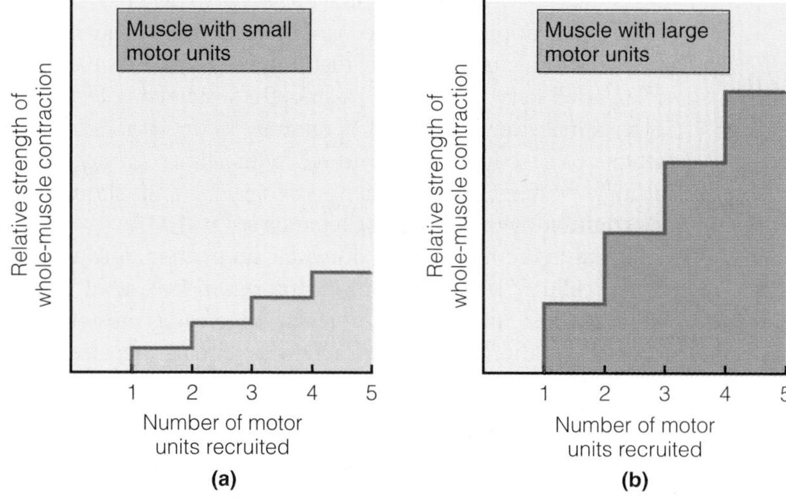

Figure 8-15 **Comparison of Motor Unit Recruitment in Muscles with Small Motor Units and Muscles with Large Motor Units** (a) Small incremental increases in strength of contraction occur during motor unit recruitment in muscles with small motor units because only a few additional fibers are called into play as each motor unit is recruited. (b) Large incremental increases in strength of contraction occur during motor unit recruitment in muscles with large motor units because so many additional fibers are stimulated with the recruitment of each additional motor unit.

Figure 8-16 **Summation and Tetanus** (a) If a muscle fiber is restimulated after it has completely relaxed, the second twitch is the same magnitude as the first twitch. (b) If a muscle fiber is restimulated before it has completely relaxed, the second twitch is added on to the first twitch, resulting in summation. (c) If a muscle fiber is stimulated so rapidly that it does not have an opportunity to relax at all between stimuli, a maximal sustained contraction known as tetanus occurs.

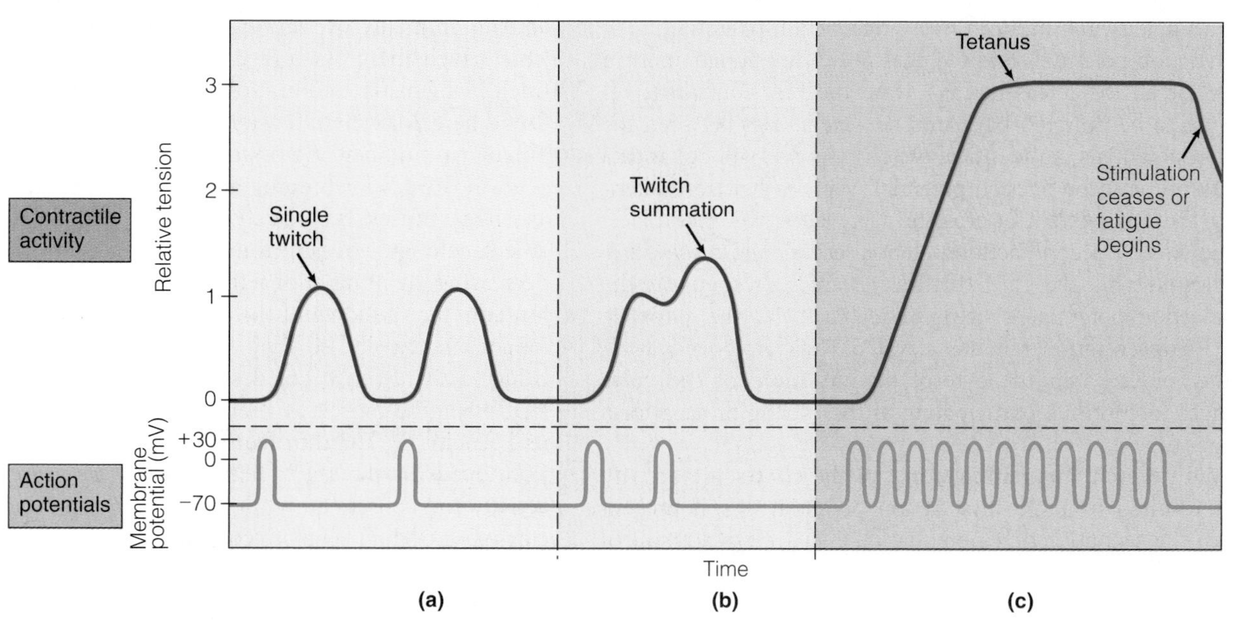

potential and recover from its refractory period before another action potential can occur. However, because the action potential and refractory period are over long before the resultant muscle twitch is completed, the muscle fiber may be restimulated while some contractile activity still exists to produce summation of the mechanical response. If the muscle fiber is stimulated so rapidly that it does not have a chance to relax at all between stimuli, a smooth, sustained contraction of maximal strength known as **tetanus** occurs (Fig. 8-16c). A tetanic contraction is usually three to four times stronger than a single twitch. (This normal physiological tetanus should not be confused with the disease tetanus; see p. 103.)

What is the mechanism of twitch summation and tetanus at the cellular level? The tension produced by a contracting muscle fiber increases as a result of greater cross-bridge cycling. As the frequency of action potentials increases, the resultant tension development increases until a maximum tetanic contraction is achieved. Sufficient Ca^{2+} is released in response to a single action potential to interact with all of the troponin within the cell. As a result, all the cross bridges are free to participate in the contractile response. How then, can repetitive action potentials bring about a greater contractile response? The difference depends on how long sufficient Ca^{2+} is available.

The cross bridges will remain active and continue to cycle as long as sufficient Ca^{2+} is present to keep the troponin-tropomyosin complex away from the cross-bridge binding sites on actin. Each troponin-tropomyosin complex spans a distance of seven actin molecules. Thus, binding of one Ca^{2+} ion to one troponin molecule leads to the uncovering of only seven cross-bridge binding sites on the thin filament.

As soon as Ca^{2+} is released in response to an action potential, the sarcoplasmic reticulum starts pumping Ca^{2+} back into the lateral sacs. As the cytosolic Ca^{2+} concentration declines with the reuptake of Ca^{2+} by the lateral sacs, less Ca^{2+} is present to bind with troponin, so some of the troponin-tropomyosin complexes slip back into their blocking positions. Consequently, not all of the cross-bridge binding sites remain available to participate in the cycling process during a single twitch induced by a single action potential.

If action potentials and twitches occur far enough apart in time for all the released Ca^{2+} from the first contractile response to be pumped back into the lateral sacs between the action potentials, an identical twitch response will occur as a result of the second action potential. With both action potentials, the same extent of cross-bridge cycling will take place. If, however, a second action potential occurs and more Ca^{2+} is released while the Ca^{2+} that was released in response to the first action potential is being taken back up, the cytosolic Ca^{2+} concentration remains elevated. This prolonged availability of Ca^{2+} in the cytosol permits more of the cross bridges to continue participating in the cycling process for a longer time.

With twitch summation, some of the contractile activity that resulted from the first action potential is still present when the second action potential takes place. As a result of another spurt of Ca^{2+} release in response to the second action potential, the magnitude of cross-bridge cycling and tension development increase correspondingly. As the frequency of action potentials increases, the duration of elevated cytosolic Ca^{2+} concentration increases, and contractile activity likewise increases until a maximum tetanic contraction is reached. With tetanus, the maximum number of cross-bridge binding sites remain uncovered so that cross-bridge cycling, and consequently tension development, are at their peak.

Because skeletal muscle must be stimulated by motor neurons to contract, the nervous system plays a key role in regulating the strength of contraction. The two main factors subject to control to accomplish gradation of contraction are the *number of motor units stimulated* and the *frequency of their stimulation.* The areas of the brain responsible for directing motor activity use a combination of tetanic contractions and precisely timed shifts of asynchronous motor unit recruitment to execute smooth rather than jerky contractions.

Additional factors not directly under nervous control also influence the tension developed during contraction (Table 8-2). Among these is the length of the fiber at the onset of contraction, to which we now turn our attention.

There is an optimal muscle length at which maximal tension can be developed upon a subsequent contraction.

A relationship exists between the length of the muscle before the onset of contraction and the tetanic tension that each contracting fiber can subsequently develop at that length. For every muscle there is an **optimal length (l_o)** at which maximal force can be achieved upon a subsequent tetanic contraction. The tension that can be achieved during tetanus at the optimal muscle length is greater than the tetanic tension that can be achieved when the contraction begins with the muscle less than or greater than its optimal length. This **length-tension relationship** can be explained by the sliding-filament mechanism of muscle contraction. At l_o when maximum tension can be developed (point A in Fig. 8-17), the thin filaments optimally overlap the regions of the thick filaments from which the cross bridges project. The central region of thick filaments is void of cross bridges; only myosin tails are found here. Maximum tension can be developed at l_o because a maximal number of cross-bridge sites are accessible to the actin molecules for binding and bending. At greater lengths, as when a muscle is passively stretched (point B), the thin filaments are pulled out from between the thick filaments, decreasing the number of actin sites available for cross-bridge binding; that is, some of the actin sites and cross bridges no longer "match up" so they "go unused." When less cross-bridge activity can occur, less tension can be developed. In fact, when the muscle is stretched to about 70% longer than its l_o (point C), the thin filaments are completely pulled out from between the thick filaments so that no cross-bridge activity and consequently no contraction can occur. If a muscle is shorter than l_o before contraction (point D), less tension can be developed for four reasons:

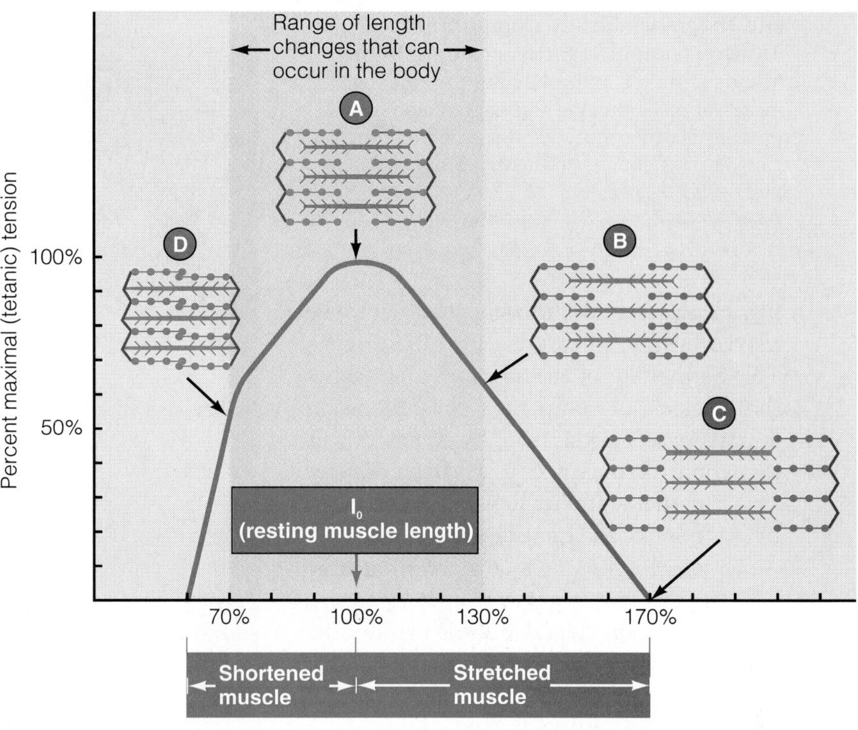

Figure 8-17 **Length-Tension Relationship** Maximal tetanic contraction can be achieved when a muscle fiber is at its optimal length (l_o) before the onset of contraction because of optimal overlap of thick-filament cross bridges and thin-filament cross-bridge binding sites (point A). The percentage of maximal tetanic contraction that can be achieved decreases when the muscle fiber is longer or shorter than l_o prior to contraction. When it is longer, fewer thin-filament binding sites are accessible for binding with thick-filament cross bridges because the thin filaments are pulled out from between the thick filaments (points B and C). When the fiber is shorter, fewer thin-filament binding sites are exposed to thick-filament cross bridges because the thin filaments overlap (point D). Also, further shortening and tension development are impeded as the thick filaments become forced against the Z lines (point D). In the body, the resting muscle length is at l_o. Furthermore, because of restrictions imposed by skeletal attachments, muscles cannot vary beyond 30% of their l_o in either direction (the range screened in light green). At the outer limits of this range, muscles are still able to achieve about 50% of their maximal tetanic contraction.

1. The thin filaments from the opposite sides of the sarcomere become overlapped, decreasing the number of actin sites exposed to the cross bridges.
2. The thick filaments become forced against the Z lines, so further shortening is impeded.
3. Besides these two mechanical factors, at muscle lengths less than 80% of l_o, not as much Ca^{2+} is released during excitation-contraction coupling for reasons unknown. Consequently, fewer actin sites are uncovered for participation in cross-bridge activity.
4. Furthermore, fewer of actin's cross-bridge binding sites are uncovered because, by an unknown mechanism, the ability of Ca^{2+} to bind to troponin and pull the troponin-tropomyosin complex aside is reduced at shorter muscle lengths.

The extremes in muscle length that prevent the development of tension occur only under experimental conditions, when a muscle is removed and stimulated at various lengths. In the body the muscles are so positioned that their relaxed length is approximately their optimal length; thus, they are capable of achieving near-maximal tetanic contraction most of the time. Because of limitations imposed by attachment to the skeleton, a muscle cannot be stretched or shortened more than 30% of its resting optimal length, and usually it deviates much less than 30% from normal length. Even at the outer limits (130% and 70% of l_o), the muscles are still able to generate half their maximum tension.

The factors we have discussed thus far that influence how much tension can be developed by a contracting muscle fiber—the frequency of stimulation and the muscle length at the onset of contraction—can vary from contraction to contraction. Other determinants of muscle fiber tension—the metabolic capability of the fiber relative to resistance to fatigue and the thickness of the fiber—do not vary from contraction to contraction but depend on the fiber type and can be modified over a period of time. We will consider these other factors in the next section on skeletal muscle metabolism and fiber types after we complete our discussion of skeletal muscle mechanics.

The two primary types of contraction are isotonic and isometric.

Tension is produced internally within the sarcomeres, the **contractile component** of the muscle, as a result of cross-bridge activity and the resultant sliding of filaments. However, the sarcomeres are not attached directly to the bones. Instead, the tension generated by these contractile elements must be transmitted to the bone via the connective tissue and tendons before the bone can be moved. Connective tissue, as well as other components of the muscle, such as intracellular elastic proteins, exhibits a certain degree of passive elasticity. These noncontractile tissues are referred to as the **series-elastic component** of the muscle; they behave like a stretchy spring placed between the internal tension-generat-

Figure 8-18 Relationship between the Contractile Component and the Series-Elastic Component in Transmission of Muscle Tension to Bone Muscle tension is transmitted to the bone by means of the stretching and tightening of the muscle's elastic connective tissue and tendon as a result of sarcomere shortening brought about by cross-bridge cycling.

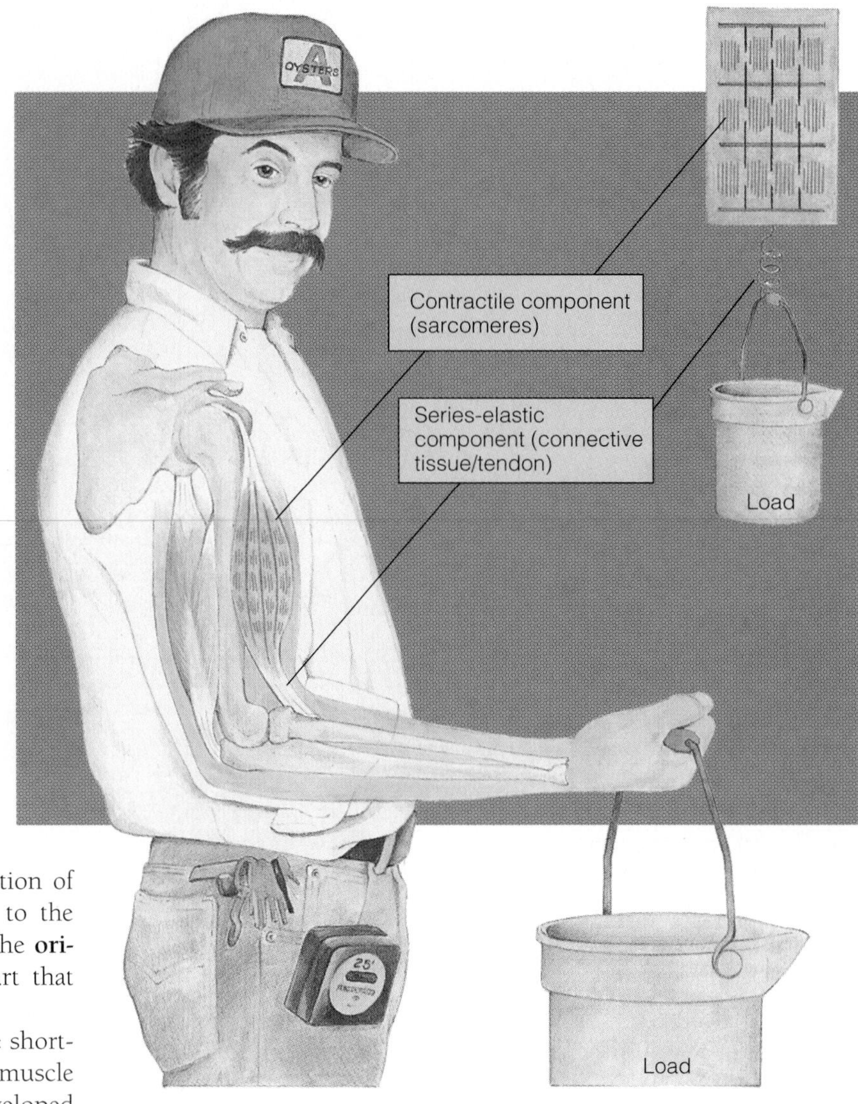

ing elements and the bone that is to be moved against an external load (Fig. 8-18). Shortening of the sarcomeres stretches the series-elastic component. Muscle tension is transmitted to the bone by means of this tightening of the series-elastic component. This externally applied tension is responsible for moving the bone against a load.

Typically, a muscle is attached to at least two different bones across a joint by means of tendons that extend from each end of the muscle (Fig. 8-19). When the muscle shortens during contraction, the position of the joint is changed as one bone is moved in relation to the other—for example, *flexion* of the elbow joint by contraction of the biceps muscle and *extension* of the elbow by contraction of the triceps. The end of the muscle attached to the more stationary part of the skeleton is called the **origin,** and the end attached to the skeletal part that moves is referred to as the **insertion.**

Not all muscle contractions result in muscle shortening and movement of bones, however. For a muscle to shorten during contraction, the tension developed in the muscle must exceed the forces that oppose movement of the bone to which the muscle's insertion is attached. In the case of elbow flexion, the opposing force, or **load,** is the weight of an object being lifted. When you flex your elbow without lifting any external object, there is still a load, albeit a minimal one—the weight of your forearm being moved against the force of gravity.

There are two primary types of contraction, depending on whether the muscle changes length during contraction. In an **isotonic contraction,** muscle tension remains constant as the muscle changes length. In an **isometric contraction,** the muscle is prevented from shortening, so tension develops at constant muscle length. The same internal events occur in both isotonic and isometric contractions: the tension-generating contractile process is turned on by muscle excitation; the cross bridges start cycling; and filament sliding shortens the sarcomeres, which stretches the series-elastic component to exert tension on the bone at the site of the muscle's insertion.

Considering your biceps as an example, assume that you are going to lift an object. When the tension developing in your biceps becomes great enough to overcome the weight of the object in your hand, you can lift the object, with the whole muscle shortening in the process. Because the weight of the object does not change as it is lifted, the muscle tension remains constant throughout the period of shortening. This is an *isotonic* (literally, "constant tension") contraction. Isotonic contractions are used for body movements and for moving external objects. What happens if you try to lift an object too heavy for you (that is, if the tension you are capable of developing in your arm muscles is less than that required to lift the load)? In this case, the muscle cannot shorten and lift the object but remains at constant length in spite of the development of tension, so an *isometric* ("constant length") contraction occurs. In addition to occurring when the load is too great, isometric contractions also take place when the tension developed in the muscle is deliberately less than that needed to move the load. In this case, the goal is to keep the muscle at fixed length although it is capable of developing more tension. These submaximal isometric contractions are important for maintaining posture (such as keeping the legs stiff while standing) and for supporting objects in a fixed position. During a given movement, a muscle may shift between isotonic and isometric contractions. For example, when you pick up a book to read, your biceps undergoes an isotonic contraction while the

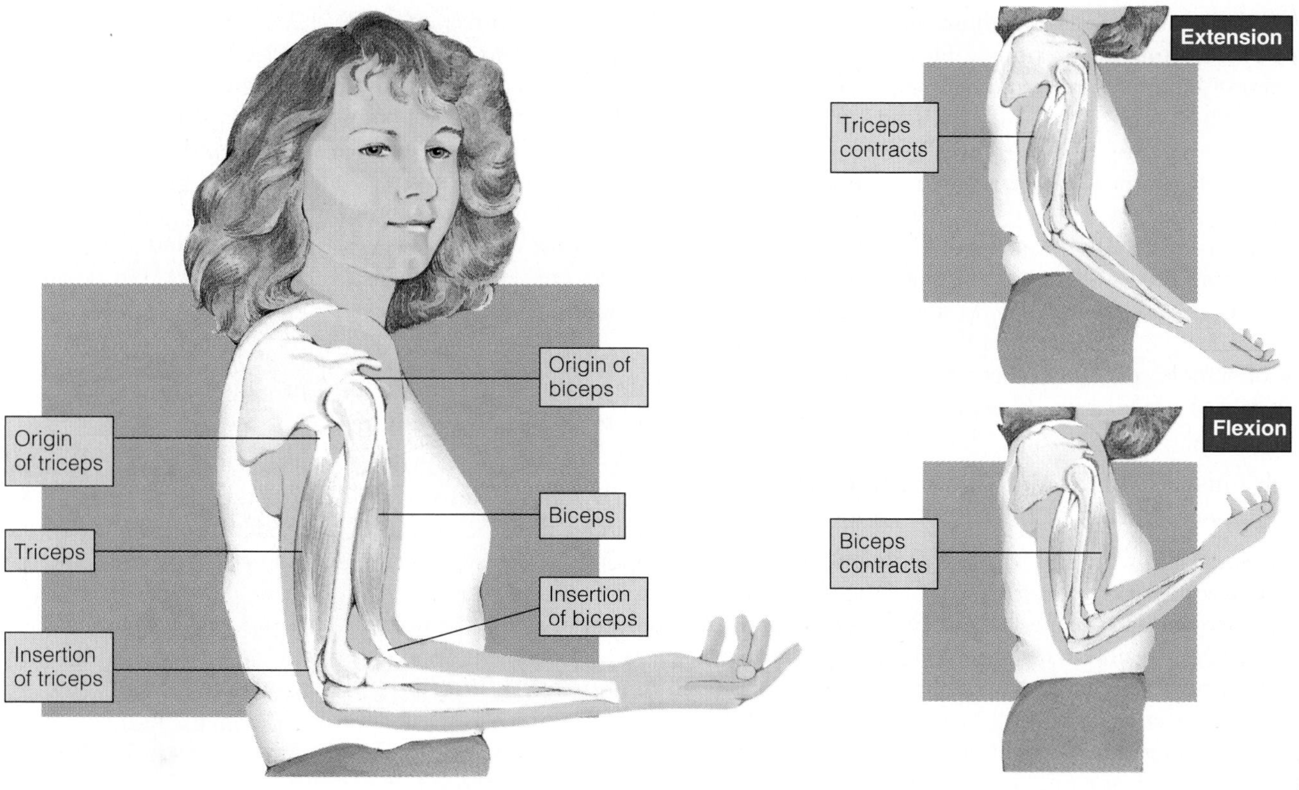

Figure 8-19 Extension and Flexion of the Elbow Joint

book is being lifted, but the contraction becomes isometric as you stop to hold the book in front of you.

There are actually two types of isotonic contraction—**concentric** and **eccentric.** In both cases the muscle changes length at constant tension. With concentric contractions, however, the muscle shortens, whereas with eccentric contractions the muscle lengthens because it is being stretched by an external force while contracting. With an eccentric contraction, the contractile activity is resisting the stretch. An example is lowering a load to the ground. During this action, the muscle fibers in the biceps are lengthening but are still contracting in opposition to being stretched. This tension supports the weight of the object.

The body is not limited to pure isotonic and isometric contractions. Muscle length and tension frequently vary throughout a range of motion. Think about drawing a bow and arrow. The tension of the biceps muscle continuously increases to overcome the progressively increasing resistance as the bow is stretched further. At the same time, the muscle progressively shortens as the bow is drawn farther back. Such a contraction occurs at neither constant tension nor constant length.

Some skeletal muscles do not attach to bones at both ends but still produce movement. For example, the muscles of the tongue are not attached at the free end. Isotonic contractions of the tongue muscles maneuver the free unattached portion of the tongue to facilitate speech and eating. The external eye muscles attach to the skull at their origin but to the eye at their insertion. Isotonic contractions of these muscles produce the eye movements that enable us to track moving objects, read, and so on. A few skeletal muscles that are completely unattached to bone actually prevent movement. These are the voluntarily controlled rings of skeletal muscles known as **sphincters** that guard the exit of urine and feces from the body by isotonically contracting.

The velocity of shortening is related to the load.

The load is also an important determinant of the **velocity,** or speed, of shortening (Fig. 8-20). The greater the load, the

Figure 8-20 **Load-Velocity Relationship** The velocity of shortening decreases as the load increases.

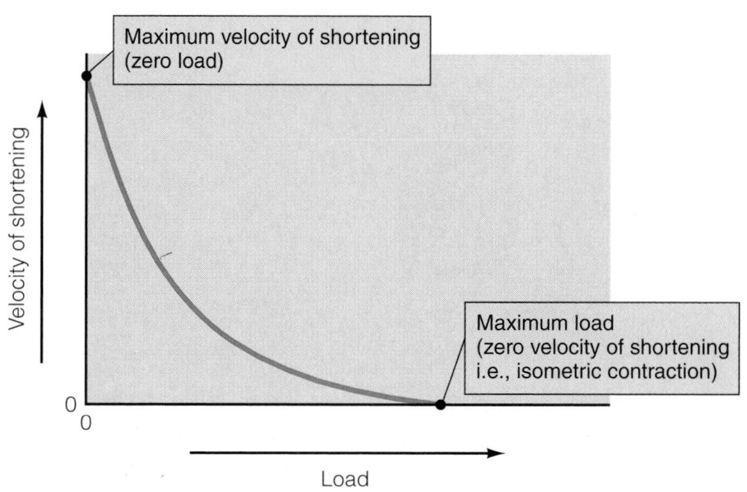

lower the velocity at which a single muscle fiber (or a constant number of contracting fibers within a muscle) shortens during an isotonic tetanic contraction. The velocity of shortening is maximal when there is no external load, progressively decreases with an increasing load, and falls to zero (no shortening—isometric contraction) when the load cannot be overcome by maximal tetanic tension. You have frequently experienced this load-velocity relationship. You can lift light objects requiring little muscle tension quickly, whereas you can lift very heavy objects only slowly, if at all. This relationship between load and shortening velocity is a fundamental property of muscle, presumably because it takes the cross bridges longer to stroke against a greater load.

Although muscles can accomplish work, much of the energy is converted to heat.

Muscle accomplishes work in a physical sense only when an object is moved. **Work** is defined as force times distance. **Force** can be equated to the muscle tension required to overcome the load (the weight of the object). The amount of work accomplished by a contracting muscle therefore depends on how much an object weighs and how far it is moved. In an isometric contraction when no object is moved, the muscle contraction's efficiency as a producer of external work is zero. All energy consumed by the muscle during the contraction is converted to heat. In an isotonic contraction, the muscle's efficiency is about 25%. Of the energy consumed by the muscle during the contraction, 25% is realized as external work whereas the remaining 75% is converted to heat. Much of this heat is not really wasted in a physiological sense because it is used in maintaining the body temperature. In fact, shivering, a form of involuntarily induced skeletal muscle contraction, is a well-known mechanism for increasing heat production on a cold day. Heavy exercise on a hot day, on the other hand, may overheat the body, because the normal heat-loss mechanisms may be unable to compensate for this increase in heat production (chapter 17).

Interactive units of skeletal muscles, bones, and joints form lever systems.

Skeletal muscles are attached to bones across joints, forming lever systems. A **lever** is a rigid structure capable of moving around a pivot point known as a **fulcrum.** In the body, the bones function as levers, the joints serve as fulcrums, and the skeletal muscles provide the force to move the bones. The portion of a lever between the fulcrum and the point where an upward force is applied is called the **power arm;** the portion between the fulcrum and the downward force exerted by a load is known as the **load arm** (Fig. 8-21a).

The most common type of lever system in the body is exemplified by flexion of the elbow joint. Skeletal muscles, such as the biceps whose contraction flexes the elbow joint, consist of many parallel (side-by-side) tension-generating fibers that can exert a large force at their insertion but shorten

Figure 8-21 **Lever Systems of Muscles, Bones, and Joints** (a) Schematic representation of the most common type of lever system in the body, showing the location of the fulcrum, the upward force, the downward force, the power arm, and the load arm. (b) Flexion of the elbow joint as an example of lever action in the body. Note that the lever ratio (length of the power arm to length of the load arm) is 1:7 (5 cm:35 cm), which amplifies the distance and velocity of movement seven times [distance moved by the muscle (extent of shortening) = 1 cm, distance moved by the hand = 7 cm; velocity of muscle shortening = 1 cm/unit of time, hand velocity = 7 cm/unit of time], but at the expense of the muscle having to exert seven times the force of the load (muscle force = 35 kg, load = 5 kg).

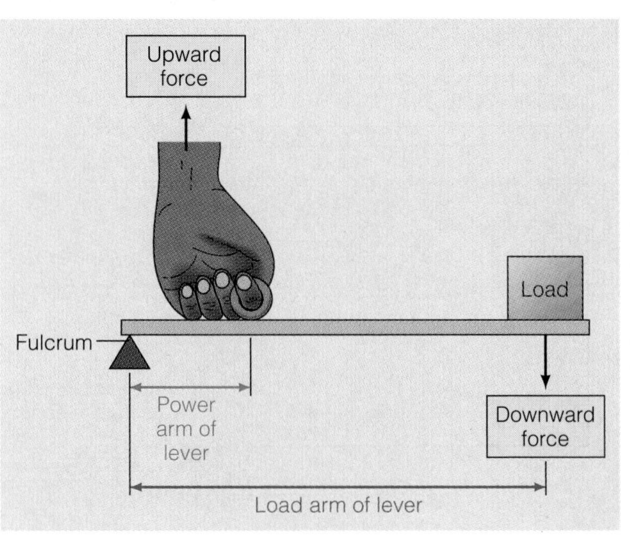

(a)

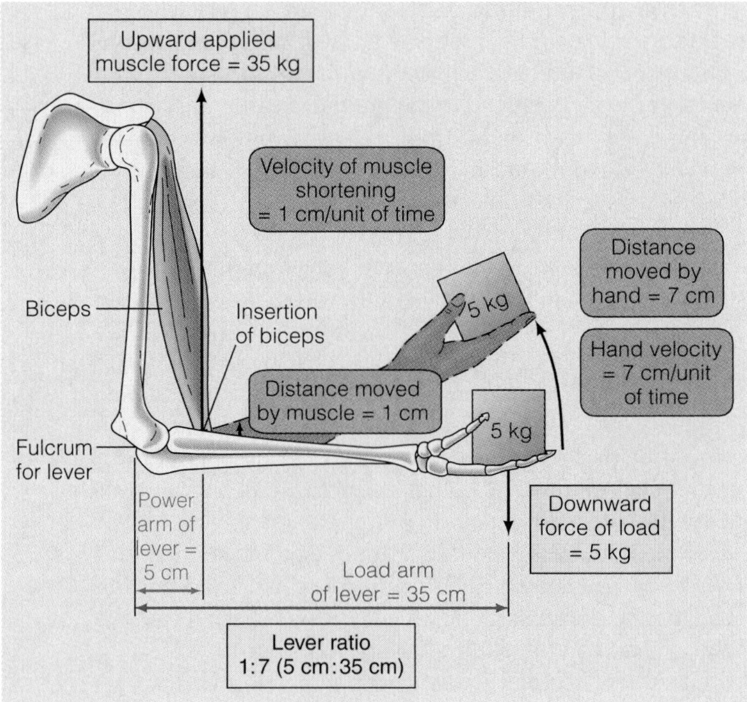

(b)

only a small distance and at relatively slow velocity. The lever system of the elbow joint amplifies the slow, short movements of the biceps to produce more rapid movements of the hand that cover a greater distance. Consider how an object weighing 5 kg is lifted by the hand (Fig. 8-21b). When the biceps contracts, it exerts an upward force at the point where it inserts on the forearm bone about 5 cm away from the elbow joint, the fulcrum. Thus, the power arm of this lever system is 5 cm long. The length of the load arm, the distance from the elbow joint to the hand, averages 35 cm. In this case, the load arm is seven times longer than the power arm, which enables the load to be moved a distance seven times greater than the shortening distance of the muscle (while the biceps shortens a distance of 1 cm, the hand moves the load a distance of 7 cm) and at a velocity seven times greater (the hand moves 7 cm during the same length of time the biceps shortens 1 cm). The disadvantage of this lever system is that at its insertion the muscle must exert a force seven times greater than the load. The product of the length of the power arm times the upward force applied must equal the product of the length of the load arm times the downward force exerted by the load. Because the load arm times the downward force is 35 cm × 5 kg, the power arm times the upward force must be 5 cm × 35 kg (the force that must be exerted by the muscle to be in mechanical equilibrium). Thus, skeletal muscles typically work at a mechanical disadvantage in that they must exert a considerably greater force than the actual load to be moved. Nevertheless, the amplification of velocity and distance afforded by the lever arrangement enables muscles to move loads faster over greater distances than would otherwise be possible. This amplification provides valuable maneuverability and speed.

Skeletal Muscle Metabolism and Fiber Types

Muscle fibers have alternate pathways for forming ATP.

Three different steps in the contraction-relaxation process require ATP:

1. Splitting of ATP by myosin ATPase provides the energy for the power stroke of the cross bridge.
2. Binding (but not splitting) of a fresh molecule of ATP to myosin permits detachment of the bridge from the actin filament at the end of a power stroke so that the cycle can be repeated. This ATP is subsequently split to provide energy for the next stroke of the cross bridge, as indicated in step 1.
3. The active transport of Ca^{2+} back into the sarcoplasmic reticulum during relaxation depends on energy derived from the breakdown of ATP.

Because ATP is the only energy source that can be directly used for these activities, ATP must constantly be supplied for contractile activity to continue. Only limited stores of ATP are immediately available in muscle tissue, but three pathways supply additional ATP as needed during muscle contraction: (1) transfer of a high-energy phosphate from creatine phosphate to ADP; (2) oxidative phosphorylation (the citric acid cycle and electron transport system); and (3) glycolysis.

Creatine phosphate is the first energy storehouse tapped at the onset of contractile activity (Fig. 8-22 ⓐ). Like ATP, creatine phosphate contains a high-energy phosphate group, which can be donated directly to ADP to form ATP. Just as energy is released when the terminal phosphate bond in ATP is split, similarly energy is released when the bond between phosphate and creatine is broken. The energy released from the hydrolysis of creatine phosphate, along with the phosphate, can be donated directly to ADP to form ATP. This reaction, which is catalyzed by the muscle cell enzyme **creatine kinase,** is reversible; energy and phosphate from ATP can be transferred to creatine to form creatine phosphate:

$$\text{creatine phosphate} + \text{ADP} \xrightleftharpoons{\text{creatine kinase}} \text{creatine} + \text{ATP}$$

As the energy reserves are built up in a resting muscle, the increased concentration of ATP favors the transfer of the high-energy phosphate group to creatine phosphate, in accordance with the law of mass action (see p. 447). Thus, most energy is stored in muscle in creatine phosphate pools. A rested muscle contains about five times as much creatine phosphate as ATP. At the onset of contraction, when the meager reserves of ATP are rapidly utilized, the reaction is reversed. Additional ATP is quickly formed by the transfer of energy and phosphate from creatine phosphate to ADP. Because only one enzymatic reaction is involved in this energy transfer, ATP can be formed rapidly (within a fraction of a second) by using creatine phosphate.

Thus, creatine phosphate is the first source for supplying additional ATP when exercise begins. Muscle ATP levels actually remain fairly constant early in contraction, but creatine phosphate stores become depleted. In fact, short bursts of high-intensity contractile effort, such as high jumps or sprints, are supported primarily by ATP derived at the expense of creatine phosphate. Other energy systems do not have a chance to become operable before the activity is over. If the energy-dependent contractile activity is to be continued, the muscle shifts to the alternate pathways of oxidative phosphorylation and glycolysis to form ATP. These multistepped pathways require time to pick up their rates of ATP formation to match the increased demands for energy, time that has been provided by the immediate supply of energy from the one-step creatine phosphate system.

Oxidative phosphorylation (see p. 32) takes place within the muscle mitochondria if sufficient O_2 is present. This pathway is fueled by glucose or fatty acids, depending on the intensity and duration of the activity (Fig. 8-22 ⓑ). Although it provides a rich yield of thirty-six ATP molecules for each glucose molecule processed, oxidative phosphorylation is relatively slow because of the number of steps involved, and it necessitates a constant supply of O_2 and nutrient fuel.

During light exercise (such as walking) to moderate exercise (such as jogging or swimming), muscle cells are able to form sufficient amounts of ATP through oxidative phosphorylation to keep pace with the modest energy demands of the contractile machinery for prolonged periods of time. To sustain ongoing oxidative phosphorylation, the exercising muscles depend on delivery of adequate O_2 and nutrient supplies via the circulatory system to maintain their activity. Activity that can be supported in this way is known as **endurance-type exercise,** or **aerobic** ("with O_2") **exercise.**

The O_2 required for oxidative phosphorylation is primarily delivered by the blood. Increased O_2 is made available to muscles during exercise by several mechanisms: deeper, more rapid breathing brings in more O_2; the heart contracts more rapidly and forcefully to pump more oxygenated blood to the tissues; more blood is diverted to the exercising muscles by dilation of the blood vessels supplying them; and the hemoglobin molecules that carry the O_2 in the blood release more O_2 in exercising muscles. (These mechanisms are discussed further in later chapters.) Furthermore, some types of muscle fibers have an abundance of **myoglobin,** which is similar to hemoglobin. Myoglobin can store small amounts of O_2, but more importantly, it increases the rate of O_2 transfer from the blood into muscle fibers.

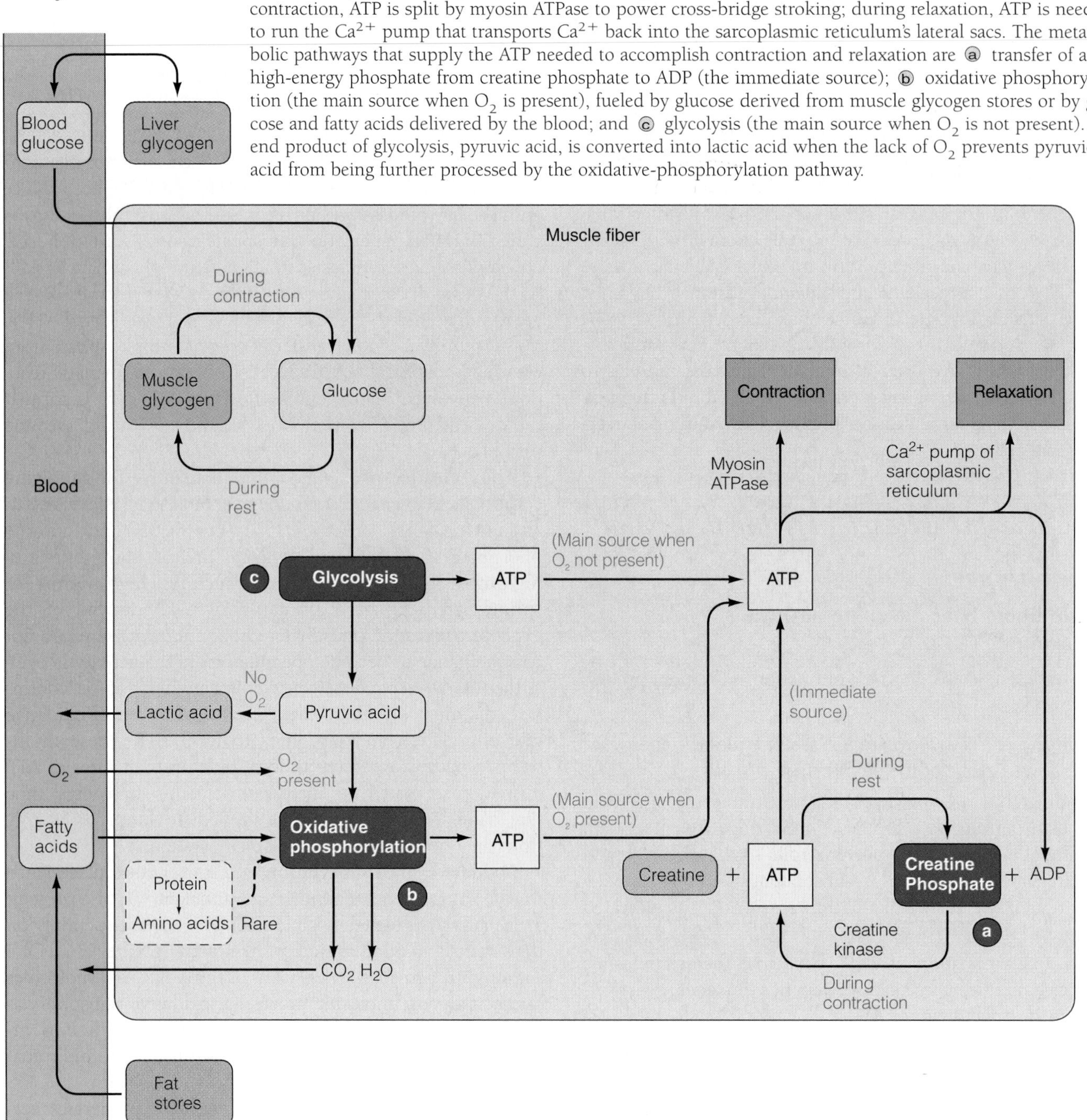

Figure 8-22 **Metabolic Pathways Producing ATP Utilized during Muscle Contraction and Relaxation** During muscle contraction, ATP is split by myosin ATPase to power cross-bridge stroking; during relaxation, ATP is needed to run the Ca^{2+} pump that transports Ca^{2+} back into the sarcoplasmic reticulum's lateral sacs. The metabolic pathways that supply the ATP needed to accomplish contraction and relaxation are ⓐ transfer of a high-energy phosphate from creatine phosphate to ADP (the immediate source); ⓑ oxidative phosphorylation (the main source when O_2 is present), fueled by glucose derived from muscle glycogen stores or by glucose and fatty acids delivered by the blood; and ⓒ glycolysis (the main source when O_2 is not present). The end product of glycolysis, pyruvic acid, is converted into lactic acid when the lack of O_2 prevents pyruvic acid from being further processed by the oxidative-phosphorylation pathway.

Glucose and fatty acids, ultimately derived from ingested food, are also delivered to the muscle cells by the blood. Excess ingested nutrients not immediately utilized are stored in the liver as glycogen (chains of glucose) and in adipose tissue as fat. Additionally, muscle cells are able to store limited quantities of glucose in the form of glycogen.

There are cardiovascular limits to the amount of O_2 that can be delivered to a muscle. In near-maximal contractions, the blood vessels that course through the muscle are compressed almost closed by the powerful contraction, severely limiting the O_2 available to the muscle fibers. Furthermore, even when O_2 is available, the relatively slow oxidative-phosphorylation system may not be able to produce ATP rapidly enough to meet the muscle's needs during intense activity. When O_2 delivery or oxidative phosphorylation cannot keep pace with the demand for ATP formation as the intensity of exercise increases, the muscle fibers rely increasingly on **glycolysis** (Fig. 8-22 ⓒ) to generate ATP (see p. 28). The chemical reactions of glycolysis yield products for ultimate entry into the oxidative-phosphorylation pathway, but glycolysis can also proceed alone in the absence of further processing of its products by oxidative phosphorylation. During glycolysis, a glucose molecule is broken down into two **pyruvic acid** molecules, yielding two ATP molecules in the process. Pyruvic acid can be further degraded by oxidative phosphorylation to extract more energy. However, glycolysis alone has two advantages over the oxidative-phosphorylation pathway: (1) glycolysis can form ATP in the absence of O_2 (operating *anaerobically,* that is, "without O_2"), and (2) it can proceed more rapidly than oxidative phosphorylation because it requires fewer steps. Although glycolysis extracts considerably fewer ATP molecules from each nutrient molecule processed, it can proceed so much more rapidly that it can outproduce oxidative phosphorylation over a given period of time if sufficient glucose is present.

Even though anaerobic glycolysis provides a means of performing intense exercise when the O_2 delivery/oxidative-phosphorylation capacity is exceeded, using this pathway has two consequences. First, large amounts of nutrient fuel must be processed, because glycolysis is much less efficient than oxidative phosphorylation in converting nutrient energy into the energy of ATP. (Glycolysis yields a net of two ATP molecules for each glucose molecule degraded, whereas the oxidative-phosphorylation pathway can extract thirty-six molecules of ATP from each glucose molecule; see p. 33.) Muscle cells are able to store limited quantities of glucose in the form of glycogen, but anaerobic glycolysis rapidly depletes the muscle's glycogen supplies. Second, the end product of anaerobic glycolysis, pyruvic acid, is converted to **lactic acid** when it cannot be further processed by the oxidative-phosphorylation pathway. Lactic acid accumulation has been implicated in the muscle soreness that occurs during the time that intense exercise is actually taking place. (The delayed-onset pain and stiffness that begin the day after unaccustomed muscular exertion, however, are probably caused by reversible structural damage.) Furthermore, lactic acid picked up by the blood is responsible for the metabolic acidosis accompanying intense exercise. Both depletion of energy reserves and the fall in muscle pH caused by lactic acid accumulation are believed to play a role in the onset of muscle fatigue, a topic to which we next turn our attention. Therefore, **high-intensity exercise,** or **anaerobic exercise,** can be sustained for only a short duration in contrast to the body's prolonged ability to sustain aerobic activities.

Fatigue has multiple causes.

Contractile activity in a particular skeletal muscle cannot be maintained at a given level indefinitely. Eventually, the tension in the muscle declines as fatigue sets in. There are apparently three different types of fatigue: muscle fatigue, neuromuscular fatigue, and central fatigue.

Muscle fatigue occurs when an exercising muscle can no longer respond to stimulation with the same degree of contractile activity. The underlying causes of muscle fatigue are unclear. The primary implicated factors include (1) accumulation of lactic acid, which may inhibit key enzymes in the energy-producing pathways or excitation-contraction coupling process, and (2) depletion of energy reserves. The time of onset of fatigue varies with the type of muscle fiber, some fibers being more resistant to fatigue than others, and the intensity of the exercise, with more rapid onset of fatigue associated with high-intensity activities.

Evidence suggests that the limiting factor during fast, powerful activities may be at the neuromuscular junction. With **neuromuscular fatigue,** the active motor neurons are not able to synthesize acetylcholine rapidly enough to sustain chemical transmission of action potentials from the motor neurons to the muscles.

Central fatigue, also known as **psychological fatigue,** occurs when the CNS no longer adequately activates the motor neurons supplying the working muscles. The person slows down or stops exercising even though the muscles are still able to perform. During strenuous exercise, central fatigue may stem from discomfort associated with the activity; it takes strong motivation (a will to win) to deliberately persevere when in pain. In less strenuous activities, central fatigue may bring about reduced physical performance in association with boredom and monotony (such as assembly-line work) or tiredness (lack of sleep). The mechanisms involved in central fatigue are poorly understood.

Increased oxygen consumption is necessary to recover from exercise.

A person continues to breathe deeply and rapidly for a period of time after exercising. The necessity for the elevated O_2 uptake during recovery from exercise is due to a variety of factors. The best known is repayment of an **oxygen debt** that was incurred during exercise, when contractile activity was being supported by ATP derived from nonoxidative sources such as creatine phosphate and anaerobic glycolysis. Oxygen is needed for recovery of the energy systems. During exercise, the creatine phosphate stores of active muscles are reduced, lactic acid may accumulate, and glycogen stores may be tapped; the extent of these effects depends on the intensity and duration of the activity. During the recovery period, fresh supplies of ATP are formed by oxidative phosphorylation using the newly acquired O_2, which is provided by the sustained increase in respiratory activity after exercise has stopped. Most of this ATP is used to resynthesize creatine phosphate to restore its reserves. This can be accomplished in a matter of a few minutes. Any accumulated lactic acid is converted back into pyruvic acid, part of which is used by the oxidative-phosphorylation system for ATP production. The remainder of the pyruvic acid is converted back into glucose by the liver. Most of this glucose in turn is used to replenish the glycogen stores drained from the muscles and liver during exercise. These biochemical transformations involving pyruvic acid require O_2 and take several hours for completion. Thus, as the O_2 debt is repaid, the creatine phosphate system is restored, lactic acid is removed, and glycogen stores are at least partially replenished. Unrelated to increased O_2 uptake is the need to restore nutrients after grueling exercise, such as marathon races, in which glycogen stores are severely depleted. In such cases, long-term recovery can take a day or more because the exhausted energy stores require nutrient intake for full replenishment. Therefore, depending on the type and duration of activity, recovery can be complete within a few minutes or can require more than a day.

Table 8-3
Characteristics of Skeletal Muscle Fibers

Characteristic	*Type of Fiber* Slow-Oxidative (Type I)	Fast-Oxidative (Type IIa)	Fast-Glycolytic (Type IIb)
Myosin-ATPase activity	Low	High	High
Speed of contraction	Slow	Fast	Fast
Resistance to fatigue	High	Intermediate	Low
Oxidative-phosphorylation capacity	High	High	Low
Enzymes for anaerobic glycolysis	Low	Intermediate	High
Mitochondria	Many	Many	Few
Capillaries	Many	Many	Few
Myoglobin content	High	High	Low
Color of fiber	Red	Red	White
Glycogen content	Low	Intermediate	High
Fiber diameter	Small	Intermediate	Large
Intensity of contraction	Low	Intermediate	High

Part of the extra O_2 uptake during recovery is not directly related to the repayment of energy stores but instead is the result of a general metabolic disturbance following exercise. For example, the local increase in muscle temperature arising from heat-generating contractile activity speeds up the rate of all chemical reactions in the muscle tissue, including those dependent on O_2. Likewise, the secretion of epinephrine, a hormone that increases O_2 consumption by the body, is elevated during exercise. Until the circulating level of epinephrine returns to its preexercise state, O_2 uptake will be increased above normal.

There are three types of skeletal muscle fibers based on differences in ATP hydrolysis and synthesis.

Based on their biochemical capacities, there are three major types of muscle fibers (Table 8-3):

1. ***Slow-oxidative (type I) fibers***
2. ***Fast-oxidative (type IIa) fibers***
3. ***Fast-glycolytic (type IIb) fibers***

As their names imply, the two main differences between these fiber types are their speed of contraction (slow or fast) and the type of enzymatic machinery they primarily use for ATP formation (oxidative or glycolytic). Fast fibers have higher myosin ATPase (ATP-splitting) activity than slow fibers. The higher the ATPase activity, the more rapidly ATP is split and the faster the rate at which energy is made available for cross-bridge cycling. The result is a fast twitch, compared to the slower twitches of those fibers that split ATP more slowly. Thus, two factors determine the speed with which a muscle contracts: the load (load-velocity relationship) and the myosin ATPase activity of the contracting fibers (fast or slow twitch).

Fiber types also differ in their ATP-synthesizing ability. Those with a greater capacity to form ATP are more resistant to fatigue. Some fibers are better equipped for oxidative phosphorylation, whereas others rely primarily on anaerobic glycolysis for synthesizing ATP. Because oxidative phosphorylation yields considerably more ATP from each nutrient molecule processed, it does not readily deplete energy stores. Furthermore, it does not result in lactic acid accumulation. Oxidative types of muscle fibers are therefore more resistant to fatigue than are glycolytic fibers.

Other related characteristics distinguishing these three fiber types are summarized in Table 8-3. As would be expected, the oxidative fibers, both slow and fast, contain an abundance of mitochondria, the organelles that house the enzymes involved in the oxidative-phosphorylation pathway. Because adequate oxygenation is essential to support this pathway, these fibers are richly supplied with capillaries. Oxidative fibers also have a high myoglobin content. Myoglobin not only helps support oxidative fibers' O_2 dependency, but it also imparts a red color to them, just as oxygenated hemoglobin is responsible for the red color of arterial blood. Accordingly, these muscle fibers are referred to as **red fibers.**

In contrast, the fast fibers specialized for glycolysis contain few mitochondria but have a high content of glycolytic enzymes instead. Also, to supply the large amounts of glucose necessary for glycolysis, they contain an abundance of stored glycogen. Because the glycolytic fibers need relatively less O_2 to function, they receive only a meager capillary supply compared with the oxidative fibers. The glycolytic fibers contain very little myoglobin and therefore are pale in color, so they are sometimes called **white fibers.** (The most readily observable comparison between red and white fibers is the dark and white meat in poultry.) In addition to their higher myosin ATPase content, these fibers are also larger in diameter than the slow-oxidative fibers because of a greater abundance of actin and myosin filaments. With these additional power-generating filaments, the fast-glycolytic fibers are capable of rapidly producing large amounts of tension, but only for a short duration because of their dependence on glycolysis for energy production.

Fast-oxidative fibers share characteristics with each of the other two types. They have high ATPase activity like the fast-glycolytic fibers and high oxidative capacity like the slow-oxidative fibers. They contract more rapidly than the slow-oxidative fibers and can maintain the contraction for a longer period of time than the fast-glycolytic fibers. However, because their rate of ATP production by oxidative phosphorylation cannot keep pace with the high rate of ATP splitting, they rely partially on glycolysis and are more prone to fatigue than the slow-oxidative fibers.

In humans, most of the muscles contain a mixture of all three fiber types; the percentage of each type is largely determined by the type of activity for which the muscle is specialized. Accordingly, a high proportion of slow-oxidative fibers are found in muscles specialized for maintaining low-intensity contractions for long periods of time without fatigue, such as the muscles of the back and legs that support the body's weight against the force of gravity. A preponderance of fast-glycolytic fibers are found in the arm muscles, which are adapted for performing rapid, forceful movements such as lifting heavy objects.

The percentage of these various fibers not only differs between muscles within an individual but also varies considerably among individuals. Most of us have an average of about 50% each of fast and slow fibers. Those genetically endowed with a higher percentage of the fast-glycolytic fibers are good candidates for power and sprint events, whereas those with a greater proportion of slow-oxidative fibers are more likely to be successful in endurance activities such as marathon races. Of course, success in any event will depend on many factors other than genetic endowment, such as the extent and type of training and the level of dedication.

Muscle fibers adapt considerably in response to the demands placed on them.

Not only is the nerve supply to a skeletal muscle essential for initiating contraction, but the motor neurons supplying a skeletal muscle are also important in maintaining the muscle's integrity and chemical composition. Different types of exercise produce different patterns of neuronal discharge to the muscle involved. Depending on the pattern of neural activity, long-term adaptive changes occur in the muscle fibers, enabling them to respond most efficiently to the types of demands placed on the muscle. Two types of changes can be induced in muscle fibers: changes in their ATP-synthesizing capacity and changes in their diameter. Regular endurance (aerobic) exercise, such as long-distance jogging or swimming, induces metabolic changes within the oxidative fibers, which are the ones primarily recruited during aerobic exercise. These changes enable the muscles to use O_2 more efficiently. For example, mitochondria, the organelles that house the enzymes involved in the oxidative-phosphorylation pathway, increase in number in the oxidative fibers. In addition, the number of capillaries supplying blood to these fibers increases. Muscles so adapted are better able to endure prolonged activity without fatiguing, but they do not change in size.

The actual size of the muscles can be increased by regular bouts of anaerobic, short-duration, high-intensity resistance training, such as weight lifting. The resulting muscle enlargement comes primarily from an increase in diameter **(hypertrophy)** of the fast-glycolytic fibers that are called into play during such powerful contractions. Most of the fiber thickening is a consequence of increased synthesis of myosin and actin filaments, which permits a greater opportunity for cross-bridge interaction and subsequently increases the muscle's contractile strength. The resultant bulging muscles are better adapted to activities that require intense strength for brief periods, but endurance has not been improved. Furthermore, **hyperplasia** (an increase in the number of muscle cells) is believed to contribute to muscle enlargement to a small extent. Muscle cells are unable to divide by mitotic cell division, but experimental evidence suggests that greatly enlarged fibers can split lengthwise down the middle, resulting in an increase in the number of fibers.

Men's muscle fibers are thicker, and accordingly, their muscles are larger and stronger than those of women, even without weight training, because of the actions of testosterone, a steroid hormone secreted primarily in males. Testosterone promotes the synthesis and assembly of myosin and actin and is responsible for the naturally larger muscle mass of men. This fact has led some athletes, both males and females, to the dangerous practice of taking this or closely related steroids to increase their athletic performance. (See the boxed feature on p. 246, ● A Closer Look at Exercise Physiology.)

Apparently, all of the muscle fibers within a single motor unit are of the same fiber type. This pattern usually is established early in life, but the two types of fast-twitch fibers are interconvertible, depending on training efforts. Regular endurance activities can convert fast-glycolytic fibers into fast-oxidative fibers, whereas fast-oxidative fibers can be shifted to fast-glycolytic fibers in response to power events such as weight training. Slow and fast fibers are not interconvertible, however. Adaptive changes that take place in skeletal muscle gradually reverse to their original state over a period of months if the regular exercise program that induced these changes is discontinued.

A CLOSER LOOK AT EXERCISE PHYSIOLOGY

Are Athletes Who Use Steroids to Gain Competitive Advantage Really Winners or Losers?

The testing of athletes for drugs and the much publicized exclusion from competition of those found to be using substances outlawed by sports federations have stirred considerable controversy. One such group of drugs are **anabolic androgenic steroids** (*anabolic* means "buildup of tissues"; *androgenic* means "male producing"; *steroids* are a class of hormone). These agents are closely related to testosterone, the natural male sex hormone, which is responsible for promoting the increased muscle mass characteristic of males.

Although their use is outlawed (possession of anabolic steroids without a prescription became a federal offense in 1991), these agents are taken by many athletes who specialize in power events such as weight lifting and sprinting in the hopes of increasing muscle mass and, accordingly, muscle strength. Both male and female athletes have resorted to using these substances in an attempt to gain a competitive edge. Anabolic steroids are also taken by bodybuilders. There are an estimated 1 million anabolic steroid abusers in the United States. Unfortunately, the use of these agents has spread into our nation's high schools. In a 1989 survey of 17,000 high school youths by the National Institute of Drug Abuse, 5% of the respondents reported current or past steroid use to "build muscles" or to "improve appearance."

Studies have confirmed that steroids can increase muscle mass when used in large amounts and coupled with heavy exercise. One reputable study demonstrated an average 8.9-pound gain of lean muscle in bodybuilders who used steriods during a ten-week period. Anecdotal evidence suggests that some steroid users have added as much as 40 pounds of muscle in a year.

The adverse effects of these drugs, however, outweigh any benefits derived. In females, who normally lack potent androgenic hormones, anabolic steroid drugs not only promote "male-type" muscle mass and strength but also "masculinize" the users in other ways, such as by inducing growth of facial hair and by lowering the voice. More importantly, in both males and females, these agents adversely affect the reproductive and cardiovascular systems and the liver, may have an impact on behavior, and may be addictive.

Adverse Effects on the Reproductive System

In males, testosterone secretion and sperm production by the testes are normally controlled by hormones from the anterior pituitary gland. In negative-feedback fashion, testosterone inhibits secretion of these controlling hormones, so that a constant circulating level of testosterone is maintained. The anterior pituitary is similarly inhibited by androgenic steroids taken as a drug. As a result, because the testes do not receive their normal stimulatory input from the anterior pituitary, testosterone secretion and sperm production decrease and the testes atrophy.

In females, inhibition of the anterior pituitary by the androgenic drugs results in repression of the hormonal output that controls ovarian function. The result is failure to ovulate, menstrual irregularities, and decreased secretion of "feminizing" female sex hormones. Their decline results in diminution in breast size and other female characteristics.

Adverse Effects on the Cardiovascular System

Use of anabolic steroids induces several cardiovascular changes that increase the risk of developing atherosclerosis, which in turn is associated with an increased incidence of heart attacks and strokes (see p. 296). Among these adverse cardiovascular effects are (1) a reduction in high density lipoproteins (HDL), the "good" cholesterol carriers that help remove cholesterol from the body, and (2) an elevation in blood pressure. Damage to the heart muscle itself has also been demonstrated in animal studies.

Adverse Effects on the Liver

Liver dysfunction is common with high steroid intake, because the liver, which normally inactivates steroid hormones and prepares them for urinary excretion, is overloaded by the excess steroid intake.

Adverse Effects on Behavior

Although the evidence is still controversial, anabolic steroid use appears to promote aggressive, even hostile behavior—so-called roid rages.

Addictive Effects

A troubling new concern is the addiction to anabolic steroids of some who abuse these drugs. In one study involving face-to-face interviews, 14% of steroid users were judged on the basis of their responses to be addicted. In another survey using anonymous, self-administered questionnaires, 57% of steroid users qualified as being addicted. This apparent tendency to become chemically dependent on steroids is alarming because the potential for adverse effects on health increases with long-term, heavy use, the kind of use that would be expected in someone hooked on the drug.

Thus, for health reasons, without even taking into account the legal and ethical issues, people should not use anabolic steroids. However, the problem appears to be worsening. Currently, the international black market for anabolic steroids is estimated at $1 billion per year.

Although training can induce changes in muscle fibers' metabolic support systems, whether a fiber is fast or slow twitch depends on the fiber's nerve supply. Slow-twitch fibers are supplied by motor neurons that exhibit a low-frequency pattern of electrical activity, whereas fast-twitch fibers are innervated by motor neurons that display intermittent rapid bursts of electrical activity. Experimental switching of motor neurons supplying slow muscle fibers with those supplying fast fibers results in a gradual reversal of the speed at which these fibers contract.

At the other extreme, if a muscle is not used, its actin and myosin content decreases, its fibers become smaller, and the muscle accordingly decreases in mass **(atrophies)** and becomes weaker. Muscle atrophy can result in two ways. **Disuse atrophy** occurs when a muscle is not used for a long period of time even though the nerve supply is intact, as when a cast or brace must be worn or during prolonged bed confinement. **Denervation atrophy** occurs after the nerve supply to a muscle is lost. If the muscle is stimulated electrically until innervation can be reestablished, such as during regeneration of a severed peripheral nerve, atrophy can be diminished but not entirely prevented. Contractile activity itself obviously plays an important role in preventing atrophy; however, poorly understood factors released from active nerve endings, perhaps packaged with the ACh vesicles, apparently contribute to the integrity and growth of muscle tissue.

When a muscle is damaged, limited repair is possible, even though muscle cells cannot divide mitotically to replace damaged cells. A small population of **myoblasts** (*myo* means "muscle," *blast* means "former"), the same undifferentiated cells that formed the muscle during embryonic development, remains close to the muscle surface. When a muscle fiber is damaged, it can be replaced by fusion of a group of these myoblasts to form a large multinucleated cell, which immediately begins to synthesize and assemble the intracellular machinery characteristic of the muscle. With extensive injury, this limited mechanism is not adequate to completely replace all of the lost fibers. In that case, the remaining fibers often hypertrophy to compensate.

Transplantation of myoblasts provides one of two glimmers of hope for victims of **muscular dystrophy,** a hereditary pathological condition characterized by progressive degeneration of contractile elements, which are ultimately replaced by fibrous tissue. (See the boxed feature on p. 249, ◆ Concepts, Challenges, and Controversies.)

With age a person experiences a slow, continuous reduction in skeletal muscle mass and an accompanying loss of strength. These losses are attributable to the following age-related changes:

- A reduction in the number of skeletal muscle fibers. Up to 30% of the fibers in skeletal muscles may be lost by age 80. Many of the lost fibers are replaced by fat tissue, with islands of fat often developing between muscle cells in older individuals. The loss of muscle fibers reduces the size of individual motor units in skeletal muscle. As a consequence, more motor units must be recruited to move a particular weight, so moving the weight requires increased effort in elderly persons.
- A reduction in diameter in the remaining skeletal muscle fibers resulting from a loss of tension-generating myofibrils. The fast-twitch glycolytic fibers seem to atrophy earlier than the slow-twitch, oxidative fibers do. Consequently, even though capacity for quick, powerful contractions declines with advancing age, the slow, prolonged contractions used for maintaining posture are not affected significantly, if at all, until very late in life.
- A reduction in the ability of muscle to adapt to exercise. Some of the loss of muscle mass is nothing more than disuse atrophy—people in general become less physically active as they get older. In fact, the best defense against age-related muscle atrophy and loss of strength is regular exercise. Even modest increases in physical activity have been shown to improve muscle mass and strength in senior citizens in their eighties and beyond. However, even a regular exercise program cannot completely prevent or reverse the age-related decline in skeletal muscle ability, because a given exercise regimen will produce less pronounced adaptive changes in the skeletal muscles of an older individual than in a younger person. The reason for this reduced adaptive ability is unclear, but it may result from impairment in the synthesis of new muscle proteins.
- A reduction in nerve activity to skeletal muscles. The age-related decline in skeletal muscle mass and strength is partially associated with age-related changes in the nervous system. There is a progressive loss of motor neurons with advancing age, as well as a reduction in the synthesis of acetylcholine by these neurons. The loss of motor neuron innervation contributes to muscle fiber atrophy, and lowered levels of acetylcholine reduce the efficiency of muscle stimulation. Furthermore, a reduction in the release of recently discovered neuropeptide Y from the motor neuron terminals may play a role (see page 210).

Control of Motor Movement

Many inputs influence motor unit output.

Particular patterns of motor unit output are responsible for motor activity, ranging from maintenance of posture and balance to stereotypical locomotor movements, such as walking, to individual, highly skilled motor activity, such as gymnastics. Control of any motor movement, no matter what its level of complexity, depends on converging input to the motor neurons of specific motor units. The motor neurons in turn trigger contraction of the muscle fibers within their respective motor units by means of the events that occur at the neuromuscular junction. Three levels of input control motor neuron output (▬ Fig. 8-23):

1. Input from afferent neurons, usually through intervening interneurons, at the level of the spinal cord—that is, spinal reflexes (see p. 149).
2. Input from the primary motor cortex. Fibers originating from cell bodies of pyramidal cells within the primary motor cortex (see p. 123) descend directly without synaptic interruption to terminate on motor neurons (or on local interneurons that terminate on motor neurons). These fibers make up the **corticospinal** (or **pyramidal**) **motor system.**
3. Input from the **multineuronal** (or **extrapyramidal**) **motor system.** The pathways composing this system include a number of synapses that involve many regions of the brain. The final link in multineuronal pathways is the brain stem, especially the reticular formation, which in turn is influenced by motor regions of the cortex, the cerebellum, and the basal nuclei. In addition, the motor cortex itself is interconnected with the thalamus as well as with

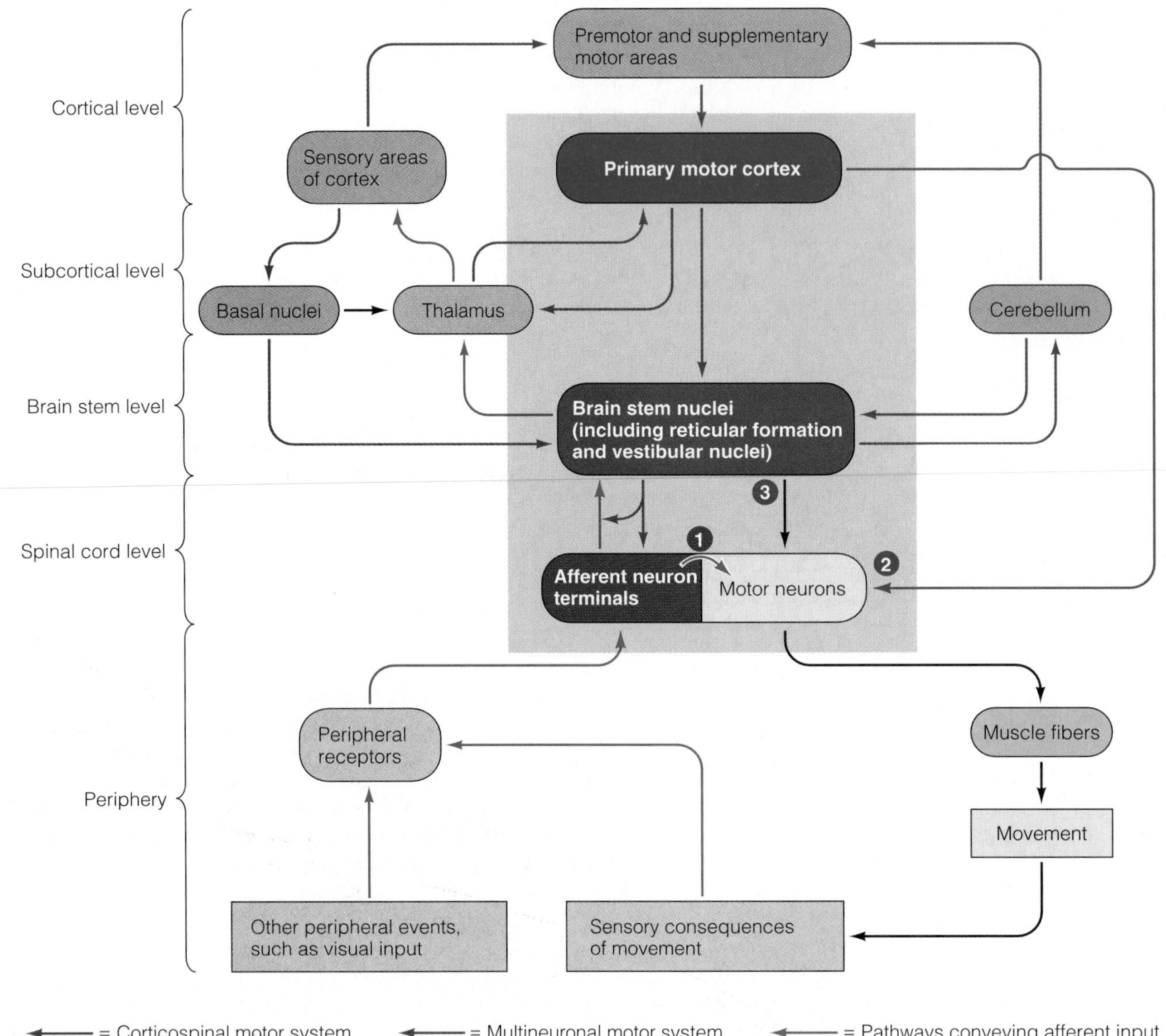

Figure 8-23 **Motor Control** The CNS is constantly apprised of muscle length and tension and other peripheral events via pathways conveying afferent input, so it can program coordinated, purposeful skeletal muscle activity. Motor movement is controlled by input to the motor neurons from ① afferent neuron terminals at the level of the spinal cord, ② the primary motor cortex, via the corticospinal motor system, and ③ brain stem nuclei, which serve as the final link in the complex multineuronal motor system involving many regions of the brain.

premotor and supplementary motor areas. The only brain regions that directly influence motor neurons are the primary motor cortex and brain stem; the other involved brain regions indirectly regulate motor activity by adjusting motor output from the motor cortex and brain stem. A number of complex interactions take place between these various brain regions; the most important are represented in Figure 8-23. (See chapter 5 for further discussion of the specific roles and interactions of these brain regions.)

The corticospinal system primarily mediates performance of fine, discrete, voluntary movements of the hands and fingers, such as those required for doing intricate needlework. Premotor and supplementary motor areas, with input from the cerebrocerebellum, plan the voluntary motor command that is issued to the appropriate motor neurons by the primary motor cortex through this descending system. The multineuronal system, in contrast, is primarily concerned with regulation of overall body posture involving involuntary movements of large muscle groups of the trunks and limbs. Considerable complex interaction and overlapping of function exist between these two systems. For example, to voluntarily manipulate your fingers to do needlework, you subconsciously assume a particular posture of your arms that enables you to hold your work.

Some of the inputs converging on motor neurons are excitatory, whereas others are inhibitory. Coordinated movement depends on an appropriate balance of activity in these inputs. If an inhibitory system originating in the brain stem is disrupted, muscles become hyperactive (increased muscle tone;

CONCEPTS, CHALLENGES, AND CONTROVERSIES

Muscular Dystrophy: When One Small Step Is a Big Deal

Hope of treatment is on the horizon for **muscular dystrophy,** a fatal muscle-wasting disease that primarily strikes boys and unrelentlessly leads to their premature death before age twenty. Muscular dystrophy encompasses a variety of hereditary pathological conditions, which have in common a progressive degeneration of contractile elements and their replacement by fibrous tissue. The gradual muscle wasting in muscular dystrophy is characterized by progressive weakness over a period of years, usually resulting in death from respiratory failure when the respiratory muscles become too weak to function adequately or from heart failure when the heart becomes too weak to beat.

The disease is caused by a recessive genetic defect on the X sex chromosome, of which males have only one copy. (Males have XY sex chromosomes; females have XX sex chromosomes). If a male inherits from either parent an X chromosome bearing the defective dystrophic gene, he is destined to develop the disease, which affects one out of every 3,500 boys worldwide. Females, on the other hand, must inherit a dystrophic-carrying gene from both parents before acquiring the condition, a much rarer occurrence.

The defective gene responsible for *Duchenne muscular dystrophy,* the most common and most devastating form of the disease, was pinpointed in 1986. The gene normally produces **dystrophin,** a large protein found in the plasma membrane of muscle cells. Dystrophin is linked with the regulation of Ca^{2+} flow into muscle cells through special Ca^{2+} "leak" channels. Dystrophic muscles are characterized by a lack of dystrophin. Even though this protein represents only 0.002% of the total amount of skeletal muscle protein, its presence is crucial. Its absence appears to permit a constant leakage of Ca^{2+} into the muscle cells through unregulated Ca^{2+} channels. This Ca^{2+} presumably activates proteases, protein-snipping enzymes that harm the muscle fibers. The resultant damage leads to the muscle wasting and ultimate fibrosis characterizing the disorder.

Although the disease is still untreatable and fatal, two different lines of research offer hope for the first time to the victims of muscular dystrophy—a cell-transplant and a gene-therapy approach. The cell-transplant technique involves the injection of myoblasts into the dystrophic muscles. **Myoblasts** ("muscle formers") are the undifferentiated cells that fuse to form the large, multinucleated skeletal muscle cells during embryonic development. After development, a small group of these immature myoblasts remains close to the muscle surface. Even though muscle cells cannot divide mitotically to replace damaged cells, limited repair of a damaged muscle is possible through the activity of these myoblasts. A group of myoblasts can fuse together to form a new skeletal muscle cell to replace a damaged cell. When the loss of muscle cells is extensive, however, as in muscular dystrophy, this limited mechanism is not adequate to replace all of the lost fibers.

A new therapeutic approach for muscular dystrophy involves the transplantation of dystrophin-producing myoblasts harvested from muscle biopsies of healthy donors into the patient's dwindling muscles. The still-experimental technique was attempted for the first time in 1990 by injection of myoblasts into one big toe of a Duchenne patient. The encouraging results: the patient could wriggle his toe more vigorously. The healthy myoblasts fused with the boy's dystrophic muscle cells, and the donated cell's normal genes produced the deficient dystrophin protein. As a result, the treated muscle increased in mass and strength. This early success was followed by more extensive procedures in which solutions containing billions of healthy myoblasts were injected into sixty-nine major muscle groups in the legs and buttocks of twenty-one boys with Duchenne's muscular dystrophy. Preliminary data showed a 43% increase in strength in the treated muscles, with 38% staying the same and 19% continuing to lose strength. Although the interpretation of these results is somewhat controversial, most specialists in the field remain optimistic that myoblast transplantation offers great promise for staving off the gradual weakening and certain death faced by those suffering from muscular dystrophy. Even though this technique cannot be applied to all weakened muscles, the hope is to salvage the muscles essential for breathing and walking so that these patients can lead longer, more normal lives.

An alternative therapeutic approach being sought is a possible "gene fix." One group of researchers have succeeded in inserting normal copies of the human dystrophin gene into the thigh muscles of mice. The gene was incorporated into about 1% of the thigh muscle cells of the mice and, importantly, produced dystrophin, which was inserted in its proper location in the cells' plasma membrane. If the technique can be improved to increase the number of treated cells that take up the introduced genes, it offers hope for an eventual gene fix for muscular dystrophy.

These steps toward an eventual treatment mean that hopefully one day the afflicted boys will be able to take steps on their own instead of being destined to wheelchairs and early death.

augmented limb reflexes) because of the unopposed activity in excitatory inputs to motor neurons, a condition known as **spastic paralysis.** In contrast, loss of excitatory input, such as that accompanying destruction of descending excitatory pathways exiting the primary motor cortex, brings about **flaccid paralysis** (muscles relaxed; inability to voluntarily contract muscles, although reflex activity is still present). Damage to the primary motor cortex on one side of the brain, as with a stroke, leads to flaccid paralysis on the opposite half of the body (**hemiplegia,** or paralysis of one side of the body). Disruption of all descending pathways, as in traumatic severance of the spinal cord, is accompanied by flaccid paralysis below the level of the damaged region—**quadriplegia** (paralysis of all four limbs) in upper spinal cord damage and **paraplegia** (paralysis of the legs) in lower spinal cord injury. Destruction of motor neurons—either their cell bodies or efferent fibers—causes flaccid paralysis and lack of reflex responsiveness in the affected muscles. Damage to the cerebellum or basal nuclei does not result in paralysis but instead in uncoordinated, clumsy activity and inappropriate patterns of movement. These regions normally smooth out activity initiated voluntarily. Damage to higher cortical regions involved in planning motor activity results in the inability to establish appropriate motor commands to accomplish desired goals.

Muscle spindles and the Golgi tendon organs provide afferent information essential for controlling skeletal muscle activity.

Coordinated, purposeful skeletal muscle activity depends on afferent input from a variety of sources. At a simple level, afferent signals indicating that your finger is touching a hot stove triggers reflex contractile activity in appropriate arm muscles to withdraw the hand from the injurious stimulus. At a more complex level, if you are going to catch a ball, the motor systems of your brain must program sequential motor commands that will move and position your body correctly for the catch, using predictions of the ball's direction and rate of movement provided by visual input. Many muscles acting simultaneously or alternately at different joints are called into play to shift your body's location and position rapidly, while maintaining your balance in the process. It is critical to have ongoing input about your body position with respect to the surrounding environment, as well as the position of your various body parts in relationship to each other. This information is necessary for establishing a neuronal pattern of activity to perform the desired movement. Your CNS must know the starting position of your body to appropriately program muscle activity. Further, it must be constantly apprised of the progression of movement it has initiated so that it can make adjustments as needed. Your brain receives this information, which is known as *proprioceptive* input (see p. 124), from receptors in your eyes, joints, vestibular apparatus, and skin, as well as from the muscles themselves. You can demonstrate your joint and muscle proprioceptive receptors in action by closing your eyes and bringing the tips of your right and left index fingers together at any point in space. You can do so without seeing where your hands are because your brain is informed of the position of your hands and other body parts at all times by afferent input from the joint and muscle receptors.

Two types of muscle receptors—*muscle spindles* and *Golgi tendon organs*—monitor changes in muscle length and tension. This information is used in two ways: (1) to apprise motor areas of the brain of muscle length and tension and (2) to control muscle length and tension in negative-feedback fashion by means of local spinal reflexes. Muscle length is monitored by muscle spindles, whereas changes in muscle tension are detected by Golgi tendon organs. Both of these receptor types are activated by muscle stretch, but they are designed to convey different types of information. Let us see how.

Muscle spindles, which are distributed throughout the fleshy part of a skeletal muscle, consist of collections of specialized muscle fibers known as **intrafusal fibers** (*fusus* means "spindle"), which lie within spindle-shaped connective tissue capsules parallel to the "ordinary" **extrafusal fibers** (Fig. 8-24). Unlike an ordinary skeletal muscle fiber,

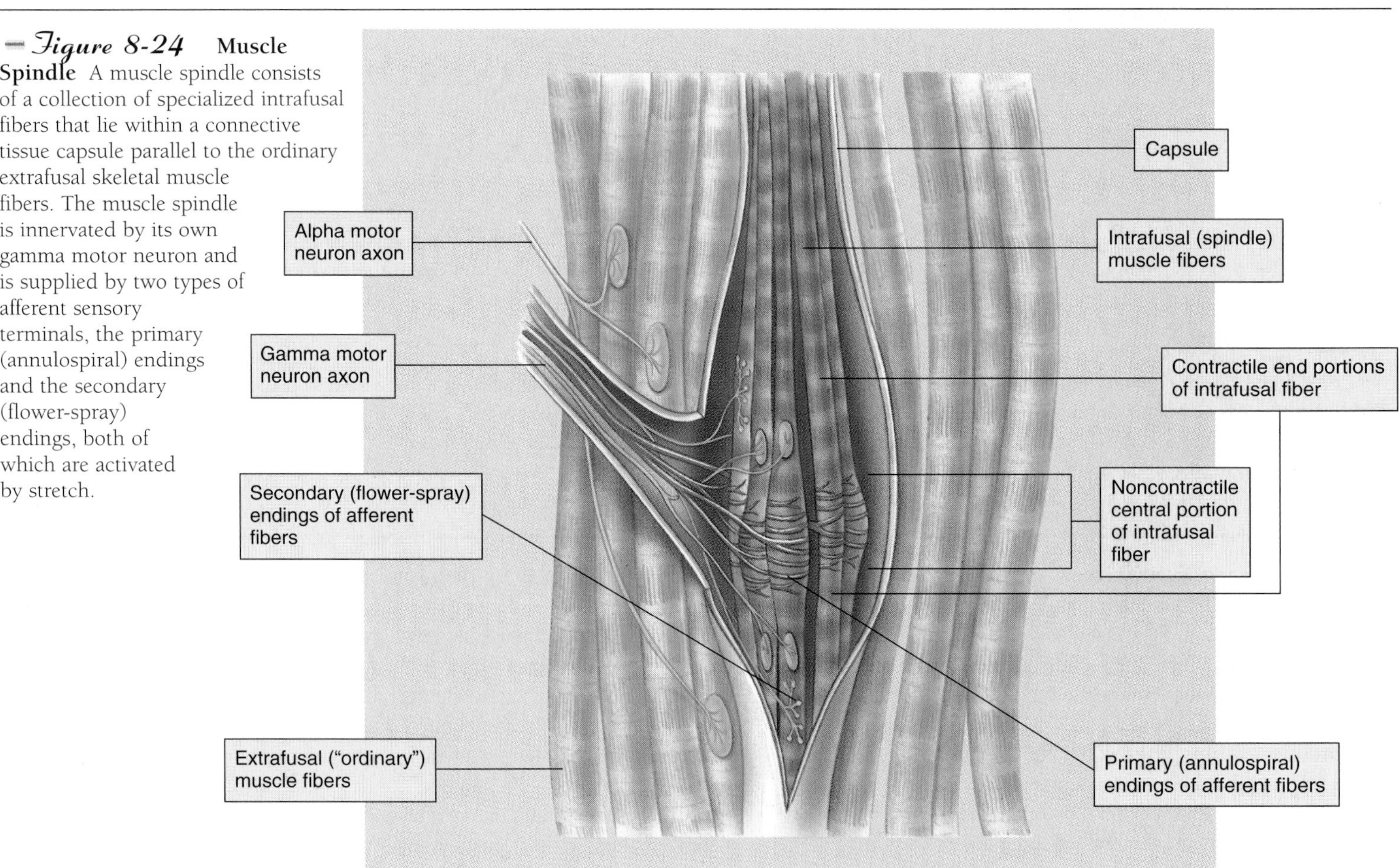

Figure 8-24 **Muscle Spindle** A muscle spindle consists of a collection of specialized intrafusal fibers that lie within a connective tissue capsule parallel to the ordinary extrafusal skeletal muscle fibers. The muscle spindle is innervated by its own gamma motor neuron and is supplied by two types of afferent sensory terminals, the primary (annulospiral) endings and the secondary (flower-spray) endings, both of which are activated by stretch.

which contains contractile elements (myofibrils) throughout its entire length, an intrafusal fiber has a noncontractile central portion, with the contractile elements being limited to both ends. Each muscle spindle has its own private efferent and afferent nerve supply. The efferent neuron that innervates a muscle spindle's intrafusal fibers is known as a **gamma motor neuron,** whereas the motor neurons that supply the ordinary extrafusal fibers are designated as **alpha motor neurons.** Two types of afferent sensory endings terminate on the intrafusal fibers and serve as muscle spindle receptors, both of which are activated by stretch. The **primary (annulospiral) endings** are wrapped around the central portion of the intrafusal fibers; they detect changes in the length of the fibers during stretching as well as the speed with which it occurs. The **secondary (flower-spray)** endings, which are clustered at the end segments of many of the intrafusal fibers, are sensitive only to changes in length.

Whenever the whole muscle is passively stretched, the intrafusal fibers within its muscle spindles are likewise stretched, increasing the rate of firing in the afferent nerve fibers whose sensory endings terminate on the stretched spindle fibers. The afferent neuron directly synapses on the alpha motor neuron that innervates the extrafusal fibers of the same muscle, resulting in contraction of that muscle (▬ Fig. 8-25a, pathway ① → ②). This **stretch reflex** serves as a local negative-feedback mechanism to resist any passive changes in muscle length so that optimal resting length can be maintained.

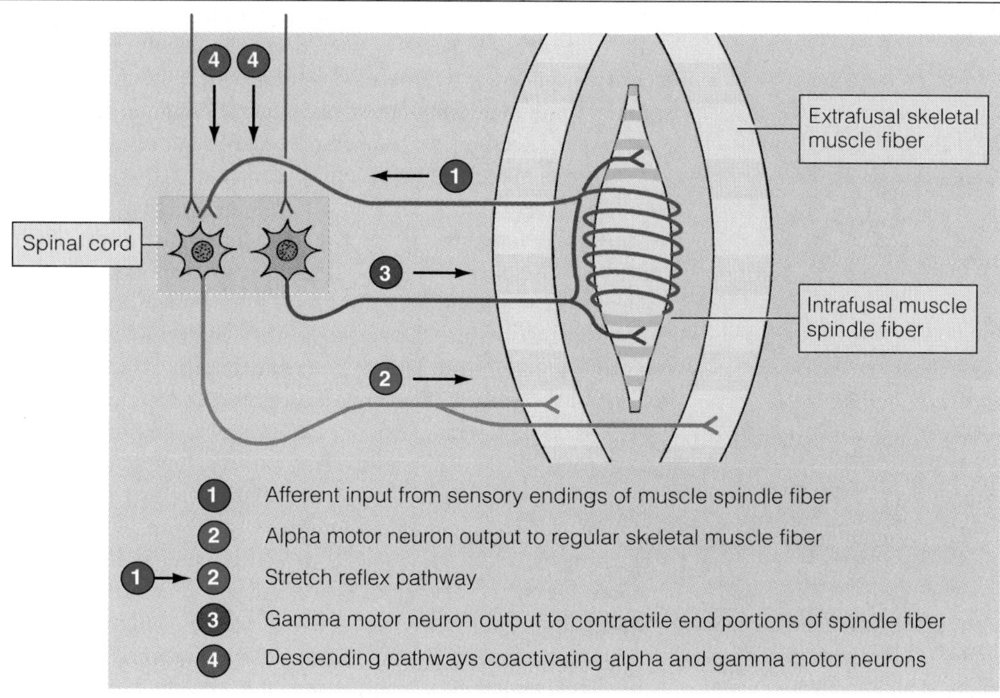

(a)

Relaxed muscle; spindle fiber sensitive to stretch of muscle

(b)

Contracted muscle; slackened spindle fiber not sensitive to stretch of muscle

(c)

Contracted muscle; contracted spindle fiber sensitive to stretch of muscle

(d)

▬ *Figure 8-25* **Muscle Spindle Function** (a) Pathways involved in the monosynaptic stretch reflex and coactivation of alpha and gamma motor neurons. (b) Status of a muscle spindle when the muscle is relaxed. (c) Status of a muscle spindle when the muscle is contracted upon alpha motor neuron stimulation. (d) Status of a muscle spindle when both the muscle and muscle spindle are contracted upon alpha and gamma motor neuron coactivation.

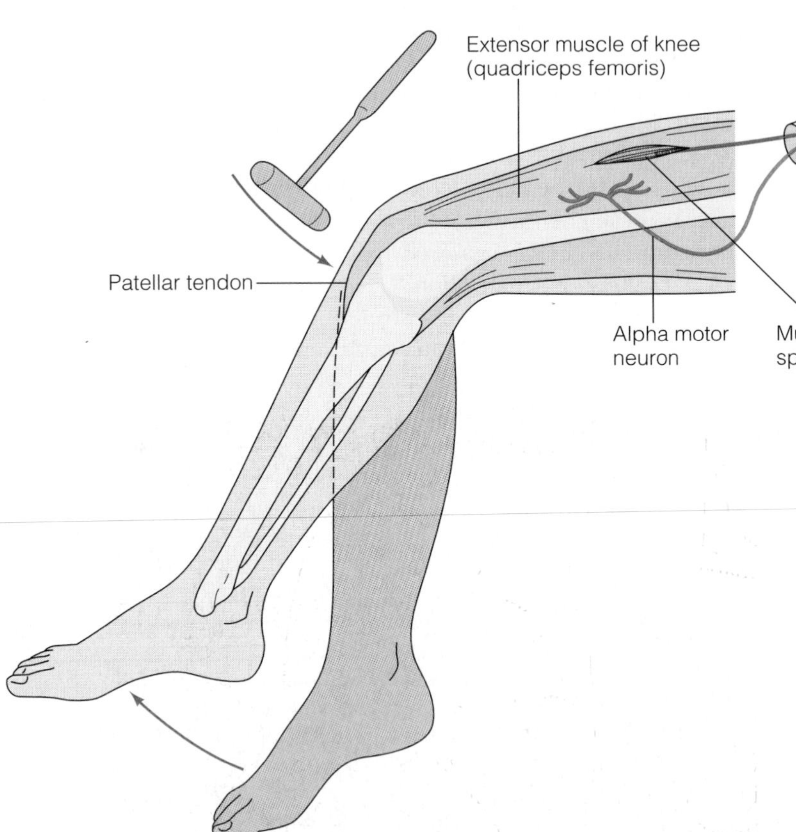

Figure 8-26 **Patellar Tendon Reflex (a Stretch Reflex)**
Tapping the patellar tendon with a rubber mallet stretches the muscle spindles in the quadriceps femoris muscle. The resultant monosynaptic stretch reflex results in contraction of this extensor muscle, causing the characteristic knee-jerk response.

The classic example of the stretch reflex is the **patellar tendon,** or **knee-jerk, reflex** (Fig. 8-26). The extensor muscle of the knee is the *quadriceps femoris,* which forms the anterior (front) portion of the thigh and is attached just below the knee to the tibia (shinbone) by the *patellar tendon.* Tapping this tendon with a rubber mallet passively stretches the quadriceps muscle, activating its spindle receptors. The resultant stretch reflex brings about contraction of this extensor muscle, causing the knee to extend and raise the foreleg in the well-known knee-jerk fashion. This test is routinely performed as a preliminary assessment of nervous system function. A normal knee jerk indicates to a physician that a number of neural and muscular components—muscle spindle, afferent input, motor neurons, efferent output, neuromuscular junctions, and the muscles themselves—are functioning normally. It also indicates the presence of an appropriate balance of excitatory and inhibitory input to the motor neurons from higher brain levels. Muscle jerks may be absent or depressed with loss of higher-level excitatory inputs or may be greatly exaggerated with loss of inhibitory input to the motor neurons from higher brain levels.

The primary purpose of the stretch reflex is to resist the tendency for the passive stretch of extensor muscles caused by gravitational forces when a person is standing upright. Whenever the knee joint tends to buckle because of gravity, the quadriceps muscle is stretched. The resultant enhanced contraction of this extensor muscle brought about by the stretch reflex quickly straightens out the knee, holding the limb extended so that the person remains standing.

Gamma motor neurons initiate contraction of the muscular end regions of intrafusal fibers (Fig. 8-25a, pathway ③). This contractile response is too weak to have any influence on whole-muscle tension, but it does have an important localized effect on the muscle spindle itself. If there were no compensating mechanisms, shortening of the whole muscle by alpha motor neuron stimulation of extrafusal fibers would cause slack in the spindle fibers so that they would be less sensitive to stretch and therefore not as effective as muscle-length detectors (Figs. 8-25b and c). **Coactivation** of the gamma motor neuron system along with the alpha motor neuron system during reflex and voluntary contractions (Fig. 8-25a, pathway ④) takes the slack out of the spindle fibers as the whole muscle shortens, permitting these receptor structures to maintain their high sensitivity to stretch over a wide range of muscle lengths. When gamma motor neuron stimulation triggers simultaneous contraction of both end muscular portions of an intrafusal fiber, the noncontractile central portion is pulled in opposite directions, tightening this region and taking out the slack (Fig. 8-25d). Whereas the extent of alpha motor neuron activation depends on the intended strength of the motor response, the extent of simultaneous gamma motor neuron activity to the same muscle depends on the anticipated distance of shortening.

In contrast to muscle spindles, which lie within the belly of the muscle, **Golgi tendon organs** are located in the tendons of the muscle, where they are able to respond to changes in the muscle's externally applied tension rather than to changes in its length. Because a number of factors determine the tension developed in the whole muscle during contraction (for example, frequency of stimulation or length of the muscle at the onset of contraction), it is essential that motor control systems be apprised of the tension actually achieved so that adjustments can be made if necessary.

The Golgi tendon organs consist of endings of afferent fibers entwined within bundles of connective tissue fibers that make up the tendon. When the extrafusal muscle fibers contract, the resultant pull on the tendon tightens the connective tissue bundles, which in turn increase the tension exerted on the bone to which the tendon is attached. In the process, the entwined Golgi organ afferent receptor endings are stretched, causing the afferent fibers to fire; the frequency of firing is directly related to the tension developed.

The afferent information is sent to the brain. In addition, other branches of the afferent neuron arising from the Golgi tendon organ inhibit, by means of an interneuron, the alpha motor neurons of the same muscle. This reflex is apparently protective in nature. When the tension becomes great enough, the high level of inhibitory input from the activated Golgi tendon organs counterbalances excitatory inputs to the alpha motor neurons. This inhibitory response halts further contraction and brings about sudden reflex relaxation, thus helping prevent damage to muscle or tendon from excessive, tension-developing muscle contractions.

Smooth and Cardiac Muscle

Smooth and cardiac muscle share some basic properties with skeletal muscle.

The two other types of muscle—smooth muscle and cardiac muscle—share some basic properties with skeletal muscle, but each also displays unique characteristics (Table 8-4). The three muscle types have several features in common. First, they all have a specialized contractile apparatus made up of thin actin filaments that slide relative to stationary thick myosin filaments in response to a rise in cytosolic Ca^{2+} to accomplish contraction. Second, they all directly use ATP as the energy source for cross-bridge cycling. However, the structure and organization of fibers within these different muscle types vary, as do their mechanisms of excitation and the means by which excitation and contraction are coupled. Furthermore, there are important distinctions in the contractile response itself. We will spend the remainder of this chapter highlighting unique features of smooth and cardiac muscle as compared with skeletal muscle, reserving a more detailed discussion of their function for chapters devoted to organs containing these muscle types.

Smooth muscle cells are small and unstriated.

The majority of smooth muscle cells are found in the walls of hollow organs and tubes. Their contraction exerts pressure on and regulates the forward movement of the contents of these structures. Both smooth and skeletal muscle cells are elongated, but in contrast to their large, cylinder-shaped, multinucleated skeletal muscle counterparts, smooth muscle cells are spindle-shaped, have a single nucleus, and are considerably smaller (2 to 10 μm in diameter and 50 to 400 μm long). Also unlike skeletal muscle cells, a single smooth muscle cell does not extend the full length of a muscle. Instead, groups of smooth muscle cells are typically arranged in sheets (Fig. 8-27a).

Three types of filaments are found in a smooth muscle cell: (1) thick myosin filaments, which are longer than those found in skeletal muscle; (2) thin actin filaments, which contain tropomyosin but lack the regulatory protein troponin; and (3), unique to smooth muscle, filaments of intermediate size, which do not directly participate in the contractile process but serve as part of the cytoskeletal framework that supports the shape of the cell. Smooth muscle filaments do not appear to form myofibrils and are not arranged in the sarcomere pat-

Figure 8-27 Microscopic View of Smooth Muscle Cells **(a)** Low-power light micrograph of smooth muscle cells. Note the spindle shape and single, centrally located nucleus. **(b)** Electron micrograph of smooth muscle cells at 14,000× magnification. Note the presence of dense bodies and lack of banding.

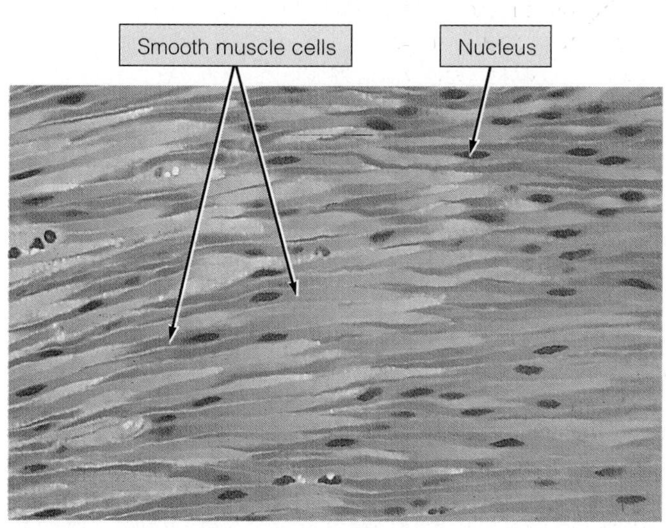

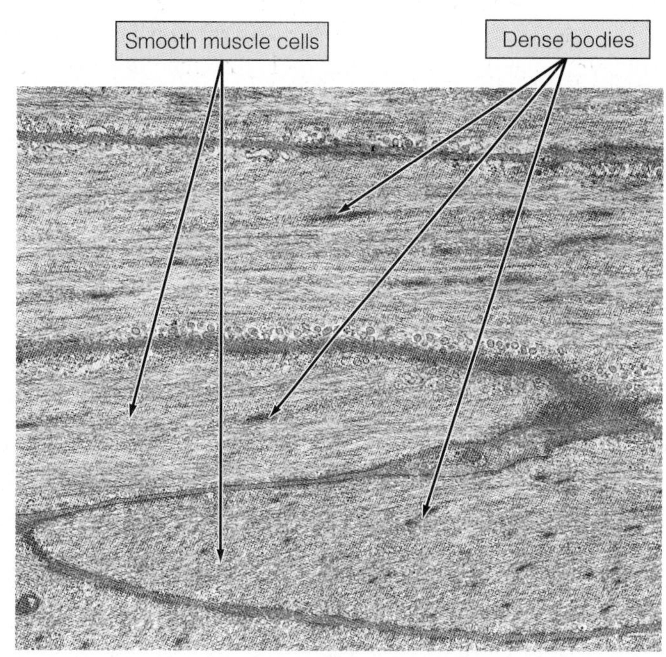

Table 8-4 Comparison of Muscle Types

Characteristic	Type of Muscle: *Skeletal*	*Multiunit Smooth*	*Single-Unit Smooth*	*Cardiac*
Location	Attached to skeleton	Large blood vessels, eye, and hair follicles	Walls of hollow organs in digestive, reproductive, and urinary tracts and in small blood vessels	Heart only
Function	Movement of body in relation to external environment	Varies with structure involved	Movement of contents within hollow organs	Pumps blood out of heart
Mechanism of contraction	Sliding-filament mechanism	Sliding-filament mechanism	Sliding-filament mechanism	Sliding-filament mechanism
Innervation	Somatic nervous system (alpha motor neurons)	Autonomic nervous system	Autonomic nervous system	Autonomic nervous system
Level of control	Under voluntary control; also subject to subconscious regulation	Under involuntary control	Under involuntary control	Under involuntary control
Initiation of contraction	Neurogenic	Neurogenic	Myogenic (pacemaker activity and slow-wave potentials)	Myogenic (pacemaker activity)
Role of nervous stimulation	Initiates contraction; accomplishes gradation	Initiates contraction; contributes to gradation	Modifies contraction; can excite or inhibit; contributes to gradation	Modifies contraction; can excite or inhibit; contributes to gradation
Modifying effect of hormones	No	Yes	Yes	Yes
Presence of thick myosin and thin actin filaments	Yes	Yes	Yes	Yes
Striated due to orderly arrangement of filaments	Yes	No	No	Yes
Presence of troponin and tropomyosin	Yes	Tropomyosin only	Tropomyosin only	Yes
Presence of T tubules	Yes	No	No	Yes

tern found in skeletal muscle. Thus, smooth muscle cells do not display the banding or striation found in skeletal muscle, giving rise to the term *smooth* for this muscle type. Lacking sarcomeres, smooth muscle does not have Z lines as such, but **dense bodies** containing the same protein constituent found in Z lines are present (Fig. 8-27b). Dense bodies are positioned throughout the smooth muscle cell as well as being attached to the internal surface of the plasma membrane. The dense bodies are held in place by a scaffold of intermediate filaments. The actin filaments are anchored to the dense bodies. Considerably more actin is present in smooth muscle cells than in skeletal muscle cells, with 10 to 15 thin filaments for each thick myosin filament in smooth muscle and 2 thin filaments for each thick filament in skeletal muscle. The thick- and thin-filament contractile units are oriented slightly diagonally from side to side within the smooth muscle cell in an elongated, diamond-shaped lattice, rather than running parallel with the long axis as myofibrils do in skeletal muscle (■ Fig. 8-28a). Relative sliding of the thin filaments past the thick filaments during contraction causes the filament lattice to reduce in length and expand from side to side. As a result, the whole cell shortens and bulges out between the points where the thin filaments are attached to the inner surface of the plasma membrane (Fig. 8-28b).

Table 8-4 Comparison of Muscle Types (continued)

Characteristic	Type of Muscle: Skeletal	Multiunit Smooth	Single-Unit Smooth	Cardiac
Level of development of sarcoplasmic reticulum	Well developed	Poorly developed	Poorly developed	Moderately developed
Cross bridges turned on by Ca^{2+}	Yes	Yes	Yes	Yes
Source of increased cytosolic Ca^{2+}	Sarcoplasmic reticulum	Extracellular fluid and sarcoplasmic reticulum	Extracellular fluid and sarcoplasmic reticulum	Extracellular fluid and sarcoplasmic reticulum
Site of Ca^{2+} regulation	Troponin in thin filaments	Myosin in thick filaments	Myosin in thick filaments	Troponin in thin filaments
Mechanism of Ca^{2+} action	Physically repositions troponin-tropomyosin complex to uncover actin cross-bridge binding sites	Chemically brings about phosphorylation of myosin cross bridges so they can bind with actin	Chemically brings about phosphorylation of myosin cross bridges so they can bind with actin	Physically repositions troponin-tropomyosin complex
Presence of gap junctions	No	Yes (very few)	Yes	Yes
ATP used directly by contractile apparatus	Yes	Yes	Yes	Yes
Myosin ATPase activity; speed of contraction	Fast or slow, depending on type of fiber	Very slow	Very slow	Slow
Means by which gradation accomplished	Varying number of motor units contracting (motor unit recruitment) and frequency at which they're stimulated (twitch summation)	Varying number of muscle fibers contracting and varying cytosolic Ca^{2+} concentration in each fiber by autonomic and hormonal influences	Varying cytosolic Ca^{2+} concentration through myogenic activity and influences of the autonomic nervous system, hormones, mechanical stretch, and local metabolites	Varying length of fiber (depending on extent of filling of the heart chambers) and varying cytosolic Ca^{2+} concentration through autonomic, hormonal, and local metabolite influence
Presence of tone in absence of external stimulation	No	No	Yes	No
Clear-cut length-tension relationship	Yes	No	No	Yes

Smooth muscle cells are turned on by Ca^{2+}-dependent phosphorylation of myosin.

If the thin filaments of smooth muscle cells do not contain troponin, and tropomyosin does not block actin's cross-bridge binding sites, what prevents actin and myosin from binding at the cross-bridges in the resting state, and how is cross-bridge activity switched on in the excited state? Smooth muscle myosin is able to interact with actin only when the myosin is *phosphorylated* (that is, has a phosphate group attached to it). During excitation, the increased cytosolic Ca^{2+} acts as an intracellular messenger, initiating a chain of biochemical events that results in phosphorylation of myosin (▬ Fig. 8-29). Smooth muscle Ca^{2+} binds with **calmodulin,** an intracellular protein found in most cells that is structurally similar to troponin (see p. 69). This Ca^{2+}-calmodulin complex binds to and activates another protein, **myosin kinase,** which in turn phosphorylates myosin. Phosphorylated myosin then binds with actin so that cross-bridge cycling can begin. When Ca^{2+} is removed, myosin is dephosphorylated (the phosphate is removed) and can no longer interact with actin, so the muscle relaxes. Thus, smooth muscle is triggered to contract by a rise in cytosolic Ca^{2+}, similar to what happens in skeletal muscle. In smooth muscle, however, Ca^{2+} ultimately turns on the cross bridges by inducing a *chemical* change in myosin in the *thick* filaments, whereas in skeletal muscle it exerts its effects by

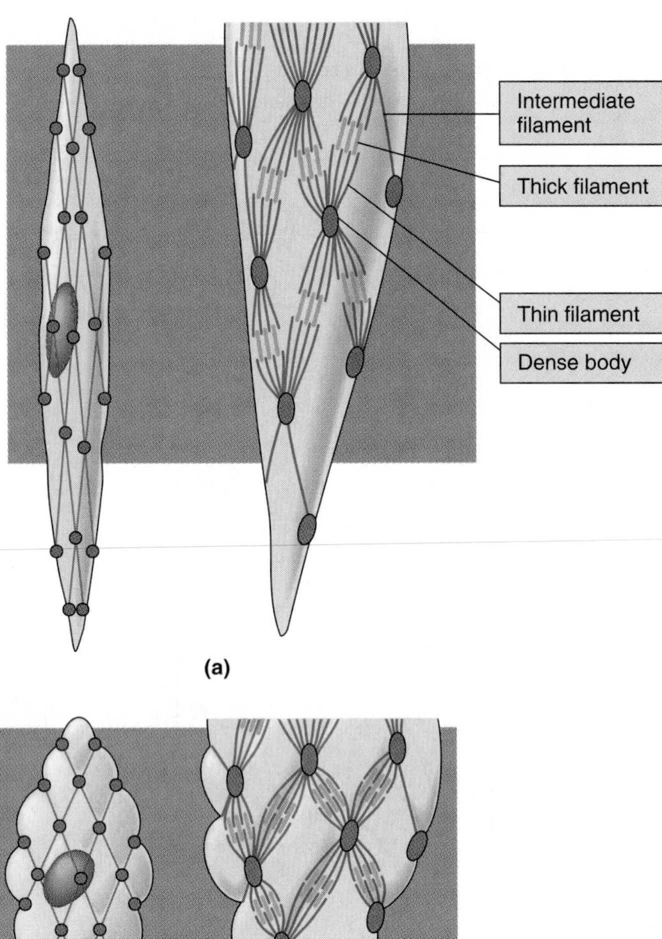

(b)

Figure 8-28 Schematic Representation of the Arrangement of Thick and Thin Filaments in a Smooth Muscle Cell in Contracted and Relaxed States (a) Relaxed smooth muscle cell. (b) Contracted smooth muscle cell.

Figure 8-29 Calcium Activation of Myosin in Smooth Muscle

Ca^{2+}
Calmodulin → Ca^{2+}-calmodulin
Inactive myosin kinase → Active myosin kinase
P_i
Inactive myosin → Phosphorylated myosin (can bind with actin)

invoking a *physical* change at the *thin* filaments (Fig. 8-30). Recall that in skeletal muscle, Ca^{2+} moves troponin and tropomyosin from their blocking position, so that actin and myosin are free to bind with each other.

The means by which excitation brings about an increase in cytosolic Ca^{2+} concentration in smooth muscle cells also differs from that for skeletal muscle. A smooth muscle cell has no T tubules and a poorly developed sarcoplasmic reticulum. The increased cytosolic Ca^{2+} that triggers the contractile response comes from two sources: most Ca^{2+} enters from the ECF, but some is released intracellularly from the sparse sarcoplasmic reticulum stores. Unlike their role in skeletal muscle cells, voltage-gated dihydropyridine receptors in the plasma membrane of smooth muscle cells function as Ca^{2+} channels. When these surface membrane channels are opened, Ca^{2+} enters down its concentration gradient from the ECF. The entering Ca^{2+} triggers the opening of Ca^{2+} channels in the sarcoplasmic reticulum, so that small additional amounts of Ca^{2+} are released from this meager source. Because smooth muscle cells are so much smaller in diameter than skeletal muscle fibers, the preponderant Ca^{2+} influx from the ECF is able to influence cross-bridge activity, even in the central portions of the cell, without the necessity of an elaborate T tubule–sarcoplasmic reticulum mechanism. Relaxation is accomplished by removal of Ca^{2+} as it is actively transported out across the plasma membrane and back into the sarcoplasmic reticulum.

Most groups of smooth muscle tissue are capable of self-excitation.

We still have not addressed the question of how smooth muscle becomes excited to contract; that is, what opens the Ca^{2+} channels in the plasma membrane and sarcoplasmic reticulum? Smooth muscle is grouped into two categories—multiunit and single-unit smooth muscle—based on differences in how the muscle fibers become excited. **Multiunit smooth muscle** exhibits properties partway between skeletal muscle and single-unit smooth muscle. As the name implies, a multiunit smooth muscle consists of multiple discrete units that function independently of each other and must be separately stimulated by nerves to contract, similar to skeletal muscle motor units. Thus, contractile activity in both skeletal muscle and multiunit smooth muscle is **neurogenic** ("nerve-produced"). Whereas skeletal muscle is innervated by the voluntary somatic nervous system (motor neurons), multiunit (as well as single-unit) smooth muscle is supplied by the involuntary autonomic nervous system. Multiunit smooth muscle is found (1) in the walls of large blood vessels; (2) in large airways to the lungs; (3) in the muscle of the eye that adjusts the lens for near or far vision; (4) in the iris of the eye, which alters the size of the pupil to adjust the amount of light entering the eye; and (5) at the base of hair follicles, contraction of which causes "goose bumps."

Most smooth muscle is of the **single-unit** variety. It is alternatively called **visceral smooth muscle** because it is found in the walls of the hollow organs or viscera (for example, the digestive, reproductive, and urinary tracts and small blood vessels). The term *single-unit smooth muscle* derives from the fact that the muscle fibers that make up this type of muscle become excited and contract as a single unit. The muscle fibers in single-unit smooth muscle are electrically linked by gap junctions (see p. 52). When an action potential occurs anywhere within a sheet of single-unit smooth muscle, it is quickly propagated via these special points of electrical contact throughout the entire group of interconnected cells, which then contract as a single coordinated unit. Such a group of interconnected muscle cells that function electrically and mechanically as a unit is known as a **functional syncytium** (*syn* means "several"; *cyt* means "cell").

Thinking about the role of the uterus during the process of labor will help you appreciate the significance of this arrangement. Muscle cells composing the uterine wall act as a functional syncytium. They repetitively become excited and contract as a unit during labor, exerting a series of coordinated "pushes" that are eventually responsible for delivering the baby. Independent, uncoordinated contractions of individual muscle cells in the uterine wall would not exert the uniformly applied pressure needed to expel the baby. A similar situation applies for single-unit smooth muscle elsewhere in the body.

Single-unit smooth muscle is **self-excitable** rather than requiring nervous stimulation for contraction. Clusters of specialized smooth muscle cells within a functional syncytium display spontaneous electrical activity; that is, they are able to undergo action potentials without any external stimulation. In contrast to the other excitable cells we have been discussing (such as neurons, skeletal muscle fibers, and multiunit smooth muscle), the self-excitable cells of single-unit smooth muscle do not maintain a constant resting potential. Instead, their membrane potential inherently fluctuates without any influence by factors external to the cell.

Two major types of spontaneous depolarizations displayed by self-excitable cells are pacemaker activity and slow-wave potentials. In **pacemaker activity** (■ Fig. 8-31a), the membrane potential gradually depolarizes on its own because of shifts in passive ionic fluxes accompanying automatic changes in channel permeability. When the membrane has depolarized to threshold, an action potential is initiated. After repo-

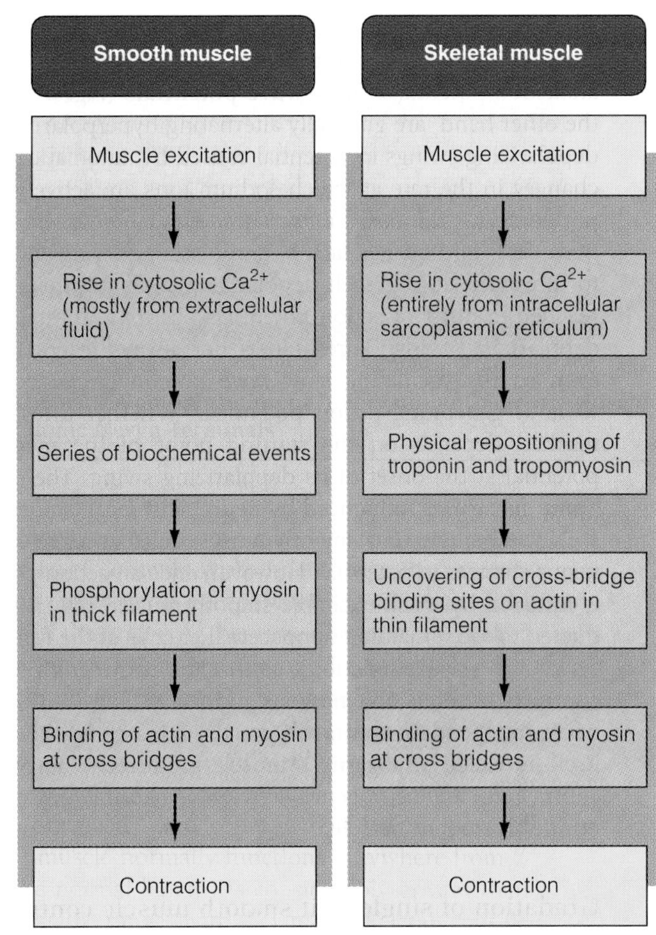

■ *Figure 8-30* **Comparison of the Role of Calcium in Bringing about Contraction in Smooth Muscle and Skeletal Muscle**

■ *Figure 8-31* **Self-Generated Electrical Activity in Smooth Muscle** (a) In pacemaker activity, the membrane gradually depolarizes to threshold on a regular periodic basis without any nervous stimulation. These regular depolarizations cyclically trigger self-induced action potentials. (b) In slow-wave potentials, the membrane gradually undergoes self-induced hyperpolarizing and depolarizing swings in potential. A burst of action potentials occurs if a depolarizing swing brings the membrane to threshold.

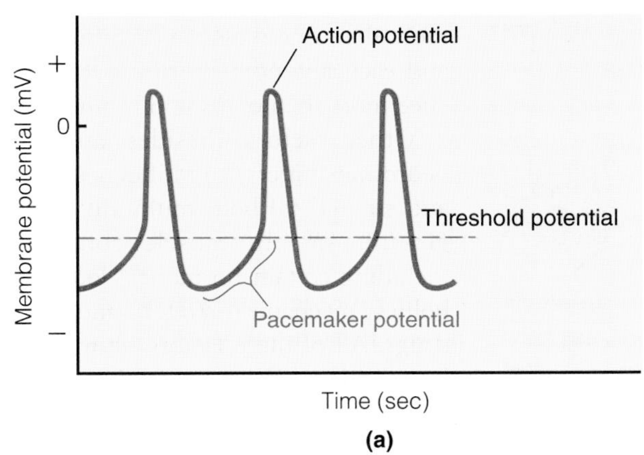

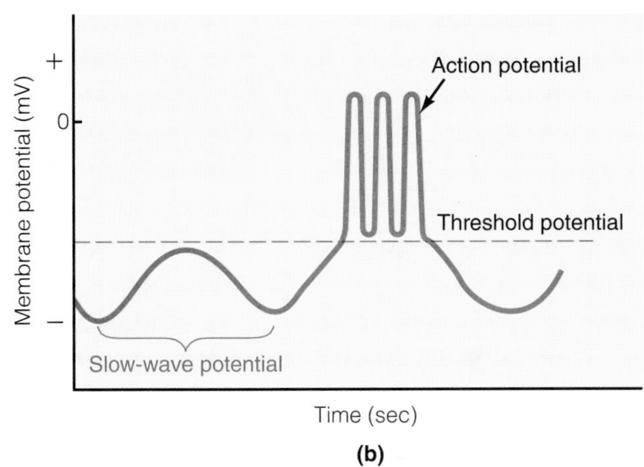

because each cross-bridge cycle uses up one molecule of ATP. Smooth muscle is therefore an economical contractile tissue, making it well suited for long-term sustained, contractions with little energy consumption and without fatigue. Unlike the rapidly changing demands placed on our skeletal muscles as we maneuver through and manipulate our external environment, our smooth muscle activities are geared for long-term duration and slower adjustments to change.

Because of its slowness and the less-ordered arrangement of its filaments, smooth muscle has often been mistakenly viewed as a poorly developed version of skeletal muscle. Actually, smooth muscle is just as highly specialized for the demands placed on it—that is, being able to economically maintain tension for prolonged periods without fatigue and being able to accommodate considerable variations in the volume of contents it encloses with little change in tension. It is an extremely adaptive, efficient tissue.

Nutrient and O_2 delivery are generally adequate to support the smooth muscle contractile process. Smooth muscle can use a wide variety of nutrient molecules for ATP production. There are no energy storage pools comparable to creatine phosphate in smooth muscle; they are not necessary. Oxygen delivery is usually adequate to keep pace with the low rate of oxidative phosphorylation needed to provide ATP for the energy-efficient smooth muscle. If necessary, anaerobic glycolysis can sustain adequate ATP production if O_2 supplies are diminished.

Cardiac muscle blends features of both skeletal and smooth muscle.

Cardiac muscle, found only in the heart, shares structural and functional characteristics with both skeletal and single-unit smooth muscle. Like skeletal muscle, cardiac muscle is striated, with its thick and thin filaments highly organized into a regular banding pattern. Cardiac thin filaments contain troponin and tropomyosin, which constitute the site of Ca^{2+} action in turning on cross-bridge activity, as in skeletal muscle. Also similar to skeletal muscle, cardiac muscle has a clear-cut length-tension relationship. Like the oxidative skeletal muscle fibers, cardiac muscle cells have an abundance of mitochondria and myoglobin. They also possess T tubules and a moderately well-developed sarcoplasmic reticulum.

As in smooth muscle, Ca^{2+} enters the cytosol from both the ECF and the sarcoplasmic reticulum during cardiac excitation. Ca^{2+} entry from the ECF through voltage-gated dihydropyridine receptors alias Ca^{2+} channels in the T tubule membrane triggers the release of Ca^{2+} intracellularly from the sarcoplasmic reticulum. Like single-unit smooth muscle, the heart displays pacemaker (but not slow-wave) activity, initiating its own action potentials without any external influence. Cardiac cells are interconnected by gap junctions that enhance the spread of action potentials throughout the heart, just as in single-unit smooth muscle. Also similarly, the heart is innervated by the autonomic nervous system, which, along with certain hormones and local factors, can modify the rate and strength of contraction.

Unique to cardiac muscle, the cardiac fibers are joined together in a branching network, and its action potentials have a much longer duration at peak reversed potential before repolarizing. Further details and the importance of cardiac muscle's features are addressed in the next chapter.

Chapter in Perspective: Focus on Homeostasis

The skeletal muscles comprise the muscular system itself. Cardiac and smooth muscle are part of the organs that comprise other body systems. Cardiac muscle is found only in the heart, which is part of the circulatory system. Smooth muscle is found in the walls of hollow organs and tubes, including the blood vessels in the circulatory system, airways in the respiratory system, bladder in the urinary system, stomach and intestines in the digestive system, and uterus and ductus deferens (the duct that provides a route of exit for sperm from the testes) in the reproductive system.

Contraction of skeletal muscles accomplishes movement of the body parts in relation to each other and movement of the whole body in relation to the external environment. Thus, these muscles permit us to move through and manipulate our external environment. At a very general level, some of these movements are aimed at maintaining homeostasis, such as moving the body toward food or away from harm. Examples of more specific homeostatic functions accomplished by skeletal muscles include the chewing and swallowing of food for further breakdown in the digestive system into usable energy-producing nutrient molecules (the mouth and throat muscles are all skeletal muscles) and the process of breathing to obtain O_2 and eliminate CO_2 (the respiratory muscles are all skeletal muscles). Generation of heat by contracting skeletal muscles also serves as the major source of heat production in the maintenance of body temperature. The skeletal muscles further accomplish many nonhomeostatic activities that enable us to work and play so that we may contribute to society and enjoy ourselves.

All of the other systems of the body, except the immune (defense) system, depend on their nonskeletal muscle components to enable them to accomplish their homeostatic functions. For example, contraction of cardiac muscle in the heart pushes life-sustaining blood forward into the blood vessels, and contraction of smooth muscle in the stomach and intestines pushes the ingested food through the digestive tract at a rate appropriate for the digestive juices secreted along the route to break down the food into usable units.

Chapter Summary

Structure of Skeletal Muscle

Muscle cells are specialized for contraction. There are three types of muscle: skeletal, smooth, and cardiac. Skeletal muscles are made up of bundles of long, cylindrical muscle cells known as muscle fibers, wrapped in connective tissue. Muscle fibers are packed with myofibrils, with each myofibril consisting of alternating, slightly overlapping stacked sets of thick and thin filaments. This arrangement leads to a skeletal muscle fiber's striated microscopic appearance. Thick filaments are composed of the protein myosin. Cross bridges made up of the myosin molecules' globular heads project from each thick filament. Thin filaments are composed primarily of the protein actin, which has the ability to bind and interact with the myosin cross bridges to bring about contraction. However, two other proteins, tropomyosin and troponin, lie across the surface of the thin filament to prevent this cross-bridge interaction in the resting state.

Molecular Basis of Skeletal Muscle Contraction

Excitation of a skeletal muscle fiber by its motor neuron brings about contraction through a series of events that results in the thin filaments sliding closer together between the thick filaments. This sliding-filament mechanism of muscle contraction is switched on by the release of Ca^{2+} from the lateral sacs of the sarcoplasmic reticulum. Calcium release occurs in response to the spread of a muscle fiber action protential into the central portions of the fiber by means of the T tubules. Released Ca^{2+} binds to the troponin-tropomyosin complex of the thin filament, causing a slight repositioning of the complex to uncover actin's cross-bridge binding sites. After the exposed actin attaches to a myosin cross bridge, the molecular interaction between actin and myosin releases the energy within the myosin head that was stored from the prior splitting of ATP by the myosin ATPase site. This released energy powers cross-bridge stroking. During a power stroke, an activated cross bridge bends toward the center of the thick filament, "rowing" in the thin filament to which it is attached. With the addition of a fresh ATP molecule to the myosin cross bridge, myosin and actin detach, the cross bridge returns to its original shape, and the cycle is repeated. Repeated cycles of cross-bridge activity slide the thin filaments inward step by step. When there is no longer a local action potential, the lateral sacs actively take up the Ca^{2+}, troponin and tropomyosin slip back into their blocking position, and relaxation occurs. The entire contractile response lasts about 100 times longer than the action potential.

Skeletal Muscle Mechanics

Gradation of whole-muscle contraction can be accomplished by (1) varying the number of muscle fibers contracting within the muscle and (2) varying the tension developed by each contracting fiber. The greater the number of active muscle fibers, the greater the whole-muscle tension. The number of fibers contracting depends on (1) the size of the muscle (the number of muscle fibers present); (2) the extent of motor unit recruitment (how many motor neurons supplying the muscle are active); and (3) the size of each motor unit (how many muscle fibers are activated simultaneously by a single motor neuron).

Also, the greater the tension developed by each contracting fiber, the stronger the contraction of the whole muscle. Two readily variable factors having an effect on the fiber tension are (1) the frequency of stimulation, which determines the extent of twitch summation, and (2) the length of the fiber before the onset of contraction. Twitch summation refers to the increase in tension accompanying repetitive stimulation of the muscle fiber. After undergoing an action potential, the muscle cell membrane recovers from its refractory period and is able to be restimulated again while some contractile activity triggered by the first action potential still remains. As a result, the contractile responses (twitches) induced by the two rapidly successive action potentials are able to sum, increasing the tension developed by the fiber. If the muscle fiber is stimulated so rapidly that it does not have a chance to start relaxing between stimuli, a smooth, sustained maximal (maximal for the fiber at that length) contraction known as tetanus takes place.

The tension developed upon a tetanic contraction also depends on the length of the fiber at the onset of contraction. At the optimal length (l_o), which is the resting muscle length, there is maximal opportunity for cross-bridge interaction due to optimal overlap of thick and thin filaments; thus, the greatest tension can be developed. At lengths shorter or longer than l_o, less tension can be developed upon contraction, primarily because a portion of the cross bridges are unable to participate.

The two primary types of muscle contraction—isometric (constant length) and isotonic (constant tension)—depend on the relationship between muscle tension and the load. If tension is less than the load, the muscle cannot shorten and lift the object but remains at constant length, producing an isometric contraction. In an isotonic contraction, the tension exceeds the load so the muscle can shorten and lift the object, maintaining constant tension throughout the period of shortening.

The bones, muscles, and joints form lever systems. The most common type amplifies the velocity and distance of muscle shortening to increase the speed and range of motion of the body part moved by the muscle. This increased maneuverability is accomplished at the expense of the muscle having to exert considerably more force than the load.

Skeletal Muscle Metabolism and Fiber Types

Three biochemical pathways furnish the ATP needed for muscle contraction: (1) the transfer of high-energy phosphates from stored creatine phosphate to ADP, providing the first source of ATP at the onset of exercise; (2) oxidative phosphorylation, which efficiently extracts large amounts of ATP from nutrient molecules if sufficient O_2 is available to support this system; and (3) glycolysis, which can synthesize ATP in the absence of O_2 but uses large amounts of stored glycogen and produces lactic acid in the process.

There are three types of muscle fibers, classified by the pathways they use for ATP synthesis (oxidative or glycolytic) and the rapidity with which they split ATP and subsequently contract (slow twitch or fast twitch): slow-oxidative fibers, fast-oxidative fibers, and fast-glycolytic fibers.

Control of Motor Movement

Control of any motor movement depends on the level of activity in the presynaptic inputs that converge on the motor neurons supplying various muscles. These inputs come from three sources: (1) spinal-reflex pathways, which originate with afferent neurons; (2) the corticospinal motor system, which originates at the primary motor cortex and is concerned primarily with discrete, intricate movements of the hands; and (3) the multineuronal motor system, which originates in the brain stem and is mostly involved with postural adjustments and involuntary movements of the trunk and limbs. The final motor output from the brain stem is influenced by the cerebellum, basal nuclei, and cerebral cortex.

Establishment and adjustment of motor commands depend on continuous afferent input, especially feedback about changes in muscle length (monitored by muscle spindles) and muscle tension (monitored by Golgi tendon organs).

Smooth and Cardiac Muscle

The thick and thin filaments of smooth muscle are not arranged in an orderly pattern, so the fibers are not striated. Cytosolic Ca^{2+}, which enters from the extracellular fluid as well as being released from sparse intracellular stores, activates cross-bridge cycling by initiating a series of biochemical reactions that result in phosphorylation of the myosin cross bridges to enable them to bind with actin. Multiunit smooth muscle is neurogenic, requiring stimulation of individual muscle fibers by its autonomic nerve supply to trigger contraction. Single-unit smooth muscle is myogenic; it is able to initiate its own contraction without any external influence as a result of spontaneous depolarizations to threshold potential brought about by automatic shifts in ionic fluxes. Once an action potential is initiated within a single-unit smooth muscle cell, this electrical activity spreads by means of gap junctions to the surrounding cells within the functional syncytium, so that the entire sheet becomes excited and contracts as a unit. The autonomic nervous system as well as hormones and local metabolites can modify the rate and strength of the self-induced contractions. Smooth muscle contractions are energy efficient, enabling this type of muscle to economically sustain long-term contractions without fatigue. This economy, coupled with the fact that single-unit smooth muscle is able to exist at a variety of lengths with little change in tension, makes single-unit smooth muscle ideally suited for its task of forming the walls of distensible hollow organs.

Cardiac muscle is found only in the heart. It has highly organized striated fibers like skeletal muscle. Like single-unit smooth muscle, some cardiac muscle fibers are capable of generating action potentials, which are spread throughout the heart with the aid of gap junctions.

Review Exercises

Objective Questions (Answers on p. E–7.)

1. Upon completion of an action potential in a muscle fiber, the contractile activity initiated by the action potential ceases. (True or false?)
2. The velocity at which a muscle shortens depends entirely on the ATPase activity of its fibers. (True or false?)
3. When a skeletal muscle is maximally stretched, it can develop maximal tension upon contraction because the actin filaments can slide in a maximal distance. (True or false?)
4. A pacemaker potential always initiates an action potential. (True or false?)
5. A slow-wave potential always initiates an action potential. (True or false?)
6. Smooth muscle can develop tension even when considerably stretched because the thin filaments still overlap with the long thick filaments. (True or false?)
7. A(n) ____________ contraction is an isotonic contraction in which the muscle shortens, whereas the muscle lengthens in a(n) ____________ isotonic contraction.
8. ____________ motor neurons supply extrafusal muscle fibers whereas intrafusal fibers are innervated by ____________ motor neurons.
9. The two types of atrophy are ____________ and ____________.
10. Which of the following provide(s) direct input to alpha motor neurons? (Indicate all correct answers.)
    a. primary motor cortex
    b. brain stem
    c. cerebellum
    d. basal nuclei
    e. spinal-reflex pathways
11. Which of the following is *not* involved in bringing about muscle relaxation?
    a. reuptake of Ca^{2+} by the sarcoplasmic reticulum
    b. no more ATP
    c. no more action potential
    d. removal of ACh at the end plate by acetylcholinesterase
    e. filaments sliding back to their resting position
12. Match the following (with reference to skeletal muscle):
    1. Ca^{2+}
    2. T tubule
    3. ATP
    4. lateral sac of the sarcoplasmic reticulum
    5. myosin
    6. troponin-tropomyosin complex
    7. actin

    (a) cyclically binds with the myosin cross bridges during contraction
    (b) has ATPase activity
    (c) supplies energy for the power stroke of a cross bridge
    (d) rapidly transmits the action potential to the central portion of the muscle fiber
    (e) stores Ca^{2+}
    (f) pulls the troponin-tropomyosin complex out of its blocking position
    (g) prevents actin from interacting with myosin when the muscle fiber is not excited

13. Using the answer code at the right, indicate what happens in the banding pattern during contraction:

____1. thick myofilament	(a) remains the same size during contraction
____2. thin myofilament	(b) decreases in length (shortens) during contraction
____3. A band	
____4. I band	
____5. H zone	
____6. sarcomere	

Essay Questions

1. Describe the levels of organization in a skeletal muscle.
2. What is responsible for the striated appearance of skeletal muscles? Describe the arrangement of thick and thin filaments that gives rise to the banding pattern.
3. What is the functional unit of skeletal muscle?
4. Describe the composition of thick and thin filaments.
5. Describe the sliding-filament mechanism of muscle contraction. How do cross-bridge power strokes bring about shortening of the muscle fiber?
6. Compare the excitation-contraction coupling process in skeletal muscle with that in smooth muscle.
7. By what means can gradation of skeletal muscle contraction be accomplished?
8. What is a motor unit? Compare the size of motor units in finely controlled muscles with those specialized for coarse, powerful contractions. Describe motor unit recruitment.
9. Explain the phenomenon of twitch summation and tetanus.
10. What effect does a skeletal muscle fiber's length at the onset of contraction have on the strength of the subsequent contraction?
11. Compare isotonic and isometric contractions.
12. Describe the role of each of the following in powering skeletal muscle contraction: ATP, creatine phosphate, oxidative phosphorylation, and glycolysis. Distinguish between aerobically and anaerobically supported exercise.
13. Compare the three types of skeletal muscle fibers.
14. What are the roles of the corticospinal system and multineuronal system in the control of motor movement?
15. Describe the structure and function of muscle spindles and Golgi tendon organs.
16. Distinguish between multiunit and single-unit smooth muscle.
17. Differentiate between neurogenic and myogenic muscle activity.
18. How can smooth muscle contraction be graded?
19. Compare the contractile speed and relative energy expenditure of skeletal muscle with that of smooth muscle.
20. In what ways is cardiac muscle functionally similar to skeletal muscle and to single-unit smooth muscle?

Quantitative Exercises (Solutions on p. E–7.)

1. Consider two individuals each throwing a softball, one a weekend athlete and the other a professional pitcher.
   a. Given the following information, calculate the velocity of the ball as it leaves the amateur's hand:
      - The distance from his shoulder socket (humeral head) to the ball is 70 cm.
      - The distance from his humeral head to the points of insertion of the muscles moving his arm forward (we must simplify here since the shoulder is such a complex joint) is 9 cm.
      - The velocity of muscle shortening is 2.6 m/sec.

   b. The professional pitcher throws the ball 85 miles per hour. If his points of insertion are also 9 cm from the humeral head and the distance from his humeral head to the ball is 90 cm, how much faster did the professional pitcher's muscles shorten compared to the amateur's?
2. The velocity at which a muscle shortens is related to the force that it can generate in the following way:

$$v = b(F_o - F)/(F + a)$$

   (Hoppensteadt and Peskin 1992)[1], where v is the velocity of shortening, and F_o can be thought of as an "upper load limit," or the maximum force a muscle can generate against a load. The parameter a is inversely proportional to the cross-bridge cycling rate, and b is proportional to the number of sarcomeres in line in a muscle. Draw the load (force)-velocity curve predicted by this equation by plotting the points $F=0$ and $F=F_o$. Values of v are on the vertical axis; and values of F are on the horizontal axis; a, b and F_o are constants.
   a. Notice that the curve generated from this equation is the same as that in Fig. 8-20, p. 239. Why does the curve have this shape; that is, what does the shape of the curve tell you about muscle performance in general?
   b. What happens to the load (force)-velocity curve when F_o is increased? When the cross-bridge cycling rate is increased? When the size of the muscle is increased? How will each of these changes affect the performance of the muscle?

---

[1]Hoppensteadt, F. C. and C. S. Peskin. *Mathematics in Medicine and the Life Sciences* (equation 9.1.1, p. 199). Springer-Verlag 1992.

Points to Ponder

(Explanations on p. E–7.)

1. Why does regular aerobic exercise provide more cardiovascular benefit than weight training does? (Hint: The heart responds to the demands placed on it in a way similar to skeletal muscle.)
2. If the biceps muscle of a child inserts 4 cm from the elbow and the length of the arm from the elbow to the hand is 28 cm, how much force must the biceps generate in order for the child to lift an 8 kg stack of books with one hand?
3. Put yourself in the position of the scientists who discovered the sliding-filament mechanism of muscle contraction by considering what molecular changes must be involved to account for the observed alterations in the banding pattern during contraction. If you were comparing a relaxed and contracted muscle fiber under a high-power light microscope (see Fig. 8-3a, p. 224), how could you determine that the thin filaments do not change in length during muscle contraction? You cannot see or measure a single thin filament at this magnification. (Hint: What landmark in the banding pattern represents each end of the thin filament? If these landmarks are the same distance apart in a relaxed and contracted fiber, then the thin filaments must not change in length.)
4. What type of off-the-snow training would you recommend for a competitive downhill skier versus a competitive cross-country skier? What adaptive skeletal muscle changes would you hope to accomplish in the athletes in each case?
5. A deadly toxin has turned out to be good news for sufferers of a number of painful, disruptive neuromuscular diseases known categorically as *dystonias*. These conditions are characterized by spasms (excessive muscle-contracting activity) that result in involuntary twisting or abnormal postures, depending on the body part affected. For example, painful neck spasms that twist the head to one side occur as a result of *spasmodic torticollis,* the most common dystonia. The Food and Drug Administration has already approved the treatment of some forms of dystonia with botulinum toxin, the toxin responsible for fatal botulism food poisoning. The therapeutic dose, however, is considerably below the amount of toxin needed to induce even mild symptoms of botulism poisoning. Explain how this toxin could be a useful therapy for dystonias. (Hint: See the effect of botulinum toxin on page 217.)
6. ***Clinical Consideration*** Jason W. is waiting impatiently for the doctor to finish removing the cast from his leg, which he broke the last day of school six weeks ago. Summer vacation is half over, and he hasn't been able to swim, play softball, or participate in any of his favorite sports. When the cast is finally off, Jason's excitement is replaced with concern when he sees that the injured limb is noticeably smaller in diameter than his normal leg. What is the explanation for this reduction in size? How can the leg be restored to its normal size and functional ability?

9

Cardiac Physiology

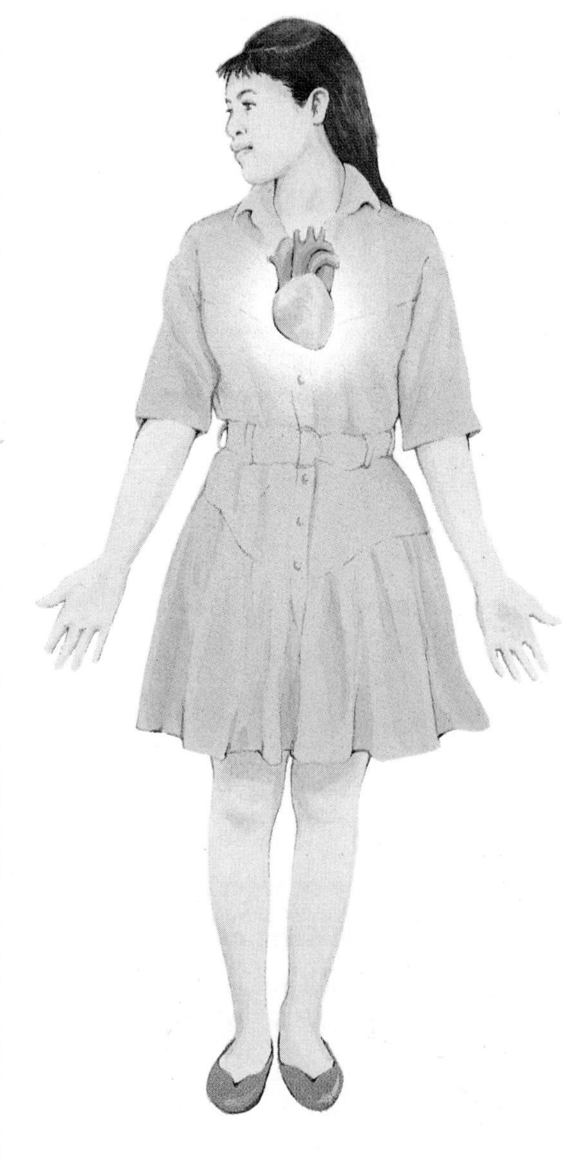

Body systems maintain homeostasis

HOMEOSTASIS
The circulatory system contributes to homeostasis by transporting O_2, CO_2, wastes, electrolytes and hormones from one part of the body to another

Homeostasis is essential for survival of cells

CELLS
Cells need a constant supply of O_2 and nutrients delivered to them by the circulatory system, which also carries away CO_2 and other wastes, in order to power life-sustaining cellular activities by the chemical reaction:

$$\text{Food} + O_2 \rightarrow CO_2 + H_2O + \text{Energy}$$

Cells make up body systems

The maintenance of homeostasis depends on essential materials such as O_2 and nutrients being continually picked up from the external environment and delivered to the cells and on waste products being continually removed. Homeostasis also depends on the transfer of hormones, which are important regulatory chemical messengers, from their site of production to their site of action. The **circulatory system,** which contributes to homeostasis by serving as the body's transport system, consists of the heart, blood vessels, and blood.

All body tissues constantly depend on the life-supporting blood flow provided to them by the contraction or beating of the **heart.** The heart drives the blood through the blood vessels for delivery to the tissues in sufficient amounts, whether the body is at rest or engaging in vigorous exercise.

Chapter Contents At a Glance

Introduction

From just a matter of days following conception until death, the beat goes on. In fact, throughout an average human life span, the heart contracts about 3 billion times, never stopping to rest except for a fraction of a second between beats. Within about three weeks after conception, even before the mother can confirm she is pregnant, the heart of the developing embryo starts to function. It is believed to be the first organ to become functional. At this time the human embryo is only a few millimeters long, about the size of a capital letter on this page.

Why does the heart develop so early, and why is it so crucial throughout life? It is because the circulatory system is the transport system of the body. A human embryo, having very little yolk available as food, depends on the prompt establishment of a circulatory system that can interact with the maternal circulation to pick up and distribute to the developing tissues the supplies so critical for survival and growth. Thus begins the story of the circulatory system, which continues throughout life to be a vital pipeline for transporting materials on which the cells of the body are absolutely dependent.

The **circulatory system** consists of three basic components:

1. The **heart** serves as the pump that imparts pressure to the blood to establish the pressure gradient needed for blood to flow to the tissues. Blood, like all liquids, flows down a pressure gradient from an area of higher pressure to an area of lower pressure.
2. The **blood vessels** serve as the passageways through which blood is directed and distributed from the heart to all parts of the body and subsequently returned to the heart (chapter 10).
3. The **blood** serves as the transport medium within which materials being transported are dissolved or suspended (chapter 11).

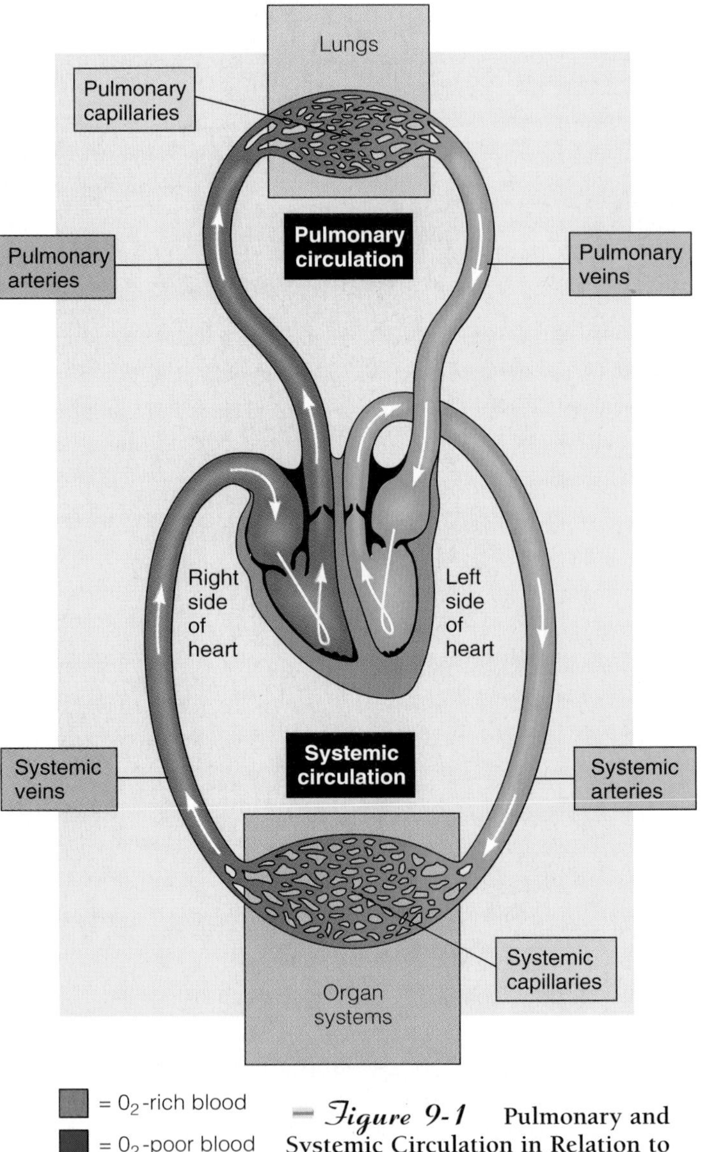

Figure 9-1 **Pulmonary and Systemic Circulation in Relation to the Heart** The circulatory system consists of two separate vascular loops: the pulmonary circulation, which carries blood between the heart and lungs, and the systemic circulation, which carries blood between the heart and organ systems.

The blood travels continuously through the circulatory system to and from the heart through two separate vascular (blood vessel) loops, both originating and terminating at the heart (Fig. 9-1). The **pulmonary** (lung) **circulation** consists of a closed loop of vessels carrying blood between the heart and lungs, whereas the **systemic circulation** consists of a circuit of vessels carrying blood between the heart and other organ systems.

Anatomical Considerations

The heart is located in the middle of the chest cavity.

The heart is a hollow muscular organ about the size of a clenched fist. It is located in the **thoracic** (chest) **cavity** approximately midline between the **sternum** (breastbone) anteriorly and the **vertebrae** (backbone) posteriorly (Fig. 9-2a). The midline location of the heart brings up a potentially confusing point. Place your hand over your heart as if to recite the Pledge of Allegiance. Where did you place your hand? People usually place their hand on the left side of the chest, even though the heart is actually in the middle of the chest. The heart has a broad base at the top and tapers to a pointed tip known as the **apex** at the bottom. It is situated at an angle under the sternum so that its base lies predominantly to the right and the apex to the left of the sternum. When the heart beats, especially when it contracts forcefully, the apex actually thumps against the inside of the chest wall on the left side. Because we become aware of the beating heart through the apex beat on the left side of the chest, we tend to think—erroneously—that the entire heart is on the left.

The fact that the heart is positioned between two bony structures, the sternum and vertebrae, makes it possible to manually drive blood out of the heart when it is not pumping effectively by rhythmically depressing the sternum (Fig. 9-2b). This maneuver compresses the heart between the sternum and vertebrae so that blood is squeezed out as if the heart were beating. In many instances, this *external cardiac compression,* which is part of **cardiopulmonary resuscitation (CPR),**

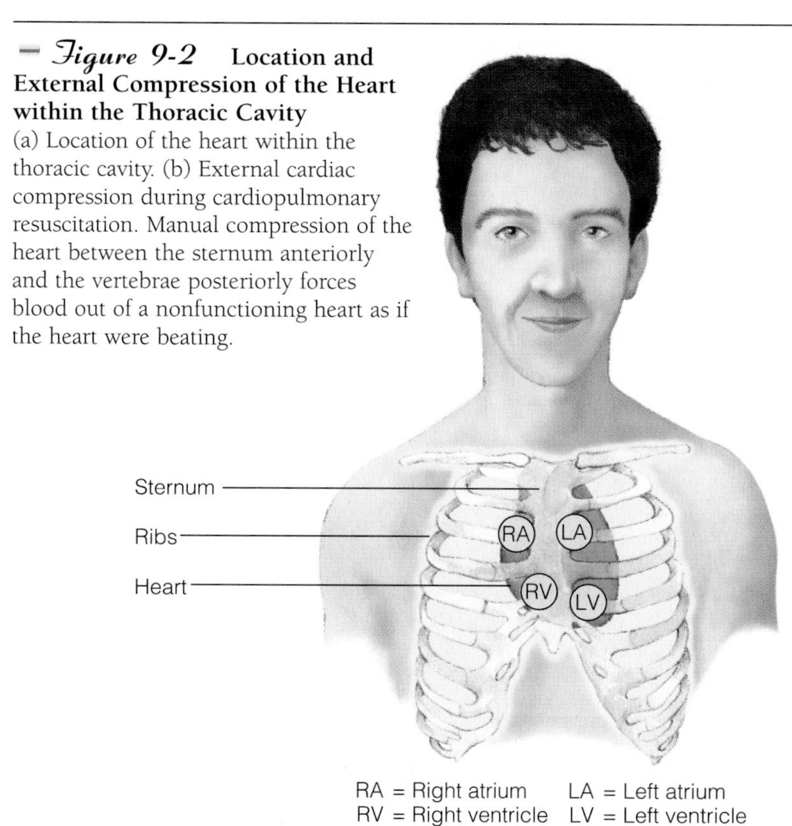

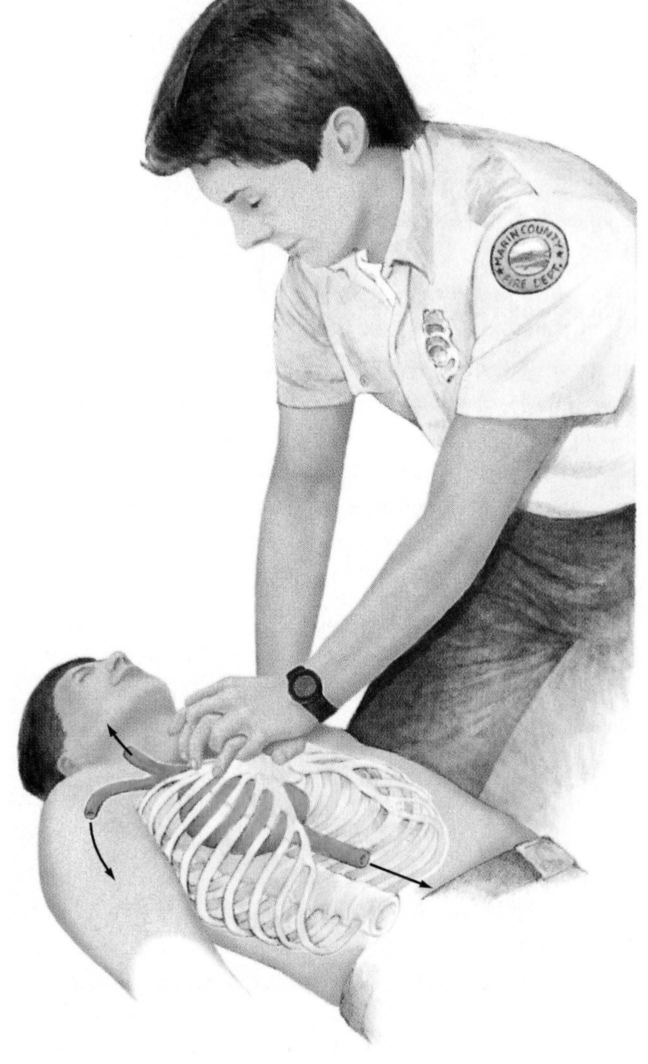

Figure 9-2 Location and External Compression of the Heart within the Thoracic Cavity
(a) Location of the heart within the thoracic cavity. (b) External cardiac compression during cardiopulmonary resuscitation. Manual compression of the heart between the sternum anteriorly and the vertebrae posteriorly forces blood out of a nonfunctioning heart as if the heart were beating.

serves as a lifesaving measure until appropriate therapy can be instituted to restore the heart to normal function.

The heart is a dual pump.

Even though anatomically the heart is a single organ, the right and left sides of the heart function as two separate pumps. The heart is divided into right and left halves and has four chambers, an upper and a lower chamber within each half (■ Fig. 9-3a). The upper chambers, the **atria** (**atrium,** singular), receive blood returning to the heart and transfer it to the lower chambers, the **ventricles,** which pump the blood from the heart. The vessels that return blood from the tissues to the atria are **veins,** and those that carry blood away from the ventricles to the tissues are **arteries.** The two halves of the heart are separated by the **septum,** a continuous muscular partition that prevents mixture of blood from the two sides of the heart. This separation is extremely important, because the right half of the heart is receiving and pumping O_2-poor blood while the left side of the heart receives and pumps O_2-rich blood.

Let us examine how the heart functions as a dual pump by tracing a drop of blood through one complete circuit (Fig. 9-3a and b). Blood returning from the systemic circulation enters the right atrium via large veins known as the **venae cavae.** The drop of blood entering the right atrium has

■ *Figure 9-3* **Blood Flow through and Pump Action of the Heart** (a) Blood flow through the heart. (b) Dual pump action of the heart. The right side of the heart receives O_2 -poor blood from the systemic circulation and pumps it into the pulmonary circulation. The left side of the heart receives O_2 -rich blood from the pulmonary circulation and pumps it into the systemic circulation. (c) Comparison of the thickness of the right and left ventricular walls. Note that the left ventricular wall is much thicker than the right wall.

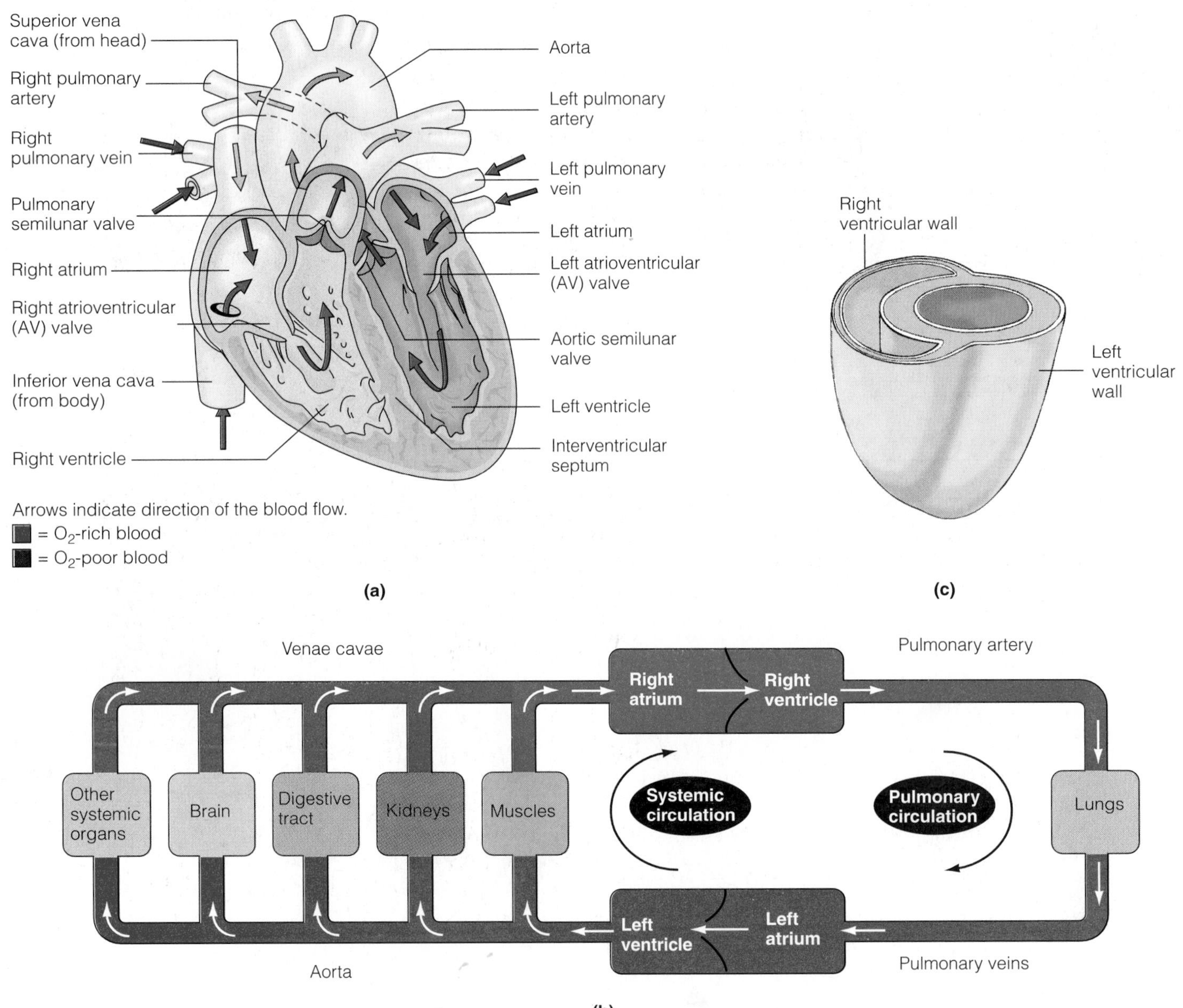

returned from the body tissues, where O_2 has been extracted from it and CO_2 has been added to it. This partially deoxygenated blood flows from the right atrium into the right ventricle, which pumps it out through the **pulmonary artery** to the lungs. Thus, the *right side of the heart pumps blood into the pulmonary circulation.* Within the lungs, the drop of blood loses its extra CO_2 and picks up a fresh supply of O_2 before being returned to the left atrium via the **pulmonary veins.** This O_2-rich blood returning to the left atrium subsequently flows into the left ventricle, the pumping chamber that propels the blood to all body systems except the lungs; that is, *the left side of the heart pumps blood into the systemic circulation.* The large artery carrying blood away from the left ventricle is the **aorta.** Major arteries branch from the aorta to supply the various tissues of the body.

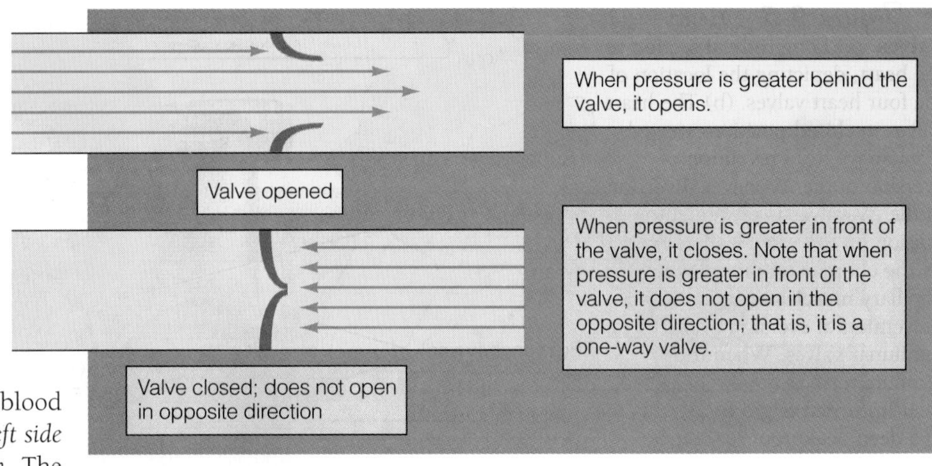
When pressure is greater behind the valve, it opens.
Valve opened
When pressure is greater in front of the valve, it closes. Note that when pressure is greater in front of the valve, it does not open in the opposite direction; that is, it is a one-way valve.
Valve closed; does not open in opposite direction

Figure 9-4 **Mechanism of Valve Action**

In contrast to the pulmonary circulation, in which all the blood flows through the lungs, the systemic circulation may be viewed as a series of parallel pathways. Part of the blood pumped out by the left ventricle goes to the muscles, part to the kidneys, part to the brain, and so on. Thus, the output of the left ventricle is distributed so that each part of the body receives a fresh blood supply; the same arterial blood does not pass from tissue to tissue. Accordingly, the drop of blood we are tracing goes to only one of the systemic tissues. Tissue cells take O_2 from the blood and use it to oxidize nutrients for energy production; in the process, the tissue cells form CO_2 as a waste product that is added to the blood (see p. 33). The drop of blood, now partially depleted of O_2 content and increased in CO_2 content, returns to the right side of the heart, which once again will pump it to the lungs. One circuit is complete. (See the boxed feature on p. 272, ◆ Concepts, Challenges, and Controversies.)

Both sides of the heart simultaneously pump equal amounts of blood. The volume of O_2-poor blood being pumped to the lungs by the right side of the heart soon becomes the same volume of O_2-rich blood being delivered to the tissues by the left side of the heart. The pulmonary circulation is a low-pressure, low-resistance system, whereas the systemic circulation is a high-pressure, high-resistance system. Therefore, even though the right and left sides of the heart pump the same amount of blood, the left side performs more work, because it pumps an equal volume of blood at a higher pressure into a higher-resistance system. Accordingly, the heart muscle on the left side is much thicker than the muscle on the right side, making the left side a stronger pump (Fig. 9-3c).

Heart valves ensure that the blood flows in the proper direction through the heart.

Blood flows through the heart in one fixed direction from veins to atria to ventricles to arteries. The presence of four one-way heart valves ensures this unidirectional flow of blood. The valves are positioned so that they open and close passively because of pressure differences, similar to a one-way door (Fig. 9-4). A forward pressure gradient (that is, a greater pressure behind the valve) forces the valve open, much as you open a door by pushing on one side of it, whereas a backward pressure gradient (that is, a greater pressure in front of the valve) forces the valve closed, just as you apply pressure to the opposite side of the door to close it. Note that a backward gradient can force the valve closed but cannot force it to swing open in the opposite direction; that is, heart valves are not like swinging, saloon-type doors.

Two of the heart valves, the **right** and **left atrioventricular (AV) valves,** are positioned between the atrium and the ventricle on the right and left sides, respectively (Fig. 9-5a). These valves allow blood to flow from the atria into the ventricles during ventricular filling (when atrial pressure exceeds ventricular pressure), but prevent the backflow of blood from the ventricles into the atria during ventricular emptying (when ventricular pressure greatly exceeds atrial pressure). If the rising ventricular pressure did not force the AV valves to close as the ventricles contracted to empty, much of the blood would inefficiently be forced back into the atria and veins instead of being pumped into the arteries. The right AV valve is also called the **tricuspid valve** (*tri* means "three") because it consists of three cusps or leaflets (Fig. 9-5b). Likewise, the left AV valve, which consists of two cusps, is often called the **bicuspid valve** (*bi* means "two") or, alternatively, the **mitral valve** (because of its physical resemblance to a mitre, or bishop's headgear).

The edges of the AV valve leaflets are fastened by tough, thin, fibrous cords of tendinous-type tissue, the **chordae tendineae,** which prevent the valves from being everted, that is, from being forced by the high ventricular pressure to open in the opposite direction into the atria. These cords extend from the edges of each cusp and attach to small, nipple-shaped **papillary muscles** (*papilla* means "nipple"), which protrude from the inner surface of the ventricular walls. When the ventricles contract, the papillary muscles also contract, pulling downward on the chordae tendineae. This pulling exerts tension on the closed AV valve cusps to hold them in position, thus helping them remain tightly sealed in the face of a strong backward pressure gradient (Fig. 9-5c).

The spread of cardiac excitation is coordinated to ensure efficient pumping.

Once initiated in the SA node, an action potential spreads throughout the rest of the heart. For efficient cardiac function, the spread of excitation should satisfy three criteria:

1. *Atrial excitation and contraction should be complete before the onset of ventricular contraction.* Complete ventricular filling requires that atrial contraction precede ventricular contraction. During the period of cardiac relaxation, the AV valves are open, so that venous blood entering the atria continues to flow directly into the ventricles. Almost 80% of ventricular filling occurs by this means prior to atrial contraction. When the atria do contract, additional blood is squeezed into the ventricles to complete ventricular filling. Ventricular contraction then occurs to eject blood from the heart into the arteries. If the atria and ventricles were to contract simultaneously, the AV valves would be closed immediately because ventricular pressures would greatly exceed atrial pressures. The ventricles have much thicker walls and, accordingly, can generate more pressure. Atrial contraction would be unproductive because the atria could not squeeze blood into the ventricles through closed valves. Therefore, to ensure complete filling of the ventricles—to obtain the remaining 20% of ventricular filling that occurs during atrial contraction—it is imperative that the atria become excited and contract before ventricular excitation and contraction.
2. *Excitation of cardiac muscle fibers should be coordinated to ensure that each heart chamber contracts as a unit to accomplish efficient pumping.* If the muscle fibers in a heart chamber were to become excited and contract randomly rather than contracting simultaneously in a coordinated fashion, they would be unable to eject blood. A smooth, uniform ventricular contraction is essential to squeeze out the blood. As an analogy, assume that you have a basting syringe full of water. If you merely poke a finger here or there into the rubber bulb of the syringe, you will not eject much water. However, if you compress the bulb in a smooth, coordinated fashion, you can squeeze out the water. In a similar manner, contraction of isolated cardiac muscle fibers is not successful in pumping blood. Such random, uncoordinated excitation and contraction of the cardiac cells is known as **fibrillation.** Ventricular fibrillation rapidly causes death because the heart is not able to pump blood into the arteries. This condition can often be corrected by **electrical defibrillation,** in which a very strong electrical current is applied on the chest wall. When this current reaches the heart, it essentially stimulates (depolarizes) all parts of the heart simultaneously. Usually, the first part of the heart to recover is the SA node, which takes over pacemaker activity, once again initiating impulses that trigger the synchronized contraction of the remainder of the heart.
3. *The pair of atria and pair of ventricles should be functionally coordinated so that both members of the pair contract simultaneously.* This permits synchronized pumping of blood into the pulmonary and systemic circulation.

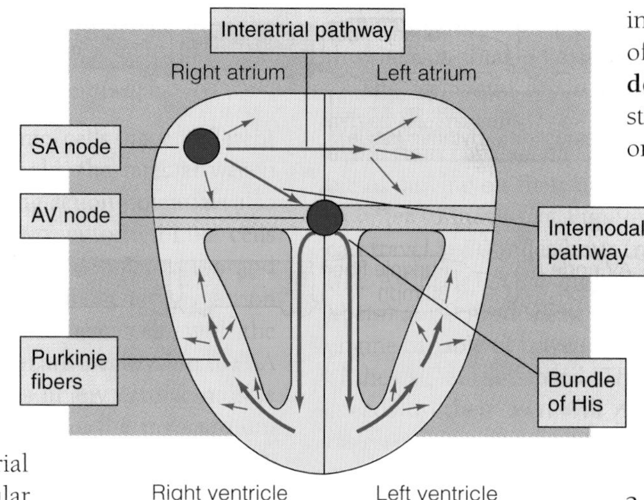

Figure 9-14 **Spread of Cardiac Excitation** An action potential initiated at the SA node first spreads throughout both atria. Its spread is facilitated by two specialized atrial conduction pathways, the interatrial and internodal pathways. The AV node is the only point where an action potential can spread from the atria to the ventricles. From the AV node, the action potential spreads rapidly throughout the ventricles, hastened by a specialized ventricular conduction system consisting of the bundle of His and Purkinje fibers.

The normal spread of cardiac excitation is carefully orchestrated to ensure that these criteria are met and the heart functions efficiently (Fig. 9-14).

Atrial excitation An action potential originating in the SA node first spreads throughout both atria, primarily from cell to cell via gap junctions. In addition several poorly delineated, specialized conduction pathways hasten conduction of the impulse through the atria (Fig. 9-11):

- The **interatrial pathway** extends from the SA node within the right atrium to the left atrium. Because this pathway rapidly transmits the action potential from the SA node to the pathway's termination in the left atrium, a wave of excitation can spread across the gap junctions throughout the left atrium at the same time a similar spread is being accomplished throughout the right atrium. This ensures that both atria become depolarized to contract more or less simultaneously.
- The **internodal pathway** extends from the SA node to the AV node. The AV node is the only point of electrical contact between the atria and ventricles; in other words, because the atria and ventricles are structurally connected by electrically nonconductive fibrous tissue, the only way an action potential in the atria can spread to the ventricles is by passing through the AV node. The internodal conduction pathway directs the spread of an action potential originating at the SA node to the AV node to ensure sequential contraction of the ventricles following atrial contraction.

Transmission between the atria and the ventricles The action potential is conducted relatively slowly through the AV node. This slowness is advantageous because it allows time for complete ventricular filling to occur. The impulse is delayed about 0.1 second (the **AV nodal delay**), which enables the atria to become completely depolarized and to contract, emptying their contents into the ventricles, before ventricular depolarization and contraction occur.

Ventricular excitation Following the AV nodal delay, the impulse travels rapidly down the bundle of His and throughout the ventricular myocardium via the Purkinje fibers. The network of fibers in this ventricular conduction system is specialized for rapid propagation of action potentials. Its presence hastens and coordinates the spread of ventricular excitation to ensure that the ventricles contract as a unit. Although this system carries the action potential rapidly to a large number of cardiac muscle cells, it does not terminate on every cell. The impulse quickly spreads from the excited cells to the remainder of the ventricular muscle cells by means of gap junctions.

The ventricular conduction system is more highly organized and more important than the interatrial and internodal conduction pathways. Because the ventricular mass is so much larger than the atrial mass, it is crucial that a rapid conduction system be present to hasten the spread of excitation in the ventricles. If the entire ventricular depolarization process depended on the cell-to-cell spread of the impulse via gap junctions, the ventricular tissue immediately adjacent to the AV node would become excited and contract before the impulse had even passed to the apex of the heart. This, of course, would not allow efficient pumping. The rapid conduction of the action potential down the bundle of His and its swift, diffuse distribution throughout the Purkinje network lead to almost simultaneous activation of the ventricular myocardial cells in both ventricular chambers, which ensures a single, smooth, coordinated contraction that can efficiently eject blood into both the systemic and pulmonary circulations at the same time.

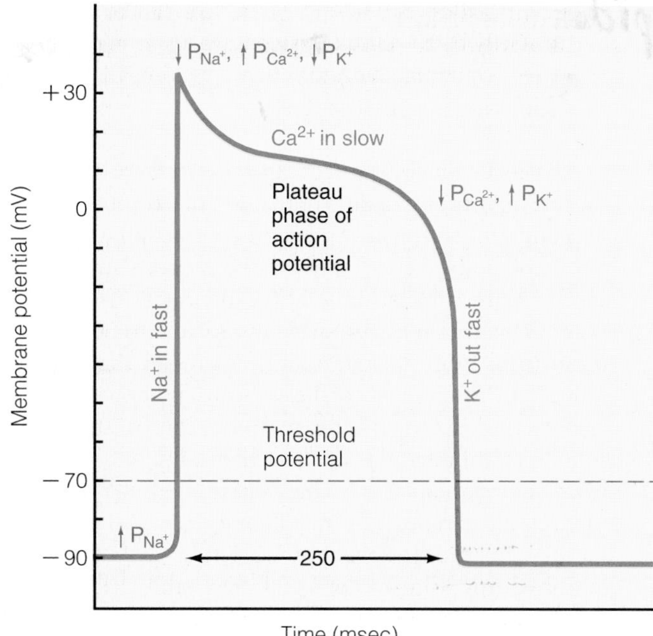

Figure 9-15 **Action Potential in Contractile Cardiac Muscle Cells** The action potential in cardiac contractile cells differs considerably from the action potential in cardiac autorhythmic cells (compare with Fig. 9-10). The membrane potential of cardiac contractile cells remains at a resting potential of −90 mV until excited. Similar to most excitable cells, the rising phase of the action potential is caused by a fast Na^+ influx and the falling phase by a fast K^+ efflux. Unique to cardiac contractile cells, the membrane potential is maintained near the peak of the action potential for several hundred milliseconds. This plateau phase of the action potential results from a slow influx of Ca^{2+} coupled with a marked decrease in K^+ permeability.

The action potential of contractile cardiac muscle cells shows a characteristic plateau.

The action potential in contractile cardiac muscle cells, although initiated by the nodal pacemaker cells, varies considerably in ionic mechanisms and shape from the SA node potential (compare Figs. 9-10 and 9-15). Unlike autorhythmic cells, the membrane of contractile cells remains essentially at rest at about −90 mV until excited by electrical activity propagated from the pacemaker. Once the membrane of a ventricular myocardial contractile cell is excited, an action potential is generated by a complicated interplay of permeability changes and membrane potential changes as follows (Fig. 9-15):

1. During the rising phase of the action potential, the membrane potential rapidly becomes reversed to a positive value of +30 mV as a result of an explosive increase in membrane permeability to Na^+ and a subsequent massive Na^+ influx. Thus far, the process is the same as in neurons and skeletal muscle cells (see p. 87). The Na^+ permeability then rapidly plummets to its low resting value, but, unique to these cardiac muscle cells, the membrane potential is maintained at this positive level for several hundred milliseconds, producing a *plateau phase* of the action potential. In contrast, the short action potential of neurons and skeletal muscle cells lasts less than a millisecond.
2. The sudden change in voltage occurring during the rising phase of the action potential brings about two voltage-dependent permeability changes that are responsible for maintaining this plateau: activation of "slow" Ca^{2+} channels and a marked decrease in K^+ permeability. Opening of the Ca^{2+} channels results in a slow, inward diffusion of Ca^{2+} because Ca^{2+} is in greater concentration in the ECF. This continued influx of positively charged Ca^{2+} prolongs the positivity inside the cell and is primarily responsible for the plateau portion of the action potential. This effect is enhanced by the concomitant decrease in K^+ permeability. The resultant reduction in outflux of positively charged K^+ prevents rapid repolarization of the membrane and thus contributes to prolongation of the plateau phase.
3. The rapid falling phase of the action potential results from inactivation of the Ca^{2+} channels and activation of K^+ channels. The decrease in Ca^{2+} permeability diminishes the slow, inward movement of positive Ca^{2+}, whereas the sudden increase in K^+ permeability simultaneously promotes rapid outward diffusion of positive K^+. Thus, rapid

repolarization at the end of the plateau is accomplished primarily by K^+ efflux, which once again makes the inside of the cell more negative than the outside and restores membrane potential to resting.

The mechanism by which an action potential in a cardiac muscle fiber brings about contraction of that fiber has features in common with the excitation-contraction coupling processes of both skeletal muscle and smooth muscle (Fig. 9-16). As in skeletal muscle, the presence of a local action potential within the T tubules causes Ca^{2+} to be released into the cytosol from the intracellular stores in the sarcoplasmic reticulum. In contrast to skeletal muscle, however, the voltage-gated dihydropyridine receptors in cardiac muscle T tubule membrane do not mechanically open foot protein Ca^{2+} channels in the sarcoplasmic reticulum membrane (see p. 230). Instead, cardiac muscle's T tubular dihydropyridine receptors, like smooth muscle's surface membrane dihydropyridine receptors, serve as Ca^{2+} channels that are opened in response to local membrane depolarization (see p. 256). Thus, unlike in skeletal muscle, Ca^{2+} diffuses into the cytosol across the T tubule membrane during a cardiac action potential. Similar to smooth muscle, this entering Ca^{2+} triggers release of Ca^{2+} from the sarcoplasmic reticulum. In contrast to skeletal muscle cells, Ca^{2+} also diffuses into the cytosol across the surface plasma membrane from the ECF during a cardiac action potential. This entering Ca^{2+} triggers even further release of Ca^{2+} from the sarcoplasmic reticulum. This extra supply of Ca^{2+} is not only the major factor responsible for the prolongation of the cardiac action potential but is also responsible for the subsequent lengthening of the period of cardiac contraction, which lasts about three times longer than a single skeletal muscle fiber contraction. This increased contractile time ensures adequate time to eject the blood.

As in skeletal muscle, the role of Ca^{2+} within the cytosol is to bind with the troponin-tropomyosin complex and physically pull it aside so that cross-bridge cycling and contraction can take place (see p. 226). However, unlike skeletal muscle, in which sufficient Ca^{2+} is always released to turn on all of the cross bridges, in cardiac muscle the extent of cross-bridge activity varies with the amount of cytosolic Ca^{2+}. Removal of Ca^{2+} from the cytosol by active pumps in both the plasma membrane and sarcoplasmic reticulum restores the blocking action of troponin and tropomyosin, so that contraction ceases and the heart muscle relaxes.

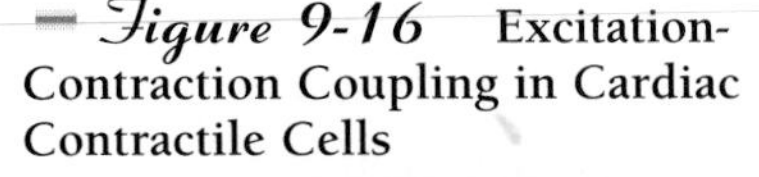

Figure 9-16 **Excitation-Contraction Coupling in Cardiac Contractile Cells**

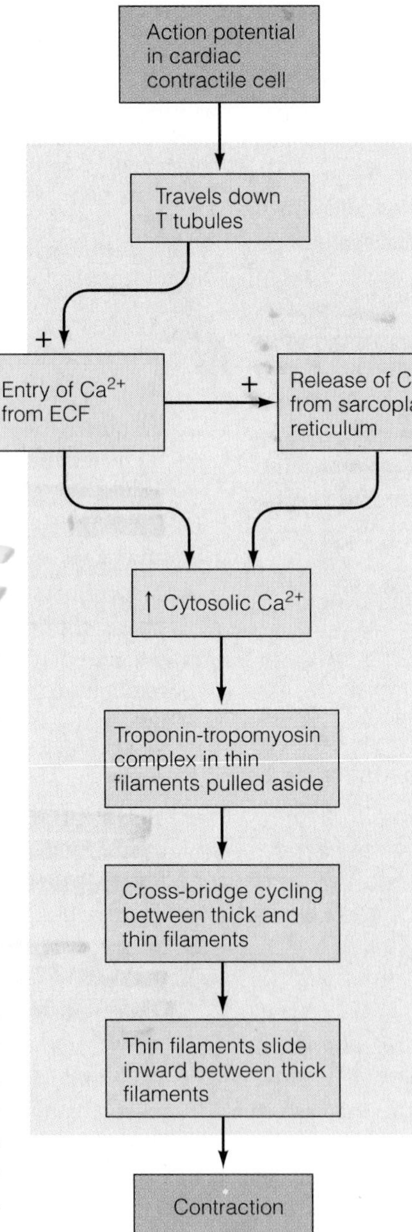

It is not surprising that changes in the ECF concentration of K^+ and Ca^{2+} can have profound effects on the heart. Abnormal levels of K^+ are most important clinically, followed to a lesser extent by Ca^{2+} imbalances. Changes in K^+ concentration in the ECF alter the K^+ concentration gradient between the ICF and ECF, thereby altering the rate of K^+ flux across the membrane. Normally, there is substantially more K^+ inside the cells than in the ECF, but with elevated ECF K^+ levels, this gradient is reduced. Associated with this change is a reduction in "resting" potential (that is, the membrane is less negative on the inside than normal because less K^+ leaves). Among the consequences is a tendency to develop ectopic foci as well as cardiac arrhythmias. Also, because the magnitude of voltage change from the reduced "resting" state to the peak of the action potential is less than from the normal "resting" state to the peak, the resultant diminution of the action potentials' intensity causes the heart to become weak, flaccid, and dilated. At the extreme, with K^+ levels elevated two to three times the normal value, the weakened heart may actually stop pumping.

A rise in ECF Ca^{2+} concentration, on the other hand, augments the strength of cardiac contraction by prolonging the plateau phase of the action potential and by increasing the cytosolic concentration of Ca^{2+}. The heart tends to contract spastically, with little time to rest between contractions. Some drugs that alter cardiac function do so by influencing Ca^{2+} movement across the myocardial cell membranes. For example, Ca^{2+} blocking agents, such as *verapamil,* block Ca^{2+} influx during an action potential, thereby reducing the force of cardiac contraction. Other drugs, such as *digitalis,* increase cardiac contractility by inducing an accumulation of cytosolic Ca^{2+}.

Tetanus of cardiac muscle is prevented by a long refractory period.

Like other excitable tissues, cardiac muscle has a refractory period. During the refractory period, which occurs immediately after the initiation of an action potential, an excitable membrane's responsiveness is totally abolished, making it impossible for another action potential to be generated. In skeletal muscle, the refractory period is very short compared with the duration of the resultant contraction, so the fiber can be restimulated again before the first contraction is complete to produce summation of contractions. Rapidly repetitive stimulation that does not allow the muscle fiber to relax between stimulations results in a sustained, maximal contraction known as tetanus (see Fig. 8–16, p. 235). In contrast, cardiac muscle has a long refractory period that lasts about 250 msec because of the prolonged action potential. This is almost as long as the period of contraction initiated by the action potential; a cardiac muscle fiber contraction averages about 300 msec in duration (▬ Fig. 9-17). Consequently, cardiac muscle cannot be restimulated until contraction is almost over, making summation of contractions and tetanus of cardiac muscle impossible. This is a valuable protective mechanism, because the pumping of blood requires alternate periods of contraction (emptying) and relaxation (filling). A prolonged tetanic contraction would prove fatal. The heart chambers could not be filled and emptied again.

The chief factor responsible for the long refractory period is inactivation, during the prolonged plateau phase, of the Na^+ channels that were activated during the initial Na^+ influx of the rising phase. Not until the membrane recovers from this inactivation process (when the membrane has already repolarized to resting), can the Na^+ channels be activated once again to begin another action potential.

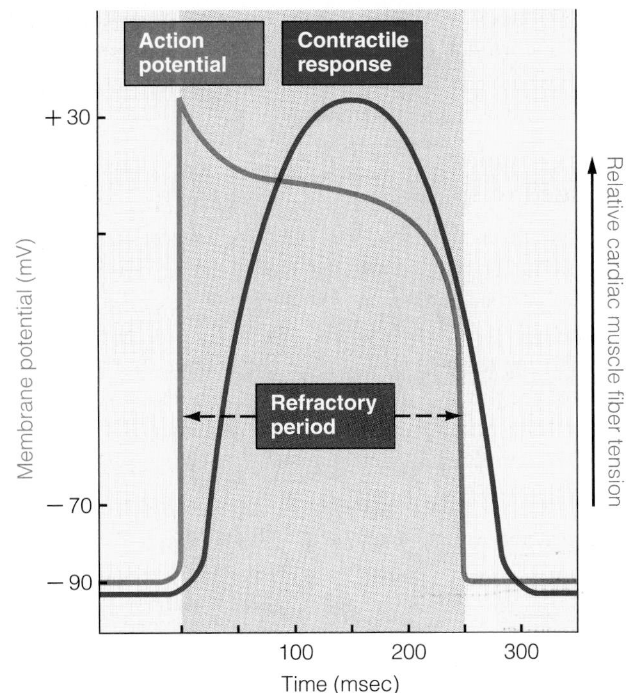

▬ *Figure 9-17* **Relationship of an Action Potential and the Refractory Period to the Duration of the Contractile Response in Cardiac Muscle**

The ECG is a record of the overall spread of electrical activity through the heart.

The electrical currents generated by cardiac muscle during depolarization and repolarization (see p. 83) spread into the tissues surrounding the heart and are conducted through the body fluids. A small portion of this electrical activity reaches the body surface, where it can be detected using recording electrodes. The record produced is an **electrocardiogram,** or **ECG.** (Originally, the term EKG was used, because this technique was developed by a German-speaking scientist, William Einthoven, and "kardia" is the word for *heart* in German.) Three important points should be remembered when considering what an ECG actually represents:

1. An ECG is a recording of that portion of the electrical activity induced in the body fluids by the cardiac impulse that reaches the surface of the body, not a direct recording of the actual electrical activity of the heart.
2. The ECG is a complex recording representing the overall spread of activity throughout the heart during depolarization and repolarization. It is not a recording of a single action potential in a single cell at a single point in time. The record at any given time represents the sum of electrical activity in all of the cardiac muscle cells, some of which may be undergoing action potentials while others may not yet be activated. For example, immediately after firing of the SA node, the atrial cells are undergoing action potentials while the ventricular cells are still at rest. At a later point, the electrical activity will have spread to the ventricular cells while the atrial cells will be repolarizing. Therefore, the overall pattern of cardiac electrical activity varies with time as the impulse passes throughout the heart.
3. The recording represents comparisons in voltage detected by electrodes at two different points on the body surface, not the actual potential. For example, the ECG does not record a potential at all when the ventricular muscle is either completely depolarized or completely repolarized; both electrodes are "viewing" the same potential, so no difference in potential between the two electrodes is recorded.

The exact pattern of electrical activity recorded from the body surface depends on the orientation of the recording electrodes. Electrodes may be loosely thought of as "eyes" that "see" electrical activity and translate it into a visible recording, the ECG record. Whether an upward deflection or downward deflection is recorded is determined by the orientation of electrodes with respect to the current flow in the heart. For example, the spread of excitation across the heart is seen differently from the right arm than from the left foot, and both of these are seen differently than a recording directly over the heart. Even though the same electrical events are occurring in the heart, different waveforms representing the same electrical activity result when this activity is recorded by electrodes at different points on the body.

To provide standard comparisons, ECG records routinely consist of twelve conventional electrode systems, or leads. When an electrocardiograph machine is connected between recording electrodes at two points on the body, the specific arrangement of each pair of connections is called a **lead.** The twelve different leads each record electrical activity in the heart from different locations—six different electrical arrangements from the limbs and six chest leads at various sites around the heart. The same twelve leads are routinely used in all ECG recordings to provide a common basis for comparison and for recognizing deviations from normal (Fig. 9-18).

Various components of the ECG record can be correlated to specific cardiac events.

Interpretation of the wave configurations recorded from each lead depends on a thorough knowledge of the sequence of the spread of cardiac excitation and the position of the heart relative to the placement of the electrodes. A normal ECG exhibits three distinct waveforms: the P wave, the QRS complex, and the T wave (Fig. 9-19). (The letters do not signify anything other than the orderly sequence of the waves. Einthoven simply started in the middle of the alphabet when naming the waves.)

- The **P wave** represents atrial depolarization.
- The **QRS complex** represents ventricular depolarization.
- The **T wave** represents ventricular repolarization.

The following important points about the ECG record should also be noted:

1. Firing of the SA node does not generate sufficient electrical activity to reach the surface of the body, so no wave is recorded for SA nodal depolarization. Therefore, the first recorded wave, the P wave, occurs when the impulse spreads across the atria.
2. In a normal ECG, there is no separate wave for atrial repolarization. The electrical activity associated with atrial repolarization normally occurs simultaneously with ventricular depolarization and is masked by the QRS complex.
3. The P wave is much smaller than the QRS complex because the atria have a much smaller muscle mass than the ventricles and consequently generate less electrical activity.
4. There are three times when no current is flowing in the heart musculature and the ECG remains at baseline:

   a. During the AV nodal delay. This delay is represented by the interval of time between the end of the P wave and

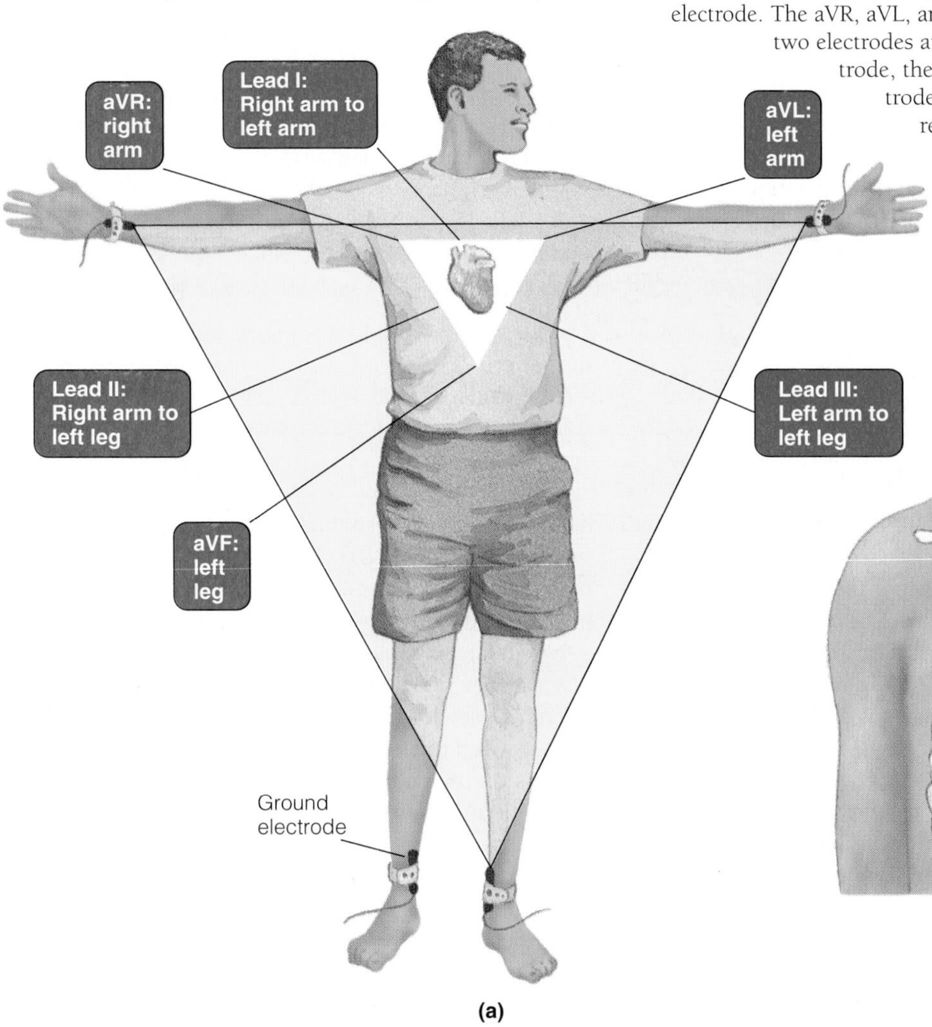

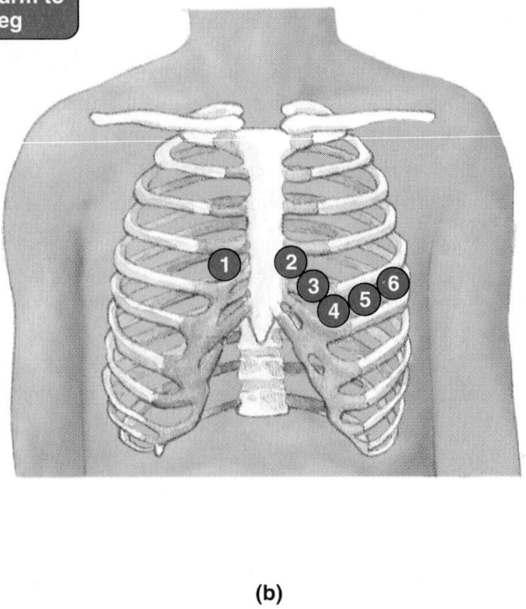

Figure 9-18 **Electrocardiogram Leads** (a) Limb leads. The six limb leads include leads I, II, III, aVR, aVL, and aVF. Leads I, II, and III are bipolar leads because two recording electrodes are used. The tracing records the *difference* in potential between the two electrodes. For example, lead I records the difference in potential detected at the right arm and left arm. The electrode placed on the right leg serves as a ground and is not a recording electrode. The aVR, aVL, and aVF leads are unipolar leads. Even though two electrodes are used, only the actual potential under one electrode, the exploring electrode, is recorded. The other electrode is set at zero potential and serves as a neutral reference point. For example, aVR records the potential reaching the right arm in comparison to the rest of the body. (b) Chest leads. The six chest leads, V_1 through V_6, are also unipolar leads. The exploring electrode mainly records the electrical potential of the cardiac musculature immediately beneath the electrode in six different locations surrounding the heart.

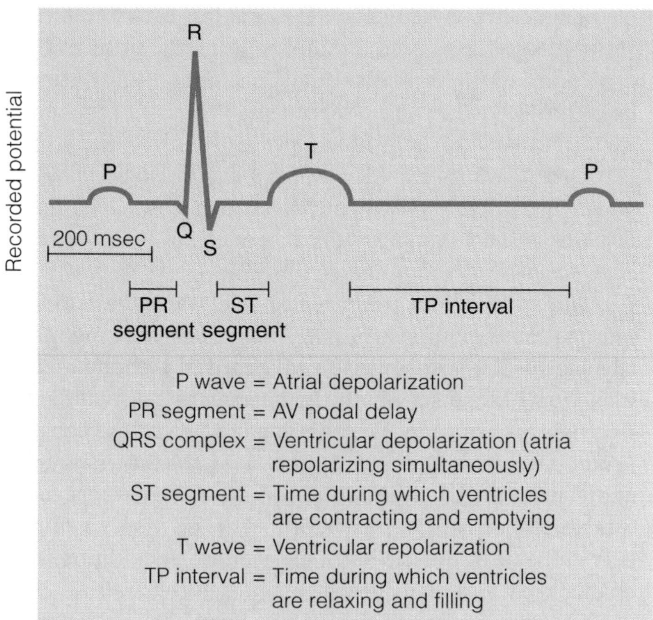

Figure 9-19 **Electrocardiogram Waveforms in Lead II**

the onset of the QRS wave; this interval is known as the **PR segment.** (It is called the PR segment rather than the PQ segment because the Q deflection is small and sometimes absent, whereas the R deflection is the dominant wave of the complex.) Current is flowing through the AV node, but the magnitude is too small to be detected by the ECG electrodes.

b. When the ventricles are completely depolarized and the cardiac contractile cells are undergoing the plateau phase of their action potential before they repolarize again, represented by the **ST segment.** This segment is the interval between QRS and T; it coincides with the time during which ventricular activation is complete and the ventricles are contracting and emptying.

c. When the heart muscle is completely at rest and ventricular filling is taking place, after the T wave and before the next P wave; this time segment is called the **TP interval.**

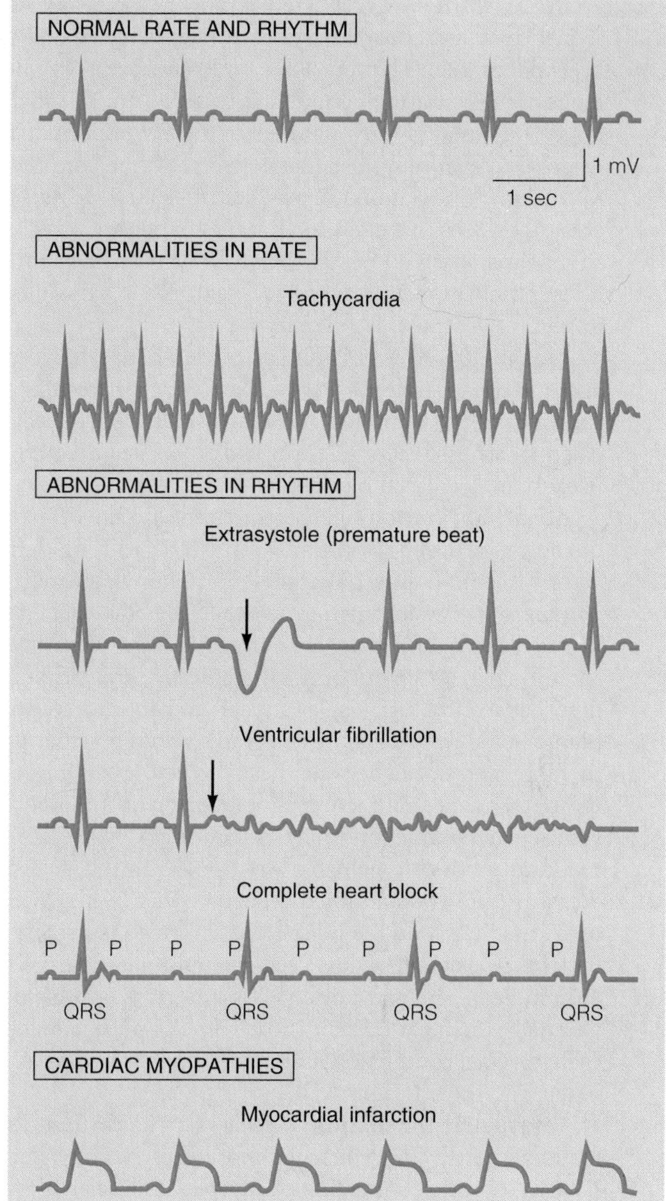

Figure 9-20 **Representative Heart Conditions Detectable through Electrocardiography**

The ECG can be useful in diagnosing abnormal heart rates, arrhythmias, and damage of heart muscle.

Because electrical activity triggers mechanical activity, abnormal electrical patterns are usually accompanied by abnormal contractile activity of the heart. Thus, evaluation of ECG patterns can provide useful information about the status of the heart. The principal deviations from normal that can be ascertained through electrocardiography are (1) abnormalities in rate, (2) abnormalities in rhythm, and (3) cardiac myopathies (Fig. 9-20).

Abnormalities in rate The distance between two consecutive QRS complexes on an ECG record is calibrated to the beat-to-beat heart rate. A rapid heart rate of more than 100 beats per minute is known as **tachycardia** (*tachy* means "fast"), whereas a slow heart rate of fewer than 60 beats per minutes is referred to as **bradycardia** (*brady* means "slow").

Abnormalities in rhythm Rhythm refers to the regularity of the ECG waves. Any variation from the normal rhythm and sequence of excitation of the heart is termed an **arrhythmia.** It may result from the presence of ectopic foci, alterations in SA node pacemaker activity, or interference with conduction. Oftentimes heart rate is also altered. *Extrasystoles* or *premature beats* originating from an ectopic focus are common deviations from normal rhythm. Other abnormalities in rhythm easily detected on an ECG include atrial flutter, atrial fibrillation, ventricular fibrillation, and heart block.

Atrial flutter is characterized by a rapid but regular sequence of atrial depolarizations at rates between 200 to 380 beats per minute. The ventricles rarely keep pace with the racing atria. Because the conducting tissue's refractory period is longer than that of the atrial muscle, the AV node is unable to respond to every impulse that converges on it from the atria. Maybe only one out of every two or three atrial impulses successfully passes through the AV node to the ventricles. Such a situation is referred to as a *2:1* or *3:1 rhythm*. The fact that not every atrial impulse reaches the ventricle in atrial flutter is important, because it precludes a rapid ventricular rate of more than 200 beats per minute. Such a high rate would not allow adequate time for ventricular filling between beats. In such a case, the output of the heart would be reduced to the extent that loss of consciousness or even death could result because of decreased blood flow to the brain.

Atrial fibrillation is characterized by rapid, irregular, uncoordinated depolarizations of the atria with no definite P waves. Accordingly, atrial contractions are chaotic and asynchronized. Since impulses reach the AV node erratically, the ventricular rhythm is also very irregular. The QRS complexes are normal in shape but occur sporadically. Variable lengths of time between ventricular beats are available for ventricular filling. Some ventricular beats come so close together that little filling can occur between beats. When less filling occurs, the subsequent contraction is weaker. In fact, some of the ventricular contractions may be too weak to eject enough blood to produce a palpable wrist pulse. In this situation, if the heart rate is determined directly, either by the apex beat or via the ECG, and the pulse rate is taken concurrently at the wrist, the heart rate will exceed the pulse rate. Such a difference in heart rate and pulse rate is known as a **pulse deficit.** Normally, the heart rate coincides with the pulse rate, because each cardiac contraction initiates a pulse wave as it ejects blood into the arteries.

Ventricular fibrillation is a very serious rhythmic abnormality in which the ventricular musculature exhibits uncoordinated, chaotic contractions. Multiple impulses travel erratically in all directions around the ventricles. The ECG tracing in ventricular fibrillations is very irregular with no detectable pattern or rhythm. The ventricles are ineffectual as pumps when contractions are so disorganized. If circulation is not restored in less than four minutes through external cardiac compression or electrical defibrillation, irreversible brain damage occurs and death is imminent.

Another type of arrhythmia, **heart block,** arises from defects in the cardiac conducting system. The atria still beat regularly, but the ventricles occasionally fail to be stimulated and thus do not contract following atrial contraction. Impulses between the atria and ventricles can be blocked to varying degrees. In some forms of heart block, only every second or third atrial impulse is passed to the ventricles. This is known as *2:1* or *3:1 block,* which can be distinguished from the 2:1 or 3:1 rhythm associated with atrial flutter by the rates involved. In heart block, the atrial rate is normal but the ventricular rate is considerably below normal, whereas in atrial flutter, the atrial rate is very high in accompaniment with a normal or above-normal ventricular rate. *Complete heart block* is characterized by complete dissociation between atrial and ventricular activity, with impulses from the atria not being conducted to the ventricles at all. The atrial beat continues to be governed by the SA node, but the ventricles generate their own impulses at a rate much slower than the atria. On the ECG, the P waves exhibit a normal rhythm. The QRS and T waves also occur regularly but at a much slower rate than the P waves and are completely independent of P wave rhythm.

Cardiac myopathies Abnormal ECG waves are also important in the recognition and assessment of **cardiac myopathies** (damage of the heart muscle). **Myocardial ischemia** refers to inadequate blood supply to the heart tissue. Actual death, or **necrosis,** of heart muscle cells, usually caused by blockage of a blood vessel supplying that area of the heart, is termed **acute myocardial infarction,** commonly known as a **heart attack.** Abnormal QRS waveforms can be seen when a portion of the heart muscle becomes necrotic. In addition to ECG changes, because damaged heart muscle cells release characteristic enzymes into the blood, the level of these enzymes in the blood provides a further index of the extent of myocardial damage.

Interpretation of an ECG is a complex task requiring extensive knowledge and training. The foregoing discussion is not intended to make you an ECG expert by any means but to give you an appreciation of the ways in which the ECG can be used as a diagnostic tool, as well as to present an overview of some of the more common abnormalities of heart function. (For a further use of the ECG, see the accompanying boxed feature, A Closer Look at Exercise Physiology.)

Mechanical Events of the Cardiac Cycle

The heart alternately contracts to empty and relaxes to fill.

The cardiac cycle consists of alternate periods of **systole** (contraction and emptying) and **diastole** (relaxation and filling). The atria and ventricles go through separate cycles of systole and diastole. Contraction occurs as a result of the spread of excitation across the heart, whereas relaxation follows the subsequent repolarization of the cardiac musculature. The following discussion correlates various events that occur concurrently during the cardiac cycle, including ECG features, pressure changes, volume changes, valve activity, and heart sounds. Reference to Figure 9-21 will facilitate this discussion. Only the events on the left side of the heart will be described, but keep in mind that identical events are occurring on the right side of the heart, except that the pressures are lower. Our discussion will begin and end with ventricular diastole to complete one full cardiac cycle.

During early ventricular diastole, the atrium is still also in diastole. This stage corresponds to the TP interval on the ECG—the interval after ventricular repolarization and before another atrial depolarization. Because of the continuous inflow of blood from the venous system into the atrium, atrial

A CLOSER LOOK AT EXERCISE PHYSIOLOGY

The What, Who, and When of Stress Testing

Stress tests, or **graded exercise tests,** are conducted primarily to aid in diagnosing or quantifying heart or lung disease and to evaluate the functional capacity of asymptomatic individuals. The tests are usually given on motorized treadmills or bicycle ergometers (stationary, variable-resistance bicycles). Workload intensity (how hard the subject is working) is adjusted by progressively increasing the speed and incline of the treadmill or by progressively increasing the pedaling frequency and resistance on the bicycle. The test starts at a low intensity and continues until a prespecified workload is achieved, physiological symptoms occur, or the subject is too fatigued to continue.

During diagnostic testing, the patient is monitored with an ECG, and blood pressure is taken each minute. A test is considered positive if ECG abnormalities occur (such as ST segment depression, inverted T waves, or dangerous arrhythmias) or if physical symptoms such as chest pain develop. A test that is interpreted as positive in a person who does not have heart disease is called a false positive test. In men, false positives occur only about 10% to 20% of the time, so the diagnostic stress test for men has a *specificity* of 80% to 90%. Women have a greater frequency of false positive tests with a corresponding lower specificity of about 70%.

The *sensitivity* of a test means that those individuals with disease are correctly identified and there are few false negatives. The sensitivity of the stress test is reported to be 60% to 80%; that is, if 100 individuals with heart disease were tested, 60 to 80 would be correctly identified, but 20 to 40 would have a false negative test. Although stress testing is now an important diagnostic tool, it is just one of several tests used to determine the presence of coronary artery disease.

Stress tests are also conducted on individuals not suspected of having heart or lung disease to determine their present functional capacity. These functional tests are administered in the same way as diagnostic tests, but they are conducted by exercise physiologists and a physician need not be present. These tests are used to establish safe exercise prescriptions, to aid athletes in establishing optimal training programs, and as research tools to evalute the effectiveness of a particular training regimen. Functional stress testing is becoming more prevalent as more people are joining hospital- or community-based wellness programs for disease prevention.

---

pressure slightly exceeds ventricular pressure even though both chambers are relaxed (point 1 in Fig. 9-21). Because of this pressure differential, the AV valve is open, and blood flows directly from the atrium into the ventricle throughout ventricular diastole (heart A in Fig. 9-21). As a result, the ventricular volume slowly continues to rise even before atrial contraction takes place (point 2). Late in ventricular diastole, the SA node reaches threshold and fires. The impulse spreads throughout the atria, which is recorded on the ECG as the P wave (point 3). Atrial depolarization brings about atrial contraction, which squeezes more blood into the ventricle, causing a rise in the atrial pressure curve (point 4). The excitation-contraction coupling process is taking place during the short delay between the P wave and the rise in atrial pressure. The corresponding rise in ventricular pressure (point 5) that occurs simultaneous to the rise in atrial pressure is due to the additional volume of blood added to the ventricle by atrial contraction (point 6 and heart B). Throughout atrial contraction, atrial pressure still slightly exceeds ventricular pressure, so the AV valve remains open.

Ventricular diastole ends at the onset of ventricular contraction. By this time, atrial contraction and ventricular filling are completed. The volume of blood in the ventricle at the end of diastole (point 7) is known as the **end-diastolic volume (EDV),** which averages about 135 ml. No more blood will be added to the ventricle during this cycle. Therefore, the end-diastolic volume is the maximum amount of blood that the ventricle will contain during this cycle.

Following atrial excitation, the impulse passes through the AV node and specialized conduction system to excite the ventricle. Simultaneously, atrial contraction is occurring. By the time ventricular activation is complete, atrial contraction is already accomplished. The QRS complex represents this ventricular excitation (point 8), which induces ventricular contraction. The ventricular pressure curve sharply increases shortly after the QRS complex, signaling the onset of ventricular systole (point 9). The slight delay between the QRS complex and the actual onset of ventricular systole is the time required for the excitation-contraction coupling process to occur. As ventricular contraction begins, ventricular pressure immediately exceeds atrial pressure. This backward pressure differential forces the AV valve closed (point 9).

After ventricular pressure exceeds atrial pressure and the AV valve has closed, the ventricular pressure must continue to increase before it exceeds aortic pressure to open the aortic valve. Therefore, between closure of the AV valve and opening of the aortic valve, there is a brief period of time when the ventricle remains a closed chamber (point 10). Because all valves are closed, no blood can enter or leave the ventricle during this time. This interval is termed the period of **isovolumetric ventricular contraction** (*isovolumetric* means "constant volume and length") (heart C). Because no blood enters or leaves the ventricle, the ventricular chamber remains at constant volume and the muscle fibers remain at constant length. This isovolumetric condition is similar to an isometric contraction in skeletal muscle. During the period of isovolumetric ventricular contraction, ventricular pressure continues to increase as the volume remains constant (point 11).

When ventricular pressure exceeds aortic pressure (point 12), the aortic valve is forced open and ejection of blood

Figure 9-21
Cardiac Cycle This graph depicts various events that occur concurrently during the cardiac cycle. Follow each horizontal strip across to see the changes that take place in the electrocardiogram; aortic, ventricular, and atrial pressures; ventricular volume; and heart sounds throughout the cycle. Late diastole, one full systole and diastole (one full cardiac cycle), and another systole are shown for the left side of the heart. Follow each vertical strip downward to see what happens simultaneously with each of these factors during each phase of the cardiac cycle. See the text (pp. 284–287) for a detailed explanation of the circled numbers. The sketches of the heart illustrate the flow of O_2-poor (dark blue) and O_2-rich (bright red) blood in and out of the ventricles during the cardiac cycle.

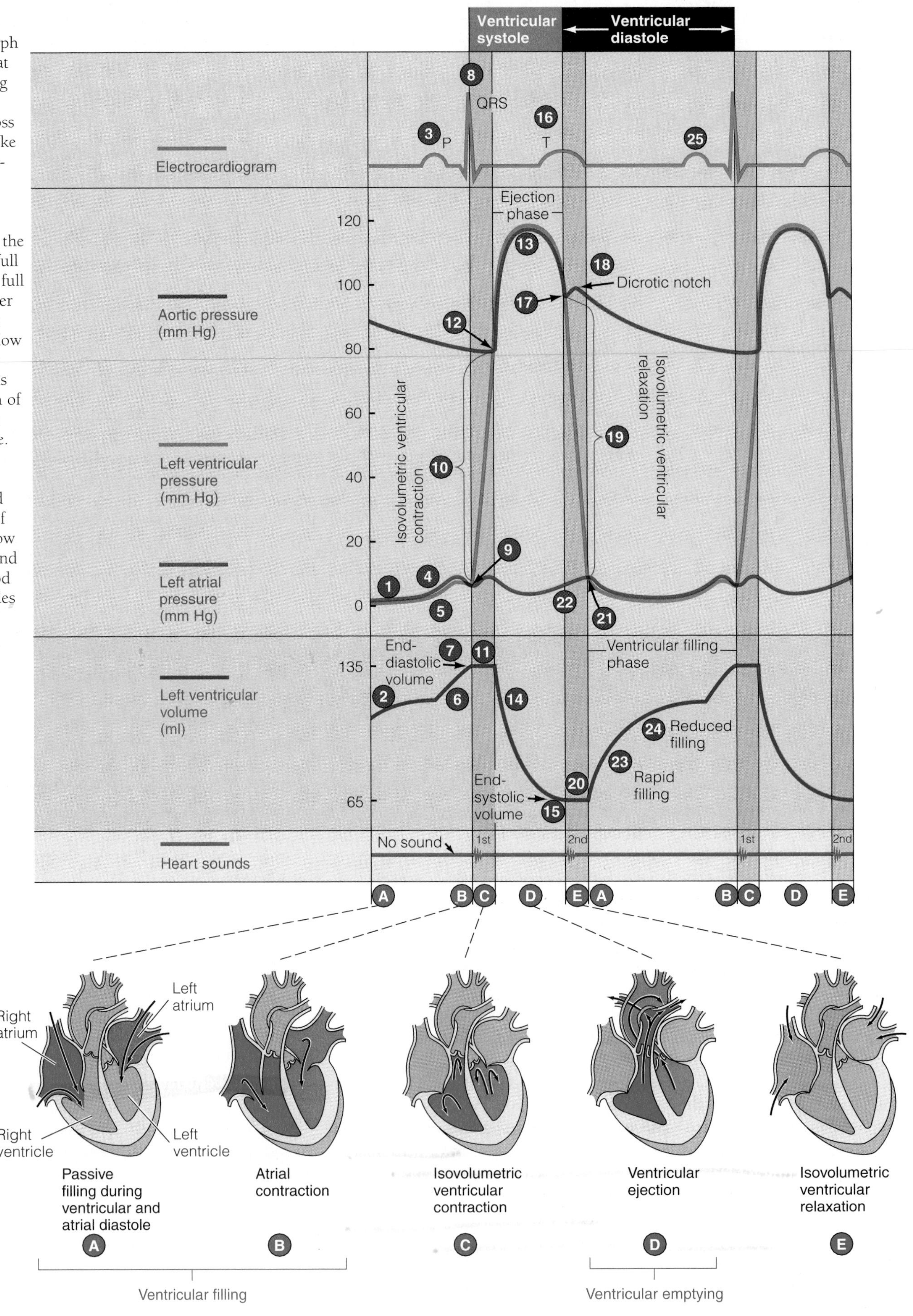

begins (heart D). The aortic pressure curve rises as blood is forced into the aorta from the ventricle faster than blood is draining off into the smaller vessels at the other end (point 13). The ventricular volume decreases substantially as blood is rapidly pumped out (point 14). Ventricular systole includes both the period of isovolumetric contraction and the ventricular ejection phase.

The ventricle does not empty completely during ejection. Normally, only about half of the blood contained within the ventricle at the end of diastole is pumped out during the subsequent systole. The amount of blood remaining in the ventricle at the end of systole when ejection is complete is known as the **end-systolic volume (ESV)**, which averages about 65 ml (point 15). This is the least amount of blood that the ventricle will contain during this cycle.

The amount of blood pumped out of each ventricle with each contraction is known as the **stroke volume (SV)**; it is equal to the end-diastolic volume minus the end-systolic volume; in other words, the difference between the volume of blood in the ventricle before contraction and the volume after contraction is the amount of blood ejected during the contraction. In our example, the end-diastolic volume is 135 ml, the end-systolic volume is 65 ml, and the stroke volume is 70 ml.

The T wave signifies ventricular repolarization occurring at the end of ventricular systole (point 16). As the ventricle starts to relax upon repolarization, ventricular pressure falls below aortic pressure and the aortic valve closes (point 17). Closure of the aortic valve produces a disturbance or notch on the aortic pressure curve known as the **dicrotic notch** (point 18). No more blood leaves the ventricle during this cycle because the aortic valve has closed. The AV valve is not yet open, however, because ventricular pressure still exceeds atrial pressure, so no blood can enter the ventricle from the atrium. Therefore, all valves are once again closed for a brief period of time known as **isovolumetric ventricular relaxation** (point 19 and heart E). The muscle fiber length and chamber volume (point 20) remain constant. No blood leaves or enters as the ventricle continues to relax and the pressure steadily falls. When the ventricular pressure falls below the atrial pressure, the AV valve opens (point 21) and ventricular filling occurs once again. Ventricular diastole includes both the period of isovolumetric ventricular relaxation and the ventricular filling phase.

Atrial repolarization and ventricular depolarization occur simultaneously, so the atria are in diastole throughout ventricular systole. Blood continues to flow from the pulmonary veins into the left atrium. As this incoming blood pools in the atrium, atrial pressure rises continuously (point 22). When the AV valve opens at the end of ventricular systole, the blood that accumulated in the atrium during ventricular systole pours rapidly into the ventricle (heart A again). Ventricular filling thus occurs rapidly at first (point 23) because of the increased atrial pressure resulting from the accumulation of blood in the atria. Then ventricular filling slows down (point 24) as the accumulated blood has already been delivered to the ventricle, and atrial pressure starts to fall. During this period of reduced filling, blood continues to flow from the pulmonary veins into the left atrium and through the open AV valve into the left ventricle. During late ventricular diastole, when ventricular filling is proceeding slowly, the SA node fires again (point 25), and the cardiac cycle starts over.

It is significant that much of ventricular filling occurs early in diastole during the rapid-filling phase. During times of rapid heart rate, the length of diastole is reduced to a much greater extent than is the length of systole. For example, if the heart rate increases from 75 to 180 beats per minute, the duration of diastole decreases about 75%, from 500 msec to 125 msec. This greatly reduces the time available for ventricular relaxation and filling. However, because much of ventricular filling is accomplished during early diastole, filling is not seriously impaired during periods of increased heart rate, such as during exercise (— Fig. 9-22). There is a limit, however, to how rapidly the heart can beat without decreasing the period of diastole to the point that ventricular filling is severely impaired. At heart rates greater than 200 beats per minute, diastolic time is too short to allow adequate ventricular filling. With inadequate filling, the resultant cardiac output is deficient. Normally, ventricular rates do not exceed 200 beats per minute because the relatively long refractory period of the AV node will not allow impulses to be conducted to the ventricles more frequently than this.

Two heart sounds associated with valve closures can be heard during the cardiac cycle.

Two major heart sounds normally can be heard with a stethoscope during the cardiac cycle. The **first heart sound** is low-pitched, soft, and relatively long—often said to sound like "lub." The **second heart sound** has a higher pitch and is shorter and sharper—often said to sound like "dup." Thus, one normally hears "lub-dup-lub-dup-lub-dup . . . ." The first

— *Figure 9-22* **Ventricular Filling Profiles during Normal and Rapid Heart Rates** Because much of ventricular filling occurs early in diastole during the rapid-filling phase, filling is not seriously impaired when diastolic time is reduced as a result of an increase in heart rate.

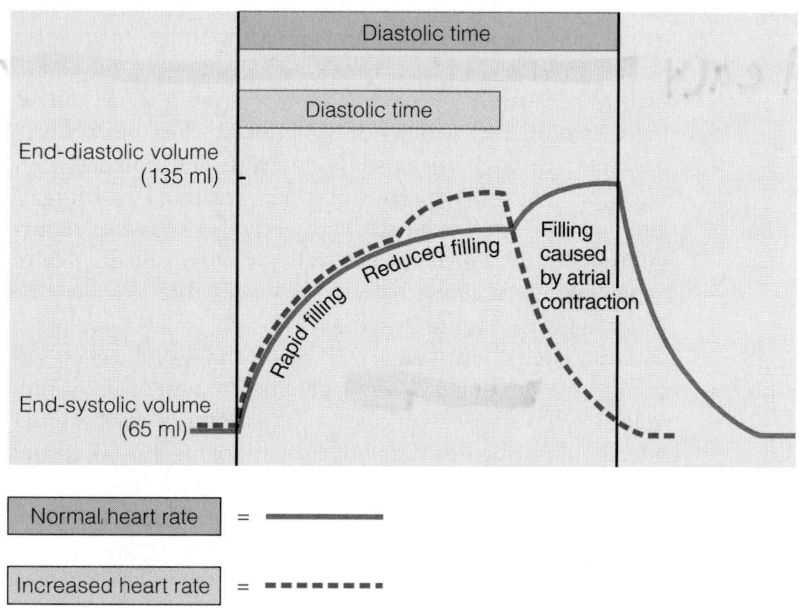

heart sound is associated with closure of the AV valves, whereas the second sound is associated with closure of the semilunar valves. Opening of valves does not produce any sound. The sounds are caused by vibrations set up within the walls of the ventricles and major arteries during valve closure, not by the valves snapping shut. Because closure of the AV valves occurs at the onset of ventricular contraction, when ventricular pressure first exceeds atrial pressure, the first heart sound signals the onset of ventricular systole. Closure of the semilunar valves occurs at the onset of ventricular relaxation, as the left and right ventricular pressures fall below the aortic and pulmonary artery pressures, respectively. The second heart sound, therefore, signals the onset of ventricular diastole.

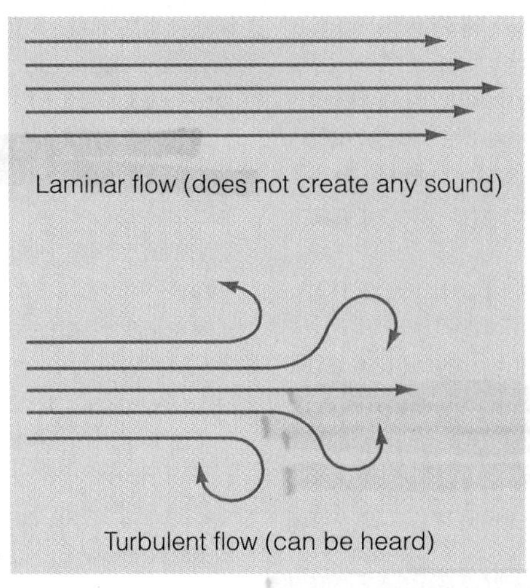

Figure 9-23 **Comparison of Laminar and Turbulent Flow**

Turbulent blood flow produces heart murmurs.

Abnormal heart sounds, or **murmurs,** are usually (but not always) associated with cardiac disease. Murmurs not involving heart pathology, so-called **functional murmurs,** are more common in young people.

Blood normally flows in a *laminar* fashion; that is, layers of the fluid slide smoothly over each other. Laminar flow does not produce any sound. When blood flow becomes turbulent, however, a sound can be heard (Fig. 9-23). Such an abnormal sound is due to vibrations created in the surrounding structures by the turbulent flow.

The most common cause of turbulence is valve malfunction, either a stenotic or an insufficient valve. A **stenotic valve** is a stiff, narrowed valve that does not open completely. Blood must be forced through the constricted opening at tremendous velocity, resulting in turbulence that produces an abnormal whistling sound similar to the sound produced when you force air rapidly through narrowed lips to whistle.

An **insufficient valve** is one that cannot close completely, usually because the valve edges are scarred and do not fit together properly. Turbulence is produced when blood flows backward through the insufficient valve and collides with blood moving in the opposite direction, creating a swishing or gurgling murmur. Such backflow of blood is known as **regurgitation.** An insufficient heart valve is often called a **leaky valve,** because it allows blood to leak back through at a time when the valve should be closed.

Most often, both valvular stenosis and insufficiency are caused by **rheumatic fever,** an autoimmune ("immunity against self") disease triggered by a streptococcus bacterial infection. Antibodies formed against toxins produced by these bacteria interact with many of the body's own tissues, resulting in immunological damage. The heart valves are among the most susceptible tissues in this regard. Large, hemorrhagic, fibrinous lesions form along the inflamed edges of an affected heart valve, causing the valve to become thickened, stiff, and scarred. Sometimes the leaflet edges permanently adhere to each other. Depending on the extent and specific nature of the lesions, the valve may become either stenotic or insufficient or some degree of both.

The valve involved and the type of defect can usually be detected by the *location* and *timing* of the murmur. Each heart valve may be heard best at a specific location on the chest. Noting the location at which a murmur is loudest helps the diagnostician determine which valve is involved. The timing of the murmur refers to the part of the cardiac cycle during which the murmur is heard. Recall that the first heart sound signals the onset of ventricular systole and the second heart sound signals the onset of ventricular diastole. Thus, a murmur occurring between the first and second heart sounds (lub-murmur-dup, lub-murmur-dup) signifies a **systolic murmur.** A **diastolic murmur,** on the other hand, occurs between the second and first heart sound (lub-dup-murmur, lub-dup-murmur). The sound of the murmur characterizes it as either a stenotic (whistling) murmur or an insufficient (swishy) murmur. Armed with these facts, one can determine the cause of a valvular murmur (Table 9-2). As an example, a whistling murmur (denoting a stenotic valve) occurring between the first and second heart sounds (denoting a systolic murmur) signifies the presence of stenosis in a valve that should be open during systole. It could be either the aortic or the pulmonary semilunar valve through which blood is being ejected. Identifying which of these valves is stenotic is accomplished by determining the location over which the murmur is best heard. The main concern with heart murmurs, of course, is not the murmur itself but the accompanying detrimental circulatory consequences caused by the defect.

Cardiac Output and Its Control

Cardiac output depends on the heart rate and the stroke volume.

Cardiac output (CO) is the volume of blood pumped by *each ventricle* per minute (not the total amount of blood pumped by the heart). During any period of time, the volume of blood flowing through the pulmonary circulation is equivalent to the volume flowing through the systemic circulation. Therefore, the cardiac output from each ventricle normally is identical, although on a beat-to-beat basis, minor variations may occur. The two determinants of cardiac output are *heart rate* (beats per minute) and *stroke volume* (volume of blood pumped per beat or stroke).

The average resting heart rate is 70 beats per minute, established by SA node rhythmicity, and the average resting stroke volume is 70 ml per beat, producing an average cardiac output of 4,900 ml/min, or close to 5 liters/min:

Table 9-2 **Timing and Type of Murmur Associated with Various Heart Valve Disorders**

Pattern Heard on Auscultation	*Type of Valve Defect*	*Timing of Murmur*	*Valve Disorder*	*Comment*
Lub-whistle-dup	Stenotic	Systolic	Stenotic semilunar valve	A whistling systolic murmur signifies that a valve that should be open during systole (a semilunar valve) does not open completely.
Lub-dup-whistle	Stenotic	Diastolic	Stenotic AV valve	A whistling diastolic murmur signifies that a valve that should be open during diastole (an AV valve) does not open completely.
Lub-swish-dup	Insufficient	Systolic	Insufficient AV valve	A swishy systolic murmur signifies that a valve that should be closed during systole (an AV valve) does not close completely.
Lub-dup-swish	Insufficient	Diastolic	Insufficient semilunar valve	A swishy diastolic murmur signifies that a valve that should be closed during diastole (a semilunar valve) does not close completely.

$$\begin{aligned} \text{cardiac output} &= \text{heart rate} \times \text{stroke volume} \\ \text{CO} &= 70 \text{ beats/min} \times 70 \text{ ml/beat} \\ &= 4{,}900 \text{ ml/min} \approx 5 \text{ liters/min} \end{aligned}$$

Because the body's total blood volume averages 5 to 5.5 liters, each half of the heart pumps the equivalent of the entire blood volume each minute. In other words, each minute the right ventricle normally pumps 5 liters of blood through the lungs, and the left ventricle pumps 5 liters of blood through the systemic circulation. At this rate, each half of the heart would pump about 2.5 million liters of blood in just one year. Yet this is only the resting cardiac output! During exercise the cardiac output can increase to 20 to 25 liters per minute, and outputs as high as 40 liters per minute have been recorded in trained athletes during heavy exercise. The difference between the cardiac output at rest and the maximum volume of blood the heart is capable of pumping per minute is known as the **cardiac reserve.** How can cardiac output vary so tremendously, depending on the demands of the body? You can readily answer this question by thinking about how your own heart pounds rapidly (increased heart rate) and forcefully (increased stroke volume) when you engage in strenuous physical activities (need for increased cardiac output). Thus, the regulation of cardiac output depends on the control of both heart rate and stroke volume, topics that will be discussed next.

Heart rate is determined primarily by autonomic influences on the SA node.

The SA node is normally the pacemaker of the heart because it has the fastest spontaneous rate of depolarization to threshold. Recall that this automatic gradual reduction of membrane potential between beats is due to a complex interplay of ion movements involving a reduction in K^+ permeability, a constant Na^+ permeability, and an increased Ca^{2+} permeability. When the SA node reaches threshold, an action potential is initiated that spreads throughout the heart, inducing the heart to contract or have a "heartbeat." This happens about 70 times per minute, setting the average heart rate at 70 beats per minute.

The heart is innervated by both divisions of the autonomic nervous system, which can modify the rate (as well as the strength) of contraction, even though nervous stimulation is not required to initiate contraction. The parasympathetic nerve to the heart, the **vagus nerve**, primarily supplies the atrium, especially the SA and AV nodes. Parasympathetic innervation of the ventricles is sparse. The cardiac sympathetic nerves also supply the atria, including the SA and AV nodes, and richly innervate the ventricles as well.

Parasympathetic and sympathetic stimulation have the following effects on the heart (▮ Table 9-3).

Effect of parasympathetic stimulation on the heart

- The parasympathetic nervous system's influence on the SA node is to decrease the heart rate (— Fig. 9-24). Acetylcholine released upon increased parasympathetic activity increases the permeability of the SA node to K^+ by slowing the closure of K^+ channels. As a result, the rate at which spontaneous action potentials are initiated is reduced through a twofold effect:
  1. Enhanced K^+ permeability hyperpolarizes the SA node membrane because more positive potassium ions leave than normal, making the inside even more negative. Because the "resting" potential starts even farther away from threshold, it takes longer to reach threshold.
  2. The enhanced K^+ permeability induced by vagal stimulation also opposes the automatic reduction in K^+ permeability that is responsible for initiating the gradual depolarization of the membrane to threshold. This countering effect decreases the rate of spontaneous depolarization, prolonging the time required to drift to threshold. Therefore, the SA node reaches threshold and fires less frequently, decreasing the heart rate.

Table 9-3 Effects of the Autonomic Nervous System on the Heart and Structures That Influence the Heart

Area Affected	*Effect of Parasympathetic Stimulation*	*Effect of Sympathetic Stimulation*
SA node	Decreases the rate of depolarization to threshold; decreases the heart rate	Increases the rate of depolarization to threshold; increases the heart rate
AV node	Decreases excitability; increases the AV nodal delay	Increases excitability; decreases the AV nodal delay
Ventricular conduction pathway	No effect	Increases excitability; hastens conduction through the bundle of His and Purkinje cells
Atrial muscle	Decreases contractility; weakens contraction	Increases contractility; strengthens contraction
Ventricular muscle	No effect	Increases contractility; strengthens contraction
Adrenal medulla (an endocrine gland)	No effect	Promotes adrenomedullary secretion of epinephrine, a hormone that augments the sympathetic nervous system's actions on the heart
Veins	No effect	Increases venous return, which increases the strength of cardiac contraction through the Frank-Starling mechanism

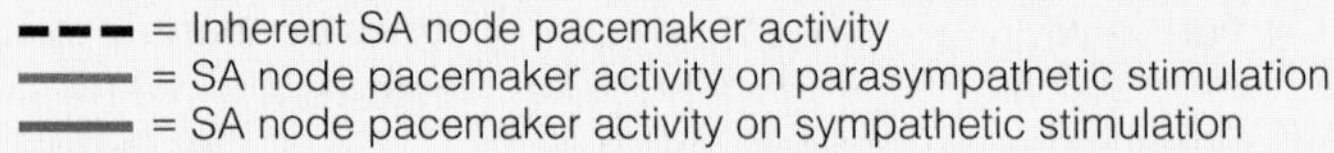

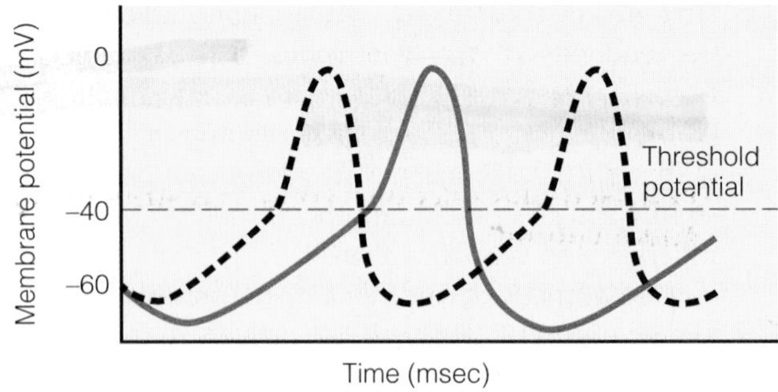

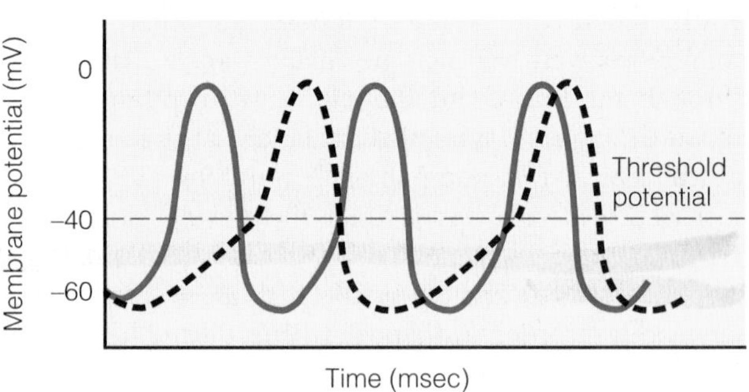

(a)

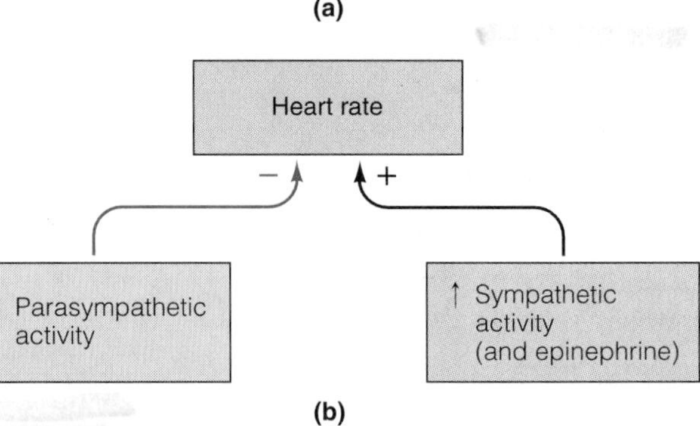

(b)

Figure 9-24 Autonomic Control of SA Node Activity and Heart Rate (a) Autonomic influence on SA node potential. Parasympathetic stimulation decreases the rate of SA nodal depolarization so that the membrane reaches threshold more slowly and has fewer action potentials, whereas sympathetic stimulation increases the rate of depolarization of the SA node so that the membrane reaches threshold more rapidly and has more frequent action potentials. (b) Control of heart rate by the autonomic nervous system. Because each SA node action potential ultimately leads to a heartbeat, increased parasympathetic activity decreases the heart rate, whereas increased sympathetic activity increases the heart rate.

- Parasympathetic influence on the AV node decreases the node's excitability, prolonging transmission of impulses to the ventricles even longer than the usual AV nodal delay. This effect is brought about by increasing K^+ permeability, which hyperpolarizes the membrane, thereby retarding the initiation of excitation in the AV node.
- Parasympathetic stimulation of the atrial contractile cells shortens the action potential, an effect that is believed to be caused by a reduction in the slow inward current carried by Ca^{2+}; that is, the plateau phase is reduced. As a result, atrial contraction is weakened.
- The parasympathetic system has little effect on ventricular contraction due to the sparsity of parasympathetic innervation to the ventricles.

Thus, the heart is more "leisurely" under parasympathetic influence—it beats less rapidly, the time between atrial and ventricular contraction is stretched out, and atrial contraction is weaker. These actions are appropriate considering that the parasympathetic system controls heart action in quiet, relaxed situations when the body is not demanding an enhanced cardiac output.

Effect of sympathetic stimulation on the heart

- In contrast, the sympathetic nervous system, which controls heart action in emergency or exercise situations, when there is a need for greater blood flow, speeds up the heart rate through its effect on the pacemaker tissue. The main effect of sympathetic stimulation on the SA node is to increase its rate of depolarization so that threshold is reached more rapidly (Fig. 9-24 and Table 9-3). Norepinephrine released from the sympathetic nerve endings appears to decrease K^+ permeability by accelerating inactivation of the K^+ channels. With fewer positive potassium ions leaving, the inside of the cell becomes less negative, creating a depolarizing effect. This swifter drift to threshold under sympathetic influence permits a greater frequency of action potentials and a correspondingly more rapid heart rate.
- Sympathetic stimulation of the AV node reduces the AV nodal delay by increasing conduction velocity, presumably by enhancing the slow, inward Ca^{2+} current.
- Similarly, sympathetic stimulation speeds up the spread of the action potential throughout the specialized conduction pathway.
- In the atrial and ventricular contractile cells, both of which have an abundance of sympathetic nerve endings, sympathetic stimulation increases contractile strength so that the heart beats more forcefully and squeezes out more blood. This effect is brought about by increasing Ca^{2+} permeability, which enhances the slow Ca^{2+} influx and intensifies Ca^{2+} participation in the excitation-contraction coupling process.

The overall effect of sympathetic stimulation on the heart, therefore, is to improve its effectiveness as a pump by increasing the heart rate, decreasing the delay between atrial and ventricular contraction, decreasing conduction time throughout the heart, and increasing the force of contraction; that is, sympathetic stimulation "revs up" the heart.

Thus, as is typical of the autonomic nervous system, parasympathetic and sympathetic effects on heart rate are antagonistic (oppose each other). At any given moment, the heart rate will be determined largely by the existing balance between the inhibitory effects of the vagus nerve and the stimulatory effects of the cardiac sympathetic nerves. Under resting conditions, parasympathetic discharge is dominant. In fact, if all autonomic nerves to the heart were blocked, the resting heart rate would increase from its average value of 70 beats per minute to about 100 beats per minute, which is the inherent rate of the SA node's spontaneous discharge when not subjected to any nervous influence. (We use 70 beats per minute as the normal rate of SA node discharge because this is the average rate under normal conditions in the body.) Alterations in the heart rate beyond this resting level in either direction can be accomplished by shifting the balance of autonomic nervous stimulation. Heart rate is increased by simultaneously increasing sympathetic and decreasing parasympathetic activity; a reduction in heart rate is brought about by a concurrent rise in parasympathetic activity and decline in sympathetic activity. The relative level of activity in these two autonomic branches to the heart in turn is primarily coordinated by the *cardiovascular control center* located in the brain stem.

Although autonomic innervation is the primary means by which heart rate is regulated, other factors affect it as well. The most important of these is epinephrine, a hormone that is secreted into the blood from the adrenal medulla upon sympathetic stimulation and that acts on the heart in a manner similar to norepinephrine to increase the heart rate. Epinephrine therefore reinforces the direct effect that the sympathetic nervous system has on the heart.

Stroke volume is determined by the extent of venous return and by sympathetic activity.

The other component that determines the cardiac output is stroke volume, the amount of blood pumped out by each ventricle during each beat. Two types of controls influence stroke volume: (1) *intrinsic control* related to the extent of venous return and (2) *extrinsic control* related to the extent of sympathetic stimulation of the heart. Both factors increase stroke volume by increasing the strength of contraction of the heart (■ Fig. 9-25). Let us examine each of these factors in more detail to see how they influence the stroke volume.

Increased end-diastolic volume results in increased stroke volume.

As more blood is returned to the heart, the heart pumps out more blood, but the relationship is not quite as simple as it appears, because the heart does not eject all the blood it contains. The direct correlation between end-diastolic volume

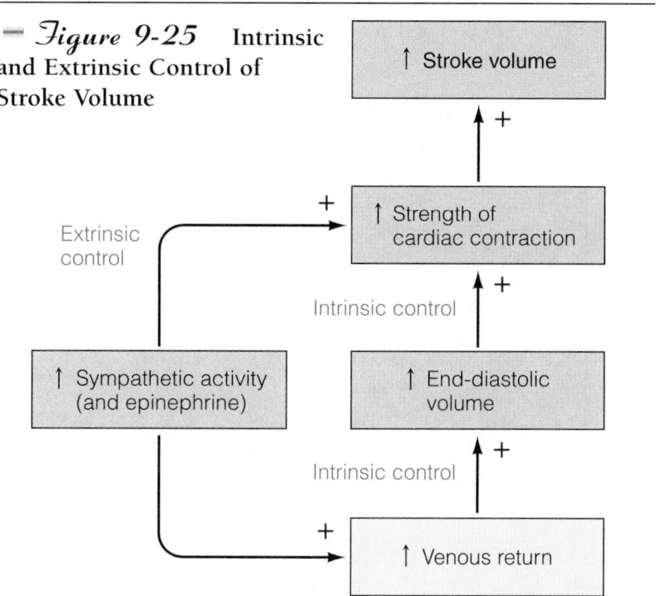

■ *Figure 9-25* **Intrinsic and Extrinsic Control of Stroke Volume**

and stroke volume constitutes the **intrinsic control** of stroke volume, which refers to the heart's inherent ability to vary the stroke volume. This intrinsic control depends on the length-tension relationship of cardiac muscle, which is similar to that of skeletal muscle. For skeletal muscle, the resting-muscle length is approximately the optimal length at which maximal tension can be developed during a subsequent contraction. When the skeletal muscle is longer or shorter than this optimal length, the subsequent contraction is weaker (see Fig. 8–17, p. 237). For cardiac muscle, the resting cardiac muscle fiber length is less than optimal length. Therefore, an increase in cardiac muscle fiber length, by moving closer to the optimal length, increases the contractile tension of the heart on the following systole (Fig. 9-26).

What causes cardiac muscle fibers to vary in length before contraction? Skeletal muscle length can vary before contraction due to the positioning of the skeletal parts to which the muscle is attached, but cardiac muscle is not attached to any bones. The main determinant of cardiac muscle fiber length is the degree of diastolic filling. An analogy is a balloon filled with water—the more water you put in, the larger the balloon becomes and the more it is stretched. Likewise, the greater the extent of diastolic filling, the larger the end-diastolic volume and the more the heart is stretched. The more the heart is stretched, the longer the initial cardiac fiber length before contraction. The increased length results in a greater force on the subsequent cardiac contraction and, consequently, a greater stroke volume. This intrinsic relationship between end-diastolic volume and stroke volume is known as the **Frank-Starling law of the heart.** Stated simply, the law says that the heart normally pumps all the blood returned to it; increased venous return results in increased stroke volume. In Figure 9-26, assume that the end-diastolic volume increases from point A to point B. You can see that this increase in end-diastolic volume is accompanied by a corresponding increase in stroke volume from point A^1 to point B^1. The extent of filling is referred to as the **preload,** because it is the workload imposed on the heart before contraction begins.

Unlike skeletal muscle, the length-tension curve of cardiac muscle normally does not have a descending limb. That is, within physiological limits, cardiac muscle does not get stretched beyond its optimal length to the point that contractile strength diminishes with further stretching.

The built-in relationship matching stroke volume with venous return has two important advantages. First, one of the most important functions served by this intrinsic mechanism is equalization of output between the right and left sides of the heart, so that the blood pumped out by the heart is equally distributed between the pulmonary and systemic circulation. If, for example, the right side of the heart ejects a larger stroke volume, more blood enters the pulmonary circulation, so venous return to the left side of the heart is increased accordingly. The increased end-diastolic volume of the left side of the heart causes it to contract more forcefully, so it too pumps out a larger stroke volume. In this way, equality of output of the two ventricular chambers is maintained. If such equalization did not happen, excessive damming of blood would occur in the venous system preceding the ventricle with the lower output.

Second, when a larger cardiac output is needed, such as during exercise, venous return is increased through action of the sympathetic nervous system and other mechanisms to be described in the next chapter. The resultant increase in end-diastolic volume automatically increases stroke volume correspondingly. Because exercise also increases heart rate, these two factors act together to increase the cardiac output so that more blood can be delivered to the exercising muscles.

Figure 9-26 **Intrinsic Control of Stroke Volume (Frank-Starling Curve)** The cardiac muscle fiber's length, which is determined by the extent of venous filling, is normally less than the optimal length for developing maximal tension. Therefore, an increase in end-diastolic volume (that is, an increase in venous return), by moving the cardiac muscle fiber length closer to optimal length, increases the contractile tension of the fibers on the next systole. A stronger contraction squeezes out more blood. Thus, as more blood is returned to the heart and the end-diastolic volume increases, the heart automatically pumps out a correspondingly larger stroke volume.

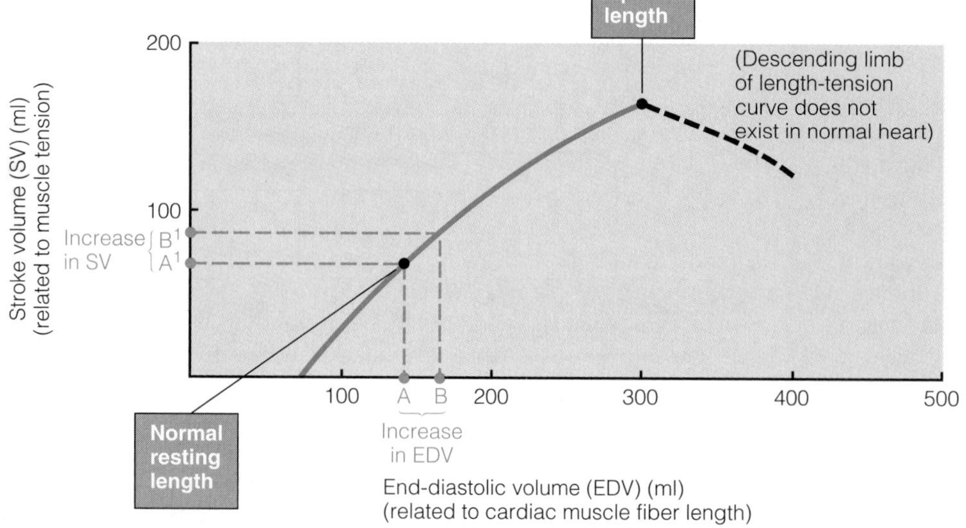

The contractility of the heart is increased by sympathetic stimulation.

In addition to intrinsic control, stroke volume is also subject to **extrinsic control** by factors originating outside the heart, the most important of which are actions of the cardiac sympathetic nerves and epinephrine (Table 9-3). Sympathetic stimulation and epinephrine enhance the heart's **contractility,** which refers to the strength of contraction at any given end-diastolic volume; in other words, the heart contracts more forcefully and squeezes out a greater percentage of the blood it contains on sympathetic stimulation, leading to more complete ejection. This increased contractility is due to the increased Ca^{2+} influx triggered by norepinephrine and epinephrine. The extra cytosolic Ca^{2+} allows the myocardial fibers to generate more force through greater cross-bridge cycling than they would without sympathetic influence. Normally, the end-diastolic volume is 135 ml and the end-systolic volume is 65 ml for a stroke volume of 70 ml (■ Fig. 9-27a). Under sympathetic influence, for the same end-diastolic volume of 135 ml, the end-systolic volume might be 35 ml and the stroke volume 100 ml (Fig. 9-27b). In effect, sympathetic stimulation shifts the Frank-Starling curve to the left (■ Fig. 9-28). Depending on the extent of sympathetic stimulation, the curve can be shifted to varying degrees, up to a maximal increase in contractile strength of about 100% greater than normal.

Sympathetic stimulation increases stroke volume not only by strengthening cardiac contractility but also by enhancing venous return (Fig. 9-27c). Sympathetic stimulation constricts the veins, which squeezes more blood forward from the veins to the heart, increasing the end-diastolic volume and subsequently increasing the stroke volume even further.

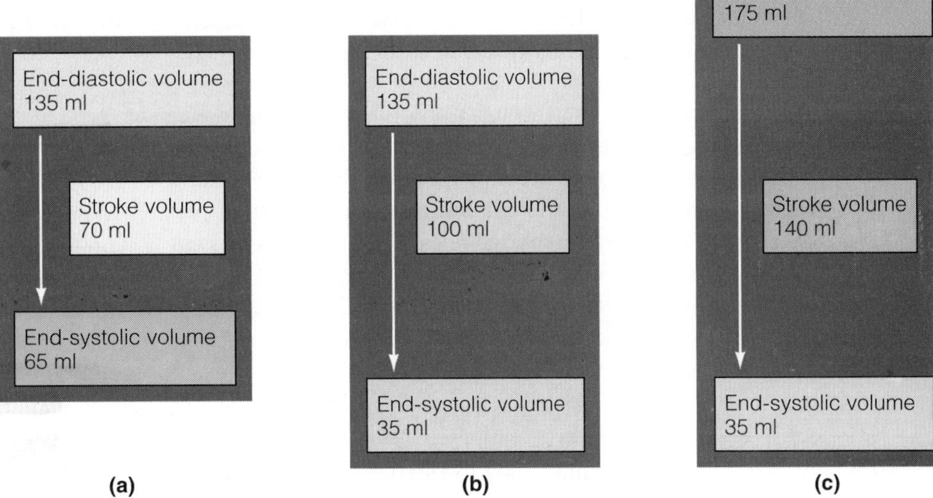

■ *Figure 9-27* **Effect of Sympathetic Stimulation on Stroke Volume** (a) Normal stroke volume. (b) Stroke volume during sympathetic stimulation. (c) Stroke volume with combination of sympathetic stimulation and increased end-diastolic volume.

The strength of cardiac muscle contraction and, accordingly, the stroke volume can thus be graded by (1) varying the initial length of the muscle fibers, which in turn depends on the degree of ventricular filling before contraction (intrinsic control); and (2) varying the extent of sympathetic stimulation (extrinsic control). This is in contrast to gradation of skeletal muscle. In skeletal muscle, twitch summation and recruitment of motor units are employed to produce variable strength of muscle contraction, but these mechanisms are not applicable to cardiac muscle. Twitch summation is impossible because of the long refractory period, and recruitment of motor units is not possible because the heart muscle cells are arranged into functional syncytia instead of distinct motor units that can be discretely activated.

All the factors that determine the cardiac output by influencing the heart rate or stroke volume are summarized in ■ Figure 9-29. Note that sympathetic stimulation increases

■ *Figure 9-28* **Shift of the Frank-Starling Curve to the Left by Sympathetic Stimulation** For the same end-diastolic volume (point A), there is a larger stroke volume (from point B to point C) upon sympathetic stimulation as a result of increased contractility of the heart. The Frank-Starling curve is shifted to the left to variable degrees, depending on the extent of sympathetic stimulation.

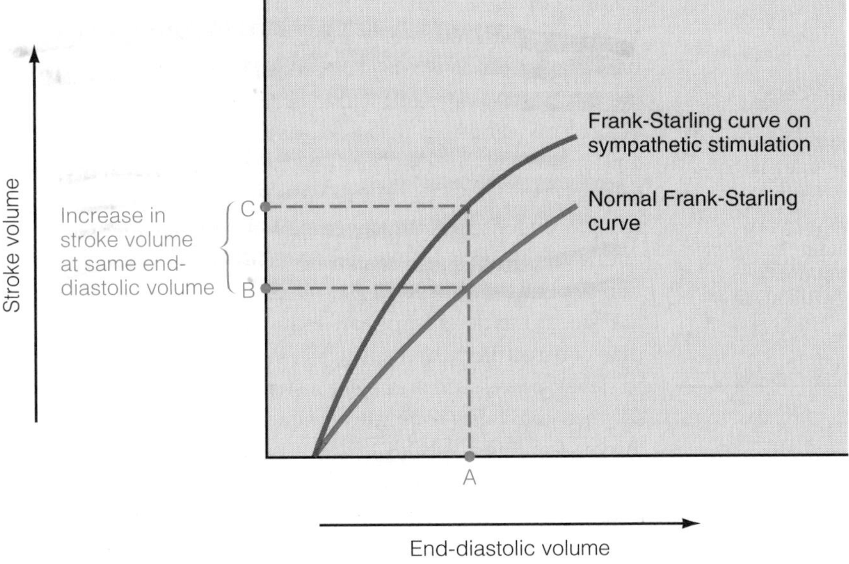

■ *Figure 9-29* **Control of Cardiac Output** Since cardiac output equals heart rate times stroke volume, this figure is a composite of Figure 9-24 (control of heart rate) and Figure 9-25 (control of stroke volume).

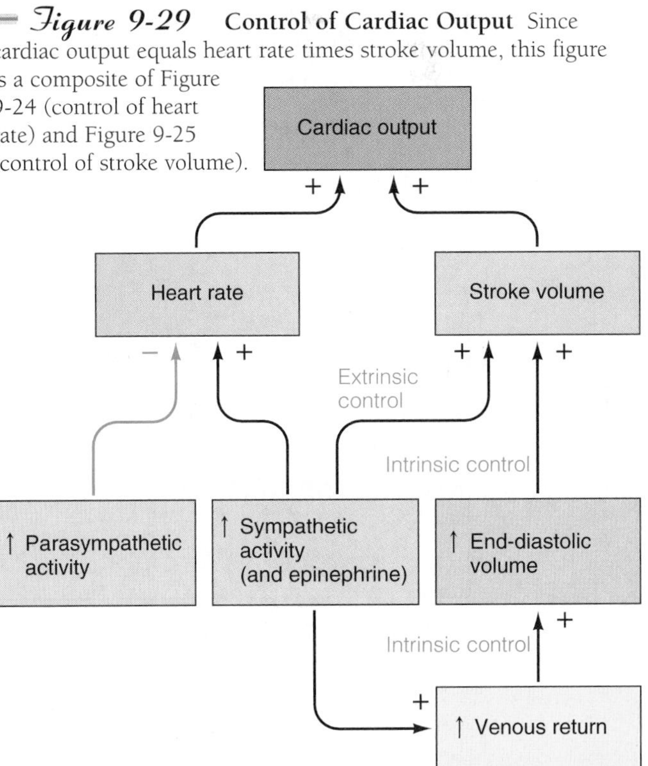

the cardiac output by increasing both heart rate and stroke volume. Sympathetic activity to the heart increases, for example, during exercise when the working skeletal muscles need increased delivery of O_2-laden blood to support their high rate of ATP consumption.

The ability of the heart to meet physiological demands for increased cardiac output declines as a person gets older; that is, the cardiac reserve decreases with aging. Researchers have recently discovered that this age-related decline is due in part to a reduction in norepinephrine release from the heart's sympathetic nerves. A dwindling of Ca^{2+} channels in the sympathetic terminals is believed to be responsible for the diminished norepinephrine release. As with neurotransmitters elsewhere, norepinephrine release depends on the entry of Ca^{2+} into the terminals through voltage-gated Ca^{2+} channels. As these channels are lost in the older heart, less norepinephrine is released when the sympathetic nerve fibers to the heart are activated, resulting in a reduction in cardiac responsiveness.

Figure 9-30 **Frank-Starling Curve in Heart Failure** (a) Shift of the Frank-Starling curve downward and to the right in a failing heart. Because its contractility is decreased, the failing heart pumps out a smaller stroke volume at the same end-diastolic volume than a normal heart does. (b) Compensations for heart failure. Reflex sympathetic stimulation shifts the Frank-Starling curve of a failing heart to the left, increasing the contractility of the heart toward normal. A compensatory increase in end-diastolic volume as a result of blood-volume expansion further increases the strength of contraction of the failing heart. Operating at a longer cardiac muscle fiber length, a compensated failing heart is able to eject a normal stroke volume.

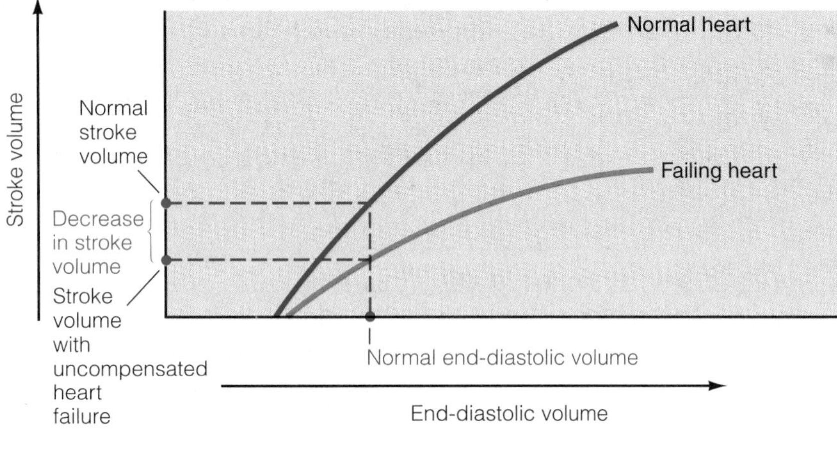

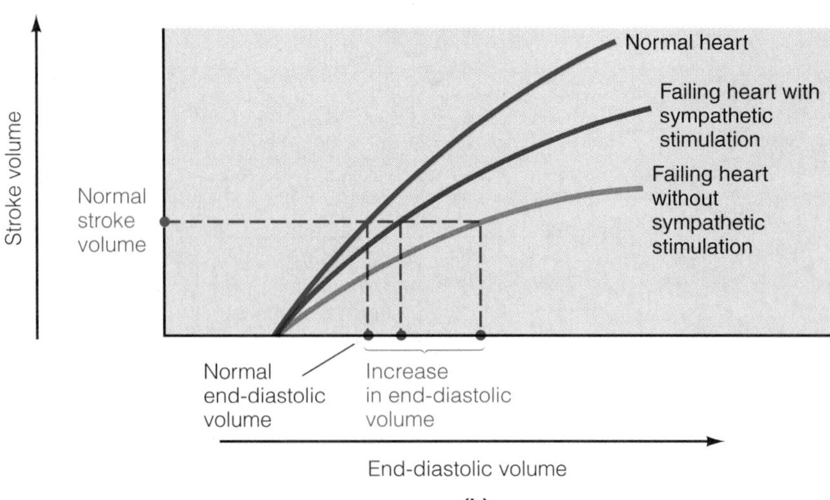

High blood pressure increases the workload of the heart.

When the ventricles contract, they must generate sufficient pressure to exceed the blood pressure in the major arteries in order to force open the semilunar valves. The arterial blood pressure is referred to as the **afterload** because it is the workload imposed on the heart after the contraction has begun. If the arterial blood pressure is chronically elevated (high blood pressure) or if the exit valve is stenotic, the ventricle has to generate more pressure to eject blood. For example, instead of generating the normal pressure of 120 mm Hg, the ventricular pressure may need to go as high as 400 mm Hg to force blood through a narrowed aortic valve.

The heart may be able to compensate for a sustained increase in afterload by enlarging (through hypertrophy of the cardiac muscle fibers; see p. 273). This enables it to contract more forcefully and maintain a normal stroke volume despite the impedance to ejection. A diseased heart or a heart weakened with age may not be able to compensate completely, however; in that case, heart failure ensues. Even if the heart is initially able to compensate for a chronic increase in afterload, the sustained extra workload placed on the heart can eventually cause pathological changes in the heart that lead to heart failure. In fact, a chronically elevated afterload is one of the two major factors that cause heart failure, a topic to which we now turn our attention.

The contractility of the heart is decreased in heart failure.

Heart failure refers to the inability of the cardiac output to keep pace with the body's demands for supplies and removal of wastes. Either one or both ventricles may fail. When a failing ventricle is unable to pump out all of the blood returned to it, the veins behind the failing ventricle become congested with blood. Heart failure may occur for a variety of reasons, but the two most common are (1) damage to the heart muscle as a result of a heart attack or impaired circulation to the cardiac muscle and (2) prolonged pumping against a chronically increased afterload, as with a stenotic semilunar valve or a sustained elevation in blood pressure.

The prime defect in heart failure is a decrease in cardiac contractility; that is, the intrinsic ability of the heart to develop pressure and eject a stroke volume is reduced so that the heart operates on a lower length-tension curve (Fig. 9-30a). The Frank-Starling curve is shifted downward and to the right such that for a given end-diastolic volume, a failing heart will pump out a smaller stroke volume than a normal healthy heart.

Two major compensatory measures help restore the stroke volume to normal in the early stages of heart failure. First, sympathetic activity to the heart is reflexly increased, which increases the contractility of the heart toward normal (Fig. 9-30b). Sympathetic stimulation can only help compensate for a limited period of time, however. The heart becomes less responsive to norepinephrine after prolonged exposure, and furthermore, the norepinephrine stores in the sympathetic nerve terminals in the heart become depleted. Second, when cardiac output is reduced, the kidneys, in a compensatory attempt to improve their reduced blood flow, retain extra salt and water in the body during urine formation to expand the blood volume. The increase in circulating blood volume increases the end-diastolic volume. The resultant stretching of the cardiac muscle fibers enables the weakened heart to pump out a normal stroke volume (Fig. 9-30b). The heart is now pumping out the blood returned to it but is operating at a longer cardiac muscle fiber length.

As the disease progresses and the contractility of the heart deteriorates further, the heart reaches a point at which it is no longer able to pump out a normal stroke volume (that is, cannot pump out all of the blood returned to it) despite compensatory measures. *Backward failure* occurs as blood that cannot enter and be pumped out by the heart continues to dam up in the venous system. *Forward failure* occurs simultaneously as the heart fails to pump an adequate amount of blood forward to the tissues because the stroke volume becomes progressively smaller. The congestion in the venous system is the reason this condition is sometimes termed **congestive heart failure.**

Left-sided failure has more serious consequences than right-sided failure. Backward failure of the left side leads to pulmonary edema (excess tissue fluid in the lungs) because blood dams up in the lungs. This fluid accumulation in the lungs reduces exchange of O_2 and CO_2 between the air and blood in the lungs, leading to reduced arterial oxygenation and an elevation of acid-forming CO_2 in the blood. In addition, one of the more serious consequences of left-sided forward failure is an inadequate blood flow to the kidneys, which causes a twofold problem. First, vital kidney function is depressed, and second, the kidneys retain even more salt and water in the body during urine formation as they attempt to expand the plasma volume even further to improve their reduced blood flow. Excessive fluid retention further deteriorates the already existing problems of venous congestion. Treatment of congestive heart failure therefore includes measures that reduce salt and water retention and increase urinary output as well as drugs that enhance the contractile ability of the weakened heart—digitalis, for example.

Nourishing the Heart Muscle

The heart receives most of its own blood supply through the coronary circulation during diastole.

Although all the blood passes through the heart, the heart muscle is unable to extract O_2 or nutrients from the blood within its chambers for two reasons. First, the watertight endocardial lining does not permit blood to pass from the chamber into the myocardium. Second, the heart walls are too thick to permit diffusion of O_2 and other supplies from the blood in the chamber to the individual cardiac cells. Therefore, like other tissues of the body, heart muscle must receive blood through blood vessels, specifically by means of the **coronary circulation.** The coronary arteries branch from the aorta just beyond the aortic valve (see Fig. 9-35), and the coronary veins empty into the right atrium.

The heart muscle receives most of its blood supply during diastole. Blood flow to the heart muscle cells is substantially reduced during systole for two reasons. First, the major branches of the coronary arteries are compressed by the contracting myocardium, and second, the entrance to the coronary vessels is partially blocked by the open aortic valve. Thus, most coronary arterial flow (about 70%) occurs during diastole, driven by the aortic blood pressure, with flow declining as aortic pressure drops. Only about 30% of coronary arterial flow occurs during systole (Fig. 9-31). The limited time for coronary blood flow becomes especially important during rapid heart rates, when diastolic time is substantially reduced. Just when increased demands are placed on the heart to pump more rapidly, it has less time to provide O_2 and nourishment to its own musculature to accomplish the increased workload.

Nevertheless, under normal circumstances the heart muscle does receive adequate blood flow to support its activities, even during exercise, when the rate of coronary blood flow increases up to five times its resting rate. Increased delivery of blood to the cardiac cells is accomplished primarily by vasodilation, or enlargement, of the coronary vessels, which allows more blood to flow through them, especially during diastole. The increased coronary blood flow is imperative to meet the heart's increased O_2 requirements because, unlike most other tissues, the heart is unable to remove much additional O_2 from the blood passing through its vessels to support increased metabolic activities. Most other tissues under resting conditions extract only about 25% of the O_2 available

Figure 9-31 **Coronary Blood Flow** Most coronary blood flow occurs during diastole because the coronary vessels are compressed almost completely closed during systole.

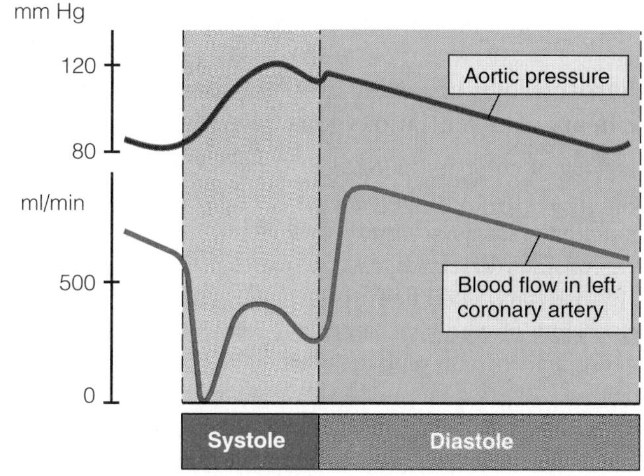

from the blood flowing through them, leaving a considerable O_2 reserve that can be drawn on when a tissue has increased O_2 needs; that is, the tissue can immediately increase the O_2 available to it by removing a greater percentage of O_2 from the blood passing through it. In contrast, the heart, even under resting conditions, removes up to 65% of the O_2 available in the coronary vessels, far more than is withdrawn by other tissues. This leaves little O_2 in reserve in the coronary blood should cardiac O_2 demands increase. Therefore, the primary means by which more O_2 can be made available to the heart muscle is by increasing coronary blood flow.

Coronary blood flow is adjusted primarily in response to changes in the heart's O_2 requirements. The major link that coordinates coronary blood flow with myocardial O_2 needs is *adenosine,* which is formed from adenosine triphosphate (ATP) during cardiac metabolic activity. Increased formation and release of adenosine from the cardiac cells occur (1) when there is a cardiac O_2 deficit or (2) when cardiac activity is increased and the heart accordingly requires more O_2 and is using more ATP as an energy source. The released adenosine induces dilation of the coronary blood vessels, thereby allowing more O_2-rich blood to flow to the more active cardiac cells to meet their increased O_2 demand (■ Fig. 9-32). This matching of O_2 delivery with O_2 needs is critical because the heart muscle depends on oxidative processes to generate energy. The heart cannot obtain sufficient ATP through anaerobic metabolism.

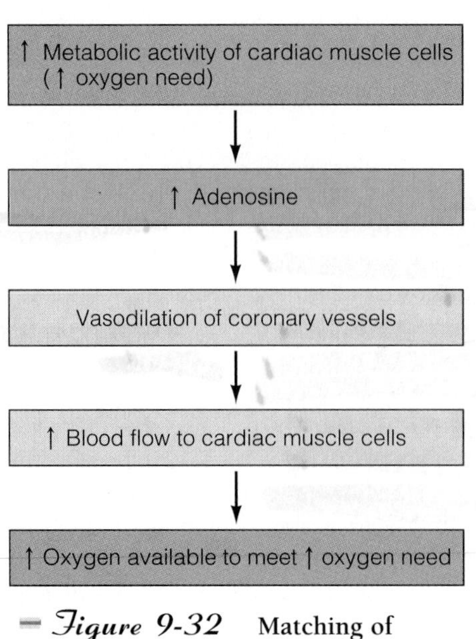

■ *Figure 9-32* **Matching of Coronary Blood Flow to the Oxygen Need of Cardiac Muscle Cells**

Although the heart has little ability to support its energy needs by means of anaerobic metabolism and must rely heavily on its O_2 supply, it can tolerate wide variations in its nutrient supply. The heart primarily uses free fatty acids and, to a lesser extent, glucose and lactate as fuel sources, depending on their availability. Because the cardiac muscle is remarkably adaptable and can shift metabolic pathways to use whatever nutrient is available, the primary danger of insufficient coronary blood flow is not fuel shortage but O_2 deficiency.

Atherosclerotic coronary artery disease can deprive the heart of essential oxygen.

Adequacy of coronary blood flow is relative to the heart's O_2 demands at any given moment. In the normal heart, coronary blood flow increases correspondingly as O_2 demands rise. With **coronary artery disease,** however, it may not be possible for coronary blood flow to keep pace with rising O_2 needs. A given rate of coronary blood flow may be adequate at rest but insufficient upon physical exertion or other stressful situations.

Complications of coronary artery disease, including *heart attacks* (see p. 284), make it the single leading cause of death in the United States. Coronary artery disease can cause *myocardial ischemia* (insufficient circulation of oxygenated blood through the coronary circulation to maintain aerobic metabolism in the heart muscle) by three mechanisms: (1) profound vascular spasm of the coronary arteries; (2) the formation of atherosclerotic plaques; and (3) thromboembolism. We will discuss each of these in turn.

Vascular spasm is an abnormal spastic constriction that transiently narrows the coronary vessels; it is most often triggered by exposure to cold, physical exertion, or anxiety. Vascular spasms are associated with the early stages of coronary artery disease. The condition is reversible and usually of insufficient duration to produce damage to the cardiac muscle. Recent evidence suggests that reduced O_2 availability in the coronary vessels causes the release of **platelet-activating factor (PAF)** from the endothelium (lining) of the vessels. PAF, which exerts a variety of biological actions, was named for its first discovered effect. Among its effects other than activating platelets, PAF, once released from the endothelium, diffuses to the underlying vascular smooth muscle and causes it to contract, bringing about vascular spasm.

Atherosclerosis is a progressive, degenerative arterial disease that leads to occlusion (gradual blockage) of affected vessels, thereby reducing blood flow through them. The initial stage of atherosclerosis is characterized by the accumulation beneath the blood vessel lining (endothelium) of excessive amounts of cholesterol-rich lipid derived from the blood. This accumulation leads to the formation of a **fatty streak.** The disease progresses as smooth muscle cells within the blood vessel wall migrate from the muscular layer of the blood vessel to a position on top of the lipid accumulation, just beneath the endothelium. Here the smooth muscle cells continue to divide and enlarge, producing **atheromas,** which are benign (noncanerous) tumors of smooth muscle cells within the blood vessel walls. Together the lipid-rich core and overlying smooth muscle form a **plaque** (■ Fig. 9-33). The plaque progressively bulges into the lumen of the vessel as it continues to develop. A thickening plaque interferes with nutrient exchange for cells of the involved arterial wall, leading to degeneration of the wall in the vicinity of the plaque. The damaged area is invaded by fibroblasts (scar tissue–forming cells), which form a collagen-rich (see p. 51) connective tissue cap over the plaque. (*Sclerosis* means excessive growth of fibrous connective tissue, hence the term *atherosclerosis* for this condition characterized by atheromas and sclerosis, along with abnormal lipid accumulation.) In the later stages of the disease, Ca^{2+} often precipitates in the plaque. A vessel so afflicted becomes hard and poorly distensible, a condition dubbed "hardening of the arteries."

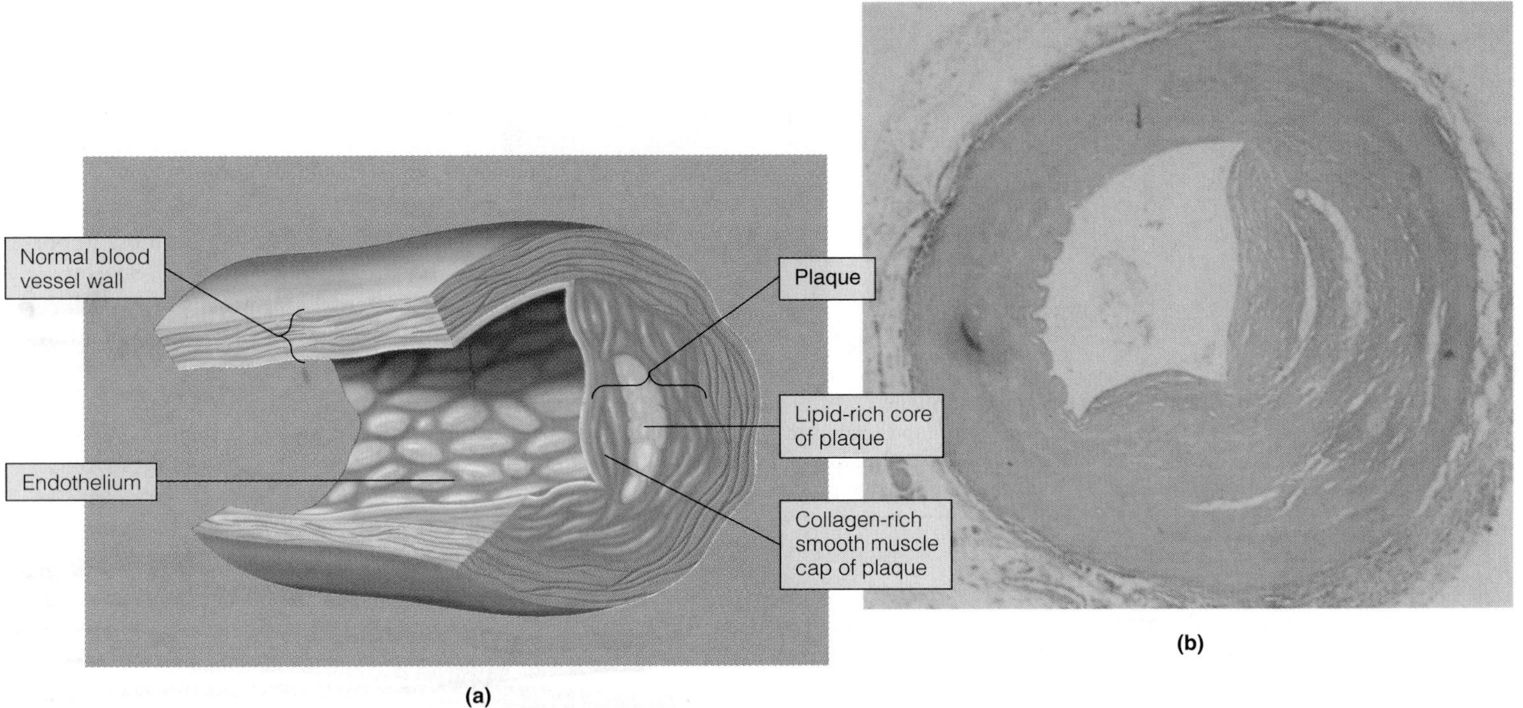

Figure 9-33 **Atherosclerotic Plaque** (a) Schematic representation of the components of a plaque. (b) Photomicrograph of a severe atherosclerotic plaque in a coronary vessel.

Atherosclerosis attacks arteries throughout the body, but the most serious consequences involve damage to the vessels of the brain and heart. In the brain, atherosclerosis is the prime cause of strokes, whereas in the heart, it brings about myocardial ischemia and its complications. The following are potential complications of coronary atherosclerosis:

1. Gradual enlargement of the protruding plaque continues to narrow the vessel lumen and progressively diminishes coronary blood flow, triggering increasingly frequent bouts of transient myocardial ischemia as the ability to match blood flow with cardiac O_2 needs becomes more limited. Although the heart cannot normally be "felt," pain is associated with myocardial ischemia. Such cardiac pain, known as **angina pectoris** ("pain of the chest"), can be felt beneath the sternum and is often referred to the left shoulder and down the left arm (see p. 148). The symptoms of angina pectoris recur whenever cardiac O_2 demands become too great in relation to the coronary blood flow—for example, during exertion or emotional stress. The pain is thought to result from stimulation of cardiac nerve endings by the accumulation of lactic acid when the heart shifts to its limited ability to perform anaerobic metabolism. The ischemia associated with the characteristically brief anginal attacks is usually temporary and reversible and can be relieved by rest, administration of vasodilator drugs such as *nitroglycerin,* or both. Nitroglycerin brings about coronary vasodilation by being metabolically converted to *nitric oxide,* which in turn relaxes the vascular smooth muscle.

2. The enlarging atherosclerotic plaque can break through the weakened endothelial lining that covers it, exposing blood to the underlying collagen in the collagen-rich connective tissue cap of the plaque. Blood platelets (formed elements of the blood involved in plugging vessel defects and in clot formation) normally do not adhere to smooth, healthy vessel linings. However, when platelets come into contact with collagen at the site of vessel damage, they stick to the site and contribute to the formation of a blood clot. Such an abnormal clot attached to a vessel wall is known as a **thrombus.** The thrombus may enlarge gradually until it completely blocks the vessel at that site, or the continued flow of blood past the thrombus may break it loose from its attachment. Such a freely floating clot, or **embolus,** may completely plug a smaller vessel as it flows downstream (Fig. 9-34). Thus, through **thromboembolism,** atherosclerosis can result in a gradual or sudden occlusion of a coronary vessel (or any other vessel).

3. When a coronary vessel is completely plugged, the cardiac tissue served by the vessel soon dies from O_2 deprivation and a heart attack occurs, unless the area can be supplied with blood from nearby vessels. Sometimes a deprived area is fortunate enough to receive blood from more than one pathway. **Collateral circulation** exists when small terminal branches from adjacent blood vessels nourish the same area. These accessory vessels cannot develop suddenly following an abrupt blockage but may be lifesaving if already developed. Such alternate vascular pathways often develop over a period of time when an atheroscle-

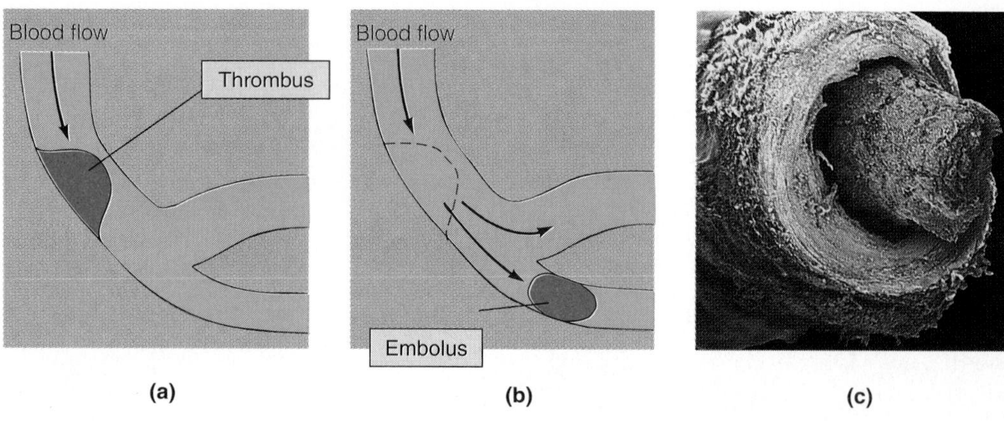

Figure 9-34 **Consequences of Thromboembolism** (a) A thrombus may enlarge gradually until it completely occludes the vessel at that site. (b) A thrombus may break loose from its attachment, forming an embolus that may completely occlude a smaller vessel downstream. (c) Scanning electron micrograph of a vessel completely occluded by a thromboembolic lesion.

rotic constriction progresses slowly, or they may be induced by sustained demands on the heart through a regular aerobic exercise program.

In the absence of collateral circulation, the extent of the damaged area during a heart attack depends on the size of the blocked vessel. The larger the vessel occluded, the greater the area deprived of its blood supply. As Figure 9-35 illustrates, a blockage at point A in the coronary circulation would cause more extensive damage than would a blockage at point B. Because there are only two major coronary arteries, complete blockage of either one of these main branches results in extensive myocardial damage. Left coronary artery blockage is most devastating because this vessel is responsible for supplying 85% of the cardiac tissue. A heart attack has four possible outcomes: immediate death, delayed death from complications, full functional recovery, or recovery with impaired function (Table 9-4).

The amount of "good" cholesterol versus "bad" cholesterol in the blood is linked to atherosclerosis.

The cause of atherosclerosis is still not entirely clear. Certain high-risk factors have been associated with an increased incidence of atherosclerosis and coronary heart disease. Included among them are genetic predisposition, obesity, advanced age, smoking, hypertension, diabetes mellitus, lack of exercise, nervous tension, and, most significantly, excess cholesterol levels in the blood.

There are two sources of cholesterol for the body: (1) dietary intake of cholesterol, with animal products such as egg yolk, red meats, and butter being especially rich in this lipid (animal fats contain cholesterol, whereas plant fats do not), and (2) manufacture of cholesterol by many organs within the body, particularly the liver. Because of the body's ability to synthesize cholesterol, there is not a direct correlation between the amount of cholesterol ingested and the cholesterol levels in the blood, although modest reductions in blood cholesterol can be accomplished by lowering the intake of animal fats. For some individuals, drugs may be necessary to satisfactorily lower blood cholesterol levels.

Actually, it is not the total blood cholesterol level but the amount of cholesterol bound to various plasma protein carriers that appears to be most important with regard to the risk of developing atherosclerotic heart disease. Because cholesterol is a lipid, it is not very soluble in blood. Most cholesterol in the blood is attached to specific plasma protein carriers in the form of lipoprotein complexes, which are soluble in blood. There are three major lipoproteins, named for their density of protein as compared to lipid: (1) **high-density lipoproteins (HDL),** which contain the most protein and least cholesterol; (2) **low-density lipoproteins (LDL),** which contain less protein and more cholesterol; and (3) **very-low-density lipoproteins (VLDL),** which contain the least protein and most lipid, but the lipid they carry is neutral fat, not cholesterol. Cholesterol carried in LDL complexes has been termed "bad" cholesterol, because cholesterol is transported *to* the cells, including those lining the blood vessel walls, by means of LDL. In contrast, cholesterol carried in HDL complexes has been dubbed "good" choles-

Figure 9-35 **Extent of Myocardial Damage as a Function of the Size of the Occluded Vessel**

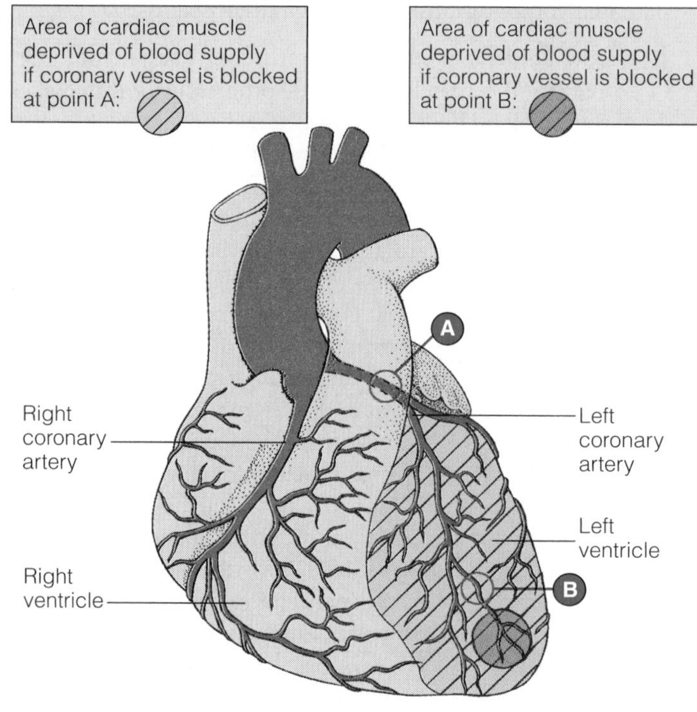

Table 9-4 Possible Outcomes of Acute Myocardial Infarction (Heart Attack)

Immediate Death	*Delayed Death from Complications*	*Full Functional Recovery*	*Recovery with Impaired Function*
Acute cardiac failure occurring because the heart is too weakened to pump effectively to support the body tissues Fatal ventricular fibrillation brought about by damage to the specialized conducting tissue or induced by O_2 deprivation	Fatal rupture of the dead, degenerating area of the heart wall Slowly progressing congestive heart failure occurring because the weakened heart is unable to pump out all the blood returned to it	Replacement of the damaged area with a strong scar, accompanied by enlargement of the remaining normal contractile tissue to compensate for the lost cardiac musculature	Persistence of permanent functional defects, such as bradycardia or conduction blocks, caused by destruction of irreplaceable autorhythmic or conductive tissues

terol, because HDL removes cholesterol *from* the cells and transports it to the liver for partial elimination from the body.

Unlike most lipids, cholesterol is not used as metabolic fuel by cells. Instead, it serves as an essential component of plasma membranes. In addition, a few special cell types use cholesterol as a precursor for the synthesis of secretory products, such as steroid hormones and bile salts. Although most cells are capable of synthesizing some of the cholesterol needed for their own plasma membranes, they cannot manufacture sufficient amounts and therefore must rely on supplemental cholesterol being delivered by the blood. This additional cholesterol is supplied either by the diet or by cells that specialize in cholesterol synthesis, especially liver cells.

Cells accomplish cholesterol uptake from the blood by synthesizing receptor proteins specifically capable of binding LDL and inserting these receptors into the cells' plasma membranes (— Fig. 9-36). When an LDL particle binds to one of the membrane receptors, the cell engulfs the particle by endocytosis (see p. 25). Within the cell, lysosomal enzymes break down the LDL to free the cholesterol, making it available to the cell for synthesis of new cellular membrane. If too much free cholesterol accumulates in the cell, there is a shutdown of both the synthesis of LDL receptor proteins (so that less cholesterol is taken up) and the cell's own cholesterol synthesis (so that less new cholesterol is made). Faced with a cholesterol shortage, on the other hand, the cell makes more LDL receptors so that it can engulf more cholesterol from the blood.

The maintenance of a blood-borne cholesterol supply to the cells involves an interaction between dietary cholesterol and the synthesis of cholesterol by the liver. When the amount of dietary cholesterol is increased, hepatic (liver) syn-

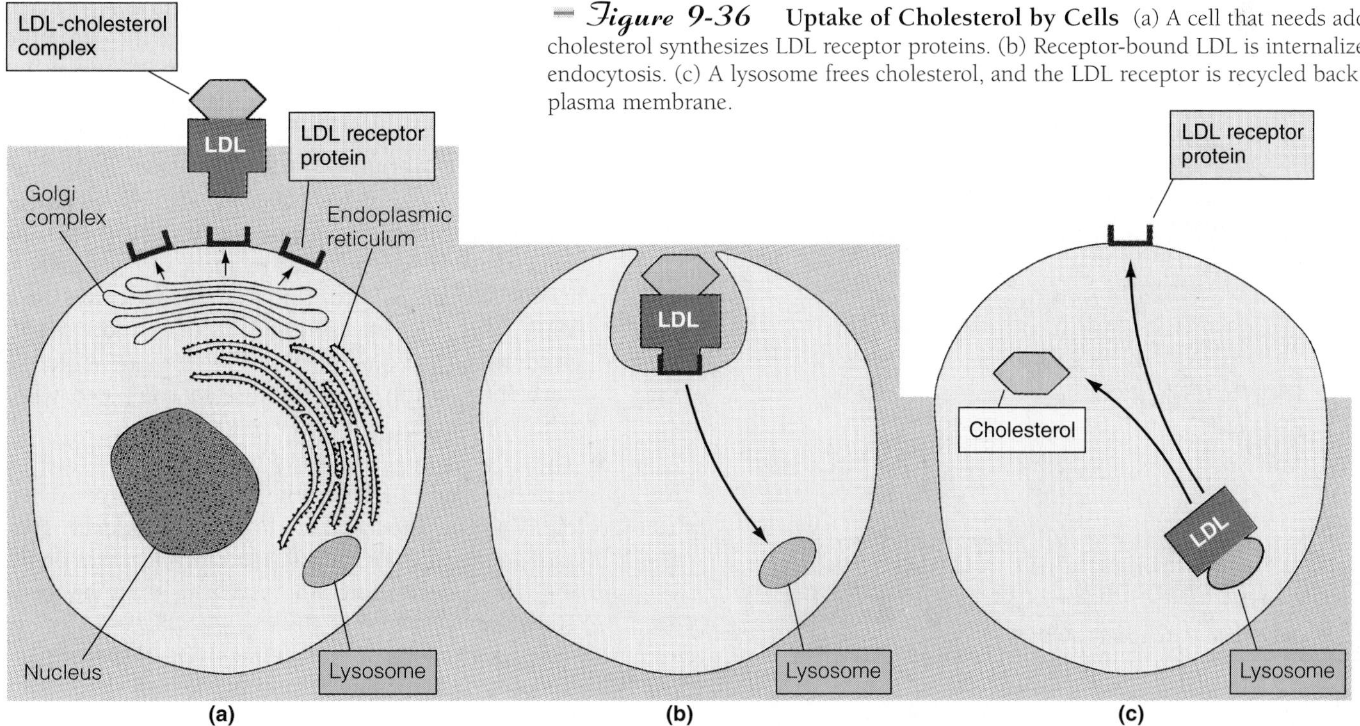

— *Figure* 9-36 **Uptake of Cholesterol by Cells** (a) A cell that needs additional cholesterol synthesizes LDL receptor proteins. (b) Receptor-bound LDL is internalized by endocytosis. (c) A lysosome frees cholesterol, and the LDL receptor is recycled back to the plasma membrane.

thesis of cholesterol is turned off because cholesterol in the blood directly inhibits a hepatic enzyme essential for cholesterol synthesis. Thus, as more cholesterol is ingested, less is produced by the liver. Conversely, when cholesterol intake from food is reduced, the liver synthesizes more of this lipid because the inhibitory effect of cholesterol on the crucial hepatic enzyme is removed. In this way, the blood concentration of cholesterol is maintained at a fairly constant level despite changes in cholesterol intake; thus, it is difficult to significantly reduce cholesterol levels in the blood by decreasing cholesterol intake.

In contrast to LDL complexes, which transport cholesterol to the cells, HDL removes cholesterol from cells and transports it to the liver. The liver in turn secretes cholesterol as well as cholesterol-derived bile salts into the bile. Bile enters the intestinal tract, where bile salts participate in the digestive process. Most of the secreted cholesterol and bile salts are subsequently reabsorbed from the intestinal tract into the blood to be recycled to the liver, but the cholesterol molecules not reclaimed by absorption are eliminated in the feces.

Figure 9-37 Role of the Liver in Cholesterol Metabolism The letters in the illustration correspond to the following steps:

(a) The liver extracts cholesterol (carried by HDL) from the blood.
(b) The liver converts most of the cholesterol into bile salts.
(c) The liver secretes bile salts and cholesterol into the bile, which empties into the intestinal lumen.
(d) Part of the cholesterol and bile salts is eliminated in the feces.
(e) Some of the bile salts are reabsorbed into the blood and recycled to the liver.
(f) The liver synthesizes new cholesterol, which is carried by LDL within the blood away from the liver to other cells.

Saturated fatty acids and polyunsaturated fatty acids are known to influence these pathways at the designated points.

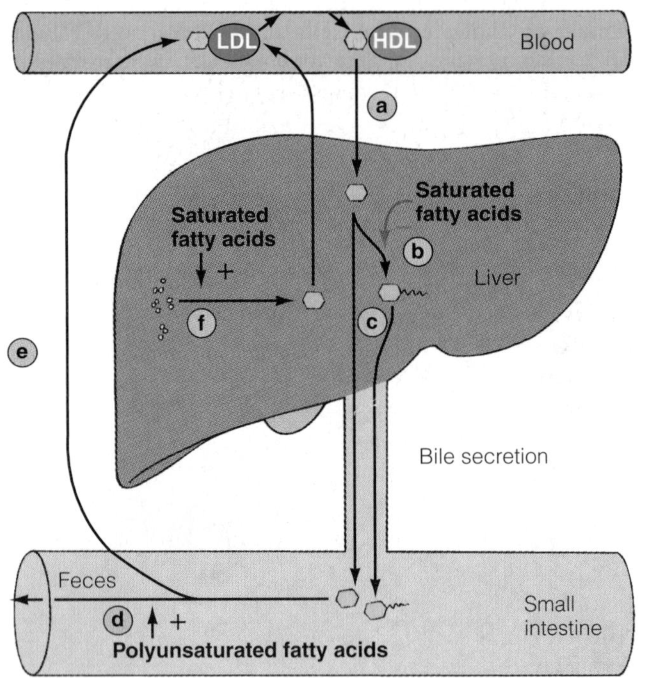

Obviously, the liver has a central role in cholesterol metabolism. At any given time, the liver may be manufacturing new cholesterol, extracting old cholesterol from the blood and secreting it into the bile, or converting old cholesterol into bile salts, which likewise are secreted into the bile (Fig. 9-37). The first process is a mechanism for adding cholesterol to the blood to supplement dietary intake. The other two pathways lead to a net loss of cholesterol from the body. Thus, the liver has a primary role in determining total blood cholesterol levels, and the interplay between LDL and HDL determines the traffic flow of cholesterol between the liver and the other cells. Whenever these mechanisms are altered, blood cholesterol levels may be affected in such a way as to influence the individual's predisposition to atherosclerosis. The following are specific examples:

- Evidence suggests that the propensity toward developing atherosclerosis substantially increases with elevated levels of LDL. In one hereditary disease, afflicted individuals lack the genes for making LDL receptor proteins. Because their cells cannot take up LDL from the blood, the blood concentration of these cholesterol-loaded lipoproteins becomes greatly elevated. Such persons have an extraordinary tendency toward developing atherosclerosis early in life, and many die in their youth from coronary artery disease. Recent studies suggest that even in individuals with normal LDL receptors, atherosclerosis may develop if they have aberrant levels of the sole protein component of LDL, **apolipoprotein B-100 (apo-B).** This is the portion of the LDL complex that interacts with the receptors.
  - Recent studies demonstrate that not all LDL is equally bad. LDL that has been oxidized by **free radicals** (very unstable, electron-deficient particles that are highly reactive) is more likely than nonoxidized LDL to promote the development of atherosclerotic plaques. In related investigations, antioxidant vitamins that prevent LDL oxidation, such as *vitamin E, vitamin C,* and *beta-carotene,* have been shown to slow plaque deposition.
  - Furthermore, size of the LDL particles is important. LDL particles that are physically very small are not removed from the blood by the liver but instead remain in the body, where they can accumulate within the arterial walls. Studies have revealed that individuals with predominantly small LDL have a threefold risk of developing coronary artery disease compared with those who have predominantly large LDL.
- The risk of atherosclerosis is inversely related to the concentration of HDL in the blood; that is, elevated levels of HDL are associated with a low incidence of atherosclerotic heart disease. In fact, a more accurate predictor of the risk of developing atherosclerosis than the total blood cholesterol level is the blood *HDL-cholesterol/total cholesterol ratio.* The higher the HDL-cholesterol concentration in relationship to the total blood cholesterol level, the lower the risk. Some other factors known to influence

atherosclerotic risk can be related to HDL levels; for example, cigarette smoking lowers HDL, and the HDL level is higher in individuals who exercise regularly. Moreover, premenopausal women, who have a lower incidence of atherosclerotic heart disease than their male counterparts, have a higher concentration of HDL, presumably because of some influence of the female sex hormone, estrogen. After production of estrogen ceases at menopause, the incidence of coronary heart disease in women parallels that in men.

- As with LDL, however, there are several types of HDL particles. Whereas HDL that contains **apolipoprotein A-I** is the "good" HDL, HDL that contains **apolipoprotein A-II** is not so good. Evidence suggests that the less desirable HDL not only may promote plaque formation but also interferes with the good cholesterol's work.
- Varying the intake of dietary fatty acids may alter total blood cholesterol levels by influencing one or more of the mechanisms involving cholesterol balance (Fig. 9-37). The blood cholesterol level tends to be raised by ingestion of saturated fatty acids found predominantly in animal fats and tropical plant oils, such as palm oil and coconut oil. These fatty acids stimulate the synthesis of cholesterol and inhibit its conversion to bile salts. On the other hand, polyunsaturated fatty acids, the predominant fatty acids of most plants, tend to reduce blood cholesterol levels by enhancing the elimination of both cholesterol and cholesterol-derived bile salts in the feces. Furthermore, dietary soluble fiber supplements such as oat bran and psyllium seed husks have been shown to reduce the level of blood cholesterol, particularly LDL-cholesterol, by physically interfering with its absorption from the intestine. Thus, dietary manipulations can reduce the cholesterol-related risk of atherosclerosis, but not just by reducing cholesterol intake.
- Investigators recently identified an **atherosclerosis susceptibility**, or **ATHS, gene.** Individuals with this gene, who have up to a threefold increased risk of suffering a heart attack, develop an atherogenic ("atherosclerosis-producing") lipoprotein profile characterized by low HDL and high LDL concentrations in the blood, among other features.
- Other researchers recently discovered another cholesterol carrier, **lipoprotein(a).** The blood concentration of lipoprotein(a) varies nearly 1,000-fold among individuals, with the tendency for higher or lower concentrations being an inherited trait. A positive correlation has been found between higher blood concentrations of lipoprotein(a) and the incidence of atherosclerosis.
- Some studies suggest that particular types of immune cells may aid and abet plaque formation. For example, one type of white blood cell leaves the blood and settles down permanently in the developing fatty streak. These cells voraciously phagocytize (see p. 25) the oxidized LDL in the fatty streak until they become greatly engorged **foam cells,** which constitute a major component of the lipid core of the plaque.
- Some researchers believe that high **triglyceride** levels in the blood also represent a high risk factor for coronary artery disease. Triglycerides, a form of neutral fat carried by VLDL particles, are the major type of dietary fat as well as the storage form of fat in the body. Studies suggest that high blood triglyceride levels are associated with low HDL levels and increased amounts of the dangerous small LDL, perhaps all resulting from abnormal breakdown of VLDL.

In addition to cholesterol and its carrier types, scientists have identified other factors that may contribute to the development of atherosclerosis, as exemplified by the following:

- Under further study is a recent finding that an *increased iron level* in the blood is positively correlated with the incidence of atherosclerosis and coronary artery disease.
- Still other investigations suggest that one type of *herpesvirus* may play a role in the development of atherosclerosis. Various herpesviruses are already known to cause fever blisters, sexually transmitted genital sores, and a flu-like illness. Researchers suspect that herpes simplex type 1 virus may infect the endothelial cells that line blood vessels and promote vascular changes that set the stage for the later development of atherosclerosis.

As you can see, the relationship between atherosclerosis, cholesterol, and other environmental and genetic factors is far from clear. Much research concerning this complex disease is currently in progress, because the incidence of atherosclerosis is so high and its consequences are potentially fatal.

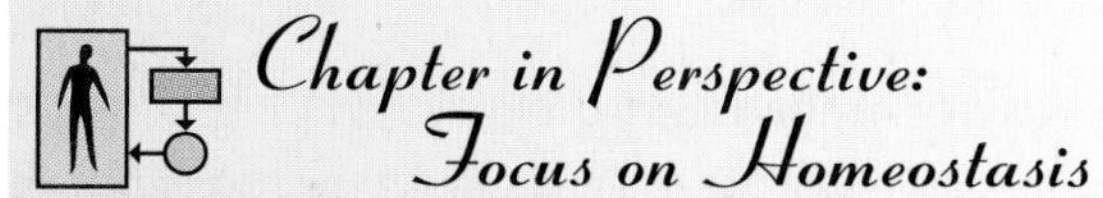

Chapter in Perspective: Focus on Homeostasis

Survival depends on continual delivery of needed supplies to all of the cells throughout the body and on ongoing removal of wastes generated by the cells. Furthermore, regulatory chemical messengers, such as hormones, must be transported from their site of production to their site of action, where they control a variety of activities, most of which are directed toward maintaining a stable internal environment.

The circulatory system contributes to homeostasis by serving as the body's transport system. It provides a means of rapidly moving materials from one part of the body to another. Without the circulatory system, materials would not get where they need to go to support life-sustaining activities nearly rapidly enough. For example, O_2 would take months to years to diffuse from the surface of the body to internal organs, yet O_2 and other substances can be picked up by the blood and delivered to all the cells in a few seconds through the heart's swift pumping action.

The heart serves as a dual pump to continuously circulate blood between the lungs, where O_2 is picked up, and the other body tissues, which use O_2 to support their energy-generating chemical reactions. As blood is pumped through the various tissues, other substances besides O_2 are also exchanged between the blood and tissues. For example, the blood picks up nutrients as it flows through the digestive organs, and other tissues remove nutrients from the blood as it flows through them.

Although all the body tissues constantly depend on the life-supporting blood flow provided to them by the heart, the heart itself is quite an independent organ. It is able to take care of many of its own needs without any outside influence. Contraction of this magnificent muscle is self-generated through a carefully orchestrated interplay of changing ionic permeabilities. Local mechanisms within the heart ensure that blood flow to the cardiac muscle normally meets the heart's need for O_2. In addition, the heart has built-in capabilities to vary its strength of contraction, depending on the amount of blood returned to it. The heart does not act entirely autonomously, however. It is innervated by the autonomic nervous system and is influenced by the hormone epinephrine, both of which can vary the rate and contractility of the heart, depending on the body's needs for blood delivery. Furthermore, as with all tissues, the cells that compose the heart depend on the other body systems for maintenance of a stable internal environment in which they can survive and function.

Chapter Summary

Anatomical Considerations

The heart is basically a dual pump that provides the driving pressure for blood flow through the pulmonary and systemic circulations. The heart has four chambers: each half of the heart consists of an atrium, or venous input chamber, and a ventricle, or arterial output chamber. Four heart valves direct the blood in the proper direction and prevent it from flowing in the reverse direction. The heart is self-excitable, initiating its own rhythmic contractions. Contraction of the spirally arranged cardiac muscle fibers produces a wringing effect important for efficient pumping. Also important for efficient pumping is the fact that the muscle fibers in each chamber act as a functional syncytium, contracting as a coordinated unit.

Electrical Activity of the Heart

The cardiac impulse originates at the SA node, the pacemaker of the heart, which has the fastest rate of spontaneous depolarization to threshold. Once initiated, the action potential spreads throughout the right and left atria, partially facilitated by specialized conduction pathways but mostly by cell-to-cell spread of the impulse through gap junctions. The impulse passes from the atria into the ventricles through the AV node, the only point of electrical contact between these chambers. The action potential is delayed briefly at the AV node, ensuring that atrial contraction precedes ventricular contraction to allow complete ventricular filling. The impulse then travels rapidly down the interventricular septum via the bundle of His and is rapidly dispersed throughout the myocardium by means of the Purkinje fibers. The remainder of the ventricular cells are activated by cell-to-cell spread of the impulse through gap junctions. Thus, the atria contract as a single unit, followed after a brief delay by a synchronized ventricular contraction.

The action potentials of contractile cardiac muscle fibers exhibit a prolonged positive phase, or plateau, accompanied by a prolonged period of contraction, which ensures adequate ejection time. This plateau is primarily due to activation of slow Ca^{2+} channels. Because a long refractory period occurs in conjunction with this prolonged plateau phase, summation and tetanus of cardiac muscle are impossible, thereby ensuring the alternate periods of contraction and relaxation essential for pumping of blood.

The spread of electrical activity throughout the heart can be recorded from the surface of the body. This record, the ECG, can provide useful information about the status of the heart.

Mechanical Events of the Cardiac Cycle

The cardiac cycle consists of three important events:

1. The generation of electrical activity as the heart autorhythmically depolarizes and repolarizes.
2. Mechanical activity consisting of alternate periods of systole (contraction and emptying) and diastole (relaxation and filling), which are initiated by the rhythmical electrical cycle.
3. Directional flow of blood through the heart chambers, guided by valvular opening and closing induced by pressure changes that are generated by mechanical activity.

Valve closing gives rise to two normal heart sounds. The first heart sound is caused by closure of the atrioventricular (AV) valves and signals the onset of ventricular systole. The second heart sound is due to closure of the aortic and pulmonary valves at the onset of diastole.

As seen in Figure 9-21 on p. 286, the atrial pressure curve remains low throughout the entire cardiac cycle, with only minor fluctuations (normally varying between 0 and 8 mm Hg). The aortic pressure curve remains high the entire time, with moderate fluctuations (normally varying between a systolic pressure of 120 mm Hg and a diastolic pressure of 80 mm Hg). The ventricular pressure curve fluctuates dramatically because ventricular pressure must be below the low atrial pressure during diastole to allow the AV valve to open so filling can take place, and it must be above the high aortic pressure during systole to force the aortic valve open to allow emptying to occur. Therefore, ventricular pressure normally varies from 0 mm Hg during diastole to slightly more than 120 mm Hg during systole.

Defective valve function produces turbulent blood flow, which is audible as a heart murmur. Abnormal valves may be either stenotic and not open completely or insufficient and not close completely.

Cardiac Output and Its Control

Cardiac output, the volume of blood ejected by each ventricle each minute, is determined by the heart rate times the stroke volume. Heart rate is varied by altering the balance of parasympathetic and sympathetic influence on the SA node. Parasympathetic stimulation slows the heart rate, and sympathetic stimulation speeds it up.

Stroke volume depends on (1) the extent of ventricular filling, with an increased end-diastolic volume resulting in a larger stroke volume by means of the length-tension relationship (intrinsic control), and (2) the extent of sympathetic stimulation, with increased sympathetic stimulation resulting in increased contractility of the heart, that is, increased strength of contraction and increased stroke volume at a given end-diastolic volume (extrinsic control).

Nourishing the Heart Muscle

Cardiac muscle is supplied with oxygen and nutrients by blood delivered to it by the coronary circulation, not by blood within the heart chambers. Most coronary blood flow occurs during diastole, because the coronary vessels are compressed by the contracting heart muscle during systole. Coronary blood flow is normally varied to keep pace with cardiac oxygen needs.

Coronary blood flow may be compromised by the development of atherosclerotic plaques, which can lead to ischemic heart disease ranging in severity from mild chest pain on exertion to fatal heart attacks. The exact cause of atherosclerosis is unclear, but apparently the ratio of cholesterol carried in the plasma by high-density lipoproteins (HDL) compared to cholesterol carried by low-density lipoproteins (LDL) is an important factor.

Review Exercises

Objective Questions (Answers on p. E–8.)

1. Adjacent cardiac muscle cells are joined end to end at specialized structures known as __________, which contain two types of membrane junctions—__________ and __________.
2. __________ is an abnormally slow heart rate whereas __________ is a rapid heart rate.
3. The link that coordinates coronary blood flow with myocardial oxygen needs is __________.
4. The left ventricle is a stronger pump than the right ventricle because more blood is needed to supply the body tissues than to supply the lungs. (True or false?)
5. The heart lies in the left half of the thoracic cavity. (True or false?)
6. The only point of electrical contact between the atria and ventricles is the fibrous skeletal rings. (True or false?)
7. The atria and ventricles each act as a functional syncytium. (True or false?)
8. Which of the following is the proper sequence of cardiac excitation?
   a. SA node → AV node → atrial myocardium → bundle of His → Purkinje fibers → ventricular myocardium.
   b. SA node → atrial myocardium → AV node → bundle of His → ventricular myocardium → Purkinje fibers.
   c. SA node → atrial myocardium → ventricular myocardium → AV node → bundle of His → Purkinje fibers.
   d. SA node → atrial myocardium → AV node → bundle of His → Purkinje fibers → ventricular myocardium.
9. What percentage of ventricular filling is normally accomplished before atrial contraction begins?
   a. 0%
   b. 20%
   c. 50%
   d. 80%
   e. 100%
10. Sympathetic stimulation of the heart:
    a. increases the heart rate.
    b. increases the contractility of the heart muscle.
    c. shifts the Frank-Starling curve to the left.
    d. Both (a) and (b) above are correct.
    e. All of the above are correct.
11. Circle the correct choice in each instance to complete the statements: During ventricular filling, ventricular pressure must be (greater than/less than) atrial pressure, while during ventricular ejection, ventricular pressure must be (greater than/less than) aortic pressure. Atrial pressure is always (greater than/less than) aortic pressure. During isovolumetric ventricular contraction and relaxation, ventricular pressure is (greater than/less than) atrial pressure and (greater than/less than) aortic pressure.
12. Circle the correct choice in each instance to complete the statement: The first heart sound is associated with closure of the (AV/semilunar) valves and signals the onset of (systole/diastole), whereas the second heart sound is associated with closure of the (AV/semilunar) valves and signals the onset of (systole/diastole).
13. Match the following:

___1. receives O_2-poor blood from the venae cavae	(a) AV valves
___2. prevent backflow of blood from the ventricles to the atria	(b) semilunar valves
___3. pumps O_2-rich blood into the aorta	(c) left atrium
___4. prevent backflow of blood from the arteries into the ventricles	(d) left ventricle
___5. pumps O_2-poor blood into the pulmonary artery	(e) right atrium
___6. receives O_2-rich blood from the pulmonary veins	(f) right ventricle
___7. permit AV valves to function as one-way valves	(g) chordae tendineae and papillary muscles

14. Match the following:

___ 1. characterized by uncoordinated excitation and contraction of the cardiac cells	(a) heart block
___ 2. caused by AV nodal damage	(b) ectopic focus
___ 3. an overly irritable area that takes over pacemaker activity	(c) ventricular fibrillation

Essay Questions

1. What are the three basic components of the circulatory system?
2. Trace a drop of blood through one complete circuit of the circulatory system.
3. Describe the location and function of the four heart valves. What prevents eversion of each of these valves?
4. What are the three layers of the heart wall? Describe the distinguishing features of the structure and arrangement of cardiac muscle cells. What are the two specialized types of cardiac muscle cells?
5. Why is the SA node the pacemaker of the heart?
6. Describe the normal spread of cardiac excitation. What is the significance of the AV nodal delay? Why is the ventricular conduction system important?
7. Compare the changes in membrane potential associated with an action potential in a nodal pacemaker cell with those in a myocardial contractile cell. What is responsible for the plateau phase?
8. Why is tetanus of cardiac muscle impossible? Why is this advantageous?
9. Draw and label the waveforms of a normal ECG. What electrical event does each component of the ECG represent?
10. Describe the mechanical events (that is, pressure changes, volume changes, valve activity, and heart sounds) that occur during the cardiac cycle. Correlate these mechanical events with the changes that take place in electrical activity.
11. Distinguish between a stenotic and an insufficient valve.
12. Define the following: end-diastolic volume, end-systolic volume, stroke volume, heart rate, cardiac output, and cardiac reserve.
13. Discuss autonomic nervous system control of heart rate.
14. Describe the intrinsic and extrinsic control of stroke volume.
15. By what means is the heart muscle provided with blood? Why does the heart receive most of its own blood supply during diastole?
16. What are the pathological changes and consequences of coronary artery disease?
17. Discuss the sources, transport, and elimination of cholesterol in the body. Distinguish between "good" cholesterol and "bad" cholesterol.

Quantitative Exercises (Solutions on p. E–8.)

1. During heavy exercise, the cardiac output can increase to as much as 35 liters/min. If stroke volume could not increase above the normal value of 70 ml, what heart rate would be necessary to achieve this cardiac output? Is such a heart rate physiologically possible?
2. How much blood remains in the heart after systole if the stroke volume is 85 ml and the end-diastolic volume is 125 ml?

Points to Ponder

(Explanations on p. E–8.)

1. The stroke volume ejected on the next heartbeat after a premature beat is usually larger than normal. Can you explain why? (Hint: At a given heart rate, the interval between a premature beat and the next normal beat is longer than the interval between two normal beats.)
2. Trained athletes usually have lower resting heart rates than normal (for example, 50 beats/min in an athlete compared to 70 beats/min in a sedentary individual). Considering that the resting cardiac output is 5,000 ml/min in both trained athletes and sedentary individuals, what is responsible for the bradycardia of trained athletes?
3. A characteristic murmur accompanies patent ductus arteriosus (see the boxed feature on p. 272). A harsh blowing murmur can be heard throughout the cardiac cycle, but becomes much more intense during ventricular systole and much less intense during ventricular diastole. What would account for the waxing and waning of the murmur with each beat of the heart?
4. Through what regulatory mechanisms is a transplanted heart, which does not have any innervation, able to adjust the cardiac output to meet the body's changing needs?
5. There are two branches of the bundle of His, the right and left bundle branches, each of which travels down its respective side of the ventricular septum (see Fig. 9-11). Occasionally, conduction through one of these branches becomes blocked (so-called *bundle-branch block*). In this case, the wave of excitation spreads out from the terminals of the intact branch and eventually depolarizes the whole ventricle, but the normally stimulated ventricle becomes completely depolarized a considerable time before the ventricle on the side of the defective bundle branch. For example, if the left bundle branch is blocked, the right ventricle will be completely depolarized two to three times more rapidly than the left ventricle. What effect would this defect have on the heart sounds?
6. ***Clinical Consideration*** On physical exam, Rachel B.'s heart rate was rapid and very irregular. Furthermore, her heart rate determined directly by listening to her heart with a stethoscope exceeded the pulse rate taken concurrently at her wrist. No definite P waves could be detected on Rachel's ECG. The QRS complexes were normal in shape but occurred sporadically. Based on these findings, what is the most likely diagnosis of Rachel's condition? Explain why the condition is characterized by a rapid, irregular heartbeat. Would cardiac output be seriously impaired by this condition? Why or why not? What accounts for the pulse deficit?

10 The Blood Vessels and Blood Pressure

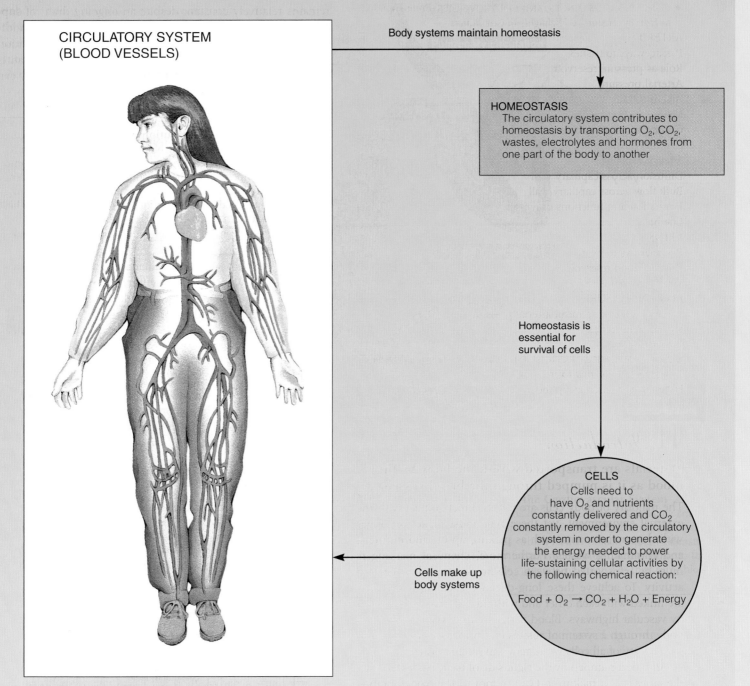

CIRCULATORY SYSTEM
(BLOOD VESSELS)

Body systems maintain homeostasis

HOMEOSTASIS
The circulatory system contributes to homeostasis by transporting O_2, CO_2, wastes, electrolytes and hormones from one part of the body to another

Homeostasis is essential for survival of cells

CELLS
Cells need to have O_2 and nutrients constantly delivered and CO_2 constantly removed by the circulatory system in order to generate the energy needed to power life-sustaining cellular activities by the following chemical reaction:
Food + O_2 → CO_2 + H_2O + Energy

Cells make up body systems

The **circulatory system** contributes to homeostasis by serving as the body's transport system. The blood vessels transport and distribute blood pumped through them by the heart to meet the body's needs for O_2 and nutrient delivery, waste removal, and hormonal signaling. The highly elastic **arteries** transport blood from the heart to the tissues and serve as a pressure reservoir to continue driving blood forward when the heart is relaxing and filling. The **mean arterial blood pressure** is closely regulated to ensure adequate blood delivery to the tissues. The amount of blood that flows through a given tissue depends on the caliber of the highly muscular **arterioles** that supply the tissue. Arteriolar caliber is subject to control so that the distribution of the cardiac output can be constantly readjusted to best serve the body's needs at the moment. The thin-walled, pore-lined **capillaries** are the actual site of exchange between the blood and the surrounding tissues. The highly distensible **veins** return the blood from the tissues to the heart and also serve as a blood reservoir.

not 100 mm Hg), because arterial pressure remains closer to diastolic than to systolic pressure for a longer portion of each cardiac cycle. At resting heart rate, about two-thirds of the cardiac cycle is spent in diastole and only one-third in systole. As an analogy, if a race car traveled 80 miles per hour (mph) for 40 minutes and 120 mph for 20 minutes, its average speed would be 93 mph, not the halfway value of 100 mph. Similarly, a good approximation of the mean arterial pressure can be determined using the following formula:

$$\text{mean arterial pressure} = \text{diastolic pressure} + 1/3 \text{ pulse pressure}$$

$$\text{at } 120/80, \text{ mean arterial pressure} = 80 \text{ mm Hg} + (1/3)\ 40 \text{ mm Hg} = 93 \text{ mm Hg}$$

It is this mean arterial pressure, not the systolic or diastolic pressures, that is monitored and regulated by blood pressure reflexes to be described later in the chapter.

Arterial pressure—whether it be systolic, diastolic, pulse or mean—is essentially the same throughout the arterial tree (Fig. 10-9). Because arteries offer little resistance to flow, only a negligible amount of pressure energy is lost in them due to friction.

Figure 10-9 Pressures throughout the Systemic Circulation Left ventricular pressure swings between a low pressure of 0 mm Hg during diastole to a high pressure of 120 mm Hg during systole. Arterial blood pressure, which fluctuates between a peak systolic pressure of 120 mm Hg and a low diastolic pressure of 80 mm Hg each cardiac cycle, is of the same magnitude throughout the large arteries. Because of the arterioles' high resistance, the pressure drops precipitously and the systolic-to-diastolic swings in pressure are converted to a nonpulsatile pressure when blood flows through the arterioles. The pressure continues to decline but at a slower rate as blood flows through the capillaries and venous system.

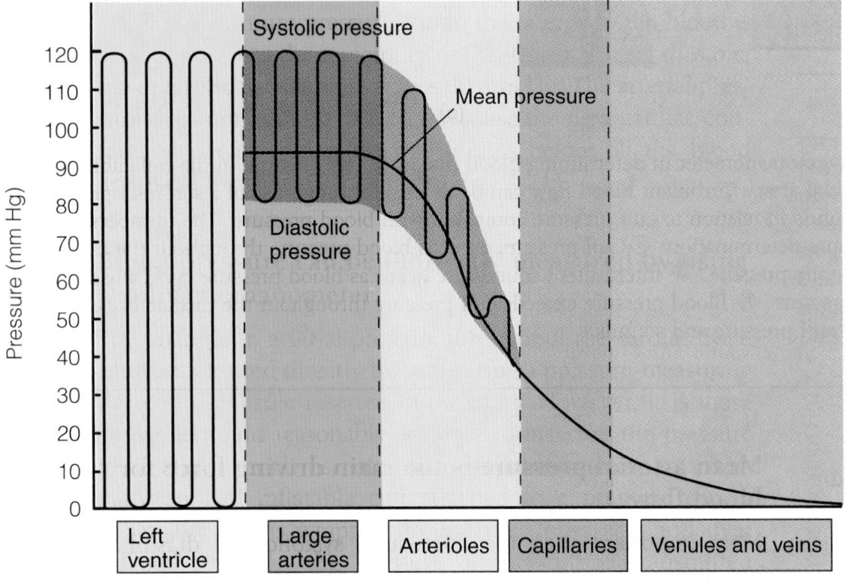

Arterioles

Arterioles are the major resistance vessels.

When an artery reaches the organ it is supplying, it branches into numerous arterioles, whose radii are small enough to offer considerable resistance to flow. In fact, the arterioles are the major resistance vessels in the vascular tree. (Even though the capillaries have smaller radii than the arterioles, we will see later how collectively the capillaries do not offer as much resistance to flow as the arteriolar level of the vascular tree does.) In contrast to the low resistance of the arteries, the high degree of arteriolar resistance causes a marked drop in mean pressure as the blood flows through these vessels. On average, the pressure falls from 93 mm Hg, the mean arterial pressure (the pressure of the blood entering the arterioles), to 37 mm Hg, the pressure of the blood leaving the arterioles and entering the capillaries (Fig. 10-9). This decline in pressure helps establish the pressure differential that encourages the flow of blood from the heart to the various organs downstream. If no pressure drop occurred in the arterioles, the pressure at the end of the arterioles (that is, at the beginning of the capillaries) would be equal to the mean arterial pressure. No pressure gradient would exist to drive the blood from the heart to the tissue capillary beds. In addition to causing a marked drop in mean blood pressure at the arteriolar level, arteriolar resistance is also responsible for converting the pulsatile systolic-to-diastolic pressure swings in the arteries into the nonfluctuating pressure present in the capillaries.

The radii (and, accordingly, the resistances) of arterioles supplying individual organs can be adjusted independently to accomplish two functions: (1) to variably distribute the cardiac output among the systemic organs, depending on the body's momentary needs, and (2) to help regulate arterial blood pressure. We will discuss the mechanisms involved in adjusting arteriolar resistance before considering how such adjustments are important in accomplishing these two functions.

Unlike arteries, arteriolar walls contain very little elastic connective tissue. However, they do have a thick layer of smooth muscle that is richly innervated by sympathetic nerve fibers. The smooth muscle is also sensitive to many local chemical changes and to a few circulating hormones. The smooth muscle layer runs circularly around the arteriole (Fig. 10-10a) so when it contracts, the vessel's circumference (and its radius) becomes smaller, thus increasing resistance and decreasing the flow through that vessel. **Vasoconstriction** is the term applied to such narrowing of a vessel (Fig. 10-10c). **Vasodilation** refers to enlargement in the circumference and radius of a vessel as a result of relaxation of its smooth muscle layer (Fig. 10-10d). Vasodilation leads to decreased resistance and increased flow through that vessel.

Arteriolar smooth muscle normally displays a state of partial constriction known as **vascular tone**, which

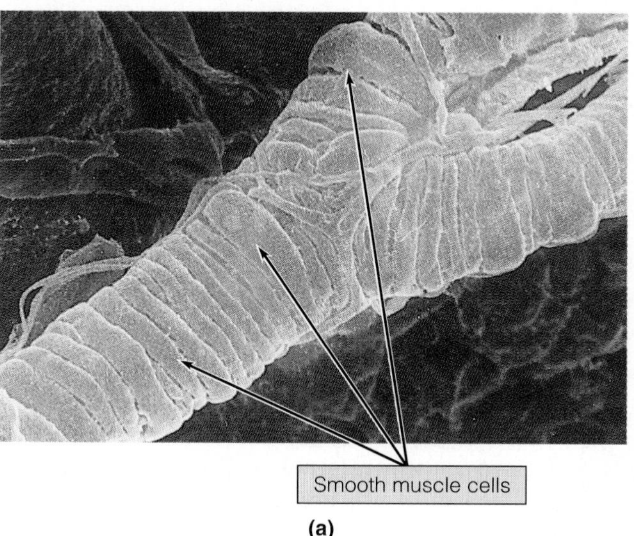

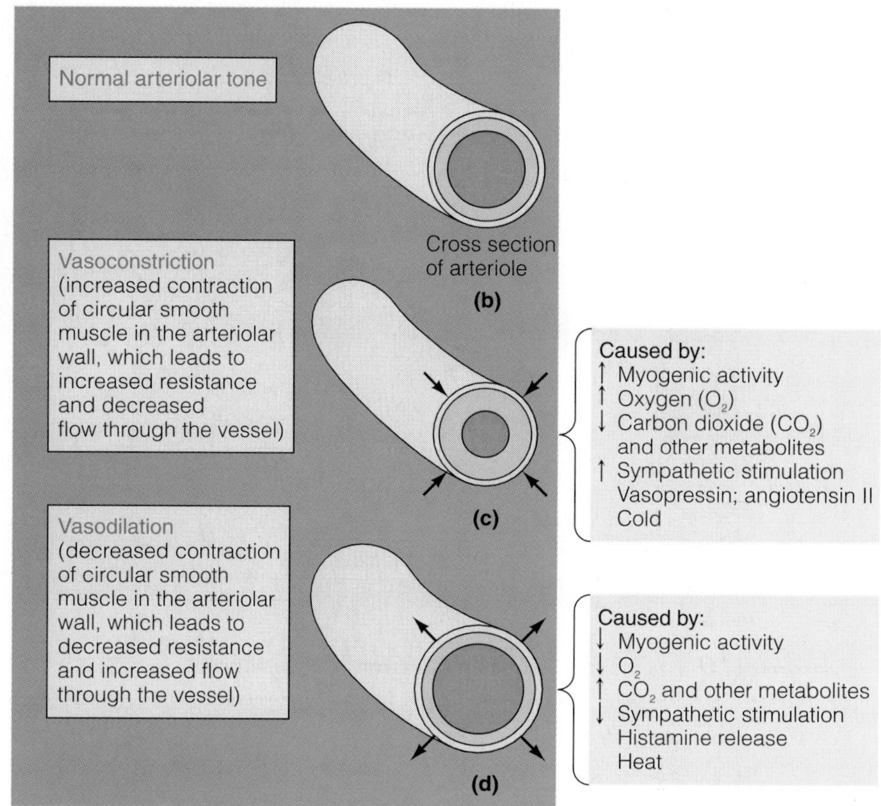

Figure 10-10 **Arteriolar Vasoconstriction and Vasodilation** (a) A scanning electron micrograph of an arteriole showing how the smooth muscle cells run circularly around the vessel wall. (b) Schematic representation of an arteriole in cross section showing normal arteriolar tone. (c) Outcome of and factors causing arteriolar vasoconstriction. (d) Outcome of and factors causing arteriolar vasodilation.

establishes a baseline of arteriolar resistance (Fig. 10-10b). Two factors are responsible for vascular tone. First, arteriolar smooth muscle has considerable myogenic activity; that is, its membrane potential fluctuates without any neural or hormonal influences, leading to self-induced contractile activity (see p. 258). Second, the sympathetic fibers supplying most arterioles continually release norepinephrine, which further enhances the vascular tone.

This ongoing tonic activity makes it possible to either increase or decrease the level of contractile activity to accomplish vasoconstriction or vasodilation, respectively. Were it not for tone, it would be impossible to reduce the tension in an arteriolar wall to accomplish vasodilation; only varying degrees of vasoconstriction would be possible.

A variety of factors can influence the level of contractile activity in arteriolar smooth muscle, thereby substantially changing resistance to flow in these vessels. These factors fall into two categories: *local (intrinsic) controls*, which are important in matching blood flow to the metabolic needs of the specific tissues in which they occur, and *extrinsic controls,* which are important in blood pressure regulation.

Local control of arteriolar radius is important in determining the distribution of cardiac output so that blood flow is matched with the tissues' metabolic needs.

The fraction of the total cardiac output delivered to each organ is not always constant; it varies, depending on the demands for blood at the time. The amount of the cardiac output received by each organ is determined by the number and caliber of the arterioles supplying that area. Recall that $F = \Delta P/R$. Because blood is delivered to all tissues at the same mean arterial pressure, the driving force for flow is identical for each organ. Therefore, differences in flow to various organs are completely determined by differences in the extent of vascularization and by differences in resistance offered by the arterioles supplying each organ. On a moment-to-moment basis, the distribution of cardiac output can be varied by differentially adjusting arteriolar resistance in the various vascular beds.

As an analogy, consider a pipe carrying water with a number of adjustable valves located throughout its length (Fig. 10-11). Assuming that water pressure in the pipe is constant, differences in the amount of water flowing into a beaker under each valve depend entirely on which valves are open and to what extent. No water enters beakers under closed valves (high resistance), and more water flows into beakers under valves that are opened completely (low resistance) than into beakers under valves that are only partially opened (moderate resistance). Similarly, more blood flows to areas whose arterioles offer the least resistance to its passage. During exercise, for example, not only is cardiac output increased, but because of vasodilation in skeletal muscle and in the heart, a greater percentage of the pumped blood is diverted to these organs to support their increased metabolic activity. Simultaneously, blood flow to the digestive tract and kidneys is reduced as a result of arteriolar vasoconstriction in

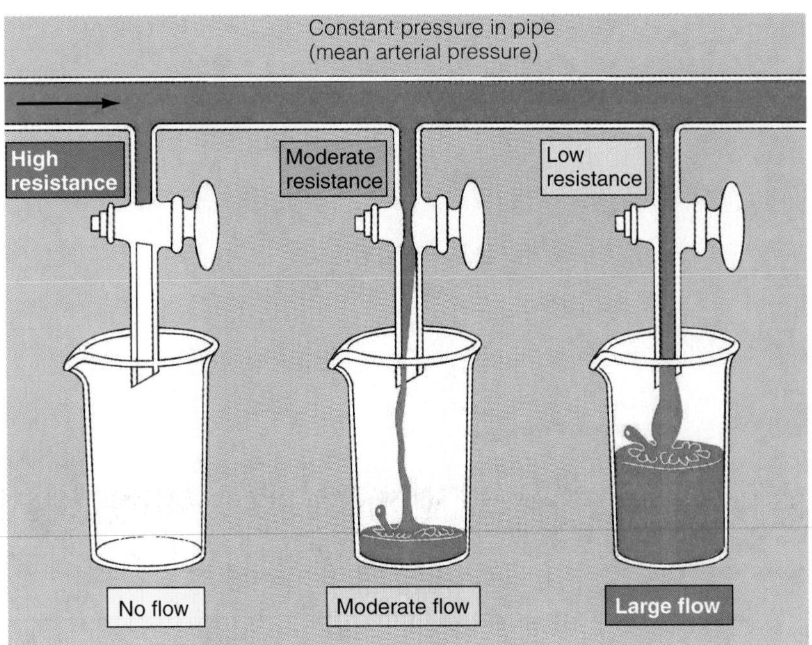

Figure 10-11 **Flow Rate as a Function of Resistance**

these organs (Fig. 10-12). Only the blood supply to the brain remains remarkably constant no matter what activity the person is engaged in, be it vigorous physical activity, intense mental concentration, or sleep. Although the *total* blood flow to the brain remains constant, new techniques demonstrate that differences in regional blood flow occur within the brain in close correlation with local neural activity patterns (see p. 116).

Local (intrinsic) controls are changes within a tissue that alter the radii of the vessels and hence adjust blood flow through the tissue by directly affecting the smooth muscle of the tissue's arterioles. Local influences may be either chemical or physical in nature. Local chemical influences on arteriolar radius include (1) local metabolic changes and (2) histamine release. Local physical influences include (1) local application of heat or cold and (2) myogenic responses to stretch. Let us examine the role and mechanism of each of these local influences.

Local chemical influences

1. *Local metabolic changes.* The influence of local metabolic changes on arteriolar radius is important in matching the blood flow through a tissue with the tissue's metabolic needs. Local metabolic controls are especially important in skeletal muscle and in the heart, the tissues whose metabolic activity and need for blood supply normally vary most extensively, and in the brain, whose overall metabolic activity and need for blood supply remain constant. Local controls help maintain the constancy of blood flow to the brain.

Arterioles lie within the tissue they are supplying and can be acted on by local factors within the tissue. During increased metabolic activity, such as when a skeletal muscle is contracting during exercise, the local concentrations of a number of the tissue's chemicals change. For example, the local O_2 concentration decreases as the actively metabolizing cells use up more O_2 to support oxidative phosphorylation for ATP production (see p. 32). This and other local chemical changes produce local arteriolar dilation by triggering relaxation of the arteriolar smooth muscle in the vicinity. Local arteriolar vasodilation subsequently increases blood flow to that particular area, a response called **active hyperemia** (*hyper* means "above normal"; *emia* means "blood"). When cells are more active metabolically, they need more blood to bring in O_2 and nutrients and to remove metabolic wastes. The increased blood flow meets these increased local needs.

Conversely, when a tissue, such as a relaxed muscle, is less active metabolically and thus has reduced needs for blood delivery, the resultant local chemical changes (for example, increased local O_2 concentration) bring about local arteriolar vasoconstriction and a subsequent reduction in blood flow to the area. Thus, local metabolic changes can adjust blood flow as needed without involving nerves or hormones.

Exactly which local chemical changes are responsible for these "selfish" local adjustments in arteriolar caliber that match a tissue's blood flow with its metabolic needs is still uncertain. The following local chemical factors have been shown experimentally to produce relaxation of arteriolar smooth muscles:

- *Decreased O_2*
- *Increased CO_2*—More CO_2 is generated as a by-product during the stepped-up pace of oxidative phosphorylation that accompanies increased activity.
- *Increased acid*—More carbonic acid is generated from the increased CO_2 produced as the metabolic activity of a cell increases. Also, lactic acid accumulates if the glycolytic pathway is used for ATP production (see p. 33).
- *Increased K^+*—Repeated action potentials that outpace the ability of the Na^+-K^+ pump to restore the resting concentration gradients (see p. 87) result in an increase in K^+ in the tissue fluid of an actively contracting muscle or a more active region of the brain.
- *Increased osmolarity*—Osmolarity (the concentration of osmotically active solutes) increases during elevated cellular metabolism because of increased formation of osmotically active particles.
- *Adenosine release*—Especially in cardiac muscle, adenosine is released in response to increased metabolic activity or O_2 deprivation (see p. 296).

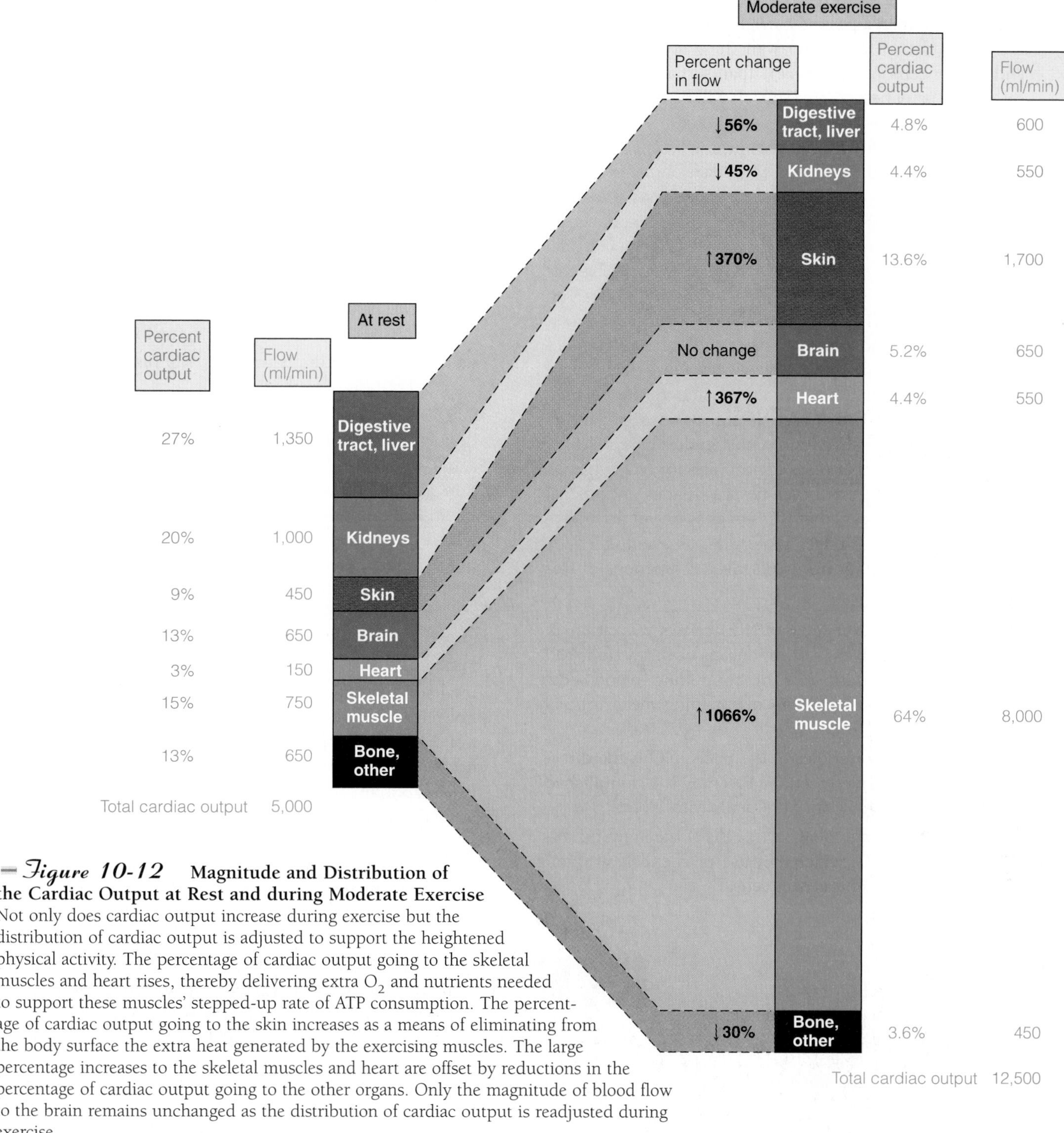

Figure 10-12 Magnitude and Distribution of the Cardiac Output at Rest and during Moderate Exercise
Not only does cardiac output increase during exercise but the distribution of cardiac output is adjusted to support the heightened physical activity. The percentage of cardiac output going to the skeletal muscles and heart rises, thereby delivering extra O_2 and nutrients needed to support these muscles' stepped-up rate of ATP consumption. The percentage of cardiac output going to the skin increases as a means of eliminating from the body surface the extra heat generated by the exercising muscles. The large percentage increases to the skeletal muscles and heart are offset by reductions in the percentage of cardiac output going to the other organs. Only the magnitude of blood flow to the brain remains unchanged as the distribution of cardiac output is readjusted during exercise.

- *Prostaglandin release*—Prostaglandins are local chemical messengers derived from fatty acid chains within the plasma membrane. Their production and release is poorly understood (see p. 716).

The relative contributions of these various chemical changes (and possibly others) in the local metabolic control of blood flow in systemic arterioles is still being investigated.

Also under investigation is the mechanism by which local metabolic factors change the contractile state of arteriolar smooth muscle. These factors do not appear to act directly on vascular smooth muscle as once was thought. Instead, it has become increasingly clear that the **endothelial cells**, the single layer of specialized epithelial cells that line the lumen of all blood vessels, release chemical mediators that play a key role in the local regulation of arteriolar caliber. Until recently, endothelial cells were regarded as little more than a passive barrier between the blood and the remainder of the vessel wall. It is now known that endothelial cells are active participants in a variety of vessel-related activities, some of which will be described elsewhere (▮ Table 10-2). Among these functions, endothelial cells release locally acting chemical messengers in response to chemical changes in their environment (such as a reduction in O_2) or physical changes (such as stretching of the vessel wall). These local chemical mediators act on the underlying smooth muscle to alter its state of contraction. Among the best studied of these local vasoactive ("acting on vessels") mediators is **endothelial-derived relaxing factor** (**EDRF**), which causes local arteriolar vasodilation by inducing relaxation of arteriolar smooth muscle in the vicinity. It does so by inhibiting the entry of contraction-inducing Ca^{2+} into these smooth muscle cells (see p. 226). EDRF has been identified as **nitric oxide** (**NO**), a small, short-lived molecule. Studies have revealed an astonishing number of biological roles for NO, which is produced in numerous other tissues besides endothelial cells. In fact, it appears that NO serves as one of the body's most important messenger molecules, as exemplified by the following range of newly discovered functions of this chemical:

- As mentioned, NO (alias EDRF) causes relaxation of arteriolar smooth muscle. By means of this action, NO plays an important role in controlling blood flow through the tissues and in maintaining mean arterial blood pressure.
- By dilating the arterioles of the penis, NO is the direct mediator of penile erection. Erection is accomplished by rapid engorgement of the penis with blood.
- Macrophages, large phagocytic cells of the immune system, produce NO, which they use as "chemical warfare" against bacteria and cancer cells.
- NO interferes with platelet function and blood clotting at sites of vessel damage.
- NO serves as a novel type of neurotransmitter in the brain and elsewhere. Unlike classical neurotransmitters, NO is not stored in vesicles and released by exocytosis, and it does not bind with receptors on its target cell. NO is synthesized on demand in the neuron terminal, then diffuses out of the terminal and into the adjacent target cell, where it brings about its effect.
- NO plays a role in the changes underlying memory (see p. 139).
- By promoting relaxation of digestive tract smooth muscle, NO helps regulate peristalsis, a type of contraction that pushes digestive tract contents forward.
- Evidence is accumulating that NO may also play a role in relaxation of skeletal muscle.

The endothelial cells release other important chemicals besides EDRF/NO. **Endothelin**, another endothelial vasoactive substance, causes arteriolar smooth muscle contraction and is one of the most potent vasoconstrictors yet identified. Still other chemicals released from the endothelium in response to chronic changes in blood flow to an organ trigger long-term vascular changes that permanently influence blood flow to a region. Some chemicals, for example, are thought to stimulate *angiogenesis* (new vessel growth).

2. *Local histamine release.* Histamine is another local chemical mediator that influences arteriolar smooth muscle, but it is not released in response to local metabolic changes and is not derived from the endothelial cells. Although histamine normally does not participate in the control of blood flow, it is important in certain pathological conditions. Histamine is synthesized and stored within special connective tissue cells in many tissues and in certain types of circulating white blood cells. When tissues are injured or during allergic reactions, histamine is released in the damaged region. By promoting relaxation of arteriolar smooth muscle, histamine is the major cause of vasodilation in an injured area. The resultant increase in blood flow into the area is responsible for the redness and contributes to the swelling associated with inflammatory responses.

Local physical influences

1. *Local heat or cold application.* Heat application, by causing localized arteriolar vasodilation, is a useful therapeutic agent for promoting increased blood flow to an area. Conversely, applying ice packs to an inflamed area produces vasoconstriction, which reduces swelling by counteracting histamine-induced vasodilation.
2. *Myogenic responses to stretch.* Evidence indicates that arteriolar smooth muscle responds to being passively stretched by myogenically increasing its tone, thereby acting to resist the initial passive stretch. Conversely, a decrease in arteriolar stretching is accompanied by a reduction in myogenic vessel tone. Endothelial-derived vasoactive substances may also contribute to these mechanically induced responses.

Table 10-2
Functions of Endothelial Cells

Functions of Endothelial Cells
Line the blood vessels and heart chambers; serve as a physical barrier between the blood and the remainder of the vessel wall
Secrete vasoactive substances, such as endothelial-derived relaxing factor, in response to local chemical and physical changes; these substances cause relaxation (vasodilation) or contraction (vasoconstriction) of the underlying smooth muscle
Secrete substances that stimulate new vessel growth and proliferation of smooth-muscle cells in vessel walls
Participate in the exchange of materials between the blood and surrounding tissues across capillaries through vesicular transport (see p. 63)
Influential in formation of platelet plugs; clotting, and clot dissolution (see chapter 11)
Participate in the determination of capillary permeability by contracting to vary the size of the pores between adjacent endothelial cells

The extent of passive stretch varies with the volume of blood delivered to the arterioles from the arteries. An increase in mean arterial pressure drives more blood forward into the arterioles and stretches them further, whereas arterial occlusion blocks blood flow into the arterioles and reduces arteriolar stretch.

Myogenic responses, coupled with metabolically induced responses, appear to be important in reactive hyperemia and pressure autoregulation.

a. *Reactive hyperemia.* When the blood supply to a region is completely occluded, arterioles in the region dilate due to (1) myogenic relaxation, which occurs in response to the diminished stretch accompanying no blood flow, and (2) changes in local chemical composition. Many of the same chemical changes occur in a blood-deprived tissue that occur during metabolically induced active hyperemia. When a tissue's blood supply is blocked, O_2 levels decrease in the deprived tissue; the tissue continues to consume O_2, but no fresh supplies are being delivered. Meanwhile, the concentrations of CO_2, acid, and other metabolites rise. Even though their production does not increase as it does when a tissue is more active metabolically, these substances accumulate in the tissue when the normal amounts produced are not "washed away" by blood.

   After the occlusion is removed, blood flow to the previously deprived tissue is transiently much higher than normal because the arterioles are widely dilated. This postocclusion increase in blood flow, called **reactive hyperemia,** can take place in any tissue. Such a response is beneficial for rapidly restoring the local chemical composition to normal. Of course, prolonged blockage of blood flow leads to irreversible tissue damage. You can demonstrate reactive hyperemia yourself by wrapping a rubber band around one of your fingers about an inch from the end for a minute. On removal, you will note that your fingertip turns red as the blood flow increases.

b. *Pressure autoregulation.* When mean arterial pressure falls (for example, because of hemorrhage or a weakened heart), the driving force is reduced, so blood flow to tissues decreases. This is a milder version of what happens during vessel occlusion. The resultant changes in local metabolites and the reduced stretch in the arterioles collectively bring about arteriolar dilation to restore tissue blood flow to normal despite the reduced driving pressure. On the negative side, widespread arteriolar dilation reduces the mean arterial pressure still further, which aggravates the problem. Conversely, in the presence of sustained elevations in mean arterial pressure (hypertension), local chemical and myogenic influences triggered by the initial increased flow of blood to tissues bring about an increase in arteriolar tone and resistance. This greater degree of vasoconstriction subsequently reduces tissue blood flow toward normal despite the elevated blood pressure (Fig. 10-13). **Pressure autoregulation** is the term applied to these local arteriolar mechanisms that are aimed at keeping tissue blood flow fairly constant in spite of rather wide deviations in mean arterial driving pressure.

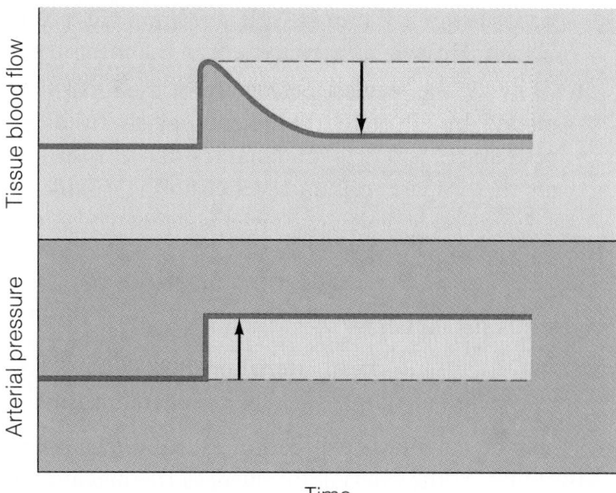

Figure 10-13 **Pressure Autoregulation of Tissue Blood Flow**
Even though blood flow through a tissue immediately increases in response to a rise in arterial pressure, the tissue blood flow is reduced gradually as a result of pressure autoregulation within the tissue, despite a sustained increase in arterial pressure.

Active hyperemia, reactive hyperemia, and histamine release all deliberately increase blood flow to a particular tissue for a specific purpose by inducing local arteriolar vasodilation. In contrast, pressure autoregulation is a means by which each tissue resists alterations in its own blood flow secondary to changes in mean arterial pressure by making appropriate adjustments in arteriolar radius.

Extrinsic control of arteriolar radius is primarily important in the regulation of arterial blood pressure.

Extrinsic control of arteriolar radius includes both neural and hormonal influences, with the effects of the sympathetic nervous system being the most important. Sympathetic nerve fibers supply arteriolar smooth muscle everywhere except in the brain. A certain level of ongoing sympathetic activity contributes to vascular tone. Increased sympathetic activity produces generalized arteriolar vasoconstriction, whereas decreased sympathetic activity leads to generalized arteriolar vasodilation.

These widespread changes in arteriolar resistance bring about changes in mean arterial blood pressure. The formula $F = \Delta P/R$ applies to the entire circulation as well as to a single vessel:

- F: Looking at the circulatory system as a whole, flow (F) through all the vessels is equal to the cardiac output.
- ΔP: The pressure gradient (ΔP) for the entire circulation is the mean arterial pressure. ΔP equals the difference in pressure between the beginning and the end of the circulatory system. Because the beginning pressure is the mean arterial pressure as the blood leaves the left ventricle at an average of 93 mm Hg and the end pressure in the

right atrium is 0 mm Hg, $\Delta P = 93$ mm Hg $- 0$ mm Hg $=$ 93 mm Hg, which is equivalent to the mean pressure.

- R: By far the greatest percentage of the total resistance (R) offered by all of the peripheral vessels (**total peripheral resistance**) is due to arteriolar resistance, because arterioles are the primary resistance vessels in the vascular tree.

Therefore, for the entire system, rearranging

$$F = \Delta P/R \quad \text{to} \quad \Delta P = F \times R$$

gives us the equation

mean arterial pressure =
cardiac output × total peripheral resistance

Thus, the extent of total peripheral resistance offered collectively by all the arterioles influences the mean arterial blood pressure immensely. A dam provides an analogy to this relationship. At the same time it restricts the flow of water downstream, a dam increases the pressure upstream by elevating the water level in the reservoir behind the dam. Similarly, generalized, sympathetically induced vasoconstriction reflexly reduces blood flow downstream to the tissue cells while elevating the upstream mean arterial pressure, thereby increasing the main driving force for blood flow to all the organs.

These effects seem to be counterproductive. Why increase the driving force for flow to the organs by increasing arterial blood pressure while reducing flow to the organs by narrowing the vessels supplying them? In effect, the sympathetically induced arteriolar responses help maintain the appropriate driving pressure head to all organs. The extent to which each organ actually receives blood flow is determined by local arteriolar adjustments that override the sympathetic constrictor effect. If all arterioles were dilated, blood pressure would fall substantially, so there would not be an adequate driving force for blood flow. An analogy is the pressure head for water in the pipes in your home. If the water pressure is adequate, you can selectively obtain satisfactory water flow at any of the faucets by turning the appropriate handle to the open position. If the water pressure in the pipes is too low, however, you cannot obtain satisfactory flow at any faucet, even if you turn the handle to the maximally open position. Thus, tonic sympathetic activity constricts most vessels (with the exception of those in the brain) to help maintain a pressure head on which organs can draw as needed through local mechanisms that control arteriolar radius.

Skeletal and cardiac muscles have the most powerful local control mechanisms with which to override generalized sympathetic vasoconstriction. For example, if you are pedaling a bicycle, the increased activity in the skeletal muscles of your legs induces overriding local vasodilation in those particular muscles, despite the generalized sympathetic vasoconstriction that accompanies exercise. As a result, more blood flows through your leg muscles but not through your inactive arm muscles. In the heart, coronary vasodilation actually accompanies sympathetic stimulation. The sympathetically induced increase in heart rate and stroke volume elevates cardiac metabolic activity, triggering a profound, locally mediated vasodilation that overpowers the weaker sympathetic vasoconstrictor effect.

The norepinephrine released from sympathetic nerve endings combines with α adrenergic receptors (see p. 210) on arteriolar smooth muscle to bring about vasoconstriction. Cerebral (brain) arterioles are the only ones that do not have α receptors, so no vasoconstriction occurs in the brain. It is important that cerebral arterioles not be reflexly constricted by neural influences, since brain blood flow must remain constant to meet the brain's continual need for O_2, no matter what is going on elsewhere in the body. Cerebral vessels are almost entirely controlled by local mechanisms that maintain a constant blood flow to support a constant level of brain metabolic activity. In fact, reflex vasconstrictor activity in the remainder of the cardiovascular system is aimed at maintaining an adequate pressure head for blood flow to the vital brain and heart.

Thus, sympathetic activity contributes in an important way to the maintenance of mean arterial pressure, assuring an adequate driving force for blood flow to the brain at the expense of organs and tissues that can better withstand reduced blood flow. Other tissues that really need additional blood, such as active muscles (including active heart muscle), obtain it through local controls that override the sympathetic effect.

There is no significant level of parasympathetic innervation to arterioles, with the exception of the abundant parasympathetic vasodilator supply to the arterioles of the penis and clitoris. The rapid, profuse vasodilation induced by parasympathetic stimulation in these organs (by means of promoting NO release) is largely responsible for accomplishing erection. Vasodilation elsewhere is produced by decreasing sympathetic vasoconstrictor activity below its tonic level.[1] When mean arterial pressure becomes elevated above normal, reflex reduction in sympathetic vasoconstrictor activity accomplishes generalized arteriolar vasodilation to help restore the driving pressure down toward normal.

The main region of the brain responsible for adjusting sympathetic output to the arterioles is the **cardiovascular control center** in the medulla of the brain stem. This is the integrating center for blood pressure regulation. Several other brain regions also influence blood distribution, the most notable being the hypothalamus, which, as part of its temperature-regulating function, controls blood flow to the skin to adjust heat loss to the environment.

In addition to neural reflex activity, several hormones also extrinsically influence arteriolar radius. These hormones include the adrenal medullary hormones *epinephrine* and *norepinephrine,* which generally reinforce the sympathetic nervous system in most tissues, as well as *vasopressin* and *angiotensin II,* which are important in the control of fluid balance.

Sympathetic stimulation of the adrenal medulla causes this endocrine gland to release epinephrine and norepinephrine. Adrenal medullary norepinephrine combines with the same α receptors as sympathetically released norepinephrine to produce generalized vasoconstriction. However, epinephrine, the

---

[1]A portion of the skeletal muscle fibers in some species are supplied by sympathetic cholinergic (ACh-releasing) fibers that bring about vasodilation in anticipation of exercise. However, the existence of such sympathetic vasodilator fibers in humans remains questionable.

more abundant of the adrenal medullary hormones, combines with both β_2 and α receptors but has a greater affinity for the β_2 receptors. Activation of β_2 receptors produces vasodilation, but not all tissues have β_2 receptors; they are most abundant in the arterioles of the heart and skeletal muscles. During sympathetic discharge, the released epinephrine combines with the β_2 receptors in the heart and skeletal muscle to reinforce local vasodilatory mechanisms in these tissues. Arterioles in digestive organs and kidneys, equipped only with α receptors, do not respond to epinephrine. Therefore, the arterioles of these organs, affected only by norepinephrine during generalized sympathetic discharge, undergo more profound vasoconstriction than those in heart and skeletal muscle do (▮ Table 10-3).

The two other hormones that extrinsically influence arteriolar tone are vasopressin and angiotensin II. Vasopressin is primarily involved in maintaining water balance by regulating the amount of water the kidneys retain for the body during urine formation. Angiotensin II is part of a hormonal pathway, the *renin-angiotensin-aldosterone pathway,* which is important in the regulation of the body's salt balance. This pathway promotes salt conservation during urine formation and also leads to water retention because salt exerts a water-holding osmotic effect in the ECF. Thus, both of these hormones play important roles in maintaining fluid balance in the body, which, in turn, is an important determinant of plasma volume and blood pressure. In addition, both vasopressin and angiotensin II are potent vaoconstrictors. Their role in this regard is especially important during hemorrhage. A sudden loss of blood reduces the plasma volume, which triggers increased secretion of both of these hormones to help restore plasma volume. Their vasoconstrictor effect also helps maintain blood pressure in the face of the abrupt loss of plasma volume. (The functions and control of these hormones are discussed more thoroughly in later chapters.)

This completes our discussion of the various factors that affect total peripheral resistance, the most important of which are controlled adjustments in arteriolar radius. These factors are summarized in ▬ Figure 10-14.

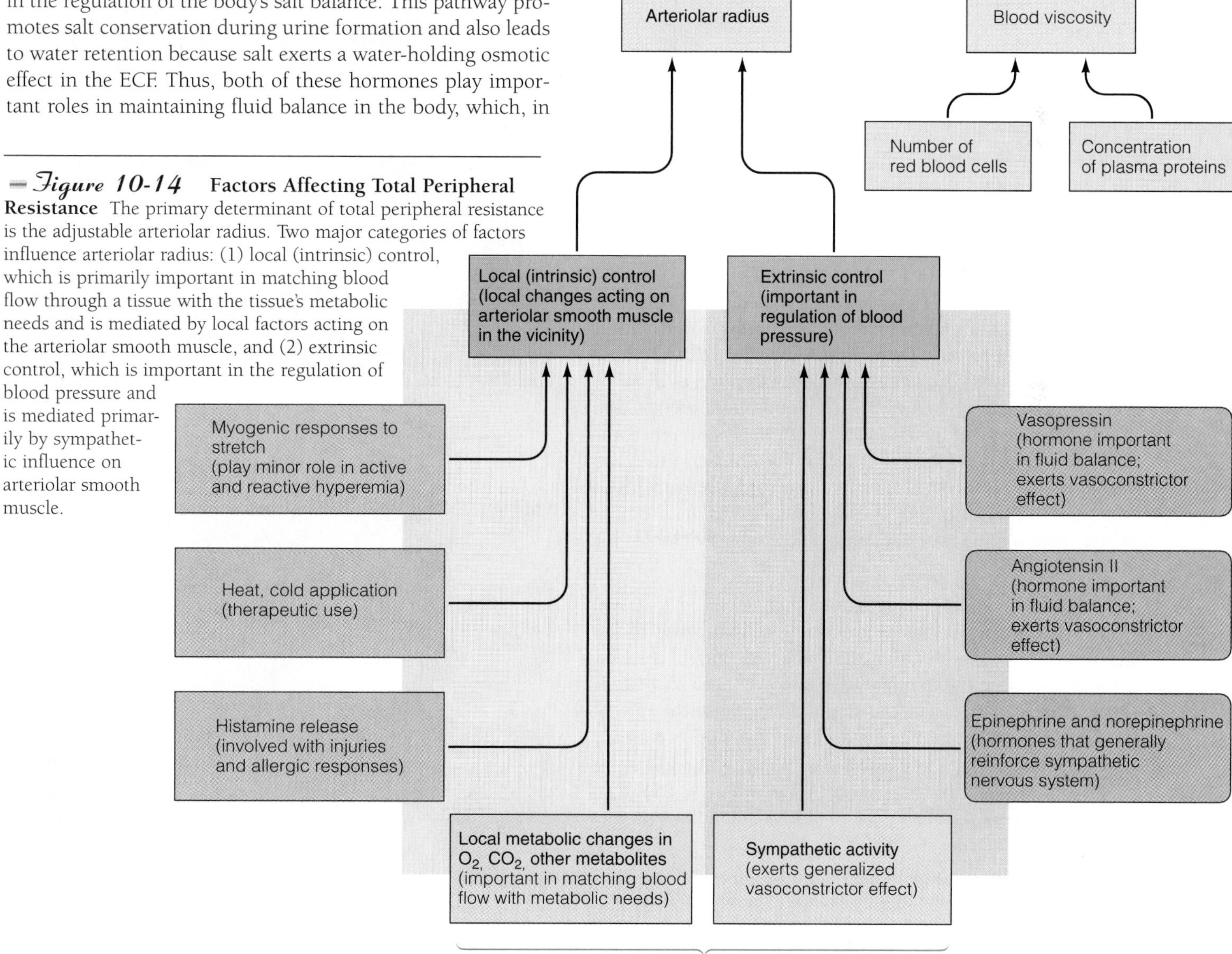

▬ ***Figure 10-14*** **Factors Affecting Total Peripheral Resistance** The primary determinant of total peripheral resistance is the adjustable arteriolar radius. Two major categories of factors influence arteriolar radius: (1) local (intrinsic) control, which is primarily important in matching blood flow through a tissue with the tissue's metabolic needs and is mediated by local factors acting on the arteriolar smooth muscle, and (2) extrinsic control, which is important in the regulation of blood pressure and is mediated primarily by sympathetic influence on arteriolar smooth muscle.

Table 10-3 Arteriolar Smooth Muscle Adrenergic Receptors

Characteristic	Receptor Type: α	Receptor Type: β_2
Location of the receptor	All arteriolar smooth muscle except in the brain	Arteriolar smooth muscle in the heart and skeletal muscles
Chemical mediator that binds with the receptor	Norepinephrine from sympathetic fibers and the adrenal medulla Epinephrine from the adrenal medulla (less affinity for this receptor)	Epinephrine from the adrenal medulla (greater affinity for this receptor)
Arteriolar smooth muscle response when the receptor is activated by binding with the chemical mediator	Vasoconstriction	Vasodilation

Capillaries

Capillaries are ideally suited to serve as sites of exchange.

Capillaries, the sites for exchange[2] of materials between the blood and tissues, branch extensively to bring blood within the reach of every cell. There are no carrier-mediated transport systems across capillaries, with the exception of those in the brain that play a role in the blood-brain barrier (see p. 119). Exchange of materials across capillary walls is accomplished primarily by the process of diffusion. Capillaries are ideally suited to enhance diffusion in accordance with Fick's law of diffusion (see p. 54). They minimize diffusion distances while maximizing surface area and time available for exchange as follows:

1. Diffusing molecules have only a short distance to travel between the blood and surrounding cells because of the thin capillary wall and small capillary diameter, coupled with the close proximity of each and every cell to a capillary. This short distance is important because the rate of diffusion slows down as the diffusion distance increases.
   a. Capillary walls are very thin (1 μm in thickness; in comparison, the diameter of a human hair is 100 μm). Capillaries are composed of only a single layer of flattened endothelial cells—essentially the lining of the other vessel types. No smooth muscle or connective tissue is present (Fig. 10-15a).
   b. Each capillary is so narrow (7 μm average diameter) that red blood cells (8 μm diameter) have to squeeze through single file (Fig. 10-15b). Consequently, plasma contents are either in direct contact with the inside of the capillary wall or are only a short diffusing distance from it.

[2]Actually, some exchange takes place across the other microcirculatory vessels, especially the postcapillary venules. The entire vasculature is a continuum and does not abruptly change from one vascular type to another. When the term *capillary exchange* is used, it tacitly refers to all exchange at the microcirculatory level, the majority of which occurs across the capillaries.

Figure 10-15 Capillary Anatomy (a) Electron micrograph of a cross section of a capillary. Note that the capillary wall consists of a single layer of endothelial cells. The nucleus of one of the endothelial cells is shown. (b) Photograph of a capillary bed. Note that the capillaries are so narrow that the red blood cells must pass through single file.

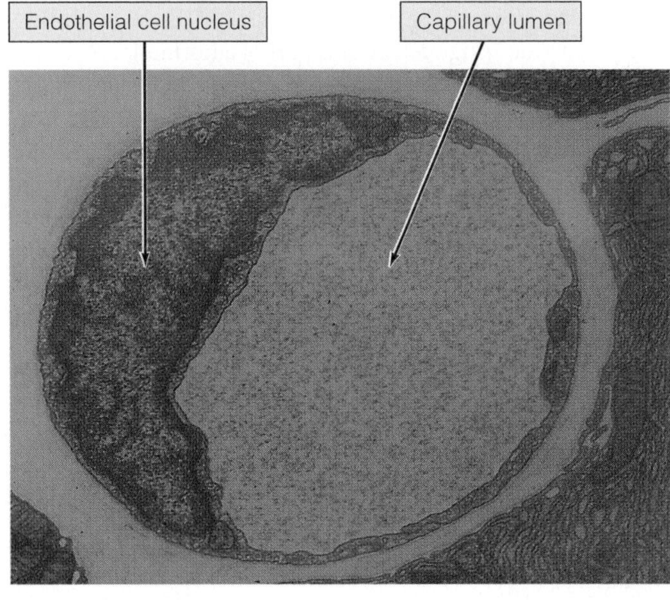

(a)

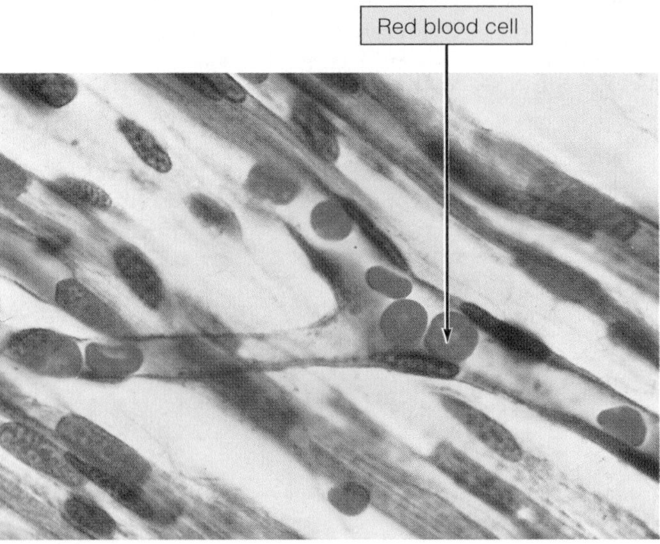

(b)

c. Because of extensive capillary branching, it is estimated that no cell is farther than 0.01 cm (4/1,000 inch) from a capillary.

2. Because capillaries are distributed in such incredible numbers (estimates range from 10 to 40 billion capillaries), a tremendous total surface area is available for exchange (an estimated 600 m^2). In spite of this large number of capillaries, at any point in time they contain only 5% of the total blood volume (250 ml out of a total of 5,000 ml). As a result, a small volume of blood is exposed to an extensive surface area. If all the capillary surfaces were stretched out in a flat sheet and the volume of blood contained within the capillaries were spread over the top, this would be roughly equivalent to spreading a half pint of paint over the floor of a high school gymnasium. Imagine how thin the paint layer would be!

3. Blood flows more slowly in the capillaries than elsewhere in the circulatory system. The extensive capillary branching is responsible for this slow velocity of blood flow through the capillaries. Let us clarify a potentially confusing point. The term "flow" can be used in two different contexts—the *flow rate,* which refers to the *volume* of blood flowing through a given segment of the circulatory system per unit of time (this is the flow we have been talking about in relation to the pressure gradient and resistance), and *velocity of flow,* which refers to the *speed* with which blood flows forward through a given segment of the circulatory system. Because the circulatory system is a closed system, the volume of blood flowing through any level of the system must equal the cardiac output. For example, if the heart pumps out 5 liters of blood per minute and 5 liters of blood per minute return to the heart, then 5 liters of blood per minute must flow through the arteries, arterioles, capillaries, and veins. Therefore, the flow rate is the same at all levels of the circulatory system.

However, the velocity with which blood flows through the different segments of the vascular tree varies because velocity of flow is inversely proportional to the total cross-sectional area of all the vessels at any given level of the circulatory system. Even though the cross-sectional area of each capillary is extremely small compared to that of the large aorta, the total cross-sectional area of all the capillaries added together is about 1,300 times greater than the cross-sectional area of the aorta because there are so many capillaries (Table 10-1). Accordingly, blood slows considerably as it passes through the capillaries (Fig. 10-16). This slow velocity allows adequate time for exchange of nutrients and metabolic end products between blood and tissues, which is the sole purpose of the entire circulatory system. As the capillaries rejoin to form veins, the total cross-sectional area is once again reduced, and the velocity of blood flow increases as blood returns to the heart.

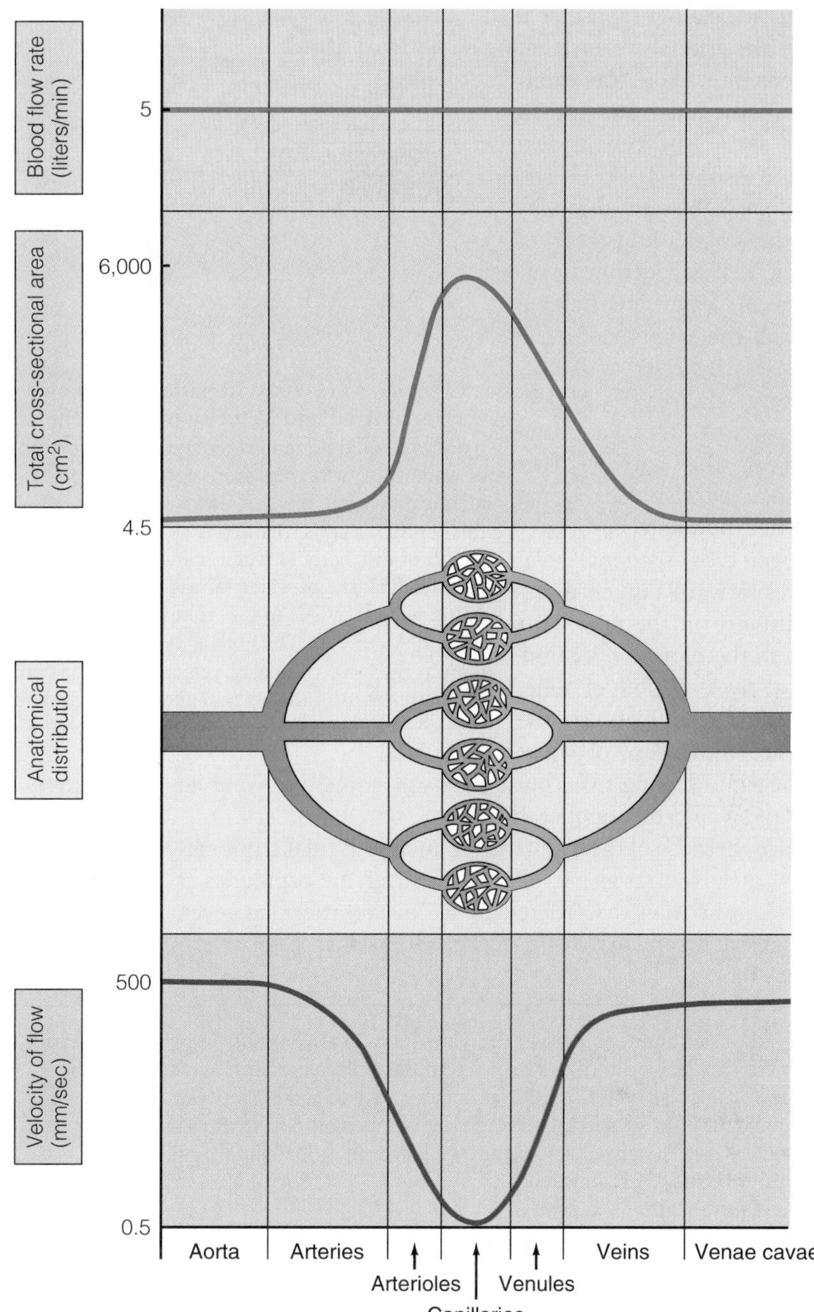

Figure 10-16 Comparison of Blood Flow Rate and Velocity of Flow in Relation to Total Cross-Sectional Area The blood flow rate (red curve) is identical through all levels of the circulatory system and is equal to the cardiac output (5 liters/min at rest). The velocity of flow (purple curve) varies throughout the vascular tree and is inversely proportional to the total cross-sectional area (green curve) of all the vessels at a given level. Note that the velocity of flow is slowest in the capillaries, which have the largest total cross-sectional area.

As an analogy, consider a river (the arterial system) that widens into a lake (the capillaries), then narrows into a river again (the venous system) (Fig. 10-17). The flow rate is the same throughout the length of this body of water, that is, identical volumes of water are flowing past all the points along the bank of the river and lake. However, the velocity of flow is slower in the wide lake than in the narrow river because the identical volume of water, now spread out over a larger cross-sectional area, moves forward a much shorter distance in the wide lake than in the narrow river during a given period of time. You could readily observe the forward movement of water in the swift-flowing river, but the forward motion of water in the lake would be unnoticeable.

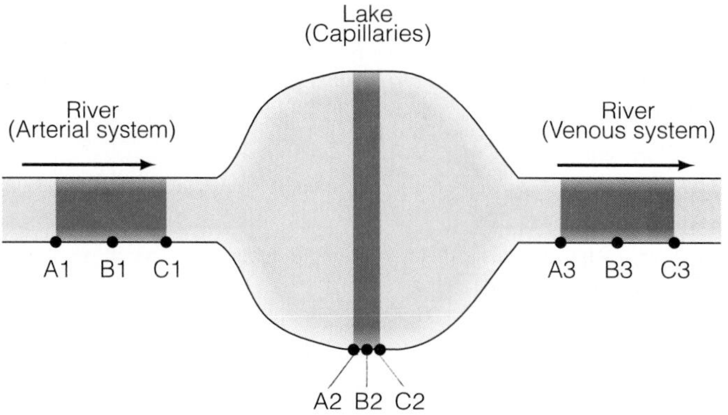

Figure 10-17 **Relationship between Total Cross-Sectional Area and Velocity of Flow** The three dark blue areas represent equal volumes of water. During one minute, this volume of water moves forward from points A to points C. Therefore, an identical volume of water flows past points B1, B2, and B3 during this minute; that is, the flow rate is the same at all points along the length of this body of water. However, during that minute, the identical volume of water moves forward a much shorter distance in the wide lake (A2 to C2) than in the much narrower river (A1 to C1 and A3 to C3). Thus, velocity of flow is much slower in the lake than in the river. Similarly, velocity of flow is much slower in the capillaries than in the arterial and venous systems.

Also, because of the capillaries' tremendous total cross-sectional area, the resistance offered by all of the capillaries is much lower than that offered by all of the arterioles, even though each capillary has a smaller radius than each arteriole. For this reason, the arterioles contribute more to total peripheral resistance. Furthermore, arteriolar caliber (and, accordingly resistance) is subject to control whereas capillary caliber cannot be adjusted.

Water-filled pores in the capillary wall permit passage of small, water-soluble substances that cannot cross the endothelial cells themselves.

Diffusion across capillary walls also depends on the walls' permeability to the materials being exchanged. The endothelial cells forming the capillary walls fit together in jigsaw-puzzle fashion, but the closeness of the fit varies considerably between organs. In most capillaries, narrow, water-filled clefts, or **pores,** are present at the junctions between the cells. (Fig. 10-18). These pores permit passage of water-soluble substances. Lipid-soluble substances, such as O_2 and CO_2, can readily pass through the endothelial cells themselves by dissolving in the lipid

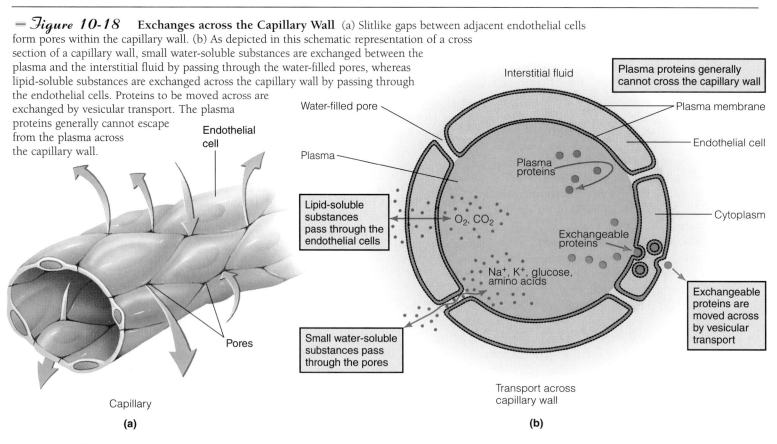

Figure 10-18 **Exchanges across the Capillary Wall** (a) Slitlike gaps between adjacent endothelial cells form pores within the capillary wall. (b) As depicted in this schematic representation of a cross section of a capillary wall, small water-soluble substances are exchanged between the plasma and the interstitial fluid by passing through the water-filled pores, whereas lipid-soluble substances are exchanged across the capillary wall by passing through the endothelial cells. Proteins to be moved across are exchanged by vesicular transport. The plasma proteins generally cannot escape from the plasma across the capillary wall.

Endothelial cell

Pores

Capillary

(a)

Interstitial fluid

Plasma proteins generally cannot cross the capillary wall

Water-filled pore

Plasma membrane

Endothelial cell

Plasma

Plasma proteins

Lipid-soluble substances pass through the endothelial cells

O_2, CO_2

Cytoplasm

Exchangeable proteins

Na^+, K^+, glucose, amino acids

Exchangeable proteins are moved across by vesicular transport

Small water-soluble substances pass through the pores

Transport across capillary wall

(b)

bilayer barrier. The size of the capillary pores varies from organ to organ. At one extreme, the endothelial cells in brain capillaries are joined by tight junctions so that pores are nonexistent. These junctions prevent transcapillary passage of materials between the cells and thus constitute part of the protective blood-brain barrier. In most tissues, small, water-soluble substances such as ions, glucose, and amino acids can readily pass through the water-filled clefts, but large, non-lipid-soluble materials such as plasma proteins are excluded from passage. At the other extreme, liver capillaries have such large pores that even proteins pass through readily. This is adaptive, because the liver's functions include synthesis of plasma proteins and the metabolism of protein-bound substances such as cholesterol. These proteins must all pass through the liver's capillary walls. The leakiness of various capillary beds is therefore a function of how tightly the endothelial cells are joined, which varies according to the different organs' needs.

The capillary wall has traditionally been considered a passive sieve, like a brick wall with permanent gaps in the mortar acting as pores. Recent studies, however, suggest that even under normal conditions, changes in endothelial cells are involved in the active regulation of capillary membrane permeability; that is, in response to appropriate signals, the "bricks" can readjust themselves to vary the size of the holes. Thus, the degree of leakiness does not necessarily remain constant for a given capillary bed. For example, researchers speculate that histamine increases capillary permeability by triggering contractile responses in endothelial cells to widen the intercellular gaps. This is not a muscular contraction, because no smooth muscle cells are present in capillaries. It is due to an actin-myosin contractile apparatus in the nonmuscular capillary cells. These large gaps make the affected capillary wall leakier so that normally retained plasma proteins escape into the surrounding tissue, where they exert an osmotic effect. Along with histamine-induced vasodilation, the resultant additional local fluid retention contributes to inflammatory swelling.

Vesicular transport also plays a limited role in the passage of materials across the capillary wall. Large non-lipid-soluble molecules such as proteinaceous hormones that must be exchanged between the blood and surrounding tissues are transported from one side of the capillary wall to the other in endocytotic-exocytotic vesicles (see p. 62).

Many capillaries are not open under resting conditions.

The branching and reconverging arrangement within capillary beds varies somewhat, depending on the tissue. Capillaries typically branch either directly from an arteriole or from a thoroughfare channel known as a **metarteriole,** which runs between an arteriole and a venule. Likewise, capillaries may rejoin at either a venule or a metarteriole (Fig. 10-19). Unlike the true capillaries within a capillary bed, metarterioles are sparsely surrounded by wisps of spiraling smooth muscle cells. These cells also form **precapillary sphincters,** each of which consists of a ring of smooth muscle around the entrance to a capillary as it arises from a metarteriole.[3] Precapillary sphincters are not innervated, but they have a high degree of myogenic tone and are sensitive to local metabolic changes. They act as stopcocks to control blood flow through the particular capillary that each one guards. Arterioles perform a similar function for a small group of capillaries. Capillaries themselves have no smooth muscle, so they cannot actively participate in the regulation of their own blood flow.

Generally, tissues that are more metabolically active have a greater density of capillaries. Muscles, for example, have relatively more capillaries than their tendinous attachments. Only about 10% of the precapillary sphincters in a resting muscle are open at any moment, however, so blood is flowing through only about 10% of the muscle's capillaries. As chemical concentrations start to change in a region of the muscle tissue supplied by closed-down capillaries, the precapillary sphincters and arterioles in the region relax. Restoration of the chemical concentrations to normal as a result of increased blood flow to that region removes the impetus for vasodilation, so the precapillary sphincters close once again and the arterioles return to normal tone. In this way, blood flow through any given capillary is often intermittent as a result of

Figure 10-19 **Capillary Bed** Capillaries branch either directly from an arteriole or from a metarteriole, a thoroughfare channel between an arteriole and venule. Capillaries rejoin at either a venule or a metarteriole. Metarterioles are surrounded by smooth muscle cells, which also form precapillary sphincters that encircle capillaries as they arise from a metarteriole.

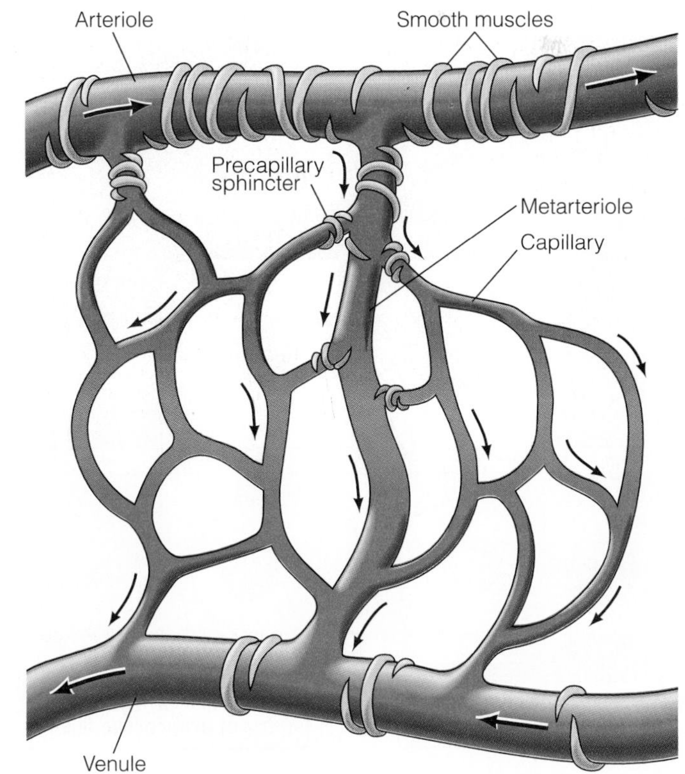

[3]Although generally accepted, the existence of precapillary sphincters in humans has not been conclusively established.

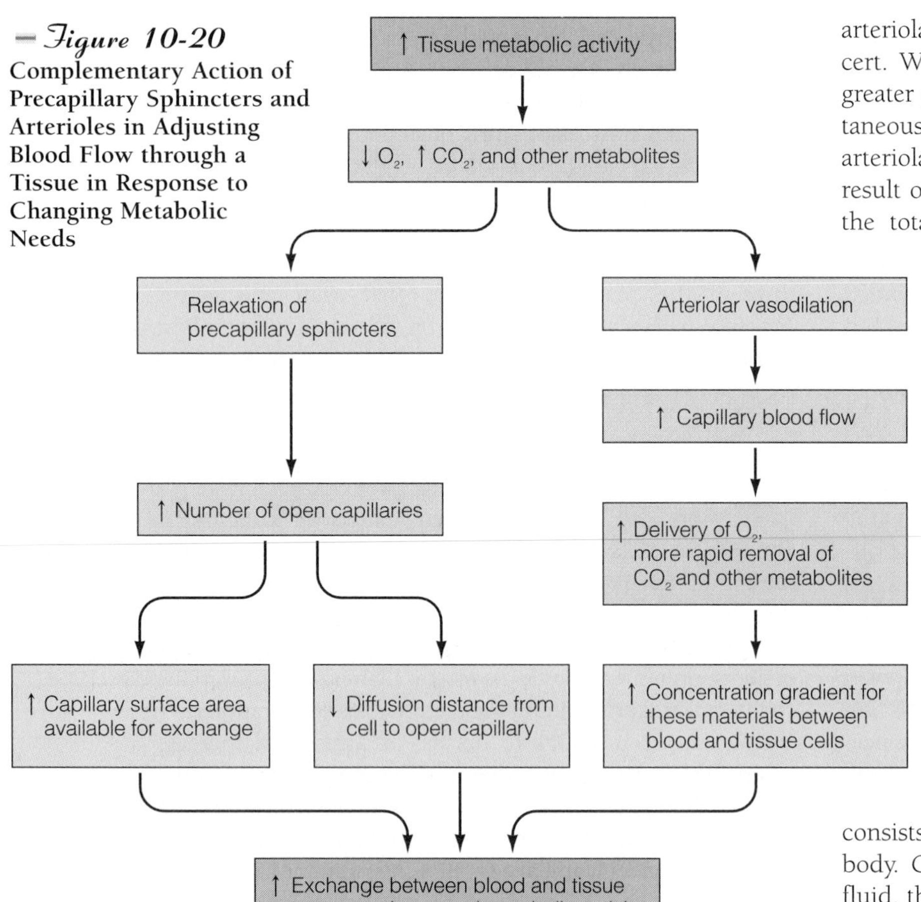

Figure 10-20 Complementary Action of Precapillary Sphincters and Arterioles in Adjusting Blood Flow through a Tissue in Response to Changing Metabolic Needs

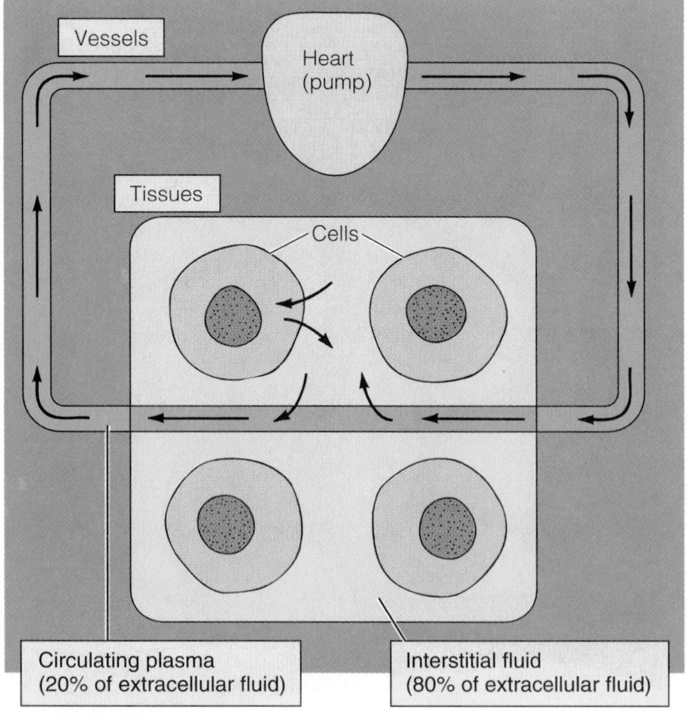

Figure 10-21 Interstitial Fluid Acting as an Intermediary between Blood and Cells

arteriolar and precapillary sphincter action working in concert. When the muscle as a whole becomes more active, a greater percentage of the precapillary sphincters relax, simultaneously opening up more capillary beds, while concurrent arteriolar vasodilation increases total flow to the organ. As a result of more blood flowing through more open capillaries, the total volume and surface area available for exchange increase, and the diffusion distance between the cells and an open capillary decreases (Fig. 10-20). Thus, blood flow through a particular tissue (assuming a constant blood pressure) is regulated by (1) the degree of resistance offered by the arterioles in the organ, controlled by sympathetic activity and local factors; and (2) the number of open capillaries, controlled by action of the same local metabolic factors on precapillary sphincters.

Diffusion across the capillary walls is important in solute exchange.

Exchanges between blood and the tissue cells are not made directly. Interstitial fluid, the true internal environment in immediate contact with the cells, acts as the go-between (Fig. 10-21). Only 20% of the ECF circulates as plasma. The remaining 80% consists of interstitial fluid, which bathes all the cells in the body. Cells exchange materials directly with the interstitial fluid, the type and extent of exchange being governed by the properties of the cellular plasma membranes. Movement across the plasma membrane may involve either passive diffusion or carrier-mediated transport. In contrast, only passive exchanges occur across the capillary wall between the plasma and interstitial fluid. Because of the permeability of the capillary walls, exchange is so thorough that the interstitial fluid takes on essentially the same composition as the incoming arterial blood, with the exception of the large plasma proteins that usually do not escape from the blood. Therefore, when we speak of exchanges between blood and tissue cells, we tacitly include interstitial fluid as a passive intermediary.

Exchanges between blood and surrounding tissues across the capillary walls are accomplished by (1) passive diffusion down concentration gradients, the primary mechanism for exchange of individual solutes; and (2) bulk flow, a process that accomplishes the totally different function of determining the distribution of the ECF volume between the vascular and interstitial fluid compartments. We will examine each of these mechanisms in more detail, starting with diffusion.

Because there are no carrier-mediated transport systems in capillary walls, solutes cross primarily by diffusion down concentration gradients. The chemical composition of arterial blood is carefully regulated to maintain the concentrations of individual solutes at levels that will promote each solute's movement in the appropriate direction across the capillary walls. The reconditioning organs primarily contribute to this homeostatic process, continuously adding nutrients and O_2 and removing CO_2 and other wastes as blood passes through them. Meanwhile, cells are constantly using up supplies and generating metabolic wastes. Diffusion of each solute continues independently as a result of the concentration difference for that solute between the blood and surrounding cells (Fig. 10-22). This process repeats itself continuously. As cells use up O_2 and glucose, the blood constantly brings in fresh supplies of these vital materials, maintaining concentration gradients that favor the net diffusion of these substances from blood to cells. Simultaneously, ongoing net diffusion of CO_2 and other metabolic wastes from cells to blood is maintained by the continual production of these wastes at the cellular level and their constant removal from the tissue level by the circulating blood.

Because the capillary wall does not limit the passage of any constituent except plasma proteins, the extent of exchanges for each solute is independently determined by the magnitude of its concentration gradient between the blood and surrounding tissues. As cells increase their level of activity, they use up more O_2 and produce more CO_2, among other things. This creates larger concentration gradients for O_2 and CO_2 between cells and blood, so more O_2 diffuses out of the blood into the cells and more CO_2 proceeds in the opposite direction to help support the increased metabolic activity.

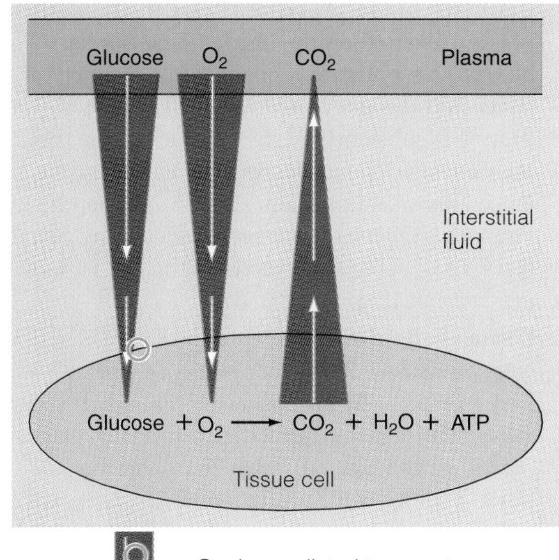

Figure 10-22 **Independent Exchange of Individual Solutes down Their Own Concentration Gradients across the Capillary Wall**

Bulk flow across the capillary wall is important in extracellular fluid distribution.

The second means by which exchange is accomplished across capillary walls is bulk flow. A volume of protein-free plasma actually filters out of the capillary, mixes with the surrounding interstitial fluid, and is subsequently reabsorbed. This process is called **bulk flow** because the various constituents of the fluid are moving together in bulk, or as a unit, in contrast to the discrete diffusion of individual solutes down concentration gradients. The capillary wall acts like a sieve, with the fluid moving through its water-filled pores. When pressure inside the capillary exceeds pressure on the outside, fluid is pushed out through the pores in a process known as **ultrafiltration.** The majority of the plasma proteins are retained on the inside during this process because of the pores' filtering effect, although a few do escape. Since all other constituents in the plasma are dragged along as a unit with the volume of fluid leaving the capillary, the filtrate is essentially a protein-free plasma. When inward-driving pressures exceed outward pressures across the capillary wall, net inward movement of fluid from the interstitial fluid compartment into the capillaries takes place through the pores, a process known as **reabsorption.**

Bulk flow occurs because of differences in the hydrostatic and colloid osmotic pressures between the plasma and interstitial fluid. Even though pressure differences exist between plasma and surrounding fluid elsewhere in the circulatory system, only the capillaries have pores that allow fluids to pass through. Four forces influence fluid movement across the capillary wall (Fig. 10-23):

Figure 10-23 **Bulk Flow across the Capillary Wall** Schematic representation of ultrafiltration and reabsorption as a result of imbalances in the forces acting across the capillary wall.

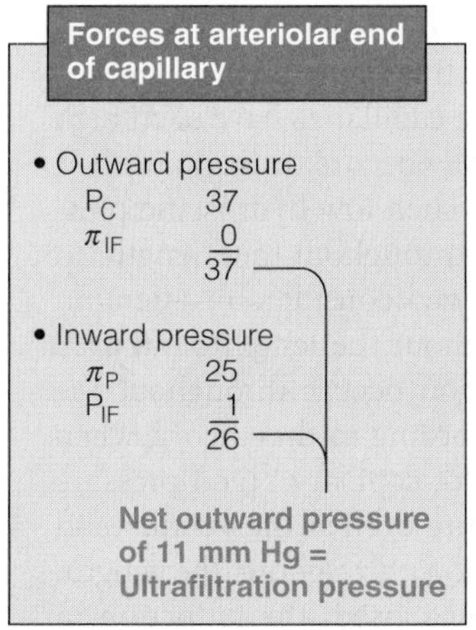

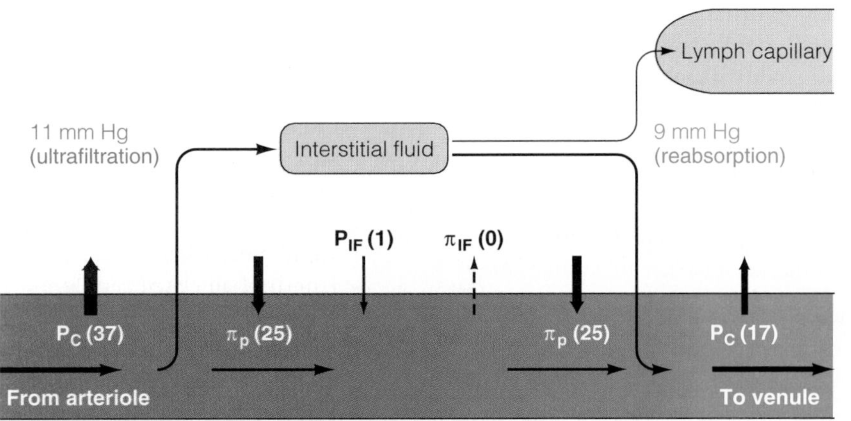

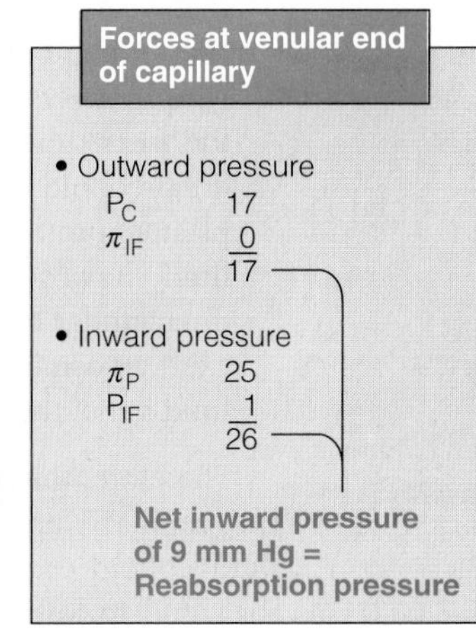

All values are given in mm Hg.

Blood capillary

1. **Capillary blood pressure (P_c)** is the fluid or hydrostatic pressure exerted on the inside of the capillary walls by the blood. This pressure tends to force fluid *out of* the capillaries into the interstitial fluid. Mean blood pressure has dropped substantially by the level of the capillaries because of frictional losses in pressure in the high-resistance arterioles upstream. On the average, the hydrostatic pressure is 37 mm Hg at the arteriolar end of a tissue capillary and has declined even further to 17 mm Hg at the venular end (Fig. 10-9, p. 314).
2. **Plasma-colloid osmotic pressure (π_P)**, also known as *oncotic pressure,* is a force caused by the colloidal dispersion (see p. A–8) of plasma proteins; it encourages fluid movement *into* the capillaries. Since the plasma proteins remain in the plasma rather than entering the interstitial fluid, a protein-concentration difference exists between the plasma and the interstitial fluid. Accordingly, there is also a water-concentration difference between these two regions. The plasma has a higher protein concentration and a lower water concentration than does the interstitial fluid. This difference exerts an osmotic effect (see p. 55) that tends to move water from the area of higher water concentration in the interstitial fluid to the area of lower water concentration (or higher protein concentration) in the plasma. Thus, the plasma proteins may be thought of as "attracting" water, although this is not actually the underlying force involved. The other plasma constituents do not exert an osmotic effect because they readily pass through the capillary wall, so their concentrations are equal in the plasma and interstitial fluid. The plasma-colloid osmotic pressure averages 25 mm Hg.
3. **Interstitial fluid hydrostatic pressure (P_{IF})** is the fluid pressure exerted on the outside of the capillary wall by the interstitial fluid. This pressure tends to force fluid *into* the capillaries. Because of the difficulties encountered in measuring interstitial fluid hydrostatic pressure, the actual value of the pressure is a controversial issue. It is either at, slightly above, or slightly below atmospheric pressure. For purposes of illustration, we will say it is 1 mm Hg.
4. **Interstitial fluid–colloid osmotic pressure (π_{IF})** is another force that does not normally contribute significantly to bulk flow. The small fraction of plasma proteins that leak across the capillary walls into the interstitial spaces are normally returned to the blood by means of the lymphatic system. Therefore, the protein concentration in the interstitial fluid is extremely low, and the interstitial fluid–colloid osmotic pressure is very close to zero. If plasma proteins pathologically leak into the interstitial fluid, however, as they do when histamine widens the intercellular clefts during tissue injury, the leaked proteins exert an osmotic effect that tends to promote movement of fluid *out of* the capillaries into the interstitial fluid.

Therefore, the two pressures that tend to force fluid out of the capillary are capillary blood pressure and interstitial fluid–colloid osmotic pressure. The two opposing pressures that tend to force fluid into the capillary are plasma-colloid osmotic pressure and interstitial fluid hydrostatic pressure.

We are now prepared to analyze the fluid movement that occurs across a capillary wall because of imbalances in these opposing physical forces (Fig. 10-23). Net exchange at a given point across the capillary wall can be calculated using the following equation:

$$\text{net exchange pressure} = \underset{\text{(outward pressure)}}{(P_C + \pi_{IF})} - \underset{\text{(inward pressure)}}{(\pi_P + P_{IF})}$$

A positive net exchange pressure (when the outward pressure exceeds the inward pressure) represents an ultrafiltration pressure. A negative net exchange pressure (when the inward pressure exceeds the outward pressure) represents a reabsorption pressure.

At the arteriolar end of the capillary, the outward pressure totals 37 mm Hg, whereas the inward pressure totals 26 mm Hg, for a net outward pressure of 11 mm Hg. Ultrafiltration takes place at the beginning of the capillary as this outward pressure gradient forces a protein-free filtrate through the capillary pores.

By the time the venular end of the capillary is reached, the capillary blood pressure has dropped, but the other pressures have remained essentially constant. At this point the outward pressure has fallen to a total of 17 mm Hg, whereas the total inward pressure is still 26 mm Hg, for a net inward pressure of 9 mm Hg. Reabsorption of fluid takes place as this inward pressure gradient forces fluid back into the capillary at its venular end. Ultrafiltration and reabsorption, collectively known as bulk flow, are thus due to a shift in the balance between the passive physical forces acting across the capillary wall. No active forces or local energy expenditures are involved in the bulk exchange of fluid between the plasma and surrounding interstitial fluid. With only minor contributions from the interstitial fluid, ultrafiltration occurs at the beginning of the capillary because capillary blood pressure exceeds plasma-colloid osmotic pressure, whereas by the end of the capillary, reabsorption takes place because blood pressure has fallen below osmotic pressure.

It is important to realize that we have taken "snapshots" at two points—at the beginning and at the end—in a hypothetical capillary. Actually, blood pressure gradually diminishes along the length of the capillary, so that progressively diminishing quantities of fluid are filtered out in the first half of the vessel and progressively increasing quantities of fluid are reabsorbed in the last half (■ Fig. 10-24). Even this situation is idealized. The pressures used in this figure are average values and controversial at that. Some capillaries have such high blood pressure that filtration actually occurs throughout their entire length, whereas others have such low hydrostatic pressure that reabsorption takes place throughout their length. In fact, a recent theory that has received considerable attention is that net filtration occurs throughout the length of all *open* capillaries, whereas net reabsorption occurs throughout the length of all *closed* capillaries. According to this theory, when the precapillary sphincter is relaxed, capillary blood pressure exceeds the plasma osmotic pressure even at the venular end of the capillary, promoting filtration throughout the length. When the precapillary sphincter is closed, the reduction in blood flow through the capillary diminishes capillary blood

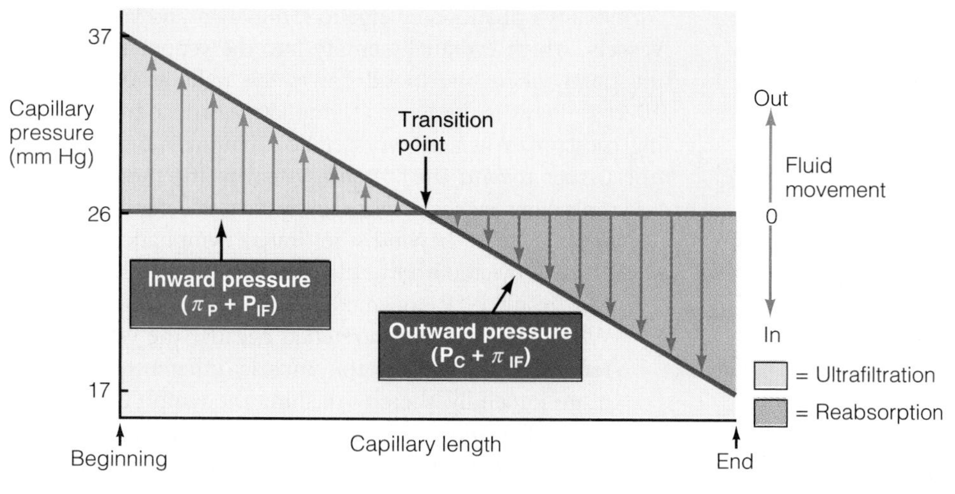

Figure 10-24 Net Filtration and Net Reabsorption along the Vessel Length The inward pressure ($\pi_p + P_{IF}$) remains constant throughout the length of the capillary whereas the outward pressure ($P_c + \pi_{IF}$) progressively declines throughout the capillary's length. In the first half of the vessel, where the declining outward pressure still exceeds the constant inward pressure, progressively diminishing quantities of fluid are filtered out (depicted by the upward red arrows). In the last half of the vessel, progressively increasing quantities of fluid are reabsorbed (depicted by the downward blue arrows) as the declining outward pressure falls farther below the constant inward pressure.

pressure below the plasma osmotic pressure even at the beginning of the capillary, so reabsorption takes place all along the capillary. Whichever mechanism is involved, the net effect is the same. A protein-free filtrate exits the capillaries and is ultimately reabsorbed.

Bulk flow does not play an important role in the exchange of individual solutes between blood and tissues, because the quantity of solutes moved across the capillary wall by bulk flow is extremely small compared to the much larger transfer of solutes by diffusion. If ultrafiltration and reabsorption are not important in the exchange of nutrients and wastes, then of what significance is bulk fluid exchange across the capillary wall? The answer is that it plays an extremely important role in regulating the distribution of ECF between the plasma and interstitial fluid. Maintenance of proper arterial blood pressure depends in part on an appropriate volume of circulating blood. If plasma volume is reduced (for example, by hemorrhage), blood pressure falls. The resultant lowering of capillary blood pressure alters the balance of forces across the capillary walls. Because the net outward pressure is decreased while the net inward pressure remains unchanged, extra fluid is shifted from the interstitial compartment into the plasma as a result of reduced filtration and increased reabsorption. The extra fluid soaked up from the interstitial fluid provides additional fluid for the plasma, temporarily compensating for the loss of blood. Meanwhile, reflex mechanisms acting on the heart and blood vessels (to be described later) also come into play to help maintain blood pressure until long-term mechanisms, such as thirst and reduction of urinary output, can restore the fluid volume to completely compensate for the loss.

Conversely, if the plasma volume becomes overexpanded, as with excessive fluid intake, the resultant elevation in capillary blood pressure forces extra fluid from the capillaries into the interstitial fluid, temporarily relieving the expanded plasma volume until the excess fluid can be eliminated from the body by long-term measures, such as increased urinary output.

These internal fluid shifts between the two ECF compartments occur automatically and immediately whenever the balance of forces acting across the capillary walls is changed; they provide a temporary mechanism to help keep the plasma volume fairly constant. In the process of restoring the plasma volume to an appropriate level, the interstitial fluid volume fluctuates, but it is much more important that the plasma volume be maintained at a constant level to ensure that the circulatory system functions effectively.

The lymphatic system is an accessory route by which interstitial fluid can be returned to the blood.

Even under normal circumstances, slightly more fluid is filtered out of the capillaries into the interstitial fluid than is reabsorbed from the interstitial fluid back into the plasma. The average net ultrafiltration pressure is 11 mm Hg, whereas the net reabsorption pressure is only 9 mm Hg (Fig. 10-23). The extra fluid filtered out as a result of this filtration-reabsorption imbalance is picked up by the **lymphatic system.** This system consists of an extensive network of one-way vessels that provide an accessory route by which fluid can be returned from the interstitial fluid to the blood. Small, blind-ended terminal lymph vessels **(initial lymphatics)** permeate almost every tissue of the body (Fig. 10-25a). The endothelial cells forming the walls of initial lymphatics slightly overlap, with their overlapping edges being free instead of attached to the surrounding cells. This arrangement creates valvelike openings in the vessel wall (Fig. 10-25b). Fluid pressure on the outside of the vessel pushes the edges inward, opening the valves and permitting interstitial fluid to enter. Once interstitial fluid enters a lymphatic vessel, it is called **lymph.** Fluid pressure on the inside forces the overlapping edges together, closing the valves so that lymph does not escape. These lymphatic valvelike openings are much larger than the pores in blood capillaries. Consequently, large particulates in the interstitial fluid, such as escaped plasma proteins and bacteria, can gain access to initial lymphatics but are excluded from blood capillaries.

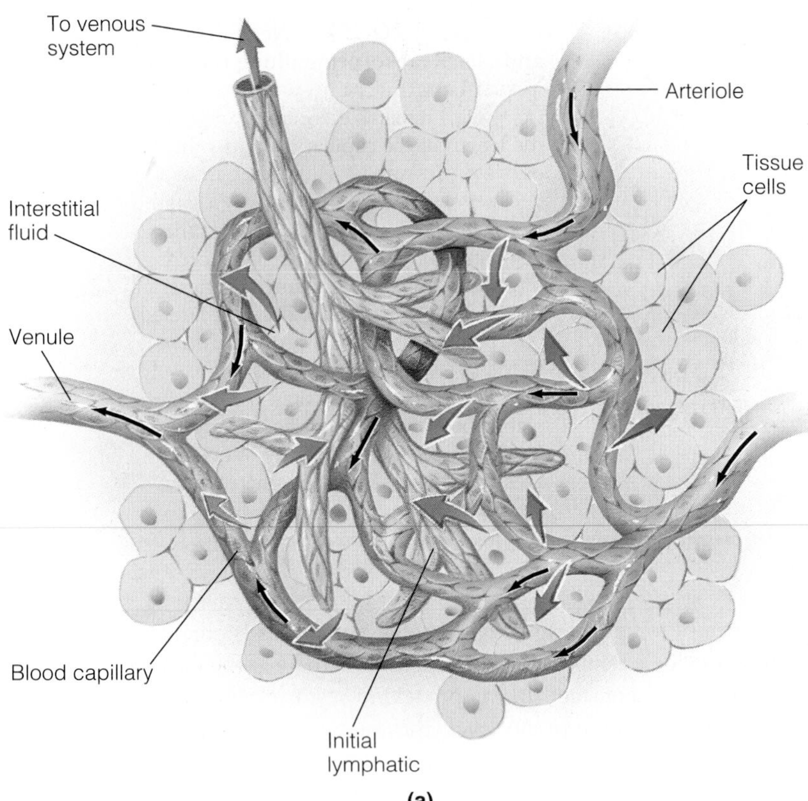
To venous system
Arteriole
Tissue cells
Interstitial fluid
Venule
Blood capillary
Initial lymphatic
(a)

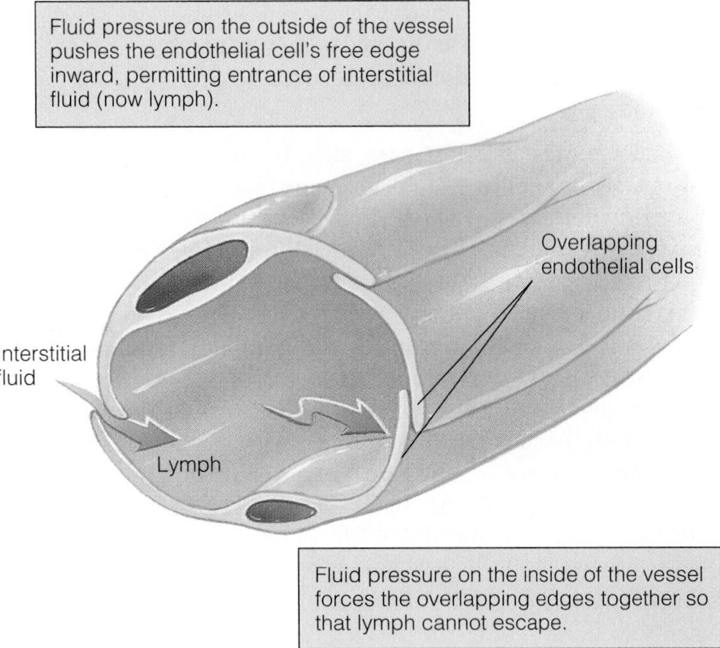
Fluid pressure on the outside of the vessel pushes the endothelial cell's free edge inward, permitting entrance of interstitial fluid (now lymph).
Overlapping endothelial cells
Interstitial fluid
Lymph
Fluid pressure on the inside of the vessel forces the overlapping edges together so that lymph cannot escape.
(b)

Figure 10-25 **Initial Lymphatics** (a) Relationship between initial lymphatics and blood capillaries. Blind-ended initial lymphatics pick up excess fluid filtered by blood capillaries and return it to the venous system in the chest. (b) Arrangement of endotheial cells in an initial lymphatic. Note that the overlapping edges of the endothelial cells create valvelike openings in the vessel wall.

Initial lymphatics converge to form larger and larger **lymph vessels,** which eventually empty into the venous system near the point where the blood enters the right atrium (Fig. 10-26a). Because there is no "lymphatic heart" to provide driving pressure, you may wonder how lymph is directed from the tissues toward the venous system in the thoracic cavity. Lymph flow is accomplished by two mechanisms. First, lymph vessels beyond the initial lymphatics are surrounded by smooth muscle, which contracts rhythmically as a result of myogenic initiation of action potentials. When this muscle is stretched because the vessel is distended with lymph, the muscle inherently contracts more forcefully, thereby pushing the lymph through the vessel. This intrinsic "lymph pump" is the major force for propelling lymph. Stimulation of lymphatic smooth muscle by the sympathetic nervous system further increases the pumping activity of the lymph vessels. Second, since lymph vessels lie between skeletal muscles, contraction of these muscles squeezes the lymph out of the vessels. One-way valves spaced at intervals within the lymph vessels direct the flow of lymph toward its venous outlet in the chest.

Following are the most important functions of the lymphatic system:

- *Return of excess filtered fluid.* Normally, capillary filtration exceeds reabsorption by about 3 liters per day (20 liters filtered, 17 liters reabsorbed) (Fig. 10-26b). Yet the entire blood volume is only 5 liters, and only 2.75 liters of that is plasma. (Blood cells make up the remainder of the blood volume.) With an average cardiac output, 7,200 liters of blood pass through the capillaries daily under resting conditions (more when cardiac output increases). Even though only a small fraction of the filtered fluid is not reabsorbed by the blood capillaries, the cumulative effect of this process being repeated with every heartbeat results in the equivalent of more than the entire plasma volume being left behind in the interstitial fluid each day. Obviously, this fluid must be returned to the circulating plasma, and this task is accomplished by the lymph vessels. The average rate of flow through the lymph vessels is 3 liters per day, compared with 7,200 liters per day through the circulatory system.
- *Defense against disease.* The lymph percolates through **lymph nodes** located en route within the lymphatic system. Passage of this fluid through the lymph nodes is an important aspect of the body's defense mechanism against disease. For example, bacteria picked up from the interstitial fluid are destroyed by special phagocytic cells located within the lymph nodes.
- *Transport of absorbed fat.* The lymphatic system is important in the absorption of fat from the digestive tract. The end products of the digestion of dietary fats are packaged by cells lining the digestive tract into fatty particles that are too large to gain access to the blood capillaries but can easily enter the initial lymphatics.
- *Return of filtered protein.* Most capillaries permit leakage of some plasma proteins during filtration. These proteins cannot readily be reabsorbed back into the blood capillaries

but can easily gain access to the initial lymphatics. If the proteins were allowed to accumulate in the interstitial fluid rather than being returned to the circulation via the lymphatics, the interstitial fluid–colloid osmotic pressure (an outward pressure) would progressively increase while the plasma-colloid osmotic pressure (an inward pressure) would progressively fall. As a result, filtration forces would gradually increase and reabsorption forces would gradually decrease, resulting in progressive accumulation of fluid in the interstitial spaces at the expense of loss of plasma volume.

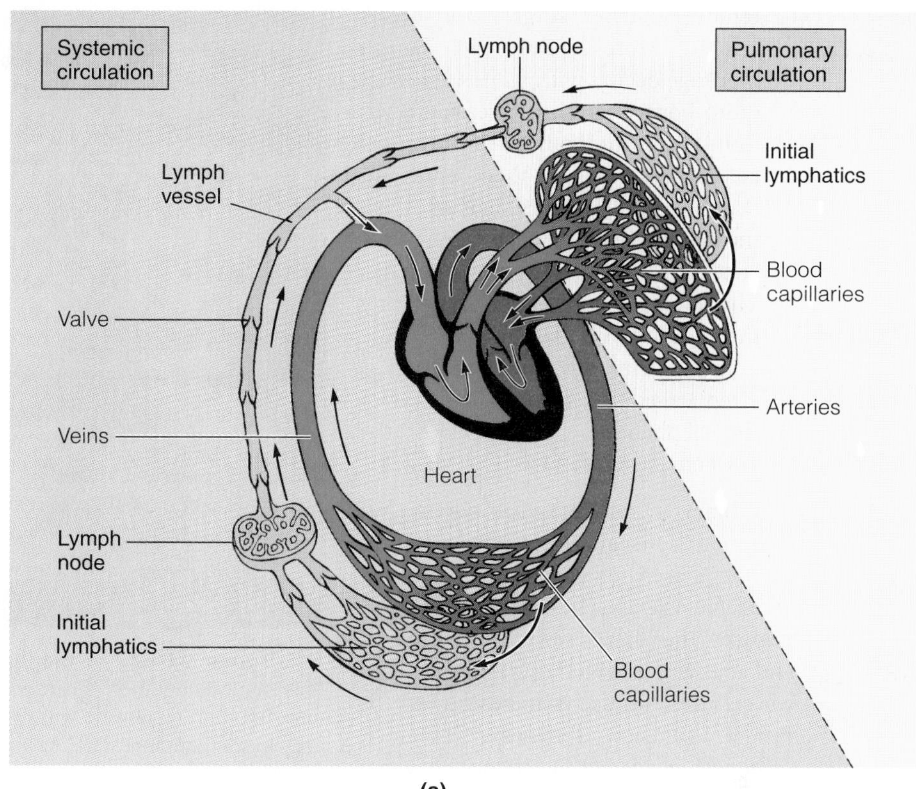

(a)

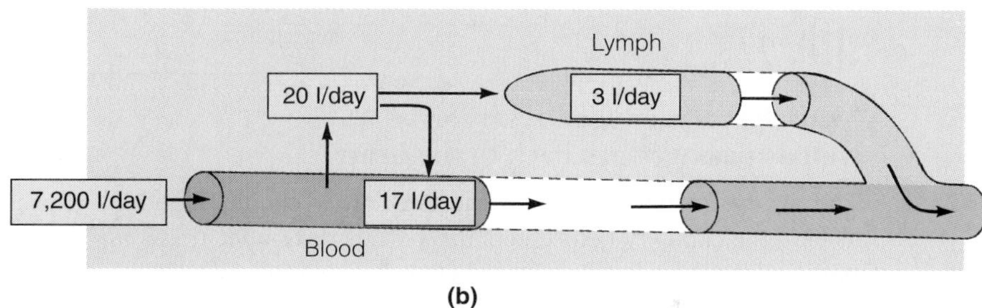

(b)

Figure 10-26 **Lymphatic System** (a) Lymph empties into the venous system near its entrance to the right atrium. (b) Lymph flow averages 3 liters per day whereas blood flow averages 7,200 liters per day.

Edema occurs when too much interstitial fluid accumulates.

Occasionally, excessive interstitial fluid does accumulate when one of the physical forces acting across the capillary walls becomes abnormal for some reason. Swelling of the tissues because of excess interstitial fluid is known as **edema.** The causes of edema can be grouped into four general categories:

1. *A reduced concentration of plasma proteins* causes a decrease in plasma-colloid osmotic pressure. Such a drop in the major inward pressure allows excess fluid to be filtered out while less than normal amounts of fluid are reabsorbed; hence extra fluid remains in the interstitial spaces. Edema caused by a decreased concentration of plasma proteins can arise in several different ways: excessive loss of plasma proteins in the urine caused by kidney disease; reduced synthesis of plasma proteins as a result of liver disease (the liver synthesizes almost all plasma proteins); a diet deficient in protein; or significant loss of plasma proteins from large burned surfaces.
2. *Increased permeability of the capillary walls* allows more plasma proteins than usual to pass from the plasma into the surrounding interstitial fluid—for example, via histamine-induced widening of the capillary pores during tissue injury or allergic reactions. The resultant fall in plasma-colloid osmotic pressure decreases the effective inward pressure while the resultant rise in interstitial fluid–colloid osmotic pressure caused by the excess protein in the interstitial fluid increases the effective outward force. This imbalance contributes in part to the localized edema associated with injuries (for example, blisters) and allergic responses (for example, hives).
3. *Increased venous pressure,* as when blood dams up in the veins, is accompanied by an increased capillary blood pressure, since the capillaries drain into the veins. This elevation in outward pressure across the capillary walls is largely responsible for the edema seen with congestive heart failure (see p. 295). Regional edema can also occur because of localized restriction of venous return. An example is the swelling often occurring in the legs and feet during pregnancy. The enlarged uterus compresses the major veins that drain the lower extremities as these vessels enter the abdominal cavity. The resultant damming of blood in these veins causes a rise in blood pressure in the capillaries of the legs and feet, which promotes regional edema of the lower extremities.
4. *Blockage of lymph vessels* produces edema because the excess filtered fluid is retained in the interstitial fluid rather than being returned to the blood through the lymphatics. The protein accumulation in the interstitial fluid compounds the problem through its osmotic effect. Local lymph blockage can occur, for example, in the arms of women whose major lymphatic drainage channels from the arm have been blocked as a result of lymph node

removal during surgery for breast cancer. More widespread lymph blockage occurs with *filariasis,* a mosquito-borne parasitic disease that is found predominantly in tropical coastal regions. In this condition, small, threadlike filaria worms infect the lymph vessels, where their presence prevents proper lymph drainage. The affected body parts, particularly the scrotum and extremities, become grossly edematous. The condition is often called *elephantiasis* because of the elephantlike appearance of the swollen extremities (Fig. 10-27).

Whatever the cause of edema, an important consequence is a reduction in exchange of materials between the blood and cells. As excess interstitial fluid accumulates, the distance between the blood and cells across which nutrients, O_2, and wastes must diffuse is increased, so the rate of diffusion decreases. Therefore, cells within edematous tissues may not be adequately supplied.

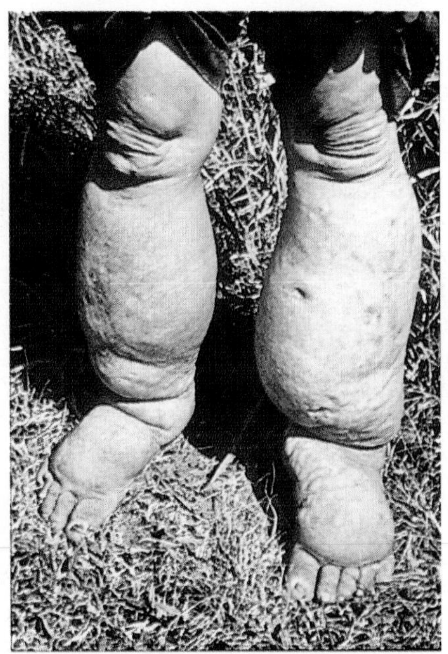

Figure 10-27 Elephantiasis
This tropical condition is caused by a mosquito-borne parasitic worm that invades the lymph vessels. As a result of the interference with lymph drainage, the affected body parts, usually the extremities, become grossly edematous, appearing elephantlike.

Veins

Veins serve as a blood reservoir as well as passageways back to the heart.

The venous system completes the circulatory circuit. Blood leaving the capillary beds enters the venous system for transport back to the heart. Veins have large radii, so they offer little resistance to flow. Furthermore, because the total cross-sectional area of the venous system gradually decreases as smaller veins converge into progressively fewer but larger vessels, the velocity of blood flow increases as the blood approaches the heart.

In addition to serving as low-resistance passageways to return blood from the tissues to the heart, systemic veins also serve as a *blood reservoir.* Because of their storage capacity, veins are often referred to as **capacitance vessels.** Veins have much thinner walls with less smooth muscle than do arteries. Since collagen fibers are considerably more abundant than elastin fibers in venous connective tissue, veins have very little elasticity, in contrast to arteries. Also, unlike arteriolar smooth muscle, venous smooth muscle has little inherent myogenic tone. Because of these features, veins are highly distensible, or stretchable, and have little elastic recoil. They easily distend to accommodate additional volumes of blood with only a small increase in venous pressure. Arteries stretched by an excess volume of blood recoil because of the elastic fibers in their walls, driving the blood forward. Veins containing an extra volume of blood simply stretch to accommodate the additional blood without tending to recoil. In this way veins serve as a blood reservoir; that is, when demands for blood are low, the veins can store extra blood in reserve because of their passive distensibility. Under resting conditions, the veins contain more than 60% of the total blood volume (Fig. 10-28). When the stored blood is needed, such as during exercise, extrinsic factors (soon to be described) drive the extra blood from the veins to the heart so that it can be pumped to the tissues. Increased venous return induces an increased cardiac stroke volume in accordance with the Frank Starling law of the heart (see p. 332). If too much blood pools in the veins instead of being returned to the heart, cardiac output is abnormally diminished. Thus, a delicate balance exists between the capacity of the veins, the extent of venous return, and the cardiac output. We will now turn our attention to the factors that affect venous capacity and contribute to venous return.

Venous return is enhanced by a number of extrinsic factors.

Venous capacity (the volume of blood that the veins can accommodate) depends on the distensibility of the vein walls (how much they can stretch to hold blood) and the influence of any externally applied pressure squeezing inwardly on the veins. At a constant blood volume, as venous capacity increases, more blood remains in the veins instead of being returned to the heart. Such venous storage decreases the **effective circulating volume.** Conversely, when venous capacity decreases, more blood is returned to the heart and continues circulating. Thus, changes in venous capacity directly influence the magnitude of venous return, which in turn is an important (although not the only) determinant of

Figure 10-28 Percentage of Total Blood Volume in Different Parts of the Circulatory System

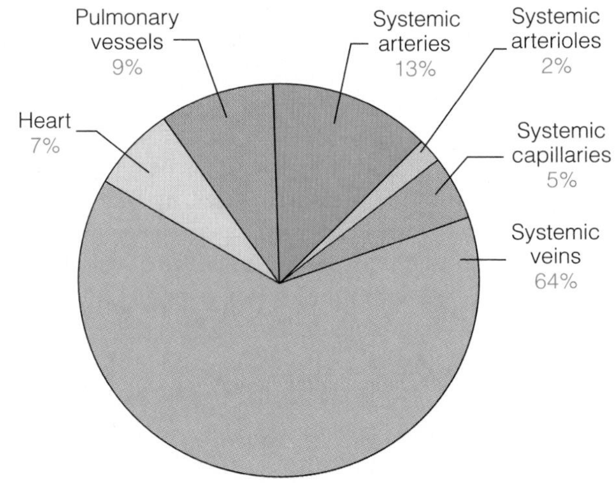

effective circulating blood volume. The magnitude of the total blood volume is also influenced on a short-term basis by passive shifts in bulk flow between the vascular and interstitial fluid compartments and on a long-term basis by factors that control total ECF volume, such as salt and water balance.

Venous return refers to the volume of blood entering each atrium per minute from the veins. Recall that the magnitude of flow through a vessel is directly proportional to the pressure gradient. Much of the driving pressure imparted to the blood by cardiac contraction has been lost by the time the blood reaches the venous system because of frictional losses along the way, especially during passage through the high-resistance arterioles. By the time the blood enters the venous system, mean pressure averages only 17 mm Hg (Fig. 10-9, p. 314). However, since atrial pressure is near 0 mm Hg, a small but adequate driving pressure still exists to promote the flow of blood through the large-radius, low-resistance veins. If atrial pressure becomes pathologically elevated, as in the presence of a leaky AV valve, the venous-to-atrial pressure gradient is decreased, reducing venous return and causing blood to dam up in the venous system (one cause of congestive heart failure; see p. 295).

In addition to the driving pressure imparted by cardiac contraction, five other factors enhance venous return: sympathetically induced venous vasoconstriction, skeletal muscle activity, the effect of venous valves, respiratory activity, and the effect of cardiac suction (Fig. 10-29). Most of these secondary factors affect venous return by influencing the pressure gradient between the veins and the heart. We will examine each in turn.

Effect of sympathetic activity on venous return Veins are not very muscular and have little inherent tone, but venous smooth muscle is abundantly supplied with sympathetic nerve fibers. Sympathetic stimulation produces venous vasoconstriction, which modestly elevates venous pressure; this, in turn, increases the pressure gradient to drive more blood from the veins into the right atrium. The veins normally have such a large diameter that the moderate vasoconstriction accompanying sympathetic stimulation has little effect on resistance to flow. Even when constricted, the veins still have a relatively large diameter and are still low-resistance vessels.

In addition to mobilizing the stored blood, venous vasoconstriction enhances venous return by decreasing venous capacity. With the filling capacity of the veins reduced, less blood draining from the capillaries remains in the veins but continues to flow instead toward the heart.

It is important to recognize the different outcomes of vasoconstriction in arterioles and veins. Arteriolar vasoconstriction *reduces* flow through these vessels because of their increased resistance (less blood can enter and flow through a narrowed arteriole), whereas venous vasoconstriction *increases* flow through these vessels because of their decreased capacity (narrowing of veins squeezes out more of the blood that is already present in the veins, thus increasing blood flow through these vessels).

Figure 10-29 **Factors That Facilitate Venous Return**

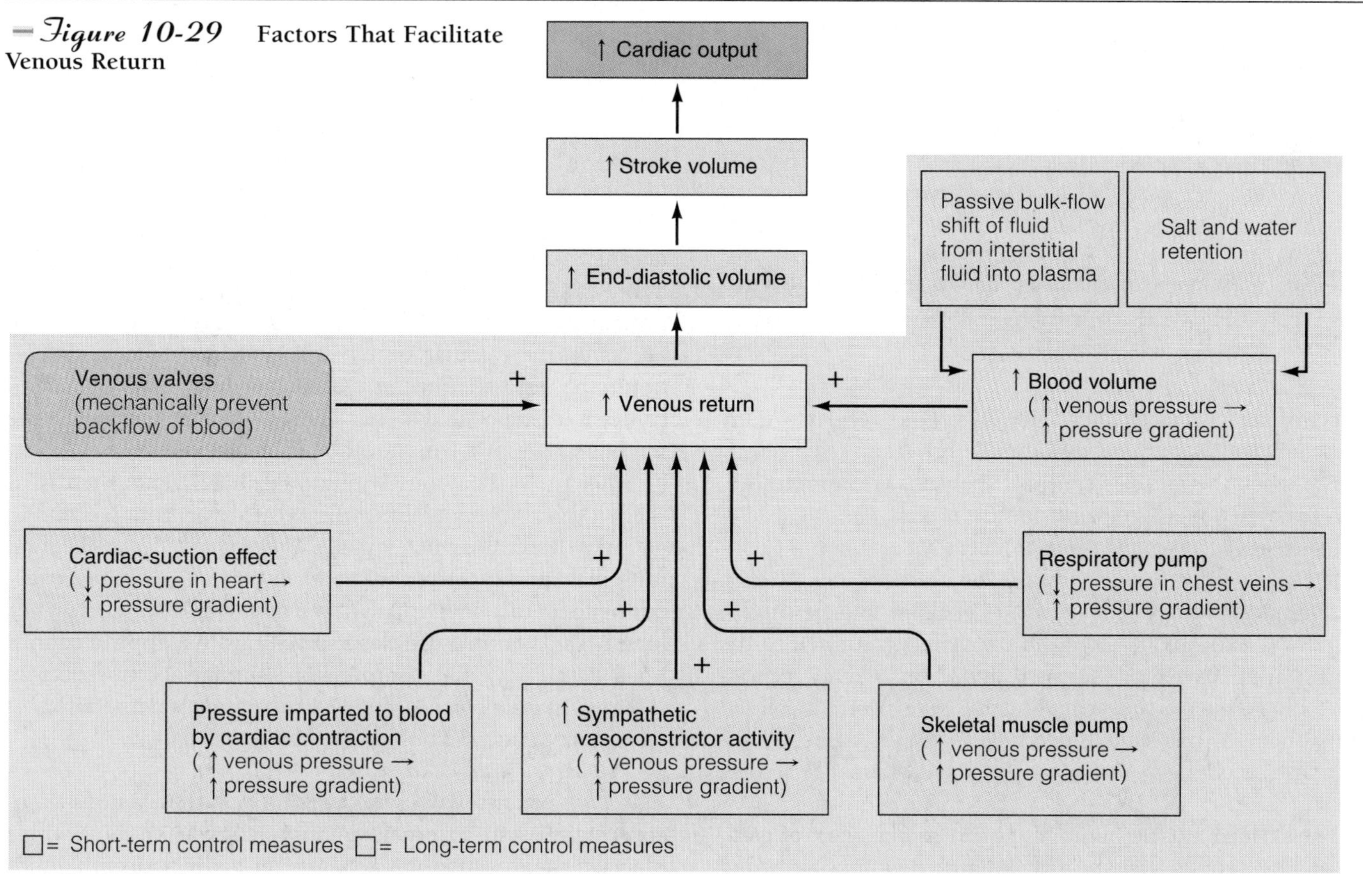

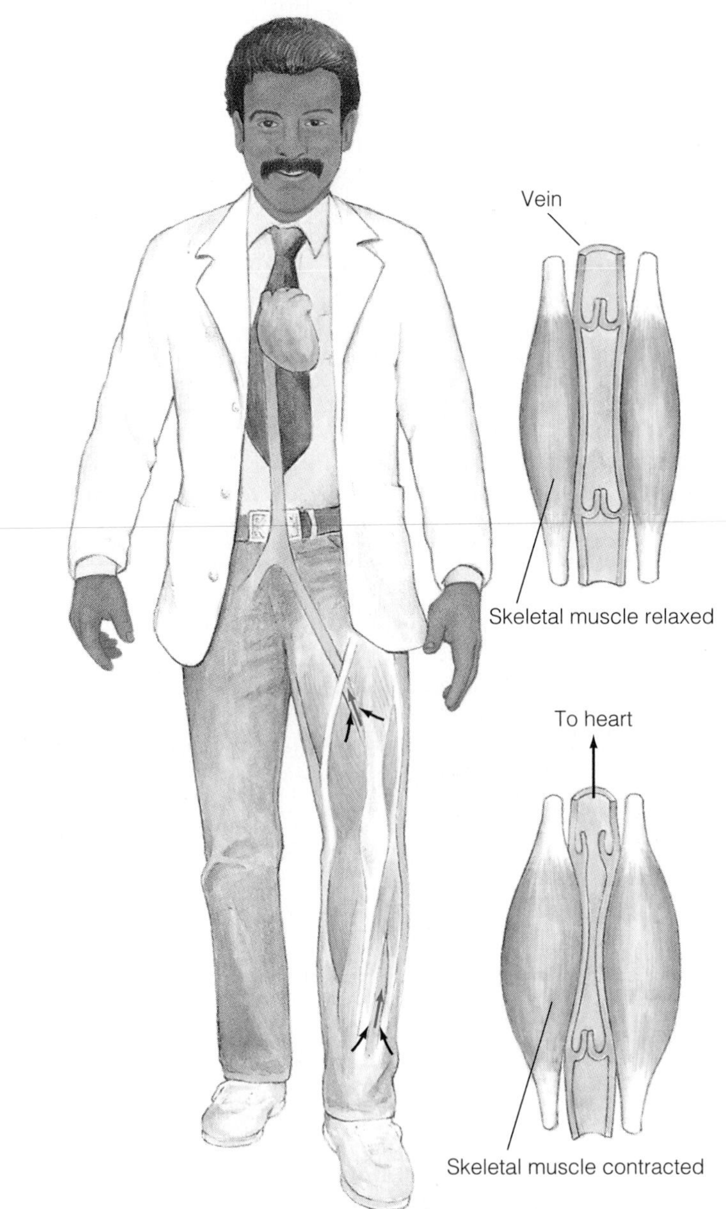

Figure 10-30 **Skeletal Muscle Pump Enhancing Venous Return**

Effect of skeletal muscle activity on venous return Many of the large veins in the extremities lie between skeletal muscles so when the muscles contract, the veins are compressed. This external venous compression decreases venous capacity and increases venous pressure, in effect squeezing fluid contained in the veins forward toward the heart (Fig. 10-30). This pumping action, known as the **skeletal muscle pump,** is one way extra blood stored in the veins is returned to the heart during exercise. Increased muscular activity pushes more blood out of the veins and into the heart. Increased sympathetic activity and the resultant venous vasoconstriction also accompany exercise, further enhancing venous return.

The skeletal muscle pump also counters the effect of gravity on the venous system. The average pressures provided thus far for various regions of the vascular tree are for a person in the horizontal position. When a person is lying down, the force of gravity is uniformly applied, so it does not have to be taken into consideration. When a person stands up, however, gravitational effects are not uniform. In addition to the usual pressure that results from cardiac contraction, vessels below the level of the heart are subjected to pressure caused by the weight of the column of blood extending from the heart to the level of the vessel (Fig. 10-31). There are two important consequences of this increased pressure. First, the distensible veins yield under the increased hydrostatic pressure, further expanding so that their capacity is increased. Even though the arteries are subjected to the same gravitational effects, they do not expand like the veins because arteries are not nearly as distensible. Much of the blood entering from the capillaries tends to pool in the expanded lower-leg veins instead of returning to the heart. Because venous return is reduced, cardiac output is decreased and the effective circulating volume is reduced. Second, the marked increase in capillary blood pressure resulting from the effect of gravity causes excessive fluid to filter out of capillary beds in the lower extremities, producing localized edema (that is, swollen feet and ankles).

Two compensatory measures normally counteract these gravitational effects. First, the resultant fall in mean arterial pressure that occurs when a person moves from a lying-down to an upright position triggers sympathetically induced venous vasoconstriction, which drives some of the pooled blood forward. Second, the skeletal muscle pump "interrupts" the column of blood by completely emptying given vein segments intermittently so that a particular portion of a vein is not subjected to the weight of the entire venous column from the heart to its level (Figs. 10-30 and 10-32). Reflex venous vasoconstriction cannot completely compensate for gravitational effects without the assistance of skeletal muscle activity. Therefore, when a person stands still for a long time, blood flow to the brain is reduced because of the decline in effective circulating volume, despite reflexes aimed at maintaining mean arterial pressure. Reduced flow of blood to the brain, in turn, leads to fainting, which returns the person to a horizontal position, thereby eliminating the gravitational effects on the vascular system and enabling effective circulation to be restored. For this reason, it is counterproductive to try to hold someone who has fainted upright. Fainting is the remedy to the problem, not the problem itself.

Since the skeletal muscle pump facilitates venous return, it is advisable to move around when you are on your feet and to get up periodically when you are working at a desk. The mild muscular activity "gets the blood moving." It is further recommended that individuals who must be on their feet for long periods of time use elastic stockings that apply a continuous gentle external compression, similar to the effect of skeletal muscle contraction, to further counter the effect of gravitational pooling of blood in the leg veins.

Effect of venous valves on venous return Venous vasoconstriction and external venous compression both drive blood in the direction of the heart. Yet if you squeeze a fluid-

Figure 10-31 Effect of Gravity on Venous Pressure
(a) In an upright adult, the blood in the vessels extending between the heart and foot is equivalent to a 1.5 m column of blood. The pressure exerted by this column of blood as a result of the effect of gravity is 90 mm Hg. The pressure imparted to the blood by the heart has declined to about 10 mm Hg in the lower-leg veins because of frictional losses in preceding vessels. The pressure caused by gravity (90 mm Hg) added to the pressure imparted by the heart (10 mm Hg) produces a venous pressure of 100 mm Hg in the ankle and foot veins. Similarly, the capillaries in the region are subjected to these same gravitational effects. (b) Because of the increased pressure caused by the gravitational effect, blood pools in the distended veins, resulting in decreased venous return. Filtration also increases across the capillary walls, resulting in swollen ankles and feet, unless compensatory measures are able to counteract the effect of gravity.

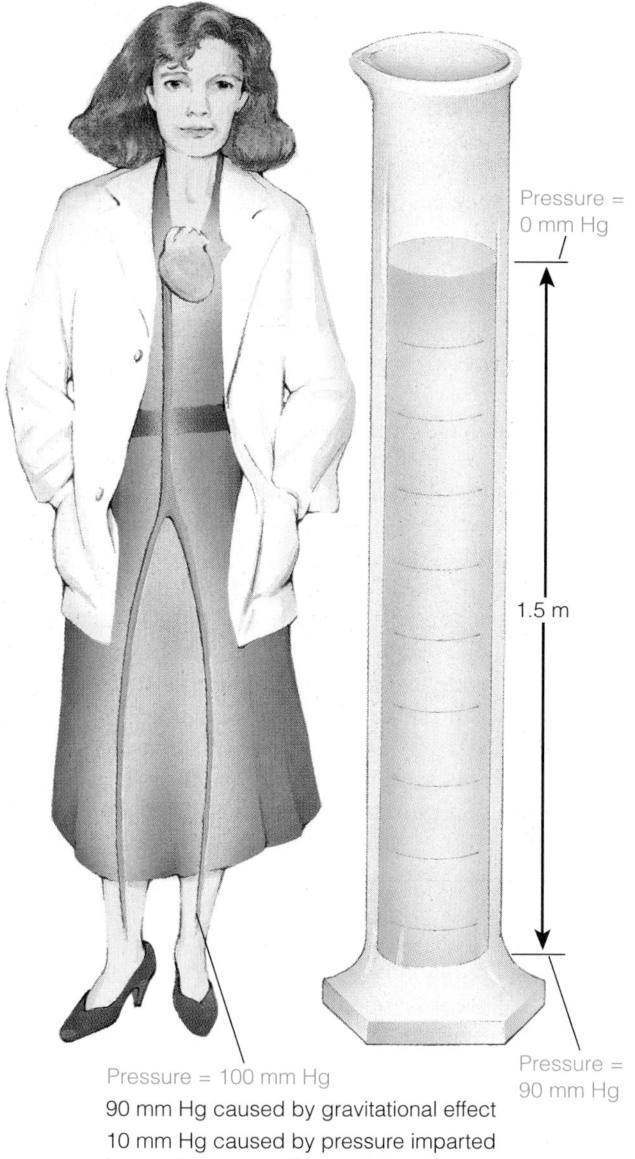

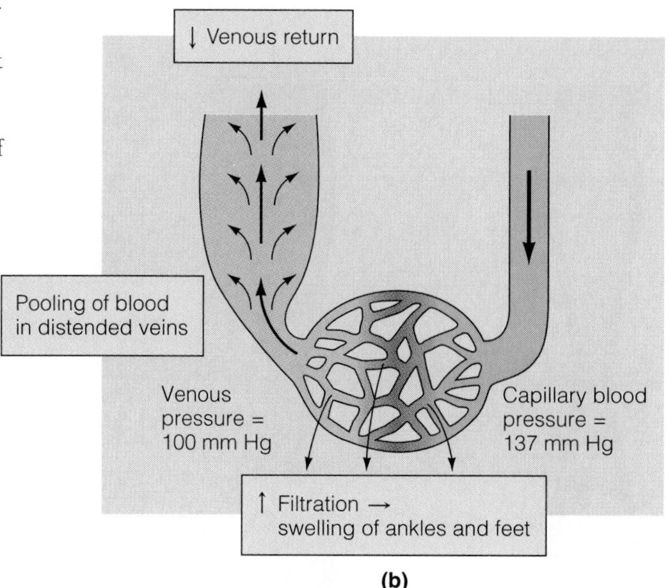

Figure 10-32 Effect of Contraction of the Skeletal Muscles of the Legs in Counteracting the Effects of Gravity
Contraction of skeletal muscles (as in walking) completely empties given vein segments, interrupting the column of blood that must be supported by the lower veins.

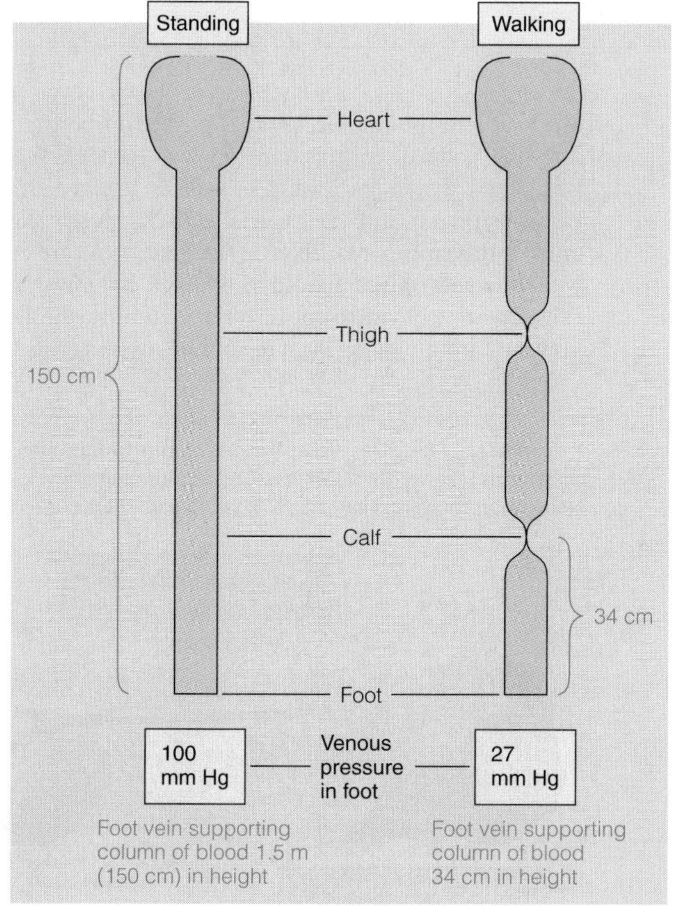

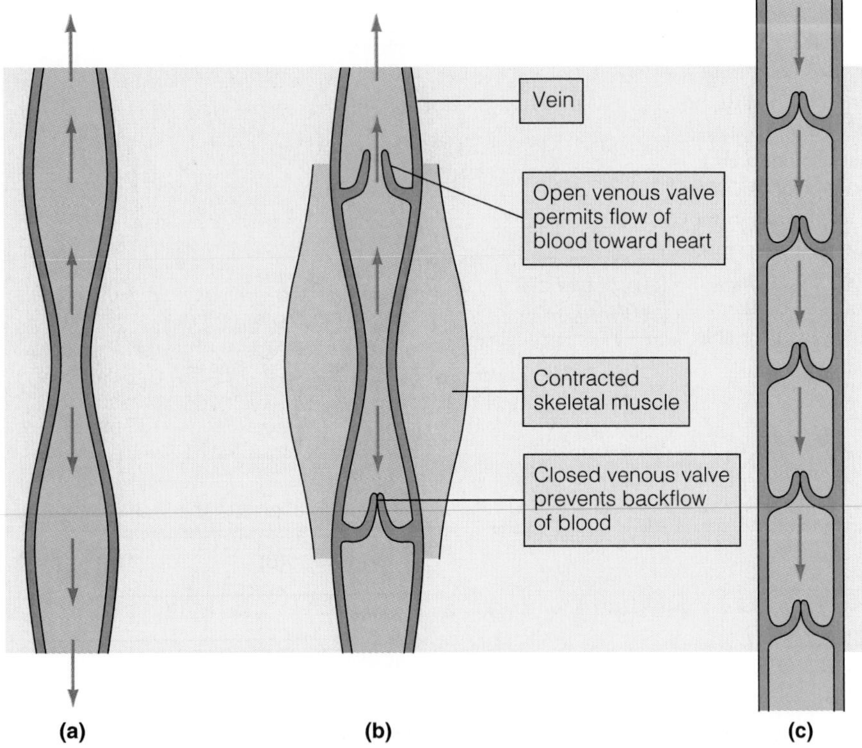

Figure 10-33 **Function of Venous Valves** (a) Fluid is pushed in both directions when a tube is squeezed in the middle. (b) Venous valves permit the flow of blood only toward the heart. (c) Venous valves play a role in counteracting gravitational effects by minimizing the backflow of blood and temporarily supporting portions of the column of blood.

filled tube in the middle, fluid is pushed in both directions from the point of constriction (Fig. 10-33a). Why, then, isn't blood driven backward as well as forward by venous vasoconstriction and the skeletal muscle pump? Blood can only be driven forward because the large veins are equipped with one-way valves spaced at 2 to 4 cm intervals; these valves permit blood to move forward toward the heart but prevent it from moving back toward the tissues (Fig. 10-33b). These venous valves also play a role in counteracting the gravitational effects of upright posture by helping minimize the backflow of blood that tends to occur when a person stands up and by temporarily supporting portions of the column of blood when the skeletal muscles are relaxed (Fig. 10-33c).

Varicose veins occur when the venous valves become incompetent and can no longer support the column of blood above them. Individuals predisposed to this condition usually have an inherited overdistensibility and weakness of their vein walls. Aggravated by frequent, prolonged standing, the veins become so distended as blood pools in them that the edges of the valves are no longer able to meet to form a seal. Varicosed superficial leg veins become visibly overdistended and tortuous. Contrary to what might be expected, the chronic pooling of blood in the pathologically distended veins does not reduce cardiac output because of a compensatory increase in total circulating blood volume. Instead, the most serious consequence of varicosed veins is the possibility of abnormal clot formation in the sluggish, pooled blood. Particularly dangerous is the risk that these clots may break loose and block small vessels elsewhere, especially the pulmonary capillaries.

Effect of respiratory activity on venous return As a result of respiratory activity, the pressure within the chest cavity averages 5 mm Hg less than atmospheric pressure. As the venous system returns blood to the heart from the lower regions of the body, it travels through the chest cavity, where it is exposed to this subatmospheric pressure. Because the venous system in the limbs and abdomen is subjected to normal atmospheric pressure, an externally applied pressure gradient exists between the lower veins (at atmospheric pressure) and the chest veins (at 5 mm Hg less than atmospheric pressure). This pressure difference squeezes blood from the lower veins to the chest veins, promoting increased venous return (Fig. 10-34). This mech-

Figure 10-34 **Respiratory Pump Enhancing Venous Return** As a result of respiratory activity, the pressure surrounding the chest veins is lower than the pressure surrounding the veins in the extremities and abdomen. This establishes an externally applied pressure gradient on the veins that drives blood toward the heart.

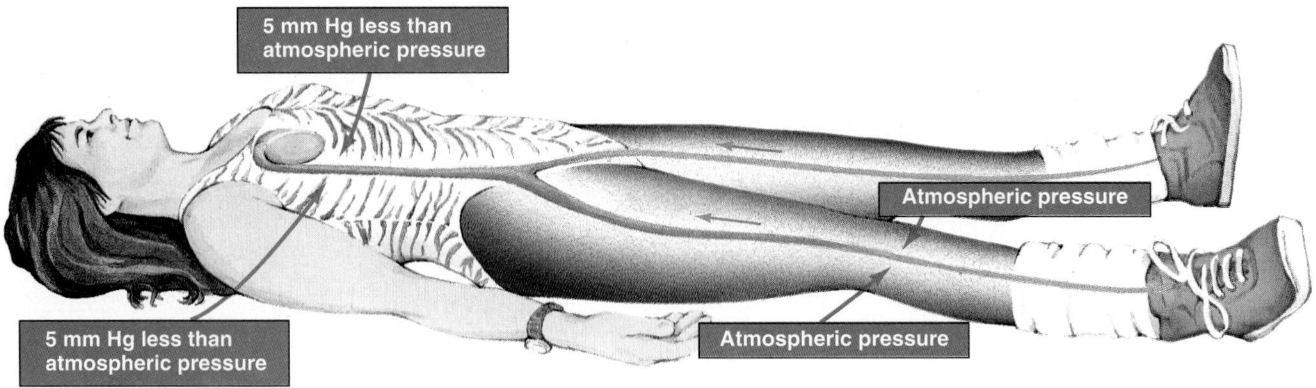

anism of facilitating venous return is known as the **respiratory pump** because it results from respiratory activity. Increased respiratory activity as well as the effects of the skeletal muscle pump and venous vasoconstriction all enhance venous return during exercise.

Effect of cardiac suction on venous return The extent of cardiac filling does not depend entirely on factors affecting the veins. The heart plays a role in its own filling. During ventricular contraction, the AV valves are drawn downward, enlarging the atrial cavities. As a result, the atrial pressure transiently drops below 0 mm Hg, thus increasing the vein-to-atria pressure gradient so that venous return is enhanced. In addition, the rapid expansion of the ventricular chambers during ventricular relaxation appears to create a transient negative pressure in the ventricles so that blood is "sucked in" from the atria and veins; that is, the negative ventricular pressure increases the vein-to-atria-to-ventricle pressure gradient, further enhancing venous return. Thus, the heart functions as a "suction pump" to facilitate cardiac filling.

Blood Pressure

Regulation of mean arterial blood pressure is accomplished by controlling cardiac output, total peripheral resistance, and blood volume.

Mean arterial blood pressure is the main driving force for propelling blood to the tissues. This pressure must be closely regulated for two reasons. First, it must be high enough to ensure sufficient driving pressure; without this pressure, the brain and other tissues will not receive adequate flow, no matter what local adjustments are made in the resistance of the arterioles supplying them. Second, the pressure must not be so high that it creates extra work for the heart and increases the risk of vascular damage and possible rupture of small blood vessels.

Elaborate mechanisms involving the integrated action of the various components of the circulatory system and other body systems are vital in the regulation of this all-important mean arterial pressure (Fig. 10-35). Remember from an

Figure 10-35 Determinants of Mean Arterial Blood Pressure Note that this figure is basically a composite of Figure 9-29, Control of Cardiac Output; Figure 10-14, Factors Affecting Total Peripheral Resistance; and Figure 10-29, Factors That Facilitate Venous Return. See the text for a discussion of the circled numbers.

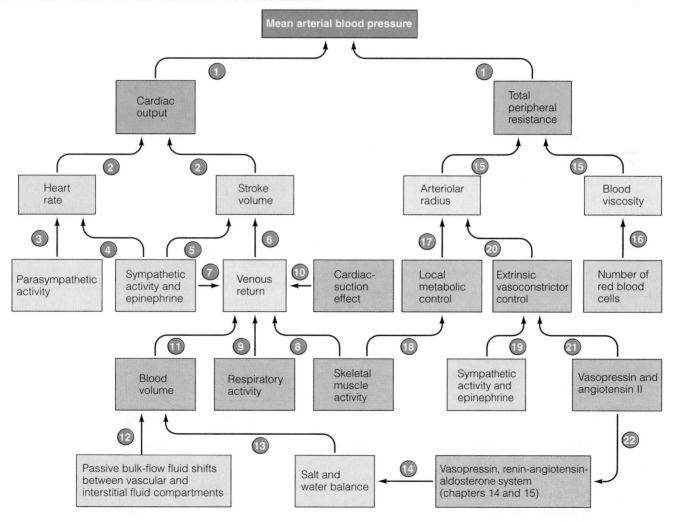

earlier discussion that the two determinants of mean arterial pressure are cardiac output and total peripheral resistance:

mean arterial pressure =
cardiac output × total peripheral resistance

(Do not confuse this equation, which indicates the *determinants* of mean arterial pressure, namely, the magnitude of both the cardiac output and total peripheral resistance, with the equation used to *calculate* mean arterial pressure, namely, mean arterial pressure = diastolic pressure + ⅓ pulse pressure.)

Recall that a number of factors, in turn, determine cardiac output (Fig. 9-29, p. 293) and total peripheral resistance (Fig. 10-14, p. 321). Thus, one can quickly appreciate the complexity of blood pressure regulation. Let's work our way through Figure 10-35, reviewing all the factors that have an effect on mean arterial blood pressure. Even though we've covered all these factors before, it is useful to pull them all together. The circled numbers in the text correspond to the numbers in the figure and indicate the portion of the figure being discussed.

- Mean arterial pressure depends on cardiac output and total peripheral resistance (① on Fig. 10-35).
- Cardiac output depends on heart rate and stroke volume ②.
- Heart rate depends on the relative balance of parasympathetic activity ③, which decreases heart rate, and sympathetic activity and epinephrine ④, which increase heart rate.
- Stroke volume increases in response to sympathetic activity ⑤ (extrinsic control of stroke volume).
- Stroke volume also increases as venous return increases ⑥ (intrinsic control of stroke volume by means of the Frank-Starling law of the heart).
- Venous return is enhanced by sympathetically induced venous vasoconstriction ⑦, the skeletal-muscle pump ⑧, the respiratory pump ⑨, and cardiac suction ⑩.
- The effective circulating blood volume also influences how much blood is returned to the heart ⑪. The blood volume depends in the short term on the magnitude of passive bulk-flow fluid shifts between the plasma and interstitial fluid across the capillary walls ⑫. In the long term, the blood volume depends on salt and water balance ⑬, which are hormonally controlled by the renin-angiotensin-aldosterone system and vasopressin, respectively ⑭.
- The other major determinant of mean arterial blood pressure, total peripheral resistance, depends on the radius of all arterioles as well as blood viscosity ⑮. The major factor determining blood viscosity is the number of red blood cells ⑯. However, arteriolar radius is the more important factor determining total peripheral resistance.
- Arteriolar radius is influenced by local (intrinsic) metabolic controls that match blood flow with metabolic needs ⑰. For example, local changes that take place in active skeletal muscles cause local arteriolar vasodilation and increased blood flow to these muscles ⑱.
- Arteriolar radius is also influenced by sympathetic activity ⑲, an extrinsic control mechanism that causes arteriolar vasoconstriction ⑳ to increase total peripheral resistance and mean arterial blood pressure.
- Arteriolar radius is also extrinsically controlled by the hormones vasopressin and angiotensin II, which are potent vasoconstrictors ㉑ as well as being important in salt and water balance ㉒.

Altering any of the pertinent factors that influence blood pressure will produce a change in blood pressure unless a compensatory change in another variable keeps the blood pressure constant. Blood flow to any given tissue depends on the driving force of the mean arterial pressure and on the degree of vasoconstriction of the tissue's arterioles. Because mean arterial pressure depends on the cardiac output and the degree of arteriolar vasoconstriction, if the arterioles in one tissue dilate, the arterioles in other tissues will have to constrict to maintain an adequate arterial blood pressure to provide a driving force to push blood not only to the vasodilated tissue but also to the brain, which depends on a constant blood supply. Thus, the cardiovascular variables must be continuously juggled to maintain a constant blood pressure in spite of tissues' varying needs for blood.

Mean arterial pressure is constantly monitored by **baroreceptors** (pressure sensors) within the circulatory system. When deviations from normal are detected, multiple reflex responses are initiated to return the arterial pressure to its normal value. *Short-term* (within seconds) adjustments are accomplished by alterations in cardiac output and total peripheral resistance, mediated by means of autonomic nervous system influences on the heart, veins, and arterioles. *Long-term* (requiring minutes to days) control involves adjusting total blood volume by restoring normal salt and water balance through mechanisms that regulate urine output and thirst (chapters 14 and 15). The magnitude of the total blood volume, in turn, has a profound effect on cardiac output and mean arterial pressure. Let us now turn our attention to the short-term mechanisms involved in the ongoing regulation of this pressure.

The baroreceptor reflex is the most important mechanism for short-term regulation of blood pressure.

Any change in mean blood pressure triggers an autonomically mediated **baroreceptor reflex** that influences the heart and blood vessels to adjust cardiac output and total peripheral resistance in an attempt to restore blood pressure to normal. Like any reflex, the baroreceptor reflex includes a receptor, an afferent pathway, an integrating center, an efferent pathway, and effector organs.

The most important receptors involved in moment-to-moment regulation of blood pressure, the **carotid sinus** and **aortic arch baroreceptors,** are mechanoreceptors sensitive to changes in both mean arterial pressure and pulse pressure. Their responsiveness to fluctuations in pulse pressure enhances their sensitivity as pressure sensors, because small changes in systolic or diastolic pressure may alter the pulse pressure without changing the mean pressure. These barore-

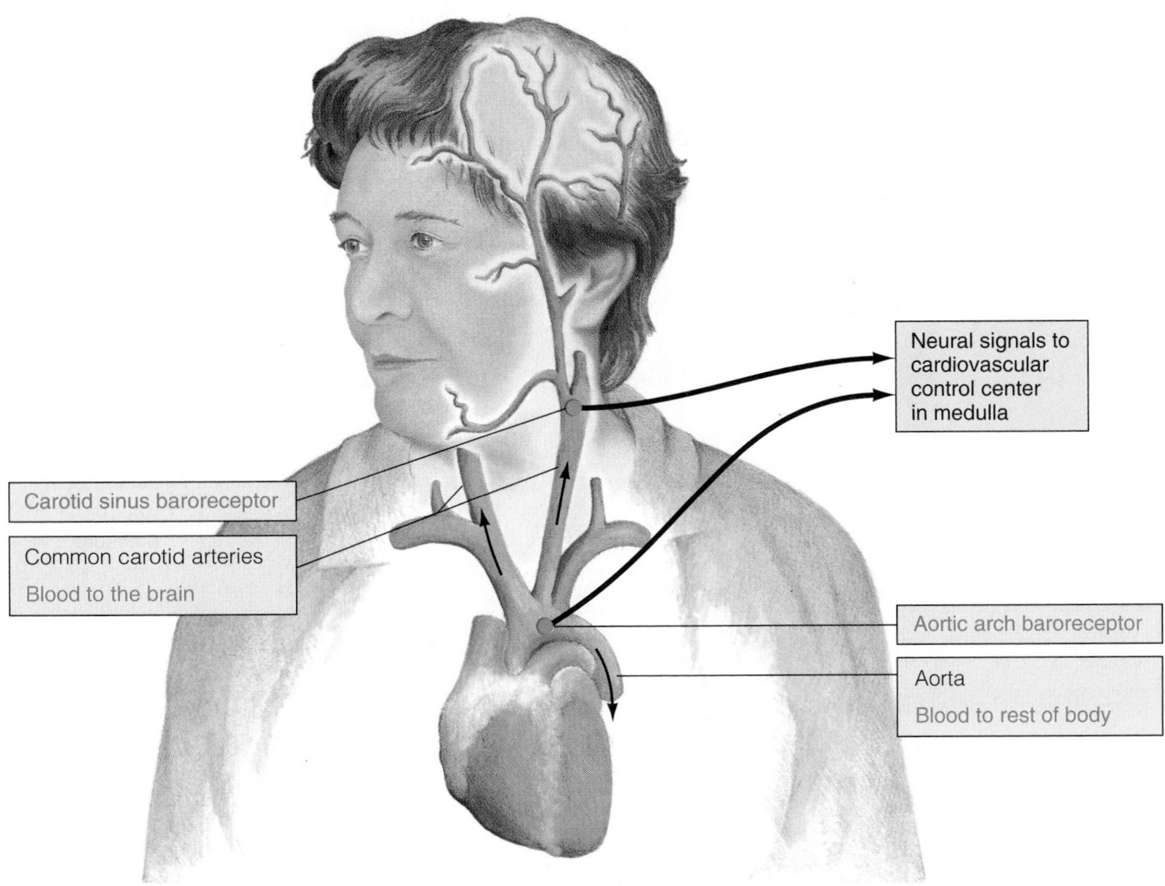

Figure 10-36 Location of the Arterial Baroreceptors The arterial baroreceptors are strategically located to monitor the mean arterial blood pressure in the arteries that supply blood to the brain (cartoid sinus baroreceptor) and to the rest of the body (aortic arch baroreceptor).

ceptors are strategically located (Fig. 10-36) to provide critical information about arterial blood pressure in the vessels leading to the brain (the carotid sinus baroreceptor) and in the major arterial trunk before it gives off branches that supply the rest of the body (the aortic arch baroreceptor).

The baroreceptors constantly provide information about blood pressure; in other words, they continuously generate action potentials in response to the ongoing pressure within the arteries. When arterial pressure (either mean or pulse pressure) increases, the receptor potential of these baroreceptors increases, thus increasing the rate of firing in the corresponding afferent neurons. Conversely, when blood pressure decreases, the rate of firing generated in the afferent neurons by the baroreceptors decreases (Fig. 10-37).

The integrating center that receives the afferent impulses about the state of arterial pressure is the **cardiovascular control center,**[4] located in the medulla within the brain stem. The efferent pathway is the autonomic nervous system. The

[4]The cardiovascular control center is sometimes divided into cardiac and vasomotor centers, which are occasionally further classified into smaller subdivisions, such as cardioacceleratory and cardioinhibitory centers and vasoconstrictor and vasodilator areas. Since these regions are highly interconnected and functionally interrelated, we will refer to them collectively as the cardiovascular control center.

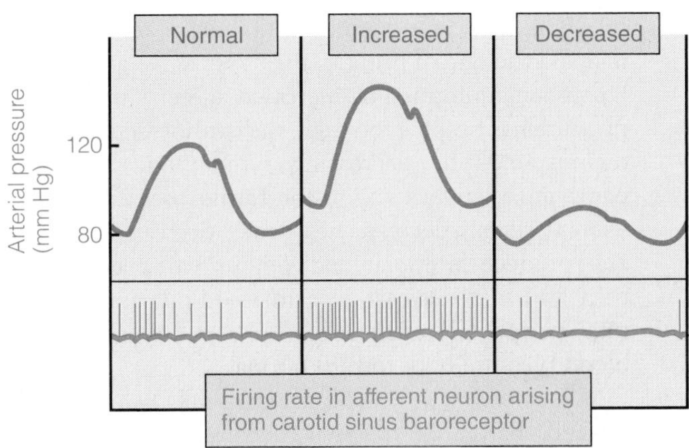

Figure 10-37 Firing Rate in the Afferent Neurons from the Carotid Sinus Baroreceptor in Relation to the Magnitude of Mean Arterial Pressure

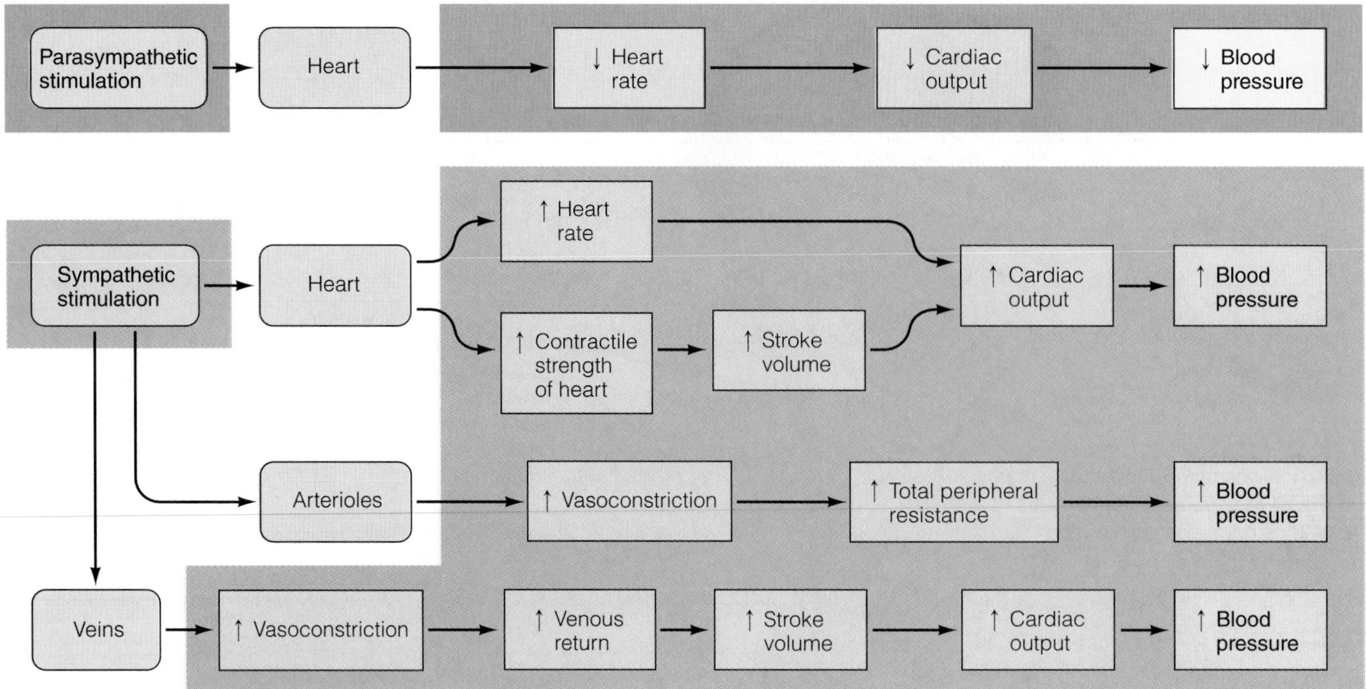

Figure 10-38 **Summary of the Effects of the Parasympathetic and Sympathetic Nervous Systems on Factors That Influence the Mean Arterial Blood Pressure**

cardiovascular control center alters the ratio between sympathetic and parasympathetic activity to the effector organs (the heart and blood vessels). To show how autonomic changes alter arterial blood pressure, Fig. 10-38 provides a review of the major effects of parasympathetic and sympathetic stimulation on the heart and blood vessels.

Let us fit all the pieces of the baroreceptor reflex together now by tracing the reflex activity that occurs to compensate for an elevation or fall in blood pressure. If for any reason arterial pressure becomes elevated above normal (Fig. 10-39a), the carotid sinus and aortic arch baroreceptors increase the rate of firing in their respective afferent neurons. Upon being informed by increased afferent firing that arterial pressure has become too high, the cardiovascular control center responds by decreasing sympathetic and increasing parasympathetic activity to the cardiovascular system. These efferent signals decrease heart rate, decrease stroke volume, and produce arteriolar and venous vasodilation, which in turn lead to a decrease in cardiac output and a decrease in total peripheral resistance, with a subsequent decrease in blood pressure back toward normal.

Conversely, when blood pressure falls below normal (Fig. 10-39b), baroreceptor activity decreases, inducing the cardiovascular center to increase sympathetic cardiac and vasoconstrictor nerve activity while decreasing its parasympathetic output. This efferent pattern of activity leads to an increase in heart rate and stroke volume coupled with arteriolar and venous vasoconstriction. These changes result in an increase in both cardiac output and total peripheral resistance, producing an elevation in blood pressure back toward normal.

Other reflexes and responses influence blood pressure.

Besides the baroreceptor reflex, whose sole function is blood pressure regulation, several other reflexes and responses influence the cardiovascular system even though they primarily are concerned with the regulation of other body functions. Some of these other influences deliberately move arterial pressure away from its normal value temporarily, overriding the baroreceptor reflex to accomplish a particular goal. These factors include the following:

1. Left atrial volume receptors and hypothalamic osmoreceptors are primarily important in water and salt balance in the body; thus, they affect the long-term regulation of blood pressure by controlling the plasma volume.
2. Chemoreceptors located in the carotid and aortic arteries, in close association with but distinct from the baroreceptors, are sensitive to low O_2 or high acid levels in the blood. These chemoreceptors' main function is to reflexly increase respiratory activity to bring in more O_2 or to blow off more acid-forming CO_2, but they also reflexly increase blood pressure by sending excitatory impulses to the cardiovascular center.
3. Cardiovascular responses associated with certain behaviors and emotions are mediated through the cerebral cortex–hypothalamic pathway and appear to be preprogrammed. These responses include the widespread changes in cardiovascular activity accompanying the generalized sympathetic fight-or-flight response, the characteristic marked increase in heart rate and blood pressure

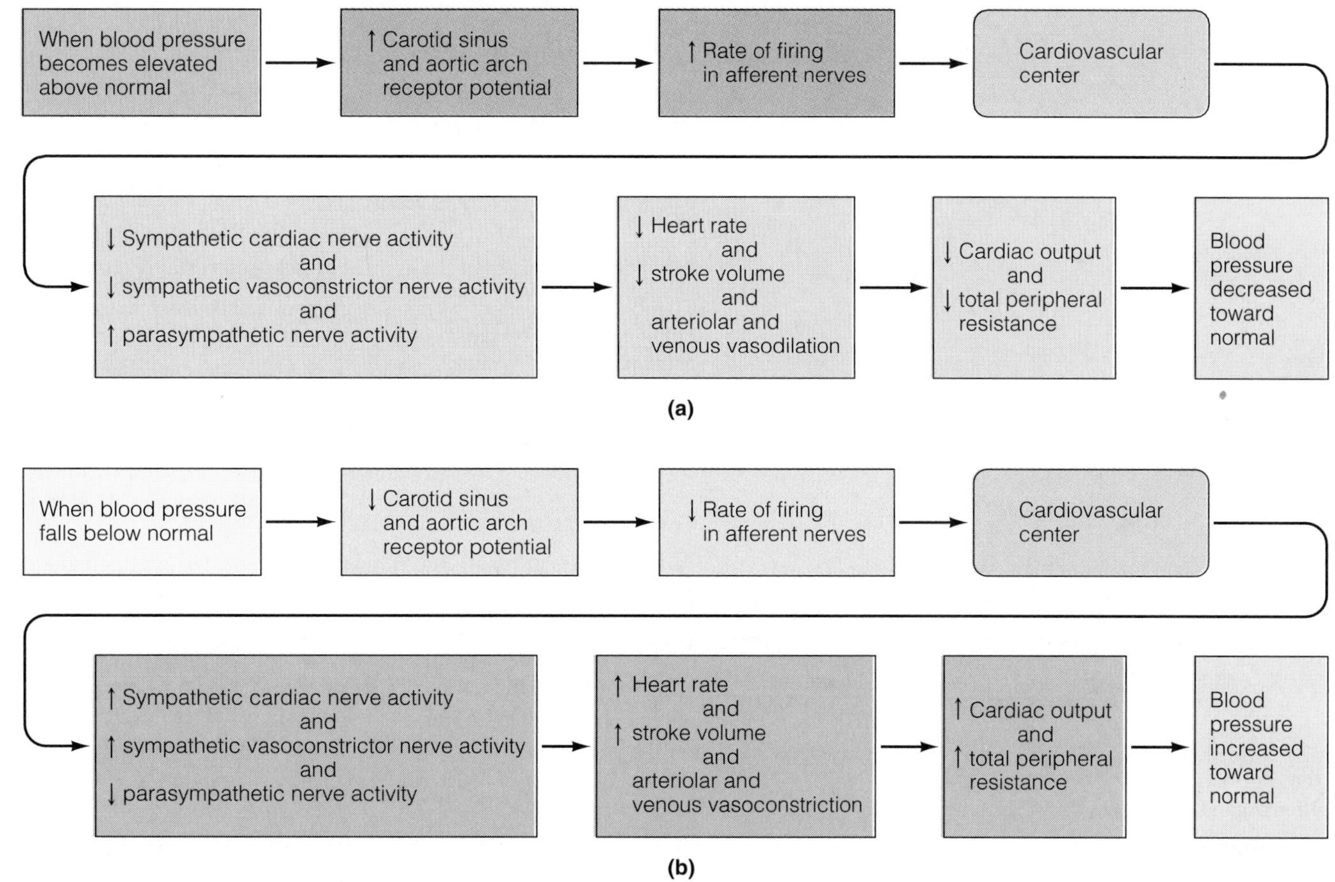

Figure 10-39 **Baroreceptor Reflexes to Restore the Blood Pressure to Normal** (a) Baroreceptor reflex in response to an elevation in blood pressure. (b) Baroreceptor reflex in response to a fall in blood pressure.

associated with sexual orgasm, and the localized cutaneous vasodilation characteristic of blushing.

4. Pronounced cardiovascular changes accompany exercise, including a substantial increase in skeletal muscle blood flow; a significant increase in cardiac output; a fall in total peripheral resistance (because of widespread vasodilation in skeletal muscles despite generalized arteriolar vasoconstriction in most organs); and a modest increase in mean arterial pressure (Table 10-4). Evidence suggests that discrete exercise centers yet to be identified induce the appropriate cardiac and vascular changes at the onset of exercise or even in anticipation of exercise. These effects are then reinforced by afferent inputs to the medullary cardiovascular center from chemoreceptors in exercising muscles as well as by local mechanisms important in maintaining vasodilation in active muscles. The baroreceptor reflex further modulates these cardiovascular responses.
5. Hypothalamic control over cutaneous (skin) arterioles for the purpose of temperature regulation takes precedence over control that the cardiovascular center has over these same vessels for the purpose of blood pressure regulation. As a result, blood pressure can fall when the skin vessels are widely dilated to eliminate excess heat from the body, even though the baroreceptor responses are calling for cutaneous vasoconstriction to help maintain adequate total peripheral resistance.
6. Vasoactive substances released from the endothelial cells play a role in the regulation of blood pressure. For example, EDRF/NO normally exerts an ongoing vasodilatory effect.
7. Recent studies suggest that numerous neurotransmitters from various regions of the brain may play a role in the control of blood pressure. Their importance and relations to the traditional pathways for blood pressure regulation are currently unclear, as are any possible roles they may play in the poorly understood condition of hypertension.

Hytertension is a serious national public health problem, but its causes are largely unknown.

Sometimes blood pressure control mechanisms do not function properly or are unable to completely compensate for changes that have taken place. Blood pressure may be above the normal range (**hypertension** if above 140/90 mm Hg) or below normal (**hypotension** if less than 100/60 mm Hg). Hypotension in its extreme form is *circulatory shock.* We will first examine hypertension and then conclude this chapter with a discussion of hypotension and shock.

Table 10-4 Cardiovascular Changes during Exercise

Cardiovascular Variable	*Change*	*Comment*
Heart rate	Increases	Occurs as a result of increased sympathetic and decreased parasympathetic activity to the SA node
Venous return	Increases	Occurs as a result of sympathetically induced venous vasoconstriction and increased activity of the skeletal muscle pump and respiratory pump
Stroke volume	Increases	Occurs both as a result of increased venous return by means of the Frank-Starling mechanism (unless diastolic filling time is significantly reduced due to a high heart rate) and as a result of a sympathetically induced increase in myocardial contractility
Cardiac output	Increases	Occurs as a result of increases in both heart rate and stroke volume
Blood flow to active skeletal muscles and heart muscle	Increases	Occurs as a result of locally controlled arteriolar vasodilation, which is reinforced by the vasodilatory effects of epinephrine and overpowers the weaker sympathetic vasoconstrictor effect
Blood flow to the brain	Unchanged	Occurs because sympathetic stimulation has no effect on brain arterioles; local control mechanisms maintain constant cerebral blood flow whatever the circumstances
Blood flow to the skin	Increases	Occurs because the hypothalamic temperature control center induces vasodilation of skin arterioles; increased skin blood flow brings heat produced by exercising muscles to the body surface where the heat can be lost to the external environment
Blood flow to the digestive system, kidneys, and other organs	Decreases	Occurs as a result of generalized sympathetically induced arteriolar vasoconstriction
Total peripheral resistance	Decreases	Occurs because resistance in the skeletal muscles, heart, and skin decreases to a greater extent than resistance in the other organs increases
Mean arterial blood pressure	Increases (modest)	Occurs because cardiac output increases to a greater extent than total peripheral resistance decreases

A definite cause for hypertension can be established in only 10% of the cases. Hypertension that occurs secondary to another primary problem is called **secondary hypertension.** The causes of secondary hypertension fall into four categories:

1. *Cardiovascular hypertension* is usually associated with chronically elevated total peripheral resistance caused by atherosclerosis (hardening of the arteries; see p. 296).
2. *Renal (kidney) hypertension* may occur as a result of either of two kidney defects: partial occlusion of the renal arteries or diseases of the kidney tissue itself.
   a. Atherosclerotic lesions protruding into the lumen of a renal artery or external compression of the vessel by a tumor may reduce blood flow through the kidney. The kidney responds by initiating the hormonal pathway involving angiotensin II. This pathway promotes salt and water retention during urine formation, thus increasing the blood volume to compensate for the reduced renal blood flow. Recall that angiotensin II is also a powerful vasoconstrictor. Although these two effects (increased blood volume and angiotensin-induced vasoconstriction) are compensatory mechanisms to improve blood flow through the narrowed renal artery, they also are responsible for elevating the arterial pressure as a whole.
   b. Renal hypertension also occurs if the kidneys are diseased and unable to eliminate the normal salt load. Salt retention induces water retention, which expands the plasma volume and leads to hypertension.
3. *Endocrine hypertension* arises from at least two different endocrine disorders: pheochromocytoma and Conn's syndrome.
   a. A *pheochromocytoma* is an adrenal medullary tumor that secretes excessive epinephrine and norepinephrine. Abnormally elevated levels of these hormones induce increased cardiac output and generalized peripheral vasoconstriction, both of which contribute to the hypertension characteristic of this disorder.
   b. *Conn's syndrome* is associated with increased production of aldosterone by the adrenal cortex. This hormone is part of the hormonal pathway responsible for salt and water retention by the kidneys (the renin-angiotensin-aldosterone pathway). Once again, the excessive salt and water load in the body caused by the increased aldosterone levels leads to an elevation in blood pressure.
4. *Neurogenic hypertension* occurs secondary to neural lesions.
   a. The problem may be erroneous blood pressure control caused by a defect in the cardiovascular control center or in the baroreceptors.

b. Neurogenic hypertension may also occur as a compensatory response to a reduction in cerebral blood flow—for example, because a major cerebral vessel is compressed by a tumor. In response to a decrease in brain blood flow, reflexes are initiated that elevate the blood pressure in an attempt to provide sufficient driving pressure to adequately supply the O_2-dependent brain tissue with blood.

The underlying cause is unknown in the remaining 90% of hypertension cases. Such hypertension is known as **primary (essential** or **idiopathic) hypertension.** Primary hypertension is undoubtedly a catchall category for elevated blood pressure caused by a variety of unknown causes rather than a single disease entity. There is a strong genetic tendency to develop primary hypertension, which can be hastened or worsened by contributing factors such as obesity, stress, smoking, and excessive ingestion of salt. Consider the following range of potential causes for primary hypertension that are currently being investigated:

- *Defects in salt management by the kidneys.* Disturbances in kidney function too minor to produce outward signs of renal disease could nevertheless insidiously lead to gradual accumulation of salt and water in the body, resulting in progressive elevation of arterial pressure.
- *Plasma membrane abnormalities such as defective Na^+-K^+ pumps.* Such defects, by altering the electrochemical gradient across plasma membranes, could change the excitability and contractility of the heart and the smooth muscle in blood vessel walls in such a way as to lead to high blood pressure. In addition, the Na^+-K^+ pump is critical to salt management by the kidneys. A genetic defect in the Na^+-K^+ pump of hypertensive-prone laboratory rats was the first gene-hypertension link to be discovered.
- *Variation in the gene that encodes for angiotensinogen.* Angiotensinogen is part of the hormonal pathway that produces the potent vasoconstrictor angiotensin II and promotes salt and water retention. One variant of the gene in humans appears to be associated with a higher incidence of hypertension. Researchers speculate that the suspect version of the gene leads to a slight excess production of angiotensinogen, thus increasing activity of this blood pressure–raising pathway. This is the first gene-hypertension link discovered in humans.
- *Endogenous digitalis-like substances.* Such substances act similarly to the drug digitalis (see p. 280) to increase cardiac contractility as well as constrict blood vessels and reduce salt elimination in the urine, all of which could cause chronic hypertension.
- *Abnormalities in EDRF/NO or other locally acting vasoactive chemicals.* For example, a shortage of NO has been discovered in the blood vessel walls of some hypertensive patients, leading to an impaired ability to accomplish blood pressure–lowering vasodilation.
- Recent experimental evidence suggests that hypertension may result from a malfunction of the vasopressin-secreting cells of the hypothalamus. Vasopressin is a potent vasoconstrictor and also promotes water retention.
- *Physical pressure on the cardiovascular control center by an overlying artery.* One neurosurgeon, in a limited number of operations, has successfully reduced high blood pressure by moving an enlarged loop of artery that pulsated against the medullary brain tissue.

Whatever the underlying defect, once initiated, hypertension appears to be self-perpetuating. Constant exposure to elevated blood pressure predisposes vessel walls to the development of atherosclerosis, which further elevates blood pressure.

The baroreceptors do not respond to bring the blood pressure back to normal during hypertension because they adapt, or are "reset," to operate at a higher level. In the presence of chronically elevated blood pressure, the baroreceptors still function to regulate blood pressure, but they maintain it at a higher mean pressure.

Hypertension imposes stresses on both the heart and the blood vessels. The heart has an increased workload because it is pumping against an increased total peripheral resistance, whereas blood vessels may be damaged by the high internal pressure, particularly when the vessel wall is weakened by the degenerative process of atherosclerosis. Complications of hypertension include congestive heart failure caused by the heart's inability to pump continuously against a sustained elevation in arterial pressure (see p. 294), strokes caused by rupture of brain vessels, and heart attacks caused by rupture of coronary vessels. Spontaneous hemorrhage due to bursting of small vessels elsewhere in the body may also occur but with less serious consequences; an example is the rupture of blood vessels in the nose, resulting in nosebleeds. Another serious complication of hypertension is renal failure caused by progressive impairment of blood flow through damaged renal blood vessels. Furthermore, retinal damage caused by changes in the blood vessels supplying the eyes may result in progressive loss of vision. Until complications occur, hypertension is symptomless because the tissues are adequately supplied with blood. Therefore, unless blood pressure measurements are made on a routine basis, the condition can go undetected until a precipitous complicating event results. When one becomes aware of these potential complications of hypertension and considers that 25% of all adults in America are estimated to be afflicted with chronic elevated blood pressure, one can appreciate the magnitude of this national health problem.

Once hypertension is detected, therapeutic intervention can reduce the course and severity of the problem. Cornerstones of treatment include reduction in salt intake and administration of diuretics (drugs that enhance urine excretion) to reduce the salt and water load in the body, thereby decreasing plasma volume. In addition, other antihypertensive drugs reduce total peripheral resistance by manipulating some aspect of autonomic function to promote arteriolar vasodilation. No matter what the original cause, agents that reduce the plasma volume or total peripheral resistance (or both) will decrease the blood pressure toward normal.

A CLOSER LOOK AT EXERCISE PHYSIOLOGY

The Ups and Downs of Hypertension and Exercise

When blood pressure is up, one way to bring it down is to increase the level of physical activity. An impressive epidemiological study that followed a group of 14,998 male graduates from Harvard for sixteen to fifty years found that participating in collegiate sports, climbing as many as fifty stairs a day, walking five blocks a day, or performing light sports activities was not protective against the development of primary hypertension. Participating in vigorous sports such as running, swimming, handball, tennis, and cross-country skiing, on the other hand, was protective against the development of hypertension, even if other risk factors were present. Other studies have supported the finding that participation in aerobic activities is protective against the development of hypertension.

A logical question to ask is whether exercise can be used as a therapy to reduce hypertension once it has already developed. Antihypertensive medication is available to lower blood pressure in severely hypertensive patients, but sometimes undesirable side effects occur. The side effects of diuretics include electrolyte imbalances, inability to handle glucose normally, and increased blood cholesterol levels. The side effects of drugs that manipulate total peripheral resistance, such as beta-blocker drugs, include increased blood triglyceride levels, lower HDL-cholesterol levels (the "good" form of cholesterol), weight gain, sexual dysfunction, and depression.

Patients with mild hypertension, arbitrarily defined as a diastolic blood pressure between 90 and 100 mm Hg and a systolic pressure of 160 mm Hg, pose a dilemma for physicians. The risks of taking the drugs may outweigh the benefits gained from lowering the blood pressure. Because of the drug therapy's possible side effects, nondrug treatment of mild hypertension may be most beneficial. The most common non-drug therapies are weight reduction, salt restriction, and exercise. Although losing weight will almost always reduce blood pressure, research has shown that weight reduction programs usually result in the loss of only twelve pounds, and the overall long-term success in keeping the weight off is only about 20%. Salt restriction is beneficial for many hypertensives, but adherence to a low-salt diet is difficult for many people because fast foods and foods prepared in restaurants usually contain high amounts of salt. Some recent studies employing exercise as a therapeutic tool have shown that blood pressure in cases of mild to moderate hypertension has been decreased in many individuals regardless of whether salt was restricted or weight was lowered. The preponderance of evidence in the literature suggests that moderate aerobic exercise performed three times per week for fifteen to sixty minutes is a beneficial therapy in most cases of mild to moderate hypertension. It is wise, therefore, to include a regular aerobic exercise program in conjunction with weight loss and salt reduction to optimally reduce high blood pressure with little or no drug intervention.

Furthermore, a regular aerobic exercise program can be employed to help reduce high blood pressure. (See the accompanying boxed feature, A Closer Look at Exercise Physiology.)

Inadequate sympathetic activity is responsible for dizziness or fainting accompanying transient hypotension.

Hypotension, or low blood pressure, occurs either when there is a disproportion between vascular capacity and blood volume or when the heart is too weak to impart sufficient driving pressure to the blood.

Two common situations in which hypotension occurs transiently are orthostatic hypotension and emotional fainting. Both are due to inadequate sympathetic activity. **Orthostatic (postural) hypotension** is a transient hypotensive condition resulting from insufficient compensatory responses to the gravitational shifts in blood that occur when a person moves from a horizontal to a vertical position, especially following prolonged bed rest. When a person moves from lying down to standing up, pooling of blood in the leg veins as a result of gravity causes a reduction in venous return and a subsequent decrease in stroke volume, leading to a fall in cardiac output and blood pressure. This fall in blood pressure is normally detected by the baroreceptors, which initiate immediate compensatory responses to restore blood pressure to its proper level. When a long-bedridden patient first starts to rise, however, these reflex compensatory adjustments are temporarily lost or reduced because of disuse. Sympathetic control of the leg veins is inadequate, so when the patient first stands up, blood pools in the lower extremities. The resultant orthostatic hypotension and decrease in blood flow to the brain are responsible for the dizziness or actual fainting that occurs. Because postural compensatory mechanisms are depressed during prolonged bed confinement, patients sometimes are put on a tilt table so that they can be moved gradually from a horizontal to an upright position. This allows the body to adjust slowly to the gravitational shifts in blood.

Transient hypotension because of emotional stress can also cause dizziness or fainting. In this situation, higher brain centers act on the cardiovascular center to inappropriately decrease sympathetic output to the vasculature. The resultant loss of vascular tone precipitates widespread arteriolar vasodilation that leads to a decrease in total peripheral resistance. Furthermore, widespread arteriolar vasodilation causes pooling of blood in the capillaries so that venous return is decreased and cardiac output is subsequently reduced. Thus, hypotension is brought about by a decrease in both total peripheral resistance and cardiac output. As the blood pressure falls, the person becomes light-headed or faints because blood flow to the brain becomes inadequate. If the person faints or lies down, cardiac output is rapidly improved as the pooled blood quickly returns to the heart. The inappropriate autonomic nerve activity is only transient; when it is discontinued, vasoconstrictor tone is quickly restored. The person normally recovers quite rapidly with no aftereffects. Even though fainting in reaction to a highly emotional situation is considered to be an inappropriate response, there is speculation that it may actually be adaptive, analogous to the "playing dead" ploy displayed by some animals. It temporarily relieves the person from the responsibility of coping with the emotionally charged situation.

Circulatory shock can become irreversible.

When blood pressure falls so low that adequate blood flow to the tissues can no longer be maintained, the condition known as **circulatory shock** occurs. Circulatory shock is categorized into four main parts (Fig. 10-40):

1. *Hypovolemic shock* is induced by a fall in blood volume, which occurs either directly through severe hemorrhage or indirectly through loss of fluids derived from the plasma (for example, severe diarrhea, excessive urinary losses, or extensive sweating).
2. *Cardiogenic shock* is due to a weakened heart's failure to pump blood adequately.
3. *Vasogenic shock* is caused by widespread vasodilation triggered by the presence of vasodilator substances. There are two types of vasogenic shock: septic and anaphylactic. *Septic shock,* which may accompany massive infections, is due to vasodilator substances released from the infective agents. Similarly, extensive histamine release accompanying severe allergic reactions can cause widespread vasodilation in *anaphylactic shock.*
4. *Neurogenic shock* also involves generalized vasodilation but not by means of the release of vasodilator substances. In this case, loss of sympathetic vascular tone leads to generalized vasodilation, similar to emotional hypotension but more pronounced and prolonged. This undoubtedly is responsible for the shock accompanying crushing injuries when blood loss has not been sufficient to cause hypovolemic shock. Deep, excruciating pain apparently inhibits sympathetic vasoconstrictor activity.

We will now examine the consequences of and compensations for shock, using hemorrhage as an example (Fig. 10-41). This figure may look intimidating, but we will work through it step by step. It is an important example that pulls together many of the principles discussed in this chapter. As

Figure 10-40 **Causes of Circulatory Shock** Circulatory shock, which occurs when mean arterial blood pressure falls so low that adequate blood flow to the tissues can no longer be maintained, may result from (1) extensive loss of blood volume (hypovolemic shock), (2) failure of the heart to pump blood adequately (cardiogenic shock), or (3) widespread arteriolar vasodilation induced by toxic or allergic vasodilator substances (vasogenic shock), or (4) neurally defective vasoconstrictor tone (neurogenic shock).

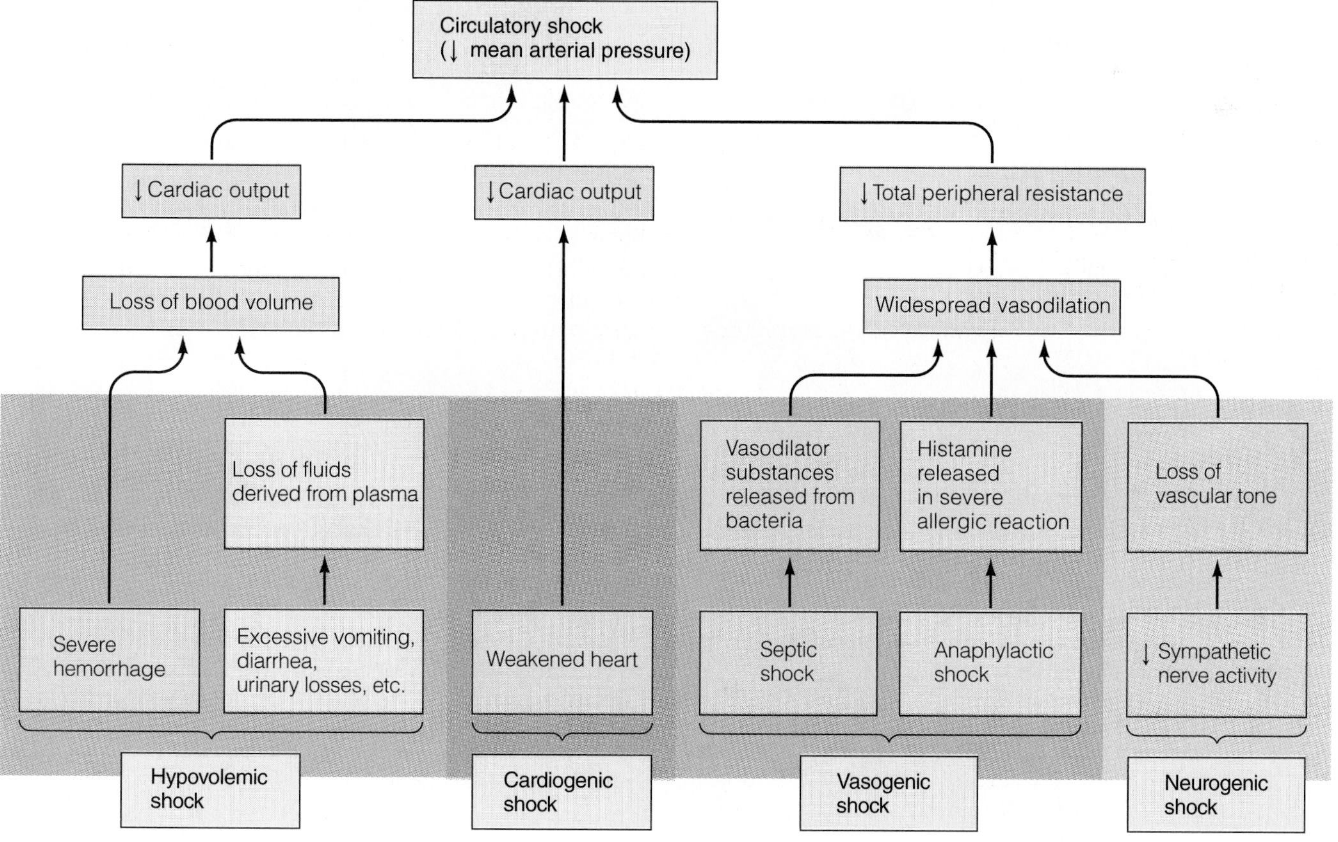

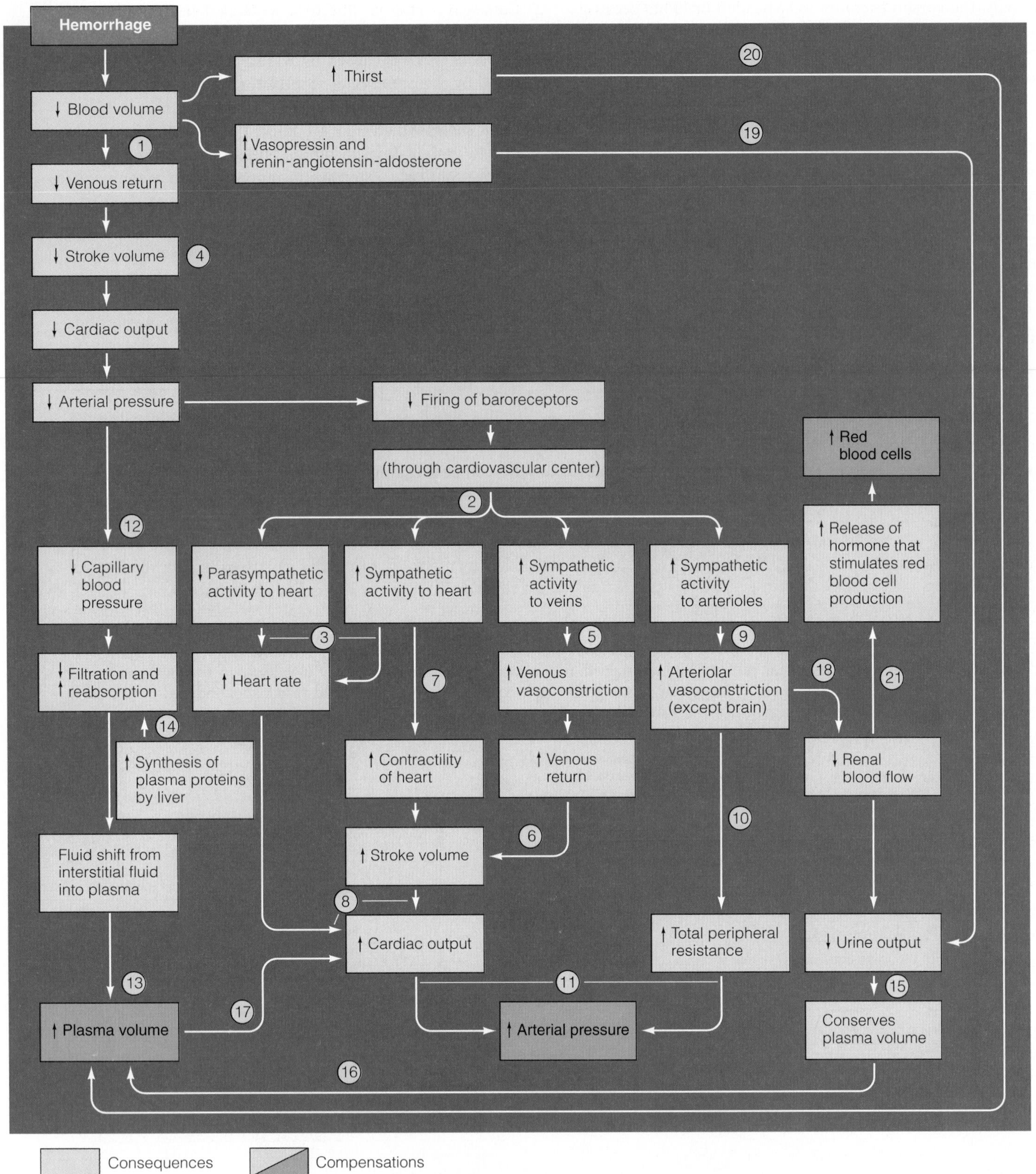

Figure 10-41 **Consequences and Compensations of Hemorrhage** The reduction in blood volume resulting from hemorrhage leads to a fall in arterial pressure. (Note the blue boxes representing consequences of hemorrhage). A series of compensations ensue (light pink boxes) that ultimately restore plasma volume, arterial pressure, and the number of red blood cells toward normal (dark pink boxes). Refer to the text (p. 347) for an explanation of the circled numbers and a detailed discussion of the compensations.

before, the circled numbers in the text correspond to the numbers in the figure and indicate the portion of the figure being discussed.

- Following severe loss of blood, the resultant reduction in circulating blood volume leads to a decrease in venous return ① and a subsequent fall in cardiac output and arterial blood pressure (Note the blue boxes, which indicate consequences of hemorrhage.)
- Compensatory measures immediately attempt to maintain adequate blood flow to the brain. (Note the pink boxes, which indicate compensations for hemorrhage.)
- The baroreceptor reflex response to the fall in blood pressure brings about increased sympathetic and decreased parasympathetic activity to the heart ②. The result is an increase in heart rate ③ to offset the reduced stroke volume ④ brought about by the loss of blood volume. With severe fluid loss, the pulse is weak because of the reduced stroke volume but rapid because of the increased heart rate.
- As a result of increased sympathetic activity to the veins, generalized venous vasoconstriction occurs ⑤, increasing venous return by means of the Frank-Starling mechanism ⑥.
- Simultaneously, sympathetic stimulation of the heart increases the heart's contractility ⑦ so that it beats more forcefully and ejects a greater volume of blood, likewise increasing the stroke volume.
- The increase in heart rate and increase in stroke volume collectively lead to an increase in cardiac output ⑧.
- Sympathetically induced generalized arteriolar vasoconstriction ⑨ leads to an increase in total peripheral resistance ⑩.
- Together, the increase in cardiac output and total peripheral resistance bring about a compensatory increase in arterial pressure ⑪.
- The original fall in arterial pressure is also accompanied by a fall in capillary blood pressure ⑫, which results in fluid shifts from the interstitial fluid into the capillaries to expand the plasma volume ⑬. This response is sometimes termed **autotransfusion**, because it restores the plasma volume as a transfusion does.
- This ECF fluid shift is enhanced by plasma-protein synthesis by the liver during the next few days following hemorrhage ⑭. The plasma proteins exert a colloid osmotic pressure to attract and retain extra fluid in the plasma.
- Urinary output is reduced, thereby conserving water that normally would have been lost from the body ⑮. This additional fluid retention helps to expand the reduced plasma volume ⑯. Expansion of plasma volume further augments the increase in cardiac output brought about by the baroreceptor reflex ⑰. Reduction in urinary output results from decreased renal blood flow caused by compensatory renal arteriolar vasoconstriction ⑱. The reduced plasma volume also triggers increased secretion of the hormone vasopressin and activation of the salt- and water-conserving renin-angiotensin-aldosterone hormonal pathway, which brings about a further reduction in urinary output ⑲.
- Increased thirst is also stimulated by a fall in plasma volume ⑳. The resultant increased fluid intake contributes to restoration of plasma volume.
- Over a longer course of time (a week or more), lost red blood cells are replaced through increased red blood cell production triggered by a reduction in O_2 delivery to the kidneys ㉑.

These compensatory mechanisms are often insufficient in the face of substantial fluid loss. Even if they are able to maintain an adequate blood pressure level, the short-term measures cannot continue indefinitely. Ultimately, fluid volume must be replaced from the outside through drinking, transfusion, or a combination of both. Blood supply to kidneys, digestive tract, skin, and other organs can be compromised to maintain blood flow to the brain only so long before organ damage begins to occur. A point may be reached at which blood pressure continues to drop rapidly because of tissue damage, despite vigorous therapy. This condition is frequently termed **irreversible shock,** in contrast to **reversible shock,** which can be corrected by compensatory mechanisms and effective therapy.

Although the exact mechanism underlying irreversibility is not currently known, there are many logical possibilities that could contribute to the unrelenting, progressive circulatory deterioration that characterizes irreversible shock. Metabolic acidosis arises when lactic acid production increases as blood-deprived tissues resort to anaerobic metabolism. Acidosis deranges the enzymatic systems responsible for energy production, limiting the capability of the heart and other tissues to produce ATP. Prolonged depression of kidney function results in electrolyte imbalances that may lead to cardiac arrhythmias. The blood-deprived pancreas releases a chemical that is toxic to the heart **(myocardial toxic factor),** further weakening the heart. Vasodilator substances build up within ischemic organs, inducing local vasodilation that overrides the generalized reflex vasoconstriction. As cardiac output progressively declines because of the heart's diminishing effectiveness as a pump and total peripheral resistance continues to fall, hypotension becomes increasingly more severe. This causes further cardiovascular failure, which leads to a further decline in blood pressure. Thus, when shock progresses to the point that the cardiovascular system itself starts to fail, a vicious positive-feedback cycle ensues that ultimately results in death.

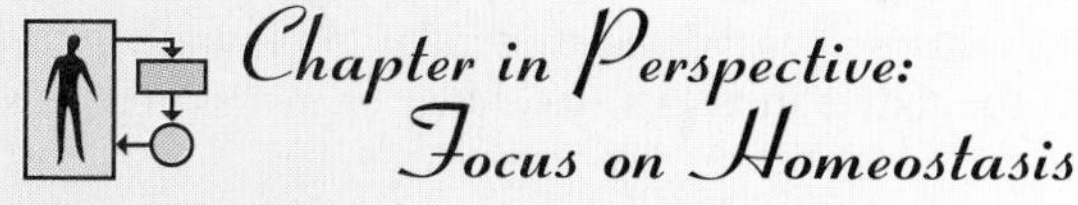

Chapter in Perspective: Focus on Homeostasis

Homeostatically, the blood vessels serve as passageways to transport blood to and from the cells for the purpose of O_2 and nutrient delivery, waste removal, distribution of fluid and electrolytes, and hormonal signaling, among other things. Cells soon die if deprived of their blood supply, with brain cells succumbing within four minutes. Blood is constantly recycled and reconditioned as it travels through the various organs by means of the vascular highways; hence, the body needs only a very small volume of blood to maintain the appropriate chemical composition of the entire internal fluid environment on which the cells depend for their survival. For example, O_2 is continually picked up by the blood in the lungs and constantly delivered to all the body cells.

The smallest of blood vessels, the capillaries, are the actual site of exchange between the blood and surrounding cells. Capillaries bring homeostatically maintained blood within 0.01 cm of every cell in the body; this proximity is critical because beyond a few centimeters, materials cannot diffuse rapidly enough to support life-sustaining activities. Oxygen that would take months to years to diffuse from the lungs to all the cells of the body is continuously delivered at the "doorstep" of every cell, where diffusion can efficiently accomplish short local exchanges between the capillaries and surrounding cells. Likewise, hormones must be rapidly transported through the circulatory system from their sites of production in endocrine glands to their sites of action in other parts of the body because they could not diffuse nearly rapidly enough to effectively exert their controlling effects, many of which are aimed toward maintaining homeostasis.

The remainder of the circulatory system is designed to transport blood to and from the capillaries. The arteries and arterioles distribute blood pumped by the heart to the capillaries for life-sustaining exchanges to take place, and the venules and veins collect blood from the capillaries and return it to the heart, where the process is repeated.

Chapter Summary

Introduction

Materials can be exchanged between various parts of the body and with the external environment by means of the blood vessel network that transports blood to and from all tissues. Organs that replenish nutrient supplies and remove metabolic wastes from the blood receive a greater percentage of the cardiac output than is warranted by their metabolic needs. These "reconditioning" organs can better tolerate reductions in blood supply than can organs that receive blood solely for the purpose of meeting their own metabolic needs. The brain is especially vulnerable to reductions in its blood supply. Therefore, the maintenance of adequate flow to this vulnerable organ is a high priority in circulatory function.

Blood flows in a closed loop between the heart and the tissues. The arteries transport blood from the heart throughout the body. The arterioles regulate the amount of blood that flows through each organ. The capillaries are the actual site where materials are exchanged between the blood and surrounding tissue. The veins return the blood from the tissues to the heart.

The flow rate of blood through a vessel is directly proportional to the pressure gradient and inversely proportional to the resistance. The higher pressure at the beginning of a vessel is established by the pressure imparted to the blood by cardiac contraction. The lower pressure at the end is due to frictional losses as flowing blood rubs against the vessel wall. Resistance, the hindrance to blood flow through a vessel, is influenced most by the vessel's radius. Resistance is inversely proportional to the fourth power of the radius, so small changes in radius profoundly influence flow. As the radius increases, resistance decreases and flow increases.

Arteries

Arteries are large-radius, low-resistance passageways from the heart to the tissues; they also serve as a pressure reservoir. Because of their elasticity, arteries expand to accommodate the extra volume of blood pumped into them by cardiac contraction and then recoil to continue driving the blood forward when the heart is relaxing.

Systolic pressure is the peak pressure exerted by the ejected blood against the vessel walls during cardiac systole. Diastolic pressure is the minimum pressure in the arteries when blood is draining off into the vessels downstream during cardiac diastole.

The average driving pressure throughout the cardiac cycle is the mean arterial pressure, which can be estimated using the following formula: mean arterial pressure = diastolic pressure + ⅓ pulse pressure.

Arterioles

Arterioles are the major resistance vessels. Their high resistance produces a large drop in mean pressure between the arteries and capillaries. This decline enhances blood flow by contributing to the pressure differential between the heart and the tissues. Tone, a baseline of contractile activity, is maintained in arterioles at all times. Arteriolar vasodilation, an expansion of arteriolar caliber above tonic level, decreases resistance and increases blood flow through the vessel, whereas vasoconstriction, a narrowing of the vessel, increases resistance and decreases flow.

Arteriolar caliber is subject to two types of control mechanisms: local (intrinsic) controls and extrinsic controls. Local controls involve local chemical changes associated with changes in the level of metabolic activity in a tissue; these controls act directly on the arteriolar smooth muscle in the vicinity to induce changes in the caliber of the arterioles supplying the tissue. By adjusting the resistance to blood flow in this manner, the local control mechanism adjusts blood flow to the tissue to match the momentary metabolic needs of the tissue. Adjustments in arteriolar caliber can be accomplished independently in different tissues by local control factors. Such adjustments are important in determining the distribution of cardiac output.

Extrinsic control is accomplished primarily by sympathetic nerve influence and to a lesser extent by hormonal influence over arteriolar smooth muscle. Extrinsic controls are important in maintaining mean arterial blood pressure. Arterioles are richly supplied with sympathetic nerve fibers, whose increased activity produces generalized vasoconstriction and a subsequent increase in mean arterial pressure. Decreased sympathetic activity produces generalized arteriolar vasodilation, which lowers mean arterial pressure. These extrinsically controlled adjustments of arteriolar caliber help maintain the appropriate pressure head for driving blood forward to the tissues.

Capillaries

The thin-walled, small-radius, extensively branched capillaries are ideally suited to serve as sites of exchange between the blood and surrounding tissues. Anatomically, the surface area for exchange is maximized and diffusion distance is minimized in the capillaries. Furthermore, because of their large total cross-sectional area, the velocity of blood flow through capillaries is relatively slow, providing adequate time for exchanges to take place.

Two types of passive exchanges—diffusion and bulk flow—take place across capillary walls. Individual solutes are exchanged primarily by diffusion down concentration gradients. Lipid-soluble substances pass directly through the single layer of endothelial cells lining a capillary, whereas water-soluble substances pass through water-filled pores between the endothelial cells. Plasma proteins generally do not escape.

Imbalances in physical pressures acting across capillary walls are responsible for bulk flow of fluid through the pores back and forth between the plasma and interstitial fluid. Fluid is forced out of the first portion of the capillary (ultrafiltration), where outward pressures (mainly capillary blood pressure) exceed inward pressures (mainly plasma-colloid osmotic pressure). Fluid is returned to the capillary along its last half, when outward pressures fall below inward pressures. The reason for the shift in balance down the length of the capillary is the continuous decline in capillary blood pressure while the plasma-colloid osmotic pressure remains constant. Bulk flow is responsible for the distribution of extracellular fluid between the plasma and the interstitial fluid.

Normally, slightly more fluid is filtered than is reabsorbed. The extra fluid, any leaked proteins, and tissue contaminants such as bacteria are picked up by the lymphatic system. Bacteria are destroyed as lymph passes through the lymph nodes en route to being returned to the venous system.

Veins

Veins are large-radius, low-resistance passageways for return of blood from the tissues to the heart. Additionally, they can accommodate variable volumes of blood and therefore act as a blood reservoir. The capacity of veins to hold blood can change markedly with little change in venous pressure. Veins are thin-walled, highly distensible vessels that can passively stretch to store a larger volume of blood.

The primary force responsible for venous flow is the pressure gradient between the veins and atrium (that is, what remains of the driving pressure imparted to the blood by cardiac contraction). Venous return is enhanced by sympathetically induced venous vasoconstriction and by external compression of the veins resulting from contraction of surrounding skeletal muscles, both of which drive blood out of the veins. One-way venous valves ensure that blood is driven toward the heart and prevented from flowing back toward the tissues. Venous return is also enhanced by the respiratory pump and the cardiac-suction effect. Respiratory activity produces a less-than-atmospheric pressure in the chest cavity, thus establishing an external pressure gradient that encourages flow from the lower veins that are exposed to atmospheric pressure to the chest veins that empty into the heart. In addition, slightly negative pressures created within the atria during ventricular systole and within the ventricles during ventricular diastole exert a suctioning effect that further enhances venous return and facilitates cardiac filling.

Blood Pressure

Regulation of mean arterial pressure depends on control of its two main determinants, cardiac output and total peripheral resistance. Control of cardiac output in turn depends on regulation of heart rate and stroke volume, whereas total peripheral resistance is determined primarily by the degree of arteriolar vasoconstriction. Short-term regulation of blood pressure is accomplished primarily by the baroreceptor reflex. Carotid sinus and aortic arch baroreceptors continuously monitor mean arterial pressure. When they detect a deviation from normal, they signal the medullary cardiovascular center, which responds by adjusting autonomic output to the heart and blood vessels to restore the blood pressure to normal. Long-term control of blood pressure involves maintenance of proper plasma volume through the kidneys' control of salt and water balance.

Blood pressure can be abnormally high (hypertension) or abnormally low (hypotension). Severe sustained hypotension resulting in generalized inadequate blood delivery to the tissues is known as circulatory shock.

Review Exercises

Objective Questions (Answers on p. E–9.)

1. In general, the parallel arrangement of the vascular system enables each organ to receive its own separate arterial blood supply. (True or false?)
2. More blood flows through the capillaries during cardiac systole than during diastole. (True or false?)
3. The capillaries contain only 5% of the total blood volume at any point in time. (True or false?)
4. The same volume of blood passes through the capillaries in a minute as passes through the aorta, even though the velocity of blood flow is much slower in the capillaries. (True or false?)
5. Because there are no carrier transport systems in capillary walls, all capillaries are equally permeable. (True or false?)
6. Because of gravitational effects, venous pressure in the lower extremities is greater when a person is standing up than when the person is lying down. (True or false?)

7. Which of the following functions is (are) attributable to arterioles? (Indicate all correct answers.)
   a. responsible for a significant decline in mean pressure, which helps establish the driving pressure gradient between the heart and tissues
   b. site of exchange of materials between the blood and surrounding tissues
   c. main determinant of total peripheral resistance
   d. determine the pattern of distribution of cardiac output
   e. play a role in regulating mean arterial blood pressure
   f. convert the pulsatile nature of arterial blood pressure into a smooth, nonfluctuating pressure in the vessels further downstream
   g. act as a pressure reservoir
8. Using the answer code below, indicate what kind of compensatory changes occur in the factors in question to restore the blood pressure to normal in response to hypovolemic hypotension resulting from severe hemorrhage:

   (a) increased
   (b) decreased
   (c) no effect

   ___ 1. rate of afferent firing generated by the carotid sinus and aortic arch baroreceptors
   ___ 2. sympathetic output by the cardiovascular center
   ___ 3. parasympathetic output by the cardiovascular center
   ___ 4. heart rate
   ___ 5. stroke volume
   ___ 6. cardiac output
   ___ 7. arteriolar radius
   ___ 8. total peripheral resistance
   ___ 9. venous radius
   ___ 10. venous return
   ___ 11. urinary output
   ___ 12. fluid retention within the body
   ___ 13. fluid movement from the interstitial fluid into the plasma across the capillaries
9. Using the answer code below, indicate whether the following factors increase or decrease venous return:

   (a) increases venous return
   (b) decreases venous return
   (c) has no effect on venous return

   ___ 1. sympathetically induced venous vasoconstriction
   ___ 2. skeletal muscle activity
   ___ 3. gravitational effects on the venous system
   ___ 4. respiratory activity
   ___ 5. increased atrial pressure associated with a leaky AV valve
   ___ 6. ventricular pressure change associated with diastolic recoil

Essay Questions

1. Compare blood flow through reconditioning organs and through organs that do not recondition the blood.
2. Discuss the relationships among flow rate, pressure gradient, and vascular resistance. What is the major determinant of resistance to flow?
3. Describe the structure and major functions of each segment of the vascular tree.
4. How do the arteries serve as a pressure reservoir?
5. Describe the indirect technique of measuring arterial blood pressure by means of a sphygmomanometer.
6. Define vasoconstriction and vasodilation.
7. Discuss the local and extrinsic controls that regulate arteriolar resistance.
8. What is the primary means by which individual solutes are exchanged across the capillary walls? What forces are responsible for bulk flow across the capillary walls? Of what importance is bulk flow?
9. How is lymph formed? What are the functions of the lymphatic system?
10. Define edema and discuss its possible causes.
11. How do veins serve as a blood reservoir?
12. Compare the effect of vasoconstriction on the rate of blood flow in arterioles and veins.
13. Discuss the factors that determine mean arterial pressure.
14. Review the effects on the cardiovascular system of parasympathetic and sympathetic stimulation.
15. Differentiate between secondary hypertension and primary hypertension. What are the potential consequences of hypertension?
16. Define circulatory shock. What are its consequences and compensations? What is irreversible shock?

Quantitative Exercises (Solutions on p. E–9.)

1. Recall that the flow rate of blood equals the pressure gradient divided by the total peripheral resistance of the vascular system. The conventional unit of resistance in physiological systems is expressed in PRU (peripheral resistance unit), which is defined as (1 liter/min)/(1 mm Hg). At rest, Tom's total peripheral resistance is about 20 PRU. Last week while playing racquetball, his cardiac output increased to 30 liters/min and his mean arterial pressure increased to 120 mm Hg. What was his total peripheral resistance during the game?
2. Systolic pressure rises as a person ages. By age 85, an average male has a systolic pressure of 180 mm Hg and a diastolic pressure of 90 mm Hg.
   a. What is the mean arterial pressure of this average 85-year-old male?
   b. From your knowledge of capillary dynamics, predict the result at the capillary level of this age-related change in mean arterial pressure if no homeostatic mechanisms were operating. (Recall that mean arterial pressure is about 93 mm Hg at age 20.)

3. Compare the flow rates in the systemic circulation and the pulmonary circulation of an individual with the following measurements:

   systemic mean arterial pressure = 95 mm Hg

   systemic resistance = 19 PRU

   pulmonary mean arterial pressure = 20 mm Hg

   pulmonary resistance = 4 PRU

4. Which of the following changes would increase the resistance in an arteriole?
   a. a longer length
   b. a smaller caliber
   c. increased sympathetic stimulation
   d. increased blood viscosity
   e. all of the above

Points to Ponder

(Explanations on p. E–9.)

1. During coronary bypass surgery, a piece of vein is removed from the patient's leg and surgically attached within the coronary circulatory system so that blood is detoured around an occluded coronary artery segment. For an extended period of time following surgery, why must the patient wear an elastic support stocking on the limb from which the vein was removed?
2. Assume a person has a blood pressure recording of 125/77:
   a. What is the systolic pressure?
   b. What is the diastolic pressure?
   c. What is the pulse pressure?
   d. What is the mean arterial pressure?
   e. Would any sound be heard when the pressure in an external cuff around the arm was 130 mm Hg? (Yes or no?)
   f. Would any sound be heard when cuff pressure was 118 mm Hg?
   g. Would any sound be heard when cuff pressure was 75 mm Hg?
3. A classmate who has been standing still for several hours working on a laboratory experiment suddenly faints. What is the probable explanation? What would you do if the person next to him tried to get him up?
4. A drug applied to a piece of excised arteriole causes the vessel to relax, but an isolated piece of arteriolar muscle stripped from the other layers of the vessel fails to respond to the same drug. What is the probable explanation?
5. Explain how each of the following antihypertensive drugs would lower arterial blood pressure:
   a. drugs that block alpha-adrenergic receptors (for example, *phentolamine*)
   b. drugs that block beta-adrenergic receptors (for example, *propranolol*) (Hint: More important than the action of these drugs on the heart and blood vessels in reducing arterial blood pressure is their inhibitory effect on the hormonal pathway involving angiotensin II that promotes salt and water conservation.)
   c. drugs that directly relax arteriolar smooth muscle (for example, *hydralazine*)
   d. diuretic drugs that increase urinary output (for example, *furosemide*)
   e. drugs that block release of norepinephrine from sympathetic endings (for example, *guanethidine*)
   f. drugs that act on the brain to reduce sympathetic output (for example, *clonidine*)
   g. drugs that block Ca^{2+} channels (for example, *verapamil*)
   h. drugs that interfere with the production of angiotensin II (for example, *captopril*)
6. ***Clinical Consideration*** Li-Ying C. has just been diagnosed as having hypertension secondary to a *pheochromocytoma,* a tumor of the adrenal medulla that secretes excessive epinephrine. Explain how this condition leads to secondary hypertension by describing the effect that excessive epinephrine would have on the various factors that determine arterial blood pressure.

11 The Blood

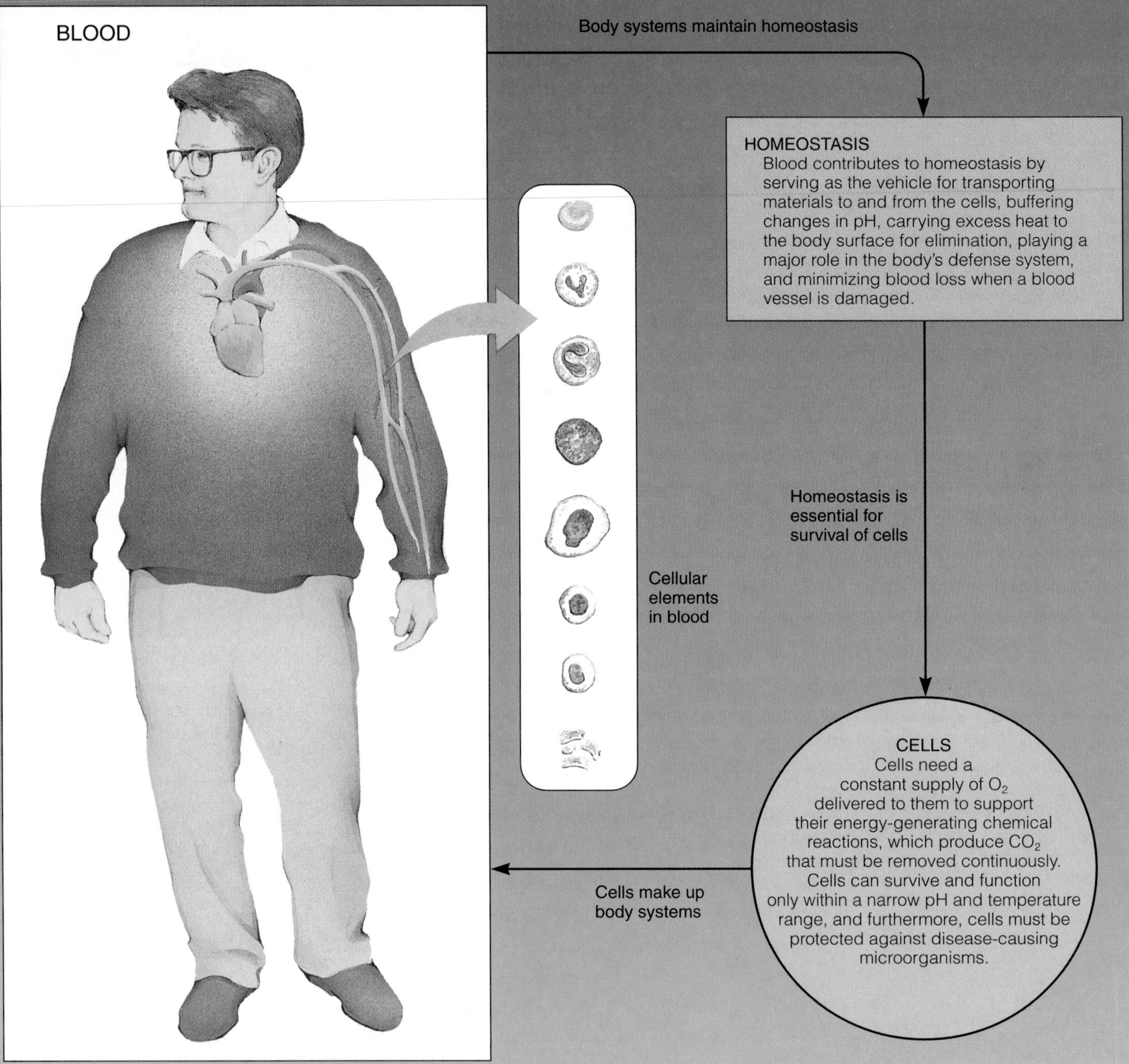

BLOOD

Body systems maintain homeostasis

HOMEOSTASIS
Blood contributes to homeostasis by serving as the vehicle for transporting materials to and from the cells, buffering changes in pH, carrying excess heat to the body surface for elimination, playing a major role in the body's defense system, and minimizing blood loss when a blood vessel is damaged.

Homeostasis is essential for survival of cells

Cellular elements in blood

CELLS
Cells need a constant supply of O_2 delivered to them to support their energy-generating chemical reactions, which produce CO_2 that must be removed continuously. Cells can survive and function only within a narrow pH and temperature range, and furthermore, cells must be protected against disease-causing microorganisms.

Cells make up body systems

Blood is the vehicle for long-distance, mass transport of materials between the cells and the external environment or between the cells themselves. Such transport is essential for maintenance of homeostasis. Blood consists of a complex liquid **plasma** in which the cellular elements—*erythrocytes, leukocytes,* and *platelets*—are suspended.

Erythrocytes (red blood cells or **rbcs)** are essentially plasma membrane–enclosed bags of **hemoglobin** that transport O_2 and, to a lesser extent, CO_2 in the blood. **Leukocytes (white blood cells** or **wbcs)**, the immune system's mobile defense units, are transported in the blood to sites of injury or invasion by disease-causing microorganisms.

Given the importance of blood, it is imperative that mechanisms exist to minimize blood loss when a vessel is injured. **Platelets (thrombocytes)** are important in **hemostasis**, the stopping of bleeding from an injured vessel.

Chapter Contents At a Glance

Introduction

Blood represents about 8% of total body weight and has an average volume of 5 liters in women and 5.5 liters in men. It consists of three types of specialized cellular elements, *erythrocytes, leukocytes,* and *platelets,* suspended in the complex liquid *plasma* (Table 11-1). The constant movement of blood as it flows through the blood vessels keeps its cellular elements rather evenly dispersed within the plasma. However, if a sample of whole blood is placed in a test tube and treated to prevent clotting, the heavier cellular elements slowly settle to the bottom and the lighter plasma rises to the top. This process can be hastened by centrifugation, which rapidly packs the cells in the bottom of the tube (Fig. 11-1). Because over 99% of the cells are erythrocytes, the **hematocrit,** or **packed cell volume,** essentially represents the percentage of total blood volume occupied by erythrocytes. Plasma accounts for the remaining volume. The hematocrit averages 42% for women and slightly higher, 45%, for men, with the average volume occupied by plasma being 58% for women and 55% for men. The white blood cells and platelets, which are colorless and less dense than red cells, are packed in a thin, cream-colored layer, the "buffy coat," on top of the packed red cell column. They represent less than 1% of the total blood volume. We will first consider the properties of the largest portion of the blood, the plasma, before turning our attention to the cellular elements.

Plasma

Many of the functions of plasma are carried out by plasma proteins.

Plasma, being a liquid, is composed of 90% water, which serves as a medium for materials being carried in the blood. Also, because water has a high capacity to hold heat, plasma is able to absorb and distribute much of the heat generated metabolically within tissues while the temperature of the blood itself undergoes only small changes. Heat energy not needed to maintain body temperature is eliminated to the environment as the blood travels close to the surface of the skin.

A large number of organic and inorganic substances are dissolved in the plasma. The most plentiful organic constituents by weight are the plasma proteins, which compose

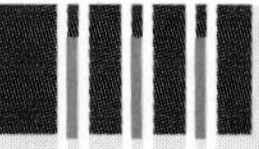

Table 11-1
Blood Constituents and Their Functions

Constituent	*Functions*
Plasma	
Water	Transport medium; carries heat
Electrolytes	Membrane excitability; osmotic distribution of fluid between the extracellular and intracellular fluid; buffering of pH changes
Nutrients, wastes, gases, hormones	Transported in blood; the blood gas CO_2 plays a role in acid-base balance
Plasma proteins	In general, exert an osmotic effect that is important in the distribution of extracellular fluid between the vascular and interstitial compartments; buffering of pH changes
Albumins	Transport many substances; make the greatest contribution to colloid osmotic pressure
Globulins	
Alpha and beta	Transport many substances; clotting factors; inactive precursor molecules
Gamma	Antibodies
Fibrinogen	Inactive precursor for the fibrin meshwork of a clot
Cellular elements	
Erythrocytes	Transport O_2 and CO_2 (mainly O_2)
Leukocytes	
Neutrophils	Phagocytes that engulf bacteria and debris
Eosinophils	Attack parasitic worms; important in allergic reactions
Basophils	Release histamine, which is important in allergic reactions, and heparin, which helps clear fat from the blood and may function as an anticoagulant
Monocytes	In transit to become tissue macrophages
Lymphocytes	
B lymphocytes	Production of antibodies
T lymphocytes	Cell-mediated immune responses
Platelets	Hemostasis

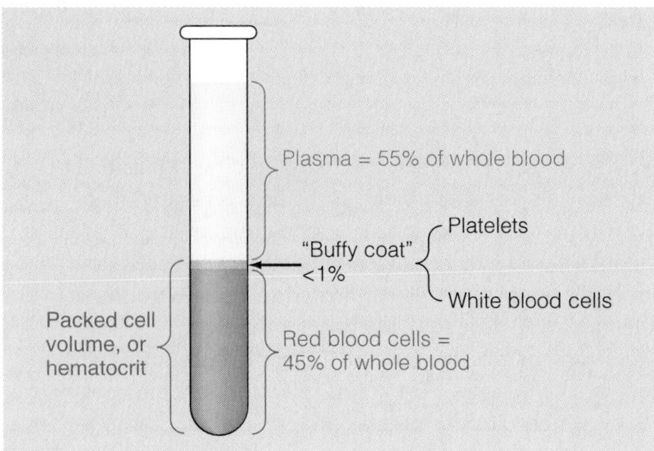

Figure 11-1 **Hematocrit** The values given are for men. The average hematocrit for women is 42%, with plasma occupying 58% of the blood volume.

6% to 8% of plasma's total weight. Inorganic constituents account for approximately 1% of plasma weight. The most abundant electrolytes (ions) in the plasma are Na^+ and Cl^-, which are the components of common salt. There are lesser amounts of HCO_3^-, K^+, Ca^{2+}, and others. The most notable functions of these extracellular fluid (ECF) ions are their roles in membrane excitability, osmotic distribution of fluid between the ECF and cells, and buffering of pH changes; these functions are discussed elsewhere. The remaining small percentage of plasma is occupied by nutrients (for example, glucose, amino acids, lipids, and vitamins); waste products (creatinine, bilirubin, and nitrogenous substances such as urea); dissolved gases (O_2 and CO_2); and hormones. Most of these substances are merely being transported in the plasma. For example, endocrine glands secrete hormones into the plasma, which transports these chemical messengers to their sites of action.

The **plasma proteins** are the one group of plasma constituents not present just for the ride. These important components normally remain in the plasma, where they perform many valuable functions. Because they are the largest of the plasma constituents, plasma proteins usually do not exit through the narrow pores in the capillary walls. Also, unlike other plasma constituents that are dissolved in the plasma water, the plasma proteins exist in a colloidal dispersion (see p. A–8).

There are three groups of plasma proteins—*albumins, globulins,* and *fibrinogen*—which are classified according to their various physical and chemical properties. The functions of the plasma proteins are elaborated on elsewhere in the text, but the following list illustrates the wide range of these functions:

- By virtue of their presence as a colloidal dispersion in the plasma and their absence in the interstitial fluid, plasma proteins establish an osmotic gradient between blood and interstitial fluid. This colloid osmotic pressure is the primary force responsible for preventing excessive loss of plasma from the capillaries into the interstitial fluid and thus helps maintain plasma volume (see p. 328).
- Plasma proteins are partially responsible for the plasma's capacity to buffer changes in pH (see p. 533).
- Plasma proteins contribute to blood viscosity, but erythrocytes are far more important in this regard (see p. 308).
- Plasma proteins are not normally used as metabolic fuels, but in a state of starvation they can be degraded to provide energy for cells.
- In addition to these general functions, each type of plasma protein performs important specific tasks:

1. **Albumins,** the most abundant of the plasma proteins, bind many substances (for example, bilirubin, bile salts, and penicillin) for transport through the plasma and contribute most extensively to the colloid osmotic pressure by virtue of their numbers.
2. There are three subclasses of **globulins: alpha (α), beta (β),** and **gamma (τ).**
   a. Specific alpha and beta globulins bind and transport a number of substances in the plasma, such as thyroid hormone, cholesterol, and iron.
   b. Many of the factors involved in the process of blood clotting, which will be described shortly, are alpha or beta globulins.
   c. Inactive precursor protein molecules, which are activated as needed by specific regulatory inputs, belong to the alpha-globulin group (for example, the alpha globulin angiotensinogen is activated to angiotensin, which plays an important role in the regulation of salt balance in the body, see p. 485).
   d. The gamma globulins are the immunogolobulins (antibodies), which are crucial to the body's defense mechanism (see p. 386).
3. **Fibrinogen** is a key factor in the blood-clotting process.

The plasma proteins generally are synthesized by the liver, with the exception of the gamma globulins, which are produced by lymphocytes, one of the types of white blood cells.

Erythrocytes

The structure of erythrocytes is well suited to their primary function of oxygen transport in the blood.

Each milliliter of blood contains about 5 billion **erythrocytes (red blood cells)** on average, commonly reported clinically in a **red blood cell count** as 5 million cells per cubic millimeter (mm^3). Erythrocytes are flat, disc-shaped cells indented in the middle on both sides, like a doughnut with a flattened center instead of a hole (they are biconcave discs 8 μm in diameter, 2 μm thick at the outer edges, and 1 μm thick in the center) (Fig. 11-2). This unique shape contributes in two ways to the efficiency with which erythrocytes perform their main function of O_2 transport in the blood. First, the biconcave shape provides a larger surface area for diffusion of O_2 across the membrane than a spherical cell of the same volume would provide. Second, the thinness of the cell enables O_2 to diffuse rapidly between the exterior and innermost regions of the cell.

Another feature of erythrocytes that facilitates their transport function is the flexibility of their membrane, which enables them to travel through the narrow, tortuous capillaries to deliver their O_2 cargo at the tissue level without rupturing in the process. Red blood cells, whose diameter is normally 8 μm, are able to deform amazingly as they squeeze single file through capillaries as narrow as 3 μm in diameter.

The most important feature that enables erythrocytes to transport O_2 is the **hemoglobin** they contain. A hemoglobin molecule consists of two parts: (1) the **globin portion,** a protein made up of four highly folded polypeptide chains, and (2) four iron-containing, nonprotein nitrogenous groups known as **heme** groups, each of which is bound to one of the polypeptides (Fig. 11-3). Each of the four iron atoms can combine reversibly with one molecule of O_2; thus, each hemoglobin molecule can pick up four O_2 passengers. Because O_2 is poorly soluble in the plasma, 98.5% of the O_2 carried in the blood is bound to hemoglobin. Hemoglobin is a pigment (that is, it is naturally colored). Because of its iron content, it appears reddish when combined with O_2 and bluish when deoxygenated. Thus, fully oxygenated arterial blood is red in color, and venous blood, which has lost some of its O_2 load at the tissue level, has a bluish cast.

In addition to carrying O_2, hemoglobin can also combine with the following:

1. *Carbon dioxide.* Hemoglobin contributes to the transport of this gas from the tissues back to the lungs.
2. *The acidic hydrogen-ion portion* (H^+) *of ionized carbonic acid,* which is generated at the tissue level from CO_2. Hemoglobin buffers this acid so that it minimally alters the pH of the blood.
3. *Carbon monoxide (CO).* This gas is not normally in the blood but, if inhaled, preferentially occupies the O_2-binding sites on hemoglobin, causing carbon monoxide poisoning.

Therefore, hemoglobin plays the key role in O_2 transport while contributing significantly to CO_2 transport and the buffering capacity of blood.

To maximize its hemoglobin content, a single erythrocyte is stuffed with several hundred million hemoglobin molecules to the exclusion of almost everything else. Erythrocytes contain no nucleus, organelles, or ribosomes. These structures are extruded during the cell's development to make room for more hemoglobin. Thus, a red blood cell is mainly a plasma membrane–enclosed sac full of hemoglobin.

Only a few crucial, nonrenewable enzymes remain within a mature erythrocyte: these are glycolytic enzymes and carbonic anhydrase. The **glycolytic enzymes** are necessary for generating the energy needed to fuel the active transport mechanisms involved in maintaining proper ionic concentrations within the cell. Ironically, even though erythrocytes are the vehicles for transport of O_2 to all other tissues of the body, they themselves cannot use the O_2 they are carrying for energy production. Erythrocytes, lacking the mitochondria that house the enzymes for oxidative phosphorylation, must rely entirely on glycolysis for ATP formation (see p. 28).

The other important enzyme within red blood cells, **carbonic anhydrase,** is critical in CO_2 transport. This enzyme cat-

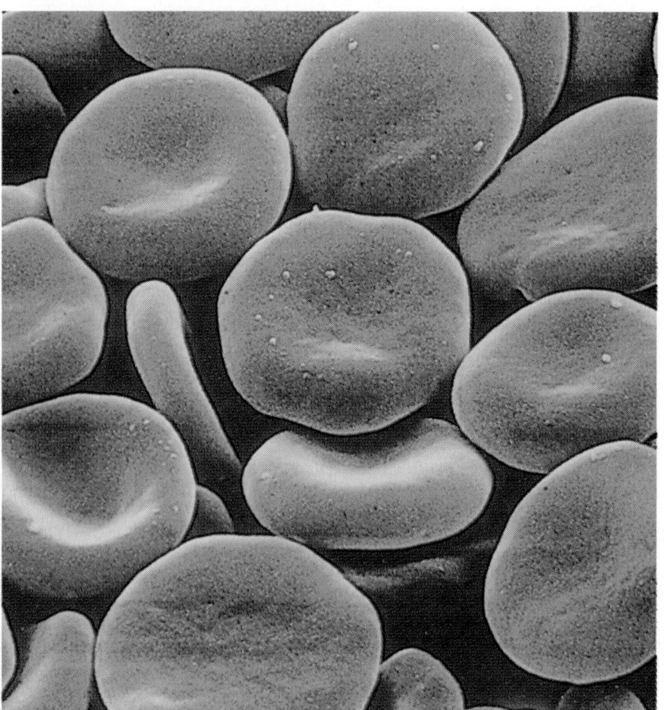

Figure 11-2 **Anatomical Characteristics of Erythrocytes** Appearance of erythrocytes under a scanning electron microscope. Note their biconcave shape.

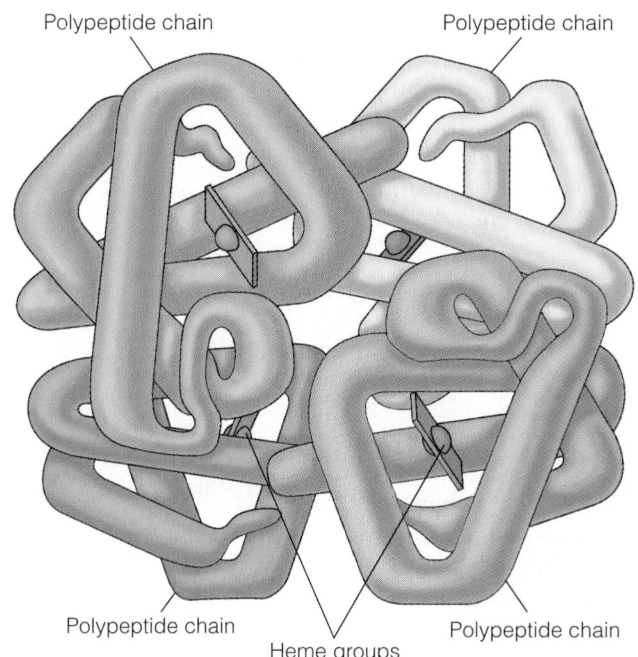

Figure 11-3 **Hemoglobin Molecule** A hemoglobin molecule consists of four highly folded polypeptide chains (the globin portion) and four iron-containing heme groups.

alyzes a key reaction that ultimately leads to the conversion of metabolically produced CO_2 into **bicarbonate ion (HCO_3^-)**, which is the primary form in which CO_2 is transported in the blood. Thus, erythrocytes contribute to CO_2 transport in two ways—by means of its carriage on hemoglobin and its carbonic anhydrase–induced conversion to HCO_3^-.

The bone marrow continuously replaces worn-out erythrocytes.

Each of us has a total of 25 to 30 trillion red blood cells streaming through our blood vessels at any given time (100,000 times more than the entire population of the United States)! Yet, these vital gas transport vehicles are short-lived and must be replaced at the average rate of 2 to 3 million cells per second. The price erythrocytes pay for their generous content of hemoglobin to the exclusion of the usual specialized intracellular machinery is a shortened life span. Without DNA and RNA, red blood cells cannot synthesize proteins for cellular repair, growth, and division or for renewal of enzyme supplies. Equipped only with initial supplies synthesized before extrusion of their nucleus, organelles, and ribosomes, erythrocytes are able to survive an average of only 120 days, in contrast to nerve and muscle cells, which last a person's entire life. During its short life span of four months, each erythrocyte travels about 700 miles as it circulates through the vasculature. As a red blood cell ages, its nonreparable plasma membrane becomes fragile and prone to rupture as the cell squeezes through tight spots in the vascular system.

Most old red blood cells meet their final demise in the **spleen,** because this organ's narrow, winding capillary network is a tight fit for these fragile cells. The spleen lies in the upper left part of the abdomen. In addition to removing most of the old erythrocytes from circulation, the spleen has a limited ability to store healthy erythrocytes in its pulpy interior; serves as a reservoir site for platelets; and contains an abundance of lymphocytes, a type of white blood cell.

Because erythrocytes cannot divide to replenish their own numbers, the old ruptured cells must be replaced by new cells produced in an erythrocyte factory—the **bone marrow**—which is the soft, highly cellular tissue that fills the internal cavities of bones. The bone marrow normally generates new red blood cells, a process known as **erythropoiesis,** at the amazing rate of 2 to 3 million per second to keep pace with the demolition of old cells.

During intrauterine development, erythrocytes are produced first by the yolk sac and then by the developing liver and spleen, until the bone marrow is formed and takes over erythrocyte production exclusively. In children most bones are filled with *red bone marrow* that is capable of blood cell production. As a person matures, however, fatty *yellow bone marrow* that is incapable of erythropoiesis gradually replaces red marrow, which remains only in the sternum (breastbone), vertebrae (backbone), ribs, base of the skull, and upper ends of the long limb bones. Red marrow not only produces red blood cells but is the ultimate source for leukocytes and platelets as well. Undifferentiated **pluripotent stem cells** reside in the red marrow, where they continuously divide and differentiate to give rise to each of the types of blood cells. The different types of cells, along with the stem cells, are intermingled in the red marrow at various stages of development. The mature cells are released into the rich supply of capillaries that permeate the red marrow. Regulatory factors act on the *hemopoietic* ("blood-producing") marrow to govern the type and number of cells generated and discharged into the blood. Of the blood cells, the mechanism for regulating red blood cell production is the best understood. We will consider it now.

Figure 11-4 Control of Erythropoiesis When less O_2 is delivered to the kidneys, they secrete the hormone erythropoietin into the blood. Erythropoietin in turn stimulates erythropoiesis (erthrocyte production) by the bone marrow. The additional circulating erythrocytes increase the O_2-carrying capacity of the blood, thus relieving the initial stimulus that triggered erythropoietin secretion.

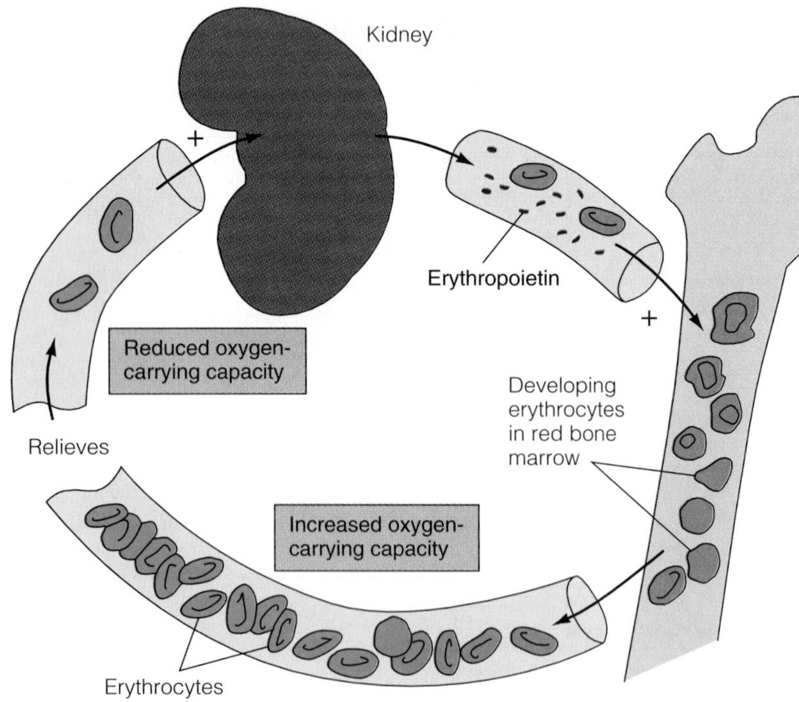

Erythropoiesis is controlled by erythropoietin from the kidneys.

The number of circulating erythrocytes normally remains fairly constant, indicative that erythropoiesis must be closely regulated. Because O_2 transport in the blood is the erythrocytes' primary function, you might logically suspect that the primary stimulus for increased erythrocyte production would be reduced O_2 delivery to the tissues. You would be correct, but low O_2 levels do not stimulate erythropoiesis by acting directly on the red bone marrow. Instead, reduced O_2 delivery to the kidneys stimulates them to secrete the hormone **erythropoietin** into the blood, and this hormone in turn stimulates erythropoiesis by the bone marrow (Fig. 11-4). Erythropoietin acts on derivatives of undifferentiated stem

Blood Doping: Is More of a Good Thing Better?

Exercising muscles require a continual supply of O_2 for generating energy to sustain endurance activities (see p. 33). Blood doping is a technique designed to temporarily increase the O_2-carrying capacity of the blood in an attempt to gain a competitive advantage. Blood doping involves removing blood from an athlete, then promptly reinfusing the plasma but freezing the red blood cells for reinfusion one to seven days before a competitive event. One to four units of blood (one unit equals 450 ml) are usually withdrawn at three- to eight-week intervals before the competition. In the periods between blood withdrawals, increased erythropoietic activity restores the red blood cell count to a normal level.

Reinfusion of the stored red blood cells temporarily increases the red blood cell count and hemoglobin level above normal. Theoretically, blood doping would benefit endurance athletes by improving the blood's O_2-carrying capacity. If too many red cells were infused, however, performance could suffer because the increased blood viscosity would decrease blood flow.

Early research into athletic performance after blood doping produced contradictory results. More recent research, however, indicates that, in a standard laboratory exercise test, athletes may realize a 5% to 13% increase in aerobic capacity; a reduction in heart rate during exercise compared to the rate during the same exercise in the absence of blood doping; improved performance; and reduced lactic acid levels in the blood. (Lactic acid is produced when muscles resort to less efficient anaerobic glycolysis for energy production—see page 33.)

Blood doping, although probably effective, is illegal in both collegiate athletics and Olympic competition for ethical and medical reasons. These prohibitive regulations are very difficult to enforce, however. Current procedures permit only the testing of urine, not blood, for the use of illegal drugs among athletes. Blood doping cannot be detected by means of a urine sample. The only way of exposing the practice of blood doping is through witnesses or self-admission.

The recent development of synthetic erythropoietin magnifies the problem of blood doping. Injection of synthetic erythropoietin stimulates red blood cell production and thus temporarily increases the O_2-carrying capacity of the blood. Use of erythropoietin is more convenient for the athlete and minimizes the number of witnesses involved, thereby reducing the probability of disclosure.

cells that are already committed to becoming red blood cells, stimulating their proliferation and maturation into mature erythrocytes. This increased erythropoietic activity elevates the number of circulating red blood cells, thereby increasing the O_2-carrying capacity of the blood and restoring O_2 delivery to the tissues to normal. Once normal O_2 delivery to the kidneys is achieved, erythropoietin secretion is turned down until needed once again. In this way, erythrocyte production is normally balanced against destruction or loss of these cells so that O_2-carrying capacity in the blood remains fairly constant. (See the accompanying boxed feature, ◆ A Closer Look at Exercise Physiology.) In response to severe loss of erythrocytes, as in hemorrhage or abnormal destruction of young circulating erythrocytes, the rate of erythropoiesis can be increased to more than six times the normal level.

The preparation of an erythrocyte for its departure from the marrow involves several steps, such as synthesis of hemoglobin and extrusion of the nucleus and organelles. Cells closest to maturity need a few days to be "finished off" and released into the blood in response to erythropoietin, and less-developed and newly proliferated cells may take up to several weeks to reach maturity. Therefore, the time required for complete replacement of lost red blood cells depends on how many are needed to return the number to normal. (When you donate blood, your circulating erythrocyte supply is replenished in less than a week.) When demands for red blood cell production are high (for example, following hemorrhage), the bone marrow may release large numbers of immature erythrocytes, known as **reticulocytes,** into the blood to quickly meet the need. These immature cells can be recognized by staining techniques that make visible the residual ribosomes and organelle remnants that have not yet been extruded. Their presence above the normal level of 0.5% to 1.5% of the total number of circulating erythrocytes is indicative of a high rate of erythropoietic activity. At very rapid rates, more than 30% of the circulating red cells may be at the immature reticulocyte stage.

The gene that directs erythropoietin synthesis has recently been identified and cloned so this hormone can now be produced in abundance in a laboratory. This capability has had a profound impact on the need for blood transfusions. For example, transfusion of a surgical patient's own predonated blood, coupled with administration of erythropoietin to stimulate further red blood cell production, has reduced the use of donor blood by as much as 50% in some hospitals. (See the boxed feature on p. 358, ◆ Concepts, Challenges, and Controversies, for an update on other alternatives to whole-blood transfusions under investigation.)

In addition to erythropoietin, which is the primary regulatory factor that adjusts the ongoing rate of red blood cell production as needed to keep erythrocyte levels constant, testosterone, the major male sex hormone, increases the basal rate of erythropoiesis. This hormone is undoubtedly responsible, at least in part, for the normally larger hematocrit in males compared to females. The additional red blood cells in males provide extra O_2-carrying capacity to meet the needs of the larger muscle mass in men, for which testosterone is also responsible.

destined to become leukocytes eventually differentiate into various committed cell lines and proliferate under the influence of appropriate stimulating factors. Granulocytes and monocytes are produced only in the bone marrow, which releases these mature leukocytes into the blood. Lymphocytes are originally derived from precursor cells in the bone marrow, but most new lymphocytes are actually produced by already-existing lymphocytes residing in the **lymphoid** (lymphocyte-containing) **tissues,** such as the lymph nodes and tonsils.

The total number of leukocytes normally ranges from 5 to 10 million cells per milliliter of blood, with an average of 7 million cells/ml, expressed as an average **white blood cell count** of 7,000/mm^3. Leukocytes are the least numerous of the cellular elements in the blood (about 1 white blood cell for every 700 red blood cells), not because fewer are produced but because they are merely in transit while in the blood. Normally, approximately two-thirds of the circulating leukocytes are granulocytes, mostly neutrophils, whereas one-third are agranulocytes, predominantly lymphocytes (Table 11–2). However, the total number of white cells and the percentage of each type may vary considerably to meet changing defense needs. Depending on the type and extent of assault the body is combating, different types of leukocytes are selectively produced at varying rates. Chemical messengers arising from invaded or damaged tissues or from activated leukocytes themselves govern the rates of production of the various leukocytes. Specific hormones analogous to erythropoietin are required to direct the differentiation and proliferation of each cell type. Some of these hormones have been identified and can be produced in the laboratory; an example is **granulocyte colony–stimulating factor,** which stimulates increased replication and release of granulocytes, especially neutrophils, from the bone marrow. This achievement has opened up the ability to administer these hormones as a powerful new therapeutic tool to bolster a person's normal defense against infection or cancer.

Examining the functions of the various types of leukocytes provides clues as to what conditions normally prompt their production. Among the granulocytes, **neutrophils** are phagocytic specialists. They invariably are the first defenders on the scene of bacterial invasion and, accordingly, are very important in inflammatory responses. Furthermore, they scavenge to clean up debris. As might be expected in view of these functions, an increase in circulating neutrophils **(neutrophilia)** typically accompanies acute bacterial infections. In fact, a differential white blood cell count (a determination of the proportion of each type of leukocyte present) can be useful in making an immediate, reasonably accurate prediction of whether an infection, such as pneumonia or meningitis, is of bacterial or viral origin. Obtaining a definitive answer as to the causative agent by culturing a sample of the infected tissue's fluid takes several days. Since an elevated neutrophil count is highly indicative of bacterial infection, it is appropriate to initiate antibiotic therapy long before the true causative agent is actually known. (Bacteria generally succumb to antibiotics, whereas viruses do not.)

Eosinophils are specialists of another type. An increase in circulating eosinophils **(eosinophilia)** is associated with allergic conditions (such as asthma and hay fever) and with internal parasite infestations (for example, worms). Eosinophils obviously cannot engulf a much larger parasitic worm, but they do attach to the worm and secrete substances that kill it.

Basophils are the least numerous and most poorly understood of the leukocytes. They are quite similar structurally and functionally to **mast cells,** which never circulate in the blood but instead are dispersed in the connective tissue throughout the body. It was once believed that basophils became mast cells by migrating from the circulatory system, but researchers have shown that basophils arise from the bone marrow whereas mast cells are derived from precursor cells located in the connective tissue. Both basophils and mast cells synthesize and store *histamine* and *heparin,* powerful chemical substances that can be released upon appropriate stimulation. Histamine release is important in allergic reactions, whereas heparin hastens the removal of fat particles from the blood following a fatty meal. Heparin can also prevent blood clotting (coagulation), but whether it plays a physiological role as an anticoagulant is still being debated.

The number of circulating granulocytes is adjusted in two ways. First, the vast majority of the mature granulocytes are stored within the bone marrow, providing a readily available pool of these cells that can be released into the blood when the need arises in response to unknown signals. Second, the actual rate of production of each type of these cells can be varied as needed. Once released into the blood from the bone marrow, a granulocyte usually remains in transit in the blood for less than a day before leaving the blood vessels to enter the tissues, where it survives another three to four days unless it dies sooner in the line of duty.

Among the agranulocytes, **monocytes,** like neutrophils, are destined to become professional phagocytes. They emerge from the bone marrow while still immature and circulate for only a day or two before settling down in various tissues throughout the body. At their new residences, monocytes continue to mature and greatly enlarge, becoming the large tissue phagocytes known as **macrophages.** A macrophage's life span may range from months to years unless it is

Table 11-2
Typical Human Blood Cell Count

Total erythrocytes = 5,000,000,000 cells/ml blood
Red blood cell count = 5,000,000/mm^3

Total leukocytes = 7,000,000 cells/ml blood
White blood cell count = 7,000/mm^3

Differential white blood cell count (percentage distribution of types of leukocytes)

Polymorphonuclear granulocytes		*Mononuclear agranulocytes*	
Neutrophils	60–70%	Lymphocytes	25–33%
Eosinophils	1–4%	Monocytes	2–6%
Basophils	0.25–0.5%		

Total platelets = 250,000,000/ml blood
Platelet count = 250,000/mm^3

destroyed sooner while performing its phagocytic activity. A phagocytic cell can ingest only a limited amount of foreign material before it succumbs itself.

Lymphocytes provide immune defense against targets for which they are specifically programmed. There are two types of lymphocytes, B lymphocytes and T lymphocytes. **B lymphocytes** produce **antibodies,** which circulate in the blood. An antibody binds with and marks for destruction (by phagocytosis or other means) the specific kinds of foreign matter, such as bacteria, that induced production of the antibody. **T lymphocytes** do not produce antibodies; instead they directly destroy their specific target cells, a process known as a **cell-mediated immune response.** The target cells of T lymphocytes include body cells invaded by viruses and cancer cells. Lymphocytes have life spans estimated at 100 to 300 days. During this period, the majority of them continually recycle among the lymphoid tissues, lymph, and blood, spending only a few hours at a time in the blood. Therefore, only a small proportion of the total lymphocytes are in transit in the blood at any given moment.

By unknown mechanisms, the number of circulating lymphocytes is frequently elevated in association with a chronic infection. In **infectious mononucleosis,** not only does the number of lymphocytes in the blood increase, but many of the lymphocytes are atypical in structure. This condition, which is caused by the *Epstein-Barr virus,* is characterized by pronounced fatigue, a mild sore throat, and low-grade fever. Full recovery usually requires a month or more.

Even though circulating-leukocyte levels may vary, changes in these levels are normally controlled and adjusted according to the body's needs. However, abnormalities in leukocyte production can occur that are not subject to regulatory mechanisms; that is, either too few or too many leukocytes can be produced. The bone marrow can greatly slow down or even stop its production of white blood cells when it is exposed to certain toxic physical agents (such as radiation) or chemical agents (such as benzene and anticancer drugs). As you might predict, the most serious consequence is the reduction in professional phagocytes (neutrophils and macrophages), which leads to a notable reduction in the body's defense capabilities against invading microorganisms. The only defense still available when the bone marrow fails is the immune capabilities of the lymphocytes produced by the lymphoid organs.

A recent discovery may hold a key to a cure for a host of blood and immune disorders as well as some types of cancer and genetic diseases. Pluripotent stem cells, the source of all blood cells, have now been isolated. The search has been arduous because stem cells represent less than 0.1% of all cells in the bone marrow. Although a great deal of research remains to be done, investigators are hopeful that stem cells can be cultured, genetically manipulated as needed, and transplanted into patients as a new form of treatment for many catastrophic diseases, such as aplastic anemia, sickle-cell anemia, AIDS, and leukemia, to name a few.

Surprisingly, one of the major consequences of **leukemia,** a cancerous condition that involves uncontrolled proliferation of white blood cells, is inadequate defense capabilities against foreign invasion. In leukemia, the white blood cell count may reach as high as 500,000/mm^3, compared with the normal 7,000/mm^3, but because the majority of these cells are abnormal or immature, they are incapable of performing their normal defense functions. Another devastating consequence of leukemia is displacement of the other blood cell lines in the bone marrow. This results in anemia because of a reduction in erythropoiesis and in internal bleeding because of a deficit of platelets. Platelets play a critical role in preventing bleeding from the myriad tiny breaks that normally occur in small blood vessel walls. Consequently, overwhelming infections or hemorrhage are the most common causes of death in leukemic patients. The next section will examine the platelets' role in greater detail to show how they normally minimize the threat of hemorrhage.

Platelets and Hemostasis

Platelets are cell fragments derived from megakaryocytes.

In addition to erythrocytes and leukocytes, **platelets** are a third type of cellular element present in the blood. Platelets are not whole cells but small cell fragments (about 2–4 μm in diameter) that are shed off the outer edges of extraordinarily large (up to 60 μm in diameter) bone marrow–bound cells known as **megakaryocytes** (Fig. 11-9). A single megakaryocyte typically produces about 1,000 platelets. Megakaryocytes are derived from the same undifferentiated stem cells that give rise to the erythrocytic and leukocytic cell lines. Platelets are essentially detached vesicles containing portions of megakaryocyte cytoplasm wrapped in plasma membrane.

An average of 250,000,000 platelets are normally present in each milliliter of blood (range of 150,000–350,000/mm^3). Platelets remain functional for an average of ten days, at which time they are removed from circulation by the tissue

Figure 11-9 Photomicrograph of a Megakaryocyte Shedding Off Platelets

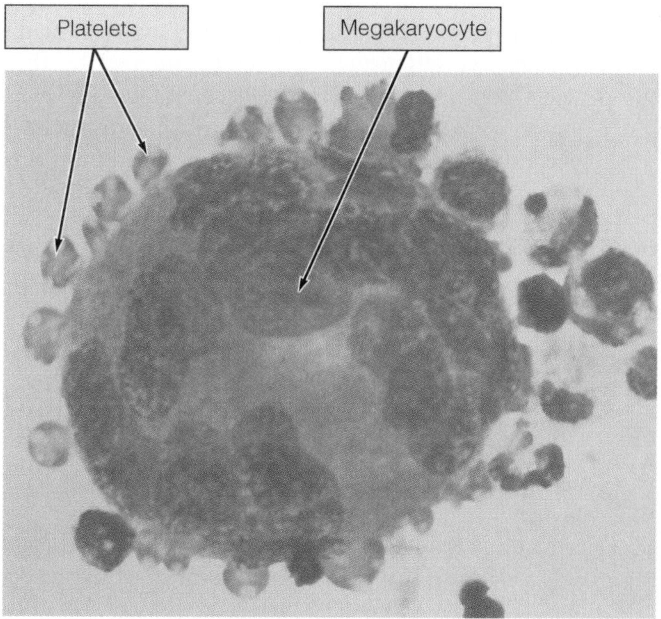

macrophages, especially those in the spleen and liver, and are replaced by newly released platelets from the bone marrow. The recently identified hormone **thrombopoietin** increases the number of megakaryocytes in the bone marrow and stimulates each megakaryocyte to produce more platelets. The factors that control thrombopoietin secretion are currently unknown.

Platelets do not leave the blood as white blood cells do, but at any given time about one-third of them are in storage in blood-filled spaces in the spleen. These stored platelets can be released from the spleen into the circulating blood as needed (for example, during hemorrhage) by sympathetically induced splenic contraction.

Because platelets are cell fragments, they lack nuclei. However, they are equipped with organelles and cytosolic enzyme systems for generating energy and synthesizing secretory products, which they store in numerous granules dispersed throughout the cytosol. Furthermore, platelets contain high concentrations of actin and myosin, which enable them to contract. Their secretory and contractile abilities are important in hemostasis, a topic to which we now turn.

Hemostasis prevents blood loss from damaged small vessels.

Hemostasis is the arrest of bleeding from a broken blood vessel—that is, the stopping of hemorrhage. (Be sure not to confuse this with the term *homeostasis.*) For bleeding to take place from a vessel, there must be a break in the vessel wall, and the pressure inside the vessel must be greater than the pressure outside it to force the blood out through the defect. The body's inherent hemostatic mechanisms normally are adequate to seal defects and stop loss of blood through small damaged capillaries, arterioles, and venules. These small vessels are frequently ruptured by the minor traumas of everyday life; such traumas are the most common source of bleeding, although we often are not even aware that any damage has taken place. The hemostatic mechanisms normally keep blood loss from these minor vascular traumas to a minimum.

The much rarer occurrence of bleeding from medium- to large-size vessels usually cannot be stopped by the body's hemostatic mechanisms alone. Bleeding from a severed artery is more profuse and therefore more dangerous than venous bleeding because the outward driving pressure is greater in the arteries (that is, arterial blood pressure is considerably higher than venous pressure). First aid measures for a severed artery include application over the wound of external pressure of greater magnitude than the arterial blood pressure to temporarily halt the bleeding until the torn vessel can be surgically closed. Hemorrhage from a traumatized vein can often be stopped simply by elevating the bleeding body part to reduce gravity's effects on pressure in the vein (see p. 334). If the accompanying drop in venous pressure is not sufficient to stop the bleeding, mild external compression is usually adequate.

Hemostasis involves three major steps: (1) *vascular spasm,* (2) *formation of a platelet plug,* and (3) *blood coagulation (clotting).* Platelets obviously play a major part in forming a platelet plug, but they contribute significantly to the other two steps as well.

Vascular spasms reduce blood flow through an injured vessel.

A cut or torn blood vessel immediately constricts as a result of both an inherent vascular response to injury and sympathetically induced vasoconstriction. This constriction slows blood flow through the defect and thus minimizes blood loss. Also, as the opposing endothelial (inner) surfaces of the vessel are pressed together by this initial **vascular spasm,** they become sticky and adhere to each other, further sealing off the damaged vessel. These physical measures alone cannot completely prevent further blood loss, but they are important in minimizing blood flow through the break in the vessel until the other hemostatic measures are able to actually plug up the defect.

Platelets aggregate to form a plug at a vessel defect.

Platelets normally do not adhere to the smooth endothelial surface of blood vessels, but when this lining is disrupted because of vessel injury, platelets attach to the exposed collagen, which is a fibrous protein present in the underlying connective tissue. Once platelets start aggregating at the site of the defect, they release several important chemicals from their storage granules (— Fig. 11-10). Among these chemicals is **adenosine diphosphate (ADP),** which causes the surface of nearby circulating platelets to become sticky, so that they adhere to the first layer of aggregated platelets. These newly aggregated platelets release more ADP, which causes more platelets to pile on, and so on; thus, a plug of platelets is rapidly built up at the defect site in a positive-feedback fashion.

This aggregating process is reinforced by the formation of a chemical messenger, **thromboxane A_2,** from a component of the platelet plasma membrane upon contact with collagen. Thromboxane A_2 is closely related to the prostaglandins, a group of locally functioning chemical messengers found extensively throughout the body (see p. 716). These local messengers are fatty acid derivatives formed from arachidonic acid, one of the fatty acids found in membrane phospholipids. Thromboxane A_2 directly promotes platelet aggregation and further enhances it indirectly by triggering the release of even more ADP from the platelet granules.

Given the self-perpetuating nature of platelet aggregation, why is the platelet plug limited to the site of vessel injury once it is initiated? (In other words, why doesn't the platelet plug continue to develop and expand over the surface of the adjacent normal vessel lining?) A key reason why this does not happen is that the normal endothelium releases **prostacyclin,** a chemical that profoundly inhibits platelet aggregation. Thus, the platelet plug is limited to the defect and does not spread to normal vascular tissue (Fig. 11-10).

The aggregated platelet plug not only physically seals the break in the vessel but also performs three other important roles. First, the actin-myosin protein complex within the aggregated platelets contracts to compact and strengthen

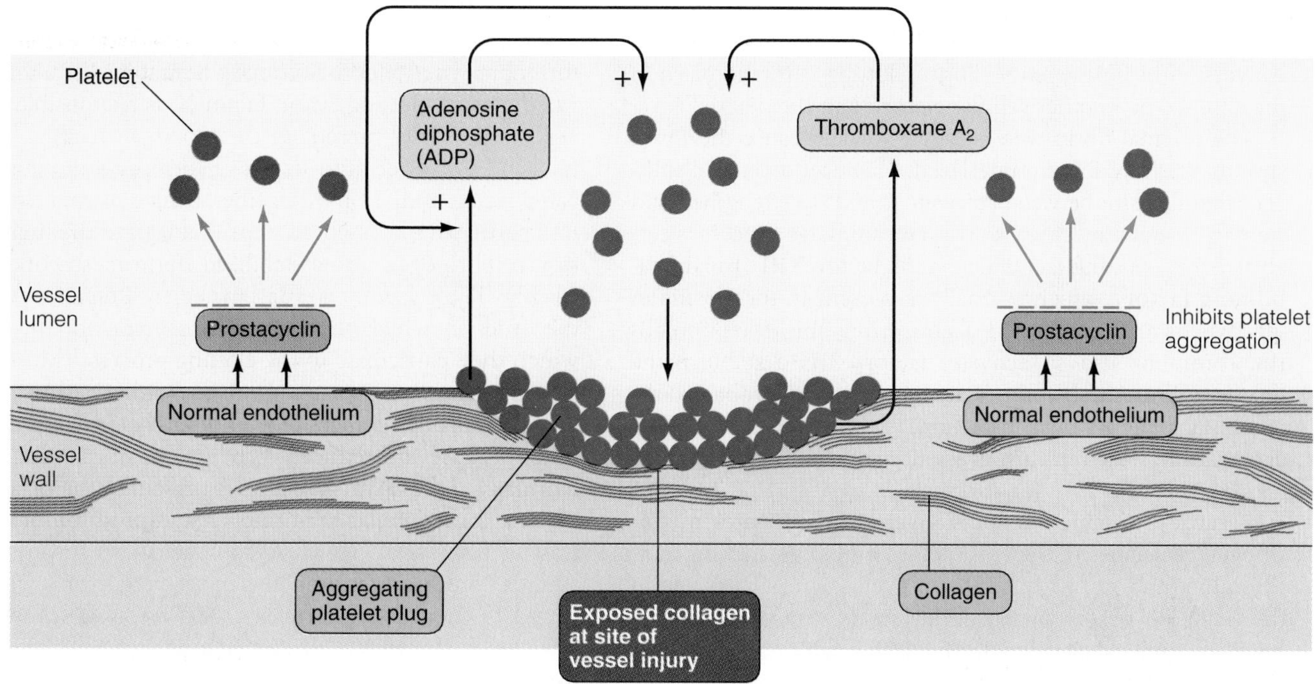

Figure 11-10 **Formation of a Platelet Plug** Platelets aggregate at a vessel defect through several positive-feedback mechanisms involving the release of adenosine diphosphate (ADP) and thromboxane A_2 from platelets that stick to exposed collagen at the site of the injury. Platelets are prevented from aggregating at the adjacent normal vessel lining by the release of prostacyclin from the endothelial cells that form the vessel lining.

what was originally a fairly loose plug. Second, the chemicals released from the platelet plug include several powerful vasoconstrictors (serotonin, epinephrine, and thromboxane A_2), which induce profound constriction of the affected vessel to reinforce the initial, self-induced vascular spasm. Third, the platelet plug releases other chemicals that enhance blood coagulation, the next step of hemostasis. Although the platelet-plugging mechanism alone is often sufficient to seal the myriad minute tears in capillaries and other small vessels that occur many times daily, larger holes in the vessels require the formation of a blood clot to completely stop the bleeding.

A triggered chain reaction involving clotting factors in the plasma results in blood coagulation.

Blood coagulation, or **clotting,** is the transformation of blood from a liquid into a solid gel. Formation of a clot on top of the platelet plug strengthens and supports the plug, reinforcing the seal over a break in a vessel. Furthermore, as blood in the vicinity of the vessel defect solidifies, it can no longer flow. Coagulation is the body's most powerful hemostatic mechanism, and it is required to stop bleeding from all but the most minute defects.

The ultimate step in clot formation is the conversion of **fibrinogen,** a large, soluble plasma protein produced by the liver and normally always present in the plasma, into **fibrin,** an insoluble, threadlike molecule. The conversion into fibrin is catalyzed by the enzyme **thrombin** at the site of the vessel injury.

Fibrin molecules adhere to the damaged vessel surface, forming a loose, netlike meshwork that traps the cellular elements of the blood. The resultant mass, or **clot,** typically appears red because of the abundance of trapped red blood cells, but the foundation of the clot is formed from fibrin derived from the plasma (Fig. 11-11). Except for platelets,

Figure 11-11 **Erythrocytes Trapped in the Fibrin Meshwork of a Clot**

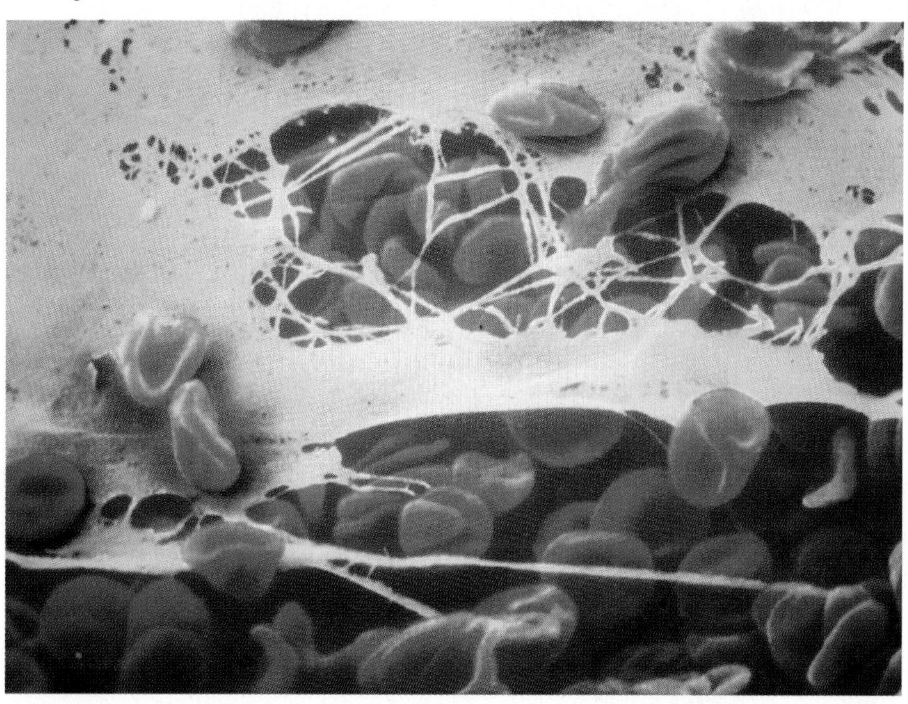

which play an important role in ultimately bringing about the conversion of fibrinogen to fibrin, clotting can take place in the absence of all other cellular elements in the blood.

The original fibrin web is rather weak because the fibrin strands are only loosely interlaced. However, chemical linkages rapidly form between adjacent strands to strengthen and stabilize the clot meshwork. This cross-linkage process is catalyzed by a clotting factor known as **factor XIII (fibrin-stabilizing factor),** which normally is present in the plasma in inactive form. In addition to ① converting fibrinogen into fibrin, thrombin also ② activates factor XIII to stabilize the resultant fibrin meshwork, ③ acts in a positive-feedback fashion to facilitate its own formation, and ④ enhances platelet aggregation, which in turn is essential to the clotting process (▬ Fig. 11-12).

Because thrombin's action converts the ever-present fibrinogen molecules in the plasma into a blood-stanching clot, thrombin must normally be absent from the plasma except in the vicinity of vessel damage. Otherwise, blood would always be coagulated, a situation incompatible with life. How can thrombin normally be absent from the plasma, yet be readily available to trigger fibrin formation when a vessel is injured? The solution lies in thrombin's existence in the plasma in the form of an inactive precursor called **prothrombin.**

The next question is, what converts prothrombin into thrombin when blood clotting is desirable? Yet another activated plasma clotting factor, **factor X,** is responsible; factor X itself is normally present in the blood in inactive form and must be converted into its active form by still another activated factor, and so on. Altogether, twelve plasma clotting factors participate in essential steps that lead to the final conversion of fibrinogen into a stabilized fibrin meshwork (▬ Fig. 11-13). These factors are designated by Roman numerals in the order in which they were discovered, not the order in which they participate in the clotting process.[1] Most of these clotting factors are plasma proteins synthesized by the liver. One of the consequences of liver disease is prolonged clotting time because of reduced production of clotting factors. Normally, these factors are always present in the plasma in an inactive form, similar to fibrinogen and prothrombin. In contrast to fibrinogen, which is converted into insoluble fibrin strands, prothrombin and the other precursors, when converted to their active form, act as proteolytic (protein-splitting) enzymes, which activate another specific factor in the clotting sequence. Once the first factor in the sequence is activated, it in turn activates the next factor, and so on, in a series of sequential reactions known as a **cascade,** until thrombin catalyzes the final conversion of fibrinogen into fibrin. Several of these steps require the presence of plasma Ca^{2+} and **platelet factor 3 (PF3),** a phospholipid secreted by the aggregated platelet plug. Thus, platelets also contribute to clot formation.

▬ *Figure 11-12* **Roles of Thrombin in Hemostasis** Thrombin, which is a component of the clotting cascade, plays multiple roles in hemostasis: ① conversion of fibrinogen to fibrin; ② activation of the factor that stabilizes the fibrin meshwork of the clot; ③ positive-feedback activation of more prothrombin into thrombin; and ④ enhancement of platelet aggregation. In positive-feedback fashion, aggregated platelets secrete PF3, which stimulates the clotting cascade that results in thrombin activation.

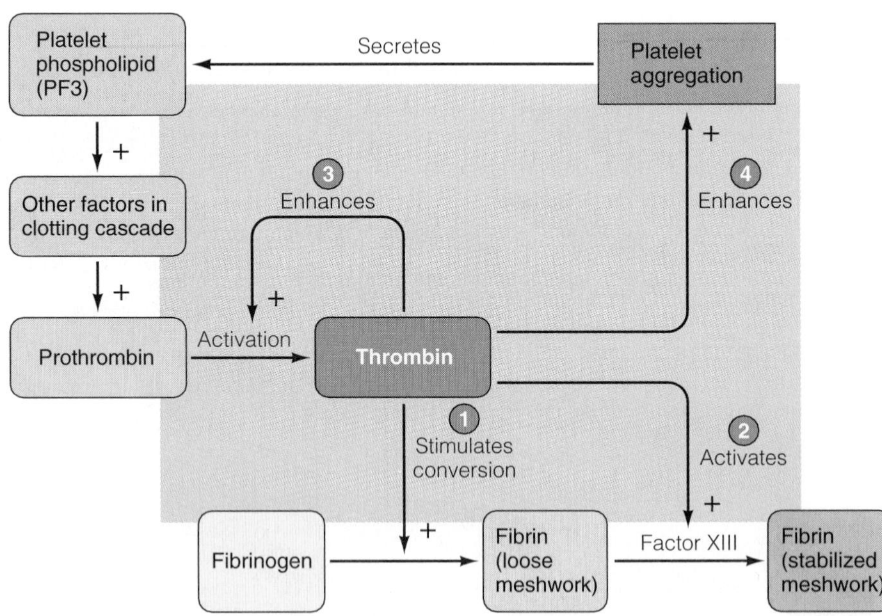

The clotting cascade may be triggered by the *intrinsic pathway* or the *extrinsic pathway:*

- The **intrinsic pathway** precipitates clotting within damaged vessels as well as clotting of blood samples in test tubes. All elements necessary to bring about clotting by means of the intrinsic pathway are present in the blood. This pathway, which involves seven separate steps (shown in blue in Fig. 11-13), is set off when **factor XII (Hageman factor)** is activated by coming into contact with either exposed collagen in an injured vessel or a foreign surface such as a glass test tube. Remember that exposed collagen also initiates platelet aggregation. Thus, formation of a platelet plug and the chain reaction leading to clot formation are simultaneously set in motion when a vessel is damaged. Furthermore, these complementary hemostatic mechanisms reinforce each other. The aggregated platelets secrete PF3, which is essential for the clotting cascade that in turn enhances further platelet aggregation (Figs. 11-12 and ▬ 11-14).
- The **extrinsic pathway** takes a shortcut and requires only four steps (shown in gray in Fig. 11-13). This pathway, which requires contact with tissue factors external to the blood, initiates

[1]The term *factor VI* is no longer used. What once was considered to be a separate factor VI has now been determined to be an activated form of factor V.

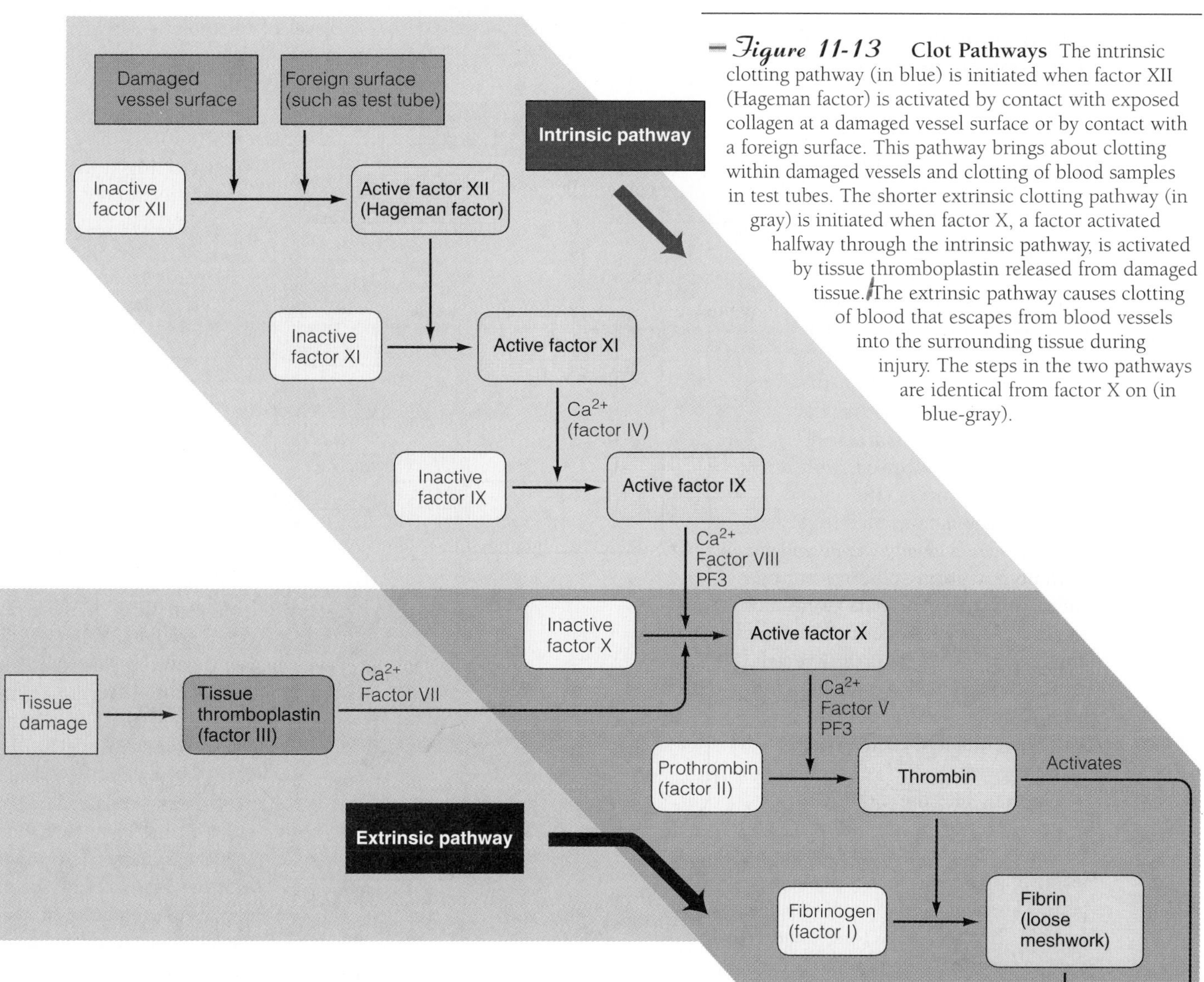

Figure 11-13 **Clot Pathways** The intrinsic clotting pathway (in blue) is initiated when factor XII (Hageman factor) is activated by contact with exposed collagen at a damaged vessel surface or by contact with a foreign surface. This pathway brings about clotting within damaged vessels and clotting of blood samples in test tubes. The shorter extrinsic clotting pathway (in gray) is initiated when factor X, a factor activated halfway through the intrinsic pathway, is activated by tissue thromboplastin released from damaged tissue. The extrinsic pathway causes clotting of blood that escapes from blood vessels into the surrounding tissue during injury. The steps in the two pathways are identical from factor X on (in blue-gray).

clotting of blood that has escaped into the tissues. When a tissue is traumatized, it releases a protein complex known as **tissue thromboplastin.** Tissue thromboplastin directly activates factor X, thereby bypassing all preceding steps of the intrinsic pathway. From this point on, the two pathways are identical.

The intrinsic and extrinsic mechanisms usually operate simultaneously. When tissue injury involves rupture of vessels, the intrinsic mechanism stops blood in the injured vessel, whereas the extrinsic mechanism clots the blood that escaped into the tissue before the vessel was sealed off. Typically, clot formation is fully developed in three to six minutes.

Once a clot is formed, contraction of the platelets trapped within the clot shrinks the fibrin meshwork, pulling the edges of the damaged vessel closer together. During **clot retraction,** fluid is squeezed from the clot. This fluid, which is essentially plasma minus fibrinogen and other clotting precursors that have been removed during the clotting process, is called **serum.**

Although a clotting process that involves so many steps might seem inefficient, the advantage is the amplification accomplished during many of the steps. One molecule of an activated factor can activate perhaps a hundred molecules of the next factor in the sequence, each of which can activate many more molecules of the next factor, and so on. In this way, large numbers of the final factors involved in clotting are rapidly activated as a result of the initial activation of only a few molecules in the beginning step of the sequence. How then is the clotting process, once initiated, confined to the site of vessel injury? If the activated clotting factors were allowed to circulate, they would induce inappropriate widespread clotting that would plug up vessels throughout the body. Fortunately, after participating in the local clotting process, the massive number of factors activated in the vicinity of vessel injury are rapidly inactivated by enzymes and other factors present in the plasma or tissue.

Fibrinolytic plasmin dissolves clots and prevents inappropriate clot formation.

A clot is not meant to be a permanent solution to vessel injury. It is a transient device to stop bleeding until the vessel can be repaired. The aggregated platelets secrete a chemical that is at least partially responsible for the invasion of fibroblasts ("fiber-formers") from the surrounding connective tissue into the wounded area of the vessel. Fibroblasts form a scar at the vessel defect. Simultaneous with the healing process, the clot, which is no longer needed to prevent hemorrhage, is slowly dissolved by a fibrinolytic (fibrin-splitting) enzyme called **plasmin.** If clots were not removed after they performed their hemostatic function, the vessels, especially the small ones that endure tiny ruptures on a regular basis, would eventually become obstructed by clots.

Plasmin, like the clotting factors, is a plasma protein produced by the liver and present in the blood in an inactive precursor form, **plasminogen.** Plasmin is activated cascade fashion by many factors, among them factor XII (Hageman factor), which also triggers the chain reaction leading to clot formation (■ Fig. 11-15). When a clot is being formed, activated plasmin becomes trapped in the clot and subsequently dissolves it by slowly breaking down the fibrin meshwork. Phagocytic white blood cells gradually remove the products of clot dissolution. You have observed the slow removal of blood that has clotted after escaping into the tissue layers of your skin following an injury. The "black-and-blue marks" of bruised skin result from deoxygenated clotted blood within the skin; this blood is eventually cleared by plasmin action, followed by the phagocytic cleanup crew.

In addition to removing clots that are no longer needed, plasmin functions continually to prevent clots from forming inappropriately. Throughout the vasculature, small amounts of fibrinogen are constantly being converted into fibrin, triggered by unknown mechanisms. Clots do not develop, however, because the fibrin is quickly disposed of by plasmin that is activated by **tissue plasminogen activator (tPA)** derived from the tissues, especially the lungs. Normally, the low level of fibrin formation is counterbalanced by a low level of fibrinolytic activity, so inappropriate clotting does not occur. Only when a vessel is damaged do additional factors precipitate the explosive chain reaction that leads to more extensive fibrin formation and results in local clotting at the site of injury.

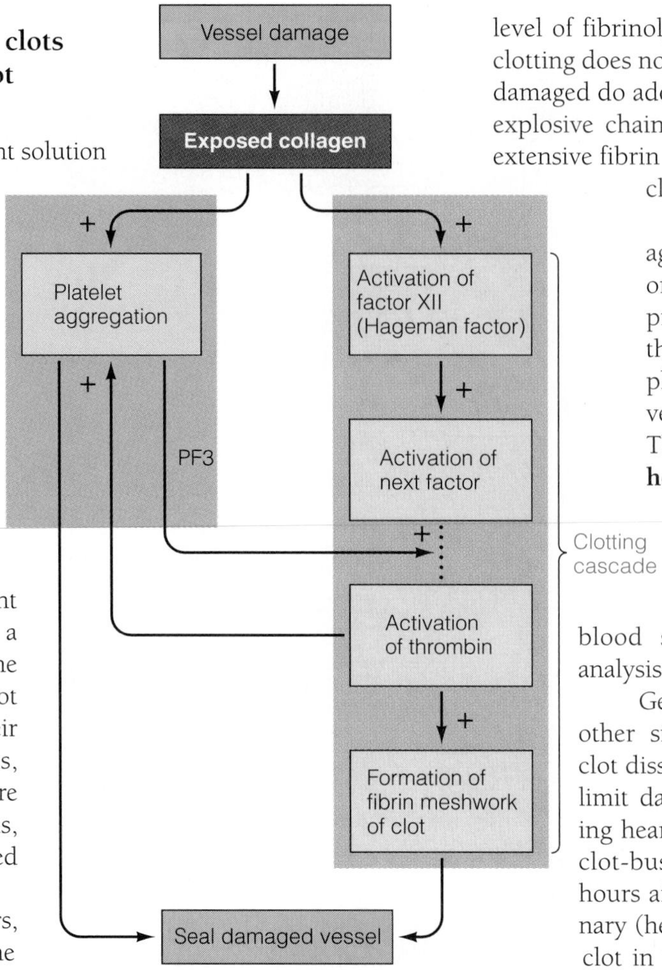

■ *Figure 11-14* **Concurrent Platelet Aggregation and Clot Formation** Exposed collagen at the site of vessel damage simultaneously initiates platelet aggregation and the clotting cascade. These two hemostatic mechanisms positively reinforce each other as they seal the damaged vessel.

Naturally occurring anticoagulants that interfere with one or more steps in the clotting process are known to exist in the body. Whether they play a physiological role in clot prevention is still being debated. The best known of these is **heparin,** which is present in basophils and mast cells. It is used extensively as an anticoagulant drug as well as to prevent clotting of blood samples drawn for clinical analysis.

Genetically engineered tPA and other similar chemicals that trigger clot dissolution are frequently used to limit damage to cardiac muscle during heart attacks. Administration of a clot-busting drug within the first hours after a clot has blocked a coronary (heart) vessel often dissolves the clot in time to restore blood flow to the cardiac muscle supplied by the blocked vessel before the muscle dies of O_2 deprivation. In recent years, tPA and related drugs have also been used successfully to dissolve a stroke-causing clot within a cerebral (brain) blood vessel, thus minimizing the loss of irreplaceable brain tissue following a stroke.

Inappropriate clotting is responsible for thromboembolism.

In spite of protective measures, clots occasionally form in intact vessels. An abnormal intravascular clot attached to a vessel wall is known as a **thrombus,** and free floating clots are called **emboli.** Several factors, acting independently or simultaneously, can cause *thromboembolism:* (1) roughened vessel surfaces associated with atherosclerosis can lead to thrombus formation (see p. 297); (2) imbalances in the clotting-anticlotting systems can likewise trigger clot formation; (3) slow-moving blood is more apt to clot, probably because small quantities of fibrin are formed and allowed to accumulate in the stagnant blood, for example, in blood pooled in varicosed leg veins; and (4) widespread clotting is occasionally triggered by the release of tissue thromboplastin into the blood from large amounts of traumatized tissue. Similar widespread clotting can occur in **septicemic shock,** in which bacteria or their toxins initiate the clotting cascade.

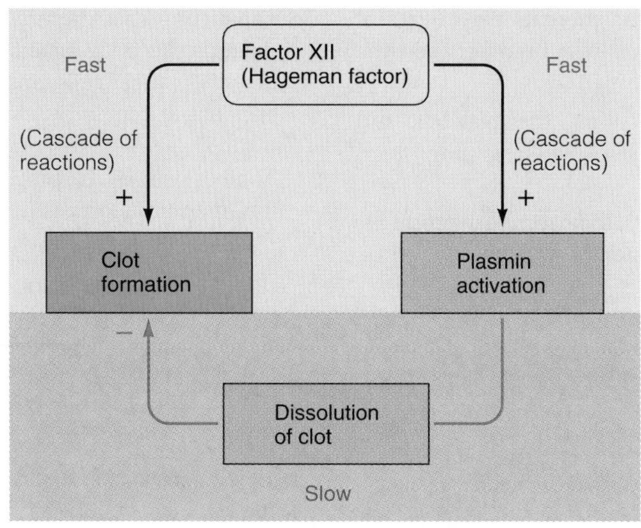

Figure 11-15 Role of Factor XII in Clot Formation and Dissolution Activation of factor XII (Hageman factor) simultaneously initiates a fast cascade of reactions that result in clot formation and a fast cascade of reactions that result in plasmin activation. Plasmin, which is trapped in the clot, subsequently slowly dissolves the clot. This action removes the clot when it is no longer needed after the vessel has been repaired.

Hemophilia is the primary condition responsible for excessive bleeding.

In contrast to inappropriate clot formation in intact vessels, the opposite hemostatic disorder is failure of clots to form promptly in injured vessels, resulting in life-threatening hemorrhage from even relatively mild traumas. The most common cause of excessive bleeding is **hemophilia,** which is caused by a deficiency of one of the factors in the clotting cascade. Although a deficiency of any of the clotting factors could block the clotting process, 80% of all hemophiliacs lack the genetic ability to synthesize factor VIII.

In contrast to the more profuse bleeding that accompanies defects in the clotting mechanism, individuals having a deficiency of platelets continuously develop hundreds of small, confined hemorrhagic areas throughout the body tissues as blood is permitted to leak from tiny breaks in the small blood vessels before coagulation takes place. Platelets normally are the primary sealers of these ever-occurring minute ruptures. In the skin of a platelet-deficient person, the diffuse capillary hemorrhages are visible as small, purplish blotches, giving rise to the term **thrombocytopenia purpura** ("the purple of thrombocyte deficiency") for this condition. (Recall that the term "thrombocytes" is an alternate name for platelets.)

Vitamin K deficiency can also cause a bleeding tendency. Vitamin K, commonly referred to as the blood-clotting vitamin, is essential for normal clot formation. Researchers recently figured out vitamin K's role in the clotting process. In a complex sequence of biochemical events, free energy released when vitamin K combines with O_2 is ultimately used in the activation processes of the clotting cascade.

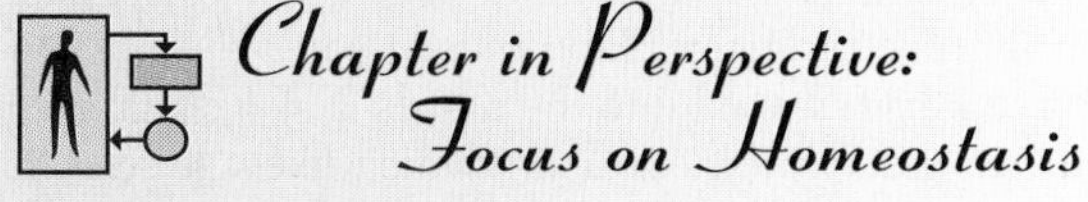

Chapter in Perspective: Focus on Homeostasis

Blood contributes to homeostasis in a variety of ways. First, the composition of the interstitial fluid, the true internal environment that surrounds and directly exchanges materials with the cells, depends on the composition of the blood plasma. Because of the thorough exchange that occurs between the interstitial and vascular compartments, the interstitial fluid has the same composition as the plasma with the exception of plasma proteins, which cannot escape through the capillary walls. Thus, the blood serves as the vehicle for rapid, long-distance mass transport of materials to and from the cells, and the interstitial fluid serves as the go-between.

Homeostasis depends on the blood carrying materials such as O_2 and nutrients to the cells as rapidly as the cells consume these supplies and carrying materials such as metabolic wastes away from the cells as rapidly as the cells produce these products. It also depends on the blood carrying hormonal messengers from their site of production to their distant site of action. Once a substance enters the blood, it can be transported throughout the body within seconds, whereas diffusion of the substance over long distances in a large multicellular organism such as a human would take months to years—a situation incompatible with life. Diffusion can, however, effectively accomplish short local exchanges of materials between the blood and surrounding cells through the intervening interstitial fluid.

The blood has special transport capabilities that enable it to move its cargo efficiently throughout the body. For example, life-sustaining O_2 is poorly soluble in water, but the blood is equipped with O_2-carrying specialists, the erythrocytes (red blood cells), which are stuffed full of heomoglobin, a complex molecule that transports O_2. Likewise, homeostatically important water-insoluble hormonal messengers are shuttled in the blood by plasma protein carriers.

Specific components of the blood perform the following additional homeostatic activities that are unrelated to blood's transport function:

- The blood helps maintain the proper pH in the internal environment by buffering changes in the acid-base load of the body.
- The blood helps maintain body temperature by absorbing heat produced by heat-generating tissues such as contracting skeletal muscles and distributing it throughout the body. Excess heat is carried by the blood to the body surface for elimination to the external environment.
- The electrolytes in the plasma are important in membrane excitability, which in turn forms the basis of nerve and muscle function.
- The electrolytes in the plasma are also important in the osmotic distribution of fluid between the extracellular and

intracellular fluid, and the plasma proteins play a critical role in the distribution of extracellular fluid between the plasma and interstitial fluid.

- Through their hemostatic functions, the platelets and clotting factors minimize the loss of life-sustaining blood following vessel injury.
- The leukocytes (white blood cells), their secretory products, and certain types of plasma proteins, such as antibodies, constitute the immune defense system. This system defends the body against invading disease-causing agents, destroys cancer cells, and paves the way for wound healing and tissue repair by clearing away debris from dead or injured cells. These actions indirectly contribute to homeostasis by helping to keep the organs that directly maintain homeostasis healthy.

Chapter Summary

Plasma

The 5- to 5.5-liter volume of blood in an adult consists of 42% to 45% erythrocytes, less than 1% leukocytes and platelets, and 55% to 58% plasma. The percentage of whole-blood volume occupied by erythrocytes is known as the hematocrit.

Plasma is a complex liquid that serves as a transport medium for substances being carried in the blood. All plasma constituents are freely diffusible across the capillary walls except the plasma proteins, which remain in the plasma and perform a variety of functions.

Erythrocytes

Erythrocytes (red blood cells) are specialized for their primary function of O_2 transport in the blood. They do not contain a nucleus, organelles, or ribosomes but instead are packed full of hemoglobin, which is an iron-containing molecule that can loosely, reversibly bind with O_2. Because O_2 is poorly soluble in blood, hemoglobin is indispensable for O_2 transport. Hemoglobin also contributes to CO_2 transport and buffering of blood by reversibly binding with CO_2 and H^+.

Unable to replace cell components, erythrocytes are destined to a short life span of about 120 days. Undifferentiated pluripotent stem cells in the red bone marrow give rise to all cellular elements of the blood. Erythrocyte production (erythropoiesis) by the marrow normally keeps pace with the rate of erythrocyte loss to keep the red cell count constant. Erythropoiesis is stimulated by erythropoietin, a hormone secreted by the kidneys in response to reduced O_2 delivery.

Leukocytes

Leukocytes (white blood cells) are the defense corps of the body. They attack foreign invaders, destroy abnormal cells that arise in the body, and clean up cellular debris. There are five types of leukocytes, each with a different task: (1) Neutrophils, the phagocytic specialists, are important in engulfing bacteria and debris. (2) Eosinophils specialize in attacking parasitic worms and play a key role in allergic responses. (3) Basophils release two chemicals: histamine, which is also important in allergic responses, and heparin, which helps clear fat particles from the blood. (4) Monocytes, upon leaving the blood, set up residence in the tissues and greatly enlarge to become the large tissue phagocytes known as macrophages. (5) Lymphocytes provide immune defense against bacteria, viruses, and other targets for which they are specifically programmed. Their defense tools include the production of antibodies and cell-mediated immune responses.

Leukocytes are present in the blood only while in transit from their site of production and storage in the bone marrow (and also in the lymphoid tissues in the case of the lymphocytes) to their site of action in the tissues. At any given time, the majority of the leukocytes are out in the tissues on surveillance missions or performing actual combative activities. All leukocytes have a limited life span and must be replenished by ongoing differentiation and proliferation of precursor cells. The total number and percentage of each of the different types of leukocytes produced vary depending on the momentary defense needs of the body.

Platelets and Hemostasis

Platelets are cell fragments derived from large megakaryocytes in the bone marrow. Platelets play an important role in hemostasis, the arrest of bleeding from an injured vessel. The three main steps in hemostasis are (1) vascular spasm, (2) platelet plugging, and (3) clot formation. Vascular spasm reduces blood flow through an injured vessel, whereas aggregation of platelets at the site of vessel injury quickly plugs the defect. Platelets start to aggregate upon contact with exposed collagen in the damaged vessel wall.

Clot formation (blood coagulation) reinforces the platelet plug and converts blood in the vicinity of a vessel injury into a nonflowing gel. The majority of factors necessary for clotting are always present in the plasma in inactive precursor form. When a vessel is damaged, exposed collagen initiates a cascade of reactions involving successive activation of these clotting factors, ultimately converting fibrinogen into fibrin. Fibrin, an insoluble threadlike molecule, is laid down as the meshwork of the clot; the meshwork in turn entangles blood cells to complete clot formation. Blood that has escaped into the tissues is also coagulated upon exposure to tissue thromboplastin, which likewise sets the clotting process into motion. When no longer needed, clots are dissolved by plasmin, a fibrinolytic factor also activated by exposed collagen.

Review Exercises

Objective Questions (Answers on p. E–10.)

1. Blood can absorb metabolic heat while undergoing only small changes in temperature. (True or false?)
2. Hemoglobin can carry only O_2. (True or false?)
3. Erythrocytes originate from the same undifferentiated stem cells as leukocytes and platelets. (True or false?)
4. Erythrocytes are unable to utilize the O_2 they contain for their own ATP formation. (True or false?)
5. White blood cells spend the majority of their time in the blood. (True or false?)
6. The body's defense capabilities are reduced in leukemia even though there are an excessive number of leukocytes. (True or false?)
7. Which type of leukocyte is produced primarily in lymphoid tissue? ____________
8. The majority of plasma clotting factors are synthesized by the ____________ .
9. Which of the following is *not* a function served by the plasma proteins?
   a. facilitate retention of fluid in the blood vessels
   b. play an important role in blood clotting
   c. bind and transport certain hormones in the blood
   d. transport O_2 in the blood
   e. serve as antibodies
   f. contribute to the buffering capacity of the blood
10. Which of the following is *not* triggered by exposed collagen?
    a. vascular spasm
    b. platelet aggregation
    c. activation of the clotting cascade
    d. activation of plasminogen
11. Match the following (an answer may be used more than once):

___ 1. causes platelets to aggregate in positive-feedback fashion	(a) prostacyclin
___ 2. activates prothrombin	(b) plasmin
___ 3. fibrinolytic enzyme	(c) ADP
___ 4. inhibits platelet aggregation	(d) fibrin
___ 5. first factor activated in intrinsic clotting pathway	(e) thrombin
___ 6. forms the meshwork of the clot	(f) factor X
___ 7. stabilizes the clot	(g) factor XII
___ 8. activates fibrinogen	(h) factor XIII
___ 9. activated by tissue thromboplastin	

12. Match the following blood abnormalities with their causes:

___1. deficiency of intrinsic factor	(a) hemolytic anemia
___2. insufficient amount of iron to synthesize adequate hemoglobin	(b) aplastic anemia
___3. destruction of bone marrow	(c) iron-deficiency anemia
___4. abnormal loss of blood	(d) hemorrhagic anemia
___5. tumorlike condition of bone marrow	(e) pernicious anemia
___6. inadequate erythropoietin secretion	(f) renal anemia
___7. excessive rupture of circulating erythrocytes	(g) primary polycythemia
___8. associated with living at high altitudes	(h) secondary polycythemia

Essay Questions

1. What is the average blood volume?
2. What is the normal percentage of blood occupied by erythrocytes and by plasma? What is the hematocrit? What is the buffy coat?
3. What is the composition of plasma?
4. List the three major groups of plasma proteins and state their functions.
5. Describe the structure and functions of erythrocytes.
6. Why are erythrocytes able to survive for only about 120 days?
7. Describe the process and control of erythropoiesis.
8. Compare the structure and functions of the five types of leukocytes.
9. Discuss the derivation of platelets.
10. Describe the three steps of hemostasis, including a comparison of the intrinsic and extrinsic pathways by which the clotting cascade is triggered.
11. Compare plasma and serum.
12. What factors normally prevent inappropriate coagulation in the vasculature?

Quantitative Exercises (Solutions on p. E–10.)

1. The normal concentration of hemoglobin in blood (as measured clinically) is 15 g/100 ml of blood.
   a. Given that one mole of hemoglobin weighs 66,000 grams, what is the concentration of hemoglobin in mM?
   b. Each hemoglobin molecule can bind four molecules of O_2. What is the concentration of O_2 bound to hemoglobin at maximal saturation (in mM)?
   c. Given that 1 mole of an ideal gas occupies 22.4 liters, what is the maximal carrying capacity of normal blood for O_2 (usually expressed in ml of O_2/liter of blood)?
2. Assume that the blood sample in Figure 11-5b, p. 359 is from a patient with hemorrhagic anemia. Given a normal blood

volume of 5 liters, a normal red-blood-cell concentration of 5 billion/ml, and a red blood cell production rate of 3 million cells/second, how long will it take the body to return the hematocrit to normal?

3. Note that in the blood sample in Figure 11-5c from a patient with polycythemia, the hematocrit has increased to 70%. An increased hematocrit increases blood viscosity, which in turn increases total peripheral resistance and increases the workload on the heart. The effect of hematocrit (*h*) on relative blood viscosity (v, viscosity relative to that of water) is given approximately by the following equation:

$$v = 1.5 \times \exp(2h)$$

Note that in this equation ***h*** is the hematocrit as a fraction, not a percent. Given a normal hematocrit of 0.40, what percent increase in viscosity would result from the polycythemia in Figure 11-5c? What percent change would this cause in total peripheral resistance?

Points to Ponder

(Explanations on p. E–10.)

1. Why do you think it is important to closely monitor blood cell counts in cancer patients who are being treated with chemotherapeutic drugs designed to destroy rapidly multiplying cells, such as cancer cells?
2. A person has a hematocrit of 62. Can you conclude from this finding that the person has polycythemia? Explain.
3. There are several different forms of hemoglobin. *Hemoglobin A* is normal adult hemoglobin. The abnormal form *hemoglobin S* causes erythrocytes to warp into fragile, sickle-shaped cells. Fetal red blood cells contain *hemoglobin F,* the production of which stops soon after birth. Now researchers are trying to goad the genes that direct hemoglobin F synthesis back into action as a means of treating sickle-cell anemia. Explain how turning on these fetal genes could be a useful remedy.
4. Low on the list of popular animals are vampire bats, leeches, and ticks, yet these animals may someday indirectly save your life. Scientists are currently examining the "saliva" of these blood-sucking creatures in search of new chemicals that might limit cardiac muscle damage in heart attack victims. What do you suspect the nature of these sought-after chemicals is?
5. With the screening methods currently employed by blood banks, about 1 out of 225,000 units of blood used for transfusions may harbor HIV, the virus that causes AIDS (see p. 399). HIV-contaminated blood that slips through the screening process comes largely from recently infected donors. The screening tests now employed only detect antibodies against HIV, which do not appear in the blood for nearly a month following HIV infection. Therefore, a window of about a month exists during which donated blood may be infectious but still pass the screening process. An estimated 10 people are infected with HIV annually by this means.

   Tests are available for detecting HIV earlier than the blood banks' antibody-based method, such as by screening for the presence of a specific protein on the surface of HIV. These tests are costly, however, and currently are used only for research. If blood banks employed these more sensitive tests, they could detect about half of the HIV-contaminated blood that currently goes to patients. The estimated cost of implementing these more expensive tests is somewhere between $70 million and $200 million. Do you think the health care system should assume this additional financial burden to prevent 4 or 5 cases of transfusion-delivered HIV infection per year?
6. ***Clinical Consideration*** Linda P. has just been diagnosed as having pneumonia. Her white blood cell count is 7,200/mm^3, with 67% of the white blood cells being neutrophils. It will take several days to obtain a definitive answer as to the causative agent by culturing a sample of discharges from her respiratory system. Based on the white blood cell count, do you think that Linda should be given antibiotics immediately, long before the causative agent is actually known? Are antibiotics likely to combat her infection?

12

The Body Defenses

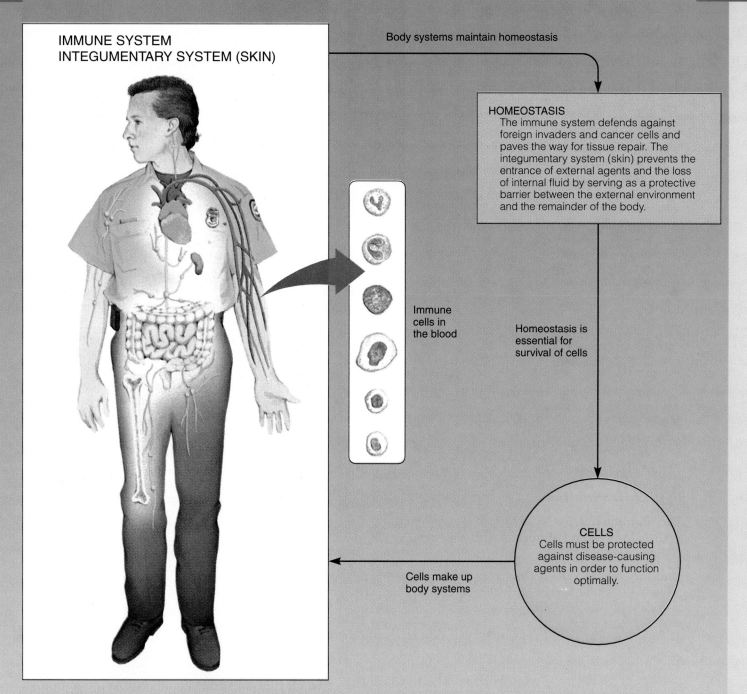

IMMUNE SYSTEM
INTEGUMENTARY SYSTEM (SKIN)

Body systems maintain homeostasis

HOMEOSTASIS
The immune system defends against foreign invaders and cancer cells and paves the way for tissue repair. The integumentary system (skin) prevents the entrance of external agents and the loss of internal fluid by serving as a protective barrier between the external environment and the remainder of the body.

Immune cells in the blood

Homeostasis is essential for survival of cells

CELLS
Cells must be protected against disease-causing agents in order to function optimally.

Cells make up body systems

Humans are constantly coming into contact with external agents that could be harmful if they gained entry into the body. The most serious of these are disease-causing microorganisms. In the event that bacteria or viruses gain entry, the body is equipped with a complex, multifaceted internal defense system—the **immune system**—which provides continual protection against invasion by foreign agents. The body surfaces that are exposed to the external environment, such as the **skin,** serve as a first line of defense to resist penetration by foreign microorganisms.

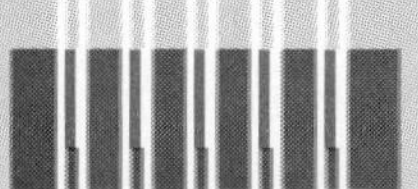

Chapter Contents At a Glance

Introduction

The immune defense system provides protection against foreign and abnormal cells and removes cellular debris.

Immunity refers to the body's ability to resist or eliminate potentially harmful foreign materials or abnormal cells. The following activities are attributable to the immune defense system, which plays a key role in recognizing and either destroying or neutralizing materials within the body that are foreign to the "normal self:"

1. Defense against invading **pathogens** (disease-producing microorganisms such as viruses and bacteria).
2. Removal of "worn-out" cells (such as aged red blood cells) and tissue debris (for example, tissue damaged by trauma or disease). The latter is essential for wound healing and tissue repair.
3. Identification and destruction of abnormal or mutant cells that have originated in the body. This function, termed *immune surveillance,* is the primary internal defense mechanism against cancer.
4. Inappropriate immune responses that lead either to *allergies,* which occur when the body turns against a normally harmless environmental chemical entity, or to *autoimmune diseases,* which happen when the defense system erroneously produces antibodies against itself, leading to destruction of a particular type of the body's own cells.
5. Rejection of tissue cells of foreign origin, which constitutes the major obstacle to successful organ transplantation.

Pathogenic bacteria and viruses are the major targets of the immune defense system.

The primary foreign enemies against which the immune system defends are bacteria and viruses. **Bacteria** are nonnucleated, single-celled microorganisms self-equipped with all machinery essential for their own survival and reproduction. Pathogenic bacteria that invade the body induce tissue damage and produce disease largely by releasing enzymes or toxins that physically injure or functionally disrupt affected cells and organs. The disease-producing power of a pathogen is known as its **virulence.**

In contrast to bacteria, **viruses,** are not self-sustaining cellular entities. They consist only of nucleic acids (DNA or RNA) enclosed by a protein coat. Because they lack cellular machinery for energy production and protein synthesis, viruses are unable to carry out metabolism and reproduce unless they invade a **host cell** (a body cell of the infected individual) and take over the cellular biochemical facilities for their own purposes. Not only do viruses sap the host cell's energy resources, but the viral nucleic acids also direct the host cell to synthesize proteins needed for viral replication. The effect of viral invasion and replication on the host cell varies with different types of viruses.

Viruses can lead to cellular damage or death in four general ways: (1) depletion of essential cellular components by the virus; (2) under the dictatorship of the virus, cellular production of substances toxic to the cell; (3) transformation of normal host cells into cancer cells; and (4) incorporation of the virus into the cell, resulting in the destruction of the cell by the body's own defense mechanisms because it is no longer recognized as a "normal-self" cell.

Leukocytes are the effector cells of the immune defense system.

The cells responsible for the various immune defense strategies are the leukocytes (white blood cells) and their derivatives, briefly reviewed as follows (see pp. 360–363):

1. Neutrophils are highly mobile phagocytic specialists that engulf and destroy unwanted materials.

2. Eosinophils secrete chemicals that destroy parasitic worms and are involved in allergic manifestations.
3. Basophils release histamine and heparin and also are involved in allergic manifestations.
4. Lymphocytes
   a. B lymphocytes are transformed into plasma cells, which secrete antibodies that indirectly lead to the destruction of foreign material.
   b. T lymphocytes are responsible for cell-mediated immunity involving direct destruction of virus-invaded cells and mutant cells through nonphagocytic means.
5. Monocytes are transformed into macrophages, which are large, tissue-bound phagocytic specialists.

A given leukocyte is present in the blood only transiently. Most of the leukocytes are out in the tissues on defense missions. As a result, the immune system's effector cells are widely dispersed throughout the body and are able to defend in any location.

Almost all leukocytes originate from common precursor stem cells in the bone marrow and are subsequently released into the blood. The only exception is lymphocytes, which arise in part from lymphocyte colonies in various lymphoid tissues that were originally populated by cells derived from the bone marrow (see p. 362). **Lymphoid tissues** refer collectively to the tissues that store, produce, or process lymphocytes. These include the lymph nodes, spleen, thymus, tonsils, adenoids, appendix, bone marrow, and aggregates of lymphoid tissue in the lining of the digestive tract called **Peyer's patches** or **gut-associated lymphoid tissue (GALT)** (Fig. 12-1). Lymphoid tissues are strategically located to intercept invading microorganisms before they have a chance to spread very far. For example, the lymphocytes populating the *tonsils* and *adenoids* are situated advantageously to respond to inhaled microbes, whereas microorganisms invading through the digestive system immediately encounter lymphocytes in the *appendix* and *GALT*. Potential pathogens that gain access to the lymph are filtered through *lymph nodes*, where they are exposed to lymphocytes as well as to macrophages that line the lymphatic passageways. The *spleen*, the largest of the lymphoid tissues, performs immune functions on the blood similar to those the lymph nodes perform on the lymph. Through actions of its lymphocyte and macrophage population, the spleen clears the blood that passes through it of microorganisms and other foreign matter and also removes worn-out red blood cells (see p. 356). The *thymus* and *bone marrow* play important roles in processing T and B lymphocytes, respectively, to prepare them to carry out their specific immune strategies. Table 12-1 summarizes

Figure 12-1
Lymphoid Tissues
The lymphoid tissues, which are dispersed throughout the body, store, produce, or process lymphocytes.

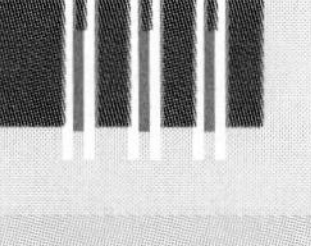

Table 12-1
Functions of Lymphoid Tissues

Lymphoid Tissue	*Function*
Bone marrow	Origin of all blood cells Site of maturational processing for B lymphocytes
Lymph nodes, tonsils, adenoids, appendix, gut-associated lymphoid tissue	Exchange lymphocytes with the lymph (remove, store, produce, and add them) Resident lymphocytes produce antibodies and sensitized T cells, which are released into the lymph Resident macrophages remove microbes and other particulate debris from the lymph
Spleen	Exchanges lymphocytes with the blood (removes, stores, produces, and adds them) Resident lymphocytes produce antibodies and sensitized T cells, which are released into the blood Resident macrophages remove microbes and other particulate debris, most notably worn-out red blood cells, from the blood Stores a small percentage of red blood cells, which can be added to the blood by splenic contraction as needed
Thymus	Site of maturational processing for T lymphocytes Secretes the hormone thymosin

the major functions of the various lymphoid tissues, some of which were described in chapter 11 and others of which are yet to be discussed in this chapter.

Immune responses may be either nonspecific or specific.

Immune responses are classified as nonspecific or specific immune responses, depending on the degree of selectivity of the defense mechanism. **Nonspecific immune responses** are inherent defense responses that nonselectively defend against foreign or abnormal material of any type, even upon initial exposure to it. Such responses provide a first line of defense against a wide range of threatening factors, including infectious agents, chemical irritants, and tissue injury accompanying mechanical trauma and burns. **Specific immune responses,** on the other hand, are selectively targeted against particular foreign material to which the body has previously been exposed. These specific responses are mediated by lymphocytes, which, upon subsequent exposure to the same offending agent, recognize and discriminately defend against it. We will first examine the nonspecific responses before turning our attention to the specific responses.

Nonspecific Immune Responses

Nonspecific defenses include inflammation, interferon, natural killer cells, and the complement system.

Nonspecific defenses that come into play whether or not there has been prior experience with the offending agent include the following:

1. *Inflammation,* a nonspecific response to tissue injury in which the phagocytic specialists—neutrophils and macrophages—play a major role, along with supportive input from other immune cell types.
2. *Interferon,* a family of proteins that nonspecifically defend against viral infection.
3. *Natural killer cells,* a special class of lymphocytelike cells that spontaneously and relatively nonspecifically lyse (rupture) and thereby destroy virus-infected host cells and cancer cells.
4. *The complement system,* a group of inactive plasma proteins that, when sequentially activated, bring about destruction of foreign cells by attacking their plasma membranes.

The complement system can nonspecifically be called into play by the presence of any foreign invader. It may also be activated by antibodies produced as part of the specific immune response to a particular microorganism. The fact that the complement system is involved in both nonspecific and specific defense mechanisms illustrates an important point. The various components of the immune system are highly interactive and interdependent, making the system highly sophisticated and effective but also complex and difficult to sort out. The most significant cooperative relationships known to exist among the immune effector cells will be pointed out as the various components of the immune system are discussed separately. As further reference, Appendix D summarizes the major immune functions and interactions addressed in this chapter.

Inflammation is a nonspecific response to foreign invasion or tissue damage.

Inflammation refers to an innate, nonspecific series of highly interrelated events that are set into motion in response to foreign invasion, tissue damage, or both. The ultimate goal of inflammation is to bring to the invaded or injured area phagocytes and plasma proteins that can (1) isolate, destroy, or inactivate the invaders; (2) remove debris; and (3) prepare for subsequent healing and repair. The overall inflammatory response is remarkably similar no matter what the triggering event (be it bacterial invasion, chemical injury, or mechanical trauma), although some subtle differences may be evident, depending on the injurious agent or the site of damage. The following sequence of events typically occurs during the inflammatory response. As an example, we will use bacterial entry into a break in the skin.

Defense by resident tissue macrophages Upon bacterial invasion through a break in the external barrier of skin, the macrophages already present in the area immediately begin phagocytizing the foreign microbes. Although the resident macrophages are usually not present in sufficient numbers to meet the challenge alone, they defend against infection during the first hour or so, before other mechanisms can be mobilized. Macrophages are usually rather stationary, gobbling debris and contaminants that come their way, but when necessary, they become mobile and migrate to sites of battle against invaders.

Localized vasodilation Almost immediately upon microbial invasion, arterioles within the area dilate, increasing blood flow to the site of injury. This localized vasodilation is primarily induced by histamine that has been released in the area of tissue damage from mast cells, a type of connective tissue–bound cell similar to circulating basophils. Increased local delivery of blood brings to the site more phagocytic leukocytes and plasma proteins, both of which are crucial to the defense response.

Increased capillary permeability Released histamine also increases the capillaries' permeability by enlarging the capillary pores (the spaces between the endothelial cells, see p. 324) so that plasma proteins that normally are prevented from leaving the blood can escape into the inflamed tissue.

Localized edema As the leaked plasma proteins accumulate in the interstitial fluid, they exert a colloid osmotic pressure. This elevation in local interstitial fluid osmotic pressure, coupled with the increased capillary blood pressure caused by increased local blood flow, favors enhanced filtration and reduced reabsorption of fluid across the involved capillaries. The end result of this shift in fluid balance is localized edema (see p. 331). Thus, the familiar swelling that accompanies inflammation is due to histamine-induced vascular changes. Likewise, the other well-known gross manifestations of inflammation such as redness and heat are largely attributable to the enhanced flow of warm arterial blood to the damaged tissue. Pain is caused both by local distention within the swollen tissue and by the direct effect of locally produced substances on the receptor endings of afferent neurons that supply the area. Keep in mind that these observable characteristics of the inflammatory process (swelling, redness, heat, and pain) are coincidental to the primary purpose of the vascular changes in the injured area—to increase the number of leukocytic phagocytes and crucial plasma proteins in the area (Fig. 12-2).

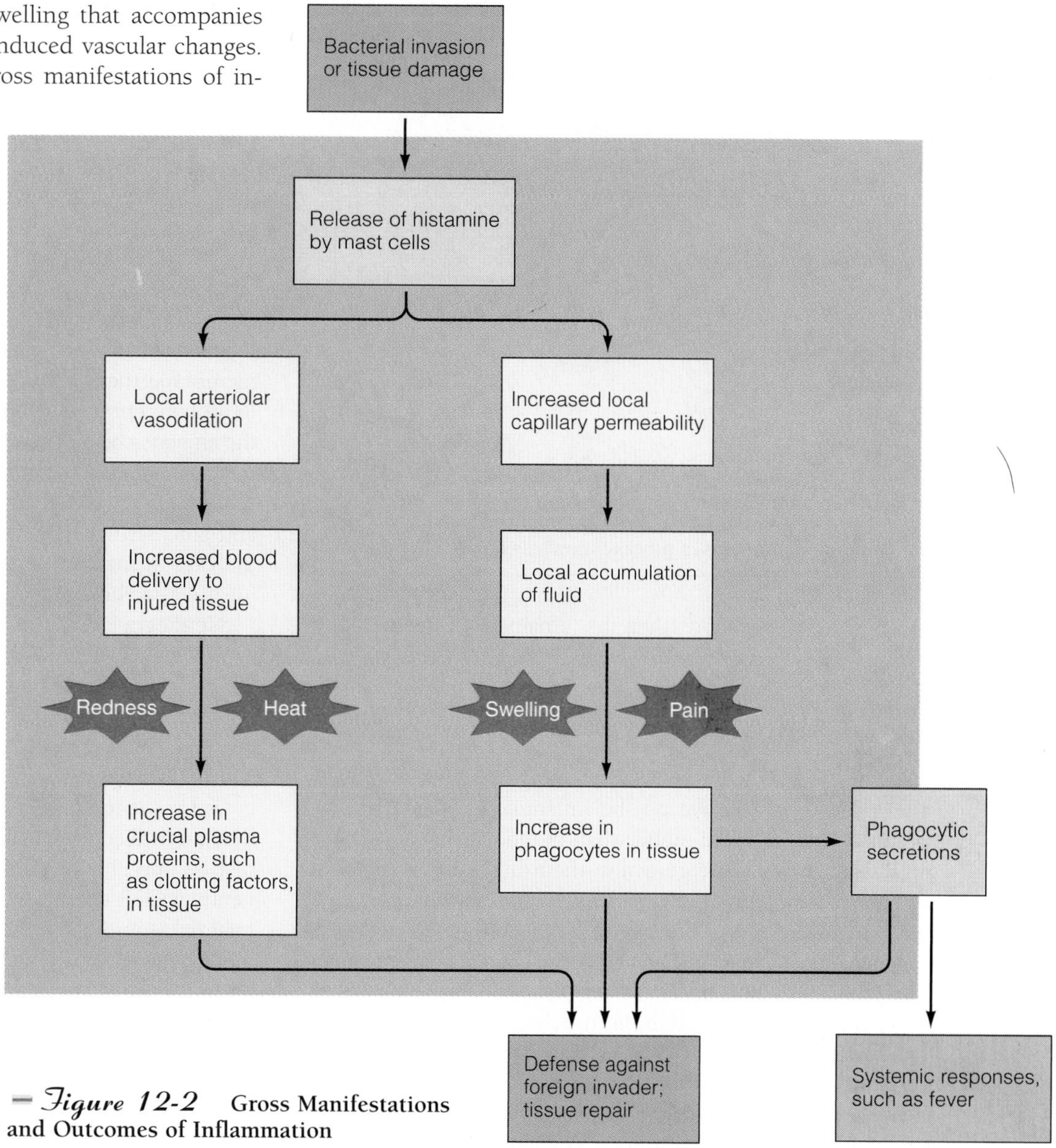

Figure 12-2 **Gross Manifestations and Outcomes of Inflammation**

Walling off of the inflamed area The leaked plasma proteins most critical to the immune response are those involved in the complement system and kinin system (to be described later) as well as clotting and anticlotting factors (see p. 365). Upon exposure to tissue thromboplastin in the injured tissue and to specific chemicals secreted by phagocytes on the scene, fibrinogen, the final factor in the clotting system, is converted into fibrin. Fibrin forms interstitial fluid clots in the spaces around the bacterial invaders and damaged cells. This walling off of the injured region from the surrounding tissues prevents or at least delays the spread of bacterial invaders and their toxic products. Later, the more slowly activated anticlotting factors gradually dissolve the clots after they are no longer needed. A few bacteria, such as streptococci, produce enzymes that act on plasminogen, an inactive plasma protein precursor, to convert it into plasmin, a proteolytic enzyme that dissolves fibrin clots. This action disrupts the walling-off process and allows the streptococcal organisms to spread extensively.

Emigration of leukocytes Within an hour after the injury, the involved area is teeming with leukocytes that have exited from the vessels. Neutrophils are the first to arrive, followed during the next eight to twelve hours by the slower-moving monocytes. The latter swell and mature into macrophages during another eight-to-twelve-hour period.

Leukocyte emigration from the blood into the tissues involves the processes of margination, diapedesis, amoeboid movement, and chemotaxis. **Margination** refers to the sticking of blood-borne leukocytes, especially neutrophils and monocytes, to the inner endothelial lining of capillaries in the affected tissue. Cell adhesion molecules (CAMs), special plasma membrane proteins that protrude from the outer surface of cells (see p. 50), are important in the process of margination. *Selectins,* one type of CAM that protrudes from the endothelial lining, cause leukocytes that are flowing by in the bloodstream to slow down and roll along the interior of the vessel, much as the nap of a carpet slows down a child's rolling toy car. This slowing down allows the leukocytes enough time to check for local activating factors—"SOS signals" from nearby injured or infected tissues. These activating factors cause the leukocytes to adhere firmly to the endothelial lining by means of interaction with another type of CAM, the *integrins.*

Soon the adhered leukocytes start exiting by a mechanism known as **diapedesis.** Assuming amoebalike behavior (see p. 40), an adhered leukocyte pushes a long, narrow projection

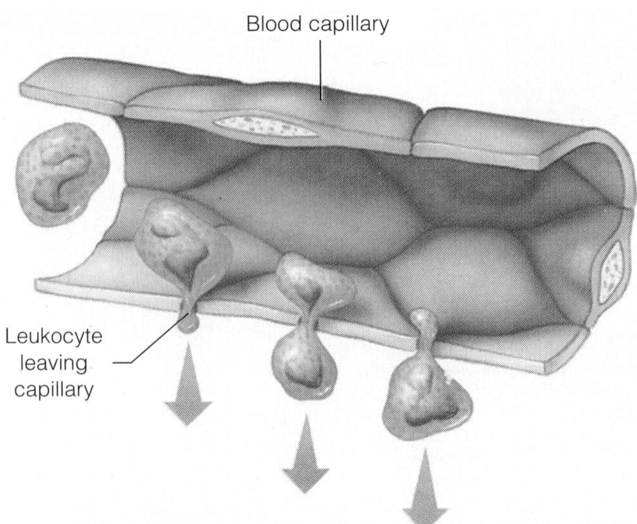

Figure 12-3 **Leukocyte Emigration from the Blood** Leukocytes emigrate from the blood into the tissues by assuming amoebalike behavior and squeezing through the capillary pores, a process known as diapedesis.

through a capillary pore; then the remainder of the cell flows forward into the projection (Fig. 12-3). In this way, the leukocyte is able to wriggle its way in about a minute through the capillary pore—even though it is much larger than the pore. Outside the vessel, the leukocyte moves in an amoeboid fashion toward the site of tissue damage and bacterial invasion. Neutrophils arrive on the inflammatory scene earliest because they are more mobile than monocytes.

Phagocytic cells are guided in their direction of migration by attraction to certain chemical mediators, or **chemotaxins,** released at the site of damage. This process is referred to as *chemotaxis*. Binding of chemotaxins with protein receptors on the plasma membrane of a phagocytic cell increases Ca^{2+} entry into the cell. Calcium, in turn, switches on the cellular contractile apparatus that leads to amoebalike crawling. Because the concentration of chemotaxins progressively increases toward the site of injury, phagocytic cells move unerringly toward this site along a chemotaxin concentration gradient.

Leukocyte proliferation Resident tissue macrophages as well as leukocytes that exited from the blood and migrated to the inflammatory site are soon joined by new phagocytic recruits from the bone marrow. Within a few hours after the onset of the inflammatory response, the number of neutrophils in the blood may increase up to four to five times that of normal. This increase is due partly to transfer into the blood of large numbers of preformed neutrophils stored in the bone marrow and partly to increased production of new neutrophils by the bone marrow. A slower-commencing but longer-lasting increase in monocyte production by the bone marrow also occurs, making larger numbers of these macrophage precursor cells available. In addition, multiplication of resident macrophages adds to the pool of these important immune cells. Proliferation of new neutrophils, monocytes, and macrophages and mobilization of stored neutrophils are stimulated by various chemical mediators released from the inflamed region.

Leukocytic destruction of bacteria Neutrophils and macrophages clear the inflamed area of infectious and toxic agents as well as tissue debris; this clearing action is the primary function of the inflammatory response. They accomplish this by both phagocytic and nonphagocytic means.

Phagocytosis involves the engulfment and intracellular degradation (breakdown) of foreign particles and tissue debris. Macrophages can engulf a bacterium in less than 0.01 second. Phagocytic cells contain an abundance of lysosomes, which are organelles filled with hydrolytic enzymes (the same type that degrade ingested food in the digestive tract). After a phagocyte has internalized targeted material, a lysosome fuses with the membrane that encloses the engulfed matter and releases its hydrolytic enzymes within the confines of the vesicle, where the enzymes begin breaking down the entrapped material (see p. 25). Many of the resultant breakdown products are low-molecular-weight organic compounds, such as amino acids and glucose, which can be used by the phagocyte itself or, in the case of macrophages, released into the extracellular fluid for use elsewhere in the body.

Nondegradable inorganic particles that have been engulfed (for example, dust, metal particles, or tattoo dyes) usually are retained indefinitely within the phagocytic cell. In fact, certain bacteria, most notably those causing tuberculosis, are likewise engulfed but not destroyed because they are resistant to lysosomal chemicals. These organisms may survive within phagocytes for many years and produce no obvious effects because they are walled off within a fibrous capsule. They cause overt disease only if allowed to escape. Such a more-or-less permanently walled-off structure in which nondestructible offending material is imprisoned is known as a **granuloma.**

Phagocytes eventually die due to accumulation of toxic by-products from foreign particle degradation or inadvertent release of destructive lysosomal chemicals into the cytosol. Neutrophils usually succumb after phagocytizing from five to twenty-five bacteria, whereas macrophages survive much longer and can engulf up to a hundred or more bacteria. Indeed, the longer-lived macrophages even clear the area of dead neutrophils in addition to other tissue debris. The **pus** that forms in an infected wound is a collection of these phagocytic cells, both living and dead; necrotic (dead) tissue liquified by lysosomal enzymes released from the phagocytes; and bacteria.

Obviously, phagocytes must be able to distinguish between normal cells and foreign or abnormal cells before accomplishing their destructive mission. Otherwise they could not selectively engulf and destroy only unwanted materials. Several different selective procedures enable phagocytes to "recognize" targets for destruction. First, dead tissue and many foreign materials have surface characteristics that differ from normal body cells. For example, the surface roughness accompanying traumatic injury to cells increases the likelihood of the cellular debris being phagocytized. Second, foreign particles are deliberately marked for phagocytic ingestion by being coated with chemical mediators generated by the immune system. Such body-produced chemicals that make bacteria more susceptible to phagocytosis are known as **opsonins.** The most important opsonins are antibodies and one of the activated proteins of the complement system.

An opsonin enhances phagocytosis by linking the foreign cell to a phagocytic cell (Fig. 12-4). One portion of an opsonin molecule binds nonspecifically to the surface of an invading bacterium, whereas another portion of the opsonin molecule binds to receptor sites specific for it on the phagocytic cell's plasma membrane. This link ensures that the bacterial victim does not have a chance to "get away" before the phagocyte can perform its lethal attack.

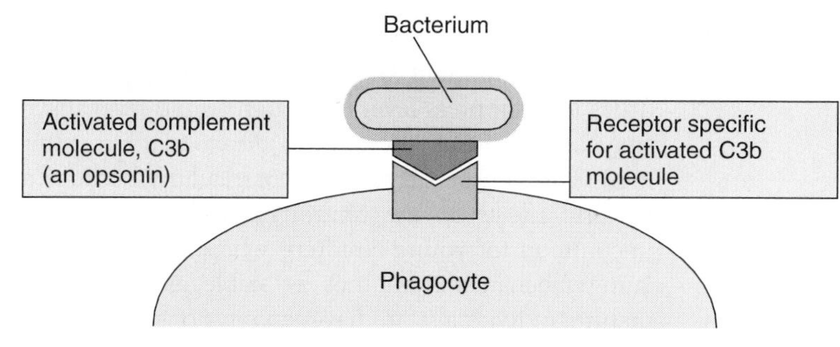

Figure 12-4 **Mechanism of Opsonin Action** One of the activated complement molecules, C3b, links a foreign cell, such as a bacterium, and a phagocytic cell by nonspecifically binding with the surface of the foreign cell and specifically binding with plasma membrane receptors on the surface of the phagocyte. This link ensures that the foreign victim does not escape before it can be engulfed by the phagocyte.

Mediation of the inflammatory response by phagocyte-secreted chemicals Microbe-stimulated phagocytes release many chemicals that function as mediators of the inflammatory response. These chemical mediators induce a broad range of interrelated immune activities, varying from local responses to the systemic manifestations that accompany microbe invasion. The following are among the most important functions of phagocytic secretions:

1. Some of the chemicals, which are highly destructive, directly kill microbes that have not been phagocytized. For example, macrophages secrete *nitric oxide,* a multipurpose chemical that is toxic to nearby microbes (see p. 318). As a more subtle means of destruction, neutrophils secrete **lactoferrin**, a protein that tightly binds with iron, making it unavailable for use by invading bacteria. Bacterial multiplication depends on high concentrations of available iron. Therefore, phagocytes are able to destroy foreign invaders both by phagocytosis and by nonphagocytic means.
2. Phagocytic secretions stimulate the release of histamine from mast cells in the vicinity. Histamine, in turn, induces the local vasodilation and increased vascular permeability that accompany inflammation.
3. Some phagocytic chemical mediators trigger both the clotting and anticlotting systems to first enhance the walling-off process and then facilitate the gradual dissolution of the fibrous clot after it is no longer needed.
4. Other secretions split **kininogens,** which are inactive precursor plasma proteins synthesized by the liver, into active **kinins.** In particular, **kallikrein,** which is released by neutrophils, can accomplish this activation. Once kinins are generated by kallikrein, they augment a variety of inflammatory events. Specifically, kinins (a) stimulate several key steps in the complement system; (b) reinforce the vascular changes induced by histamine; (c) activate nearby pain receptors, and are thus partially responsible for the soreness associated with inflammation; and (d) act as powerful chemotaxins to induce phagocyte migration into the affected area. In positive-feedback fashion, the newly arriving neutrophils release kallikrein, which can generate still more kinins that chemotactically entice more neutrophils to join the battle (Fig. 12-5).

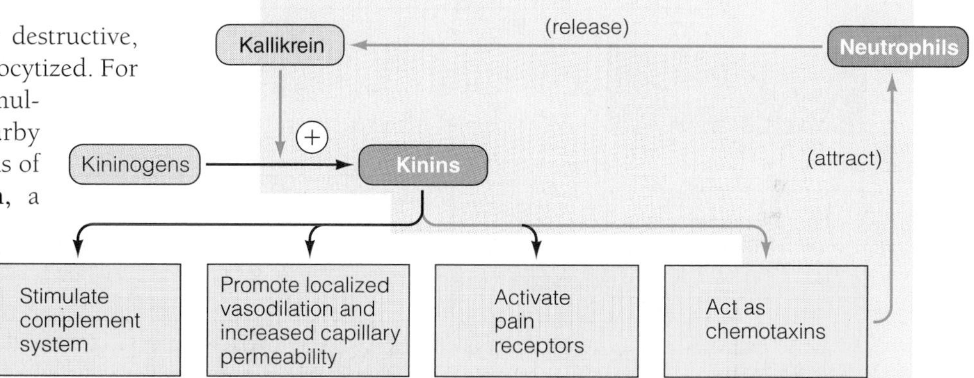

Figure 12-5 **Relationship between the Kinin System and Neutrophils in an Inflammatory Response** The green arrows indicate a positive-feedback loop between the kinin system and neutrophils during an inflammatory response. Neutrophils release kallikrein, which converts kininogens, inactive plasma proteins, into kinins. Active kinins augment a variety of inflammatory events, including acting as chemotaxins to attract more neutrophils to the area.

5. One chemical in particular released by phagocytes, **endogenous pyrogen (EP),** induces the development of fever. This response occurs especially when the invading organisms have spread into the bloodstream. Endogenous pyrogen is believed to cause release within the hypothalamus of *prostaglandins,* locally acting chemical messengers that "turn up" the hypothalamic "thermostat" that regulates body temperature. The function of the resultant elevation in body temperature in fighting infection remains unclear. The fact that fever is such a common systemic manifestation of inflammation suggests that the higher body temperature plays an important beneficial role in the overall inflammatory response. Recent evidence strongly supports this supposition. For example, higher temperatures augment the process of phagocytosis and increase the rate at which the many enzyme-dependent inflammatory activities proceed. One theory proposes that an elevated body temperature increases bacterial requirements for iron while reducing the plasma concentration of iron. According to this theory, the resultant iron deficiency interferes with bacterial multiplication. Resolving the con-

troversial issue of whether a fever can be beneficial is extremely important in view of the widespread use of drugs that suppress fever. Although a mild fever may possibly be beneficial, there is no doubt that an extremely high fever can be detrimental, particularly because of its harmful effects on the central nervous system. It is not uncommon for young children, whose temperature-regulating mechanisms are not as stable as those of more mature individuals, to have convulsions in association with high fevers.

6. One phagocyte secretory product decreases the plasma concentration of iron by altering iron metabolism within the liver, spleen, and other tissues, thus reducing the amount of iron available to support bacterial multiplication.
7. The same phagocytic secretion also stimulates **granulopoiesis,** the synthesis and release of neutrophils and other granulocytes by the bone marrow. This effect is especially prominent in response to bacterial infections.
8. This particular phagocytic mediator also stimulates the release of **acute-phase proteins** from the liver. This collection of proteins, which have not yet been sorted out by scientists, exert a multitude of wide-ranging effects associ-

Figure 12-6 **Phagocyte Responsibilities** Phagocytes not only destroy foreign particles or damaged cells by phagocytosis but also secrete chemical mediators that destroy the targeted material by nonphagocytic means and augment many aspects of inflammation and other immune processes. The pink boxes represent final outcomes of phagocytic actions.

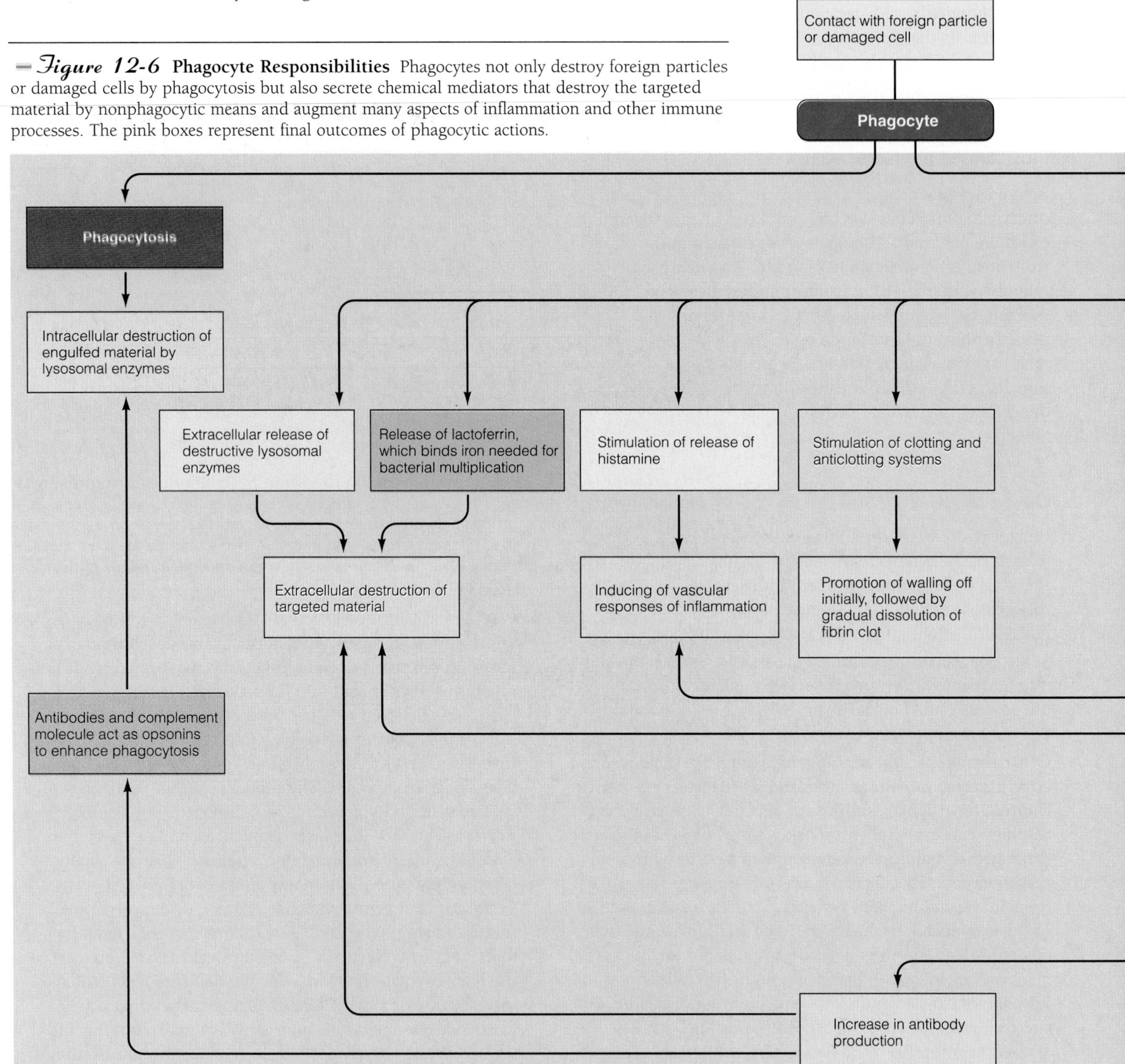

ated with the inflammatory process, tissue repair, and immune cell activities.

The latter three effects (plasma iron reduction, enhanced granulopoiesis, and release of acute-phase proteins) are all attributed to **leukocyte endogenous mediator (LEM),** a chemical mediator secreted by macrophages. Evidence suggests that LEM and EP are the same substance or at least very closely related.

9. Another effect of phagocytic secretions is to enhance the proliferation and differentiation of both B and T lymphocytes, which, in turn, are responsible for antibody production and cell-mediated immunity, respectively. Specifically, **interleukin 1 (IL-1),** a secretory product released by macrophages, is responsible for this effect on lymphocytes. Amazingly, IL-1 is identical to (or closely related to) EP and LEM. Apparently, the same chemical substance is responsible for a diverse array of effects throughout the body, all of which are geared toward defending the body against infection or tissue injury. In fact, the release of EP/LEM/IL-1 can be elicited by stressful situations not associated with microbe invasion (for example, during endurance exercise). Accordingly, this mediator may be part of a generalized nonspecific protective response.

This list of events that are augmented by chemicals secreted by phagocytes is not complete, but it serves to illustrate the diversity and complexity of responses elicited by these mediators. As will soon become evident, there are other important macrophage-lymphocyte interactions that are not dependent on the release of chemicals from phagocytic cells. Thus, the effect that phagocytes, especially macrophages, ultimately have on microbial invaders far exceeds their "engulf and destroy" tactics (Fig. 12-6).

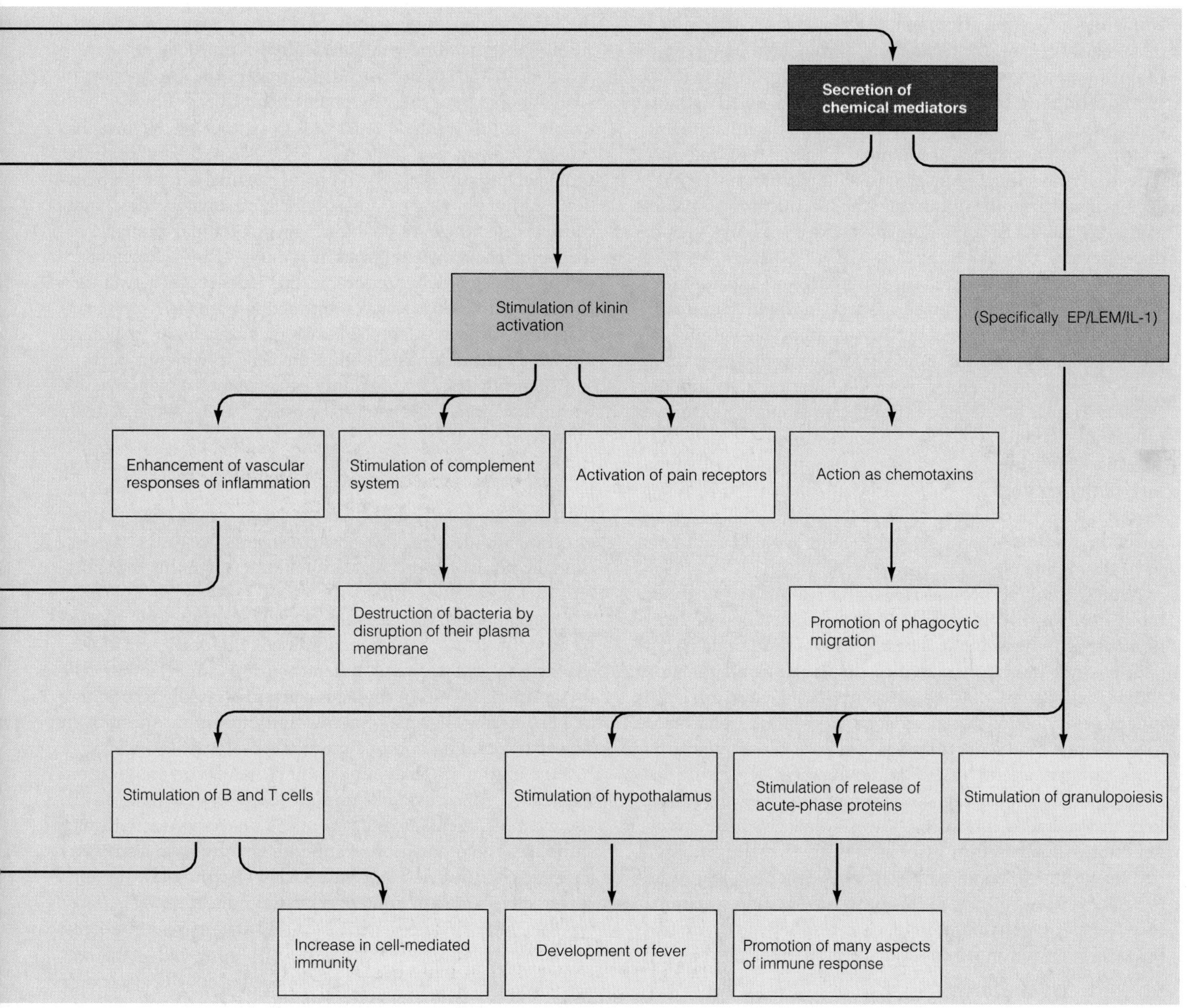

Tissue repair The ultimate purpose of the inflammatory process is to isolate and destroy injurious agents and to clear the area for tissue repair. In some tissues (for example, skin, bone, and liver), the healthy organ-specific cells surrounding the injured area undergo cell division to replace the lost cells, often accomplishing perfect repair. In nonregenerative tissues such as nerve and muscle, however, lost cells are replaced by **scar tissue.** Fibroblasts, a type of connective tissue cell, start to divided rapidly in the vicinity and secrete large quantities of the protein collagen (see p. 51), which fills in the region vacated by the lost cells and results in the formation of scar tissue. Even in a tissue as readily replaceable as the skin, scar formation sometimes takes place when complex underlying structures, such as hair follicles and sweat glands, are permanently destroyed by deep wounds.

Salicylates and glucocorticoid drugs suppress the inflammatory response.

Numerous drugs can suppress the inflammatory process; the most effective are the *salicylates* and related compounds (aspirin-type drugs) and *glucocorticoids* (drugs similar to the steroid hormone cortisol, which is secreted by the adrenal cortex; see p. 663). Salicylates interfere with the inflammatory response by decreasing histamine release, resulting in a reduction in swelling, redness, and pain. Furthermore, salicylates reduce fever by inhibiting the production of prostaglandins, the local mediators of endogenous pyrogen–induced fever.

Glucocorticoids, which are potent anti-inflammatory drugs, suppress almost every aspect of the inflammatory response. In addition, they destroy lymphocytes within lymphoid tissue and reduce antibody production. These therapeutic agents are useful for treating undesirable immune responses, such as allergic reactions (for example, poison ivy rash and asthma) and the inflammation associated with arthritis. Unfortunately, however, by suppressing inflammatory and other immune responses that localize and eliminate bacteria, such therapy also reduces the body's ability to resist infection. For this reason, glucocorticoids should be administered discriminately.

What about the normal role of the adrenal cortical hormone cortisol in this regard? Is cortisol secretion counterproductive to the immune defense system? Traditionally, cortisol has not been considered to display anti-inflammatory activity at normal blood concentrations. Instead, anti-inflammatory action has been attributed only to pharmacological levels of glucocorticoids (that is, at blood concentrations brought about by the administration of cortisol-like drugs that are higher than the normal physiological range). Recent evidence, however, suggests that cortisol, whose secretion is increased in response to any stressful situation, does exert anti-inflammatory activity even at normal physiological levels. According to this proposal, the anti-inflammatory effect of cortisol modulates stress-activated immune responses, preventing them from overshooting and thus protecting us against damage by potentially overreactive defense mechanisms.

Interferon transiently inhibits multiplication of viruses in most cells.

Besides the inflammatory response, another nonspecific defense mechanism is the release of **interferon** from virus-infected cells. Interferon is not a single type of molecule but a family of similar proteins. Interferon briefly provides nonspecific resistance to viral infections by transiently interfering with replication of the same or unrelated viruses in other host cells. In fact, interferon was named for its ability to "interfere" with viral replication.

When a virus invades a cell, the presence of viral nucleic acid induces the cell's genetic machinery to synthesize interferon, which is secreted into the extracellular fluid. Interferon acts as a "whistle-blower," forewarning healthy cells of potential viral attack and helping them prepare to resist such an attack. Once released from a virus-infected cell, interferon binds with receptors on the plasma membranes of healthy neighboring cells or even distant cells that it reaches through the bloodstream, signaling these cells to prepare for the possibility of impending viral attack. Interferon does not have a direct antiviral effect; instead, it triggers the production of virus-blocking enzymes by potential host cells. Binding with interferon induces these other cells to synthesize enzymes that can break down viral messenger RNA (see p. B–7) and inhibit protein synthesis, both of which are essential for viral replication. Although viruses are still able to invade these forewarned cells, they are unable to govern cellular protein synthesis for their own replication (Fig. 12-7). Researchers recently learned that interferon exerts its effect by activating a heretofore unknown signal transduction pathway (see p. 64) involving a newly identified class of intracellular chemicals known as **Janus kinases** within the cells to which it binds.

The newly synthesized inhibitory enzymes remain inactive within the tipped-off potential host cell unless it is actually invaded by a virus, at which time the enzymes are activated by the presence of viral nucleic acid. This activation requirement protects the cell's own messenger RNA and protein-synthesizing machinery from unnecessary inhibition by these enzymes should viral invasion not occur. Because activation can take place only during a limited time span, this is a short-term defense mechanism.

Interferon is released nonspecifically from any cell infected by any virus and, in turn, can induce temporary self-protective activity against many different viruses in any other cells that it reaches. Thus, it provides a general, rapidly responding defense strategy against viral invasion until more specific but slower-responding immune mechanisms come into play.

In addition to facilitating inhibition of viral replication, interferon reinforces other immune activities (Table 12-2). For example, it enhances macrophage phagocytic activity and stimulates the production of antibodies. Interferon also exerts anticancer as well as antiviral effects. Fortunately, its anticancer effects are not limited to virally induced cancers. Most types of human cancer are not caused by viruses. Interferon markedly enhances the actions of cell-killing cells—the *natural killer cells* and a special type of T lymphocyte, *cytotoxic T cells*—which attack and destroy both virus-infected cells and cancer cells. Furthermore, interferon itself slows cell division and suppresses tumor growth.

The discovery of interferon's antiviral and anticancer effects led to great excitement and anticipation about its potential role as a miracle therapeutic weapon against enemies ranging from the common cold virus to fatal invasive cancers. For a quarter of a century after its discovery in 1957, however, it was impossible to collect human interferon in sufficient quantities to make it feasible to study its therapeutic possibilities. (Interferon is species specific; each species of higher animals produces a type of interferon capable of conferring protection only to other members of the same species.) During the 1980s, technological advances enabled researchers to introduce human interferon genes into bacterial genes so that these bacteria produce human interferon. Through this **recombinant DNA technology,** bacteria can be turned into interferon "factories" for large-scale commercial production of this valuable immune agent. With an unlimited supply of interferon available, much research was undertaken to determine interferon's potential effectiveness in treating viral diseases and cancer. However, many investigative studies were not as encouraging as hoped. Although interferon has not lived up to earlier expectations that it would be a miracle drug, it has been approved by the Food and Drug Administration for some forms of cancer, including one previously fatal, rare form of *leukemia* and the AIDS-associated *Kaposi's sarcoma,* as well as for *genital warts* caused by *papillomavirus.* Interferon is also the first drug approved for treating *multiple sclerosis.* As clinical studies continue, the list is expected to grow. The research thrust is toward treatment rather than prevention, since it would not be economically feasible to maintain a continuous preventive level of this substance in the population because of the transient nature of interferon's defense capabilities.

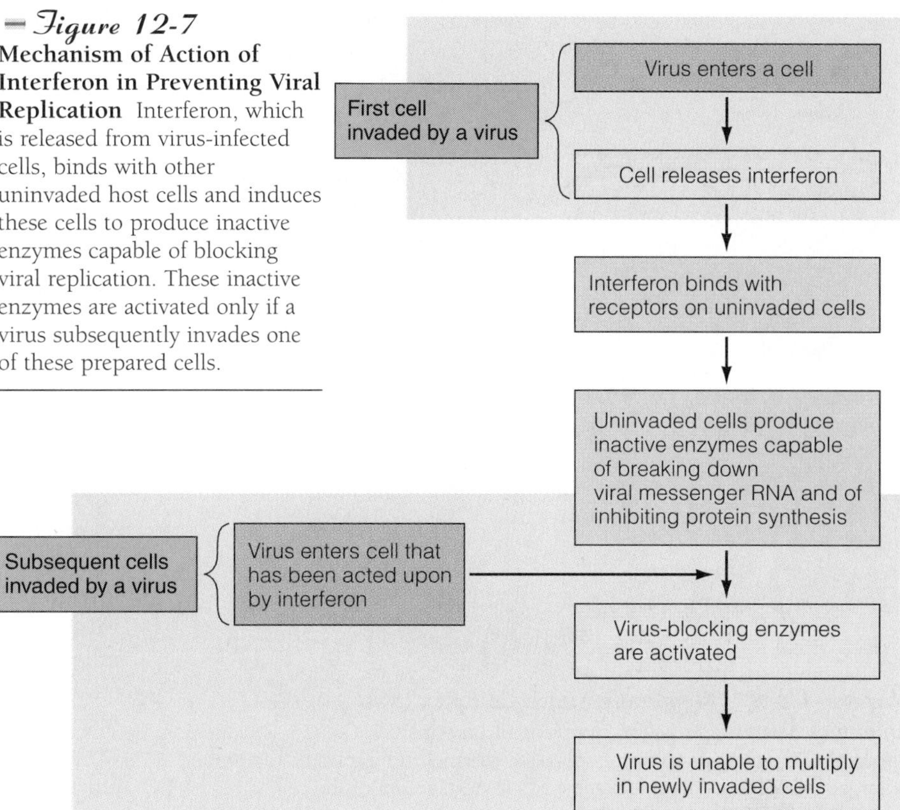

First cell invaded by a virus
Virus enters a cell
Cell releases interferon
Interferon binds with receptors on uninvaded cells
Uninvaded cells produce inactive enzymes capable of breaking down viral messenger RNA and of inhibiting protein synthesis
Subsequent cells invaded by a virus
Virus enters cell that has been acted upon by interferon
Virus-blocking enzymes are activated
Virus is unable to multiply in newly invaded cells

Figure 12-7
Mechanism of Action of Interferon in Preventing Viral Replication Interferon, which is released from virus-infected cells, binds with other uninvaded host cells and induces these cells to produce inactive enzymes capable of blocking viral replication. These inactive enzymes are activated only if a virus subsequently invades one of these prepared cells.

Natural killer cells destroy virus-infected cells and cancer cells upon first exposure to them.

Natural killer cells are naturally occurring, lymphocytelike cells that nonspecifically destroy virus-infected cells and cancer cells by directly lysing their membranes upon first exposure to them. Their mode of action and major targets are similar to those of cytotoxic T cells, but the latter can fatally attack only the specific types of virus-infected cells and cancer cells to which they have been previously exposed. Furthermore, following exposure, cytotoxic T cells require a maturation period before they are capable of launching their lethal assault. The natural killer cells provide an immediate, nonspecific defense against virus-invaded cells and cancer cells before the more specific and more abundant cytotoxic T cells become functional.

The complement system kills microorganisms directly both on its own and in conjunction with antibodies and also augments the inflammatory response.

The **complement system** is another defense mechanism brought into play nonspecifically in response to invading organisms. It can also be triggered by antibodies as part of the specific immune strategy. In fact, the system derives its name

Table 12-2
Functions of Interferon

Functions	*Antiviral Effect*	*Anticancer Effect*
Released by virus-infected cells and transiently interferes with viral replication in other host cells by inducing them to produce enzymes that destroy viral messenger RNA and inhibit virus-directed protein synthesis	✓	
Enhances macrophage phagocytic activity	✓	✓
Stimulates production of antibodies	✓	
Enhances actions of natural killer cells and cytotoxic T cells	✓	✓
Slows cell division and suppresses tumor growth		✓

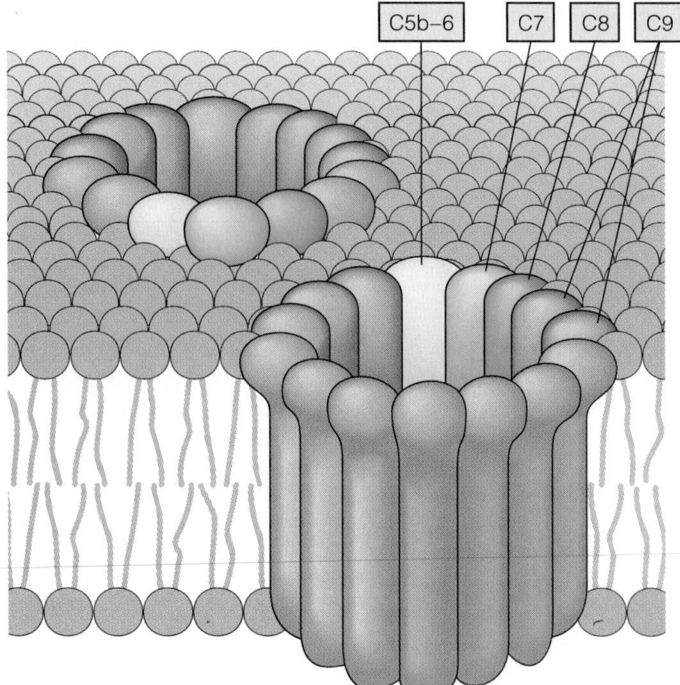

Figure 12-8 **Membrane-Attack Complex (MAC) of the Complement System** Activated complement proteins C5, C6, C7, C8, and a number of C9s aggregate to form a porelike channel in the plasma membrane of the target cell. The resultant leakage leads to destruction of the cell.

from the fact that it "complements" the action of antibodies, being the primary mechanism activated by antibodies to kill foreign cells.

In the same tradition as the clotting and anticlotting systems and the kinin system, the complement system consists of plasma proteins that are produced by the liver and circulate in the blood in inactive form. Once the first component, C1, is activated, it activates the next component, C2, and so on, in a sequential cascade of activation reactions. The five final components, C5 through C9, assemble into a large, doughnut-shaped protein complex, the **membrane-attack complex (MAC),** which attacks the surface membrane of nearby microorganisms by imbedding itself so that a large channel is created through the microbial surface membrane (Fig. 12-8). This hole-punching technique makes the membrane extremely leaky; the resulting osmotic flux of water into the victim cell causes it to swell and burst. This complement-induced lysis is the major means of directly killing microbes without phagocytizing them.

Table 12-3
Functions of the Complement System

Complement components C5 through C9, when activated, form a membrane-attack complex that punches lethal holes in the target cell

Other activated complement components augment the inflammatory process by:

- Serving as chemotaxins
- Acting as opsonins
- Promoting localized vasodilation and increased capillary permeability
- Stimulating histamine release from mast cells in the vicinity
- Activating kinins

The powerful complement cascade can be set into motion in two ways: (1) by exposure to particular carbohydrate chains (mannose) present on the surfaces of microorganisms but not found on human cells (the **alternate pathway,** a nonspecific immune response) and (2) by exposure to antibodies produced against a specific foreign invader (the **classical pathway,** a specific immune response). In either case, activation of the complement system brings about direct lysis of the invader and reinforcement of other general inflammatory tactics.

Unlike the other cascade systems, in which the sole function of the various components is activation of the next precursor in the sequence, several of the activated proteins in the complement cascade perform additional important functions on their own (Table 12-3). Besides the direct destruction of foreign cells accomplished by the membrane-attack complex, various other activated complement components augment the inflammatory process by (1) serving as chemotaxins, which attract and guide professional phagocytes to the site of complement activation (that is, the site of microbial invasion); (2) acting as opsonins by binding with microbes and thereby enhancing their phagocytosis; (3) promoting vasodilation and increased vascular permeability to increase blood flow to the invaded area; (4) stimulating the release of histamine from mast cells in the vicinity, which in turn enhances the local vascular changes characteristic of inflammation; and (5) activating kinins, which further reinforce inflammatory reactions.

What restricts the activated complement system's destructive tactics to undesirable cells, such as invading bacteria? Several of the activated components in the cascade are very unstable. Because these unstable components are able to perpetuate the sequence only in the immediate vicinity in which they are activated before they decompose, the complement attack is confined to the surface membrane of the microbe whose presence initiated activation of the system. Nearby host cells are thus spared from lytic attack.

Specific Immune Responses: General Concepts

Specific immune responses include antibody-mediated immunity accomplished by B lymphocyte derivatives and cell-mediated immunity accomplished by T lymphocytes.

A specific immune response is a selective attack aimed at limiting or neutralizing a particular offending target for which the body has been specially prepared following prior exposure to it. There are two classes of specific immune responses: **antibody-mediated**, or **humoral, immunity,** involving the production of antibodies by B lymphocyte derivatives known as *plasma cells;* and **cell-mediated immunity,** involving the production of *activated T lymphocytes,* which directly attack unwanted cells.

B and T lymphocytes (B and T cells) have different life histories and, more importantly, different properties and functions. Both types of lymphocytes, like all blood cells, are derived from common stem cells in the bone marrow. Whether a lymphocyte and all of its progeny are destined to be B or T cells depends on the site of final maturation and differentiation of the original cell in the lineage (Fig. 12-9). During fetal life and early childhood, some of the immature lymphocytes migrate through the blood to the thymus, where they undergo further processing to become T lymphocytes. The **thymus** is a lymphoid tissue located midline within the chest cavity above the heart in the space between the lungs (Fig. 12-1). Lymphocytes that mature without benefit of "thymic education" become B lymphocytes. B lymphocytes were first discovered in birds, where the maturational processing takes place in a gut-related lymphoid tissue unique to birds, the bursa of Fabricius; hence the name B cells. In humans, the site of B cell maturation and differentiation is uncertain, although it is generally assumed to be the bone marrow.

Upon being released into the blood from either the bone marrow or the thymus, mature B and T cells take up residence and establish lymphocyte colonies in the peripheral lymphoid tissues. Here, upon appropriate stimulation, they undergo cell division to produce new generations of either B or T cells, depending on their ancestry. After early childhood, most new lymphocytes are derived from these peripheral lymphocyte colonies rather than from the bone marrow.

The role of the thymus remained obscure until recently because its removal from an adult had no obvious effect. Because most of the migration and differentiation of T cells occurs early in development, the thymus gradually atrophies and becomes less important as the individual matures. It does, however, continue to produce **thymosin,** a hormone important in maintaining the T cell lineage. Thymosin enhances proliferation of new T cells within the peripheral lymphoid tissues and augments the immune capabilities of existing T cells. Recent evidence indicates that secretion of thymosin decreases after about thirty to forty years of age. This decline has been implicated as a contributing factor in aging. It is further speculated that diminishing T cell capacity with advancing age might somehow be linked to the increased susceptibility to viral infections and cancer that occurs as a person ages. T cells play an especially important role in defense against viruses and virus-induced cancer.

Lymphocytes are able to specifically recognize and selectively respond to an almost limitless variety of foreign agents as well as cancer cells. The recognition and response processes are different for B and T cells. In general, B cells recognize free-existing foreign invaders such as bacteria and their toxins and a few viruses, which they combat by secreting antibodies specific for the invaders. T cells specialize in recognizing and destroying body cells gone awry, including virus-infected cells and cancer cells.

Each of us has an estimated total of 2 trillion lymphocytes, which, if aggregated together in a mass, would be comparable to the size of the brain or liver. At any one time, the majority of these lymphocytes are concentrated in the various strategically located lymphoid tissues, but both B and T cells continually circulate among the lymph, blood, and body tissues, where they remain on constant surveillance.

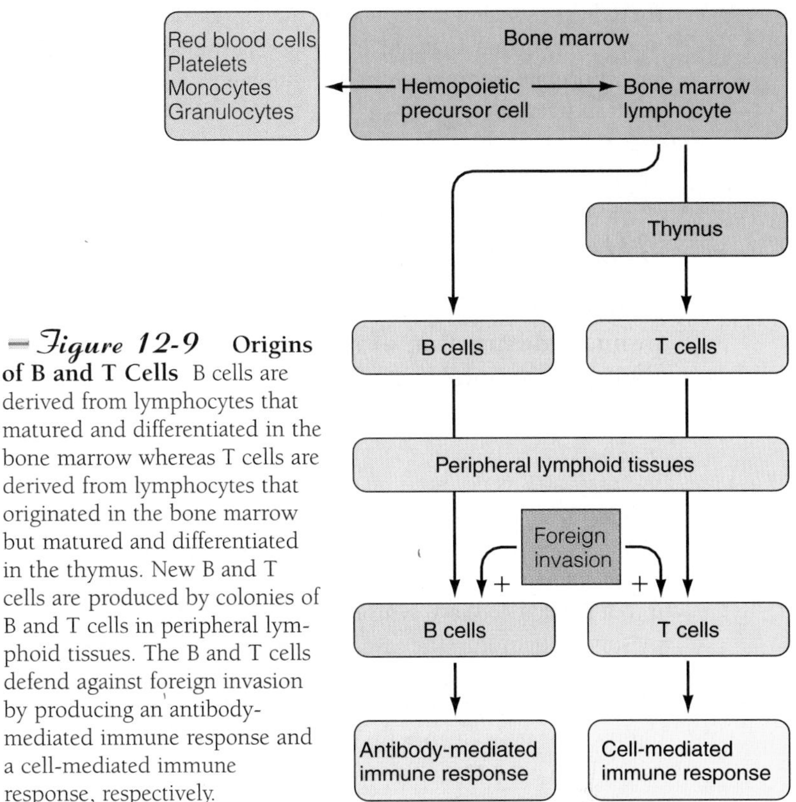

***Figure 12-9* Origins of B and T Cells** B cells are derived from lymphocytes that matured and differentiated in the bone marrow whereas T cells are derived from lymphocytes that originated in the bone marrow but matured and differentiated in the thymus. New B and T cells are produced by colonies of B and T cells in peripheral lymphoid tissues. The B and T cells defend against foreign invasion by producing an antibody-mediated immune response and a cell-mediated immune response, respectively.

An antigen induces an immune response against itself.

Both B and T cells must be able to specifically recognize unwanted cells and other material to be destroyed or neutralized as being distinct from the body's own normal cells. The presence of antigens enables lymphocytes to make this distinction. An **antigen** is a large, complex molecule that triggers a specific immune response against itself when it gains entry into the body. In general, the more complex a molecule, the greater its antigenicity. Foreign proteins are the most common antigens because of their size and structural complexity, although other macromolecules, such as large polysaccharides, can also act as antigens. Antigens may exist as isolated molecules, such as bacterial toxins, or they may be an integral part of a multimolecular structure, such as being present on the surface of an invading foreign microbe. A complex antigen may have many **antigenic determinant sites,** each capable of stimulating the production of and interacting with different antibodies. For simplicity, we will consider each antigenic determinant site to be a separate antigen.

Many low-molecular-weight organic substances that are not antigenic by themselves can become antigenic if they attach to body proteins. Such small molecules are known as **haptens.** Antibodies developed against the hapten-protein combination can react in the future against the hapten alone should it be reintroduced to the body. Examples of haptens include poison ivy toxin, various drugs (such as penicillin), and other agents that are otherwise harmless but can elicit inappropriate immune responses known as allergies in sensitized individuals.

B Lymphocytes: Antibody-Mediated Immunity

Antibodies amplify the inflammatory response to promote destruction of the antigen that stimulated their production.

Each B and T cell has receptors on its surface for binding with one particular type of the multitude of possible antigens. In the case of B cells, binding with antigen induces the cell to differentiate into a **plasma cell,** which produces antibodies that are able to combine with the specific type of antigen that stimulated the antibodies' production. During differentiation into a plasma cell, a B lymphocyte swells as the rough endoplasmic reticulum (the site for synthesis of proteins to be exported) greatly expands (■ Fig. 12-10). Because antibodies are proteins, plasma cells essentially become prolific protein factories, producing up to 2,000 antibody molecules per second. So great is the commitment of a plasma cell's protein-synthesizing machinery to antibody production that it is unable to maintain protein synthesis for its own viability and growth. Consequently, it dies after a brief five- to seven-day, highly productive life span.

Antibodies are secreted into the blood or lymph, depending on the location of the activated plasma cells, but all antibodies eventually gain access to the blood, where they are known as **gamma globulins,** or **immunoglobulins.** Antibodies are grouped into the following five subclasses based on differences in their biological activity:

- **IgM** immunoglobulin serves as the B cell surface receptor for antigen attachment and is secreted in the early stages of plasma cell response.
- **IgG,** the most abundant immunoglobulin in the blood, is produced copiously when the body is subsequently exposed to the same antigen.

  Together, IgM and IgG antibodies are responsible for most specific immune responses against bacterial invaders and a few types of viruses.
- **IgE** helps protect against parasitic worms and is the antibody mediator for common allergic responses, such as hay fever, asthma, and hives.
- **IgA** immunoglobulins are found in secretions of the digestive, respiratory, and genitourinary systems, as well as in milk and tears.
- **IgD** is present on the surface of many B cells, but its function is uncertain.

It is important to note that this classification is based on different ways in which antibodies function. It does not imply that there are only five different antibodies. Within each functional subclass, there are millions of different antibodies, each able to bind only with a specific antigen.

Antibody proteins of all five subclasses are composed of four interlinked polypeptide chains—two long, heavy chains and two short, light chains—arranged in the shape of a Y (■ Fig. 12-11). Characteristics of the arm regions of the Y determine with what antigen the antibody can bind (that is, the *specificity* of the antibody). Properties of the tail portion of the antibody, on the other hand, determine the *functional properties* of the antibody (what the antibody does once it binds with antigen). An antibody has two identical antigen-binding sites, one at the tip of each arm. These **antigen-binding fragments (Fab)** are unique for each different antibody, so that each antibody can interact only with an antigen that specifically matches it, much like a lock and key. The tremendous variation in the antigen-binding fragments of different antibodies is responsible for the extremely large number of unique antibodies that are capable of binding specifically with millions of different antigens.

In contrast to these variable Fab regions at the arm tips, the tail portion of every antibody within each immunoglobulin subclass is identical. The tail, the antibody's so-called **constant (Fc) region,** contains binding sites for particular mediators of antibody-induced activities, which vary among the different subclasses. In fact, differences in the constant region are the basis for distinguishing between the different immunoglobulin subclasses. For example, the constant tail region of IgG antibodies, when activated by antigen binding in the Fab region, binds with phago-

■ *Figure 12-10* **Comparison of an Unactivated B Cell and a Plasma Cell** Electron micrograph of (a) an unactivated B cell, or small lymphocyte, and (b) a plasma cell. A plasma cell is an activated B cell. It is filled with an abundance of rough endoplasmic reticulum distended with antibody molecules. (Contributed by Dr. Dorothea Zucker-Franklin, New York University Medical Center.)

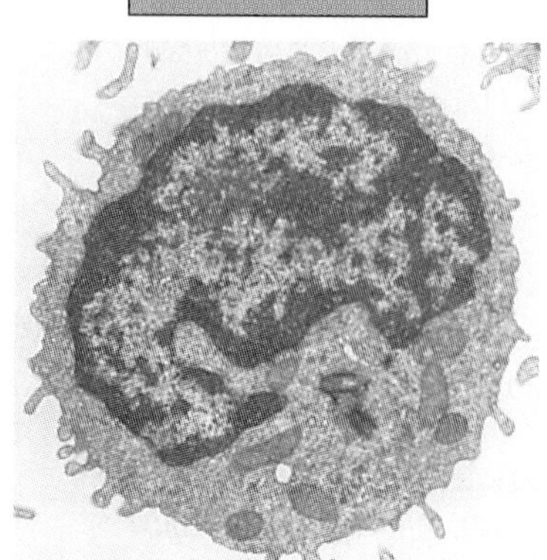

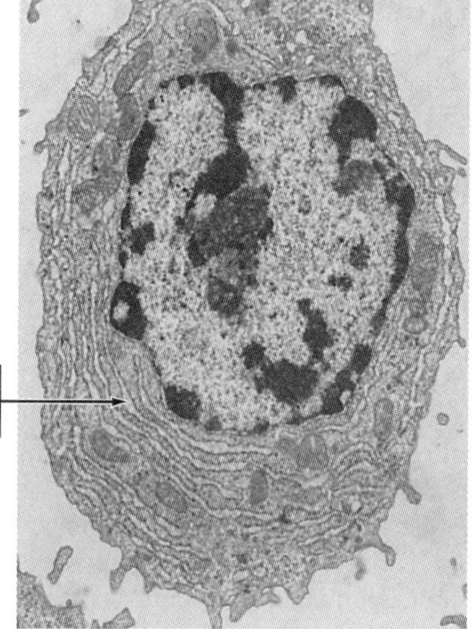

cytic cells and serves as an opsonin to enhance phagocytosis. In comparison, the constant tail region of IgE antibodies attaches to mast cells and basophils, even in the absence of antigen. When the appropriate antigen or hapten gains entry to the body and binds with the attached antibodies, this triggers the release of histamine from the affected mast cells and basophils. Histamine, in turn, induces the allergic manifestations that follow.

Immunoglobulins cannot directly destroy foreign organisms or other unwanted materials upon binding with antigens on their surfaces. Instead, antibodies exert their protective influence in one of two general ways: physical hindrance of antigens or amplification of nonspecific immune responses (Fig. 12-12).

Antibodies can physically hinder some antigens from exerting their detrimental effects. For example, by combining with bacterial toxins, antibodies can prevent these harmful chemicals from interacting with susceptible cells. This process is known as **neutralization.** Similarly, antibodies are able to bind with surface antigens on some types of viruses, preventing these viruses from entering cells, where they could exert their damaging effects. Sometimes multiple antibody molecules can cross-link numerous antigen molecules into chains or lattices of antigen-antibody complexes. The process in which foreign cells, such as bacteria or mismatched transfused red blood cells, bind together in such a clump is known as **agglutination.** When linked antigen-antibody complexes involve soluble antigens, such as tetanus toxin, the lattice can become so large that it precipitates out of solution. (**Precipitation** is the process in which a substance separates from a solution.) Within the body, these physical-hindrance mechanisms play only a minor protective role against invading agents. However, the tendency for certain antigens to agglutinate or precipitate upon forming large complexes with antibodies specific for them is useful clinically and experimentally for detecting the presence of particular antigens or antibodies. Pregnancy diagnosis tests, for example, employ this principle to detect the presence in the urine of a hormone secreted soon after conception.

Antibodies' most important function by far is to profoundly augment the nonspecific immune responses already initiated by the invaders. Antibodies mark or identify foreign material as targets for actual destruction by the complement system, phagocytes, or killer cells while enhancing the activity of these other defense systems as follows:

1. *Activation of the complement system.* When an appropriate antigen binds with an antibody, receptors on the tail portion of the antibody are able to bind with and activate C1, the first component of the complement system. This sets off the cascade of events leading to formation of the membrane-attack complex, which is specifically directed at the membrane of the invading cell that bears the antigen that initiated the activation process. In fact, antibody is the most powerful activator of the complement system. The biochemical attack subsequently unleashed against the invader's membrane is the most important mechanism by which antibodies exert their protective influence. Furthermore, various activated complement components enhance virtually every aspect of the inflammatory process. Note that the same complement system is activated by an antigen-antibody complex regardless of the type of antigen. Although the binding of antigen to antibody is highly specific, the outcome, which is determined by the antibody's constant tail region, is identical for all activated antibodies within a given subclass; for example, all IgG antibodies activate the same complement system.
2. *Enhancement of phagocytosis.* As mentioned previously, antibodies, especially IgG, act as opsonins. The tail portion of an antigen-bound IgG antibody is able to bind with a receptor on the surface of a phagocyte and subsequently promote the phagocytosis of the antigen-containing victim attached to the antibody.
3. *Stimulation of killer (K) cells.* The binding of antibody to antigen can also induce attack of the antigen-bearing cell by a **killer (K) cell.** K cells are similar to natural killer cells except that K cells require the target cell to be coated with antibodies before they can destroy it by lysing its plasma membrane. K cells have receptors for the constant tail portion of antibodies.

In these ways, antibodies, though unable to directly destroy invading bacteria or other undesirable material, bring about destruction of the antigens to which they are specifically attached by amplifying other nonspecific lethal defense mechanisms.

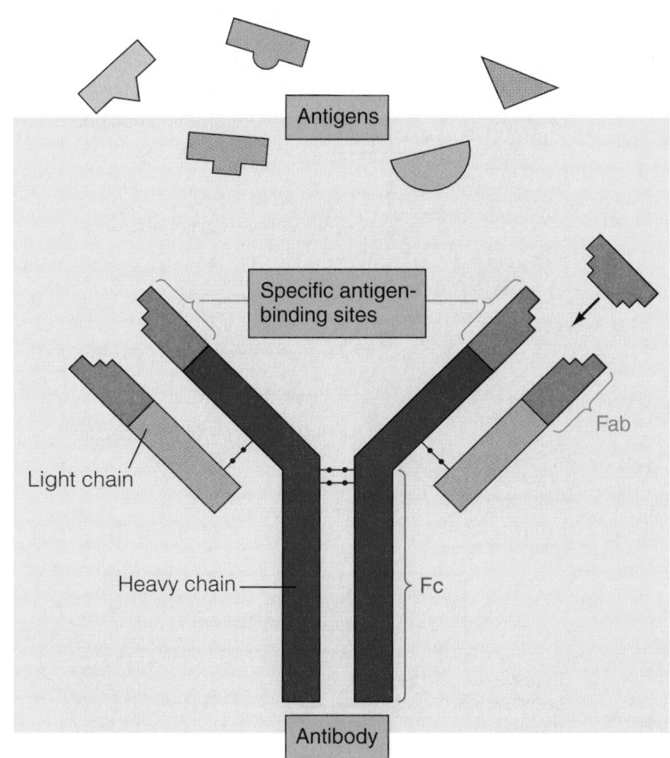

Figure 12-11 **Antibody Structure** An antibody is Y-shaped. It is able to bind only with the specific antigen that "fits" its antigen-binding sites (Fab) on the arm tips. The tail region (Fc) binds with particular mediators of antibody-induced activities.

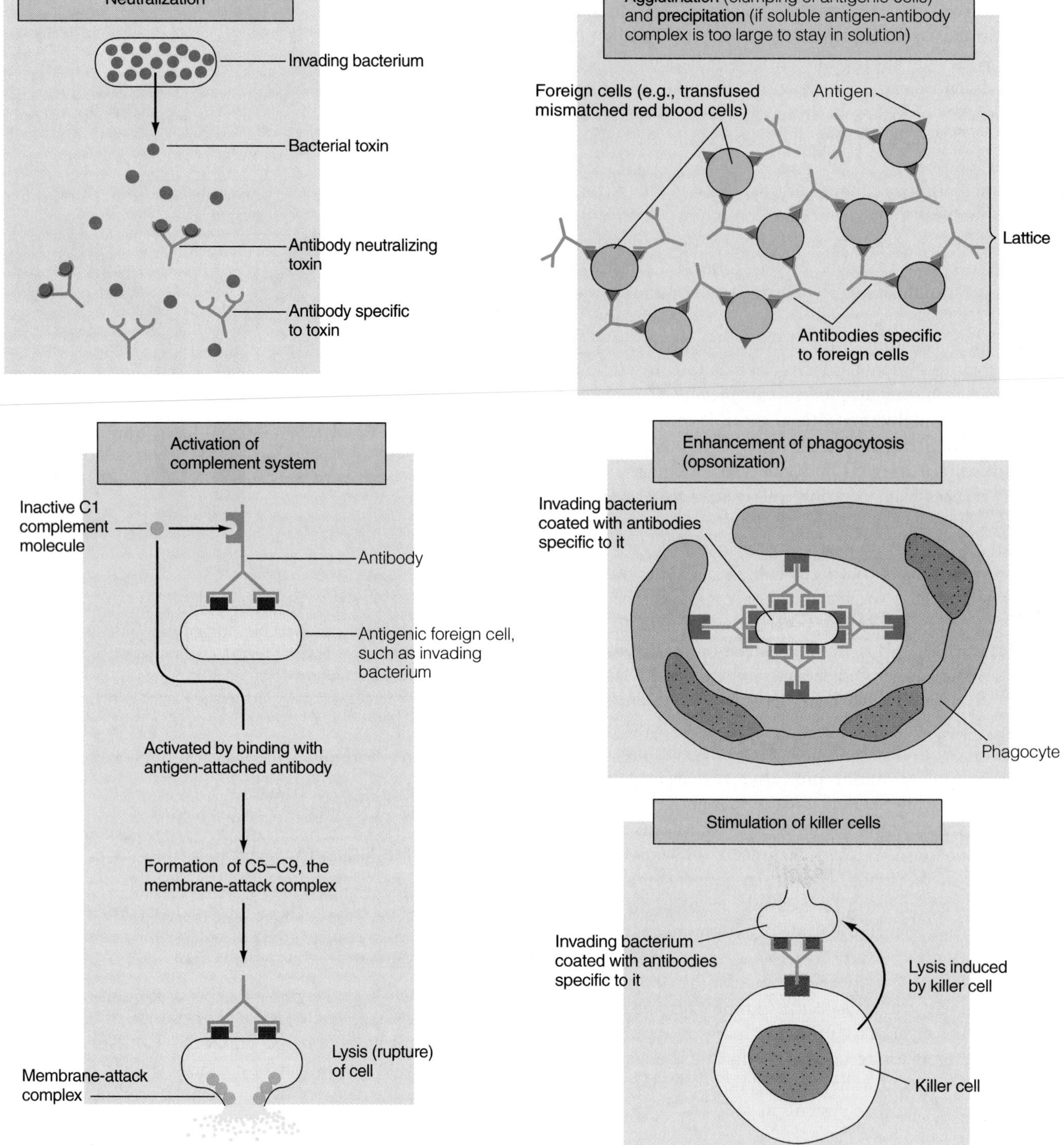

Figure 12-12 **How Antibodies Help Eliminate Invading Microbes** Antibodies physically hinder antigens through (1) neutralization or (2) agglutination and precipitation. Antibodies amplify nonspecific immune responses by (1) activating the complement system, (2) enhancing phagocytosis by acting as opsonins, and (3) stimulating killer cells.

Occasionally, an overzealous antigen-antibody response can inadvertently cause damage to normal cells as well as to invading foreign cells. Typically, antigen-antibody complexes, formed in response to foreign invasion, are removed by phagocytic cells after having revved up nonspecific defense strategies. If large numbers of these complexes are continuously produced, however, the phagocytes are unable to clear away all of the immune complexes formed. Antigen-antibody complexes that are not removed continue to activate the complement system, among other things. Excessive amounts of activated complement and other inflammatory agents may "spill over," damaging the surrounding normal cells as well as the unwanted cells. Furthermore, destruction is not necessarily restricted to the initial site of inflammation. Antigen-antibody complexes may freely circulate and become trapped in the kidneys, joints, brain, small vessels of the skin, and elsewhere, causing widespread inflammation and tissue damage. Such damage produced by immune complexes is referred to as an **immune-complex disease,** which can be a complicating outcome of bacterial, viral, or parasitic infection.

More insidiously, immune-complex disease can also occur as a result of overzealous inflammatory activity prompted by the presence of immune complexes formed by "self-antigens" (proteins synthesized by the person's own body) and antibodies erroneously produced against them. *Rheumatoid arthritis* is brought about in this way.

Each antigen stimulates a different clone of B lymphocytes to produce antibodies.

Consider the diversity of foreign molecules that a person can potentially encounter during a lifetime. Yet each B lymphocyte is preprogrammed to respond to only one of these millions of different antigens. Other antigens cannot combine with the same B cell and induce it to secrete different antibodies. The astonishing implication is that each of us is equipped with millions of different preformed B lymphocytes, at least one for every possible antigen that we might ever encounter—including those specific for synthetic substances that do not exist in nature.

According to early immunological theory, antibodies were believed to be "made to order" whenever a foreign antigen gained entry to the body. In contrast, the currently accepted **clonal selection theory** proposes that diverse B lymphocytes are produced during fetal development, each capable of synthesizing antibody against a particular antigen before ever being exposed to it. All offspring of a particular ancestral B lymphocyte form a family of identical cells, or a **clone,** which is committed to producing the same specific antibody. B cells remain dormant, not actually secreting their particular antibody product until (or unless) they come into contact with the appropriate antigen. When an antigen gains entry to the body, it "selects" (that is, activates) the particular clone of B cells that bear receptors on their surface uniquely specific for that antigen, hence the term "clonal selection theory" (Fig. 12-13).

Figure 12-13 **Clonal Selection Theory** (a) The B cell clone specific to the antigen proliferates and differentiates into plasma cells and memory cells. (b) Plasma cells secrete antibodies that bind with free antigen (antigen not attached to B cells). (c) Memory cells expand the specific clone and are primed and ready for subsequent exposure to the same antigen.

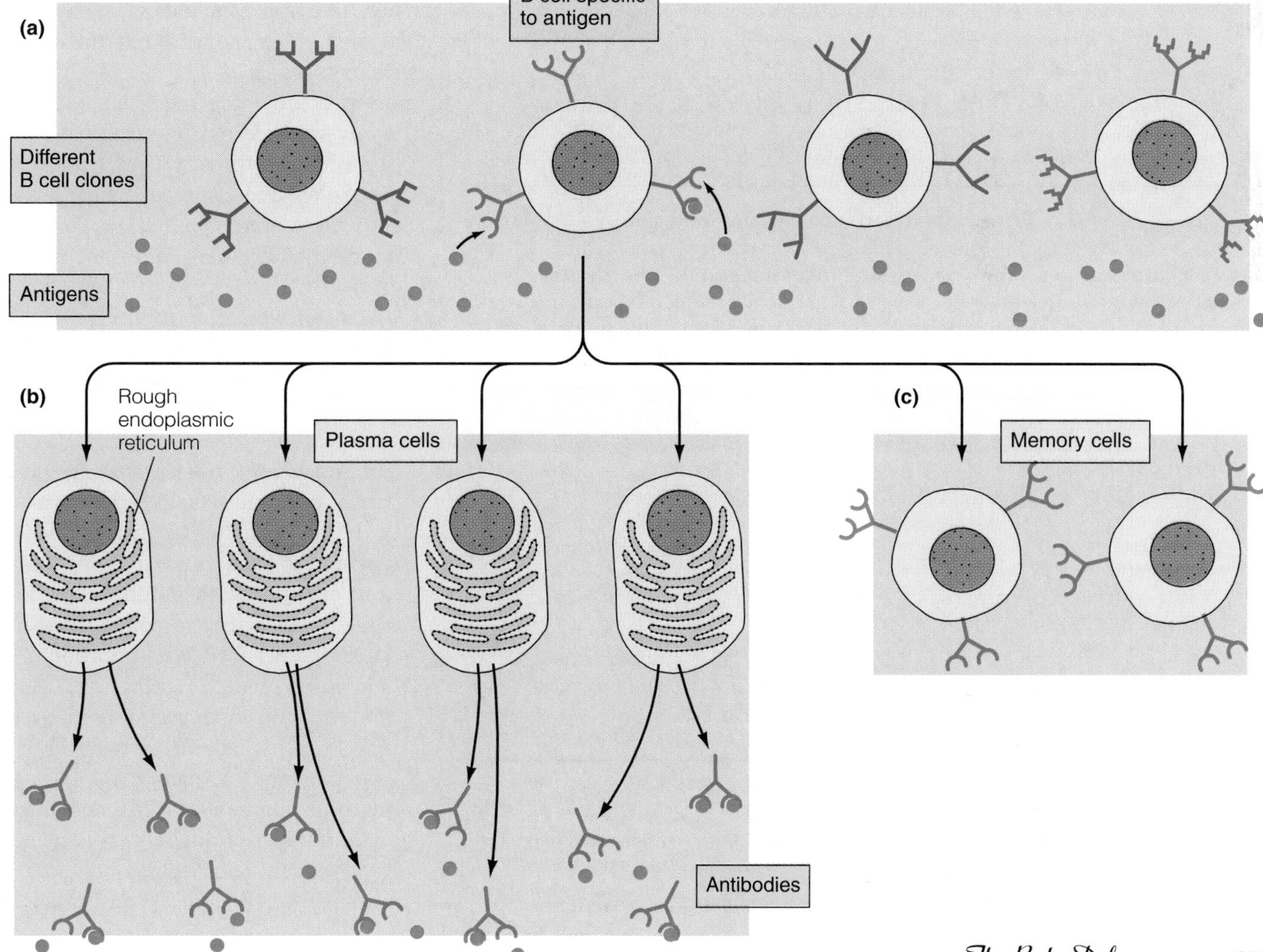

The first antibodies produced by a newly formed B cell are IgM immunoglobulins, which are inserted into the cell's plasma membrane rather than being secreted. Here they serve as receptor sites for binding with a specific kind of antigen, almost like "advertisements" for the kind of antibody the cell can produce. Binding of the appropriate antigen to a B cell amounts to "placing an order" for the manufacture and secretion of large quantities of that particular antibody.

Antigen binding causes the activated B cell clone to multiply and differentiate into two cell types—*plasma cells* and *memory cells*. Most progeny are transformed into plasma cells, which are prolific producers of customized antibodies that contain the same antigen-binding sites as the surface receptors. However, plasma cells switch to the production of IgG antibodies, which are secreted rather than remaining membrane bound. In the blood, the secreted antibodies combine with invading free (not bound to lymphocytes) antigen, marking it for destruction by the complement system, phagocytic ingestion, or other means.

Not all of the new B lymphocytes produced by the specifically activated clone differentiate into antibody-secreting plasma cells. A small proportion of them become **memory cells**, which do not participate in the current immune attack against the antigen but instead remain dormant and expand the specific clone. Should the person ever be exposed to the same antigen again, these memory cells are primed and ready for even more immediate action than were the original lymphocytes in the clone.

During initial contact with a microbial antigen, the antibody response is delayed for several days until plasma cells are formed and does not reach its peak for a couple of weeks (■ Fig. 12-14). This response is known as the **primary response.** Meanwhile, symptoms characteristic of the particular microbial invasion persist until either the invader succumbs to the mounting specific immune attack against it or the infected individual dies. After reaching the peak, the antibody levels gradually decline over a period of time, although some circulating antibody from this primary response may persist for a prolonged period. Long-term protection against the same antigen, however, is primarily attributable to the memory cells. If the same antigen ever reappears, the long-lived memory cells launch a more rapid, more potent, and longer-lasting **secondary response** than occurred during the primary response. This swifter, more powerful immune attack is frequently adequate to prevent or minimize overt infection upon subsequent exposures to the same microbe, forming the basis of long-term immunity against a specific disease.

The original antigenic exposure that induces the formation of memory cells can occur through either actually having the disease or being vaccinated (■ Fig. 12-15). During vaccination, the individual is deliberately exposed to a pathogen that has been stripped of its disease-inducing capability but can still induce antibody formation against it. (See the boxed feature on p. 392, ♦Concepts, Challenges, and Controversies.)

Even though each of us has essentially the same original pool of different B lymphocyte clones, the pool gradually becomes appropriately biased to respond most efficiently to each individual's particular antigenic environment. Those clones specific for antigens to which an individual is never exposed remain dormant for life, whereas those specific for antigens in the individual's environment typically become expanded and enhanced through the formation of highly responsive memory cells.

Memory cells are not formed for some diseases, so no lasting immunity is conferred by an initial exposure, as in the case of "strep throat." The course and severity of the disease are the same each time a person is reinfected with a microbe that the immune system does not "remember," regardless of the number of prior exposures.

Considering the millions of different antigens against which each of us has the potential to actively produce antibodies, how is it possible for an individual to have such a tremendous diversity of B lymphocytes, each capable of producing a different antibody? Antibodies are proteins that are produced in accordance with a nuclear DNA blueprint. Because all cells of the body, including the antibody-producing cells, contain the same nuclear DNA, it is hard to imagine how enough DNA could be packaged within the nuclei of every cell to code for the millions of different antibodies (a different portion of the genetic code being used by each B cell clone), along with all of the other genetic instructions used by other cells. It is now known that only a relatively small number of gene fragments actually code for antibody synthesis, but these fragments are cut, reshuffled, and spliced in a vast number of different combinations during B cell development. Each different combination gives rise to a unique B cell clone. Even further diversification of

■ *Figure 12-14* **Primary and Secondary Immune Responses** (a) Primary response on first exposure to a microbial antigen. (b) Secondary response on subsequent exposure to the same microbial antigen. Note that the primary response does not peak for a couple of weeks, whereas the secondary response peaks in a week. Also, the magnitude of the secondary response is 100 times that of the primary response. (The relative antibody response is in the logarithmic scale.)

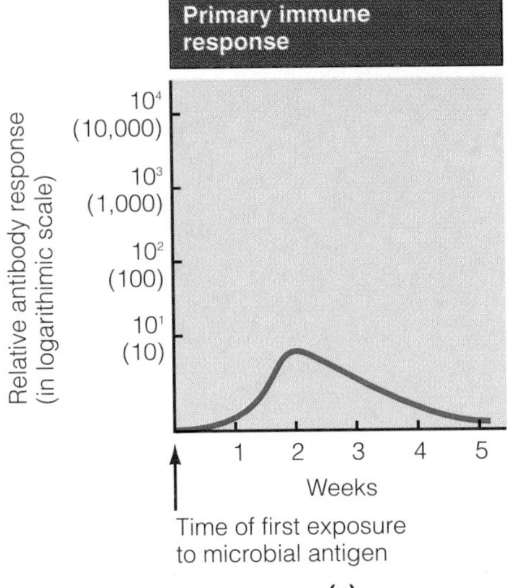

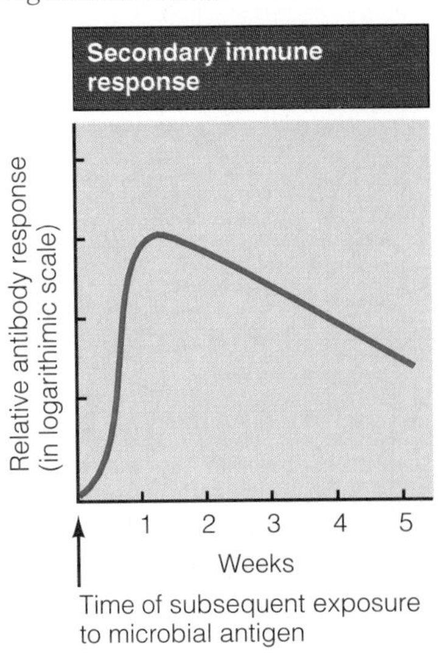

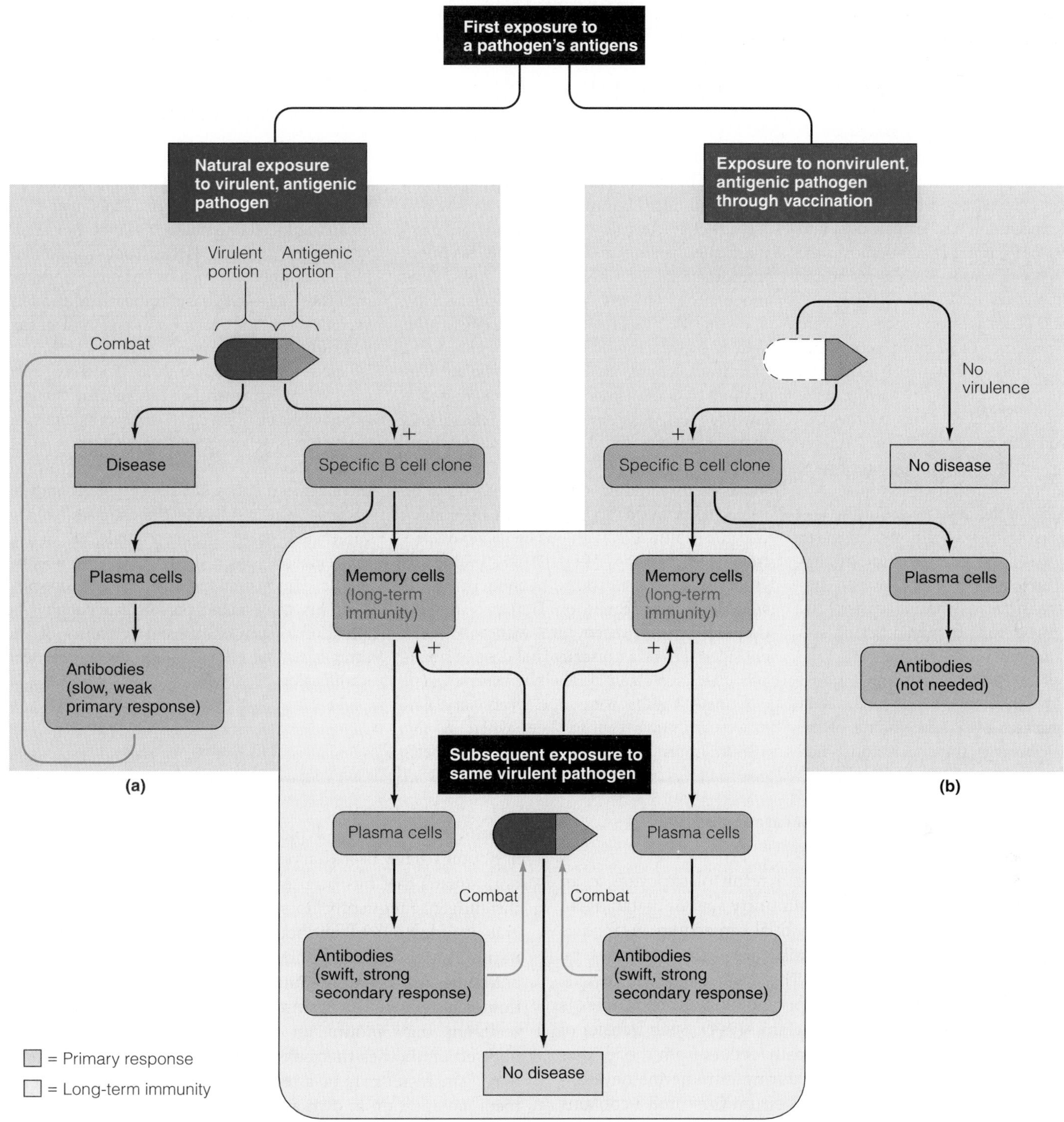

Figure 12-15 **Means of Acquiring Long-Term Immunity** Long-term immunity against a pathogen can be acquired through having the disease or being vaccinated against it. (a) Exposure to a virulent (disease-producing) pathogen. (b) Vaccination with a modified pathogen that is no longer virulent (that is, can no longer produce disease) but is still antigenic. In both cases, long-term memory cells are produced that mount a swift, secondary response that prevents or minimizes symptoms on a subsequent natural exposure to the same virulent pathogen.

antibody genes is subsequently accomplished by somatic mutation (see p. B–14). The antibody genes of already-formed B cells are highly prone to mutations in the region that codes for the variable antigen-binding sites on the antibodies. Each different mutant cell in turn gives rise to a new clone. Thus, the great diversity of antibodies is made possible by the reshuffling of a small set of gene fragments during B cell development as well as by further somatic mutation in already-formed B cells. In this way, a huge antibody repertoire is possible using only a modest share of the genetic blueprint.

CONCEPTS, CHALLENGES, AND CONTROVERSIES

Vaccination: A Victory Over Many Dreaded Diseases

Modern society has come to hope and even expect that vaccines can be developed to protect us from almost any dreaded infectious disease. This expectation has been brought into sharp focus by our current frustration over the inability to date to develop a successful vaccine against HIV, the virus that causes AIDS.

Nearly 2,500 years ago, our ancestors were aware of the existence of immune protection. Writing about a plague that struck Athens in 430 B.C., Thucydides observed that the same person was never attacked twice by this disease. However, the ancients did not understand the basis of this protection, so they were unable to manipulate it to their advantage.

Early attempts at deliberately acquiring lifelong protection against smallpox, a dreaded disease that was highly infectious and frequently fatal (up to 40% of the sick died), consisted of intentionally exposing oneself by coming into direct contact with a person suffering from a milder form of the disease. The hope was to protect against a future fatal bout of smallpox by deliberately inducing a mild case of the disease. By the beginning of the seventeenth century, this technique had evolved into using a needle to extract small amounts of pus from active smallpox pustules (the fluid-filled bumps on the skin, which leave a characteristic depressed scar or "pox" mark after healing) and introducing this infectious material into healthy individuals. This inoculation process was accomplished by applying the pus directly to slight cuts in the skin or by inhaling dried pus.

Edward Jenner, an English physician, was the first to demonstrate that immunity against cowpox, a disease similar to but less serious than smallpox, could also protect humans against smallpox. Having observed that milkmaids who acquired cowpox seemed to be protected from smallpox, Jenner in 1796 inoculated a healthy boy with pus he had extracted from cowpox boils. After the boy recovered, Jenner (not being restricted by modern ethical standards of research on human subjects) deliberately inoculated him with what was considered to be a normally fatal dose of smallpox infectious material. The boy survived.

Jenner's results were not taken seriously, however, until a century later when, in the 1880s, Louis Pasteur, the first great experimental immunologist, extended Jenner's technique. Pasteur demonstrated that the disease-inducing capability of organisms could be greatly reduced (attenuated) so that they could no longer produce disease but would still induce antibody formation when introduced into the body—the basic principle of modern vaccines. His first vaccine was against anthrax, a deadly disease of sheep and cows. Pasteur isolated and heated anthrax bacteria, then injected these attenuated organisms into a group of healthy sheep. A few weeks later at a gathering of fellow scientists, Pasteur injected these vaccinated sheep as well as a group of unvaccinated sheep with fully potent anthrax bacteria. The result was dramatic—all of the vaccinated sheep survived while all of the unvaccinated sheep died. Pasteur's notorious public demonstrations such as this, coupled with his charismatic personality, caught the attention of physicians and scientists of the time, sparking the development of modern immunology.

Active immunity is self-generated; passive immunity is "borrowed."

The production of antibodies as a result of exposure to an antigen is referred to as **active immunity** against that antigen. A second way in which an individual can acquire antibodies is by the direct transfer of antibodies actively formed by another person (or animal). The immediate "borrowed" immunity conferred upon receipt of preformed antibodies is known as **passive immunity** (Table 12-4). Such transfer of antibodies of the IgG class normally occurs from the mother to the fetus across the placenta during intrauterine development. In addition, a mother's colostrum (first milk) contains IgA antibodies that provide further protection for breast-fed babies. Passively transferred antibodies are usually broken down in less than a month, but meanwhile the newborn is provided important immune protection (essentially the same as its mother's) until it can begin actively mounting its own immune responses. Antibody-synthesizing ability does not develop for about a month after birth.

Passive immunity is sometimes employed clinically to provide immediate protection or to bolster resistance against an extremely virulent infectious agent or potentially lethal toxin to which a person has been exposed (for example, rabies virus, tetanus toxin in nonimmunized individuals, and poisonous snake venom). Typically, the administered preformed antibodies have been harvested from another source (often nonhuman) that has been exposed to an attenuated form of the antigen. Frequently, horses or sheep are used in the deliberate production of antibodies to be collected for passive immunizations. Although injection of serum containing these antibodies **(antiserum** or **antitoxin)** is beneficial in providing immediate protection against the specific disease or toxin, the recipient may mount an immune response against the injected antibodies themselves because they are foreign proteins. The result may be a severe allergic reaction to the treatment, a condition known as **serum sickness.**

Natural immunity is actually a special case of actively acquired immunity.

Certain antibodies were once thought to occur naturally in the blood. Antibodies associated with blood types are the classic example of "natural antibodies." The surface membranes of human erythrocytes contain inherited antigens that vary depending on blood type. With the major blood group system, the **ABO system,** the erythrocytes of individuals with type A blood contain A antigens; those with type B blood contain B antigens; those with type AB blood have both A and B antigens; and those with type O blood do not have any A or B red blood cell surface antigens. Antibodies against erythrocyte antigens not present on the body's own erythrocytes begin to appear in human plasma after a person is about six months of age. Accordingly, the plasma of type A blood contains anti-B antibodies; type B blood contains anti-A antibodies; no antibodies related to the ABO system are present in type AB blood; and both anti-A and anti-B antibodies are present in type O blood. Typically, one would expect antibody production against A or B antigen to be induced only if blood containing the alien antigen were injected into the body. However, high levels of these antibodies are found in the plasma of individuals who have never been exposed to a different type blood. Consequently, these were considered to be naturally occurring antibodies, that is, produced without any known exposure to the antigen. It is now known that individuals are unknowingly exposed at an early age to small amounts of A- and B-like antigens associated with common intestinal bacteria. Antibodies produced against these foreign antigens coincidentally also interact with a nearly identical foreign-blood-group antigen, even upon first exposure to it.

If a person is administered blood of an incompatible type, two different antigen-antibody interactions take place. By far the more serious consequences arise from the effect of the antibodies in the recipient's plasma on the incoming donor erythrocytes. The effect of the donor's antibodies on the recipient's erythrocyte-bound antigens is less important unless a large amount of blood is transfused, because the donor's antibodies are so diluted by the recipient's plasma that little red blood cell damage takes place in the recipient.

Antibody interaction with erythrocyte-bound antigen may result in agglutination (clumping) or hemolysis (rupture) of the attacked red blood cells. Agglutination and hemolysis of donor red blood cells by antibodies in the recipient's plasma can lead to a sometimes fatal **transfusion reaction** (— Fig. 12-16). Agglutinated clumps of incoming donor cells can plug small blood vessels. In addition, one of the most lethal consequences of mismatched transfusions is acute kidney failure caused by the release of large amounts of hemoglobin from damaged donor erythrocytes. If the free hemoglobin in the plasma rises above a critical level, it will precipitate in the kidneys and block the urine-forming structures, leading to acute kidney shutdown.

Because type O individuals do not have any A or B antigens, their erythrocytes will not be attacked by either anti-A or anti-B antibodies, so they are considered to be **universal donors.** Their blood can be transfused into persons of any blood type. However, type O individuals can receive only type O blood because the anti-A and anti-B antibodies present in their plasma will attack either A or B antigens in incoming blood. In contrast, type AB individuals are called **universal recipients.** Lacking both anti-A and anti-B antibodies, they can accept donor blood of any type, although they can donate blood only to other AB persons. Because their erythrocytes possess both A and B antigens, their cells would be attacked if transfused into individuals with antibodies against either of these antigens.

The terms *universal donor* and *universal recipient* are somewhat misleading, however. In addition to the ABO system, numerous other erythrocyte antigens and plasma antibodies can cause transfusion reactions, the most important of which is the Rh factor. Individuals who possess the **Rh factor** (an erythrocyte antigen first observed in rhesus monkeys, hence the designation *Rh*) are said to have *Rh-positive* blood, whereas those lacking the Rh factor are considered to be *Rh-negative.* In contrast to the ABO system, no naturally occurring anti-

Table 12-4 Active versus Passive Immunity

Characteristic	*Active Immunity*	*Passive Immunity*
Exposure to the antigen required for immunity to develop?	Yes, either through natural exposure during an infection or through artificial exposure during vaccination	No
Source of circulating antibodies	Antibodies self-generated on exposure to the antigen	Antibodies "borrowed"—that is, produced from another source and transferred to the individual, either naturally across the placenta or through the milk from the mother to the offspring, or artificially through injection of antibodies harvested from a vaccinated animal
Injection required?	Injection of attenuated antigen required for artificial active immunity	Injection of borrowed antibodies required for artificial passive immunity
Time to develop resistance to antigen exposure	Several weeks (primary response to the antigen) to several days (secondary response to the antigen) required for antibody production	Immediate on acquisition of borrowed antibodies
Duration of resistance	Long—may be lifelong	Short—less than a month
Indication for artificial acquisition of this type of immunity	Well before potential exposure to a virulent pathogen to allow enough time for the appropriate B cell clone to respond; boosters as needed to maintain active immunity	Immediately after exposure to an extremely virulent virus or a potentially lethal toxin when protection is needed immediately (rather than waiting for the more slowly responding active immunity to develop)

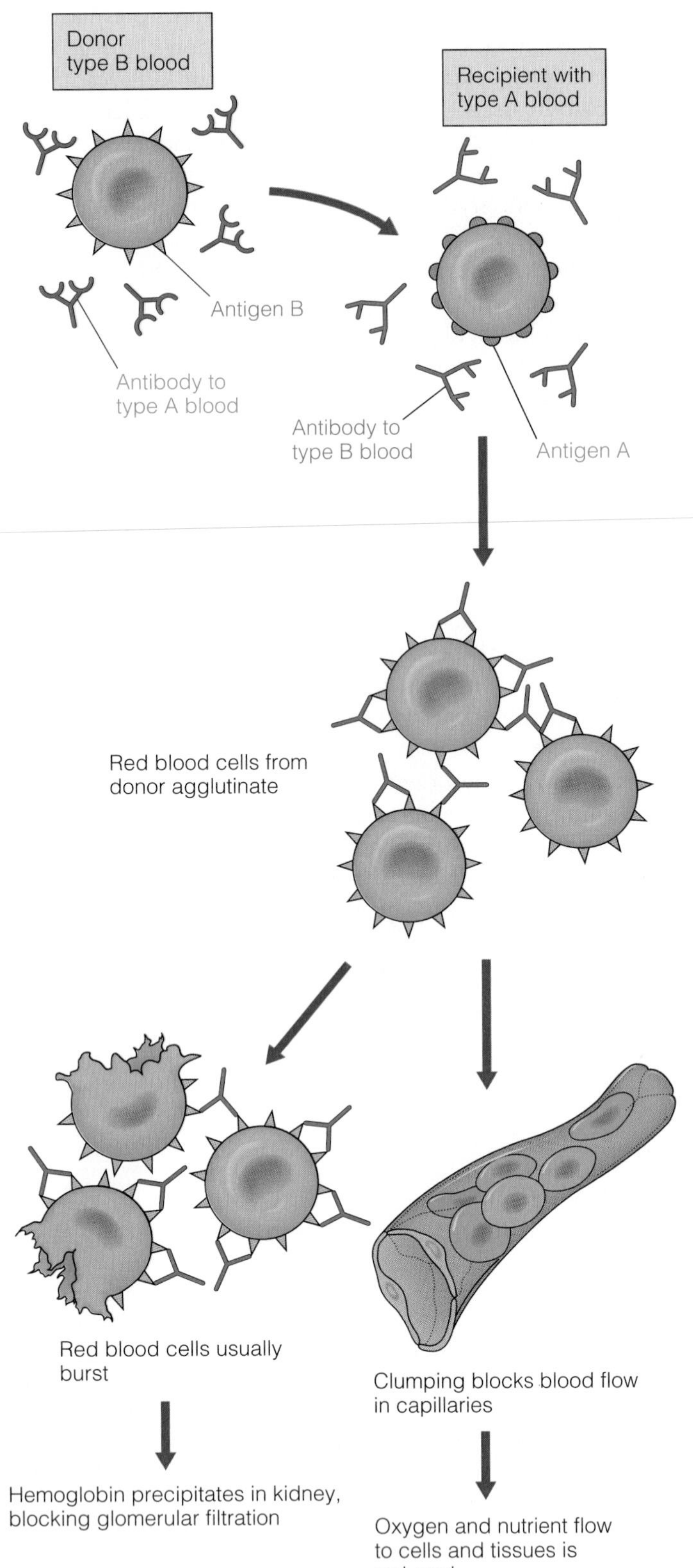
Donor
type B blood
Recipient with
type A blood
Antigen B
Antibody to
type A blood
Antibody to
type B blood
Antigen A
Red blood cells from
donor agglutinate
Red blood cells usually
burst
Clumping blocks blood flow
in capillaries
Hemoglobin precipitates in kidney,
blocking glomerular filtration
Oxygen and nutrient flow
to cells and tissues is
reduced

Figure 12-16 **Transfusion Reaction** A transfusion reaction resulting from type B blood being transfused into a recipient with type A blood.

bodies develop against the Rh factor. Anti-Rh antibodies are produced only by Rh-negative individuals when (and if) they are first exposed to the foreign Rh antigen present in Rh-positive blood. A subsequent transfusion of Rh-positive blood could produce a transfusion reaction in such a sensitized Rh-negative person. Rh-positive individuals, in contrast, never produce antibodies against the Rh factor that they themselves possess. Therefore, Rh-negative individuals should be administered only Rh-negative blood, whereas Rh-positive persons can safely receive either Rh-negative or Rh-positive blood. The Rh factor is of particular medical importance when an Rh-negative mother develops antibodies against the erythrocytes of an Rh-positive fetus she is carrying, a condition known as **erythroblastosis fetalis,** or **hemolytic disease of the newborn.**

Except in extreme emergencies, it is safest to individually cross-match blood before a transfusion is undertaken even though the ABO and Rh typing is already known, because there are approximately twelve other minor human erythrocyte antigen systems. Compatibility is determined by mixing the red blood cells from the potential donor with plasma from the recipient. If no clumping occurs, the blood is considered to be adequately matched for transfusion.

In addition to being an important consideration in transfusions, the various blood group systems are also of legal importance in disputed paternity cases, because the erythrocyte antigens are inherited. In recent years, however, DNA "finger-printing" has become a more definitive test.

Lymphocytes respond only to antigens that have been processed and presented to them by macrophages.

B cells typically cannot perform their task of antibody production without assistance from macrophages and, in most cases, from T cells as well (Fig. 12-17). Relevant B cell clones are not able to recognize and produce antibodies in response to "raw" foreign antigens entering the body; a B cell clone must be formally "introduced" to the antigen before it will react to it. Invading organisms or other antigens are first engulfed by macrophages, which cluster around the appropriate B cell clone and handle the formal introduction. During phagocytosis, the macrophage processes the raw antigen intracellularly and then "presents" the processed antigen by exposing it on the outer surface of the macrophage's plasma membrane in such a way that the adjacent B cells can recognize and be activated by it. When a macrophage engulfs a foreign microbe, it digests the microbe into antigenic peptides (small protein fragments). Each antigenic peptide is then bound to an **MHC molecule,** which is synthesized within the endoplasmic reticulum–Golgi complex. An MHC molecule has a deep groove into which a variety of antigenic peptides can bind, depending on what the macrophage has engulfed. Loading of the antigenic peptide onto an MHC molecule takes place in a newly discovered specialized organelle within antigen-presenting cells, the **compartment for peptide loading,** or **CPL.** The MHC molecule then transports the bound antigen to the cell surface where it is presented to passing lymphocytes. In addition, these antigen-presenting macrophages secrete **interleukin 1,** a multipurpose chemical mediator that enhances the differentiation and proliferation of the now-activated B cell clone. Interleukin 1 (alias endogenous pyro-

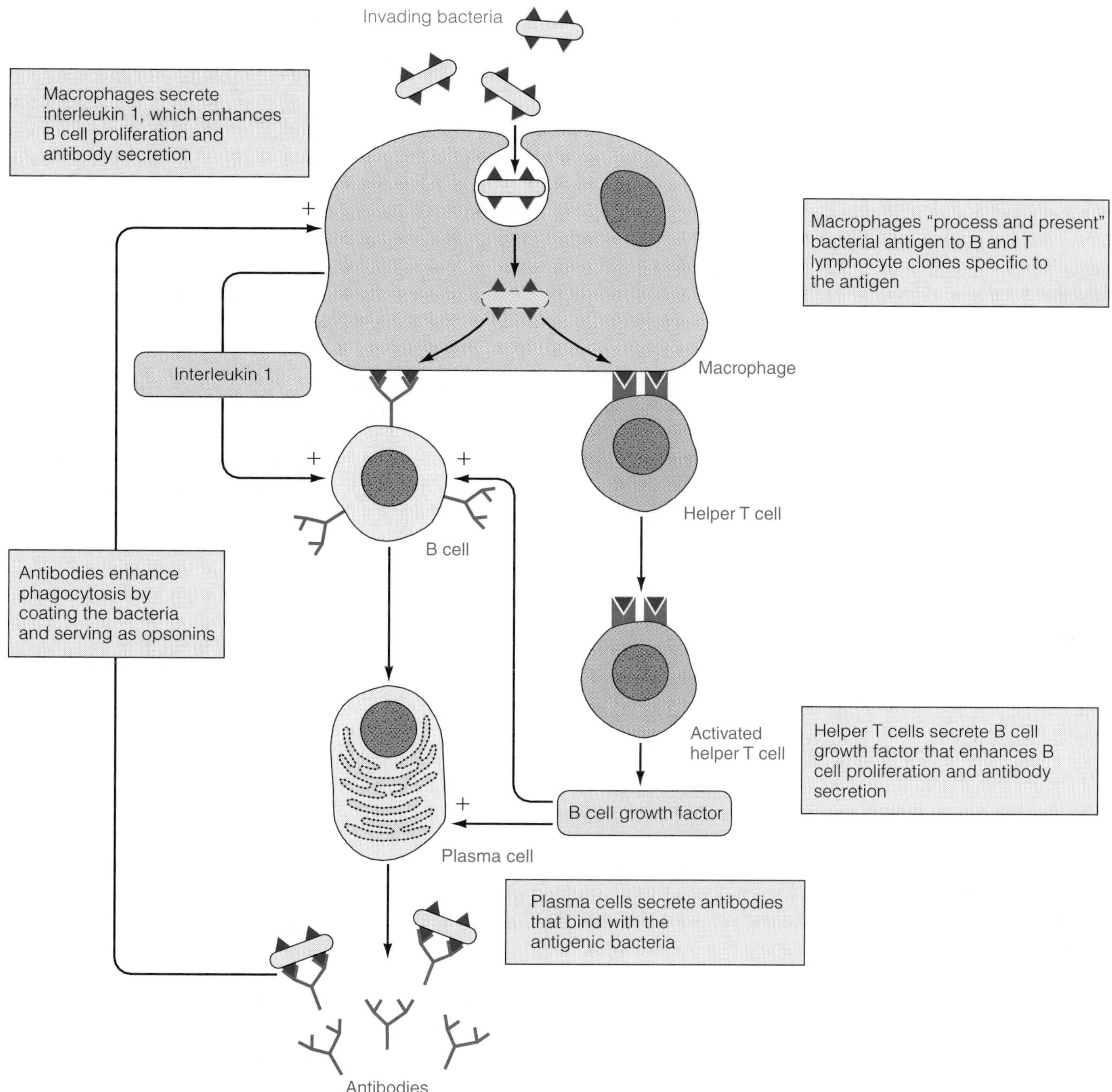
Invading bacteria
Macrophages secrete interleukin 1, which enhances B cell proliferation and antibody secretion
+
Macrophages "process and present" bacterial antigen to B and T lymphocyte clones specific to the antigen
Interleukin 1
Macrophage
+
+
Helper T cell
B cell
Antibodies enhance phagocytosis by coating the bacteria and serving as opsonins
Activated helper T cell
Helper T cells secrete B cell growth factor that enhances B cell proliferation and antibody secretion
+
B cell growth factor
Plasma cell
Plasma cells secrete antibodies that bind with the antigenic bacteria
Antibodies

Figure 12-17 **Synergistic Interactions among Macrophages, B Cells, and Helper T Cells** B and T cells cannot react to a newly entering foreign antigen until the antigen has been processed and presented to them by macrophages. These antigen-presenting cells also secrete interleukin 1, which stimulates proliferation of the activated B cells. These B cells are transformed into plasma cells, which produce antibodies against the antigen. Activated helper T cells secrete B cell growth factor, which further stimulates B cell proliferation and antibody production. The antibodies not only lead to the demise of the foreign antigen but also serve as opsonins to enhance phagocytosis by the macrophages.

gen/leukocyte endogenous mediator) is also largely responsible for the fever and malaise accompanying many infections. In collaborative fashion, activated lymphocytes secrete antibodies that, among other things, enhance further phagocytic activity.

Many antigens are similarly presented to T cells. One specialized class of T lymphocytes, called helper T cells, help B cells upon being activated by macrophage-presented antigen. The helper T cells secrete a chemical mediator, **B cell growth factor,** which further contributes to B cell function in concert with the interleukin 1 secreted by macrophages. Therefore, mutually supportive interactions among macrophages, B cells, and helper T cells synergistically reinforce the phagocyte-antibody immune attack against the foreign intruder. ▮Table 12-5 summarizes the nonspecific and specific immune strategies that defend against bacterial invasion.

Table 12-5 **Nonspecific and Specific Immune Responses to Bacterial Invasion**

Nonspecific Immune Mechanisms	*Specific Immune Mechanisms*
Inflammation	Processing and presenting of bacterial antigen by macrophages to B cells specific to the antigen
Engulfment of invading bacteria by resident tissue macrophages	Proliferation and differentiation of the activated B cell clone into plasma cells and memory cells
Histamine-induced vascular responses to enhance delivery of increased blood flow to the area, bringing in additional immune-effector cells and plasma proteins	Secretion by plasma cells of customized antibodies, which specifically bind to invading bacteria
Walling off of the invaded area by a fibrin clot	Enhancement by interleukin 1 secreted by macrophages
Emigration of neutrophils and monocytes/macrophages to the area to engulf and destroy foreign invaders and to remove cellular debris	Enhancement by helper T cells, which have been activated by the same bacterial antigen processed and presented to them by macrophages
Secretion by phagocytic cells of chemical mediators, which enhance both nonspecific and specific immune responses and induce local and systemic symptoms associated with an infection	Binding of antibodies to invading bacteria and enhancement of nonspecific mechanisms that lead to the bacteria's destruction
Nonspecific activation of the complement system	Action as opsonins to enhance phagocytic activity
Formation of a hole-punching membrane-attack complex that lyses bacterial cells	Activation of lethal complement system
Enhancement of many steps of inflammation	Stimulation of killer cells, which directly lyse bacteria
	Persistence of memory cells capable of responding more rapidly and more forcefully should the same bacteria be encountered again

T Lymphocytes: Cell-Mediated Immunity

The three types of T cells are specialized to kill virus-infected host cells and to help or suppress other immune cells.

As important as B lymphocytes and their antibody products are in specific defense against invading bacteria and other foreign material, they represent only half of the body's specific immune defense corps. The T lymphocytes are equally important in defense against most viral infections and also play an important regulatory role in immune mechanisms (Table 12-6). Whereas B cells and antibodies defend against conspicuous invaders in the extracellular fluid, T cells defend against covert invaders that hide out inside cells where antibodies and the complement system cannot reach them. Unlike B cells, which secrete antibodies that can attack antigen at long distances, T cells do not secrete antibodies. Instead, they must be in direct contact with their targets, a process known as *cell-mediated immunity.* T cells release chemicals that destroy targeted cells with which they make contact, such as virus-infected cells and cancer cells.

Like B cells, T cells are clonal and exquisitely antigen-specific. On its plasma membrane, each T cell bears unique receptor proteins, similar although not identical to the surface receptors on B cells. Unlike B cells, T cells are activated by foreign antigen only when it is present on the surface of a cell that also carries a marker of the individual's own identity; that is, both foreign antigens and **self-antigens** must be present on a cell's surface before a T cell can bind with it (with one important exception being whole transplanted foreign cells). It is during thymic education that T cells learn to recognize foreign antigens only in combination with the individual's own tissue antigens, a lesson that is passed on to all T cells' future progeny. The importance of this dual antigen requirement and the nature of the self-antigens will be described shortly.

A delay of a few days generally follows exposure to the appropriate antigen before **sensitized,** or **activated, T cells** are prepared to launch a cell-mediated immune attack. When exposed to a specific antigen combination, cells of the complementary T cell clone proliferate and differentiate for several days, yielding large numbers of activated T cells that carry out various cell-mediated responses. There are three subpopulations of T cells, depending on their roles when activated by antigen:

1. **Cytotoxic T cells (killer T cells** or **CD 8 cells),** which destroy host cells bearing foreign antigen, such as body cells invaded by viruses, cancer cells, and transplanted cells.
2. **Helper T cells (CD4 cells),** which enhance the development of antigen-stimulated B cells into antibody-secreting cells, enhance activity of the appropriate cytotoxic and suppressor T cells, and activate macrophages.
3. **Suppressor T cells,** which suppress both B cell antibody production and cytotoxic and helper T cell activity.

The vast majority of the billions of T lymphocytes are believed to be of the helper or suppressor varieties, which do not directly participate in the immune destruction of invading pathogens. Collectively, these subpopulations are referred to as **regulatory T cells,** because they modulate the activities of B cells and cytotoxic T cells as well as their own activities and those of macrophages.

Like B cells, not all activated T cell progeny become effector T cells. A small proportion of them remain dormant, serving as a pool of memory T cells that are primed and ready to respond even more swiftly and vigorously should the same foreign antigen ever reappear within a body cell. T cells, even activated ones, generally have long life spans, in contrast to B cells, which rapidly work themselves to death producing antibodies once they have been converted into plasma cells upon antigen stimulation. Thus, immunity for cell-mediated responses is similar to that for antibody responses, but it is generally of longer duration. Passive cell-mediated immunity against a particular pathogen can also be conferred through administration of activated T lymphocytes that have been harvested from another individual or animal who possesses active immunity against the particular microbe.

Exposure to antigen frequently activates both the B and T cell mechanisms simultaneously. Just as the regulatory T cells can facilitate or suppress the secretion of antibody by B cells, antibodies may either enhance or block the ability of cytotoxic T cells to destroy a victim cell, depending on the circumstances. Most of the effects exerted by lymphocytes on other immune cells (such as other lymphocytes and macrophages) are mediated by means of the secretion of chemical messengers. All chemicals other than antibodies that are secreted by leukocytes are collectively called **cytokines,** the majority of which are produced by T cells. Unlike antibodies, cytokines do not interact directly with the antigen responsible for inducing their production. The roles of specific cytokines will be described in the following discussion of each of the T cell subpopulations.

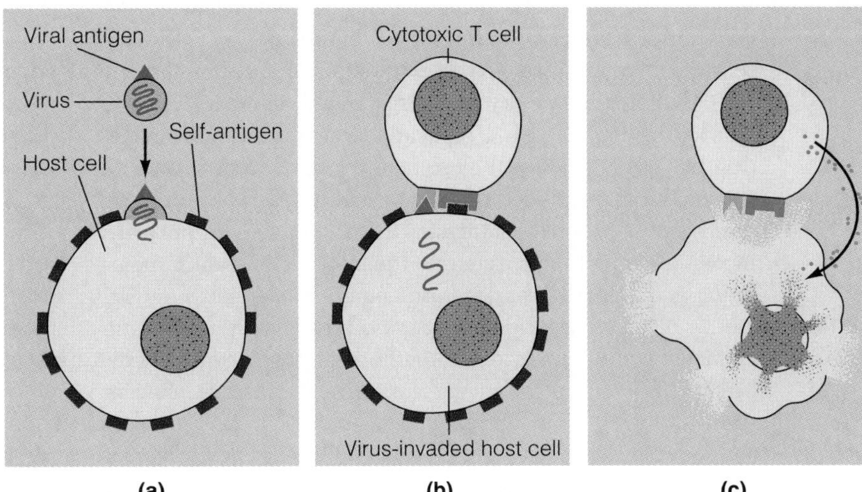

Figure 12-18 **A Cytotoxic T Cell Lysing a Virus-Invaded Host Cell** (a) When a virus invades a host cell, the viral antigen is displayed on the surface of the host cell alongside the cell's self-antigen. (b) The cytotoxic T cell recognizes and binds with a specific foreign antigen (viral antigen) in association with the self-antigen. (c) The cytotoxic T cell releases chemicals that destroy the attacked cell before the virus can enter the nucleus and start to replicate.

Cytotoxic T cells The targets of cytotoxic T cells most frequently are host cells infected with viruses. When a virus invades a body cell, as it must to survive, the envelope of antigenic proteins surrounding the virus are incorporated into the host cell's surface membrane (Fig. 12-18). To attack the intracellular virus, cytotoxic T cells must destroy the infected host cell in the process. Cytotoxic T cells of the clone specific

Table 12-6 **B versus T Lymphocytes**

Characteristic	*B Lymphocytes*	*T Lymphocytes*
Ancestral origin	Bone marrow	Bone marrow
Site of maturational processing	Bone marrow	Thymus
Receptors for antigens	Antibodies inserted in the plasma membrane serve as surface receptors; highly specific	Surface receptors present but differing from antibodies; highly specific
Bind with	Extracellular antigens such as bacteria, free viruses, and other circulating foreign material	Foreign antigen in association with self-antigen, such as virus-infected cells
Antigen must be processed and presented by macrophages	Yes	Yes
Types of active cells	Plasma cells	Cytotoxic T cells, helper T cells, suppressor T cells
Formation of memory cells	Yes	Yes
Type of immunity	Antibody-mediated immunity	Cell-mediated immunity
Secretory product	Antibodies	Cytokines
Function	Help eliminate free foreign invaders by enhancing nonspecific immune responses against them; provide immunity against most bacteria and a few viruses	Lyse virus-infected cells and cancer cells; provide immunity against most viruses and fungi and a few bacteria; aid B cells in antibody production
Life span	Short	Long

for this particular virus recognize and bind to the viral antigens and self-antigens on the surface of the infected cell. Thus sensitized by viral antigen, a cytotoxic T cell either directly kills the victim cell by releasing chemicals that lyse the attacked cell before viral replication can begin or indirectly destroys the infected cell by signaling it to commit suicide.

The direct means by which cytotoxic T cells as well as natural killer cells destroy a targeted cell is by releasing **perforin** molecules, which penetrate into the target cell's surface membrane and join together to form porelike channels (Fig. 12-19). This technique of killing a cell by punching holes in its membrane is similar to the method employed by the membrane-attack complex of the complement cascade. Recent studies suggest that cytotoxic T cells also indirectly bring about death of infected host cells by producing chemicals that induce these virus-infected cells to self-destruct, a process known as *apoptosis* (see p. 26). The virus released upon destruction of the host cell is directly destroyed in the extracellular fluid by phagocytic cells, neutralizing antibodies, and the complement system. Meanwhile, the cytotoxic T cell, which has not been harmed in the process, can move on to kill other infected host cells. The surrounding healthy cells replace the lost cells by means of cell division. Usually, not many of the host cells have to be destroyed to halt a viral infection. If the virus has had a chance to multiply, however, with replicated virus leaving the original cell and spreading to other host cells, so many of the host cells may be sacrificed by the cytotoxic T cell defense mechanism that serious malfunction may ensue.

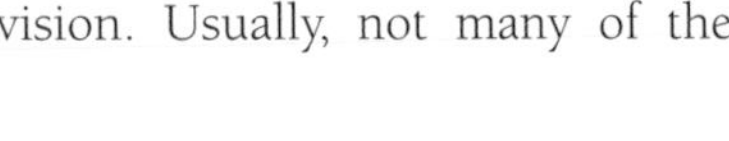

Recall that other nonspecific defense mechanisms also come into play to combat viral infections, most notably natural killer cells, interferon, macrophages, and the complement system. As usual, an intricate web of interplay exists among the immune defenses that are launched against viral invaders (Table 12-7).

The usual method of destroying virus-infected host cells is not appropriate for the nervous system. If cytotoxic T cells destroyed virus-infected neurons, the lost cells could not be replaced because neurons cannot reproduce. Fortunately, virus-infected neurons are spared from extermination by the immune system, but how then are neurons protected from viruses? Immunologists long thought that the only antiviral defenses for neurons were those aimed at free viruses in the extracellular fluid. Surprising new research has revealed, however, that antibodies not only target viruses for destruction in the extracellular fluid but can also eliminate viruses inside neurons. It is unclear whether antibodies actually enter

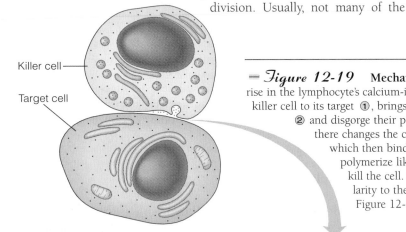

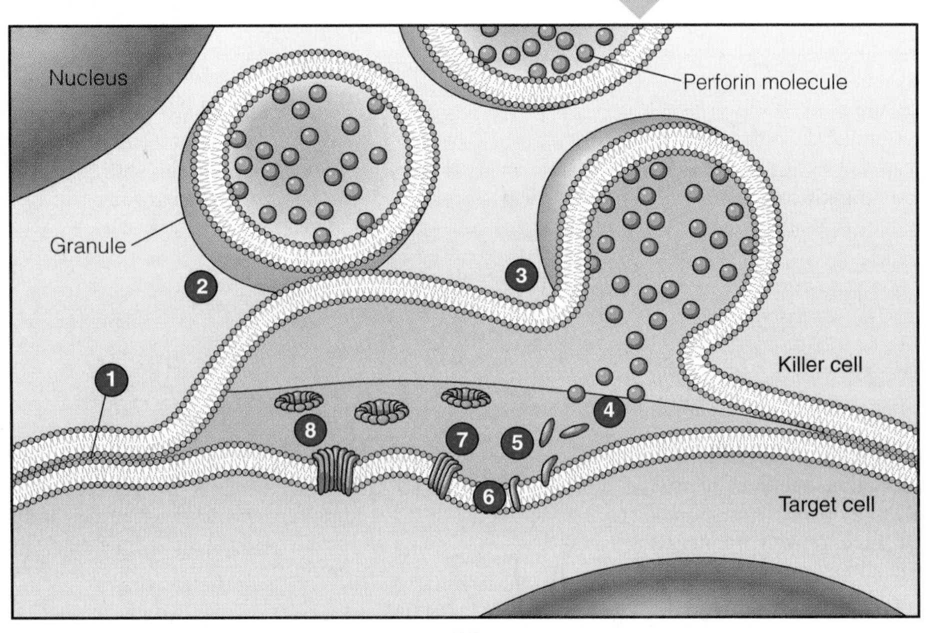

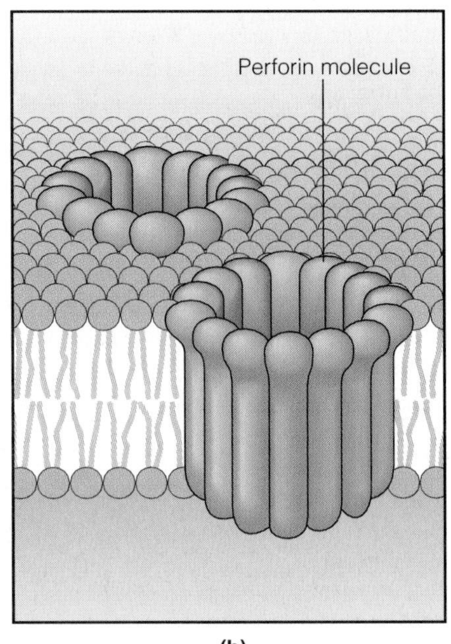

Figure 12-19 **Mechanism of Killing by Killer Cells** (a) Details of the killing process. A rise in the lymphocyte's calcium-ion level, apparently triggered by the receptor-mediated binding of the killer cell to its target ①, brings about exocytosis, in which the granules fuse with the cell membrane ② and disgorge their perforin ③ into the small intercellular space abutting the target. Calcium there changes the conformation of the individual perforin molecules, or monomers ④, which then bind to the target cell membrane ⑤ and insert into it ⑥. The monomers polymerize like staves of a barrel ⑦ to form pores ⑧ that admit water and salts and kill the cell. (b) Enlargement of perforin-formed pores in a target cell. Note the similarity to the membrane-attack complex formed by complement molecules (see Figure 12-8).

the neurons and interfere directly with viral replication (neurons have been shown to take up antibodies near their synaptic endings) or bind with the surface of nerve cells and trigger intracellular changes that stop viral replication. The fact that some viruses, such as the herpesvirus, persist for years in nerve cells, occasionally "flaring up" to produce symptoms, demonstrates that the antibodies' intraneuronal mechanism does not provide a foolproof antiviral defense for neurons.

Helper T cells In contrast to cytotoxic T cells, helper T cells are not killer cells. Instead, helper T cells secrete cytokines that "help," or augment, nearly all aspects of the immune response. The following are among the best known of these T cell chemical messengers:

1. As noted earlier, helper T cells secrete *B cell growth factor,* which enhances the antibody-secreting ability of the activated B cell clone. Antibody secretion is greatly reduced in the absence of helper T cells, even though T cells themselves do not produce antibodies.
2. Helper T cells similarly secrete **T cell growth factor,** also known as **interleukin 2 (IL-2),** which augments the activity of cytotoxic T cells, suppressor T cells, and even other helper T cells responsive to the invading antigen. In typical interplay fashion, interleukin 1 secreted by macrophages not only enhances the activity of both the appropriate B and T cell clones but also stimulates the secretion of interleukin 2 by the activated helper T cells. (The 16 known interleukins that mediate interactions between various leukocytes—*interleukin* means "between leukocytes"—were numbered in the order of their discovery.)
3. Some chemicals secreted by T cells act as *chemotaxins* to lure more neutrophils and macrophages-to-be to the invaded area.
4. Once macrophages are attracted to the area, **macrophage-migration inhibition factor,** another important cytokine released from helper T cells, keeps these large phagocytic cells in the region by inhibiting their outward migration. As a result, a great number of chemotactically attracted macrophages accumulate in the infected area. This factor also confers greater phagocytic power on the gathered macrophages. These so-called **angry macrophages** have more powerful destructive ability. They are especially important in defending against the bacteria that cause tuberculosis, because such microbes are able to survive simple phagocytosis by nonactivated macrophages.
5. Some interleukins secreted by helper T cells activate eosinophils and promote the development of IgE antibodies for defense against parasitic worms.

Helper T cells are by far the most numerous of the T cells, making up 60% to 80% of circulating T cells. Because of the important role these cells play in "turning on" the full power of all of the other activated lymphocytes and macrophages, helper T cells may constitute the immune system's "master switch." It is for this reason that **acquired immune deficiency syndrome (AIDS),** caused by the **human immunodeficiency virus (HIV),** is so devastating to the immune defense system. The AIDS virus selectively invades helper T cells, destroying or incapacitating the cells that normally orchestrate much of the immune response (Fig. 12-20). The virus also invades macrophages, further crippling the immune system, and sometimes enters brain cells as well, leading to the dementia (severe impairment of intellectual capacity) noted in some AIDS victims.

Recent studies have demonstrated the existence of two subsets of helper T cells—**T helper 1 (TH1) cells** and **T helper 2 (TH2) cells.** TH1 and TH2 cells augment different patterns of immune responses by secreting different types of cytokines. TH1 cells rally a cell-mediated (cytotoxic T cell) response, which is appropriate for infections with intracellular microbes, such as viruses, whereas TH2 cells promote humoral immunity by B cells and rev up eosinophil activity for defense against parasitic worms. Helper T cells produced in the thymus are in a "naive" state until they encounter the antigen they are primed to recognize. Whether a naive helper T cell becomes a TH1 or TH2 cell depends on which

Table 12-7
Defenses against Viral Invasion

When the virus is free in the extracellular fluid,

Macrophages:

Destroy the free virus by phagocytosis.

Process and present the viral antigen to both B and T cells.

Secrete interleukin 1, which activates B and T cell clones specific to the viral antigen.

Plasma cells derived from B cells specific to the viral antigen secrete antibodies that:

Neutralize the virus to prevent its entry into a host cell.

Activate the complement cascade that directly destroys the free virus and enhances phagocytosis of the virus by acting as an opsonin.

When the virus has entered a host cell (which it must do to survive and multiply, with the replicated viruses leaving the original host cell to enter the extracellular fluid in search of other host cells),

Interferon:

Is secreted by virus-infected cells.

Binds with and prevents viral replication in other host cells.

Enhances the killing power of macrophages, natural killer cells, and cytotoxic T cells.

Natural killer cells:

Nonspecifically lyse virus-infected host cells.

Cytotoxic T cells:

Are specifically sensitized by the viral antigen; lyse the infected host cells before the virus has a chance to replicate.

Helper T cells:

Secrete cytokines, which enhance cytotoxic T cell activity and B cell antibody production.

When a virus-infected cell is destroyed, the free virus is released into the extracellular fluid, where it is attacked directly by macrophages, antibodies, and the activated complement components.

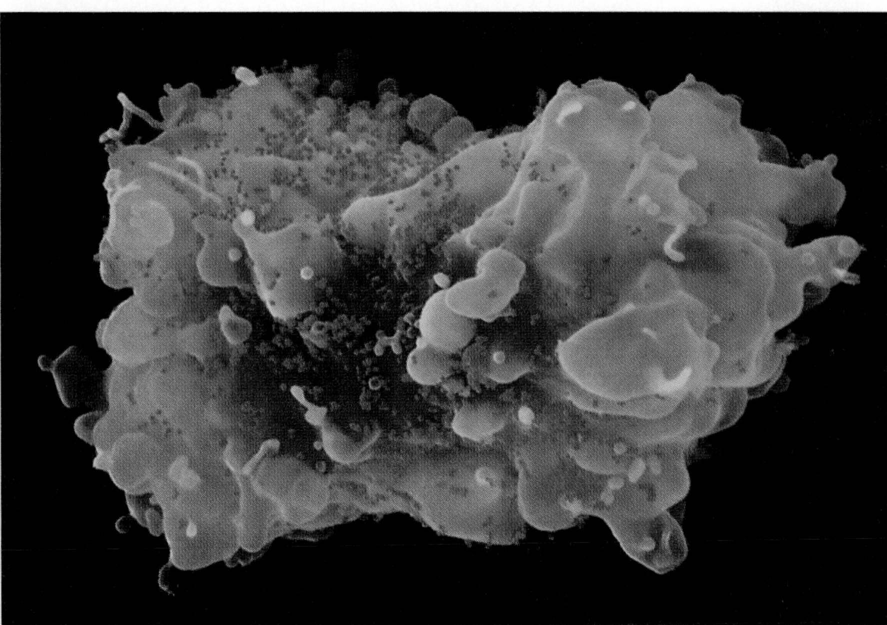

Figure 12-20 **AIDS Virus** Human immunodeficiency virus (HIV) (in purple), the AIDS-causing virus, on a helper T lymphocyte, HIV's primary target.

cytokines are secreted by the macrophage as it presents the antigen to the naive T cell. **Interleukin 12 (IL-12)** drives a naive T cell specific for the antigen to become a TH1 cell, whereas **interleukin 4 (IL-4)** favors the development of a naive cell into a TH2 cell. Thus, by means of macrophage secretions, the nonspecific immune system can influence the whole tenor of the specific immune response.

Suppressor T cells Much less is known about suppressor T cells than about the other T cell subpopulations. Suppressor T cells do not have to be presented with antigen to become active. They apparently limit immune reactions in a "check and balance" relationship with the other lymphocytes. Whereas B cells, cytotoxic T cells, and especially helper T cells reinforce each other's immune activities, suppressor T cells limit the responses of all the other immune cells. In negative-feedback fashion, helper T cells prod suppressor T cells into action; the suppressor T cells, in turn, inhibit the helper T cells and the other cells revved up for duty by helper T cell influence. Because of this feedback loop, the immune response tends to be self-limiting. Such an inhibitory effect by the suppressor T cells helps prevent excessive immune reactions that might be detrimental to the body. Suppressor T cells generally increase in number more slowly in response to a viral infection than the cytotoxic and helper T cells do, so the suppressor cells help shut down the immune response after it has already served its purpose.

The immune system is normally tolerant of self-antigens.

Suppressor T cells probably also play an important role in preventing the immune system from attacking the person's own tissues, a phenomenon known as **tolerance.** Presumably, during the genetic "cut, shuffle, and paste process" that goes on during lymphocyte development, some B and T cells would by chance be formed that could react against the body's own tissue antigens. If these lymphocyte clones were allowed to function, they would destroy the individual's own body. Fortunately, the immune system normally does not produce antibodies or activated T cells against its own body antigens but instead directs its destructive tactics only at foreign antigens. At least five different mechanisms appear to be involved in tolerance:

1. *Clonal deletion.* In response to continuous exposure to body antigens early in development, lymphocyte clones specifically capable of attacking these self-antigens in some cases are permanently destroyed. New findings demonstrate that the thymus triggers apoptosis (cellular suicide) of immature T cells that would react to the body's own proteins. Without ever having had a chance to leave the thymus, these self-destructed T cells are gobbled up by macrophages. This is the major mechanism by which tolerance is developed.
2. *Clonal anergy.* A backup to clonal deletion, **clonal anergy,** has recently been identified. The premise of this mechanism is that a T cell must receive two specific simultaneous signals to be activated (turned on), one from its compatible antigen and a stimulatory cosignal molecule known as **B7** found only on the surface of an antigen-presenting cell. Both signals are present for foreign antigens, which are introduced to T cells by antigen-presenting cells such as macrophages. Once a T cell is turned on by finding its matching antigen in accompaniment with the cosignal, the cell no longer needs the cosignal to interact with other cells. For example, an activated cytotoxic T cell can destroy any virus-invaded cell that bears the viral antigen even though the infected cell does not possess the cosignal. In contrast, these dual signals—antigen plus cosignal—never are present for self-antigens because these antigens are not handled by cosignal-bearing antigen-presenting cells. The first exposure to a single signal from a self-antigen turns *off* the compatible T cell, rendering the cell unresponsive to further exposure to the antigen instead of spurring the cell to proliferate. This reaction is referred to as clonal anergy (*anergy* means "lack of energy") because T cells are being inactivated (that is, "become lazy") rather than activated by their antigens.
3. *Inhibition by suppressor T cells.* Some lymphocyte clones specific for the body's own tissues that are not eliminated during early development are inhibited throughout life by suppressor T cells.
4. *Antigen sequestering.* Some self-molecules are normally hidden from the immune system because they never come into direct contact with the extracellular fluid in which the

immune cells and their products circulate. An example of such a segregated antigen is thyroglobulin, a complex protein sequestered within the hormone-secreting structures of the thyroid gland.

5. *Granting of immune privilege.* A few tissues, most notably the testes and the eyes, are considered to be immune privileged because they escape immune attack even when transplanted in an unrelated individual. Scientists recently discovered that the cellular plasma membranes in these immune privileged tissues possess a specific molecule that triggers apoptosis of approaching activated lymphocytes that could attack the tissues.

Occasionally, the immune system fails to make the distinction between self-antigens and foreign antigens, unleashing its deadly powers against one or more of the body's own tissues. A condition in which the immune system fails to recognize and tolerate self-antigens associated with particular tissues is known as an **autoimmune disease,** of which myasthenia gravis is an example. People with this condition erroneously produce antibodies against the acetylcholine receptors on their own skeletal muscle fibers (see p. 217).

Autoimmune diseases may arise from a number of different causes:

1. A reduction in suppressor T cell activity or an imbalance in the ratio of suppressor to helper T cells specific for self-antigens may be responsible for the development of some autoimmune conditions.
2. Normal self-antigens may be modified by factors such as drugs, environmental chemicals, viruses, or genetic mutations so that they are no longer recognized and tolerated by the immune system.
3. Exposure of normally inaccessible self-antigens sometimes induces an immune attack against these antigens. Since the immune system is usually never exposed to hidden self-antigens, it does not "learn" to tolerate them. Inadvertent exposure of these normally inaccessible antigens to the immune system because of tissue disruption caused by injury or disease can lead to a rapid immune attack against the affected tissue, just as if these self-proteins were foreign invaders. *Hashimoto's disease,* which involves the production of antibodies against thryoglobulin and the destruction of the thyroid gland's hormone-secreting capacity, is one such example.
4. Exposure of the immune system to a foreign antigen structurally almost identical to a self-antigen may induce the production of antibodies or activated T lymphocytes that not only interact with the foreign antigen but also cross-react with the closely similar body antigen. An example is the streptococcal bacteria responsible for "strep throat." The bacteria possess antigens that are structurally very similar to self-antigens in the tissue covering the heart valves of some individuals, in which case the antibodies produced against the streptococcal organisms may also bind with this heart tissue. The resultant inflammatory response is responsible for the heart valve lesions associated with *rheumatic fever.*

The major histocompatibility complex is the code for surface membrane–enclosed self-antigens unique for each individual.

What is the nature of the self-antigens that the immune system learns to recognize as markers of a person's own cells? These self-antigens are plasma membrane–bound glycoproteins (proteins with sugar attached) known as **MHC molecules** because their synthesis is directed by a group of genes called the **major histocompatibility complex** or **MHC.** These are the same MHC molecules that escort engulfed foreign antigen to the cell surface for presentation by antigen-presenting cells. The MHC genes are the most variable ones in humans. More than a hundred different MHC molecules have been identified in human tissue, but each individual has a code for only three to six of these possible antigens. Because of the tremendous number of different combinations possible, the exact pattern of MHC molecules varies from one individual to another, much like a "biochemical fingerprint," except in identical twins, who have the same MHC self-antigens.

MHC molecules alone on a cell surface signal to immune cells, "Leave me alone; I'm one of you." T cells typically bind with MHC self-antigens only when they are in association with a foreign antigen, such as a viral protein, also displayed on the cell surface. In the case of cytotoxic T cells, the outcome of this binding is destruction of the infected body cell. Since cytotoxic T cells do not bind to MHC self-antigens in the absence of foreign antigen, normal body cells are protected from lethal immune attack.

T cells do bind with MHC antigens present on the surface of *transplanted cells* in the absence of foreign viral antigen. The ensuing destruction of the transplanted cells is responsible for rejection of transplanted or grafted tissues. Presumably, some of the recipient's T cells "mistake" the MHC antigens of the donor cells for a closely resembling combination of a conventional viral foreign antigen complexed with the recipient's MHC self-antigens.

To minimize the rejection phenomenon, the tissues of donor and recipient are matched according to MHC antigens as closely as possible. Therapeutic procedures to suppress the immune system then follow. In the past, the primary immunosuppressive tools included radiation therapy and drugs aimed at destroying the actively multiplying lymphocyte populations, plus anti-inflammatory drugs that suppressed the growth of all lymphoid tissue. However, these measures not only suppressed the T cells that were primarily responsible for rejecting transplanted tissue but also depleted the antibody-secreting B cells. Unfortunately, the treated individual was left with little specific immune protection against bacterial and viral infections. In recent years, new therapeutic agents have become extremely useful in selectively depressing T cell–mediated immune activity while leaving B cell humoral immunity essentially intact. For example, *cyclosporin* blocks interleukin 2, the cytokine secreted by helper T cells that is required for expansion of the selected cytotoxic T cell clone. Furthermore, a new technique under investigation holds promise of completely preventing rejection of transplanted tissues even from an unmatched donor. This technique involves the use of tailor-made antibodies that block two specific facets of the rejection process. If proven to be safe and effective, the technique will have a tremendous impact on tissue transplantation.

The major histocompatibility (*histo* means "tissue"; *compatibility* means "ability to get along") complex was so named because these genes and the self-antigens they encode were first discerned in relation to tissue typing (similar to blood typing), which is done to obtain the most compatible matches for tissue grafting and transplantation. It is important to realize, however, that transfer of tissue from one individual to another does not normally occur in nature. The natural function of MHC antigens lies in their ability to direct the responses of T cells, not in their artificial role in the rejection of transplanted tissue.

Each individual has two main classes of MHC-encoded glycoproteins that are differentially recognized by cytotoxic T and helper T cells (— Fig. 12-21). Cytotoxic T cells are able to respond to foreign antigen only in association with **class I MHC glycoproteins,** which are found on the surface of virtually all nucleated body cells. **Class II MHC glycoproteins,** which are recognized by helper T cells, are restricted to the surface of a few special types of immune cells, such as B cells, cytotoxic T cells, and macrophages. The class I and II markers serve as signposts to guide cytotoxic and helper T cells to the precise cellular locations where their immune capabilities can be most effective. The binding of suppressor T cells is less well understood.

Because cytotoxic T cells cannot dispose of free viruses, bacteria, or other antigens circulating within the host's body fluids, it would be inefficient for this subpopulation of T cells to recognize and bind with such extracellular antigens. To carry out their role of dealing with pathogens that have invaded host cells, it is appropriate that cytotoxic T cells bind only with cells of the organism's own body that have been infected by viruses—that is, with foreign antigen in association with self-antigen. Since any nucleated body cell can be invaded by viruses, essentially all cells display class I MHC glycoproteins, enabling cytotoxic T cells to attack any invaded host cell. In contrast, helper T cells can bind with foreign antigen only when it is found on the surfaces of immune cells with which it interacts—namely, those bearing class II glycoproteins. These include the macrophages, which "present" antigen to helper T cells, and B and T cells, whose activity is enhanced by helper T cells. The capabilities of helper T cells would be squandered if these cells were able to bind with body cells other than these special immune cells. Thus, specific binding requirements for the various T cells help ensure the appropriate T cell responses.

Immune surveillance against cancer cells involves an interplay among cytotoxic T cells, natural killer cells, macrophages, and interferon.

Besides destruction of virus-infected host cells, another important function generally attributed to the T cell system is its role in recognizing and destroying newly arisen, potentially cancerous tumor cells before they have a chance to multiply and spread, a process known as **immune surveillance.** Any normal cell may be transformed into a cancer cell if mutations occur within its genes responsible for controlling cell division and growth. Such mutations may occur by chance alone or, more frequently, by exposure to **carcinogenic** (cancer-causing) factors such as ionizing radiation, certain environmental chemicals, certain viruses, or physical irritants.

Cellular multiplication and growth are normally under strict control, but the regulatory mechanisms are largely unknown. Cell multiplication in an adult is generally

— *Figure 12-21* **Distinctions between Class I and Class II Major Histocompatibility Complex (MHC) Glycoproteins** Specific binding requirements for cytotoxic and helper T cells ensure that these cells bind only with the targets that they can influence. Cytotoxic T cells can recognize and bind with foreign antigen only when the antigen is in association with class I MHC glycoproteins, which are found on the surface of all body cells. This requirement is met when a virus invades a body cell, whereupon the cell is destroyed by the cytotoxic T cells. Helper T cells, which enhance the activities of other immune cells, can recognize and bind with foreign antigen only when it is in association with class II MHC glycoproteins, which are found only on the surface of these other immune cells.

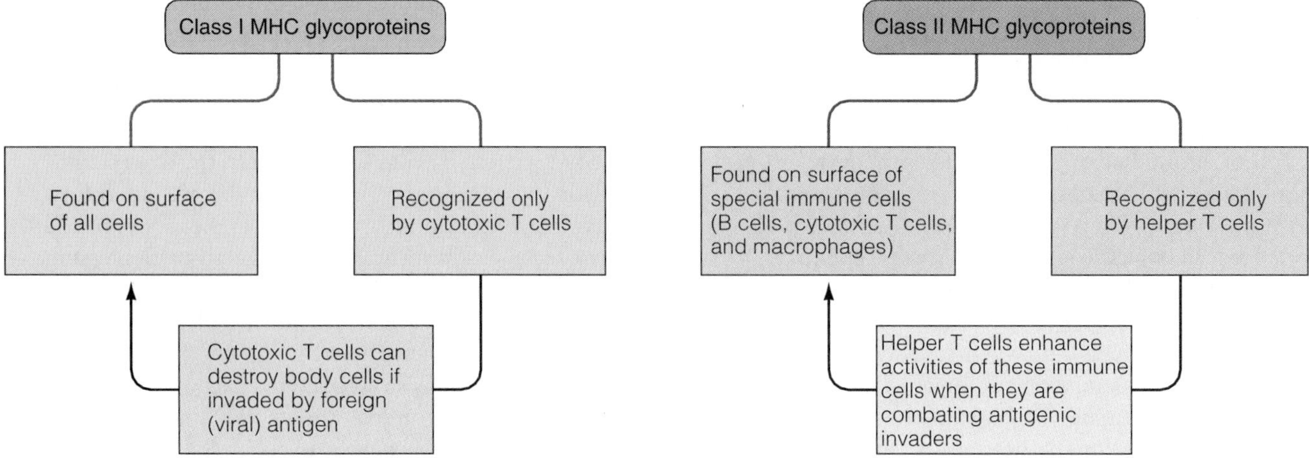

restricted to the replacement of lost cells. Furthermore, cells normally respect their own place and space in the body's society of cells. If a cell that has been transformed into a tumor cell manages to escape destruction, however, it defies the normal controls on its proliferation and position. Unrestricted multiplication of a single tumor cell results in a **tumor** that consists of a clone of cells identical to the original mutated cell.

If the mass is slow growing, stays put in its original location, and does not infiltrate into the surrounding tissue, it is considered a **benign** tumor. In contrast, the transformed cell may multiply rapidly and form an invasive mass that lacks the altruistic behavior characteristic of normal cells. Such invasive tumors are known as **malignant** tumors, commonly referred to as **cancer.** Malignant tumor cells usually do not adhere well to the neighboring normal cells, with the result that often some of the cancer cells break away from the parent tumor. These "emigrant" cancer cells are transported through the blood to new territories, where they continue to proliferate, forming multiple malignant tumors. **Metastasis** is the term applied to this spread of cancer to other parts of the body.

If a malignant tumor is detected early, before it has metastasized, it can be removed surgically. Once cancer cells have dispersed and seeded multiple cancerous sites, surgical elimination of the malignancy is impossible. In this case, agents that interfere with rapidly dividing and growing cells, such as certain chemotherapeutic drugs, are employed in an attempt to destroy the malignant cells. Unfortunately, these agents are also detrimental to normal body cells, especially rapidly proliferating cells such as blood cells and the cells lining the digestive tract.

Untreated cancer is eventually fatal in most cases for several interrelated reasons. The uncontrollably growing malignant mass crowds out normal cells by vigorously competing with them for space and nutrients, yet the cancer cells are unable to take over the functions of the cells they are destroying. Cancer cells typically remain immature and do not become specialized, often resembling embryonic cells instead (■ Fig. 12-22). Such dedifferentiated malignant cells lack the ability to perform the specialized functions of the normal cell type from which they mutated. Affected organs gradually become disrupted to the point that they are no longer able to perform their life-sustaining functions, and death results.

Even though many body cells undergo mutations throughout a person's lifetime, most of these mutations do not result in malignancy for three reasons:

1. Only a fraction of the mutations involve loss of control over the cell's growth and multiplication. More frequently, other facets of cellular function are altered.
2. Evidence suggests that a cell usually becomes cancerous only after an accumulation of multiple independent mutations. A single mutation generally is not sufficient. This requirement could contribute at least in part to the much higher incidence of cancer in older individuals, in whom mutations have had more time to accumulate in a single cell lineage. Alternatively, a few cancers have been shown to be caused by tumor viruses, which permanently alter particular DNA sequences of the cells they invade.
3. Potentially cancerous cells that do arise are usually destroyed by the immune system early in their development. Presumably, the immune system recognizes cancer cells because they bear new and different surface antigens alongside the cell's normal self-antigens, because of either genetic mutation or invasion by a tumor virus.

Immune surveillance against cancer depends on an interplay among three types of immune cells—cytotoxic T cells, natural killer cells, and macrophages—as well as interferon. Not only are all three of these immune cell types able to attack and destroy cancer cells directly, but all of them also secrete interferon. Interferon in turn inhibits multiplication of cancer

■ *Figure 12-22* **Comparison of Normal and Cancerous Cells in the Large Respiratory Airways** The normal cells display numerous specialized cilia, which constantly contract in whiplike motion to sweep debris and microorganisms from the respiratory airways so they do not gain entrance to the deeper portions of the lungs. The cancerous cells are not ciliated, so they are unable to perform this specialized defense task.

A CLOSER LOOK AT EXERCISE PHYSIOLOGY

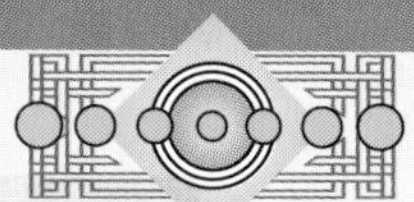

Exercise: A Help or Hindrance to Immune Defense?

For years people who engage in moderate exercise regimens have claimed they have fewer colds when they are in good aerobic condition. In contrast, elite athletes and their coaches have often complained about the number of upper respiratory infections that the athletes seem to contract at the height of their competitive seasons. The results of recent scientific studies lend support to both of these claims. The impact of exercise on immune defense depends on the intensity of the exercise.

Animal studies have shown that high-intensity exercise after experimentally induced infection results in more severe infection. Moderate exercise performed prior to infection or to tumor implantation, on the other hand, results in less severe infection and slower tumor growth in experimental animals.

Recent studies on humans lend further support to the hypothesis that exhaustive exercise suppresses immune defense whereas moderate exercise stimulates the immune system. A survey of 2,300 runners competing in the 1987 Los Angeles Marathon indicated that those who trained more than sixty miles a week had twice the number of respiratory infections of those who trained less than twenty miles a week in the two months preceding the race. In another study, ten elite athletes were asked to run on a treadmill for three hours at the same pace they would run in competition. Blood tests after the run indicated that natural killer cell activity had decreased by 25% to 50%, and this decrease lasted for six hours. The runners also showed a 60% increase in the stress hormone cortisol, which is known to suppress immunity. Other studies have shown that athletes have lower resting salivary IgA levels compared with control subjects and that their respiratory mucosal immunoglobulins are decreased after prolonged exhaustive exercise. These results suggest a lower resistance to respiratory infection following high-intensity exercise. Because of these results, researchers in the field recommend that athletes keep exposure to respiratory viruses to a minimum by avoiding crowded places or anyone with a cold or flu for the first six hours after strenuous competition.

On the other hand, a study evaluating the effects of a moderate exercise program in which a group of women walked forty-five minutes a day, five days a week, for fifteen weeks found that the walkers' antibody levels and natural killer cell activity increased throughout the exercise program. Other studies using moderate exercise on stationary bicycles in subjects over the age of sixty-five showed increases in natural killer cell activity as large as those found in young people.

Unfortunately, the few studies conducted on those infected with human immunodeficiency virus (HIV, the AIDS virus) have not found an improvement in immune function with exercise. The studies have shown that HIV-positive patients can gain strength through resistance training and improve psychological well-being through exercise and that they suffer no detrimental effects from moderate exercise.

---

gic response appears within about twenty minutes after a sensitized individual is exposed to an allergen, whereas in **delayed hypersensitivity**, the reaction is not generally manifested until a day or so following exposure. The difference in timing is due to the different mediators involved. A particular allergen may activate either a B cell or a T cell response. Immediate allergic reactions involve B cells and are elicited by antibody interactions with an allergen; delayed reactions involve T cells and the more slowly responding process of cell-mediated immunity against the allergen. Let us examine the causes and consequences of each of these reactions in more detail.

Immediate hypersensitivity In immediate hypersensitivity, the antibodies involved and the events that ensue upon exposure to an allergen differ from the typical antibody-mediated response to bacteria. The most common allergens that provoke immediate hypersensitivities are pollen grains, bee stings, penicillin, certain foods, molds, dust, feathers, and animal fur. (Actually, people allergic to cats are not allergic to the fur itself. The true allergen is in the cat's saliva, which is deposited on the fur during licking. Likewise, people are not allergic to dust or feathers per se, but to tiny mites that inhabit the dust or feathers and eat the scales constantly being shed from the skin.) For unclear reasons, these allergens bind to and elicit the synthesis of IgE antibodies rather than the IgG antibodies associated with bacterial antigens. When an individual with an allergic tendency is first exposed to a particular allergen, compatible helper T cells secrete **interleukin 4 (IL-4)**, which prods compatible B cells to synthesize IgE antibodies specific for the allergen. During this initial **sensitization** period, no symptoms are evoked, but memory cells are formed that are primed for a more powerful response on subsequent reexposure to the same allergen.

In contrast to the antibody-mediated response elicited by bacterial antigens, IgE antibodies do not freely circulate. Instead, their tail portions attach to *mast cells* and *basophils.* Recall that mast cells are connective tissue–bound "cousins" of circulating basophils, both of which produce and store an arsenal of chemicals, such as histamine, in preformed granules (see p. 362). Mast cells are most plentiful in regions that come into contact with the external environment, such as the skin, the outer surface of the eyes, and the linings of the respiratory system and digestive tract. Binding of an appropriate allergen with the outreached arm regions of the IgE antibodies that are lodged tail first in a mast cell or basophil triggers the rupture of the cell's preformed granules. As a result, histamine and other chemical mediators spew forth into the surrounding tissue.

A single mast cell (or basophil) may be coated with a number of different IgE antibodies, each able to bind with a different allergen. Thus, the mast cell can be triggered to release its chemical products by any one of a number of different allergens (Fig. 12-25). These released chemicals are responsible for the reactions that characterize immediate hypersensitivity. The following are among the most important chemicals released during immediate allergic reactions:

1. **Histamine,** which brings about vasodilation and increased capillary permeability as well as increased mucus production.
2. **Slow-reactive substance of anaphylaxis (SRS-A),** which induces prolonged and profound contraction of smooth muscle, especially of the small respiratory airways.
3. **Eosinophil chemotactic factor,** which specifically attracts eosinophils to the area. Interestingly, eosinophils release enzymes that inactivate SRS-A and may also inhibit histamine, perhaps serving as an "off switch" to limit the allergic response.

Symptoms vary depending on the site, allergen, and mediators involved. Most frequently, the reaction is localized to the body site in which the IgE-bearing cells first come into contact with the allergen. If the reaction is limited to the upper respiratory passages after a person inhales an allergen such as ragweed pollen, the released chemicals bring about the symptoms characteristic of **hay fever**—for example, nasal congestion caused by histamine-induced localized edema and sneezing and runny nose caused by increased mucus secretion in response to local irritation. If the reaction is concentrated primarily within the bronchioles (the small respiratory airways that lead to the tiny air sacs within the lungs), **asthma** results. Contraction of the smooth muscle in the walls of the bronchioles narrows or constricts these passageways, making breathing difficult. Localized swelling in the skin because of allergy-induced histamine release causes **hives.** An allergic reaction in the digestive tract in response to an ingested allergen can lead to diarrhea.

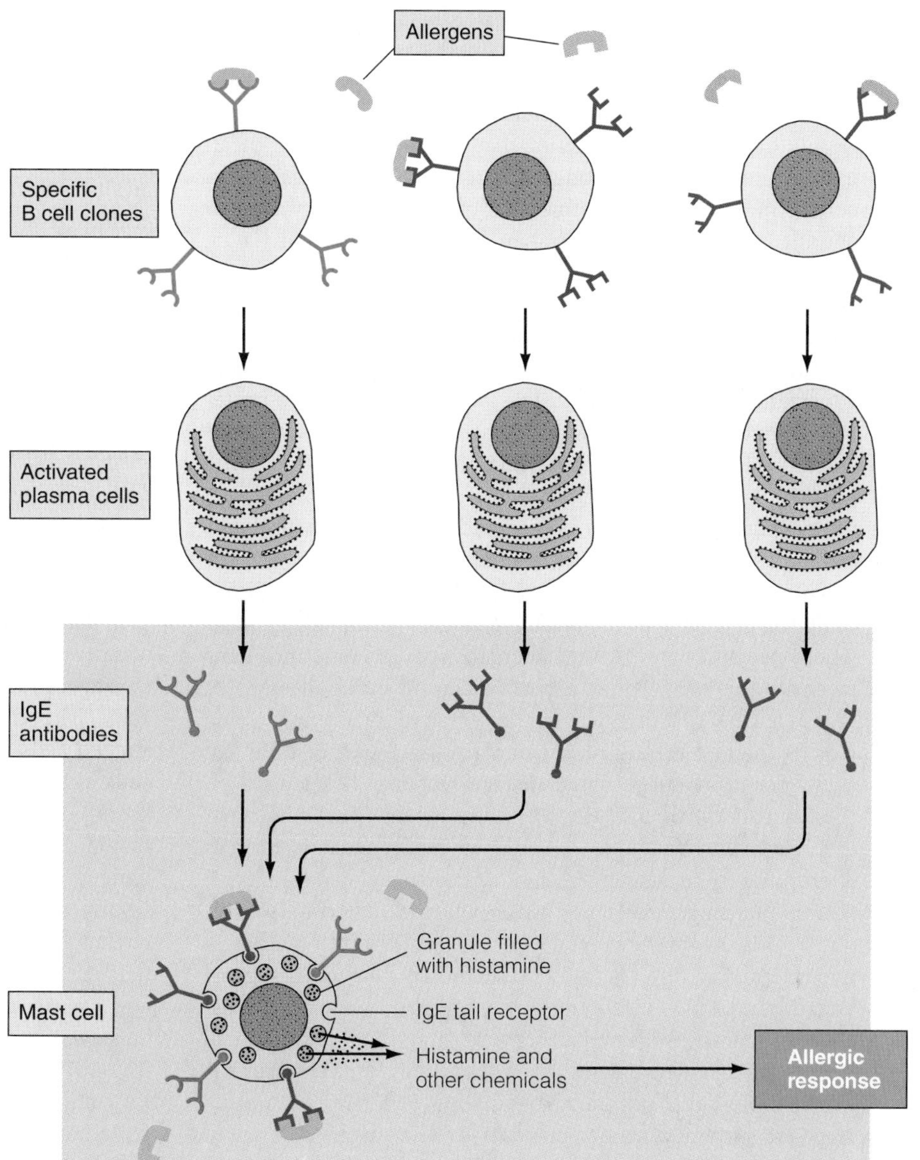

Figure 12-25 **Role of IgE Antibodies and Mast Cells in Immediate Hypersensitivity** B cell clones are converted into plasma cells, which secrete IgE antibodies on contact with the allergen for which they are specific. The Fc tail portion of all IgE antibodies, regardless of the specificity of their Fab arm regions, binds to receptor proteins specific for IgE tails on mast cells and basophils. Unlike B cells, each mast cell bears a variety of antibody surface receptors for binding different allergens. When an allergen combines with an IgE receptor specific for it on the surface of a mast cell, the mast cell releases histamine and other chemicals by exocytosis. These chemicals elicit the allergic response.

Treatment of localized immediate allergic reactions with antihistamines often offers only partial relief of the symptoms, because some of the manifestations are invoked by other chemical mediators not blocked by these drugs. For example, antihistamines are not particularly effective in treating asthma, the most serious symptoms of which are invoked by SRS-A. Adrenergic drugs (which mimic the sympathetic nervous system) are helpful through their vasoconstrictor-bronchodilator actions in counteracting the effects of both histamine and SRS-A. Anti-inflammatory drugs such as cortisol derivatives may be necessary to block the inflammatory response.

A life-threatening systemic reaction can occur if the allergen becomes blood-borne or if very large amounts of chemicals are released from the localized site into the circulation. When large amounts of these chemical mediators gain access to the blood, the extremely serious systemic (involving the entire body) reaction known as **anaphylactic shock** occurs. Severe hypotension (see p. 341) that can lead to circulatory failure results from widespread vasodilation and a massive shift of plasma fluid into the interstitial spaces as a result of a generalized increase in capillary permeability. Concurrently, pronounced bronchiolar constriction occurs and can lead to respiratory failure. The victim may suffocate because of an inability to move air through the narrowed airways. Unless countermeasures, such as injection of a vasoconstrictor-bronchodilator drug, are undertaken immediately, anaphylactic shock is frequently fatal. This reaction is why even a single bee sting or a single dose of penicillin can be so dangerous in individuals sensitized to these allergens.

Although the immediate hypersensitivity response differs considerably from the typical IgG antibody response to bacterial infections, it is strikingly similar to the immune response elicited by parasitic worms. Shared characteristics of the immune reactions to allergens and parasitic worms include the production of IgE antibodies and increased basophil and eosinophil activity. This finding has led to the proposal that harmless allergens somehow trigger an immune response designed to fight worms. Mast cells are concentrated in areas where parasitic worms (and allergens) could come into contact with the body. Parasitic worms can penetrate the skin or digestive tract or can attach to the digestive tract lining. Some worms migrate through the lungs during a part of their life cycle. Scientists suspect that the IgE response helps ward off these invaders through the following means. The inflammatory response in the skin could wall off parasitic worms attempting to burrow in. Coughing and sneezing could expel worms that migrated to the lungs. Diarrhea could help flush out worms before they could penetrate or attach to the digestive tract lining. Interestingly, epidemiological studies suggest that the incidence of allergies in a country rises as the presence of parasites decreases. Thus, superfluous immediate hypersensitivity responses to normally harmless allergens might represent a pointless marshaling of a honed immune response system "with nothing better to do" in the absence of parasitic worms.

Delayed hypersensitivity Some allergens invoke delayed hypersensitivity, a T cell–mediated immune response, rather than an immediate, B cell–IgE antibody response. Among these allergens are poison ivy toxin and certain chemicals to which the skin is frequently exposed, such as cosmetics and household cleaning agents. Most commonly, the response is characterized by a delayed skin eruption that reaches its peak intensity one to three days following contact with an allergen to which the T system has previously been sensitized. To illustrate, poison ivy toxin is a hapten that may bind with skin proteins with which it comes into contact. The toxin itself does not harm the skin upon contact, but it activates T cells specific for the toxin, including formation of a memory component. Upon subsequent exposure to the toxin, activated T cells diffuse into the skin within a day or two, combining with the poison ivy toxin that is present. The resultant interaction gives rise to the tissue damage and discomfort typically associated with the condition. The best relief is obtained from application of anti-inflammatory preparations, such as those containing cortisol derivatives.

External Defenses

The body's defenses against foreign microbes are not limited to the intricate, interrelated immune mechanisms that destroy the microorganisms that have actually invaded the body. In addition to the internal immune defense system, the body is equipped with external defense mechanisms designed to prevent microbial penetration wherever body tissues are exposed to the external environment. The most obvious external defense is the **skin**, or **integument**, which covers the outside of the body.

The skin consists of an outer protective epidermis and an inner, connective tissue dermis.

The skin, which is the largest organ of the body, not only serves as a mechanical barrier between the external environment and the underlying tissues but is dynamically involved in defense mechanisms and other important functions as well. The skin consists of two layers, an outer *epidermis* and an inner *dermis* (Fig. 12-26).

The **epidermis** consists of numerous layers of epithelial cells. The inner epidermal layers are composed of cube-shaped cells that are living and rapidly dividing, whereas the cells in the outer layers are dead and flattened. The epidermis has no direct blood supply. Its cells are nourished only by diffusion of nutrients from a rich vascular network in the underlying dermis. The newly forming cells in the inner layers constantly push the older cells closer to the surface, farther and farther from their nutrient supply. This, coupled with the fact that the outer layers are continuously subjected to pressure and "wear and tear," causes these older cells to die and become flattened. Epidermal cells are tightly bound together by spot desmosomes (see p. 51), which interconnect with intracellular keratin filaments (see p. 41) to form a strong, cohesive covering. During maturation of a keratin-producing cell, keratin filaments progressively accumulate and cross-link with each other within the cytoplasm. As the outer cells die, this fibrous keratin core remains, forming flattened, hardened scales that provide a tough, protective **keratinized layer.** As the scales of the outermost keratinized layer slough or flake off through abrasion, they are continuously replaced by means of cell division in the deeper epidermal layers. The rate of cell division, and consequently the thickness of this keratinized layer, varies in different regions of the body. It is thickest in the areas where the skin is subjected to the most pressure, such as the bottom of the feet.

The keratinized layer is airtight, fairly waterproof, and impervious to most substances. It serves to resist passage in both directions between the body and the external environment. For example, it minimizes loss of water and other vital constituents from the body. This protective layer's value in holding in body fluids becomes obvious in severe burns. Not only can bacterial infections occur in the unprotected underlying tissue, but even more serious are the systemic consequences of loss of body water and plasma proteins, which escape from the exposed, burned surface. The resultant circulatory disturbances can be life-threatening.

Likewise, the skin barrier impedes passage into the body of most materials that come into contact with the body surface, including bacteria and toxic chemicals. In many instances, the skin modifies compounds that come into contact with it. For example, epidermal enzymes are able to convert many potential carcinogens into harmless compounds. Some materials, however, especially lipid-soluble substances, are able to penetrate intact skin through the lipid bilayers of the plasma membranes of the epidermal cells. Drugs that can be absorbed by the skin, such as estrogen and nicotine, are sometimes administered in the form of a cutaneous "patch" impregnated with the drug.

Scientists recently developed a promising new technique of "injecting" non-lipid-soluble drugs through the skin without a needle. Low-frequency ultrasound waves—sound waves at pitches higher than the human ear can detect—are applied over a skin patch impregnated with a large-molecule, non-lipid-soluble drug such as insulin. The sound waves cause the O_2 and CO_2 present in the skin to form bubbles. These ultrasound-induced bubbles nudge aside some of the lipid bilayer molecules in the outer, dead epidermal cells. The subsequent collapse of these bubbles creates temporary paths big enough to permit the drug to slip painlessly through the usually impenetrable skin. When treatment ends, the original alignment of the lipid bilayer molecules is restored, and the temporary tunnels produced by ultrasound disappear. (In contrast to the low-frequency ultrasound waves used in this technique, the higher-frequency ultrasound waves used to "film" a developing fetus during pregnancy do not affect skin permeability.)

Under the epidermis is the **dermis**, a connective tissue layer that contains many elastin fibers (for stretch) and collagen fibers (for strength), as well as an abundance of blood vessels and specialized nerve endings. The dermal blood vessels not only supply both the dermis and epidermis but also play a major role in temperature regulation. The caliber of these vessels, and hence the volume of blood flowing through them, is subject to control to vary the amount of heat exchange between these skin surface vessels and the external environment (chapter 17). Receptors at the peripheral endings of afferent nerve fibers in the dermis detect pressure, temperature, pain, and other somatosensory input. Efferent nerve endings in the dermis control blood vessel caliber, hair erection, and secretion by the skin's exocrine glands.

Special infoldings of the epidermis into the underlying dermis form the skin's exocrine glands—the sweat glands and

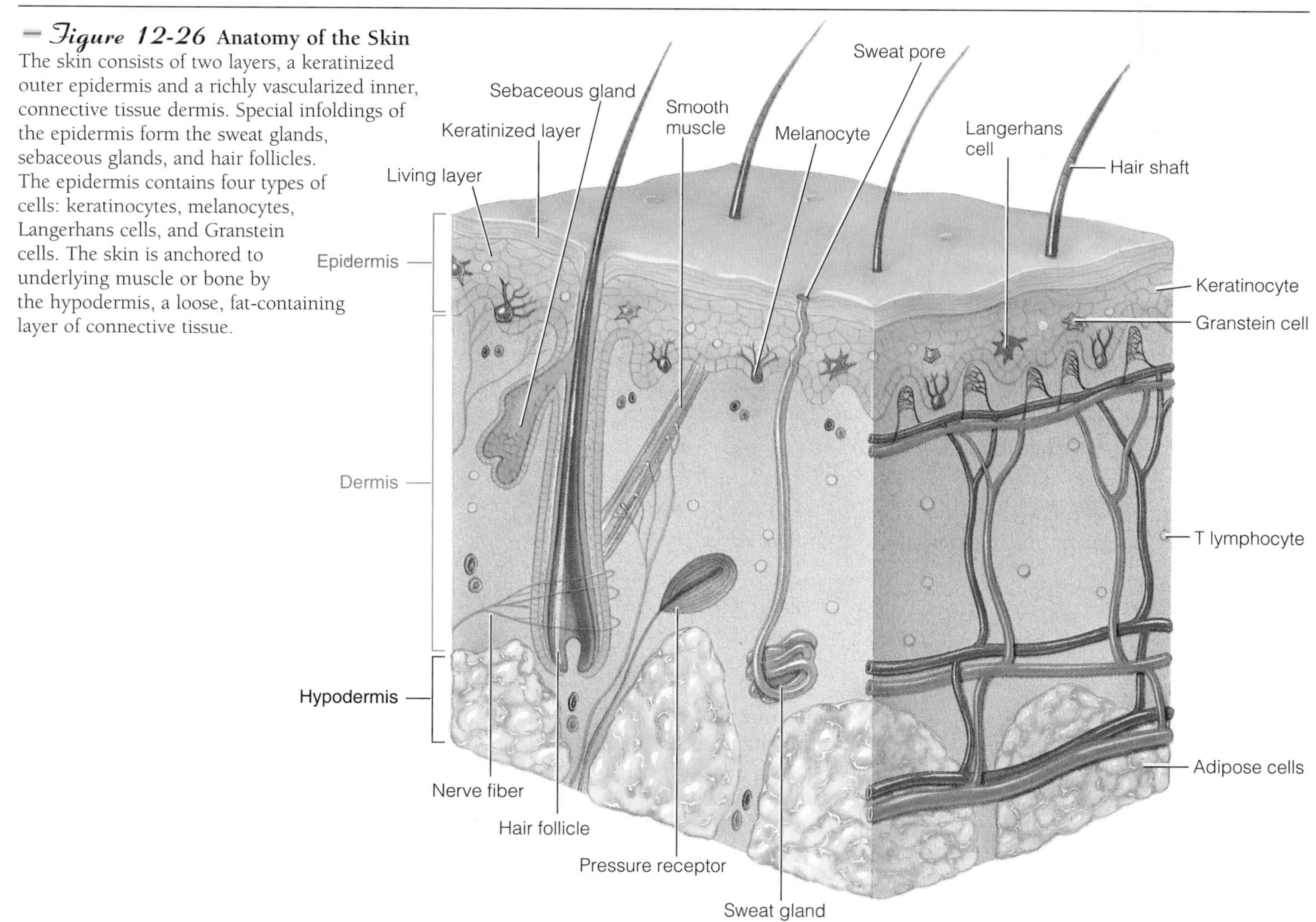

***Figure 12-26* Anatomy of the Skin**
The skin consists of two layers, a keratinized outer epidermis and a richly vascularized inner, connective tissue dermis. Special infoldings of the epidermis form the sweat glands, sebaceous glands, and hair follicles. The epidermis contains four types of cells: keratinocytes, melanocytes, Langerhans cells, and Granstein cells. The skin is anchored to underlying muscle or bone by the hypodermis, a loose, fat-containing layer of connective tissue.

sebaceous glands—as well as the hair follicles. **Sweat glands,** which are located over the majority of the body, release a dilute salt solution through small openings, the sweat pores, onto the surface of the body. Evaporation of this sweat cools the skin and is important in temperature regulation. The amount of sweat produced is subject to regulation and depends on the environmental temperature, the amount of heat-generating muscular activity, and various emotional factors (for example, a person often sweats when nervous). A special type of sweat gland located in the axilla (armpit) and pubic region produces a protein-rich sweat that supports the growth of surface bacteria, which give rise to a characteristic odor. In contrast, most sweat, as well as the secretions from the sebaceous glands, contains chemicals that are generally highly toxic to bacteria.

The cells of the **sebaceous glands** produce an oily secretion known as **sebum** that is released into adjacent hair follicles. From there the oily sebum flows to the surface of the skin, oiling both the hairs and the outer keratinized layers of the skin to help waterproof them and prevent them from drying and cracking. Insufficient protection by sebum is evidenced by chapped hands or lips. The sebaceous glands are particularly active during adolescence, causing the oily skin prominent in teenagers.

Each **hair follicle** is lined by special keratin-producing cells, which secrete keratin and other proteins that form the hair shaft. Hairs increase the sensitivity of the skin's surface to tactile (touch) stimuli. In some lower species, this function is more exquisitely fine-tuned. For example, the whiskers on a cat are extremely sensitive in this regard. An even more important role of hair in lower species is heat conservation, but this function is not significant in us relatively hairless humans. Like hair, the nails are another special keratinized product derived from living epidermal structures, the nail beds.

The skin is anchored to the underlying tissue (muscle or bone) by the **hypodermis** (*hypo* means "below"), also known as **subcutaneous tissue** (*sub* means "under," *cutaneous* means "skin"), a loose layer of connective tissue. Most fat cells in the body are housed within the hypodermis. These subcutaneous fat deposits throughout the body are collectively referred to as **adipose tissue.**

Specialized cells in the epidermis produce keratin and melanin and participate in immune defense.

The epidermis contains four distinct resident cell types—*melanocytes, keratinocytes, Langerhans cells,* and *Granstein cells*—plus transient T lymphocytes that are scattered throughout the epidermis and dermis. Each of these resident cell types performs specialized functions.

Melanocytes produce the pigment **melanin,** which they disperse to surrounding skin cells. The amount and type of melanin, which can vary among black, brown, yellow, and red pigments, are responsible for the different shades of skin color of the various races. Melanin is produced through complex biochemical pathways in which the melanocyte enzyme *tyrosinase* plays a key role. Recent studies have demonstrated that most people, regardless of skin color, have enough tyrosinase that, if fully functional, it could result in enough melanin to make their skin very black. In those with lighter skin, however, two genetic factors prevent this melanocyte enzyme from functioning at full capacity: (1) much of the tyrosinase produced is in an inactive form, and (2) various inhibitors that block tyrosinase action are produced. As a result, less melanin is produced. In addition to hereditary determination of melanin content, the amount of this pigment can be increased transiently in response to exposure to ultraviolet light rays from the sun. This additional melanin, the outward appearance of which constitutes a "tan," performs the protective function of absorbing harmful ultraviolet light rays.

The most abundant epidermal cells are the **keratinocytes,** which, as the name implies, are specialists in keratin production. As they die, they form the outer protective keratinized layer. They are also responsible for generating hair and nails. A surprising, recently discovered function is that keratinocytes are also important immunologically. They secrete interleukin 1 (a product also secreted by macrophages), which influences the maturation of T cells that tend to localize in the skin. Interestingly, the epithelial cells of the thymus have been shown to bear anatomical, molecular, and functional similarities to those of the skin. Apparently, some post-thymic steps in T cell maturation take place in the skin under keratinocyte guidance.

The two other epidermal cell types also play a role in immunity. Both **Langerhans cells,** which migrate to the skin from the bone marrow, and **Granstein cells,** the most recently discovered and least understood type of epidermal cells, serve as antigen-presenting cells. Langerhans cells present antigen to helper T cells, thereby facilitating their responsiveness to skin-associated antigens. In contrast, it appears that Granstein cells interact with suppressor T cells, probably serving as a "brake" on skin-activated immune responses. It is significant that Langerhans cells are more susceptible to damage by ultraviolet radiation (as from the sun) than are Granstein cells. Loss of Langerhans cells as a result of exposure to ultraviolet radiation can detrimentally lead to a predominant suppressor signal rather than the normally dominant helper signal, leaving the skin more vulnerable to microbial invasion and cancer cells.

The various epidermal components of the immune system are collectively termed **skin-associated lymphoid tissue** or **SALT** (— Fig. 12-27). Recent research suggests that the skin probably plays an even more elaborate role in specific immune defense than described here. This is appropriate because the skin serves as the major interface with the external environment.

In addition, the epidermis synthesizes vitamin D in the presence of sunlight. The cell type that produces vitamin D is undetermined. Vitamin D, which is derived from a precursor molecule closely related to cholesterol, promotes the absorption of Ca^{2+} from the digestive tract into the blood. Dietary supplements of vitamin D are usually required because typically the skin is not exposed to sufficient sunlight to produce adequate amounts of this essential chemical.

Protective measures within body cavities that communicate with the external environment discourage pathogen invasion into the body.

The human body's defense system must guard against entry of potential pathogens not only through the outer surface of the body but also through the internal cavities that communicate directly with the external environment—namely, the digestive system, the genitourinary system, and the respiratory system. These systems employ various strategies to destroy microorganisms entering through these routes.

Saliva secreted into the mouth at the entrance of the digestive system contains an enzyme that lyses certain ingested bacteria. "Friendly" bacteria that live on the back of the tongue convert food-derived nitrate into nitrite, which is swallowed. Acidification of nitrite on reaching the highly acidic stomach generates nitric oxide, which is toxic to a variety of microorganisms. Furthermore, many of the surviving bacteria that are swallowed are killed directly by the strongly acidic gastric juice that they encounter in the stomach. Farther down the tract, the intestinal lining is endowed with gut-associated lymphoid tissue. These defensive mechanisms are not 100% effective, however. Some bacteria do manage to survive and reach the large intestine (the last portion of the digestive tract), where they continue to flourish. Surprisingly, this normal microbial population provides a natural barrier against infection within the lower intestine. These harmless resident flora competitively suppress the growth of potential pathogens that have managed to escape the antimicrobial measures of earlier parts of the digestive tract. Occasionally, orally administered antibiotic therapy against one infection within the body may actually induce another infection in the intestinal tract. By knocking out some of the normal intestinal flora, an antibiotic may permit an antibiotic-resistant pathogenic species to overgrow.

Within the genitourinary (reproductive and urinary) system, would-be invaders encounter hostile conditions in the acidic urine and acidic vaginal secretions. The genitourinary organs also produce a sticky mucus, which, like flypaper, entraps small invading particles. Subsequently, the particles are either engulfed by phagocytes or are swept out as the organ empties (for example, they are flushed out with urine flow).

The respiratory system is likewise equipped with several important defense mechanisms against inhaled particulate matter. The respiratory system is the largest surface of the body that comes into direct contact with the increasingly polluted external environment. The surface area of the respiratory system that is exposed to the air is thirty times that of the skin. Larger airborne particles are filtered out of the inspired air by hairs at the entrance of the nasal passages. Lymphoid tissues, the *tonsils* and *adenoids,* provide immunological protection against inspired pathogens near the beginning of the respiratory system. Farther down in the respiratory airways, millions of tiny hairlike projections known as cilia (see p. 37) constantly beat in an outward direction. The respiratory airways are coated with a layer of thick, sticky mucus secreted by epithelial cells within the airway lining. This mucus sheet, laden with any inspired particulate debris (such as dust) that adheres to it, is constantly moved upward to the throat by ciliary action. This moving "staircase" of mucus is known as the **mucus escalator.** The dirty mucus is either expectorated (spit out) or in most cases swallowed without the person even being aware of it; any undigestible foreign particulate matter is subsequently eliminated in the feces. Besides keeping the lungs clean, this mechanism is an important defense against bacterial infection, because many bacteria enter the body on dust particles. Also contributing to defense against respiratory infections are IgA antibodies secreted in the mucus. In addition, an abundance of phagocytic specialists called the **alveolar macrophages** scavenge within

Figure 12-27 **Skin-Associated Lymphoid Tissue (SALT)**

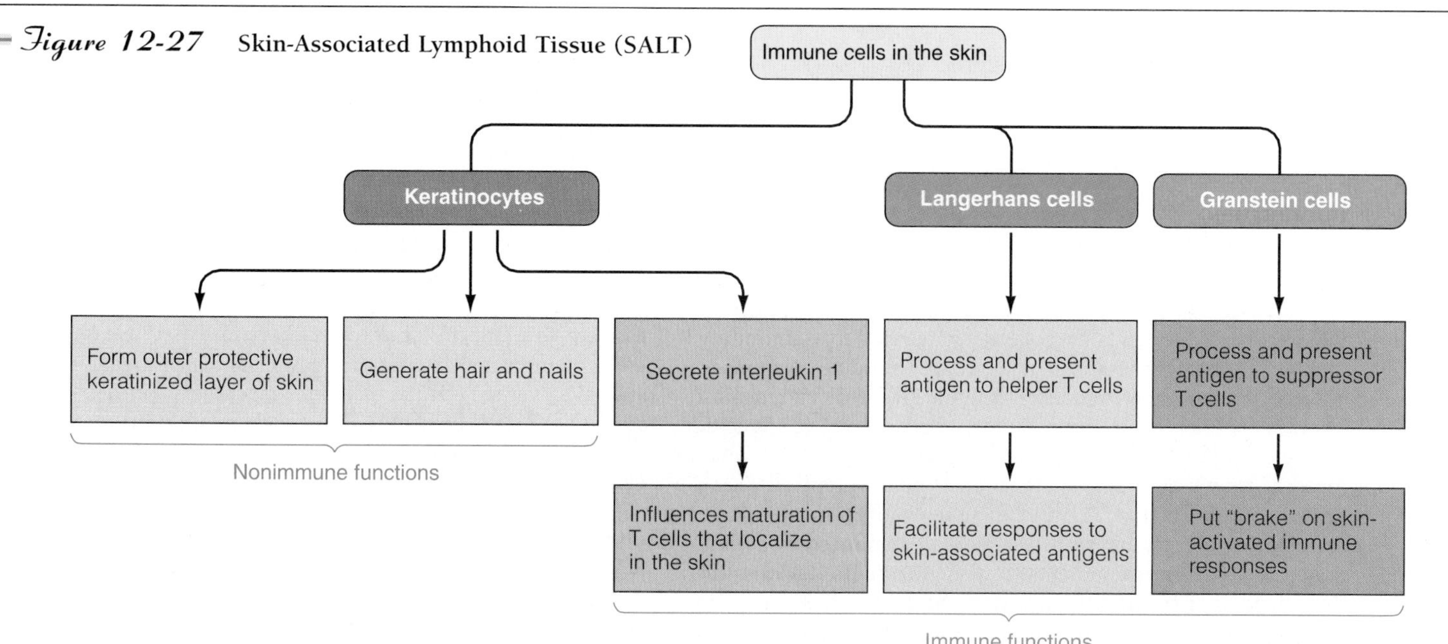

the air sacs (alveoli) of the lungs. Further respiratory defenses include coughs and sneezes. These commonly experienced reflex mechanisms involve forceful outward expulsion of material in an attempt to remove irritants from the trachea (*coughs*) or nose (*sneezes*).

Cigarette smoking suppresses these normal respiratory defenses. The smoke from a single cigarette can paralyze the cilia for several hours, with repeated exposure eventually leading to ciliary destruction. Failure of ciliary activity to sweep out a constant stream of particulate-laden mucus enables inspired carcinogens to remain in contact with the respiratory airways for prolonged periods. Furthermore, cigarette smoke incapacitates alveolar macrophages. Not only do particulates in cigarette smoke overwhelm the macrophages, but certain components of cigarette smoke have a direct toxic effect on the macrophages, reducing their ability to engulf foreign material. In addition, noxious agents in tobacco smoke irritate the mucous linings of the respiratory tract, resulting in excess mucus production, which may partially obstruct the airways. "Smoker's cough" is an attempt to dislodge this excess stationary mucus. These and other direct toxic effects on lung tissue lead to the increased incidence of lung cancer and chronic respiratory diseases associated with cigarette smoking. Air pollutants include some of the same substances found in cigarette smoke and can similarly affect the respiratory system. We will examine the respiratory system in greater detail in the next chapter.

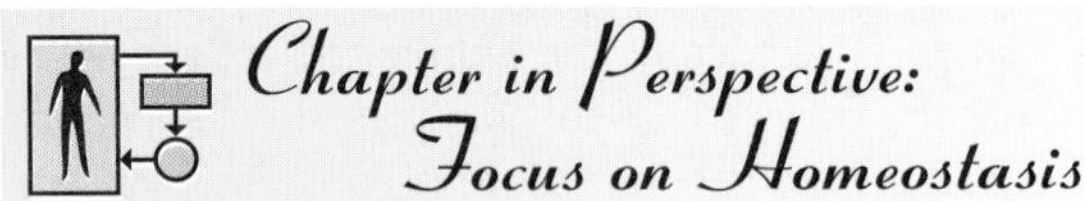

Chapter in Perspective: Focus on Homeostasis

We could not survive beyond early infancy were it not for the body's defense mechanisms. These mechanisms resist and eliminate potentially harmful foreign agents with which we continuously come into contact in our hostile external environment and also destroy abnormal cells that often arise within the body. Homeostasis can be optimally maintained, and thus life sustained, only if the body cells are not physically injured or functionally disrupted by pathogenic microorganisms or are not replaced by abnormally functioning cells, such as traumatized cells or cancer cells. The immune defense system—a complex, multifaceted, interactive network of leukocytes, their secretory products, and plasma proteins—contributes indirectly to homeostasis by keeping other cells alive so that they can perform their specialized activities to maintain a stable internal environment. The immune system protects the other healthy cells from foreign agents that have gained entrance to the body, eliminates newly arisen cancer cells, and clears away dead and injured cells to pave the way for replacement with healthy new cells.

The skin contributes indirectly to homeostasis by serving as a protective barrier between the external environment and the remainder of the body cells. It helps prevent harmful foreign agents such as pathogens and toxic chemicals from entering the body and helps prevent the loss of precious internal fluids from the body. The skin also contributes directly to homeostasis by helping maintain body temperature by means of the sweat glands and adjustments in skin blood flow. The amount of heat carried to the body surface for dissipation to the external environment is determined by the volume of warmed blood flowing through the skin.

Other systems that have internal cavities in contact with the external environment, such as the digestive, genitourinary, and respiratory systems, also have defense capabilities to prevent harmful external agents from entering the body through these avenues.

Chapter Summary

Introduction

Foreign invaders and newly arisen mutant cells are immediately confronted with multiple interrelated defense mechanisms aimed at destroying and eliminating anything that is not part of the normal self. These mechanisms, collectively referred to as immunity, include both nonspecific and specific immune responses. Nonspecific immune responses nonselectively defend against foreign material even upon initial exposure to it. Specific immune responses are selectively targeted against particular invaders for which the body has been specially prepared after a prior exposure. Leukocytes and their derivatives are the major effector cells of the immune system and are reinforced by a number of different plasma proteins. Leukocytes are produced in the bone marrow, then circulate transiently in the blood. They spend most of their time, however, on defense missions in the tissues. Some lymphocytes are also produced, differentiated, and perform their defense activities within lymphoid tissues strategically located at likely points of foreign infiltration.

The most common invaders are bacteria and viruses. Bacteria are self-sustaining, single-celled organisms, which produce disease by virtue of the destructive chemicals they release. Viruses are protein-coated nucleic acid particles, which invade host cells and take over the cellular metabolic machinery for their own survival to the detriment of the host cell. In addition to defending against these microbes and mutant cells, the immune cells also clean up cellular debris, preparing the way for tissue repair.

Nonspecific Immune Responses

Nonspecific immune responses, which form a first line of defense against atypical cells (foreign, mutant, or injured cells) even upon first exposure to them, include inflammation, interferon, natural killer cells, and the complement system.

Inflammation is a nonspecific response to foreign invasion or tissue damage mediated largely by the professional phagocytes (neutrophils and monocytes-turned-macrophages) and their secretions. The phagocytic cells destroy foreign and damaged cells both by phagocytosis and by the release of lethal chemicals. Histamine-induced vasodilation and increased permeability of local vessels at the site of invasion or injury permit enhanced delivery of more phagocytic leukocytes and inactive plasma protein precursors crucial to the inflammatory process, such as clotting factors and components of the complement system. These vascular changes are also largely responsible for the observable local manifestations of inflammation—swelling, redness, heat, and pain.

Interferon is nonspecifically released by virus-infected cells and transiently inhibits viral multiplication in other cells to which it binds. Interferon further exerts anticancer effects by slowing division and growth of tumor cells as well as by enhancing the power of killer cells.

Natural killer cells nonspecifically lyse and destroy virus-infected cells and cancer cells on first exposure to them.

Upon being activated by locally released factors or microbes themselves at the site of invasion, the complement system directly destroys the foreign invaders by lysing their membranes and also augments other aspects of the inflammatory process.

Specific Immune Responses

Following initial exposure to a microbial invader, specific components of the immune system become specially prepared to selectively attack the particular foreigner. Not only is the immune system able to recognize foreign molecules as different from self-molecules—so that destructive immune reactions are not unleashed against the body itself—but it can also distinguish between millions of different foreign molecules. The cells of the specific immune system, the lymphocytes, are each uniquely equipped with surface membrane receptors that are able to bind lock-and-key fashion with only one specific complex foreign molecule, which is known as an antigen. The tremendous variation in antigen-detecting ability between different lymphocytes arises from the shuffling around of a few different gene segments, coupled with a high incidence of somatic mutation, during lymphocyte development. Those lymphocytes produced by chance that are able to attack the body's own antigen-bearing cells are eliminated or suppressed so that they are prevented from functioning. In this way, the body is able to "tolerate" (not attack) its own antigens. The major surface antigens of all nucleated cells are known as MHC molecules, which are coded for by the major histocompatibility complex (MHC), a group of genes with DNA sequences unique for each individual.

There are two broad classes of specific immune responses: antibody-mediated immunity and cell-mediated immunity. In both instances, the ultimate outcome of a particular lymphocyte binding with a specific antigen is destruction of the antigen, but the effector cells, stimuli, and tactics involved are different. Plasma cells derived from B lymphocytes (B cells) are responsible for antibody-mediated immunity, whereas T lymphocytes (T cells) accomplish cell-mediated immunity. B cells develop from a lineage of lymphocytes that originally matured within the bone marrow. The T cell lineage arises from lymphocytes that migrated from the bone marrow to the thymus to complete their maturation.

After being activated by antigen associated with a foreign invader, a lymphocyte (either a B cell or a T cell, depending on the characteristics of the antigen) rapidly proliferates, producing a clone of its own kind that can specifically wage battle against the invader. Some of the newly developed lymphocytes do not participate in the attack but become memory cells that lie in waiting, ready to launch a swifter and more forceful attack should the same foreigner ever invade the body again. B cells and T cells have different targets because their requirements for antigen recognition differ. Each B cell recognizes specific free extracellular antigen that is not associated with cell-bound self-antigens, such as that found on the surface of bacteria. Appropriately, the activated B cell differentiates into a plasma cell, which is specialized to secrete freely circulating antibodies that besiege the freely existing invading bacteria (or other foreign substance) that induced their production. Antibodies do not directly destroy the foreign material. Instead, they intensify lethal nonspecific immune mechanisms already called into play by the foreign invasion. Antibodies activate the complement system, enhance phagocytosis, and stimulate killer cells.

T cells, in contrast, have a dual binding requirement of foreign antigen in association with MHC molecules on the surface of one of the body's own cells. Two different types of host cells meet this requirement: (1) virally invaded host cells and (2) other immune cells with foreign antigen attached on their surface. The presence of different classes of self-antigens on the surface of these foreign antigen–bearing host cells causes three different types of T cells to differentially interact with them: (1) Cytotoxic T cells are able to bind only with virus-infected host cells, whereupon they release toxic substances that kill the infected cell. (2) Helper T cells can only bind with other T cells, B cells, and macrophages that have encountered foreign antigen. Subsequently, helper T cells enhance the immune powers of these other effector cells by secreting specific chemical mediators. (3) Suppressor T cells suppress both T and B cells that have been activated by antigen, thereby preventing the immune system from overresponding and potentially damaging normal host cells. Such differential activation of the various types of lymphocytes assures that the appropriate specific immune response ensues to dispose of the particular enemy efficiently. Moreover, B cells, the various T cells, and macrophages reinforce each other's defense strategies, primarily by releasing a number of important secretory products.

In a process known as immune surveillance, natural killer cells, cytotoxic T cells, macrophages, and the interferon that they collectively secrete normally eradicate newly arisen cancer cells before they have a chance to spread.

Immune Diseases

Occasionally, through a deficiency of B or T cells, the immune system fails to defend normally against bacterial or viral infections, respectively. In contrast, in certain instances the immune system becomes overzealous. In autoimmune disease, the immune system erroneously turns against one of the person's own tissues that it no longer recognizes and tolerates as self. With immune-complex diseases, body tissues are inadvertently destroyed as an overabundance of antigen-antibody complexes activates excessive quantities of lethal complement, which destroys surrounding normal cells as well as the antigen. Allergies occur when the immune system inappropriately launches a symptom-producing, body-damaging attack against an allergen, a normally harmless environmental antigen.

External Defenses

The body surfaces exposed to the outside environment—both the outer covering of skin and the linings of internal cavities that communicate with the external environment—serve not only as mechanical barriers to deter would-be pathogenic invaders but also play an active role in thwarting entry of bacteria and other unwanted materials.

The skin consists of two layers: an outer vascular, keratinized epidermis and an inner, connective tissue dermis. The epidermis contains four cell types: melanocytes, keratinocytes, Langerhans cells, and Granstein cells. Melanocytes produce a pigment, melanin, the color and amount of which is responsible for the varying shades of skin color. Melanin protects the skin by absorbing harmful ultraviolet radiation. The most abundant cells are the keratinocytes, producers of the tough keratin that forms the outer protective layer of the skin. This physical barrier discourages bacteria and other harmful environmental agents from entering the body and prevents water and other valuable body substances from escaping. Keratinocytes further serve immunologically by secreting interleukin 1, which enhances post-thymic T cell maturation within the skin. Langerhans cells and Granstein cells also function in specific immunity by presenting antigen to helper T cells and suppressor T cells, respectively.

The dermis contains (1) blood vessels, which nourish the skin and play an important role in regulating body temperature; (2) sensory nerve endings, which provide information about the external environment; and (3) several exocrine glands and hair follicles, which are formed by specialized invaginations of the overlying epithelium. The skin's exocrine glands include sebaceous glands, which produce sebum, an oily substance that softens and waterproofs the skin, and sweat glands, which produce cooling sweat. Hair follicles produce hairs, the distribution and function of which are minimal in humans. Additionally, the skin synthesizes vitamin D in the presence of sunlight.

Besides the skin, the other main routes by which potential pathogens enter the body are (1) the digestive system, which is defended by an antimicrobial salivary enzyme, destructive acidic gastric secretions, gut-associated lymphoid tissue, and harmless colonic resident flora; (2) the genitourinary system, which is protected by destructive acidic and particle-entrapping mucus secretions; and (3) the respiratory system, whose defense depends on alveolar macrophage activity and on secretion of a sticky mucus that traps debris, which is subsequently swept out by ciliary action. Other respiratory defenses include nasal hairs, which filter out large inspired particles; reflex cough and sneeze mechanisms, which expel irritant materials from the trachea and nose, respectively; and the tonsils and adenoids, which defend immunologically.

Review Exercises

Objective Questions (Answers on p. E–11.)

1. The complement system can only be activated by antibodies. (True or false?)
2. Specific immune responses are accomplished by neutrophils. (True or false?)
3. Damaged tissue is always replaced by scar tissue. (True or false?)
4. Active immunity against a particular disease can be acquired only by actually having the disease. (True or false?)
5. A secondary response has a more rapid onset, is more potent, and has a longer duration than a primary response. (True or false?)
6. The complement system's ____________ forms a doughnut-shaped complex that imbeds in a microbial surface membrane, causing osmotic lysis of the victim cell.
7. ____________ is a collection of phagocytic cells, necrotic tissue, and bacteria.
8. ____________ refers to the localized response to microbial invasion or tissue injury that is accompanied by swelling, heat, redness, and pain.
9. A chemical that enhances phagocytosis by serving as a link between a microbe and the phagocytic cell is known as a(n) ____________.
10. ____________ refer collectively to all of the chemical messengers other than antibodies secreted by lymphocytes.
11. Which of the following statements concerning leukocytes is (are) *incorrect?*
    a. Monocytes are transformed into macrophages.
    b. T lymphocytes are transformed into plasma cells that secrete antibodies.
    c. Neutrophils are highly mobile phagocytic specialists.
    d. Basophils release histamine.
    e. Lymphocytes arise in part from lymphoid tissues.
12. Using the answer code below, indicate whether the following characteristics of the specific immune system apply to antibody-mediated immunity or cell-mediated immunity:
    (a) antibody-mediated immunity
    (b) cell-mediated immunity
    (c) both antibody-mediated and cell-mediated immunity

    ___ 1. involves secretion of antibodies
    ___ 2. mediated by B cells
    ___ 3. mediated by T cells
    ___ 4. accomplished by thymic-educated lymphocytes
    ___ 5. triggered by the binding of specific antigens to complementary lymphocytic receptors
    ___ 6. involves formation of memory cells in response to initial exposure to an antigen
    ___ 7. primarily aimed against virus-infected host cells
    ___ 8. protects primarily against bacterial invaders
    ___ 9. directly destroys targeted cells
    ___10. involved in rejection of transplanted tissue
    ___11. requires binding of a lymphocyte to a free extracellular antigen
    ___12. requires dual binding of a lymphocyte with both foreign antigen and self-antigens present on the surface of a host cell

13. Using the answer code below, indicate whether the following characteristics apply to the epidermis or dermis:

(a) epidermis

(b) dermis

___1. is the inner layer of skin

___2. has layers of epithelial cells that are dead and flattened

___3. has no direct blood supply

___4. contains sensory nerve endings

___5. contains keratinocytes

___6. contains melanocytes

___7. contains rapidly dividing cells

___8. is mostly connective tissue

14. Match the following:

___1. a family of proteins that nonspecifically defend against viral infection

___2. a response to tissue injury in which neutrophils and macrophages play a major role

___3. a group of plasma proteins that, when activated, bring about destruction of foreign cells by attacking their plasma membranes

___4. lymphocytelike entities that spontaneously lyse tumor cells and virus-infected host cells

(a) complement system

(b) natural killer cells

(c) interferon

(d) inflammation

Essay Questions

1. Distinguish between bacteria and viruses.
2. Summarize the functions of each of the lymphoid tissues.
3. Distinguish between specific and nonspecific immune responses.
4. Compare the life history of B cells and T cells.
5. What is an antigen?
6. Describe the structure of an antibody. List and describe the five subclasses of immunoglobulins.
7. In what ways do antibodies exert their effect?
8. Describe the clonal selection theory.
9. Compare the functions of B cells and T cells. What are the roles of the three types of T cells?
10. Summarize the functions of macrophages in immune defense.
11. What mechanisms are believed to be involved in tolerance?
12. What is the importance of class I and class II MHC glycoproteins?
13. Describe the factors that contribute to immune surveillance against cancer cells.
14. Distinguish among immune deficiency disease, autoimmune disease, immune-complex disease, immediate hypersensitivity, and delayed hypersensitivity.
15. What are the immune functions of the skin?

Quantitative Exercises (Solutions on p. E–11.)

1. As a result of the nonspecific immune response to an infection, for example from a cut on the skin, capillary walls near the site of infection become very permeable to plasma proteins that normally remain in the blood. These proteins diffuse into the interstitial fluid, raising the interstitial fluid-colloid osmotic pressure. This increased colloid osmotic pressure causes fluid to leave the circulation and accumulate in the tissue, forming a welt. This process is referred to as the *wheal response*. The wheal response is mediated in part by histamine secreted from mast cells in the area of infection. The histamine binds to receptors, called *H-1 receptors*, on capillary endothelial cells. The histamine signal is transduced via a second-messenger pathway involving phospholipase C (see p. 69). In response to this signal, the capillary endothelial cells contract (via internal actin-myosin interaction), which causes a widening of the intercellular gaps (pores) between the capillary endothelial cells (see p. 324). In addition, substance P (see p. 162) also contributes to pore widening. Plasma proteins can pass through these widened pores and leave the capillaries. Looking at Fig. 10-23, p. 327, compare the magnitude of the wheal response (that is, the extent of localized edema) if π_{IF} were raised (a) from 0 mm Hg to 5 mm Hg and (b) from 0 mm Hg to 10 mm Hg. In both cases, compare the net exchange pressure (NEP) at the arteriolar end of the capillary, the venular end of the capillary, and the average NEP. (Assume the other forces acting across the capillary wall remain unchanged.)

Points to Ponder

(Explanations on p. E–12.)

1. Compare the defense mechanisms that come into play in response to bacterial and viral pneumonia.
2. Why does the fact that HIV (the AIDS virus) frequently mutates make it difficult to develop a vaccine against this virus?
3. What impact would failure of the thymus to develop embryonically have on the immune system after birth?
4. Medical researchers are currently working on ways to "teach" the immune system to view foreign tissue as "self." What useful clinical application will the technique have?
5. When someone looks at you, are the cells of your body they are viewing dead or alive?
6. ***Clinical consideration*** Heather L., who has Rh-negative blood, has just given birth to her first child, who has Rh-positive blood. Both mother and baby are fine, but the doctor administers an Rh immunoglobulin preparation so that any future Rh-positive babies Heather has will not suffer from hemolytic disease of the newborn (see p. 394). During gestation (pregnancy), fetal and maternal blood do not mix. Instead, materials are exchanged between these two circulatory systems across the placenta, a special organ that develops during gestation from both maternal and fetal structures (see p. 737). Red blood cells are unable to cross the placenta but antibodies can cross. During the birthing process, a small amount of the infant's blood may enter the maternal circulation. Why did Heather's first-born child not have hemolytic disease of the newborn; that is, why didn't maternal antibodies against the Rh factor attack the fetal Rh-positive red blood cells during gestation? Why would any subsequent Rh-positive babies Heather might carry be likely to develop hemolytic disease of the newborn if she were not treated with Rh immunoglobulin? How would administration of Rh immunoglobulin immediately following Heather's first pregnancy with an Rh-positive child prevent hemolytic disease of the newborn in a subsequent pregnancy with another Rh-positive child? Similarly, why must Rh immunoglobulin be administered to Heather following the birth of every Rh-positive child she bears? Suppose Heather were not treated with Rh immunoglobulin following the birth of her first Rh-positive child and a second Rh-positive child developed hemolytic disease of the newborn. Would administration of Rh immunoglobulin to Heather immediately following the second birth prevent this condition in a third Rh-positive child? Why or why not?

13 The Respiratory System

RESPIRATORY SYSTEM

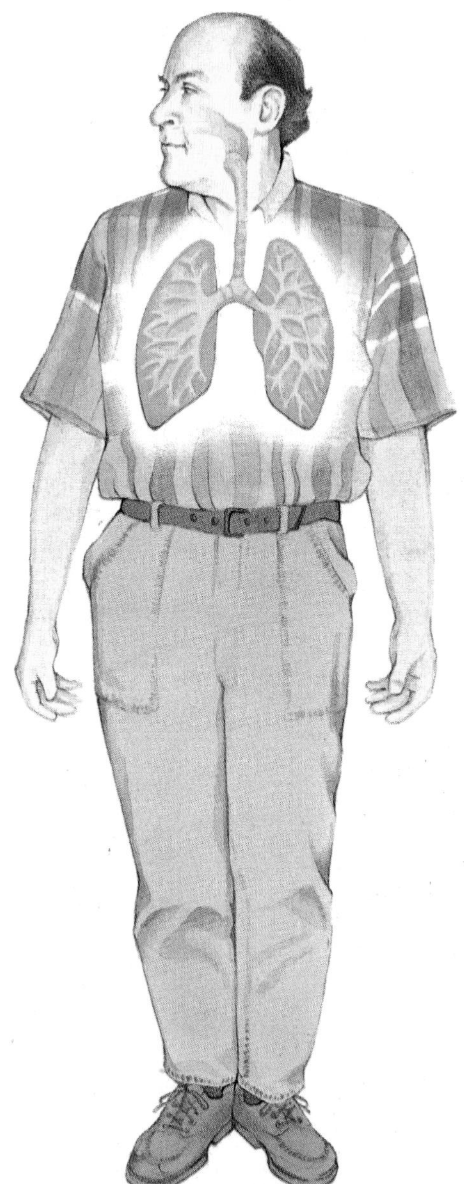

Body systems maintain homeostasis

HOMEOSTASIS

The respiratory system contributes to homeostasis by obtaining O_2 from and eliminating CO_2 to the external environment. It helps regulate the pH of the internal environment by adjusting the rate of removal of acid-forming CO_2.

Homeostasis is essential for survival of cells

CELLS

Cells need a constant supply of O_2 delivered to them to support their energy-generating chemical reactions, which produce CO_2 that must be removed continuously. Furthermore, CO_2 generates carbonic acid with which the body must continuously deal in order to maintain the proper pH in the internal environment. Cells can survive only within a narrow pH range.

Cells make up body systems

Energy is essential for sustaining life-supporting cellular activities, such as protein synthesis and active transport across plasma membranes. The cells of the body need a continual supply of O_2 to support their energy-generating chemical reactions. The CO_2 produced during these reactions must be eliminated from the body at the same rate it is produced to prevent dangerous fluctuations in pH (that is, to maintain the acid-base balance), because CO_2 generates carbonic acid.

Respiration involves the sum of the processes that accomplish ongoing passive movement of O_2 from the atmosphere to the tissues to support cellular metabolism, as well as the continual passive movement of metabolically produced CO_2 from the tissues to the atmosphere. The **respiratory system** contributes to homeostasis by exchanging O_2 and CO_2 between the atmosphere and the blood. The blood transports O_2 and CO_2 between the respiratory system and tissues.

Chapter Contents At a Glance

Introduction

The respiratory system does not participate in all steps of respiration.

The primary function of respiration is to obtain O_2 for use by the body's cells and to eliminate the CO_2 the cells produce. Most people think of respiration as the process of breathing in and breathing out. In physiology, however, respiration has a much broader meaning. **Internal** or **cellular respiration** refers to the intracellular metabolic processes carried out within the mitochondria, which use O_2 and produce CO_2 during the derivation of energy from nutrient molecules (see p. 33). The **respiratory quotient (R.Q.)**, the ratio of CO_2 produced to O_2 consumed, varies depending on the foodstuff consumed. When carbohydrate is being used, the R.Q. is 1; that is, for every molecule of O_2 consumed, one molecule of CO_2 is produced: $C_6H_{12}O_6 + 6O_2 \rightarrow 6CO_2 + 6H_2O + \text{ATP}$. For fat utilization, the R.Q. is 0.7; for protein, it is 0.8. On a typical American diet consisting of a mixture of these three nutrients, resting O_2 consumption averages about 250 ml/min, and CO_2 production averages about 200 ml/min, for an average R.Q. of 0.8:

$$\text{R.Q.} = \frac{CO_2 \text{ produced}}{O_2 \text{ consumed}} = \frac{200 \text{ ml/min}}{250 \text{ ml/min}} = 0.8$$

External respiration refers to the entire sequence of events involved in the exchange of O_2 and CO_2 between the external environment and the cells of the body. External respiration, the topic of this chapter, encompasses four steps (▬ Fig. 13-1):

1. Air is alternately moved in and out of the lungs so that exchange of air can occur between the atmosphere (external environment) and the air sacs **(alveoli)** of the lungs. This exchange is accomplished by the mechanical act of **breathing,** or **ventilation.** The rate of ventilation is regulated so that the flow of air between the atmosphere and the alveoli is adjusted according to the body's metabolic needs for O_2 uptake and CO_2 removal.
2. Oxygen and CO_2 are exchanged between air in the alveoli and blood within the pulmonary (*pulmonary* means "lung") capillaries by the process of diffusion.
3. Oxygen and CO_2 are transported by the blood between the lungs and the tissues.
4. Exchange of O_2 and CO_2 takes place between the tissues and the blood by the process of diffusion across the systemic (tissue) capillaries.

The respiratory system does not accomplish all the steps of respiration; it is involved only with ventilation and the exchange of O_2 and CO_2 between the lungs and blood (steps 1 and 2). The circulatory system carries out the remaining steps.

The respiratory system additionally performs the following nonrespiratory functions:

- It provides a route for water loss and heat elimination. Inspired atmospheric air is humidified and warmed by the respiratory airways before it is expired. Moistening of inspired air is essential to prevent the alveolar linings from drying out. Oxygen and CO_2 cannot diffuse through dry membranes.
- It enhances venous return (see the "respiratory pump," p. 337).
- It contributes to the maintenance of normal acid-base balance by altering the amount of H^+-generating CO_2 exhaled (see p. 534).
- It enables speech, singing, and other vocalization.
- It defends against inhaled foreign matter (see p. 411).
- It removes, modifies, activates, or inactivates various materials passing through the pulmonary circulation. All blood returning to the heart from the tissues must pass through the lungs before being returned to the systemic circulation. The lungs, therefore, are uniquely situated to partially or completely remove specific materials that have been added to the blood at the tissue level before they have a chance to reach other parts of the body by means of the arterial system. For example, prostaglandins, a collection of chemical messengers released in numerous tissues to

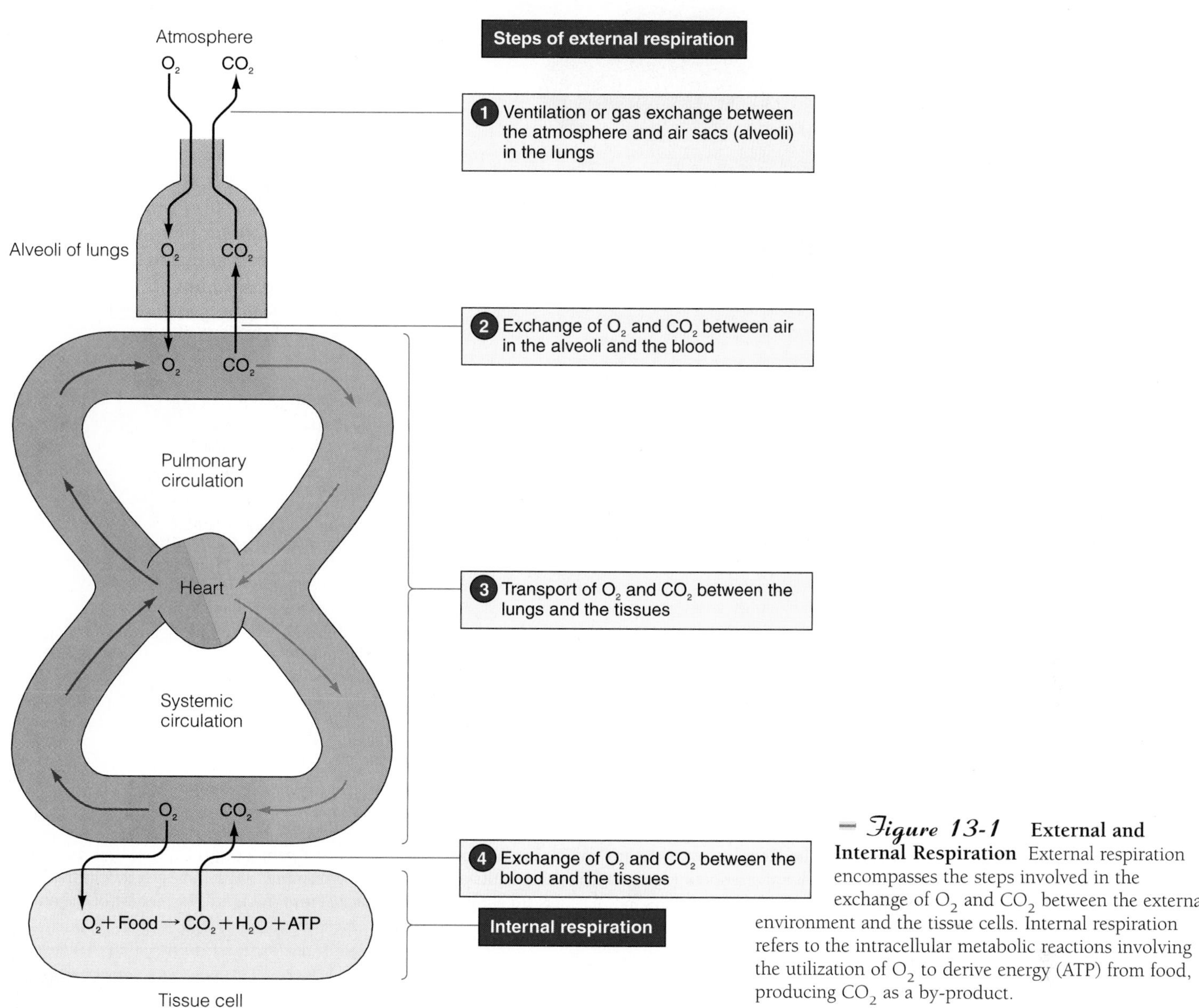
Atmosphere
O_2 CO_2
Steps of external respiration
1 Ventilation or gas exchange between the atmosphere and air sacs (alveoli) in the lungs
Alveoli of lungs
O_2 CO_2
2 Exchange of O_2 and CO_2 between air in the alveoli and the blood
O_2 CO_2
Pulmonary circulation
Heart
3 Transport of O_2 and CO_2 between the lungs and the tissues
Systemic circulation
O_2 CO_2
4 Exchange of O_2 and CO_2 between the blood and the tissues
O_2 + Food → CO_2 + H_2O + ATP
Internal respiration
Tissue cell

Figure 13-1 **External and Internal Respiration** External respiration encompasses the steps involved in the exchange of O_2 and CO_2 between the external environment and the tissue cells. Internal respiration refers to the intracellular metabolic reactions involving the utilization of O_2 to derive energy (ATP) from food, producing CO_2 as a by-product.

mediate particular local responses (see p. 716), may spill into the blood, but they are inactivated during passage through the lungs so that they cannot exert systemic effects. On the other hand, the lungs activate angiotensin II, a hormone that plays an important role in regulating the concentration of Na^+ in the extracellular fluid (see p. 485).

- The nose, a part of the respiratory system, serves as the organ of smell (see p. 196).

The respiratory airways conduct air between the atmosphere and alveoli.

The **respiratory system** includes the respiratory airways leading into the lungs, the lungs themselves, and the structures of the thorax (chest) involved in producing movement of air through the airways into and out of the lungs. The **respiratory airways** are tubes that carry air between the atmosphere and the alveoli, the latter being the only site where exchange of gases can take place between air and blood. The airways (Fig. 13-2a) begin with the **nasal passages (nose).** The nasal passages open into the **pharynx (throat),** which serves as a common passageway for both the respiratory and the digestive systems. Two tubes lead from the pharynx—the **trachea (windpipe),** through which air is conducted to the lungs, and the **esophagus,** the tube through which food passes to the stomach. Air normally enters the pharynx through the nose, but it can enter by the mouth as well when the nasal passages are congested; that is, you can breathe through your mouth when you have a cold. Because the pharynx serves as a common passageway for food and air, reflex mechanisms exist to close off the trachea during swallowing so that food enters the esophagus and not the airways. The esophagus remains closed except during swallowing to prevent air from entering the stomach during breathing.

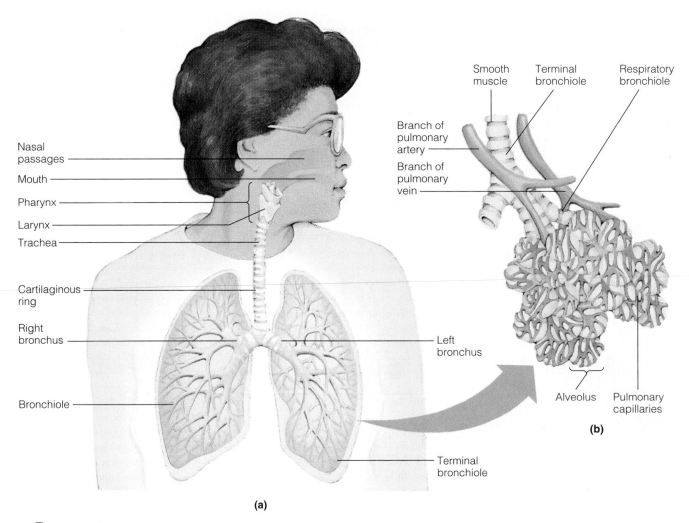

Smooth muscle
Terminal bronchiole
Respiratory bronchiole
Branch of pulmonary artery
Branch of pulmonary vein
Nasal passages
Mouth
Pharynx
Larynx
Trachea
Cartilaginous ring
Right bronchus
Left bronchus
Bronchiole
Terminal bronchiole
Alveolus
Pulmonary capillaries
(b)
(a)

Figure 13-2 **Anatomy of the Respiratory System** (a) The respiratory airways. (b) Enlargement of the alveoli (air sacs) at the terminal end of the airways. Most alveoli are clustered in grapelike arrangements at the end of the terminal bronchioles. A few individual alveoli bud off laterally along the last portion of the airways. Because gas exchange can take place across them, these smallest of the airways are termed *respiratory bronchioles.*

Located at the entrance of the trachea is the **larynx,** or **voice box,** the anterior protrusion of which forms the "Adam's apple." The **vocal folds,**[1] two bands of elastic tissue that lie across the opening of the larynx, can be stretched and positioned in different shapes by laryngeal muscles (Fig. 13-3). As air is moved past the taut vocal folds, they vibrate to produce the many different sounds of speech. The lips, tongue, and soft palate modify the sounds into recognizable sound patterns. During swallowing, the vocal folds assume a function not related to speech; they are brought into tight apposition to each other to close off the entrance to the trachea.

Beyond the larynx, the trachea divides into two main branches, the right and left **bronchi,** which enter the right and left lungs, respectively. Within each lung, the bronchus continues to branch into progressively narrower, shorter, and more numerous airways, much like the branching of a tree. The smaller branches are known as **bronchioles.** Clustered at the ends of the terminal bronchioles are the alveoli, the tiny air sacs where gas exchange between air and blood takes place (Fig. 13-2b).

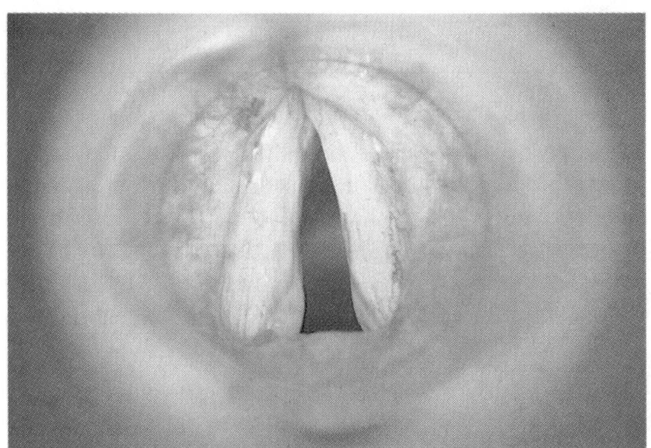

Figure 13-3 **Vocal Folds** Photograph of the vocal folds as viewed from above at the laryngeal opening.

To permit airflow in and out of the gas-exchanging portions of the lungs, the continuum of conducting airways from

[1]The commonly used term *vocal cords* has been replaced by the more descriptive term *vocal folds* in recent years.

the entrance through the terminal bronchioles to the alveoli must remain open. The trachea and larger bronchi are fairly rigid, nonmuscular tubes encircled by a series of cartilaginous rings that prevent the tubes' compression. The smaller bronchioles have no cartilage to hold them open. Their walls contain smooth muscle that is innervated by the autonomic nervous system and is sensitive to certain hormones and local chemicals. These factors, by varying the degree of contraction of bronchiolar smooth muscle and hence the caliber of these small terminal airways, are able to regulate the amount of air passing between the atmosphere and each cluster of alveoli.

The gas-exchanging alveoli are small, thin-walled, inflatable air sacs encircled by a jacket of pulmonary capillaries.

The lungs are ideally structured for their function of gas exchange. Recall that according to Fick's law of diffusion (see p. 54), the shorter the distance through which diffusion must take place, the greater the rate of diffusion. Also, the greater the surface area across which diffusion can take place, the greater the rate of diffusion.

The alveoli are clusters of thin-walled, inflatable, grapelike sacs at the terminal branches of the conducting airways (Fig. 13-4b). The alveolar walls consist of a single layer of flattened **Type I alveolar cells** (Fig. 13-4a). The walls of the dense network of pulmonary capillaries encircling each alveolus are also only one cell-layer thick. The interstitial space between an alveolus and the surrounding capillary network forms an extremely thin barrier, with only 0.5 µm separating the air in the alveoli from the blood in the pulmonary capillaries. (A sheet of tracing paper is about 50 times thicker than this air-to-blood barrier.) The thinness of this barrier facilitates gas exchange.

Furthermore, the alveolar air-blood interface presents a tremendous surface area for exchange. The lungs contain

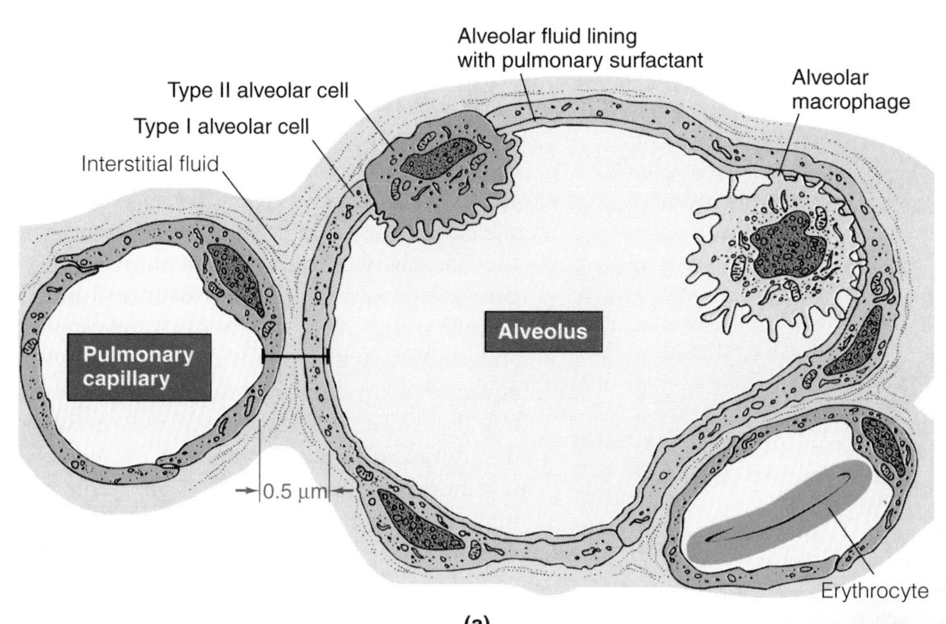

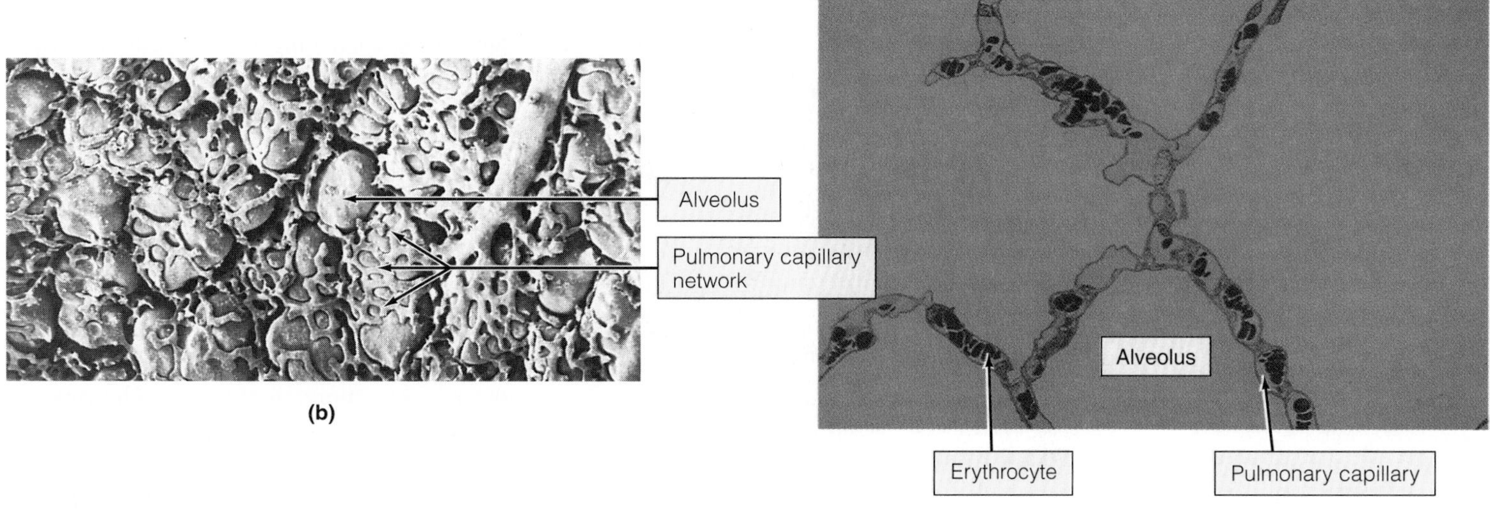

Figure 13-4 **Alveolus and Associated Pulmonary Capillaries** (a) A schematic representation of a detailed electron microscope view of an alveolus and surrounding capillaries. A single layer of flattened Type I alveolar cells forms the alveolar walls. Type II alveolar cells embedded within the alveolar wall secrete pulmonary surfactant. Wandering alveolar macrophages are found within the alveolar lumen. (The size of the cells and respiratory membrane is exaggerated compared to the size of the alveolar and pulmonary capillary lumens. The diameter of an alveolus is actually about 600 times larger than the intervening space between air and blood.) (b) A scanning electron micrograph of alveoli showing the rich capillary network surrounding them. [Source: From *Tissues and Organs: A Text-Atlas of Scanning Electron Microscopy* by Richard G. Kessel and Randy H. Kardon (New York: W. H. Freeman and Company, 1979). Copyright by and reprinted with permission from the authors.] (c) A transmission electron micrograph showing several alveoli and the close relationship of the capillaries surrounding them.

about 300 million alveoli, each about 300 μm (0.3 mm) in diameter. So dense are the pulmonary capillary networks that each alveolus is encircled by an almost continuous sheet of blood (Figs. 13-4b and c). The total surface area thus exposed between alveolar air and pulmonary capillary blood is about 75 m^2 (about the size of a tennis court). In contrast, if the lungs consisted of a single hollow chamber of the same dimensions instead of being divided into myriad alveolar units, the total surface area would be only about 0.01 m^2.

In addition to the thin, wall-forming Type I cells, the alveolar epithelium also contains **Type II alveolar cells** (Fig. 13-4a), which secrete **pulmonary surfactant,** a phospholipoprotein complex that facilitates lung expansion (to be described later). Also present within the lumen of the air sacs are the defensive alveolar macrophages (see p. 411).

Minute **pores of Kohn** present in the alveolar walls permit airflow between adjacent alveoli, a process known as **collateral ventilation.** These passageways are especially important in allowing fresh air to enter an alveolus whose terminal conducting airway is blocked because of disease.

The lungs occupy much of the thoracic cavity.

There are two **lungs,** each divided into several lobes and each supplied by one of the bronchi. The lung tissue itself consists of the series of highly branched airways, the alveoli, the pulmonary blood vessels, and large quantities of elastic connective tissue. The only muscle within the lungs is the smooth muscle in the walls of the arterioles and the walls of the bronchioles, both of which are subject to control. There is no muscle within the alveolar walls to cause them to inflate and deflate during the breathing process. Instead, changes in lung volume are brought about through changes in the dimensions of the thorax.

The lungs occupy most of the volume of the **thoracic (chest) cavity,** the only other structures in the chest being the heart and associated vessels, the esophagus, the thymus, and some nerves. The outer chest wall **(thorax)** is formed by twelve pairs of curved **ribs,** which join the **sternum** (breastbone) anteriorly and the **thoracic vertebrae** (backbone) posteriorly. The rib cage provides bony protection for the lungs and heart. The **diaphragm,** which forms the floor of the thoracic cavity, is a large, dome-shaped sheet of skeletal muscle that completely separates the thoracic cavity from the abdominal cavity. It is penetrated only by the esophagus and blood vessels traversing between the thoracic and abdominal cavities. The thoracic cavity is enclosed at the neck by muscles and connective tissue. The only communication between the thorax and the atmosphere is through the respiratory airways into the alveoli. Like the lungs, the chest wall contains considerable amounts of elastic connective tissue.

Figure 13-5 **Pleural Sac** (a) Pushing a lollipop into a water-filled balloon produces a relationship analogous to that between each double-walled, closed pleural sac and the lung that it surrounds and separates from the thoracic wall. (b) Schematic representation of the relationship of the pleural sac to the lungs and thorax. One layer of the pleural sac, the *visceral pleura,* closely adheres to the surface of the lung (*viscus* means "organ") then reflects back on itself to form another layer, the *parietal pleura,* which lines the interior surface of the thoracic wall (*paries* means "wall"). The relative size of the pleural cavity between these two layers is grossly exaggerated for the purpose of visualization.

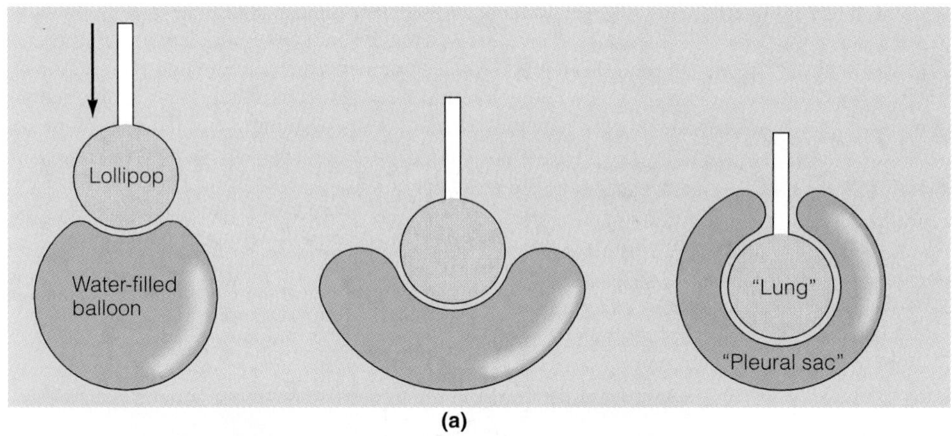

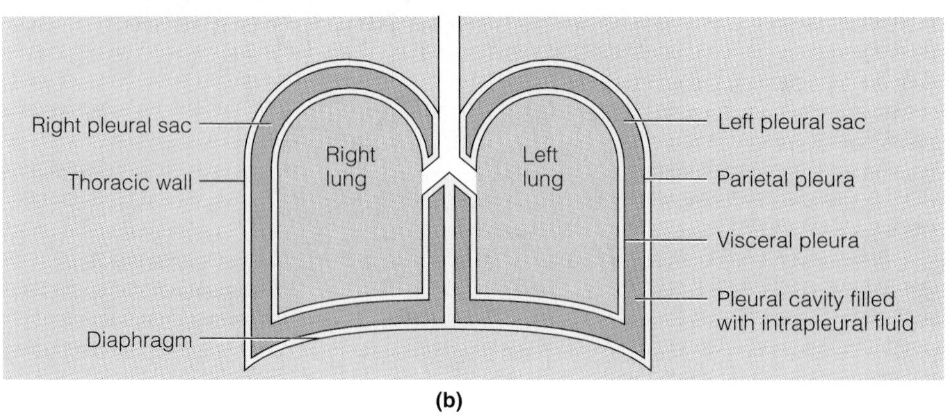

A pleural sac separates each lung from the thoracic wall.

Separating each lung from the thoracic wall and other surrounding structures is a double-walled, closed sac called the **pleural sac** (Fig. 13-5). The dimensions of the **pleural cavity** within the pleural sac are greatly exaggerated in the illustration to aid visualization; in reality the layers of the pleural sac are in close contact with one another. The surfaces of the pleura secrete a thin **intrapleural fluid** (*intra* means "within"), which lubricates the pleural surfaces as they slide past each other during respiratory movements. **Pleurisy,** an inflammation of the pleural sac, is accompanied by painful breathing because each inflation and each deflation of the lungs cause a "friction rub."

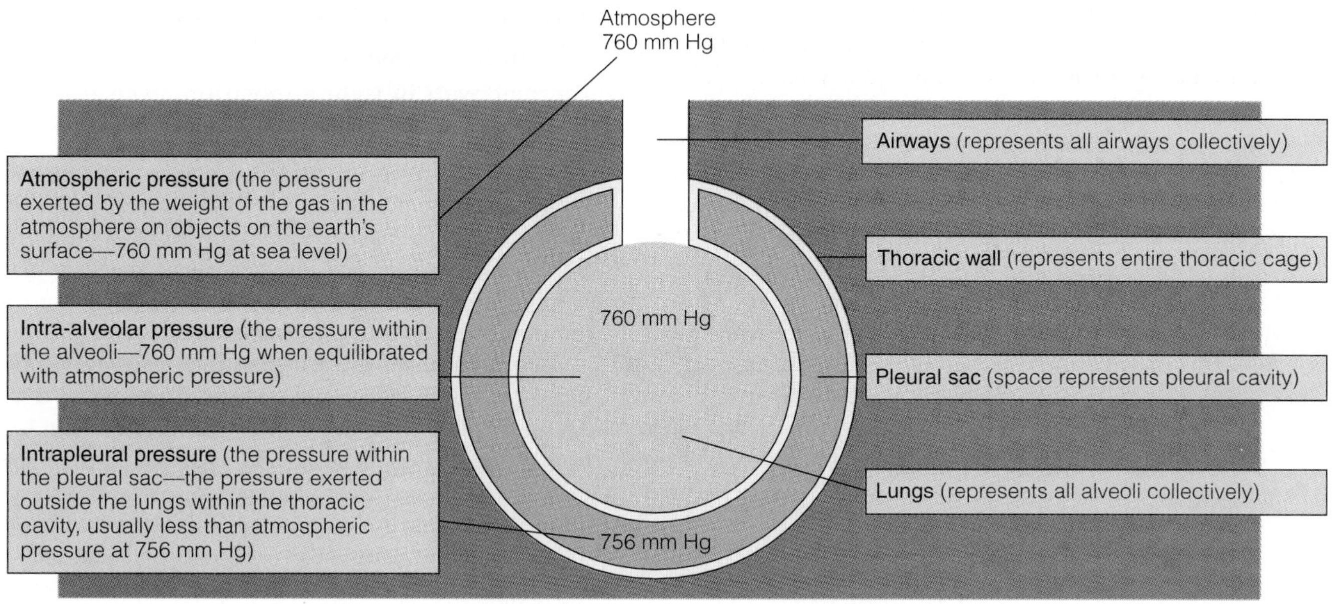
Atmosphere
760 mm Hg
Airways (represents all airways collectively)
Atmospheric pressure (the pressure exerted by the weight of the gas in the atmosphere on objects on the earth's surface—760 mm Hg at sea level)
Thoracic wall (represents entire thoracic cage)
760 mm Hg
Intra-alveolar pressure (the pressure within the alveoli—760 mm Hg when equilibrated with atmospheric pressure)
Pleural sac (space represents pleural cavity)
Intrapleural pressure (the pressure within the pleural sac—the pressure exerted outside the lungs within the thoracic cavity, usually less than atmospheric pressure at 756 mm Hg)
Lungs (represents all alveoli collectively)
756 mm Hg

Figure 13-6 **Pressures Important in Ventilation**

Respiratory Mechanics

Interrelationships among atmospheric, intra-alveolar, and intrapleural pressures are important in respiratory mechanics.

Air tends to move from a region of higher pressure to a region of lower pressure, that is, down a **pressure gradient.** Air flows in and out of the lungs during the act of breathing by moving down alternately reversing pressure gradients established between the alveoli and the atmosphere by cyclical respiratory muscle activity. Three different pressure considerations are important in ventilation (Fig. 13-6):

1. **Atmospheric (barometric) pressure** is the pressure exerted by the weight of the air in the atmosphere on objects on the earth's surface. At sea level it equals 760 mm Hg (Fig. 13-7). Atmospheric pressure diminishes with increasing altitude above sea level as the column of air above the earth's surface correspondingly decreases. Minor fluctuations in atmospheric pressure occur at any height because of changing weather conditions (that is, when barometric pressure is rising or falling).
2. **Intra-alveolar pressure,** also known as **intrapulmonary pressure,** is the pressure within the alveoli. Because the alveoli communicate with the atmosphere through the conducting airways, air quickly flows down its pressure gradient any time intra-alveolar pressure differs from atmospheric pressure; the airflow continues until the two pressures equilibrate (become equal).
3. **Intrapleural pressure** is the pressure within the pleural sac. Also known as **intrathoracic pressure,** it is the pressure exerted outside the lungs within the thoracic cavity. The intrapleural pressure is usually less than atmospheric pressure, averaging 756 mm Hg at rest. Just as blood pres-

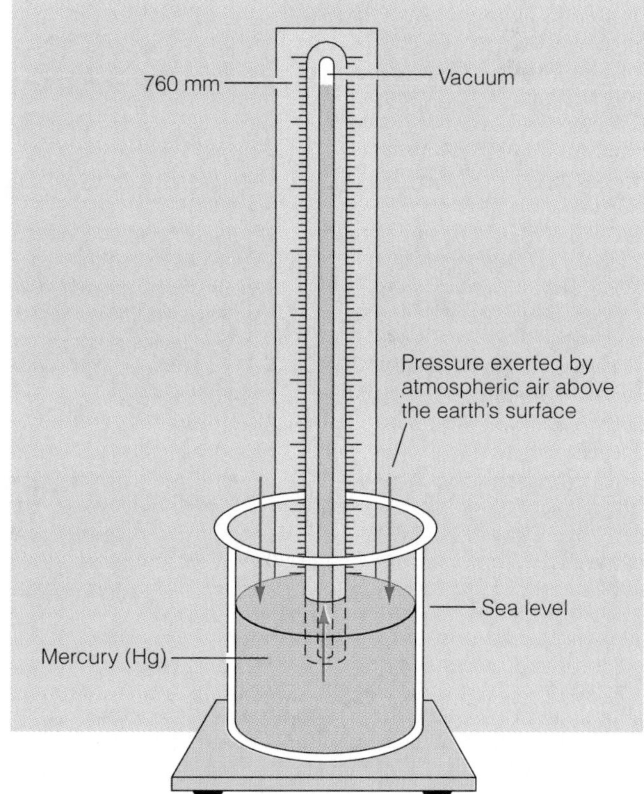
760 mm
Vacuum
Pressure exerted by atmospheric air above the earth's surface
Sea level
Mercury (Hg)

Figure 13-7 **Atmospheric Pressure** The pressure exerted on objects by the atmospheric air above the earth's surface at sea level can push a column of mercury to a height of 760 mm. Therefore, atmospheric pressure at sea level is considered to be 760 mm Hg.

sure is recorded using atmospheric pressure as a reference point (that is, a systolic blood pressure of 120 mm Hg is 120 mm Hg greater than the atmospheric pressure of 760 mm Hg or in reality 880 mm Hg), 756 mm Hg is sometimes referred to as a pressure of −4 mm Hg, although there is really no such thing as an *absolute* negative pressure. A pressure of −4 mm Hg is just negative when compared with the normal atmospheric pressure of 760 mm Hg. To avoid confusion, we will use absolute positive values throughout our discussion of respiration.

Intrapleural pressure does not equilibrate with atmospheric or intra-alveolar pressure, because there is no direct communication between the pleural cavity and either the atmosphere or the lungs. Since the pleural sac is a closed sac with no openings, air cannot enter or leave despite any pressure gradients that might exist between it and surrounding regions.

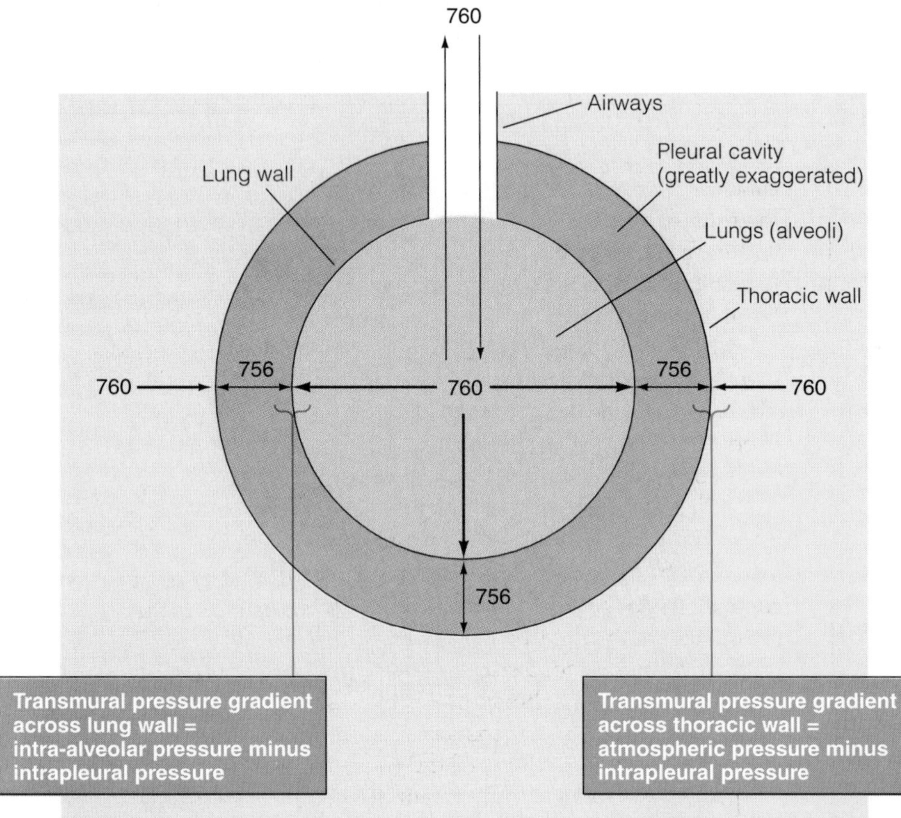

Figure 13-8 **Transmural Pressure Gradient** Across the lung wall, the intra-alveolar pressure of 760 mm Hg pushes outward, while the intrapleural pressure of 756 mm Hg pushes inward. This 4 mm Hg difference in pressure constitutes a transmural pressure gradient that pushes out on the lungs, stretching them to fill the larger thoracic cavity. Across the thoracic wall, the atmospheric pressure of 760 mm Hg pushes inward, while the intrapleural pressure of 756 mm Hg pushes outward. This 4 mm Hg difference in pressure constitutes a transmural pressure gradient that pushes inward and compresses the thoracic wall.

The intrapleural fluid's cohesiveness and the transmural pressure gradient hold the lungs and thoracic wall in tight apposition, even though the lungs are smaller than the thorax.

The thoracic cavity is larger than the unstretched lungs because the thoracic wall grows more rapidly than the lungs during development. However, two forces—the intrapleural fluid's cohesiveness and the transmural pressure gradient—hold the thoracic wall and lungs in close apposition, stretching the lungs to fill the larger thoracic cavity.

The polar water molecules in the intrapleural fluid resist being pulled apart because of their attraction to each other. The resultant cohesiveness of the intrapleural fluid tends to hold the pleural surfaces together. Thus, the intrapleural fluid can be considered very loosely as a "stickiness" or "glue" between the lining of the thoracic wall and the lung. Have you ever tried to pull apart two smooth surfaces held together by a thin layer of liquid, such as two wet glass slides? If so, you know that the two surfaces act as if they are stuck together by the thin layer of water. Even though you can easily slip the slides back and forth relative to reach other (just as the intrapleural fluid facilitates movement of the lungs against the interior surface of the chest wall), you can pull the slides apart only with great difficulty, because the molecules within the intervening liquid resist being separated. This relationship is partly responsible for the fact that changes in thoracic dimension are always accompanied by corresponding changes in lung dimension; that is, when the thorax expands, the lungs, being stuck to the thoracic wall by virtue of the intrapleural fluid's cohesiveness, do likewise.

An even more important reason that the lungs follow the movements of the chest wall is the **transmural pressure gradient** that exists across the lung wall (*trans* means "across"; *mural* means "wall") (Fig. 13-8). The intra-alveolar pressure, equilibrated with atmospheric pressure at 760 mm Hg, is greater than the intrapleural pressure of 756 mm Hg, so a greater pressure is pushing outward than is pushing inward across the lung wall. This net outward pressure differential, the transmural pressure gradient, pushes out on the lungs, stretching, or distending, them. Because of this pressure gradient, the lungs are always forced to expand to fill the thoracic cavity.

A similar transmural pressure gradient exists across the thoracic wall. The atmospheric pressure pushing inward on the thoracic wall is greater than the intrapleural pressure pushing outward on this same wall, so the chest wall tends to be "squeezed in" or compressed compared to what it would be in an unrestricted state. The effect of the transmural pressure gradient across the lung wall is much more pronounced, however, because the highly

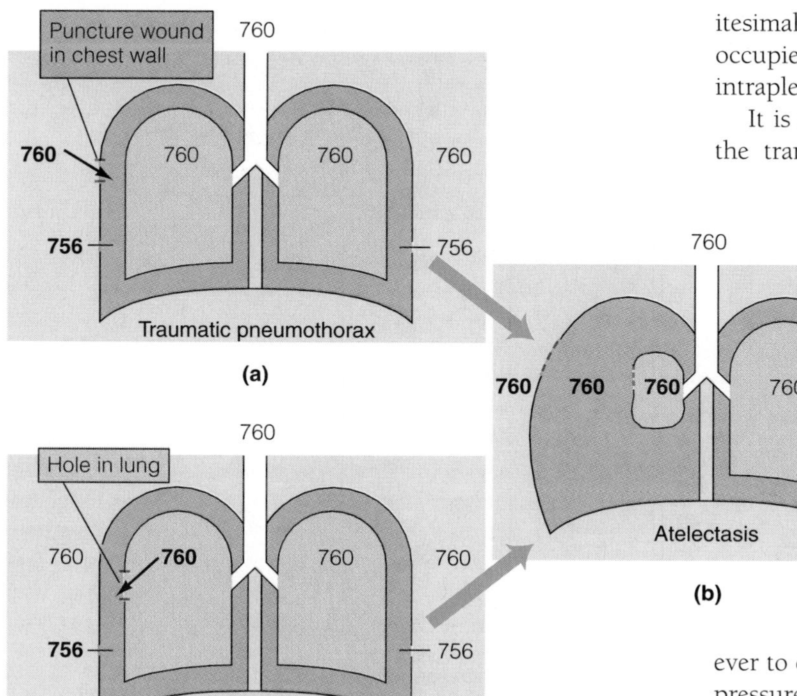

Figure 13-9 **Pneumothorax** (a) Traumatic pneumothorax. A puncture in the chest wall permits air from the atmosphere to flow down its pressure gradient and enter the pleural cavity, abolishing the transmural pressure gradient. (b) When the transmural pressure gradient is abolished, the lung collapses to its unstretched size (atelectasis), and the chest wall springs outward. (c) Spontaneous pneumothorax. A hole in the lung wall permits air to move down its pressure gradient and enter the pleural cavity from the lungs, abolishing the transmural pressure gradient. As with traumatic pneumothorax, the lung collapses to its unstretched size.

distensible lungs are influenced by this modest pressure differential to a much greater extent than is the more rigid thoracic wall.

Because neither the thoracic wall nor the lungs are in their natural position when they are held in apposition to each other, they constantly try to assume their own inherent dimensions. The stretched lungs have a tendency to pull inward away from the thoracic wall, whereas the compressed thoracic wall tends to move outward away from the lungs. The transmural pressure gradient and intrapleural fluid's cohesiveness, however, prevent these structures from pulling away from each other except to the slightest degree. Even so, the resultant ever-so-slight expansion of the pleural cavity is sufficient to drop the pressure in this cavity by 4 mm Hg, bringing the intrapleural pressure to the subatmospheric level of 756 mm Hg. This pressure drop occurs because the pleural cavity is filled with fluid, which cannot expand to fill the slightly larger volume. Therefore, a vacuum exists in the infinitesimal space in the slightly expanded pleural cavity not occupied by intrapleural fluid, producing a small drop in intrapleural pressure below atmospheric pressure.

It is important to recognize the interrelationship between the transmural pressure gradient and the subatmospheric intrapleural pressure. The lungs are stretched and the thorax is compressed because a transmural pressure gradient exists across their walls due to the presence of a subatmospheric intrapleural pressure. The intrapleural pressure in turn is subatmospheric because the stretched lungs and compressed thorax tend to pull away from each other, slightly expanding the pleural cavity and dropping the intrapleural pressure below atmospheric pressure.

If the intrapleural pressure were ever to equilibrate with atmospheric pressure, the transmural pressure gradient would be abolished. As a result, the lungs and thorax would separate and assume their own inherent dimensions. This is exactly what happens if air is permitted to enter the pleural cavity, a condition known as **pneumothorax** ("air in the chest"). (The intrapleural fluid's cohesiveness cannot hold the lungs and thoracic wall in apposition in the absence of the transmural pressure gradient.)

Normally, air does not enter the pleural cavity because there is no communication between the cavity and either the atmosphere or the alveoli. However, if the chest wall is punctured (for example, by a stab wound or a broken rib), air flows down its pressure gradient from the higher atmospheric pressure and rushes into the pleural space (Fig. 13-9a). Intrapleural and intra-alveolar pressure are now both equilibrated with atmospheric pressure, so a transmural pressure gradient no longer exists across either the lung wall or the chest wall. With no force present to stretch the lung, it collapses to its unstretched size, a condition known as **atelectasis** (Fig. 13-9b). The thoracic wall likewise springs outward to its unrestricted dimensions, but this has much less serious consequences than collapse of the lung. Similarly, pneumothorax and lung collapse can occur if air enters the pleural cavity through a hole in the lung produced, for example, by a disease process (Fig. 13-9c).

Bulk flow of air into and out of the lungs occurs because of cyclical intra-alveolar pressure changes brought about indirectly by respiratory muscle activity.

Because air flows down a pressure gradient, the intra-alveolar pressure must be less than atmospheric pressure for air to flow into the lungs during inspiration. Similarly, the intra-alveolar pressure must be greater than atmospheric pressure for air to flow out of the lungs during expiration. Intra-alveolar pressure can be changed by altering the volume of the lungs, in accordance with Boyle's law. **Boyle's law** states that

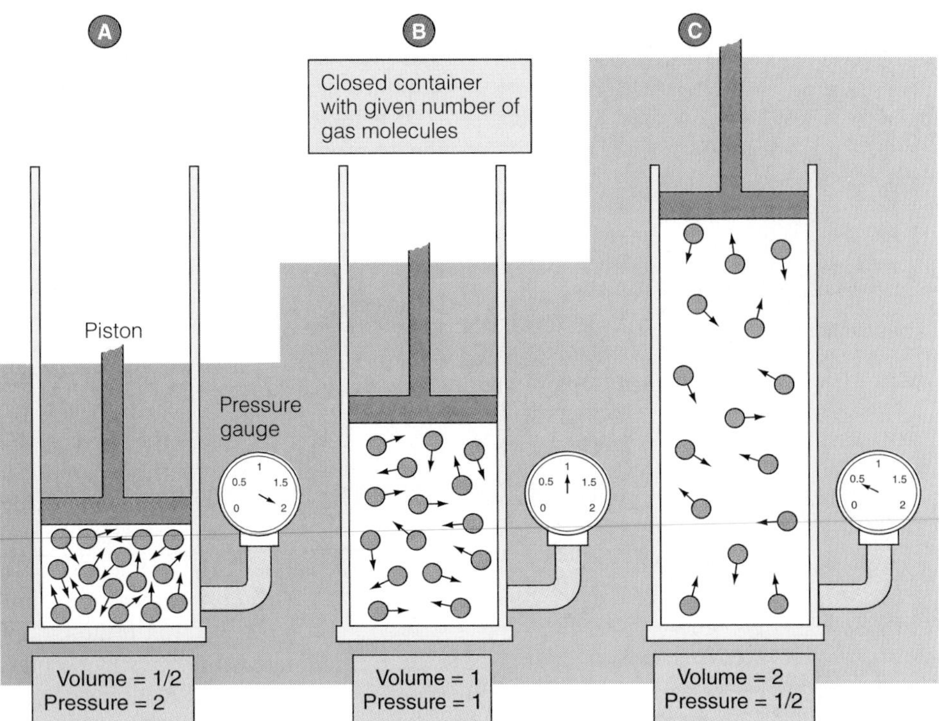

Figure 13-10 **Boyle's Law** Each container has the same number of gas molecules. Given the random motion of gas molecules, the likelihood of a gas molecule striking the interior wall of the container and exerting pressure varies inversely with the volume of the container at any constant temperature. The gas in container Ⓑ exerts more pressure than the same gas in larger container Ⓒ but less pressure than the same gas in smaller container Ⓐ. This relationship is stated as Boyle's law: $P_1V_1 = P_2V_2$. As the volume of a gas increases, the pressure of the gas decreases proportionately; conversely, the pressure increases proportionately as the volume decreases.

at any constant temperature, the pressure exerted by a gas varies inversely with the volume of the gas (Fig. 13-10); that is, as the volume of a gas increases, the pressure exerted by the gas decreases proportionately, and conversely, the pressure increases proportionately as the volume decreases.

The respiratory muscles that accomplish breathing do not act directly on the lungs to change their volume. Instead, these muscles change the volume of the thoracic cavity, causing a corresponding change in lung volume because the thoracic wall and lungs are linked together by the intrapleural fluid's cohesiveness and the transmural pressure gradient.

Let us follow the changes that occur during one respiratory cycle—that is, one breath in **(inspiration)** and out **(expiration).** Before the beginning of inspiration, the respiratory muscles are relaxed, no air is flowing, and intra-alveolar pressure is equal to atmospheric pressure. The major **inspiratory muscles**—the muscles that contract to accomplish an inspiration during quiet breathing—include the *diaphragm* and *external intercostal muscles* (Fig. 13-11 and Table 13-1). At the onset of inspiration, these muscles are stimulated to contract, resulting in enlargement of the tho-

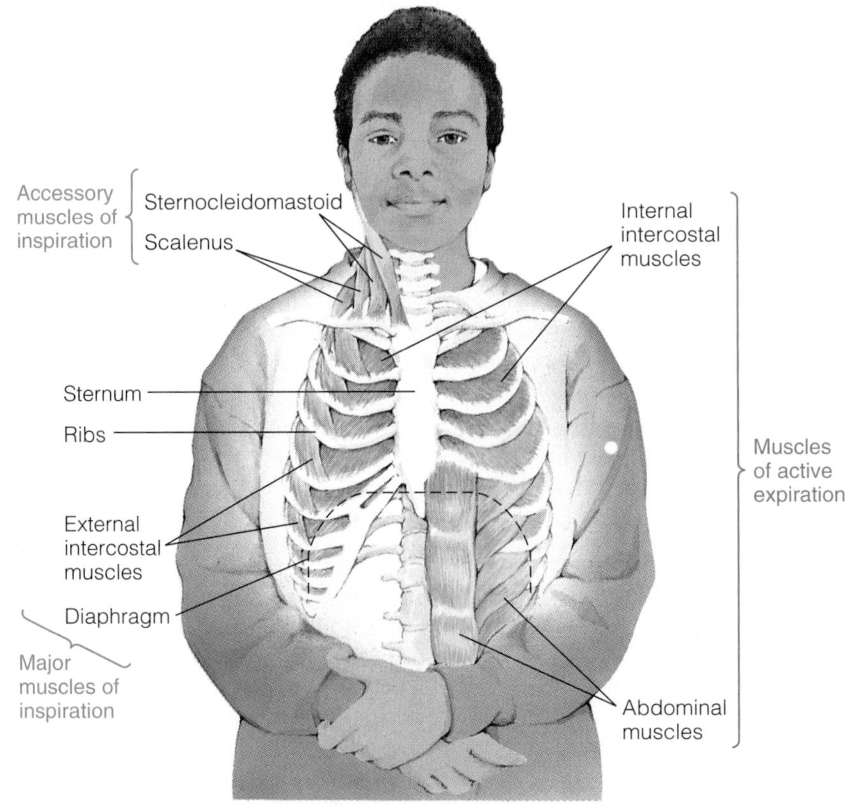

Figure 13-11 **Anatomy of the Respiratory Muscles**

Table 13-1 Actions of the Respiratory Muscles

Muscles	Result of Muscle Contraction	Timing of Stimulation to Contract
Inspiratory muscles		
Diaphragm	Descends downward, increasing the vertical dimension of the thoracic cavity	Every inspiration; the primary muscle of inspiration
External intercostal muscles	Elevate ribs upward and outward, enlarging the thorax in both the front-to-back and side-to-side dimensions	Every inspiration; play a secondary complementary role to the primary action of the diaphragm
Neck muscles (scalenus, sternocleiodomastoid)	Raise the sternum and elevate the first two ribs, enlarging the upper portion of the thoracic cavity	Only during forceful inspiration; accessory inspiratory muscles
Expiratory muscles		
Abdominal muscles	Increase the intra-abdominal pressure, which exerts an upward force on the diaphragm to decrease the vertical dimension of the thoracic cavity	Only during active (forced) expiration
Internal intercostal muscles	Flatten the thorax by pulling the ribs downward and inward, decreasing the front-to-back and side-to-side dimensions of the thoracic cavity	Only during active (forced) expiration

racic cavity. The major inspiratory muscle is the diaphragm, a sheet of skeletal muscle that forms the floor of the thoracic cavity and is innervated by the **phrenic nerve.** The relaxed diaphragm assumes a dome shape that protrudes upward into the thoracic cavity. When the diaphragm contracts upon stimulation by the phrenic nerve, it descends downward, enlarging the volume of the thoracic cavity by increasing its vertical dimension (■ Fig. 13-12a). The abdominal wall, if relaxed, can be seen to bulge outward during inspiration as the descending diaphragm pushes the abdominal contents downward and forward. Seventy-five percent of the enlargement of the thoracic cavity during quiet inspiration is attributed to contraction of the diaphragm.

Two sets of **intercostal** (*inter* means "between"; *costa* means "rib") **muscles** lie between the ribs—the external intercostal muscles lie on top of the internal intercostal muscles. Whereas contraction of the diaphragm enlarges the thoracic cavity in the vertical dimension, contraction of the **external intercostal muscles,** whose fibers run downward and forward between adjacent ribs, enlarges the thoracic cavity in both the lateral (side-to-side) and anteroposterior (front-to-back) dimensions. When the external intercostals contract, they elevate the ribs and subsequently the sternum upward and outward (Figs. 13-12a and b). **Intercostal nerves** activate these intercostal muscles.

Before inspiration, at the end of the preceding expiration, intra-alveolar pressure is equal to atmospheric pressure so no air is flowing into or out of the lungs (■ Fig. 13-13a). As the thoracic cavity enlarges, the lungs are also forced to expand to fill the larger thoracic cavity. As the lungs enlarge, the intra-alveolar pressure drops because the same number of air molecules now occupy a larger lung volume. In a typical inspiratory excursion, the intra-alveolar pressure drops 1 mm Hg to 759 mm Hg (Fig. 13-13b). Since the intra-alveolar pressure is now less than atmospheric pressure, air flows into the lungs down the pressure gradient from higher to lower pressure. Air continues to enter the lungs until no further gradient exists—that is, until intra-alveolar pressure equals atmospheric pressure. Thus, lung expansion is not caused by movement of air into the lungs; instead, air flows into the lungs because of the fall in intra-alveolar pressure brought about by lung expansion.

During inspiration, the intrapleural pressure falls to 754 mm Hg as a result of expansion of the thorax. The resultant increase in the transmural pressure gradient during inspiration ensures that the lungs are stretched to fill the expanded thoracic cavity.

Deeper inspirations (more air breathed in) can be accomplished by contracting the diaphragm and external intercostal muscles more forcefully and by bringing the **accessory inspiratory muscles** into play to further enlarge the thoracic cavity. Contraction of these accessory muscles, which are located in the neck (Fig. 13-11 and Table 13-1), raises the sternum and elevates the first two ribs, enlarging the upper portion of the thoracic cavity. As the thoracic cavity increases even further in volume than under resting conditions, the lungs likewise expand even more, dropping the intra-alveolar pressure even further. Consequently, a larger inward flow of air occurs before equilibration with atmospheric pressure is achieved; that is, a deeper breath occurs.

At the end of inspiration, the inspiratory muscles relax. The diaphragm assumes its original dome-shaped position when it relaxes; the elevated rib cage falls because of gravity when the external intercostals relax; and the chest wall and stretched lungs recoil to their preinspiratory size because of their elastic properties, much as a stretched balloon would upon release (Fig. 13-12c). As the lungs recoil and become smaller in volume, the intra-alveolar pressure rises, because the greater number of air molecules contained within the larger lung volume at the end of inspiration are now compressed into a smaller volume. In a resting expiration, the intra-alveolar pressure increases about 1 mm Hg above

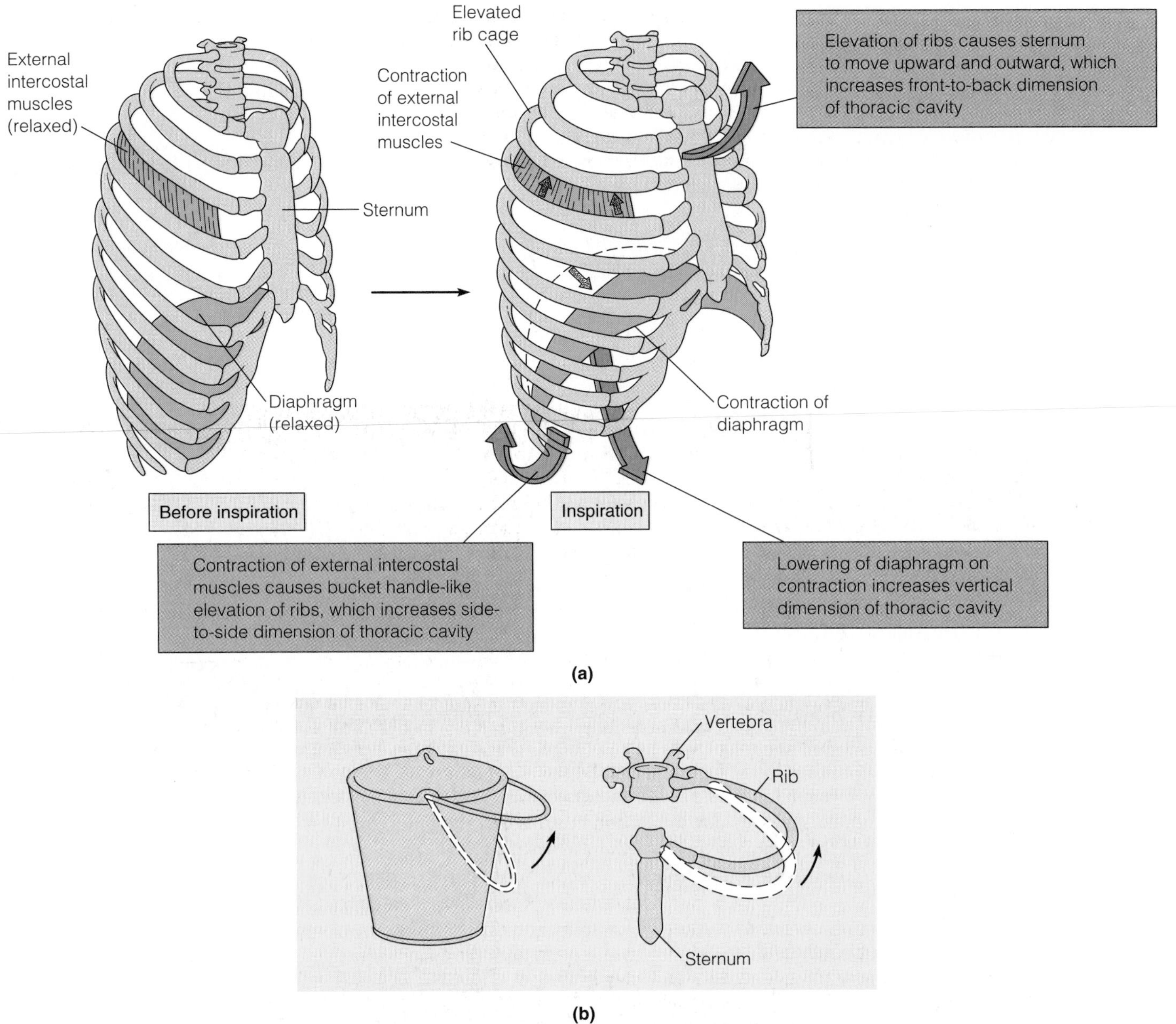

Figure 13-12 Respiratory Muscle Activity during Inspiration and Expiration (a) Inspiration, during which the diaphragm descends on contraction, increasing the vertical dimension of the thoracic cavity. Contraction of the external intercostal muscles elevates the ribs and subsequently the sternum to enlarge the thoracic cavity from front-to-back and side-to-side. (b) The rib elevations produced by contraction of the external intercostal muscles are similar to the lifting of a bucket handle. Notice that elevating the bucket handle also moves it *outward.* (c) Quiet passive expiration, during which the diaphragm relaxes, reducing the volume of the thoracic cavity from its peak inspiratory size. As the external intercostal muscles relax, the elevated rib cage falls because of the force of gravity. This also reduces the volume of the thoracic cavity. (d) Active expiration, during which contraction of the abdominal muscles increases the intra-abdominal pressure, exerting an upward force on the diaphragm. This reduces the vertical dimension of the thoracic cavity further than it is reduced during quiet passive expiration. Contraction of the internal intercostal muscles decreases the front-to-back and side-to-side dimensions by flattening the ribs and sternum.

atmospheric level to 761 mm Hg (Fig. 13-13c). Air now leaves the lungs down its pressure gradient from high intra-alveolar pressure to lower atmospheric pressure. Outward flow of air ceases when intra-alveolar pressure becomes equal to atmospheric pressure and a pressure gradient no longer exists. Figure 13-14 summarizes the intra-alveolar and intrapleural pressure changes that take place during one respiratory cycle.

During quiet breathing, expiration is normally a *passive* process, since it is accomplished by elastic recoil of the lungs on relaxation of the inspiratory muscles, with no muscular exertion or energy expenditure required. In contrast, inspiration is *always active,* because it is brought about only by contraction of inspiratory muscles at the expense of energy utilization. To empty the lungs more completely and more rapidly than is accomplished during quiet breathing, as dur-

Contraction of internal intercostal muscles flattens ribs and sternum, further reducing side-to-side and front-to-back dimensions of thoracic cavity

Relaxation of external intercostal muscles

Relaxation of diaphragm

Passive expiration

Return of diaphragm, ribs, and sternum to resting position on relaxation of inspiratory muscles restores thoracic cavity to preinspiratory size

(c)

Contraction of internal intercostal muscles

Contraction of abdominal muscles

Position of relaxed abdominal muscles

Active expiration

Contraction of abdominal muscles causes diaphragm to be pushed upward, further reducing vertical dimension of thoracic cavity

(d)

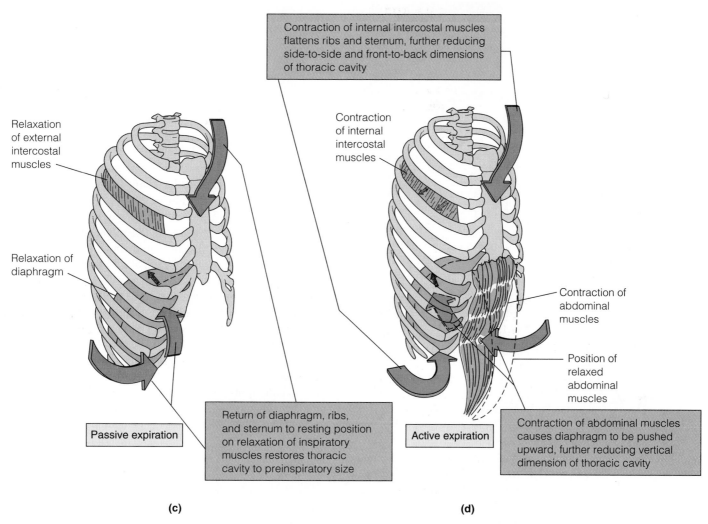

Figure 13-12 **Respiratory Muscle Activity during Inspiration and Expiration (continued)**

Figure 13-13 **Changes in Lung Volume and Intra-alveolar Pressure during Inspiration and Expiration** (a) Before inspiration, at the end of the preceding expiration. Intra-alveolar pressure is equilibrated with atmospheric pressure and no air is flowing. (b) Inspiration. As the lungs increase in volume during inspiration, the intra-alveolar pressure decreases, establishing a pressure gradient that favors the flow of air into the alveoli from the atmosphere; that is, an inspiration occurs. (c) Expiration. As the lungs recoil to their preinspiratory size upon relaxation of the inspiratory muscles, the intra-alveolar pressure increases, establishing a pressure gradient that favors the flow of air out of the alveoli into the atmosphere; that is, an expiration occurs.

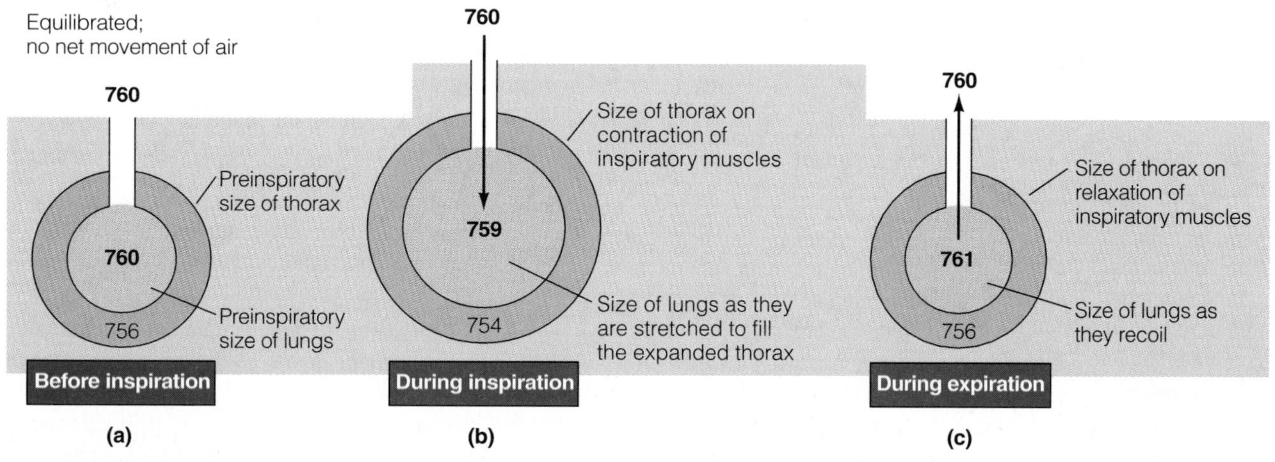

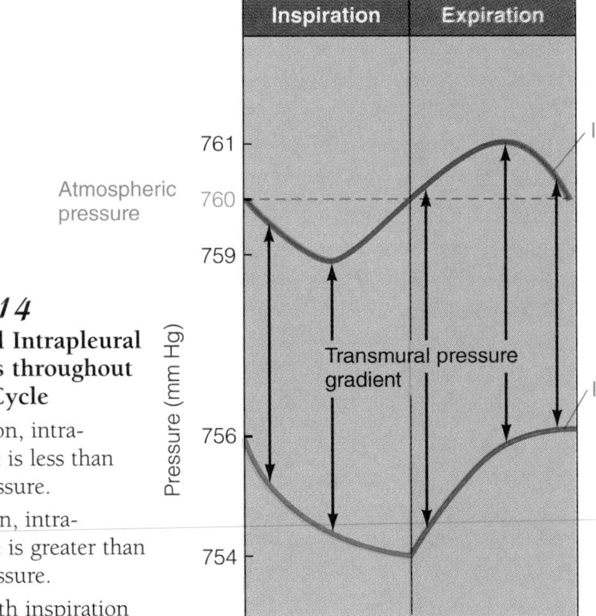

Figure 13-14

Intra-alveolar and Intrapleural Pressure Changes throughout the Respiratory Cycle

- During inspiration, intra-alveolar pressure is less than atmospheric pressure.
- During expiration, intra-alveolar pressure is greater than atmospheric pressure.
- At the end of both inspiration and expiration, intra-alveolar pressure is equal to atmospheric pressure, because the alveoli are in direct communication with the atmosphere and air continues to flow down its pressure gradient until the two pressures equilibrate.
- Throughout the respiratory cycle, the intrapleural pressure is lower than the intra-alveolar pressure. Thus, a transmural pressure gradient always exists, and the lung is always stretched to some degree, even during expiration.

ing the deeper breaths accompanying exercise, expiration does become active. The intra-alveolar pressure must be increased even further above atmospheric pressure than can be accomplished by simple relaxation of the inspiratory muscles and elastic recoil of the lungs. To produce such a **forced,** or **active, expiration, expiratory muscles** must contract to further reduce the volume of the thoracic cavity and lungs (Figs. 13-11 and 13-12 and Table 13-1). The most important expiratory muscles are (unbelievable as it may seem at first) the *muscles of the abdominal wall.* As the abdominal muscles contract, the resultant increase in intra-abdominal pressure exerts an upward force on the diaphragm, pushing it further up into the thoracic cavity than its relaxed position, thus decreasing the vertical dimension of the thoracic cavity even more (Fig. 13-12d). The other expiratory muscles are the **internal intercostal muscles,** whose contraction pulls the ribs downward and inward, flattening the chest wall and further decreasing the size of the thoracic cavity; this action is just the opposite of that of the external intercostal muscles (Fig. 13-12d).

As the active contraction of the expiratory muscles further reduces the volume of the thoracic cavity, the lungs also become further reduced in volume because they do not have to be stretched as much to fill the smaller thoracic cavity; that is, they are permitted to recoil to an even smaller volume. The intra-alveolar pressure increases further as the air in the lungs is confined within this smaller volume. The differential between intra-alveolar and atmospheric pressure is even greater now than during passive expiration, so more air leaves down the pressure gradient before equilibration is achieved. In this way, the lungs are emptied more completely during forceful, active expiration than during quiet, passive expiration.

During forceful expiration, the intrapleural pressure exceeds atmospheric pressure, but the lungs do not collapse. Because the intra-alveolar pressure is also increased correspondingly, a transmural pressure gradient still exists across the walls of the lungs, keeping them stretched to fill the thoracic cavity. For example, if the pressure within the thorax increases 10 mm Hg, the intrapleural pressure becomes 766 mm Hg and the intra-alveolar pressure becomes 770 mm Hg—still a 4 mm Hg pressure difference.

Because the diaphragm is the major inspiratory muscle and its relaxation also causes expiration, paralysis of the intercostal muscles alone does not seriously influence quiet breathing. Disruption of diaphragm activity caused by nerve or muscle disorders, however, leads to respiratory paralysis. Fortunately, the phrenic nerve arises from the spinal cord in the neck region (cervical segments 3, 4, and 5) and then descends to the diaphragm at the floor of the thorax, instead of arising from the thoracic region of the cord as might be expected. For this reason, individuals completely paralyzed below the neck as a result of traumatic severance of the spinal cord are still able to breathe, even though they have lost use of all other skeletal muscles in their trunk and limbs.

Airway resistance becomes an especially important determinant of airflow rates when the airways are narrowed by disease processes.

Thus far we have discussed airflow in and out of the lungs as a function of the magnitude of the pressure gradient between the alveoli and the atmosphere, with the pressure gradient changing through alterations in the dimensions of the thoracic cavity and subsequently of the lungs. However, just as flow of blood through the blood vessels depends not only on the pressure gradient but also on the resistance to the flow offered by the vessels, so it is with airflow:

$$F = \frac{\Delta P}{R}$$

where

F = airflow rate
ΔP = difference between atmospheric and intra-alveolar pressure (pressure gradient)
R = resistance of airways, determined by their radii

The primary determinant of resistance to airflow is the radius of the conducting airways. We ignored airway resistance in our preceding discussion of pressure gradient–induced airflow rates because, in a healthy respiratory system, the radius of the conducting system is sufficiently large that resistance remains extremely low. Therefore, the pressure gradient between the alveoli and the atmosphere is usually the primary factor determining the airflow rate. Indeed, the airways normally offer such low resistance that only very small pressure gradients of 1 to 2 mm Hg need be created to achieve adequate rates of airflow in and out of the lungs. (By comparison, it would take a pressure gradient 250 times greater to move air through a smoker's pipe than through the respiratory airways at the same flow rate.)

Normally, modest adjustments in airway size can be accomplished by autonomic nervous system regulation to suit the body's needs. Parasympathetic stimulation, which occurs in quiet, relaxed situations when the demand for airflow is not high, promotes bronchiolar smooth muscle contraction, which increases airway resistance by producing **bronchoconstriction** (a reduction in bronchiolar caliber). In contrast, sympathetic stimulation and to a greater extent its associated hormone, epinephrine, bring about **bronchodilation** (an increase in bronchiolar caliber) and decreased airway resistance by promoting bronchiolar smooth muscle relaxation (Table 13-2). Thus, during periods of sympathetic domination, when increased demands for O_2 uptake are actually or potentially placed on the body, bronchodilation occurs to ensure that the pressure gradients established by respiratory muscle activity are able to achieve maximum airflow rates with minimum resistance. Because of this bronchodilator action, epinephrine or similar drugs are useful therapeutic tools to counteract airway constriction in patients with bronchial spasms.

Resistance becomes an extremely important impediment to airflow when airway lumens become abnormally narrowed as a result of disease. We have all transiently experienced the effect that increased airway resistance has on breathing when we have a cold. We know how difficult it is to produce an adequate airflow rate through a "stuffy nose" when the nasal passages are narrowed as a result of swelling and mucus accumulation.

More serious is **chronic obstructive pulmonary disease (COPD)**, a group of lung diseases characterized by increased airway resistance resulting from the narrowing of the lumen of the lower airways. When airway resistance increases, a larger pressure gradient must be established to maintain even a normal airflow rate. For example, if resistance is doubled by narrowing of airway lumens, ΔP must be doubled through increased respiratory muscle exertion to induce the same flow rate of air in and out of the lungs as a normal individual accomplishes during quiet breathing. Accordingly, patients with COPD have to work harder to breathe.

Chronic obstructive pulmonary disease encompasses three chronic (long-term) diseases: asthma, chronic bronchitis, and emphysema. In **asthma,** airway obstruction is due to (1) profound constriction of the smaller airways caused by allergy-induced spasm of the smooth muscle in the walls of these airways (see p. 407); (2) plugging of the airways by excess secretion of a very thick mucus; and (3) thickening of the walls of the airways due to inflammation and histamine-induced edema (see p. 331).

Chronic bronchitis is a long-term inflammatory condition of the lower respiratory airways, generally triggered by frequent exposure to irritating cigarette smoke, polluted air, or allergens. In response to the chronic irritation, the airways become narrowed because of prolonged edematous thickening of the airway linings, coupled with overproduction of a thick mucus. Despite frequent coughing associated with the chronic irritation, the plugged mucus often cannot be satisfactorily removed, especially since the ciliary mucus escalator is immobilized by the irritants (see p. 411). Pulmonary bacterial infections frequently occur, because the accumulated mucus serves as an excellent medium for bacterial growth.

Emphysema is characterized by collapse of the smaller airways and a breakdown of alveolar walls. This irreversible condition can arise in two different ways. Most commonly, emphysema results from excessive release of destructive enzymes such as *trypsin* from alveolar macrophages in response to chronic exposure to cigarette smoke or other inhaled chemical irritants. The lungs are normally protected from damage by these enzymes by α_1-*antitrypsin,* a protein

Table 13-2 **Factors Affecting Airway Resistance**

Status of Airways	*Effect on Resistance*	*Factors Producing the Effect*
Bronchoconstriction	↓ radius, ↑ resistance to airflow	*Pathological factors:* Allergy-induced spasm of the airways caused by: Slow-reactive substance of anaphylaxis Histamine Physical blockage of the airways caused by: Excess mucus Edema of the walls Airway collapse *Physiological control factors:* Neural control: parasympathetic stimulation Local chemical control: ↓ CO_2 concentration
Bronchodilation	↑ radius, ↓ resistance to airflow	*Pathological factors:* none *Physiological control factors:* Neural control: sympathetic stimulation (minimal effect) Hormonal control: epinephrine Local chemical control: ↑ CO_2 concentration

that inhibits trypsin. Excessive secretion of these destructive enzymes in response to chronic irritation, however, can overwhelm the protective capability of α_1-antitrypsin so that these enzymes destroy not only foreign materials but lung tissue as well. Loss of lung tissue leads to the breakdown of alveolar walls and collapse of small airways, the characteristics of emphysema. Less frequently, emphysema arises from a genetic inability to produce α_1-antitrypsin so that the lung tissue has no protection from trypsin. The unprotected lung tissue gradually disintegrates under the influence of even small amounts of macrophage-released enzymes in the absence of chronic exposure to inhaled irritants.

When airway resistance is increased as a result of chronic obstructive lung disease of any type, expiration is more difficult to accomplish than inspiration. The smaller airways, lacking the cartilaginous rings that hold the larger airways open, are held open by the same transmural pressure gradient that distends the alveoli. Expansion of the thoracic cavity during inspiration indirectly dilates the airways even further than their expiratory dimensions, similar to alveolar expansion, so airway resistance is lower during inspiration than during expiration. In a normal individual, the airway resistance is always so low that the slight variation occurring between inspiration and expiration is not noticeable. When airway resistance has substantially increased, however, such as during an asthmatic attack, the difference between inspiration and expiration is quite noticeable. Thus, an asthmatic has more difficulty expiring than inspiring, giving rise to the characteristic "wheeze" as air is forced out through the narrowed airways.

In normal individuals, the smaller airways collapse and further outflow of air is halted only at very low lung volumes (Fig. 13-

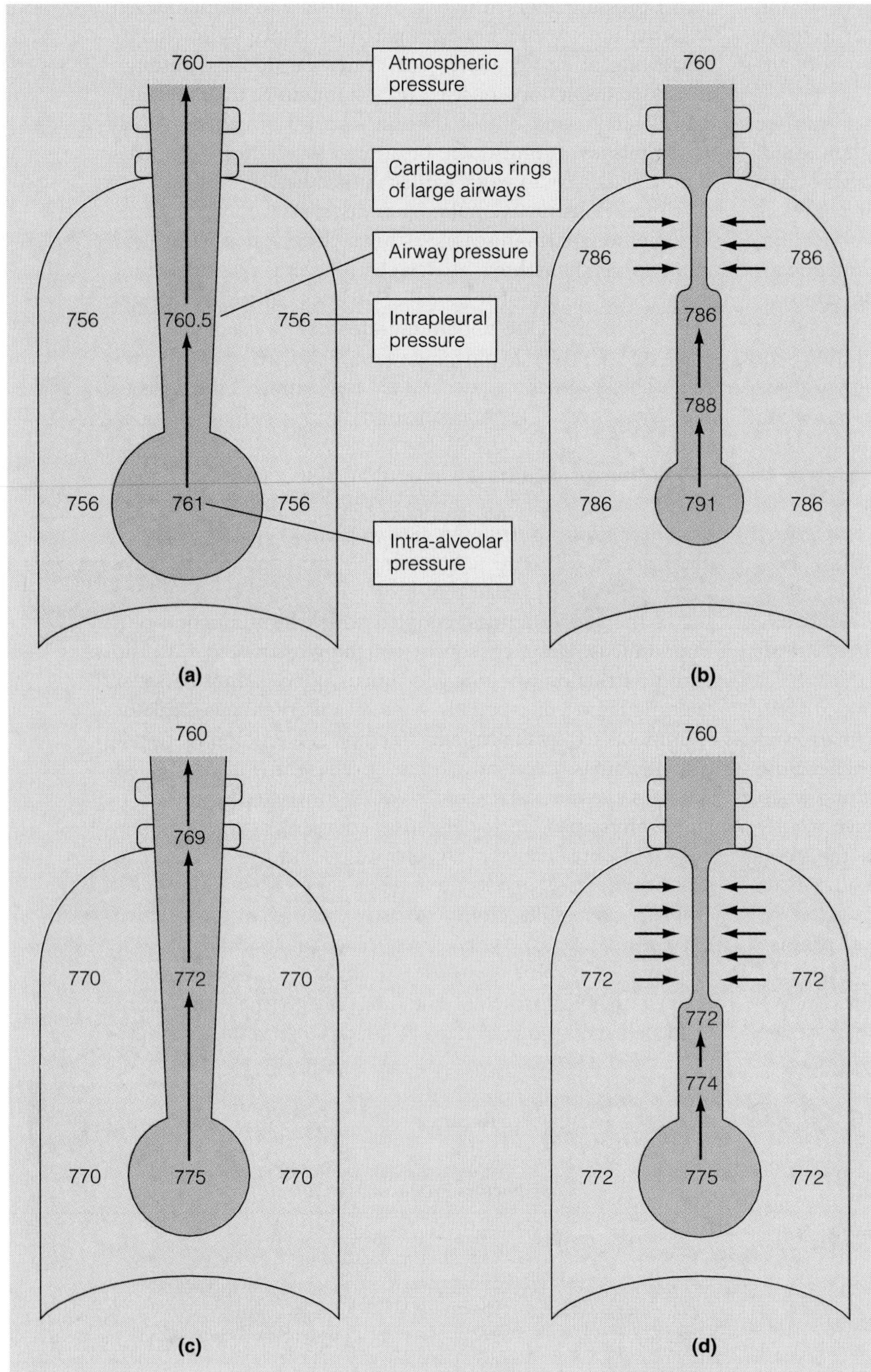

Figure 13-15 Airway Collapse during Forced Expiration (a) Normal quiet breathing, during which airway resistance is low, so there is little frictional loss of pressure within the airways. Intrapleural pressure remains less than airway pressure throughout the length of the airways, so the airways remain open. (b) Maximal forced expiration, during which both intra-alveolar and intrapleural pressures are markedly increased. When frictional losses cause the airway pressure to fall below the surrounding elevated intrapleural pressure, the small nonrigid airways are compressed closed, blocking further expiration of air through the airway. In normal individuals, this occurs only at very low lung volumes. (c) Routine exercise. Even though intrapleural pressure is elevated during the active expiration accompanying routine vigorous activity, the airway pressure does not drop below the intrapleural pressure until the level at which the airways are held open by cartilaginous rings, so airway collapse does not occur. (d) Obstructive lung disease. Premature airway collapse occurs for two reasons: (1) the pressure drop along the airways is magnified as a result of increased airway resistance, and (2) the intrapleural pressure is higher than normal due to the loss, as in emphysema, of lung tissue that is responsible for the lung's tendency to recoil and pull away from the thoracic wall. Excessive air trapped in the alveoli behind the compressed bronchiolar segments reduces the amount of gas exchanged between the alveoli and the atmosphere. Therefore, less alveolar air is "freshened" with each breath when airways collapse at higher lung volumes in patients with obstructive lung disease.

15). Because of this airway collapse, the lungs can never be emptied completely.

Local controls act on the smooth muscle of the airways and arterioles to maximally match blood flow to airflow.

We have been talking about overall airway resistance in the lungs, but like arteriolar smooth muscle, bronchiolar smooth muscle is sensitive to local changes within its immediate environment, particularly to local CO_2 levels. If an alveolus is receiving too little airflow (ventilation) in comparison to its blood flow (perfusion), CO_2 levels will increase in the alveolus and surrounding tissue as more CO_2 is dropped off by the blood than is being exhaled into the atmosphere. This local increase in CO_2 acts directly on the bronchiolar smooth muscle involved to induce the airway supplying the underaerated alveolus to relax. The resultant decrease in airway resistance leads to an increased airflow (for the same ΔP) to the involved alveolus, so its airflow now matches its blood supply (Fig. 13-16). The converse is also true. A localized decrease in CO_2 associated with an alveolus that is receiving too much air for its blood supply directly increases contractile activity of the airway smooth muscle involved, causing the airway supplying this overaerated alveolus to constrict. The result is a reduction in airflow to the overaerated alveolus.

A similar locally induced effect on pulmonary vascular smooth muscle also takes place simultaneously to maximally match blood flow to airflow. Just as in the systemic circulation, distribution of the cardiac output to different alveolar capillary networks can be controlled by adjusting the resistance to blood flow through specific pulmonary arterioles. If the blood flow is greater than the airflow to a given alveolus, the O_2 level in the alveolus and surrounding tissues will fall below normal as more O_2 than usual is extracted from the alveolus by the overabundance of blood. The local decrease in O_2 concentration causes vasoconstriction of the pulmonary arteriole supplying this particular capillary bed; the result is a reduction in blood flow to match the smaller airflow. Conversely, an increase in alveolar O_2 concentration caused by a mismatched large airflow and small blood flow brings about pulmonary vasodilation, which increases blood flow to match the larger airflow. Note that the local effect of O_2 on pulmonary arteriolar smooth muscle is appropriately just the opposite of its effect on systemic arteriolar smooth muscle (Table 13-3). In the systemic circulation, a decrease in O_2 in a tissue causes localized vasodilation to increase blood flow to the deprived area, and vice versa, which is important in matching blood supply to local metabolic needs (see p. 316).

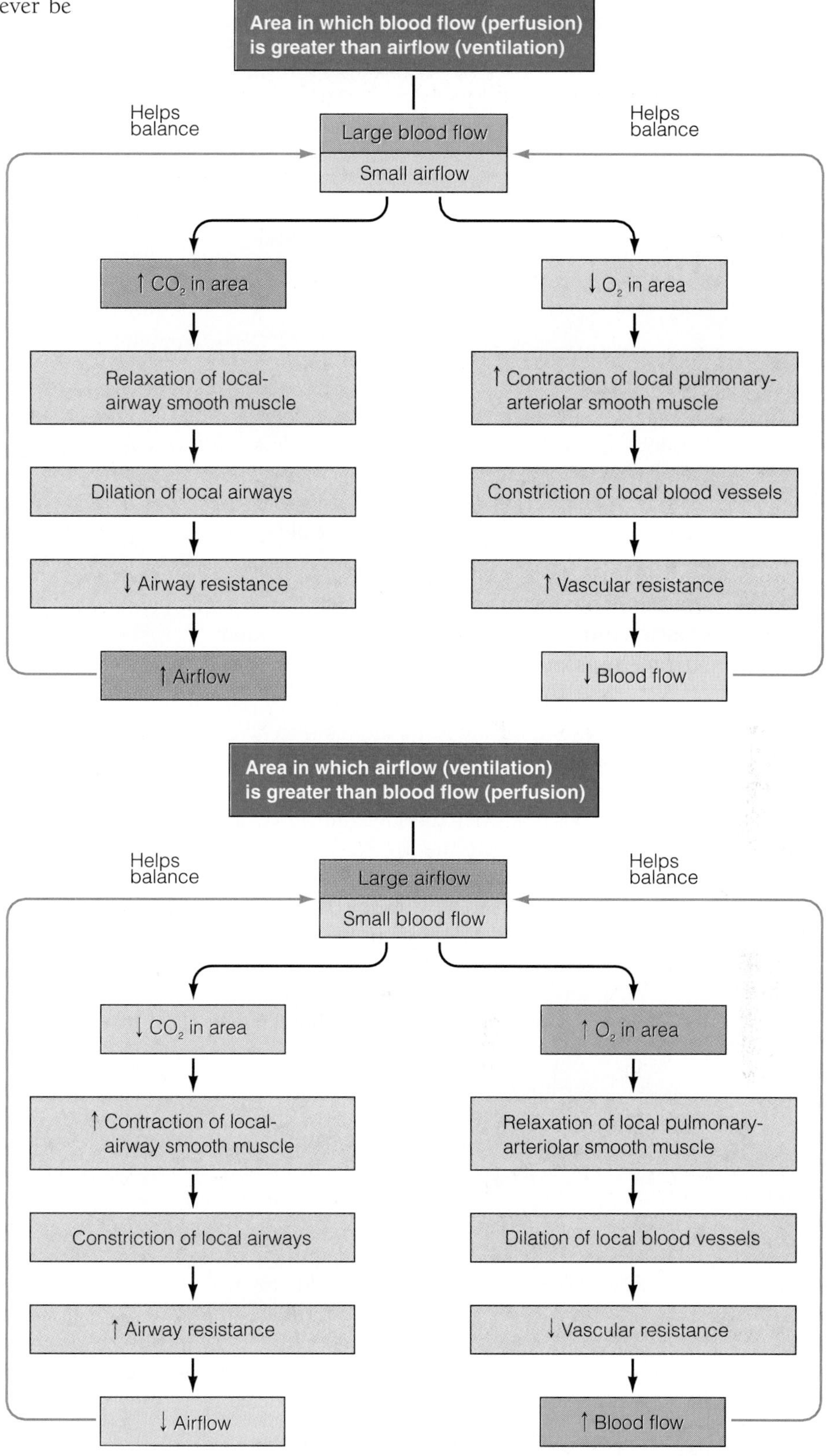

Figure 13-16 **Local Controls to Match Airflow and Blood Flow to an Area of the Lung**

The two mechanisms for matching airflow and blood flow function concurrently, so normally very little air or blood is wasted in the lung. Some regional differences in ventilation

Table 13-3
Effects of Local Changes in O_2 on the Pulmonary and Systemic Arterioles

Vessels	Effect of a Local Change in O_2	
	Decreased O_2	*Increased O_2*
Pulmonary arterioles	Vasoconstriction	Vasodilation
Systemic arterioles	Vasodilation	Vasoconstriction

and perfusion exist from the top to the bottom of the lung because of gravitational effects. Nevertheless, airflow and blood flow at a particular alveolar interface are usually matched to the greatest extent possible by means of these local controls to accomplish efficient exchange of O_2 and CO_2.

Elastic behavior of the lungs is due to elastic connective tissue fibers and alveolar surface tension.

You have learned that during the respiratory cycle, the lungs alternately expand during inspiration and recoil during expiration. What properties of the lungs enable them to behave like balloons, being stretchable and then snapping back to their resting position when the stretching forces are removed? Two interrelated concepts are involved in pulmonary elasticity: elastic recoil and compliance.

Elastic recoil refers to how readily the lungs rebound after having been stretched. It is responsible for the lungs returning to their preinspiratory volume when the inspiratory muscles relax at the end of inspiration.

Compliance refers to how much effort is required to stretch or distend the lungs; it is analogous to how hard you have to work to blow up a balloon. (By comparison, 100 times more distending pressure is required to inflate a child's toy balloon than to inflate the lungs.) Specifically, compliance is a measure of the magnitude of change in lung volume accomplished by a given change in the transmural pressure gradient, the force that stretches the lungs. A highly compliant lung stretches further for a given increase in the pressure difference than does a less compliant lung. Stated another way, the lower the compliance of the lungs, the larger the transmural pressure gradient that must be created during inspiration to produce normal lung expansion. In turn, a greater-than-normal transmural pressure gradient during inspiration can be achieved only by making the intrapleural pressure more subatmospheric than usual. This is accomplished by greater expansion of the thorax through more vigorous contraction of the inspiratory muscles. Therefore, the less compliant the lungs, the more work required to produce a given degree of inflation. A poorly compliant lung is referred to as a "stiff" lung, because it lacks normal stretchability. Respiratory compliance can be decreased by a number of factors, such as replacement of normal lung tissue with fibrous connective tissue as a result of breathing in asbestos fibers or similar irritants.

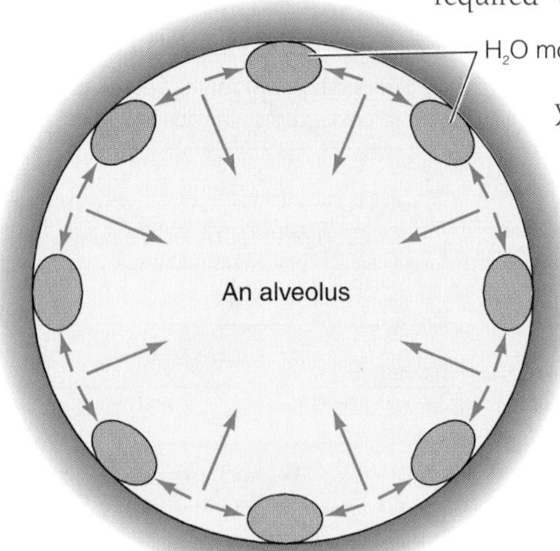

Figure 13-17 Alveolar Surface Tension The attractive forces between the water (H_2O) molecules in the liquid film that lines the alveolus are responsible for surface tension. Because of its surface tension, an alveolus (1) resists being stretched, (2) tends to be reduced in surface area or size, and (3) tends to recoil after being stretched.

Pulmonary elastic behavior depends mainly on two factors: highly elastic connective tissue in the lungs and alveolar surface tension. Pulmonary connective tissue contains large quantities of elastin fibers (see p. 51). Not only do these fibers exhibit elastic properties themselves, but they are arranged into a meshwork that amplifies their elastic behavior, much like the threads in a piece of stretch-knit fabric. The entire piece of fabric (or lung) is stretchier and tends to bounce back to its original shape more than the individual threads (elastin fibers) of which the fabric is woven.

An even more important factor influencing elastic behavior of the lungs is the **alveolar surface tension** displayed by the thin liquid film that lines each alveolus. At an air-water interface, the water molecules at the surface are more strongly attracted to other surrounding water molecules than to the air above the surface. This unequal attraction produces a force known as surface tension at the surface of the liquid. Surface tension is responsible for a twofold effect. First, the liquid layer resists any force that increases its surface area; that is, it opposes expansion of the alveolus because the surface water molecules oppose being pulled apart. Accordingly, the greater the surface tension, the less compliant the lungs. Second, the liquid surface area tends to become as small as possible because the surface water molecules, being preferentially attracted to each other, try to get as close together as possible. Thus, the surface tension of the liquid lining an alveolus tends to reduce the size of the alveolus, squeezing in on the air within it (Fig. 13-17). This property, along with the rebound of the stretched elastin fibers, is responsible for the lungs' elastic recoil back to their preinspiratory size when inspiration is over.

Pulmonary surfactant decreases surface tension and contributes to lung stability.

The cohesive forces between water molecules are so strong that if the alveoli were lined with water alone, the surface tension would be so great that the lungs would collapse; the recoil force attributable to the elastin fibers and high surface tension would exceed the opposing stretching force of the transmural pressure gradient. Furthermore, the lungs would be very poorly compliant, so exhausting muscular efforts would be required to accomplish stretching and inflation of the alveoli.

Law of LaPlace:
Magnitude of inward-directed pressure (P) in a bubble (alveolus) = $\frac{2 \times \text{Surface tension (T)}}{\text{Radius (r) of bubble (alveolus)}}$

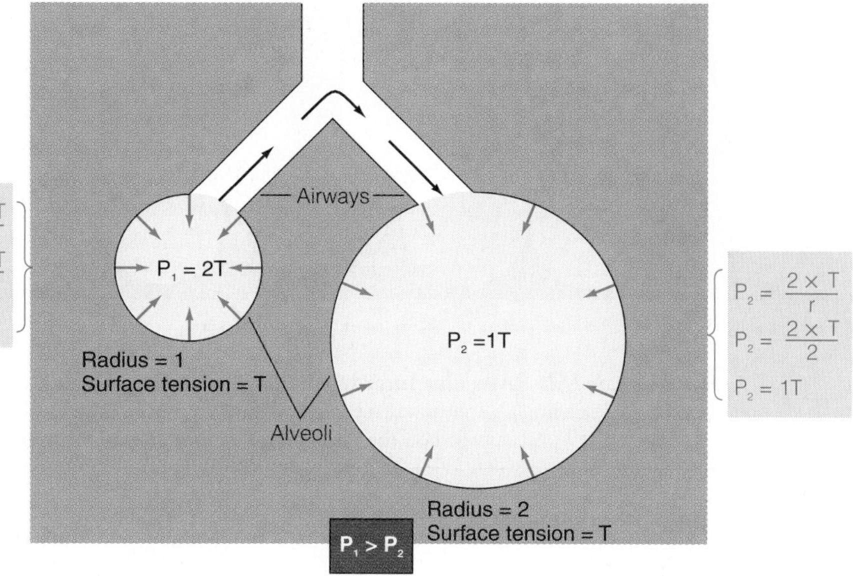

The tremendous surface tension of pure water is normally counteracted by *pulmonary surfactant,* a complex mixture of lipids and proteins secreted by the Type II alveolar cells (Fig. 13-4a). Pulmonary surfactant intersperses between the water molecules in the fluid lining the alveoli and lowers the alveolar surface tension because the cohesive force between a water molecule and an adjacent pulmonary surfactant molecule is very low. By lowering the alveolar surface tension, pulmonary surfactant provides two important benefits: (1) it increases pulmonary compliance, thus reducing the work of inflating the lungs; and (2) it reduces the lungs' tendency to recoil, so that they do not collapse as readily.

Pulmonary surfactant's role in reducing the alveoli's tendency to recoil, thereby discouraging alveolar collapse, is important in helping maintain lung stability. The division of the lung into myriad tiny air sacs provides the advantage of a tremendously increased surface area for exchange of O_2 and CO_2, but it also presents the problem of maintaining the stability of all these alveoli. Recall that the pressure generated by alveolar surface tension is directed inward, squeezing in on the air in the alveoli. If the alveoli are visualized as spherical bubbles, according to **LaPlace's law,** the magnitude of the inward-directed collapsing pressure is directly proportional to the surface tension and inversely proportional to the radius of the bubble:

$$P = \frac{2T}{r}$$

where

P = inward-directed collapsing pressure
T = surface tension
r = radius of bubble (alveolus)

Because the collapsing pressure is inversely proportional to the radius, the smaller the alveolus, the smaller its radius and the greater its tendency to collapse at a given surface tension. Accordingly, if two alveoli of unequal size but the same surface tension are connected by the same terminal airway, the smaller alveolus—because it generates a larger collapsing pressure—has a tendency to collapse and empty its air into the larger alveolus (Fig. 13-18a). Small alveoli normally do not collapse and blow up larger alveoli, how-

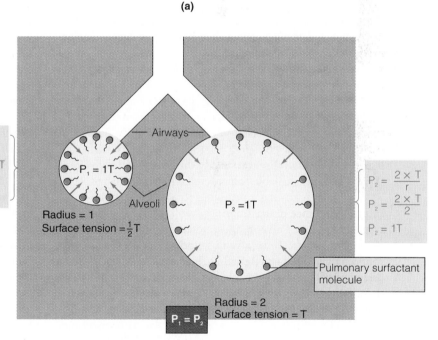

***Figure 13-18* Role of Pulmonary Surfactant in Counteracting the Tendency for Small Alveoli to Collapse into Larger Alveoli**
(a) According to the law of LaPlace, if two alveoli of unequal size but the same surface tension are connected by the same terminal airway, the smaller alveolus—because it generates a larger inward-directed collapsing pressure—has a tendency (without pulmonary surfactant) to collapse and empty its air into the larger alveolus. (b) Pulmonary surfactant reduces the surface tension of a smaller alveolus more than that of a larger alveolus. This reduction in surface tension offsets the effect of the smaller radius in determining the inward-directed pressure. Consequently, the collapsing pressures of the small and large alveoli are comparable. Therefore, in the presence of pulmonary surfactant, a small alveolus does not collapse and empty its air into the larger alveolus.

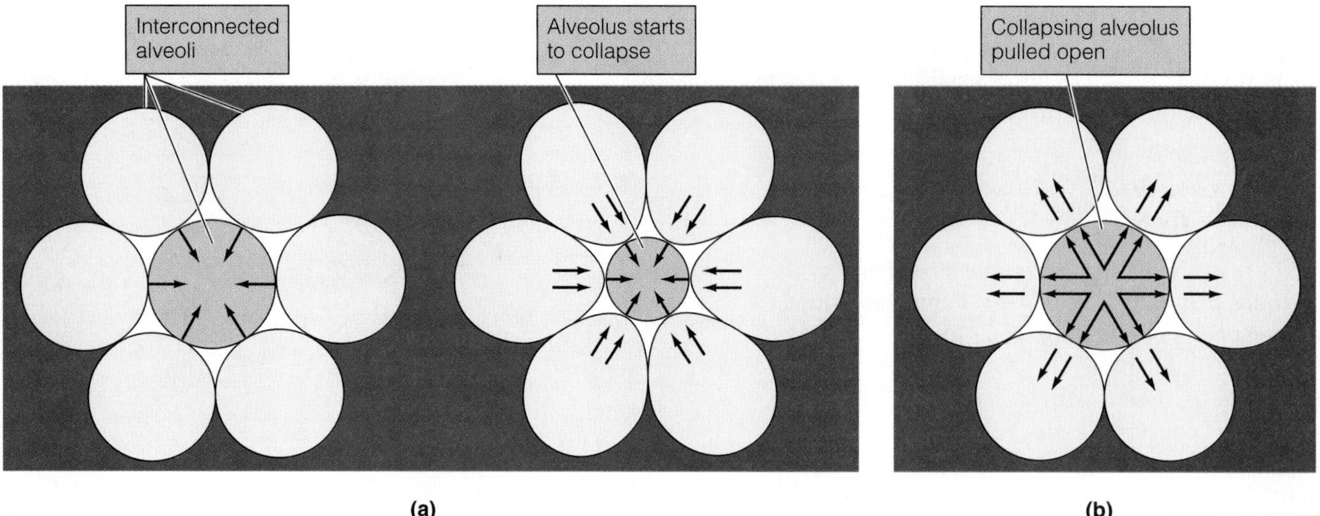

Figure 13-19 Alveolar Interdependence (a) When an alveolus (in pink) in a group of interconnected alveoli starts to collapse, the surrounding alveoli (in cream) are stretched by the collapsing alveolus. (b) As the neighboring alveoli recoil in resistance to being stretched, they pull outward on the collapsing alveolus. This expanding force pulls the collapsing alveolus open.

ever, because pulmonary surfactant reduces the surface tension of small alveoli more than that of larger alveoli. Pulmonary surfactant decreases surface tension to a greater degree in small alveoli than in larger alveoli because the surfactant molecules are closer together in the smaller alveoli. The larger an alveolus, the more spread out are its surfactant molecules and the less effect they have on reducing surface tension. The surfactant-induced lower surface tension of small alveoli offsets the effect of their smaller radii in determining the inward-directed pressure. Therefore, the presence of surfactant causes the collapsing pressure of small alveoli to become comparable to that of larger alveoli and minimizes the tendency for small alveoli to collapse and empty their contents into larger alveoli (Fig. 13-18b). Pulmonary surfactant therefore helps stabilize the sizes of the alveoli and helps keep them open and available to participate in gas exchange.

A second factor that contributes to alveolar stability is the **interdependence** of neighboring alveoli. Each alveolus is surrounded by other alveoli and interconnected with them by connective tissue. If an alveolus starts to collapse, the surrounding alveoli are stretched as their walls are pulled in the direction of the caving-in alveolus (Fig. 13-19a). In turn, these neighboring alveoli, by recoiling in resistance to being stretched, exert expanding forces on the collapsing alveolus and thereby help keep it open (Fig. 13-19b). This phenomenon, which can be likened to a stalemated "tug of war" between adjacent alveoli, is termed *interdependence.*

The opposing forces acting on the lung (that is, the forces keeping the alveoli open and the countering forces that promote alveolar collapse) are summarized in Table 13-4.

New findings point to an important role for pulmonary surfactant in the lung defense system in addition to its traditional role of decreasing surface tension and thereby helping to maintain alveolar stability. Studies suggest that the protein component of pulmonary surfactant enhances phagocytosis of bacteria and viruses by alveolar macrophages and assists the ciliary mucus escalator in the respiratory airways (see p. 411).

A deficiency of pulmonary surfactant is responsible for newborn respiratory distress syndrome.

The developing fetal lungs normally do not have the ability to synthesize pulmonary surfactant until late in pregnancy. Especially in an infant born prematurely, pulmonary surfactant may be insufficient to reduce the alveolar surface tension to manageable levels. The resultant collection of symptoms that develop are referred to as **newborn respiratory distress syndrome.** Very strenuous inspiratory efforts are required to overcome the high surface tension in an attempt to inflate the poorly compliant lungs. Adding to the dilemma, the work of breathing is further increased because the alveoli, in the absence of surfactant, tend to collapse almost completely during each expiration. It is more difficult (requires a greater transmural pressure differential) to expand a collapsed alveolus by a given volume than to increase an already partially expanded alveolus by the same volume. The situation is analogous to blowing up a new balloon. It takes more effort to blow in that first breath of air when starting to blow up a new balloon than to blow additional breaths into the already partially expanded balloon. With newborn respiratory distress syndrome, it is as though the infant must start blowing up a new balloon with every breath. Lung expansion may require transmural pressure gradients of 20 to 30 mm Hg (compared to the normal 4 to 6 mm Hg) to overcome the tendency of surfactant-deprived alveoli to collapse.

The problem is compounded by the fact that the newborn's muscles are still weak. The respiratory distress associated with surfactant deficiency may soon lead to death as breathing efforts become exhausting or inadequate to support sufficient gas exchange.

This life-threatening condition affects 30,000 to 50,000 newborns, primarily premature infants, each year in the United States. Until the surfactant-secreting cells mature sufficiently, the condition is treated by surfactant replacement. Before exogenous surfactant became available, the only therapy was to force air into the baby's lungs at greater-than-atmospheric, or "positive," pressure. By artificially increasing the atmospheric pressure, a sufficient pressure gradient could be established to drive air into the lungs.

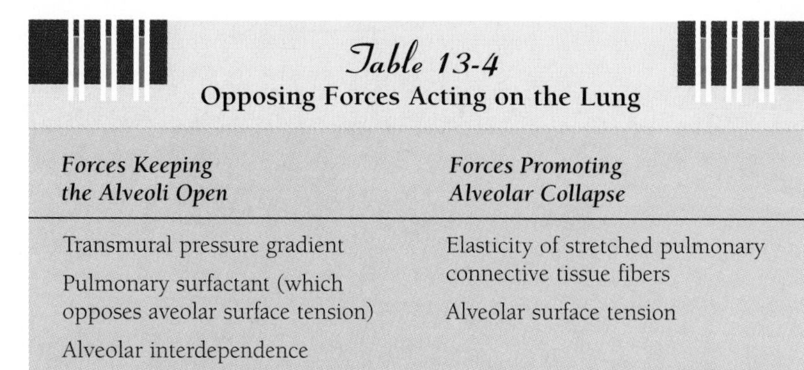

Table 13-4
Opposing Forces Acting on the Lung

Forces Keeping the Alveoli Open	*Forces Promoting Alveolar Collapse*
Transmural pressure gradient	Elasticity of stretched pulmonary connective tissue fibers
Pulmonary surfactant (which opposes aveolar surface tension)	Alveolar surface tension
Alveolar interdependence	

The work of breathing normally requires only about 3% of total energy expenditure.

During normal quiet breathing, the respiratory muscles must work during inspiration to expand the lungs against their elastic forces and to overcome airway resistance, whereas expiration is a passive process. Normally, the lungs are highly compliant and airway resistance is low, so only about 3% of the total energy expended by the body is used to accomplish quiet breathing. The work of breathing may be increased in four different situations:

1. *When pulmonary compliance is decreased,* more work is required to expand the lungs.
2. *When airway resistance is increased,* more work is required to achieve the greater pressure gradients necessary to overcome the resistance so that adequate airflow can occur.
3. *When elastic recoil is decreased,* passive expiration may be inadequate to expel the volume of air normally exhaled during quiet breathing. Thus, the abdominal muscles must work to aid in emptying the lungs, even when the person is at rest.
4. *When there is a need for increased ventilation,* such as during exercise, more work is required to accomplish both a greater depth of breathing (a larger volume of air moving in and out with each breath) and a faster rate of breathing (more breaths per minute).

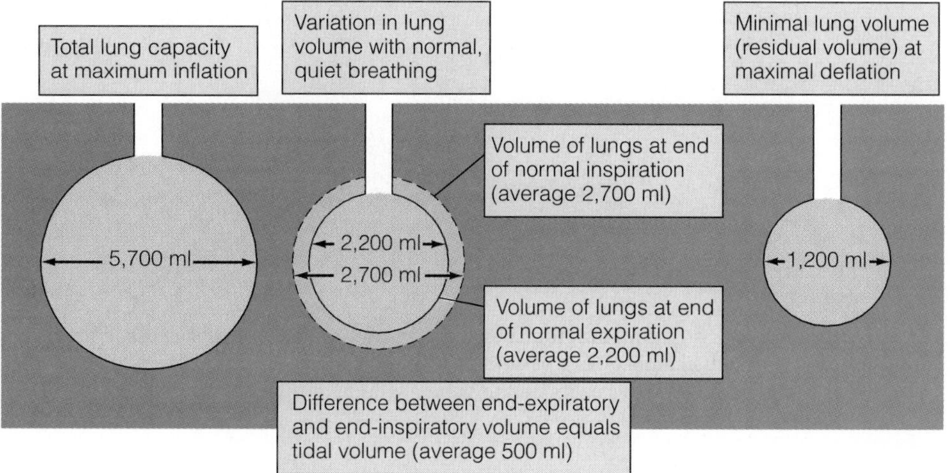

Values are average for a healthy young adult male; values for females are somewhat lower.

Figure 13-20 **Normal Range and Extremes of Lung Volume in an Adult Male**

During strenuous exercise, the amount of energy required to power pulmonary ventilation may increase up to 25-fold. However, because total energy expenditure by the body is increased up to 15- to 20-fold during heavy exercise, the energy used to accomplish the increased ventilation still represents only about 5% of total energy expended. In contrast, in patients with poorly compliant lungs or obstructive lung disease, the energy required for breathing even at rest may increase to 30% of total energy expenditure. In such cases, the individual's exercise ability is severely limited, as breathing itself becomes an exhausting exercise.

Normally, the lungs contain about 2 to 2.5 liters of air during the respiratory cycle but can be filled to more than 5.5 liters or emptied to about 1 liter.

On average, in healthy young adults, the maximum amount of air that the lungs can hold is about 5.7 liters in males (4.2 liters in females). Anatomical build, age, the distensibility of the lungs, and the presence or absence of respiratory disease affect this total lung capacity. Normally, during quiet breathing, the lungs are not anywhere near maximally inflated nor are they deflated to their minimum volume. Thus, the lungs normally remain moderately inflated throughout the respiratory cycle. At the end of a normal quiet expiration, the lungs still contain about 2,200 ml of air. During each typical breath under resting conditions, about 500 ml of air are inspired and the same quantity is expired, so during quiet breathing the lung volume varies between 2,200 ml at the end of expiration to 2,700 ml at the end of inspiration (Fig. 13-20). During maximal expiration, lung volume can be decreased to 1,200 ml in males (1,000 ml in females), but the lungs can never be completely deflated because the small airways collapse during forced expirations at low lung volumes, blocking further outflow of air (Fig. 13-15b).

An important outcome of not being able to empty the lungs completely is that, even during maximal expiratory efforts, gas exchange can still continue between blood flowing through the lungs and the remaining alveolar air. Instead of the wide fluctuations that would occur in O_2 uptake and CO_2 removal by the blood if the lungs were to completely fill and empty with each breath, the gas content of the blood leaving the lungs for delivery to the tissues normally remains remarkably constant throughout the respiratory cycle. Furthermore, recall that it takes less effort to inflate a partially inflated alveolus than a totally collapsed one.

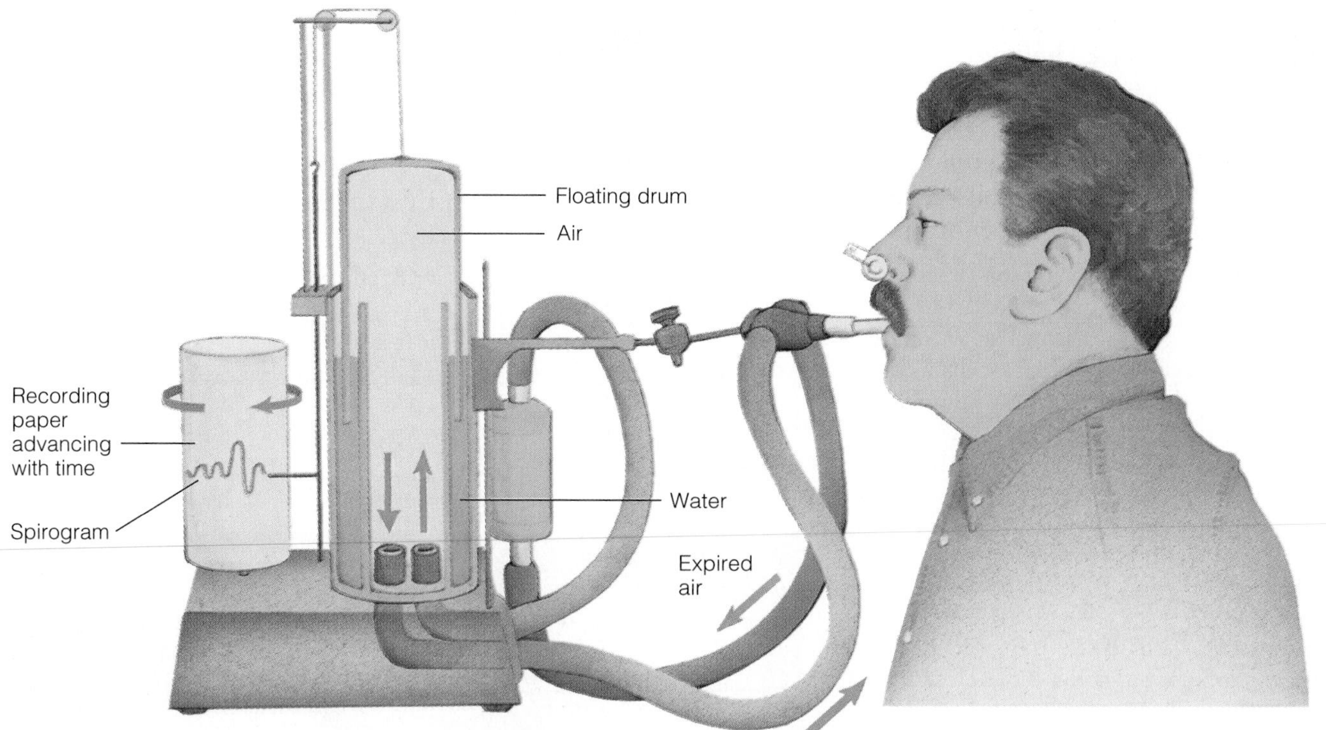

Figure 13-21 **A Spirometer** A spirometer is a device that measures the volume of air breathed in and out; it consists of an air-filled drum floating in a water-filled chamber. As a person breathes air in and out of the drum through a connecting tube, the resultant rise and fall of the drum are recorded as a spirogram, which is calibrated to the magnitude of the volume change.

The changes in lung volume that occur with different respiratory efforts can be measured using a **spirometer.** Basically, a spirometer consists of an air-filled drum floating in a water-filled chamber. As the person breathes air in and out of the drum through a tube connecting the mouth to the air chamber, the drum rises and falls in the water chamber (Fig. 13-21). This rise and fall can be recorded as a **spirogram,** which is calibrated to volume changes. The pen records inspiration as an upward deflection and expiration as a downward deflection.

Figure 13-22 is a hypothetical example of a spirogram in a healthy young adult male. Generally, the values are lower for females. The following lung volumes and lung capacities (a lung capacity is the sum of two or more lung volumes) can be determined:

- **Tidal volume (TV).** The volume of air entering or leaving the lungs during a single breath. Average value under resting conditions = 500 ml.
- **Inspiratory reserve volume (IRV).** The extra volume of air that can be maximally inspired over and above the typical resting tidal volume. The IRV is accomplished by maximal contraction of the diaphragm, external intercostal muscles, and accessory inspiratory muscles. Average value = 3,000 ml.
- **Inspiratory capacity (IC).** The maximum volume of air that can be inspired at the end of a normal quiet expiration (IC = IRV + TV). Average value = 3,500 ml.
- **Expiratory reserve volume (ERV).** The extra volume of air that can be actively expired by maximal contraction of the expiratory muscles beyond that normally passively expired at the end of a typical resting tidal volume. Average value = 1,000 ml.
- **Residual volume (RV).** The minimum volume of air remaining in the lungs even after a maximal expiration. Average value = 1,200 ml. The residual volume cannot be measured directly with a spirometer because this volume of air does not move in and out of the lungs. It can be determined indirectly, however, through gas dilution

Figure 13-22 **Normal Spirogram of a Healthy Young Adult Male**

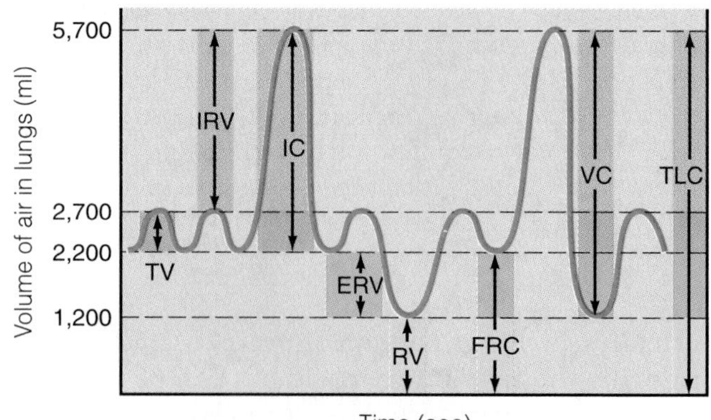

TV = Tidal volume (500 ml)
IRV = Inspiratory reserve volume (3,000 ml)
IC = Inspiratory capacity (3,500 ml)
ERV = Expiratory reserve volume (1,000 ml)
RV = Residual volume (1,200 ml)
FRC = Functional residual capacity (2,200 ml)
VC = Vital capacity (4,500 ml)
TLC = Total lung capacity (5,700 ml)

techniques involving inspiration of a known quantity of a harmless tracer gas such as helium.

- **Functional residual capacity (FRC).** The volume of air in the lungs at the end of a normal passive expiration (FRC = ERV + RV). Average value = 2,200 ml.
- **Vital capacity (VC).** The maximum volume of air that can be moved out during a single breath following a maximal inspiration. The subject first inspires maximally, then expires maximally (VC = IRV + TV + ERV). The VC represents the maximum volume change possible within the lungs (Fig. 13-23). It is rarely used because the maximal muscle contractions involved become exhausting, but it is useful in ascertaining the functional capacity of the lungs. Average value = 4,500 ml.
- **Total lung capacity (TLC).** The maximum volume of air that the lungs can hold (TLC = VC + RV). Average value = 5,700 ml.
- **Forced expiratory volume in one second (FEV_1).** The volume of air that can be expired during the first second of expiration in a VC determination. Usually, FEV_1 is about 80% of VC; that is, normally 80% of the air that can be forcibly expired from maximally inflated lungs can be expired within 1 second. This measurement gives an indication of the maximal airflow rate that is possible from the lungs.

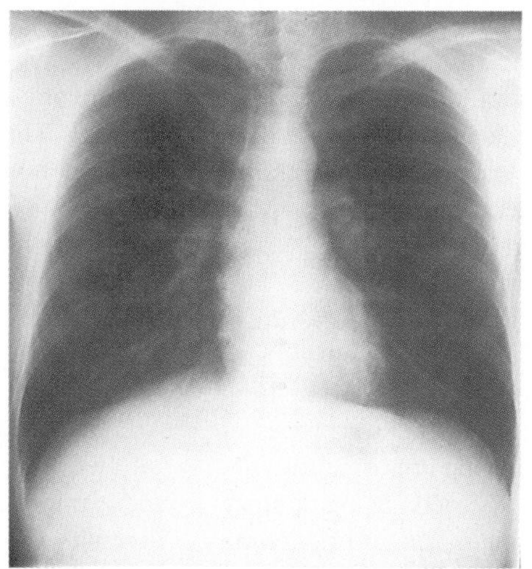

(a)

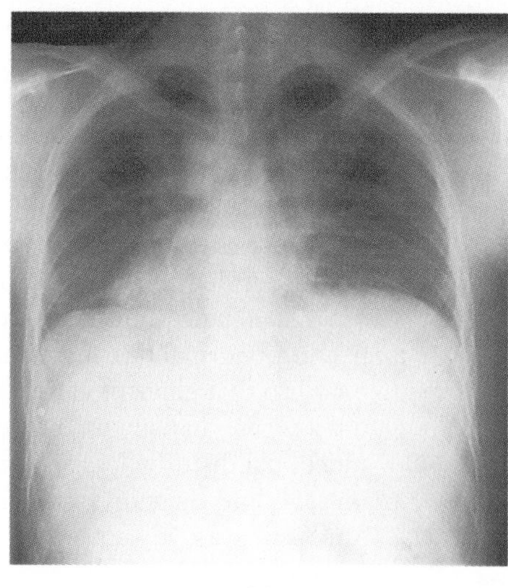

(b)

Figure 13-23 **X Rays of Lungs Showing Maximum Volume Change** (a) Maximum volume of the lungs at maximum inspiration. (b) Minimum volume of the lungs at maximum expiration. The difference between these two volumes is the vital capacity, which is the maximum volume of air that can be moved out during a single breath following a maximum inspiration.

Measurement of the lungs' various volumes and capacities is of more than pure academic interest, because such determinations provide a useful tool to the diagnostician in various respiratory disease states. Two general categories of respiratory dysfunction yield abnormal results during spirometry—*obstructive lung disease* and *restrictive lung disease* (Fig. 13-24). However, you should not have the impression that

Figure 13-24 **Abnormal Spirograms Associated with Obstructive and Restrictive Lung Diseases** (a) Spirogram in obstructive lung disease. Since a patient with obstructive lung disease experiences more difficulty emptying the lungs than filling them, the total lung capacity (TLC) is essentially normal, but the functional residual capacity (FRC) and the residual volume (RV) are elevated as a result of the additional air trapped in the lungs following expiration. Because the RV is increased, the vital capacity (VC) is reduced. With more air remaining in the lungs, less of the TLC is available to be used in exchanging air with the atmosphere. Another common finding is a markedly reduced FEV_1 since the airflow rate is reduced by the airway obstruction. Even though both the VC and the FEV_1 are reduced, the FEV_1 is reduced more markedly than is the VC. As a result, the FEV_1/VC% is much lower than the normal 80%; that is, much less than 80% of the reduced VC can be blown out during the first second. (b) Spirogram in restrictive lung disease. In this disease the lungs are less compliant than normal. Total lung capacity, inspiratory capacity, and VC are reduced, since the lungs cannot be expanded as normal. The percentage of the VC that can be exhaled within 1 second is the normal 80% or an even higher percentage, because air can flow freely in the airways. Therefore, the FEV_1/VC% is particularly useful in distinguishing between obstructive and restrictive lung disease. Also, in contrast to obstructive lung disease, the RV is usually normal in restrictive lung disease.

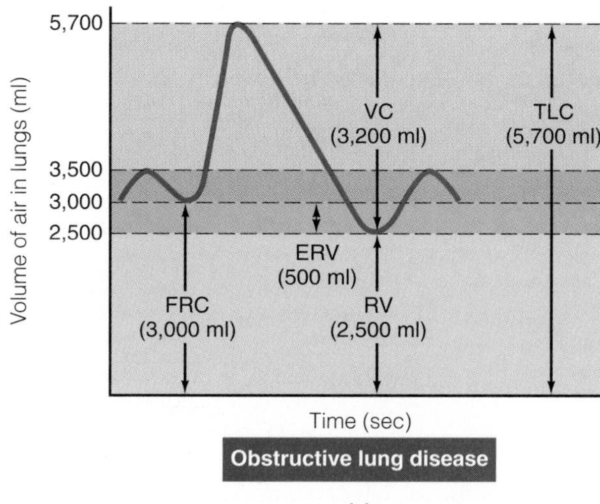

(a)

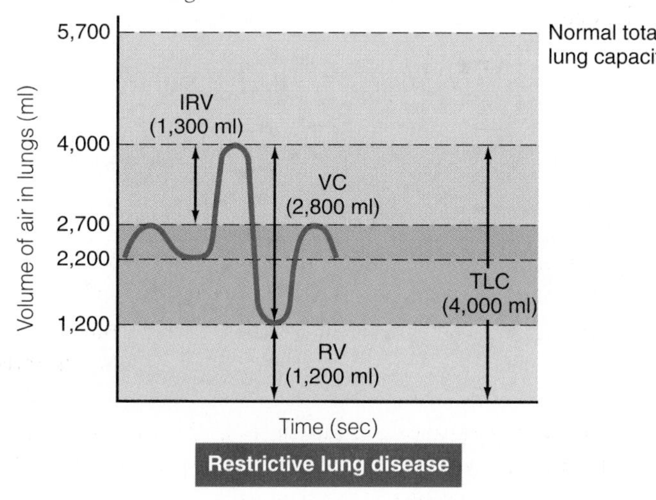

(b)

these are the only categories of respiratory dysfunction nor that spirometry is the only pulmonary function test. Other conditions affecting respiratory function include (1) diseases impairing diffusion of O_2 and CO_2 across the pulmonary membranes; (2) reduced ventilation because of mechanical failure, as with neuromuscular disorders affecting the respiratory muscles, or because of depression of the respiratory control center due to alcohol, drugs, or other chemicals; (3) failure of adequate pulmonary blood flow; or (4) ventilation/perfusion abnormalities involving a poor matching of air and blood so that efficient gas exchange cannot occur. Some lung diseases are actually a complex mixture of different types of functional disturbances. To determine what abnormalities are present, the diagnostician relies on a variety of respiratory function tests in addition to spirometry, including X-ray examination, blood gas determinations, and tests to measure the diffusion capacity of the alveolar capillary membrane.

Alveolar ventilation is less than pulmonary ventilation because of the presence of dead space.

Various changes in lung volume represent only one factor in the determination of **pulmonary,** or **minute, ventilation,** which is the volume of air breathed in and out in one minute. The other important factor is **respiratory rate,** which averages 12 breaths per minute.

$$\underset{\text{(ml/min)}}{\text{pulmonary ventilation}} = \underset{\text{(ml/breath)}}{\text{tidal volume}} \times \underset{\text{(breaths/min)}}{\text{respiratory rate}}$$

At an average tidal volume of 500 ml/breath and a respiratory rate of 12 breaths per minute, pulmonary ventilation is 6,000 ml, or 6 liters, of air breathed in and out in one minute under resting conditions. For a brief period of time, a healthy young adult male can voluntarily increase his total pulmonary ventilation 25-fold, to 150 liters/min. To increase pulmonary ventilation, both tidal volume and respiratory rate increase, but depth of breathing is increased more than frequency of breathing.

When increasing pulmonary ventilation, it is usually more advantageous to have a greater increase in tidal volume than in respiratory rate because of the presence of anatomical dead space. Not all of the inspired air gets down to the site of gas exchange in the alveoli. Part of it remains in the conducting airways, where it is not available for gas exchange. The volume of the conducting passages in an adult averages about 150 ml. This volume is considered to be **anatomical dead space** because air within these conducting airways is useless for exchange purposes. Anatomical dead space has a pronounced effect on the efficiency of pulmonary ventilation. In effect, even though 500 ml of air are moved in and out with each breath, only 350 ml are actually exchanged between the atmosphere and the alveoli because of the 150 ml occupying the anatomical dead space (Fig. 13-25).

Since the amount of atmospheric air that reaches the alveoli and is actually available for exchange with the blood is more important than the total amount breathed in and out, **alveolar ventilation**—the volume of air exchanged between the atmosphere and the alveoli per minute—is more important than pulmonary ventilation. In determining alveolar ventilation, the amount of wasted air moved in and out through the anatomical dead space must be taken into account, as follows:

$$\text{alveolar ventilation} = (\text{tidal volume} - \text{dead-space volume}) \times \text{respiratory rate}$$

With average resting values,

$$\text{alveolar ventilation} = (500\ \text{ml/breath} - 150\ \text{ml dead-space volume}) \times 12\ \text{breaths/min} = 4{,}200\ \text{ml/min}$$

Thus, with quiet breathing, alveolar ventilation is 4,200 ml/min, whereas pulmonary ventilation is 6,000 ml/min.

To understand how important dead-space volume is in determining the magnitude of alveolar ventilation, examine the effect of various breathing patterns on alveolar ventilation in Table 13-5. If a person deliberately breathes deeply (for example, a tidal volume of 1,200 ml) and slowly (for example, a respiratory rate of 5 breaths/min), pulmonary ventilation is 6,000 ml/min, the same as during quiet breathing at rest, but alveolar ventilation increases to 5,250 ml/min compared to the resting rate of 4,200 ml/min. In contrast, if a person deliberately breathes shallowly (for example, a tidal volume of 150 ml) and rapidly (a frequency of 40 breaths/min), pulmonary ventilation would still be 6,000 ml/min; however, alveolar ventilation would be 0 ml/min. In effect, the person

Table 13-5 Effect of Different Breathing Patterns on Alveolar Ventilation

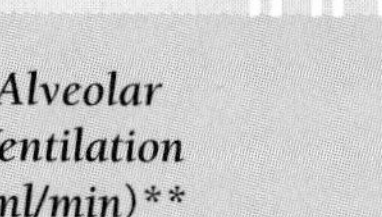

Breathing Pattern	*Tidal Volume (ml/breath)*	*Respiratory Rate (breaths/min)*	*Dead-Space Volume (ml)*	*Pulmonary Ventilation (ml/min)**	*Alveolar Ventilation (ml/min)***
Quiet breathing at rest	500	12	150	6,000	4,200
Deep, slow breathing	1,200	5	150	6,000	5,250
Shallow, rapid breathing	150	40	150	6,000	0

*Equals tidal volume × respiratory rate.
**Equals (tidal volume − dead-space volume) × respiratory rate.

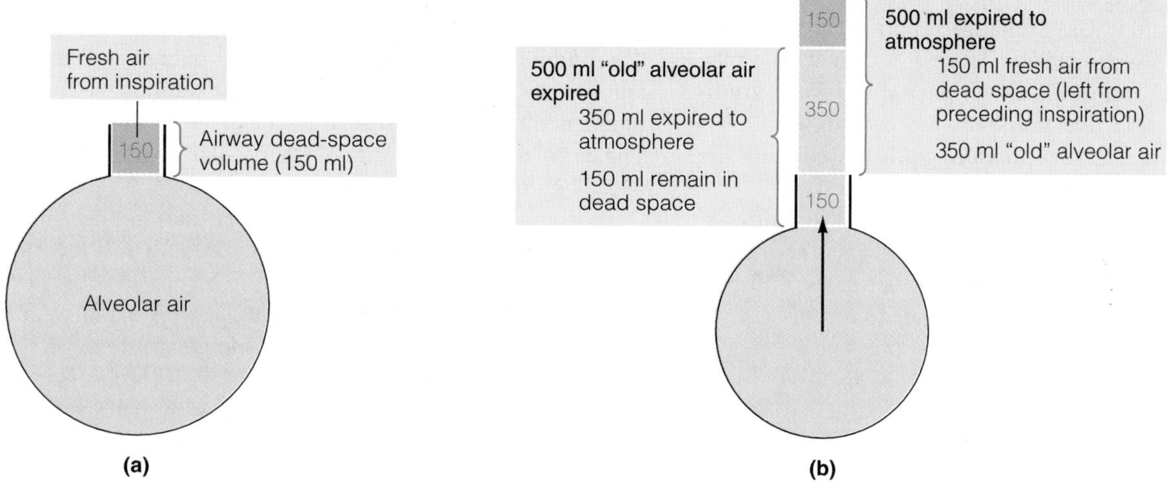

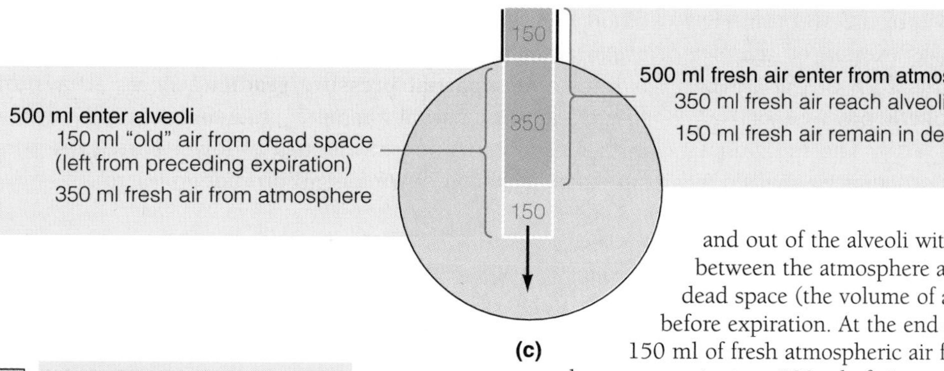

Figure 13-25 **Effect of Dead-Space Volume on Exchange of Tidal Volume between the Atmosphere and the Alveoli** Even though 500 ml of air move in and out between the atmosphere and the respiratory system and 500 ml move in and out of the alveoli with each breath, only 350 ml are actually exchanged between the atmosphere and the alveoli because of the presence of anatomical dead space (the volume of air in the respiratory airways). (a) After inspiration, before expiration. At the end of inspiration, the respiratory airways are filled with 150 ml of fresh atmospheric air from the inspiration. (b) During expiration. During the subsequent expiration, 500 ml of air are expired to the atmosphere. The first 150 ml expired are the fresh air that was retained in the airways and never used. The remaining 350 ml are "old" alveolar air that has participated in gas exchange with the blood. During the same expiration, 500 ml of gas also leave the alveoli. The first 350 ml are expired to the atmosphere; the other 150 ml of old alveolar air never reach the outside but remain in the conducting airways. (c) During inspiration. On the next inspiration, 500 ml of gas enter the alveoli. The first 150 ml to enter the alveoli are the old alveolar air that remained in the dead space during the preceding expiration. The other 350 ml entering the alveoli are fresh air inspired from the atmosphere. Simultaneously, 500 ml of air enter from the atmosphere. The first 350 ml of atmospheric air reach the alveoli; the other 150 ml remain in the conducting airways to be expired without benefit of being exchanged with the blood, as the cycle repeats itself.

would only be drawing air in and out of the anatomical dead space without any atmospheric air being exchanged with the alveoli, where it could be useful. The individual could voluntarily maintain such a breathing pattern for only a few minutes before losing consciousness, at which time normal breathing would resume.

The value of reflexly bringing about a larger increase in depth of breathing than in rate of breathing when pulmonary ventilation is increased during exercise should now be apparent. It is the most efficient means of elevating alveolar ventilation. When tidal volume is increased, the entire increase goes toward elevating alveolar ventilation, while an increase in respiratory rate does not go entirely toward increasing alveolar ventilation. When respiratory rate is increased, the frequency with which air is wasted in the dead space is also increased, because a portion of *each* breath must move in and out of the dead space. As needs vary, ventilation is normally adjusted to a tidal volume and respiratory rate that meet those needs most efficiently in terms of energy cost.

We have assumed that all the atmospheric air entering the alveoli participates in exchanges of O_2 and CO_2 with pulmonary blood. However, the match between air and blood is not always perfect, because not all alveoli are equally ventilated with air and perfused with blood. Any ventilated alveoli that do not participate in gas exchange with blood because they are inadequately perfused are considered **alveolar dead space.** In normal persons, alveolar dead space is quite small and of little importance, but it can be increased to even lethal levels in several types of pulmonary disease.

Gas Exchange

Gases move down partial pressure gradients.

The ultimate purpose of breathing is to provide a continual supply of fresh O_2 for pickup by the blood and to constantly remove CO_2 unloaded from the blood. The blood acts as a transport system for O_2 and CO_2 between the lungs and tissues, with the tissue cells extracting O_2 from the blood and eliminating CO_2 into it. Gas exchange at both the pulmonary-capillary and the tissue-capillary levels involves simple passive diffusion of O_2 and CO_2 down **partial pressure gradients.** There are no active transport mechanisms for these gases. Let us see what partial pressure gradients are and how they are established.

Atmospheric air is a mixture of gases; typical dry air contains about 79% nitrogen (N_2) and 21% O_2, with almost negligible percentages of CO_2, H_2O vapor, other gases, and pollutants. Altogether, these gases exert a total atmospheric pressure of 760 mm Hg at sea level. This total pressure is equal to the sum of the pressures that each gas in the mixture partially contributes. The pressure exerted by a particular gas is directly proportional to the percentage of that gas in the total air mixture. Every gas molecule, no matter what its size, exerts the same amount of pressure; for example, a N_2 molecule exerts the same pressure as an O_2 molecule. Since 79% of the air consists of N_2 molecules, 79% of the 760 mm Hg atmospheric pressure, or 600 mm Hg, is exerted by the N_2 molecules. Similarly, since O_2 represents 21% of the atmosphere, 21% of the 760 mm Hg atmospheric pressure, or 160 mm Hg, is exerted by O_2 (■ Fig. 13-26). The individual pressure exerted independently by a particular gas within a mixture of gases is known as its **partial pressure**, designated by P_{gas}. Thus, the partial pressure of O_2 in atmospheric air, $\mathbf{P_{O_2}}$, is normally 160 mm Hg. The atmospheric partial pressure of CO_2, $\mathbf{P_{CO_2}}$, is negligible at 0.03 mm Hg.

Gases dissolved in a liquid such as blood or another body fluid are also considered to exert a partial pressure. The amount of a gas that will dissolve in the blood depends on the solubility of the gas in blood and on the partial pressure of the gas in the alveolar air to which the blood is exposed. Because the solubilities of O_2 and CO_2 in blood remain constant, the amount of O_2 and CO_2 dissolved in the pulmonary capillary blood is directly proportional to the alveolar P_{O_2} and P_{CO_2}. The alveolar partial pressure of a particular gas can be thought of as "holding" that gas in solution in the blood.

If, as in the case with O_2, the alveolar partial pressure of a gas is *higher* than the partial pressure of that gas in the blood entering the pulmonary capillaries, the higher alveolar partial pressure drives more O_2 into the blood. Oxygen diffuses from the alveoli and dissolves in the blood until blood P_{O_2} becomes equal to alveolar P_{O_2}. Conversely, if the alveolar partial pressure of a gas is *lower* than its partial pressure in the entering blood—the situation that exists for CO_2—the lower alveolar partial pressure permits some of the CO_2 to escape from solution (that is, to no longer be dissolved) in the blood. As CO_2 comes out of solution, it diffuses into the alveoli until blood P_{CO_2} equilibrates with alveolar P_{CO_2}. Such a difference in partial pressure between pulmonary blood and alveolar air is known as a **partial pressure gradient.** A gas always diffuses down its partial pressure gradient from the area of higher partial pressure to the area of lower partial pressure, similar to diffusion down a concentration gradient.

■ *Figure 13-26* **Concept of Partial Pressures** The partial pressure exerted by each gas in a mixture equals the total pressure times the fractional composition of the gas in the mixture.

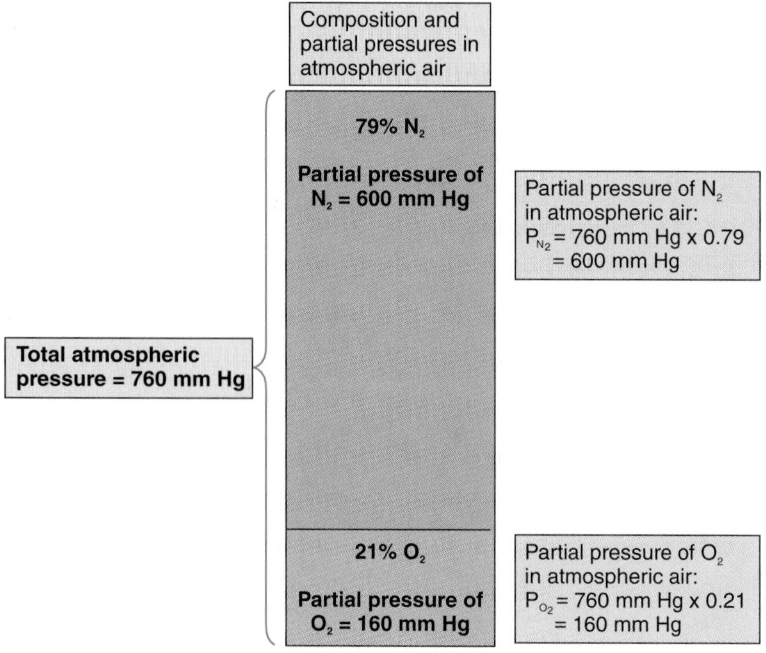

Oxygen enters and CO_2 leaves the blood in the lungs passively down partial pressure gradients.

Alveolar air is not of the same composition as inspired atmospheric air for two reasons. First, as soon as atmospheric air enters the respiratory passages, it becomes saturated with H_2O by exposure to the moist airways. Like any other gas, water vapor exerts a partial pressure. At body temperature, the partial pressure of H_2O vapor is 47 mm Hg. Humidification of inspired air in effect "dilutes" the partial pressure of the inspired gases by 47 mm Hg, because the sum of the partial pressures must total the atmospheric pressure of 760 mm Hg. In moist air, $P_{H_2O} = 47$ mm Hg, $P_{N_2} = 563$ mm Hg, and $P_{O_2} = 150$ mm Hg.

Second, alveolar P_{O_2} is also lower than atmospheric P_{O_2} because fresh inspired air is mixed with the large volume of old air that remained in the lungs and dead space at the end of the preceding expiration (the functional residual capacity). Only about one-seventh of the total alveolar air is replaced by fresh atmospheric air with each normal breath. Thus, at the end of inspiration, less than 15% of the air in the alveoli is fresh air. As a result of humidification and the small turnover of alveolar air, the average alveolar P_{O_2} is 100 mm Hg, compared to the atmospheric P_{O_2} of 160 mm Hg.

It is logical to think that alveolar P_{O_2} would increase during inspiration with the arrival of fresh air and would decrease during expiration. Only small fluctuations of a few mm Hg occur, however, for two reasons. First, only a small proportion of the total alveolar air is exchanged with each breath. The relatively small volume of high-P_{O_2} air that is inspired is quickly mixed with the much larger volume of retained alveolar air, which has a lower P_{O_2}. Thus, the O_2 in the inspired air can only slightly elevate the total alveolar P_{O_2}. Even this potentially small elevation of P_{O_2} is diminished for another reason. Oxygen is continually moving by passive diffusion down its partial pressure gradient from the alveoli into the blood. The O_2 arriving in the alveoli in the newly inspired air simply replaces the O_2 diffusing out of the alveoli into the pulmonary capillaries. Therefore, the alveolar P_{O_2} remains relatively constant at about 100 mm Hg throughout the respiratory cycle. Because the pulmonary blood P_{O_2} equilibrates with the alveolar P_{O_2}, the P_{O_2} of the blood likewise remains fairly constant at this same value. Accordingly, the amount of O_2 in the blood available to the tissues varies only slightly during the respiratory cycle.

A similar situation exists in reverse for CO_2. Carbon dioxide, which is continually produced by the body tissues as a metabolic waste product, is constantly added to the blood at the level of the systemic capillaries. In the pulmonary capillaries, CO_2 diffuses down its partial pressure gradient from the blood into the alveoli and is subsequently removed from the body during expiration. Like O_2, alveolar P_{CO_2} remains fairly constant throughout the respiratory cycle but at a lower value of 40 mm Hg. Ventilation constantly replenishes alveolar P_{O_2}, keeping it relatively high, and constantly removes CO_2, keeping alveolar P_{CO_2} relatively low. Thus, the appropriate partial pressure gradients between the alveoli and blood are maintained to ensure that O_2 enters the blood and CO_2 leaves the blood.

The blood entering the pulmonary capillaries is systemic venous blood pumped to the lungs through the pulmonary arteries. This blood, having just returned from the body tissues, is relatively low in O_2, with a P_{O_2} of 40 mm Hg, and is relatively high in CO_2, with a P_{CO_2} of 46 mm Hg. As this blood flows through the pulmonary capillaries, it is exposed to alveolar air (Fig. 13-27). Since the alveolar P_{O_2} at 100 mm Hg is higher than the P_{O_2} of 40 mm Hg in the blood entering the lungs, O_2 diffuses down its partial pressure gradient from the alveoli into the blood until no further gradient exists. As the blood leaves the pulmonary capillaries, it has a P_{O_2} equal to alveolar P_{O_2} at 100 mm Hg. The partial pressure gradient for CO_2 is in the opposite direction. Blood entering the pulmonary capillaries has a P_{CO_2} of 46 mm Hg, whereas alveolar P_{CO_2} is only 40 mm Hg. Carbon

Figure 13-27 **Oxygen and CO_2 Exchange across Pulmonary and Systemic Capillaries Caused by Partial Pressure Gradients** Alveolar P_{O_2} remains relatively high and alveolar P_{CO_2} remains relatively low because a portion of the alveolar air is exchanged for fresh atmospheric air with each breath. In contrast, the systemic venous blood entering the lungs is relatively low in O_2 and high in CO_2, having given up O_2 and picked up CO_2 at the systemic capillary level. This establishes partial pressure gradients between the alveolar air and pulmonary capillary blood that induce the passive diffusion of O_2 into the blood and CO_2 out of the blood until the blood and alveolar partial pressures become equal. The blood leaving the lungs is thus relatively high in O_2 and low in CO_2 compared to the partial pressures in the O_2-consuming, CO_2-producing tissue cells to which it is delivered. Consequently, partial pressure gradients for gas exchange at the tissue level favor the passive movement of O_2 out of the blood into the cells to support their metabolic requirements and also favor the simultaneous transfer of CO_2 into the blood. The blood then returns to the lungs to once again fill up on O_2 and dump off CO_2. The systemic arterial P_{O_2} and P_{CO_2} normally remain essentially constant, having equilibrated with the alveolar partial pressures, which remain essentially constant. In contrast, the systemic venous P_{O_2} and P_{CO_2} vary, depending on the level of metabolic activity.

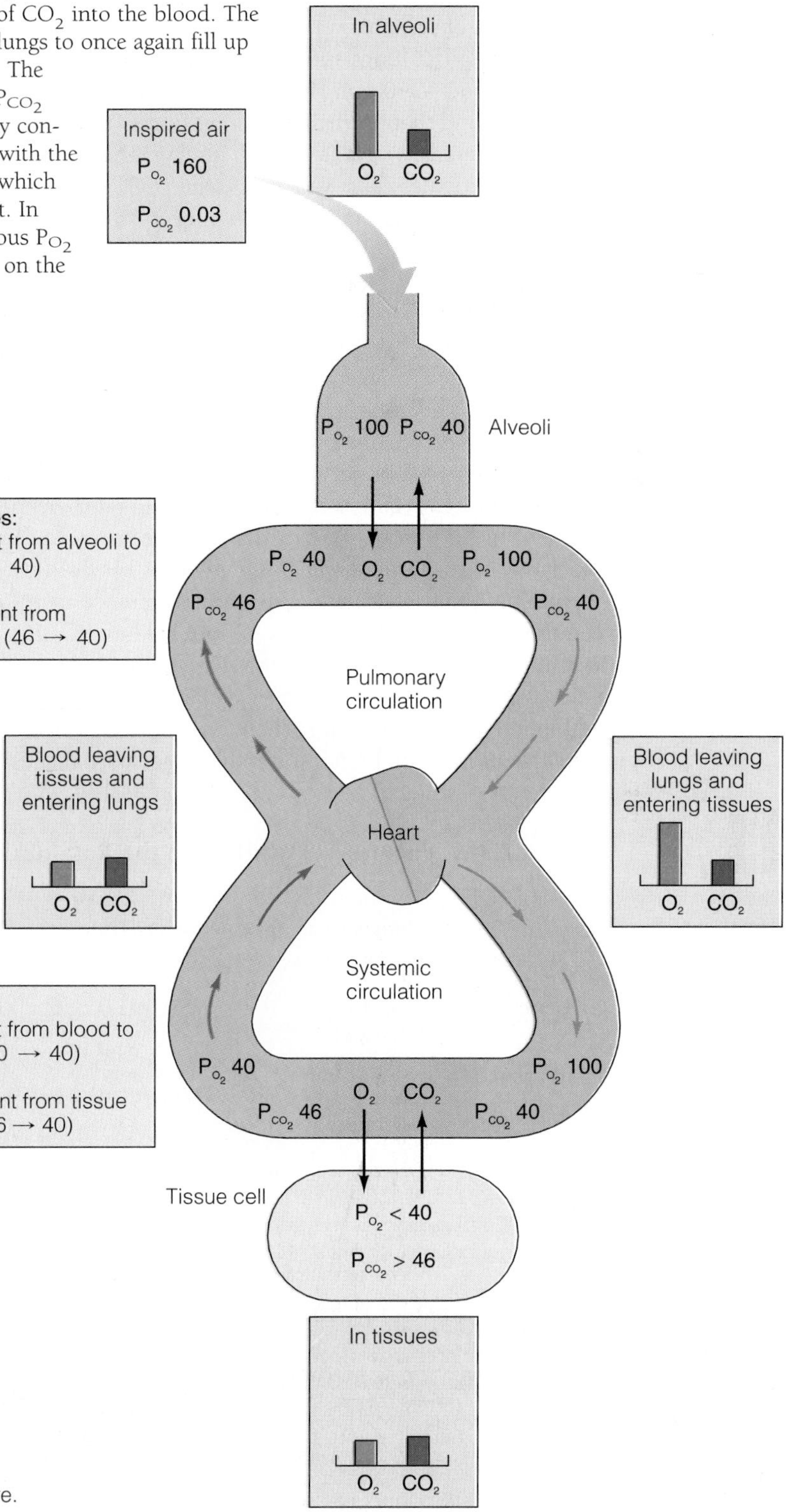

Across pulmonary capillaries:
O_2 partial pressure gradient from alveoli to blood = 60 mm Hg (100 → 40)

CO_2 partial pressure gradient from blood to alveoli = 6 mm Hg (46 → 40)

Across systemic capillaries:
O_2 partial pressure gradient from blood to tissue cell = 60 mm Hg (100 → 40)

CO_2 partial pressure gradient from tissue cell to blood = 6 mm Hg (46 → 40)

Numbers are mm Hg pressure.

dioxide diffuses from the blood into the alveoli until blood P_{CO_2} equilibrates with alveolar P_{CO_2}. Thus, the blood leaving the pulmonary capillaries has a P_{CO_2} of 40 mm Hg. As the blood passes through the lungs, it picks up O_2 and gives up CO_2 simply by diffusion down partial pressure gradients that exist between the blood and the alveoli. After leaving the lungs, the blood, which now has a P_{O_2} of 100 mm Hg and a P_{CO_2} of 40 mm Hg, is returned to the heart to be subsequently pumped out to the body tissues as systemic arterial blood.

Note that blood returning to the lungs from the tissues still contains O_2 (P_{O_2} of systemic venous blood = 40 mm Hg) and that blood leaving the lungs still contains CO_2 (P_{CO_2} of systemic arterial blood = 40 mm Hg). The additional O_2 carried in the blood beyond that normally given up to the tissues represents an immediately available O_2 reserve that can be tapped by the tissue cells whenever their O_2 demands increase. The CO_2 remaining in the blood even after passage through the lungs plays an important role in the acid-base balance of the body, because CO_2 generates carbonic acid. Furthermore, arterial P_{CO_2} is important in driving respiration. This mechanism will be described later.

The amount of O_2 picked up in the lungs matches the amount extracted and used by the tissues. When the tissues metabolize more actively (for example, during exercise), more O_2 is extracted from the blood at the tissue level, reducing the systemic venous P_{O_2} even lower than 40 mm Hg—for example, to a P_{O_2} of 30 mm Hg. When this blood returns to the lungs, a larger-than-normal P_{O_2} gradient exists between the newly entering blood and alveolar air. The difference in P_{O_2} between the alveoli and blood is now 70 mm Hg (alveolar P_{O_2} of 100 mm Hg and blood P_{O_2} of 30 mm Hg), compared to the normal P_{O_2} gradient of 60 mm Hg (alveolar P_{O_2} of 100 mm Hg and blood P_{O_2} of 40 mm Hg). Therefore, more O_2 diffuses from the alveoli into the blood down the larger partial pressure gradient before blood P_{O_2} equals alveolar P_{O_2}. This additional transfer of O_2 into the blood replaces the increased amount of O_2 consumed, so O_2 uptake matches O_2 use even when O_2 consumption increases. At the same time that more O_2 is diffusing from the alveoli into the blood because of the increased partial pressure gradient, ventilation is stimulated so that O_2 enters the alveoli more rapidly from the atmosphere to replace the O_2 diffusing into the blood.

Similarly, the amount of CO_2 given up to the alveoli from the blood matches the amount of CO_2 picked up at the tissues. Once again, an increase in ventilation associated with increased activity assures that increased amounts of CO_2 delivered to the alveoli are blown off to the atmosphere.

Factors other than the partial pressure gradient influence the rate of gas transfer.

We have been discussing diffusion of O_2 and CO_2 between the blood and the alveoli as if these gases' partial pressure gradients were the sole determinants of their rates of diffusion. Recall that, according to Fick's law of diffusion, the rate of diffusion of a gas through a sheet of tissue also depends on the surface area and thickness of the membrane through which the gas is diffusing and on the diffusion coefficient of the particular gas (■ Table 13-6). Normally, changes in the rate of gas exchange are determined primarily by changes in partial pressure gradients between the blood and alveoli, because these other factors are relatively constant under resting conditions.

During exercise, however, the surface area available for exchange can be physiologically increased to enhance the rate of gas transfer. During resting conditions, some of the pulmonary capillaries are typically closed, because the normally low pressure of the pulmonary circulation is inadequate to keep all of the capillaries open. During exercise, when the pulmonary blood pressure is raised as a result of increased

Table 13-6 **Factors That Influence the Rate of Gas Transfer across the Alveolar Membrane**

Factor	*Influence on the Rate of Gas Transfer Across the Alveolar Membrane*	*Comments*
Partial pressure gradients of O_2 and CO_2	Rate of transfer ↑ as partial pressure gradient ↑	Major determinant of the rate of transfer
Surface area of the alveolar membrane	Rate of transfer ↑ as surface area ↑	Surface area remains constant under resting conditions Surface area ↑ during exercise as more pulmonary capillaries open up when the cardiac output increases and the alveoli expand as breathing becomes deeper Surface area ↓ with pathological conditions such as emphysema and atelectasis
Thickness of the barrier separating the air and blood across the alveolar membrane	Rate of transfer ↓ as thickness ↑	Thickness normally remains constant Thickness ↑ with pathological conditions such as pulmonary edema, pulmonary fibrosis, and pneumonia
Diffusion coefficient (solubility of the gas in the membrane)	Rate of transfer ↑ as diffusion coefficient ↑	Diffusion coefficient for CO_2 is 20 times that of O_2, offsetting the smaller partial pressure gradient for CO_2; therefore, approximately equal amounts of CO_2 and O_2 are transferred across the membrane

cardiac output, many of the previously closed pulmonary capillaries are forced open. This increases the surface area of blood available for exchange. Furthermore, the alveolar membranes are stretched further than normal during exercise because of the larger tidal volumes (deeper breathing). Such stretching increases the alveolar surface area and decreases the thickness of the alveolar membrane. Collectively, these changes expedite gas exchange during exercise.

On the other hand, several pathological conditions can markedly reduce the pulmonary surface area and, in turn, decrease the rate of gas exchange. Most notably, surface area is reduced in emphysema because many of the alveolar walls are lost, resulting in larger but fewer chambers (Fig. 13-28). Loss of surface area for exchange is likewise associated with atelectic regions of the lung and also results when part of the lung tissue is surgically removed—for example, in the treatment of lung cancer.

Inadequate gas exchange can also occur when the thickness of the barrier separating the air and blood is pathologically increased. As the thickness increases, the rate of gas transfer decreases, because a gas takes longer to diffuse through the greater thickness. Thickness increases in (1) *pulmonary edema,* an excess accumulation of interstitial fluid between the alveoli and pulmonary capillaries caused by pulmonary inflammation or left-sided congestive heart failure (see p. 295); (2) *pulmonary fibrosis* involving replacement of delicate lung tissue with thick fibrous tissue in response to certain chronic irritants; and (3) *pneumonia,* which is characterized by inflammatory fluid accumulation within or around the alveoli. Most commonly, pneumonia is due to bacterial or viral infection of the lungs, but it may also arise from accidental aspiration (breathing in) of food, vomitus, or chemical agents.

The rate of gas transfer is directly proportional to the diffusion coefficient (D), a constant value related to the solubility of a particular gas in the lung tissues and to its molecular weight ($D \propto sol \sqrt{mw}$). The diffusion coefficient for CO_2 is 20 times that of O_2 because CO_2 is much more soluble in body tissues than O_2 is. The rate of CO_2 diffusion across the respiratory membranes is therefore 20 times more rapid than that of O_2 for a given partial pressure gradient. This difference in diffusion coefficients is normally offset by the difference in partial pressure gradients that exist for O_2 and CO_2 across the alveolar capillary membrane. The CO_2 partial pressure gradient is 6 mm Hg (P_{CO_2} of 46 mm Hg in the blood; P_{CO_2} of 40 mm Hg in the alveoli), compared to the O_2 gradient of 60 mm Hg (P_{O_2} of 100 mm Hg in the alveoli; P_{O_2} of 40 mm Hg in the blood).

Normally, approximately equal amounts of O_2 and CO_2 are exchanged—a respiratory quotient's worth. Even though a given volume of blood spends three-fourths of a second passing through the pulmonary capillary bed, P_{O_2} and P_{CO_2} are usually both equilibrated with alveolar partial pressures by the time the blood has traversed only one-third the length of the pulmonary capillaries. This means that the lung normally has enormous diffusion reserves, a fact that becomes extremely important during heavy exercise. The time the blood spends in transit in the pulmonary capillaries is decreased as pulmonary blood flow increases with the greater cardiac output that accompanies exercise. Even when less time is available for exchange, blood P_{O_2} and P_{CO_2} are normally able to equilibrate with alveolar levels because of the lungs' diffusion reserves.

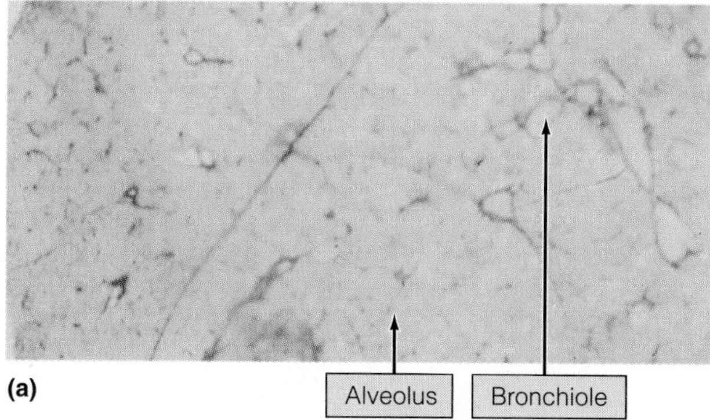

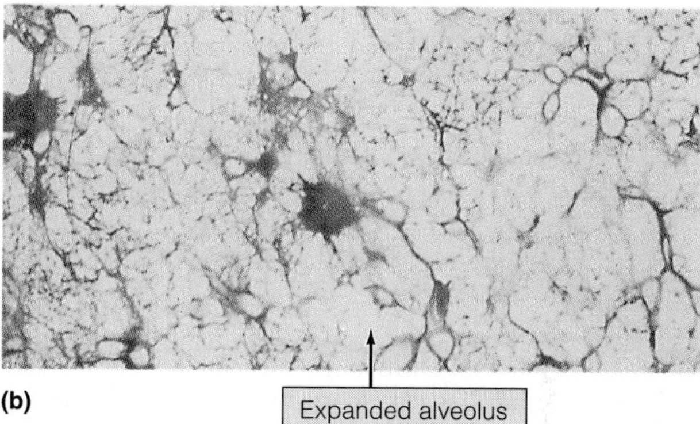

Figure 13-28 **Comparison of Normal and Emphysematous Lung Tissue** (a) A thin section of lung tissue from a normal individual. Each of the smallest clear spaces is an alveolar lumen. (b) A thin section of lung tissue from a patient with emphysema. These are photographs of the actual tissue, unmagnified. Note the loss of alveolar walls in the emphysematous lung tissue, resulting in larger but fewer alveolar chambers.

In a diseased lung in which diffusion is impeded because the surface area is decreased or the blood-air barrier is thickened, O_2 transfer is usually more seriously impaired than CO_2 transfer because of the larger CO_2 diffusion coefficient. By the time the blood reaches the end of the pulmonary capillary network, it is more likely to have equilibrated with alveolar P_{CO_2} than with alveolar P_{O_2}, because CO_2 can diffuse more rapidly through the respiratory barrier. In milder conditions, diffusion of both O_2 and CO_2 might remain adequate at rest, but during exercise, when pulmonary transit time is decreased, the blood gases, especially O_2, may not have completely equilibrated with the alveolar gases before the blood leaves the lungs.

Gas exchange across the systemic capillaries also occurs down partial pressure gradients.

Just as they do at the pulmonary capillaries, O_2 and CO_2 move between the systemic capillary blood and the tissue cells by

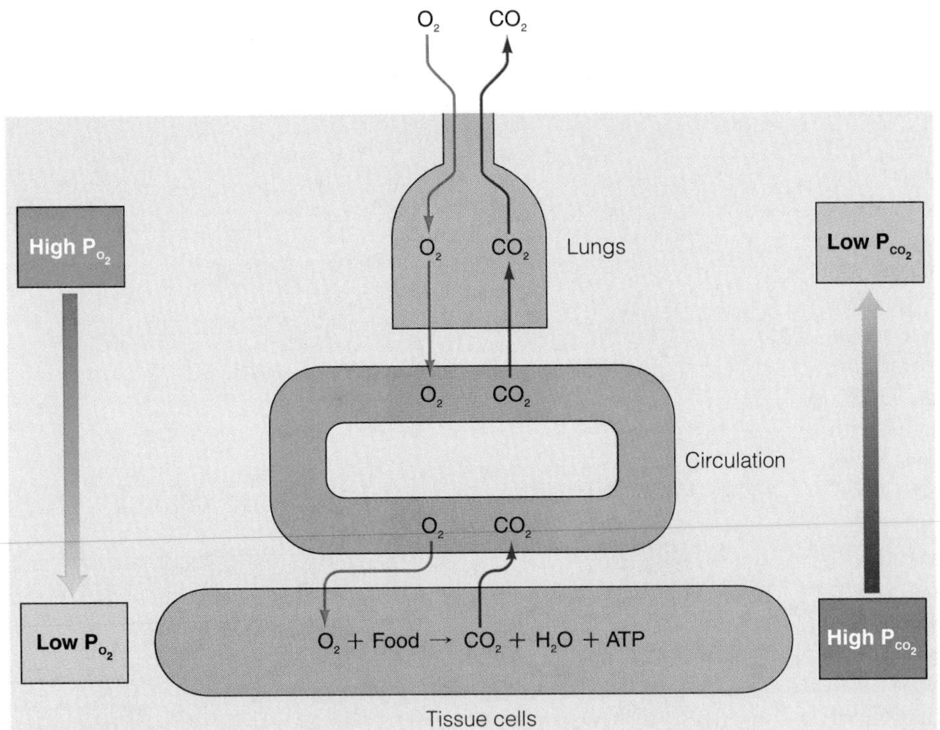

Figure 13-29 **Net Diffusion Gradients for O_2 and CO_2 between the Lungs and Tissues** As a result of alveolar ventilation constantly bringing in fresh supplies of O_2 and the cells constantly using O_2, a net gradient exists for diffusion of O_2, first from the alveoli to the blood and then from the blood to the tissue cells. A net gradient exists in reverse for diffusion of CO_2, first from the tissue cells to the blood and then from the blood to the alveoli. This gradient is due to the continual production of CO_2 at the tissue level and the continual removal of CO_2 from the alveoli by alveolar ventilation.

simple passive diffusion down partial pressure gradients. Refer again to Figure 13-27. The arterial blood that reaches the systemic capillaries is essentially the same blood that left the lungs by means of the pulmonary veins, because the only two places in the entire circulatory system at which gas exchange can take place are the pulmonary capillaries and the systemic capillaries. The arterial P_{O_2} is 100 mm Hg and the arterial P_{CO_2} is 40 mm Hg, the same as alveolar P_{O_2} and P_{CO_2}.

The cells constantly consume O_2 and produce CO_2 through oxidative metabolism. Cellular P_{O_2} averages about 40 mm Hg and P_{CO_2} about 46 mm Hg, although these values are highly variable, depending on the level of cellular metabolic activity. Oxygen moves by diffusion down its partial pressure gradient from the entering systemic capillary blood (P_{O_2} = 100 mm Hg) into the adjacent cells (P_{O_2} = 40 mm Hg) until equilibrium is reached. Therefore, the P_{O_2} of venous blood leaving the systemic capillaries is equal to the tissue P_{O_2} at an average of 40 mm Hg. The reverse situation exists for CO_2. Carbon dioxide rapidly diffuses out of the cells (P_{CO_2} = 46 mm Hg) into the entering capillary blood (P_{CO_2} = 40 mm Hg) down the partial pressure gradient created by the ongoing production of CO_2. Transfer of CO_2 continues until blood P_{CO_2} equilibrates with tissue P_{CO_2}.[2] Accordingly, the blood leaving the systemic capillaries has an average P_{CO_2} of 46 mm Hg. This systemic venous blood, which is relatively low in O_2 (P_{O_2} = 40 mm Hg) and relatively high in CO_2 (P_{CO_2} = 46 mm Hg), returns to the heart and is subsequently pumped to the lungs as the cycle repeats itself.

The more actively a tissue is metabolizing, the lower the cellular P_{O_2} falls and the higher the cellular P_{CO_2} rises. As a consequence of the larger blood-to-cell partial pressure gradients, more O_2 diffuses from the blood into the cells and more CO_2 moves in the opposite direction before blood P_{O_2} and P_{CO_2} achieve equilibrium with the surrounding cells. Thus, the amount of O_2 transferred to the cells and the amount of CO_2 carried away from the cells depend on the rate of cellular metabolism.

Note that net diffusion of O_2 occurs first between the alveoli and the blood and then between the blood and the tissues due to the O_2 partial pressure gradients created by continuous utilization of O_2 in the cells and continuous replenishment of fresh alveolar O_2 provided by alveolar ventilation. Net diffusion of CO_2 occurs in the reverse direction, first between the tissues and the blood and then between the blood and the alveoli, due to the CO_2 partial pressure gradients created by continuous production of CO_2 in the cells and the continuous removal of alveolar CO_2 through the process of alveolar ventilation (Fig. 13-29).

Gas Transport

Most O_2 in the blood is transported bound to hemoglobin.

Oxygen picked up by the blood at the lungs must be transported to the tissues for cell use. Conversely, CO_2 produced at the cellular level must be transported to the lungs for elimination.

Oxygen is present in the blood in two forms: physically dissolved and chemically bound to hemoglobin (Table 13-7). Very little O_2 is physically dissolved in the plasma water because O_2 is poorly soluble in body fluids. The amount dissolved is directly proportional to the P_{O_2} of the blood; the higher the P_{O_2}, the more O_2 dissolved. At a normal arterial P_{O_2} of 100 mm Hg, only 3 ml of O_2 can dissolve in 1 liter of blood. Thus, only 15 ml of O_2/min can be dissolved in the normal pulmonary blood flow of 5 liters/min (the resting cardiac output). Even under resting conditions, the cells consume 250 ml of O_2/min, and this may increase up to 25-fold during strenuous exercise. To deliver the O_2 required by the tissues even at rest, the cardiac output would have to be 83.3

[2]Actually, the partial pressures of the systemic blood gases never completely equilibrate with tissue P_{O_2} and P_{CO_2}. Because the cells are constantly consuming O_2 and producing CO_2, the tissue P_{O_2} is always slightly less than the P_{O_2} of the blood leaving the systemic capillaries, and the tissue P_{CO_2} always slightly exceeds the systemic venous P_{CO_2}.

liters/min if O_2 could only be transported in dissolved form. Obviously, there must be an additional mechanism for transporting O_2 to the tissues. This mechanism is *hemoglobin (Hb)*. Only 1.5% of the O_2 in the blood is dissolved; the remaining 98.5% is transported in combination with Hb. *The O_2 bound to Hb does not contribute to the P_{O_2} of the blood;* thus, blood P_{O_2} is not a measure of the total O_2 content of the blood but only of the portion of O_2 that is dissolved.

Hemoglobin, an iron-bearing protein molecule contained within the red blood cells, has the ability to form a loose, easily reversible combination with O_2 (see p. 355). When not combined with O_2, Hb is referred to as **reduced hemoglobin;** when combined with O_2, it is called **oxyhemoglobin (HbO_2)**:

$$\underset{\text{reduced hemoglobin}}{Hb + O_2} \rightleftharpoons \underset{\text{oxyhemoglobin}}{HbO_2}$$

We need to answer several important questions about the role of Hb in O_2 transport. What determines whether O_2 and Hb are combined or dissociated (separated)? Why does Hb combine with O_2 in the lungs and release O_2 at the tissues? How can a variable amount of O_2 be released at the tissue level, depending on the level of tissue activity? How can we talk about O_2 transfer between blood and surrounding tissues in terms of O_2 partial pressure gradients when 98.5% of the O_2 is bound to Hb and thus does not contribute to the P_{O_2} of the blood at all?

The P_{O_2} is the primary factor determining the percent hemoglobin saturation.

Each of the four atoms of iron within the heme portions of a hemoglobin molecule is able to combine with an O_2 molecule, so each Hb molecule can carry up to four molecules of O_2. Hemoglobin is considered *fully saturated* when all of the Hb present is carrying its maximum O_2 load. The **percent hemoglobin (% Hb) saturation,** a measure of the extent to which the Hb present is combined with O_2, can vary from 0% to 100%.

The most important factor determining the % Hb saturation is the P_{O_2} of the blood, which in turn is related to the concentration of O_2 physically dissolved in the blood. According to the **law of mass action,** if the concentration of one of the substances involved in a reversible reaction is increased, the reaction is driven toward the opposite side. Conversely, if the concentration of one of the substances is decreased, the reaction is driven toward that side. Applying this law to the reversible reaction involving Hb and O_2 ($Hb + O_2 \rightleftharpoons HbO_2$), when the blood P_{O_2} is increased, as it is in the pulmonary capillaries, the reaction is driven toward the right side of the equation, resulting in increased formation of HbO_2 (increased % Hb saturation). When the blood P_{O_2} is decreased, as it is in the systemic capillaries, the reaction is driven toward the left side of the equation and oxygen is released from Hb as HbO_2 dissociates (decreased % Hb saturation). Thus, because of the difference in P_{O_2} at the lungs and other tissues, Hb automatically "loads up" on O_2 in the lungs, where fresh supplies of O_2 are continually being provided by ventilation, and "unloads" it in the tissues, which are constantly using up O_2.

The relationship between blood P_{O_2} and % Hb saturation is not linear, however, a point that is very important physiologically. Doubling the partial pressure does not double the % Hb saturation. Rather, the relationship between these variables is depicted by an S-shaped curve known as the **O_2-Hb dissociation** (or **saturation**) **curve** (Fig. 13-30). Note that

Table 13-7

Methods of Gas Transport in the Blood

Gas	*Method of Transport in Blood*	*Percentage Carried in This Form*
O_2	Physically dissolved	1.5
	Bound to hemoglobin	98.5
CO_2	Physically dissolved	10
	Bound to hemoglobin	30
	As bicarbonate (HCO_3^-)	60

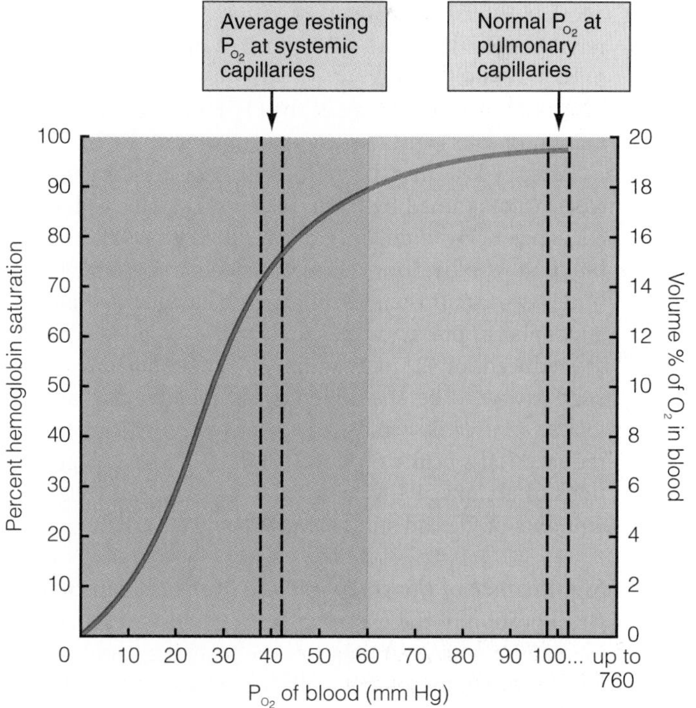

Figure 13-30 **Oxygen-Hemoglobin (O_2-Hb) Dissociation (Saturation) Curve** The percent hemoglobin saturation (the scale on the left side of the graph) depends on the P_{O_2} of the blood. The relationship between these two variables is depicted by an S-shaped curve with a plateau region between a blood P_{O_2} of 60 and 100 mm Hg and a steep portion between 0 and 60 mm Hg. Another way of expressing the effect of blood P_{O_2} on the amount of O_2 bound with hemoglobin is the volume percent of O_2 in the blood (ml of O_2 bound with hemoglobin in each 100 ml of blood). That relationship is represented by the scale on the right side of the graph.

at the upper end, between a blood P_{O_2} of 60 and 100 mm Hg, the curve flattens off, or plateaus. Within this pressure range, a rise in P_{O_2} produces only a small increase in the extent to which Hb is bound with O_2. In contrast, in the P_{O_2} range of 0 to 60 mm Hg, a small change in P_{O_2} results in a large change in the extent to which Hb is combined with O_2, as depicted by the steep lower part of the curve. Both the upper plateau and lower steep portion of the curve have physiological significance.

Significance of the plateau portion of the O_2-Hb curve The plateau portion of the curve is in the blood P_{O_2} range that exists at the pulmonary capillaries where O_2 is being loaded onto Hb. The systemic arterial blood leaving the lungs, having equilibrated with alveolar P_{O_2}, normally has a P_{O_2} of 100 mm Hg. Looking at the O_2-Hb curve, note that at a blood P_{O_2} of 100 mm Hg, Hb is 97.5% saturated. Therefore, the Hb in the systemic arterial blood normally is almost fully saturated.

If the alveolar P_{O_2} and consequently the arterial P_{O_2} fall below normal, there is little reduction in the total amount of O_2 transported by the blood until the P_{O_2} falls below 60 mm Hg because of the plateau region of the curve. If the arterial P_{O_2} falls 40%, from 100 to 60 mm Hg, the concentration of dissolved O_2 as reflected by the P_{O_2} is likewise reduced 40%. At a blood P_{O_2} of 60 mm Hg, however, the % Hb saturation is still remarkably high at 90%. Accordingly, the total O_2 content of the blood is only slightly decreased despite the 40% reduction in P_{O_2} because Hb is still carrying an almost-full load of O_2, and, as mentioned before, the vast majority of O_2 is transported by Hb rather than being dissolved. On the other hand, even if the blood P_{O_2} is greatly increased, say, to 600 mm Hg by breathing pure O_2, very little additional O_2 is added to the blood. A small extra amount of O_2 dissolves, but the % Hb saturation can be maximally increased by only another 2.5%, to 100% saturation. Therefore, in the P_{O_2} range between 60 and 600 mm Hg or even higher, there is only a 10% difference in the amount of O_2 carried by Hb. Thus, the plateau portion of the O_2-Hb curve provides a good margin of safety in O_2-carrying capacity of the blood.

Arterial P_{O_2} may be reduced because of pulmonary diseases accompanied by inadequate ventilation or defective gas exchange or by circulatory disorders that result in inadequate blood flow to the lungs. It may also fall in healthy individuals under two circumstances: (1) at high altitudes, where the total atmospheric pressure and hence the P_{O_2} of the inspired air are reduced, or (2) in O_2-deprived environments at sea level, such as would be encountered if someone were accidentally locked in a vault. Unless the arterial P_{O_2} becomes markedly reduced (falls below 60 mm Hg) in either pathological conditions or abnormal environmental circumstances, near-normal amounts of O_2 can still be carried to the tissues.

Significance of the steep portion of the O_2-Hb curve The steep portion of the curve between 0 and 60 mm Hg is in the blood P_{O_2} range that exists at the systemic capillaries, where O_2 is being unloaded from Hb. In the systemic capillaries, the blood equilibrates with the surrounding tissue cells at an average P_{O_2} of 40 mm Hg. Note on Figure 13-30 that at a P_{O_2} of 40 mm Hg, the % Hb saturation is 75%. The blood arrives in the tissue capillaries at a P_{O_2} of 100 mm Hg with 97.5% Hb saturation. Since Hb can only be 75% saturated at the P_{O_2} of 40 mm Hg in the systemic capillaries, nearly 25% of the HbO_2 must dissociate, yielding reduced Hb and O_2. This released O_2 is free to diffuse down its partial pressure gradient from the red blood cells through the plasma and interstitial fluid into the tissue cells.

The Hb in the venous blood returning to the lungs is still normally 75% saturated. If the tissue cells are metabolizing more actively, the P_{O_2} of the systemic capillary blood falls (for example, from 40 to 20 mm Hg) because the cells are consuming O_2 more rapidly. Note on the curve that this 20 mm Hg drop in P_{O_2} decreases the % Hb saturation from 75% to 30%; that is, about 45% more of the total HbO_2 than normal gives up its O_2 for tissue use. The normal 60 mm Hg drop in P_{O_2} from 100 to 40 mm Hg in the systemic capillaries causes about 25% of the total HbO_2 to unload its O_2. In comparison, a further drop in P_{O_2} of only 20 mm Hg results in an additional 45% of the total HbO_2 unloading its O_2 because the O_2 partial pressures in this range are operating in the steep portion of the curve. In this range, only a small drop in systemic capillary P_{O_2} can automatically make large amounts of O_2 immediately available to meet the O_2 needs of more actively metabolizing tissues. As much as 85% of the Hb may give up its O_2 to actively metabolizing cells during strenuous exercise. In addition to this more thorough withdrawal of O_2 from the blood, even more O_2 is made available to actively metabolizing cells, such as exercising muscles, by circulatory and respiratory adjustments that increase the flow rate of oxygenated blood through the active tissues.

By acting as a storage depot, hemoglobin promotes the net transfer of O_2 from the alveoli to the blood.

We still have not really clarified the role of Hb in gas exchange. Because blood P_{O_2} depends entirely on the concentration of *dissolved* O_2, we could ignore the O_2 bound to Hb in our earlier discussion of O_2 being driven from the alveoli to the blood by a P_{O_2} gradient. However, Hb does play a crucial role in permitting the transfer of large quantities of O_2 before blood P_{O_2} equilibrates with the surrounding tissues (— Fig. 13-31). Hemoglobin does so by acting as a "storage depot" for O_2, removing the O_2 from solution as soon as it enters the blood from the alveoli. Because only dissolved O_2 contributes to the P_{O_2}, the O_2 stored in Hb cannot contribute to blood P_{O_2}. When systemic venous blood enters the pulmonary capillaries, its P_{O_2} is considerably lower than the alveolar P_{O_2}, so O_2 immediately diffuses into the blood, raising the blood P_{O_2}. As soon as the P_{O_2} of the blood increases, the percentage of Hb that can bind with O_2 likewise increases, as indicated by the O_2-Hb curve. Consequently, most of the O_2 that has diffused into the blood combines with Hb and no longer contributes to blood P_{O_2}. As O_2 is removed from solution by combining with Hb, blood P_{O_2} falls to about the same level it was when the blood entered the lungs, despite the fact that the total quantity of O_2 in the blood actually has increased. Since the blood P_{O_2} is once again considerably

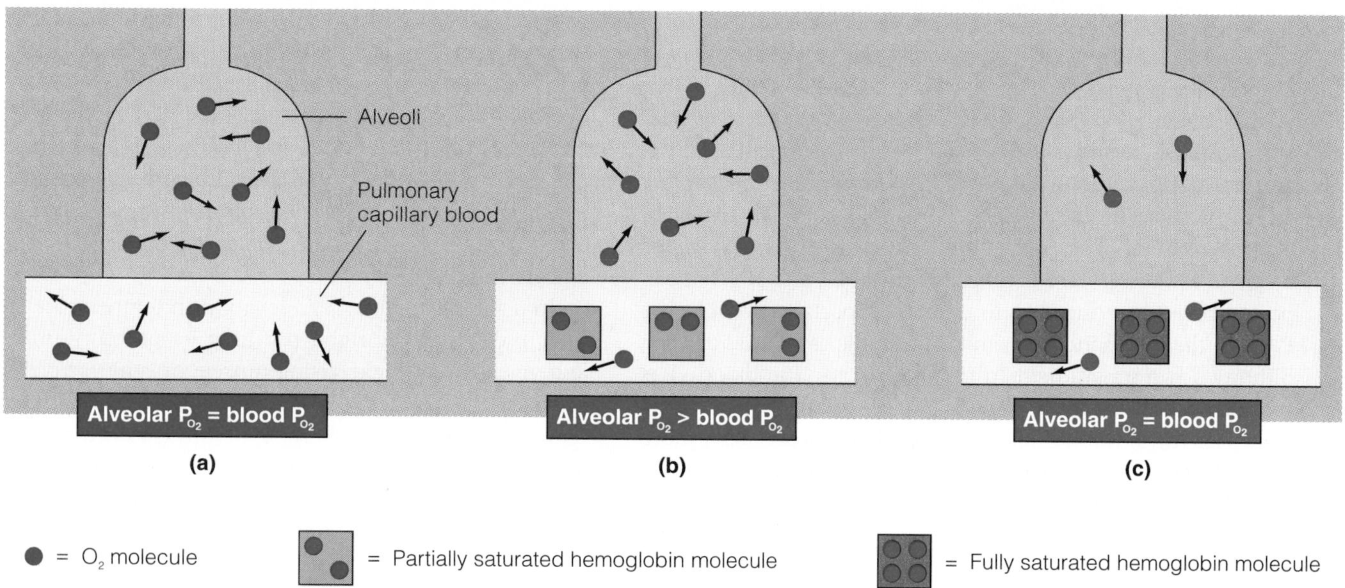

Figure 13-31 **Hemoglobin Facilitating a Large Net Transfer of O_2 by Acting as a Storage Depot to Keep P_{O_2} Low** (a) In the hypothetical situation in which no hemoglobin is present in the blood, the alveolar P_{O_2} and the pulmonary capillary blood P_{O_2} are at equilibrium. (b) Hemoglobin has been added to the pulmonary capillary blood. As the Hb starts to bind with O_2, it removes O_2 from solution. Since only dissolved O_2 contributes to blood P_{O_2}, the blood P_{O_2} falls below that of the alveoli, even though the same number of O_2 molecules are present in the blood as in part (a). By "soaking up" some of the dissolved O_2, Hb favors the net diffusion of more O_2 down its partial pressure gradient from the alveoli to the blood. (c) Hemoglobin is fully saturated with O_2, and the alveolar and blood P_{O_2} are at equilibrium again. The blood P_{O_2} resulting from dissolved O_2 is equal to the alveolar P_{O_2}, despite the fact that the total O_2 content in the blood is much greater than in part (a) when blood P_{O_2} was equal to alveolar P_{O_2} in the absence of Hb.

below alveolar P_{O_2}, more O_2 diffuses from the alveoli into the blood, only to be soaked up by Hb again.

Even though we have considered this process in stepwise fashion for clarity, net diffusion of O_2 from alveoli to blood occurs continuously until Hb becomes as completely saturated with O_2 as it can be at that particular P_{O_2}. At a normal P_{O_2} of 100 mm Hg, Hb is 97.5% saturated. Thus, by soaking up O_2, Hb keeps blood P_{O_2} low and prolongs the existence of a partial pressure gradient so that a large net transfer of O_2 into the blood can take place. Not until Hb can store no more O_2 (that is, Hb is maximally saturated for that P_{O_2}) does all of the O_2 transferred into the blood remain dissolved and directly contribute to the P_{O_2}. Only at this time does the blood P_{O_2} rapidly equilibrate with the alveolar P_{O_2} and bring further O_2 transfer to a halt, but this point is not reached until Hb is already loaded to the maximum extent possible. Once the blood P_{O_2} equilibrates with the alveolar P_{O_2}, no further O_2 transfer can take place, no matter how little or how much total O_2 has already been transferred.

The reverse situation occurs at the tissue level. Since the P_{O_2} of blood entering the systemic capillaries is considerably higher than the P_{O_2} of the surrounding tissue, O_2 immediately diffuses from the blood into the tissues, lowering blood P_{O_2}. When blood P_{O_2} falls, Hb is forced to unload some of its stored O_2 because the % Hb saturation is reduced. As the O_2 released from Hb dissolves in the blood, the blood P_{O_2} increases and is once again above the P_{O_2} of the surrounding tissues. This favors further movement of O_2 out of the blood, despite the fact that the total quantity of O_2 in the blood has already been reduced. Only when Hb is no longer able to release any more O_2 into solution (when Hb is unloaded to the greatest extent possible for the P_{O_2} existing at the systemic capillaries) can blood P_{O_2} become as low as in the surrounding tissue. At this time, further transfer of O_2 ceases. Hemoglobin, because it bears a large quantity of stored O_2 that can be liberated by a slight reduction in P_{O_2} at the systemic capillary level, permits the transfer of tremendously more O_2 from the blood into the cells than would be possible in its absence.

Thus, Hb plays an important role in the *total quantity* of O_2 that the blood can pick up in the lungs and drop off in the tissues. If Hb levels are reduced to one-half of normal, as in a severely anemic patient (see p. 358), the O_2-carrying capacity of the blood is reduced by 50% even though the arterial P_{O_2} is the normal 100 mm Hg with 97.5% Hb saturation. Only half as much Hb is available to be saturated, emphasizing once again how critical the presence of Hb is in determining the total amount of O_2 that can be picked up at the lungs and made available to the tissues.

Increased CO_2, acidity, temperature, and 2,3-diphosphoglycerate shift the O_2-Hb dissociation curve to the right.

Even though the primary factor determining the % Hb saturation is the P_{O_2} of the blood, other factors can affect the affinity, or bond strength, between Hb and O_2 and, accordingly, can shift the O_2-Hb curve (that is, change the % Hb saturation at a given P_{O_2}). These other factors are CO_2, acidity, temperature, and 2,3-diphosphoglycerate, which we will examine separately. The O_2-Hb dissociation curve with which you are already familiar (Fig. 13-30) is a typical curve at normal arterial CO_2 and acidity levels, normal body temperature, and normal 2,3-diphosphoglycerate concentration.

An increase in P_{CO_2} shifts the O_2-Hb curve to the right (Fig. 13-32). The % Hb saturation still depends on the P_{O_2}, but for any given P_{O_2}, the amount of O_2 and Hb that can be combined is reduced. This effect is important, because the P_{CO_2} of the blood increases in the systemic capillaries as CO_2 diffuses down its gradient from the cells into the blood. The presence of this additional CO_2 in the blood in effect decreases the affinity of Hb for O_2, so Hb unloads even more O_2 at the tissue level than it would if the reduction in P_{O_2} in the systemic capillaries were the only factor affecting % Hb saturation.

An increase in acidity also shifts the curve to the right. Because CO_2 generates carbonic acid (H_2CO_3), the blood becomes more acidic at the systemic capillary level as it picks up CO_2 from the tissues. The resultant reduction in Hb affinity for O_2 in the presence of increased acidity aids in releasing even more O_2 at the tissue level for a given P_{O_2}. In actively metabolizing cells, such as exercising muscles, not only is more carbonic acid–generating CO_2 produced, but lactic acid also may be produced if the cells resort to anaerobic metabolism (see p. 243). The resultant local elevation of acid in the working muscles facilitates further unloading of O_2 in the very tissues that have the greatest O_2 need.

The influence of CO_2 and acid on the release of O_2 is known as the **Bohr effect.** Both CO_2 and the hydrogen-ion (H^+) component of acids are able to combine reversibly with Hb at sites other than the O_2-binding sites. The result is an alteration in the molecular structure of Hb that reduces its affinity for O_2.

In a similar manner, an elevation in temperature shifts the O_2-Hb curve to the right, resulting in more unloading of O_2 at a given P_{O_2}. An exercising muscle or other actively metabolizing cell produces heat. The resultant local elevation in temperature enhances O_2 release from Hb for use by the more active tissues.

Thus, increases in CO_2, acidity, and temperature at the tissue level, all of which are associated with increased cellular metabolism and increased O_2 consumption, enhance the effect of a drop in P_{O_2} in facilitating the release of O_2 from Hb. These effects are largely reversed at the pulmonary level, where the extra acid-forming CO_2 is blown off and the local environment is cooler. Appropriately, therefore, Hb has a higher affinity for O_2 in the pulmonary capillary environment, thus enhancing the effect of the elevation in P_{O_2} in loading O_2 onto Hb.

The preceding changes take place in the *environment* of the red blood cells, but a factor *inside* the red blood cells can also affect the degree of O_2-Hb binding: **2,3-diphosphoglycerate (DPG).** This erythrocyte constituent, which is produced during red blood cell metabolism, can bind reversibly with Hb and reduce its affinity for O_2, just as CO_2 and H^+ do. Thus, an increased level of DPG, like the other factors, shifts the O_2-Hb curve to the right, enhancing O_2 unloading as the blood flows through the tissues. DPG production by red blood cells gradually increases whenever Hb in the arterial blood is chronically undersaturated—that is, when arterial HbO_2 is below normal. This condition may occur in individuals living at high altitudes or in those suffering from certain types of circulatory or respiratory diseases or anemia. By promoting the liberation of O_2 from Hb at the tissue level, an increased amount of DPG helps maintain O_2 availability for tissue use under circumstances associated with decreased arterial O_2 supply. However, unlike the other factors—which normally are present only at the tissue level and thus shift the O_2-Hb curve to the right only in the systemic capillaries, where the shift is advantageous in unloading O_2—DPG is present in the red blood cells throughout the circulatory system and, accordingly, shifts the curve to the right to the same degree in both the tissues and the lungs. As a result, DPG decreases the ability to load O_2 at the pulmonary level, which is the negative side of increased DPG production.

Figure 13-32 **Effect of Increased P_{CO_2}, H^+, Temperature, 2,3-Diphosphoglycerate, and Carbon Monoxide on the O_2-Hb Curve** Increased P_{CO_2}, acid, temperature, and 2,3-diphosphoglycerate, as found at the tissue level, shift the O_2-Hb curve to the right. As a result, less O_2 and Hb can be combined at a given P_{O_2}, so that more O_2 is unloaded from Hb for use by the tissues. Carbon monoxide poisoning, on the other hand, shifts the O_2-Hb curve to the left so that less O_2 is unloaded from Hb at the tissue level for a given P_{O_2}.

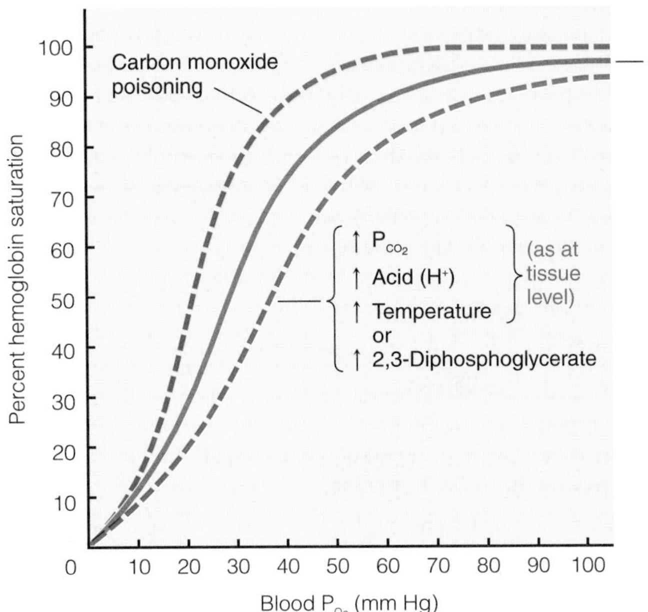

Oxygen-binding sites on hemoglobin have a much higher affinity for carbon monoxide than for O_2.

Carbon monoxide (CO) and O_2 compete for the same binding sites on Hb, but the affinity of Hb for CO is 240 times that of the bond strength between Hb and O_2. The combination of CO and Hb is known as **carboxyhemoglobin (HbCO).** Because Hb preferentially latches onto CO, the presence of even small amounts of CO can tie up a disproportionately large share of Hb, making the latter unavailable for O_2 transport. Even though the Hb concentration and P_{O_2} are normal, the O_2 content of the blood is seriously reduced. If enough CO is present, the cells die from O_2 deprivation. Adding to the toxicity of CO, the presence of HbCO shifts the Hb-O_2 curve to the *left* (Fig. 13-32); thus, the already limited quantity of O_2-bearing Hb is unable to unload as much of its O_2 at the tissue level for a given P_{O_2}.

Fortunately, CO is not a normal constituent of inspired air. It is a poisonous gas produced during the incomplete combustion (burning) of carbon products such as automobile gasoline, coal, wood, and tobacco. Carbon monoxide is especially dangerous because it is so insidious. If CO is being produced in a closed environment so that its concentration continues to increase (for example, in a parked car with the motor running and windows closed), it can reach lethal levels without the victim ever being aware of the danger. Carbon monoxide is not detectable because it is odorless, colorless, tasteless, and nonirritating. Furthermore, for reasons to be described later, the victim has no sensation of breathlessness and makes no attempt to increase ventilation, even though the cells are O_2-starved.

The majority of CO_2 is transported in the blood as bicarbonate.

When arterial blood flows through the tissue capillaries, CO_2 diffuses down its partial pressure gradient from the tissue cells into the blood. Carbon dioxide is transported in the blood in three ways: (1) physically dissolved, (2) bound to Hb, and (3) as bicarbonate (Fig. 13-33 and Table 13-7).

As with dissolved O_2, the amount of CO_2 *physically dissolved* in the blood depends on the P_{CO_2}. Because CO_2 is more soluble than O_2 in the blood, a greater proportion of the total CO_2 in the blood is physically dissolved compared to O_2. Even so, only 10% of the blood's total CO_2 content is carried this way at the normal systemic venous P_{CO_2} level.

Another 30% of the CO_2 combines with Hb to form **carbamino hemoglobin ($HbCO_2$).** Carbon dioxide binds with the globin portion of Hb in contrast to O_2, which combines with the heme portions. Reduced Hb has a greater affinity for CO_2 than does HbO_2. The unloading of O_2 from Hb in the tissue capillaries therefore facilitates the picking up of CO_2 by Hb.

By far the most important means of CO_2 transport is as **bicarbonate (HCO_3^-)**, with 60% of the CO_2 being converted into HCO_3^- by the following chemical reaction, which takes place within the red blood cells:

$$CO_2 + H_2O \overset{\text{carbonic anhydrase}}{\rightleftharpoons} H_2CO_3 \rightleftharpoons H^+ + HCO_3^-$$

Figure 13-33 **Carbon Dioxide Transport in the Blood** Carbon dioxide (CO_2) picked up at the tissue level is transported in the blood to the lungs in three ways: physically dissolved ①, bound to hemoglobin (Hb) ②, and as bicarbonate ion (HCO_3^-) ③. Hemoglobin is present only in the red blood cells, as is carbonic anhydrase, the enzyme that catalyzes the production of HCO_3^-. The H^+ generated during the production of HCO_3^- also binds to Hb. Bicarbonate moves by facilitated diffusion down its concentration gradient out of the red blood cell into the plasma, and chloride (Cl^-) moves by means of the same passive carrier into the red blood cell down the electrical gradient created by the outward diffusion of HCO_3^-.

The reactions that occur at the tissue level are reversed at the pulmonary level, where CO_2 diffuses out of the blood to enter the alveoli.

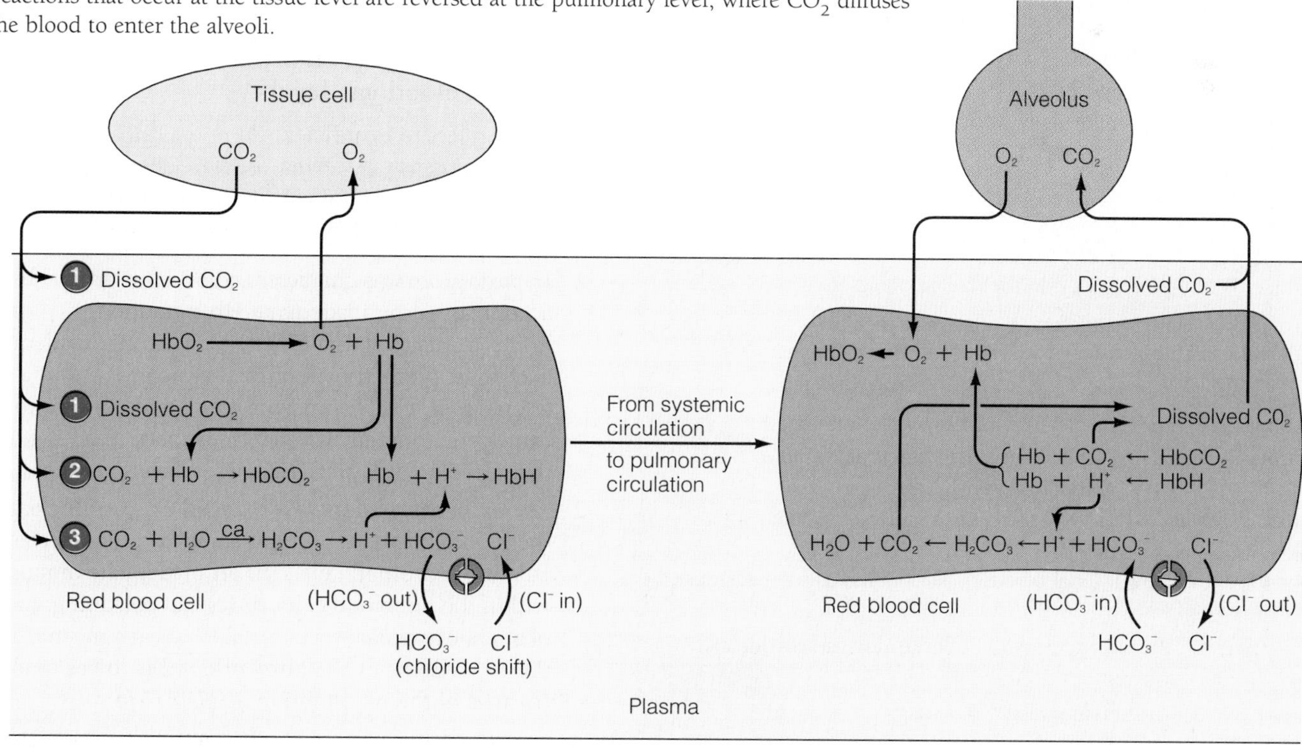

In the first step, CO_2 combines with H_2O to form **carbonic acid (H_2CO_3).** This reaction can occur very slowly in the plasma, but it proceeds swiftly within the red blood cells because of the presence of the erythrocyte enzyme **carbonic anhydrase,** which catalyzes (speeds up) the reaction. As is characteristic of acids, some of the carbonic acid molecules spontaneously dissociate into hydrogen ions (H^+) and bicarbonate ions (HCO_3^-). The one carbon and two oxygen atoms of the original CO_2 molecule are thus present in the blood as an integral part of HCO_3^-. This is beneficial because HCO_3^- is more soluble in the blood than CO_2.

As this reaction proceeds, HCO_3^- and H^+ start to accumulate within the red blood cells in the systemic capillaries. The red cell membrane has a HCO_3^--Cl^- carrier that passively facilitates the diffusion of these ions in opposite directions across the membrane. The membrane is relatively impermeable to H^+. Consequently, HCO_3^- but not H^+ diffuses down its concentration gradient out of the erythrocytes into the plasma. Since HCO_3^- is a negatively charged ion, the outflux of HCO_3^- unaccompanied by a comparable outward diffusion of positively charged ions creates an electrical gradient (see p. 55). Chloride ions (Cl^-), the dominant plasma anions, diffuse into the red blood cells down this electrical gradient to restore electric neutrality. This inward shift of Cl^- in exchange for the outflux of CO_2-generated HCO_3^- is known as the **chloride (Cl^-) shift.**

The vast majority of the accumulated H^+ within the erythrocytes following the dissociation of H_2CO_3 becomes bound to Hb. As is the case with CO_2, reduced Hb has a greater affinity for H^+ than HbO_2 does. Therefore, the unloading of O_2 once again facilitates the pickup of CO_2-generated H^+ by Hb. Because only free dissolved H^+ contributes to the acidity of a solution, the venous blood would be considerably more acidic than the arterial blood if it were not for Hb mopping up most of the H^+ generated at the tissue level.

The fact that removal of O_2 from Hb increases the ability of Hb to pick up CO_2 and CO_2-generated H^+ is known as the **Haldane effect.** The Haldane effect and Bohr effect work in synchrony to facilitate O_2 liberation and the uptake of CO_2 and CO_2-generated H^+ at the tissue level. Increased CO_2 and H^+ cause increased O_2 release from Hb by means of the Bohr effect; increased O_2 release from Hb in turn causes increased CO_2 and H^+ uptake by Hb through the Haldane effect. The entire process is very efficient. Reduced Hb must be carried back to the lungs to refill on O_2 anyway. Meanwhile, after O_2 is released, Hb picks up new passengers—CO_2 and H^+—that are going in the same direction to the lungs.

The reactions that occur at the tissue level as CO_2 enters the blood from the tissues are reversed once the blood reaches the lungs and CO_2 leaves the blood to enter the alveoli (Fig. 13-33).

Figure 13-34 summarizes the transport and exchange of O_2 and CO_2 between the atmosphere and the cells of the body; that is, it summarizes the process of external respiration.

Table 13-8
Miniglossary of Clinically Important Respiratory States

Apnea Transient cessation of breathing

Asphyxia O_2 starvation of tissues, caused by a lack of O_2 in the air, respiratory impairment, or inability of the tissues to utilize O_2

Cyanosis Blueness of the skin resulting from insufficiently oxygenated blood in the arteries

Dyspnea Difficult or labored breathing

Eupnea Normal breathing

Hypercapnia Excess CO_2 in the arterial blood

Hyperpnea Increased pulmonary ventilation that matches increased metabolic demands, as in exercise

Hyperventilation Increased pulmonary ventilation in excess of metabolic requirements, resulting in decreased P_{CO_2} and respiratory alkalosis

Hypocapnia Below-normal CO_2 in the arterial blood

Hypoventilation Underventilation in relation to metabolic requirements, resulting in increased P_{CO_2} and respiratory acidosis

Hypoxia Insufficient O_2 at the cellular level

- ***Anemic hypoxia*** Reduced O_2-carrying capacity of the blood
- ***Circulatory hypoxia*** Too little oxygenated blood delivered to the tissues; also known as stagnant hypoxia
- ***Histotoxic hypoxia*** Inability of the cells to utilize O_2 available to them
- ***Hypoxic hypoxia*** Low arterial blood P_{O_2} accompanied by inadequate Hb saturation

Respiratory arrest Permanent cessation of breathing (unless clinically corrected)

Suffocation O_2 deprivation as a result of an inability to breathe oxygenated air

Various respiratory states are characterized by abnormal blood gas levels.

Hypoxia refers to insufficient O_2 at the cellular level. (Table 13-8 is a glossary of terms used to describe various states associated with respiratory abnormalities.) There are four general categories of hypoxia:

1. *Hypoxic hypoxia* is characterized by a low arterial blood P_{O_2} accompanied by inadequate Hb saturation. It is caused by (a) a respiratory malfunction involving inadequate gas exchange, typified by a normal alveolar P_{O_2} but a reduced arterial P_{O_2}, or (b) exposure to high altitude or to a suffocating environment where atmospheric P_{O_2} is reduced so that alveolar and arterial P_{O_2} are likewise reduced.
2. *Anemic hypoxia* refers to a reduced O_2-carrying capacity of the blood. It can be brought about by (a) a decrease in circulating red blood cells, (b) an inadequate amount of Hb within the red blood cells, or (c) CO poisoning. In all cases of anemic hypoxia, the arterial P_{O_2} is at a normal level, but the O_2 content of the arterial blood is lower than normal because of the reduction in available Hb.
3. *Circulatory hypoxia* arises when too little oxygenated blood is delivered to the tissues. Circulatory hypoxia can be

restricted to a limited area as a result of a local vascular spasm or blockage. On the other hand, the body may experience circulatory hypoxia in general as a result of congestive heart failure or circulatory shock. The arterial P_{O_2} and O_2 content are typically normal, but too little oxygenated blood reaches the cells.

4. In *histotoxic hypoxia,* O_2 delivery to the tissues is normal, but the cells are unable to use the O_2 available to them. The classic example is *cyanide poisoning.* Cyanide blocks cellular enzymes essential for internal respiration.

Hyperoxia, an above-normal arterial P_{O_2}, cannot occur when a person is breathing atmospheric air at sea level.

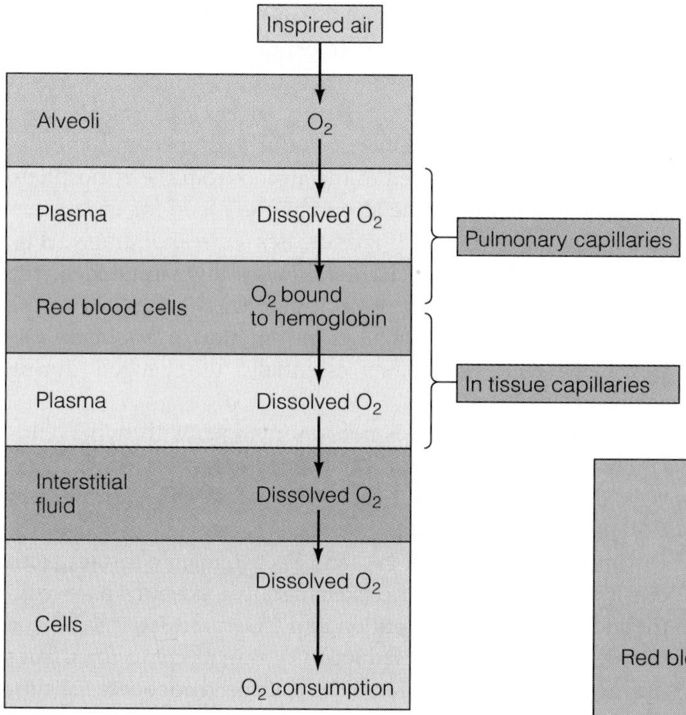

Figure 13-34 **The Paths of O_2 and CO_2 during External Respiration** (a) The path of O_2 between the atmosphere and tissue cells. (b) The path of CO_2 between the tissue cells and red blood cells in the tissue capillaries. (c) The path of CO_2 between the red blood cells in the pulmonary capillaries and the atmosphere.

Inspired air
Alveoli — O_2
Plasma — Dissolved O_2
Red blood cells — O_2 bound to hemoglobin
Plasma — Dissolved O_2
Interstitial fluid — Dissolved O_2
Cells — Dissolved O_2 — O_2 consumption
Pulmonary capillaries
In tissue capillaries
(a)

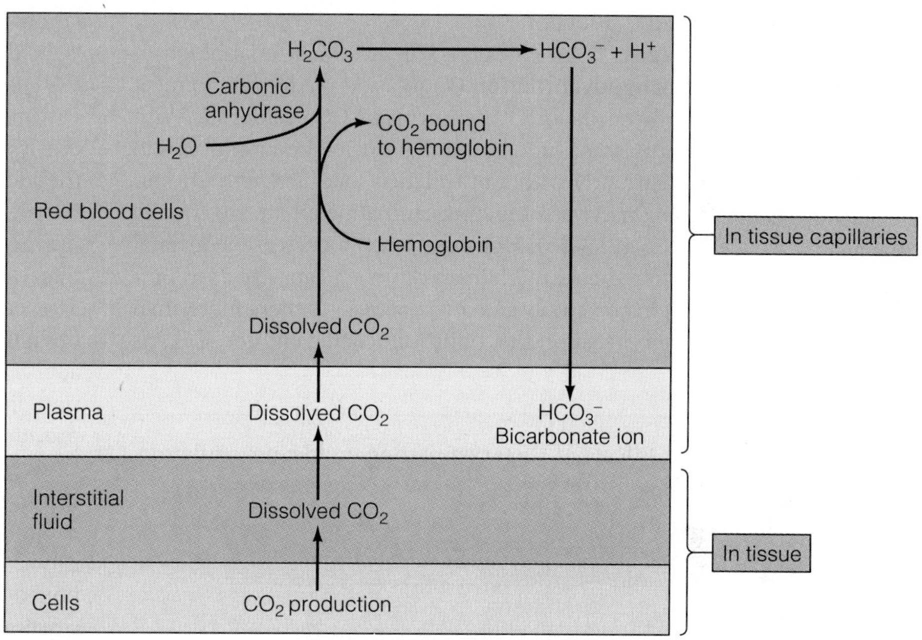

Red blood cells — H_2CO_3 → $HCO_3^- + H^+$; Carbonic anhydrase; H_2O; CO_2 bound to hemoglobin; Hemoglobin; Dissolved CO_2
Plasma — Dissolved CO_2 — HCO_3^- Bicarbonate ion
Interstitial fluid — Dissolved CO_2
Cells — CO_2 production
In tissue capillaries
In tissue
(b)

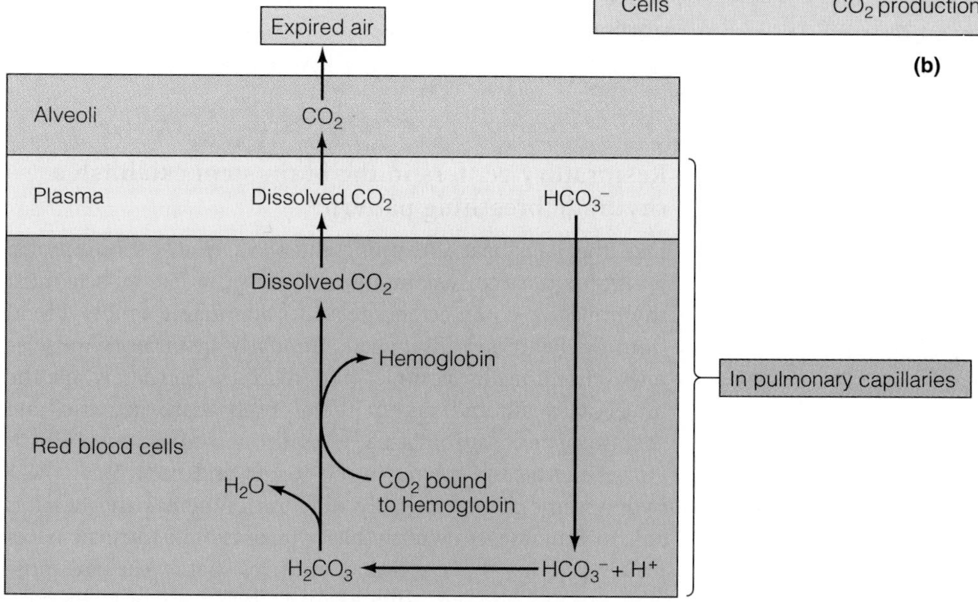

Expired air
Alveoli — CO_2
Plasma — Dissolved CO_2 — HCO_3^-
Red blood cells — Dissolved CO_2; Hemoglobin; CO_2 bound to hemoglobin; H_2O; H_2CO_3 ← $HCO_3^- + H^+$
In pulmonary capillaries
(c)

However, breathing supplemental O_2 can increase alveolar and consequently arterial P_{O_2}. Because a greater percentage of the inspired air is O_2, a greater percentage of the total pressure of the inspired air is attributable to the O_2 partial pressure, so more O_2 dissolves in the blood before arterial P_{O_2} equilibrates with alveolar P_{O_2}. Even though arterial P_{O_2} is increased, the *total* blood O_2 content is not significantly increased because Hb is nearly fully saturated at the normal arterial P_{O_2}. In certain pulmonary diseases associated with a reduced arterial P_{O_2}, however, breathing supplemental O_2 can be beneficial in establishing a larger alveoli-to-blood driving gradient, thereby improving the arterial P_{O_2}. On the other hand, far from being advantageous, a markedly elevated arterial P_{O_2} can be dangerous. If the arterial P_{O_2} is too high, **oxygen toxicity** can occur. Even though the total O_2 content of the blood is only slightly increased, some cells can be damaged by exposure to a high P_{O_2}. In particular, brain damage and damage to the retina, causing blindness, are problems associated with O_2 toxicity. Therefore, O_2 therapy must be administered cautiously.

Hypercapnia refers to excess CO_2 in the arterial blood; it is caused by **hypoventilation** (ventilation inadequate to meet the metabolic needs for O_2 delivery and CO_2 removal). With most lung diseases, CO_2 accumulation in the arterial blood occurs concurrently with an O_2 deficit, because both O_2 and CO_2 exchange between the lungs and atmosphere are equally affected (Fig. 13-35). However, when a decrease in arterial P_{O_2} is due to reduced pulmonary diffusing capacity, as in pulmonary edema or emphysema, O_2 transfer suffers more than CO_2 transfer because the diffusion coefficient for CO_2 is twenty times that of O_2. As a result, hypoxic hypoxia occurs much more readily than hypercapnia in these circumstances.

Figure 13-35 **Effects of Hyperventilation and Hypoventilation on Arterial P_{O_2} and P_{CO_2}**

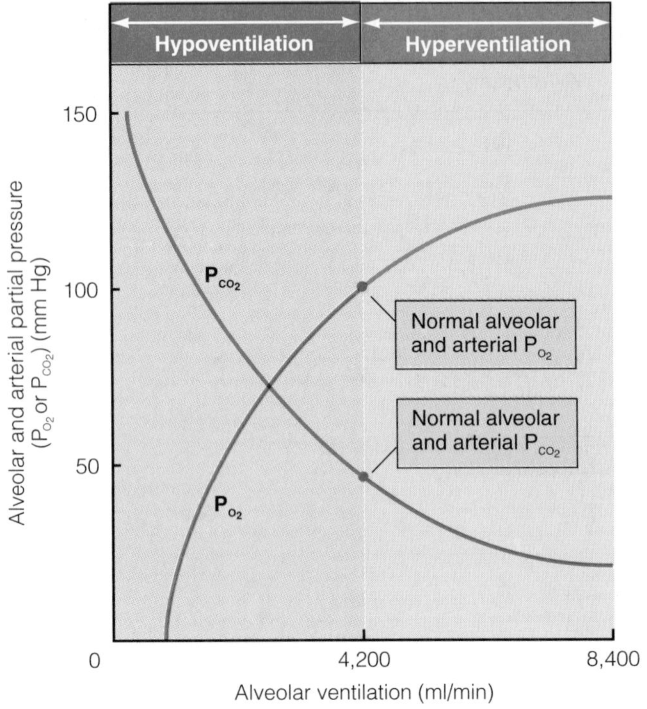

Hypocapnia, below-normal arterial P_{CO_2} levels, is brought about by hyperventilation. **Hyperventilation** occurs when a person "overbreathes," that is, when the rate of ventilation is in excess of the body's metabolic needs for CO_2 removal so that CO_2 is blown off to the atmosphere more rapidly than it is produced in the tissues and arterial P_{CO_2} falls. Hyperventilation can be triggered by anxiety states, fever, and aspirin poisoning. Alveolar P_{O_2} increases during hyperventilation as more fresh O_2 is delivered to the alveoli from the atmosphere than is extracted from the alveoli by the blood for tissue consumption, and arterial P_{O_2} increases correspondingly (Fig. 13-35). However, since Hb is almost fully saturated at the normal arterial P_{O_2}, very little additional O_2 is added to the blood. Except for the small extra amount of dissolved O_2, blood O_2 content remains essentially un-changed during hyperventilation.

Increased ventilation is not synonymous with hyperventilation. Increased ventilation that matches an increased metabolic demand, such as the increased need for O_2 delivery and CO_2 elimination during exercise, is termed **hyperpnea.** During exercise, alveolar P_{O_2} and P_{CO_2} remain constant, with the increased atmospheric exchange just keeping pace with the increased O_2 consumption and CO_2 production.

The consequences of reduced O_2 availability to the tissues during hypoxia are apparent. The cells need adequate O_2 supplies to sustain their energy-generating metabolic activities. The consequences of abnormal blood CO_2 levels are less obvious. Changes in blood CO_2 concentration primarily affect acid-base balance. Hypercapnia results in an elevated production of carbonic acid. The subsequent generation of excess H^+ produces an acidic condition termed *respiratory acidosis.* Conversely, less-than-normal amounts of H^+ are generated through carbonic acid formation in conjunction with hypocapnia. The resultant alkalotic (less acidic than normal) condition is called *respiratory alkalosis* (chapter 15). (See the boxed feature on p. 456, ◆Concepts, Challenges, and Controversies.)

Control of Respiration

Respiratory centers in the brain stem establish a rhythmic breathing pattern.

Like the heartbeat, breathing must occur in a continuous, cyclical pattern to sustain life processes. Cardiac muscle must rhythmically contract and relax to alternately empty blood from the heart and fill it again. Similarly, inspiratory muscles must rhythmically contract and relax to alternately fill the lungs with air and empty them. Both these activities are accomplished automatically without conscious effort. However, the underlying mechanisms and control of these two systems are remarkably different. Whereas the heart is able to generate its own rhythm by means of its intrinsic pacemaker activity, the respiratory muscles, being skeletal muscles, require nervous stimulation to bring about their contraction. The rhythmic pattern of breathing is established by cyclical neural activity to the respiratory muscles. In other words, the pacemaker activity that establishes the rhythmicity of breathing resides in the respiratory control centers in the brain, not in the lungs or respiratory muscles themselves. The nerve supply to the heart, not being necessary to initiate the heartbeat, only serves to modify the rate and strength of cardiac contraction. In contrast, the nerve supply to the respiratory system is absolutely essential in maintaining breathing and in reflexly adjusting the level of ventilation to match changing needs for O_2 uptake and CO_2 removal. Furthermore, unlike cardiac activity, which is not subject to voluntary control, respiratory activity can be voluntarily modified to accomplish speaking, singing, whistling, playing a wind instrument, or holding one's breath while swimming.

Neural control of respiration involves three distinct components: (1) the factors responsible for generating the alternating inspiration/expiration rhythm, (2) the factors that regulate the magnitude of ventilation (that is, the rate and depth of breathing) to match body needs, and (3) the factors that modify respiratory activity to serve other purposes. The latter modifications may be either voluntary, as in the breath control required for speech, or involuntary, as in the respiratory maneuvers involved in a cough or sneeze.

Respiratory control centers housed in the brain stem are responsible for generating the rhythmic pattern of breathing. The primary respiratory control center, the **medullary respiratory center,** consists of several aggregations of neuronal cell bodies within the medulla that provide output to the respiratory muscles. In addition, there are two other respiratory centers higher in the brain stem in the pons—the **apneustic center** and **pneumotaxic center.** These pontine centers influence the output from the medullary respiratory center (Fig. 13-36). Exactly how these various regions interact to establish respiratory rhythmicity is unclear, but the following factors are believed to contribute.

Inspiratory and expiratory neurons in the medullary center We rhythmically breathe in and out during quiet breathing because of alternate contraction and relaxation of the inspiratory muscles, namely, the diaphragm and external intercostal muscles, supplied by the phrenic nerve and intercostal nerves, respectively. The cell bodies for the neuronal fibers composing these nerves are located in the spinal cord. Impulses originating in the medullary center terminate on these motor neuron cell bodies (Fig. 13-37). When these motor neurons are activated, they in turn stimulate the inspiratory muscles, leading to inspiration; when these neurons are not firing, the inspiratory muscles relax and expiration takes place.

The medullary respiratory center consists of two neuronal clusters known as the dorsal respiratory group and the ventral respiratory group (Fig. 13-36). The **dorsal respiratory**

Figure 13-36 **Respiratory Control Centers in the Brain Stem**

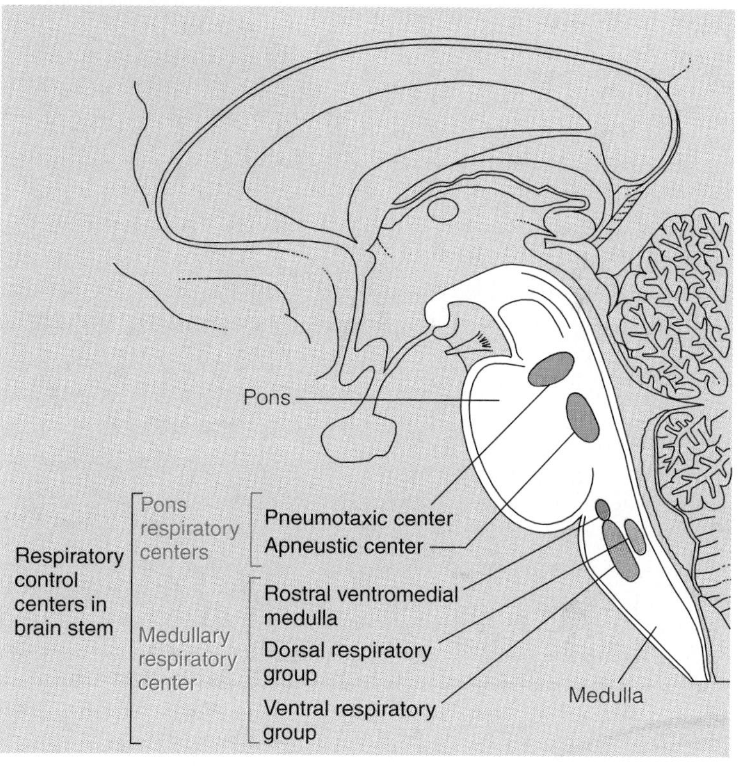

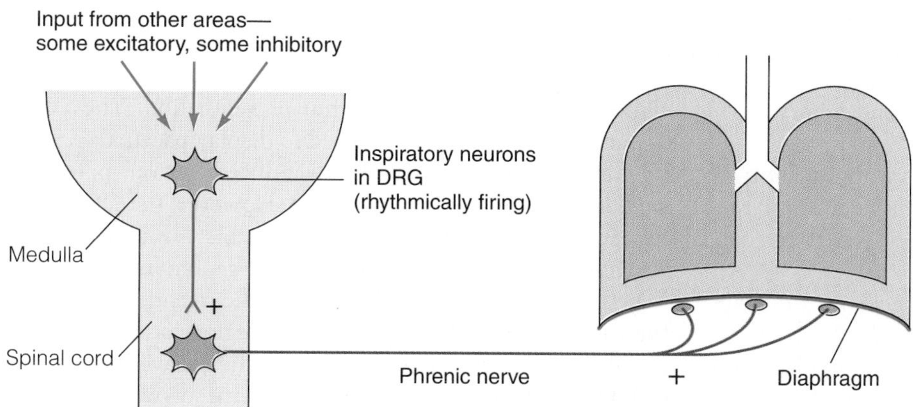

Figure 13-37 **Schematic Representation of Medullary Dorsal Respiratory Group (DRG) Control of Inspiration** Inspiration takes place when the inspiratory neurons are firing and activating the motor neurons that supply the inspiratory muscles. Expiration takes place when the inspiratory neurons cease firing, so that the motor neurons supplying the inspiratory muscles are no longer activated.

CONCEPTS, CHALLENGES, AND CONTROVERSIES

Effects of Heights and Depths on the Body

Our bodies are optimally equipped for existence at normal atmospheric pressure. Ascent into mountains high above sea level or descent into the depths of the ocean can have adverse effects on the body.

Effects of High Altitude on the Body

The atmospheric pressure progressively declines as altitude increases. At 18,000 ft above sea level, the atmospheric pressure is only 380 mm Hg—half of its normal sea level value. Since the proportion of O_2 and N_2 in the air remains the same, the P_{O_2} of inspired air at this altitude is 21% of 380 mm Hg, or 80 mm Hg, with alveolar P_{O_2} being even lower at 45 mm Hg. At any altitude above 10,000 ft, the arterial P_{O_2} falls into the steep portion of the O_2-Hb curve, below the safety range of the plateau region. As a result, the % Hb saturation in the arterial blood declines precipitously with further increases in altitude.

People who rapidly ascend to altitudes of 10,000 ft or more experience symptoms of **acute mountain sickness** attributable to hypoxic hypoxia and the resultant hypocapnia-induced alkalosis. The increased ventilatory drive to obtain more O_2 causes respiratory alkalosis because acid-forming CO_2 is blown off more rapidly than it is produced. Symptoms of mountain sickness include fatigue, nausea, loss of appetite, labored breathing, rapid heart rate (triggered by hypoxia as a compensatory measure to increase circulatory delivery of available O_2 to the tissues), and nerve dysfunction characterized by poor judgment, dizziness, and incoordination.

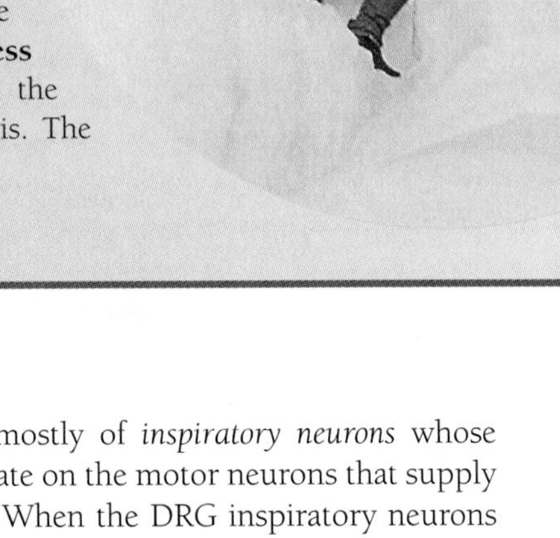

Despite these acute responses to high altitude, millions of people live at elevations above 10,000 ft, with some villagers even residing in the Andes at altitudes greater than 16,000 ft. How do they live and function normally? They do so through the process of **acclimatization.** When a person remains at high altitude, the acute compensatory responses of increased ventilation and increased cardiac output are gradually replaced over a period of days by more slowly developing compensatory measures that permit adequate oxygenation of the tissues and restoration of normal acid-base balance. Red blood cell (rbc) production is increased, stimulated by erythropoietin in response to reduced O_2 delivery to the kidneys (see p. 356). The rise in the number of rbcs increases the O_2-carrying capacity of the blood. Hypoxia also promotes the synthesis of DPG within the rbcs, so that O_2 is unloaded from Hb more easily at the tissues. The num-

group (DRG) consists mostly of *inspiratory neurons* whose descending fibers terminate on the motor neurons that supply the inspiratory muscles. When the DRG inspiratory neurons fire, inspiration takes place; when they cease firing, expiration occurs. Expiration is brought to an end as the inspiratory neurons once again reach threshold and fire.

The DRG has important interconnections with the **ventral respiratory group (VRG).** The VRG is composed of *inspiratory neurons* and *expiratory neurons,* both of which remain inactive during normal quiet breathing. This region is called into play by the DRG as an "overdrive" mechanism during periods when demands for ventilation are increased. It is especially important in active expiration. No impulses are generated in the descending pathways from the expiratory neurons during quiet breathing. Only during active expiration do the expiratory neurons stimulate the motor neurons supplying the expiratory muscles (the abdominal and internal intercostal muscles). Furthermore, the VRG inspiratory neurons, when stimulated by the DRG, rev up inspiratory activity when demands for ventilation are high.

Until recently, the DRG was generally thought to be responsible for the basic rhythm of ventilation. The DRG's inspiratory neurons were believed to display pacemaker activity, repetitively undergoing self-induced action potentials similar to the SA node of the heart. However, the DRG is no longer thought to be directly responsible for initiating the rhythmic pattern of breathing. Instead, generation of respiratory rhythm is now widely believed to lie outside the DRG in the **rostral ventromedial medulla,** a region located near the upper (head) end of the VRG. Researchers have identified a network of neurons in this region, some of which display pacemaker activity and all of which are believed to play a role in the rhythm-generating process. The interplay between these neurons that is ultimately responsible for the rhythmicity of ber of capillaries within the tissues is increased, reducing the distance that O_2 must diffuse from the blood to reach the cells. Furthermore, acclimatized cells are able to use O_2 more efficiently through an increase in the number of mitochondria, the energy organelles (see p. 27). The kidneys restore the arterial pH to nearly normal by conserving acid that normally would have been lost in the urine.

These compensatory measures are not without undesirable tradeoffs. For example, the greater number of circulating rbcs increases blood viscosity (makes the blood "thicker"), thereby increasing resistance to blood flow. As a result, the heart has to work harder to pump blood through the vessels (see p. 360).

Effects of Deep-Sea Diving on the Body

When a deep-sea diver descends underwater, the body is exposed to greater than atmospheric pressure. Pressure rapidly increases with sea depth as a result of the weight of the water. Pressure is already doubled by about 30 ft below sea level. The air provided by scuba equipment is delivered to the lungs at these high pressures. Recall that (1) the amount of a gas in solution is directly proportional to the partial pressure of the gas and (2) air is composed of 79% N_2. Nitrogen is poorly soluble in body tissues, but the high P_{N_2} that occurs during deep-sea diving causes more of this gas than normal to dissolve in the body tissues. The small amount of N_2 dissolved in the tissues at sea level has no known effect, but as more N_2 dissolves at greater depths, **nitrogen narcosis**, or **"raptures of the deep,"** ensues. Nitrogen narcosis is believed to result from a reduction in the excitability of neurons due to the highly lipid-soluble N_2 dissolving in their lipid membranes. At 150 ft under water, divers experience a feeling of euphoria and become drowsy, similar to the effect of having a few cocktails. At lower depths, divers become weak and clumsy, and at 350 to 400 ft, they lose consciousness. *Oxygen toxicity* resulting from the high P_{O_2} is another possible detrimental effect of descending deep under water.

Another problem associated with deep-sea diving occurs during ascent. If a diver who has been submerged long enough for a significant amount of N_2 to become dissolved in the tissues suddenly ascends to the surface, the rapid reduction in P_{N_2} causes N_2 to quickly come out of solution and form bubbles of gaseous N_2 in the body. The consequences depend on the amount and location of the bubble formation. This condition is called **decompression sickness** or **"the bends,"** the latter term arising because the victim often bends over in pain. Decompression sickness can be prevented by slow ascent to the surface or by gradual decompression in a decompression tank so that the excess N_2 can slowly escape through the lungs without bubble formation.

breathing remains to be identified. The rate at which the inspiratory neurons rhythmically fire is driven by synaptic input from the rostral ventromedial medulla and from elsewhere in the body. Thus, the on-off nature of the respiratory cycle is very complex and incompletely understood.

Influences from the pneumotaxic and apneustic centers The pontine centers exert "fine-tuning" influences over the medullary center to help produce normal, smooth inspirations and expirations. The pneumotaxic center sends impulses to the DRG that help "switch off" the inspiratory neurons, thereby limiting the duration of inspiration. In contrast, the apneustic center prevents the inspiratory neurons from being switched off, thus providing an extra boost to the inspiratory drive. In this check-and-balance system, the pneumotaxic center is dominant over the apneustic center, helping to bring inspiration to a halt and allowing expiration to occur normally. Without the pneumotaxic brakes, the breathing pattern consists of prolonged inspiratory gasps abruptly interrupted by very brief expirations. This abnormal pattern of breathing is known as **apneusis;** hence, the center responsible for this type of breathing is the apneustic center. Apneusis may occur in certain types of severe brain damage.

Hering-Breuer reflex When the tidal volume is large (greater than 1 liter), as during exercise, the **Hering-Breuer reflex** is triggered to prevent overinflation of the lungs. **Pulmonary stretch receptors** located within the smooth muscle layer of the airways are activated by the stretching of the lungs at large tidal volumes. Action potentials from these stretch receptors travel through afferent nerve fibers to the medullary center and inhibit the inspiratory neurons. This negative feedback from the highly stretched lungs themselves helps cut inspiration short before the lungs become overinflated.

Carbon dioxide–generated hydrogen-ion concentration in the brain extracellular fluid is normally the primary regulator of the magnitude of ventilation.

No matter how much O_2 is extracted from the blood or how much CO_2 is added to it at the tissue level, the P_{O_2} and P_{CO_2} of the systemic arterial blood leaving the lungs are held remarkably constant, indicative of the fact that arterial blood gas content is subject to precise regulation. Arterial blood gases are maintained within the normal range by varying the magnitude of ventilation to match the body's needs for O_2 uptake and CO_2 removal. If more O_2 is extracted from the alveoli and more CO_2 is dropped off by the blood because the tissues are metabolizing more actively, ventilation is increased correspondingly to bring in more fresh O_2 and blow off more CO_2.

The medullary respiratory center receives inputs that provide information about the body's needs for gas exchange. It responds by sending appropriate signals to the motor neurons supplying the respiratory muscles to adjust the rate and depth of ventilation to meet those needs. The two most obvious signals to increase ventilation are a decreased arterial P_{O_2} or an increased arterial P_{CO_2}. Intuitively, you would suspect that if O_2 levels in the arterial blood declined or if CO_2 accumulated, ventilation would be stimulated to obtain more O_2 or to eliminate the excess CO_2. These two factors do indeed influence the magnitude of ventilation, but not to the same degree nor through the same pathway. Also, a third chemical factor, H^+, has a notable influence on the level of respiratory activity. We will examine the role of each of these important chemical factors in the control of ventilation (Table 13-9).

Role of decreased arterial P_{O_2} in regulating ventilation Arterial P_{O_2} is monitored by **peripheral chemoreceptors** known as the **carotid bodies** and **aortic bodies,** which are located at the bifurcation of the common carotid arteries and in the arch of the aorta, respectively (Fig. 13-38). These chemoreceptors, which respond to specific changes in the chemical content of the arterial blood that bathes them, are distinctly different from the carotid sinus and aortic arch baroreceptors located in the same vicinity. The latter, being important in the regulation of systemic arterial blood pressure, monitor pressure changes rather than chemical changes.

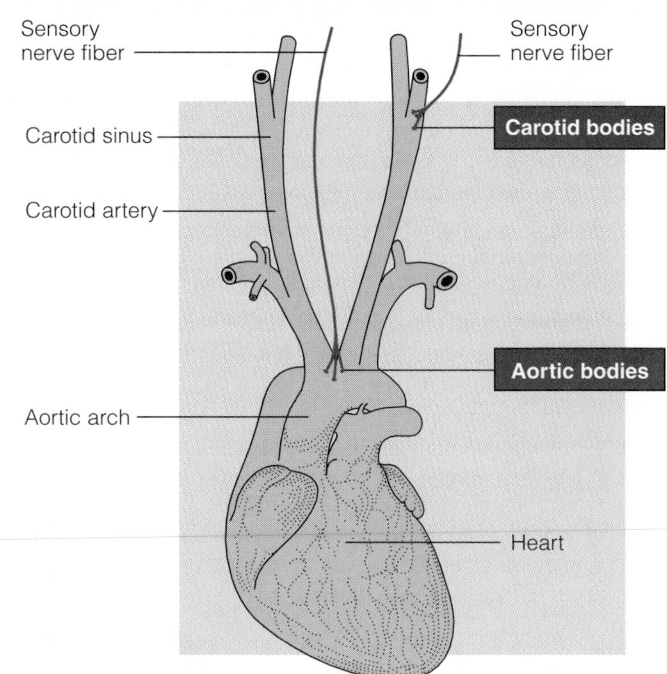

Figure 13-38 **Location of the Peripheral Chemoreceptors** The carotid bodies are located in the carotid sinus, and the aortic bodies are located in the aortic arch.

The peripheral chemoreceptors are not sensitive to modest reductions in arterial P_{O_2}. The arterial P_{O_2} must fall below 60 mm Hg (>40% reduction) before the peripheral chemoreceptors respond by sending afferent impulses to the medullary inspiratory neurons, thereby reflexly increasing ventilation. Because arterial P_{O_2} only falls below 60 mm Hg in the unusual circumstances of severe pulmonary disease or reduced atmospheric P_{O_2}, it does not play a role in the normal ongoing regulation of respiration. This fact might seem surprising at first thought, since one of the primary functions of ventilation is to provide sufficient O_2 for uptake by the blood. However, there is no need to increase ventilation until the arterial P_{O_2} falls below 60 mm Hg because of the margin of safety in % Hb saturation afforded by the plateau portion of the O_2-Hb curve. Hemoglobin is still 90% saturated at an arterial P_{O_2} of 60 mm Hg, but the % Hb saturation drops precipitously when the P_{O_2} falls below this level. Therefore,

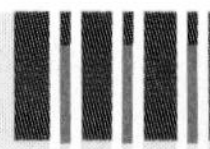

Table 13-9 **Influence of Chemical Factors on Respiration**

Chemical Factor	*Effect on the Peripheral Chemoreceptors*	*Effect on the Central Chemoreceptors*
↓ P_{O_2} in the arterial blood	Stimulates only when the arterial P_{O_2} has fallen to the point of being life-threatening (<60 mm Hg); an emergency mechanism	Directly depresses the central chemoreceptors and the respiratory center itself when <60 mm Hg
↑ P_{CO_2} in the arterial blood	Weakly stimulates	Strongly stimulates; is the dominant control of ventilation
(↑ H^+ in the brain ECF)		(Levels >70–80 mm Hg directly depress the respiratory center and central chemoreceptors)
↑ H^+ in the arterial blood	Stimulates; important in acid-base balance	Does not affect; cannot penetrate the blood-brain barrier

reflex stimulation of respiration by the peripheral chemoreceptors serves as an important emergency mechanism in dangerously low arterial P_{O_2} states. Indeed, this reflex mechanism is a lifesaver, because a low arterial P_{O_2} tends to directly depress the respiratory center, as it does all the rest of the brain (Fig. 13-39). Except for the peripheral chemoreceptors, the level of activity in all nervous tissue becomes reduced in the face of O_2 deprivation. Were it not for stimulatory intervention of the peripheral chemoreceptors when the arterial P_{O_2} falls threateningly low, a vicious cycle ending in cessation of breathing would ensue. Direct depression of the respiratory center by the markedly low arterial P_{O_2} would further reduce ventilation, leading to an even greater fall in arterial P_{O_2}, which would even further depress the respiratory center until ventilation ceased and death occurred.

Since the peripheral chemoreceptors respond to the P_{O_2} of the blood, *not* the total O_2 content of the blood, O_2 content in the arterial blood can fall to dangerously low or even fatal levels without the peripheral chemoreceptors ever responding to reflexly stimulate respiration. Remember that only physically dissolved O_2 contributes to blood P_{O_2}. The total O_2 content in the arterial blood can be reduced in anemic states, in which O_2-carrying Hb is reduced, or in CO poisoning, when the Hb is preferentially bound to this molecule rather than to O_2. In both cases, arterial P_{O_2} is normal so respiration is not stimulated, even though O_2 delivery to the tissues may be so reduced that the person dies from cellular O_2 deprivation.

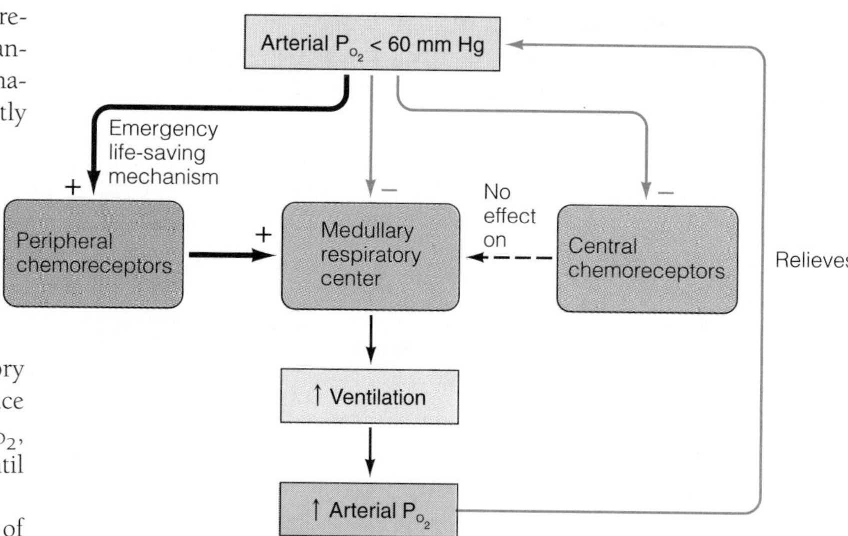

Figure 13-39 **Effect of Threateningly Low Arterial P_{O_2} (<60 mm Hg) on Ventilation**

Role of increased arterial P_{CO_2} in regulating ventilation In contrast to arterial P_{O_2}, which does not contribute to the minute-to-minute regulation of respiration, arterial P_{CO_2} is the most important input regulating the magnitude of ventilation under resting conditions. This role is appropriate, because changes in alveolar ventilation have an immediate and pronounced effect on arterial P_{CO_2}, whereas changes in ventilation have little effect on % Hb saturation and O_2 availability to the tissues until the arterial P_{O_2} falls by more than 40%. Even slight alterations from normal in arterial P_{CO_2} induce a significant reflex effect on ventilation. An increase in arterial P_{CO_2} reflexly stimulates the respiratory center, with the resultant increase in ventilation promoting elimination of the excess CO_2 to the atmosphere. Conversely, a fall in arterial P_{CO_2} reflexly reduces the respiratory drive. The subsequent decrease in ventilation allows metabolically produced CO_2 to accumulate so that P_{CO_2} can be returned to normal.

Surprisingly, given the key role of arterial P_{CO_2} in regulating respiration, there are no important receptors that monitor arterial P_{CO_2} per se. The carotid and aortic bodies are only weakly responsive to changes in arterial P_{CO_2}, so they play only a minor role in reflexly stimulating ventilation in response to an elevation in arterial P_{CO_2}. More important in linking changes in arterial P_{CO_2} to compensatory adjustments in ventilation are the **central chemoreceptors,** located in the medulla in the vicinity of the respiratory center. These central chemoreceptors do not monitor CO_2 itself; however, they are sensitive to changes in CO_2-induced H^+ concentration in the brain extracellular fluid (ECF) that bathes them.

Movement of materials across the brain capillaries is restricted by the blood-brain barrier (see p. 119). Because this barrier is readily permeable to CO_2, any increase in arterial P_{CO_2} causes a similar rise in brain ECF P_{CO_2} as CO_2 diffuses down its pressure gradient from the cerebral blood vessels into the brain ECF. The increased P_{CO_2} within the brain ECF causes a corresponding increase in the concentration of H^+ according to the law of mass action as it applies to this reaction: $CO_2 + H_2O \rightleftharpoons H_2CO_3 \rightleftharpoons H^+ + HCO_3^-$. An elevation in H^+ concentration in the brain ECF directly stimulates the central chemoreceptors, which in turn increase ventilation by stimulating the respiratory center through synaptic connections (Fig. 13-40). As the excess CO_2 is subsequently blown off, the arterial P_{CO_2} and the P_{CO_2} and H^+ concentration of the brain ECF are returned to normal. Conversely, a decline in arterial P_{CO_2} below normal is paralleled by a fall in P_{CO_2} and H^+ in the brain ECF, the result of which is a central chemoreceptor–mediated decrease in ventilation. As CO_2 produced by cellular metabolism is consequently allowed to accumulate, arterial P_{CO_2} and P_{CO_2} and H^+ of the brain ECF are restored toward normal.

Unlike CO_2, H^+ is not readily able to permeate the blood-brain barrier, so H^+ present in the plasma cannot gain access to the central chemoreceptors. Accordingly, the central chemoreceptors are responsive only to H^+ generated within the brain ECF itself as a result of CO_2 entry. Thus, the major mechanism controlling ventilation under resting conditions is specifically aimed at regulating the brain ECF H^+ concentration, which in turn is a direct reflection of the arterial P_{CO_2}. Unless there are extenuating circumstances such as reduced availability of O_2 in the inspired air, arterial P_{O_2} is coincidentally also maintained at its normal value by the brain ECF H^+ ventilatory driving mechanism.

The powerful influence of the central chemoreceptors on the respiratory center is responsible for your inability to deliberately hold your breath for more than about a minute. While

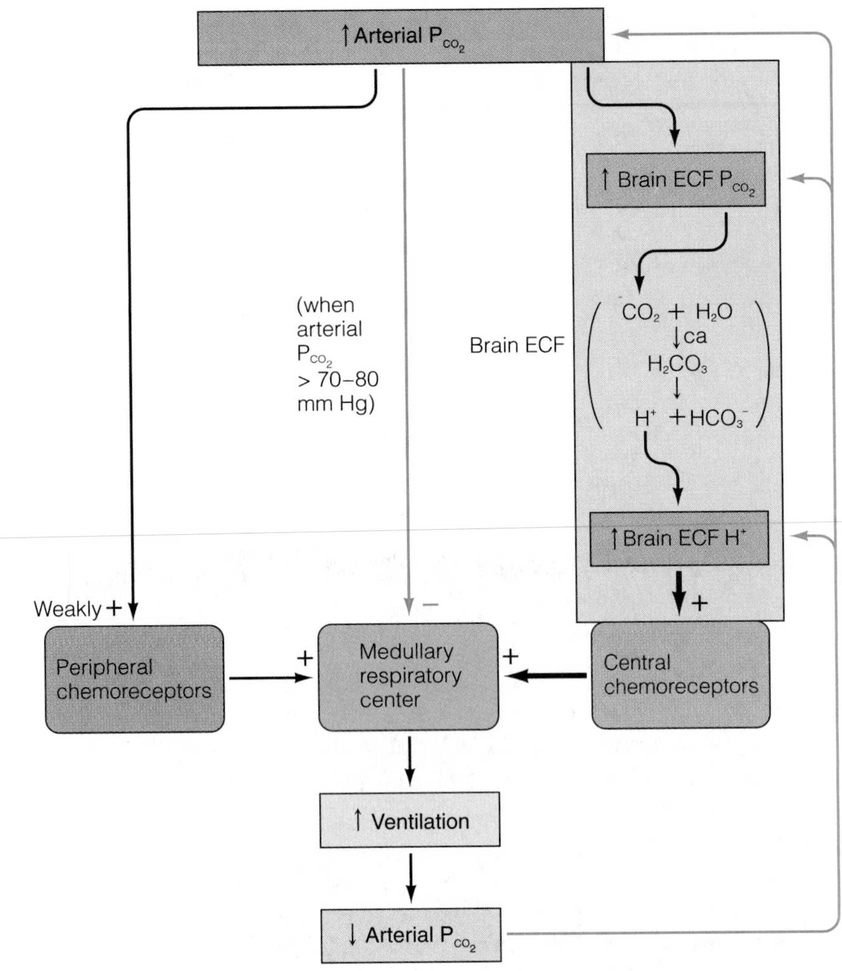

Figure 13-40 Effect of Increased Arterial P_{CO_2} on Ventilation

you hold your breath, metabolically produced CO_2 continues to accumulate in your blood and subsequently to build up the H^+ concentration in your brain ECF. Finally, the increased P_{CO_2}-H^+ stimulant to respiration becomes so powerful that central-chemoreceptor excitatory input overrides voluntary inhibitory input to respiration, so breathing resumes despite deliberate attempts to prevent it. Breathing resumes long before arterial P_{O_2} falls to the threateningly low levels that trigger the peripheral chemoreceptors. Therefore, you cannot deliberately hold your breath long enough to create a dangerously high level of CO_2 or low level of O_2 in the arterial blood.

In contrast to the normal reflex stimulatory effect of the increased P_{CO_2}-H^+ mechanism on respiratory activity, very high levels of CO_2 directly depress the entire brain, including the respiratory center, just as very low levels of O_2 do. Up to a P_{CO_2} of 70–80 mm Hg, progressively higher P_{CO_2} levels induce correspondingly more vigorous respiratory efforts in an attempt to blow off the excess CO_2. A further increase in P_{CO_2} beyond 70–80 mm Hg, however, does not further increase ventilation but actually depresses the respiratory neurons. For this reason, CO_2 must be removed and O_2 supplied in closed environments such as closed-system anesthesia machines, submarines, or space capsules. Otherwise, CO_2 could reach lethal levels, not only because of its depressant effect on respiration but also because of the resultant severe state of respiratory acidosis.

During prolonged hypoventilation caused by certain types of chronic lung disease, an elevated P_{CO_2} occurs simultaneously with a markedly reduced P_{O_2}. In most cases, the elevated P_{CO_2} (acting via the central chemoreceptors) and the reduced P_{O_2} (acting via the peripheral chemoreceptors) are *synergistic;* that is, the combined stimulatory effect on respiration exerted by these two inputs together is greater than the sum of their independent effects. However, some patients with severe chronic lung disease lose their sensitivity to an elevated arterial P_{CO_2}. In the presence of a prolonged increase in H^+ generation in the brain ECF as a result of long-standing CO_2 retention, enough HCO_3^- may cross the blood-brain barrier to buffer, or "neutralize," the excess H^+. The additional HCO_3^- combines with the excess H^+, removing it from solution so that it no longer contributes to free H^+ concentration. By raising the brain ECF HCO_3^- concentration, the brain ECF H^+ concentration is restored to normal despite the fact that arterial P_{CO_2} and brain ECF P_{CO_2} remain high. The central chemoreceptors are no longer aware of the elevated P_{CO_2} because the brain ECF H^+ is normal. Since the central chemoreceptors no longer reflexly stimulate the respiratory center in response to the elevated P_{CO_2}, the drive to eliminate CO_2 is blunted in such patients; that is, their level of ventilation is abnormally low considering their high arterial P_{CO_2}. In these patients, the hypoxic drive to ventilation becomes their primary respiratory stimulus, in contrast to normal individuals, in whom the arterial P_{CO_2} level is the dominant factor governing the magnitude of ventilation. Ironically, administering O_2 to such patients to relieve the hypoxic condition can markedly depress their drive to breathe by elevating the arterial P_{O_2} and removing the primary driving stimulus for respiration. Because of this danger, O_2 therapy must be administered cautiously in patients with long-term pulmonary diseases.

Role of increased arterial H^+ concentration in regulating ventilation Changes in arterial H^+ concentration cannot influence the central chemoreceptors because H^+ does not readily cross the blood-brain barrier. However, the aortic and carotid body peripheral chemoreceptors are highly responsive to fluctuations in arterial H^+ concentration, in contrast to their weak sensitivity to deviations in arterial P_{CO_2} and their unresponsiveness to arterial P_{O_2} until it falls 40% below normal.

Any change in arterial P_{CO_2} brings about a corresponding change in the H^+ concentration of the blood as well as of the brain ECF. These CO_2-induced H^+ changes in the arterial blood are detected by the peripheral chemoreceptors; the result is reflex stimulation of ventilation in response to an increase in arterial H^+ concentration and depression of ventilation in association with a decrease in arterial H^+ concentration. However, these changes in ventilation mediated by the peripheral chemoreceptors are far less important than the powerful central-chemoreceptor mechanism in adjusting ventilation in response to changes in CO_2-generated H^+ concentration.

The peripheral chemoreceptors do play a major role in adjusting ventilation in response to alterations in arterial H^+ concentration unrelated to fluctuations in P_{CO_2}. In many situations, even though P_{CO_2} is normal, arterial H^+ concentration is changed by the addition or loss of non-carbonic acid from the body. For example, arterial H^+ concentration increases during diabetes mellitus because excess H^+-generating keto acids are abnormally produced and added to the blood. A rise in arterial H^+ concentration reflexly stimulates ventilation by means of the peripheral chemoreceptors. Conversely, the peripheral chemoreceptors reflexly suppress respiratory activity in response to a fall in arterial H^+ concentration resulting from nonrespiratory causes. Changes in ventilation by this mechanism are extremely important in regulating the acid-base balance of the body. By changing the magnitude of ventilation, the amount of H^+-generating CO_2 that is eliminated can be varied. The resultant adjustment in the amount of H^+ added to the blood from CO_2 can compensate for the nonrespiratory-induced abnormality in arterial H^+ concentration that first elicited the respiratory response (Fig. 13-41).

Exercise profoundly increases ventilation, but the mechanisms involved are unclear.

Alveolar ventilation may increase up to 20-fold during heavy exercise to keep pace with the increased demand for O_2 uptake and CO_2 output. (Table 13-10 highlights the changes in O_2- and CO_2-related variables that occur during exercise.) The cause of increased ventilation during exercise is still largely speculative. It would seem logical that

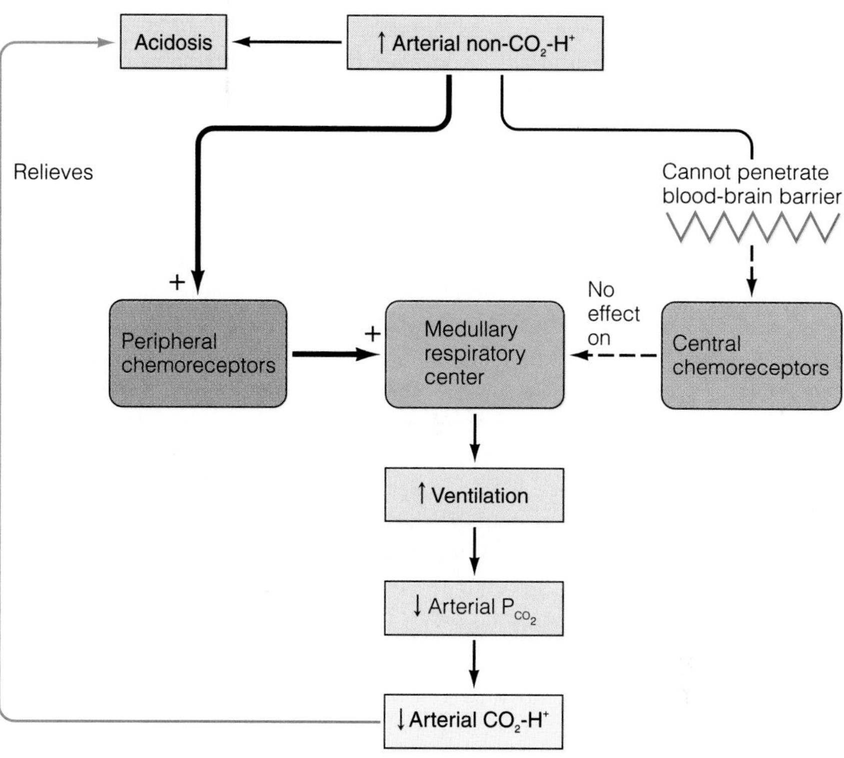

Figure 13-41 **Effect of Increased Arterial Non-Carbonic Acid–Generated Hydrogen Ion (Non-CO_2-H^+) on Ventilation**

Table 13-10 **Oxygen and Carbon Dioxide–Related Variables during Exercise**

O_2- or CO_2-Related Variable	*Change*	*Comment*
O_2 utilization	Marked increase	Active muscles are oxidizing nutrient molecules more rapidly to meet their increased energy needs
CO_2 production	Marked increase	More actively metabolizing muscles produce more CO_2
Alveolar ventilation	Marked increase	By mechanisms not completely understood, alveolar ventilation keeps pace with or even slightly exceeds the increased metabolic demands during exercise
Arterial P_{O_2} **Arterial P_{CO_2}**	Normal or slight ↑ Normal or slight ↓	Despite a marked increase in O_2 utilization and CO_2 production during exercise, alveolar ventilation keeps pace with or even slightly exceeds the stepped-up rate of O_2 consumption and CO_2 production
O_2 delivery to muscles	Marked increase	Although arterial P_{O_2} remains normal, O_2 delivery to muscles is greatly increased by the increased blood flow to exercising muscles accomplished by increased cardiac output coupled with local vasodilation of active muscles
O_2 extraction by muscles	Marked increase	Increased utilization of O_2 lowers the P_{O_2} at the tissue level, which results in more O_2 unloading from hemoglobin; this is enhanced by ↑ P_{CO_2}, ↑H^+ and ↑ temperature
CO_2 removal from muscles	Marked increase	The increased blood flow to exercising muscles removes the excess CO_2 produced by these more actively metabolizing tissues
Arterial H^+ concentration		
Mild to moderate exercise	Normal	Since carbonic acid–generating CO_2 is held constant in arterial blood, arterial H^+ concentration does not change
Heavy exercise	Modest increase	In heavy exercise, when muscles resort to anaerobic metabolism, lactic acid is added to the blood

changes in the "big three" chemical factors—decreased P_{O_2}, increased P_{CO_2}, and increased H^+—could account for the increase in ventilation. This does not appear to be the case, however.

- Despite the marked increase in O_2 utilization during exercise, arterial P_{O_2} does not decrease but remains normal or may actually increase slightly. This is because the increase in alveolar ventilation keeps pace with or even slightly exceeds the stepped-up rate of O_2 consumption.
- Likewise, despite the marked increase in CO_2 production during exercise, arterial P_{CO_2} does not increase but remains normal or decreases slightly. This is because the extra CO_2 is removed as rapidly or even more rapidly than it is produced as a result of the increase in ventilation.
- During mild or moderate exercise, H^+ concentration does not increase because H^+-generating CO_2 is held constant. During heavy exercise, H^+ concentration does increase somewhat because of the release of H^+-generating lactic acid into the blood as a result of anaerobic metabolism in the exercising muscles. Even so, the elevation in H^+ concentration resulting from lactic acid formation is not sufficient to account for the large increase in ventilation accompanying exercise.

Some investigators argue that the constancy of the three chemical regulatory factors during exercise is evidence that ventilatory responses to exercise are actually being controlled by these factors—particularly by P_{CO_2}, because it is normally the dominant control during resting conditions. According to this reasoning, how else could alveolar ventilation be increased in exact proportion to CO_2 production, thereby keeping the P_{CO_2} constant? This proposal, however, cannot account for the observation that during heavy exercise, alveolar ventilation may be increased relatively more than the increase in CO_2 production, thereby actually causing a slight decline in P_{CO_2}. Also, ventilation increases abruptly at the onset of exercise (within seconds), long before changes in arterial blood gases could become important influences on the respiratory center (which requires a matter of minutes).

Researchers have suggested that a number of other factors, including the following, play a role in the ventilatory response to exercise:

1. *Reflexes originating from body movements.* Joint and muscle receptors excited during muscle contraction reflexly stimulate the respiratory center, abruptly increasing ventilation. Even passive movement of the limbs (for example, someone else alternately flexing and extending a person's knee) may increase ventilation several-fold through activation of these receptors, even though no actual exercise is occurring. Thus, the mechanical events of exercise are believed to play an important role in coordinating respiratory activity with the increased metabolic requirements of the active muscles.
2. *Increase in body temperature.* Much of the energy generated during muscle contraction is converted to heat rather than to actual mechanical work. Heat-loss mechanisms such as sweating are frequently unable to keep pace with the increased heat production that accompanies increased physical activity, so body temperature often increases slightly during exercise. Since an increase in body temperature stimulates ventilation, this exercise-related heat production undoubtedly contributes to the respiratory response to exercise. For the same reason, increased ventilation often accompanies a fever.
3. *Epinephrine release.* The adrenal medullary hormone epinephrine also stimulates ventilation. The level of circulating epinephrine is increased during exercise in response to the sympathetic nervous system discharge that accompanies increased physical activity.
4. *Impulses from the cerebral cortex.* Especially at the onset of exercise, the motor areas of the cerebral cortex are believed to simultaneously stimulate the medullary respiratory neurons and activate the motor neurons of the exercising muscles. This is similar to the cardiovascular adjustments initiated by the motor cortex at the onset of exercise. In this way, the motor region of the brain calls forth increased ventilatory and circulatory responses to support the increased physical activity that it is about to orchestrate. These anticipatory adjustments are unusual in that regulatory steps are taken *before* any homeostatic factors have actually changed. In the usual case, regulatory adjustments take place *after* a factor has been altered in an attempt to restore homeostasis.

None of these factors or combinations of factors are fully satisfactory in explaining the abrupt and profound effect exercise has on ventilation, nor can they completely account for the high degree of correlation between respiratory activity and the body's needs for gas exchange during exercise. (For a discussion of how measurement of O_2 consumption during exercise can be used to determine a person's maximum work capacity, see the accompanying boxed feature, A Closer Look at Exercise Physiology.)

Ventilation can be influenced by factors unrelated to the need to supply O_2 or remove CO_2.

Respiratory rate and depth can be modified for reasons other than the need to supply O_2 or remove CO_2. Protective reflexes such as sneezing and coughing temporarily govern respiratory activity in an effort to expel irritant materials from the respiratory passages. Inhalation of particularly noxious agents frequently triggers immediate cessation of ventilation. Pain originating anywhere in the body reflexly stimulates the respiratory center (for example, one "gasps" with pain). Involuntary modification of breathing also occurs during the expression of various emotional states, such as laughing, crying, sighing, and groaning. The emotionally induced modifications are mediated through connections between the limbic system in the brain (which is responsible for emotions) and the respiratory center. In addition, the respiratory center is reflexly inhibited during swallowing, when the airways are closed to prevent food from entering the lungs.

Human beings also have considerable voluntary control over ventilation. Voluntary control of breathing is accom-

A CLOSER LOOK AT EXERCISE PHYSIOLOGY

How to Find Out How Much Work You're Capable of Doing

The best single predictor of a person's work capacity is the determination of the maximum volume of O_2 the person is capable of using per minute to oxidize nutrient molecules for energy production. *Maximal O_2 consumption,* or *max VO_2*, is measured by having the person engage in exercise, usually on a treadmill or bicycle ergometer (a stationary bicycle with variable resistance). The workload is incrementally increased until the person becomes exhausted. Expired air samples collected during the last minutes of exercise, when O_2 consumption is at a maximum because the person is working as hard as possible, are analyzed for the percentages of O_2 and CO_2 they contain. Furthermore, the volume of air expired is measured. Equations are then employed to determine the amount of O_2 consumed, taking into account the percentages of O_2 and CO_2 in the inspired air, the total volume of air expired, and the percentages of O_2 and CO_2 in the exhaled air.

Maximal O_2 consumption depends on three systems. The respiratory system is essential for ventilation and exchange of O_2 and CO_2 between the air and blood in the lungs. The circulatory system is required to deliver O_2 to the working muscles. Finally, the muscles must have the oxidative enzymes available to use the O_2 once it has been delivered.

Regular aerobic exercise can improve max VO_2 by making the heart and respiratory system more efficient, thereby delivering more O_2 to the working muscles. Exercised muscles themselves become better equipped to use O_2 once it is delivered. The number of functional capillaries increases, as do the number and size of mitochondria, which contain the oxidative enzymes.

Maximal O_2 consumption is measured in liters per minute and then converted into milliliters per kilogram of body weight per minute so that large and small people can be compared. As would be expected, athletes have the highest values for maximal O_2 consumption. The max VO_2 for male cross-country skiers has been recorded to be as high as 94 ml O_2/kg/min. Distance runners maximally consume between 65 and 85 ml O_2/kg/min, and football players have max VO_2 values between 45 and 65 ml O_2/kg/min, depending on the position they play. Sedentary young men maximally consume between 25 and 45 ml O_2/kg/min. Female values for max VO_2 are 20% to 25% lower than for males when expressed as ml/kg/min of total body weight. The difference in max VO_2 between females and males is only 8% to 10% when expressed as ml/kg/min of lean body weight, however, because females generally have a higher percentage of body fat (the female sex hormone estrogen promotes fat deposition).

Available norms are used to classify people as being low, fair, average, good, or excellent in aerobic capacity for their age group. Exercise physiologists use max VO_2 measurements to prescribe or adjust training regimens to help people achieve their optimal level of aerobic conditioning.

plished by the cerebral cortex, which does not act on the respiratory center in the brain stem but instead sends impulses directly to the motor neurons in the spinal cord that supply the respiratory muscles. We can voluntarily hyperventilate ("overbreathe") or, at the other extreme, hold our breath, but only for a brief period of time. The resulting chemical changes in the arterial blood directly and reflexly influence the respiratory center, which in turn overrides the voluntary input to the motor neurons of the respiratory muscles. Other than these extreme forms of deliberately controlling ventilation, we also control our breathing to perform such voluntary acts as speaking, singing, and whistling.

During apnea, a person subconsciously "forgets to breathe," whereas during dyspnea, a person consciously feels that ventilation is inadequate.

Apnea is the transient cessation of ventilation with the expectation that breathing will resume spontaneously. The condition is called **respiratory arrest** if breathing does not resume. Because ventilation is normally decreased and the central chemoreceptors are less sensitive to the arterial P_{CO_2} drive during sleep, especially REM sleep (see p. 143), apnea is most likely to occur during this time. Victims of **sleep apnea** may stop breathing for a few seconds or up to one or two minutes as many as 500 times a night. Mild forms of sleep apnea are not dangerous unless the sufferer has pulmonary or circulatory disease, the consequences of which can be compounded by recurrent bouts of apnea.

In exaggerated cases of sleep apnea, the victim may be unable to recover from an apneic period, and death results. This is the case in **sudden infant death syndrome (SIDS)**, or "crib death." With this tragic form of sleep apnea, an otherwise healthy two- or four-month-old infant is found dead in his or her crib for no apparent reason. The underlying cause of SIDS is the subject of intense investigation. Most evidence suggests that the baby "forgets to breathe" as a result of the immaturity of the respiratory control mechanisms, either in the brain stem or in the chemoreceptors that monitor the body's respiratory status. For example, on autopsy, more than half the victims have been shown to have poorly developed carotid bodies, the more important of the peripheral chemoreceptors. Recent studies have shown that certain risk factors make babies more vulnerable to SIDS. Among them are sleeping position (a higher incidence of SIDS is associated with sleeping on the abdomen rather than on the back or side) and exposure to nicotine during fetal life or after birth. Infants whose mothers smoked during pregnancy or who breathe cigarette smoke in the home are three times more likely to die of SIDS than those not exposed to smoke.

In contrast to sleep apnea, in which the victim unconsciously stops breathing, people who have **dyspnea** have the subjective sensation that they are not getting enough air; that is, they feel "short of breath." Dyspnea is the mental anguish associated with the unsatiated desire for more adequate ventilation. It often accompanies the labored breathing characteristic of obstructive lung disease or the pulmonary edema associated with congestive heart failure. In contrast, during exercise a person can breathe very hard without experiencing the sensation of dyspnea, because such exertion is not accompanied by a sense of anxiety over the adequacy of ventilation. Surprisingly, dyspnea is not directly related to chronic elevation of arterial P_{CO_2} or reduction of P_{O_2}. The subjective feeling of air hunger may occur even when alveolar ventilation and the blood gases are normal. Some individuals experience dyspnea when they *perceive* that they are short of air even though this is not actually the case, such as when they are in a crowded elevator.

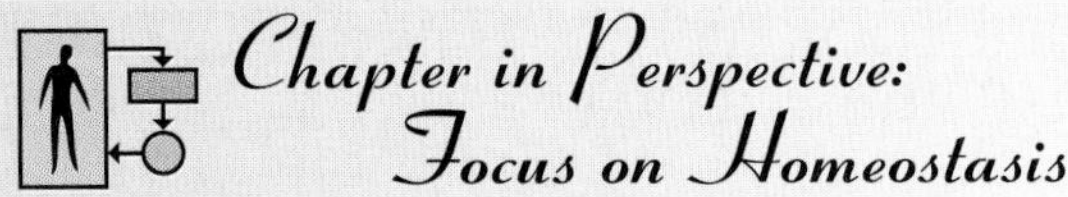

The respiratory system contributes to homeostasis by obtaining O_2 from and eliminating CO_2 to the external environment. All body cells ultimately need an adequate supply of O_2 to use in oxidizing nutrient molecules to generate ATP. Brain cells, which are especially dependent on a continual supply of O_2, die if deprived of O_2 for more than four minutes. Even cells that can resort to anaerobic ("without O_2") metabolism for energy production, such as strenuously exercising muscles, can do so only transiently by incurring an O_2 debt that ultimately must be repaid (see p. 243).

As a result of these energy-yielding metabolic reactions, large quantities of CO_2 are produced that must be eliminated from the body. Because CO_2 and H_2O form carbonic acid, adjustments in the rate of CO_2 elimination by the respiratory system are important in the regulation of acid-base balance in the internal environment. Cells can survive only within a narrow pH range.

Chapter Summary

Introduction

Internal respiration refers to the intracellular metabolic reactions that utilize O_2 and produce CO_2 during energy-yielding oxidation of nutrient molecules. External respiration encompasses the various steps involved in the transfer of O_2 and CO_2 between the external environment and tissue cells. The respiratory and circulatory systems function together to accomplish external respiration.

The respiratory system accomplishes exchange of air between the atmosphere and the lungs through the process of ventilation. Exchange of O_2 and CO_2 between the air in the lungs and the blood in the pulmonary capillaries takes place across the extremely thin walls of the air sacs, or alveoli. Respiratory airways conduct air from the atmosphere to this gas-exchanging portion of the lungs. The lungs are housed within the closed compartment of the thorax, the volume of which can be changed by contractile activity of surrounding respiratory muscles.

Respiratory Mechanics

Ventilation, or breathing, is the process of cyclically moving air in and out of the lungs, so that old alveolar air that has already participated in exchange of O_2 and CO_2 with the pulmonary capillary blood can be exchanged for fresh atmospheric air. Ventilation is mechanically accomplished by alternately shifting the direction of the pressure gradient for airflow between the atmosphere and the alveoli through the cyclical expansion and recoil of the lungs. Alternate contraction and relaxation of the inspiratory muscles (primarily the diaphragm) indirectly produce periodic inflation and deflation of the lungs by cyclically expanding and compressing the thoracic cavity, with the lungs passively following its movements.

Because energy is required for contraction of the inspiratory muscles, inspiration is an active process, but expiration is passive during quiet breathing because it is accomplished by elastic recoil of the lungs on relaxation of inspiratory muscles at no energy expense. For more forceful active expiration, contraction of the expiratory muscles (namely, the abdominal muscles) further decreases the size of the thoracic cavity and lungs, which further increases the intra-alveolar–to–atmospheric pressure gradient. The larger the gradient between the alveoli and the atmosphere in either direction, the larger the airflow rate, because air continues to flow until the intra-alveolar pressure equilibrates with atmospheric pressure.

Besides being directly proportional to the pressure gradient, airflow rate is also inversely proportional to airway resistance. Because airway resistance, which depends on the caliber of the conducting airways, is normally very low, airflow rate usually depends primarily on the pressure gradient established between the alveoli and the atmosphere. If airway resistance is pathologically increased by chronic obstructive pulmonary disease, the pressure gradient must be correspondingly increased by more vigorous respiratory muscle activity to maintain a normal airflow rate.

The lungs can be stretched to varying degrees during inspiration and then recoil to their preinspiratory size during expiration because of their elastic behavior. Pulmonary compliance refers to the distensibility of the lungs—how much they stretch in response to a given change in the transmural pressure gradient, the stretching force exerted across the lung wall. Elastic recoil refers to the phenomenon of the lungs snapping back to their resting position during expiration. Pulmonary elastic behavior depends on the elastic connective tissue meshwork within the lungs and on alveolar surface tension/pulmonary surfactant interaction. Alveolar surface tension, which is due to the attractive forces between the surface water molecules in the liquid film lining each alveolus, tends to resist the alveolus being stretched upon inflation (decreases compliance) and tends to return it back to a smaller surface area during deflation (increases lung rebound). If the alveoli were lined by water alone, the surface tension would be so great that the lungs would be poorly compliant and would tend to collapse. Type II alveolar cells secrete pulmonary surfactant, a phospholipoprotein that intersperses between the water molecules and lowers the alveolar surface tension, thereby increasing the compliance of the lungs and counteracting the tendency for alveoli to collapse.

The lungs can be filled to over 5.5 liters upon maximal inspiratory effort or emptied to about 1 liter upon maximal expiratory effort. Normally, however, the lungs operate at "half-full." The lung volume typically varies from about 2 to 2.5 liters as an average tidal volume of 500 ml of air is moved in and out with each breath.

The amount of air moved in and out of the lungs in one minute, the pulmonary ventilation, is equal to tidal volume times respiratory rate. However, not all of the air moved in and out is available for O_2 and CO_2 exchange with the blood because part of it occupies the conducting airways, known as the anatomical dead space. Alveolar ventilation, the volume of air exchanged between the atmosphere and the alveoli in one minute, is a measure of the air actually available for gas exchange with the blood. Alveolar ventilation equals (tidal volume minus the dead space volume) times respiratory rate.

Gas Exchange

Oxygen and CO_2 move across body membranes by passive diffusion down partial pressure gradients. Net diffusion of O_2 occurs first between the alveoli and the blood and then between the blood and the tissues as a result of the O_2 partial pressure gradients created by continuous utilization of O_2 in the cells and continuous replenishment of fresh alveolar O_2 provided by ventilation. Net diffusion of CO_2 occurs in the reverse direction, first between the tissues and the blood and then between the blood and the alveoli, as a result of the CO_2 partial pressure gradients created by continuous production of CO_2 in the cells and continuous removal of alveolar CO_2 through the process of ventilation.

Gas Transport

Because O_2 and CO_2 are not very soluble in the blood, they must be transported primarily by mechanisms other than simply being physically dissolved. Only 1.5% of the O_2 is physically dissolved in the blood, with 98.5% chemically bound to hemoglobin (Hb). The primary factor that determines the extent to which Hb and O_2 are combined (the % Hb saturation) is the P_{O_2} of the blood. The relationship between blood P_{O_2} and % Hb saturation is such that in the P_{O_2} range found in the pulmonary capillaries, Hb is still almost fully saturated even if the blood P_{O_2} falls as much as 40%; this provides a margin of safety by ensuring near-normal O_2 delivery to the tissues despite a substantial reduction in arterial P_{O_2}. On the other hand, in the P_{O_2} range found in the systemic capillaries, large increases in Hb unloading occur in response to a small local decline in blood P_{O_2} associated with increased cellular metabolism; thus, more O_2 is provided to match the increased tissue needs.

Carbon dioxide picked up at the systemic capillaries is transported in the blood by three methods: (1) 10% is physically dissolved; (2) 30% is bound to Hb; and (3) 60% is in the form of bicarbonate (HCO_3^-). The erythrocyte enzyme carbonic anhydrase catalyzes the conversion of CO_2 to HCO_3^- according to the reaction: $CO_2 + H_2O \rightleftharpoons H_2CO_3 \rightleftharpoons H^+ + HCO_3^-$. The generated H^+ binds to Hb. These reactions are all reversed in the lungs as CO_2 is eliminated to the alveoli.

Control of Respiration

Ventilation involves two distinct aspects, both of which are subject to neural control: (1) rhythmic cycling between inspiration and expiration and (2) regulation of the magnitude of ventilation, which in turn depends on control of respiratory rate and depth of tidal volume. Respiratory rhythm is primarily established by a complex neuronal network that displays pacemaker activity and drives the inspiratory neurons located in the respiratory control center in the medulla of the brain stem. When these inspiratory neurons fire, impulses ultimately reach the inspiratory muscles to bring about inspiration. When the inspiratory neurons cease firing, the inspiratory muscles relax and expiration takes place. If active expiration is to occur, the expiratory muscles are activated by output from the medullary expiratory neurons at this time. This basic rhythm is smoothed out by a balance of activity in the apneustic and pneumotaxic centers located higher in the brain stem in the pons. The apneustic center prolongs inspiration, whereas the more powerful pneumotaxic center limits inspiration.

Three chemical factors play a role in determining the magnitude of ventilation: the P_{CO_2}, P_{O_2}, and H^+ concentration of the arterial blood. The dominant factor in the minute-to-minute regulation of ventilation is the arterial P_{CO_2}. An increase in arterial P_{CO_2} is the most potent chemical stimulus for increasing ventilation. Changes in arterial P_{CO_2} alter ventilation primarily by bringing about corresponding changes in the brain ECF H^+ concentration, to which the central chemoreceptors are exquisitely sensitive. The peripheral chemoreceptors are responsive to an increase in arterial H^+ concentration, which likewise reflexly brings about increased ventilation. The resultant adjustment in arterial H^+-generating CO_2 is important in maintaining the acid-base balance of the body. The peripheral chemoreceptors also reflexly stimulate the respiratory center in response to a marked reduction in arterial P_{O_2} (<60 mm Hg). This response serves as an emergency mechanism to increase respiration when the arterial P_{O_2} levels fall below the safety range provided by the plateau portion of the O_2-Hb curve.

Review Exercises

Objective Questions (Answers on p. E–12.)

1. Breathing is accomplished by alternate contraction and relaxation of muscles within the lung tissue. (True or false?)
2. The alveoli normally empty completely during maximal expiratory efforts. (True or false?)
3. Alveolar ventilation does not always increase when pulmonary ventilation increases. (True or false?)

4. Oxygen and CO_2 have equal diffusion coefficients. (True or false?)
5. Hemoglobin has a higher affinity for O_2 than for any other substance. (True or false?)
6. Rhythmicity of breathing is brought about by pacemaker activity displayed by the respiratory muscles. (True or false?)
7. The expiratory neurons send impulses to the motor neurons controlling the expiratory muscles during normal quiet breathing. (True or false?)
8. The three forces that tend to keep the alveoli open are __________, __________, and __________.
9. The two forces that promote alveolar collapse are __________ and __________.
10. __________ is a measure of the magnitude of change in lung volume accomplished by a given change in the transmural pressure gradient.
11. __________ refers to the phenomenon of the lungs snapping back to their resting size after having been stretched.
12. __________ is the erythrocytic enzyme responsible for catalyzing the conversion of CO_2 into HCO_3^-.
13. Which of the following reactions take(s) place at the pulmonary capillaries?
    a. $Hb + O_2 \rightarrow HbO_2$
    b. $CO_2 + H_2O \rightarrow H_2CO_3 \rightarrow H^+ + HCO_3^-$
    c. $Hb + CO_2 \rightarrow HbCO_2$
    d. $HbH \rightarrow Hb + H^+$
14. Using the answer code below, indicate which chemoreceptors are being described:
    (a) peripheral chemoreceptors
    (b) central chemoreceptors
    (c) both peripheral and central chemoreceptors
    (d) neither peripheral nor central chemoreceptors
    ____ 1. stimulated by an arterial P_{O_2} of 80 mm Hg
    ____ 2. stimulated by an arterial P_{O_2} of 55 mm Hg
    ____ 3. directly depressed by an arterial P_{O_2} of 55 mm Hg
    ____ 4. weakly stimulated by an elevated arterial P_{CO_2}
    ____ 5. strongly stimulated by an elevated brain ECF H^+ induced by an elevated arterial P_{CO_2}
    ____ 6. stimulated by an elevated arterial H^+ concentration
15. Indicate the O_2 and CO_2 partial pressure relationships that are important in gas exchange by circling > (greater than), < (less than), or = (equal to) as appropriate in each of the following statements;
    a. P_{O_2} in blood entering the pulmonary capillaries is (>, <, or =) P_{O_2} in the alveoli.
    b. P_{CO_2} in blood entering the pulmonary capillaries is (>, <, or =) P_{CO_2} in the alveoli.
    c. P_{O_2} in the alveoli is (>, <, or =) P_{O_2} in blood leaving the pulmonary capillaries.
    d. P_{CO_2} in the alveoli is (>, <, or =) P_{CO_2} in blood leaving the pulmonary capillaries.
    e. P_{O_2} in blood leaving the pulmonary capillaries is (>, <, or =) P_{O_2} in blood entering the systemic capillaries.
    f. P_{CO_2} in blood leaving the pulmonary capillaries is (>, <, or =) P_{CO_2} in blood entering the systemic capillaries.
    g. P_{O_2} in blood entering the systemic capillaries is (>, <, or =) P_{O_2} in the tissue cells.
    h. P_{CO_2} in blood entering the systemic capillaries is (>, <, or =) P_{CO_2} in the tissue cells.
    i. P_{O_2} in the tissue cells is (>, <, or approximately =) P_{O_2} in blood leaving the systemic capillaries.
    j. P_{CO_2} in the tissue cells is (>, <, or approximately =) P_{CO_2} in blood leaving the systemic capillaries.
    k. P_{O_2} in blood leaving the systemic capillaries is (>, <, or =) P_{O_2} in blood entering the pulmonary capillaries.
    l. P_{CO_2} in blood leaving the systemic capillaries is (>, <, or =) P_{CO_2} in blood entering the pulmonary capillaries.

Essay Questions

1. Distinguish between internal and external respiration. List the steps in external respiration.
2. Describe the components of the respiratory system. What is the site of gas exchange?
3. Compare atmospheric, intra-alveolar, and intrapleural pressures.
4. Why are the lungs normally stretched even during expiration?
5. Explain why air enters the lungs during inspiration and leaves during expiration.
6. Why is inspiration normally active and expiration normally passive?
7. Why does airway resistance become an important determinant of airflow rates in chronic obstructive pulmonary disease?
8. Explain pulmonary elasticity in terms of elastic recoil and compliance. What are the source and function of pulmonary surfactant?
9. Define the various lung volumes and capacities.
10. Compare pulmonary ventilation and alveolar ventilation. What is the consequence of anatomical and alveolar dead space?
11. What determines the partial pressures of a gas in air and in blood?
12. List the methods of O_2 and CO_2 transport in the blood.
13. What is the primary factor that determines the percent hemoglobin saturation? What is the significance of the plateau and the steep portions of the O_2-Hb dissociation curve?
14. How does hemoglobin promote the net transfer of O_2 from the alveoli to the blood?
15. Explain the Bohr and Haldane effects.
16. Define the following: hypoxic hypoxia, anemic hypoxia, circulatory hypoxia, histotoxic hypoxia, hypercapnia, hypocapnia, hyperventilation, hypoventilation, hyperpnea, apnea, and dyspnea.
17. What are the locations and functions of the three respiratory control centers? Distinguish between the DRG and the VRG.
18. What factors contribute to rhythmicity of breathing?

Quantitative Exercises (Solutions on p. E–12.)

1. The two curves in Figure 13-35 show the partial pressures for O_2 and CO_2 at various alveolar ventilation rates. These curves can be calculated from the following two equations:

$$P_{AO_2} = P_{IO_2} - (V_{O_2}/V_A)\ 863\ \text{mm Hg}$$

$$P_{ACO_2} = (V_{CO_2}/V_A)\ 863\ \text{mm Hg}$$

In these equations, P_{AO_2} = the partial pressure of O_2 in the alveoli, P_{ACO_2} = the partial pressure CO_2 in the alveoli, P_{IO_2} = the partial pressure of O_2 in the inspired air, V_{O_2} = the rate of O_2 consumption by the body, V_{CO_2} = the rate of CO_2 production by the body, V_A = the rate of alveolar ventilation, and 863 mm Hg is a constant that accounts for atmospheric pressure and temperature.

John is in training for a marathon tomorrow and just ate a meal of pasta (assume this is pure carbohydrate, which is metabolized with an RQ of 1). His alveolar ventilation rate is 3.0 liters/min and he is consuming O_2 at a rate of 300 ml/min. What is the value of John's P_{ACO_2}?

2. Assume you are flying in an airplane that is cruising at 18,000 feet, where the pressure outside the plane is 380 mm Hg.
   a. Calculate the partial pressure of O_2 in the air outside the plane, ignoring water vapor pressure.
   b. If the plane depressurized, what would be the value of your P_{AO_2}? Assume that the ratio of your O_2 consumption to ventilation was not changed (that is, equaled 0.06) and note that under these conditions the constant in the equation that accounts for atmospheric pressure and temperature decreases from 863 mm Hg to 431.5 mm Hg.
   c. Calculate your P_{ACO_2}, assuming that your CO_2 production and ventilation rates remained unchanged at 200 ml/min and 4.2 liters/min, respectively.
3. A student has a tidal volume of 350 ml. While breathing at a rate of 12 breaths/min, her alveolar ventilation is 80% of her minute ventilation. What is her anatomical dead space volume?

Points to Ponder

(Explanations on p. E–13.)

1. Why is it important that airplane interiors are pressurized (that is, the pressure is maintained at sea level atmospheric pressure even though the atmospheric pressure surrounding the plane is substantially lower)? Explain the physiological value of using O_2 masks if the pressure in the airplane interior cannot be maintained.
2. Would hypercapnia accompany the hypoxia produced in each of the following situations? Explain why or why not.
   a. cyanide poisoning
   b. pulmonary edema
   c. restrictive lung disease
   d. high altitude
   e. severe anemia
   f. congestive heart failure
   g. obstructive lung disease
3. If a person lives 1 mile above sea level at Denver, Colorado, where the atmospheric pressure is 630 mm Hg, what would the P_{O_2} of the inspired air be once it is humidified in the respiratory airways before it reaches the alveoli?
4. Based on what you know about the control of respiration, explain why it is dangerous to voluntarily hyperventilate to lower the arterial P_{CO_2} before going underwater. The purpose of the hyperventilation is to stay under longer before P_{CO_2} rises above normal and drives the swimmer to surface for a breath of air.
5. If a person whose alveolar membranes are thickened by disease has an alveolar P_{O_2} of 100 mm Hg and an alveolar P_{CO_2} of 40 mm Hg, which of the following values of systemic arterial blood gases are most likely to exist?
   a. P_{O_2} = 105 mm Hg P_{CO_2} = 35 mm Hg
   b. P_{O_2} = 100 mm Hg P_{CO_2} = 40 mm Hg
   c. P_{O_2} = 90 mm Hg P_{CO_2} = 45 mm Hg

   If the person is administered 100% O_2, will the arterial P_{O_2} increase, decrease, or remain the same? Will the arterial P_{CO_2} increase, decrease, or remain the same?
6. ***Clinical Consideration*** Keith M., a former heavy cigarette smoker, has severe emphysema. What effect does this condition have on his airway resistance? How does this change in airway resistance influence Keith's inspiratory and expiratory efforts? Describe how his respiratory muscle activity and intra-alveolar pressure changes compare to normal to accomplish a normal tidal volume. How would his spirogram compare to normal? What influence would Keith's condition have on gas exchange in his lungs? What blood gas abnormalities are likely to be present? Would it be appropriate to administer O_2 to Keith to relieve his hypoxic condition?

14

The Urinary System

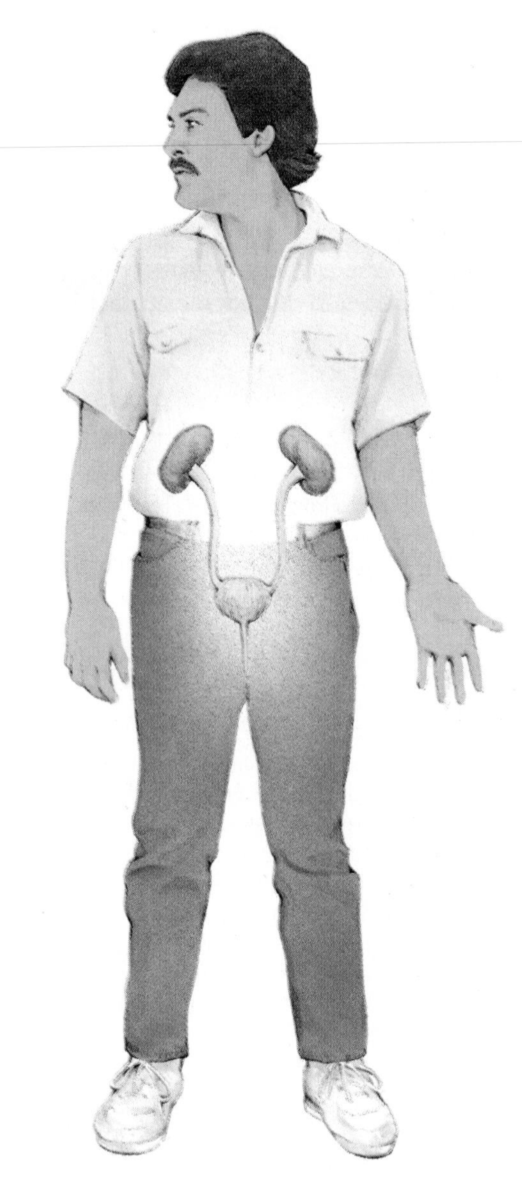

Body systems maintain homeostasis

HOMEOSTASIS
The urinary system contributes to homeostasis by helping regulate the volume, electrolyte composition and pH of the internal environment and by eliminating metabolic waste products.

Homeostasis is essential for survival of cells

CELLS
The concentrations of salt, acids, and other electrolytes must be closely regulated because even small changes can have a profound impact on cell function. Also, the waste products that are continually generated by cells as they perform life-sustaining chemical reactions must be removed because these wastes are toxic if allowed to accumulate.

Cells make up body systems

The survival and proper functioning of cells depend on the maintenance of stable concentrations of salt, acids, and other electrolytes in the internal fluid environment. Cell survival also depends on the continual removal of toxic metabolic wastes produced by the cells as they perform life-sustaining chemical reactions. The **kidneys** play a major role in maintaining homeostasis by regulating the concentration of many of the plasma constituents, especially the electrolytes and water, and by eliminating all the metabolic wastes (except CO_2, which is removed by the lungs). As plasma repeatedly filters through the kidneys, they retain constituents of value for the body and eliminate the undesirable or excess materials in the **urine.** Of special importance is the kidneys' ability to regulate the volume and osmolarity (solute concentration) of the internal fluid environment by controlling salt and water balance. Also critical is their ability to help regulate pH by controlling elimination of acid and base in the urine.

Chapter Contents At a Glance

Introduction

The kidneys perform a variety of functions aimed at maintaining homeostasis.

There would be no animal life-forms on dry land today if it were not for the development of kidneys (or comparable organs). The simplest forms of life live in an external environment of fixed composition, the sea. Likewise, the individual cells of more complex multicellular organisms are able to function and survive only in a fluid environment of essentially constant composition similar to the sea. This salty internal fluid environment is the extracellular fluid (ECF) that bathes all the cells of the body and must be homeostatically maintained.

Simple marine organisms have no influence over the fluid environment in which they live. Exchanges between a simple life-form and its aqueous environment do not alter the external environment's composition because the volume of the organism's body is so small compared to the tremendous volume of the sea in which the organism lives. In contrast, in a complex land animal, exchanges between the cells and the ECF could notably alter the composition of this small, private internal fluid environment were it not for mechanisms that maintain its stability. To become separated from a watery environment of fixed composition and be free to move about in a dry and ever-changing external environment, land animals have internalized their own bit of sealike water and are equipped with mechanisms to maintain its constancy.

To a large extent, terrestrial animals are able to live on dry land independent of the sea because of their kidneys, the organs that, in concert with the hormonal and neural inputs that control their function, are primarily responsible for maintaining the stability of ECF volume and electrolyte composition. By adjusting the quantity of water and various plasma constituents that are either conserved for the body or eliminated in the urine, the kidneys are able to maintain water and electrolyte balance within the very narrow range compatible with life, despite wide variations in intake and losses of these constituents through other avenues.

When there is a surplus of water or a particular electrolyte such as salt (NaCl) in the ECF, the kidneys can eliminate the excess in the urine. If there is a deficit, the kidneys cannot actually provide additional quantities of the depleted constituent, but they can limit the urinary losses of the material in short supply and thus conserve it until more of the depleted substance can be ingested. Accordingly, the kidneys can compensate more efficiently for excesses than for deficits, as is further reflected by the fact that in some instances the kidneys cannot completely halt the loss of a particular valuable substance in the urine, even though the substance may be in short supply. A prime example is the case of a H_2O deficit. Even if a person is not consuming any H_2O, the kidneys are obligated to put out about half a liter of H_2O in the urine each day to accomplish another major role as the body's "cleaners."

In addition to the kidneys' important regulatory role in maintaining fluid and electrolyte balance, they are the primary route for elimination of potentially toxic metabolic wastes and foreign compounds from the body. These wastes cannot be eliminated in solid form; they must be excreted in solution, obligating the kidneys to produce a minimum volume of around 500 ml of waste-filled urine per day. Because the H_2O eliminated in the urine is derived from the blood plasma, a person stranded without H_2O is eventually obligated to urinate himself or herself to death by depleting the plasma volume to a fatal level as H_2O is inexorably removed to accompany the wastes. Fortunately, except under such extreme circumstances, the kidneys are able to maintain stability in the internal fluid environment despite the usual variations in intake of fluids and electrolytes.

Not only are the kidneys able to adjust for wide variations in ingestion of H_2O, salt, and other electrolytes, but they also make adjustments in the urinary output of these ECF constituents to compensate for their abnormal losses through heavy sweating, vomiting, diarrhea, or hemorrhage. Thus, urine composition varies widely as the kidneys adjust for differences in intake as well as losses of various substances in an attempt to maintain the ECF within the narrow limits compatible with life.

The following are the specific functions performed by the kidneys, most of which are directed toward preserving the constancy of the internal fluid environment:

1. *Maintaining H_2O balance in the body.*
2. *Regulating the quantity and concentration of most ECF ions,* including Na^+, Cl^-, K^+, HCO_3^-, Ca^{2+}, Mg^{2+}, SO_4^{2-}, PO_4^{3-}, and H^+. Even minor fluctuations in the ECF concentrations of some of these electrolytes can have profound influences. For example, changes in the ECF concentration of K^+ can potentially lead to fatal cardiac dysfunction.
3. *Maintaining proper plasma volume,* thereby contributing significantly to the long-term regulation of arterial blood pressure. This function is accomplished through the kidneys' regulatory role in salt and H_2O balance.
4. *Helping maintain the proper acid-base balance* of the body by adjusting urinary output of H^+ and HCO_3^-.
5. *Maintaining the proper osmolarity* (concentration of solutes) of body fluids, primarily through regulation of H_2O balance.
6. *Excreting (eliminating) the end products (wastes) of bodily metabolism* such as urea, uric acid, and creatinine. If allowed to accumulate, these wastes are toxic, especially to the brain.
7. *Excreting many foreign compounds* such as drugs, food additives, pesticides, and other exogenous nonnutritive materials that have gained entrance to the body.
8. *Secreting erythropoietin,* a hormone that stimulates red blood cell production (see p. 356).
9. *Secreting renin,* an enzymatic hormone that triggers a chain reaction important in the process of salt conservation by the kidneys.
10. *Converting vitamin D into its active form.*

The kidneys form the urine; the remainder of the urinary system is ductwork that carries the urine to the outside.

The **urinary system** consists of the urine-forming organs—the **kidneys**— and the structures that carry the urine from the kidneys to the outside for elimination from the body (Fig.

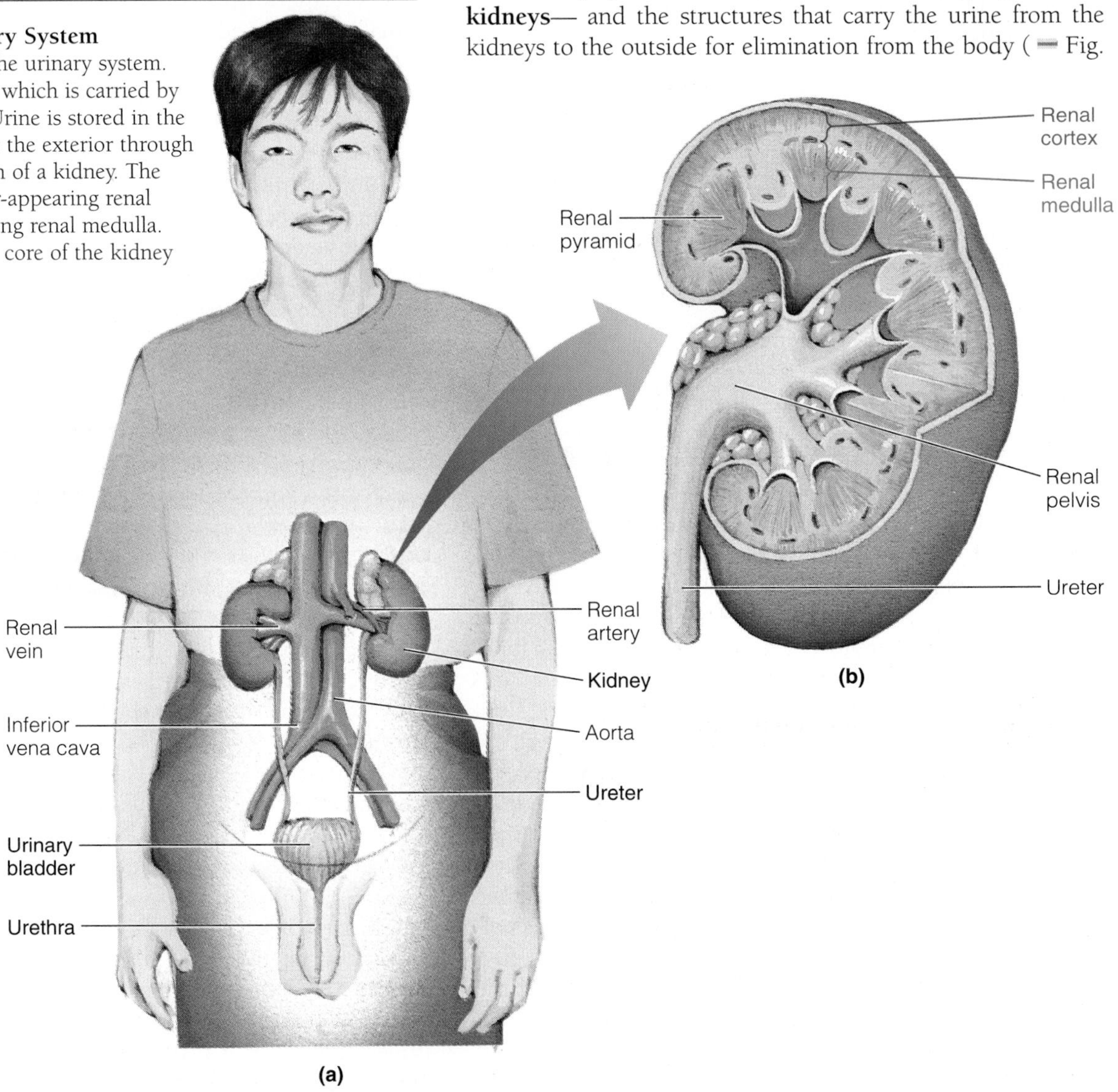

Figure 14-1 **The Urinary System**
(a) Location of the components of the urinary system. The pair of kidneys form the urine, which is carried by the ureters to the urinary bladder. Urine is stored in the bladder and periodically emptied to the exterior through the urethra. (b) Longitudinal section of a kidney. The kidney consists of an outer granular-appearing renal cortex and an inner striated-appearing renal medulla. The renal pelvis at the medial inner core of the kidney collects urine after it is formed.

14-1a). The kidneys are a pair of bean-shaped organs that lie in the back of the abdominal cavity, one on each side of the vertebral column slightly above the waistline. Each kidney is supplied by a **renal artery** and a **renal vein,** which, respectively, enter and leave the kidney at the medial indentation that gives this organ its beanlike form. The kidney acts on the plasma flowing through it to produce urine, conserving materials to be retained in the body and eliminating unwanted materials into the urine.

After urine is formed, it drains into a central collecting cavity, the **renal pelvis,** located at the medial inner core of each kidney (Fig. 14-1b). From there urine is channeled into the **ureter,** a smooth muscle–walled duct that exits at the medial border in close proximity to the renal artery and vein. There are two ureters, one carrying urine from each kidney to the single urinary bladder.

The **urinary bladder,** which temporarily stores urine, is a hollow, distensible sac whose volume can be adjusted by varying the contractile state of the smooth muscle within its walls. Periodically, urine is emptied from the bladder to the outside through another tube, the **urethra.** The urethra in females is straight and short, passing directly from the neck of the bladder to the outside (— Fig. 14-2; see also Fig. 20–2). In males the urethra is much longer and follows a curving course from the bladder to the outside, passing through both the prostate gland and the penis (Figs. 14-1a and 14-2b; see also Fig. 20–1). The male urethra serves the dual function of providing both a route for elimination of urine from the bladder and a passageway for semen from the reproductive organs. The prostate gland lies below the neck of the bladder and completely encircles the urethra. Prostatic enlargement, which often occurs during middle to older age, can partially or completely occlude the urethra, thereby impeding the flow of urine.

The parts of the urinary system beyond the kidneys merely serve as ductwork to transport urine to the outside. Once formed by the kidneys, urine is not altered in composition or volume as it moves downstream through the remainder of the urinary system.

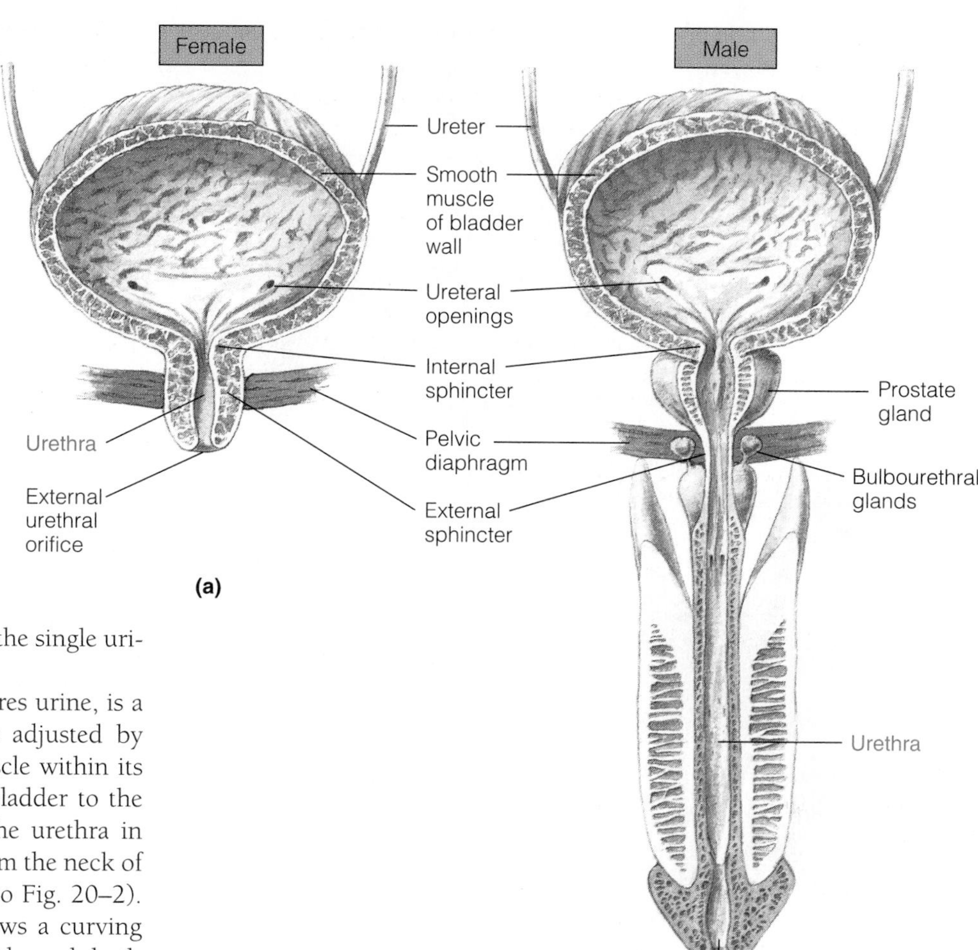

— *Figure 14-2* **Comparison of the Urethra in Females and Males** (a) In females, the urethra is straight and short. (b) In males, the much longer urethra passes through the prostate gland and penis.

The nephron is the functional unit of the kidney.

Each kidney is composed of about 1 million microscopic functional units known as **nephrons,** which are bound together by connective tissue. Recall that a functional unit is the smallest unit within an organ capable of performing all of that organ's functions. Because the primary function of the kidneys is to produce urine and, in so doing, maintain constancy in the ECF composition, a nephron is the smallest unit capable of urine formation.

The arrangement of nephrons within the kidneys gives rise to two distinct regions—an outer granular-appearing region, the **renal cortex,** and an inner region of striated-appearing triangles, the **renal pyramids,** which collectively compose the **renal medulla** (Fig. 14-1b).

Knowledge of the structural arrangement of an individual nephron is essential for understanding the distinction between the cortical and medullary regions of the kidney and, more importantly, for understanding renal function. Each nephron consists of a *vascular component* and a *tubular component,* both of which are intimately related structurally and functionally (— Fig. 14-3). The dominant portion of the vascular component is the **glomerulus,** a ball-like tuft of capillaries through which part of the water and solutes are filtered from the blood passing through. This filtered fluid, which is almost identical in composition to the plasma, then passes through the tubular component of the nephron, where it is modified by various transport processes that convert it into urine.

On entering the kidney, the renal artery systematically subdivides to ultimately form many small vessels known as **afferent arterioles,** one of which supplies each nephron. The afferent arteriole delivers blood to the glomerular capillaries, which rejoin to form another arteriole, the **efferent arteriole,** through which blood that was not filtered into the tubular

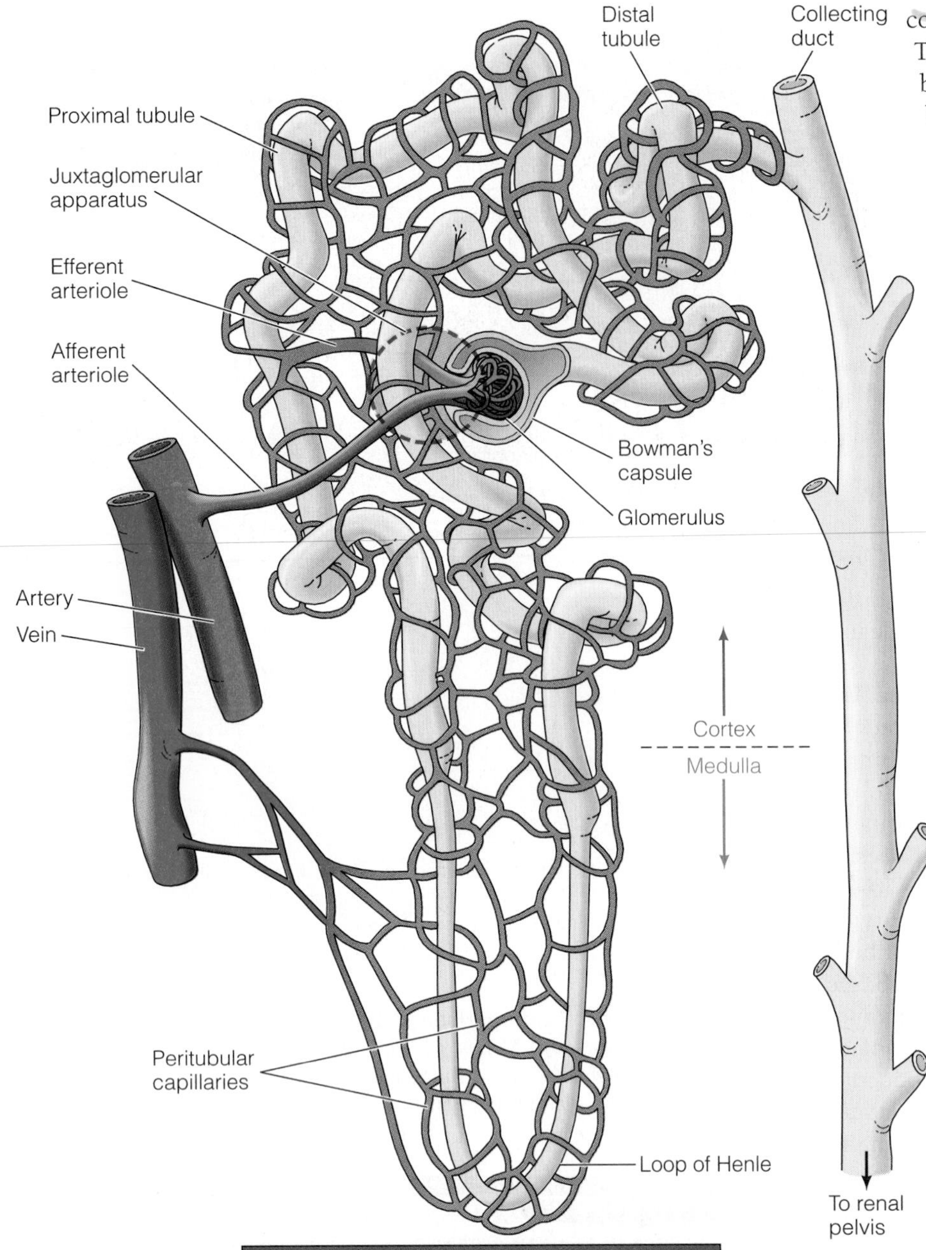

Overview of Functions of Parts of a Nephron

Vascular component

- **Afferent arteriole**—carries blood to the glomerulus
- **Glomerulus**—a tuft of capillaries that filters a protein-free plasma into the tubular component
- **Efferent arteriole**—carries blood from the glomerulus
- **Peritubular capillaries**—supply the renal tissue; involved in exchanges with the fluid in the tubular lumen

Combined vascular/tubular component

- **Juxtaglomerular apparatus**—secretes substances involved in the control of kidney function

Tubular component

- **Bowman's capsule**—collects the glomerular filtrate
- **Proximal tubule**—uncontrolled reabsorption and secretion of selected substances occur here
- **Loop of Henle**—establishes an osmotic gradient in the renal medulla that is important in the kidney's ability to produce urine of varying concentration
- **Distal tubule and collecting duct**—variable, controlled reabsorption of Na^+ and H_2O and secretion of K^+ and H^+occur here; fluid leaving the collecting duct is urine, which enters the renal pelvis

Figure 14-3 A Nephron

component leaves the glomerulus (Fig. 14-4). The efferent arterioles are the only arterioles in the body that drain from capillaries. Typically, arterioles break up into capillaries that rejoin to form venules. At the glomerular capillaries, no O_2 or nutrients are extracted from the blood for use by the kidney tissues nor are waste products picked up from the surrounding tissue. Thus, arterial blood enters the glomerular capillaries through the afferent arteriole, and arterial blood leaves the glomerulus through the efferent arteriole.

The efferent arteriole quickly subdivides into a second set of capillaries, the **peritubular capillaries**, which supply the renal tissue with blood and are important in exchanges between the tubular system and blood during conversion of the filtered fluid into urine. These peritubular capillaries, as their name implies (*peri* means "around"), are intertwined around the tubular system. The peritubular capillaries rejoin to form venules that ultimately drain into the renal vein, by which blood leaves the kidney.

The tubular component of each nephron is a hollow, fluid-filled tube formed by a single layer of epithelial cells. Even though the tubule is continuous from its beginning in close proximity to the glomerulus to its ending at the renal pelvis, it is arbitrarily divided into various segments based on differences in structure and function that occur along its length (Figs. 14-3 and 14-5). The tubular component begins with **Bowman's capsule**, an expanded, double-walled invagination that cups around the glomerulus to collect the fluid filtered from the glomerular capillaries. From Bowman's capsule, the filtered fluid passes into the **proximal tubule**, which lies entirely within the cortex and is highly coiled or convoluted throughout much of its course. The next segment, the **loop of Henle**, forms a sharp U-shaped or hairpin loop that dips into the renal medulla. The *descending limb* of Henle's loop plunges from the cortex into the medulla; the *ascending limb* traverses back up into the cortex. The ascending limb returns to the glomerular region of its own nephron, where it passes through the fork formed by the afferent and efferent arterioles. Both the tubular and vascular cells at this point are specialized to form the **juxtaglomerular apparatus** (*juxta* means "next to"), a structure that lies next to the glomerulus and plays an important role in regulating kidney function. Beyond the juxtaglomerular apparatus, the tubule once again becomes highly coiled to form the **distal tubule**, which also lies entirely within the cortex. The distal tubule empties into a **collecting duct** or **tubule**, with each collecting duct draining fluid from up to eight separate nephrons. Each collecting duct plunges down through the medulla to empty its fluid contents (which have now been converted into urine) into the renal pelvis.

There are two types of nephrons—**cortical nephrons** and **juxtamedullary nephrons**—distinguished by the location and length of some of their structures (Fig. 14-5). (Note the distinction between *juxtamedullary* nephrons and *juxtaglomerular* apparatus.) All nephrons originate in the cortex, but the glomeruli of cortical nephrons lie in the outer layer of the cortex, whereas the glomeruli of juxtamedullary nephrons lie in the inner layer of the cortex, adjacent to the medulla. The presence of all glomeruli and associated Bowman's capsules in the cortex is responsible for the region's granular appearance. These two nephron types differ most markedly in their loops of Henle. The hairpin loop of cortical nephrons dips only slightly into the medulla. In contrast, the loop of juxtamedullary nephrons plunges through the entire depth of the medulla.

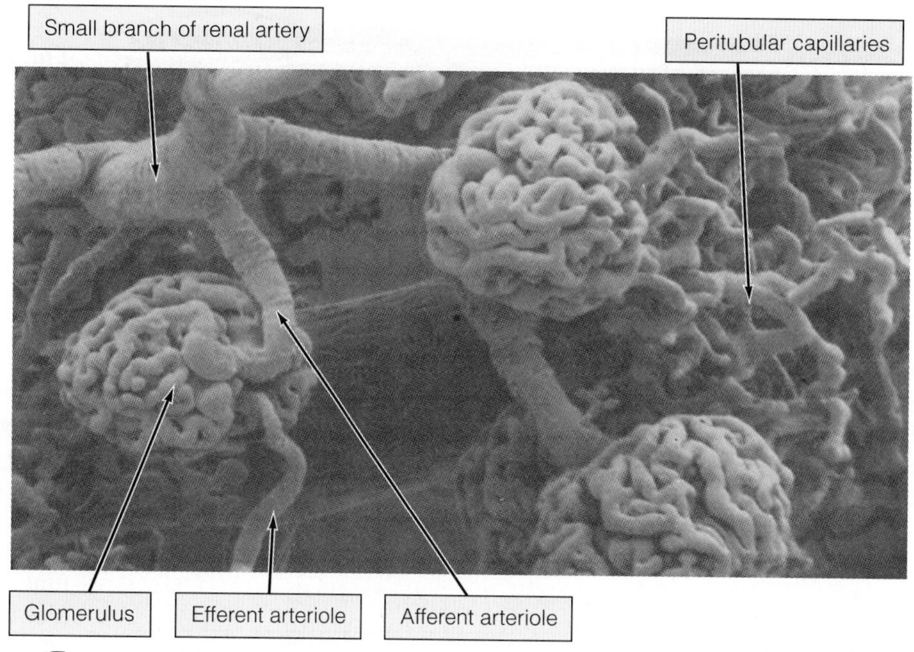

Figure 14-4 Scanning Electron Micrograph of a Glomerulus and Associated Arterioles [Source: From *Tissues and Organs: A Text-Atlas of Scanning Electron Microscopy* by Richard G. Kessel and Randy H. Kardon (New York: W. H. Freeman and Company, 1979). Copyright by and reprinted with permission of the authors.]

Juxtamedullary nephron: long-looped nephron important in establishing the medullary vertical osmotic gradient

Cortical nephron: most abundant type of nephron

Distal tubule
Glomerulus
Bowman's capsule
Proximal tubule
Distal tubule
Cortex
Medulla
Proximal tubule
Descending limb of loop of Henle
Collecting duct
Loop of Henle
Other nephrons emptying into the same collecting duct
Thick portion of ascending limb of loop of Henle
Vasa recta
Thin portion of ascending limb of loop of Henle
To renal pelvis
For better visualization, the nephrons are grossly exaggerated in size, and the peritubular capillaries have been omitted, except for the vasa recta.

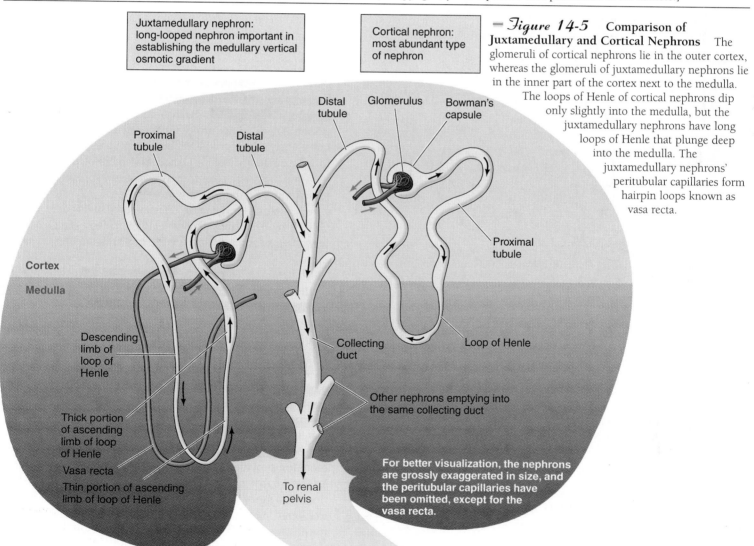

Figure 14-5 Comparison of Juxtamedullary and Cortical Nephrons The glomeruli of cortical nephrons lie in the outer cortex, whereas the glomeruli of juxtamedullary nephrons lie in the inner part of the cortex next to the medulla. The loops of Henle of cortical nephrons dip only slightly into the medulla, but the juxtamedullary nephrons have long loops of Henle that plunge deep into the medulla. The juxtamedullary nephrons' peritubular capillaries form hairpin loops known as vasa recta.

Furthermore, the peritubular capillaries of juxtamedullary nephrons form hairpin vascular loops known as **vasa recta** ("straight vessels"), which run in close association with the loops of Henle. As they course through the medulla, the collecting ducts of both cortical and juxtamedullary nephrons run parallel to the ascending and descending limbs of the juxtamedullary nephrons' loops of Henle and vasa recta. This parallel arrangement of tubules and vessels creates the medullary tissue's striated appearance. More importantly, as you will see, this arrangement, coupled with the permeability and transport characteristics of the long loops of Henle and vasa recta, plays a key role in the kidneys' ability to produce urine of varying concentrations, depending on the needs of the body. About 80% of the nephrons in humans are of the cortical type. Species with greater urine-concentrating abilities than humans, such as the desert rat, have a greater proportion of juxtamedullary nephrons.

Figure 14-6 **Basic Renal Processes** Anything filtered or secreted but not reabsorbed is excreted in the urine and lost from the body. Anything filtered and subsequently reabsorbed or not filtered at all enters the venous blood and is saved for the body.

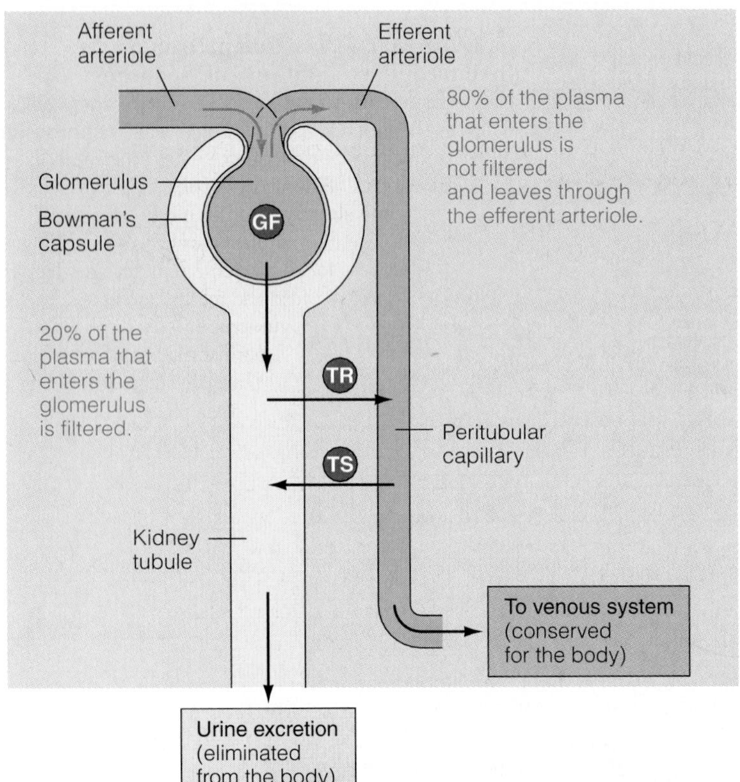

Afferent arteriole
Efferent arteriole
80% of the plasma that enters the glomerulus is not filtered and leaves through the efferent arteriole.
Glomerulus
Bowman's capsule
GF
20% of the plasma that enters the glomerulus is filtered.
TR
TS
Peritubular capillary
Kidney tubule
To venous system (conserved for the body)
Urine excretion (eliminated from the body)
GF = Glomerular filtration—nondiscriminant filtration of a protein-free plasma from the glomerulus into Bowman's capsule
TR = Tubular reabsorption—selective movement of filtered substances from the tubular lumen into the peritubular capillaries
TS = Tubular secretion—selective movement of nonfiltered substances from the peritubular capillaries into the tubular lumen

The three basic renal processes are glomerular filtration, tubular reabsorption, and tubular secretion.

Three basic processes are involved in the formation of urine: *glomerular filtration, tubular reabsorption,* and *tubular secretion.* To aid in visualizing the relationships among these renal processes, it is useful to unwind the nephron schematically, as in Figure 14-6.

As blood flows through the glomerulus, filtration of protein-free plasma occurs through the glomerular capillaries into Bowman's capsule. Normally, about 20% of the plasma that enters the glomerulus is filtered. This process, known as **glomerular filtration,** is the first step in urine formation. On the average, 180 liters (about 47.5 gallons) of glomerular filtrate (filtered fluid) are formed collectively through all the glomeruli each day. Considering that the average plasma volume in an adult is 2.75 liters, this means that the entire plasma volume is filtered by the kidneys about 65 times per day. If everything filtered were to pass out in the urine, the total plasma volume would be urinated in less than half an hour! This does not happen, however, because the kidney tubules and peritubular capillaries are intimately related throughout their lengths, so that materials can be transferred between the fluid inside the tubules and the blood within the peritubular capillaries.

As the filtrate flows through the tubules, substances of value to the body are returned to the peritubular capillary plasma. This selective movement of substances from inside the tubule (the tubular lumen) into the blood is referred to as **tubular reabsorption.** Reabsorbed substances are not lost from the body in the urine but are carried instead by the peritubular capillaries to the venous system and then to the heart to be recirculated again. Of the 180 liters of plasma filtered per day, 178.5 liters on the average are reabsorbed, with the remaining 1.5 liters passing into the renal pelvis to be eliminated as urine. In general, substances that need to be conserved by the body are selectively reabsorbed, whereas unwanted substances that need to be eliminated remain in the urine.

The third renal process, **tubular secretion,** which refers to the selective transfer of substances from the peritubular capillary blood into the tubular lumen, provides a second route for substances to enter the renal tubules from the blood. The first means by which substances move from the plasma into the tubular lumen is by glomerular filtration. However, only about 20% of the plasma flowing through the glomerular capillaries is filtered into Bowman's capsule; the remaining 80% flows on through the efferent arteriole into the peritubular capillaries. A few substances may be discriminately transferred by tubular secretion from the plasma in the peritubular capillaries into the tubular lumen. Tubular secretion provides a mechanism for more rapidly eliminating selected substances from the plasma by extracting an additional quantity of a particular substance from the 80% of unfiltered plasma in the peritubular capillaries and adding it to the quantity of the substance already present in the tubule as a result of filtration.

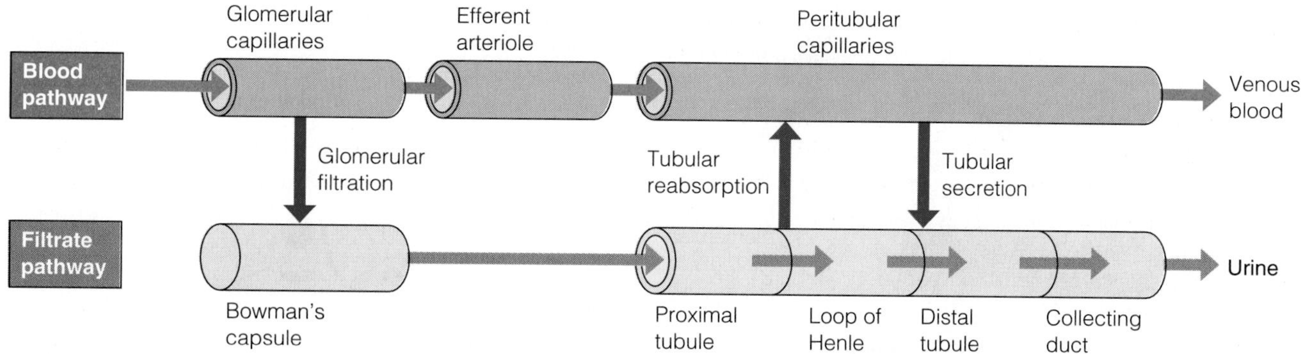

Figure 14-7 **Pathways Traveled by Blood and Filtrate as Urine is Formed in the Nephron**

Urine excretion refers to the elimination of substances from the body in the urine. It is not really a separate process but rather is the result of the first three processes. All plasma constituents that gain access to the tubules—that is, are filtered or secreted—but are not reabsorbed, remain in the tubules and pass into the renal pelvis to be excreted as urine and eliminated from the body (Figs. 14-6 and 14-7). (Do not confuse *excretion* with *secretion.*) Note that anything filtered and subsequently reabsorbed or not filtered at all enters the venous blood from the peritubular capillaries and thus is conserved for the body, despite passage through the kidneys.

Glomerular filtration is largely an indiscriminate process. With the exception of blood cells and plasma proteins, all constituents within the blood—H_2O, nutrients, electrolytes, wastes, and so on—are nonselectively filtered. The highly discriminating tubular processes then go to work on the filtrate to return to the blood a fluid of the composition and volume necessary to maintain the constancy of the internal fluid environment. The unwanted filtered material is left behind in the tubular fluid to be excreted as urine. Glomerular filtration can be thought of as pushing a portion of plasma, with all of its essential components as well as those that need to be eliminated from the body, onto a tubular "conveyor belt" that terminates at the renal pelvis, which is the collecting point for urine within the kidney. All plasma constituents that enter this conveyor belt and are not subsequently returned to the plasma by the end of the line are spilled out of the kidney as urine. It is up to the tubular system to salvage by reabsorption the filtered materials that need to be preserved for the body while leaving behind substances that need to be excreted. In addition, some substances are not only filtered but are also secreted onto the tubular conveyor belt, so that the amounts of these substances excreted in the urine are greater than the amounts that were filtered. For many substances, these renal processes are subject to physiological control. Thus, the kidneys handle each constituent in the plasma in a characteristic manner by a particular combination of filtration, reabsorption, and secretion.

The kidneys act only on the plasma, yet ECF consists of both plasma and interstitial fluid. The interstitial fluid is actually the true internal fluid environment of the body, because it is the only component of the ECF that comes into direct contact with the cells. However, because of the free exchange between plasma and interstitial fluid across the capillary walls (with the exception of plasma proteins), interstitial fluid composition reflects the composition of plasma. Thus, by performing their regulatory and excretory roles on the plasma, the kidneys maintain the proper interstitial fluid environment for optimal cell function. Most of the remainder of this chapter will be devoted to considering how the basic renal processes are accomplished and the mechanisms by which they are carefully regulated to help maintain homeostasis.

Glomerular Filtration

The glomerular membrane is more than 100 times more permeable than capillaries elsewhere.

Fluid filtered from the glomerulus into Bowman's capsule must pass through the three layers that make up the **glomerular membrane** (Fig. 14-8): (1) the wall of the glomerular capillaries, (2) an acellular gelatinous layer known as the basement membrane, and (3) the inner layer of Bowman's capsule. Collectively, these layers function as a fine molecular sieve that retains the blood cells and plasma proteins but permits H_2O and solutes of small molecular dimension to filter through. The glomerular capillary wall, which consists of a single layer of flattened endothelial cells, is perforated by many large pores, or *fenestrae,* that render it more than 100 times more permeable to H_2O and solutes than capillaries elsewhere in the body. A *basement membrane* composed of collagen and glycoproteins is sandwiched between the glomerulus and Bowman's capsule. The collagen provides structural strength, and the glycoproteins discourage the filtration of small plasma proteins. Even though the larger plasma proteins cannot be filtered because they cannot fit through the capillary pores, the pores are actually large enough to permit passage of albumin, the smallest of plasma proteins. The glycoproteins, however, being strongly negatively charged, repel albumin and other plasma proteins because the latter are also negatively charged. Therefore, plasma proteins are almost completely excluded from the filtrate, with less than 1% of the albumin molecules escaping into Bowman's capsule. Some renal diseases characterized by excessive albumin in the urine (albuminuria) are believed to be due to disruption of the negative charges within the glomerular membrane, which causes

Sieve

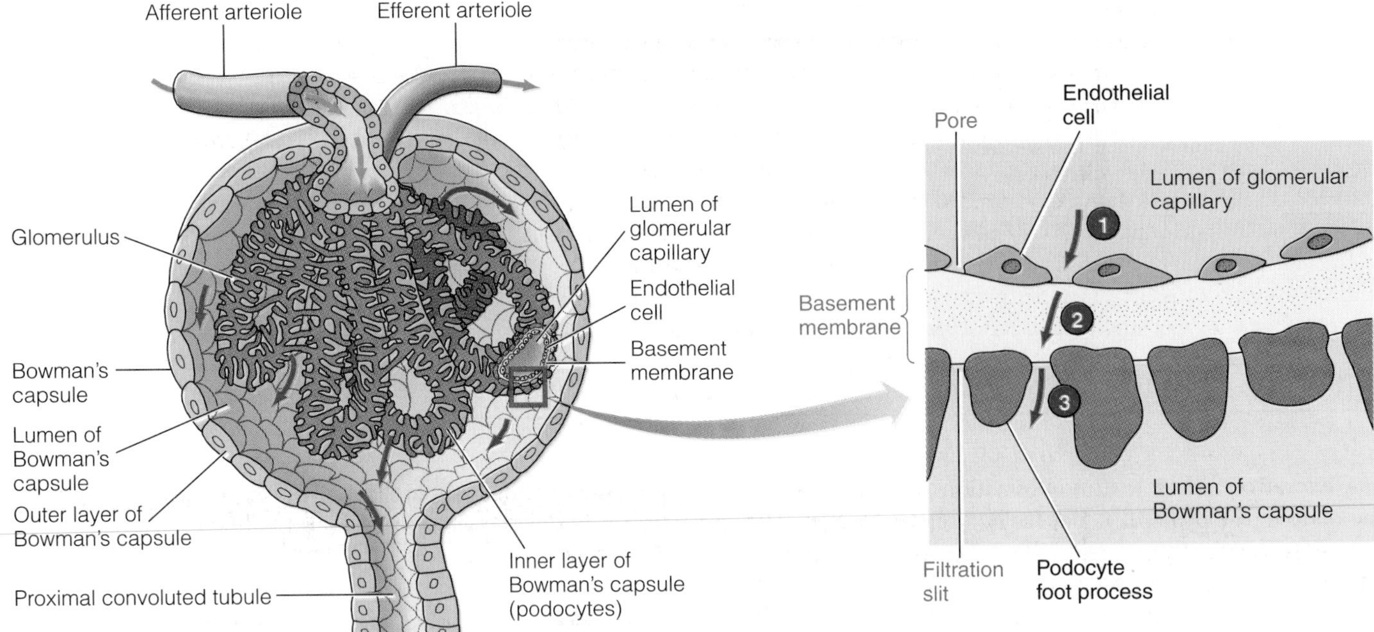

Figure 14-8 Layers of the Glomerular Membrane
To be filtered, a substance must pass through ① the pores between the endothelial cells of the glomerular capillary, ② an acellular basement membrane, and ③ the filtration slits between the foot processes of the podocytes of the inner layer of Bowman's capsule.

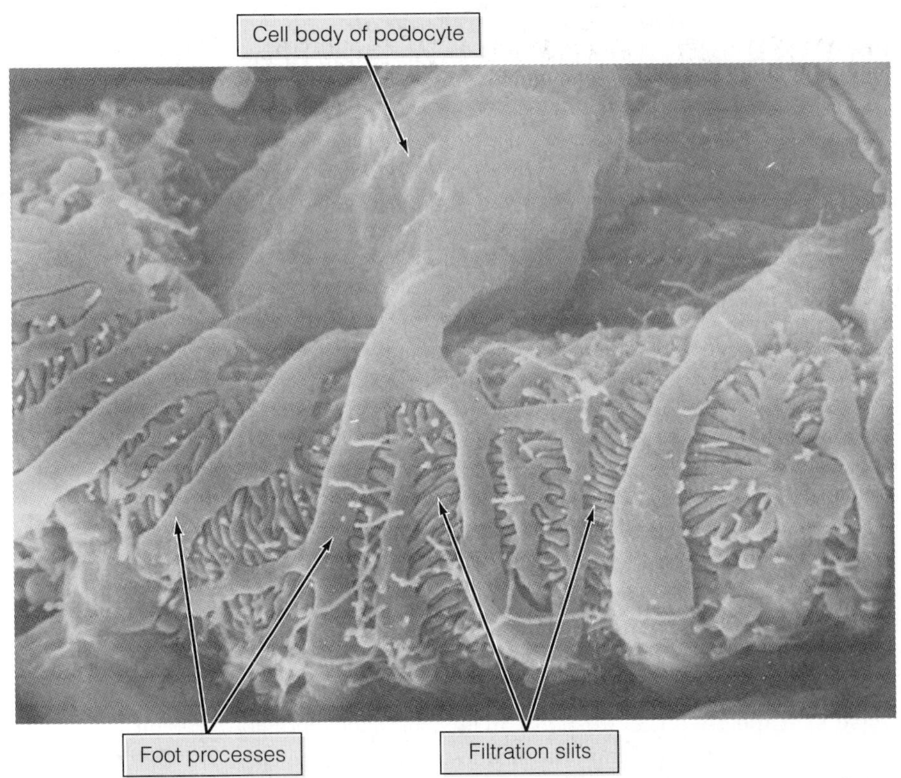

Figure 14-9 Bowman's Capsule Podocytes with Foot Processes and Filtration Slits Note the filtration slits between adjacent foot processes on this scanning electron micrograph. The podocytes and their foot processes encircle the glomerular capillaries. [Source: From *Tissues and Organs: A Text-Atlas of Scanning Electron Microscopy* by Richard G. Kessel and Randy H. Kardon (New York: W. H. Freeman and Company, 1979). Copyright by and reprinted with permission of the authors.]

the membrane to be more permeable to albumin even though the size of the pores remains constant.

The final layer of the glomerular membrane, which is the inner layer of Bowman's capsule, consists of **podocytes**, octopuslike cells that encircle the glomerular tuft. Each podocyte bears many elongated foot processes (*podo* means "foot") that interdigitate with foot processes of adjacent podocytes (Fig. 14-9), much as you slide your fingers between each other as you cup your hands around a ball. The narrow slits between adjacent foot processes, known as **filtration slits**, provide a pathway through which fluid exiting the glomerular capillaries can enter the lumen of Bowman's capsule. Thus, the route that filtered substances take across the glomerular membrane is completely extracellular—first through capillary pores, then through the acellular basement membrane, and finally through capsular filtration slits (Fig. 14-8).

The glomerular capillary blood pressure is the major force responsible for inducing glomerular filtration.

To accomplish glomerular filtration, a force must be present to drive a portion of the plasma in the glomerulus through the openings in the glomerular membrane. No active transport mechanisms or local energy expenditures are involved in moving fluid from the plasma across the glomerular membrane into Bowman's capsule. Passive physical forces similar to those acting across capillaries elsewhere are responsible for glomerular filtration. Because the glomerulus is a tuft of capillaries, the same principles of fluid dynamics that are responsible for ultrafiltration across other capillaries apply (see p. 327), except for two important differences: (1) the glomerular capillaries are much more permeable than capillaries elsewhere, so more fluid is filtered for a given filtration pressure, and (2) the balance of forces across the glomerular membrane is such that filtration occurs throughout the entire length of the capillaries. In contrast, the balance of forces in other capillaries shifts, so that filtration occurs in the beginning portion of the vessel but reabsorption occurs toward the vessel's end.

Three physical forces are involved in glomerular filtration (■ Table 14-1): (1) glomerular capillary blood pressure, (2) plasma-colloid osmotic pressure, and (3) Bowman's capsule hydrostatic pressure. The glomerular capillary blood pressure is the fluid pressure exerted by the blood within the glomerular capillaries. It ultimately depends on contraction of the heart (the source of energy that produces glomerular filtration) and the resistance to blood flow offered by the afferent and efferent arterioles. The glomerular capillary blood pressure, at an estimated average value of 55 mm Hg, is higher than capillary blood pressure elsewhere, because the diameter of the afferent arteriole is larger than that of the efferent arteriole. Since blood can more readily enter the glomerulus through the wide afferent arteriole than it can leave through the narrower efferent arteriole, glomerular capillary blood pressure is maintained high as a result of blood damming up in the glomerular capillaries. Furthermore, because of the high resistance offered by the efferent arterioles, blood pressure does not have the same tendency to decrease along the length of the glomerular capillaries as it does along other capillaries. This elevated, nondecremental glomerular blood pressure tends to push fluid out of the glomerulus into Bowman's capsule along the glomerular capillaries' entire length, and it is the major force responsible for producing glomerular filtration.

Whereas glomerular capillary blood pressure *favors* filtration, the two other forces acting across the glomerular membrane (plasma-colloid osmotic pressure and Bowman's capsule hydrostatic pressure) *oppose* filtration. Plasma-colloid osmotic pressure is caused by the unequal distribution of plasma proteins across the glomerular membrane. Because plasma proteins cannot be filtered, they are present in the glomerular capillaries but are absent in Bowman's capsule. Accordingly, the concentration of H_2O is higher in Bowman's capsule than in the glomerular capillaries. The resultant tendency for H_2O to move by osmosis down its own concentration gradient from Bowman's capsule into the glomerulus opposes glomerular filtration. This opposing osmotic force averages 30 mm Hg, which is slightly higher than across other capillaries. It is higher because considerably more H_2O is filtered out of the glomerular blood, so the concentration of plasma proteins is higher than elsewhere.

The fluid in Bowman's capsule exerts a hydrostatic (fluid) pressure that is estimated to be about 15 mm Hg. This pressure, which tends to push fluid out of Bowman's capsule, opposes the filtration of fluid from the glomerulus into Bowman's capsule.

As can be seen in Table 14-1, there is an imbalance in the forces acting across the glomerular membrane. The total force favoring filtration is attributable to the glomerular capillary blood pressure at 55 mm Hg. The total of the two forces opposing filtration is 45 mm Hg. The net difference favoring filtration (10 mm Hg of pressure) is referred to as the **net filtration pressure.** This modest pressure is responsible for forcing large volumes of fluid from the blood through the highly permeable glomerular membrane.

The actual rate of filtration, the **glomerular filtration rate (GFR),** depends not only on the net filtration pressure but also on how much glomerular surface area is available for penetration and how permeable the glomerular membrane is (that is, how "holey" it is). These properties of the glomerular membrane are collectively referred to as the **filtration coefficient (K_f).** Accordingly:

$$GFR = K_f \times \text{net filtration pressure}$$

Normally, about 20% of the plasma that enters the glomerulus is filtered at the net filtration pressure of 10 mm Hg, producing collectively through all glomeruli 180 liters of glomerular filtrate each day for an average GFR of 125 ml/min in males and 160 liters of filtrate per day for an average GFR of 115 ml/min in females.

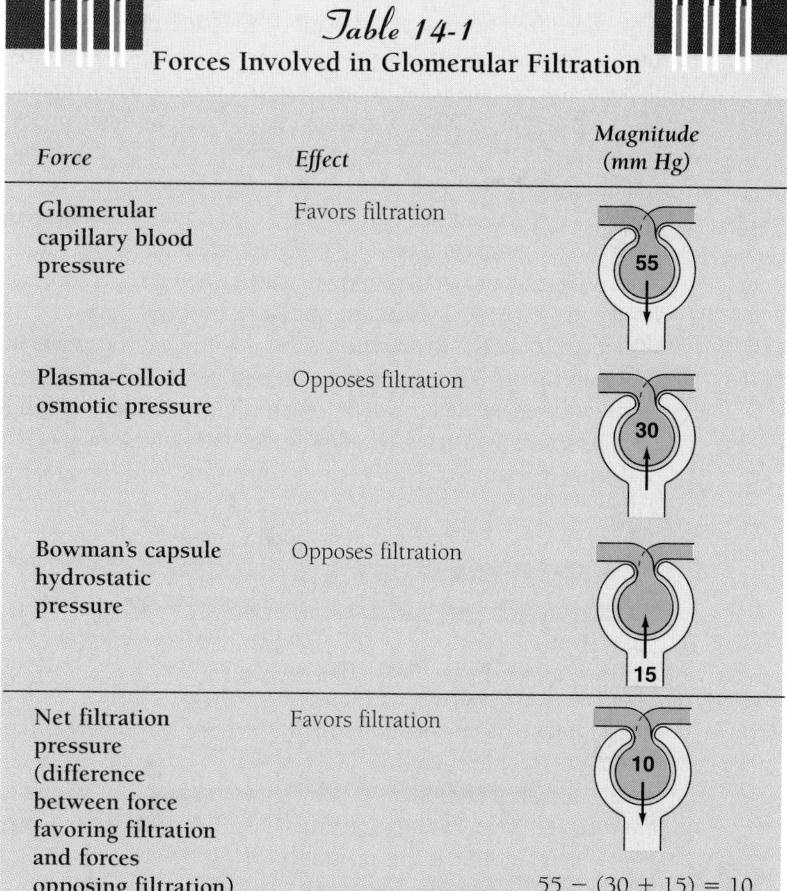

Table 14-1
Forces Involved in Glomerular Filtration

Force	Effect	Magnitude (mm Hg)
Glomerular capillary blood pressure	Favors filtration	55
Plasma-colloid osmotic pressure	Opposes filtration	30
Bowman's capsule hydrostatic pressure	Opposes filtration	15
Net filtration pressure (difference between force favoring filtration and forces opposing filtration)	Favors filtration	10 55 − (30 + 15) = 10

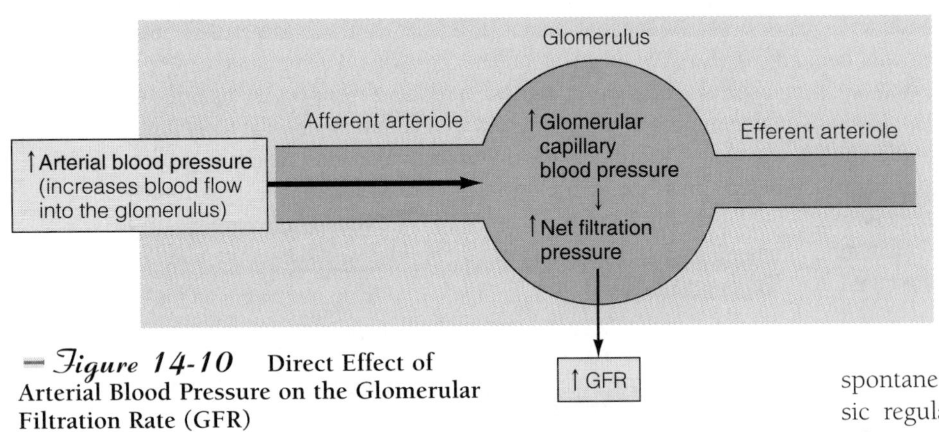

Figure 14-10 **Direct Effect of Arterial Blood Pressure on the Glomerular Filtration Rate (GFR)**

The most common factor resulting in a change in the GFR is an alteration in the glomerular capillary blood pressure.

Because the net filtration pressure responsible for inducing glomerular filtration is simply due to an imbalance of opposing physical forces between the glomerular capillary plasma and Bowman's capsule fluid, alterations in any of these physical forces can affect the GFR. We will examine the effect that changes in each of these physical forces have on the GFR.

Plasma-colloid osmotic pressure and Bowman's capsule hydrostatic pressure are not subject to regulation and, under normal conditions, do not vary substantially. However, they can change pathologically and thus inadvertently affect the GFR. Because plasma-colloid osmotic pressure opposes filtration, a decrease in plasma protein concentration, by reducing this pressure, leads to an increase in the GFR. An uncontrollable reduction in plasma protein concentration might occur, for example, in severely burned patients who lose a large quantity of protein-rich, plasma-derived fluid through the exposed burned surface of their skin. Conversely, in situations in which the plasma-colloid osmotic pressure is elevated, such as in cases of dehydrating diarrhea, the GFR is reduced. Bowman's capsule hydrostatic pressure can become uncontrollably elevated and filtration subsequently can decrease in the presence of a urinary tract obstruction, such as a kidney stone or prostatic enlargement. A damming up of fluid behind the obstruction elevates capsular hydrostatic pressure.

Unlike plasma-colloid osmotic pressure and Bowman's capsule hydrostatic pressure—which may be uncontrollably altered in various disease states, thereby inadvertently altering the GFR—glomerular capillary blood pressure can be controlled to adjust the GFR to suit the body's needs. Assuming that all other factors remain constant, the magnitude of the glomerular capillary blood pressure depends on the rate of blood flow in each of the glomeruli, which in turn is determined largely by the magnitude of the mean systemic arterial blood pressure and the resistance offered by the afferent arterioles. GFR is controlled by two mechanisms, both of which are directed at adjusting glomerular blood flow by regulating the caliber and thus the resistance of the afferent arteriole. These are (1) autoregulation, which is aimed at preventing spontaneous changes in GFR, and (2) extrinsic sympathetic control, which is aimed at long-term regulation of arterial blood pressure.

Autoregulation of GFR Because the arterial blood pressure is the force that drives blood into the glomerulus, the glomerular capillary blood pressure and, accordingly, the GFR would increase in direct proportion to an increase in arterial pressure if everything else remained constant (— Fig. 14-10). Similarly, a fall in arterial blood pressure would be accompanied by a decline in GFR. Such spontaneous changes in GFR are largely prevented by intrinsic regulatory mechanisms initiated by the kidneys themselves, a process known as **autoregulation** (*auto* means "self"). The kidneys can, within limits, maintain a constant blood flow in the glomerular capillaries (and thus a constant glomerular capillary blood pressure and a stable GFR) despite changes in the driving arterial pressure. They do so primarily by altering afferent arteriolar caliber, thereby adjusting resistance to flow through these vessels. For example, if the GFR increases as a direct result of a rise in arterial pressure, the net filtration pressure and GFR can be reduced to normal by constriction of the afferent arteriole (— Fig. 14-11a), which

Figure 14-11 **Adjustments of Afferent Arteriole Caliber to Alter the GFR** (a) Arteriolar adjustment to reduce the GFR. (b) Arteriolar adjustment to increase the GFR.

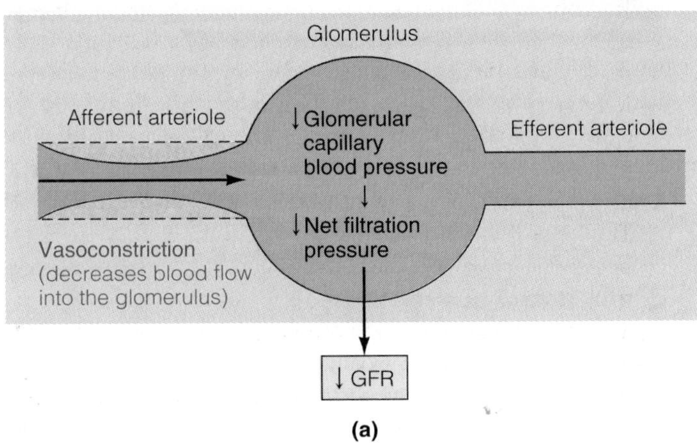

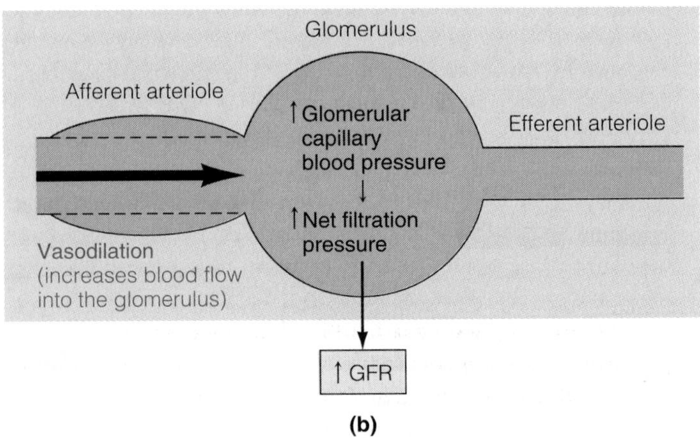

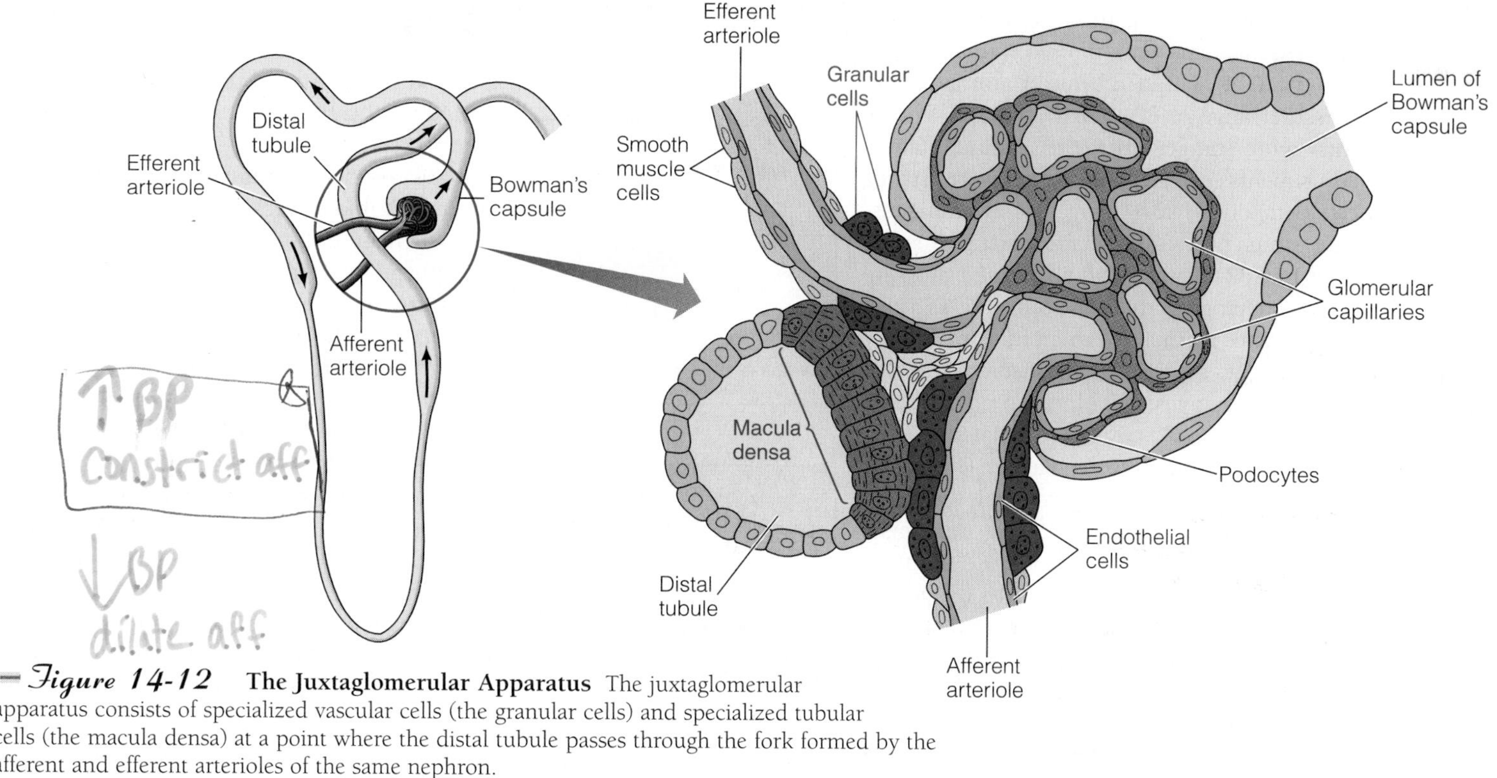

Figure 14-12 **The Juxtaglomerular Apparatus** The juxtaglomerular apparatus consists of specialized vascular cells (the granular cells) and specialized tubular cells (the macula densa) at a point where the distal tubule passes through the fork formed by the afferent and efferent arterioles of the same nephron.

decreases the flow of blood into the glomerulus. This local adjustment lowers the glomerular blood pressure and the GFR to normal. Conversely, when GFR falls in the presence of a decline in arterial pressure, glomerular pressure can be increased to normal by vasodilation of the afferent arteriole, which allows more blood to enter despite the reduction in driving pressure (Fig. 14-11b). The resultant buildup of glomerular blood volume increases glomerular blood pressure, which in turn brings the GFR back up to normal.

The exact mechanisms responsible for accomplishing autoregulatory responses still elude complete understanding. Currently, two intrarenal mechanisms are thought to contribute to autoregulation: (1) a myogenic mechanism, which responds to changes in pressure within the nephron's vascular component, and (2) a tubuloglomerular feedback mechanism, which senses changes in flow through the nephron's tubular component.

The **myogenic mechanism** is a common property of vascular smooth muscle. Arteriolar vascular smooth muscle contracts inherently in response to the stretch accompanying increased pressure within the vessel. Accordingly, the afferent arteriole automatically constricts on its own when it is stretched because of an increased arterial driving pressure. This response helps limit blood flow into the glomerulus to normal despite the elevated arterial pressure. Conversely, inherent relaxation of an unstretched afferent arteriole when pressure within the vessel is reduced increases blood flow into the glomerulus despite the fall in arterial pressure.

The **tubuloglomerular feedback mechanism** involves the *juxtaglomerular apparatus*, which is the specialized combination of tubular and vascular cells in the region of the nephron where the tubule, after having bent back on itself, passes through the angle formed by the afferent and efferent arterioles as they join the glomerulus (Figs. 14-3 and 14-12). Within the arteriolar walls at their point of contact with the tubule, the smooth muscle cells are specialized to form **granular cells,** so called because they contain numerous secretory granules. Specialized tubular cells in this region are collectively known as the **macula densa.** The macula densa cells detect changes in the rate at which fluid is flowing past them through the tubule. If the GFR is increased secondary to an elevation in arterial pressure, more fluid than normal is filtered and reaches the distal tubule (Fig. 14-13). In response, the macula densa cells trigger the release of vasoactive chemicals from the juxtaglomerular apparatus, which in turn cause the afferent arteriole to constrict, reducing glomerular blood flow and returning GFR to normal. The exact nature of these locally released vasoactive chemicals is currently unclear. Several chemicals have been identified, some of which are vasoconstrictors (for example, endothelin) and others vasodilators (for example, bradykinin), but their exact contributions are yet to be determined. In the opposite situation, when the macula densa cells detect that the flow rate of fluid through the tubules is low because of a spontaneous decline in GFR accompanying a fall in arterial pressure, these cells induce afferent arteriolar vasodilation by altering the rate of secretion of the relevant vasoactive chemicals. The resultant increase in glomerular flow rate restores the GFR to normal. Thus, by means of the juxtaglomerular apparatus, the tubule of a nephron is able to monitor the rate of movement of fluid through it and adjust the GFR accordingly. This tubuloglomerular feedback mechanism is initiated by the tubule to help each nephron regulate the rate of filtration through its own glomerulus.

The myogenic and tubuloglomerular feedback mechanisms work in unison to autoregulate the GFR within the mean arterial blood pressure range of 80 to 180 mm Hg. Within this wide range, intrinsic autoregulatory adjustments of afferent arteriolar resistance can compensate for changes in arterial pressure, thus preventing inappropriate fluctuations in GFR, even though glomerular pressure tends to change in the same direction as arterial pressure. Normal mean arterial pressure is 93 mm Hg, so this range encompasses the transient changes in blood pressure that accompany daily activities unrelated to the need for the kidneys to regulate H_2O and salt excretion, such as the normal elevation in blood pressure accompanying exercise. Autoregulation is important because unintentional shifts in GFR could lead to dangerous imbalances of fluid, electrolytes, and wastes. Since at least a certain portion of the filtered fluid is always excreted, the amount of fluid excreted in the urine is automatically increased as the GFR increases. If it were not for autoregulation, the GFR would increase and H_2O and solutes would be lost needlessly as a result of the rise in arterial pressure accompanying heavy exercise. On the other hand, if the GFR were too low, the kidneys would not be able to eliminate sufficient quantities of wastes, excess electrolytes, and other materials that should be excreted. Autoregulation thus greatly blunts the direct effect that changes in arterial pressure would otherwise have on GFR and subsequently on H_2O, solute, and waste excretion.

When changes in mean arterial pressure fall outside the autoregulatory range, these mechanisms are unable to compensate. Therefore, dramatic changes in mean arterial pressure (<80 mm Hg or >180 mm Hg) directly cause the glomerular capillary pressure and, accordingly, the GFR to decrease or increase in proportion to the change in arterial pressure.

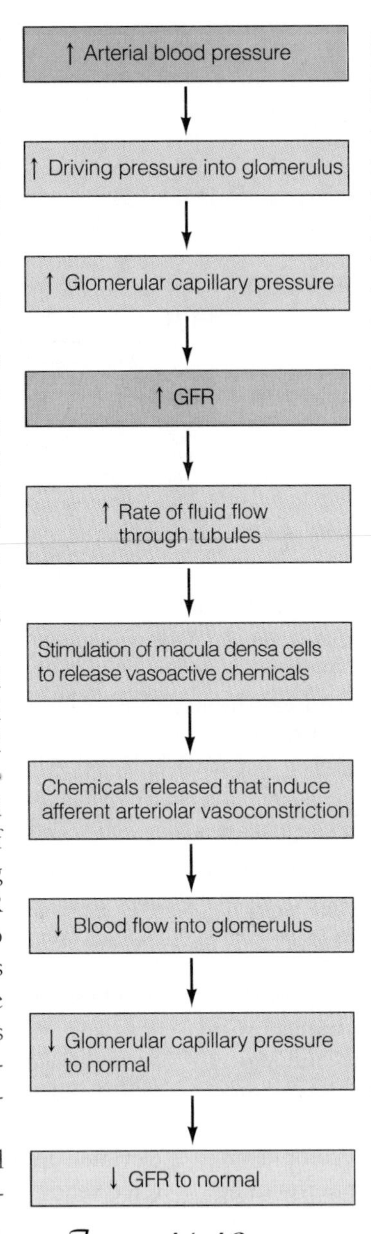

Figure 14-13
Tubuloglomerular Feedback Mechanism of Autoregulation

Extrinsic sympathetic control of the GFR In addition to the intrinsic autoregulatory mechanisms designed to keep the GFR constant in the face of fluctuations in arterial blood pressure, the GFR can be *changed on purpose*—even when the mean arterial blood pressure is within the autoregulatory range—by extrinsic-control mechanisms that override the autoregulatory responses. Extrinsic control of GFR, which is mediated by sympathetic nervous system input to the afferent arterioles, is aimed at the regulation of arterial blood pressure. The parasympathetic nervous system does not exert any influence on the kidneys.

If plasma volume is decreased—for example, because of hemorrhage—the resultant fall in arterial blood pressure is detected by the arterial carotid sinus and aortic arch baroreceptors (see p. 338), which initiate neural reflexes to increase blood pressure toward normal. These reflex responses are coordinated by the cardiovascular control center in the brain stem and are mediated primarily through increased sympathetic activity to the heart and blood vessels. Although the resultant increase in both cardiac output and total peripheral resistance helps to increase blood pressure toward normal, the plasma volume is still reduced. In the long term, the plasma volume must be restored to normal. One of the compensations for a depleted plasma volume is a reduction in urine output so that more fluid than normal is conserved for the body. This reduction in urine output is accomplished in part by a reduction in the GFR; if less fluid is filtered, less fluid is available to be excreted.

No new mechanism is required to decrease the GFR. It is reduced as a result of the baroreceptor reflex response to a fall in blood pressure (Fig. 14-14). During this reflex, sympathetically induced vasoconstriction occurs in the majority of the arterioles throughout the body as a compensatory mechanism to increase total peripheral resistance. Among the arterioles that constrict in response to the baroreceptor reflex are the afferent arterioles carrying blood to the glomeruli. The afferent arterioles are innervated with sympathetic vasoconstrictor fibers to a far greater extent than the efferent arterioles. When the afferent arterioles constrict as a result of increased sympathetic activity, less blood flows into the glomeruli than normal, causing the glomerular capillary blood pressure to fall (Fig. 14-11a). The resultant decrease in GFR in turn leads to a reduction in urine volume. In this way, some of the H_2O and salt that would have been lost in the urine are saved for the body, helping in the long term to restore the plasma volume to normal so that the short-term cardiovascular adjustments that have been made are no longer necessary. Other mechanisms, such as increased tubular reabsorption of H_2O and salt as well as increased thirst (described more thoroughly elsewhere), also contribute to the long-term maintenance of blood pressure, despite a loss of plasma volume, by helping to restore plasma volume.

Conversely, if blood pressure is elevated (for example, because of an expansion of plasma volume following ingestion of excessive fluid), the opposite responses occur. When the baroreceptors detect a rise in blood pressure, sympathetic vasoconstrictor activity to the arterioles, including the renal afferent arterioles, is reflexly reduced, allowing afferent arteriolar vasodilation to occur. As more blood enters the glomeruli through the dilated afferent arterioles, glomerular capillary blood pressure rises, increasing the GFR (Fig. 14-11b). As more fluid is filtered, more fluid is available to be eliminated in the urine. Contributing to the increase in urine volume is a hormonally adjusted reduction in the tubular reabsorption of H_2O and salt. By these two renal mechanisms—increased glomerular filtration and decreased tubular reabsorption of H_2O and salt—urine volume is increased and the excess fluid is eliminated from the body. A reduction in thirst and fluid intake also contributes to restoring an elevated blood pressure to normal.

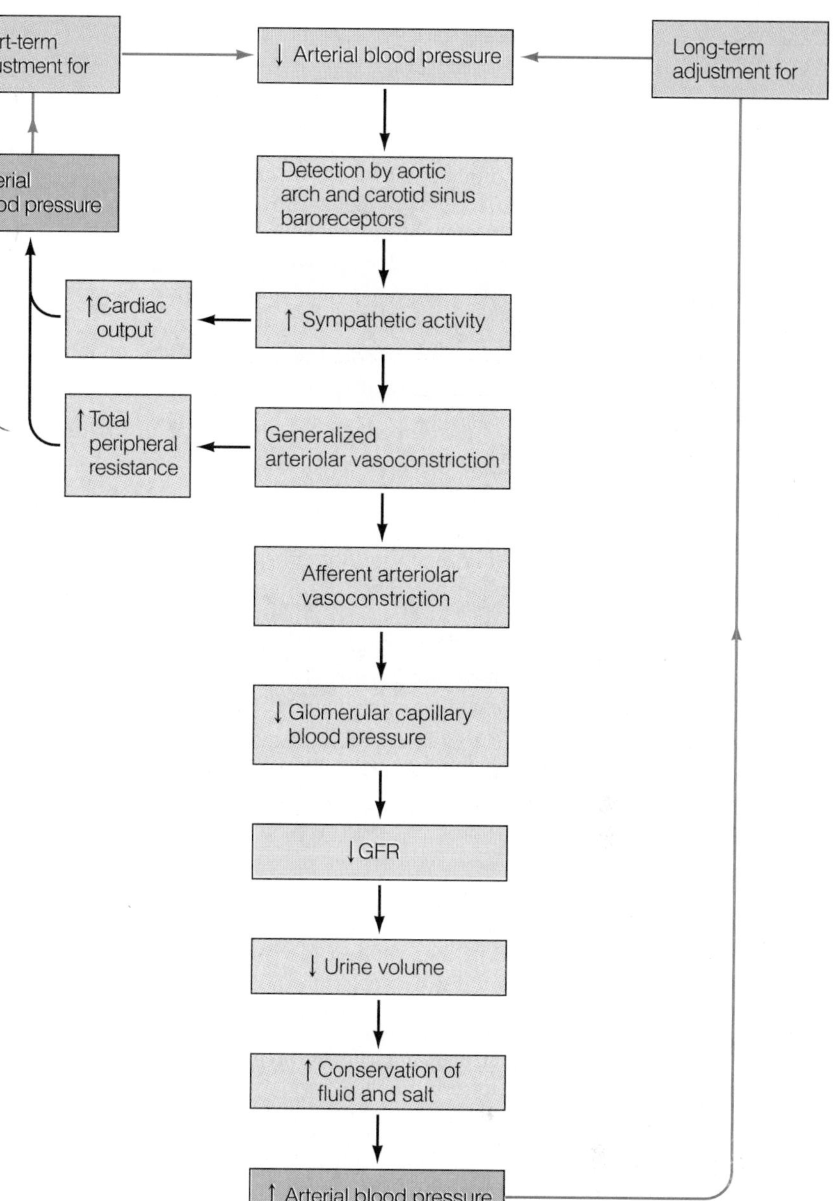

Figure 14-14 **Baroreceptor Reflex Influence on the GFR in Long-Term Regulation of Arterial Blood Pressure**

The GFR can be influenced by changes in the filtration coefficient.

Thus far we have discussed changes in the GFR as a result of changes in the net filtration pressure. The rate of glomerular filtration, however, is dependent on the filtration coefficient (K_f) as well as on the net filtration pressure. For years K_f was considered to be a constant, except in disease situations in which the glomerular membrane becomes leakier than usual. Exciting new research to the contrary indicates that K_f is subject to change under physiological control. Both factors on which K_f depends—the surface area and the permeability of the glomerular membrane—can be modified by contractile activity within the membrane.

The surface area available for filtration within the glomerulus is represented by the inner surface of the glomerular capillaries that comes into contact with blood. Each tuft of glomerular capillaries is held together by **mesangial cells.** These cells also function as phagocytes and contain contractile elements (that is, actinlike filaments). Contraction of these mesangial cells closes off a portion of the filtering capillaries, thereby reducing the surface area available for filtration within the glomerular tuft. When the net filtration pressure remains unchanged, this reduction in K_f leads to a decrease in GFR. Evidence suggests that sympathetic stimulation causes the mesangial cells to contract, thus providing a second mechanism (besides promoting afferent arteriolar vasoconstriction) by which sympathetic activity can decrease the GFR. In addition, several hormones and local chemical mediators suspected or known to be involved in the control of other renal mechanisms, such as tubuloglomerular feedback or tubular reabsorption, have been shown to influence mesangial cell contractile activity.

The podocytes also possess actinlike contractile filaments, whose contraction or relaxation can, respectively, decrease or increase the number of filtration slits open in the inner membrane of Bowman's capsule by changing the shapes and proximities of the foot processes (Fig. 14-15). The number of slits is a determinant of permeability; the more open slits, the greater the permeability. Contractile activity of the podocytes, which in turn affects permeability and the K_f, appears to be under physiological control.

As the full extent of various regulatory mechanisms is becoming understood, it is evident that control of glomerular filtration (both through adjustments in afferent arteriolar resistance and K_f) and control of tubular reabsorption are complexly interrelated.

The kidneys normally receive 20% to 25% of the cardiac output.

At the average net filtration pressure and K_f, 20% of the plasma that enters the kidneys is converted into glomerular filtrate. Thus, at an average GFR of 125 ml/min, the total renal plasma flow must average about 625 ml/min. Because 55% of whole blood consists of plasma (that is, hematocrit = 45), the total flow of blood through the kidneys averages 1,140 ml/min. This quantity is about 22% of the total cardiac output of 5 liters (5,000 ml)/min, although the kidneys compose less than 1% of total body weight.

The kidneys need to receive such a seemingly disproportionate share of the cardiac output because they must continuously perform their regulatory and excretory functions on the huge volumes of plasma delivered to them to maintain

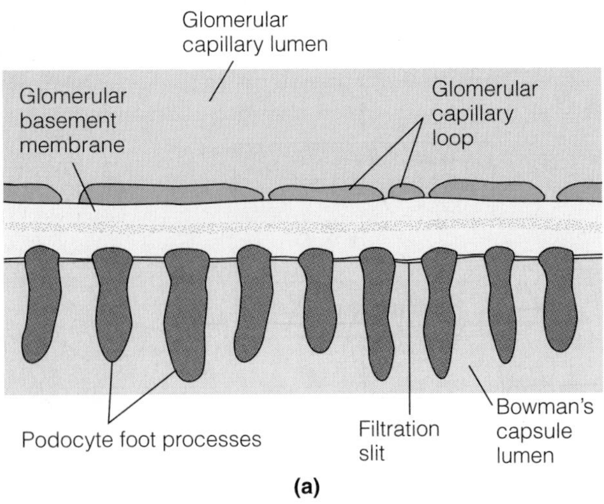

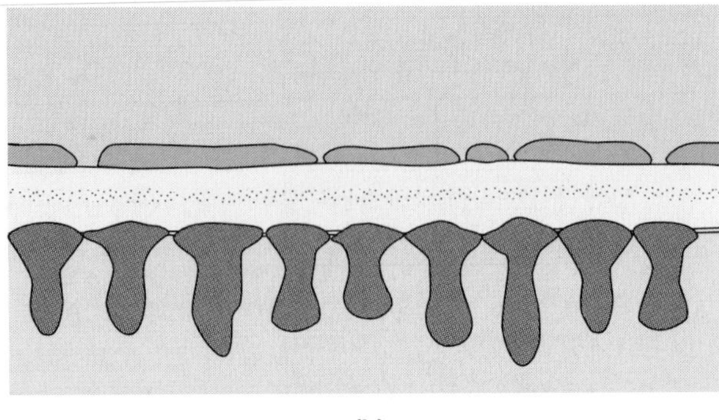

Figure 14-15 Change in the Number of Open Filtration Slits Caused by Podocyte Relaxation and Contraction (a) Podocyte relaxation causes the bases of the foot processes to narrow, thereby increasing the number of fully open intervening filtration slits spanning a given area. (b) Podocyte contraction causes the foot processes to flatten and thus decreases the number of intervening filtration slits.

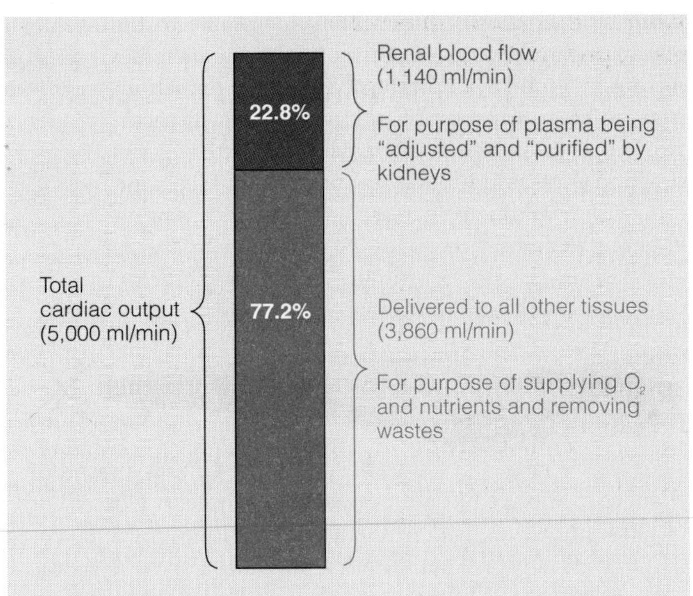

Figure 14-16 Percentage of the Cardiac Output Distributed to the Kidneys

stability in the internal fluid environment. Most of the blood goes to the kidneys not to supply the renal tissue but to be adjusted and purified by the kidneys. On the average, 20% to 25% of the blood pumped out by the heart each minute "goes to the cleaners" instead of serving its normal purpose of exchanging materials with the tissues (Fig. 14-16). Only by continuously processing such a large proportion of the blood are the kidneys able to precisely regulate the volume and electrolyte composition of the internal environment and adequately eliminate the large quantities of metabolic waste products that are constantly produced.

Tubular Reabsorption

Tubular reabsorption is tremendous, highly selective, and variable.

All plasma constituents except the proteins are nondiscriminantly filtered together through the glomerular capillaries. In addition to waste products and excess materials that need to be eliminated from the body, the filtered fluid also contains nutrients, electrolytes, and other substances that the body cannot afford to lose in the urine. Indeed, through the ongoing process of glomerular filtration, greater quantities of these materials are filtered per day than are even present in the entire body. It is important that the essential materials that are filtered be returned to the blood by the process of *tubular reabsorption,* the discrete transfer of substances from the tubular lumen into the peritubular capillaries.

Tubular reabsorption is a highly selective process. All constituents except plasma proteins are at the same concentration in the glomerular filtrate as in the plasma. In most cases, the quantity of each material that is reabsorbed is the amount required to maintain the proper composition and volume of the internal fluid environment. In general, the tubules have a high reabsorptive capacity for substances needed by the body and little or no reabsorptive capacity for substances of no value (Table 14-2). Accordingly, only a small percentage, if any, of filtered plasma constituents that are useful to the body are present in the urine, most having been reabsorbed and returned to the blood. Only excess amounts of essential materials such as electrolytes are excreted in the urine. For the essential plasma constituents regulated by the kidneys, absorptive capacity may vary depending on the body's needs. In contrast, a large percentage of filtered waste products is present in the urine. These wastes, which are useless or even potentially harmful to the body if allowed to accumulate, are not reabsorbed to any extent. Instead, they remain in the tubules to be eliminated in the urine. As H_2O and other valuable constituents are reabsorbed, the waste products remaining in the tubular fluid become highly concentrated.

Considering the magnitude of glomerular filtration, the extent of tubular reabsorption is tremendous: the tubules typically reabsorb 99% of the filtered H_2O (47 gallons/day), 100% of the filtered sugar (2.5 pounds/day), and 99.5% of the filtered salt (0.36 pounds/day).

Tubular reabsorption involves transepithelial transport.

Throughout its entire length, the tubule is one cell-layer thick and is in close proximity to a surrounding peritubular capillary (— Fig. 14-17). Adjacent tubular cells do not come into contact with each other except where they are joined by tight junctions (see p. 52) at their lateral edges near their *luminal membranes,* which face the tubular lumen. Interstitial fluid lies in the gaps between adjacent cells—the **lateral spaces**—as well as between the tubules and capillaries. The *basolateral membrane* faces the interstitial fluid at the base and lateral edges of the cell.

The tight junctions largely prevent substances except H_2O from moving *between* the cells, so materials must pass *through* the cells to leave the tubular lumen and gain entry to the blood. To be reabsorbed, a substance must traverse five distinct barriers (Fig. 14-17):

Step 1. It must leave the tubular fluid by crossing the luminal membrane of the tubular cell.

Step 2. It must pass through the cytosol from one side of the tubular cell to the other.

Step 3. It must traverse the basolateral membrane of the tubular cell to enter the interstitial fluid.

Step 4. It must diffuse through the interstitial fluid.

Step 5. It must penetrate the capillary wall to enter the blood plasma.

This entire sequence of steps is known as **transepithelial** ("across the epithelium") **transport.**

There are two types of tubular reabsorption—**passive reabsorption** and **active reabsorption**—depending on whether local energy expenditure is required for transfer of a particular substance. In passive reabsorption, *all* steps in the transepithelial transport of a substance from the tubular lumen to the plasma are passive; that is, no energy is expended for the substance's net movement, which occurs down electrochemical or osmotic gradients (see p. 55). On the other hand, a substance is said to be actively reabsorbed if any one of the steps in the sequence requires energy, even if the four other steps are passive. With active reabsorption, net movement of the substance from the tubular lumen to the plasma occurs against an electrochemical gradient. Substances that are actively reabsorbed are of particular importance to the body, such as glucose, amino acids, and other organic nutrients, as well as Na^+ and other electrolytes such as PO_4^{3-}. Rather than specifically describing the reabsorptive

Table 14-2

Fate of Various Substances Filtered by the Kidneys

Substance	*Average Percentage of Filtered Substance Reabsorbed*	*Average Percentage of Filtered Substance Excreted*
Water	99	1
Sodium	99.5	0.5
Glucose	100	0
Urea (*a waste product*)	50	50
Phenol (*a waste product*)	0	100

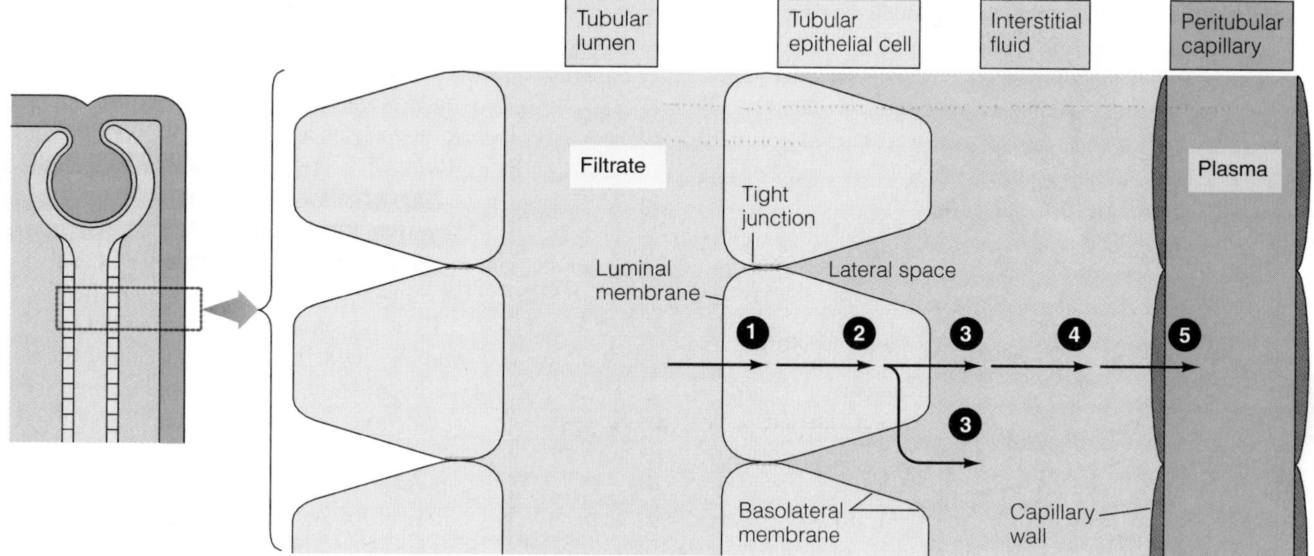

— *Figure 14-17* **Steps of Transepithelial Transport** To be reabsorbed (move from the filtrate to the plasma), a substance must traverse five distinct barriers: ① across the luminal cell membrane; ② through the cytosol; ③ across the basolateral cell membrane; ④ through the interstitial fluid; and ⑤ across the capillary wall.

process for each of the many filtered substances that are returned to the plasma, we will provide illustrative examples of the general mechanisms involved after first highlighting the unique and important case of Na^+ reabsorption.

An energy-dependent Na^+-K^+ ATPase transport mechanism in the basolateral membrane is essential for Na^+ reabsorption.

Sodium reabsorption is unique and complex. Eighty percent of the total energy requirement of the kidneys is used for Na^+ transport, indicative of the importance of this process. Unlike most filtered solutes, Na^+ is reabsorbed throughout the tubule, but to varying extents in different regions. Of the Na^+ filtered, 99.5% is normally reabsorbed, of which on average 67% is reabsorbed in the proximal tubule, 25% in the loop of Henle, and 8% in the distal and collecting tubules. Sodium reabsorption plays different important roles in each of these segments, as will become apparent as our discussion continues.

- Sodium reabsorption in the proximal tubule plays a pivotal role in the reabsorption of glucose, amino acids, H_2O, Cl^-, and urea.
- Sodium reabsorption in the loop of Henle, along with Cl^- reabsorption, plays a critical role in the kidneys' ability to produce urine of varying concentrations and volumes, depending on the body's need to conserve or eliminate H_2O.
- Sodium reabsorption in the distal portion of the nephron is variable and subject to hormonal control, being important in the regulation of ECF volume. It is also linked in part to K^+ secretion and H^+ secretion.

The active step in Na^+ reabsorption involves the energy-dependent Na^+-K^+ ATPase carrier located in the tubular cell's basolateral membrane (— Fig. 14-18). This carrier is the same one that is present in all cells and actively extrudes Na^+ from the cell (see p. 61). As this basolateral pump transports Na^+ out of the tubular cell into the lateral space, it keeps the intracellular Na^+ concentration low while it simultaneously builds up the concentration of Na^+ in the lateral space; that is, it moves Na^+ *against* a concentration gradient. Because the intracellular Na^+ concentration is kept low by basolateral-pump activity, a concentration gradient is established that favors the diffusion of Na^+ from its higher concentration in the tubular lumen across the luminal border through Na^+ channels into the tubular cell. Once within the cell, the Na^+ is actively extruded to the lateral space by the basolateral pump. Sodium continues to diffuse down a concentration gradient from its high concentration in the lateral space into the surrounding interstitial fluid and finally into the peritubular capillary blood. Thus, net transport of Na^+ from the tubular lumen into the blood occurs at the expense of energy.

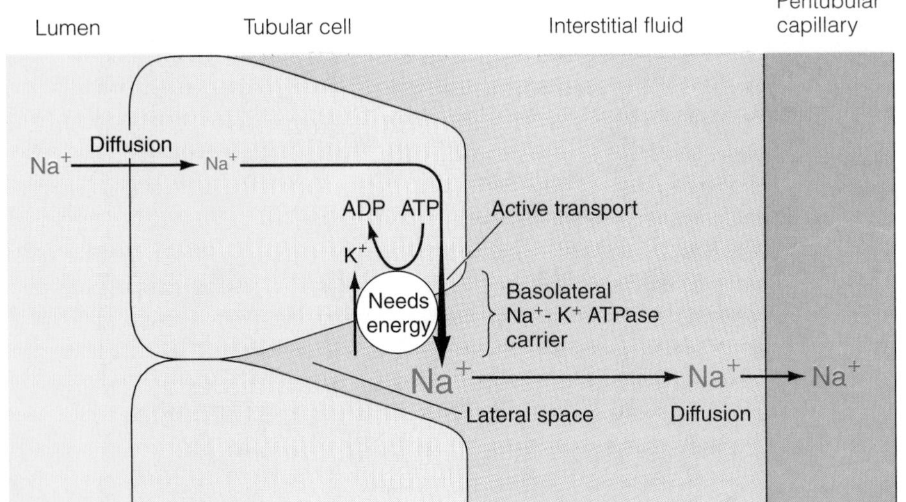

— *Figure 14-18* **Sodium Reabsorption** The basolateral Na^+-K^+ ATPase carrier actively transports Na^+ from the tubular cell into the interstitial fluid within the lateral space. This process establishes a concentration gradient for diffusion of Na^+ from the lumen into the tubular cell and from the lateral space into the peritubular capillary, accomplishing net transport of Na^+ from the tubular lumen into the blood at the expense of energy.

Aldosterone stimulates Na^+ reabsorption in the distal and collecting tubules; atrial natriuretic peptide inhibits it.

In the proximal tubule and loop of Henle, a constant percentage of the filtered Na^+ is reabsorbed regardless of the **Na^+ load** (*total amount* of Na^+ in the body fluids, *not the concentration* of Na^+ in the body fluids). The reabsorption of a small percentage of the filtered Na^+ is subject to hormonal control in the distal portion of the tubule. The extent of this controlled reabsorption is inversely related to the magnitude of the Na^+ load in the body. If there is too much Na^+, little of this controlled Na^+ is reabsorbed; instead, it is lost in the urine, thereby removing excess Na^+ from the body. On the other hand, if Na^+ is depleted, most or all of this controlled Na^+ is reabsorbed, conserving for the body Na^+ that otherwise would be lost in the urine. The most important and best known hormonal system involved in the regulation of Na^+ is the **renin-angiotensin-aldosterone system,** which stimulates Na^+ reabsorption in the distal and collecting tubules.

The Na^+ load in the body is reflected by the ECF volume. Sodium and its accompanying anion Cl^- account for more than 90% of the ECF's osmotic activity. (NaCl is common table salt.) Recall that osmotic pressure can be thought of loosely as a force that attracts and holds H_2O (see p. 57). When the Na^+ load is above normal and the ECF's osmotic activity is therefore increased, the extra Na^+ "holds" extra H_2O, expanding the ECF volume. Conversely, when the Na^+ load is below normal, thereby decreasing ECF osmotic activity, less H_2O than normal can be held in the ECF, so the ECF volume is reduced. Since plasma is a component of the ECF, the most important consequence of a change in ECF volume is the corresponding change in blood pressure accompanying expansion (↑ blood pressure) or reduction (↓ blood pressure) of the plasma volume.

The granular cells of the juxtaglomerular apparatus (Fig. 14-12) secrete a hormone, **renin,** into the blood in response to a fall in NaCl/ECF volume/blood pressure. This function is in addition to the role the juxtaglomerular apparatus plays in autoregulation, and renin is distinct from the local vasoactive chemicals that influence glomerular blood flow. These interrelated signals for increased renin secretion are all indicative of the necessity to expand the plasma volume to increase the arterial pressure to normal on a long-term basis. Through a complex series of events, increased renin secretion brings about increased Na^+ reabsorption by the distal portion of the tubule. Chloride always passively follows Na^+ down the electrical gradient established by sodium's active movement. The ultimate benefit of this salt retention is that it osmotically induces H_2O retention, which helps restore the plasma volume and blood pressure.

Let us examine the mechanism by which renin secretion ultimately leads to increased Na^+ reabsorption (Fig. 14-19). Once secreted into the blood, renin acts as an enzyme to activate **angiotensinogen** into **angiotensin I.** Angiotensinogen is a plasma protein synthesized by the liver and always present in the plasma in high concentration. On passing through the lungs via the pulmonary circulation, angiotensin I is converted into **angiotensin II** by **angiotensin-converting enzyme (ACE),** which is abundant in the pulmonary capillaries. Angiotensin II is the primary stimulus for the secretion of the hormone **aldosterone** from the adrenal cortex. The adrenal cortex is an endocrine gland that produces several different hormones, each of which is secreted in response to different stimuli.

Among its actions, aldosterone increases Na^+ reabsorption by the distal and collecting tubules. It does so by promoting

Figure 14-19 **Renin-Angiotensin-Aldosterone System** The kidneys secrete the hormone renin in response to a reduction in NaCl/ECF volume/arterial blood pressure. Renin activates angiotensinogen, a plasma protein produced by the liver, into angiotensin I. Angiotensin I is converted into angiotensin II by angiotensin-converting enzyme produced in the lungs. Angiotensin II stimulates the adrenal cortex to secrete the hormone aldosterone, which stimulates Na^+ reabsorption by the kidneys. The resultant retention of Na^+ exerts an osmotic effect that holds more H_2O in the ECF. Together the conserved Na^+ and H_2O help correct the original stimuli that activated this renin-angiotensin-aldosterone system. Angiotensin II also exerts other effects that help rectify the original stimuli.

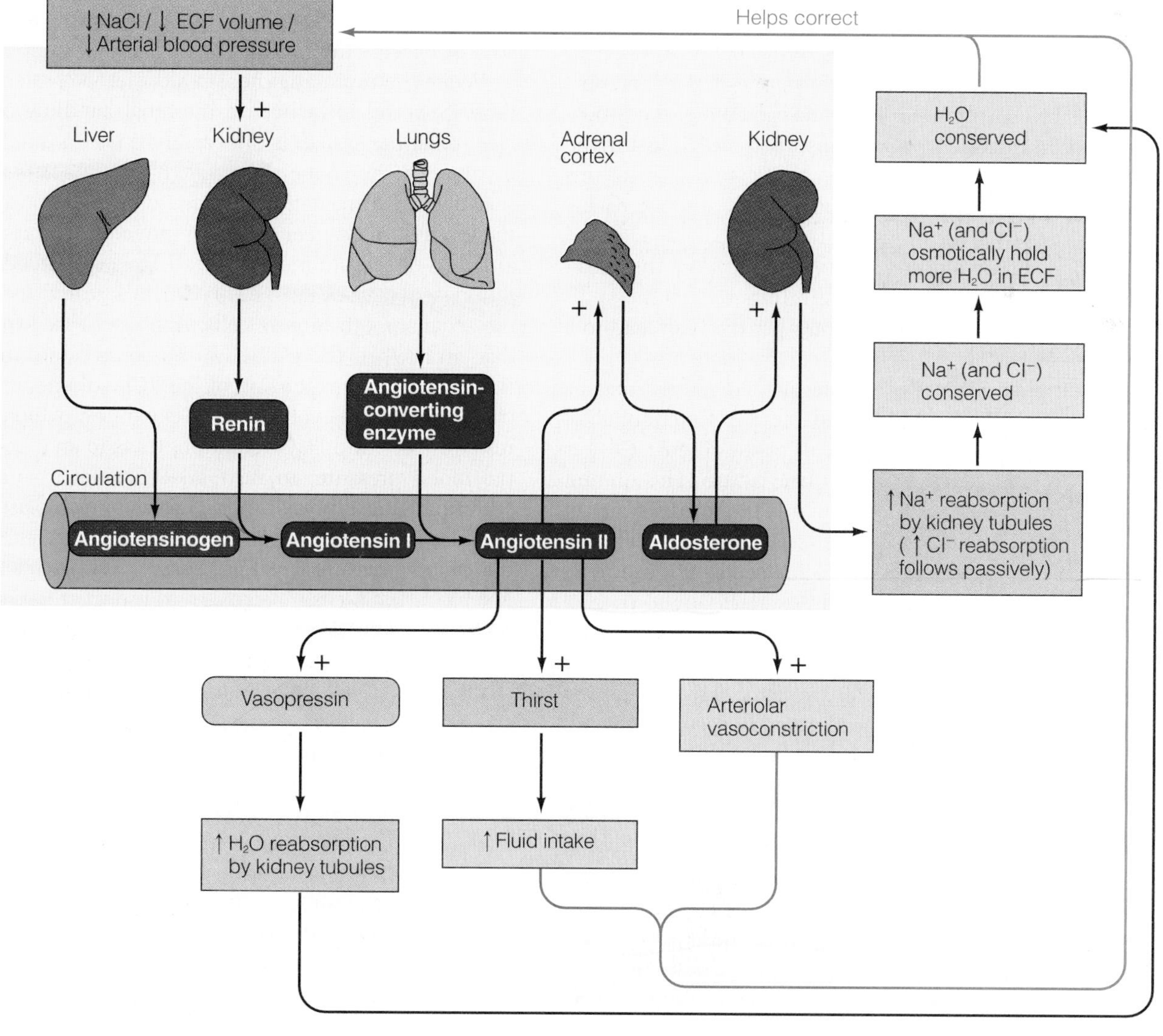

the insertion of additional Na^+ channels into the luminal membranes and additional Na^+-K^+ ATPase carriers into the basolateral membranes of the distal and collecting tubular cells. The net result is a greater passive inward flux of Na^+ into the tubular cells from the lumen and increased active pumping of Na^+ out of the cells into the plasma—that is, an increase in Na^+ reabsorption. Chloride (Cl^-) follows passively along the electrical gradient produced by active Na^+ reabsorption.

The renin-angiotensin-aldosterone system thus promotes salt retention and a resultant H_2O retention and elevation of arterial blood pressure. Acting in negative-feedback fashion, this system alleviates the factors that triggered the initial release of renin—namely, salt depletion, plasma-volume reduction, and decreased arterial blood pressure. In addition to stimulating aldosterone secretion, angiotensin II is also a potent constrictor of arterioles, thereby directly increasing blood pressure by increasing total peripheral resistance (see p. 320). Furthermore, it stimulates thirst (increasing fluid intake) and stimulates vasopressin (a hormone that increases H_2O retention by the kidneys), both of which contribute to plasma-volume expansion and elevation of arterial pressure.

The opposite situation exists when the Na^+ load, ECF and plasma volume, and arterial blood pressure are above normal. Under these circumstances, renin secretion is inhibited. Consequently, because angiotensinogen is not activated to angiotensin I and II, aldosterone secretion is not stimulated. Without aldosterone, the small aldosterone-dependent portion of Na^+ reabsorption in the distal segments of the tubule does not occur. Instead, this nonreabsorbed Na^+ is lost in the urine. In the absence of aldosterone, the ongoing loss of this small percentage of filtered Na^+ can rapidly remove excess Na^+ from the body. Even though only about 8% of the filtered Na^+ is dependent on aldosterone for reabsorption, this small loss, multiplied manyfold as the entire plasma volume is filtered through the kidneys many times per day, can lead to a sizeable loss of Na^+.

In the complete absence of aldosterone, 20 g of salt may be excreted per day. With maximum aldosterone secretion, all of the filtered Na^+ (and, accordingly, all of the filtered Cl^-) is reabsorbed, so salt excretion in the urine is zero. The amount of aldosterone secreted, and consequently the relative amount of salt conserved versus salt excreted, usually varies between these extremes, depending on the body's needs. For example, an "average" salt consumer typically excretes about 10 g of salt per day in the urine, a heavy salt consumer excretes more, and an individual who has lost considerable salt during heavy sweating has a lower urinary salt excretion. By varying the amount of renin and aldosterone secreted in accordance with the salt-determined fluid load in the body, the kidneys are able to finely adjust the amount of salt conserved or eliminated. In doing so, they maintain the salt load and ECF volume/arterial blood pressure at a relatively constant level in spite of wide variations in salt consumption and abnormal losses of salt-laden fluid.

It should not be surprising that some cases of hypertension (high blood pressure) are due to abnormal increases in renin-angiotensin-aldosterone activity. This system is also responsible in part for the fluid retention and edema accompanying congestive heart failure. Because of the failing heart, the cardiac output is reduced and arterial blood pressure is low in spite of a normal or even expanded plasma volume. When a fall in blood pressure is due to a failing heart rather than a reduction in the salt/fluid load in the body, the salt- and fluid-retaining reflexes triggered by the low blood pressure are inappropriate. Sodium excretion may fall to virtually zero despite continued salt ingestion and accumulation in the body. The resultant expansion of the ECF produces edema and exacerbates the congestive heart failure as the weakened heart is unable to pump the additional plasma volume.

Because their salt-retaining mechanisms are being inappropriately triggered by the reduction in blood pressure, patients with congestive heart failure are placed on low-salt diets. Often, they are treated with **diuretics,** which are therapeutic agents that cause *diuresis* (increased urinary output) and thus promote loss of fluid from the body. Many of these drugs function by inhibiting tubular reabsorption of Na^+. As more Na^+ is excreted, more H_2O is also lost from the body, thus helping to remove the excess ECF. **ACE inhibitor drugs** which block the action of angiotensin-converting enzyme (ACE), are also beneficial in the treatment of congestive heart failure as well as certain cases of hypertension. By blocking the generation of angiotensin II, ACE inhibitors halt the ultimate salt- and fluid-conserving actions and arteriolar constrictor effects of the renin-angiotensin-aldosterone system.

While the renin-angiotensin-aldosterone system exerts the most powerful influence on the renal handling of Na^+, this Na^+-retaining system is opposed by a Na^+-losing system that involves the hormone **atrial natriuretic peptide (ANP)** and several similar, recently identified natriuretic hormones from the brain. (*Natriuretic* means "inducing excretion of large amounts of sodium in the urine.") The heart, in addition to its pump action, produces ANP, which is stored in granules in the atrial smooth muscle cells. ANP is released from the cardiac atria when these smooth muscle cells are mechanically stretched due to expansion of the ECF volume, including the circulating plasma volume. This expansion, which occurs as a result of Na^+ and H_2O retention, leads to an increase in arterial blood pressure. In turn, ANP promotes natriuresis and accompanying diuresis, thus decreasing the plasma volume, and also exerts effects on the cardiovascular system to lower the blood pressure (— Fig. 14-20). The primary action of ANP is to inhibit Na^+ reabsorption in the distal parts of the nephron, thus increasing Na^+ excretion in the urine. ANP further increases Na^+ excretion in the urine by inhibiting two steps of the Na^+-conserving renin-angiotensin-aldosterone system. ANP inhibits renin secretion by the kidneys and acts on the adrenal cortex to inhibit aldosterone secretion. ANP also promotes natriuresis and accompanying diuresis by increasing the GFR through dilation of the afferent arterioles, leading to an increase in glomerular capillary blood pressure, and by causing relaxation of the glomerular mesangial cells, leading to an increase in K_f. As more salt and water are filtered, more salt and water are excreted in the urine. Besides its indirect effect in lowering blood pressure by reducing the Na^+ load and hence the fluid load in the body, ANP also directly lowers blood pressure by decreasing the cardiac output and reducing peripheral vascular resistance by means of inhibiting sympathetic nervous activity to the heart and blood vessels.

The relative contributions of ANP and possibly other salt-losing, blood pressure–lowering factors in the maintenance of salt and H_2O balance and blood pressure regulation are presently being intensely investigated. This subject is not purely of academic interest, because it is likely that derangements of this system will be found to contribute to hypertension. For example, a deficiency of a counterbalancing natriuretic system could theoretically cause long-term hypertension, since the powerful Na^+-conserving system would be unopposed. Even if the natriuretic factors are not found to contribute to the development of hypertension or congestive heart failure, they might prove to be useful in treating these circulatory disorders by reducing the Na^+ load in the body.

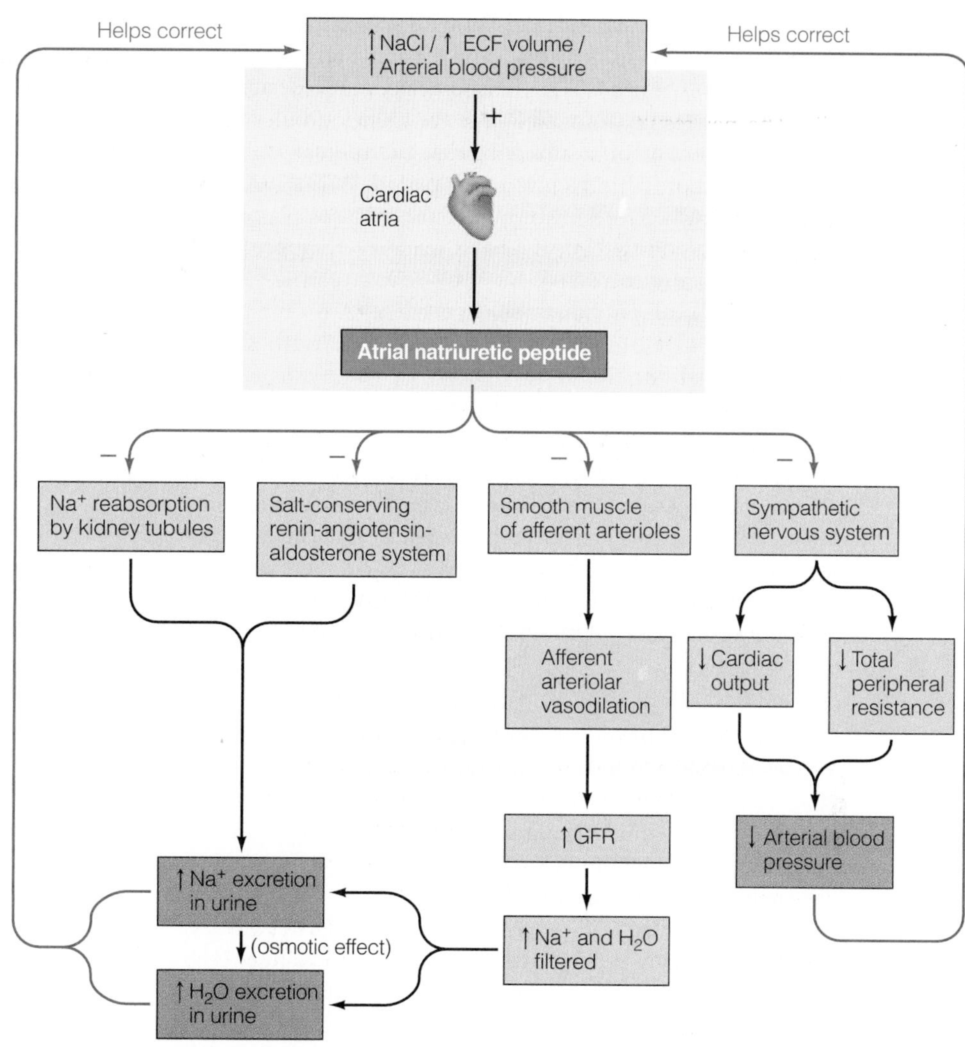

Figure 14-20 **Atrial Natriuretic Peptide** The atria secrete the hormone atrial natriuretic peptide in response to being stretched as a result of Na^+ retention, expansion of the ECF volume, and increase in arterial blood pressure. Atrial natriuretic peptide in turn promotes natriuretic, diuretic, and hypotensive effects to help correct the original stimuli that resulted in its release.

Glucose and amino acids are reabsorbed by Na^+-dependent secondary active transport.

Large quantities of nutritionally important organic molecules such as glucose and amino acids are filtered each day. Because these substances normally are completely reabsorbed back into the blood by energy- and Na^+-dependent mechanisms located in the proximal tubule, none of these materials are usually excreted in the urine. This rapid and thorough reabsorption early in the tubules protects against the loss of these important organic nutrients.

Even though glucose and amino acids are actively moved uphill against their concentration gradients from the tubular lumen into the blood until their concentration in the tubular fluid is virtually zero, no energy is directly used to operate the glucose or amino acid carriers. Glucose and amino acids are transferred by means of secondary active transport; a specialized cotransport carrier simultaneously transfers both Na^+ and the specific organic molecule from the lumen into the cell (see p. 62). The lumen-to-cell Na^+ concentration gradient maintained by the energy-consuming basolateral Na^+-K^+ ATPase pump drives this cotransport system and pulls the organic molecule against its concentration gradient without the direct expenditure of energy. Because the overall process of glucose and amino acid reabsorption depends on the use of energy, these organic molecules are considered to be actively reabsorbed, even though energy is not used directly to transport them across the membrane. In essence, glucose and amino acids get a "free ride" at the expense of energy already used in the reabsorption of Na^+. Secondary active transport requires the presence of Na^+ in the lumen; without Na^+ the cotransport carrier is inoperable. Once transported into the tubular cells, glucose and amino acids passively diffuse down their concentration gradients across the basolateral membrane into the plasma, facilitated by a non-energy-dependent carrier.

With the exception of Na^+, actively reabsorbed substances exhibit a tubular maximum.

No Tm for Na

All actively reabsorbed substances bind with plasma membrane carriers that transfer them across the membrane against a concentration gradient. Each carrier is specific for the types of substances it can transport; for example, the glucose cotransport carrier cannot transport amino acids, or vice versa. Since a limited number of each carrier type are present in the cells lining the tubules, there is an upper limit on the quantity of a particular substance that can be actively transported from the tubular fluid in a given period of time. The maximum reabsorption rate is reached when all the carriers specific for a particular substance are fully "occupied" or saturated (see p. 58), so that they cannot handle any additional passengers at that time. This transport maximum, designated as the **tubular maximum**, or **T_m**, in the kidney tubules is the maximum amount of a substance that the tubular cells can actively transport within a given time period. With the exception of Na^+, all actively reabsorbed substances display a T_m. (Sodium does not display a T_m because aldosterone promotes the synthesis of more active Na^+-K^+ ATPase carriers in the distal and collecting tubular cells as needed.) Any quantity of a substance filtered beyond its T_m fails to be reabsorbed, and escapes instead into the urine.

The plasma concentrations of some but not all substances that display carrier-limited reabsorption are regulated by the kidneys. How can the kidneys regulate some actively reabsorbed substances but not others, when the renal tubules limit the quantity of each of these substances that can be reabsorbed and returned to the plasma? We will compare glucose, a substance that has a T_m but is not regulated by the kidneys, with phosphate, a T_m-limited substance that is regulated by the kidneys.

Glucose reabsorption The normal plasma concentration of glucose is 100 mg of glucose/100 ml of plasma. Because glucose is freely filterable at the glomerulus, it passes into Bowman's capsule at the same concentration it has in the plasma. Accordingly, 100 mg of glucose are present in every 100 ml of plasma filtered. With 125 ml of plasma normally being filtered each minute (average GFR = 125 ml/min), 125 mg of glucose pass into Bowman's capsule with this filtrate every minute. The quantity of any substance filtered per minute, known as its **filtered load**, can be calculated as follows:

$$\text{filtered load of a substance} = \text{plasma concentration of the substance} \times \text{GFR}$$

$$\begin{aligned}\text{filtered load of glucose} &= 100 \text{ mg}/100 \text{ ml} \times 125 \text{ ml/min} \\ &= 125 \text{ mg/min}\end{aligned}$$

At a constant GFR, the filtered load of glucose is directly proportional to the plasma glucose concentration. Doubling the plasma glucose concentration to 200 mg/100 ml doubles the filtered load of glucose to 250 mg/min, and so on (Fig. 14-21).

The T_m for glucose averages 375 mg/min; that is, the glucose carrier mechanism is capable of actively reabsorbing up to 375 mg of glucose per minute before it reaches it maximum transport capacity. At a normal plasma glucose concentration of 100 mg/100 ml, the 125 mg of glucose filtered per minute can readily be reabsorbed by the glucose carrier mechanism, because the filtered load is well below the T_m for glucose. Ordinarily, therefore, no glucose appears in the urine because all of the filtered glucose is reabsorbed. Not until the filtered load of glucose exceeds 375 mg/min is the T_m reached. When more glucose is filtered per minute than can be reabsorbed because the T_m is exceeded, the maximum amount is reabsorbed, while the rest remains in the filtrate to be excreted. Accordingly, the plasma glucose concentration must be greater than 300 mg/100 ml—more than three times the normal value—before glucose starts spilling into the urine.

The plasma concentration at which the T_m of a particular substance is reached and the substance first starts appearing in the urine is known as the **renal threshold.** At the normal T_m of 375 mg/min and GFR of 125 ml/min, the renal threshold for glucose is 300 mg/100 ml. Beyond the T_m, reabsorption remains constant at its maximum rate, and any further increase in the filtered load is accompanied by a directly proportional increase in the amount of the substance excreted. For example, at a plasma glucose concentration of 400 mg/100 ml, the filtered load of glucose is 500 mg/min, 375 mg/min of which can be reabsorbed (a T_m's worth) and 125 mg/min of which are excreted in the urine. At a plasma glucose concentration of 500 mg/100 ml, the filtered load is 625 mg/min, still only 375 mg/min can be reabsorbed, and 250 mg/min spill into the urine (Fig. 14-21).

The plasma glucose concentration can become extremely high in diabetes mellitus, an endocrine disorder involving deficiency of insulin, a pancreatic hormone. This hormone is important in facilitating the transport of glucose into many of the body's cells. In insulin deficiency, the glucose that cannot gain entry into the cells remains in the plasma, elevating the plasma glucose concentration. Consequently, although glucose does not normally appear in the urine, it is found in the urine of persons with diabetes when the plasma glucose concentration exceeds the renal threshold, even though there has been no change in renal function.

What happens when the plasma glucose concentration falls below normal? The renal tubules, of course, reabsorb all of the filtered glucose, because the glucose reabsorptive capacity is far from being exceeded. The kidneys cannot do anything to raise a low plasma glucose level to normal. They simply return all of the filtered glucose to the plasma.

Thus, the kidneys do not influence the plasma glucose concentration over a wide range of values from abnormally low levels up to three times the normal level. Because the T_m for glucose is well above the normal filtered load, the kidneys usually conserve all the glucose, thereby protecting against the loss of this important nutrient in the urine. The kidneys do not regulate glucose because they do not maintain glucose at some specific plasma concentration; instead, this concentration is normally regulated by endocrine and liver mechanisms, with the kidneys merely maintaining whatever plasma glucose concentration is set by these other mechanisms (except when excessively high levels overwhelm the kidney's reabsorptive capacity). The same general principle holds true for other organic plasma nutrients, such as amino acids and water-soluble vitamins.

PO_4^{3-} Phosphate reabsorption The kidneys do directly contribute to the regulation of many electrolytes, such as phosphate (PO_4^{3-}) and calcium (Ca^{2+}), because the renal thresholds of these inorganic ions equal their normal plasma concentrations. We will use PO_4^{3-} as an example. Our diets are generally rich in PO_4^{3-}, but because the tubules can reabsorb up to the normal plasma concentration's worth of PO_4^{3-} and no more, the excess ingested PO_4^{3-} is quickly spilled into the urine, restoring the plasma concentration to normal. The greater the amount of PO_4^{3-} ingested beyond the body's needs, the greater the amount excreted. In this way, the kidneys maintain the desired plasma PO_4^{3-} concentration while eliminating any excess PO_4^{3-} ingested.

Unlike the reabsorption of organic nutrients, the reabsorption of PO_4^{3-} and Ca^{2+} is also subject to hormonal control. Parathyroid hormone can alter the renal thresholds for PO_4^{3-} and Ca^{2+}, thus adjusting the quantity of these electrolytes conserved, depending on the body's momentary needs (chapter 19).

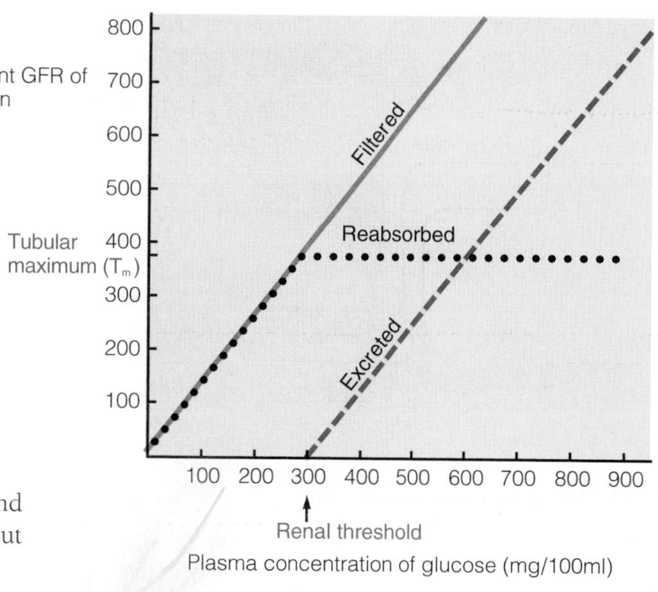

Plasma concentration of substance × GFR = Amount of substance filtered

For example, 100 mg glucose/100 ml plasma × 125 ml plasma filtered/min = 125 mg glucose filtered/min

Plasma Concentration (mg/100ml)	GFR (ml/min)	Amount Filtered (mg/min)	T_m (mg/min)	Amount Reabsorbed (mg/min)	Amount Excreted (mg/min)
a. 80	125	100	375	100	0
b. 100	125	125	375	125	0
c. 200	125	250	375	250	0
d. 300	125	375	375	375	0
e. 400	125	500	375	375	125
f. 500	125	625	375	375	250

***Figure 14-21* Renal Handling of Glucose as a Function of Plasma Glucose Concentration** At a constant GFR, the quantity of glucose filtered per minute is directly proportional to the plasma concentration of glucose. All of the filtered glucose can be reabsorbed up to the tubular maximum (T_m), the maximum amount of glucose the tubular cells can actively transport per minute. Urinary excretion of glucose does not occur until the amount of glucose filtered per minute exceeds the T_m. At that point, the maximum amount of glucose is reabsorbed (a T_m's worth), and the rest remains in the filtrate to be excreted in the urine. The renal threshold is the plasma concentration at which the T_m is reached and glucose first starts appearing in the urine.

Active Na^+ reabsorption is responsible for the passive reabsorption of Cl^-, H_2O, and urea.

Not only is the secondary active reabsorption of glucose and amino acids linked to the basolateral Na^+-K^+ pump, but the passive reabsorption of Cl^-, H_2O, and urea, also depends on this active Na^+ reabsorption mechanism.

Chloride reabsorption The negatively charged chloride ions are passively reabsorbed down the electrical gradient created by the active reabsorption of the positively charged sodium ions. The amount of Cl^- reabsorbed is determined by the rate of active Na^+ reabsorption instead of being directly controlled by the kidneys.

Water reabsorption Water is passively reabsorbed by osmosis throughout the length of the tubule. Of the H_2O filtered, 80% is obligatorily reabsorbed in the proximal tubules and loops of Henle, osmotically following solute reabsorption. This reabsorption occurs regardless of the H_2O load in the body and is not subject to regulation. Variable amounts of the remaining 20% are reabsorbed in the distal portions of the tubule; the extent of reabsorption in the distal and collecting tubules is subject to direct hormonal control, depending on the body's state of hydration. During the process of reabsorption, H_2O passes through recently discovered **water channels,** or **aquaporins,** formed by specific plasma membrane proteins in the tubular cells. Different types of water channels are present in various parts of the nephron. The water channels in the proximal tubule are always open, accounting for the high H_2O permeability of this region. The channels in the distal parts of the nephron, on the other hand, are regulated by the hormone *vasopressin,* accounting for the variable H_2O reabsorption in this region.

The driving force for H_2O reabsorption in the proximal tubule is a compartment of hypertonicity in the lateral spaces between the tubular cells that is established by the active extrusion of Na^+ by the basolateral pump (Fig. 14-22). As a result of this pump activity, the concentration of Na^+ rapidly diminishes in the tubular fluid and tubular cells while it simultaneously increases in the localized region within the lateral spaces. This osmotic gradient induces the passive net flow of H_2O from the lumen into the lateral spaces, either through the cells or intercellularly through "leaky" tight junctions. The accumulation of fluid in the lateral spaces results in a buildup of hydrostatic (fluid) pressure, which flushes H_2O out of the lateral spaces into the interstitial fluid and finally into the peritubular capillaries.

This return of filtered H_2O to the plasma is enhanced by the fact that the plasma-colloid osmotic pressure is greater in the peritubular capillaries than elsewhere. The concentration of plasma proteins, which is responsible for the plasma-colloid osmotic pressure, is elevated in the blood entering the peritubular capillaries because of the extensive filtration of

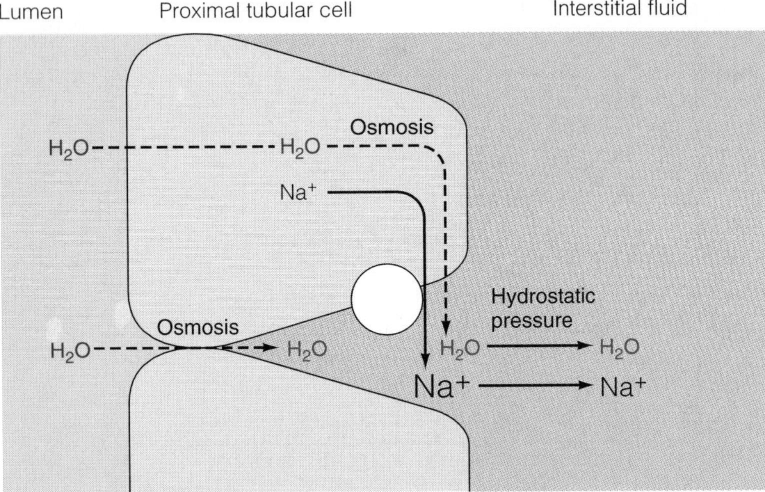

Figure 14-22 Water Reabsorption in the Proximal Tubule The force for H_2O reabsorption is the compartment of hypertonicity in the lateral spaces established by active extrusion of Na^+ by the basolateral pump. The dashed arrows show the direction of osmotic movement of H_2O.

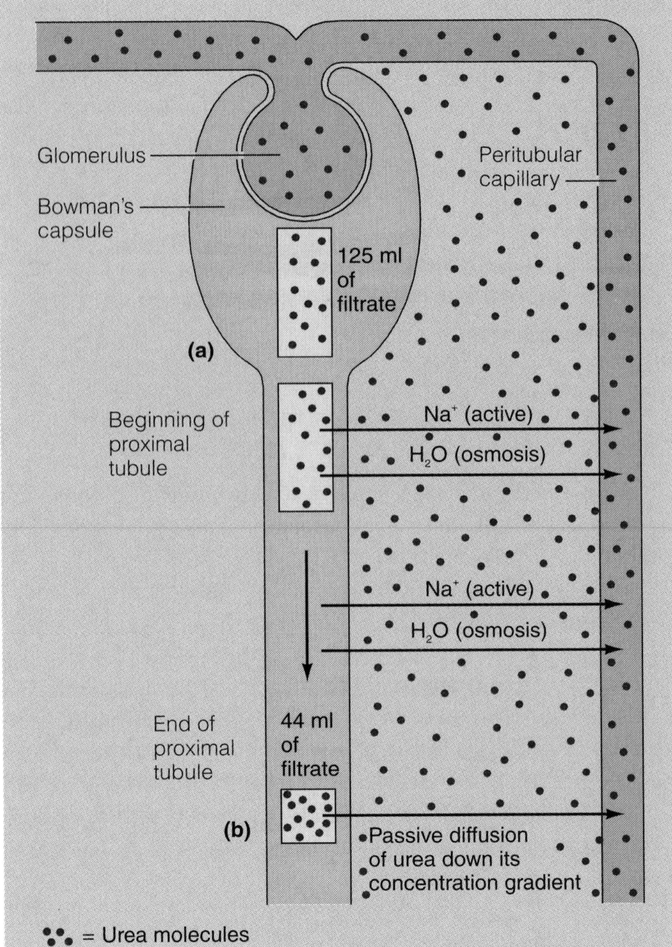

Figure 14-23 Passive Reabsorption of Urea at the End of the Proximal Tubule (a) In Bowman's capsule and at the beginning of the proximal tubule, urea is at the same concentration as in the plasma and surrounding interstitial fluid. (b) By the end of the proximal tubule, 65% of the original filtrate has been reabsorbed, leading to concentration of the filtered urea in the remaining filtrate. This establishes a concentration gradient favoring the passive reabsorption of urea.

H_2O through the glomerular capillaries upstream. The plasma proteins left behind in the glomerulus are concentrated into a smaller volume of plasma H_2O, increasing the plasma-colloid osmotic pressure of the unfiltered blood that leaves the glomerulus and enters the peritubular capillaries. This force tends to "pull" H_2O into the peritubular capillaries, simultaneous with the "push" of the hydrostatic pressure in the lateral spaces that drives H_2O toward the capillaries. By these means, 65% of the filtered H_2O—117 liters per day—is passively reabsorbed by the end of the proximal tubule. Neither the proximal tubule nor indeed any other portion of the tubule directly requires energy for this tremendous reabsorption of H_2O.

Another 15% of the filtered H_2O is obligatorily reabsorbed from the loop of Henle. Adjustable reabsorption of the remaining 20% of the filtered H_2O is accomplished in the distal and collecting tubules under control of vasopressin. The mechanisms responsible for H_2O reabsorption beyond the proximal tubule will be described later.

Urea reabsorption In addition to Cl^- and H_2O, the passive reabsorption of urea is also indirectly linked to active Na^+ reabsorption. Urea is a waste product resulting from the breakdown of protein. The osmotically induced reabsorption of H_2O in the proximal tubule secondary to active Na^+ reabsorption produces a concentration gradient for urea that favors the passive reabsorption of this nitrogenous waste as follows (Fig. 14-23). Because of extensive reabsorption of H_2O in the proximal tubule, the original 125 ml/min of filtrate are progressively reduced, until only 44 ml/min of fluid remain in the lumen by the end of the proximal tubule (with 65% of the H_2O in the original filtrate, or 81 ml/min, having been reabsorbed). Substances that have been filtered but not reabsorbed become progressively more concentrated in the tubular fluid as H_2O is reabsorbed while they are left behind. Urea is one such substance. Urea's concentration as it is filtered at the glomerulus is identical to its concentration in the plasma entering the peritubular capillaries. The quantity of urea present within the 125 ml of filtered fluid at the beginning of the proximal tubule, however, is concentrated almost threefold in the small volume of only 44 ml left at the end of the proximal tubule. As a result, the urea concentration within the tubular fluid becomes considerably greater than the plasma urea concentration in the adjacent capillaries. Therefore, a concentration gradient is created for urea to passively diffuse from the tubular lumen into the peritubular capillary plasma. Since the walls of the proximal tubules are only moderately permeable to urea, about 50% of the filtered urea is passively reabsorbed by this means.

Even though only half of the filtered urea is eliminated from the plasma with each pass through the nephrons, this removal rate is adequate. Only in impaired kidney function, when much less than half of the urea is removed, does the urea concentration in the plasma become elevated. An elevated urea level was one of the first chemical characteristics to be identified in the plasma of patients with severe renal failure. Accordingly, clinical measurement of **blood urea nitrogen (BUN)** came into use as a crude assessment of kidney function. It is now known that the most serious consequences of renal failure are not attributable to the retention of urea, which itself is not especially toxic, but rather to the accumulation of other substances that are not adequately excreted because of their failure to be properly secreted—most notably H^+ and K^+. Health professionals still often refer to renal failure as **uremia** ("urea in the blood"), indicative of the presence of excess urea in the blood, even though urea retention is not this condition's major threat.

In general, unwanted waste products are not reabsorbed.

The other filtered waste products besides urea, such as *phenol* and *creatinine,* are likewise concentrated in the tubular fluid as H_2O leaves the filtrate to enter the plasma, but they are not passively reabsorbed as urea is. Urea molecules, being the smallest of waste products, are the only wastes able to be passively reabsorbed as a result of this concentrating effect. Even though the other wastes are also concentrated in the tubular fluid, they are unable to leave the lumen down their concentration gradients to be passively reabsorbed because they are unable to permeate the tubular wall. Therefore, the waste products, failing to be reabsorbed, generally remain in the tubules and are excreted in the urine in a highly concentrated form. This excretion of metabolic wastes is not subject to physiological control. When renal function is normal, however, the excretory processes proceed at a satisfactory rate even though they are not controlled.

Tubular Secretion

The most important secretory processes are those for H^+, K^+, and organic ions.

By providing a second route of entry into the tubules for selected substances, *tubular secretion,* the discrete transfer of substances from the peritubular capillaries into the tubular lumen, may be viewed as a supplemental mechanism that hastens the elimination of these compounds from the body. Anything that gains entry to the tubular fluid, whether by glomerular filtration or tubular secretion, and fails to be reabsorbed is eliminated in the urine.

Tubular secretion involves transepithelial transport just as tubular reabsorption does, but now the steps are reversed. As with reabsorption, tubular secretion may be active or passive. The most important substances secreted by the tubules are hydrogen ion (H^+), potassium (K^+), and organic anions and cations, many of which are compounds foreign to the body.

Hydrogen-ion secretion Renal H^+ secretion is extremely important in the regulation of the acid-base balance in the body. Hydrogen ion can be added to the filtered fluid by being secreted by the proximal, distal, and collecting tubules. The extent of H^+ secretion depends on the acidity of the body fluids. When the body fluids are too acidic, H^+ secretion increases. Conversely, H^+ secretion is reduced when the H^+ concentration in the body fluids is too low. (See chapter 15 for further detail.)

Potassium-ion secretion Potassium ion is an example of a substance that is selectively moved in opposite directions in different parts of the tubule; it is actively reabsorbed in the proximal tubule and actively secreted in the distal and collecting tubules. Potassium-ion reabsorption early in the tubule occurs in a constant, unregulated fashion, whereas K^+ secretion later in the tubule is variable and subject to regulation. Normally, a quantity equivalent to about 10% to 15% of the filtered K^+ is excreted in the urine. However, the filtered K^+ is almost completely reabsorbed in the proximal tubule, so most of the K^+ appearing in the urine is derived from controlled K^+ secretion in the distal portions of the nephron rather than from filtration.

During K^+ depletion, K^+ secretion in the distal portions of the nephron is reduced to a minimum, so only the small percentage of filtered K^+ that escapes reabsorption in the proximal tubule is excreted in the urine. In this way, K^+ that normally would have been lost in the urine is conserved for the body. On the other hand, when plasma K^+ levels are elevated, K^+ secretion is adjusted so that just enough K^+ is added to the filtrate for elimination to reduce the plasma K^+ concentration to normal. Thus K^+ secretion, not the filtration or reabsorption of K^+, is varied in a controlled fashion to regulate the rate of K^+ excretion and maintain the desired plasma K^+ concentration.

Potassium-ion secretion in the distal and collecting tubules is coupled to Na^+ reabsorption by means of the energy-dependent basolateral Na^+-K^+ pump (Fig. 14-24). This

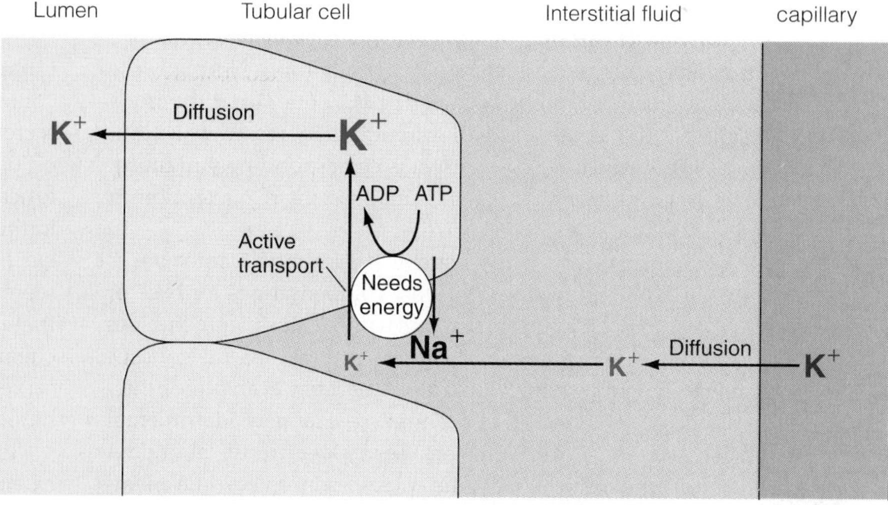

Figure 14-24 **Potassium-Ion Secretion** The basolateral pump simultaneously transports Na^+ into the lateral space and K^+ into the tubular cell.

pump not only moves Na^+ out into the lateral space but also transports K^+ into the tubular cells. The resultant high intracellular K^+ concentration favors the net diffusion of K^+ from the cells into the tubular lumen. Movement across the luminal membrane occurs passively through the large number of K^+ channels present in this barrier. By keeping the interstitial fluid concentration of K^+ low as it transports K^+ into the tubular cells from the surrounding interstitial fluid, the basolateral pump encourages the passive diffusion of K^+ out of the peritubular capillary plasma into the interstitial fluid. Potassium exiting the plasma in this manner is subsequently pumped into the cells, from which it diffuses into the lumen. In this way, the basolateral pump actively induces the net secretion of K^+ from the peritubular capillary plasma into the tubular lumen.

Several factors are able to alter the rate of K^+ secretion, the most important being the hormone aldosterone, which stimulates K^+ secretion by the tubular cells late in the nephron simultaneous to enhancing these cells' reabsorption of Na^+. An elevation in plasma K^+ concentration directly stimulates the adrenal cortex to increase its output of aldosterone, which in turn promotes the secretion and ultimate urinary excretion and elimination of the excess K^+. Conversely, a decline in plasma K^+ concentration causes a reduction in aldosterone secretion and a corresponding decrease in aldosterone-stimulated renal K^+ secretion.

Note that a rise in plasma K^+ concentration directly stimulates aldosterone secretion by the adrenal cortex, whereas a fall in plasma Na^+ concentration stimulates aldosterone secretion by means of the complex renin-angiotensin pathway. Thus, aldosterone secretion can be stimulated by two separate pathways (■ Fig. 14-25). No matter what the stimulus, however, increased aldosterone secretion always promotes simultaneous Na^+ reabsorption and K^+ secretion. For this reason, K^+ secretion can be inadvertently stimulated as a result of increased aldosterone activity brought about by Na^+ depletion, ECF volume reduction, or a fall in arterial blood pressure totally unrelated to K^+ balance. The resultant inappropriate loss of K^+ can lead to K^+ deficiency.

Another factor that can inadvertently alter the magnitude of K^+ secretion is the acid-base status of the body. The basolateral pump in the distal portions of the nephron can secrete either K^+ or H^+ in exchange for reabsorbed Na^+. An increased rate of secretion of either K^+ or H^+ is accompanied by a decreased rate of secretion of the other ion. Normally the kidneys secrete a preponderance of K^+, but when the body fluids are too acidic and H^+ secretion is increased as a compensatory measure, K^+ secretion is correspondingly reduced. This reduced secretion leads to inadvertent K^+ retention in the body fluids.

Except in the overriding circumstances of K^+ imbalances inadvertently induced during renal compensations for Na^+ or ECF volume deficits or acid-base imbalances, the kidneys usually exert a fine degree of control over plasma K^+ concentration. This is extremely important, because even minor fluctuations in plasma K^+ concentration can have detrimental consequences. Potassium ion, being the most abundant cation in the intracellular fluid, plays a key role in the membrane electrical activity of excitable tissues. Both increases or decreases in the plasma (ECF) K^+ concentration can alter the intracellular-to-extracellular K^+ concentration gradient, which in turn can change the resting membrane potential. A rise in ECF K^+ concentration leads to a reduction in resting potential and a subsequent increase in excitability, especially of heart muscle. This cardiac overexcitability can lead to a rapid heart rate and even fatal cardiac arrhythmias. Conversely, a fall in ECF K^+ concentration results in hyperpolarization of nerve and muscle cell membranes, which reduces their excitability. The manifestations of ECF K^+ depletion are skeletal muscle weakness, diarrhea and abdominal distention caused by smooth muscle dysfunction, and abnormalities in cardiac rhythm and impulse conduction.

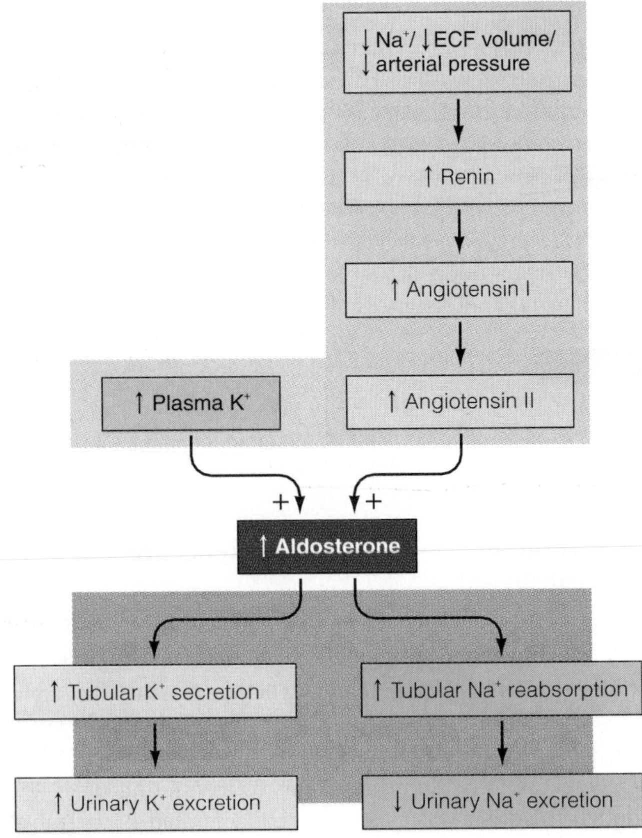

■ *Figure 14-25* **Dual Control of Aldosterone Secretion by K^+ and Na^+**

Organic anion and cation secretion The proximal tubule contains two distinct types of secretory carriers, one for the secretion of organic anions and a separate system for secretion of organic cations. These systems serve several important functions. First, by adding more of a particular type of organic ion to the quantity that has already gained entry to the tubular fluid by means of glomerular filtration, these organic secretory pathways facilitate the excretion of these substances. Included among these organic ions are certain blood-borne chemical messengers such as prostaglandins that, having served their purpose, need to be rapidly removed from the blood so that their biological activity is not unduly prolonged.

Second, in some important instances, organic ions are extensively but not irreversibly bound to plasma proteins.

Because they are attached to plasma proteins, these substances cannot be filtered through the glomeruli. Tubular secretion facilitates the elimination of these nonfilterable organic ions in the urine. Even though a given organic ion is largely bound to plasma proteins, a small percentage of these ions always exists in free or unbound form in the plasma. Removal of this free organic ion by secretion permits the "unloading" of some of the bound ion, which is then free to be secreted. This, in turn, encourages the unloading of even more organic ion, and so on.

Third, and most important is the ability of the organic-ion secretory systems to eliminate many foreign compounds from the body. The organic-ion systems can secrete a large number of different organic ions, both those produced endogenously (within the body) and those foreign organic ions that have gained access to the body fluids. This nonselectivity permits these organic-ion secretory systems to hasten the removal of many foreign organic chemicals, including food additives, environmental pollutants (for example, pesticides), drugs, and other non-nutritive organic substances that have gained entrance to the body.

The liver plays an important role in this regard. Many foreign organic compounds are not ionic in their original form, so they cannot be secreted by the organic-ion systems. The liver converts these foreign substances into an anionic form that facilitates their secretion by the organic-anion system and thus accelerates their elimination.

The rate of excretion of foreign organic compounds is not subject to control. Although the relatively nonselective organic-ion secretory systems enhance the removal of these substances from the body, this mechanism is not subject to physiological adjustments.

Many drugs, such as penicillin, are eliminated from the body by means of the proximal tubule organic-ion secretory systems. To keep the plasma concentration of these drugs at effective levels, the dosage has to be repeated on a regular, frequent basis to keep pace with the rapid removal of these compounds in the urine.

This completes our discussion of the reabsorptive and secretory processes that occur across the proximal and distal portions of the nephron. These processes are summarized in Table 14-3.

Urine Excretion and Plasma Clearance

On the average, 1 milliliter of urine is excreted per minute.

Typically, of the 125 ml of plasma filtered per minute, 124 ml/min are reabsorbed, so the final quantity of urine formed averages 1 ml/min. Thus, of the 180 liters filtered per day, 1.5 liters of urine are excreted.

Urine contains high concentrations of various waste products plus variable amounts of the substances regulated by the kidneys, with any excess quantities having spilled into the urine. Useful substances are conserved by reabsorption, so they do not appear in the urine.

Table 14-3
Summary of Transport across Proximal and Distal Portions of the Nephron

Reabsorption	*Secretion*
Proximal Tubule	
67% of filtered Na^+ actively reabsorbed; not subject to control; Cl^- follows passively	Variable H^+ secretion, depending on acid-base status of body
All filtered glucose and amino acids reabsorbed by secondary active transport; not subject to control	Organic-ion secretion; not subject to control
Variable amounts of filtered PO_4^{3-} and other electrolytes reabsorbed; subject to control	
65% of filtered H_2O osmotically reabsorbed; not subject to control	
50% of filtered urea passively reabsorbed; not subject to control	
All filtered K^+ reabsorbed; not subject to control	
Distal Tubule and Collecting Duct	
Variable Na^+ reabsorption, controlled by aldosterone; Cl^- follows passively	Variable H^+ secretion, depending on acid-base status of body
Variable H_2O reabsorption, controlled by vasopressin	Variable K^+ secretion, controlled by aldosterone

A relatively small change in the quantity of filtrate reabsorbed can bring about a large change in the volume of urine formed. For example, a reduction of less than 1% in the total reabsorption rate, from 124 to 123 ml/min, increases the urinary excretion rate by 100%, from 1 to 2 ml/min.

Plasma clearance refers to the volume of plasma cleared of a particular substance per minute.

Compared to the plasma entering the kidneys through the renal arteries, the plasma leaving the kidneys through the renal veins lacks the materials that were left behind to be eliminated in the urine. By excreting substances in the urine, the kidneys clean or "clear" the plasma flowing through them of these substances. For any substance, its **plasma clearance** is defined as the volume of plasma that is completely cleared of that substance by the kidneys per minute.[1] It does not refer

[1]Actually, plasma clearance is an artificial concept, because when a particular substance is excreted in the urine, that substance's concentration in the plasma as a whole is uniformly decreased as a result of thorough mixing in the circulatory system. However, it is useful for comparative purposes to consider clearance in effect as the volume of plasma that would have contained the total quantity of the substance that the kidneys excreted in one minute; that is, the hypothetical volume of plasma completely cleared of that substance per minute.

to the *amount of the substance* removed but to the *volume of plasma* from which that amount was removed. Plasma clearance is actually a more useful measure than urine excretion; it is more important to know what effect urine excretion has on removing materials from the body fluids than to know the volume and composition of the discarded urine. Plasma clearance expresses the kidneys' effectiveness in removing various substances from the internal fluid environment.

Plasma clearance can be calculated for any plasma constituent as follows:

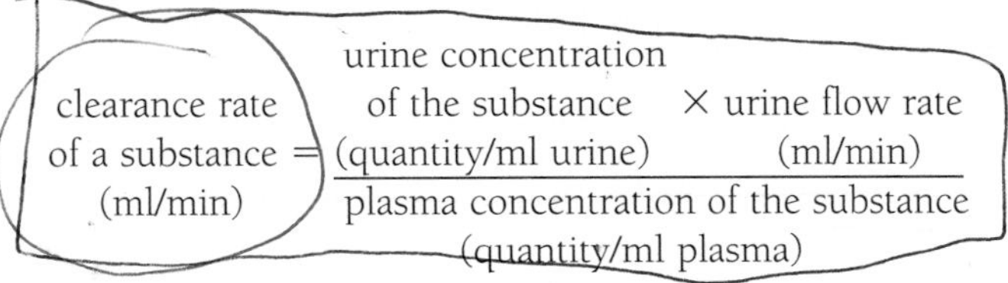

$$\begin{array}{c}\text{clearance rate}\\ \text{of a substance}\\ \text{(ml/min)}\end{array} = \frac{\begin{array}{c}\text{urine concentration}\\ \text{of the substance}\\ \text{(quantity/ml urine)}\end{array} \times \begin{array}{c}\text{urine flow rate}\\ \text{(ml/min)}\end{array}}{\begin{array}{c}\text{plasma concentration of the substance}\\ \text{(quantity/ml plasma)}\end{array}}$$

The plasma clearance rate varies for different substances, depending on how the kidneys handle each substance.

If a substance is filtered but not reabsorbed or secreted, its plasma clearance rate equals the GFR Assume that a plasma constituent, substance X, is freely filterable at the glomerulus but is not reabsorbed or secreted. As 125 ml/min of plasma are filtered and subsequently reabsorbed, the quantity of substance X originally contained within the 125 ml is left behind in the tubules to be excreted. Thus, 125 ml of plasma are cleared of substance X each minute (Fig. 14-26a). (Of the 125 ml/min of plasma filtered, 124 ml/min of the filtered fluid are returned through the process of reabsorption to the plasma minus substance X, thus clearing this 124 ml/min of substance X. In addition, the 1 ml/min of fluid lost in the urine is replaced in the long term by an equivalent volume of ingested H_2O that is already clear of substance X. Therefore, 125 ml of plasma cleared of substance X are, in effect, returned to the plasma for every 125 ml of plasma filtered per minute.)

There is no endogenous chemical with the characteristics of substance X. All substances naturally present in the plasma, even wastes, are reabsorbed or secreted to some extent. However, **inulin** (do not confuse with insulin), a harmless foreign carbohydrate produced by onions and garlic, is freely filtered and not reabsorbed or secreted—an ideal substance X. Inulin can be injected and its plasma clearance determined as a clinical means of ascertaining the GFR. Since all glomerular filtrate formed is cleared of inulin, the volume of plasma cleared of inulin per minute equals the volume of plasma filtered per minute—that is, the GFR.

Although the determination of inulin plasma clearance is accurate and straightforward, it is not very convenient because inulin must be infused continuously throughout the determination to maintain a constant plasma concentration. Therefore, the plasma clearance of an endogenous substance, **creatinine,** is often used instead to give a rough estimate of the GFR. Creatinine, an end product of muscle metabolism, is produced at a relatively constant rate. It is freely filtered and not reabsorbed but is slightly secreted. Accordingly, creatinine clearance is not a completely accurate reflection of the GFR, but it does provide a close approximation and can be more readily determined than inulin clearance.

If a substance is filtered and reabsorbed but not secreted, its plasma clearance rate is always less than the GFR Some or all of a reabsorbable substance that has been filtered is returned to the plasma. Because less than the filtered volume of plasma will have been cleared of the substance, the plasma clearance rate of a reabsorbable substance is always less than the GFR. For example, the plasma clearance for glucose is normally zero. All of the filtered glucose is reabsorbed along with the rest of the returning filtrate, so none of the plasma is cleared of glucose (Fig. 14-26b).

For a substance that is partially reabsorbed, such as urea, only part of the filtered plasma is cleared of that substance. With about 50% of the filtered urea being passively reabsorbed, only half of the filtered plasma, or 62.5 ml, is cleared of urea each minute (Fig. 14-26c).

If a substance is filtered and secreted but not reabsorbed, its plasma clearance rate is always greater than the GFR Tubular secretion allows the kidneys to clear certain materials from the plasma more efficiently. Only 20% of the plasma entering the kidneys is filtered. The remaining 80% passes unfiltered into the peritubular capillaries. The only means by which this unfiltered plasma can be cleared of any substance during this trip through the kidneys before being returned to the general circulation is by the process of secretion. An example is H^+. Not only will the plasma that is filtered be cleared of nonreabsorbable H^+, but the plasma from which H^+ is secreted will also be cleared of H^+. For example, if the quantity of H^+ that is secreted is equivalent to the quantity of H^+ present in 25 ml of plasma, the clearance rate for H^+ will be 150 ml/min at the normal GFR of 125 ml/min. Every minute 125 ml of plasma will lose its H^+ through the process of filtration and failure of reabsorption, and 25 more ml of plasma will lose its H^+ through the process of secretion. The plasma clearance for a secreted substance is always greater than the GFR (Fig. 14-26d).

Just as inulin can be used clinically to determine the GFR, the plasma clearance of another foreign compound, the organic anion **para-aminohippuric acid (PAH),** can be used to measure renal plasma flow. Like inulin, PAH is freely filterable and nonreabsorbable. It differs, however, in that all of the PAH in the plasma that escapes filtration is secreted from the peritubular capillaries. Thus, PAH is removed from *all* of the plasma that flows through the kidneys—both from the plasma that is filtered and subsequently reabsorbed without its PAH, and from the unfiltered plasma that continues on in the peritubular capillaries and loses its PAH by means of active secretion into the tubules. Because all of the plasma that flows through the kidneys is cleared of PAH, the plasma clearance for PAH is a reasonable estimate of the rate of plasma flow through the kidneys. Typically, renal plasma flow averages 625 ml/min, for a renal blood flow (plasma, plus blood cells) of 1,140 ml/min—over 20% of the cardiac output.

Knowing PAH clearance (renal plasma flow) and inulin clearance (GFR), you can easily determine the **filtration frac-**

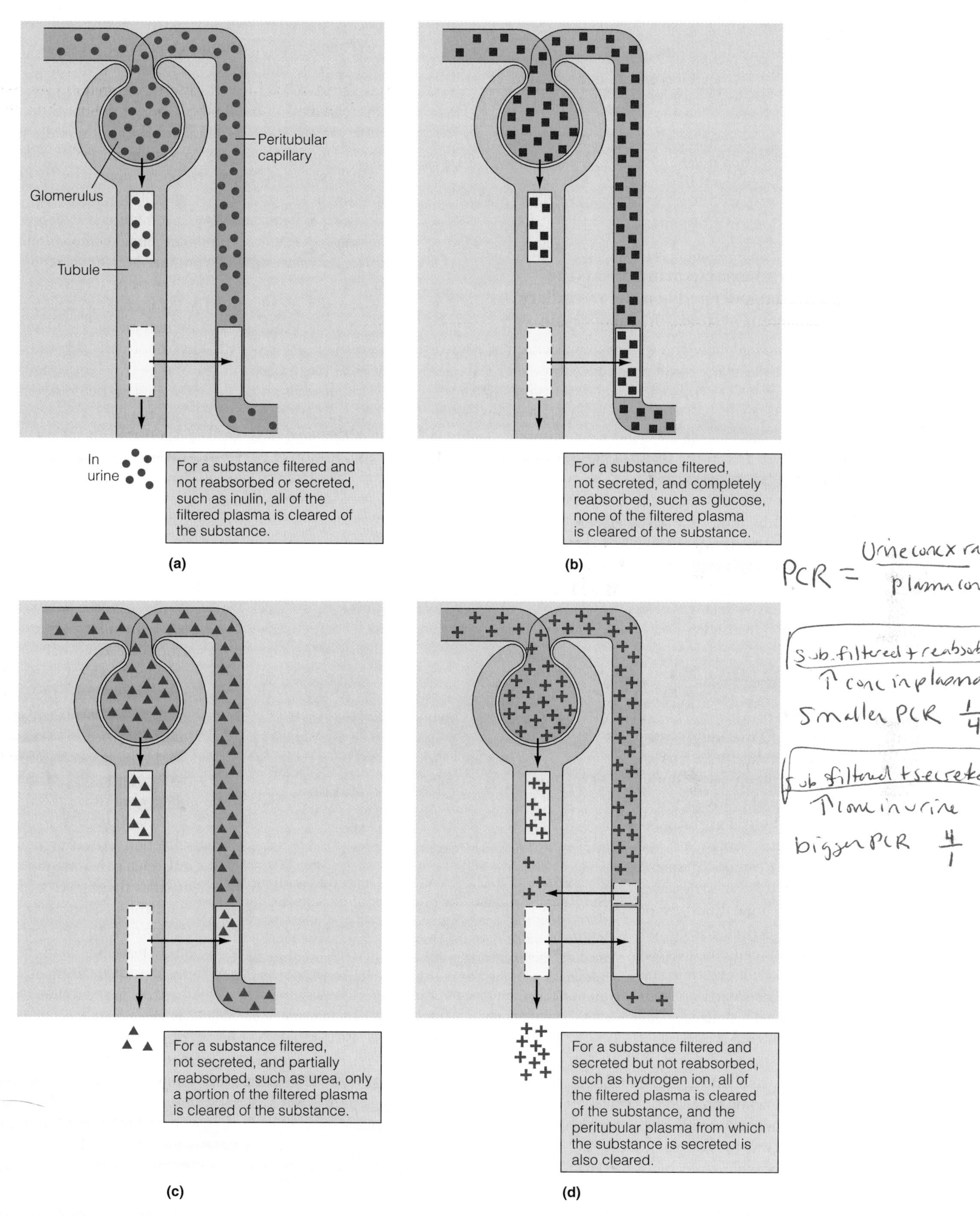

Peritubular capillary
Glomerulus
Tubule
In urine
For a substance filtered and not reabsorbed or secreted, such as inulin, all of the filtered plasma is cleared of the substance.
(a)
For a substance filtered, not secreted, and completely reabsorbed, such as glucose, none of the filtered plasma is cleared of the substance.
(b)
For a substance filtered, not secreted, and partially reabsorbed, such as urea, only a portion of the filtered plasma is cleared of the substance.
(c)
For a substance filtered and secreted but not reabsorbed, such as hydrogen ion, all of the filtered plasma is cleared of the substance, and the peritubular plasma from which the substance is secreted is also cleared.
(d)

Figure 14-26 Plasma Clearance for Substances Handled in Different Ways by the Kidneys

tion, the fraction of the plasma flowing through the glomeruli that is filtered into the tubules:

$$\text{filtration fraction} = \frac{\text{GFR (plasma inulin clearance)}}{\text{renal plasma flow (plasma PAH clearance)}}$$

$$= \frac{125 \text{ ml/min}}{625 \text{ ml/min}} = 20\%$$

Thus, 20% of the plasma that enters the glomeruli is typically filtered.

The ability to excrete urine of varying concentrations depends on the medullary countercurrent system and vasopressin.

Having considered how the kidneys deal with a variety of solutes in the plasma, we will now concentrate on renal handling of plasma H_2O. The ECF osmolarity (solute concentration) depends on the relative amount of H_2O compared to solute. At normal fluid balance and solute concentration, the body fluids are said to be **isotonic** at an osmolarity of 300 milliosmols/liter (mosm/liter) (see p. A-7). If there is too much H_2O relative to the solute load, the body fluids are **hypotonic,** which means that they are too dilute at an osmolarity less than 300 mosm/liter. On the other hand, if a H_2O deficit exists relative to the solute load, the body fluids are too concentrated or are **hypertonic**, having an osmolarity greater than 300 mosm/liter.

Generally speaking, the osmolarity of the ECF is uniform throughout the body. Knowing that the driving force for H_2O reabsorption throughout the entire length of the tubules is an osmotic gradient between the tubular lumen and surrounding interstitial fluid, you would expect, based on osmotic considerations, that the kidneys could not excrete urine more or less concentrated than the body fluids. Indeed, this would be the case if the interstitial fluid surrounding the tubules in the kidneys were identical in osmolarity to the remaining body fluids. Water reabsorption would proceed only until the tubular fluid equilibrated osmotically with the interstitial fluid, and there would be no way to eliminate excess H_2O when the body fluids were hypotonic or to conserve H_2O in the presence of hypertonicity.

Fortunately, a large vertical osmotic gradient is uniquely maintained in the interstitial fluid of the medulla of each kidney. The concentration of the interstitial fluid progressively increases from the cortical boundary down through the depth of the renal medulla until it reaches a maximum of 1,200 mosm/liter in humans at the junction with the renal pelvis (Fig. 14-27). This vertical osmotic gradient remains constant regardless of the fluid balance of the body.

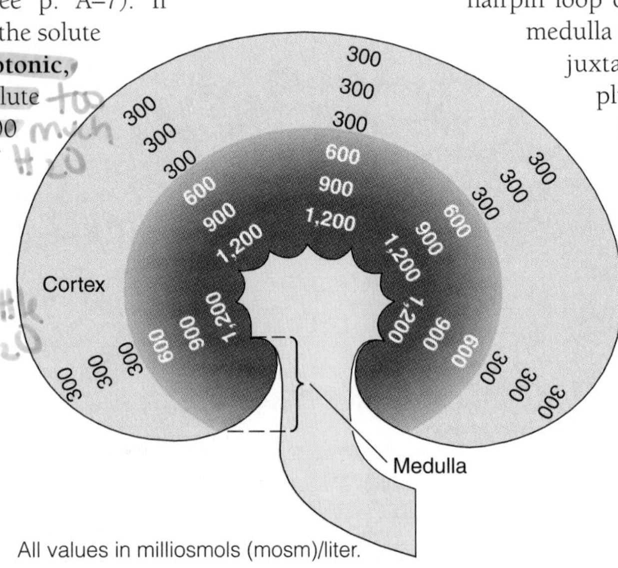

Figure 14-27 **Vertical Osmotic Gradient in the Renal Medulla** The osmolarity of the interstitial fluid throughout the renal cortex is isotonic at 300 mosm/liter, but the osmolarity of the interstitial fluid in the renal medulla increases progressively from 300 mosm/liter at the boundary with the cortex to a maximum of 1,200 mosm/liter at the junction with the renal pelvis.

The presence of this gradient enables the kidneys to produce urine that ranges in concentration from 100 to 1,200 mosm/liter, depending on the body's state of hydration. When the body is in ideal fluid balance, 1 ml/min of isotonic urine is formed. When the body is overhydrated (too much H_2O), the kidneys are able to produce a large volume of dilute urine (up to 25 ml/min and hypotonic at 100 mosm/liter), thus eliminating the excess H_2O in the urine. Conversely, the kidneys are able to put out a small volume of concentrated urine (down to 0.3 ml/min and hypertonic at 1,200 mosm/liter) when the body is dehydrated (too little H_2O), thus conserving H_2O for the body.

Unique anatomical arrangements and complex functional interactions between the various nephron components present in the renal medulla are responsible for the establishment and utilization of the vertical osmotic gradient. Recall that the hairpin loop of Henle dips only slightly into the medulla in the cortical nephrons, but in the juxtamedullary nephrons, the loop plunges through the entire depth of the medulla so that the tip of the loop lies near the renal pelvis (Fig. 14-5). Also, the vasa recta of the juxtamedullary nephrons follow the same deep hairpin loop as the long loop of Henle. Flow in both the long loops of Henle and the vasa recta is considered countercurrent because the flow in the two closely adjacent limbs of the loop is in opposite directions. Also running through the medulla in the descending direction only on their way to the renal pelvis are the collecting tubules that serve both types of nephrons. This arrangement, coupled with the permeability and transport characteristics of these tubular segments, plays a key role in the kidneys' ability to produce urine of varying concentrations, depending on the body's needs for water conservation or elimination. Briefly, the juxtamedullary nephrons' long loops of Henle *establish the vertical osmotic gradient,* their vasa recta *prevent the dissolution of this gradient* while providing blood to the renal medulla, and the collecting tubules of all nephrons *use the gradient,* in conjunction with the hormone vasopressin, to produce urine of varying concentrations. Collectively, this entire functional organization is known as the **medullary countercurrent system.** We will examine each of its facets in greater detail.

The long loops of Henle of juxtamedullary nephrons establish the medullary vertical osmotic gradient by means of countercurrent multiplication We will follow the filtrate through a long-looped nephron to see how this structure establishes a vertical osmotic gradient in the medulla. Immediately after the filtrate is formed, uncontrolled osmotic reabsorption of filtered H_2O occurs in the proximal tubule secondary to active Na^+ reabsorption. As a result, by the end of the proximal tubule, about 65% of the filtrate has been reabsorbed, but the 35% remaining in the tubular lumen still has the same osmolarity as the body fluids. Therefore, the fluid entering the loop of Henle is still isotonic. An additional 15% of the filtered H_2O is obligatorily reabsorbed from the loop of Henle during the establishment and maintenance of the vertical osmotic gradient, with the osmolarity of the tubular fluid being altered in the process.

The following functional distinctions between the descending limb of a long Henle's loop (which carries fluid from the proximal tubule down into the depths of the medulla) and the ascending limb (which carries fluid up and out of the medulla into the distal tubule) are critical to the establishment of the incremental osmotic gradient in the medullary interstitial fluid.

The *descending limb:*

1. Is highly permeable to H_2O.
2. tubule that does not do so).

The *ascending limb:*

1. Actively transports NaCl out of the tubular lumen into the surrounding interstitial fluid.
2. Is always impermeable to H_2O, so salt leaves the tubular fluid without H_2O osmotically following along.

The close proximity and countercurrent flow of the two limbs allow important interactions to occur between them. Even though the flow of fluids is continuous through the loop of Henle, we will visualize what happens step by step, much like an animated film run so slowly that each individual frame can be viewed.

- *Initial scene* (— *Fig. 14-28a*) Before the vertical osmotic gradient is established, the medullary interstitial fluid concentration is uniformly 300 mosm/liter, as is the remainder of the body fluids.
- *Step 1 (Fig. 14-28b)* The active salt pump in the ascending limb is able to transport NaCl out of the lumen until the surrounding interstitial fluid is 200 mosm/liter more concentrated than the tubular fluid in this limb. When the ascending limb pump starts actively extruding salt, the medullary interstitial fluid becomes hypertonic. Water cannot follow osmotically from the ascending limb because this limb is impermeable to H_2O. However, net diffusion of H_2O does occur from the descending limb into the interstitial fluid. The tubular fluid entering the descending limb from the proximal tubule is isotonic. Because the descending limb is highly permeable to H_2O, net diffusion of H_2O occurs by osmosis out of the descending limb into the more concentrated interstitial fluid. The passive movement of H_2O out of the descending limb continues until the osmolarities of the fluid in the descending limb and interstitial fluid become equilibrated. Thus, the tubular fluid entering the loop of Henle immediately starts to become more concentrated as it loses H_2O. At equilibrium, the osmolarity of the ascending limb fluid is 200 mosm/liter and the osmolarities of the interstitial fluid and descending limb fluid are equal at 400 mosm/liter.
- *Step 2 (Fig. 14-28c)* If we now advance the entire column of fluid in the loop of Henle several "frames," a mass of 200 mosm/liter fluid exits from the top of the ascending limb into the distal tubule, and a new mass of isotonic fluid at 300 mosm/liter enters the top of the descending limb from the proximal tubule. At the bottom of the loop, a comparable mass of 400 mosm/liter fluid from the descending limb moves forward around the tip into the ascending limb, placing it opposite a 400 mosm/liter region in the descending limb. Note that the 200 mosm/liter concentration difference has been lost at both the top and the bottom of the loop.
- *Step 3 (Fig. 14-28d)* The ascending limb pump again transports NaCl out while H_2O passively leaves the descending limb until a 200 mosm/liter difference is reestablished between the ascending limb and both the interstitial fluid and descending limb at each horizontal level. Note, however, that the concentration of tubular fluid is progressively increasing in the descending limb and progressively decreasing in the ascending limb.
- *Step 4 (Fig. 14-28e)* As the tubular fluid is advanced still further, the 200 mosm/liter concentration gradient is disrupted once again at all horizontal levels.
- *Step 5 (Fig. 14-28f)* Again, active extrusion of NaCl from the ascending limb, coupled with the net diffusion of H_2O out of the descending limb, reestablishes the 200 mosm/liter gradient at each horizontal level.
- *Steps 6 and on (Fig. 14-28g)* As the fluid flows slightly forward again and this stepwise process continues, the fluid in the descending limb becomes progressively more hypertonic until it reaches a maximum concentration of 1,200 mosm/liter at the bottom of the loop, four times the normal concentration of body fluids. Because the interstitial fluid always achieves equilibrium with the descending limb, an incremental vertical concentration gradient ranging from 300 to 1,200 mosm/liter is likewise established in the medullary interstitial fluid. In contrast, the concentration of the tubular fluid progressively decreases in the ascending limb as salt is pumped out but H_2O is unable to follow. In fact, the tubular fluid even becomes hypotonic as it leaves the ascending limb to enter the distal tubule at a concentration of 100 mosm/liter, one-third the normal concentration of body fluids.

Note that although a gradient of only 200 mosm/liter exists between the ascending limb and the surrounding fluids at each medullary horizontal level, a much larger vertical gradient exists from the top to the bottom of the medulla. Even

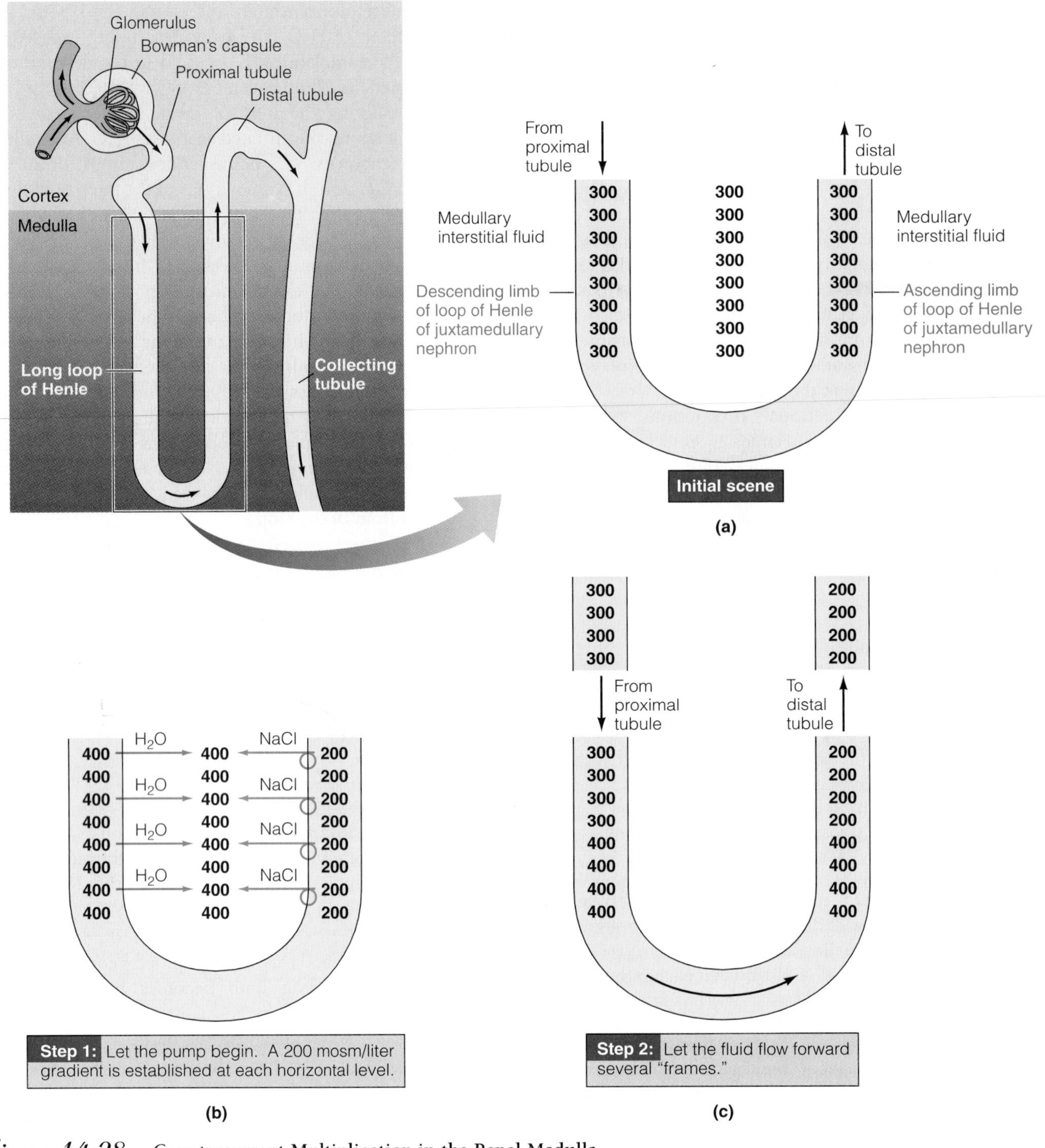

Figure 14-28 Countercurrent Multiplication in the Renal Medulla

though the ascending limb pump can generate a gradient of only 200 mosm/liter, this effect is multiplied into a large vertical gradient because of the countercurrent flow within the loop. This concentrating mechanism accomplished by the loop of Henle is known as **countercurrent multiplication.**

We have artificially described countercurrent multiplication in a "stop-and-flow," stepwise fashion to facilitate understanding. It is important to realize that once the incremental medullary gradient is established, it remains constant because of the continuous flow of fluid coupled with the ongoing ascending limb active transport activity and accompanying descending limb passive fluxes.

If you consider only what happens to the tubular fluid as it flows through the loop of Henle, the whole process seems to be an exercise in futility. The isotonic fluid that enters the loop becomes progressively more concentrated as it flows down the descending limb, achieving a maximum concentration of 1,200 mosm/liter, only to become progressively more dilute as it flows up the ascending limb, finally leaving the loop at a minimum concentration of 100 mosm/liter. What is the

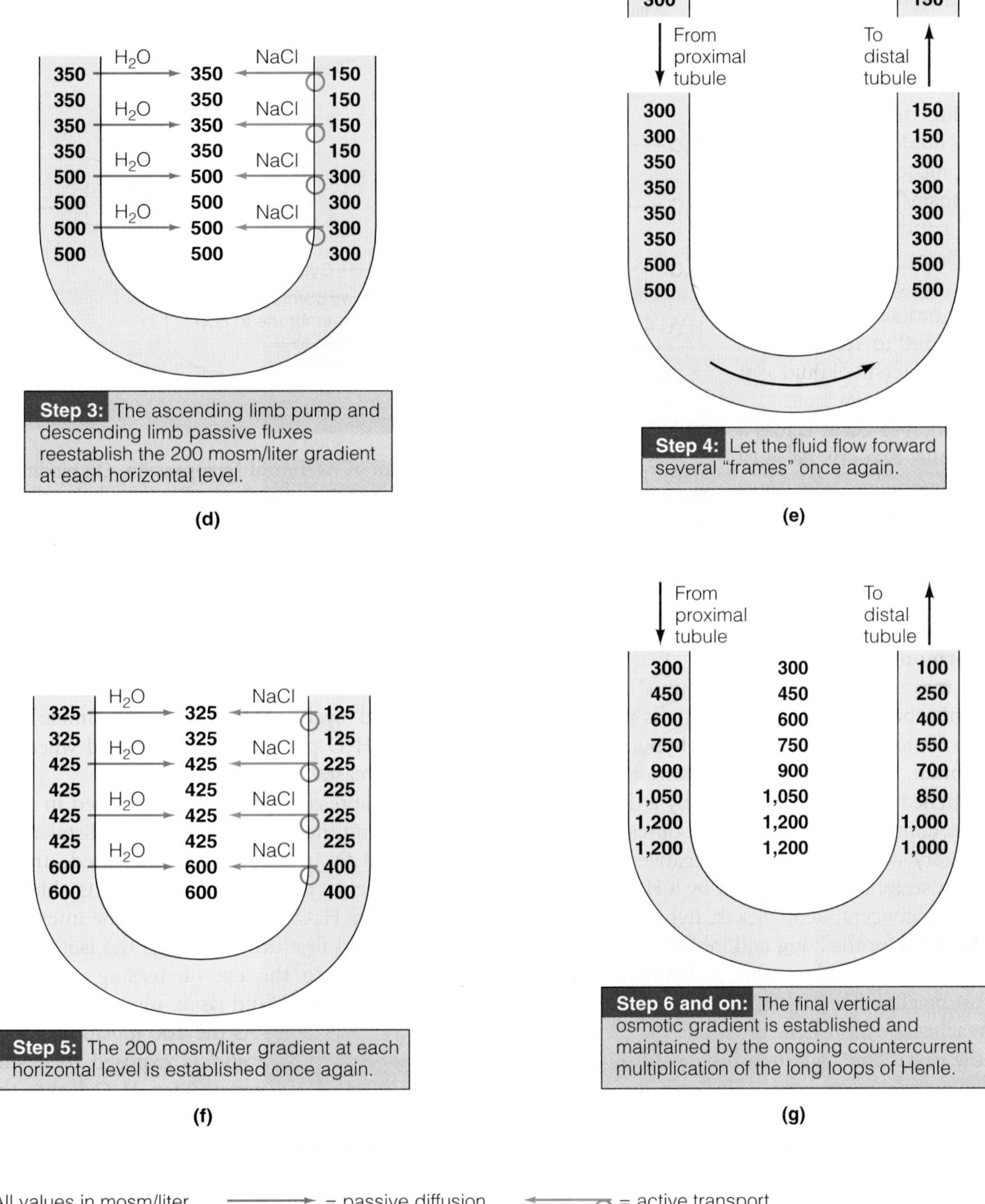

point of concentrating the fluid fourfold and then turning around and diluting it until it leaves at one-third the concentration at which it entered? Such a mechanism offers two benefits. First, it establishes a vertical osmotic gradient in the medullary interstitial fluid. This gradient, in turn, is used by the collecting ducts to concentrate the tubular fluid so that a urine *more concentrated* than normal body fluids can be excreted. Second, the fact that the fluid is hypotonic as it enters the distal portions of the tubule enables the kidneys to excrete a urine *more dilute* than normal body fluids. Let us see how.

The medullary vertical osmotic gradient permits excretion of urine of differing concentrations by means of vasopressin-controlled, variable H_2O reabsorption from the final tubular segments Following obligatory H_2O reabsorption from the proximal tubule (65% of the filtered H_2O) and loop of Henle (15% of the filtered H_2O), 20% of the filtered H_2O remains in the lumen to enter the distal and collecting tubules for variable reabsorption that is under hormonal control. This is still a large volume of filtered H_2O subject to regulated reabsorption; 20% × GFR (180 liters/day) =

hormone that triggers a series of events ultimately leading to increased secretion of aldosterone from the adrenal cortex. Aldosterone increases Na^+ reabsorption from the distal portions of the tubule, thus correcting for the original reduction in Na^+/ECF volume/blood pressure.

The other electrolytes actively reabsorbed by the tubules, such as PO_4^{3-} and Ca^{2+}, have their own independently functioning carrier systems. Because these carriers, like the organic-nutrient cotransport carriers, can become saturated, each exhibits a maximal carrier-limited transport capacity, or T_m. Once the filtered load of an actively reabsorbed substance exceeds the T_m, reabsorption proceeds at a constant maximal rate, with the additional filtered quantity of the substance being excreted in the urine.

Tubular Secretion

By means of the process of tubular secretion, the kidney tubules are able to selectively add some substances to the quantity already filtered. Secretion of substances hastens their excretion in the urine. The most important secretory systems are for (1) H^+, which is important in the regulation of acid-base balance; (2) K^+, which keeps the plasma K^+ concentration at an appropriate level to maintain normal membrane excitability in muscles and nerves; and (3) organic ions, which accomplishes more efficient elimination of foreign organic compounds from the body.

Urine Excretion and Plasma Clearance

Of the 125 ml/min filtered in the glomeruli, normally only 1 ml/min remains in the tubules to be excreted as urine. Only wastes and excess electrolytes not wanted by the body are left behind, dissolved in a given volume of H_2O to be eliminated in the urine. Because the excreted material is removed or "cleared" from the plasma, the term plasma clearance refers to the volume of plasma being cleared of a particular substance each minute by means of renal activity.

The kidneys are able to excrete urine of varying volumes and concentrations to either conserve or eliminate H_2O, depending on whether the body has a H_2O deficit or excess, respectively. The kidneys are able to produce urine ranging from 0.3 ml/min at 1,200 mosm/liter to 25 ml/min at 100 mosm/liter by reabsorbing variable amounts of H_2O from the distal portions of the nephron. This variable reabsorption is made possible by the establishment of a vertical osmotic gradient ranging from 300 to 1,200 mosm/liter in the medullary interstitial fluid by means of the loop of Henle countercurrent system and urea recycling between the collecting tubule and Henle's loops. This vertical osmotic gradient to which the hypotonic (100 mosm/liter) tubular fluid is exposed as it passes through the distal portions of the nephron establishes a passive driving force for progressive reabsorption of H_2O from the tubular fluid, but the actual extent of H_2O reabsorption depends on the amount of vasopressin (antidiuretic hormone) secreted. Vasopressin increases the permeability of the distal and collecting tubules to H_2O; they are impermeable to H_2O in its absence. Vasopressin secretion increases in response to a H_2O deficit, and H_2O reabsorption increases accordingly. Vasopressin secretion is inhibited in response to a H_2O excess, thereby reducing H_2O reabsorption. In this way, adjustments in vasopressin-controlled H_2O reabsorption help correct any fluid imbalances.

Once formed, urine is propelled by peristaltic contractions through the ureters from the kidneys to the urinary bladder for temporary storage. The bladder can accommodate up to 250 to 400 ml of urine before stretch receptors within its wall initiate the micturition reflex. This reflex causes involuntary emptying of the bladder by simultaneous bladder contraction and opening of both the internal and external urethral sphincters. Micturition can transiently be voluntarily prevented until a more opportune time for bladder evacuation by deliberate tightening of the external sphincter and surrounding pelvic diaphragm.

Review Exercises

Objective Questions (Answers on p. E-14.)

1. Part of the kidneys' energy supply is used to accomplish glomerular filtration. (True or false?)
2. Sodium reabsorption is under hormonal control throughout the length of the tubule. (True or false?)
3. Glucose and amino acids are reabsorbed by secondary active transport. (True or false?)
4. Solute excretion is always accompanied by comparable H_2O excretion. (True or false?)
5. Water excretion can occur without comparable solute excretion. (True or false?)
6. The functional unit of the kidneys is the ________.
7. ________ is the only ion actively reabsorbed in the proximal tubule and actively secreted in the distal and collecting tubules.
8. The minimum volume of obligatory H_2O loss that must accompany the excretion of wastes each day is ________ ml.
9. Indicate whether each of the following factors would (a) increase or (b) decrease the GFR, if everything else remained constant.
   ___1. a rise in Bowman's capsule pressure resulting from ureteral obstruction by a kidney stone
   ___2. a fall in plasma protein concentration resulting from loss of these proteins from a large burned surface
   ___3. a dramatic fall in arterial blood pressure following severe hemorrhage (<80 mm Hg)
   ___4. afferent arteriolar vasoconstriction
   ___5. tubuloglomerular feedback response to a reduction in tubular flow rate
   ___6. myogenic response of an afferent arteriole stretched as a result of an increased driving blood pressure
   ___7. sympathetic activity to the afferent arterioles
   ___8. contraction of mesangial cells
   ___9. contraction of podocytes

10. Which of the following filtered substances is normally *not* present in the urine at all?
   a. Na^+
   b. PO_4^{3-}
   c. urea
   d. H^+
   e. glucose

11. Reabsorption of which of the following substances is *not* linked in some way to active Na^+ reabsorption?
   a. glucose
   b. PO_4^{3-}
   c. H_2O
   d. urea
   e. Cl^-

In questions 12–14, indicate the proper sequence through which fluid flows as it traverses the structures in question by writing the identifying letters in the proper order in the blanks.

12. a. ureter __ __ __ __ __
   b. kidney
   c. urethra
   d. bladder
   e. renal pelvis

13. a. efferent arteriole __ __ __ __ __ __
   b. peritubular capillaries
   c. renal artery
   d. glomerulus
   e. afferent arteriole
   f. renal vein

14. a. loop of Henle __ __ __ __ __ __ __
   b. collecting duct
   c. Bowman's capsule
   d. proximal tubule
   e. renal pelvis
   f. distal tubule
   g. glomerulus

15. Using the answer code below, indicate what the osmolarity of the tubular fluid is at each of the designated points in a nephron:
   (a) isotonic (300 mosm/liter)
   (b) hypotonic (100 mosm/liter)
   (c) hypertonic (1,200 mosm/liter)
   (d) ranging from hypotonic to hypertonic (100 mosm/liter to 1,200 mosm/liter)

   ___ 1. Bowman's capsule
   ___ 2. end of proximal tubule
   ___ 3. tip of Henle's loop of juxtamedullary nephron (at the bottom of the U turn)
   ___ 4. end of Henle's loop of juxtamedullary nephron (before entry into distal tubule)
   ___ 5. end of collecting duct

Essay Questions

1. List the functions of the kidneys.
2. Describe the anatomy of the urinary system. Describe the components of a nephron.
3. Describe the three basic renal processes; indicate how they relate to urine excretion. Distinguish between *secretion* and *excretion*.
4. Discuss the forces involved in glomerular filtration. What is the average GFR?
5. How is GFR regulated as part of the baroreceptor reflex?
6. Why do the kidneys receive a seemingly disproportionate share of the cardiac output? What percentage of renal blood flow is normally filtered?
7. List the steps in transepithelial transport.
8. Distinguish between active and passive reabsorption.
9. Describe all of the tubular transport processes that are linked to the basolateral Na^+-K^+ ATPase carrier.
10. Describe the renin-angiotensin-aldosterone system. What are the source and function of atrial natriuretic peptide?
11. To what do the terms *tubular maximum* (T_m) and *renal threshold* refer? Compare two substances that display a T_m, one that is and one that is not regulated by the kidneys.
12. What is the importance of tubular secretion? What are the most important secretory processes?
13. What is the average rate of urine formation?
14. Define plasma clearance.
15. What is responsible for the presence of a vertical osmotic gradient in the medullary interstitial fluid? Of what importance is this gradient?
16. Discuss the function of vasopressin.
17. Describe the transfer of urine to, the storage of urine in, and the emptying of urine from the bladder.

Quantitative Exercises (Solutions on p. E–14.)

1. Two patients are voiding protein in their urine. To determine whether or not this proteinuria is indicative of a serious problem , a physician injects small amounts of inulin and PAH into each patient. Recall that inulin is freely filtered and neither secreted nor reabsorbed in the nephron and that PAH at this concentration is completely removed from the blood by tubular secretion. The data collected are given in the following table, where $[x]_u$ is the concentration of substance x (either inulin or PAH) in the urine (in mM); $[x]_p$ is the concentration of x in the plasma, and v_u is the flow rate of urine (in ml/min).

Patient	$[I]_u$	$[I]_p$	$[PAH]_u$	$[PAH]_p$	v_u
1	25	2	186	3	10
2	31	1.5	300	4.5	6

   a. Calculate each patient's GFR and renal plasma flow.
   b. Calculate the renal blood flow for each patient, assuming both have a hematocrit of 0.45.
   c. Calculate the filtration fraction for each patient
   d. Which of the values calculated for each patient are within the normal range? Which values are abnormal? What could be causing these deviations from normal?

2. What is the filtered load of sodium if inulin clearance is 125 ml/min and the sodium concentration in plasma is 145mM?
3. Calculate a patient's rate of urine production, given that his inulin clearance is 125 ml/min and his urine and plasma concentrations of inulin are 300 mg/liter and 3 mg/liter, respectively.
4. If the urine concentration of a substance is 7.5 mg/ml of urine, its plasma concentration is 0.2 mg/ml of plasma, and the urine flow rate is 2 ml/min, what is the clearance rate of the substance? Is the substance being reabsorbed or secreted by the kidneys?

Points to Ponder

(Explanations on p. E–15.)

1. The long-looped nephrons of animals adapted to survive with minimal water consumption, such as desert rats, have relatively much longer loops of Henle than humans have. Of what benefit would these longer loops be?
2. If the plasma concentration of substance X is 200 mg/100 ml and the GFR is 125 ml/min, the filtered load of this substance is ______________ .

   If the T_m for substance X is 200 mg/min, how much of the substance will be reabsorbed at a plasma concentration of 200 mg/100 ml and a GFR of 125 ml/min? ______________ How much of substance X will be excreted? ______________
3. Conn's syndrome is an endocrine disorder brought about by a tumor of the adrenal cortex that secretes excessive aldosterone in uncontrolled fashion. Based on what you know about the functions of aldosterone, describe what the most prominent features of this condition would be.
4. Due to a mutation, a child was born with an ascending limb of Henle that was water permeable. What would be the minimum/maximum urine osmolarities (in units of mosm/liter) the child could produce?
   a. 100/300
   b. 300/1,200
   c. 100/100
   d. 1,200/1,200
   e. 300/300
5. An accident victim suffers permanent damage of the lower spinal cord and is paralyzed from the waist down. Describe what governs bladder emptying in this individual.
6. ***Clinical Consideration*** Marcus T. has noted a gradual decrease in his urine flow rate and is now experiencing difficulty in initiating micturition. He needs to urinate frequently, and often he feels as if his bladder is not empty even though he has just urinated. Analysis of Marcus's urine reveals no abnormalities. Are his urinary tract symptoms most likely caused by kidney disease, a bladder infection, or prostate enlargement?

Fluid and Acid-Base Balance 15

SYSTEMS OF MAJOR IMPORTANCE IN MAINTAINING FLUID AND ACID-BASE BALANCE

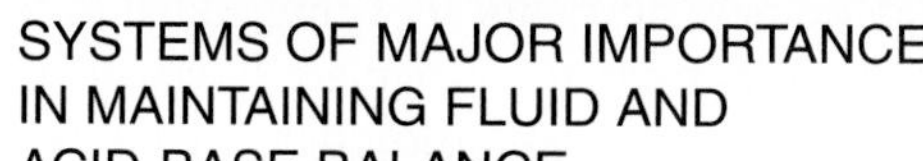

Body systems maintain homeostasis
HOMEOSTASIS
The kidneys, in conjunction with hormones involved in salt and water balance, are responsible for maintaining the volume and osmolarity of the extracellular fluid (internal environment). The kidneys, along with the respiratory system and chemical buffer systems in the body fluids, also contribute to homeostasis by maintaining the proper pH in the internal environment.
Homeostasis is essential for survival of cells
$CO_2 + H_2O \overset{ca}{\rightleftharpoons} H_2CO_3 \rightleftharpoons H^+ + HCO_3^-$
Chemical buffer systems in body fluids
CELLS
The volume of circulating blood must be maintained to help ensure adequate pressure to drive life-sustaining blood to the cells.
The osmolarity of fluid surrounding the cells must be closely regulated to prevent detrimental osmotic movement of water between the cells and ECF.
The pH of the internal environment must remain stable because pH changes alter neuromuscular excitability and enzyme activity, among other serious consequences.
Cells make up body systems

Homeostasis depends on maintaining a balance between the input and the output of all constituents present in the internal fluid environment. Regulation of **fluid balance** involves two separate components: *control of ECF volume,* of which circulating plasma volume is a part, and *control of ECF osmolarity* (solute concentration). The kidneys control ECF volume by maintaining **salt balance** and control ECF osmolarity by maintaining **water balance.** The kidneys maintain this balance by adjusting the output of salt and water in the urine as needed to compensate for variable input and abnormal losses of these constituents.

Similarly, the kidneys contribute to the maintenance of **acid-base balance** by adjusting the urinary output of hydrogen ion (acid) and bicarbonate ion (base) as needed. Also contributing to acid-base balance are the lungs, which can adjust their rate of excretion of hydrogen-ion-generating CO_2, and the chemical buffer systems in the body fluids.

Chapter Contents At a Glance

Fluid Balance

Input must equal output if balance is to be maintained.

The cells of complex multicellular organisms are able to survive and function only within a very narrow range of composition of the extracellular fluid (ECF), the internal fluid environment that bathes them. The quantity of any particular substance in the ECF is considered to be a readily available internal **pool.** The amount of the substance in the pool may be increased either by transferring more in from the external environment (most commonly by ingestion) or by metabolically producing it within the body (▬ Fig. 15-1). Substances may be removed from the body by excretion to the outside or by being used up in a metabolic reaction. If the quantity of a substance is to remain stable within the body, its input by means of ingestion or metabolic production must be balanced by an equal output by means of excretion or metabolic consumption. This relationship, known as the **balance concept,** is extremely important in the maintenance of homeostasis. Not all input and output pathways are applicable for every body fluid constituent. For example, salt is not synthesized or consumed by the body, so the stability of salt concentration in the body fluids depends entirely on a balance between salt ingestion and salt excretion.

The ECF pool can further be altered by transferring a particular ECF constituent into storage within the cells or bones. If the body as a whole has a surplus or deficit of a particular stored substance, the storage site can be expanded or partially depleted to maintain the ECF concentration of the substance within homeostatically prescribed limits. For example, following absorption of a meal, when more glucose is entering the plasma than is being consumed by the cells, the surfeit of glucose can be temporarily stored in muscle and liver cells in the form of glycogen. This storage depot can then be tapped between meals as necessary to maintain the plasma glucose level when no new nutrients are being added to the blood by eating. It is important to recognize, however, that internal storage capacity is limited. Although an internal exchange between the ECF and a storage depot can temporarily restore the plasma concentration of a particular substance to normal, in the long run any excess or deficit of that constituent must be compensated for by appropriate adjustments in total body input or output.

Another possible internal exchange between the pool and the remainder of the body is the reversible incorporation of certain plasma constituents into more complex molecular structures. For example, iron is incorporated into hemoglobin within the red blood cells during their synthesis but is released intact back into the body fluids when the red cells degenerate. This process differs from metabolic consumption of a substance, in which the substance is irretrievably converted into another form—for example, glucose converted into CO_2 plus H_2O plus energy. It also differs from storage in that the latter serves no purpose other than storage, whereas reversible incorporation into a more complex structure serves a specific purpose.

When total body input of a particular substance equals its total body output, a **stable balance** exists. When the gains via input for a substance exceed its losses via output, a **positive balance** exists. The result is an increase in the total amount of the substance in the body. In contrast, when the losses for a substance exceed its gains, a **negative balance** exists and the total amount of the substance in the body decreases.

Changing the magnitude of any of the input or output pathways for a given substance can alter its plasma concentration. To maintain homeostasis, any change in input must be balanced by a corresponding change in output (for example, increased salt intake must be matched by a corresponding increase in salt output in the urine), and, conversely, increased losses must be compensated for by increased intake. Thus, maintenance of a stable balance necessitates control. However, not all input and output pathways are regulated to maintain balance. Generally, input of various plasma constituents is poorly controlled or not controlled at all. We frequently ingest salt and H_2O, for example, not because we *need* them but because we *want* them, so the intake of salt and H_2O is highly variable. Likewise, hydrogen ion (H^+) is uncontrollably generated internally and added to the body fluids. Salt, H_2O, and H^+ can also be lost to the external environment to varying degrees through the digestive tract (vomiting), skin (sweating), and elsewhere without regard for salt, H_2O, or H^+ balance in the body. Compensatory adjustments in the urinary excretion of these substances are responsible for maintaining the body fluid's volume and salt and acid

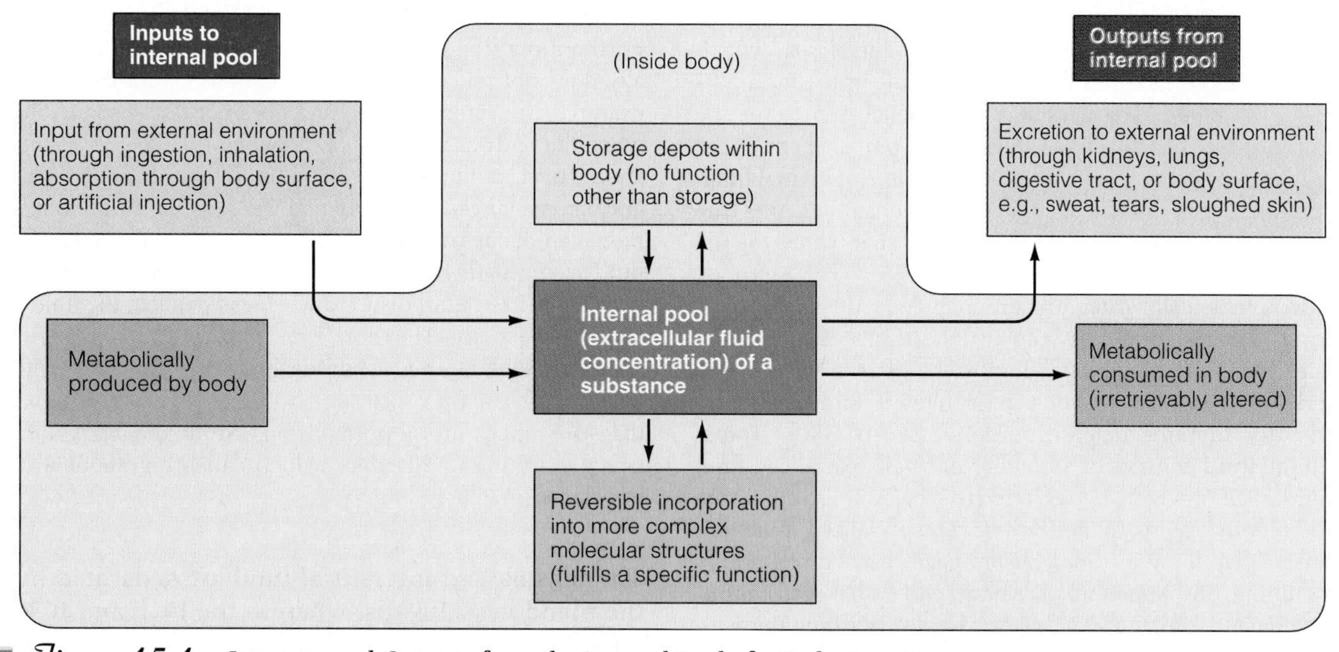

Figure 15-1 Inputs to and Outputs from the Internal Pool of a Body Constituent

composition within the extremely narrow homeostatic range compatible with life despite the wide variations in input and unregulated losses of these plasma constituents. This chapter will be devoted to the regulation of fluid balance (maintenance of salt and H_2O balance) and acid-base balance (maintenance of H^+ balance).

Body water is distributed between the intracellular and extracellular fluid compartments.

Water is by far the most abundant component of the human body, constituting 60% of body weight on average but ranging from 40% to 80%. The H_2O content of an individual remains fairly constant over a period of time, largely because of the kidneys' efficiency in regulating H_2O balance, but the percentage of body H_2O varies from person to person. The reason for the wide range in body H_2O among individuals is the variability in the amount of adipose tissue (fat) different people have. Adipose tissue has a low H_2O content compared to other tissues. Plasma, as you might suspect, is more than 90% H_2O. Even the soft tissues such as skin, muscles, and internal organs consist of 70% to 80% H_2O. The relatively drier skeleton is only 22% H_2O. Fat, however, is the driest tissue of all, having only 10% H_2O content. Accordingly, a high body H_2O content is associated with leanness and a low body H_2O content with obesity, since a larger proportion of the body is composed of relatively dry fat in overweight individuals.

The percentage of body H_2O is also influenced by the sex and age of the individual. Women have a lower body H_2O content than men, primarily because the female sex hormone, estrogen, promotes fat deposition in the breasts, buttocks, and elsewhere. This not only gives rise to the typical female figure, but also endows women with a higher proportion of adipose tissue and, therefore, a lower body H_2O content. The percentage of body H_2O also decreases progressively with age.

Body H_2O is distributed between two major fluid compartments: the fluid within the cells, *intracellular fluid (ICF)*, and the fluid surrounding the cells, *extracellular fluid (ECF)* (Table 15-1). (The terms "H_2O" and "fluid" are commonly used interchangeably. Although this usage is not entirely accurate because it ignores the solutes in the body fluids, it is acceptable when discussing the total volume of the fluids, since the major proportion of these fluids consists of H_2O.) The ICF compartment comprises about two-thirds of the total body H_2O. Even though each cell contains its own unique mixture of constituents, these trillions of minute fluid compartments are sufficiently similar to be considered collectively as one large fluid compartment. The remaining one-third of

Table 15-1
Classification of Body Fluid

Compartment	*Volume of Fluid (in Liters)*	*Percentage of Body Fluid*	*Percentage of Body Weight*
Total body fluid	42	100%	60%
Intracellular fluid (ICF)	28	67	40
Extracellular fluid (ECF)	14	33	20
Plasma	2.8	6.6 (20% of ECF)	4
Interstitial fluid	11.2	26.4 (80% of ECF)	16
Lymph	Negligible	Negligible	Negligible
Transcellular fluid	Negligible	Negligible	Negligible

the body H_2O found in the ECF compartment is further subdivided into plasma and interstitial fluid. The *plasma*, which makes up about one-fifth of the ECF volume, is the fluid portion of the blood. The *interstitial fluid*, which represents the other four-fifths of the extracellular fluid compartment, is the fluid that lies in the spaces between the cells. Interstitial fluid, sometimes also known as *tissue fluid*, constitutes the true internal environment in that it is the fluid that bathes the tissue cells.

Two other major categories are included in the ECF compartment: lymph and transcellular fluid. *Lymph* is fluid being returned from the interstitial fluid to the plasma by means of the lympathic system, where it is filtered through lymph nodes for immune defense purposes (see p. 330). **Transcellular fluid** consists of a number of small specialized fluid volumes, all of which are secreted by specific cells into a particular body cavity to perform some specialized function. Transcellular fluid includes *cerebrospinal fluid* (surrounding, cushioning, and nourishing the brain and spinal cord); *intraocular fluid* (maintaining the shape of and nourishing the eye); *synovial fluid* (lubricating and serving as a shock absorber for the joints); *pericardial, intrapleural*, and *peritoneal fluids* (lubricating movements of the heart, lungs, and intestines, respectively); and the *digestive juices* (digesting ingested foods). Although these fluids are extremely important functionally, they represent an insignificant fraction of the total body H_2O. Furthermore, the transcellular compartment as a whole usually does not reflect changes in the body's fluid balance. For example, the cerebrospinal fluid does not decrease in volume when the body as a whole is experiencing a negative H_2O balance. This is not to say that these fluid volumes never change. Localized changes in a particular transcellular fluid compartment can occur pathologically (such as too much intraocular fluid accumulating in the eyes of persons with glaucoma), but such a localized fluid disturbance does not affect the fluid balance of the body. Therefore, the transcellular compartment can usually be ignored when dealing with problems of fluid balance. The main exception to this generalization is when digestive juices are abnormally lost from the body during heavy vomiting or diarrhea, which can bring about a fluid imbalance.

The plasma and interstitial fluid are separated by the blood vessel walls, whereas the ECF and ICF are separated by cellular plasma membranes.

Several barriers separate the body fluid compartments, limiting the movement of H_2O and solutes between the various compartments to differing degrees. The two components of the ECF—plasma and interstitial fluid—are separated by the walls of the blood vessels. However, H_2O and all plasma constituents with the exception of plasma proteins are continuously and freely exchanged between the plasma and the interstitial fluid by passive means across the thin, pore-lined capillary walls. Accordingly, plasma and interstitial fluid are nearly identical in composition, except that interstitial fluid lacks plasma proteins. Any change in one of these ECF compartments is quickly reflected in the other compartment because they are constantly mixing.

In contrast to the very similar composition of the vascular and interstitial fluid compartments, the composition of the ECF differs considerably from that of the ICF (Fig. 15-2). Each cell is surrounded by a highly selective plasma membrane that permits passage of certain materials while excluding others. Movement through the membrane barrier occurs by both passive and active means and may be highly discriminating. Among the major differences between the ECF and ICF are (1) the presence of cellular proteins in the ICF that are unable to permeate the enveloping membranes to leave the cells and (2) the unequal distribution of Na^+ and K^+ and their attendant anions as a result of the action of the membrane-bound Na^+-K^+ ATPase pump that is present in all cells. This pump actively transports Na^+ out of and K^+ into cells; for this reason, Na^+ is the primary ECF cation, and K^+ is primarily found in the ICF. This unequal distribution of Na^+ and K^+, coupled with differences in membrane permeability to these ions, is responsible for the electrical properties of cells, including the initiation and propagation of action potentials in excitable tissues (see chapters 3 and 4).

Figure 15-2 **Ionic Composition of the Major Body Fluid Compartments**

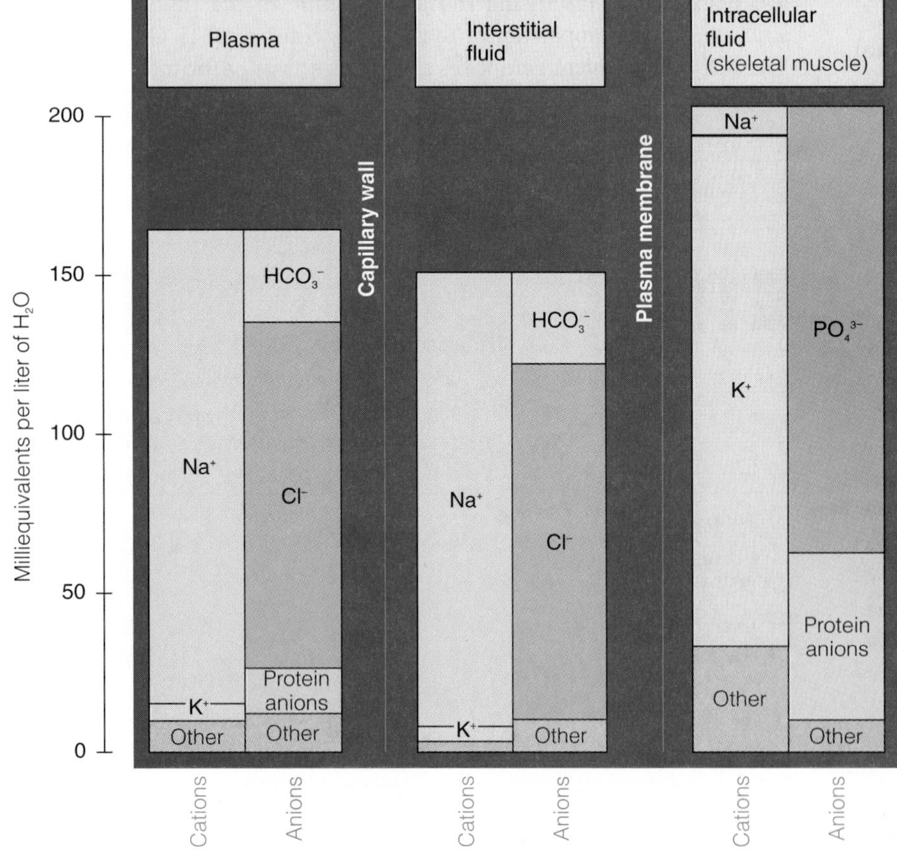

Except for the extremely small, electrically imbalanced portion of the total intracellular and extracellular ions that are involved in membrane potential, the majority of the ECF and ICF ions are electrically balanced. In the ECF, Na^+ is accompanied primarily by the anion Cl^- (chloride) and to a lesser extent by HCO_3^- (bicarbonate). The major intracellular anions are PO_4^{3-} (phosphate) and the negatively charged proteins trapped within the cell.

Although all cells' plasma membranes display selective permeability, all cells are freely permeable to H_2O. The movement of H_2O between the plasma and the interstitial fluid across capillary walls is governed by relative imbalances between capillary blood pressure (a fluid, or hydrostatic, pressure) and colloid osmotic pressure (see p. 328). In contrast, the net transfer of H_2O between the interstitial fluid and the ICF across the cellular plasma membranes occurs as a result of osmotic effects alone. The hydrostatic pressures of the interstitial fluid and ICF are both extremely low and fairly constant.

Fluid balance is maintained by regulating ECF volume and ECF osmolarity.

Extracellular fluid serves as an intermediary between the cells and the external environment. All exchanges of H_2O and other constituents between the ICF and the external world must occur through the ECF. Water added to the body fluids always enters the ECF compartment first, and fluid always leaves the body by way of the ECF.

Plasma is the only fluid that can be directly acted on to control its volume and composition. However, because of the free exchange across the capillary walls, if the volume and composition of the plasma are regulated, the volume and composition of the interstitial fluid bathing the cells are likewise regulated. Thus, any control mechanism that operates on the plasma in effect regulates the entire ECF. The ICF in turn is influenced by changes in the ECF to the extent permitted by the permeability of the membrane barriers surrounding the cells.

The factors regulated to maintain fluid balance in the body are ECF volume and ECF osmolarity. Although regulation of these two factors is closely interrelated, both being dependent on the relative NaCl and H_2O load in the body, the reason that they are closely controlled are significantly different:

1. *Extracellular fluid volume* must be closely regulated to help *maintain blood pressure.* Maintenance of *salt balance* is of primary importance in the long-term regulation of ECF volume.
2. *Extracellular fluid osmolarity* must be closely regulated to *prevent swelling or shrinking of the cells.* Maintenance of *water balance* is of primary importance in the regulation of ECF osmolarity.

We will examine each of these factors in more detail (Table 15-2).

Control of ECF volume is important in the long-term regulation of blood pressure.

A reduction in ECF volume causes a fall in arterial blood pressure by decreasing the plasma volume. Conversely, a rise in ECF volume increases the arterial blood pressure by expanding the plasma volume. Two compensatory measures come into play to transiently adjust the blood pressure until the ECF volume can be restored to normal:

1. Baroreceptor reflex mechanisms alter both cardiac output and total peripheral resistance through autonomic nervous system effects on the heart and blood vessels (see p. 338). These immediate cardiovascular responses are designed to minimize the effect a deviation in circulating volume has on blood pressure.
2. Fluid shifts occur temporarily and automatically between the plasma and interstitial fluid. A reduction in plasma volume is partially compensated for by a shift of fluid out of the interstitial compartment into the blood vessels, thereby expanding the circulating plasma volume at the

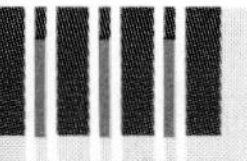
Table 15-2 **Summary of the Regulation of ECF Volume and Osmolarity**

Regulated Variable	*Need to Regulate the Variable*	*Outcomes if the Variable Is Not Normal*	*Mechanism for Regulating the Variable*
ECF volume	Important in the long-term control of arterial blood pressure	↓ECF volume → ↓arterial blood pressure ↑ECF volume → ↑arterial blood pressure	Maintenance of salt balance; salt osmotically "holds" H_2O, so the Na^+ load determines the ECF volume. Accomplished primarily by aldosterone-controlled adjustments in urinary Na^+ excretion
ECF osmolarity	Important to prevent detrimental osmotic movement of H_2O between the ECF and ICF	↓ECF osmolarity (hypotonicity) → H_2O enters the cell → cells swell ↑ECF osmolarity (hypertonicity) → H_2O leaves the cells → cells shrink	Maintenance of free H_2O balance. Accomplished primarily by vasopressin-controlled adjustments in excretion of H_2O in the urine

expense of the interstitial compartment. Conversely, when the plasma volume is too large, much of the excess fluid is shifted into the interstitial compartment. These shifts occur immediately and automatically as a result of changes in the balance of hydrostatic and osmotic forces acting across the capillary walls that arise when plasma volume deviates from normal (see p. 329).

These two measures provide temporary relief to help keep the blood pressure fairly constant, but they are not designed to be long-term solutions. Furthermore, these short-term compensatory measures are limited in their ability to minimize a change in blood pressure. If the plasma volume is too inadequate, the blood pressure remains too low no matter how vigorous the pump action of the heart, how constricted the resistance vessels, or what proportion of interstitial fluid shifts into the blood vessels. Conversely, if the plasma volume is greatly overexpanded, blood pressure cannot be restored down to normal even with maximum dilation of the resistance vessels and other short-term measures.

It is important, therefore, that other compensatory measures come into play in the long run to restore the ECF volume to normal. This responsibility for long-term regulation of blood pressure rests with the kidneys and the thirst mechanism, which control urinary output and fluid intake, respectively. In so doing, they accomplish needed fluid exchanges between the ECF and the external environment to regulate the body's total fluid volume. Accordingly, they have an important long-term influence on arterial blood pressure.

Control of salt balance is primarily important in regulating ECF volume.

Sodium and its attendant anions account for more than 90% of the ECF's osmotic activity. Because osmotic activity can be equated with "water-holding power," the total *Na^+ load* (the total quantity of NaCl, not its concentration) in the ECF determines the total amount of H_2O that will be osmotically retained in the ECF. The total mass of Na^+ salts in the ECF therefore determines the ECF's volume, and, appropriately, regulation of ECF volume depends primarily on controlling salt balance.

Table 15-3
Daily Salt Balance

Salt Input		*Salt Output*	
Avenue	*Amount (g/day)*	*Avenue*	*Amount (g/day)*
Ingestion	10.5	Obligatory loss in sweat and feces	0.5
		Controlled excretion in urine	10.0
Total input	10.5	Total output	10.5

To maintain salt balance at a set level, salt input must equal salt output, thus preventing salt accumulation or deficit in the body. The only avenue for salt input is ingestion, which typically is well in excess of the body's need for replacement of obligatory salt losses. A half gram of salt per day is adequate to replace the small amounts of salt usually lost in the feces and sweat. In our example of a typical daily salt balance (Table 15-3), salt intake is 10.5 g per day. (The average American salt intake is 10 to 15 g per day, although many people are consciously reducing their salt intake.)

Since we typically consume salt in excess of our needs, it is obvious that salt intake in humans is not well controlled. Carnivores (meat eaters) and omnivores (eaters of meat and plants, like humans), which naturally get sufficient salt in fresh meat (meat contains an abundance of salt-rich ECF), normally do not manifest a physiological appetite to seek additional salt. In contrast, herbivores (plant eaters), which lack salt naturally in their diets, develop a salt hunger and will travel miles to a salt lick. Humans generally have a hedonistic rather than a regulatory appetite for salt; we consume salt because we like it rather than because we have a physiological need, except in the unusual circumstance of severe salt depletion caused by a deficiency of aldosterone, the salt-conserving hormone.

The excess ingested salt must be excreted in the urine to maintain salt balance. The three avenues for salt output are obligatory loss of salt in sweat and feces and controlled excretion of salt in the urine (Table 15-3). The total amount of sweat produced is unrelated to salt balance, being determined instead by factors that control body temperature. Aldosterone can reduce the sweat's salt content, however, thus helping to conserve salt in a hot environment. The small salt loss in the feces is not subject to control. Except when sweating heavily or during diarrhea, the body normally uncontrollably loses only about 0.5 g of salt per day. This amount is actually the only salt that normally needs to be replaced by salt intake. Because salt consumption is typically far in excess of the meager amount needed to compensate for uncontrolled losses, the kidneys precisely excrete the excess salt in the urine to maintain salt balance. In our example, 10 g of salt are eliminated in the urine per day so that total salt output exactly equals salt input. By regulating the rate of urinary salt excretion (that is, by regulating the rate of Na^+ excretion, with Cl^- following along), the kidneys normally keep the total Na^+ mass in the ECF constant despite any notable changes in dietary intake of salt or unusual losses through sweating, diarrhea, or other means. As a reflection of keeping the total Na^+ mass in the ECF constant, the ECF volume in turn is maintained within the narrowly prescribed limits essential for normal circulatory function.

Deviations in ECF volume accompanying changes in salt load are responsible for triggering renal compensatory responses that quickly bring the Na^+ load and ECF volume back into line. Sodium is freely filtered at the glomerulus and actively reabsorbed, but it is not secreted by the tubules, so the amount of Na^+ excreted in the urine represents the amount of Na^+ that is filtered but is not subsequently reabsorbed:

$$Na^+ \text{ excreted} = Na^+ \text{ filtered} - Na^+ \text{ reabsorbed}$$

The kidneys accordingly adjust the amount of salt excreted by controlling two processes: (1) the glomerular filtration rate (GFR) and (2) more importantly, the tubular reabsorption of Na^+.

Control of the amount of Na^+ filtered through regulation of the GFR The amount of Na^+ filtered is equal to the plasma Na^+ concentration times the GFR. At any given plasma Na^+ concentration, any alteration in the GFR will correspondingly alter the amount of Na^+ filtered. Thus, control of the GFR can adjust the amount of Na^+ filtered each minute.

The GFR is deliberately changed to alter the amount of salt and fluid filtered as part of the general baroreceptor reflex response to a change in blood pressure (see Fig. 14-14), p. 481). The afferent arterioles that supply the renal glomeruli are constricted as part of the generalized vasoconstriction aimed at elevating a reduction in blood pressure. As a result of reduced blood flow into the glomeruli, GFR decreases and, accordingly, the amount of Na^+ and accompanying fluid that are filtered decreases. Consequently, excretion of salt and fluid is diminished. The conserved salt and fluid that otherwise would have been filtered and excreted help minimize the reduction in fluid volume and contribute to long-term restitution of blood pressure. Conversely, an elevation in ECF volume and arterial blood pressure is reflexly countered by a baroreceptor reflex response that leads to an increase in GFR, which in turn results in enhanced salt and fluid excretion. The elimination of extra salt and fluid that otherwise would have been conserved helps relieve the expanded plasma volume.

Note that adjustments in the amount of salt filtered are accomplished as part of the general blood pressure–regulating reflexes. Changes in Na^+ load in the body are not sensed as such; instead they are monitored indirectly through the effect that Na^+ ultimately has on blood pressure via its role in determining the ECF volume. Fittingly, baroreceptors that monitor fluctuations in blood pressure are responsible for bringing about adjustments in the amounts of Na^+ filtered and eventually excreted.

Control of the amount of Na^+ reabsorbed through the renin-angiotensin-aldosterone system The amount of Na^+ reabsorbed also depends on regulatory systems that play an important role in controlling blood pressure. Although Na^+ is reabsorbed throughout the tubule's length, only its reabsorption in the distal portions of the tubule is subject to control. The main factor controlling the extent of Na^+ reabsorption in the distal and collecting tubule is the powerful renin-angiotensin-aldosterone system, which promotes Na^+ reabsorption and thereby Na^+ retention. Sodium retention in turn promotes osmotic retention of H_2O and the subsequent expansion of plasma volume and elevation of arterial blood pressure. Appropriately, this Na^+ conserving system is activated by a reduction in NaCl/ECF volume/arterial blood pressure (see Fig. 14-19, p. 485).

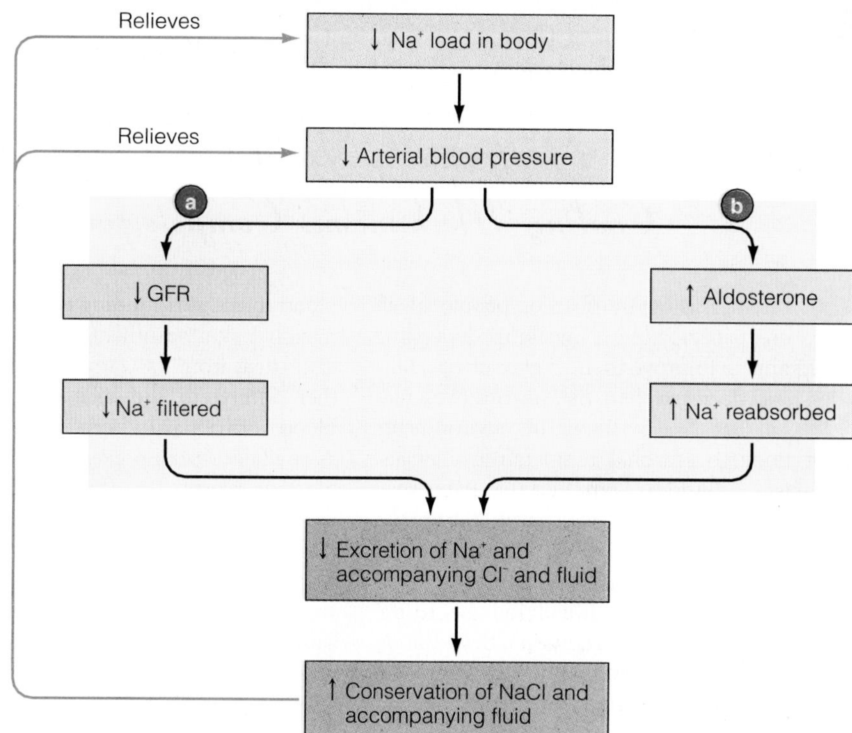

Figure 15-3 **Dual Effect of a Fall in Arterial Blood Pressure on Renal Handling of Na^+**

It should be apparent that control of GFR and Na^+ reabsorption are highly interrelated and that both are intimately tied in with long-term regulation of ECF volume as reflected by the blood pressure. Specifically, a fall in arterial blood pressure brings about a twofold effect in the renal handling of Na^+ (Fig. 15-3): (1) a reflex reduction in the GFR to decrease the amount of Na^+ filtered and (2) a hormonally adjusted increase in the amount of Na^+ reabsorbed. Together these effects reduce the amount of Na^+ excreted, thereby conserving for the body the Na^+ and accompanying H_2O necessary to compensate for the fall in arterial pressure. (See the boxed feature on p. 522, A Closer Look at Exercise Physiology.) Conversely, a rise in arterial blood pressure brings about (1) increases in the amount of Na^+ filtered, and (2) a reduction in renin-aldosterone activity, which decreases salt (and fluid) reabsorption. Together these actions increase salt (and fluid) excretion, thereby eliminating the extra fluid that was expanding the plasma volume and increasing the arterial pressure.

Control of ECF osmolarity prevents changes in ICF volume.

Maintenance of fluid balance in the body depends on the regulation of both ECF volume and ECF osmolarity. Whereas regulation of ECF volume is important in the long-term control of blood pressure, regulation of ECF osmolarity is important in preventing changes in cell volume.

A CLOSER LOOK AT EXERCISE PHYSIOLOGY

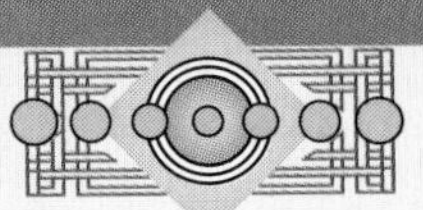

A Potentially Fatal Clash: When Exercising Muscles and Cooling Mechanisms Compete for an Inadequate Plasma Volume

An increasing number of people of all ages are participating in walking or jogging programs to improve their level of physical fitness and decrease their risk of cardiovascular disease. For people living in environments that undergo seasonal temperature changes, exercising outdoors can become dangerous due to fluid loss during the transition from the cool days of spring to the hot, humid days of summer. If exercise intensity is not modified until the participant gradually adjusts to the hotter environmental conditions, dehydration and salt loss can indirectly lead to heat cramps, heat exhaustion, or ultimately heat stroke and death.

Acclimatization refers to the gradual adaptations the body makes in order to maintain long-term homeostasis in response to a prolonged physical change in the surrounding environment, such as a change in temperature. When a person exercises in the heat without gradually adapting to the hotter environment, the body faces a terrible dilemma. During exercise, large amounts of blood must be delivered to the muscles to supply O_2 and nutrients and to remove the wastes that accumulate from their high rate of activity. Exercising muscles also produce heat. To maintain the body temperature in the face of this extra heat, blood flow to the skin is increased, so that heat from the warmed blood can be lost through the skin to the surrounding environment. If the environmental temperature is hotter than the body temperature, heat cannot be lost from the blood to the surrounding environment despite maximal skin vasodilation. Instead, the body gains heat from its warmer surroundings, further adding to the dilemma. Because extra blood is diverted to both the muscles and the skin when a person exercises in the heat, less blood is returned to the heart, and the heart pumps less blood per beat in accordance with the Frank-Starling mechanism (see p. 292). Therefore the heart must beat faster than it would in a cool environment to deliver the same amount of blood per minute. The increased rate of cardiac pumping further contributes to heat production.

The sweat rate also increases so that evaporative cooling can take place to help maintain the body temperature during periods of excessive heat gain. In an unacclimatized person, maximal sweat rate is about 1.5 liters per hour. During sweating, water-retaining salt as well as water is lost. The resulting loss of plasma volume through sweating further depletes the blood supply available for muscular exercise and for cooling through skin vasodilation.

The heart has a maximum rate at which it can pump. If exercise continues at a high intensity and this maximal rate is reached, the exercising muscles win the contest for blood supply. The body responds by constricting the skin arterioles, sacrificing cooling to maintain cardiac output and blood pressure. If exercise continues, body heat continues to rise, and heat exhaustion (rapid, weak pulse, hypotension, profuse sweating, and disorientation) or heat stroke (failure of the temperature-control center in the hypothalamus; hot, dry skin; extreme confusion or unconsciousness; and possibly death) can occur. In fact, every year people die of heat stroke in marathons run during hot, humid weather.

On the other hand, if a person exercises in the heat for two weeks at reduced, safe intensities, the body makes the following adaptations so that after acclimatization the person can do the same amount of work as was possible in a cool environment: (1) The plasma volume is increased by as much as 12%. The expansion of plasma volume provides sufficient blood to both supply the exercising muscles and direct blood to the skin for cooling. (2) The person begins sweating at a lower temperature so that the body does not get so hot before cooling begins. (3) The sweat rate increases as much as three times to 4 liters per hour with a more even distribution over the body. This increase in evaporative cooling reduces the need for cooling by skin vasodilation. (4) The sweat becomes more dilute so that less salt is lost in the sweat. The retained salt exerts an osmotic effect to hold water in the body and help maintain circulating plasma volume. These adaptations take fourteen days and will occur only if the person exercises in the heat. Being patient until these changes take place will enable the person to exercise safely throughout the summer months.

The **osmolarity** of a fluid is a measure of the concentration of the individual solute particles dissolved in it. The higher the osmolarity, the higher the concentration of solutes or, to look at it differently, the lower the concentration of H_2O. Water tends to move by osmosis down its own concentration gradient from an area of lower solute (higher H_2O) concentration to an area of higher solute (lower H_2O) concentration.

Osmosis occurs across the cellular plasma membranes only when there is a difference in concentration of nonpenetrating solutes between the ECF and ICF. Solutes that can penetrate a barrier separating two fluid compartments quickly become equally distributed between the two compartments and thus do not contribute to osmotic differences. The osmotic activity across the capillary wall is due to the unequal distribution of plasma proteins. Plasma proteins are present only in the plasma (as their name implies), because they are too large to penetrate the capillary walls and enter the interstitial fluid. All other solutes are in essentially the same concentration in the plasma and interstitial fluid, so they do not contribute to any unequal distribution of H_2O that would induce osmosis across the capillary wall.

In contrast, osmotic activity across the cellular plasma membranes is directly related to any differences in ionic concentration between the ECF and ICF. Plasma proteins play no role in the osmosis of H_2O across the cell membranes because they are absent in both the interstitial fluid and the ICF that are separated by the cell membranes.

Sodium and its attendant anions, being by far the most abundant solutes in the ECF in terms of numbers of particles, account for the vast majority of the ECF's osmotic activity. In contrast, K^+ and its accompanying intracellular anions are responsible for the ICF's osmotic activity. Even though small amounts of Na^+ and K^+ passively diffuse across the plasma membrane all the time, these ions behave as if they are nonpenetrating because of Na^+-K^+ pump activity. Any Na^+ that passively diffuses down its electrochemical gradient into the cell is promptly pumped back outside, so the result is the same as if Na^+ were barred from the cells. The same holds true in reverse for K^+; it in effect remains trapped within the cells. The resulting unequal distribution of Na^+ and K^+ and their accompanying anions between the ECF and the ICF is responsible for the osmotic activity of these two fluid compartments.

Normally, the osmolarities of the ECF and ICF are the same, because the total concentration of K^+ and other effectively nonpenetrating solutes inside the cells is equal to the total concentration of Na^+ and other effectively nonpenetrating solutes in the fluid surrounding the cells. Even though the nonpenetrating solutes in the ECF and ICF differ, their concentrations are normally identical, and it is the number (not the nature) of the unequally distributed particles per volume that determines the fluid's osmolarity. Because the osmolarities of the ECF and ICF are normally equal, no net movement of H_2O usually occurs into or out of the cells. Therefore, cell volume normally remains constant, because no H_2O osmotically enters or leaves the cells.

However, any circumstance that results in a loss or gain of *free* H_2O (that is, loss or gain of H_2O that is not accompanied by comparable solute deficit or excess) leads to changes in ECF osmolarity. If there is a deficit of free H_2O in the ECF, the solutes become too concentrated, and the ECF osmolarity becomes abnormally high (that is, becomes *hypertonic*) (see p. 496). If there is excess free H_2O in the ECF, the solutes become too dilute, and the ECF osmolarity becomes abnormally low (that is, becomes *hypotonic*). When the ECF osmolarity changes with respect to the ICF osmolarity, osmosis takes place, with H_2O either leaving or entering the cells, depending, respectively, on whether the ECF is more or less concentrated than the ICF.

The osmolarity of the ECF must therefore be regulated to prevent these undesirable shifts of H_2O into or out of the cells. As far as the ECF itself is concerned, the concentration of its solutes does not really matter. However, it is crucial that ECF osmolarity be maintained within very narrow limits to prevent the cells from swelling (by osmotically gaining H_2O from the ECF) or shrinking (by osmotically losing H_2O to the ECF).

We will examine the fluid shifts that occur between the ECF and the ICF when the ECF osmolarity becomes hypertonic or hypotonic relative to the ICF. Then we will consider how water balance and subsequently ECF osmolarity are normally maintained to minimize detrimental changes in cell volume.

ECF hypertonicity Hypertonicity of the ECF, or the excessive concentration of ECF solutes, is usually associated with **dehydration,** or a negative free H_2O balance. Dehydration with accompanying hypertonicity can be brought about in three major ways: (1) insufficient H_2O intake, such as might occur during desert travel or might accompany difficulty in swallowing; (2) excessive H_2O loss, such as might occur during heavy sweating, vomiting, or diarrhea (even though both H_2O and solutes are lost during these conditions, relatively more H_2O is lost, so the remaining solutes become more concentrated); and (3) **diabetes insipidus,** a disease characterized by a deficiency of *vasopressin (antidiuretic hormone),* the hormone that increases the permeability of the distal and collecting tubules to H_2O and thus enhances water conservation by reducing urinary output of water. In diabetes insipidus, the kidneys cannot conserve H_2O because they are unable to reabsorb H_2O from the distal portions of the nephron in the absence of vasopressin. Such patients typically produce up to 20 liters of very dilute urine per day, compared to the normal average of 1.5 liters per day. Unless H_2O intake keeps pace with this tremendous loss of H_2O in the urine, the person quickly dehydrates. Patients complain that they spend an extraordinary amount of time day and night going to the bathroom and getting drinks. Fortunately, they can be treated with replacement vasopressin administered by nasal spray.

On rare occasions, the ECF becomes hypertonic in the absence of dehydration because of the abnormal accumulation of osmotically active solutes that do not normally contribute significantly to ECF osmotic activity. This occurs, for example, with the high blood urea levels in uremia (kidney failure).

Whenever the ECF compartment becomes hypertonic, H_2O moves out of the cells by osmosis into the more concentrated ECF until the ICF osmolarity equilibrates with the ECF. Thus, although the H_2O deficit occurs first in the ECF, the H_2O deficit quickly becomes equally distributed between the ECF and ICF as a result of the osmotic shift of H_2O out of the cells. The cells shrink as H_2O leaves them. (See the box feature on p. 524, ● Concepts, Challenges, and Controversies.) Of particular concern is the fact that considerable shrinking of brain neurons causes disturbances in brain function, which can be manifested as mental confusion and irrationality in moderate cases and can bring about possible delirium, convulsions, or coma in more severe hypertonic conditions. Rivaling the neural symptoms in seriousness are the circulatory disturbances that arise from a reduction in plasma volume in association with dehydration. Circulatory problems may range from a slight reduction in blood pressure to circulatory shock and death.

Other more common symptoms become apparent, even in mild cases of dehydration. For example, dry skin and sunken eyeballs are indications of loss of H_2O from the underlying soft tissues, and the tongue becomes dry and parched because of suppressed salivary secretion.

ECF hypotonicity Hypotonicity of the ECF is usually associated with **overhydration;** that is, excess free H_2O is present. When a positive free H_2O balance exists, the ECF is less concentrated (more dilute) than normal. Usually, any surplus free H_2O is promptly excreted in the urine, so hypotonicity generally does not occur. However, hypotonicity can arise in three ways: (1) Patients with renal failure who are unable to

CONCEPTS, CHALLENGES, AND CONTROVERSIES

Breaching the Blood-Brain Barrier

In the early 1980s, Neil Shay, a thirty-six-year-old cross-country truck driver, received a grim diagnosis. While making a delivery run, Shay suddenly became disoriented; he didn't know where he was or where he was going. By the time he checked into a nearby hospital, his mental fog was coupled with a searing headache. There he learned the bad news: the diagnosis was CNS lymphoma, a rare type of cancer.

Shay had three malignant tumors in his brain about the size of golf balls. None of the standard approaches for treating cancer seemed to be available in his case. The tumors were too scattered and extensive to remove surgically, and they could not be eradicated by radiation therapy. Although lymphomas elsewhere in the body frequently succumb to chemotherapy, these cancer-killing drugs could not penetrate the staunch blood-brain barrier (see p. 119), so they were unable to reach the life-threatening brain tumors. Shay was given less than a year to live.

More than a decade later, however, Shay is still alive, and free of brain cancer, thanks to the ingenuity of neurosurgeon Edward Neuwelt. Using his knowledge of physiology, Neuwelt developed a pioneering method of getting chemotherapeutic drugs through the blood-brain barrier to treat the brain cancer. Applying his idea for the first time on Shay, Neuwelt threaded a catheter into one of his patient's carotid arteries, the vessels that supply the brain with blood. For thirty seconds, he pumped a concentrated solution of mannitol, a type of sugar, through the catheter, making the brain-bound blood very hypertonic. The sugar treatment was followed immediately by an infusion of potent chemotherapeutic drugs, and this time, unlike ever before, these cancer-killing agents were able to get through the blood-brain barrier.

How did the pretreatment with sugar accomplish this feat? The endothellial cells that form the walls of the brain capillaries are sealed together by tight junctions, which normally prevent substances from passing between the cells. Flooding the brain with hypertonic blood osmotically drew water from these endothelial cells. As the cells shrank, they pulled apart slightly. The chemotherapeutic drugs that followed the mannitol were able to pass through this transient breach in the blood-brain barrier and reach the cancer cells in Shay's brain. The endothelial cells quickly regained water and returned to their normal state as soon as fresh isotonic blood reached the brain, but meanwhile the life-saving drug had snuck through the barrier. The technique has been used on other brain lymphoma patients, over half of whom are alive today.

Nevertheless, this approach is not the final answer to penetrating the blood-brain barrier with chemotherapeutic and other drugs. For one thing, it is not risk-free. During the brief time the blood-brain barrier is breached, other chemicals that are usually barred from the brain can also enter. These undesirable chemicals include blood-borne hormones, such as epinephrine, that can act as neurotransmitters and set off unwanted nervous activity. Some patients undergoing this treatment have suffered seizures and developed more long-lasting neurological problems. Also, blood-borne infectious agents and toxins could enter the brain through the passages opened up for the chemotherapeutic drugs.

Other investigators are looking for alternative ways to smuggle neuropharmaceuticals across an undisturbed blood-brain barrier by capitalizing on the normal means by which substances get into the brain. As neuroscientists learn more about the brain, the possibility of developing drugs to treat a variety of brain disorders is continuing to expand. Yet these new neuropharmaceuticals are useless unless they can get into the brain. With passage between the endothelial cells blocked, substances normally get through the blood-brain barrier in one of two ways: (1) Lipid-soluble substances that must get into the brain, such as life-sustaining O_2 and regulatory steroid hormones, simply passively diffuse through the lipid plasma membranes of the brain endothelial cells. (2) Non-lipid-soluble substances that the brain needs, such as glucose for energy production and amino acids for protein synthesis, are transported across the brain endothelial cells by highly selective membrane-bound carriers. Twelve different specific transport systems have been identified to date.

Some foreign chemicals are already known to cross the blood-brain barrier. For example, agents such as nicotine, alcohol, and cocaine exert their effects so quickly because they are lipid soluble and pass through the blood-brain barrier with ease. Likewise, most of the drugs that are useful in treating mental illness, such as antidepressants, tranquilizers, and sedatives, are at least partially lipid soluble.

Scientists are now looking for ways to get non-lipid-soluble drugs across the barrier. These important compounds, which have largely been excluded from the brain, include antibiotics, chemotherapeutic agents, neurotransmitters, neuropeptides, nerve growth factor, and many hormones. New approaches being investigated include (1) "lipidizing" a water-soluble medication by chemically linking it to a fat group, (2) tricking one of the membrane-bound carriers to transport foreign therapeutic passengers, and (3) inducing endocytosis of drugs by brain endothelial cell membranes.

excrete a dilute urine become hypotonic when they consume relatively more H_2O than solutes. (2) Hypotonicity can occur transiently in healthy people if H_2O is rapidly ingested to such an excess that the kidneys are unable to respond quickly enough to eliminate the extra H_2O. (3) Hypotonicity can occur when excess H_2O without solute is retained in the body as a result of inappropriate secretion of vasopressin. Vasopressin is normally secreted in response to a H_2O deficit, which is relieved by increasing H_2O reabsorption in the distal portion of the nephrons. However, vasopressin secretion, and therefore hormonally controlled tubular H_2O reabsorption, can be increased in response to pain, acute infections, trauma, and other stressful situations, even when there is no H_2O deficit in the body. The increase in vasopressin secretion and resultant H_2O retention elicited in response to stress are appropriate in anticipation of potential blood loss in the stressful situation; the extra retained H_2O could minimize the effect a loss of blood volume would have on blood pressure.

However, because modern-day stressful situations generally are not accompanied by blood loss, the increased vasopressin secretion is inappropriate as far as the body's fluid balance is concerned. The reabsorption and retention of too much H_2O lead to dilution of the body's solutes.

Whichever way it is brought about, excess free H_2O retention first dilutes the ECF compartment, making it hypotonic. The resultant difference in osmotic activity between the ECF and ICF induces H_2O to move by osmosis from the more dilute ECF into the cells, with the cells swelling as H_2O moves into them osmotically. Again, the H_2O excess originally present in the ECF rapidly becomes equally distributed between the ECF and ICF as H_2O osmotically enters the cells. Like the shrinking of cerebral neurons, pronounced swelling of brain cells also leads to brain dysfunction. Symptoms include confusion, irritability, lethargy, headache, dizziness, vomiting, drowsiness, and, in severe cases, even convulsions, coma, and death. Nonneural symptoms of overhydration include weakness caused by the swelling of muscle cells and circulatory disturbances, including hypertension and edema, caused by expansion of the plasma volume.

The condition of overhydration, hypotonicity, and cellular swelling resulting from excess free H_2O retention is known as **water intoxication.** It should not be confused with the fluid retention that occurs with excess salt retention. In the latter case, the ECF is still isotonic because the increase in salt is matched by a corresponding increase in H_2O. Because the interstitial fluid is still isotonic, there is no osmotic gradient to drive the extra H_2O into the cells. The excess salt and H_2O burden is therefore confined to the ECF compartment, with circulatory consequences being the most important concern. In the case of water intoxication, the body's H_2O content is increased relative to the salt content, and the excess H_2O becomes evenly distributed between the ECF and ICF compartments. In addition to any circulatory disturbances, symptoms caused by cellular swelling become a problem.

Isotonic fluid gain or loss Let us contrast the situations of hypertonicity and hypotonicity with what happens as a result of isotonic fluid gain or loss. An example of an isotonic fluid gain is therapeutic intravenous administration of an isotonic solution, such as isotonic saline. When the isotonic fluid is injected into the ECF compartment, the ECF volume increases, but the concentration of ECF solutes remains unchanged; in other words, the ECF is still isotonic. Because the ECF's osmolarity has not changed, the ECF and ICF are still in osmotic equilibrium, so no net fluid shift occurs between the two compartments. The ECF compartment has increased in volume without causing a shift of H_2O into the cells. Thus, unless trying to correct an osmotic imbalance, intravenous fluid therapy should be isotonic to prevent fluctuations in intracellular volume and possible neural symptoms.

Similarly, in the case of an isotonic fluid loss such as occurs in hemorrhage, the loss is confined to the ECF with no corresponding loss of fluid from the ICF. Fluid does not shift out of the cells because the ECF remaining within the body is still isotonic, so there is no osmotic gradient to draw H_2O out of the cells. Of course, many other mechanisms come into play to counteract the loss of blood, but the ICF compartment is not directly affected by the loss.

Thus, when the ECF and ICF are in osmotic equilibrium, no net movement of H_2O into or out of the cells occurs regardless of whether the ECF volume increases or decreases. Shifts in H_2O between the ECF and ICF take place only when the ECF becomes more or less concentrated than the cells and this usually arises from a loss or gain, respectively, of free H_2O.

Control of water balance by means of vasopressin and thirst is of primary importance in regulating ECF osmolarity.

Because cells, especially the brain neurons, do not function properly when they are either shrunk or swollen, it is important that ECF osmolarity be closely regulated to prevent osmotic fluid shifts between the ECF and ICF. Control of H_2O balance is crucial for regulating ECF osmolarity. Because increases in free H_2O cause the ECF to become too dilute and deficits of free H_2O cause the ECF to become too concentrated, the osmolarity of the ECF must be immediately corrected by restoring stable H_2O balance to avoid osmotically induced fluxes of H_2O into or out of the cells.

To maintain a stable H_2O balance, H_2O input must equal H_2O output. In a person's typical daily H_2O balance (▮Table 15-4), a little more than a liter of H_2O is added to the body by drinking liquids. Surprisingly, an amount almost equal to that is obtained from eating solid food. Recall that muscles consist of about 75% H_2O; meat is therefore 75% H_2O because it is animal muscle. Likewise, fruits and vegetables consist of 60% to 90% H_2O. Therefore, people normally obtain almost as much H_2O from solid foods as from the liquids they drink. The third source of H_2O input is metabolically produced H_2O. Chemical reactions within the cells convert food and O_2 into energy, producing CO_2 and H_2O in the process. This **metabolic H_2O** produced during cellular metabolism and released into the ECF averages about 350 ml/day. The average H_2O intake from these three sources thus

Table 15-4
Daily Water Balance

Water Input		*Water Output*	
Avenue	*Quantity (ml/day)*	*Avenue*	*Quantity (ml/day)*
Fluid intake	1,250	Insensible loss (from lungs and nonsweating skin)	900
H_2O in food intake	1,000	Sweat	100
Metabolically produced H_2O	350	Feces	100
		Urine	1,500
Total input	2,600	Total Output	2,600

totals 2,600 ml/day. Another source of H_2O often employed therapeutically is intravenous infusion of fluid.

On the output side of the H_2O balance tally, a person loses close to a liter of H_2O daily without being aware of it. This so-called **insensible loss** occurs from the lungs and nonsweating skin. During the process of respiration, inspired air becomes saturated with H_2O within the airways. This H_2O is lost when the moistened air is subsequently expired. Normally, we are not aware of this H_2O loss, but it can be recognized on cold days, when the H_2O vapor condenses so that we can "see our breath." The other insensible loss is the continual loss of H_2O from the skin even in the absence of sweating. Water molecules are able to diffuse through the cells of the skin and evaporate without being noticed. Fortunately, the skin is fairly waterproof because of its keratinized exterior layer, which protects against a much greater loss of H_2O by this avenue (see p. 408). When this protective surface layer is lost, such as when a person has extensive burns, fluid loss from the burned surface can increase to the point of causing serious fluid-balance problems.

Sensible loss (loss of which the person is aware) of H_2O from the skin occurs through sweating, which represents another avenue of H_2O output. At an air temperature of 68°F, an average of 100 ml of H_2O is lost daily through sweating. Loss of water from sweating can vary substantially, of course, depending on the environmental temperature and humidity and the degree of physical activity; it may range from zero up to as much as several liters per hour in very hot weather.

Another passageway for H_2O loss from the body is through the feces. Normally, only about 100 ml of H_2O are lost via this route each day. During the process of fecal formation in the large intestine, most of the H_2O is absorbed out of the digestive tract lumen into the blood, thereby conserving fluid and solidifying the digestive tract's contents for elimination. Additional losses of H_2O can occur from the digestive tract through vomiting or diarrhea.

By far the most important output mechanism is urine excretion, with 1,500 ml of urine being produced daily on the average. The total H_2O output is 2,600 ml/day, the same as the volume of H_2O input in our example. This balance is not by chance. Normally, H_2O input matches H_2O output so that the H_2O in the body remains in balance.

Of the many sources of H_2O input and output, only two can be regulated to maintain H_2O balance. On the intake side, thirst influences the amount of fluid ingested, and on the output side, the kidneys can adjust the magnitude of urine formation. Control of H_2O output in the urine is the most important mechanism in the control of H_2O balance. Some of the other factors are regulated but not for the purpose of maintaining H_2O balance. Food intake is subject to regulation to maintain energy balance, whereas control of sweating is important in the maintenance of body temperature. Metabolic H_2O production and insensible losses are completely unregulated.

Control of water output in the urine by vasopressin Fluctuations in ECF osmolarity caused by imbalances between H_2O input and output are quickly compensated for by adjusting the urinary excretion of H_2O without changing the usual excretion of salt; that is, H_2O reabsorption and excretion are partially dissociated from solute reabsorption and excretion, so the amount of free H_2O retained or eliminated can be varied to quickly restore ECF osmolarity to normal. Adjustments in free H_2O reabsorption and excretion are accomplished through changes in vasopressin secretion (see p. 500). Throughout most of the nephron, H_2O reabsorption is important in regulating ECF volume because salt reabsorption is accompanied by comparable H_2O reabsorption. In the distal portions of the tubule, however, variable free H_2O reabsorption can take place without comparable salt reabsorption because of the presence of a vertical osmotic gradient in the renal medulla to which the collecting tubule is exposed. Vasopressin increases the permeability of this late portion of the tubule to H_2O. Depending on the amount of vasopressin present, the amount of free H_2O reabsorbed can be adjusted as necessary to restore ECF osmolarity to normal.

Control of water input by thirst **Thirst** is the subjective feeling that drives one to ingest H_2O. A **thirst center** is located in the hypothalamus in close proximity to the vasopressin-secreting cells. Thirst increases H_2O input, whereas vasopressin decreases H_2O output by reducing urine production.

Regulation of vasopressin secretion and thirst The hypothalamic control centers that regulate vasopressin secretion (and thus urinary output) and thirst (and thus drinking) act in concert. Vasopressin secretion and thirst are both stimulated by a free H_2O deficit and suppressed by a free H_2O excess. Thus, appropriately, the same circumstances that call for a reduction in urinary output to conserve body H_2O also give rise to the sensation of thirst to replenish body H_2O. The predominant excitatory input for both vasopressin secretion and thirst comes from **hypothalamic osmoreceptors** located near the vasopressin-secreting cells and thirst center. These osmoreceptors monitor the osmolarity of the fluid surrounding them, which in turn reflects the concentration of the entire internal fluid environment. As the osmolarity increases (too little H_2O) and the need for H_2O conservation increases, vasopressin secretion and thirst are both stimulated (■ Fig. 15-4). As a result, reabsorption of H_2O in the distal and collecting tubules is increased so that urinary output is reduced and H_2O is conserved, while H_2O intake is simultaneously encouraged. These actions restore depleted H_2O stores, thus relieving the hypertonic condition by diluting the solutes to normal concentration. On the other hand, H_2O excess, as manifested by a reduction in ECF osmolarity, brings about increased urinary output (through a decrease in vasopressin secretion) as well as suppression of thirst, which together reduce the water load in the body.

Even though the major stimulus for vasopressin secretion and thirst is an increase in ECF osmolarity, the vasopressin-secreting cells and thirst center are both influenced to a moderate extent by changes in ECF volume mediated by input from the **left atrial volume receptors.** Located in the left atrium, these volume receptors monitor the blood pressure, which is a reflection of the ECF volume. In response to a major reduction in ECF volume and arterial pressure, as during hemorrhage, the left atrial volume receptors reflexly stimulate both vasopressin secretion and thirst. The outpouring of vasopressin and increased thirst lead to decreased urine output and increased fluid intake, respectively. Furthermore, vasopressin, at the circulating levels elicited by a large decline in ECF volume and arterial pressure, exerts a potent vasoconstrictor effect on arterioles (thus giving rise to its name), in addition to having an effect on the kidney tubules. Both by helping expand the ECF and plasma volume and by increasing the total peripheral resistance, vasopressin helps relieve the low blood pressure that elicited the vasopressin secretion. Conversely, vasopressin and thirst are both inhibited when the ECF/plasma volume and arterial blood pressure are elevated. The resultant suppression of H_2O intake, coupled with the elimination of the excess ECF/plasma volume in the urine, helps to restore the blood pressure to normal.

Recall that low ECF/plasma volume and low arterial blood pressure also reflexly increase aldosterone secretion. The resultant increase in Na^+ reabsorption ultimately leads to osmotic retention of H_2O, expansion of ECF volume, and an increase in arterial blood pressure. In fact, aldosterone-controlled Na^+ reabsorption is the most important factor in regulating ECF volume, with the vasopressin and thirst mechanism playing only a supportive role. Left atrial volume receptor input to vasopressin secretion and the thirst center become important only in response to fairly serious changes in ECF/plasma volume and arterial pressure. Osmoreceptor input is normally the dominant factor controlling vasopressin secretion and thirst.

Yet another stimulus for increasing both thirst and vasopressin is angiotensin II (Table 15-5). When the renin-angiotension-aldosterone mechanism is activated to conserve Na^+, angiotensin II, in addition to stimulating aldosterone secretion, acts directly on the brain to give rise to the urge to drink and concurrently stimulates vasopressin to enhance renal H_2O reabsorption (see p. 486). The resultant increased H_2O intake and decreased urinary output help to correct the reduction in ECF volume that triggered the renin-angiotensin-aldosterone system.

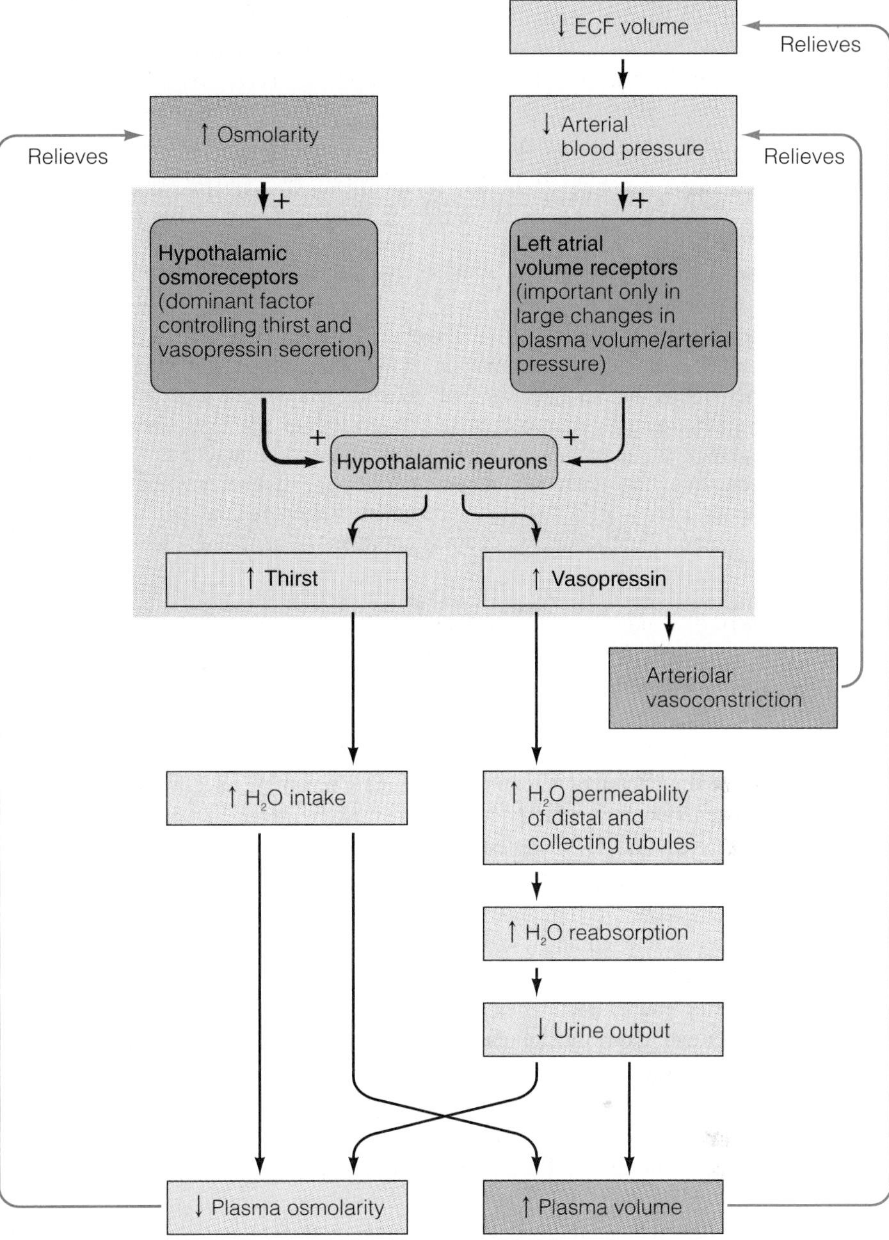

Figure 15-4 **Control of Increased Vasopressin Secretion and Thirst during a H_2O Deficit**

Several factors affect vasopressin secretion but not thirst. As described earlier, vasopressin is stimulated by stress-related inputs such as pain, fear, and trauma that have nothing directly to do with maintaining H_2O balance. In fact, H_2O retention as a result of the inappropriate secretion of vasopressin can bring about a hypotonic H_2O imbalance. On the other hand, alcohol inhibits vasopressin secretion and can lead to ECF hypertonicity by promoting excessive free H_2O excretion.

One stimulus that promotes thirst but not vasopressin secretion is a direct effect of dryness of the mouth. Nerve endings in the mouth are directly stimulated by dryness, which causes an intense sensation of thirst that can often be relieved merely by moistening the mouth even though no H_2O is actually ingested. A dry mouth can exist when salivation is suppressed by factors unrelated to the body's H_2O content, such as nervousness, excessive smoking, or certain drugs.

Factors that affect vasopressin secretion or thirst but have nothing directly to do with the body's need for H_2O are usually short-lived. It is important to recognize that the dominant, long-standing control of vasopressin and thirst *is* directly correlated with the body's state of H_2O—namely, by the status of ECF osmolarity and, to a lesser extent, by ECF volume.

In addition, there is evidence of some kind of "oral H_2O metering," at least in animals. A thirsty animal will rapidly drink only enough H_2O to satisfy its H_2O deficit. It stops drinking before the ingested H_2O has had time to be absorbed

Table 15-5 Factors Controlling Vasopressin Secretion and Thirst

Factor	Effect on Vasopressin Secretion	Effect on Thirst	Comment
↑ECF osmolarity	↑	↑	Major stimulus for vasopressin secretion and thirst
↓ECF volume	↑	↑	Important only in large changes in ECF volume/arterial blood pressure
Angiotensin II	↑	↑	Part of dominant pathway for promoting compensatory salt and H_2O retention when ECF volume/arterial blood pressure are reduced
Pain, fear, trauma and other stress-related inputs	Inappropriate ↑ unrelated to body's H_2O balance	No effect	Promotes excess H_2O retention and ECF hypotonicity (resultant H_2O retention of potential value in maintaining arterial blood pressure in case of blood loss in the stressful situation)
Alcohol	Inappropriate ↓ unrelated to body's H_2O balance	No effect	Promotes excess H_2O loss and ECF hypertonicity
Dry mouth	No effect	↑	Nerve endings in the mouth that ultimately give rise to the sensation of thirst are directly stimulated by dryness

from the digestive tract and actually return the ECF compartment to normal. Exactly what factors are involved in signaling that enough H_2O has been consumed is still uncertain. It might be a learned anticipatory response based on past experience. This mechanism seems to be less effective in humans, because we frequently drink more than is necessary to meet the needs of our bodies or, conversely, may not drink enough to make up a deficit.

In fact, even though the thirst mechanism exists to control H_2O intake, fluid consumption by humans is often influenced more by habit and sociological factors than by the need to regulate H_2O balance. Thus, even though H_2O intake is critical in maintaining fluid balance, it is not precisely controlled in humans, who err especially on the side of excess H_2O consumption. We usually drink when we are thirsty, but we often drink even when we are not thirsty because, for example, we are on a coffee break. With H_2O intake being inadequately controlled and indeed even contributing to H_2O imbalances in the body, the primary factor involved in maintaining H_2O balance is urinary output regulated by the kidneys. Accordingly, vasopressin-controlled H_2O reabsorption is of primary importance in regulating ECF osmolarity.

Just as vasopressin secretion is the major factor controlling ECF osmolarity but plays a supportive role in controlling ECF volume, aldosterone secretion is the major factor controlling ECF volume but plays a supportive role in controlling ECF osmolarity. Researches have recently learned that increased osmolarity accompanying H_2O deprivation acts directly on the adrenal cortex to reduce aldosterone secretion, thus decreasing renal Na^+ reabsorption and increasing urinary excretion of Na^+. It is proposed that this controlling mechanism, which is in addition to the renin-angiotensin pathway and the direct action of plasma K^+ concentration on aldosterone secretion (see Fig. 14-25, p. 492), may be important in reducing a detrimental osmotic flux of H_2O out of the cells in the face of prolonged H_2O deprivation and elevated ECF osmolarity.

Acid-Base Balance

Acids liberate free hydrogen ions whereas bases accept them.

Acid-base balance actually refers to the precise regulation of **free** (that is, unbound) **hydrogen-ion (H^+)** concentration in the body fluids. To indicate the concentration of a chemical, its symbol is enclosed in brackets [ ]. Thus, $[H^+]$ designates H^+ concentration.

Acids are a special group of hydrogen-containing substances that *dissociate,* or separate, when in solution to liberate free H^+ and anions (negatively charged ions). Many other substances (for example, carbohydrates) also contain hydrogen, but they are not classified as acids because the hydrogen is tightly bound within their molecular structure and is never liberated as free H^+.

A strong acid has a greater tendency to dissociate in solution than a weak acid does; that is, a greater percentage of a strong acid's molecules separates into free H^+ and anions. Hydrochloric acid (HCl) is an example of a strong acid; every HCl molecule dissociates into free H^+ and Cl^- (chloride) when dissolved in H_2O. With a weaker acid such as carbonic acid (H_2CO_3), only a portion of the molecules dissociate in solution into H^+ and HCO_3^- (bicarbonate anions). The remaining H_2CO_3 molecules remain intact. Since only the free hydrogen ions contribute to the acidity of a solution, H_2CO_3 is a weaker acid than HCl because H_2CO_3 does not yield as many free hydrogen ions per number of acid molecules present in solution (Fig. 15-5).

The extent of dissociation for a given acid is always constant; that is, when in solution, the same proportion of a particular acid's molecules always separate to liberate free H^+, with the other portion always remaining intact. The constant degree of dissociation for a particular acid (in this example H_2CO_3) is expressed by its **dissociation constant (K)** as follows:

$$[H^+]\ [HCO_3^-]/[H_2CO_3] = K$$

- $[H^+]\ [HCO_3^-]$ represents the concentration of ions resulting from H_2CO_3 dissociation.
- $[H_2CO_3]$ represents the concentration of intact (undissociated) H_2CO_3.

The dissociation constant varies for different acids.

A **base** is a substance that can combine with a free H^+ and thus remove it from solution. A strong base is able to bind H^+ more readily than a weak base can.

The pH designation is used to express hydrogen-ion concentration.

The $[H^+]$ in the ECF is normally 4×10^{-8} or 0.00000004 equivalents per liter. The concept of pH has been developed to express $[H^+]$ more conveniently. Specifically, **pH** equals the logarithm (log) to the base 10 of the reciprocal of the hydrogen-ion concentration:

$$pH = \log 1/[H^+]$$

Two important points should be noted about this formula:

1. Because $[H^+]$ is in the denominator, *a high $[H^+]$ corresponds to a low pH, and a low $[H^+]$ corresponds to a high pH.* The greater the $[H^+]$, the larger the number by which 1 must be divided, and the lower the pH.
2. *Every unit change in pH actually represents a tenfold change in $[H^+]$* because of the logarithmic relationship. A log to the base 10 indicates how many times 10 must be multiplied by itself to produce a given number. For example, the log of 10=1, whereas the log of 100=2. Ten must be multiplied by itself twice to yield 100 (10×10=100). Numbers less than 10 have logs less than 1. Numbers between 10 and 100 have logs between 1 and 2, and so on. Accordingly, each unit of change in pH is indicative of a tenfold change in $[H^+]$. For example, a solution with a pH of 7 has a $[H^+]$ 10 times less than that of a solution with a pH of 6 (1 pH-unit difference) and 100 times less than that of a solution with a pH of 5 (2 pH-unit difference).

The pH of pure H_2O is 7.0, which is considered to be chemically neutral. An extremely small proportion of H_2O molecules dissociate into hydrogen ions and hydroxyl (OH^-) ions. Because OH^- has the ability to bind with H^+ to once again form a H_2O molecule, it is considered to be basic. Since an equal number of acidic hydrogen ions and basic hydroxyl ions are formed, H_2O is neutral, being neither acidic nor basic. Solutions having a pH less than 7.0 contain a higher $[H^+]$ than pure H_2O and are considered to be **acidic.** Conversely, solutions having a pH value greater than 7.0 have a lower $[H^+]$ and are considered to be **basic** or **alkaline** (Fig. 15-6a). Fig. 15-7 compares the pH values of common solutions.

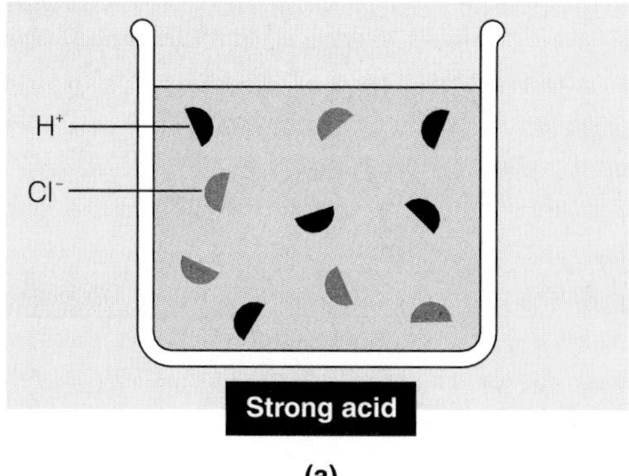

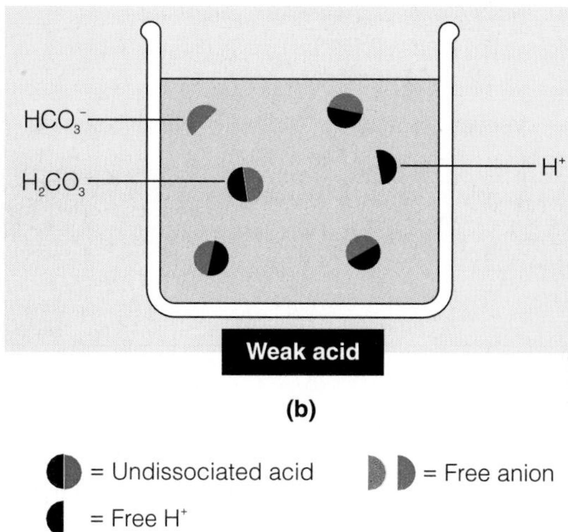

Figure 15-5 **Comparison of a Strong and a Weak Acid** (a) Five molecules of a strong acid. A strong acid such as HCl (hydrochloric acid) completely dissociates into free H^+ and anions in solution. (b) Five molecules of a weak acid. A weak acid such as H_2CO_3 (carbonic acid) only partially dissociates into free H^+ and anions in solution.

The pH of arterial blood is normally 7.45 and the pH of venous blood is 7.35, for an average blood pH of 7.4. The pH of venous blood is slightly lower (more acidic) than that of arterial blood because of H^+ generated by the formation of H_2CO_3 from CO_2 picked up at the tissue capillaries. **Acidosis** exists whenever the blood pH falls below 7.35, whereas **alkalosis** occurs when the blood pH is above 7.45 (Fig. 15-6b). Note that the reference point for determining the body's acid-base status is not the chemically neutral pH of 7.0 but the normal plasma pH of 7.4. Thus, a plasma pH of 7.2 is considered acidotic even though in chemistry a pH of 7.2 is considered basic.

Death occurs if arterial pH falls outside the range of 6.8 to 8.0 for more than a few seconds because an arterial pH of less than 6.8 or greater than 8.0 is not compatible with life. Obviously, therefore, $[H^+]$ in the body fluids must be carefully regulated.

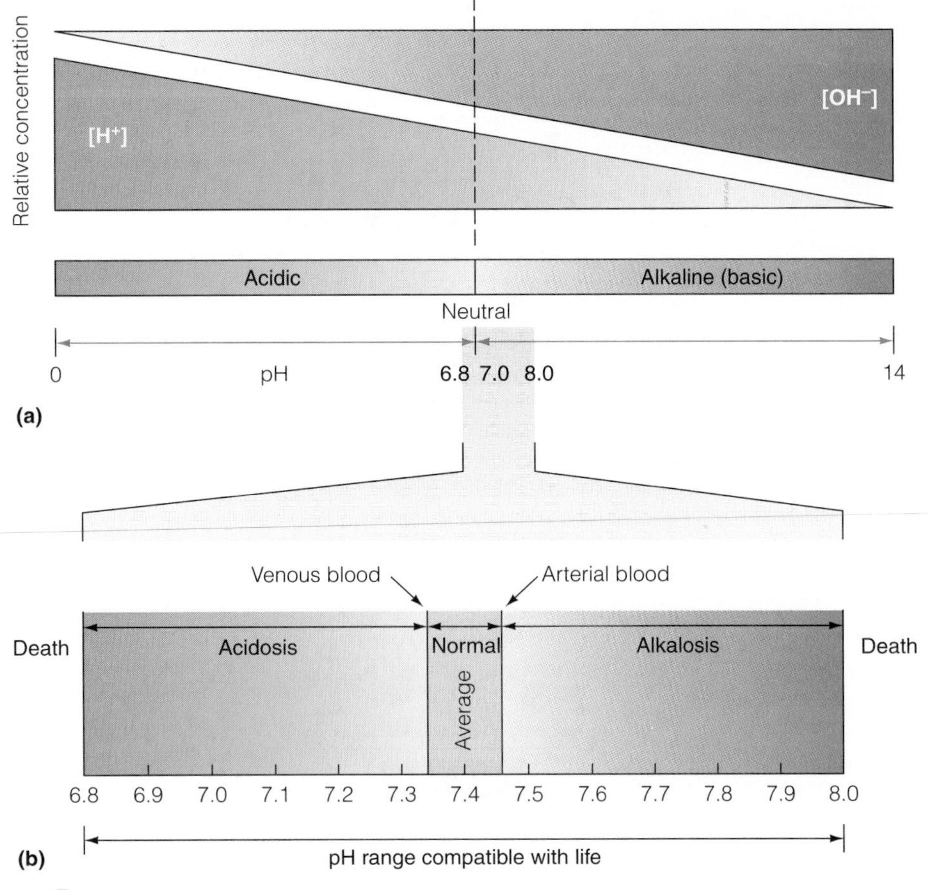

Figure 15-6 pH Considerations in Chemistry and Physiology
(a) Relationship of pH to the relative concentrations of H^+ and base (OH^-) under chemically neutral, acidic, and alkaline conditions. (b) Plasma pH range under normal, acidosis, and alkalosis conditions.

Fluctuations in hydrogen-ion concentration have profound effects on body chemistry.

Only a narrow pH range is compatible with life because even small changes in $[H^+]$ have dramatic effects on normal cell function. The prominent consequences of fluctuations in $[H^+]$ include the following:

- Changes in excitability of nerve and muscle cells are among the major clinical manifestations of pH abnormalities. The major clinical effect of increased $[H^+]$ (acidosis) is depression of the central nervous system. Acidotic patients become disoriented and, in more severe cases, eventually die in a state of coma. In contrast, the major clinical effect of decreased $[H^+]$ (alkalosis) is overexcitability of the nervous system, first the peripheral nervous system and later the central nervous system. Peripheral nerves become so excitable that they fire even in the absence of normal stimuli. Such overexcitability of the afferent (sensory) nerves give rise to abnormal "pins and needles" tingling sensations. On the other hand, overexcitability of efferent (motor) nerves brings about muscle twitches and, in more pronounced cases, severe muscle spasms. Death may occur in extreme alkalosis as spasm of the respiratory muscles seriously impairs breathing. Alternatively, severely alkalotic patients may die of convulsions resulting from overexcitability of the central nervous system. In less serious situations, CNS overexcitability is manifested as extreme nervousness.
- Hydrogen-ion concentration exerts a marked influence on enzyme activity. Even slight deviations in $[H^+]$ alter the shape and activity of protein molecules. Since enzymes are proteins, a shift in the body's acid-base balance disturbs the normal pattern of metabolic activity catalyzed by these enzymes. Some cellular chemical reactions are accelerated; others are depressed.
- Changes in $[H^+]$ influence K^+ levels in the body. When reabsorbing Na^+ from the filtrate, the renal tubular cells secrete either K^+ or H^+ in exchange (see p. 492). Normally, they secrete a preponderance of K^+ compared to H^+. Because of the intimate relationship between secretion of H^+ and K^+ by the kidneys, an increased rate of secretion of one of these ions is accompanied by a decreased rate of secretion of the other. For example, if more H^+ than normal is eliminated by the kidneys, as occurs when the body fluids become acidotic, less K^+ than usual can be excreted. The resultant K^+ retention can affect cardiac function, among other detrimental consequences.

Hydrogen ions are continually being added to the body fluids as a result of metabolic activities.

As with any other constituent, to maintain a constant $[H^+]$ in the body fluids, input of hydrogen ions must be balanced by an equal output. On the input side, only a small amount of acid capable of dissociating to release H^+ is taken in with food, such as the weak citric acid found in oranges. Most H^+ in the body fluids is generated internally from metabolic activities. Normally, H^+ is continually being added to the body fluids from the three following sources:

1. *Carbonic acid formation.* The major source of H^+ is through H_2CO_3 formation from metabolically produced CO_2. Cellular oxidation of nutrients yields energy, with CO_2 and H_2O as end products. Catalyzed by the enzyme *carbonic anhydrase* (ca), CO_2 and H_2O form H_2CO_3, which then partially dissociates to liberate free H^+ and HCO_3^-:

$$CO_2 + H_2O \overset{ca}{\rightleftharpoons} H_2CO_3 \rightleftharpoons H^+ + HCO_3^-$$

This reaction is reversible because it can proceed in either direction, depending on the concentrations of the substances involved as dictated by the *law of mass action* (see p. 447). Within the systemic capillaries, the CO_2 level in the blood increases as metabolically produced CO_2 enters from the tissues. This drives the reaction to the acid side, generating H^+ as well as HCO_3^- in the process. In

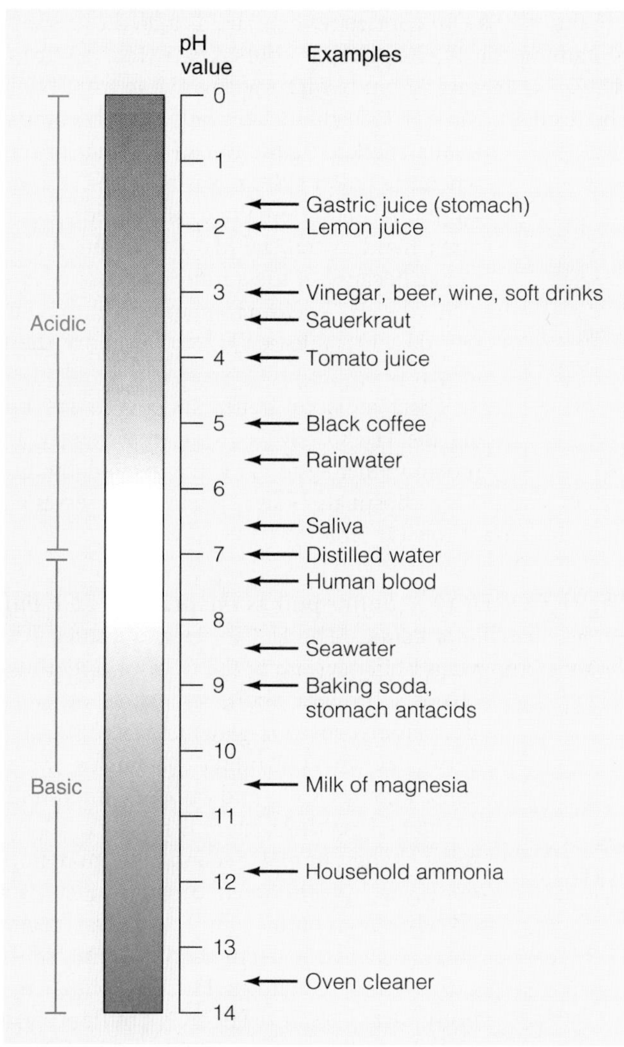

Figure 15-7 **Comparison of pH Values of Common Solutions**

the lungs, the reaction is reversed: CO_2 diffuses from the blood flowing through the pulmonary capillaries into the alveoli (air sacs), from which it is expired to the atmosphere. The resultant reduction in CO_2 in the blood drives the reaction toward the CO_2 side. Hydrogen ion and HCO_3^- form H_2CO_3, which rapidly decomposes into CO_2 and H_2O once again. The CO_2 is exhaled while the hydrogen ions generated at the tissue level are incorporated into H_2O molecules.

When the respiratory system is able to keep pace with the rate of metabolism, there is no net gain or loss of H^+ in the body fluids from metabolically produced CO_2. When the rate of CO_2 removal by the lungs does not match the rate of CO_2 production at the tissue level, however, the resultant accumulation or deficit of CO_2 in the body leads to an excess or shortage, respectively, of free H^+ in the body fluids.

2. *Inorganic acids produced during the breakdown of nutrients.* Dietary proteins and other ingested nutrient molecules that are found abundantly in meat contain a large quantity of sulfur and phosphorus. When these molecules are broken down, sulfuric acid and phosphoric acid are produced as by-products. Being moderately strong acids, these two inorganic acids dissociate to a large extent, liberating free H^+ into the body fluids. In contrast, the breakdown of fruits and vegetables produces bases that, to some extent, neutralize the acids derived from protein metabolism. Generally, however, more acids than bases are produced during the breakdown of ingested food, leading to an excess of these acids.
3. *Organic acids resulting from intermediary metabolism.* Numerous organic acids are produced during normal intermediary metabolism (see p. 35). For example, fatty acids are produced during fat metabolism, and lactic acid is produced by muscles during heavy exercise. These acids partially dissociate to yield free H^+.

Hydrogen-ion generation therefore normally goes on continuously as a result of ongoing metabolic activities. Furthermore, in certain disease states, additional acids may be produced that further contribute to the total body pool of H^+. For example, in diabetes mellitus, large quantities of keto acids may be produced as a result of abnormal fat metabolism. Some types of acid-producing medications may also add to the total load of H^+ that must be handled by the body. Thus, input of H^+ is unceasing, highly variable, and essentially unregulated.

The crux of H^+ balance is maintaining the normal alkalinity of the ECF (pH 7.4) despite this constant onslaught of acid. The generated free H^+ must be largely removed from solution while in the body and ultimately must be eliminated from the body so that the pH of the body fluids can remain within the narrow range compatible with life. Mechanisms must also exist to compensate rapidly for the occasional situation in which the ECF becomes too alkaline.

Three lines of defense against changes in $[H^+]$ operate to maintain the $[H^+]$ of body fluids at a nearly constant level despite unregulated input: (1) the *chemical buffer systems,* (2) the *respiratory mechanism of pH control,* and (3) the *renal mechanism of pH control.* We will look at each of these methods.

Chemical buffer systems act as the first line of defense against changes in hydrogen-ion concentration.

A **chemical buffer system** is a mixture in a solution of two (or perhaps three) chemical compounds that minimize pH changes when either an acid or a base is added to or removed from the solution. A buffer system consists of a pair of substances involved in a reversible reaction—one substance that can yield free H^+ as the $[H^+]$ starts to fall and another that can bind with free H^+ (thus removing it from solution) when $[H^+]$ starts to rise. An important example of such a buffer system is the carbonic acid: bicarbonate (H_2CO_3:HCO_3^-) buffer pair, which is involved in the following reversible reaction:

$$H^+ + HCO_3^- \rightleftharpoons H_2CO_3$$

When a strong acid such as HCl is added to an unbuffered solution, all the dissociated H^+ remains free in the solution

(Fig. 15-8a). In contrast, when HCl is added to a solution containing the H_2CO_3:HCO_3^- buffer pair (Fig. 15-8b), the HCO_3^- immediately binds with the free H^+ to form H_2CO_3. The resultant weak H_2CO_3 dissociates only slightly compared to the marked reduction in pH that occurred when the buffer system was not present and the additional H^+ remained unbound. On the other hand, when the pH of the solution starts to rise because of the addition of base or loss of acid, the H^+-yielding member of the buffer pair, H_2CO_3, releases H^+ to minimize the rise in pH.

There are four buffer systems in the body: (1) the H_2CO_3:HCO_3^- buffer system, (2) the protein buffer system, (3) the hemoglobin buffer system, and (4) the phosphate buffer system. Each serves a different important role (Table 15-6).

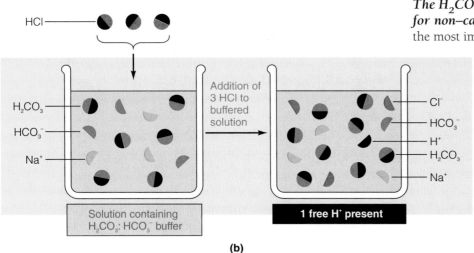

Figure 15-8 **Action of Chemical Buffers** (a) Addition of HCl to an unbuffered solution. All of the added hydrogen ions (H^+) remain free and contribute to the acidity of the solution. (b) Addition of HCl to a buffered solution. Bicarbonate ions (HCO_3^-), the basic member of the buffer pair, bind with some of the added H^+ and remove them from solution so that they do not contribute to its acidity.

The H_2CO_3:HCO_3^- buffer pair is the primary ECF buffer for non–carbonic acids The H_2CO_3:HCO_3^- buffer pair is the most important buffer system in the ECF for buffering pH changes brought about by causes other than fluctuations in CO_2-generated H_2CO_3. It is a very effective ECF buffer system for two reasons. First, H_2CO_3 and HCO_3^- are abundant in the ECF, so this system is readily available to resist changes in pH. Second, and more importantly, each component of this buffer pair is closely regulated. The kidneys regulate HCO_3^-, and the respiratory system regulates CO_2, which generates H_2CO_3. Thus, in the body the H_2CO_3:HCO_3^- buffer system includes involvement of CO_2 via the following reaction with which you are already familiar:

$$H^+ + HCO_3^- \rightleftharpoons H_2CO_3 \rightleftharpoons CO_2 + H_2O$$

When new H^+ is added to the plasma from any source other than CO_2 (for example through lactic acid released into the ECF from exercising muscles), the preceding reaction is driven toward the right side of the equation. As the extra H^+ binds with HCO_3^-, it no longer contributes to the acidity of the body fluids, so the rise in $[H^+]$ is abated. In the converse situation, when the plasma $[H^+]$ occasionally falls below normal for some reason other than a change in CO_2 (such as the loss of plasma-derived HCl in the gastric juices during vomiting), the reaction is driven toward the left side of the equation. Dissolved CO_2 and H_2O in the plasma form H_2CO_3, which generates additional H^+ to make up for the H^+ deficit. In so doing, the H_2CO_3:HCO_3^- buffer system resists the fall in $[H^+]$.

This buffer system cannot buffer changes in pH induced by fluctuations in H_2CO_3. A buffer system cannot buffer itself. Consider, for example, the situation in which the plasma $[H^+]$ is elevated because of CO_2 retention associated with a respiratory problem. The rise in CO_2 drives the reaction to the left according to the law of mass action, resulting in an elevation in $[H^+]$. The increase in $[H^+]$ occurs as a result of

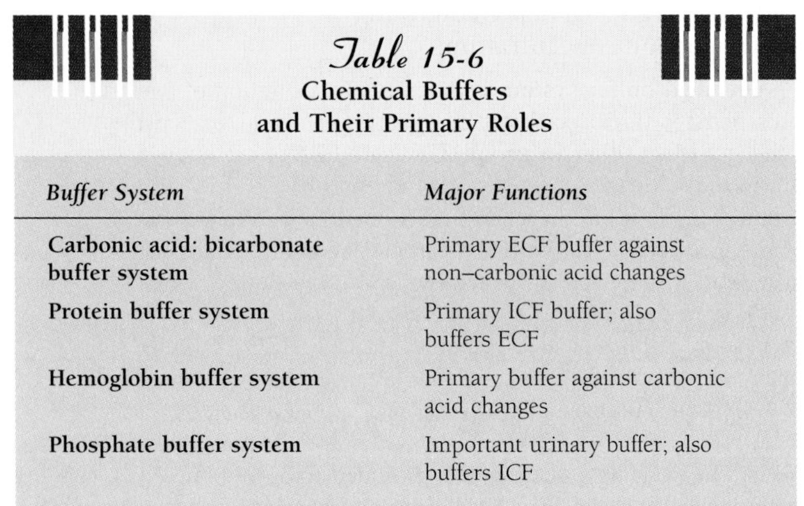

Table 15-6
Chemical Buffers and Their Primary Roles

Buffer System	*Major Functions*
Carbonic acid: bicarbonate buffer system	Primary ECF buffer against non–carbonic acid changes
Protein buffer system	Primary ICF buffer; also buffers ECF
Hemoglobin buffer system	Primary buffer against carbonic acid changes
Phosphate buffer system	Important urinary buffer; also buffers ICF

the reaction being driven to the *left* because of an increase in CO_2, so the elevated $[H^+]$ cannot drive the reaction to the *right* to buffer the increase in $[H^+]$. Only if the increase in $[H^+]$ is brought about by some mechanism other than CO_2 accumulation can this buffer system be shifted to the CO_2 side of the equation and be effective in reducing the $[H^+]$. Likewise, in the opposite situation, the H_2CO_3:HCO_3^- buffer system cannot compensate for a reduction in $[H^+]$ caused by a deficit of CO_2 by generating more H^+-yielding H_2CO_3 when the problem in the first place is a shortage of H_2CO_3-forming CO_2. Other mechanisms to be described shortly, are available for resisting fluctuations in pH caused by changes in CO_2 levels.

The relationship between $[H^+]$ and the members of a buffer pair can be expressed according to the **Henderson-Hasselbalch equation,** which, for the H_2CO_3:HCO_3^- buffer system, is as follows:

$$pH = pK + \log [HCO_3^-]/[H_2CO_3]$$

Although you do not need to know the mathematical manipulations involved, it will be helpful for you to understand how this formula is derived. Recall that the dissociation constant K for H_2CO_3 acid is

$$[H^+]\ [HCO_3^-]/[H_2CO_3] = K$$

and that the relationship between pH and $[H^+]$ is

$$pH = \log 1/[H^+]$$

Then, by solving the dissociation constant formula for $[H^+]$ (that is, $[H^+] = K \times [H_2CO_3]/[HCO_3^-]$) and replacing this value for $[H^+]$ in the pH formula, one comes up with the Henderson-Hasselbalch equation.

Practically speaking, $[H_2CO_3]$ is a direct reflection of the concentration of dissolved CO_2, henceforth referred to as $[CO_2]$, because most of the CO_2 in the plasma is converted into H_2CO_3. (The dissolved CO_2 concentration is equivalent to P_{CO_2}, as described in the chapter on respiration.) Therefore, the equation becomes

$$pH = pK + \log [HCO_3^-]/[CO_2]$$

The pK is the logarithm of 1/K and, like K, always remains a constant for any given acid. For H_2CO_3, the pK is 6.1. Since the pK is always a constant, changes in pH are associated with changes in the ratio between $[HCO_3^-]$ and $[CO_2]$. Normally, the ratio between $[HCO_3^-]$ and $[CO_2]$ in the ECF is 20 to 1; that is, there is 20 times more HCO_3^- than CO_2. Plugging this ratio into our formula:

$$\begin{aligned} pH &= pK + \log [HCO_3^-]/[CO_2] \\ &= 6.1 + \log 20/1 \end{aligned}$$

The log of 20 is 1.3. Therefore pH = 6.1 + 1.3 = 7.4, which is the normal pH of plasma.

When the ratio of $[HCO_3^-]$ to $[CO_2]$ increases above 20/1, pH increases. Accordingly, either a rise in $[HCO_3^-]$ or a fall in $[CO_2]$, both of which increase the $[HCO_3^-]/[CO_2]$ ratio if the other component remains constant, shifts the acid-base balance toward the alkaline side. In contrast, when the $[HCO_3^-]/[CO_2]$ ratio decreases below 20/1, pH decreases toward the acid side. This can occur either if the $[HCO_3^-]$ decreases or if the $[CO_2]$ increases while the other component remains constant. Because $[HCO_3^-]$ is regulated by the kidneys and $[CO_2]$ by the lungs, the pH of the plasma can be shifted up and down by kidney and lung influences. The kidneys and lungs regulate pH (and thus free $[H^+]$) largely by controlling plasma $[HCO_3^-]$ and $[CO_2]$, respectively, to restore their ratio to normal. Accordingly,

$$pH \propto \frac{[HCO_3^-]\ \text{controlled by kidney function}}{[CO_2]\ \text{controlled by respiratory function}}$$

Because of this relationship, not only do both the kidneys and lungs normally participate in pH control but renal or respiratory dysfunction can also induce acid-base imbalances by altering the $[HCO_3^-]/[CO_2]$ ratio.

The protein buffer system is primarily important intracellularly The most plentiful buffers of the body fluids are the proteins, including the intracellular proteins and the plasma proteins. Proteins are excellent buffers because they contain both acidic and basic groups that can give up or take up H^+. Quantitatively, the protein system is most important in buffering changes in $[H^+]$ in the ICF because of the sheer abundance of the intracellular proteins. A more limited number of plasma proteins reinforces the H_2CO_3:HCO_3^- system in extracellular buffering.

The hemoglobin buffer system buffers hydrogen ion generated from carbonic acid Hemoglobin (Hb) buffers the H^+ generated from metabolically produced CO_2 in transit between the tissues and the lungs. At the systemic capillary level, CO_2 continuously diffuses into the blood from the tissue cells where it is being produced. The greatest percentage of this CO_2 forms H_2CO_3, which partially dissociates into H^+ and HCO_3^-. Simultaneously, some of the oxyhemoglobin (HbO_2), releases O_2, which diffuses into the tissues. Reduced (unoxygenated) Hb has a greater affinity for H^+ than HbO_2 does. Therefore, most of the H^+ generated from CO_2 at the tissue level becomes bound to reduced Hb and no longer contributes to the acidity of the body fluids:

$$H^+ + Hb \rightleftharpoons HHb$$

At the lungs the reactions are reversed. As Hb picks up O_2 diffusing from the alveoli into the red blood cells, the affinity of Hb for H^+ is decreased, so H^+ is released. This liberated H^+ combines with HCO_3^- to yield H_2CO_3, which in turn produces CO_2, which is exhaled. Meanwhile, the hydrogen has been reincorporated into neutral H_2O molecules. Were it not for Hb, the blood would become much too acidic after picking up CO_2 at the tissues. With the tremendous buffering capacity of the Hb system, venous blood is only slightly more acidic than arterial blood despite the large volume of H^+-generating CO_2 carried in the venous blood.

The phosphate buffer system is an important urinary buffer The phosphate buffer system is composed of an acid phosphate salt (NaH_2PO_4) that can donate a free H^+ when the $[H^+]$ falls and a basic phosphate salt (Na_2HPO_4) that can accept a free H^+ when the $[H^+]$ rises. Basically, this buffer pair can alternately switch a H^+ for a Na^+ as demanded by the $[H^+]$:

$$Na_2HPO_4 + H^+ \rightleftharpoons NaH_2PO_4 + Na^+$$

Even though the phosphate pair is a good buffer, its concentration in the ECF is rather low, so it is not very important as an ECF buffer. Because phosphates are more abundant within the cells, this system contributes significantly to intracellular buffering, being rivaled only by the more plentiful intracellular proteins. Even more importantly, the phosphate system serves as an excellent urinary buffer. Humans normally consume more phosphate than needed. The excess phosphate filtered through the kidneys is not reabsorbed but remains in the tubular fluid to be excreted (because the renal threshold for phosphate is exceeded; see p. 488). This excreted phosphate buffers the urine as it is being formed by removing from solution the H^+ secreted into the tubular fluid. None of the other body fluid buffer systems are present in the tubular fluid to play a role in buffering urine during its formation. Most or all of the filtered HCO_3^- and CO_2 (alias H_2CO_3) are reabsorbed, whereas Hb and plasma proteins are not even filtered.

All chemical buffer systems act immediately, within fractions of a second, to minimize changes in pH. When $[H^+]$ is altered, the involved buffer systems' reversible chemical reactions are shifted at once in favor of compensating for the change in $[H^+]$. Accordingly, the buffer systems are the *first line of defense* against changes in $[H^+]$ since they are the first mechanism to respond.

Through the mechanism of buffering, most hydrogen ions seem to "disappear" from the body fluids between the time of their generation and their elimination. It must be emphasized, however, that none of the chemical buffer systems actually *eliminates* H^+ from the body. These ions are merely removed from solution by being incorporated within one of the members of the buffer pair, thus preventing the hydrogen ions from contributing to the body fluids' acidity. Because each buffer system has a limited capacity to "soak up" H^+, the H^+ that is unceasingly produced must ultimately be removed from the body. If H^+ were not eventually eliminated, soon all the body fluid buffers would already be bound with H^+ and there would be no further buffering ability.

The respiratory and renal mechanisms of pH control actually eliminate acid from the body instead of merely suppressing it, but they respond more slowly than the chemical buffer systems. We will now turn our attention to these other defenses against changes in acid-base balance.

The respiratory system regulates hydrogen-ion concentration by controlling the rate of CO_2 removal from the plasma through adjustments in pulmonary ventilation.

The respiratory system plays an important role in acid-base balance through its ability to alter pulmonary ventilation and, consequently, to alter the excretion of H^+-generating CO_2. Because of this ability, it should not be surprising that the level of respiratory activity is governed at least in part by the arterial $[H^+]$ (Table 15-7). When arterial $[H^+]$ increases, the respiratory center in the brain stem is reflexly stimulated (see p. 460) to increase pulmonary ventilation (the rate at which gas is exchanged between the lungs and the atmosphere). As the rate and depth of breathing increase, more CO_2 than usual is blown off, so less H_2CO_3 than normal is added to the body fluids. Since CO_2 forms acid, the removal of CO_2 in essence removes acid from the body. Conversely, when arterial $[H^+]$ falls, pulmonary ventilation is reduced. As a result of slower, shallower breathing, metabolically produced CO_2 diffuses from the cells into the blood faster than it is removed from the blood by the lungs, so higher-than-usual amounts of acid-forming CO_2 accumulate in the blood, thus restoring $[H^+]$ toward normal.

The lungs are extremely important in maintaining the $[H^+]$ of the plasma. Every day they remove from the body fluids what amounts to 100 times more H^+ derived from car-

Table 15-7 **Respiratory Adjustments to Acidosis and Alkalosis Induced by Nonrespiratory Causes**

	Acid-Base Status		
Respiratory Compensations	*Normal (pH 7.4)*	*Nonrespiratory acidosis (pH 7.1)*	*Nonrespiratory alkalosis (pH 7.7)*
Spirogram records at various pHs			
Respiratory rate	Normal	↑	↓
Tidal volume	Normal	↑	↓
Ventilation	Normal	↑	↓
Rate of CO_2 removal	Normal	↑	↓
Rate of carbonic-acid formation	Normal	↓	↑
Rate of H^+ generation from CO_2	Normal	↓	↑

bonic acid than the kidneys remove from non–carbonic acid sources. Furthermore, the respiratory system, through its ability to regulate arterial [CO_2], can adjust the amount of H^+ added to the body fluids from this source as needed to restore pH toward normal when fluctuations in [H^+] from non–carbonic acid sources occur.

Respiratory regulation acts at a moderate speed, coming into play only when the chemical buffer systems alone are unable to minimize [H^+] changes. When deviations in [H^+] occur, the buffer systems respond immediately, whereas adjustments in ventilation require a few minutes to be initiated. If a deviation in [H^+] is not swiftly and completely corrected by the buffer systems, the respiratory system comes into action a few minutes later, thus serving as the *second line of defense* against changes in [H^+].

Of course, when changes in [H^+] occur because of fluctuations in [CO_2] that arise from respiratory abnormalities, the respiratory mechanism cannot contribute at all to the control of pH; for example, if acidosis exists because of CO_2 accumulation caused by lung disease, the impaired lungs cannot possibly compensate for the acidosis by increasing the rate of CO_2 removal. The buffer systems (other than the H_2CO_3:HCO_3^- pair) plus renal regulation are the only mechanisms available for defending against respiratory-induced acid-base abnormalities.

The kidneys contribute powerfully to control of acid-base balance by controlling both hydrogen-ion and bicarbonate concentrations in the blood.

The kidneys are the *third line of defense* against changes in [H^+] in the body fluids; they require hours to days to compensate for changes in body fluid pH, compared to the immediate responses of the buffer systems and the few-minute delay before the respiratory system responds. However, the kidneys are the most potent acid-base regulatory mechanism; not only can they vary removal of H^+ from any source, but they can also variably conserve or eliminate HCO_3^-, depending on the acid-base status of the body. For example, during renal compensation for acidosis, for each H^+ excreted in the urine, a new HCO_3^- is added to the plasma to buffer, by means of the H_2CO_3:HCO_3^- system, yet another H^+ that still remains in the body fluids. By simultaneously removing acid (H^+) from and adding base (HCO_3^-) to the body fluids, the kidneys are able to restore the pH toward normal more effectively than the lungs, which can adjust only the amount of H^+-forming CO_2 in the body.

Also contributing to the kidneys' acid-base regulatory potency is their ability to return the pH almost exactly to normal. In contrast, when some nonrespiratory abnormality has altered the [H^+], the respiratory system alone can return the pH to only 50% to 75% of the way toward normal; this is because the driving force governing the compensatory ventilatory response is diminished as the pH moves toward normal. As an example, ventilation is increased in response to a rise in arterial [H^+], but as the [H^+] is gradually reduced because of the stepped-up removal of H^+-forming CO_2, the ventilatory response is also gradually reduced. In comparison, the kidneys continue to respond to a change in pH until compensation is essentially complete.

The kidneys control the pH of the body fluids by adjusting three interrelated factors: (1) H^+ excretion, (2) HCO_3^- excretion, and (3) ammonia (NH_3) secretion.

Hydrogen-ion excretion Acids are continuously being added to the body fluids as a result of metabolic activities, yet the generated H^+ must not be allowed to accumulate. Although the body's buffer systems can resist changes in pH by removing H^+ from solution, the persistent production of acidic metabolic products would eventually overwhelm the limits of this buffering capacity. The constantly generated H^+ must therefore ultimately be eliminated from the body. The lungs are able to remove only carbonic acid through the elimination of CO_2. The task of eliminating H^+ derived from sulfuric, phosphoric, lactic and other acids rests with the kidneys.

Almost all of the excreted H^+ enters the urine by means of secretion. Recall that the filtration rate of H^+ equals plasma [H^+] times GFR. Since plasma [H^+] is extremely low (less than in pure H_2O except during extreme acidosis, when the pH falls below 7.0), the filtration rate of H^+ is likewise extremely low. This minute amount of filtered H^+ is excreted in the urine. However, the majority of excreted H^+ gains entry into the tubular fluid by being secreted. The proximal, distal, and collecting tubules all secrete H^+. Since the kidneys normally excrete H^+, urine is usually acidic, having an average pH of 6.0. Not only do the kidneys continuously eliminate the normal amount of H^+ that is constantly being produced from non–carbonic acid sources, but they can also alter their rate of H^+ secretion to compensate for changes in [H^+] arising from abnormalities in the concentration of carbonic acid.

The magnitude of H^+ secretion depends on a direct effect of the plasma's acid-base status on the kidneys' tubular cells (— Fig. 15-9). No neural or hormonal control is involved. When the [H^+] of the plasma passing through the peritubular capillaries is elevated above normal, the tubular cells respond by secreting greater-than-usual amounts of H^+ from the plasma into the tubular fluid to be excreted in the urine.

— *Figure 15-9* **Control of the Rate of Tubular H^+ Secretion**

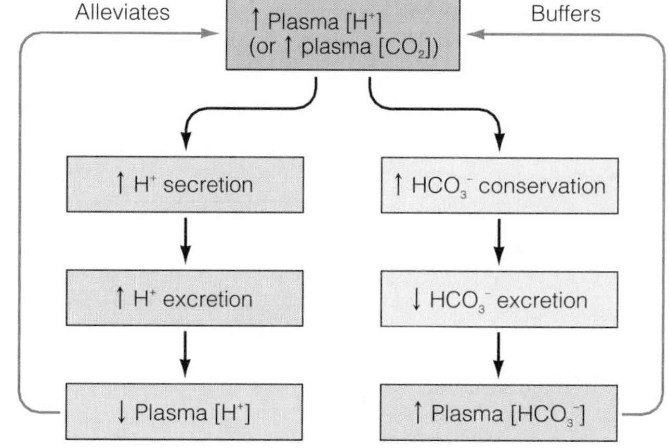

Conversely, when plasma $[H^+]$ is lower than normal, the kidneys conserve H^+ by reducing its secretion and subsequent excretion in the urine. The kidneys are unable to raise plasma $[H^+]$ by reabsorbing more of the filtered H^+ because there are no reabsorptive mechanisms for H^+. The only way the kidneys can reduce H^+ excretion is by secreting less H^+.

The H^+ secretory process begins in the tubular cells with CO_2 that has come from three sources: CO_2 that has diffused into the tubular cells from either the plasma or the tubular fluid or CO_2 that has been metabolically produced within the tubular cells. Under the influence of carbonic anhydrase, CO_2 and H_2O form H_2CO_3, which dissociates into H^+ and HCO_3^-. An energy-dependent carrier in the luminal membrane then transports H^+ out of the cell into the tubular lumen. In part of the nephron, this carrier transports Na^+ derived from the glomerular filtrate in the opposite direction, so H^+ secretion and Na^+ reabsorption are partially linked.

Because the chemical reactions for H^+ secretion begin with CO_2, the rate at which they proceed is influenced by $[CO_2]$. When plasma $[CO_2]$ increases, these reactions proceed more rapidly and the rate of H^+ secretion speeds up (Fig. 15-9). Conversely, the rate of H^+ secretion is reduced when plasma $[CO_2]$ falls below normal. This is especially important in renal compensations for acid-base abnormalities involving a change in H_2CO_3 caused by respiratory dysfunction. The kidneys can therefore adjust H^+ excretion to compensate for changes in both carbonic and non–carbonic acids.

Bicarbonate excretion Before being eliminated by the kidneys, the H^+ generated from non–carbonic acids is buffered to a large extent by plasma HCO_3^-. Appropriately, therefore, renal handling of acid-base balance also involves the adjustment of HCO_3^- excretion, depending on the H^+ load in the plasma (Fig. 15-9).

The kidneys regulate plasma $[HCO_3^-]$ by two interrelated mechanisms: (1) variable reabsorption of the filtered HCO_3^- back into the plasma and (2) variable addition of new HCO_3^- to the plasma. Both of these mechanisms are inextricably linked with H^+ secretion by the kidney tubules. Every time a H^+ is secreted into the tubular fluid, a HCO_3^- is simultaneously transferred into the peritubular capillary plasma. Whether a filtered HCO_3^- is reabsorbed or a new HCO_3^- is added to the plasma in accompaniment with H^+ secretion depends on whether filtered HCO_3^- is present in the tubular fluid to react with the secreted H^+.

Bicarbonate is freely filtered, but because the luminal membranes of the tubular cells are impermeable to the filtered HCO_3^-, it cannot diffuse into these cells. Therefore, reabsorption of HCO_3^- must occur indirectly (– Fig. 15-10). Hydrogen ion secreted into the tubular fluid combines with the filtered HCO_3^- to form H_2CO_3. Under the influence of carbonic anhydrase, which is present on the surface of the luminal membrane, H_2CO_3 decomposes into CO_2 and H_2O within the filtrate. Unlike HCO_3^-, CO_2 can easily penetrate the tubular cell membranes. Within the cells, CO_2 and H_2O, under the influence of intracellular carbonic anhydrase, form H_2CO_3, which dissociates into H^+ and HCO_3^-. Because HCO_3^- can permeate the tubular cells' basolateral membrane, it passively diffuses out of the cells and into the peritubular capillary plasma. Meanwhile, the generated H^+ is actively secreted. Since the disappearance of a HCO_3^- from the tubular fluid is coupled with the appearance of another HCO_3^- in the plasma, a HCO_3^- has, in effect, been "reabsorbed." Even though the HCO_3^- entering the plasma is not the same HCO_3^- that was filtered, the net result is the same as if HCO_3^- were directly reabsorbed.

Normally, slightly more hydrogen ions are secreted into the tubular fluid than bicarbonate ions are filtered. Accordingly, all of the filtered HCO_3^- is usually absorbed because secreted H^+ is available in the tubular fluid to combine with it to form highly reabsorbable CO_2. The vast majority of the secreted H^+ combines with HCO_3^- and is not excreted because it is "used up" in the process of HCO_3^- reabsorption. However, the slight excess of secreted H^+ that is not matched by filtered HCO_3^- is excreted in the urine. This normal H^+ excretion rate keeps pace with the normal rate of non–carbonic acid H^+ production.

Secretion of H^+ that is *excreted* is coupled with the *addition of new* HCO_3^- to the plasma, in contrast to the secreted H^+ that is coupled with HCO_3^- *reabsorption* and is *not excreted,* instead being incorporated into H_2O molecules. When all of the filtered HCO_3^- has been reabsorbed and additional secreted H^+ is generated by the dissociation of H_2CO_3, the HCO_3^- produced by this reaction diffuses into the plasma as a "new" HCO_3^-. It is termed "new" because its appearance in the plasma is not associated with reabsorption of

– ***Figure 15-10*** **Hydrogen-Ion Secretion Coupled with Bicarbonate Reabsorption** Since the disappearance of a filtered HCO_3^- from the tubular fluid is coupled with the appearance of another HCO_3^- in the plasma, HCO_3^- is considered to have been "reabsorbed."

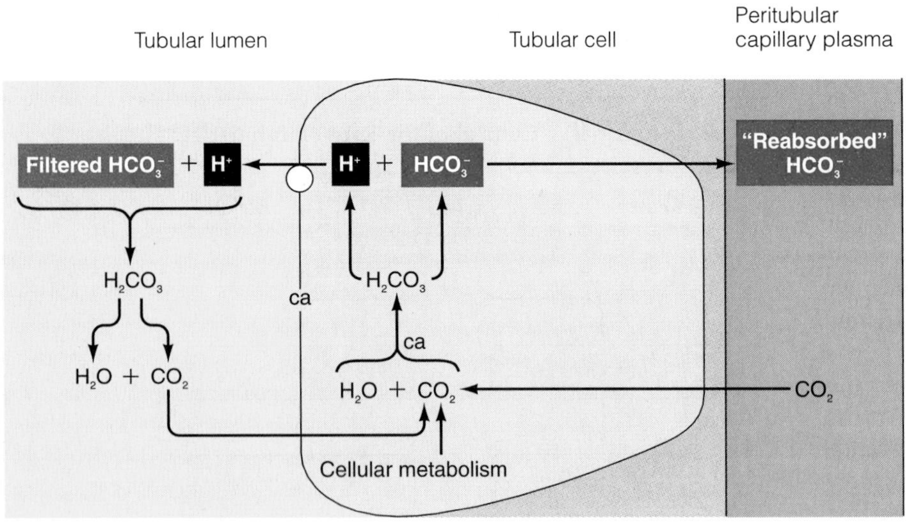

ca = Carbonic anhydrase

Table 15-8 Summary of Renal Responses to Acidosis and Alkalosis

Acid-Base Abnormality	*H^+ Secretion*	*H^+ Excretion*	*HCO_3^- Reabsorption and Addition of New HCO_3^- to Plasma*	*HCO_3^- Excretion*	*pH of Urine*	*Compensatory Change in Plasma pH*
Acidosis	↑	↑	↑	Normal (zero; all filtered is reabsorbed)	Acidic	Alkalinization toward normal
Alkalosis	↓	↓	↓	↑	Alkaline	Acidification toward normal

filtered HCO_3^- (Fig. 15-11). Meanwhile, the secreted H^+ combines with urinary buffers, especially basic phosphate (HPO_4^{2-}) and is excreted.

When the plasma [H^+] is elevated during acidosis, more H^+ is secreted than normal. At the same time, less HCO_3^- is filtered than normal because more of the plasma HCO_3^- is used up in buffering the excess H^+ as HCO_3^- plus H^+ is converted to H_2CO_3. This greater-than-usual inequity between filtered HCO_3^- and secreted H^+ has two consequences. First, more of the secreted H^+ is excreted in the urine because more hydrogen ions are entering the tubular fluid at a time when fewer of them are needed to reabsorb the reduced quantities of filtered HCO_3^-. In this way, extra H^+ is eliminated from the body, making the urine more acidic than normal. Second, because the excretion of H^+ is linked with the addition of new HCO_3^- to the plasma, more HCO_3^- than usual enters the plasma that is passing through the kidneys. This additional HCO_3^- is available to buffer the excess H^+ present in the body.

In the opposite situation of alkalosis, the rate of H^+ secretion diminishes while the rate of HCO_3^- filtration increases compared to normal. When the plasma [H^+] is below normal, a smaller proportion of the HCO_3^- pool is tied up buffering H^+, so the plasma [HCO_3^-] is elevated above normal. As a result, the rate of HCO_3^- filtration is correspondingly increased. Not all of the filtered HCO_3^- is reabsorbed, because bicarbonate ions are in excess of secreted hydrogen ions in the tubular fluid and HCO_3^- cannot be reabsorbed without first reacting with H^+. The excess HCO_3^- is left in the tubular fluid to be excreted in the urine, thus reducing the plasma [HCO_3^-] while causing the urine to become alkaline.

In short, when plasma [H^+] is increased above normal during *acidosis,* renal compensation includes the following (Table 15-8):

1. Increased secretion and subsequent increased excretion of H^+ in the urine, thereby eliminating the excess H^+ and decreasing plasma [H^+].
2. Reabsorption of all of the filtered HCO_3^-, plus addition of new HCO_3^- to the plasma, resulting in increased plasma [HCO_3^-].

When plasma [H^+] is reduced below normal during *alkalosis,* renal responses include the following:

1. Decreased secretion and subsequent reduced excretion of H^+ in the urine, resulting in conservation of H^+ and increased plasma [H^+].
2. Incomplete reabsorption of filtered HCO_3^- and subsequent increased excretion of HCO_3^-, resulting in reduction of plasma [HCO_3^-].

Figure 15-11 **Hydrogen-Ion Secretion and Excretion Coupled with the Addition of New HCO_3^- to the Plasma** Secreted H^+ does not combine with filtered HPO_4^{2-} and is not subsequently excreted until all of the filtered HCO_3^- has been "reabsorbed," as depicted in Figure 15-10. Once all of the filtered HCO_3^- has combined with secreted H^+, further secreted H^+ is excreted in the urine, primarily in association with urinary buffers such as basic phosphate. Excretion of H^+ is coupled with the appearance of new HCO_3^- in the plasma. The "new" HCO_3^- represents a net gain rather than being merely a replacement for filtered HCO_3^-.

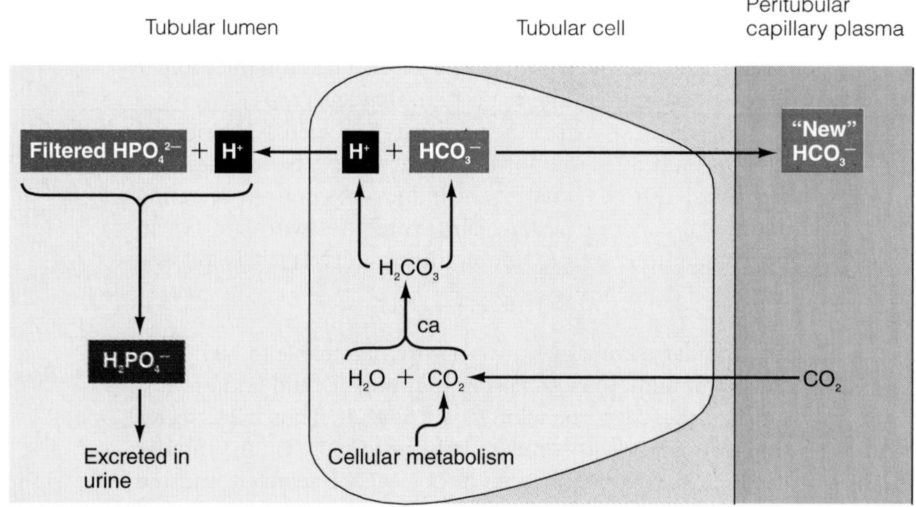

Note that to compensate for acidosis, the kidneys acidify the urine (by getting rid of extra H^+) and alkalinize the plasma (by conserving HCO_3^-) to bring the pH to normal. In the opposite case—alkalosis—the kidneys make the urine alkaline (by eliminating excess HCO_3^-) while acidifying the plasma (by conserving H^+).

Ammonia secretion The energy-dependent H^+ carriers in the tubular cells are capable of secreting H^+ against a concentration gradient until the tubular fluid (urine) becomes 800 times more acidic than the plasma. At this point, further H^+ secretion ceases, because the gradient becomes too great for the secretory process to continue. It is impossible for the kidneys to acidify the urine beyond a gradient-limited urinary pH of 4.5. If left unbuffered as free H^+, only about 1% of the excess H^+ typically excreted daily would produce a urinary pH of this magnitude at normal urine flow rates, and elimination of the other 99% of the usually secreted H^+ load would be prevented, a situation that would be intolerable. For H^+ secretion to proceed, the majority of secreted H^+ must be buffered in the tubular fluid so that it does not exist as free H^+ and, accordingly, does not contribute to the tubular acidity.

Bicarbonate cannot buffer urinary H^+ as it does the ECF, because HCO_3^- is not excreted in the urine simultaneously with H^+. (Whichever of these substances is in excess in the plasma is excreted in the urine.) There are, however, two important urinary buffers: (1) filtered phosphate buffers and (2) secreted **ammonia (NH_3)**. Normally, secreted H^+ is first buffered by the phosphate buffer system, which is in the tubular fluid because excess ingested phosphate has been filtered but not reabsorbed. The basic member of the phosphate buffer pair binds with secreted H^+. Basic phosphate is present in the tubular fluid by virtue of dietary excess, not because of any deliberate mechanism to buffer secreted H^+. When H^+ secretion is high, the buffering capacity of urinary phosphates is exceeded, but the kidneys cannot respond by excreting more basic phosphate. Only the quantity of phosphate reabsorbed, not the quantity excreted, is subject to control. As soon as all of the basic phosphate ions that are coincidentally excreted have "soaked up" H^+, the acidity of the tubular fluid quickly rises as more H^+ ions are secreted. Without additional buffering capacity from another source, H^+ secretion would soon be halted abruptly as the free $[H^+]$ in the tubular fluid quickly rose to the critical limiting level.

When acidosis exists, the tubular cells secrete NH_3 into the tubular fluid once the normal urinary phosphate buffers are saturated. This NH_3 enables the kidneys to continue secreting additional H^+ ions because NH_3 combines with free H^+ in the tubular fluid to form **ammonium ion (NH_4^+)** as follows:

$$NH_3 + H^+ \rightarrow NH_4^+$$

The tubular membranes are not very permeable to NH_4^+, so the ammonium ions remain in the tubular fluid and are lost in the urine, each one taking a H^+ with it. Thus, NH_3 secretion during acidosis serves to buffer excess H^+ in the tubular fluid, so that large amounts of H^+ can be secreted into the urine before the pH falls to the limiting value of 4.5. Were it not for NH_3 secretion, the extent of H^+ secretion would be limited to whatever phosphate buffering capacity coincidentally happened to be present as a result of dietary excess.

In contrast to the phosphate buffers, which are in the tubular fluid because they have been filtered but not reabsorbed, NH_3 is deliberately synthesized from the amino acid *glutamine* within the tubular cells, from which it readily diffuses down its concentration gradient into the tubular fluid. The rate of NH_3 secretion is controlled by a direct effect on the tubular cells of the amount of excess H^+ to be transported in the urine. When an individual has been acidotic for more than two or three days, the rate of NH_3 production increases substantially. This extra NH_3 provides additional buffering capacity to allow H^+ secretion to continue after the normal phosphate buffering capacity is overwhelmed during renal compensation for acidosis.

Acid-base imbalances can arise from either respiratory dysfunction or metabolic disturbances.

Deviations from normal acid-base status are divided into four general categories, depending on the source and direction of the abnormal change in $[H^+]$. These categories are respiratory acidosis, respiratory alkalosis, metabolic acidosis, and metabolic alkalosis.

Because of the relationship between $[H^+]$ and the concentrations of the members of a buffer pair, changes in $[H^+]$ are reflected by changes in the ratio of $[HCO_3^-]$ to $[CO_2]$. Recall that the normal ratio is 20/1. Using the Henderson-Hasselbalch equation and with pK being 6.1 and the log of 20 being 1.3, normal pH = 6.1 + 1.3 = 7.4. Determinations of $[HCO_3^-]$ and $[CO_2]$ provide more meaningful information about the underlying factors responsible for a particular acid-base status than do direct measurements of $[H^+]$ alone. The following rules of thumb apply when examining acid-base imbalances *before any compensations take place:*

1. A change in pH that has a respiratory cause will be associated with an abnormal $[CO_2]$, giving rise to a change in carbonic acid-generated H^+. In contrast, a pH deviation of metabolic origin will be associated with an abnormal $[HCO_3^-]$ as a result of the participation of HCO_3^- in buffering abnormal amounts of H^+ generated from non–carbonic acids.
2. Anytime the $[HCO_3^-]/[CO_2]$ ratio falls below 20/1, an acidosis exists. The log of any number lower than 20 is less than 1.3 and, when added to the pK of 6.1, will yield an acidotic pH below 7.4. Anytime the ratio exceeds 20/1, an alkalosis exists. The log of any number greater than 20 is more than 1.3 and, when added to the pK of 6.1, will yield an alkalotic pH above 7.4.

Putting these two points together:

- *Respiratory acidosis* has a ratio of less than 20/1 arising from an increase in $[CO_2]$.
- *Respiratory alkalosis* has a ratio greater than 20/1 because of a decrease in $[CO_2]$.
- *Metabolic acidosis* has a ratio of less than 20/1 associated with a fall in $[HCO_3^-]$.

- *Metabolic alkalosis* has a ratio greater than 20/1 arising from an elevation in [HCO_3^-].

We will examine each of these categories separately in more detail, paying particular attention to possible causes and the compensations that occur. The "balance beam" concept, presented in Fig. 15-12 in conjunction with the Henderson-Hasselbalch equation, will help you better visualize the contributions of the lungs and kidneys to the causes and compensations of various acid-base disorders. The normal situation is represented in Figure 15-12a.

Respiratory acidosis **Respiratory acidosis** is the result of abnormal CO_2 retention arising from *hypoventilation* (see p. 454). As less-than-normal amounts of CO_2 are lost through the lungs, the resultant increase in H_2CO_3 formation and dissociation leads to an elevated [H^+]. Possible causes include lung disease, depression of the respiratory center by drugs or disease, nerve or muscle disorders that reduce respiratory muscle ability, or (transiently) even the simple act of holding one's breath.

In uncompensated respiratory acidosis (Fig. 15-12b), [CO_2] is elevated (in our example, it is doubled) while [HCO_3^-] is normal, so the ratio is 20/2 (10/1) and pH is reduced. Let us clarify a potentially confusing point. You might wonder why when [CO_2] is elevated and drives the reaction $CO_2 + H_2O \rightleftharpoons H_2CO_3 \rightleftharpoons H^+ + HCO_3^-$ to the right, we say that [H^+] becomes elevated but [HCO_3^-] remains normal although the same quantities of H^+ and HCO_3^- are produced when CO_2-generated H_2CO_3 dissociates. The answer lies in the fact that normally the [HCO_3^-] is 600,000 times the [H^+]. For every one hydrogen ion and 600,000 bicarbonate ions present in the ECF, the generation of one additional H^+ and one HCO_3^- doubles the [H^+] (a 100% increase) but only increases the [HCO_3^-] 0.00017% (from 600,000 to 600,001 ions). Therefore, an elevation in [CO_2] brings about a pronounced increase in [H^+], but [HCO_3^-] remains essentially normal.

Compensatory measures act to restore pH to normal. The chemical buffers immediately take up additional H^+, but the respiratory mechanism is usually not able to respond with compensatory increased ventilation because an impairment in respiratory activity is the problem in the first place. Thus, the kidneys are most important in compensating for respiratory acidosis. They conserve all of the filtered HCO_3^- and add new HCO_3^- to the plasma while simultaneously secreting and, accordingly, excreting more H^+.

As a result, HCO_3^- stores in the body become elevated. In our example (Fig. 15-12c), the plasma [HCO_3^-] is doubled, so the [HCO_3^-]/[CO_2] ratio is 40/2 rather than 20/2 as it was in the uncompensated state. A ratio of 40/2 is equivalent to a normal 20/1 ratio, so the pH is once again the normal 7.4. Enhanced renal conservation of HCO_3^- has fully compensated for CO_2 accumulation, thus restoring the pH to normal, although both the [CO_2] and the [HCO_3^-] are now distorted. Note that maintenance of normal pH depends on preserving a normal ratio between [HCO_3^-] and [CO_2], no matter what the absolute values of each of these buffer components are. (Compensation is never fully complete because the pH can be restored close to but not precisely to normal. In our examples, however, we assume full compensation for ease in mathematical calculations. Also bear in mind that the values used are only representative. Deviations in pH actually occur over a range and the degree to which compensation can be accomplished varies.)

Respiratory alkalosis The primary defect in **respiratory alkalosis** is excessive loss of CO_2 from the body as a result of *hyperventilation*. When pulmonary ventilation increases out of proportion to the rate of CO_2 production, too much CO_2 is blown off. Consequently, less H_2CO_3 is formed and [H^+] decreases.

Possible causes of respiratory alkalosis include fever, anxiety, and aspirin poisoning, all of which excessively stimulate ventilation without regard to the status of O_2, CO_2, or H^+ in the body fluids. Respiratory alkalosis also occurs as a result of physiological mechanisms at high altitude. When the low concentration of O_2 in the arterial blood reflexly stimulates ventilation in an attempt to obtain more O_2 for the body, too much CO_2 is blown off in the process, inadvertently leading to an alkalotic state (see p. 456).

Looking at the biochemical abnormalities in uncompensated respiratory alkalosis (Fig. 15-12d), the increase in pH reflects a reduction in [CO_2] (half the normal value in our example) while the [HCO_3^-] remains normal. This yields an alkalotic ratio of 20/0.5, which is comparable to 40/1.

Compensatory measures act to shift the pH back toward normal. The chemical buffer systems liberate H^+ to diminish the severity of the alkalosis. As plasma [CO_2] and [H^+] fall below normal because of excessive ventilation, two of the normally potent stimuli for driving ventilation are removed. This effect tends to "put brakes" on the extent to which some nonrespiratory related factor such as fever or anxiety can overdrive ventilation. Therefore, hyperventilation does not continue completely unabated. If the situation continues for a few days, the kidneys compensate by conserving H^+ and excreting more HCO_3^-. If, as in our example (Fig. 15-12e), the HCO_3^- stores are reduced by half by loss of HCO_3^- in the urine, the [HCO_3^-]/[CO_2] ratio becomes 10/0.5, equivalent to the normal 20/1. Therefore, the pH is restored to normal by reducing the HCO_3^- load to compensate for the CO_2 loss.

Metabolic acidosis **Metabolic acidosis** (also known as **nonrespiratory acidosis**) encompasses all types of acidosis besides that caused by excess CO_2 in the body fluids. In the uncompensated state (Fig. 15-12f), metabolic acidosis is always characterized by a reduction in plasma [HCO_3^-] (in our example it is halved), while [CO_2] remains normal, producing an acidotic ratio of 10/1. The problem may arise from excessive loss of HCO_3^--rich fluids from the body or from an accumulation of non–carbonic acids. In the latter case, plasma HCO_3^- is used up in the process of buffering the additional H^+.

This type of acid-base disorder is the one more frequently encountered. The following are the most common causes.

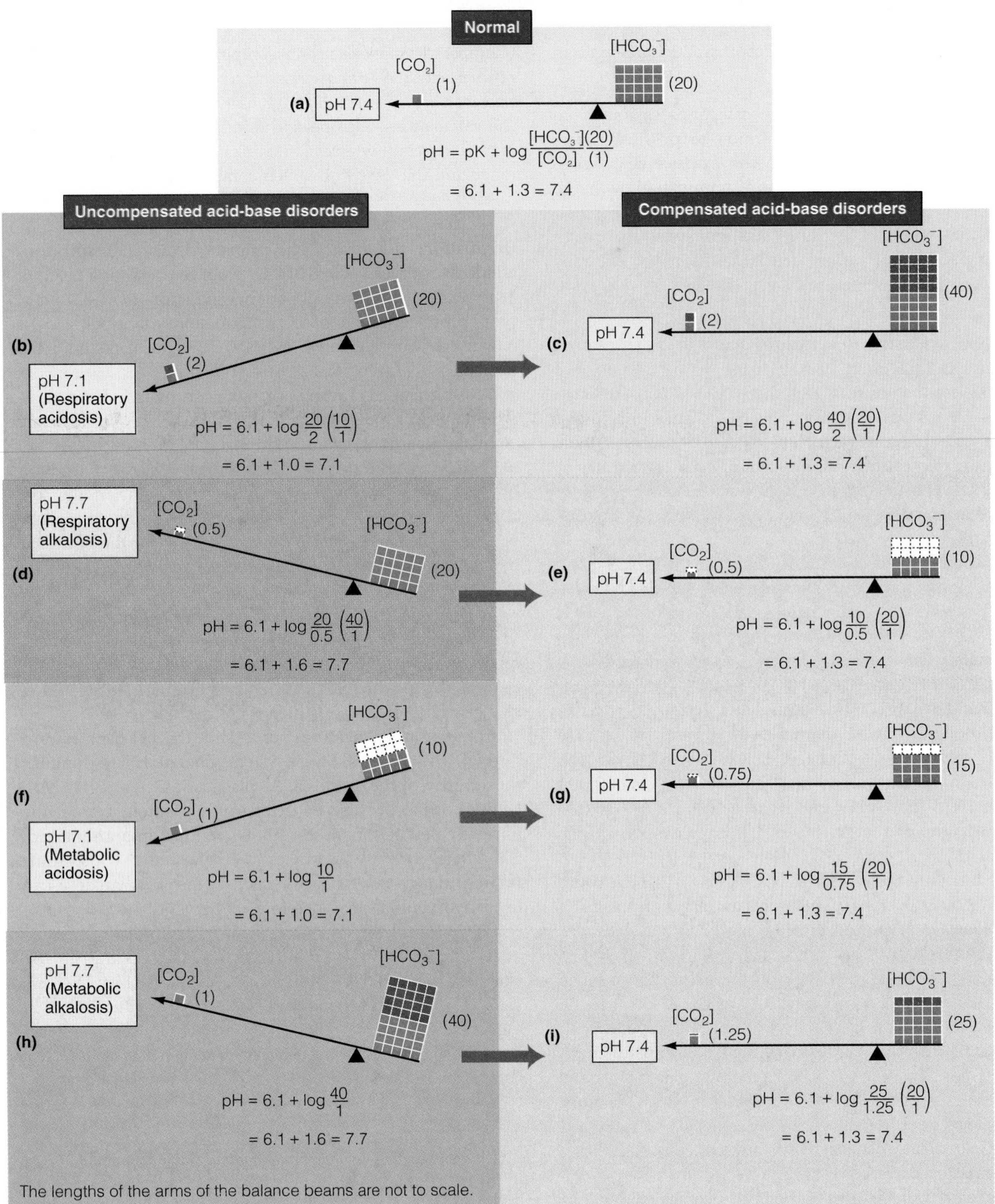

Figure 15-12 **Schematic Representation of the Relationship of HCO_3^- and CO_2 Concentrations to pH in Various Acid-Base Statuses** (a) Normal acid-base balance. The $[HCO_3^-]/[CO_2]$ ratio is 20/1. (b) Uncompensated respiratory acidosis. The $[HCO_3^-]/[CO_2]$ ratio is reduced (20/2) because CO_2 has accumulated. (c) Compensated respiratory acidosis. Compensatory retention of HCO_3^- to balance the CO_2 accumulation restores the $[HCO_3^-]/[CO_2]$ ratio to a normal equivalent (40/2). (d) Uncompensated respiratory alkalosis. The $[HCO_3^-]/[CO_2]$ ratio is increased (20/0.5) due to a reduction in CO_2. (e) Compensated respiratory alkalosis. Compensatory elimination of HCO_3^- to balance the CO_2 deficit restores the $[HCO_3^-]/[CO_2]$ ratio to a normal equivalent (10/0.5). (f) Uncompensated metabolic acidosis. The $[HCO_3^-]/[CO_2]$ ratio is reduced (10/1) due to a HCO_3^- deficit. (g) Compensated metabolic acidosis. Conservation of HCO_3^-, which partially makes up for the HCO_3^- deficit, and a compensatory reduction in CO_2 restore the $[HCO_3^-]/[CO_2]$ to a normal equivalent (15/0.75). (h) Uncompensated metabolic alkalosis. The $[HCO_3^-]/[CO_2]$ ratio is increased (40/1) due to excess HCO_3^-. (i) Compensated metabolic alkalosis. Elimination of some of the extra HCO_3^- and a compensatory increase in CO_2 restore the $[HCO_3^-]/[CO_2]$ ratio to a normal equivalent (25/1.25).

1. *Severe diarrhea.* During the process of digestion, a HCO_3^--rich digestive juice is normally secreted into the digestive tract and is subsequently reabsorbed back into the plasma when digestion is completed. During diarrhea, this HCO_3^- is lost from the body rather than being reabsorbed. The reduction in plasma $[HCO_3^-]$ without a corresponding reduction in $[CO_2]$ lowers the pH. Because of the loss of HCO_3^-, less HCO_3^- is available to buffer H^+, leading to more free H^+ in the body fluids. Looking at the situation differently, loss of HCO_3^- shifts the $H^+ + HCO_3^- \rightleftharpoons CO_2 + H_2O$ reaction to the left to compensate for the HCO_3^- deficit, increasing $[H^+]$ above normal in the process.
2. *Diabetes mellitus.* Abnormal fat metabolism resulting from the inability of cells to preferentially use glucose in the absence of insulin results in the formation of excess keto acids whose dissociation causes an increase in plasma $[H^+]$.
3. *Strenuous exercise.* When muscles resort to anaerobic glycolysis during strenuous exercise (see p. 243), excess lactic acid is produced, leading to a rise in plasma $[H^+]$.
4. *Uremic acidosis.* In severe renal failure (uremia), the kidneys are unable to rid the body of even the normal amounts of H^+ generated from the non–carbonic acids formed by the body's ongoing metabolic processes, so H^+ starts to accumulate in the body fluids. Also, the kidneys are unable to conserve an adequate amount of HCO_3^- to be used for buffering the normal acid load. The resultant fall in $[HCO_3^-]$ without a concomitant reduction in $[CO_2]$ is correlated with a decline in pH to an acidotic level.

Except in uremic acidosis, metabolic acidosis is compensated for by both respiratory and renal mechanisms as well as by chemical buffers. The buffers take up extra H^+, the lungs blow off additional H^+-generating CO_2, and the kidneys excrete more H^+ and conserve more HCO_3^-. In our example (Fig. 15-12g), these compensatory measures restore the ratio to normal by reducing $[CO_2]$ to 75% of normal and by raising $[HCO_3^-]$ halfway back toward normal (up from 50% to 75% of the normal value). This brings the ratio to 15/0.75 (equivalent to 20/1).

Note that in compensating for metabolic acidosis, the lungs deliberately displace $[CO_2]$ from normal in an attempt to restore $[H^+]$ toward normal. Whereas in respiratory-induced acid-base disorders, an abnormal $[CO_2]$ is the *cause* of the $[H^+]$ imbalance, in metabolic acid-base disorders, $[CO_2]$ is intentionally shifted from normal as an important *compensation* for the $[H^+]$ imbalance.

When kidney disease is the cause of metabolic acidosis, complete compensation is not possible because the renal mechanism is not available for pH regulation. Recall that the respiratory system is capable of compensating only up to 75% of the way toward normal. Uremic acidosis is very serious, because the kidneys cannot help to restore the pH all the way to normal.

Metabolic alkalosis **Metabolic alkalosis** is a reduction in plasma $[H^+]$ caused by a relative deficiency of non–carbonic acids. This acid-base disturbance is associated with an increase in $[HCO_3^-]$, which, in the uncompensated state, is not accompanied by a change in $[CO_2]$. In our example (Fig. 15-12h), $[HCO_3^-]$ is doubled, producing an alkalotic ratio of 40/1. This condition arises most commonly from the following:

1. *Vomiting,* which results in abnormal loss of H^+ from the body as a result of loss of acidic gastric juices. Hydrochloric acid is secreted into the stomach lumen during the process of digestion. Bicarbonate is added to the plasma during gastric HCl secretion. This HCO_3^- is neutralized by H^+ as the gastric secretions are eventually reabsorbed back into the plasma, so normally there is no net addition of HCO_3^- to the plasma from this source. However, when this acid is lost from the body during vomiting, not only is plasma $[H^+]$ decreased, but also reabsorbed H^+ is no longer available to neutralize the extra HCO_3^- added to the plasma during gastric HCl secretion. Thus, loss of HCl in effect increases plasma $[HCO_3^-]$. (On the other hand, with "deeper" vomiting, HCO_3^- in the digestive juices secreted into the upper intestine may be lost in the vomitus, resulting in an acidosis instead of an alkalosis.)
2. *Ingestion of alkaline drugs,* such as when baking soda ($NaHCO_3$, which dissociates in solution into Na^+ and HCO_3^-) is used as a self-administered remedy for the treatment of gastric hyperacidity. By neutralizing excess acid in the stomach, HCO_3^- relieves the symptoms of stomach irritation and heartburn, but when more HCO_3^- than needed is ingested, the extra HCO_3^- is absorbed from the digestive tract and increases plasma $[HCO_3^-]$. The extra HCO_3^- binds with some of the free H^+ normally present in the plasma from non–carbonic acid sources, leading to a reduction in free $[H^+]$. (In contrast, commercially prepared alkaline products for the treatment of gastic hyperacidity are not absorbed from the digestive tract to any extent and therefore do not alter the acid-base status of the body.)

In metabolic alkalosis, the chemical buffer systems immediately liberate H^+ and ventilation is reduced so that extra H^+-generating CO_2 is retained in the body fluids. If the condition persists for several days, the kidneys conserve H^+ and excrete the excess HCO_3^- in the urine. The resultant compensatory increase in $[CO_2]$ (up 25% in our example—Fig. 15-12i) and the partial reduction in $[HCO_3^-]$ (75% of the way back down toward normal in our example) together restore the $[HCO_3^-]/[CO_2]$ ratio back to the equivalent of 20/1 at 1.25/25.

It should be obvious that an individual's acid-base status cannot be assessed on the basis of pH alone. Uncompensated acid-base abnormalities can readily be distinguished on the basis of deviations of either $[CO_2]$ or $[HCO_3^-]$ from normal (Table 15-9). When compensation has been accomplished and the pH is essentially normal, determinations of $[CO_2]$ and $[HCO_3^-]$ can reveal the presence of an acid-base disorder. Note, however, that in both compensated respiratory acidosis and compensated metabolic alkalosis, $[CO_2]$ and $[HCO_3^-]$ are both above normal. With respiratory acidosis,

Table 15-9 **Summary of $[CO_2]$, $[HCO_3^-]$, and pH in Uncompensated and Compensated Acid-Base Abnormalities**

Acid-Base Status	*pH*	*$[CO_2]$ (compared to normal)*	*$[HCO_3^-]$ (compared to normal)*	*$[HCO_3^-]/[CO_2]$*
Normal	Normal	Normal	Normal	20/1
Uncompensated respiratory acidosis	Decreased	Increased	Normal	20/2 (10/1)
Compensated respiratory acidosis	Normal	Increased	Increased	40/2 (20/1)
Uncompensated respiratory alkalosis	Increased	Decreased	Normal	20/0.5 (40/1)
Compensated respiratory alkalosis	Normal	Decreased	Decreased	10/0.5 (20/1)
Uncompensated metabolic acidosis	Decreased	Normal	Decreased	10/1
Compensated metabolic acidosis	Normal	Decreased	Decreased	15/0.75 (20/1)
Uncompensated metabolic alkalosis	Increased	Normal	Increased	40/1
Compensated metabolic alkalosis	Normal	Increased	Increased	25/1.25 (20/1)

the original problem is an abnormal increase in $[CO_2]$, and a compensatory increase in $[HCO_3^-]$ restores the $[HCO_3^-]/[CO_2]$ ratio to 20/1. Metabolic alkalosis, on the other hand, is characterized by an abnormal increase in $[HCO_3^-]$ in the first place; then $[CO_2]$ is deliberately increased to restore the ratio to normal. Similarly, compensated respiratory alkalosis and compensated metabolic acidosis share similar patterns of $[CO_2]$ and $[HCO_3^-]$. Respiratory alkalosis starts out with a reduction in $[CO_2]$, which is compensated by a reduction in $[HCO_3^-]$. With metabolic acidosis, $[HCO_3^-]$ falls below normal, followed by a compensatory decrease in $[CO_2]$.

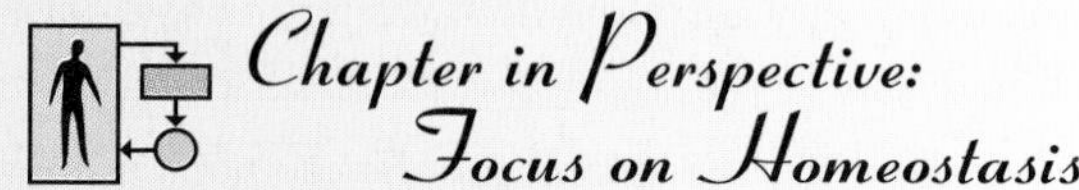

Chapter in Perspective: Focus on Homeostasis

Homeostasis depends on maintaining a balance between the input and output of all constituents present in the internal fluid environment. Regulation of fluid balance involves two separate components: control of salt balance and control of H_2O balance. Control of salt balance is primarily important in the long-term regulation of arterial blood pressure because the body's salt load affects the osmotic determination of the ECF volume, of which plasma volume is a part. An increased salt load in the ECF leads to an expansion in ECF volume, including plasma volume, which in turn causes a rise in blood pressure. Conversely, a reduction in the ECF salt load brings about a fall in blood pressure. Salt balance is maintained by constantly adjusting salt output in the urine to match unregulated, variable salt intake.

Control of H_2O balance is important in preventing changes in ECF osmolarity, which would induce detrimental osmotic shifts of H_2O between the cells and the ECF. Such shifts of H_2O into or out of the cells would cause the cells to swell or shrink, respectively. Cells, especially brain neurons, do not function normally when swollen or shrunken. Water balance is largely maintained by controlling the volume of H_2O lost in the urine to compensate for uncontrolled losses of variable volumes of H_2O from other avenues, such as through sweating or diarrhea, and for poorly regulated H_2O intake. Even though a thirst mechanism exists to control H_2O intake based on need, the amount drunk is often influenced by social custom and habit instead of thirst alone.

A balance between input and output of H^+ is also critical to maintaining the body's acid-base balance within the narrow limits compatible with life. Deviations in the internal fluid environment's pH lead to altered neuromuscular excitability, to changes in enzymatically controlled metabolic activity, and to K^+ imbalances, which can cause cardiac arrhythmias. These effects are fatal if the pH falls outside the range of 6.8 to 8.0.

Hydrogen ions are uncontrollably, continually being added to the body fluids as a result of ongoing metabolic activities, yet the ECF's pH must be kept constant at a slightly alkaline level of 7.4 for optimal body function. Like salt and H_2O balance, control of H^+ output by the kidneys is the main regulatory factor in achieving H^+ balance. Assisting the kidneys in eliminating H^+ are the lungs, which can adjust their rate of excretion of H^+-generating CO_2.

The assertion that the same input-output balance that applies to salt and H_2O also applies to H^+ homeostasis must be modified by the fact that H^+ is buffered in the body. This buffering mechanism can take up or liberate H^+, thereby transiently keeping its concentration constant within the body until its output can be brought into line with its input. Such a mechanism is not available for salt or H_2O balance.

Chapter Summary

Fluid Balance

On average, the body fluids compose 60% of total body weight. This figure varies between individuals, depending on how much fat (a low H_2O content tissue) they possess. Two-thirds of the body H_2O is found in the intracellular fluid (ICF). The remaining one-third present in the extracellular fluid (ECF) is distributed between the plasma (20% of the ECF) and the interstitial fluid (80% of the ECF).

Because all plasma constituents are freely exchanged across the capillary walls, the plasma and interstitial fluid are nearly identical in composition, except for the lack of plasma proteins in the interstitial fluid. In contrast, the ECF and ICF have markedly different compositions because the cell membrane barriers are highly selective as to what materials are transported into or out of the cells.

The essential components of fluid balance are control of ECF volume by maintaining salt balance and control of ECF osmolarity by maintaining water balance. Because of the osmotic holding power of Na^+, the major ECF cation, a change in the body's total Na^+ content brings about a corresponding change in ECF volume, including plasma volume, which, in turn, alters the arterial blood pressure in the same direction. Appropriately, changes in ECF volume and arterial blood pressure are compensated for in the long run by Na^+-regulating mechanisms. Salt intake is not controlled in humans, but control of salt output in the urine is closely regulated. Blood pressure–regulating mechanisms can vary the GFR, and accordingly the amount of Na^+ filtered, by adjusting the caliber of the afferent arterioles supplying the glomeruli. Simultaneously, blood pressure–regulating mechanisms can vary the secretion of aldosterone, the hormone that promotes Na^+ reabsorption by the renal tubules. By varying Na^+ filtration and Na^+ reabsorption, the extent of Na^+ excretion in the urine can be adjusted to regulate the plasma volume and subsequently the arterial blood pressure in the long term.

Changes in ECF osmolarity are primarily detected and corrected by the systems responsible for maintaining H_2O balance. The osmolarity of the ECF must be closely regulated to prevent osmotic shifts of H_2O between the ECF and ICF, because cell swelling or shrinking is deleterious, especially to brain neurons. Excess free H_2O in the ECF dilutes the ECF solutes, with the resultant ECF hypotonicity driving H_2O into the cells. An ECF free H_2O deficit, on the other hand, concentrates the ECF solutes, and consequently H_2O leaves the cells to enter the hypertonic ECF. To prevent these detrimental fluxes, regulation of free H_2O balance is accomplished largely by vasopressin and, to a lesser degree, by thirst. Changes in vasopressin secretion and thirst are both governed primarily by hypothalamic osmoreceptors, which monitor ECF osmolarity. The amount of vasopressin secreted determines the extent of free H_2O reabsorption by the distal portions of the nephrons, thereby determining the volume of urinary output. Simultaneously, the intensity of thirst controls the volume of fluid intake. However, because the volume of fluid drunk is often not directly correlated with the intensity of thirst, control of urinary output by vasopressin is the most important regulatory mechanism for maintaining H_2O balance.

Acid-Base Balance

Acids liberate free hydrogen ions (H^+) into solution; bases bind with free hydrogen ions and remove them from solution. Acid-base balance refers to the regulation of H^+ concentration ($[H^+]$) in the body fluids. To precisely maintain $[H^+]$, input of H^+ by means of metabolic production of acids within the body must continually be matched with H^+ output by urinary excretion of H^+ and respiratory removal of H^+-generating CO_2. Furthermore, between the time of this generation and elimination, H^+ must be buffered within the body to prevent marked fluctuations in $[H^+]$.

Hydrogen-ion concentration frequently is expressed in terms of pH, which is the logarithm of $1/[H^+]$. The normal pH of the plasma is 7.4, slightly alkaline compared to neutral H_2O, which has a pH of 7.0. A pH lower than normal (higher $[H^+]$ than normal) is indicative of a state of acidosis. A pH higher than normal (lower $[H^+]$ than normal) characterizes a state of alkalosis. Fluctuations in $[H^+]$ have profound effects on body chemistry, most notably: (1) changes in neuromuscular excitability, with acidosis depressing excitability, especially in the central nervous system, and alkalosis producing overexcitability of both the peripheral and the central nervous systems; (2) disruption of normal metabolic reactions by altering the structure and function of all enzymes; and (3) alterations in plasma $[K^+]$ brought about by H^+-induced changes in the rate of K^+ elimination by the kidneys.

The primary challenge in controlling acid-base balance is the maintenance of normal plasma alkalinity in the face of continual addition of H^+ to the plasma from ongoing metabolic activity. The three lines of defense for resisting changes in $[H^+]$ are (1) the chemical buffer systems, (2) respiratory control of pH, and (3) renal control of pH.

Chemical buffer systems, the first line of defense, each consist of a pair of chemicals involved in a reversible reaction, one that can liberate H^+ and the other that can bind H^+. A buffer pair acts immediately to minimize any changes in pH that occur by acting according to the law of mass action.

The respiratory system, constituting the second line of defense, normally eliminates the metabolically produced CO_2 so that H_2CO_3 does not accumulate in the body fluids. When the chemical buffers alone have been unable to immediately minimize a pH change, the respiratory system responds within a few minutes by altering its rate of CO_2 removal. An increase in $[H^+]$ arising from non–carbonic acid sources stimulates respiration so that more H_2CO_3-forming CO_2 is blown off, compensating for the acidosis by reducing the generation of H^+ from H_2CO_3. Conversely, a fall in $[H^+]$ depresses respiratory activity so that CO_2 and thus H^+-generating H_2CO_3 can accumulate in the body fluids to compensate for the alkalosis.

The kidneys are the third and most powerful line of defense. They require hours to days to compensate for a deviation in body fluid pH. However, they not only eliminate the normal amount of H^+ produced from non-H_2CO_3 sources, but they can also alter their rate of H^+ removal in response to changes in both non-H_2CO_3 and H_2CO_3 acids. In contrast, the lungs can adjust only H^+ generated from H_2CO_3. Furthermore, the kidneys can regulate $[HCO_3^-]$ in the body fluids as well. The kidneys compensate for acidosis by secreting excess H^+ in the urine while adding new HCO_3^- to the plasma to expand the HCO_3^- buffer pool. During alkalosis, the kidneys conserve H^+ by reducing its secretion in the urine. They also eliminate HCO_3^-, which is in excess because less HCO_3^- than usual is tied up buffering H^+ when H^+ is in short supply.

Secreted H^+ that is to be excreted in the urine must be buffered in the tubular fluid to prevent the H^+ concentration gradient from becoming so great that it prevents further H^+ secretion. Normally, H^+ is buffered by the urinary phosphate buffer pair, which is abundant in the tubular fluid because excess dietary phosphate spills into the urine to be excreted from the body. In acidosis, when all of the phosphate buffer is already used up in buffering the extra secreted H^+, the kidneys secrete NH_3 into the tubular fluid to serve as a buffer so that H^+ secretion can continue.

There are four types of acid-base imbalances: respiratory acidosis, respiratory alkalosis, metabolic acidosis, and metabolic alkalosis. Respiratory acid-base disorders originate with deviations from normal $[CO_2]$, whereas metabolic acid-base imbalances encompass all deviations in pH other than those caused by abnormal $[CO_2]$.

Review Exercises

Objective Questions (Answers on p. E–15.)

1. The only avenue by which materials can be exchanged between the cells and the external environment is the ECF. (True or false?)
2. Water is driven into the cells when the ECF volume is expanded by an isotonic fluid gain. (True or false?)
3. Salt balance in humans is poorly regulated because of our hedonistic salt appetite. (True or false?)
4. An unintentional increase in CO_2 is a cause of respiratory acidosis, but a deliberate increase in CO_2 is a compensation for metabolic alkalosis. (True or false?)
5. Secreted H^+ that is coupled with HCO_3^- reabsorption is not excreted whereas secreted H^+ that is excreted is linked with the addition of new HCO_3^- to the plasma. (True or false?)
6. The largest body fluid compartment is the ____________.
7. Specialized fluid volumes secreted by specific cells into a particular cavity within the body for a specific purpose are collectively known as ____________.
8. Of the two members of the H_2CO_3:HCO_3^- buffer system, ____________ is regulated by the lungs whereas ____________ is regulated by the kidneys.
9. Which of the following individuals would have the lowest percentage of body H_2O?
   a. a chubby baby
   b. a well-proportioned female college student
   c. a well-muscled male college student
   d. an obese, elderly woman
   e. a lean, elderly man
10. Which of the following factors does *not* increase vasopressin secretion?
    a. ECF hypertonicity
    b. alcohol
    c. stressful situations
    d. an ECF volume deficit
    e. angiotensin II
11. pH (indicate all correct answers.)
    a. equals log $1/[H^+]$.
    b. equals pK + log $[CO_2]/[HCO_3^-]$.
    c. is high in acidosis.
    d. falls lower as $[H^+]$ increases.
    e. is normal when the $[HCO_3^-]/[CO_2]$ ratio is 20/1.
12. Acidosis (indicate all correct answers.)
    a. causes overexcitability of the nervous system.
    b. exists when the plasma pH falls below 7.35.
    c. occurs when the $[HCO_3^-]/[CO_2]$ ratio exceeds 20/1.
    d. occurs when CO_2 is blown off more rapidly than it is being produced by metabolic activities.
    e. occurs when excessive HCO_3^- is lost from the body such as in diarrhea.
13. The kidney tubular cells secrete NH_3 (Indicate all correct answers.)
    a. when the urinary pH becomes too high.
    b. when the body is in a state of alkalosis.
    c. to enable further renal secretion of H^+ to occur.
    d. to buffer excess filtered HCO_3^-.
    e. when there is excess NH_3 in the body fluids.
14. Complete the following chart:

$\frac{[HCO_3^-]}{[CO_2]}$	*Uncompensated Abnormality*	*Possible Cause*	*pH*
10/1	1. ________	2. ________	3. ________
20/0.5	4. ________	5. ________	6. ________
20/2	7. ________	8. ________	9. ________
40/1	10. ________	11. ________	12. ________

Essay Questions

1. Explain the balance concept.
2. Outline the distribution of body H_2O.
3. Compare the ionic composition of plasma, interstitial fluid, and intracellular fluid.
4. What factors are regulated to maintain the body's fluid balance?
5. Why is regulation of ECF volume important? How is it regulated?

6. Why is regulation of ECF osmolarity important? How is it regulated? What are the causes and consequences of ECF hypertonicity and ECF hypotonicity?
7. Outline the sources of input and output in a daily salt balance and a daily H_2O balance. Which are subject to control to maintain the body's fluid balance?
8. Distinguish between an acid and a base.
9. What is the relationship between $[H^+]$ and pH?
10. What is the normal pH of body fluids? How does this compare to the pH of H_2O? Define acidosis and alkalosis.
11. What are the consequences of fluctuations in $[H^+]$?
12. What are the body's sources of H^+?
13. Describe the three lines of defense against changes in $[H^+]$ in terms of the mechanisms and speed of action.
14. List and indicate the functions of each of the body's chemical buffer systems.
15. What are the causes of the four categories of acid-base imbalances?
16. Why is uremic acidosis so serious?

Quantitative Exercises (Solutions on p. E–15.)

1. Given that plasma pH = 7.4, arterial P_{CO_2} = 40 mm Hg, and each mm Hg partial pressure of CO_2 is equivalent to a plasma $[CO_2]$ of 0.03 mM, what is the value of plasma $[HCO_3^-]$?
2. Death occurs if the plasma pH falls outside the range of 6.8 to 8.0 for an extended time. What is the concentration range of H^+ represented by this pH range?
3. If a person drinks 1 liter of distilled water, use the data in Table 15-1 to calculate the resulting percent increase in total body water (TBW), ICF, ECF, plasma, and interstitial fluid. Repeat the calculations for ingestion of 1 liter of isotonic NaCl. Which solution would be better at expanding plasma volume in a patient who has just hemorrhaged?

Points to Ponder

(Explanations on p. E–15.)

1. Alcoholic beverages inhibit vasopressin secretion. Given this fact, predict the effect of alcohol on the rate of urine formation. Predict the actions of alcohol on ECF osmolarity. Explain why a person still feels thirsty after excessive consumption of alcoholic beverages.
2. If a person loses 1,500 ml of salt-rich sweat and drinks 1,000 ml of water during the same time period, what will happen to vasopressin secretion? Why is it important to replace both the water and the salt?
3. If a solute that can penetrate the plasma membrane such as dextrose (a type of sugar) is dissolved in sterile water at a concentration equal to that of normal body fluids and then is injected intravenously, what would be the impact on the body's fluid balance?
4. Explain why it is safer to treat gastric hyperacidity with antacids that are poorly absorbed from the digestive tract than with baking soda, which is a good buffer for acid but is readily absorbed.
5. Which of the following reactions would occur to buffer the acidosis accompanying severe pneumonia?
   a. $H^+ + HCO_3^- \rightarrow H_2CO_3 \rightarrow CO_2 + H_2O$
   b. $CO_2 + H_2O \rightarrow H_2CO_3 \rightarrow H^+ + HCO_3^-$
   c. $H^+ + Hb \rightarrow HHb$
   d. $HHb \rightarrow H^+ + Hb$
   e. $NaH_2PO_4 + Na^+ \rightarrow Na_2HPO_4 + H^+$
6. ***Clinical Consideration*** Marilyn Y. has had pronounced diarrhea for over a week as a result of having acquired samonellosis, an intestinal bacterial infection, from improperly handled food. What impact has this prolonged diarrhea had on her fluid and acid-base balance? In what ways has Marilyn's body been trying to compensate for these imbalances?

16 The Digestive System

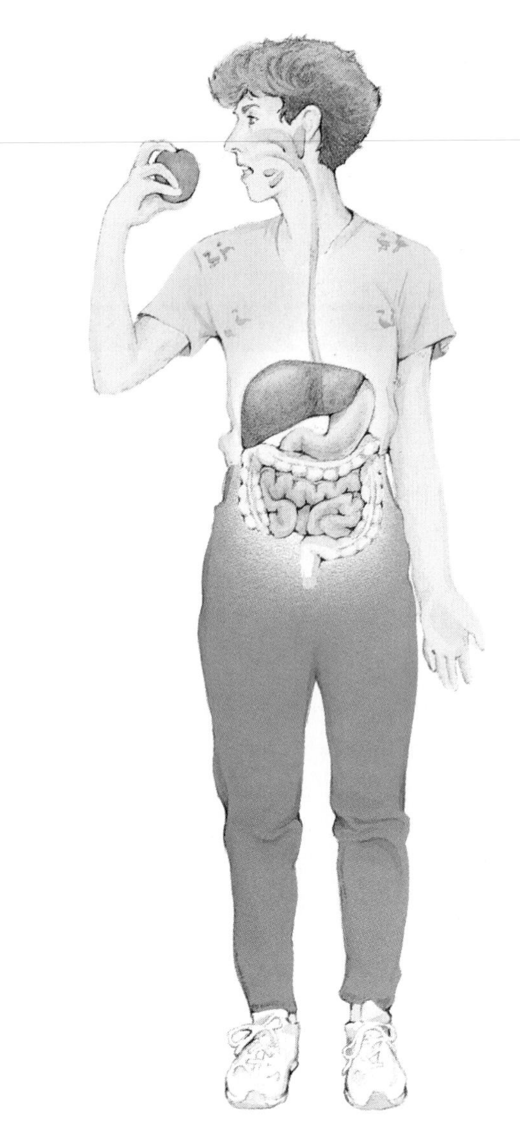

Body systems maintain homeostasis

HOMEOSTASIS
The digestive system contributes to homeostasis by transferring nutrients, water, and electrolytes from the external environment to the internal environment.

Homeostasis is essential for survival of cells

CELLS
Cells need a constant supply of nutrients to support their energy-generating chemical reactions:

Food + $O_2 \rightarrow CO_2 + H_2O$ + Energy.

Also, proper cell function depends on maintaining the availability of water and various electrolytes.

Cells make up body systems

To maintain homeostasis, nutrient molecules used for energy production must continually be replaced by the acquisition of new, energy-rich nutrients. Similarly, water and electrolytes that are constantly lost in the urine and sweat and through other avenues must be replenished on a regular basis. The **digestive system** contributes to homeostasis by transferring nutrients, water, and electrolytes from the external environment to the internal environment. The digestive system does not directly regulate the concentration of any of these constituents in the internal environment. It does not vary nutrient, water, or electrolyte uptake based on body needs (with few exceptions), but rather, it optimizes conditions for digesting and absorbing what is ingested.

Chapter Contents At a Glance

Introduction

The primary function of the **digestive system** is to transfer nutrients (after modifying them), water, and electrolytes from the food we eat into the body's internal environment. Ingested food is essential as an energy source, or "fuel," from which the cells can produce ATP to carry out their particular energy-dependent activities, such as active transport, contraction, synthesis, and secretion. Food is also a source of building supplies for the renewal and addition of body tissues.

Plants can capture the sun's energy and manufacture the organic molecules they need from inorganic compounds such as CO_2 and H_2O through the process of **photosynthesis:**

$$\underset{\text{(from sun)}}{\text{energy}} + CO_2 + H_2O \xrightarrow{\text{plant photosynthesis}} \underset{\text{molecules}}{\text{organic}} + O_2$$

Humans cannot harvest energy from sunlight directly, so they have to survive on second-hand energy by eating either plants or other animals that have eaten plants. In turn, humans use the organic molecules (in food) and O_2 to produce energy, with CO_2 and H_2O being formed as by-products in the process:

$$\underset{\text{molecules (in food)}}{\text{organic}} + O_2 \xrightarrow{\text{human cell metabolism}} \underset{\text{(for use by human cells)}}{\text{energy}} + CO_2 + H_2O$$

Note that these fundamental reactions in plants and humans (as well as other animals) are the reverse of each other, thus providing a natural balance between these forms of life. (Both plants and animals perform additional chemical reactions that involve these organic molecules as well.)

The act of eating does not automatically make the preformed organic molecules in food available to the body cells as a source of fuel or as building blocks. The food first must be digested or broken down into small, simple molecules that can be absorbed from the digestive tract into the circulatory system for distribution to the cells. Normally, about 95% of the ingested food is made available for the body's use.

We will first provide an overview of the digestive system, examining the common features of the various components of the system, before we begin a detailed tour of the tract from beginning to end.

The digestive system performs four basic digestive processes.

There are four basic digestive processes: *motility, secretion, digestion,* and *absorption.*

Motility **Motility** refers to the muscular contractions that mix and move forward the contents of the digestive tract. Like vascular smooth muscle, the smooth muscle in the walls of the digestive tract maintains a constant low level of contraction known as **tone.** Tone is important in maintaining a steady pressure on the contents of the digestive tract as well as in preventing its walls from remaining permanently stretched following distention.

Superimposed on this ongoing tonic base are two basic types of digestive motility: propulsive movements and mixing movements. *Propulsive movements* propel or push the contents forward through the digestive tract at varying speeds, with the rate of propulsion depending on the functions accomplished by the different regions; that is, food is moved forward in a given segment at an appropriate velocity to allow that segment to "do its job." For example, transit of food through the esophagus is rapid, which is appropriate because this structure merely serves as a passageway from the mouth to the stomach. In comparison, in the small intestine, the major site of digestion and absorption, the contents are moved forward slowly, allowing sufficient time for the breakdown and absorption of food to be accomplished.

Mixing movements serve a twofold function. First, by mixing food with the digestive juices, these movements promote digestion of the food. Second, they facilitate absorption by

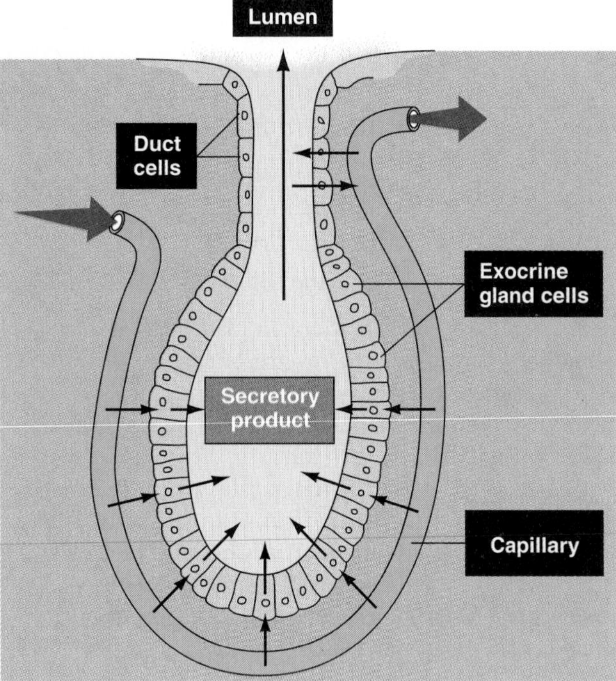

Figure 16-1 General Mode of Exocrine Gland Secretion
Exocrine gland cells extract from the plasma by both active and passive means the raw materials that they need to produce their secretory product. This product is emptied into ducts, which lead, in the case of the digestive system, to the lumen of the digestive tract. Frequently, the secretion is modified as it moves through the duct by active and passive transport mechanisms within the membranes of the cells lining the duct.

exposing all portions of the intestinal contents to the absorbing surfaces of the digestive tract.

Contraction of the smooth muscle within the walls of the digestive organs accomplishes the movement of material through most of the digestive tract, with the exceptions of the ends of the tract—the mouth through the early portion of the esophagus at the beginning and the external anal sphincter at the end—where motility involves skeletal muscle rather than smooth muscle activity. Accordingly, the acts of chewing, swallowing, and defecation have voluntary components since skeletal muscle is under voluntary control, whereas motility accomplished by smooth muscle throughout the remainder of the tract is controlled by complex involuntary mechanisms.

Secretion A number of digestive juices are secreted into the digestive tract lumen by exocrine glands (see p. 3) located along the route, each with its own specific secretory product or products. Each **digestive secretion** consists of water, electrolytes, and specific organic constituents that are important in the digestive process, such as enzymes, bile salts, or mucus. The secretory cells extract from the plasma large volumes of water and the raw materials necessary to produce their particular secretion (Fig. 16-1). Secretion of all digestive juices requires energy, both for active transport of some of the raw materials into the cell (others diffuse in passively) and for synthesis of secretory products by the endoplasmic reticulum. These exocrine cells are richly endowed with mitochondria to support the extensive energy requirement necessary for secretion. The secretions are released into the digestive tract lumen upon appropriate neural or hormonal stimulation. Normally, the digestive secretions are reabsorbed in one form or another back into the blood after their participation in digestion. Failure to do so (because of vomiting or diarrhea, for example) results in loss of this fluid that has been "borrowed" from the plasma.

Digestion **Digestion** refers to the breaking-down process whereby the structurally complex foodstuffs of the diet are converted into smaller absorbable units by the enzymes produced within the digestive system. Humans consume three different biochemical categories of energy-rich foodstuffs: *carbohydrates, proteins,* and *fats.* These large molecules are unable to cross plasma membranes intact to be absorbed from the lumen of the digestive tract into the blood or lymph. The process of digestion degrades these large food molecules into smaller nutrient molecules that can be absorbed (Table 16-1).

The simplest form of **carbohydrates** is the simple sugars or **monosaccharides** ("one-sugar" molecules), such as **glucose, fructose,** and **galactose,** very few of which are normally found in the diet. Most ingested carbohydrate is in the form of **polysaccharides** ("many-sugar" molecules), which consist of chains of interconnected glucose molecules. The most common polysaccharide consumed is **starch** derived from plant sources. Additionally, meat contains **glycogen,** the polysaccharide storage form of glucose in muscle. **Cellulose,** another dietary polysaccharide that is found in plant walls, cannot be digested into its constituent monosaccharides by the digestive juices secreted in humans; thus, it represents the undigested fiber or "bulk" of our diets. Besides polysaccharides, a lesser source of dietary carbohydrate is in the form of **disaccharides** ("two-sugar" molecules), including **sucrose** (table sugar, which consists of one glucose and one fructose molecule) and **lactose** (milk sugar made up of one glucose and one galactose molecule).

Through the process of digestion, starch, glycogen, and dissacharides are converted into their constituent monosaccharides, principally glucose with small amounts of fructose and galactose. These monosaccharides are the absorbable units for carbohydrates.

The second category of foodstuffs is **proteins,** which consist of various combinations of **amino acids** held together by peptide bonds (see p. A–12). Through the process of digestion, proteins are degraded primarily into their constituent amino acids as well as a few **small polypeptides** (several amino acids linked together by peptide bonds), both of which are the absorbable units for protein.

Fats represent the third category of foodstuffs. Most dietary fat is in the form of **triglycerides,** which are neutral fats, each consisting of a **glycerol** with three (*tri* means "three") **fatty acid** molecules attached. During digestion, two of the fatty acid molecules are split off, leaving a **monoglyceride,** a glycerol molecule with one (*mono* means "one") fatty acid molecule attached. Thus, the end products of fat digestion are monoglycerides and free fatty acids, which are the absorbable units of fat.

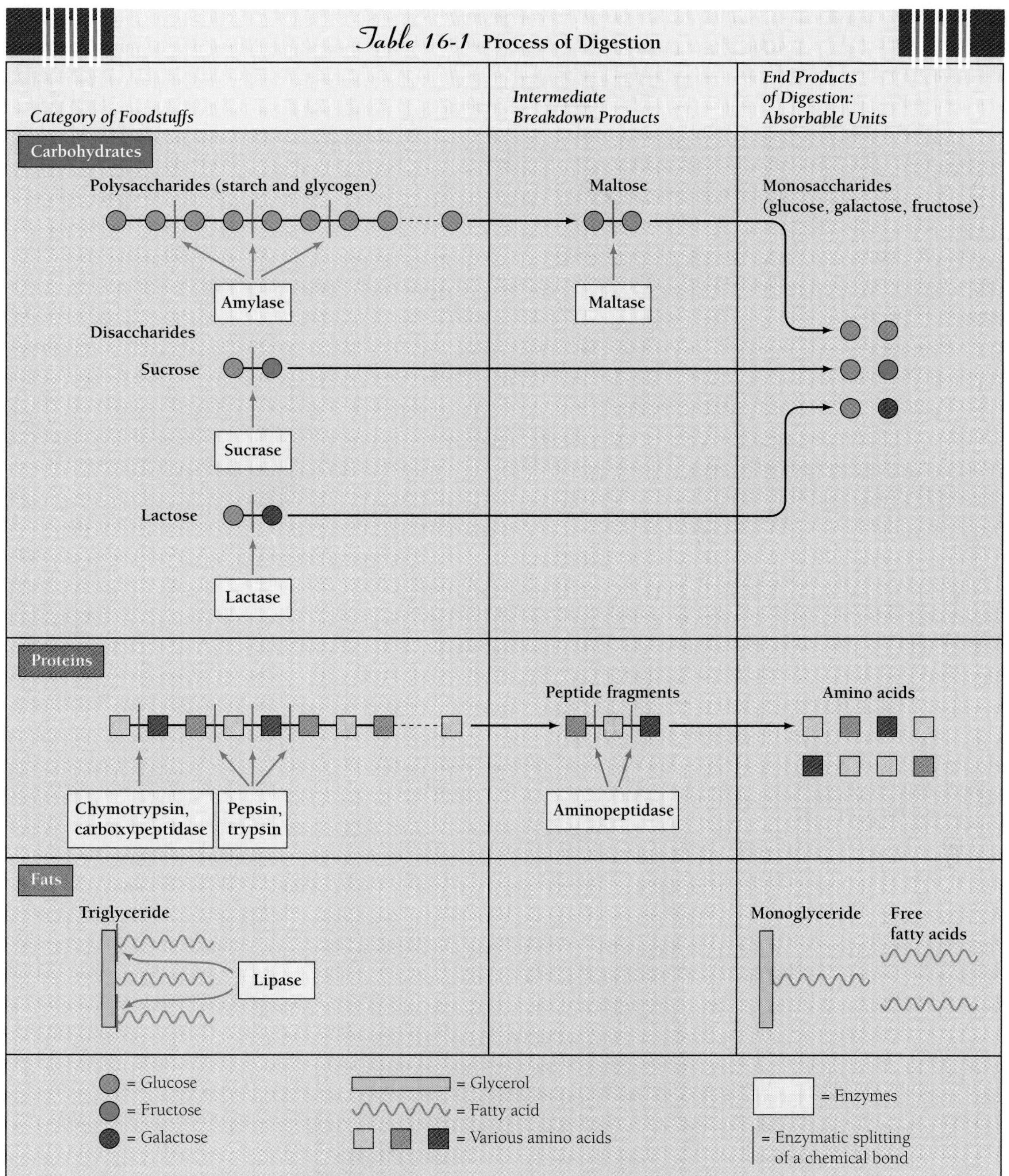

Table 16-1 **Process of Digestion**

Digestion is accomplished by enzymatic **hydrolysis** ("breakdown by water"). By adding H_2O at the bond site, enzymes in the digestive secretions break down the bonds that hold the small molecular subunits within the nutrient molecules together, thus setting the small molecules free (— Fig. 16-2 on p. 552). These small subunits were originally joined to form nutrient molecules by the removal of H_2O at the bond sites. Hydrolysis replaces the H_2O and frees the small absorbable units. Digestive enzymes are specific in the bonds they can hydrolyze. As food moves through the digestive tract,

Table 16-2 Anatomy and Functions of Components of the Digestive System

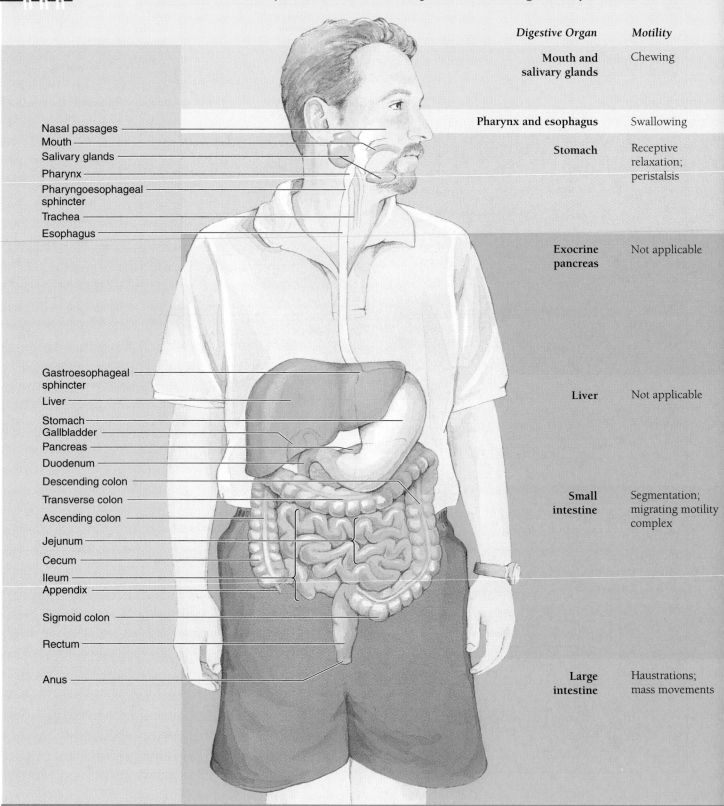

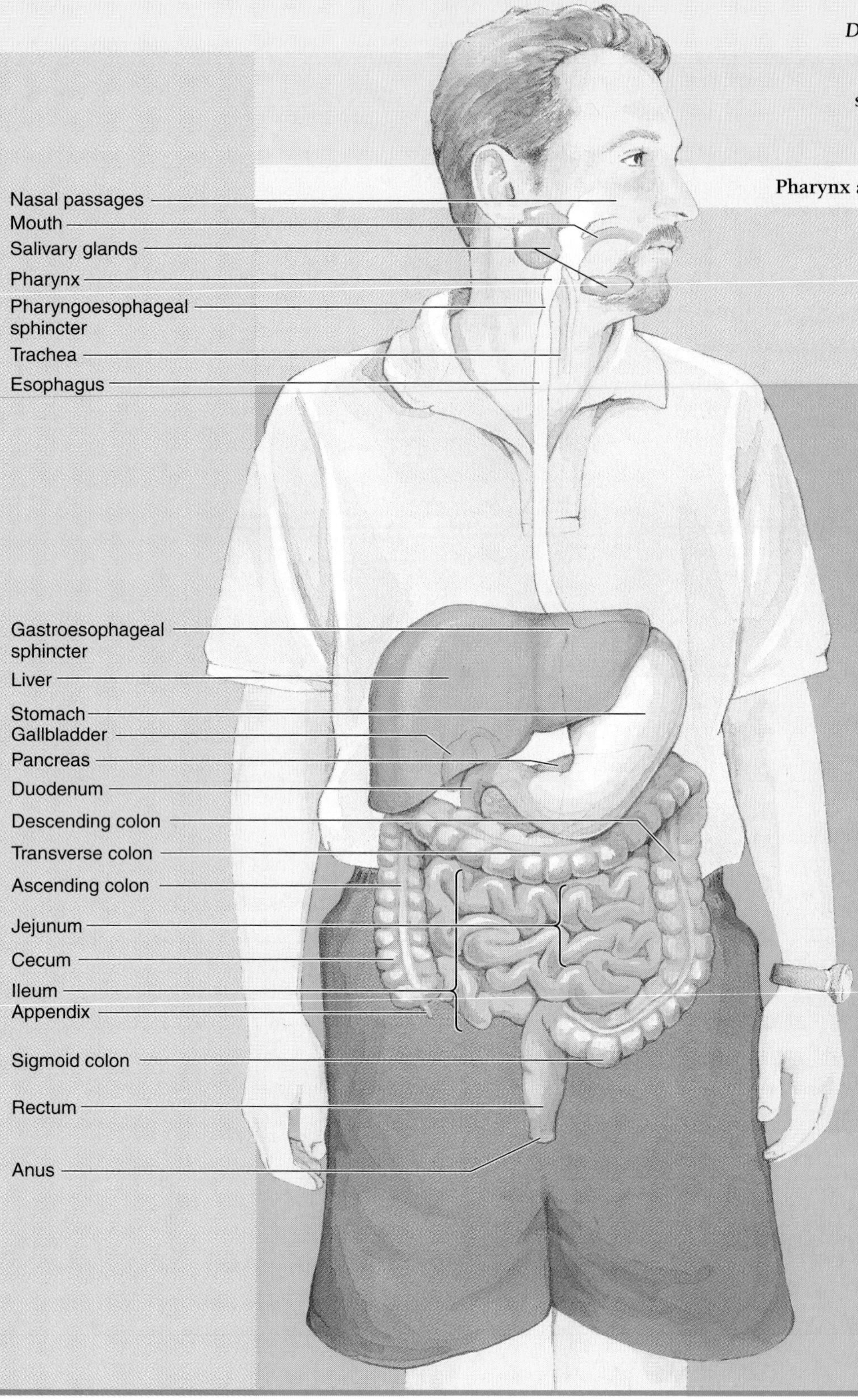

Nasal passages
Mouth
Salivary glands
Pharynx
Pharyngoesophageal sphincter
Trachea
Esophagus
Gastroesophageal sphincter
Liver
Stomach
Gallbladder
Pancreas
Duodenum
Descending colon
Transverse colon
Ascending colon
Jejunum
Cecum
Ileum
Appendix
Sigmoid colon
Rectum
Anus

Digestive Organ	*Motility*
Mouth and salivary glands	Chewing
Pharynx and esophagus	Swallowing
Stomach	Receptive relaxation; peristalsis
Exocrine pancreas	Not applicable
Liver	Not applicable
Small intestine	Segmentation; migrating motility complex
Large intestine	Haustrations; mass movements

Secretion	Digestion	Absorption
Saliva • Amylase • Mucus • Lysozyme	Carbohydrate digestion begins	No foodstuffs; a few medications—for example, nitroglycerin
Mucus	None	None
Gastric juice • HCI • Pepsin • Mucus • Intrinsic factor	Carbohydrate digestion continues in body of stomach; protein digestion begins in antrum of stomach	No foodstuffs; a few lipid-soluble substances, such as alcohol and aspirin
Pancreatic digestive enzymes • Trypsin, chymotrypsin, carboxypeptidase • Amylase • Lipase Pancreatic aqueous $NaHCO_3$ secretion	These pancreatic enzymes accomplish digestion in duodenal lumen	Not applicable
Bile • Bile salts • Alkaline secretion • Bilirubin	Bile does not digest anything, but bile salts facilitate fat digestion and absorption in duodenal lumen	Not applicable
Succus entericus • Mucus • Salt (Small intestine enzymes are not secreted but function intracellularly in the brush border—disaccharidases and aminopeptidases)	In lumen, under influence of pancreatic enzymes and bile, carbohydrate and protein digestion continue and fat digestion is completely accomplished; in brush border, carbohydrate and protein digestion completed	All nutrients, most electrolytes, and water
Mucus	None	Salt and water, converting contents to feces

it is subjected to various enzymes, each of which breaks down the food molecules even further. In this way, large food molecules are converted to simple absorbable units in a progressive, stepwise fashion as the digestive tract contents are propelled forward.

Absorption Digestion is completed and most absorption occurs in the small intestine. Through the process of **absorption,** the small absorbable units that result from digestion, along with water, vitamins, and electrolytes, are transferred from the digestive tract lumen into the blood or lymph.

As we examine the digestive tract from beginning to end, we will discuss the four processes of motility, secretion, digestion, and absorption as they take place within each digestive organ (Table 16-2).

The digestive tract and accessory digestive organs make up the digestive system.

The digestive system consists of the digestive (gastrointestinal; *gastro* means "stomach") tract plus the accessory digestive organs. The **accessory digestive organs** include the *salivary glands,* the *exocrine pancreas,* and the *biliary system,* which is composed of the *liver* and *gallbladder.* These exocrine organs are located outside the wall of the digestive tract and empty their secretions through ducts into the digestive tract lumen. They develop from outpouchings of the embryonic digestive tube and maintain their connection with the digestive tract through the ducts that are formed.

The **digestive tract** is essentially a tube about 9 m (30 feet) in length that runs through the middle of the body from the mouth to the anus. (The length is 9 m in a cadaver but only about half that in a living person because of ongoing contractions of the tract's muscular walls.) The digestive tract includes the following organs (Table 16-2): *mouth; pharynx* (throat); *esophagus; stomach; small intestine* (consisting of the *duodenum, jejunum,* and *ileum*); *large intestine* (composed of the *cecum, appendix, colon,* and *rectum*); and *anus.* It should be noted that these organs are continuous with each other and are discussed as separate entities only because of their regional modifications, which allow them to specialize in particular digestive activities.

Since the digestive tract is continuous from the mouth to the anus, the lumen of this tube, like the lumen of a straw, is continuous with the external environment. As a result, the contents within the lumen of the digestive tract are technically outside the body, just as the soda that you suck through a straw is not a part of the straw. Only after a substance has been absorbed from the lumen across the intestinal wall is it considered to have become a part of the body. This fact is important because conditions essential to the digestive process can be tolerated in the digestive tract lumen but could not be tolerated in the body proper. Consider the following examples:

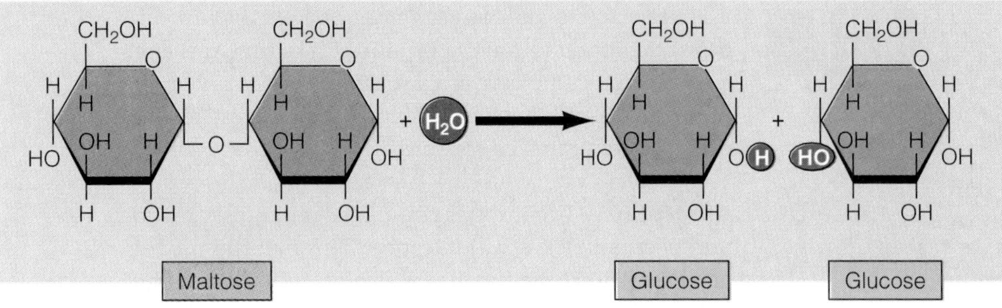

Figure 16-2 **An Example of Hydrolysis** In this example, the disaccharide maltose (the intermediate breakdown product of polysaccharides) is broken down into two glucose molecules by the addition of H_2O at the bond site.

- The pH of the stomach contents falls as low as 2 as a result of the gastric secretion of hydrochloric acid (HCl), yet in the body fluids the range of pH compatible with life is 6.8 to 8.0.
- The harsh digestive enzymes that hydrolyze food could also destroy the body's own tissues that produce them. Therefore, once they are synthesized in inactive form, these enzymes are not activated until they reach the lumen, where they actually attack the food outside the body (that is, within the lumen), thereby protecting the body tissues against self-digestion.
- The lower portion of the intestine is inhabited by millions of living microorganisms that are normally harmless and even beneficial, yet if these same microorganisms enter the body proper (as may happen with a ruptured appendix), they may be extremely harmful or even lethal.

The wall of the digestive tract has the same general structure throughout most of its length from the esophagus to the anus, with some local variations characteristic for each region. A cross section of the digestive tube (Fig. 16-3) reveals four major tissue layers. From the innermost layer of the tract outward they are the *mucosa,* the *submucosa,* the *muscularis externa,* and the *serosa.*

The **mucosa** lines the luminal surface of the digestive tract. It is divided into three layers:

- The primary component of the mucosa is a **mucous membrane,** an inner epithelial layer that serves as a protective surface as well as being modified in particular areas for secretion and absorption. The mucous membrane contains *exocrine cells* for secretion of digestive juices, *endocrine cells* for secretion of gastointestinal hormones, and *epithelial cells* specialized for absorbing digestive nutrients.
- The **lamina propria** is a thin middle layer of connective tissue on which the epithelium rests. Small blood vessels, lymph vessels, and nerve fibers pass through the lamina propria, and it houses the gut-associated lymphoid tissue (GALT), which is important in the defense against intestinal bacteria (see p. 375).
- The **muscularis mucosa** is a sparse outer layer of smooth muscle that lies adjacent to the submucosa.

The mucosal surface is generally not flat and smooth but is highly folded with many ridges and valleys that greatly increase the surface area available for absorption. The degree of folding varies in different areas of the digestive tract, being most extensive in the small intestine, where maximum absorption occurs, and least extensive in the esophagus, which merely serves as a transit tube. The pattern of surface folding can be modified by contraction of the muscularis mucosa. This is important primarily in exposing different areas of the absorptive surface to the luminal contents.

The **submucosa** ("under the mucosa") is a thick layer of connective tissue that provides the digestive tract with its distensibility and elasticity. It contains the larger blood and lymph vessels, both of which send branches inward to the mucosal layer and outward to the surrounding thick muscle layer. Also lying within the submucosa is a nerve network known as the *submucous plexus,* which helps control local activities of each gut region.

Surrounding the submucosa is the **muscularis externa,** the major smooth muscle coat of the digestive tube. In most parts of the tract, it consists of two layers: an *inner circular layer* and an *outer longitudinal layer.* The fibers of the inner smooth muscle layer (adjacent to the submucosa) run circularly around the circumference of the tube. Contraction of these circular fibers constricts or decreases the diameter of the lumen at the point of contraction. Contraction of the fibers in the outer layer, which run longitudinally along the length of the tube, accomplishes shortening of the tube. Together, contractile activity of these smooth muscle layers produces the propulsive and mixing movements. Lying between the two muscle layers is another nerve network, the *myenteric plexus,* which, along with the submucous plexus, helps to regulate local gut activity.

The outer connective tissue covering of the digestive tract is the **serosa,** which secretes a watery serous fluid that lubricates and prevents friction between the digestive organs and the surrounding viscera. Throughout much of the tract, the serosa is continuous with the **mesentery,** which suspends the digestive organs from the inner wall of the abdominal cavity like a sling (Fig. 16-3). This attachment provides relative fixation, supporting the digestive organs in proper position, while still allowing them freedom for mixing and propulsive movements. One cause of **hernias,** or protrusions of an organ through the muscular wall of the cavity that contains them, is mesenteric tearing. Such tearing allows a portion of the digestive tube to fall free from its attachment and bulge through the abdominal wall.

Regulation of digestive function is complex and synergistic.

Digestive motility and secretion are carefully regulated to maximize digestion and absorption of the ingested food. Four factors are involved in the regulation of digestive system function: (1) autonomous smooth muscle function, (2) intrinsic nerve plexuses, (3) extrinsic nerves, and (4) gastrointestinal hormones.

Autonomous smooth muscle function Like self-excitable cardiac muscle cells, some smooth muscle cells are "pacesetter" cells that do not have a constant resting potential but rather display rhythmic, spontaneous variations in membrane potential. The prominent type of self-induced electrical activity in digestive smooth muscle is **slow-wave potentials** (see p. 258), alternatively referred to as the digestive tract's **basic electrical rhythm (BER)** or **pacesetter potential.** Most researchers believe that musclelike but noncontractile cells known as the **interstitial cells of Cajal** are the pacesetter cells responsible for instigating cyclic slow-wave activity. These pacesetter cells are located at the boundaries of the muscle layers. The slow-wave potentials initiated by these cells spread to the adjacent contractile smooth muscle cells. Slow waves are not action potentials and do not directly induce muscle contraction; they are rhythmic, wavelike fluctuations in membrane potential that cyclically bring the membrane closer to or farther from threshold. These slow-wave oscillations are believed to be due to cyclical variations in the rate at which the Na^+ pump transports Na^+ out of the pacesetter cell. Should these waves reach threshold at the peaks of depolarization, a volley of action potentials is triggered at each peak, resulting in repeating, rhythmical cycles of muscle contraction.

Whether threshold is reached depends on the effect of various mechanical, nervous system, and hormonal factors that influence the "resting" potential, or the starting point around which the slow-wave rhythm oscillates. If the starting point is nearer the threshold level, as it is when food is present in the digestive tract, the depolarizing slow-wave peak reaches threshold, so action potential frequency and its accompanying contractile activity increase. Conversely, if the starting point is farther from threshold, as when no food is present, there is less likelihood of reaching threshold, so action potential frequency is lowered and contractile activity is reduced.

Like cardiac muscle, sheets of smooth muscle cells are connected by gap junctions (see p. 52), which serve as points of low electrical resistance so that electrical activity initiated in a digestive tract pacesetter cell can spread to adjacent smooth muscle cells. If threshold is reached and action potentials are triggered, the whole muscle sheet behaves like a functional syncytium, becoming excited and contracting as a unit. If threshold is not achieved, the oscillating electrical activity continues to sweep across the muscle without being accompanied by contractile activity.

Figure 16-3 **Layers of the Digestive Tract Wall** The digestive tract wall consists of four major layers: from the innermost out, they are the mucosa, submucosa, muscularis externa, and serosa.

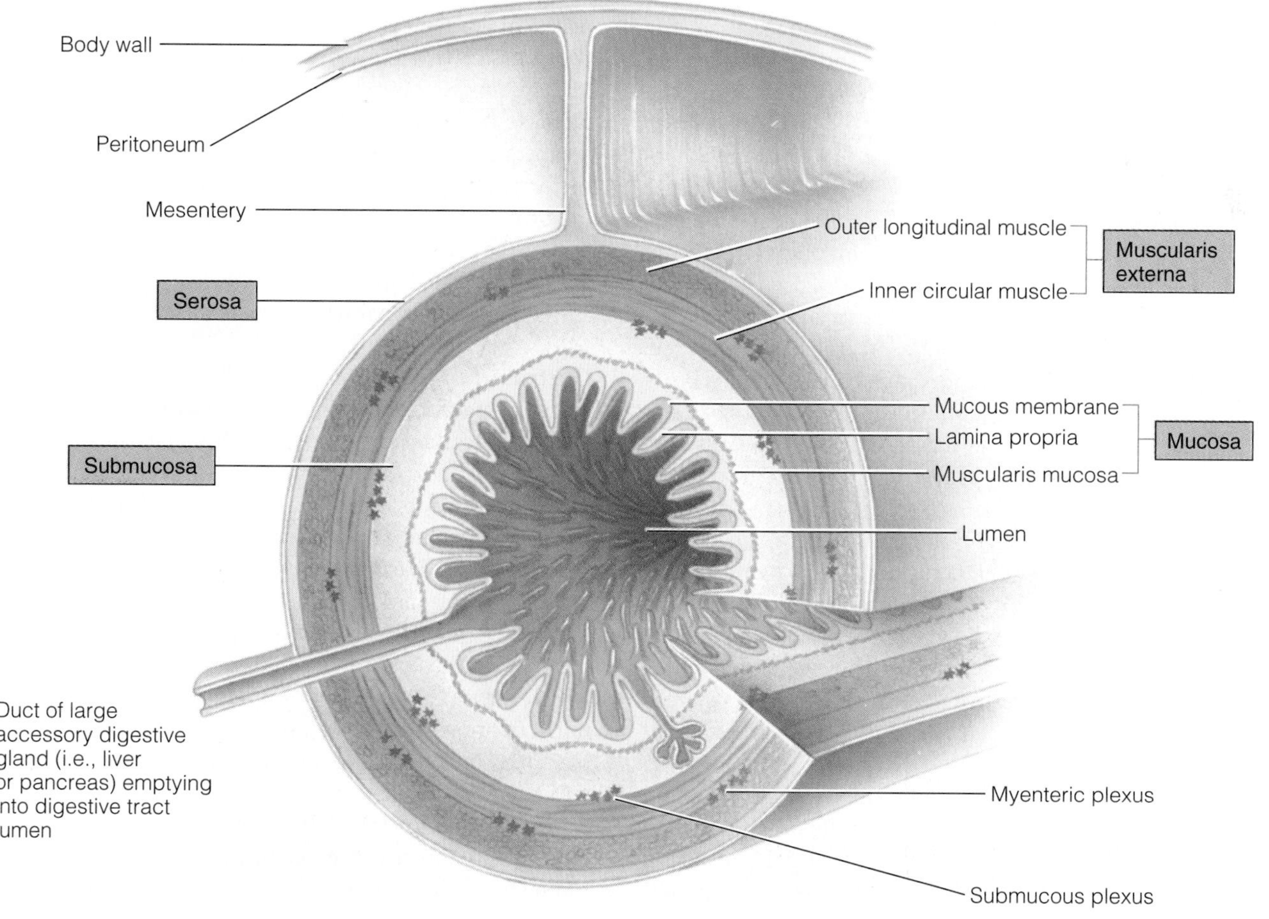

The rate (frequency) of rhythmic digestive contractile activities, such as peristalsis in the stomach, segmentation in the small intestine, and haustrations in the large intestine, depends on the inherent rate established by the involved pacesetter cells. (Specific details about these rhythmic contractions will be discussed when we examine the organs involved.) The intensity of these contractions depends on the number of action potentials that occur when the slow-wave potential reaches threshold, which in turn depends on how long threshold is sustained. At threshold, voltage-gated (see p. 84) Ca^{2+} channels are activated, resulting in Ca^{2+} influx into the smooth muscle cell. The greater the number of action potentials, the higher the cytosolic Ca^{2+} concentration, the greater the cross-bridge activity, and the stronger the contraction. Other factors that influence contractile activity also do so by altering the cytosolic Ca^{2+} concentration. Thus, the level of contractility can range from low-level tone to vigorous mixing and propulsive movements by varying the cytosolic Ca^{2+} concentration.

Intrinsic nerve plexuses The second factor involved in the regulation of digestive tract function is the **intrinsic nerve plexuses.** A nerve plexus is an interconnecting network of nerve cells. Two major networks of nerve fibers form the plexuses of the digestive tract: the **myenteric (Auerbach's) plexus,** which is located between the longitudinal and circular smooth muscle layers (*myo* means "muscle," *enteric* means "intestine," in reference to the location of this plexus between the two muscle layers of the intestine), and the **submucous (Meissner's) plexus,** which is located in the submucosa. These two plexuses are known as intrinsic plexuses because they are located entirely within the digestive tract wall. They run the entire length from the esophagus to the anus. Thus, unlike any other organ system, the digestive tract has its own intramural ("within wall") nervous system, which contains as many neurons as the spinal cord and endows the tract with a considerable degree of self-regulation. Together, these two plexuses are often termed the **enteric nervous system.**

The intrinsic plexuses influence all facets of digestive tract activity. Various types of neurons are present in the intrinsic plexuses. Some are sensory neurons, which possess receptors that respond to specific local stimuli in the digestive tract. Other local neurons innervate the smooth muscle cells and exocrine and endocrine cells of the digestive tract to directly affect digestive tract motility, secretion of digestive juices, and secretion of gastrointestinal hormones. As with the central nervous system, these input and output neurons of the enteric nervous system are linked by interneurons. Some of the output neurons are excitatory and some are inhibitory. For example, neurons that release *acetylcholine* as a neurotransmitter promote contraction of digestive tract smooth muscle whereas the neurotransmitters *nitric oxide* and *vasoactive intestinal peptide* act in concert to cause its relaxation. These intrinsic nerve networks are primarily responsible for coordinating local activity within the digestive tract. To illustrate, if a large piece of food gets stuck in the esophagus, local contractile responses coordinated by the intrinsic plexuses are initiated to push the food forward. Intrinsic nerve activity can in turn be influenced by the extrinsic nerves.

Extrinsic nerves The **extrinsic nerves** are the nerves that originate outside the digestive tract and innervate the various digestive organs—namely, nerve fibers from both branches of the autonomic nervous system. The autonomic nerves influence digestive tract motility and secretion either by modifying ongoing activity in the intrinsic plexuses, altering the level of gastrointestinal hormone secretion, or, in some instances, acting directly on the smooth muscle and glands.

Recall that, in general, the sympathetic and parasympathetic nerves supplying any given tissue exert opposing actions on that tissue. The sympathetic system, which dominates in fight-or-flight situations, tends to inhibit or slow down digestive tract contraction and secretion. This action is appropriate considering that digestive processes are not of highest priority when the body is faced with an emergency or threat from the external environment. The parasympathetic nervous system, on the other hand, dominates in quiet, relaxed situations, when general maintenance types of activities such as digestion can proceed optimally. Accordingly, the parasympathetic nerve fibers supplying the digestive tract, which arrive primarily by way of the vagus nerve, tend to increase smooth muscle motility and promote secretion of digestive enzymes and hormones.

In addition to being called into play during generalized sympathetic or parasympathetic discharge, the autonomic nerves supplying the digestive system can be discretely activated to modify only digestive activity. One of the major purposes of specific activation of extrinsic innervation is the coordination of activity between different regions of the digestive system; for example, the act of chewing food reflexly increases not only salivary secretion but also stomach, pancreatic, and liver secretion via vagal reflexes in anticipation of the arrival of food. Another purpose of specific activation is the provision of a pathway by which factors outside the digestive system can influence digestion, as, for example, with the vagally mediated increase in digestive juices that occurs in anticipation of a meal when a person sees or smells food.

Gastrointestinal hormones The fourth factor influencing digestive tract activity is hormonal control. Tucked within the mucosa of certain regions of the digestive tract are endocrine gland cells that release hormones into the blood upon appropriate stimulation. These **gastrointestinal hormones** are carried through the blood to other areas of the digestive tract, where they exert either excitatory or inhibitory influences on smooth muscle and exocrine gland cells. In feedforward fashion (see p. 12), they also act on the endocrine cells of the pancreas to influence the secretion of pancreatic hormones, which play a key role in the uptake and storage of absorbed nutrient molecules. Gastrointestinal hormones are released primarily in response to specific local changes in the luminal contents (such as the presence of protein, fat, or acid), acting either directly on the endocrine gland cells or indirectly through the intrinsic plexuses or extrinsic autonomic nerves. Interestingly, many of these same hormones are released from neurons in the brain, where they act as neurotransmitters and neuromodulators. During embryonic development, certain cells of the developing neural tissue migrate to the digestive system, where they become endocrine cells.

Receptor activation alters digestive activity through neural reflexes and hormonal pathways.

The wall of the digestive tract contains three different types of sensory receptors that respond to local chemical or mechanical changes in the digestive tract: (1) *chemoreceptors* sensitive to chemical components within the lumen; (2) *mechanoreceptors* (pressurereceptors) sensitive to stretch or tension within the wall; and (3) *osmoreceptors* sensitive to the osmolarity of the luminal contents. Stimulation of these receptors elicits neural reflexes or secretion of hormones, both of which alter the level of activity in the digestive system's effector cells. These effector cells include smooth muscle cells (for modifying motility), exocrine gland cells (for controlling secretion of digestive juices), and endocrine gland cells (for varying secretion of gastrointestinal and pancreatic hormones) (Fig. 16-4).

Receptor activation may elicit two types of neural reflexes—short reflexes and long reflexes. When the intrinsic nerve networks influence local motility or secretion in response to specific local stimulation, such a reflex, in which all elements of the reflex are located within the wall of the digestive tract itself, is known as a **short reflex.** Extrinsic autonomic nervous activity can be superimposed on the local controls to modify smooth muscle and glandular responses, either to correlate activity between different regions of the digestive system or to modify digestive system activity in response to external influences. Because the autonomic reflexes involve long pathways between the central nervous system and digestive system, they are known as **long reflexes.** In addition to these neural reflexes, digestive system activity is also coordinated by the secretion of gastrointestinal hormones, which are triggered directly by local changes in the digestive tract or by short or long reflexes.

From this overview, it should be apparent that regulation of gastrointestinal function is very complex, being influenced by many synergistic, interrelated pathways designed to ensure that the appropriate responses occur to digest and absorb the ingested food. Nowhere else in the body is such a level of overlapping control exercised.

We are now going to take a "tour" of the digestive tract, beginning with the mouth and ending with the anus. We will examine the four basic digestive processes of motility, secretion, digestion, and absorption at each digestive organ along the way. Table 16-2 summarizes these activities and serves as a useful reference throughout the remainder of the chapter.

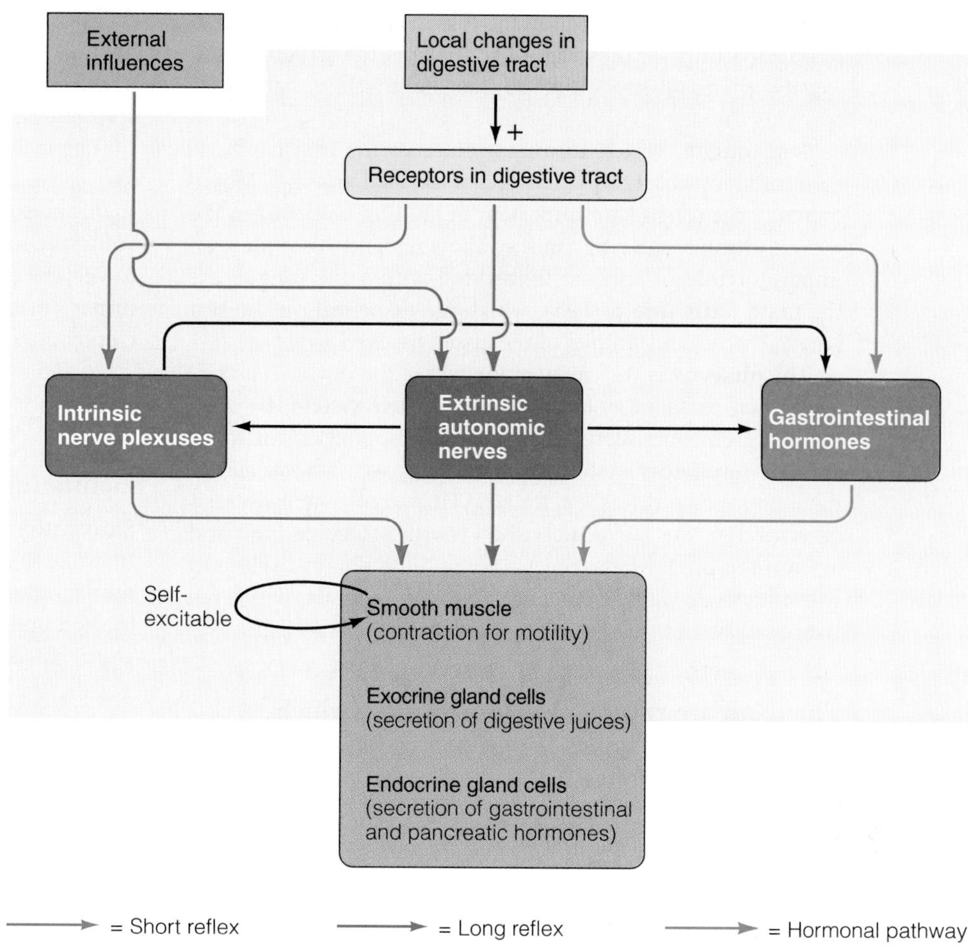

Figure 16-4 **Summary of Pathways Controlling Digestive System Activities**

Mouth

The oral cavity is the entrance to the digestive tract.

Entry to the digestive tract is through the **mouth** or **oral cavity.** The opening is formed by the muscular **lips,** which help procure, guide, and contain the food in the mouth. The lips also serve nondigestive functions; they are important in speech (articulation of many sounds depends on a particular lip formation) and as a sensory receptor in interpersonal relationships (for example, as in kissing).

The **palate,** which forms the arched roof of the oral cavity, separates the mouth from the nasal passages. Its presence allows breathing and chewing or sucking to take place simultaneously. Embryologically, the palate is derived from projections that grow inward from the jaws on both sides and fuse in the midline of the oral cavity. Failure of these projections to unite properly results in a **cleft palate,** which can interfere with sucking, eating, and speech if not surgically corrected. Toward the front of the mouth, the palate is made of bone, forming what is known as the **hard palate.** There is no bone in the portion of the palate toward the rear of the mouth; this region is called the **soft palate.** Hanging down from the soft palate in the rear of the throat is a dangling projection, the

uvula, which plays an important role in sealing off the nasal passages during swallowing. (The uvula is the structure you elevate when you say "ahhh" so that the physician can better see your throat.)

The **tongue,** which forms the floor of the oral cavity, is composed of voluntarily controlled skeletal muscle. Movements of the tongue are important in guiding food within the mouth during chewing and swallowing and also play an important role in speech. Embedded within the tongue are the **taste buds** (see p. 195), which are dispersed on the soft palate, throat, and linings of the cheeks as well.

The **pharynx** is the cavity at the rear of the throat. It acts as a common passageway for both the digestive system (by serving as the link between the mouth and esophagus, for food) and the respiratory system (by providing access between the nasal passages and trachea, for air). This arrangement necessitates mechanisms (to be described shortly) to guide food and air into the proper passageways beyond the pharynx. Housed within the side walls of the pharynx are the **tonsils,** which are lymphoid tissues that are part of the body's defense team.

The teeth are responsible for chewing, which breaks up food, mixes it with saliva, and stimulates digestive secretions.

The first step in the digestive process is **mastication** or **chewing,** the motility of the mouth that involves the slicing, tearing, grinding, and mixing of ingested food by the **teeth.** The teeth are firmly embedded in and protrude from the jawbones. The exposed portion of a tooth is covered by **enamel,** the hardest structure of the body. Enamel is formed prior to the tooth's eruption by special cells that are lost as the tooth erupts. Since it cannot be regenerated after the tooth has erupted, any defects ("cavities") that develop in the enamel must be patched by artificial "fillings," or else the surface will continue to erode into the underlying living pulp.

The upper and lower teeth normally fit together when the jaws are closed. This **occlusion** allows food to be ground and crushed between the tooth surfaces. When the teeth do not make proper contact with each other, they cannot accomplish their normal cutting and grinding action adequately. Such **malocclusion** results from abnormal positioning of the teeth and is often caused either by overcrowding of teeth too large for the available jaw space or by one jaw being displaced in relation to the other. In addition to ineffective chewing, malocclusion can cause abnormal wearing of affected tooth surfaces and dysfunction and pain of the **temporomandibular joint (TMJ),** where the jawbones articulate with each other. Malocclusions can often be corrected by applying braces, which exert prolonged gentle pressure against the teeth to move them gradually to the desired position.

The purposes of chewing are (1) to grind and break food up into smaller pieces to facilitate swallowing; (2) to mix food with saliva; and (3) to stimulate the taste buds. The latter not only gives rise to the pleasurable subjective sensation of taste but also, in feedforward fashion, reflexly increases salivary, gastric, pancreatic, and bile secretion to prepare for the arrival of food.

The act of chewing can be voluntary, but most chewing during a meal is a rhythmic reflex brought about by activation of the skeletal muscles of the jaws, lips, cheeks, and tongue in response to the pressure of food against the oral tissues.

The teeth can exert forces much greater than those necessary to eat ordinary food. For example, the molars in an adult man can exert a crushing force of up to 200 pounds, which is sufficient to crack a hard nut, but ordinarily these powerful forces are not used. In fact, the degree of occlusion is more important than the force of the bite in determining the efficiency of chewing.

Saliva begins carbohydrate digestion but plays more important roles in oral hygiene and in facilitating speech.

Saliva, the secretion associated with the mouth, is produced by three major pairs of salivary glands—the **sublingual, submandibular,** and **parotid glands** (— Fig. 16-5)—that are located outside of the oral cavity and discharge saliva through short ducts into the mouth. In addition, minor salivary glands, the **buccal glands,** are located in the mucosa lining the cheeks.

Saliva is composed of about 99.5% H_2O and 0.5% electrolytes and protein. As is typical of exocrine gland secretion, the process of salivation occurs in two stages. First, the glandular portion of the salivary glands, the *acini,* produces a primary secretion with an electrolyte (Na^+, Cl^-, K^+, and

— *Figure 16-5* **Salivary Glands** Three major pairs of salivary glands (parotid, sublingual, and submandibular) are located in and around the oral cavity and empty into the mouth via small ducts.

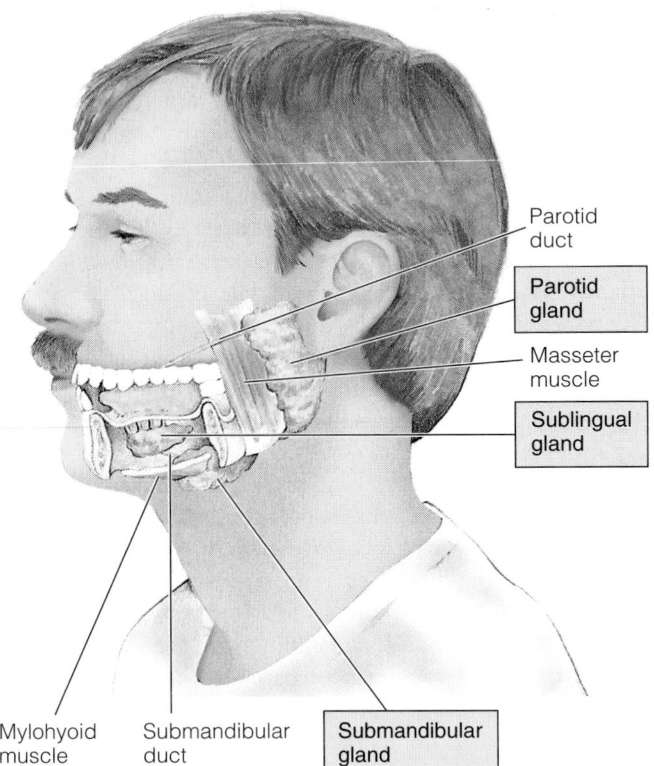

HCO_3^-) composition similar to that of the plasma. Second, as the primary secretion flows through the salivary ducts, the duct cells reabsorb Na^+ and Cl^- from the saliva and add more K^+ and HCO_3^- to it. As a result, the salivary NaCl (salt) concentration is considerably below that in the plasma, a fact believed important in the perception of salty tastes. Similarly, discrimination of sweet tastes is presumably enhanced by the absence of glucose in the saliva. The most important salivary proteins—*amylase, mucus,* and *lysozyme*—contribute to the functions of saliva as follows:

1. Saliva begins digestion of carbohydrate in the mouth through action of **salivary amylase,** an enzyme that breaks polysaccharides down into **maltose,** a disaccharide consisting of two glucose molecules.
2. Saliva facilitates swallowing by moistening food particles, thereby holding them together, and by providing lubrication through the presence of **mucus,** which is thick and slippery.
3. Saliva exerts some antibacterial action by means of a two-fold effect— first by **lysozyme,** an enzyme that lyses, or destroys, certain bacteria, and second by rinsing away material that may serve as a food source for bacteria.
4. Saliva serves as a solvent for molecules that stimulate the taste buds. Only molecules in solution can react with taste bud receptors. You can demonstrate this for yourself: dry your tongue and then drop some sugar on it—you cannot taste the sugar until it is moistened.
5. Saliva aids speech by facilitating movements of the lips and tongue. It is difficult to talk when the mouth feels dry.
6. Saliva plays an important role in oral hygiene by helping to keep the mouth and teeth clean. The constant flow of saliva helps to flush away food residues, shed epithelial cells, and foreign particles. Saliva's contribution in this regard is apparent to anyone who has experienced a foul taste in the mouth when salivation is suppressed for a while, such as during a fever or states of prolonged anxiety.
7. Bicarbonate buffers in the saliva neutralize acids in food as well as acids produced by bacteria in the mouth, thereby helping to prevent **dental caries** (cavities).

In spite of these many functions, saliva is not essential for the digestion and absorption of foods, because enzymes produced by the pancreas and small intestine can complete the digestion of food even in the absence of salivary and gastric secretion. The main problems associated with diminished salivary secretion, a condition known as **xerostomia,** are difficulty in chewing and swallowing, inarticulate speech unless frequent sips of water are taken when talking, and a rampant increase in dental caries.

The continuous low level of salivary secretion can be increased by simple and conditioned reflexes.

On the average, about 1 to 2 liters of saliva are secreted per day, ranging from a continuous spontaneous basal rate of 0.5 ml/min to a maximum flow rate of about 5 ml/min in response to a potent stimulus such as sucking on a lemon. The continuous spontaneous secretion of saliva, even in the absence of apparent stimuli, is due to constant low-level stimulation by the parasympathetic nerve endings that terminate in the salivary glands. This basal secretion is important in keeping the mouth and throat moist at all times.

In addition to this continuous, low-level secretion, salivary secretion may be enhanced by two different types of salivary reflexes (Fig. 16-6): (1) the simple, or unconditioned salivary reflex and (2) the acquired, or conditioned, salivary reflex. The **simple,** or **unconditioned, salivary reflex** occurs when chemoreceptors and pressurereceptors within the oral cavity respond to the presence of food. On activation, these receptors initiate impulses in afferent nerve fibers that carry the information to the **salivary center,** which is located in the medulla of the brain stem, as are all the brain centers that control digestive activities. The salivary center in turn sends impulses via the extrinsic autonomic nerves to the salivary glands to promote increased salivation. Dental procedures promote salivary secretion in the absence of food in the mouth because these manipulations activate pressurereceptors in the mouth.

With the **acquired,** or **conditioned, salivary reflex,** salivation occurs without oral stimulation. Just thinking about, seeing, smelling, or hearing the preparation of pleasant food initiates salivation through this reflex. All of us have experienced such "mouth watering" in anticipation of something delicious to eat. This reflex is a learned response based on previous experience. Inputs that arise outside the mouth and are mentally associated with the pleasure of eating act through the cerebral cortex to stimulate the medullary salivary center.

The salivary center controls the degree of salivary output by means of the autonomic nerves that supply the salivary glands. Unlike the autonomic nervous system elsewhere in

Figure 16-6 **Control of Salivary Secretion**

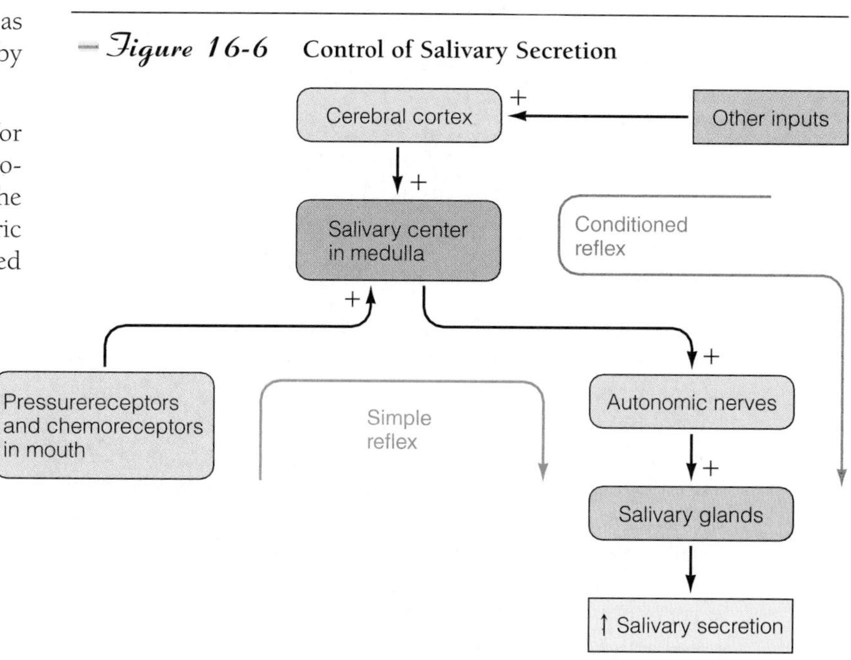

the body, sympathetic and parasympathetic responses in the salivary glands are not antagonistic. Both sympathetic and parasympathetic stimulation increase salivary secretion, but the quantity, characteristics, and mechanisms are different. Parasympathetic stimulation, which exerts the dominant role in salivary secretion, produces a prompt and abundant flow of watery saliva that is rich in enzymes. Sympathetic stimulation, on the other hand, produces a much smaller volume of thick saliva that is rich in mucus. Because sympathetic stimulation elicits a smaller volume of saliva, the mouth feels drier than usual during circumstances when the sympathetic system is dominant, such as stress situations. Thus, people experience a dry feeling in the mouth when they are nervous about giving a speech.

Salivary secretion is the only digestive secretion entirely under neural control. All other digestive secretions are regulated by both nervous system reflexes and hormones.

Digestion in the mouth is minimal, and no absorption of nutrients occurs.

Digestion in the mouth involves the hydrolysis of polysaccharides into disaccharides by amylase. However, most digestion by this enzyme is accomplished in the body of the stomach after the food mass and saliva have been swallowed. Acid inactivates amylase, but in the center of the food mass, where stomach acid has not yet reached, this salivary enzyme continues to function for several more hours.

No absorption of foodstuff occurs from the mouth. Importantly, some therapeutic agents can be absorbed by the oral mucosa, a prime example being a vasodilator drug, *nitroglycerin,* which is used by certain cardiac patients to relieve anginal attacks (see p. 297) associated with myocardial ischemia (see p. 284).

Pharynx and Esophagus

Swallowing is a sequentially programmed all-or-none reflex.

The motility associated with the pharynx and esophagus is **swallowing,** or **deglutition.** Most of us think of swallowing as the limited act of moving food out of the mouth into the esophagus. However, swallowing actually refers to the entire process of moving food from the mouth through the esophagus into the stomach.

Swallowing is initiated when a **bolus,** or ball of food, is voluntarily forced by the tongue to the rear of the mouth into the pharynx. The pressure of the bolus in the pharynx stimulates pharyngeal pressurereceptors, which send afferent impulses to the **swallowing center** located in the medulla. The swallowing center then reflexly activates in the appropriate sequence the muscles that are involved in swallowing. Swallowing is an example of a sequentially programmed all-or-none reflex in which multiple responses are triggered in a specific timed sequence; that is, a number of highly coordinated activities are initiated in a regular pattern over a period of time to accomplish the act of swallowing. Swallowing is initiated voluntarily, but once it is initiated, it cannot be stopped. Perhaps you have experienced this when a large piece of hard candy inadvertently slipped to the rear of your throat, triggering an unintentional swallow.

During the oropharyngeal stage of swallowing, food is directed into the esophagus and is prevented from entering the wrong passageways.

Swallowing is arbitrarily divided into two stages: the oropharyngeal stage and the esophageal stage. The **oropharyngeal stage** lasts about 1 second and consists of moving the bolus from the mouth through the pharynx and into the esophagus. When the bolus enters the pharynx during swallowing, it must be directed into the esophagus and prevented from entering the other openings that communicate with the pharynx. In other words, food must be prevented from reentering the mouth, from entering the nasal passages, and from entering the trachea. All of this is accomplished by the following coordinated activities (— Fig. 16-7):

- Food is prevented from reentering the mouth during swallowing by the position of the tongue against the hard palate.
- The uvula is elevated and lodges against the back of the throat, sealing off the nasal passage from the pharynx so that food does not enter the nose.
- Food is prevented from entering the trachea primarily by elevation of the larynx and tight closure of the vocal folds across the laryngeal opening, or **glottis.** The first portion of the trachea is the *larynx,* or *voice box,* across which are stretched the *vocal folds* (see p. 420). During swallowing, the vocal folds serve a purpose unrelated to speech. Contraction of laryngeal muscles aligns the vocal folds in tight apposition to each other, thus sealing the glottis entrance. Also, the bolus tilts a small flap of cartilaginous tissue, the **epiglottis** (*epi* means "upon"), backward down over the closed glottis as further protection from food entering the respiratory airways.
- The individual does not attempt futile respiratory efforts when the respiratory passages are temporarily sealed off during swallowing because the swallowing center briefly inhibits the nearby respiratory center.
- With the larynx and trachea sealed off, pharyngeal muscles contract to force the bolus into the esophagus.

The esophagus is guarded by sphincters at both ends.

The **esophagus** is a fairly straight muscular tube that extends between the pharynx and stomach (Table 16-2). Lying for the most part in the thoracic cavity, it penetrates the diaphragm and joins the stomach in the abdominal cavity a few centimeters below the diaphragm. Occasionally, a portion of the stomach herniates through the esophageal hiatus and protrudes into the thoracic cavity, a condition known as a **hiatal hernia.**

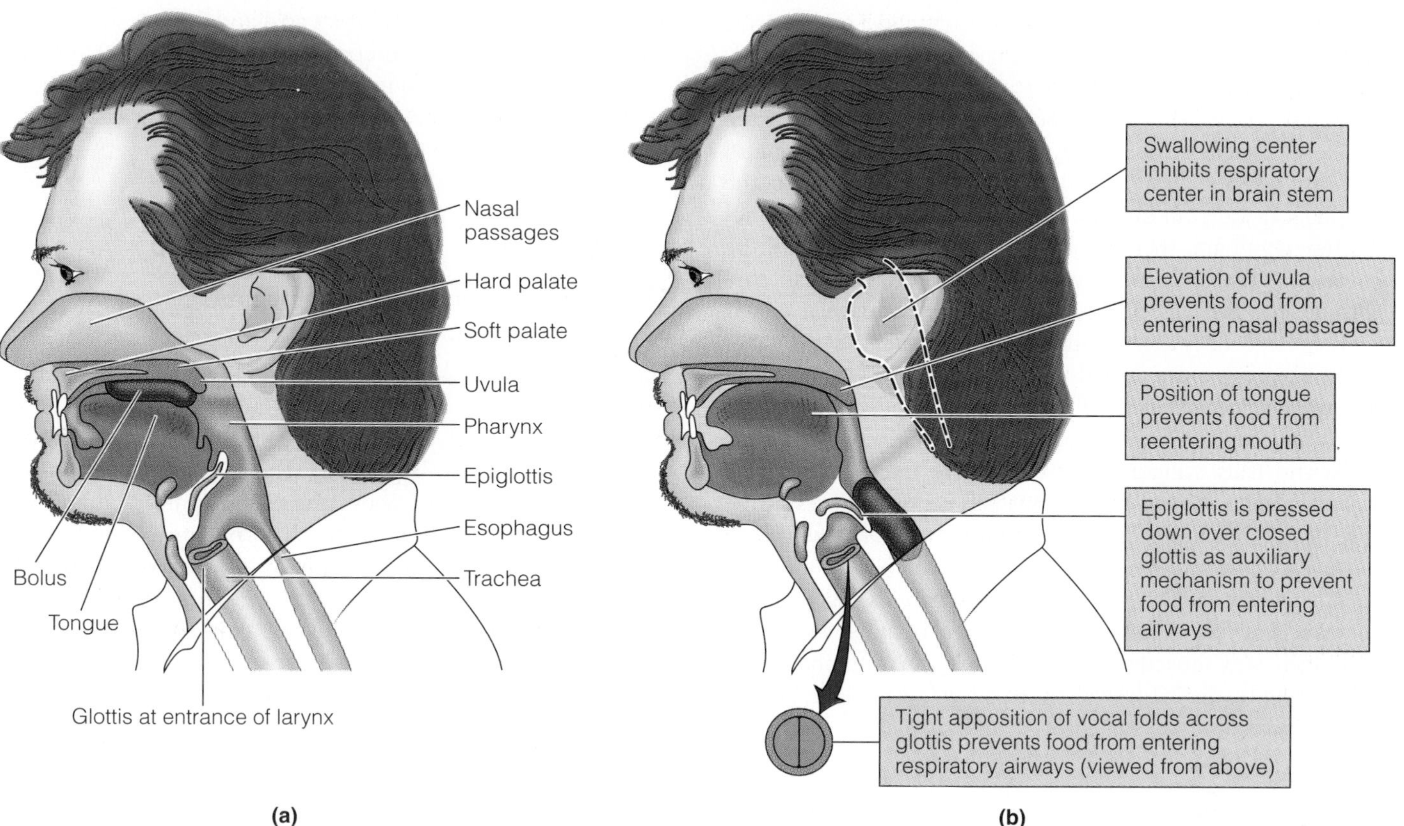

Figure 16-7 **Oropharyngeal Stage of Swallowing** (a) Position of the oropharyngeal structures at rest. (b) Changes that occur during the oropharyngeal stage of swallowing to prevent the bolus of food from entering the wrong passageways.

The esophagus is guarded at both ends by sphincters. A sphincter is a ringlike muscular structure that, when closed, prevents passage through the tube it guards. The upper esophageal sphincter is the **pharyngoesophageal sphincter,** and the lower sphincter is the **gastroesophageal sphincter.**

Since the esophagus is exposed to subatmospheric intrapleural pressure as a result of respiratory activity (see p. 423), a pressure gradient exists between the atmosphere and the esophagus. Accordingly, if the entrance to the esophagus were not closed, air would enter the esophagus as well as the trachea with each breath. Except during a swallow, the pharyngoesophageal sphincter keeps the entrance to the esophagus closed to prevent large volumes of air from entering the esophagus and stomach during breathing. Instead, the air is directed only into the respiratory airways. Were it not for the pharyngoesophageal sphincter, the digestive tract would be subjected to large volumes of gas, which would lead to excessive **eructation** (burping). In contrast to most sphincters, passive elastic tensions in the walls of the pharyngoesophageal sphincter cause the esophagus to be closed when this sphincter is relaxed. During swallowing, this sphincter contracts, opening the sphincter and allowing the bolus to pass into the esophagus. Once the bolus has entered the esophagus, the pharyngoesophageal sphincter closes, the respiratory airways are opened, and breathing resumes. The oropharyngeal stage is complete, and about 1 second has passed since the swallow was first voluntarily initiated.

Peristaltic waves push the food through the esophagus.

The **esophageal stage** of the swallow now begins. The swallowing center initiates a **primary peristaltic wave** that sweeps from the beginning to the end of the esophagus, forcing the bolus ahead of it through the esophagus to the stomach. **Peristalsis** refers to ringlike contractions of the circular smooth muscle that move progressively forward with a stripping motion, pushing the bolus ahead of the contraction (Fig. 16-8). Thus, propulsion of food through the esophagus is an

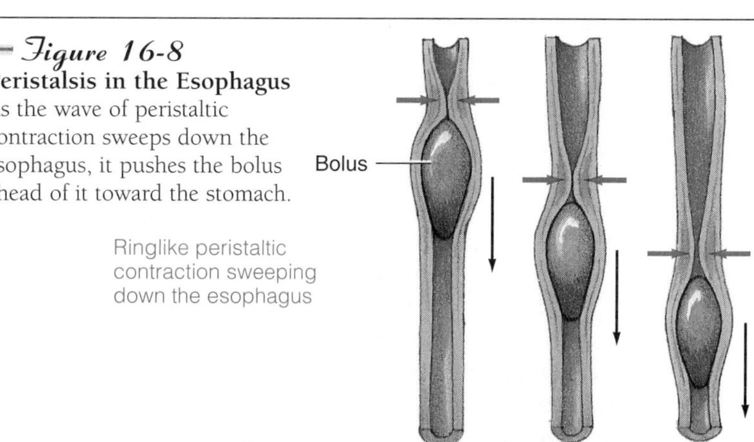

Figure 16-8
Peristalsis in the Esophagus
As the wave of peristaltic contraction sweeps down the esophagus, it pushes the bolus ahead of it toward the stomach.

active process that does not rely on gravity. Food can be pushed to the stomach even during a headstand. The peristaltic wave takes about 5 to 9 seconds to reach the lower end of the esophagus. Progression of the wave is controlled by the swallowing center, with innervation by means of the vagus.

Liquids, not being held up by friction against the esophageal wall, fall quickly to the lower esophageal sphincter by gravity and then must wait about 5 seconds until the primary peristaltic wave finally arrives before they can pass through the gastroesophageal sphincter.

If a large or sticky swallowed bolus, such as a bite of peanut butter sandwich, fails to be carried along to the stomach by the primary wave of peristalsis, the lodged bolus distends the esophagus, stimulating pressurereceptors within its walls and thus initiating a second, more forceful peristaltic wave that is mediated by the intrinsic nerve plexuses at the level of the distention. These **secondary peristaltic waves** do not involve the swallowing center, nor is the person aware of their occurrence. Distention of the esophagus also reflexly increases salivary secretion. The trapped bolus is eventually dislodged and moved forward through the combination of lubrication by the extra swallowed saliva and the forceful secondary peristaltic waves.

The gastroesophageal sphincter prevents reflux of gastric contents.

Except during swallowing, the gastroesophageal sphincter remains contracted to maintain a barrier between the stomach and esophagus, thus reducing the possibility of reflux of acidic gastric contents into the esophagus. If gastric contents do flow back into the esophagus in spite of the sphincter, the acidity of these contents irritates the esophagus, causing the esophageal discomfort known as **heartburn.** (The heart itself is not involved at all.)

As the peristaltic wave sweeps down the esophagus, the gastroesophageal sphincter relaxes reflexly so that the bolus can pass into the stomach. After the bolus has entered the stomach, the gastroesophageal sphincter again contracts.

In a condition known as **achalasia**, the lower esophageal sphincter fails to relax during swallowing and instead contracts more vigorously. Food accumulates in the esophagus, which causes the esophagus to become enormously distended as the food's passage into the stomach is greatly delayed. People with this condition are prone to aspiration pneumonia because of the increased likelihood that some of the detained meal in the esophagus may spill into the pharynx and accidentally be aspirated (sucked) into the lungs. The underlying defect is apparently the result of damage to the myenteric nerve plexus in the region of the gastroesophageal sphincter.

Esophageal secretion is entirely protective.

Esophageal secretion is entirely mucus. In fact, mucus is secreted throughout the length of the digestive tract. By providing lubrication for passage of food, esophageal mucus lessens the likelihood that the esophagus will be damaged by any sharp edges in the newly entering food. Furthermore, it protects the esophageal wall from acid and enzymes in gastric juice if gastric reflux should occur.

The entire transit time in the pharynx and esophagus averages a mere 6 to 10 seconds, too short a time for any digestion or absorption to occur in this region.

Stomach

The stomach stores food and begins protein digestion.

The **stomach** is a J-shaped saclike chamber lying between the esophagus and small intestine. It is arbitrarily divided into three sections based on anatomical, histological, and functional distinctions (Fig. 16-9). The **fundus** is the portion of the stomach that lies above the esophageal opening. The middle or main part of the stomach is the **body.** The smooth muscle layers in the fundus and body are relatively thin, but the lower portion of the stomach, the **antrum,** has much heavier musculature. This difference in muscle thickness plays an important role in gastric motility in these two regions, as you will see shortly. There are also glandular differences in the mucosa of these regions, as will be described later. The terminal portion of the stomach consists of the **pyloric sphincter,** which acts as a barrier between the stomach and the upper part of the small intestine, the duodenum.

The stomach performs several functions. The most important function is to store ingested food until it can be emptied into the small intestine at a rate appropriate for optimal digestion and absorption. It takes hours to digest and absorb a meal that was consumed in only a matter of minutes. Because the small intestine is the primary site for this digestion and absorption, it is important that the stomach store the food and meter it into the duodenum at a rate that does not exceed the small intestine's capacities. A second function of the stomach is to secrete hydrochloric acid (HCl) and enzymes that begin protein digestion. Finally, through the stomach's mixing movements, the ingested food is pulverized and mixed with gastric secretions to produce a thick liquid mixture known as **chyme.**

Stomach motility is complex and subject to multiple regulatory inputs.

There are four aspects of gastric motility: (1) gastric filling, (2) gastric storage, (3) gastric mixing, and (4) gastric emptying.

Gastric filling When empty, the stomach has a volume of about 50 ml, but it can expand to a capacity of about 1 liter (1,000 ml) during a meal. Accommodation of such a 20-fold change in volume would create tension in the walls of the stomach and greatly increase intragastric pressure were it not for two factors: (1) the basic plasticity of the stomach smooth muscle and (2) the receptive relaxation of the stomach as it fills.

Plasticity refers to the ability of smooth muscle to maintain a constant tension over a wide range of lengths, unlike skeletal and cardiac muscle, which exhibit clear-cut length-tension relationships (see p. 236). Thus, when the stomach smooth muscle fibers are stretched by gastric filling, they yield without an increase in muscle tension. Beyond a certain level of stretch, however, a stretch-activated contraction can be superimposed on this passive plastic behavior. Substantial stretch depolarizes the pacesetter cells, bringing them closer to resting potential so that the slow-wave potential is able to reach threshold and initiate contractile activity.

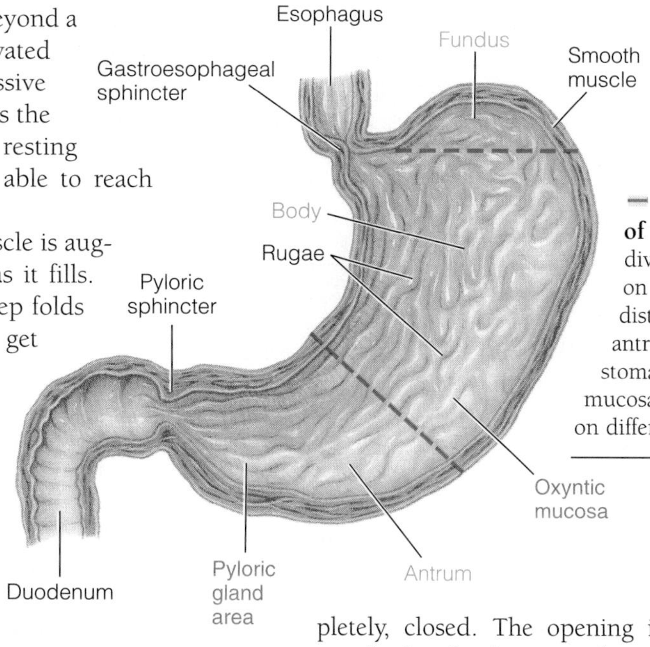

Figure 16-9 Anatomy of the Stomach The stomach is divided into three sections based on structural and functional distinctions—the fundus, body, and antrum. The mucosal lining of the stomach is divided into the oxyntic mucosa and the pyloric gland area based on differences in glandular secretion.

This basic yielding behavior of smooth muscle is augmented by reflex relaxation of the stomach as it fills. The interior of the stomach is thrown into deep folds known as **rugae.** During a meal, the folds get smaller and flatten out as the stomach relaxes slightly with each mouthful, much like the gradual expansion of a collapsed ice bag as it is being filled. This reflex relaxation of the stomach as it is receiving food is called **receptive relaxation;** it enhances the stomach's ability to accommodate the extra volume of food with little rise in stomach pressure. Of course, if more than 1 liter of food is consumed, the stomach becomes overdistended and the person experiences discomfort. Receptive relaxation is triggered by the act of eating and is mediated by the vagus nerve.

Gastric storage Recall that some smooth muscle cells are capable of rhythmic, autonomous, partial depolarization. One such group of these pacesetter cells is located in the upper fundus region of the stomach. These cells generate slow-wave potentials that sweep down the length of the stomach toward the pyloric sphincter at a rate of three per minute. This rhythmic pattern of spontaneous depolarizations—the basic electrical rhythm, or BER, of the stomach—occurs continuously and may or may not be accompanied by contraction of the stomach's circular smooth muscle layer. Depending on the level of excitability in the smooth muscle, it may be brought to threshold by this flow of current and undergo action potentials, which in turn initiate contractions recognized as peristaltic waves that sweep over the stomach in pace with the BER at a rate of three per minute.

Once initiated, the peristaltic wave spreads over the fundus and body to the antrum and pyloric sphincter. Because the muscle layers are thin in the fundus and body, the peristaltic contractions in this region are weak. When the waves reach the antrum, they become much stronger and more vigorous because the muscle there is much thicker.

Since only feeble mixing movements occur in the body and fundus, food emptied into the stomach from the esophagus is stored in the relatively quiet body without being mixed. The fundic area usually does not store food but contains only a pocket of gas. Food is gradually fed from the body into the antrum, where mixing does take place.

Gastric mixing The strong antral peristaltic contractions are responsible for the mixing of food with gastric secretions to produce chyme. Each antral peristaltic wave propels chyme forward toward the pyloric sphincter. Tonic contraction of the pyloric sphincter normally keeps it almost, but not completely, closed. The opening is large enough for water and other fluids to pass through with ease but too small for the thicker chyme to pass through except when a strong peristaltic antral contraction pushes it through. Even then, of the 30 ml of chyme that the antrum can hold, usually only a few milliliters of antral contents are forced into the duodenum with each peristaltic wave. Before more chyme can be squeezed out, the peristaltic wave reaches the pyloric sphincter and causes it to contract more forcefully, sealing off the exit and blocking further passage into the duodenum. The bulk of the antral chyme that was being propelled forward but failed to be pushed into the duodenum is abruptly halted at the closed sphincter and is tumbled back into the antrum, only to be propelled forward and tumbled back again as the new peristaltic wave advances (Fig. 16-10). This tossing back and forth, called **retropulsion,** accomplishes thorough mixing of the chyme in the antrum.

Gastric emptying In addition to being responsible for gastric mixing, the antral peristaltic contractions provide the driving force for gastric emptying. The amount of chyme that escapes into the duodenum with each peristaltic wave before the pyloric sphincter closes tightly depends largely on the strength of peristalsis. The intensity of antral peristalsis can vary markedly under the influence of different signals from both the stomach and the duodenum; thus gastric emptying is regulated by both gastric and duodenal factors. These factors influence the stomach's excitability by slightly depolarizing or hyperpolarizing the gastric smooth muscle. This excitability in turn is a determinant of the degree of antral peristaltic activity. The greater the excitability, the more frequently the BER will generate action potentials, the greater the degree of peristaltic activity in the antrum, and the faster the rate of gastric emptying (Table 16-3).

Factors in the stomach that influence the rate of gastric emptying. The main gastric factor that influences the strength of contraction is the amount of chyme in the stomach. Other things being equal, the stomach empties at a rate proportional

Figure 16-10 Gastric Emptying and Mixing as a Result of Antral Peristaltic Contractions (a) Gastric emptying. A peristaltic contraction originates in the upper fundus and sweeps down toward the pyloric sphincter, becoming more vigorous as it reaches the thick-muscled antrum. As the strong antral peristaltic contraction propels the chyme forward, a small portion of chyme is pushed through the partially open sphincter into the duodenum. The stronger the antral contraction, the more chyme is emptied with each contractile wave. (b) Gastric mixing. When the peristaltic contraction reaches the pyloric sphincter, the sphincter is tightly closed and no further emptying takes place. When chyme that was being propelled forward hits the closed sphincter, it is tossed back into the antrum. Mixing of chyme is accomplished as chyme is propelled forward and tossed back into the antrum with each peristaltic contraction.

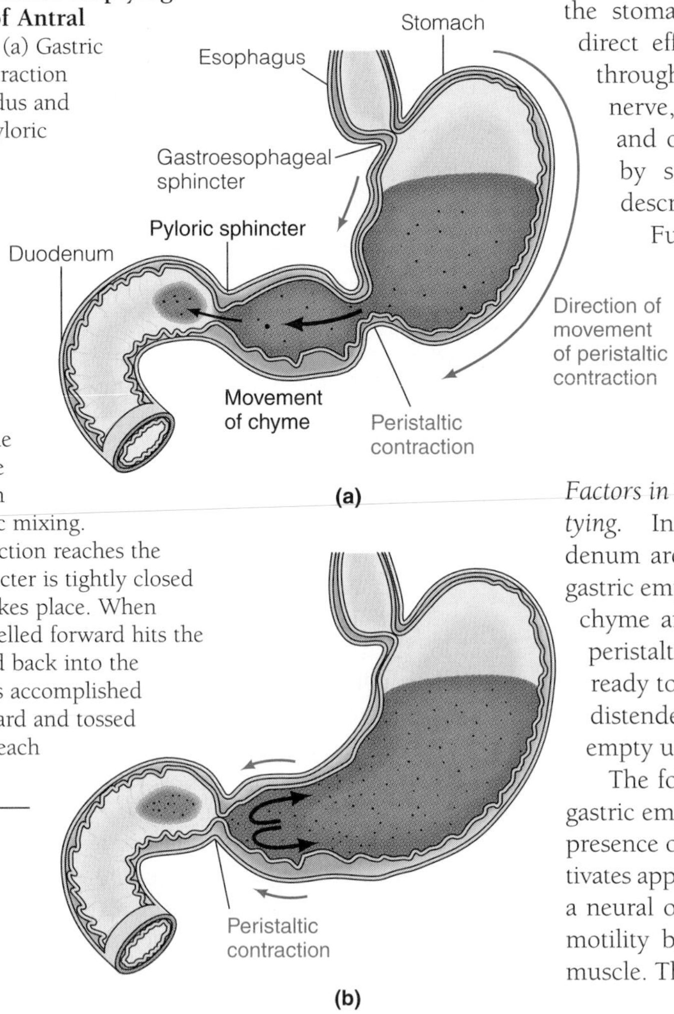

to the volume of chyme in it at any given time. Distention of the stomach triggers increased gastric motility through a direct effect of stretch on the smooth muscle as well as through involvement of the intrinsic plexuses, the vagus nerve, and the stomach hormone *gastrin.* (The control and other functions of this hormone, which is secreted by special endocrine cells in the antrum, will be described later.)

Furthermore, the degree of fluidity of the chyme in the stomach influences gastric emptying. The stomach contents must be converted into a finely divided, thick liquid form before emptying. The sooner the appropriate degree of fluidity can be achieved, the more rapidly the contents are ready to be evacuated.

Factors in the duodenum that influence the rate of gastric emptying. In spite of these gastric influences, factors in the duodenum are of primary importance in controlling the rate of gastric emptying. The duodenum must be ready to receive the chyme and can act to delay gastric emptying by reducing peristaltic activity in the stomach until the duodenum is ready to accommodate more chyme. Even if the stomach is distended and its contents are in a liquid form, it cannot empty until the duodenum is ready to deal with the chyme.

The four most important duodenal factors that influence gastric emptying are *fat, acid, hypertonicity,* and *distention.* The presence of one or more of these stimuli in the duodenum activates appropriate duodenal receptors, thereby triggering either a neural or a hormonal response that puts brakes on gastric motility by reducing the excitability of the gastric smooth muscle. The subsequent reduction in antral peristaltic activity

Table 16-3 Factors Regulating Gastric Motility and Emptying

Factors	*Mode of Regulation*	*Effects on Gastric Motility and Emptying*
Within the stomach		
Volume of chyme	Distention has a direct effect on gastric smooth muscle excitability, as well as acting through the intrinsic plexuses, the vagus nerve, and gastrin	Increased volume stimulates motility and emptying
Degree of fluidity	Direct effect; contents must be in a fluid form to be evacuated	Increased fluidity allows more rapid emptying
Within the duodenum		
Presence of fat, acid, hypertonicity, or distention	Initiates the enterogastric reflex or triggers the release of enterogastrones (cholecystokinin, secretin, gastric inhibitory peptide)	These factors in the duodenum inhibit further gastric motility and emptying until the duodenum has coped with factors already present
Outside the digestive system		
Emotion	Alters autonomic balance	Stimulates or inhibits motility and emptying
Intense pain	Increases sympathetic activity	Inhibits motility and emptying
Decreased glucose utilization in the hypothalamus	Increases vagal activity	Stimulates motility; accompanied by hunger pangs

Pregame Meal: What's In and What's Out?

Many coaches and athletes believe intensely in special food rituals before a competitive event. For example, a football team may always breakfast on steak before a game. Another may always include bananas in their pregame meal. Do these rituals work?

Many studies have been done to determine the effect of the pregame meal on athletic performance. Although laboratory studies have shown that substances such as caffeine improve endurance, no food substance that will greatly enhance performance has been identified. The athlete's prior training is the most important determinant of performance. Even though no particular food confers a special benefit before an athletic contest, some food choices can actually hinder the competitors. For example, a meal of steak is high in fat and could take so long to digest that it might impair the football team's performance and thus should be avoided. On the other hand, food rituals that do not impair performance, such as eating bananas, but give the athletes a morale boost or extra confidence are harmless and should be respected. People may attach special meanings to eating certain foods, and their faith in these practices can make the difference between winning and losing a game.

The greatest benefit of the pregame meal is to prevent hunger during competition. Because the stomach can take from one to four hours to empty, an athlete should eat at least three to four hours before competition begins. Excessive quantities of food should not be consumed before competition. Food that remains in the stomach during competition may cause nausea and possibly vomiting. This condition can be aggravated by nervousness, which slows digestion and delays gastric emptying by means of the sympathetic nervous system.

The best choices are foods that are high in carbohydrate and low in fat and protein. High-carbohydrate foods are recommended because they are emptied from the stomach more quickly than fat or protein is. Carbohydrates do not inhibit gastric emptying by means of enterogastrone release, whereas fat and protein do. Fats in particular delay gastric emptying and are slowly digested. Metabolic processing of proteins yields nitrogenous wastes such as urea whose osmotic activity draws water from the body and increases urine volume, both of which are undesirable during an athletic event. Good choices for a pregame meal include breads, pasta, rice, potatoes, gelatins, and fruit juices. Not only will these complex carbohydrates be emptied from the stomach if consumed one to four hours before a competitive event, but they also will help maintain the blood glucose level during the event.

Although it might seem logical to consume something sugary immediately before a competitive event to provide an "energy boost," beverages and foods high in sugar should be avoided because they trigger insulin release. Insulin is the hormone that enhances glucose entry into most body cells. Once the person begins exercising, insulin sensitivity increases (see p. 678), which results in a decrease in the plasma glucose level. A lowered plasma glucose level induces feelings of fatigue and an increased use of muscle glycogen stores, which can limit performance in endurance events such as the marathon. Therefore, sugar consumption just before a competition can actually impair performance instead of giving the sought-after energy boost.

Within an hour of competition, it is best for athletes to drink only plain water to ensure adequate hydration.

slows down the rate of gastric emptying. The *neural response* is mediated through both the intrinsic nerve plexuses (short reflex) and the autonomic nerves (long reflex). Collectively, these reflexes are called the **enterogastric reflex.** The *hormonal response* involves the release from the duodenal mucosa of several hormones collectively known as **enterogastrones.** These hormones are transported by the blood to the stomach, where they inhibit antral contractions to reduce gastric emptying. Three of these enterogastrones have been clearly identified: **secretin, cholecystokinin (CCK),** and **gastric inhibitory peptide (GIP).** Secretin was the first hormone discovered (in 1902). Because it was a secretory product that entered the blood, it was termed *secretin.* The name *cholecystokinin* derives from the fact that this same hormone is also responsible for contraction of the bile-containing gallbladder (*chole* means "bile"; *cysto* means "bladder"; and *kinin* means "contraction"). The name *gastric inhibitory peptide* is self-explanatory; it is a peptide hormone that inhibits the stomach.

Let us examine why it is important that each of these stimuli in the duodenum (fat, acid, hypertonicity, and distention) delays gastric emptying (acting through the enterogastric reflex or one of the enterogastrones).

- *Fat.* Fat is digested and absorbed more slowly than the other nutrients. Furthermore, fat digestion and absorption take place only within the lumen of the small intestine. Therefore, when fat is already present in the duodenum, further gastric emptying of more fatty stomach contents into the duodenum is prevented until the small intestine has processed the fat already there. In fact, fat is the most potent stimulus for inhibition of gastric motility. This is evident when one compares the rate of emptying of a high-fat meal (after six hours some of a bacon-and-eggs meal may still be in the stomach) with that of a protein and carbohydrate meal (a meal of lean meat and potatoes may empty in three hours). (For a discussion of the pregame meal before participation in an athletic event, see the accompanying boxed feature, A Closer Look at Exercise Physiology.)
- *Acid.* Since the stomach secretes hydrochoric acid (HCl), highly acidic chyme is emptied into the duodenum, where it is neutralized by sodium bicarbonate ($NaHCO_3$) secreted into the duodenal lumen from the pancreas. Unneutralized acid irritates the duodenal mucosa and inactivates the pancreatic digestive enzymes that are secreted into the

duodenal lumen. Appropriately, therefore, unneutralized acid in the duodenum inhibits further emptying of acidic gastric contents until complete neutralization can be accomplished.

- *Hypertonicity.* As molecules of protein and starch are digested in the duodenal lumen, large numbers of amino acid and glucose molecules are released. If absorption of these amino acid and glucose molecules does not keep pace with the rate at which protein and carbohydrate digestion proceeds, these large numbers of molecules remain in the chyme and increase the osmolarity of the duodenal contents. Osmolarity depends on the number of molecules present, not on their size, and one protein molecule may be split into several hundred amino acid molecules, each of which has the same osmotic activity as the original protein molecule. The same holds true for one large starch molecule, which yields many smaller but equally osmotically active glucose molecules. Since water is freely diffusable across the duodenal wall, water enters the duodenal lumen from the plasma as the duodenal osmolarity rises. Large volumes of water entering the intestine from the plasma lead to intestinal distention, and, more importantly, circulatory disturbances ensue because of the reduction in plasma volume. To prevent these effects, gastric emptying is reflexly inhibited when the osmolarity of the duodenal contents starts to rise. Thus, the amount of food entering the duodenum for further digestion into a multitude of additional osmotically active particles is reduced until absorption processes have had an opportunity to catch up.
- *Distention.* Too much chyme in the duodenum inhibits the emptying of even more gastric contents, thus allowing the distended duodenum time to cope with the excess volume of chyme it already contains before it receives an additional quantity.

Peristaltic contractions occur in the empty stomach before the next meal.

After a meal is finally emptied from the stomach, there are no more gastric factors to enhance gastric excitability, so peristaltic contractions eventually die out and the stomach remains quiet for a while. In conjunction with the sensation of hunger before the next meal, however, peristaltic contractions begin again, sweeping over the nearly empty antrum. This arousal of stomach motility appears to be mediated by increased parasympathetic activity, perhaps brought about by the hypothalamus in response to a fall in hypothalamic glucose utilization as the time for the next meal approaches. One of the leading theories of why we get hungry also involves decreased use of glucose by the hypothalamus (chapter 17). An individual may experience the sensation of hunger pangs when these peristaltic contractions are occurring, but the contractions themselves are not responsible for the sensation. Rather, both the sensation of hunger and the increased peristaltic activity are triggered simultaneously by the reduced amount of glucose being metabolized by the brain.

Emotions can influence gastric motility.

Other factors unrelated to digestion, such as emotions, can also alter gastric motility by acting through the autonomic nerves to influence the degree of gastric smooth muscle excitability. Even though the effect of emotions on gastric motility varies from one individual to another and is not always predictable, sadness and fear generally tend to decrease motility, whereas anger and aggression tend to increase it. In addition to emotional influences, intense pain from any part of the body tends to inhibit motility, not just in the stomach but throughout the digestive tract. This response is brought about by increased sympathetic activity and a corresponding decrease in parasympathetic activity.

The body of the stomach does not actively participate in the act of vomiting.

Vomiting, or **emesis**, the forceful explulsion of gastric contents out through the mouth, is generally perceived as being caused by abnormal gastric motility. However, vomiting is not accomplished by reverse peristalsis, as might be predicted. Actually, the stomach itself does not actively participate in the act of vomiting. The stomach, the esophagus, the gastroesophageal sphincter, and the pyloric sphincter are all relaxed during vomiting. The major force for expulsion comes, surprisingly, from contraction of the respiratory muscles—namely, the diaphragm (the major inspiratory muscle) and the abdominal muscles (the muscles of active expiration).

Vomiting begins with a deep inspiration and closure of the glottis. The contracting diaphragm descends downward on the stomach while simultaneous contraction of the abdominal muscles compresses the abdominal cavity, increasing the intra-abdominal pressure and forcing the abdominal viscera upward. As the flaccid stomach is squeezed between the diaphragm from above and the compressed abdominal cavity from below, the gastric contents are forced into the esophagus and out through the mouth. The glottis is closed, so vomited material does not enter the respiratory airways. Also, the uvula is elevated to close off the nasal cavity.

Often when vomitus first enters the esophagus, the pharyngoesophageal sphincter remains closed so that no gastric contents enter the mouth. Distention of the esophagus by the vomitus induces secondary peristaltic waves that force the gastric contents back into the stomach. This cycle repeats itself as the contents are immediately squeezed up into the esophagus again. This is the act of *retching* or *heaves*. After a series of heaves, when the pressure becomes great enough, the person thrusts out the jaw, pulling open the pharyngoesophageal sphincter. The gastric contents are then forced through the esophagus, through the pharyngoesophageal sphincter, and out through the mouth. During this time, the duodenum contracts strongly, which may force some of the intestinal contents back into the stomach and out with the vomitus. The vomited material may therefore be stained with the yellowish bile that has entered the small intestine from the liver and gallbladder.

The vomiting cycle may be repeated several times until the stomach is emptied. Vomiting is usually preceded by profuse salivation, sweating, rapid heart rate, and the sensation of nausea, all of which are characteristic of a generalized discharge of the autonomic nervous system.

This complex act of vomiting is coordinated by a **vomiting center** in the medulla. It is currently unclear whether this center consists of a circumscribed cluster of neurons or a more diffuse network of interacting neurons. Nausea, retching, and vomiting can be initiated by afferent input to the vomiting center from a number of receptors throughout the body. The causes of vomiting include the following:

- Tactile (touch) stimulation of the back of the throat, which is one of the most potent stimuli. For example, sticking a finger in the back of the throat or even the presence of a tongue depressor or dental instrument in the back of the mouth is sufficient stimulation to cause gagging and even vomiting in some people.
- Irritation or distention of the stomach and duodenum.
- Elevated intracranial pressure, such as that caused by cerebral hemorrhage. Thus, vomiting following a head injury is considered a bad sign; it suggests swelling or bleeding within the cranial cavity.
- Rotation or acceleration of the head producing dizziness, such as occurs in motion sickness.
- Intense pain arising from a variety of organs, such as that accompanying passage of a kidney stone.
- Chemical agents, including drugs or noxious substances that initiate vomiting (that is, **emetics**) either by acting in the upper portions of the gastrointestinal tract or by stimulating chemoreceptors in a specialized **chemoreceptor trigger zone** adjacent to the vomiting center in the brain. Activation of this zone triggers the vomiting reflex. For example, the chemotherapeutic agents used in the treatment of cancer often cause vomiting by acting on the chemoreceptor trigger zone.
- Psychogenic vomiting induced by emotional factors, including those accompanying nauseating sights and odors and anxiety before taking an examination or in other stressful situations.

With excessive vomiting, the body experiences large losses of secreted fluids and acids that normally would be reabsorbed. The resultant reduction in plasma volume can lead to dehydration and circulatory problems, while the loss of acid from the stomach can lead to metabolic alkalosis (see p. 541).

Vomiting is not always detrimental, however. Limited vomiting brought about by irritation of the digestive tract can provide a useful service in removing noxious material from the stomach rather than allowing it to be retained and absorbed. In fact, emetics are frequently given in the case of accidental ingestion of a poison to quickly remove the offending substance from the body.

Gastric pits are the source of gastric digestive secretions.

Each day the stomach secretes about 2 liters of gastric juice. The cells responsible for gastric secretion are located in the lining of the stomach, the gastric mucosa, which is divided into two distinct areas: (1) the **oxyntic mucosa,** which lines the body and fundus, and (2) the **pyloric gland area (PGA),** which lines the antrum (Fig. 16-11a). The mucosal gland

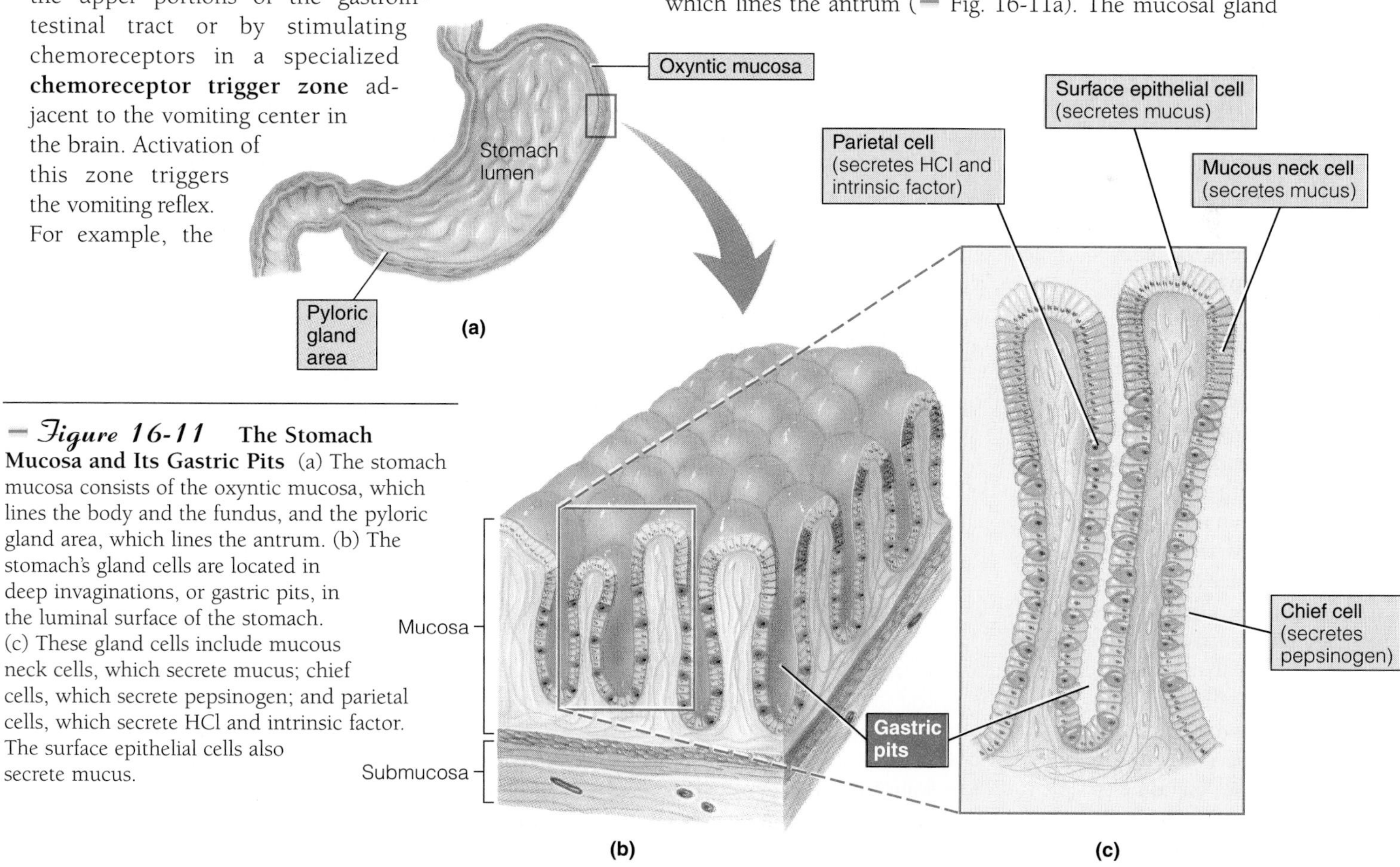

Figure 16-11 **The Stomach Mucosa and Its Gastric Pits** (a) The stomach mucosa consists of the oxyntic mucosa, which lines the body and the fundus, and the pyloric gland area, which lines the antrum. (b) The stomach's gland cells are located in deep invaginations, or gastric pits, in the luminal surface of the stomach. (c) These gland cells include mucous neck cells, which secrete mucus; chief cells, which secrete pepsinogen; and parietal cells, which secrete HCl and intrinsic factor. The surface epithelial cells also secrete mucus.

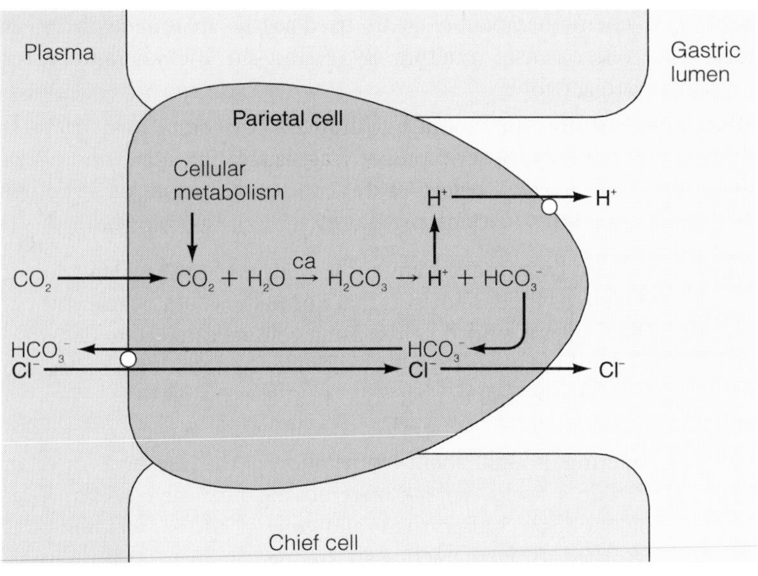

Figure 16-12 Mechanism of HCl Secretion
The stomach's parietal cells actively secrete H^+ and Cl^- by the actions of two separate pumps. The secreted H^+ is derived from H_2CO_3 that is generated within the cell from CO_2 that is either metabolically produced in the cell or diffuses in from the plasma. The secreted Cl^- is transported into the parietal cell from the plasma. The HCO_3^- generated from H_2CO_3 dissociation is transported into the plasma in exchange for the secreted Cl^-.

cells are found in **gastric pits,** which are invaginations or deep pockets in the luminal surface of the stomach (Fig. 16-11b and c). Three types of secretory cells are found in the walls of the pits in the oxyntic mucosa. The entrance, or neck, of the gastric pit is lined by **mucous neck cells,** which secrete a thin, watery *mucus.* (*Mucous* is the adjective; *mucus* is the noun.) The deeper portions of the pit are lined by **chief cells,** which secrete the enzyme precursor *pepsinogen,* and **parietal** (or **oxyntic**) **cells,** which secrete *HCl* and *intrinsic factor.* The parietal cells are located on the outer wall of the gastric pits and do not come into contact with the pit lumen (Fig. 16-11c). (*Parietal* means "wall," a reference to the location of these cells. *Oxyntic* means "sharp," a reference to these cells' potent HCl secretory product.) Although the parietal cells are separated from the lumen of the gastric pit by the chief cells, they transmit their HCl secretion into the lumen through fine channels, or **canaliculi,** that pass between the chief cells.

Between the gastric pits, the gastric mucosa is covered by surface epithelial cells, which secrete a thick, viscous, alkaline mucus that forms a visible layer several millimeters thick over the surface of the mucosa.

The mucous neck cells rapidly divide and serve as the parent cells of all new cells of the gastric mucosa. The daughter cells that result from cell division either migrate out of the pit to become surface epithelial cells or migrate down into the deeper parts of the pit to differentiate into chief or parietal cells. Through this activity, the entire stomach mucosa is replaced about every three days.

The gastric pits of the PGA primarily secrete mucus and a small amount of pepsinogen; no acid is secreted in this area, in contrast to the oxyntic mucosa. More importantly, endocrine cells in the PGA secrete the hormone *gastrin* into the blood. Thus, the most important gastric digestive secretions produced within the body and fundus are HCl, pepsinogen, mucus, and intrinsic factor, which are released into the gastric lumen. On the other hand, the most important product of the PGA is the hormone gastrin, which is released into the blood. We will examine each of these secretory products in more detail.

Hydrochloric acid secretion The parietal cells actively secrete HCl into the lumen of the gastric pits, which in turn empty into the lumen of the stomach. The pH of the luminal contents falls as low as 2 as a result of this HCl secretion. Hydrogen ion (H^+) and chloride ion (Cl^-) are actively transported by separate pumps in the parietal cell's plasma membrane. Hydrogen ion is actively transported against a tremendous concentration gradient, with the H^+ concentration being as much as 3 to 4 million times greater in the lumen than in the blood. Because of the high energy expenditure needed to move H^+ against such a large gradient, parietal cells have an abundance of mitochondria, the energy organelles. Chloride is also actively secreted but against a much smaller concentration gradient of only 1½ times.

The secreted H^+ is not transported from the plasma but is derived instead from metabolic processes within the parietal cell (Fig. 16-12). Whenever a H^+ is secreted, neutrality of the interior of the cell is maintained by generation of a new H^+ from carbonic acid (H_2CO_3) to replace the secreted H^+. The parietal cells contain an abundance of the enzyme carbonic anhydrase (ca). In the presence of carbonic anhydrase, H_2O readily combines with CO_2, which either has been produced within the parietal call by metabolic processes or has diffused in from the blood. The combination of H_2O and CO_2 results in the formation of H_2CO_3, which partially dissociates to yield H^+ and HCO_3^-:

$$CO_2 + H_2O \xrightarrow{ca} H_2CO_3 \rightarrow H^+ + HCO_3^-$$

The generated H^+ replaces the one secreted. The generated HCO_3^- is moved into the plasma by the same carrier that actively transports Cl^- from the plasma into the gastric lumen, similar to the Cl^- shift that occurs in red blood cells (see p. 452). This exchange of HCO_3^- for Cl^- maintains electrical neutrality in the plasma during HCl secretion.

Although HCl does not actually digest anything and is not absolutely essential to gastrointestinal function, it does perform several functions that assist digestion. Hydrochloric acid (1) activates the enzyme precursor pepsinogen to an active enzyme, pepsin, and provides an acid medium that is optimal for pepsin activity; (2) aids in the breakdown of connective tissue and muscle fibers, thereby reducing large food particles into smaller particles; and (3) along with salivary lysozyme, kills most of the microorganisms ingested with the food, although some do escape and continue to grow and multiply in the large intestine.

Pepsinogen secretion The major digestive constituent of gastric secretion is **pepsinogen,** an inactive enzymatic molecule synthesized and packaged by the endoplasmic reticulum and Golgi complex of the chief cells. Pepsinogen is stored in the chief cell's cytoplasm within secretory vesicles known as **zymogen granules,** from which it is released by exocytosis (see p. 25) upon appropriate stimulation. When pepsinogen is secreted into the gastric lumen, HCl cleaves off a small fragment of the molecule, converting it to the active form of the enzyme, **pepsin** (— Fig. 16-13). Once formed, pepsin acts on other pepsinogen molecules to produce more pepsin. A mechanism such as this, whereby an active form of an enzyme activates other molecules of the same enzyme, is referred to as an **autocatalytic** ("self-activating") **process.**

Pepsin initiates protein digestion by splitting certain amino acid linkages in proteins to yield peptide fragments (small amino acid chains); it works most effectively in the acid environment provided by HCl. Because pepsin can digest protein, it must be stored and secreted in an inactive form so that it does not digest the cells in which it is formed. (The primary structural component of cells is protein.) Therefore, pepsin is maintained in the inactive form of pepsinogen until it reaches the gastric lumen, where it is activated by HCl.

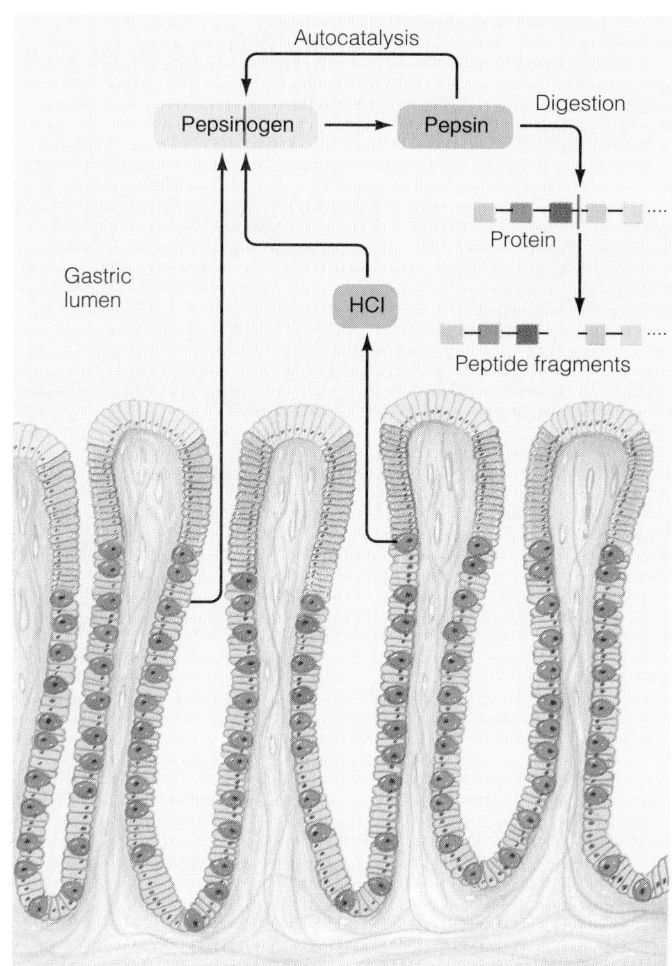

— *Figure 16-13* **Pepsinogen Activation in the Stomach Lumen**
In the lumen, hydrochloric acid (HCl) activates pepsinogen to its active form, pepsin, by cleaving off a small fragment. Once activated, pepsin autocatalytically activates more pepsinogen and begins protein digestion. Secretion of pepsinogen in the inactive form prevents it from digesting the protein structures of the cells in which it is produced. Its activation process does not begin until it reaches the lumen and comes into contact with HCl secreted by a separate cell in the gastric pit.

Mucus secretion The surface of the gastric mucosa is covered by a layer of mucus, which is derived from the surface epithelial cells and mucous neck cells. This mucus serves as a protective barrier against several forms of potential injury to the gastric mucosa:

- By virtue of its lubricating properties, mucus protects the gastric mucosa against mechanical injury.
- It helps protect the stomach wall from self-digestion because pepsin is inhibited when it comes in contact with the mucus layer coating the stomach lining. (However, mucus does not affect pepsin activity in the lumen, where digestion of dietary protein proceeds without interference.)
- Being alkaline, mucus helps protect against acid injury by neutralizing HCl in the vicinity of the gastric lining, but it does not interfere with the function of HCl in the lumen.

Intrinsic factor secretion **Intrinsic factor,** another secretory product of the parietal cells in addition to HCl, is important in the absorption of vitamin B_{12}. Only when in combination with intrinsic factor can this vitamin be absorbed by a special transport mechanism, presumably endocytosis, in the terminal portions of the ileum. Vitamin B_{12} is essential for the normal formation of red blood cells. In the absence of intrinsic factor, vitamin B_{12} fails to be absorbed, so erythrocyte production is defective, and *pernicious anemia* results (see p. 358). Pernicious anemia is typically caused by an autoimmune attack (see p. 401) against the parietal cells. This condition is treated by regular injections of vitamin B_{12}, thus bypassing the defective digestive-tract absorptive mechanism.

Occasionally, the oxyntic mucosa atrophies or degenerates. With the loss of chief and parietal cells, the stomach is unable to secrete pepsinogen, HCl, and intrinsic factor. Although pepsin and acid normally begin protein digestion in the stomach, they are not absolutely essential to protein digestion. If necessary, the pancreatic and small intestinal enzymes can completely digest proteins. The most detrimental consequence of **gastric mucosal atrophy** is the loss of intrinsic factor and the subsequent development of pernicious anemia.

Gastrin secretion Special endocrine cells, the **G cells,** located in the pyloric gland area (PGA) of the stomach secrete the hormone **gastrin** into the blood upon appropriate stimulation. After being carried by the blood back to the body and fundus of the stomach, gastrin stimulates the parietal and chief cells, thereby promoting secretion of a highly acidic gas-

Table 16-4 Stimulation of Gastric Secretion

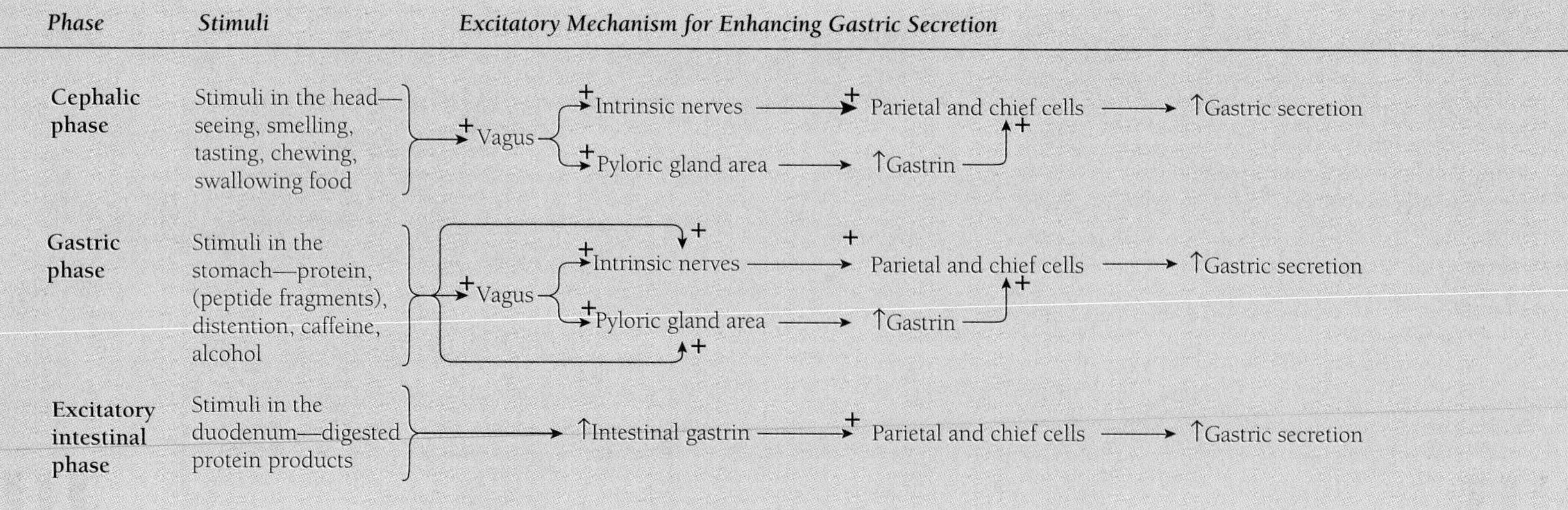

Phase	Stimuli	Excitatory Mechanism for Enhancing Gastric Secretion
Cephalic phase	Stimuli in the head—seeing, smelling, tasting, chewing, swallowing food	+ Vagus → + Intrinsic nerves → + Parietal and chief cells → ↑Gastric secretion; + Vagus → + Pyloric gland area → ↑Gastrin → + Parietal and chief cells
Gastric phase	Stimuli in the stomach—protein, (peptide fragments), distention, caffeine, alcohol	+ Vagus → + Intrinsic nerves → + Parietal and chief cells → ↑Gastric secretion; + Vagus → + Pyloric gland area → ↑Gastrin → + Parietal and chief cells; stimuli also act directly (+) on intrinsic nerves and pyloric gland area
Excitatory intestinal phase	Stimuli in the duodenum—digested protein products	→ ↑Intestinal gastrin → + Parietal and chief cells → ↑Gastric secretion

tric juice. Gastrin is also *trophic* (growth-promoting) to the mucosa of the stomach and small intestine, thereby maintaining their secretory capabilities.

Control of gastric secretion involves three phases.

The rate of gastric secretion can be influenced by (1) factors arising before food ever reaches the stomach; (2) factors resulting from the presence of food in the stomach; and (3) factors in the duodenum after food has left the stomach. Accordingly, gastric secretion is divided into three phases—the cephalic, gastric, and intestinal phases (Table 16-4).

Cephalic phase The cephalic phase of gastric secretion refers to the increased secretion of HCl and pepsinogen that occurs in feedforward fashion in response to stimuli acting in the head (*cephalic* refers to "head") even before food reaches the stomach. Thinking about, tasting, smelling, chewing, and swallowing food increase gastric secretion by means of vagal nerve activity in two ways. First, vagal stimulation of the intrinsic plexuses promotes increased secretion of HCl and pepsinogen by the secretory cells. Second, vagal stimulation of the PGA causes the release of gastrin, which in turn further enhances secretion of HCl and pepsinogen.

Gastric phase The gastric phase of gastric secretion occurs when food actually reaches the stomach. Stimuli acting in the stomach—namely, *protein,* especially peptide fragments; *distention; caffeine;* or *alcohol*—increase gastric secretion by means of overlapping efferent pathways. For example, protein in the stomach, the most potent stimulus, initiates short local reflexes in the intrinsic nerve plexuses to stimulate the secretory cells. Furthermore, protein initiates long reflexes so that extrinsic vagal fibers to the stomach are activated. Vagal activity further enhances intrinsic nerve stimulation of the secretory cells and triggers the release of gastrin. Protein also directly stimulates the release of gastrin. Gastrin in turn is a powerful stimulus for further acid and pepsinogen secretion. Through these synergistic and overlapping pathways, protein induces the secretion of a highly acidic, pepsin-rich gastric juice, which continues the digestion of the protein that first initiated the process.

When the stomach is distended with protein-rich food that needs to be digested, these secretory responses are appropriate. Caffeine and, to lesser extent, alcohol also stimulate the secretion of a highly acidic gastric juice, even when no food is present. This unnecessary acid can irritate the linings of the stomach and duodenum. For this reason, persons with ulcers or gastric hyperacidity should avoid caffeinated and alcoholic beverages.

Intestinal phase The third phase of gastric secretion, the intestinal phase, encompasses the factors originating in the small intestine that influence gastric secretion. The intestinal phase has both an excitatory and an inhibitory component (Fig. 16-14).

To a limited extent, the presence of the products of protein digestion in the duodenum stimulates further gastric secretion by triggering the release of an **intestinal gastrin** that is carried by the blood to the stomach. This is the *excitatory component* of the intestinal phase of gastric secretion. It is as if the small intestine, on noting the arrival of protein fragments from the stomach, offers the stomach a "helping hand" in digesting the protein by enhancing gastric secretion.

The *inhibitory component* of the intestinal phase of gastric secretion is dominant over the excitatory component, however. The inhibitory component is important in helping to shut off the flow of gastric juices as chyme begins to be emptied into the small intestine, a topic to which we now turn our attention.

Gastric secretion gradually decreases as food empties from the stomach into the intestine.

We now know what factors turn on gastric secretion before and during a meal, but how is the flow of gastric juices shut off as chyme begins to be emptied from the stomach into the small intestine? Gastric secretion is gradually reduced by three different means as the stomach empties (Table 16-5):

- As the meal is gradually emptied into the duodenum, the major stimulus for enhanced gastric secretion—the presence of protein in the stomach—is withdrawn.
- After foods leave the stomach and gastric juices accumulate to such an extent that gastric pH falls very low, gastric secretion is inhibited because a high concentration of H^+ directly inhibits the PGA from releasing gastrin. As gastrin secretion declines, the most potent stimulant of gastric secretion is withdrawn.
- The same stimuli that inhibit gastric motility (fat, acid, hypertonicity, or distention in the duodenum brought about by stomach emptying) inhibit gastric secretion as well; the enterogastric reflex and the enterogastrones suppress the gastric secretory cells while they simultaneously reduce the excitability of the gastric smooth muscle cells. This inhibitory response is the inhibitory component of the intestinal phase of gastric secretion (Fig. 16-14). It is more powerful than the excitatory component mediated by intestinal gastrin.

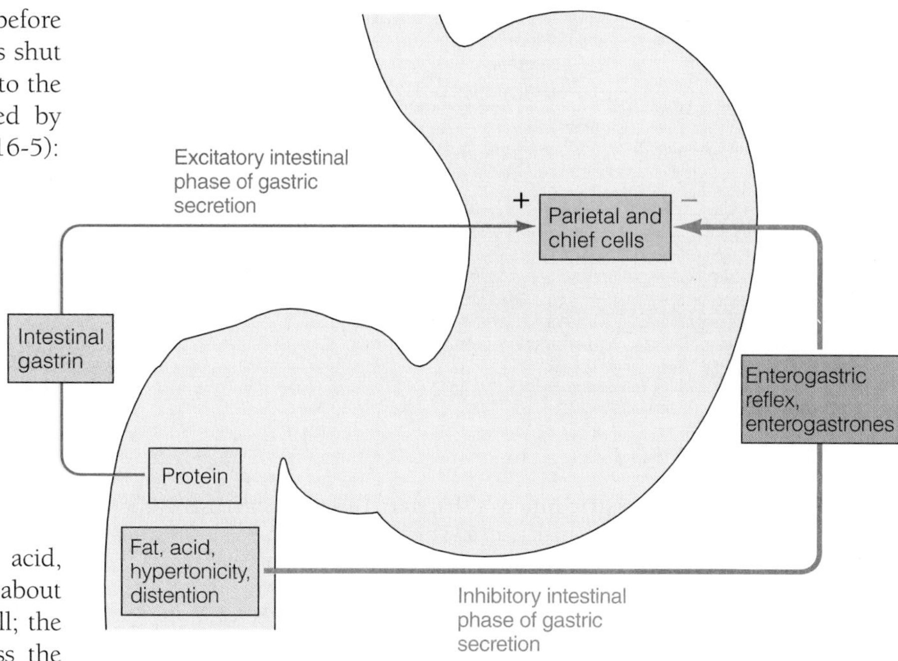

Figure 16-14 **Excitatory and Inhibitory Components of the Intestinal Phase of Gastric Secretion** The intestine influences gastric secretion by the parietal and chief cells in two ways. In the excitatory component of the intestinal phase of gastric secretion, protein in the duodenum triggers the release of intestinal gastrin, which stimulates gastric secretion. In the inhibitory component of the intestinal phase of gastric secretion, fat, acid, hyptertonicity, and distention in the duodenum inhibit gastric secretion by means of the enterogastric reflex and enterogastrones. The inhibitory component is more powerful than the excitatory component.

The stomach lining is protected from gastric secretions by the gastric mucosal barrier.

How can the stomach contain strong acid contents and proteolytic enzymes without destroying itself? We already learned that mucus provides a protective coating. In addition, other

Table 16-5 **Inhibition of Gastric Secretion**

Region	*Stimuli*	*Inhibitory Mechanism for Reducing Gastric Secretion*		
Body and antrum	Removal of protein and distention as the stomach empties	– Intrinsic nerves		↓ Gastric secretion
		– Vagus		
		– Pyloric gland area →	↓ Gastrin	
Antrum	Accumulation of acid	– Pyloric gland area →	↓ Gastrin →	↓ Gastric secretion
Duodenum (inhibitory intestinal phase of gastric secretion)	Fat Acid Hypertonicity Distention	+ Enterogastric reflex		↓ Gastric secretion
		↑ Enterogastrones (cholecystokinin, secretin, gastric inhibitory peptide)		

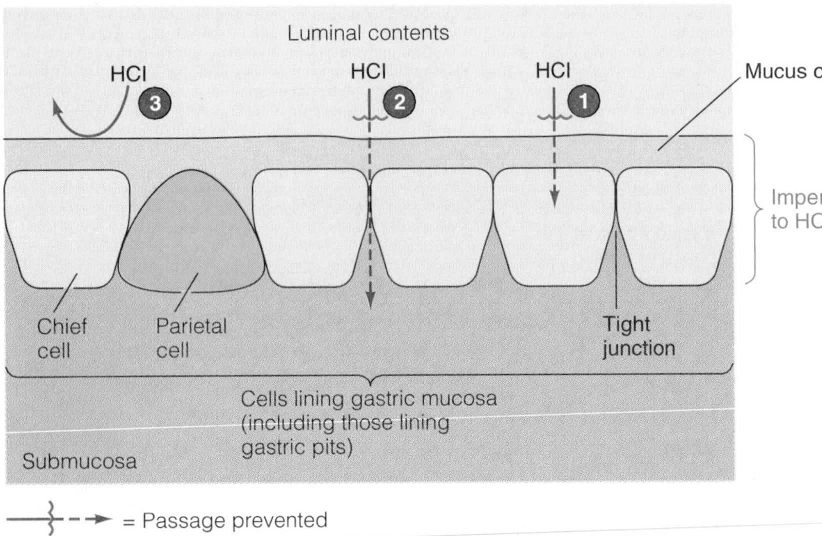

Figure 16-15 **Gastric Mucosal Barrier** The gastric mucosal barrier encompasses the following factors that enable the stomach to contain acid without injuring itself: the luminal membranes of the gastric mucosal cells are impermeable to H^+ so that HCl cannot penetrate into the cells ①, and they are joined by tight junctions that prevent HCl from penetrating between the cells ②. A mucus coating over the gastric mucosa offers further protection ③.

barriers to mucosal acid damage are provided by the mucosal lining itself (▬ Fig. 16-15). First, the luminal membranes of the gastric mucosal cells are almost impermeable to H^+, so acid cannot penetrate *into* the cells and cause cellular damage. Furthermore, the lateral edges of these cells are joined together near their luminal borders by tight junctions (see p. 52), so acid cannot diffuse *between* the cells from the lumen into the underlying submucosa. The properties of the gastric mucosa that enable the stomach to contain acid without injuring itself constitute the **gastric mucosal barrier.** These protective mechanisms are further enhanced by the fact that the entire stomach lining is replaced every three days. Because of rapid mucosal turnover, cells are usually replaced before they are exposed to the wear and tear of harsh gastric conditions long enough to suffer damage.

In spite of the protection provided by mucus, by the gastric mucosal barrier, and by the frequent turnover of cells, the barrier occasionally is broken, and the gastric wall is injured by its acidic and enzymatic contents. When this occurs, an erosion, or **peptic ulcer,** of the stomach wall results. (Excessive gastric reflux into the esophagus and dumping of excessive acidic gastric contents into the duodenum can lead to peptic ulcers in these locations as well.)

Until recently, the exact cause of ulcers was unknown, but in a surprising discovery in the early 1990s, the bacterium *Helicobacter pylori* was pinpointed as the cause of more than 80% of all peptic ulcers. Thirty percent of the population harbors *H. pylori.* Those with this "slow" bacterium have a 3 to 12 times greater risk of developing an ulcer within 10 to 20 years of acquiring the infection than those without the bacterium. They are also at increased risk of developing stomach cancer.

For years, scientists had overlooked the possibility that ulcers could be triggered by a infectious agent because bacteria typically cannot survive in a strongly acidic environment such as the stomach lumen. An exception, *H. pylori* exploits several strategies to survive in this hostile environment. First, these organisms are motile, being equipped with four to six flagella (see p. 37) (▬ Fig. 16-16), which enable them to tunnel through and take up residence under the stomach's thick mucus layer. Here, they are protected from the highly acidic gastric contents. Furthermore, *H. pylori* preferentially settles in the antrum, which has no acid-producing parietal cells, although HCl from the upper portions of the stomach does reach the antrum. Also, these bacteria produce *urease,* an enzyme that breaks down urea, an end product of protein metabolism, into ammonia (NH_3) and CO_2. Ammonia serves as a buffer (see p. 538) to neutralize stomach acid locally in the vicinity of the *H. pylori.*

H. pylori contributes to ulcer formation by secreting toxins that cause a persistent inflammation, or *chronic superficial gastritis,* at the site they colonize. This inflammatory response apparently weakens the gastric mucosal barrier.

Alone or in conjunction with this infectious culprit, other factors are known to contribute to ulcer formation. Frequent exposure to some chemicals can break the gastric mucosal barrier; the most important of these are ethyl alcohol and

Figure 16-16 **Helicobacter pylori** *Helicobacter pylori,* the bacterium responsible for most cases of peptic ulcers, possesses flagella that enable it to tunnel beneath the protective mucus layer that coats the stomach lining.

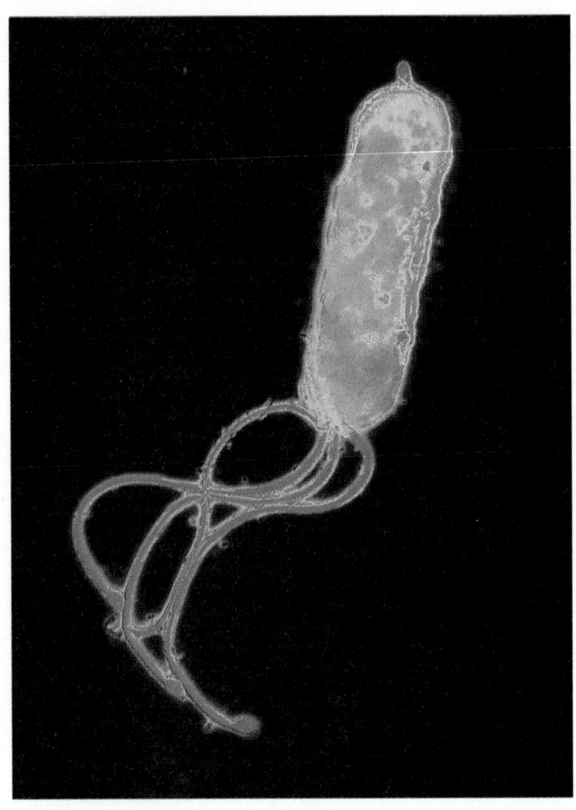

nonsteroidal anti-inflammatory drugs (NSAIDs), such as aspirin, ibuprofen, or more potent medications for the treatment of arthritis or other chronic inflammatory processes. The barrier frequently breaks in patients with preexisting debilitating conditions, such as severe injuries or infections. The persistence of stressful situations is frequently associated with ulcer formation, presumably because the emotional response to the stress can bring about excessive stimulation of gastric secretion.

When the gastric mucosal barrier is broken, acid and pepsin diffuse into the mucosa and underlying submucosa with serious pathophysiological consequences (▬ Fig. 16-17). Acid triggers the release of *histamine,* a potent acid stimulant that is produced and stored in large amounts in the submucosa. Released histamine stimulates secretion of more acid, which can diffuse back into the submucosa to stimulate further histamine release, triggering more acid release, and so on, thus establishing a vicious cycle. (Histamine is not believed to play a role in the normal control of gastric secretion.) The surface erosion or ulcer progressively enlarges as increasing levels of acid and pepsin continue to damage the stomach wall. Two of the most serious consequences of ulcers are (1) hemorrhage resulting from damage to submucosal capillaries and (2) perforation, or complete erosion through the stomach wall, resulting in the escape of potent gastric contents into the abdominal cavity.

With the discovery of the infectious component of the majority of ulcers, antibiotics, which can completely cure the condition, are now the treatment of choice. For the two decades prior to the discovery of *H. pylori,* one of the leading treatments for ulcers was antihistamine (cimetidine) that specifically blocks H-2 receptors, the type of receptors that bind histamine released from the stomach. These receptors differ from H-1 receptors that bind the histamine involved in allergic respiratory disorders. Accordingly, traditional antihistamines used for respiratory allergies (such as hay fever and asthma) are not effective against ulcers, nor is cimetidine useful for respiratory problems. Before the discovery of the role these unique histamine receptors play in ulcer formation, the treatment of ulcers was aimed at either neutralizing stomach acidity through the use of antacids or removing factors known to enhance gastric secretion. Accordingly, the treatment included (1) a bland diet void of caffeine and alcohol, (2) cutting the vagus nerve supply to the stomach, and (3) removal of the stomach antrum to eliminate the source of gastrin. Even though symptoms subsided, the condition often recurred because the underlying cause was not eradicated.

Carbohydrate digestion continues in the body of the stomach, whereas protein digestion begins in the antrum.

Two separate digestive processes take place within the stomach. In the body of the stomach, food remains in a semisolid mass, because peristaltic contractions in this region are too weak for mixing to occur. Because food is not mixed with gastric secretions in the body of the stomach, very little protein digestion occurs here. Acid and pepsin are able to attack only the surface of the food mass. In the interior of the mass, however, carbohydrate digestion continues under the influence of salivary amylase. Even though acid inactivates salivary amylase, the unmixed interior of the food mass is free of acid.

Digestion by the gastric juice itself is accomplished in the antrum of the stomach, where the food is thoroughly mixed with HCl and pepsin, thereby initiating protein digestion.

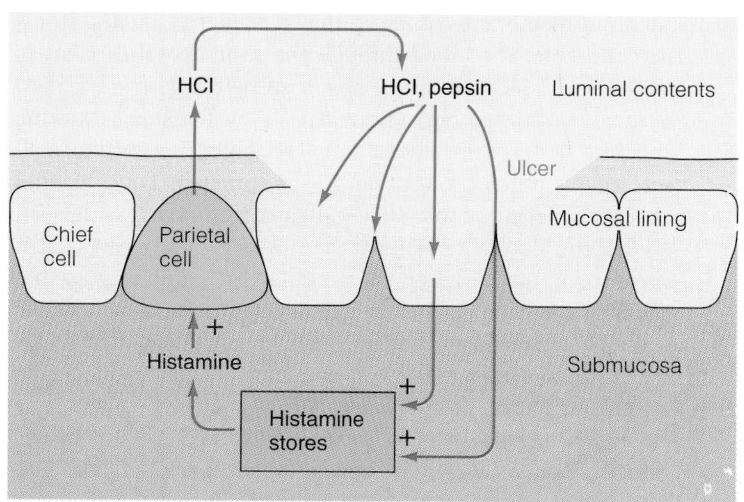

▬ *Figure 16-17* **Ulcer Formation** When acid and pepsin are able to break through a weakened or overwhelmed gastric mucosal barrier, the acid stimulates the release of histamine that is stored in the submucosa. Histamine in turn stimulates the parietal cells to secrete more acid, which diffuses through the broken barrier to trigger the release of more histamine, as the vicious cycle continues. An ulcer is formed and progressively enlarges as the acid and pepsin continue to erode the gastric mucosa.

The stomach absorbs alcohol and aspirin but no food.

No food or water is absorbed into the blood from the stomach mucosa. Carbohydrate and protein digestion have not been completed in the stomach. These large, partially digested substances are not lipid soluble, so they cannot penetrate the cell membranes. Furthermore, no special transport mechanisms are present in the stomach wall to facilitate the absorption of these nutrients. Fat digestion has not even begun in the stomach. Although it is theoretically possible for fat molecules, which are lipid soluble, to pass through the lipid portions of the cell membranes that line the stomach, this does not occur to any appreciable extent because dietary fat in the gastric lumen separates into large fat droplets that float in the chyme. This reduces the possibility that fat molecules will come into contact with the gastric mucosa. Ingested electrolytes such as Na^+ and Ca^{2+} exist as non-lipid-soluble ions that can be absorbed only through use of special transport systems, which exist in the small intestine but not in the stomach. Likewise, H_2O exchange does not take place across the gastric mucosa because the stomach is impermeable to H_2O.

Even though none of the ingested food is absorbed from the stomach, two noteworthy nonnutrient substances are absorbed directly by the stomach—*ethyl alcohol* and *aspirin.* Alcohol is lipid soluble to a degree, so it can diffuse through the lipid membranes of the epithelial cells that line the stomach and enter the blood through the submucosal capillaries. Yet although alcohol can be absorbed by the gastric mucosa, it can be absorbed even more rapidly by the small intestine mucosa, because the surface area for absorption in the small intestine is much greater than in the stomach. Thus, alcohol absorption occurs more slowly if gastric emptying is delayed so that the alcohol remains in the stomach longer. Since fat is the most potent duodenal stimulus for inhibiting gastric motility, consumption of fat-rich foods (for example, whole milk or pizza) before or during alcohol ingestion delays gastric emptying and prevents the alcohol from producing its effects as rapidly.

Another category of substances absorbed by the gastric mucosa includes weak acids, most notably acetylsalicylic acid (aspirin). In the highly acidic environment of the stomach lumen, weak acids are almost totally unionized; that is, the H^+ and associated anion of the acid are bound together. In an unionized form, these weak acids are lipid soluble, so they can be absorbed quickly by crossing the plasma membranes of the epithelial cells that line the stomach. Most other drugs are not absorbed until they reach the small intestine, so they do not begin to take effect as quickly.

Pancreatic and Biliary Secretions

When gastric contents are emptied into the small intestine, they are mixed not only with juice secreted by the small intestine mucosa but also with the secretions of the exocrine pancreas and liver that are emptied into the duodenal lumen. We will discuss the roles of each of these accessory digestive organs before we examine the contributions of the small intestine itself.

The pancreas is a mixture of exocrine and endocrine tissue.

The **pancreas** is an elongated gland that lies behind and below the stomach, above the first loop of the duodenum (Fig. 16-18). It is a mixed gland that contains both exocrine and endocrine tissue. The predominant exocrine portion consists of grapelike clusters of secretory cells that form sacs known as **acini,** which connect to ducts that eventually empty into the duodenum. The smaller endocrine portion consists of isolated islands of endocrine tissue, the **islets of Langerhans,** which are dispersed throughout the pancreas. The most important hormones secreted by the islet cells are insulin and glucagon (chapter 19). The exocrine and endocrine pancreas have nothing in common except that they share the same location.

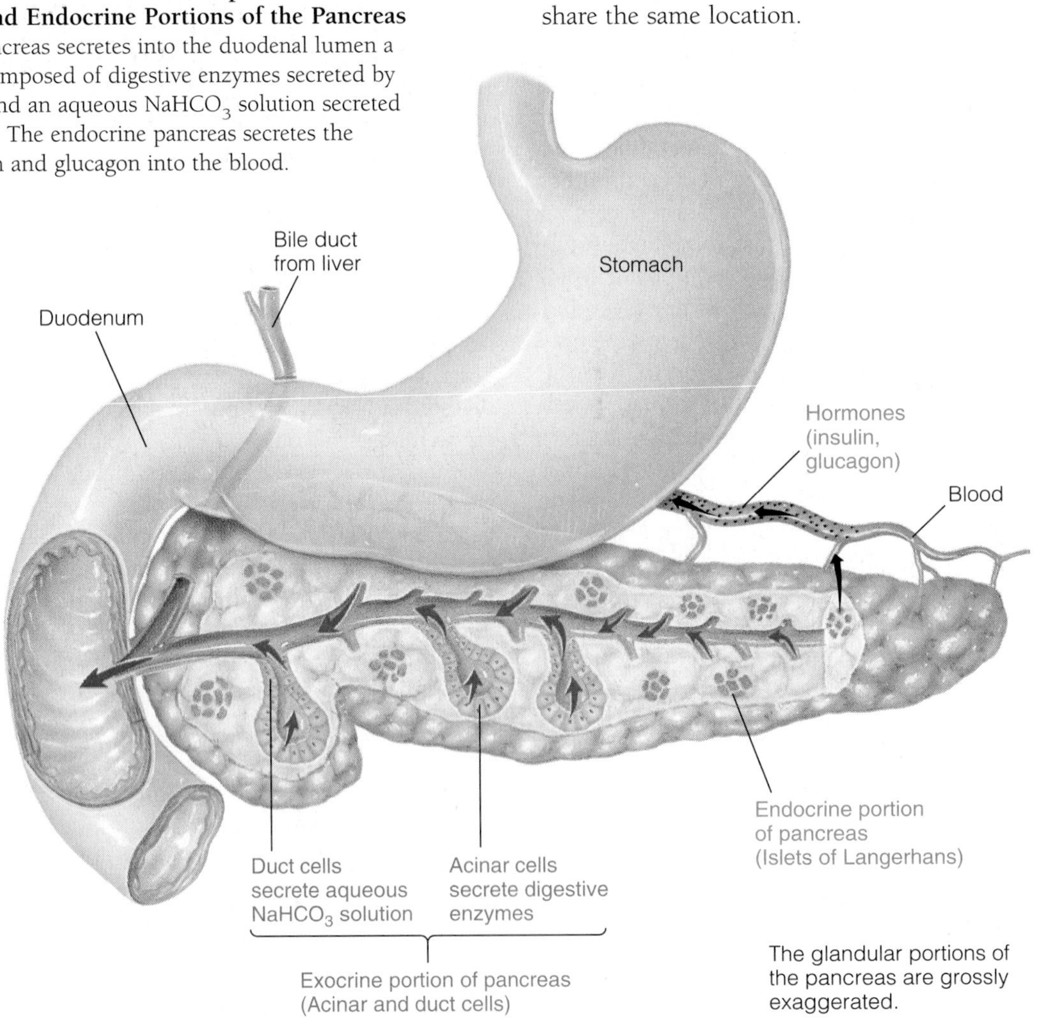

Figure 16-18 **Schematic Representation of the Exocrine and Endocrine Portions of the Pancreas**
The exocrine pancreas secretes into the duodenal lumen a digestive juice composed of digestive enzymes secreted by the acinar cells and an aqueous $NaHCO_3$ solution secreted by the duct cells. The endocrine pancreas secretes the hormones insulin and glucagon into the blood.

The exocrine pancreas secretes digestive enzymes and an aqueous alkaline fluid.

The **exocrine pancreas** secretes a pancreatic juice consisting of two components—a potent *enzymatic secretion* and an *aqueous* (watery) *alkaline secretion* that is rich in sodium bicarbonate ($NaHCO_3$). The pancreatic enzymes are actively secreted by the *acinar cells.* The aqueous $NaHCO_3$ component is actively secreted by the *duct cells* that line the early portion of the pancreatic ducts, and it then is modified as it passes down the ducts.

Like pepsinogen, pancreatic enzymes are synthesized by the endoplasmic reticulum and Golgi complex of the acinar cells, then are stored within zymogen granules and released by exocytosis as needed. The acinar cells secrete three different types of pancreatic enzymes that are capable of digesting all three categories of foodstuffs. These pancreatic enzymes are important because they are capable of almost completely digesting food in the absence of all other digestive secretions. The three types of pancreatic enzymes are (1) **proteolytic enzymes,** which are involved in protein digestion; (2) **pancreatic amylase,** which contributes to carbohydrate digestion in a way similar to salivary amylase; and (3) **pancreatic lipase,** the only enzyme important in fat digestion.

Pancreatic proteolytic enzymes The three major proteolytic enzymes secreted by the pancreas are *trypsinogen, chymotrypsinogen,* and *procarboxypeptidase,* each of which is secreted in an inactive form. When **trypsinogen** is secreted into the duodenal lumen, it is activated to its active enzyme form, **trypsin,** by **enterokinase,** an enzyme embedded in the luminal border of the cells that line the duodenal mucosa. Trypsin then autocatalytically activates more trypsinogen. Like pepsinogen, trypsinogen must remain inactive within the pancreas to prevent this proteolytic enzyme from digesting the cells in which it is formed. Trypsinogen remains inactive, therefore, until it reaches the duodenal lumen, where enterokinase triggers the activation process, which then proceeds autocatalytically. As further protection, the pancreatic tissue also produces a chemical known as **trypsin inhibitor,** which blocks trypsin's actions should spontaneous activation of trypsinogen inadvertently occur within the pancreas.

Chymotrypsinogen and **procarboxypeptidase,** the other pancreatic proteolytic enzymes, are converted by trypsin to their active forms, **chymotrypsin** and **carboxypeptidase,** respectively, within the duodenal lumen. Thus, once enterokinase has activated some of the trypsin, trypsin is then responsible for the remainder of the activation process.

Each of these proteolytic enzymes attacks different peptide linkages. The end products that result from this action are a mixture of amino acids and small peptide chains. Mucus secreted by the intestinal cells provides protection against digestion of the small intestine wall by the activated proteolytic enzymes.

Pancreatic amylase Like salivary amylase, pancreatic amylase plays an important role in carbohydrate digestion by converting polysaccharides into disaccharides. Amylase is secreted in the pancreatic juice in an active form because active amylase does not present a danger to the secretory cells.

Pancreatic lipase Pancreatic lipase is extremely important because it is the only enzyme secreted throughout the entire digestive system that can accomplish digestion of fat. Pancreatic lipase hydrolyzes dietary triglycerides into monoglycerides and free fatty acids, which are the absorbable units of fat. Like amylase, lipase is secreted in its active form because there is no risk of pancreatic self-digestion by lipase.

When pancreatic enzymes are deficient, digestion of food is incomplete. Because the pancreas is the only significant source of lipase, pancreatic enzyme deficiency results in serious maldigestion of fats. The principal clinical manifestation of pancreatic exocrine insufficiency is **steatorrhea,** or excessive undigested fat in the feces. Up to 60% to 70% of the ingested fat may be excreted in the feces. Digestion of protein and carbohydrates is impaired to a lesser degree because salivary, gastric, and small intestinal enzymes contribute to the digestion of these two foodstuffs.

Pancreatic aqueous alkaline secretion Pancreatic enzymes function best in a neutral or slightly alkaline environment, yet the highly acidic gastric contents are emptied into the duodenal lumen in the vicinity of pancreatic enzyme entry into the duodenum. It is imperative that the acidic chyme be quickly neutralized in the duodenal lumen, not only to allow optimal functioning of the pancreatic enzymes but also to prevent acid damage to the duodenal mucosa. Therefore, the alkaline ($NaHCO_3$-rich) fluid secreted by the pancreas into the duodenal lumen serves the important function of neutralizing the acidic chyme as the latter is emptied into the duodenum from the stomach. This aqueous $NaHCO_3$ secretion is by far the largest component of pancreatic secretion. The volume of pancreatic secretion ranges between 1 and 2 liters per day, depending on the types and degree of stimulation.

Pancreatic exocrine secretion is hormonally regulated to maintain neutrality of the duodenal contents and to optimize digestion.

Pancreatic exocrine secretion is regulated primarily by hormonal mechanisms. A small amount of parasympathetically induced pancreatic secretion occurs during the cephalic phase of digestion, with a further token increase occurring during the gastric phase in response to gastrin. However, the predominant stimulation of pancreatic secretion occurs during the intestinal phase of digestion when chyme is in the small intestine. The release of the two major enterogastrones, secretin and cholecystokinin (CCK), in response to chyme in the duodenum plays the central role in the control of pancreatic secretion (▬ Fig. 16-19).

Of the factors that stimulate enterogastrone release, the primary stimulus specifically for secretin release is acid in the duodenum. Secretin in turn is carried by the blood to the pancreas, where it stimulates the duct cells to markedly increase their secretion of a $NaHCO_3$-rich aqueous fluid into the duodenum. Even though other stimuli may cause the

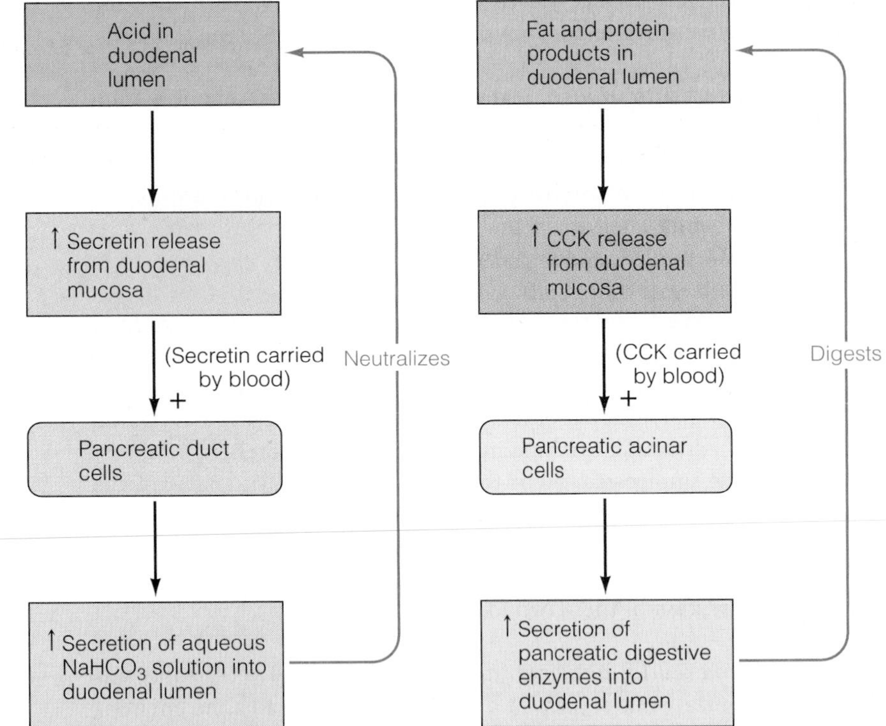

Figure 16-19 **Hormonal Control of Pancreatic Exocrine Secretion**

release of secretin, it is appropriate that the most potent stimulus is acid because secretin promotes the alkaline pancreatic secretion that neutralizes the acid. This mechanism provides a control system for maintaining neutrality of the chyme in the intestine. The amount of secretin released is proportional to the amount of acid that enters the duodenum, so the amount of $NaHCO_3$ secreted parallels the duodenal acidity.

Cholecystokinin, on the other hand, is important in the regulation of pancreatic digestive enzyme secretion. The main stimulus for release of CCK from the duodenal mucosa is the presence of fat and to a lesser extent protein products. The circulatory system transports CCK to the pancreas where it stimulates the pancreatic acinar cells to increase digestive enzyme secretion. Among these enzymes are lipase and the proteolytic enzymes, which appropriately bring about further digestion of the fat and protein that initiated the response and also help digest carbohydrate. In contrast to fat and protein, carbohydrate does not have any direct influence on pancreatic digestive enzyme secretion.

All three types of pancreatic digestive enzymes are packaged together in the zymogen granules, so all the pancreatic enzymes are released together on exocytosis of the granules. Therefore, even though the *total amount* of enzymes released varies depending on the type of meal consumed (the most being secreted in response to fat), the *proportion* of enzymes released does not vary on a meal-to-meal basis. That is, a high-protein meal does not cause the release of a greater proportion of proteolytic enzymes. Evidence suggests, however, that long-term adjustments in the proportion of the types of enzymes produced may occur as an adaptive response to a prolonged change in diet. For example, a greater proportion of proteolytic enzymes are produced with a long-term switch to a high-protein diet. Cholecystokinin may play a role in pancreatic digestive enzyme adaptation to changes in diet.

Just as gastrin is trophic to the stomach and small intestine, CCK and secretin exert trophic effects on the exocrine pancreas to maintain its integrity.

The liver performs various important functions including bile production.

Besides pancreatic juice, the other secretory product that is emptied into the duodenal lumen is **bile.** The **biliary system** includes the *liver,* the *gallbladder,* and associated ducts.

The **liver** is the largest and most important metabolic organ in the body; it can be viewed as the body's major "biochemical factory." Its importance to the digestive system is its secretion of *bile salts,* but the liver performs a wide variety of other functions, including the following:

1. Metabolic processing of the major categories of nutrients (carbohydrates, proteins, and lipids) after their absorption from the digestive tract.
2. Detoxification or degradation of body wastes and hormones as well as drugs and other foreign compounds.
3. Synthesis of plasma proteins, including those necessary for the clotting of blood and those that transport steroid and thyroid hormones and cholesterol in the blood.
4. Storage of glycogen, fats, iron, copper, and many vitamins.
5. Activation of vitamin D, which the liver accomplishes in conjunction with the kidneys.
6. Removal of bacteria and worn-out red blood cells, thanks to its resident macrophages (see p. 362).
7. Excretion of cholesterol and bilirubin, the latter being a breakdown product derived from the destruction of worn-out red blood cells.

Given this wide range of complex functions, there is amazingly little specialization of cells within the liver. Each liver cell, or **hepatocyte** (*hepato* means "liver"; *cyte* means "cell"), appears to be able to perform the same wide variety of metabolic and secretory tasks, with the exception of the phagocytic activities carried out by the resident macrophages, which are known as **Kupffer cells.** The specialization comes from the highly developed organelles within each hepatocyte.

To carry out these wide-ranging tasks, the anatomical organization of the liver permits each hepatocyte to be in direct contact with blood from two sources: venous blood coming directly from the digestive tract and arterial blood coming from the aorta. Venous blood enters the liver by means of a unique and complex vascular connection between the digestive tract and the liver that is known as the **hepatic portal system** (Fig. 16-20). The veins draining the digestive tract do not directly join the inferior vena cava, the large vein that returns blood to the heart. Instead, the veins from the stomach and intestine enter the hepatic portal vein, which carries the products absorbed from the digestive tract directly to the liver for processing, storage, or detoxification before they gain access to the general circulation. Within the liver, the portal vein once again breaks up into a capillary network (the liver *sinusoids*) to permit exchange between the blood and hepatocytes before draining into the hepatic vein, which joins the inferior vena cava. The hepatocytes also are provided with fresh arterial blood, which supplies their oxygen and delivers blood-borne metabolites for hepatic processing.

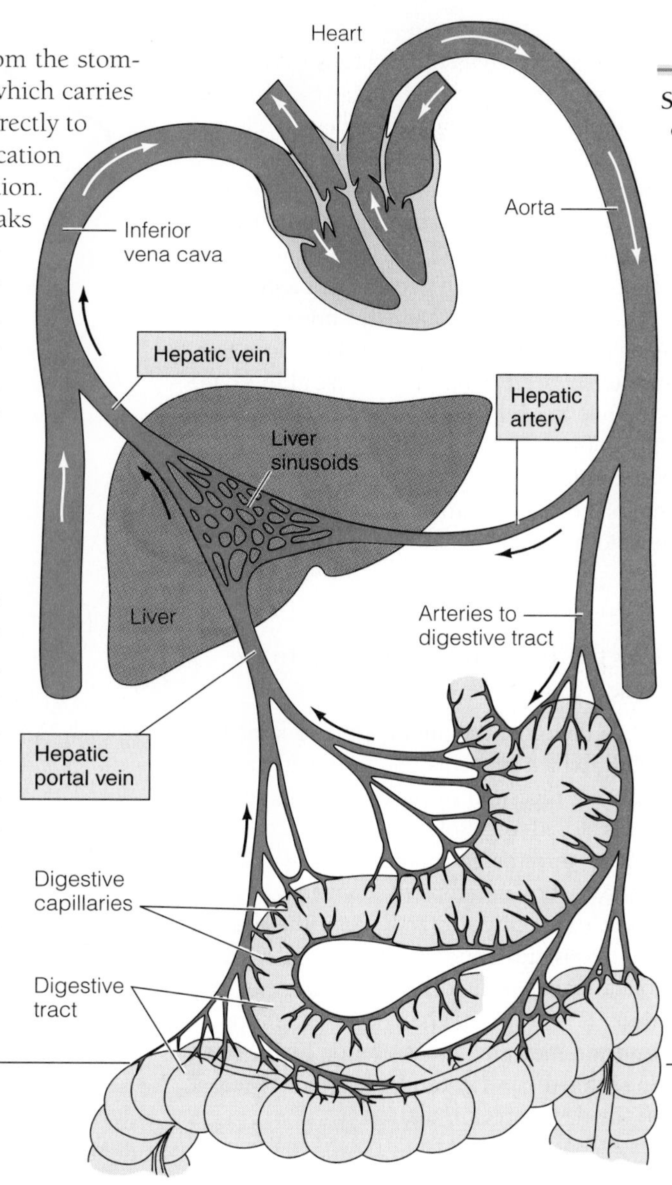

Figure 16-20 Schematic Representation of Liver Blood Flow The liver receives blood from two sources: (1) venous blood draining the digestive tract is carried by the hepatic portal vein to the liver for processing and storage of newly absorbed nutrients, and (2) arterial blood, which provides the liver's O_2 supply and carries blood-borne metabolites for hepatic processing, is delivered by the hepatic artery. Blood leaves the liver via the hepatic vein.

The liver lobules are delineated by vascular and bile channels.

The liver is organized into functional units known as **lobules,** which are hexagonal arrangements of tissue surrounding a central vein, like a six-sided angel food cake with the hole representing the central vein (Fig. 16-21a). At the outer edge of each "slice" of the lobule are three vessels: a branch of the hepatic artery, a branch of the portal vein, and a bile duct. Blood from the branches of both the hepatic artery and the portal vein flows from the periphery of the lobule into large, expanded capillary spaces called

Figure 16-21 Anatomy of the Liver
(a) Hepatic lobule. (b) Wedge of a hepatic lobule.

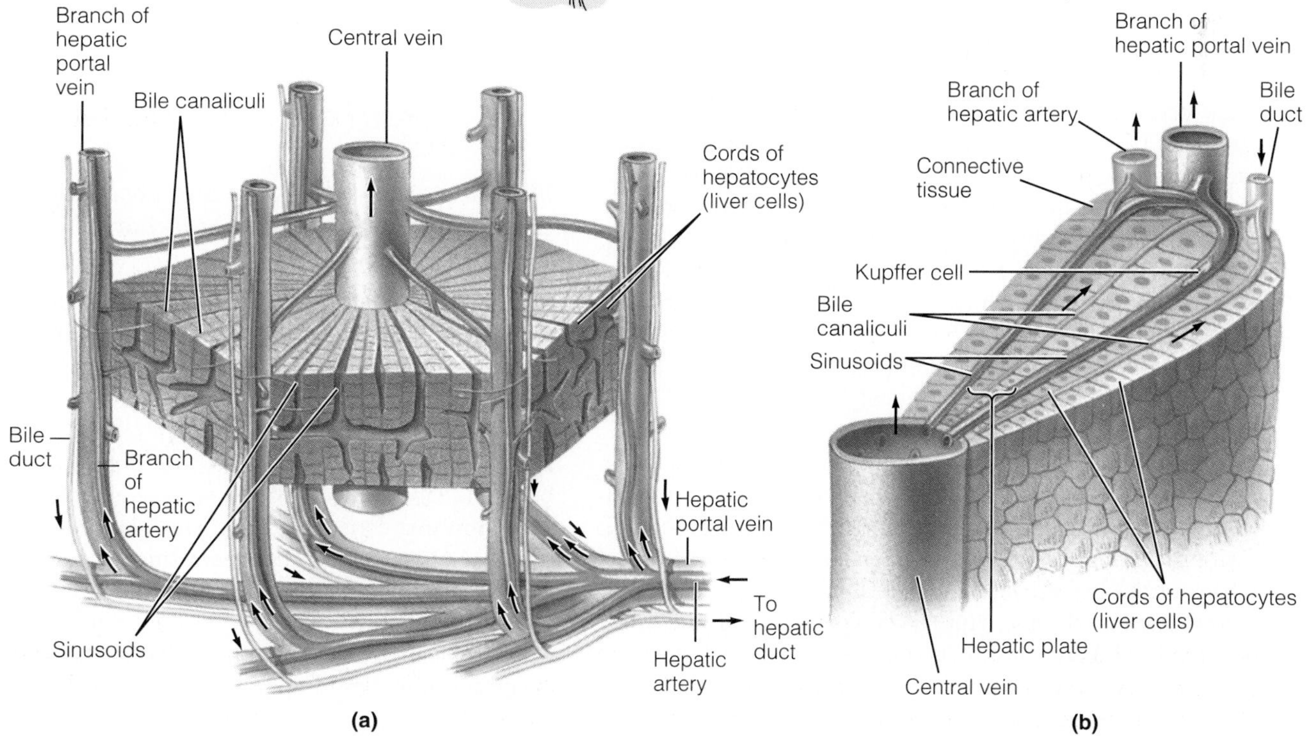

sinusoids, which run between rows of liver cells to the central vein like spokes on a bicycle wheel (Fig. 16-21b). The Kupffer cells line the sinusoids and engulf and destroy old red blood cells and bacteria that pass through in the blood. The hepatocytes are arranged between the sinusoids in plates two cell layers thick, so that each lateral edge faces a sinusoidal pool of blood. The central veins of all the liver lobules converge to form the hepatic vein, which carries the blood away from the liver. The thin bile-carrying channel, a **bile canaliculus,** runs between the cells within each hepatic plate. Hepatocytes continuously secrete bile into these thin channels, which carry the bile to a bile duct at the periphery of the lobule. The bile ducts from the various lobules converge to eventually form the common bile duct, which transports the bile from the liver to the duodenum. Each hepatocyte is in contact with a sinusoid on one side and a bile canaliculus on the other side.

Bile is secreted by the liver and is diverted to the gallbladder between meals.

The opening of the bile duct into the duodenum is guarded by the **sphincter of Oddi**, which prevents bile from entering the duodenum except during digestion of meals (Fig. 16-22). When this sphincter is closed, most of the bile secreted by the liver is diverted back up into the **gallbladder,** a small saclike structure tucked beneath but not directly connected to the liver. Thus, bile is not transported directly from the liver to the gallbladder. The bile is subsequently stored and concentrated in the gallbladder between meals. After a meal, bile enters the duodenum as a result of the combined effects of gallbladder emptying and increased bile secretion by the liver. The amount of bile secreted per day ranges from 250 ml to 1 liter, depending on the degree of stimulation.

Bile salts are recycled through the enterohepatic circulation.

Bile consists of an *aqueous alkaline fluid* similar to the pancreatic $NaHCO_3$ secretion as well as several organic constituents, including *bile salts, cholesterol, lecithin,* and *bilirubin.* The organic constituents are derived from hepatocyte activity, whereas the water, $NaHCO_3$, and other inorganic salts are added by the duct cells. Even though bile does not contain any digestive enzymes, it is important for the digestion and absorption of fats, primarily through the activity of the bile salts.

Bile salts are derivatives of cholesterol. They are actively secreted into the bile and eventually enter the duodenum along with the other biliary constituents. Following their participation in fat digestion and absorption, most bile salts are reabsorbed back into the blood by special active transport mechanisms located in the terminal ileum, the last portion of the small intestine. From here the bile salts are returned by means of the hepatic portal system to the liver, which resecretes them into the bile. This recycling of bile salts (and some of the other biliary constituents) between the small intestine and liver is referred to as the **enterohepatic circulation** (*entero* means "intestine"; *hepatic* means "liver") (Fig. 16-22).

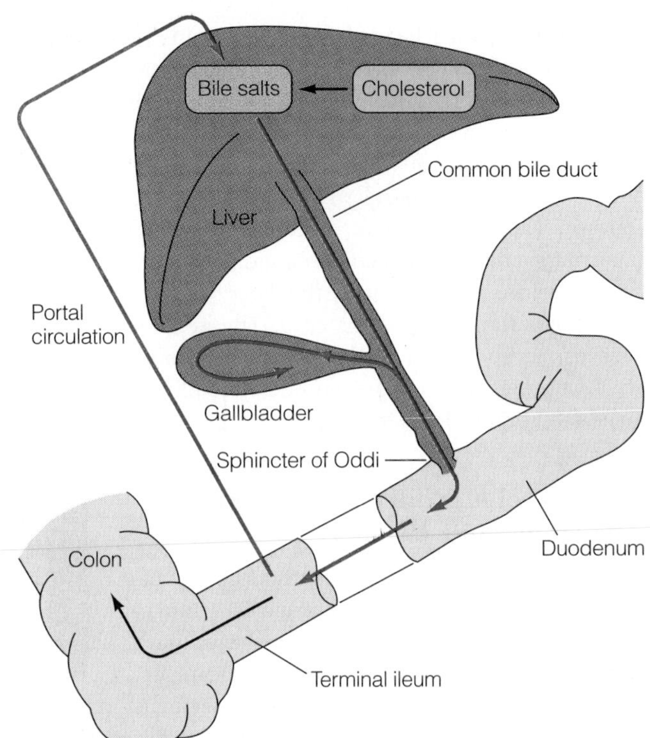

Figure 16-22 **Enterohepatic Circulation of Bile Salts** The majority of bile salts are recycled between the liver and small intestine through the enterohepatic circulation, designated by the blue arrows. After participating in fat digestion and absorption, most bile salts are reabsorbed by active transport in the terminal ileum and returned through the hepatic portal vein to the liver, which resecretes them in the bile.

The total amount of bile salts in the body averages about 3 to 4 g, yet 3 to 15 g of bile salts may be emptied into the duodenum in a single meal. Obviously, the bile salts must be recycled many times per day. Usually, only about 5% of the secreted bile escapes into the feces daily. These lost bile salts are replaced by new bile salts synthesized by the liver; thus, the size of the pool of bile salts is kept constant.

Bile salts aid fat digestion and absorption through their detergent action and micellar formation, respectively.

Bile salts aid fat digestion through their detergent action (emulsification) and facilitate fat absorption through their participation in the formation of micelles. Both functions are related to the structure of bile salts.

Detergent action of bile salts **Detergent action** refers to bile salts' ability to convert large fat globules into a **lipid emulsion** that consists of many small fat droplets suspended in the aqueous chyme, thus increasing the surface area available for attack by pancreatic lipase. To digest fat, lipase must come into direct contact with the triglyceride molecule. Because fat molecules are not soluble in water, they tend to aggregate into large droplets in the aqueous environment of the small intestine lumen. If bile salts did not emulsify these

Figure 16-23 Schematic Structure and Function of Bile Salts (a) Schematic representation of the structure of bile salts and their adsorption on the surface of a small fat droplet. A bile salt consists of a lipid-soluble portion that dissolves in the fat droplet and a negatively charged, water-soluble portion that projects from the surface of the droplet. (b) Formation of a lipid emulsion through the action of bile salts. When a large fat droplet is broken up into smaller fat droplets by intestinal contractions, bile salts adsorb on the surface of the small droplets, creating "shells" of negatively charged, water-soluble bile salt components that cause the fat droplets to repel each other. This action holds the fat droplets apart and prevents them from recoalescing, thereby increasing the surface area of exposed fat available for digestion by pancreatic lipase.

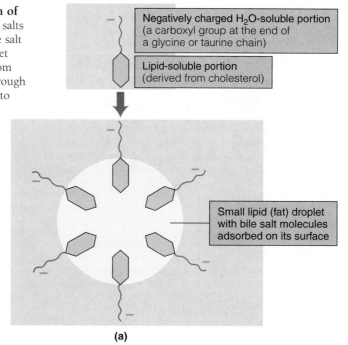

Negatively charged H_2O-soluble portion (a carboxyl group at the end of a glycine or taurine chain)

Lipid-soluble portion (derived from cholesterol)

Small lipid (fat) droplet with bile salt molecules adsorbed on its surface

(a)

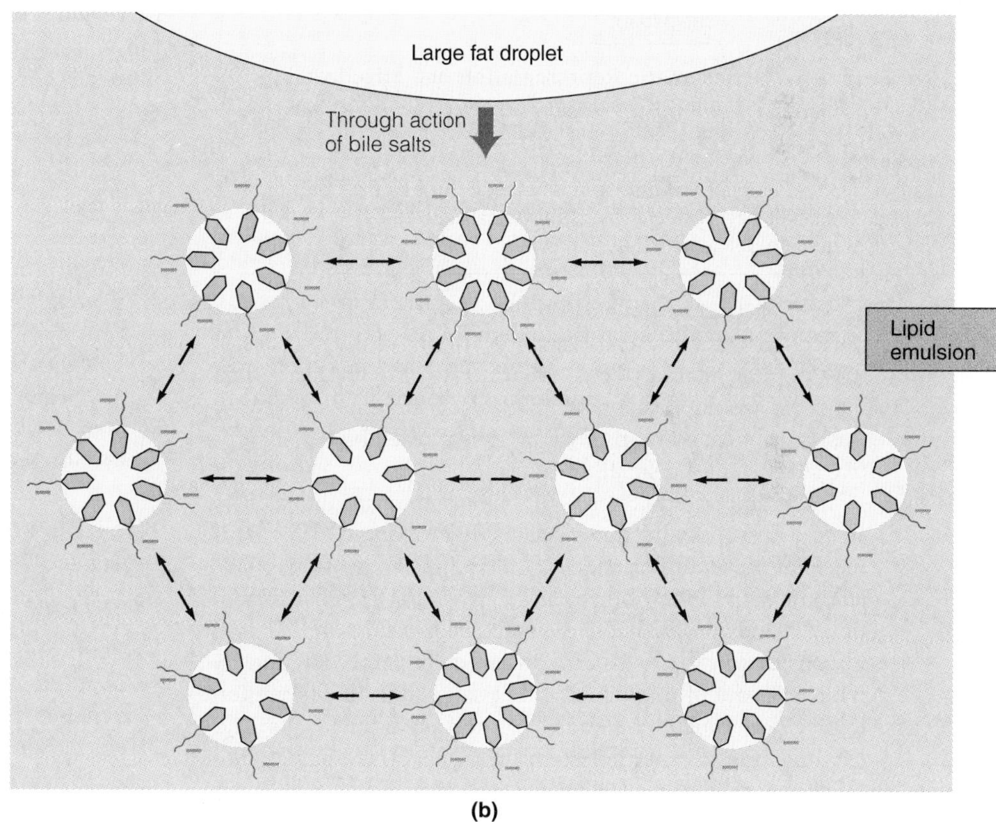

(b)

large droplets, lipase could act on the lipids only at the surface of the large droplets, and triglyceride digestion would be greatly prolonged.

Bile salts exert a detergent action similar to that of the detergent you use to break up grease when you wash dishes. A bile salt molecule contains a lipid-soluble portion (a steroid derived from cholesterol) plus a negatively charged, water-soluble portion. Bile salts *adsorb* on the surface of a fat droplet; that is, the lipid-soluble portion of the bile salt dissolves in the fat droplet, leaving the charged water-soluble portion projecting from the surface of the droplet (Fig. 16-23a). Intestinal mixing movements break up large fat droplets into smaller ones. These small droplets would quickly recoalesce were it not for bile salts adsorbing on their surface and creating a "shell" of water-soluble negative charges on the surface of each little droplet. Because like charges repel, these negatively charged groups on the droplet surfaces cause the fat droplets to repel each other (Fig. 16-23b). This electrical repulsion prevents the small droplets from recoalescing into large fat droplets and thus produces a lipid emulsion that increases the surface area available for lipase action. The increased surface area is extremely important in accomplishing rapid completion of fat digestion; without bile salts, digestion of fat would be very slow.

Micellar formation Bile salts—along with cholesterol and lecithin, which are also constituents of the bile—play an important role in facilitating fat absorption through micellar formation. Like bile salts, lecithin has both a lipid-soluble and a water-soluble portion, whereas cholesterol is almost totally insoluble in water. In a **micelle,** the bile salts and lecithin aggregate in small clusters with their fat-soluble portions huddled together in the middle to form a hydrophobic ("water fearing") core while their water-soluble portions form an outer hydrophilic ("water-loving") shell (Fig. 16-24). A micellar aggregate is about one-millionth the size of an emulsified lipid droplet. Micelles, being water soluble by virtue of their hydrophilic shells, can dissolve water-insoluble (and hence lipid-soluble) substances in their lipid-soluble cores. Micelles thus provide a handy vehicle for carrying water-insoluble substances through the watery luminal contents. The most important lipid-soluble substances thus carried are the products of fat digestion (monoglyerides and free fatty acids) as well as fat-soluble vitamins, which are all trans-

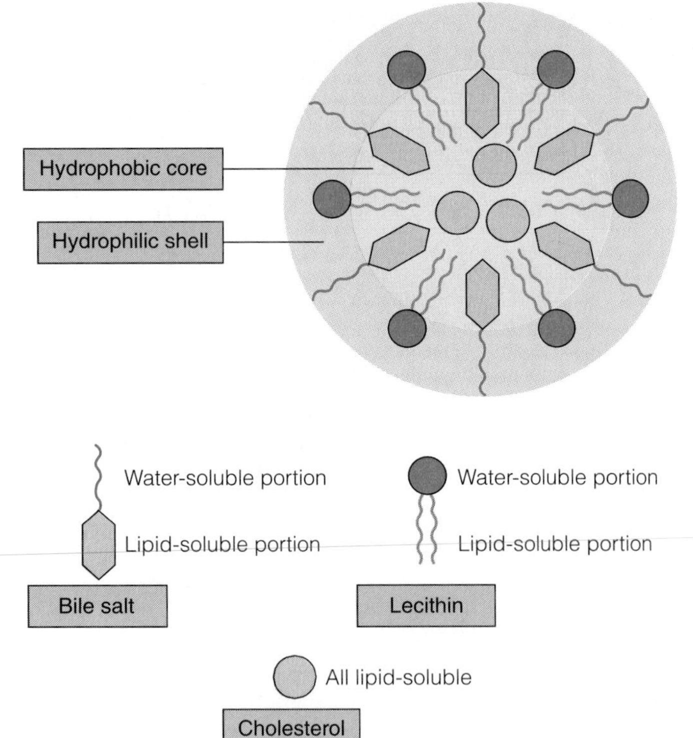

Figure 16-24 **Schematic Representation of a Micelle** Bile constituents (bile salts, lecithin, and cholesterol) aggregate to form micelles that consist of a hydrophilic (water-soluble) shell and a hydrophobic (lipid-soluble) core. Because the outer shell of a micelle is water soluble, the products of fat digestion, which are not water soluble, can be carried through the watery luminal contents to the absorptive surface of the small intestine by dissolving in the micelle's lipid-soluble core.

ported to their sites of absorption by means of the micelles. If they did not hitch a ride in the water-soluble micelles, these nutrients would float on the surface of the aqueous chyme (just as oil floats on top of water), never reaching the absorptive surfaces of the small intestine.

In addition, cholesterol, a highly water-insoluble substance, dissolves in the micelle's hydrophobic core. This mechanism is important in cholesterol homeostasis. The amount of cholesterol that can be carried in micellar formation depends on the relative amount of bile salts and lecithin in comparison to cholesterol. When cholesterol secretion by the liver is out of proportion to bile salt and lecithin secretion (either too much cholesterol or too little bile salts and lecithin), the excess cholesterol in the bile precipitates into microcrystals that can aggregate into **gallstones.** One treatment of cholesterol-containing gallstones involves ingestion of bile salts to increase the bile salt pool in an attempt to dissolve the cholesterol stones. Only about 75% of gallstones are derived from cholesterol, however. The other 25% are made up of abnormal precipitates of another bile constituent, bilirubin.

Bilirubin is a waste product excreted in the bile.

Bilirubin, the other major constituent of bile, does not play a role in digestion at all but instead is one of the few waste products excreted in the bile. Bilirubin is the primary bile pigment derived from the breakdown of worn-out red blood cells. The typical life-span of a red blood cell in the circulatory system is 120 days. Worn-out red blood cells are removed from the blood by the macrophages that line the liver sinusoids and reside in other areas in the body. Bilirubin is the end product resulting from the degradation of the heme (iron-containing) portion of the hemoglobin contained within these old red blood cells. This bilirubin is extracted from the blood by the hepatocytes and is actively excreted into the bile.

Bilirubin is a yellow pigment that gives bile its yellow color. Within the intestinal tract, this pigment is modified by bacterial enzymes, giving rise to the characteristic brown color of the feces. When bile secretion does not occur, as when the bile duct is completely obstructed by a gallstone, the feces are grayish white. A small amount of bilirubin is normally reabsorbed by the intestine back into the blood, and when it is eventually excreted in the urine, it is largely responsible for the urine's yellow color. The kidneys are unable to excrete bilirubin until after it has been modified during its passage through the liver and intestine.

If bilirubin is formed more rapidly than it can be excreted, it accumulates in the body and causes **jaundice.** Patients with this condition appear yellowish, with this color being seen most easily in the whites of their eyes. Jaundice can be brought about in three different ways:

1. *Prehepatic* (the problem occurs "before the liver"), or *hemolytic, jaundice* is due to excessive breakdown (hemolysis) of red blood cells, which results in the liver being presented with more bilirubin than it is capable of excreting.
2. *Hepatic* (the problem is the "liver") *jaundice* occurs when the liver is diseased and is unable to deal with even the normal load of bilirubin.
3. *Posthepatic* (the problem occurs "after the liver"), or *obstructive, jaundice* occurs when the bile duct is obstructed, such as by a gallstone, so that bilirubin cannot be eliminated in the feces.

Bile salts are the most potent stimulus for increased bile secretion.

Bile secretion may be increased by chemical, hormonal, and neural mechanisms:

- *Chemical mechanism (bile salts).* Any substance that increases bile secretion by the liver is called a **choleretic.** The most potent choleretic is bile salts themselves. Between meals bile is stored in the gallbladder, but during a meal bile is emptied into the duodenum as the gallbladder contracts. After bile salts participate in fat digestion and absorption, they are reabsorbed and returned by the enterohepatic circulation to the liver, where they act as potent choleretics to stimulate further bile secretion. Therefore, during a meal, when bile salts are needed and being used, bile secretion by the liver is enhanced.
- *Hormonal mechanism (secretin).* Besides increasing the aqueous $NaHCO_3$ secretion by the pancreas, secretin also stimulates an aqueous alkaline bile secretion by the liver ducts without any corresponding increase in bile salts.

- *Neural mechanism (vagus nerve).* Vagal stimulation of the liver plays a minor role in bile secretion during the cephalic phase of digestion, promoting an increase in liver bile flow before food ever reaches the stomach or intestine.

The gallbladder stores and concentrates bile between meals and empties during meals.

Even though the factors just described increase bile secretion by the liver during and after a meal, bile secretion by the liver occurs continuously. Between meals the secreted bile is shunted into the gallbladder, where it is stored and concentrated. Active transport of salt out of the gallbladder, with water following osmotically, results in a 5 to 10 times concentration of the organic constituents. Since the gallbladder stores this concentrated bile, it is the primary site for precipitation of concentrated bile constituents into gallstones. Fortunately, the gallbladder does not play an essential digestive role, so its removal as a treatment for gallstones or other gallbladder disease presents no particular problem. The bile secreted between meals is stored instead in the common bile duct, which becomes dilated.

During digestion of a meal, when chyme reaches the small intestine, the presence of food, especially fat products, in the duodenal lumen triggers the release of CCK. This hormone stimulates contraction of the gallbladder and relaxation of the sphincter of Oddi, so bile is discharged into the duodenum, where it appropriately aids in the digestion and absorption of the fat that initiated the release of CCK.

Hepatitis and cirrhosis are the most common liver disorders.

Hepatitis is an inflammatory disease of the liver that results from a variety of causes, including viral infection or exposure to toxic agents. With viral hepatitis, the viruses replicate in the nuclei of the hepatocytes, bringing about injury, inflammation, and even death of the infected cells. Toxic hepatitis occurs as a result of injury and destruction of liver cells from abusive exposure to toxic chemicals or drugs, including alcohol, carbon tetrachloride, and certain tranquilizers. Hepatitis ranges in severity from mild, reversible symptoms to acute massive liver damage, with possible imminent death resulting from acute hepatic failure.

Repeated or prolonged hepatic inflammation, usually in association with chronic alcoholism, can lead to **cirrhosis,** a condition in which damaged hepatocytes are permanently replaced by connective tissue. Liver tissue has the ability to regenerate, normally undergoing a gradual turnover of cells. If part of the hepatic tissue is destroyed, the lost tissue can be replaced by an increase in the rate of cell division. A circulating factor is apparently responsible for regulating liver cell proliferation, although the nature and mechanism of this regulatory factor remain a mystery. There is a limit, however, to how rapidly hepatocytes can be replaced. In addition to hepatocytes, dispersed between the hepatic plates are a small number of fibroblasts (connective tissue cells) that form a supporting framework for the liver. If the liver is repeatedly exposed to toxic substances such as alcohol so often that new hepatocytes cannot be generated rapidly enough to replace the damaged cells, the sturdier fibroblasts take advantage of the situation and overproduce. This extra connective tissue leaves little space for the hepatocytes' regrowth. Because nodules of hepatocytes that do manage to regenerate are often less functional and are frequently separated from an adequate blood supply by the thick bands of connective tissue, even the new hepatocyte growth may die. Thus, as cirrhosis develops slowly over time, active liver tissue is gradually reduced, leading eventually to chronic liver failure. Symptoms of hepatic malfunction may not appear until 70% to 80% of the liver tissue is destroyed, demonstrating that, just as with the kidneys, our bodies are endowed with a generous reserve of hepatic tissue.

Small Intestine

The **small intestine** is the site where most digestion and absorption take place. No further digestion is accomplished after the luminal contents pass beyond the small intestine, and no further absorption of nutrients occurs, although the large intestine does absorb small amounts of salt and water. The small intestine is a tube about 6.3 m (21 ft) long with a small diameter of 2.5 cm (1 in). It lies coiled within the abdominal cavity, extending between the stomach and the large intestine. It is arbitrarily divided into three segments—the **duodenum** [the first 20 cm (8 in)], the **jejunum** [2.5 m (8 ft)], and the **ileum** [3.6 m (12 ft.)].

Segmentation contractions mix and slowly propel the chyme.

Segmentation, the small intestine's primary method of motility, both mixes and slowly propels the chyme. Segmentation consists of oscillating, ringlike contractions of the circular smooth muscle along the length of the small intestine; between the contracted segments are relaxed areas containing a small bolus of chyme. The contractile rings occur every few centimeters, dividing the small intestine into segments like a chain of sausages. These contractile rings do not sweep along the length of the intestine as peristaltic waves do. Rather, after a brief period of time, the contracted segments relax, and ringlike contractions appear in the previously relaxed areas (■ Fig. 16-25). The new contraction forces the chyme in a previously relaxed segment to move in both directions into the now relaxed adjacent segments. A newly relaxed segment therefore receives chyme from both the contracting segment immediately ahead of it and the one immediately behind it. Shortly thereafter, the areas of contraction and relaxation alternate again. In this way, the chyme is chopped, churned, and thoroughly mixed. These contractions can be compared to squeezing a pastry tube with your hands to mix the contents. This mixing serves the dual functions of mixing the chyme with the digestive juices secreted into the small intes-

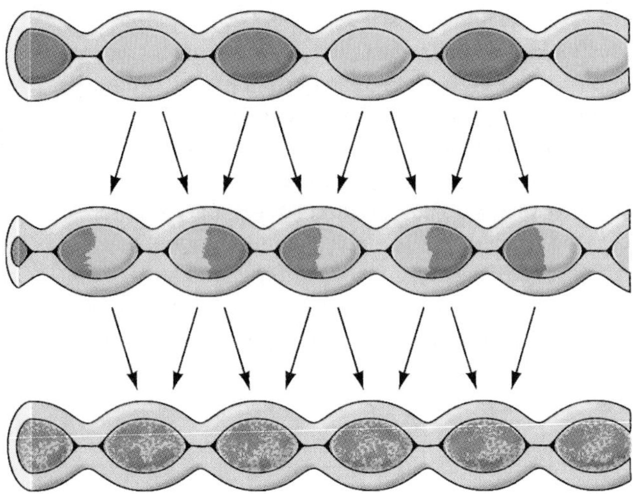

Figure 16-25 Segmentation Segmentation consists of ringlike contractions along the length of the small intestine. Within a matter of seconds, the contracted segments relax and the previously relaxed areas contract. These oscillating contractions thoroughly mix the chyme within the small intestine lumen.

tine lumen and exposing all of the chyme to the absorptive surfaces of the small intestine mucosa.

Segmentation contractions are initiated by the small intestine's pacesetter cells, which produce a basic electrical rhythm (BER) similar to the gastric BER responsible for peristalsis in the stomach. If the small intestine BER brings the circular smooth muscle layer to threshold, segmentation contractions are induced, with the frequency of segmentation following the frequency of the BER.

The circular smooth muscle's degree of responsiveness and thus the intensity of segmentation contractions can be influenced by distention of the intestine, by the hormone gastrin, and by extrinsic nerve activity. All of these factors influence the excitability of the small intestine smooth muscle cells by moving the "resting" potential around which the BER oscillates closer to or farther from the threshold. Segmentation is slight or absent between meals but becomes very vigorous immediately after a meal. Both the duodenum and the ileum start to segment simultaneously when the meal first enters the small intestine. The duodenum starts to segment primarily in response to local distention caused by the presence of chyme. Segmentation of the empty ileum, on the other hand, appears to be brought about by gastrin, which is secreted in response to the presence of chyme in the stomach, a mechanism known as the **gastroileal reflex.** Extrinsic nerves can modify the strength of these contractions. Parasympathetic stimulation enhances segmentation, whereas sympathetic stimulation depresses segmental activity.

Segmentation not only accomplishes mixing but also is the primary factor responsible for slowly moving chyme through the small intestine. How can this be when each segmental contraction propels chyme both forward and backward? The chyme slowly progresses forward because the frequency of segmentation declines along the length of the small intestine. The pacesetter cells in the duodenum spontaneously depolarize at a faster rate than those farther down the tract, with the pacesetter cells in the terminal ileum exhibiting the slowest rate of spontaneous depolarization. Segmentation contractions occur in the duodenum at a rate of 12 per minute, compared to only 9 per minute in the terminal ileum. Because segmentation occurs with greater frequency in the upper part of the small intestine than in the lower part, more chyme on average is pushed forward than is pushed backward. As a result, chyme is moved very slowly from the upper to the lower part of the small intestine, being shuffled back and forth to accomplish thorough mixing and absorption in the process. This slow propulsive mechanism is advantageous because it allows ample time for the digestive and absorptive processes to take place. The contents usually take 3 to 5 hours to move through the small intestine.

The migrating motility complex sweeps the intestine clean between meals.

When most of the meal has been absorbed, segmentation contractions cease and are replaced between meals by the **migrating motility complex**, or **"intestinal housekeeper."** This between-meal motility consists of weak, repetitive peristaltic waves that move a short distance down the intestine before dying out. The waves start at the stomach and migrate down the intestine; that is, each new peristaltic wave is initiated at a site a little farther down the small intestine. These short peristaltic waves take about 100 to 150 minutes to gradually migrate from the stomach to the end of the small intestine, with each contraction "sweeping" any remnants of the preceding meal plus mucosal debris and bacteria forward toward the colon, just like a good "intestinal housekeeper." After the end of the small intestine is reached, the cycle begins again and continues to repeat itself until the next meal. It is speculated that the unconfirmed hormone **motilin** might play a role in regulating the migrating motility complex. When the next meal arrives, segmental activity is triggered again, and the migrating motility complex ceases.

The ileocecal juncture prevents contamination of the small intestine by colonic bacteria.

At the juncture between the small and large intestines, the last portion of the ileum empties into the cecum (Fig. 16-26). Two factors contribute to this region's ability to act as a barrier between the small and large intestines. First, the anatomical arrangement is such that valvelike folds of tissue protrude from the ileum into the lumen of the cecum. When the ileal contents are pushed forward, this **ileocecal valve** is easily pushed open, but the folds of tissue are forcibly closed when the cecal contents attempt to move backward. Second, the smooth muscle within the last several centimeters of the ileal wall is thickened, forming a sphincter that is under neural and hormonal control. Most of the time this **ileocecal sphincter** remains at least mildly constricted. Pressure on the cecal side of the sphincter causes it to contract more forcibly; distention of the ileal side causes the sphincter to relax, a reaction that is mediated by the intrinsic plexuses in the area. In this way, the ileocecal sphincter prevents the bacteria-laden

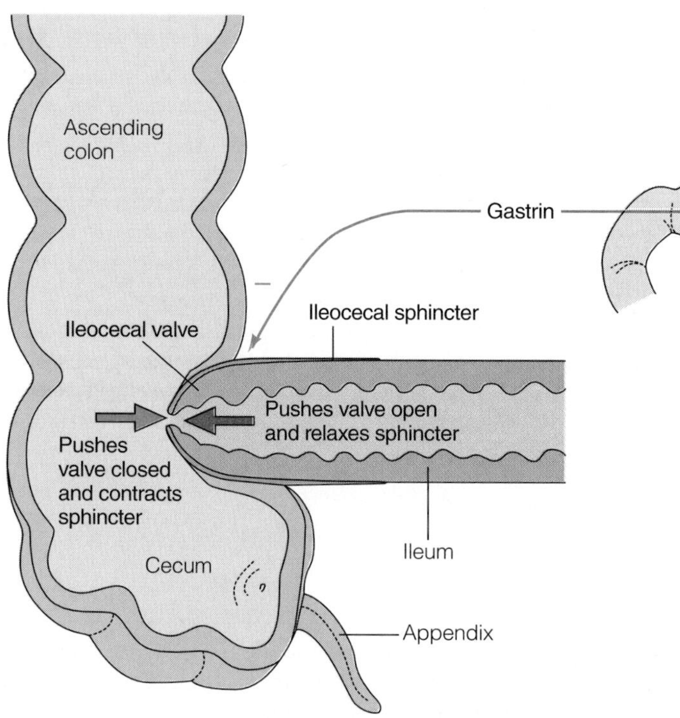

Figure 16-26 **Control of the Ileocecal Valve/Sphincter** The juncture between the ileum and large intestine consists of the ileocecal valve, which is surrounded by thickened smooth muscle, the ileocecal sphincter. Pressure on the cecal side pushes the valve closed and contracts the sphincter, preventing the bacteria-laden colonic contents from contaminating the nutrient-rich small intestine. The valve/sphincter opens and allows ileal contents to enter the large intestine in response to pressure on the ileal side of the valve and to the hormone gastrin secreted as a new meal enters the stomach.

contents of the large intestine from contaminating the small intestine and at the same time allows the ileal contents to pass into the colon. If the colonic bacteria were to gain access to the nutrient-rich small intestine, they would multiply rapidly. Relaxation of the sphincter is enhanced through the release of gastrin at the onset of a meal, when increased gastric activity is taking place. This relaxation allows the undigested fibers and unabsorbed solutes from the preceding meal to be moved forward as the new meal enters the tract.

Small intestine secretions do not contain any digestive enzymes.

Each day the exocrine gland cells located in the small intestine mucosa secrete into the lumen about 1.5 liters of an aqueous salt and mucus solution known as the **succus entericus** (*succus* means "juice"; *entericus* means "of the intestine"). No digestive enzymes are secreted into this intestinal juice. The small intestine does synthesize digestive enzymes, but they act intracellularly within the borders of the epithelial cells that line the lumen instead of being secreted directly into the lumen.

Are any functions served by this small intestine secretion, since no digestive enzymes are involved? The mucus in the secretion provides protection and lubrication. Furthermore, this aqueous secretion provides an abundance of H_2O to participate in the enzymatic digestion of food. Recall that digestion involves hydrolysis—bond breakage by reaction with H_2O—which proceeds most efficiently when all the reactants are in solution.

The regulation of small intestine secretion is not clearly understood. Secretion of succus entericus does increase after a meal. The most effective stimulus for secretion appears to be local stimulation of the small intestine mucosa by the presence of chyme.

Digestion in the small intestine lumen is accomplished by pancreatic enzymes, whereas the small intestine enzymes act intracellularly.

Digestion within the small intestine lumen is accomplished by the pancreatic enzymes, with fat digestion being enhanced by bile secretion. As a result of pancreatic enzymatic activity, fats are completely reduced to their absorbable units of monoglycerides and free fatty acids, proteins are broken down into small peptide fragments and some amino acids, and carbohydrates are reduced to disaccharides and some monosaccharides. Thus, fat digestion is completed within the small intestine lumen, but carbohydrate and protein digestion have not been brought to completion.

Special actin-stiffened, hairlike projections on the luminal surface of the small intestine epithelial cells form the **brush border** (see p. 40), which contains three different categories of enzymes: (1) **enterokinase,** which activates the pancreatic enzyme trypsinogen; (2) the **disaccharidases (maltase, sucrase,** and **lactase),** which complete carbohydrate digestion by hydrolyzing the remaining disaccharides (maltose, sucrose, and lactose, respectively) into their constituent monosaccharides; and (3) the **aminopeptidases,** which hydrolyze the small peptide fragments into their amino acid components, thereby completing protein digestion. Thus, carbohydrate and protein digestion are completed intracellularly within the confines of the brush border. (▮ Table 16-6 provides a summary of the digestive processes for the three major categories of nutrients.)

A fairly common disorder, **lactose intolerance,** involves a deficiency of lactase, the disaccharidase specific for the digestion of lactose, or milk sugar. Most children under four years of age have adequate lactase, but this may be gradually lost so that in many adults, lactase activity is diminished or absent. When milk or dairy products (except those in which lactose has already been digested by bacterial action during processing, such as some kinds of cheese and yogurt) are consumed by an individual with lactase deficiency, the undigested lactose remains in the lumen and has several related consequences. First, accumulation of undigested lactose creates an osmotic gradient that draws H_2O into the intestinal lumen. Second, bacteria residing in the large intestine possess lactose-splitting ability, so they eagerly attack the lactose as an energy source, producing large quantities of CO_2 and methane gas in the process. Distention of the intestine by both fluid and gas

Table 16-6 Digestive Processes for the Three Major Categories of Nutrients

Nutrients	Enzymes for Digesting Nutrient	Source of Enzymes	Site of Action of Enzymes	Action of Enzymes	Absorbable Units of Nutrients
Carbohydrate	Amylase	Salivary glands	Mouth and body of stomach	Hydrolyzes polysaccharides to disaccharides	
		Exocrine pancreas	Small intestine lumen		
	Disaccharidases (maltase, sucrase, lactase)	Small intestine epithelial cells	Small intestine brush border	Hydrolyzes disaccharides to monosaccharides	Monosaccharides, especially glucose
Protein	Pepsin	Stomach chief cells	Stomach antrum	Hydrolyzes protein to peptide fragments	
	Trypsin, chymotrypsin, carboxypeptidase	Exocrine pancreas	Small intestine lumen	Attack different peptide fragments	
	Aminopeptidases	Small intestine epithelial cells	Small intestine brush border	Hydrolyzes peptide fragments to amino acids	Amino acids and a few small peptides
Fat	Lipase	Exocrine pancreas	Small intestine lumen	Hydrolyzes triglycerides to fatty acids and monoglycerides	
	Bile salts (not an enzyme)	Liver	Small intestine lumen	Emulsify large fat globules for attack by pancreatic lipase	Fatty acids and monoglycerides

produces pain (cramping) and diarrhea. Depending on the extent of the lactase deficiency and the quantity of lactose ingested, symptoms can very from mild abdominal discomfort to severe dehydrating diarrhea.

The small intestine is remarkably well adapted for its primary role in absorption.

All products of carbohydrate, protein, and fat digestion, as well as most of the ingested electrolytes, vitamins, and water, are normally absorbed by the small intestine indiscriminately. Usually, only the absorption of calcium and iron is adjusted to the body's needs. Thus, the more food that is consumed, the more that will be digested and absorbed, as those who are trying to control their weight are all too painfully aware.

Most absorption occurs in the duodenum and jejunum; very little occurs in the ileum, not because the ileum does not have absorptive capacity but because most absorption has already been accomplished before the intestinal contents reach the ileum. The small intestine has an abundant reserve absorptive capacity. In fact, about 50% of the small intestine can be removed with little interference to absorption—with one exception. If the terminal ileum is removed, vitamin B_{12} and bile salts are not properly absorbed because the specialized transport mechanisms for these two substances are located only in this region. All other substances can be absorbed throughout the length of the small intestine.

The mucous lining of the small intestine is remarkably well adapted for its special absorptive function for two reasons: (1) it has a very large surface area, and (2) the epithelial cells in this lining possess a variety of specialized transport mechanisms. The following special modifications of the small intestine mucosa greatly increase the surface area available for absorption (Fig. 16-27):

- The inner surface of the small intestine is thrown into circular folds that are visible to the naked eye and increase the surface area threefold.
- Projecting from this folded surface are microscopic fingerlike projections known as **villi**, which give the lining a velvety appearance and increase the surface area by another 10 times (Fig. 16-28). The surface of each villus is covered by epithelial cells interspersed occasionally with mucous cells.
- Even smaller hairlike projections known as **microvilli** (or the *brush border*) arise from the luminal surface of these epithelial cells, increasing the surface area another 20-fold. Each epithelial cell has as many as 3,000 to 6,000 of these microvilli, which are visible only with an electron microscope. It is within the membrane of this brush border that the small intestine enzymes perform their functions.

Altogether, the folds, villi, and microvilli provide the small intestine with a luminal surface area 600 times greater than it would have been if it were a tube of the same length and

Figure 16-27 Small Intestine Absorptive Surface
(a) Gross structure of the small intestine. (b) One of the circular folds of the small intestine mucosa, which collectively increase the absorptive surface area threefold. (c) Microscopic fingerlike projection known as a villus. Collectively, the villi increase the surface area another tenfold. (d) Electron microscope view of a villus epithelial cell, depicting the presence of microvilli on its luminal border; the microvilli increase the surface area another 20-fold. Altogether, these surface modifications increase the small intestine's absorptive surface area 600-fold.

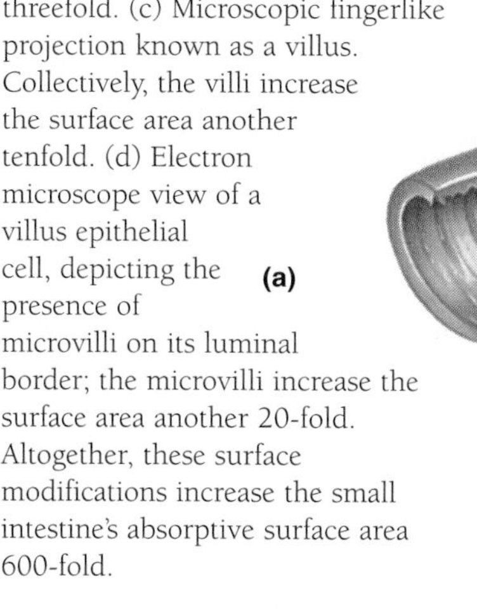

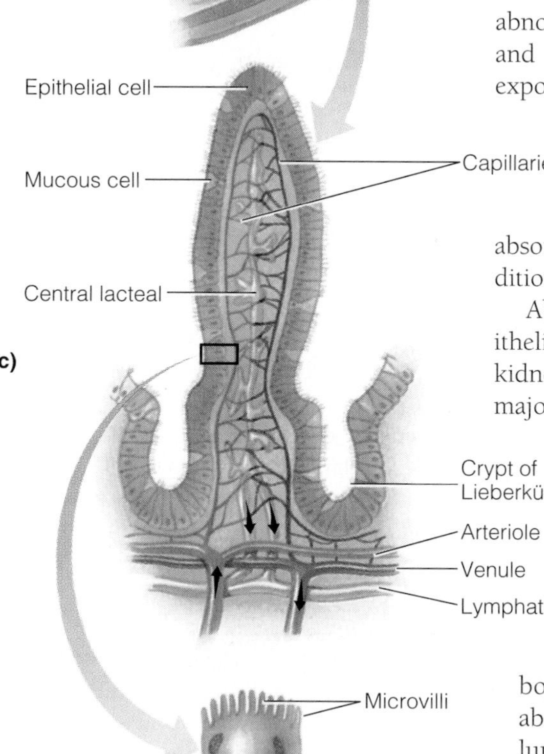

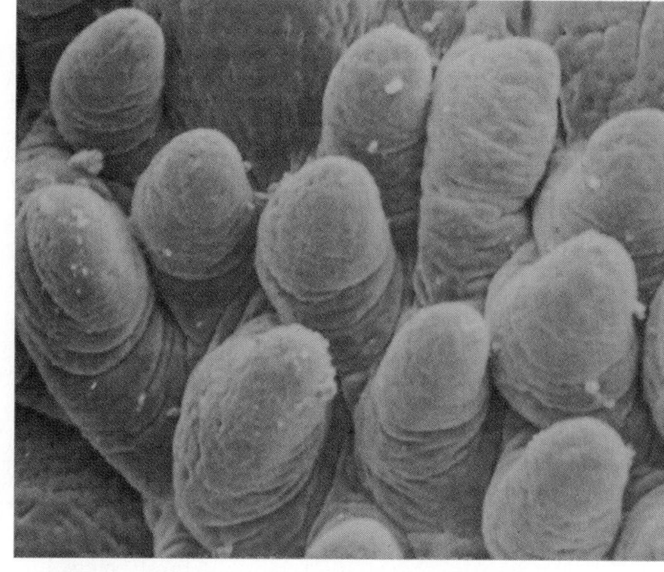

Figure 16-28 Scanning Electron Micrograph of Villi Projecting from the Small Intestine Mucosa

diameter lined by a flat surface. In fact, if the surface area of the small intestine were spread out flat, it would cover an entire tennis court.

Malabsorption (impairment of absorption) may be caused by damage to or reduction of the surface area of the small intestine. One of the most common causes is **gluten enteropathy.** In this condition, the individual's small intestine is abnormally sensitive to gluten, a protein constituent of wheat and some other grains. Through an unknown mechanism, exposure to gluten damages the intestinal villi: the normally luxuriant array of villi is reduced, the mucosa becomes flattened, and the brush border becomes short and stubby (Fig. 16-29). Because this loss of villi decreases the surface area available for absorption, absorption of all nutrients is impaired. The condition is treated by eliminating gluten from the diet.

Absorption across the digestive tract wall involves transepithelial transport similar to movement of material across the kidney tubules (see p. 483). Each villus has the following major components (Fig. 16-27c):

- Epithelial cells that cover the surface of the villus. The epithelial cells are joined at their lateral borders by tight junctions, which limit passage of luminal contents between the cells, although the tight junctions in the small intestine are "leakier" than those in the stomach. Within their luminal brush borders, these epithelial cells possess carriers for absorption of specific nutrients and electrolytes from the lumen as well as the intracellular digestive enzymes that complete carbohydrate and protein digestion.
- A connective tissue core. This core is formed by the lamina propria.
- A capillary network. Each villus is supplied by an arteriole that breaks up into a capillary network within the villus. The capillaries rejoin to form a venule that drains away from the villus.
- A terminal lymphatic vessel known as a **central lacteal.**

During the process of absorption, digested substances enter the capillary network or the central lacteal. To be absorbed, a substance must pass completely through the epithelial cell,

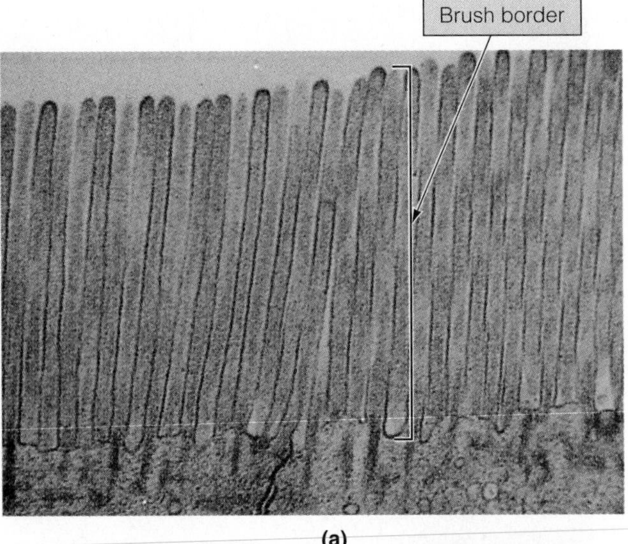

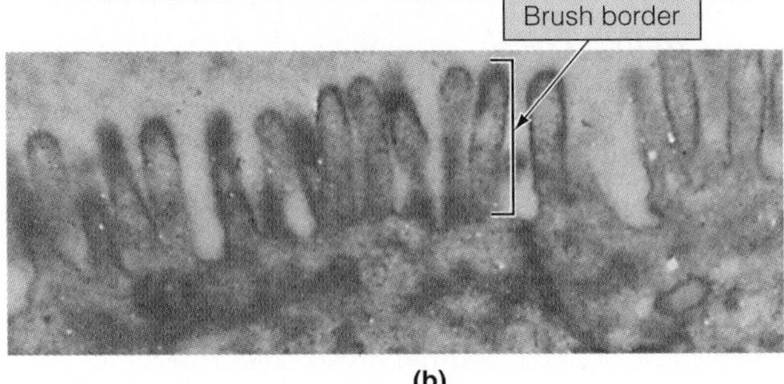

Figure 16-29 Reduction in the Brush Border with Gluten Enteropathy (a) Electron micrograph of the brush border of a small intestine epithelial cell in a normal individual. (b) Electron micrograph of the short, stubby brush border of a small intestine epithelial cell in a patient with gluten enteropathy.

diffuse through the interstitial fluid within the connective tissue core of the villus, and then cross the wall of a capillary or lymph vessel. Like renal transport, intestinal absorption may be an active or passive process, with active absorption involving energy expenditure during at least one of the steps in the transepithelial transport process.

The mucosal lining experiences rapid turnover.

Dipping down into the mucosal surface between the villi are shallow invaginations known as the **crypts of Lieberkühn** (Fig. 16-27c). Unlike the gastric pits, these intestinal crypts do not secrete digestive enzymes, but they do secrete water and electrolytes, which, along with the mucus secreted by the cells on the villus surface, constitute the succus entericus. Furthermore, the crypts function as "nurseries." The epithelial cells lining the small intestine slough off and are replaced at a rapid rate as a result of high mitotic activity in the crypts. New cells that are continually being produced at the bottom of the crypts migrate up the villi and, in the process, push off the older cells at the tips of the villi into the lumen. In this manner, more than 100 million intestinal cells are shed per minute. The entire trip from crypt to tip averages about 3 days, so the epithelial lining of the small intestine is replaced approximately every 3 days. Because of this high rate of cell division, the cells within the crypts are very sensitive to damage by radiation and anticancer drugs, both of which may inhibit cell division.

The cells undergo several changes as they migrate up the villus. The concentration of brush border enzymes increases and the capacity for absorption improves, so the cells at the tip of the villus have the greatest digestive and absorptive capability. Just at their peak, these cells are pushed off by the newly migrating cells. Thus, the luminal contents are constantly exposed to cells that are optimally equipped to complete the digestive and absorptive functions efficiently. Furthermore, just as in the stomach, the rapid turnover of cells in the small intestine is essential because of the harsh luminal conditions. Cells exposed to the abrasive and corrosive luminal contents are easily damaged and cannot live for long, so they must be continually replaced by a fresh supply of newborn cells.

The old cells that are sloughed off into the lumen are not entirely lost to the body. These cells are digested, with the cell constituents being absorbed into the blood and reclaimed for synthesis of new cells, among other things.

Special mechanisms facilitate absorption of most nutrients.

We will now turn our attention to the mechanisms through which the specific dietary constituents are normally absorbed.

Salt and water absorption Sodium may be absorbed both passively and actively. When the electrochemical gradient favors movement of Na^+ from the lumen to the blood, passive diffusion of Na^+ can occur *between* the intestinal epithelial cells through the "leaky" tight junctions into the interstitial fluid within the villus. Movement of Na^+ *through* the cells is energy dependent and involves at least two different carriers, similar to the process of Na^+ reabsorption across the kidney tubules (see p. 484). Sodium may passively enter the epithelial cells at their luminal surface by itself, or it may be cotransported with glucose or amino acid. It is then actively pumped out of the cell at the basolateral border into the interstitial fluid in the lateral spaces between the cells where they are not joined by tight junctions. From the interstitial fluid, Na^+ eventually diffuses into the capillaries.

As with the renal tubules in the early portion of the nephron, the absorption of Cl^-, H_2O, glucose, and amino acids from the small intestine is linked to this energy-dependent Na^+ absorption. Chloride passively follows down the electrical gradient created by Na^+ absorption and can be actively absorbed as well if needed. Most H_2O absorption in the digestive tract depends on the active carrier that pumps Na^+ into the lateral spaces, resulting in a concentrated area of high osmotic pressure in that localized region between the cells, similar to the situation in the kidneys (see p. 489). This localized high osmotic pressure induces H_2O to move from the lumen through the cell (and possibly from the lumen through the leaky tight junction) into the lateral space. Water entering the space reduces the osmotic pressure but raises the hydrostatic (fluid) pressure. As a result, H_2O is flushed out of the lateral space into the interior of the villus, where it is picked up by the capillary network. Meanwhile, more Na^+ is pumped into the lateral space to encourage more H_2O absorption.

Carbohydrate absorption Dietary carbohydrate is presented to the small intestine for absorption mainly in the forms of the disaccharides maltose (the product of polysaccharide digestion), sucrose, and lactose (Table 16-1 and Fig. 16-30). The disac-

Figure 16-30 **Carbohydrate Digestion and Absorption** The dietary polysaccharides starch and glycogen are converted into the disaccharide maltose through the action of salivary and pancreatic amylase. Maltose and the dietary disaccharides lactose and sucrose are converted to their respective monosaccharides by the disaccharidases (maltase, lactase, and sucrase) located in the brush borders of the small intestine epithelial cells. The monosaccharides glucose and galactose are absorbed into the interior of the cell and eventually enter the blood by means of Na^+- and energy-dependent secondary active transport. The monosaccharide fructose is absorbed into the blood by passive facilitated diffusion.

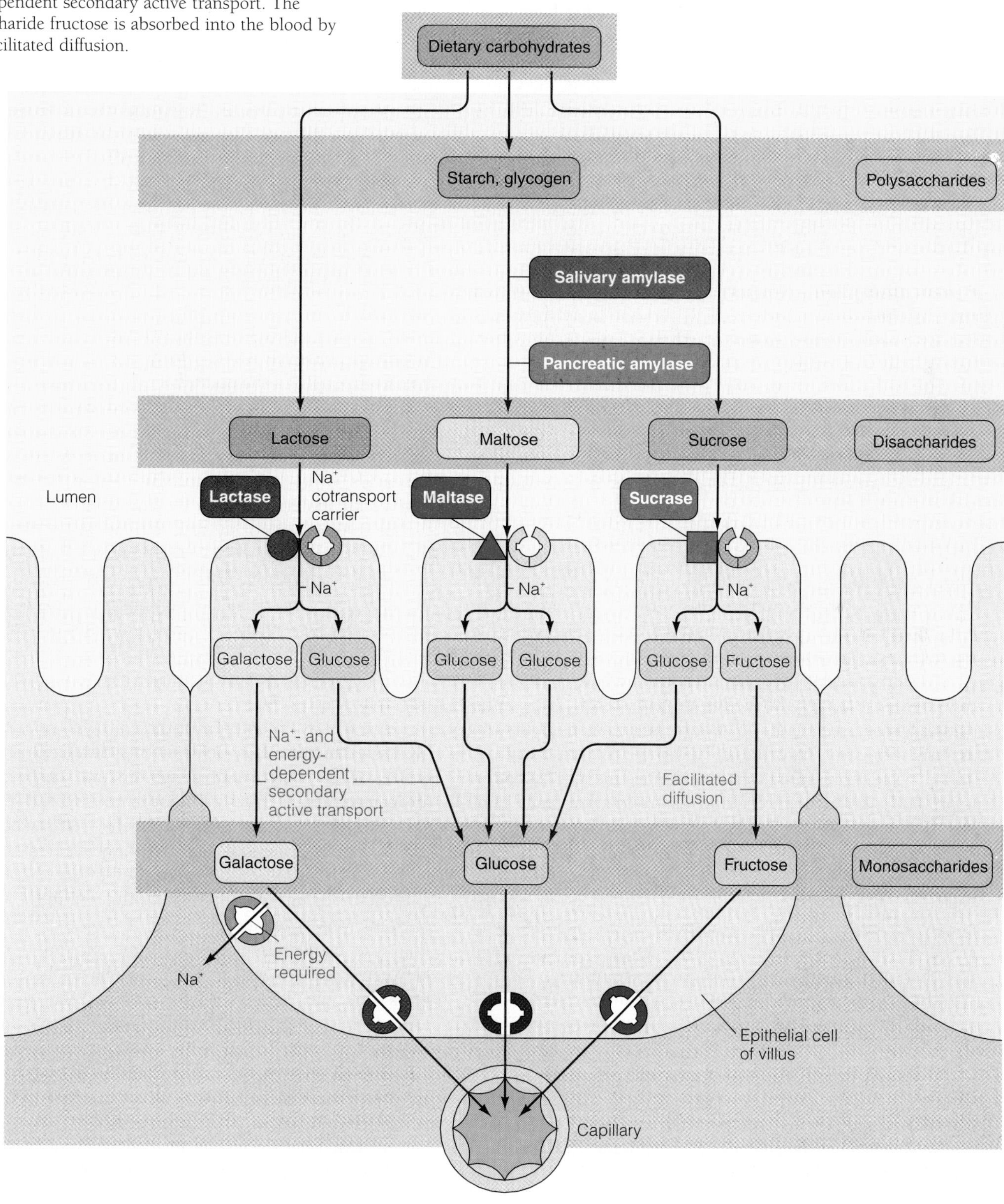

charidases located in the brush borders of the small intestine cells further reduce these disaccharides into the absorbable monosaccharide units of glucose, galactose, and fructose.

Glucose and galactose are both absorbed by *secondary active transport*, in which cotransport carriers on the luminal border transport both the monosaccharide and Na^+ from the lumen into the interior of the intestinal cell. The operation of these cotransport carriers, which do not directly use energy themselves, depends on the Na^+ concentration gradient established by the energy-consuming basolateral Na^+-K^+ pump (see p. 62). Glucose (or galactose), having been concentrated in the cell by the cotransport carriers, leaves the cell down its concentration gradient by means of a passive carrier in the basolateral border to enter the blood within the villus. In addition to glucose being absorbed through the cells by means of the cotransport carrier, recent evidence suggests that a significant amount of glucose crosses the epithelial barrier through "leaky" tight junctions between the epithelial cells. Fructose is absorbed into the blood solely by facilitated diffusion (passive carrier-mediated transport; see p. 59).

Protein absorption Not only are ingested proteins digested and absorbed, but endogenous ("within the body") proteins that have entered the digestive tract lumen from the three following sources are digested and absorbed as well:

1. Digestive enzymes, all of which are proteins, that have been secreted into the lumen.
2. Proteins within the cells that are pushed off from the villi into the lumen during the process of mucosal turnover.
3. Small amounts of plasma proteins that normally leak from the capillaries into the digestive tract lumen.

About 20 to 40 g of endogenous protein enter the lumen each day from these three sources. This quantity can amount to more than half of the protein presented to the small intestine for digestion and absorption. All endogenous proteins must be digested and absorbed along with the dietary proteins to prevent depletion of the body's protein stores. The amino acids absorbed from both food and the endogenous protein are used primarily to synthesize new protein in the body.

The protein presented to the small intestine for absorption is primarily in the form of amino acids and a few small peptide fragments (— Fig. 16-31). Amino acids are absorbed across the intestinal cells by secondary active transport, similar to glucose and galactose absorption. Thus, glucose, galactose, and amino acids all get a "free ride" in on the energy expended for Na^+ transport. Small peptides gain entry by means of a different carrier and are broken down into their constituent amino acids by the aminopeptidases in the brush borders or by intracellular peptidases. Like monosaccharides, amino acids enter the capillary network within the villus.

Thus, absorption of the digestion end products of both carbohydrates and proteins involves special carrier-mediated transport systems that require energy expenditure and Na^+ cotransport, and both categories of end products are absorbed into the blood.

Fat absorption Fat absorption is quite different from carbohydrate and protein absorption because the insolubility of fat in water presents a special problem. Fat must be transferred from the watery chyme through the watery body fluids even though it is not water soluble. Therefore, fat must undergo a series of physical and chemical transformations to circumvent this problem during its digestion and absorption (— Fig. 16-32).

When the stomach contents are emptied into the duodenum, the ingested fat is aggregated into large, oily triglyceride droplets that float in the chyme. Recall that through the bile salts' detergent action in the small intestine, the large droplets are dispersed into a lipid emulsification of small droplets, thereby exposing a much greater surface area of fat for digestion by pancreatic lipase. The products of lipase digestion (monoglycerides and free fatty acids) are also not very water soluble, so very little of these end products of fat digestion can diffuse through the aqueous chyme to reach the absorptive lining. However, biliary components facilitate absorption of these fatty end products through formation of micelles. Remember that micelles are water-soluble particles that can carry the end products of fat digestion within their lipid-soluble interiors. Once these micelles reach the luminal membranes of the epithelial cells, the monoglycerides and free fatty acids passively diffuse from the micelles through the lipid component of the epithelial cell membranes to enter the interior of these cells. As these fat products leave the micelles and are absorbed across the epithelial cell membranes, the micelles are able to pick up more monoglycerides and free fatty acids, which have been produced from digestion of other triglyceride molecules in the fat emulsion.

Bile salts continuously repeat their fat-solubilizing function down the length of the small intestine until all the fat is absorbed. Then the bile salts themselves are reabsorbed in the terminal ileum by special active transport. This is an efficient process, because relatively small amounts of bile salts can facilitate digestion and absorption of large amounts of fat, with each bile salt performing its ferrying function repeatedly before it is reabsorbed.

Once within the interior of the epithelial cells, the monoglycerides and free fatty acids are resynthesized into triglycerides. These triglycerides congolomerate into droplets and are coated with a layer of lipoprotein (synthesized by the endoplasmic reticulum of the epithelial cell), which renders the fat droplets water soluble. The large, coated fat droplets, known as **chylomicrons,** are extruded by exocytosis from the epithelial cells into the interstitial fluid within the villus. The chylomicrons subsequently enter the central lacteals rather than the capillaries because of the structural differences between these two vessels. Capillaries have a basement membrane (an outer layer of polysaccharides) that prevents chylomicrons from entering, but the lymph vessels do not have this barrier. Thus, fat can be absorbed into the lymphatics but not directly into the blood. (Fatty acids with short- or medium-length carbon chains can enter the blood, but very few of these are eaten in the normal diet.)

The actual transfer of monoglycerides and free fatty acids from the chyme across the cell membranes of the intestinal

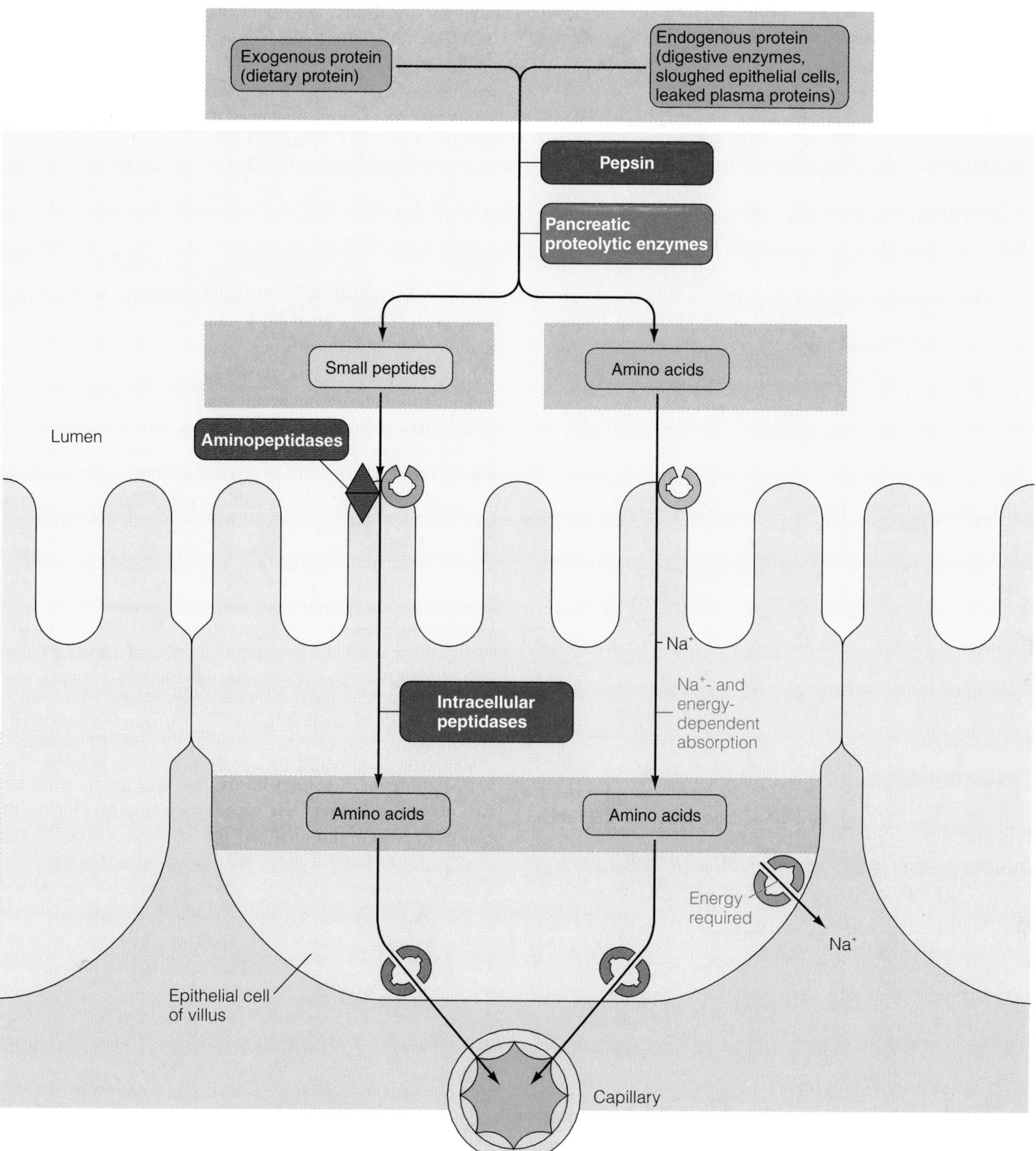

Figure 16-31 Protein Digestion and Absorption Dietary and endogenous proteins are hydrolyzed to their constituent amino acids and a few small peptide fragments by gastric pepsin and the pancreatic proteolytic enzymes. Amino acids are absorbed into the small intestine epithelial cells and eventually enter the blood by means of Na^+- and energy-dependent secondary active transport. Various amino acids are transported by carriers specific for them. The small peptides, which are absorbed by a different type of carrier, are broken down into their amino acids by aminopeptidases in the epithelial cells' brush borders or by intracellular peptidases.

epithelial cells is a passive process, because the lipid-soluble fatty end products merely dissolve in and pass through the lipid portions of the membrane. Fat absorption is therefore said to be a passive process. However, the overall sequence of events necessary for fat absorption does require energy. For example, bile salts are actively secreted by the liver, and the resynthesis of triglycerides and formation of chylomicrons within the epithelial cells are active processes.

Vitamin absorption Water-soluble vitamins are primarily absorbed with water, whereas fat-soluble vitamins are carried in the micelles and absorbed passively with the end products of fat digestion. Absorption of some of the vitamins can also be accomplished by carriers, if necessary. Vitamin B_{12} is unique in that it must be in combination with gastric intrinsic factor for absorption by special transport in the terminal ileum.

Iron and calcium absorption In contrast to the almost complete, unregulated absorption of other ingested electrolytes, the absorption of dietary iron and calcium may not be complete because it is subject to regulation that depends on the body's needs for these electrolytes.

Iron is essential for hemoglobin production. The normal iron intake is typically 15–20 mg/day, yet a man usually absorbs about 0.5–1 mg/day into the blood, and a woman takes up slightly more at 1.0–1.5 mg/day (women's need for iron is greater because of the periodic loss of iron in the menstrual blood flow).

Two main steps are involved in the absorption of iron into the blood: (1) absorption of iron from the lumen into the intestinal epithelial cells and (2) absorption of iron from the epithelial cells into the blood (Fig. 16-33).

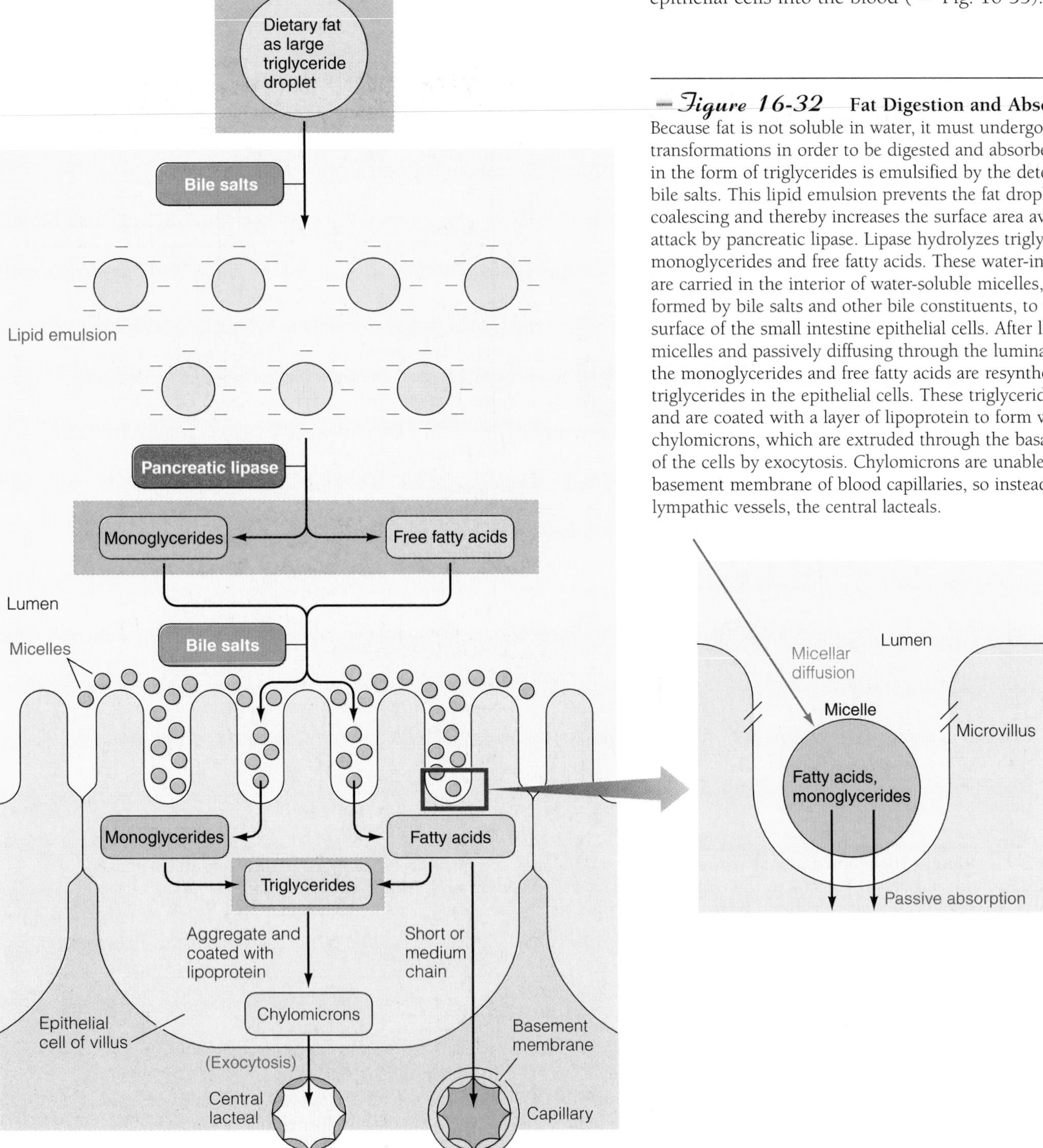

Figure 16-32 **Fat Digestion and Absorption**
Because fat is not soluble in water, it must undergo a series of transformations in order to be digested and absorbed. Dietary fat in the form of triglycerides is emulsified by the detergent action of bile salts. This lipid emulsion prevents the fat droplets from coalescing and thereby increases the surface area available for attack by pancreatic lipase. Lipase hydrolyzes triglycerides into monoglycerides and free fatty acids. These water-insoluble products are carried in the interior of water-soluble micelles, which are formed by bile salts and other bile constituents, to the luminal surface of the small intestine epithelial cells. After leaving the micelles and passively diffusing through the luminal membranes, the monoglycerides and free fatty acids are resynthesized into triglycerides in the epithelial cells. These triglycerides aggregate and are coated with a layer of lipoprotein to form water-soluble chylomicrons, which are extruded through the basal membrane of the cells by exocytosis. Chylomicrons are unable to cross the basement membrane of blood capillaries, so instead they enter the lymphatic vessels, the central lacteals.

Iron is actively transported from the lumen into the epithelial cells, with women having about four times more active transport sites for iron than men. The extent to which ingested iron is taken up by the epithelial cells depends on the type of iron consumed (ferrous iron, Fe^{2+}, is absorbed more easily than ferric iron, Fe^{3+}) and the presence of other substances in the lumen that can either promote or reduce iron absorption. For example, ascorbic acid (vitamin C) increases iron absorption, primarily by reducing ferric to ferrous iron. Phosphate and oxalate, on the other hand, combine with ingested iron to form insoluble iron salts that cannot be absorbed.

After active absorption into the small intestine epithelial cells, iron has two possible fates:

1. Iron that is immediately needed for the production of red blood cells is absorbed into the blood for delivery to the bone marrow, the site of red blood cell production. Iron is transported in the blood by a plasma protein carrier known as **transferrin.** The hormone responsible for stimulating red blood cell production, erythropoietin (see p. 356), is believed to also enhance iron absorption from the intestinal cells into the blood. The absorbed iron is then used in the synthesis of hemoglobin for the newly produced red blood cells.
2. Iron that is not immediately needed remains stored within the epithelial cells in a granular form called **ferritin,** which cannot be absorbed into the blood. If the blood level of iron is too high, excess iron may be dumped from the blood into this unabsorbable pool of ferritin in the intestinal epithelial cells. Iron stored as ferritin is lost in the feces within three days as the epithelial cells containing these granules are sloughed off during mucosal regeneration. Large amounts of iron in the feces give them a dark, almost black color.

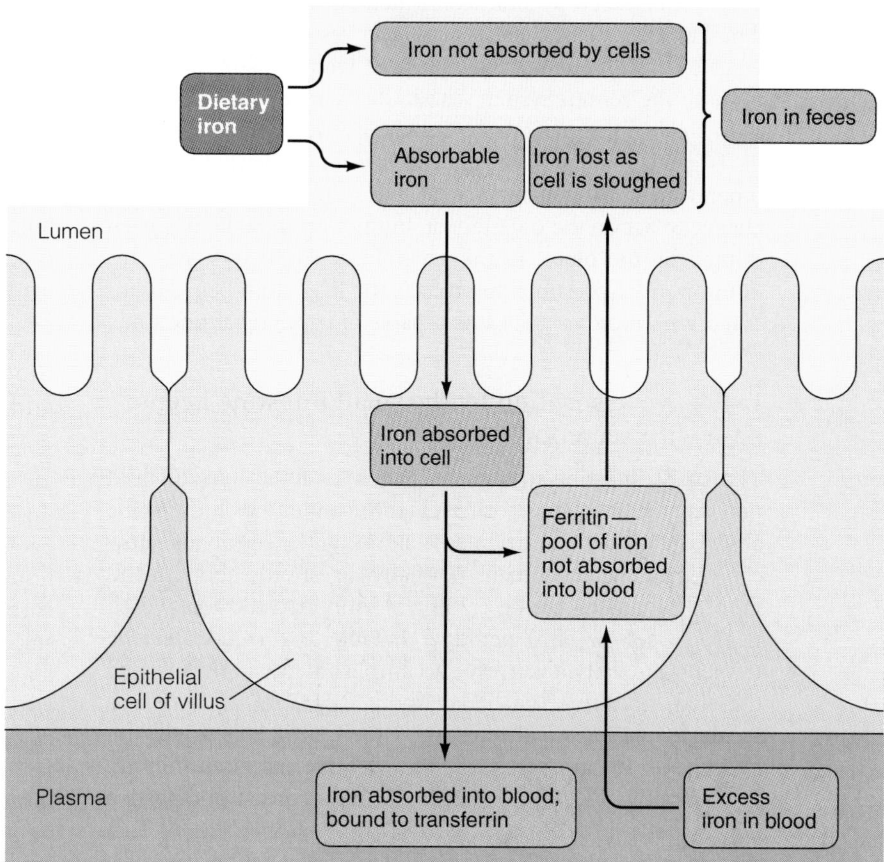

Figure 16-33 **Iron Absorption** Not all ingested iron is in a form that can be absorbed. Dietary iron that is absorbed into the small intestine epithelial cells and is immediately needed for red blood cell production is transferred into the blood and carried to the bone marrow bound to transferrin, a plasma protein carrier. Absorbed dietary iron that is not immediately needed is stored in the epithelial cells as ferritin, which cannot be transferred into the blood. Excess iron in the blood can be dumped into the ferritin pool. This unused iron, along with dietary iron that was not absorbed, is lost in the feces as the ferritin-containing epithelial cells are sloughed.

The amount of calcium (Ca^{2+}) absorbed is also regulated. Absorption of Ca^{2+} is accomplished partly by passive diffusion but mostly by active transport. Vitamin D greatly stimulates this active transport. Vitamin D can exert this effect only after it has been activated in the liver and kidneys, a process that is enhanced by parathyroid hormone. Appropriately, secretion of parathyroid hormone increases in response to a fall in Ca^{2+} concentration in the blood. Normally, of the average 1,000 mg of Ca^{2+} taken in daily, only about two-thirds is absorbed in the small intestine, with the rest passing out in the feces.

Most absorbed nutrients immediately pass through the liver for processing.

The venules that leave the small intestine villi, along with those from the remainder of the digestive tract, empty into the portal vein, which carries the blood to the liver. Consequently, anything absorbed into the digestive capillaries first must pass through the hepatic biochemical factory before entering the general circulation. Thus, the products of carbohydrate and protein digestion as well as the electrolytes and H_2O are channeled into the liver, where many of these products are subjected to immediate metabolic processing. Furthermore, harmful substances that may have been absorbed are detoxified by the liver before gaining access to the general circulation. After passing through the portal circulation, the venous blood from the digestive system is emptied into the vena cava and returned to the heart to be distributed throughout the body, carrying glucose and amino acids for use by the tissues.

Fat, which cannot penetrate the intestinal capillaries, is picked by the central lacteal and enters the lymphatic system instead, thereby bypassing the hepatic portal system. Contractions of the villi, presumably accomplished by the muscularis mucosa, periodically compress the central lacteal and "milk" the lymph out of this vessel. Increased contractions of the villi are known to occur following a meal and are perhaps mediated by an unconfirmed hormone from the duodenal mucosa, **villikinin.** The lymph vessels eventually converge to form the *thoracic duct,* a large lymph vessel that empties into the venous system within the chest. By this means, fat ultimately gains access to the circulatory system. The absorbed fat is carried by the systemic circulation to the liver and to other tissues of the body. Therefore, the liver does have a chance to act on the digested fat, but not until the fat has been diluted by the blood in the general circulatory system. This dilution of fat presumably protects the liver from being inundated with more fat than it is capable of handling at one time.

Extensive absorption by the small intestine keeps pace with secretion.

The small intestine normally absorbs about 9 liters of fluid per day in the form of H_2O and solutes, including the absorbable units of nutrients, vitamins, and electrolytes. How can that be, when humans normally ingest only about 1,250 ml of fluid and consume 1,250 g of solid food (80% of which is H_2O—see p. 525) per day? Table 16-7 illustrates the tremendous daily absorptive accomplishments performed by the small intestine. Each day about 9,500 ml of H_2O and solutes enter the small intestine. Note that of this 9,500 ml, only 2,500 ml are ingested from the external environment. The remaining 7,000 ml (7 liters) of fluid consist of digestive juices that are essentially derived from plasma. Recall that plasma is the ultimate source of digestive secretions because the secretory cells extract the necessary raw materials for their secretory product from the plasma. Considering that the entire plasma volume is only about 2.75 liters, it is obvious that absorption must closely parallel secretion to prevent the plasma volume from falling sharply. Of the 9,500 ml of fluid entering the small intestine lumen per day, about 95%, or 9,000 ml of fluid, is normally absorbed by the small intestine back into the plasma, with only 500 ml of the small intestine contents passing on into the colon. Thus, the digestive juices are not lost from the body. After the constituents of the juices are secreted into the digestive tract lumen and perform their function, they are returned to the plasma. The only secretory product that escapes from the body is bilirubin, a waste product that must be eliminated.

Table 16-7
Volumes Absorbed by the Small and Large Intestine per Day

Volume entering the small intestine per day			
Sources	Ingested	Food Eaten	1,250 g*
		Fluid drunk	1,250 ml
	Secreted from the plasma	Saliva	1,500 ml
		Gastric juice	2,000 ml
		Pancreatic juice	1,500 ml
		Bile	500 ml
		Intestinal juice	1,500 ml
			9,500 ml
Volume absorbed by the small intestine per day			9,000 ml
Volume entering the colon from the small intestine per day			500 ml
Volume absorbed by the colon per day			350 ml
Volume of feces eliminated from the colon per day			150 g*

*One milliliter of H_2O weighs 1 g. Therefore, because a high percentage of food and feces is H_2O, we can roughly equate grams of food or feces with milliliters of fluid.

Biochemical balance among the stomach, pancreas, and small intestine is normally maintained.

Since the secreted juices are normally absorbed back into the plasma, the acid-base balance of the body is not altered by digestive processes. When secretion and absorption do not parallel each other, however, acid-base abnormalities can result. Figure 16-34 is a summary of the biochemical balance that normally exists among the stomach, pancreas, and small intestine. The arterial blood entering the stomach contains Cl^-, CO_2, H_2O, and Na^+, among other things. During HCl secretion, the gastric parietal cells extract Cl^-, CO_2, and H_2O from the plasma (the CO_2 and H_2O being essential for H^+ secretion) and add HCO_3^- to it (the HCO_3^- being formed in the process of generating H^+). The HCO_3^- diffuses into the plasma to replace the secreted Cl^- and to electrically balance the Na^+ in the plasma. Plasma Na^+ levels are not altered by gastric secretory processes. Because HCO_3^- is an alkaline ion, the venous blood leaving the stomach is more alkaline than the arterial blood delivered to it. The overall acid-base balance of the body is not altered, however, because the pancreatic duct cells extract a comparable amount of HCO_3^- (along with Na^+) from the plasma to neutralize the acidic gastric chyme as it is emptied into the small intestine. Within the intestinal lumen, the alkaline pancreatic $NaHCO_3$ secretion neutralizes the gastric HCl secretion, yielding NaCl and H_2CO_3. The latter molecules form Na^+ and Cl^- plus CO_2 and H_2O, respectively. All four of these constituents (Na^+, Cl^-, CO_2 and H_2O) are absorbed by the intestinal epithelium into the plasma. Note that these are exactly the same constituents present in the arterial blood entering the stomach. Thus, through these interactions, the body normally does not experience a net gain or loss of acid or base during digestion.

Diarrhea results in loss of fluid and electrolytes.

When vomiting or diarrhea occur, these normal neutralization processes cannot take place. We have already described vomiting in the section on gastric motility. The other common digestive tract disturbance that can lead to a loss of fluid and an acid-base imbalance is **diarrhea.** This condition is characterized by passage of a highly fluid fecal matter, often with increased frequency of defecation. Just as with vomiting, the effects of diarrhea can be either beneficial or harmful. Diarrhea is beneficial when rapid emptying of the intestine

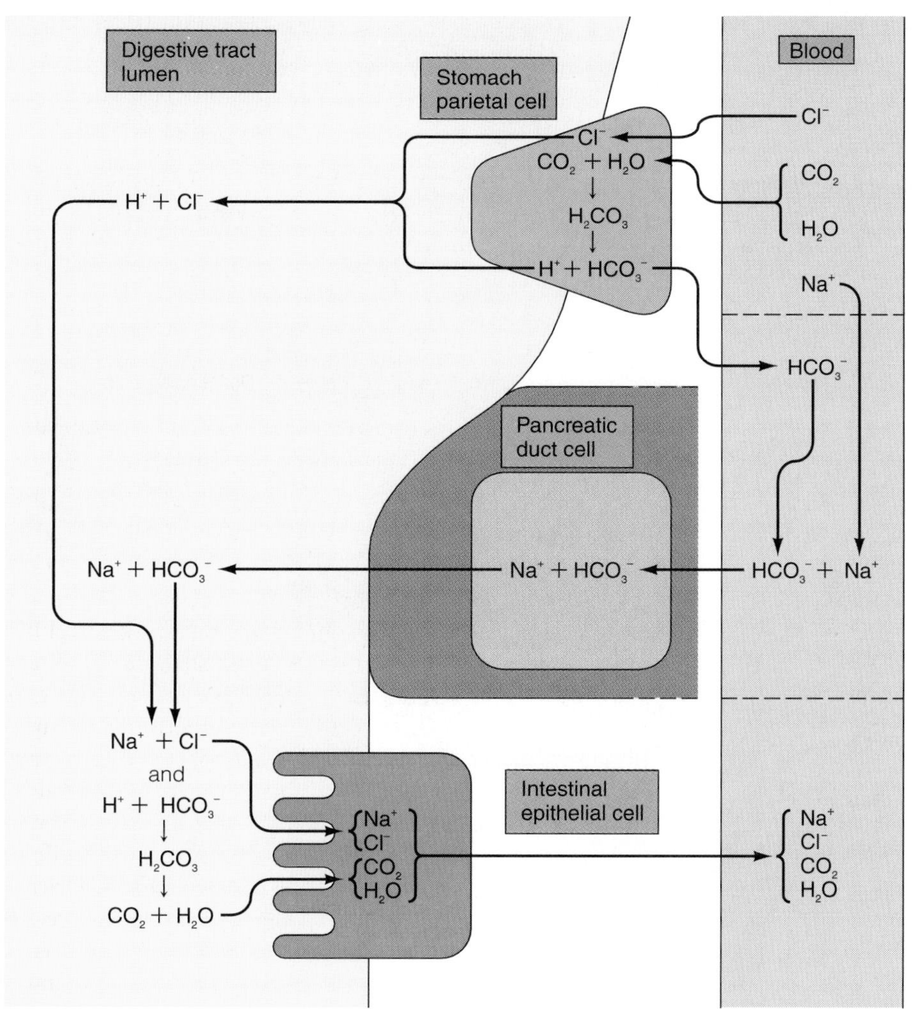

Figure 16-34 **Biochemical Balance among the Stomach, Pancreas, and Small Intestine** When digestion and absorption proceed normally, no net loss or gain of acid or base or other chemicals from the body fluids occurs as a consequence of digestive secretions. The parietal cells of the stomach extract Cl^-, CO_2, and H_2O from and add HCO_3^- to the blood during HCl secretion. The pancreatic duct cells extract the HCO_3^- as well as Na^+ from the blood during $NaHCO_3$ secretion. Within the small intestine lumen, pancreatic $NaHCO_3$ neutralizes gastric HCl to form NaCl and H_2CO_3, which decomposes into CO_2 and H_2O. Subsequently, the intestinal cells absorb Na^+, Cl^-, CO_2 and H_2O into the blood, thereby replacing the constituents that were extracted from the blood during gastric and pancreatic secretion.

hastens the elimination of harmful material from the body. However, not only are some of the ingested materials lost, but some of the secreted materials that normally would have been reabsorbed are lost as well. Excessive loss of intestinal contents causes dehydration, loss of nutrient material, and metabolic acidosis resulting from the loss of HCO_3^- (see p. 541). The abnormal fluidity of the feces in diarrhea usually occurs because the intestine is unable to absorb fluid as extensively as normal. This extra unabsorbed fluid passes out in the feces.

The following are the causes of diarrhea:

1. The most common cause of diarrhea is excessive intestinal motility, which arises either from local irritation of the gut wall caused by bacterial or viral infection of the intestine or from emotional stress. Rapid transit of the intestinal contents does not allow sufficient time for adequate absorption of fluid to occur.
2. Diarrhea also occurs when excess osmotically active particles, such as those found in lactase deficiency, are present in the digestive tract lumen. These particles cause excessive fluid to enter and be retained in the lumen, thus increasing the fluidity of the feces.
3. Toxins of the bacterium *Vibrio cholera* (the causative agent of cholera) and certain other microorganisms promote the secretion of excessive amounts of fluid by the small intestine mucosa, resulting in profuse diarrhea. Diarrhea produced in response to toxins from infectious agents is the leading cause of death of small children in developing nations. Fortunately, a low-cost, effective *oral rehydration therapy* that takes advantage of the intestine's glucose cotransport carrier is saving the lives of millions of children. (See the boxed feature on p. 594, ◆Concepts, Challenges, and Controversies.)

Large Intestine

The large intestine is primarily a drying and storage organ.

The **large intestine** consists of the colon, cecum, appendix, and rectum (Fig. 16-35). The **cecum** forms a blind-ended pouch below the junction of the small and large intestines at the ileocecal valve. The small, fingerlike projection at the bottom of the cecum is the **appendix,** a lymphoid tissue that houses lymphocytes (see p. 375). The **colon,** which makes

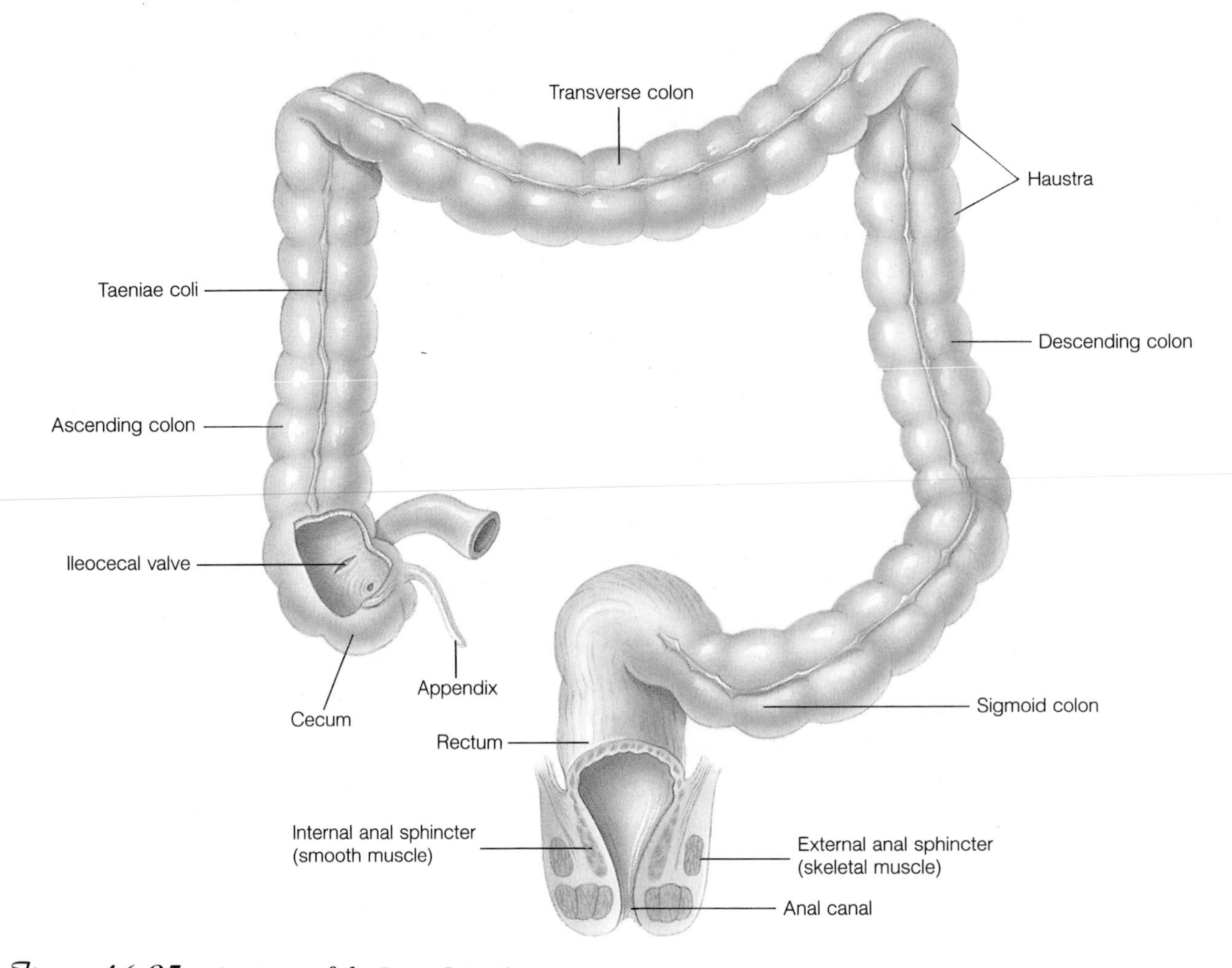

Figure 16-35 **Anatomy of the Large Intestine**

up most of the large intestine, is not coiled like the small intestine but consists of three relatively straight portions—the *ascending colon,* the *transverse colon,* and the *descending colon.* The terminal portion of the descending colon becomes S-shaped, forming the *sigmoid colon* (*sigmoid* means "S-shaped"), then straightens out to form the **rectum** (*rectum* means "straight").

The outer longitudinal smooth muscle layer does not completely surround the large intestine. Instead, it consists only of three separate, conspicuous, longitudinal bands of muscle, the **taeniae coli,** which run the length of the large intestine. These taeniae coli are shorter than the underlying circular smooth muscle and mucosal layers would be if the latter were stretched out flat. Because of this, the underlying layers are gathered into pouches or sacs called **haustra,** much as the material of a full skirt is gathered at the narrower waistband. The haustra are not merely passive permanent gathers, however; they actively change location as a result of contraction of the circular smooth muscle layer.

The colon normally receives about 500 ml of chyme from the small intestine each day. Since most digestion and absorption have been accomplished in the small intestine, the contents delivered to the colon consist of undigestible food residues (such as cellulose), unabsorbed biliary components, and the remaining fluid. The colon extracts more H_2O and salt from the contents. What remains to be eliminated is known as **feces.** The primary function of the large intestine is to store this fecal material before defecation. Cellulose and other indigestible substances in the diet provide bulk and help maintain regular bowel movements by contributing to the volume of the colonic contents.

Haustral contractions slowly shuffle the colonic contents back and forth while mass movements propel colonic contents long distances.

Most of the time, movements of the large intestine are slow and nonpropulsive, as is appropriate for its absorptive and storage functions. The colon's primary method of motility is **haustral contractions,** or **haustrations,** initiated by the autonomous rhythmicity of colonic smooth muscle cells. These contractions, which throw the large intestine into haustra, are similar to small intestine segmentations but occur much less frequently. Thirty minutes may elapse between haustral contractions, whereas segmentation contractions in the small intestine occur at rates of between 9 and 12 per minute. The location of the haustral sacs gradually changes as a relaxed segment that has formed a sac slowly contracts while a previously contracted area simultaneously relaxes to form a new sac. These movements are nonpropulsive; they slowly shuffle the contents in a back-and-forth mixing movement that exposes the colonic contents to the absorptive mucosa. Haustral contractions are largely controlled by locally mediated reflexes involving the intrinsic plexuses.

Because of this slow colonic movement, bacteria have time to grow and accumulate in the large intestine. In contrast, in the small intestine the contents are normally moved through too rapidly for bacterial growth to occur. Not all ingested bacteria are destroyed by salivary lysozyme and gastric HCl, so the surviving bacteria continue to thrive in the large intestine. Most colonic microorganisms are harmless in this location.

Three to four times a day, generally after meals, a marked increased in motility takes place during which large segments of the ascending and transverse colon contract simultaneously, driving the feces one-third to three-fourths of the length of the colon in a few seconds. These massive contractions, appropriately called **mass movements,** drive the colonic contents into the distal portion of the large intestine, where material is stored until defecation occurs.

When food enters the stomach, mass movements occur in the colon primarily by means of the **gastrocolic reflex,** which is mediated from the stomach to the colon by gastrin and by the extrinsic autonomic nerves. In many people, this reflex is most evident after the first meal of the day and is often followed by the urge to defecate. Thus, when a new meal enters the digestive tract, reflexes are initiated to move the existing contents farther along down the tract to make way for the incoming food. The gastroileal reflex moves the remaining small intestine contents into the large intestine, and the gastrocolic reflex pushes the colonic contents into the rectum, triggering the defecation reflex.

Feces are eliminated by the defecation reflex.

When mass movements of the colon move fecal material into the rectum, the resultant distention of the rectum stimulates stretch receptors in the rectal wall, thus initiating the **defecation reflex.** This reflex causes the **internal anal sphincter** (which is composed of smooth muscle) to relax and the rectum and sigmoid colon to contract more vigorously. If the **external anal sphincter** (which is composed of skeletal muscle) is also relaxed, defecation occurs. Being skeletal muscle, the external anal sphincter is under voluntary control. The initial distention of the rectal wall is accompanied by the conscious urge to defecate. If circumstances are unfavorable for defecation, voluntary tightening of the external anal sphincter can prevent defecation in spite of the defecation reflex. If defecation is delayed, the distended rectal wall gradually relaxes, and the urge to defecate subsides until the next mass movement propels more feces into the rectum, once again distending the rectum and triggering the defecation reflex. During periods of nonactivity, both anal sphincters remain contracted to ensure fecal continence.

When defecation does occur, it is usually assisted by voluntary straining movements that involve simultaneous contraction of the abdominal muscles and a forcible expiration against a closed glottis. This maneuver brings about a large increase in intra-abdominal pressure, which assists in elimination of the feces.

Constipation occurs when the feces become too dry.

If defecation is delayed too long, **constipation** may result. When the colonic contents are retained for longer periods of time than normal, more than the usual amount of H_2O is absorbed, causing the feces to become hard and dry. Normal variations in frequency of defecation among individuals range from after every meal to up to once a week. When the frequency is delayed beyond what is normal for a particular individual, constipation and its attendant symptoms may occur. These symptoms include abdominal discomfort, dull headache, loss of appetite sometimes accompanied by nausea, and mental depression. Contrary to popular belief, these symptoms are not caused by toxins absorbed from the retained fecal material. Although some potentially toxic substances are produced by bacterial metabolism in the colon, these substances normally pass through the portal system and are removed by the liver before they can reach the systemic circulation. Instead, the symptoms associated with constipation seem to be due to prolonged distention of the large intestine, particularly the rectum, because these sensations are promptly alleviated following relief from distention.

Possible causes for delayed defecation that might lead to constipation include (1) ignoring the urge to defecate; (2) decreased colon motility accompanying aging, emotion, or a low-bulk diet; (3) obstruction of fecal movement in the large bowel caused by a local tumor or colonic spasm; and (4) impairment of the defecation reflex, such as through injury of the nerve pathways involved.

If hardened fecal material becomes lodged in the appendix, it may obstruct normal circulation and mucus secretion in this narrow, blind-ended appendage. This blockage leads to inflammation of the appendix, or **appendicitis.** The appendix often becomes swollen and filled with pus, and the inflamed tissue may die as a result of local circulatory interference. If not surgically removed, the diseased appendix may rupture, spewing its infectious contents into the abdominal cavity.

Large intestine secretion is protective in nature.

The large intestine does not secrete any digestive enzymes. None are needed because digestion is completed before chyme ever reaches the colon. Colonic secretion consists of an alkaline (HCO_3^-) mucus solution, whose function is to protect the large intestine mucosa from mechanical and chemical injury. The mucus provides lubrication to facilitate passage of the feces, whereas the HCO_3^- neutralizes irritating acids produced by local bacterial fermentation. Secretion increases in response to mechanical and chemical stimulation of the

CONCEPTS, CHALLENGES, AND CONTROVERSIES

Oral Rehydration Therapy: Sipping a Simple Solution Saves Lives

Diarrhea-inducing microoganisms such as *Vibrio cholera,* which causes cholera (see p. 70), are the leading cause of death in children under five worldwide. The problem is especially pronounced in developing countries, refugee camps, and elsewhere where poor sanitary conditions encourage the spread of the microorganisms and medical supplies and health care personnel are scarce. Fortunately, a low-cost, easily obtainable, uncomplicated remedy—oral rehydration therapy—has been developed to combat potentially fatal diarrhea. This treatment exploits the secondary active transport carriers located at the luminal border of the villus epithelial cells.

Let us examine the pathophysiology of life-threatening diarrhea and then see how simple oral rehydration therapy can save lives. During digestion of a meal, the crypt cells of the small intestine normally secrete succus entericus, a salt and mucus solution, into the lumen. These cells actively transport Cl^- into the lumen, promoting the parallel passive transport of Na^+ and H_2O from the blood into the lumen. The fluid provides the watery environment needed for the enzymatic breakdown of the ingested nutrients into absorbable units. Glucose and amino acids, the absorbable units of dietary carbohydrates and proteins, respectively, are absorbed by secondary active transport. This absorption mechanism utilizes Na^+-glucose (or amino acid) cotransport carriers located at the basolateral membrane of the villus epithelial cells (see p. 62). In addition, separate active Na^+ carriers not linked with nutrient absorption transfer Na^+, passively accompanied by Cl^- and H_2O, from the lumen into the blood. The net result of these various carrier activities is absorption of the secreted salt and H_2O along with the digested nutrients. Normally, absorption of salt and H_2O exceeds their secretion so that not only are the secreted fluids salvaged but additional ingested salt and H_2O are absorbed as well.

Cholera and most diarrhea-inducing microbes cause diarrhea by stimulating the secretion of Cl^- and/or impairing the absorption of Na^+. As a result, more fluid is secreted from the blood into the lumen than is subsequently transferred back into the blood. The excess fluid is lost in the feces, producing the watery stool characteristic of diarrhea. More importantly, the loss of fluids and electrolytes that came from the blood leads to dehydration. The subsequent reduction in effective circulating plasma volume can cause death in a matter of days or even hours, depending on the severity of the fluid loss.

About fifty years ago, physicians learned that replacing the lost fluids and electrolytes intravenously saves the lives of most diarrhea victims. In many parts of the world, however, adequate facilities, equipment, and personnel are not available to administer intravenous rehydration therapy. Consequently, millions of children still succumbed to diarrhea annually.

In 1966, researchers learned that the Na^+-glucose cotransport carrier system is not affected by diarrhea-causing microbes. This discovery led to the development of **oral rehydration therapy (ORT).** When both Na^+ and glucose are present in the lumen, the cotransport carrier transports them both from the lumen into the villus epithelial cells, from which they enter the blood. Since H_2O osmotically follows the absorbed Na^+, ingestion of a glucose and salt solution promotes the uptake of fluid into the blood from the intestinal tract without the need for intravenous replacement of fluids.

colonic mucosa mediated by short reflexes and parasympathetic innervation.

No digestion takes place within the large intestine because there are no digestive enzymes. However, the colonic bacteria do digest some of the cellulose and use it for their own metabolism.

The large intestine absorbs salt and water, converting the luminal contents into feces.

Some absorption takes place within the colon but not to the same extent as in the small intestine. Because the luminal surface of the colon is fairly smooth, it has considerably less absorptive surface area than the small intestine. Furthermore, no specialized transport mechanisms are present in the colonic mucosa for absorption of glucose or amino acids, as there are in the small intestine. When excessive small intestine motility delivers the contents to the colon before absorption of nutrients has been completed, the colon is unable to absorb these materials, and they are lost in diarrhea.

The colon normally absorbs salt and H_2O. Sodium is actively absorbed, Cl^- follows passively down the electrical gradient, and H_2O follows osmotically. Bacteria in the colon synthesize some vitamins that the colon is capable of absorbing, but this is normally not a significant contribution, except in the case of vitamin K.

Through absorption of salt and H_2O, a firm fecal mass is formed. Of the 500 ml of material entering the colon per day from the small intestine, the colon normally absorbs about 350 ml, leaving 150 g of feces to be eliminated from the body each day (Table 16-7). This fecal material normally consists of 100 g of H_2O and 50 g of solid, including undigested cellulose, bilirubin, bacteria, and small amounts of salt. Thus, contrary to popular thinking, the digestive tract is not a major excretory passageway for eliminating wastes from the body. The main waste product excreted in the feces is bilirubin. The other fecal constituents are unabsorbed food residues and bacteria, which were never actually a part of the body.

Intestinal gases are absorbed or expelled.

Occasionally, instead of fecal material passing from the anus, intestinal gas, or **flatus**, passes out. This gas is derived primarily from two sources: (1) swallowed air—as much as 500

The first proof of ORT's life-saving ability in the field came in 1971 when several million refugees poured into India from war-ravaged Bangladesh. Of the thousands of refugees who fell victim to cholera and other diarrheal diseases, more than 30% died because of the scarcity of sterile fluids and needles for intravenous therapy. In one refugee camp, however, under the supervision of a group of scientists who had been experimenting with oral rehydration therapy, families were taught to administer ORT to diarrhea victims, most of whom were small children. The scarce intravenous solutions were reserved for those unable to drink. Death due to diarrhea was reduced to 3% in this camp, compared with a tenfold higher mortality among refugees elsewhere.

Based on this evidence, the World Health Organization (WHO) started aggressively promoting ORT. Packets of dry ingredients for ORT are now manufactured locally in more than sixty countries. The WHO estimates that about 30% of the world's children who contract diarrhea are treated with the prepackaged mixture or home-prepared versions. In the United States, commercially prepared oral solutions are widely available at pharmacies and supermarkets. An estimated 1 million children worldwide are saved annually as a result of ORT.

Since the inception of ORT, scientists have been capitalizing on other physiological principles to hasten fluid uptake from the digestive tract. For example, adding amino acids to the ORT formula further increases fluid absorption by exploiting the Na^+–amino acid cotransport carriers as well as the Na^+-glucose cotransport carriers. Furthermore, certain amino acids hasten replacement of intestinal cells damaged by diarrhea-inducing agents.

Another approach has been the substitution of starches and proteins for simple glucose. As these nutrient molecules are digested, they yield large numbers of glucose and amino acid molecules, which accelerate the uptake of nutrients, Na^+ and H_2O from the lumen. Simply adding more glucose or amino acids themselves to the solution would not speed up fluid absorption. The ORT formula should be isotonic. Making the solution hypertonic by adding more glucose or amino acids would be counterproductive: the extra osmotically active particles would actually cause H_2O to move from the blood into the hypertonic lumen. Recall that each large starch or protein molecule exerts the same osmotic effect as each smaller glucose or amino acid molecule (see p. 564). An isotonic solution of starch and protein molecules is gradually digested into several hundred times as many constituent glucose and amino acid molecules, respectively. Since each of these small absorbable units is transferred from the lumen by means of a cotransport carrier, ultimately dragging Na^+ and H_2O with it, an isotonic starch and protein solution promotes much greater fluid absorption than an isotonic glucose and amino acid solution does. Of course, net absorption occurs only if the rate of cotransport carrier uptake keeps pace with the rate at which the starch and protein are digested into absorbable units. Otherwise, the luminal contents would become hypertonic and promote counterproductive osmotic flux of H_2O from the blood into the lumen.

Researchers are experimenting with prepackaged ORT formulas using precooked foods such as rice powder instead of glucose. Furthermore, scientists have devised "recipes" for ORT that can be prepared at home using readily available foods that contain both starch and protein. An example is rice. If every family in the world knew how to prepare a food-based ORT, deaths from diarrhea would plummet to almost zero. Use of food-based ORT is limited for now, however, because of the challenge of educating families worldwide on the preparation and use of this simple, low-cost, life-saving technique.

ml of air may be swallowed during a meal— and (2) gas produced by bacterial fermentation in the colon. The presence of gas percolating through the luminal contents gives rise to gurgling sounds known as **borborygmi.** Eructation (burping) removes most of the swallowed air from the stomach, but some passes on into the intestine. Usually, very little gas is present in the small intestine, because the gas is either quickly absorbed or passes on into the colon. Most gas in the colon is due to bacterial activity, with the quantity and nature of the gas depending on the type of food eaten and the characteristics of the colonic bacteria. Some foods, such as beans, contain types of carbohydrates for which humans lack digestive enzymes. These fermentable carbohydrates enter the colon, where they are attacked by gas-producing bacteria. Much of the gas entering or forming in the large intestine is absorbed through the intestinal mucosa. The remainder is expelled through the anus.

To accomplish selective expulsion of gas when fecal material is also present in the rectum, the abdominal muscles and external anal sphincter are voluntarily contracted simultaneously. When contraction of the abdominal muscles raises the pressure sufficiently against the contracted anal sphincter, the pressure gradient forces air out at a high velocity through a slitlike anal opening that is too narrow for solid feces to escape. This passage of air at high velocity causes the edges of the anal opening to vibrate, giving rise to the characteristic low-pitched sound accompanying passage of gas.

Overview of the Gastrointestinal Hormones

Throughout our discussion of digestion, we have repeatedly mentioned different functions of the four accepted gastrointestinal hormones: gastrin, secretin, cholecystokinin, and gastric inhibitory peptide. We will now fit all of these functions together so that you can appreciate the overall adaptive importance of these interactions (Table 16-8).

Chyme in the stomach, especially if it contains protein, stimulates the release of gastrin, which acts on parietal and chief cells to increase secretion of HCl and pepsinogen. These two substances in turn are of primary importance in initiating digestion of the protein that promoted their release. Furthermore, gastrin enhances gastric motility, stimulates ileal motility, relaxes the ileocecal sphincter, and induces mass movements in the colon—functions that are all aimed at keeping the contents moving through the tract upon the arrival of a new meal. Gastrin also is trophic not only to the stomach mucosa but also the small intestine mucosa, helping maintain a well-developed, functionally viable digestive tract lining. Predictably, gastrin secretion is inhibited by an accumulation of acid in the stomach and by the presence in the duodenal lumen of acid and other constituents that necessitate a delay in gastric secretion.

As the stomach empties into the duodenum, the presence of acid in the duodenum stimulates the release of secretin into the blood. Secretin performs four major interrelated functions: (1) it inhibits gastric emptying to prevent further acid from entering the duodenum until the acid that is already present is neutralized; (2) it inhibits gastric secretion to reduce the amount of acid being produced; (3) it stimulates the pancreatic duct cells to produce a large volume of aqueous $NaHCO_3$ secretion, which is emptied into the duodenum to neutralize the acid; and (4) it stimulates secretion by the liver of a $NaHCO_3$-rich bile, which likewise is emptied into the duodenum to assist in the neutralization process. Neutralization of the acidic chyme in the duodenum helps prevent damage to the duodenal walls and provides a suitable environment for the optimal functioning of the pancreatic digestive enzymes, which are inhibited by acid.

As chyme is emptied from the stomach, fat and other nutrients enter the duodenum. These nutrients, especially fat and to a lesser extent protein products, cause the release of cholecystokinin (CCK) from the duodenal mucosa. This hormone also performs several important interrelated functions: (1) it inhibits gastric motility and secretion, thereby allowing adequate time for the nutrients already in the duodenum to be digested and absorbed; (2) it stimulates the pancreatic acinar cells to increase secretion of pancreatic enzymes, which continue the digestion of these nutrients in the duodenum (this action is especially important for fat digestion, because

Table 16-8 **Source, Control, and Functions of the Major Gastrointestinal Hormones**

Hormone	*Source*	*Primary Stimulus for Secretion*	*Functions*
Gastrin	G cells of the stomach's pyloric gland area	Protein in the stomach	Stimulates secretion by the parietal and chief cells Enhances gastric motility Stimulates ileal motility Relaxes the ileocecal sphincter Induces colonic mass movements Is trophic to the stomach and small intestine mucosa
Secretin	Endocrine cells in the duodenal mucosa	Acid in the duodenal lumen	Inhibits gastric emptying Inhibits gastric secretion Stimulates aqueous $NaHCO_3$ secretion by the pancreatic duct cells Stimulates secretion of $NaHCO_3$-rich bile by the liver Is trophic to the exocrine pancreas
Cholecystokinin	Endocrine cells in the duodenal mucosa	Nutrients in the duodenal lumen, especially fat products and to a lesser extent protein products	Inhibits gastric emptying Inhibits gastric secretion Stimulates digestive enzyme secretion by the pancreatic acinar cells Causes gallbladder contraction Causes relaxation of the sphincter of Oddi Is trophic to the exocrine pancreas May cause long-term adaptive changes in the proportion of pancreatic enzymes Contributes to satiety
Gastric inhibitory peptide	Endocrine cells in the duodenal mucosa	Fat, acid, hypertonicity, glucose, and distention in the duodenum	Inhibits gastric emptying Inhibits gastric secretion Stimulates insulin secretion by the endocrine pancreas

pancreatic lipase is the only enzyme that digests fat); and (3) it causes contraction of the gallbladder and relaxation of the sphincter of Oddi so that bile is emptied into the duodenum to aid fat digestion and absorption. Bile salts' detergent action is particularly important in enabling pancreatic lipase to perform its digestive task. Once again, the multiple effects of CCK are remarkably well adapted to dealing with the fat and other nutrients whose presence in the duodenum triggered this hormone's release.

Furthermore, it is appropriate that both secretin and CCK, which have profound stimulatory effects on the exocrine pancreas, are trophic to this tissue. Recall that CCK has also been implicated in long-term adaptive changes in the proportion of pancreatic enzymes produced in response to prolonged changes in diet.

Besides facilitating the digestion of ingested nutrients, CCK is an important regulator of food intake. It plays a key role in satiety, the sensation of having had enough to eat (see p. 606).

Gastric inhibitory peptide (GIP) is the most recently recognized gastrointestinal hormone. We have already mentioned its role in inhibiting gastric motility and secretion in response to fat, acid, hypertonicity and distention in the duodenum. In addition, it has been shown to stimulate insulin release by the pancreas, so it is also called **glucose-dependent insulinotrophic peptide** (once again GIP). Again, this is remarkably adaptive. As soon as the meal is absorbed, the body has to shift its metabolic gears to use and store the newly arriving nutrients. The metabolic activities of this absorptive phase are largely under the control of insulin (see p. 679). Stimulated by the presence of a meal in the digestive tract, GIP initiates the release of insulin in anticipation of the absorption of the meal in a "feedforward" fashion. Insulin is especially important in promoting the uptake and storage of glucose. Appropriately, glucose in the duodenum has recently been shown to increase GIP secretion.

This overview of the multiple, integrated, adaptive functions of the gastrointestinal hormones provides an excellent example of the remarkable efficiency of the human body. The list of hormones that regulate digestive function will undoubtedly grow, because a number of peptides found in the digestive tract are suspected of being hormones. Of these candidate hormones, motilin and villikinin have already been mentioned.

Chapter in Perspective: Focus on Homeostasis

To maintain constancy in the internal environment, materials that are used up in the body (such as nutrients and O_2) or uncontrollably lost from the body (such as evaporative H_2O loss from the airways or salt loss in sweat) must constantly be replaced by new supplies of these materials from the external environment. All of these replacement supplies except O_2 are acquired through the digestive system. Fresh supplies of O_2 are transferred to the internal environment by the respiratory system, but all of the nutrients, H_2O, and various electrolytes needed to maintain homeostasis are acquired through the digestive system. The large, complex food that is ingested is broken down by the digestive system into small absorbable units. These small energy-rich nutrient molecules are transferred across the small intestine epithelium into the blood for delivery to the cells to replace the nutrients constantly used for ATP production and for repair and growth of body tissues. Likewise, ingested H_2O, salt, and other electrolytes are absorbed by the intestine into the blood.

Unlike most body systems, regulation of digestive system activities is not aimed at maintaining homeostasis. The quantity of nutrients and H_2O ingested is subject to control, but the quantity of ingested materials absorbed by the digestive tract is not subject to control, with few exceptions. The hunger mechanism governs food intake to help maintain energy balance (chapter 17), and the thirst mechanism controls H_2O intake to help maintain H_2O balance (chapter 15). However, we often do not heed these control mechanisms and eat and drink even when we are not hungry or thirsty. Once these materials are in the digestive tract, the digestive system does not vary its rate of nutrient, H_2O, or electrolyte uptake according to body needs (with the exception of iron and calcium); rather, it optimizes conditions for digesting and absorbing what is ingested. Truly, what you eat is what you get. The digestive system is subject to many regulatory processes, but these are not influenced by the nutritional or hydration state of the body. Instead, these control mechanisms are governed by the composition and volume of digestive tract contents so that the rate of motility and secretion of digestive juices are optimal for digestion and absorption of the ingested food.

If excess nutrients are ingested and absorbed, the extra is placed in storage, such as in adipose tissue (fat), so that the blood level of nutrient molecules is kept at a constant level. Excess ingested H_2O and electrolytes are eliminated in the urine to homeostatically maintain the blood levels of these constituents.

Chapter Summary

Introduction

The four basic digestive processes are motility, secretion, digestion, and absorption. Digestive activities are carefully regulated by synergistic autonomous, neural (both intrinsic and extrinsic), and hormonal mechanisms to assure that the ingested food is maximally made available to the body for energy production and as synthetic raw materials. The digestive tract consists of a continuous tube that runs from the mouth to the anus, with local modifications that reflect regional specializations for carrying out digestive functions. The lumen of the digestive tract is continuous with the external environment, so its contents are technically outside the body; this arrangement permits digestion of food without self-digestion occurring in the process.

Mouth, Pharynx, and Esophagus

Food enters the digestive system through the mouth, where it is chewed and mixed with saliva to facilitate swallowing. The salivary enzyme, amylase, begins the digestion of polysaccharides, a process that continues in the stomach after the food has been swallowed until amylase is eventually inactivated by the acidic gastric juice. More important than its minor digestive function, saliva is essential for articulate speech and plays an important role in dental health. Salivary secretion is controlled by a salivary center in the medulla, mediated by autonomic innervation of the salivary glands.

Following chewing, the tongue propels the bolus of food to the rear of the throat, which initiates the swallowing reflex. The swallowing center in the medulla coordinates a complex group of activities that result in closure of the respiratory passages and propulsion of food through the pharynx and esophagus into the stomach.

The esophageal secretion, mucus, is protective in nature. No nutrient absorption occurs in the mouth, pharynx, or esophagus.

Stomach

The stomach, a saclike structure located between the esophagus and small intestine, stores ingested food for variable periods of time until the small intestine is ready to process it further for final absorption. The four aspects encompassing gastric motility are gastric filling, storage, mixing, and emptying. Gastric filling is facilitated by vagally mediated receptive relaxation of the stomach musculature. Gastric storage takes place in the body of the stomach, where peristaltic contractions of the thin muscular walls are too weak to mix the contents. Gastric mixing takes place in the thick-muscled antrum as a result of vigorous peristaltic contractions. Gastric emptying is influenced by factors in both the stomach and duodenum. The volume and fluidity of chyme in the stomach tend to promote emptying of the stomach contents. The duodenal factors, which are the dominant factors controlling gastric emptying, tend to delay gastric emptying until the duodenum is ready to receive and process more chyme. The specific factors in the duodenum that delay gastric emptying by inhibiting stomach peristaltic activity are fat, acid, hypertonicity, and distention.

Carbohydrate digestion continues in the body of the stomach under the influence of the swallowed salivary amylase. Protein digestion is initiated in the antrum of the stomach, where vigorous peristaltic contractions mix the food with gastric secretions, converting it to a thick liquid mixture known as chyme. Gastric secretions into the stomach lumen include (1) HCl, which activates pepsinogen, denatures protein, and kills bacteria; (2) pepsinogen, which, once activated, initiates protein digestion; (3) mucus, which provides a protective coating to supplement the gastric mucosal barrier, enabling the stomach to contain the harsh luminal contents without self-digestion; and (4) intrinsic factor, which plays a vital role in vitamin B_{12} absorption, a constituent essential for normal red blood cell production. The stomach also secretes into the blood the hormone gastrin, which plays a dominant role in regulating gastric secretion. Histamine, a potent gastric stimulant that is not normally secreted, is released into the stomach lumen with devastating effects during ulcer formation.

Both gastric motility and gastric secretion are under complex control mechanisms, involving not only gastrin but also vagal and intrinsic nerve responses and enterogastrone hormones (secretin, cholecystokinin, and gastric inhibitory peptide) secreted from the small intestine mucosa. Regulation of the stomach is aimed at balancing the rate of gastric activity with the ability of the small intestine to handle the arrival of acidic, fat-laden contents from the stomach.

No nutrients are absorbed from the stomach.

Pancreatic and Biliary Secretions

Pancreatic exocrine secretions and bile from the liver both enter the duodenal lumen. Pancreatic secretions include (1) potent digestive enzymes from the acinar cells, which digest all three categories of foodstuff, and (2) an aqueous $NaHCO_3$ solution from the duct cells, which neutralizes the acidic contents emptied into the duodenum from the stomach. This neutralization is important to protect the duodenum from acid injury and to allow the pancreatic enzymes, which are inactivated by acid, to perform their important digestive functions. Pancreatic secretion is primarily under hormonal control, which matches the composition of the pancreatic juice with the needs in the duodenal lumen.

The liver, the body's largest and most important metabolic organ, performs many varied functions. Its contribution to digestion is the secretion of bile, which contains bile salts. Bile salts aid fat digestion through their detergent action and facilitate fat absorption through formation of water-soluble micelles that can carry the products of fat digestion to their absorptive site. Between meals, bile is stored and concentrated in the gallbladder, which is hormonally stimulated to contract and empty the bile into the duodenum during digestion of a meal. After participating in fat digestion and absorption, bile salts are reabsorbed and returned via the hepatic portal system to the liver, where they not only are resecreted but act as a potent choleretic to stimulate the secretion of even more bile. Bile also contains bilirubin, a derivative of degraded hemoglobin, which is the major excretory product in the feces.

Small Intestine

The small intestine is the main site for digestion and absorption. Segmentation, its primary motility, thoroughly mixes the food with pancreatic, biliary, and small intestinal juices to facilitate digestion; it also exposes the products of digestion to the absorptive surfaces. Between meals, the migrating motility complex sweeps the lumen clean.

The juice secreted by the small intestine does not contain any digestive enzymes. The enzymes synthesized by the small intestine act intracellularly within the brush border membranes of the epithelial cells. These enzymes complete the digestion of carbohydrates and protein before these nutrients enter the blood. The energy-dependent process of Na^+ absorption provides the driving force for Cl^-, water, glucose, and amino acid absorption. Fat digestion is accomplished entirely in the lumen of the small intestine by pancreatic lipase. Because they are not soluble in water, the products of fat digestion must undergo a series of transformations that enable them to be passively absorbed, eventually entering the lymph. The small intestine absorbs almost everything presented to it, from ingested food to digestive secretions to sloughed epithelial cells. Only a small amount of fluid and nondigestible food residue passes on to the large intestine.

The small intestine lining is remarkably adapted to its digestive and absorptive function. It is thrown into folds that bear a rich array of fingerlike projections, the villi, which are furnished with a multitude of even smaller hairlike protrusions, the microvilli. Altogether, these surface modifications tremendously increase the area available to house the membrane-bound enzymes and to accomplish both active and passive absorption. This impressive lining is replaced approximately every three days to ensure an optimally healthy and functional presence of epithelial cells in spite of harsh luminal conditions.

Large Intestine

The colon serves primarily to concentrate and store undigested food residues and biliary waste products until they can be eliminated from the body as feces. No secretion of digestive enzymes or absorption of nutrients takes place in the colon, all nutrient digestion and absorption having been completed in the small intestine. Haustral contractions slowly shuffle the colonic contents back and forth to accomplish absorption of most of the remaining fluid and electrolytes. Mass movements occur several times a day, usually following meals, propelling the feces long distances. Movement of feces into the rectum triggers the defecation reflex, which can be voluntarily prevented by contraction of the external anal sphincter if the time is inopportune for elimination. The alkaline mucus secretion of the large intestine is primarily protective in nature.

Review Exercises

Objective Questions (Answers on p. E–16.)

1. The extent of nutrient uptake from the digestive tract depends on the body's needs. (True or false?)
2. The stomach is relaxed during vomiting. (True or false?)
3. Acid cannot normally penetrate into or between the cells lining the stomach, which enables the stomach to contain acid without injuring itself. (True or false?)
4. Protein is continually lost from the body through digestive secretions and sloughed epithelial cells, which pass out in the feces. (True or false?)
5. Foodstuffs not absorbed by the small intestine are absorbed by the large intestine. (True or false?)
6. The endocrine pancreas secretes secretin and CCK. (True or false?)
7. A digestive reflex involving the autonomic nerves is known as a ____________ reflex, whereas a reflex in which all elements of the reflex arc are located within the gut wall is known as a ____________ reflex.
8. When food is broken down and mixed with gastric secretions, the resultant thick liquid mixture is known as ____________.
9. The entire lining of the small intestine is replaced approximately every ____________ days.
10. The two substances absorbed by specialized transport mechanisms located only in the terminal ileum are ____________ and ____________.
11. The most potent choleretic is ____________.
12. Which of the following is *not* a function of saliva?
    a. begins digestion of carbohydrate
    b. facilitates absorption of glucose across the oral mucosa
    c. facilitates speech
    d. exerts an antibacterial effect
    e. plays an important role in oral hygiene
13. Match the following:
    ____1. prevents reentry of food into the mouth during swallowing
    ____2. triggers the swallowing reflex
    ____3. seals off the nasal passages during swallowing
    ____4. prevents air from entering the esophagus during breathing
    ____5. closes off the respiratory airways during swallowing
    ____6. prevents gastric contents from backing up into the esophagus

    (a) closure of the pharyngoesophageal sphincter
    (b) elevation of the uvula
    (c) position of the tongue against the hard palate
    (d) closure of the gastroesophageal sphincter
    (e) bolus pushed to the rear of the mouth by the tongue
    (f) tight apposition of the vocal folds

14. Use the answer code below to identify the characteristics of the following substances:

(a) pepsin
(b) mucus
(c) HCl
(d) intrinsic factor
(e) histamine

___ 1. activates pepsinogen
___ 2. inhibits amylase
___ 3. essential for vitamin B_{12} absorption
___ 4. can act autocatalytically
___ 5. not normally secreted but a potent stimulant for acid secretion
___ 6. breaks down connective tissue and muscle fibers
___ 7. begins protein digestion
___ 8. serves as a lubricant
___ 9. kills ingested bacteria
___10. is alkaline
___11. deficient in pernicious anemia
___12. coats the gastric mucosa

Essay Questions

1. Describe the four basic digestive processes.
2. List the three categories of energy-rich foodstuffs and the absorbable units of each.
3. List the components of the digestive system. Describe the cross-sectional anatomy of the digestive tract.
4. What four general factors are involved in the regulation of digestive system function? What is the role of each?
5. Describe the types of motility in each component of the digestive tract. What factors control each type of motility?
6. State the composition of the digestive juice secreted by each component of the digestive system. Describe the factors that control each digestive secretion.
7. List the enzymes involved in the digestion of each category of foodstuff. Indicate the source and control of secretion of each of the enzymes.
8. Why are some digestive enzymes secreted in inactive form? How are they activated?
9. What absorption processes take place within each component of the digestive tract? What special adaptations of the small intestine enhance its absorptive capacity?
10. Describe the absorptive mechanisms for salt, water, carbohydrate, protein, and fat.
11. What are the contributions of the accessory digestive organs? What are the nondigestive functions of the liver?
12. Summarize the functions of each of the gastrointestinal hormones.
13. What waste product is excreted in the feces?
14. How is vomiting accomplished? What are the causes and consequences of vomiting, diarrhea, and constipation?
15. Describe the process of mucosal turnover in the small intestine.

Quantitative Exercises (Solutions on p. E–17.)

1. Suppose a lipid droplet in the gut is essentially a sphere with a diameter of 1 cm.
   a. What is the surface area to volume ratio of the droplet? *Hint:* The area of a sphere is $4\pi r^2$, and the volume is $4/3\pi r^3$.
   b. Now, suppose that this sphere were emulsified into 100 essentially equal-sized droplets. What is the average surface area to volume ratio of each droplet?
   c. How much greater is the total surface area of these 100 droplets compared to the original single droplet?
   d. How much did the total volume change as a result of emulsification?

Points to Ponder

(Explanations on p. E–17.)

1. Why do patients who have had a large portion of their stomachs removed for treatment of stomach cancer or severe peptic ulcer disease have to eat small quantities of food frequently instead of consuming three meals a day?
2. The number of immune cells in the gut-associated lymphoid tissue (GALT) is estimated to be equal to the total number of these defense cells in the rest of the body. Speculate on the adaptive significance of this extensive defense capability of the digestive system.
3. By what means would defecation be accomplished in a patient paralyzed from the waist down because of a lower spinal cord injury?
4. After bilirubin is extracted from the blood by the liver, it is conjugated (combined) with glycuronic acid by the enzyme glucuronyl transferase within the liver. Only when conjugated can bilirubin be actively excreted into the bile. For the first few days of life, the liver does not make adequate quantities of glucuronyl transferase. Explain how this transient enzyme deficiency leads to the common condition of jaundice in newborns.
5. Explain why removal of either the stomach or the terminal ileum leads to pernicious anemia.
6. ***Clinical Consideration*** Thomas W. experiences a sharp pain in his upper right abdomen after eating a high-fat meal. Also, he has noted that his feces are grayish white instead of brown. What is the most likely cause of his symptoms? Explain why each of these symptoms occurs with this condition.

Energy Balance and Temperature Regulation

17

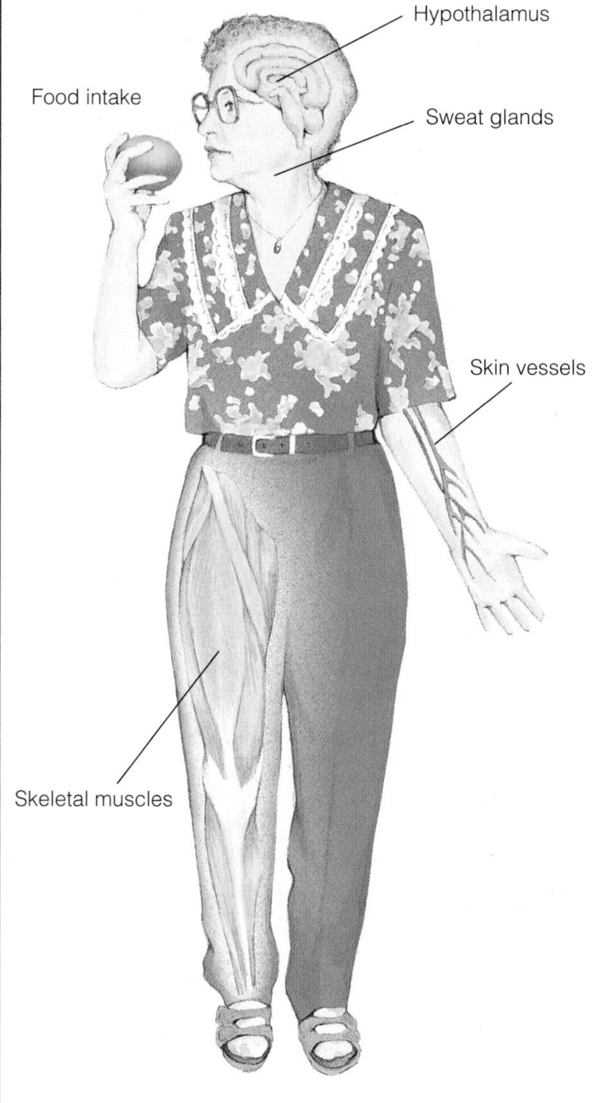

Body systems maintain homeostasis

HOMEOSTASIS

Among the factors that are homeostatically maintained are the availability of energy-rich nutrients to the cells and the temperature of the internal environment. The hypothalamus helps regulate food intake, which is of primary importance in energy balance. The hypothalamus also helps maintain body temperature. It can vary heat production by the skeletal muscles and can adjust heat loss from the skin surface by varying the amount of warmed blood flowing through the skin vessels and by controlling sweat production.

Homeostasis is essential for survival of cells

CELLS

Each cell needs energy to perform the functions essential for its own survival and to carry out its specialized contribution toward maintaining homeostasis. All energy used by cells is ultimately provided by food intake, which is regulated to maintain energy balance. Body temperature must be maintained at a fairly constant level to prevent unwanted changes in the rate of chemical reactions within cells and to prevent damage to cell proteins from heat.

Cells make up body systems

Food intake is essential to provide energy to power life-sustaining cell activities. The energy (caloric) value of ingested food must equal the body's total energy needs for body weight to remain constant. **Energy balance** and thus body weight are maintained by controlling food intake.

Energy expenditure by the body generates heat, which is important in **temperature regulation.** Humans are usually in external environments cooler than their bodies, so they must constantly generate heat internally to maintain their body temperatures. Also, they must have mechanisms to cool the body if it gains too much heat from heat-generating skeletal muscle activity or from a hot external environment. Body temperature must be regulated because the rate of cellular chemical reactions is temperature dependent and overheating damages cell proteins.

The hypothalamus is the major integrating center for maintenance of both energy balance and body temperature.

Chapter Contents At a Glance

Energy Balance

Most food energy is ultimately converted into heat in the body.

Each cell in the body needs energy to perform the functions essential for the cell's own survival (such as active transport and cellular repair) and to carry out its specialized contributions toward maintenance of homeostatic balance (such as gland secretion and muscle contraction). All energy used by cells is ultimately provided by food intake. Chemical energy locked in the bonds that hold the atoms together in nutrient molecules is released when these molecules are broken down in the body. Energy harvested from biochemical processing of ingested nutrients either is used immediately to perform biological work or is stored in the body for later use as needed during periods when food is not being digested and absorbed.

According to the *first law of thermodynamics,* energy can be neither created nor destroyed. Therefore, energy is subject to the same kind of input-output balance as are the chemical components of the body such as H_2O and salt (see p. 516).

The energy in ingested foodstuffs constitutes energy input to the body. Energy output or expenditure falls into two categories (Fig. 17-1): external work and internal work. **External work** refers to the energy expended when skeletal muscles are contracted to move external objects or to move the body in relation to the environment. **Internal work** constitutes all other forms of biological energy expenditure that do not accomplish mechanical work outside the body. Internal work encompasses two types of energy-dependent activities: (1) skeletal muscle activity used for purposes other than external work, such as the contractions associated with postural maintenance and shivering, and (2) all the energy-expending activities that must go on all the time just to sustain life. The latter include the work of pumping blood and breathing; the energy required for active transport of critical materials across plasma membranes; and the energy used during synthetic reactions essential for the maintenance, repair, and growth of cellular structures—in short, the "metabolic cost of living."

Not all energy in nutrient molecules can be harnessed to perform biological work. Energy cannot be created or destroyed, but it can be converted from one form to another. The energy in nutrient molecules that is not used to energize work is transformed into **thermal energy,** or **heat.** During biochemical processing, only about 50% of the energy in nutrient molecules is transferred to ATP; the other 50% of nutrient energy is immediately lost as heat. During ATP expenditure by the cells, another 25% of the energy derived from ingested food becomes heat. Since the body is not a heat engine, it cannot convert heat into work. Therefore, not more than 25% of nutrient energy is available to accomplish work, either external or internal. The remaining 75% is lost as heat during the sequential transfer of energy from nutrient molecules to ATP to cellular systems.

Furthermore, of the energy actually captured for use by the body, almost all expended energy eventually becomes heat. To exemplify, energy expended by the heart to pump blood is gradually changed into heat by friction as blood flows through the vessels. Likewise, energy used in the synthesis of cellular structural protein eventually appears as heat when that protein is degraded during the normal course of turnover of bodily constituents. Even in the performance of external work, skeletal muscles convert chemical energy into mechanical energy inefficiently, with as much as 75% of the expended energy being lost as heat. Thus, all energy that is liberated from ingested food but not directly used for movement of external objects or stored in fat (adipose tissue) deposits (or, in the case of growth, as protein) eventually becomes body heat. This heat is not entirely wasted energy, however, because much of it is used to maintain body temperature.

The metabolic rate is the rate of energy use.

The rate at which energy is expended by the body during both external and internal work is known as the **metabolic rate:**

$$\text{metabolic rate} = \text{energy expenditure/unit of time}$$

Since most of the body's energy expenditure eventually appears as heat, the metabolic rate is normally expressed in terms of the rate of heat production in kilocalories per hour. The basic unit of heat energy is the **calorie,** which is the amount of heat required to raise the temperature of 1 g of H_2O 1°C. This unit is too small to be convenient when discussing the human body because of the magnitude of heat involved, so the **kilocalorie** or **Calorie,** which is equivalent to 1,000 calories, is used. When nutritionists speak of "calories" in quantifying the energy content of various foods, they are actually referring to kilocalories or Calories. Four kilocalories of heat energy are released when 1 g of glucose is oxidized or "burned," whether the oxidation takes place inside or outside the body.

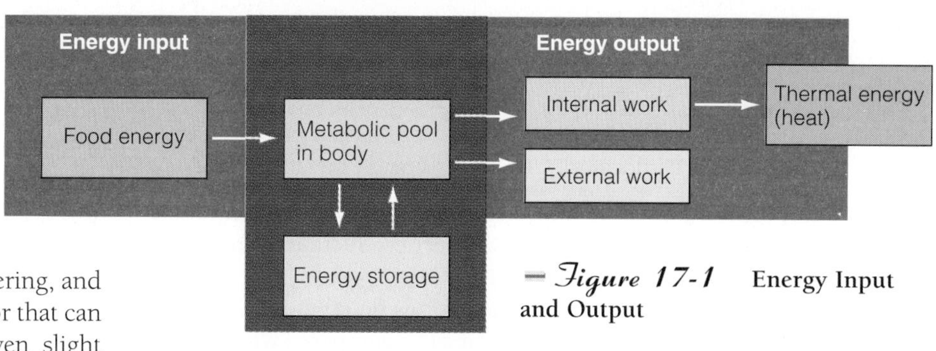

Figure 17-1 **Energy Input and Output**

The metabolic rate and consequently the amount of heat produced vary depending on a variety of factors, such as exercise, food intake, shivering, and anxiety. Increased skeletal muscle activity is the factor that can increase metabolic rate to the greatest extent. Even slight increases in muscle tone notably elevate the metabolic rate, and various levels of physical activity alter energy expenditure and heat production markedly (■ Table 17-1). For this reason, a person's metabolic rate is determined under standardized basal conditions established to control as many as possible of the variables that can alter metabolic rate. In this way, the metabolic activity necessary to maintain the basic body functions at rest can be determined. Thus, the so-called **basal metabolic rate (BMR)** is a reflection of the body's "idling speed," or the minimal waking rate of internal energy expenditure. The BMR is measured under the following specified conditions:

1. The person should be at physical rest, having refrained from exercise for at least 30 minutes to eliminate any contribution of muscular exertion to heat production.
2. The individual should be at mental rest to minimize skeletal muscle tone (people "tense up" when they are nervous) and to prevent a rise in epinephrine secretion. Stress elicits the secretion of the hormone epinephrine, which increases the metabolic rate.
3. The measurement should be performed at a comfortable room temperature so that the individual does not shiver. Shivering can markedly increase heat production because the purpose of these reflex oscillating skeletal muscle contractions is to generate heat in response to cold exposure.
4. The subject should not have eaten any food within 12 hours before the BMR determination to avoid **diet-induced thermogenesis** (*thermo* means "heat"; *genesis* means "production"), or the obligatory increase in metabolic rate that occurs as a consequence of food intake. This short-lived (less than 12-hour) rise in metabolic rate is due not to digestive activities but to the increased metabolic activity associated with the processing and storage of ingested nutrients, especially by the major biochemical factory, the liver.

The rate of heat production in BMR determinations can be measured directly or indirectly. **Direct calorimetry** involves the cumbersome procedure of placing the subject in an insulated chamber with H_2O circulating through the walls. The difference in the temperature of the H_2O entering and leaving the chamber reflects the amount of heat liberated by the subject and picked up by the H_2O as it passes through the chamber. Even though this method provides a direct measurement of heat production, it is not practical because a calorimeter chamber is costly and takes up a lot of space. Therefore, a more practical method of indirectly determining the rate of heat production was developed for widespread use. With **indirect calorimetry**, only the subject's O_2 uptake per unit of time is measured, which is a simple task using minimal equipment. Recall that:

$$\text{food} + \mathbf{O_2} \rightarrow CO_2 + H_2O + \text{energy (mostly transformed into } \mathbf{heat}\text{)}$$

Accordingly, a direct relationship exists between the volume of O_2 used and the quantity of heat produced. This relationship also depends on the type of food being oxidized. Although carbohydrates, proteins, and fats require different amounts of O_2 for their oxidation and yield different amounts of kilocalories when oxidized, an average estimate can be made of the quantity of heat produced per liter of O_2 consumed on a typical mixed American diet. This approximate value, known as the **energy equivalent of O_2**, is 4.8 kilocalories of energy liberated per liter of O_2 consumed. Using

Table 17-1
Rate of Energy Expenditure for a 70-kg Person during Different Types of Activity

Form of Activity	*Energy Expenditure (Kcal/Hour)*
Sleeping	65
Awake, lying still	77
Sitting at rest	100
Standing relaxed	105
Getting dressed	118
Typewriting	140
Walking slowly on level (2.6 mi/hr)	200
Carpentry, painting a house	240
Sexual intercourse	280
Bicycling on level (5.5 mi/hr)	304
Shoveling snow, sawing wood	480
Swimming	500
Jogging (5.3 mi/hr)	570
Rowing (20 strokes/min)	828
Walking up stairs	1,100

this method, the metabolic rate of a person consuming 15 liters/hr of O_2 can be estimated as follows:

15	liters/hr	= O_2 consumption
×4.8	kilocalories/liter	= energy equivalent of O_2
72	kilocalories/hr	= estimated metabolic rate

In this way, a simple measurement of O_2 consumption can be used to reasonably approximate heat production in the determination of metabolic rate.

Once the rate of heat production is determined under the prescribed basal conditions, it must be compared with normal values for persons of the same sex, age, height, and weight because these factors all affect the basal rate of energy expenditure. For example, a large man actually has a higher rate of heat production than a smaller man, but when expressed in terms of total surface area (which is a reflection of height and weight), the output in kilocalories per hour per square meter of surface area is normally about the same.

Thyroid hormone is the primary but not sole determinant of the rate of basal metabolism. As the level of active circulating thyroid hormone increases, the BMR increases correspondingly.

Surprisingly, the BMR is not the body's lowest metabolic rate. The rate of energy expenditure during sleep is 10% to 15% lower than the BMR, presumably because of the more complete muscle relaxation that occurs during the paradoxical stage of sleep (see p. 144).

Energy input must equal energy output to maintain a neutral energy balance.

Since energy cannot be created or destroyed, energy input must equal energy output, as represented by the following equation:

energy input = energy output

energy in food consumed = external work + internal heat production ± stored energy

There are three possible states of energy balance:

- *Neutral energy balance.* If the amount of energy in food intake exactly equals the amount of energy expended by the muscles in performing external work plus the basal internal energy expenditure that eventually appears as body heat, then energy input and output are exactly in balance, and body weight remains constant.
- *Positive energy balance.* If the amount of energy in food intake is greater than the amount of energy expended by means of external work and internal functioning, the extra energy taken in but not used is stored in the body, primarily as adipose tissue, so body weight increases.
- *Negative energy balance.* Conversely, if the energy derived from food intake is less than the body's immediate energy requirements, the body must use stored energy to supply energy needs, and, accordingly, body weight decreases.

To maintain a constant body weight (with the exception of minor fluctuations caused by changes in H_2O content), energy acquired through food intake must equal energy expenditure by the body. Because the average adult maintains a fairly constant weight over long periods of time, this implies that precise homeostatic mechanisms exist to maintain a long-term balance between energy intake and energy expenditure. Theoretically, total body energy content could be maintained at a constant level by regulating the magnitude of food intake, physical activity, or internal work and heat production. Control of food intake to match changing metabolic expenditures is the major means of maintaining a neutral energy balance. The level of physical activity is principally under voluntary control. For example, there are no regulatory factors that automatically impel an obese person to exercise; in fact, most overweight people are even less inclined to engage in physical activities than are their slimmer counterparts. There are mechanisms that can alter the degree of internal work and heat production, but these control measures are aimed primarily at regulating body temperature rather than total energy balance. Some studies suggest, however, that after several weeks of eating less or more than desired, small counteracting changes in metabolism may occur. For example, a compensatory increase in the body's efficiency of energy use in response to underfeeding may partially explain why some dieters become stuck at a plateau after having lost the first 10 or so pounds of weight fairly easily. Similarly, a compensatory reduction in the efficiency of energy use in response to overfeeding may account in part for the difficulty experienced by very thin people who are deliberately trying to gain weight. Despite the possibility of compensatory changes in metabolism, there is no question that regulation of food intake is the most important factor in the long-term maintenance of energy balance and body weight.

Food intake is controlled primarily by the hypothalamus, but the mechanisms involved are not fully understood.

Control of food intake is primarily a function of the hypothalamus. Classically, the hypothalamus is considered to house a pair of **feeding** or **appetite centers** located in the lateral (outer) regions of the hypothalamus, one on each side, and another pair of **satiety centers** located in the ventromedial (underside middle) area. The functions of these areas have been elucidated by a series of experiments that involve either destruction or stimulation of these specific regions. Stimulation of the clusters of nerve cells designated as feeding or appetite centers makes the animal hungry, driving it to eat voraciously, whereas selective destruction of these areas suppresses eating and food-intake behavior to the point that the animal starves itself to death. In contrast, stimulation of the satiety centers signals satiety, or the feeling of having had enough to eat. Consequently, the stimulated animal refuses to eat, even if previously deprived of food. As expected, destruction of this area produces the opposite effect—profound overeating and obesity because the animal never achieves a feeling of being full (Fig. 17-2). Thus, the feeding centers tell us to eat, whereas the satiety centers tell us when we have had enough. Although it is convenient to consider these specific areas as exciting and inhibiting feeding behavior, respectively, this approach is too simplistic. Other areas of the brain as well as the liver are now known to play important roles in controlling food intake, but their contributions and interrelationships with the hypothalamus remain obscure. Undoubtedly, complex systems rather than isolated centers control feeding and satiety.

Whatever the site of final integration, a major interest is determining what input factors to these integrating centers govern feeding onset and termination as well as total energy intake. Exactly what switches feeding behavior on and off is still unclear. Even though food intake is adjusted to balance changing energy expenditures over a period of time, there are no calorie receptors per se to monitor energy input, energy output, or total body energy content. A number of proposals have been set forth, each supported by some experimental findings, but none of them alone is able to account for all observed feeding behavior. Undoubtedly, control of food intake does not depend on changes in a single signal but is determined by the integration of many inputs that provide information about the body's energy status. Some information is apparently used for short-term regulation of food intake, helping to control meal size and frequency. Even so, over a 24-hour period, the energy in ingested food rarely matches energy expenditure for that day. The correlation between total caloric intake and total energy output is excellent, however, over long periods of time. As a result, the total energy content of the body—and, consequently, body weight—remains relatively constant on a long-term basis.

The following factors are among those that have been hypothesized as contributing to the control of food intake (— Fig. 17-3). The relative importance of each of these factors (and possibly others) remains unclear.

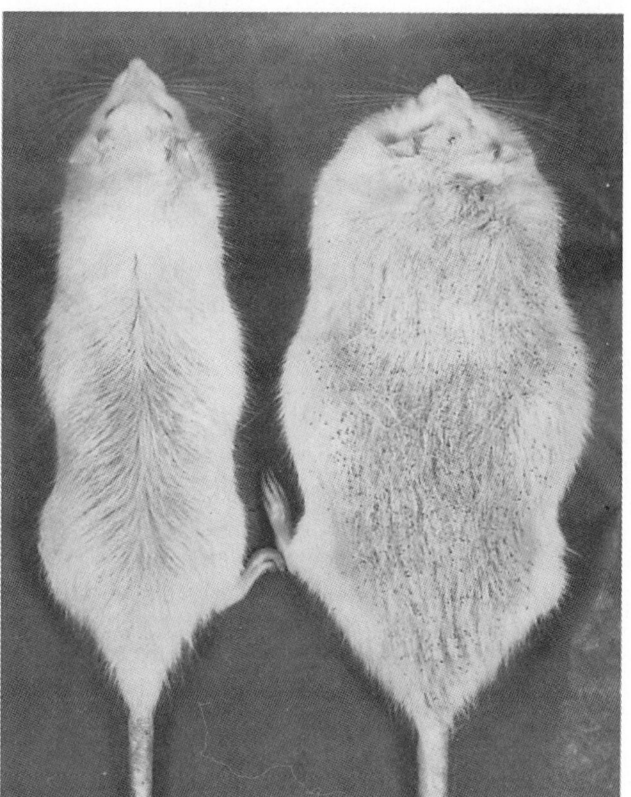

— *Figure 17-2* **Comparison of a Normal Rat with a Rat Whose Satiety Center Has Been Destroyed** Several months after destruction of the satiety center in the ventromedial area of the hypothalamus, the rat on the right had gained considerable weight as a result of overeating compared to its normal litter mate on the left. Rats sustaining lesions in this area also display less grooming behavior, accounting for the soiled appearance of the fat rat.

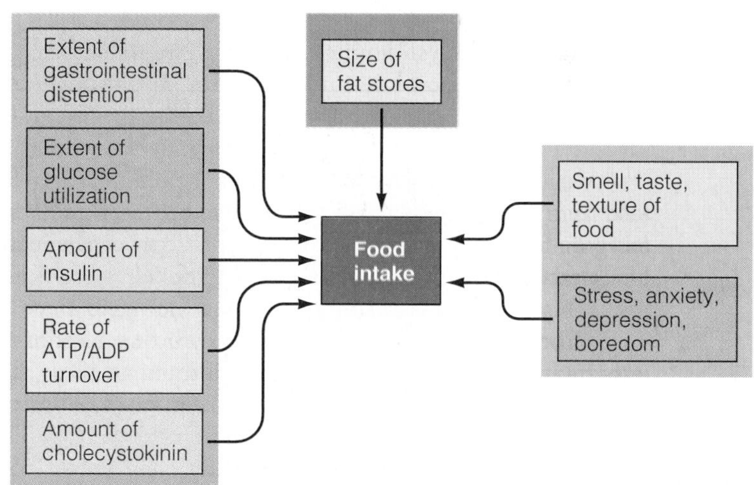

— *Figure 17-3* **Factors that Influence Food Intake**

The size of fat stores According to the **lipostatic** ("fat stability") **theory,** increased fat storage in adipose tissue signals satiety. This signal is generally considered to be responsible for the long-term matching of food intake to energy expenditure so that total body energy content remains balanced and body weight remains constant. Researchers recently discovered the blood-borne signal between fat stores and the areas of the brain controlling food intake. Adipose tissue releases the hormone **leptin** (*leptin* means "thin"), which directly or indirectly tells the brain the size of the fat stores. The amount of leptin in the blood appears to be an excellent indicator of the total amount of triglyceride fat stored in the adipose tissue. Leptin is believed to reduce feeding behavior, acting in negative-feedback fashion to regulate adipose stores and thus body weight.

The extent of gastrointestinal distention In addition to the mechanisms involved in the long-term control of body weight, other factors are believed to play a role in controlling the timing and size of meals. Early proposals suggested that cues of emptiness or fullness of the digestive tract signaled hunger or satiety, respectively. For example, stimulation of gastric stretch receptors has been shown to suppress food intake. However, neural input arising from stomach distention plays a more important role in controlling the rate of gastric emptying than in signaling satiety (see p. 564). Researchers now believe that internal blood-borne signals reflecting the depletion or availability of energy-producing substances are more important than stomach volume in controlling the initiation and cessation of eating.

The extent of glucose utilization The **glucostatic** ("glucose stability") **theory** proposes that satiety is signaled by increased glucose utilization, such as occurs during a meal when more glucose is available for use because it is being absorbed from the digestive tract. Conversely, after absorption of a meal is complete, and no new glucose is entering the blood, the resultant reduction in the cells' glucose use arouses the sensation of hunger.

In a related proposal, some researchers suggest that an increased level of insulin in the blood signals satiety. Insulin, a hormone secreted by the pancreas in response to a rise in the concentration of glucose and other nutrients in the blood following a meal, stimulates cellular uptake, utilization, and storage of glucose and other nutrients. Thus, the increase in insulin secretion that accompanies nutrient abundance and promotes increased glucose utilization is an appropriate satiety signal.

The intensity of cell power production The **ischymetric** (*ischys* means "power"; *metric* means "measure") **theory** suggests that the signal for short-term control of food intake is not a deficit or surplus of any particular major nutrient such as glucose but is linked instead to the magnitude of cellular power (ATP) production. Changes in the availability of one or all of the nutrients to a cell may result in decreases or increases in the rate of ATP/ADP turnover, which in turn could be transduced into some sort of blood-borne or neural signal of low power (hunger) or high power (satiety).

The level of cholecystokinin secretion Evidence has accumulated that **cholecystokinin (CCK),** one of the gastrointestinal hormones released from the duodenal mucosa during digestion of a meal, may be an important satiety signal. This theory could explain why we stop eating before the ingested food is actually digested, absorbed, and made available to meet the body's energy needs. We feel satisfied when adequate food to replenish the stores is in the digestive tract, even though the body's energy stores are still low.

Cholecystokinin is secreted in response to the presence of nutrients in the small intestine. Through multiple effects on the digestive system, CCK facilitates the digestion and absorption of these nutrients (see p. 596). It is appropriate that this blood-borne signal, whose rate of secretion is correlated with the amount of nutrients ingested, also contributes to the sense of being filled after a meal has been consumed but before it has actually been digested and absorbed.

Influences of neurotransmitters Various neuronal circuits within the brain are apparently involved in controlling feeding behavior. Exposure of the brain to various neurotransmitters has been shown to induce different patterns of food intake in experimental animals. For example norepinephrine and neuropeptide Y increase carbohydrate consumption whereas dopamine and serotonin suppress carbohydrate consumption. How the release of these neurotransmitters is linked to the control of food intake is unclear.

Psychosocial influences Thus far we have described possible involuntary signals that might automatically occur to control food intake. However, as with water intake, people's eating habits are also shaped by psychological and social factors. Often our decision to eat or stop eating is not determined merely by whether we are hungry or full, respectively. Frequently, we eat out of habit (eating three meals a day on schedule no matter what our status on the hunger-satiety continuum) or because of social custom (food often plays a prime role in entertainment, leisure, and business activities).

Furthermore, the amount of pleasure derived from eating can reinforce feeding behavior. Eating foods with an enjoyable taste, smell, and texture can increase appetite and food intake. This has been demonstrated in an experiment in which rats were offered their choice of highly palatable human foods. They overate by as much as 70% to 80% and became obese. When the rats returned to eating their regular monotonous but nutritionally balanced rat chow, their obesity was rapidly reversed as their food intake was controlled once again by physiological drives rather than by hedonistic urges for the tastier offerings.

Stress, anxiety, depression, and boredom have also been shown to alter feeding behavior in ways that are unrelated to energy needs in both experimental animals and humans. Thus, any comprehensive explanation of how food intake is controlled must take into account these voluntary eating acts that can reinforce or override the internal signals governing feeding behavior.

Obesity occurs when more calories are consumed than are burned up.

Obesity is defined as excessive fat content in the adipose tissue stores. The arbitrary boundary for obesity is generally considered to be greater than 20% overweight compared to normal standards. Obesity occurs when, over a period of time, more kilocalories are ingested in food than are used to support the body's energy needs, with the excessive energy being stored as triglycerides in adipose tissue. The causes of obesity are many and obscure. Some factors that may be involved include (1) disturbances of the satiety-appetite regulatory centers in the hypothalamus, (2) emotional disturbances in which overeating replaces other gratifications, (3) development of an excessive number of fat cells as a result of overfeeding, (4) certain endocrine disorders such as hypothyroidism, (5) hereditary tendencies, (6) the palatability of available food, and (7) lack of physical exercise.

Numerous studies have shown that, on the average, fat people do not eat any more than thin people. One possible explanation is that overweight persons do not overeat but "underexercise." Studies have shown that very low levels of physical activity are not accompanied by comparable reductions in food intake. Another explanation is that excess energy input occurs only during the time that obesity is actually developing. Some investigators suggest that the hypothalamic centers determining long-term satiety are "set at a higher level" in obese persons. Thus, overweight people do tend to maintain their weight, but at a higher set point than normal. Once obesity has developed, all that is required to maintain the condition is that energy input equals energy output.

According to a recent theory, obesity might have a leptin link. The gene that produces leptin, the **obese** (or **ob**) **gene,** was first identified in a strain of genetically obese mice whose body weight is three times normal on average. Their obesity arises from a failure to produce leptin as a result of a defective ob gene. When administered leptin, the fat mice ate much less and increased their metabolic rates, resulting in dramatic weight loss. Unlike obese mice, most obese people don't have a problem producing leptin. In fact, they have an overabundance of circulating leptin, reflecting their enlarged adipose stores. Scientists suspect that the problem lies with faulty leptin receptors in the brain that do not respond appropriately to the high levels of circulating leptin. Thus, the brain does not detect leptin as a satiety signal until a higher set point is achieved. Alternatively, other genetic errors not involving the ob gene and its leptin product may be at fault. Following the discovery of the ob gene, researchers have uncovered other mutated "obesity" genes in strains of overweight mice. For example, the pattern of development of obesity in mice that possess the **tubby gene** more closely parallels that of obese humans than does the obesity of mice with a mutated ob gene. Investigators have identified a gene similar to the tubby gene in humans and are now examining whether this gene is mutated in obese people.

Other studies further complicate the weight-control issue by suggesting that even though the total caloric content of food remains the same, the frequency of eating may be important in the distribution of the nutrients. In rat and monkey studies, nibblers had relatively more body protein and less body fat than animals given the same food in several large meals each day. Further studies suggest that, even though total caloric input for the day remains unchanged, foods consumed before bedtime are more likely to be put into storage. In this regard, bedtime snacks are more fattening than they would be if they were eaten earlier in the day.

The type of food ingested also appears to be important. High-fat diets are especially obesity producing for several reasons. First, fewer calories are used to convert ingested fat into stored fat than to convert ingested carbohydrate or protein into stored fat. This difference occurs because fewer energy-consuming biochemical steps are needed to process and store dietary fat than to process and store dietary carbohydrate or protein. Therefore, if the same number of calories of fat and carbohydrate or protein are ingested, more calories are "left over" after the caloric price for processing and storing the fat has been paid than remain after the carbohydrate or protein has been processed. Accordingly, calorie for calorie, fat is more fattening than carbohydrate or protein. In addition, as fat intake increases at the expense of reduced carbohydrate consumption, glucose utilization decreases. As predicted by the glucostatic theory, low glucose utilization promotes increased food intake, resulting in greater energy input before satiety is signaled. Furthermore, studies suggest that a high-fat diet may increase the set point for body weight, perhaps in part by limiting leptin's action as a satiety signal.

Thus, our knowledge about the causes and control of obesity is still rather limited, as evidenced by the number of people who are constantly trying to stabilize their weight at a more desirable level. This is important from more than an aesthetic viewpoint. It is known that obesity, especially of the android type, can predispose an individual to illness and premature death from a multitude of disease. (See the boxed feature on p. 608, A Closer Look at Exercise Physiology.)

Persons suffering from anorexia nervosa have a pathological fear of gaining weight.

The converse of obesity is generalized nutritional deficiency. The obvious causes for reduction of food intake below energy needs are lack of availability of food, interference with the swallowing or digestive mechanism, and impairment of appetite. Chronic diseases such as renal failure, cancer, and tuberculosis are commonly accompanied by a lack of appetite, which contributes to the subsequent weight loss characteristic of these "wasting" diseases. The mechanisms for this appetite suppression are obscure. Researchers have identified a protein called **cachetin** (*cachexia* refers to a chronic, wasting condition), which is released by the immune system and is believed to contribute to the development of emaciation in many patients with chronic, debilitating diseases or cancer.

Another poorly understood disorder in which lack of appetite is a prominent feature is **anorexia nervosa.** Patients with this disorder, most commonly adolescent girls and young women, have a morbid fear of becoming fat. As a result of having a distorted body image, they tend to visualize themselves as being much heavier than they actually are. Because they have an aversion to food, they eat very little and consequently lose considerable weight, perhaps even starving themselves to death. Other characteristics of the condition include altered secretion of many hormones, absence of menstrual periods, and low body temperature. It is unclear whether these symptoms occur secondarily as a result of general malnutrition or arise independently of the eating disturbance as a part of a primary hypothalamic malfunction. Many investigators think the underlying problem may be psychological rather than biological. Some experts suspect that anorexics may suffer from addiction to endogenous opiates, self-produced morphine-like substances (see p. 164) that are thought to be released during prolonged starvation.

Temperature Regulation

Internal core temperature is homeostatically maintained at 100°F.

Humans are usually in environments cooler than their bodies, so they must constantly generate heat internally to maintain body temperature. Heat production ultimately depends on the oxidation of metabolic fuel derived from food.

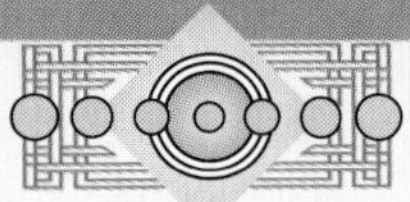

What the Scales Don't Tell You

Body composition refers to the percentage of body weight that is composed of lean tissue and adipose tissue. The assessment of body composition is an important component in evaluating a person's health status. The age-height-weight tables used by insurance companies can be misleading for determining healthy body weight. Many athletes, for example, would be considered overweight by these charts. A football player may be 6 feet 5 inches tall and weigh 300 pounds, but have only 12% body fat. This player's extra weight is muscle, not fat, and therefore is not a detriment to his health. A sedentary person, on the other hand, may be normal on the height-weight charts but have 30% body fat. This person should maintain body weight while increasing muscle mass and decreasing fat. Ideally, men should have 15% fat or less and women should have 20% fat or less.

The most accurate method for assessing body composition is underwater weighing. This technique is based on the fact that lean tissue is denser than water and fat tissue is less dense than water. (You can readily demonstrate this for yourself by dropping a piece of lean meat and a piece of fat into a glass of water; the lean meat will sink and the fat will float.) The most common method of underwater weighing requires the person to expel all the air from his or her lungs and then completely submerge in a tank of water while sitting in a swing that is attached to a scale. The results are used to determine body density using equations that take into consideration the density of water, the difference between the person's weight in air and underwater, and the residual volume of air remaining in the lungs. Because of the difference in density between lean and fat tissue, people who have more fat have a lower density and weigh relatively less underwater than in air compared to their lean counterparts. Body composition is then determined by means of an equation that correlates percentage fat with body density.

Another common way to assess body composition is skinfold thickness. Because approximately half of the body's total fat content is located just beneath the skin, total body fat can be estimated from measurements of skinfold thickness taken at various sites on the body. Skinfold thickness is determined by pinching up a fold of skin at one of the designated sites and measuring its thickness by means of a caliper, a hinged instrument that fits over the fold and is calibrated to measure thickness. Mathematical equations specific for the person's age and sex can be used to predict the percentage of fat from the skinfold-thickness scores. A major criticism of skinfold assessments is that accuracy depends on the investigator's skill.

There are different ways to be fat, and one way is more dangerous than the other. Obese patients can be classified into two categories—*android,* a male-type of adipose tissue distribution, and *gynoid,* a female-type distribution—based on the anatomical distribution of adipose tissue measured as the ratio of waist circumference to hip circumference. Android obesity is characterized by abdominal fat distribution (people shaped as "apples"), whereas gynoid obesity is characterized by fat distribution in the hips and thighs (people shaped as "pears"). Both sexes can display either android or gynoid obesity.

Android obesity is associated with a number of disorders, including insulin resistance, type II (adult-onset) diabetes mellitus, excess blood lipid levels, high blood pressure, coronary heart disease, and stroke. Gynoid obesity is not associated with high risk for these diseases. Because android obesity is associated with increased risk for disease, it is most important for apple-shaped overweight individuals to reduce their fat stores.

Exercise physiologists often assess body composition as an aid in prescribing and evaluating exercise programs. Exercise generally reduces the percentage of body fat and, by increasing muscle mass, increases the percentage of lean tissue.

Research on the success of weight-reduction programs indicates that it is very difficult for people to lose weight, but when weight loss occurs, it is from the areas of increased stores. Recent interesting research has shed some light on the problems of obesity. Studies indicate that the resting metabolic rate may be an inherited trait. In one study, at three months of age, babies of obese parents showed 20% less energy expenditure than babies of lean parents. Also, diet-induced thermogenesis (DIT), which is the increase in metabolism that follows consumption of carbohydrate or protein, has been shown to be lower in those who have been obese since childhood. In other words, people who have been obese since childhood are very efficient at storing the excess calories they ingest. Lean people may metabolize more of the calories they ingest, with the energy being given off as heat.

Even after obese people reduce their weight to normal levels, their DIT remains lower than someone of the same weight who has always been lean. This would be an admirable physiological trait in times of food deprivation, but in times of food abundance, low DIT and a low metabolic rate can predispose an individual to obesity. A person with these traits has to eat less than his or her lean counterpart to maintain normal weight.

Because very-low-calorie diets are difficult to maintain, an alternative to severely cutting caloric intake to lose weight is to increase energy expenditure through physical exercise. An aerobic exercise program further helps reduce the risk of the disorders associated with android obesity and helps to reduce fat stores.

Because cellular function is sensitive to fluctuations in internal temperature, humans homeostatically maintain body temperature at a level that is optimal for cellular metabolism to proceed in a stable fashion. Even moderate elevations of body temperature begin to cause nerve malfunction and irreversible protein denaturation. Most people suffer convulsions when the internal body temperature reaches about 106°F (41°C); 110°F (43.3°C) is considered the upper limit compatible with life. On the other hand, most of the body's tissues can transiently withstand substantial cooling. This characteristic is useful during cardiac surgery when the heart must be stopped. The patient's body temperature is deliberately lowered. The cooled tissues need less nourishment than they do at normal body temperature because of their pronounced reduction in metabolic activity. The lower O_2 need of cooled tissues also accounts for the occasional survival of drowning victims who have been submerged in icy water considerably longer than one can normally survive without O_2.

Normal body temperature has traditionally been considered 98.6°F (37°C). However, a recent study indicates that normal body temperature varies among individuals and varies throughout the day, ranging from 96.0°F in the morning to 99.9°F in the evening, with an overall average of 98.2°F. These values are considered normal for temperatures taken orally (by mouth). Furthermore, there is no one body temperature because the temperature varies from organ to organ. From a thermoregulatory viewpoint, the body may conveniently be viewed as a *central core* surrounded by an *outer shell*. The temperature within the inner core, which consists of the abdominal and thoracic organs, the central nervous system, and the skeletal muscles, generally remains fairly constant. It is this internal **core temperature** that is subject to precise regulation to maintain its homeostatic constancy. The core tissues function best at a relatively constant temperature of around 100°F. The skin and subcutaneous fat constitute the outer shell. In contrast to the constant high temperature in the core, the temperature within the shell is generally cooler and may vary substantially. For example, skin temperature may fluctuate between 68° and 104°F without damage. In fact, as you will see, the temperature of the skin is deliberately varied as a control measure to help maintain the core's thermal constancy.

We are accustomed to thinking in terms of oral, rectal, or axillary (under the armpit) temperature because these are easy sites for monitoring body temperature. The oral and axillary temperatures are comparable, while rectal temperature averages about 1°F higher. Also recently available is a temperature-monitoring instrument that scans the heat generated by the eardrum and converts this temperature into an oral equivalent. However, none of these measurements is an absolute indication of the internal core temperature, which averages about 100°F. Even though the core temperature is held relatively constant, several factors cause it to vary slightly:

1. Most people's core temperature normally varies about 1.8°F (1°C) during the day, with the lowest level occurring early in the morning before rising (6 to 7 A.M. and the highest point occurring in late afternoon (5 to 7 P.M.). This variation is due to an innate biological rhythm or "biological clock."
2. Women also experience a monthly rhythm in core temperature in connection with their menstrual cycle. The core temperature averages 0.9°F (0.5°C) higher during the last half of the cycle from the time of ovulation to menstruation. This mild sustained elevation in temperature was once thought to be caused by the increased secretion of progesterone, one of the ovarian hormones, during this period, but this is no longer believed to be the case. The actual cause is still undetermined.
3. The core temperature increases during exercise because of the tremendous increase in heat production by the contracting muscles. During hard exercise, the core temperature may increase to as much as 104°F (40°C). This temperature would be considered to be a fever in a resting person, but it is normal during strenuous exercise.
4. Because the temperature-regulating mechanisms are not 100% effective, the core temperature may vary slightly with exposure to extremes of temperature. For example, the core temperature may fall several degrees in cold weather or rise a degree or so in hot weather.

Thus, the core temperature can vary at the extremes between about 96° to 104°F but usually deviates less than a few degrees. This relative constancy is made possible by multiple thermoregulatory mechanisms coordinated by the hypothalamus.

Heat gain must balance heat loss to maintain a stable core temperature.

The core temperature is a reflection of the body's total heat content. To maintain a constant total heat content and thus a stable core temperature, heat input to the body must balance heat output (Fig. 17-4). *Heat input* occurs by way of heat gain from the external environment and internal heat production, the latter being the most important source of heat for the body. Recall that most of the body's energy expenditure ultimately appears as heat. This heat is important in the maintenance of core temperature. In fact, usually more heat is generated than is required to maintain the body temperature at a normal level, so the excess heat must be eliminated from the body. *Heat output* occurs by way of heat loss from exposed body surfaces to the external environment.

Balance between heat input and output is frequently disturbed by (1) changes in internal heat production for purposes unrelated to regulation of body temperature, most notably by exercise, which markedly increases heat production, and (2) changes in the external environmental temperature that influence the degree of heat gain or heat loss that occurs between the body and its surroundings. To maintain body temperature within narrow limits in spite of changes in metabolic heat production and changes in environmental temperature, compensatory adjustments must take place in heat-loss and heat-gain mechanisms.

Figure 17-4 **Heat Input and Output**

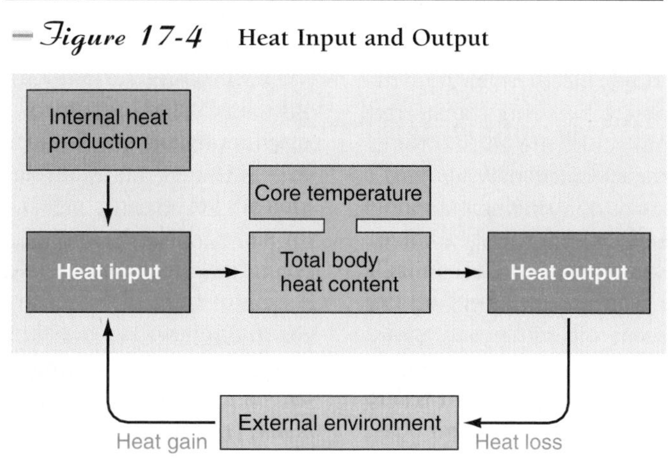

If the core temperature starts to fall, heat production is increased and heat loss is minimized so that normal temperature can be restored. Conversely, if the temperature starts to rise above normal, it can be corrected by increasing heat loss while simultaneously reducing heat production.

We will now elaborate on the means by which heat gains and losses can be adjusted to maintain body temperature, starting with a discussion of the methods of heat exchange between the body and its surroundings.

Heat exchange between the body and the environment takes place by radiation, conduction, convection, and evaporation.

All heat loss or heat gain between the body and the external environment must take place between the body surface and its surroundings. The same physical laws of nature that govern heat transfer between inanimate objects also control the transfer of heat between the body surface and the environment. The temperature of an object may be thought of as a measure of the concentration of heat within the object. Accordingly, heat always moves down its concentration gradient; that is, down a **thermal gradient** from a warmer to a cooler region.

The body uses four mechanisms of heat transfer: *radiation, conduction, convection,* and *evaporation.* **Radiation** is the emission of heat energy from the surface of a warm body in the form of **electromagnetic waves,** or **heat waves,** which travel through space (➡ Fig. 17-5a). When radiant energy strikes an object and is absorbed, the energy of the wave motion is transformed into heat within the object. The human body both emits (source of heat loss) and absorbs (source of heat gain) radiant energy. Whether the body loses or gains heat by radiation depends on the difference in temperature between the skin surface and the surfaces of various other objects in the body's environment. Because net transfer of heat by radiation is always from warmer objects to cooler ones, the body gains heat by radiation from objects warmer than the skin surface, such as the sun, a radiator, or burning logs. On the other hand, the body loses heat by radiation to objects in its environment whose surfaces are cooler than the surface of the skin, such as building walls, furniture, or trees. On the average, humans lose close to half of their heat energy through radiation.

Conduction is the transfer of heat between objects of differing temperatures that are in *direct contact* with each other (Fig. 17-5b). Heat moves down its thermal gradient from the warmer to the cooler object by being transferred from molecule to molecule. All molecules are constantly in vibratory motion, with warmer molecules moving faster than cooler ones. When molecules of differing heat content touch each other, the faster moving, warmer molecule agitates the cooler molecule into more rapid motion, thereby warming up the cooler molecule. During this process, the original warmer molecule loses some of its thermal energy as it slows down and cools off a bit. Therefore, given enough time, the temperature of the two touching objects eventually equalizes.

The rate of heat transfer by conduction depends on the *temperature difference* between the touching objects and the *thermal conductivity* of the substances involved (that is, how easily heat is conducted by the molecules of the substances). Heat can be lost or gained by conduction when the skin is in contact with a good conductor. When you hold a snowball, for example, your hand becomes cold because heat moves by conduction from your hand to the snowball. Conversely, when you apply a heating pad to a body part, the part is warmed up as heat is transferred directly from the pad to the body.

Similarly, you either lose or gain heat by conduction to the layer of air in direct contact with your body. The direction of heat transfer depends on whether the air is cooler or warmer, respectively, than your skin. Only a small percentage of total heat exchange between the skin and environment takes place by conduction alone, however, because air is not a very good conductor of heat. (This is why swimming pool water at 80°F feels cooler than air at the same temperature; heat is conducted more rapidly from the body surface into the water, which is a good conductor, than into the air, which is a poor conductor.)

Convection refers to the transfer of heat energy by *air (or H_2O) currents.* As the body loses heat by conduction to the surrounding cooler air, the air in immediate contact with the skin is warmed. Because warm air is lighter (less dense) than cool air, the warmed air rises while cooler air moves in next to the skin to replace the vacating warm air. The process is then repeated (Figure 17-5c). These air movements, known as convection currents, help carry heat away from the body. If it were not for convection currents, no further heat could be dissipated from the skin by conduction once the temperature of the layer of air immediately around the body equilibrated with skin temperature.

The combined conduction-convection process of dissipating heat from the body is enhanced by forced movement of air across the body surface, either by external air movements, such as those caused by the wind or a fan, or by movement of the body through the air, such as during bicycle riding. Because forced air movement sweeps away the air warmed by conduction and replaces it with cooler air more rapidly, a greater total amount of heat can be carried away from the body over a given time period. Thus, wind makes us feel cooler on hot days, and windy days in the winter are more chilling than calm days at the same cold temperature. For this reason, weather forecasters have developed the concept of *wind chill factor.*

Evaporation is the final method of heat transfer used by the body. When water evaporates from the skin surface, the heat required to transform water from a liquid to a gaseous state is absorbed from the skin, thereby cooling the body (Fig. 17-5d). Evaporative heat loss makes you feel cooler when your bathing suit is wet than when it is dry. Evaporative heat loss occurs continually from the linings of the respiratory airways and from the surface of the skin. Heat is continuously lost through the H_2O vapor in the expired air as a result of the air's humidification during its passage through the respiratory system. Similarly, because the skin is not completely waterproof, H_2O molecules constantly diffuse through the skin and

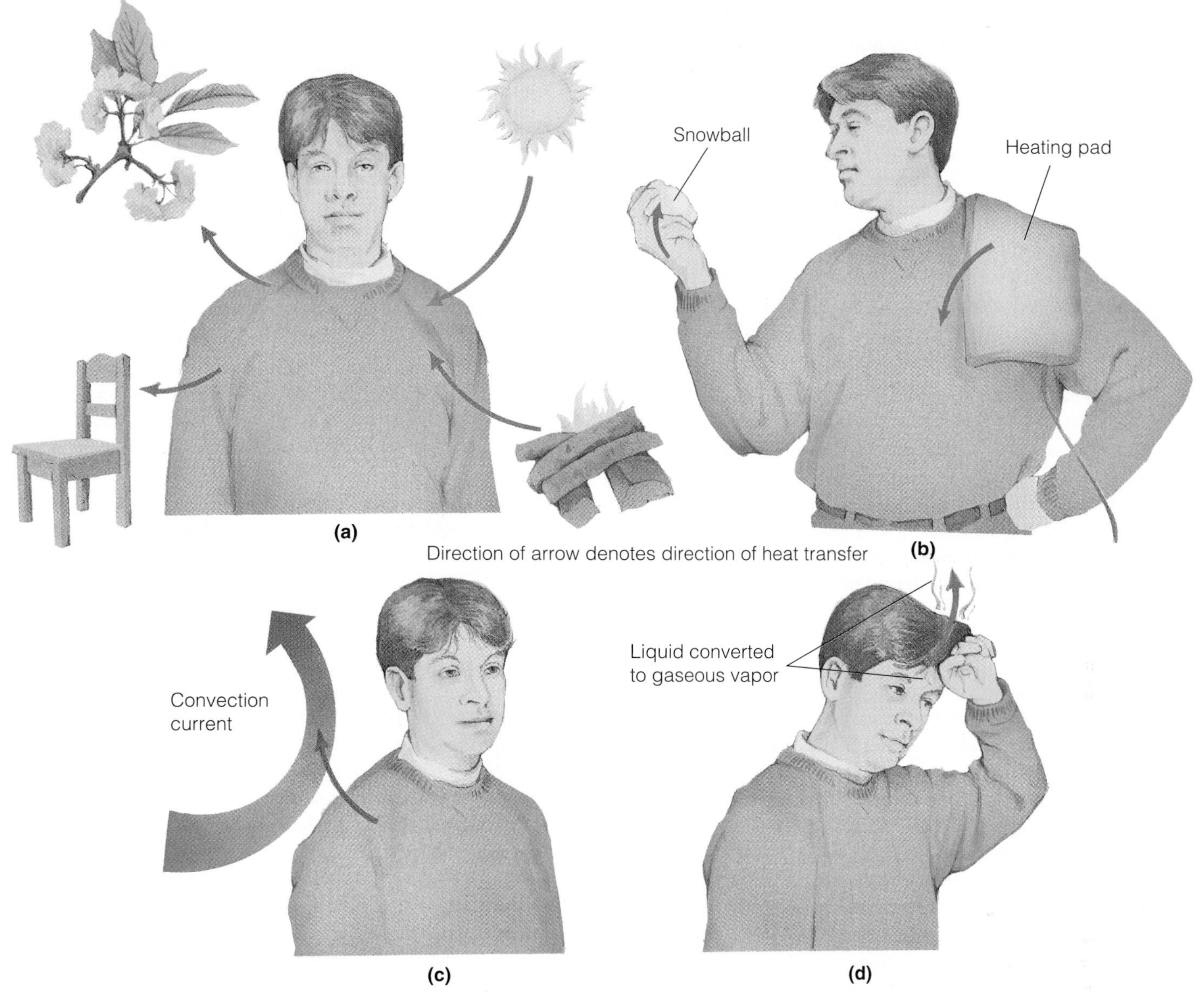

Figure 17-5 **Mechanisms of Heat Transfer** (a) Radiation—the transfer of heat energy from a warmer object to a cooler object in the form of electromagnetic waves ("heat waves"), which travel through space. (b) Conduction—the transfer of heat from a warmer object to a cooler object that is in direct contact with the warmer one. The heat is transferred through the movement of thermal energy from molecule to adjacent molecule. (c) Convection—the transfer of heat energy by air currents. Cool air warmed by the body through conduction rises and is replaced by more cool air. This process is enhanced by the forced movement of air across the body surface. (d) Evaporation—conversion of a liquid such as sweat into a gaseous vapor, a process that requires heat (the heat of vaporization), which is absorbed from the skin.

evaporate. This ongoing evaporation from the skin is completely unrelated to the sweat glands. These passive evaporative heat-loss processes are not subject to physiological control and go on even in very cold weather, when the problem is one of conserving body heat.

Sweating, on the other hand, is an active evaporative heat-loss process under sympathetic nervous control. The rate of evaporative heat loss can be deliberately adjusted by means of sweating, which is an important homeostatic mechanism to eliminate excess heat as needed. In fact, when the environmental temperature exceeds the skin temperature, sweating is the *only* avenue for heat loss, because the body is gaining heat by radiation and conduction under these circumstances.

Sweat is a dilute salt solution that is actively extruded to the surface of the skin by sweat glands dispersed all over the body. Sweat must be evaporated from the skin for heat loss to occur. If sweat merely drips from the surface of the skin or is wiped away, no heat loss is accomplished. The most important factor determining the extent of evaporation of sweat is the *relative humidity* of the surrounding air (the percentage of H_2O vapor actually present in the air compared to the greatest amount that the air can possibly hold at that temperature; for example, a relative humidity of 70% means that the air contains 70% of the H_2O vapor it is capable of holding). When the relative humidity is high, the air is already almost fully saturated with H_2O, so it has limited ability to take up additional moisture from the skin. Thus, little evaporative heat loss can occur on hot, humid days. The sweat glands continue to secrete, but the sweat simply remains on the skin or drips off, instead of evaporating and producing a cooling effect. As a measure of the discomfort associated with combined heat and high humidity, meteorologists have devised the *temperature-humidity index.*

The hypothalamus integrates a multitude of thermosensory inputs from both the core and the surface of the body.

The hypothalamus serves as the body's thermostat. The home thermostat keeps track of the temperature in a room and triggers a heating mechanism (the furnace) or a cooling mechanism (the air conditioner) as necessary to maintain the room temperature at the indicated setting. Similarly, the hypothalamus, as the body's thermoregulatory integrating center, receives afferent information about the temperature in various regions of the body and initiates extremely complex, coordinated adjustments in heat-gain and heat-loss mechanisms as necessary to correct any deviations in core temperature from the "normal setting." The hypothalamic thermostat is far more sensitive than your home thermostat. The hypothalamus is able to respond to changes in blood temperature as small as 0.01°C. The hypothalamus's degree of response to deviations in body temperature is finely matched so that precisely enough heat is lost or generated as needed to restore the temperature to normal.

To make the appropriate adjustments in the delicate balance between the heat-loss mechanisms and the opposing heat-producing and heat-conserving mechanisms, the hypothalamus must be continuously apprised of both the skin temperature and the core temperature by means of specialized temperature-sensitive receptors called **thermoreceptors.** *Peripheral thermoreceptors* monitor skin temperature throughout the body and transmit information about changes in surface temperature to the hypothalamus. The core temperature is monitored by *central thermoreceptors,* which are located in the hypothalamus itself as well as elsewhere in the central nervous system and the abdominal organs.

Two centers for temperature regulation have been identified in the hypothalamus. The *posterior region* is activated by cold and subsequently triggers reflexes that mediate heat production and heat conservation. The *anterior region,* which is activated by warmth, initiates reflexes that mediate heat loss. Let us examine the means by which the hypothalamus fulfills its thermoregulatory functions (▬ Fig. 17-6).

Shivering is the primary involuntary means of increasing heat production.

The body can gain heat as a result of internal heat production generated by metabolic activity or from the external environment if the latter is warmer than body temperature. Because body temperature usually is higher than environmental temperature, metabolic heat production is the primary source of body heat. In a resting person, most body heat is produced by the thoracic and abdominal organs as a result of ongoing, cost-of-living metabolic activities. Above and beyond this basal level, the rate of metabolic heat production can be variably increased primarily by changes in skeletal muscle activity or to a lesser extent by certain hormonal actions. Thus, changes in skeletal muscle activity constitute the major way heat gain is controlled for temperature regulation.

In response to a fall in core temperature caused by exposure to cold, the hypothalamus takes advantage of the fact that increased skeletal muscle activity generates more heat. Acting through descending pathways that terminate on the motor neurons controlling the body's skeletal muscles, the hypothalamus first gradually increases skeletal muscle tone. (Muscle tone refers to the constant level of tension within the muscles.) Soon, shivering begins. **Shivering** consists of rhythmic, oscillating skeletal muscle contractions that occur at a rapid rate of ten to twenty per second. This mechanism is very effective in increasing heat production; all of the energy liberated during these muscle tremors is converted to heat because no external work is accomplished. Within a matter of seconds to minutes, internal heat production may increase two- to fivefold as a result of shivering.

Frequently, reflex changes in skeletal muscle activity are augmented by increased voluntary, heat-producing actions such as bouncing up and down or hand clapping. Such behavioral responses appear to share neural systems in common with the involuntary physiological responses. The hypothalamus and the limbic system of which it is a part (see p. 132) are extensively involved with controlling motivated behavior. Therefore, one should not think of the hypothalamic homeostatic control systems as operating only at the subconscious level.

Although reflex and voluntary changes in muscle activity are the major means of increasing the rate of heat production, **nonshivering (chemical) thermogenesis** also plays a role in thermoregulation. In most experimental animals, chronic cold exposure brings about an increase in metabolic heat production that is independent of muscle contraction, appearing instead to involve changes in heat-generating chemical activity. In humans, nonshivering thermogenesis is most important in newborns, because they lack the ability to shiver. Nonshivering thermogenesis is mediated by the hormones epinephrine and thyroid hormone, both of which increase heat production by stimulating fat metabolism. Newborns have deposits of a special type of adipose tissue known as **brown fat,** which is especially capable of converting chemical energy into heat. The role of nonshivering thermogenesis in adults remains controversial.

In the opposite situation—an elevation in core temperature caused by heat exposure—two mechanisms are employed to reduce heat-producing skeletal muscle activity: muscle tone is reflexly reduced, and voluntary movement is curtailed. When the air becomes very warm, people often complain that it is "too hot even to move." These responses are not as effective at reducing heat production during heat exposure as are the muscular responses that increase heat production during cold exposure for two reasons. First, because muscle tone is normally quite low, the capacity to reduce it

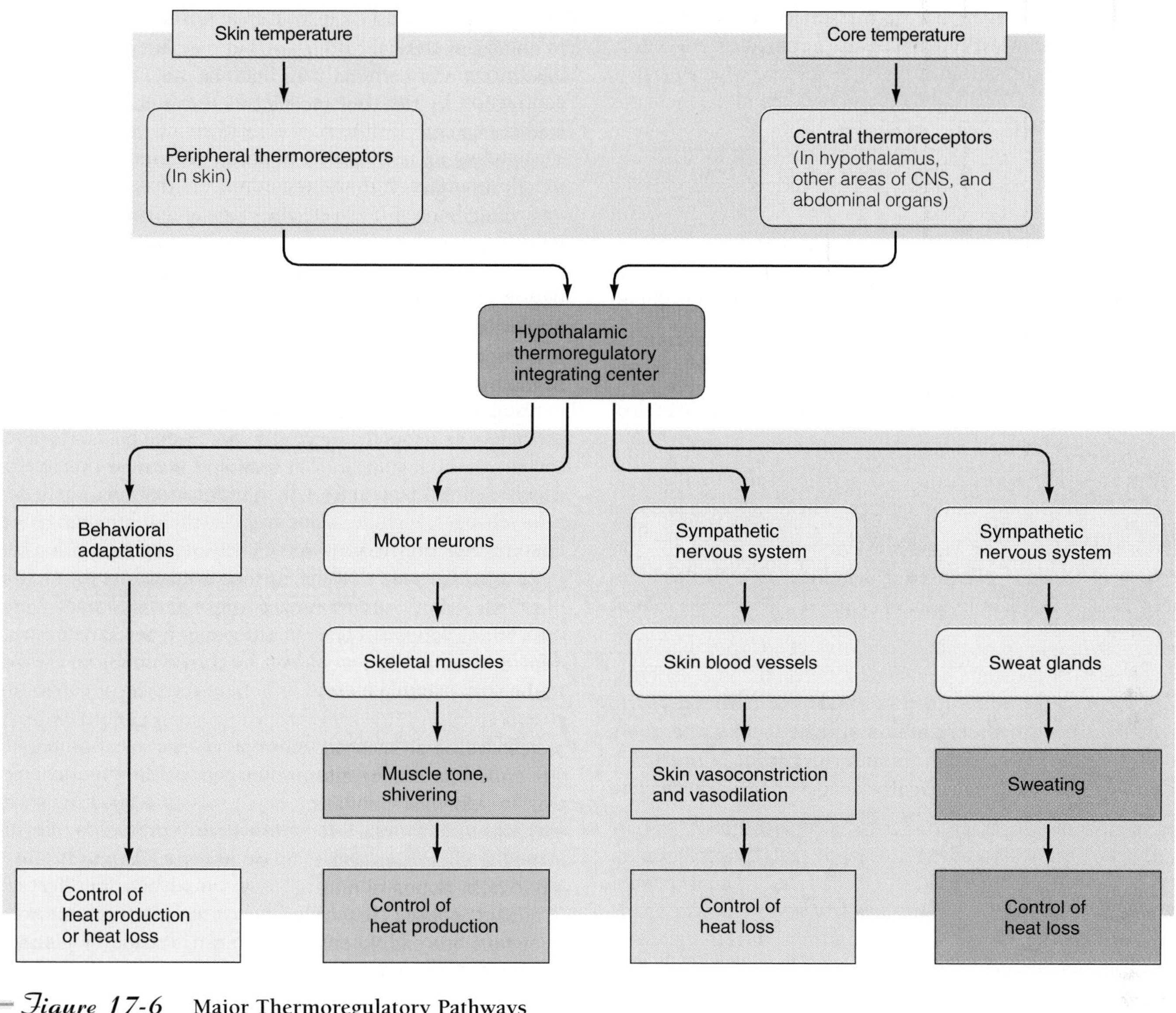

Figure 17-6 Major Thermoregulatory Pathways

further is limited. Second, the elevated body temperature tends to increase the rate of metabolic heat production because the temperature has a direct effect on the rate of chemical reactions.

The magnitude of heat loss can be adjusted by varying the flow of blood through the skin.

Heat-loss mechanisms are also subject to control, again largely by the hypothalamus. When we are hot, we want to increase heat loss to the environment; when we are cold, we want to decrease heat loss. The amount of heat lost to the environment by radiation and conduction-convection is largely determined by the temperature gradient between the skin and the external environment. The body's central core is a heat-generating chamber in which the temperature must be maintained at approximately 100°F. Surrounding the core is an insulating shell through which heat exchanges between the body and the external environment take place. In an attempt to maintain a constant core temperature, the insulative capacity and temperature of the shell can be adjusted to vary the temperature gradient between the skin and external environment, thereby influencing the extent of heat loss.

The insulative capacity of the shell can be varied by controlling the amount of blood flowing through the skin. Blood flow to the skin serves two functions. First, it provides a nutritive blood supply to the skin. Second, as blood is pumped to the skin from the heart, it has been heated in the central core and carries this heat to the skin. Most of the flow of blood through the skin is for purposes of temperature regulation; at normal room temperature, 20 to 30 times more blood flows through the skin than is needed to meet the skin's nutritional needs.

In the process of thermoregulation, skin blood flow can vary tremendously, from 400 ml/min up to 2,500 ml/min. The more blood that reaches the skin from the warm core, the closer the skin's temperature is to the core temperature. The skin's blood vessels diminish the effectiveness of the skin as an insulator by carrying heat to the surface, where it can be lost from the body by radiation and conduction-convection. Accordingly, vasodilation of the skin vessels, which permits increased flow of heated blood through the skin, increases heat loss or, if the environmental temperature is above the core temperature, reduces heat gain. Conversely, vasoconstriction of the skin vessels, which reduces blood flow through the skin, decreases heat loss by keeping the warm blood in the central core, where it is insulated from the external environment. Cold, relatively bloodless skin provides excellent insulation between the core and the environment. However, the skin is not a perfect insulator, even with maximum vasoconstriction. Despite minimal blood flow to the skin, some heat can still be transferred by conduction from the deeper organs to the skin surface and then can be lost from the skin to the environment.

These skin vasomotor responses are coordinated by the hypothalamus by means of sympathetic nervous system output. Increased sympathetic activity to the skin vessels produces heat-conserving vasoconstriction in response to cold exposure, whereas decreased sympathetic activity produces heat-losing vasodilation of the skin vessels in response to heat exposure.

The hypothalamus simultaneously coordinates heat-production mechanisms and heat-loss and heat-conservation mechanisms to regulate core temperature homeostatically.

Let us now pull together the coordinated adjustments in heat production as well as heat loss and heat conservation in response to exposure to either a cold or a hot environment (Fig. 17-6 and Table 17-2). In response to cold exposure, the posterior region of the hypothalamus directs increased heat production such as by shivering, while simultaneously decreasing heat loss (that is, conserving heat) by skin vasoconstriction and other measures.

Because there is a limit to the body's ability to reduce skin temperature through vasoconstriction, even maximum vasoconstriction is not sufficient to prevent excessive heat loss when the external temperature falls too low. Accordingly, other measures must be instituted to further reduce heat loss. In animals with dense fur or feathers, the hypothalamus, acting through the sympathetic nervous system, brings about contraction of the tiny muscles at the base of the hair or feather shafts to lift the hair or feathers off the skin surface. This puffing up traps a layer of poorly conductive air between the skin surface and the environment, thus increasing the insulating barrier between the core and the cold air and reducing heat loss. Even though the hair-shaft muscles contract in humans in response to cold exposure, this heat-retention mechanism is ineffective because of the low density and fine texture of most human body hair. The result instead is useless *goosebumps*.

After maximum skin vasoconstriction has been achieved as a result of exposure to cold, further heat dissipation in humans can be prevented only by behavioral adaptations, such as postural changes that reduce as much as possible the exposed surface area from which heat can escape. These postural changes include maneuvers such as hunching over, clasping the arms in front of the chest, or curling up in a ball. Putting on warmer clothing further insulates the body from too much heat loss. Clothing entraps layers of poorly conductive air between the skin surface and the environment, thereby diminishing loss of heat by conduction from the skin to the cold external air and curtailing the flow of convection currents.

Under the opposite circumstance—heat exposure—the anterior part of the hypothalamus reduces heat production by decreasing skeletal muscle activity and promotes increased heat loss by inducing skin vasodilation. When even maximal skin vasodilation is inadequate to rid the body of excess heat, sweating is brought into play to accomplish further heat loss through evaporation. In fact, if the air temperature rises above the temperature of maximally vasodilated skin, the temperature gradient reverses itself so that heat is gained from the environment. Sweating is the only means of heat loss under these conditions.

Humans sweat all over their bodies, but many species either lack or have limited distribution of sweat glands; for example, dogs have sweat glands only on their paw pads. These species must employ alternative methods, such as pant-

Table 17-2 **Coordinated Adjustments in Response to Cold or Heat Exposure**

In Response to Cold Exposure (Coordinated by the Posterior Hypothalamus)		*In Response to Heat Exposure (Coordinated by the Anterior Hypothalamus)*	
Increased Heat Production	*Decreased Heat Loss (Heat Conservation)*	*Decreased Heat Production*	*Increased Heat Loss*
Increased muscle tone	Skin vasoconstriction	Decreased muscle tone	Skin vasodilation
Shivering	Postural changes to reduce exposed surface area (hunching shoulders, etc.)*	Decreased voluntary exercise*	Sweating
Increased voluntary exercise*	Warm clothing*		Cool clothing*
Nonshivering thermogenesis			

*Behavioral adaptations.

ing, to enhance heat loss during heat exposure. The hypothalamus induces a shallow, rapid breathing pattern that permits large volumes of air to move over the hot, moist tongue and respiratory airways without leading to imbalances in the blood concentration of O_2 and CO_2. The resultant increase in evaporative heat loss from the respiratory tract provides a cooling effect.

In addition to sweating, humans employ voluntary measures, such as using fans, wetting the body, drinking cold beverages, and wearing cool clothing, to further enhance heat loss. Contrary to popular belief, wearing light-colored, loose clothing is cooler than being nude. Naked skin absorbs almost all of the radiant energy that strikes it, whereas light-colored clothing reflects almost all of the radiant energy that falls on it. Thus, if light-colored clothing is loose and thin enough to permit convection currents and evaporative heat loss to occur, wearing it is actually cooler than going without any clothes at all.

Skin vasomotor activity is highly effective in controlling heat loss in environmental temperatures between the upper 60s and mid 80s. This range, within which core temperature can be kept constant by vasomotor responses without calling supplementary heat-production or heat-loss mechanisms into play, is called the **thermoneutral zone.** When external air temperature falls below the lower limits of the ability of skin vasoconstriction to reduce heat loss further, the major burden of maintaining core temperature is borne by increased heat production, especially shivering. At the other extreme, when external air temperature exceeds the upper limits of the ability of skin vasodilation to increase heat loss further, sweating becomes the dominant factor in maintaining core temperature.

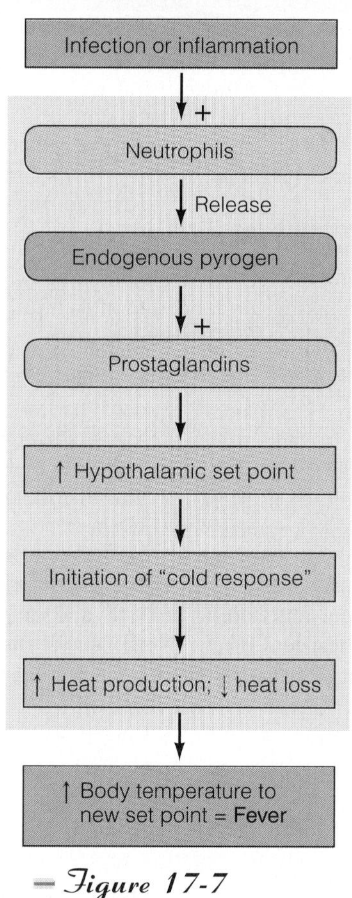

Figure 17-7
Fever Production

During a fever, the hypothalamic thermostat is "reset" at an elevated temperature.

Fever refers to an elevation in body temperature as a result of infection or inflammation. In response to microbial invasion, certain white blood cells release a chemical known as **endogenous pyrogen,** which, among its many infection-fighting effects (see p. 379), acts on the hypothalamic thermoregulatory center to raise the setting of the thermostat (Fig. 17-7). The hypothalamus now maintains the temperature at the new set level instead of maintaining normal body temperature. If, for example, endogenous pyrogen raises the set point to 102°F (as recorded orally), the hypothalamus senses that the normal prefever temperature is too cold, so it initiates the cold-response mechanisms to elevate the temperature to 102°F. Shivering is initiated to rapidly increase heat production, while skin vasoconstriction is brought about to rapidly reduce heat loss, both of which drive the temperature upward. These events account for the sudden cold chills often experienced at the onset of a fever. Because the person feels cold, he or she may put on more blankets as a voluntary mechanism that helps elevate body temperature by conserving body heat. Once the new temperature is achieved, body temperature is regulated as normal in response to cold and heat but at a higher setting. Thus, fever production in response to an infection is a deliberate outcome and is not due to a breakdown of thermoregulatory mechanisms. Although the physiological significance of a fever is still unclear, many medical experts believe that a rise in body temperature has a beneficial role in fighting infection. A fever augments the inflammatory response and may interfere with bacterial multiplication (see p. 379).

Endogenous pyrogen raises the set point of the hypothalamic thermostat during fever production by triggering the local release of *prostaglandins,* which are local chemical mediators that act directly on the hypothalamus. Aspirin reduces a fever by inhibiting the synthesis of prostaglandins. Aspirin does not lower the temperature in a nonfebrile person, because prostaglandins are not present in the hypothalamus in appreciable quantities in the absence of endogenous pyrogen.

The exact molecular cause of a fever "breaking" naturally is unknown, although it presumably results from a reduction in pyrogen release or decreased prostaglandin synthesis. When the hypothalamic set point is restored to normal, the temperature at 102°F (in this example) is too high. The heat-response mechanisms are instituted to cool down the body. Skin vasodilation occurs and sweating commences. The person feels hot and throws off extra covers. The gearing up of these heat-loss mechanisms by the hypothalamus reduces the temperature to normal.

Hyperthermia can occur unrelated to infection.

Hyperthermia denotes any elevation in body temperature above the normally accepted range. The term *fever* is usually reserved for an elevation in temperature caused by resetting of the hypothalamic set point by the release of endogenous pyrogen during infection or inflammation; hyperthermia refers to all other imbalances between heat gain and heat loss that increase body temperature. Hyperthermia has a variety of causes, some of which are normal and harmless, others pathological and fatal.

The most common cause of hyperthermia is sustained exercise. As a physical consequence of the tremendous heat load generated by exercising muscles, body temperature rises during the initial stage of exercise because heat gain exceeds heat

CONCEPTS, CHALLENGES, AND CONTROVERSIES

The Extremes of Heat and Cold Can Be Fatal

Heat exhaustion refers to a state of collapse, usually manifested by fainting, that is caused by reduced blood pressure brought about as a result of overtaxing the heat-loss mechanisms. Extensive sweating reduces cardiac output by depleting the plasma volume, and pronounced skin vasodilation causes a drop in total peripheral resistance. Since blood pressure is determined by cardiac output times total peripheral resistance, blood pressure falls, an insufficient amount of blood is delivered to the brain, and fainting takes place. Thus, heat exhaustion is a consequence of overactivity of the heat-loss mechanisms rather than a breakdown of these mechanisms. Because the heat-loss mechanisms have been very active, body temperature is only mildly elevated in heat exhaustion. By forcing the cessation of activity when the heat-loss mechanisms are no longer able to cope with heat gain through exercise or a hot environment, heat exhaustion serves as a safety valve to help prevent the more serious consequences of heat stroke.

Heat stroke is an extremely dangerous situation that arises from the complete breakdown of the hypothalamic thermoregulatory systems. Heat exhaustion may progress into heat stroke if the heat-loss mechanisms continue to be overtaxed. Heat stroke is more likely to occur upon overexertion during a prolonged exposure to a hot, humid environment. The elderly, in whom thermoregulatory responses are generally slower and less efficient, are particularly vulnerable to heat stroke during prolonged, stifling heat waves. So too are individuals who are taking certain common tranquilizers, because these drugs interfere with the hypothalamic thermoregulatory centers' neurotransmitter activity.

The most striking feature of heat stroke is a lack of compensatory heat-loss measures, such as sweating, in the face of a rapidly rising body temperature. No sweating occurs despite a markedly elevated body temperature because the hypothalamic thermoregulatory control centers are not functioning properly and cannot initiate heat-loss mechanisms. During the development of heat stroke, body temperature starts to climb as the heat-loss mechanisms are eventually overwhelmed by prolonged, excessive heat gain. Once the core temperature reaches the point at which the hypothalamic temperature-control centers are damaged by the heat, the body temperature rapidly rises even higher because of the complete shutdown of heat-loss mechanisms. Furthermore, as the body temperature increases, the rate of metabolism increases correspondingly because higher temperatures speed up the rate of all chemical reactions; the result is even greater heat production. This positive-feedback state sends the temperature spiraling upward. Heat stroke is a very dangerous situation that is rapidly fatal if untreated. Even with treatment to halt and reverse the rampant rise in body temperature, there is still a high rate of mortality; the rate of permanent disability in survivors is also high because of irreversible damage caused by the high internal heat.

At the other extreme, the body can be harmed by cold exposure in two ways: frostbite and generalized hypothermia. **Frostbite** involves excessive cooling of a particular part of the body to the point where tissue in that area is damaged. If exposed tissues actually freeze, tissue damage occurs as a result of disruption of the cells by formation of ice crystals or by lack of liquid water. **Hypothermia,** a fall in body temperature, occurs when generalized cooling of the body exceeds the ability of the normal heat-producing and heat-conserving regulatory mechanisms to match the excessive heat loss. As hypothermia sets in, the rate of all metabolic processes slows down because of the declining temperature. Higher cerebral functions are the first to be affected by body cooling, leading to loss of judgment, apathy, disorientation, and tiredness, all of which diminish the cold victim's ability to initiate voluntary mechanisms to reverse the falling body temperature. As body temperature continues to plummet, depression of the respiratory center occurs, reducing the ventilatory drive so that breathing becomes slow and weak. Activity of the cardiovascular system also is gradually reduced. The heart is slowed and cardiac output decreased. Disturbances of cardiac rhythm occur, eventually leading to ventricular fibrillation and death.

---

loss (Fig. 17-8). The elevation in core temperature reflexly triggers heat-loss mechanisms (skin vasodilation and sweating), which eliminate the discrepancy between heat production and heat loss. As soon as the heat-loss mechanisms are stepped up sufficiently to equalize heat production, the core temperature stabilizes at a level slightly above the set point despite continued heat-producing exercise. Thus, during sustained exercise, body temperature initially rises, then is maintained at the higher level as long as the exercise continues.

A completely different way in which hyperthermia can be brought about is by excessive heat production in connection with abnormally high circulating levels of epinephrine or thyroid hormone that result from dysfunctions of the adrenal medulla or thyroid gland, respectively. Both of these hor-

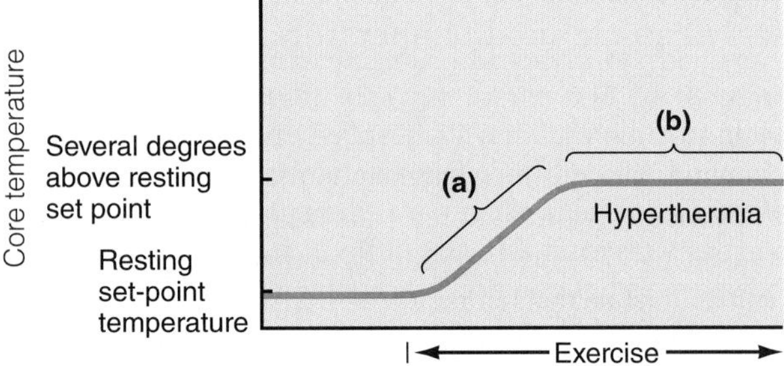

Figure 17-8 Hyperthermia in Sustained Exercise
(a) At the onset of exercise, the rate of heat production initially exceeds the rate of heat loss so the core temperature rises. (b) When heat-loss mechanisms are reflexly increased sufficiently to equalize the elevated heat production, the core temperature stabilizes slightly above the resting point for the duration of the exercise.

mones elevate the core temperature by increasing the overall rate of metabolic activity and heat production.

Another way in which hyperthermia occurs is malfunction of the hypothalamic control centers. Certain brain lesions, for example, destroy the normal regulatory capacity of the hypothalamic thermostat. When the thermoregulatory mechanisms are not functional, lethal hyperthermia may occur very rapidly. Normal metabolism produces enough heat to kill a person in less than five hours if the heat-loss mechanisms are completely shut down. In addition to brain lesions, a breakdown in the function of hypothalamic thermoregulation can also occur upon exposure to severe, prolonged heat stress. Similarly, the body can be harmed by extreme cold exposure. (See the accompanying boxed feature, ◆ Concepts, Challenges, and Controversies.)

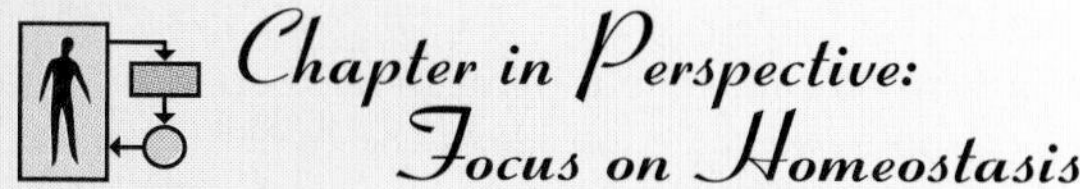

Chapter in Perspective: Focus on Homeostasis

Because energy can be neither created nor destroyed, input must equal output in the case of both the body's total energy balance and its heat-energy balance in order for body weight and body temperature, respectively, to remain constant. If total energy input exceeds total energy output, the extra energy is stored in the body, and body weight increases. Similarly, if the input of heat energy exceeds its output, body temperature increases. Conversely, if output exceeds input, body weight decreases or body temperature falls. The hypothalamus is the major integrating center for the maintenance of both a constant total energy balance (and thus a constant body weight) and a constant heat-energy balance (and thus a constant body temperature).

Body temperature, which is one of the homeostatically regulated factors of the internal environment, must be maintained within narrow limits because the structure and reactivity of the chemicals that compose the body are temperature sensitive. Deviations in body temperature outside a limited range result in protein denaturation and death of the individual if the temperature rises too high, or metabolic slowing and death if the temperature falls too low.

Body weight, in contrast, varies widely among individuals. Only the extremes of imbalances between total energy input and output become incompatible with life. For example, in the face of insufficient energy input in the form of ingested food during prolonged starvation, the body resorts to breaking down muscle protein to meet its needs for energy expenditure once the adipose stores are depleted. Body weight dwindles due to this self-cannibalistic mechanism until death finally occurs as a result of loss of heart muscle, among other things. At the other extreme, when the food energy consumed greatly exceeds the energy expended, the extra energy input is stored as adipose tissue, and body weight increases. The resultant gross obesity can also lead to heart failure. Not only must the heart work harder to pump blood to the excess adipose tissue, but obesity also predisposes the individual to atherosclerosis and heart attacks (see p. 296).

Chapter Summary

Energy Balance

Energy input to the body in the form of food energy must equal energy output, because energy cannot be created or destroyed. Energy output or expenditure includes (1) external work performed by skeletal muscles to accomplish movement of an external object or movement of the body through the external environment and (2) internal work, which consists of all other energy-dependent activities that do not accomplish external work, including active transport, smooth and cardiac muscle contraction, glandular secretion, and protein synthesis. Only about 25% of the chemical energy in food is harnessed to do biological work. The rest is immediately converted to heat. Furthermore, all of the energy expended to accomplish internal work is eventually converted into heat, and 75% of the energy expended by working skeletal muscles is lost as heat. Therefore, most of the energy in food ultimately appears as body heat. The metabolic rate, which is energy expenditure per unit of time, is measured in kilocalories of heat produced per hour.

For a neutral energy balance, the energy in ingested food must equal energy expended in performing work. If more food is consumed than energy expended, the extra energy is stored in the body, primarily as adipose tissue, so body weight increases. On the other hand, if more energy is burned than is available in the food, body energy stores are used to support energy expenditure, so body weight decreases. Usually, body weight remains fairly constant over a prolonged period of time (except during growth) because food intake is adjusted to match energy expenditure on a long-term basis. Food intake is controlled primarily by the hypothalamus by means of complex, poorly understood regulatory mechanisms in which hunger and satiety are important components.

Temperature Regulation

The body can be thought of as a heat-generating core (internal organs, CNS, and skeletal muscles) surrounded by a shell of variable insulating capacity (the skin). The skin exchanges heat energy with the external environment, with the direction and amount of heat transfer depending on the environmental temperature and the momentary insulating capacity of the shell. The four physical means by which heat is exchanged between the body and the external environment are (1) radiation (net movement of heat energy via electromagnetic waves); (2) conduction (exchange of heat energy by direct contact); (3) convection (transfer of heat energy by means of air currents); and (4) evaporation (extraction of heat energy from the body by the heat-requiring conversion of liquid H_2O to H_2O vapor). Because heat energy moves from warmer to cooler objects, radiation, conduction, and convection can be channels for either heat loss or heat gain, depending on whether surrounding objects are cooler or warmer, respectively, than the body surface. Normally, they are avenues for heat loss, along with evaporation resulting from sweating.

To prevent serious cellular malfunction, the core temperature must be held constant at about 100°F (equivalent to an average oral temperature of 98.2°F) by continuously balancing heat gain and heat loss despite changes in environmental temperature and variation in internal heat production. This thermoregulatory balance is controlled by the hypothalamus. The hypothalamus is apprised of the skin temperature by peripheral thermoreceptors and of the core temperature by central thermoreceptors, the most important of which are located in the hypothalamus itself. The primary means of heat gain is heat production by metabolic activity, the biggest contributor being skeletal muscle contraction. Heat loss is adjusted by sweating and by controlling to the greatest extent possible the temperature gradient between the skin and the surrounding environment. The latter is accomplished by regulating the caliber of the skin's blood vessels. Vasoconstriction of the skin vessels reduces the flow of warmed blood through the skin so that skin temperature falls. The layer of cool skin between the core and the environment increases the insulating barrier between the warm core and the external air. Conversely, skin vasodilation brings more warmed blood through the skin so that skin temperature approaches the core temperature, thus reducing the insulative capacity of the skin.

Upon exposure to cool surroundings, the core temperature starts to fall as heat loss increases due to the larger-than-normal skin-to-air temperature gradient. The hypothalamus responds to reduce the heat loss by inducing skin vasoconstriction while simultaneously increasing heat production through heat-generating shivering. Conversely, in response to a rise in core temperature (resulting either from excessive internal heat production accompanying exercise or from excessive heat gain upon exposure to a hot environment), the hypothalamus triggers heat-loss mechanisms, such as skin vasodilation and sweating, while simultaneously decreasing heat production, such as by reducing muscle tone. In both cold and heat responses, voluntary behavioral actions also contribute importantly to maintenance of thermal homeostasis.

A fever occurs when endogenous pyrogen released from white blood cells in response to infection raises the hypothalamic set point. An elevated core temperature develops as the hypothalamus initiates cold-response mechanisms to raise the core temperature to the new set point.

Review Exercises

Objective Questions (Answers on p. E–17.)

1. If more food energy is consumed than is expended, the excess energy is lost as heat. (True or false?)
2. All of the energy within nutrient molecules can be harnessed to perform biological work. (True or false?)
3. Each liter of O_2 contains 4.8 kilocalories of heat energy. (True or false?)
4. A body temperature greater than 98.2°F is always indicative of a fever. (True or false?)
5. Core temperature is relatively constant, but skin temperature can vary markedly. (True or false?)
6. Sweat that drips off the body has no cooling effect. (True or false?)
7. Production of "goosebumps" in response to cold exposure has no value in regulating body temperature. (True or false?)
8. The posterior region of the hypothalamus triggers shivering and skin vasoconstriction. (True or false?)
9. The primary means of involuntarily increasing heat production is ____________.
10. Increased heat production independent of muscle contraction is known as ____________.
11. The only means of heat loss when the environmental temperature exceeds the core temperature is ____________.
12. Which of the following statements concerning heat exchange between the body and the external environment is *incorrect?*
    a. Heat gain is primarily by means of internal heat production.
    b. Radiation serves as a means of heat gain but not of heat loss.
    c. Heat energy always moves down its concentration gradient from warmer to cooler objects.
    d. The temperature gradient between the skin and the external air is subject to control.
    e. Very little heat is lost from the body by conduction alone.
13. Which of the following statements concerning fever production is *incorrect?*
    a. Endogenous pyrogen is released by white blood cells in response to microbial invasion.
    b. The hypothalamic set point is elevated.
    c. The hypothalamus initiates cold-response mechanisms to increase the body temperature.
    d. Prostaglandins appear to mediate the effect.
    e. The hypothalamus is not effective in regulating body temperature during a fever.

14. Using the answer code below, indicate which mechanism of heat transfer is being described:

(a) radiation
(b) conduction
(c) convection
(d) evaporation

___ 1. sitting on a cold metal chair
___ 2. sunbathing on the beach
___ 3. a gentle breeze
___ 4. sitting in front of a fireplace
___ 5. sweating
___ 6. riding in a car with the windows open
___ 7. lying on an electric blanket
___ 8. sitting in a wet bathing suit
___ 9. fanning yourself
___10. immersion in cool water

Essay Questions

1. Differentiate between external and internal work.
2. Define metabolic rate and basal metabolic rate. Explain the process of indirect calorimetry.
3. Describe the three states of energy balance.
4. By what means is energy balance primarily maintained?
5. List the sources of heat input and output for the body.
6. What are the two hypothalamic centers for temperature regulation?
7. Discuss the compensatory measures that occur in response to a fall in core temperature as a result of cold exposure and in response to a rise in core temperature as a result of heat exposure.

Quantitative Exercises (Solutions on p. E–18.)

1. The basal metabolic rate (BMR) is a measure of how much energy the body consumes to maintain its "idling speed." The normal BMR = about 72 kcal/hr (see p. 604). The vast majority of this energy is converted to heat. Our thermoregulatory systems function to eliminate this heat so as to keep body temperature constant. If our bodies were not able to lose this heat, our temperature would rise until we boiled (of course a person would die before reaching that temperature). It is relatively easy to calculate how long it would take to reach the hypothetical boiling point. If an amount of energy ΔU is put into a liquid of mass m, the temperature change ΔT (in °C) is given by:

$$\Delta T = \Delta U/m \times C$$

In this equation, C is the specific heat of the liquid. For water, C = 1.0 kcal/kg-°C. Use this information to calculate how long it would take for the heat from the BMR to boil your body fluids (assume 42 liters of water in your body and a starting point of normal body temperature at 37°C). When exercising maximally, a person consumes about 1,000 kcal/hr. How long would it take to boil in this case?

Points to Ponder

(Explanations on p. E–18.)

1. Explain how drugs that selectively inhibit CCK increase feeding behavior in experimental animals.
2. What advice would you give an overweight friend who asks for your help in designing a safe, sensible, inexpensive program for losing weight?
3. Why is it dangerous to engage in heavy exercise on a hot, humid day?
4. Describe the avenues for heat loss in a person soaking in a hot bath.
5. Consider the difference between you and a fish in a local pond with regard to control of body temperature. Humans are *thermoregulators;* they are able to maintain a remarkably constant, rather high internal body temperature despite the body's exposure to a wide range of environmental temperatures. To maintain thermal homeostasis, humans physiologically manipulate mechanisms within their bodies to adjust heat production, heat conservation, and heat loss. In contrast, fish are *thermoconformers;* their body temperatures conform to the temperature of their surroundings. Thus, their body temperatures vary capriciously with changes in the environmental temperature. Even though fish produce heat, they cannot physiologically regulate internal heat production, nor can they control heat exchange with their environment to maintain a constant body temperature when the temperature in their surroundings rises or falls. Knowing this, do you think fish run a fever when they have a systemic infection? Why or why not?
6. ***Clinical Consideration*** Michael F., a drowning victim, was pulled from the icy water by rescuers 15 minutes after he fell through the thin ice on which he was skating. Michael is now alert and recuperating in the hospital. How can you explain his "miraculous" survival even though he was submerged for 15 minutes and irreversible brain damage, soon followed by death, normally occurs if the brain is deprived of its critical O_2 supply for more than four or five minutes?

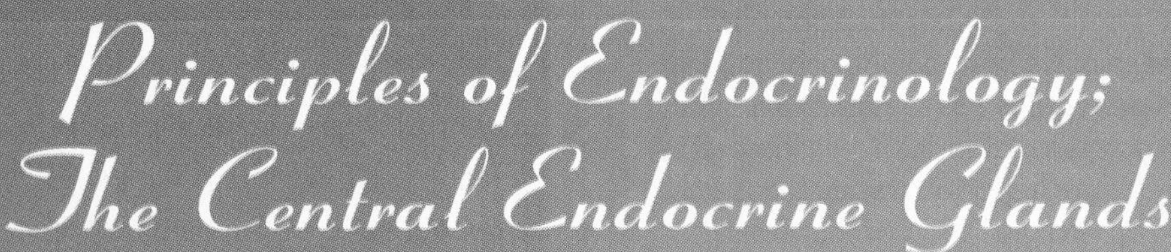

18 Principles of Endocrinology; The Central Endocrine Glands

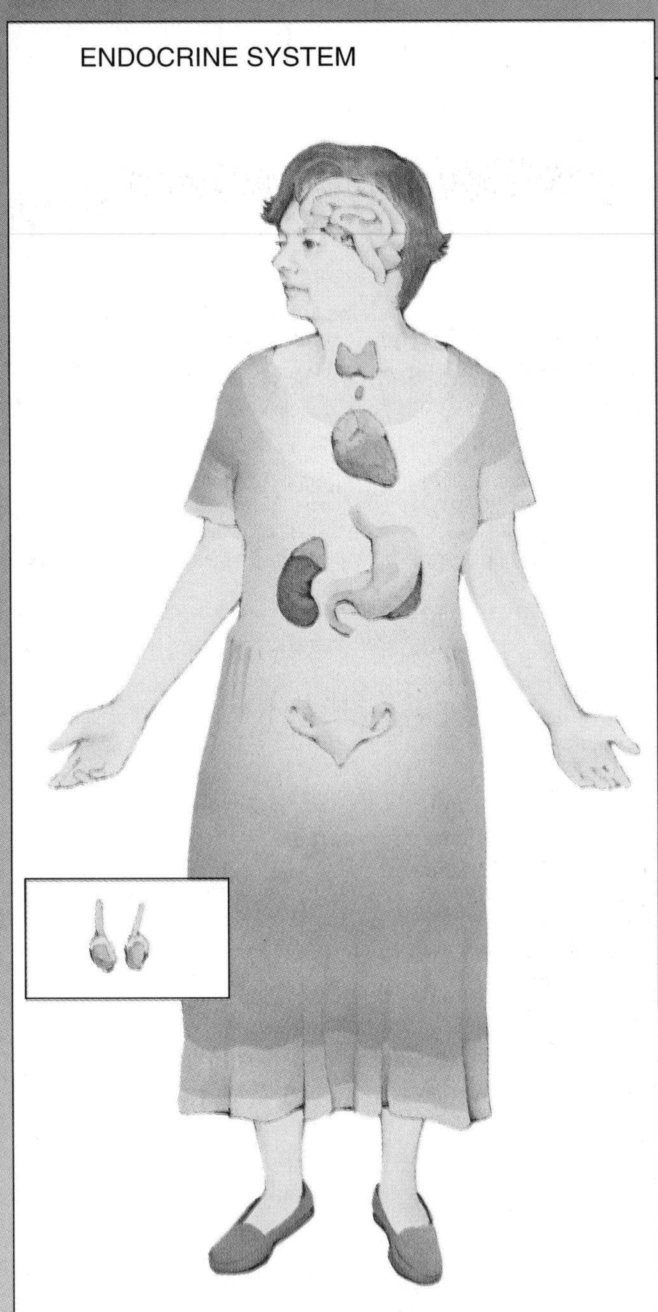

Body systems maintain homeostasis

HOMEOSTASIS
The endocrine system, one of the body's two major control systems, secretes hormones which act on their target cells to regulate the blood concentrations of nutrient molecules, water, salt and other electrolytes, among other homeostatic activities. Hormones also play a key role in controlling growth and reproduction and in stress adaptation.

Homeostasis is essential for survival of cells

CELLS
Cells need a constant supply of nutrients to support their energy-generating chemical reactions. Normal cell function also depends on a proper balance of water and various electrolytes.

Cells make up body systems

The **endocrine system** generally regulates activities that require duration rather than speed. Endocrine glands release **hormones,** blood-borne chemical messengers that act on target cells typically located a long distance from the endocrine gland. Most of the target cells' activities under the control of hormones are directed toward maintaining homeostasis. The central endocrine glands include the pineal gland, hypothalamus, and pituitary gland. The **pineal gland,** a part of the brain, secretes a hormone important in establishing the body's biological rhythms. The **hypothalamus,** also a part of the brain, and the **posterior pituitary gland** act as a unit to release hormones essential for maintaining water balance and for giving birth and breast-feeding. The hypothalamus also secretes regulatory hormones that control the hormonal output of the anterior pituitary gland. The **anterior pituitary gland** secretes six hormones, which in turn largely control the hormonal output of several peripheral endocrine glands. One anterior pituitary hormone, growth hormone, promotes growth and influences nutrient homeostasis.

Chapter Contents At a Glance

General Principles of Endocrinology

Hormones are secreted by widely dispersed endocrine glands and exert a variety of regulatory effects throughout the body.

The **endocrine system** is composed of the ductless endocrine glands (see p. 4) that are scattered throughout the body (Fig. 18-1). Even though the endocrine glands for the most part are not connected anatomically, they constitute a system in a functional sense. They all accomplish their functions by secreting hormones into the blood, and many functional interactions take place among the various endocrine glands. Once secreted, a hormone travels in the blood to its distant target cells, where it regulates or directs a particular function. **Endocrinology** is the study of the homeostatic chemical adjustments and other activities accomplished by hormones.

Recall that neurosecretory neurons release their chemical messengers, *neurohormones,* into the blood (see p. 66). In contrast, ordinary neurons release their chemical messengers, neurotransmitters, into a synaptic cleft. Like hormones, neurohormones are distributed by the blood to the target cells. Thus, neurosecretory neurons are considered part of the endocrine system. The general term "hormone" will tacitly include both blood-borne hormonal and neurohormonal messengers.

Even though hormones are distributed throughout the body by the blood, only specific target cells are able to respond to each hormone because only the target cells possess receptors (particular plasma membrane proteins) for binding

Figure 18-1 The Endocrine System

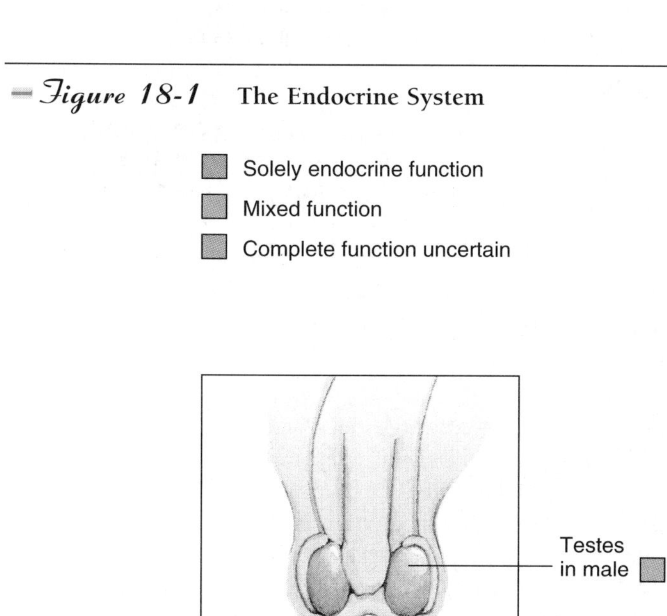

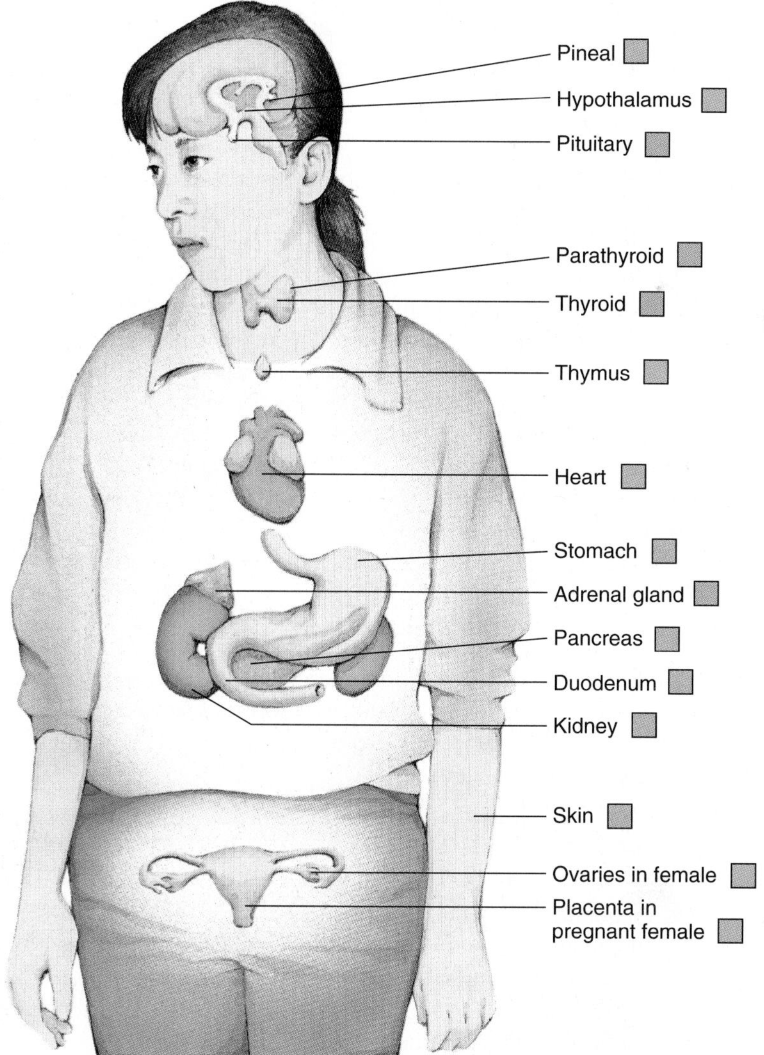

with the particular hormone. Binding of a hormone with its specific target cell receptors initiates a chain of events within the target cells to bring about the hormone's final effect. Some hormones have a single target cell type; others have many.

The endocrine system is one of the body's two major control systems, the other being the nervous system, with which you are already familiar (chapters 4–7). Recall that the endocrine and nervous systems have different modes of action and different realms of authority, yet they also have much in common and interact extensively (see pp. 110–112). The nervous and endocrine systems are specialized for controlling different types of activities. In general, the nervous system is responsible for coordinating rapid, precise responses and is especially important in mediating interactions with the external environment. The endocrine system, on the other hand, primarily controls activities that require duration rather than speed, including the following:

1. Regulating organic metabolism and H_2O and electrolyte balance, which are important collectively in maintaining a constant internal environment.
2. Inducing adaptive changes to help the body cope with stressful situations.
3. Promoting smooth, sequential growth and development.
4. Controlling reproduction.
5. Regulating red blood cell production.
6. Along with the autonomic nervous system, controlling and integrating both circulation and the digestion and absorption of food.

The sole function of some hormones is regulating the production and secretion of another hormone. A hormone that has as its primary function the regulation of hormone secretion by another endocrine gland is classified functionally as a **tropic hormone.** For example, the only function of thyroid-stimulating hormone (TSH) from the anterior pituitary gland is the regulation of thyroid hormone secretion by the thyroid gland. A **nontropic hormone,** on the other hand, primarily exerts its effects on nonendocrine target tissues. Thyroid hormone, which increases the rate of O_2 consumption and the metabolic activity of almost every cell in the body, is an example of a nontropic hormone.

The following points add to the complexity of the endocrine system:

- A single endocrine gland may produce multiple hormones. The anterior pituitary, for example, secretes six different hormones; each is under different control mechanisms and has different functions, some being tropic and others having nontropic effects.
- A single hormone may be secreted by more than one endocrine gland. For example, the hypothalamus and pancreas both secrete the hormone somatostatin.
- Frequently, a single hormone has more than one type of target cell and therefore can induce more than one type of effect. As an example, vasopressin promotes H_2O reabsorption by the kidney tubules as well as vasoconstriction of arterioles throughout the body.
- A single target cell may be influenced by more than one hormone. Some cells contain an array of receptors for responding in different ways to different hormones. To illustrate, insulin promotes the conversion of glucose into glycogen within liver cells by stimulating one particular hepatic enzyme, whereas another hormone, glucagon, enhances the degradation of glycogen into glucose within liver cells by activating yet another hepatic enzyme.
- The same chemical messenger may be either a hormone or a neurotransmitter, depending on its source and mode of delivery to the target cell. Norepinephrine, which is secreted as a hormone by the adrenal medulla and released as a neurotransmitter from sympathetic postganglionic nerve fibers, is a prime example.
- Some organs are exclusively endocrine in function (they specialize in hormonal secretion alone, the anterior pituitary and thyroid glands being examples), whereas other organs of the endocrine system perform nonendocrine functions in addition to secreting hormones. For example, the testes produce sperm and also secrete the male sex hormone testosterone. Other examples of mixed organs are the ovaries, digestive tract, pancreas, kidneys, and even the brain. In each case except the brain, mixed function occurs because the organ houses nonendocrine tissue plus isolated clusters of endocrine cells that migrated to the organ during embryonic development. The endocrine function of the brain derives from the presence of neurosecretory neurons. There are no endocrine cells as such in the brain.

Because hormones are widely dispersed through the circulatory system, those that have multiple target cell types are able to coordinate the activities of various tissues toward a common goal. Tissues such as muscle, liver, and fat all respond to hormones in a fashion that ensures that their contributions to overall intermediary metabolism are acting in concert.

Endocrine systems also provide temporal (time) coordination of function. This is particularly apparent in the endocrine control of reproductive cycles, such as the menstrual cycle, in which normal function requires highly specific patterns of change in the secretion of various hormones.

This has been a brief overview of the general functions of the endocrine system. ▮Table 18-1 on p. 624 summarizes the most important specific functions of the major hormones. Some of these hormones have been introduced elsewhere and will not be discussed further; these are the gastrointestinal hormones (chapter 16), the renal hormones (erythropoietin in chapter 11 and renin in chapter 14), atrial natriuretic peptide from the heart (chapter 15), and thymosin (chapter 12). The remainder of the hormones will be described in greater detail in this and the next two chapters.

As extensive as Table 18-1 appears to be, it leaves out a variety of "candidate" or potential hormones that have not fully qualified as hormones, either because their characteristics do not quite fit the classic definition of a hormone or because they have been discovered so recently that their hormonal status has not yet been conclusively documented. The table also excludes the hormones secreted by the effector cells of the defense system (white blood cells and macrophages) and a variety of recently revealed and poorly understood growth factors that promote growth of specific tissues, such as *epidermal growth factor* and *nerve growth factor.* In addition, the table is undoubtedly incomplete in ways as yet unknown. First, it is more than likely that additional hormones will be discovered; the list of hormones still continues to grow as a result of the probing techniques of endocrinologists. Second, an exciting finding is that hormones with well-established functions sometimes exert other, quite unrelated functions as well. As an example, researchers determined that antidiuretic hormone, which controls H_2O retention by the kidneys during the formation of urine, also exerts vasoconstrictor effects on arteriolar smooth muscle, giving rise to its more commonly used alternative name of vasopressin. More recently, vasopressin has also been implicated as playing roles in fever, learning, memory, and behavior.

Hormones are chemically classified into three categories: peptides, amines, and steroids.

Hormones are not all similar chemically, but instead fall into three distinct classes according to their biochemical structure (Table 18-2 on p. 626): (1) peptides and proteins, (2) amines, and (3) steroids. The first two categories are both amino acid derivatives. The **peptide** and **protein hormones** consist of specific amino acids arranged in a chain of varying length; the shorter chains are peptides and the longer ones are categorized as proteins. For convenience, we will refer to this entire category as peptides. The majority of hormones fall into this class, including those secreted by the hypothalamus, anterior pituitary, posterior pituitary, pineal gland, pancreas, parathyroid gland, gastrointestinal tract, kidneys, liver, thyroid C cells, and heart. The **amines** are derived from the amino acid *tyrosine* and include the hormones secreted by the thyroid gland and adrenal medulla. The adrenomedullary hormones are specifically known as *catecholamines.* The **steroids,** which include the hormones secreted by the adrenal cortex and gonads, as well as most placental hormones, are neutral lipids derived from cholesterol.

Minor differences in chemical structure between hormones within each category often result in profound differences in biological response. For example, note the subtle difference between testosterone, the male sex hormone responsible for inducing the development of masculine characteristics, and estradiol, the predominant form of estrogen, which is the feminizing female sex hormone (Fig. 18-2).

The structural classification of hormones is of more than biochemical interest. The chemical properties of a hormone, most notably, its solubility, determine the means by which the hormone is synthesized, stored, and secreted; the way it is transported in the blood; and the mechanism by which it exerts its effects at the target cell. The following differences in the solubility of the various types of hormones are critical to their function (Table 18-2, p. 626):

1. All peptides and catecholamines are hydrophilic (water-loving) and lipophobic (lipid-fearing); that is, they are highly H_2O soluble and have low lipid solubility.

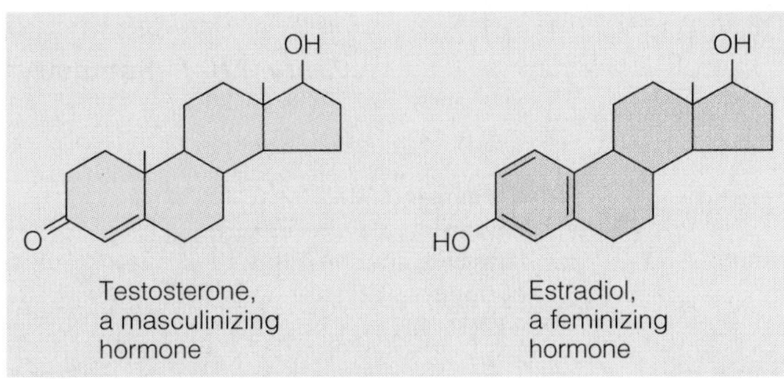

Figure 18-2 **Comparison of Testosterone and Estradiol**

2. All steroid and thyroid hormones are lipophilic (lipid-loving) and hydrophobic (water-fearing); that is, they have high lipid solubility and are poorly soluble in H_2O.

The mechanisms of hormone synthesis, storage, and secretion vary according to the class of hormone.

Because of their chemical differences, the means by which the various classes of hormones are synthesized, stored, and secreted differ as follows:

Peptide hormones Peptide hormones are synthesized by the same method used for the manufacture of any protein that is to be exported (see p. 21). Because they are destined to be released from the endocrine cell, the synthesized hormones must be segregated from intracellular proteins by being sequestered in a membrane-enclosed compartment until they are secreted. Briefly, the synthesis of peptide hormones requires the following steps:

1. Large precursor proteins, or **preprohormones,** are synthesized by ribosomes on the rough endoplasmic reticulum. They then migrate to the Golgi complex in membrane-enclosed vesicles that pinch off from the smooth endoplasmic reticulum.
2. During their journey through the endoplasmic reticulum and Golgi complex, the large preprohormone precursor molecules are pruned first to **prohormones** and finally to **active hormones.** The peptide "scraps" that are left over as a large preprohormone molecule is cleaved to form the classic hormone are often stored and cosecreted along with the hormone. This raises the possibility that these other peptides may also exert biological effects that differ from the traditional hormonal product; that is, the cell may actually be secreting multiple hormones, but the functions of the other peptide products are for the most part unknown. A known example is the cleavage of the large precursor molecule *pro-opiomelanocortin* into three active products: *adrenocorticotropic hormone (ACTH), melanocyte-stimulating hormone (MSH),* and *β-endorphin.*
3. The Golgi complex concentrates the finished hormones, then packages them into secretory vesicles that are pinched off and stored in the cytoplasm until an appropriate signal triggers their secretion. By storing peptide hor-

Table 18-1 Summary of the Major Hormones

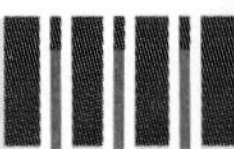

Endocrine Gland	*Hormones*	*Target Cells*	*Major Functions of Hormones*
Hypothalamus	Releasing and inhibiting hormones (TRH, CRH, GnRH, GHRH, GHIH, PRF, PIH)	Anterior pituitary	Controls release of anterior pituitary hormones
Posterior pituitary (hormones stored in)	Vasopressin (antidiuretic hormone)	Kidney tubules	Increases H_2O reabsorption
		Arterioles	Produces vasoconstriction
	Oxytocin	Uterus	Increases contractility
		Mammary glands (breasts)	Causes milk ejection
Anterior pituitary	Thyroid-stimulating hormone (TSH)	Thyroid follicular cells	Stimulates T_3 and T_4 secretion
	Adrenocorticotropic hormone (ACTH)	Zona fasciculata and zona reticularis of adrenal cortex	Stimulates cortisol secretion
	Growth hormone	Bone; soft tissues	Essential but not solely responsible for growth; stimulates growth of bones and soft tissues; metabolic effects include protein anabolism, fat mobilization, and glucose conservation.
		Liver	Stimulates somatomedin secretion
	Follicle-stimulating hormone (FSH)	Females: ovarian follicles	Promotes follicular growth and development; stimulates estrogen secretion
		Males: seminiferous tubules in testes	Stimulates sperm production
	Luteinizing hormone (LH) (interstitial cell–stimulating hormone—ICSH)	Females: ovarian follicle and corpus luteum	Stimulates ovulation, corpus luteum development, and estrogen and progesterone secretion
		Males: interstitial cells of Leydig in testes	Stimulates testosterone secretion
	Prolactin	Females: mammary glands	Promotes breast development; stimulates milk secretion
		Males	Uncertain
Thyroid gland follicular cells	Tetraiodothyronine (T_4 or thyroxine); triiodothyronine (T_3)	Most cells	Increases the metabolic rate; essential for normal growth and nerve development
Thyroid gland C cells	Calcitonin	Bone	Decreases plasma calcium concentration
Adrenal cortex			
Zona glomerulosa	Aldosterone (mineralocorticoid)	Kidney tubules	Increases Na^+ reabsorption and K^+ secretion
Zona fasciculata and zona reticularis	Cortisol (glucocorticoid)	Most cells	Increases blood glucose at the expense of protein and fat stores; contributes to stress adaptation
	Androgens (dehydroepiandrosterone)	Females: bone and brain	Responsible for the pubertal growth spurt and sex drive in females
Adrenal medulla	Epinephrine and norepinephrine	Sympathetic receptor sites throughout the body	Reinforces the sympathetic nervous system; contributes to stress adaptation and blood pressure regulation
Endocrine pancreas (islets of Langerhans)	Insulin (β cells)	Most cells	Promotes cellular uptake, utilization, and storage of absorbed nutrients

Endocrine Gland	*Hormones*	*Target Cells*	*Major Functions of Hormones*
Endocrine pancreas (continued)	Glucagon (α cells)	Most cells	Important for the maintenance of nutrient levels in the blood during the postabsorptive state
	Somatostatin (D cells)	Digestive system	Inhibits digestion and absorption of nutrients
		Pancreatic islet cells	Inhibits secretion of all pancreatic hormones
Parathyroid gland	Parathyroid hormone (PTH)	Bone, kidneys, intestine	Increases plasma calcium concentration; decreases plasma phosphate concentration; stimulates vitamin D activation
Gonads			
Female: ovaries	Estrogen (estradiol)	Female sex organs; body as a whole	Promotes follicular development; responsible for development of secondary sex characteristics; stimulates uterine and breast growth
		Bone	Promotes closure of the epiphyseal plate
	Progesterone	Uterus	Prepares for pregnancy
Male: testes	Testosterone	Male sex organs; body as a whole	Stimulates sperm production; responsible for development of secondary sex characteristics; promotes sex drive
		Bone	Enhances pubertal growth spurt; promotes closure of the epiphyseal plate
Testes and ovaries	Inhibin	Anterior pituitary	Inhibits secretion of follicle-stimulating hormone
Pineal gland	Melatonin	Brain; anterior pituitary; reproductive organs; immune system; possibly others	Entrains body's biological rhythm with external cues; believed to inhibit gonadotropins; initiation of puberty possibly caused by a reduction in melatonin secretion; acts as an antioxidant; enhances immunity
Placenta	Estrogen (estriol); progesterone	Female sex organs	Help maintain pregnancy; prepare breasts for lactation
	Chorionic gonadotropin	Ovarian corpus luteum	Maintains corpus luteum of pregnancy
Kidneys	Renin (→ angiotensin)	Zona glomerulosa of adrenal cortex (acted on by angiotensin, which is activated by renin)	Stimulates aldosterone secretion
	Erythropoietin	Bone marrow	Stimulates erythrocyte production
Stomach	Gastrin	Digestive tract exocrine glands and smooth muscles; pancreas; liver; gallbladder	Control of motility and secretion to facilitate digestive and absorptive processes
Duodenum	Secretin; cholecystokinin; gastric inhibitory peptide		
Liver	Somatomedins	Bone; soft tissues	Promotes growth
Skin	Vitamin D	Intestine	Increases absorption of ingested calcium and phosphate
Thymus	Thymosin	T lymphocytes	Enhances T lymphocyte proliferation and function
Heart	Atrial natriuretic peptide	Kidney tubules	Inhibits Na^+ reabsorption

Table 18-2 Chemical Classification of Hormones

Properties	*Peptides*	*Amines* — *Catecholamines*	*Amines* — *Thyroid Hormone*	*Steroids*
Structure	Chains of specific amino acids, for example: Cys^1–s–s–Cys^6–Pro^7–Arg^8–Gly^9NH_2; Tyr^2, Asn^5, Phe^3 — Gln^4 (vasopressin)	Tyrosine derivative, for example: (epinephrine)	Iodinated tyrosine derivative, for example: (thyroxine, T_4)	Cholesterol derivative, for example: (cortisol)
Solubility	Hydrophilic (lipophobic)	Hydrophilic (lipophobic)	Lipophilic (hydrophobic)	Lipophilic (hydrophobic)
Synthesis	In rough endoplasmic reticulum; packaged in Golgi complex	In cytosol	In colloid, an inland extracellular site	Stepwise modification of cholesterol molecule in various intracellular compartments
Storage	Large amounts in secretory granules	In chromaffin granules	In colloid	Not stored; cholesterol precursor stored in lipid droplets
Secretion	Exocytosis of granules	Exocytosis of granules	Endocytosis of colloid	Simple diffusion
Transport in blood	As free hormone	Half bound to plasma proteins	Mostly bound to plasma proteins	Mostly bound to plasma proteins
Receptor site	Surface of target cell	Surface of target cell	Inside target cell	Inside target cell
Mechanism of action	Channel changes or activation of second messenger system to alter activity of preexisting proteins that produce the effect	Activation of second messenger system to alter activity of preexisting proteins that produce the effect	Activation of specific genes to produce new proteins that produce the effect	Activation of specific genes to produce new proteins that produce the effect
Hormones of this type	All hormones from the hypothalamus, anterior pituitary, posterior pituitary, pineal gland, pancreas, parathyroid gland, gastrointestinal tract, kidneys, liver, thyroid C cells, heart	Only hormones from the adrenal medulla	Only hormones from the thyroid follicular cells	Hormones from the adrenal cortex and gonads plus most placental hormones (vitamin D is steroidlike)

mones in a readily releasable form, the gland can respond rapidly to any demands for increased secretion without the necessity of first increasing hormone synthesis.

4. Upon appropriate stimulation, the secretory vesicles fuse with the plasma membrane and release their contents to the outside by the process of exocytosis (see p. 25). Such secretion usually does not go on continuously; it is triggered only by specific stimuli. The secreted hormone is subsequently picked up by the blood for distribution.

Steroid hormones The following steps are performed by all *steroidogenic* (steroid-producing) cells to produce and release their hormonal product:

1. Cholesterol is the common precursor for all steroid hormones. Although steroidogenic cells synthesize some cholesterol on their own, most of this raw material is derived from the low-density lipoproteins (LDL) that have been internalized into the cell and degraded by lysosomal enzymes to liberate free cholesterol (see p. 299). The uptake and degradation of LDL are regulated so that more cholesterol is made available to steroidogenic cells at times of increased need for steroid hormones. Furthermore, unused cholesterol may be chemically modified and stored in large amounts as lipid droplets within steroidogenic cells. Conversion of this major storage form of cholesterol into free cholesterol for use in steroid-hormone produc-

tion is also subject to control. Thus, the provision of free cholesterol for use by steroidogenic cells can be closely coordinated with the body's overall need for the hormone product.

2. Synthesis of the various steroid hormones from cholesterol requires a series of enzymatic reactions that modify the basic cholesterol molecule—for example, by varying the type and position of side groups attached to the cholesterol framework or the degree of saturation within the steroid ring (Fig. 18-3). Each of the conversions from cholesterol to a specific steroid hormone requires the assistance of a number of enzymes that are limited to certain steroidogenic organs. Accordingly, each steroidogenic organ is able to produce only the steroid hormone or hormones for which it has a complete set of appropriate enzymes. For example, a key enzyme necessary for the production of cortisol is found only in the adrenal cortex, so no other steroidogenic organ is able to produce this hormone. Each of the enzymes necessary for the conversion of cholesterol into a steroid hormone is localized in a specific intracellular compartment, such as the mitochondria or endoplasmic reticulum. The steroid molecule is therefore shuttled back and forth by unknown means between different compartments within the steroidogenic cell for step-by-step modification until the final secretory product is formed.

3. Unlike peptide hormones, steroid hormones are not stored after their formation. Once formed, the lipid-soluble steroid hormones immediately diffuse through the steroidogenic cell's lipid plasma membrane to enter the blood. Only the hormone precursor cholesterol is stored in significant quantities within steroidogenic cells. Accordingly, the rate of steroid-hormone secretion is controlled entirely by the rate of hormone synthesis. In contrast, peptide-hormone secretion is controlled primarily by regulating the release of presynthesized, stored hormone.

4. Following their secretion into the blood, some steroid hormones undergo further interconversions within the blood or other organs, where they are converted into more potent or different hormones.

Amines The amine hormones—thyroid hormone and adrenomedullary catecholamines—have unique synthetic and secretory pathways that will be thoroughly described when

Figure 18-3 **Steroidogenic Pathways for the Major Steroid Hormones** All steroid hormones are produced through a series of enzymatic reactions that modify cholesterol molecules, such as by varying the side groups attached to them. Each steroidogenic organ can produce only those steroid hormones for which it has a complete set of the enzymes needed to appropriately modify cholesterol. For example, the testes have the enzymes necessary to convert cholesterol into testosterone (male sex hormone) whereas the ovaries possess the enzymes needed to yield progesterone and the various estrogens (female sex hormones).

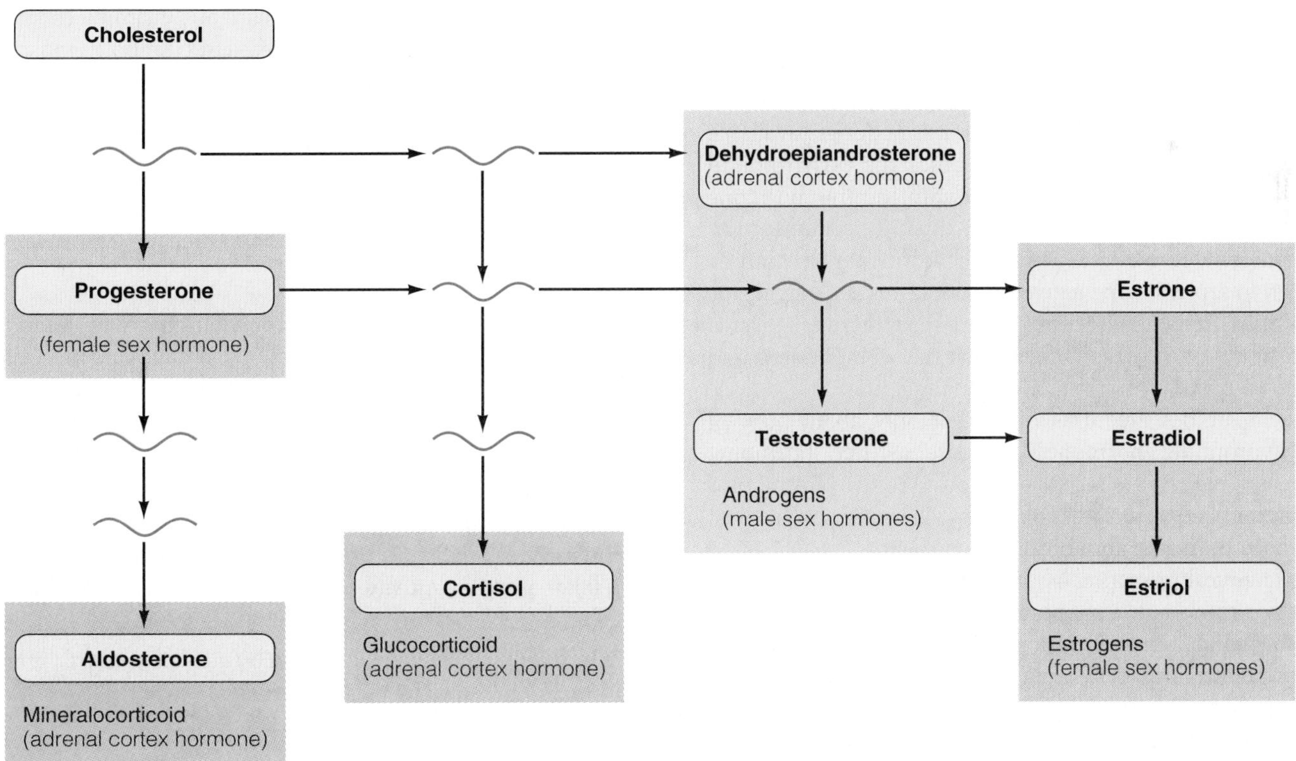

each of these hormones is specifically addressed. However, the amines have the following features in common:

1. They are derived from the naturally occurring amino acid tyrosine.
2. None of the enzymes directly involved in the synthesis of either of these hormone types are located in organelle compartments within the secretory cells.
3. Both types of amines are stored until they are secreted.

Water-soluble hormones are transported dissolved in the plasma, whereas lipid-soluble hormones are largely transported bound to plasma proteins.

All hormones are carried by the blood, but they are not all transported in the same manner. The hydrophilic (water-soluble) peptide hormones are transported simply dissolved in the plasma. However, lipophilic (lipid-soluble) steroids and thyroid hormone, which are poorly soluble in water, cannot dissolve in the aqueous plasma in sufficient quantities to account for their known plasma concentrations. Instead, the majority of the lipophilic hormones circulate in the blood to their target cells reversibly bound to plasma proteins. Some are bound to specific plasma proteins that are designed to carry only one type of hormone, whereas other plasma proteins, such as albumin, nondiscriminately pick up any "hitchhiking" hormone.

In the case of some hormones, 1% or less of the total hormone remains unbound. This figure is noteworthy because only the small, unbound, freely dissolved fraction of a lipophilic hormone is biologically active (that is, free to cross capillary walls and bind with target cell receptors to exert an effect). Once a hormone has interacted with a target cell, it is rapidly inactivated or removed so that it is no longer available to interact with another target cell. Since the carrier-bound hormone is in dynamic equilibrium with the free hormone pool, the bound form of steroid and thyroid hormones provides a large reserve of these lipophilic hormones that can be called on to replenish the active free pool. It is the magnitude of the small free effective pool rather than the total plasma concentration of a particular hormone that is monitored and adjusted to maintain normal endocrine function.

Catecholamines are unusual in that only about 50% of these hydrophilic hormones circulate as free hormone, whereas the other 50% are loosely bound to the plasma protein albumin. Because catecholamines are water soluble, the importance of this protein binding is unclear.

The chemical properties of a hormone dictate not only the means by which it is transported in the blood but also the means by which it can be artificially introduced into the blood for therapeutic purposes. Because the digestive system does not secrete enzymes that can digest steroid and thyroid hormones, these hormones, such as the sex steroids contained in birth control pills, can be absorbed intact from the digestive tract into the blood when taken orally. None of the other types of hormones can be taken orally because they would be attacked and converted into inactive fragments by protein-digesting enzymes. Therefore, these hormones must be administered when necessary by nonoral routes; for example, insulin deficiency (diabetes mellitus) is treated by daily injections of insulin.

Hormones generally produce their effect by altering intracellular protein activity.

Hormones must bind with target cell receptors specific for them in order to induce their effect. However, the location of the receptors within the target cell and the mechanism by which binding of the hormone with the receptors induces a response vary, depending on the hormone's solubility characteristics. Hormones can be grouped into two categories based on the location of their receptors (Table 18-2):

1. The hydrophilic peptides and catecholamines, which are poorly soluble in lipid, are not able to pass through the lipid membrane barriers of their target cells. Instead they bind with specific receptors located on the *outer plasma membrane surface* of the target cell.
2. The lipophilic steroids and thyroid hormone easily pass through the surface membrane to bind with specific receptors located *inside* the target cell.

Each interaction between a particular hormone and a target cell receptor produces a highly characteristic target cell response that differs for different hormones and differs between different target cells influenced by the same hormone. For example, one of the adrenomedullary catecholamines, epinephrine, through its ubiquitous distribution and target cell specialization, simultaneously produces such diverse effects as contraction of vascular smooth muscle, relaxation of respiratory airway smooth muscle, and breakdown of glycogen (stored glucose) in the liver.

Even though hormones elicit a wide variety of biological responses, all hormones ultimately influence their target cells by altering the cell's protein activity. Hormones exert an effect on their target cell's proteins through three general means: (■ Fig. 18-4 and Table 18-2):

1. A few hydrophilic hormones, upon binding with a target cell's surface receptors, bring about changes in the cell's permeability (either opening or closing channels to one or more ions) by *altering the conformation (shape) of adjacent channel-forming proteins already present in the membrane.*
2. Most surface-binding hydrophilic hormones function by *activating second messenger systems* within the target cell. This activation directly *alters the activity of preexisting intracellular proteins, usually enzymes,* to produce the desired effect.
3. All lipophilic hormones function by *activating specific genes in the target cell to cause the formation of new intracellular proteins,* which in turn produce the desired effect.

Even though all hormones ultimately influence their target cells by altering intracellular protein activity, the onset of action varies depending primarily on the mode employed. Hormones that act through a second messenger system to alter a preexisting enzyme's activity elicit full action within a few minutes. In contrast, hormonal responses that require the synthesis of new protein may take up to several hours before any action is initiated. Let us examine each of the two major postreceptor (after the hormone binds with the receptor) events in more detail.

Postreceptor events: hydrophilic hormones Because second messenger systems in general, and cyclic AMP (cAMP) in particular, play such a central role in hydrophilic hormone activity, it is worthwhile to briefly review the steps (see Fig. 3-22, p. 69):

1. Binding of the extracellular first messenger, the hydrophilic hormone, to its surface membrane receptor activates by means of G protein intermediaries the membrane-bound enzyme adenylyl cyclase, which is located on the cytoplasmic side of the plasma membrane.
2. Activated adenylyl cyclase converts intracellular ATP to cAMP, the intracellular second messenger.
3. Cyclic AMP triggers a preprogrammed series of biochemical steps that bring about a change in the shape and function of specific preexisting enzymatic proteins in the cell.
4. These altered enzymatic proteins are responsible for bringing about a change in cell activity. The resultant change is the target cell's ultimate physiological response to the hormone.

Once the hormone is removed, cAMP is converted to inactive cAMP by a specific chemical within the cytoplasm, and the intracellular message is "erased."

The nature of the preexisting enzymatic proteins whose activity is ultimately modified by a second messenger varies in different target cells. Thus, various target cells respond very differently to the universal mechanism of hormonally induced changes in their cAMP levels. Cyclic AMP can "turn on" (or "turn off") different cellular events depending on the different kinds of enzyme activity that are ultimately modified in the different target cells.

Most but not all hydrophilic hormones use cAMP as their second messenger. A few are known to use intracellular Ca^{2+} as the second messenger; for others, the second messenger is still unknown.

Postreceptor events: lipophilic hormones All lipophilic hormones (steroids and thyroid hormone) produce their effects in their target cells by enhancing the synthesis of new enzymatic or structural proteins. Their effects result from stimulation of the cell's genes as follows (Fig. 18-5):

1. Free lipophilic hormone (hormone not bound with its plasma protein carrier) diffuses through the plasma membrane of the target cell and binds with its specific receptor inside the cell. Most lipophilic hormone receptors are located in the nucleus.
2. Each receptor has a specific region for binding with its hormone and another region for binding with DNA. In the absence of the hormone, the receptor exists in a multiprotein aggregate in association with proteins of the **molecular chaperone machinery.** These chaperone proteins normally function elsewhere in the cell to ensure that newly produced proteins are folded into their correct shape. (The chaperones prevent "illicit liaisons" between various parts of the protein chain during folding.) The chaperone proteins are believed to have additional functions when in association with steroid receptors. For example, they may stabilize the receptor's hormone-binding site so that the site is accessible to the hormone instead of being tucked in a fold out of reach of the hormone. Also, the chaperone proteins are believed to cover the DNA-binding portion of the receptor. Binding of a lipophilic hormone with its receptor triggers release of the receptor from the chaperone proteins and subsequent exposure of the receptor's DNA-binding site.
3. Once the hormone is bound to the receptor, the hormone-receptor complex binds with DNA at a specific attachment site on DNA known as the **hormone response element (HRE).** Different steroid hormones and thyroid hormone, once bound with their respective receptors, attach at different HREs on DNA. For example, the estrogen-receptor complex binds at DNA's estrogen-response element.
4. Binding of the hormone-receptor complex with DNA ultimately "turns on" specific genes within the target cell.

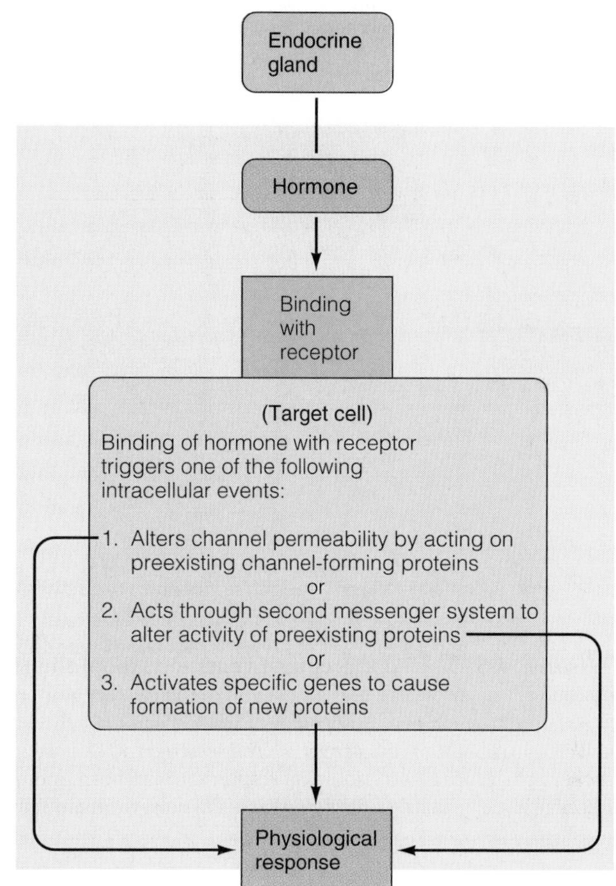

Figure 18-4 **Three General Means by Which Hormones Elicit Biological Responses**

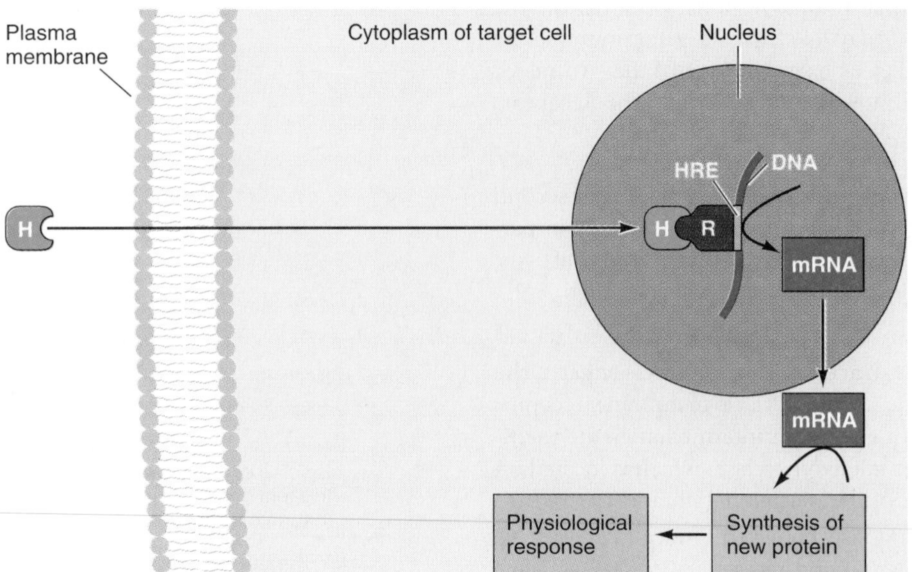

Figure 18-5 **Postreceptor Events for Lipophilic Hormones** A lipophilic hormone diffuses through the plasma and nuclear membranes of its target cell and binds with a nuclear receptor specific for it. The hormone-receptor complex in turn binds with the hormone response element, a segment of DNA specific for the hormone-receptor complex. DNA binding activates specific genes, which produce complementary messenger RNA. Messenger RNA leaves the nucleus and directs the synthesis of new proteins, which accomplish the target cell's ultimate physiological response to the hormone. [Source: Adapted with permission from George A. Hedge, Howard D. Colby, and Robert L. Goodman, *Clinical Endocrine Physiology* (Philadelphia: W. B. Saunders Company, 1987), Figure 1-9, p. 20.]

5. The activated genes direct the synthesis of new cell protein by producing complementary messenger RNA, which enters the cytoplasm and binds to a ribosome, the "workbench" that mediates the assembly of new proteins (p. B–7).
6. The newly synthesized protein produces the target cell's ultimate physiological response to the hormone.

As a result of this mechanism, different genes are activated by different lipophilic hormones, resulting in different biological effects.

Despite the differences in the molecular events induced by hydrophilic and lipophilic hormones, the two hormone types have several characteristics in common that have important implications for hormonal actions:

- First, the actions of hormones are greatly amplified at the target cell. Hormones are greatly diluted by the blood and thus must exert their effect at incredibly low concentrations—as low as 1 picogram (10^{-12} gram; 1 millionth of a millionth of a gram) per ml—as opposed to the much higher localized concentration of neurotransmitter at the target cell during neural communication. Interaction of one hormonal molecule with its receptor can result in the formation of many active protein products that ultimately carry out the physiological effect. For example, one peptide hormone results in the production of numerous cAMP messengers, each in turn activating many latent enzymes (see p. 70). Similarly, one steroid-hormone–activated gene induces the formation of many messenger RNA molecules, each of which is used to make many enzymes.
- A second common feature of hormone action is that hormones regulate the rates of existing reactions instead of initiating new reactions. Enzymes under hormonal control usually show some activity even in the absence of the hormone.
- Finally, hormone action is relatively slow and prolonged. In general, lipophilic hormones, which act by increasing protein synthesis, take longer to produce an effect than hydrophilic hormones, which need only to activate a preformed enzyme. Once an enzyme is activated, it no longer depends on the presence of the hormone. As a result, a hormone's effect usually lasts for some time after its withdrawal.

The effective plasma concentration of a hormone is normally regulated by changes in its rate of secretion.

The primary function of hormones is the regulation of various homeostatic activities. Because hormones' effects are proportional to their concentrations in the blood, it follows that these concentrations must be subject to control according to homeostatic need. The plasma concentration of free, biologically active hormone—and thus the hormone's availability to its receptors—depends on several factors (Fig. 18-6): (1) the hormone's rate of secretion into the blood by the

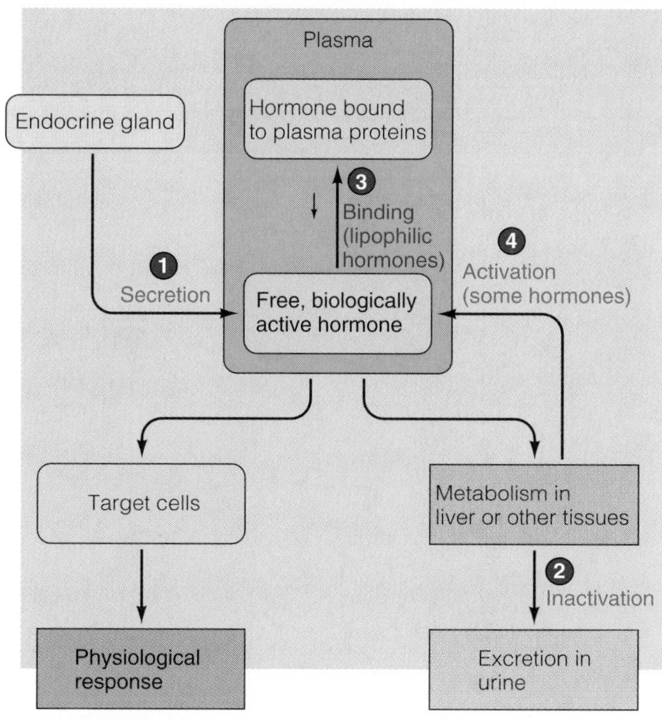

Figure 18-6 Factors Affecting the Plasma Concentration of Free, Biologically Active Hormone The plasma concentration of free, biologically active hormone, which can interact with its target cells to produce a physiological response, depends on the hormone's rate of secretion by the endocrine gland ①, its rate of metabolic inactivation and excretion ②, its extent of binding to plasma proteins (for lipophilic hormones) ③, and its rate of metabolic activation (for a few hormones) ④.

endocrine gland, (2) its rate of removal from the blood by metabolic inactivation and excretion in the urine, (3) for lipophilic hormones, its extent of binding to plasma proteins, and (4) for a few hormones, its rate of metabolic activation.

Normally, the effective plasma concentration of a hormone is regulated by appropriate adjustments in the rate of its secretion. Endocrine glands do not secrete their hormones at a constant rate; the secretion rates of all hormones vary subject to control, often by a combination of several complex mechanisms. The regulatory system for each hormone will be considered in detail in the subsequent sections. As you proceed through the discussion of the endocrine system, note the following general mechanisms of controlling secretion; they are common to many different hormones.

Negative-feedback control Negative feedback is a prominent feature of hormonal control systems (see p. 11). Stated simply, *negative feedback exists when the output of a system opposes a change in input.* Negative feedback maintains the plasma concentration of a hormone at a given level, similar to the way in which a home heating system maintains the room temperature at a given set point. Control of hormonal secretion provides some classic physiological examples of negative feedback. For example (Fig. 18-7), when the plasma concentration of free circulating thyroid hormone falls below a given "set point," the anterior pituitary secretes thyroid-stimulating hormone (TSH), which stimulates the thyroid to increase its secretion of thyroid hormone. Thyroid hormone in turn inhibits further secretion of TSH by the anterior pituitary. Negative feedback ensures that once thyroid gland secretion has been "turned on" by TSH, it will not continue unabated but instead will be "turned off" when the appropriate level of free circulating thyroid hormone has been achieved. Thus, the effect of a particular hormone's actions can bring about inhibition of its own secretion. The feedback loops often become quite complex.

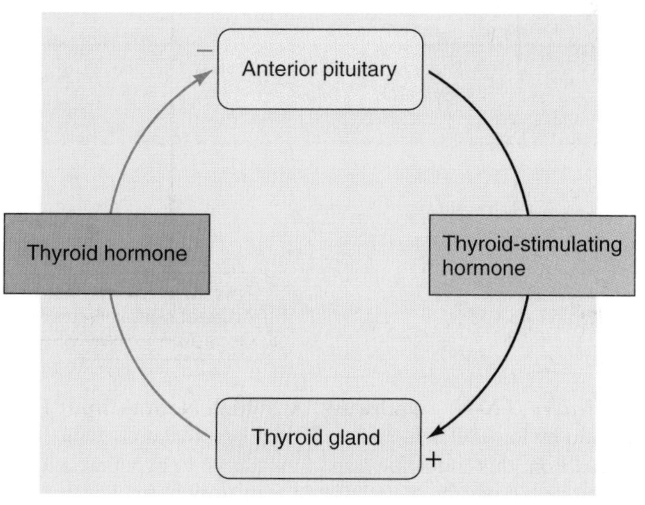

Figure 18-7 Negative-Feedback Control

Neuroendocrine reflexes Many endocrine control systems involve **neuroendocrine reflexes,** which include neural as well as hormonal components. The purpose of such reflexes is to produce a sudden increase in hormone secretion (that is, "turn up the thermostat setting") in response to a specific stimulus, frequently a stimulus external to the body. The nervous system can influence hormonal secretion by four different means, each of which will be elaborated upon when the

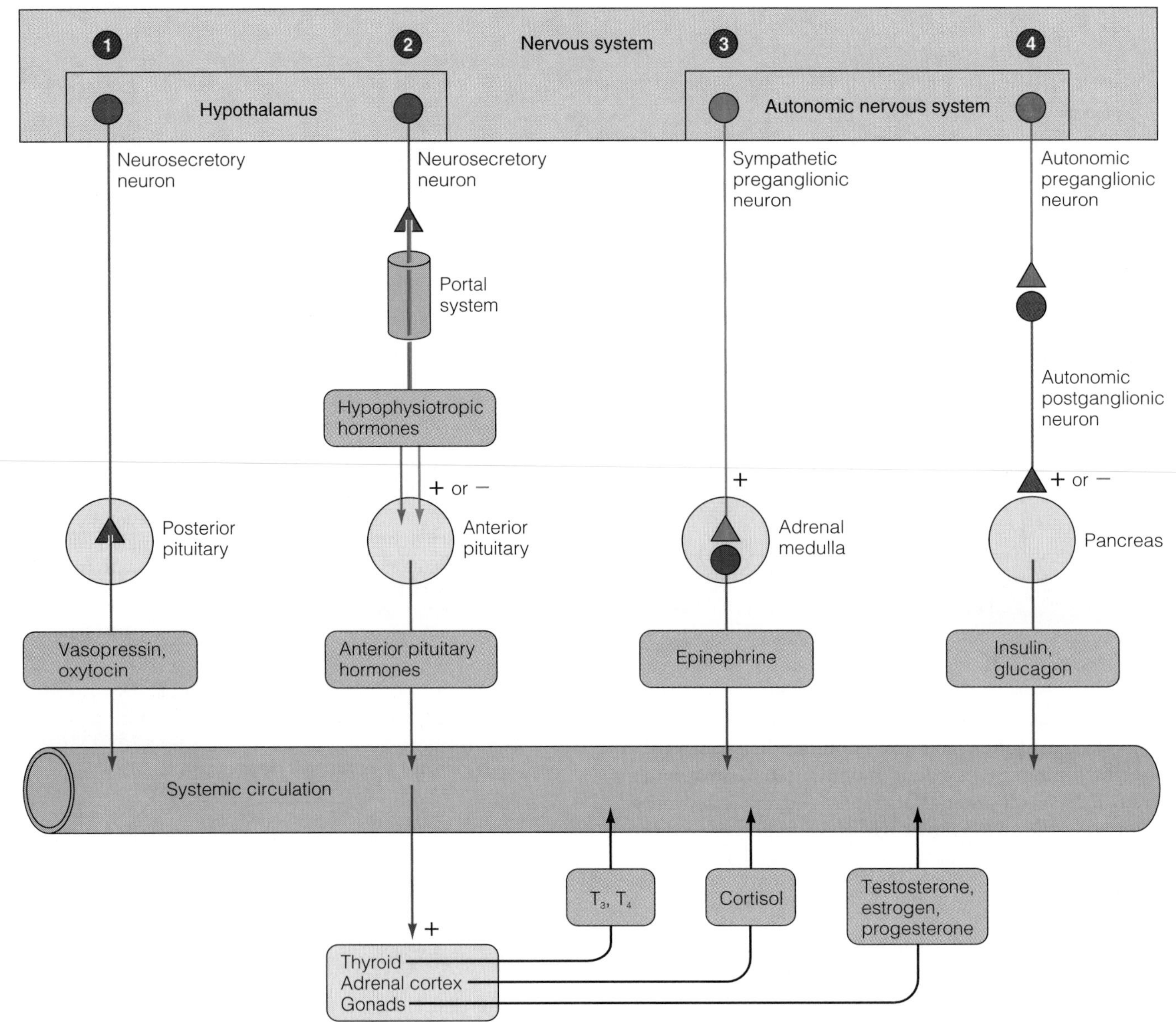

Figure 18-8 **Pathways by Which Neural Input Influences Hormonal Secretion** The nervous system influences hormonal secretion by four different means: ① the hypothalamus actually produces the hormones that are stored in the posterior pituitary and released from this endocrine gland in response to hypothalamic stimulation. ② The release of hormones produced by the anterior pituitary is controlled in part by hypothalamic regulatory hormones that reach the anterior pituitary by a special vascular portal system. The anterior pituitary hormones in turn control the release of a number of other hormones from other endocrine glands. ③ The release of hormones produced by the adrenal medulla, which consists of modified sympathetic postganglionic neurons, is entirely under control of sympathetic preganglionic neurons. ④ Postganglionic autonomic neurons are among many regulatory inputs to the endocrine cells of the pancreas. Gastrointestinal hormones are influenced by the autonomic nervous system in a similar way.

specific hormones involved are discussed (Fig. 18-8). In some instances, neural input to the endocrine gland is the only factor regulating secretion of the hormone. For example, secretion of epinephrine by the adrenal medulla is solely under the control of the sympathetic nervous system. Some endocrine control systems, on the other hand, include both feedback control, which maintains a constant basal level of the hormone, and neuroendocrine reflexes, which cause sudden bursts in secretion in response to a sudden increased need for the hormone, such as increased secretion of cortisol by the adrenal cortex during a stress response.

Diurnal (ciradian) rhythms Although hormone secretion rates are usually regulated by some form of negative feedback, this does not imply that they are always maintained at a constant level. Instead, the secretion rates of all hormones rhythmically fluctuate up and down as a function of time. The most common endocrine rhythm is the **diurnal** ("day-night"), or **circadian** ("around a day"), **rhythm,** which is characterized by repetitive oscillations in hormone levels that are very regular and have a frequency of one cycle every 24 hours. This rhythmicity appears to be caused by endogenous oscillators similar to the self-paced respiratory neurons in the brain stem that are responsible for the rhythmic motions of breathing. Unlike the rhythmicity of breathing, however, endocrine rhythms are locked on, or "entrained," to external cues, called **zeitgebers** ("time givers"), such as the light-dark cycle or the activity cycle; that is, the inherent 24-hour cycles of peak and ebb of hormone secretion are set to "march in step" with cycles of light/activity and dark/inactivity. For example, cortisol secretion rises during the night, reaching its peak secretion in the morning before a person rises, then falls throughout the day to its lowest level at bedtime (Fig. 18-9). Inherent hormonal rhythmicity and the entrainment of zeitgebers are not accomplished by the endocrine organs themselves, but instead are the result of the central nervous system changing the set point of these organs. Negative-feedback control mechanisms operate to maintain whatever set point is established for that time of day. Some endocrine cycles operate on time scales other than a circadian rhythm—some much shorter than a day and some much longer. A well-known example of the latter is the monthly menstrual cycle.

Periodicity is not unique to the endocrine system. Humans have similar biological clocks for other bodily functions such as temperature regulation (see p. 609). The master biological clock that serves as the pacemaker for many circadian rhythms in the body is the **suprachiasmatic nucleus (SCN),** which consists of a cluster of nerve cell bodies in the hypothalamus above the optic chiasm, the point at which part of the nerve fibers from each eye cross to the opposite half of the brain (see p. 180). (*Supra* means "above;" *chiasm* means "cross.") The self-induced rhythmic firing of the SCN neurons plays a major role in establishing many of the body's inherent 24-hour rhythms. The SCN works in conjunction with the *pineal gland* and its hormonal product *melatonin* to synchronize the various circadian rhythms with the 24-hour day/night cycle. The mechanism and purpose served by these rhythmic fluctuations in biological activity are unclear, but the effect of upsetting them is well known by people who experience **jet lag** when their inherent rhythm is out of step with external cues. We will discuss further the SCN, the pineal gland, and melatonin, a hormone that has functions beyond those related to timekeeping, after we complete our general overview of endocrinology.

The effective plasma concentration of a hormone can be influenced by the hormone's transport, metabolism, and excretion.

Even though the effective plasma concentration of a hormone is normally regulated by appropriate adjustments in the rate of secretion, alterations in the transport, metabolism, or excretion of a hormone can also influence the size of its effective pool, sometimes inappropriately (Fig. 18-6). For example, since the liver synthesizes plasma proteins, liver disease may result in abnormal endocrine activity because of a change in the balance between free and bound pools of certain hormones.

Eventually, all hormones are metabolized by enzyme-mediated reactions that modify the hormonal structure in some way. In most cases this results in inactivation of the hormone. Hormone metabolism is not always a mechanism for removal of used hormones, however. In some cases, a hormone is *activated* by metabolism; that is, the hormone's product has greater activity than the original hormone. For example, after the thyroid hormone thyroxine is secreted, it is converted to a more powerful hormone by enzymatic removal of one of the iodine atoms it contains. Usually, the rate of such hormone activation is itself under hormonal control.

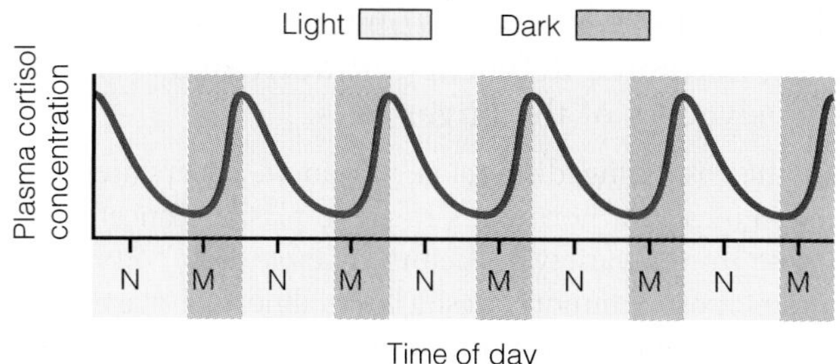

Figure 18-9 **Diurnal Rhythm of Cortisol Secretion**
[Source: Adapted with permission from George A. Hedge, Howard D. Colby, and Robert L. Goodman, *Clinical Endocrine Physiology* (Philadelphia: W. B. Saunders Company, 1987), Figure 1-13, p. 28.]

The liver is the most common site for metabolic hormonal inactivation (or activation, if that be the case), but some hormones are also metabolized in the kidneys, blood, or target cells. Hormonal inactivation and excretion are not subject to control. The primary means of eliminating hormones and their metabolites from the blood is by urinary excretion. Because the liver and kidneys are important in the removal of hormones from the blood by means of metabolic inactivation and urinary excretion, patients with liver or kidney disease may suffer from excess activity of certain hormones solely as a result of reduced hormone elimination. On the other hand, when liver and kidney function are normal, measurement of urinary concentrations of hormones and their metabolites provides a useful noninvasive means of assessing endocrine function, because the rate of excretion of these products in the urine is a direct reflection of their rate of secretion by the endocrine glands.

The amount of time after a hormone is secreted before it is inactivated and the means by which this is accomplished differ for different classes of hormones. In general, the hydrophilic peptides and catecholamines are easy targets for blood and tissue enzymes, so they remain in the blood only briefly (a few minutes to a few hours) before being enzymatically inactivated. In the case of some peptide hormones, the target cell actually engulfs the bound hormone by endocytosis and degrades it intracellularly. In contrast, binding of lipophilic hormones to plasma proteins renders them less vulnerable to metabolic inactivation and prevents them from escaping into the urine. Therefore, lipophilic hormones are removed from the plasma much more slowly. They may persist in the blood for hours (steroids) or up to a week (thyroid hormone). In general, lipophilic hormones undergo a series of reactions that reduces their biological activity and enhances their H_2O solubility so that they can be freed of their plasma protein carriers and be eliminated in the urine.

Endocrine disorders are attributable to hormonal excess, hormonal deficiency, or decreased responsiveness of the target cells.

From the preceding discussion, it should be apparent that abnormalities in a hormone's effective plasma concentration can arise from a variety of factors (Table 18-3). Endocrine disorders most commonly result from abnormal plasma concentrations of a hormone due to inappropriate rates of secretion—that is, too little hormone secreted **(hyposecretion)** or too much hormone secreted **(hypersecretion).** Occasionally, endocrine dysfunction arises because target cell responsiveness to the hormone is abnormally low, even though the plasma concentration of the hormone is normal.

Hyposecretion If an endocrine gland is secreting too little of its hormone because of an abnormality within that gland, the condition is referred to as *primary hyposecretion.* If, on the other hand, the endocrine gland is normal but is secreting too little hormone because of a deficiency of its tropic hormone, the condition is known as *secondary hyposecretion.* The following are among the many different factors (each listed with an example) that may be responsible for hormone deficiency: (1) genetic (inborn absence of an enzyme that catalyzes synthesis of the hormone); (2) dietary (lack of iodine, which is necessary for synthesis of thyroid hormone); (3) chemical or toxic (certain insecticide residues may destroy the adrenal cortex); (4) immunologic (autoimmune antibodies may cause self-destruction of the person's own thyroid tissue); (5) other disease processes (cancer or tuberculosis may coincidentally destroy endocrine glands); (6) *iatrogenic* (physician-induced, such as surgical removal of a thyroid tumor); and (7) *idiopathic* (meaning the cause is not known).

The most common method of treating hormone hyposecretion is administration of a hormone that is the same as (or similar to, such as from another species) the one that is deficient or missing. Such replacement therapy is straightforward in theory, but the hormone's source and means of administration involve some practical problems. The sources of hormone preparation for clinical use include (1) endocrine tissues from domestic livestock; (2) placental tissue and urine of pregnant women; (3) laboratory synthesis of hormones; and (4) "hormone factories," or bacteria into which genes coding for the production of human hormones have been introduced. The method of choice for a given hormone is determined largely by its structural complexity and degree of species specificity.

Hypersecretion Like hyposecretion, hypersecretion by a particular endocrine gland is designated as primary or secondary depending on whether the defect lies in that gland or is due to excessive stimulation from the outside, respectively. Hypersecretion may be caused by (1) tumors that ignore the normal regulatory input and continuously secrete excess hormone and (2) immunologic factors, such as excessive stimulation of the thyroid gland by an abnormal antibody that mimics the action of TSH, the thyroid tropic hormone. Excessive levels of a particular hormone may also arise from substance abuse, such as the outlawed practice among athletes of using certain steroids that increase muscle mass by promoting protein synthesis in muscle cells (see p. 246).

There are several ways of treating hormonal hypersecretion. If a tumor is the culprit, it may be surgically removed or destroyed with radiation treatment. In some instances, hypersecretion can be limited by drugs that block hormone synthesis or inhibit hormone secretion. Sometimes the condition may be treated by giving drugs that inhibit the action of the hormone without actually reducing the excess hormone secretion.

Abnormal target cell responsiveness Endocrine dysfunction can also occur as a result of target cells not responding adequately to the hormone, even though the effective plasma concentration of a hormone is normal. This inadequate responsiveness may be caused, for example, by an inborn lack of receptors for the hormone, such as is seen in *testicular feminization syndrome.* In this condition, receptors for testosterone, a masculinizing hormone produced by the male testes, are not produced because of a specific genetic defect. Although adequate testosterone is available, masculinization does not take place, just as if no testosterone were present. Abnormal responsiveness may also occur if the target cells for a particular hormone lack an enzyme essential to carrying out the response.

Table 18-3
Means by Which Endocrine Disorders Can Arise

Too Little Hormone Activity	*Too Much Hormone Activity*
Too little hormone secreted by the endocrine gland (hyposecretion)*	Too much hormone secreted by the endocrine gland (hypersecretion)*
Increased removal of the hormone from the blood	Reduced plasma protein binding of the hormone (too much free, biologically active hormone)
Abnormal tissue responsiveness to the hormone	Decreased removal of the hormone from the blood
Lack of target cell receptors	Decreased inactivation
Lack of an enzyme essential to the target cell response	Decreased excretion

*Most common causes of endocrine dysfunction.

The responsiveness of a target cell to its hormone can be varied by regulating the number of its hormone-specific receptors.

In contrast to endocrine dysfunction caused by *unintentional* receptor abnormalities, the target cell receptors for a particular hormone can be *deliberately altered* as a result of physiological control mechanisms. A target cell's response to a hormone is correlated with the number of the cell's receptors occupied by molecules of that hormone, which in turn depends on the number of receptors present in the target cell for that hormone as well as on the plasma concentration of the hormone. Thus, the response of a target cell to a given plasma concentration can be fine-tuned up or down by varying the number of receptors available for hormone binding.

As an illustration, when the plasma concentration of insulin is chronically elevated, the total number of target cell receptors for insulin is reduced as a direct result of the effect that an elevated level of insulin has on the insulin receptors. This phenomenon, known as **down regulation,** constitutes an important locally acting negative-feedback mechanism that prevents the target cells from overreacting to the high concentration of insulin; that is, the target cells are *desensitized* to insulin, helping to blunt the effect of insulin hypersecretion. Down regulation of insulin is accomplished by the following mechanism. The binding of insulin to its surface receptors induces endocytosis of the hormone-receptor complex, which is subsequently attacked by intracellular lysosomal enzymes. This internalization serves a twofold purpose: it provides a pathway for degradation of the hormone, and it also plays a role in regulating the number of receptors available for binding on the target cell's surface. At high plasma insulin concentrations, the number of surface receptors for insulin is gradually reduced as a result of the accelerated rate of receptor internalization and degradation brought about by increased hormonal binding. The rate of synthesis of new receptors within the endoplasmic reticulum and their insertion in the plasma membrane do not keep pace with their rate of destruction. Over time, this self-induced loss of target cell receptors for insulin results in a reduction in the target cell's sensitivity to the elevated hormone concentration.

A given hormone's effects are influenced not only by the concentration of the hormone itself but also by the concentrations of other hormones that interact with it. Because hormones are widely distributed through the blood, target cells may be exposed simultaneously to many different hormones, giving rise to numerous complex hormonal interactions on target cells. Hormones frequently alter the receptors for other kinds of hormones as part of their normal physiological activity. A hormone can influence the activity of another hormone at a given target cell in one of three ways: permissiveness, synergism, and antagonism. With **permissiveness,** one hormone must be present in adequate amounts for the full exertion of another hormone's effect. In essence, the first hormone, by enhancing a target cell's responsiveness to another hormone, "permits" this other hormone to exert its full effect. For example, thyroid hormone increases the number of receptors for epinephrine in epinephrine's target cells, thereby increasing the effectiveness of epinephrine. Epinephrine is only marginally effective in the absence of thyroid hormone. **Synergism** occurs when the actions of several hormones are complementary and their combined effect is greater than the sum of their separate effects. An example is the synergistic action of follicle-stimulating hormone and testosterone, both of which are required to maintain the normal rate of sperm production. Synergism probably results from each hormone's influence on the number or affinity of receptors for the other hormone.

Antagonism occurs when one hormone causes the loss of another hormone's receptors, reducing the effectiveness of the second hormone. To illustrate, progesterone (a hormone secreted during pregnancy that *decreases* contractions of the uterus) inhibits uterine responsiveness to estrogen (another hormone secreted during pregnancy that *increases* uterine contractions). Progesterone, by causing loss of estrogen receptors on uterine smooth muscle, prevents estrogen from exerting its excitatory effects during pregnancy and thus keeps the uterus a quiet (noncontracting) environment suitable for the developing fetus.

Pineal Gland

The **pineal gland,** a tiny, pine-cone–shaped structure located in the center of the brain (see Fig. 18-1), secretes the hormone **melatonin.** (Do no confuse *melatonin* with the skin-darkening pigment, *melanin*—see p. 410). Although melatonin was discovered in 1959, investigators have only recently begun to unravel its many functions. One of melatonin's most widely accepted roles is helping to keep the body's inherent circadian rhythms in synchrony with the light/dark cycle. Melatonin is the hormone of darkness. Melatonin secretion increases up to 10-fold during the darkness of night and then falls to low levels during the light of day. The eyes cue the pineal gland about the absence or presence of light by means of a neural pathway that passes through the suprachiasmatic nucleus. This pathway is distinct from the neural systems that result in vision perception. Fluctuations in melatonin secretion in turn help entrain the body's biological rhythms with the external light/dark cues. A new approach to treating patients, **chronotherapy** (*chrono* means "time"), involves synchronizing the timing of medications with the body's rhythms to enhance the drug's desired therapeutic effects and minimize its adverse side effects.

Other proposed roles of melatonin besides regulating the body's biological clock include the following:

- Melatonin induces a natural sleep without the side effects that accompany hypnotic sedatives.
- Melatonin is believed to inhibit the hormones that stimulate reproductive activity. Initiation of puberty is possibly caused by a reduction in melatonin secretion.
- In a related role, in some species, seasonal fluctuations in melatonin secretion associated with changes in the number of daylight hours are important triggers for seasonal breeding, migration, and hibernation.
- In another related role, melatonin is being used on a clinical-trial basis as a method of birth control because at high levels it shuts down ovulation (egg release). A male contraceptive utilizing melatonin to stop sperm production is also under development.
- Melatonin appears to be a very effective **antioxidant,** a defense tool against biologically damaging free radicals. *Free radicals* are very unstable electron-deficient particles that are highly reactive and destructive. Free radicals have been implicated in several chronic diseases such as

Tinkering with our Biological Clocks

Research shows that the hectic pace of modern life, stress, noise, pollution, and the irregular work schedules many workers follow can upset internal rhythms, illustrating how a healthy external environment affects our own internal environment—and our health.

Dr. Richard Restak, a neurologist and author, notes that the "usual rhythms of wakefulness and sleep . . . seem to exert a stabilizing effect on our physical and psychological health." The greatest disrupter of our natural circadian rhythms, says Restak, is the variable work schedule, surprisingly common in the United States and other industrialized countries. Today, one out of every four working men and one out of every six working women has a variable work schedule—shifting frequently between day and night work. In many industries, workers are on the job day and night to make optimal use of equipment and buildings. As a result, more restaurants and stores are open twenty-four hours a day and more health care workers must be on duty at night to care for accident victims.

To spread the burden, many companies that maintain shifts round the clock alter their workers' schedules. One week, employees work the day shift. the next week, they're shifted to the "graveyard shift" from 12:00 midnight to 8:00 A.M. The next week, they work the night shift from 4:00 P.M. to 12 midnight. Many shift workers report they that feel tired most of the time and have trouble staying awake at the job. Their work performance suffers because of fatigue and general malaise. When workers arrive home for bed, they're exhausted but can't sleep, because they're trying to doze off at a time when the body is trying to wake them up. Unfortunately, weekly changes in schedule never permit workers' internal alarm clocks to fully adjust. Most people, in fact, require four to fourteen days to adjust to a new schedule.

Workers on alternating shifts suffer more ulcers, insomnia, irritability, depression, and tension than workers on unchanging shifts. Their lives are never the same. To make matters worse, tired, irritable workers whose judgment is impaired by fatigue pose a threat to society as a whole. Consider an example.

At 4:00 A.M. in the control room of the Three Mile Island Nuclear Reactor in Pennsylvania, three operators made the first mistake in a series of errors that led to the worst nuclear accident in U.S. history. The operators did not notice warning lights and failed to observe that a crucial valve had remained open. When the morning shift operators entered the control room the next day, they quickly discovered the problems, but it was too late. Pipes in the system had burst, sending radioactive steam and water into the air and into two buildings. John Gofman and Arthur Tamplin, two radiation health experts, estimate that the radiation released from the accident will cause at least 300 and possibly as many as 900 additional fatal cases of cancer in the residents living near the troubled reactor, although other "experts" (especially in the nuclear industry) contest these projections, saying the accident will have an unnoticeable effect. Whatever the outcome, the 1979 accident at Three Mile Island cost several billion dollars to clean up.

Late in April 1986, another nuclear power plant went amok. This accident, in the former Soviet Union, was far more severe. In the early hours of the morning, two engineers were testing the reactor. They deactivated key safety systems in violation of standard operational protocol. This single error in judgment (possibly due to fatigue) led to the largest and most costly nuclear accident in the history of the world. Steam built up inside the reactor and blew the roof off the containment building. A thick cloud of radiation rose skyward and then spread throughout Europe and the world. While workers battled to cover the molten radioactive core that spewed radiation into the sky, the whole world watched in horror.

The Chernobyl disaster, like the accident at Three Mile Island, may have been the result of workers operating at a time unsuitable for clear thinking. One has to wonder how many plane crashes, auto accidents, and acts of medical malpractice can be traced to judgment errors resulting from our insistence on working against the inherent body rhythms.

Thanks to studies of biological rhythms, researchers are finding ways to reset the biological clocks, which could help lessen the misery and suffering of shift workers and could increase the performance of the graveyard shift workers. For instance, one simple measure is to put shift workers on three-week cycles to give their clocks time to adjust. And instead of shifting workers from daytime to graveyard shifts, transfer them forward, rather than backward (for example, from a daytime to a nighttime shift). It's a much easier adjustment. Bright lights can also be used to reset the biological clock as well. It's a small price to pay for a healthy workforce and a safer society. Furthermore, use of supplemental melatonin, the hormone that sets the internal clock to march in step with environmental cycles, may prove to be useful in resetting the body's clock when that clock is out of synchrony with external cues.

coronary artery disease (see p. 300) and cancer and are believed to contribute to the aging process. Melatonin as an antioxidant scavenges free radicals, thus reducing their detrimental impact on the body.

- Evidence suggests that melatonin may slow the aging process, perhaps by removing free radicals or by other means.
- Melatonin appears to enhance immunity and has been shown to reverse some of the age-related shrinkage of the thymus, the source of T lymphocytes (see p. 385), in old experimental animals.

Because of melatonin's many proposed roles, use of supplemental melatonin for a variety of conditions is very promising. However, most researchers are cautious about recommending supplemental melatonin until its effectiveness as a drug is further substantiated. Meanwhile, many people are turning to melatonin as a health food supplement; as such, it is not regulated by the Food and Drug Administration for safety and effectiveness. The two biggest self-prescribed uses of melatonin are to prevent jet lag and as a sleep aid.

As the timekeeper that sets the body's internal clock to coincide with external cycles, melatonin is taken by some travelers and shift workers to "reset" the body's clock to ward off jet lag and related symptoms associated with the body's biological rhythms being out of synchrony with external cues. (See the accompanying boxed feature, ◆ Concepts, Challenges, and Controversies.)

As a sleep aid, melatonin sales are especially high among senior citizens. This observation is in keeping with the fact that melatonin secretion declines steadily during adulthood. This decline occurs because calcium deposits *(pineal sand)* gradually accumulate in the pineal gland, diminishing its melatonin production. This age-related reduction in melatonin secretion may be one reason that many elderly people have difficulty getting a good night's sleep.

Hypothalamus and Pituitary

The pituitary gland consists of anterior and posterior lobes.

The **pituitary gland,** or **hypophysis,** is a small endocrine gland located in a bony cavity at the base of the brain just below the hypothalamus (Fig. 18-10). If you point one finger between your eyes and another finger toward one of your ears, the imaginary point where these lines would intersect is about the location of your pituitary. The pituitary is connected to the hypothalamus by a thin stalk, the **infundibulum,** which contains nerve fibers and small blood vessels.

The pituitary has two anatomically and functionally distinct lobes, the **posterior pituitary** and the **anterior pituitary.** The posterior pituitary, being derived embryonically from an outgrowth of the brain, is composed of nervous tissue and thus is also termed the **neurohypophysis.** The anterior pituitary, in contrast, consists of glandular epithelial tissue derived embryonically from an outpouching that buds off from the roof of the mouth. Accordingly, the anterior pituitary is also known as the **adenohypophysis** (*adeno* means "glandular"). The anterior and posterior pituitary have nothing more in common than their location. The posterior pituitary is connected to the hypothalamus by a neural pathway, whereas the anterior pituitary is connected to the hypothalamus by a vascular link.

In some species, the adenohypophysis also includes a third, well-defined intermediate lobe, but humans lack this lobe. In lower vertebrates, the intermediate lobe secretes several **melanocyte-stimulating hormones** or **MSHs,** which regulate skin coloration by controlling the dispersion of granules containing the pigment **melanin.** By causing variable skin darkening in certain amphibians, reptiles, and fishes, MSHs play a vital role in the camouflage of these species. In humans, a small amount of MSH is secreted by the anterior pituitary. The function of this MSH, if any, is still unclear. It is not involved in the differences in the amount of melanin deposited in the skin of various races, nor is it associated with the process of tanning. However, excessive MSH activity does cause darkening of the skin. Some evidence implicates MSH in humans in the totally different role of influencing the excitability of the nervous system, perhaps playing a role in improving memory and learning.

The hypothalamus and posterior pituitary form a neurosecretory system that secretes vasopressin and oxytocin.

The release of hormones from both the posterior and the anterior pituitary is directly controlled by the hypothalamus, but the nature of the relationship is entirely different. The hypothalamus and posterior pituitary form a neuroendocrine system that consists of a population of neurosecretory neurons whose cell bodies lie in two well-defined clusters in the hypo-

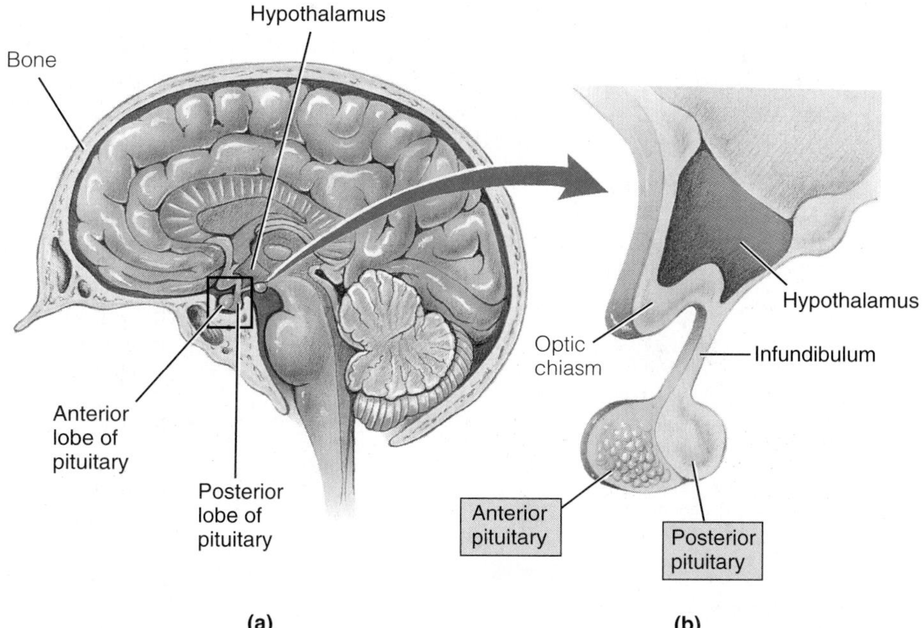

Figure 18-10 Anatomy of the Pituitary Gland (a) Relation of the pituitary gland to the hypothalamus and to the remainder of the brain. (b) Schematic enlargement of the pituitary gland and its connection to the hypothalamus.

thalamus (the **supraoptic** and **paraventricular nuclei**) and whose axons pass down through the connecting stalk to terminate on capillaries in the posterior pituitary (Fig. 18-11). The posterior pituitary consists of these neuronal terminals plus glial-like supporting cells called **pituicytes.** Functionally as well as anatomically, the posterior pituitary is simply an extension of the hypothalamus. The posterior pituitary does not actually produce any hormones. It simply stores and, upon appropriate stimulation, releases into the blood two small peptide hormones (actually, neurohormones), *vasopressin* and *oxytocin,* which are synthesized by the neuronal cell bodies in the hypothalamus. Both of these hydrophilic peptides are made in both the supraoptic and the paraventricular nuclei, but a single neuron is capable of producing only one of these hormones. The synthesized hormones are packaged in secretory granules that are transported down the cytoplasm of the axon (see p. 37) to be stored in the neuronal terminals within the posterior pituitary. Each terminal stores either vasopressin or oxytocin but not both. Thus, these hormones can be released independently as needed. Upon stimulatory input to the hypothalamus, either vasopressin or oxytocin is released into the blood from the posterior pituitary by exocytosis of the appropriate secretory granules. This hormonal release is triggered in response to action potentials that originate in the hypothalamic cell body and sweep down the axon to the neuronal terminal in the posterior pituitary. As in any other neuron, action potentials are generated in these neurosecretory neurons in response to synaptic input to their cell bodies.

Figure 18-11 **Relationship of the Hypothalamus and Posterior Pituitary** The paraventricular and supraoptic nuclei both contain neurons that produce vasopressin and oxytocin. The hormone, either vasopressin or oxytocin depending on the neuron, is synthesized in the neuronal cell body in the hypothalamus and travels down the axon to be stored in the neuronal terminals within the posterior pituitary. The stored hormone is released into the systemic blood upon excitation of the neuron.

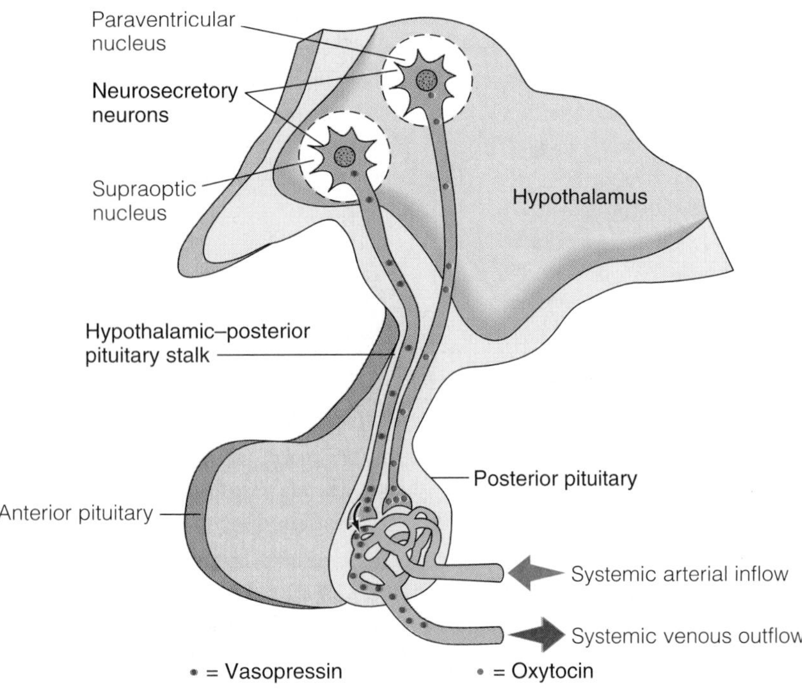

The actions of vasopressin and oxytocin are briefly summarized here to make our endocrine story complete. They are described more thoroughly elsewhere—vasopressin in chapters 14 and 15 and oxytocin in chapter 20.

Vasopressin Vasopressin (antidiuretic hormone, ADH) has two major effects that correspond to its two names: (1) it enhances the retention of H_2O by the kidneys (an antidiuretic effect), and (2) it causes contraction of arteriolar smooth muscle (a vessel pressor effect). The first effect is of greater physiological importance. Under normal conditions, vasopressin is the primary endocrine factor that regulates urinary H_2O loss and overall H_2O balance. In contrast, typical levels of vasopressin play only a minor role in regulating blood pressure by means of the hormone's pressor effect.

The major control for hypothalamic-induced release of vasopressin from the posterior pituitary is input from hypothalamic osmoreceptors, which increase vasopressin secretion in response to a rise in plasma osmolarity. A less powerful input from the left atrial volume receptors increases vasopressin secretion in response to a fall in ECF volume and arterial blood pressure (see p. 526). The osmoreceptor and volume receptor mechanisms both function as negative-feedback systems to resist changes in ECF osmolarity and volume, respectively, by adjusting the H_2O load in the body through variable vasopressin activity. (See the boxed feature on p. 640, A Closer Look at Exercise Physiology.)

Oxytocin Oxytocin stimulates contraction of the uterine smooth muscle to aid in expulsion of the baby during childbirth, and it promotes ejection of the milk from the mammary glands (breasts) during breast-feeding. Appropriately, oxytocin secretion is increased by reflexes that originate within the birth canal during childbirth and by reflexes that are triggered when the infant suckles the breast.

The anterior pituitary secretes six established hormones, many of which are tropic to other endocrine glands.

Unlike the posterior pituitary, which releases hormones that are synthesized by the hypothalamus, the anterior pituitary itself synthesizes the hormones that it releases into the blood. Different cell populations within the anterior pituitary produce and secrete six established peptide hormones. The actions of each of these hormones will be dealt with in detail in subsequent sections. For now, a brief statement of their primary effects will be given to provide a rationale for their nomenclature (Fig. 18-12):

1. **Growth hormone (GH, somatotropin),** the primary hormone responsible for regulating overall body growth, is also important in intermediary metabolism.
2. **Thyroid-stimulating hormone (TSH, thyrotropin)** stimulates secretion of thyroid hormone and growth of the thyroid gland.

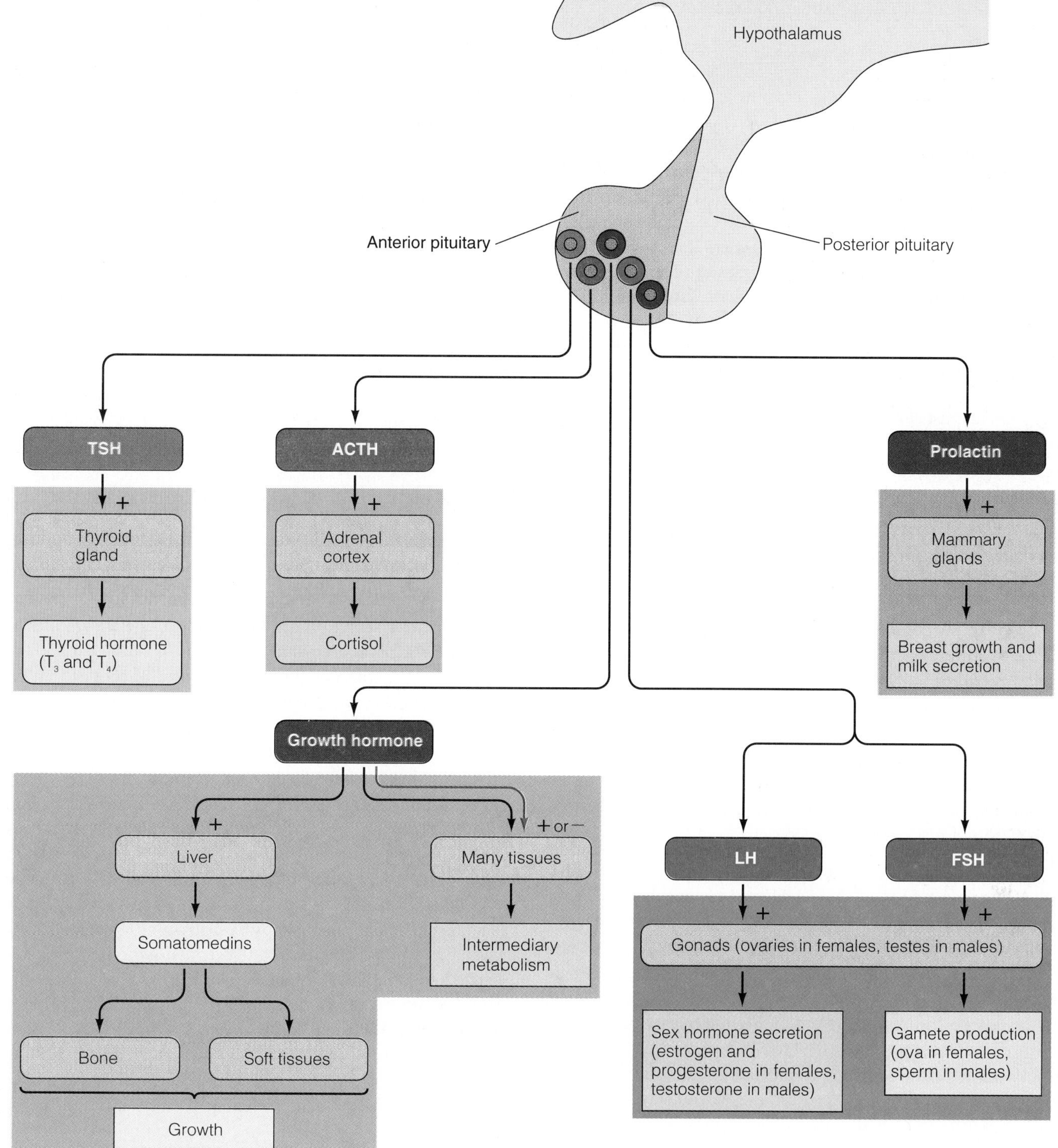

Figure 18-12 **Functions of the Anterior Pituitary Hormones** Five different endocrine cell types produce the six anterior pituitary hormones—TSH, ACTH, growth hormone, LH and FSH (produced by the same cell type), and prolactin—which exert a wide range of effects throughout the body.

3. **Adrenocorticotropic hormone (ACTH, adrenocorticotropin)** stimulates cortisol secretion by the adrenal cortex and promotes growth of the adrenal cortex.
4. **Follicle-stimulating hormone (FSH)** has different functions in males and females. In females it stimulates growth and development of ovarian follicles, within which the ova, or eggs, develop. Furthermore, FSH promotes secretion of the hormone estrogen by the ovaries. In males FSH is required for sperm production.
5. **Luteinizing hormone (LH)** also functions differently in females and males. In females LH is responsible for ovulation, luteinization (that is, the formation of a postovulatory hormone-secreting corpus luteum in the ovary), and regulation of ovarian secretion of the female sex hor-

A CLOSER LOOK AT EXERCISE PHYSIOLOGY

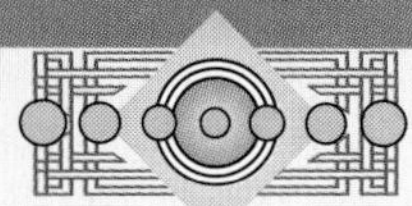

The Endocrine Response to the Challenge of Combined Heat and Marching Feet

When one exercises in a hot environment, maintenance of plasma volume becomes a critical homeostatic concern. Exercise in the heat results in the loss of large amounts of fluid through sweating. Simultaneously, blood is needed for shunting to the skin for cooling and for increased blood flow to nourish the working muscles. There must also be adequate venous return to maintain cardiac output. The hypothalamus-posterior pituitary neurosecretory system responds to these multiple, conflicting needs for fluid by releasing water-conserving vasopressin in an attempt to reduce urinary fluid loss to preserve plasma volume.

Studies have generally shown that exercise in heat stimulates vasopressin release, which results in decreased urinary fluid loss. In one study conducted during an eighteen-mile road march in heat, the participants' average urine output dropped to 134 ml (the normal urine output during the same time period would be about twice that much), while sweat loss averaged 4 liters. Overhydration prior to exercise appears to decrease the intensity of this response, suggesting that increased vasopressin release is related to plasma osmolarity. If fluid loss is not adequately replaced, plasma osmolarity increases. When the hypothalamic osmoreceptors detect this hypertonic condition, they promote increased secretion of vasopressin from the posterior pituitary. Some investigators believe, however, that increased vasopressin release results from other factors, such as changes in blood pressure or in renal blood flow. Regardless of the mechanism, vasopressin release is an important physiologic response to exercise in heat.

mones, estrogen and progesterone. In males the same hormone stimulates the interstitial cells of Leydig in the testes to secrete the male sex hormone, testosterone, giving rise to its alternate name of **interstitial cell–stimulating hormone (ICSH).**

6. **Prolactin (PRL)** enhances breast development and milk production in females. Its function in males is uncertain, although evidence indicates that it may induce the production of testicular LH receptors.

Because they each regulate the secretion of another specific endocrine gland, TSH, ACTH, FSH, and LH are all tropic hormones. More specifically, FSH and LH are collectively referred to as **gonadotropins** because they control secretion of the sex hormones by the gonads (ovaries and testes). Since growth hormone has recently been shown to exert its growth-promoting effects indirectly by stimulating the release of the liver hormones, the somatomedins, it too is sometimes categorized as a tropic hormone. Among the anterior pituitary hormones, prolactin is the only one that does not stimulate secretion of another hormone. Of the tropic hormones, FSH, LH, and growth hormone differ from TSH and ACTH in that the former exert nontropic functions in addition to stimulating secretion of other hormones.

Hypothalamic releasing and inhibiting hormones are delivered to the anterior pituitary by the hypothalamic-hypophyseal portal system to control anterior pituitary hormone secretion.

None of the anterior pituitary hormones are secreted at a constant rate. Even though each of these hormones has a unique control system, there are some common regulatory patterns. The two most important factors that regulate anterior pituitary hormone secretion are (1) hypothalamic hormones and (2) feedback by target gland hormones.

Because the anterior pituitary secretes hormones that control the secretion of various other hormones, it has long had the undeserved title of "master gland." It is now known that the release of each of the anterior pituitary hormones is largely controlled by still other hormones produced by the hypothalamus and that secretion of these regulatory neurohormones in turn is controlled by a variety of neural and hormonal inputs to the hypothalamic neurosecretory cells (Fig. 18-13).

The secretion of each of the anterior pituitary hormones is stimulated or inhibited by one or more of the seven generally accepted hypothalamic **hypophysiotropic** (*hypophysis* means "pituitary"; *tropic* means "regulating") **hormones,** which are listed in Table 18-4. Depending on their actions, these small peptides are called *releasing hormones* or *inhibiting hormones.* (The term *factor* rather than *hormone* is used for any substance whose structure is not yet known. A substance may qualify as a full-fledged hormone only if it has been isolated and chemically identified.) In each case, the primary action of the hormone is apparent from its name. For example, **thyrotropin-releasing hormone (TRH)** stimulates the release of TSH (alias thyrotropin) from the anterior pituitary, whereas **prolactin-inhibiting hormone (PIH)** inhibits the release of prolactin from the anterior pituitary. Although it was originally speculated that there was a neat one-to-one correspondence—one hypophysiotropic hormone for each anterior pituitary hormone—it is now clear that many of the hypothalamic hormones have more than one effect, with their names indicating only the function that was initially attributed to them. Moreover, a single anterior pituitary hormone may be regulated by two or more hypophysiotropic hormones, which may even exert opposing effects. For example, **growth hormone–releasing hormone (GHRH)** stimulates

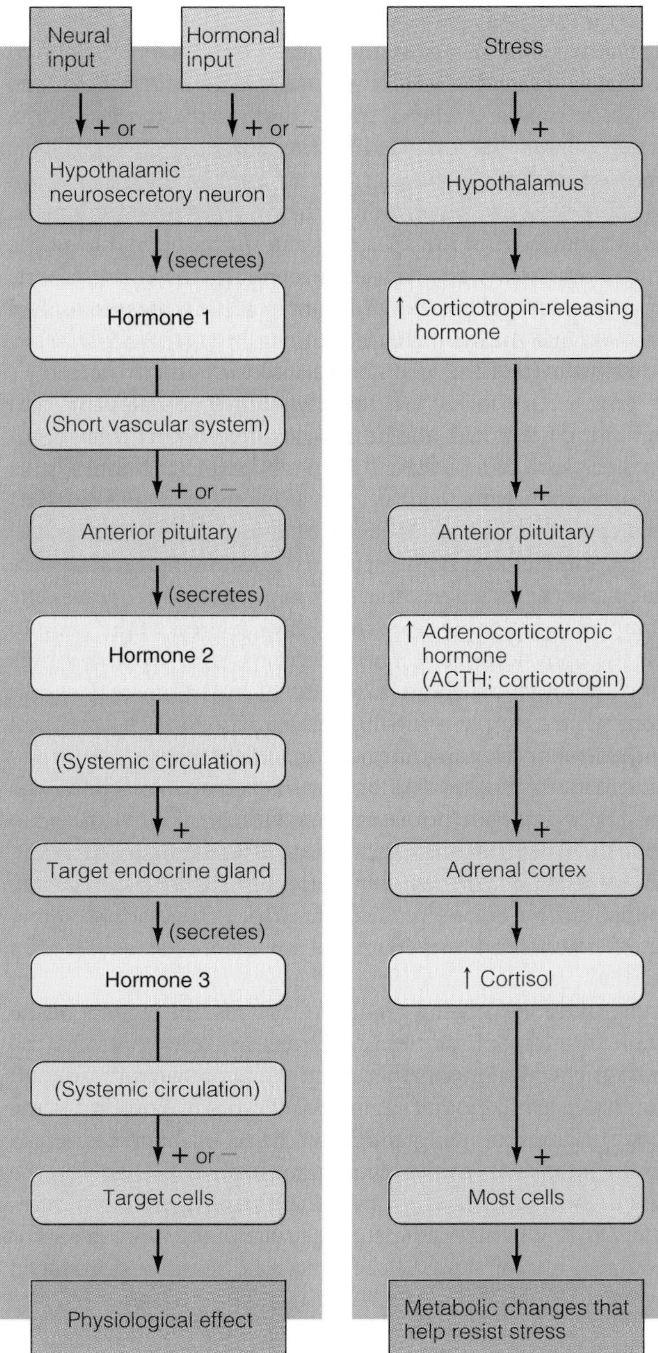

Figure 18-13 **Hierarchical Chain of Command in Endocrine Control** The hypothalamic hypophysiotropic hormone (hormone 1) controls the output of the anterior pituitary tropic hormone (hormone 2). This tropic hormone in turn regulates the secretion of the target endocrine gland's hormone (hormone 3), which exerts the final physiological effect. The pathway leading to cortisol secretion provides a specific example of this endocrine chain of command.

growth hormone secretion, whereas **growth hormone–inhibiting hormone (GHIH),** also known as **somatostatin,** inhibits it. The output of the anterior pituitary growth hormone—secreting cells (that is, the rate of growth hormone secretion) in response to two such opposing inputs depends on the relative concentrations of these hypothalamic hormones as well as on the intensity of other regulatory inputs. This is analogous to the output of a nerve cell (that is, the rate of action potential propagation) being dependent on the relative magnitude of excitatory and inhibitory synaptic inputs (EPSPs and IPSPs) to it (see p. 98).

Chemical messengers that are identical in structure to the hypothalamic releasing and inhibiting hormones and to vasopressin are produced in many areas of the brain outside the hypothalamus. Instead of being released into the blood, these messengers act locally as neurotransmitters and neuromodulators in these extrahypothalamic sites. They are thought to modulate a variety of functions that range from motor activity (TRH) to libido (GnRH) to learning (vasopressin). Thus, they provide a further example of the multiplicity of use of chemical messengers.

The hypothalamic regulatory hormones reach the anterior pituitary by means of a unique vascular link. In contrast to the direct neural connection between the hypothalamus and posterior pituitary, the anatomical and functional link between the hypothalamus and anterior pituitary is an unusual capillary-to-capillary connection, the **hypothalamic-hypophyseal portal system.** A portal system is a vascular arrangement in which venous blood flows directly from one capillary bed through a connecting vessel to another capillary bed without passing through the systemic circulation. The largest and best-known portal system is the hepatic portal system, which drains intestinal venous blood directly into the liver for immediate processing of absorbed nutrients (see

Table 18-4
Major Hypophysiotropic Hormones

Hormone	*Effect on the Anterior Pituitary*
Thyrotropin-releasing hormone (TRH)	Stimulates release of TSH (thyrotropin) and prolactin
Corticotropin-releasing hormone (CRH)	Stimulates release of ACTH (corticotropin)
Gonadotropin-releasing hormone (GnRH)	Stimulates release of FSH and LH (gonadotropins)
Growth hormone–releasing hormone (GHRH)	Stimulates release of growth hormone
Growth hormone–inhibiting hormone (GHIH)	Inhibits release of growth hormone and TSH
Prolactin-releasing factor (PRF)	Stimulates release of prolactin
Prolactin-inhibiting hormone (PIH)	Inhibits release of prolactin

p. 574). Although much smaller, the hypothalamic-hypophyseal portal system is no less important because it provides a critical link between the brain and much of the endocrine system. It begins in the base of the hypothalamus with a group of capillaries that recombine into small portal vessels, which pass down through the connecting stalk into the anterior pituitary. Here they branch to form most of the anterior pituitary capillaries, which in turn drain into the systemic venous system (Fig. 18-14).

Note that almost all of the blood supply to the anterior pituitary must first pass through the hypothalamus. Since materials can be exchanged between the blood and surrounding tissue only at the capillary level, the hypothalamic-hypophyseal portal system provides a route where releasing and inhibiting hormones can be picked up at the hypothalamus and delivered immediately and directly to the anterior pituitary at relatively high concentrations, completely bypassing the systemic circulation. The hypophysiotropic hormones would be considerably more diluted if they had to travel to the anterior pituitary by means of the usual systemic circulatory route. This type of chemical messenger distribution is intermediate between a classic endocrine system, in which the hormone is released into the general circulation, and a neuronal system, in which the neurotransmitter is released into a confined space (that is, a synapse).

The axons of the neurosecretory neurons that produce the hypothalamic regulatory hormones terminate on the capillaries at the origin of the portal system. These hypothalamic neurons secrete their hormones in the same way as the hypothalamic neurons that produce vasopressin and oxytocin. The hormone is synthesized in the cell body and then transported to the axon terminal. It is stored there until its release into an adjacent capillary when, upon appropriate stimulation, an action potential is generated in the neuron. The major difference is that the hypophysiotropic hormones are released into the portal vessels, which deliver them to the anterior pituitary, where they control the release of anterior pituitary hormones into the systemic circulation. In contrast, the hypothalamic hormones stored in the posterior pituitary are themselves released into the systemic circulation (Fig. 18-15).

Knowing that the secretion of anterior pituitary hormones is largely controlled by the hypothalamic releasing and inhibiting hormones, the next logical question is, what regulates the secretion of these hypophysiotropic hormones? Like other neurons, the neurons secreting these regulatory hormones receive abundant input of information (both neural and hormonal and both excitatory and inhibitory) that they must integrate. Studies are still in progress to unravel the complex neural input from many diverse areas of the brain to the hypophysiotropic secretory neurons. Some of these inputs carry information about a variety of environmental conditions. One example is the marked increase in the secretion of corticotropin-releasing hormone (CRH) in response to stressful situations (Fig. 18-13). Numerous neural connections also exist between the hypothalamus and the portions of the brain concerned with emotions (the limbic system—see p. 132). Thus, secretion of hypophysiotropic hormones is greatly influenced by emotions. The menstrual irregularities sometimes experienced by women who are emotionally upset are a common manifestation of this relationship.

In addition to being regulated by different regions of the brain, the hypophysiotropic neurons are also controlled by various chemical inputs that reach the hypothalamus through the blood. Unlike other regions of the brain, portions of the hypothalamus are not guarded by the blood-brain barrier, so the hypothalamus can easily monitor chemical changes in the blood. In some instances, hypophysiotropic secretion is influenced by the immediate metabolic state of the individual. For example, one or more of the metabolic responses produced by growth hormone (such as an elevated blood glucose level) negatively feed back to the hypothalamus to control growth hormone secretion. The more common blood-borne factors that influence hypothalamic neurosecretion are the negative-feedback effects of either anterior pituitary or target-gland hormones, to which we now turn our attention.

Figure 18-14 Vascular Link between the Hypothalamus and Anterior Pituitary Hypothalamic capillaries, which pick up the hypophysiotropic hormones, rejoin to form the hypothalamic-hypophyseal portal system. This vascular link passes to the anterior pituitary where it branches into the anterior pituitary capillaries. The hypophysiotropic hormones leave the blood across the anterior pituitary capillaries and control the release of anterior pituitary hormones, which enter these capillaries for distribution throughout the body.

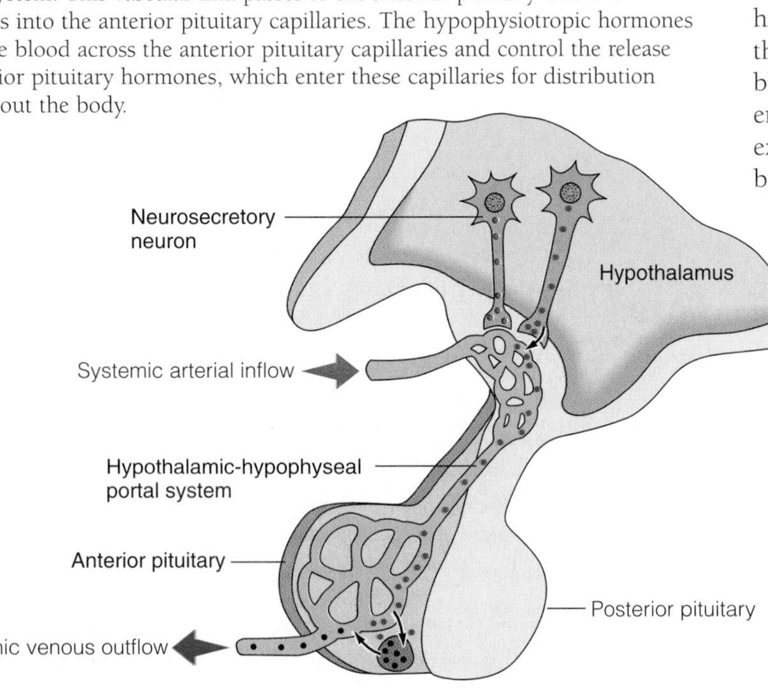

In general, feedback by target-gland hormones strives to maintain relatively constant rates of anterior pituitary hormone secretion.

In most cases, hypophysiotropic hormones initiate a three-hormone sequence: (1) hypophysiotropic hormone, (2) anterior pituitary tropic hormone, and (3) peripheral target–endocrine gland hormone (Fig. 18-13). With one exception, in addition to producing its physiological effects, the target-gland hormone also acts to suppress secretion of the tropic hormone that is driving it. This suppression, designated **long-loop negative feedback,** is accomplished by the target-gland hormone acting either directly on the pituitary itself or on the release of hypothalamic hormones, which in turn regulate anterior pituitary function (Fig. 18-16). As an example, consider the CRH-ACTH-cortisol system. Hypothalamic CRH (corticotropin-releasing hormone) stimulates the anterior pituitary to secrete ACTH (adrenocorticotropic hormone, alias corticotropin), which in turn stimulates the adrenal cortex to secrete cortisol. The final hormone in the system, cortisol, inhibits the hypothalamus to reduce CRH secretion and also reduces the sensitivity of the ACTH-secreting cells to CRH by acting directly on the anterior pituitary. Through this double-barreled approach, cortisol exerts negative-feedback control to stabilize its own plasma concentration. If plasma cortisol levels start to rise above a prescribed set level, cortisol suppresses its own further secretion by its inhibitory actions at the hypothalamus and anterior pituitary. This mechanism ensures that once a hormonal system is activated, its secretion does not continue unabated. If plasma cortisol levels fall below the desired set point, cortisol's inhibitory actions at the hypothalamus and anterior pituitary are reduced, so the driving forces for cortisol secretion (CRH-ACTH) increase accordingly.

Similarly, the other target-gland hormones act by means of long-loop negative feedback. The goal of such feedback is to maintain relatively constant levels of the target-gland hormone. Remember, however, that the diurnal rhythms are superimposed on this type of stabilizing negative-feedback regulation. Furthermore, other controlling inputs may break through the negative-feedback control to alter hormone secretion (that is, change the set level) at times of special need.

The one exception to the negative-feedback relationship just described is the long-loop positive-feedback effect of estrogen on LH secretion. This causes an extremely dramatic rise in LH secretion that is responsible for triggering ovulation. This relationship will be discussed further in chapter 20 (see p. 727).

Besides the physiologically more important long-loop feedback mechanism just described, evidence exists for **short-loop negative feedback,** which refers to a pituitary hormone's action at the hypothalamus to inhibit the release of its stimulatory hypophysiotropic hormone (Fig. 18-16). For example, prolactin, which is the only anterior pituitary hormone that does not participate in a three-hormone sequence, is believed to act directly on the hypothalamus to influence secretion of hypophysiotropic hormones that control prolactin secretion. It is possible that anterior pituitary tropic hormones employ similar short-loop feedback on their respective hypophysiotropic neurons. The recent discovery that a substantial portion of the blood flowing from the anterior pituitary may actually go back to the hypothalamus

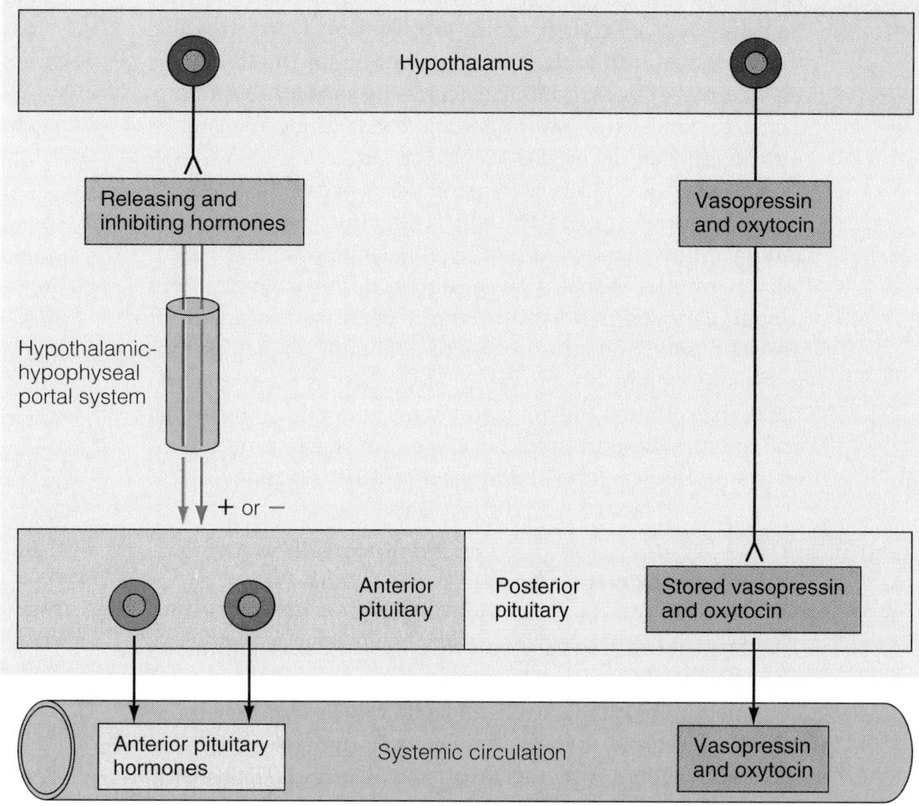

Figure 18-15 **Comparison of the Hypothalamic Relationship to the Anterior Pituitary and Posterior Pituitary** Hypothalamic releasing and inhibiting hormones pass through the hypothalamic-hypophyseal portal system to control the release into the systemic circulation of hormones produced by the anterior pituitary. The hypothalamus itself produces vasopressin and oxytocin, which are stored in the posterior pituitary and released into the systemic circulation on hypothalamic stimulation.

Figure 18-16 **Negative Feedback in Hypothalamic–Anterior Pituitary Control Systems**

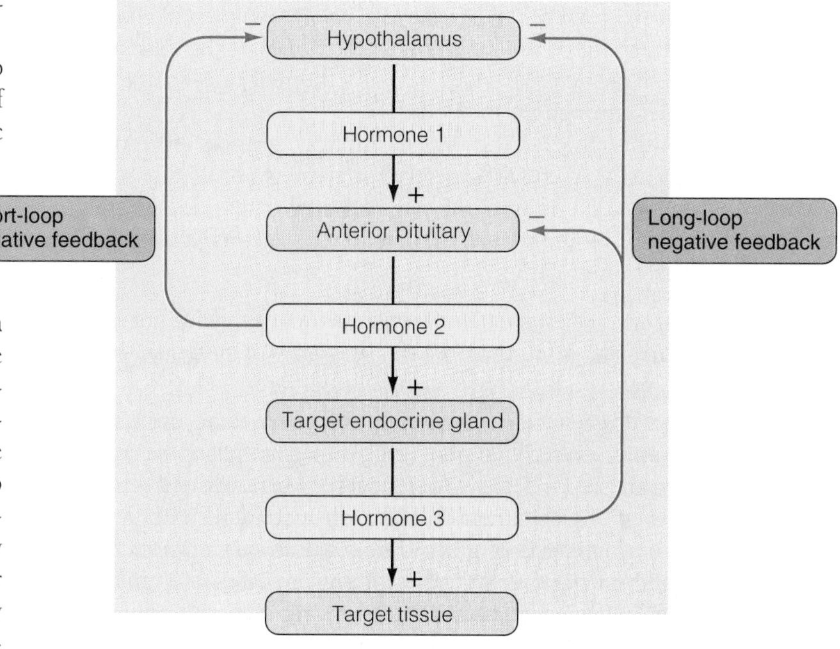

before entering the general circulation suggests that this may be the case. Such retrograde flow of blood makes it plausible that pituitary hormones may reach the brain in sufficiently high concentrations to influence releasing-hormone secretion. The actual existence and importance of such mechanisms in humans remain uncertain.

In addition, other hormones *outside* a particular sequence may also exert important influences, either stimulatory or inhibitory, on the secretion of hypothalamic or anterior pituitary hormones within a given sequence. For example, even though estrogen is not in the direct chain of command for prolactin secretion, this sex steroid notably enhances prolactin secretion by the anterior pituitary. This is but one example of the common phenomenon in the endocrine system that one seemingly unrelated hormone can have pronounced effects on the secretion or actions of another hormone.

Endocrine Control of Growth

The detailed functions and control of all the anterior pituitary hormones except growth hormone are discussed elsewhere in conjunction with the peripheral target tissues that they influence; for example, thyroid-stimulating hormone is covered in the next chapter with the discussion of the thyroid gland. Accordingly, growth hormone is the only anterior pituitary hormone that will be elaborated on at this time.

Growth depends on growth hormone but is influenced by other factors as well.

In growing children, continuous net protein synthesis occurs under the influence of growth hormone as the body steadily gets larger. Weight gain alone is not synonymous with growth, because weight gain may occur as a result of retention of excess H_2O or fat without true structural growth of tissues. Growth requires net synthesis of proteins and includes lengthening of the long bones (the bones of the extremities) as well as increases in the size and number of cells in the soft tissues throughout the body.

Although, as the name implies, growth hormone is absolutely essential for growth, it alone is not wholly responsible for determining the rate and final magnitude of growth in a given individual. The following factors also have an impact on growth:

- *Genetic determination* of an individual's maximum growth capacity. Attainment of this full growth potential further depends on the other factors listed here.
- *An adequate diet,* including sufficient total protein and ample essential amino acids to accomplish the protein synthesis necessary for growth. Malnourished children never achieve their full growth potential. The growth-stunting effects of inadequate nutrition are most profound when they occur in infancy. In severe cases, the child may be locked in to irreversible stunting of body growth and brain development. About 70% of the total growth of the brain occurs in the first two years of life. On the other hand, one cannot exceed his or her genetically determined maximum by eating more than an adequate diet. The excess food intake produces obesity instead of growth.
- *Freedom from chronic disease and stressful environmental conditions.* Stunting of growth under adverse circumstances is due in large part to the prolonged stress-induced secretion of cortisol from the adrenal cortex. Cortisol exerts several potent antigrowth effects, such as promoting protein breakdown, inhibiting growth in the long bones, and blocking the secretion of growth hormone. Even though sickly or stressed children do not grow well, if the condition is rectified before adult size is achieved, they can rapidly catch up to their normal growth curve through a remarkable spurt in growth.
- *A normal milieu of growth-influencing hormones.* In addition to the absolutely essential growth hormone, other hormones, including thyroid hormone, insulin, and the sex hormones, play secondary roles in the promotion of growth.

The rate of growth is not continuous nor are the factors responsible for promoting growth the same throughout the growth period. *Fetal growth* seems to be largely independent of hormonal control, with the size at birth being determined principally by genetic and environmental factors. Hormonal factors begin to play an important role in the regulation of growth after birth. Genetic and nutritional factors also strongly affect growth during this period.

Children display two periods of rapid growth—a *postnatal growth spurt* during their first two years of life and a *pubertal growth spurt* during adolescence (Fig. 18-17). From the age of two until puberty, the *rate* of linear growth progressively declines, even though the child is still growing. Before puberty there is little sexual difference in height or weight. During puberty, a marked acceleration in linear growth takes place because of the lengthening of the long bones. Puberty begins at about age eleven in girls and thirteen in boys and lasts for several years in both sexes. The mechanisms responsible for the pubertal growth spurt are not clearly understood. Apparently, both genetic and hormonal factors are involved. Some evidence indicates that growth hormone secretion is

Figure 18-17 Normal Growth Curve

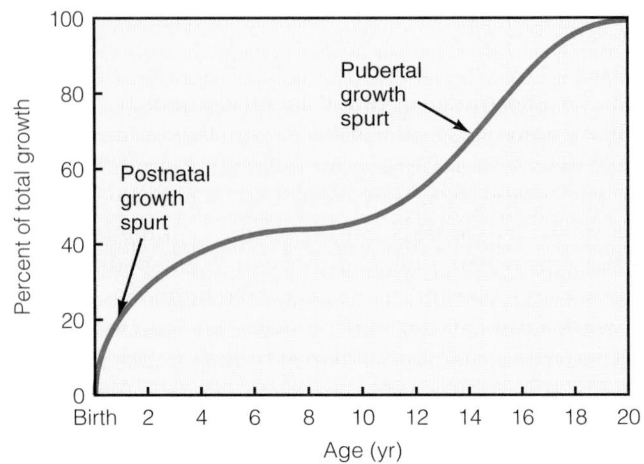

elevated during puberty and thus may contribute to the acceleration of growth during this time. Furthermore, **androgens** ("male" sex hormones), whose secretion increases dramatically at puberty, also contribute to the pubertal growth spurt by promoting protein synthesis and bone growth. The potent androgen from the male testes, testosterone, is of greatest importance in promoting a sharp increase in height in adolescent boys, whereas the less potent adrenal androgens from the adrenal gland, which also manifest a sizable increase in secretion during adolescence, are most likely important in the female pubertal growth spurt. Although estrogen secretion by the ovaries also begins during puberty, it is unclear what role this "female" sex hormone may play in the pubertal growth spurt in girls. There is no doubt that testosterone and estrogen both ultimately act on bone to halt its further growth so that full adult height is attained by the end of adolescence.

Growth hormone is essential for growth, but it also exerts metabolic effects not related to growth.

Growth hormone is the most abundant hormone produced by the anterior pituitary, even in adults in whom growth has already ceased. The continued high secretion of growth hormone beyond the growing period implies that this hormone must have important influences other than on growth. Its growth-promoting effects are fairly well understood. Its metabolic actions not related to growth are known, but their physiological role remains somewhat nebulous.

Metabolic actions unrelated to growth Growth hormone increases fatty acid levels in the blood by enhancing the breakdown of triglyceride fat stored in adipose tissue, and it increases blood glucose levels by decreasing glucose uptake by muscles. Muscles use the mobilized fatty acids instead of glucose as a metabolic fuel. Thus, the overall metabolic effect of growth hormone is to mobilize fat stores as a major energy source while conserving glucose for glucose-dependent tissues such as the brain. The brain can use only glucose as its metabolic fuel, yet nervous tissue cannot store glycogen (stored glucose) to any extent. This metabolic pattern is suitable for maintaining the body during prolonged fasting or other situations when the body's energy needs exceed available glucose stores.

Growth-promoting actions on soft tissues When tissues are responsive to its growth-promoting effects, growth hormone stimulates growth of both soft tissues and the skeleton. Growth of soft tissues is accomplished by (1) increasing the number of cells **(hyperplasia)** by stimulating cell division and (2) increasing the size of cells **(hypertrophy)** by favoring synthesis of proteins, the main structural component of cells.

Growth hormone stimulates almost all aspects of protein synthesis while it simultaneously inhibits protein degradation. It promotes the uptake of amino acids (the raw materials for protein synthesis) by cells, decreasing blood amino acid levels in the process. Furthermore, it stimulates the cellular machinery responsible for accomplishing protein synthesis according to the cell's genetic code.

Growth-promoting actions on bones Growth of the long bones resulting in increased height is the most dramatic effect of growth hormone. **Bone** is a living tissue. Being a form of connective tissue, it consists of cells and an extracellular organic matrix that is produced by the cells. The bone cells that produce the organic matrix are known as **osteoblasts** ("bone-formers"). The organic matrix is composed of collagen fibers (see p. 51) in a mucopolysaccharide-rich semisolid gel referred to as *ground substance*. This matrix has a rubbery consistency and is responsible for the tensile strength of bone (the resilience of bone to breakage when tension is applied). Bone is made hard by precipitation of calcium phosphate crystals within the matrix. These inorganic crystals provide the bone with compressional strength (the ability of bone to hold its shape when squeezed or compressed). If bones were composed entirely of inorganic crystals, they would be brittle, like pieces of chalk. Bones have structural strength approaching that of reinforced concrete, yet they are not brittle and are much lighter in weight as a result of the structural blending of an organic scaffolding hardened by inorganic crystals. **Cartilage** is similar to bone, except that living cartilage is not calcified.

A long bone basically consists of a fairly uniform cylindrical shaft, the **diaphysis,** with a flared articulating knob at either end, an **epiphysis.** In a growing bone, the diaphysis is separated at each end from the epiphysis by a layer of cartilage known as the **epiphyseal plate** (— Fig. 18-18a). The central cavity of the bone is filled with bone marrow, which is the site of blood cell production (see p. 356).

Growth in *thickness* of bone is achieved by the addition of new bone on top of the already existing bone on the outer surface. This growth occurs through activity of osteoblasts within the **periosteum,** a connective tissue sheath that covers the outer bone surface. As new bone is being deposited by osteoblast activity on the external surface, other cells within the bone, the **osteoclasts** ("bone-breakers"), dissolve the bony tissue on the inner surface adjacent to the marrow cavity. In this way, the marrow cavity is enlarged to keep pace with the increase in the circumference of the bone shaft.

Growth in *length* of long bones is accomplished by a different mechanism than growth in thickness. Bones grow in length as a result of proliferation of the cartilage cells in the epiphyseal plates (Fig. 18-18b). During growth, new cartilage cells **(chondrocytes)** are produced through cell division on the outer edge of the plate adjacent to the epiphysis. As new chondrocytes are being formed on the epiphyseal border, the older cartilage cells toward the diaphyseal border are enlarging. This combination of proliferation of new cartilage cells and hypertrophy of maturing chondrocytes causes the epiphyseal plate to temporarily increase in width. This thickening of the intervening cartilaginous plate causes the bony epiphysis to be pushed farther away from the diaphysis. Soon the matrix surrounding the oldest hypertrophied cartilage becomes calcified. Since cartilage lacks its own capillary network, the survival of cartilage cells depends on diffusion of nutrients and O_2 through the ground substance, a process that is prevented by the deposition of calcium salts. As a result, the old nutrient-deprived cartilage cells on the diaphyseal border die. As osteoclasts clear away the dead chondrocytes and the calcified matrix that imprisoned them, the area is invaded by osteoblasts, which swarm upward from the diaphysis, trailing their capillary supply with them. These new tenants lay down bone around the persisting remnants of disintegrating cartilage until the inner region of cartilage on the diaphyseal side of the plate is entirely replaced by bone. When this **ossification** ("bone-forming") process is completed, the bone on the diaphyseal side has increased in length, and the epiphyseal plate has been returned to its original thickness. The cartilage that has been replaced by bone on the diaphyseal end of the plate is equivalent in thickness to the new cartilaginous growth on the epiphyseal end of the plate. Thus, bone growth is made possible by the growth and death of cartilage, which acts like a "spacer" to push the epiphysis farther out while it provides a framework for future formation of bone on the end of the diaphysis.

As the extracellular matrix produced by an osteoblast becomes calcified, the osteoblast like its chondrocyte predecessor becomes entombed by the matrix that it has deposited around itself. Unlike chondrocytes, however, osteoblasts trapped within a calcified matrix do not die because they are supplied by nutrients transported to them through small canals that the osteoblasts themselves form by sending out cytoplasmic processes around which the bony matrix is deposited. Thus, within the final bony product, a network of permeating tunnels radiates from each entrapped osteoblast, serving as a lifeline system for nutrient delivery and waste removal. The entrapped osteoblasts, which are now called **osteocytes,** retire from active bone-forming duty, since their imprisonment prevents them from laying down any new bone. However, they are involved in the hormonally regulated exchange of calcium between bone and the blood. This exchange is under the control of parathyroid hormone (to be discussed in the next chapter; see p. 688), not growth hormone.

Growth hormone promotes growth of bone in both thickness and length. It stimulates the proliferation of epiphyseal cartilage, thereby making space for more bone formation, and also stimulates osteoblast activity. Growth hormone is able to

Figure 18-18 **Anatomy and Growth of Long Bones** (a) Anatomy of long bones. (b) Two sections of the same epiphyseal plate at different times, depicting the growth in length of long bones.

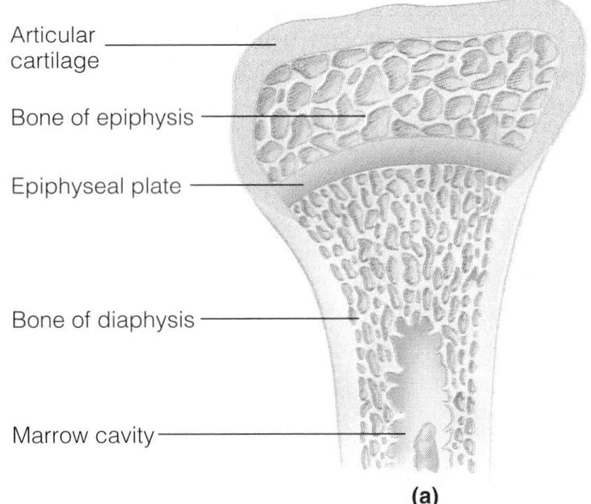

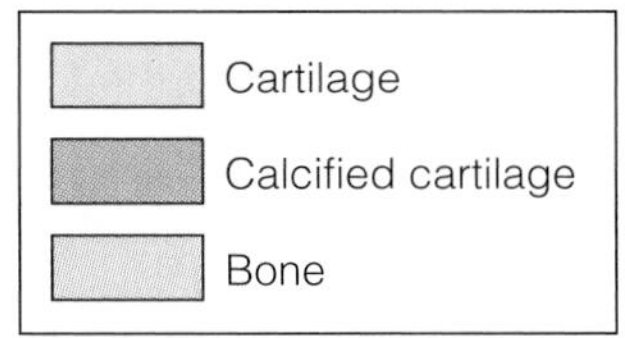

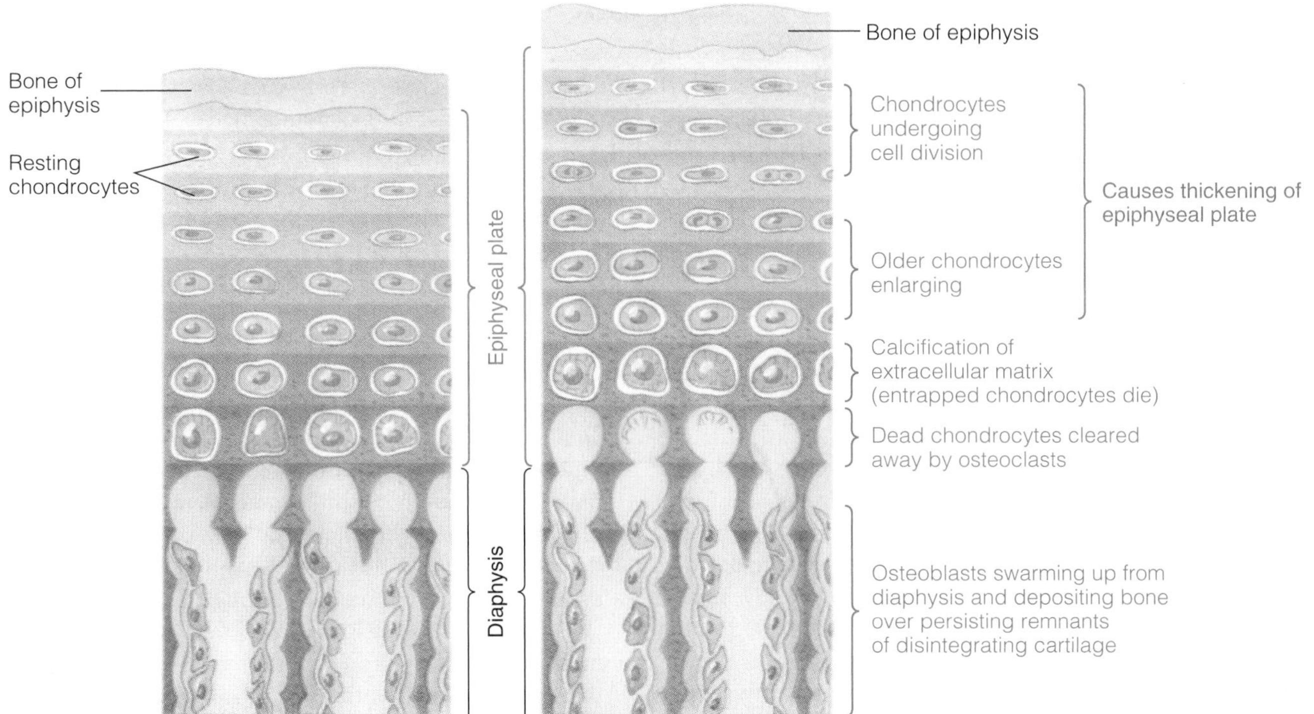

promote lengthening of long bones as long as the epiphyseal plate remains cartilaginous, or is "open." At the end of adolescence, under the influence of the sex hormones, these plates completely ossify, or "close," so that the bones can grow no further in length despite the presence of growth hormone. Thus, after the plates are closed, the individual does not grow any taller.

Growth hormone exerts its growth-promoting effects indirectly by stimulating somatomedins.

Growth hormone does not act directly on its target cells to bring about its growth-producing actions (increased cell division, enhanced protein synthesis, and bone growth). These effects are directly brought about by peptide mediators known as **somatomedins,** whose synthesis is stimulated by growth hormone. These peptides are also referred to as **insulin-like growth factors (IGF)** because they are structurally and functionally similar to insulin. Two somatomedins—IGF I and IGF II—have been identified.

The major site of somatomedin production is the liver, which releases this peptide product into the blood. However, somatomedin production has also been demonstrated in a variety of other tissues. It has been proposed that somatomedins produced locally in target tissues may act through paracrine means (see p. 66) for at least some of the growth hormone–induced effects. Such a mechanism could account for the fact that blood levels of growth hormone are no higher, and indeed circulating somatomedin levels are lower, during the first several years of life compared to adult values, even though growth is quite rapid during the postnatal period. Local production of somatomedins in target tissues may possibly be more important than delivery of blood-borne somatomedins during this time.

To complicate the situation further, the production and activity of somatomedins are also controlled by a number of factors other than growth hormone, including age and nutritional status. As a result, changes in circulating somatomedin levels do not always coincide with changes in growth hormone secretion. For example, fasting decreases IGF I levels even though it increases growth hormone secretion. Furthermore, the two different somatomedins are not always secreted in parallel. A dramatic increase in IGF I levels accompanies the moderate increase in growth hormone at puberty, which may, of course, be an important factor in the pubertal growth spurt. In contrast, IGF II concentrations do not appear to increase during puberty. These few examples serve to illustrate the confusing state of knowledge regarding the role and control of these important growth hormone mediators.

Growth hormone secretion is regulated by two hypophysiotropic hormones.

Two antagonistic regulatory hormones from the hypothalamus are involved in control of growth hormone secretion: growth hormone–releasing hormone (GHRH), which is stimulatory, and growth hormone–inhibiting hormone (GHIH or somatostatin), which is inhibitory (Fig. 18-19). (Note the distinction between *somatotropin,* alias growth hormone; *somatomedin,* the liver hormone that directly mediates the effects of growth hormone; and *somatostatin,* which inhibits growth hormone secretion.) Any factor that increases growth

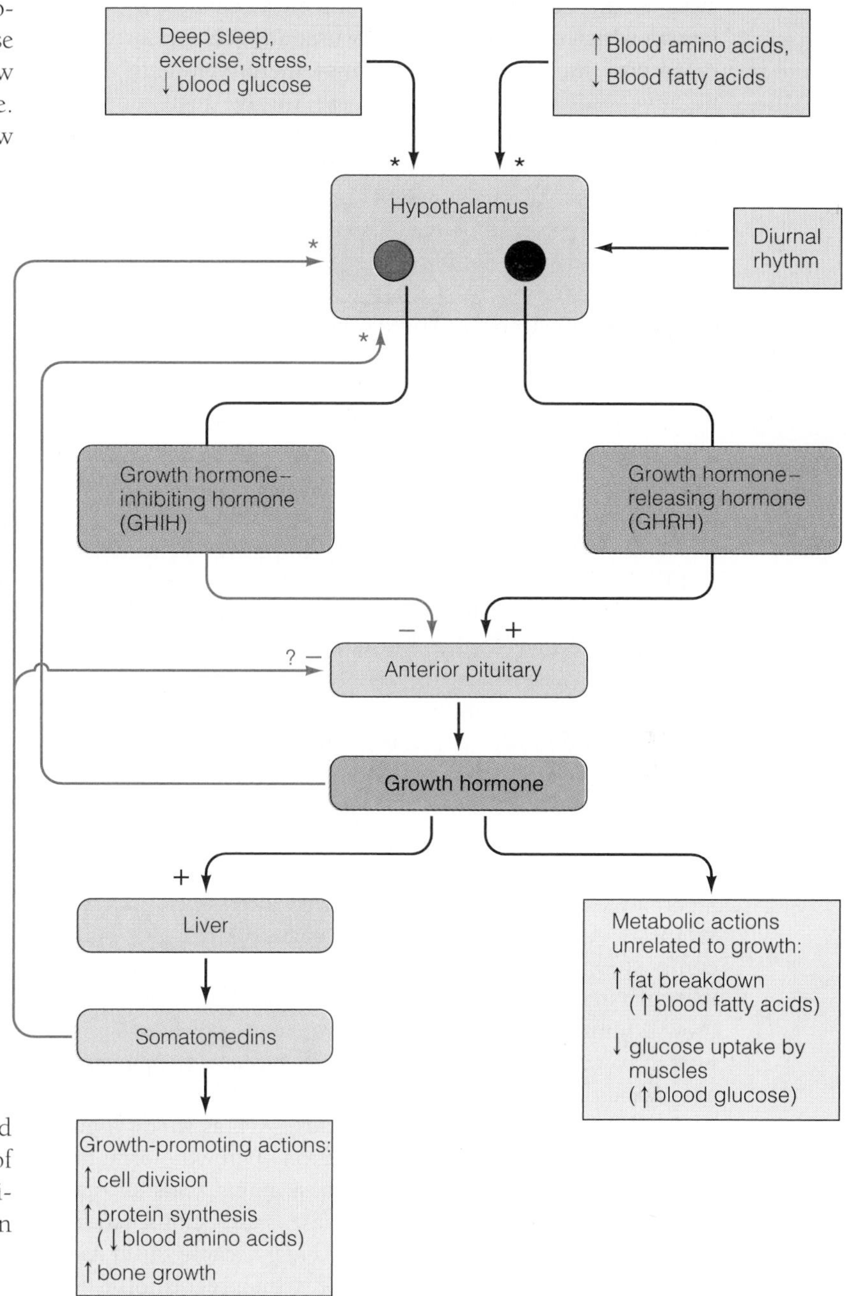

Figure 18-19 **Control of Growth Hormone Secretion**

hormone secretion could theoretically do so either by stimulating GHRH release or by inhibiting GHIH release. It is not known which of these pathways is used in each specific case.

As with the other hypothalamus–anterior pituitary axes, negative-feedback loops participate in the regulation of growth hormone secretion. Both growth hormone and the somatomedins inhibit pituitary secretion of growth hormone, presumably by stimulating GHIH release from the hypothalamus. The somatomedins may also exert direct effects on the anterior pituitary to inhibit the effects of GHRH on growth hormone release.

Interacting with or overriding the basic negative-feedback control system are a number of factors that influence growth hormone secretion. Growth hormone secretion displays a well-characterized diurnal rhythm. Through most of the day, growth hormone levels tend to be low and fairly constant. Approximately one hour after the onset of deep sleep, however, growth hormone secretion markedly increases up to five times the daytime value, then rapidly drops over the next several hours.

Superimposed on this diurnal undulation in growth hormone secretion are further bursts in secretion that occur in response to exercise, stress, and hypoglycemia (low blood glucose), the major stimuli for increased secretion. The benefit of increased growth hormone secretion during these situations when energy demands outstrip the body's glucose reserves is presumably the conservation of glucose for the brain and the provision of fatty acids as an alternative energy source for muscle.

Because growth hormone uses up fat stores and promotes synthesis of body proteins, it encourages a change in body composition away from adipose deposition toward an increase in muscle protein. Accordingly, the increase in growth hormone secretion that accompanies exercise may at least in part mediate exercise's effects in reducing the percentage of body fat while increasing the lean body mass.

An abundance of blood amino acids, such as after a high-protein meal, also enhances the secretion of growth hormone, which in turn promotes the use of these amino acids for protein synthesis. Growth hormone release is also stimulated by a decline in the level of blood fatty acids. Because of the fat-mobilizing actions of growth hormone, such regulation contributes to the maintenance of fairly constant blood fatty acid levels.

Note that the known regulatory inputs for growth hormone secretion are aimed at adjusting the levels of glucose, amino acids, and fatty acids in the blood. There are no known growth-related signals that influence growth hormone secretion. The whole issue of what really controls growth is complicated by the fact that growth hormone levels during early childhood, a period of quite rapid linear growth, are similar to those found in normal adults. As mentioned earlier, the poorly understood control of somatomedin activity may be important in this regard. Another related question is, why aren't tissues still responsive to growth hormone's growth-promoting effects in adulthood? We know that we do not grow any taller after adolescence because the epiphyseal plates have closed, but why don't the soft tissues continue to grow through hypertrophy and hyperplasia under the influence of growth hormone? One speculation is that the levels of growth hormone may only be high enough to produce its growth-promoting effects during the bursts in secretion that occur in deep sleep. It is interesting to note that time spent in deep sleep is greatest in infancy and gradually declines with age. Still, even as we age, we spend some time in deep sleep, yet we do not gradually get larger. Further research will be needed to unravel some of these mysteries.

Abnormal growth hormone secretion results in aberrant growth patterns.

Diseases related to both deficiencies and excesses of growth hormone can occur. The effects on the pattern of growth are much more pronounced than the metabolic consequences.

Growth hormone deficiency Growth hormone deficiency may be caused by a pituitary defect (lack of growth hormone) or occur secondarily to hypothalamic dysfunctions (lack of GHRH). Hyposecretion of growth hormone in a child results in **dwarfism.** The predominant feature is short stature caused by retarded skeletal growth. Less obvious characteristics include poorly developed musculature (reduction in muscle protein synthesis) and excess subcutaneous fat (less fat mobilization).

In addition, growth may be thwarted because the tissues fail to respond normally to growth hormone. An example is **Laron dwarfism,** which is characterized by abnormal growth hormone receptors that are unresponsive to the hormone. The symptoms of this condition resemble those of severe growth hormone deficiency even though blood levels of growth hormone are actually high. In some instances, growth hormone levels are adequate and target cell responsiveness is normal, but somatomedins are deficient. African pygmies are an interesting example. Their short stature is attributable to a genetic deficit of the most potent of the somatomedins.

The onset of growth hormone deficiency in adulthood after growth is already complete produces relatively few symptoms. Growth hormone—deficient adults tend to have reduced muscle strength (less muscle protein) as well as decreased bone density (less osteoblast activity during ongoing bone remodeling). (See the accompanying boxed feature, ◆Concepts, Challenges, and Controversies.)

Growth hormone excess Hypersecretion of growth hormone is most often caused by a tumor of the growth hormone–producing cells of the anterior pituitary. The symptoms that result depend on the age of the individual when the abnormal secretion begins. If overproduction of growth hormone begins in childhood before the epiphyseal plates are closed, the principal manifestation of the disorder is a rapid growth in height without distortion of body proportions. Appropriately, this condition is known as **gigantism.** If not treated by removal of the tumor or by drugs that block the effect of growth hormone, the individual may reach a height of 8 feet or more. All of the soft tissues grow correspondingly, so the body is still well proportioned.

CONCEPTS, CHALLENGES, AND CONTROVERSIES

Growth and Youth in a Bottle?

Unlike most hormones, growth hormone's structure and activity differ among species. As a result, growth hormone from animal sources is ineffective in the treatment of growth hormone deficiency in humans. In the past, the only source of human growth hormone was pituitary glands from human cadavers. This supply was never adequate, and it has been removed from the market because of fear of viral contamination. Recently, however, an unlimited supply of human growth hormone has become available through the technique of genetic engineering. The gene that directs the synthesis of growth hormone in humans has been introduced into bacteria, converting the bacteria into "factories" that synthesize human growth hormone.

Even though genetic engineering has succeeded in providing an adequate supply of growth hormone, the solution has given rise to problems for the medical community regarding the circumstances under which synthetic growth hormone treatment is appropriate. Synthetic human growth hormone has currently been approved by the Food and Drug Administration only for use in growth hormone–deficient children.

Growth hormone therapy is also being considered for the treatment of other disorders including (1) promoting faster healing of the skin in severely burned patients; (2) counteracting the wasting, or loss of body mass, that accompanies chronic, debilitating diseases; and (3) treating obesity. In preliminary studies using growth hormone in obese patients, the subjects did not lose weight, but their bodies became leaner (more muscle, less fat) during the treatment.

Another group for whom replacement growth hormone therapy may be beneficial is the elderly. Scientists suspect that in many people growth hormone production may start to dwindle after age forty. This decline may contribute to some of the characteristic signs of aging, such as decreased muscle mass (growth hormone promotes synthesis of proteins, including muscle protein), increased fat deposition (growth hormone promotes leanness by mobilizing fat stores for use as an energy source), and thinner, sagging skin (growth hormone promotes proliferation of skin cells).

Several studies in the early 1990s suggested that some of these consequences of aging may be partially reversed or counteracted through the use of synthetic growth hormone in people over age sixty who have a measurable growth hormone deficit. Elderly growth hormone–deficient experimental subjects who were provided with supplemental growth hormone demonstrated an increase in muscle mass, a reduction in fatty tissue, and a thickening of the skin. Even though these early studies were exciting, scientists caution that synthetic growth hormone should not be viewed as a potential cure for old age. More recent results have been more discouraging. For example, despite an increase in lean body mass, many individuals do not experience an increase in muscle strength. Also, when growth hormone supplementation is used for an extended time or in large doses, detrimental side effects include an increased likelihood of diabetes, kidney stones, high blood pressure, and carpal tunnel syndrome. (*Carpal tunnel syndrome* is a wrist disorder—*carpal* means "wrist"—involving thickening and narrowing of the tunnel through which the nerve supply to the hand muscles passes.) Furthermore, synthetic growth hormone is costly and must be injected on a regular basis. For these reasons, many investigators no longer view synthetic growth hormone as a potential "fountain of youth." Instead, they are hopeful that it can be used in a more limited way to strengthen muscle and bone sufficiently in the many elderly who have growth hormone deficits to help reduce the incidence of bone-breaking falls that often lead to disability and loss of independence.

An ethical dilemma is whether the drug should be made legally available for others who have normal growth hormone levels but desire the product's growth-promoting actions for cosmetic or athletic reasons, such as normally growing teenagers who wish to attain even greater stature. The drug is already being used illegally by some athletes and bodybuilders.

Hesitancy in making the drug more readily available stems in part from the fact that synthetic growth hormone is a double-edged hormone. Although it has positive effects such as promoting growth and muscle mass, it also has negative effects, such as its potential troubling side effects. Furthermore, a recent British study revealed that supplemental growth hormone therapy in children who do not lack the hormone causes redistribution of the body's fat and protein. The investigators compared two groups of otherwise healthy six- to eight-year-olds who were among the shortest for their age. One group consisted of children who were receiving growth hormone, the other of children who were not. At the end of six months, the children taking the synthetic hormone had outpaced the untreated group in growth by more than 1.5 inches per year. However, the untreated children added both muscle and fat as they grew, whereas the treated children became unusually muscular and lost up to 76% of their body fat. The loss of fat became especially obvious in their faces and limbs, giving them a rawboned, gangly appearance. It is unclear what long-term effects—either deleterious or desirable—these dramatic changes in body composition might have. Scientists also express concern that these readily observable physical changes may be accompanied by more subtle abnormalities in organs and cells. Thus, the debate about whether to make growth hormone available to normal but short children is likely to continue until further investigation demonstrates if it is safe and appropriate.

If growth hormone hypersecretion occurs after adolescence when the epiphyseal plates have already closed, further growth in height is prevented. Under the influence of excess growth hormone, however, the bones become thicker and the soft tissues, especially connective tissue and skin, proliferate. This disproportionate growth pattern produces a disfiguring condition known as **acromegaly** (*acro* means "extremity"; *megaly* means "large"). Bone thickening is most obvious in the extremities and face. A marked coarsening of the features to an almost apelike appearance gradually develops as the jaws

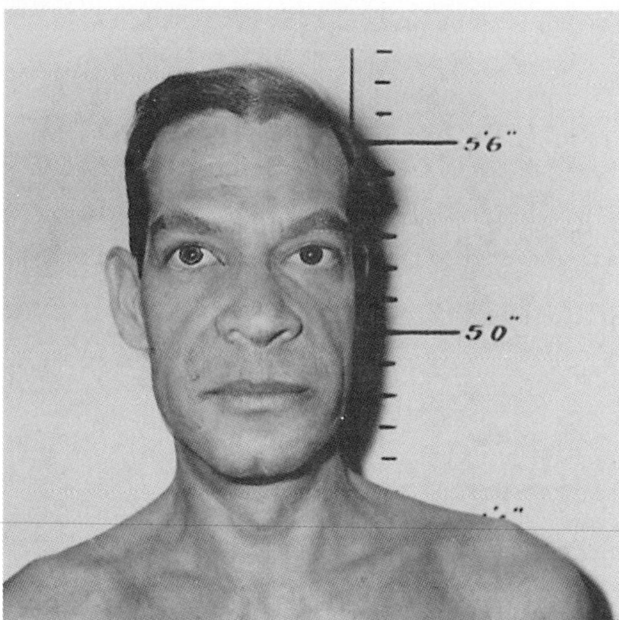

Figure 18-20 Patient with Acromegaly Note the prominent cheekbones and jaw caused by thickening of the facial bones and skin.

and cheekbones become more prominent because of the thickening of the facial bones and the skin (Fig. 18-20). The hands and feet enlarge, and the fingers and toes become greatly thickened. Peripheral nerve disorders often occur as a result of entrapment of nerves by overgrowth of connective tissue or bone or both.

Visual disturbances may accompany either gigantism or acromegaly. This is not due to excessive growth hormone per se but rather to the fact that a growth hormone–secreting tumor causes enlargement of the anterior pituitary gland, which lies in close proximity to the optic chiasm, the point at which the nerve fibers passing from the eyes cross over on their way to the visual cortex of the brain (Fig. 18-10b).

Other hormones besides growth hormone are essential for normal growth.

Several other hormones in addition to growth hormone contribute in special ways to overall growth.

- *Thyroid hormone* is essential for growth but is not itself directly responsible for producing growth-promoting effects. It plays a permissive role in promoting skeletal growth; the actions of growth hormone are fully manifested only in the presence of adequate amounts of thyroid hormone. As a result, growth is severely stunted in hypothyroid children, but hypersecretion of thyroid hormone does not cause excessive growth.
- *Insulin* is believed to be an important growth-promoting factor, as is evidenced by a high correlation between insulin levels and growth. Growth failure often accompanies insulin deficiency, and hyperinsulinism is frequently associated with excessive growth. Since insulin promotes protein synthesis, its growth-promoting effects should not be surprising. However, insulin's growth-promoting effects may also arise from a different mechanism other than its direct effect on protein synthesis. Insulin is structurally similar to the somatomedins and may interact with the somatomedin (IGF I) receptor, which is very similar to the insulin receptor.
- *Androgens,* which are believed to play an important role in the pubertal growth spurt, are powerful stimulants of protein synthesis in many organs. Androgens stimulate linear growth, promote weight gain, and increase muscle mass. The most potent androgen, testicular testosterone, is responsible for the development of the heavier musculature in men as compared to women. These androgenic growth-promoting effects depend on the presence of growth hormone. Androgens have virtually no effect on body growth in the absence of growth hormone, but in its presence, they synergistically enhance linear growth. Although androgens stimulate growth, they ultimately stop further growth by promoting closure of the epiphyseal plates.
- *Estrogens,* like androgens, ultimately terminate linear growth by stimulating complete conversion of the epiphyseal plates to bone. However, the effects of estrogen on growth prior to bone maturation are not well understood. Some studies suggest that large doses of estrogen may even inhibit further body growth.

Several factors contribute to the average height differences between men and women. First, since puberty occurs about two years earlier in girls than in boys, on the average boys have two more years of prepubertal growth than girls. As a result, boys are usually several inches taller than girls at the start of their respective growth spurts. Second, boys experience a greater androgen-induced growth spurt than girls before their respective gonadal steroids seal their long bones from further growth; this results in greater heights in men than in women on the average. Third, recent evidence suggests that androgens "imprint" the brains of males during development, giving rise to a "masculine" secretory pattern of growth hormone characterized by higher cyclical peaks, which are speculated to contribute to the greater height of males.

In addition to these hormones that exert overall effects on body growth, a number of poorly understood peptide *growth factors* have been identified that stimulate mitotic activity of specific tissues (for example, epidermal growth factor).

Chapter in Perspective: Focus on Homeostasis

The endocrine system is one of the body's two major control systems, the other being the nervous system. Through its relatively slowly acting hormonal messengers, the endocrine system generally regulates activities that require duration rather than speed. Most of these activities are directed toward maintaining homeostasis, as exemplified by the following:

- Hormones help maintain the proper concentration of nutrients in the internal environment by directing chemical reactions involved in the cellular uptake, storage, and release of these molecules. Furthermore, the rate at which these nutrients are metabolized is controlled in large part by the endocrine system.

- The control of H_2O balance, which is responsible for maintaining ECF osmolarity and proper cell volume, is largely accomplished by hormonal regulation of H_2O reabsorption by the kidneys during urine formation.
- Salt balance, which is important in the maintenance of ECF volume and arterial blood pressure, is achieved by hormonally controlled adjustments in salt reabsorption by the kidneys during urine formation. Likewise, hormones act on various target cells to maintain the plasma concentration of calcium and other electrolytes.
- The endocrine system orchestrates a wide range of adjustments that help the body maintain homeostasis in response to stressful situations.
- The endocrine and nervous systems work in concert to control the circulatory and digestive systems, which in turn carry out important homeostatic activities.

Unrelated to homeostasis, hormones direct the growing process and control most aspects of the reproductive system.

Chapter Summary

General Principles of Endocrinology

Hormones are long-distance chemical messengers secreted by the ductless endocrine glands into the blood, which transports them to specific target sites where they regulate or direct a particular function by altering protein activity within the target cells. Even though hormones are able to reach all tissues via the blood, they exert their effects only at their target cells because these cells alone have unique receptors for binding the hormone. The endocrine system is especially important in regulating fuel metabolism, H_2O and electrolyte balance, growth, and reproduction.

Hormones are grouped into three categories—based on differences in their mode of synthesis, storage, secretion, transport in the blood, and interaction with target cells—peptides, steroids, and amines, the latter including thyroid hormone and adrenomedullary catecholamines. Peptides and catecholamines are hydrophilic; steroid and thyroid hormones are lipophilic. Hydrophilic hormones are synthesized and packaged for export by the endoplasmic reticulum/Golgi complex route, stored in secretory vesicles, and released by exocytosis on appropriate stimulation. They dissolve freely in the blood for transport to their target cells where they bind with surface membrane receptors. Upon binding, a hydrophilic hormone triggers a chain of intracellular events by means of a second messenger system that ultimately alters preexisting cellular proteins, usually enzymes, which exert the effect leading to the target cell's response to the hormone.

Steroids are synthesized by modifications of stored cholesterol by means of enzymes specific for each steroidogenic tissue. Steroids are not stored in the endocrine cells. Being lipophilic, they diffuse out through the lipid membrane barrier as soon as they are synthesized. Control of steroids is directed at their synthesis. Lipophilic steroids and thyroid hormone are both transported in the blood largely bound to carrier plasma proteins, with only free, unbound hormone being biologically active. Lipophilic hormones readily enter through the lipid membrane barriers of their target cells and bind with nuclear receptors. Hormonal binding activates the synthesis of new intracellular proteins that carry out the hormone's effect on the target cell.

The effective plasma concentration of each hormone is normally controlled by regulated changes in the rate of hormone secretion. Secretory output of endocrine cells is influenced by one or more of three different types of direct regulatory inputs: (1) neural input, which increases hormone secretion in response to a specific need and also governs diurnal variations in secretion; (2) input from another hormone, which involves either stimulatory input from a tropic hormone or inhibitory input from a target cell hormone in negative-feedback fashion; and (3) changes in the plasma concentration of an organic nutrient or an electrolyte being regulated by the hormone, also acting in negative-feedback fashion. In all cases, the regulated secretory output is designed to accomplish particular adjustments needed to maintain homeostasis, promote growth, or control reproduction.

Endocrine dysfunction arises when too much or too little of any particular hormone is secreted. Symptoms are generally referable to exaggerated or insufficient target cell responses normally controlled by the hormone.

Pineal Gland

The pineal gland secretes the hormone melatonin, the study of which is in its infancy. In response to input from the eyes, melatonin secretion rhythmically fluctuates with the light/dark cycle, decreasing in the light and increasing in the dark. Melatonin's proposed roles include (1) synchronizing the body's natural circadian rhythms with external cues such as the light/dark cycle; (2) promoting sleep; (3) influencing reproductive activity, including the onset of puberty; (4) acting as an antioxidant to remove damaging free radicals; and (5) enhancing immunity.

Hypothalamus and Pituitary

The pituitary gland consists of two distinct lobes, the posterior pituitary and the anterior pituitary. The hypothalamus, a portion of the brain, secretes nine peptide hormones; two are stored in the posterior pituitary, and seven are carried through a special vascular link to the anterior pituitary, where they either stimulate or inhibit the release of particular anterior pituitary hormones.

The posterior pituitary is essentially a neural extension of the hypothalamus. Two small peptide hormones, vasopressin and oxytocin, are synthesized within the cell bodies of neurosecretory neurons located in the hypothalamus, from which they pass down the axon to be stored in nerve terminals within the posterior pituitary. These hormones are independently released from the posterior pituitary into the blood in response to action potentials originating in the hypothalamus.

The anterior pituitary secretes six different peptide hormones that it produces itself. With the exception of prolactin, which stimulates milk secretion, the five other anterior pituitary hormones stimulate and maintain other endocrine tissues; that is, they are tropic. The sole function of thyroid-stimulating hormone (TSH) is to stimulate secretion of thyroid hormone, and that of adrenocorticotropic hormone (ACTH) is to stimulate secretion of cortisol by the adrenal cortex. The gonadotropic hormones—follicle-stimulating hormone (FSH) and luteinizing hormone (LH)—stimulate production of gametes (eggs and sperm) as well as secretion of sex hormones. Growth hormone stimulates growth and exerts metabolic effects as well.

The anterior pituitary releases its hormones into the blood at the bidding of releasing and inhibiting hormones from the hypothalamus. The hypothalamus in turn is influenced by a variety of neural and hormonal controlling inputs. Both the hypothalamus and the anterior pituitary are inhibited in negative-feedback fashion by the product of the target endocrine gland in the hypothalamus/anterior pituitary/target-gland axis.

Endocrine Control of Growth

Growth hormone promotes growth indirectly by stimulating the liver's production of somatomedins, which act directly on bone and soft tissues to cause growth. The growth hormone/somatomedin pathway causes growth by stimulating protein synthesis, cell division, and lengthening and thickening of bones. Growth hormone also directly exerts metabolic effects unrelated to growth on the liver, adipose tissue, and muscle, such as conservation of carbohydrates and mobilization of fat stores.

Growth hormone secretion by the anterior pituitary is regulated in negative-feedback fashion by two hypothalamic hormones, growth hormone–releasing hormone and growth hormone–inhibiting hormone. Growth hormone levels are not highly correlated with periods of rapid growth. The primary signals for increased growth hormone secretion are related to metabolic needs rather than growth, namely, deep sleep, stress, exercise, and low blood glucose levels.

Review Exercises

Objective Questions (Answers on p. E–18.)

1. One endocrine gland may secrete more than one hormone. (True or false?)
2. One hormone may influence more than one type of target cell. (True or false?)
3. All endocrine glands are exclusively endocrine in function. (True or false?)
4. A single target cell may be influenced by more than one hormone. (True or false?)
5. Each steroidogenic organ has all of the enzymes necessary to produce any steroid hormone. (True or false?)
6. Hyposecretion or hypersecretion of a specific hormone can occur even though its endocrine gland is perfectly normal. (True or false?)
7. Inhibition of the anterior pituitary by a target-gland hormone is known as short-loop negative feedback. (True or false?)
8. Growth hormone levels in the blood are no higher during the early childhood growing years than during adulthood. (True or false?)
9. A hormone that has as its primary function the regulation of another endocrine gland is classified functionally as a ____________ hormone.
10. Self-induced reduction in the number of receptors for a specific hormone is known as ____________.
11. Activity within the cartilaginous layer of bone known as the ____________ is responsible for linear growth of long bones.
12. Indicate the relationships among the hormones in the hypothalamic/anterior pituitary/adrenal cortex system by using the following answer code to identify which hormone belongs in each blank:

    (a) cortisol
    (b) ACTH
    (c) CRH

    (1) ____________ from the hypothalamus stimulates the secretion of (2) ____________ from the anterior pituitary. (3) ____________ in turn stimulates the secretion of (4) ____________ from the adrenal cortex. In negative-feedback fashion, (5) ____________ inhibits secretion of (6) ____________ and furthermore reduces the sensitivity of the anterior pituitary to (7) ________.

Essay Questions

1. List the overall functions of the endocrine system.
2. Compare the three categories of hormones in terms of chemical structure; mechanisms of synthesis, storage, and secretion; transport in the blood; and interaction with target cells.
3. By what means is the plasma concentration of a hormone normally regulated?
4. List and briefly state the functions of the posterior pituitary hormones.
5. List and briefly state the functions of the anterior pituitary hormones.
6. Compare the relationship between the hypothalamus and posterior pituitary with the relationship between the hypothalamus and anterior pituitary. Describe the role of the hypothalamic-hypophyseal portal system and the hypothalamic releasing and inhibiting hormones.
7. Describe the actions of growth hormone that are unrelated to growth. What are growth hormone's growth-promoting actions? What is the role of somatomedins?
8. Discuss the control of growth hormone secretion.

Points to Ponder

(Explanations on p. E–19.)

1. A new supervisor at a local hospital decides to rotate the nursing staff to a different shift every week so that one group of employees is not always "stuck" on an undesirable shift. From a physiological viewpoint, do you think this proposal is advisable?
2. Would you expect the concentration of hypothalamic releasing and inhibiting hormones in a systemic venous blood sample to be higher, lower, or the same as the concentration of these hormones in a sample of hypothalamic-hypophyseal portal blood?
3. A patient displays symptoms of excess cortisol secretion. What factors could be measured in a blood sample to determine whether the condition is caused by a defect at the hypothalamic/anterior pituitary level or the adrenal cortex level?
4. Why would males with testicular feminization syndrome be unusually tall?
5. A black market for growth hormone abuse already exists among weight lifters and other athletes. What actions of growth hormone would induce a full-grown athlete to take supplemental doses of this hormone? What are the potential detrimental side effects?
6. ***Clinical Consideration*** At age 18, 8-foot Anthony O. was diagnosed with gigantism due to a pituitary tumor. The condition was treated by surgical removal of his pituitary gland. What hormonal replacement therapy do you think Anthony's physician prescribed following this procedure?

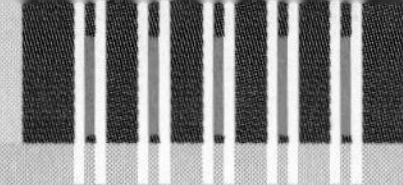

19 The Peripheral Endocrine Glands

ENDOCRINE SYSTEM

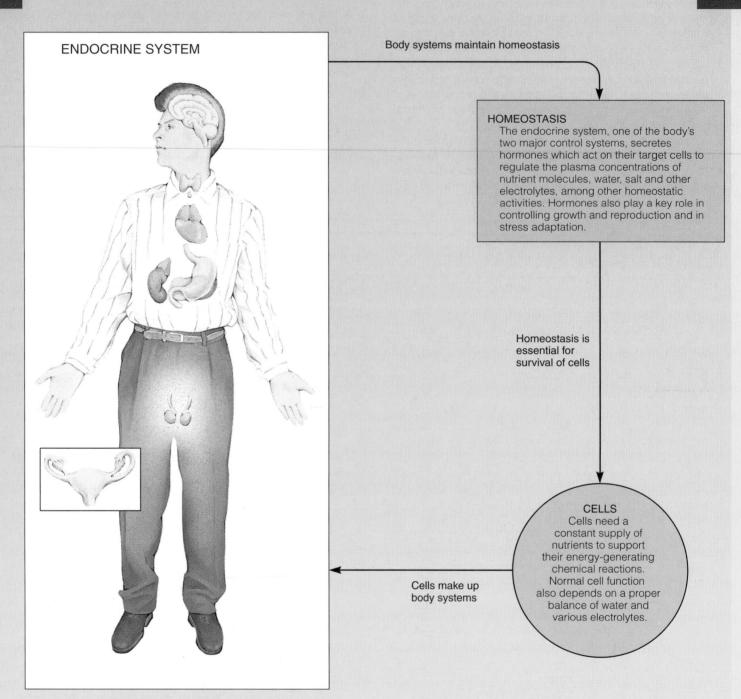

Body systems maintain homeostasis

HOMEOSTASIS
The endocrine system, one of the body's two major control systems, secretes hormones which act on their target cells to regulate the plasma concentrations of nutrient molecules, water, salt and other electrolytes, among other homeostatic activities. Hormones also play a key role in controlling growth and reproduction and in stress adaptation.

Homeostasis is essential for survival of cells

CELLS
Cells need a constant supply of nutrients to support their energy-generating chemical reactions. Normal cell function also depends on a proper balance of water and various electrolytes.

Cells make up body systems

The endocrine system, by means of the blood-borne hormones it secretes, generally regulates activities that require duration rather than speed. The peripheral endocrine glands include the **thyroid gland,** which controls the body's basal metabolic rate; the **adrenal glands,** which secrete hormones important in the metabolism of nutrient molecules, adaptation to stress, and maintenance of salt balance; the **endocrine pancreas,** which secretes hormones important in the metabolism of nutrient molecules; and the **parathyroid glands,** which secrete a hormone important in Ca^{2+} metabolism.

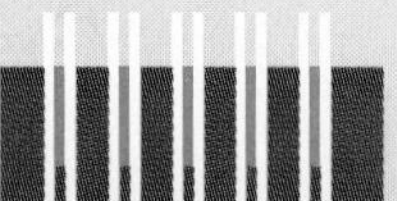

Chapter Contents At a Glance

Thyroid Gland

The major thyroid hormone secretory cells are organized into colloid-filled spheres.

The **thyroid gland** consists of two lobes of endocrine tissue joined in the middle by a narrow portion of the gland, giving it a bow tie–shaped appearance (Fig. 19-1a). The gland is even located in the appropriate place for a bow tie, lying over the trachea just below the larynx. The major thyroid secretory cells are arranged into hollow spheres, each of which forms a functional unit called a **follicle.** Consequently, these secretory cells are often referred to as **follicular cells.** On a microscopic section (Fig. 19-1b), the follicles appear as rings of follicular cells enclosing an inner lumen filled with **colloid,** a substance that serves as an "inland" extracellular storage site for thyroid hormones.

The chief constituent of the colloid is a large, complex molecule known as **thyroglobulin,** within which are incorporated the thyroid hormones in their various stages of synthesis. The follicular cells produce two iodine-containing hormones derived from the amino acid tyrosine: **tetraiodothyronine (T_4 or thyroxine)** and **triiodothyronine (T_3).** The prefixes *tetra* and *tri* and the subscripts 4 and 3 denote the number of iodine atoms incorporated into each of these hormones. These two hormones, collectively referred to as **thyroid hormone,** are important regulators of overall basal metabolic rate.

Interspersed in the interstitial spaces between the follicles is another secretory cell type, the **C cells,** so called because they secrete the peptide hormone **calcitonin,** which plays a role in calcium metabolism. Calcitonin is not related in any way to the two other major thyroid hormones. We will restrict our discussion of thyroid hormones to the secretions of the follicular cells, deferring coverage of calcitonin until a later section dealing with endocrine control of calcium balance.

Figure 19-1 Anatomy of the Thyroid Gland (a) Gross anatomy of the thyroid gland, anterior view. The thyroid gland lies over the trachea just below the larynx and consists of two lobes connected by a thin strip called the isthmus. (b) Light-microscope appearance of the thyroid gland. The thyroid gland is composed primarily of colloid-filled spheres enclosed by a single layer of follicular cells.

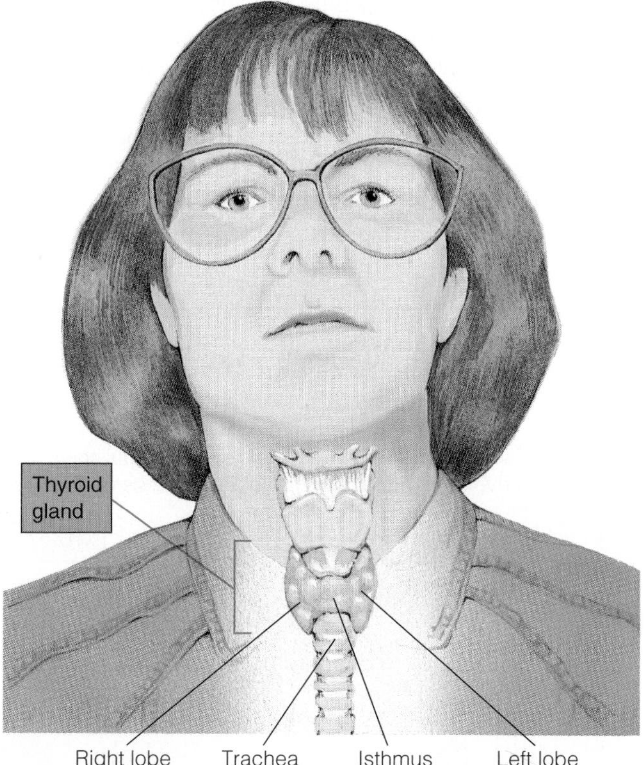

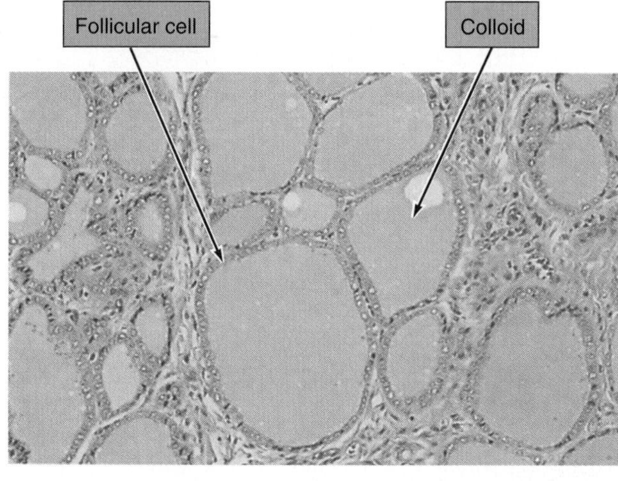

All of the steps of thyroid hormone synthesis occur on the large thyroglobulin molecule, which subsequently stores the hormones.

The basic ingredients for thyroid hormone synthesis are tyrosine and iodine, both of which must be taken up from the blood by the follicular cells. Tyrosine, an amino acid, is synthesized in sufficient amounts by the body, so it is not an essential dietary requirement. The iodine needed for thyroid hormone synthesis, on the other hand, must be obtained from dietary intake. The synthesis, storage, and secretion of thyroid hormone involve the following steps:

1. All steps of thyroid hormone synthesis take place on the thyroglobulin molecules within the colloid. Thyroglobulin itself is produced by the endoplasmic reticulum/Golgi complex of the thyroid follicular cells. Tyrosine becomes incorporated in the much larger thyroglobulin molecules as the latter are being produced. Once produced, tyrosine-containing thyroglobulin is exported from the follicular cells into the colloid by exocytosis (step 1 in Figure 19-2).
2. The thyroid captures iodine from the blood and transfers it into the colloid by means of a very active "iodine pump" or "iodine-trapping mechanism"—the powerful, energy-requiring carrier proteins located in the outer membranes of the follicular cells (step 2). Almost all of the iodine in the body is moved against its concentration gradient to become trapped in the thyroid for the purpose of thyroid hormone synthesis. Iodine serves no other purpose in the body.
3. Within the colloid, iodine is quickly attached to a tyrosine within the thyroglobulin molecule. Attachment of one iodine to tyrosine yields **monoiodotyrosine (MIT)** (step 3a). Attachment of two iodines to tyrosine yields **diiodotyrosine (DIT)** (step 3b).
4. Next, a coupling process occurs between the iodinated tyrosine molecules to form the thyroid hormones. Coupling of two DITs (each bearing two iodine atoms) yields **tetraiodothyronine** (**T_4** or **thyroxine**), the four-iodine form of thyroid hormone (step 4a). Coupling of one MIT (with one iodine) and one DIT (with two iodines) yields **triiodothyronine** or **T_3** (with three iodines) (step 4b). Coupling does not occur between two MIT molecules.

Because these reactions occur within the thyroglobulin molecule, all the products remain attached to this large protein. Thyroid hormones remain stored in this form in the colloid until they are split off and secreted. It is estimated that sufficient thyroid hormone to supply the body's needs for several months is normally stored in the colloid.

Figure 19-2 **Synthesis, Storage, and Secretion of Thyroid Hormone** Tyrosine-containing TGB produced within the thyroid follicular cells is transported into the colloid by exocytosis ①. Iodine is actively transported from the blood into the colloid by the follicular cells ②. Attachment of one iodine to tyrosine within the TGB molecule yields MIT (3a); attachment of two iodines yields DIT (3b). Coupling of two DITs yields T_4 (4a); coupling of one MIT and one DIT yields T_3 (4b). On appropriate stimulation, the thyroid follicular cells engulf a portion of TGB-containing colloid by phagocytosis ⑤. Lysosomes attack the engulfed vesicle and split the iodinated products from TGB ⑥. T_3 and T_4 diffuse into the blood (7a). MIT and DIT are deiodinated, and the freed iodine is recycled for synthesis of more hormone (7b).

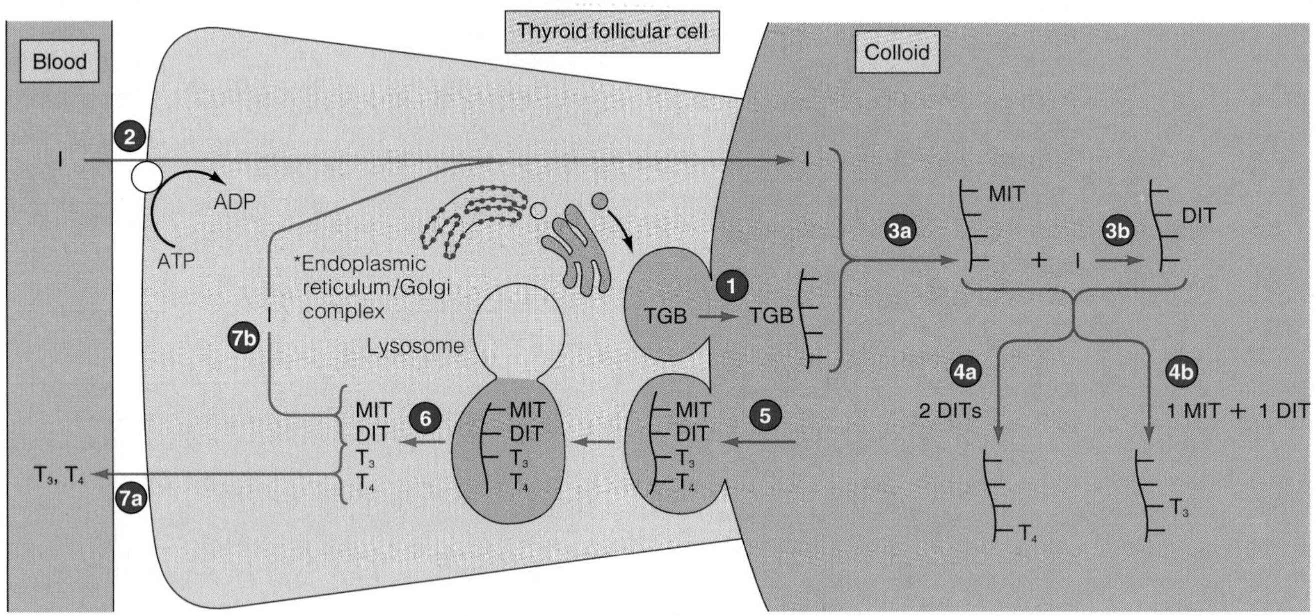

TGB = Thyroglobulin
I = Iodine
MIT = Monoiodotyrosine
DIT = Diiodotyrosine
T_3 = Triiodothyronine
T_4 = Tetraiodothyronine (thyroxine)

* Organelles not drawn to scale. Endoplasmic reticulum/Golgi complex are proportionally too small.

The follicular cells phagocytize thyroglobulin-laden colloid to accomplish thyroid hormone secretion.

The release of the thyroid hormones into the systemic circulation requires a rather complex process for two reasons. First, before their release, T_4 and T_3 are still bound within the thyroglobulin molecule. Second, these hormones are stored at an inland extracellular site, the follicular lumen; before they can enter the blood vessels that course through the interstitial spaces, they must be transported completely across the follicular cells. The process of thyroid hormone secretion essentially involves the follicular cells "biting off" a piece of colloid, breaking the thyroglobulin molecule down into its component parts, and "spitting out" the freed T_4 and T_3 into the blood. Upon appropriate stimulation for thyroid hormone secretion, the follicular cells internalize a portion of the thyroglobulin-hormone complex by phagocytizing a piece of colloid (step 5 of Fig. 19-2). Within the cells, the membrane-enclosed droplets of colloid coalesce with lysosomes, whose enzymes split off the biologically active thyroid hormones, T_4 and T_3, as well as the inactive iodotyrosines, MIT and DIT (step 6). The thyroid hormones, being very lipophilic, pass freely through the outer membranes of the follicular cells and into the blood (step 7a). The MIT and DIT are of no endocrine value. The follicular cells contain an enzyme that swiftly removes the iodine from MIT and DIT, allowing the freed iodine to be recycled for synthesis of more hormone (step 7b). This highly specific enzyme will remove iodine only from the worthless MIT and DIT, not the valuable T_4 or T_3.

Most of the secreted T_4 is converted into T_3 outside the thyroid.

About 90% of the secretory product released from the thyroid gland is in the form of T_4, yet T_3 is about four times more potent in its biological activity. However, most of the secreted T_4 is converted into T_3, or *activated,* by being stripped of one of its iodines in the liver and kidneys. About 80% of the circulating T_3 is derived from secreted T_4 that has been peripherally stripped. Therefore, T_3 is the major biologically active form of thyroid hormone at the cellular level, even though the thyroid gland secretes mostly T_4.

For the most part, both T_4 and T_3 are transported bound to specific plasma proteins.

Once released into the blood, the highly lipophilic thyroid hormone molecules very quickly bind with several plasma proteins. Less than 1% of the T_3 and less than 0.1% of the T_4 remain in the unbound (free) form. This is remarkable considering that only the free portion of the total thyroid hormone pool has access to the target cell receptors and thus is able to exert an effect.

Three different plasma proteins are important in thyroid hormone binding: *thyroxine-binding globulin* selectively binds only thyroid hormones—55% of the circulating T_4 and 65% of the T_3—even though its name specifies only "thyroxine" (T_4); *albumin* nonselectively binds many lipophilic hormones, including 10% of the T_4 and 35% of the T_3; and *thyroxine-binding prealbumin* binds the remaining 35% of the T_4.

Thyroid hormone is the primary determinant of the body's overall metabolic rate and is also important for bodily growth and normal development and function of the nervous system.

Virtually every tissue in the body is affected either directly or indirectly by thyroid hormone. The effects of T_3 and T_4 can be grouped into several overlapping categories.

Effect on metabolic rate Thyroid hormone increases the body's overall basal metabolic rate or "idling speed" (see p. 603). This hormone is the most important regulator of the body's rate of O_2 consumption and energy expenditure under resting conditions.

Compared to other hormones, the action of thyroid hormone is "sluggish." Only after a delay of several hours is the metabolic response to thyroid hormone detectable, and the maximal response is not evident for several days. The duration of the response is also quite long, partially because thyroid hormone is not rapidly degraded but also because the response continues to be expressed for days or even weeks after the plasma thyroid hormone concentrations have returned to normal.

Calorigenic effect Closely related to thyroid hormone's overall metabolic effect is its **calorigenic** ("heat-producing") effect. Increased metabolic activity results in increased heat production.

Effect on intermediary metabolism In addition to increasing the general metabolic rate, thyroid hormone modulates the rates of many specific reactions involved in fuel metabolism. The effects of thyroid hormone on the metabolic fuels are multifaceted; not only can it influence both the synthesis and degradation of carbohydrate, fat, and protein, but small or large amounts of the hormone may induce opposite effects. For example, the conversion of glucose to glycogen, the storage form of glucose, is facilitated by small amounts of thyroid hormone, but the reverse—the breakdown of glycogen into glucose—occurs with large amounts of the hormone. Similarly, adequate amounts of thyroid hormone are essential for the protein synthesis needed for normal bodily growth, yet protein degradation effects predominate at high doses. In general, at abnormally high plasma levels of thyroid hormone, such as in thyroid hypersecretion, the overall effect is to favor consumption rather than storage of fuel, as manifested by depletion of liver glycogen stores, depletion of fat stores, and muscle wasting resulting from protein degradation. (The main structural component of cells is protein. Muscle cells are particularly rich in structural protein because they are packed full of contractile elements made of protein—that is, the actin and myosin filaments.)

Sympathomimetic effect Any action similar to one produced by the sympathetic nervous system is known as a **sym-**

pathomimetic ("sympathetic-mimicking") **effect.** Thyroid hormone increases target cell responsiveness to catecholamines (epinephrine and norepinephrine), the chemical messengers used by the sympathetic nervous system and its hormonal reinforcements from the adrenal medulla. Thyroid hormone presumably accomplishes this permissive action by causing a proliferation of specific catecholamine target cell receptors (see p. 635). Because of this action, many of the effects observed when thyroid hormone secretion is elevated are similar to those that accompany activation of the sympathetic nervous system (a sympathomimetic effect).

Effect on the cardiovascular system Through its effect of increasing the heart's responsiveness to circulating catecholamines, thyroid hormone increases heart rate and force of contraction, thus increasing cardiac output. In addition, in response to the heat load generated by the calorigenic effect of thyroid hormone, peripheral vasodilation occurs to carry the extra heat to the body surface for elimination to the environment (see p. 614).

Effect on growth and the nervous system Thyroid hormone is essential for normal growth. The growth-promoting effect of thyroid hormone seems to be secondary to its effects on growth hormone. Thyroid hormone not only stimulates growth hormone secretion but also promotes the effects of growth hormone (or somatomedins) on the synthesis of new structural proteins and on skeletal growth. Thyroid-deficient children have stunted growth that is reversible with thyroid replacement therapy. Unlike excess growth hormone, however, excess thyroid hormone does not result in excessive growth.

Thyroid hormone plays a crucial role in the normal development of the nervous system, especially the CNS, an effect impeded in children with thyroid deficiency from birth. Thyroid hormone is also essential for normal CNS activity in adults. Abnormal thyroid hormone levels are associated with behavioral changes. Furthermore, the conduction velocity of peripheral nerves varies directly with the availability of thyroid hormone.

Thyroid hormone is regulated by the hypothalamus-pituitary-thyroid axis.

Thyroid-stimulating hormone (TSH), the thyroid tropic hormone from the anterior pituitary, is the most important physiological regulator of thyroid hormone secretion (Fig. 19-3). Almost every step of thyroid hormone synthesis and release is stimulated by TSH.

In addition to enhancing thyroid hormone secretion, TSH is responsible for maintaining the structural integrity of the thyroid gland. In the absence of TSH, the thyroid atrophies (decreases in size) and secretes its hormones at a very low rate. Conversely, it undergoes hypertrophy (increase in the size of each follicular cell) and hyperplasia (increase in the number of follicular cells) in response to excess TSH stimulation.

The hypothalamic **thyrotropin-releasing hormone (TRH),** in tropic fashion, "turns on" TSH secretion by the anterior pituitary, whereas thyroid hormone, in negative-feedback fashion, "turns off" TSH secretion. In the hypothalamus-pituitary-thyroid axis, inhibition is exerted primarily at the level of the anterior pituitary. Like the other negative-feedback loops, the one between thyroid hormone and TSH tends to maintain a stable thyroid hormone output.

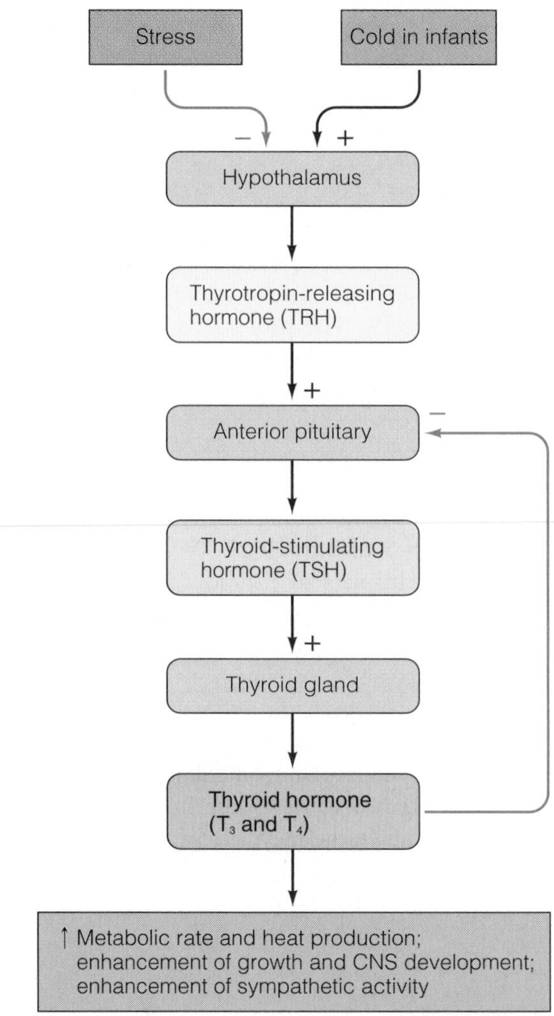

Figure 19-3 **Regulation of Thyroid Hormone Secretion**

Day-to-day regulation of free thyroid hormone levels is apparently accomplished by negative feedback between the thyroid and anterior pituitary whereas long-range adjustments are mediated by the hypothalamus. Unlike most of the other hormonal systems, the hormones in the thyroid axis in an adult normally do not undergo sudden, wide swings in secretion. The relatively steady rate of thyroid hormone secretion is in keeping with the sluggish, long-lasting responses that this hormone induces; there would be no adaptive value in suddenly increasing or decreasing plasma thyroid hormone levels. The only known factor that increases TRH secretion (and, accordingly, TSH and thyroid hormone secretion) is exposure to cold in infants; this is a highly adaptive mechanism in newborns. The dramatic increase in heat-producing thyroid hormone secretion is thought to contribute to the maintenance of body temperature in the face of the abrupt drop in surrounding temperature at birth, as the infant goes from the mother's warm body to the cooler environmental air. A similar TSH response to cold exposure does not occur in adults, although it makes sense physiologically and does occur in many types of experimental animals.

Various types of stress are known to inhibit TSH and thyroid hormone secretion, presumably through neural influences on the hypothalamus, although the adaptive importance of this inhibition is unclear.

Abnormalities of thyroid function include both hypothyroidism and hyperthyroidism.

Normal thyroid function is referred to as **euthyroidism.** Abnormalities of thyroid function are among the most common of all endocrine disorders. They fall into two major categories—*hypothyroidism* and *hyperthyroidism*—reflecting deficient and excess thyroid hormone secretion, respectively. A number of specific causes can give rise to each of these conditions (Table 19-1). Whatever the cause, the consequences of too little or too much thyroid hormone secretion are largely predictable, based on a knowledge of the functions of thyroid hormone.

Hypothyroidism **Hypothyroidism** can result (1) from primary failure of the thyroid gland itself; (2) secondary to a deficiency of TRH, TSH, or both; or (3) from an inadequate dietary supply of iodine. The symptoms of hypothyroidism are largely referable to a reduction in overall metabolic activity. Among other things, a patient with hypothyroidism has a reduced basal metabolic rate; displays poor tolerance of cold (lack of the calorigenic effect); has a tendency to gain excessive weight (not burning fuels at a normal rate); is easily fatigued (lower energy production); has a slow, weak pulse (caused by a reduction in the rate and strength of cardiac contraction and a lowered cardiac output); and exhibits slow reflexes and slow mentation (because of the effect on the nervous system), characterized by diminished alertness, slow speech, and poor memory.

Another notable characteristic is an edematous condition caused by infiltration of the skin with complex, water-retaining carbohydrate molecules, presumably as a result of altered metabolism. The resultant puffy appearance, primarily of the face, hands, and feet, is known as **myxedema.** In fact, the term *myxedema* is often used as a synonym for hypothyroidism in an adult because of the prominence of this symptom.

If an individual has hypothyroidism from birth, a condition known as **cretinism** develops. Because adequate levels of thyroid hormone are essential for normal growth and CNS development, cretinism is characterized by dwarfism and mental retardation as well as other general symptoms of thyroid deficiency. The mental retardation is preventable if replacement therapy is started promptly, but it is not reversible once it has developed for a few months after birth, even with later treatment with thyroid hormone. For this reason, an important public health measure is the recent establishment of neonatal screening programs in which congenital hypothyroidism can be detected from a single drop of a newborn's blood.

Treatment of hypothyroidism, with one exception, consists of replacement therapy through the administration of exogenous thyroid hormone. The exception is hypothyroidism caused by iodine deficiency, in which the remedy is adequate dietary iodine.

Hyperthyroidism The most common cause of **hyperthyroidism** is **Grave's disease**, an autoimmune disease in which the body erroneously produces **thyroid-stimulating immunoglobulin (TSI),** an antibody whose target is the TSH receptors on the thyroid cells. Thyroid-stimulating immunoglobulin stimulates both secretion and growth of the thyroid in a manner similar to TSH. Unlike TSH, however, TSI is not subject to negative-feedback inhibition by thyroid hormone, so thyroid secretion and growth continue unchecked (Fig. 19-4). Less frequently, hyperthyroidism occurs secondary to excess TRH or TSH or in association with a hypersecreting thyroid tumor.

As expected, the hyperthyroid patient has an elevated basal metabolic rate. The resultant increase in heat production leads to excessive perspiration and poor tolerance of heat. In spite of the increased appetite and food intake that occur in response to the increased metabolic demands, body weight typically falls because the body is burning fuel at an abnormally rapid rate. Net degradation of endogenous carbohydrate, fat, and protein stores occurs. The resultant loss of

Table 19-1 Types of Thyroid Dysfunctions

Thyroid Dysfunction	*Cause*	*Plasma Concentrations of Relevant Hormones*	*Goiter Present?*
Hypothyroidism	Primary failure of thyroid gland	↓ T_3 and T_4, ↑ TSH	Yes
	Secondary to hypothalamic or anterior pituitary failure	↓ T_3 and T_4, ↓ TRH and/or ↓ TSH	No
	Lack of dietary iodine	↓ T_3 and T_4, ↑ TSH	Yes
Hyperthyroidism	Abnormal presence of thyroid-stimulating immunoglobulin(TSI) (Grave's disease)	↑ T_3 and T_4, ↓ TSH	Yes
	Secondary to excess hypothalamic or anterior pituitary secretion	↑ T_3 and T_4, ↑ TRH and/or ↑ TSH	Yes
	Hypersecreting thyroid tumor	↑ T_3 and T_4, ↓ TSH	No

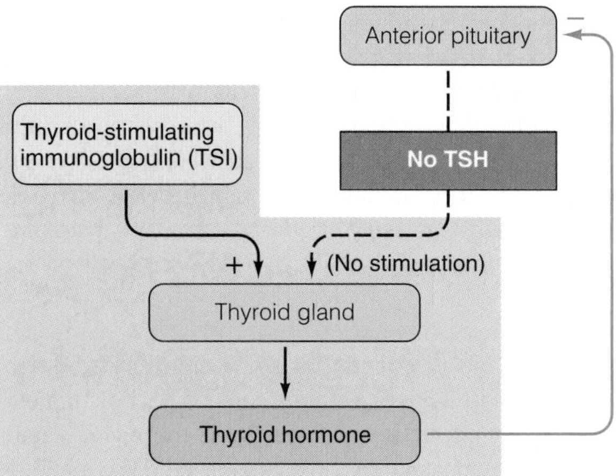

Figure 19-4 Role of Thyroid-Stimulating Immunoglobulin in Grave's Disease Thyroid-stimulating immunoglobulin, an antibody erroneously produced in the autoimmune condition of Grave's disease, binds with the TSH receptors on the thyroid gland and continuously stimulates thyroid hormone secretion outside the normal negative-feedback control system.

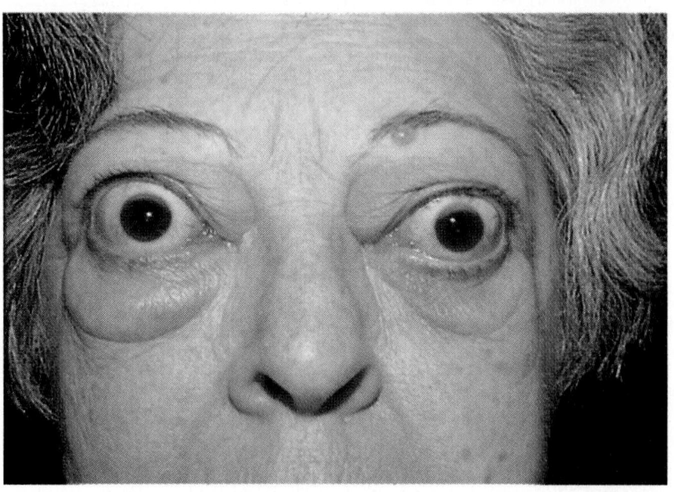

Figure 19-5 Patient Displaying Exophthalmos Abnormal fluid retention behind the eyeballs causes them to bulge forward.

skeletal muscle protein results in weakness. Various cardiovascular abnormalities are associated with hyperthyroidism, caused both by the direct effects of thyroid hormone and by its interactions with catecholamines. Heart rate and strength of contraction may increase so much that the individual has palpitations (an unpleasant awareness of the heart's activity). In severe cases, the heart may fail to meet the body's metabolic demands in spite of increased cardiac output. Nervous system involvement is manifested by an excessive degree of mental alertness to the point where the patient is irritable, tense, anxious, and excessively emotional.

A prominent feature of Grave's disease but not of the other types of hyperthyroidism is **exophthalmos** (bulging eyes) (Fig. 19-5). Complex, water-retaining carbohydrates are deposited behind the eyes, although why this happens is still being debated. The resultant fluid retention behind the eyes pushes the eyeballs forward so that they bulge from their bony orbit. The eyeballs may bulge so far that the lids cannot completely close, in which the case the eyes become dry, irritated, and prone to corneal ulceration. Even after correction of the hyperthyroid condition, these troublesome eye symptoms may persist.

Three general methods of treatment are available for suppressing excess thyroid hormone secretion: surgical removal of a portion of the oversecreting thyroid gland; administration of radioactive iodine, which, after being concentrated in the thyroid gland by the iodine pump, selectively destroys thyroid glandular tissue; and use of antithyroid drugs that specifically interfere with thyroid hormone synthesis.

A goiter may or may not accompany either hypothyroidism or hyperthyroidism.

A **goiter** refers to an enlarged thyroid gland. Because of the location of the thyroid over the trachea, a goiter is readily palpable and usually highly visible (Fig. 19-6). A goiter will occur whenever there is excessive stimulation of the thyroid gland by either TSH or TSI. Note from Table 19-1 that a goiter may accompany hypothyroidism or hyperthyroidism, but it need not be present in either condition. Knowing the hypothalamus-pituitary-thyroid axis and feedback control, we can predict which types of thyroid dysfunction will be accompanied by a goiter. We will consider hypothyroidism first.

- Hypothyroidism secondary to hypothalamic or anterior pituitary failure will not be accompanied by a goiter because the thyroid gland is not being adequately stimulated, let alone excessively stimulated.
- With hypothyroidism caused by thyroid gland failure or lack of iodine, a goiter does develop because the circulating level of thyroid hormone is so low that there is little negative-feedback inhibition on the anterior pituitary, and TSH secretion is therefore elevated. TSH acts on the thyroid to increase the size and number of follicular cells and to increase their rate of secretion. If the thyroid cells are incapable of secreting hormone because of a lack of a critical enzyme or lack of iodine, no amount of TSH will be able to induce these cells to secrete T_3 and T_4. However, TSH can still promote hypertrophy and hyperplasia of the thyroid with a consequent paradoxical enlargement of the gland (that is, a goiter), even though the gland is still underproducing.

Similarly, a goiter may or may not accompany hyperthyroidism:

- Excessive TSH secretion resulting from a hypothalamic or anterior pituitary defect would obviously be accompanied by a goiter and excess T_3 and T_4 secretion because of overstimulation of thyroid growth. Since the thyroid gland in this circumstance is also capable of responding to excess TSH with increased hormone secretion, hyperthyroidism is present with this goiter.
- In Grave's disease, a hypersecreting goiter occurs because TSI promotes growth of the thyroid as well as enhancing secretion of thyroid hormone. Because the high levels of circulating T_3 and T_4 inhibit the anterior pituitary, TSH

secretion itself is low. In all other cases when a goiter is present, TSH levels are elevated and are directly responsible for excessive growth of the thyroid.

- Hyperthyroidism resulting from overactivity of the thyroid in the absence of overstimulation, such as that caused by an uncontrolled thyroid tumor, is not accompanied by a goiter. The spontaneous secretion of excessive amounts of T_3 and T_4 inhibits TSH, so there is no stimulatory input to promote growth of the thyroid.

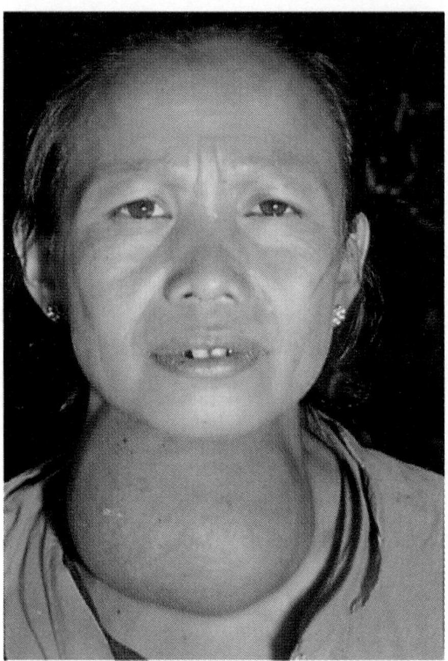

Figure 19-6 Patient with a Goiter

Adrenal Glands

Each adrenal gland consists of an outer, steroid-secreting adrenal cortex and an inner, catecholamine-secreting adrenal medulla.

There are two **adrenal glands**, one embedded above each kidney in a capsule of fat (*adrenal* means "next to the kidney") (Fig. 19-7a). Each adrenal is actually composed of two endocrine organs, one surrounding the other. The outer layers composing the **adrenal cortex** secrete a variety of steroid hormones; the inner portion, the **adrenal medulla**, secretes catecholamines. Derived embryologically from different structures, the adrenal cortex and medulla secrete hormones belonging to different chemical categories, whose functions, mechanisms of action, and regulation are entirely different.

The adrenal cortex secretes mineralocorticoids, glucocorticoids, and sex hormones.

About 80% of the adrenal gland is composed of the cortex, which consists of three different layers or zones: the **zona glomerulosa**, the outermost layer; the **zona fasciculata**, the

Figure 19-7 Anatomy of the Adrenal Glands
(a) Location and structure of the adrenal glands. (b) Layers of the adrenal cortex.

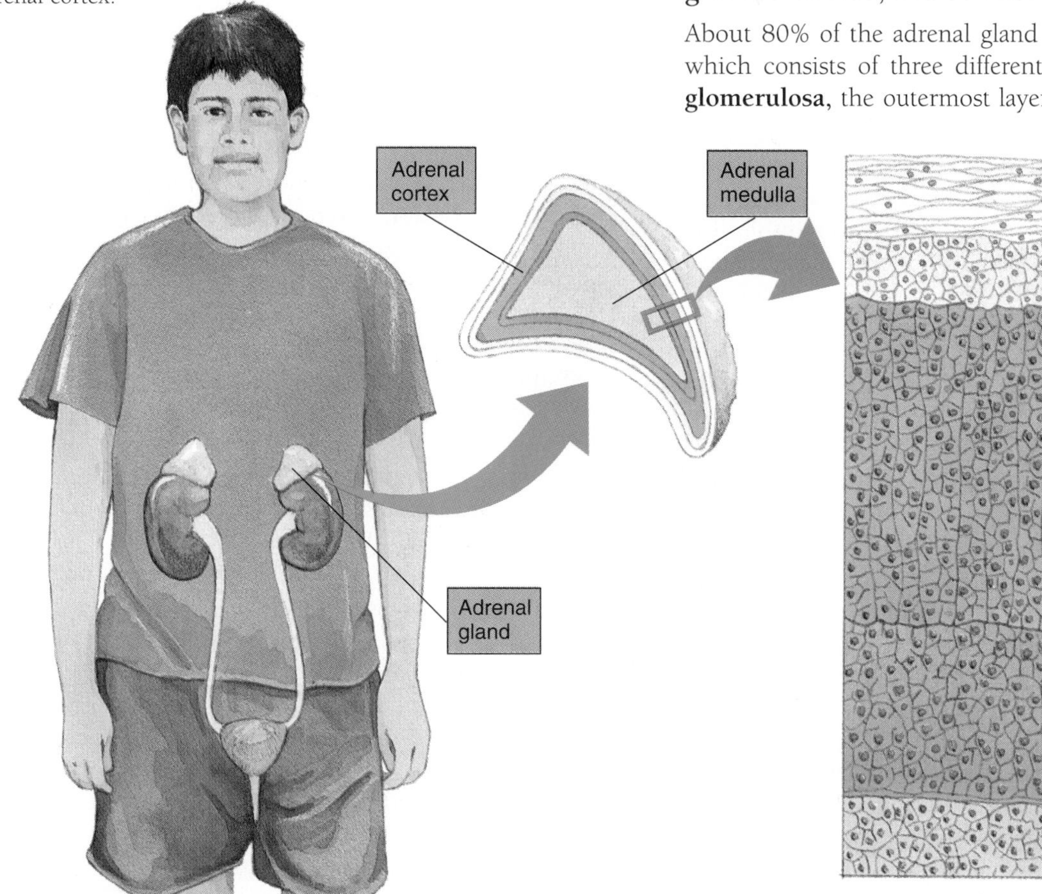

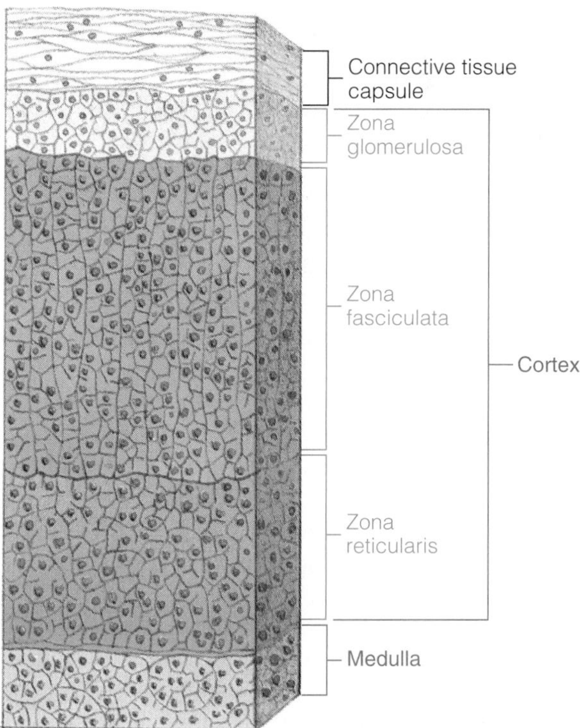

middle and largest portion; and the **zona reticularis,** the innermost zone (Fig. 19-7b). The adrenal cortex produces a number of different **adrenocortical hormones,** all of which are steroids derived from the common precursor molecule, cholesterol. Slight variations in structure confer different functional capabilities on the various adrenocortical hormones. On the basis of their primary actions, the adrenal steroids can be divided into three categories: (1) **mineralocorticoids,** mainly *aldosterone,* which influence mineral (electrolyte) balance; (2) **glucocorticoids,** primarily *cortisol,* which play a major role in glucose metabolism as well as in protein and lipid metabolism; and (3) **sex hormones** identical or similar to those produced by the gonads (testes in males, ovaries in females).

The production of adrenocortical and other steroid hormones requires a series of steps in which the cholesterol molecule undergoes various enzymatic modifications. The different functional types of adrenal steroids are produced in anatomically distinct portions of the adrenal cortex. This zonation is the result of differential distribution of the enzymes that are required to catalyze the different biosynthetic pathways leading to the formation of each of these steroids. Of the two major adrenocortical hormones, aldosterone is produced exclusively in the zona glomerulosa, while cortisol synthesis is limited to the two inner layers of the cortex, with the zona fasciculata being the major source of this glucocorticoid. No other steroidogenic tissues have the capability of producing either mineralocorticoids or glucocorticoids. In contrast, the adrenal sex hormones, also produced by the two inner cortical zones, are produced in far greater abundance in the gonads.

Being lipophilic, the adrenocortical hormones are all carried in the blood extensively bound to plasma proteins. About 60% of circulating aldosterone is protein bound, primarily to nonspecific albumin. Cortisol is 98% bound, mostly to a plasma protein specific for it called **corticosteroid-binding globulin (transcortin).** Likewise, 98% of the adrenocortical sex hormones are bound, in this case exclusively to albumin.

Mineralocorticoids' major effects are on electrolyte balance and blood pressure homeostasis.

The actions and regulation of the primary adrenocortical mineralocorticoid, **aldosterone,** are described thoroughly elsewhere (chapters 14 and 15). To highlight aldosterone activity, its principal site of action is on the distal and collecting tubules of the kidney, where it promotes Na^+ retention and enhances K^+ elimination during the formation of urine. The promotion of Na^+ retention by aldosterone secondarily induces osmotic retention of H_2O, thereby causing expansion of the ECF volume, which is important in the long-term regulation of blood pressure.

Mineralcorticoids are *essential for life.* Without aldosterone, a person rapidly dies from circulatory shock because of the marked fall in plasma volume caused by excessive losses of H_2O-holding Na^+. With most other hormonal deficiencies, death is not imminent, even though a chronic hormonal deficiency may eventually lead to a premature death.

Aldosterone secretion is increased by (1) activation of the renin-angiotensin-aldosterone system by factors related to a reduction in Na^+ and a fall in blood pressure and (2) direct stimulation of the adrenal cortex by a rise in plasma K^+ concentration (see Fig. 14-25, p. 492). In addition to its effect on aldosterone secretion, angiotensin promotes growth of the zona glomerulosa, in a manner similar to the effect of TSH on the thyroid. The adrenal tropic hormone ACTH primarily affects the inner cortical zones and has only a weak effect in stimulating aldosterone secretion. Thus, unlike cortisol regulation, the regulation of aldosterone secretion is largely independent of anterior pituitary control.

Glucocorticoids exert metabolic effects and have an important role in adaptation to stress.

Cortisol, the primary glucocorticoid, plays an important role in carbohydrate, protein, and fat metabolism; exhibits significant permissive actions for other hormonal activities; and helps people resist stress.

Metabolic effects The overall effect of cortisol's metabolic actions is to increase the concentration of blood glucose at the expense of protein and fat stores. Specifically, cortisol performs the following functions:

- It stimulates hepatic **gluconeogenesis** (*gluco* means "glucose"; *neo* means "new"; *genesis* means "production"), which refers to the conversion of noncarbohydrate sources (namely, amino acids) into carbohydrate within the liver. Between meals or during periods of fasting, when no new nutrients are being absorbed into the blood for utilization and storage, the glycogen (stored glucose) in the liver tends to become depleted as it is broken down to release glucose into the blood. Gluconeogenesis is an important factor in replenishing hepatic glycogen stores and thus in maintaining normal blood glucose levels between meals. This is essential because the brain can use only glucose as its metabolic fuel, yet nervous tissue cannot store glycogen to any extent. The concentration of glucose in the blood must therefore be maintained at an appropriate level to adequately supply the glucose-dependent brain with nutrients.
- It inhibits glucose uptake and use by many tissues, but not the brain, thus sparing glucose for use by the brain, which absolutely requires it as a metabolic fuel. This action contributes to the increase in blood glucose concentration brought about by gluconeogenesis.
- It stimulates protein degradation in many tissues, especially muscle. By breaking down a portion of muscle proteins into their constituent amino acids, cortisol increases the blood amino acid concentration. These mobilized amino acids are available for use in gluconeogenesis or wherever else they are needed, such as for repair of damaged tissue or synthesis of new cellular structures.
- It facilitates lipolysis (*lysis* means "breakdown"), the breakdown of lipid (fat) stores in adipose tissue, thus releasing free fatty acids into the blood. The mobilized fatty

acids are available as an alternative metabolic fuel for tissues that can use this energy source in lieu of glucose, thereby conserving glucose for the brain.

Permissive actions Cortisol is extremely important for its permissiveness (see p. 635). For example, cortisol must be present in adequate amounts to permit the catecholamines to induce vasoconstriction. A person lacking cortisol, if untreated, may go into circulatory shock in a stressful situation that demands immediate widespread vasoconstriction.

Role in adaptation to stress Cortisol plays a key role in adaptation to stress. **Stress** refers to the generalized, nonspecific response of the body to any factor that overwhelms, or threatens to overwhelm, the body's compensatory abilities to maintain homeostasis. Contrary to popular usage, the agent inducing the response is correctly called a *stressor,* whereas *stress* refers to the state induced by the stressor. The following types of noxious stimuli illustrate the range of factors that can induce a stress response: *physical* (trauma, surgery, intense heat or cold); *chemical* (reduced O_2 supply, acid-base imbalance); *physiological* (heavy exercise, hemorrhagic shock, pain); *psychological* or *emotional* (anxiety, fear, sorrow); and *social* (personal conflicts, change in lifestyle). Stress of any kind is one of the major stimuli for increased cortisol secretion.

Although cortisol's precise role in adapting to stress is not known, a speculative but plausible explanation might be as follows. A primitive human or an animal wounded or faced with a life-threatening situation must forgo eating. A cortisol-induced shift away from protein and fat stores in favor of expanded carbohydrate stores and increased availability of blood glucose would help protect the brain from malnutrition during the imposed fasting period. Also, the amino acids liberated by protein degradation would provide a readily available supply of building blocks for tissue repair should physical injury occur. Thus, an increased pool of glucose, amino acids, and fatty acids is available for use as needed.

Other physiological actions Cortisol is also known to alter mood and behavior. The underlying mechanisms are unclear.

Anti-inflammatory and immunosuppressive effects When cortisol or synthetic cortisol-like compounds are administered to yield higher than physiological concentrations of glucocorticoids (that is, *pharmacological levels*), not only are all of the metabolic effects increased in magnitude, but several important new actions not evidenced at normal physiological levels are seen. The most noteworthy of glucocorticoids' pharmacological effects are anti-inflammatory and immunosuppressive effects (see p. 382). (Although these actions are traditionally considered to occur only at pharmacological levels, recent studies suggest that cortisol may exert anti-inflammatory effects even at normal physiological levels.) Synthetic glucocorticoids have been developed that maximize the anti-inflammatory and immunosuppressive effects of these steroids while minimizing the metabolic effects.

Administration of large amounts of glucocorticoid inhibits almost every step of the inflammatory response, making these steroids effective drugs in treating conditions in which the inflammatory response itself has become a destructive process, such as rheumatoid arthritis. It is important to recognize that glucocorticoids used in this manner do not affect the underlying disease process; they merely suppress the body's response to the disease. Because glucocorticoids also exert multiple inhibitory effects on the overall immune process, such as "knocking out of commission" the white blood cells responsible for antibody production and destruction of foreign cells, these agents have also proved useful in the management of various allergic disorders and in the prevention of organ transplant rejections.

When these steroids are employed therapeutically, they should be used only when warranted and then only sparingly for several important reasons. First, because they suppress the normal inflammatory and immune responses that form the backbone of the body's defense system, a glucocorticoid-treated individual has limited ability to resist infections. Second, in addition to the anti-inflammatory and immunosuppressive effects readily exhibited at pharmacological levels, other less desirable effects may also be observed with prolonged exposure to supraphysiological concentrations of glucocorticoids. These effects include development of gastric ulcers, high blood pressure, atherosclerosis, and menstrual irregularities. Third, high levels of exogenous glucocorticoids act in negative-feedback fashion to suppress the hypothalamus-pituitary axis that drives normal glucocorticoid secretion and maintains the integrity of the adrenal cortex. Prolonged suppression of this axis can lead to irreversible atrophy of the cortisol-secreting cells of the adrenal gland and thus to permanent inability of the body to produce its own cortisol.

Cortisol secretion is directly regulated by ACTH.

Cortisol secretion by the adrenal cortex is regulated by a long-loop negative-feedback system involving the hypothalamus and anterior pituitary (▬ Fig. 19-8). Adrenocorticotropic hormone (ACTH) from the anterior pituitary stimulates the adrenal cortex to secrete cortisol. ACTH is derived from a large precursor molecule, **pro-opiomelanocortin**, produced within the endoplasmic reticulum of the anterior pituitary's ACTH-secreting cells (see p. 622). Prior to secretion, this large precursor is pruned into ACTH and several other biologically active peptides, namely, *melanocyte-stimulating hormone (MSH)* (see p. 637) and a morphine-like substance, *β-endorphin* (see p. 164). The possible significance of these multiple secretory products from a single precursor molecule will be addressed later.

Being tropic to the zona fasciculata and zona reticularis, ACTH stimulates both the growth and the secretory output of these two inner layers of the cortex. In the absence of adequate amounts of ACTH, these layers shrink considerably, and cortisol secretion is drastically reduced. Recall that angiotensin, not ACTH, is responsible for maintaining the size of the zona glomerulosa. Like the actions of TSH on the thyroid gland, ACTH enhances many steps in the synthesis of cortisol.

The ACTH-producing cells in turn secrete only at the command of corticotropin-releasing hormone (CRH) from the hypothalamus. The feedback control loop is completed by cortisol's inhibitory actions on CRH and ACTH secretion by the hypothalamus and anterior pituitary, respectively.

The negative-feedback system for cortisol maintains a relatively constant average level of cortisol secretion that is punctuated by alternating bursts of modest secretion separated by "silent" periods of little or no secretion. The amount of cortisol secreted with each burst does not vary much from one secretory episode to the next. However, the total amount of cortisol secreted during a given period of time can be changed by varying the *frequency* of secretory bursts. Reminiscent of summation in skeletal muscle contraction, one burst "adds on" to the previous burst, increasing the plasma concentration of cortisol beyond that achieved by a single burst. Thus, the mean plasma concentration of cortisol can be elevated by stimulating the secretory bursts to occur more frequently (Fig. 19-9).

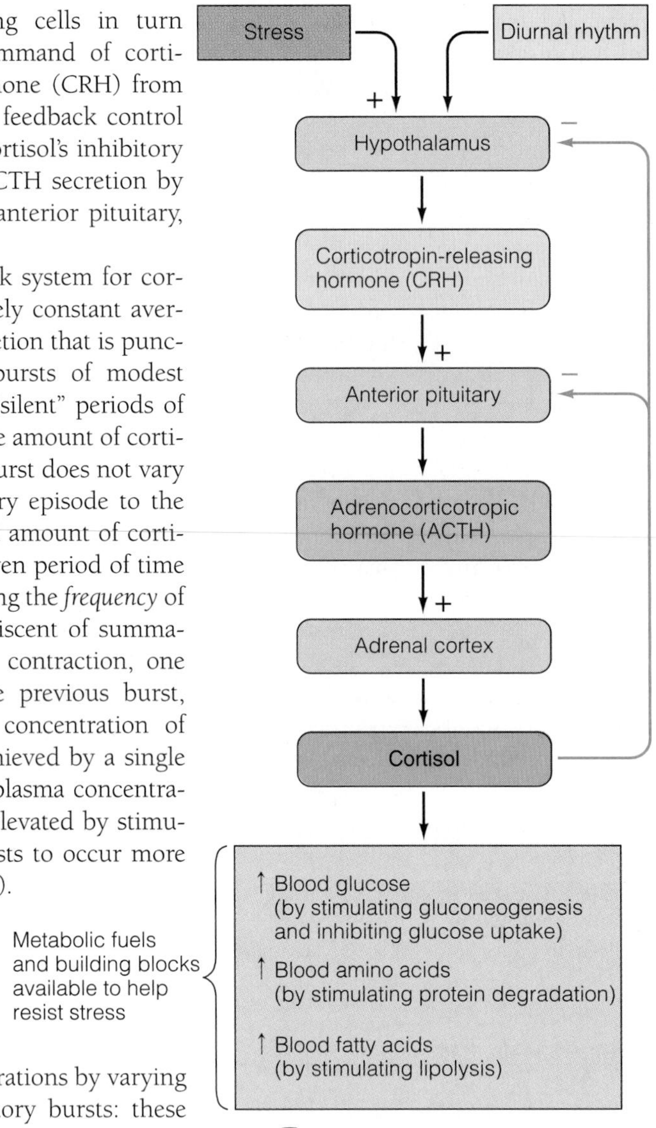

Superimposed on the basic negative-feedback control system are two additional factors that influence plasma cortisol concentrations by varying the frequency of secretory bursts: these are *diurnal rhythm* and *stress.* Recall that there is a characteristic diurnal rhythm in plasma cortisol concentration, with the highest level occurring in the morning and the lowest level at night (see Fig. 18-9, p. 633). This diurnal rhythm, which is intrinsic to the hypothalamus-pituitary control system, is related primarily to the sleep-wake cycle. The peak and low levels are reversed in a person who works at night and sleeps during the day. Such time-dependent variations in secretion are of more than academic interest for several reasons. First, it is important clinically to know at what time of day a blood sample was taken when interpreting the significance of a particular value. Second, the linking of cortisol secretion to day-night activity patterns raises serious questions about the common practice of swing shifts at work (that is, constantly switching day and night shifts among employees). Third, because cortisol helps a person resist stress, increasing attention is being given to the time of day various surgical procedures are performed.

The other major factor that is independent of, and in fact can override, the stabilizing negative-feedback control is stress. Dramatic increases in cortisol secretion, mediated by the central nervous system through enhanced activity of the CRH-ACTH system, occur in response to all kinds of stressful situations. The magnitude of the increase in plasma cortisol concentration is generally proportional to the intensity of the stressful stimulation; a greater increase in cortisol levels is evoked in response to severe stress than to mild stress.

Stress
Diurnal rhythm
+
Hypothalamus
–
Corticotropin-releasing hormone (CRH)
+
Anterior pituitary
–
Adrenocorticotropic hormone (ACTH)
+
Adrenal cortex
Cortisol
Metabolic fuels and building blocks available to help resist stress
↑ Blood glucose (by stimulating gluconeogenesis and inhibiting glucose uptake)
↑ Blood amino acids (by stimulating protein degradation)
↑ Blood fatty acids (by stimulating lipolysis)

Figure 19-8 **Control of Cortisol Secretion**

The adrenal cortex secretes both male and female sex hormones in both sexes.

In both sexes, the adrenal cortex produces both *androgens,* or "male" sex hormones, and *estrogens,* or "female" sex hormones. Because the enzymes required for the production of these sex hormones are found in very low concentrations in the adrenocortical cells, androgens and estrogens are normally produced in very small quantities from this source. The main site of production for the sex hormones is the gonads: the testes for androgens (of which testosterone is the most powerful and most abundant) and the ovaries for estrogens. Accordingly, males have a preponderance of circulating androgens, whereas in females, estrogens are predominant. However, no hormones are unique to either males or females (except those from the placenta during pregnancy) because small amounts of the sex hormone of the opposite sex are produced by the adrenal cortex in both sexes.

Under normal circumstances, the adrenal androgens and estrogens are not sufficiently abundant or powerful to induce masculinizing or feminizing effects, respectively. The only adrenal sex hormone that has any biological importance is the androgen **dehydroepiandrosterone (DHEA).** The testes' primary androgen product is the potent testosterone, but the most abundant adrenal androgen is the much weaker DHEA. Adrenal DHEA is overpowered by testicular testosterone in males but is of physiological significance in females, who otherwise lack androgens. This adrenal androgen is responsible for androgen-dependent processes in the female such as growth of pubic and axillary (armpit) hair, enhancement of the pubertal growth spurt, and development and maintenance of the female sex drive.

In addition to controlling cortisol secretion, ACTH (not the pituitary gonadotropic hormones) controls adrenal androgen secretion. In general, cortisol and DHEA output by the adrenal cortex parallel each other. However, adrenal androgens feed back outside the hypothalamus-pituitary-adrenal cortex loop. Instead of inhibiting CRH, DHEA inhibits gonadotropin-releasing hormone, just as testicular androgens do. Furthermore, sometimes adrenal androgen and cortisol output diverge from each other—for example, at the time of puberty, adrenal androgen secretion undergoes a marked surge, but cortisol secretion does not change. This enhanced secretion initiates the development of androgen-dependent processes in females. In males the same thing is accomplished primarily by testicular androgen secretion, which is also aroused at puberty. The nature of the pubertal inputs to the adrenals and gonads is still unresolved.

The surge in DHEA secretion that begins at puberty peaks between the ages of 25 and 30. After 30, DHEA secretion slowly tapers off until by the age of 60 the plasma DHEA concentration is less than 15% of its peak level. Some scientists suspect that the age-related decline of DHEA and other hormones such as human growth hormone (see p. 649) and melatonin (see p. 637) plays a role in some of the problems associated with getting older. Early studies with DHEA replacement therapy demonstrated some physical improvement, such as an increase in lean muscle mass and a decrease in fat, but the most pronounced effect was a marked increase in psychological well-being and an improved ability to cope with stress. Advocates for DHEA replacement therapy do not suggest that maintaining youthful levels of this hormone will be a fountain of youth (that is, it is not going to extend the life span), but they do propose that it may help people feel and act younger as they age. Other scientists caution that evidence supporting DHEA as an anti-aging therapy is still sparse. Also, they are concerned about the use of DHEA supplementation until it has been thoroughly studied for possible deleterious side effects.

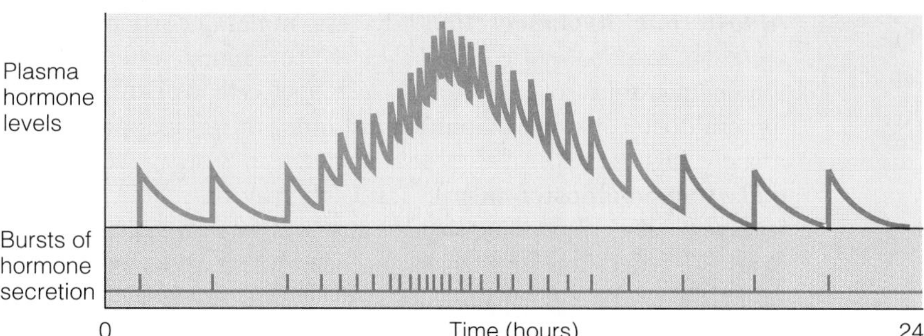

Figure 19-9 Changes in Plasma Cortisol Concentration Caused by Changes in the Frequency of Bursts of Hormone Secretion
[Source: Adapted with permission from George A. Hedge, Howard D. Colby, and Robert L. Goodman, *Clinical Endocrine Physiology* (Philadelphia: W. B. Saunders Company, 1987), Figure 4-4, p. 80.]

The adrenal cortex may secrete too much or too little of any one of its hormones.

Although uncommon, there are a number of different disorders of adrenocortical function. Excessive secretion may occur with any of the three categories of adrenocortical hormones. Accordingly, three main patterns of symptoms resulting from hyperadrenalism can be distinguished, depending on which hormone type is in excess: aldosterone hypersecretion, cortisol hypersecretion, and adrenal androgen hypersecretion (Table 19-2).

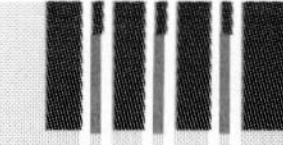

Table 19-2 Major Adrenocortical Abnormalities

Abnormality	*Condition*	*Cause*	*Symptoms*
Excess aldosterone	Conn's syndrome (primary hyperaldosteronism)	Hypersecreting tumor of the zona glomerulosa	Hypernatremia; hypokalemia; hypertension
	Secondary hyperaldosteronism	Inappropriately high activity of the renin-angiotensin system	
Excess cortisol	Cushing's syndrome	Excess CRH and/or ACTH caused by hypothalamic or anterior pituitary disease; hypersecreting tumor of the inner layers of the adrenal cortex; ACTH-secreting tumor in the lung	Glucose excess; protein shortage; abnormal fat distribution
Excess androgen	Adrenogenital syndrome	Lack of an enzyme in the cortisol steroidogenic pathway	Inappropriate masculinization in all but adult males
Deficient aldosterone and cortisol	Addison's disease (primary adrenocortical insufficiency)	Destruction or idiopathic atrophy of the adrenal cortex	Related to aldosterone deficiency: hyperkalemia; hyponatremia; hypotension (if severe enough, fatal)
Deficient cortisol	Secondary adrenocortical insufficiency	Insufficient ACTH caused by hypothalamic or anterior pituitary failure	Related to cortisol deficiency: poor response to stress; hypoglycemia; lack of permissiveness for many metabolic activities

Aldosterone hypersecretion Excess mineralocorticoid secretion may be caused by (1) a hypersecreting adrenal tumor made up of aldosterone-secreting cells (**primary hyperaldosteronism** or **Conn's syndrome**) or (2) inappropriately high activity of the renin-angiotensin system **(secondary hyperaldosteronism).** The latter may be produced by any number of conditions that cause a chronic reduction in arterial blood flow to the kidneys, thereby excessively activating the renin-angiotensin-aldosterone system. An example is atherosclerotic narrowing of the renal arteries.

The symptoms of both primary and secondary hyperaldosteronism are related to the exaggerated effects of aldosterone—namely, excessive Na^+ retention (**hypernatremia)** and K^+ depletion **(hypokalemia).** Also, high blood pressure (hypertension) is generally present, at least partially because of excessive Na^+ and fluid retention.

Cortisol hypersecretion Excessive cortisol secretion **(Cushing's syndrome)** can be caused by (1) overstimulation of the adrenal cortex by excessive amounts of CRH and/or ACTH; (2) adrenal tumors that uncontrollably secrete cortisol independent of ACTH; or (3) ACTH-secreting tumors located in places other than the pituitary, most commonly in the lung. Whatever the cause, the prominent characteristics of this syndrome are related to the exaggerated effects of glucocorticoid, with the main symptoms being reflections of excessive gluconeogenesis. When too many amino acids are converted into glucose, the body suffers from combined glucose excess (high blood glucose) and protein shortage. Because the resultant hyperglycemia and glucosuria (glucose in the urine) mimic diabetes mellitus, the condition is sometimes referred to as adrenal diabetes. For reasons that are unclear, some of the extra glucose is deposited as body fat in locations characteristic for this disease, typically in the abdomen and face and above the shoulder blades. The abnormal fat distributions in the latter two locations are descriptively called a "moon-face" and a "buffalo hump," respectively. The appendages, in contrast, remain thin.

Besides the effects attributable to excessive glucose production, other effects arise from the widespread mobilization of amino acids from body proteins for use as glucose precursors. Loss of muscle protein leads to muscle weakness and fatigue. The protein-poor, thin skin of the abdomen becomes overstretched by the excessive underlying fat deposits. As a result, the subdermal tissues rupture, causing the formation of irregular, reddish-purple linear streaks. The weakness of blood vessel walls caused by depletion of structural protein results in an excessive tendency to bruise and to form ecchymoses (small patches of subcutaneous bleeding). Formation of collagen, a major structural protein of connective tissue, is depressed. This interferes with scar formation, so wounds heal poorly. Furthermore, loss of the proteinaceous collagen framework of bone leads to weakness of the skeleton, so fractures may result from little or no apparent injury.

Adrenal androgen hypersecretion Excess adrenal androgen secretion, a masculinizing condition, is more common than the extremely rare feminizing condition of excess adrenal estrogen secretion. Either condition is referred to as **adrenogenital syndrome,** emphasizing the pronounced effects that excessive adrenal sex hormones have on the genitalia and associated sexual characteristics.

The symptoms that result from excess androgen secretion depend on the sex of the individual and the age when the hyperactivity first begins. Because androgens exert masculinizing effects, a woman with this disease tends to develop a male pattern of body hair, a condition referred to as **hirsutism.** She usually also acquires other male secondary sexual characteristics, such as deepening of the voice and more muscular arms and legs. The breasts become smaller, and menstruation may cease as a result of androgen suppression of the woman's hypothalamus-pituitary-ovarian pathway for her own female sex hormone secretion.

Female infants born with adrenogenital syndrome manifest male-type external genitalia because excessive androgen secretion occurs early enough during fetal life to induce development of their genitalia along male lines, similar to the development of males under the influence of testicular androgen. The clitoris, which is the female homologue of the male penis, enlarges under androgen influence and takes on a penile-like appearance so that in some cases it is difficult at first to determine the child's sex. This condition is known as **male pseudohermaphroditism.** (A true hermaphrodite has the gonads of both sexes.)

Excessive adrenal androgen secretion in prepubertal boys causes them to prematurely develop male secondary sexual characteristics—for example, deep voice, beard, enlarged penis, and sex drive. This condition is referred to as **precocious pseudopuberty** to differentiate it from true puberty, which occurs as a result of increased testicular activity. In precocious pseudopuberty, the androgen secretion from the adrenal cortex is not accompanied by sperm production or any other gonadal activity because the testes are still in their nonfunctional prepubertal state.

Overactivity of adrenal androgens in adult males has no apparent effect because of the already existing male sex characteristics.

The adrenogenital syndrome is most commonly caused by an inherited enzymatic defect in the cortisol steroidogenic pathway. The pathway for synthesis of androgens branches from the normal biosynthetic pathway for cortisol (see Fig. 18-3, p. 627). When an enzyme specifically essential for synthesis of cortisol is deficient, the result is decreased secretion of cortisol. The decline in cortisol secretion removes the negative-feedback effect on the hypothalamus and anterior pituitary so that levels of CRH and ACTH increase considerably (■ Fig. 19-10). The defective adrenal cortex is incapable of responding to this increased ACTH secretion with cortisol output and instead shunts more of its cholesterol precursor into the androgen pathway. The result is excess DHEA production. This excess androgen does not inhibit ACTH but rather inhibits the gonadotropins. Because gamete production is not stimulated in the absence of gonadotropins, individuals with adrenogenital syndrome are sterile. Of course, the victims also exhibit symptoms of cortisol deficiency.

The symptoms of adrenal virilization, sterility, and cortisol deficiency are all reversed by glucocorticoid therapy. Administration of exogenous glucocorticoid replaces the cortisol deficit and, more dramatically, also inhibits the hypothalamus and pituitary so that ACTH secretion is suppressed. Once ACTH secretion is reduced, the profound stimulation of the adrenal cortex ceases, and androgen secretion declines markedly. Removal of the large quantities of adrenal androgens from the circulation allows masculinizing characteristics to gradually recede and normal gonadotropin secretion to resume. Without understanding how these hormonal systems are related, it would be very difficult to comprehend how glucocorticoid administration could dramatically reverse symptoms of masculinization and sterility.

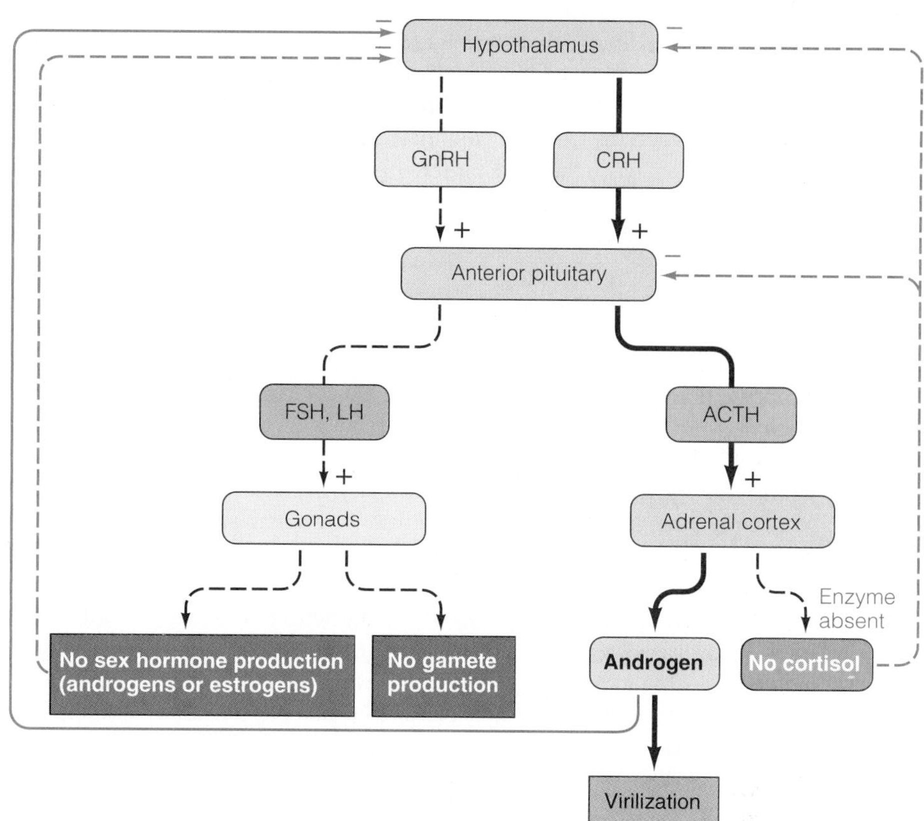

Figure 19-10 **Hormonal Interrelationships in Adrenogenital Syndrome** The adrenocortical cells that are supposed to produce cortisol produce androgens instead because of a deficiency of a specific enzyme essential for cortisol synthesis. Because no cortisol is secreted to act in negative-feedback fashion, CRH and ACTH levels are elevated. The adrenal cortex responds to the increased ACTH by further increasing androgen secretion. The excess androgen produces virilization and inhibits the gonadotropin pathway, with the result that the gonads stop producing sex hormones and gametes.

Adrenocortical insufficiency If one adrenal gland is nonfunctional or removed, the other healthy organ can take over the function of both through hypertrophy and hyperplasia. Therefore, both glands must be affected before adrenocortical insufficiency occurs.

In **primary adrenocortical insufficiency,** also known as **Addison's disease,** all layers of the adrenal cortex are undersecreting. This condition is most commonly due to autoimmune destruction of the gland by erroneous production of adrenal cortex–attacking antibodies. **Secondary adrenocortical insufficiency** may occur because of a pituitary or hypothalamic abnormality, resulting in insufficient ACTH secretion. In Addison's disease, both aldosterone and cortisol are deficient, whereas in the secondary form of the condition, only cortisol is deficient, because aldosterone secretion does not depend on ACTH stimulation.

The symptoms associated with aldosterone deficiency in Addison's disease are the most threatening. If severe enough, the condition is fatal because aldosterone is essential for life. However, the loss of adrenal function may develop slowly and insidiously so that aldosterone secretion may be subnormal but not totally lacking. Patients with aldosterone deficiency display K^+ retention **(hyperkalemia)** caused by reduced K^+ loss in the urine and Na^+ depletion **(hyponatremia)** caused by excessive urinary loss of Na^+. The former results in disturbances in cardiac rhythm. The latter results in a fall in ECF volume, including a reduction in circulating blood volume, which in turn leads to low blood pressure (hypotension).

Symptoms associated with cortisol deficiency are as would be expected; poor response to stress, hypoglycemia (low blood glucose) caused by reduced gluconeogenic activity, and lack of permissive action for many metabolic activities. With the primary form of the disease, hyperpigmentation (darkening of the skin) resulting from excessive secretion of ACTH is also seen. Because the pituitary is normal, the decline in cortisol secretion brings about an uninhibited elevation in ACTH output. Recall that both ACTH and melanocyte-stimulating hormone (MSH) are produced from the same large precursor molecule, pro-opiomelanocortin. As a result, MSH activity also becomes elevated when the blood levels of ACTH are very high.

The catecholamine-secreting adrenal medulla is a modified sympathetic postganglionic neuron.

The adrenal medulla is actually a modified part of the sympathetic nervous system. A sympathetic pathway consists of two neurons in sequence—a *preganglionic neuron* originating in the CNS, whose axonal fiber terminates on a second peripherally located *postganglionic neuron,* which in turn terminates on the effector organ (see p. 204). The neurotransmitter released by sympathetic postganglionic fibers is norepinephrine, which interacts locally with the innervated organ by binding with specific target receptors known as *adrenergic receptors.*

The adrenal medulla is composed of modified postganglionic sympathetic neurons. Unlike ordinary postganglionic sympathetic neurons, those in the adrenal medulla do not possess axonal fibers that terminate on effector organs. Instead, the ganglionic cell bodies within the adrenal medulla release their chemical transmitter directly into the circulation upon stimulation by the preganglionic fiber (see Fig. 7-3, p. 209). In this case, the transmitter qualifies as a hormone instead of a neurotransmitter. Like sympathetic fibers, the adrenal medulla does release norepinephrine, but its most abundant secretory output is a similar chemical messenger known as **epinephrine.** Both epinephrine and norepinephrine belong to the chemical class of *catecholamines,* which are derived from the amino acid tyrosine (see p. 623). Epinephrine is the same as norepinephrine except that it has a methyl group added to it.

Catecholamine synthesis is accomplished almost entirely within the cytosol of the adrenomedullary secretory cells, with only one step taking place within hormonal storage granules. Once produced, epinephrine and norepinephrine are stored in **chromaffin granules,** which are similar to the transmitter storage vesicles found in sympathetic nerve endings. Accordingly, adrenomedullary tissue is often called *chromaffin tissue.* The chromaffin granules have a very efficient, active transport system for catecholamine uptake; as a result, the concentration of epinephrine in the chromaffin granules is at least 25,000 times greater than that in the cytosol. Segregation of catecholamines in chromaffin granules protects them from being destroyed by cytosolic enzymes during storage. Furthermore, packaging of these hormones into granules is a prerequisite for their export from the adrenomedullary cells. Catecholamines are secreted into the circulation by exocytosis of chromaffin granules; their release is analogous to the release mechanism for secretory vesicles that contain stored peptide hormones or the release of norepinephrine at sympathetic postganglionic terminals.

Of the total adrenomedullary catecholamine output, epinephrine accounts for 80% and norepinephrine for 20%. Whereas epinephrine is produced exclusively by the adrenal medulla, the bulk of norepineprhine in the body is produced by sympathetic postganglionic fibers. Adrenomedullary norepinephrine is generally secreted in quantities too small to exert significant effects on target cells. Therefore, for practical purposes, we can assume that norepinephrine effects are predominantly mediated directly by the sympathetic nervous system and that epinephrine effects are brought about exclusively by the adrenal medulla.

Epinephrine and norepinephrine have differing affinities for three distinctive receptor types: α, β_1, and β_2 adrenergic receptors (see p. 210). Some target cells have only α receptors, some have only β_2 receptors, and some have both α and β_2 receptors; β_1 receptors are found almost exclusively in the heart. In general, the responses elicited by activation of α and β_1 receptors are excitatory in nature, whereas the responses to stimulation of β_2 receptors are inhibitory.

Neither catecholamine is totally specific for particular receptor types, but each has markedly different affinities. Norepinephrine binds predominantly with α and β_1 receptors located in the vicinity of postganglionic sympathetic-fiber terminals. Hormonal epinephrine, which is able to reach all α and β_1 receptors by means of its circulatory distribution, interacts with these receptors with approximately the same potency as neurotransmitter norepinephrine. In addition, epinephrine activates β_2 receptors, over which the sympathetic system exerts little influence. Epinephrine is at least 10 times more potent than norepinephrine at β_2 receptors. In fact, many of the essentially epinephrine-exclusive β_2 receptors are located at tissues not even supplied by the sympathetic system but reached by epinephrine through the blood. An example is skeletal muscle, where epinephrine exerts metabolic effects such as promoting the breakdown of stored glycogen.

Epinephrine and norepinephrine exert similar effects in many tissues, with epinephrine generally reinforcing sympathetic activity. For example, epinephrine and norepinephrine both accelerate the heart rate by binding with cardiac β_1 receptors, and both induce generalized arteriolar vasoconstriction of the skin, digestive tract, and kidneys through their mutual α receptor activation. Yet there are some important differences in response that can be explained on the basis of differential activation of different receptor types. As an example, epinephrine, through its exclusive β_2 receptor activation, brings about vasodilation of the blood vessels that supply skeletal muscles and the heart. This effect is in addition to its generalized vasoconstrictor effect mediated by α-receptor stimulation. Also, epinephrine is able to exert some unique effects, such as its metabolic effects, because it reaches places not supplied by sympathetic fibers.

It is important to realize, however, that epinephrine functions only at the bidding of the sympathetic nervous system, which is solely responsible for stimulating its secretion from the adrenal medulla. Epinephrine secretion always accompanies a generalized sympathetic discharge, so sympathetic activity indirectly exerts control over the actions performed by epinephrine. By having the more versatile circulating epinephrine at its call, the sympathetic system has a means of reinforcing its own neurotransmitter effects plus a way of exerting additional actions on tissues that it does not directly innervate.

Epinephrine reinforces the sympathetic nervous system and exerts additional metabolic effects as well.

Adrenomedullary hormones are not essential for life, but virtually all organs in the body are affected by these catecholamines. They play important roles in stress responses, regulation of arterial blood pressure, and control of fuel metabolism. Together, the sympathetic nervous system and adrenomedullary epinephrine mobilize the body's resources to support peak physical exertion in the face of impending danger. The sympathetic and epinephrine actions constitute a fight-or-flight response that prepares the individual to combat an enemy or flee from danger. The following sections discuss epinephrine's major effects, which it accomplishes in collaboration with the sympathetic transmitter norepinephrine or performs alone to complement direct sympathetic response.

Effects on organ systems The sympathetic system and epinephrine both exert widespread effects on organ systems that are ideally suited for fight-or-flight responses (see p. 208). Under the influence of epinephrine and the sympathetic system, the rate and strength of cardiac contraction increase, resulting in increased cardiac output, and their generalized vasoconstrictor effects increase total peripheral resistance. Together, these effects cause an increase in arterial blood pressure, thus ensuring an appropriate driving pressure to force blood to the organs that are most vital for meeting the emergency. Meanwhile, vasodilation of coronary and skeletal muscle blood vessels induced by epinephrine and local metabolic factors shifts blood to the heart and skeletal muscles from other vasoconstricted regions of the body.

Because of their profound influence on the heart and vasculature, the adrenomedullary catecholamines and sympathetic system also play an important role in the ongoing maintenance of arterial blood pressure.

Epinephrine (but not norepinephrine) dilates the respiratory airways to reduce the resistance encountered in moving air in and out of the lungs. Epinephrine also reduces digestive activity and inhibits bladder emptying, both activities that can be "put on hold" during a fight-or-flight situation.

Metabolic effects Epinephrine exerts some important metabolic effects, even at blood hormone concentrations lower than those required for eliciting the cardiovascular responses. In general, epinephrine prompts the mobilization of stored carbohydrate and fat to provide immediately available energy for use as needed to fuel muscular work. Specifically, epinephrine increases the blood glucose level by several different mechanisms. First, it stimulates both hepatic (liver) gluconeogenesis and **glycogenolysis,** the latter referring to the breakdown of stored glycogen into glucose, which is released into the blood. Epinephrine also stimulates glycogenolysis in skeletal muscles. Because of the difference in enzyme content between liver and muscle, however, muscle glycogen cannot be converted directly to glucose. Instead, lactic acid is released into the blood as a result of the breakdown of muscle glycogen. This lactic acid is removed from the blood by the liver and converted into glucose, so epinephrine's actions on skeletal muscle indirectly contribute to an increase in blood glucose levels. Epinephrine and the sympathetic system may further add to this hyperglycemic effect by inhibiting the secretion of insulin, the pancreatic hormone primarily responsible for removal of glucose from the blood, and by stimulating glucagon, another pancreatic hormone that promotes hepatic glycogenolysis and gluconeogenesis. In addition to increasing blood glucose levels, epinephrine also increases the blood fatty acids level by promoting lipolysis.

Epinephrine's metabolic effects are appropriate for fight-or-flight situations. The elevated levels of glucose and fatty acids provide additional fuel to power the muscular movement required by the situation and also assure adequate nourishment for the brain during the crisis when no new nutrients are being consumed. Muscles can use fatty acids for energy production, but the brain cannot.

Because of its other widespread actions, epinephrine also increases the overall metabolic rate. Under the influence of epinephrine, many tissues metabolize at a faster rate. For example, the work of the heart and respiratory muscles is increased, and the pace of liver metabolism is stepped up. Thus, epinephrine as well as thyroid hormone can increase the metabolic rate.

Other effects Epinephrine affects the central nervous system to promote a state of arousal and increased CNS alertness. This permits "quick thinking" to help cope with the impending emergency. Many drugs that are used as stimulants or sedatives probably exert their effects by altering catecholamine levels in the CNS.

Both epinephrine and norepinephrine cause sweating, which helps the body rid itself of extra heat generated by increased muscular activity. Also, epinephrine acts on smooth muscles within the eyes to dilate the pupil and flatten the lens. These actions adjust the eyes for more encompassing vision so that the whole threatening scene can be quickly viewed.

Sympathetic stimulation of the adrenal medulla is solely responsible for epinephrine release.

Catecholamine secretion by the adrenal medulla is controlled entirely by sympathetic input to the gland. When the sympathetic system is activated under conditions of fear or stress, it simultaneously triggers a surge of adrenomedullary catecholamine release, flooding the circulation with up to 300 times the normal concentration of epinephrine. Although a number of different factors have been shown to influence adrenal catecholamine secretion, they all act by increasing preganglionic sympathetic impulses to the adrenal medulla. Among the major factors that stimulate increased adrenomedullary output are a variety of stressful conditions such as physical or psychological trauma, hemorrhage, illness, exercise, hypoxia (low arterial O_2), cold exposure, and hypoglycemia (low blood glucose). The amount of epinephrine released depends on the type and intensity of the stressful stimulus.

Adrenomedullary dysfunction is very rare.

Adrenomedullary hyposecretion is not a recognized clinical entity. No adverse effects have been attributed to a deficiency of epinephrine, presumably not because a deficiency never occurs but because the majority of epinephrine's functions can be duplicated by activation of the sympathetic nervous system alone.

The only catecholamine disorder is a **pheochromocytoma,** a rarely occurring catecholamine-secreting tumor. These tumors are usually, but not always, located in the adrenal medulla. Pheochromocytomas may contain up to 20 times more catecholamines per gram of tissue than normal medullary tissue, with release of these hormones not being subject to neural control. The symptoms of this condition are directly attributable to the actions of excessive amounts of catecholamines, the most common of which are high blood pressure, rapid heart rate, palpitations, excessive sweating and high blood glucose.

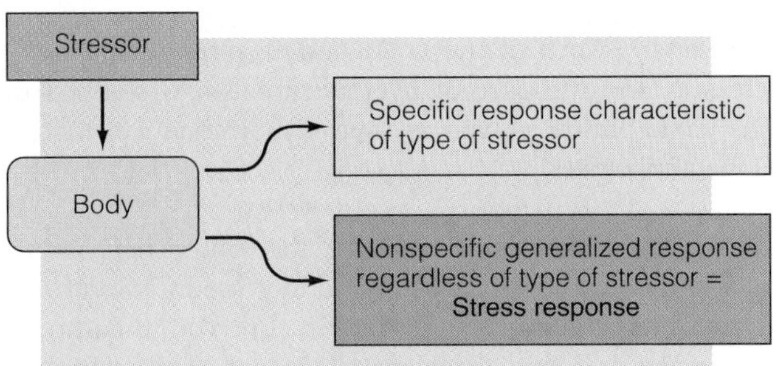

Figure 19-11 Action of a Stressor on the Body

The stress response is a generalized, nonspecific pattern of neural and hormonal reactions to any situation that threatens homeostasis.

Since both components of the adrenal gland play an extensive role in responding to stress, this is an appropriate place to pull together the various major factors involved in the stress response. Recall that a variety of noxious physical, chemical, physiological, and psychosocial stimuli that threaten to overwhelm the body's compensatory ability to maintain homeostasis can elicit a stress response. Different stressors may produce some specific responses characteristic of that stressor; for example, the body's specific response to cold exposure is shivering and skin vasoconstriction, whereas the specific response to bacterial invasion includes increased phagocytic activity and antibody production. In addition to their specific response, however, all stressors also produce a similar nonspecific, generalized response regardless of the type of stressor (Fig. 19-11).

Dr. Hans Selye was the first to recognize this commonality of responses to noxious stimuli in what he called the **general adaptation syndrome.** When a stressor is recognized, both nervous and hormonal responses are called into play to bring about defensive measures to cope with the emergency. The result is a state of intense readiness and mobilization of biochemical resources.

To appreciate the value of the multifaceted stress response, imagine a primitive cave dweller who has just seen a large wild beast lurking in the shadows. The major neural response to such a stressful stimulus is generalized activation of the sympathetic nervous system. The resultant increase in cardiac output and ventilation as well as the diversion of blood from vasoconstricted regions of suppressed activity, such as the digestive tract and kidneys, to the more active vasodilated skeletal muscles and heart prepare the body for a fight-or-flight response. Simultaneously, the sympathetic system calls forth hormonal reinforcements in the form of a massive outpouring of epinephrine from the adrenal medulla. Epinephrine strengthens sympathetic responses and reaches places not innervated by the sympathetic system to perform additional functions, such as mobilizing carbohydrate and fat stores.

Besides epinephrine, a number of other hormones are involved in the overall stress response (Table 19-3). The predominant hormonal response is activation of the CRH-ACTH-cortisol system. Recall that cortisol's role in helping the body cope with stress is presumed to be related to its metabolic effects. Cortisol breaks down fat and protein stores while expanding carbohydrate stores and increasing the availability of blood glucose. A logical assumption is that the increased pool of glucose, amino acids, and fatty acids is available for use as needed, such as to sustain nourishment to the brain and provide building blocks for repair of damaged tissues.

Besides the effects of cortisol in the hypothalamus-pituitary-adrenal cortex axis, there is evidence that ACTH may play a role in resisting stress. ACTH is one of several peptides that facilitate learning and behavior. Thus, it is possible that an increase in ACTH during psychosocial stress might help the body cope more readily with similar stressors in the future by facilitating the learning of appropriate behavioral responses. Furthermore, ACTH is not released alone from its anterior pituitary storage vesicles. Pruning of the large proopiomelanocortin precursor molecule yields not only ACTH but also morphinelike β-endorphin and similar compounds. These compounds are cosecreted with ACTH upon stimulation by CRH during stress. It has been hypothesized that β-endorphin, as a potent endogenous opiate (see p. 164), might exert a role in mediating analgesia (reduction of pain perception) should physical injury be inflicted during stress. It is further speculated that these cosecreted peptides have possible roles in learning, mood alterations, and appetite sup-

Table 19-3
Major Hormonal Changes during the Stress Response

Hormone	*Change*	*Purpose Served*
Epinephrine	↑	Reinforces the sympathetic nervous system to prepare the body for "fight or flight" Mobilizes carbohydrate and fat energy stores; increases blood glucose and blood fatty acids
CRH-ACTH-cortisol	↑	Mobilizes energy stores and metabolic building blocks for use as needed; increases blood glucose, blood amino acids, and blood fatty acids ACTH facilitates learning and behavior β-endorphin cosecreted with ACTH may mediate analgesia
Glucagon	↑	Act in concert to increase blood glucose and blood fatty acids
Insulin	↓	
Renin-angiotensin-aldosterone	↑	Conserve salt and H_2O to expand the plasma volume; help sustain blood pressure when acute loss of plasma volume occurs Angiotensin II and vasopressin cause arteriolar vasoconstriction to increase blood pressure Vasopressin facilitates learning
Vasopressin	↑	

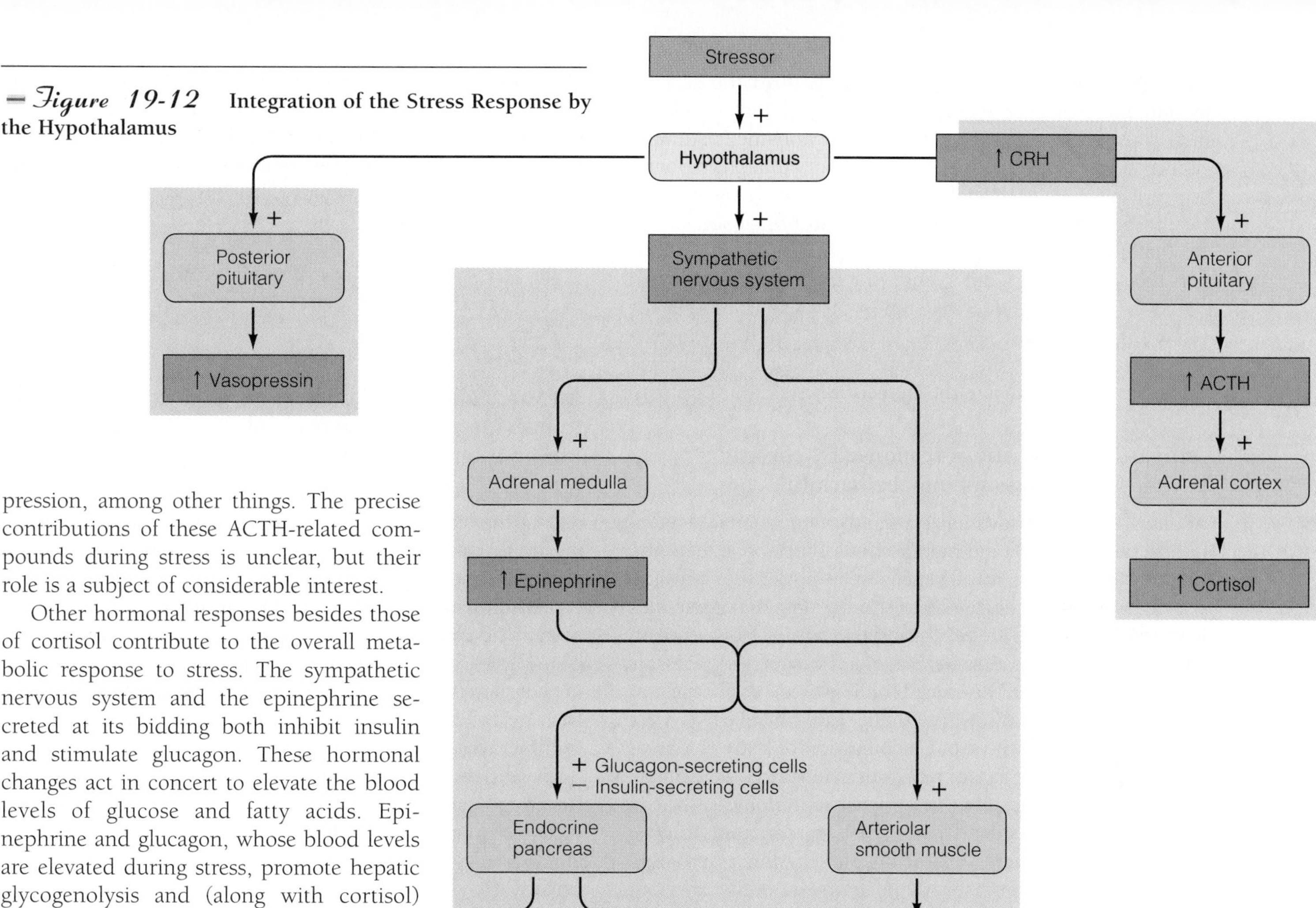

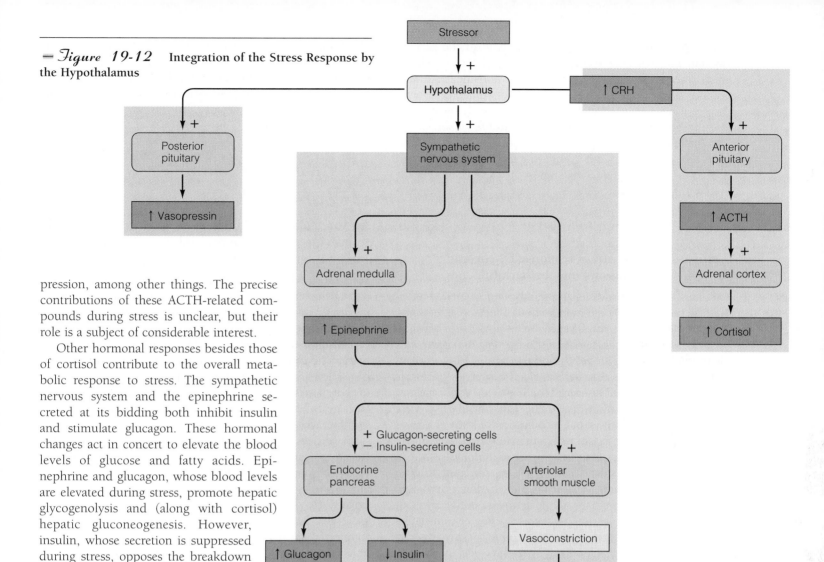

Figure 19-12 Integration of the Stress Response by the Hypothalamus

pression, among other things. The precise contributions of these ACTH-related compounds during stress is unclear, but their role is a subject of considerable interest.

Other hormonal responses besides those of cortisol contribute to the overall metabolic response to stress. The sympathetic nervous system and the epinephrine secreted at its bidding both inhibit insulin and stimulate glucagon. These hormonal changes act in concert to elevate the blood levels of glucose and fatty acids. Epinephrine and glucagon, whose blood levels are elevated during stress, promote hepatic glycogenolysis and (along with cortisol) hepatic gluconeogenesis. However, insulin, whose secretion is suppressed during stress, opposes the breakdown of liver glycogen stores. All of these effects contribute to increasing the concentration of blood glucose. The primary stimulus for insulin secretion is a rise in blood glucose; in turn, a primary effect of insulin is to lower blood glucose. If it were not for the deliberate inhibition of insulin during the stress response, the hyperglycemia caused by stress would stimulate secretion of glucose-lowering insulin. As a result, the elevation in blood glucose could not be sustained. Stress-related hormonal responses also promote a release of fatty acids from fat stores, because lipolysis is favored by epinephrine, glucagon, and cortisol but opposed by insulin.

In addition to the hormonal changes that mobilize energy stores during stress, other hormones are simultaneously called into play to sustain blood volume and blood pressure during the emergency. The sympathetic system and epinephrine have major responsibilities in acting directly on the heart and blood vessels to improve circulatory function. In addition, the renin-angiotensin-aldosterone system is activated as a consequence of a sympathetically induced reduction of blood supply to the kidneys (see p. 484). Vasopressin secretion is also increased during stressful situations. Collectively, these hormones expand the plasma volume by promoting retention of salt and H_2O. Presumably, the enlarged plasma volume serves as a protective measure to help sustain blood pressure should acute loss of plasma fluid occur through hemorrhage or heavy sweating during the impending period of danger. Vasopressin and angiotensin also have direct vasopressor effects, which would be of benefit in maintaining an adequate arterial pressure in the face of acute blood loss. Vasopressin is further believed to facilitate learning, which has implications for future adaptation to stress.

The multifaceted stress response is coordinated by the hypothalamus.

All the individual responses to stress just described are either directly or indirectly influenced by the hypothalamus (Fig. 19-12). The hypothalamus receives input concerning physical and emotional stressors from virtually all areas of the brain and from many receptors throughout the body. In response, the hypothalamus directly activates the sympathetic nervous system, secretes CRH to stimulate ACTH and cortisol release, and triggers the release of vasopressin. Sympathetic stimulation in turn brings about the secretion of epinephrine, with which it has a conjoined effect on the pancreatic secretion of insulin and glucagon. Furthermore, vasoconstriction of the renal afferent arterioles by the catecholamines indirectly triggers the secretion of renin by reducing the flow of oxygenated blood through the kidneys. Renin in turn sets in motion the renin-angiotensin-aldosterone mechanism. In this way, the hypothalamus integrates the responses of both the sympathetic nervous system and the endocrine system during stress.

Activation of the stress response by chronic psychosocial stressors may be harmful.

Acceleration of cardiovascular and respiratory activity, retention of salt and H_2O, and mobilization of metabolic fuels and building blocks can be of benefit in response to a physical stressor, such as an athletic competition. Most of the stressors in our everyday lives are psychosocial in nature, however, and yet they induce these same magnified responses. Stressors such as anxiety about an exam, conflicts with loved ones, or impatience while sitting in a traffic jam can elicit a stress response. Although the rapid mobilization of body resources is appropriate in the face of real or threatened physical injury, it is generally inappropriate in response to nonphysical stress. If no extra energy is demanded, no tissue damaged, and no blood lost, body stores are being broken down and fluid retained needlessly, probably to the detriment of the emotionally stressed individual. In fact, there is strong circumstantial evidence for a link between chronic exposure to psychosocial stressors and the development of pathological conditions such as atherosclerosis and high blood pressure, although no definitive cause-and-effect relationship has been ascertained. As a result of "unused" stress responses, could hypertension result from too much sympathetic vasoconstriction? From too much salt and H_2O retention? From too much vasopressin and angiotensin pressor activity? A combination of these? Other factors? Recall that these same diseases can develop with prolonged exposure to pharmacological levels of glucocorticoids. Could long-standing lesser elevations of cortisol, such as might occur in the face of continual psychosocial stressors, do the same thing, only more slowly? Considerable work remains to be done to evaluate the contributions that the stressors in our everyday lives make toward disease production.

Endocrine Control of Fuel Metabolism

All three classes of nutrient molecules can be used to provide cellular energy and, to a large extent, can be interconverted.

We have just discussed the metabolic changes that are elicited during the stress response. Now we will concentrate on the metabolic patterns that occur in the absence of stress, including the hormonal factors that govern this normal metabolism.

Metabolism refers to all the chemical reactions that occur within the cells of the body. Those reactions involving the degradation, synthesis and transformation of the three classes of energy-rich organic molecules—protein, carbohydrate, and fat—are collectively known as **intermediary metabolism** or **fuel metabolism** (▮ Table 19-4).

During the process of digestion, large nutrient molecules **(macromolecules)** are broken down into their smaller absorbable subunits as follows: proteins are converted into amino acids, complex carbohydrates into monosaccharides (mainly glucose), and triglycerides (dietary fats) into monoglycerides and free fatty acids. These absorbable units are transferred from the digestive tract lumen into the blood, either directly or by way of the lymph (chapter 16).

These organic molecules are constantly exchanged between the blood and the cells of the body (— Fig. 19-13). The chemical reactions in which the organic molecules participate within the cells are categorized into two metabolic processes: anabolism and catabolism. **Anabolism** refers to the buildup or synthesis of larger organic macromolecules from the small organic molecular subunits. Anabolic reactions generally require the input of energy in the form of ATP. These

Table 19-4 **Summary of Reactions in Fuel Metabolism**

Metabolic Process	*Reaction*	*Consequence*
Glycogenesis	Glucose → Glycogen	↓ Blood glucose
Glycogenolysis	Glycogen → Glucose	↑ Blood glucose
Gluconeogenesis	Amino acids → Glucose	↑ Blood glucose
Protein synthesis	Amino acids → Protein	↓ Blood amino acids
Protein degradation	Protein → Amino acids	↑ Blood amino acids
Fat synthesis (lipogenesis or triglyceride synthesis)	Fatty acids and glycerol → Triglycerides	↓ Blood fatty acids
Fat breakdown (lipolysis or triglyceride degradation)	Triglycerides → Fatty acids and glycerol	↑ Blood fatty acids

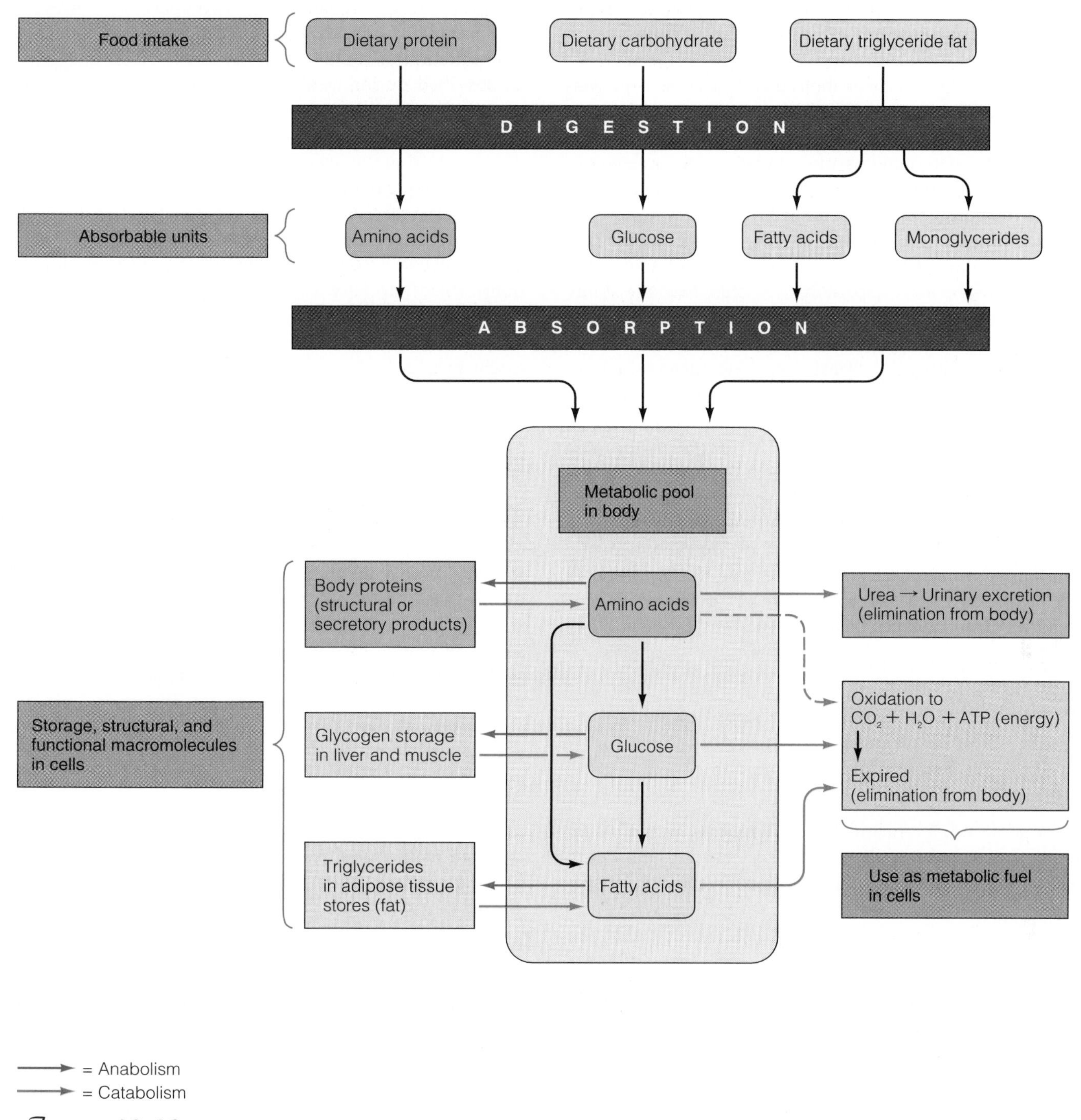

Figure 19-13 **Summary of the Major Pathways Involving Organic Nutrient Molecules**

reactions result in either (1) the manufacture of materials needed by the cell, such as cellular structural proteins or secretory products, or (2) storage of excess ingested nutrients not immediately needed for energy production or needed as cellular building blocks. Storage is in the form of glycogen (the storage form of glucose) or fat reservoirs. **Catabolism,** on the other hand, refers to the breakdown, or degradation, of large, energy-rich organic molecules within cells. Catabolism encompasses two levels of breakdown: (1) hydrolysis (see p. 549) of large cellular organic macromolecules into their smaller subunits, similar to the process of digestion except that the reactions take place within the cells of the body instead of within the digestive tract lumen (for example, release of glucose by the catabolism of stored glycogen), and (2) oxidation of the smaller subunits, such as glucose, to yield energy for ATP production (see p. 33). Indeed, even the structural components of cells represent stored energy, albeit an "expensive" energy source, because they contain energy-rich proteins that can be cannibalized if necessary to yield energy. As an alternative to energy production, the smaller, multipotential organic subunits derived from intracellular hydrolysis may be released into the blood. These mobilized glucose, fatty acid, and amino acid molecules can then be used as needed for energy production or cellular synthesis elsewhere in the body.

In an adult, the rates of anabolism and catabolism are generally in balance, so the adult body remains in a dynamic steady state and appears unchanged even though the organic molecules that determine its structure and function are continuously being turned over. During growth, anabolism exceeds catabolism.

In addition to being able to resynthesize catabolized organic molecules back into the same type of molecules, many cells of the body, especially liver cells, have the ability to convert most types of small organic molecules into other types—as in, for example, the transformation of amino acids into glucose or fatty acids. Because of these interconversions, adequate nourishment can be provided by a wide range of molecules present in different types of foods. There are limits, however. **Essential nutrients,** such as the essential amino acids and vitamins, cannot be formed in the body by conversion from another type of organic molecule.

The major fate of both ingested carbohydrates and fats is catabolism to yield energy. Amino acids are predominantly used for protein synthesis, but can be used to supply energy after being converted to carbohydrate or fat. Thus, all three categories of foodstuff can be used as fuel, and excesses of any foodstuff can be deposited as stored fuel.

The brain must be continuously supplied with glucose, even between meals when no new nutrients are being taken up from the digestive tract.

At a superficial level, fuel metabolism appears relatively simple: the amount of nutrients in the diet must be sufficient to meet the body's needs for energy production and cellular synthesis. This simple relationship is complicated, however, by two important considerations. First, dietary fuel intake is intermittent, not continuous. As a result, excess energy must be absorbed during meals and stored for use during fasting periods between meals, when dietary sources of metabolic fuel are not available (▮Table 19-5). Excess circulating glucose is stored in the liver and muscle as *glycogen,* a large molecule consisting of interconnected glucose molecules. Because glycogen is a relatively small energy reservoir, less than a day's energy needs can be stored in this form. Once the liver and muscle glycogen stores are "filled up," additional glucose is transformed into fatty acids and glycerol, which are used to synthesize *triglycerides* (glycerol with three fatty acids attached), primarily in adipose tissue (fat) and to a lesser extent in muscle. Excess circulating fatty acids derived from dietary intake also become incorporated into triglycerides. Excess circulating amino acids not needed for protein synthesis are not stored as extra protein but are converted to glucose and fatty acids, which ultimately end up being stored as triglycerides. Thus, the major site of energy storage for excess nutrients of all three classes is adipose tissue. Normally, sufficient triglyceride is stored to provide energy for about two months, more so in an overweight individual. Consequently, during any prolonged period of fasting, the fatty acids released from triglyceride catabolism serve as the primary source of energy for most tissues. The catabolism of stored triglycerides frees glycerol as well as fatty acids, but quantitatively speaking, the fatty acids are far more important. Catabolism of stored fat yields 90% fatty acids and 10% glycerol by weight. Glycerol can be converted to glucose by the liver and contributes in a small way to maintaining blood glucose during a fast.

As a third energy reservoir, a substantial amount of energy is stored as *structural protein,* primarily in muscle, the most abundant protein mass in the body. Protein is not the first

Table 19-5 Stored Metabolic Fuel in the Body

Metabolic Fuel	*Circulating Form*	*Storage Form*	*Major Storage Site*	*Percentage of Total Body Energy Content (and Calories*)*	*Reservoir Capacity*	*Role*
Carbohydrate	Glucose	Glycogen	Liver, muscle	1% (1,500 calories)	Less than a day's worth of energy	First energy source; essential for the brain
Fat	Free fatty acids	Triglycerides	Adipose tissue	77% (143,000 calories)	About 2 months' worth of energy	Primary energy reservoir; energy source during a fast
Protein	Amino acids	Body proteins	Muscle	22% (41,000 calories)	Death results long before capacity is utilized because of structural and functional impairment	Source of glucose for the brain during a fast; last resort to meet other energy needs

*Actually refers to kilocalories; see p. 602.

choice to tap as an energy source, however, because it serves other essential functions; in contrast, the glycogen and triglyceride reservoirs serve solely as energy depots.

The second factor complicating fuel metabolism is that the brain normally depends on the delivery of adequate blood glucose as its sole source of energy. Consequently, it is essential that the blood glucose concentration be maintained above a critical level. The blood glucose concentration is typically 100 mg glucose/100 ml plasma and is normally maintained within the narrow limits of 70–110 mg/100 ml. Liver glycogen is an important reservoir for maintaining blood glucose levels during a short fast. However, liver glycogen is depleted relatively rapidly, so during a longer fast, other mechanisms must be used to ensure that the energy requirements of the glucose-dependent brain are met. First, when new dietary glucose is not entering the blood, tissues not obligated to use glucose shift their metabolic gears to burn fatty acids instead, thus sparing glucose for the brain. Fatty acids are made available by catabolism of triglyceride stores as an alternative energy source for non-glucose-dependent tissues. Second, amino acids can be converted to glucose by gluconeogenesis, whereas fatty acids cannot. Thus, once glycogen stores are depleted despite glucose sparing, new glucose supplies for the brain are provided by the catabolism of body proteins and conversion of the freed amino acids into glucose.

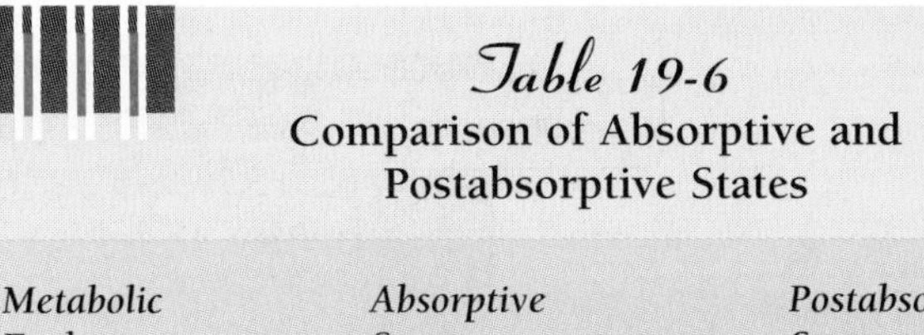

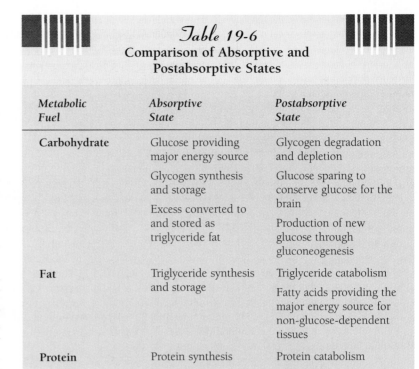

Table 19-6
Comparison of Absorptive and Postabsorptive States

Metabolic Fuel	*Absorptive State*	*Postabsorptive State*
Carbohydrate	Glucose providing major energy source Glycogen synthesis and storage Excess converted to and stored as triglyceride fat	Glycogen degradation and depletion Glucose sparing to conserve glucose for the brain Production of new glucose through gluconeogenesis
Fat	Triglyceride synthesis and storage	Triglyceride catabolism Fatty acids providing the major energy source for non-glucose-dependent tissues
Protein	Protein synthesis Excess converted to and stored as triglyceride fat	Protein catabolism Amino acids used for gluconeogenesis

Metabolic fuels are stored during the absorptive state and are mobilized during the postabsorptive state.

From the preceding discussion, it should be obvious that the disposition of organic molecules depends on the body's metabolic state. There are two functional metabolic states related to eating and fasting cycles—the absorptive state and the postabsorptive state, respectively (Table 19-6). Following a meal, ingested nutrients are being absorbed and entering the blood during the **absorptive,** or **fed, state.** During this time, glucose is plentiful and serves as the major energy source. Very little of the absorbed fat and amino acids is used for energy during the absorptive state because most cells prefer to use glucose when it is available. Extra nutrients not immediately used for energy or structural repairs are channeled into storage as glycogen or triglycerides.

The average meal is completely absorbed in about four hours. Therefore, on a typical three-meals-a-day diet, no nutrients are being absorbed from the digestive tract during late morning and late afternoon and throughout the night. These times constitute the **postabsorptive,** or **fasting, state.** During this state, endogenous energy stores are mobilized to provide energy, while gluconeogenesis and glucose sparing are used to maintain the blood glucose at an adequate level to nourish the brain. The synthesis of protein and fat is curtailed. Instead, stores of these organic molecules are catabolized for glucose formation and energy production, respectively. Carbohydrate synthesis does occur through gluconeogenesis, but the utilization of glucose for energy is greatly reduced.

Note that the blood concentration of nutrients does not fluctuate markedly between the absorptive and postabsorptive states. During the absorptive state, the glut of absorbed nutrients is swiftly removed from the blood and placed into storage; during the postabsorptive state, these stores are catabolized to maintain the blood concentrations at levels necessary to sustain tissue energy demands.

As already described, during these alternating metabolic states, various tissues play different roles. The liver plays the primary role in maintenance of normal blood glucose levels. It stores glycogen when excess glucose is available, releases glucose into the blood when needed, and is the principal site for metabolic interconversions such as gluconeogenesis. Adipose tissue serves as the primary energy storage site and is important in the regulation of fatty acid levels in the blood. Muscle is the primary site of amino acid storage and is the major energy user. The brain normally can use only glucose as an energy source, yet it does not store glycogen, making it mandatory that blood glucose levels be maintained.

Several other organic intermediates play a lesser role as energy sources—namely, glycerol, lactic acid, and ketone bodies. As mentioned earlier, *glycerol* derived from triglyceride hydrolysis (it is the backbone to which the fatty acid chains are attached) can be converted to glucose by the liver. Similarly, *lactic acid,* which is produced by the incomplete catabolism of glucose via glycolysis in muscle (Fig. 19-14; see also p. 243), can also be converted to glucose in the liver. **Ketone bodies** are a group of compounds produced by the liver during glucose sparing. Unlike other tissues, when the liver uses fatty acids as an energy source, it oxidizes them only

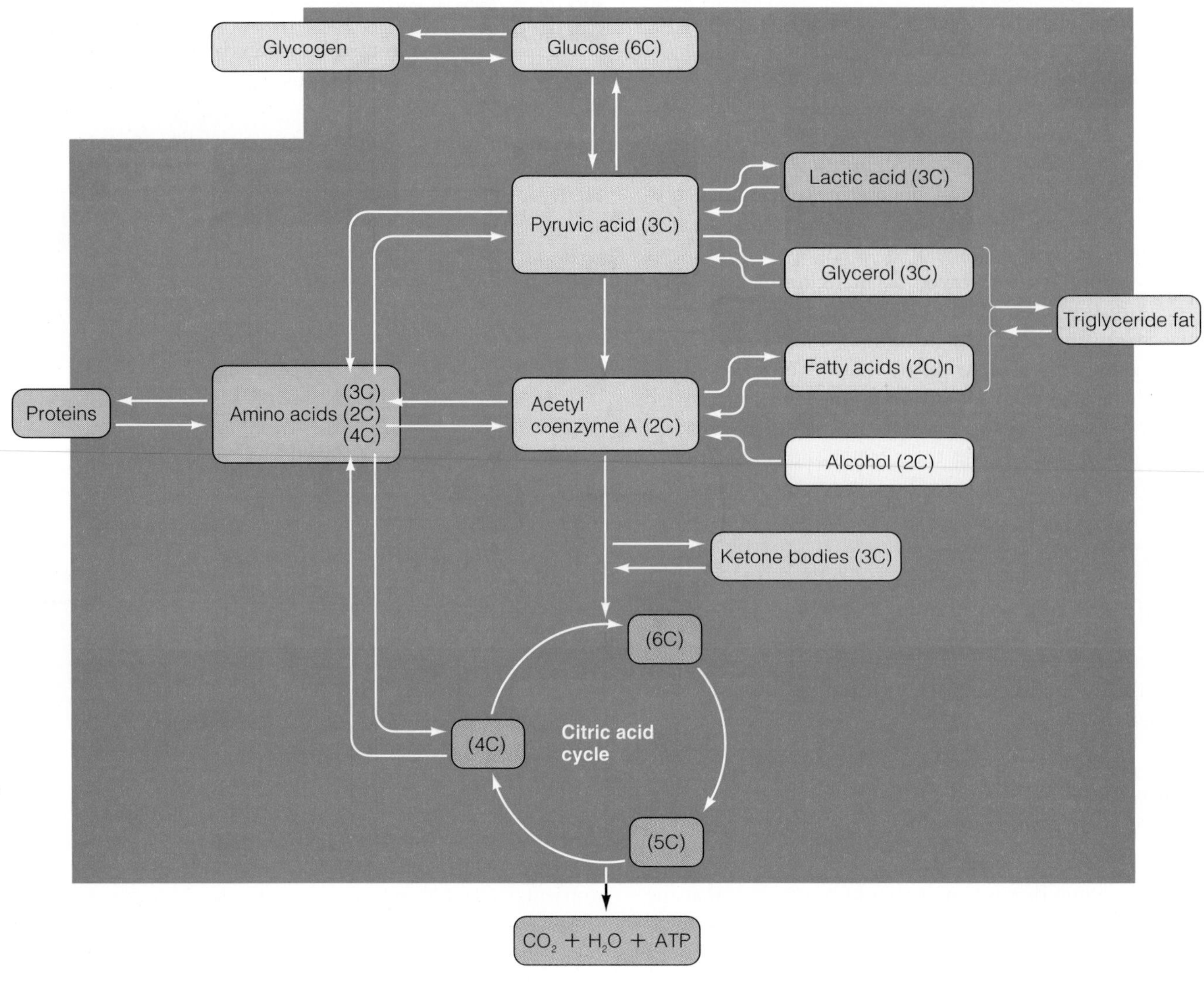

Figure 19-14 **Major Pathways in Energy Production by, Storage of, and Interconversions between Organic Nutrient Molecules**

to acetyl coenzyme A (acetyl CoA), which it is unable to process through the citric acid cycle for further energy extraction. Thus, the liver does not degrade fatty acids all the way to CO_2 and H_2O for maximum energy release. Instead, it partially extracts the available energy and converts the remaining energy-bearing acetyl CoA molecules into ketone bodies, which it releases into the blood. Ketone bodies serve as an alternative energy source for tissues capable of oxidizing them further by means of the citric acid cycle.

During long-term starvation, the brain starts using ketones instead of glucose as a major energy source. Since death resulting from starvation is usually due to protein wasting rather than hypoglycemia (low blood glucose), prolonged survival without any caloric intake requires that gluconeogenesis be kept to a minimum while the energy needs of the brain are not compromised. A sizable portion of cell protein can be catabolized without serious cellular malfunction, but a point is finally reached at which a cannibalized cell is no longer able to function adequately. To ward off the fatal point of failure so long as possible during prolonged starvation, the brain starts using ketones as a major energy source, correspondingly decreasing its use of glucose. Use by the brain of this fatty acid "table scrap" left over from the liver's "meal" limits the necessity of mobilizing body proteins for glucose production to nourish the brain. Both of the major metabolic adaptations to prolonged starvation—a decrease in protein catabolism and use of ketones by the brain—are attributable to the high levels of ketones in the blood at the time. Utilization of ketones by the brain occurs only when the blood ketone level is high. The high blood levels of ketones also directly inhibit protein degradation in muscle. Thus, ketones are responsible for sparing body proteins while satisfying the brain's energy needs.

The pancreatic hormones, insulin and glucagon, are most important in regulating fuel metabolism.

How does the body "know" when to shift its metabolic gears from one of net anabolism and nutrient storage to one of net catabolism and glucose sparing? The flow of organic nutrients along metabolic pathways is influenced by a variety of hormones, including insulin, glucagon, epinephrine, cortisol, and growth hormone. Under most circumstances, the pancreatic hormones, insulin and glucagon, are the dominant hormonal regulators that shift the metabolic pathways back and forth from net anabolism to net catabolism and glucose sparing, depending on whether the body is in a state of feasting or fasting, respectively.

The **pancreas** is an organ composed of both exocrine and endocrine tissues. The exocrine portion of the pancreas secretes a watery alkaline solution and digestive enzymes through the pancreatic duct into the digestive tract lumen. Scattered throughout the pancreas between the exocrine cells are clusters, or "islands," of endocrine cells known as the **islets of Langerhans** (see Fig. 16-18, p. 572). The most abundant pancreatic endocrine cell type is the **β (beta) cell,** the site of *insulin* synthesis and secretion. Next most important are the **α (alpha) cells,** which produce *glucagon.* The **D (delta) cells** are the pancreatic site of *somatostatin* synthesis, whereas the least common islet cells, the **PP cells,** secrete *pancreatic polypeptide.*

The most important pancreatic hormones in the regulation of fuel metabolism are insulin and glucagon. Accordingly, we will pay the most attention to these two pancreatic hormones, giving only a brief highlight now of the other two endocrine pancreatic secretory products. **Somatostatin** is also produced by the hypothalamus, where it serves to inhibit the secretion of growth hormone and TSH. Furthermore, somatostatin is produced by cells lining the digestive tract, where it is suspected of acting locally as a paracrine to inhibit most digestive processes. Pancreatic somatostatin also exerts a variety of inhibitory effects on the digestive system, the overall effect of which is to inhibit digestion of nutrients and to decrease nutrient absorption. Somatostatin is released from the pancreatic D cells in direct response to an increase in blood glucose and blood amino acids during absorption of a meal. By exerting its inhibitory effects, pancreatic somatostatin acts in negative-feedback fashion to put the brakes on the rate at which the meal is being digested and absorbed, thereby preventing excessive plasma levels of nutrients. In addition, pancreatic somatostatin may play an important role in the local regulation of pancreatic hormone seretion. The secretion of insulin, glucagon, pancreatic polypeptide, and somatostatin itself is decreased by the local presence of somatostatin, but the physiological importance of such paracrine function has not been determined.

Little is known about **pancreatic polypeptide.** Its effects seem to be concerned primarily with inhibiting gastrointestinal function. Like somatostatin, pancreatic polypeptide appears to have no direct effects on carbohydrate, protein, or fat metabolism.

Insulin lowers blood glucose, amino acid, and fatty acid levels and promotes anabolism of these small nutrient molecules.

Insulin has important effects on carbohydrate, fat, and protein metabolism. It lowers the blood levels of glucose, fatty acids, and amino acids and promotes their storage. As these nutrient molecules enter the blood during the absorptive state, insulin promotes their cellular uptake and conversion into glycogen, triglycerides, and protein, respectively. Insulin exerts its many effects either by altering transport of specific blood-borne nutrients into cells or by altering the activity of the enzymes involved in specific metabolic pathways.

Actions on carbohydrates The maintenance of blood glucose homeostasis is a particularly important function of the pancreas. Circulating glucose concentrations are determined by the balance that exists among the following processes (Fig. 19-15): glucose absorption from the digestive tract; transport

Figure 19-15 **Factors Affecting Blood Glucose Concentration**

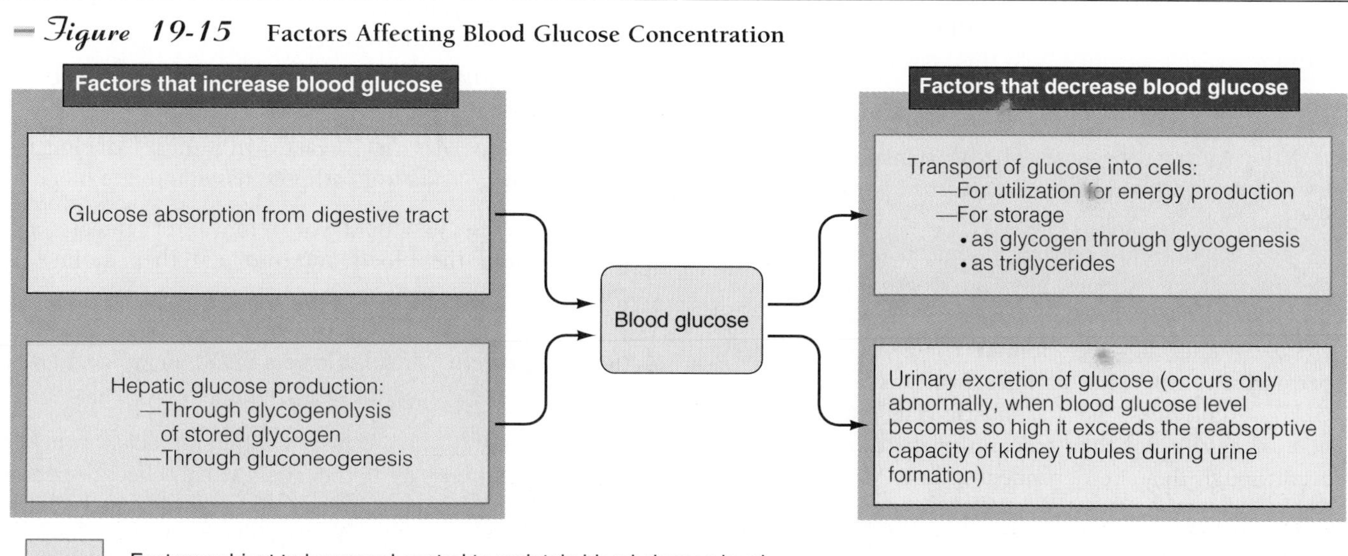

of glucose into cells; cellular (primarily hepatic) glucose production; and (abnormally) urinary excretion of glucose. Insulin exerts a fourfold effect to lower blood glucose levels and promote carbohydrate storage:

1. Insulin facilitates glucose transport into most cells. Glucose transport between the blood and cells is accomplished by means of plasma membrane carriers known as **glucose transporters (GLUT).** Six forms of glucose transporters have been identified, named in the order they were discovered—GLUT-1, GLUT-2, and so on. These glucose transporters all accomplish passive facilitated diffusion of glucose across the plasma membrane (see p. 59) and are distinct from the Na^+-glucose cotransport carriers responsible for secondary active transport of glucose across the kidney and intestinal epithelia (see pp. 62, 487, and 586). Each member of the GLUT family performs slightly different functions. For example, *GLUT-1* transports glucose across the blood-brain barrier, *GLUT-2* transfers into the adjacent bloodstream the glucose that has entered the kidney and intestinal cells by means of the cotransport carriers, and *GLUT-3* is the main transporter of glucose into neurons. The glucose transporter responsible for the majority of glucose uptake by most cells of the body is *GLUT-4,* which operates only at the bidding of insulin. Glucose molecules cannot readily penetrate most cell membranes in the absence of insulin, making most tissues highly dependent on insulin for uptake of glucose from the blood and for its subsequent use. GLUT-4 is especially abundant in the tissues that account for the bulk of glucose uptake from the blood during the absorptive state, namely, skeletal muscle and adipose tissue cells. GLUT-4 is the only type of glucose transporter that is responsive to insulin. Unlike the other types of GLUT molecules, which are always present in the plasma membranes at the sites where they perform their functions, GLUT-4 is excluded from the plasma membrane in the absence of insulin. Insulin promotes glucose uptake by the phenomenon of **transporter recruitment.** Insulin-dependent cells maintain a pool of intracellular vesicles containing GLUT-4. Insulin induces these vesicles to move to the plasma membrane and fuse with it, thus inserting GLUT-4 molecules into the plasma membrane. In this way, increased insulin secretion promotes a rapid, 10- to 30-fold increase in glucose uptake by insulin-dependent cells. When insulin secretion decreases, these glucose transporters are retrieved from the membrane and returned to the intracellular pool.

   Several tissues are not dependent on insulin for their glucose uptake—namely, the brain, working muscles, and the liver. The brain, which requires a constant supply of glucose for its minute-to-minute energy needs, is freely permeable to glucose at all times by means of GLUT-1 and GLUT-3 molecules. Skeletal muscle cells are not dependent on insulin for their glucose uptake during exercise, even though they are dependent at rest. Muscle contraction triggers the insertion of GLUT-4 into the plasma membranes of exercising muscle cells in the absence of insulin. This fact is important in the management of diabetes mellitus (insulin deficiency), as will be described later. The liver is also not dependent on insulin for glucose uptake because it does not use GLUT-4; however, insulin does enhance the metabolism of glucose by the liver by stimulating the first step in glucose metabolism, the phosphorylation of glucose to form glucose-6-phosphate. The phosphorylation of glucose as it enters the cell keeps the intracellular concentration of "plain" glucose low so that a gradient favoring the facilitated diffusion of glucose into the cell is maintained.
2. Insulin stimulates **glycogenesis,** the production of glycogen from glucose, in both skeletal muscle and the liver.
3. Insulin inhibits glycogenolysis, the breakdown of glycogen into glucose. By inhibiting the degradation of glycogen, it likewise favors carbohydrate storage and decreases glucose output by the liver.
4. Insulin further decreases hepatic glucose output by inhibiting gluconeogenesis, the conversion of amino acids into glucose in the liver. Insulin accomplishes this in two ways: by decreasing the amount of amino acids in the blood that are available to the liver for gluconeogenesis, and by inhibiting the hepatic enzymes required for the conversion of amino acids into glucose.

Thus, insulin decreases the concentration of blood glucose by promoting the cells' uptake of glucose from the blood for utilization and storage, while simultaneously blocking the two mechanisms by which the liver releases glucose into the blood (glycogenolysis and gluconeogenesis). Insulin is the only hormone capable of lowering the blood glucose level.

Actions on fat Insulin exerts multiple effects to lower blood fatty acids and promote triglyceride storage:

1. It increases the transport of glucose into adipose tissue cells by means of GLUT-4 recruitment. Glucose serves as a precursor for the formation of fatty acids and glycerol, which are the raw materials for triglyceride synthesis.
2. It activates enzymes that catalyze the production of fatty acids from glucose derivatives.
3. It promotes the entry of fatty acids from the blood into adipose tissue cells.
4. It inhibits lipolysis (fat breakdown), thus reducing the release of fatty acids from adipose tissue into the blood.

Collectively, these actions favor removal of glucose and fatty acids from the blood and promote their storage as trigylcerides.

Actions on protein Insulin lowers blood amino acid levels and enhances protein synthesis through several effects:

1. It promotes the active transport of amino acids from the blood into muscles and other tissues. This effect decreases the circulating amino acid level and provides the building blocks for protein synthesis within the cells.

2. It increases the rate of amino acid incorporation into protein by stimulating the cells' protein-synthesizing machinery.
3. It inhibits protein degradation.

The collective result of these actions is a protein anabolic effect. For this reason, insulin is essential for normal growth.

In short, insulin stimulates biosynthetic pathways that lead to increased glucose utilization, increased carbohydrate and fat storage, and increased protein synthesis. In so doing, this hormone lowers the blood glucose, fatty acid, and amino acid levels. This metabolic pattern is characteristic of the absorptive state. Indeed, insulin secretion rises during this state and is responsible for shifting metabolic pathways to net anabolism.

When insulin secretion is low, the opposite effects occur. The rate of glucose entry into cells is reduced, and net catabolism rather than net synthesis of glycogen, triglycerides, and protein occurs. This pattern is reminiscent of the postabsorptive state; indeed, insulin secretion is reduced during the postabsorptive state. However, the other major pancreatic hormone, glucagon, also plays an important role in shifting from absorptive to postabsorptive metabolic patterns, as will be described later.

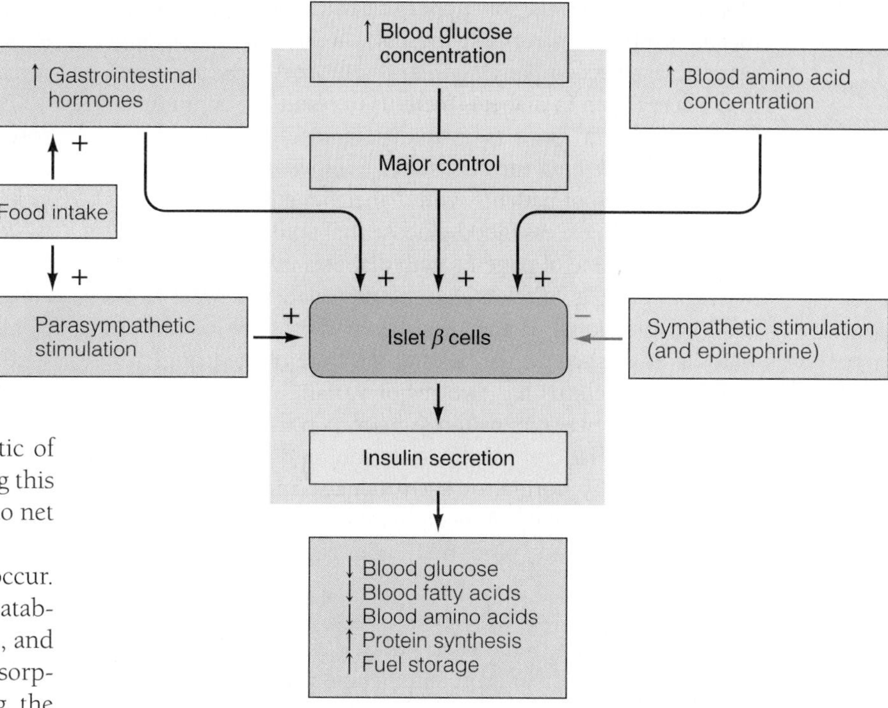

Figure 19-16 **Factors Controlling Insulin Secretion**

The primary stimulus for increased insulin secretion is an increase in blood glucose concentration.

The primary control of insulin secretion is a direct negative-feedback system between the pancreatic β cells and the concentration of glucose in the blood flowing to them. An elevated blood glucose level, such as occurs during absorption of a meal, directly stimulates synthesis and release of insulin by the β cells. The increased insulin in turn reduces the blood glucose to normal while it promotes utilization and storage of this nutrient. Conversely, a fall in blood glucose below normal, such as occurs during fasting, directly inhibits insulin secretion. Lowering the rate of insulin secretion shifts metabolism from the absorptive to the postabsorptive pattern. Thus, this simple negative-feedback system is able to maintain a relatively constant supply of glucose to the tissues without requiring the participation of nerves or other hormones.

In addition to plasma glucose concentration, other inputs are involved in the regulation of insulin secretion as follows (Fig. 19-16):

- An elevated blood amino acid level, such as occurs following ingestion of a high-protein meal, directly stimulates the β cells to increase insulin secretion. In negative-feedback fashion, the increased insulin enhances the entry of these amino acids into the cells, lowering the blood amino acid level while promoting protein synthesis.
- The major gastrointestinal hormones secreted by the digestive tract in response to the presence of food, especially gastric inhibitory peptide (see p. 597), stimulate pancreatic insulin secretion in addition to having direct regulatory effects on the digestive system. Through this control, insulin secretion is increased in "feedforward," or anticipatory, fashion even before nutrient absorption increases the concentration of glucose and amino acids in the blood.
- The autonomic nervous system also directly influences insulin secretion. The islets are richly innervated by both parasympathetic (vagal) and sympathetic nerve fibers. The increase in parasympathetic activity that occurs in response to food in the digestive tract stimulates insulin release. This, too, is a feedforward response in anticipation of nutrient absorption. In contrast, sympathetic stimulation and the concurrent increase in epinephrine both inhibit insulin secretion. The reduction in insulin allows the blood glucose level to increase, an appropriate response to the circumstances under which generalized sympathetic activation occurs—namely, stress (fight or flight) and exercise. In both of these situations, extra fuel is needed for increased muscle activity.

The symptoms of diabetes mellitus are characteristic of an exaggerated postabsorptive state.

Diabetes mellitus is by far the most common of all endocrine disorders. The acute symptoms of diabetes mellitus are attributable to inadequate insulin action. Because insulin is the only hormone capable of lowering blood glucose levels, one of the most prominent features of diabetes mellitus is elevated blood glucose levels, or **hyperglycemia.** *Diabetes* literally means "syphon" or "running through," a reference to the large urine volume accompanying this condition. A large urine volume occurs both in diabetes mellitus (a result of insulin insufficiency) and in diabetes insipidus (a result of vasopressin deficiency). *Mellitus* means "sweet"; *insipidus* means "tasteless." The urine of patients with diabetes mellitus acquires its sweetness from excess blood glucose that spills into the urine, whereas the urine of patients with diabetes insipidus contains no sugar, so it is tasteless. (Aren't you glad you were not a health professional at the time when these two conditions were distinguished on the basis of the taste of the urine?)

Diabetes mellitus has two major variants, differing in the capacity for pancreatic insulin secretion: *Type I diabetes,* characterized by a lack of insulin secretion, and *Type II diabetes,* characterized by normal or even increased insulin secretion but reduced sensitivity of insulin's target cells to its presence. (See the boxed feature on p. 682, ♦Concepts, Challenges, and Controversies.)

The acute consequences of diabetes mellitus can be grouped according to the effects of inadequate insulin action on carbohydrate, fat, and protein metabolism (▬Fig. 19-17). The figure looks overwhelming, but the numbers correspond to the numbers in the following discussion to help you work your way through this complex disease step-by-step.

Since the postabsorptive metabolic pattern is induced by low insulin activity, the changes that occur in diabetes mellitus are an exaggeration of this state, with the exception of hyperglycemia. In the usual fasting state, the blood glucose level is slightly below normal. Hyperglycemia, the hallmark of diabetes mellitus, arises from reduced glucose uptake by cells, coupled with increased output of glucose from the liver (① in Fig. 19-17). Hepatic output of glucose increases as the glucose-yielding processes of glycogenolysis and gluconeogenesis proceed unchecked in the absence of insulin. Because most of the body's cells are unable to use glucose without the assistance of insulin, an ironic extracellular glucose excess occurs coincident with an intracellular glucose deficiency—"starvation in the midst of plenty." Even though the non-insulin-dependent brain is adequately nourished during diabetes mellitus, further consequences of the disease lead to brain dysfunction, as you will see shortly.

When the blood glucose rises to the level that the amount of glucose filtered exceeds the tubular cells' capacity for reabsorption, glucose appears in the urine *(glucosuria)* ②. Glucose in the urine exerts an osmotic effect that draws H_2O with it, producing an osmotic diuresis characterized by *polyuria* (frequent urination) ③. The excess fluid lost from the body leads to dehydration ④, which in turn can ultimately lead to peripheral circulatory failure because of the marked reduction in blood volume ⑤. Circulatory failure, if uncorrected, can lead to death because of low cerebral blood flow ⑥ or secondary renal failure due to inadequate filtration pressure ⑦. Furthermore, cells lose water as the body becomes dehydrated as a result of an osmotic shift of water from the cells into the hypertonic extracellular fluid ⑧. Brain cells are especially sensitive to shrinking so that nervous system malfunction ensues ⑨ (see p. 523). Another characteristic symptom of diabetes mellitus is *polydipsia* (excessive thirst) ⑩, which is actually a compensatory mechanism to counteract the dehydration.

The story is not complete. In the face of intracellular glucose deficiency, appetite is stimulated, leading to *polyphagia* (excessive food intake) ⑪. In spite of the increased food intake, however, progressive weight loss occurs as a result of the effects of insulin deficiency on fat and protein metabolism. Triglyceride synthesis decreases while lipolysis increases, resulting in large-scale mobilization of fatty acids from triglyceride stores ⑫. The increased blood fatty acids are used to a large extent by the cells as an alternative energy source. Increased liver utilization of fatty acids results in the release of excessive ketone bodies into the blood, causing *ketosis* ⑬. Since the ketone bodies include several different acids such as acetoacetic acid that result from the incomplete breakdown of fat during hepatic energy production, this developing ketosis leads to progressive metabolic acidosis ⑭. Acidosis depresses the brain and, if severe enough, can lead to diabetic coma and death ⑮.

A compensatory measure for metabolic acidosis is increased ventilation to blow off extra acid-forming CO_2 ⑯. Exhalation of one of the ketone bodies, acetone, causes a "fruity" breath odor. Sometimes, because of this odor, a patient collapsed in a diabetic coma is unfortunately mistaken by passersby for a "wino" passed out in a state of drunkenness. (This situation illustrates the merits of medical-alert identification tags.) Persons with Type I diabetes are much more prone to develop ketosis than are Type II diabetics.

The effects of a lack of insulin on protein metabolism result in a net shift toward protein catabolism. The net breakdown of muscle proteins leads to wasting and weakness of skeletal muscles ⑰ and, in child diabetics, a reduction in overall growth. Reduced amino acid uptake coupled with increased protein degradation results in excess amino acids in the blood ⑱. The increased circulating amino acids can be used for additional gluconeogenesis, which further aggravates the hyperglycemia ⑲.

As you can readily appreciate from this overview, diabetes mellitus is a complicated disease that can lead to disturbances in carbohydrate, fat, and protein metabolism and in fluid and acid-base balance. It can also have repercussions on the circulatory system, kidneys, respiratory system, and nervous system.

In addition to these potential consequences of untreated diabetes, which can be explained on the basis of insulin's short-term metabolic effects, numerous long-range complications of this disease frequently occur after 15 to 20 years despite treatment to prevent the short-term effects. These chronic complications, which account for the shorter life expectancy of diabetics, primarily involve degenerative disorders of the vasculature and nervous system. Cardiovascular lesions are the most common cause of premature death in diabetics. Heart disease and strokes occur with greater incidence than in nondiabetics. Because vascular lesions often develop in the kidneys and retinas of the eyes, kidney failure is a very common long-term complication of diabetes, and diabetes is the second leading cause of blindness in the United States. Impaired delivery of blood to the extremities may cause these

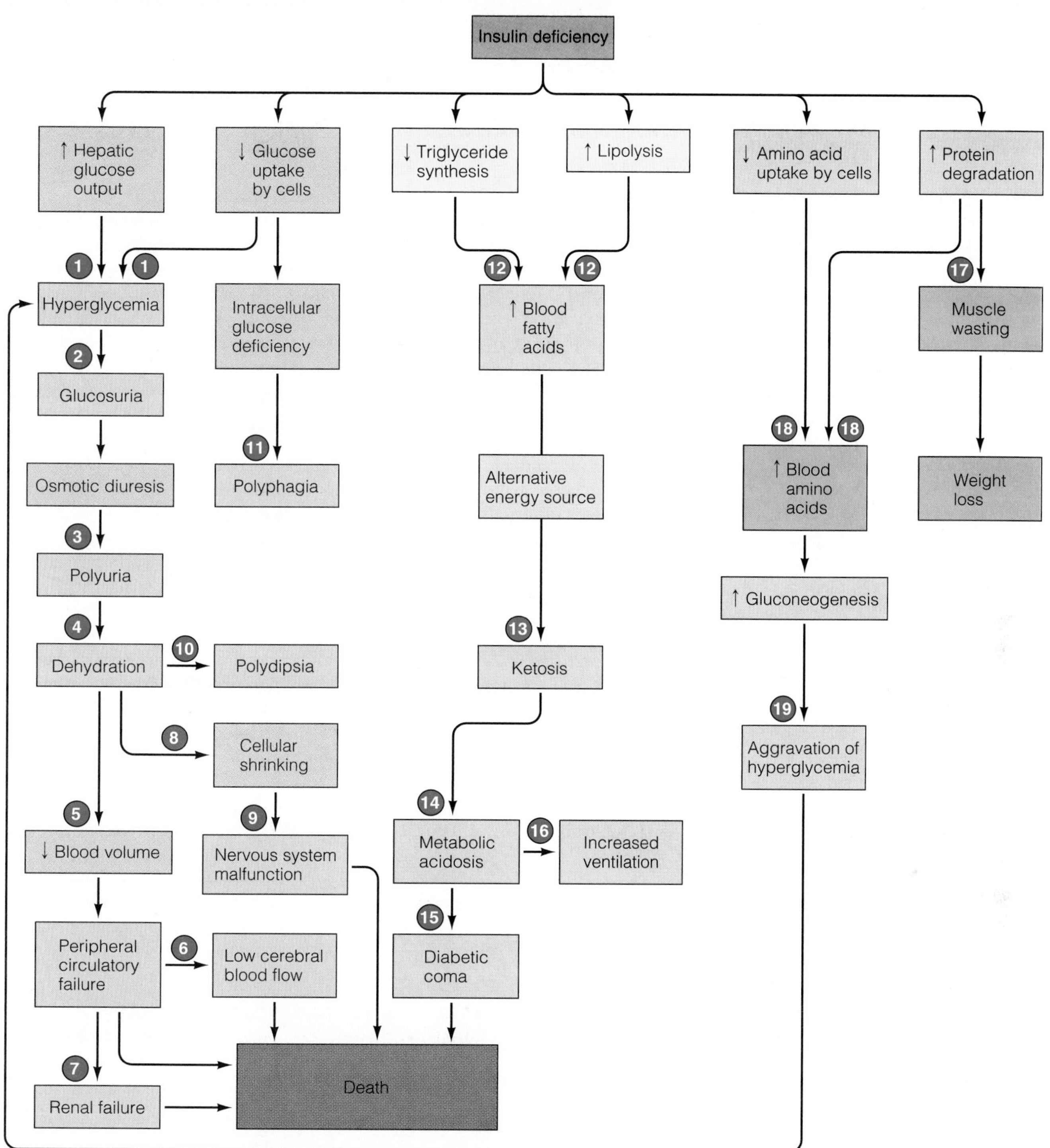

Figure 19-17 Acute Effects of Diabetes Mellitus The acute consequences of diabetes mellitus can be grouped according to the effects of inadequate insulin action on carbohydrate, fat, and protein metabolism. These effects ultimately cause death through a variety of pathways. See p. 680 for an explanation of the circled numbers.

tissues to become gangrenous, and toes or even whole limbs may have to be amputated as a result. In addition to the problems related to the circulatory system, degenerative lesions in nerves lead to multiple neuropathies that result in dysfunction of the brain, spinal cord, and peripheral nerves.

Regular exposure of tissues to excess blood glucose over a prolonged time leads to tissue alterations responsible for the development of these long-range vascular and neural degenerative complications. Thus, the best management for diabetes mellitus is to continuously keep blood glucose levels within normal limits to diminish the incidence of these chronic abnormalities. However, the blood glucose levels of diabetic patients on traditional therapy typically fluctuate over a broader range than normal, exposing their tissues to a moderately elevated blood glucose during a portion of each day.

Diabetics and Insulin: Some Have It and Some Don't

There are two distinct types of diabetes mellitus (see the accompanying table). **Type I (insulin-dependent** or **juvenile-onset) diabetes mellitus,** which accounts for 10% to 20% of all cases of diabetes, is characterized by a lack of insulin secretion. In **Type II (non-insulin-dependent** or **maturity-onset) diabetes mellitus,** insulin secretion may be normal or even increased, but insulin's target cells are less sensitive than normal to this hormone. Although either type can first be manifested at any age, Type I has a greater prevalence in children, whereas the onset of Type II more generally occurs in adulthood, giving rise to the age-related designations of the two conditions. Genetic as well as environmental factors appear to be important in the development of both types of diabetes mellitus.

Type I diabetes is an autoimmune process involving the erroneous, selective destruction of the pancreatic β cells by inappropriately activated T lymphocytes (see p. 401). Because Type I diabetics suffer a total or near-total lack of insulin secretion by their pancreatic β cells, they require exogenous insulin for survival. This dependence is the basis for the alternative name insulin-dependent diabetes mellitus for this form of the disease.

Type II diabetics, on the other hand, do secrete insulin in varying amounts. In fact, insulin levels may be normal or even exceed those in nondiabetics. The basic problem in Type II diabetes is not lack of insulin but reduced sensitivity of insulin's target cells to its presence, usually because of down-regulation (see p. 635) of insulin receptors in association with obesity. Chronic overeating by an obese person results in the secretion of increased amounts of insulin to maintain the blood glucose at normal levels by putting the excess nutrients in storage. In response to chronic hyperinsulinemia, the number of insulin receptors gradually becomes reduced over time. The resultant decrease in sensitivity to insulin in obese but otherwise normal individuals is overcome by secretion of additional insulin. In this way, the excess nutrients are stored despite the decreased availability of insulin receptors, so blood glucose homeostasis is maintained. In obese diabetes-prone individuals, however, the sustained overtaxing of the pancreas by chronic excessive nutrient intake eventually exceeds the reserve secretory capacity of the genetically weak β cells. Even though insulin secretion may be normal or somewhat elevated, symptoms of insulin insufficiency develop because the amount of insulin is still inadequate to prevent significant hyperglycemia in the presence of excess nutrient absorption.

The symptoms in Type II diabetes are usually slower in onset and less severe than in Type I diabetes. Whereas Type I diabetics are permanently insulin dependent, dietary control and weight reduction may be all that is necessary to completely reverse the symptoms in Type II diabetics. Therefore, Type II diabetes is alternatively known as non-insulin-dependent diabetes. As the magnitude of insulin secretion decreases in connection with reduced caloric intake, the number of insulin receptors gradually returns to normal and so too does target-tissue responsiveness to insulin. Exercise is also useful in the management of both types of diabetes, because working muscles are not insulin dependent. Exercising muscles take up and use some of the excess glucose in the blood, thus reducing the overall need for insulin.

Occasionally, Type II diabetics are administered drugs that remove excess nutrients from the blood by driving the weakened β cells to secrete more insulin than they do on their own. These therapeutic agents are referred to as *oral hypoglycemic drugs* since they can be taken orally to bring about a reduction in blood glucose. In contrast, insulin must be administered by injection because this peptide hormone would be digested by proteolytic enzymes if it were to be swallowed. Although oral hypoglycemics alleviate the symptoms of diabetes, their use is controversial because they overwork already weakened β cells; the end result may be the complete "burnout" of these cells so that they are no longer able to produce insulin at all. In such a case, the previously non-insulin-dependent patient would have to be placed on insulin therapy for life. Weight reduction and exercise are safer and more permanent solutions.

Comparison of Type I and Type II Diabetes Mellitus

Characteristic	*Type I Diabetes*	*Type II Diabetes*
Level of insulin secretion	None or almost none	May be normal or exceed normal
Typical age of onset	Childhood	Adulthood
Percentage of diabetics	10–20%	80–90%
Basic defect	Destruction of β cells	Reduced sensitivity of insulin's target cells
Associated with obesity?	No	Usually
Genetic and environmental factors important in precipitating overt disease?	Yes	Yes
Speed of development of symptoms	Rapid	Slow
Development of ketosis	Common if untreated	Rare
Treatment	Insulin injections; dietary management	Dietary control and weight reduction; occasionally oral hypoglycemic drugs

Treatment of diabetes has traditionally depended on insulin injections (for Type I), dietary management, weight control, and exercise. Because exercising muscles take up and use some of the excess glucose in the blood, the overall need for insulin is reduced. Current research on several fronts may dramatically change the approach to diabetic therapy. New treatments on the horizon that promise more precise control of blood glucose levels include an implanted, glucose-sensitive, insulin-releasing "artificial pancreas" and pancreatic islet transplants. Scientists have developed several types of devices that isolate donor islet cells from the recipient's immune system. The islet cells are enclosed by a selectively permeable membrane that permits passage of nutrients, O_2, electrolytes, and insulin but excludes white blood cells, antibodies, and other immune effectors. In experimental diabetic animals, such encapsulated islet cells have been shown to secrete insulin in precise relation to the blood glucose, but some technical problems must be resolved before the technique is available for use in humans. Such immunoisolation of islet cells would permit use of xenografts (grafts from other animals; see p. 6), thus circumventing the shortage of human donor cells. Pig islet cells would be an especially good source because porcine insulin is nearly identical to human insulin and these cells are readily available from slaughterhouses. Furthermore, recent advances in understanding the underlying molecular defects in diabetes have even made investigators hopeful that safe therapies may be developed within this decade to prevent new cases of diabetes.

Insulin excess causes brain-starving hypoglycemia.

Let us now look at the opposite of diabetes mellitus, insulin excess, which is characterized by **hypoglycemia** (low blood glucose) and can occur in two different ways. First, insulin excess can occur in a diabetic patient when two much insulin has been injected for the person's caloric intake and exercise level, resulting in so-called **insulin shock.** Second, an abnormally high blood insulin level may occur in a nondiabetic individual who has a β cell tumor or whose β cells are overresponsive to glucose. The true incidence of the latter condition, so-called **reactive hypoglycemia,** is a subject of intense controversy, because laboratory measurements to confirm the presence of low blood glucose during the time of symptoms have not been performed in the majority of people who have been diagnosed as having the condition. In mild cases, the symptoms of hypoglycemia, such as tremor, fatigue, sleepiness, and inability to concentrate, are nonspecific. Because these symptoms could also be attributable to emotional problems or other factors, a definitive diagnosis based on symptoms alone is impossible to make.

The consequences of insulin excess are primarily manifestations of the effects of hypoglycemia on the brain. Recall that the brain relies on a continuous supply of blood glucose for its nourishment and that glucose uptake by the brain does not depend on insulin. With insulin excess, more glucose than necessary is driven into the other insulin-dependent cells of the body. The result is a lowering of the blood glucose level so that not enough glucose is left in the blood to be delivered to the brain. The brain literally starves in the presence of hypoglycemia. The symptoms, therefore, are primarily referable to depressed brain function, which, if severe enough, may rapidly progress to unconsciousness and death. Persons with overresponsive β cells usually do not become sufficiently hypoglycemic to manifest these more serious consequences, but they do show milder symptoms of depressed CNS activity.

The treatment of hypoglycemia depends on the cause. In the case of a diabetic with insulin overdose, something sugary should be taken at the first indication of a hypoglycemic attack. Prompt treatment of severe hypoglycemia is imperative if brain damage is to be prevented. Note that a diabetic can lose consciousness and die from either diabetic ketoacidotic coma caused by prolonged insulin deficiency or acute hypoglycemia caused by insulin shock. Fortunately, the other accompanying signs and symptoms differ sufficiently between the conditions to enable medical caretakers to administer appropriate therapy, either insulin or glucose. For example, ketoacidotic coma is accompanied by deep, labored breathing (a compensatory measure for the metabolic acidosis) and fruity breath (due to exhalation of ketone bodies), whereas insulin shock is not.

Ironically, even though reactive hypoglycemia is characterized by a low blood glucose level, persons with this disorder are treated by limiting their intake of sugar and other glucose-yielding carbohydrates to prevent their β cells from overresponding to a high glucose intake. With low carbohydrate intake, the blood glucose does not rise as much during the absorptive state. Since blood glucose elevation is the primary regulator of insulin secretion, the β cells are not stimulated as much with a low-carbohydrate meal as with a typical meal. Accordingly, reactive hypoglycemia is less likely to occur. Giving a symptomatic individual with reactive hypoglycemia something sugary, as is done with a hypoglycemic diabetic, temporarily alleviates the symptoms, because the blood glucose level is transiently restored to normal so that the brain's energy needs are once again satisfied. As soon as the extra glucose triggers further insulin release, however, the situation is merely aggravated.

Glucagon in general opposes the actions of insulin.

Even though insulin plays a central role in controlling the metabolic adjustments between the absorptive and postabsorptive states, the secretory product of the pancreatic islet α cells, **glucagon,** is also very important. Many physiologists view the insulin-secreting β cells and the glucagon-secreting α cells as a coupled endocrine system whose combined secretory output is a major factor in the regulation of fuel metabolism.

Glucagon affects many of the same metabolic processes that are influenced by insulin, but in most cases glucagon's actions are opposite to those of insulin. The major site of action of glucagon is the liver, where it exerts a variety of effects on carbohydrate, fat, and protein metabolism.

Actions on carbohydrate The overall effects of glucagon on carbohydrate metabolism result in an increase in hepatic glucose production and release and thus an increase in blood glucose levels. Glucagon exerts its hyperglycemic effects by decreasing glycogen synthesis, promoting glycogenolysis, and stimulating gluconeogenesis.

Actions on fat Glucagon also antagonizes the actions of insulin with regard to fat metabolism by promoting fat breakdown and inhibiting triglyceride synthesis. Glucagon enhances hepatic ketone production **(ketogenesis)** by promoting the conversion of fatty acids to ketone bodies. Thus, the blood levels of fatty acids and ketones increase under glucagon's influence.

Actions on protein Glucagon inhibits hepatic protein synthesis and promotes degradation of hepatic protein. Stimulation of gluconeogenesis further contributes to glucagon's catabolic effect on hepatic protein metabolism. Glucagon promotes protein catabolism in the liver, but it does not have any significant effect on blood amino acid levels because it does not affect muscle protein, the major protein store in the body.

Glucagon secretion is increased during the postabsorptive state.

Considering the catabolic effects of glucagon on the body's energy stores, you would be correct in assuming that glucagon secretion is increased during the postabsorptive state and decreased during the absorptive state, just the opposite of insulin secretion. In fact, insulin is sometimes referred to as a "hormone of feasting" and glucagon as a "hormone of fasting." Insulin tends to put nutrients in storage when their blood levels are high, such as following a meal, whereas glucagon promotes catabolism of nutrient stores between meals to keep up the blood nutrient levels, especially blood glucose.

Like insulin secretion, the major factor regulating glucagon secretion is a direct effect of the blood glucose concentration on the endocrine pancreas. In this case, the pancreatic α cells increase glucagon secretion in response to a fall in blood glucose. The hyperglycemic actions of this hormone tend to restore the blood glucose level to normal. Conversely, an increase in blood glucose concentration, such as occurs after a meal, inhibits glucagon secretion, which likewise tends to restore the blood glucose level to normal. Thus, there is a direct negative-feedback relationship between blood glucose concentration and the α cells' rate of secretion, but it is in the opposite direction of the effect of blood glucose on the β cells; in other words, an elevated blood glucose level inhibits glucagon secretion but stimulates insulin secretion, whereas a fall in blood glucose level leads to increased glucagon secretion and decreased insulin secretion (Fig. 19-18). Since glucagon raises blood glucose and insulin decreases blood glucose, the changes in secretion of these pancreatic hormones in response to deviations in blood glucose work together homeostatically to restore blood glucose levels to normal. Similarly, a fall in blood fatty acid concentration directly stimulates glucagon output and inhibits insulin output by the pancreas, both of which are negative-feedback control mechanisms to restore the blood fatty acid level to normal.

The opposite effects exerted by the blood concentrations of glucose and fatty acids on the pancreatic α and β cells are appropriate for regulating the circulating levels of these nutrient molecules, because the actions of insulin and glucagon on carbohydrate and fat metabolism oppose one another. The effect of blood amino acid concentration on the secretion of these two hormones is a different story. A rise in blood amino acid concentration stimulates *both* glucagon and insulin secretion. Why this seeming paradox, since glucagon does not exert any effect on blood amino acid concentration? The identical effect of high blood amino acid levels on both glucagon and insulin secretion makes sense if you consider the concomitant effects these two hormones have on blood glucose levels (Fig. 19-19). If during absorption of a protein-rich meal the rise in blood amino acids stimulated only insulin secretion, hypoglycemia might result. Because little carbohydrate is available for absorption following consumption of a high-protein meal, the amino acid–induced increase in insulin secretion would drive too much glucose into the cells, causing a sudden, inappropriate drop in the blood glucose level. However, the simultaneous increase in glucagon secretion elicited by elevated blood amino acid levels increases hepatic glucose production. Since the hyperglycemic effects of glucagon counteract the hypoglycemic actions of insulin, the net result is maintenance of normal blood glucose levels (and prevention of hypoglycemic starvation of the brain) during absorption of a meal that is high in protein but low in carbohydrates.

Glucagon excess can aggravate the hyperglycemia of diabetes mellitus.

No known clinical abnormalities are attributable to glucagon deficiency or excess per se. However, diabetes mellitus is frequently accompanied by excess glucagon secretion, because insulin is required for glucose to gain entry into the α cells, where it can

Figure 19-18 Complementary Interactions of Glucagon and Insulin

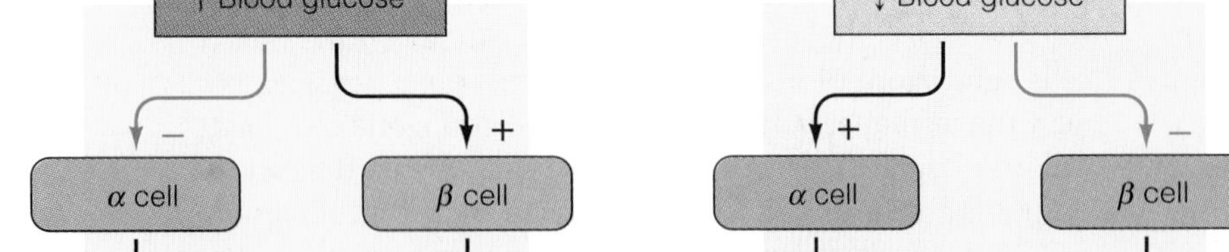
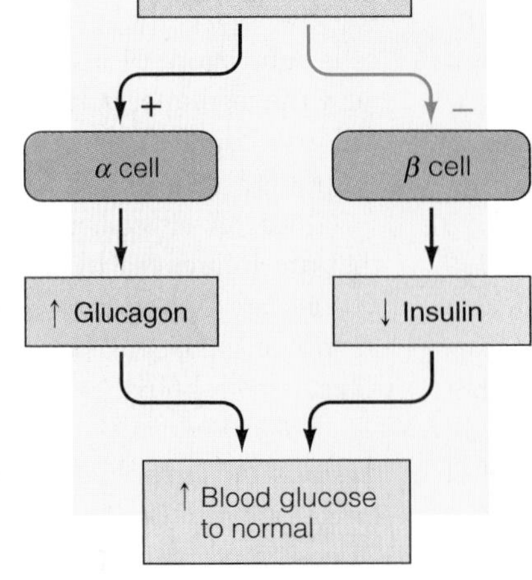

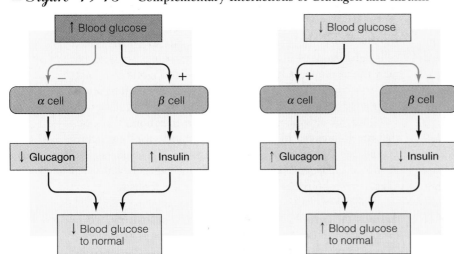

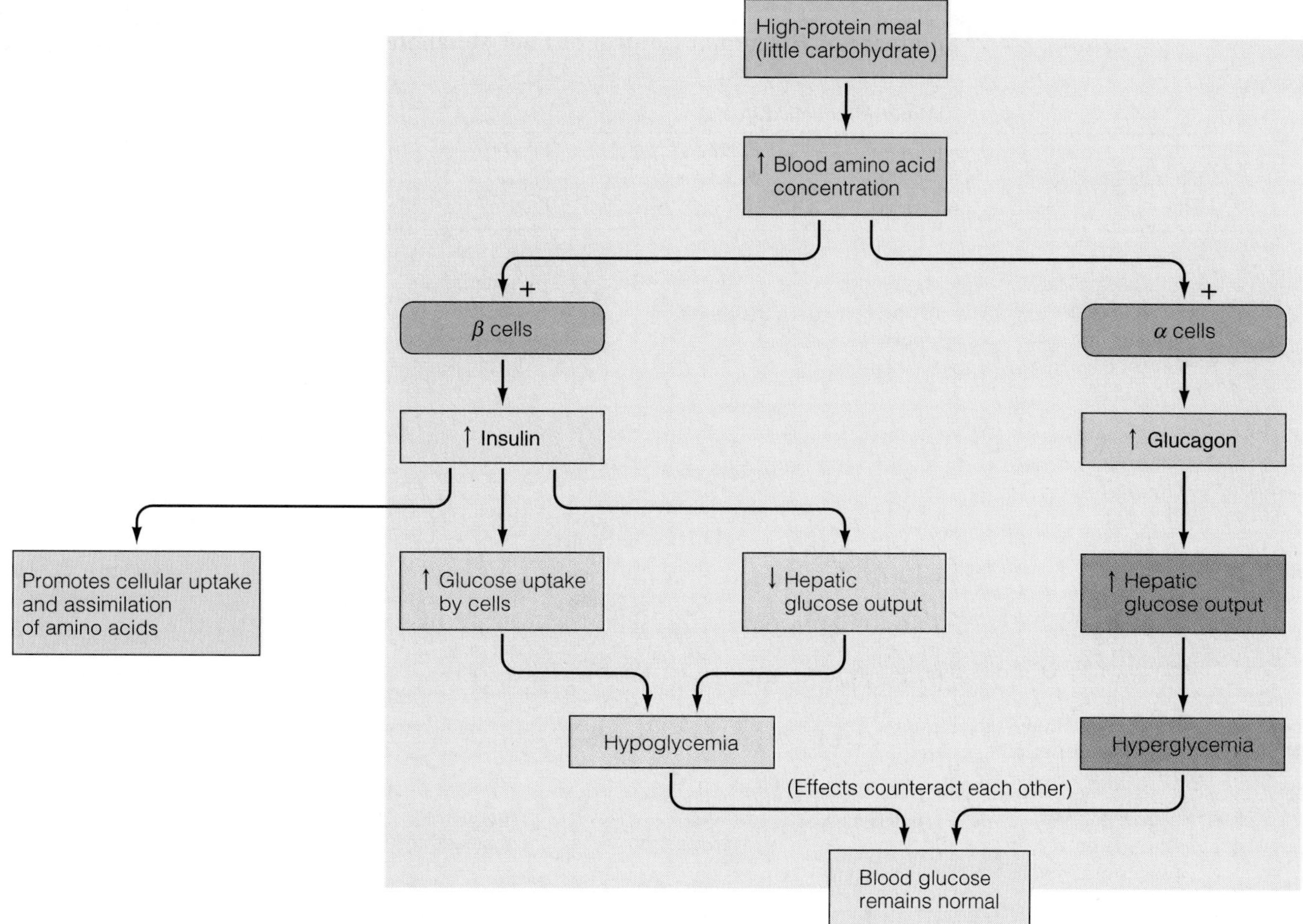

Figure 19-19 **Counteracting Actions of Glucagon and Insulin on Blood Glucose during Absorption of a High-Protein Meal**

exert control over glucagon secretion. As a result, diabetics frequently have a high rate of glucagon secretion concurrent with their insulin insufficiency because the elevated blood glucose is not able to inhibit glucagon secretion as it normally would. Since glucagon is a hormone that raises blood glucose, its excess intensifies the hyperglycemia of diabetes mellitus. For this reason, some insulin-dependent diabetics respond best to a combination of insulin and somatostatin therapy. By inhibiting glucagon secretion, somatostatin indirectly helps to achieve better reduction of the elevated blood glucose concentration than can be accomplished by insulin therapy alone.

Epinephrine, cortisol, growth hormone, and thyroid hormone also exert direct metabolic effects.

The pancreatic hormones are the most important regulators of normal fuel metabolism. However, several other hormones exert direct metabolic effects even though control of their secretion is keyed to factors other than transitions in metabolism between feasting and fasting states (Table 19-7).

The stress hormones, epinephrine and cortisol, both increase blood levels of glucose and fatty acids through a variety of metabolic effects. In addition, cortisol mobilizes amino acids by promoting protein catabolism. Neither of these hormones plays important roles in the regulation of fuel metabolism under resting conditions, but they are important for the metabolic responses to stress. Cortisol also appears to contribute to the maintenance of blood glucose concentration during long-term starvation.

Growth hormone has protein anabolic effects in muscle. In fact, this is one of its growth-promoting features. Although growth hormone can elevate the blood levels of glucose and fatty acids, it is normally of little importance to the overall regulation of fuel metabolism. Deep sleep, stress, exercise, and severe hypoglycemia stimulate growth hormone secretion, possibly to provide fatty acids as an energy source and spare glucose for the brain under these circumstances. Growth hormone, like cortisol, appears to contribute to the maintenance of blood glucose concentrations during starvation.

The effects of thyroid hormone on intermediary metabolism are decidedly mixed. It stimulates both anabolic and catabolic actions as well as the overall metabolic rate. Catabolic effects tend to dominate at higher thyroid hormone levels. However, changes in thyroid hormone secretion are usually not important for fuel homeostasis for two reasons. First, control of thyroid hormone secretion is not directed toward maintenance of nutrient levels in the blood. Second,

Table 19-7 **Summary of Hormonal Control of Fuel Metabolism**

	Major Metabolic Effects				Control of Secretion	
Hormone	*Effect on Blood Glucose*	*Effect on Blood Fatty Acids*	*Effect on Blood Amino Acids*	*Effect on Muscle Protein*	*Major Stimuli for Secretion*	*Primary Role in Metabolism*
Insulin	↓ +Glucose uptake +Glycogenesis −Glycogenolysis −Gluconeogenesis	↓ +Triglyceride synthesis −Lipolysis	↓ +Amino acid uptake	↑ +Protein synthesis −Protein degradation	↑ Blood glucose ↑ Blood amino acids	Primary regulator of absorptive and postabsorptive cycles
Glucagon	↑ +Glycogenolysis +Gluconeogenesis −Glycogenesis	↑ +Lipolysis −Triglyceride synthesis	No effect	No effect	↓ Blood glucose ↑ Blood amino acids	Regulation of absorptive and postabsorptive cycles in concert with insulin; protection against hypoglycemia
Epinephrine	↑ +Glycogenolysis +Gluconeogenesis −Insulin secretion +Glucagon secretion	↑ +Lipolysis	No effect	No effect	Sympathetic stimulation during stress and exercise	Provision of energy for emergencies and exercise
Cortisol	↑ +Gluconeogenesis −Glucose uptake by tissues other than brain; glucose sparing	↑ +Lipolysis	↑ +Protein degradation	↓ +Protein degradation	Stress	Mobilization of metabolic fuels and building blocks during adaptation to stress
Growth hormone	↑ −Glucose uptake by muscles; glucose sparing	↑ +Lipolysis	↓ +Amino acid uptake	↓ +Protein synthesis −Protein degradation +Synthesis of DNA and RNA	Deep sleep Stress Exercise Hypoglycemia	Promotion of growth; normally little role in metabolism; mobilization of fuels plus glucose sparing in extenuating circumstances

the onset of thyroid hormone action is too slow to have any significant effect on the rapid adjustments required to maintain normal blood levels of nutrients. During starvation, less T_4 is converted to the more potent T_3. This shift in the pattern of circulating thyroid hormone is probably in large part responsible for the reduction in basal metabolic rate that usually occurs with prolonged fasting.

Note that, with the exception of the anabolic effects of growth hormone on protein metabolism, all of the metabolic actions of these other hormones are opposite to those of insulin. Insulin alone is able to reduce blood glucose and blood fatty acid levels, whereas glucagon, epinephrine, cortisol, and growth hormone all increase blood levels of these nutrients. Because of this, these other hormones are considered **insulin antagonists.** Thus, the main reason diabetes mellitus has such devastating metabolic consequences is that no other control mechanism is available to pick up the slack to promote anabolism when insulin activity is insufficient, so the catabolic reactions promoted by other hormones are allowed to proceed unchecked. The only exception is protein anabolism stimulated by growth hormone.

Endocrine Control of Calcium Metabolism

Plasma calcium must be closely regulated to prevent changes in neuromuscular excitability.

Besides regulating the concentration of organic nutrient molecules in the blood by manipulation of anabolic and catabolic pathways, the endocrine system also regulates the plasma concentration of a number of inorganic electrolytes. As you already know, aldosterone controls Na^+ and K^+ concentrations in the ECF. Three other hormones—*parathyroid hormone, calcitonin,* and *vitamin D*—control calcium (Ca^{2+}) and phosphate (PO_4^{3-}) metabolism. These hormonal agents concern themselves with regulation of plasma Ca^{2+}, and in the process, plasma PO_4^{3-} is also maintained. Plasma Ca^{2+} concentration is one of the most tightly controlled variables in the body. The need for the precise regulation of plasma Ca^{2+} stems from its critical influence on so many body activities.

About 99% of the Ca^{2+} in the body is in crystalline form within the skeleton and teeth. Of the remaining 1%, about 0.9% is found intracellularly within the soft tissues; less than 0.1% is present in the ECF. Approximately half of the plasma Ca^{2+} either is bound to plasma proteins and therefore restricted to the plasma or is complexed with PO_4^{3-} and not free to participate in chemical reactions. The other half of the plasma Ca^{2+} is freely diffusible and can readily pass into the interstitial fluid and interact with the cells. The free Ca^{2+} in the plasma and interstitial fluid is considered to be a single pool. Only this free Ca^{2+} is biologically active and subject to regulation; it constitutes less than one-thousandth of the total Ca^{2+} in the body.

This small, freely diffusible fraction of ECF Ca^{2+} plays a vital role in a number of essential activities, including the following:

1. *Neuromuscular excitability.* Even minor variations in the concentration of free ECF Ca^{2+} can have a profound and immediate impact on the sensitivity of excitable tissues. A fall in free Ca^{2+} results in overexcitability of nerves and muscles; conversely, a rise in free Ca^{2+} depresses neuromuscular excitability. These effects result from the influence of Ca^{2+} on membrane permeability to Na^+. A decrease in free Ca^{2+} increases Na^+ permeability, with the resultant influx of Na^+ moving the resting potential closer to threshold. Consequently, in the presence of **hypocalcemia** (low blood Ca^{2+}), excitable tissues may be brought to threshold by normally ineffective physiological stimuli, so that skeletal muscles discharge and contract (go into spasm) "spontaneously" (in the absence of normal stimulation). If severe enough, spastic contraction of the respiratory muscles results in death by asphyxiation. **Hypercalcemia** (elevated blood Ca^{2+}), on the other hand, is also life-threatening because it causes cardiac arrhythmias accompanied by generalized depression of neuromuscular excitability.
2. *Excitation-contraction coupling in cardiac and smooth muscle.* Entry of ECF Ca^{2+} into cardiac and smooth muscle cells, resulting from increased Ca^{2+} permeability in response to an action potential, triggers the contractile mechanism. Calcium is also necessary for excitation-contraction coupling in skeletal muscle fibers, but in this case the Ca^{2+} is released from intracellular Ca^{2+} stores in response to an action potential. A significant part of the increase in cytosolic Ca^{2+} in cardiac muscle cells is also derived from internal stores.

   Note that a *rise in cytosolic Ca^{2+}* within a muscle cell causes contraction, whereas an *increase in free ECF Ca^{2+}* decreases neuromuscular excitability and reduces the likelihood of contraction occurring. Unless this point is kept in mind, it is difficult to understand why low plasma Ca^{2+} levels induce muscle hyperactivity when Ca^{2+} is necessary to switch on the contractile apparatus. We are talking about two different Ca^{2+} pools, which exert different effects.
3. *Stimulus-secretion coupling.* The entry of Ca^{2+} into secretory cells, which results from increased permeability to Ca^{2+} in response to appropriate stimulation, triggers the release of the secretory product by exocytosis. This process is important for the secretion of neurotransmitters by nerve cells and for peptide and catecholamine hormone secretion by endocrine cells.
4. *Maintenance of tight junctions between cells.* Calcium forms part of the intercellular cement that holds particular cells tightly together.
5. *Clotting of blood.* Calcium serves as a cofactor in several steps of the cascade of reactions that lead to clot formation.

Because of the profound effects of deviations in free Ca^{2+}, especially on neuromuscular excitability, the plasma concentration of this electrolyte is regulated with extraordinary precision. In addition, intracellular Ca^{2+} (not the free ECF Ca^{2+}) serves as a second messenger in many cells and is involved in cell motility and cilia action. Finally, the Ca^{2+} in bone and teeth is essential for the structural and functional integrity of these tissues.

Control of calcium metabolism includes regulation of both calcium homeostasis and calcium balance.

Maintenance of the proper plasma concentration of free Ca^{2+} differs from regulation of Na^+ and K^+ in two important regards. Sodium and K^+ homeostasis is maintained primarily by regulating the urinary excretion of these electrolytes so that controlled output matches uncontrolled input. In the case of Ca^{2+}, in contrast, not all the ingested Ca^{2+} is absorbed from the digestive tract; instead, the extent of absorption is hormonally controlled and is dependent on the Ca^{2+} status of the body. In addition, bone serves as a large Ca^{2+} reservoir that can be drawn on to maintain the free plasma Ca^{2+} concentration within the narrow limits compatible with life should dietary intake become too low. Exchange of Ca^{2+} between the ECF and bone is also subject to control. Similar inhouse stores are not available for Na^+ and K^+.

Regulation of Ca^{2+} metabolism depends on hormonal control of exchanges between the ECF and three other compartments: bone, kidneys, and intestine. Accordingly, control of Ca^{2+} metabolism encompasses two aspects. First, regulation of **calcium homeostasis** involves the immediate adjustments required to maintain a *constant free plasma Ca^{2+} concentration* on a minute-to-minute basis. This is largely accomplished by rapid exchanges between the bone and ECF and to a lesser extent by modifications in urinary excretion of Ca^{2+}. Second, regulation of **calcium balance** involves the more

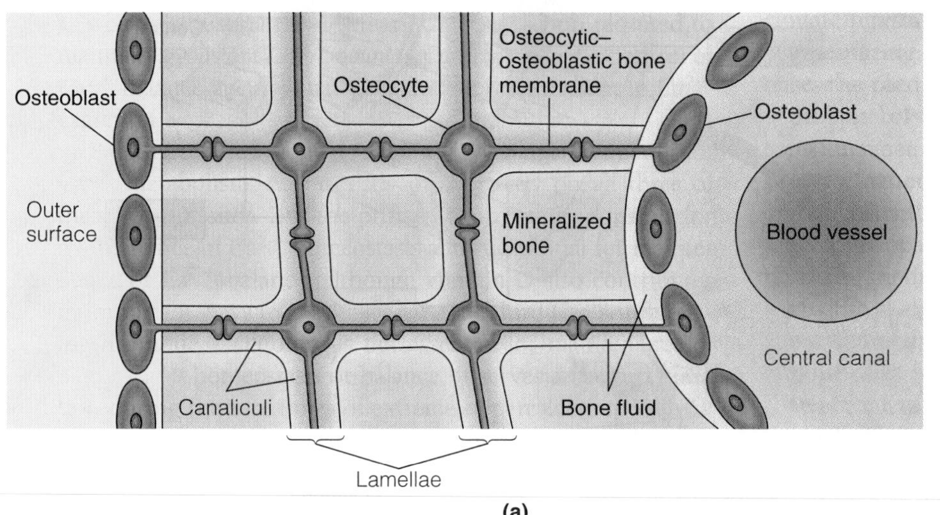

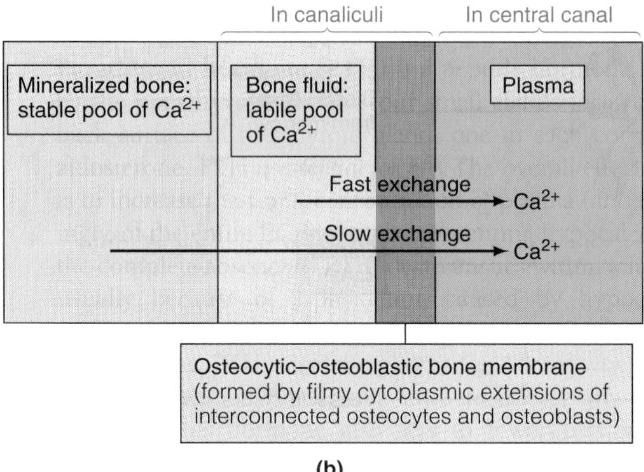

Figure 19-21 Relationship of Mineralized Bone, Bone Cells, Bone Fluid, and the Blood (a) Schematic representation of the osteocytic-osteoblastic bone membrane. The entombed osteocytes and surface osteoblasts are interconnected by long cytoplasmic processes that extend from these cells and connect to each other within the canaliculi. This interconnecting cell network, the osteocytic-osteoblastic bone membrane, separates the mineralized bone from the plasma in the central canal. Bone fluid lies between the membrane and the mineralized bone. (b) Schematic representation of fast and slow exchange of Ca^{2+} between the bone and the plasma. In a fast exchange, Ca^{2+} is moved from the labile pool in the bone fluid into the plasma by means of PTH-activated Ca^{2+} pumps in the osteocytic-osteoblastic bone membrane. In a slow exchange, Ca^{2+} is moved from the stable pool in the mineralized bone into the plasma by means of PTH-induced dissolution of the bone.

Under conditions of chronic hypocalcemia, such as might occur with dietary Ca^{2+} deficiency, PTH influences the slow exchange of Ca^{2+} between the bone itself and the ECF by promoting actual localized dissolution of bone. It does so by stimulating osteoclasts to gobble up bone, increasing the formation of more osteoclasts, and transiently inhibiting the bone-forming activity of the osteoblasts. Bone contains such a great abundance of Ca^{2+} in comparison to the plasma (more than 1,000 times as much) that even when PTH promotes increased bone resorption, there are no immediate discernible effects on the skeleton because such a minute amount of bone is affected. Yet the negligible amount of Ca^{2+} "borrowed" from the bone bank can be lifesaving in terms of restoring the free plasma Ca^{2+} level to normal. The borrowed Ca^{2+} is then redeposited in the bone at another time when Ca^{2+} supplies are more abundant. Meanwhile, the plasma Ca^{2+} level has been maintained without sacrificing the integrity of the bone. However, prolonged excess PTH secretion over months or years is eventually evidenced by the formation of cavities throughout the skeleton that are filled with very large, overstuffed osteoclasts.

When PTH promotes dissolution of the calcium phosphate crystals in bone to harvest their Ca^{2+} content, both Ca^{2+} and PO_4^{3-} are released into the plasma. An elevation in plasma PO_4^{3-} is undesirable, but PTH deals with this dilemma by its actions on the kidneys.

Actions on kidneys Parathyroid hormone stimulates Ca^{2+} conservation and promotes PO_4^{3-} elimination by the kidneys during the formation of urine. Under the influence of PTH, the kidneys are able to reabsorb more of the filtered Ca^{2+}, so less Ca^{2+} escapes into the urine. This effect increases the plasma Ca^{2+} level and decreases urinary Ca^{2+} losses. (It would be futile to dissolve bone to obtain more Ca^{2+} only to lose it in the urine).

At the same time that it stimulates renal Ca^{2+} reabsorption, PTH decreases PO_4^{3-} reabsorption, thus increasing urinary PO_4^{3-} excretion. As a result, PTH causes a fall in plasma PO_4^{3-} levels at the same time it increases Ca^{2+} concentrations.

This PTH-induced removal of extra PO_4^{3-} from the body fluids is essential for preventing reprecipitation of the Ca^{2+} freed from the bone. Because of the solubility characteristics of calcium phosphate salt, the product of the plasma concentration of Ca^{2+} times the plasma concentration of PO_4^{3-} must remain an approximately constant value. Therefore, an inverse relationship exists between the plasma concentrations of Ca^{2+} and PO_4^{3-}; for example, when the plasma PO_4^{3-} level rises, some of the plasma Ca^{2+} is forced back into the bone through hydroxyapatite crystal formation, thus reducing the plasma Ca^{2+} level and keeping the calcium phosphate product constant. This inverse relationship occurs because the free ions in the ECF are in equilibrium with the bone crystals.

Recall that both Ca^{2+} and PO_4^{3-} are released from the bone when PTH promotes bone dissolution. Because PTH is secreted only when plasma Ca^{2+} falls below normal, the released Ca^{2+} is needed to restore plasma Ca^{2+} to normal, yet the released PO_4^{3-} tends to increase plasma PO_4^{3-} levels above normal. If the plasma PO_4^{3-} level were allowed to rise above normal, some of the plasma Ca^{2+} would have to be redeposited back in the bone along with the PO_4^{3-} in order to keep the calcium phosphate product constant. This redeposition of Ca^{2+} would lower plasma Ca^{2+}, just the opposite of the desired effect. Therefore, PTH acts on the kidneys to decrease the reabsorption of PO_4^{3-} by the renal tubules. This increases the urinary excretion of PO_4^{3-} and lowers its plasma concentration, even though extra PO_4^{3-} is being released from the bone into the blood. Such action prevents the self-defeating redeposition of released Ca^{2+} back into the bone.

The third important action of PTH on the kidneys (besides increasing Ca^{2+} reabsorption and decreasing PO_4^{3-} reabsorption) is to enhance the activation of vitamin D by the kidneys.

Action on the intestine Although PTH has no direct effect on the intestine, it indirectly increases both Ca^{2+} and PO_4^{3-} reabsorption from the small intestine by means of its role in vitamin D activation. This vitamin, in turn, directly increases intestinal absorption of Ca^{2+} and PO_4^{3-}.

The primary regulator of PTH secretion is the plasma concentration of free calcium.

All the effects of PTH are aimed at raising the plasma Ca^{2+} levels. Appropriately, PTH secretion is increased in response to a fall in plasma Ca^{2+} concentration and decreased by a rise in plasma Ca^{2+} levels. The secretory cells of the parathyroid glands are directly and exquisitely sensitive to changes in free plasma Ca^{2+}. Since PTH regulates plasma Ca^{2+} concentration, this relationship forms a simple negative-feedback loop for controlling PTH secretion without involving any nervous or other hormonal intervention (Fig. 19-22).

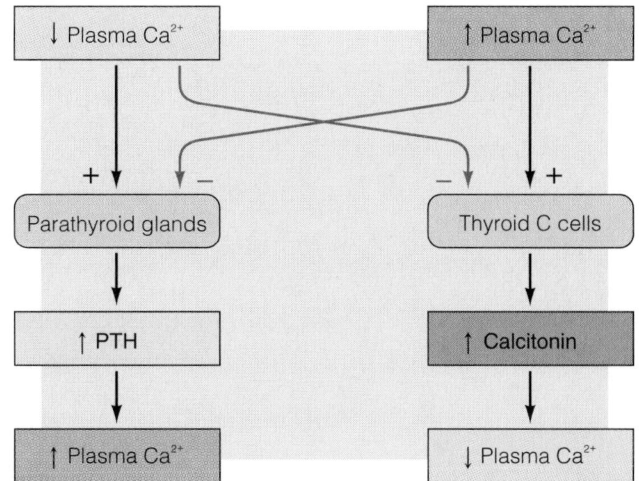

Figure 19-22 **Negative-Feedback Loops Controlling Parathyroid Hormone (PTH) and Calcitonin Secretion**

Calcitonin lowers the plasma calcium concentration but is not important in the normal control of calcium metabolism.

Calcitonin, the hormone produced by the C cells of the thyroid gland, also exerts an influence on plasma Ca^{2+} levels. Like, PTH, calcitonin has two effects on bone, but in this case, both effects *decrease* plasma Ca^{2+} levels. First, on a short-term basis, calcitonin decreases Ca^{2+} movement from the bone fluid into the plasma. Second, on a long-term basis, calcitonin decreases bone resorption by inhibiting the activity of osteoclasts. The suppression of bone resorption results in decreased plasma PO_4^{3-} levels as well as a reduced plasma Ca^{2+} concentration. The hypocalcemic and hypophosphatemic effects of calcitonin are due entirely to this hormone's actions on bone. It has no effect on the kidneys or intestine.

As with PTH, the primary regulator of calcitonin release is the free plasma Ca^{2+} concentration, but in contrast to its effect on PTH release, an increase in plasma Ca^{2+} stimulates calcitonin secretion and a fall in plasma Ca^{2+} inhibits calcitonin secretion (Fig. 19-22). Since calcitonin reduces plasma Ca^{2+} levels, this system constitutes a second simple negative-feedback control over plasma Ca^{2+} concentration, one that is opposed to the PTH system.

Most evidence suggests, however, that calcitonin plays little or no role in the normal control of Ca^{2+} or PO_4^{3-} metabolism. Although calcitonin will protect against hypercalcemia, this condition rarely occurs under normal circumstances. Moreover, neither thyroid removal nor calcitonin-secreting tumors alter circulating levels of Ca^{2+} or PO_4^{3-}, implying that this hormone is not normally essential to the maintenance of Ca^{2+} or PO_4^{3-} homeostasis. Calcitonin may, however, play a role in protecting skeletal integrity when there is a large Ca^{2+} demand, such as during pregnancy or breast-feeding.

Vitamin D is actually a hormone that increases calcium absorption in the intestine.

The final factor involved in the regulation of Ca^{2+} metabolism is **cholecalciferol**, or **vitamin D,** a steroidlike compound that is essential for Ca^{2+} absorption in the intestine. Strictly speaking, vitamin D should be considered a hormone because it can be produced in the skin from a precursor related to cholesterol (7-dehydrocholesterol) on exposure to sunlight. It is subsequently released into the blood to act at a distant target site, the intestine. The skin, therefore, is actually an endocrine gland and vitamin D a hormone. Traditionally, however, this chemical messenger has been considered a vitamin for two reasons. First, it was originally discovered and isolated from a dietary source and tagged as a vitamin. Second, even though the skin would be an adequate source of vitamin D if it were exposed to sufficient sunlight, indoor dwelling and clothing in response to cold weather and social customs preclude significant exposure of the skin to sunlight most of the time. At least part of the essential vitamin D must therefore be derived from dietary sources.

Regardless of its source, vitamin D is biologically inactive when it first enters the blood from either the skin or the digestive tract. It must be activated by two sequential biochemical alterations that involve the addition of two hydroxyl (−OH)

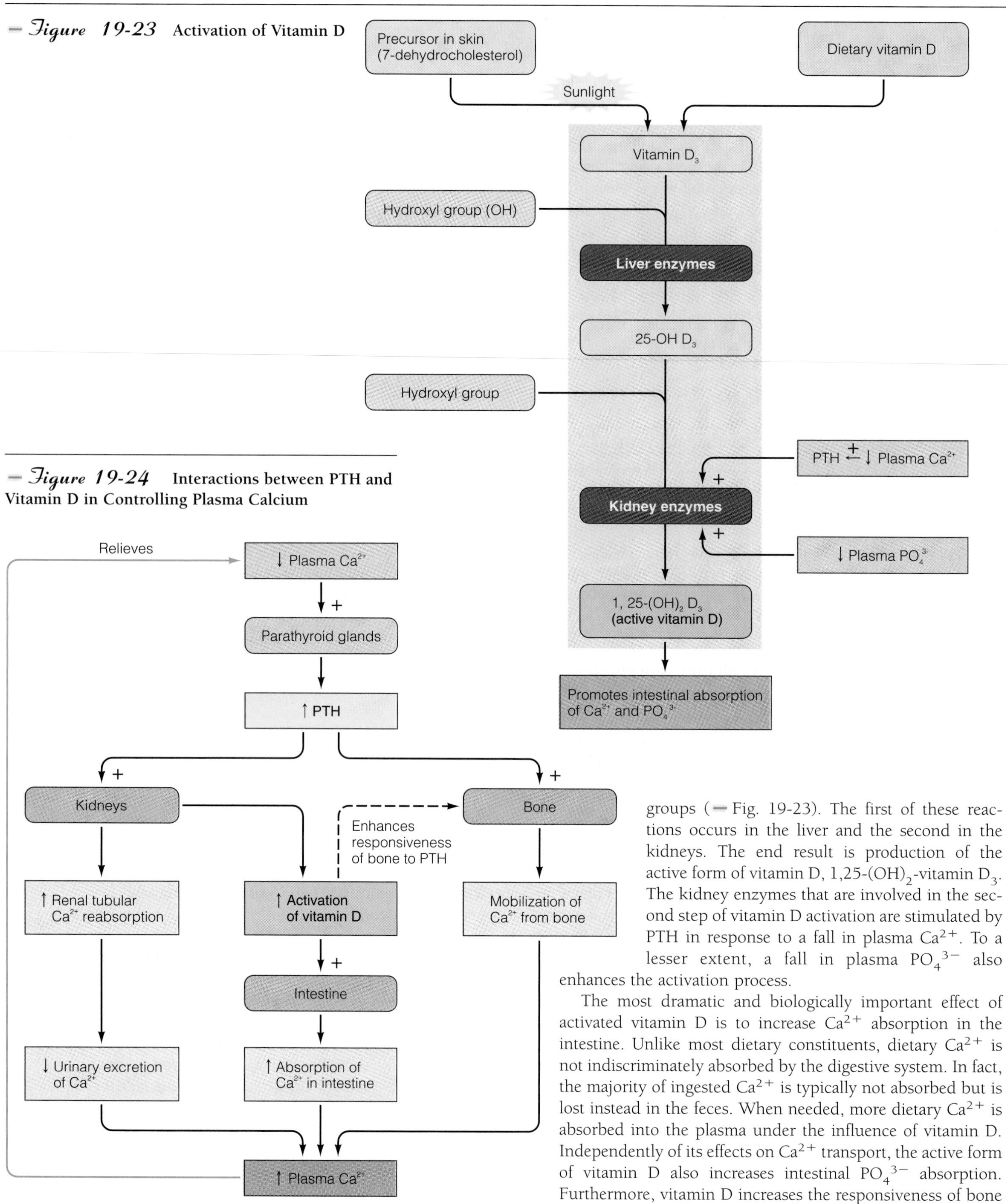

Figure 19-23 Activation of Vitamin D

Figure 19-24 Interactions between PTH and Vitamin D in Controlling Plasma Calcium

groups (Fig. 19-23). The first of these reactions occurs in the liver and the second in the kidneys. The end result is production of the active form of vitamin D, 1,25-$(OH)_2$-vitamin D_3. The kidney enzymes that are involved in the second step of vitamin D activation are stimulated by PTH in response to a fall in plasma Ca^{2+}. To a lesser extent, a fall in plasma PO_4^{3-} also enhances the activation process.

The most dramatic and biologically important effect of activated vitamin D is to increase Ca^{2+} absorption in the intestine. Unlike most dietary constituents, dietary Ca^{2+} is not indiscriminately absorbed by the digestive system. In fact, the majority of ingested Ca^{2+} is typically not absorbed but is lost instead in the feces. When needed, more dietary Ca^{2+} is absorbed into the plasma under the influence of vitamin D. Independently of its effects on Ca^{2+} transport, the active form of vitamin D also increases intestinal PO_4^{3-} absorption. Furthermore, vitamin D increases the responsiveness of bone to PTH. Thus, vitamin D and PTH have a closely interdependent relationship (Fig. 19-24).

Parathyroid hormone is principally responsible for controlling Ca^{2+} homeostasis, because the actions of vitamin D are too sluggish for it to contribute substantially to the minute-to-minute regulation of plasma Ca^{2+} concentration. However, both PTH and vitamin D are essential to Ca^{2+} balance, the process that ensures that, over the long term, Ca^{2+} input into the body is equivalent to Ca^{2+} output. When dietary Ca^{2+} intake is reduced, the resultant transient fall in plasma Ca^{2+} level stimulates PTH secretion. The increased PTH has two effects that are important for the maintenance of Ca^{2+} balance: (1) it stimulates Ca^{2+} reabsorption by the kidneys, thereby decreasing Ca^{2+} output, and (2) it activates vitamin D, which increases the efficiency of uptake of ingested Ca^{2+}. Since PTH also promotes bone resorption, a substantial loss of bone minerals occurs if Ca^{2+} intake is reduced for a prolonged period, even though bone is not directly involved in maintaining Ca^{2+} input and output in balance.

Phosphate metabolism is controlled by the same mechanisms that regulate calcium metabolism.

Plasma PO_4^{3-} concentration is not as tightly controlled as plasma Ca^{2+} concentration. Phosphate is regulated directly by vitamin D and indirectly by the plasma Ca^{2+}-PTH feedback loop. To illustrate, a fall in plasma PO_4^{3-} concentration **(hypophosphatemia)** exerts a twofold effect to help restore the circulating PO_4^{3-} level to normal (Fig. 19-25). First, because of the inverse relationship between the PO_4^{3-} and Ca^{2+} concentrations in the plasma, a fall in plasma PO_4^{3-} causes an increase in plasma Ca^{2+}, which directly suppresses PTH secretion. In the presence of reduced PTH, PO_4^{3-} reabsorption by the kidneys increases, returning plasma PO_4^{3-} concentration toward normal. Second, a fall in plasma PO_4^{3-} also results in increased activation of vitamin D, which then promotes PO_4^{3-} absorption in the intestine. This further helps to alleviate the initial hypophosphatemia.

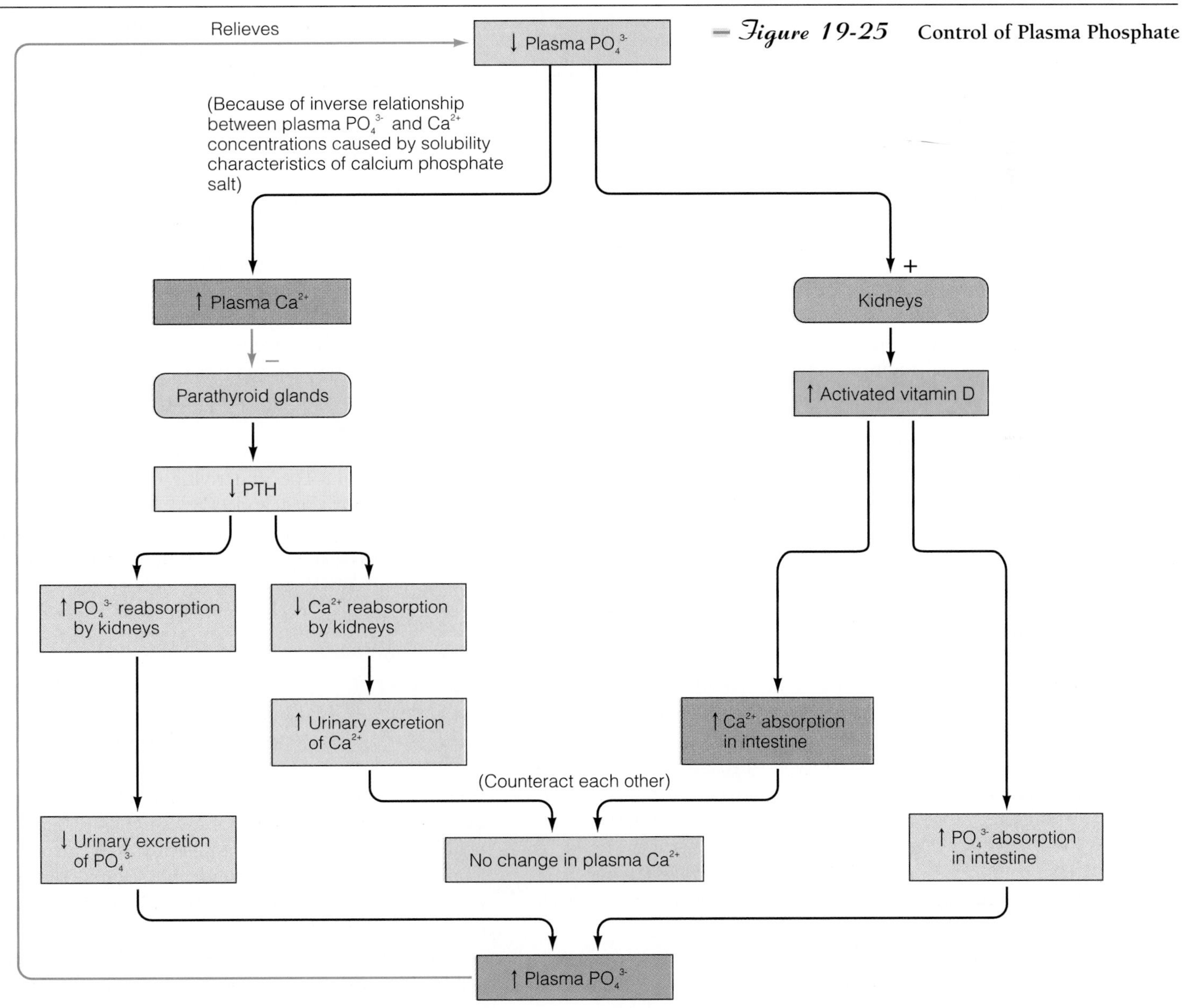

Figure 19-25 **Control of Plasma Phosphate**

Note that Ca^{2+} balance is not compromised by these changes. Although the increase in activated vitamin D stimulates Ca^{2+} absorption, the concurrent fall in PTH produces a compensatory increase in urinary Ca^{2+} excretion because less of the filtered Ca^{2+} is reabsorbed.

Disorders in calcium metabolism may arise from abnormal levels of parathyroid hormone or vitamin D.

The primary disorders that affect Ca^{2+} metabolism are too much or too little PTH or a deficiency of vitamin D.

PTH hypersecretion Excess PTH secretion, or **hyperparathyroidism,** which is usually due to a hypersecreting tumor in one of the parathyroid glands, is characterized by hypercalcemia and hypophosphatemia. The affected individual can be asymptomatic or symptoms can be severe, depending on the magnitude of the problem. The following are among the possible consequences:

- Hypercalcemia reduces the excitability of muscle and nervous tissue, leading to muscle weakness and neurological disorders, including decreased alertness, poor memory, and depression. Cardiac disturbances may also occur.
- Excessive mobilization of Ca^{2+} and PO_4^{3-} from skeletal stores leads to thinning of bone, which may result in skeletal deformities and increased incidence of fractures.
- An increased incidence of Ca^{2+}-containing kidney stones occurs because the excess quantity of Ca^{2+} being filtered through the kidneys may precipitate and form stones. These stones may impair renal function. Passage of the stones through the ureters causes extreme pain. Because of these potential multiple consequences, hyperparathyroidism has been called a disease of "bones, stones, and abdominal groans."
- To further account for the "abdominal groans," hypercalcemia can cause digestive disorders such as peptic ulcers, nausea, and constipation.

PTH hyposecretion Because of the parathyroid glands' close anatomical relation to the thyroid, the most common cause of deficient PTH secretion, or **hypoparathyroidism,** used to be inadvertent removal of the parathyroid glands (before their existence was known) during surgical removal of the thyroid gland (for treatment of thyroid disease). If all of the parathyroid tissue was removed, these patients died, of course, because PTH is essential for life. In fact, physicians were puzzled why some patients died soon after thyroid removal even though there were no apparent surgical complications. Now that the location and importance of the parathyroid glands have been discovered, surgeons are careful to leave parathyroid tissue during thyroid removal. Rarely, PTH hyposecretion occurs as a result of an autoimmune attack against the parathyroid glands.

Hypoparathyroidism leads to hypocalcemia and hyperphosphatemia. The symptoms are primarily referable to increased neuromuscular excitability caused by the reduction in the level of free plasma Ca^{2+}. In the complete absence of parathyroid hormone, death is imminent because of hypocalcemic spasm of respiratory muscles. With a relative deficiency rather than a complete absence of PTH, milder symptoms of increased neuromuscular excitability are evident. Muscle cramps and twitches derive from spontaneous activity in the motor nerves, while tingling and pins-and needles sensations result from spontaneous activity in the sensory nerves. Mental changes include irritability and paranoia.

Vitamin D deficiency The major consequence associated with vitamin D deficiency is impaired intestinal absorption of Ca^{2+}. In the face of reduced Ca^{2+} uptake, PTH maintains the plasma Ca^{2+} level at the expense of the bones. As a result, the bone matrix is not properly mineralized because Ca^{2+} salts are not available for deposition. The demineralized bones become soft and deformed, bowing to the pressures of weight bearing, especially in children. This condition is known as **rickets** in children and **osteomalacia** in adults.

Osteoporosis The rates of bone formation and resorption are about equal throughout most of adult life, so total bone mass usually remains fairly constant during this period. Exceptions include minor hormonally induced fluctuations in bone mass to maintain Ca^{2+} homeostasis or mechanically induced adjustments in bone mass in response to changes in compressional weight bearing (for example, taking up exercise or being confined to bed). Such is not the case, however, during the first 20 and last 20 years of an average life span. During the first two decades of life, when growth is occurring, bone deposition exceeds bone resorption under the influence of growth hormone (see p. 646). In contrast, by 50 to 60 years of age, bone resorption often exceeds bone formation. The result is a reduction in bone mass known as **osteoporosis.** This condition is characterized by a diminished laying down of organic matrix as a result of reduced osteoblast activity and/or increased osteoclast activity rather than abnormal bone calcification. The underlying cause of osteoporosis is uncertain. Plasma Ca^{2+} and PO_4^{3-} levels are normal, as are PTH and vitamin D concentrations. Osteoporosis occurs with greatest frequency in postmenopausal women, suggesting that estrogen withdrawal plays a role. (See the accompanying boxed feature, A Closer Look at Exercise Physiology.)

Chapter in Perspective: Focus on Homeostasis

A number of peripherally located endocrine glands play key roles in maintaining homeostasis, primarily by means of their regulatory influences over the rate of various metabolic reactions and over electrolyte balance. These endocrine glands all secrete hormones in response to specific stimuli. The hormones in turn exert effects that act in negative-feedback fashion to resist the change that induced their secretion, thus maintaining stability in the internal environment. The specific contributions of the peripheral endocrine glands to homeostasis include the following:

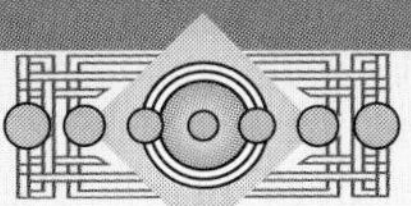

Osteoporosis: The Bane of Brittle Bones

Osteoporosis, a decrease in bone density resulting from reduced deposition of the bone's organic matrix, is a major health problem that affects 25 million people in the United States. The condition is especially prevalent among postmenopausal women. After menopause, women start losing 1% or more bone density each year. Skeletons of elderly women are typically only 50% to 80% as dense as at their peak at about age 35, while elderly men's skeletons retain 80% to 90% of their youthful density.

Osteoporosis is responsible for the greater incidence of bone fractures among women over the age of 50 than among the population at large. Because bone mass is reduced, the bones are more susceptible to fracture in response to a fall, blow, or lifting action that normally would not strain stronger bones. Osteoporosis is the underlying cause of approximately 1.5 million fractures each year, of which 530,000 are vertebral fractures and 227,000 are hip fractures. The cost of rehabilitation is in excess of $10 billion per year. The cost in pain, suffering, and loss of independence is not measurable. One-half of all American women have spinal pain and deformity by age 75.

There appear to be two types of osteoporosis, caused by different mechanisms. Type I osteoporosis affects women soon after menopause and is characterized by vertebral crush fractures or fractures of the arm just above the wrist. It is hypothesized that these fractures occur as a result of the reduction in bone density that accompanies the estrogen deficiency of menopause. Diminished estrogen levels are associated with increased activity of the bone-dissolving osteoclasts and reduced activity of the bone-building osteoblasts. Type II osteoporosis occurs in men as well as women, although it affects females twice as often as males. It is characterized by hip fractures as well as fractures at other sites. Because Type II osteoporosis occurs later in life, the decreased ability to absorb Ca^{2+} associated with advancing age may play a key role in the development of the condition, although estrogen deficiency probably also contributes, accounting for the higher incidence in women.

Estrogen replacement therapy, Ca^{2+} supplementation, and a regular weight-bearing exercise program have been the most common therapeutic approaches used to minimize or reverse bone loss. However, estrogen therapy has been linked to an increased risk of breast cancer, and Ca^{2+} alone has not been as effective in halting bone thinning as was once hoped. The Food and Drug Administration has recently approved two new drugs for the treatment of osteoporosis, and several other potential osteoporosis drugs are in the experimental stage. The two approved drugs are alendronate and calcitonin in a nasal-spray form. *Alendronate* is the first nonhormonal osteoporosis drug. It appears to work by blocking osteoclasts' bone-destroying actions. *Calcitonin,* the thyroid C cell hormone that slows osteoclast activity, had been used in the past to treat advanced osteoporosis, but it had to be injected daily, a deterrent to patient compliance. Now calcitonin is available in a more patient-friendly nasal spray.

Nearing approval for use in treating osteoporosis is *slow-release sodium fluoride,* which promotes the rebuilding of bone by mechanisms that are as yet unclear. Even though sodium fluoride was known to promote bone deposition, it caused troubling side effects, such as making the bone more susceptible to fracture (the opposite of the desired effect) and causing peptic ulcers. The slow-release formula appears to avoid these side effects. Another osteoporosis drug available in other parts of the world but not approved in the United States is *calcitriol,* a form of vitamin D that helps the body absorb Ca^{2+}. A problematic side effect of this agent is increased risk of kidney stones.

Despite advances in osteoporosis therapy, treatment is still often less than satisfactory, and prevention is by far the best approach to managing this disease. Development of strong bones to begin with before menopause through a good Ca^{2+}-rich diet and adequate exercise appears to be the best preventive measure. A large reservoir of bone at midlife may delay the clinical manifestations of osteoporosis in later life. Continued physical activity throughout life appears to retard or prevent bone loss, even in the elderly.

It is well documented that osteoporosis can result from disuse—that is, from reduced mechanical loading of the skeleton. Space travel has clearly shown that lack of gravity results in a decrease in bone density. Studies of athletes, on the other hand, demonstrate that physical activity increases bone density. Within groups of athletes, bone density correlates directly with the load that the bone must bear. If one looks at athletes' femurs (thigh bones), the greatest bone density is found in weight lifters, followed in order by throwers, runners, soccer players, and finally swimmers. In fact, the bone density of swimmers does not differ from that of nonathletic controls. Swimming does not place any strain on bones. The bone density in the playing arm of male tennis players has been found to be as much as 35% greater than in their other arm; female tennis players have been found to have 28% greater density in their playing arm than in their other arm. One study found that very mild activity in nursing home patients, whose average age was 82 years, not only slowed bone loss but even resulted in bone buildup over a 36-month period. Thus, exercise is a good defense against osteoporosis.

The exact mechanism by which exercise increases bone mass is unknown. According to one proposal, exercise places strain on bone, which causes changes in electrical potential that induce bone formation.

A gene has recently been identified that influences a person's risk of developing osteoporosis. Researchers have found that the gene that codes for vitamin D receptors comes in two varieties. One produces receptors that result in higher bone density than the other. For those with the less efficient bone-building gene, a combination of adequate dietary Ca^{2+} intake and exercise is especially important to promote development of strong bones and overcome a genetic predisposition for osteoporosis.

- Two closely related hormones secreted by the thyroid gland, tetraiodothyronine (T_4) and triiodothyronine (T_3), increase the overall metabolic rate. Not only does this action influence the rate at which nutrient molecules and O_2 within the internal environment are used by the cells, but it also produces heat, which is a contributing factor to the control of body temperature. T_4 and T_3 also stimulate a number of specific metabolic reactions, most of which influence the plasma concentrations of nutrient molecules.
- The adrenal cortex secretes three classes of hormones. Aldosterone, the primary mineralocorticoid, is essential for Na^+ and K^+ balance. Because of Na^+'s osmotic effect, Na^+ balance is critical to maintaining the proper ECF volume and arterial blood pressure. This action is essential for life. Without aldosterone's Na^+- and H_2O-conserving effect, so much plasma volume would be lost in the urine that death would quickly ensue. Maintenance of K^+ balance is essential for homeostasis because changes in extracellular K^+ have a profound impact on neuromuscular excitability, thus jeopardizing normal functioning of the heart, among other detrimental effects.
- Cortisol, the primary glucocorticoid secreted by the adrenal cortex, increases the plasma concentrations of glucose, fatty acids, and amino acids above normal. Although these actions disrupt the maintenance of stable concentrations of these molecules in the internal environment, they contribute to homeostasis indirectly by making the molecules readily available as energy sources or building blocks for tissue repair to help the body adapt to stressful situations.
- The sex hormones secreted by the adrenal cortex do not contribute to homeostasis.
- The major hormone secreted by the adrenal medulla, epinephrine, generally reinforces activities of the sympathetic nervous system. It contributes to homeostasis directly by its role in blood pressure regulation. Epinephrine also contributes to homeostasis indirectly by helping prepare the body for peak physical responsiveness in fight-or-flight situations. This includes increasing the plasma concentrations of glucose and fatty acids above normal to provide additional energy sources to support increased physical activity.
- The two major hormones secreted by the endocrine pancreas, insulin and glucagon, are important in shifting metabolic pathways between the absorptive and post-absorptive states to maintain the appropriate plasma levels of nutrient molecules.
- Parathyroid hormone from the parathyroid glands is critical to the maintenance of the plasma concentration of Ca^{2+}. PTH is essential for life because of Ca^{2+}'s effect on neuromuscular excitability. In the absence of PTH, death rapidly occurs due to asphyxiation resulting from pronounced spasms of the respiratory muscles.

Chapter Summary

Thyroid Gland

The thyroid gland contains two types of endocrine secretory cells: (1) follicular cells, which produce the iodine-containing hormones, T_4 (thyroxine or tetraiodothyronine) and T_3 (triiodothyronine), collectively known as thyroid hormone, and (2) C cells, which synthesize a Ca^{2+}-regulating hormone, calcitonin.

Thyroid hormone is the primary determinant of the overall metabolic rate of the body. By accelerating the metabolic rate of most tissues, it increases heat production. Thyroid hormone also enhances the actions of the chemical mediators of the sympathetic nervous system. Through this and other means, thyroid hormone indirectly increases cardiac output. Finally, thyroid hormone is essential for normal growth as well as the development and function of the nervous system.

Thyroid hormone secretion is regulated by a negative-feedback system between hypothalamic TRH, anterior pituitary TSH, and thyroid gland T_3 and T_4. The feedback loop maintains thyroid hormone levels relatively constant. Cold exposure in newborn infants is the only input to the hypothalamus known to be effective in increasing TRH and thereby thyroid hormone secretion.

Adrenal Glands

Each of the pair of adrenal glands consists of two separate endocrine organs—an outer steroid-secreting adrenal cortex and an inner catecholamine-secreting adrenal medulla. The adrenal cortex secretes three different categories of steroid hormones: mineralocorticoids (primarily aldosterone), glucocorticoids (primarily cortisol), and adrenal sex hormones (primarily the weak androgen, dehydroepiandrosterone).

Aldosterone regulates Na^+ and K^+ balance and is important for blood pressure homeostasis, which is accomplished secondarily as a result of the osmotic effect of Na^+ in maintaining the plasma volume, a lifesaving effect. Control of aldosterone secretion is related to Na^+ and K^+ balance and blood pressure regulation and is not influenced by ACTH.

Cortisol helps regulate fuel metabolism and is important in stress adaptation. It increases the blood levels of glucose, amino acids, and fatty acids and spares glucose for use by the glucose-dependent brain. The mobilized organic molecules are available for use as needed for energy or for repair of injured tissues. Cortisol secretion is regulated by a negative-feedback loop involving hypothalamic CRH and pituitary ACTH. The most potent stimulus for increasing activity of the CRH/ACTH/cortisol axis is stress.

Dehydroepiandrosterone is responsible for the sex drive and growth of pubertal hair in females.

The adrenal medulla is composed of modified sympathetic postganglionic neurons, which secrete the catecholamine epinephrine into the blood in response to sympathetic stimulation.

For the most part, epinephrine reinforces the sympathetic system in its general systemic "fight-or-flight" responses and in its maintenance of arterial blood pressure. Epinephrine also exerts important metabolic effects, namely, increasing blood glucose and blood fatty acids. The primary stimulus for increased adrenomedullary secretion is activation of the sympathetic system by stress.

Endocrine Control of Fuel Metabolism

Intermediary or fuel metabolism refers collectively to the synthesis (anabolism), breakdown (catabolism), and transformations of the three classes of energy-rich organic nutrients—carbohydrate, fat, and protein—within the body. Glucose and fatty acids derived respectively from carbohydrates and fats are primarily used as metabolic fuels, whereas amino acids derived from proteins are primarily used for the synthesis of structural and enzymatic proteins.

During the absorptive state following a meal, the excess absorbed nutrients not immediately needed for energy production or protein synthesis are stored to a limited extent as glycogen in the liver and muscle but mostly as triglycerides in adipose tissue. During the postabsorptive state between meals when no new nutrients are entering the blood, the glycogen and triglyceride stores are catabolized to release nutrient molecules into the blood. If necessary, body proteins are degraded to release amino acids for conversion into glucose. It is essential to maintain the blood glucose concentration above a critical level even during the postabsorptive state, because the brain depends on blood-delivered glucose as its energy source. Tissues not dependent on glucose switch to fatty acids as their metabolic fuel, sparing glucose for the brain.

These shifts in metabolic pathways between the absorptive and postabsorptive state are hormonally controlled. The most important hormone in this regard is insulin. Insulin is secreted by the β cells of the islets of Langerhans, the endocrine portion of the pancreas. The other major pancreatic hormone, glucagon, is secreted by the α cells of the islets. Insulin is an anabolic hormone; it promotes the cellular uptake of glucose, fatty acids, and amino acids and enhances their conversion into glycogen, triglycerides, and proteins, respectively. In so doing, it lowers the blood concentrations of these small organic molecules. Insulin secretion is increased during the absorptive state, primarily by a direct effect of an elevated blood glucose on the β cells, and is largely responsible for directing the organic traffic into cells during this state.

Glucagon mobilizes the energy-rich molecules from their stores during the postabsorptive state. Glucagon, which is secreted in response to a direct effect of a fall in blood glucose on the pancreatic α cells, in general opposes the actions of insulin.

Endocrine Control of Calcium Metabolism

Changes in the concentration of free, diffusible plasma Ca^{2+}, the biologically active form of this ion, produce profound and life-threatening effects, most notably on neuromuscular excitability. Hypercalcemia reduces excitability, whereas hypocalcemia brings about overexcitability of nerves and muscles. If the overexcitability is severe enough, fatal spastic contractions of respiratory muscles can occur.

Three hormones regulate the plasma concentration of Ca^{2+} (and concurrently regulate PO_4^{3-})—parathyroid hormone (PTH), calcitonin, and vitamin D. PTH, whose secretion is directly increased by a fall in plasma Ca^{2+} concentration, acts on bone, kidneys, and the intestine to raise the plasma Ca^{2+} concentration. In so doing, it is essential for life by preventing the fatal consequences of hypocalcemia. The specific effects of PTH on bone are to promote Ca^{2+} movement from the bone fluid into the plasma in the short term and to promote localized dissolution of bone by enhancing activity of the osteoclasts (bone-dissolving cells) in the long term. Dissolution of the calcium phosphate bone crystals releases PO_4^{3-} as well as Ca^{2+} into the plasma. Parathyroid hormone acts on the kidneys to enhance the reabsorption of filtered Ca^{2+}, thereby reducing the urinary excretion of Ca^{2+} and increasing its plasma concentration. Simultaneously, PTH reduces renal PO_4^{3-} reabsorption, in this way increasing PO_4^{3-} excretion and lowering plasma PO_4^{3-} levels. This is important because a rise in plasma PO_4^{3-} would force the deposition of some of the plasma Ca^{2+} back into the bone. Furthermore, PTH facilitates the activation of vitamin D, which in turn stimulates Ca^{2+} and PO_4^{3-} absorption from the intestine.

Vitamin D can be synthesized from a cholesterol derivative in the skin when exposed to sunlight, but frequently this endogenous source is inadequate, so vitamin D must be supplemented by dietary intake. From either source, vitamin D must be activated first by the liver and then by the kidneys (the site of PTH regulation of vitamin D activation) before it can exert its effect on the intestine.

Calcitonin, a hormone produced by the C cells of the thyroid gland, is the third factor that regulates Ca^{2+}. In negative-feedback fashion, calcitonin is secreted in response to an increase in plasma Ca^{2+} concentration and acts to lower plasma Ca^{2+} levels by inhibiting activity of bone osteoclasts. Calcitonin is unimportant except during the rare condition of hypercalcemia.

Review Exercises

Objective Questions (Answers on p. E–19).

1. The response to thyroid hormone is detectable within a few minutes after its secretion. (True or false?)
2. "Male" sex hormones are produced in both males and females by the adrenal cortex. (True or false?)
3. Adrenal androgen hypersecretion is usually due to a deficit of an enzyme crucial to cortisol synthesis. (True or false?)
4. Excess glucose and amino acids as well as fatty acids can be stored as triglycerides. (True or false?)
5. Insulin is the only hormone that can lower blood glucose levels. (True or false?)
6. The most life-threatening consequence of hypocalcemia is its impact on reducing blood clotting. (True or false?)

7. All ingested Ca^{2+} is indiscriminately absorbed in the intestine. (True or false?)
8. The $Ca_3(PO_4)_2$ bone crystals form a labile pool from which Ca^{2+} can rapidly be extracted under the influence of PTH. (True or false?)
9. The lumen of the thyroid follicle is filled with ____________, the chief constituent of which is a large, complex glycoprotein known as ____________.
10. The common large precursor molecule that yields ACTH, MSH, and β-endorphin is known as ____________.
11. ____________ refers to the conversion of glucose into glycogen. ____________ refers to the conversion of glycogen into glucose. ____________ refers to the conversion of amino acids into glucose.
12. The three major tissues that are not dependent on insulin for their glucose uptake are ____________, ____________, and ____________.
13. The three compartments with which ECF Ca^{2+} is exchanged are ____________, ____________, and ____________.
14. Which of the following hormones does *not* exert a direct metabolic effect?
    a. epinephrine
    b. growth hormone
    c. aldosterone
    d. cortisol
    e. thyroid hormone
15. Which of the following are characteristic of the postabsorptive state?
    a. glycogenolysis
    b. gluconeogenesis
    c. lipolysis
    d. glycogenesis
    e. protein synthesis
    f. triglyceride synthesis
    g. protein degradation
    h. increased insulin secretion
    i. increased glucagon secretion
    j. glucose sparing
16. Indicate the primary circulating form and storage form of each of the three classes of organic nutrients:

	Primary Circulating Form	*Primary Storage Form*
Carbohydrate	1. ____________	2. ____________
Fat	3. ____________	4. ____________
Protein	5. ____________	6. ____________

Essay Questions

1. Describe the steps of thyroid hormone synthesis.
2. What are the effects of T_3 and T_4? Which is the more potent? What is the source of most circulating T_3?
3. Describe the regulation of thyroid hormone.
4. Discuss the causes and symptoms of both hypothyroidism and hyperthyroidism. When does a goiter occur?
5. What hormones are secreted by the adrenal cortex? What are the functions and control of each of these hormones?
6. Discuss the causes and symptoms of each type of adrenocortical dysfunction.
7. What is the relationship of the adrenal medulla to the sympathetic nervous system? What are the functions of epinephrine? How is epinephrine release controlled?
8. Define stress. Describe the neural and hormonal responses to a stressor.
9. Define fuel metabolism, anabolism, and catabolism.
10. Distinguish between the absorptive and postabsorptive states with regard to the dispensation of nutrient molecules.
11. Name the two major cell types of the islets of Langerhans, and indicate the hormonal product of each.
12. Compare the functions and control of insulin secretion with those of glucagon secretion.
13. What are the consequences of diabetes mellitus? Distinguish between Type I and Type II diabetes mellitus.
14. Why must plasma Ca^{2+} be closely regulated?
15. Discuss the contributions of parathyroid hormone, calcitonin, and vitamin D to Ca^{2+} metabolism. Describe the source and control of each of these hormones.
16. Discuss the major disorders in Ca^{2+} metabolism.

Points to Ponder

(Explanations on p. E–19.)

1. Iodine is naturally present in salt water and is abundant in soil along the coastal regions. Fish and shellfish living in the ocean and plants grown in coastal soil take up iodine from their environment. Fresh water does not contain iodine, and the soil becomes more iron poor the farther inland. Knowing this, explain why the midwestern United States was once known as an *endemic goiter belt* because of the high incidence of goiter in this region. Why is this region no longer an endemic goiter belt even though the soil is still iodine poor?
2. Explain why people who work night shifts adapt better to this routine when exposed to bright light during the night and darkness during the day.
3. Why would an infection tend to increase the blood glucose level of a diabetic individual?
4. Tapping the facial nerve at the angle of the jaw in a patient with moderate hyposecretion of a particular hormone elicits a characteristic grimace on that side of the face. What endocrine abnormality could give rise to this so-called *Chvostek's sign?*
5. Soon after a technique to measure plasma Ca^{2+} levels was developed in the 1920s, physicians observed that hypercalcemia accompanied a broad range of cancers. Early researchers proposed that malignancy-associated hypercalcemia arose from metastatic (see p. 403) tumor cells that invaded and destroyed bone, releasing Ca^{2+} into the blood. This conceptual framework was overturned when physicians noted that

hypercalcemia was often found in the absence of bone lesions. Furthermore, cancer patients often manifested hypophosphatemia in addition to hypercalcemia. This finding led investigators to suspect that the tumors might be producing a PTH-like substance. Explain how they reached this conclusion. In 1987, this substance was identified and named *parathyroid hormone–related peptide (PTHrP),* which binds to and activates PTH receptors.

6. ***Clinical Consideration*** Najma G. sought medical attention after her menstrual periods ceased and she started getting excessive facial hair. Also, she had been thirstier than usual and urinated more frequently. A clinical evaluation revealed that Najma was hyperglycemic. Her physician told her that she had an endocrine disorder dubbed "diabetes of bearded ladies." Based on her symptoms and your knowledge of the endocrine system, what underlying defect do you think is responsible for Najma's condition?

20 The Reproductive System

REPRODUCTIVE SYSTEM

Body systems maintain homeostasis

HOMEOSTASIS
The reproductive system does not contribute to homeostasis but is essentiial for perpetuation of the species.

Homeostasis is essential for survival of cells

CELLS

Cells make up body systems

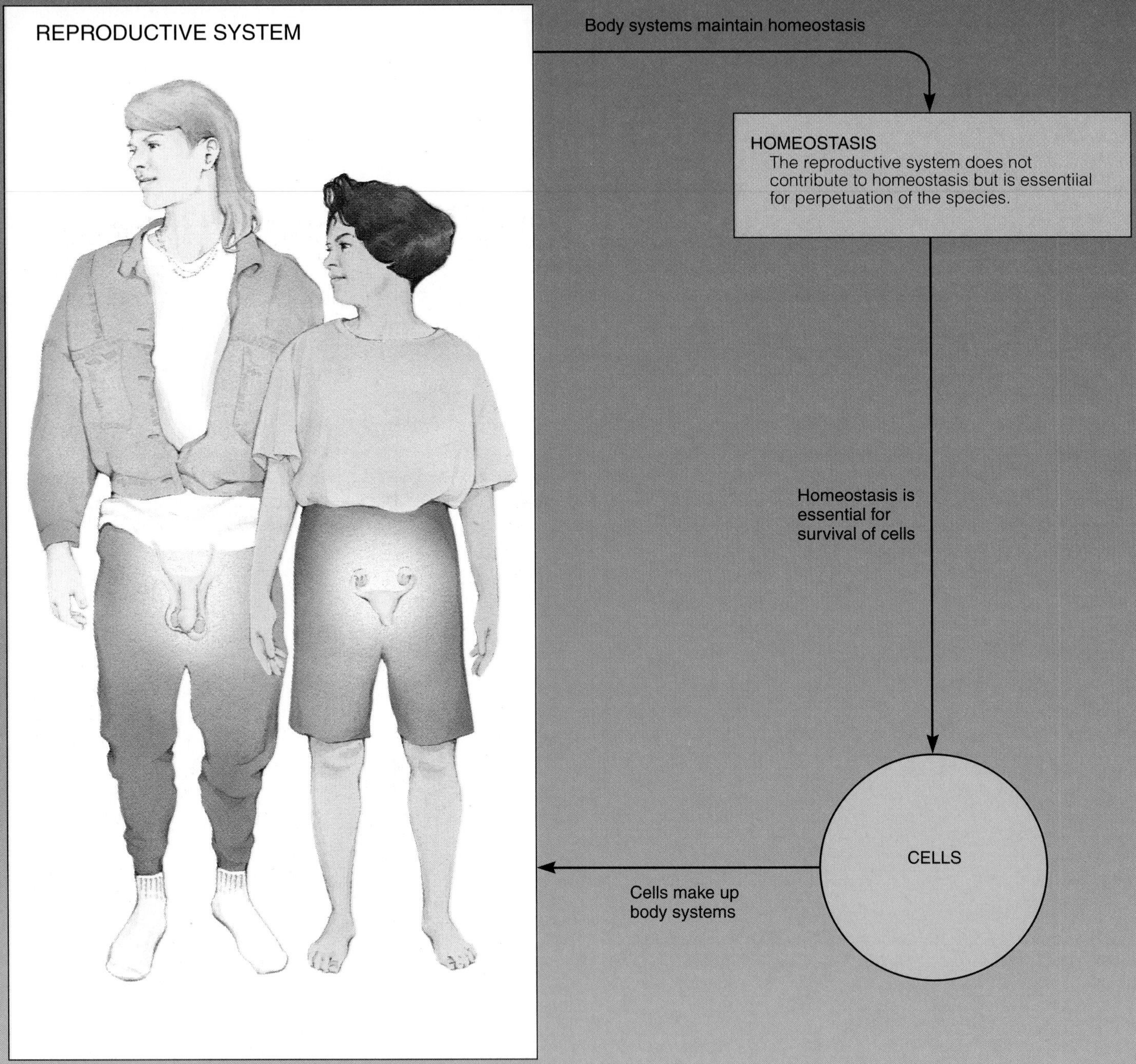

Normal functioning of the **reproductive system** is not aimed toward homeostasis and is not necessary for survival of an individual, but it is essential for survival of the species.Only through reproduction can the complex genetic blueprint of each species survive beyond the lives of individual members of the species.

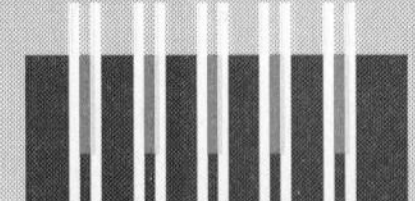

Chapter Contents At a Glance

Introduction

The reproductive system is not essential for survival of the individual, but it is necessary for survival of the species and has a profound impact on a person's life.

The central theme of this book has been the physiological processes aimed at maintaining homeostasis to ensure survival of the individual. We are now going to disembark from this theme to discuss the reproductive system, which primarily serves the purpose of perpetuating the species. Normal functioning of the reproductive system is not aimed toward homeostasis and is not necessary for survival of an individual, but it is essential for survival of the species. Only through reproduction can the complex genetic blueprint of each species survive beyond the lives of individual members of the species.

Even though the reproductive system does not contribute to homeostasis and is not essential for survival of an individual, it still plays an important role in a person's life. For example, the manner in which people relate as sexual beings contributes in significant ways to psychosocial behavior and has important influences on how people view themselves and how they interact with others. Reproductive function also has a profound effect on society. The universal organization of societies into family units provides a stable environment that is conducive for perpetuating our species. On the other hand, the population explosion and its resultant drain on dwindling resources have recently led to worldwide concern with the means by which reproduction can be limited.

Reproductive capability depends on an intricate relationship among the hypothalamus, anterior pituitary, reproductive organs, and target cells of the sex hormones. In addition to these basic biological processes, sexual behavior and attitudes are deeply influenced by emotional factors and the sociocultural mores of the society in which the individual lives. We will concentrate on the basic sexual and reproductive functions that are under nervous and hormonal control and will not examine the psychological and social ramifications of sexual behavior.

The reproductive system includes the gonads and reproductive tract.

Reproduction depends on the union of male and female **gametes (reproductive,** or **germ, cells)**, each with a half set of chromosomes, to form a new individual with a full, unique set of chromosomes. Unlike the other body systems, which are essentially identical in the two sexes, the reproductive systems of males and females are remarkably different, befitting their different roles in the reproductive process. The male and female **reproductive systems** are designed to enable union of genetic material from the two sexual partners, and the female system is equipped to house and nourish the offspring to the developmental point at which it can survive independently in the external environment.

The **primary reproductive organs**, or **gonads**, consist of a pair of **testes** in the male and a pair of **ovaries** in the female. In both sexes, the mature gonads perform the dual function of (1) producing gametes **(gametogenesis)**, that is, **spermatozoa (sperm)** in the male and **ova (eggs)** in the female, and (2) secreting sex hormones, specifically, **testosterone** in males and **estrogen** and **progesterone** in females.

In addition to the gonads, the reproductive system in each sex includes a **reproductive tract** encompassing a system of ducts that are specialized to transport or house the gametes after they are produced, plus **accessory sex glands** that empty their supportive secretions into these passageways. In females, the *breasts* are also considered accessory reproductive

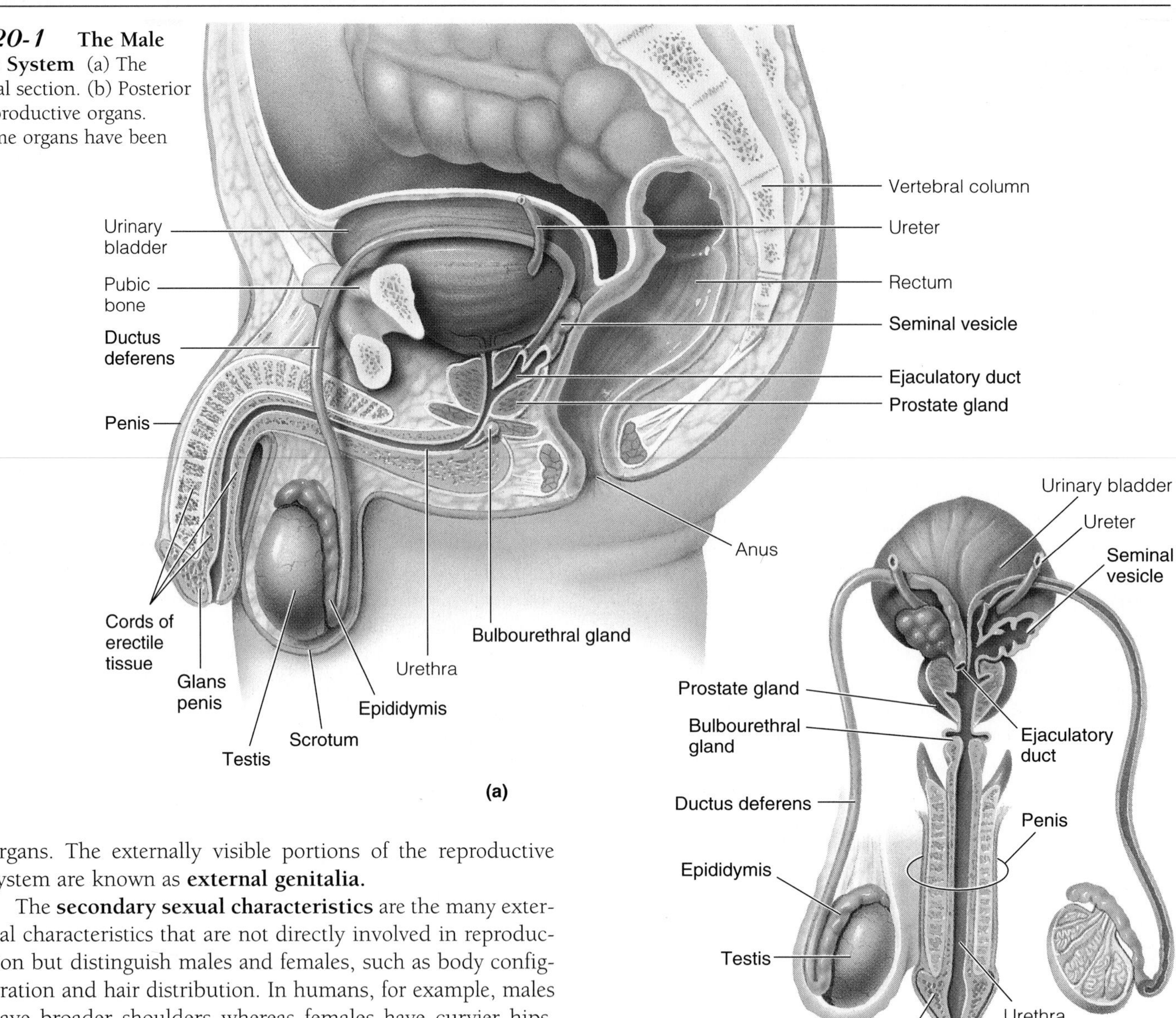

Figure 20-1 The Male Reproductive System (a) The pelvis in sagittal section. (b) Posterior view of the reproductive organs. Portions of some organs have been removed.

organs. The externally visible portions of the reproductive system are known as **external genitalia.**

The **secondary sexual characteristics** are the many external characteristics that are not directly involved in reproduction but distinguish males and females, such as body configuration and hair distribution. In humans, for example, males have broader shoulders whereas females have curvier hips, and males have beards while females do not. Testosterone in the male and estrogen in the female are responsible for the development and maintenance of these characteristics. Progesterone has no influence on secondary sexual characteristics. Even though growth of axillary and pubic hair at puberty is promoted in both sexes by androgens—testosterone in males and adrenocortical dehydroepiandrosterone in females (see p. 664)—this hair growth is not a secondary sexual characteristic because both sexes display this feature. Thus, testosterone and estrogen alone are responsible for the nonreproductive distinguishing features. In some species, the secondary sexual characteristics are of great importance in courting and mating behavior; for example, the rooster's headdressing attracts the female's attention, and the stag's antlers are useful to ward off other males. In humans, the differentiating marks between males and females do serve to attract the opposite sex, but attraction is also strongly influenced by the complexities of human society and cultural behavior.

The essential reproductive functions of the male are (1) production of sperm *(spermatogenesis)* and (2) delivery of sperm to the female. The sperm-producing organs, the testes, are suspended outside the abdominal cavity in a skin-covered sac, the **scrotum,** which lies within the angle between the legs. The male reproductive system is designed to deliver sperm to the female reproductive tract in a liquid vehicle, *semen,* which is conducive to sperm viability. The major male accessory sex glands, whose secretions provide the bulk of the semen, are the *seminal vesicles, prostate gland,* and *bulbourethral glands* (Figure 20-1). The **penis** is the organ used to deposit semen in the female. Sperm exit the testes through the *epididymis, ductus (vas) deferens, ejaculatory duct,* and *urethra,* the latter being a canal that runs the length of the penis.

The female's role in reproduction is more complicated than the male's. The essential female reproductive functions include (1) production of ova *(oogenesis);* (2) reception of sperm; (3) transport of the sperm and ovum to a common site for union *(fertilization,* or *conception);* (4) maintenance of the

Figure 20-2 **The Female Reproductive System** (a) The pelvis in sagittal section. (b) Posterior view of the reproductive organs. (c) Perineal view of the external genitalia.

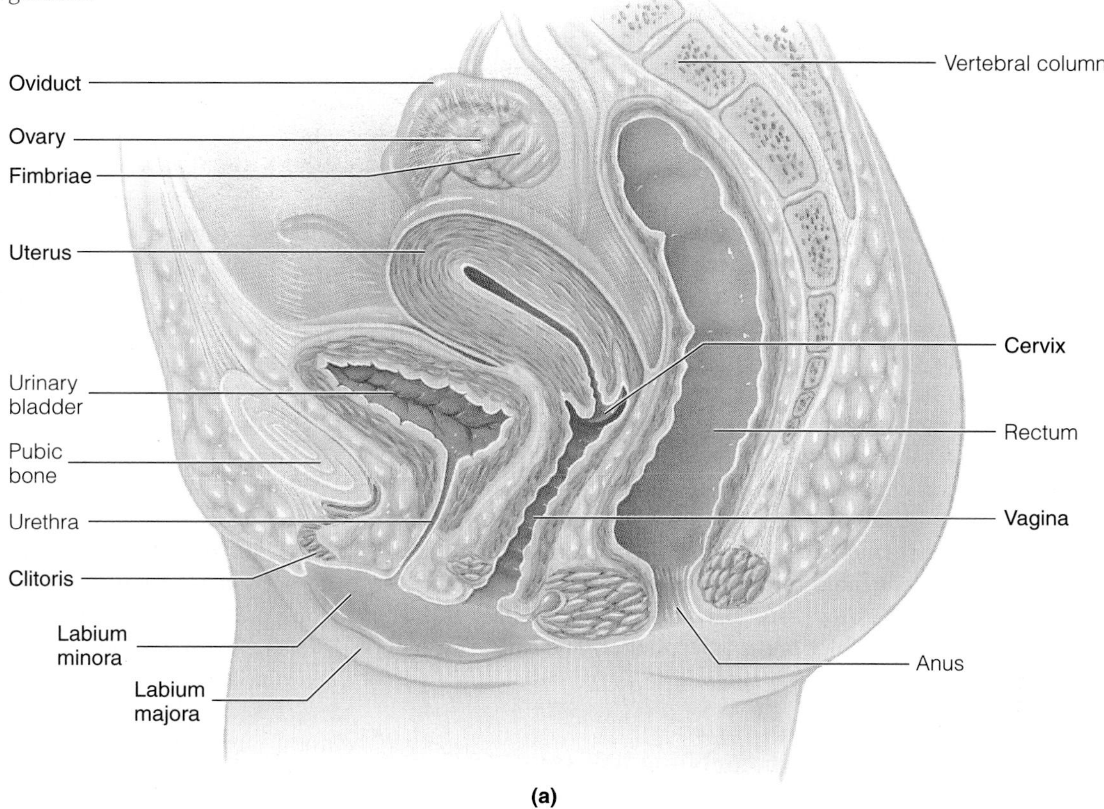

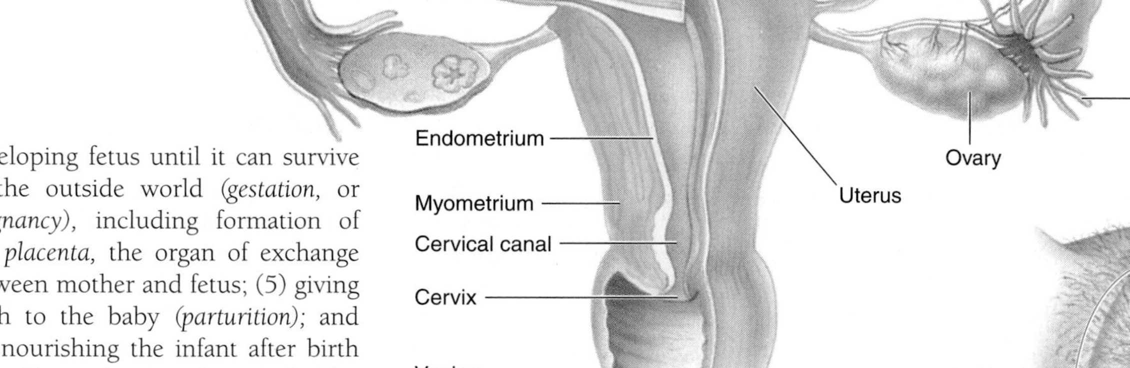

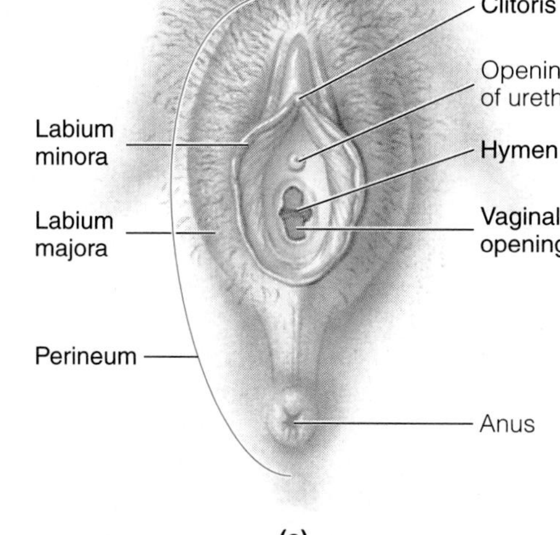

developing fetus until it can survive in the outside world *(gestation,* or *pregnancy),* including formation of the *placenta,* the organ of exchange between mother and fetus; (5) giving birth to the baby *(parturition);* and (6) nourishing the infant after birth by milk production *(lactation).* The product of fertilization is known as an **embryo** during the first two months of intrauterine development when tissue differentiation is taking place. Beyond this time, the developing living being is recognizable as human and is known as a **fetus** during the remainder of gestation. Although no further tissue differentiation takes place during fetal life, it is a time of tremendous tissue growth and maturation.

The ovaries and female reproductive tract lie within the pelvic cavity (Figure 20-2a and b). The female reproductive tract consists of two **oviducts (uterine**, or **Fallopian**,

tubes), which pick up ova upon ovulation and serve as the site for fertilization; the thick-walled hollow **uterus,** which is primarily responsible for maintaining the fetus during its development and expelling it at the end of pregnancy; and the **vagina,** a muscular, expansible tube connecting the uterus to the external environment. The lowest portion of the uterus, the **cervix,** projects into the vagina and contains a single small opening, the **cervical canal.** Sperm are deposited in the vagina by the penis during sexual intercourse. The cervical canal serves as a pathway for sperm through the uterus to the site of fertilization in the oviduct, and when greatly dilated during parturition, it serves as the passageway for delivery of the baby from the uterus.

The **vaginal opening** is located in the **perineal region** between the urethral opening anteriorly and the anal opening posteriorly (Fig. 20-2c). It is partially covered by a thin mucous membrane, the **hymen,** which can be physically disrupted in a variety of ways, including by the first sexual intercourse. The vaginal and urethral openings are surrounded laterally by two pairs of skin folds, the **labia minora** and **labia majora.** The smaller labia minora are located medially to the more prominent labia majora. The **clitoris,** a small erotic structure composed of tissue identical to the penis, lies at the anterior end of the folds of the labia minora. The female external genitalia are collectively referred to as the **vulva.**

Reproductive cells each contain a half set of chromosomes.

The DNA molecules that carry the cell's genetic code are not randomly crammed into the nucleus but are precisely organized into **chromosomes** (see p. B-3). Each chromosome consists of a different DNA molecule that contains a unique set of genes. **Somatic** (body) **cells** contain forty-six chromosomes (the **diploid number**), which can be sorted into twenty-three pairs on the basis of various distinguishing features. Chromosomes composing a matched pair are termed **homologous chromosomes,** one member of each pair having been derived from the individual's maternal parent and the other member from the paternal parent. Gametes (that is, sperm and eggs) contain only one member of each homologous pair for a total of twenty-three chromosomes (the **haploid number**).

Gametogenesis is accomplished by meiosis.

Most cells in the human body have the ability to reproduce themselves, a process important in growth, replacement, and repair of tissues. Cell division involves two components: division of the nucleus and division of the cytoplasm. Nuclear division in somatic cells is accomplished by **mitosis.** In mitosis, the chromosomes replicate (make duplicate copies of themselves), then the identical chromosomes are separated so that a complete set of genetic information (that is, a diploid number of chromosomes) is distributed to each of the two new daughter cells. Nuclear division in the specialized case of gametes is accomplished by **meiosis,** in which only a half set of genetic information (that is, a haploid number of chromosomes) is distributed to each of four new daughter cells (see Fig. B-11).

During meiosis, a specialized diploid germ cell undergoes one chromosome replication followed by two nuclear divisions. In the first meiotic division, the replicated chromosomes do not separate into two individual, identical chromosomes but remain joined together. The doubled chromosomes sort themselves into homologous pairs, and the pairs separate so that each of two daughter cells receives a half set of doubled chromosomes. During the second meiotic division, the doubled chromosomes within each of the two daughter cells separate and are distributed into two cells, yielding four daughter cells, each containing a half set of chromosomes, a single member of each pair. During this process, the maternally and paternally derived chromosomes of each homologous pair are distributed to the daughter cells in random assortments containing one member of each chromosome pair without regard for its original derivation. That is, not all of the mother-derived chromosomes go to one daughter cell and the father-derived chromosomes to the other cell. More than 8 million (2^{23}) different mixtures of the twenty-three paternal and maternal chromosomes are possible. This genetic mixing provides novel new combinations of chromosomes.

Thus, sperm and ova each have a unique haploid number of chromosomes. When fertilization takes place, a sperm and ovum fuse to form the start of a new individual with forty-six chromosomes, one member of each chromosomal pair having been inherited from the mother and the other member from the father (▬ Fig. 20-3).

The sex of an individual is determined by the combination of sex chromosomes.

Whether individuals are destined to be males or females is a genetic phenomenon determined by the sex chromosomes they possess. As the twenty-three chromosome pairs are separated during meiosis, each sperm or ovum receives only one member of each chromosome pair. Twenty-two of the chromosome pairs are **autosomal chromosomes** that code for general human characteristics as well as for specific traits such as eye color. The remaining pair of chromosomes are the **sex chromosomes,** of which there are two genetically different types—a larger **X chromosome** and a smaller **Y chromosome. Sex determination** depends on the combination of sex chromosomes: **genetic males** have both an X and a Y sex chromosome; **genetic females** have two X sex chromosomes. Thus, the genetic difference responsible for all of the anatomical and functional distinctions between males and females is the single Y chromosome. Males have it; females do not.

As a result of meiosis during gametogenesis, all chromosome pairs are separated so that each daughter cell contains only one member of each pair, including the sex chromosome pair. When the XY sex chromosome pair separates during sperm formation, half the sperm receive an X chromosome and the other half a Y chromosome. In contrast, during oogenesis, every ovum receives an X chromosome because separation of the XX sex chromosome pair yields only X chromosomes. During fertilization, combination of an X-bearing sperm with an X-bearing ovum produces a genetic female, XX, whereas union of a Y-bearing sperm with an X-bearing ovum results in a genetic male, XY. Thus, genetic sex is determined at the time of conception and depends on which type of sex chromosome is contained within the fertilizing sperm.

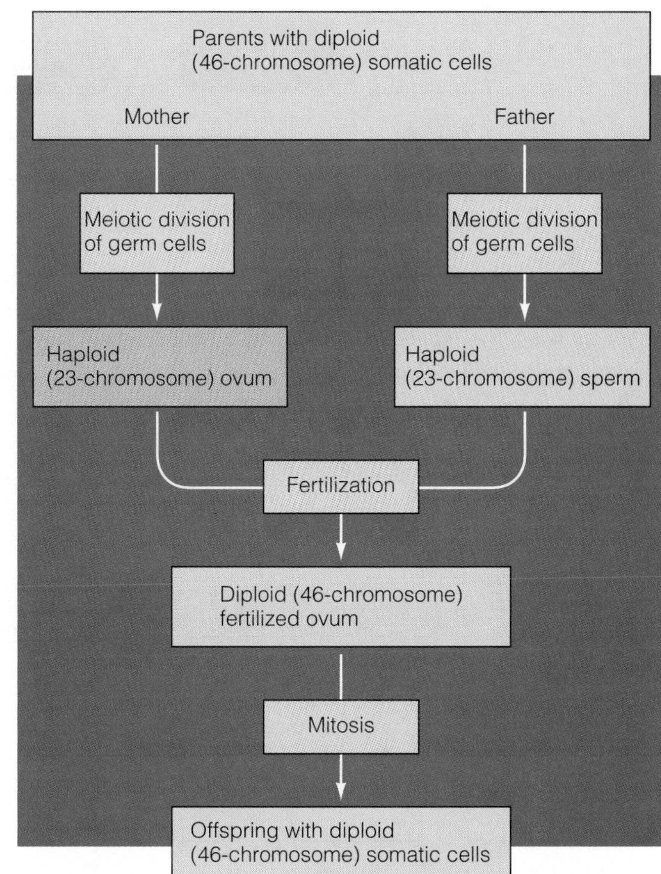

Figure 20-3 **Chromosomal Distribution in Sexual Reproduction**

Sex differentiation along male or female lines depends on the presence or absence of masculinizing determinants during critical periods of embryonic development.

Differences between males and females exist at three levels: genetic, gonadal, and phenotypic (anatomical) sex (Fig. 20-4). **Genetic sex,** which depends on the combination of sex chromosomes at the time of conception, in turn determines **gonadal sex,** that is, whether testes or ovaries develop. The presence or absence of a Y chromosome determines gonadal differentiation. For the first month and a half of gestation, all embryos have the potential to differentiate along either male or female lines, because the developing reproductive tissues of both sexes are identical and indifferent. Gonadal specificity appears during the seventh week of intrauterine life when the indifferent gonadal tissue of a genetic male begins to differentiate into testes under the influence of the **sex-determining region** of the Y chromosome **(SRY),** the single gene that is responsible for sex determination. This gene triggers a chain of reactions that leads to physical development of a male. The sex-determining region of the Y chromosome "masculinizes" the gonads (induces their development into testes) by stimulating production of **H-Y antigen** by primitive gonadal cells. H-Y antigen, which is a specific plasma membrane protein found only in males, directs differentiation of the gonads into testes. Because genetic females lack the SRY gene and consequently do not produce H-Y antigen, their gonadal cells never receive a signal for testicular formation, so the undifferentiated gonadal tissue starts developing during the ninth week into ovaries instead.

Phenotypic sex, the apparent anatomical sex of an individual, depends on the genetically determined gonadal sex. **Sexual differentiation** refers to the embryonic development of the external genitalia and reproductive tract along either male or female lines. As with the undifferentiated gonads, embryos of both sexes have the potential to develop either male or female reproductive tracts and external genitalia. Differentiation into a male-type reproductive system is induced by **androgens,** which are masculinizing hormones secreted by the developing testes. Testosterone is the most potent androgen. The absence of these testicular hormones in female fetuses results in the development of a female-type reproductive system. By ten to twelve weeks of gestation, the sexes can easily be distinguished by the anatomical appearance of the external genitalia.

Male and female external genitalia develop from the same embryonic tissue. In both sexes, the undifferentiated external genitals consist of a *genital tubercle,* paired *urethral folds* surrounding a urethral groove, and, more laterally, *genital (labioscrotal) swellings* (Fig. 20-5). The **genital tubercle** gives rise to exquisitely sensitive erotic tissue—in males the **glans penis** (the cap at the distal end of the penis) and in females the clitoris. The major distinctions between the glans penis and clitoris are the smaller size of the clitoris and the penetration of the glans penis by the urethral opening. The urethra is the tube through which urine is transported from the bladder to the outside and also serves in males as a passageway for exit of semen through the penis to the outside. In males, the **urethral folds** fuse around the urethral groove to form the penis, which encircles the urethra. The **genital swellings** similarly fuse to form the scrotum and **prepuce,** a fold of skin that extends over the end of the penis and more or less completely covers the glans penis. In females, the urethral folds and genital swellings do not fuse at midline but develop instead into the labia minora and labia majora, respectively. The urethral groove remains open, providing access to the interior through the urethral opening and vaginal orifice.

Although the male and female external genitalia develop from the same undifferentiated embryonic tissue, this is not the case with the reproductive tracts. Two primitive duct systems—the Wolffian ducts and the Müllerian ducts—develop in all embryos. In males, the reproductive tract develops from

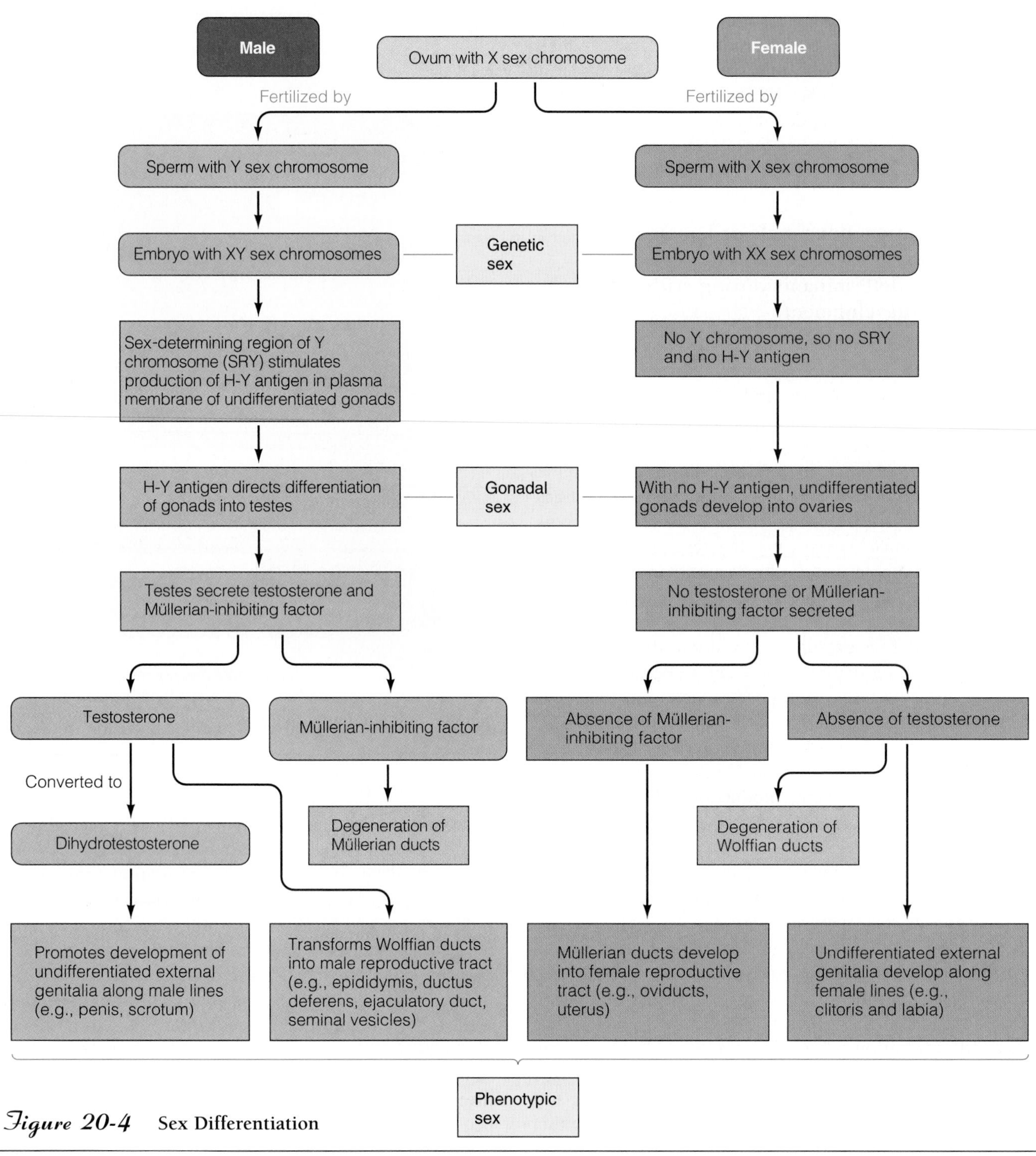

Figure 20-4 **Sex Differentiation**

the **Wolffian ducts** and the Müllerian ducts degenerate, whereas in females the **Müllerian ducts** differentiate into the reproductive tract and the Wolffian ducts regress. Since both duct systems are present before sexual differentiation occurs, the early embryo has the potential to develop either a male or a female reproductive tract. Development of the reproductive tract along male or female lines is determined by the presence or absence of two hormones secreted by the fetal testes—*testosterone* and *Müllerian-inhibiting factor* (Fig. 20-4). A hormone released by the placenta, *human chorionic gonadotropin,* appears to be the stimulus for this early testicular secretion. Testosterone induces development of the Wolffian ducts into the male reproductive tract (epididymis, ductus deferens, ejaculatory duct, and seminal vesicles). This hormone, after being converted into **dihydrotestosterone (DHT),** is also responsible for differentiating the external genitalia into the penis and scrotum. Meanwhile, Müllerian-inhibiting factor causes regression of the Müllerian ducts. In the absence of testosterone and Müllerian-inhibiting factor in females, the Wolffian ducts regress, the Müllerian ducts

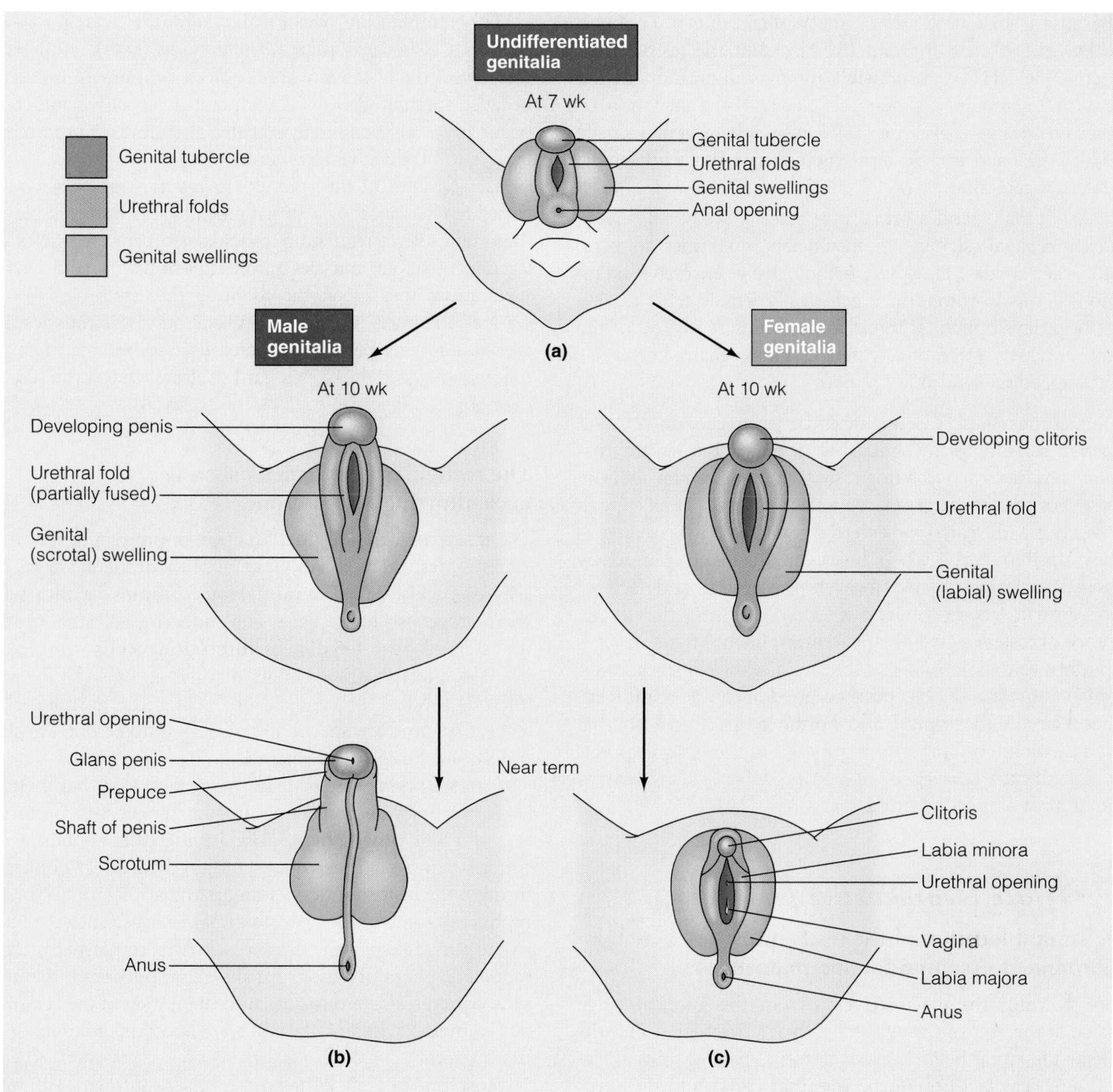

Figure 20-5 **Sexual Differentiation of the External Genitalia** (a) Undifferentiated stage (7 weeks). (b) Male development. (c) Female development.

develop into the female reproductive tract (oviducts, uterus, and vagina), and the external genitalia differentiate into the clitoris and labia.

Note that the indifferent embryonic reproductive tissue passively develops into a female structure unless actively acted on by masculinizing factors. In the absence of male testicular hormones, a female reproductive tract and external genitalia develop regardless of the genetic sex of the individual. Ovaries do not even need to be present for feminization of the fetal genital tissue. Such a control pattern for determining sex differentiation is appropriate considering that fetuses of both sexes are exposed to high concentrations of female sex hormones throughout gestation. If female sex hormones exerted influence over the development of the reproductive tract and external genitalia, all fetuses would be feminized.

In the usual case, genetic sex and sex differentiation are compatible; that is, a genetic male appears to be a male anatomically and functions as a male, and the same compatibility holds true for females. Occasionally, however, discrepancies occur between genetic and anatomical sexes because of errors in sex differentiation, as the following examples illustrate:

- If testes in a genetic male fail to properly differentiate and secrete hormones, the result is the development of an apparent anatomical female in a genetic male, who, of course, will be sterile.

- Because testosterone acts on the Wolffian ducts to convert them into a male reproductive tract but the testosterone derivative DHT is responsible for masculinization of the external genitalia, a genetic deficiency of the enzyme that converts testosterone into DHT results in a genetic male with testes and a male reproductive tract but with female external genitalia.
- The adrenal gland normally secretes a weak androgen, *dehydroepiandrosterone,* in insufficient quantities to masculinize females. However, pathologically excessive secretion of this hormone in a genetically female fetus during critical developmental stages imposes differentiation of the reproductive tract and genitalia along male lines (see adrenogenital syndrome, p. 666).

Sometimes these discrepancies between genetic sex and apparent sex are not recognized until puberty, when the discovery produces a psychologically traumatic gender identity crisis. For example, a masculinized genetic female with ovaries but with male-type external genitalia may be reared as a boy until puberty, when breast enlargement (caused by estrogen secretion by the awakening ovaries) and lack of beard growth (caused by lack of testosterone secretion in the absence of testes) signal an apparent problem. Therefore, it is important to diagnose any problems in sexual differentiation in infancy. Once a sex has been assigned, it can be reinforced, if necessary, with surgical and hormonal treatment so that psychosexual development can proceed as normally as possible. Less dramatic cases of inappropriate sex differentiation often appear as sterility problems.

Male Reproductive Physiology

The scrotal location of the testes provides a cooler environment essential for spermatogenesis.

Embryonically, the testes develop from the gonadal ridge located at the rear of the abdominal cavity. In the last months of fetal life, they begin a slow descent, passing out of the abdominal cavity through the **inguinal canal** into the scrotum, one testis dropping into each pocket of the scrotal sac. Testosterone from the fetal testes is responsible for inducing descent of the testes into the scrotum. Although the time is somewhat variable, descent is usually complete by the seventh month of gestation. As a result, descent is complete in 98% of full-term baby boys, but in a substantial percentage of premature male infants, the testes are still within the inguinal canal at birth. In most instances of retained testes, descent occurs naturally before puberty or can be encouraged with administration of testosterone. Rarely, a testis remains undescended into adulthood, a condition known as **cryptorchidism** ("hidden testis"). Following descent of the testes into the scrotum, the opening in the abdominal wall through which the inguinal canal passes closes snugly around the sperm-carrying duct and blood vessels that traverse between each testis and the abdominal cavity. Incomplete closure or rupture of this opening permits abdominal viscera to slip through, resulting in an **inguinal hernia.**

The temperature within the scrotum averages several degrees Celsius less than normal body (core) temperature. Descent of the testes into this cooler environment is essential because spermatogenesis is temperature sensitive and cannot occur at normal body temperature. Therefore, a cryptorchid is unable to produce viable sperm.

The position of the scrotum in relation to the abdominal cavity can be varied by a spinal reflex mechanism that plays an important role in regulating testicular temperature. Reflex contraction of scrotal muscles upon exposure to a cold environment raises the scrotal sac to bring the testes closer to the warmer abdomen. Conversely, relaxation of the muscles upon exposure to heat permits the scrotal sac to become more pendulous, moving the testes farther from the warm core of the body.

The testicular Leydig cells secrete masculinizing testosterone.

The testes perform the dual function of producing sperm and secreting testosterone. About 80% of the testicular mass consists of highly coiled **seminiferous tubules,** within which spermatogenesis takes place. The endocrine cells that produce testosterone—the **Leydig,** or **interstitial, cells**—are located in the connective tissue (interstitial tissue) between the seminiferous tubules (see Fig. 20-6b). Thus, the portions of the testes that produce sperm and secrete testosterone are structurally and functionally distinct.

Testosterone is a steroid hormone derived from a cholesterol precursor molecule, as are the female sex hormones, estrogen and progesterone. The Leydig cells contain a high concentration of the enzymes required to direct cholesterol through the testosterone-yielding pathway (see p. 627). Once produced, some of the testosterone is secreted into the blood, where it is transported, primarily bound to plasma proteins, to its target sites of action. A substantial portion of the newly synthesized testosterone goes into the lumen of the seminiferous tubules, where it plays an important role in sperm production.

Most but not all of testosterone's actions are ultimately directed toward ensuring delivery of sperm to the female. The effects of testosterone can be grouped into five categories: (1) effects on the reproductive system before birth; (2) effects on sex-specific tissues after birth; (3) other reproduction-related effects; (4) effects on secondary sexual characteristics; and (5) nonreproductive actions (Table 20-1).

Effects on the reproductive system before birth Before birth, testosterone secretion by the fetal testes is responsible for masculinizing the reproductive tract and external genitalia and for promoting descent of the testes into the scrotum, as already described. After birth, testosterone secretion ceases, and the testes and remainder of the reproductive system remain small and nonfunctional until puberty.

Effects on sex-specific tissues after birth **Puberty** refers to the period of arousal and maturation of the previously nonfunctional reproductive system, culminating in attainment of sexual maturity and the ability to reproduce. Its onset usually occurs sometime between the ages of ten and fourteen; on the average it begins about two years earlier in females than in males. Usually lasting three to five years, puberty encompasses a complex sequence of endocrine, physical, and behavioral events. **Adolescence** is a broader concept that refers to the entire transition period between childhood and adulthood, not just sexual maturation.

At puberty, the Leydig cells start secreting testosterone once again, and spermatogenesis is initiated in the seminiferous tubules for the first time. Testosterone is responsible for growth and maturation of the entire male reproductive system. Under the influence of the pubertal surge in testosterone secretion, the testes enlarge and become capable of spermatogenesis, the accessory sex glands enlarge and become secretory, and the penis and scrotum enlarge. Ongoing testosterone secretion is essential for spermatogenesis and for maintaining a mature male reproductive tract throughout adulthood. Following **castration** (surgical removal of the testes) or testicular failure caused by disease, the sex organs regress in size and function.

Other reproduction-related effects Testosterone is responsible for development of sexual libido at puberty and helps to maintain the sex drive in the adult male. Stimulation of this behavior by testosterone is important for facilitating delivery of sperm to females. In humans, libido is also influenced by many interacting social and emotional factors. Once libido has developed, testosterone is no longer absolutely required for its maintenance. Castrated males often remain sexually active but at a reduced level.

In another reproduction-related function, testosterone participates in the normal negative-feedback control of gonadotropin hormone secretion by the anterior pituitary, a topic that will be covered more thoroughly later.

Effects on secondary sexual characteristics All male secondary sexual characteristics depend on testosterone for their development and maintenance. These nonreproductive male characteristics induced by testosterone include (1) the male pattern of hair growth (for example, beard and chest hair and, in genetically predisposed men, baldness); (2) a deep voice caused by enlargement of the larynx and thickening of the vocal cords; (3) thick skin; and (4) the male body configuration (for example, broad shoulders and heavy arm and leg musculature) as a result of protein deposition. A male castrated before puberty (a **eunuch**) does not mature sexually nor does he develop secondary sexual characteristics.

Nonreproductive actions Testosterone exerts several important effects not related to reproduction. It has a general protein anabolic (synthesis) effect and promotes bone growth, thus contributing to the more muscular physique of males and to the pubertal growth spurt. Ironically, testosterone not only stimulates bone growth but eventually prevents further growth by sealing the growing ends of the long bones (that is, ossifying, or "closing," the epiphyseal plates—see p. 647). Testosterone also stimulates oil secretion by the sebaceous glands. This effect is most striking during the adolescent surge of testosterone secretion, predisposing the young man to develop acne.

In animals, testosterone induces aggressive behavior, but whether it influences human behavior other than in the area of sexual behavior is an unresolved issue. Even though some athletes and bodybuilders who take testosterone-like anabolic androgenic steroids to increase muscle mass have been observed to display more aggressive behavior (see p. 246), it is unclear to what extent general behavioral differences between the sexes are hormonally induced or are a result of social conditioning.

Once initiated at puberty, testosterone secretion and spermatogenesis occur continuously throughout the male's life. Testicular efficiency gradually declines after forty-five to fifty years of age, however, even though men in their seventies and beyond may continue to enjoy an active sex life and some even father a child at this late age. The gradual diminution in circulating testosterone levels and in sperm production is not caused by a decrease in stimulation of the testes but probably arises instead from degenerative changes associated with aging that occur in the small testicular blood ves-

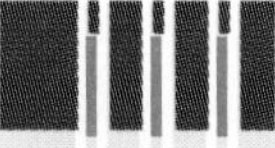

Table 20-1
Effects of Testosterone

Effects Before Birth

Masculinizes the reproductive tract and external genitalia
Promotes descent of the testes into the scrotum

Effects on Sex-Specific Tissues

Promotes growth and maturation of the reproductive system at puberty
Essential for spermatogenesis
Maintains the reproductive tract throughout adulthood

Other Reproductive Effects

Develops the sex drive at puberty
Controls gonadotropin hormone secretion

Effects On Secondary Sexual Characteristics

Induces the male pattern of hair growth (e.g., beard)
Causes the voice to deepen because of thickening of the vocal cords
Promotes muscle growth responsible for the male body configuration

Nonreproductive Actions

Exerts a protein anabolic effect
Promotes bone growth at puberty and then closure of the epiphyseal plates
May induce aggressive behavior

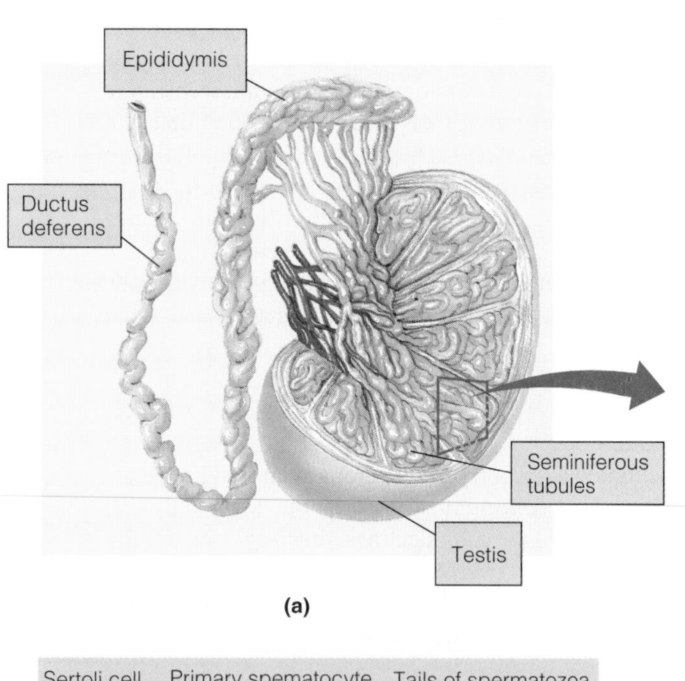

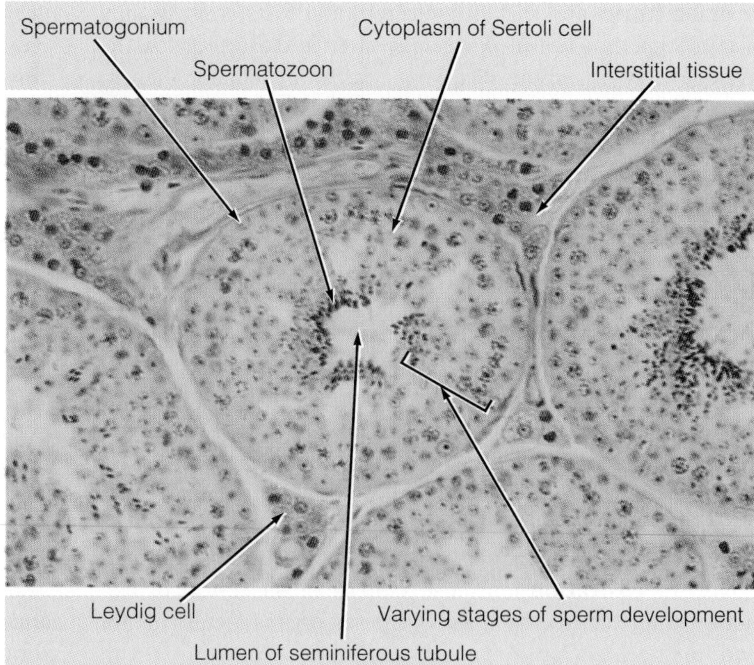

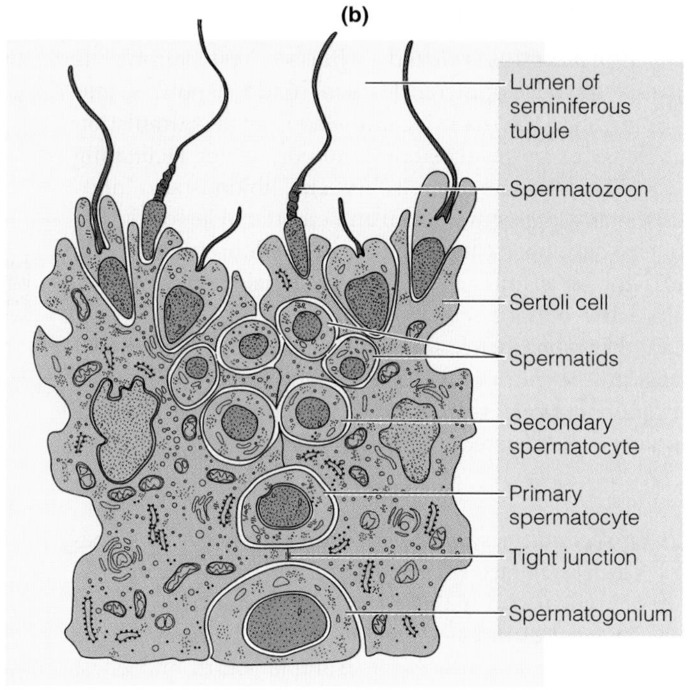

Figure 20-6 Testicular Anatomy Depicting the Site of Spermatogenesis (a) Longitudinal section of a testis showing the location and arrangement of the seminiferous tubules, the sperm-producing portion of the testis. (b) Light micrograph of a cross section of a seminiferous tubule. The undifferentiated germ cells (the spermatogonia) lie in the periphery of the tubule, and the differentiated spermatozoa are in the lumen, with the various stages of sperm development in between. (c) Scanning electron micrograph of a cross section of a seminiferous tubule. [Source: From *Tissues and Organs: A Text-Atlas of Scanning Electron Microscopy* by Richard G. Kessel and Randy H. Kardon (New York: W. H. Freeman and Company, 1979). Copyright by and reprinted with permission from the authors.] (d) Relationship of the Sertoli cells to the developing sperm cells.

sels. This gradual decline is often termed "male menopause," although it is not deliberately programmed as is female menopause. Some physicians have begun prescribing testosterone replacement therapy for their older patients who suffer from low libido. Opponents of this practice fear that supplemental testosterone, even in smaller doses than used for bodybuilding, may increase the risk of prostate cancer and heart disease.

Spermatogenesis yields an abundance of highly specialized, mobile sperm.

About 250 m (800 feet) of sperm-producing seminiferous tubules are packed within the testes (Fig. 20-6a). Two functionally important cell types are present in these tubules: *germ cells,* most of which are in various stages of sperm development, and *Sertoli cells,* which provide crucial support for spermatogenesis (Fig. 20-6b, c, and d). **Spermatogenesis** is a complex process by which relatively undifferentiated primordial germ cells, the **spermatogonia** (each of which contains a diploid complement of forty-six chromosomes), proliferate and are converted into extremely specialized, motile spermatozoa (sperm), each bearing a randomly distributed haploid set of twenty-three chromosomes.

Microscopic examination of a seminiferous tubule reveals layers of germ cells in an anatomical progression of sperm development, starting with the least differentiated in the outer layer and moving inward through various stages of division to the lumen, where the highly differentiated sperm are ready for exit from the testis (Fig. 20-6b, c and d). Spermatogenesis takes sixty-four days for development from a spermatogonium to a mature sperm. Up to several hundred million sperm may reach maturity daily. Spermatogenesis encompasses three major stages: *mitotic proliferation, meiosis,* and *packaging* (Fig. 20-7).

Mitotic proliferation Spermatogonia located in the outermost layer of the tubule continuously divide mitotically, with all new cells bearing the full complement of forty-six chromosomes that are identical to those of the parent cell. Such proliferation provides a continual supply of new germ cells. Following mitotic division of a spermatogonium, one of the daughter cells remains at the outer edge of the tubule as an undifferentiated spermatogonium, thus maintaining the germ cell line. The other daughter cell starts moving toward the lumen while undergoing the various steps required to form sperm, which will be released into the lumen. In humans, the sperm-forming daughter cell divides mitotically twice more to form four identical **primary spermatocytes.** After the last mitotic division, the primary spermatocytes enter a resting phase during which the chromosomes are duplicated and the doubled strands remain together in preparation for the first meiotic division.

Meiosis During meiosis, each primary spermatocyte (with forty-six doubled chromosomes) forms two **secondary spermatocytes** (each with twenty-three doubled chromosomes) during the first meiotic division, finally yielding four **sper-**

Figure 20-7 Spermatogenesis

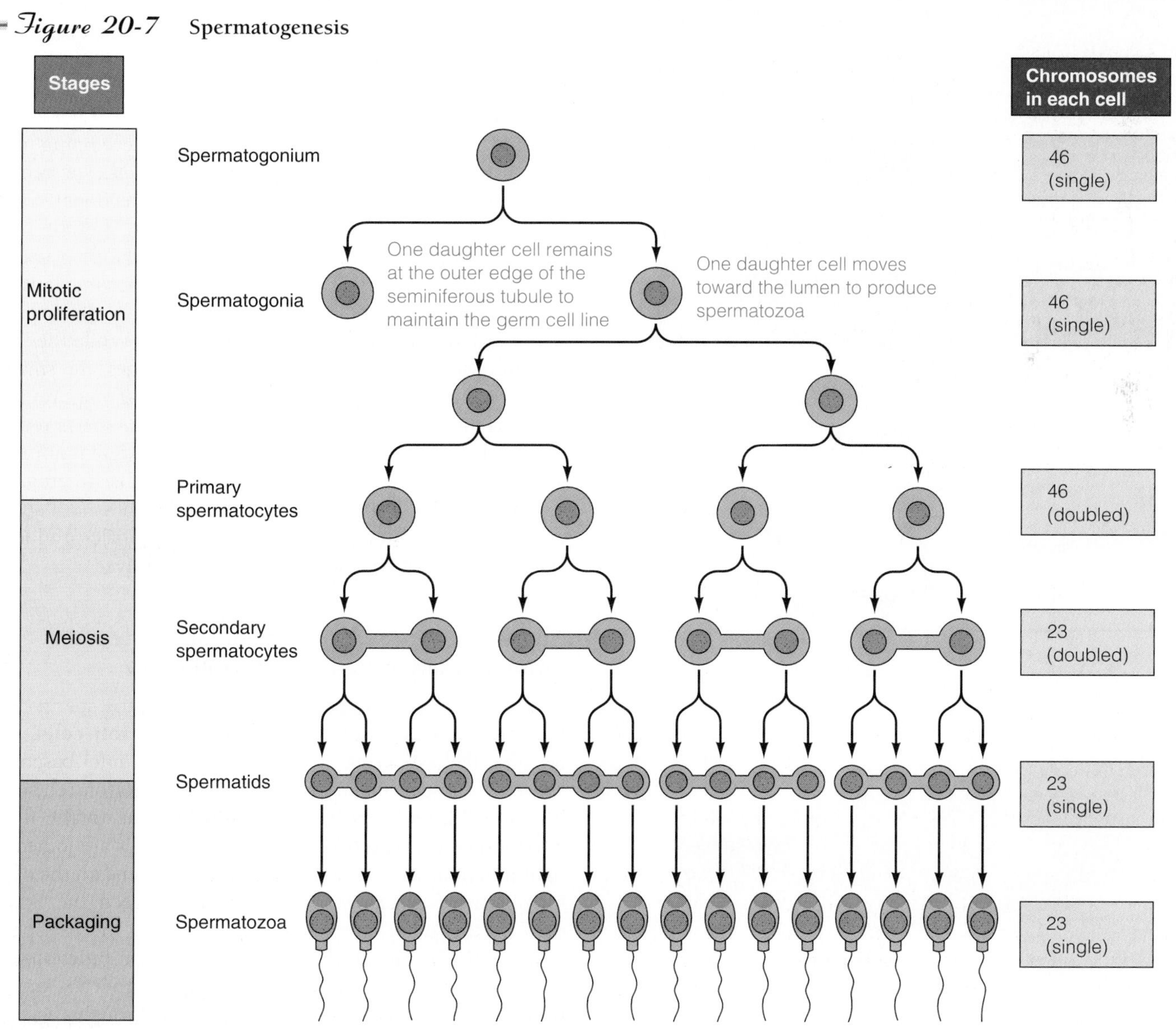

matids (each with twenty-three single chromosomes) as a result of the second meiotic division.

No further division takes place beyond this stage of spermatogenesis. Each spermatid is remodeled into a single spermatozoon. Because each sperm-producing spermatogonium mitotically produces four primary spermatocytes and each primary spermatocyte meiotically yields four spermatids (spermatozoa-to-be), the spermatogenic sequence in humans can theoretically produce sixteen spermatozoa each time a spermatogonium initiates this process. Usually, however, some cells are lost at various stages, so the efficiency of productivity is rarely this high.

Figure 20-8 **Anatomy of a Spermatozoon** (a) A phase-contrast photomicrograph of human spermatozoa. (b) Schematic representation of a spermatozoon in "frontal" view. (c) Longitudinal section of the head portion of a spermatozoon in "side" view.

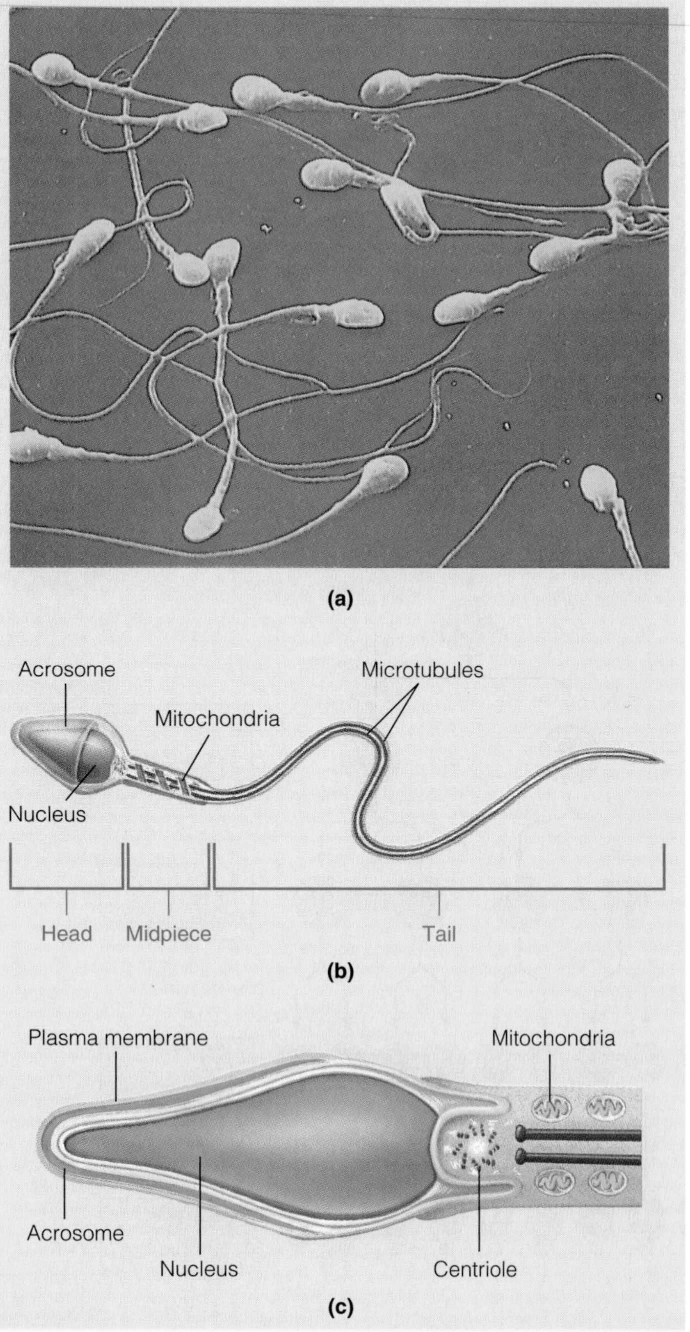

Packaging Even after meiosis, spermatids still resemble undifferentiated spermatogonia structurally, except for their half complement of chromosomes. Production of extremely specialized, mobile spermatozoa from spermatids requires extensive remodeling, or packaging, of cellular elements, a process known as **spermiogenesis.** Sperm are essentially "stripped-down" cells in which most of the cytosol and any organelles not needed for the task of delivering the sperm's genetic information to an ovum have been extruded. A **spermatozoon** has four parts (Fig. 20-8): a head, an acrosome, a midpiece, and a tail. The **head** consists primarily of the nucleus, which contains the sperm's complement of genetic information. The **acrosome,** an enzyme-filled vesicle at the tip of the head, is used as an "enzymatic drill" for penetrating the ovum. The acrosome is formed by aggregation of vesicles produced by the endoplasmic reticulum/Golgi complex before these organelles are discarded. Mobility for the spermatozoon is provided by a long, whiplike **tail** that grows out of one of the centrioles. Movement of the tail, which occurs as a result of relative sliding of its constituent microtubules (see p. 37), is powered by energy generated by the mitochondria concentrated within the **midpiece** of the sperm.

Until sperm maturation is complete, the developing germ cells arising from a single primary spermatocyte remain joined by cytoplasmic bridges. These connections, which result from incomplete cytoplasmic division, permit cytoplasm to be exchanged among the four developing sperm. This linkage is important because the X chromosome, but not the Y chromosome, contains genes that code for cellular products essential for development of the sperm. (While the large X chromosome contains several thousand genes, the small Y chromosome has only about fifteen, the most important of which is the SRY gene.) During meiosis, half the sperm receive an X and the other half a Y chromosome. If it were not for the sharing of cytoplasm so that all the haploid cells are provided with the products coded for by X chromosomes until sperm development is complete, the Y-bearing, male-producing sperm would not be able to develop and survive.

Throughout their development, sperm remain intimately associated with Sertoli cells.

In addition to the spermatogonia and developing sperm cells, the seminiferous tubules also house the **Sertoli cells.** The Sertoli cells form a ring that extends from the outer basement membrane to the lumen of the tubule. Each Sertoli cell spans the entire distance from the outer membrane to the fluid-filled lumen (Fig. 20-6d). Adjacent Sertoli cells are joined by tight junctions (see p. 52) at a point slightly beneath the outer membrane. Spermatogonia are tucked between the Sertoli cells at the outer perimeter of the tubule in the spaces between the basement membrane and the tight junctions.

During spermatogenesis, developing sperm cells arising from spermatogonial mitotic activity pass through the tight junctions, which transiently separate to make a path for them, then migrate toward the lumen in intimate association with the adjacent Sertoli cells. The cytoplasm of the Sertoli cells envelops the migrating germ cells, which remain buried within these cytoplasmic recesses throughout their development.

The supportive Sertoli cells perform the following functions essential for spermatogenesis:

1. The tight junctions between adjacent Sertoli cells form a **blood-testes barrier.** Because this barrier prevents blood-borne substances from passing between the cells to gain entry to the lumen of the seminiferous tubule, only selected molecules that are able to pass through the Sertoli cells reach the intratubular fluid. As a result, the composition of the intratubular fluid varies considerably from that of the blood. The unique composition of this fluid that bathes the germ cells is believed to be critical for later stages of sperm development. The blood-testes barrier also prevents the antibody-producing cells in the extracellular fluid from reaching the tubular sperm factory, thus preventing the formation of antibodies against the highly differentiated spermatozoa.
2. Since the secluded developing sperm cells do not have direct access to blood-borne nutrients, the Sertoli cells provide nourishment for them.
3. The Sertoli cells have an important phagocytic function. They engulf the cytoplasm extruded from the spermatids during their remodeling and destroy defective germ cells that fail to successfully complete all stages of spermatogenesis.
4. The Sertoli cells secrete into the lumen **seminiferous tubule fluid,** which "flushes" the released sperm from the tubule into the epididymis for storage and further processing.
5. An important component of this Sertoli secretion is **androgen-binding protein.** As the name implies, this protein binds androgens (that is, testosterone), thus maintaining a very high level of this hormone within the lumen of the seminiferous tubules. This high local concentration of testosterone is essential for sustaining sperm production. Androgen-binding protein is necessary to retain testosterone within the lumen because this steroid hormone is lipid soluble and could easily diffuse across the plasma membranes and leave the lumen.
6. The Sertoli cells are the site of action for control for spermatogenesis by both testosterone and follicle-stimulating hormone (FSH). The Sertoli cells themselves release another hormone, *inhibin,* which acts in negative-feedback fashion to regulate FSH secretion.

The two anterior pituitary gonadotropic hormones, LH and FSH, control testosterone secretion and spermatogenesis.

The testes are controlled by the two gonadotropic hormones secreted by the anterior pituitary, **luteinizing hormone (LH)** and **follicle-stimulating hormone (FSH),** which are named for their functions in females (see p. 639). These hormones act on separate components of the testes (Fig. 20-9).

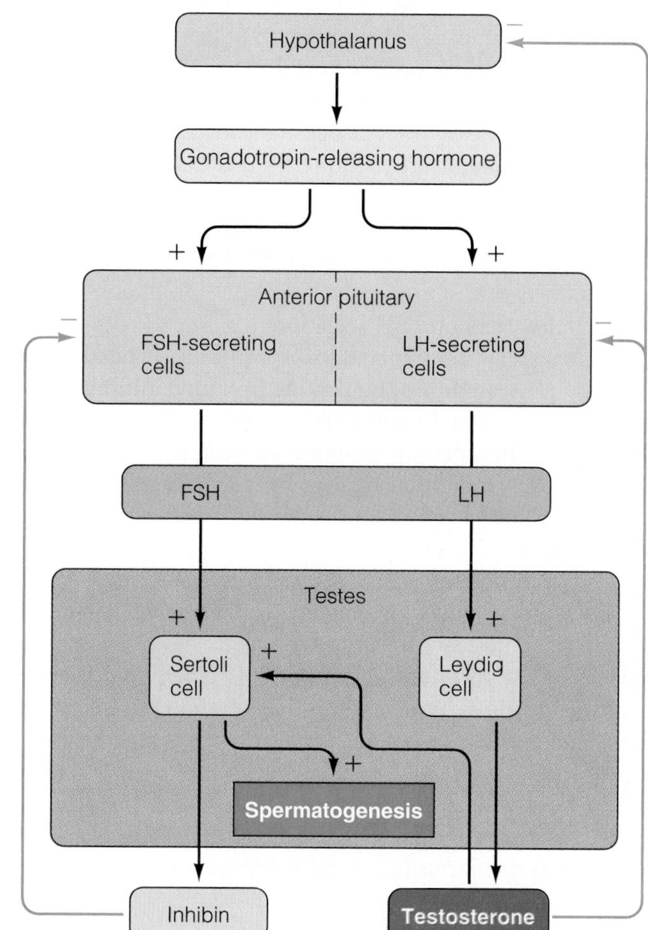

Figure 20-9 **Control of Testicular Function**

Luteinizing hormone acts on the Leydig (interstitial) cells to regulate testosterone secretion, accounting for its alternative name in males—*interstitial cell–stimulating hormone (ICSH).* Follicle-stimulating hormone acts on the seminiferous tubules, specifically the Sertoli cells, to enhance spermatogenesis. (There is no alternative name for FSH in males.) Secretion of both LH and FSH from the anterior pituitary is stimulated in turn by a single hypothalamic hormone, **gonadotropin-releasing hormone (GnRH)** (see p. 641).

Once every two to three hours, GnRH is released from the hypothalamus in secretory bursts, with no secretion occurring in between. Since GnRH stimulates the gonadotropic hormone secretory cells in the anterior pituitary, this pulsatile pattern of hypothalamic secretion results in similar episodic bursts in LH and FSH secretion.

Even though GnRH stimulates both LH and FSH secretion, the blood concentrations of these two gonadotropic hormones do not always parallel each other for two reasons. First, LH is removed from the blood more rapidly between the secretory bursts than is the more slowly metabolized FSH, so the pulsatile variations in blood levels of LH are much more pronounced than are those of FSH. Second, two other regulatory factors besides GnRH—*testosterone* and *inhibin*—differentially influence the secretory rate of LH and FSH. Testosterone, the product of LH stimulation of the Leydig cells, acts in negative-feedback fashion to inhibit LH secretion in two ways. The predominant negative-feedback effect of testosterone is to decrease the episodes of GnRH release by acting on the hypothalamus, thus indirectly decreasing both LH and FSH release by the anterior pituitary. In addition, testosterone acts directly on the anterior pituitary to reduce the responsiveness of the LH secretory cells to GnRH. The latter action explains why testosterone exerts a greater inhibitory effect on LH secretion than on FSH secretion.

The testicular inhibitory signal specifically directed at controlling FSH secretion is the peptide hormone **inhibin,** which is secreted by the Sertoli cells. Inhibin is believed to act directly on the anterior pituitary to inhibit FSH secretion. This feedback inhibition of FSH by a Sertoli cell product is appropriate, because FSH stimulates spermatogenesis by acting on the Sertoli cells.

Both testosterone and FSH play critical roles in controlling spermatogenesis, each exerting its effect by acting on the Sertoli cells. Testosterone is essential for both mitosis and meiosis of the germ cells, whereas FSH is required for spermatid remodeling. Testosterone concentration is much higher in the testes than in the blood because a substantial portion of this hormone produced locally by the Leydig cells is retained in the intratubular fluid complexed with androgen-binding protein secreted by the Sertoli cells. Only this high concentration of testicular testosterone is adequate to sustain sperm production.

Gonadotropin-releasing hormone activity increases at puberty.

Even though the fetal testes secrete testosterone, which directs masculine development of the reproductive system, after birth the testes become quiescent until puberty. During the prepubertal period, LH and FSH are not secreted at adequate levels to stimulate any significant testicular activity. The prepubertal delay in the onset of reproductive capability allows time for the individual to mature physically (though not necessarily psychologically) enough to handle childrearing. (This physical maturation is especially important in the female, whose body must support the developing fetus.)

The pubertal process is initiated by an increase in GnRH activity sometime between eight and twelve years of age. Early in puberty, GnRH secretory pulses occur only at night, causing brief nocturnal increases in LH secretion and accordingly, testosterone secretion. The duration of episodic GnRH secretion gradually increases as puberty progresses until the adult pattern of GnRH, FSH, LH, and testosterone secretion is established. Under the influence of the rising levels of testosterone during puberty, the physical changes that encompass the secondary sexual characteristics and reproductive maturation become evident.

The low level of GnRH activity during the prepubertal period appears to be due to active inhibition of GnRH release by both hormonal and neural mechanisms. Before puberty, the hypothalamus is extremely sensitive to the negative-feedback actions of testosterone, so the very small amounts of testosterone produced by the prepubertal testes are able to inhibit GnRH release. At puberty, the hypothalamus becomes less sensitive to feedback inhibition by testosterone. Because the low levels of testosterone no longer suppress the hypothalamus, GnRH and gonadotropic hormone levels rise. Also, during the prepubertal period, the hypothalamic GnRH cells are subjected to direct neural inhibition, which is similarly removed at puberty.

The factors responsible for removing these inhibitory mechanisms, thus initiating puberty in humans, remain a mystery. Three proposals, none of which is fully satisfactory, have been put forth. One possibility is a preprogrammed, age-related reduction in inhibitory activity. A second suggestion links attainment of a critical body weight or percentage of body fat to increased GnRH release. The leading proposal focuses on a potential role for the hormone *melatonin,* which is secreted by the *pineal gland* within the brain (see p. 635). Melatonin, whose secretion decreases during exposure to the light and increases during exposure to the dark, has an anti-gonadotropic effect in many species. Light striking the eyes inhibits the nerve pathways that are responsible for stimulating melatonin secretion. In many seasonally breeding species, the overall decrease in melatonin secretion in connection with longer days and shorter nights initiates the mating season. Some researchers suggest that an observed reduction in the overall rate of melatonin secretion at puberty in humans—particularly during the night, when the peaks in GnRH secretion first occur—is the trigger for the onset of puberty.

Recall that the pineal gland is also believed to have extremely widespread influences in addition to its probable link with the reproductive system. For example, it may serve as the timekeeper that synchronizes biological rhythms in hormone secretion and body temperature with external cues. Other proposed roles for melatonin include promoting sleep, acting as a powerful antioxidant, and enhancing immunity.

The ducts of the reproductive tract store and concentrate sperm and increase their motility and fertility as well.

The remainder of the male reproductive system (besides the testes) is designed to deliver sperm to the female reproductive tract. Essentially, it consists of (1) a tortuous pathway of tubes that transport sperm from the testes to the outside of the body; (2) several glands, which contribute secretions that are important to the viability and motility of the sperm; and (3) the penis, which is designed to penetrate and deposit the sperm within the vagina of the female.

A comma-shaped **epididymis** is loosely attached to the rear surface of each testis (Figs. 20-1 and 20-6a). After sperm are produced in the seminiferous tubules, they are swept into the epididymis as a result of the pressure created by the continual secretion of tubular fluid by the Sertoli cells. The epididymal ducts from each testis converge to form a large, thick-walled, muscular duct called the **ductus (vas) deferens.** The ductus deferens from each testis passes up out of the scrotal sac and runs back through the inguinal canal into the abdominal cavity, where it eventually empties into the urethra at the neck of the bladder (Fig. 20-1). The urethra carries sperm out of the penis during ejaculation, the forceful expulsion of semen from the body.

These ducts perform several important functions (▮ Table 20-2). The epididymis and ductus deferens serve as the sperm's exit route from the testis. As they leave the testis, the sperm are capable of neither movement nor fertilization. They gain both capabilities during their passage through the epididymis. This maturational process is stimulated by the testosterone retained within the tubular fluid bound to androgen-binding protein. Sperm's capacity to fertilize is enhanced even further by exposure to secretions of the female reproductive tract. This enhancement of sperm's capacity in the male and female reproductive tracts is known as **capacitation.** The epididymis also concentrates the sperm a hundredfold by absorbing most of the fluid that enters from the seminiferous tubules. The maturing sperm are slowly moved through the epididymis into the ductus deferens by rhythmic contractions of the smooth muscle in the walls of these tubes. The ductus deferens serves as an important site for sperm storage. Because the tightly packed sperm are relatively inactive and their metabolic needs are accordingly low, they can be stored in the ductus deferens for many days, even though they have no nutrient blood supply and are nourished only by simple sugars present in the tubular secretions.

In a **vasectomy,** a common sterilization procedure in males, a small segment of each ductus deferens (alias vas deferens, hence the term *vasectomy*) is surgically removed after it passes from the testis but before it enters the inguinal canal, thus blocking the exit of sperm from the testes. The sperm that build up behind the tied-off testicular end of the severed ductus are removed by phagocytosis. Although this procedure blocks sperm exit, it does not interfere with testosterone activity because the Leydig cells secrete testosterone into the blood, not through the ductus deferens. Thus, there should be no diminution of testosterone-dependent masculinity or libido following a vasectomy.

The accessory sex glands contribute the bulk of the semen.

Several accessory sex glands—the seminal vesicles and prostate—empty their secretions into the duct system before it joins the urethra (Fig. 20-1). A pair of saclike *seminal vesicles* empty into the last portion of the two ductus deferens, one on each side. The short segment of duct that passes beyond the entry point of the seminal vesicle to join the urethra constitutes the *ejaculatory duct.* The *prostate* is a large single gland that completely surrounds the ejaculatory ducts and urethra. In a significant number of men, prostatic enlargement occurs in middle to older age. Difficulty in urination is often encountered as the enlarging prostate impinges on the portion of the urethra that passes through the prostate. Another pair of accessory sex glands, the *bulbourethral glands,* drain into the urethra after it has passed through the prostate just before it enters the penis. Numerous mucus-secreting glands are also located along the length of the urethra.

During ejaculation, the accessory sex glands contribute secretions that provide support for the continuing viability of

Table 20-2 **Location and Functions of the Components of the Male Reproductive System**

Component	*Number and Location*	*Functions*
Testis	Pair; located in the scrotum, a skin-covered sac suspended within the angle between the legs	Produce sperm Secrete testosterone
Epididymis and ductus deferens	Pair; one epididymis attached to the rear of each testis; one ductus deferens travels from each epididymis up out of the scrotal sac through the inguinal canal and empties into the urethra at the neck of the bladder	Serve as the sperm's exit route from the testis Serve as the site for maturation of the sperm for motility and fertility Concentrate and store the sperm
Seminal vesicle	Pair; both empty into the last portion of the ductus deferens, one on each side	Supply fructose to nourish the ejaculated sperm Secrete prostaglandins that stimulate motility to help transport the sperm within the male and female Provide the bulk of the semen Provide precursors for the clotting of semen
Prostate gland	Single; completely surrounds the urethra at the neck of the bladder	Secretes an alkaline fluid that neutralizes the acidic vaginal secretions Triggers clotting of the semen to keep the sperm in the vagina during penis withdrawal
Bulbourethral gland	Pair; both empty into the urethra, one on each side, just before the urethra enters the penis	Secrete mucus for lubrication

the sperm inside the female reproductive tract. These secretions constitute the bulk of the **semen,** which consists of a mixture of accessory sex gland secretions, sperm, and mucus. Sperm make up only a small percentage of the total ejaculated fluid.

Although the accessory sex gland secretions are not absolutely essential for fertilization, they do make contributions that greatly facilitate the fertilization process (Table 20-2):

- The **seminal vesicles** (1) supply fructose, which serves as the primary energy source for ejaculated sperm; (2) secrete *prostaglandins,* which stimulate contractions of the smooth muscle in both the male and female reproductive tracts, thereby helping to transport sperm from their storage site in the male to the site of fertilization in the female oviduct; (3) provide more than half the semen, which helps to wash the sperm into the urethra and also dilutes the thick mass of sperm, thus enabling them to develop motility; and (4) secrete fibrinogen, a precursor of fibrin, which forms the meshwork of a clot (see p. 365).
- The **prostate gland** (1) secretes an alkaline fluid that neutralizes the acidic vaginal secretions, an important

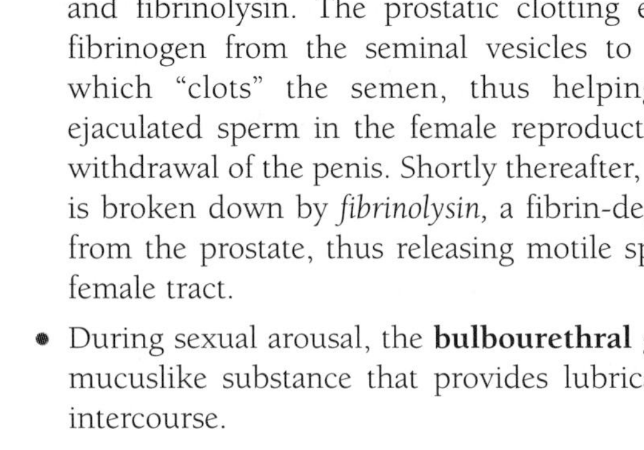

function because sperm are more viable in a slightly alkaline environment, and (2) provides clotting enzymes and fibrinolysin. The prostatic clotting enzymes act on fibrinogen from the seminal vesicles to produce fibrin, which "clots" the semen, thus helping to keep the ejaculated sperm in the female reproductive tract during withdrawal of the penis. Shortly thereafter, the seminal clot is broken down by *fibrinolysin,* a fibrin-degrading enzyme from the prostate, thus releasing motile sperm within the female tract.
- During sexual arousal, the **bulbourethral glands** secrete a mucuslike substance that provides lubrication for sexual intercourse.

Table 20-3
Known or Suspected Actions of Prostaglandins

Body System Activity	*Actions of Prostaglandins*
Reproductive system	Promote sperm transport by action on smooth muscle in the male and female reproductive tracts
	Important in menstruation
	Play a role in ovulation
	Contribute to preparation of the maternal portion of the placenta
	Contribute to parturition
Respiratory system	Some promote bronchodilation, others bronchoconstriction
Urinary system	Increase the renal blood flow
	Increase excretion of water and salt
Digestive system	Inhibit HCl secretion by the stomach
	Stimulate intestinal motility
Nervous system	Influence neurotransmitter release and action
	Act at the hypothalamic "thermostat" to increase body temperature
Endocrine system	Enhance cortisol secretion
	Influence tissue responsiveness to hormones in many instances
Circulatory system	Influence platelet aggregation
Fat metabolism	Inhibit fat breakdown
Defense system	Promote many aspects of inflammation, including fever and development of pain

Prostaglandins are ubiquitous, locally acting chemical messengers.

Although **prostaglandins** were first identified in the semen and were believed to be of prostate gland origin (hence their name, even though they are actually secreted into the semen by the seminal vesicles), their production and actions are by no means limited to the reproductive system. These 20-carbon fatty acid derivatives are among the most ubiquitous chemical messengers in the body. They are produced in virtually all tissues from arachidonic acid, a fatty acid constituent of the phospholipids within the plasma membrane. Prostaglandins (and other closely related arachidonic acid derivatives that are often included for convenience in the category of prostaglandins, namely, *prostacyclins, thromboxanes,* and *leukotrienes*) are among the most biologically active compounds known. Upon appropriate stimulation, arachidonic acid is split from the plasma membrane by a membrane-bound enzyme and then is converted into the appropriate prostaglandin, which acts locally within or near its site of production. After prostaglandins act, they are rapidly inactivated by local enzymes before they gain access to the blood, or if they do reach the circulatory system, they are swiftly degraded on their first pass through the lungs so that they are not dispersed through the systemic arterial system.

Prostaglandins are designated as belonging to one of three groups—PGA, PGE, or PGF—according to structural variations in the five-carbon ring that they contain at one end (Fig. 20-10). Within each group, prostaglandins are further identified by the number of double bonds present in the two side chains that project from the ring structure (for example, PGE_1 has one double bond and PGE_2 has two double bonds).

Prostaglandins exert a bewildering variety of effects. Not only are slight variations in prostaglandin structure accompanied by profound differences in biological action, but the same prostaglandin molecule may even exert opposite effects in different tissues. Besides enhancing sperm transport in semen, these abundant chemical messengers are known or suspected to exert other actions in the female reproductive system and in the respiratory, urinary, digestive, nervous, and endocrine systems, in addition to having effects on platelet aggregation, fat metabolism, and inflammation (Table 20-3).

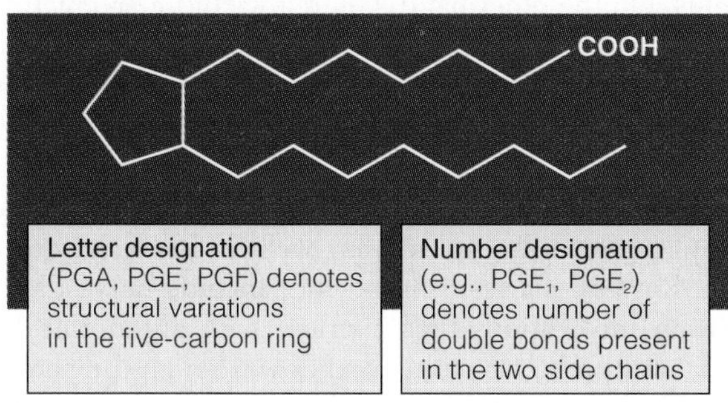
COOH
Letter designation
(PGA, PGE, PGF) denotes
structural variations
in the five-carbon ring
Number designation
(e.g., PGE_1, PGE_2)
denotes number of
double bonds present
in the two side chains

Figure 20-10 **Structure and Nomenclature of Prostaglandins**

As prostaglandins' various actions are better understood, new ways of manipulating them therapeutically are becoming available. A classic example is the use of aspirin, which blocks the conversion of arachidonic acid into prostaglandins, for fever reduction and pain relief. Prostaglandin action is also therapeutically inhibited in the treatment of premenstrual symptoms and menstrual cramping. Furthermore, specific prostaglandins have been medically administered in such diverse situations as inducing labor, treating asthma, and treating gastric ulcers.

Sexual Intercourse between Males and Females

Ultimately, union of male and female gametes in humans requires delivery of semen into the female vagina through the **sex act**, also known as **sexual intercourse, coitus,** or **copulation.**

The male sex act is characterized by erection and ejaculation.

The *male sex act* involves two components: (1) **erection,** or hardening of the normally flaccid penis to permit its entry into the vagina, and (2) **ejaculation,** or forceful expulsion of semen into the urethra and out of the penis (Table 20-4). In addition to these strictly reproduction-related components, the **sexual response cycle** also encompasses broader physiological responses that can be divided into four phases:

1. The *excitement phase,* which includes erection accompanied by testicular vasocongestion (engorgement with blood) and heightened sexual awareness.
2. The *plateau phase,* which is characterized by intensification of these responses, plus more generalized body responses, such as steadily increasing heart rate, blood pressure, respiratory rate, and muscle tension.
3. The *orgasmic phase,* which includes ejaculation as well as other responses that culminate the mounting sexual excitement and are collectively experienced as an intense physical pleasure.
4. The *resolution phase,* which returns the genitalia and body systems to their prearousal state.

The human sexual response is a multicomponent experience that, in addition to these physiological phenomena, also encompasses emotional, psychological, and sociological factors. We will examine only the physiological aspects of sex.

Erection is not caused by contraction of skeletal muscles within the penis, as might be expected, but by engorgement of the penis with blood. The penis consists almost entirely of **erectile tissue** made up of three columns of spongelike vascular spaces extending the length of the organ (Fig. 20-1). In the absence of sexual excitation, the erectile tissues contain little blood, because the arterioles that supply these vascular chambers are constricted. As a result, the penis remains small and flaccid. During sexual arousal, these arterioles reflexly dilate and the erectile tissue fills with blood, causing the penis to enlarge both in length and width and to become more rigid. A larger buildup of blood and further enhancement of erection are achieved by a reduction in venous outflow. The veins that drain the erectile tissue are compressed as a result of engorgement and expansion of the vascular spaces brought about by increased arterial inflow. These local vascular responses—reflexly increased arterial inflow and mechanically reduced venous outflow—transform the penis into a hardened, elongated organ capable of penetrating the vagina.

Table 20-4 **Components of the Male Sex Act**

Components of the Male Sex Act	*Definition*	*Accomplished By*
Erection	Hardening of the normally flaccid penis to permit its entry into the vagina	Engorgement of the penis erectile tissue with blood as a result of marked parasympathetically induced vasodilation of the penile arterioles and mechanical compression of the veins.
Ejaculation		
Emission phase	Emptying of sperm and accessory sex gland secretions (semen) into the urethra	Sympathetically induced contraction of the smooth muscle in the walls of the ducts and accessory sex glands
Expulsion phase	Forceful expulsion of semen from the penis	Motor neuron–induced contraction of the skeletal muscles at the base of the penis

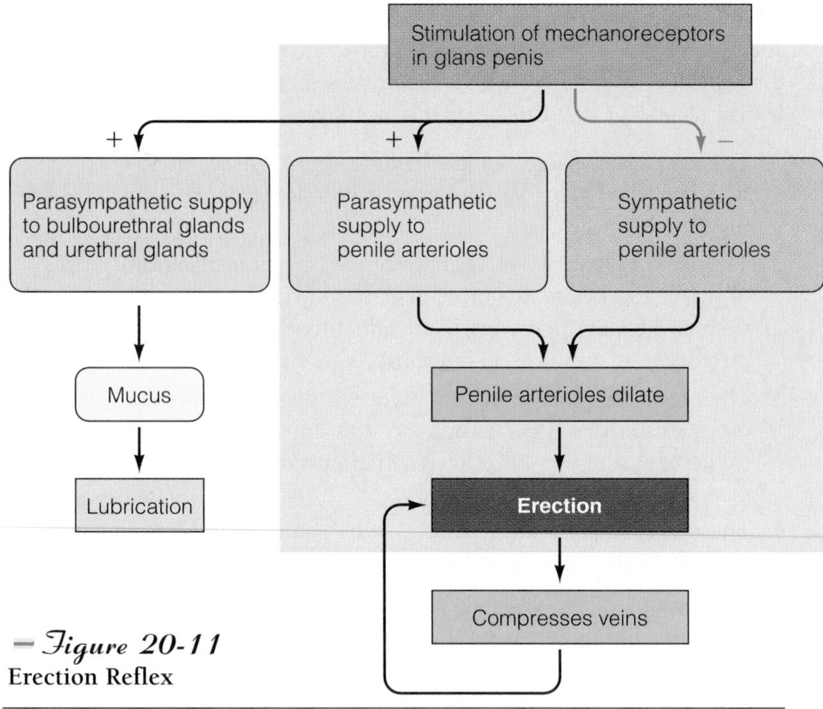

Figure 20-11
Erection Reflex

The erection reflex is a spinal reflex triggered by stimulation of highly sensitive mechanoreceptors located in the glans penis, which caps the tip of the penis. Tactile stimulation of the glans reflexly triggers increased parasympathetic vasodilator activity and decreased sympathetic vasoconstrictor activity to the penile arterioles; the result is rapid, pronounced vasodilation of these arterioles and an ensuing erection (Fig. 20-11). This parasympathetically induced vasodilation is the major instance of direct parasympathetic control over blood vessel caliber in the body. Recent evidence suggests that parasympathetic stimulation brings about relaxation of penile arteriolar smooth muscle by means of nitric oxide (alias endothelial-derived relaxing factor), which is known to cause arteriolar vasodilation in response to local tissue changes elsewhere in the body (see p. 318). Arterioles are typically supplied only by sympathetic nerves, with increased sympathetic activity producing vasoconstriction and decreased sympathetic activity resulting in vasodilation. Concurrent parasympathetic stimulation and sympathetic inhibition of penile arterioles accomplish vasodilation more rapidly and in greater magnitude than is possible in other arterioles supplied only by sympathetic nerves. Through this efficient means of rapidly increasing blood flow into the penis, complete erection can be accomplished in as little as 5 to 10 seconds. At the same time, parasympathetic impulses promote secretion of lubricating mucus from the bulbourethral glands and the urethral glands in preparation for coitus.

This basic spinal reflex can be either facilitated or inhibited by higher brain centers through descending pathways that also terminate on the autonomic nerves supplying the penile arterioles. As an example of facilitation, psychic stimuli, such as viewing something sexually exciting, can induce an erection in the complete absence of tactile stimulation of the penis. On the other hand, failure to achieve an erection in spite of appropriate stimulation **(impotence)** may occur as a result of inhibition of the erection reflex by higher brain centers. In fact, impotence is more commonly caused by psychological influences than by actual physical limitations. A man who becomes overly anxious about his ability to perform the sex act may well be on his way to failure. Physical causes of impotence include nerve damage, certain medications that interfere with autonomic function, and problems with blood flow through the penis.

The second component of the male sex act is *ejaculation*. Like erection, ejaculation is accomplished by a spinal reflex. The same types of tactile and psychic stimuli that induce erection cause ejaculation when the level of excitation intensifies to a critical peak. The overall ejaculatory response occurs in two phases: emission and expulsion. First, sympathetic impulses cause sequential contraction of smooth muscles in the prostate, reproductive ducts, and seminal vesicles. This contractile activity delivers prostatic fluid, then sperm, and finally seminal vesicle fluid (collectively, *semen*) into the urethra. This phase of the ejaculatory reflex is known as **emission.** During this time, the sphincter at the neck of the bladder is tightly closed to prevent semen from entering the bladder and urine from being expelled along with the ejaculate through the urethra. Second, the filling of the urethra with semen triggers nerve impulses that activate a series of skeletal muscles at the base of the penis. Rhythmic contractions of these muscles occur at 0.8-second intervals and increase the pressure within the penis, forcibly expelling the semen through the urethra to the exterior. This is the **expulsion** phase of ejaculation.

The rhythmic contractions that occur during semen expulsion are accompanied by involuntary rhythmic throbbing of pelvic muscles and peak intensity of the overall body responses that were climbing during the earlier phases. Heavy breathing, a heart rate of up to 180 beats per minute, marked generalized, skeletal muscle contraction, and heightened emotions are characteristic. These pelvic and overall systemic responses that culminate the sex act are associated with an intense pleasure characterized by a feeling of release and complete gratification, an experience known as **orgasm.**

During the resolution phase following orgasm, sympathetic vasoconstrictor impulses slow the inflow of blood into the penis, causing the erection to subside. A deep relaxation ensues, often accompanied by a feeling of fatigue. Muscle tone returns to normal while the cardiovascular and respiratory systems return to their prearousal level of activity. Once ejaculation has occurred, a temporary refractory period of variable duration ensues before sexual stimulation can trigger another erection. Males are therefore unable to experience multiple orgasms within a matter of minutes as females sometimes do.

The volume and sperm content of the ejaculate depend on the length of time between ejaculations. The average volume of semen is 2.75 ml, ranging from 2 to 6 ml, the higher volumes following periods of continence. An average human ejaculate contains about 180 million sperm (66 million/ml), but some ejaculates contain as many as 400 million sperm.

Environmental "Estrogens": Bad News for the Reproductive System

Unknowingly, we have been polluting our environment with estrogenic chemicals as an unintended side effect of industrialization during the last fifty years. Sometimes known as **xenoestrogens** (*xeno* means "foreign"), these environmental contaminants mimic or alter the action of estrogen, the feminizing steroid hormone produced by the female ovaries. Though not yet conclusive, laboratory and field studies suggest that xenoestrogens might be responsible for some disturbing trends in reproductive health problems, such as falling sperm counts in males and an increased incidence of breast cancer in females.

Estrogenic pollutants are everywhere in our environment. They contaminate our food, drinking water, and air. Proved xenoestrogens include (1) certain weed killers and insecticides, (2) some detergent breakdown products, (3) petroleum by-products found in car exhaust, (4) a common food preservative used to retard rancidity, and (5) softeners that make plastics flexible. These plastic softeners are commonly found in food packaging and can readily leach into fluids with which they come in contact. They are among the most plentiful industrial contaminants in our environment.

Scientists are only beginning to identify and understand the implications for reproductive health of the myriad synthetic chemicals that have become such an integral part of modern societies. They suspect that estrogen-mimicking chemicals may underlie a spectrum of reproductive disorders that have been on the rise during the past fifty years—the same time period during which large amounts of these pollutants have been introduced into our environment. Following are examples of problems in male reproductive function that may be circumstantially linked to xenoestrogen exposure:

- *Falling sperm counts.* The average sperm count has fallen from 113 million sperm per milliliter of semen in 1940 to 66 million/ml in 1990. Making matters worse, the volume of a single ejaculate has declined from 3.40 ml to 2.75 ml. This means that men on average are now ejaculating less than half the number of sperm as men did fifty years ago—a drop from more than 380 million sperm to about 180 million sperm per ejaculate. Furthermore, the number of motile sperm has also dipped. Importantly, the sperm count has not declined in the less polluted areas of the world during the same time period.
- *Increased incidence of testicular cancer.* Cases of testicular cancer have tripled since 1940, and the rate continues to climb.
- *Rising number of male reproductive tract abnormalities at birth.* The incidence of *cryptorchidism* (undescended testis) nearly doubled from the 1950s to the 1970s. The number of cases of *hypospadia,* a malformation of the penis, more than doubled between the mid-1960s and the mid-1980s. Hypospadia results when the urethral fold fails to fuse closed during the development of a male fetus.
- *Evidence of gender bending in animals.* Some fish and wild animal populations that have been severely exposed to environmental estrogens—such as those living in or near water heavily polluted with hormone-mimicking chemical wastes—display a high rate of grossly impaired reproductive systems. Examples include male fish that are hermaphrodites (possessing both male and female reproductive parts) and male alligators with abnormally small penises. Presumably, excessive estrogen exposure is emasculating these populations.

Xenoestrogens are also implicated in the rising incidence of breast cancer in females. Breast cancer is 25% to 30% more prevalent now than in the 1940s. Many of the established risk factors for breast cancer, such as starting to menstruate earlier than usual and undergoing menopause later than usual, are associated with an elevation in the total lifetime exposure to estrogen. Since increased exposure to natural estrogen bumps up the risk for breast cancer, prolonged exposure to environmental estrogens might be contributing to the rising prevalence of this malignancy among women (and men too).

Although many feminizing pollutants act by mimicking estrogen, others exert their effect by blocking the male's own androgens from accomplishing their intended effects. Antiandrogenic action demasculinizes males, producing an outcome similar to that brought about by estrogenic agents.

(See the accompanying boxed feature, ◆Concepts, Challenges, and Controversies.) Both quantity and quality of the sperm are important determinants of fertility. A man is considered clinically infertile if his sperm concentration falls below 20 million/ml of semen. Even though only one spermatozoon actually fertilizes the ovum, large numbers of accompanying sperm are needed to provide sufficient acrosomal enzymes to break down the barriers surrounding the ovum until the victorious sperm penetrates into the ovum's cytoplasm. The quality of sperm also must be taken into account when assessing the fertility potential of a semen sample. The presence of substantial numbers of sperm with abnormal motility or structure, such as sperm with distorted tails, reduces the chances of fertilization.

The female sexual cycle parallels that of males in many ways.

Both sexes experience the same four phases of the sexual cycle—excitement, plateau, orgasm, and resolution. Furthermore, the physiological mechanisms responsible for orgasm are fundamentally the same in males and females.

The excitement phase in females can be initiated by either physical or psychological stimuli. Tactile stimulation of the clitoris and surrounding perineal area is an especially powerful sexual stimulus. These stimuli trigger spinal reflexes that bring about parasympathetically induced vasodilation of arterioles throughout the vagina and external genitalia. The resultant inflow of blood becomes evident as swelling of the labia and erection of the clitoris. The latter—like its male homologue, the penis—is composed largely of erectile tissue. Vasocongestion of the vaginal capillaries forces fluid out of the vessels into the vaginal lumen. This fluid, which is the first positive indication of sexual arousal, serves as the primary lubricant for intercourse. Additional lubrication is provided by the mucus secretions from the male and by mucus released during sexual arousal from glands located at the outer opening of the vagina. Also during the excitement phase in the female, the nipples become erect and the breasts enlarge as a result of vasocongestion. In addition, the majority of women show a *sex flush* during this time, which is caused by increased blood flow through the skin.

During the plateau phase, the changes initiated during the excitement phase intensify, while systemic responses similar to those in the male (such as increased heart rate, blood pressure, respiratory rate, and muscle tension) occur. Further vasocongestion of the lower third of the vagina during this time reduces its inner capacity so that it tightens around the thrusting penis, thus heightening tactile sensation for the male. Simultaneously, the uterus raises upward, lifting the cervix and enlarging the upper two-thirds of the vagina. This ballooning, or **tenting effect,** creates a space for ejaculate deposition.

If erotic stimulation continues, the sexual response culminates in orgasm as sympathetic impulses trigger rhythmic contractions of the pelvic musculature at 0.8-second intervals, the same rate as in males. The contractions occur most intensely in the engorged lower third of the vaginal canal; hence this region is often termed the **orgasmic platform.** Systemic responses identical to those of the male orgasm also occur. In fact, the orgasmic experience in females parallels that of males with two exceptions. First, there is no female counterpart to ejaculation. Second, females do not become refractory following an orgasm, so they can respond immediately to continued erotic stimulation and achieve multiple orgasms. If stimulation continues, the sexual intensity only diminishes to the plateau level following orgasm and can quickly be brought to a peak again. Women have been known to achieve as many as twelve successive orgasms in this manner.

During resolution, pelvic vasocongestion and the systemic manifestations gradually subside. As with males, this is a time of great physical relaxation for females.

Female Reproductive Physiology

Complex cycling characterizes female reproductive physiology.

Female reproductive physiology is much more complex than male reproductive physiology. Unlike the continuous sperm production and essentially constant testosterone secretion characteristic of the male, release of ova is intermittent, and secretion of female sex hormones displays wide cyclical swings. The tissues influenced by these sex hormones also undergo cyclical changes, the most obvious of which is the monthly menstrual cycle. During each cycle, the female reproductive tract is prepared for the fertilization and implantation of an ovum released from the ovary at ovulation. If fertilization does not occur, the cycle repeats itself. If fertilization does occur, the cycles are interrupted while the female system adapts to nurture and protect the newly conceived human being until it has developed into an individual capable of living outside the maternal environment. Furthermore, the female continues her reproductive duties after birth by producing milk (lactation) for the baby's nourishment. Thus, the female reproductive system is characterized by complex cycles that are interrupted only by more complex changes should pregnancy ensue.

The ovaries, as the primary female reproductive organs, perform the dual function of producing ova (oogenesis) and secreting the female sex hormones, estrogen and progesterone. These hormones act together to promote fertilization of the ovum and to prepare the female reproductive system for pregnancy. Estrogen in the female is responsible for many functions similar to those carried out by testosterone in the male, such as maturation and maintenance of the entire female reproductive system and establishment of female secondary sexual characteristics. In general, the actions of estrogen are important to preconception events. Estrogen is essential for ova maturation and release, development of physical characteristics that are sexually attractive to males, and transport of sperm from the vagina to the site of fertilization in the oviduct. Furthermore, estrogen contributes to breast development in anticipation of lactation. The other ovarian steroid, progesterone, is important in preparing a suitable environment for nourishing a developing embryo/fetus and for contributing to the breasts' ability to produce milk.

As in males, reproductive capability begins at puberty in females, but unlike males, who have reproductive potential through life, female reproductive potential ceases during middle age at **menopause.**

Chromosome division in oogenesis parallels that in spermatogenesis, but there are major qualitative and quantitative sexual differences in gametogenesis.

Oogenesis contrasts sharply with spermatogenesis in several important aspects, even though the identical steps of chromosome replication and division take place during gamete production in both sexes. The undifferentiated primordial germ cells in the fetal ovaries, the **oogonia** (comparable to the spermatogonia), divide mitotically to give rise to 6 to 7 million oogonia by the fifth month of gestation, at which time mitotic proliferation ceases. During the last part of fetal life, the oogonia begin the early steps of the first meiotic division but do not complete it. Known now as **primary oocytes,** they contain forty-six replicated chromosomes, which are gathered into homologous pairs but do not separate. The primary oocytes remain in this state of meiotic arrest for years until they are prepared for ovulation.

Before birth, each primary oocyte is surrounded by a single layer of **granulosa cells** to form a **primary follicle.** Oocytes that fail to be incorporated into follicles degenerate, and at birth only about 2 million primary follicles remain, each containing a single primary oocyte capable of producing a single ovum. No new oocytes or follicles appear after birth; the follicles already present in the ovaries at birth serve as a reservoir from which all ova throughout the reproductive life of a female must arise. Of these follicles, only about 400 will mature and release ova. The pool of primary follicles present at birth gives rise to an ongoing trickle of developing follicles. Once it starts to develop, a follicle is destined for one of two fates: it will reach maturity and ovulate, or it will degenerate to form scar tissue, a process known as **atresia.** Until puberty, all of the follicles that start to develop undergo atresia in the early stages without ever ovulating. Even for the first few years after puberty, many of the cycles are **anovulatory** (that is, no ovum is released). Of the initial pool of follicles, 99.98% never ovulate but instead undergo atresia at some stage in development. By menopause, which occurs on average in a woman's early fifties, few, if any, primary follicles remain, having either already ovulated or become atretic. From this point on, the woman's reproductive capacity ceases.

This limited gamete potential, which is already determined at birth in females, is in sharp contrast to the continual process of spermatogenesis in males, who have the potential to produce several hundred million sperm in a single day. Furthermore, there is considerable chromosome wastage in oogenesis compared with spermatogenesis.

The primary oocyte within a primary follicle is still a diploid cell that contains forty-six doubled chromosomes. From puberty until menopause, a portion of the resting pool of follicles starts developing into **secondary (antral) follicles** on a cyclical basis. The number of developing follicles at any time is roughly proportional to the size of the pool, but the mechanisms that determine which follicles in the pool will develop during a given cycle are unknown. Development of a secondary follicle is characterized by growth of the primary oocyte and by expansion and differentiation of the surrounding cell layers. Oocyte enlargement is due to a buildup of cytoplasmic materials that will be needed by the early embryo. Just before ovulation, the primary oocyte, whose nucleus has been in meiotic arrest for years, completes its first meiotic division. This division yields two daughter cells, each receiving a set of twenty-three doubled chromosomes, analogous to the formation of secondary spermatocytes (— Fig. 20-12). However, almost all of the cytoplasm remains with one of the daughter cells, now called the **secondary oocyte,** which is destined to become the ovum. The chromosomes of the other daughter cell together with a small share of cytoplasm form the **first polar body.** In this way, the ovum-to-be loses half of its chromosomes to form a haploid gamete but retains all of its nutrient-rich cytoplasm. The nutrient-poor polar body soon degenerates.

— *Figure 20-12* **Oogenesis** Compare with Figure 20-7, spermatogenesis.

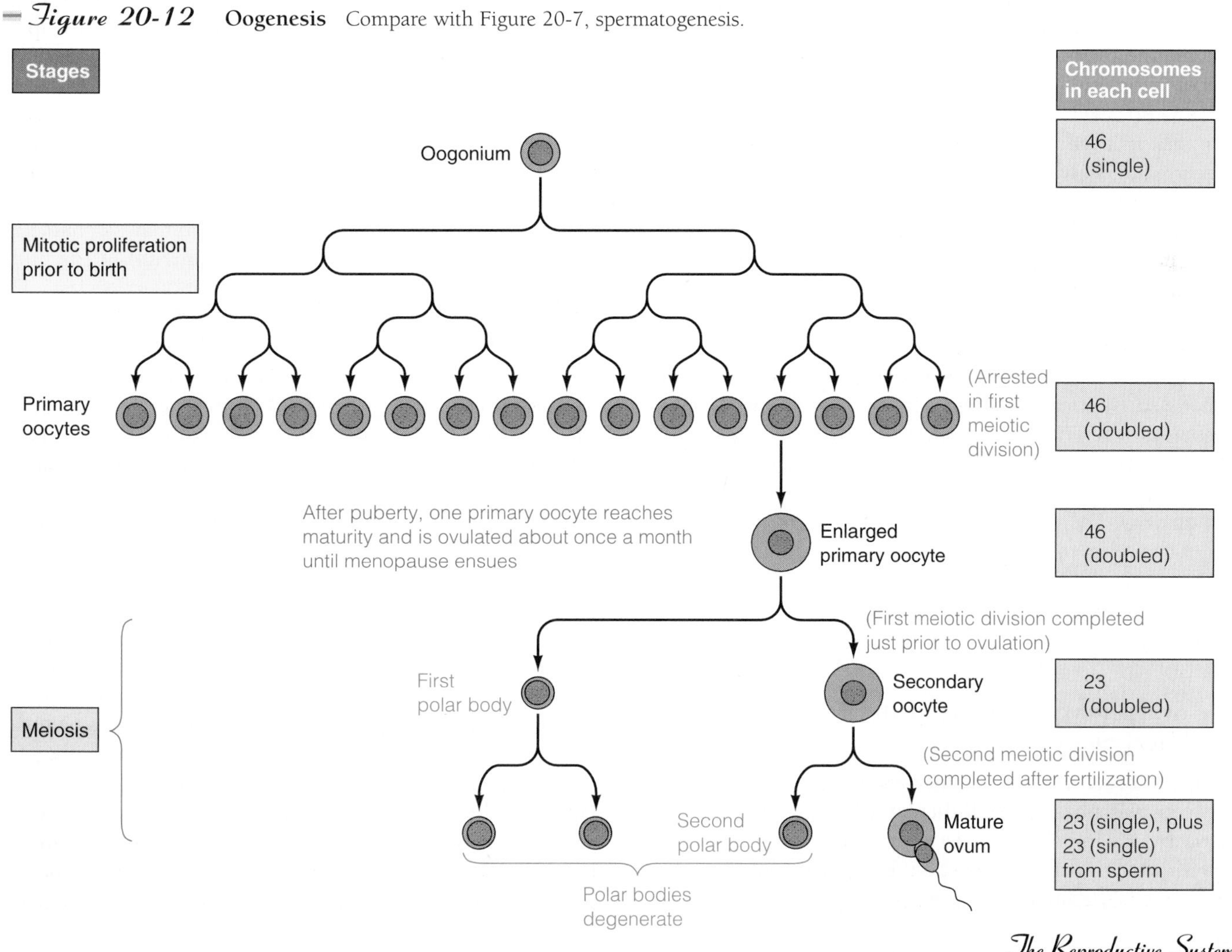

It is actually the secondary oocyte, and not the mature ovum, that is ovulated and fertilized, but common usage refers to the developing female gamete as an ovum even in its primary and secondary oocyte stages. Sperm entry into the secondary oocyte is needed to trigger the second meiotic division. Oocytes that are not fertilized never complete this final division. During this division, a half set of chromosomes along with a thin layer of cytoplasm is extruded as the **second polar body.** The other half set of twenty-three unpaired chromosomes remains behind in what is now the **mature ovum.** These twenty-three maternal chromosomes unite with the twenty-three paternal chromosomes of the penetrating sperm to complete fertilization. If the first polar body has not already degenerated, it too undergoes the second meiotic division at the same time the fertilized secondary oocyte is dividing its chromosomes.

Thus, the steps involved in chromosome distribution during oogenesis parallel those of spermatogenesis except that the cytoplasmic distribution and time span for completion sharply differ. Just as four haploid spermatids are produced by each primary spermatocyte, four haploid daughter cells are produced by each primary oocyte (if the first polar body does not degenerate before it completes the second meiotic division). In spermatogenesis, each daughter cell develops into a highly specialized, motile spermatozoon unencumbered by unessential cytoplasm and organelles, its only destiny being to supply half of the genes for a new individual. In oogenesis, however, of the four daughter cells, only the one destined to become the ovum receives cytoplasm. This uneven distribution of cytoplasm is important, because the ovum, in addition to providing half the genes, provides all of the cytoplasmic components needed to support early development of the fertilized ovum. The large, relatively undifferentiated ovum contains numerous nutrients, organelles, and structural and enzymatic proteins. The three other cytoplasm-scarce daughter cells, the polar bodies, rapidly degenerate, their chromosomes being deliberately wasted.

Note also the considerable difference in time required for completion of spermatogenesis and oogenesis. It takes about two months for a spermatogonium to develop into fully remodeled spermatozoa. In contrast, development of an oogonium (present before birth) to a mature ovum requires anywhere from eleven years (beginning of ovulatory cycles at onset of puberty) to fifty years (end of ovulation at onset of menopause). There is considerable speculation that the older age of ova released by women in their late thirties and forties accounts for the higher incidence of genetic abnormalities, such as Down's syndrome, in children born to women in this age range.

The ovarian cycle consists of alternating follicular and luteal phases.

After the onset of puberty, the ovary constantly alternates between two phases: the **follicular phase,** which is dominated by the presence of *maturing follicles,* and the **luteal phase,** which is characterized by the presence of the *corpus luteum* (to be described shortly). This cycle is normally interrupted only by pregnancy and is finally terminated by menopause. The average ovarian cycle lasts twenty-eight days, but this varies among women and among cycles in any particular woman. The follicle operates in the first half of the cycle to produce a mature egg ready for ovulation at midcycle. The corpus luteum takes over during the second half of the cycle to prepare the female reproductive tract for pregnancy if fertilization of the released egg occurs.

At any given time throughout the cycle, a portion of the primary follicles are starting to develop. However, only those that do so during the follicular phase, when the hormonal milieu is right to promote their maturation, continue beyond the early stages of development. The others, lacking hormonal support, undergo atresia. During follicular development, as the primary oocyte is synthesizing and storing materials for future use if fertilized, important changes are taking place in the cells surrounding the reactivated oocyte in preparation for the egg's release from the ovary (▬ Fig. 20-13). First, the single layer of *granulosa cells* in a primary follicle proliferates to form several layers that surround the oocyte. These granulosa cells secrete a thick, gel-like material that covers the oocyte and separates it from the surrounding granulosa cells. This intervening membrane is known as the **zona pellucida.** At the same time, specialized ovarian connective tissue cells at the edge of the growing follicle proliferate and differentiate to form an outer layer of **thecal cells.** The thecal and granulosa cells, collectively known as **follicular cells,** function as a unit to secrete estrogen. Of the three physiologically important estrogens—estradiol, estrone, and estriol—**estradiol** is the principal ovarian estrogen.

The hormonal environment that exists during the follicular phase promotes enlargement and development of the follicular cells' secretory capacity, converting the primary follicle into a secondary, or antral, follicle capable of estrogen secretion. This stage of follicular development is characterized by the formation of a fluid-filled **antrum** in the midst of the granulosa cells (Figs. 20-13 and ▬ 20-14). The follicular fluid originates partially from transudation (passage through capillary pores) of plasma and partially from follicular cell secretions. As the follicular cells start producing estrogen, some of this hormone is secreted into the blood for distribution throughout the body. However, a portion of the estrogen collects in the hormone-rich antral fluid.

The oocyte has reached full size by the time the antrum begins to form. The shift to an antral follicle initiates a period of rapid follicular growth. During this time, the follicle increases in size from a diameter of less than 1 mm to 12–16 mm shortly before ovulation. Part of the follicular growth is due to continued proliferation of the granulosa and thecal cells, but most is due to a dramatic expansion of the antrum. As the follicle grows, estrogen is produced in increasing quantities. One of the follicles usually grows more rapidly than the others, developing into a **mature (preovulatory, tertiary,** or **Graafian) follicle** within about fourteen days after the onset of follicular development. The antrum occupies most of the space in a mature follicle. The oocyte, surrounded by the zona pellucida and a single layer of granulosa cells, is displaced

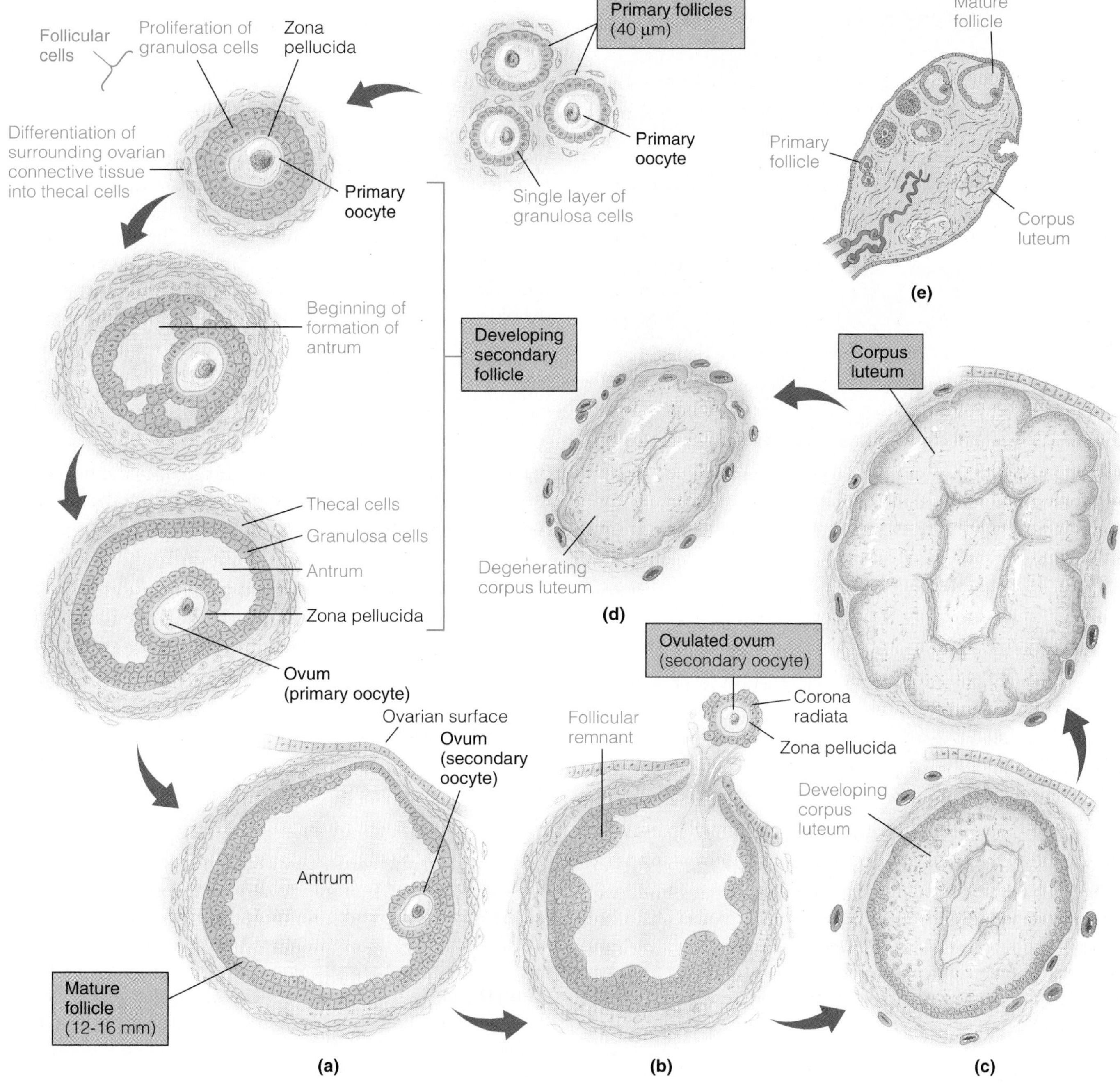

Figure 20-13 **Development of the Follicle, Ovulation, and Formation of the Corpus Luteum** (a) Stages in follicular development from a primary follicle through a mature follicle. (b) Rupture of a mature follicle and release of an ovum (secondary oocyte) at ovulation. (c) Formation of a corpus luteum from the old follicular cells following ovulation. (d) Degeneration of the corpus luteum if the released ovum is not fertilized. (e) Ovary (actual size), showing development of a follicle, ovulation, and formation and degeneration of a corpus luteum.

asymmetrically at one side of the growing follicle in a little mound that protrudes into the antrum.

The greatly expanded mature follicle bulges on the ovarian surface, creating a thin area that ruptures to release the oocyte at **ovulation.** Rupture of the follicle is facilitated by the release from the follicular cells of enzymes that digest the connective tissue in the follicular wall. The bulging wall is thus weakened so that it balloons out even further to the point that it can no longer contain the rapidly expanding follicular contents.

Just before ovulation, the oocyte completes its first meiotic division. The ovum (secondary oocyte), still surrounded by its tightly adhering zona pellucida and granulosa cells (now called the **corona radiata,** meaning "radiating crown") is swept out of the ruptured follicle into the abdominal cavity by

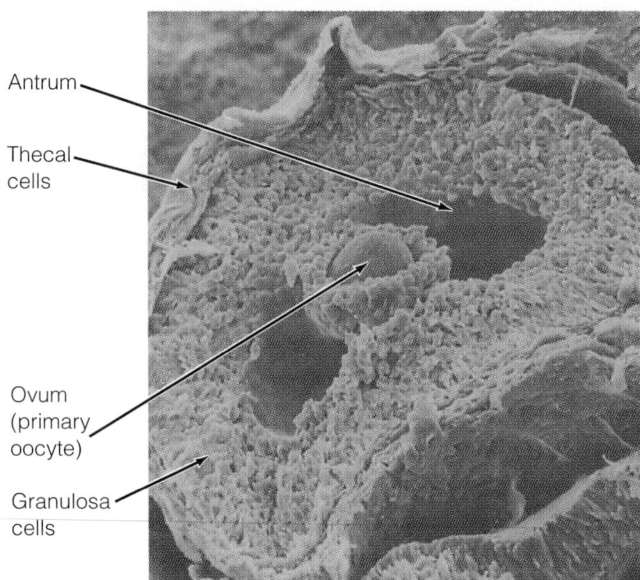

Figure 20-14 Scanning Electron Micrograph of a Developing Secondary Follicle

the leaking antral fluid (Fig. 20-13b). The released ovum is quickly drawn into the oviduct, where fertilization may or may not take place.

The other developing follicles that failed to reach maturation and ovulate undergo degeneration, never to be reactivated. Occasionally, two (or perhaps more) follicles reach maturation and ovulate at about the same time. If both are fertilized, **fraternal twins** result. Because fraternal twins arise from separate ova fertilized by separate sperm, they share no more in common than any other two siblings except for the same birth date. **Identical twins,** on the other hand, develop from a single fertilized ovum that completely divides into two separate, genetically identical embryos at a very early stage in development.

Rupture of the follicle at ovulation signals the end of the follicular phase and ushers in the luteal phase. The ruptured follicle that is left behind in the ovary following release of the ovum undergoes a rapid change. The granulosa and thecal cells remaining in the remnant follicle first collapse into the emptied antral space that has been partially filled by clotted blood. These old follicular cells soon undergo a dramatic structural transformation to form the **corpus luteum,** in a process called **luteinization** (Fig. 20-13c). The follicular-turned-luteal cells hypertrophy and are converted into very active steroidogenic (steroid hormone–producing) tissue. Abundant storage of cholesterol, the steroid precursor molecule, in lipid droplets within the corpus luteum gives this tissue a yellowish appearance, hence its name (*corpus* means "body"; *luteum* means "yellow"). The corpus luteum becomes highly vascularized as blood vessels from the thecal region invade the luteinizing granulosa. These changes are appropriate for the corpus luteum's function, which is to secrete abundant quantities of progesterone along with lesser amounts of estrogen into the blood. Estrogen secretion in the follicular phase followed by progesterone secretion in the luteal phase is essential for preparing the uterus to be a suitable site for implantation of a fertilized ovum. The corpus luteum becomes fully functional within four days after ovulation, but it continues to increase in size for another four or five days. If the released ovum is not fertilized and does not implant, the corpus luteum degenerates within fourteen days after its formation (Fig. 20-13d). The luteal cells degenerate and are phagocytized, the vascular supply is withdrawn, and connective tissue rapidly fills in to form a fibrous tissue mass known as the **corpus albicans** ("white body"). The luteal phase is now over, and one ovarian cycle is complete. A new wave of follicular development, which begins when degeneration of the old corpus luteum is completed, signals the onset of a new follicular phase.

If fertilization and implantation do take place, the corpus luteum continues to grow and produce increasing quantities of progesterone and estrogen instead of degenerating. Now called the *corpus luteum of pregnancy,* this ovarian structure persists until the end of pregnancy. It provides the hormones essential for the maintenance of pregnancy until the developing placenta is able to take over this crucial function.

The ovarian cycle is regulated by complex hormonal interactions among the hypothalamus, anterior pituitary, and ovarian endocrine units.

The ovary has two related endocrine units: the estrogen-secreting follicle during the first half of the cycle and the corpus luteum, which secretes both progesterone and estrogen, during the last half of the cycle. These units are sequentially triggered by complex cyclical hormonal relationships among the hypothalamus, anterior pituitary, and these two ovarian endocrine units.

As in the male, gonadal function in the female is directly controlled by the anterior pituitary gonadotropic hormones, follicle-stimulating hormone (FSH) and luteinizing hormone (LH). These hormones, in turn, are regulated by pulsatile episodes of hypothalamic gonadotropin-releasing hormone (GnRH) and feedback actions of gonadal hormones. Differing from the male, however, control of the female gonads is complicated by the cyclical nature of ovarian function. For example, the effects of FSH and LH on the ovaries depend on the stage of the ovarian cycle. Also in contrast to the male, FSH is not strictly responsible for gametogenesis, nor is LH solely responsible for gonadal hormone secretion. We will consider control of follicular function, ovulation, and the corpus luteum separately, using Figure 20-15 as a means of integrating the various concurrent and sequential activities that take place throughout the cycle.

Control of follicular function The factors that initiate follicular development are poorly understood. The early stages of preantral follicular growth and oocyte maturation do not require gonadotropic stimulation. Hormonal support is required, however, for antrum formation, further follicular development, and estrogen secretion. Estrogen, FSH, and LH are all needed. Antrum formation is induced by FSH. Both

Figure 20-15 Correlation between Hormonal Levels and Cyclical Ovarian and Uterine Changes During the follicular phase (the first half of the ovarian cycle), the ovarian follicle secretes estrogen under the influence of FSH, LH, and estrogen itself. The low but rising levels of estrogen (1) inhibit FSH secretion, which declines during the last part of the follicular phase, and (2) incompletely suppress tonic LH secretion, which continues to rise throughout the follicular phase. When the follicular output of estrogen reaches its peak, the high levels of estrogen trigger a surge in LH secretion at midcycle. This LH surge brings about ovulation of the mature follicle. Estrogen secretion plummets when the follicle meets its demise at ovulation.

The old follicular cells are transformed into the corpus luteum, which secretes progesterone as well as estrogen during the luteal phase (the last half of the ovarian cycle). Progesterone strongly inhibits both FSH and LH, which continue to decrease throughout the luteal phase. The corpus luteum degenerates in about two weeks if the released ovum has not been fertilized and implanted in the uterus. Progesterone and estrogen levels sharply decrease when the corpus luteum degenerates, removing the inhibitory influences on FSH and LH. As these anterior pituitary hormone levels start to rise again upon the withdrawal of inhibition, they begin to stimulate the development of a new batch of follicles as a new follicular phase is ushered in.

Concurrent uterine phases reflect the influences of the ovarian hormones on the uterus. Early in the follicular phase, the highly vascularized, nutrient-rich endometrial lining is sloughed off (the uterine menstrual phase). This sloughing results from the withdrawal of estrogen and progesterone when the old corpus luteum degenerated at the end of the preceding luteal phase. Late in the follicular phase, the rising levels of estrogen cause the endometrium to thicken (the uterine proliferative phase). After ovulation, progesterone from the corpus luteum brings about vascular and secretory changes in the estrogen-primed endometrium to produce a suitable environment for implantation (the uterine secretory, or progestational, phase). When the corpus luteum degenerates, a new ovarian follicular phase and uterine menstrual phase begin.

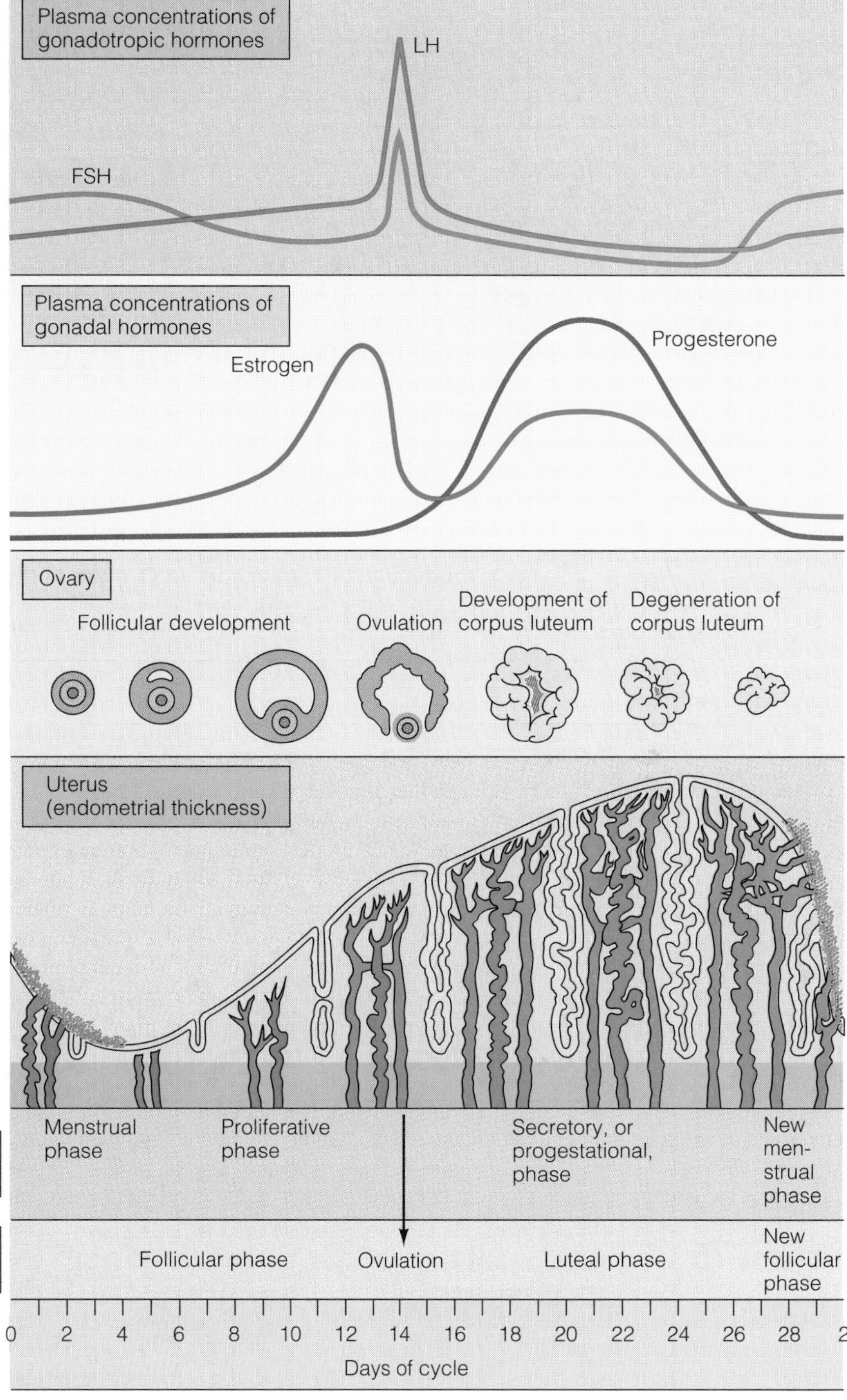

FSH and estrogen stimulate proliferation of the granulosa cells. Both LH and FSH are required for synthesis and secretion of estrogen by the follicle, but these hormones act on different cells and at a different steps in the estrogen production pathway (Fig. 20-16). Both granulosa and thecal cells participate in estrogen production. The conversion of cholesterol into estrogen requires a number of sequential steps, the last of which is conversion of androgens into estrogens (see Fig. 18-3, p. 627). Thecal cells readily produce androgens but have limited capacity to convert them into estrogens. Granulosa cells, on the other hand, are readily able to convert androgens into estrogens but are unable to produce androgens in the first place. Luteinizing hormone acts on the thecal cells to stimulate androgen production, whereas FSH acts on the granulosa cells to promote the conversion of thecal androgens (which diffuse into the granulosa cells from the thecal cells) into estrogens. Because low basal levels of FSH are sufficient to promote this final conversion to estrogen, the rate of estrogen secretion by the follicle primarily depends on the circulating level of LH, which continues to rise during the

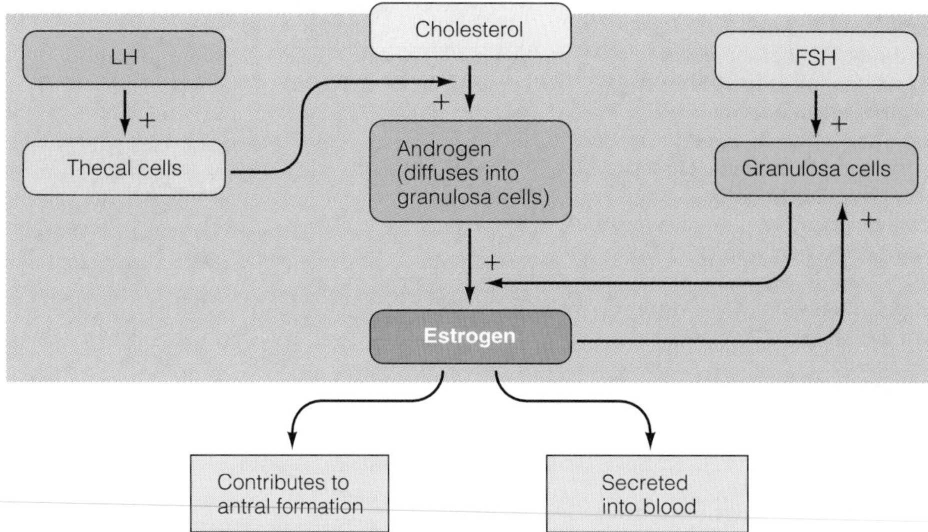

Figure 20-16 Production of Estrogen by an Ovarian Follicle Stimulated by LH, thecal cells convert cholesterol into androgen, which diffuses into the granulosa cells. The granulosa cells, stimulated by FSH, in turn convert this androgen into estrogen. Part of the estrogen remains in the ovarian follicle and contributes to antral formation, while the other part is secreted into the blood.

Figure 20-17 Feedback Control of FSH and Tonic LH Secretion during the Follicular Phase

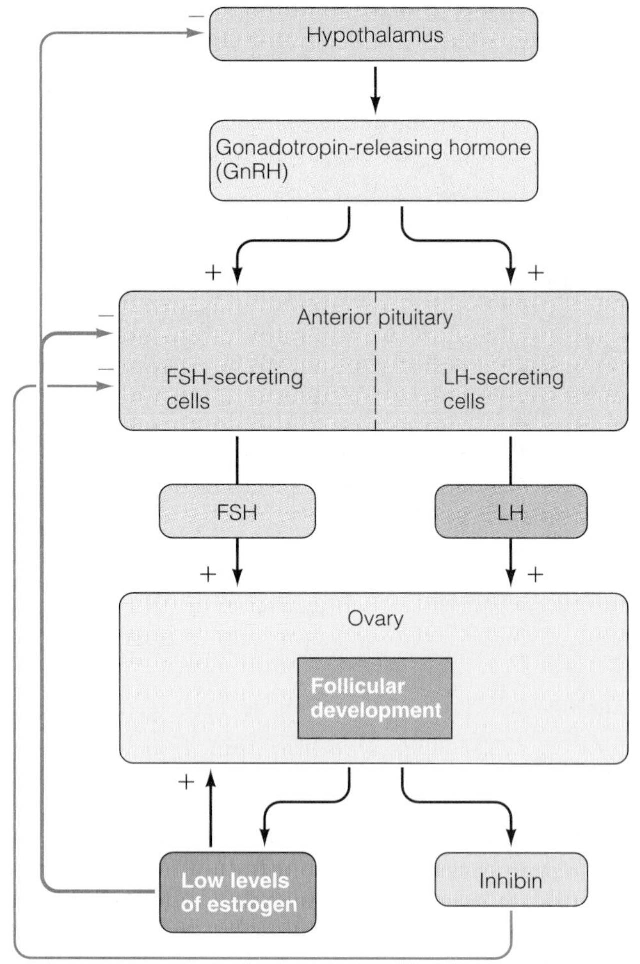

follicular phase. Furthermore, as the follicle continues to grow, more estrogen is produced simply because more estrogen-producing follicular cells are present.

Part of the estrogen produced by the growing follicle is secreted into the blood and is responsible for the steadily increasing plasma estrogen levels during the follicular phase. The remainder of the estrogen remains within the follicle, contributing to the antral fluid and stimulating further granulosa cell proliferation.

The secreted estrogen, in addition to acting on sex-specific tissues such as the uterus, inhibits the hypothalamus and anterior pituitary in negative-feedback fashion (Fig. 20-17). The low but rising levels of estrogen characterizing the follicular phase act directly on the hypothalamus to inhibit GnRH secretion, thus suppressing GnRH-prompted release of FSH and LH from the anterior pituitary. However, estrogen's primary effect is directly on the pituitary itself. Estrogen reduces the sensitivity to GnRH of the cells that produce gonadotropic hormones, especially the FSH-producing cells.

This differential sensitivity of FSH- and LH-producing cells induced by estrogen is at least in part responsible for the fact that the plasma FSH level, unlike the plasma LH concentration, declines during the follicular phase as the estrogen level rises (Fig. 20-15). Another contributing factor to the fall in FSH during the follicular phase is secretion of *inhibin* by the follicular cells. Inhibin preferentially inhibits FSH secretion by acting at the anterior pituitary, just as it does in the male. The decline in FSH secretion brings about atresia of all but the single most mature of the developing follicles.

In contrast to FSH, LH secretion continues to rise slowly during the follicular phase despite inhibition of GnRH (and thus, indirectly, LH) secretion. This seeming paradox is due to the fact that estrogen alone cannot completely suppress **tonic** (low-level, ongoing) **LH secretion;** both estrogen and progesterone are required to completely inhibit tonic LH secretion. Since progesterone does not appear until the luteal phase of the cycle, the basal level of circulating LH slowly increases during the follicular phase under incomplete inhibition by estrogen alone.

Control of ovulation Ovulation and subsequent luteinization of the ruptured follicle are triggered by an abrupt, massive increase in LH secretion. This **LH surge** brings about four major changes in the follicle:

1. It halts estrogen synthesis by the follicular cells.
2. It reinitiates meiosis in the oocyte of the developing follicle, apparently by blocking release of an *oocyte maturation-inhibiting substance* produced by the granulosa cells. This substance is believed to be responsible for arresting meiosis in the primary oocytes once they are wrapped within granulosa cells in the fetal ovary.
3. It triggers production of locally acting prostaglandins, which induce ovulation by promoting vascular changes that cause rapid swelling of the follicle while inducing enzymatic digestion of the follicular wall. Together these actions lead to rupture of the weakened wall that covers the bulging follicle.
4. It causes differentiation of follicular cells into luteal cells. Because the LH surge triggers both ovulation and luteinization, formation of the corpus luteum automatically follows ovulation. Thus, the midcycle burst in LH secretion is a dramatic point in the cycle; it terminates the follicular phase and initiates the luteal phase.

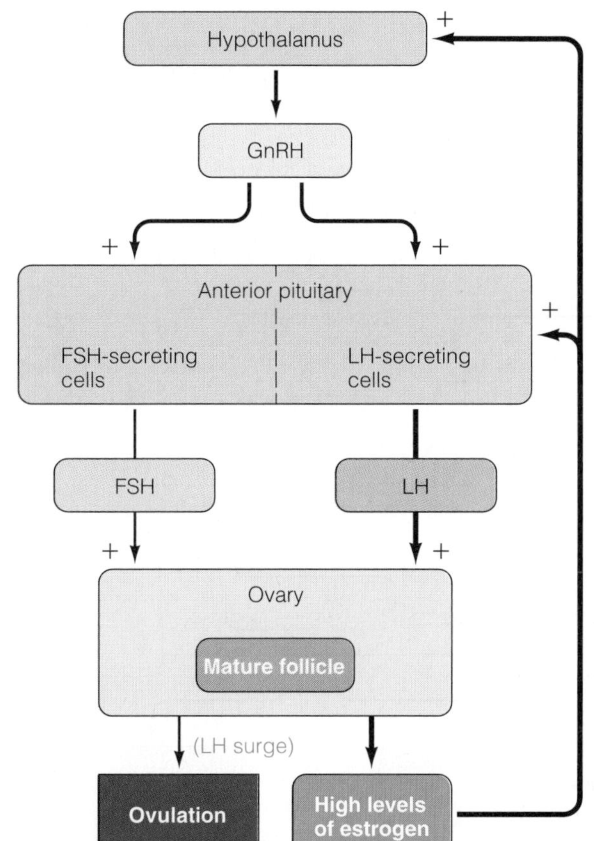

Figure 20-18 **Control of the LH Surge at Ovulation**

The two different modes of LH secretion—the tonic secretion of LH responsible for promoting ovarian hormone secretion and the LH surge that causes ovulation—not only occur at different times and produce different effects on the ovaries but also are controlled by different mechanisms. Tonic LH secretion is partially suppressed by the inhibitory action of the low, rising levels of estrogen during the follicular phase and is completely suppressed by the increasing levels of progesterone during the luteal phase. Since tonic LH secretion stimulates both estrogen and progesterone secretion, this is a typical negative-feedback control system.

In contrast, the LH surge is triggered by a *positive-feedback* effect. Whereas the low, rising levels of estrogen early in the follicular phase *inhibit* LH secretion, the high level of estrogen that occurs during peak estrogen secretion late in the follicular phase (Fig. 20-15) *stimulates* LH secretion and initiates the LH surge (Fig. 20-18). Thus, LH enhances estrogen production by the follicle, and the resultant peak estrogen concentration stimulates LH secretion. The high plasma concentration of estrogen acts directly on the hypothalamus to increase the frequency of GnRH pulses, thereby increasing both LH and FSH secretion. It also acts directly on the anterior pituitary to specifically increase the sensitivity of LH-secreting cells to GnRH. The latter effect accounts for the much greater surge in LH secretion compared to FSH secretion at midcycle. There is no known role for the modest midcycle surge in FSH that accompanies the pronounced and pivotal LH surge. Because only a mature, preovulatory follicle, not follicles in earlier stages of development, is capable of secreting sufficiently high levels of estrogen to trigger the LH surge, ovulation is not induced until a follicle has reached the proper size and degree of maturation. In a way, then, the follicle lets the hypothalamus know when it is ready to be stimulated to ovulate. The LH surge lasts for only one to two days at midcycle, just before ovulation.

Control of the corpus luteum Luteinizing hormone "maintains" the corpus luteum; that is, after triggering development of the corpus luteum, LH stimulates ongoing steroid hormone secretion by this ovarian structure. Under the influence of LH, the corpus luteum secretes both progesterone and estrogen, with progesterone being its most abundant hormonal product. The plasma progesterone level increases for the first time during the luteal phase. No progesterone is secreted during the follicular phase (except for a slight output of progesterone by the about-to-rupture follicle under the influence of the LH surge). Therefore, the follicular phase is dominated by estrogen and the luteal phase by progesterone (Fig. 20-15).

A transitory drop in the level of circulating estrogen occurs at midcycle as the estrogen-secreting follicle meets its demise at ovulation. The estrogen level climbs again during the luteal phase because of the corpus luteum's activity, although it does not reach the same peak as during the follicular phase. What keeps the moderately high estrogen level during the luteal

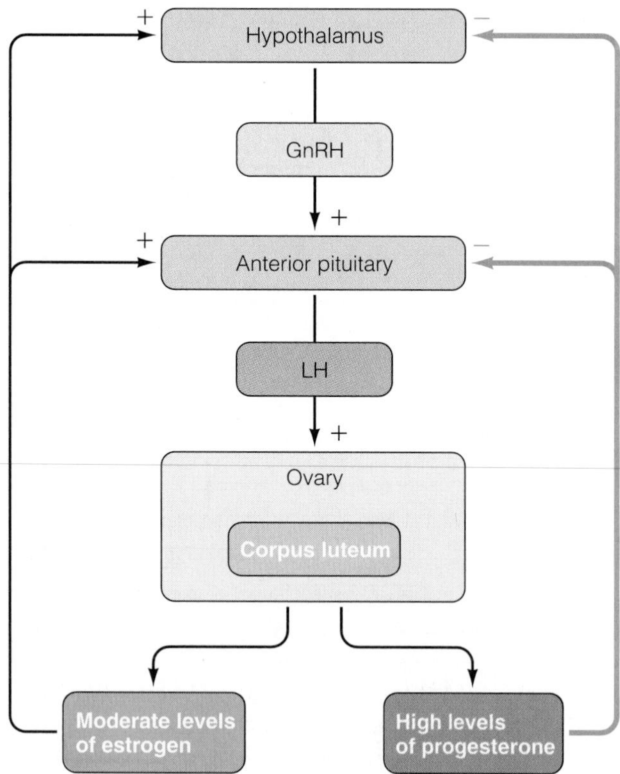

Figure 20-19 **Feedback Control during the Luteal Phase**

phase from triggering another LH surge? Progesterone. Even though a high level of estrogen stimulates LH secretion, progesterone, which dominates the luteal phase, powerfully inhibits LH secretion as well as FSH secretion (Fig. 20-19). Inhibition of FSH and LH by progesterone prevents new follicular maturation and ovulation during the luteal phase. Under progesterone's influence, the reproductive system is gearing up to support the just-released ovum, should it be fertilized, instead of preparing other ova for release. No inhibin is secreted by the luteal cells.

The corpus luteum functions for two weeks, then degenerates if fertilization does not occur. The mechanisms responsible for degeneration of the corpus luteum are not fully understood. The declining level of circulating LH, driven down by inhibitory actions of progesterone, undoubtedly contributes to the corpus luteum's downfall. Prostaglandins and estrogen released by the luteal cells themselves may play a role. Demise of the corpus luteum terminates the luteal phase and sets the stage for a new follicular phase. As the corpus luteum degenerates, plasma progesterone and estrogen levels fall rapidly because these hormones are no longer being produced (Fig. 20-15). Withdrawal of the inhibitory effects of these hormones on the hypothalamus allows FSH and tonic LH secretion to modestly increase once again. Under the influence of these gonadotropic hormones, another batch of primary follicles is induced to mature as a new follicular phase begins.

The uterine changes that occur during the menstrual cycle reflect hormonal changes during the ovarian cycle.

The fluctuations in circulating levels of estrogen and progesterone that occur during the ovarian cycle induce profound changes in the uterus, giving rise to the **menstrual,** or **uterine, cycle.** Because it reflects hormonal changes that occur during the ovarian cycle, the menstrual cycle averages twenty-eight days, as does the ovarian cycle, although there is considerable variation from this mean even in normal adults. This variability is primarily a reflection of differing lengths of the follicular phase; the duration of the luteal phase is fairly constant. The outward manifestation of the cyclical changes that occur in the uterus is the menstrual bleeding that takes place once during each menstrual cycle (that is, once a month). Less obvious changes take place throughout the cycle, however, as the uterus is prepared for implantation should a released ovum be fertilized, then is stripped clean of its prepared lining (menstruation) if implantation does not occur, only to repair itself and start preparing for the ovum that will be released during the next cycle.

We will briefly examine the influences of estrogen and progesterone on the uterus and then consider the effects of cyclical fluctuations of these hormones on uterine structure and function. The uterus consists of two main layers: the **myometrium,** the outer smooth muscle layer; and the **endometrium,** the inner lining that contains numerous blood vessels and glands. Estrogen stimulates growth of both the myometrium and the endometrium. It also induces the synthesis of progesterone receptors in the endometrium. Thus, progesterone is able to exert an effect on the endometrium only after it has been "primed" by estrogen. Progesterone acts on the estrogen-primed endometrium to convert it into a hospitable and nutritious lining suitable for implantation of a fertilized ovum. Under the influence of progesterone, the endometrial connective tissue becomes loose and edematous as a result of an accumulation of electrolytes and water, which facilitates implantation of the fertilized ovum. Progesterone further prepares the endometrium to sustain an early-developing embryo by inducing the endometrial glands to secrete and store large quantities of glycogen and by causing tremendous growth of the endometrial blood vessels. Progesterone also reduces the contractility of the uterus to provide a quiet environment for implantation and embryonic growth.

The menstrual cycle consists of three phases: the *menstrual phase,* the *proliferative phase,* and the *secretory,* or *progestational, phase.* The **menstrual phase** is the most overt phase, being characterized by discharge of blood and endometrial debris from the vagina. By convention, the first day of menstruation is considered the start of a new cycle. It coincides with termination of the ovarian luteal phase and onset of the follicular phase. As the corpus luteum degenerates because fertilization and implantation of the ovum released during the preceding cycle did not take place, circulating levels of estrogen and progesterone drop precipitously. Since the net effect of estrogen and progesterone is preparation of the endometrium for implantation of a fertilized ovum, withdrawal of these steroids deprives the highly vascular, nutrient-rich uterine lining of its hormonal support. The fall in ovarian hormone levels also stimulates release of a uterine prostaglandin that causes vasoconstriction of the endometrial vessels, thus disrupting the blood supply to the endometrium. The subsequent reduction in O_2 delivery causes death of the endometrium, including its blood vessels. The resulting bleeding through the disintegrating vessels flushes the dying endometrial tissue into the uterine lumen. The entire uterine lining sloughs during each menstrual period except for a deep, thin layer of epithelial cells and glands from which the endometrium will regenerate. The same local uterine prostaglandin also stimulates mild rhythmic contractions of the uterine myometrium. These contractions help expel the blood and endometrial debris from the uterine cavity out through the vagina as **menstrual flow.** Excessive uterine contractions caused by the overproduction of prostaglandin are responsible for the menstrual cramps **(dysmenorrhea)** experienced by some women.

The average blood loss during a single menstrual period is 50 to 150 ml. Blood that seeps slowly through the degenerating endometrium clots within the uterine cavity, then is acted on by fibrinolysin, a fibrin dissolver that breaks down the fibrin that forms the meshwork of the clot. Therefore, blood in the menstrual flow usually does not clot because it has already clotted and the clot has been dissolved before it passes out of the vagina. When blood flows rapidly through the leaking vessels, however, it may not be exposed to sufficient fibrinolysin, so blood clots may appear when the menstrual flow is most profuse. In addition to the blood and endometrial debris, large numbers of leukocytes are found in the menstrual flow. These white blood cells play an important defense role in helping the raw endometrium resist infection.

Menstruation typically lasts for about five to seven days after degeneration of the corpus luteum, coinciding in time with the early portion of the ovarian follicular phase (Fig. 20-15). Withdrawal of estrogen and progesterone upon degeneration of the corpus luteum leads simultaneously to sloughing of the endometrium (menstruation) and development of new follicles in the ovary under the influence of rising gonadotropic hormone levels. The drop in gonadal hormone secretion removes inhibitory influences from the hypothalamus and anterior pituitary, so FSH and LH secretion rise and a new follicular phase begins. After five to seven days under the influence of FSH and LH, the newly growing follicles are secreting sufficient quantities of estrogen to induce repair and growth of the endometrium.

Thus, menstrual flow ceases, and the **proliferative phase** of the uterine cycle begins concurrent with the last portion of the ovarian follicular phase as the endometrium starts to repair itself and proliferate under the influence of estrogen from the newly growing follicles. When the menstrual flow ceases, a thin endometrial layer less than 1 mm thick remains. Estrogen stimulates proliferation of epithelial cells, glands, and blood vessels in the endometrium, increasing this lining to a thickness of 3 to 5 mm. The estrogen-dominant proliferative phase lasts from the end of menstruation to ovulation. Peak estrogen levels trigger the LH surge responsible for ovulation.

Following ovulation, when a new corpus luteum is formed, the uterus enters the **secretory,** or **progestational, phase,** which coincides in time with the ovarian luteal phase. The corpus luteum secretes large amounts of progesterone and estrogen. Progesterone acts on the thickened, estrogen-primed endometrium to convert it to a richly vascularized, glycogen-filled tissue. This period is called either the secretory phase, because the endometrial glands are actively secreting glycogen, or the progestational ("before pregnancy") phase, in reference to the development of a lush endometrial lining capable of supporting an early embryo. If fertilization and implantation do not occur, the corpus luteum degenerates and a new follicular phase and menstrual phase begin once again. (For the effects of exercise, see the boxed feature on p. 730, A Closer Look at Exercise Physiology.)

Fluctuating estrogen and progesterone levels produce cyclical changes in cervical mucus.

Hormonally induced changes also take place in the cervix during the ovarian cycle. Under the influence of estrogen during the follicular phase, the mucus secreted by the cervix becomes abundant, clear, and thin. This change, which is most pronounced when estrogen is at its peak and ovulation is approaching, facilitates passage of sperm through the cervical canal. After ovulation, under the influence of progesterone from the corpus luteum, the mucus becomes thick and sticky, essentially forming a plug across the cervical opening. This plug constitutes an important defense mechanism by preventing bacteria that might threaten a pregnancy should conception have occurred from entering the uterus from the vagina. Sperm also cannot penetrate this thick mucus barrier.

Pubertal changes in females are similar to those in males, but menopausal changes are unique to females.

Regular menstrual cycles are absent in both young and aging females, but for different reasons.

Pubertal changes The female reproductive system does not become active until puberty. Unlike the fetal testes, the fetal ovaries do not need to be functional, because feminization of the female reproductive system automatically takes place in the absence of fetal testosterone secretion without the presence of female sex hormones. The female reproductive system remains quiescent from birth until puberty, which occurs at about twelve years of age when hypothalamic GnRH activity increases for the first time. As in the male, the mechanisms responsible for the onset of puberty are not clearly understood but are believed to involve the pineal gland and melatonin secretion.

GnRH begins stimulating the release of the anterior pituitary gonadotropic hormones, which in turn stimulate ovarian activity. The resultant secretion of estrogen by the activated ovaries induces growth and maturation of the female reproductive tract as well as development of the female secondary sexual characteristics. Estrogen's prominent action in the lat-

A CLOSER LOOK AT EXERCISE PHYSIOLOGY

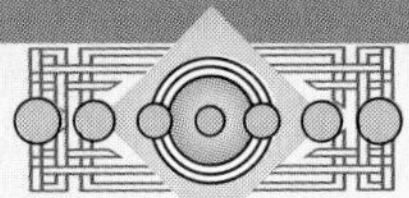

Menstrual Irregularities: When Cyclists and Other Female Athletes Do Not Cycle

Since the 1970s, as women in growing numbers have participated in a variety of sports requiring vigorous training regimens, researchers have become increasingly aware that many women experience changes in their menstrual cycles as a result of athletic participation. These changes are referred to as *athletic menstrual cycle irregularity (AMI)*. The menstrual cycle dysfunction can vary in severity from amenorrhea (cessation of menstrual periods) to oligomenorrhea (cycles at irregular or infrequent intervals) to cycles that are normal in length but are anovulatory (no ovulation) or that have a short or inadequate luteal phase.

In early research studies using surveys and questionnaires to determine the prevalence of the problem, the frequency of these sport-related disorders varied from 2% to 51%. In contrast, the rate of occurrence of menstrual cycle dysfunction in females of reproductive age in the general population is 2% to 5%. A major problem of using surveys to determine the frequency of menstrual cycle irregularity is the questionable accuracy of recall of menstrual periods. Furthermore, without blood tests to determine hormone levels throughout the cycle, a woman would not know whether she was anovulatory or had a shortened luteal phase. Studies in which hormone levels have been determined through the menstrual cycle have demonstrated that seemingly normal cycles in athletes frequently have a short luteal phase (less than two days long with low progesterone levels).

In a study conducted to determine whether strenuous exercise spanning two menstrual cycles would induce menstrual disorders, twenty-eight initially untrained college women with documented ovulation and luteal adequacy served as subjects. The women participated in an eight-week exercise program in which they initially ran four miles per day and progressed to ten miles per day by the fifth week. They were expected to participate daily in three and one-half hours of moderate-intensity sports. Only four women had normal menstrual cycles during the training. Abnormalities that occurred as a result of training included abnormal bleeding, delayed menstrual periods, abnormal luteal function, and loss of LH surge. All women returned to normal cycles within six months after training. The results of this study suggest that the frequency of AMI with strenuous exercise may be much greater than indicated by questionnaire alone. In other studies using low-intensity exercise regimens, AMI was much less frequent.

The mechanisms of AMI are unknown at present, although studies have implicated rapid weight loss, decreased percentage of body fat, dietary insufficiencies, prior menstrual dysfunction, stress, age at onset of training, and the intensity of training as factors that play a role. Epidemiologists have shown that if girls participate in vigorous sports before **menarche** (the first menstrual period), menarche is delayed. On the average, athletes have their first period when they are about three years older than their nonathletic counterparts. Furthermore, females who participate in sports before menarche seem to have a higher frequency of AMI throughout their athletic careers than those who begin to train after menarche. Hormonal changes that have been found in female athletes include (1) severely depressed FSH levels, (2) elevated LH levels, (3) low progesterone during the luteal phase, (4) low estrogen levels in the follicular phase, and (5) an FSH/LH environment totally unbalanced as compared to age-matched nonathletic women. The preponderance of evidence indicates that cycles return to normal once vigorous training is stopped.

The major problem associated with athletic amenorrhea is a reduction in bone-mineral density. Studies have shown that the mineral density in the vertebrae of the lower spine of those with athletic amenorrhea is lower than in athletes with normal menstrual cycles and lower than in age-matched nonathletes. However, amenorrheic runners have higher bone-mineral density than amenorrheic nonathletes, presumably because the mechanical stimulus of exercise helps to retard bone loss. Studies have shown that amenorrheic athletes are at higher risk for stress-related fractures than athletes with normal menstrual cycles. One study, for example, found stress fractures in six of eleven amenorrheic runners but in only one of six runners with normal menstrual cycles. The mechanism for bone loss is probably the same as is found in postmenopausal osteroporosis—lack of estrogen (see p. 695). The problem is serious enough that an amenorrheic athlete should discuss the possibility of estrogen replacement therapy with her physician.

There may be some positive benefits of athletes' menstrual dysfunction. A recent epidemiologic study to determine if the long-term reproductive and general health of women who had been college athletes differed from that of college nonathletes showed that former athletes had less than half the lifetime occurrence rate of cancers of the reproductive system and half the breast cancer occurrence compared to nonathletes. Because these are hormone-sensitive cancers, the delayed menarche and lower estrogen levels found in women athletes may play a key role in decreasing the risk of cancer of the reproductive system and breast.

ter regard is to promote fat deposition in strategic locations, such as the breasts, buttocks, and thighs, giving rise to the typical curvaceous female figure. Enlargement of the breasts at puberty is due primarily to fat deposition in the breast tissue and not to functional development of the mammary glands. Three other pubertal changes in females—growth of axillary and pubic hair, the pubertal growth spurt, and development of libido—are attributable to a spurt in adrenal androgen secretion at puberty, not to estrogen. The pubertal rise in estrogen does close the epiphyseal plates, however, halting further growth in height, similar to the effect of testosterone in males.

Menopausal changes The cessation of a woman's menstrual cycles at menopause sometime between the age of forty-five and fifty-five years is attributable to the limited supply of ovarian follicles present at birth. Once this reservoir is depleted, ovarian cycles, and hence menstrual cycles, cease. Thus, the termination of reproductive potential in a middle-aged woman is "preprogrammed" at her own birth. Evolutionarily, menopause may have developed as a mechanism to prevent pregnancy in women beyond the time that they could likely rear a child before their own death.

Menopause is preceded by a period of progressive ovarian failure characterized by increasingly irregular cycles and dwindling estrogen levels. This entire period of transition from sexual maturity to cessation of reproductive capability is known as the **climacteric.** Ovarian estrogen production declines from as much as 300 mg per day to essentially nothing. Postmenopausal women are not completely devoid of estrogen, however, because adipose tissue, the liver, and the adrenal cortex continue to produce up to 20 mg of estrogen per day. Nevertheless, the loss of ovarian estrogen brings about a host of physical and emotional changes. The absence of ovarian estrogen is responsible for the physical postmenopausal changes that occur in the reproductive tract (in addition to cessation of ovarian and menstrual cycles). These include vaginal dryness, which can cause discomfort during sex, and gradual atrophy of the genital organs. However, postmenopausal women still have a sex drive because of their adrenal androgens.

Since estrogen has widespread physiological actions beyond the reproductive system, the dramatic loss of ovarian estrogen that accompanies menopause has impacts on other body systems, most notably the skeleton and the cardiovascular system:

- Estrogen helps build strong bones, thus shielding premenopausal women from the bone-thinning condition of osteoporosis (see p. 695). The postmenopausal reduction in estrogen leads to increased activity of the bone-dissolving osteoclasts and diminished activity of the bone-building osteoblasts. The result is a decrease in bone density and a greater incidence of bone fractures.
- Estrogen exerts cardioprotective effects that give females a fifteen- to twenty-year advantage over males in evading atherosclerotic coronary artery disease. The incidence of coronary artery disease and its accompanying heart attacks rapidly climbs following menopause. Coronary artery disease is the leading cause of death in women, accounting for over half of the deaths in females each year. Estrogen helps reduce the risk of heart attacks in several different ways. First, estrogen inhibits the development of atherosclerosis through its beneficial effect on cholesterol metabolism. Postmenopausal estrogen deficiency leads to a decrease in the production of HDL-cholesterol (the "good" cholesterol) and an increase in LDL-cholesterol (the "bad" cholesterol) (see p. 298). Second, estrogen acts as an antioxidant that helps protect the endothelial cells from the ravages of free radicals that characterize the early stages of atherosclerotic coronary artery disease (see p. 300). In addition, estrogen promotes arteriolar vasodilation, which helps improve coronary blood flow and prevents the vessel spasms that accompany early coronary atherosclerosis.
- Estrogen also helps modulate the actions of epinephrine and norepinephrine on the arteriolar walls. The menopausal diminution of estrogen leads to unstable control of blood flow, especially in the skin vessels. Transient increases in the flow of warm blood through these superficial vessels are responsible for the **"hot flashes"** that frequently accompany menopause. Vasomotor stability is gradually restored in postmenopausal women so that hot flashes eventually cease.

It is unclear to what extent emotional symptoms associated with waning ovarian function, such as depression and irritability, are caused by estrogen withdrawal or are psychological reactions to the implications of menopause.

About 20% of menopausal and postmenopausal women in the United States take supplemental estrogen to counter the effects of their natural decline in estrogen. Those on hormone replacement therapy enjoy stronger bones, a reduced risk of heart attacks, fewer hot flashes, and less mood swings than those not using estrogens. However, estrogen therapy increases the risk of breast cancer, a detrimental side effect that can be diminished to some extent by taking estrogen in combination with progesterone.

Supplemental estrogen therapy is the most commonly employed hormone replacement therapy utilized to restore youthful levels of hormones that naturally decline with age. Others include melatonin (see p. 636), human growth hormone (see p. 649), DHEA (see p. 665), and testosterone (see p. 710). Physicians are increasingly prescribing, and patients are increasingly requesting, such hormone supplementation in the hopes of helping the body age better. Proponents of this "youth preservation" therapy claim that one must maintain youthful levels of these hormones to maintain a youthful level of health. Those who oppose the practice are concerned about the troubling, sometimes serious, side effects that may occur.

Males do not experience a complete gonadal failure as females do for two reasons. First, a male's germ cell supply is unlimited because of continuing mitotic activity of the spermatogonia. Second, gonadal hormone secretion in males is not inextricably dependent on gametogenesis, as it is in females. If female sex hormones were produced by separate tissues unrelated to those responsible for gametogenesis, as they are in the male, cessation of estrogen and progesterone secretion would not automatically accompany termination of oogenesis.

The oviduct is the site of fertilization.

You have now learned about the events that take place if fertilization does not occur. Because the primary function of the reproductive system is, of course, reproduction, we will next turn our attention to the sequence of events that ensue when this function is accomplished. (See the boxed feature on p. 732, ◆ Concepts, Challenges, and Controversies.)

CONCEPTS, CHALLENGES, AND CONTROVERSIES

Reproduction-related Technology: Taking the Making of Life into our Own Hands

Implicit in our ever-increasing understanding of the mechanisms underlying physiological functions is our growing ability to correct or compensate for defective functions or to manipulate various functions to our advantage. Most advances in the area of biomedical technology have been welcomed as new ways to save, prolong, or enhance human lives. However, one area of new technological control over biological processes has aroused considerable moral, ethical, and legal controversy—namely, our growing power to determine the very existence of new individuals. Technological advances that influence reproductive ability have opened up many new avenues for those desiring to control their potential progeny through artificial means. A brief listing of capabilities already available or under development will indicate how extensive our control over the future generation is:

- A variety of methods of *contraception* are available to avoid pregnancy by preventing the sperm and egg from joining or, once united, by preventing the fertilized egg from implanting in the uterus where it could develop into another human being.
- An unwanted pregnancy can be terminated by several different methods of *abortion* that remove the developing individual from its supportive uterine environment.
- An unborn child can be unobtrusively "viewed" in its mother's womb through *ultrasound* techniques. Furthermore, a sample of its cells can be extracted from the uterus through *amniocentesis* or *placental biopsy* and analyzed to determine the presence or absence of numerous genetic disorders. If defects are discovered, the parents have to make the difficult choice of terminating the pregnancy or knowingly bringing into the world a child who will be physically or mentally impaired.
- Sperm can be frozen and stored in *sperm banks* where a woman can "shop" for a "father" with particular traits for her future child; the woman can then undergo *artificial insemination* with the chosen sperm and have a normal pregnancy and birth without active participation or even knowledge of the male.
- Infertile women who fail to ovulate (do not release eggs) can be given *"fertility pills,"* which are hormones that promote ovulation.
- Infertile women who have a mechanical blockage in the pathway where the egg and sperm unite can still have their own child through *in vitro fertilization,* often dubbed the "test-tube baby" technique. After hormonally promoting multiple ovulation, the physician collects the eggs through a small incision in the woman's abdomen, then incubates them with sperm donated by the father-to-be. If fertilization takes place in this test-tube environment, the fertilized egg is placed in the mother's uterus, where it develops as would a naturally fertilized egg.
- In vitro fertilization often results in fertilization of more than one egg. Viable unused *embryos can be frozen* for future use.
- Because many externally fertilized eggs fail to implant when inserted in the uterus, several variations on in vitro fertilization have been developed. With *gamete intra-Fallopian transfer (GIFT),* unfertilized eggs and sperm are mixed and inserted into the oviduct (Fallopian tube). Thus, this procedure bypasses a mechanical blockage in the upper oviduct while permitting fertilization and implantation to proceed naturally. With *zygote intra-Fallopian transfer (ZIFT),* a zygote resulting from in vitro fertilization is placed in the oviduct rather than in the uterus so that it can leisurely descend before implanting as zygotes normally do. Using GIFT and ZIFT techniques, infertility clinics have reported

Fertilization, the union of male and female gametes, normally occurs in the **ampulla,** the upper third of the oviduct (Fig. 20-2). Thus, both the ovum and the sperm must be transported from their gonadal site of production to the ampulla (Fig. 20-20).

Ovum transport to the oviduct At ovulation the ovum is released into the abdominal cavity, but it is quickly picked up by the oviduct. The dilated end of the oviduct cups around the ovary and contains **fimbriae,** fingerlike projections that contract in a sweeping motion to guide the released ovum into the oviduct (Figs. 20-2b and 20-20). Furthermore, the fimbriae are lined by cilia, which are fine hairlike projections that beat in waves toward the interior of the oviduct, further assuring the ovum's passage into the oviduct (see p. 37). Within the oviduct, the ovum is rapidly propelled by peristaltic contractions and ciliary action to the ampulla.

Conception can take place during a very limited time span in each cycle (the **fertile period**). If not fertilized, the ovum begins to disintegrate within 12 to 24 hours and is subsequently phagocytized by cells that line the reproductive tract. Fertilization must therefore occur within 24 hours after ovulation, when the ovum is still viable. Sperm typically survive about 48 hours but can survive up to five days in the female reproductive tract, so sperm deposited within five days before ovulation to 24 hours after ovulation may be able to fertilize the released ovum, although these times are subject to considerable variation.

Occasionally, an ovum fails to be transported into the oviduct and remains instead in the peritoneal cavity. Rarely, such an ovum gets fertilized, resulting in an **ectopic (abdominal) pregnancy,** in which the fertilized egg implants in the rich vascular supply to the digestive organs rather than in its usual site in the uterus. If this unusual pregnancy proceeds to implantation rates two to three times those of ordinary in vitro fertilization procedures.

- The sperm of some men are "lazy" or lack the necessary acrosomal enzymes to penetrate the formidable barriers surrounding an egg. Working under powerful magnification and using a thin needle, an infertility specialist can overcome this problem by inserting a single spermatozoon directly into an egg, a technique known as *microinjection.*
- Another technique under development to facilitate sperm penetration of an egg is *zona drilling.* In this procedure, a tiny hole is drilled in the zona pellucida, the outer membrane surrounding an unfertilized egg. When the prepared egg is mixed with sperm, the hole serves as a ready-made passageway for entry of sperm. This new technology offers hope for couples with sperm insufficiencies or zona abnormalities.
- Although sperm and embryos can be successfully frozen and stored, unfertilized eggs have not been able to withstand similar processing. However, researchers have accomplished a major step toward eventual success in *egg freezing.* When this technique is perfected, a young woman facing the loss of ovarian function because of surgery or chemotherapy for cancer could opt to collect and store eggs for future use before the ovary-destroying procedure is performed. A more controversial use of the egg-freezing technique might be its employment by women who wish to postpone childrearing for professional or other reasons. A woman's fertility declines sharply and the risk of having children with genetic abnormalities such as Down's syndrome rises after age forty. Evidence indicates that the age of the eggs, not the reproductive tract, is responsible for the decline in fertility and rise in fetal abnormalities. Accordingly, if egg freezing were available, a woman could collect healthy eggs during her most fertile years and put them on ice for later use.
- It is technically possible, although legally questionable, for a couple unable to have children because of the woman's inability to bear a child to "rent" another woman's uterus to carry their child. The *surrogate mother* can be artificially inseminated with the father's sperm, or an in vitro fertilized egg from the couple can be introduced into the "rented" uterus.
- In vitro fertilization techniques and *preimplantation genetic testing* make possible the screening of early embryos for potential genetic defects before implantation. Many genetic abnormalities, such as sickle-cell anemia and cystic fibrosis, can be detected in the genetic code from the moment of fertilization. Because an early embryo can tolerate the loss of a cell without impairing normal development, it is possible to remove a single cell from an eight-cell embryo and analyze it for genetic defects before implanting it in the female reproductive tract.

While these new technologies are being hailed as modern miracles, they are also making us tread on new moral, ethical, and legal ground. What rights do the partners who created a frozen embryo have, for example, or what rights does the frozen embryo itself have? Because legislation has not kept pace with technological capabilities, many ambiguities are being decided through precedent-setting lawsuits. For example, a woman and her ex-husband recently engaged in a highly controversial legal battle over whether she could attempt to become pregnant after their divorce using frozen embryos that the couple had created while still married. The court ruled that she could not use the embryos against his will. In another case, the court ruled that frozen embryos could not inherit the estate left by a wealthy, childless couple who died in a plane crash, leaving the embryos as their only direct "heirs." Many researchers, physicians, regulators, and concerned citizens advocate the establishment of a permanent national ethics advisory board that could promote public discussion of ethical issues and offer ethical guidelines as innovative techniques push us further into controversial, uncharted territory.

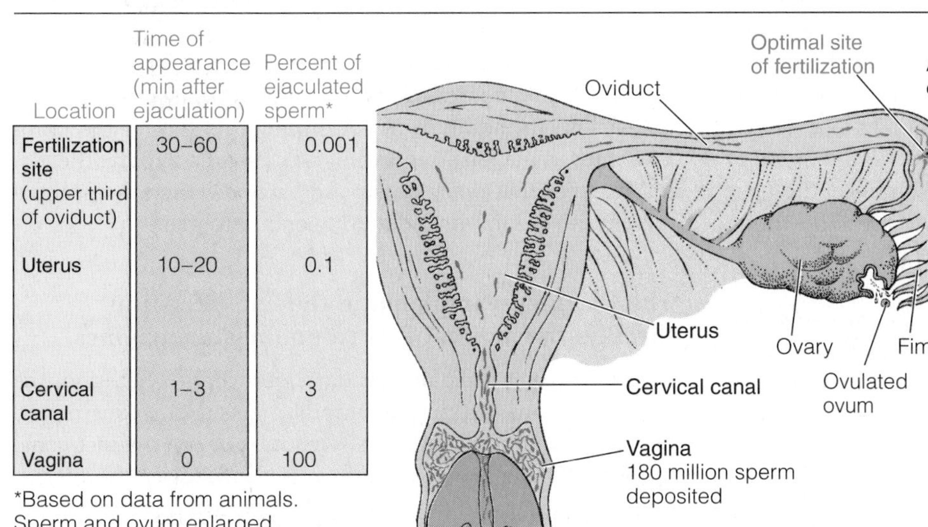

Location
Time of appearance (min after ejaculation)
Percent of ejaculated sperm*
Fertilization site (upper third of oviduct)
30–60
0.001
Uterus
10–20
0.1
Cervical canal
1–3
3
Vagina
0
100
*Based on data from animals.
Sperm and ovum enlarged.
Optimal site of fertilization
Oviduct
Ampulla of oviduct
Sperm surrounding ovum
Uterus
Ovary
Fimbria
Ovulated ovum
Cervical canal
Vagina
180 million sperm deposited
Penis

Figure 20-20 **Sperm and Ovum Transport to the Site of Fertilization**

term, the baby must be delivered surgically, because the normal vaginal exit is not available to it.

Sperm transport to the oviduct Once sperm are deposited in the vagina at ejaculation, they must travel through the cervical canal, through the uterus, and then up to the egg in the upper third of the oviduct (Fig. 20-20). The first sperm arrive in the oviduct within a half hour after ejaculation. Even though sperm are mobile by means of whiplike contractions of their tails, 30 minutes is much too soon for a sperm's own mobility to transport it to the site of fertilization. To accomplish this formidable journey, sperm need the help of the female reproductive tract. The first hurdle is passage through the cervical canal. The cervical mucus is too thick to permit sperm penetration throughout most of the cycle because of high progesterone or low estrogen levels. Only when estrogen levels are high, as occurs in the presence of a mature follicle about to ovulate, does the cervical mucus become thin and watery enough to permit sperm to penetrate. Sperm migrate up the cervical canal under their own power. The canal remains penetrable for only two or three days during each cycle, around the time of ovulation.

Once the sperm have entered the uterus, contractions of the myometrium churn them around in "washing-machine" fashion. This action quickly disperses the sperm throughout the uterine cavity. When sperm reach the oviduct, they must travel upward against the predominant action of the cilia lining this passageway, most of which beat downward to propel the nonmotile egg toward the uterus. Sperm transport in the oviduct is facilitated by antiperistaltic (upward) contractions of the oviduct smooth muscle, perhaps assisted by tracts of cilia that beat upward in the opposite direction of the main ciliary current. These myometrial and oviduct contractions that facilitate sperm transport are induced by the high estrogen level that exists just prior to ovulation, perhaps aided by seminal prostaglandins. Furthermore, new research indicates that ova are not passive partners in conception. Mature eggs have been shown to release an as-yet-unidentified chemical that attracts sperm and causes them to propel themselves toward the waiting female gamete.

Even around ovulation time when sperm can penetrate the cervical canal, of the several hundred million sperm deposited in a single ejaculate, only a few thousand make it to the oviduct (Fig. 20-20). The fact that only a very small percentage of the deposited sperm ever reach their destination is one reason why sperm concentration must be so high (> 20 million/ml of semen) for a man to be fertile. The other reason is that the acrosomal enzymes of many sperm are needed to break down the barriers surrounding the ovum (Fig. 20-21).

Fertilization The tail of the sperm is used to maneuver for final penetration of the ovum. To fertilize an ovum, a sperm must first pass through the corona radiata and zona pellucida surrounding it. The acrosomal enzymes, which are exposed as the acrosomal membrane disrupts on contact with the corona radiata, enable the sperm to tunnel a path through these protective barriers (Fig. 20-22). Sperm are able to penetrate the zona pellucida only after binding with specific receptor

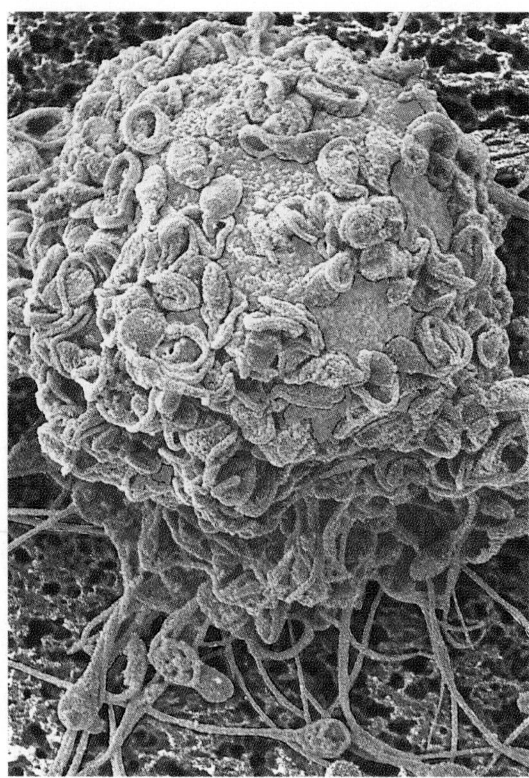

Figure 20-21 **Scanning Electron Micrograph of Sperm Amassed at the Surface of an Ovum**

sites on the surface of this layer. (Only sperm of the same species are able to bind to these receptors and pass through.) The first sperm to reach the ovum itself fuses with the plasma membrane of the ovum (actually a secondary oocyte), triggering a chemical change in the ovum's surrounding membrane that makes this outer layer impenetrable to the entry of any more sperm. This phenomenon is known as **block to polyspermy** ("many sperm").

The head of the fused sperm is gradually pulled into the ovum's cytoplasm by a growing cone that engulfs it. The sperm's tail is frequently lost in this process, but it is the head that carries the crucial genetic information. Penetration of the sperm into the cytoplasm triggers the final meiotic division of the secondary oocyte. Within an hour, the sperm and egg nuclei fuse. In addition to contributing its half of the chromosomes to the fertilized ovum, now called a **zygote,** the victorious sperm also activates ovum enzymes that are essential for the early embryonic developmental program.

The blastocyst implants in the endometrium through the action of its trophoblastic enzymes.

During the first three to four days following fertilization, the zygote remains within the ampulla, because a constriction between the ampulla and the remainder of the oviduct canal prevents further movement of the zygote toward the uterus. The zygote is not idle during this time, however. It is rapidly undergoing a number of mitotic cell divisions to form a solid

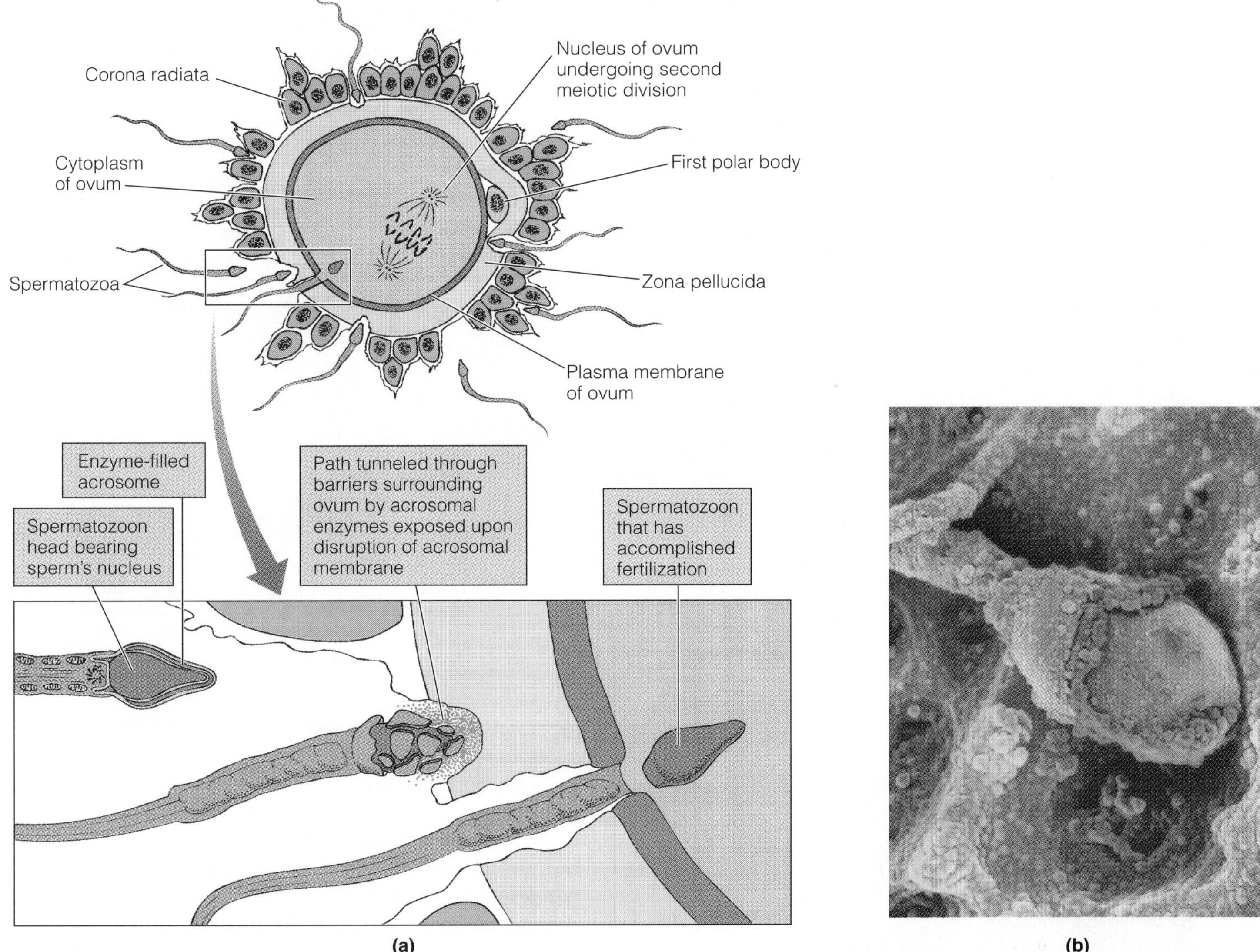

Figure 20-22 **Process of Fertilization** (a) Schematic representation of sperm tunneling the barriers surrounding the ovum. (b) Scanning electron micrograph of a spermatozoon in which the acrosomal membrane has been disrupted and the acrosomal enzymes (in red) are exposed.

ball of cells called the **morula** (Fig. 20-23). Meanwhile, the rising levels of progesterone from the newly developing corpus luteum that formed after ovulation stimulate release of glycogen from the endometrium into the reproductive tract lumen for use as energy by the early embryo. The nutrients stored in the cytoplasm of the ovum can sustain the product of conception for less than a day. The concentration of secreted nutrients increases more rapidly in the small confines of the ampulla than in the uterine lumen. About three to four days after ovulation, progesterone is being produced in sufficient quantities to relax the oviduct constriction, thus permitting the morula to be rapidly propelled into the uterus by oviductal peristaltic contractions and ciliary activity. The temporary delay before the developing embryo passes into the uterus allows sufficient nutrients to accumulate in the uterine lumen to support the embryo until implantation can take place. If the morula arrives prematurely, it dies.

When the morula descends to the uterus, it floats freely within the uterine cavity for another three to four days, living on the endometrial secretions and continuing to divide. During the first six to seven days after ovulation, while the developing embryo is in transit in the oviduct and floating in the uterine lumen, the uterine lining is simultaneously being prepared for implantation under the influence of luteal-phase progesterone. During this time, the uterus is in its secretory or progestational phase, storing up glycogen and becoming richly vascularized.

Occasionally, the morula fails to descend into the uterus and continues instead to develop and implant in the lining of the oviduct. This leads to a **tubal pregnancy**, which must be terminated. Such a pregnancy can never succeed, because the oviduct cannot expand as the uterus does to accommodate the growing embryo. The first warning of a tubal pregnancy is pain caused by stretching of the oviduct by the growing

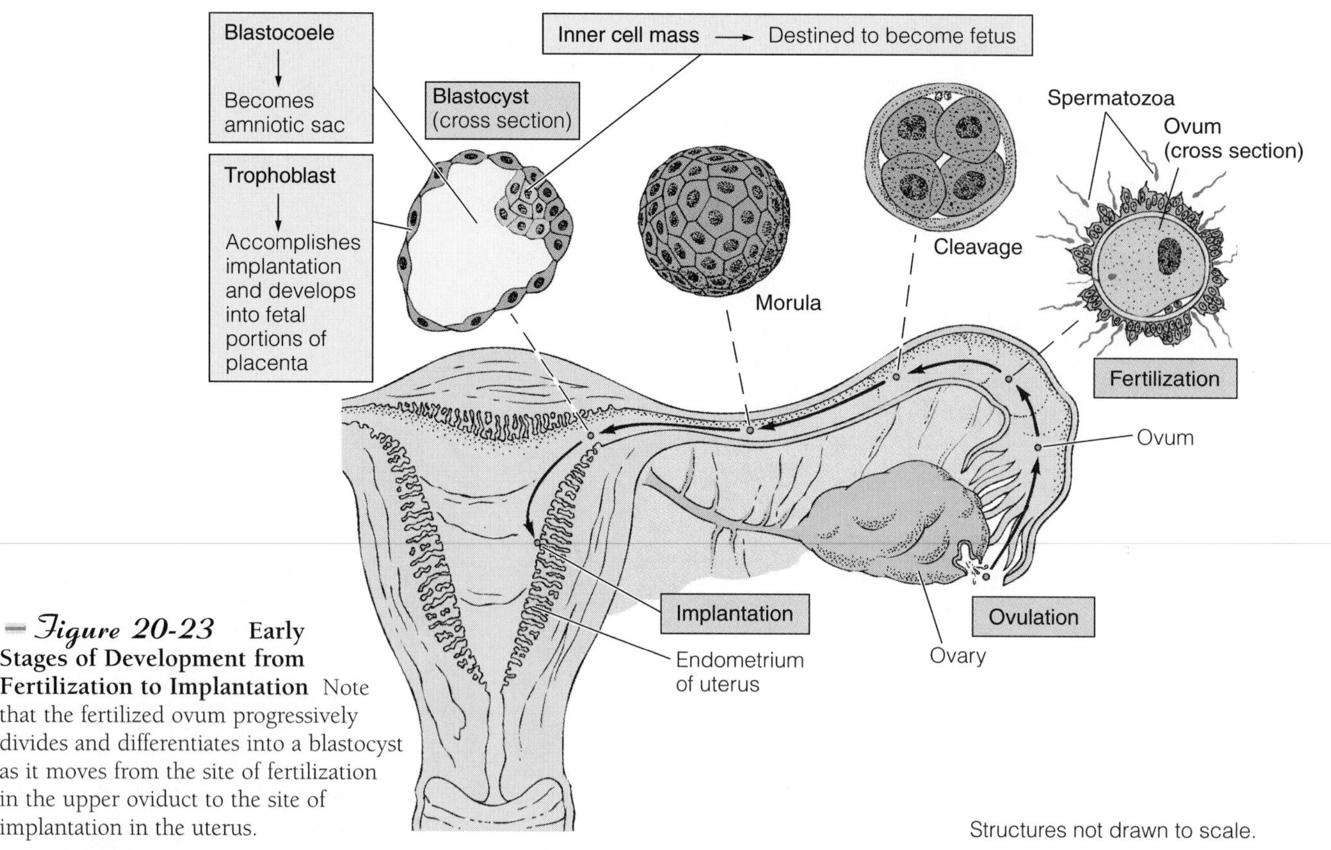

Figure 20-23 **Early Stages of Development from Fertilization to Implantation** Note that the fertilized ovum progressively divides and differentiates into a blastocyst as it moves from the site of fertilization in the upper oviduct to the site of implantation in the uterus.

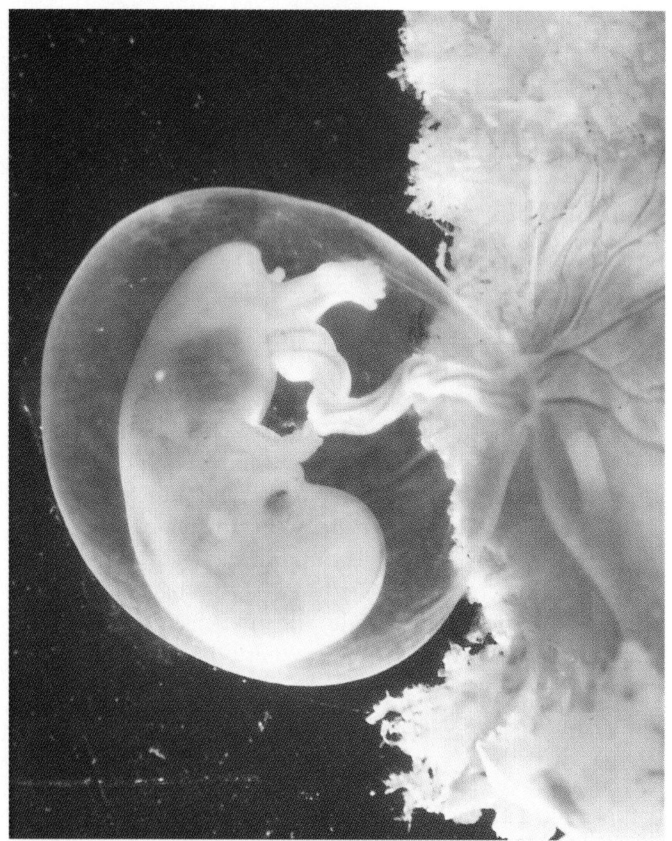

Figure 20-24 **A Human Fetus Surrounded by the Amniotic Sac** The fetus is near the end of the first trimester of development.

embryo. If not removed, the enlarging embryo will rupture the oviduct, causing possibly lethal hemorrhage.

In the usual case, by the time the endometrium is suitable for implantation (about a week after ovulation), the morula has descended to the uterus and continued to proliferate and differentiate into a *blastocyst* capable of implantation. Thus, the week's delay after fertilization and before implantation allows time for both the endometrium and the developing embryo to prepare for implantation.

A **blastocyst** is a single-layered sphere of cells encircling a fluid-filled cavity, with a dense mass of cells grouped together at one side (Fig. 20-23). This dense mass, called the **inner cell mass,** is destined to become the embryo/fetus itself. The remainder of the blastocyst will never be incorporated into the fetus but will serve a supportive role during intrauterine life. The thin outermost layer, the **trophoblast,** is responsible for accomplishing implantation, after which it develops into the fetal portion of the placenta. The fluid-filled cavity, the **blastocoele,** will become the fluid within the **amniotic sac** which surrounds and cushions the fetus throughout gestation (Fig. 20-24).

When the blastocyst is ready to implant, its surface becomes sticky. By this time the endometrium is ready to accept the early embryo. The blastocyst adheres to the uterine lining on its inner-cell-mass side (Fig. 20-25a). **Implantation** begins when the trophoblastic cells overlying the inner cell mass release protein-digesting enzymes upon contact with the endometrium. These enzymes digest pathways between the endometrial cells, thus permitting finger-like cords of trophoblastic cells to penetrate into the depths of

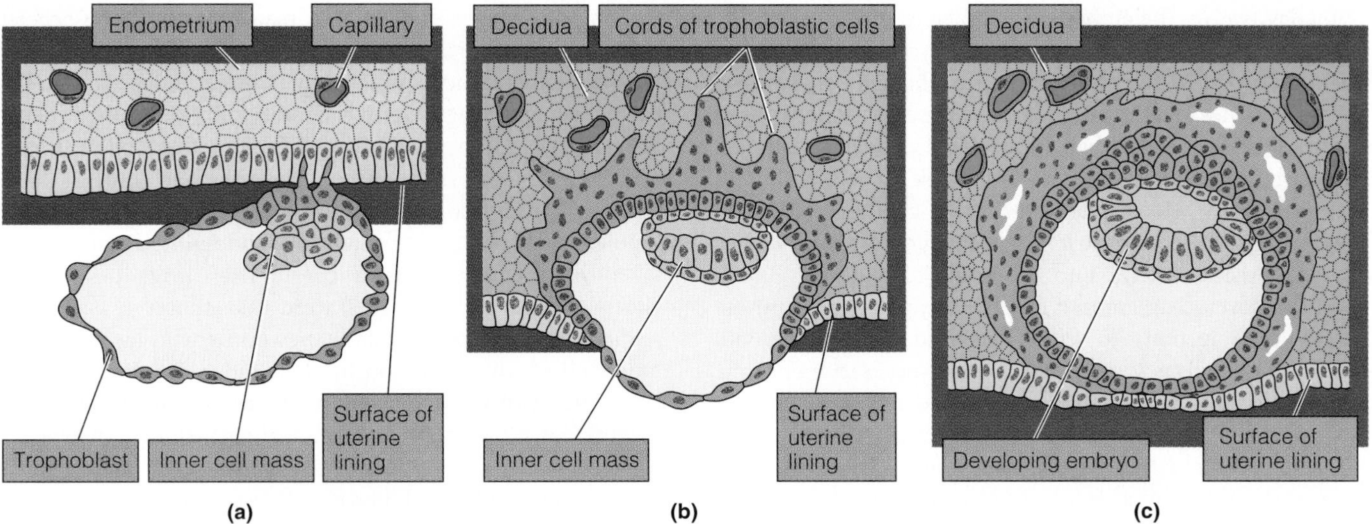

Figure 20-25 **Implantation of the Blastocyst** (a) The free-floating blastocyst adheres to the endometrial lining. (b) Cords of trophoblastic cells tunnel into the endometrium, carving out a hole for the blastocyst. The boundaries between the cells in the advancing trophoblastic tissue disintegrate. (c) When implantation is finished, the blastocyst is completely buried in the endometrium.

the endometrium, where they continue to digest uterine cells (Figure 20-25b). Through its cannibalistic actions, the trophoblast performs the dual functions of (1) accomplishing implantation as it carves out a hole in the endometrium for the blastocyst and (2) making metabolic fuel and raw materials available for the developing embryo as the advancing trophoblastic projections break down the nutrient-rich endometrial tissue.

Stimulated by the invading trophoblast, the endometrial tissue at the site of contact undergoes dramatic changes that enhance its ability to support the implanting embryo. In response to a chemical messenger released by the blastocyst, the underlying endometrial cells secrete prostaglandins, which act to locally increase vascularization, produce edema, and enhance nutrient storage. The endometrial tissue so modified at the implantation site is called the **decidua.** It is into this super-rich decidual tissue that the blastocyst becomes embedded. After the blastocyst burrows into the decidua by means of trophoblastic activity, a layer of endometrial cells covers over the surface of the hole, completely burying the blastocyst within the uterine lining (Figure 20-25c). The trophoblastic layer continues to digest the surrounding decidual cells, providing energy for the embryo until the placenta develops.

The placenta is the organ of exchange between maternal and fetal blood.

The glycogen stores in the endometrium are only sufficient to nourish the embryo during its first few weeks. To sustain the growing embryo/fetus for the duration of its intrauterine life, the **placenta,** a specialized organ of exchange between the maternal and fetal blood, is rapidly developed (Fig. 20-26). The placenta is derived from both trophoblastic and decidual tissue.

Figure 20-26 **Placentation** Schematic representation of interlocking maternal and fetal structures that form the placenta.

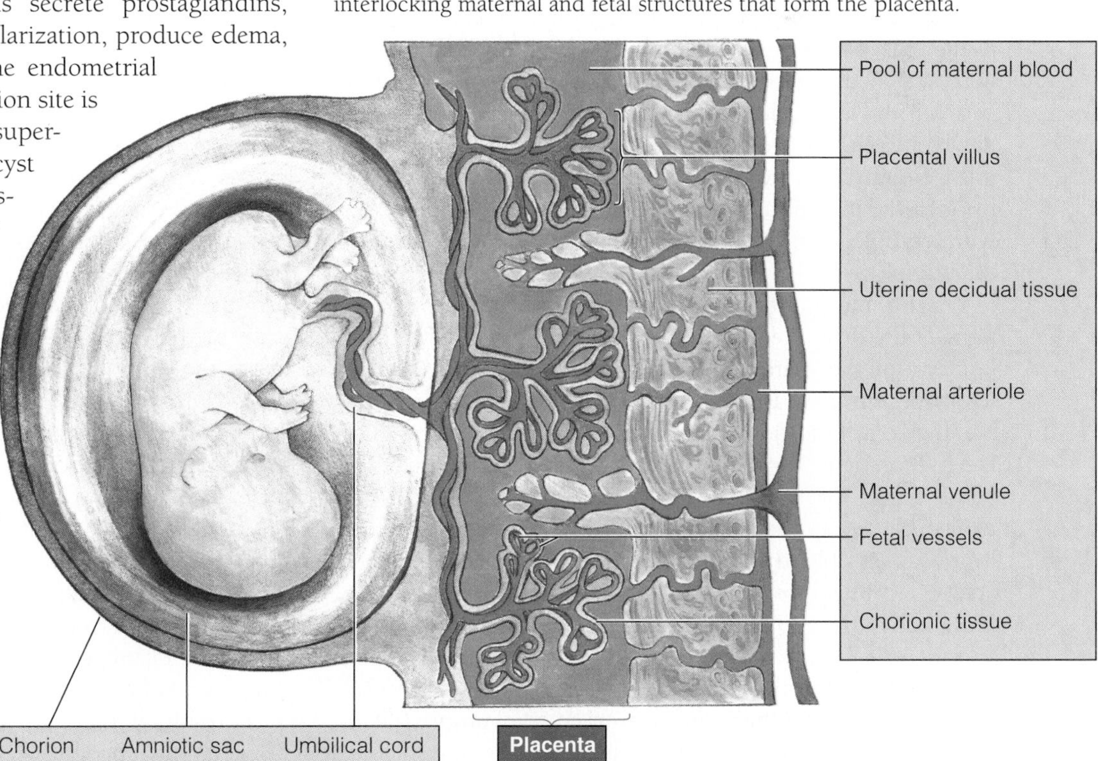

By day twelve, the embryo is completely embedded in the decidua. By this time the trophoblastic layer is two cell layers thick and is called the **chorion.** As the chorion continues to release enzymes and expand, it forms an extensive network of cavities within the decidua. As decidual capillary walls are eroded by the expanding chorion, these cavities fill with maternal blood, which is prevented from clotting by an anticoagulant produced by the chorion. Fingerlike projections of chorionic tissue extend into the pools of maternal blood. Soon the developing embryo sends out capillaries into these chorionic projections to form **placental villi.** Some villi extend completely across the blood-filled spaces to anchor the fetal portion of the placenta to the endometrial tissue, but most simply project into the pool of maternal blood.

Each placental villus contains embryonic (later fetal) capillaries surrounded by a thin layer of chorionic tissue, which separates the embryonic/fetal blood from the maternal blood in the intervillus spaces. There is no actual mingling of maternal and fetal blood, but the barrier between them is extremely thin. To visualize this relationship, think of your hands (the fetal capillary blood vessels) in rubber gloves (the chorionic tissue) immersed in water (the pool of maternal blood). Only the rubber gloves separate your hands from the water. In the same way, only the thin chorionic tissue (plus the capillary wall of the fetal vessels) separates the fetal and maternal blood. It is across this extremely thin barrier that all materials are exchanged between these two bloodstreams. This entire system of interlocking maternal (decidual) and fetal (chorionic) structures makes up the placenta.

Even though not fully developed, the placenta is well established and operational by five weeks after implantation. By this time, the heart of the developing embryo is pumping blood into the placental villi as well as to the embryonic tissues. Throughout gestation, fetal blood continuously traverses between the placental villi and the circulatory system of the fetus by means of the **umbilical artery** and **umbilical vein,** which are wrapped within the **umbilical cord,** a lifeline between the fetus and the placenta (Fig. 20-26). The maternal blood within the placenta is continuously replaced as fresh blood enters through the uterine arterioles, percolates through the intervillus spaces, where it exchanges substances with fetal blood in the surrounding villi, then exits through the uterine vein.

During intrauterine life, the placenta performs the functions of the digestive system, the respiratory system, and the kidneys for the "parasitic" fetus. This is not to say that the fetus does not have these organ systems, but rather that they are unable to (and do not need to) function within the uterine environment. Nutrients and O_2 diffuse from the maternal blood across the thin placental barrier into the fetal blood, while CO_2 and other metabolic wastes simultaneously diffuse from the fetal blood into the maternal blood. The nutrients and O_2 brought to the fetus in the maternal blood are acquired by the mother's digestive and respiratory systems, and the CO_2 and wastes transferred into the maternal blood are eliminated by the mother's lungs and kidneys, respectively. Thus, the mother's digestive tract, respiratory system, and kidneys serve the fetus's needs as well as her own.

Some substances traverse the placental barrier by special mediated transport systems in the placental membranes, whereas others move across by simple diffusion. Unfortunately, many drugs, environmental pollutants, other chemical agents, and microorganisms in the mother's bloodstream are also able to cross the placental barrier, and some of them may be harmful to the developing fetus. Individuals born limbless as a result of exposure to thalidomide, a tranquilizer taken by pregnant women before this drug's devastating effects on the growing fetus were known, serve as a grim reminder of this fact. Similarly, newborns who have become "addicted" during gestation by their mother's use of an abusive drug such as heroin suffer withdrawal symptoms after birth. Even more common chemical agents such as aspirin, alcohol, and agents in cigarette smoke can reach the fetus and have adverse effects. Likewise, fetuses can acquire AIDS before birth if their mothers are infected with the virus. A pregnant women should therefore be very cautious about potentially harmful exposure from any source.

In addition to serving as an organ of exchange, the placenta in some way acts as a protective barrier to prevent the embryo from being immunologically rejected by the mother. The embryo is actually a "foreigner" because it is half derived from genetically different paternal chromosomes. Furthermore, the placenta becomes a temporary endocrine organ during pregnancy, a topic to which we now turn.

Hormones secreted by the placenta play a critical role in the maintenance of pregnancy.

The placenta has the remarkable capacity to secrete a number of peptide and steroid hormones essential for the maintenance of pregnancy. The most important are *human chorionic gonadotropin, estrogen,* and *progesterone* (Table 20-5). Serving as the major endocrine organ of pregnancy, the placenta is unique among endocrine tissues in two regards. First, it is a transient tissue. Second, secretion of its hormones is not subject to extrinsic control, in contrast to the stringent, often complex mechanisms that regulate the secretion of other hormones. Instead, the type and rate of placental hormone secretion depend primarily on the stage of pregnancy.

One of the first events following implantation is secretion by the developing chorion of **human chorionic gonadotropin (hCG),** a peptide hormone that prolongs the life span of the corpus luteum. Recall that during the ovarian cycle, the corpus luteum degenerates and the highly prepared, luteal-dependent uterine lining sloughs if fertilization and implantation do not occur. When fertilization does occur, the implanted blastocyst saves itself from being flushed out in menstrual flow by producing hCG. This hormone, which is functionally similar to LH, stimulates and maintains the corpus luteum so that it does not degenerate. Now called the **corpus luteum of pregnancy,** this ovarian endocrine unit grows even larger and produces increasingly greater amounts of estrogen and progesterone for an additional ten weeks until the placenta takes over secretion of these steroid hormones. As a result of the persistence of circulating estrogen and progesterone, the thick, pulpy endometrial tissue is maintained instead of sloughing. Accordingly, menstruation ceases during pregnancy.

Stimulation by hCG is necessary to maintain the corpus luteum of pregnancy because LH, which maintains the corpus luteum during the normal luteal phase of the uterine cycle, is suppressed through feedback inhibition by the high levels of progesterone. Suppression of the anterior pituitary hormones by the high progesterone levels also precludes further follicular maturation and ovulation for the duration of the pregnancy.

The maintenance of a normal pregnancy depends on high concentrations of progesterone and estrogen. Thus, production of hCG is critical during the first trimester to maintain ovarian output of these hormones. In a male fetus, hCG also stimulates the precursor Leydig cells in the fetal testes to secrete testosterone, which masculinizes the developing reproductive tract.

The secretion rate of hCG increases rapidly during early pregnancy to save the corpus luteum from demise. Peak secretion of hCG occurs about sixty days after the end of the last menstrual period (Fig. 20-27). By the tenth week of pregnancy, hCG output declines to a low rate of secretion that is maintained for the duration of gestation. The fall in hCG occurs at a time when the corpus luteum is no longer needed for its steroidal hormone output because the placenta has begun to secrete substantial quantities of estrogen and progesterone. The corpus luteum of pregnancy partially regresses as hCG secretion dwindles, but it is not converted into scar tissue until after delivery of the baby.

Human chorionic gonadotropin is eliminated from the body in the urine, where its detection forms the basis of pregnancy diagnosis tests. This hormone can be detected in the urine as early as the first month of pregnancy, about two weeks after the first missed menstrual period. Since this is before the growing embryo can be detected by physical examination, the test permits early confirmation of pregnancy.

A frequent early clinical sign of pregnancy is **morning sickness,** a daily bout of nausea and vomiting that often occurs in the morning but can take place at any time of day. Since this condition usually appears shortly after implantation and coincides with the time of peak hCG production, it is speculated that this early placental hormone may trigger the symptoms, perhaps by acting on the chemoreceptor trigger zone in the vomiting center (see p. 565).

A logical question is why the developing placenta does not start producing estrogen and progesterone in the first place instead of secreting hCG, which in turn stimulates the corpus luteum to secrete these two critical hormones. The answer is that, for different reasons, the placenta is unable to produce sufficient quantities of either estrogen or progesterone in the first trimester of pregnancy. In the case of estrogen, the placenta does not have all of the enzymes needed for complete synthesis of this hormone. Estrogen synthesis requires a complex interaction between the placenta and the fetus (Fig. 20-28). The placenta is able to convert the androgen hormone produced by the fetal adrenal cortex, dehydroepiandrosterone (DHEA), into estrogen. The placenta is unable to produce estrogen until the fetus has developed to the point that its

Table 20-5 Placental Hormones

Hormone	*Function*
Human chorionic gonadotropin	Maintains the corpus luteum of pregnancy Stimulates secretion of testosterone by the developing testes in XY embryos
Estrogen ***(also secreted by the corpus luteum of pregnancy)***	Stimulates growth of the myometrium, increasing uterine strength for parturition Helps prepare the mammary glands for lactation
Progesterone ***(also secreted by the corpus luteum of pregnancy)***	Suppresses uterine contractions to provide a quiet environment for the fetus Promotes formation of a cervical mucus plug to prevent uterine contamination Helps prepare the mammary glands for lactation
Human chorionic somatomammotropin	Helps prepare the mammary glands for lactation Believed to reduce maternal utilization of glucose so that greater quantities of glucose may be shunted to the fetus
Relaxin ***(also secreted by corpus luteum of pregnancy)***	Softens the cervix in preparation for cervical dilation at parturition Loosens the connective tissue between the pelvic bones in preparation for parturition

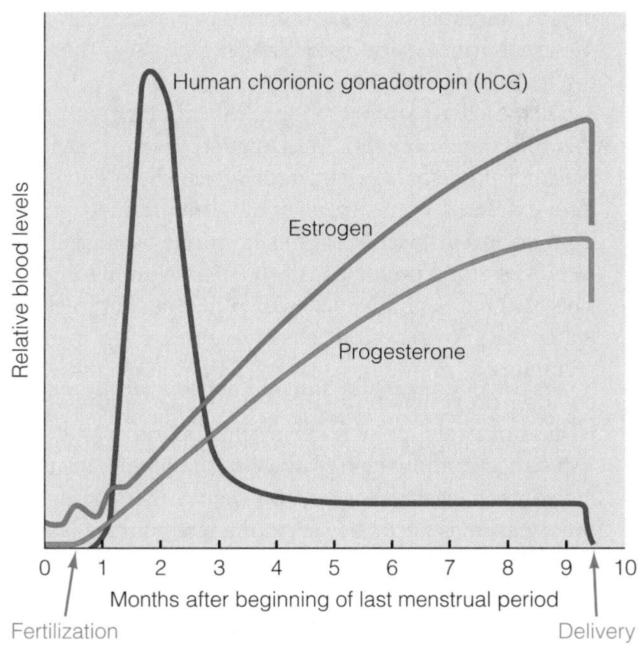

Figure 20-27 **Secretion Rates of Placental Hormones**

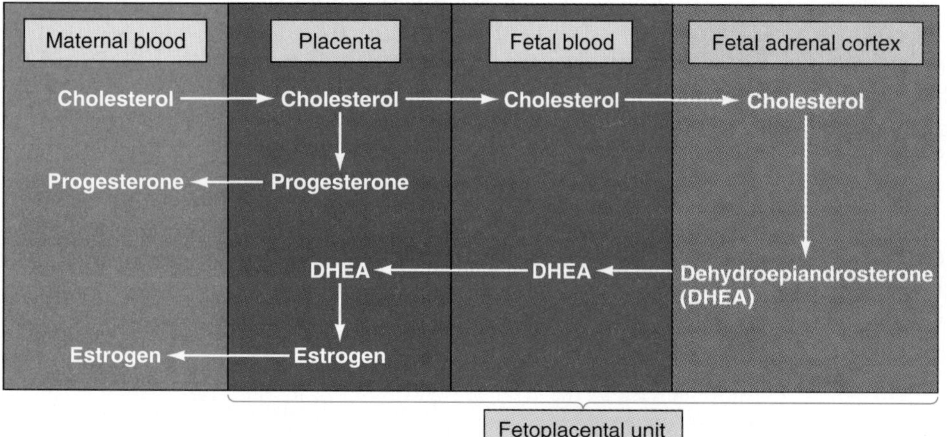

Figure 20-28 **Secretion of Estrogen and Progesterone by the Placenta** The placenta secretes increasing quantities of progesterone and estrogen into the maternal blood after the first trimester. The placenta itself has the ability to convert cholesterol into progesterone but lacks some of the enzymes necessary to convert cholesterol into estrogen. However, the placenta can convert DHEA derived from cholesterol in the fetal adrenal cortex into estrogen when DHEA reaches the placenta by means of the fetal blood.

adrenal cortex is secreting DHEA into the blood. The placenta extracts DHEA from the fetal blood and converts it into estrogen, which it then secretes into the maternal blood. Because both fetal and placental tissues participate in this biosynthetic pathway, estrogen is said to be produced by the **fetoplacental unit.** The primary estrogen synthesized by this unit is **estriol,** in contrast to the main estrogen product of the ovaries, estradiol. Consequently, measurement of estriol levels in the maternal urine can be used clinically to assess the viability of the fetus.

In the case of progesterone, the placenta can synthesize this hormone soon after implantation. Even though the early placenta possesses the enzymes necessary to convert cholesterol extracted from the maternal blood into progesterone, it does not produce much of this hormone because the amount of progesterone produced is proportional to placental weight. The placenta is simply too small in the first ten weeks of pregnancy to produce sufficient quantities of progesterone to maintain the endometrial tissue. The notable increase in circulating progesterone in the last seven months of gestation reflects placental growth during this period.

Estrogen stimulates growth of the myometrium, which increases in size throughout pregnancy. The stronger uterine musculature is needed to expel the fetus during labor. Estriol also promotes development of the ducts within the mammary glands through which milk will be ejected during lactation. Progesterone performs various roles throughout pregnancy. Its primary function is to prevent miscarriage by suppressing contractions of the uterine myometrium. Progesterone also promotes formation of a mucus plug in the cervical canal to prevent vaginal contaminants from reaching the uterus. Finally, placental progesterone stimulates the development of milk glands in the breasts in preparation for lactation.

Maternal body systems respond to the increased demands of gestation.

The period of **gestation (pregnancy)** is about thirty-eight weeks from conception (forty weeks from the end of the last menstrual period). During gestation, the embryo/fetus continues to grow and develop to the point of being able to leave its maternal life support system. Meanwhile, a number of physical changes take place within the mother to accommodate the demands of the pregnancy. The most obvious change is uterine enlargement. The uterus expands and increases in weight more than 20 times, exclusive of its contents. The breasts enlarge and develop the capability of producing milk. Body systems other than the reproductive system also make needed adjustments. The volume of blood increases by 30%, and the cardiovascular system responds to the increasing demands of the growing placental mass. The weight gain experienced during pregnancy is due only in part to the weight of the fetus. The remainder is primarily caused by the increased weight of the uterus, including the placenta, and the increased blood volume. Respiratory activity is increased by about 20% to handle the additional fetal requirements for O_2 utilization and CO_2 removal. Urinary output increases, and the kidneys excrete the additional wastes from the fetus. The increased metabolic demands of the growing fetus result in increased nutritional requirements for the mother. In general, the fetus takes what it needs from the mother, even if this leaves the mother with a nutritional deficit. For example, if the mother does not consume sufficient Ca^{2+} in her diet, Ca^{2+} will be mobilized from the maternal bones to ensure adequate calcification of the fetal bones. (See the accompanying boxed feature, A Closer Look at Exercise Physiology.)

Parturition is accomplished by a positive-feedback cycle.

Parturition (labor, delivery, or **birth)** requires (1) dilation of the cervical canal to accommodate passage of the fetus from the uterus through the vagina and to the outside and (2) contractions of the uterine myometrium that are sufficiently strong to expel the fetus. During the first two trimesters of gestation, the uterus remains relatively quiet because of the inhibitory effect of the high levels of progesterone on the uterine musculature. During the last trimester, the uterus becomes progressively more excitable so that mild contractions **(Braxton-Hicks contractions)** are experienced with increasing strength and frequency. Sometimes these contractions become regular enough to be mistaken as the onset of labor, a phenomenon called "false labor."

Several other events take place near the end of pregnancy in preparation for parturition. The cervix begins to soften (or "ripen") as a result of the dissociation of its connective tissue fibers. This cervical softening is believed to be caused by

A CLOSER LOOK AT EXERCISE PHYSIOLOGY

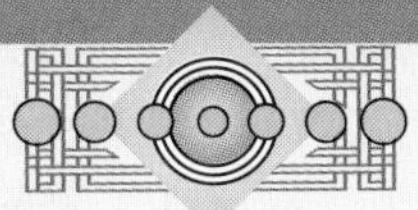

Exercise and Pregnancy: A Compatible or Contraindicated Combination?

Researchers have established that a regular exercise program is beneficial for optimal health, and a large number of young women participate in both competitive and noncompetitive exercise. Many of these women do not want to interrupt their exercise programs during pregnancy.

The questions most often asked about exercise during pregnancy are (1) How does exercise influence the course of pregnancy and the growing fetus, and (2) How does pregnancy alter the capacity to exercise?

One of the most serious concerns about exercise during pregnancy involves the intensity of the exercise and the resultant hyperthermia and redistribution of blood flow. Hyperthermia is not uncommon during high-intensity exercise (see p. 615), and body temperatures of 40°C have been recorded in long-distance runners. Pregnancy might compound this problem of elevated body temperature because the ratio of surface area to body mass becomes smaller due to the presence of a growing fetus. Thus, it might be more difficult to dissipate body heat. The fetus may also be an additional source of heat. However, reports indicate that peak core body temperatures reached during exercise decrease more rapidly in late pregnancy than in the nonpregnant state. These reports also indicate that the temperatures reached during exercise decrease more rapidly as pregnancy progresses, suggesting that the dissipation of body heat during pregnancy is somehow enhanced.

During exercise, blood flow is redistributed to the working muscles and to the skin (for cooling) at the expense of the uterine area. It is still unknown whether uterine blood flow during exercise is maintained at a level that is sufficient for the fetus. However, noninvasive ultrasound investigations found that intervillus blood flow in healthy pregnant women was not decreased by moderate exercise.

Several reports have indicated that a woman who engages in moderate exercise on a regular basis will have a better course of pregnancy than a sedentary woman. There is no evidence that moderate exercise results in harm to the fetus. Indeed, a recent study demonstrated that babies of women who exercised regularly throughout pregnancy were bigger at birth than those of inactive mothers, and bigger babies tend to be healthier. Exercise during pregnancy should involve large muscle groups, be aerobic, and not require delicate balance. Swimming is suggested as an ideal exercise. The cool water keeps the internal temperature down, and the buoyancy provided by the water tends to offset gravity, preventing joint injury. This type of exercise also reduces the chance of the pregnant uterus being in a position to compress the descending aorta and disrupt blood flow to the uterus, as might be the case in supine exercise or stationary bicycling. Exercise in water may also decrease swelling in the lower legs.

It is recommended that exercise during pregnancy be moderate, with exercising heart rates kept below 160 beats per minute. New strenuous exercise programs should not be started during the first or third trimesters, but a gradual program may be started during the second trimester.

A proper exercise program during pregnancy may result in improved aerobic and muscular fitness, easier labor and recovery from labor, and an enhanced feeling of psychological well-being. Exercise generally can be resumed about eight weeks after delivery or as advised by a physician.

---

relaxin, a peptide hormone produced by the corpus luteum of pregnancy and by the placenta. Relaxin also "relaxes" the pelvic bones. Meanwhile, the fetus shifts downward (the baby "drops") and is normally oriented so that the head is in contact with the cervix in preparation for exiting through the birth canal. In a **breech birth**, any part of the body other than the head approaches the birth canal first.

Rhythmic, coordinated contractions, usually painless at first, begin at the onset of true labor. As labor progresses, the contractions occur with increasing frequency and intensity and are accompanied by increasing discomfort. These strong, rhythmic contractions force the fetus against the cervix, resulting in dilation of the cervix. Then, after having dilated the cervix sufficiently for passage of the fetus, these contractions force the fetus out through the birth canal.

The exact factors responsible for triggering this change in uterine contractility and thus initiating parturition are not clearly established, although endocrine factors are believed to be the most important. According to the leading proposal, a dramatic, progressive increase in the concentration of myometrial receptors for oxytocin, probably induced by the increasing levels of estrogen during pregnancy, is ultimately responsible for initiating labor. **Oxytocin** is a peptide hormone that is produced by the hypothalamus, stored in the posterior pituitary, and released into the blood from the posterior pituitary upon nervous stimulation by the hypothalamus (see p. 638). Oxytocin is a powerful uterine muscle stimulant and is known to play the key role in the progression of labor. However, this hormone was discounted as serving as the trigger to parturition because the circulating levels of oxytocin remain constant prior to the onset of labor. The discovery that uterine responsiveness to oxytocin is 100 times greater at term than in nonpregnant women (because of the increased concentration of myometrial oxytocin receptors) gives rise to speculation that labor is initiated when the oxytocin receptor concentration reaches a critical threshold level that permits the onset of strong, coordinated contractions in response to ordinary levels of circulating oxytocin.

Furthermore, one new study suggests that the uterus itself synthesizes oxytocin. Investigators have noted a steady increase throughout gestation in activity of a gene in the uterus of pregnant rats that codes for oxytocin synthesis. This oxytocin-coding activity of the uterine gene peaks at 150 times normal levels just before the onset of labor. Therefore,

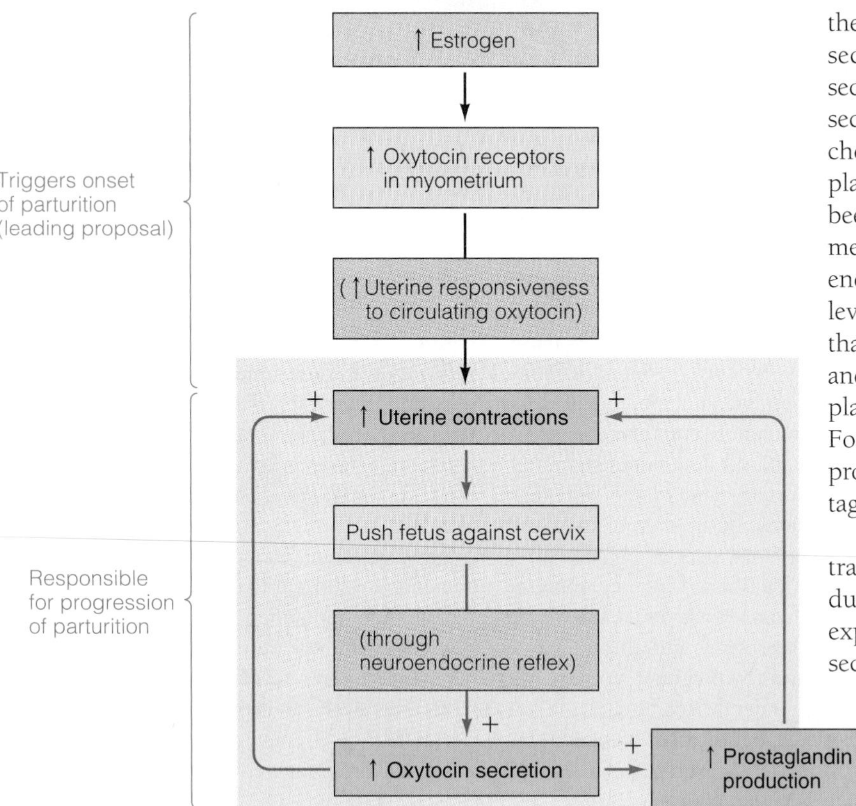

Figure 20-29 **Positive Hormonal Feedback during Parturition**

the local level of oxytocin in the uterus is considerably higher than the blood concentration of oxytocin at the onset of labor. What triggers oxytocin production in the uterus and what role this local oxytocin plays in the onset of labor are presently unclear.

The idea that maternal factors determine the onset of labor has recently been challenged by the finding that in sheep at least the trigger for parturition comes not from the mother but from the fetus. When the fetal sheep's hypothalamus reaches a critical level of maturation, it signals the anterior pituitary to increase ACTH secretion (see p. 639). ACTH in turn causes increased cortisol secretion by the adrenal cortex. Upon reaching the placenta by means of the fetal blood, fetal cortisol promotes the conversion of progesterone into estrogen, which is secreted into the maternal blood. Since progesterone inhibits uterine contractility and estrogen enhances it, this fetal-induced shift in circulating maternal hormones brings about the uterine contractions associated with the onset of labor. Whether this mechanism is also operable in humans remains to be determined. In fact, this and the oxytocin receptor mechanism may both contribute to the onset of parturition in complementary fashion.

An alternative proposal recently set forth is the idea of a **"placental clock"** that ticks out a predetermined length of time until parturition. According to this proposal, the timing of parturition is established early in pregnancy, with delivery occurring at the endpoint of an ongoing maturational process that extends throughout the entire gestation. The ticking of the placental clock is believed to be measured by placental secretion of *corticotropin-releasing hormone (CRH)*. CRH secreted by the hypothalamus regulates anterior pituitary secretion of ACTH, but CRH is also secreted by the placental chorion into the maternal circulation. CRH levels in maternal plasma rise as the pregnancy progresses. Researchers have been able to accurately predict the timing of parturition by measuring the maternal plasma levels of CRH as early as the end of the first trimester of pregnancy. Higher than normal levels were associated with premature deliveries while lower than normal levels were indicative of late deliveries. These and other data suggest that a critical level of maternal CRH of placental origin may act directly as a trigger for parturition. For example, CRH has been shown to stimulate the release of prostaglandin from the placental decidua. Decidual prostaglandin is a potent stimulant of uterine muscle contractions.

Once triggered by whatever mechanism, myometrial contractions progressively increase in frequency, strength, and duration throughout labor until the uterine contents are expelled. At the beginning of labor, contractions lasting thirty seconds or less occur about every twenty-five to thirty minutes; by the end, they last sixty to ninety seconds and occur every two to three minutes.

Myometrial contractions incessantly increase as labor progresses because a positive-feedback cycle involving oxytocin and prostaglandin ensues (Fig. 20-29). Each uterine contraction begins at the top of the uterus and sweeps downward, forcing the fetus toward the cervix. Pressure of the fetus against the cervix accomplishes two things. First, the fetal head pushing against the softened cervix acts as a wedge to dilate the cervical canal. Second, cervical stretch stimulates the release of oxytocin through a neuroendocrine reflex. Stimulation of receptors in the cervix in response to fetal pressure produces a neural signal that travels up the spinal cord to the hypothalamus, which in turn triggers oxytocin release from the posterior pituitary. This additional oxytocin promotes more powerful uterine contractions. As a result, the fetus is pushed more forcefully against the cervix, stimulating the release of even more oxytocin, and so on. This cycle is reinforced as oxytocin stimulates prostaglandin production by the decidua. As a powerful myometrial stimulant, prostaglandin further enhances uterine contractions. Oxytocin secretion, prostaglandin production, and uterine contractions continue to increase in positive-feedback fashion throughout labor until the pressure on the cervix is relieved by delivery.

Labor is divided into three stages: (1) cervical dilation, (2) delivery of the baby, and (3) delivery of the placenta (Fig. 20-30). At the onset of labor or sometime during the first stage, the membranes surrounding the amniotic sac, or "bag of waters," rupture. As the amniotic fluid escapes out of the vagina, it helps to lubricate the birth canal. During the first stage, the cervix is forced to dilate to accommodate the diameter of the baby's head, usually to a maximum of 10 cm. This stage is the longest, lasting from several hours to as long as twenty-four hours in a first pregnancy. If another part of the fetus's body other than the head is oriented against the cervix, it is generally less effective than the head as a wedge. The head

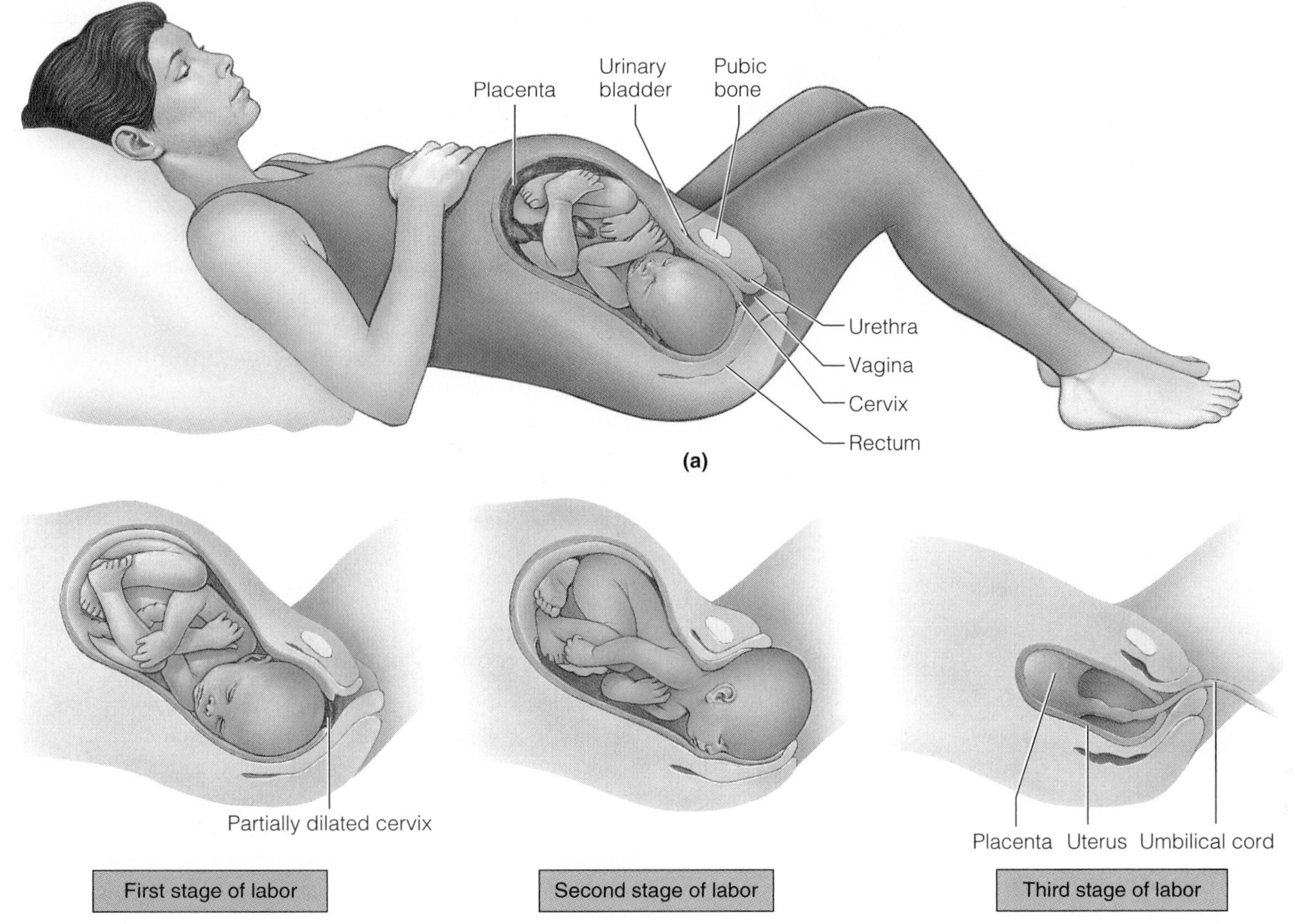

Figure 20-30 **Stages of Labor** (a) Position of the fetus near the end of pregnancy. (b) First stage of labor: cervical dilation. (c) Second stage of labor: delivery of the baby. (d) Third stage of labor: delivery of the placenta.

has the largest diameter of the baby's body. If the baby approaches the birth canal feet first, the cervix may not be dilated sufficiently by the feet to permit passage of the head. Without medical intervention in such a case, the baby's head would remain stuck behind the too narrow cervical opening.

The second stage of labor, the actual birth of the baby, begins once cervical dilation is complete. When the infant begins to move through the cervix and vagina, stretch receptors in the vagina activate a neural reflex that triggers contractions of the abdominal wall in synchrony with the uterine contractions. These abdominal contractions greatly increase the force pushing the baby through the birth canal. The mother can help deliver the infant by voluntarily contracting the abdominal muscles at this time in unison with each uterine contraction (that is, "push" with each "labor pain"). Stage two is usually much shorter than the first stage and lasts thirty to ninety minutes. The infant is still attached to the placenta by the umbilical cord at birth. The cord is tied and severed with the stump shriveling up in a few days to form the **umbilicus (navel).**

Shortly after delivery of the baby, a second series of uterine contractions causes the placenta to separate from the myometrium and be expelled through the vagina. Delivery of the placenta, or **afterbirth,** constitutes the third stage of labor, which is typically the shortest stage, being completed within fifteen to thirty minutes after the baby is born. After the placenta is expelled, continued contractions of the myometrium constrict the uterine blood vessels supplying the site of placental attachment to prevent hemorrhage.

After delivery, the uterus shrinks to its pregestational size, a process known as **involution,** which takes four to six weeks to complete. During involution, the remaining endometrial tissue that was not expelled with the placenta gradually disintegrates and sloughs off, producing a vaginal discharge called **lochia** that continues for three to six weeks following parturition. After this period, the endometrium is restored to its nonpregnant state.

Involution occurs largely because of the precipitous fall in circulating estrogen and progesterone when the placental source of these steroids is lost at delivery. The process is facilitated in mothers who breast-feed their infants because of the oxytocin released in response to suckling. In addition to playing an important role in lactation, this periodic nursing-induced postpartum release of oxytocin promotes myometrial contractions that help maintain uterine muscle tone, thus enhancing involution. Involution is usually complete in about four weeks in nursing mothers but takes about six weeks in those who do not breast-feed.

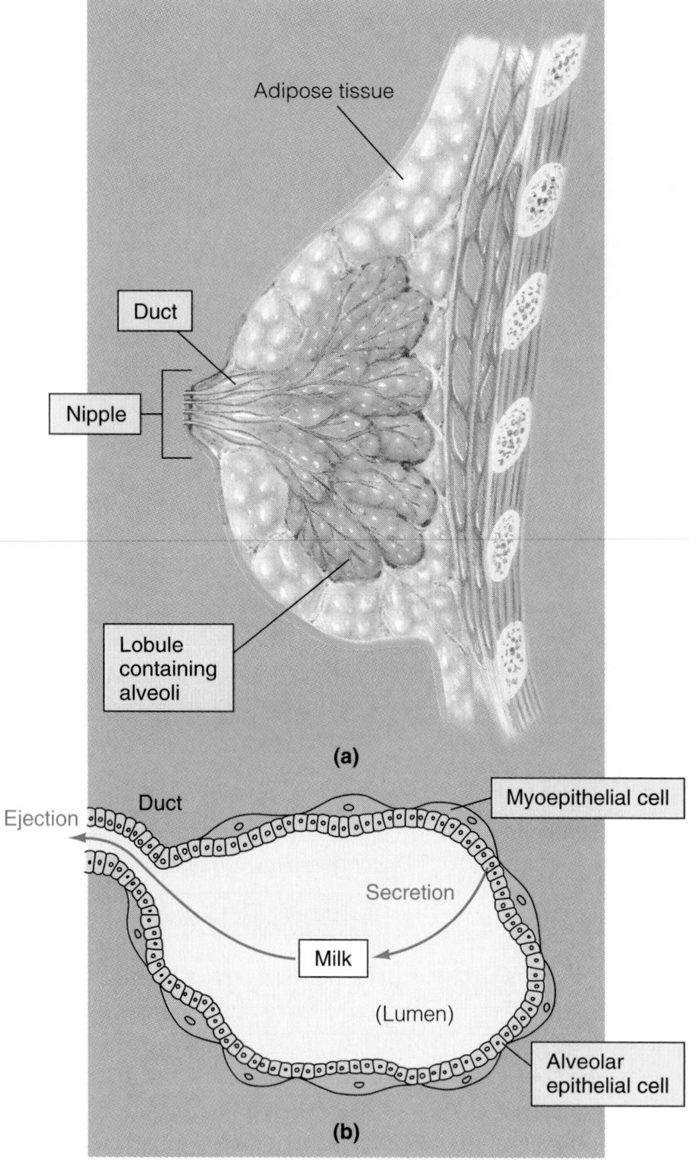

Figure 20-31 **Mammary Gland Anatomy**
(a) Internal structure of the mammary gland, lateral view.
(b) Schematic representation of the microscopic structure of an alveolus within the mammary gland. The alveolar epithelial cells secrete milk into the lumen. Contraction of the surrounding myoepithelial cells ejects the secreted milk out through the duct.

Postpartum depression may be due to low levels of CRH.

For weeks following parturition, many new mothers experience **postpartum depression** (*post* means "after"; *partum* means "birth"). These so-called afterbaby blues have been variably attributed to exhaustion associated with the demands of a new baby, abrupt loss of the high levels of placental estrogen and progesterone, and other unproven speculations. Now researchers have found evidence for a specific physiological cause of postpartum depression. During pregnancy, levels of CRH of placental origin in the maternal bloodstream rise to as much as three times the nonpregnant CRH level. In addition to CRH's proposed role in determining the length of gestation and the timing of parturition, scientists postulate that the stress hormone cortisol ultimately secreted in response to the elevated CRH helps the mother cope with the stress of pregnancy and parturition (see p. 663).

When the placenta is gone following delivery, maternal CRH levels plummet. Earlier studies have shown that CRH helps combat depression. A new investigation demonstrated that the CRH levels of some new mothers were as low as those observed in some forms of clinical depression. The mothers with the lowest levels evidenced the most postpartum depression. The researchers suspect that CRH is lower than normal in new mothers because the placental CRH disrupts the normal feedback mechanism. During pregnancy, the elevated placental-derived CRH inhibits CRH release from the hypothalamus. After delivery, the hypothalamus takes a variable amount of time to recover from the prolonged suppression of its CRH output during gestation. Until the hypothalamus rebounds and starts secreting normal amounts of CRH once again, the woman is apt to experience depression.

Lactation requires multiple hormonal inputs.

The female reproductive system supports the new being from the moment of its conception through gestation and continues to nourish it during its early life outside the supportive uterine environment. Milk (or its equivalent) is essential for survival of the newborn. Accordingly, during gestation the **mammary glands,** or **breasts,** are prepared for **lactation (milk production).**

The breasts in nonpregnant females are composed mostly of adipose tissue and a rudimentary duct system. The size of the breasts is determined by the amount of adipose tissue, which has nothing to do with the ability to produce milk. Under the hormonal environment present during pregnancy, the mammary glands develop the internal glandular structure and function necessary for milk production. A breast capable of lactating consists of a network of progressively smaller ducts that branch out from the nipple and terminate in lobules (Fig. 20-31a). Each *lobule* is made up of a cluster of saclike epithelial-lined *alveoli* that constitute the milk-producing glands. Milk is synthesized by the epithelial cells, then secreted into the alveolar lumen, which is drained by a milk-collecting duct that transports the milk to the surface of the nipple (Fig. 20-31b).

During pregnancy, the high concentration of estrogen promotes extensive duct development, whereas the high level of progesterone stimulates abundant alveolar-lobular formation. Elevated concentrations of **prolactin** (an anterior pituitary hormone stimulated by the rising levels of estrogen) and **human chorionic somatomammotropin** (a peptide hormone produced by the placenta) also contribute to mammary gland development by inducing the synthesis of enzymes needed for milk production.

Most of these changes in the breasts occur during the first half of gestation, so the mammary glands are fully capable of producing milk by the middle of pregnancy. However, milk secretion does not occur until parturition. The high estrogen and progesterone concentrations during the last half of pregnancy prevent lactation by blocking prolactin's stimulatory action on milk secretion. Prolactin is the primary stimulant of milk secretion. Thus, even though the high levels of placental steroids induce the development of the milk-producing machinery in the breasts, they prevent these glands from becoming operational until the baby is born and milk is needed. The abrupt decline in estrogen and progesterone that occurs with loss of the placenta at parturition initiates lactation. (The functions of estrogen and progesterone during gestation and lactation as well as throughout the reproductive life of females are summarized in ▮Table 20-6.)

Once milk production begins after delivery, two hormones are critical for maintaining lactation: (1) *prolactin,* which acts on the alveolar epithelium to promote secretion of milk, and (2) *oxytocin,* which produces **milk ejection.** The latter refers to the forced expulsion of milk from the lumen of the alveoli out through the ducts. Release of both of these hormones is stimulated by a neuroendocrine reflex triggered by suckling (▬Fig. 20-32). Milk cannot be directly sucked out of the alveolar lumen by the infant. Instead, milk must be actively squeezed out of the alveoli into the ducts and hence toward the nipple by contraction of specialized **myoepithelial cells** (musclelike epithelial cells) that surround each alveolus (Fig. 20-31b). Suckling of the breast by the infant stimulates sensory nerve endings in the nipple, resulting in initiation of action potentials that travel up the spinal cord to the hypothalamus. Thus activated, the hypothalamus triggers a burst of oxytocin release from the posterior pituitary. Oxytocin in turn stimulates contraction of the myoepithelial cells in the breasts to bring about milk ejection, or "milk letdown." Milk letdown continues only as long as the infant continues to nurse. In this way, the milk ejection reflex ensures that milk exits the breasts only when and in the amount needed by the baby. Even though the alveoli may be full of milk, the milk

▬ *Figure 20-32* **Suckling Reflexes**

Suckling
↓ +
Mechanoreceptors in nipple
↓ +
Hypothalamus

Nervous pathway
↓ +
Posterior pituitary
↓
↑ Oxytocin
↓ +
Contraction of myoepithelial cells surrounding alveoli
↓
Milk ejection

↓ Prolactin-inhibiting hormone or ↑ prolactin-releasing hormone (?)
↓ +
Anterior pituitary
↓
↑ Prolactin
↓ +
↑ Milk secretion

Table 20-6
Actions of Estrogen and Progesterone

Estrogen

Effects on Sex-Specific Tissues

Essential for egg maturation and release

Stimulates growth and maintenance of the entire female reproductive tract

Stimulates granulosa cell proliferation, which leads to follicle maturation

Thins the cervical mucus to permit sperm penetration

Enhances transport of the sperm to the oviduct by stimulating upward contractions of the uterus and oviduct

Stimulates growth of the endometrium and myometrium

Induces synthesis of progesterone receptors in the endometrium and, during gestation, of myometrial receptors for oxytocin

Other Reproductive Effects

Promotes development of secondary sexual characteristics

Controls GnRH and gonadotropin secretion

- Low levels inhibit secretion
- High levels responsible for triggering the LH surge

Stimulates duct development in the breasts during gestation

Inhibits milk-secreting actions of prolactin during gestation

Nonreproductive Effects

Promotes fat deposition

Increases bone density; closes the epiphyseal plates

Reduces blood cholesterol (incidence of atherosclerosis lower in females until after menopause)

Exerts vascular effects (deficiency produces "hot flashes" at menopause)

Progesterone

Prepares a suitable environment for nourishment of a developing embryo/fetus

Promotes formation of a thick mucus plug in the cervical canal

Inhibits hypothalamic GnRH and gonadotropin secretion

Stimulates alveolar development in the breasts during gestation

Inhibits the milk-secreting actions of prolactin during gestation

Inhibits uterine contractions during gestation

cannot be released without oxytocin. The reflex can become conditioned to stimuli other than suckling, however. For example, the infant's cry can trigger milk letdown, thus causing a spurt of milk to leak from the nipples. On the other hand, psychological stress, acting through the hypothalamus, can easily inhibit milk ejection. For this reason, a positive attitude toward breast-feeding and a relaxed environment are essential for successful breast-feeding.

Suckling not only triggers oxytocin release but also stimulates prolactin secretion. Prolactin output by the anterior pituitary is controlled by two hypothalamic secretions: **prolactin-inhibiting hormone (PIH)** and **prolactin-releasing factor (PRF).** PIH has been shown to be dopamine, which also serves as a neurotransmitter in the brain. The chemical nature of PRF has not been identified with certainty, but scientists suspect that PRF is oxytocin secreted by the hypothalamus into the hypothalamic-hypophyseal portal system to stimulate prolactin secretion by the anterior pituitary (see p. 641). This role of oxytocin is distinct from the roles of oxytocin produced by the hypothalamus and stored in the posterior pituitary. The oxytocin released by the posterior pituitary plays a key role in parturition by causing powerful uterine muscle contractions and also brings about milk ejection by stimulating contraction of the myoepithelial cells of lactating breasts. Throughout most of the female's life, PIH is the dominant influence, so prolactin concentrations normally remain low. During lactation, a burst in prolactin secretion occurs each time the infant suckles. Afferent impulses initiated in the nipple upon suckling are carried by the spinal cord to the hypothalamus. This reflex ultimately leads to prolactin release by the anterior pituitary, although it is unclear whether this outcome is accomplished by inhibition of PIH or stimulation of PRF secretion or both. Prolactin then acts on the alveolar epithelium to promote secretion of milk to replenish the milk lost from the alveoli by milk ejection (Fig. 20-32).

Concurrent stimulation by suckling of both milk ejection and milk production ensures that the rate of milk synthesis keeps pace with the baby's needs for milk. The more the infant nurses, the more milk is removed by letdown and the more milk is produced.

In addition to prolactin, which is the most important factor controlling synthesis of milk, at least four other hormones are essential for their permissive role in milk production: these are cortisol, insulin, parathyroid hormone, and growth hormone.

Among its actions, growth hormone enhances milk production by increasing the mammary gland's uptake from the blood of nutrients needed for the synthesis of milk. Some dairy farmers have been treating their cows with a recently available, genetically engineered version of cow growth hormone, alias *bovine somatotropin (BST),* because BST has been shown to increase a cow's milk production by 10% to 25%. Milk from BST-treated cows is nutritionally equivalent to that of milk from untreated cows. The Food and Drug Administration claims that BST poses no health risk to humans because BST is structurally different from human growth hormone and is not physiologically active in humans. Consumers opposed to the practice dumped gallons of milk outside supermarkets across the United States when the unlabeled BST product first arrived in the dairy case. Opponents are concerned not only about the potential for unknown health risks in humans from consumption of BST but also about the known increased incidence of mastitis (infection of the udder, or mammary gland) in BST-treated cows. As farmers use more antibiotics to treat the mastitis, the risk that antibiotics may slip past routine screening measures and enter the nation's milk supply increases. Antibiotic residues in milk pose a threat to individuals allergic to these antibiotics and furthermore might encourage bacterial resistance in humans to these antibiotics.

Breast-feeding is advantageous to both the infant and the mother.

Milk is composed of water, triglyceride fat, the carbohydrate lactose (milk sugar), a number of proteins, vitamins, and the minerals calcium and phosphate. In addition to supplying nutrients, milk contains a host of immune cells, antibodies, and other chemicals that help protect the infant against infection until it can mount an effective immune response on its own a few months after birth. **Colostrum,** the milk produced for the first five days postpartum, contains lower concentrations of fat and lactose but higher concentrations of immunoprotective components. All human babies acquire some passive immunity during gestation through the passage of antibodies across the placenta from the mother to the fetus (see p. 392). These antibodies are short-lived, however, and often do not persist until the infant can fend for itself immunologically. Breast-fed babies gain additional protection during this vulnerable period through a variety of mechanisms:

- Breast milk contains an abundance of immune cells—both B and T lymphocytes, macrophages, and neutrophils (see p. 374)—that produce antibodies and destroy pathogenic microorganisms outright. These cells are especially plentiful in colostrum.
- **Secretory IgA,** a special type of antibody, is present in great amounts in breast milk. Secretory IgA consists of two IgA antibody molecules (see p. 386) joined together with a so-called secretory component that helps protect the antibodies from destruction by the infant's acidic gastric juice and digestive enzymes. The collection of IgA antibodies a breast-fed baby receives is specifically targeted against the particular pathogens in the environment of the mother—and, accordingly, of the infant as well. Appropriately, therefore, these antibodies provide protection against the infectious microbes that the infant is most likely to encounter.
- Some components in mother's milk, such as mucus, adhere to potentially harmful microorganisms, thus preventing them from attaching to and crossing the intestinal mucosa.
- *Lactoferrin* is a breast-milk constituent that thwarts growth of harmful bacteria by decreasing the availability of iron, a mineral needed for multiplication of these pathogens (see p. 379).

- *Bifidus factor* in breast milk, on the other hand, promotes multiplication of the nonpathogenic microorganism *Lactobacillus bifidus* in the infant's digestive tract. Growth of this harmless bacterium helps crowd out potentially harmful bacteria.
- Other components in breast milk promote maturation of the baby's digestive system so that it is less vulnerable to diarrhea-causing bacteria and viruses.
- Still other factors in breast milk hasten the development of the infant's own immune capabilities.

Thus, breast milk helps protect infants from disease in a variety of ways.

Infants who are bottle fed on a formula made from cow's milk or another substitute do not have the protective advantage provided by human milk, and accordingly, have a higher incidence of infections of the digestive tract, respiratory tract, and ears than do breast-fed babies. Also, the digestive system of a newborn is better equipped to handle human milk than cow milk–derived formula, so bottle-fed babies tend to have more digestive upsets.

Breast-feeding is also advantageous for the mother. Oxytocin release triggered by nursing hastens uterine involution. In addition, suckling suppresses the menstrual cycle by inhibiting LH and FSH secretion, probably through inhibition of GnRH. Lactation, therefore, prevents ovulation and serves as a means of preventing pregnancy (although it is not 100% effective as a means of contraception), thus permitting all the mother's resources to be directed toward the newborn instead of being shared with a new embryo.

When the infant is weaned, two mechanisms contribute to the cessation of milk production. First, without suckling, prolactin secretion is not stimulated, thus removing the primary stimulus for continued milk synthesis and secretion. Also because of the lack of suckling, milk letdown does not occur in the absence of oxytocin release. Since milk production does not immediately shut down, milk accumulates in the alveoli, causing engorgement of the breasts. The resultant pressure buildup acts directly on the alveolar epithelial cells to suppress further milk production. Cessation of lactation at weaning therefore results from a lack of suckling-induced stimulation of both prolactin and oxytocin secretion.

Table 20-7
Average Failure Rate of Various Contraceptive Techniques

Contraceptive Method	*Average Failure Rate (Annual Pregnancies/ 100 Women)*
None	90
Natural (rhythm) methods	20–30
Coitus interruptus	23
Chemical contraceptives	20
Barrier methods	10–15
Oral contraceptives	2–2.5
Implanted contraceptives	1
Intrauterine device	4

Contraceptive techniques act by blocking sperm transport, ovulation, or implantation.

Couples wishing to engage in sexual intercourse but avoid pregnancy have a number of methods of **contraception** ("against conception") available. These methods, which range in effectiveness (Table 20-7) and ease of use, act by blocking one of three major steps in the reproductive process: sperm transport to the ovum, ovulation, or implantation.

Blockage of sperm transport to the ovum

- *Natural contraception* or the *rhythm method* of birth control relies on abstinence from intercourse during the woman's fertile period. The woman can predict when ovulation is to occur based on keeping careful records of her menstrual cycles. This technique is only partially effective because of variability in cycles. The time of ovulation can be determined more precisely by recording body temperature each morning before getting up. Body temperature rises slightly about a day after ovulation has taken place. The temperature rhythm method is not useful in determining when it is safe to engage in intercourse before ovulation, but it can be helpful in determining when it is safe to resume sex after ovulation. This technique is also employed by couples as a method of fertility. The rise in body temperature occurs during the time period that the ovulated ovum may still be viable and capable of being fertilized.

  One group of scientists is working on developing a rapid, sensitive, at-home test using a few drops of urine to detect (1) a rise in estrogen, a "red-light" signal that ovulation will occur in about four days and intercourse should be avoided (if contraception is the goal) and (2) a rise in progesterone, a "green-light" signal that ovulation is past and sexual activity can be resumed.
- *Coitus interruptus* involves withdrawal of the penis from the vagina before ejaculation occurs. This method is only moderately effective, however, because timing is difficult and some sperm may pass out of the urethra prior to ejaculation.
- *Chemical contraceptives,* such as spermicidal ("sperm-killing") jellies, foams, creams, and suppositories, when inserted into the vagina are toxic to sperm for about an hour following application.
- *Barrier methods* mechanically prevent sperm transport to the oviduct. For males, the *condom* is a thin, strong rubber or latex sheath placed over the erect penis prior to ejaculation to prevent sperm from entering the vagina. For females, the *diaphragm* is a flexible rubber dome that is inserted through the vagina and positioned over the cervix to block sperm entry into the cervical canal. It is held in

position by lodging snugly against the vaginal wall. The diaphragm must be fitted by a trained professional and must be left in place for at least six hours but no longer than twenty-four hours after intercourse. Barrier methods are often used in conjunction with spermicidal agents for increased effectiveness. The *cervical cap* is a recently developed alternative to the diaphragm. Smaller than a diaphragm, the cervical cap, which is coated with a film of spermicide, cups over the cervix and is held in place by suction.

The *contraceptive sponge* is a nonprescription polyurethane (clear plastic) device that is placed in the vagina next to the cervix. It acts as a physical deterrent to sperm transport and also releases a spermicidal agent for up to twenty-four hours. The sponge has been implicated as the underlying cause of toxic shock syndrome in a small number of this product's users.

The *female condom* (or *vaginal pouch*) is the latest barrier method developed. It is a 7-inch-long, polyurethane, cylindrical pouch that is closed on one end and open on the other end, with a flexible ring at both ends. The ring at the closed end of the device is inserted into the vagina and fits over the cervix, similar to a diaphragm. The ring at the open end of the pouch is positioned outside the vagina over the external genitalia.

- *Sterilization,* which involves surgical disruption of either the ductus deferens *(vasectomy)* in men or the oviduct *(tubal ligation)* in women, is considered to be a permanent method of preventing sperm and ovum from uniting.

Prevention of ovulation

- *Oral contraceptives,* or *birth control pills,* prevent ovulation by suppressing gonadotropin secretion. These pills, which contain synthetic estrogen-like and progesterone-like steroids, are taken for three weeks, either in combination or in sequence, and then are withdrawn for one week. These steroids, like the natural steroids produced during the ovarian cycle, inhibit GnRH and thus FSH and LH secretion. As a result, follicle maturation and ovulation do not take place, so conception is impossible. The endometrium responds to the exogenous steroids by thickening and developing secretory capacity, just as it would to the natural hormones. When these synthetic steroids are withdrawn after three weeks, the endometrial lining sloughs and menstruation occurs, as it normally would upon degeneration of the corpus luteum. In addition to blocking ovulation, oral contraceptives also prevent pregnancy by increasing the viscosity of cervical mucus, which makes sperm penetration more difficult, and by decreasing muscular contractions in the female reproductive tract, which reduces sperm transport to the oviduct. Oral contraceptives are available only by prescription. They have been shown to increase the risk of intravascular clotting, especially in women who also smoke tobacco.
- A new approach to contraception is a *long-acting subcutaneous* ("under the skin") *implantation* of synthetic progesterone, which acts similarly to oral contraceptives by blocking ovulation and thickening the cervical mucus to prevent sperm transport. Unlike oral contraceptives, however, which must be taken on a regular basis, this new contraceptive, once implanted, is effective for five years. Six matchstick-sized capsules containing the steroid are inserted under the skin in the inner arm above the elbow. The capsules slowly release the synthetic progesterone at a nearly steady rate for five years, thus sustaining their contraceptive effect for a prolonged period. Of concern is a preliminary finding that these long-lasting implantations may exert toxic effects on the nervous system. Similar in action but on a shorter time scale are injectable time-released synthetic female sex hormones that exert contraceptive effects for a month or three months, depending on the product.

Blockage of implantation

- Blockage of implantation is most commonly accomplished by insertion of a small *intrauterine device (IUD)* into the uterus by a physician. The IUD's mechanism of action is not completely understood, although most evidence suggests that the presence of this foreign object in the uterus induces a local inflammatory response that prevents implantation of a fertilized ovum. Although the IUD is a convenient birth control method because it does not require ongoing attention by the user, it is no longer as popular as it once was because of reported complications associated with its use, the most serious of which are pelvic inflammatory disease, permanent infertility, and uterine perforation. Today's IUD models are much safer than the earlier models, however, so these complications are now rare.
- Implantation can also be blocked by the so-called *morning-after pill,* which is a different type of oral contraceptive than the usual birth control pill. The high-estrogen morning-after pill is taken during the early luteal phase within a few days after conception may have occurred. It prevents implantation by inducing premature degeneration of the corpus luteum so that the developing endometrium's hormonal support is withdrawn. Because of side effects, such as nausea and vomiting, and because of the increased risks of cardiovascular disease associated with high doses of estrogen, this contraceptive method is not used on a routine basis. It is beneficial for one-time usage in special circumstances, however, such as for rape victims who may have conceived.

Future possibilities

- On the horizon are improved varieties of currently available contraceptive techniques, such as hormone-releasing IUDs that last for five years and hormone-releasing subcutaneous implants using fewer rods that are easier to insert and remove.
- Scientists hope to have another form of hormonal manipulation available by the turn of the century—doughnut-

shaped vaginal rings that release synthetic progesterone alone or in combination with estrogen.

- A future birth control technique is **immunocontraception**—the use of vaccines that prod the immune system to produce antibodies targeted against a particular protein critical to the reproductive process. The contraceptive effects of the vaccines are expected to last about a year. For example, in the testing stage is a vaccine that induces the formation of antibodies against human chorionic gonadotropin so that this essential corpus luteum–supporting hormone is not effective should pregnancy occur.
- Some researchers are exploring ways to block the union of sperm and egg by interfering with a specific interaction that normally occurs between the male and female gametes. For example, under study are chemicals introduced into the vagina that trigger premature release of the acrosomal enzymes, thus depriving the sperm of a means to fertilize an ovulated egg. Other investigators are working on a vaccine against the protein in a sperm's head that normally binds to the receptor sites on the zona pellucida surrounding the egg.
- Some scientists are seeking ways to manipulate hormones in males to block spermatogenesis. Although a "male birth control pill" remains a distant possibility, a more imminent prospect is injection of synthetic androgen to shut down the GnRH, gonadotropin, sperm production pathway. Alternatively, male vaccines against GnRH or FSH would stop sperm production. Since blocking GnRH action would hinder testosterone secretion as well, supplemental testosterone injections would be needed.
- Still another possibility under investigation is manipulation of the anterior pituitary secretion of FSH and LH by GnRH-like drugs. The use of these drugs as contraceptives in both females and males is being explored.
- Another outlook for male contraception is chemical sterilization designed to be reversed, unlike surgical sterilization, which is considered irreversible. In this experimental technique, a nontoxic polymer is injected into the ductus deferens, where the chemical interferes with the sperms' fertilizing capabilities. Flushing of the polymer from the ductus deferens by a solvent reverses the contraceptive effect.

Termination of unwanted pregnancies

- When contraceptive practices fail or are not used and an unwanted pregnancy results, women sometimes turn to *abortion* to terminate the pregnancy. More than half of the approximately 6.4 million pregnancies in the United States each year are unintended, and about 1.6 million of them end with an abortion. Although surgical removal of an embryo/fetus is legal in the United States, the practice of abortion is fraught with emotional, ethical, and political controversy.
- Adding to the controversy is the discovery by a French pharmaceutical in 1980 of an "abortion pill," *RU 486,* now officially called *mifepristone,* which terminates an early pregnancy by chemical interference rather than by standard surgical procedures. RU 486 (named after Roussel-Uclaf, the company that developed the drug) is a progesterone antagonist. It binds tightly with the progesterone receptors on the target cells but does not evoke progesterone's usual effects and prevents progesterone from binding and acting. Deprived of progesterone activity, the highly developed endometrial tissue sloughs off, carrying the implanted embryo with it. RU 486 is typically given in conjunction with a prostaglandin that induces uterine contractions to help expel the endometrium and embryo. Even though RU 486 has been in use since 1988 in France and more recently in other European countries, the makers of the drug were unwilling to seek approval for marketing it in the United States, largely because of the heated controversy surrounding the abortion issue. In 1994, however, Roussel-Uclaf negotiated a licensing agreement with the Population Council, a New York–based nonprofit contraceptive research group. The Population Council instituted a two-year study in support of eventual approval of RU 486 by the U.S. Food and Drug Administration. Plans call for the product to be manufactured for local use by a U.S. pharmaceutical company selected by the Population Council.

The end is a new beginning.

Reproduction is an appropriate way to end our discussion of physiology from cells to systems. The single cell resulting from the union of male and female gametes divides mitotically and differentiates into a multicellular individual made up of a number of different organ systems that interact cooperatively to maintain homeostasis (that is, stability in the internal environment). All the life-supporting homeostatic processes introduced throughout this book begin all over again at the start of a new life.

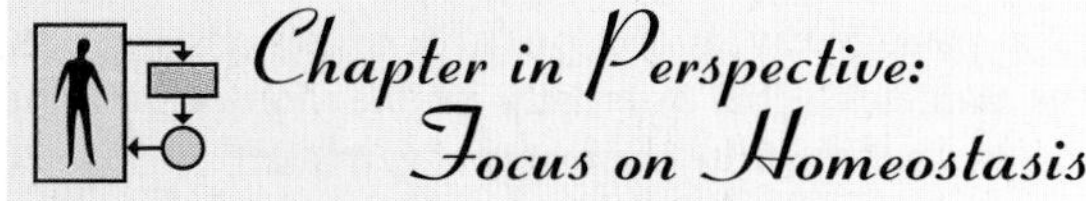

The reproductive system is unique in that it is not essential for homeostasis or for survival of the individual, but it is essential for sustaining the thread of life from generation to generation. Reproduction depends on the union of male and female gametes (reproductive cells), each with a half set of chromosomes, to form a new individual with a full, unique set of chromosomes. Unlike the other body systems, which are essentially identical in the two sexes, the reproductive systems of males and females are remarkably different, befitting their different roles in the reproductive process.

The male system is designed to continuously produce huge numbers of motile spermatozoa that are delivered to the female during the sex act. Male gametes must be produced in abundance for two reasons: (1) only a small percentage of them survive the hazardous journey through the female reproductive tract to the site of fertilization, and (2) the cooperative effort of many spermatozoa is required to break down the barriers surrounding the female gamete (ovum or egg) to enable one spermatozoon to penetrate and unite with the ovum.

The female reproductive system undergoes complex changes on a monthly cyclical basis. During the first half of the cycle, a single nonmotile ovum is prepared for release. During the second half, the reproductive system is geared toward preparing a suitable environment for supporting the ovum if fertilization (union with a spermatozoon) occurs. If fertilization does not occur, the prepared supportive environment within the uterus sloughs off, and the cycle starts over again as a new ovum is prepared for release. If fertilization occurs, the female reproductive system adjusts to support growth and development of the new individual until it can survive on its own on the outside.

There are three important parallels in the male and female reproductive systems, even though they differ considerably in structure and function. First, the same set of undifferentiated reproductive tissues in the embryo can develop into either a male or a female system, depending on the presence or absence, respectively, of male-determining factors. Second, the same hormones—namely, hypothalamic GnRH and anterior pituitary FSH and LH—control reproductive function in both sexes. In both cases, gonadal steroids and inhibin act in negative-feedback fashion to control hypothalamic and anterior pituitary output. Third, the same events take place in the developing gamete's nucleus during sperm formation and egg formation, although males produce millions of sperm in one day, whereas females produce only about 400 ova in a lifetime.

Chapter Summary

Introduction

Both sexes produce gametes (reproductive cells), sperm in males and ova (eggs) in females, each of which bears one member of each of the twenty-three pairs of chromosomes present in human cells. Union of a sperm and an ovum at fertilization results in the beginning of a new individual with twenty-three complete pairs of chromosomes, half from the father and half from the mother.

The reproductive system is anatomically and functionally distinct in males and females. Males produce sperm and deliver them into the female. Females produce ova, accept sperm delivery, and provide a suitable environment for supporting development of a fertilized ovum until the new individual can survive on its own in the external world. In both sexes, the reproductive system consists of (1) a pair of gonads, testes in males and ovaries in females, which are the primary reproductive organs that produce the gametes and secrete sex hormones, and (2) a reproductive tract composed of a system of ducts and associated glands that provide a passageway and supportive secretions, respectively, for the gametes. The externally visible portions of the reproductive system constitute the external genitalia.

Sex determination is a genetic phenomenon dependent on the combination of sex chromosomes at the time of fertilization, an XY combination being a genetic male and an XX combination a genetic female. Sex differentiation refers to the embryonic development of the gonads, reproductive tract, and external genitalia along male or female lines, which gives rise to the apparent anatomical sex of the individual. In the presence of masculinizing factors, a male reproductive system develops; in their absence, a female system develops.

Male Reproductive Physiology

Spermatogenesis (sperm production) occurs in the highly coiled seminiferous tubules within the testes. Leydig cells located in the interstitial spaces between these tubules secrete the male sex hormone testosterone into the blood. Testosterone is secreted before birth to masculinize the developing reproductive system; then its secretion ceases until puberty, at which time it begins once again and continues throughout life. Testosterone is responsible for maturation and maintenance of the entire male reproductive tract, for development of secondary sexual characteristics, and for stimulating libido.

The testes are regulated by the anterior pituitary hormones, luteinizing hormone (LH) and follicle-stimulating hormone (FSH). These gonadotropic hormones in turn are under control of hypothalamic gonadotropin-releasing hormone (GnRH). Testosterone secretion is regulated by LH stimulation of the Leydig cells, and, in negative-feedback fashion, testosterone inhibits gonadotropin secretion. Spermatogenesis requires both testosterone and FSH. Testosterone stimulates the mitotic and meiotic divisions required to transform the undifferentiated diploid germ cells, the spermatogonia, into undifferentiated haploid spermatids. The remodeling of spermatids into highly specialized motile spermatozoa is stimulated by FSH. A spermatozoon consists only of a DNA-packed head bearing an enzyme-filled acrosome at its tip for penetrating the ovum, a midpiece containing the metabolic machinery for energy production, and a whiplike motile tail. Also present in the seminiferous tubules are Sertoli cells, which protect, nurse, and enhance the germ cells throughout their development. Sertoli cells also secrete inhibin, a hormone that inhibits FSH secretion, thus completing the negative-feedback loop.

The still immature sperm are flushed out of the seminiferous tubules into the epididymis by fluid secreted by the Sertoli cells. The epididymis and ductus deferens store and concentrate the sperm and increase their motility and fertility prior to ejaculation. During ejaculation, the sperm are mixed with secretions released by the accessory glands, which contribute the bulk of the semen. The seminal vesicles supply fructose for energy and prostaglandins, which promote smooth muscle motility in both the male and female reproductive tracts to enhance sperm transport. The prostate gland contributes an alkaline fluid for neutralizing the acidic vaginal secretions. The bulbourethral glands release lubricating mucus.

Sexual Intercourse between Males and Females

The male sex act consists of erection and ejaculation, which are part of a much broader systemic, emotional response that typifies the male sexual response cycle. Erection is a hardening of the normally flaccid penis that enables it to penetrate the female vagina. Erection is accomplished by marked vasocongestion of the penis brought about by reflexly induced vasodilation of the arterioles supplying the penile erectile tissue. When sexual excitation reaches a critical peak, ejaculation occurs. It consists of two stages: (1) emission, the emptying of semen (sperm and accessory sex gland secretions) into the urethra, and (2) expulsion of semen from the penis. The latter is accompanied by a set of characteristic systemic responses and intense pleasure referred to as orgasm.

Females experience a sexual cycle similar to males, with both having excitation, plateau, orgasmic, and resolution phases. The major differences are that women do not ejaculate and they are capable of multiple orgasms. During the female sexual response, the outer third of the vagina constricts to grip the penis while the inner two-thirds expands to create space for sperm deposition.

Female Reproductive Physiology

In the nonpregnant state, female reproductive function is controlled by a complex, cyclical negative-feedback control system between the hypothalamus (GnRH), anterior pituitary (FSH and LH), and ovaries (estrogen, progesterone, and inhibin). During pregnancy, placental hormones become the main controlling factors.

The ovaries perform the dual and interrelated functions of oogenesis (producing ova) and secreting estrogen and progesterone. Two related ovarian endocrine units sequentially accomplish these functions: the follicle and the corpus luteum. Oogenesis and estrogen secretion take place within an ovarian follicle during the first half of each reproductive cycle (the follicular phase). At approximately midcycle, the maturing follicle releases a single ovum (ovulation). The empty follicle is then converted into a corpus luteum, which produces progesterone as well as estrogen during the last half of the cycle (the luteal phase). This endocrine unit is responsible for preparing the uterus as a suitable site for implantation should the released ovum be fertilized. If fertilization and implantation do not occur, the corpus luteum degenerates. The consequent withdrawal of hormonal support for the highly developed uterine lining causes it to disintegrate and slough, producing menstrual flow. Simultaneously, a new follicular phase is initiated. Menstruation ceases and the uterine lining (endometrium) repairs itself under the influence of rising estrogen levels from the newly maturing follicle.

If fertilization does take place, it occurs in the oviduct as the released egg and sperm deposited in the vagina are both transported to this site. The fertilized ovum begins to divide mitotically. Within a week it grows and differentiates into a blastocyst capable of implantation. Meanwhile, the endometrium has become richly vascularized and stocked with stored glycogen under the influence of luteal phase progesterone. It is into this especially prepared lining that the blastocyst implants by means of enzymes released by the blastocyst's outer layer. These enzymes digest the nutrient-rich endometrial tissue, accomplishing the dual function of carving out a hole in the endometrium for implantation of the blastocyst while at the same time releasing nutrients from the endometrial cells for use by the developing embryo.

Following implantation, an interlocking combination of fetal and maternal tissues, the placenta, develops. The placenta is the organ of exchange between the maternal and fetal blood and also acts as a transient, complex endocrine organ that secretes a number of hormones essential for pregnancy. Human chorionic gonadotropin, estrogen, and progesterone are the most important of these hormones.

At parturition, rhythmic contractions of increasing strength, duration, and frequency accomplish the three stages of labor: dilation of the cervix, birth of the baby, and delivery of the placenta (afterbirth). Once the contractions are initiated at the onset of labor, a positive-feedback cycle is established that progressively increases their force. As contractions push the fetus against the cervix, secretion of oxytocin, a powerful uterine muscle stimulant, is reflexly increased. The extra oxytocin causes stronger contractions, giving rise to even more oxytocin release, and so on. This positive-feedback cycle progressively intensifies until cervical dilation and delivery are accomplished.

During gestation, the breasts are specially prepared for lactation. The elevated levels of placental estrogen and progesterone, respectively, promote development of the ducts and alveoli in the mammary glands. Prolactin stimulates the synthesis of enzymes essential for milk production by the alveolar epithelial cells. However, the high gestational level of estrogen and progesterone prevents prolactin from promoting milk production. Withdrawal of the placental steroids at parturition initiates lactation. Lactation is sustained by suckling, which triggers the release of oxytocin and prolactin. Oxytocin causes milk ejection by stimulating the myoepithelial cells surrounding the alveoli to squeeze the secreted milk out through the ducts. Prolactin stimulates the production of more milk to replace the milk ejected as the baby nurses.

Review Exercises

Objective Questions (Answers on p. E–20.)

1. It is possible for a genetic male to have the anatomical appearance of a female. (True of false?)
2. Testosterone secretion essentially ceases from birth until puberty. (True or false?)
3. The pineal gland secretes more melatonin during the light than during the dark. (True or false?)
4. Females do not experience erection. (True or false?)
5. Most of the lubrication for sexual intercourse is provided by the female. (True or false?)
6. If a follicle fails to reach maturity during one ovarian cycle, it can finish maturing during the next cycle. (True or false?)
7. Low but rising levels of estrogen inhibit tonic LH secretion whereas high levels of estrogen stimulate the LH surge. (True or false?)

8. Spermatogenesis takes place within the ______________ of the testes, stimulated by the hormones ______________ and ______________.

9. During estrogen production by the follicle, the ______________ cells under the influence of the hormone ______________ produce androgens, and the ______________ cells under the influence of the hormone ______________ convert these androgens into estrogens.

10. The source of estrogen and progesterone during the first ten weeks of gestation is the ______________.

11. Detection of ______________ in the urine is the basis of pregnancy diagnosis tests.

12. Which of the following statements concerning chromosomal distribution is *incorrect?*
    a. All human somatic cells contain twenty-three chromosomal pairs for a total diploid number of forty-six chromosomes.
    b. Each gamete contains twenty-three chromosomes, one member of each chromosomal pair.
    c. During meiotic division, the members of the chromosome pairs regroup themselves into the original combinations derived from the individual's mother and father for separation into haploid gametes.
    d. Sex determination depends on the combination of sex chromosomes, an XY combination being a genetic male, XX a genetic female.
    e. The sex chromosome content of the fertilizing sperm determines the sex of the offspring.

13. When the corpus luteum degenerates,
    a. circulating levels of estrogen and progesterone rapidly decline.
    b. FSH and LH secretion start to rise as the inhibitory effects of the gonadal steroids are withdrawn.
    c. the endometrium sloughs.
    d. Both a and b are correct.
    e. All of the above are correct.

14. Match the following:

___ 1. secrete(s) prostaglandins	(a) epididymis and ductus deferens
___ 2. increase(s) motility and fertility of sperm	(b) prostate gland
___ 3. secrete(s) an alkaline fluid	(c) seminal vesicles
___ 4. provide(s) fructose	(d) bulbourethral glands
___ 5. storage site for sperm	(e) penis
___ 6. concentrate(s) the sperm a hundredfold	
___ 7. secrete(s) fibrinogen	
___ 8. provide(s) clotting enzymes	
___ 9. contain(s) erectile tissue	

15. Using the answer code below, indicate when each event takes place during the ovarian cycle:
    (a) occurs during the follicular phase
    (b) occurs during the luteal phase
    (c) occurs during both the follicular and luteal phases

    ___ 1. development of antral follicles
    ___ 2. secretion of estrogen
    ___ 3. secretion of progesterone
    ___ 4. menstruation
    ___ 5. repair and proliferation of the endometrium
    ___ 6. increased vascularization and glycogen storage in the endometrium

Essay Questions

1. What are the primary reproductive organs, gametes, sex hormones, reproductive tract, accessory sex glands, external genitalia, and secondary sexual characteristics in males and females?
2. List the essential reproductive functions of the male and of the female.
3. Discuss the differences between males and females with regard to genetic, gonadal, and phenotypic sex.
4. What parts of the male and female reproductive systems develop from each of the following: genital tubercle, urethral folds, genital swellings, Wolffian ducts, and Müllerian ducts?
5. Discuss the source and functions of testosterone.
6. Describe the three major stages of spermatogenesis. Discuss the functions of each part of a spermatozoon. What are the roles of Sertoli cells?
7. Discuss the control of testicular function.
8. Compare the sex act in males and females.
9. Compare oogenesis with spermatogenesis.
10. Describe the events of the follicular and luteal phases of the ovarian cycle. Correlate the phases of the uterine cycle with those of the ovarian cycle.
11. How are the ovum and spermatozoa transported to the site of fertilization? Describe the process of fertilization.
12. Describe the process of implantation and placenta formation.
13. What are the functions of the placenta? What hormones does the placenta secrete?
14. What is the role of human chorionic gonadotropin?
15. What is the leading proposal for the mechanism that initiates parturition? What are the stages of labor? What is the role of oxytocin?
16. Describe the hormonal factors that play a role in lactation.
17. Summarize the actions of estrogen and progesterone.

Points to Ponder

(Explanations on p. E–20.)

1. The hypothalamus releases GnRH in pulsatile bursts once every two to three hours, with no secretion occurring in between. A promising line of research for a new method of contraception involves administration of GnRH-like drugs. In what way could such drugs act as contraceptives when GnRH is the hypothalamic hormone that triggers the chain of events leading to ovulation? (*Hint:* The anterior pituitary is "programmed" to respond only to the normal pulsatile pattern of GnRH.)
2. Occasionally, testicular tumors composed of interstitial cells of Leydig may secrete up to 100 times the normal amount of testosterone. When such a tumor develops in young children, they grow up much shorter than their genetic potential. Explain why. What other symptoms would be present?
3. What type of sexual dysfunction might arise in men taking drugs that inhibit sympathetic nervous system activity as part of the treatment for high blood pressure?
4. Explain the physiological basis for administering a posterior pituitary extract to induce or facilitate labor.
5. The symptoms of menopause are sometimes treated with supplemental estrogen and progesterone. Why wouldn't treatment with GnRH or FSH and LH also be effective?
6. ***Clinical Consideration*** Maria A., who is in her second month of gestation, has been experiencing severe abdominal cramping. Her physician has diagnosed her condition as a *tubal pregnancy:* The developing embryo is implanted in the oviduct instead of in the uterine endometrium. Why must this pregnancy be surgically terminated?

Appendix A

A Review of Chemical Principles

BY SPENCER SEAGER

Weber State College

Appendix A Contents At a Glance

The chemical nature of all matter makes an understanding of chemistry helpful in many areas of study, including human physiology. This appendix contains a brief discussion of some basic chemical concepts that you are encouraged to examine as needed while you study the material in the textbook.

Atoms, Elements, Compounds, and Molecules

All matter is made up of tiny particles called **atoms.** These particles are too small to be seen individually, even with the most powerful electron microscopes available today. However, the work of generations of scientists has led to an understanding of many characteristics of atoms that help us explain and understand the behavior of matter.

Even though they are tiny, atoms are composed of three types of smaller particles. **Protons** and **neutrons** are particles of nearly identical mass, and they make up the nucleus of an atom. Protons carry a positive charge, whereas neutrons have no charge. **Electrons,** the third type of particle found in atoms, move rapidly around the central nucleus (Fig. A-1). Electrons have a much smaller mass than protons and neutrons and are negatively charged. The charge of a proton exactly matches that of an electron, but it is opposite in sign. In all atoms, the number of protons in the nucleus is equal to the number of electrons moving around the nucleus, so their charges balance and the atoms are neutral.

A pure substance that contains only one type of atom is called an **element.** A pure sample of the element carbon contains only carbon atoms, even though the atoms might be arranged in a form called diamond or in a form called graphite.

Pure substances called **compounds** contain more than one type of atom. Pure water, for example, is a compound that contains atoms of hydrogen and atoms of oxygen in a 2-to-1 ratio, regardless of whether the water is in the form of liquid, solid (ice), or vapor (steam). A **molecule** is the smallest unit of a pure substance that has the properties of that substance and is capable of a stable, independent existence. For exam-

Figure A-1 **The Atom** The atom consists of two regions. The central nucleus contains protons and neutrons and makes up 99.9% of the mass. Surrounding the nucleus is the electron cloud, where the electrons move rapidly around the nucleus. (Figure not drawn to scale.)

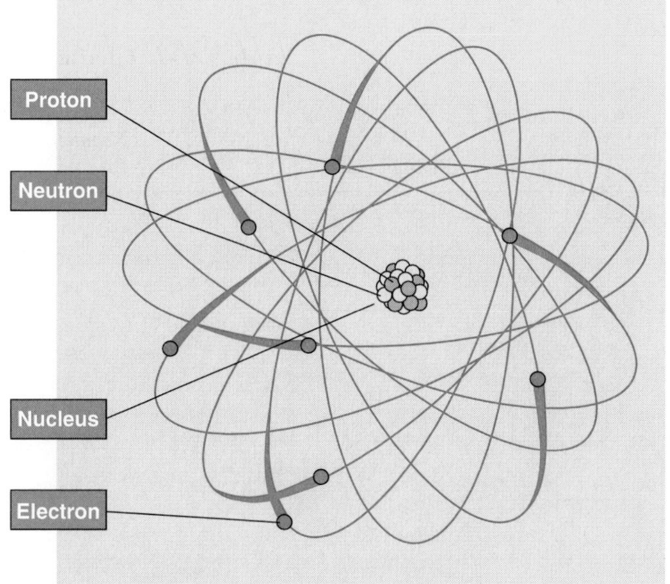

ple, a molecule of water consists of two atoms of hydrogen and one atom of oxygen, held together by chemical bonds.

Exactly what are we talking about when we refer to a "type" of atom? That is, what makes carbon, hydrogen, and oxygen atoms different? The answer is the number of protons in the nucleus. Regardless of where they are found, all hydrogen atoms have one proton in the nucleus, all carbon atoms have six, and all oxygen atoms have eight. Of course, these numbers also represent the number of electrons moving around each nucleus, because the number of electrons and number of protons in an atom are equal. The number of protons in the nucleus of an atom of an element is called the **atomic number** of the element.

As expected, tiny atoms have tiny masses. For example, the actual mass of a hydrogen atom is 1.67×10^{-24} g, that of a carbon atom is 1.99×10^{-23} g, and that of an oxygen atom is 2.66×10^{-23}g. These very small numbers are inconvenient to work with in calculations, so a system of relative masses has been developed. These relative masses simply compare the actual masses of the atoms with each other. Suppose the actual masses of two people were determined to be 45.50 kg and 113.75 kg. Their relative masses are determined by dividing each mass by the smaller mass of the two: 45.50/45.50 = 1.00 and 113.75/45.50 = 2.50. Thus, the relative masses of the two people are 1.00 and 2.50; these numbers simply express the fact that the mass of the heavier person is 2.50 times that of the other person. The relative masses of atoms are called **atomic masses,** or **atomic weights,** and are given in atomic mass units *(amu)*. In this system, hydrogen atoms, the least massive of all atoms, have an atomic weight of 1.01 amu. The atomic weight of carbon atoms is 12.01 amu, and that of oxygen atoms is 16.00 amu. Thus, oxygen atoms have a mass about 16 times that of hydrogen atoms. Table A-1 gives the atomic weights and some other characteristics of the elements that are most important physiologically.

Chemical Bonds

Since all matter is made up of particles called atoms, we must conclude that the atoms somehow are held together to form matter. The forces holding atoms together are called **chemical bonds.** Not all chemical bonds are formed in the same way, but all of them involve the electrons of atoms. It is now understood that the electrons of each atom have energies and other characteristics that allow them to be classified into groupings called **shells.** In general, electrons will belong to the lowest energy shell possible, but the specific shells of an atom have maximum capacities that cannot be exceeded. For example, the first, or lowest, energy shell of every atom can contain a maximum of only two electrons, while the second, or next highest, energy shell can contain a maximum of eight electrons. Different atoms have different numbers of electrons in the various shells. Hydrogen atoms have only one electron, so it is in the first shell. Helium atoms have two electrons, which are both in the first shell and fill it. Carbon atoms have six electrons, two in the first shell and four in the second shell, while the eight electrons of oxygen are arranged with two in the first shell and six in the second shell.

Having filled electron shells provides an energy benefit. That is, the average energy per electron is lower for the second shell when the shell holds the maximum number of eight electrons instead of any number from one to seven. This energy benefit leads to a general statement about the electronic behavior of atoms: *atoms tend to undergo processes that result in a filled outermost electron shell.* Thus, it is the electrons of the outer or higher energy shell that determine the bonding characteristics of an atom.

Consider sodium atoms (Na) and chlorine atoms (Cl) (— Fig. A-2). Sodium atoms have eleven electrons: two in the first shell, eight in the second shell, and one in the third

Table A-1 Characteristics of Selected Elements

Name	*Symbol*	*Number of Protons*	*Atomic Number*	*Atomic Weight (amu)*
Hydrogen	H	1	1	1.01
Carbon	C	6	6	12.01
Nitrogen	N	7	7	14.01
Oxygen	O	8	8	16.00
Sodium	Na	11	11	22.99
Magnesium	Mg	12	12	24.31
Phosphorus	P	15	15	30.97
Sulfur	S	16	16	32.06
Chlorine	Cl	17	17	35.45
Potassium	K	19	19	39.10
Calcium	Ca	20	20	40.08

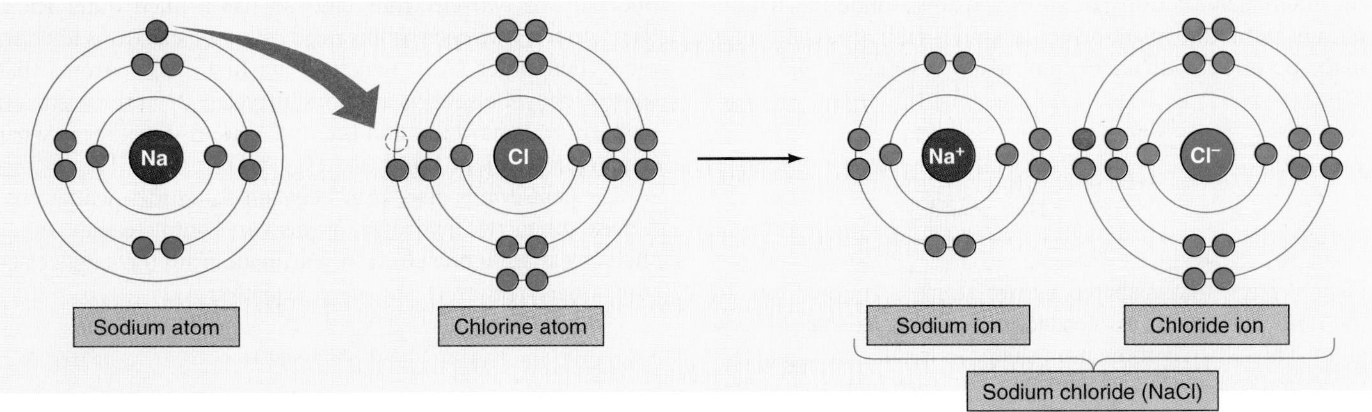

Figure A-2 **Ions and Ionic Bonds** Sodium (Na) and chlorine (Cl) atoms both have partially filled outermost shells. Therefore, sodium tends to give up its lone electron in the outer shell to chlorine, thus filling chlorine's outer shell. As a result, sodium becomes a positively charged ion, and chlorine becomes a negatively charged ion known as chloride. The oppositely charged ions attract each other, forming an ionic bond.

shell. Chlorine atoms have seventeen electrons: two in the first shell, eight in the second shell, and seven in the third shell. Because eight electrons are required to fill the second and third shells, sodium atoms have one electron more than is needed to provide a filled second shell, while chlorine atoms have one less electron than is needed to fill the third shell. Each sodium atom can lose an electron to a chlorine atom, leaving each sodium with ten electrons, eight of which are in the second shell, which is now the outer shell occupied by electrons. By accepting one electron, each chlorine atom has a total of eighteen electrons, with eight of them in the third, or outer, shell.

As a result of giving up and accepting electrons, the sodium atoms and chlorine atoms have achieved filled outer shells, but now each atom is unbalanced electrically. While each sodium now has ten electrons, it still has eleven protons in the nucleus and a net electrical charge of $+1$. Similarly, each chlorine now has eighteen electrons, but only seventeen protons. Thus, each chlorine has a -1 charge. Charged atoms such as these are called **ions.** Positively charged ions are called **cations,** whereas negatively charged ions are called **anions.** Since opposite charges attract, sodium ions (Na^+) and charged chlorine atoms, now called *chloride* ions (Cl^-), are attracted toward each other. It is this attraction, known as an **ionic bond,** that bonds the ions together in the compound **sodium chloride, NaCl,** which is common table salt. A sample of sodium chloride actually contains sodium and chloride ions in a three-dimensional geometric arrangement called a crystal lattice. The ions of opposite charge occupy alternate sites within the lattice (Fig. A-3).

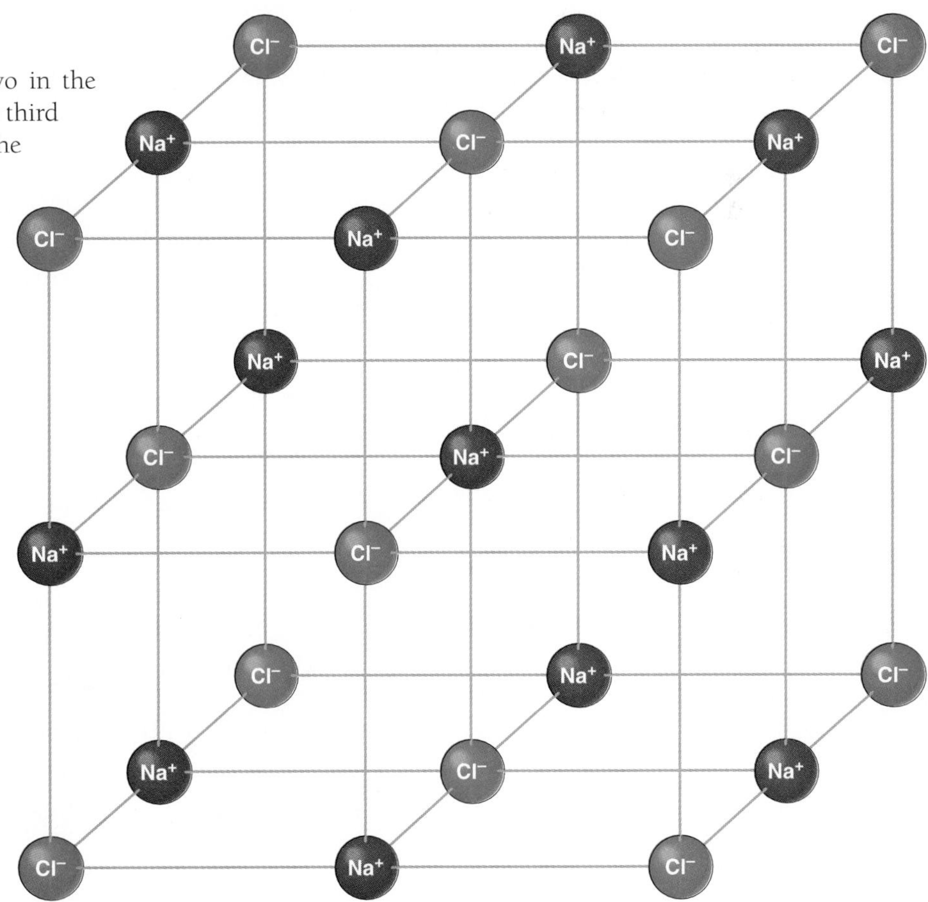

Figure A-3 **Crystal Lattice for Sodium Chloride (Table Salt)**

It is not energetically favorable for an atom to give up or accept more than three electrons. In spite of this, carbon atoms, which have four electrons in their outer shell, are known to form compounds. This observation led scientists to propose another bonding mechanism for atoms. Atoms that would have to lose or gain four or more electrons to achieve outer-shell stability usually bond by *sharing* electrons. Thus, a carbon atom can share its four outer electrons with the four electrons of four hydrogen atoms, as shown in Equation A-1;

the outer-shell electrons are shown as dots around the symbol of each atom. The resulting compound is methane, CH_4, a gas made up of individual CH_4 molecules:

$$\cdot\dot{\underset{\cdot}{C}}\cdot + 4\cdot H \rightarrow H:\overset{H}{\underset{H}{\ddot{\underset{\cdot\cdot}{C}}}}:H \qquad \text{Eq. A-1}$$

(shared electron pairs; shared electron pairs)

Each electron that is shared by two atoms is counted toward the number of electrons needed to fill the outer shell of each atom. Thus, each carbon atom shares four pairs, or eight electrons, and so has eight in its outer shell. Each hydrogen shares one pair, or two electrons, and so has a filled outer shell. (Remember, hydrogen atoms need only two electrons to complete their outer shell, which is the first shell.) Atoms that share a pair of electrons are both attracted toward the shared pair. This mutual attraction bonds the atoms together in what is called a **covalent bond** (Fig. A-4).

Covalent bonds also form between some identical atoms. For example, two hydrogen atoms can complete their outer shells by sharing one electron pair made from the single electrons of each atom, as shown in Equation A-2:

$$H\cdot + \cdot H \rightarrow H:H \qquad \text{Eq. A-2}$$

Figure A-4 **A Covalent Bond** A covalent bond is formed when atoms that share a pair of electrons are both attracted toward the shared pair.

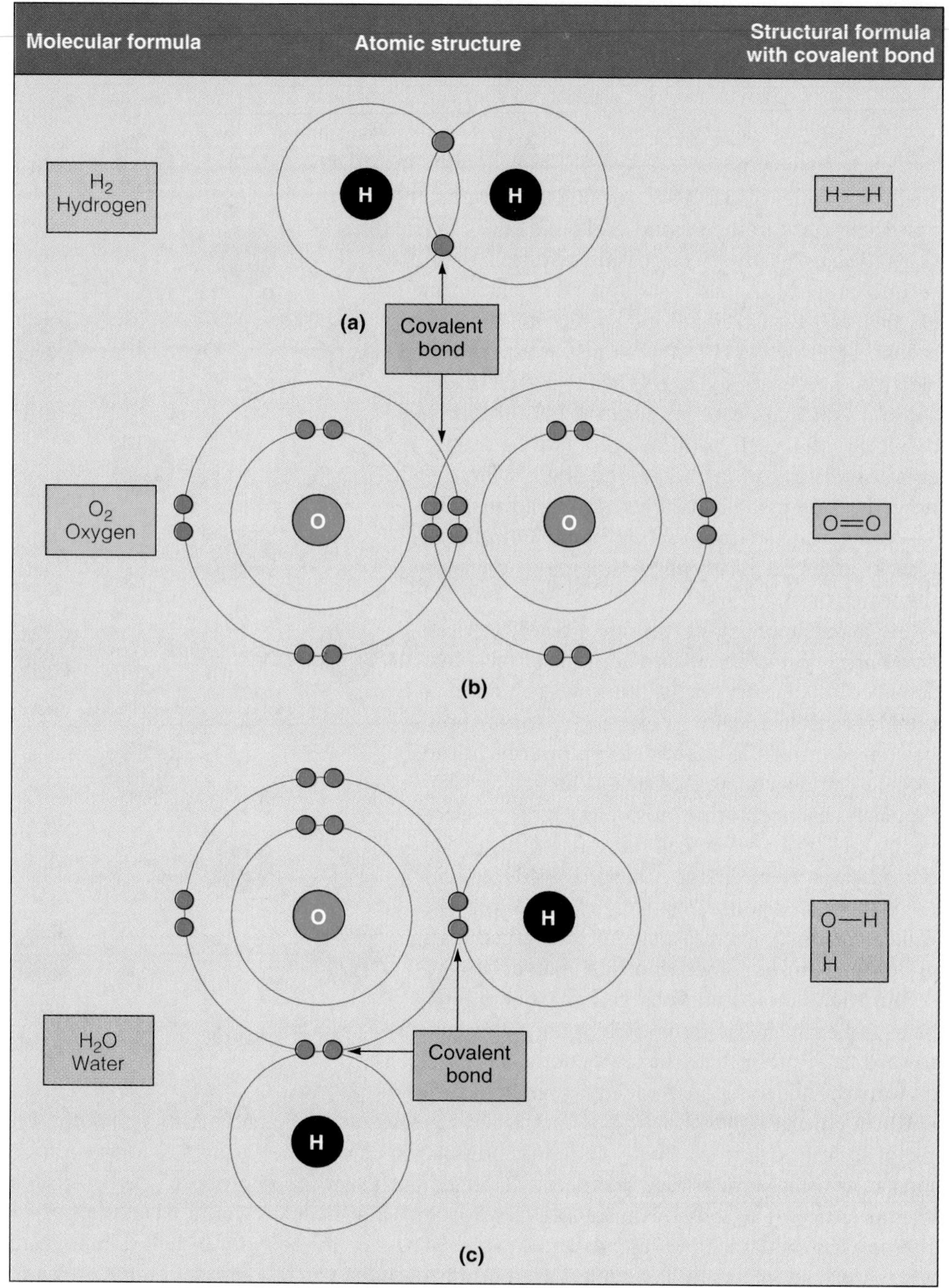

Thus, hydrogen gas consists of individual H_2 molecules (Fig. A-4a). (A subscript following a chemical symbol indicates the number of that type of atom present in the molecule.) Several other nonmetallic elements also exist as molecules because covalent bonds form between identical atoms; examples include oxygen (O_2) (Fig. A-4b), chlorine (Cl_2), and nitrogen (N_2).

One of the most familiar covalently bonded substances is water (H_2O) (Fig. A-4c). Equation A-3 represents the formation of its covalent bonds:

$$\begin{matrix} H\cdot \\ H\cdot \end{matrix} + \cdot\ddot{\underset{\cdot}{O}}: \rightarrow H:\ddot{\underset{\cdot\cdot}{O}}: \atop \quad H \qquad \text{Eq. A-3}$$

The water molecule is sometimes represented as

$$\begin{matrix} H - O \\ \quad | \\ \quad H \end{matrix}$$

where the nonshared electron pairs are not shown and the covalent bonds, or shared pairs, are represented by dashes. The water molecule is a good example of a **polar molecule;** that is, a molecule in which the electrons are not distributed uniformly (Fig. A-5). Polar molecules result because shared electrons are unequally attracted by the atoms that share them. When the atoms sharing an electron pair are identical, the electrons are attracted equally by both atoms and so are shared equally; such molecules are **nonpolar.** Examples of molecules containing equally shared electrons are H_2, O_2, and N_2. When the sharing atoms are not identical, the shared pair of electrons is pulled closer to one atom than to the other. The side of the molecule to which the electrons are pulled is electrically negative compared to the other side. In water molecules, the oxygen atom pulls shared electrons more strongly than do the hydrogen atoms. Thus, the oxygen side of a water molecule is more negative than the hydrogen sides. The electron distribution is not uniform, and the water molecule is polar. Atoms of oxygen, nitrogen, phosphorus, and chlorine strongly attract electrons when they are bonded to other atoms.

Polar molecules are attracted to other polar molecules. In water, for example, an attraction exists between the positive hydrogen ends of some molecules and the negative oxygen ends of others. Hydrogen is not a part of all polar molecules, but when it is covalently bonded to an atom that strongly attracts electrons to form a covalent molecule, the attraction of the positive (hydrogen) end of the polar molecule to the negative end of another polar molecule is called a **hydrogen bond** (Fig. A-6). Thus, the polar attractions of water molecules to each other are an example of hydrogen bonding.

Chemical Reactions

Processes in which chemical bonds are broken and/or formed are called **chemical reactions.** Reactions are represented by equations in which the reacting substances **(reactants)** are written on the left, the produced substances **(products)** are written on the right, and an arrow points from the reactants to the products. These conventions are illustrated in Equation A-4:

$$\underset{\text{reactants}}{A + B} \rightarrow \underset{\text{products}}{C + D} \qquad \text{Eq. A-4}$$

Let's look at a specific example, the combustion of methane gas, CH_4, as shown in Equation A-5:

$$CH_4 + 2O_2 \rightarrow CO_2 + 2H_2O \qquad \text{Eq. A-5}$$

According to this equation, one molecule of methane gas reacts with two molecules of oxygen gas to produce one molecule of carbon dioxide gas and two molecules of water vapor.

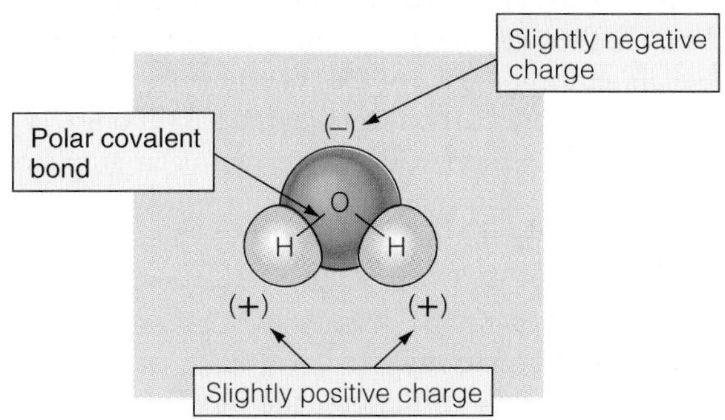

Figure A-5 **A Polar Molecule** A water molecule is an example of a polar molecule, in which the distribution of shared electrons is not uniform. Because the oxygen atom pulls the shared electrons more strongly than the hydrogen atoms do, the oxygen side of the molecule is slightly negatively charged, and the hydrogen sides are slightly positively charged.

Figure A-6 **A Hydrogen Bond** A hydrogen bond is formed by the attraction of a positively charged hydrogen end of a polar molecule to the negatively charged end of another polar molecule.

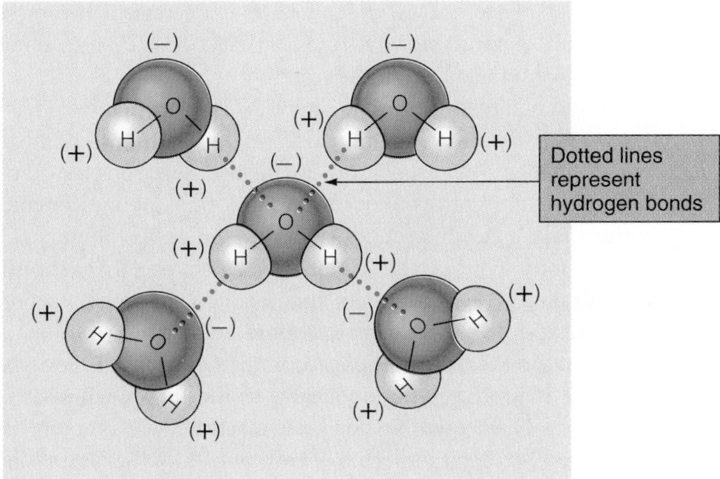

Coefficients such as the 2s to the left of the O_2 and H_2O are used so that the total number of each type of atom is the same on the left and right sides of the equation. Thus, the preceding reaction has one carbon atom on the left (in one CH_4) and one on the right (in one CO_2), four hydrogen atoms on the left (in one CH_4) and four on the right (in two H_2Os), and four oxygen atoms on the left (in two O_2s) and four on the right (in one CO_2 and two H_2Os). Equations in which the same number of each type of atom appears on both sides are called **balanced equations.**

Under appropriate conditions, the products of a reaction can be changed back to the reactants. For example, carbon dioxide gas dissolves in and reacts with water to form carbonic acid, H_2CO_3:

$$CO_2 + H_2O \rightarrow H_2CO_3 \qquad \text{Eq. A-6}$$

Carbonic acid is not very stable, however, and as soon as some is formed, part of it decomposes to give carbon dioxide and water:

$$H_2CO_3 \rightarrow CO_2 + H_2O \qquad \text{Eq. A-7}$$

Reactions that go in both directions are called **reversible reactions.** They are usually represented by double arrows pointing in both directions:

$$CO_2 + H_2O \rightleftharpoons H_2CO_3 \qquad \text{Eq. A-8}$$

Theoretically, every reaction is reversible. Often, however, conditions are such that a reaction, for all practical purposes, goes only in one direction; such a reaction is called **irreversible.** For example, in order for the combustion reaction of methane (Eq. A-5) to be reversible, the gaseous CO_2 and water vapor products would have to remain in the vicinity of the reaction site (otherwise, there is no way they could get together to react). In practice, these gaseous products leave the reaction site and have no chance to recombine. Thus, the ordinary combustion of methane is considered irreversible. Some other irreversible reactions are those that take place when an explosion occurs and when an egg is fried.

The rates (speeds) of chemical reactions are influenced by a number of factors, of which catalysts are one of the most important. A **catalyst** is a substance that speeds up a reaction without being used up in the reaction.

Living organisms produce catalysts called **enzymes.** These enzymes exert amazing influence on the rates of chemical reactions that take place in the organisms. Reactions that take weeks or even months to occur under normal laboratory conditions take place in seconds under the influence of enzymes in the body. One of the fastest-acting enzymes is **carbonic anhydrase,** which catalyzes the reaction between carbon dioxide and water to form carbonic acid. This reaction is important in the transport of carbon dioxide from tissue cells, where it is produced metabolically, to the lungs, where it is excreted. The equation for the reaction was shown in Equation A-6. Each molecule of carbonic anhydrase catalyzes the conversion of 36 million CO_2 molecules per minute. Enzymes are important in essentially every chemical reaction that takes place in living organisms.

Formulas, Equations, and the Mole

We have already discussed the concept of atomic weights. The idea of comparing masses by using relative values is also useful in discussing molecules. Since molecules are made up of atoms, the relative mass of a molecule is simply the sum of the relative masses (atomic weights) of the atoms found in the molecule. The relative masses of molecules are called **molecular masses** or **molecular weights.** The molecular weight of water, H_2O, is thus the sum of the atomic weights of two hydrogen atoms and one oxygen atom, or 1.01 amu + 1.01 amu + 16.00 amu = 18.02 amu.

Not all compounds exist in the form of molecules. Ionically bonded substances such as sodium chloride consist of three-dimensional arrangements of sodium ions (Na^+) and chloride ions (Cl^-) in a 1-to-1 ratio. The formulas for ionic compounds reflect only the ratio of the ions in the compound and should not be interpreted in terms of molecules. Thus, the formula for sodium chloride, NaCl, indicates that the ions combine in a 1-to-1 ratio. It is convenient to apply the concept of relative masses to ionic compounds even though they do not exist as molecules. The **formula weight** for such compounds is defined as the sum of the atomic weights of the atoms found in the formula. Thus, the formula weight of magnesium chloride, an ionic compound with the formula $MgCl_2$, is equal to the sum of the atomic weights of one magnesium atom and two chlorine atoms, or 24.31 amu + 35.45 amu + 35.45 amu+= 95.21 amu.

As we have seen, chemical reactions can be represented by equations and discussed in terms of numbers of molecules, atoms, and ions reacting with each other. However, it is much more convenient to describe reactions in terms of amounts of reactants and products that can readily be measured in units such as grams. Using the mole concept makes such descriptions possible. A **mole** (abbreviated *mol*) of a pure element or compound is the amount of material contained in a sample of the pure substance that has a mass in grams equal to the substance's atomic weight (for elements) or the molecular weight or formula weight (for compounds). Thus, 1 mole of potassium, K, would be a sample of the element with a mass of 39.10 g. Similarly, a mole of water, H_2O, would have a mass of 18.02 g, and a mole of sodium chloride, NaCl, would be a sample with a mass of 58.44 g.

The fact that atomic weights, molecular weights, and formula weights are relative masses leads to a fundamental characteristic of moles. One mole of hydrogen atoms has a mass of 1.01 g, and 1 mole of oxygen atoms has a mass of 16.00 g. The ratio of atomic weights for the two elements is 16.00/1.01, the same as the ratio of the masses of 1 mole of each element: 16.00 g/1.01 g. Therefore, 1 mole of hydrogen contains exactly the same number of hydrogen atoms as the number of oxygen atoms in 1 mole of oxygen. Thus, it is possible and sometimes useful to think of a mole as a specific number of particles. This number, called **Avogadro's number,** is equal to 6.02×10^{23}.

Solutions

Most chemical reactions in the body take place between reactants that have dissolved to form solutions. **Solutions** are homogenous mixtures containing a relatively large amount of one substance called the **solvent** and smaller amounts of one or more substances called **solutes.** Salt water, for example, contains mostly water, which is thus the solvent, and a smaller amount of salt, which is the solute. Water is the solvent in most solutions found in the human body.

When ionic solutes are dissolved in water to form solutions, the resulting solution will conduct electricity. This is not true for most covalently bonded solutes. For example, a salt-water solution conducts electricity, but a sugar-water solution does not. When salt dissolves in water, the solid lattice of Na^+ and Cl^- is broken down, and the individual ions are separated and distributed uniformly throughout the solution. These mobile, charged ions conduct electricity through the solution. Solutes that form ions in solution and conduct electricity are called **electrolytes.** Some very polar covalent molecules also behave this way. When sugar dissolves, however, individual covalently bonded sugar molecules leave the solid and become uniformly distributed throughout the solution. These uncharged molecules cannot conduct a current. Solutes that do not form conductive solutions are called **nonelectrolytes.**

The amount of solute dissolved in a specific amount of solution can vary. For example, a salt-water solution might contain 1 g of salt in 100 ml of solution, or it could contain 10 g of salt in 100 ml of solution. Both solutions are salt-water solutions, but they have different concentrations of solute. The **concentration** of a solution indicates the relationship between the amount of solute and the amount of solvent or the amount of solution. Concentrations can be given in a number of different units.

Concentrations given in terms of **molarity** (abbreviated *M*) give the number of moles of solute in exactly 1 liter of solution. Thus, a half molar (0.5 M) solution of NaCl would contain one-half mole, or 29.22 g, of NaCl in each liter of solution.

Concentrations given in terms of **molality** (abbreviated *m*) give the number of moles of solute dissolved in 1,000 g (1 kg) of solvent. Thus, a half molal (0.5 m) solution of NaCl would contain NaCl and water in the ratio of 29.22 g NaCl to each 1,000 g (or 1,000 ml) of water.

Note the difference between molarity and molality. Molarity gives the amount of solute in a specific amount of solution, whereas molality gives the amount of solute dissolved in a specific amount of solvent.

When the solute is an electrolyte, it is sometimes useful to express the concentration of the solution in a unit that gives information about the amount of ionic charge in the solution. This is done by expressing concentration in terms of **normality** (abbreviated *N*). The normality of a solution gives the number of equivalents of solute in exactly 1 liter of solution. An **equivalent** of an electrolyte is the amount that produces 1 mole of positive (or negative) charges when it dissolves. The number of equivalents of an electrolyte can be calculated by multiplying the number of moles of electrolyte by the total number of positive charges produced when one formula unit of the electrolyte dissolves. Consider sodium chloride (NaCl) and calcium chloride ($CaCl_2$) as examples. The ionization reactions for one formula unit of each solute are:

$$NaCl \rightarrow Na^+ + Cl^- \qquad \text{Eq. A–9}$$

$$CaCl_2 \rightarrow Ca^{2+} + 2Cl^- \qquad \text{Eq. A–10}$$

Thus, 1 mole of NaCl produces 1 mole of positive charges (Na^+) and so contains 1 equivalent:

$$(1 \text{ mole NaCl})(1) = 1 \text{ equivalent}$$

where the number 1 used to multiply the 1 mole of NaCl came from the +1 charge on Na^+.

One mole of $CaCl_2$ produces 1 mole of Ca^{2+}, which is 2 moles of positive charge. Thus, 1 mole of $CaCl_2$ contains 2 equivalents:

$$(1 \text{ mole } CaCl_2)(2) = 2 \text{ equivalents}$$

where the number 2 used in the multiplication came from the +2 charge on Ca^{2+}.

If two solutions were made such that one contained 1 mole of NaCl per liter and the other contained 1 mole of $CaCl_2$ per liter, the NaCl solution would contain 1 equivalent of solute per liter and would be 1 normal (1 N). The $CaCl_2$ solution would contain 2 equivalents of solute per liter and would be 2 normal (2 N).

Another expression of concentration frequently used in physiology is **osmolarity** (abbreviated *osm*), which indicates the total *number* of solute particles in a liter of solution instead of the relative weights of the specific solutes. The osmolarity of a solution is the product of M and n, where *n* is the number of moles of solute particles obtained when 1 mole of solute dissolves. Because nonelectrolytes such as glucose do not dissociate in solution, n = 1 and the osmolarity (n times M) is equal to the molarity of the solution. For electrolyte solutions, the osmolarity exceeds the molarity by a factor equal to the number of ions produced upon dissociation of each molecule in solution. For example, because a NaCl molecule dissociates into two ions, Na^+ and Cl^-, the osmolarity of a 1 M solution of NaCl is 2×1 M = 2 osm.

Suspensions and Colloids

Suspensions and *colloids,* like solutions, consist of two or more components, with much more of one component than the others. In solutions, the component present in largest amount is called the solvent; in suspensions and colloids, it is called the **dispersing medium.** In solutions, the components present in smaller amounts are called solutes; in suspensions and colloids, they are called **dispersed phases.** An important difference between solutions and suspensions/colloids is the size of the particles dissolved or dispersed. In solutions, solute particles are ions or small molecules. Dispersed-phase particles are much larger than ions or small molecules. When the dispersed-phase particles are no more than about 100 times the size of the largest solution solute particles, the suspension is called a **colloid** or **colloidal dispersion.** The dispersed-phase particles of colloids generally do not settle out. They are small enough to be kept in suspension by the constant buffeting they receive from the motion of dispersing-medium molecules. All dispersed-phase particles of colloids carry electrical charges of the same sign. Thus, they repel each other and, despite their collisions with other dispersed-phase particles, do not get together to form larger particles that would settle. When dispersed-phase particles are larger than those in colloids, the dispersed phase will settle out. Such mixtures are usually called **suspensions.**

The relatively large size of dispersed-phase particles in colloids and suspensions causes them to scatter light. As a result, colloids and suspensions appear cloudy. Solutions are clear because the dissolved solute particles are too small to scatter light. In the body, the dispersing medium of most colloids is water, and the colloids are liquids. However, colloids can occur in all three states, depending on the state of the dispersing medium.

Inorganic and Organic Chemicals

Chemicals are commonly classified into two categories: inorganic and organic. The original criterion used for this classification was the origin of the chemicals. Those that came from living or once-living sources were *organic,* and those that came from other sources were *inorganic.* Today, the basis for classification is the element carbon. **Organic** chemicals are generally those that contain carbon. All others are classified as **inorganic.** A few carbon-containing chemicals are also classified as inorganic; the most common are pure carbon in the form of diamond and graphite, carbon dioxide (CO_2), carbon monoxide (CO), carbonates such as limestone ($CaCO_3$), bicarbonates such as baking soda ($NaHCO_3$), and cyanides such as sodium cyanide (NaCN).

The unique ability of carbon atoms to bond to each other and form networks of carbon atoms results in an interesting fact. Even though organic chemicals are required to contain one specific element, carbon, millions of these compounds have been identified. Some were isolated from natural plant or animal sources, and many have been synthesized in laboratories. Inorganic chemicals include all the other 108 elements and their compounds; the number of known inorganic chemicals is estimated to be about 250,000.

Another result of carbon's ability to bond to itself is the large size of some organic molecules. Molecules classified as organic range in size from methane, CH_4, a small, simple molecule with one carbon atom, to molecules such as DNA that contain as many as a million carbon atoms. Large molecules such as this are often called **macromolecules.** Macromolecules include many naturally occurring molecules such as DNA as well as many molecules that are synthetically produced. In this latter category are numerous substances that are widely used today, including synthetic textiles, (for example, nylon, dacron, and orlon) and plastics (for example, lucite, plexiglass, and teflon). Another general name given to these materials is **polymer,** which means "many units," reflecting the fact that polymeric macromolecules are made by the bonding together of a large number of smaller molecules.

Acids, Bases, and Salts

Acids, bases, and salts are among the most common and important compounds studied in chemistry. Both inorganic and organic compounds fit into these three categories. Until late in the nineteenth century, these substances were classified on the basis of such properties as taste or by the color changes induced in certain dyes. Acids taste sour, bases bitter, and salts salty. Litmus, a dye, is red in the presence of acids and blue in the presence of bases. These and other observations led to the correct conclusions that acids and bases are chemical opposites and that salts are produced when acids and bases react with each other. Today, acids and bases are defined in more precise ways.

In 1887, Swedish chemist Svante Arrhenius proposed a theory defining acids and bases. He said that an *acid* is any substance that will dissociate, or break apart, when dissolved in water and in the process release a hydrogen ion, H^+. Similarly, *bases* are substances that dissociate when dissolved in water and in the process release a hydroxide ion, OH^-. Hydrogen chloride (HCl) and sodium hydroxide (NaOH) are examples of Arrhenius acids and bases; their dissociations in water are represented in Equations A-11 and A-12, respectively:

$$HCl \rightarrow H^+ + Cl^- \qquad \text{Eq. A-11}$$

$$NaOH \rightarrow Na^+ + OH^- \qquad \text{Eq. A-12}$$

Note that the hydrogen ion is a bare proton, the nucleus of a hydrogen atom. Also note that both HCl and NaOH would behave as electrolytes.

Arrhenius did not know that free hydrogen ions cannot exist in water. It is now believed that they covalently bond to water molecules to form hydronium ions, as shown in Equation A-13:

$$H^+ + :\ddot{\underset{\underset{H}{|}}{O}} - H \rightarrow \left[H - \ddot{\underset{\underset{H}{|}}{O}} - H \right]^+ \qquad \text{Eq. A-13}$$

In 1923, Johannes Brønsted in Denmark and Thomas Lowry in England proposed an acid-base theory that took this behavior into account. They defined an **acid** as any hydrogen-containing substance that donates a proton (hydrogen ion) to another substance and a **base** as any substance that accepts a proton. According to these definitions, the acidic behavior of HCl given in Equation A-11 is rewritten as shown in Equation A-14:

$$HCl + H_2O \rightleftharpoons H_3O^+ + Cl^- \qquad \text{Eq. A-14}$$

Note that this reaction is shown to be reversible, and the hydronium ion is represented as H_3O^+.

In Equation A-14, the HCl acts as an acid in the forward (left-to-right) reaction, while water acts as a base. In the reverse reaction (right-to-left), the hydronium ion gives up a proton and thus is an acid, while the chloride ion, Cl^-, accepts the proton and so is a base. It is still a common practice to use equations such as A-11 to simplify the representation of the dissociation of an acid, even though it is recognized that equations like A-14 are more correct.

At room temperature, **inorganic salts** are crystalline solids that contain the positive ion (cation) of an Arrhenius base such as NaOH and the negative ion (anion) of an acid such as HCl. Salts can be produced by mixing solutions of appropriate acids and bases, allowing a neutralization reaction to occur. In **neutralization reactions,** the acid and base react to form a salt and water. Most salts that form are water-soluble and can be recovered by evaporating the water. Equations A-15 and A-16 are neutralization reactions:

$$HCl + NaOH \rightarrow NaCl + H_2O \qquad \text{Eq. A–15}$$

$$H_2SO_4 + Cu(OH)_2 \rightarrow CuSO_4 + 2H_2O \qquad \text{Eq. A–16}$$

When acids or bases are used as solutes in solutions, the concentrations can be expressed as normalities just as they were earlier for salts. An equivalent of acid is the amount that gives up 1 mole of H^+ in solution. Thus, 1 mole of HCl is also 1 equivalent, but 1 mole of H_2SO_4 is 2 equivalents. Bases are described in a similar way, but an equivalent is the amount of base that gives 1 mole of OH^-.

Functional Groups of Organic Molecules

The study of organic compounds is simplified by the concept of functional groups. All organic compounds can be classified according to the functional group or groups they contain. **Functional groups** are specific combinations of atoms that generally react in the same way, regardless of the number of carbon atoms in the molecule to which they are attached. For example, all *aldehydes* contain a functional group that contains one carbon atom, one oxygen atom, and one hydrogen atom covalently bonded in a specific way:

$$(-\overset{\overset{\displaystyle O}{\|}}{C}-H)$$

The carbon atom in an aldehyde group forms a single covalent bond with the hydrogen atom and a **double bond** (a bond in which two covalent bonds are formed between the same atoms, designated by a double line between the atoms) with the oxygen atom. The aldehyde group is attached to the rest of the molecule by a single covalent bond extending to the left of the carbon atom. Most reactions of aldehydes involve this group, so most aldehyde reactions are the same regardless of the size and nature of the rest of the molecule to which the aldehyde group is attached. Reactions of physiological importance often occur between two functional groups or between one functional group and a small molecule such as water.

Carbohydrates

Carbohydrates are organic compounds of tremendous biological and commercial importance. They are widely distributed in nature and include such familiar substances as starch, table sugar, and cellulose. Carbohydrates have four important functions in living organisms: they provide energy, they supply carbon atoms for the synthesis of cell components, they serve as a stored form of chemical energy, and they form part of the structural elements of some cells.

Carbohydrates contain carbon, hydrogen, and oxygen. They acquired their name because most of them contain these three elements in an atomic ratio of one carbon to two hydrogens to one oxygen. This ratio suggests that the general formula is CH_2O and that the compounds are simply carbon hydrates or carbohydrates. It is now known that they are not hydrates of carbon, but the name persists. All carbohydrates have a large number of functional groups per molecule. The most common functional groups in carbohydrates are *alcohol, ketone* and *aldehyde*—

$$\underset{\text{alcohol}}{(-OH)},\ \underset{\text{ketone}}{(-\overset{\overset{\displaystyle O}{\|}}{C}-)},\ \underset{\text{aldehyde}}{(-\overset{\overset{\displaystyle O}{\|}}{C}-H)}$$

or functional groups formed by reactions between pairs of these three.

The simplest carbohydrates are simple sugars, also called **monosaccharides.** As their name indicates, they consist of single (*mono* means "one") units called saccharides. The molecular structure of *glucose,* an important monosaccharide, is shown in Figure A-7a. In solution, most glucose molecules assume the ring form shown in Figure A-7b. Other common monosaccharides are *fructose, galactose,* and *ribose.*

Disaccharides (*di* means "two") are sugars formed by a reaction between two monosaccharide molecules. Some common examples of disaccharides are *sucrose* (common table sugar) and *lactose* (milk sugar). Sucrose molecules are formed from one glucose and one fructose molecule. Lactose molecules each contain one galactose and one glucose unit.

The large number of functional groups on carbohydrate molecules makes it possible for large numbers of simple car-

Figure A-7 **Forms of Glucose** (a) Chain. (b) Ring.

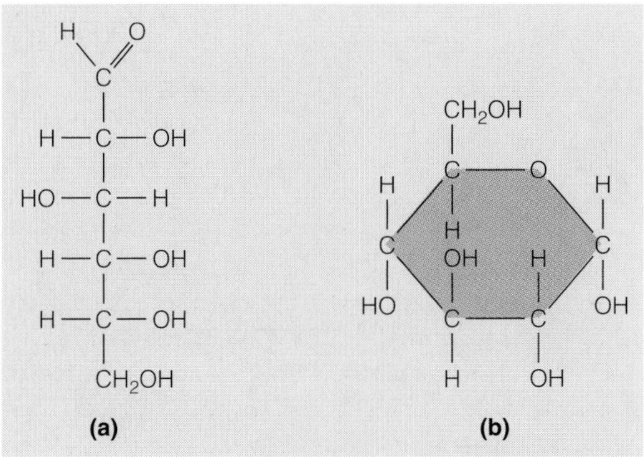

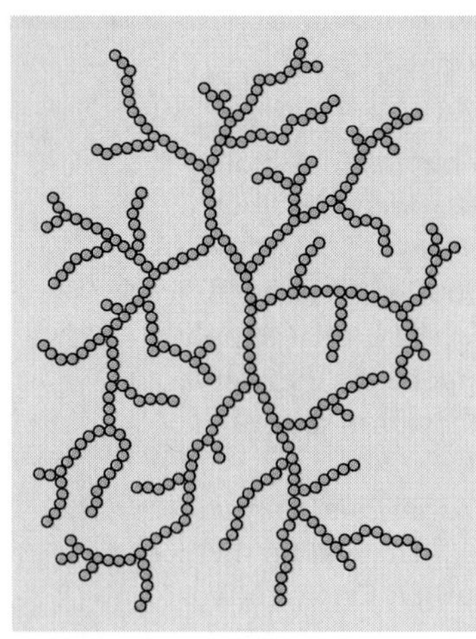

Figure A-8 **A Simplified Representation of Glycogen** Each circle represents a glucose molecule.

bohydrate molecules to bond together and form long chains and branched networks. These substances are called **polysaccharides,** a name that indicates that they contain many saccharide units (*poly* means "many"). Three common polysaccharides that are made up entirely of glucose units are glycogen, starch, and cellulose.

Glycogen is a storage carbohydrate found in animals. It is a highly branched polysaccharide that averages a branch every eight to twelve glucose units. The structure of glycogen is represented in Figure A-8, where each circle represents one glucose unit.

Starch, a storage carbohydrate of plants, consists of two fractions, amylose and amylopectin. Amylose consists of long, essentially unbranched chains of glucose units. Amylopectin is a highly branched network of glucose units averaging twenty-four to thirty glucose units per branch. Thus, it is less highly branched than glycogen. *Cellulose,* a structural carbohydrate of plants, exists in the form of long, unbranched chains of glucose units.

The bonding between the glucose units of cellulose is slightly different than the bonding between the glucose units of glycogen and starch. Humans have digestive enzymes that catalyze the breaking (hydrolysis) of the glucose-to-glucose bonds in starch but lack the necessary enzymes to hydrolyze cellulose glucose-to-glucose bonds. Thus, starch is a food for humans, but cellulose is not.

Lipids

The group of compounds called lipids is made up of substances with widely different compositions and molecular structures. Unlike carbohydrates, which are classified on the basis of their molecular structure, substances are classified as lipids on the basis of their solubility. **Lipids** are compounds that are insoluble in water but soluble in nonpolar solvents. Thus, lipids are the waxy, greasy, or oily compounds found in plants and animals. Lipids repel water, a useful characteristic of the protective wax coatings found on some plants. Fats and oils are energy-rich and have relatively low densities. These properties account for the use of fats and oils as stored energy in plants and animals. Still other lipids occur as structural components, especially in cellular membranes.

Simple lipids contain just two types of components, fatty acids and alcohols. **Fatty acid molecules** consist of a hydrocarbon chain with a *carboxylic acid* functional group (—COOH) on the end. The hydrocarbon chain can be of variable length, but natural fatty acids always contain an even number of carbon atoms. The hydrocarbon chain can also contain one or more double bonds between carbon atoms. Fatty acids with no double bonds are called **saturated fatty acids,** whereas those with double bonds are called **unsaturated fatty acids.** The more double bonds present, the higher the degree of unsaturation. The most common alcohol found in simple lipids is **glycerol** (glycerin), a three-carbon alcohol that has three alcohol functional groups (—OH).

Simple lipids called fats and oils are formed by a reaction between the carboxylic acid group of three fatty acids and the three alcohol groups of glycerol. The resulting lipid is called a **triglyceride** or **triacylglycerol.** Such lipids are classified as fats or oils on the basis of their melting points. *Fats* are solids at room temperature, whereas *oils* are liquids. Their melting points depend on the degree of unsaturation of the fatty acids of the triglyceride. The melting point goes down with increasing degree of unsaturation. Thus, oils contain more unsaturated fatty acids than do fats. Examples of the components of fats and oils and a typical triglyceride molecule are shown in Figure A-9.

When triglycerides form, a molecule of water is released as each fatty acid reacts with glycerol. Adipose tissue in the body contains triglycerides. When the body uses adipose tissue as an energy source, the triglycerides react with water to release free fatty acids into the blood. The fatty acids can be used as an immediate energy source by many organs. In the liver, free

Figure A-9 **Triglyceride Components and Structure**

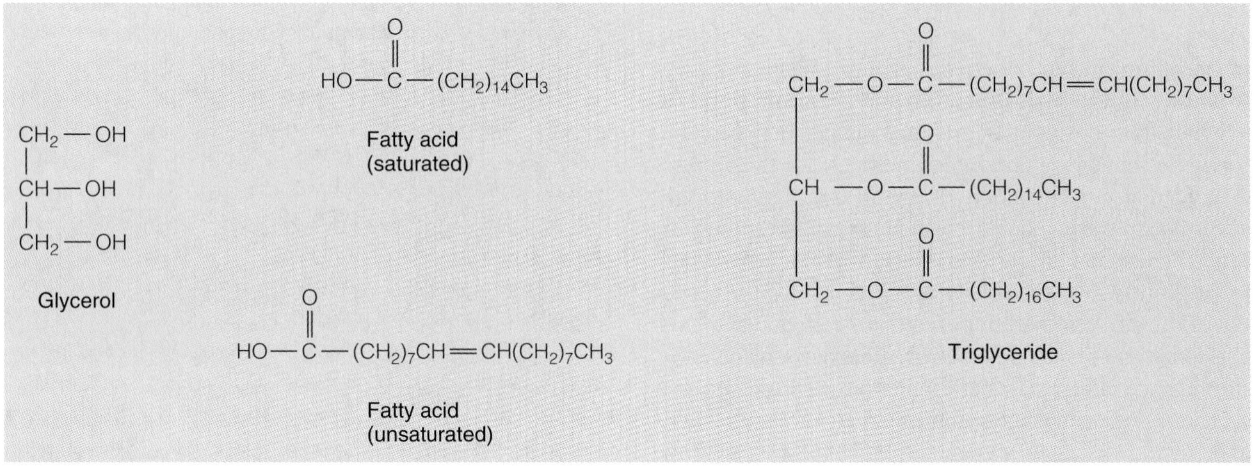

Glycerol:

$CH_2—OH$
$CH—OH$
$CH_2—OH$

Fatty acid (saturated):

$HO—C(=O)—(CH_2)_{14}CH_3$

Fatty acid (unsaturated):

$HO—C(=O)—(CH_2)_7CH=CH(CH_2)_7CH_3$

Triglyceride:

$CH_2—O—C(=O)—(CH_2)_7CH=CH(CH_2)_7CH_3$
$CH—O—C(=O)—(CH_2)_{14}CH_3$
$CH_2—O—C(=O)—(CH_2)_{16}CH_3$

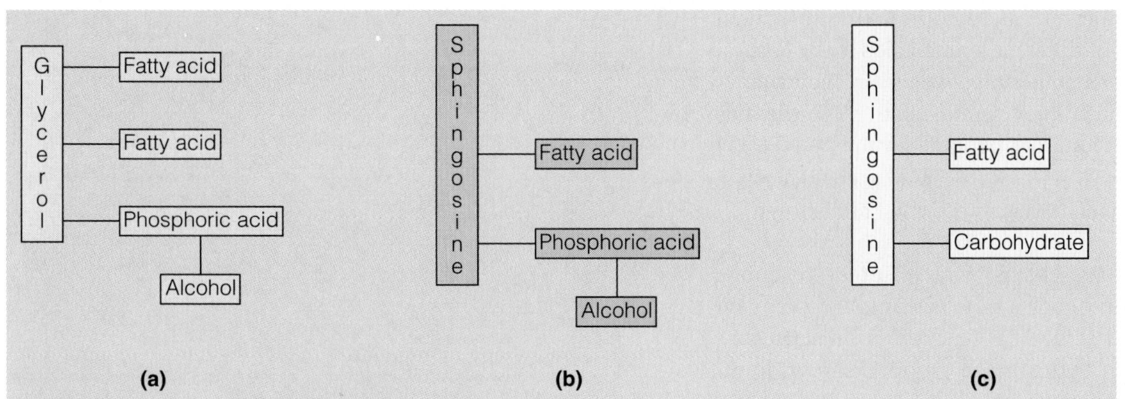

Figure A-10 **Examples of Complex Lipids** (a) A phosphoglyceride. (b) A sphingolipid (sphingosine is an alcohol). (c) A glycolipid.

fatty acids are converted into compounds called **ketone bodies.** Two of the ketone bodies are acids and one is the ketone called acetone.

Complex lipids contain more than two types of components. The different complex lipids usually contain three or more of the following components: glycerol, fatty acids, phosphoric acid, an alcohol other than glycerol, and a carbohydrate. Those that contain phosphoric acid are called **phospholipids.** Figure A-10 contains representations of a few complex lipids; it emphasizes the components but does not give details of the molecular structures.

Steroids are lipids that have a unique structural feature consisting of a fused carbon ring system containing three six-membered rings and a single five-membered ring (Fig. A-11). Different steroids possess this characteristic ring structure but have different functional groups and carbon chains attached.

Cholesterol, a steroidal alcohol, is the most abundant steroid in the human body. It is a component of cell membranes and is used by the body to produce other important steroids that include bile salts, male and female sex hormones, and adrenocortical hormones. The structures of cholesterol and cortisol, an important adrenocortical hormone, are given in Figure A-12.

Proteins

The name *protein* is derived from the Greek word *proteios,* which means "of first importance." It is certainly an appropriate term for these very important biological compounds. Proteins are indispensable components of all living things, where they play crucial roles in all biological processes.

Figure A-11 **The Steroid Ring System** (a) Detailed. (b) Simplified.

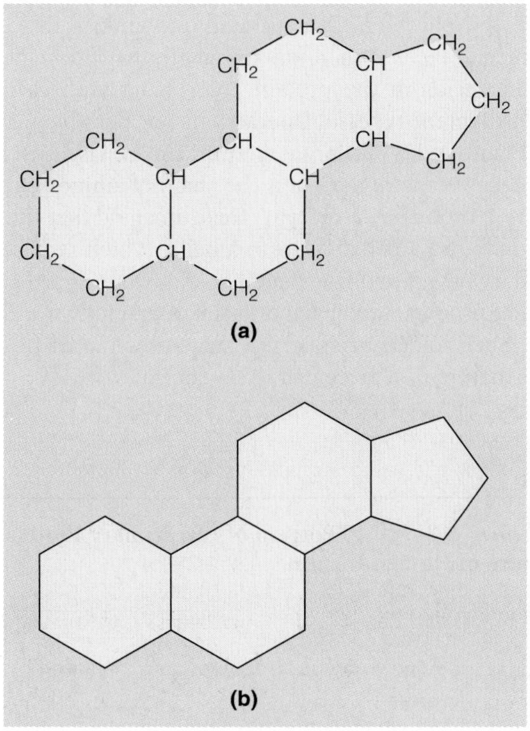

Figure A-12 **Examples of Steroidal Compounds**

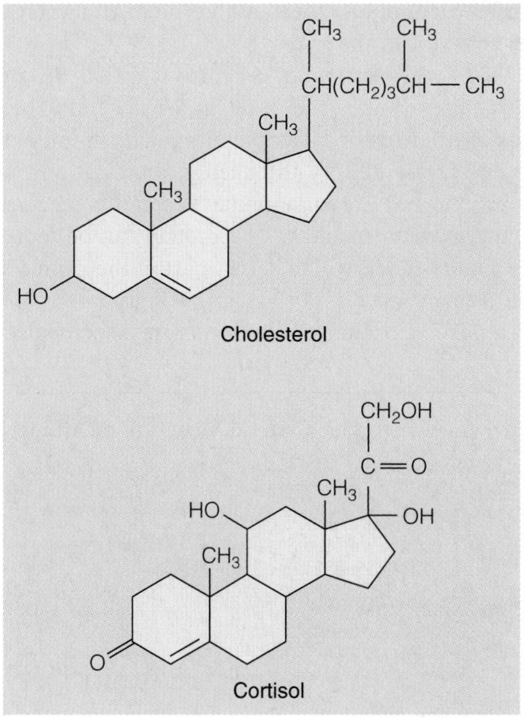

Cholesterol

Cortisol

Proteins are macromolecules made up of subunits called **amino acids.** Hundreds of different amino acids, both natural and synthetic, are known, but only twenty are commonly found in natural proteins. Each amino acid molecule has three important parts: an amino functional group ($—NH_2$), a carboxyl functional group ($—COOH$), and a characteristic side chain or R group. These components are shown in Figure A-13.

Amino acids form long chains as a result of reactions between the amino group of one amino acid and the carboxyl group of another amino acid. This reaction is illustrated in Equation A-17, in which the carboxyl group is shown in an expanded form for clarity:

$$H_2N—\underset{\underset{CH_3}{|}}{CH}—\overset{\overset{O}{\|}}{C}—OH + H_2N—CH_2—\overset{\overset{O}{\|}}{C}—OH \rightarrow$$

$$H_2N—\underset{\underset{CH_3}{|}}{CH}—\overset{\overset{O}{\|}}{C}—NH—CH_2—\overset{\overset{O}{\|}}{C}—OH + H_2O \qquad \text{(peptide bond: C—NH)}$$

Eq. A-17

Notice that after the two molecules react, the ends of the product still have an amino group and a carboxyl group that can react to extend the chain length. The covalent bond formed in the reaction is called a **peptide bond** (Fig. A-14).

On a molecular scale, proteins are immense molecules. Their size can be illustrated by comparing a glucose molecule to a molecule of hemoglobin, a protein. Glucose has a molecular weight of 180 amu and a molecular formula of $C_6H_{12}O_6$. Hemoglobin, a relatively small protein, has a molecular weight of 65,000 amu and a molecular formula of $C_{2952}H_{4664}O_{832}N_{812}S_8Fe_4$. The many atoms in a protein are not arranged in a random way. In fact, proteins have a high degree of structural organization that plays an important role in their behavior in the body.

The first level of protein structure is called the **primary structure.** It is simply the order in which amino acids are bonded together to form the protein chain. Amino acids are frequently represented by three-letter abbreviations, such as Gly for glycine and Arg for arginine. When this practice is followed, the primary structure of a protein can be represented as in Figure A-15, which shows part of the primary structure of human insulin, or as in Figure A-16a, which depicts a portion of the primary structure of hemoglobin.

Figure A-13 **The General Structure of Amino Acids**

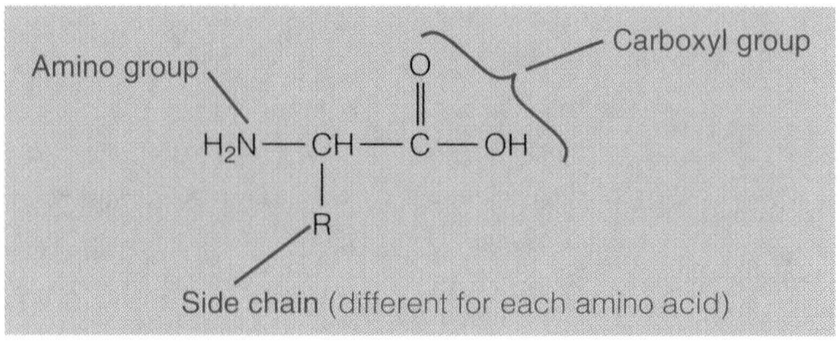

Amino group
$H_2N—\underset{\underset{R}{|}}{CH}—\overset{\overset{O}{\|}}{C}—OH$
Carboxyl group
Side chain (different for each amino acid)

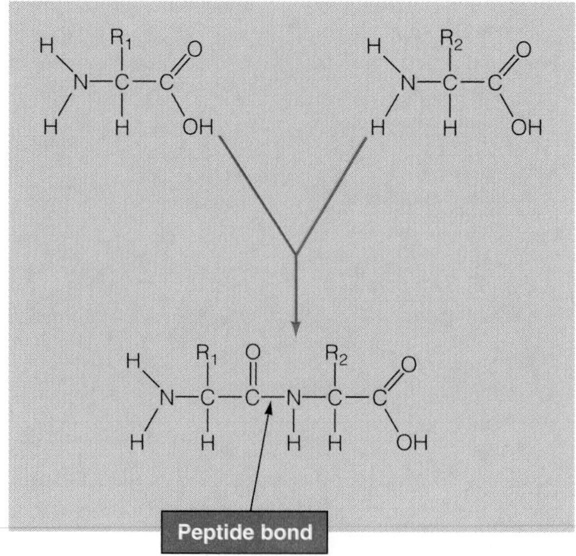

Figure A-14 **A Peptide Bond** In forming a peptide bond, the carboxyl group of one amino acid reacts with the amino group of another amino acid.

The second level of protein structure, called the **secondary structure,** results when hydrogen bonding occurs between the amino hydrogen of one amino acid in the primary chain and the carboxyl oxygen

$$(—\overset{\overset{O}{\|}}{C}—)$$

of another amino acid in the same or another chain. When the hydrogen bonding occurs between amino acids in the same chain, the chain assumes a coiled, helical shape called the alpha (α) helix, which is by far the most common secondary structure found in natural proteins (Fig. A-16b).

The third level of structure in proteins is the **tertiary structure.** It results when functional groups of the side chains of amino acids in the protein chain react with each other. Several different types of interactions are possible, as shown in Figure A-17. Tertiary structures can be visualized by letting a length of wire represent the chain of amino acids in the primary structure of a protein. Next, imagine that the wire is wound around a pencil to form a helix, which represents the secondary structure. The pencil is removed, and the helical structure is now folded back on itself or carefully wadded into a ball. Such folded or spherical structures represent the tertiary structure of a protein (Fig. A-16c).

Figure A-15 **A Portion of the Primary Protein Structure of Human Insulin**

Thr—Lys—Pro—Thr—Tyr—Phe—Phe—Gly—Arg— · · · · ·

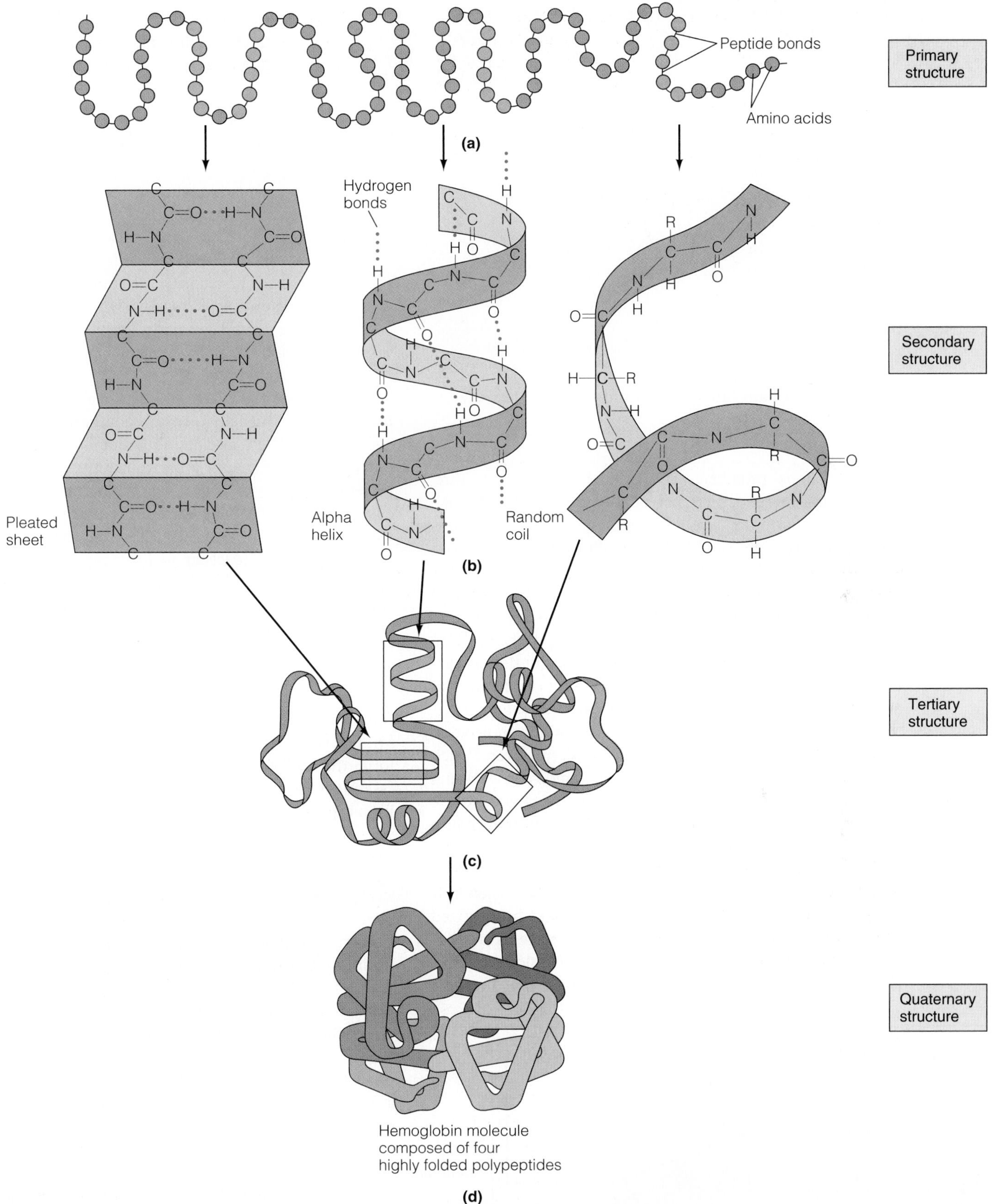

Figure A-16 **Levels of Protein Structure** Proteins can have four levels of structure. (a) The primary structure is a particular sequence of amino acids bonded in a chain. (b) At the secondary level, hydrogen bonding occurs between various amino acids within the chain, causing the chain to assume a particular shape. The most common secondary protein structure in the body is the alpha helix. (c) The tertiary structure is formed by the folding of the secondary structure into a functional three-dimensional configuration. (d) Some proteins form a fourth level of structure composed of several polypeptides, as exemplified by hemoglobin.

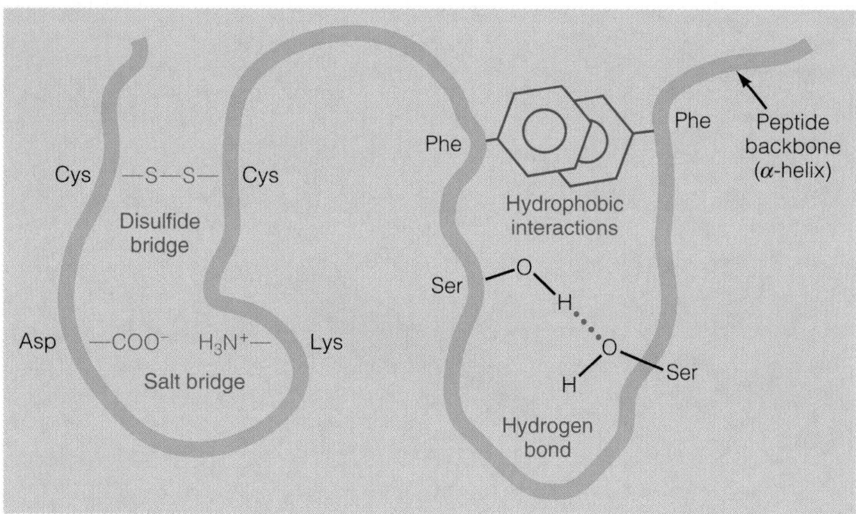

Figure A-17 **Side Chain Interactions Leading to the Tertiary Protein Structure**

All functional proteins exist in at least a tertiary structure. In some instances, several polypeptides interact with each other to form a fourth level of protein structure, the **quaternary structure.** For example, hemoglobin is made up of four highly folded polypeptide chains (the *globin* portion), along with four iron-containing *heme* groups (Fig. A-16d). One of the important functions of proteins is to serve as enzymes that catalyze the many essential chemical reactions of the body. In addition to catalyzing reactions, proteins can undergo reactions themselves. Two of the most important are hydrolysis and denaturation. Notice that according to Equation A-17, the formation of peptide bonds releases water molecules. Under appropriate conditions, it is possible to reverse such reactions by adding water to the peptide bonds and breaking them. **Hydrolysis** ("breakdown by H_2O") reactions of this type convert large proteins into smaller fragments or even into individual amino acids. **Denaturation** of proteins occurs when the bonds holding a protein chain in its characteristic tertiary or secondary conformation are broken. When this happens, the protein chain takes on a random, disorganized conformation. Denaturation can result when proteins are subjected to heating, treatment with specific chemicals such as alcohol or heavy metal ions, or extremes of pH. In some instances, denaturation is accompanied by coagulation or precipitation, as illustrated by the changes that occur in the white of an egg as it is fried.

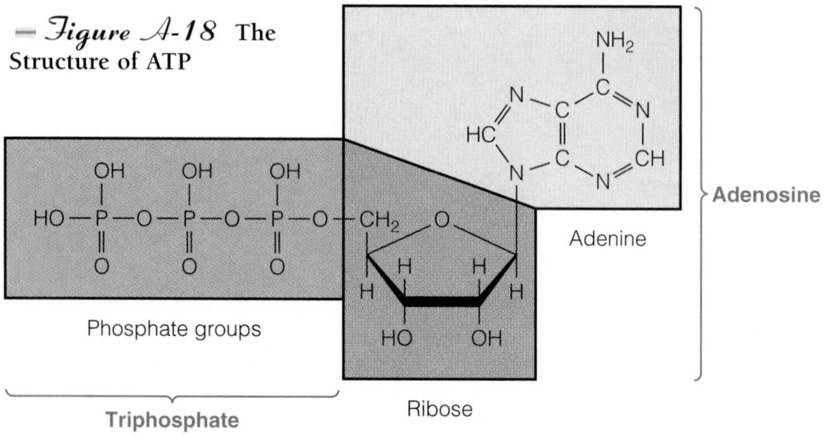

Nucleic Acids

High-molecular-weight macromolecules called **nucleic acids** are the compounds that enable genetic information to be stored in living cells and passed on to future generations. These important biomolecules are classified into two categories: **ribonucleic acids (RNA)** and **deoxyribonucleic acids (DNA).** Deoxyribonucleic acids are found primarily in the nuclei of cells, and ribonucleic acids are found primarily in the cytoplasm that surrounds cell nuclei.

Both types of nucleic acid are made up of units called **nucleotides,** which in turn are composed of three simpler components. Each nucleotide contains an organic nitrogenous base, a sugar, and a phosphate group. The three components are chemically bonded together with the sugar molecule lying between the base and the phosphate. In RNA, the sugar is ribose, whereas in DNA it is deoxyribose. When nucleotides bond together to form nucleic acid chains, the bonding is between the phosphate of one nucleotide and the sugar of another. Thus, the resulting nucleic acids consist of chains of alternating phosphates and sugar molecules, with a base molecule extending out of the chain from each sugar molecule (see Fig. B-1, p. B-2).

The chains of nucleic acid assume structural features somewhat like those found in proteins. DNA occurs in the form of two chains that mutually coil around one another to form the well-known double helix. Some RNA occurs in essentially straight chains, while in other types the chain forms specific loops or helices.

High-Energy Biomolecules

Certain molecules in the body store energy that is released during the metabolism of foods and make it available to the parts of the cells where it is needed to do specific cellular work. The primary substance that performs this function is **adenosine triphosphate,** or **ATP.** The structure of this molecule is shown in Figure A-18. Energy is stored in the phosphate bonds of this molecule and is released to the cells when a phosphate bond reacts with water to form adenosine diphosphate (ADP) and inorganic phosphate (P_i) (Eq. A-18).

$$ATP \rightarrow ADP + P_i + \text{energy for use by cell} \qquad \text{Eq. A-16}$$

More energy is released when the second phosphate reacts with water in the same way and ADP is converted to adenosine monophosphate (AMP).

Under the influence of an enzyme, AMP can be converted to a cyclical form called **cyclic AMP** or **cAMP,** which affects the activities of a number of enzymes involved in important reactions in the body.

Figure A-18 **The Structure of ATP**

Appendix B

Storage, Replication, and Expression of Genetic Information

Appendix B Contents At a Glance

DEOXYRIBONUCLEIC ACID (DNA) AND CHROMOSOMES
DNA structure; genes
Chromosomes and chromatin
PROTEIN SYNTHESIS
Complementary base pairing
Types of RNA
Steps of protein synthesis
CELL DIVISION
Mitosis
Meiosis
MUTATIONS

Deoxyribonucleic Acid (DNA) and Chromosomes

The nucleus perpetuates the genetic blueprint and serves as the control center of the cell.

The nucleus of the cell houses **deoxyribonucleic acid (DNA),** the genetic blueprint that is unique for each individual. This genetic material serves two essential functions. First, DNA contains "instructions" for assembling the structural and enzymatic proteins of the cell. Cellular enzymes in turn control the formation of other cellular structures and also determine the functional activity of the cell by regulating the rate at which metabolic reactions proceed. The nucleus serves as the cell's control center by directly or indirectly controlling almost all cell activities through the role its DNA plays in governing protein synthesis. Since cells make up the body, the DNA code determines the structure and the function of the body as a whole. The DNA that an organism possesses not only dictates whether the organism is a human, a toad, or a pea but also determines the unique physical and functional characteristics of that individual, all of which ultimately depend on the proteins produced under DNA control. Second, by replicating (making copies of itself), DNA perpetuates the genetic blueprint within all new cells formed within the body and is responsible for passing on genetic information from parents to children. We will first examine the coding mechanism used by DNA and then turn our attention to the means by which DNA replicates itself and controls protein synthesis.

Deoxyribonucleic acid is a double helix composed of nucleotides arranged in a particular sequence unique for each individual.

Deoxyribonucleic acid is a huge molecule, composed in humans of millions of nucleotides arranged into two long, paired strands that spiral around each other to form a double helix. Each **nucleotide** has three components: (1) a *nitrogenous base,* a ring-shaped organic molecule containing nitrogen; (2) a five-carbon ring-shaped sugar molecule, which in the case of DNA is *deoxyribose;* and (3) a phosphate group. Nucleotides are joined end to end by linkages between the sugar of one nucleotide and the phosphate group of the adjacent nucleotide to form a long polynucleotide ("many nucleotide") strand with a sugar-phosphate backbone and bases projecting out one side (Fig. B-1). There are four different bases in DNA—the double ringed bases **adenine (A)** and **guanine (G)** and the single-ringed bases **cytosine (C)** and **thymine (T).** The two polynucleotide strands within a DNA molecule are wrapped around each other and oriented so that their bases all project to the interior of the helix. The strands are held together by weak hydrogen bonds (Fig. A-6, see p. A-5) formed between the bases of adjoining strands. Base pairing is highly specific: adenine pairs only with thymine and guanine pairs only with cytosine (Fig. B-2).

The composition of the repetitive sugar-phosphate backbones that form the "sides" of the DNA "ladder" is identical for every molecule of DNA, but the sequence of the linked bases that form the "rungs" varies among different DNA molecules. The particular sequence of bases in a DNA molecule serves as "instructions," or a "code," that dictates the assembly of amino acids into a given order for the synthesis of specific **polypeptides** (chains of amino acids linked by peptide bonds; see p. A-12). A **gene** is a stretch of DNA that codes for the synthesis of a particular polypeptide. Polypeptides, in turn, are folded into a three-dimensional configuration to form a functional protein. Not all portions of a DNA molecule code for structural or enzymatic proteins. Some stretches of DNA code for proteins that regulate genes. Other segments

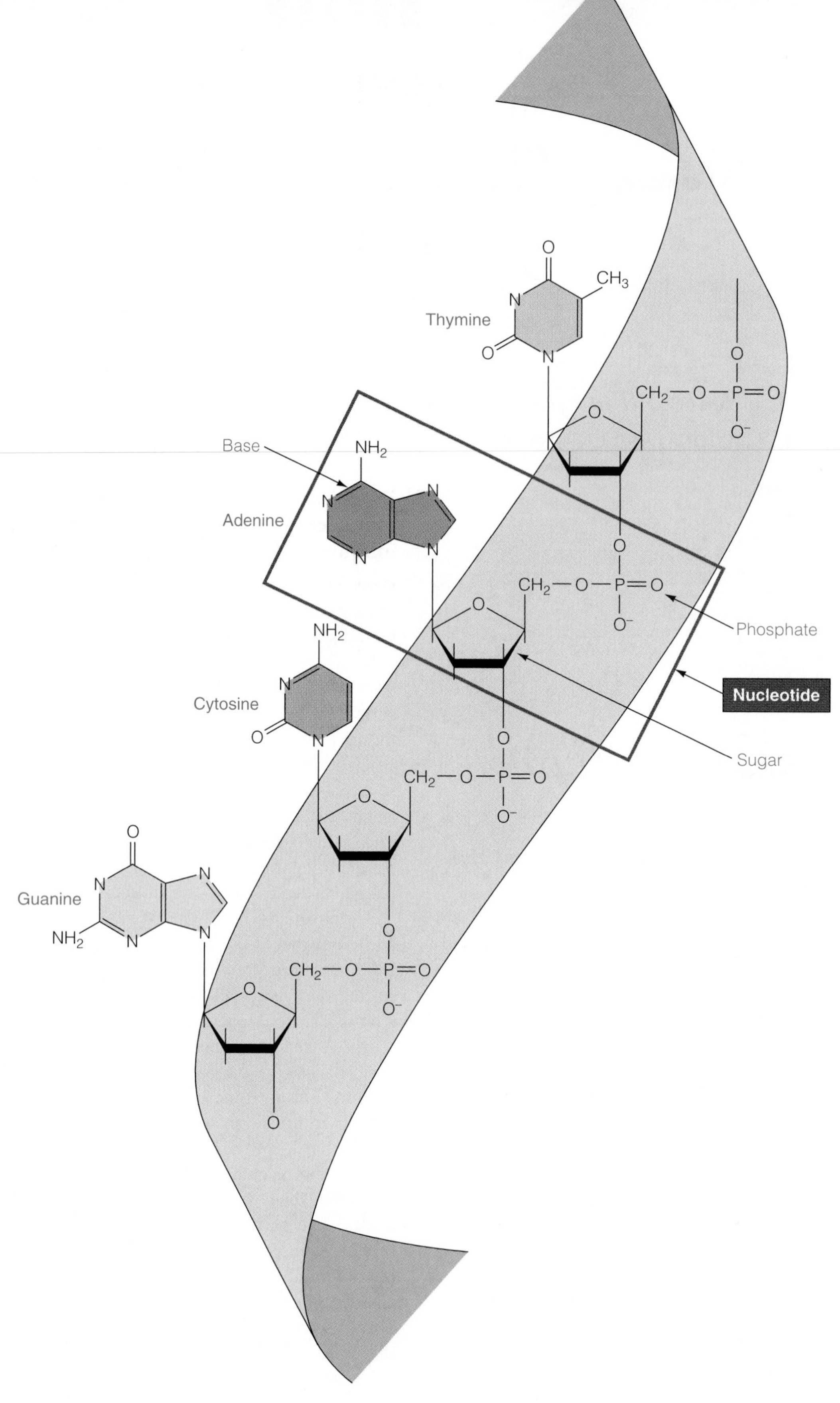

Figure B-1 **Polynucleotide Strand** Sugar-phosphate bonds link adjacent nucleotides together to form a polynucleotide strand with bases projecting to one side. The sugar-phosphate backbone is identical in all polynucleotides, but the sequence of the bases varies.

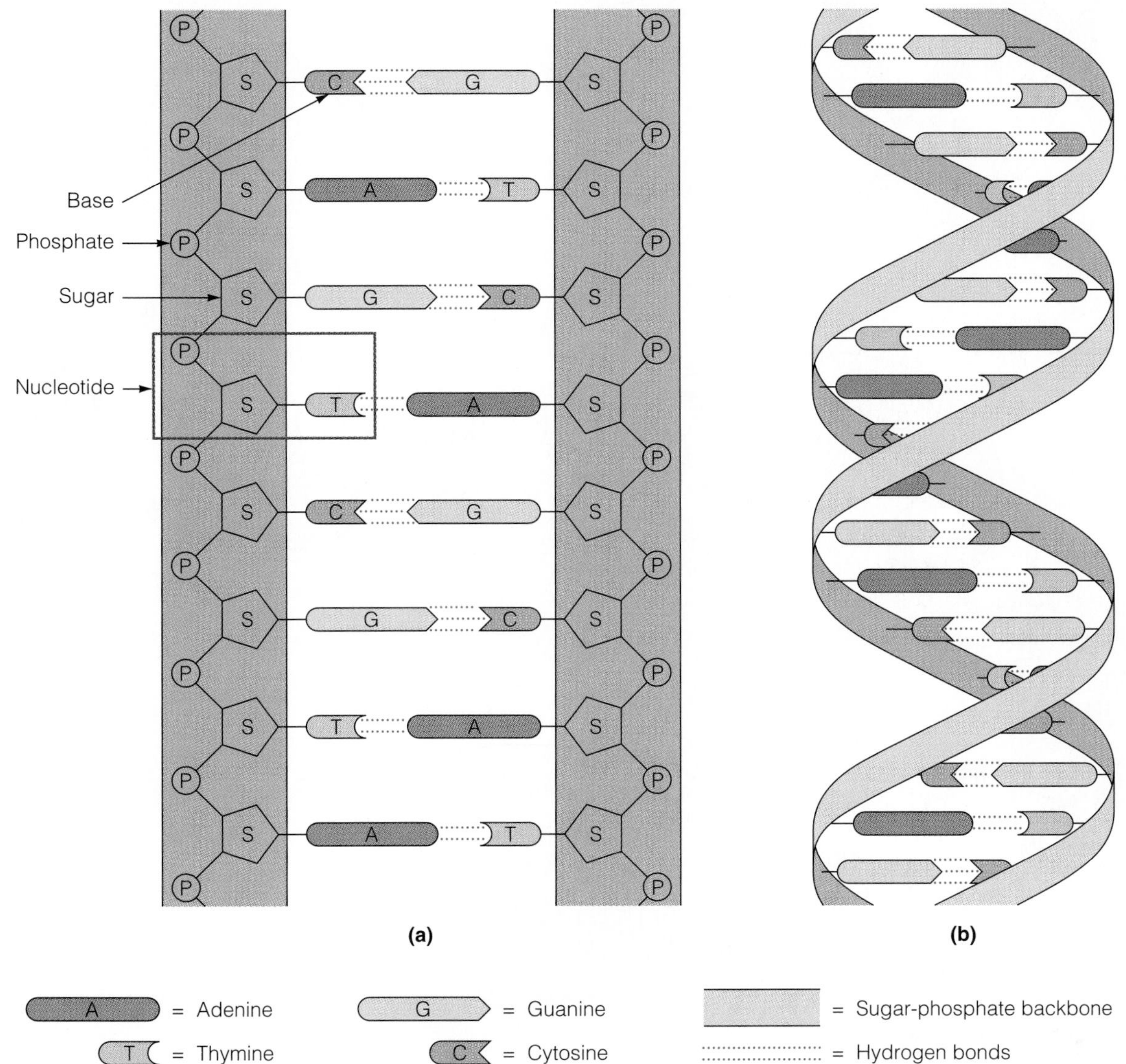

Figure B-2 **Complementary Base Pairing in DNA** (a) Two polynucleotide strands held together by weak hydrogen bonds formed between the bases of adjoining strands—adenine always paired with thymine and guanine always paired with cytosine. (b) Arrangement of the two bonded polynucleotide strands of a DNA molecule into a double helix.

appear to be important in organizing and packaging DNA within the nucleus. Still other regions are "nonsense" base sequences that have no apparent significance.

Deoxyribonucleic acid is precisely packaged within the nucleus.

The DNA molecules within each human cell, if lined up end to end, would extend more than 2 m (2,000,000 μm), yet these molecules are packed into a nucleus that is only 5 μm in diameter. These molecules are not randomly crammed into the nucleus but are precisely organized into **chromosomes.** Each chromosome consists of a different DNA molecule and contains a unique set of genes. **Somatic** (body) **cells** contain 46 chromosomes (the **diploid number**), which can be sorted into 23 pairs on the basis of various distinguishing features. Chromosomes composing a matched pair are termed **homologous chromosomes,** one member of each pair having been derived from the individual's maternal parent and the other member from the paternal parent. **Germ** (reproductive) **cells** (that is, sperm and eggs) contain only one member of each homologous pair for a total of 23 chromosomes (the **haploid number**). Union of a sperm and an egg results in a new diploid cell with 46 chromosomes, consisting of a set of 23 chromosomes from the mother and another set of 23 from the father.

The packaging and compression of DNA molecules into discrete chromosomal units are accomplished in part by nuclear proteins associated with DNA. Two classes of proteins—histone and nonhistone proteins—bind with DNA. **Histones** form bead-shaped bodies that play a key role in packaging DNA into its chromosomal structure. The **nonhistones** are believed to be important in gene regulation. The complex formed between the DNA and its associated proteins is known as **chromatin.** The long threads of DNA within a

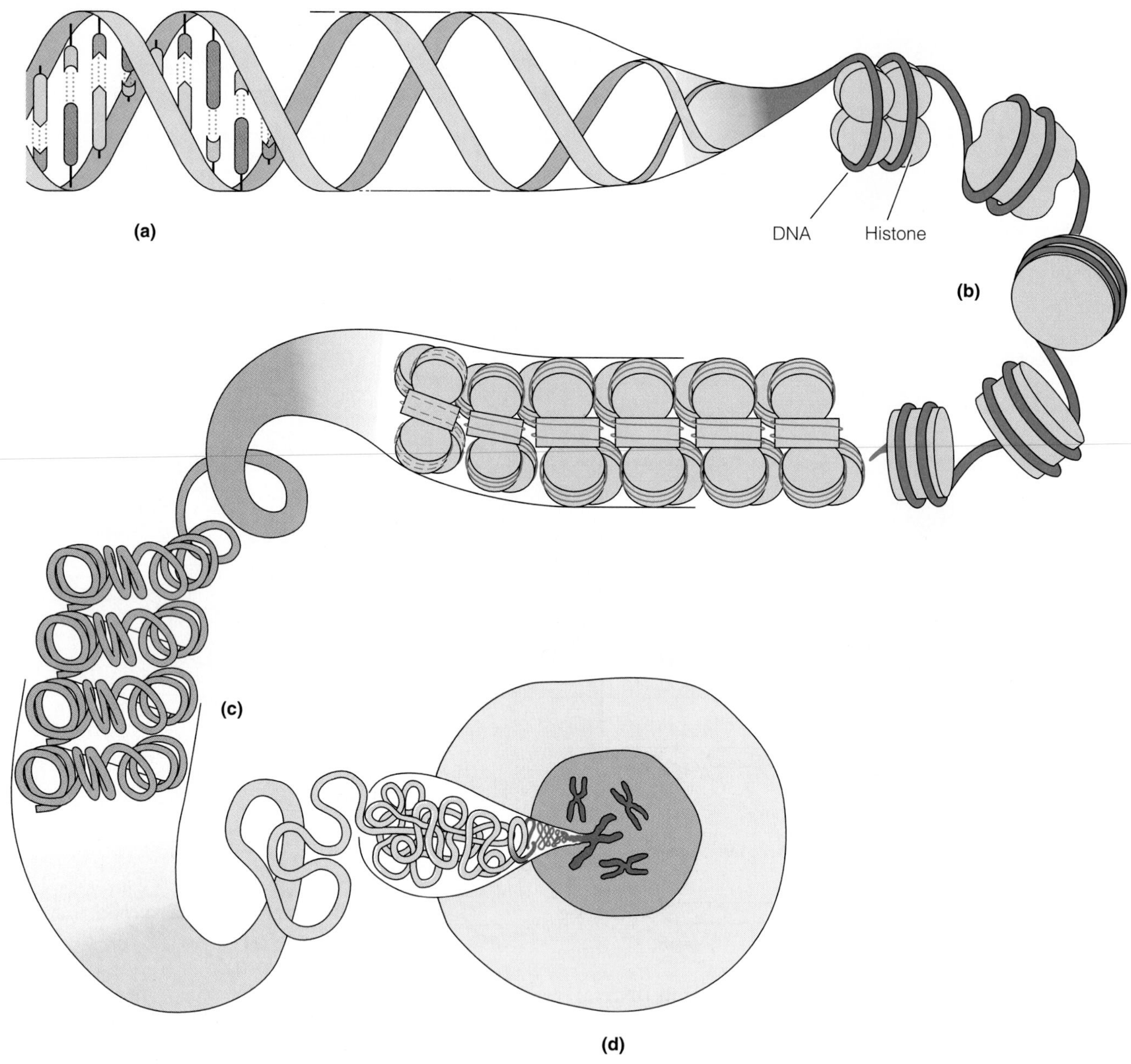

Figure B-3 Levels of Organization of DNA (a) Double helix of a DNA molecule. (b) DNA molecule wound around histone proteins, forming a "beads-on-a-string" structure. (c) Further folding and supercoiling of the DNA-histone complex. (d) Rodlike chromosomes, the most condensed form of DNA, which are visible in the cell's nucleus during cell division.

chromosome are wound around histones at regular intervals, thus compressing a given DNA molecule to about one-sixth its fully extended length. This "beads-on-a-string" structure is further folded and supercoiled into higher and higher levels of organization to further condense DNA into rodlike chromosomes that are readily visible by means of a light microscope during cell division (Fig. B-3). When the cell is not dividing, the chromosomes partially "unravel" or decondense to a less compact form of chromatin that is indistinct under a light microscope but appears as thin strands and clumps with an electron microscope. The decondensed form of DNA is its working form; that is, it is the form used as a template for protein assembly.

Protein Synthesis

Complementary base pairing serves as the foundation for both DNA replication and the initial step of protein synthesis.

During replication, the two DNA strands "unzip" as the weak bonds between the paired bases are enzymatically broken. New nucleotides present within the nucleus pair with the exposed bases from each strand (Fig. B-4). New adenine-bearing nucleotides pair with exposed thymine-bearing nucleotides in an old strand, and new guanine-bearing nucleotides pair with exposed cytosine-bearing nucleotides in

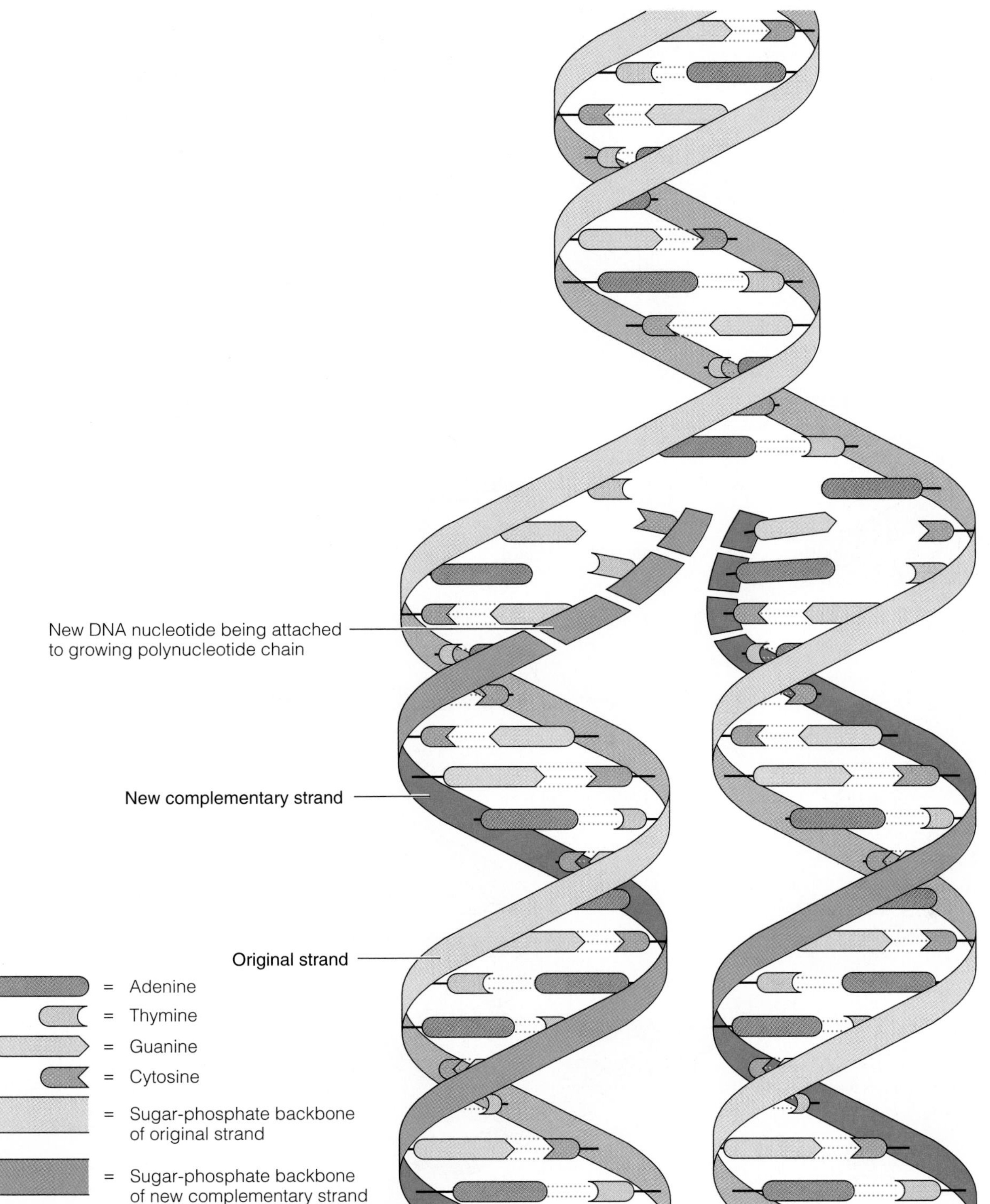

Figure B-4 **Complementary Base Pairing during DNA Replication** During DNA replication, the DNA molecule is unzipped, and each old strand directs the formation of a new strand; the result is two identical double-helix DNA molecules.

an old strand. This complementary base pairing is initiated at one end of the two old strands and proceeds in an orderly fashion to the other end. The new nucleotides attracted to and thus aligned in a prescribed order by the old nucleotides are sequentially joined by sugar-phosphate linkages to form two new strands that are complementary to each of the old strands. This replication process results in two complete double-stranded DNA molecules, one strand within each molecule having come from the original DNA molecule and one strand having been newly formed by complementary base pairing. These two DNA molecules are both identical to the original DNA molecule, with the "missing" strand in each of the original separated strands having been produced as a result of the imposed pattern of base pairing. This replication

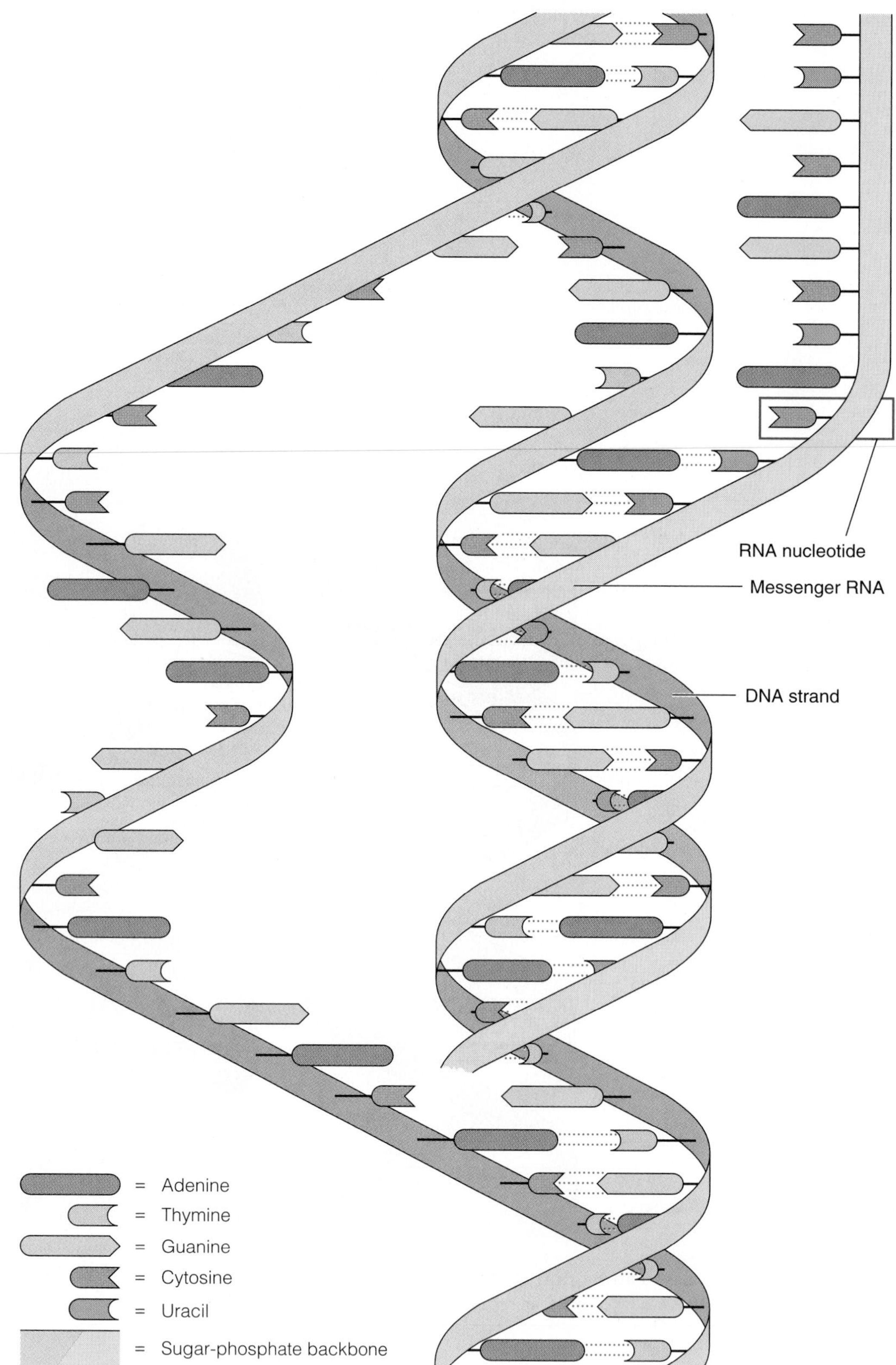

Figure B-5 **Complementary Base Pairing during DNA Transcription** During DNA transcription, a messenger RNA molecule is formed as RNA nucleotides are assembled by complementary base pairing at a given segment of one strand of an unzipped DNA molecule (that is, a gene).

process, which occurs only during cell division, is essential for ensuring the perpetuation of the genetic code in both of the new daughter cells. The duplicate copies of DNA are separated and evenly distributed to the two halves of the cell before it divides.

At other times, when DNA is not replicating in preparation for cell division, it serves as a blueprint for dictating cellular protein synthesis. How is this accomplished when DNA is sequestered within the nucleus and protein synthesis is carried out by ribosomes within the cytoplasm? Several types of another nucleic acid, **ribonucleic acid (RNA),** serve as the "go-between." Ribonucleic acid differs structurally from DNA in three regards: (1) the five-carbon sugar in RNA is *ribose* instead of deoxyribose, the only difference between them being the presence in ribose of a single oxygen atom that is absent in deoxyribose; (2) RNA contains the closely related base **uracil** instead of thymine, with the three other bases being the same as in DNA; and (3) RNA is single-stranded and not self-replicating. All RNA molecules are produced in the nucleus using DNA as a template or mold, then exit the nucleus through openings in the nuclear membrane known as **nuclear pores.** These pores are large enough for passage of RNA molecules but preclude passage of the much larger DNA molecules.

The DNA instructions for assembling a particular protein coded in the base sequence of a given gene are "transcribed" into a molecule of **messenger RNA (mRNA).** The segment of the DNA molecule to be copied uncoils, and the base pairs separate to expose the particular sequence of bases in the gene. In any given gene, only one of the DNA strands is used as a template for transcribing RNA, with the copied strand varying for different genes along the same DNA molecule. The beginning and end of a gene within a DNA strand are designated by particular base sequences that serve as "start" and "stop" signals. **Transcription** is accomplished by complementary base pairing of free RNA nucleotides with their DNA counterparts in the exposed gene (▬ Fig. B-5). The same pairing rules apply except that uracil, the RNA nucleotide substitute for thymine, pairs with adenine in the exposed DNA nucleotides. As soon as the RNA nucleotides pair with their DNA counterparts, sugar-phosphate bonds are formed to join the nucleotides together into a single-stranded RNA molecule that is released from DNA once transcription is complete. The original conformation of DNA is then restored. The RNA strand is much shorter than a DNA strand, because only a one-gene segment of DNA is transcribed into a single RNA molecule. The length of the finished RNA transcript varies, depending on the size of the gene. Within its nucleotide base sequence, this RNA transcript contains instructions for assembling a particular protein. Note that the message is coded in a base sequence that is *complementary to, not identical to,* the original DNA code.

Messenger RNA delivers the final coded message to the ribosomes for **translation** into a particular amino acid sequence to form a given protein. Thus, genetic information flows from DNA (which can replicate itself) through RNA to protein. This is accomplished first by *transcription* of the DNA code into a complementary RNA code, followed by *translation*

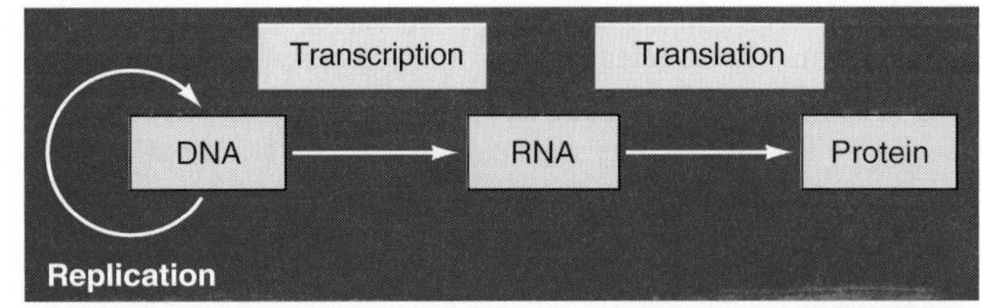

▬ *Figure B-6* **Flow of Genetic Information from DNA through RNA to Protein by Transcription and Translation**

of the RNA code into a specific protein (▬ Fig. B-6). The structural and functional characteristics of the cell as determined by its protein composition can be varied, subject to control, depending on which genes are "switched on" to produce mRNA.

Free nucleotides present in the nucleus cannot be randomly joined together to form either DNA or RNA strands, because the enzymes required to link together the sugar and phosphate components of nucleotides are active only when bound to DNA. This ensures that DNA, mRNA, and protein assembly occur only according to genetic plan.

Three forms of RNA participate in protein synthesis.

Besides messenger RNA, two other forms of RNA are required for translation of the genetic message into cellular protein: ribosomal RNA and transfer RNA. Messenger RNA carries the coded message from nuclear DNA to a cytoplasmic ribosome, where it directs the synthesis of a particular protein. **Ribosomal RNA (rRNA)** is an essential component of *ribosomes,* the "workbenches" for protein synthesis. Ribosomal RNA "reads" the base-sequence code of mRNA and translates it into the appropriate amino acid sequence during protein synthesis. **Transfer RNA (tRNA)** transfers the appropriate amino acids in the cytosol to their designated site in the amino acid sequence of the protein under construction.

Twenty different amino acids are used to construct proteins, yet only four different nucleotide bases are used to code for these twenty amino acids. In the "genetic dictionary," each different amino acid is specified by a **triplet code** that consists of a specific sequence of three bases in the DNA nucleotide chain. For example, the DNA sequence ACA (adenine, cytosine, adenine) specifies the amino acid cysteine, whereas the sequence ATA specifies the amino acid tyrosine. Each DNA triplet code is transcribed into mRNA as a complementary code word, or **codon,** consisting of a sequenced order of the three bases that pair with the DNA triplet. For example, the DNA triplet code ATA is transcribed as UAU (uracil, adenine, uracil) in mRNA.

Sixty-four different DNA triplet combinations (and, accordingly, sixty-four different mRNA codon combinations) are possible using the four different nucleotide bases (4^3). Of these possible combinations, sixty-one code for specific amino acids and the remaining three serve as "stop signals." A stop signal acts as a "period" at the end of a "sentence" that specifies the amino acid sequence in a particular protein; that is, ribosomal RNA releases the finished polypeptide product when it reaches a stop codon. Because sixty-one triplet codes each specify a particular amino acid and there are twenty different amino acids, a given amino acid may be specified by more than one base-triplet combination. For example, tyrosine is specified by the DNA sequence ATG as well as by ATA. In addition, one DNA triplet code, TAC (mRNA codon sequence AUG) functions as a "start signal" in addition to specifying the amino acid methionine. This code marks the place on mRNA where translation is to begin so that the message is started at the correct end and thus reads in the right

Figure B-7 **Ribosomal Assembly and Protein Translation** (a) On binding with a messenger RNA (mRNA) molecule, the small ribosomal subunit joins with the large subunit to form a functional ribosome. A transfer RNA (tRNA), charged with its specific amino acid passenger, binds to mRNA by means of complementary base pairing between the tRNA anticodon and the first mRNA codon positioned in the first ribosomal binding site. (b) Another tRNA molecule attaches to the next codon on mRNA positioned in the second ribosomal binding site. (c) The amino acid from the first tRNA is linked to the amino acid on the second tRNA. The first tRNA detaches. (d) The mRNA molecule shifts forward one codon (a distance of a three-base sequence). Another charged tRNA moves in to attach with the next codon on mRNA, which has now moved into the second ribosomal binding site. The amino acids from the tRNA in the first ribosomal site are linked with the amino acid in the second site. This process continues, with the polypeptide chain continuing to grow, until a stop codon is reached and the polypeptide chain is released.

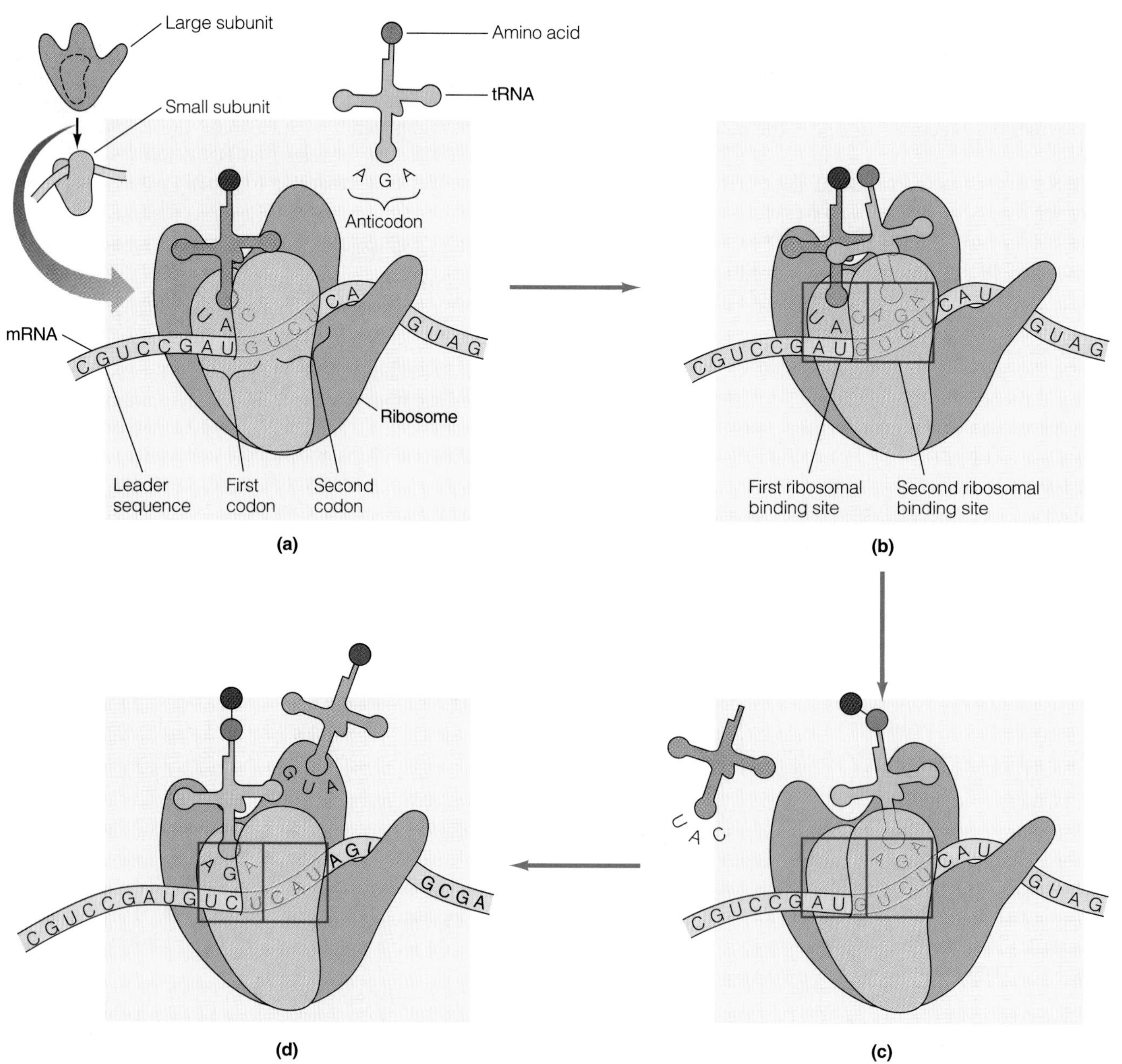

direction. Interestingly, the same genetic dictionary is used universally; a given three-base code stands for the same amino acid in all living things, including microorganisms, plants, and animals.

The three steps of protein synthesis are initiation, elongation, and termination.

A **ribosome** brings together all components that participate in protein synthesis—mRNA, tRNA, and amino acids—and provides the enzymes and energy required for linking the amino acids together. The nature of the protein synthesized by a given ribosome is determined by the mRNA message that is being translated. Each mRNA serves as a code for only one particular polypeptide.

A ribosome is an rRNA-protein structure organized into two subunits of unequal size. Only when a protein is being synthesized are these subunits brought together (— Fig. B-7a). During assembly of a ribosome, an mRNA molecule attaches to the smaller of the ribosomal subunits by means of a *leader sequence,* a section of mRNA that precedes the start codon. The small subunit with mRNA attached then binds to a large subunit to form a complete, functional ribosome. When the two subunits unite, a groove is formed that accommodates the mRNA molecule as it is being translated.

Free amino acids in the cytoplasm are not able to "recognize" and bind directly with their specific codons in mRNA. Transfer RNA is required to bring the appropriate amino acid to its proper codon. Even though tRNA is single-stranded, as are all RNA molecules, it is folded back onto itself into a T shape with looped ends (— Fig. B-8). The open-ended stem portion recognizes and binds to a specific amino acid. There are at least twenty different varieties of tRNA, each able to bind with only one of the twenty different kinds of amino acids. A tRNA is said to be "charged" when it is carrying its passenger amino acid. The loop end of a tRNA opposite the amino acid binding site contains a sequence of three exposed bases, known as the **anticodon,** which is complementary to the mRNA codon that specifies the amino acid being carried. Through complementary base pairing, a tRNA can bind with mRNA and insert its amino acid into the protein under construction only at the site designated by the codon for the amino acid. For example, the tRNA molecule that binds with tyrosine bears the anticodon AUA, which can pair only with the mRNA codon UAU, which specifies tyrosine. This dual binding function of tRNA molecules ensures that the correct amino acids are delivered to mRNA for assembly in the order specified by the genetic code. Transfer RNA can only bind with mRNA at a ribosome, so protein assembly does not occur except in the confines of a ribosome.

— *Figure B-8* **Structure of a tRNA Molecule** (a) The open end attaches to free amino acids. (b) The anticodon loop attaches to a complementary mRNA codon.

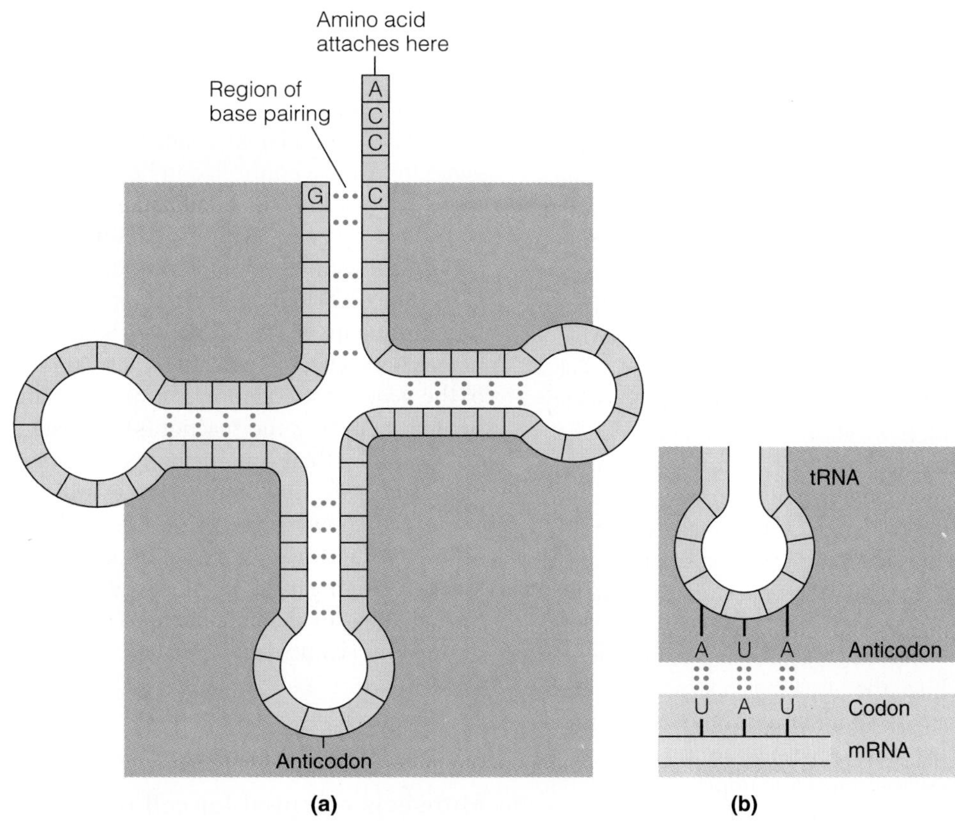

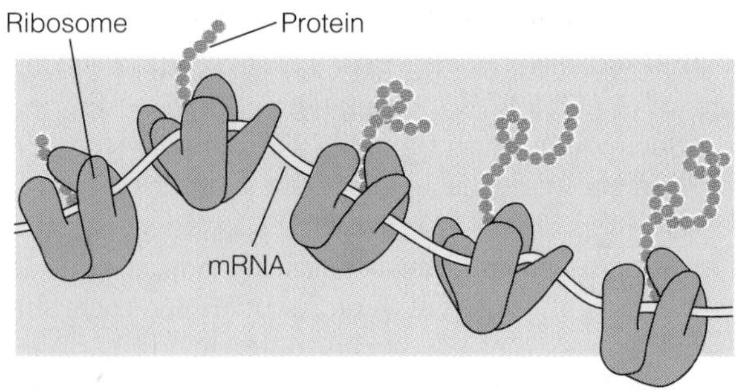

Figure B-9 **A Polyribosome** A polyribosome is formed by numerous ribosomes simultaneously translating mRNA.

Protein synthesis is initiated when a charged tRNA molecule bearing the anticodon specific for the start codon binds at this site on mRNA. A second charged tRNA bearing the anticodon specific for the next codon in the mRNA sequence then occupies the site next to the first tRNA (Fig. B-7b). At any given time, a ribosome can accommodate only two tRNA molecules bound to adjacent codons. Through enzymatic action, a peptide bond is formed between the two amino acids that are linked to the stems of the adjacent tRNA molecules (Fig. B-7c). The linkage is subsequently broken between the first tRNA and its amino acid passenger, leaving the second tRNA with a chain of two amino acids. The uncharged tRNA molecule (that is, the one minus its amino acid passenger) is released from mRNA. The ribosome then moves along the mRNA molecule by precisely three bases, a distance of one codon, so that the tRNA bearing the chain of two amino acids is moved into the number one ribosomal site for tRNA. Then, an incoming charged tRNA with a complementary anticodon for the third codon in the mRNA sequence occupies the number two ribosomal site that was vacated by the second tRNA (Fig. B-7d). The chain of two amino acids subsequently binds with and is transferred to the third tRNA to form a chain of three amino acids. Through repetition of this process, amino acids are subsequently added one at a time to a growing polypeptide chain in the order designated by the mRNA codon sequence as the ribosomal translation machinery moves stepwise along the mRNA molecule one codon at a time. This process is rapid. Up to ten to fifteen amino acids can be added per second. Elongation of the polypeptide chain continues until the ribosome reaches a stop codon in the mRNA molecule, at which time the polypeptide is released. The polypeptide is then folded and modified into a full-fledged protein. The ribosomal subunits dissociate and are free to reassemble into another ribosome for translation of other mRNA molecules.

Protein synthesis is energetically expensive. Attachment of each new amino acid to the growing polypeptide chain requires a total investment of splitting four high-energy phosphate bonds—two to charge tRNA with its amino acid, one to bind tRNA to the ribosomal-mRNA complex, and one to move the ribosome forward one codon.

A number of copies of a given protein can be produced from a single mRNA molecule before the latter is chemically degraded. As one ribosome moves forward along the mRNA molecule, a new ribosome attaches at the starting point on mRNA and also starts translating the message. Attachment of many ribosomes to a single mRNA molecule results in a *polyribosome*. Multiple copies of the identical protein are produced as each ribosome moves along and translates the same message (Fig. B-9). The released proteins are used within the cytosol, except for the few that move into the nucleus through the nuclear pores.

In contrast to the cytosolic polyribosomes, recall that ribosomes directed to bind with the rough endoplasmic reticulum (ER) feed their growing polypeptide chains into the ER lumen (see p. 21). The resultant proteins are subsequently packaged for export out of the cell or for replacement of membrane components within the cell.

Control of gene activity and protein transcription are incompletely understood.

Since each somatic cell in the body has the identical DNA blueprint, you might assume that they would all produce the same proteins. This is not the case, however, because different cell types are able to transcribe different sets of genes and thus synthesize different sets of structural and enzymatic proteins. For example, only red blood cells are able to synthesize hemoglobin, even though all body cells carry the DNA instructions for hemoglobin synthesis. Only about 7% of the DNA sequences in a typical cell are ever transcribed into mRNA for ultimate expression as specific proteins.

Control of gene expression involves gene regulatory proteins that activate ("switch on") or repress ("switch off") the genes that code for specific proteins within a given cell. Various DNA segments that do not code for structural and enzymatic proteins code for synthesis of these regulatory proteins. The molecular mechanisms by which these regulatory genes in turn are controlled in human cells are only beginning to be understood. In some instances, regulatory proteins are controlled by **gene-signaling factors** that bring about differential gene activity among various cells to accomplish specialized tasks. The largest group of known gene-signaling factors in humans is the hormones. Some hormones exert their homeostatic effect by selectively altering the transcription rate of the genes that code for enzymes that are in turn responsible for catalyzing the reaction(s) regulated by the hormone. For example, the hormone cortisol promotes the breakdown of fat stores by stimulating synthesis of the enzyme that catalyzes the conversion of stored fat into its component fatty acids. In other cases, gene action appears to be time-specific; that is, certain genes are expressed only at a certain developmental stage in the individual. This is especially important during embryonic development.

Cell Division

Mitosis is essential for cell reproduction.

Most cells in the human body have the ability to reproduce themselves, a process important in growth, replacement, and repair of tissues. The rate at which cells divide is highly variable. Cells within the deeper layers of the intestinal lining divide every few days to replace cells that are continually sloughed off the surface of the lining into the lumen of the digestive tract. In this way, the entire intestinal lining is replaced about every three days. At the other extreme are nerve cells, which permanently lose the ability to divide beyond a certain period of fetal growth and development. Consequently, when nerve cells are lost through trauma or disease, they cannot be replaced. In between these two extremes are cells that divide infrequently except when needed to replace damaged or destroyed tissue. The factors that control the rate of cell division remain obscure.

Recall that cell division involves two components: nuclear division and cytoplasmic division **(cytokinesis).** Nuclear division in somatic cells is accomplished by **mitosis,** in which a complete set of genetic information (that is, a diploid number of chromosomes) is distributed to each of two new daughter cells.

A cell capable of dividing alternates between periods of mitosis and nondivision. The interval of time between cell division is known as **interphase.** Since mitosis takes less than an hour to complete, the vast majority of cells in the body at any given time are in interphase.

Replication of DNA and growth of the cell take place during interphase in preparation for mitosis. Although mitosis is a continuous process, it displays four distinct phases: *prophase, metaphase, anaphase,* and *telophase* (■ Fig. B-10a).

- **Prophase:**
  1. Chromatin condenses and becomes microscopically visible as chromosomes. The condensed duplicate strands of DNA, known as **sister chromatids,** remain joined together within the chromosome at a point called the **centromere** (■ Fig. B-11, p. B-14).
  2. Cells contain a pair of **centrioles,** short cylindrical structures that form the mitotic spindle during cell division (see Fig. 2-1, p. 21). The centriole pair divides, and the daughter centrioles move to opposite ends of the cell, where they assemble between them a mitotic spindle made up of microtubules (see p. 38).
  3. The membrane surrounding the nucleus starts to break down.
- **Metaphase:**
  1. The nuclear membrane completely disappears.
  2. The 46 chromosomes, each consisting of a pair of sister chromatids, align themselves at the midline, or equator, of the cell. Each chromosome becomes attached to the spindle by means of several spindle fibers that extend from the centriole to the centromere of the chromosome.
- **Anaphase:**
  1. The centromeres split, converting each pair of sister chromatids into two identical chromosomes, which separate and move toward opposite poles of the spindle. Motor proteins (see p. 37) are responsible for pulling the chromosomes along the spindle fibers toward the poles.
  2. At the end of anaphase, an identical set of 46 chromosomes is present at each of the poles, for a transient total of 92 chromosomes in the soon-to-be-divided cell.
- **Telophase:**
  1. The cytoplasm divides through formation and gradual tightening of an actin contractile ring at the midline of the cell, thus forming two separate daughter cells, each with a full diploid set of chromosomes (see Fig. 2-23a, p. 39).
  2. The spindle fibers disassemble.
  3. The chromosomes uncoil to their decondensed chromatin form.
  4. A nuclear membrane reforms in each new cell.

Cell division is complete with the end of telophase. Each of the new cells now enters interphase.

Meiosis is essential for formation of reproductive cells.

Nuclear division in the specialized case of germ cells is accomplished by **meiosis,** in which only a half set of genetic information (that is, a haploid number of chromosomes) is distributed to each daughter cell. Meiosis differs from mitosis in several important regards (Fig. B-10b). Specialized diploid germ cells undergo one chromosome replication followed by two nuclear divisions to produce four haploid germ cells.

- **Meiosis I:**
  1. During prophase of the first meiotic division, the members of each homologous pair of chromosomes line up side by side to form a **tetrad,** which is a group of four sister chromatids with two identical chromatids within each member of the pair.
  2. The process of crossing over occurs during this period, when the maternal copy and the paternal copy of each chromosome are paired. **Crossing over** involves a physical exchange of chromosome material between nonsister chromatids within a tetrad (■ Fig. B-12, p. B-14). This process yields new chromosome combinations, thus contributing to genetic diversity.
  3. During metaphase, the 23 tetrads line up at the equator.
  4. At anaphase, homologous chromosomes, each consisting of a pair of sister chromatids joined at the centromere, separate and move toward opposite poles. Maternally and paternally derived chromosomes migrate to opposite poles in random assortments of one member of each chromosome pair without regard for its original derivation. This genetic mixing provides novel new combinations of chromosomes.
  5. During the first telophase, the cell divides into two cells. Each cell contains 23 chromosomes consisting of two sister chromatids.
- **Meiosis II:**
  1. Following a brief interphase in which no further replication occurs, the 23 unpaired chromosomes line up at the equator, the centromeres split, and the sister chromatids separate for the first time into independent chromosomes that move to opposite poles.

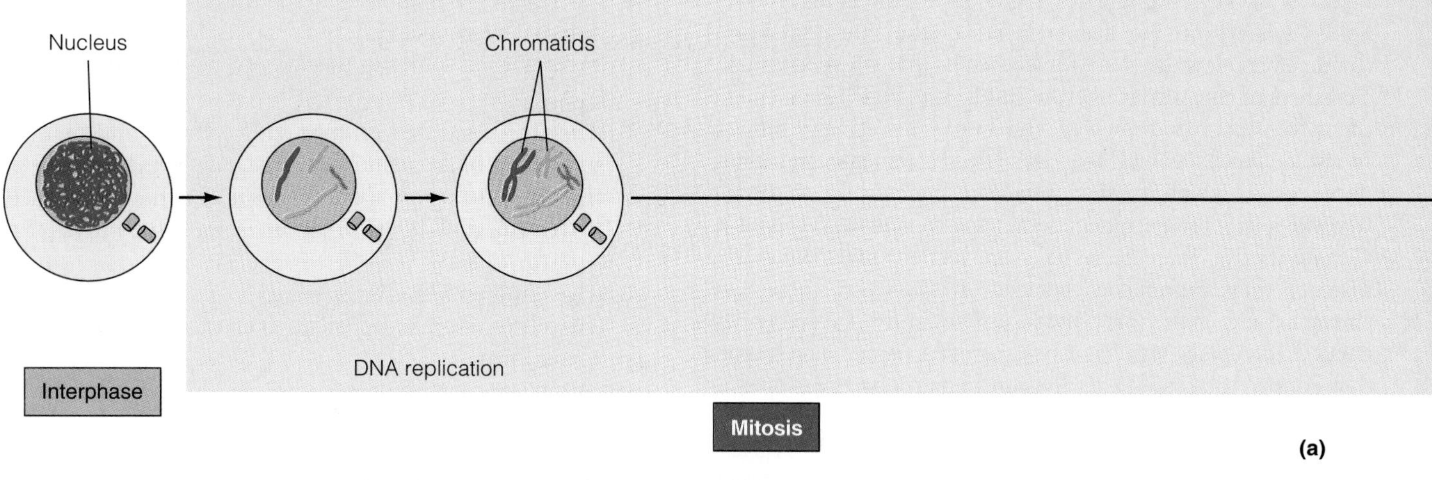

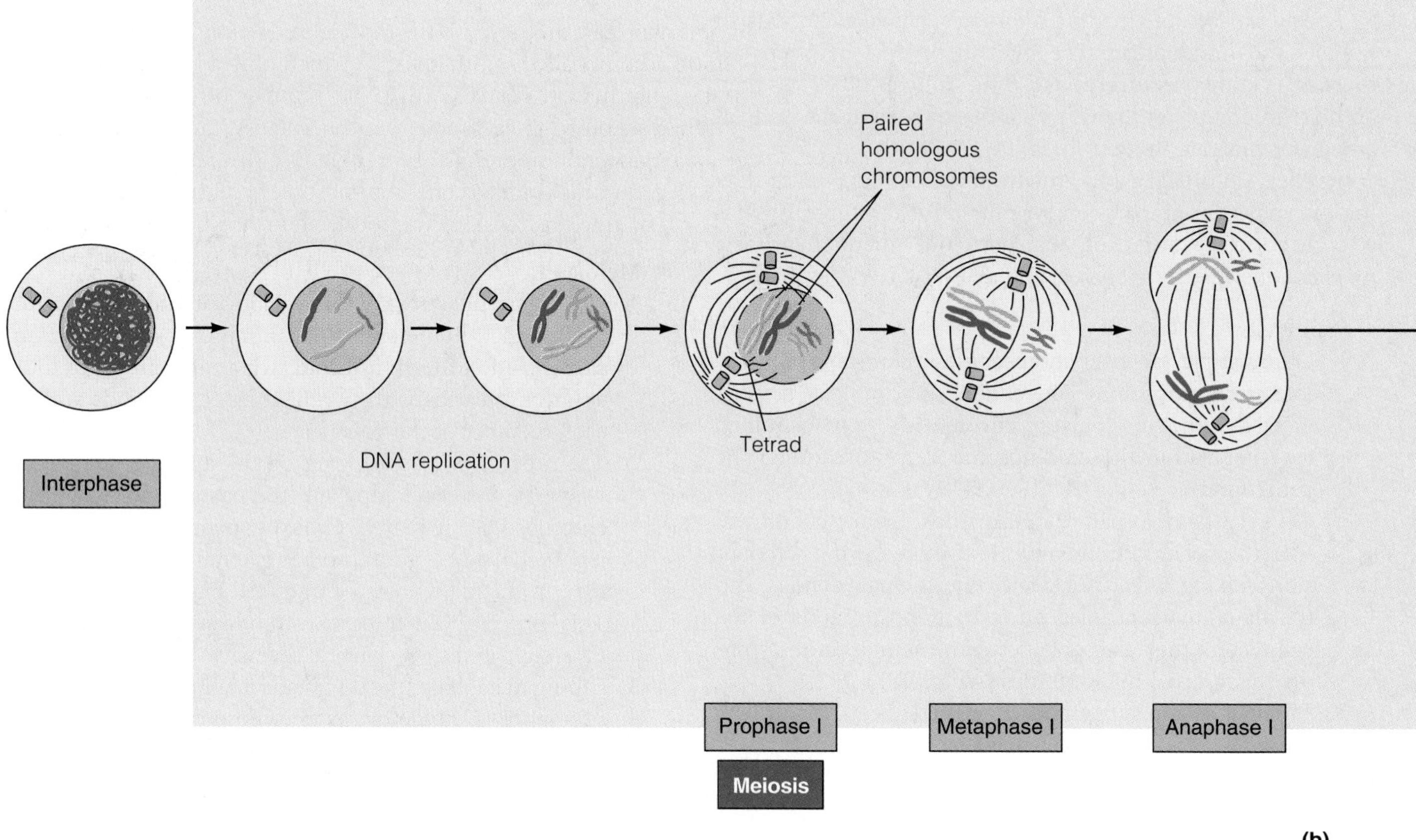

Figure B-10 **A Comparison of Events in Mitosis and Meiosis** (a) Mitosis. (b) Meiosis.

2. During cytokinesis, each of the daughter cells derived from the first meiotic division forms two new daughter cells. The end result is four daughter cells, each containing a haploid set of chromosomes.

Union of a haploid sperm and haploid egg results in a zygote (fertilized egg) that contains the diploid number of chromosomes. Development of a new multicellular individual from the zygote is accomplished by mitosis and cell differentiation. Since DNA is normally faithfully replicated in its entirety during each mitotic division, all cells in the body possess an identical aggregate of DNA molecules. Structural and functional variations between different cell types result from differential gene expression.

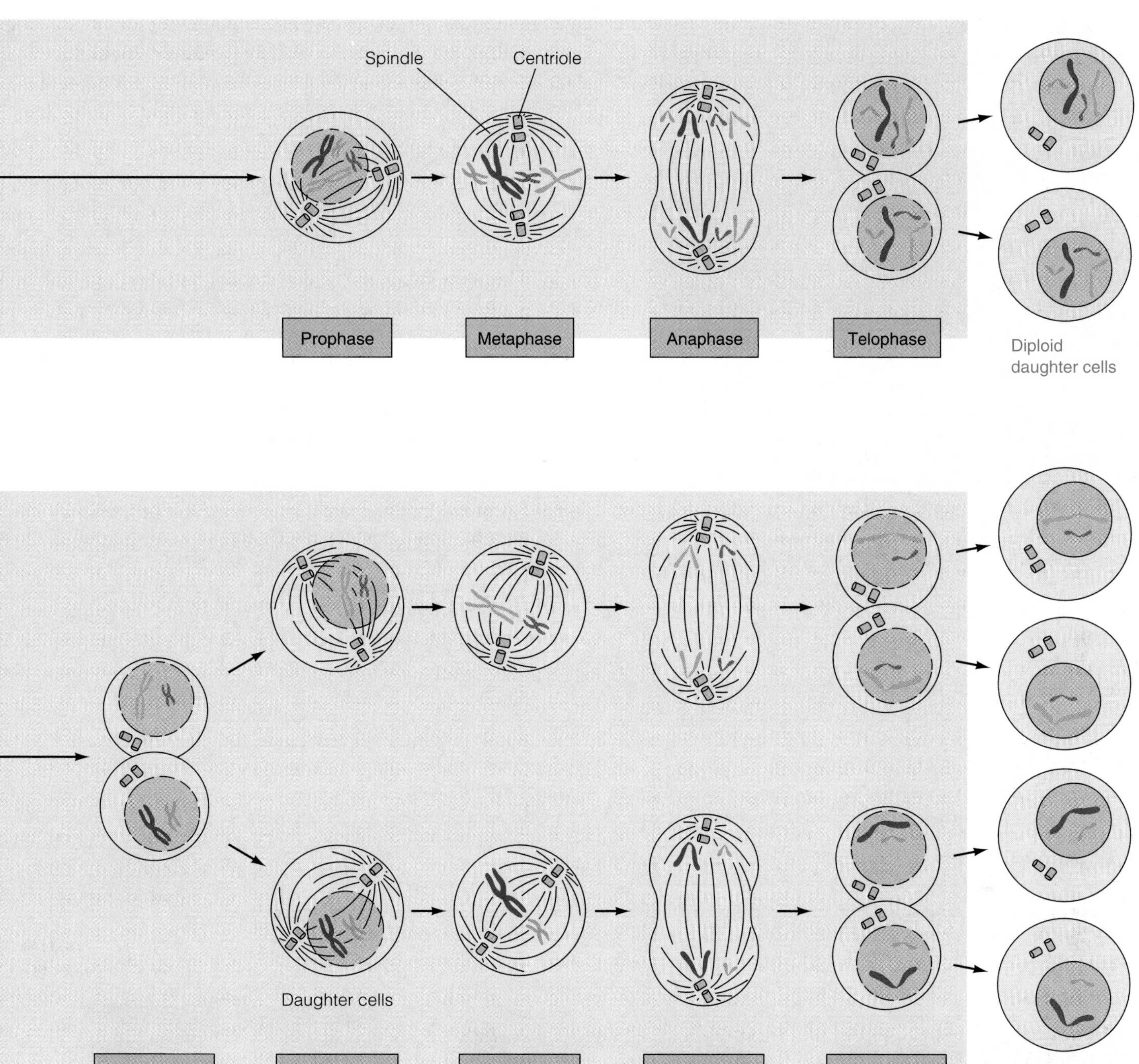

Mutations

Mutations can be harmless, deleterious, fatal, or beneficial.

It is estimated that about 10^{16} cell divisions take place in the body during the course of a person's lifetime to accomplish growth, repair, and normal cell turnover. Because more than 3 billion nucleotides must be replicated during each cell division, it is no wonder that "copying errors" occasionally occur. Any change in the DNA sequence is known as a **point (gene) mutation.** A point mutation arises when a base is inadvertently substituted, added, or deleted during the replication process.

When a base is inserted in the wrong position during DNA replication, the mistake can often be corrected by a built-in

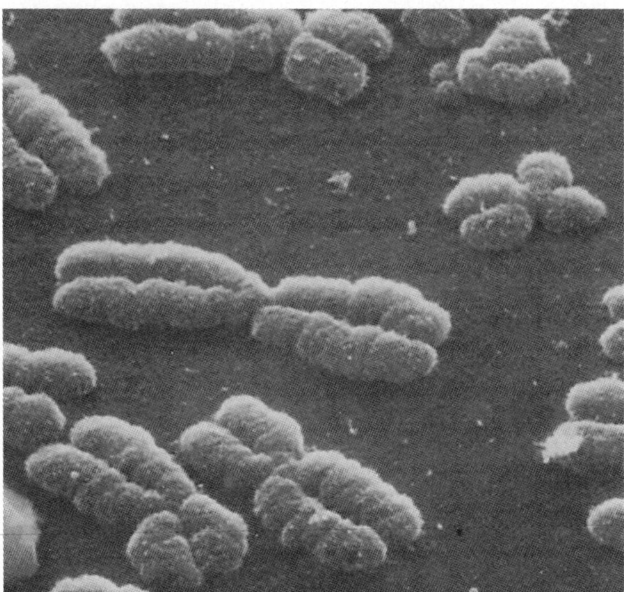

Figure B-11 **A Scanning Electron Micrograph of Human Chromosomes from a Dividing Cell** The replicated chromosomes appear as double structures, with identical sister chromatids joined in the middle at a common centromere.

"proofreading" system. Repair enzymes remove the newly replicated strand back to the defective segment at which time normal base pairing resumes to resynthesize a corrected strand. Not all mistakes can be corrected, however.

Mutations can arise spontaneously by chance alone or they can be induced by **mutagens,** which are factors that increase the rate at which mutations take place. Mutagens include various chemical agents as well as ionizing radiation such as X rays and atomic radiation. Mutagens promote mutations either by chemically altering the DNA base code through a variety of mechanisms or by interfering with the repair enzymes so that abnormal base segments cannot be cut out.

Depending on the location and nature of a change in the genetic code, a given mutation may (1) have no noticeable effect if it does not alter a critical region of a cellular protein; (2) adversely alter cell function if it impairs the function of a crucial protein; (3) be incompatible with the life of the cell, in which case the cell dies and the mutation is lost with it; or (4) in rare cases, prove beneficial if a more efficient structural or enzymatic protein results. If a mutation occurs in a body cell (a **somatic mutation**), the outcome will be reflected as an alteration in all future copies of the cell in the affected individual, but it will not be perpetuated beyond the life of the individual. If, on the other hand, a mutation occurs in a sperm- or egg-producing cell **(germ cell mutation),** the genetic alteration may be passed on to succeeding generations.

In most instances, cancer results from multiple somatic mutations that occur over a course of time within DNA segments known as **proto-oncogenes.** Proto-oncogenes are normal genes whose coded products are important in the regulation of cell growth and division. These genes have the potential of becoming overzealous **oncogenes** ("cancer genes"), which induce the uncontrolled cell proliferation characteristic of cancer. Proto-oncogenes can become cancer producing as a result of several sequential mutations in the gene itself or by changes in adjacent regions that regulate the proto-oncogenes. Less frequently, tumor viruses become incorporated in the DNA blueprint and act as oncogenes.

Figure B-12 **Crossing Over** (a) During prophase I of meiosis, each homologous pair of chromosomes lines up side by side to form a tetrad. (b) Physical exchange of chromosome material occurs between nonsister chromatids. (c) As a result of this crossing over, new combinations of genetic material are formed within the chromosomes.

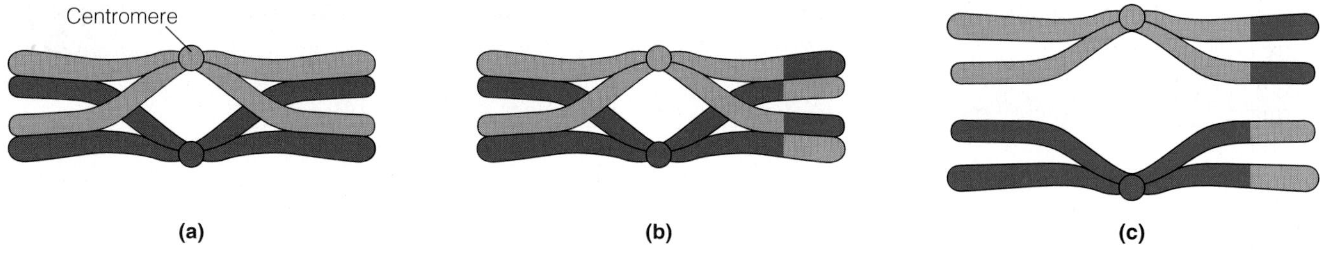

Appendix C

Principles of Quantitative Reasoning

BY KIM E. COOPER AND JOHN D. NAGY

Arizona State University

Introduction

Historically, as a branch of science matures, it typically becomes more precise and usually more quantitative. This trend is becoming increasingly true of biology and especially of physiology. Most students, however, are uncomfortable with quantitative reasoning. Students are usually quite capable of doing the mechanical manipulations of mathematics but have trouble translating back and forth between words, concepts, and equations. This appendix is meant to help you become more comfortable working with equations.

Why Are Equations Useful?

A great deal of what we do in science involves establishing functional relations between variables of interest (for example, blood pressure and heart rate, transport rate and concentration gradient). Equations are simply a compact and exact way of expressing such relationships. The tools of mathematics then allow us to draw conclusions systematically from these relationships. Mathematics is a very powerful set of tools or, more generally, a very powerful way of thinking. Mathematics allows you to think extremely precisely, and therefore clearly, about complex relationships. Equations and quantitative notions are the keys to that precision. For example, a quantitative comparison of the predictions of a theory against the results of measurement forms the basis of statistics and much of the process of hypothesis testing on which science is based. A scientific conclusion without adequate quantification and statistical backing may be little more than an impression or opinion.

It may seem odd to say that mathematics allows you to think more clearly about complex ideas. People unfamiliar with mathematical thinking often complain that even simple relationships produce complicated equations and that complex relationships are mathematically intractable. Certainly, many basic concepts require considerable mathematical expertise to be handled properly, but such concepts are in fact not simple. More commonly, however, many simple equations are seen as complex because many students are poorly trained in how to think about equations.

How to Think about an Equation

In this section we will take the first, and often overlooked, step in thinking quantitatively. How do we begin to think about some new equation presented to us? We start by becoming comfortable with the "meaning" of an equation. This step is absolutely necessary if you are to use an equation properly. As a specific example, consider the Nernst equation (see p. 73) for potassium. Here are several forms you will find in various books; they all say essentially the same thing:

$$E_{K^+} = (RT/zF)\log\{[K^+]_{out}/\ [K^+]_{in}\}$$
$$E_{K^+} = (RT/zF)\ 2.303\ \log\{[K^+]_{out}/\ [K^+]_{in}\}$$
$$E_{K^+} = (61\ \text{mV}/z)\ \log\{[K^+]_{out}/\ [K^+]_{in}\}$$

For many students these equations may be like meaningless strings of symbols. What are these equations trying to tell us? What do they represent? The following four steps may help you become comfortable with any new equation. Try them with the Nernst equation.

1. Be sure you can define the symbols and give dimensions and units. Check the equation for dimensional consistency.

   One of the first steps is to figure out which symbols represent the variables of interest and which are simple constants. In this case, all the symbols are constants except two.

   E_{K^+} is the Nernst (equilibrium) potential for potassium. It represents the concentration gradient (force of diffusion) on a mole of potassium ions. E_{K^+} has the dimensions of a voltage and is usually given in units of mV. This dimension is used so that the concentration gradient is expressed in the same dimensions as the other force acting on the ions, that is, the electrical gradient. Using the same dimensions makes it possible to compare the two forces.

   $[K^+]$ represents the concentration of potassium. With the subscript "out," this symbol refers to the concentration of potassium outside the cell. With the subscript "in," this symbol refers to the concentration of potassium inside the cell. $[K^+]$ has dimensions of concentration and is usually expressed in units of mM (millimolars; millimols/liter).

2. Identify the dependent and independent variables. Try to find normal values and ranges for the variables. Before continuing, we should define "dependent" and "independent" variables. Remember that equations represent rela-

tionships between variables. Whenever you hear the word *relationship,* think of a graph, like this, for example:

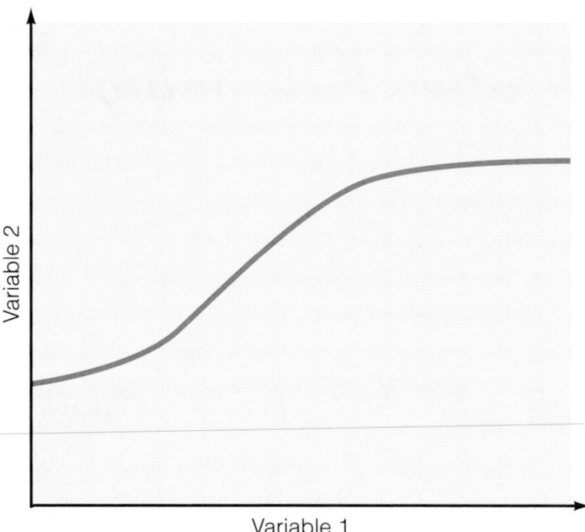

Graphs are often a good way to represent relationships and therefore equations. What this graph says is that the value of variable 2 depends on the value of variable 1. Thus, for any value of variable 1, the corresponding value for variable 2 can be determined from the graph. In other words, variable 1 determines the value of variable 2. Because variable 2 depends on variable 1, we call variable 2 the *dependent variable.* Variable 1, on the other hand, is independent of variable 2, so we call variable 1 the *independent variable.* There can be any number of dependent and independent variables.

How do you determine which variables are dependent and which are independent? The answer usually depends on cause and effect: "effects" are dependent variables and "causes" are independent. For example, we know (see chapter 10, p. 320) that mean arterial pressure (MAP) is the product of cardiac output (CO) and total peripheral resistance (TPR); that is:

$$\text{MAP} = \text{CO} \times \text{TPR}$$

MAP is on the left-hand side of this equation because we think of mean arterial pressure as a result of cardiac output and total peripheral resistance. Or, to put this another way, mean arterial pressure is a function of cardiac output and total peripheral resistance. As a cause-effect relationship, to think of mean arterial pressure somehow "causing" cardiac output to be a certain value seems backward. Therefore, MAP is the effect, the dependent variable, and we place it on the left-hand side of the equality symbol. Conversely, CO and TPR are the causes, the independent variables, and we put them on the right.

In our Nernst equation example, the independent variables are the concentrations. The dependent variable is the Nernst potential because we think of the potential as being a result of the ion concentrations. We also know that E_{K+} is about −90mV, and $[K^+]_{out}$ and $[K^+]_{in}$ are about 5 mM and 150 mM, respectively.

3. Identify the constants and know their numerical values:

   R is the gas constant. It has dimensions of energy per mole per degree of temperature and the value of 8.31 joules/kelvin · mole. It is also convenient to note that a joule = volt × coulomb.

   T is temperature, with the dimension of temperature being in units of kelvins. Normal body temperature is around 37°C [= 308 kelvins (K)].

   z is the valence of the ion. Valence is the charge on an ion, including the sign. For potassium, z = +1.

   F is Faraday's constant, which has dimensions of charge per mole, units of coulombs per mole, and a value of 96,500 C/mol.

   Refer back to the Nernst equations given on the preceding page. Note that the constants just defined appear in the first two equations, but not the third. In the third equation, the quantity RT/F has already been evaluated for you, as follows:

$$RT/F = [(8.31\ V \cdot C/K \cdot mol)(308K)]/(96{,}500\ C/mol) = 26.5\ mV$$

   We multiply this value by 2.303 to convert the natural logarithm to the base 10 logarithm. Note that 26.5 mV × 2.303 = 61.1 mV.

4. State the equation in words. Summarize it in a few sentences so that someone can understand what it is about. Don't just say the names of the symbols.

   Just saying the names of the symbols would be equivalent to saying the following: The Nernst potential is given by a constant times the logarithm of the ratio of the ion concentrations. This is certainly true but does nothing to aid our intuition. A preferable statement would be: The Nernst equation allows us to calculate the force pushing ions into or out of a cell via diffusion. This is valuable because we can compare this force to the force moving ions in and out via the membrane voltage and see which is larger and hence in what direction the ions will actually move. The force is expressed in electrical units so we can compare it directly with the membrane voltage. The constants convert from concentration to electrical units.

Only when you understand what an equation means, will you be able to use it to answer questions. The next section gives you some guidance in taking this next step.

How to Think with an Equation

Before you can use an equation to help you think, you need to develop a few basic skills. Luckily, these skills are not difficult to learn.

1. Be sure you know the algebraic rules for manipulating variables within any function (√, exp, log, etc.) involved.

   In the case of the Nernst equation, the tricky function is the logarithm. You should consult a college-level algebra book if you are hazy on the rules of working with logs or any other function. For instance, it is useful to know that:

$$\log\{A\} = -\log\{1/A\}$$

and that the log operation is undone by taking it to the power of 10; that is:

$$10^{\log\{A\}} = A$$

2. Be able to solve for any variable in terms of the others.

   Given just three variables (E_{K^+}, $[K^+]_{in}$, $[K^+]_{out}$), only a few types of questions can be asked. Two of the three variables must be given, and you must solve for the third. If the two concentrations are given, then the formula is already set to give you the Nernst potential. If the Nernst potential and one concentration are given, however, you must be able to solve for the other concentration. See if you can do this and obtain the two following equations:

$$[K^+]_{out} = [K^+]_{in} 10^{(E_{K^+}/61\text{ mV})}$$

$$[K^+]_{in} = [K^+]_{out} 10^{(-E_{K^+}/61\text{ mV})}$$

3. Be able to sketch, at least approximately, the dependence of any variable on any other variable. The ability to do this is exceedingly valuable. Sketching helps you generate insight about equations; it helps you understand what an equation means. Therefore, sketching helps you understand the solution, as well as solve the problem. If you apply this technique consistently, you may find equations far simpler to handle than you previously suspected. In addition, be sure you can relate your sketch to experimental measurements and physiological situations.

   For example, we can draw the relationships between the Nernst potential for potassium and the external potassium ion concentration predicted by the equations as follows:

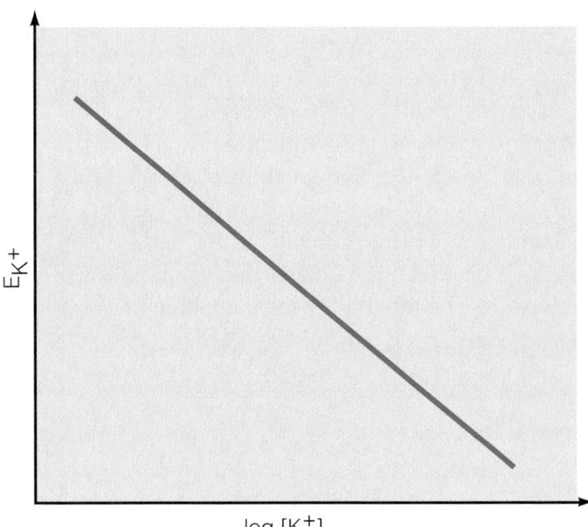

This sketch makes clear that the Nernst potential, which can be measured physiologically, should decrease linearly as the log $[K^+]_{out}$ increases, which can be controlled experimentally. Therefore, this sketch suggests an experiment: vary $[K^+]_{out}$. If E_{K^+} does not decrease linearly with increasing log $[K^+]_{out}$, then we would have a flaw in our understanding. The Nernst equation would not describe the real situation, as we think it should. Scientific advances are almost always heralded by such contradictions.

4. Be able to combine several equations to find new relationships. Combining separate pieces of information is always useful. In fact, some scientists have argued that this activity is all scientists ever do. To integrate knowledge for yourself, you must be able to combine the various relations you learn about into new combinations. This allows you to solve increasingly complex problems.

   As an example, consider the following relation:

$$I_{K^+} = G_{K^+}(V_m - E_{K^+})$$

   This equation describes the number of potassium ions flowing across a membrane if both a concentration gradient (E_{K^+}) and an electrical gradient (V_m) are present. This equation can be combined with the Nernst equation for potassium to answer the following question. Suppose V_m, G_{K^+}, and $[K^+]_{in}$ are fixed. What would the external potassium ion concentration have to be such that there is no net flux of potassium ions across the membrane? No net flux implies that $I_{K^+} = 0$. But, if $G_{K^+} > 0$, $I_{K^+} = 0$ only when the membrane voltage equals the Nernst potential ($V_m = E_{K^+}$). To answer the question, then, we set $E_{K^+} = V_m$ in the Nernst equation and solve it for $[K^+]_{out}$.

5. Know the equation's underlying assumptions and limits of validity. Every equation comes from some underlying theory or set of observations and, therefore, has some limited range of validity and rests on certain assumptions. Failure to understand this simple point often leads students to apply equations outside their realm of applicability. In that case, even though the math is done correctly, the results will be incorrect.

   In the case of the Nernst equation, things are fairly simple. This equation is derived from a very powerful theory known as equilibrium thermodynamics, and hence it has very wide applicability. As another example, consider enzyme kinetics. The rate at which an enzyme catalyzes a reaction (*v*) is related to the concentration of substrate on which the enzyme works (*[S]*) by an equation called the Michaelis-Menton relationship. The graph of this relationship looks like this:

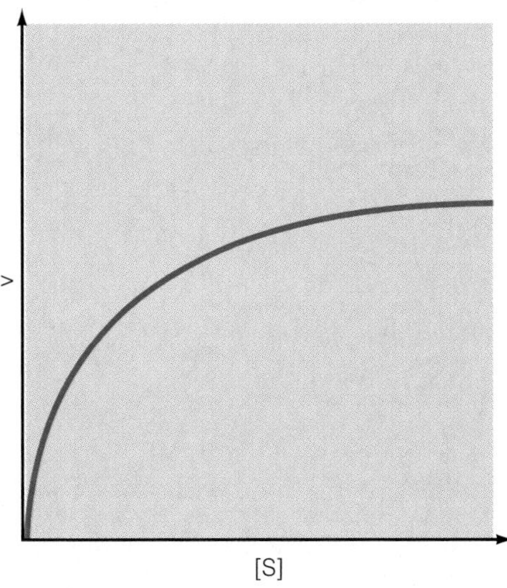

However, some real enzymes, lactate dehydrogenase, for example, do not have this relationship between reaction velocity and substrate concentration. Instead, those quantities relate as follows:

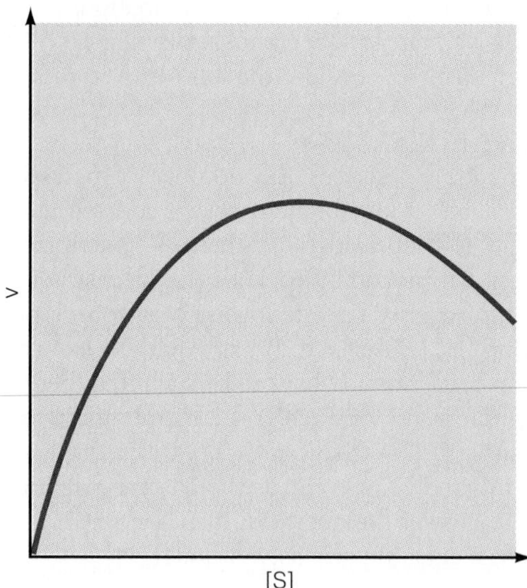

At high substrate concentrations, the enzyme actually is inhibited by too much substrate. The Michaelis-Menton theory, which works well at low [S], is invalid at higher [S] for such enzymes.

An Approach to Problem Solving

The final step is to apply these skills to solve a problem. As an example, calculate the concentration of potassium that must exist inside a cell if $E_{K^+} = -95$ mV and the interstitial fluid has a potassium concentration of 4 mM. Try using the following procedure to solve this problem:

1. Get a clear picture of what is being asked. State it out loud or write it down.
   The question asks for the concentration of potassium in the cell, that is, $[K^+]_{in}$.
2. Determine what you need to know to answer the question:
   To answer this, you need to know E_{K^+} and $[K^+]_{out}$.
3. Determine what information is given. Is it sufficient? Are other relevant facts or relationships not stated in the problem? Specifically, do you need any other equations?
   E_{K^+} is given explicitly in the problem, but you have to translate the words to realize that $[K^+]_{out} = 4$ mM.
4. Manipulate the equation algebraically so that the unknown is on the left-hand side and everything else is on the right-hand side.
   We now solve the Nernst equation for $[K^+]_{in}$. This was done on p. C–3:

$$[K^+]_{in} = [K^+]_{out}\ 10^{(-E_{K^+}/61\ \text{mV})}$$

   Substituting for the values given, we obtain:

$$\begin{aligned}[K^+]_{in} &= 4\ \text{mM}\ 10^{(95\text{mV}/61\ \text{mV})} \\ &= 4\ \text{mM}\ 10^{(1.56)} \\ &= 4\ \text{mM}\ (36.3) \\ &= 145\ \text{mM}\end{aligned}$$

5. Is the answer dimensionally correct? Do not skip this step. It will tell you immediately if something went wrong.
   Yes, the answer is in mM, which are the proper dimension and unit.
6. Does the answer make sense?
   Yes the value is not alarmingly low or high. Also, since the potassium ion is positively charged, if the potassium concentration outside the cell is lower than that inside the cell, then the interior of the cell would have to be negative at passive equilibrium, which it is.

Apply the approach to problem solving we have just outlined to the quantitative questions in the chapters. When you start out, apply the approach formally and carefully. For example, go through each step and write everything out as we have done for the Nernst equation. After a time, you may not need to be so formal. Also, as you progress, you will develop your own style and approach to problem solving. Be prepared to spend some time and patience on some of the problems. Not every answer will be immediately apparent. This situation is normal. If you run into difficulties, relax, return to this appendix for guidance, and work through the problem again carefully. Do not go immediately to the answers if you are having difficulty with a problem. It may ease the frustration, but you will be cheating yourself of a valuable learning experience. Besides, being able to solve challenging problems has its own rewards.

Appendix D

Control, Functions, and Interactions of Immune Response Effectors

Appendix D Control, Functions, and Interactions of Immune Response Effectors

Immune Response Effectors	*Description*	*Controlling Mechanism(s)*	*Function(s)*
Acute-phase proteins[a]	Collection of proteins from the liver	Stimulated by LEM	Exert multitude of wide-ranging effects associated with inflammation, tissue repair, and immune cell activity
Angry macrophages[b]	Large tissue-bound phagocytic cells	Stimulated by MIF	Display superpower phagocytic ability
Antibodies[a]	Special plasma proteins secreted by plasma cells; same as gamma globulins or immunoglobulins	Production and release stimulated by presence of "matching" antigen, which might be any large, complex molecule that triggers an immune attack against itself	Amplify nonspecific immune responses against the antigen that stimulated their production In some instances, physically hinder antigen by neutralization or agglutination
Autoantibodies[a]	Antibody proteins against self-antigens (own body proteins)	Breakdown of mechanisms for tolerance of self-antigens	Attack one or more of body's own tissues
B lymphocytes (B cells) (*see also* **Plasma cells**)[b]	White blood cells	Activated by macrophage-presented antigen; stimulated by helper T cells, IL-1, interferon, B cell growth factor, and helper T cells	Converted into plasma cells, which produce antibodies Secrete lymphokines
Basophils[b]	White blood cells	Binding of allergen to lgE antibodies coating the basophils	Release histamine; involved in immediate allergic reactions
B cell growth factor[a]	Lymphokine secreted by helper T cells	Release stimulated by macrophage-presented antigen and IL-1	Enhances antibody secretion by activated B cell clone
Blocking antibodies[a]	Antibody proteins	Production induced by cancer cells	Protect harmful cancer cells from attack by cytotoxic T cells
Complement system[a]	Group of inactive plasma proteins that can be sequentially activated	Activated by antibody (classical pathway) and by carbohydrate chains on surface of foreign invaders (alternate pathway); several steps stimulated by kinins	Destroys foreign cells by punching holes in their plasma membranes Facilitates every step of inflammatory process (e.g., various components serve as opsonins, stimulate secretion of histamine, or activate kinins)
Cortisol[a]	Steroid hormone secreted by adrenal cortex	Stimulated by stress	Anti-inflammatory action at pharmacological levels and possibly at physiological levels, the latter of which may prevent stress-activated immune responses from overshooting
Chemotaxins[a]	Local chemical mediators	Released from a variety of sources at site of inflammation	Attract phagocytes to site of inflammation

[a]Chemical mediator.
[b]Effector cell.

Immune Response Effectors	*Description*	*Controlling Mechanism(s)*	*Function(s)*
Cytokines[a]	Chemical mediators, other than antibodies secreted by lymphocytes (including B and T cell growth factor and MIF)	Released by presence of macrophage-presented antigen	Enhance all activity of B and T cells Act as chemotaxins Retain macrophages in inflamed area
Cytotoxic T cells[b]	Form of T lymphocytes	Stimulated by interferon, IL-1, and helper T cells; activated by macrophage-presented antigen	Destroy virally infected cells or cancer cells Secrete interferon Mediate delayed hypersensitivity
Endogenous pyrogen (EP)[a]	Chemical mediator released by monocytes/macrophages; identical to LEM and IL-1	Released in presence of microbes	Causes development of fever
Eosinophil chemotactic factor[a]	Chemical mediator in allergic responses	Released by mast cells or basophils activated by allergen	Attracts eosinophils to site of allergen contact
Eosinophils[b]	White blood cells	Stimulated by eosinophil chemotactic factor and presence of internal parasites	Involved in allergic manifestations Destroy parasitic worms
Fibrinogen (inactive); fibrin (active)[a]	Final plasma protein in clotting cascade	Activated by tissue thromboplastin in injured area and chemical mediators released from microbe-stimulated phagocytes	Forms interstitial fluid clots to wall off injured or invaded area
Fibroblasts[b]	Connective tissue cells	Start dividing in area vacated by lost tissue cells	Form scar tissue
Gamma globulin (*see* Antibodies[a]**)**			
Helper T cells[b]	Form of T lymphocyte	Activated by macrophage-presented antigen; stimulated by IL-1	Enhance activity of B cells, cytotoxic T cells, and suppressor T cells through secretion of B cell growth factor and T cell growth factor (IL-2) Activate macrophages
Histamine[a]	Chemical mediator released from mast cells and basophils	Stimulated by chemical mediators released from microbe-activated phagocytes, complement system, and allergens	Important in inflammation Important in immediate allergic reactions Acts on microcirculation to cause localized vasodilation and increased capillary permeability
Immunoglobulin (*see* Antibodies)[a]			
Interferon[a]	Family of proteins	Released from cells invaded by virus	Protects noninvaded cells from viral invasion Increases killing ability of cytotoxic T cells, NK cells, and phagocytes Stimulates production of antibodies Slows cell division and suppresses tumor growth; exerts anticancer effect
Interleukin 1 (IL-1)[a]	Chemical mediator released by macrophages; identical to EP and LEM	Released in presence of microbes	Enhances proliferation and differentiation of both B and T lymphocytes

[a]Chemical mediator.
[b]Effector cell.

Immune Response Effectors	*Description*	*Controlling Mechanism(s)*	*Function(s)*
Interleukin 2 (IL-2)[a]	Lymphokine secreted by helper T cells; identical to T cell growth factor	Released by presence of macrophage-presented antigen; stimulated by IL-1	Augments activity of all T cells
Kallikrein[a]	Chemical mediator released from neutrophils	Released in presence of microbes	Activates kinins
Killer (K) cells[b]	Lymphocytelike cells	Stimulated by antibody on surface of target cell	Lyse plasma membrane of target cell
Kininogen (inactive); kinin (active)[a]	Plasma protein	Stimulated by kallikrein released from neutrophils and by complement component	Stimulates complement system Enhances vasodilation and increases capillary permeability in inflammation Activates nearby pain receptors Acts as powerful chemotaxin
Lactoferrin[a]	Chemical mediator secreted by neutrophils	Stimulated by presence of microbes	Binds iron so it is unavailable for essential use by invading bacteria
Leukocyte endogenous mediator (LEM)[a]	Chemical mediator released by monocytes/macrophages; identical to EP and IL-1	Released in presence of microbes	Decreases plasma iron Stimulates granulopoiesis Increases release of acute-phase proteins
Lymphocytes[b]	White blood cells—two types: B and T lymphocytes	Stimulated by products specific for each type of cell (*see* specific cells)	Responsible for specific immune defense
Macrophage migration inhibition factor (MIF)[a]	Lymphokine secreted by helper T cells	Released by presence of macrophage-presented antigen	Prevents macrophages from migrating out of inflamed area Increases phagocytic power of macrophages (activates angry macrophages)
Macrophages[b]	Large, tissue-bound phagocytic specialists	Stimulated by presence of microbes; enhanced by interferon; activated by helper T cells; stimulated by antibodies on surface of foreign material; attracted by chemotaxins; linked to target cell by opsonin	Important in nonspecific defense; first line of defense Phagocytize foreign material and cellular debris Secrete chemical mediators that exert a variety of effects (*see* Phagocytic secretions) Clear way for tissue repair by removing debris Secrete interferon Process and present antigen to B and T cells, thereby assisting in antibody formation and T cell sensitization
Memory cells[b]	B and T cells produced during initial infection but remaining dormant until subsequent invasion by same antigen	Stimulated by subsequent invasion of same antigen	Expand activated clone Launch swifter, more powerful attack on subsequent exposure
Monocytes[b]	White blood cells	Various factors stimulate their derivatives, the macrophages (*see* Macrophages)	Converted into large, tissue-bound macrophages after leaving the blood
Natural killer (NK) cells[b]	Lymphocytelike cells	Stimulated by interferon	Nonspecifically lyse virus-infected cells and tumor cells on first exposure to them Secrete interferon

[a]Chemical mediator.
[b]Effector cell.

Immune Response Effectors	*Description*	*Controlling Mechanism(s)*	*Function(s)*
Neutrophils[b]	White blood cells	Stimulated by presence of microbes, by antibodies on surface of foreign material, and by circulating antibodies; attracted by chemotaxins; linked to target cell by opsonins	Important in nonspecific defense Important in inflammatory process; highly mobile phagocytic specialists Release chemicals that exert a variety of effects (*see* Phagocytic secretions)
Opsonins[a]	Antibodies and one of activated proteins of complement system	Produced in response to factors that stimulate antibody secretion (*see* Antibodies) and complement activation (*see* Complement system)	Enhance phagocytosis by linking foreign cell to phagocytic cell
Phagocytic secretions (EP/LEM/IL-1 and others)[a]	Chemical mediators released from neutrophils or macrophages	Released in presence of microbes	Directly kill microbes extracellularly Trigger clotting and anticlotting systems Activate kinins Induce development of fever Decrease plasma iron Stimulate granulopoiesis Stimulate secretion of acute-phase proteins Enhance proliferation and differentiation of both B and T cells
Plasma cells[b]	Activated form of B lymphocyte	Specific B cells that match invading antigen proliferate and are converted into antibody-secreting plasma cells; stimulated by B cell growth factor, helper T cells, and interferon	Secrete customized antibodies specific to invading antigen
Plasminogen (inactive); plasmin (active)[a]	Final plasma protein in anticlotting cascade	Activated by chemical mediators released from microbe-stimulated phagocytes	Slowly dissolves interstitial fluid clot formed during inflammatory response
Prostaglandin[a]	Fatty acid–derived local chemical mediator	Released in hypothalamus by EP	Raises set point of hypothalamic thermostat to produce fever
Slow reactive substance of anaphylaxis (SRS-A)[a]	Chemical mediator released from mast cells and basophils in allergic reactions	Released by binding of allergen to lgE antibodies coating mast cells and basophils	Induced profound, prolonged constriction of small airways Increases capillary permeability
Suppressor T cells[b]	Form of T lymphocyte	Stimulated by helper T cells	Limit immune reaction in check-and-balance fashion
T cell growth factor[a]	Lymphokine secreted by helper T cells	Release stimulated by macrophage-presented antigen and IL-1	Augments activity of all T cells
Thymosin[a]	Collection of hormones produced by thymus	Unknown	Enhances proliferation of T cell colonies Enhances immune capabilities of existing T cells
T lymphocytes (T cells) (*see* **Cytotoxic T cells, Helper T cells, and Suppressor T cells)**[b]	White blood cells	Activated by macrophage-presented antigen; stimulated by interferon, IL-1, and helper T cells	Responsible for cell-mediated immunity (kills cells through nonphagocytic means)

[a]Chemical mediator.
[b]Effector cell.

Appendix E

Answers to End-of-Chapter Objective Questions, Quantitative Exercises, and Points to Ponder

CHAPTER 1 Homeostasis: The Foundation of Physiology

OBJECTIVE QUESTIONS

(Questions on p. 14.)

1. e
2. b
3. c
4. T
5. F (Unicellular organisms and cells in a multicellular organism all perform similar basic functions essential for the cell's survival.)
6. T
7. muscle tissue, nervous tissue, epithelial tissue, connective tissue
8. secretion
9. exocrine, endocrine, hormones
10. intrinsic, extrinsic
11. 1.d, 2.g, 3.a, 4.e, 5.b, 6.j, 7.h, 8.i, 9.c, 10.f

POINTS TO PONDER

(Questions on p. 15.)

1. The respiratory system eliminates internally produced CO_2 to the external environment. A decrease in CO_2 in the internal environment brings about a reduction in respiratory activity (that is, slower, shallower breathing) so that CO_2 produced within the body is allowed to accumulate instead of being blown off as rapidly as normal to the external environment. The extra CO_2 retained in the body increases the CO_2 levels in the internal environment to normal.
2. (b) below normal, (c) elevated, (b) be too acidic
3. b
4. immune defense system
5. When a person is engaged in strenuous exercise, the temperature-regulating center in the brain will bring about widening of the blood vessels of the skin. The resultant increased blood flow through the skin will carry the extra heat generated by the contracting muscles to the body surface, where it can be lost to the surrounding environment.
6. *Clinical Consideration* Loss of fluids threatens the maintenance of proper plasma volume and blood pressure. Loss of acidic digestive juices threatens the maintenance of the proper pH in the internal fluid environment. The urinary system will help restore the proper plasma volume and pH by reducing the amount of water and acid eliminated in the urine. The respiratory system will help restore the pH by adjusting the rate of removal of acid-forming CO_2. Adjustments will be made in the circulatory system to help maintain blood pressure despite fluid loss. Increased thirst will encourage increased fluid intake to help restore plasma volume. These compensatory changes in the urinary, respiratory, and circulatory systems, as well as the sensation of thirst, will all be regulated by the two control systems, the nervous and endocrine systems. Furthermore, the endocrine system will make internal adjustments to help maintain the concentration of nutrients in the internal environment even though no new nutrients are being absorbed from the digestive system.

CHAPTER 2 Cellular Physiology

OBJECTIVE QUESTIONS

(Questions on p. 44.)

1. plasma membrane
2. deoxyribonucleic acid (DNA), nucleus
3. organelles, cytosol, cytoskeleton
4. endoplasmic reticulum, Golgi complex
5. oxidative, hydrogen peroxide
6. adenosine triphosphate (ATP)
7. F (Human cells cannot be seen by the unaided eye; they are about ten times smaller than the smallest point visible to the naked eye.)
8. F (Lysosomal enzymes are less potent in the cytosol than in the acidic environment within the lysosome. Furthermore, cells can usually tolerate the limited damage accompanying the inadvertent rupture of a few lysosomes because most cell parts are renewable.)
9. T
10. 1.b, 2.a, 3.b
11. 1.b, 2.c, 3.c, 4.a, 5.b, 6.c, 7.a, 8.c

QUANTITATIVE EXERCISES

(Questions on p. 44.)

1. b
2. 24 moles O_2/day × 6 moles ATP/mole O_2 = 144 moles ATP/day
   144 moles ATP/day × 507g ATP/mole = 73,000g ATP/day
   1,000 g/2.2 lb = 73,000 g/x lb
   1,000 x = 160,600
   x = approximately 160 lb
3. 144 mol/day (7,300 cal/mol) = 1,051,200 cal/day (1,051 kilocal/day)
4. About 2/3 of the water in the body is intracellular. Since a person's mass is about 60% water, for a 150-pound (68 kg) person,

   68 kg(0.6)(2/3) = 27.6 kg

   is the mass of water. Assume that 1 ml of body water weighs 1 gm. Then the total volume in the person's cell is about 27.6 L. The volume of an average cell is

   $$\frac{4}{3}\pi(1 \times 10^{-3}\ \text{cm})^3 \approx 4.9 \times 10^{-9}\ \text{cm}^3$$
   $$= 4.2 \times 10^{-9}\ \text{ml}$$

   So, the number of cells in a 68 kg person is about

   $$27.65\ \frac{\text{L}}{\text{ml}}\left(\frac{1{,}000\ \text{ml}}{1\ \text{L}}\right)\left(\frac{1\ \text{cell}}{4\times10^{-9}\ \text{ml}}\right)$$
   $$= 6.9125 \times 10^{12}\text{cells/ml}$$
5. $150\ \text{mg}\left(\frac{1\ \text{ml}}{0.015\ \text{mg}}\right) = 10{,}000\ \text{ml}\ (10\ \text{L})$

POINTS TO PONDER

(Questions on p. 45.)

1. lysosomes
2. With cyanide poisoning, the cellular activities that depend on ATP expenditure could not continue, such as synthesis of new chemical compounds, membrane transport, and mechanical work. The resultant inability of the heart to pump blood and failure of the respiratory muscles to accomplish breathing would lead to imminent death.
3. catalase
4. ATP is required for muscle contraction. Muscles are able to store limited supplies of nutrient fuel for use in the generation of ATP. During anaerobic exercise, muscles generate ATP from these nutrient stores by means of glycolysis, which yields two molecules of ATP per glucose molecule processed. During aerobic exercise, muscles can generate ATP by means of oxidative phosphorylation, which yields thirty-six molecules of ATP per glucose molecule processed. Because glycolysis inefficiently generates ATP from nutrient fuels, it rapidly depletes the muscle's limited stores of fuel, and ATP can no longer be produced to sustain the muscle's contractile activity. Aerobic exercise, on the other hand, can be sustained for prolonged periods. Not only does oxidative phosphorylation use far less nutrient fuel to generate ATP, but it can be supported by nutrients delivered to the muscle by means of the blood instead of relying on stored fuel in the muscle. Intense anaerobic exercise outpaces the ability to deliver supplies to the muscle by the blood, so the muscle must rely on stored fuel and inefficient glycolysis, thus limiting anaerobic exercise to brief periods of time before energy sources are depleted.
5. skin.
   The mutant keratin weakens the skin cells of patients with epidermolysis bullosa so that the skin blisters in response to even a light touch.
6. ***Clinical Consideration*** Some hereditary forms of male sterility involving nonmotile sperm have been traced to defects in the cytoskeletal components of the sperm's flagella. These same individuals usually also have long histories of recurrent respiratory tract disease because the same types of defects are present in their respiratory cilia, which are unable to clear mucus and inhaled particles from the respiratory system.

CHAPTER 3 The Plasma Membrane and Membrane Potential

OBJECTIVE QUESTIONS

(Questions on p. 78.)

1. T
2. F (The hydrophobic regions correspond to the light space; the hydrophilic regions correspond to the two dark layers.)
3. T
4. F (Only 20% of the membrane potential is *directly* generated by the $Na^+ - K^+$ pump. The other 80% is caused by the passive diffusion of K^+ and Na^+ down concentration gradients. The $Na^+ - K^+$ pump *indirectly* contributes to membrane potential by maintaining these concentration gradients.)
5. pinocytosis, phagocytosis, endocytosis
6. opening or closing specific membrane channels, activating an intracellular second messenger
7. G protein
8. negative, positive
9. 1.b, 2.a, 3.b, 4.a, 5.c, 6.b, 7.a, 8.b
10. 1.a, 2.a, 3.b, 4.a, 5.b, 6.a, 7.b
11. 1.c, 2.b, 3.a, 4.a, 5.c, 6.b, 7.c, 8.a, 9.b

QUANTITATIVE EXERCISES

(Questions on p. 79.)

1. $E = \frac{61\ \text{mV}}{z} \log \frac{c_o}{c_i}$

   a. $\frac{61\text{mV}}{2} \log \frac{1\times10^{-3}}{100\times10^{-3}} = +\ 122\ \text{mV}$

   b. $\frac{61\text{mV}}{-1} \log \frac{110\times10^{-3}}{10\times10^{-3}} = -\ 63.5\ \text{mV}$

2. $I_x = G_x(V_m - E_x)$

$$E_{Na^+} = 61\text{mV} \log \frac{145\text{ mM}}{15\text{mM}} = 60.1\text{ mV}$$

a. $= 1\text{ nS } (-70\text{ mV} - 60.1\text{ mV})$
$= 1\text{ nS } (-130\text{ mV})$
$= -130\text{ pA }$ (A = amperes)

b. Entering

c. With concentration gradient; with electrical gradient

3. $$V_m = \frac{G_{Na^+}}{G_T} E_{Na^+} + \frac{G_{K^+}}{G_T} E_{K^+}$$

a. $G_T = 1\text{ nS} + 5.3\text{ nS} = 6.3\text{ nS}$;
$$V_m = \frac{1}{6.3} 59.1\text{ mV} + \frac{5.3}{6.3}(-94.4\text{ mV})$$
$= 9.4\text{ mV} - 79.4\text{ mV} = -70\text{ mV}$

b. $E_{K^+} = 0$ mV; $V_m = 9.4$ mV; i.e., large depolarization

POINTS TO PONDER

(Questions on p. 80.)

1. c.
As Na^+ moves from side 1 to side 2 down its concentration gradient, Cl^- remains on side 1, unable to permeate the membrane. The resultant separation of charges produces a membrane potential, negative on side 1 due to unbalanced chloride ions and positive on side 2 due to unbalanced sodium ions. Sodium does not continue to move to side 2 until its concentration gradient is dissipated because of the development of an opposing electrical gradient.
2. more positive.
Because the electrochemical gradient for Na^+ is inward, the membrane potential would become more positive as a result of an increased influx of Na^+ into the cell if the membrane were more permeable to Na^+ than to K^+. (Indeed, this is what happens during the rising phase of an action potential once threshold potential is reached—see chapter 4).
3. d. active transport.
Leveling off of the curve designates saturation of a carrier molecule, so carrier-mediated transport is involved. The graph indicates that active transport is being utilized instead of facilitated diffusion because the concentration of the substance in the intracellular fluid is greater than the concentration in the extracellular fluid at all points until after the transport maximum is reached. Thus, the substance is being moved *against* a concentration gradient, so active transport must be the method of transport being used.
4. vesicular transport.
The maternal antibodies in the infant's digestive tract lumen are taken up by the intestinal cells by pinocytosis and are extruded on the opposite side of the cell into the interstitial fluid by exocytosis. The antibodies are picked up from the intestinal interstitial fluid by the blood supply to the region.
5. accelerate.
During an action potential, Na^+ enters and K^+ leaves the cell. Repeated action potentials would eventually "run down" the Na^+ and K^+ concentration gradients were it not for the Na^+-K^+ pump returning the Na^+ that entered back to the outside and the K^+ that left back to the inside. Indeed, the rate of pump activity is accelerated by the increase in both ICF Na^+ and ECF K^+ concentrations that occurs as a result of action potential activity, thus hastening the restoration of the concentration gradients.
6. ***Clinical Consideration*** As Cl^- is secreted by the intestinal cells into the intestinal tract lumen, Na^+ follows passively along the established electrical gradient. Water passively accompanies this salt (Na^+ and Cl^-) secretion by osmosis. The toxin produced by the cholera pathogen prevents the normal inactivation of cAMP in intestinal cells so that cAMP levels rise. An increase in cAMP opens the Cl^- channels in the luminal membranes of these cells. Increased secretion of Cl^- and the subsequent passively induced secretion of Na^+ and water are responsible for the severe diarrhea that characterizes cholera.

CHAPTER 4 Neuronal Physiology

OBJECTIVE QUESTIONS

(Questions on p. 106.)

1. T
2. F (The membrane is returned to resting potential after it reaches the peak of an action potential as a result of K^+ efflux. The Na^+-K^+ pump gradually restores the original concentration gradients following an action potential by pumping out the Na^+ that entered and pumping in the K^+ that left during the action potential.)
3. F (Even though K^+ leaves the cell during an action potential, there is still much more K^+ inside than outside the cell following the action potential because only an extremely small percentage of the total intracellular K^+ leaves.)
4. F (Presynaptic neurons can either excite or inhibit postsynaptic neurons by means of generating EPSPs or IPSPs, respectively. However, postsynaptic neurons cannot generate EPSPs or IPSPs in presynaptic neurons.)
5. refractory period
6. axon hillock
7. synapse
8. temporal summation
9. spatial summation
10. convergence, divergence
11. 1.b, 2.a, 3.a, 4.b, 5.b, 6.a

QUANTITATIVE EXERCISES

(Questions on p. 107.)

1. **a.** 0.6 m (1 sec/0.7 m) = 0.8571 sec
b. 0.6 m (1 sec/120 m) = 0.005 sec
c. unmyelinated: 0.8591 sec
myelinated: 0.007 sec

d. unmyelinated: 0.8621 sec
myelinated: 0.01 sec

2. Total conduction time for the single axon is 1/60 sec. Let v m/sec be the unknown conduction velocity for the three neurons. Our equation for the total conduction time then is:

$$\frac{1}{60}\text{ sec} = \left(\frac{1}{v} \times 1\text{ m}\right) + 0.002\text{ sec}$$

Solving for *v*, we obtain

v m/sec = 1 m/(1/60 sec. − 0.002 sec) = 68.18 m/sec

3. $$25 \times 10^{-3}\text{ V}\left[\frac{3.3\ \mu\text{S/cm}^2(240\ \mu\text{S /cm}^2)}{(3.3 + 240)\ \mu\text{S/cm}^2}\right]\log\frac{240(145)}{3.3(4)}$$

$$= 25 \times 10^{-3}(11.1361)\text{V} \times \mu\text{S/cm}^2 = 0.2784\mu\text{A/cm}^2$$

POINTS TO PONDER
(Questions on p. 107.)

1. c.
The action potentials would stop as they met in the middle. As the two action potentials moving toward each other both reached the middle of the axon, the two adjacent patches of membrane in the middle would be in a refractory period so further propagation of either action potential would be impossible.
2. A subthreshold stimulus would transiently depolarize the membrane but not sufficiently to bring the membrane to threshold, so no action potential would occur. Because a threshold stimulus would bring the membrane to threshold, an action potential would occur. An action potential of the same magnitude and duration would occur in response to a suprathreshold stimulus as to a threshold stimulus. Because of the all-or-none law, a stimulus larger than that necessary to bring the membrane to threshold would not produce a larger action potential. (The magnitude of the stimulus is coded in the *frequency* of action potentials generated in the neuron, not the *size* of the action potentials.)
3. The hand could be pulled away from the hot stove by flexion of the elbow accomplished by summation of EPSPs at the cell bodies of the neurons controlling the biceps muscle, thus bringing these neurons to threshold. The subsequent action potentials generated in these neurons would stimulate contraction of the biceps. Simultaneous contraction of the triceps muscle, which would oppose the desired flexion of the elbow, could be prevented by generation of IPSPs at the cell bodies of the neurons controlling this muscle. These IPSPs would keep the triceps neurons from reaching threshold and firing so that the triceps would not be stimulated to contract.

   The arm could deliberately be extended despite a painful finger prick by voluntarily generating EPSPs to override the reflex IPSPs at the neuronal cell bodies controlling the triceps while simultaneously generating IPSPs to override the reflex EPSPs at the neuronal cell bodies controlling the biceps.
4. Treatment for Parkinson's disease is aimed toward restoring dopamine activity in the basal nuclei. However, this treatment may lead to excessive dopamine activity in otherwise normal areas of the brain that also utilize dopamine as a neurotransmitter. Excessive dopamine activity in a particular region of the brain (the limbic system) is believed to be among the causes of schizophrenia. Therefore, symptoms of schizophrenia sometimes occur as a side effect during treatment for Parkinson's disease.
5. An EPSP, being a graded potential, spreads decrementally from its site of initiation in the postsynaptic neuron. If presynaptic neuron A (near the axon hillock of the postsynaptic cell) and presynaptic neuron B (on the opposite side of the postsynaptic cell body) both initiate EPSPs of the same magnitude and frequency, the EPSPs from A will be of greater strength when they reach the axon hillock than will the EPSPs from B. An EPSP from B will decrease more in magnitude as it travels farther before reaching the axon hillock, the region of lowest threshold and thus the site of action potential initiation. Temporal summation of the larger EPSPs from A may bring the axon hillock to threshold and initiate an action potential in the postsynaptic neuron, whereas temporal summation of the weaker EPSPs from B at the axon hillock may not be sufficient to bring this region to threshold. Thus, the proximity of a presynaptic neuron to the axon hillock can bias its influence on the postsynaptic cell.
6. ***Clinical Consideration*** Initiation and propagation of action potentials would not occur in nerve fibers acted upon by local anesthetic because blockage of Na^+ channels by the local anesthetic would prevent the massive opening of voltage-gated Na^+ channels at threshold potential. As a result, pain impulses (action potentials in nerve fibers that carry pain signals) would not be initiated and propagated to the brain and reach the level of conscious awareness.

CHAPTER *5* The Central Nervous System

OBJECTIVE QUESTIONS
(Questions on p. 153.)

1. F (The major function of CSF is to serve as a shock-absorbing fluid to prevent brain damage during sudden, jarring movements of the head.)
2. F (The brain cannot produce ATP under anaerobic conditions.)
3. F (Damage to the left hemisphere brings about paralysis and loss of sensation on the right side of the body because of fiber crossover.)
4. T
5. F (The left hemisphere specializes in verbal and analytical skills whereas the right side excels in artistic and musical ability.)
6. F (The brain has considerable plasticity; the organizational pattern of sensory and motor regions of the cortex is subject to constant subtle modifications based on use-dependent competition for cortical space.)
7. habituation

8. consolidation
9. dorsal, ventral
10. receptor, afferent pathway, integrating center, efferent pathway, effector
11. 1.a, 2.c, 3.a and b, 4.b, 5.a, 6.c, 7.c
12. 1.c, 2.e, 3.a, 4.b, 5.f, 6.d
13. 1.d, 2.c, 3.f, 4.e, 5.a, 6.b

POINTS TO PONDER

(Questions on p. 154.)

1. Only the left hemisphere has language ability. When sharing of information between the two hemispheres is prevented as a result of severance of the corpus callosum, visual information presented only to the right hemisphere cannot be verbally identified by the left hemisphere, because the left hemisphere is unaware of the information. However, the information can be recognized by nonverbal means, of which the right hemisphere is capable.
2. Insulin excess drives too much glucose into insulin-dependent cells so that the blood glucose falls below normal and insufficient glucose is delivered to the non-insulin-dependent brain. Therefore, the brain, which depends on glucose as its energy source, does not receive adequate nourishment.
3. c.
   A severe blow to the back of the head is most likely to traumatize the visual cortex in the occipital lobe.
4. Salivation when seeing or smelling food, striking the appropriate letter on the keyboard when typing, and many of the actions involved in driving a car are conditioned reflexes. You undoubtedly will have many other examples.
5. Strokes occur when a portion of the brain is deprived of its vital O_2 and glucose supply because the cerebral blood vessel supplying the area either is blocked by a clot or has ruptured. Although a clot-dissolving drug could be helpful in restoring blood flow through a cerebral vessel blocked by a clot, such a drug would be detrimental in the case of a ruptured cerebral vessel sealed by a clot. Dissolution of a clot sealing a ruptured vessel would lead to renewed hemorrhage through the vessel and exacerbation of the problem.
6. ***Clinical Consideration*** The deficits following the stroke—numbness and partial paralysis on the upper right side of the body and inability to speak—are indicative of damage to the left somatosensory cortex and left primary motor cortex in the regions devoted to the upper part of the body plus Broca's area.

CHAPTER *6* The Peripheral Nervous System: Afferent Division; Special Senses

OBJECTIVE QUESTIONS

(Questions on p. 200.)

1. transduction
2. adequate stimulus
3. T
4. F (Only afferent information that reaches conscious levels of the brain is considered to be sensory information.)
5. T
6. T
7. T
8. F (Displacement of the round window just dissipates pressure. Displacement of the basilar membrane and the resultant activation of the organ of Corti's hair cells generate neural impulses that are perceived as sound sensations.)
9. T
10. F (Each taste receptor responds in varying degrees to all four primary tastes.)
11. F (Adaptation to odors involves some sort of CNS adaptation, not olfactory receptor adaptation.)
12. 1.a, 2.b, 3.c, 4.c, 5.c, 6.a, 7.b, 8.b
13. 1.f, 2.h, 3.l, 4.d, 5.i, 6.e, 7.b, 8.j, 9.a, 10.g, 11.c, 12.k

QUANTITATIVE EXERCISES

(Questions on p. 201.)

1. The slow pain pathway takes about (1.3 m) (1 sec/12 m) = 0.1083 sec. The fast pathway takes (1.3 m) (1 sec/30 m) = 0.0433 sec. The difference is 0.1083 sec −0.0433 sec = 0.065 sec = 65 msec.
2. **a.** The amount of light entering the eye is proportional, approximately, to the area of the open pupil. Recall that the area of a circle is πr^2. Let r be the pupil radius and A_1 be the original pupil area. Halving the diameter also halves the radius, so the new pupil area is

   $$\pi\left(\frac{1}{2}r^2\right) = \frac{1}{4}\pi r^2 = \frac{1}{4}A_1$$

   Therefore, the amount of light allowed into the eye is a quarter of what it was originally.

   b. The area of a rectangle is *hw*, where *h* is the height and *w* the width. Halving either dimension halves the area, and hence the amount of light allowed into the eye.

   c. The cat's pupil can be considered more precise. Think about the coarse and fine adjustments on a microscope. Fine adjustment translates rotations of the knob into much smaller movement of the stage than does coarse adjustment.
3. **a.** Solve the following for I:

   $$\beta = (10\text{ dB})\log_{10}(I/I_0)$$
   $$I = I_0 10^{B/10}\text{W/m}^2$$

   Therefore,

   $$I_i = 10^{-12}(10^{20/10}) = 10^{-12}(10^2) = 10^{-10}\text{W/m}^2;$$
   $$I_{ii} = 10^{-12}(10^{70/10}) = 10^{-12}(10^7) = 10^{-5}\text{W/m}^2;$$
   $$I_{iii} = 10^{-12}(10^{120/10}) = 10^{-12}(10^{12}) = 1\text{W/m}^2;$$
   $$I_{iv} = 10^{-12}(10^{170/10}) = 10^{-12}(10^{17}) = 10^5\text{W/m}^2$$

b. Because of the logarithm in the definition of decibel, the sound intensity increases exponentially with respect to sound level. This fact should be clear from the definition of dB solved for I. This result implies that the human ear performs well throughout an enormous range of sound intensities.

POINTS TO PONDER

(Questions on p. 202.)

1. Pain is a conscious warning that tissue damage is occurring or about to occur. A patient unable to feel pain because of a nerve disorder does not consciously take measures to withdraw from painful stimuli and thus prevent more serious tissue damage.
2. Pupillary dilation (mydriasis) can be deliberately induced by ophthalmic instillation of either an adrenergic drug (such as epinephrine or related compound) or a cholinergic blocking drug (such as atropine or related compounds). Adrenergic drugs produce mydriasis by causing contraction of the sympathetically supplied radial (dilator) muscle of the iris. Cholinergic blocking drugs cause pupillary dilation by blocking parasympathetic activity to the circular (constrictor) muscle of the iris so that action of the adrenergically controlled radial muscle of the iris is unopposed.
3. The defect would be in the left optic tract or optic radiation.
4. Fluid accumulation in the middle ear in accompaniment with middle ear infections impedes the normal movement of the tympanic membrane, ossicles, and oval window in response to sound. All of these structures vibrate less vigorously in the presence of fluid, thereby causing temporary hearing impairment. Chronic fluid accumulation in the middle ear is sometimes relieved by surgical implantation of drainage tubes in the eardrum. Hearing is restored to normal as the fluid drains to the exterior. Usually, the tubes "fall out" as the eardrum heals and pushes out the foreign object.
5. The sense of smell is reduced when you have a cold, even though the cold virus does not directly adversely affect the olfactory receptor cells, because odorants are not able to reach the receptor cells as readily when the mucous membranes lining the nasal passageways are swollen and excess mucus is present.
6. ***Clinical Consideration*** Syncope most frequently occurs as a result of inadequate delivery of blood carrying sufficient oxygen and glucose supplies to the brain. Possible causes include circulatory disorders such as impaired pumping of the heart or low blood pressure; respiratory disorders resulting in poorly oxygenated blood; anemia, in which the oxygen-carrying capacity of the blood is reduced; or low blood glucose resulting from improper endocrine management of blood glucose levels. Vertigo, on the other hand, typically results from a dysfunction of the vestibular apparatus, arising, for example, from viral infection or trauma, or abnormal neural processing of vestibular information, as, for example, with a brain tumor.

CHAPTER 7 The Peripheral Nervous System: Efferent Division

OBJECTIVE QUESTIONS

(Questions on p. 219.)

1. T
2. F (Action potentials are transmitted on a one-to-one basis at neuromuscular junctions, but one action potential in a presynaptic neuron does not bring about one action potential in a postsynaptic neuron at a synapse. A postsynaptic neuron is brought to threshold and has an action potential only upon summation of EPSPs.)
3. c
4. c
5. sympathetic, parasympathetic
6. adrenal medulla
7. 1.a, 2.b, 3.a, 4.b, 5.a, 6.a, 7.b
8. 1.b, 2.b, 3.a, 4.a, 5.b, 6.b, 7.a

QUANTITATIVE EXERCISES

(Questions on p. 219.)

1. $$t = \frac{x^2}{2D} = \frac{(200\text{ nm})^2}{2\times10^{-5}\text{ cm}^2/\text{sec}}$$
$$= \frac{4\times10^{-14}\text{m}^2\cdot\text{sec}}{2\times10^{-5}\text{ cm}^2}\left(\frac{10^4\text{ cm}^2}{\text{m}^2}\right) = 20\ \mu\text{sec}$$

POINTS TO PONDER

(Questions on p. 220.)

1. By promoting arteriolar constriction, epinephrine administered in conjunction with local anesthetics reduces blood flow to the region and thus helps the anesthetic stay in the region instead of being carried away by the blood.
2. No. Atropine blocks the effect of acetylcholine at muscarinic receptors but does not affect nicotinic receptors. Nicotinic receptors are present on the motor end plates of skeletal muscle fibers.
3. The voluntarily controlled external urethral sphincter is composed of skeletal muscle and supplied by the somatic nervous system.
4. By interfering with normal acetylcholine activity at the neuromuscular junction, α bungarotoxin leads to skeletal muscle paralysis, with death ultimately occurring as a result of an inability to contract the diaphragm and breathe.
5. If the motor neurons that control the respiratory muscles, especially the diaphragm, are destroyed by poliovirus or amyotropic lateral sclerosis, the person is unable to breathe and dies (unless breathing is assisted by artificial means).
6. ***Clinical Consideration*** Drugs that block β_1 receptors are useful for prolonged treatment of angina pectoris because they interfere with sympathetic stimulation of the heart during exercise or emotionally stressful situations. By preventing increased cardiac metabolism and

thus an increased need for oxygen delivery to the cardiac muscle during these situations, beta blockers can reduce the frequency and severity of angina attacks.

CHAPTER 8 Muscle Physiology

OBJECTIVE QUESTIONS

(Questions on p. 262.)

1. F (The contractile response lasts about 100 times longer than the action potential.)
2. F (The velocity of shortening depends on the magnitude of the load as well as on the ATPase activity of its fibers.)
3. F (When a muscle is maximally stretched, the thin filaments are completely pulled out from between the thick filaments so that no cross-bridge activity and consequently no contraction can occur. Maximal tension is achieved when the muscle is at its optimal length and the thin filaments optimally overlap the thick filaments.)
4. T
5. F (Slow-wave potentials initiate action potentials only when the automatic depolarizing swing in potential reaches threshold. Whether threshold is reached depends on the starting point of the membrane potential at the onset of its depolarizing swing.)
6. T
7. concentric, eccentric
8. alpha, gamma
9. denervation atrophy, disuse atrophy
10. a, b, e
11. b
12. 1.f, 2.d, 3.c, 4.e, 5.b, 6.g, 7.a
13. 1.a, 2.a, 3.a, 4.b, 5.b, 6.b

QUANTITATIVE EXERCISES

(Questions on p. 263.)

1. **a.** For the weekend athlete, the lever ratio is 70 cm/9 cm. So, the velocity at the end of the arm is 2.6 m/sec (70/9) = 20.2 m/sec (about 45 mph).
   b. For the professional ballplayer, the lever ratio is 90 cm/9 cm. So
   $10x = 85$ mph
   $x = 8.5$ mi/hr(1,609 m/mi) (1 hr/3,600 sec)
   $= 3.8$ m/sec.
2. The force-velocity curve is as follows:

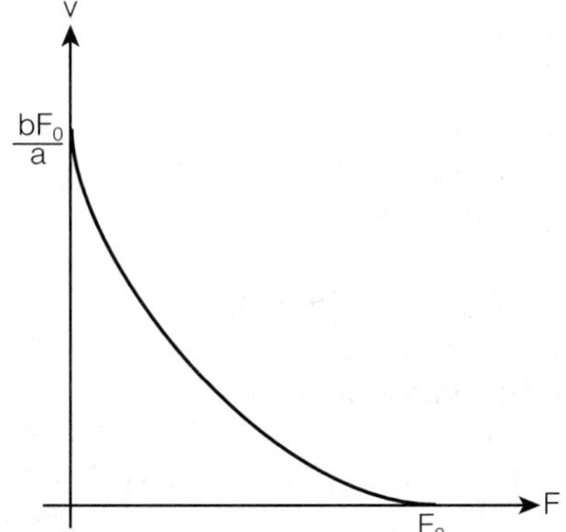

   a. The shape of the curve indicates that it takes time to develop force, and that the greater the force developed, the more time is needed.
   b. The maximum velocity will not change when F_0 is increased, but the muscle is able to lift heavier loads, or to generate more force. The maximum load will not increase when the cross-bridge cycling rate increases, but the muscle will be able to lift lighter loads faster. If the muscle increases in size, b increases, and the entire curve shifts up with respect to the v axis.

POINTS TO PONDER

(Questions on p. 264.)

1. By placing increased demands on the heart to sustain increased delivery of O_2 and nutrients to working skeletal muscles, regular aerobic exercise induces changes in cardiac muscle that enable it to use O_2 more efficiently, such as increasing the number of capillaries supplying blood to the heart muscle. Intense exercise of short duration, such as weight training, in contrast, does not induce cardiac efficiency. Because this type of exercise relies on anaerobic glycolysis for ATP formation, no demands are placed on the heart for increased delivery of blood to the working muscles.
2. The power arm of the lever is 4 cm, and the load arm is 28 cm for a lever ratio of 1:7 (4 cm:28 cm). Thus, to lift an 8 kg stack of books with one hand, the child must generate an upward applied force in the biceps muscle of 56 kg. (With a lever ratio of 1:7, the muscle must exert seven times the force of the load; 7×8 kg $= 56$ kg.)
3. The length of the thin filaments is represented by the distance between a Z line and the edge of the adjacent H zone. This distance remains the same in a relaxed and contracted myofibril, leading to the conclusion that the thin filaments do not change in length during muscle contraction.
4. Regular bouts of anaerobic, short-duration, high-intensity resistance training would be recommended for competitive downhill skiing. By promoting hypertrophy of the fast glycolytic fibers, such exercise better adapts the muscles to activities that require intense strength for brief periods, such as a swift, powerful descent downhill. In contrast, regular aerobic exercise would be more beneficial for competitive cross-country skiers. Aerobic exercise induces metabolic changes within the oxidative fibers that enable the muscles to use O_2 more efficiently. These changes, which include an increase in mitochondria and capillaries within the oxidative fibers, adapt the muscles to better endure the prolonged activity of cross-country skiing without fatiguing.
5. Botulinum toxin blocks the release of acetylcholine from the terminal button in response to an action potential in the motor neuron. Dystonias are believed to occur as a result of an abnormality in the basal nuclei, leading to excessive release of acetylcholine from motor neuron

terminals at the affected muscles. Injection of minute doses of botulinum toxin in the affected area can be used to reduce acetylcholine activity to normal levels so that the excessive, disruptive muscle-contracting activity can be curbed. Because the effect is temporary, patients require periodic injections. This treatment provides relief from the symptoms but does not address the underlying central nervous system disorder.

6. ***Clinical Consideration*** The muscles in the immobilized leg have undergone disuse atrophy. The physician or physical therapist can prescribe regular resistance-type exercises that specifically use the atrophied muscles to help restore them to their normal size.

CHAPTER 9 Cardiac Physiology

OBJECTIVE QUESTIONS

(Questions on p. 303.)

1. intercalated discs, desmosomes, gap junctions
2. bradycardia, tachycardia
3. adenosine
4. F (The left ventricle is a stronger pump because it pumps the same quantity of blood as the right ventricle does but must pump into the high-pressure, high-resistance systemic circulation whereas the right ventricle pumps into the low-pressure, low-resistance pulmonary circulation.)
5. F (The heart is located approximately midline within the thoracic cavity.)
6. F (The only point of electrical contact between the atria and ventricles is the AV node.)
7. T
8. d
9. d
10. e
11. less than, greater than, less than, greater than, less than
12. AV, systole, semilunar, diastole
13. 1.e, 2.a, 3.d, 4.b, 5.f, 6.c, 7.g
14. 1.c, 2.a, 3.b

QUANTITATIVE EXERCISES

(Questions on p. 304.)

1. CO = HR × SV
   35 L/min = HR × 0.07L
   HR = (35 L/min)/(0.07 L) = 500 beats/min

This rate is not physiologically possible.

2. ESV = EDV − SV
   = 125 ml − 85 ml
   = 40 ml

POINTS TO PONDER

(Questions on p. 304.)

1. Because, at a given heart rate, the interval between a premature beat and the next normal beat is longer than the interval between two normal beats, the heart fills for a longer period of time following a premature beat before the next period of contraction and emptying begins. Because of the longer filling time, the end-diastolic volume is larger, and, according to the Frank-Starling law of the heart, the subsequent stroke volume will also be correspondingly larger.
2. Trained athletes' hearts are stronger and can pump blood more efficiently so that the resting stroke volume is larger than in an untrained person. For example, if the resting stroke volume of a strong-hearted athlete is 100 ml, a resting heart rate of only 50 beats/minute produces a normal resting cardiac output of 5,000 ml/minute. An untrained individual with a resting stroke volume of 70 ml, in contrast, must have a heart rate of about 70 beats/minute to produce a comparable resting cardiac output.
3. The direction of flow through a patent ductus arteriosus is the reverse of the flow that occurs through this vascular connection during fetal life. With a patent ductus arteriosus, some of the blood present in the aorta is shunted into the pulmonary artery because, after birth, the aortic pressure is greater than the pulmonary artery pressure. This abnormal blood flow produces a "machinery murmur," which lasts throughout the cardiac cycle but is more intense during systole and less intense during diastole. Thus, the murmur waxes and wanes with each beat of the heart, sounding somewhat like a washing machine as the agitator rotates back and forth. The murmur is present throughout the cardiac cycle because a pressure differential between the aorta and pulmonary artery is present during both systole and diastole. The murmur is more intense during systole because more blood is diverted through the patent ductus arteriosus as a result of the greater pressure differential between the aorta and pulmonary artery during ventricular systole than during ventricular diastole. Typically, the systolic aortic pressure is 120 mm Hg, and the systolic pulmonary arterial pressure is 24 mm Hg, for a pressure differential of 96 mm Hg. By contrast, the diastolic aortic pressure is normally 80 mm Hg, and the diastolic pulmonary arterial pressure is 8 mm Hg, for a pressure differential of 72 mm Hg.
4. A transplanted heart that does not have any innervation adjusts the cardiac output to meet the body's changing needs by means of both intrinsic control (the Frank-Starling mechanism) and extrinsic hormonal influences, such as the effect of epinephrine on the rate and strength of cardiac contraction.
5. In left bundle-branch block, the right ventricle becomes completely depolarized more rapidly than the left ventricle. As a result, the right ventricle contracts before the left ventricle, and the right AV valve is forced closed prior to closure of the left AV valve. Because the two AV valves do not close in unison, the first heart sound is "split"; that is, two distinct sounds in close succession can be detected as closure of the left valve lags behind closure of the right valve.
6. ***Clinical Consideration*** The most likely diagnosis is atrial fibrillation. This condition is characterized by

rapid, irregular, uncoordinated depolarizations of the atria. Many of these depolarizations reach the AV node at a time when it is not in its refractory period, thus bringing about frequent ventricular depolarizations and a rapid heartbeat. However, since impulses reach the AV node erratically, the ventricular rhythm and thus the heartbeat are also very irregular as well as being rapid.

Ventricular filling is only slightly reduced despite the fact that the fibrillating atria are unable to pump blood because most ventricular filling occurs during diastole prior to atrial contraction. Because of the erratic heartbeat, variable lengths of time are available between ventricular beats for ventricular filling. However, the majority of ventricular filling occurs early in ventricular diastole after the AV valves first open, so even though the filling period may be shortened, the extent of filling may be near normal. Only when the ventricular filling period is very short is ventricular filling substantially reduced.

Cardiac output, which depends on stroke volume and heart rate, usually is not seriously impaired with atrial fibrillation. Because ventricular filling is only slightly reduced during most cardiac cycles, stroke volume, as determined by the Frank-Starling mechanism, is likewise only slightly reduced. Only when the ventricular filling period is very short and the cardiac muscle fibers are operating on the lower end of their length-tension curve is the resultant ventricular contraction weak. When the ventricular contraction becomes too weak, the ventricles eject a small or no stroke volume. During most cardiac cycles, however, the slight reduction in stroke volume is often offset by the increased heart rate, so that cardiac output is usually near normal. Furthermore, if the mean arterial blood pressure falls because the cardiac output does decrease, increased sympathetic stimulation of the heart brought about by the baroreceptor reflex helps restore cardiac output to normal by shifting the Frank-Starling curve to the left.

On those cycles when ventricular contractions are too weak to eject enough blood to produce a palpable wrist pulse, if the heart rate is determined directly, either by the apex beat or via the ECG, and the pulse rate is taken concurrently at the wrist, the heart rate will exceed the pulse rate, producing a pulse deficit.

CHAPTER *10* The Blood Vessels and Blood Pressure

OBJECTIVE QUESTIONS

(Questions on p. 349.)

1. T
2. F (Blood flow is continuous through the capillaries. The driving force for continued blood flow to the tissues during cardiac diastole is provided by the elastic recoil of the arterial walls that had been stretched by blood during cardiac systole.)
3. T
4. T
5. F (Capillaries vary in pore size.)
6. T
7. a, c, d, e, f
8. 1.b, 2.a, 3.b, 4.a, 5.a, 6.a, 7.b, 8.a, 9.b, 10.a, 11.b, 12.a, 13.a
9. 1.a, 2.a, 3.b, 4.a, 5.b, 6.a

QUANTITATIVE EXERCISES

(Questions on p. 350.)

1. (120 mm Hg)/(30 L/min) = 4 PRU
2. **a.** 90 mm Hg + (180 mm Hg − 90 mm Hg)/3 = 120 mm Hg

   b. Since the other forces acting across the capillary wall, such as plasma colloid osmotic pressure, typically do not change with age, one would suspect fluid loss from the capillaries into the tissues as a result of the increase in capillary blood pressure.
3. systemic: (95 mm Hg)/(19 PRU) = 95 mm Hg/(19 mm Hg/L/min) = 5 L/min
   pulmonary: (20 mm Hg)/(4 PRU) = 5 L/min
4. **e.** all of the above

POINTS TO PONDER

(Questions on p. 351.)

1. An elastic support stocking increases external pressure on the remaining veins in the limb to produce a favorable pressure gradient that promotes venous return to the heart and minimizes swelling that would result from fluid retention in the extremity.
2. **a.** 125 mm Hg

   b. 77 mm Hg

   c. 48 mm Hg; (125 mm Hg − 77 mm = 48 mm Hg)

   d. 93 mm Hg; [77 + ⅓(48) = 77 + 16 = 93 mm Hg]

   e. No; no blood would be able to get through the brachial artery, so no sound would be heard.

   f. Yes; blood would flow through the brachial artery when the arterial pressure was between 118 and 125 mm Hg and would not flow through when the arterial pressure fell below 118 mm Hg. The turbulence created by this intermittent blood flow would produce sounds.

   g. No; blood would flow continuously through the brachial artery in smooth, laminar fashion, so no sound would be heard.
3. The classmate has apparently fainted because of insufficient blood flow to the brain as a result of pooling of blood in the lower extremities brought about by standing still for a prolonged time during the laboratory experiment. When the person faints and assumes a horizontal position, the pooled blood will quickly be returned to the heart, improving cardiac output and blood flow to his brain. Trying to get the person up would be counterproductive, so the classmate trying to get him up should be advised to let him remain lying down until he recovers on his own.
4. The drug is apparently causing the arteriolar smooth muscle to relax by causing the release of a local vasoac-

tive chemical mediator from the endothelial cells that induces relaxation of the underlying smooth muscle.

5. **a.** Since activation of alpha-adrenergic receptors in vascular smooth muscle brings about vasoconstriction, blockage of alpha-adrenergic receptors reduces vasoconstrictor activity, thereby lowering the total peripheral resistance and arterial blood pressure.
   b. Since drugs that block beta-adrenergic receptors inhibit the pathway that promotes salt and water conservation (the renin-angiotensin-aldosterone system), less salt and water are retained, thus reducing the plasma volume and arterial blood pressure.
   c. Drugs that directly relax arteriolar smooth muscle lower arterial blood pressure by promoting arteriolar vasodilation and reducing total peripheral resistance.
   d. Diuretic drugs reduce the plasma volume, thereby lowering arterial blood pressure, by increasing urinary output. Salt and water that normally would have been retained in the plasma are excreted in the urine.
   e. Since sympathetic activity promotes generalized arteriolar vasoconstriction, thereby increasing total peripheral resistance and arterial blood pressure, drugs that block the release of norepinephrine from sympathetic endings lower blood pressure by preventing this sympathetic vasoconstrictor effect.
   f. Similarly, drugs that act on the brain to reduce sympathetic output lower blood pressure by preventing the effect of sympathetic activity on promoting arteriolar vasoconstriction and the resultant increase in total peripheral resistance and arterial blood pressure.
   g. Drugs that block Ca^{2+} channels reduce the entry of Ca^{2+} into the vascular smooth muscle cells from the ECF in response to excitatory input. Since the level of contractile activity in vascular smooth muscle cells depends on their cytosolic Ca^{2+} concentration, drugs that block Ca^{2+} channels reduce the contractile activity of these cells by reducing Ca^{2+} entry and lowering their cytosolic Ca^{2+} concentration. Total peripheral resistance and, accordingly, arterial blood pressure are decreased as a result of reduced arteriolar contractile activity.
   h. Drugs that interfere with the production of angiotensin II block activation of the hormonal pathway that promotes salt and water conservation (the renin-angiotensin-aldosterone system). As a result, more salt and water are lost in the urine, and less fluid is retained in the plasma. The resultant reduction in plasma volume lowers the arterial blood pressure.
6. ***Clinical Consideration*** The abnormally elevated levels of epinephrine found with a pheochromocytoma bring about secondary hypertension by (1) increasing the heart rate; (2) increasing cardiac contractility, which increases stroke volume; (3) causing venous vasoconstriction, which increases venous return and subsequently stroke volume by means of the Frank-Starling mechanism; and (4) causing arteriolar vasoconstriction, which increases total peripheral resistance. Increased heart rate and stroke volume both lead to increased cardiac output. Increased cardiac output and increased total peripheral resistance both lead to increased arterial blood pressure.

CHAPTER *11* The Blood

OBJECTIVE QUESTIONS

(Questions on p. 371.)

1. T
2. F (Hemoglobin normally carries CO_2 and H^+ from the tissues to the lungs. It also has a high affinity for carbon monoxide.)
3. T
4. T
5. F (White blood cells are in the blood only while in transit from their site of production and storage in the bone marrow or lymphoid tissues to their site of action in the tissues.)
6. T
7. lymphocytes
8. liver
9. d
10. a
11. 1.c, 2.f, 3.b, 4.a, 5.g, 6.d, 7.h, 8.e, 9.f
12. 1.e, 2.c, 3.b, 4.d, 5.g, 6.f, 7.a, 8.h

QUANTITATIVE EXERCISES

(Questions on p. 371.)

1. **a.** (15 g)/(100 ml) = (150 g/L)
   (150 g/L) $\times$ (1 mole/66 $\times$ 10^3 g) = 2.27 mM
   b. (2.27 mM) $\times$ (4 O_2/Hb) = 9.09 mM
   c. (9.09 $\times$ 10^{-3} moles O_2/L blood) $\times$
   (22.4 L O_2/1 mole O_2) = 204 ml O_2/L blood
2. Normal blood contains 5 $\times$ 10^9 rbcs/ml.
   Normal blood volume is 5 L.
   Thus, a normal person has (5 $\times$ 10^9 rbcs/ml) $\times$ (5,000 ml) = 25 $\times$ 10^{12} rbcs.
   The normal hematocrit (Ht) is 45% while the anemic has a Ht of 30%. This represents a loss of 1/3 of the rbcs, that is, 8.3 $\times$ 10^{12} rbcs. If rbcs are produced at a rate of 3 $\times$ 10^6 rbcs/sec, then the time to reestablish the Ht is:
   8.3 $\times$ 10^{12} rbcs/(3 $\times$ 10^6 rbcs/sec) = 2.77 $\times$ 10^6 sec = 32 days
   Thus, it takes about a month to replace a hemorrhagic loss of rbcs of this magnitude.
3. v = 1.5 $\times$ exp(2h); calculate v for h = 0.4 and h = 0.7.
   When h = 0.4, v = 1.5 $\times$ exp (0.8) = 3.3.
   When h = 0.7, v = 1.5 $\times$ exp (1.4) = 6.1.
   6.1/3.3 = 1.84, that is, an 84% increase in viscosity. Since resistance is directly proportional to viscosity, the resistance will also increase by 84%.

POINTS TO PONDER

(Questions on p. 372.)

1. Because blood cells have a short life span, they must continuously be replaced through rapid multiplication

of precursor stem cells. These rapidly multiplying stem cells may be destroyed inadvertently by chemotherapeutic drugs designed to destroy rapidly multiplying cancer cells. It is important, therefore, to closely monitor blood cell counts to ascertain that the number of functional precursor cells is still adequate to maintain satisfactory levels of circulating blood cells.

2. No, you cannot conclude that a person with a hematocrit of 62 definitely has polycythemia. With 62% of the whole-blood sample consisting of erythrocytes (normal being 45%), the number of erythrocytes compared to the plasma volume is definitely elevated. However, the person may have polycythemia, in which the number of erythrocytes is abnormally high, or may be dehydrated, in which case a normal number of erythrocytes is concentrated in a smaller than normal plasma volume.
3. If the genes that direct fetal hemoglobin F synthesis could be reactivated in a patient with sickle-cell anemia, a portion of the abnormal hemoglobin S that causes the erythrocytes to warp into defective sickle-shaped cells would be replaced by "healthy" hemoglobin F, thus sparing a portion of the rbcs from premature rupture. Hemoglobin F would not completely replace hemoglobin S because the gene for synthesis of hemoglobin S would still be active.
4. Most heart attack deaths are attributable to the formation of abnormal clots that prevent normal blood flow. The sought-after chemicals in the "saliva" of blood-sucking creatures are agents that break up or prevent the formation of these abnormal clots.

   Although genetically engineered tissue plasminogen activator (tPA) is already being used as a clot-busting drug (see p. 368), this agent brings about degradation of fibrinogen as well as fibrin. Thus, even though the life-threatening clot in the coronary circulation is dissolved, the fibrinogen supplies in the blood are depleted for up to 24 hours until new fibrinogen is synthesized by the liver. If the patient sustains a ruptured vessel in the interim, insufficient fibrinogen might be available to form a blood-staunching clot. For example, many patients treated with tPA suffer hemorrhagic strokes within 24 hours of treatment due to incomplete sealing of a ruptured cerebral vessel. Therefore, scientists are searching for better alternatives to combat abnormal clot formation by examining the naturally occurring chemicals produced by blood-sucking creatures that permit them to suck a victim's blood without the blood clotting.
5. This question asks for your opinion, so there is no "right" answer. So far, the decision makers in our health care system have felt that the additional financial burden of the more expensive testing of our nation's blood supply for HIV is not warranted.
6. ***Clinical Consideration*** Since the white blood cell count is within the normal range, the patient's pneumonia is most likely not caused by a bacterial infection. Bacterial infections are typically accompanied by an elevated total white blood cell count and an increase in percentage of neutrophils. Therefore, the pneumonia is probably caused by a virus. Because antibiotics are more useful in combating bacterial than viral infections, antibiotics are not likely to be useful in combating this patient's pneumonia.

CHAPTER *12* The Body Defenses

OBJECTIVE QUESTIONS

(Questions on p. 414.)

1. F (The complement system can also be activated by the presence of any foreign invader.)
2. F (Specific immune responses are accomplished by lymphocytes.)
3. F (In tissues capable of regeneration, lost cells are replaced by cell division of surrounding, healthy, organ-specific cells so that perfect repair is accomplished.)
4. F (Active immunity can also be acquired through vaccination.)
5. T
6. membrane-attack complex
7. pus
8. inflammation
9. opsonin
10. cytokines
11. b
12. 1.a, 2.a, 3.b, 4.b, 5.c, 6.c, 7.b, 8.a, 9.b, 10.b, 11.a, 12.b
13. 1.b, 2.a, 3.a, 4.b, 5.a, 6.a, 7.a, 8.b
14. 1.c, 2.d, 3.a, 4.b

QUANTITATIVE EXERCISES

(Questions on p. 415.)

1. NEP = net outward pressure − net inward pressure
   NEP = $(P_C + \pi_{IF}) - (P_{IF} + \pi_P)$
   Note for this problem, $(P_{IF} + \pi_P)$ = (25 mm Hg + 1 mm Hg) = 26 mm Hg, is constant for all cases.

Condition	*Arteriolar end NEP*	*Venular end NEP*	*Average NEP*
Normal (π_{IF} = 0 mm Hg)	(37 + 0) − 26 = +11 mm Hg	(17 + 0) − 26 = −9 mm Hg	(+11 − 9)/2 = +1 mm Hg (outward)
a. Mild edema (π_{IF} = 5 mm Hg)	(37 + 5) − 26 = +16 mm Hg	(17 + 5) − 26 = −4 mm Hg	(+16 − 4)/2 = +6 mm Hg (outward)
b. Extreme edema (π_{IF} = 10 mm Hg)	(37 + 10) − 26= +21 mm Hg	(17 + 10) − 26 = +1 mm Hg	(+21 + 1)/2 = +11 mm Hg (outward)

POINTS TO PONDER

(Questions on p. 415.)

1. See p. 396 for a summary of immune responses to bacterial invasion and p. 399 for a summary of defenses against viral invasion.
2. A vaccine against a particular microbe can be effective only if it induces formation of antibodies and/or activated T cells against a stable antigenic determinant site that is present on all microbes of this type. Because HIV frequently mutates, it has not been possible to produce a vaccine against it. Specific immune responses induced by vaccination against one form of HIV may prove to be ineffective against a slightly modified version of the virus.
3. Failure of the thymus to develop embryonically would lead to an absence of T lymphocytes and no cell-mediated immunity after birth. This outcome would seriously compromise the individual's ability to defend against viral invasion and cancer.
4. Researchers are currently working on ways to "teach" the immune system to view foreign tissue as "self" as a means of preventing the immune systems of organ-transplant patients from rejecting the foreign tissue while leaving the patients' immune defense capabilities fully intact. The immunosuppressive drugs currently being used to prevent transplant rejection cripple the recipients' immune defense systems and leave the patients more vulnerable to microbial invasion.
5. The skin cells that are visible on the body's surface are all dead.
6. ***Clinical Consideration*** Heather's first-born Rh-positive child did not have hemolytic disease of the newborn because the fetal and maternal blood did not mix during gestation. Consequently, Heather did not produce any maternal antibodies against the fetus's Rh factor during gestation.

   Because a small amount of the infant's blood likely entered the maternal circulation during the birthing process, Heather would produce antibodies against the Rh factor as she was first exposed to it at that time. During any subsequent pregnancies with Rh-positive fetuses, Heather's maternal antibodies against the Rh factor could cross the placental barrier and bring about destruction of fetal erythrocytes.

   If, however, any Rh factor that accidentally mixed with the maternal blood during the birthing process were immediately tied up by Rh immunoglobulin administered to the mother, the Rh factor would not be available to induce maternal antibody production. Thus, no anti-Rh antibodies would be present in the maternal blood to threaten the rbcs of an Rh-positive fetus in a subsequent pregnancy. (The exogenously administered Rh immunoglobulin, being a passive form of immunity, is short-lived. In contrast, the active immunity that would result if Heather were exposed to Rh factor would be long-lived due to the formation of memory cells.)

   Rh immunoglobulin must be administered following the birth of every Rh-positive child Heather bears to sop up any Rh factor before it can induce antibody production. Once an immune attack against Rh factor is launched, subsequent treatment with Rh immunoglobulin will not reverse the situation. Thus, if Heather were not treated with Rh immunoglobulin following the birth of a first Rh-positive child and a second Rh-positive child developed hemolytic disease of the newborn, administration of Rh immunoglobulin following the second birth would not prevent the condition in a third Rh-positive child. Nothing could be done to eliminate the maternal antibodies already present.

CHAPTER *13* The Respiratory System

OBJECTIVE QUESTIONS

(Questions on p. 465.)

1. F (The respiratory muscles directly change the size of the thoracic cavity, not the lungs. There are no muscles in the lungs, except for the smooth muscle in the walls of the airways and blood vessels.)
2. F (A volume of air, known as the residual volume, remains in the lungs after a maximal expiration because air is trapped in the alveoli as a result of compression of the small nonrigid airways by the high intrapleural pressure accompanying the forceful expiration.)
3. T
4. F (The diffusion coefficient for CO_2 is 20 times that of O_2.)
5. F (Hemoglobin has a much higher affinity for CO than for O_2.)
6. F (Rhythmicity of breathing is brought about by pacemaker activity of a neuronal network within the rostral ventromedial medulla that drives the inspiratory neurons, not by the respiratory muscles themselves.)
7. F (Impulses from the expiratory neurons to the expiratory muscles occur only during active expiration, not during normal quiet breathing.)
8. transmural pressure gradient, pulmonary surfactant action, alveolar interdependence
9. pulmonary elasticity, alveolar surface tension
10. compliance
11. elastic recoil
12. carbonic anhydrase
13. a
14. 1.d, 2.a, 3.b, 4.a, 5.b, 6.a
15. a.<, b.>, c. =, d. =, e. = ,f. =, g.>, h.<, i. approximately =, j. approximately =, k. =, l. =

QUANTITATIVE EXERCISES

(Questions on p. 467.)

For general reference for questions 1 and 2:

$$P_{AO_2} = P_{IO_2} - (V_{O_2}/V_A) \times 863 \text{ mm Hg}$$

$$P_{ACO_2} = (V_{CO_2}/V_A) \times 863 \text{ mm Hg}$$

1. $V_A = 3$ L/min
$V_{O_2} = 0.3$ L/min, RQ = 1, therefore $V_{CO_2} = 0.3$ L/min
$P_{ACO_2} = (0.3 \text{ L/min}/3 \text{ L/min}) \times 863$ mm Hg
$= 86.3$ mm Hg
2. a. 380 mm Hg $\times$ 0.21 = 79.8 mm Hg

b. $P_{AO_2} = 79.8 \text{ mm Hg} - (0.06) \times 431.5 \text{ mm Hg}$
$= 79.8 \text{ mm Hg} - 25.8 \text{ mm Hg}$
$= 54 \text{ mm Hg}$

c. $P_{ACO_2} = (0.2 \text{ L/min}/4.2 \text{ L/min}) \times 431.5 \text{ mm Hg}$
$= 20.5 \text{ mm Hg}$

3. TV = 350 ml, BR = 12/min, $V_A = 0.8 \times V_E$, DS = ?
$V_A = BR \times (TV - DS)$
$V_E = BR \times TV$
$0.8 = V_A/V_E = [BR \times (TV - DS)]/(BR \times TV)$
$= 1 - (DS/TV)$
$0.8 = 1 - (DS/350 \text{ ml})$
$DS/350 \text{ ml} = 0.2$
$DS = 0.2(350 \text{ ml}) = 70 \text{ ml}$

POINTS TO PONDER

(Questions on p. 467.)

1. Total atmospheric pressure decreases with increasing altitude, yet the percentage of O_2 in the air remains the same. At an altitude of 30,000 feet, the atmospheric pressure is only 226 mm Hg. Since 21% of atmospheric air consists of O_2, the P_{O_2} of inspired air at 30,000 feet is only 47.5 mm Hg, and alveolar P_{O_2} is even lower at about 20 mm Hg. At this low P_{O_2}, hemoglobin is only about 30% saturated with O_2—much too low to sustain tissue needs for O_2.
   The P_{O_2} of inspired air can be increased by two means when flying at high altitude. First, by pressurizing the plane's interior to a pressure comparable to that of atmospheric pressure at sea level, the P_{O_2} of inspired air within the plane is 21% of 760 mm Hg, or the normal 160 mm Hg. Accordingly, alveolar and arterial P_{O_2} and percent hemoglobin saturation are likewise normal. In the emergency situation of failure to maintain internal cabin pressure, breathing pure O_2 can raise the P_{O_2} considerably above that accomplished by breathing normal air. When a person is breathing pure O_2, the entire pressure of inspired air is attributable to O_2. For example, with a total atmospheric pressure of 226 mm Hg at an altitude of 30,000 feet, the P_{O_2} of inspired pure O_2 is 226 mm Hg, which is more than adequate to maintain normal arterial hemoglobin saturation.
2. a. Hypercapnia would not accompany the hypoxia associated with cyanide poisoning. In fact, CO_2 levels decline because oxidative metabolism is blocked by the tissue poisons so that CO_2 is not being produced.
   b. Hypercapnia could but may not accompany the hypoxia associated with pulmonary edema. Pulmonary diffusing capacity is reduced in pulmonary edema, but O_2 transfer suffers more than CO_2 transfer because the diffusion coefficient for CO_2 is 20 times that for O_2. As a result, hypoxia occurs much more readily than hypercapnia in these circumstances. Hypercapnia does occur, however, when pulmonary diffusing capacity is severely impaired.
   c. Hypercapnia would accompany the hypoxia associated with restrictive lung disease because ventilation is inadequate to meet the metabolic needs for both O_2 delivery and CO_2 removal. Both O_2 and CO_2 exchange between the lungs and atmosphere are equally affected.
   d. Hypercapnia would not accompany the hypoxia associated with high altitude. In fact, arterial P_{CO_2} levels actually decrease. One of the compensatory responses in acclimatization to high altitudes is reflex stimulation of ventilation as a result of the reduction in arterial P_{O_2}. This compensatory hyperventilation to obtain more O_2 blows off too much CO_2 in the process, so arterial P_{CO_2} levels decline below normal.
   e. Hypercapnia would not accompany the hypoxia associated with severe anemia. Reduced O_2-carrying capacity of the blood has no influence on blood CO_2 content, so arterial P_{CO_2} levels are normal.
   f. Hypercapnia would exist in accompaniment with circulatory hypoxia associated with congestive heart failure. Just as the diminished blood flow fails to deliver adequate O_2 to the tissues, it also fails to remove sufficient CO_2.
   g. Hypercapnia would accompany the hypoxic hypoxia associated with obstructive lung disease because ventilation would be inadequate to meet the metabolic needs for both O_2 delivery and CO_2 removal. Both O_2 and CO_2 exchange between the lungs and atmosphere would be equally affected.
3. $P_{O_2} = 122$ mm Hg
$0.21(\text{atmospheric pressure} - \text{partial pressure of } H_2O)$
$= 0.21(630 \text{ mm Hg} - 47 \text{ mm Hg})$
$= 0.21(583 \text{ mm Hg}) = 122 \text{ mm Hg}$
4. Voluntarily hyperventilating before going underwater lowers the arterial P_{CO_2} but does not increase the O_2 content in the blood. Because the P_{CO_2} is below normal, the person can hold his or her breath longer than usual before the arterial P_{CO_2} increases to the point that he or she is driven to surface for a breath. Therefore, the person can stay underwater longer. The risk, however, is that the O_2 content of the blood, which was normal, not increased, before going underwater, continues to fall. Therefore, the O_2 level in the blood can fall dangerously low before the CO_2 level builds to the point of driving the person to take a breath. Low arterial P_{O_2} does not stimulate respiratory activity until it has plummeted to 60 mm Hg. Meanwhile, the person may lose consciousness and drown due to inadequate O_2 delivery to the brain. If the person does not hyperventilate so that both the arterial P_{CO_2} and O_2 content are normal before going underwater, the buildup of CO_2 will drive the

person to the surface for a breath before the O_2 levels fall to a dangerous point.

5. c.

The arterial P_{O_2} will be less than the alveolar P_{O_2}, and the arterial P_{CO_2} will be greater than the alveolar P_{CO_2}. Because pulmonary diffusing capacity is reduced, arterial P_{O_2} and P_{CO_2} do not equilibrate with alveolar P_{O_2} and P_{CO_2}.

If the person is administered 100% O_2, the alveolar P_{O_2} will increase, and the arterial P_{O_2} will increase accordingly. Even though arterial P_{O_2} will not equilibrate with alveolar P_{O_2}, it will be higher than when the person is breathing atmospheric air.

The arterial P_{CO_2} will remain the same whether the person is administered 100% O_2 or is breathing atmospheric air. The alveolar P_{CO_2} and thus the blood-to-alveolar P_{CO_2} gradient are not changed by breathing 100% O_2 because the P_{CO_2} in atmospheric air and 100% O_2 are both essentially zero (P_{CO_2} in atmospheric air = 0.3 mm Hg).

6. ***Clinical Consideration*** Emphysema is characterized by a collapse of smaller respiratory airways and a breakdown of alveolar walls. Because of the collapse of smaller airways, airway resistance is increased with emphysema. As with other chronic obstructive pulmonary diseases, expiration is impaired to a greater extent than inspiration because airways are naturally dilated slightly more during inspiration than expiration as a result of the greater transmural pressure gradient during inspiration. Because airway resistance is increased, a patient with emphysema must produce larger-than-normal intra-alveolar pressure changes to accomplish a normal tidal volume. Unlike quiet breathing in a normal person, the accessory inspiratory muscles (neck muscles) and the muscles of active expiration (abdominal muscles and internal intercostal muscles) must be brought into play to inspire and expire a normal tidal volume of air.

The spirogram would be characteristic of chronic obstructive pulmonary disease. Since the patient experiences more difficulty emptying the lungs than filling them, the total lung capacity would be essentially normal, but the functional residual capacity and the residual volume would be elevated as a result of the additional air trapped in the lungs following expiration. Because the residual volume is increased, the inspiratory capacity and vital capacity will be reduced. Also, the FEV_1 will be markedly reduced since the airflow rate is decreased by the airway obstruction. The FEV_1-to-vital capacity ratio will be much lower than the normal 80%.

Because of the reduced surface area for exchange as a result of a breakdown of alveolar walls, gas exchange would be impaired. Therefore, arterial P_{CO_2} would be elevated and arterial P_{O_2} reduced compared to normal.

Ironically, administering O_2 to this patient to relieve his hypoxic condition would markedly depress his drive to breathe by elevating the arterial P_{O_2} and removing the primary driving stimulus for respiration. Because of this danger, O_2 therapy should either not be administered or administered extremely cautiously.

CHAPTER *14* The Urinary System

OBJECTIVE QUESTIONS

(Questions on p. 512.)

1. F (Glomerular filtration is a passive process, requiring no energy expenditure by the kidneys.)
2. F (Sodium reabsorption is not under hormonal control in the proximal tubule and loop of Henle. Aldosterone increases sodium reabsorption in the distal and collecting tubules.)
3. T
4. T
5. T
6. nephron
7. potassium
8. 500
9. 1.b, 2.a, 3.b, 4.b, 5.a, 6.b, 7.b, 8.b, 9.b
10. e
11. b
12. b, e, a, d, c
13. c, e, d, a, b, f
14. g, c, d, a, f, b, e
15. 1.a, 2.a, 3.c, 4.b, 5.d

QUANTITATIVE EXERCISES

(Questions on p. 513.)

1.

	Patient 1	*Patient 2*
GFR	125 ml/min	124 ml/min
RPF	620 ml/min	400 ml/min
RBF	1,127 ml/min	727 ml/min
FF	0.20	0.31

All of patient 1's values are within the normal range. Patient 2's GFR is normal, but he has a low renal plasma flow and a high filtration fraction. Therefore, his GFR is too high for that RPF. This might imply enlarged filtration slits or a "leaky" glomerulus in general. The low RPF may imply low renal blood pressure, perhaps from a partially blocked renal artery.

2. filtered load = GFR × plasma concentration

$(0.125\ \text{liter/min}) \times (145\ \text{mmol/liter})$

$= 18.125\ \text{mmol/min}$

3. $GFR = (U \times [I]_U/[I]_B$

$U = (GFR \times [I]_B)/[I]_U = (125\ \text{ml/min})(3\ \text{mg/L})/(300\ \text{mg/L}) = 1.25\ \text{ml/min}$

4.

$$\text{clearance rate of a substance} = \frac{\text{urine concentration of the substance} \times \text{urine flow rate}}{\text{plasma concentration of the substance}} = \frac{7.5\ \text{mg/ml} \times 2\ \text{ml/min}}{0.2\ \text{mg/ml}} = 75\ \text{ml/min}$$

Since a clearance rate of 75 ml/min is less than the average GFR of 125 ml/min, the substance is being reabsorbed.

POINTS TO PONDER

(Questions on p. 514.)

1. The longer loops of Henle in desert rats (known as kangaroo rats) permit a greater magnitude of countercurrent multiplication and thus a larger medullary vertical osmotic gradient. As a result, these rodents are able to produce urine that is concentrated up to an osmolarity of almost 6,000 mosm/liter, which is five times more concentrated than maximally concentrated human urine at 1,200 mosm/liter. Because of this tremendous concentrating ability, kangaroo rats never have to drink; the H_2O produced metabolically within their cells during oxidation of foodstuff (food + O_2 yields CO_2 + H_2O + energy) is sufficient for their needs.
2. **a.** 250 mg/min filtered
   filtered load of substance = plasma concentration of substance × GFR
   filtered load of substance = 200 mg/100 ml × 125 ml/min
   = 250 mg/min
   b. 200 mg/min reabsorbed
   A T_m's worth of the substance will be reabsorbed.
   c. 50 mg/min excreted
   amount of substance excreted = amount of substance filtered − amount of substance reabsorbed
   = 250 mg/min − 200 mg/min = 50 mg/min
3. Aldosterone stimulates Na^+ reabsorption and K^+ secretion by the renal tubules. Therefore, the most prominent features of Conn's syndrome (hypersecretion of aldosterone) are hypernatremia (elevated Na^+ levels in the blood) caused by excessive Na^+ reabsorption, hypophosphatemia (below normal K^+ levels in the blood) caused by excessive K^+ secretion, and hypertension (elevated blood pressure) caused by excessive salt and water retention.
4. e. 300/300.
   If the ascending limb were permeable to water, it would not be possible to establish a vertical osmotic gradient in the interstitial fluid of the renal medulla, nor would the ascending-limb fluid become hypotonic before entering the distal tubule. As the ascending limb pumped NaCl into the interstitial fluid, water would osmotically follow, so both the interstitial fluid and the ascending limb would remain isotonic at 300 mosm/liter. With the tubular fluid entering the distal tubule being 300 mosm/liter instead of the normal 100 mosm/liter, it would not be possible to produce a urine with an osmolarity less than 300 mosm/liter. Likewise, in the absence of the medullary vertical osmotic gradient, it would not be possible to produce a urine more concentrated than 300 mosm/liter, no matter how much vasopressin was present.
5. Because the descending pathways between the brain and the motor neurons supplying the external urethral sphincter and pelvic diaphragm are no longer intact, the accident victim can be longer voluntarily control micturition. Therefore, bladder emptying in this individual will be governed entirely by the micturition reflex.
6. ***Clinical Consideration*** prostate enlargement

CHAPTER *15* Fluid and Acid-Base Balance

OBJECTIVE QUESTIONS

(Questions on p. 544.)

1. T
2. F (With an isotonic fluid gain, there is no change in ECF osmolarity to induce a fluid shift between the ECF and the cells, so the fluid gain is confined to the ECF compartment.)
3. F (Salt balance is carefully regulated. Despite wide variations in salt intake, a stable salt balance is maintained by controlling the urinary excretion of salt.)
4. T
5. T
6. intracellular fluid
7. transcellular fluid
8. $[H_2CO_3]$, $[HCO_3^-]$
9. d
10. b
11. a, d, e
12. b, e
13. c
14. 1. metabolic acidosis, 2. diabetes mellitus, 3. pH = 7.1, 4. respiratory alkalosis, 5. anxiety, 6. pH = 7.7, 7. respiratory acidosis, 8. pneumonia, 9. pH = 7.1, 10. metabolic alkalosis, 11. vomiting, 12. pH = 7.7

QUANTITATIVE EXERCISES

(Questions on p. 545.)

1. $pH = 6.1 + \log [HCO_3^-]/(0.03 \text{ mM/mm Hg} \times 40 \text{ mm Hg})$
   $7.4 = 6.1 + \log [HCO_3^-]/1.2 \text{ mM}$
   $\log [HCO_3^-]/1.2 \text{ mM} = 7.4 - 6.1 = 1.3$
   $[HCO_3^-] = 1.2 \text{ mM} \times (10^{1.3}) = 24 \text{ mM}$
2. $pH = -\log [H^+]$, $[H^+] = 10^{-pH}$
   $[H^+] = 10^{-6.8} = 158$ nM for pH = 6.8
   $[H^+] = 10^{-8.0} = 10$ nM for pH = 8.0
3. Note that distilled water is permeable across all barriers, so it will distribute equally among all compartments. However, the saline does not enter cells, so it will stay in the ECF. The resultant distributions are summarized in the chart on p. E-16. Clearly, saline is better at expanding the plasma volume.

POINTS TO PONDER

(Questions on p. 545.)

1. The rate of urine formation increases when alcohol inhibits vasopressin secretion and the kidneys are unable to reabsorb water from the distal and collecting tubules. Because extra free water that normally would have been reabsorbed from the distal parts of the tubule is lost from the body in the urine, the body becomes dehydrated and the ECF osmolarity increases following

Ingested fluid	Compartment	Size of compartment before ingestion	Size of compartment after ingestion	% increase in size of compartment after ingestion
Distilled water	TBW	42 L	43 L	2%
	ICF (2/3 TBW)	28 L	28.667 L	2%
	ECF (1/3 TBW)	14 L	14.333 L	2%
	plasma (20% ECF)	2.8 L	2.866 L	2%
	ISF (80% ECF)	11.2 L	11.466	2%
Saline	TBW	42 L	43 L	2%
	ICF	28 L	28 L	0%
	ECF	14 L	15 L	7%
	plasma	2.8 L	3 L	7%
	ISF	11.2	12 L	7%

alcohol consumption. That is, more fluid is lost in the urine than is consumed in the alcoholic beverage as a result of alcohol's action on vasopressin. Thus, the imbibing person experiences a water deficit and still feels thirsty, despite the recent fluid consumption.

2. If a person loses 1,500 ml of salt-rich sweat and drinks 1,000 ml of water without replacing the salt during the same time period, there will still be a volume deficit of 500 ml, and the body fluids will have become hypotonic (the remaining salt in the body will be diluted by the ingestion of 1,000 ml of free H_2O). As a result, the hypothalamic osmoreceptors (the dominant input) will signal the vasopressin-secreting cells to *decrease* vasopressin secretion and thus increase urinary excretion of the extra free water that is making the body fluids too dilute. Simultaneously, the left atrial volume receptors will signal the vasopressin-secreting cells to *increase* vasopressin secretion to conserve water during urine formation and thus help to relieve the volume deficit. These two conflicting inputs to the vasopressin-secreting cells are counterproductive. This is why it is important to replace both water and salt following heavy sweating or abnormal loss of other salt-rich fluids. If salt is replaced along with water intake, the ECF osmolarity remains close to normal, and the vasopressin-secreting cells receive signals only to increase vasopressin secretion to help restore the ECF volume to normal.
3. When a dextrose solution equal in concentration to that of normal body fluids is injected intravenously, the ECF volume is expanded but the ECF and ICF are still osmotically equal. Therefore, no net movement of water occurs between the ECF and ICF. When the dextrose enters the cell and is metabolized, however, the ECF becomes hypotonic as this solute leaves the plasma. If the excess free water is not excreted in the urine rapidly enough, water will move into the cells by osmosis.
4. Because baking soda ($NaHCO_3$) is readily absorbed from the digestive tract, treatment of gastric hyperacidity with baking soda can lead to metabolic alkalosis as too much HCO_3^- is absorbed. Treatment with antacids that are poorly absorbed is safer because these products remain in the digestive tract and do not produce an acid-base imbalance.
5. c.
   The hemoglobin buffer system buffers carbonic acid–generated hydrogen ion. In the case of respiratory acidosis accompanying severe pneumonia, the $H^+ + Hb \rightarrow HHb$ reaction will be shifted toward the HHb side, thus removing some of the extra free H^+ from the blood.
6. ***Clinical Consideration*** The resultant prolonged diarrhea will lead to dehydration and metabolic acidosis due, respectively, to excessive loss in the feces of fluid and $NaHCO_3$ that normally would have been absorbed into the blood.
   Compensatory measures for dehydration have included increased vasopressin secretion, resulting in increased water reabsorption by the distal and collecting tubules and a subsequent reduction in urine output. Simultaneously, fluid intake has been encouraged by increased thirst. The metabolic acidosis has been combatted by removal of excess H^+ from the ECF by the HCO_3^- member of the H_2CO_3:HCO_3^- buffer system, by increased ventilation to reduce the amount of acid-forming CO_2 in the body fluids, and by the kidneys excreting extra H^+ and conserving HCO_3^-.

CHAPTER *16* The Digestive System

OBJECTIVE QUESTIONS

(Questions on p. 599.)

1. F (The digestive system does not vary nutrient, water, or electrolyte uptake based on body needs, but instead it absorbs all digestible food that is ingested.)
2. T
3. T
4. F (The protein components of digestive secretions and sloughed epithelial cells are digested and absorbed.)
5. F (The large intestine does not have the ability to absorb nutrients.)

6. F (The major hormones secreted by the endocrine pancreas are insulin and glucagon. Secretin and CCK, which are secreted by the small intestine, act on the exocrine pancreas.)
7. long, short
8. chyme
9. three
10. vitamin B_{12}, bile salts
11. bile salts
12. b
13. 1.c, 2.e, 3.b, 4.a, 5.f, 6.d
14. 1.c, 2.c, 3.d, 4.a, 5.e, 6.c, 7.a, 8.b, 9.c, 10.b, 11.d, 12.b

QUANTITATIVE EXERCISES

(Questions on p. 600.)

1. a. $r = 0.5$ cm, area $= 4\pi(0.25) = \pi$ cm^2
   volume $= (4\pi/3)(0.5)^3 = 0.5236$ cm^3
   area/volume = 6
   b. The new spheres now have a volume of 5.236×10^{-3} cm^3, so the average radius is therefore 0.1077 cm. The area of a sphere with that radius is $4\pi(0.1077 \text{ cm})^2 = 0.1458$ cm^2
   area/volume = 27.8
   c. Area emulsified/area droplet $= (100)(0.1458 \text{ cm}^2)/\pi \text{ cm}^2 = 4.64$.
   Thus, the total surface area of all 100 emulsified droplets is 4.64 times the area of the original larger lipid droplet.
   d. Volume emulsified/volume droplet $= (100)(5.236 \times 10^{-3} \text{ cm}^3)/0.5236 \text{ cm}^3 = 1.0$.
   Thus, the total volume did not change as a result of emulsification, as would be expected because the total volume of the lipid is conserved during emulsification. The volume originally present in the large droplet is divided up among the 100 emulsified droplets.

POINTS TO PONDER

(Questions on p. 600.)

1. Patients who have had their stomachs removed must eat small quantities of food frequently instead of consuming the typical three meals a day because they have lost the ability to store food in the stomach and meter it into the small intestine at an optimal rate. If a person without a stomach consumed a large meal that entered the small intestine all at once, the luminal contents would quickly become too hypertonic as digestion of the large nutrient molecules into a multitude of small, osmotically active, absorbable units outpaced the more slowly acting process of absorption of these units. As a consequence of this increased luminal osmolarity, water would enter the small intestine lumen from the plasma by osmosis, resulting in circulatory disturbances as well as intestinal distention. To prevent this "dumping syndrome" from occurring, the patient must "feed" the small intestine only small amounts of food at a time so that absorption of the digestive end products can keep pace with their rate of production. The person has to consciously take over metering the delivery of food into the small intestine because the stomach is no longer present to assume this responsibility.
2. The gut-associated lymphoid tissue launches an immune attack against any pathogenic (disease-causing) microorganisms that enter the readily accessible digestive tract and escape destruction by salivary lysozyme or gastric HCl. This action defends against entry of these potential pathogens into the body proper. The large number of immune cells in the gut-associated lymphoid tissue is adaptive as a first line of defense against foreign invasion when considering that the surface area of the digestive tract lining represents the largest interface between the body proper and the external environment.
3. Defecation would be accomplished entirely by the defecation reflex in a patient paralyzed from the waist down because of lower spinal cord injury. Voluntary control of the external anal sphincter would be impossible because of interruption in the descending pathway between the primary motor cortex and the motor neuron supplying this sphincter.
4. When insufficient glucuronyl transferase is available in the neonate to conjugate all of the bilirubin produced during erythrocyte degradation with glycuronic acid, the extra unconjugated bilirubin cannot be excreted into the bile. Therefore, this extra bilirubin remains in the body, giving rise to mild jaundice in the newborn.
5. Removal of the stomach leads to pernicious anemia because of the resultant lack of intrinsic factor, which is necessary for absorption of vitamin B_{12}. Removal of the terminal ileum leads to pernicious anemia because this is the only site where vitamin B_{12} can be absorbed.
6. ***Clinical Consideration*** A person whose bile duct is blocked by a gallstone experiences a painful "gallbladder attack" after eating a high-fat meal because the ingested fat triggers the release of cholecystokinin, which stimulates gallbladder contraction. As the gallbladder contracts and bile is squeezed into the blocked bile duct, the duct becomes distended prior to the blockage. This distention is painful.
   The feces are grayish white because no bilirubin-containing bile enters the digestive tract when the bile duct is blocked. Bilirubin, when acted on by bacterial enzymes, is responsible for the brown color of feces, which are grayish white in its absence.

CHAPTER 17 Energy Balance and Temperature Regulation

OBJECTIVE QUESTIONS

(Questions on p. 618.)

1. F (The excess energy is stored in the body, primarily as adipose tissue.)
2. F (Only about 25% of the chemical energy in nutrient molecules is harnessed to do biological work.)

3. F (Oxygen does not contain heat energy. For each liter of O_2 consumed in oxidizing nutrient molecules, 4.8 kilocalories of heat are liberated from the food on the average.)
4. F (Body temperature normally varies several degrees Fahrenheit.)
5. T
6. T
7. T
8. T
9. shivering
10. nonshivering thermogenesis
11. sweating
12. b
13. e
14. 1.b, 2.a, 3.c, 4.a, 5.d, 6.c, 7.b, 8.d, 9.c, 10.b

QUANTITATIVE EXERCISES

(Questions on p. 619.)

1. From physics we know that $\Delta T(°C) = \Delta U/(C \times m)$.
   Also note that $\Delta U/t$ = BMR; i.e., the rate of using energy is the basal metabolic rate. *m* represents the mass of body fluid; for a typical person, this is 42 L.
   $(42\ L) \times (1\ kg/L) = 42\ kg$
   $C = 1.0\ kcal/(kg—°C)$
   Given that water boils at 100°C and normal body temperature is 37°C, we need to change the temperature by 63°C. Thus:

   $t = (\Delta T \times C \times m)/BMR = (63°C)[1.0\ kcal/(kg—°C)] (42\ kg)/(75\ kcal/hr) = 35\ hr$

   At the higher metabolic rate during exercise:
   $t = (63°C)[1.0\ kcal/(kg—°C)](42\ kg)/(1{,}000\ kcal/hr) = 2.6\ hr$

POINTS TO PONDER

(Questions on p. 619.)

1. Evidence suggests that CCK serves as a satiety signal. It is believed to serve as a signal to stop eating when enough food has been consumed to meet the body's energy needs, even though the food is still in the digestive trace. Therefore, when drugs that inhibit CCK release are administered to experimental animals, the animals overeat because this satiety signal is not released.
2. Don't go on a "crash diet." Be sure to eat a nutritionally balanced diet that provides all essential nutrients, but reduce total caloric intake, especially by cutting down on high-fat foods. Spread out consumption of the food throughout the day instead of just eating several large meals. Avoid bedtime snacks. Burn more calories through a regular exercise program.
3. Engaging in heavy exercise on a hot day is dangerous because of problems arising from trying to eliminate the extra heat generated by the exercising muscles. First, there will be conflicting demands for distribution of the cardiac output—temperature-regulating mechanisms will trigger skin vasodilation to promote heat loss from the skin surface, whereas metabolic changes within the exercising muscles will induce local vasodilation in the muscles to match the increased metabolic needs with increased blood flow. Further exacerbating the problem of conflicting demands for blood flow is the loss of effective circulating plasma volume resulting from the loss of a large volume of fluid through another important cooling mechanism, sweating. Therefore, it is difficult to maintain an effective plasma volume and blood pressure and simultaneously keep the body from overheating when engaging in heavy exercise in the heat, so heat exhaustion is likely to ensue.
4. When a person is soaking in a hot bath, loss of heat by radiation, conduction, convection, and evaporation is limited to the small surface area of the body exposed to the cooler air. Heat is being gained by conduction at the larger skin surface area exposed to the hotter water.
5. The thermoconforming fish would not run a fever when it has a systemic infection because it has no mechanisms for regulating internal heat production or for controlling heat exchange with its environment. The fish's body temperature varies capriciously with the external environment no matter whether it has a systemic infection or not. It is not able to maintain body temperature at a "normal" set point or an elevated set point (i.e., a fever).
6. ***Clinical Consideration*** Cooled tissues need less nourishment than they do at normal body temperature because of their pronounced reduction in metabolic activity. The lower O_2 need of cooled tissues accounts for the occasional survival of drowning victims who have been submerged in icy water considerably longer than one could normally survive without O_2.

CHAPTER 18 Principles of Endocrinology; The Central Endocrine Glands

OBJECTIVE QUESTIONS

(Questions on p. 652.)

1. T
2. T
3. F (Some endocrine glands exert nonendocrine effects in addition to secreting hormones.)
4. T
5. F (Each steroidogenic organ has a limited set of enzymes for producing only a given type or types of steroid hormones.)
6. T
7. F (Inhibition of the anterior pituitary and hypothalamus by a target-organ hormone is known as long-loop negative feedback. Short-loop negative feedback involves inhibition of the hypothalamus by an anterior pituitary hormone.)

8. T
9. tropic
10. down regulation
11. epiphyseal plate
12. 1.c, 2.b, 3.b, 4.a, 5.a, 6.c, 7.c

POINTS TO PONDER

(Questions on p. 653.)

1. It is not advisable to rotate the nursing staff on different shifts every week because such a practice would disrupt the individuals' natural circadian rhythms. Such disruption can affect physical and psychological health and have a negative impact on work performance.
2. The concentration of hypothalamic releasing and inhibiting hormones would be considerably lower (in fact, almost nonexistent) in a systemic venous blood sample compared to the concentration of these hormones in a sample of hypothalamic-hypophyseal portal blood. These hormones are secreted into the portal blood for local delivery between the hypothalamus and anterior pituitary. Any portion of these hormones picked up by the systemic blood at the anterior pituitary capillary level is greatly diluted by the much larger total volume of systemic blood compared to the extremely small volume of blood within the portal vessel.
3. If CRH and/or ACTH is elevated in accompaniment with the excess cortisol secretion, the condition is secondary to a defect at the hypothalamic/anterior pituitary level. If CRH and ACTH levels are below normal in accompaniment with the excess cortisol secretion, the condition is due to a primary defect at the adrenal cortex level, with the excess cortisol inhibiting the hypothalamus and anterior pituitary in negative-feedback fashion.
4. Males with testicular feminization syndrome would be unusually tall because of the inability of testosterone to promote closure of the epiphyseal plates of the long bones in the absence of testosterone receptors.
5. Full-grown athletes sometimes illegally take supplemental doses of growth hormone because it promotes increased skeletal muscle mass through its protein anabolic effect. However, excessive growth hormone can have detrimental side effects, such as possibly causing diabetes or high blood pressure.
6. ***Clinical Consideration*** Hormonal replacement therapy following pituitary gland removal should include thyroid hormone (the thyroid gland will not produce sufficient thyroid hormone in the absence of TSH) and glucocorticoid (because of the absence of ACTH), especially in stress situations. If indicated, male or female sex hormones can be replaced, even though these hormones are not essential for survival. For example, testosterone in males plays an important role in libido. Growth hormone and prolactin need not be replaced because their absence will produce no serious consequences in this individual. Vasopressin may have to be replaced if insufficient quantities of this hormone are picked up by the blood at the hypothalamus in the absence of the posterior pituitary.

CHAPTER *19* The Peripheral Endocrine Organs

OBJECTIVE QUESTIONS

(Questions on p. 697.)

1. F (Only after a delay of several hours is the metabolic response to thyroid hormone first detectable, and the maximal response is not evident for several days.)
2. T
3. T
4. T
5. T
6. F (The most life-threatening consequence of hypocalcemia is overexcitability of nerves and muscles.)
7. F (Absorption of ingested Ca^{2+} from the intestine is controlled by vitamin D.)
8. F (The labile pool of Ca^{2+} is the bone fluid. The stable pool of Ca^{2+} is the bone crystals.)
9. colloid, thyroglobulin
10. pro-opiomelanocortin
11. glycogenesis, glycogenolysis, gluconeogenesis
12. brain, working muscles, liver
13. bone, kidneys, digestive tract
14. c
15. a, b, c, g, i, j
16. 1. glucose, 2. glycogen, 3. free fatty acids, 4. triglycerides, 5. amino acids, 6. body proteins

POINTS TO PONDER

(Questions on p. 698.)

1. The midwestern United States is no longer an endemic goiter belt even though the soil is still iodine poor because individuals living in this region obtain iodine from iodine-supplemented nutrients, such as iodinated salt, and from seafood and other naturally iodine-rich foods shipped from coastal regions.
2. Melatonin helps synchronize the body's inherent biological rhythms with external cues such as the light/dark cycles. Melatonin secretion increases during the dark of night and decreases during the light of day. For people who work the night shift, their activity (work)/inactivity (sleep) cycles are out of sync with the normal biological rhythms entrained to the natural light/dark cycles. Exposure to bright light during the night and darkness during the day artificially helps to synchronize the workers' biological rhythms with their unnatural work/sleep cycles.
3. An infection elicits the stress response, which brings about increased secretion of cortisol and epinephrine, both of which increase the blood glucose level. This can become a problem in managing the blood glucose level of a diabetic patient. When the blood glucose is elevated too high, the patient can reduce it by injecting addi-

tional insulin, or, preferably, by reducing carbohydrate intake and/or exercising to use up some of the extra blood glucose. In a normal individual, the check-and-balance system between insulin and the other hormones that oppose insulin's actions helps to maintain the blood glucose within reasonable limits during the stress response.

4. The presence of Chvostek's sign is due to increased neuromuscular excitability caused by moderate hyposecretion of parathyroid hormone.
5. If malignancy-associated hypercalcemia arose from metastatic tumor cells that invaded and destroyed bone, both hypercalcemia and hyperphosphatemia would result as calcium phosphate salts were released from the destructed bone. The fact that hypophosphatemia, not hyperphosphatemia, often accompanies malignancy-associated hypercalcemia led investigators to rule out bone destruction as the cause of the hypercalcemia. Instead, they suspected that the tumors produced a substance that mimics the actions of PTH in promoting concurrent hypercalcemia and hypophosphatemia.
6. ***Clinical Consideration*** "Diabetes of bearded ladies" is descriptive of both excess cortisol and excess adrenal androgen secretion. Excess cortisol secretion causes hyperglycemia and glucosuria. Glucosuria promotes osmotic diuresis, which leads to dehydration and a compensatory increased sensation of thirst. All these symptoms—hyperglycemia, glucosuria, polyuria, and polydipsia—mimic diabetes mellitus. Excess adrenal androgen secretion in females promotes masculinizing characteristics, such as beard growth. Simultaneous hypersecretion of both cortisol and adrenal androgen most likely occurs secondary to excess CRH/ACTH secretion, because ACTH stimulates both cortisol and androgen production by the adrenal cortex.

CHAPTER *20* The Reproductive System

OBJECTIVE QUESTIONS

(Questions on p. 751.)

1. T
2. T
3. F (Melatonin secretion decreases during exposure to light.)
4. F (The clitoris becomes erect during sexual arousal.)
5. T
6. F (A follicle that fails to reach maturity undergoes atresia; that is, it degenerates and forms scar tissue.)
7. T
8. seminiferous tubules, FSH, testosterone
9. thecal, LH, granulosa, FSH
10. corpus luteum of pregnancy
11. human chorionic gonadotropin
12. c
13. e
14. 1.c, 2.a, 3.b, 4.c, 5.a, 6.a, 7.c, 8.b, 9.e
15. 1.a, 2.c, 3.b, 4.a, 5.a, 6.b

POINTS TO PONDER

(Questions on p. 753.)

1. The anterior pituitary responds only to the normal pulsatile pattern of GnRH and does not secrete gonadotropins in response to continuous exposure to GnRH. In the absence of FSH and LH secretion, ovulation and other events of the ovarian cycle do not ensue, so continuous GnRH administration may find use as a contraceptive technique.
2. Testosterone hypersecretion in a young boy causes premature closure of the epiphyseal plates so that he stops growing before he reaches his genetic potential for height. The child would also display signs of precocious pseudopuberty, characterized by premature development of secondary sexual characteristics, such as deep voice, beard, enlarged penis, and sex drive.
3. A potentially troublesome side effect of drugs that inhibit sympathetic nervous system activity as part of the treatment for high blood pressure is the inability to carry out the sex act in males. Both divisions of the autonomic nervous system are required for the male sex act. Parasympathetic activity is essential for accomplishing erection, and sympathetic activity is important for ejaculation.
4. Posterior pituitary extract contains an abundance of stored oxytocin, which can be administered to induce or facilitate labor by increasing uterine contractility. Exogenous oxytocin is most successful in inducing labor if the woman is near term, presumably because of the increasing concentration of myometrial oxytocin receptors at that time.
5. GnRH or FSH and LH are not effective in treating the symptoms of menopause because the ovaries are no longer responsive to the gonadotropins. Thus, treatment with these hormones would not cause estrogen and progesterone secretion. In fact, GnRH, FSH, and LH levels are already elevated in postmenopausal women because of lack of negative feedback by the ovarian hormones.
6. ***Clinical Consideration*** The first warning of a tubal pregnancy is pain caused by stretching of the oviduct by the growing embryo. A tubal pregnancy must be surgically terminated because the oviduct cannot expand as the uterus does to accommodate the growing embryo. If not removed, the enlarging embryo will rupture the oviduct, causing possibly lethal hemorrhage.

Appendix F

Supplemental Reading List

The following list includes more advanced textbooks and relevant articles in lay science journals. The list is not meant to be comprehensive but is merely a starting point for those interested in pursuing specific topics in more depth. Publications in specialty science journals are not included.

General Physiology References

Berne, R. M., and M. N. Levy, eds. *Physiology.* 3d ed. St. Louis: C. V. Mosby, 1992.

Ganong, W. F. *Review of Medical Physiology,* 17th ed. Norwalk: Appleton & Lange, 1995.

Geison, G. L., ed. *Physiology in the American Context.* New York: Oxford University Press, 1987.

Guyton, A. C. *Textbook of Medical Physiology.* 9th ed. Philadelphia: W. B. Saunders, 1995.

Johnson. L. R., ed. *Essential Medical Physiology.* New York: Raven Press, 1991.

Patton, H. D., A. F. Fuchs, B. Hille, A. M. Scher, and R. Steiner, eds. *Textbook of Physiology.* 21st ed. Philadelphia: W. B. Saunders, 1989.

Vick, R. L. *Contemporary Medical Physiology.* Menlo Park, Calif.: Addison-Wesley, 1984.

Exercise Physiology References

Brooks, G. A., and T. D. Fahey, *Exercise Physiology: Human Bioenergetics and Its Applications.* New York: John Wiley & Sons, 1984.

Cody, F. W. J., ed. *Neural Control of Skilled Human Movement.* Colchester, UK: Portland Press, 1995.

DeVries, H. A. *Physiology of Exercise.* 5th ed. Dubuque, Iowa: W. C. Brown, 1994.

Harries, M., C. Williams, W. D. Stanish, and L. J. Micheli, eds. *Oxford Textbook of Sports Medicine.* New York: Oxford University Press, 1994.

Kent, M., ed. *The Oxford Dictionary of Sports Science and Medicine.* New York: Oxford University Press, 1996.

Komi, P. V., ed. *Strength and Power in Sport.* London: Blackwell Scientific Publications, 1992.

Lamb, D. R. *Physiology of Exercise: Responses and Adaptations.* 2d ed. New York: Macmillan, 1984.

McArdle, W. D., F. D. Katch, and V. L. Katch. *Exercise Physiology: Energy, Nutrition, and Human Performance.* 3d ed. Philadelphia: Lea & Febiger, 1991.

McArdle, W. D., F. D. Katch, and V. L. Katch. *Essentials of Exercise Physiology.* Baltimore: Williams & Wilkins, 1994.

Noble, B. J. *Physiology of Exercise and Sport.* St. Louis: Times Mirror, Mosby College Publications, 1991.

Rosato, F. D. *Fitness and Wellness.* 3d ed. St. Paul: West Publishing Co., 1994.

Rowell, L. B. and J. T. Shepherd, eds. *Exercise: Regulation and Integration of Multiple Systems, Handbook of Physiology: Section 12.* New York: Oxford University Press, 1996.

Ryan, A. J., and F. L. Allman, eds. *Sports Medicine.* 2d ed. San Diego: Academic Press, 1989.

Shephard, R. J., and P. D. Aistrand, eds. *Endurance in Sport.* London: Blackwell Scientific Publications, 1992.

Teitz, C. C., ed. *Scientific Foundations of Sports Medicine.* Philadelphia: Decker, 1989.

Viru, A. *Adaptation in Sports Training.* Boca Raton: CRC Press, Inc., 1995.

Whipp, B. J. and A. J. Sargeant, eds. *Physiological Determinants of Human Exercise Tolerance.* Colchester, UK: Portland Press, 1997.

Wood, S. C. and R. C. Roach, eds. *Sports and Exercise Medicine.* New York: Marcel Dekker, Inc., 1994.

Principles of Homeostasis and Regulation (CHAPTER 1)

Adolph, E. F., ed. *The Development of Homeostasis.* New York: Academic Press, 1960.

Jones, R. W. *Principles of Biological Regulation: An Introduction to Feedback Systems.* New York: Academic Press, 1973.

Langley, L. L., ed. *Homeostasis: Origin of the Concept.* Stroudsburg, Pa.: Dowden, Hutchison, & Ross, 1973.

______ . *Origins of Physiological Regulations.* New York: Academic Press, 1968.

Cellular Physiology
(CHAPTERS 2 AND 3)

Alberts, B., D. Bray, J. Lewis, M. Raff, K. Roberts, and J. Watson. *Molecular Biology of the Cell.* 3d ed. New York: Garland Publishing, 1994.

Berridge, M. J. "The Molecular Basis of Communication within the Cell." *Scientific American* 253 (October 1985).

Bretscher, M. S. "The Molecules of the Cell Membrane." *Scientific American* 253 (October 1985).

Byrne, J. H., and S. G. Schultz. *An Introduction to Membrane Transport and Bioelectricity:* Foundations of General Physiology and Electrochemical Signaling, 2d ed. New York: Raven Press, 1994.

Carafoli, E., and J. T. Penniston. "The Calcium Signal." *Scientific American* 253 (November 1985).

______. "Caveolae Caveat." *Discover* 16 (January 1995).

______. "Cell 'Caves' Harbor Clues to Diseases." *Science News* 146 (July 2, 1994).

Chien, S., ed. *Molecular Biology in Physiology.* New York: Raven Press, 1989.

Darnell, J., H. Lodish, and D. Baltimore. *Molecular Cell Biology.* 2d ed. New York: Scientific American Books, 1990.

Ezzell, C. "Sticky Situations: Picking Apart the Molecules That Glue Cells Together." *Science News* 141 (June 13, 1992).

Goodsell, D. S. *The Machinery of Life.* Berlin: Springer-Verlag, 1992.

Hanson, B. "Message in a Barrel." *Discover* 13 (June 1992).

______. "How Cells Absorb Glucose." *Scientific American* 266 (January 1992).

Jacobson, K., E. D. Sheets, and R. Simson. "Revisiting the Fluid Mosaic Model of Membranes." *Science* 268 (June 1995).

Linder, M. E., and A. G. Gilman. "G Proteins." *Scientific American* 267 (July 1992).

Matthews, G. G. *Cellular Physiology of Nerve and Muscle,* 2d ed. London: Blackwell Scientific Publications, 1991.

The Molecules of Life: Readings from Scientific American. New York: Scientific American Books, 1986.

Murray, M. "Life on the Move." *Discover* 12 (March 1991).

Neher, E., and B. Sakmann. "The Patch Clamp Technique." *Scientific American* 266 (March 1992).

Oakley, B. R. and C. E. Oakley. "Tubulin and Microtubules." *Scientific American: Science & Medicine* 2 (January/February 1995).

Palca, J. "The Promise of a Cure." *Discover* 15 (June 1994).

Rasmussen, H. "The Cycling of Calcium as an Intracellular Messenger." *Scientific American* 261 (October 1989).

Rothman, J. E. "The Compartmental Organization of the Golgi Apparatus." *Scientific American* 253 (September 1985).

Rothman, J. E. and L. Orci. "Budding Vesicles in Living Cells." *Scientific American* 274 (March 1996).

Satir, P. "Mechanism of Ciliary Movement—What's New?" *News in Physiological Sciences* 4 (August 1989).

Scott-Burden, T. "Extracellular Matrix: The Cellular Environment." *News in Physiological Sciences* 9 (June 1994).

Sharon, N. and H. Lis. "Carbohydrates in Cell Recognition." *Scientific American* 268 (January 1993).

Skou, J. C. "The Na-K Pump." *News in Physiological Sciences* 7 (June 1992).

Spiegel, A. M., T. Jones, W. Simonds, and L. Weinstein. *G Proteins.* Boca Raton: CRC Press, Inc., 1994.

Stossel, T. P. "The Machinery of Cell Crawling." *Scientific American* 271 (September 1994).

Tosteson, D. C., ed. *Membrane Transport: People and Ideas.* New York: Oxford University Press, 1989.

Travis, J. "Cell Biologists Explore 'Tiny Caves.'" *Science* 262 (November 19, 1993).

Travis, J. "What's in the Vault? An ignored cell component may often account for why chemotherapy fails." *Science News* 150 (July 27, 1996).

Welsh, M. J. and A. E. Smith. "Cystic Fibrosis." *Scientific American* 273 (December 1995).

Nervous System
(CHAPTERS 4, 5, 6, AND 7)

Abelson, P. H., E. Butz, and S. H. Snyder, eds. *Neuroscience.* Washington, D.C.: American Association for the Advancement of Science. 1985.

______. "Additional genes may affect color vision." *Science News* 147. (February 18, 1995).

"Adult Neurons: Not Too Old to Divide." *Science News* 141 (April 4, 1992).

Alkon, D. L. "Memory Storage and Neural Systems." *Scientific American* 261 (July 1989).

Alonso, A. D. C., I. Grundke-Iqbal, and K. Iqbal. "Alzheimer's disease hyperphosphorylated tau sequesters normal tau into tangles of filaments and disassembles microtubules." *Nature Medicine* 2 (July 1996).

Aoki, C., and P. Siekevitz. "Plasticity in Brain Development." *Scientific American* 259 (December 1988).

Appenzeller, O. *The Autonomic Nervous System.* 3d ed. New York: Elsevier, 1982.

Axel, R. "The Molecular Logic of Smell." *Scientific American* 273 (October 1995).

Barinaga, M. "Neurons Tap out a Code That May Help Locate Sounds." *Science* 264 (May 6, 1994).

Barinaga, M. "Dendrites Shed Their Dull Image." *Science* 268 (April 14, 1995).

Barinaga, M. "Remapping the Motor Cortex." *Science* 268 (June 23, 1995).

Bloom, F. E., and A. Lazerson. *Brain, Mind and Behavior* 2d ed. New York: W. H. Freeman, 1995.

Bloom, F. E., ed. *Handbook of Physiology, Section 1, Volume IV; Intrinsic Regulatory Systems of the Brain.* New York: Oxford University Press, 1986.

Borg, E., and S. A. Counter. "The Middle-Ear Muscles." *Scientific American* 261 (August 1989).

Bradford, H. F. *Chemical Neurobiology: An Introduction to Neurochemistry.* New York: W. H. Freeman, 1995.

______. "Brain changes linked to phantom-limb pain." *Science News* 147 (June 10, 1995).

Chalmers, D. J. "The Puzzle of Conscious Experience." *Scientific American* 273 (December 1995).

Chollar, S. "The Nerves of Some People." *Discover* 12 (February 1991).

Ciriello, J., F. R. Calaresu, L. P. Renaud, and C. Polosa. *Organization of the Autonomic Nervous System—Central and Peripheral Mechanisms.* New York: Alan R. Liss, 1987.

Damasio, A. R., and H. Damasio. "Brain and Language." *Scientific American* 267 (September 1992).

Darian-Smith, I., ed. *Handbook of Physiology, Section 1, Volume III, Parts 1 & 2: Sensory Processes.* New York: Oxford University Press, 1984.

Darson, H. and M. B. Segal, eds. *Physiology of the CSF and Blood-Brain Barriers.* Boca Raton: CRC Press, Inc., 1996.

Dowling, J. E. *Neurons and Networks: An Introduction to Neuroscience.* Boston: The Belknap Press of Harvard University Press, 1992.

Ezzell, C. "Alzheimer's Alchemy." *Science News* 141 (March 7, 1992).

Fackelmann, K. A. "Myelin on the Mend." *Science News* 137 (April 7, 1990).

Fischbach, G. D. "Mind and Brain." *Scientific American* 267 (September 1992).

Fitzgerald, M. J. T. *Neuroanatomy Basic and Applied.* London: Bailliere Tindall, 1985.

Freedman, D. H. "In the Realm of the Chemical." *Discover* 14 (June 1993).

Freeman, W. J. "The Physiology of Perception." *Scientific American* 264 (February 1991).

French, A. S. and P. H. Torkkeli. "The Basis of Rapid Adaptation in Mechanoreceptors." *News in Physiological Sciences* 9 (August 1994).

Fuller, R. W. "Neural Functions of Serotonin." *Scientific American: Science & Medicine* 2 (July/August 1995).

Glausiusz, J. "Brain, Heal Thyself." *Discover* 17 (August 1996).

Goldman-Rakic, P. S. "Working Memory and the Mind." *Scientific American* 267 (September 1992).

Goldstein, G. W., and A. L. Betz. "The Blood-Brain Barrier." *Scientific American* 255 (September 1986).

Gutin, J. C. "Good Vibrations." *Discover* 14 (June 1993).

Hamill, O. P. and D. W. McBride, Jr. "Mechanoreceptive Membrane Channels." *American Scientist* 83 (January–February 1995).

______ . "How the Nose Knows." *Discover* 13 (January 1992).

Jahn, A. F., and J. R. Santos-Sacchi, eds. *Physiology of the Ear.* New York: Raven Press, 1988.

Kandel, E. R., ed. *Handbook of Physiology, Section 1, Volume I, Parts 1 & 2: Cellular Biology of Neurons.* New York: Oxford University Press, 1977.

Kandel, E. R., and R. D. Hawkins. "The Biological Basis of Learning and Individuality." *Scientific American* 267 (September 1992).

Kandel, E. R., and J. H. Schwartz. *Principles of Neural Science.* 3d ed. New York: Elsevier, 1991.

Katzman, R., and T. Saitoh. "Advances in Alzheimer's Disease." *The FASEB Journal* 5 (March 1991).

Kimelberg, H. K., and M. D. Norenberg. "Astrocytes." *Scientific American* 260 (April 1989).

Kuffler, S. W., J. G. Nicholls, and A. R. Martin. *From Neuron to Brain.* 3d ed. Sunderland, Mass.: Sinauer Associates, 1992.

______ . "Left Brain May Serve as Language Director." *Science News* 141 (March 7, 1992).

Levitan, J. B., and L. K. Kaczmarek. *The Neuron: Cell and Molecular Biology.* New York: Oxford University Press, 1991.

Lindemann, B. "Sweet and Salty: Transduction in Taste." *News in Physiological Sciences* 10 (August 1995).

Lipkin, R. "Neural Code Breakers; What language do neurons use to communicate?" *Science News* 149 (June 22, 1996).

Livingstone, M. "Art, Illusion, and the Visual System." *Scientific American* 258 (January 1988).

Llinás, R. R., ed. *The Biology of the Brain: From Neurons to Networks.* New York: W. H. Freeman, 1995.

______ . *The Workings of the Brain: Development, Memory and Perception.* New York: Scientific American Books, 1990.

Lundberg, J. M., J. Pernow, and J. S. Lacroix. "Neuropeptide Y: Sympathetic Cotransmitter and Modulator?" *News in Physiological Sciences* 4 (February 1989).

Masland, R. H. "The Functional Architecture of the Retina." *Scientific American* 255 (December 1986).

McKean, K. "Pain." *Discover* 7 (October 1986).

McLaughlin, S. and R. F. Margolskee. "The Sense of Taste." *American Scientist* 82 (November–December 1994).

Melzack, R. "Phantom Limbs." *Scientific American* 266 (April 1992).

______ . "Memories Might Be Made of This." *Science News* 139 (May 25, 1991).

______ . "Memory in a Neuron." *Scientific American* 260 (January 1989).

Miller, S. K. "Picking Up Parkinson's Pieces." *Discover* 12 (May 1991).

______ . "Molecular Custodians Sweep Away Odorants." *Science News* 136 (December 9, 1989).

Montgomery, G. "Molecules of Memory." *Discover* 10 (December 1989).

Müller, E. E., and G. Nistico. *Brain Messengers and the Pituitary.* San Diego: Academic Press, 1989.

Nathans, J. "The Genes for Color Vision." *Scientific American* 260 (February 1989).

______ . "The Nerve-Muscle Connection." *Discover* 15 (May 1994).

Orkand, R. K. and S. C. Opava. "Glial Function in Homeostasis of the Neuronal Microenvironment." *News in Physiological Sciences* 9 (December 1994).

Ottoson, D. *Physiology of the Nervous System.* New York: Oxford University Press, 1983.

______ . "Parkinson's Progress." *Discover* 13 (March 1992).

Pennisi, E. "Microglial Madness." *Science News* 144 (December 4, 1993).

Pennisi, E. "A Molecular Whodunit: New Twists in the Alzheimer's Mystery." *Science News* 145 (January 1, 1994).

Petit, T. L., and G. O. Ivy, eds. *Neural Plasticity: A Lifespan Approach.* New York: Alan R. Liss, 1987.

Piani, D., D. B. Constam, K. Frei, and A. Fontana. "Macrophages in the Brain: Friends or Enemies." *News in Physiological Sciences* 9 (April 1994).

Pickles, J. O. *An Introduction to the Physiology of Hearing.* San Diego: Academic Press, 1988.

Plum, F. *Handbook of Physiology, Section 1, Volume V, Parts 1 & 2: Higher Functions of the Brain.* New York: Oxford University Press, 1987.

______ . "Protein Protects, Restores Neurons." *Science News* 147 (January 28, 1995).

Pujol, R., M. Eybalin, and J-L Puel. "Recent Advances in Cochlear Neurotransmission: Physiology and Pathophysiology." *News in Physiological Sciences* 10 (August 1995).

Radgetsky, P. "The Brainiest Cells Alive." *Discover* 12 (April 1991).

Raichle, M. E. "Visualizing the Mind." *Scientific American* 270 (April 1994).

Richardson, S. "The Smell Files." *Discover* 15 (August 1995).

Risau, W. "Differentiation of Blood-Brain Barrier Endothelium." *News in Physiological Sciences* 4 (August 1989).

Rock, I., ed. *The Perceptual World.* New York: W. H. Freeman, 1995.

Roses, A. D. "Apolipoprotein E and Alzheimer Disease." *Scientific American: Science & Medicine* 2 (September/ October 1995).

Roush, W. "Envisioning an Artificial Retina." *Science* 268 (May 1995).

Schnapf, J. L., and D. A. Baylor. "How Photoreceptor Cells Respond to Light." *Scientific American* 256 (April 1987).

Selkoe, D. J. "Amyloid Protein and Alzheimer's Disease." *Scientific American* 265 (November 1991).

Shatz, C. J. "The Developing Brain." *Scientific American* 267 (September 1992).

Shepherd, G. M. *Neurobiology.* New York: Oxford University Press, 1988.

Shreeve, J. "Touching the Phantom." *Discover* 14 (June 1993).

Springer, S. P., and G. Deutsch. *Left Brain, Right Brain.* New York: W. H. Freeman, 1995.

Streit, W. J. and C. A. Kincaid-Colton. "The Brain's Immune System." *Scientific American* 273 (November 1995).

Stryer, L. "The Molecules of Visual Excitation." *Scientific American* 251 (July 1987).

Stux, G. *Basics of Acupuncture* 2d ed. Berlin: Springer-Verlag, 1994.

Tuomanen, E. "Breaching the Blood-Brain Barrier." *Scientific American* 268 (February 1993).

Wang, M., and A. Freeman. *Neural Function.* Boston: Little, Brown, 1987.

Weiss, R. "Shadows of Thoughts Revealed." *Science News* 138 (November 10, 1990).

Zeki, S. "The Visual Image in Mind and Brain." *Scientific American* 267 (September 1992).

Zivin, J. A., and D. W. Choi. "Stroke Therapy." *Scientific American* 265 (July 1991).

Zuker, C. S. "A Taste of Things to Come." *Nature* 376 (August 1995).

Muscle
(CHAPTER 8)

Bourne, G. H., ed. *The Structure and Function of Muscle.* 2d ed. 4 vols. New York: Academic Press, 1972–1974.

Brooks, V. B., ed. *Handbook of Physiology, Section 1, Volume II, Parts 1 & 2: Motor Control.* New York: Oxford University Press, 1981.

Cope, Timothy C. and M. J. Pinter. "The Size Principle: Still Working After All These Years." *News in Physiological Sciences* 10 (December 1995).

Entman, M. L., and W. B. Van Winkle. *Sarcoplasmic Reticulum in Muscle Physiology.* Boca Raton, Fla.: CRC Press, 1986.

Gaesser, G. A., and G. A. Brooks. "Metabolic Bases of Excess Post-Exercise Oxygen Consumption: A Review." *Medicine and Science in Sports and Exercise* 16 (1984).

Hoberman, J. M. and C. E. Yesalis. "The History of Synthetic Testosterone." *Scientific American* 272 (February 1995).

Hochachka, P. W., ed. *Muscles as Molecular and Metabolic Machines.* Boca Raton: CRC Press, Inc., 1994.

Huxley, A. *Reflections on Muscle.* Princeton, N.J.: Princeton University Press, 1980.

Jones, D. A., and J. M. Round. *Skeletal Muscle in Health and Disease: A Textbook of Muscle Physiology.* Manchester, England: Manchester University Press, 1990.

Junge, D. *Nerve and Muscle Excitation.* 3d ed. Sunderland, Mass.: Sinauer Associates, 1992.

Keynes, R. D., and D. J. Aidley. *Nerve and Muscle.* Cambridge: Cambridge University Press, 1981.

Komi, P. V., ed. *Strength and Power in Sport.* London: Blackwell Scientific Publications, 1992.

______ . "Muscle molecules: A textbook picture." *Science News* 145 (March 5, 1994).

Peachey, L. D. and R. H. Adrian, eds. *Handbook of Physiology, Section 10: Skeletal Muscle.* New York: Oxford University Press, 1983.

Prosser, C. L. "Smooth Muscle: Diversity and Rhythmicity." *News in Physiological Sciences* 7 (June 1992).

Rios, E., and G. Pizarro. "Voltage Sensors and Calcium Channels of Excitation-Contraction Coupling." *News in Physiological Science* 3 (December 1988).

Shephard, R. J., and P. O. Aistrand, eds. *Endurance in Sport.* London: Blackwell Scientific Publications, 1992.

Squire, J. M. *Molecular Mechanisms in Muscular Contraction.* Boca Raton, Fla.: CRC Press, 1990.

"3-D Atomic View of Muscle Molecule." *Science News* 144 (July 3, 1993).

Circulatory System
(CHAPTERS 9, 10, AND 11)

Berne, R. M., ed. *Handbook of Physiology, Section 2, Volume I: The Heart.* New York: Oxford University Press, 1979.

Berne, R. M. *Cardiovascular Physiology.* 6th ed. St. Louis: C. V. Mosby, 1991.

"A Better Red." *Scientific American* 266 (February 1992).

Bevan, J. A. and R. D. Bevan. "Is Innervation a Prime Regulator of Cerebral Blood Flow?" *News in Physiological Sciences* 8 (August 1993).

______ . "Blood Flow." *Discover* 13 (December 1992).

Bohr, D. F., A. P. Somlyo, and H. V. Sparks, Jr., eds. *Handbook of Physiology, Section 2, Volume II: Vascular Smooth Muscle.* New York Oxford University Press, 1980.

Brown, M. S. and J. Goldstein. "How LDL Receptors Influence Cholesterol and Atherosclerosis." *Scientific American* 251 (November 1984).

______ . "Can lipoprotein(a) foretell heart trouble?" *Science News* 144 (November 13, 1993).

Fishman, A. P. and D. W. Richards, eds. *Circulation of the Blood: Men and Ideas.* Bethesda: American Physiological Society, 1982.

Fozzard, H. A., E. Haber, R. B. Jennings, A. M. Katz, H. E. Morgan, eds. *The Heart and Cardiovascular System, Scientific Foundations.* 2d ed. 2 vols. New York: Raven Press, 1991.

Garfein, O. B., ed. *Current Concepts in Cardiovascular Physiology.* San Diego: Academic Press, 1990.

Gewirtz, H. "The Coronary Circulation: Limitations of Current Concepts of Metabolic Control." *News in Physiological Sciences* 6 (December 1991).

Goerke, J., and A. H. Mines. *Cardiovascular Physiology* 2d ed. New York: Raven Press, 1995.

Golde, D. W. "The Stem Cell." *Scientific American* 265 (December 1991).

______ . "Hemoglobin Molecule's Secret Revealed." *Science News* 149 (March 23, 1996).

Honig, C. R. *Modern Cardiovascular Physiology.* 2d ed. Boston: Little, Brown, 1988.

______ . "Immune cells trigger attack on plaque." *Science News* 146 (October 22, 1994).

Jordan, D. and J. M. Marshall, eds. *Cardiovascular Regulation.* Colchester, UK: Portland Press, 1995.

Katz, A. M. *Physiology of the Heart.* 2d ed. New York: Raven Press, 1992.

Lawn, R. M. "Lipoprotein(a) in Heart Disease." *Scientific American* 266 (June 1992).

Little, R. C. *Physiology of the Heart and Circulation.* 4th ed. Chicago: Year Book Medical Publishers, 1989.

Lüscher, T. F. and Y. Dohi. "Endothelium-Derived Relaxing Factor and Endothelin in Hypertension." *News in Physiological Sciences* 7 (June 1992).

McHale, N. G. "Role of the Lymph Pump and Its Control." *News in Physiological Sciences* 10 (June 1995).

Mellander, S. and J. Björnberg. "Regulation of Vascular Smooth Muscle Tone and Capillary Pressure." News in Physiological Sciences 7 (June 1992).

Miller, V. M. "Interactions between Neural and Endothelial Mechanisms in Control of Vascular Tone." *News in Physiological Science* 6 (April 1991).

Mohrman, D. E., and L. J. Heller. *Cardiovascular Physiology.* 3d ed. New York: McGraw-Hill, 1991.

______ . "A New Treatment for Sickle Cells." *Discover* 14 (July 1993).

Oliwenstein, L. "Liquid Assets." *Discover* 14 (September 1993).

Opie, L. H. *The Heart: Physiology and Metabolism, Second edition.* New York: Raven Press, 1991.

Porzig, H. "Signaling Mechanisms in Erythropoiesis: New Insights." *News in Physiological Sciences* 6 (December 1991).

Radetsky, P. "The Mother of All Blood Cells." *Discover* 16 (March 1995).

Renkin, E. and C. C. Michel, eds. *Handbook of Physiology, Section 2, Volume IV, Parts 1 & 2: Microcirculation.* New York: Oxford University Press, 1984.

Richardson, S. "Fixing Hemophilia." *Discover* 16 (August 1995).

Rodgers, G. P., C. T. Noguchi, and A. N. Schechter. "Sickle Cell Anemia." *Scientific American: Science & Medicine* 1 (September/October 1994).

Schmid-Schönbein, G. W. and B. W. Zweifach. "Fluid Pump Mechanisms in Initial Lymphatics." *News in Physiological Sciences* 9 (April 1994).

______ . "Scientists hail platelet factor's promise." *Science News* 146 (October 8, 1994).

Seachrist, L. "Shocking Rhythms: How do jolts of electricity bring people back to life after heart fibrillations?" *Science News* 149 (January 27, 1996).

Shepherd, J. T., and P. M. Vanhoutte, *The Human Cardiovascular System: Facts and Concepts.* New York: Raven Press, 1979.

Shepherd, J. T., ed. *Handbook of Physiology, Section 2, volume III, Parts 1 & 2: Peripheral Circulation and Organ Blood Flow.* New York: Oxford University Press, 1983.

Smith, J. J. *Circulatory Physiology: The Essentials.* 3d ed. Baltimore: Williams & Wilkins, 1990.

Smith, J. J., and J. P. Kampine. *Circulatory Physiology.* Baltimore: Williams & Wilkins, 1984.

Snyder, S. H., and D. S. Bredt. "Biological Roles of Nitric Oxide." *Scientific American* 266 (May 1992).

Sparks, H. V. *Essentials of Cardiovascular Physiology.* Minneapolis: University of Minnesota Press, 1987.

Vanhoutte, P. M., and T. F. Luscher, eds. *Endothelium-Derived Vasoactive Factors.* Boca Raton, Fla.: CRC Press, 1990.

Winslow, R. M. "Blood substitutes—a moving target." *Nature Medicine* 1 (November 1995).

Woolf, N. and M. J. Davies. "Arterial Plaque and Thrombus Formation." *Scientific American: Science & Medicine* 1 (September/October 1994).

Immune Defense
(CHAPTER 12)

Barrett, J. T. *Textbook of Immunology.* 5th ed. St. Louis: C. V. Mosby, 1988.

Black, P. H. "Psychoneuroimmunology: Brain and Immunity," *Scientific American: Science & Medicine* 2 (November/December 1995).

Caldwell, M. "Crucibles in the Cell." *Discover* 16 (January 1995).

Edelson, R. L., and J. M. Fink. "The Immunologic Function of Skin." *Scientific American* 252 (June 1985).

______ . "Electric pulses pour drugs through skin." *Science News* 144 (November 20, 1993).

Engelhard, V. H. "How Cells Process Antigens." *Scientific American* 271 (August 1994).

Fuchs, E. "Clues to B-cell memory," *Nature Medicine* 2 (July 1996).

Greaves, M. F., J. J. T. Owen, and M. C. Raff. "T and B Lymphocytes: Origins, Properties, and Roles in Immune Responses." *Excerpta Medica.* Amsterdam: 1973.

______ . "Immune presentation: In the groove II." *Science News* 145 (March 26, 1994).

______ . "Interferons: The Future is Now." *Harvard Health Letter* 19 (September 1994).

Janeway, Jr., C. A. "How the Immune System Recognizes Invaders." *Scientific American* 269 (September 1993).

Johnson, H. M., J. K. Russell, and C. H. Pontzer. "Superantigens in Human Disease." *Scientific American* 266 (April 1992).

Johnson, H. M., F. W. Bazer, B. E. Szente, and M. A. Jarpe. "How Interferons Fight Disease." *Scientific American* 270 (May 1994).

Katz, A. M. *Physiology of the Heart, Second edition.* New York: Raven Press, 1992.

Lichtenstein, L. M. "Allergy and the Immune System." *Scientific American* 269 (September 1993).

Marrack, P., and J. Kappler. "The T Cell and Its Receptor." *Scientific American* 254 (February 1986).

Marrack, P. and J. W. Kappler. "How the Immune System Recognizes the Body." *Scientific American* 269 (September 1993).

Marsh, J. A. and M. D. Kendall. *The Physiology of Immunity.* Boca Raton: CRC Press, Inc., 1996.

Milstein, C. "From Antibody Structure to Immunological Diversification of Immune Response." *Science* 231 (March 1986).

"Overview: Tolerating Self." *Scientific American* 263 (September 1990).

Paul, W. E. "Infectious Diseases and the Immune System." *Scientific American* 269 (September 1993).

Pendick, D. "Ties that Bind." *Science News* 143 (January 1993).

Pennisi, E. "Pinning down T cell death in the thymus." *Science News* 146 (November 5, 1994).

Radesky, P. "Of Parasites & Pollens." *Discover* 14 (September 1993).

Rennie, J. "The Body Against Itself." *Scientific American* 263 (December 1990).

Romagnani, S. "T_H1 and T_H2 Subsets of CD4+ Lymphocytes." *Scientific American: Science & Medicine* 1 (May/June 1994).

______ . "A Room of Their Own." *Science News* 145 (May 21, 1994).

Schwartz, R. H. "T Cell Anergy." *Scientific American* 269 (September 1993).

______ . "The skin we're in." *Discover* 15 (November 1994).

______ . "Sounding out a better way to deliver drugs." *Science News* 148 (August 12, 1995).

Steinman, L. "Autoimmune Disease." *Scientific American* 269 (September 1993).

von Boehmer, H., and P. Kisielow. "How the Immune System Learns about Self." *Scientific American* 265 (October 1991).

Respiratory System
(CHAPTER 13)

Cherniack, N. S. and J. G. Widdicombe, eds. *Handbook of Physiology, Section 3, Volume II, Parts 1 & 2: Control of Breathing.* New York: Oxford University Press, 1986.

Crystal, R. G., J. B. West, P. J. Barnes, N. S. Cherniak, and E. R. Weibel, eds. *The Lung, Scientific Foundations.* 2 vols. New York: Raven Press, 1991.

Decramer, M. "Respiratory Muscle Interaction." *News in Physiological Sciences* 8 (June 1993).

Duffin, J., K. Ezure, and J. Lipski. "Breathing Rhythm Generation: Focus on the Rostral Ventrolateral Medulla." *News in Physiological Sciences* 10 (June 1995).

Farhi, L. E. and S. M. Tenney, eds. *Handbook of Physiology, Section 3, Volume IV: Gas Exchange.* New York: Oxford University Press, 1987.

Hlastala, M. P. and A. J. Berger, eds. *Physiology of Respiration.* New York: Oxford University Press, 1996.

Houston, C. S. "Mountain Sickness." *Scientific American* 267 (October 1992).

MacKlem, P. T. and J. Mead, eds. *Handbook of Physiology, Section 3, Volume III, Parts 1 & 2: Mechanics of Breathing.* New York: Oxford University Press, 1986.

Malthew, O. P., ed. *Respiratory Function of the Upper Airway.* New York: Dekker, 1988.

Mines, A. H. *Respiratory Physiology, Third edition.* New York: Raven Press, 1993.

Moon, R. E., R. D. Vann, and P. B. Bennett. "The Physiology of Decompression Illness." *Scientific American* 273 (August 1995).

Nicholas, T. E. "Control of Turnover of Alveolar Surfactant." *News in Physiological Sciences* 8 (February 1993).

Taylor, A. E., K. Rehder, R. E. Hyatt, and J. C. Parker. *Clinical Respiratory Physiology.* Philadelphia: W. B. Saunders, 1989.

West, J. B. *Respiratory Physiology: The Essentials.* 4th ed. Baltimore: Williams & Wilkins, 1990.

West, J. B., ed. *Respiratory Physiology: People and Ideas.* New York: Oxford University Press, 1996.

Urinary System and Acid-Base Balance
(CHAPTERS 14 AND 15)

Armstrong, D. L. and R. E. White. "Natriuretic Peptides and Receptors." *Scientific American: Science & Medicine* 1 (March/April 1994).

Brenner, B. M., and F. C. Rector. *The Kidney.* Philadelphia: W. B. Saunders, 1991.

Cowley, A. W., J. F. Liard, and D. A. Ausiello, eds. *Vasopressin.* New York: Raven Press, 1988.

Davenport, H. W. *The ABC of Acid-Base Chemistry.* 7th ed. Chicago: University of Chicago Press, 1978.
Dempster, J. A., A. N. VanHoek, and C. H. Van Os. "The Quest for Water Channels." *News in Physiological Sciences* 7 (August 1992).
Gottschalk, C. W., R. W. Berliner, and G. H. Giebisch. *Renal Physiology: People and Ideas.* New York: Oxford University Press, 1987.
Grossman, S. P. *Thirst and Sodium Appetite.* San Diego: Academic Press, 1990.
Haperin, M. L., and M. B. Goldstein. *Fluid, Electrolyte and Acid-Base Physiology,* 2d ed. Philadelphia: W. B. Saunders, 1994.
Ito, S. "Role of Nitric Oxide in Glomerular Arterioles and Macula Densa." *News in Physiological Sciences* 9 (June 1994).
Klah, S. *The Kidney and Body Fluids in Health and Disease.* New York: Plenum Medical Book Co., 1982.
Koushanpour, E., and W. Kriz. *Renal Physiology: Principles, Structure, and Function.* 2d ed. New York: Springer-Verlag, 1986.
Lang, F., G. L. Busch, H. Völkl, and D. Häussinger. "Cell Volume: A Second Message in Regulation of Cellular Function." *News in Physiological Sciences* 10 (February 1995).
Marsh, D. J. *Renal Physiology.* New York: Raven Press, 1983.
Opie, L. H. "ACE Inhibitors: Almost Too Good to be True." *Scientific American: Science & Medicine* 1 (July/August 1994).
______ . *The Regulation of Acid-Base Balance.* New York: Raven Press, 1989.
______ . *The Regulation of Potassium Balance.* New York: Raven Press, 1989.
______ . *The Regulation of Sodium and Chloride Balance.* New York: Raven Press, 1989.
Robinson, J. R. *Reflections on Renal Function.* 2d ed. London: Blackwell Scientific Publications, 1988.
Rose, B. D. *Clinical Physiology of Acid-Base and Electrolyte Disorders.* 4th ed. New York: McGraw-Hill, 1994.
Ruskoaho, H. and O. Vuolteenaho. "Regulation of Atrial Natriuretic Peptide Secretion." *News in Physiological Sciences* 8 (December 1993).
Seldin, D. W., and G. Giebisch, eds. *The Kidney: Physiology and Pathophysiology.* 2d ed. New York: Raven Press, 1992.
Seldin, D. W. and G. H. Giebisch. *Clinical Disturbances of Water Metabolism.* New York: Raven Press, 1993.
Strange, K. "Are All Cell Volume Changes the Same?" *News in Physiological Sciences* 9 (October 1994).
Sullivan, L. P., and J. J. Grantham. *Physiology of the Kidney.* 2d ed. Philadelphia: Lea & Febiger, 1982.
Valtin, H. *Renal Function: Mechanisms Preserving Fluid and Solute Balance in Health.* 2d ed. Boston: Little, Brown, 1983.
Vander, A. J. *Renal Physiology.* 5th ed. New York: McGraw-Hill, 1995.
Windhager, E. E., ed. *Handbook of Physiology, Section 8, Volumes I and II: Renal Physiology.* New York: Oxford University Press, 1992.
Wolf, G. and E. G. Neilson. "Angiotensin II as a Renal Cytokine." *News in Physiological Sciences* 9 (February 1994).

Digestive System
(CHAPTER 16)

Arias, I. M., J. L. Boyer, N. Fausto, W. B. Jakoby, D. Schachter, and D. A. Shafritz, eds. *The Liver: Biology and Pathobiology, Third edition.* New York: Raven Press, 1994.
______ . "As the Stomach Churns." *Discover* 15 (August 1994).
Blaser, M. J. "The Bacteria behind Ulcers." *Scientific American* 274 (February 1996).
______ . "Defending us from our dirty mouths." *Science News* 147 (March 18, 1995).
Cooke, H. J. "Role of the 'Little Brain' in the Gut in Water and Electrolyte Homeostasis." *The FASEB Journal* 3 (February 1989).
Davenport, H. W. *A Digest of Digestion.* 2d ed. Chicago: Year Book Medical Publishers, 1978.
Field, M. and R. A. Frizzell, eds. *Handbook of Physiology, Section 6, Volume IV: The Gastrointestinal System: Intestinal Absorption and Secretion.* New York: Oxford University Press, 1991.
Forte, J. G. and S. G. Schultz, eds. *Handbook of Physiology, Section 6, Volume III: The Gastrointestinal System: Salivary, Gastric, Pancreatic, and Hepatobiliary Secretion.* New York: Oxford University Press, 1989.
González-Gallego, J. "New Concepts in Hepatocellular Transport and Metabolism of Bilirubin." *News in Physiological Sciences* 10 (February 1995).
Granger, D. N., J. A. Barrowman, and D. R. Kvietys. *Clinical Gastrointestinal Physiology.* Philadelphia: W. B. Saunders, 1985.
Green, S. J. "Nitric Oxide in Mucosal Immunity." *Nature Medicine* 1 (June 1995).
Grélot, L. and A. D. Miller. "Vomiting–Its Ins and Outs." *News in Physiological Sciences* 9 (June 1994).
Johnson, L. R., ed. *Physiology of the Gastrointestinal Tract, 3d ed.* New York: Raven Press, 1994.
Lewis, L. D., and J. A. Williams. "Cholecystokinin: A Key Integrator of Nutrient Assimilation." *News in Physiological Sciences* 5 (August 1990).
Makhlouf, G. M. and S. G. Schultz, eds. *Handbook of Physiology, Section 6, Volume II: The Gastrointestinal System: Neural and Endocrine Biology.* New York: Oxford University Press, 1989.
______ . *Physiology of the Digestive Tract.* 5th ed. Chicago: Year Book Medical Publishers, 1982.
Richardson, S. "Tongue Bugs." *Discover* 16 (October 1995).
Schiller, L. R. "Peristalsis." *Scientific American: Science & Medicine* 1 (November/December 1994).
Sherman, J. H. *Gastrointestinal Physiology.* New York: McGraw Hill, 1992.
Wood, J. D., and S. G. Schultz, eds. *Handbook of Physiology, Section 6, Volume I, Parts 1 & 2: The Gastrointestinal System: Motility and Circulation.* New York: Oxford University Press, 1989.

Wood, J. D. "Communication between Minibrain in Gut and Enteric Immune System. *News in Physiological Sciences* 6 (April 1991).

Energy Balance and Temperature Regulation
(CHAPTER 17)

Bergh, C. and P. Södersten. "Anorexia nervosa, self-starvation and the reward of stress." *Nature Medicine* 2 (January 1996).

______ . "Fifth obesity gene found in mice," *Science News* 149 (April 27, 1996).

Gibbs, W. W. "Gaining on Fat," *Scientific American* 275 (August 1996).

Hamilton, E. M. N., E. N. Whitney, and F. S. Sizer. *Nutrition: Concepts and Controversies.* 5th ed. St. Paul: West Publishing Co., 1991.

Harris, R. B. S. "Role of Set-Point Theory in Regulation of Body Weight." *The FASEB Journal* 4 (December 1990).

Jéquier, E. "Body Weight Regulation in Humans: The Importance of Nutrient Balance." *News in Physiological Sciences* 8 (December 1993).

Lönnquist, F., P. Arner, L. Nordors, and M. Schalling. "Overexpression of the obese (ob) gene in adipose tissue of human obese subjects." *Nature Medicine* 1 (September 1995).

______ . "New chapters in the leptin tale." *Science News* 149 (March 16, 1996).

______ . "Obesity Hormone is No Magic Bullet for Weight Loss." *Environmental Nutrition* (October 1995).

______ . "Obesity Insights." *Harvard Women's Health Watch* III (October 1995).

Pennisi, E. "Fanfare over Finding First Fat Gene." *Science News* 146 (December 3, 1994).

Raloff, J. "Body temperature: Don't look for 98.6°F." *Science News* 142 (September 26, 1992).

Seachrist, L. and J. Travis. "Hormone triggers cells to turn to fat." *Science News* 148 (December 9, 1995).

Shitzer, A., and R. Eberhart, eds. *Heat Transfer in Medicine and Biology.* New York: Plenum Press, 1985.

Travis, J. "Mouse Obesity Cured by Hormone." *Science News* 148 (July 29, 1995).

Vogel, S. "The Mouse on the Left Needs Leptin." *Discover* 17 (January 1996).

Weiss, P. L. "Fat and Fiction." *Science News* 138 (September 1, 1990).

Westerterp-Plantenga, M. S., E. W. H. M. Fredix, A. B. Steffens, and H. R. Kissileff, eds. *Food Intake and Energy Expenditure.* Boca Raton: CRC Press, Inc., 1994.

______ . "Why Do We Stop Eating?" *News in Physiological Sciences* 5 (April 1990).

Endocrine and Reproductive Systems
(CHAPTERS 18, 19 AND 20)

Alexander, N. J. "Future Contraceptives." *Scientific American* 273 (September 1995).

______ . "Assisted Reproduction." *Harvard Women's Health Watch* (April 1995)

Atkinson, M. A., and N. K. Maclaren. "What Causes Diabetes?" *Scientific American* 263 (July 1990).

Bolander, F. F. *Molecular Endocrinology* 2d ed. San Diego: Academic Press, 1994.

Burger, H., and D. DeKretser. *Comprehensive Endocrinology: The Testis* 2d ed. New York: Raven Press, 1989.

Carmichael, S. W., and H. Winkler. "The Adrenal Chromaffin Cell." *Scientific American* 253 (August 1985).

Challis, J. R. G. "CRH, a placental clock and preterm labour." *Nature Medicine* 1 (May 1995).

Crapo, L. *Hormones: Messengers of Life.* New York: W. H. Freeman, 1995.

Davis, D. D. and H. L. Bradlow. "Can Environmental Estrogens Cause Breast Cancer?" *Scientific American* 273 (October 1995).

"Deciding to Be Born." *Discover* 13 (May 1992).

Diamond, J. "Why Women Change." *Discover* 17 (July 1996).

Ebadi, M., M. Samejima, and R. F. Pfeiffer. "Pineal Gland in Synchronizing and Refining Physiological Events." *News in Physiological Sciences* 8 (February 1993).

"Eggs Not Silent Partners in Conception." *Science News* 139 (April 6, 1991).

Elmer-Dewitt, P. "Making Babies." *Time.* September 30, 1991.

Evans, R. M. "The Steroid and Thyroid Hormone Receptor Superfamily." *Science* 240 (May 13, 1988).

Fackelmann, K. A. "Sex Protection: Balancing the Equation." *Science News* 141 (March 14, 1992).

Fackelmann, K. A. "Cloning Human Embryoes: Exploring the science of a controversial experiment." *Science News* 145 (February 5, 1994).

"Fetus Tells Mother: It's Time for Labor." *Science News* 140 (September 21, 1991).

Frayn, K. N. *Metabolic Regulation: A Human Perspective.* Colchester, UK: Portland Press, 1996.

Goodman, H. M. *Basic Medical Endocrinology, 2d ed.* New York: Raven Press, 1994.

Griffin, J. and S. Ojeda, eds. *Textbook of Endocrine Physiology, 3d ed.* New York: Oxford University Press, 1996.

Hedge, G. A., H. D. Colby, and R. L. Goodman, *Clinical Endocrine Physiology.* Philadelphia: W. B. Saunders, 1987.

Henry, J. P. "Biological Basis of the Stress Response." *News in Physiological Sciences* 8 (April 1993).

"Hormone May Restore Muscle in Elderly." *Science News* 138 (July 14, 1990).

Illnerová, H. "Mammalian Circadian Clock and Its Resetting." *News in Physiological Sciences* 6 (June 1991).

James, D. E. "The Mammalian Facilitative Glucose Transporter Family." *News in Physiological Sciences* 10 (April 1995).

Knobil, E., J. D. Neill, G. S. Greenwald, C. L. Markert, and D. W. Pfaff, eds. *The Physiology of Reproduction, Third edition.* New York: Raven Press, 1994.

Lacy, P. E. "Treating Diabetes with Transplanted Cells." *Scientific American* 273 (July 1995).

Lamb, J. C., IV, ed. *Physiology and Toxicology of Male Reproduction.* San Diego: Academic Press, 1988.

Lienhard, G. E., J. W. Slot, D. E. James, and M. M. Mueckler. "How Cells Absorb Glucose." *Scientific American* 266 (January 1992).

Martin, C. R. *Endocrine Physiology.* New York: Oxford University Press, 1985.

Martin, C. R. *The Dictionary of Endocrinology and Related Biomedical Sciences.* New York: Oxford University Press, 1995.

McCann, S. M., ed. *Endocrinology: People and Ideas.* New York: Oxford University Press, 1988.

McLean, M., A. Bisits, J. Davies, R. Woods, P. Lowry, and R. Smith. "A placental clock controlling the length of human pregnancy." *Nature Medicine* 1, no. 5 (May 1995).

______ . "Melatonin: 'Miracle Hormone' or Media Hype?" *The University of Texas–Houston Health Science Center Lifetime Health Letter* 7 (November 1995).

Moran, N. "Man-made chemicals and reproductive health." *Nature Medicine* 1 (September 1995).

Müller, E. E., and G. Nisticò. *Brain Messengers and the Pituitary.* San Diego: Academic Press, 1989.

"New Clues to Diabetes' Cause and Treatment." *Science News* 140 (December 21 and 28, 1991).

Newman, J. "How Breast Milk Protects Newborns." *Scientific American* 273 (December 1995).

Norman, A. W., and G. Litwack. *Hormones.* San Diego: Academic Press, 1987.

Orci, L., J. Vassali, and A. Perrelet. "The Insulin Factory." *Scientific American* 259 (September 1988).

______ . "Ovulation heralds end of fertile period." *Science News* 148 (December 16, 1995).

Pennisi, E. "Diabetes Stopped Before It Starts." *Science News* 144 (November 6, 1993).

Raloff, J. "Ecocancers: Do environmental factors underlie a breast cancer epidemic?" *Science News* 144 (July 3, 1993).

Raloff, J. "That Feminine Touch: Are men suffering from prenatal or childhood exposures to 'hormonal' toxicants?" *Science News* 145 (January 22, 1994).

Raloff, J. "Drug of Darkness: Can a pineal hormone head off everything from breast cancer to aging?" *Science News* 147 (May 13, 1995).

Raloff, J. "Beyond Estrogens: Why unmasking hormone-mimicking pollutants proves so challenging." *Science News* 148 (July 15, 1995).

Rennie, J. "Malignant Mimicry: False estrogens may cause cancer and lower sperm counts." *Scientific American* 269 (September 1993).

______ . "Reproductive equality: A male Pill?" *Science News* 149 (June 22, 1996).

Richardson, S. "Not by Testosterone Alone." *Discover* 16 (April 1995).

Sarrel, P. M., E. G. Lufkin, M. J. Oursler, and D. Keefe. "Estrogen Actions in Arteries, Bone, and Brain." *Scientific American: Science & Medicine* 1 (July/August 1994).

Schwartz, W. J. "Internal Timekeeping." *Scientific American: Science & Medicine* 3 (May/June 1996).

Serra, G. *Comprehensive Endocrinology: The Ovary.* New York: Raven Press, 1983.

Smith, D. F. "Steroid Receptors and Molecular Chaperones." *Scientific American: Science & Medicine* 2 (July/August 1995).

Smith, S. M. "Melatonin: Sleep Aid, Jet-Lag Antidote, Plus the Promise of More." *Environmental Nutrition* (November 1995).

Snyder, S. H. "The Molecular Basis of Communication between Cells." *Scientific American* 253 (October 1985).

Spera, G., A Fabbrini, L. Gnessi, and C. W. Bardin, eds. *Molecular and Cellular Biology of Reproduction.* New York: Raven Press, 1992.

Stroh, M. "The Root of Impotence: Does Nitric Oxide Hold the Key?" *Science News* 142 (July 4, 1992).

Tepperman, J. *Metabolic and Endocrine Physiology.* 5th ed. Chicago: Year Book Medical Publishers, 1987.

Travis, J. "Hormonal clock predicts premature births." *Science News* 147 (April 29, 1995).

Ulmann, A., G. Teutsch, and D. Philibert. "RU 486." *Scientific American* 262 (June 1990).

Villanúa, M. A., C. Agrasal, and A. I. Esquifino. "New Perspectives in the Research of Pineal Gland." *News in Physiological Sciences* 5 (February 1990).

Wilson, J. D., and D. W. Foster, eds. *Williams' Textbook of Endocrinology.* 8th ed. Philadelphia: W. B. Saunders, 1992.

Glossary

A band one of the dark bands that alternate with light (I) bands to create a striated appearance in a skeletal or cardiac muscle fiber when these fibers are viewed with a light microscope

absorptive state the metabolic state following a meal when nutrients are being absorbed and stored; fed state

accessory digestive organs exocrine organs outside of the wall of the digestive tract that empty their secretions through ducts into the digestive tract lumen

accessory sex glands glands that empty their secretions into the reproductive tract

accommodation the ability to adjust the strength of the lens in the eye so that both near and far sources can be focused on the retina

acetylcholine (ACh) (as'-uh-teal-KŌ-lēn) the neurotransmitter released from all autonomic preganglionic fibers, parasympathetic postganglionic fibers, and motor neurons

acetylcholinesterase (AChE) (as'-uh-teal-kō-luh-NES-tuh-rās) an enzyme present in the motor end plate membrane of a skeletal muscle fiber that inactivates acetylcholine

ACh see *acetylcholine*

AChE see *acetylcholinesterase*

acid a hydrogen-containing substance that yields a free hydrogen ion and anion on dissociation

acidosis (as-i-DŌ-sus) blood pH of less than 7.35

acini (ĀS-i-nī) the secretory component of saclike exocrine glands, such as digestive enzyme-producing pancreatic glands or milk-producing mammary glands

ACTH see *adrenocorticotropic hormone*

actin the contractile protein that forms the backbone of the thin filaments in muscle fibers

active expiration emptying of the lungs more completely than when at rest by contracting the expiratory muscles; also called *forced expiration*

active force a force that requires expenditure of cellular energy (ATP) in the transport of a substance across the plasma membrane

active reabsorption when any one of the five steps in the transepithelial transport of a substance reabsorbed across the kidney tubules requires energy expenditure

active transport active carrier-mediated transport involving transport of a substance against its concentration gradient across the plasma membrane

acuity discriminative ability; the ability to discern between two different points of stimulation

acute myocardial infarction (mī'-ō-KAR-dē-ul) death of heart muscle cells caused by disruption of their blood supply; a heart attack

adaptation a reduction in receptor potential in spite of sustained stimulation of the same magnitude

adenosine diphosphate (ADP) (uh-DEN-uh-sēn) the two-phosphate product formed from the splitting of ATP to yield energy for the cell's use

adenosine triphosphate (ATP) the body's common energy "currency," which consists of an adenosine with three phosphate groups attached; splitting of the high-energy, terminal phosphate bond provides energy to power cellular activities

adenylyl cyclase (ah-DEN-il-il sī-klās) the membrane-bound enzyme which is activated by a G protein intermediary in response to binding of an extracellular messenger with a surface membrane receptor and which in turn activates cyclic AMP, an intracellular second messenger

ADH see *vasopressin*

adipose tissue the tissue specialized for storage of triglyceride fat; found under the skin in the hypodermis

ADP see *adenosine diphosphate*

adrenal cortex (uh-DRĒ-nul) the outer portion of the adrenal gland; secretes three classes of steroid hormones: glucocorticoids, mineralcorticoids, and sex hormones

adrenal medulla (muh-DUL-uh) the inner portion of the adrenal gland; an endocrine gland that is a modified sympathetic ganglion that secretes the hormones epinephrine and norepinephrine into the blood in response to sympathetic stimulation

adrenergic fibers (ad'-ruh-NUR-jik) nerve fibers that release norepinephrine as their neurotransmitter

adrenocorticotropic hormone (ACTH) (ad-rē'-nō-kor'tuh-kō-TRŌP-ik) an anterior pituitary hormone that stimulates cortisol secretion by the adrenal cortex and promotes growth of the adrenal cortex

aerobic referring to a condition in which oxygen is available

aerobic exercise exercise that can be supported by ATP formation accomplished by oxidative phosphorylation because adequate O_2 is available to support the muscle's modest energy demands; also called *endurance-type exercise*

afferent arteriole (AF-er-ent ar-TIR-ē-ōl) the vessel that carries blood into the glomerulus of the kidney's nephron

afferent division the portion of the peripheral nervous system that carries information from the periphery to the central nervous system

afferent neuron neuron that possesses a sensory receptor at its peripheral ending and carries information to the central nervous system

after hyperpolarization (hī'-pur-pō-luh-ruh-ZĀ-shun) a slight, transient hyperpolarization that sometimes occurs at the end of an action potential

agranulocytes (ā-GRAN-yuh-lō-sīts') leukocytes that do not contain granules, including lymphocytes and monocytes

albumin (al-BEW-min) the smallest and most abundant of the plasma proteins; binds and transports many water-insoluble substances in the blood; contributes extensively to plasma-colloid osmotic pressure

aldosterone (al-dō-steer-OWN) or (al-DOS-tuh-rōn) the adrenocortical hormone that stimulates Na^+ reabsorption by the distal and collecting tubules of the kidney's nephron during urine formation.

alkalosis (al'-kuh-LŌ-sus) blood pH of greater than 7.45

allergy acquisition of an inappropriate specific immune reactivity to a normally harmless environmental substance

all-or-none law an excitable membrane either responds to a stimulus with a maximal action potential that spreads nondecrementally throughout the membrane, or it does not respond with an action potential at all

alpha cells the endocrine pancreatic cells that secrete the hormone glucagon

alpha motor neuron a motor neuron that innervates ordinary skeletal muscle fibers

alveolar surface tension (al-VĒ-ō-lur) the surface tension of the fluid lining the alveoli in the lungs; see *surface tension*

alveolar ventilation the volume of air exchanged between the atmosphere and alveoli per minute; equals (tidal volume minus dead-space volume) times respiratory rate

alveoli (al-VĒ-ō-lī) the air sacs across which O_2 and CO_2 are exchanged between the blood and air in the lungs

amines (ah-means) hormones derived from the amino acid tyrosine; includes thyroid hormone and catecholamines

amoeboid movement (uh-MĒ-boid) "crawling" movement of white blood cells, similar to the means by which amoebas move

anabolism (ah-NAB-ō-li-zum) the buildup, or synthesis, of larger organic molecules from the small organic molecular subunits

anaerobic (an'-uh-RŌ-bik) referring to a condition in which oxygen is not present

anaerobic exercise high-intensity exercise that can be supported by ATP formation accomplished by anaerobic glycolysis for brief periods of time when O_2 delivery to a muscle is inadequate to support oxidative phosphorylation

analgesic (an-al-JEE-zic) pain relieving

anatomy the study of body structure

androgen a masculinizing "male" sex hormone; includes testosterone from the testes and dehydroepiandrosterone from the adrenal cortex

anemia a reduction below normal in O_2-carrying capacity of the blood

anions (AN-ī-on) negatively charged ions that have gained one or more electrons in their outer shell

ANP see *atrial natriuretic peptide*

antagonism actions opposing each other; in the case of hormones, when one hormone causes the loss of another hormone's receptors, reducing the effectiveness of the second hormone

anterior pituitary the glandular portion of the pituitary that synthesizes, stores and secretes six different hormones: growth hormone, TSH, ACTH, FSH, LH, and prolactin

antibody an immunoglobulin produced by a specific activated B lymphocyte (plasma cell) against a particular antigen; binds with the specific antigen against which it is produced and promotes the antigenic invader's destruction by augmenting nonspecific immune responses already initiated against the antigen

antibody-mediated immunity a specific immune response accomplished by antibody production by B cells

antidiuretic hormone (an'-ti-dī-'-yū-RET-ik) see *vasopressin*

antigen a large, complex molecule that triggers a specific immune response against itself when it gains entry into the body

antioxidant a substance that helps inactivate biologically damaging free radicals

antrum (of ovary) the fluid-filled cavity formed within a developing ovarian follicle

antrum (of stomach) the lower portion of the stomach

aorta (a-OR-tah) the large vessel that carries blood from the left ventricle

aortic valve a one-way valve that permits the flow of blood from the left ventricle into the aorta during ventricular emptying but prevents the backflow of blood from the aorta into the left ventricle during ventricular relaxation

apoptosis (ā-pop-TŌ-sis) programmed cell death; deliberate self-destruction of a cell

appetite centers see *feeding centers*

aqueous humor (Ā-kwē-us) the clear watery fluid in the anterior chamber of the eye; provides nourishment for the cornea and lens

arterioles (ar-TIR-ē-ōlz) the highly muscular, high-resistance vessels, the caliber of which can be changed subject to control to determine how much of the cardiac output is distributed to each of the various tissues

artery a vessel that carries blood away from the heart

ascending tract a bundle of nerve fibers of similar function that travels up the spinal cord to transmit signals derived from afferent input to the brain

asthma an obstructive pulmonary disease characterized by profound constriction of the smaller airways caused by allergy-induced spasm of the smooth muscle in the walls of these airways

astrocyte a type of glial cell in the brain; major functions include holding the neurons together in proper spatial relationship and inducing the brain capillaries to form tight junctions important in the blood-brain barrier

atherosclerosis (ath-uh-rō-skluh-RŌ-sus) a progressive, degenerative arterial disease that leads to gradual blockage of affected vessels, thereby reducing blood flow through them

atmospheric pressure the pressure exerted by the weight of the air in the atmosphere on objects on the earth's surface; equals 760 mm Hg at sea level

ATP see *adenosine triphosphate*

ATPase an enzyme that possesses ATP-splitting ability

ATP synthetase (sin-thuh-TĀS) the enzyme within the mitochondrial inner membrane that phosphorylates ADP to ATP

atrial natriuretic peptide (ANP) (Ā-trē-al NĀ-tree-ur-eh'-tik) a peptide hormone released from the cardiac atria that promotes urinary loss of Na^+

atrioventricular (AV) node (ā'-trē-ō-ven-TRIK-yuh-lur) a small bundle of specialized cardiac cells located at the junction of the atria and ventricles that serves as the only site of electrical contact between the atria and ventricles

atrioventricular (AV) valve a one-way valve that permits the flow of blood from the atrium to the ventricle during filling of the heart but prevents the backflow of blood from the ventricle to the atrium during emptying of the heart

atrium (atria, plural) (Ā-tree-um) an upper chamber of the heart that receives blood from the veins and transfers it to the ventricle

atrophy (AH-truh-fē) decrease in mass of an organ

autoimmune disease disease characterized by erroneous production of antibodies against one of the body's own tissues

autonomic nervous system the portion of the efferent division of the peripheral nervous system that innervates smooth and cardiac muscle and exocrine glands; composed of two subdivisions, the sympathetic nervous system and the parasympathetic nervous system

autorhythmicity the ability of an excitable cell to rhythmically initiate its own action potentials

AV nodal delay the delay in impulse transmission between the atria and ventricles at the AV node to allow sufficient time for the atria to become completely depolarized and contract, emptying their contents into the ventricles, before ventricular depolarization and contraction occur

AV valve see *atrioventricular valve*

axon a single, elongated tubular extension of a neuron that conducts action potentials away from the cell body; also known as a *nerve fiber*

axon hillock the first portion of a neuronal axon plus the region of the cell body from which the axon leaves; the site of action-potential initiation in most neurons

axon terminals the branched endings of a neuronal axon, which release a neurotransmitter that influences target cells in close association with the axon terminals

baroreceptor reflex an autonomically mediated reflex response that influences the heart and blood vessels to oppose a change in mean arterial blood pressure

baroreceptors receptors located within the circulatory system that monitor blood pressure

basal metabolic rate (BĀ-sul) the minimal waking rate of internal energy expenditure; the body's "idling speed"

basal nuclei several masses of gray matter located deep within the white matter of the cerebrum of the brain; play an important inhibitory role in motor control

base a substance that can combine with a free hydrogen ion and remove it from solution

basic electrical rhythm (BER) self-induced electrical activity of the digestive tract smooth muscle

basilar membrane (BAS-ih-lar) the membrane that forms the floor of the middle compartment of the cochlea and bears the organ of Corti, the sense organ for hearing

basophils (BAY-so-fills) white blood cells that synthesize, store, and release histamine, which is important in allergic responses, and heparin, which hastens the removal of fat particles from the blood

BER see *basic electrical rhythm*

beta (β) cells the endocrine pancreatic cells that secrete the hormone insulin

bicarbonate (HCO_3^-) the anion resulting from dissociation of carbonic acid, H_2CO_3

bile salts cholesterol derivatives secreted in the bile that facilitate fat digestion through their detergent action and facilitate fat absorption through their micellar formation

biliary system (BIL-ē-air'-ē) the bile-producing system, consisting of the liver, gallbladder, and associated ducts

bilirubin (bill-eh-RŪ-bin) a bile pigment, which is a waste product derived from the degradation of hemoglobin during the breakdown of old red blood cells

bipolar neurons the nerve cells in the middle layer of the retina; synapse with the photoreceptors of the eye

blastocyst the developmental stage of the fertilized ovum by the time it is ready to implant; consists of a single-layered sphere of cells encircling a fluid-filled cavity

blood-brain barrier special structural and functional features of the brain capillaries that limit access of materials from the blood into the brain tissue

B lymphocytes (B cells) white blood cells that produce antibodies against specific targets to which they have been exposed

body of the stomach the main, or middle, part of the stomach

body system a collection of organs that perform related functions and interact to accomplish a common activity that is essential for survival of the whole body; for example, the digestive system

bone marrow the soft, highly cellular tissue that fills the internal cavities of bones and is the source of most blood cells

Bowman's capsule the beginning of the tubular component of the kidney's nephron that cups around the glomerulus and collects the glomerular filtrate as it is formed

Boyle's law (boils) at any constant temperature, the pressure exerted by a gas varies inversely with the volume of the gas

brain the most anterior, most highly developed portion of the central nervous system

brain stem the portion of the brain that is continuous with the spinal cord and that serves as an integrating link between the spinal cord and higher brain levels and controls many life-sustaining processes, such as breathing, circulation, and digestion

bronchioles (BRONG-kē-ōlz) the small, branching airways within the lungs

bronchoconstriction narrowing of the respiratory airways

bronchodilation widening of the respiratory airways

brush border the collection of microvilli projecting from the luminal border of epithelial cells lining the digestive tract and kidney tubules

buffer see *chemical buffer system*

bulbourethral glands (bul-bo-you-WREATH-ral) male accessory sex glands that secrete mucus for lubrication

bulk flow movement in bulk of a protein-free plasma across the capillary walls between the blood and surrounding interstitial fluid; encompasses ultrafiltration and reabsorption

bundle of His (hiss) a tract of specialized cardiac cells that rapidly transmits an action potential down the interventricular septum of the heart

calcitonin (kal'-suh-TŌ-nun) a hormone secreted by the thyroid C cells that lowers plasma Ca^{2+} levels

calcium balance maintenance of a constant total amount of Ca^{2+} in the body; accomplished by slowly responding adjustments in intestinal Ca^{2+} absorption and in urinary Ca^{2+} excretion

calcium homeostasis maintenance of a constant free plasma Ca^{2+} concentration; accomplished by rapid exchanges of Ca^{2+} between the bone and ECF and to a lesser extent by modifications in urinary Ca^{2+} excretion

calmodulin (kal'-MA-jew-lin) an intracellular Ca^{2+} binding protein that, upon activation by Ca^{2+}, induces a change in structure and function of another intracellular protein; especially important in smooth-muscle excitation-contraction coupling

CAMs see *cell adhesion molecules*

capillaries the thin-walled, pore-lined smallest of blood vessels, across which exchange between the blood and surrounding tissues takes place

carbonic anhydrase (an-HĪ-drās) the enzyme that catalyzes the conversion of CO_2 and H_2O into carbonic acid, H_2CO_3

cardiac cycle one period of systole and diastole

cardiac muscle the specialized muscle found only in the heart

cardiac output (CO) the volume of blood pumped by each ventricle each minute; equals stroke volume times heart rate

cardiovascular control center the integrating center located in the medulla of the brain stem that controls mean arterial blood pressure

carrier-mediated transport transport of a substance across the plasma membrane facilitated by a carrier molecule

carrier molecules membrane proteins, which, by undergoing reversible changes in shape so that specific binding sites are alternately exposed at either side of the membrane, are able to bind with and transfer particular substances unable to cross the plasma membrane on their own

cascade a series of sequential reactions that culminates in a final product, such as a clot

catabolism (kuh-TAB-ō-li-zum) the breakdown, or degradation, of large, energy-rich molecules within cells

catalase (KAT-ah-lās) an antioxidant enzyme found in peroxisomes that decomposes potent hydrogen peroxide into harmless H_2O and O_2

catecholamines (kat'-uh-KŌ-luh-means) the chemical classification of the adrenomedullary hormones

cations (KAT-ī-onz) positively charged ions that have lost one or more electrons from their outer shell

caveolae (kā-vē-Ō-lē) cavelike indentations in the outer surface of the plasma membrane that contain an abundance of membrane receptors and serve as important sites for signal transduction

C cells the thyroid cells that secrete calcitonin

cell the smallest unit capable of carrying out the processes associated with life; the basic unit of both structure and function of living organisms

cell adhesion molecules (CAMs) proteins that protrude from the surface of the plasma membrane and form loops or other appendages that the cells use to grip ahold of each other and the surrounding connective-tissue fibers

cell body the portion of a neuron that houses the nucleus and organelles

cell-mediated immunity a specific immune response accomplished by activated T lymphocytes, which directly attack unwanted cells

center a functional collection of cell bodies within the central nervous system

central chemoreceptors (kē-mō-rē-SEP-turz) receptors located in the medulla near the respiratory center that respond to changes in ECF H^+ concentration resulting from changes in arterial P_{CO_2} and adjust respiration accordingly

central lacteal (LAK-tē-ul) the initial lymphatic vessel that supplies each of the small intestinal villi

central nervous system (CNS) the brain and spinal cord

central sulcus (SUL-kus) a deep infolding of the brain surface that runs roughly down the middle of the lateral surface of each cerebral hemisphere and separates the parietal and frontal lobes

centrioles (SEN-tree-ōl) a pair of short cylindrical structures within a cell that form the mitotic spindle during cell division

cerebellum (ser'-uh-BEL-um) the portion of the brain attached at the rear of the brain stem and concerned with maintaining proper position of the body in space and subconscious coordination of motor activity

cerebral cortex the outer shell of gray matter in the cerebrum; site of initiation of all voluntary motor output and final perceptual processing of all sensory input as well as integration of most higher neural activity

cerebral hemispheres the cerebrum's two halves, which are connected by a thick band of neuronal axons

cerebrospinal fluid (ser'-uh-brō-SPĪ-nul) or (sah-REE-brō-SPĪ-nul) a special cushioning fluid that is produced by, surrounds, and flows through the central nervous system

cerebrum (SER-uh-brum) or (sah-REE-brum) the division of the brain that consists of the basal nuclei and cerebral cortex

channels small water-filled pathways through the plasma membrane; formed by membrane proteins that span the membrane and provide highly selective passage for small water-soluble substances such as ions

chemical bonds the forces holding atoms together

chemical buffer system a mixture in a solution of two or more chemical compounds that minimize pH changes when either an acid or a base is added to or removed from the solution

chemical mediator a chemical that is secreted by a cell and that influences an activity outside of the cell

chemical messenger-gated channels channels that open or close in response to the binding of a specific messenger with a membrane receptor site that is in close association with the channel

chemoreceptor (KĒ-mō-rē-sep'-tur) a sensory receptor sensitive to specific chemicals

chemotaxin (kē-mō-TAK-sin) a chemical released at an inflammatory site that attracts phagocytes to the area

chief cells the stomach cells that secrete pepsinogen

cholecystokinin (CCK) (kō'-luh-sis-tuh-kī-nun) a hormone released from the duodenal mucosa primarily in response to the presence of fat; inhibits gastric motility and secretion, stimulates pancreatic enzyme secretion, and stimulates gallbladder contraction

cholesterol a type of fat molecule that serves as a precursor for steroid hormones and bile salts and is a stabilizing component of the plasma membrane

cholinergic fibers (kō'-lin-ER-jik) nerve fibers that release acetylcholine as their neurotransmitter

chronic obstructive pulmonary disease a group of lung diseases characterized by increased airway resistance resulting from narrowing of the lumen of the lower airways; includes asthma, chronic bronchitis, and emphysema

chyme (kīm) a thick liquid mixture of food and digestive juices

cilia (SILL-ee-ah) motile, hairlike protrusions from the surface of cells lining the respiratory airways and the oviducts

ciliary body the portion of the eye that produces aqueous humor and contains the ciliary muscle

ciliary muscle a circular ring of smooth muscle within the eye whose contraction increases the strength of the lens to accommodate for near vision

circadian rhythm (sir-KĀ-dē-un) repetitive oscillations in the set point of various body activities, such as hormone levels and body temperature, that are very regular and have a frequency of one cycle every twenty-four hours, usually linked to light/dark cycles; diurnal rhythm; biological rhythm

circulatory shock when mean arterial blood pressure falls so low that adequate blood flow to the tissues can no longer be maintained

citric acid cycle a cyclical series of biochemical reactions that involves the further processing of intermediate breakdown products of nutrient molecules, resulting in the generation of carbon dioxide and the preparation of hydrogen carrier molecules for entry into the high-energy-yielding electron transport chain

CNS see *central nervous system*

cochlea (KOK-lē-uh) the snail-shaped portion of the inner ear that houses the receptors for sound

collecting tubule the last portion of tubule in the kidney's nephron that empties into the renal pelvis

colloid (KOL-oid) the thyroglobulin-containing substance enclosed within the thyroid follicles

complement system a collection of plasma proteins that are activated in cascade fashion on exposure to invading microorganisms, ultimately producing a membrane attack complex that destroys the invaders

compliance the distensibility of a hollow, elastic structure, such as a blood vessel or the lungs; a measure of how easily the structure can be stretched

concave curved in, as a surface of a lens that diverges light rays

concentration gradient a difference in concentration of a particular substance between two adjacent areas

concentric contraction an isotonic muscle contraction during which the muscle shortens

conduction transfer of heat between objects of differing temperatures that are in direct contact with each other

conduction by local current flow the means by which an action potential is propagated throughout a nonmyelinated nerve fiber; local current flow between an active and adjacent inactive area brings the inactive area to threshold, triggering an action potential in a previously inactive area

cones the eye's photoreceptors used for color vision in the light

congestive heart failure the inability of the cardiac output to keep pace with the body's needs for blood delivery, with blood damming up in the veins behind the failing heart

connective tissue tissue that serves to connect, support, and anchor various body parts; distinguished by relatively few cells dispersed within an abundance of extracellular material

contractile component the sarcomere-containing myofibrils within a muscle fiber that are capable of shortening on excitation

contractile proteins myosin and actin, whose interaction brings about shortening (contraction) of a muscle fiber

controlled variable some factor that can vary but is controlled to reduce the amount of variability and keep the factor at a relatively steady state

convection transfer of heat energy by air or water currents

convergence the converging of many presynaptic terminals from thousands of other neurons on a single neuronal cell body and its dendrites so that activity in the single neuron is influenced by the activity in many other neurons

convex curved out, as a surface in a lens that converges light rays

core temperature the temperature within the inner core of the body (abdominal and thoracic organs, central nervous system, and skeletal muscles) that is homeostatically maintained at about 100°F

cornea (KOR-nee-ah) the clear, anteriormost outer layer of the eye through which light rays pass to the interior of the eye

coronary artery disease atherosclerotic plaque formation and narrowing of the coronary arteries that supply the heart muscle

coronary circulation the blood vessels that supply the heart muscle

corpus luteum (LOO-tē-um) the ovarian structure that develops from a ruptured follicle following ovulation

cortisol (KORT-uh-sol) the adrenocortical hormone that plays an important role in carbohydrate, protein, and fat metabolism and helps the body resist stress

cranial nerves the twelve pairs of peripheral nerves, the majority of which arise from the brain stem

cross bridges the myosin molecules' globular heads that protrude from a thick filament within a muscle fiber and interact with the actin molecules in the thin filaments to bring about shortening of the muscle fiber during contraction

cyclic adenosine monophosphate (cyclic AMP or cAMP) an intracellular second messenger derived from adenosine triphosphate (ATP)

cyclic AMP see *cyclic adenosine monophosphate*

cytokines all chemicals other than antibodies that are secreted by lymphocytes

cytoplasm (SĪ-tō-plaz'-um) the portion of the cell interior not occupied by the nucleus

cytoskeleton a complex intracellular protein network that acts as the "bone and muscle" of the cell

cytosol (SĪ-tuh-sol') the semiliquid portion of the cytoplasm not occupied by organelles

cytotoxic T cells (sī'-tō-TOK-sik) the population of T cells that destroys host cells bearing foreign antigen, such as body cells invaded by viruses or cancer cells

dead-space volume the volume of air that occupies the respiratory airways as air is moved in and out and which is not available to participate in exchange of O_2 and CO_2 between the alveoli and atmosphere

dehydration a water deficit in the body

dehydroepiandrosterone (DHEA) (dē-HĪ-drō-ep-i-and-row-steer-own) the androgen (masculinizing hormone) secreted by the adrenal cortex in both sexes

dendrites projections from the surface of a neuron's cell body that carry signals toward the cell body

deoxyribonucleic acid (DNA) (dē-OK-sē-rī-bō-new-klā'-ik) the cell's genetic material, which is found within the nucleus and which provides codes for protein synthesis and serves as a blueprint for cell replication

depolarization (dē'-pō-luh-ruh-ZĀ-shun) a reduction in membrane potential from resting potential; movement of the potential from resting toward 0 mV

dermis the connective tissue layer that lies under the epidermis in the skin; contains the skin's blood vessels and nerves

descending tract a bundle of nerve fibers of similar function that travels down the spinal cord to relay messages from the brain to efferent neurons

desmosome (dez'-muh-sōm) an adhering junction between two adjacent but nontouching cells formed by the extension of filaments between the cells' plasma membranes; most abundant in tissues that are subject to considerable stretching

DHEA see *dehydroepiandrosterone*

diabetes insipidus (in-sip'-ud-us) an endocrine disorder characterized by a deficiency of vasopressin

diabetes mellitus (muh-LĪ-tus) an endocrine disorder characterized by inadequate insulin action

diaphragm (DIE-uh-fram) a dome-shaped sheet of skeletal muscle that forms the floor of the thoracic cavity; the major inspiratory muscle

diastole (dī-AS-tō-lē) the period of cardiac relaxation and filling

diencephalon (dī'-un-SEF-uh-lan) the division of the brain that consists of the thalamus and hypothalamus

diet-induced thermogenesis (DIT) (thur'-mō-JEN-uh-sus) the obligatory increase in metabolic rate that occurs as a result of the processing and storage of ingested nutrients

diffusion random collisions and intermingling of molecules as a result of their continuous thermally induced random motion

digestion the breaking-down process whereby the structurally complex foodstuffs of the diet are converted into smaller absorbable units by the enzymes produced within the digestive system

diploid number (DIP-loid) a complete set of forty-six chromosomes (twenty-three pairs), as found in all human somatic cells

distal tubule a highly convoluted tubule that extends between the loop of Henle and the collecting duct in the kidney's nephron

DIT see *diet-induced thermogenesis*

diurnal rhythm (dī-urn'-ul) repetitive oscillations in hormone levels that are very regular and have a frequency of one cycle every twenty-four hours, usually linked to the light-dark cycle; circadian rhythm; biological rhythm

divergence the diverging, or branching, of a neuron's axon terminals, so that activity in this single neuron influences the many other cells with which its terminals synapse

DNA see *deoxyribonucleic acid*

dorsal root ganglion a cluster of afferent neuronal cell bodies located adjacent to the spinal cord

down regulation a reduction in the number of receptors for (and thereby the target cells' sensitivity to) a particular hormone as a direct result of the effect that an elevated level of the hormone has on its own receptors

eccentric contraction (ex-SEN-trik) an isotonic contraction during which the muscle lengthens as it resists being stretched by an external force

ECG see *electrocardiogram*

edema (i-DĒ-muh) swelling of tissues as a result of excess interstitial fluid

EDV see *end-diastolic volume*

EEG see *electroencephalogram*

effector organs the muscles or glands that are innervated by the nervous system and that carry out the nervous system's orders to bring about a desired effect, such as a particular movement or secretion

efferent division (EF-er-ent) the portion of the peripheral nervous system that carries instructions from the central nervous system to effector organs

efferent neuron neuron that carries information from the central nervous system to an effector organ

efflux (Ē-flux) movement out of the cell

elastic recoil rebound of the lungs after having been stretched

electrical gradient a difference in charge between two adjacent areas

electrocardiogram (ECG) the graphic record of the electrical activity that reaches the surface of the body as a result of cardiac depolarization and repolarization

electrochemical gradient the simultaneous existence of an electrical gradient and concentration (chemical) gradient for a particular ion

electroencephalogram (EEG) (i-lek'-trō-in-SEF-uh-luh-gram') a graphic record of the collective postsynaptic potential activity in the cell bodies and dendrites located in the cortical layers under a recording electrode

electrolytes solutes that form ions in solution and conduct electricity

embolus (EM-bō-lus) a freely floating clot

emphysema (em'-fuh-ZĒ-muh) a pulmonary disease characterized by collapse of the smaller airways and a breakdown of alveolar walls

end-diastolic volume (EDV) the volume of blood in the ventricle at the end of diastole, when filling is complete

endocrine glands ductless glands that secrete hormones into the blood

endocytosis (en'-dō-sī-TŌ-sis) internalization of extracellular material within a cell as a result of the plasma membrane forming a pouch that contains the extracellular material, then sealing at the surface of the pouch to form a small, intracellular, membrane-enclosed vesicle with the contents of the pouch trapped inside

endogenous opiates (en-DAJ'-eh-nus ō'-pē-ātz) endorphins and enkephalins, which bind with opiate receptors and are important in the body's natural analgesic system

endogenous pyrogen (pī'-ruh-jun) a chemical released from macrophages during inflammation that acts by means of local prostaglandins to raise the set point of the hypothalamic thermostat to produce a fever

endometrium (en'-dō-MĒ-trē-um) the lining of the uterus.

endoplasmic reticulum (en'-dō-PLAZ-mik ri-TIK-yuh-lum) an organelle consisting of a continuous membranous network of fluid-filled tubules and flattened sacs, partially studded with ribosomes; synthesizes proteins and lipids for formation of new cell membrane and other cell components and manufactures products for secretion

endothelial-derived relaxing factor (EDRF) (en'-dō-THĒ-lē-ul) a local chemical mediator released from the endothelial cells lining an arteriole that diffuses locally to cause relaxation of the arteriolar smooth muscle in the vicinity

endothelium (en'-dō-THĒ-lē-um) the thin, single-celled layer of epithelial cells that lines the entire circulatory system

end plate potential (EPP) the graded receptor potential that occurs at the motor end plate of a skeletal muscle fiber in response to binding with acetylcholine

end-systolic volume (ESV) the volume of blood in the ventricle at the end of the systole, when emptying is complete

endurance-type exercise see *aerobic exercise*

enterogastrones (ent'-uh-rō-GAS-trōnz) hormones secreted by the duodenal mucosa that inhibit gastric motility and secretion; include secretin, cholecystokinin, and gastric inhibitory peptide

enterohepatic circulation (en'-tur-ō-hi-PAT-ik) the recycling of bile salts and other bile constituents between the small intestine and liver by means of the hepatic portal vein

enzyme a special protein molecule that speeds up a particular chemical reaction in the body

eosinophils (ē'-uh-SIN-uh-fils) white blood cells that are important in allergic responses and in combating internal parasite infestations

epidermis (ep'-uh-DER-mus) the outer layer of the skin, consisting of numerous layers of epithelial cells, with the outermost layers being dead and flattened

epinephrine (ep'-uh-NEF-rin) the primary hormone secreted by the adrenal medulla; important in preparing the body for "fight-or-flight" responses and in regulation of arterial blood pressure; adrenaline

epiphyseal plate (eh-pif-i-SEE-al) a layer of cartilage that separates the diaphysis (shaft) of a long bone from the epiphysis (flared end); the site of growth of bones in length before the cartilage ossifies (turns into bone)

epithelial tissue (ep'-uh-THĒ-lē-ul) a functional grouping of cells specialized in the exchange of materials between the cell and its environment; lines and covers various body surfaces and cavities and forms secretory glands

EPSP see *excitatory postsynaptic potential*

equilibrium potential the potential that exists when the concentration gradient and opposing electrical gradient for a given ion exactly counterbalance each other so that there is no net movement of the ion

erythrocytes (i-RITH-ruh-sīts) red blood cells, which are plasma membrane-enclosed bags of hemoglobin that transport O_2 and to a lesser extent CO_2 and H^+ in the blood

erythropoiesis (i-rith'-rō-poi-Ē-sus) erythrocyte production by the bone marrow

erythropoietin the hormone released from the kidneys in response to a reduction in O_2 delivery to the kidneys; stimulates the bone marrow to increase erythrocyte production

esophagus (i-SOF-uh-gus) a straight muscular tube that extends between the pharynx and stomach

estrogen feminizing "female" sex hormone

ESV see *end-systolic volume*

excitable tissue tissue capable of producing electrical signals when excited; includes nervous and muscle tissue

excitation-contraction coupling the series of events linking muscle excitation (the presence of an action potential) to muscle contraction (filament sliding and sarcomere shortening)

excitatory postsynaptic potential (EPSP) (pōst'-si-NAP-tik) a small depolarization of the postsynaptic membrane in response to neurotransmitter binding, thereby bringing the membrane closer to threshold

excitatory synapse (SIN-aps') synapse in which the postsynaptic neuron's response to neurotransmitter release is a small depolarization of the postsynaptic membrane, bringing the membrane closer to threshold

exercise physiology the study of both the functional changes that occur in response to a single session of exercise and the adaptations that occur as a result of regular, repeated exercise sessions

exocrine glands glands that secrete through ducts to the outside of the body or into a cavity that communicates with the outside

exocytosis (eks'-ō-sī-TŌ-sis) fusion of a membrane-enclosed intracellular vesicle with the plasma membrane, followed by the opening of the vesicle and the emptying of its contents to the outside

expiration a breath out

expiratory muscles the skeletal muscles whose contraction reduces the size of the thoracic cavity and allows the lungs to recoil to a smaller size, bringing about movement of air from the lungs to the atmosphere

external intercostal muscles inspiratory muscles whose contraction elevates the ribs, thereby enlarging the thoracic cavity

external environment the environment that surrounds the body

external work energy expended by contracting skeletal muscles to move external objects or to move the body in relation to the environment

extracellular fluid all of the body's fluid found outside of the cells; consists of interstitial fluid and plasma

extracellular matrix an intricate meshwork of fibrous proteins embedded in a watery, gel-like substance; secreted by local cells

extrinsic controls regulatory mechanisms initiated outside of an organ that alter the activity of the organ; accomplished by the nervous and endocrine systems

extrinsic nerves the nerves that originate outside of the digestive tract and innervate the various digestive organs

facilitated diffusion passive carrier-mediated transport involving transport of a substance down its concentration gradient across the plasma membrane

fatigue inability to maintain muscle tension at a given level despite sustained stimulation

feedforward mechanism a response designed to prevent an anticipated change in a controlled variable

feeding (appetite) centers neuronal clusters in the lateral regions of the hypothalamus that drive the individual to eat

fibrinogen (fī-BRIN-uh-jun) a large soluble plasma protein that is converted into an insoluble, threadlike molecule that forms the meshwork of a clot during blood coagulation

Fick's law of diffusion the rate of net diffusion of a substance across a membrane is directly proportional to the substance's concentration gradient, the membrane's permeability to the substance, and the surface area of the membrane and inversely proportional to the substance's molecular weight and the diffusion distance

fight-or-flight response the changes in activity of the various organs innervated by the autonomic nervous system in response to sympathetic stimulation, which collectively prepare the body for strenuous physical activity in the face of an emergency or stressful situation, such as a physical threat from the outside environment

fire when an excitable cell undergoes an action potential

first messenger an extracellular messenger, such as a hormone, that binds with a surface membrane receptor and activates an intracellular second messenger to carry out the desired cellular response

flagellum (fluh-JEL-um) the single, long, whiplike appendage that serves as the tail of a spermatozoon

follicle (of ovary) a developing ovum and the surrounding specialized cells

follicle-stimulating hormone (FSH) an anterior pituitary hormone that stimulates ovarian follicular development and estrogen secretion in females and stimulates sperm production in males

follicular cells (of ovary) (fah-LIK-you-lar) collectively, the granulosa and thecal cells

follicular cells (of thyroid gland) the cells that form the walls of the colloid-filled follicles in the thyroid gland and secrete thyroid hormone

follicular phase the phase of the ovarian cycle dominated by the presence of maturing follicles prior to ovulation

forebrain the division of the brain that consists of the diencephalon and cerebrum

Frank-Starling law of the heart intrinsic control of the heart, such that increased venous return resulting in increased end-diastolic volume leads to an increased strength of contraction and increased stroke volume; that is, the heart normally pumps out all of the blood returned to it.

free radicals very unstable electron-deficient particles that are highly reactive and destructive

frontal lobes the lobes of the cerebral cortex that lie at the top of the brain in front of the central sulcus and that are responsible for voluntary motor output, speaking ability, and elaboration of thought

FSH see *follicle-stimulating hormone*

fuel metabolism see *intermediary metabolism*

functional syncytium (sin-sish'-ē-um) a group of smooth or cardiac muscle cells that are interconnected by gap junctions and function electrically and mechanically as a single unit

functional unit the smallest component of an organ that can perform all the functions of the organ

gametes (GAM-ētz) reproductive, or germ, cells, each containing a haploid set of chromosomes; sperm and ova

gamma motor neuron a motor neuron that innervates the fibers of a muscle-spindle receptor

ganglion (GAN-glē-un) a collection of neuronal cell bodies located outside the central nervous system

ganglion cells the nerve cells in the outermost layer of the retina and whose axons form the optic nerve

gap junction a communicating junction formed between adjacent cells by small connecting tunnels that permit passage of charge-carrying ions between the cells so that electrical activity in one cell is spread to the adjacent cell

gastrin a hormone secreted by the pyloric gland area of the stomach that stimulates the parietal and chief cells to secrete a highly acidic gastric juice

gestation pregnancy

glands epithelial tissue derivatives that are specialized for secretion

glial cells (glē-ul) serve as the connective tissue of the CNS and help support the neurons both physically and metabolically; include astrocytes, oligodendrocytes, ependymal cells, and microglia

glomerular filtration (glō-MER-yū-lur) filtration of a protein-free plasma from the glomerular capillaries into the tubular component of the kidney's nephron as the first step in urine formation

glomerular filtration rate (GFR) the rate at which glomerular filtrate is formed

glomerulus (glō-MER-yū-lus) a ball-like tuft of capillaries in the kidney's nephron that filters water and solute from the blood as the first step in urine formation

glucagon (GLOO-kuh-gon) the pancreatic hormone that raises blood glucose and blood fatty acid levels

glucocorticoids (gloo'-kō-KOR-ti-koidz) the adrenocortical hormones that are important in intermediary metabolism and in helping the body resist stress; primarily cortisol

gluconeogenesis (gloo'-kō-nē'-ō-JEN-uh-sus) the conversion of amino acids into glucose

glycogen (GLĪ-kō-jen) the storage form of glucose in the liver and muscle

glycogenesis (glī'-kō-JEN-i-sus) the conversion of glucose into glycogen

glycogenolysis (glī'-kō-juh-NOL-i-sus) the conversion of glycogen to glucose

glycolysis (glī-KOL-uh-sus) a biochemical process that takes place in the cell's cytosol and involves the breakdown of glucose into two pyruvic acid molecules

GnRH see *gonadotropin-releasing hormone*

Golgi complex (GOL-jē) an organelle consisting of sets of stacked, flattened membranous sacs; processes raw materials transported to it from the endoplasmic reticulum into finished products and sorts and directs the finished products to their final destination

gonadotropin-releasing hormone (GnRH) (gō-nad'-uh-TRŌ-pin) the hypothalamic hormone that stimulates the release of FSH and LH from the anterior pituitary

gonadotropins FSH and LH; hormones that are tropic to the gonads

gonads (GŌ-nadz) the primary reproductive organs, which produce the gametes and secrete the sex hormones; testes and ovaries

G protein a membrane-bound intermediary, which, when activated upon binding of an extracellular first messenger to a surface receptor, activates the enzyme adenylyl cyclase on the intracellular side of the membrane in the cAMP second-messenger system

gradation of contraction variable magnitudes of tension produced in a single whole muscle

graded potential a local change in membrane potential that occurs in varying grades of magnitude; serves as a short-distance signal in excitable tissues

granulocytes (gran'-yuh-lō-sīts) leukocytes that contain granules, including neutrophils, eosinophils, and basophils

granulosa cells (gran'-yuh-LŌ-suh) the layer of cells immediately surrounding a developing oocyte within an ovarian follicle

gray matter the portion of the central nervous system composed primarily of densely packaged neuronal cell bodies and dendrites

growth hormone (GH) an anterior pituitary hormone that is primarily responsible for regulating overall body growth and is also important in intermediary metabolism; somatotropin

H^+ see *hydrogen ion*

haploid number (HAP-loid) the number of chromosomes found in gametes; a half set of chromosomes, one member of each pair, for a total of twenty-three chromosomes in humans

hapten a low-molecular-weight organic substance that is not antigenic by itself but can become antigenic if it attaches to body proteins

Hb see *hemoglobin*

hCG see *human chorionic gonadotropin*

heart failure an inability of the cardiac output to keep pace with the body's demands for supplies and for removal of wastes

helper T cells the population of T cells that enhances the activity of other immune response effector cells

hematocrit (hi-mat'-uh-krit) the percentage of blood volume occupied by erythrocytes as they are packed down in a centrifuged blood sample

hemoglobin (HĒ-muh-glō'-bun) a large iron-bearing protein molecule found within erythrocytes that binds with and transports most O_2 in the blood; also carries some of the CO_2 and H^+ in the blood

hemolysis (hē-MOL-uh-sus) rupture of red blood cells

hemostasis (hē'-mō-STĀ-sus) the stopping of bleeding from an injured vessel

hepatic portal system (hi-PAT-ik) a complex vascular connection between the digestive tract and liver such that venous blood from the digestive system drains into the liver for processing of absorbed nutrients before being returned to the heart

hippocampus (hip-oh-CAM-pus) the elongated, medial portion of the temporal lobe that is a part of the limbic system and is especially crucial for forming long-term memories

histamine a chemical released from mast cells or basophils that brings about vasodilation and increased capillary permeability; important in allergic responses

homeostasis (hō'-mē-ō-STĀ-sus) maintenance by the highly coordinated, regulated actions of the body systems of relatively stable chemical and physical conditions in the internal fluid environment that bathes the body's cells

hormone a long-distance chemical mediator that is secreted by an endocrine gland into the blood, which transports it to its target cells

hormone response element (HRE) the specific attachment site on DNA for a given steroid hormone and its nuclear receptor

host cell a body cell infected by a virus

HRE see *hormone response element*

human chorionic gonadotropin (hCG) (kō-rē-ON-ik gō-nad'-uh-TRŌ-pin) a hormone secreted by the developing placenta that stimulates and maintains the corpus luteum of pregnancy

hydrogen ion (H^+) the cationic portion of a dissociated acid

hydrolysis (hī-DROL-uh-sis) the digestion of a nutrient molecule by the addition of water at a bond site

hydrostatic pressure (hī-dro-STAT-ik) the pressure exerted by fluid on the walls that contain it

hypercalcemia (hī-pur-kal'-SE-mē-uh) high blood Ca^{2+} levels

hyperglycemia (hī'-pur-glī-SĒ-mē-uh) elevated blood glucose concentration

hyperplasia (hī'-pur-PLĀ-zē-uh) an increase in the number of cells

hyperpolarization an increase in membrane potential from resting potential; potential becomes even more negative than at resting potential

hypersecretion too much of a particular hormone secreted

hypertension (hī'-pur-TEN-shun) sustained, above-normal mean arterial blood pressure

hypertonic (hī'-pur-TON-ik) having an osmolarity greater than normal body fluids; more concentrated than normal

hypertrophy (hī-PUR-truh-fē) increase in the size of an organ as a result of an increase in the size of its cells

hyperventilation overbreathing; when the rate of ventilation is in excess of the body's metabolic needs for CO_2 removal

hypocalcemia (hī-pō-kal'-SĒ-mē-uh) low blood Ca^{2+} levels

hypophysiotropic hormones (hi-PŌ-fiz-ē-oh-TRO-pik) hormones secreted by the hypothalamus that regulate the secretion of anterior pituitary hormones; see also *releasing hormone* and *inhibiting hormone*

hyposecretion too little of a particular hormone secreted

hypotension (hi-po-TEN-chun) sustained, below normal mean arterial blood pressure

hypothalamic hypophyseal portal system (hī'-pō-thuh-LAM-ik hī-pō-FIZ-ē-ul) the vascular connection between the hypothalamus and anterior pituitary gland used for the pickup and delivery of hypophysiotropic hormones

hypothalamus (hī'-pō-THAL-uh-mus) the brain region located beneath the thalamus that is concerned with regulating many aspects of the internal fluid environment, such as water and salt balance and food intake; serves as an important link between the autonomic nervous system and endocrine system

hypotonic (hī'-pō-TON-ik) having an osmolarity less than normal body fluids; more dilute than normal

hypoventilation underbreathing; ventilation inadequate to meet the metabolic needs for O_2 delivery and CO_2 removal

hypoxia (hī-POK-sē-uh) insufficient O_2 at the cellular level

I band one of the light bands that alternate with dark (A) bands to create a striated appearance in a skeletal or cardiac muscle fiber when these fibers are viewed with a light microscope

IGF see *insulin-like growth factor*

immune surveillance recognition and destruction of newly arisen cancer cells by the immune system

immunity the body's ability to resist or eliminate potentially harmful foreign materials or abnormal cells

immunoglobulins (im'-ū-nō-GLOB-yū-lunz) antibodies; gamma globulins

impermeable prohibiting passage of a particular substance through the plasma membrane

implantation the burrowing of a blastocyst into the endometrial lining

inflammation an innate, nonspecific series of highly interrelated events, especially involving neutrophils, macrophages, and local vascular changes, that are set into motion in response to foreign invasion or tissue damage

influx movement into the cell

inhibin (ih-HIB-un) a hormone secreted by the Sertoli cells of the testes or by the ovarian follicles that inhibits FSH secretion

inhibiting hormone a hypothalamic hormone that inhibits the secretion of a particular anterior pituitary hormone

inhibitory postsynaptic potential (IPSP) (pōst'-si-NAP-tik) a small hyperpolarization of the postsynaptic membrane in response to neurotransmitter binding, thereby moving the membrane farther from threshold

inhibitory synapse (SIN-aps') synapse in which the postsynaptic neuron's response to neurotransmitter release is a small hyperpolarization of the postsynaptic membrane, moving the membrane farther from threshold

inorganic referring to substances that do not contain carbon; from nonliving sources

inspiration a breath in

inspiratory muscles the skeletal muscles whose contraction enlarges the thoracic cavity, bringing about lung expansion and movement of air into the lungs from the atmosphere

insulin (IN-suh-lin) the pancreatic hormone that lowers blood levels of glucose, fatty acids, and amino acids and promotes their storage

insulin-like growth factor (IGF) synonymous with somatomedins

integrating center a region that determines efferent output based on processing of afferent input

integument (in-teg̅-yuh-munt) the skin and underlying connective tissue

intercostal muscles (int-ur-kos'-tul) the muscles that lie between the ribs; see also *external intercostal muscles* and *internal intercostal muscles*

interferon (in'-tur-FĒR-on) a chemical released from virus-invaded cells that provides nonspecific resistance to viral infections by transiently interfering with replication of the same or unrelated viruses in other host cells

interleukin 1 (int-ur-loo-kin) a multipurpose chemical mediator released from macrophages that enhances B cell activity

interleukin 2 a chemical mediator secreted by helper T cells that augments the activity of all T cells

intermediary metabolism the collective set of intracellular chemical reactions that involve the degradation, synthesis, and transformation of small nutrient molecules; also known as fuel metabolism

intermediate filaments threadlike cytoskeletal elements that play a structural role in parts of the cells subject to mechanical stress

internal environment the body's aqueous extracellular environment, which consists of the plasma and interstitial fluid and which must be homeostatically maintained for the cells to make life-sustaining exchanges with it

internal intercostal muscles expiratory muscles whose contraction pulls the ribs downward and inward, thereby reducing the size of the thoracic cavity

internal respiration the intracellular metabolic processes carried out within the mitochondria that use O_2 and produce CO_2 during the derivation of energy from nutrient molecules

internal work all forms of biological energy expenditure that do not accomplish mechanical work outside of the body

interneuron neuron that lies entirely within the central nervous system and is important for integration of peripheral responses to peripheral information as well as for the abstract phenomena associated with the "mind"

interstitial fluid (in'-tur-STISH-ul) the portion of the extracellular fluid that surrounds and bathes all of the body's cells

intra-alveolar pressure (in'-truh-al-VĒ-uh-lur) the pressure within the alveoli

intracellular fluid the fluid collectively contained within all of the body's cells

intrapleural pressure (in'-truh-PLOOR-ul) the pressure within the pleural sac

intrinsic controls local control mechanisms inherent to an organ

intrinsic factor a special substance secreted by the parietal cells of the stomach that must be combined with vitamin B_{12} for this vitamin to be absorbed by the intestine; deficiency produces pernicious anemia

intrinsic nerve plexuses interconnecting networks of nerve fibers within the digestive tract wall

ion an atom that has gained or lost one or more of its electrons, so that it is not electrically balanced

IPSP see *inhibitory postsynaptic potential*

iris a pigmented smooth muscle that forms the colored portion of the eye and controls pupillary size

Islets of Langerhans (LAHNG-er-honz) the endocrine portion of the pancreas that secretes the hormones insulin and glucagon into the blood

isometric contraction (ī'-sō-MET-rik) a muscle contraction in which the development of tension occurs at constant muscle length

isotonic (ī'-sō-TON-ik) having an osmolarity equal to normal body fluids

isotonic contraction a muscle contraction in which muscle tension remains constant as the muscle fiber changes length

juxtaglomerular apparatus (juks'-tuh-glō-MER-yū-lur) a cluster of specialized vascular and tubular cells at a point where the ascending limb of the loop of Henle passes through the fork formed by the afferent and efferent arterioles of the same nephron in the kidney

keratin (CARE-uh-tin) the protein found in the intermediate filaments in skin cells that give the skin strength and help form a waterproof outer layer

killer (K) cells cells that destroy a target cell that has been coated with antibodies by lysing its membrane

kinesin (kī-NĒ-sin) the transport, or motor, protein that transports secretory vesicles along the microtubular highway within neuronal axons by "walking" along the microtubule

lactation milk production by the mammary glands

lactic acid an end product formed from pyruvic acid during the anaerobic process of glycolysis

larynx (LARE-inks) the "voice box" at the entrance of the trachea; contains the vocal cords

lateral inhibition the phenomenon in which the most strongly activated signal pathway originating from the center of a stimulus area inhibits the less excited pathways from the fringe areas by means of lateral inhibitory connections within sensory pathways

lateral sacs the expanded saclike regions of a muscle fiber's sarcoplasmic reticulum; store and release calcium, which plays a key role in triggering muscle contraction

law of mass action if the concentration of one of the substances involved in a reversible reaction is increased, the reaction is driven toward the opposite side, and if the concentration of one of the substances is decreased, the reaction is driven toward that side

left ventricle the heart chamber that pumps blood into the systemic circulation

length-tension relationship the relationship between the length of a muscle fiber at the onset of contraction and the tension the fiber can achieve on a subsequent tetanic contraction

lens a transparent, biconvex structure of the eye that refracts (bends) light rays and whose strength can be adjusted to accommodate for vision at different distances

leukocyte endogenous mediator (LEM) (LOO-kō-sīt en-DAJ-eh-nus mē'-dē-ĀT-or) a chemical mediator secreted by macrophages that is identical to endogenous pyrogen and exerts a wide array of effects associated with inflammation

leukocytes (LOO-kuh-sīts) white blood cells, which are the immune system's mobile defense units

Leydig cells (LĪ-dig) the interstitial cells of the testes that secrete testosterone

LH see *luteinizing hormone*

LH surge the burst in LH secretion that occurs at midcycle of the ovarian cycle and triggers ovulation

limbic system (LIM-bik) a functionally interconnected ring of forebrain structures that surrounds the brain stem and is concerned with emotions, basic survival and sociosexual behavioral patterns, motivation, and learning

lipid emulsion a suspension of small fat droplets held apart as a result of adsorption of bile salts on their surface

loop of Henle (HEN-lē) a hairpin loop that extends between the proximal and distal tubule of the kidney's nephron

lumen (LOO-men) the interior space of a hollow organ or tube

luteal phase (LOO-tē-ul) the phase of the ovarian cycle dominated by the presence of a corpus luteum

luteinization (loot'-ē-un-uh-ZĀ-shun) formation of a postovulatory corpus luteum in the ovary

luteinizing hormone (LH) an anterior pituitary hormone that stimulates ovulation, luteinization, and secretion of estrogen and progesterone in females and stimulates testosterone secretion in males

lymph interstitial fluid that is picked up by the lymphatic vessels and returned to the venous system, meanwhile passing through the lymph nodes for defense purposes

lymphocytes white blood cells that provide immune defense against targets for which they are specifically programmed

lymphoid tissues tissues that produce and store lymphocytes, such as lymph nodes and tonsils

lysosomes (LĪ-sō-sōmz) organelles consisting of membrane-enclosed sacs containing powerful hydrolytic enzymes that destroy unwanted material within the cell, such as internalized foreign material or cellular debris

macrophages (MAK-ruh-fājs) large, tissue-bound phagocytes

mast cells cells located within connective tissue that synthesize, store, and release histamine, as during allergic responses

mature follicle an ovarian follicle that is ready to ovulate; Graafian follicle

mean arterial blood pressure the average pressure responsible for driving blood forward through the arteries into the tissues throughout the cardiac cycle; equals cardiac output times total peripheral resistance

mechanically gated channels channels that open or close in response to stretching or other mechanical deformation

mechanistic approach explanation of body functions in terms of mechanisms of action, that is, the "how" of events that occur in the body

mechanoreceptor (meh-CAN-oh-rē-SEP-tur) or (mek'-uh-nō-rē-SEP-tur) a sensory receptor sensitive to mechanical energy, such as stretching or bending

medullary respiratory center (med-you-LAIR-ē) several aggregations of neuronal cell bodies within the medulla that provide output to the respiratory muscles and receive input important for regulating the magnitude of ventilation

meiosis (mī-Ō-sis) cell division in which the chromosomes replicate followed by two nuclear divisions so that only a half set of chromosomes is distributed to each of four new daughter cells

melanocyte stimulating hormone (MSH) (mel-ah-NŌ-sīt) a hormone produced by the anterior pituitary in humans and by the intermediate lobe of the pituitary in lower vertebrates; regulates skin coloration by controlling the dispersion of melanin granules in lower vertebrates; presumably involved with memory and learning in humans

melatonin (mel-uh-TŌ-nin) a hormone secreted by the pineal gland during darkness that helps entrain the body's biological rhythms with the external light/dark cues

membrane attack complex a collection of the five final activated components of the complement system that aggregate to form a porelike channel in the plasma membrane of an invading microorganism, with the resultant leakage leading to destruction of the invader

membrane potential a separation of charges across the membrane; a slight excess of negative charges lined up along the inside of the plasma membrane and separated from a slight excess of positive charges on the outside

memory cells B or T cells that are newly produced in response to a microbial invader, but which do not participate in the current immune response against the invader but instead remain dormant, ready to launch a swift, powerful attack should the same microorganism invade again in the future

menstrual cycle (men'-stroo-ul) the cyclical changes in the uterus in accompaniment with the hormonal changes in the ovarian cycle

menstrual phase the phase of the menstrual cycle characterized by sloughing of endometrial debris and blood out through the vagina

messenger RNA carries the transcribed genetic blueprint for synthesis of a particular protein from nuclear DNA to the cytoplasmic ribosomes where the protein synthesis takes place

metabolic acidosis (met-uh-bol'-ik) acidosis resulting from any cause other than excess accumulation of carbonic acid in the body

metabolic alkalosis (al'-kuh-LŌ-sus) alkalosis caused by a relative deficiency of non-carbonic acid

metabolic rate energy expenditure per unit of time

micelle (mī-SEL) a water-soluble aggregation of bile salts, lecithin, and cholesterol that has a hydrophilic shell and a hydrophobic core; carries the water-insoluble products of fat digestion to their site of absorption

microfilaments cytoskeletal elements made of actin molecules (as well as myosin molecules in muscle cells); play a major role in various cellular contractile systems and serve as a mechanical stiffener for microvilli

microtrabecular lattice (mī'-krō-truh-BEK'-yuh-lur) cytoskeletal element consisting of a meshwork of exceedingly fine, interlinked filaments that suspends and functionally links larger cytoskeletal elements and various organelles

microtubules cytoskeletal elements made of tubulin molecules arranged into long, slender, unbranched tubes that help maintain asymmetrical cell shapes and coordinate complex cell movements

microvilli (mī'-krō-VIL-ī) actin-stiffened, nonmotile, hairlike projections from the luminal surface of epithelial cells lining the digestive tract and kidney tubules; tremendously increase the surface area of the cell exposed to the lumen

micturition (mik-too-RISH-un) or (mik'-chuh-RISH-un) the process of bladder emptying; urination

milk ejection the squeezing out of milk produced and stored in the alveoli of the breasts by means of contraction of the myoepithelial cells that surround each alveolus

millivolts the unit used to measure membrane potential

mineralocorticoids (min'-uh-rul-ō-KOR-ti-koidz) the adrenocortical hormones that are important in Na^+ and K^+ balance; primarily aldosterone

mitochondria (mī-tō-KON-drē-uh) the energy organelles, which contain the enzymes for oxidative phosphorylation

mitosis (mī-TŌ-sis) cell division in which the chromosomes replicate before nuclear division, so that each of the two daughter cells receives a full set of chromosomes

mitotic spindle the system of microtubules assembled during mitosis along which the replicated chromosomes are directed away from each other toward opposite sides of the cell prior to cell division

molecule a chemical substance formed by the linking of atoms together; the smallest unit of a given chemical substance

monocytes (MAH-nō-sīts) white blood cells that emigrate from the blood, enlarge, and become macrophages, large tissue phagocytes

monosaccharides (mah'-nō-SAK-uh-rīdz) simple sugars, such as glucose; the absorbable unit of digested carbohydrates

motor activity movement of the body accomplished by contraction of skeletal muscles

motor end plate the specialized portion of a skeletal muscle fiber that lies immediately underneath the terminal button of the motor neuron and possesses receptor sites for binding acetylcholine released from the terminal button

motor neurons the neurons that innervate skeletal muscle and whose axons constitute the somatic nervous system

motor unit one motor neuron plus all of the muscle fibers it innervates

motor unit recruitment the progressive activation of a muscle fiber's motor units to accomplish increasing gradations of contractile strength

mucosa (mew-KŌ-sah) the innermost layer of the digestive tract that lines the lumen

multiunit smooth muscle a smooth muscle mass that consists of multiple discrete units that function independently of each other and that must be separately stimulated by autonomic nerves to contract

muscarinic receptor type of cholinergic receptor found at the effector organs of all parasympathetic postganglionic fibers

muscle fiber a single muscle cell, which is relatively long and cylindrical in shape

muscle tension see *tension*

muscle tissue a functional grouping of cells specialized for contraction and force generation

myelin (MĪ-uh-lun) an insulative lipid covering that surrounds myelinated nerve fibers at regular intervals along the axon's length; each patch of myelin is formed by a separate myelin-forming cell that wraps itself jelly-roll fashion around the neuronal axon

myelinated fibers neuronal axons covered at regular intervals with insulative myelin

myocardial ischemia (mī'-ō-KAR-dē-ul is-KĒ-mē-uh) inadequate blood supply to the heart tissue

myocardium (mī'-ō-KAR-dē-um) the cardiac muscle within the heart wall

myofibril (mī'-ō-FĪB-rul) a specialized intracellular structure of muscle cells that contains the contractile apparatus

myometrium (mī'-ō-mē-TRĒ-um) the smooth muscle layer of the uterus

myosin (MĪ-uh-sun) the contractile protein that forms the thick filaments in muscle fibers

Na^+-K^+ pump a carrier that actively transports Na^+ out of the cell and K^+ into the cell

natural killer cells naturally occurring, lymphocyte-like cells that nonspecifically destroy virus-infected cells and cancer cells by directly lysing their membranes on first exposure to them

negative balance situation in which the losses for a substance exceed its gains so that the total amount of the substance in the body decreases

negative feedback a regulatory mechanism in which a change in a controlled variable triggers a response that opposes the change, thus maintaining a relatively steady set point for the regulated factor

nephron (NEF-ron') the functional unit of the kidney; consisting of an interrelated vascular and tubular component, it is the smallest unit that can form urine

nerve a bundle of peripheral neuronal axons, some afferent and some efferent, enclosed by a connective tissue covering and following the same pathway

nervous system one of the two major control systems of the body; in general, coordinates rapid activities of the body, especially those involving interactions with the external environment

nervous tissue a functional grouping of cells specialized for initiation and transmission of electrical signals

net diffusion the difference between two opposing movements

net filtration pressure the net difference in the hydrostatic and osmotic forces acting across the glomerular membrane that favors the filtration of a protein-free plasma into Bowman's capsule

neuroendocrinology the study of the interaction between the nervous and endocrine systems

neuroglia see *glial cells*

neurohormones hormones released into the blood by neurosecretory neurons

neuromodulators (ner'ō-MA-jew-lā'-torz) chemical messengers that bind to neuronal receptors at nonsynaptic sites (that is, not at the subsynaptic membrane) and bring about long-term changes that subtly depress or enhance synaptic effectiveness

neuromuscular junction the juncture between a motor neuron and a skeletal muscle fiber

neuron (NER-on) a nerve cell, typically consisting of a cell body, dendrites, and an axon and specialized to initiate, propagate, and transmit electrical signals

neuropeptides large, slowly acting peptide molecules released from axon terminals along with classical neurotransmitters; most neuropeptides function as neuromodulators

neurotransmitter the chemical messenger that is released from the axon terminal of a neuron in response to an action potential and influences another neuron or an effector with which the neuron is anatomically linked

neutrophils (new'-truh-filz) white blood cells that are phagocytic specialists and important in inflammatory responses and defense against bacterial invasion

nicotinic receptor (nick'-o-TIN-ik) type of cholinergic receptor found at all autonomic ganglia and the motor end plates of skeletal muscle fibers

nitric oxide a recently identified local chemical mediator released from endothelial cells and other tissues; exerts a wide array of effects, ranging from causing local arteriolar vasodilation to acting as a toxic agent against foreign invaders to serving as a unique type of neurotransmitter

nociceptor (nō'-sē-SEP-tur) a pain receptor, sensitive to tissue damage

Node of Ranvier (RAN-vē-ā) the portions of a myelinated neuronal axon between the segments of insulative myelin; the axonal regions where the axonal membrane is exposed to the ECF and membrane potential exists

nonspecific immune responses inherent defense responses that nonselectively defend against foreign or abnormal material, even upon initial exposure to it; see also *inflammation, interferon, natural killer cells,* and *complement system*

nontropic hormone a hormone that exerts its effects on nonendocrine target tissues

norepinephrine (nor'-ep-uh-NEF-run) the neurotransmitter released from sympathetic postganglionic fibers; noradrenaline

nucleus(of brain) (NŪ-klē-us) a functional aggregation of neuronal cell bodies within the brain

nucleus(of cells) a distinct spherical or oval structure, which is usually located near the center of a cell and which contains the cell's genetic material, deoxyribonucleic acid (DNA)

occipital lobes (ok-sip'-ut-ul) the lobes of the cerebral cortex that are located posteriorly and are responsible for initially processing visual input

O_2-Hb dissociation curve a graphic depiction of the relationship between arterial P_{O_2} and percent hemoglobin saturation

oligodendrocytes (ol-i-gō'-DEN-drō-sītz) the myelin-forming cells of the central nervous system

oogenesis (ō'-ō-JEN-uh-sus) egg production

opsonin (op'-suh-nun) body-produced chemical that links bacteria to macrophages, thereby making the bacteria more susceptible to phagocytosis

optic nerve the bundle of nerve fibers that leave the retina, relaying information about visual input

optimal length the length before the onset of contraction of a muscle fiber at which maximal force can be developed upon a subsequent tetanic contraction

organ a distinct structural unit composed of two or more types of primary tissue organized to perform one or more particular functions; for example, the stomach

organelles (or'-gan-ELZ) distinct, highly organized, membrane-bound intracellular compartments, each containing a specific set of chemicals for carrying out a particular cellular function

organic referring to substances that contain carbon; from living or once-living sources

organism a living entity, either single-celled (unicellular) or made up of many cells (multicellular)

organ of Corti (KOR-tē) the sense organ of hearing within the inner ear that contains hair cells whose hairs are bent in response to sound waves, setting up action potentials in the auditory nerve

osmolarity (oz'-mō-LAR-ut-ē) a measure of the concentration of solute molecules in a solution

osmosis (os-MŌ-sis) movement of water across a membrane down its own concentration gradient toward the area of higher solute concentration

osteoblasts (OS-tē-ō-blasts') bone cells that produce the organic matrix of bone

osteoclasts bone cells that dissolve bone in their vicinity

otolith organs (ōt'-ul-ith) sense organs in the inner ear that provide information about rotational changes in head movement; include the utricle and saccule

oval window the membrane-covered opening that separates the air-filled middle ear from the upper compartment of the fluid-filled cochlea in the inner ear

overhydration water excess in the body

ovulation (ov'-yuh-LĀ-shun) release of an ovum from a mature ovarian follicle

oxidative phosphorylation (fos'-fōr-i-LĀ-shun) the entire sequence of mitochondrial biochemical reactions that uses oxygen to extract energy from the nutrients in food and transforms it into ATP, producing CO_2 and H_2O in the process

oxyhemoglobin (ok-si-HĒ-muh-glō-bun) hemoglobin combined with O_2

oxyntic mucosa (ok-SIN-tic) the mucosa that lines the body and fundus of the stomach; contains gastric pits lined by mucous neck cells, parietal cells, and chief cells

oxytocin (ok'-sē-TŌ-sun) a hypothalamic hormone that is stored in the posterior pituitary and stimulates uterine contraction and milk ejection

pacemaker activity self-excitable activity of an excitable cell in which its membrane potential gradually depolarizes to threshold on its own

Pacinian corpuscle (pa-SIN-ē-un) a rapidly adapting skin receptor that detects pressure and vibration

pancreas (PAN-krē-us) a mixed gland composed of an exocrine portion that secretes digestive enzymes and an aqueous alkaline secretion into the duodenal lumen and an endocrine portion that secretes the hormones insulin and glucagon into the blood

paracrine (PEAR-uh-krin) a local chemical messenger whose effect is exerted only on neighboring cells in the immediate vicinity of its site of secretion

parasympathetic nervous system (pear'-uh-sim-puh-THET-ik) the subdivision of the autonomic nervous system that dominates in quiet, relaxed situations and promotes body maintenance activities such as digestion and emptying of the urinary bladder

parathyroid glands (pear'-uh-THĪ-roid) four small glands located on the posterior surface of the thyroid gland that secrete parathyroid hormone

parathyroid hormone (PTH) a hormone that raises plasma Ca^{2+} levels

parietal cells (puh-RĪ-ut-ul) the stomach cells that secrete hydrochloric acid and intrinsic factor

parietal lobes the lobes of the cerebral cortex that lie at the top of the brain behind the central sulcus and contain the somatosensory cortex

partial pressure the individual pressure exerted independently by a particular gas within a mixture of gases

partial pressure gradient a difference in the partial pressure of a gas between two regions that promotes the movement of the gas from the region of higher partial pressure to the region of lower partial pressure

parturition (par'-too-RISH-un) delivery of a baby

passive expiration expiration accomplished during quiet breathing as a result of elastic recoil of the lungs on relaxation of the inspiratory muscles, with no energy expenditure required

passive force a force that does not require expenditure of cellular energy to accomplish transport of a substance across the plasma membrane

passive reabsorption when none of the steps in the transepithelial transport of a substance reabsorbed across the kidney tubules requires energy expenditure

pathogens (PATH-uh-junz) disease-causing microorganisms, such as bacteria or viruses

pathophysiology (path'-ō-fiz-ē-UL-uh-jē) abnormal functioning of the body associated with disease

pepsin; pepsinogen (pep-SIN-uh-jun) an enzyme secreted in inactive form by the stomach that, once activated, begins protein digestion

peptide hormones hormones that consist of a chain of specific amino acids of varying length

percent hemoglobin saturation a measure of the extent to which the hemoglobin present is combined with O_2

perception the conscious interpretation of the external world as created by the brain from a pattern of nerve impulses delivered to it from sensory receptors

peripheral chemoreceptors (kē'-mō-rē-SEP-turz) the carotid and aortic bodies, which respond to changes in arterial P_{O_2}, P_{CO_2}, and H^+ and adjust respiration accordingly

peripheral nervous system (PNS) nerve fibers that carry information between the central nervous system and other parts of the body

peristalsis (per'-uh-STOL-sus) ringlike contractions of the circular smooth muscle of a tubular organ that move progressively forward with a stripping motion, pushing the contents of the organ ahead of the contraction

peritubular capillaries (per'-i-TŪ-bū-lur) capillaries that intertwine around the tubules of the kidney's nephron; they supply the renal tissue and participate in exchanges between the tubular fluid and blood during the formation of urine

permeable permitting passage of a particular substance

permissiveness when one hormone must be present in adequate amounts for the full exertion of another hormone's effect

pernicious anemia (per-KNEE-shus) the anemia produced as a result of intrinsic factor deficiency

peroxisomes (puh-ROK'-suh-sōmz) organelles consisting of membrane-bound sacs that contain powerful oxidative enzymes that detoxify various wastes produced within the cell or foreign compounds that have entered the cell

pH the logarithm to the base 10 of the reciprocal of the hydrogen ion concentration; $pH = \log 1/[H^+]$ or $pH = -\log[H^+]$

phagocytosis (FAG-oh-sī-TŌ-sus) a type of endocytosis in which large, multimolecular, solid particles are engulfed by a cell

pharynx (FARE-inks) the back of the throat, which serves as a common passageway for the digestive and respiratory systems

phosphorylation (fos'-fōr-i-LĀ-shun) addition of a phosphate group to a molecule

photoreceptor a sensory receptor responsive to light

phototransduction the mechanism of converting light stimuli into electrical activity by the rods and cones of the eye

phrenic nerve (FREN-ik) the nerve that supplies the diaphragm

physiology (fiz-ē-OL-ō-gē) the study of body functions

pineal gland (PIN-ē-ul) a small endocrine gland located in the center of the brain that secretes the hormone melatonin

pinocytosis (pin-oh-cī-TŌ-sus) type of endocytosis in which the cell internalizes fluid

pitch the tone of a sound, determined by the frequency of vibrations (that is, whether a sound is a C or G note)

pituitary gland (pih-TWO-ih-tair-ee) a small endocrine gland connected by a stalk to the hypothalamus; consists of the anterior pituitary and posterior pituitary

placenta (plah-SEN-tah) the organ of exchange between the maternal and fetal blood; also secretes hormones that support the pregnancy

plaque a deposit of cholesterol and other lipids, perhaps calcified, in thickened, abnormal smooth muscle cells within blood vessels as a result of atherosclerosis

plasma the liquid portion of the blood

plasma cell an antibody-producing derivative of an activated B lymphocyte

plasma clearance the volume of plasma that is completely cleared of a given substance by the kidneys per minute

plasma colloid osmotic pressure (KOL-oid os-MOT-ik) the force caused by the unequal distribution of plasma proteins between the blood and surrounding fluid that encourages fluid movement into the capillaries

plasma membrane a protein-studded lipid bilayer that encloses each cell, separating it from the extracellular fluid

plasma proteins the proteins that remain within the plasma, where they perform a number of important functions; include albumins, globulins, and fibrinogen

plasticity (plas-TIS-uh-tē) the ability of portions of the brain to assume new responsibilities in response to the demands placed on it

platelets (PLATE-lets) specialized cell fragments in the blood that participate in hemostasis by forming a plug at a vessel defect

pleural sac (PLOOR-ul) a double-walled, closed sac that separates each lung from the thoracic wall

pluripotent stem cells precursor cells that reside in the bone marrow and continuously divide and differentiate to give rise to each of the types of blood cells

polycythemia (pol-i-sī-THĒ-mē-uh) excess circulating erythrocytes, accompanied by an elevated hematocrit

polysaccharides (pol'-ē-SAK-uh-rīdz) complex carbohydrates, consisting of chains of interconnected glucose molecules

positive balance situation in which the gains via input for a substance exceed its losses via output, so that the total amount of the substance in the body increases

positive feedback a regulatory mechanism in which the input and the output in a control system continue to enhance each other so that the controlled variable is progressively moved farther from a steady state

postabsorptive state the metabolic state after a meal is absorbed during which endogenous energy stores must be mobilized and glucose must be spared for the glucose-dependent brain; fasting state

posterior pituitary the neural portion of the pituitary that stores and releases into the blood on hypothalamic stimulation two hormones produced by the hypothalamus, vasopressin and oxytocin

postganglionic fiber (pōst'-gan-glē-ON-ik) the second neuron in the two-neuron autonomic nerve pathway; originates in an autonomic ganglion and terminates on an effector organ

postsynaptic neuron (pōst'-si-NAP-tik) the neuron that conducts its action potentials away from a synapse

preganglionic fiber the first neuron in the two-neuron autonomic nerve pathway; originates in the central nervous system and terminates on an autonomic ganglion

pressure gradient a difference in pressure between two regions that drives the movement of blood or air from the region of higher pressure to the region of lower pressure

presynaptic facilitation enhanced release of neurotransmitter from a presynaptic axon terminal as a result of excitation of another neuron that terminates on the axon terminal

presynaptic inhibition a reduction in the release of neurotransmitter from a presynaptic axon terminal as a result of excitation of another neuron that terminates on the axon terminal

presynaptic neuron (prē-si-NAP-tik) the neuron that conducts its action potentials toward a synapse

primary active transport a carrier-mediated transport system in which energy is directly required to operate the carrier and move the transported substance against its concentration gradient

primary follicle a primary oocyte surrounded by a single layer of granulosa cells in the ovary

primary motor cortex the portion of the cerebral cortex that lies anterior to the central sulcus and is responsible for voluntary motor output

progestational phase see *secretory phase*

prolactin (PRL) (prō-LAK-tun) an anterior pituitary hormone that stimulates breast development and milk production in females

proliferative phase the phase of the menstrual cycle during which the endometrium is repairing itself and thickening following menstruation; lasts from the end of the menstrual phase until ovulation

pro-opiomelanocortin (prō-ō-PĒ-ō-ma-LAN-oh-kor'-tin) a large precursor molecule produced by the anterior pituitary that is cleaved into adrenocorticotropic hormone, melanocyte stimulating hormone, and endorphin

proprioception (prō'-prē-ō-SEP-shun) awareness of position of body parts in relation to each other and to surroundings

prostaglandins (pros'-tuh-GLAN-dins) local chemical mediators that are derived from a component of the plasma membrane, arachidonic acid

prostate gland a male accessory sex gland that secretes an alkaline fluid, which neutralizes acidic vaginal secretions

protein kinase (KĪ-nase) an enzyme that phosphorylates and thereby induces a change in the shape and function of a particular intracellular protein

proteolytic enzymes (prōt'-ē-uh-LIT-ik) enzymes that digest protein

proximal tubule (PROKS-uh-mul) a highly convoluted tubule that extends between Bowman's capsule and the loop of Henle in the kidney's nephron

PTH see *parathyroid hormone*

pulmonary artery (PULL-mah-nair-ē) the large vessel that carries blood from the right ventricle to the lungs

pulmonary circulation the closed loop of blood vessels carrying blood between the heart and lungs

pulmonary surfactant (sur-FAK-tunt) a phospholipoprotein complex secreted by the Type II alveolar cells that intersperses between the water molecules that line the alveoli, thereby lowering the surface tension within the lungs

pulmonary valve a one-way valve that permits the flow of blood from the right ventricle into the pulmonary artery during ventricular emptying but prevents the backflow of blood from the pulmonary artery into the right ventricle during ventricular relaxation

pulmonary veins the large vessels that carry blood from the lungs to the heart

pulmonary ventilation the volume of air breathed in and out in one minute; equals tidal volume times respiratory rate

pupil an adjustable round opening in the center of the iris through which light passes to the interior portions of the eye

Purkinje fibers (pur-KIN-jē) small terminal fibers that extend from the bundle of His and rapidly transmit an action potential throughout the ventricular myocardium

pyloric gland area (PGA) (pī-LŌR-ik) the specialized region of the mucosa in the antrum of the stomach that secretes gastrin

pyloric sphincter (pī-lōr'-ik SFINGK-tur) the juncture between the stomach and duodenum

radiation emission of heat energy from the surface of a warm body in the form of electromagnetic waves

reabsorption the net movement of interstitial fluid into the capillary

receptor see *sensory receptor* or *receptor site*

receptor potential the graded potential change that occurs in a sensory receptor in response to a stimulus; generates action potentials in the afferent neuron fiber

receptor site membrane protein that binds with a specific extracellular chemical messenger, thereby bringing about a series of membrane and intracellular events that alter the activity of the particular cell

reduced hemoglobin hemoglobin that is not combined with O_2

reflex any response that occurs automatically without conscious effort; the components of a reflex arc include a receptor, afferent pathway, integrating center, efferent pathway, and effector

refraction bending of a light ray

refractory period (rē-FRAK-tuh-rē) the time period when a recently activated patch of membrane is refractory (unresponsive) to further stimulation, preventing the action potential from spreading backward into the area through which it has just passed, thereby ensuring the unidirectional propagation of the action potential away from the initial site of activation

regulatory proteins troponin and tropomyosin, which play a role in regulating muscle contraction by either covering or exposing the sites of interaction between the contractile proteins

regulatory T cells helper and suppressor T cells

releasing hormone a hypothalamic hormone that stimulates the secretion of a particular anterior pituitary hormone

renal cortex an outer granular-appearing region of the kidney

renal medulla (RĒ-nul muh-DUL-uh) an inner striated-appearing region of the kidney

renal threshold the plasma concentration at which the T_m of a particular substance is reached and the substance first starts appearing in the urine

renin (RĒ-nin) an enzymatic hormone released from the kidneys in response to a decrease in NaCl/ECF volume/arterial blood pressure; activates angiotensinogen

renin-angiotensin-aldosterone system (an'-jē-ō-TEN-sun al-dō-steer-OWN) the salt-conserving system triggered by the release of renin from the kidneys, which activates angiotensin, which stimulates aldosterone secretion, which stimulates Na^+ reabsorption by the kidney tubules during the formation of urine

repolarization (rē'-pō-luh-ruh-ZĀ-shun) return of membrane potential to resting potential following a depolarization

reproductive tract the system of ducts that are specialized to transport or house the gametes after they are produced

residual volume the minimum volume of air remaining in the lungs even after a maximal expiration

resistance hindrance of flow of blood or air through a passageway (blood vessel or respiratory airway, respectively)

respiration the sum of processes that accomplish ongoing passive movement of O_2 from the atmosphere to the tissues, as well as the continual passive movement of metabolically produced CO_2 from the tissues to the atmosphere

respiratory acidosis (as-i-DŌ-sus) acidosis resulting from abnormal retention of CO_2 arising from hypoventilation

respiratory airways the system of tubes that conducts air between the atmosphere and the alveoli of the lungs

respiratory alkalosis (al'-kuh-LŌ-sus) alkalosis caused by excessive loss of CO_2 from the body as a result of hyperventilation

respiratory rate breaths per minute

resting membrane potential the membrane potential that exists when an excitable cell is not displaying an electrical signal

reticular activating system (RAS) (ri-TIK-ū-lur) ascending fibers that originate in the reticular formation and carry signals upward to arouse and activate the cerebral cortex

reticular formation a network of interconnected neurons that runs throughout the brain stem and initially receives and integrates all synaptic input to the brain

retina the innermost layer in the posterior region of the eye that contains the eye's photoreceptors, the rods and cones

ribonucleic acid (RNA) (rī-bō-new-KLĀ-ik) a nucleic acid that exists in three forms (messenger RNA, ribosomal RNA, and transfer RNA), which participate in gene transcription and protein synthesis

ribosomes (RĪ-bō-sōms) special ribosomal RNA-protein complexes that synthesize proteins under the direction of nuclear DNA

right atrium (A-trē'-um) the heart chamber that receives venous blood from the systemic circulation

right ventricle the heart chamber that pumps blood into the pulmonary circulation

RNA see *ribonucleic acid*

rods the eye's photoreceptors used for night vision

round window the membrane-covered opening that separates the lower chamber of the cochlea in the inner ear from the middle ear

salivary amylase (AM-uh-lās') an enzyme produced by the salivary glands that begins carbohydrate digestion in the mouth and continues in the body of the stomach after the food and saliva have been swallowed

saltatory conduction (SAL-tuh-tōr'-ē) the means by which an action potential is propagated throughout a myelinated fiber, with the impulse jumping over the myelinated regions from one node of Ranvier to the next

SA node see *sinoatrial node*

sarcomere (SAR-kō-mir) the functional unit of skeletal muscle; the area between two Z lines within a myofibril

sarcoplasmic reticulum (ri-TIK-yuh-lum) a fine meshwork of interconnected tubules that surrounds a muscle fiber's myofibrils; contains expanded lateral sacs, which store calcium that is released into the cytosol in response to a local action potential

satiety centers (suh-TĪ-ut-ē) neuronal clusters in the ventromedial region of the hypothalamus that inhibit feeding behavior

saturation when all of the binding sites on a carrier molecule are occupied

Schwann cells (shwah-'n) the myelin-forming cells of the peripheral nervous system

secondary active transport a transport mechanism in which a carrier molecule for glucose or an amino acid is driven by a Na^+ concentration gradient established by the energy-dependent Na^+ pump to transfer the glucose or amino acid uphill without directly expending energy to operate the carrier

secondary follicle a developing ovarian follicle that is secreting estrogen and forming an antrum

secondary sexual characteristics the many external characteristics that are not directly involved in reproduction but that distinguish males and females

second messenger an intracellular chemical that is activated by binding of an extracellular first messenger to a surface receptor site and that triggers a preprogrammed series of biochemical events, which result in altered activity of intracellular proteins to control a particular cellular activity

secretin (si-KRĒT-'n) a hormone released from the duodenal mucosa primarily in response to the presence of acid; inhibits gastric motility and secretion and stimulates secretion of a $NaHCO_3$ solution from the pancreas and from the liver

secretion release to a cell's exterior, on appropriate stimulation, of substances that have been produced by the cell

secretory phase the phase of the menstrual cycle characterized by the development of a lush endometrial lining capable of supporting a fertilized ovum; also known as the *progestational phase*

secretory vesicles (VES-i-kuls) membrane-enclosed sacs containing proteins that have been synthesized and processed by the endoplasmic reticulum/Golgi complex of the cell and which will be released to the cell's exterior by exocytosis on appropriate stimulation

segmentation the small intestine's primary method of motility; consists of oscillating, ringlike contractions of the circular smooth muscle along the small intestine's length

self-antigens antigens that are characteristic of a person's own cells

semen (SĒ-men) a mixture of accessory sex gland secretions and sperm

semicircular canals sense organ in the inner ear that detects rotational or angular acceleration or deceleration of the head

semilunar valves (sem-Ī-LEW-nur) the aortic and pulmonary valves

seminal vesicles (VES-i-kuls) male accessory sex glands that supply fructose to ejaculated sperm and secrete prostaglandins

seminiferous tubules (sem'-uh-NIF-uh-rus) the highly coiled tubules within the testes that produce spermatozoa

sensory afferent a pathway coming into the central nervous system that carries information that reaches the level of consciousness

sensory input includes somatic sensation and special senses

sensory receptor an afferent neuron's peripheral ending, which is specialized to respond to a particular stimulus in its environment

septum the muscular partition that separates the right and left halves of the heart

series-elastic component the noncontractile portions of a skeletal muscle fiber, including the connective tissue and sarcoplasmic reticulum

Sertoli cells (sur-TŌ-lē) cells located in the seminiferous tubules that support spermatozoa during their development

signal transduction the sequence of events in which incoming signals (instructions from extracellular chemical messengers such as hormones) are conveyed to the cell's interior for execution

single-unit smooth muscle the most abundant type of smooth muscle; made up of muscle fibers that are interconnected by gap junctions so that they become excited and contract as a unit; also known as *visceral smooth muscle*

sinoatrial (SA) node (sī-nō-Ā-trē-ul) a small specialized autorhythmic region in the right atrial wall of the heart that has the fastest rate of spontaneous depolarizations and serves as the normal pacemaker of the heart

skeletal muscle striated muscle, which is attached to the skeleton and is responsible for movement of the bones in purposeful relation to one another; innervated by the somatic nervous system and under voluntary control

slow-wave potentials self-excitable activity of an excitable cell in which its membrane potential undergoes gradually alternating depolarizing and hyperpolarizing swings

smooth muscle involuntary muscle innervated by the autonomic nervous system and found in the walls of hollow organs and tubes

somatic cells (sō-MAT-ik) body cells, as contrasted with reproductive cells

somatic nervous system the portion of the efferent division of the peripheral nervous system that innervates skeletal muscles; consists of the axonal fibers of the alpha motor neurons

somatic sensation sensory information arising from the body surface, including somesthetic sensation and proprioception

somatomedins (sō'-mat-uh-MĒ-dinz) hormones secreted by the liver or other tissues, in response to growth hormone, that act directly on the target cells to promote growth

somatosensory cortex the region of the parietal lobe immediately behind the central sulcus; the site of initial processing of somesthetic and proprioceptive input

somesthetic sensations (SEW-mess-THEH-tik) awareness of sensory input such as touch, pressure, temperature, and pain from the body's surface

sound waves traveling vibrations of air that consist of regions of high pressure caused by compression of air molecules alternating with regions of low pressure caused by rarefaction of the molecules

spatial summation the summing of several postsynaptic potentials arising from the simultaneous activation of several excitatory (or several inhibitory) synapses

special senses vision, hearing, taste, and smell

specific immune responses responses that are selectively targeted against particular foreign material to which the body has previously been exposed; see also *antibody-mediated immunity* and *cell-mediated immunity*

specificity ability of carrier molecules to transport only specific substances across the plasma membrane

spermatogenesis (spur'-mat-uh-JEN-uh-sus) sperm production

sphincter (sfink-tur) a voluntarily controlled ring of skeletal muscle that controls passage of contents through an opening into or out of a hollow organ or tube

spinal reflex a reflex that is integrated by the spinal cord

spleen a lymphoid tissue in the upper left part of the abdomen that stores lymphocytes and platelets and destroys old red blood cells

state of equilibrium no net change in a system is occurring

steroids (STEER-oidz) hormones derived from cholesterol

stimulus a detectable physical or chemical change in the environment of a sensory receptor

stress the generalized, nonspecific response of the body to any factor that overwhelms, or threatens to overwhelm, the body's compensatory abilities to maintain homeostasis

stretch reflex a monosynaptic reflex in which an afferent neuron originating at a stretch-detecting receptor in a skeletal muscle terminates directly on the efferent neuron supplying the same muscle to cause it to contract and counteract the stretch

stroke volume (SV) the volume of blood pumped out of each ventricle with each contraction, or beat, of the heart

subcortical regions the brain regions that lie under the cerebral cortex, including the basal nuclei, thalamus, and hypothalamus

submucosa the connective tissue layer of the digestive tract that lies under the mucosa and contains the larger blood and lymph vessels and a nerve network

substance P the neurotransmitter released from pain fibers

subsynaptic membrane (sub-suh-NAP-tik) the portion of the postsynaptic cell membrane that lies immediately underneath a synapse and contains receptor sites for the synapse's neurotransmitter

suppressor T cells the population of T cells that suppresses the activity of the other T cells, serving to limit immune responses in check-and-balance fashion

suprachiasmatic nucleus (soup'-ra-KĪ-as mat-ik) a cluster of nerve cell bodies in the hypothalamus that serves as the master biological clock, acting as the pacemaker that establishes many of the body's circadian rhythms

surface tension the force at the liquid surface of an air-water interface resulting from the greater attraction of water molecules to the surrounding water molecules than to the air above the surface; a force that tends to decrease the area of a liquid surface and resists stretching of the surface

sympathetic nervous system the subdivision of the autonomic nervous system that dominates in emergency ("fight-or-flight") or stressful situations and prepares the body for strenuous physical activity

synapse (SIN-aps') the specialized junction between two neurons where an action potential in the presynaptic neuron influences the membrane potential of the postsynaptic neuron by means of the release of a chemical messenger that diffuses across the small cleft that separates the two neurons

synergism (SIN-er-jiz'-um) when several actions are complementary, so that their combined effect is greater than the sum of their separate effects

systemic circulation (sis-TEM-ik) the closed loop of blood vessels carrying blood between the heart and body systems

systole (SIS-tō-lē) the period of cardiac contraction and emptying

T_3 see *triiodothyronine*

T_4 see *thyroxine*

tactile (TACK-til) referring to touch

target cell receptors receptors located on a target cell that are specific for a particular chemical mediator

target cells the cells that a particular extracellular chemical messenger, such as a hormone or a neurotransmitter, influences

teleological approach (tē'-lē-ō-LA-ji-kul) explanation of body functions in terms of their particular purpose in fulfilling a bodily need, that is, the "why" of body processes

temporal lobes the lobes of the cerebral cortex that are located laterally and that are responsible for initially processing auditory input

temporal summation the summing of several postsynaptic potentials occurring very close together in time because of successive firing of a single presynaptic neuron

tension the force produced during muscle contraction by shortening of the sarcomeres, resulting in stretching and tightening of the muscle's elastic connective tissue and tendon, which transmit the tension to the bone to which the muscle is attached

terminal button a motor neuron's enlarged knoblike ending that terminates near a skeletal muscle fiber and releases acetylcholine in response to an action potential in the neuron

testosterone (tes-TOS-tuh-rōn) the male sex hormone, secreted by the Leydig cells of the testes

tetanus (TET'-n-us) a smooth, maximal muscle contraction that occurs when the fiber is stimulated so rapidly that it does not have a chance to relax at all between stimuli

thalamus (THAL-uh-mus) the brain region that serves as a synaptic integrating center for preliminary processing of all sensory input on its way to the cerebral cortex

thecal cells (THAY-kel) the outer layer of specialized ovarian connective tissue cells in a maturing follicle

thermoreceptor (thur'-mō-rē-SEP-tur) a sensory receptor sensitive to heat and cold

thick filaments specialized cytoskeletal structures within skeletal muscle that are made up of myosin molecules and interact with the thin filaments to accomplish shortening of the fiber during muscle contraction

thin filaments specialized cytoskeletal structures within skeletal muscle that are made up of actin, tropomyosin, and troponin molecules and interact with the thick filaments to accomplish shortening of the fiber during muscle contraction

thoracic cavity (thō-RAS-ik) chest cavity

threshold potential the critical potential that must be reached before an action potential is initiated in an excitable cell

thromboembolism (throm'-bō-EM-buh-liz-um) a condition characterized by the presence of thrombi and emboli in the circulatory system

thrombus an abnormal clot attached to the inner lining of a blood vessel

thymus (THIGH-mus) a lymphoid gland located midline in the chest cavity that processes T lymphocytes and produces the hormone thymosin, which maintains the T cell lineage

thyroglobulin (thī'-rō-GLOB-yuh-lun) a large, complex molecule on which all steps of thyroid hormone synthesis and storage take place

thyroid gland a bilobed endocrine gland that lies over the trachea and secretes three hormones, thyroxine and triiodothyronine, which regulate overall basal metabolic rate, and calcitonin, which contributes to control of calcium balance

thyroid hormone collectively, the hormones secreted by the thyroid follicular cells, namely, thyroxine and triiodothyronine

thyroid-stimulating hormone (TSH) an anterior pituitary hormone that stimulates secretion of thyroid hormone and promotes growth of the thyroid gland; thyrotropin

thyroxine (thī-ROCKS-in) the most abundant hormone secreted by the thyroid gland; important in the regulation of overall metabolic rate; also known as *tetraiodothyronine* or T_4

tidal volume the volume of air entering or leaving the lungs during a single breath

tight junction an impermeable junction between two adjacent epithelial cells formed by the sealing together of the cells' lateral edges near their luminal borders; prevents passage of substances between the cells

tissue (1) a functional aggregation of cells of a single specialized type, such as nerve cells forming nervous tissue; (2) the aggregate of various cellular and extracellular components that make up a particular organ, such as lung tissue

T lymphocytes (T cells) white blood cells that accomplish cell-mediated immune responses against targets to which they have been previously exposed; see also *cytotoxic T cells, helper T cells,* and *suppressor T cells*

T_m see *transport maximum*

tone the ongoing baseline of activity in a given system or structure, as in muscle tone, sympathetic tone, or vascular tone

tonic LH secretion the low-level, ongoing secretion of LH that occurs throughout most of the ovarian cycle

total peripheral resistance the resistance offered by all of the peripheral blood vessels, with arteriolar resistance contributing most extensively

trachea (TRĀ-kē-uh) the "windpipe"; the conducting airway that extends from the pharynx and branches into two bronchi, each entering a lung

tract a bundle of nerve fibers (axons of long interneurons) with a similar function within the spinal cord

transduction conversion of stimuli into action potentials by sensory receptors

transepithelial transport (tranz-ep-i-THĒ-lē-al) the entire sequence of steps involved in the transfer of a substance across the epithelium between either the renal tubular lumen or digestive tract lumen and the blood

transmural pressure gradient the pressure difference across the lung wall (intra-alveolar pressure is greater than intrapleural pressure) that stretches the lungs to fill the thoracic cavity, which is larger than the unstretched lungs

transporter recruitment the phenomenon of inserting additional transporters (carriers) for a particular substance into the plasma membrane, thereby increasing membrane permeability to the substance, in response to an appropriate stimulus

transport maximum (T_m) the maximum rate of a substance's carrier-mediated transport across the membrane when the carrier is saturated; known as *tubular maximum* in the kidney tubules

transverse tubule (T tubule) a perpendicular infolding of the surface membrane of a muscle fiber; rapidly spreads surface electric activity into the central portions of the muscle fiber

triglycerides (trī-GLIS-uh-rīdz) neutral fats composed of one glycerol molecule with three fatty acid molecules attached

triiodothyronine (T_3) (trī-ī-ō-dō-THĪ-rō-nēn) the most potent hormone secreted by the thyroid follicular cells; important in the regulation of overall metabolic rate

trophoblast (TRŌF-uh-blast') the outer layer of cells in a blastocyst that is responsible for accomplishing implantation and developing the fetal portion of the placenta

tropic hormone (TRŌ-pik) a hormone that regulates the secretion of another hormone

tropomyosin (trop'-uh-MĪ-uh-sun) one of the regulatory proteins found in the thin filaments of muscle fibers

troponin (tro-PŌ-nun) one of the regulatory proteins found in the thin filaments of muscle fibers

TSH see *thyroid-stimulating hormone*

T tubule see *transverse tubule*

tubular maximum (T_m) the maximum amount of a substance that the renal tubular cells can actively transport within a given time period; the kidney cells' equivalent of transport maximum

tubular reabsorption the selective transfer of substances from the tubular fluid into the peritubular capillaries during the formation of urine

tubular secretion the selective transfer of substances from the peritubular capillaries into the tubular lumen during the formation of urine

twitch a brief, weak contraction that occurs in response to a single action potential in a muscle fiber

twitch summation the addition of two or more muscle twitches as a result of rapidly repetitive stimulation, resulting in greater tension in the fiber than that produced by a single action potential

tympanic membrane (tim-PAN-ik) the eardrum, which is stretched across the entrance to the middle ear and which vibrates when struck by sound waves funneled down the external ear canal

Type I alveolar cells (al-VĒ-ō-lur) the single layer of flattened epithelial cells that forms the wall of the alveoli within the lungs

Type II alveolar cells the cells within the alveolar walls that secrete pulmonary surfactant

ultrafiltration the net movement of a protein-free plasma out of the capillary into the surrounding interstitial fluid

ureter (yū-RĒ-tur) a duct that transmits urine from the kidney to the bladder

urethra (yū-RĒ-thruh) a tube that carries urine from the bladder to outside the body

urine excretion the elimination of substances from the body in the urine; anything filtered or secreted and not reabsorbed is excreted

vagus nerve (VĀ-gus) the tenth cranial nerve, which serves as the major parasympathetic nerve

vasoconstriction (vā'-zō-kun-STRIK-shun) the narrowing of a blood vessel lumen as a result of contraction of the vascular circular smooth muscle

vasodilation the enlargement of a blood vessel lumen as a result of relaxation of the vascular circular smooth muscle

vasopressin (vā-zō-PRES-sin) a hormone secreted by the hypothalamus, then stored and released from the posterior pituitary; increases the permeability of the distal and collecting tubules of the kidneys to water and promotes arteriolar vasoconstriction; also known as *antidiuretic hormone (ADH)*

vaults recently discovered organelles shaped like octagonal barrels; believed to serve as transporters for messenger RNA from the nucleus to sites of protein synthesis; may be important in cellular movement.

vein a vessel that carries blood toward the heart

velocity speed

vena cava (VĒ-nah CĀV-ah), (**venae cavae,** plural; VĒ-nē cāv-ē) a large vein that empties blood into the right atrium

venous return (VĒ-nus) the volume of blood returned to each atrium per minute from the veins

ventilation the mechanical act of moving air in and out of the lungs; breathing

ventricle (VEN-tri-kul) a lower chamber of the heart that pumps blood into the arteries

vesicle (VES-i-kul) a small, intracellular, fluid-filled, membrane-enclosed sac

vesicular transport movement of large molecules or multimolecular materials into or out of the cell by means of being enclosed in a vesicle, as in endocytosis or exocytosis

vestibular apparatus (veh-STIB-yuh-lur) the component of the inner ear that provides information essential for the sense of equilibrium and for coordinating head movements with eye and postural movements; consists of the semicircular canals, utricle, and saccule

villus (villi, plural**)** (VIL-us) microscopic fingerlike projections from the inner surface of the small intestine

virulence (VIR-you-lentz) the disease-producing power of a pathogen

visceral afferent a pathway coming into the central nervous system that carries subconscious information derived from the internal viscera

visceral smooth muscle (VIS-uh-rul) see *single-unit smooth muscle*

viscosity (viss-KOS-i-tē) the friction developed between molecules of a fluid as they slide over each other during flow of the fluid; the greater the viscosity, the greater the resistance to flow

vital capacity the maximum volume of air that can be moved out during a single breath following a maximal inspiration

voltage-gated channels channels that open or close in response to changes in membrane potential

white matter the portion of the central nervous system composed of myelinated nerve fibers

Z line a flattened disc-like cytoskeletal protein that connects the thin filaments of two adjoining sarcomeres

zona fasciculata (zō-nah-fa-SICK-ū-lah-ta) the middle and largest layer of the adrenal cortex; major source of cortisol

zona glomerulosa (glō-MER-yū-lō-sah) the outermost layer of the adrenal cortex; sole source of aldosterone

zona reticularis (ri-TIK-yuh-lair-us) the innermost layer of the adrenal cortex; produces cortisol, along with the zona fasciculata

Index

Credits

2-2b K.G. Murti/Visuals Unlimited; 2-6b David M. Phillips/Visuals Unlimited; 2-9c Prof. Marcel Bessis, Science Source/Photo Researchers; 2-10b Bill Longcore/Photo Researchers; 2-16c Dr. Leonard H. Rome, UCLA School of Medicine; 2-17a,b Elizabeth R. Walker, Ph.D., Associate Professor, Department of Anatomy, School of Medicine, West Virginia University, and Dennis O. Overman, Ph.D., Associate Professor, Department of Anatomy, School of Medicine, West Virginia University; 2-20 PIR-CNRI, Science Source/Photo Researchers; 2-21a Adapted and reproduced with permission from Bruce Alberts, Dennis Bray, Julian Lewis, Martin Raff, Keith Roberts, and James D. Watson: *Molecular Biology of the Cell* (New York: Garland Publishing), Figure 10-27, p. 565; 2-21b David M. Phillips/Visuals Unlimited; 2-22 Adapted and reproduced with permission from Bruce Alberts, Dennis Bray, Julian Lewis, Martin Raff, Keith Roberts, and James D. Watson: *Molecular Biology of the Cell* (New York: Garland Publishing), Figure 10-30, p. 567; 2-23b David M. Phillips/Visuals Unlimited; 2-25b M. Abbey/Visuals Unlimited; 3-1 Don W. Fawcett/Visuals Unlimited; 4-10b David M. Phillips/Visuals Unlimited; 4-12c C. Raines/Visuals Unlimited; 4-17b Science VU/E.R. Lewis, T.E. Everhart, and Y.Y. Zeevi, University of California/Visuals Unlimited; 5-4 Nancy Kedersha, Ph.D., Research Scientist - Cell Biology, ImmunoGen, Inc.; Chapter 5 Concepts, Challenges, and Controversies box photo courtesy of Washington University School of Medicine, St. Louis; 6-15b Bill Beatty/Visuals Unlimited; 6-19b Patricia N. Farnsworth, Ph.D., Professor of Physiology and Opthalmology, University of Medicine & Dentistry of New Jersey, New Jersey Medical School; 6-22 A.L. Blum/Visuals Unlimited; 6-24c Omikron, Science Source/Photo Researchers; 6-40d Dean Hillman, Ph.D., Professor of Physiology, New York University Medical School; 7-4 Eric V. Grave, Science Source/Photo Researchers; 8-3b M. Abbey, Science Source/Photo Researchers; 8-4a Photo - Fawcett (micrograph by J. Auber), Science Source/Photo Researchers; 8-27a Dr. Brian Eyden, Science Source/Photo Researchers; 8-27b Dr. Brenda Russell, Professor of Physiology, University of Illinois at Chicago; 9-8 Photo by John Cunningham/Visuals Unlimited; 9-33b Sloop-Ober/Visuals Unlimited; 9-34c Copyright Boehringer Ingelheim International GmbH, photo Lennart Nilsson/Bonnier Alba AB; 10-5 Triarch/Visuals Unlimited; 10-10a Fawcett-Uehara-Suyama, Science Source/Photo Researchers; 10-15a R.J. Bolander, D. Fawcett/Visuals Unlimited; 10-15b From *Behold Man* (Boston: Little, Brown and Company, 1974: 63). Photo Lennart Nilsson/Bonnier Alba AB; 10-27 Science VU - Fred Marsik/Visuals Unlimited; 10-32 Adapted from Little, R.C., and Little, W.C.: *Physiology of the Heart and Circulation*, 4th ed. Copyright © 1989 by Year Book Medical Publishers, Inc. By permission of the author and Mosby-Year Book, Inc.; 11-2 David M. Phillips/Visuals Unlimited; 11-6 Stanley Flegler/Visuals Unlimited; 11-7 Erythrocytes photos courtesy of Barbara O'Connor, Teaching Supervisor, Clinical Hematology, Yale-New Haven Hospital, New Haven, Connecticut. Neutrophil, eosinophil, basophil, monocyte, lymphocyte, and platelets from Barbara O'Connor, *A Color Atlas and Instruction Manual of Peripheral Blood Cell Morphology* (Baltimore, MD: Williams & Wilkins Co., 1984). Reprinted with permission from the author and publisher; 11-9 Copyright © 1995 Discover Magazine; 11-11 Copyright Boehringer Ingelheim International GmbH, photo Lennart Nilsson/Bonnier Alba AB; 12-8 Adapted from "How Killer Cells Kill," by John Ding-E Young and Zanvil A. Cohn. Copyright © 1988 by Scientific American, Inc. All rights reserved; 12-19 Adapted from "How Killer Cells Kill," by John Ding-E Young and Zanvil A. Cohn. Copyright © 1988 by Scientific American, Inc. All rights reserved; 12-20 Copyright Boehringer International GmbH, photo Lennart Nilsson/Bonnier Alba AB; 12-22 Copyright Boehringer International GmbH, photo Lennart Nilsson/Bonnier Alba AB; 12-24 Copyright Boehringer International GmbH, photo Lennart Nilsson/Bonnier Alba AB; 13-3 Custom Medical Stock Photo; 13-4c Don Fawcett/Visuals Unlimited; 13-23a,b SIU/Visuals Unlimited; 13-28 a,b Val Vallyathan, Ph.D., Professor, Department of Pathology, School of Medicine, West Virginia University and Research Physiologist, National Institute of Occupational Safety and Health.; Chapter 13 Concepts, Challenges, and Controversies box photo p. 456 by F. Stuart Westmorland/Photo Researchers and p. 457 by Dave B. Fleetham/Visuals Unlimited; 14-15 Adapted with permission from *Federation Proceedings* 42: 3046 - 3052; 1983; 16-16 A.B. Dowsett, Science Photo Library/ Photo Researchers; 16-28 Michael Webb/Visuals Unlimited; 16-29 Thomas W. Sheehy, M.D.; Robert L. Slaughter, M.D., "The Malabsorption Syndrome," by Medcom, Inc. Reproduced by permission of Medcom, Inc.; 17-2 Wilbert E. Gladfelter, Ph.D., Professor Emeritus, Department of Physiology, School of Medicine, West Virginia University; 18-20 Lester V. Bergman and Associates, Inc.; 19-1b Elizabeth R. Walker, Ph.D., Associate Professor, Department of Anatomy, School of Medicine, West Virginia University, and Dennis O. Overman, Ph.D., Associate Professor, Department of Anatomy, School of Medicine, West Virginia University; 19-5 Lester V. Bergman and Associates, Inc.; 19-6 Lester V. Bergman and Associates, Inc.; 20-6b Elizabeth R. Walker, Ph.D., Associate Professor, Department of Anatomy, School of Medicine, West Virginia University, and Dennis O. Overman, Ph.D., Associate Professor, Department of Anatomy, School of Medicine, West Virginia University; 20-8a David M. Phillips/Visuals Unlimited; 20-14 Dr. P. Bagavandoss, Department of Biological Sciences, Kent State University; 20-21 David Scharf; 20-22b Photo Lennart Nilsson/Bonnier Alba AB, *A Child Is Born*. Copyright © 1966, 1977 Dell Publishing Company, Inc.; 20-24 Photo Lennart Nilsson/Bonnier Alba AB, *A Child Is Born*. Copyright © 1966, 1977 Dell Publishing Company, Inc.; B-11 From Christine J. Harrison et al.: "Cytogenics," Cell Genetics 35: 21-27 (1983), Figure 3B. Reprinted with permission from Dr. Christine Harrison and S. Karger AG, Basel.